# Energy Management
# Handbook
## Sixth Edition

# ENERGY MANAGEMENT
# HANDBOOK
## SIXTH EDITION

---

BY

## WAYNE C. TURNER

*SCHOOL OF INDUSTRIAL ENGINEERING AND MANAGEMENT*

*OKLAHOMA STATE UNIVERSITY*

AND

## STEVE DOTY

*COLORADO SPRINGS UTILITIES*

*COLORADO SPRINGS, COLORADO*

**THE FAIRMONT PRESS, INC.**

CRC Press
Taylor & Francis Group

Library of Congress Cataloging-in-Publication Data

Turner, Wayne C., 1942-
Energy management handbook / by Wayne C. Turner & Steve Doty. -- 6th ed.
    p. cm.
Includes bibliographical references and index.
ISBN: 0-88173-542-6 (print) — 0-88173-543-4 (electronic)
    1. Power resources--Handbooks, manuals, etc. 2. Energy conservation--
Handbooks, manuals, etc. I. Doty, Steve. II. Title.

TJ163.2.T87 2006
658.2'6--dc22

2006041263

Published by The Fairmont Press, Inc.
700 Indian Trail
Lilburn, GA 30047
tel: 770-925-9388; fax: 770-381-9865
http://www.fairmontpress.com

Distributed by Taylor & Francis Ltd.
6000 Broken Sound Parkway NW, Suite 300
Boca Raton, FL 33487, USA
E-mail: orders@crcpress.com

Distributed by Taylor & Francis Ltd.
23-25 Blades Court
Deodar Road
London SW15 2NU, UK
E-mail: uk.tandf@thomsonpublishingservices.co.uk

Printed in the United States of America
10 9 8 7 6 5 4 3 2

0-88173-542-6 (The Fairmont Press, Inc.)
0-8493-8234-3 (Taylor & Francis Ltd.)

# CONTENTS

# FOREWORD TO THE SIXTH EDITION

Since its first edition was published more than two decades ago, *Energy Management Handbook* has remained the leading reference of choice used by thousands of energy management professionals for one fundamental reason. With this new edition, Dr. Turner and Mr. Doty continue to bring readers both the cutting-edge developments they need to know about, as well as the broad scope of practical information they must have to accomplish real and significant energy cost reduction goals. No other single publication has been as influential in defining and guiding the energy management profession.

This new sixth edition builds upon and is no less essential than its predecessors. Comprehensive in scope, it provides today's energy managers with the tools they will require to meet the challenges of a new era of predicted rising energy costs and supply uncertainties—ongoing developments which seem certain to impact virtually every aspect of the cost of doing business in the decades ahead. The new edition also examines the impact of the passage and implementation of the Energy Policy Act of 2005, which puts in place new energy efficiency requirements for government facilities, as well as energy-efficiency-related tax incentives for commercial buildings.

As evidence continues to lend credence to the reality of global climate change, a growing number of businesses are seeing the "good business sense" of reducing greenhouse emissions and developing sustainable, green facilities. The sixth edition of *Energy Management Handbook* includes substantial new material on sustainability, high performance facilities and related technologies.

In many ways the evolution of *Energy Management Handbook* has paralleled that of the Association of Energy Engineers (AEE) in meeting the needs of and setting the standards for the modern energy management profession which has emerged since the 1970s. Therefore, it seems very appropriate that the publication of this important new sixth edition officially kicks off AEE's 30th anniversary celebration. As our organization completes its third decade of serving more than 8,000 members in 77 countries, it would be nearly impossible to overstate the impact that *Energy Management Handbook* has had for those we serve. The book is an official reference and preparatory text for AEE's Certified Energy Manager (CEM) program, the most widely recognized professional credential in the energy management field, having certified more than 6,000 professionals since its inception in 1981. In addition numerous large corporations have selected *Energy Management Handbook* as their official corporate energy management reference.

There is no doubt that *Energy Management Handbook* will continue its role as the indispensable reference for all energy managers who must meet the daunting energy supply and cost control challenges which lie ahead.

*Albert Thumann, P.E., C.E.M.*
*Executive Director, The Association of Energy Engineers*
*April 2006*

# Preface to the Sixth Edition

When the first introduction to *Energy Management Handbook* was written in 1982, I was in college, worried more about how to repay student loans than anything else. But this is now.

This book lives to serve its readers. In helping to edit the book, it has been my goal to keep the material fresh, pertinent and useful. My approach has been to view it from the reader's perspective, and to assure that the book provides good value. I look forward to further improvements, over time, as technologies continue to change and refine, since it will keep *me* current in the process.

As I see it, the core intentions of this book are these:

* To address an audience of practicing energy managers and persons entering this trade. It is a tool for energy managers to get their questions answered, and get things accomplished.

* To provide a resource of current, accurate, useful information in a readable format.

* To emphasize applications, and include practical examples to clarify key topics, especially savings calculations. Background development and derivations should be limited to just what is needed to support the application messages.

For this edition, there are some significant changes, including a rewrite of the automatic controls chapter, and all-new chapters on ground source heat pumps, and green building design.

As I have learned, editors check for errors and adjust grammar or format, but do not change the author's message. The primary strategy for quality in this book is to choose the very best authors, so I'll extend my personal thanks to each of them for their contributions. And I owe Wayne Turner a debt of thanks for the opportunity to contribute. I can only hope to do as well.

I am excited about contributing to the long-running success of this book, because of its potential to influence other energy professionals, and therefore the public at large. Writing is just writing, until it changes a behavior or helps someone get something accomplished—then it becomes golden!

*Steve Doty*
*Colorado Springs, CO*
*April 2006*

# CHAPTER 1

# INTRODUCTION

**DR. WAYNE C. TURNER,
REGENTS PROFESSOR**
Oklahoma State University
Stillwater, Ok.

**DR. BARNEY L. CAPEHART, PROFESSOR**
University of Florida
Gainesville, Fla.

**STEPHEN A. PARKER**
Pacific Northwest National Laboratory
Richland, WA

**STEVE DOTY**
Colorado Springs Utilities
Colorado Springs, CO

## 1.1 BACKGROUND

Mr. Al Thumann, Executive Director of the Association of Energy Engineers, said it well in the Foreword. "The energy 'roller coaster' never ceases with new turns and spirals which make for a challenging ride." Those professionals who boarded the ride in the late 70's and stayed on board have experienced several ups and downs. First, being an energy manager was like being a mother, John Wayne, and a slice of apple pie all in one. Everyone supported the concept and success was around every bend. Then, the mid-80's plunge in energy prices caused some to wonder "Do we really need to continue energy management?"

Sometime in the late 80's, the decision was made. Energy management is good business but it needs to be run by professionals. The Certified Energy Manager Program of the Association of Energy Engineers became popular and started a very steep growth curve. AEE continues to grow in membership and stature.

About the same time (late 80's), the impact of the Natural Gas Policy Act began to be felt. Now, energy managers found they could sometimes save significant amounts of money by buying "spot market" natural gas and arranging transportation. About the only thing that could be done in purchasing electricity was to choose the appropriate rate schedule and optimize parameters (power factor, demand, ratchet clauses, time of use, etc.—see Chapter 18 on energy rate schedules).

With the arrival of the Energy Policy Act of 1992, electricity deregulation moved closer to reality, creating the opportunity of purchasing electricity from wherever the best deal could be found and to wheel the electric energy through the grid. Several states moved toward electrical deregulation, with some successes. But there were also some failures that made the energy industry pause and reflect. The prospect of electric deregulation and sharing grid infrastructure caused utilities to change their business view of their portion of the grid. Investment in expanding or upgrading this infrastructure became risky business for individual utilities, and so most chose to maintain the existing grid systems they owned, with a wait-and-see approach. Through electricity trading that manipulated pricing, problems with implementation changed the electric deregulation movement trend from slow to stop. Since good business relationships are good for all, some revisions to the EPACT-92 deregulation provisions may be necessary to see greater acceptance, and to sustain the concept in practice. To regain the confidence of the consumers, a greater degree of oversight of the business practices and the sharing of the vital US grid infrastructure may be necessary. This need is further accentuated by concerns of security and reliability of our nation's electrical grid, spurred by national events in September 2001 (9-11) and August 2003 (Blackout). Even with the bumps as electricity deregulation was first tried, wider scale electric deregulation remains an exciting concept and energy managers are watching with anticipation. As new skills are learned and beneficial industry relationships are created, the prospects of larger scale deregulation will improve.

However, EPACT-1992's impact is further reaching. If utilities must compete with other producers of electricity, then they must be "lean and mean." As Mr. Thumann mentions in the Foreword, this means many of the Demand Side Management (DSM) and other conservation activities of the utilities are being cut or eliminated. The roller coaster ride goes on.

In 2005, the Bush Administration enacted the Energy Policy Act of 2005. This Act provides new opportunities and incentives for energy improvements in the country, including strong incentives for renewable energy sources and net metering. It is hoped that the tax incentives provided under this Act will become tools for the private sector to spur change with the free enterprise system. Similar in style to individual utility incentive programs, the Act's success will depend largely on the ability of private firms, such as consultants, ESCOs and

Performance Contractors, to find partnering solutions to connect the program funding mechanism and the customer points of use. EPACT-2005 also updates the federal energy improvement mandates with a newer, stricter, baseline year (2003) and a new timeline for energy reduction requirements. The federal building segment remains an excellent target for large-scale improvement, as well as setting the all-important high visibility example for private industry to follow.

The Presidential Executive Orders mentioned in Chapter 20 created the Federal Energy Management Program (FEMP) to aid the federal sector in meeting federal energy management goals. The potential FEMP savings are mammoth and new professionals affiliated with federal, as well as state and local governments have joined the energy manager ranks. However, as Congress changes complexion, the FEMP and even DOE itself may face uncertain futures. The roller coaster ride continues.

FEMP efforts are showing results. Figure 1.3 outlines the goals that have been established for FEMP and reports show that the savings are apparently on schedule to meet all these goals. As with all such programs, reporting and measuring is difficult and critical. However, that energy and money is being saved is undeniable. More important, however, to most of this book's readers are the Technology Demonstration Programs and Technology Alerts being published by the Pacific Northwest Laboratories of Battelle in cooperation with the US DOE. Both of these programs are dramatically speeding the incorporation of new technology and the Alerts are a great source of information for all energy managers. (Information is available on the WEB).

As utility DSM programs shrink, while private sector businesses and the federal government expand their needs for energy management programs, the door is opening for the ESCOs (Energy Service Companies), Shared Savings Providers, Performance Contractors, and other similar organizations. These groups are providing the auditing, energy and economic analyses, capital and monitoring to help other organizations reduce their energy consumption and reduce their expenditures for energy services. By guaranteeing and sharing the savings from improved energy efficiency and improved productivity, both groups benefit and prosper.

Throughout it all, energy managers have proven time and time again, that energy management is cost effective. Furthermore, energy management is vital to our national security, environmental welfare, and economic productivity. This will be discussed in the next section.

## 1.2  THE VALUE OF ENERGY MANAGEMENT

Business, industry and government organizations have all been under tremendous economic and environmental pressures in the last few years. Being economically competitive in the global marketplace and meeting increasing environmental standards to reduce air and water pollution have been the major driving factors in most of the recent operational cost and capital cost investment decisions for all organizations. Energy management has been an important tool to help organizations meet these critical objectives for their short term survival and long-term success.

The problems that organizations face from both their individual and national perspectives include:

• Meeting more stringent environmental quality standards, primarily related to reducing global warming and reducing acid rain.

Energy management helps improve environmental quality. For example, the primary culprit in global warming is carbon dioxide, $CO_2$. Equation 1.1, a balanced chemistry equation involving the combustion of methane (natural gas is mostly methane), shows that 2.75 pounds of carbon dioxide is produced for every pound of methane combusted. Thus, energy management, by reducing the combustion of methane can dramatically reduce the amount of carbon dioxide in the atmosphere and help reduce global warming. Commercial and industrial energy use accounts for about 45 percent of the carbon dioxide released from the burning of fossil fuels, and about 70 percent of the sulfur dioxide emissions from stationary sources.

$$CH_4 + 2\ O_2 = CO_2 + 2\ H_2O$$
$$(12 + 4*1) + 2(2*16) =$$
$$(12 + 2*16) + 2(2*1 + 16) \tag{1.1}$$

Thus, 16 pounds of methane produces 44 pounds of carbon dioxide; or 2.75 pounds of carbon dioxide is produced for each pound of methane burned.

Energy management reduces the load on power plants as fewer kilowatt hours of electricity are needed. If a plant burns coal or fuel oil, then a significant amount of acid rain is produced from the sulphur dioxide emitted by the power plant. Acid rain problems then are reduced through energy management, as are $NO_x$ problems.

Less energy consumption means less petroleum field development and subsequent on-site pollution.

Less energy consumption means less thermal pollution at power plants and less cooling water discharge. Reduced cooling requirements or more efficient satisfaction of those needs means less CFC usage and reduced ozone depletion in the stratosphere. The list could go on almost indefinitely, but the bottom line is that energy management helps improve environmental quality.

- Becoming—or continuing to be—economically competitive in the global marketplace, which requires reducing the cost of production or services, reducing industrial energy intensiveness, and meeting customer service needs for quality and delivery times.

Significant energy and dollar savings are available through energy management. Most facilities (manufacturing plants, schools, hospitals, office buildings, etc) can save according to the profile shown in Figure 1.1. Even more savings have been accomplished by some programs.

| | |
|---|---|
| √ | Low cost activities first year or two: 5 to 15% |
| √ | Moderate cost, significant effort, three to five years: 15 to 30% |
| √ | Long-term potential, higher cost, more engineering: 30 to 50% |

**Figure 1.1  Typical Savings
Through Energy Management**

Thus, large savings can be accomplished often with high returns on investments and rapid paybacks. Energy management can make the difference between profit and loss and can establish real competitive enhancements for most companies.

Energy management in the form of implementing new energy efficiency technologies, new materials and new manufacturing processes and the use of new technologies in equipment and materials for business and industry is also helping companies improve their productivity and increase their product or service quality. Often, the energy savings is not the main driving factor when companies decide to purchase new equipment, use new processes, and use new high-tech materials. However, the combination of increased productivity, increased quality, reduced environmental emissions, and reduced energy costs provides a powerful incentive for

companies and organizations to implement these new technologies.

Total Quality Management (TQM) is another emphasis that many businesses and other organizations have developed over the last decade. TQM is an integrated approach to operating a facility, and energy cost control should be included in the overall TQM program. TQM is based on the principle that front-line employees should have the authority to make changes and other decisions at the lowest operating levels of a facility. If employees have energy management training, they can make informed decisions and recommendations about energy operating costs.

- Maintaining energy supplies that are:
  — Available without significant interruption, and
  — Available at costs that do not fluctuate too rapidly.

Once again, the country is becoming dependent on imported oil. During the time of the 1979 oil price crisis, the U.S. was importing almost 50% of our total oil consumption. By 1995, the U.S. was again importing 50% of our consumption. Today (2003) we are importing even more (approximately 54%), and the price has dramatically increased. Thus, the U S is once again vulnerable to an oil embargo or other disruption of supply. The major difference is that there is a better balance of oil supply among countries friendly to the U.S. Nonetheless, much of the oil used in this country is not produced in this country. The trade balance would be much more favorable if we imported less oil.

- Helping solve other national concerns which include:
  — Need to create new jobs
  — Need to improve the balance of payments by reducing costs of imported energy
  — Need to minimize the effects of a potential limited energy supply interruption

None of these concerns can be satisfactorily met without having an energy efficient economy. Energy management plays a key role in helping move toward this energy efficient economy.

## 1.3  THE ENERGY MANAGEMENT PROFESSION

Energy management skills are important to people in many organizations, and certainly to people who perform duties such as energy auditing, facility or building management, energy and economic analysis, and maintenance. The number of companies employing professionally trained energy managers is large and growing. A partial list of job titles is given in Figure 1.2. Even though this is only a partial list, the breadth shows the robustness of the profession.

For some of these people, energy management will be their primary duty, and they will need to acquire in-depth skills in energy analysis as well as knowledge about existing and new energy using equipment and technologies. For others—such as maintenance managers—energy management skills are simply one more area to cover in an already full plate of duties and expectations. The authors are writing this *Energy Management Handbook* for both of these groups of readers and users.

Twenty years ago, few university faculty members would have stated their primary interest was energy management, yet today there are numerous faculty who prominently list energy management as their principal specialty. In 2003, there were 26 universities throughout the country listed by DOE as Industrial Assessment Centers or Energy Analysis and Diagnostic Centers. Other Universities offer coursework and/or do research in energy management but do not have one of the above centers. Finally, several professional Journals and Magazines now publish exclusively for energy managers while we know of none that existed 15 years ago.

The need for energy management in federal facilities predates the U.S. Department of Energy. Since 1973, the President and Congress have called on federal agencies to lead by example in energy conservation and management in its own facilities, vehicles and operations. Both the President and the Congress have addressed the issue of improving energy efficiency in federal facilities several times since the mid- 1970's. Each new piece of legislation and executive order has combined past experiences with new approaches in[12] an effort to promote further efficiency gains in federal agencies . The Federal Energy Management Program (FEMP) was established

in the early 1970's to coordinate federal agency reporting, analysis of energy use and to encourage energy conservation and still leads that effort today. Executive Order 13123, *Greening the Government Through Efficient Energy Management,* signed by President Clinton in June 1999, is the most recent directive for federal agencies. A brief summary of the goals of that executive order is given in Figure 1.3. In addition to the goals, Executive Order 13123 outlined several other requirements for federal agencies aimed at improving energy efficiency, reducing greenhouse gases and other emissions, increasing the use of renewable energy, and promoting federal leadership in energy management.

Like energy management itself, utility DSM programs have had their ups and downs. DSM efforts peaked in the late 80s and early 90s, and have since retrenched significantly as utility deregulation and the movement to retail wheeling have caused utilities to reduce staff and cut costs as much as possible. This short-term cost cutting is seen by many utilities as their only way to become a competitive low-cost supplier of electric power. Once their large customers have the choice of their power supplier, they want to be able to hold on to these customers by offering rates that are competitive with other producers around the country. In the meantime, the other energy services provided by the utility are being reduced or eliminated in this corporate downsizing effort.

This reduction in electric utility incentive and rebate programs, as well as the reduction in customer support, has produced a gap in energy service assistance that is being met by a growing sector of equipment supply companies and energy service consulting firms that are willing and able to provide the technical and financial assistance that many organizations previously got from their local electric utility. New business opportunities and many new jobs are being created in this shift away from utility support to energy service company support. Energy management skills are extremely important in this rapidly expanding field, and even critical to those companies that are in the business of identifying energy savings and providing a guarantee of the savings results.

---

- Plant Energy Manager
- Utility Energy Auditor
- State Agency Energy Analyst
- Consulting Energy Manager
- DSM Auditor/Manager

- Building/Facility Energy Manager
- Utility Energy Analyst
- Federal Energy Analyst
- Consulting Energy Engineer

**Figure 1.2 Typical Energy Management Job Titles**

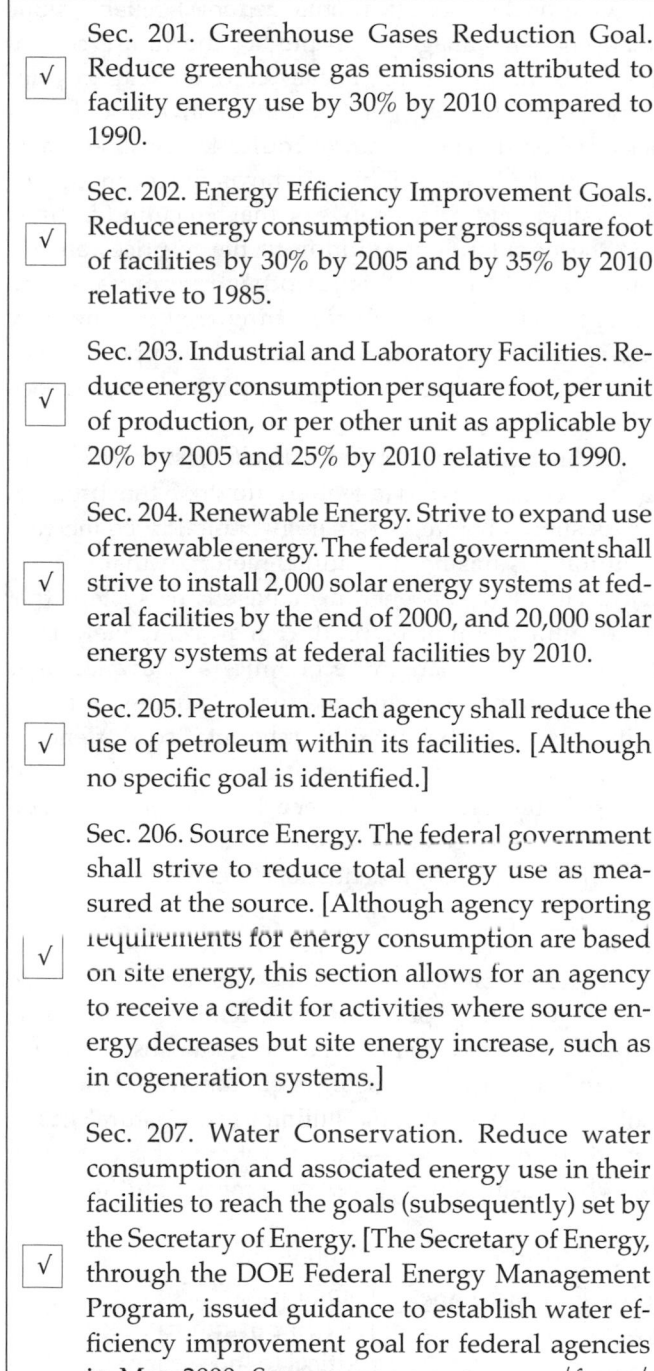

Sec. 201. Greenhouse Gases Reduction Goal. Reduce greenhouse gas emissions attributed to facility energy use by 30% by 2010 compared to 1990.

Sec. 202. Energy Efficiency Improvement Goals. Reduce energy consumption per gross square foot of facilities by 30% by 2005 and by 35% by 2010 relative to 1985.

Sec. 203. Industrial and Laboratory Facilities. Reduce energy consumption per square foot, per unit of production, or per other unit as applicable by 20% by 2005 and 25% by 2010 relative to 1990.

Sec. 204. Renewable Energy. Strive to expand use of renewable energy. The federal government shall strive to install 2,000 solar energy systems at federal facilities by the end of 2000, and 20,000 solar energy systems at federal facilities by 2010.

Sec. 205. Petroleum. Each agency shall reduce the use of petroleum within its facilities. [Although no specific goal is identified.]

Sec. 206. Source Energy. The federal government shall strive to reduce total energy use as measured at the source. [Although agency reporting requirements for energy consumption are based on site energy, this section allows for an agency to receive a credit for activities where source energy decreases but site energy increase, such as in cogeneration systems.]

Sec. 207. Water Conservation. Reduce water consumption and associated energy use in their facilities to reach the goals (subsequently) set by the Secretary of Energy. [The Secretary of Energy, through the DOE Federal Energy Management Program, issued guidance to establish water efficiency improvement goal for federal agencies in May 2000. See www.eere.energy.gov/femp/resources/waterguide.html for details.]

**Figure 1.3. Federal Agency Goals as Established by Executive Order 13123.**

Thus, the future for energy management is extremely promising. It is cost effective, it improves environmental quality, it helps reduce the trade deficit, and it helps reduce dependence on foreign fuel supplies. Energy management will continue to grow in size and importance.

## 1.4 SOME SUGGESTED PRINCIPLES OF ENERGY MANAGEMENT

(The material in this section is repeated verbatim from the first and second editions of this handbook. Mr. Roger Sant who was then director of the Energy Productivity Center of the Carnegie-Mellon Institute of Research in Arlington, VA, wrote this section for the first edition. It was unchanged for the second edition. Now, the fourth edition is being printed. The principles developed in this section are still sound. Some of the number quoted may now be a little old; but the principles are still sound. Amazing, but what was right 18 years ago for energy management is still right today. The game has changed, the playing field has moved; but the principles stay the same).

If energy productivity is an important opportunity for the nation as a whole, it is a necessity for the individual company. It represents a real chance for creative management to reduce that component of product cost that has risen the most since 1973.

Those who have taken advantage of these opportunities have done so because of the clear intent and commitment of the top executive. Once that commitment is understood, managers at all levels of the organization can and do respond seriously to the opportunities at hand. Without that leadership, the best designed energy management programs produce few results. In addition, we would like to suggest four basic principles which, if adopted, may expand the effectiveness of existing energy management programs or provide the starting point of new efforts.

The first of these is to *control the costs of the energy function or service provided, but not the Btu of energy.* As most operating people have noticed, energy is just a means of providing some service or benefit. With the possible exception of feedstocks for petrochemical production, energy is not consumed directly. It is always converted into some useful function. The existing data are not as complete as one would like, but they do indicate some surprises. In 1978, for instance, the aggregate industrial expenditure for energy was $55 billion. Thirty-five percent of that was spent for machine drive from electric motors, 29% for feedstocks, 27% for process heat, 7% for electrolytic functions, and 2% for space conditioning and light. As shown in Table 1.1, this is in blunt contrast to measuring these functions in Btu. Machine drive, for example, instead of 35% of the dollars, required only 12% of the Btu.

In most organizations it will pay to be even more specific about the function provided. For instance, evaporation, distillation, drying, and reheat are all typical of

**Table 1.1 Industrial Energy Functions by Expenditure and Btu, 1978**

| Function | Dollar Expenditure (billions) | Percent of Expenditure | Percent of Total Btu |
|---|---|---|---|
| Machine drive | 19 | 35 | 12 |
| Feedstocks | 16 | 29 | 35 |
| Process steam | 7 | 13 | 23 |
| Direct heat | 4 | 7 | 13 |
| Indirect heat | 4 | 7 | 13 |
| Electrolysis | 4 | 7 | 3 |
| Space conditioning and lighting | 1 | 1 | 1 |
| Total | 55 | 100 | 100 |

*Source:* Technical Appendix, *The Least-Cost Energy Strategy,* Carnegie-Mellon University Press, Pittsburgh, Pa., 1979, Tables 1.2.1 and 11.3.2.

the uses to which process heat is put. In some cases it has also been useful to break down the heat in terms of temperature so that the opportunities for matching the heat source to the work requirement can be utilized.

In addition to energy costs, it is useful to measure the depreciation, maintenance, labor, and other operating costs involved in providing the conversion equipment necessary to deliver required services. These costs add as much as 50% to the fuel cost.

It is the total cost of these functions that must be managed and controlled, not the Btu of energy. The large difference in cost of the various Btu of energy can make the commonly used Btu measure extremely misleading. In November 1979, the cost of 1 Btu of electricity was nine times that of 1 Btu of steam coal. Table 1.2 shows how these values and ratios compare in 2005.

One of the most desirable and least reliable skills for an energy manager is to predict the future cost of energy. To the extent that energy costs escalate in price beyond the rate of general inflation, investment paybacks will be shortened, but of course the reverse is also true. A quick glance at Table 1.2 shows the inconsistency in overall energy price changes over this period in time. Even the popular conception that energy prices always go up was not true for this period, when normalized to constant dollars. This volatility in energy pricing may account for some business decisions that appear overly conservative in establishing rate of return or payback period hurdles.

Availabilities also differ and the cost of maintaining fuel flexibility can affect the cost of the product. And as shown before, the average annual price increase of natural gas has been almost three times that of electricity. Therefore, an energy management system that controls Btu per unit of product may completely miss the effect of the changing economics and availabilities of energy alternatives and the major differences in usability of each fuel. Controlling the total cost of energy functions is much more closely attuned to one of the principal interests of the executives of an organization—controlling costs.

**NOTE:** The recommendation to control energy dollars and not Btus does not always apply. For example, tracking building energy use per year for comparison to prior years is best done with Btus since doing so negates the effect of energy price volatility. Similarly, comparing the heating use of a commercial facility against an industry segment benchmark using cost alone can yield wild results if, for example, one building uses natural gas to heat while another uses electric resistance; this is another case where using Btus yields more meaningful results.

**Table 1.2 Cost of Industrial Energy per Million Btu, 1979 and 2005**

| Fuel | Cost (1979) | Cost (1979) Adjusted to 2004 Dollars | Cost (2005) | Pct Change in 25 years (constant dollars) |
|---|---|---|---|---|
| Industrial Coal | 1.11 | 2.32 | 1.78 | <-23%> |
| Natural Gas | 2.75 | 5.75 | 8.41 | +46% |
| Residual Fuel | 2.95 | 6.17 | 8.74 | +41% |
| Distillate Fuel | 4.51 | 9.65 | 10.82 | +12% |
| Electricity | 10.31 | 21.58 | 17.57 | <-18%> |
| | | | | |
| | | Approximated as 3% inflation per year over 25 years | Source: EIA Annual Energy Outlook, 2004 dollars per million Btu | |

A second principle of energy management is to *control energy functions as a product cost, not as a part of manufacturing or general overhead.* It is surprising how many companies still lump all energy costs into one general or manufacturing overhead account without identifying those products with the highest energy function cost. In most cases, energy functions must become part of the standard cost system so that each function can be assessed as to its specific impact on the product cost.

The minimum theoretical energy expenditure to produce a given product can usually be determined en route to establishing a standard energy cost for that product. The seconds of 25-hp motor drive, the minutes necessary in a 2200°F furnace to heat a steel part for fabrication, or the minutes of 5-V electricity needed to make an electrolytic separation, for example, can be determined as theoretical minimums and compared with the actual figures. As in all production cost functions, the minimum standard is often difficult to meet, but it can serve as an indicator of the size of the opportunity.

In comparing actual values with minimum values, four possible approaches can be taken to reduce the variance, usually in this order:

1.  An hourly or daily control system can be installed to keep the function cost at the desired level.

2.  Fuel requirements can be switched to a cheaper and more available form.

3.  A change can be made to the process methodology to reduce the need for the function.

4.  New equipment can be installed to reduce the cost of the function.

The starting point for reducing costs should be in achieving the minimum cost possible with the present equipment and processes. Installing management control systems can indicate what the lowest possible energy use is in a well-controlled situation. It is only at that point when a change in process or equipment configuration should be considered. An equipment change prior to actually minimizing the expenditure under the present system may lead to oversizing new equipment or replacing equipment for unnecessary functions.

The third principle is to *control and meter only the main energy functions*—the roughly 20% that make up 80% of the costs. As Peter Drucker pointed out some time ago, a few functions usually account for a majority of the costs. It is important to focus controls on those that represent the meaningful costs and aggregate the

remaining items in a general category. Many manufacturing plants in the United States have only one meter, that leading from the gas main or electric main into the plant from the outside source. Regardless of the reasonableness of the standard cost established, the inability to measure actual consumption against that standard will render such a system useless. Submetering the main functions can provide the information not only to measure but to control costs in a short time interval. The cost of metering and submetering is usually incidental to the potential for realizing significant cost improvements in the main energy functions of a production system.

The fourth principle is to put *the major effort of an energy management program into installing controls and achieving results.* It is common to find general knowledge about how large amounts of energy could be saved in a plant. The missing ingredient is the discipline necessary to achieve these potential savings. Each step in saving energy needs to be monitored frequently enough by the manager or first-line supervisor to see noticeable changes. Logging of important fuel usage or behavioral observations are almost always necessary before any particular savings results can be realized. Therefore, it is critical that an energy director or committee have the authority from the chief executive to install controls, not just advise line management. Those energy managers who have achieved the largest cost reductions actually install systems and controls; they do not just provide good advice.

As suggested earlier, the overall potential for increasing energy productivity and reducing the cost of energy services is substantial. The 20% or so improvement in industrial energy productivity since 1972 is just the beginning. To quote the energy director of a large chemical company: "Long-term results will be much greater."

Although no one knows exactly how much we can improve productivity in practice, the American Physical Society indicated in their 1974 energy conservation study that it is theoretically possible to achieve an eightfold improvement of the 1972 energy/production ratio.[9] Most certainly, we are a long way from an economic saturation of the opportunities (see, e.g., Ref. 10). The common argument that not much can be done after a 15 or 20% improvement has been realized ought to be dismissed as baseless. Energy productivity provides an expanding opportunity, not a last resort. The chapters in this book provide the information that is necessary to make the most of that opportunity in each organization.

## References

1. *Statistical Abstract of the United States,* U.S. Government Printing Office, Washington, D.C., 1999.
2. *Energy User News,* Jan. 14, 1980.
3. JOHN G. WINGER et al., *Outlook for Energy in the United States*

*to* 1985, The Chase Manhattan Bank, New York, 1972, p 52.

4. DONELLA H. MEADOWS et al., *The Limits to Growth,* Universe Books, New York, 1972, pp. 153-154.

5. JIMMY E. CARTER, July 15, 1979, "Address to the Nation," *Washington Post,* July 16, 1979, p. A14.

6. *Monthly Energy Review,* Jan. 1980, U.S. Department of Energy, Washington, D.C., p. 16.

7. *Monthly Energy Review,* Jan. 1980, U.S. Department of Energy, Washington D.C., p. 8; *Statistical Abstract of the United States, U.S.* Government Printing Office, Washington, D.C., 1979, Table 1409; *Energy User News,* Jan. 20, 1980, p. 14.

8. American Association for the Advancement of Science, "U.S. Energy Demand: Some Low Energy Futures," *Science,* Apr. 14, 1978, p. 143.

9. American Physical Society Summer Study on Technical Aspects of Efficient Energy Utilization, 1974. Available as W.H. CARNAHAN et al., *Efficient Use of Energy, a Physics Perspective,* from NTIS PB-242-773, or in *Efficient Energy Use, Vol.* 25 of the American Institute of Physics Conference Proceedings.

10. R.W. SANT, *The Least-Cost Energy Strategy,* Carnegie-Mellon University Press, Pittsburgh, Pa., 1979

11. U.S. Congress Office of Technology Assessment (OTA). Energy Efficiency in the Federal Government: Government by Good Example? OTA-E-492, U.S. Government Printing Office, Washington D.C., May 1991.

12. U.S. Air Force. *DOD Energy Manager's Handbook* Volume 1: Installation Energy Management. Washington D.C., April 1993.

# CHAPTER 2

# EFFECTIVE ENERGY MANAGEMENT

**WILLIAM H. MASHBURN, P.E., CEM**

Professor Emeritus
Mechanical Engineering Department
Virginia Polytechnic Institute & State University
Blacksburg, Virginia

## 2.1 INTRODUCTION

Some years ago, a newspaper headline stated, "Lower energy use leaves experts pleased but puzzled." The article went on to state "Although the data are preliminary, experts are baffled that the country appears to have broken the decades-old link between economic growth and energy consumption."

For those involved in energy management, this comes as no surprise. We have seen companies becoming more efficient in their use of energy, and that's showing in the data. Those that have extracted all possible savings from downsizing, are now looking for other ways to become more competitive. Better management of energy is a viable way, so there is an upward trend in the number of companies that are establishing an energy management program. Management is now beginning to realize they are leaving a lot of money on the table when they do not instigate a good energy management plan.

With the new technologies and alternative energy sources now available, this country could possibly reduce its energy consumption by 50%—if there were no barriers to the implementation. But of course, there are barriers, mostly economic. Therefore, we might conclude that **managing energy is not a just technical challenge, but one of how to best implement those technical changes within economic limits, and with a minimum of disruption.**

Unlike other management fads that have come and gone, such as value analysis and quality circles, the need to manage energy will be permanent within our society.

There are several reasons for this:

- There is a direct economic return. Most opportunities found in an energy survey have less than a two year payback. Some are immediate, such as load shifting or going to a new electric rate schedule.

- Most manufacturing companies are looking for a competitive edge. A reduction in energy costs to manufacture the product can be immediate and permanent. In addition, products that use energy, such as motor driven machinery, are being evaluated to make them more energy efficient, and therefore more marketable. Many foreign countries where energy is more critical, now want to know the maximum power required to operate a piece of equipment.

- Energy technology is changing so rapidly that state-of-the-art techniques have a half life of ten years at the most. Someone in the organization must be in a position to constantly evaluate and update this technology.

- Energy security is a part of energy management. Without a contingency plan for temporary shortages or outages, and a strategic plan for long range plans, organizations run a risk of major problems without immediate solutions.

- Future price shocks will occur. When world energy markets swing wildly with only a five percent decrease in supply, as they did in 1979, it is reasonable to expect that such occurrences will happen again.

Those people then who choose—or in many cases are drafted—to manage energy will do well to recognize this continuing need, and exert the extra effort to become skilled in this emerging and dynamic profession.

The purpose of this chapter is to provide the fundamentals of an energy management program that can be, and have been, adapted to organizations large and small. Developing a working organizational structure may be the most important thing an energy manager can do.

## 2.2 ENERGY MANAGEMENT PROGRAM

All the components of a comprehensive energy management program are depicted in Figure 2-1. These components are the organizational structure, a policy, and plans for audits, education, reporting, and strategy. It is hoped that by understanding the fundamentals of managing energy, the energy manager can then adapt a good working program to the existing organizational structure. Each component is discussed in detail below.

## 2.3  ORGANIZATIONAL STRUCTURE

The organizational chart for energy management shown in Figure 2-1 is generic. It must be adapted to fit into an existing structure for each organization. For example, the presidential block may be the general manager, and VP blocks may be division managers, but the fundamental principles are the same. The main feature of the chart is the location of the energy manager. This position should be high enough in the organizational structure to have access to key players in management, and to have a knowledge of current events within the company. For example, the timing for presenting energy projects can be critical. Funding availability and other management priorities should be known and understood. The organizational level of the energy manager is also indicative of the support management is willing to give to the position.

### 2.3.1 Energy Manager

One very important part of an energy management program is to have top management support. More important, however, is the selection of the energy manager, who can among other things secure this support. The person selected for this position should be one with a vision of what managing energy can do for the company. Every successful program has had this one thing in common—one person who is a shaker and mover that makes things happen. The program is then built around this person.

There is a great tendency for the energy manager to become an energy engineer, or a prima donna, and attempt to conduct the whole effort alone. Much has been accomplished in the past with such individuals working alone, but for the long haul, managing the program by involving everyone at the facility is much more productive and permanent. **Developing a working organizational structure may be the most important thing an energy manager can do.**

The role and qualifications of the energy manager have changed substantially in the past few years, caused mostly by EPACT-1992 requiring certification of federal energy managers, deregulation of the electric utility industry bringing both opportunity and uncertainty, and by performance contracting requiring more business skills than engineering. In her book titled "Performance Contracting: Expanded Horizons," Shirley Hansen give the following requirements for an energy management:

- Set up an Energy Management Plan
- Establish energy records
- Identify outside assistance
- Assess future energy needs
- Identify financing sources
- Make energy recommendations
- Implement recommendations

**Figure 2.1**

- Provide liaison for the energy committee
- Plan communication strategies
- Evaluate program effectiveness

Energy management programs can, and have, originated within one division of a large corporation. The division, by example and savings, motivates people at corporate level to pick up on the program and make energy management corporate wide. Many also originate at corporate level with people who have facilities responsibility, and have implemented a good corporate facilities program. They then see the importance and potential of an energy management program, and take a leadership role in implementing one. In every case observed by the author, **good programs have been instigated by one individual who has recognized the potential, is willing to put forth the effort—in addition to regular duties—will take the risk of pushing new concepts, and is motivated by a seemingly higher calling to save energy**.

If initiated at corporate level, there are some advantages and some precautions. Some advantages are:

- More resources are available to implement the program, such as budget, staff, and facilities.

- It top management support is secured at corporate level, getting management support at division level is easier.

- Total personnel expertise throughout the corporation is better known and can be identified and made known to division energy managers.

- Expensive test equipment can be purchased and maintained at corporate level for use by divisions as needed.

- A unified reporting system can be put in place.

- Creative financing may be the most needed and the most important assistance to be provided from corporate level.

- Impacts of energy and environmental legislation can best be determined at corporate level.

- Electrical utility rates and structures, as well as effects of unbundling of electric utilities, can be evaluated at corporate level.

Some precautions are:

- Many people at division level may have already done a good job of saving energy, and are cautious about corporate level staff coming in and taking credit for their work.

- All divisions don't progress at the same speed. Work with those who are most interested first, then through the reporting system to top management give them credit. Others will then request assistance.

### 2.3.2 Energy Team

The coordinators shown in Figure 2-1 represent the energy management team within one given organizational structure, such as one company within a corporation. This group is the core of the program. The main criteria for membership should be an indication of interest. There should be a representative from the administrative group such as accounting or purchasing, someone from facilities and/or maintenance, and a representative from each major department.

This energy team of coordinators should be appointed for a specific time period, such as one year. Rotation can then bring new people with new ideas, can provide a mechanism for tactfully removing nonperformers, and involve greater numbers of people in the program in a meaningful way.

Coordinators should be selected to supplement skills lacking in the energy manager since, as pointed out above, it is unrealistic to think one energy manager can have all the qualifications outlined. So, total skills needed for the team, including the energy manager may be defined as follows:

- Have enough technical knowledge within the group to either understand the technology used by the organization, or be trainable in that technology.

- Have a knowledge of potential new technology that may be applicable to the program.

- Have planning skills that will help establish the organizational structure, plan energy surveys, determine educational needs, and develop a strategic energy management plan.

- Understand the economic evaluation system used by the organization, particularly payback and life cycle cost analysis.

- Have good communication and motivational skills since energy management involves everyone within the organization.

The strengths of each team member should be evaluated in light of the above desired skills, and their assignments made accordingly.

### 2.3.3 Employees

**Employees are shown as a part of the organizational structure, and are perhaps the greatest untapped resource in an energy management program.** A structured method of soliciting their ideas for more efficient use of energy will prove to be the most productive effort of the energy management program. A good energy manager will devote 20% of total time working with employees. Too many times employee involvement is limited to posters that say "Save Energy."

Employees in manufacturing plants generally know more about the equipment than anyone else in the facility because they operate it. They know how to make it run more efficiently, but because there is no mechanism in place for them to have an input, their ideas go unsolicited.

An understanding of the psychology of motivation is necessary before an employee involvement program can be successfully conducted. Motivation may be defined as the amount of physical and mental energy that a worker is willing to invest in his or her job. Three key factors of motivation are listed below:

• Motivation is already within people. The task of the supervisor is not to provide motivation, but to know how to release it.

• The amount of energy and enthusiasm people are willing to invest in their work varies with the individual. Not all are over-achievers, but not all are lazy either.

• The amount of personal satisfaction to be derived determines the amount of energy an employee will invest in the job.

Achieving personal satisfaction has been the subject of much research by industrial psychologists, and they have emerged with some revealing facts. For example. They have learned that most actions taken by people are done to satisfy a physical need—such as the need for food—or an emotional need—such as the need for acceptance, recognition, or achievement.

Research has also shown that many efforts to motivate employees deal almost exclusively with trying to satisfy physical needs, such as raises, bonuses, or fringe benefits. These methods are effective only for the short term, so we must look beyond these to other needs that may be sources of releasing motivation,

A study done by Heresy and Blanchard [1] in 1977 asked workers to rank job related factors listed below. The results were as follows:

1. Full appreciation for work done
2. Feeling "in" on things
3. Understanding of personal problems
4. Job security
5. Good wages
6. Interesting work
7. Promoting and growth in the company
8. Management loyalty to workers
9. Good working conditions
10. Tactful discipline of workers

This priority list would no doubt change with time and with individual companies, but the rankings of what supervisors thought employees wanted were almost diametrically opposed. They ranked good wages as first.

It becomes obvious from this that **job enrichment is a key to motivation**. Knowing this, the energy manager can plan a program involving employees that can provide job enrichment by some simple and inexpensive recognitions.

Some things to consider in employee motivation are as follows:

• There appears to be a positive relationship between fear arousal and persuasion if the fear appeals deal with topics primarily of significance to the individual; e.g., personal well being.

• The success of persuasive communication is directly related to the credibility of the source of communication and may be reduced if recommended changes deviate too far from existing beliefs and practices.

• When directing attention to conservation, display the reminder at the point of action at the appropriate time for action, and specify who is responsible for taking the action and when it should occur. Generic posters located in the work area are not effective.

• Studies have shown that pro-conservation attitudes and actions will be enhanced through associations with others with similar attitudes, such as being part of an energy committee.

• Positive effects are achieved with financial incentives if the reward is in proportion to the savings, and represents respectable increments of spendable income.

- Consumers place considerable importance on the potential discomfort in reducing their consumption of energy. Changing thermostat settings from the comfort zone should be the last desperate act for an energy manager.

- Social recognition and approval is important, and can occur through such things as the award of medals, designation of employee of the month, and selection to membership in elite sub-groups. Note that the dollar cost of such recognitions is minimal.

- The potentially most powerful source of social incentives for conservation behavior—but the least used—is the commitment to others that occurs in the course of group decisions.

Before entering seriously into a program involving employees, be prepared to give a heavy commitment of time and resources. In particular, have the resources to respond quickly to their suggestions.

## 2.4. ENERGY POLICY

A well written energy policy that has been authorized by management is as good as the proverbial license to steal. It provides the energy manager with the authority to be involved in business planning, new facility location and planning, the selection of production equipment, purchase of measuring equipment, energy reporting, and training—things that are sometimes difficult to do.

If you already have an energy policy, chances are that it is too long and cumbersome. To be effective, the policy should be short—two pages at most. Many people confuse the policy with a procedures manual. It should be bare bones, but contain the following items as a minimum:

- Objectives—this can contain the standard motherhood and flag statements about energy, but the most important is that the organization will incorporate energy efficiency into facilities and new equipment, with emphasis on life cycle cost analysis rather than lowest initial cost.

- Accountability—This should establish the organizational structure and the authority for the energy

manager, coordinators, and any committees or task groups.

- Reporting—Without authority from top management, it is often difficult for the energy manager to require others within the organization to comply with reporting requirements necessary to properly manage energy. The policy is the place to establish this. It also provides a legitimate reason for requesting funds for instrumentation to measure energy usage.

- Training—If training requirements are established in the policy, it is again easier to include this in budgets. It should include training at all levels within the organization.

Many companies, rather that a comprehensive policy encompassing all the features described above, choose to go with a simpler policy statement.

Appendices A and B give two sample energy policies. Appendix A is generic and covers the items discussed above. Appendix B is a policy statement of a multinational corporation.

## 2.5 PLANNING

Planning is one of the most important parts of the energy management program, and for most technical people is the least desirable. It has two major functions in the program. First, a good plan can be a shield from disruptions. Second, by scheduling events throughout the year, continuous emphasis can be applied to the energy management program, and will play a major role in keeping the program active.

Almost everyone from top management to the custodial level will be happy to give an opinion on what can be done to save energy. Most suggestions are worthless. It is not always wise from a job security standpoint to say this to top management. However, if you inform people—especially top management—that you will evaluate their suggestion, and assign a priority to it in your plan, not only will you not be disrupted, but may be considered effective because you do have a plan.

Many programs were started when the fear of energy shortages was greater, but they have declined into oblivion. By planning to have events periodically through the year, a continued emphasis will be placed on energy management. Such events can be training programs, audits, planning sessions, demonstrations, research projects, lectures, etc.

**The secret to a workable plan is to have people who are required to implement the plan involved in the planning process.** People feel a commitment to making things work if they have been a part of the design. This is fundamental to any management planning, but more often that not is overlooked. However, in order to prevent the most outspoken members of a committee from dominating with their ideas, and rejecting ideas from less outspoken members, a technique for managing committees must be used. A favorite of the author is the Nominal Group Technique developed at the University of Wisconsin in the late 1980's by Andre Delbecq and Andrea Van de Ven [2]. This technique consists of the following basic steps:

1. Problem definition—The problem is clearly defined to members of the group.

2. Grouping—Divide large groups into smaller groups of seven to ten, then have the group elect a recording secretary.

3. Silent generation of ideas—Each person silently and independently writes as many answers to the problem as can be generated within a specified time.

4. Round-robin listing—Secretary lists each idea individually on an easel until all have been recorded.

5. Discussion—Ideas are discussed for clarification, elaboration, evaluation and combining.

6. Ranking—Each person ranks the five most important items. The total number of points received for each idea will determine the first choice of the group.

## 2.6 AUDIT PLANNING

The details of conducting audits are discussed in a comprehensive manner in Chapter 4, but planning should be conducted prior to the actual audits. The planning should include types of audits to be performed, team makeup, and dates.

By making the audits specific rather than general in nature, much more energy can be saved. Examples of some types of audits that might be considered are:

- Tuning-Operation-Maintenance (TOM)
- Compressed air
- Motors
- Lighting
- Steam system
- Water
- Controls
- HVAC
- Employee suggestions

By defining individual audits in this manner, it is easy to identify the proper team for the audit. Don't neglect to bring in outside people such as electric utility and natural gas representatives to be team members. Scheduling the audits, then, can contribute to the events that will keep the program active.

With the maturing of performance contracting, energy managers have two choices for the energy audit process. They may go through the contracting process to select and define the work of a performance contractor, or they can set up their own team and conduct audits, or in some cases such as a corporate energy manager, performance contracting may be selected for one facility, and energy auditing for another. Each has advantages and disadvantages.

Advantages of performance contracting are:

- No investment is required of the company—other than that involved in the contracting process, which can be very time consuming.

- A minimum of in-house people are involved, namely the energy manager and financial people.

Disadvantages are:

- Technical resources are generally limited to the contracting organization.

- Performance contracting is still maturing, and many firms underestimate the work required

- The contractor may not have the full spectrum of skills needed.

- The contractor may not have an interest in low/cost no/cost projects.

Advantages of setting up an audit team are:

- The team can be selected to match equipment to be audited, and can be made up of in-house personnel, outside specialists, or best, a combination of both.

- They can identify all potential energy conservation projects, both low-cost/no-cost as well as large capital investments.

- The audit can be an excellent training tool by involving others in the process, and by adding a training component as a part of the audit.

Disadvantages of an audit team approach:

- Financing identified projects becomes a separate issue for the energy manager.

- It takes a well organized energy management structure to take full advantage of the work of the audit team.

## 2.7 EDUCATIONAL PLANNING

A major part of the energy manager's job is to provide some energy education to persons within the organization. In spite of the fact that we have been concerned with it for the past two decades, **there is still a sea of ignorance concerning energy**.

Raising the energy education level throughout the organization can have big dividends. The program will operate much more effectively if management understands the complexities of energy, and particularly the potential for economic benefit; the coordinators will be more effective is they are able to prioritize energy conservation measures, and are aware of the latest technology; the quality and quantity of employee suggestions will improve significantly with training.

Educational training should be considered for three distinct groups—management, the energy team, and employees.

### 2.7.1 Management Training

It is difficult to gain much of management's time, so subtle ways must be developed to get them up to speed. Getting time on a regular meeting to provide updates on the program is one way. When the momentum of the program gets going, it may be advantageous to have a half or one day presentation for management.

A good concise report periodically can be a tool to educate management. Short articles that are pertinent to your educational goals, taken from magazines and newspapers can be attached to reports and sent selectively. Having management be a part of a training program for either the energy team or employees, or both, can be an educational experience since we learn best when we have to make a presentation.

Ultimately, the energy manager should aspire to be a part of business planning for the organization. A strategic plan for energy should be a part of every business plan. This puts the energy manager into a position for more contact with management people, and thus the opportunity to inform and teach.

### 2.7.2 Energy Team Training

Since the energy team is the core group of the energy management program, proper and thorough training for them should have the highest priority. Training is available from many sources and in many forms.

- Self study—this necessitates having a good library of energy related materials from which coordinators can select.

- In-house training—may be done by a qualified member of the team—usually the energy manager, or someone from outside.

- Short courses offered by associations such as the Association of Energy Engineers [3], by individual consultants, by corporations, and by colleges and universities.

- Comprehensive courses of one to four weeks duration offered by universities, such the one at the University of Wisconsin, and the one being run cooperatively by Virginia Tech and N.C. State University.

For large decentralized organizations with perhaps ten or more regional energy managers, an annual two or three-day seminar can be the base for the educational program. Such a program should be planned carefully. The following suggestions should be incorporated into such a program:

- Select quality speakers from both inside and outside the organization.

- This is an opportunity to get top management support. Invite a top level executive from the organization to give opening remarks. It may be wise to offer to write the remarks, or at least to provide some material for inclusion.

- Involve the participants in workshop activities so they have an opportunity to have an input into the program. Also, provide some practical tips on energy savings that they might go back and implement immediately. One or two good ideas can sometimes pay for their time in the seminar.

- Make the seminar first class with professional speakers; a banquet with an entertaining—not technical—after dinner speaker; a manual that includes a schedule of events, biosketches of speakers, list of attendees, information on each topic presented, and other things that will help pull the whole seminar together. Vendors will contribute things for door prizes.

- You may wish to develop a logo for the program, and include it on small favors such as cups, carrying cases, etc.

### 2.7.3 Employee Training

A systematic approach for involving employees should start with some basic training in energy. This will produce a much higher quality of ideas from them. Employees place a high value on training, so a side benefit is that morale goes up. Simply teaching the difference between electrical demand and kilowatt hours of energy, and that compressed air is very expensive is a start. Short training sessions on energy can be injected into other ongoing training for employees, such as safety. A more comprehensive training program should include:

- Energy conservation in the home
- Fundamentals of electric energy
- Fundamentals of energy systems
- How energy surveys are conducted and what to look for

## 2.8 STRATEGIC PLANNING

Developing an objective, strategies, programs, and action items constitutes strategic planning for the energy management program. It is the last but perhaps the most important step in the process of developing the program, and unfortunately is where many stop. The very name "Strategic Planning" has an ominous sound for those who are more technically inclined. However, by using a simplified approach and involving the energy management team in the process, a plan can be developed using a flow chart that will define the program for the next five years.

If the team is involved in developing each of the components of objective, strategies, programs, and action items—using the Nominal Group Technique—the result will be a simplified flow chart that can be used for many purposes. First, it is a protective plan that discourages intrusion into the program, once it is established and approved. It provides the basis for resources such as funding and personnel for implementation. It projects strategic planning into overall planning by the organization, and hence legitimizes the program at top management level. By involving the implementers in the planning process, there is a strong commitment to make it work.

Appendix C contains flow charts depicting a strategic plan developed in a workshop conducted by the author by a large defense organization. It is a model plan in that it deals not only with the technical aspects of energy management but also the funding, communications, education, and behavior modification.

## 2.9 REPORTING

There is no generic form to that can be used for reporting. There are too many variables such as organization size, product, project requirements, and procedures already in existence. The ultimate reporting system is one used by a chemical company making a textile product. The Btu/lb of product is calculated on a computer system that gives an instantaneous reading. This is not only a reporting system, but one that detects maintenance problems. Very few companies are set up to do this, but many do have some type of energy index for monthly reporting.

When energy prices fluctuate wildly, the best energy index is usually based on Btus; but, when energy prices are stable, the best index is dollars. However, there are still many factors that will influence any index, such as weather, production, expansion or contraction of facilities, new technologies, etc.

The bottom line is that any reporting system has to be customized to suit individual circumstances. And, while reporting is not always the most glamorous part of managing energy, it can make a contribution to the program by providing the bottom line on its effectiveness. It is also a straight pipeline into management, and can be a tool for promoting the program.

The report is probably of most value to the one who prepares it. It is a forcing function that requires all information to be pulled together in a coherent manner. This requires much thought and analysis that might not otherwise take place.

By making reporting a requirement of the energy

policy, getting the necessary support can be easier. In many cases, the data may already be collected on a periodic basis and put into a computer. It may simply require combining production data and energy data to develop an energy index.

Keep the reporting requirements as simple as possible. The monthly report could be something as simple as adding to an ongoing graph that compares present usage to some baseline year. Any narrative should be short, with data kept in a file that can be provided for any supporting in-depth information.

**With all the above considered, the best way to report is to do it against an audit that has been performed at the facility.** One large corporation has its facilities report in this manner, and then has an award for those that complete all energy conservation measures listed on the audit.

## 2.10 OWNERSHIP

The key to a successful energy management program is within this one word—ownership. This extends to everyone within the organization. Employees that operate a machine "own" that machine. Any attempt to modify their "baby" without their participation will not succeed. They have the knowledge to make or break the attempt. Members of the energy team are not going to be interested in seeing one person—the energy manger—get all the fame and glory for their efforts. Management people that invest in energy projects want to share in the recognition for their risk taking. A corporate energy team that goes into a division for an energy audit must help put a person from the division in the energy management position, then make sure the audit belongs to the division. Below are more tips for success that have been compiled from observing successful energy management programs.

- Have a plan. A plan dealing with organization, surveys, training, and strategic planning—with events scheduled—has two advantages. It prevents disruptions by non-productive ideas, and it sets up scheduled events that keeps the program active.

- Give away—or at least share—ideas for saving energy. The surest way to kill a project is to be possessive. If others have a vested interest they will help make it work.

- Be aggressive. The energy team—after some training—will be the most energy knowledgeable group within the company. Too many management decisions are made with a meager knowledge of the effects on energy.

- Use proven technology. Many programs get bogged down trying to make a new technology work, and lose sight of the easy projects with good payback. Don't buy serial number one. In spite of price breaks and promise of vendor support, it can be all consuming to make the system work.

- Go with the winners. Not every department within a company will be enthused about the energy program. Make those who are look good through the reporting system to top management, and all will follow.

- A final major tip—ask the machine operator what should be done to reduce energy. Then make sure they get proper recognition for ideas.

## 2.11 SUMMARY

Let's now summarize by assuming you have just been appointed energy manager of a fairly large company. What are the steps you might consider in setting up an energy management program? Here is a suggested procedure.

### 2.11.1 Situation Analysis
Determine what has been done before. Was there a previous attempt to establish an energy management program? What were the results of this effort? Next, plot the energy usage for all fuels for the past two—or more—years, then project the usage, and cost, for the next five years at the present rate. This will not only help you sell your program, but will identify areas of concentration for reducing energy.

### 2.11.2 Policy
Develop some kind of acceptable policy that gives authority to the program. This will help later on with such things as reporting requirements, and need for measurement instrumentation.

### 2.11.3 Organization
Set up the energy committee and/or coordinators.

### 2.11.4 Training
With the committee involvement, develop a training plan for the first year.

### 2.11.5 Audits

Again with the committee involvement, develop an auditing plan for the first year.

### 2.11.6 Reporting

Develop a simple reporting system.

### 2.11.7 Schedule

From the above information develop a schedule of events for the next year, timing them so as to give periodic actions from the program, which will help keep the program active and visible.

### 2.11.8 Implement the program

## 2.12 CONCLUSION

Energy management has now matured to the point that it offers outstanding opportunities for those willing to invest time and effort to learn the fundamentals. It requires technical and management skills which broadens educational needs for both technical and management people desiring to enter this field. Because of the economic return of energy management, it is attractive to top management, so exposure of the energy manager at this level brings added opportunity for recognition and advancement. Managing energy will be a continuous need, so persons with this skill will have personal job security as we are caught up in the down sizing fad now permeating our society.

### References

1. Hersey, Paul and Kenneth H. Blanchard, *Management of Organizational Behavior: Utilizing Human Resources*, Harper and Row, 1970
2. Delbecq, Andre L., Andrew H. Van de Ven, and David H. Gustafson, *Group Techniques for Program Planning*, Green Briar Press, 1986.
3. Mashburn, William H., *Managing Energy Resources in Times of Dynamic Change*, Fairmont Press, 1992
4. Turner, Wayne, *Energy Management Handbook*, 2nd edition, Chapter 2, Fairmont Press, 1993.

# Appendix A

## ENERGY POLICY

Acme Manufacturing Company

Policy and Procedures Manual

Subject: Energy Management Program

### I. Policy

Energy Management shall be practiced in all areas of the Company's operation.

### II. Energy Management Program Objectives

It is the Company's objective to use energy efficiently and provide energy security for the organization for both immediate and long range by:

- Utilizing energy efficiently throughout the Company's operations.

- Incorporating energy efficiency into existing equipment and facilities, and in the selection and purchase of new equipment.

- Complying with government regulations—federal, state, and local.

- Putting in place an Energy Management Program to accomplish the above objectives.

### III. Implementation

A. Organization

The Company's Energy Management Program shall be administered through the Facilities Department.

*1. Energy Manager*

The Energy Manager shall report directly to the Vice President of Facilities, and shall have overall responsibility for carrying out the Energy Management Program.

*2. Energy Committee*

The Energy Manager may appoint and Energy Committee to be comprised of representatives from various departments. Members will serve for a specified period of time. The purpose of the Energy Committee is to advise the Energy Manager on the operation of the Energy Management Program, and to provide assistance on specific tasks when needed.

*3. Energy Coordinators*

Energy Coordinators shall be appointed to represent a specific department or division. The Energy Manager shall establish minimum qualification standards for Coordinators, and shall have joint approval authority for each Coordinator appointed.

Coordinators shall be responsible for maintaining an ongoing awareness of energy consumption and expenditures in their assigned areas. They shall recommend and implement energy conservation projects and energy management practices.

Coordinators shall provide necessary information for reporting from their specific areas.

They may be assigned on a full-time or part-time basis; as required to implement programs in their areas.

B.    Reporting

The energy Coordinator shall keep the Energy Office advised of all efforts to increase energy efficiency in their areas. A summary of energy cost savings shall be submitted each quarter to the Energy Office.

The Energy Manager shall be responsible for consolidating these reports for top management.

C. Training

The Energy Manager shall provide energy training at all levels of the Company.

IV.   Policy Updating

The Energy Manager and the Energy Advisory Committee shall review this policy annually and make recommendations for updating or changes.

## Appendix B

**POLICY STATEMENT**

Acme International Corporation is committed to the efficient, cost effective, and environmentally responsible use of energy throughout its worldwide operations. Acme will promote energy efficiency by implementing cost-effective programs that will maintain or improve the quality of the work environment, optimize service reliability, increase productivity, and enhance the safety of our workplace.

## Appendix C

**Figure 2.2**

Figure 2.4

Figure 2.3

Figure 2.6

Figure 2.5

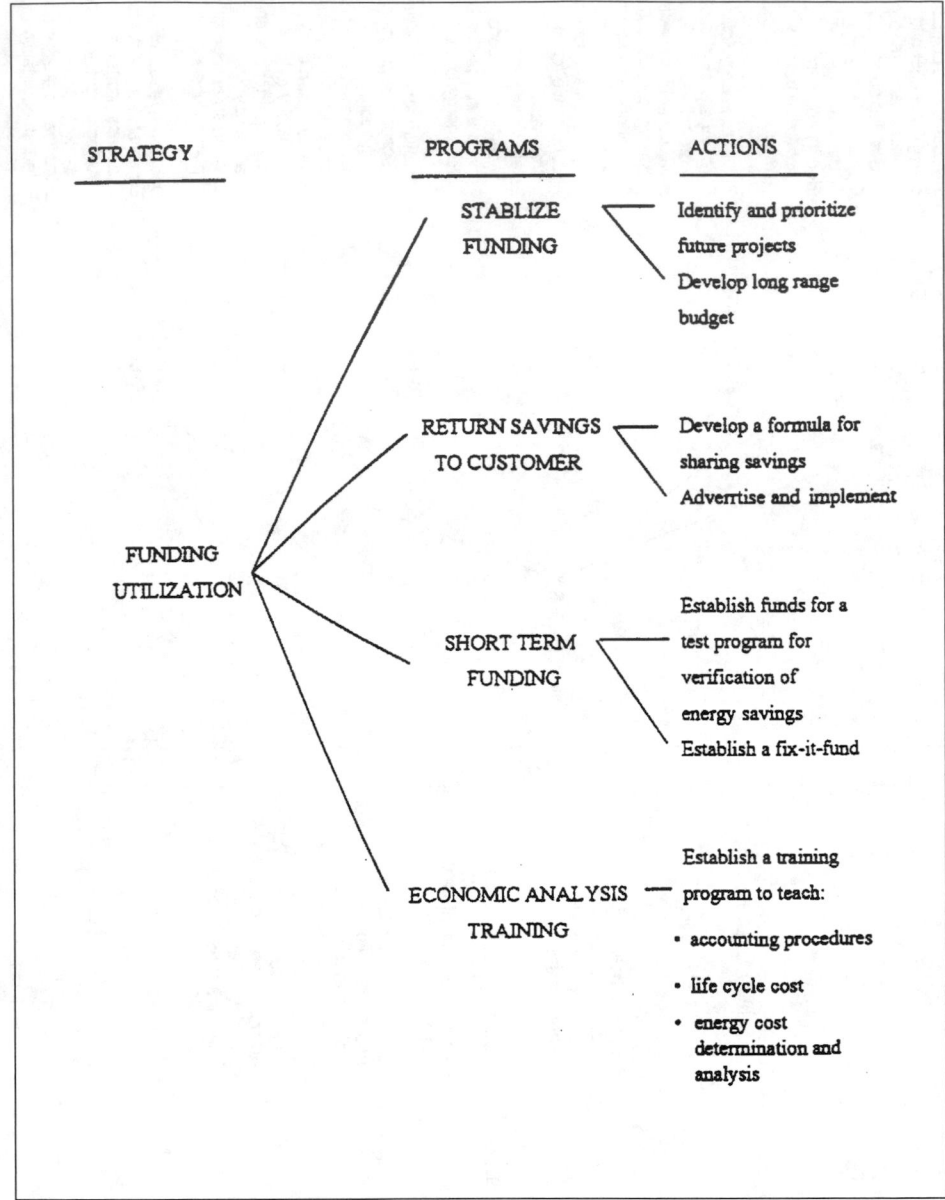

Figure 2.7

# CHAPTER 3

# ENERGY AUDITING

**BARNEY L. CAPEHART AND MARK B. SPILLER**
University of Florida — Gainesville Regional Utilities
Gainesville, FL

**SCOTT FRAZIER**
Oklahoma State University

## 3.1 INTRODUCTION

Saving money on energy bills is attractive to businesses, industries, and individuals alike. Customers whose energy bills use up a large part of their income, and especially those customers whose energy bills represent a substantial fraction of their company's operating costs, have a strong motivation to initiate and continue an on-going energy cost-control program. No-cost or very low-cost operational changes can often save a customer or an industry 10-20% on utility bills; capital cost programs with payback times of two years or less can often save an additional 20-30%. In many cases these energy cost control programs will also result in both reduced energy consumption and reduced emissions of environmental pollutants.

The energy audit is one of the first tasks to be performed in the accomplishment of an effective energy cost control program. An energy audit consists of a detailed examination of how a facility uses energy, what the facility pays for that energy, and finally, a recommended program for changes in operating practices or energy-consuming equipment that will cost-effectively save dollars on energy bills. The energy audit is sometimes called an energy survey or an energy analysis, so that it is not hampered with the negative connotation of an audit in the sense of an IRS audit. The energy audit is a positive experience with significant benefits to the business or individual, and the term "audit" should be avoided if it clearly produces a negative image in the mind of a particular business or individual.

## 3.2 ENERGY AUDITING SERVICES

Energy audits are performed by several different groups. Electric and gas utilities throughout the country offer free residential energy audits. A utility's residential energy auditors analyze the monthly bills, inspect the construction of the dwelling unit, and inspect all of the energy-consuming appliances in a house or an apartment. Ceiling and wall insulation is measured, ducts are inspected, appliances such as heaters, air conditioners, water heaters, refrigerators, and freezers are examined, and the lighting system is checked.

Some utilities also perform audits for their industrial and commercial customers. They have professional engineers on their staff to perform the detailed audits needed by companies with complex process equipment and operations. When utilities offer free or low-cost energy audits for commercial customers, they usually only provide walk-through audits rather than detailed audits. Even so, they generally consider lighting, HVAC systems, water heating, insulation and some motors.

Large commercial or industrial customers may hire an engineering consulting firm to perform a complete energy audit. Other companies may elect to hire an energy manager or set up an energy management team whose job is to conduct periodic audits and to keep up with the available energy efficiency technology.

The U.S. Department of Energy (U.S. DOE) funds a program where universities around the country operate Industrial Assessment Centers which perform free energy audits for small and medium sized manufacturing companies. There are currently 30 IACs funded by the Industrial Division of the U.S. DOE.

The Institutional Conservation Program (ICP) is another energy audit service funded by the U.S. Department of Energy. It is usually administered through state energy offices. This program pays for audits of schools, hospitals, and other institutions, and has some funding assistance for energy conservation improvements.

## 3.3 BASIC COMPONENTS OF AN ENERGY AUDIT

An initial summary of the basic steps involved in conducting a successful energy audit is provided here, and these steps are explained more fully in the sections that follow. This audit description primarily addresses the steps in an industrial or large-scale commercial audit, and not all of the procedures described in this section are required for every type of audit.

The audit process starts by collecting information about a facility's operation and about its past record of utility bills. This data is then analyzed to get a picture of how the facility uses—and possibly wastes—energy, as well as to help the auditor learn what areas to examine to reduce energy costs. Specific changes—called Energy Conservation Opportunities (ECOs)—are identified and evaluated to determine their benefits and their cost-effectiveness. These ECOs are assessed in terms of their costs and benefits, and an economic comparison is made to rank the various ECOs. Finally, an Action Plan is created where certain ECOs are selected for implementation, and the actual process of saving energy and saving money begins.

### 3.3.1 The Auditor's Toolbox

To obtain the best information for a successful energy cost control program, the auditor must make some measurements during the audit visit. The amount of equipment needed depends on the type of energy-consuming equipment used at the facility, and on the range of potential ECOs that might be considered. For example, if waste heat recovery is being considered, then the auditor must take substantial temperature measurement data from potential heat sources. Tools commonly needed for energy audits are listed below:

### Tape Measures

The most basic measuring device needed is the tape measure. A 25-foot tape measure l" wide and a 100-foot tape measure are used to check the dimensions of walls, ceilings, windows and distances between pieces of equipment for purposes such as determining the length of a pipe for transferring waste heat from one piece of equipment to the other.

### Lightmeter

One simple and useful instrument is the lightmeter which is used to measure illumination levels in facilities. A lightmeter that reads in footcandles allows direct analysis of lighting systems and comparison with recommended light levels specified by the Illuminating Engineering Society. A small lightmeter that is portable and can fit into a pocket is the most useful. Many areas in buildings and plants are still significantly over-lighted, and measuring this excess illumination then allows the auditor to recommend a reduction in lighting levels through lamp removal programs or by replacing inefficient lamps with high efficiency lamps that may not supply the same amount of illumination as the old inefficient lamps.

### Thermometers

Several thermometers are generally needed to measure temperatures in offices and other worker areas, and to measure the temperature of operating equipment. Knowing process temperatures allows the auditor to determine process equipment efficiencies, and also to identify waste heat sources for potential heat recovery programs. Inexpensive electronic thermometers with interchangeable probes are now available to measure temperatures in both these areas. Some common types include an immersion probe, a surface temperature probe, and a radiation shielded probe for measuring true air temperature. Other types of infra-red thermometers and thermographic equipment are also available. An infra-red "gun" is valuable for measuring temperatures of steam lines that are not readily reached without a ladder.

### Infrared Cameras

Infrared cameras are still rather expensive pieces of equipment. An investment of at least $25,000 is needed to have a quality infrared camera. However, these are very versatile pieces of equipment and can be used to find overheated electrical wires, connections, neutrals, circuit breakers, transformers, motors and other pieces of electrical equipment. They can also be used to find wet insulation, missing insulation, roof leaks, and cold spots. Thus, infrared cameras are excellent tools for both safety related diagnostics and energy savings diagnostics. A good rule of thumb is that if one safety hazard is found during an infrared scan of a facility, then that has paid for the cost of the scan for the entire facility. Many insurers require infrared scans of buildings for facilities once a year.

### Voltmeter

An inexpensive voltmeter is useful for determining operating voltages on electrical equipment, and especially useful when the nameplate has worn off of a piece of equipment or is otherwise unreadable or missing. The most versatile instrument is a combined volt-ohm-ammeter with a clamp-on feature for measuring currents in conductors that are easily accessible. This type of multimeter is convenient and relatively inexpensive. Any newly purchased voltmeter, or multimeter, should be a true RMS meter for greatest accuracy where harmonics might be involved.

### Clamp On Ammeter

These are very useful instruments for measuring current in a wire without having to make any live electrical connections. The clamp is opened up and put around one insulated conductor, and the meter reads the current in that conductor. New clamp on ammeters can be purchased rather inexpensively that read true RMS values. This is important because of the level of harmonics in many of our facilities. An idea of the level of harmonics in a load can be estimated from using an old non-RMS ammeter, and then a true RMS ammeter to measure the current. If there is more than a five to ten percent difference between the two readings, there is a significant harmonic content to that load.

## Wattmeter/Power Factor Meter

A portable hand-held wattmeter and power factor meter is very handy for determining the power consumption and power factor of individual motors and other inductive devices. This meter typically has a clamp-on feature which allows an easy connection to the current-carrying conductor, and has probes for voltage connections. Any newly purchased wattmeter or power factor meter, should be a true RMS meter for greatest accuracy where harmonics might be involved

## Combustion Analyzer

Combustion analyzers are portable devices capable of estimating the combustion efficiency of furnaces, boilers, or other fossil fuel burning machines. Two types are available: digital analyzers and manual combustion analysis kits. Digital combustion analysis equipment performs the measurements and reads out in percent combustion efficiency. These instruments are fairly complex and expensive.

The manual combustion analysis kits typically require multiple measurements including exhaust stack temperature, oxygen content, and carbon dioxide content. The efficiency of the combustion process can be calculated after determining these parameters. The manual process is lengthy and is frequently subject to human error.

## Airflow Measurement Devices

Measuring air flow from heating, air conditioning or ventilating ducts, or from other sources of air flow is one of the energy auditor's tasks. Airflow measurement devices can be used to identify problems with air flows, such as whether the combustion air flow into a gas heater is correct. Typical airflow measuring devices include a velometer, an anemometer, or an airflow hood. See section 3.4.3 for more detail on airflow measurement devices.

## Blower Door Attachment

Building or structure tightness can be measured with a blower door attachment. This device is frequently used in residences and in office buildings to determine the air leakage rate or the number of air changes per hour in the facility. This is often helps determine whether the facility has substantial structural or duct

leaks that need to be found and sealed. See section 3.4.2 for additional information on blower doors.

## Smoke Generator

A simple smoke generator can also be used in residences, offices and other buildings to find air infiltration and leakage around doors, windows, ducts and other structural features. Care must be taken in using this device, since the chemical "smoke" produced may be hazardous, and breathing protection masks may be needed. See section 3.4.1 for additional information on the smoke generation process, and use of smoke generators.

## Safety Equipment

The use of safety equipment is a vital precaution for any energy auditor. A good pair of safety glasses is an absolute necessity for almost any audit visit. Hearing protectors may also be required on audit visits to noisy plants or areas with high horsepower motors driving fans and pumps. Electrical insulated gloves should be used if electrical measurements will be taken, and asbestos gloves should be used for working around boilers and heaters. Breathing masks may also be needed when hazardous fumes are present from processes or materials used. Steel-toe and steel-shank safety shoes may be needed on audits of plants where heavy materials, hot or sharp materials or hazardous materials are being used. (See section 3.3.3 for an additional discussion of safety procedures.)

## Miniature Data Loggers

Miniature—or mini—data loggers have appeared in low cost models in the last five years. These are often devices that can be held in the palm of the hand, and are electronic instruments that record measurements of temperature, relative humidity, light intensity, light on/off, and motor on/off. If they have an external sensor input jack, these little boxes are actually general purpose data loggers. With external sensors they can record measurements of current, voltage, apparent power (kVA), pressure, and $CO_2$.

These data loggers have a microcomputer control chip and a memory chip, so they can be initialized and then can record data for periods of time from days to weeks. They can record data on a 24 hour a day basis, without any attention or intervention on the part of the energy auditor. Most of these data loggers interface with a digital computer PC, and can transfer data into a spreadsheet of the user's choice, or can use the software provided by the suppliers of the loggers.

Collecting audit data with these small data loggers

gives a more complete and accurate picture of an energy system's overall performance because some conditions may change over long periods of time, or when no one is present.

## Vibration Analysis Gear

Relatively new in the energy manager's tool box is vibration analysis equipment. The correlation between machine condition (bearings, pulley alignment, etc.) and energy consumption is related and this equipment monitors such machine health. This equipment comes in various levels of sophistication and price. At the lower end of the spectrum are vibration pens (or probes) that simply give real-time amplitude readings of vibrating equipment in in/sec or mm/sec. This type of equipment can cost under $1,000. The engineer compares the measured vibration amplitude to a list of vibration levels (ISO2372) and is able to determine if the vibration is excessive for that particular piece of equipment.

The more typical type of vibration equipment will measure and log the vibration into a database (on-board and downloadable). In addition to simply measuring vibration amplitude, the machine vibration can be displayed in time or frequency domains. The graphs of vibration in the frequency domain will normally exhibit spikes at certain frequencies. These spikes can be interpreted by a trained individual to determine the relative health of the machine monitored.

The more sophisticated machines are capable of trend analysis so that facility equipment can be monitored on a schedule and changes in vibration (amplitudes and frequencies) can be noted. Such trending can be used to schedule maintenance based on observations of change. This type of equipment starts at about $3,000 and goes up depending on features desired.

### 3.3.2 Preparing for the Audit Visit

Some preliminary work must be done before the auditor makes the actual energy audit visit to a facility. Data should be collected on the facility's use of energy through examination of utility bills, and some preliminary information should be compiled on the physical description and operation of the facility. This data should then be analyzed so that the auditor can do the most complete job of identifying Energy Conservation Opportunities during the actual site visit to the facility.

## Energy Use Data

The energy auditor should start by collecting data on energy use, power demand and cost for at least the previous 12 months. Twenty-four months of data might be necessary to adequately understand some types of billing methods. Bills for gas, oil, coal, electricity, etc. should be compiled and examined to determine both the amount of energy used and the cost of that energy. This data should then be put into tabular and graphic form to see what kind of patterns or problems appear from the tables or graphs. Any anomaly in the pattern of energy use raises the possibility for some significant energy or cost savings by identifying and controlling that anomalous behavior. Sometimes an anomaly on the graph or in the table reflects an error in billing, but generally the deviation shows that some activity is going on that has not been noticed, or is not completely understood by the customer.

## Rate Structures

To fully understand the cost of energy, the auditor must determine the rate structure under which that energy use is billed. Energy rate structures may go from the extremely simple ones—for example, $1.00 per gallon of Number 2 fuel oil, to very complex ones—for example, electricity consumption which may have a customer charge, energy charge, demand charge, power factor charge, and other miscellaneous charges that vary from month to month. Few customers or businesses really understand the various rate structures that control the cost of the energy they consume. The auditor can help here because the customer must know the basis for the costs in order to control them successfully.

- Electrical Demand Charges: The demand charge is based on a reading of the maximum power in kW that a customer demands in one month. Power is the rate at which energy is used, and it varies quite rapidly for many facilities. Electric utilities average the power reading over intervals from fifteen minutes to one hour, so that very short fluctuations do not adversely affect customers. Thus, a customer might be billed for demand for a month based on a maximum value of a fifteen minute integrated average of their power use.

- Ratchet Clauses: Some utilities have a ratchet clause in their rate structure which stipulates that the minimum power demand charge will be the highest demand recorded in the last billing period or some percentage (i.e., typically 70%) of the highest power demand recorded in the last year. The ratchet clause can increase utility charges for facilities during periods of low activity or where power demand is tied to extreme weather.

- Discounts/Penalties: Utilities generally provide

discounts on their energy and power rates for customers who accept power at high voltage and provide transformers on site. They also commonly assess penalties when a customer has a power factor less than 0.9. Inductive loads (e.g., lightly loaded electric motors, old fluorescent lighting ballasts, etc.) reduce the power factor. Improvement can be made by adding capacitance to correct for lagging power factor, and variable capacitor banks are most useful for improving the power factor at the service drop. Capacitance added near the loads can effectively increase the electrical system capacity. Turning off idling or lightly loaded motors can also help.

- Wastewater charges: The energy auditor also frequently looks at water and wastewater use and costs as part of the audit visit. These costs are often related to the energy costs at a facility. Wastewater charges are usually based on some proportion of the metered water use since the solids are difficult to meter. This can needlessly result in substantial increases in the utility bill for processes which do not contribute to the wastewater stream (e.g., makeup water for cooling towers and other evapo-

rative devices, irrigation, etc.). A water meter can be installed at the service main to supply the loads not returning water to the sewer system. This can reduce the charges by up to two-thirds.

Energy bills should be broken down into the components that can be controlled by the facility. These cost components can be listed individually in tables and then plotted. For example, electricity bills should be broken down into power demand costs per kW per month, and energy costs per kWh. The following example illustrates the parts of a rate structure for an industry in Florida.

Example: A company that fabricates metal products gets electricity from its electric utility at the following general service demand rate structure.

Rate structure:

| | | |
|---|---|---|
| Customer cost | = | $21.00 per month |
| Energy cost | = | $0.051 per kWh |
| Demand cost | = | $6.50 per kW per month |
| Taxes | = | Total of 8% |
| Fuel adjustment | = | A variable amount per kWh each month |

The energy use and costs for that company for a year are summarized below:

**Summary of Energy Usage and Costs**

| Month | kWh Used (kWh) | kWh Cost ($) | Demand (kW) | Demand Cost ($) | Total Cost ($) |
|---|---|---|---|---|---|
| Mar | 44960 | 1581.35 | 213 | 1495.26 | 3076.61 |
| Apr | 47920 | 1859.68 | 213 | 1495.26 | 3354.94 |
| May | 56000 | 2318.11 | 231 | 1621.62 | 3939.73 |
| Jun | 56320 | 2423.28 | 222 | 1558.44 | 3981.72 |
| Jul | 45120 | 1908.16 | 222 | 1558.44 | 3466.60 |
| Aug | 54240 | 2410.49 | 231 | 1621.62 | 4032.11 |
| Sept | 50720 | 2260.88 | 222 | 1558.44 | 3819.32 |
| Oct | 52080 | 2312.19 | 231 | 1621.62 | 3933.81 |
| Nov | 44480 | 1954.01 | 213 | 1495.26 | 3449.27 |
| Dec | 38640 | 1715.60 | 213 | 1495.26 | 3210.86 |
| Jan | 36000 | 1591.01 | 204 | 1432.08 | 3023.09 |
| Feb | 42880 | 1908.37 | 204 | 1432.08 | 3340.45 |
| Totals | 569,360 | 24,243.13 | 2,619 | 18,385.38 | 42,628.51 |
| Monthly Averages | 47,447 | 2,020.26 | 218 | 1,532.12 | 3,552.38 |

The auditor must be sure to account for all the taxes, the fuel adjustment costs, the fixed charges, and any other costs so that the true cost of the controllable energy cost components can be determined. In the electric rate structure described above, the quoted costs for a kW of demand and a kWh of energy are not complete until all these additional costs are added. Although the rate structure says that there is a basic charge of $6.50 per kW per month, the actual cost including all taxes is $7.02 per kW per month. The average cost per kWh is most easily obtained by taking the data for the twelve month period and calculating the cost over this period of time. Using the numbers from the table, one can see that this company has an average energy cost of $0.075 per kWh.

These data are used initially to analyze potential ECOs and will ultimately influence which ECOs are recommended. For example, an ECO that reduces peak demand during a month would save $7.02 per kW per month. Therefore, the auditor should consider ECOs that would involve using certain equipment during the night shift when the peak load is significantly less than the first shift peak load. ECOs that save both energy and demand on the first shift would save costs at a rate of $0.075 per kWh. Finally, ECOs that save electrical energy during the off-peak shift should be examined too, but they may not be as advantageous; they would only save at the rate of $0.043 per kWh because they are already used off-peak and there would not be any additional demand cost savings.

**Physical and Operational Data for the Facility**

The auditor must gather information on factors likely to affect energy use in the facility. Geographic location, weather data, facility layout and construction, operating hours, and equipment can all influence energy use.

- **Geographic Location/Weather Data:** The geographic location of the facility should be noted, together with the weather data for that location. Contact the local weather station, the local utility or the state energy office to obtain the average degree days for heating and cooling for that location for the past twelve months. This degree-day data will be very useful in analyzing the need for energy for heating or cooling the facility. Bin weather data would also be useful if a thermal envelope simulation of the facility were going to be performed as part of the audit.

- **Facility Layout:** Next the facility layout or plan should be obtained, and reviewed to determine the facility size, floor plan, and construction features such as wall and roof material and insulation levels, as well as door and window sizes and construction. A set of building plans could supply this information in sufficient detail. It is important to make sure the plans reflect the "as-built" features of the facility, since many original building plans do not get updated after building alterations.

- **Operating Hours:** Operating hours for the facility should also be obtained. Is there only a single shift? Are there two shifts? Three? Knowing the operating hours in advance allows some determination as to whether some loads could be shifted to off-peak times. Adding a second shift can often be cost effective from an energy cost view, since the demand charge can then be spread over a greater amount of kWh.

- **Equipment List:** Finally, the auditor should get an equipment list for the facility and review it before conducting the audit. All large pieces of energy-consuming equipment such as heaters, air conditioners, water heaters, and specific process-related equipment should be identified. This list, together with data on operational uses of the equipment allows a good understanding of the major energy-consuming tasks or equipment at the facility. As a general rule, the largest energy and cost activities should be examined first to see what savings could be achieved. The greatest effort should be devoted to the ECOs which show the greatest savings, and the least effort to those with the smallest savings potential.

The equipment found at an audit location will depend greatly on the type of facility involved. Residential audits for single-family dwellings generally involve smaller-sized lighting, heating, air conditioning and refrigeration systems. Commercial operations such as grocery stores, office buildings and shopping centers usually have equipment similar to residences, but much larger in size and in energy use. However, large residential structures such as apartment buildings have heating, air conditioning and lighting that is very similar to many commercial facilities. Business operations is the area where commercial audits begin to involve equipment substantially different from that found in residences.

Industrial auditors encounter the most complex equipment. Commercial-scale lighting, heating, air conditioning and refrigeration, as well as office business equipment, is generally used at most industrial facilities. The major difference is in the highly specialized equipment used for the industrial production processes. This can include equipment for chemical mixing and blending, metal plating and treatment, welding, plastic injection molding, paper making and printing, metal refining, electronic assembly, and making glass, for example.

### 3.3.3 Safety Considerations

Safety is a critical part of any energy audit. The audit person or team should be thoroughly briefed on safety equipment and procedures, and should never place themselves in a position where they could injure themselves or other people at the facility. Adequate safety equipment should be worn at all appropriate times. Auditors should be extremely careful making any measurements on electrical systems, or on high temperature devices such as boilers, heaters, cookers, etc. Electrical gloves or asbestos gloves should be worn as appropriate.

The auditor should be careful when examining any operating piece of equipment, especially those with open drive shafts, belts or gears, or any form of rotating machinery. The equipment operator or supervisor should be notified that the auditor is going to look at that piece of equipment and might need to get information from some part of the device. If necessary, the auditor may need to come back when the machine or device is idle in order to safely get the data. The auditor should never approach a piece of equipment and inspect it without the operator or supervisor being notified first.

### Safety Checklist

1. Electrical:
   a. Avoid working on live circuits, if possible.
   b. Securely lock off circuits and switches before working on a piece of equipment.
   c. Always keep one hand in your pocket while making measurements on live circuits to help prevent cardiac arrest.
2. Respiratory:
   a. When necessary, wear a full face respirator mask with adequate filtration particle size.
   b. Use activated carbon cartridges in the mask when working around low concentrations of noxious gases. Change the cartridges on a regular basis.
   c. Use a self-contained breathing apparatus for work in toxic environments.
3. Hearing:
   a. Use foam insert plugs while working around loud machinery to reduce sound levels up to 30 decibels.

### 3.3.4 Conducting the Audit Visit

Once the information on energy bills, facility equipment and facility operation has been obtained, the audit equipment can be gathered up, and the actual visit to the facility can be made.

### Introductory Meeting

The audit person—or team—should meet with the facility manager and the maintenance supervisor and briefly discuss the purpose of the audit and indicate the kind of information that is to be obtained during the visit to the facility. If possible, a facility employee who is in a position to authorize expenditures or make operating policy decisions should also be at this initial meeting.

### Audit Interviews

Getting the correct information on facility equipment and operation is important if the audit is going to be most successful in identifying ways to save money on energy bills. The company philosophy towards investments, the impetus behind requesting the audit, and the expectations from the audit can be determined by interviewing the general manager, chief operating officer, or other executives. The facility manager or plant manager is one person that should have access to much of the operational data on the facility, and a file of data on facility equipment. The finance officer can provide any necessary financial records (e.g., utility bills for electric, gas, oil, other fuels, water and wastewater, expenditures for maintenance and repair, etc.).

The auditor must also interview the floor supervisors and equipment operators to understand the building and process problems. Line or area supervisors usually have the best information on times their equipment is used. The maintenance supervisor is often the primary person to talk to about types of lighting and lamps, sizes of motors, sizes of air conditioners and space heaters, and electrical loads of specialized process equipment. Finally, the maintenance staff must be interviewed to find the equipment and performance problems.

The auditor should write down these people's names, job functions and telephone numbers, since it is frequently necessary to get additional information after the initial audit visit.

### Walk-through Tour

A walk-through tour of the facility or plant tour should be conducted by the facility/plant manager, and should be arranged so the auditor or audit team can see the major operational and equipment features of the facility. The main purpose of the walk-through tour is to obtain general information. More specific information should be obtained from the maintenance and operational people after the tour.

### Getting Detailed Data

Following the facility or plant tour, the auditor or audit team should acquire the detailed data on facility equipment and operation that will lead to identifying the significant Energy Conservation Opportunities (ECOs) that may be appropriate for this facility. This includes data on lighting, HVAC equipment, motors, water heating, and specialized equipment such as refrigerators, ovens, mixers, boilers, heaters, etc. This data is most easily recorded on individualized data sheets that have been prepared in advance.

### What to Look for

- **Lighting:** Making a detailed inventory of all lighting is important. Data should be recorded on numbers of each type of light fixtures and lamps, wattages of lamps, and hours of operation of groups of lights. A lighting inventory data sheet should be used to record this data. Using a lightmeter, the auditor should also record light intensity readings for each area. Taking notes on types of tasks performed in each area will help the auditor select alternative lighting technologies that might be more energy efficient. Other items to note are the areas that may be infrequently used and may be candidates for occupancy sensor controls of lighting, or areas where daylighting may be feasible.

- **HVAC Equipment:** All heating, air conditioning and ventilating equipment should be inventoried. Prepared data sheets can be used to record type, size, model numbers, age, electrical specifications or fuel use specifications, and estimated hours of operation. The equipment should be inspected to determine the condition of the evaporator and condenser coils, the air filters, and the insulation on the refrigerant lines. Air velocity measurement may also be made and recorded to assess operating efficiencies or to discover conditioned air leaks. This data will allow later analysis to examine alternative equipment and operations that would reduce energy costs for heating, ventilating, and air conditioning.

- **Electric Motors:** An inventory of all electric motors over 1 horsepower should also be taken. Prepared data sheets can be used to record motor size, use, age, model number, estimated hours of operation, other electrical characteristics, and possibly the operating power factor. Measurement of voltages, currents, and power factors may be appropriate for some motors. Notes should be taken on the use of motors, particularly recording those that are infrequently used and might be candidates for peak load control or shifting use to off-peak times. All motors over 1 hp and with times of use of 2000 hours per year or greater, are likely candidates for replacement by high efficiency motors—at least when they fail and must be replaced.

- **Water Heaters:** All water heaters should be examined, and data recorded on their type, size, age, model number, electrical characteristics or fuel use. What the hot water is used for, how much is used, and what time it is used should all be noted. Temperature of the hot water should be measured.

- **Waste Heat Sources:** Most facilities have many sources of waste heat, providing possible opportunities for waste heat recovery to be used as the substantial or total source of needed hot water. Waste heat sources are air conditioners, air compressors, heaters and boilers, process cooling systems, ovens, furnaces, cookers, and many others. Temperature measurements for these waste heat sources are necessary to analyze them for replacing the operation of the existing water heaters.

- **Peak Equipment Loads:** The auditor should particularly look for any piece of electrically powered equipment that is used infrequently or whose use could be controlled and shifted to off-peak times. Examples of infrequently used equipment include trash compactors, fire sprinkler system pumps (testing), certain types of welders, drying ovens, or any type of back-up machine. Some production machines might be able to be scheduled for off-peak. Water heating could be done off-peak if a storage system is available, and off-peak thermal storage can be accomplished for on-peak heating or cooling of buildings. Electrical measurements of voltages, currents, and wattages may be helpful. Any information which leads to a piece of equipment being used off-peak is valuable, and could result in substantial savings on electric bills. The auditor should be especially alert for those infrequent on-peak uses that might help explain anomalies on the energy demand bills.

- **Other Energy-Consuming Equipment:** Finally, an inventory of all other equipment that consumes a substantial amount of energy should be taken. Commercial facilities may have extensive computer and copying equipment, refrigeration and cooling equipment, cooking devices, printing equipment, water heaters, etc. Industrial facilities will have many highly specialized process and production operations and machines. Data on types, sizes, capacities, fuel use, electrical characteristics, age, and operating hours should be recorded for all of this equipment.

## Preliminary Identification of ECOs

As the audit is being conducted, the auditor should take notes on potential ECOs that are evident. Identifying ECOs requires a good knowledge of the available energy efficiency technologies that can accomplish the same job with less energy and less cost. For example, overlighting indicates a potential lamp removal or lamp change ECO, and inefficient lamps indicates a potential lamp technology change. Motors with high use times are potential ECOs for high efficiency replacements. Notes on waste heat sources should indicate what other heating sources they might replace, and how far away they are from the end use point. Identifying any potential ECOs during the walk-through will make it easier later on to analyze the data and to determine the final ECO recommendations.

## 3.3.5 Post-Audit Analysis

Following the audit visit to the facility, the data collected should be examined, organized and reviewed for completeness. Any missing data items should be obtained from the facility personnel or from a re-visit to the facility. The preliminary ECOs identified during the audit visit should now be reviewed, and the actual analysis of the equipment or operational change should be conducted. This involves determining the costs and the benefits of the potential ECO, and making a judgment on the cost-effectiveness of that potential ECO.

Cost-effectiveness involves a judgment decision that is viewed differently by different people and different companies. Often, Simple Payback Period (SPP) is used to measure cost-effectiveness, and most facilities want a SPP of two years or less. The SPP for an ECO is found by taking the initial cost and dividing it by the annual savings. This results in finding a period of time for the savings to repay the initial investment, without using the time value of money. One other common measure of cost-effectiveness is the discounted benefit-cost ratio. In this method, the annual savings are discounted when they occur in future years, and are added together to find the present value of the annual savings over a specified period of time.

The benefit-cost ratio is then calculated by dividing the present value of the savings by the initial cost. A ratio greater than one means that the investment will more than repay itself, even when the discounted future savings are taken into account.

Several ECO examples are given here in order to illustrate the relationship between the audit information obtained and the technology and operational changes recommended to save on energy bills.

## Lighting ECO

First, an ECO technology is selected—such as replacing an existing 400 watt mercury vapor lamp with a 325 watt multi-vapor lamp when it burns out. The cost of the replacement lamp must be determined. Product catalogs can be used to get typical prices for the new lamp—about $10 more than the 400 watt mercury vapor lamp. The new lamp is a direct screw-in replacement, and no change is needed in the fixture or ballast. Labor cost is assumed to be the same to install either lamp. The benefits—or cost savings—must be calculated next. The power savings is 400-325 = 75 watts. If the lamp operates for 4000 hours per year and electric energy costs $0.075/kWh, then the savings is (.075 kW)(4000 hr/year)($0.075/kWh) = $22.50/year. This gives a SPP = $10/$22.50/yr =.4 years, or about 5 months. This would be considered an extremely cost-effective ECO. (For illustration purposes, ballast wattage has been ignored.)

## Motor ECO

A ventilating fan at a fiberglass boat manufacturing company has a standard efficiency 5 hp motor that runs at full load two shifts a day, or 4160 hours per year. When this motor wears out, the company will have an ECO of using a high efficiency motor. A high efficiency 5 hp motor costs around $80 more to purchase than the standard efficiency motor. The standard motor is 83% efficient and the high efficiency model is 88.5% efficient. The cost savings is found by calculating (5 hp)(4160 hr/yr)(.746 kW/hp)[(1/.83) −(1/.885)]($.075/kWh) = (1162 kWh)*($0.075) = $87.15/year. The SPP = $80/$87.15/yr =.9 years, or about 11 months. This is also a very attractive ECO when evaluated by this economic measure.

The discounted benefit-cost ratio can be found once a motor life is determined, and a discount rate is selected. Companies generally have a corporate standard for the discount rate used in determining their measures used to make investment decisions. For a 10 year assumed life, and a 10% discount rate, the present worth factor is found as 6.144 (see Chapter 4, Appendix 4A). The benefit-cost ratio is found as B/C = ($87.15)(6.144)/$80 = 6.7. This is an extremely attractive benefit-cost ratio.

**Peak Load Control ECO**

A metals fabrication plant has a large shot-blast cleaner that is used to remove the rust from heavy steel blocks before they are machined and welded. The cleaner shoots out a stream of small metal balls—like shotgun pellets—to clean the metal blocks. A 150 hp motor provides the primary motive force for this cleaner. If turned on during the first shift, this machine requires a total electrical load of about 180 kW which adds directly to the peak load billed by the electric utility. At $7.02/kW/month, this costs (180 kW)*($7.02/kW/month) = $1263.60/month. Discussions with line operating people resulted in the information that the need for the metal blocks was known well in advance, and that the cleaning could easily be done on the evening shift before the blocks were needed. Based on this information, the recommended ECO is to restrict the shot-blast cleaner use to the evening shift, saving the company $15,163.20 per year. Since there is no cost to implement this ECO, the SPP = O; that is, the payback is immediate.

**3.3.6 The Energy Audit Report**

The next step in the energy audit process is to prepare a report which details the final results and recommendations. The length and detail of this report will vary depending on the type of facility audited. A residential audit may result in a computer printout from the utility. An industrial audit is more likely to have a detailed explanation of the ECOs and benefit-cost analyses. The following discussion covers the more detailed audit reports.

The report should begin with an executive summary that provides the owners/managers of the audited facility with a brief synopsis of the total savings available and the highlights of each ECO. The report should then describe the facility that has been audited, and provide information on the operation of the facility that relates to its energy costs. The energy bills should be presented, with tables and plots showing the costs and consumption. Following the energy cost analysis, the recommended ECOs should be presented, along with the calculations for the costs and benefits, and the cost-effectiveness criterion.

Regardless of the audience for the audit report, it should be written in a clear, concise and easy-to-understand format and style. The executive summary should be tailored to non-technical personnel, and technical jargon should be minimized. A client who understands the report is more likely to implement the recommended ECOs. An outline for a complete energy audit report is shown below.

**Energy Audit Report Format**

Executive Summary
> A brief summary of the recommendations and cost savings

Table of Contents

Introduction
> Purpose of the energy audit
> Need for a continuing energy cost control program

Facility Description
> Product or service, and materials flow
> Size, construction, facility layout, and hours of operation
> Equipment list, with specifications

Energy Bill Analysis
> Utility rate structures
> Tables and graphs of energy consumptions and costs
> Discussion of energy costs and energy bills

Energy Conservation Opportunities
> Listing of potential ECOs
> Cost and savings analysis
> Economic evaluation

Action Plan
> Recommended ECOs and an implementation schedule
> Designation of an energy monitor and ongoing program

Conclusion
> Additional comments not otherwise covered

**3.3.7 The Energy Action Plan**

The last step in the energy audit process is to recommend an action plan for the facility. Some companies will have an energy audit conducted by their electric utility or by an independent consulting firm, and will then make changes to reduce their energy bills. They may not spend any further effort in the energy cost control area until several years in the future when another energy audit is conducted. In contrast to this is the company which establishes a permanent energy cost control program, and assigns one person—or a team of people—to continually monitor and improve the energy efficiency and energy productivity of the company. Similar to a Total Quality Management program where a company seeks to continually improve the quality of its products, services and operation, an energy cost control program seeks continual improvement in the amount of product produced for a given expenditure for energy.

The energy action plan lists the ECOs which should be implemented first, and suggests an overall implementation schedule. Often, one or more of the recommended ECOs provides an immediate or very short payback period, so savings from that ECO—or those ECOs can be used to generate capital to pay for implementing the other ECOs. In addition, the action plan also suggests that a company designate one person as the energy monitor for the facility. This person can look at the monthly energy bills and see whether any unusual costs are occurring, and can verify that the energy savings from ECOs is really being seen. Finally, this person can continue to look for other ways the company can save on energy costs, and can be seen as evidence that the company is interested in a future program of energy cost control.

## 3.4 SPECIALIZED AUDIT TOOLS

### 3.4.1 Smoke Sources

Smoke is useful in determining airflow characteristics in buildings, air distribution systems, exhaust hoods and systems, cooling towers, and air intakes. There are several ways to produce smoke. Ideally, the smoke should be neutrally buoyant with the air mass around it so that no motion will be detected unless a force is applied. Cigarette and incense stick smoke, although inexpensive, do not meet this requirement.

Smoke generators using titanium tetrachloride ($TiCl_4$) provide an inexpensive and convenient way to produce and apply smoke. The smoke is a combination of hydrochloric acid (HCl) fumes and titanium oxides produced by the reaction of $TiCl_4$ and atmospheric water vapor. This smoke is both corrosive and toxic so the use of a respirator mask utilizing activated carbon is strongly recommended. Commercial units typically use either glass or plastic cases. Glass has excellent longevity but is subject to breakage since smoke generators are often used in difficult-to-reach areas. Most types of plastic containers will quickly degrade from the action of hydrochloric acid.

Small Teflon® squeeze bottles (i.e., 30 ml) with attached caps designed for laboratory reagent use resist degradation and are easy to use. The bottle should be stuffed with 2-3 real cotton balls then filled with about 0.15 fluid ounces of liquid $TiCl_4$. Synthetic cotton balls typically disintegrate if used with titanium tetrachloride.

This bottle should yield over a year of service with regular use. The neck will clog with debris but can be cleaned with a paper clip.

Some smoke generators are designed for short time use. These bottles are inexpensive and useful for a day of smoke generation, but will quickly degrade. Smoke bombs are incendiary devices designed to emit a large volume of smoke over a short period of time. The smoke is available in various colors to provide good visibility. These are useful in determining airflow capabilities of exhaust air systems and large-scale ventilation systems. A crude smoke bomb can be constructed by placing a stick of elemental phosphorus in a metal pan and igniting it. A large volume of white smoke will be released. This is an inexpensive way of testing laboratory exhaust hoods since many labs have phosphorus in stock.

More accurate results can be obtained by measuring the chemical composition of the airstream after injecting a known quantity of tracer gas such as sulphur hexafluoride into an area. The efficiency of an exhaust system can be determined by measuring the rate of tracer gas removal. Building infiltration/exfiltration rates can also be estimated with tracer gas.

### 3.4.2 Blower Door

The blower door is a device containing a fan, controller, several pressure gauges, and a frame which fits in the doorway of a building. It is used to study the pressurization and leakage rates of a building and its air distribution system under varying pressure conditions. The units currently available are designed for use in residences although they can be used in small commercial buildings as well. The large quantities of ventilation air limit blower door use in large commercial and industrial buildings.

An air leakage/pressure curve can be developed for the building by measuring the fan flow rate necessary to achieve a pressure differential between the building interior and the ambient atmospheric pressure over a range of values. The natural air infiltration rate of the building under the prevailing pressure conditions can be estimated from the leakage/pressure curve and local air pressure data. Measurements made before and after sealing identified leaks can indicate the effectiveness of the work.

The blower door can help to locate the source of air leaks in the building by depressurizing to 30 Pascals and searching potential leakage areas with a smoke source. The air distribution system typically leaks on both the supply and return air sides. If the duct system is located outside the conditioned space (e.g., attic, under floor, etc.), supply leaks will depressurize the building and increase the air infiltration rate; return air leaks will pressurize the building, causing air to exfiltrate. A combination of supply and return air leaks is difficult to detect without sealing off the duct system at the registers and measuring the leakage rate of the building compared to that of the unsealed duct system. The difference between the two conditions is a measure of the leakage attributable to the air distribution system.

### 3.4.3 Airflow Measurement Devices

Two types of anemometers are available for measuring airflow: vane and hot-wire. The volume of air moving through an orifice can be determined by estimating the free area of the opening (e.g., supply air register, exhaust hood face, etc.) and multiplying by the air speed. This result is approximate due to the difficulty in determining the average air speed and the free vent area. Regular calibrations are necessary to assure the accuracy of the instrument. The anemometer can also be used to optimize the face velocity of exhaust hoods by adjusting the door opening until the anemometer indicates the desired airspeed.

Airflow hoods also measure airflow. They contain an airspeed integrating manifold which averages the velocity across the opening and reads out the airflow volume. The hoods are typically made of nylon fabric supported by an aluminum frame. The instrument is lightweight and easy to hold up against an air vent. The lip of the hood must fit snugly around the opening to assure that all the air volume is measured. Both supply and exhaust airflow can be measured. The result must be adjusted if test conditions fall outside the design range.

## 3.5 INDUSTRIAL AUDITS

### 3.5.1 Introduction

Industrial audits are some of the most complex and most interesting audits because of the tremendous variety of equipment found in these facilities. Much of the industrial equipment can be found during commercial audits too. Large chillers, boilers, ventilating fans, water heaters, coolers and freezers, and extensive lighting systems are often the same in most industrial operations as those found in large office buildings or shopping centers. Small cogeneration systems are often found in both commercial and industrial facilities.

The highly specialized equipment that is used in industrial processes is what differentiates these facilities from large commercial operations. The challenge for the auditor and energy management specialist is to learn how this complex—and often unique—industrial equip-

ment operates, and to come up with improvements to the processes and the equipment that can save energy and money. The sheer scope of the problem is so great that industrial firms often hire specialized consulting engineers to examine their processes and recommend operational and equipment changes that result in greater energy productivity.

### 3.5.2 Audit Services

A few electric and gas utilities are large enough, and well-enough staffed, that they can offer industrial audits to their customers. These utilities have a trained staff of engineers and process specialists with extensive experience who can recommend operational changes or new equipment to reduce the energy costs in a particular production environment. Many gas and electric utilities, even if they do not offer audits, do offer financial incentives for facilities to install high efficiency lighting, motors, chillers, and other equipment. These incentives can make many ECOs very attractive.

Small and medium-sized industries that fall into the Manufacturing Sector—SIC 2000 to 3999, and are in the service area of one of the Industrial Assessment Centers funded by the U.S. Department of Energy, can receive free energy audits throughout this program. There are presently 30 IACs operating primarily in the eastern and mid-western areas of the U.S. These IACs are administered by the U.S. Department of Energy. A search vehicle using "Industrial Assessment Centers" will yield updated locations.

### 3.5.3 Industrial Energy Rate Structures

Except for the smallest industries, facilities will be billed for energy services through a large commercial or industrial rate category. It is important to get this rate structure information for all sources of energy—electricity, gas, oil, coal, steam, etc. Gas, oil and coal are usually billed on a straight cost per unit basis—e.g. $0.90 per gallon of #2 fuel oil. Electricity and steam most often have complex rate structures with components for a fixed customer charge, a demand charge, and an energy charge. Gas, steam, and electric energy are often available with a time of day rate, or an interruptible rate that provides much cheaper energy service with the understanding that the customer may have his supply interrupted (stopped) for periods of several hours at a time. Advance notice of the interruption is almost always given, and the number of times a customer can be interrupted in a given period of time is limited.

### 3.5.4 Process and Technology Data Sources

For the industrial audit, it is critical to get in advance as much information as possible on the specialized process equipment so that study and research can be performed to understand the particular processes being used, and what improvements in operation or technology are available. Data sources are extremely valuable here; auditors should maintain a library of information on processes and technology and should know where to find additional information from research organizations, government facilities, equipment suppliers and other organizations.

## EPRI/GRI

The Electric Power Research Institute (EPRI) and the Gas Research Institute (GRI) are both excellent sources of information on the latest technologies of using electric energy or gas. EPRI has a large number of on-going projects to show the cost-effectiveness of electro-technologies using new processes for heating, drying, cooling, etc. GRI also has a large number of projects underway to help promote the use of new cost-effective gas technologies for heating, drying, cooling, etc. Both of these organizations provide extensive documentation of their processes and technologies; they also have computer data bases to aid customer inquiries.

## U.S. DOE Industrial Division

The U.S. Department of Energy has an Industrial Division that provides a rich source of information on new technologies and new processes. This division funds research into new processes and technologies, and also funds many demonstration projects to help insure that promising improvements get implemented in appropriate industries. The Industrial Division of USDOE also maintains a wide network of contacts with government-related research laboratories such as Oak Ridge National Laboratory, Brookhaven National Laboratory, Lawrence Berkeley National Laboratory, Sandia National Laboratory, and Battelle National Laboratory. These laboratories have many of their own research, development and demonstration programs for improved industrial and commercial technologies.

## State Energy Offices

State energy offices are also good sources of information, as well as good contacts to see what kind of incentive programs might be available in the state. Many states offer programs of free boiler tune-ups, free air conditioning system checks, seminars on energy efficiency for various facilities, and other services. Most state energy offices have well-stocked energy libraries, and are also tied into other state energy research organizations, and to national laboratories and the USDOE.

## Equipment Suppliers

Equipment suppliers provide additional sources for data on energy efficiency improvements to processes. Marketing new, cost-effective processes and technologies provides sales for the companies as well as helping industries to be more productive and more economically competitive. The energy auditor should compare the information from all of the sources described above.

### 3.5.5 Conducting the Audit
### Safety Considerations

Safety is the primary consideration in any industrial audit. The possibility of injury from hot objects, hazardous materials, slippery surfaces, drive belts, and electric shocks is far greater than when conducting residential and commercial audits. Safety glasses, safety shoes, durable clothing and possibly a safety hat and breathing mask might be needed during some audits. Gloves should be worn while making any electrical measurements, and also while making any measurements around boilers, heaters, furnaces, steam lines, or other very hot pieces of equipment. In all cases, adequate attention to personal safety is a significant feature of any industrial audit.

### Lighting

Lighting is not as great a percent of total industrial use as it is in the commercial sector on the average, but lighting is still a big energy use and cost area for many industrial facilities. A complete inventory of all lighting should be taken during the audit visit. Hours of operation of lights are also necessary, since lights are commonly left on when they are not needed. Timers, Energy Management Systems, and occupancy sensors are all valuable approaches to insuring that lights that are not needed are not turned on. It is also important to look at the facility's outside lighting for parking and for storage areas.

During the lighting inventory, types of tasks being performed should also be noted, since light replacement with more efficient lamps often involves changing the color of the resultant light. For example, high pressure sodium lamps are much more efficient than mercury vapor lamps or even metal halide lamps, but they produce a yellowish light that makes fine color distinction difficult. However, many assembly tasks can still be performed adequately under high pressure sodium lighting. These typically include metal fabrication, wood product fabrication, plastic extrusion, and many others.

### Electric Motors

A common characteristic of many industries is their extensive use of electric motors. A complete inventory of all motors over 1 hp should be taken, as well as recording data on how long each motor operates during a day. For motors with substantial usage times, replacement with high-efficiency models is almost always cost effective. In addition, consideration should be given to replacement of standard drive belts with synchronous belts which transmit the motor energy more efficiently. For motors which are used infrequently, it may be possible to shift the use to off-peak times, and to achieve a kW demand reduction which would reduce energy cost.

## HVAC Systems

An inventory of all space heaters and air conditioners should be taken. Btu per hour ratings and efficiencies of all units should be recorded, as well as usage patterns. Although many industries do not heat or air condition the production floor area, they almost always have office areas, cafeterias, and other areas that are normally heated and air conditioned. For these conditioned areas, the construction of the facility should be noted—how much insulation, what are the walls and ceilings made of, how high are the ceilings. Adding additional insulation might be a cost effective ECO.

Production floors that are not air conditioned often have large numbers of ventilating fans that operate anywhere from one shift per day to 24 hours a day. Plants with high heat loads and plants in the mild climate areas often leave these ventilating fans running all year long. These are good candidates for high efficiency motor replacements. Timers or an Energy Management System might be used to turn off these ventilating fans when the plant is shut down.

## Boilers

All boilers should be checked for efficient operation using a stack gas combustion analyzer. Boiler specifications on Btu per hour ratings, pressures and temperatures should be recorded. The boiler should be varied between low-fire, normal-fire, and high-fire, with combustion gas and temperature readings taken at each level. Boiler tune-up is one of the most common, and most energy-saving operations available to many facilities. The auditor should check to see whether any waste heat from the boiler is being recovered for use in a heat recuperator or for some other use such as water heating. If not, this should be noted as a potential ECO.

## Specialized Equipment

Most of the remaining equipment encountered during the industrial audit will be the highly specialized process production equipment and machines. This equipment should all be examined and operational data taken, as well as noting hours and periods of use. All heat sources should be considered carefully as to whether they could be replaced with sources using waste heat, or whether a particular heat source could serve as a provider of waste heat to another application. Operations where both heating and cooling occur periodically—such as a plastic extrusion machine—are good candidates for reclaiming waste heat, or in sharing heat from a machine needing cooling with another machine needing heat.

## Air Compressors

Air compressors should be examined for size, operating pressures, and type (reciprocating or screw), and whether they use outside cool air for intake. Large air compressors are typically operated at night when much smaller units are sufficient. Also, screw-type air compressors use a large fraction of their rated power when they are idling, so control valves should be installed to prevent this loss. Efficiency is improved with intake air that is cool, so outside air should be used in most cases—except in extremely cold temperature areas.

The auditor should determine whether there are significant air leaks in air hoses, fittings, and in machines. Air leaks are a major source of energy loss in many facilities, and should be corrected by maintenance action. Finally, air compressors are a good source of waste heat. Nearly 90% of the energy used by an air compressor shows up as waste heat, so this is a large source of low temperature waste heat for heating input air to a heater or boiler, or for heating hot water for process use.

## 3.6 COMMERCIAL AUDITS

### 3.6.1 Introduction

Commercial audits span the range from very simple audits for small offices to very complex audits for multi-story office buildings or large shopping centers. Complex commercial audits are performed in substantially the same manner as industrial audits. The following discussion highlights those areas where commercial audits are likely to differ from industrial audits.

Commercial audits generally involve substantial consideration of the structural envelope features of the facility, as well as significant amounts of large or specialized equipment at the facility. Office buildings, shopping centers and malls all have complex building envelopes that should be examined and evaluated. Building materials, insulation levels, door and window construction, skylights, and many other envelope features must be considered in order to identify candidate ECOs.

Commercial facilities also have large capacity equipment, such as chillers, space heaters, water heaters, refrigerators, heaters, cookers, and office equipment such as computers and copy machines. Small cogeneration systems are also commonly found in commercial facilities and institutions such as schools and hospitals. Much of the equipment in commercial facilities is the same type and size as that found in manufacturing or industrial facilities. Potential ECOs would look at more efficient equipment, use of waste heat, or operational changes to use less expensive energy.

### 3.6.2 Commercial Audit Services

Electric and gas utilities, as well as many engineering consulting firms, perform audits for commercial facilities. Some utilities offer free walk-through audits for commercial customers, and also offer financial incentives for customers who change to more energy efficient equipment. Schools, hospitals and some other government institutions can qualify for free audits under the ICP program described in the first part of this chapter. Whoever conducts the commercial audit must initiate the ICP process by collecting information on the rate energy rate structures, the equipment in use at the facility, and the operational procedures used there.

### 3.6.3 Commercial Energy Rate Structures

Small commercial customers are usually billed for energy on a per energy unit basis, while large commercial customers are billed under complex rate structures containing components related to energy, rate of energy use (power), time of day or season of year, power factor, and numerous other elements. One of the first steps in a commercial audit is to obtain the rate structures for all sources of energy, and to analyze at least one to two year's worth of energy bills. This information should be put into a table and also plotted.

### 3.6.4 Conducting the Audit

A significant difference in industrial and commercial audits arises in the area of lighting. Lighting in commercial facilities is one of the largest energy costs—sometimes accounting for half or more of the entire electric bill. Lighting levels and lighting quality are extremely important to many commercial operations. Retail sales operations, in particular, want light levels that are far in excess of standard office values. Quality of light in terms of color is also a big concern in retail sales, so finding acceptable ECOs for reducing lighting costs is much more difficult for retail facilities than for office buildings. The challenge is to find new lighting technologies that allow high light levels and warm color while reducing the wattage required. New T8 and T10 fluorescent lamps, and metal halide lamp replacements for mercury vapor lamps offer these features, and usually represent cost-effective ECOs for retail sales and other facilities.

## 3.7 RESIDENTIAL AUDITS

Audits for large, multi-story apartment buildings can be very similar to commercial audits. (See section 3.6.) Audits of single-family residences, however, are generally fairly simple. For single-family structures, the energy audit focuses on the thermal envelope and the appliances such as the heater, air conditioner, water heater, and "plug loads."

The residential auditor should start by obtaining past energy bills and analyzing them to determine any patterns or anomalies. During the audit visit, the structure is examined to determine the levels of insulation, the conditions of and seals for windows and doors, and the integrity of the ducts. The space heater and/or air conditioner is inspected, along with the water heater. Equipment model numbers, age, size, and efficiencies are recorded. The post-audit analysis then evaluates potential ECOs such as adding insulation, adding double-pane windows, window shading or insulated doors, and changing to higher efficiency heaters, air conditioners, and water heaters. The auditor calculates costs, benefits, and Simple Payback Periods and presents them to the owner or occupant. A simple audit report—often in the form of a computer printout is given to the owner or occupant.

## 3.8 INDOOR AIR QUALITY

### 3.8.1 Introduction

Implementation of new energy-related standards and practices has contributed to a degradation of indoor air quality. In fact, the quality of indoor air has been found to exceed the Environmental Protection Agency (EPA) standards for outdoor air in many homes, businesses, and factories. Thus, testing for air quality problems is done in some energy audits both to prevent exacerbating any existing problems and to recommend ECOs that might improve air quality. Air quality standards for the industrial environment have been published by the American Council of Governmental Industrial Hygienists (ACGIH) in their booklet "Threshold Limit Values."

No such standards currently exist for the residential and commercial environments although the ACGIH standards are typically and perhaps inappropriately used. The EPA has been working to develop residential and commercial standards for quite some time.

### 3.8.2 Symptoms of Air Quality Problems

Symptoms of poor indoor air quality include, but are not limited to: headaches; irritation of mucous membranes such as the nose, mouth, throat, lungs; tearing, redness and irritation of the eyes; numbness of the lips, mouth, throat; mood swings; fatigue; allergies; coughing; nasal and throat discharge; and irritability. Chronic exposure to some compounds can lead to damage to internal organs such as the liver, kidney, lungs, and brain; cancer; and death.

### 3.8.3 Testing

Testing is required to determine if the air quality is acceptable. Many dangerous compounds, like carbon monoxide and methane without odorant added, are odorless and colorless. Some dangerous particulates such as asbestos fibers do not give any indication of a problem for up to twenty years after inhalation. Testing must be conducted in conjunction with pollution-producing processes to ensure capture of the contaminants. Testing is usually performed by a Certified Industrial Hygienist (CIH).

### 3.8.4 Types of Pollutants

Airstreams have three types of contaminants: particulates like dust and asbestos; gases like carbon monoxide, ozone, carbon dioxide, volatile organic compounds, anhydrous ammonia, radon, outgassing from urea-formaldehyde insulation, low oxygen levels; and biologicals like mold, mildew, fungus, bacteria, and viruses.

### 3.8.5 Pollutant Control Measures

#### Particulates

Particulates are controlled with adequate filtration near the source and in the air handling system. Mechanical filters are frequently used in return air streams, and baghouses are used for particulate capture. The coarse filters used in most residential air conditioners typically have filtration efficiencies below twenty percent. Mechanical filters called high efficiency particulate apparatus (HEPA) are capable of filtering particles as small as 0.3 microns at up to 99% efficiency. Electrostatic precipitators remove particulates by placing a positive charge on the walls of collection plates and allowing negatively charged particulates to attach to the surface. Periodic cleaning of

the plates is necessary to maintain high filtration efficiency. Loose or friable asbestos fibers should be removed from the building or permanently encapsulated to prevent entry into the respirable airstream. While conducting an audit, it is important to determine exactly what type of insulation is in use before disturbing an area to make temperature measurements.

#### Problem Gases

Problem gases are typically removed by ventilating with outside air. Dilution with outside air is effective, but tempering the temperature and relative humidity of the outdoor air mass can be expensive in extreme conditions. Heat exchangers such as heat wheels, heat pipes, or other devices can accomplish this task with reduced energy use. Many gases can be removed from the airstream by using absorbent/adsorbent media such as activated carbon or zeolite. This strategy works well for spaces with limited ventilation or where contaminants are present in low concentrations. The media must be checked and periodically replaced to maintain effectiveness.

Radon gas—Ra 222—cannot be effectively filtered due to its short half life and the tendency for its Polonium daughters to plate out on surfaces. Low oxygen levels are a sign of inadequate outside ventilation air. A high level of carbon dioxide (e.g., 1000-10,000 ppm) is not a problem in itself but levels above 1000 ppm indicate concentrated human or combustion activity or a lack of ventilation air. Carbon dioxide is useful as an indicator compound because it is easy and inexpensive to measure.

#### Microbiological Contaminants

Microbiological contaminants generally require particular conditions of temperature and relative humidity on a suitable substrate to grow. Mold and mildew are inhibited by relative humidity levels less than 50%. Air distribution systems often harbor colonies of microbial growth. Many people are allergic to microscopic dust mites. Cooling towers without properly adjusted automated chemical feed systems are an excellent breeding ground for all types of microbial growth.

#### Ventilation Rates

Recommended ventilation quantities are published by the American Society of Heating, Refrigerating, and Air-Conditioning Engineers (ASHRAE) in Standard 62, "Ventilation for Acceptable Air Quality." These ventilation rates are for effective systems. Many existing systems fail in entraining the air mass efficiently. The density of the contaminants relative to air must be considered in locating the exhaust air intakes and ventilation supply air registers.

## Liability

Liability related to indoor air problems appears to be a growing but uncertain issue because few cases have made it through the court system. However, in retrospect, the asbestos and ureaformaldehyde pollution problems discovered in the last two decades suggest proceeding with caution and a proactive approach.

## 3.9 CONCLUSION

Energy audits are an important first step in the overall process of reducing energy costs for any building, company, or industry. A thorough audit identifies and analyzes the changes in equipment and operations that will result in cost-effective energy cost reduction. The energy auditor plays a key role in the successful conduct of an audit, and also in the implementation of the audit recommendations.

## Bibliography

*Instructions For Energy Auditors, Volumes I and II*, U.S. Department of Energy, DOE/CS-0041/12&13, September, 1978. Available through National Technical Information Service, Springfield, VA.

*Energy Conservation Guide for Industry and Commerce, National Bureau of Standards Handbook* 115 and Supplement, 1978. Available through U.S. Government Printing Office, Washington, DC.

*Guide to Energy Management, Third Edition*, Capehart, B.L., Turner, W.C., and Kennedy, W.J., The Fairmont Press, Atlanta, GA, 2000.

Illuminating Engineering Society, *IES Lighting Handbook, Ninth Edition*, New York, NY, 2000.

*Total Energy Management*, A Handbook prepared by the National Electrical Contractors Association and the National Electrical Manufacturers Association, Washington, DC.

*Handbook of Energy Audits*, Thumann, Albert, Fifth Edition, The Fairmont Press, Atlanta, GA.

*Industrial Energy Management and Utilization*, Witte, Larry C., Schmidt, Philip S., and Brown, David R., Hemisphere Publishing Corporation, Washington, DC, 1988.

Threshold Limit Values for Chemical Substances and Physical Agents and Biological Exposure Indices, 1990-91 American Conference of Governmental Industrial Hygienists.

*Ventilation for Acceptable Indoor Air Quality*, ASHRAE 62-1999, American Society of Heating, Refrigerating and Air-Conditioning Engineers, Inc., 1999.

Facility Design and Planning Engineering Weather Data, Departments of the Air Force, the Army, and the Navy, 1978.

*Handbook of Energy Engineering, Fourth Edition*, Thumann, A., and Mehta, D.P., The Fairmont Press, Atlanta, GA.

# CHAPTER 4

# ECONOMIC ANALYSIS

**DR. DAVID PRATT**
Industrial Engineering and Management
Oklahoma State University
Stillwater, OK

## 4.1 OBJECTIVE

The objective of this chapter is to present a coherent, consistent approach to economic analysis of capital investments (energy related or other). Adherence to the concepts and methods presented will lead to sound investment decisions with respect to time value of money principles. The chapter opens with material designed to motivate the importance of life cycle cost concepts in the economic analysis of projects. The next three sections provide foundational material necessary to fully develop time value of money concepts and techniques. These sections present general characteristics of capital investments, sources of funds for capital investment, and a brief summary of tax considerations which are important for economic analysis. The next two sections introduce time value of money calculations and several approaches for calculating project measures of worth based on time value of money concepts. Next the measures of worth are applied to the process of making decisions when a set of potential projects are to be evaluated. The final concept and technique section of the chapter presents material to address several special problems that may be encountered in economic analysis. This material includes, among other things, discussions of inflation, non-annual compounding of interest, and sensitivity analysis. The chapter closes with a brief summary and a list of references which can provide additional depth in many of the areas covered in the chapter.

## 4.2 INTRODUCTION

Capital investment decisions arise in many circumstances. The circumstances range from evaluating business opportunities to personal retirement planning. Regardless of circumstances, the basic criterion for evaluating any investment decision is that the revenues (savings) generated by the investment must be greater than the costs incurred. The number of years over which the revenues accumulate and the comparative importance of future dollars (revenues or costs) relative to present dollars are important factors in making sound investment decisions. This consideration of costs over the entire life cycle of the investments gives rise to the name *life cycle cost* analysis which is commonly used to refer to the economic analysis approach presented in this chapter. An example of the importance of life cycle costs is shown in Figure 4.1 which depicts the estimated costs of owning and operating an oil-fired furnace to heat a 2,000-square-foot house in the northeast United States. Of particular note is that the initial costs represent only 23% of the total costs incurred over the life of the furnace. The life cycle cost approach provides a significantly better evaluation of long term implications of an investment than methods which focus on first cost or near term results.

**Figure 4.1 15-Year life cycle costs of a heating system**

Life cycle cost analysis methods can be applied to virtually any public or private business sector investment decision as well as to personal financial planning decisions. Energy related decisions provide excellent examples for the application of this approach. Such decisions include: evaluation of alternative building designs which have different initial costs, operating and maintenance costs, and perhaps different lives; evaluation of investments to improve the thermal performance of an existing building (wall or roof insulation, window glazing); or evaluation of alternative heating, ventilating, or air conditioning systems. For federal buildings, Congress and the President have mandated, through legislation and executive order, energy conservation goals that must be met using cost-effective measures. The life cycle cost approach is mandated as the means of evaluating cost effectiveness.

## 4.3 GENERAL CHARACTERISTICS OF CAPITAL INVESTMENTS

### 4.3.1 Capital Investment Characteristics

When companies spend money, the outlay of cash can be broadly categorized into one of two classifications; expenses or capital investments. Expenses are generally those cash expenditures that are routine, on-going, and necessary for the ordinary operation of the business. Capital investments, on the other hand, are generally more strategic and have long term effects. Decisions made regarding capital investments are usually made at higher levels within the organizational hierarchy and carry with them additional tax consequences as compared to expenses.

Three characteristics of capital investments are of concern when performing life cycle cost analysis. First, capital investments usually require a relatively large initial cost. "Relatively large" may mean several hundred dollars to a small company or many millions of dollars to a large company. The initial cost may occur as a single expenditure such as purchasing a new heating system or occur over a period of several years such as designing and constructing a new building. It is not uncommon that the funds available for capital investments projects are limited. In other words, the sum of the initial costs of all the viable and attractive projects exceeds the total available funds. This creates a situation known as capital rationing which imposes special requirements on the investment analysis. This topic will be discussed in Section 4.8.3.

The second important characteristic of a capital investment is that the benefits (revenues or savings) resulting from the initial cost occur in the future, normally over a period of years. The period between the initial cost and the last future cash flow is the life cycle or life of the investment. It is the fact that cash flows occur over the investment's life that requires the introduction of time value of money concepts to properly evaluate investments. If multiple investments are being evaluated and if the lives of the investments are not equal, special consideration must be given to the issue of selecting an appropriate planning horizon for the analysis. Planning horizon issues are introduced in Section 4.8.5.

The last important characteristic of capital investments is that they are relatively irreversible. Frequently, after the initial investment has been made, terminating or significantly altering the nature of a capital investment has substantial (usually negative) cost consequences. This is one of the reasons that capital investment decisions are usually evaluated at higher levels of the organizational hierarchy than operating expense decisions.

### 4.3.2 Capital Investment Cost Categories

In almost every case, the costs which occur over the life of a capital investment can be classified into one of the following categories:

- Initial Cost,
- Annual Expenses and Revenues,
- Periodic Replacement and Maintenance, or
- Salvage Value.

As a simplifying assumption, the cash flows which occur during a year are generally summed and regarded as a single end-of-year cash flow. While this approach does introduce some inaccuracy in the evaluation, it is generally not regarded as significant relative to the level of estimation associated with projecting future cash flows.

Initial costs include all costs associated with preparing the investment for service. This includes purchase cost as well as installation and preparation costs. Initial costs are usually nonrecurring during the life of an investment. Annual expenses and revenues are the recurring costs and benefits generated throughout the life of the investment. Periodic replacement and maintenance costs are similar to annual expenses and revenues except that they do not (or are not expected to) occur annually. The salvage (or residual) value of an investment is the revenue (or expense) attributed to disposing of the investment at the end of its useful life.

### 4.3.3 Cash Flow Diagrams

A convenient way to display the revenues (savings) and costs associated with an investment is a *cash flow diagram*. By using a cash flow diagram, the timing of the cash flows are more apparent and the chances of properly applying time value of money concepts are increased. With practice, different cash flow patterns can be recognized and they, in turn, may suggest the most direct approach for analysis.

It is usually advantageous to determine the time frame over which the cash flows occur first. This establishes the horizontal scale of the cash flow diagram. This scale is divided into time periods which are frequently, but not always, years. Receipts and disbursements are then located on the time scale in accordance with the problem specifications. Individual outlays or receipts are indicated by drawing vertical lines appropriately placed along the time scale. The relative magnitudes can be suggested by the heights, but exact scaling generally does not enhance the meaningfulness of the diagram. Upward directed lines indicate cash inflow (revenues or savings) while downward directed lines indicate cash outflow (costs).

Figure 4.2 illustrates a cash flow diagram. The cash flows depicted represent an economic evaluation of whether to choose a baseboard heating and window air conditioning system or a heat pump for a ranger's house in a national park [Fuller and Petersen, 1994]. The differential costs associated with the decision are:

- The heat pump costs (cash outflow) $1500 more than the baseboard system,

- The heat pump saves (cash inflow) $380 annually in electricity costs,

- The heat pump has a $50 higher annual maintenance costs (cash outflow),

- The heat pump has a $150 higher salvage value (cash inflow) at the end of 15 years,

- The heat pump requires $200 more in replacement maintenance (cash outflow) at the end of year 8.

Although cash flow diagrams are simply graphical representations of income and outlay, they should exhibit as much information as possible. During the analysis phase, it is useful to show the Minimum Attractive Rate of Return (an interest rate used to account for the time value of money within the problem) on the cash flow diagram, although this has been omitted in Figure 4.2. The requirements for a good cash flow diagram are completeness, accuracy, and legibility. The measure of a successful diagram is that someone else can understand the problem fully from it.

## 4.4  SOURCES OF FUNDS

Capital investing requires a source of funds. For large companies multiple sources may be employed.

The process of obtaining funds for capital investment is called financing. There are two broad sources of financial funding; debt financing and equity financing. Debt financing involves borrowing and utilizing money which is to be repaid at a later point in time. Interest is paid to the lending party for the privilege of using the money. Debt financing does not create an ownership position for the lender within the borrowing organization. The borrower is simply obligated to repay the borrowed funds plus accrued interest according to a repayment schedule. Car loans and mortgage loans are two examples of this type of financing. The two primary sources of debt capital are loans and bonds. The cost of capital associated with debt financing is relatively easy to calculate since interest rates and repayment schedules are usually clearly documented in the legal instruments controlling the financing arrangements. An added benefit to debt financing under current U.S. tax law (as of April 2000) is that the interest payments made by corporations on debt capital are tax deductible. This effectively lowers the cost of debt financing since for debt financing with deductible interest payments, the after-tax cost of capital is given by:

$$\text{Cost of Capital}_{\text{AFTERTAX}} =$$
$$\text{Cost of Capital}_{\text{BEFORETAX}} * (1 - \text{TaxRate})$$

where the tax rate is determine by applicable tax law.

The second broad source of funding is equity financing. Under equity financing the lender acquires an ownership (or equity) position within the borrower's organization. As a result of this ownership position, the lender has the right to participate in the financial success of the organization as a whole. The two primary sources of equity financing are stocks and retained earnings. The cost of capital associated with shares of stock is much debated within the financial community. A detailed presentation of the issues and approaches is beyond the

**Figure 4.2. Heat pump and baseboard system differential life cycle costs**

scope of this chapter. Additional reference material can be found in Park and Sharp-Bette [1990]. One issue over which there is general agreement is that the cost of capital for stocks is higher than the cost of capital for debt financing. This is at least partially attributable to the fact that interest payments are tax deductible while stock dividend payments are not.

If any subject is more widely debated in the financial community than the cost of capital for stocks, it is the cost of capital for retained earnings. Retained earnings are the accumulation of annual earnings surpluses that a company retains within the company's coffers rather than pays out to the stockholders as dividends. Although these earnings are held by the company, they truly belong to the stockholders. In essence the company is establishing the position that by retaining the earnings and investing them in capital projects, stockholders will achieve at least as high a return through future financial successes as they would have earned if the earnings had been paid out as dividends. Hence, one common approach to valuing the cost of capital for retained earnings is to apply the same cost of capital as for stock. This, therefore, leads to the same generally agreed result. The cost of capital for financing through retained earnings generally exceeds the cost of capital for debt financing.

In many cases the financing for a set of capital investments is obtained by packaging a combination of the above sources to achieve a desired level of available funds. When this approach is taken, the overall cost of capital is generally taken to be the weighted average cost of capital across all sources. The cost of each individual source's funds is weighted by the source's fraction of the total dollar amount available. By summing across all sources, a weighted average cost of capital is calculated.

**Example 1**

Determine the weighted average cost of capital for financing which is composed of:

25% loans with a before tax cost of capital of 12%/ yr and

75% retained earnings with a cost of capital of 10%/yr.

The company's effective tax rate is 34%.

Cost of Capital$_{LOANS}$ = 12% * (1 - 0.34) = 7.92%

Cost of Capital$_{RETAINED\ EARNINGS}$ = 10%

Weighted Average Cost of Capital = (0.25)*7.92% + (0.75)*10.00% = 9.48%

## 4.5 TAX CONSIDERATIONS

### 4.5.1 After Tax Cash Flows

Taxes are a fact of life in both personal and business decision making. Taxes occur in many forms and are primarily designed to generate revenues for governmental entities ranging from local authorities to the Federal government. A few of the most common forms of taxes are income taxes, ad valorem taxes, sales taxes, and excise taxes. Cash flows used for economic analysis should always be adjusted for the combined impact of all relevant taxes. To do otherwise, ignores the significant impact that taxes have on economic decision making. Tax laws and regulations are complex and intricate. A detailed treatment of tax considerations as they apply to economic analysis is beyond the scope of this chapter and generally requires the assistance of a professional with specialized training in the subject. A high level summary of concepts and techniques that concentrate on Federal income taxes are presented in the material which follows. The focus is on Federal income taxes since they impact most decisions and have relatively wide and general application.

The amount of Federal taxes due are determined based on a tax rate multiplied by a taxable income. The rates (as of April 2000) are determined based on tables of rates published under the Omnibus Reconciliation Act of 1993 as shown in Table 4.1. Depending on income range, the marginal tax rates vary from 15% of taxable income to 39% of taxable income. *Taxable income* is calculated by subtracting *allowable deductions* from *gross income*. Gross income is generated when a company sells its product or service. Allowable deductions include salaries and wages, materials, interest payments, and depreciation as well as other costs of doing business as detailed in the tax regulations.

The calculation of taxes owed and after tax cash flows (ATCF) requires knowledge of:

- Before Tax Cash Flows (BTCF), the net project cash flows before the consideration of taxes due, loan payments, and bond payments;

- Total loan payments attributable to the project, including a breakdown of principal and interest components of the payments;

- Total bond payments attributable to the project, including a breakdown of the redemption and interest components of the payments; and

- Depreciation allowances attributable to the project.

**Table 4.1 Federal tax rates based on the Omnibus Reconciliation Act of 1993**

| Taxable Income (TI) | Taxes Due | Marginal Tax Rate |
|---|---|---|
| $0 < TI ≤ $50,000 | 0.15*TI | 0.15 |
| $50,000 < TI ≤ $75,000 | $7,500+0.25(TI-$50,000) | 0.25 |
| $75,000 < TI ≤ $100,000 | $13,750+0.34(TI-$75,000) | 0.34 |
| $100,000 < TI ≤ $335,000 | $22,250+0.39(TI-$100,000) | 0.39 |
| $335,000 < TI ≤ $10,000,000 | $113,900+0.34(TI-$335,000) | 0.34 |
| $10,000,000 < TI ≤ $15,000,000 | $3,400,000+0.35(TI-$10,000,000) | 0.35 |
| $15,000,000 < TI ≤ $18,333,333 | $5,150,000+0.38(TI-$15,000,000) | 0.38 |
| $18,333,333 < TI | $6,416,667+0.35(TI-$18,333,333) | 0.35 |

Given the availability of the above information, the procedure to determine the ATCF on a year-by-year basis proceeds using the following calculation for each year:

- Taxable Income = BTCF - Loan Interest - Bond Interest - Deprecation

- Taxes = Taxable Income * Tax Rate

- ATCF = BTCF - Total Loan Payments - Total Bond Payments - Taxes

An important observation is that Depreciation reduces Taxable Income (hence, taxes) but does not directly enter into the calculation of ATCF since it is not a true cash flow. It is not a true cash flow because no cash changes hands. Depreciation is an accounting concept design to stimulate business by reducing taxes over the life of an asset. The next section provides additional information about depreciation.

## 4.5.2 Depreciation

Most assets used in the course of a business decrease in value over time. U.S. Federal income tax law permits reasonable deductions from taxable income to allow for this. These deductions are called depreciation allowances. To be depreciable, an asset must meet three primary conditions: (1) it must be held by the business for the purpose of producing income, (2) it must wear out or be consumed in the course of its use, and (3) it must have a life longer than a year.

Many methods of depreciation have been allowed under U.S. tax law over the years. Among these methods are straight line, sum-of-the-years digits, declining balance, and the accelerated cost recovery system. Descrip-

tions of these methods can be found in many references including economic analysis text books [White, et al., 1998]. The method currently used for depreciation of assets placed in service after 1986 is the Modified Accelerated Cost Recovery System (MACRS). Determination of the allowable MACRS depreciation deduction for an asset is a function of (1) the asset's property class, (2) the asset's basis, and (3) the year within the asset's recovery period for which the deduction is calculated.

Eight property classes are defined for assets which are depreciable under MACRS. The property classes and several examples of property that fall into each class are shown in Table 4.2. Professional tax guidance is recommended to determine the MACRS property class for a specific asset.

The basis of an asset is the cost of placing the asset in service. In most cases, the basis includes the purchase cost of the asset plus the costs necessary to place the asset in service (e.g., installation charges).

Given an asset's property class and its depreciable basis the depreciation allowance for each year of the asset's life can be determined from tabled values of MACRS percentages. The MACRS percentages specify the percentage of an asset's basis that are allowable as deductions during each year of an asset's recovery period. The MACRS percentages by recovery year (age of the asset) and property class are shown in Table 4.3.

## Example 2

Determine depreciation allowances during each recovery year for a MACRS 5-year property with a basis of $10,000.

Year 1 deduction: $10,000 * 20.00% = $2,000

Year 2 deduction: $10,000 * 32.00% = $3,200

**Table 4.2 MACRS property classes**

| Property Class | Example Assets |
|---|---|
| 3-Year Property | special handling devices for food<br>special tools for motor vehicle manufacturing |
| 5-Year Property | computers and office machines<br>general purpose trucks |
| 7-Year Property | office furniture<br>most manufacturing machine tools |
| 10-Year Property | tugs & water transport equipment<br>petroleum refining assets |
| 15-Year Property | fencing and landscaping<br>cement manufacturing assets |
| 20-Year Property | farm buildings<br>utility transmission lines and poles |
| 27.5-Year Residential Rental Property | rental houses and apartments |
| 31.5-Year Nonresidential Real Property | business buildings |

Year 3 deduction: $10,000 * 19.20% = $1,920

Year 4 deduction: $10,000 * 11.52% = $1,152

Year 5 deduction: $10,000 * 11.52% = $1,152

Year 6 deduction: $10,000 * 5.76% = $576

The sum of the deductions calculated in Example 2 is $10,000 which means that the asset is "fully depreciated" after six years. Though not shown here, tables similar to Table 4.3 are available for the 27.5-Year and 31.5-Year property classes. Their usage is similar to that outlined above except that depreciation is calculated monthly rather than annually.

## 4.6 TIME VALUE OF MONEY CONCEPTS

### 4.6.1 Introduction

Most people have an intuitive sense of the time value of money. Given a choice between $100 today and $100 one year from today, almost everyone would prefer the $100 today. Why is this the case? Two primary factors lead to this time preference associated with money; interest and inflation. Interest is the ability to earn a return on money which is loaned rather than consumed. By taking the $100 today and placing it in an interest bearing bank account (i.e., loaning it to the bank), one year from today an amount greater than $100 would be available for withdrawal. Thus, taking the $100 today and loaning it to earn interest, generates a sum greater than $100 one year from today and thus is preferred. The amount in excess of $100 that would be available depends upon the interest rate being paid by the bank. The next section develops the mathematics of the relationship between interest rates and the timing of cash flows.

The second factor which leads to the time preference associated with money is inflation. Inflation is a complex subject but in general can be described as a decrease in the purchasing power of money. The impact of inflation is that the "basket of goods" a consumer can buy today with $100 contains more than the "basket" the consumer could buy one year from today. This decrease in purchasing power is the result of inflation. The subject of inflation is addressed in Section 4.9.4.

### 4.6.2 The Mathematics of Interest

The mathematics of interest must account for the amount and timing of cash flows. The basic formula for studying and understanding interest calculations is:

$$F_n = P + I_n$$

where: $F_n$ = a future amount of money at the *end* of the nth year;

$P$ = a present amount of money at the beginning of the year which is n years prior to $F_n$;

## Table 4.3 MACRS percentages by recovery year and property class

| Recovery Year | 3-Year Property | 5-Year Property | 7-Year Property | 10-Year Property | 15-Year Property | 20-Year Property |
|---|---|---|---|---|---|---|
| 1 | 33.33% | 20.00% | 14.29% | 10.00% | 5.00% | 3.750% |
| 2 | 44.45% | 32.00% | 24.49% | 18.00% | 9.50% | 7.219% |
| 3 | 14.81% | 19.20% | 17.49% | 14.40% | 8.55% | 6.677% |
| 4 | 7.41% | 11.52% | 12.49% | 11.52% | 7.70% | 6.177% |
| 5 | | 11.52% | 8.93% | 9.22% | 6.93% | 5.713% |
| 6 | | 5.76% | 8.92% | 7.37% | 6.23% | 5.285% |
| 7 | | | 8.93% | 6.55% | 5.90% | 4.888% |
| 8 | | | 4.46% | 6.55% | 5.90% | 4.522% |
| 9 | | | | 6.56% | 5.91% | 4.462% |
| 10 | | | | 6.55% | 5.90% | 4.461% |
| 11 | | | | 3.28% | 5.91% | 4.462% |
| 12 | | | | | 5.90% | 4.461% |
| 13 | | | | | 5.91% | 4.462% |
| 14 | | | | | 5.90% | 4.461% |
| 15 | | | | | 5.91% | 4.462% |
| 16 | | | | | 2.95% | 4.461% |
| 17 | | | | | | 4.462% |
| 18 | | | | | | 4.461% |
| 19 | | | | | | 4.462% |
| 20 | | | | | | 4.461% |
| 21 | | | | | | 2.231% |

$I_n$ = the amount of accumulated interest over n years; and

n = the number of years between P and F.

The goal of studying the mathematics of interest is to develop a formula for $F_n$ which is expressed only in terms of the present amount P, the annual interest rate i, and the number of years n. There are two major approaches for determining the value of $I_n$; simple interest and compound interest. Under simple interest, interest is earned (charged) only on the original amount loaned (borrowed). Under compound interest, interest is earned (charged) on the original amount loaned (borrowed) plus any interest accumulated from previous periods.

### 4.6.3 Simple Interest

For simple interest, interest is earned (charged) only on the original principal amount at the rate of i% per year (expressed as i%/yr). Table 4.4 illustrates the annual calculation of simple interest. In Table 4.4 and the formulas which follow, the interest rate i is to be expressed as a decimal amount (e.g., 8% interest is expressed as 0.08).

At the beginning of year 1 (end of year 0), P dollars (e.g., $100) are deposited in an account earning i%/yr (e.g., 8%/yr or 0.08) simple interest. Under simple compounding, during year 1 the P dollars ($100) earn P*i dollars ($100*0.08 = $8) of interest. At the end of the year 1 the balance in the account is obtained by adding P dollars (the original principal, $100) plus P*i (the interest earned during year 1, $8) to obtain P+P*i ($100+$8=$108). Through algebraic manipulation, the end of year 1 balance can be expressed mathematically as P*(1+i) dollars ($100*1.08=$108).

The beginning of year 2 is the same point in time as the end of year 1 so the balance in the account is P*(1+i) dollars ($108). During year 2 the account again

**Table 4.4 The mathematics of simple interest**

| Year (t) | Amount At Beginning Of Year | Interest Earned During Year | Amount At End Of Year ($F_t$) |
|---|---|---|---|
| 0 | - | - | P |
| 1 | P | Pi | P + Pi = P (1 + i) |
| 2 | P (1 + i) | Pi | P (1+ i) + Pi = P (1 + 2i) |
| 3 | P (1 + 2i) | Pi | P (1+ 2i) + Pi = P (1 + 3i) |
| n | P (1 + (n-1)i) | Pi | P (1+ (n-1)i) + Pi = P (1 + ni) |

earns P*i dollars ($8) of interest since under simple compounding, interest is paid only on the *original* principal amount P ($100). Thus at the end of year 2, the balance in the account is obtained by adding P dollars (the original principal) plus P*i (the interest from year 1) plus P*i (the interest from year 2) to obtain P+P*i+P*i ($100+$8+$8=$116). After some algebraic manipulation, this can be written conveniently mathematically as P*(1+2*i) dollars ($100*1.16=$116).

Table 4.4 extends the above logic to year 3 and then generalizes the approach for year n. If we return our attention to our original goal of developing a formula for $F_n$ which is expressed only in terms of the present amount P, the annual interest rate i, and the number of years n, the above development and Table 4.4 results can be summarized as follows:

For Simple Interest
$$F_n = P (1+n*i)$$

**Example 3**

Determine the balance which will accumulate at the end of year 4 in an account which pays 10%/yr simple interest if a deposit of $500 is made today.

$F_n = P * (1 + n*i)$

$F_4 = 500 * (1 + 4*0.10)$

$F_4 = 500 * (1 + 0.40)$

$F_4 = 500 * (1.40)$

$F_4 = $700$

**4.6.4 Compound Interest**

For compound interest, interest is earned (charged) on the original principal amount *plus any accumulated*

*interest from previous years* at the rate of i% per year (i%/yr). Table 4.5 illustrates the annual calculation of compound interest. In the Table 4.5 and the formulas which follow, i is expressed as a decimal amount (i.e., 8% interest is expressed as 0.08).

At the beginning of year 1 (end of year 0), P dollars (e.g., $100) are deposited in an account earning i%/yr (e.g., 8%/yr or 0.08) compound interest. Under compound interest, during year 1 the P dollars ($100) earn P*i dollars ($100*0.08 = $8) of interest. Notice that this the same as the amount earned under simple compounding. This result is expected since the interest earned in previous years is zero for year 1. At the end of the year 1 the balance in the account is obtain by adding P dollars (the original principal, $100) plus P*i (the interest earned during year 1, $8) to obtain P+P*i ($100+$8=$108). Through algebraic manipulation, the end of year 1 balance can be expressed mathematically as P*(1+i) dollars ($100*1.08=$108).

During year 2 and subsequent years, we begin to see the power (if you are a lender) or penalty (if you are a borrower) of compound interest over simple interest. The beginning of year 2 is the same point in time as the end of year 1 so the balance in the account is P*(1+i) dollars ($108). During year 2 the account earns i% interest on the original principal, P dollars ($100), <u>and</u> it earns i% interest on the accumulated interest from year 1, P*i dollars ($8). Thus the interest earned in year 2 is [P+P*i]*i dollars ([$100+$8]*0.08=$8.64). The balance at the end of year 2 is obtained by adding P dollars (the original principal) plus P*i (the interest from year 1) plus [P+P*i]*i (the interest from year 2) to obtain P+P*i+[P+P*i]*i dollars ($100+$8+$8.64=$116.64). After some algebraic ma-

### Table 4.5 The Mathematics of Compound Interest

| Year (t) | Amount At Beginning Of Year | Interest Earned During Year | Amount At End Of Year ($F_t$) |
|---|---|---|---|
| 0 | - | - | P |
| 1 | P | Pi | $P + Pi$ <br> $= P(1+i)$ |
| 2 | $P(1+i)$ | $P(1+i)i$ | $P(1+i) + P(1+i)i$ <br> $= P(1+i)(1+i)$ <br> $= P(1+i)^2$ |
| 3 | $P(1+i)^2$ | $P(1+i)^2 i$ | $P(1+i)^2 + P(1+i)^2 i$ <br> $= P(1+i)^2(1+i)$ <br> $= P(1+i)^3$ |
| n | $P(1+i)^{n-1}$ | $P(1+i)^{n-1} i$ | $P(1+i)^{n-1} + P(1+i)^{n-1} i$ <br> $= P(1+i)^{n-1}(1+i)$ <br> $= P(1+i)^n$ |

nipulation, this can be written conveniently mathematically as $P*(1+i)^n$ dollars ($100*1.082 = $116.64$).

Table 4.5 extends the above logic to year 3 and then generalizes the approach for year n. If we return our attention to our original goal of developing a formula for $F_n$ which is expressed only in terms of the present amount P, the annual interest rate i, and the number of years n, the above development and Table 4.5 results can be summarized as follows:

For Compound Interest
$$F_n = P(1+i)^n$$

**Example 4**

Repeat Example 3 using compound interest rather than simple interest.

$F_n = P * (1+i)^n$

$F_4 = 500 * (1 + 0.10)^4$

$F_4 = 500 * (1.10)^4$

$F_4 = 500 * (1.4641)$

$F_4 = \$732.05$

Notice that the balance available for withdrawal is higher under compound interest ($732.05 > $700.00$). This is due to earning interest on principal plus interest rather than earning interest on just original principal. Since compound interest is by far more common in prac-tice than simple interest, the remainder of this chapter is based on <u>compound interest</u> unless explicitly stated otherwise.

**4.6.5 Single Sum Cash Flows**

Time value of money problems involving compound interest are common. Because of this frequent need, tables of compound interest time value of money factors can be found in most books and reference manuals that deal with economic analysis. The factor $(1+i)^n$ is known as the *single sum, future worth factor* or the *single payment, compound amount factor*. This factor is denoted $(F|P,i,n)$ where F denotes a future amount, P denotes a present amount, i is an interest rate (expressed as a percentage amount), and n denotes a number of years. The factor $(F|P,i,n)$ is read "to find F given P at i% for n years." Tables of values of $(F|P,i,n)$ for selected values of i and n are provided in Appendix 4A. The tables of values in Appendix 4A are organized such that the annual interest rate (i) determines the appropriate page, the time value of money factor $(F|P)$ determines the appropriate column, and the number of years (n) determines the appropriate row.

**Example 5**

Repeat Example 4 using the single sum, future worth factor.

$F_n = P * (1 + i)^n$

$F_n = P * (F|P,i,n)$

$F_4 = 500 * (F|P,10\%,4)$

$F_4 = 500 * (1.4641)$

$F_4 = 732.05$

The above formulas for compound interest allow us to solve for an unknown F given P, i, and n. What if we want to determine P with known values of F, i, and n? We can derive this relationship from the compound interest formula above:

$$F_n = P (1+i)^n$$

dividing both sides by $(1+i)^n$ yields

$$P = \frac{F^n}{(1 + i)^n}$$

which can be rewritten as

$$P = F_n (1+i)^{-n}$$

The factor $(1+i)^{-n}$ is known as the *single sum, present worth factor* or the *single payment, present worth factor*. This factor is denoted $(P|F,i,n)$ and is read "to find P given F at i% for n years." Tables of $(P|F,i,n)$ are provided in Appendix 4A.

**Example 6**

To accumulate $1000 five years from today in an account earning 8%/yr compound interest, how much must be deposited today?

$P = F_n * (1 + i)^{-n}$

$P = F_5 * (P|F,i,n)$

$P = 1000 * (P|F,8\%,5)$

$P = 1000 * (0.6806)$

$P = 680.60$

To verify your solution, try multiplying 680.60 * $(F|P,8\%,5)$. What would expect for a result? (Answer: $1000) If you're still not convinced, try building a table like Table 4.5 to calculate the year end balances each year for five years.

**4.6.6 Series Cash Flows**

Having considered the transformation of a single sum to a future worth when given a present amount and vice versa, let us generalize to a series of cash flows. The future worth of a series of cash flows is simply the sum of the future worths of each individual cash flow. Similarly, the present worth of a series of cash flows is the

sum of the present worths of the individual cash flows.

**Example 7**

Determine the future worth (accumulated total) at the end of seven years in an account that earns 5%/yr if a $600 deposit is made today and a $1000 deposit is made at the end of year two?

for the $600 deposit, n=7 (years between today and end of year 7)

for the $1000 deposit, n=5 (years between end of year 2 and end of year 7)

$F7 = 600 * (F|P,5\%,7) + 1000 * (F|P,5\%,5)$

$F7 = 600 * (1.4071) + 1000 * (1.2763)$

$F7 = 844.26 + 1276.30 = \$2120.56$

**Example 8**

Determine the amount that would have to be deposited today (present worth) in an account paying 6%/yr interest if you want to withdraw $500 four years from today and $600 eight years from today (leaving zero in the account after the $600 withdrawal).

for the $500 deposit n=4, for the $600 deposit n=8

$P = 500 * (P|F,6\%,4) + 600 * (P|F,6\%,8)$

$P = 500 * (0.7921) + 600 * (0.6274)$

$P = 396.05 + 376.44 = \$772.49$

**4.6.7 Uniform Series Cash Flows**

A uniform series of cash flows exists when the cash flows in a series occur every year and are all equal in value. Figure 4.3 shows the cash flow diagram of a uniform series of withdrawals. The uniform series has length 4 and amount 2000. If we want to determine the amount of money that would have to be deposited today to support this series of withdrawals starting one year from today, we could use the approach illustrated in Example 8 above to determine a present worth component for each individual cash flow. This approach would require us to sum the following series of factors (assuming the interest rate is 9%/yr):

$$P = 2000*(P|F,9\%,1) + 2000*(P|F,9\%,2) + 2000*(P|F,9\%,3) + 2000*(P|F,9\%,4)$$

After some algebraic manipulation, this expression can be restated as:

P =  2000*[(P | F,9%,1) + (P | F,9%,2) +
        (P | F,9%,3) + (P | F,9%,4)]

P =  2000*[(0.9174) + (0.8417) + (0.7722) + (0.7084)]

P =  2000*[3.2397] = $6479.40

**Figure 4.3. Uniform series cash flow**

Fortunately, uniform series occur frequently enough in practice to justify tabulating values to eliminate the need to repeatedly sum a series of (P | F,i,n) factors. To accommodate uniform series factors, we need to add a new symbol to our time value of money terminology in addition to the single sum symbols P and F. The symbol "A" is used to designate a uniform series of cash flows. When dealing with uniform series cash flows, the symbol A represents the amount of each annual cash flow and the n represents the number of cash flows in the series. The factor (P | A,i,n) is known as the *uniform series, present worth factor* and is read "to find P given A at i% for n years." Tables of (P | A,i,n) are provided in Appendix 4A. An algebraic expression can also be derived for the (P | A,i,n) factor which expresses P in terms of A, i, and n. The derivation of this formula is omitted here, but the resulting expression is shown in the summary table (Table 4.6) at the end of this section.

An important observation when using a (P | A,i,n) factor is that the "P" resulting from the calculation occurs one period prior to the first "A" cash flow. In our example the first withdrawal (the first "A") occurred one year after the deposit (the "P"). Restating the example problem above using a (P | A,i,n) factor, it becomes:

P = A * (P | A,i,n)

P = 2000 * (P | A,9%,4)

P = 2000 * (3.2397) = $6479.40

This result is identical (as expected) to the result using the (P | F,i,n) factors. In both cases the interpretation of the result is as follows: if we deposit $6479.40 in an account paying 9%/yr interest, we could make withdrawals of $2000 per year for four years starting one year after the initial deposit to deplete the account at the end of 4 years.

The reciprocal relationship between P and A is symbolized by the factor (A | P,i,n) and is called the *uniform series, capital recovery factor*. Tables of (A | P,i,n) are

provided in Appendix 4A and the algebraic expression for (A | P,i,n) is shown in Table 4.6 at the end of this section. This factor enables us to determine the amount of the equal annual withdrawals "A" (starting one year after the deposit) that can be made from an initial deposit of "P."

**Example 9**

Determine the equal annual withdrawals that can be made for 8 years from an initial deposit of $9000 in an account that pays 12%/yr. The first withdrawal is to be made one year after the initial deposit.

A = P * (A | P,12%,8)

A = 9000 * (0.2013)

A = $1811.70

Factors are also available for the relationships between a future worth (accumulated amount) and a uniform series. The factor (F | A,i,n) is known as the *uniform series future worth* factor and is read "to find F given A at i% for n years." The reciprocal factor, (A | F,i,n), is known as the *uniform series sinking fund* factor and is read "to find A given F at i% for n years." An important observation when using an (F | A,i,n) factor or an (A | F,i,n) factor is that the "F" resulting from the calculation occurs at the same point in time as to the last "A" cash flow. The algebraic expressions for (A | F,i,n) and (F | A,i,n) are shown in Table 4.6 at the end of this section.

**Example 10**

If you deposit $2000 per year into an individual retirement account starting on your 24th birthday, how much will have accumulated in the account at the time of your deposit on your 65th birthday? The account pays 6%/yr.

n = 42 (birthdays between 24th and 65th, inclusive)

F = A * (F | A,6%,42)

F = 2000 * (175.9505) = $351,901

**Example 11**

If you want to be a millionaire on your 65th birthday, what equal annual deposits must be made in an account starting on your 24th birthday? The account pays 10%/yr.

n = 42 (birthdays between 24th and 65th, inclusive)

A = F * (A | F,10%,42)

A = 1000000 * (0.001860) = $1860

### 4.6.8 Gradient Series

A gradient series of cash flows occurs when the value of a given cash flow is greater than the value of the previous period's cash flow by a constant amount. The symbol used to represent the constant increment is G. The factor (P|G,i,n) is known as the *gradient series, present worth factor*. Tables of (P|G,i,n) are provided in Appendix 4A. An algebraic expression can also be derived for the (P|G,i,n) factor which expresses P in terms of G, i, and n. The derivation of this formula is omitted here, but the resulting expression is shown in the summary table (Table 4.6) at the end of this section.

It is not uncommon to encounter a cash flow series that is the sum of a uniform series and a gradient series. Figure 4.4 illustrates such a series. The uniform component of this series has a value of 1000 and the gradient series has a value of 500. By convention the first element of a gradient series has a zero value. Therefore, in Figure 4.4, both the uniform series and the gradient series have length four (n=4). Like the uniform series factor, the "P" calculated by a (P|G,i,n) factor is located one period before the first element of the series (which is the zero element for a gradient series).

**Example 12**

Assume you wish to make the series of withdrawals illustrated in Figure 4.4 from an account which pays 15%/yr. How much money would you have to deposit today such that the account is depleted at the time of the last withdrawal?

This problem is best solved by recognizing that the cash flows are a combination of a uniform series of value 1000 and length 4 (starting at time=1) plus a gradient series of size 500 and length 4 (starting at time=1).

$$P = A * (P|A,15\%,4) + G * (P|G,15\%,4)$$
$$P = 1000 * (2.8550) + 500 * (3.7864)$$

$$P = 2855.00 + 1893.20 = \$4748.20$$

Occasionally it is useful to convert a gradient series to an equivalent uniform series of the same length. Equivalence in this context means that the present value (P) calculated from the gradient series is numerically equal to the present value (P) calculated from the uniform series. One way to accomplish this task with the time value of money factors we have already considered is to convert the gradient series to a present value using a (P|G,i,n) factor and then convert this present value to a uniform series using an (A|P,i,n) factor. In other words:

Figure 4.4. Combined uniform series and gradient series cash flow

$$A = [G * (P|G,i,n)] * (A|P,i,n)$$

An alternative approach is to use a factor known as the *gradient-to-uniform series conversion factor*, symbolized by (A|G,i,n). Tables of (A|G,i,n) are provided in Appendix 4A. An algebraic expression can also be derived for the (A|G,i,n) factor which expresses A in terms of G, i, and n. The derivation of this formula is omitted here, but the resulting expression is shown in the summary table (Table 4.6) at the end of this section.

### 4.6.9 Summary of Time Value of Money Factors

Table 4.6 summarizes the time value of money factors introduced in this section. Time value of money factors are useful in economic analysis because they provide a mechanism to accomplish two primary functions: (1) they allow us to replace a cash flow at one point in time with an equivalent cash flow (in a time value of money sense) at a different point in time and (2) they allow us to convert one cash flow pattern to another (e.g., convert a single sum of money to an equivalent cash flow series or convert a cash flow series to an equivalent single sum). The usefulness of these two functions when performing economic analysis of alternatives will become apparent in Sections 4.7 and 4.8 which follow.

### 4.6.10 The Concepts of Equivalence and Indifference

Up to this point the term "equivalence" has been used several times but never fully defined. It is appropriate at this point to formally define equivalence as well as a related term, indifference.

In economic analysis, "equivalence" means "the state of being equal in value." The concept is primarily applied to the comparison of two or more cash flow profiles. Specifically, two (or more) cash flow profiles are equivalent if their time value of money worths at a common point in time are equal.

Question: Are the following two cash flows equivalent at 15%/yr?
Cash Flow 1: Receive $1,322.50 two years from today
Cash Flow 2: Receive $1,000.00 today

**Table 4.6 Summary of discrete compounding time value of money factors**

| To Find | Given | Factor | Symbol | Name |
|---------|-------|--------|--------|------|
| P | F | $(1+i)^{-n}$ | $(P\,|\,F,i,n)$ | Single Payment, Present Worth Factor |
| F | P | $(1+i)^{n}$ | $(F\,|\,P,i,n)$ | Single Payment, Compound Amount Factor |
| P | A | $\dfrac{(1+i)^{n}-1}{i(1+i)^{n}}$ | $(P\,|\,A,i,n)$ | Uniform Series, Present Worth Factor |
| A | P | $\dfrac{i(1+i)^{n}}{(1+i)^{n}-1}$ | $(A\,|\,P,i,n)$ | Uniform Series, Capital Recovery Factor |
| F | A | $\dfrac{(1+i)^{n}-1}{i}$ | $(F\,|\,A,i,n)$ | Uniform Series, Compound Amount Factor |
| A | F | $\dfrac{i}{(1+i)^{n}-1}$ | $(A\,|\,F,i,n)$ | Uniform Series, Sinking Fund Factor |
| P | G | $\dfrac{1-(1+ni)(1+i)^{-n}}{i^{2}}$ | $(P\,|\,G,i,n)$ | Gradient Series, Present Worth Factor |
| A | G | $\dfrac{(1+i)^{-n}-(1+ni)}{i\left[(1+i)^{-n}-1\right]}$ | $(A\,|\,G,i,n)$ | Gradient Series, Uniform Series Factor |

**Analysis Approach 1**: Compare worths at t=0 (present worth)

PW(1) = 1,322.50*(P | F,15,2) = 1322.50*0.756147 = 1,000

PW(2) = 1,000

Answer: Cash Flow 1 and Cash Flow 2 are equivalent

**Analysis Approach 2**: Compare worths at t=2 (future worth)

FW(1) = 1,322.50

FW(2) = 1,000*(F | P,15,2) = 1,000*1.3225 = 1,322.50

Answer: Cash Flow 1 and Cash Flow 2 are equivalent

Generally the comparison (hence the determination of equivalence) for the two cash flow series in this example would be made as present worths (t=0) or future worths (t=2), but the equivalence definition holds regardless of the point in time chosen. For example:

**Analysis Approach 3**: Compare worths at t=1

W1(1)   = 1,322.50*(P | F,15,1)

      = 1,322.50*0.869565 = 1,150.00

W1(2) = 1,000*(F | P,15,1) = 1,000*1.15 = 1,150.00

Answer: Cash Flow 1 and Cash Flow 2 are equivalent

Thus, the selection of the point in time, t, at which to make the comparison is completely arbitrary. Clearly however, some choices are more intuitively appealing than others (t= 0 and t=2 in the above example).

In economic analysis, "indifference" means "to have no preference" The concept is primarily applied in the comparison of two or more cash flow profiles. Specifically, a potential investor is indifferent between two (or more) cash flow profiles if they are equivalent.

Question: Given the following two cash flows at 15%/yr which do you prefer?

Cash Flow 1: Receive $1,322.50 two years from today

Cash Flow 2: Receive $1,000.00 today

Answer: Based on the equivalence calculations above, given these two choices, an investor is indifferent.

The concept of equivalence can be used to break a large, complex problem into a series of smaller more manageable ones. This is done by taking advantage of the fact that, in calculating the economic worth of a cash flow profile, any part of the profile can be replaced by an equivalent representation without altering the worth of the profile at an arbitrary point in time.

Question: You are given a choice between (1) receiving P dollars today or (2) receiving the cash flow series illustrated in Figure 4.5. What must the value of P be for you to be indifferent between the two choices if i=12%/yr?

**Figure 4.5 A cash flow series**

Analysis Approach: To be indifferent between the choices, P must have a value such that the two alternatives are equivalent at 12%/yr. If we select t=0 as the common point in time upon which to base the analysis (present worth approach), then the analysis proceeds as follows.

PW(Alt 1) = P

Because P is already at t=0 (today), no time value of money factors are involved.

PW(Alt 2)

Step 1 - Replace the uniform series (t=3 to 7) with an equivalent single sum, $V_2$, at t=2 (one period before the first element of the series).

$V_2$ = 2,000 * (P | A,12%,5) = 2,000 * 3.6048 = 7,209.60

Step 2 - Replace the single sum $V_2$, with an equivalent value $V_0$ at t=0:

PW(Alt 2) = $V_0$ = $V_2$ * (P | F,12,2) = 7,209.60 * 0.7972 = 5,747.49

Answer: To be indifferent between the two alternatives, they must be equivalent at t=0. To be equivalent, P must have a value of $5,747.49

## 4.7    PROJECT MEASURES OF WORTH

### 4.7.1 Introduction

In this section measures of worth for investment projects are introduced. The measures are used to evaluate the attractiveness of a single investment opportunity. The measures to be presented are (1) present worth, (2) annual worth, (3) internal rate of return, (4) savings investment ratio, and (5) payback period. All but one of these measures of worth require an interest rate to calculate the worth of an investment. This interest rate is commonly referred to as the Minimum Attractive Rate of Return (MARR). There are many ways to determine a

value of MARR for investment analysis and no one way is proper for all applications. One principle is, however, generally accepted. MARR should always exceed the cost of capital as described in Section 4.4, Sources of Funds, presented earlier in this chapter.

In all of the measures of worth below, the following conventions are used for defining cash flows. At any given point in time (t = 0, 1, 2,..., n), there may exist both revenue (positive) cash flows, $R_t$, and cost (negative) cash flows, $C_t$. The net cash flow at t, $A_t$, is defined as $R_t$ - $C_t$.

### 4.7.2 Present Worth

Consider again the cash flow series illustrated in Figure 4.5. If you were given the opportunity to "buy" that cash flow series for $5,747.49, would you be interested in purchasing it? If you expected to earn a 12%/yr return on your money (MARR=12%), based on the analysis in the previous section, your conclusion would be (should be) that you are indifferent between (1) retaining your $5,747.49 and (2) giving up your $5,747.49 in favor of the cash flow series. Figure 4.6 illustrates the net cash flows of this second investment opportunity.

What value would you expect if we calculated the present worth (equivalent value of all cash flows at t=0) of Figure 4.6? We must be careful with the signs (directions) of the cash flows in this analysis since some represent cash outflows (downward) and some represent cash inflows (upward).

PW = -5747.49 + 2000*(P | A,12%,5)*(P | F,12%,2)

PW = -5747.49 + 2000*(3.6048)*(0.7972)

PW = -5747.49 + 5747.49 = $0.00

The value of zero for present worth indicates indifference regarding the investment opportunity. We would just as soon do nothing (i.e., retain our $5747.49) as invest in the opportunity.

What if the same returns (future cash inflows) where offered for a $5000 investment (t=0 outflow),

**Figure 4.6 An investment opportunity**

would this be more or less attractive? Hopefully, after a little reflection, it is apparent that this would be a more attractive investment because you are getting the same returns but paying less than the indifference amount for them. What happens if calculate the present worth of this new opportunity?

PW = -5000 + 2000*(P I A,12%,5)*(P I F,12%,2)

PW = -5000 + 2000*(3.6048)*(0.7972)

PW = -5000.00 + 5747.49 = \$747.49

The positive value of present worth indicates an attractive investment. If we repeat the process with an initial cost greater than \$5747.49, it should come as no surprise that the present worth will be negative indicating an unattractive investment.

The concept of present worth as a measure of investment worth can be generalized as follows:

Measure of Worth: Present Worth

Description: All cash flows are converted to a single sum equivalent at time zero using i=MARR.

Calculation Approach: $PW = \sum_{t=0}^{n} A_t (P|F,i,t)$

Decision Rule: If PW≥0, then the investment is attractive.

**Example 13**

Installing thermal windows on a small office building is estimated to cost \$10,000. The windows are expected to last six years and have no salvage value at that time. The energy savings from the windows are expected to be \$2525 each year for the first three years and \$3840 for each of the remaining three years. If MARR is 15%/yr and the present worth measure of worth is to be used, is this an attractive investment?

**Figure 4.7 Thermal windows investment**

The cash flow diagram for the thermal windows is shown in Figure 4.7.

PW =
−10000+2525*(P I F,15%,1)+2525*(P I F,15%,2)
+2525*(P I F,15%,3)+3840*(P I F,15%,4)+
3840*(P I F,15%,5)+3840*(P I F,15%,6)

PW =
−10000+2525*(0.8696)+2525*(0.7561)
+2525*(0.6575)+3840*(0.5718)+3840*(0.4972)+
3840*(0.4323)

PW =
−10000+2195.74+1909.15+1660.19+2195.71
+1909.25+1660.03

PW =   \$1530.07

Decision: PW≥0 (\$1530.07≥0.0), therefore the window investment is attractive.

An alternative (and simpler) approach to calculating PW is obtained by recognizing that the savings cash flows are two uniform series; one of value \$2525 and length 3 starting at t=1 and one of value \$3840 and length 3 starting at t=4.

PW =   -10000+2525*(P I A,15%,3)+3840*
(P I A,15%,3)*(P I F,15%,3)

PW =   -10000+2525*(2.2832)+3840*(2.2832)*
(0.6575) = \$1529.70

Decision: PW≥0 (\$1529.70>0.0), therefore the window investment is attractive.

The slight difference in the PW values is caused by the accumulation of round off errors as the various factors are rounded to four places to the right of the decimal point.

**4.7.3 Annual Worth**

An alternative to present worth is annual worth (AW). The annual worth measure converts all cash flows to an equivalent uniform annual series of cash flows over the investment life using i=MARR. The annual worth measure is generally calculated by first calculating the present worth measure and then multiplying this by the appropriate (A I P,i,n) factor. A thorough review of the tables in Appendix 4A or the equations in Table 4.6 leads to the conclusion that for all values of i (i>0) and n (n>0), the value of (A I P,i,n) is greater than zero. Hence, if PW>0, then AW>0;

if PW<0, then AW<0; and

if PW=0, then AW=0

because the only difference between PW and AW is multiplication by a positive, non-zero value, namely $(A \mid P,i,n)$. The decision rule for investment attractiveness for PW and AW are identical; positive values indicate an attractive investment; negative values indicate an unattractive investment; zero indicates indifference. Frequently the only reason for choosing between AW and PW as a measure of worth in an analysis is the preference of the decision maker.

The concept of annual worth as a measure of investment worth can be generalized as follows:

Measure of Worth: Annual Worth

Description: All cash flows are converted to an equivalent uniform annual series of cash flows over the planning horizon using i=MARR.

Calculation Approach: $AW = PW (A \mid P,i,n)$

Decision Rule: If AW≥0, then the investment is attractive.

**Example 14**

Reconsider the thermal window data of Example 13. If the annual worth measure of worth is to be used, is this an attractive investment?

$AW = PW (A \mid P,15\%,6)$

$AW = 1529.70 (0.2642) = \$404.15/yr$

Decision: AW ≥0 (\$404.15>0.0), therefore the window investment is attractive.

**4.7.4 Internal Rate of Return**

One of the problems associated with using the present worth or the annual worth measures of worth is that they depend upon knowing a value for MARR. As mentioned in the introduction to this section, the "proper" value for MARR is a much debated topic and tends to vary from company to company and decision maker to decision maker. If the value of MARR changes, the value of PW or AW must be recalculated to determine whether the attractiveness/unattractiveness of an investment has changed.

The internal rate of return (IRR) approach is designed to calculate a rate of return that is "internal" to the project. That is,

if IRR > MARR, the project is attractive,

if IRR < MARR, the project is unattractive,

if IRR = MARR, indifferent.

Thus, if MARR changes, no new calculations are required. We simply compare the calculated IRR for the project to the new value of MARR and we have our decision.

The value of IRR is typically determined through a trial and error process. An expression for the present worth of an investment is written without specifying a value for i in the time value of money factors. Then, various values of i are substituted until a value is found that sets the present worth (PW) equal to zero. The value of i found in this way is the IRR.

As appealing as the flexibility of this approach is, their are two major drawbacks. First, the iterations required to solve using the trial and error approach to solution can be time consuming. This factor is mitigated by the fact that most spreadsheets and financial calculators are pre-programmed to solve for an IRR value given a cash flow series. The second, and more serious, drawback to the IRR approach is that some cash flow series have more than one value of IRR (i.e., more than one value of i sets the PW expression to zero). A detailed discussion of this multiple solution issue is beyond the scope of this chapter, but can be found in White, et al. [1998], as well as most other economic analysis references. However, it can be shown that, if a cash flow series consists of an initial investment (negative cash flow at t=0) followed by a series of future returns (positive or zero cash flows for all t>0) then a unique IRR exists. If these conditions are not satisfied a unique IRR is not guaranteed and caution should be exercised in making decisions based on IRR.

The concept of internal rate of return as a measure of investment worth can be generalized as follows:

Measure of Worth: Internal Rate of Return

Description: An interest rate, IRR, is determined which yields a present worth of zero. IRR implicitly assumes the reinvestment of recovered funds at IRR.

Calculation Approach:

find IRR such that $PW = \sum_{t=0}^{n} A_t(P \mid F,IRR,t) = 0$

Important Note: Depending upon the cash flow series, multiple IRRs may exist! If the cash flow series consists of an initial investment (net negative cash flow) followed by a series of future returns (net non-negative cash flows), then a unique IRR exists.

<u>Decision Rule</u>: If IRR is unique and IRR $\geq$ MARR, then the investment is attractive.

## Example 15

Reconsider the thermal window data of Example 13. If the internal rate of return measure of worth is to be used, is this an attractive investment?

First we note that the cash flow series has a single negative investment followed by all positive returns, therefore, it has a unique value for IRR. For such a cash flow series it can also be shown that as i increases PW decreases.

From example 11, we know that for i=15%:

PW = -10000+2525*(P|A,15%,3)+3840*(P|A,15%,3)*
(P|F,15%,3)

PW = -10000+2525*(2.2832)+3840*(2.2832)*
(0.6575) = $1529.70

Because PW>0, we must increase i to decrease PW toward zero for i=18%:

PW = -10000+2525*(P|A,18%,3)+3840*
(P|A,18%,3)*(P|F,18%,3)

PW = -10000+2525*(2.1743)+3840*(2.1743)*
(0.6086) = $571.50

Since PW>0, we must increase i to decrease PW toward zero for i=20%:

PW = -10000+2525*(P|A,20%,3)+3840*
(P|A,20%,3)*(P|F,20%,3)

PW = -10000+2525*(2.1065)+3840*(2.1065)*
(0.5787) = -$0.01

Although we could interpolate for a value of i for which PW=0 (rather than -0.01), for practical purposes PW=0 at i=20%, therefore IRR=20%.

<u>Decision</u>: IRR$\geq$MARR (20%>15%), therefore the window investment is attractive.

## 4.7.5 Saving Investment Ratio

Many companies are accustomed to working with benefit cost ratios. An investment measure of worth which is consistent with the present worth measure and has the form of a benefit cost ratio is the savings investment ratio (SIR). The SIR decision rule can be derived from the present worth decision rule as follows:

Starting with the PW decision rule

$$PW \geq 0$$

replacing PW with its calculation expression

$$\sum_{t=0}^{n} A_t(P|F,i,t) \leq 0$$

which, using the relationship $A_t = R_t - C_t$, can be restated

$$\sum_{t=0}^{n} (R_t - C_t)(P|F,i,t) \geq 0$$

which can be algebraically separated into

$$\sum_{t=0}^{n} R_t(P|F,i,t) - \sum_{t=0}^{n} C_t(P|F,i,t) \geq 0$$

adding the second term to both sides of the inequality

$$\sum_{t=0}^{n} R_t(P|F,i,t) \geq \sum_{t=0}^{n} C_1(P|F,i,t) \geq 0$$

dividing both sides of the inequality by the right side term

$$\frac{\sum_{t=0}^{n} R_1(P|F,i,t)}{\sum_{t=0}^{n} C_1(P|F,i,t)} \geq 1$$

which is the decision rule for SIR.

The SIR represents the ratio of the present worth of the revenues to the present worth of the costs. If this ratio exceeds one, the investment is attractive.

The concept of savings investment ratio as a measure of investment worth can be generalized as follows:

<u>Measure of Worth</u>: Savings Investment Ratio

<u>Description</u>: The ratio of the present worth of positive cash flows to the present worth of (the absolute value of) negative cash flows is formed using i=MARR.

<u>Calculation Approach</u>: $SIR = \dfrac{\sum_{t=0}^{n} R_t(P|F,i,t)}{\sum_{t=0}^{n} C_t(P|F,i,t)}$

Decision Rule: If SIR $\geq 1$, then the investment is attractive.

**Example 16**

Reconsider the thermal window data of Example 13. If the savings investment ratio measure of worth is to be used, is this an attractive investment?

From example 13, we know that for i=15%:

$$SIR = \frac{\sum_{t=0}^{n} R_t(P\,|\,F,i,t)}{\sum_{t=0}^{n} C_t(P\,|\,F,i,t)}$$

$$SIR = \frac{2525*(P\,|\,A,15\%,3)+3840*(P\,|\,A,15\%,3)*(P\,|\,F,15\%,3)}{1000}$$

$$SIR = \frac{11529.70}{10000.00} = 1.15297$$

Decision: SIR$\geq$1.0 (1.15297>1.0), therefore the window investment is attractive.

An important observation regarding the four measures of worth presented to this point (PW, AW, IRR, and SIR) is that they are all consistent and equivalent. In other words, an investment that is attractive under one measure of worth will be attractive under each of the other measures of worth. A review of the decisions determined in Examples 13 through 16 will confirm the observation. Because of their consistency, it is not necessary to calculate more than one measure of investment worth to determine the attractiveness of a project. The rationale for presenting multiple measures which are essentially identical for decision making is that various individuals and companies may have a preference for one approach over another.

### 4.7.6 Payback Period

The payback period of an investment is generally taken to mean the number of years required to recover the initial investment through net project returns. The payback period is a popular measure of investment worth and appears in many forms in economic analysis literature and company procedure manuals. Unfortunately, all too frequently, payback period is used inappropriately and leads to decisions which focus exclusively on short term results and ignore time value of money concepts. After presenting a common form of payback period these shortcomings will be discussed.

Measure of Worth: Payback Period

Description: The number of years required to recover the initial investment by accumulating net project returns is determined.

Calculation Approach:

PBP = the smallest $m$ such that $\sum_{t=1}^{m} A_t \geq C_0$

Decision Rule: If PBP is less than or equal to a predetermined limit (often called a hurdle rate), then the investment is attractive.

Important Note: This form of payback period ignores the time value of money and ignores returns beyond the predetermined limit.

The fact that this approach ignores time value of money concepts is apparent by the fact that no time value of money factors are included in the determination of m. This implicitly assumes that the applicable interest rate to convert future amounts to present amounts is zero. This implies that people are indifferent between $100 today and $100 one year from today, which is an implication that is highly inconsistent with observable behavior.

The short-term focus of the payback period measure of worth can be illustrated using the cash flow diagrams of Figure 4.8. Applying the PBP approach above yields a payback period for investment (a) of PBP=2 (1200>1000 @ t=2) and a payback period for investment (b) of PBP=4 (1000300>1000) @ t=4). If the decision hurdle rate is 3 years (a very common rate), then investment (a) is attractive but investment (b) is not. Hopefully, it is obvious that judging (b) unattractive is not good decision making since a $1,000,000 return four years after a $1,000 investment is attractive under almost any value of MARR. In point of fact, the IRR for (b) is 465% so for any value of MARR less than 465%, investment (b) is attractive.

## 4.8 ECONOMIC ANALYSIS

### 4.8.1 Introduction

The general scenario for economic analysis is that a set of investment alternatives are available and a decision must be made regarding which ones (if any) to accept and which ones (if any) to reject. If the analysis is deterministic, then an assumption is made that cash flow amounts, cash flow timing, and MARR are known with

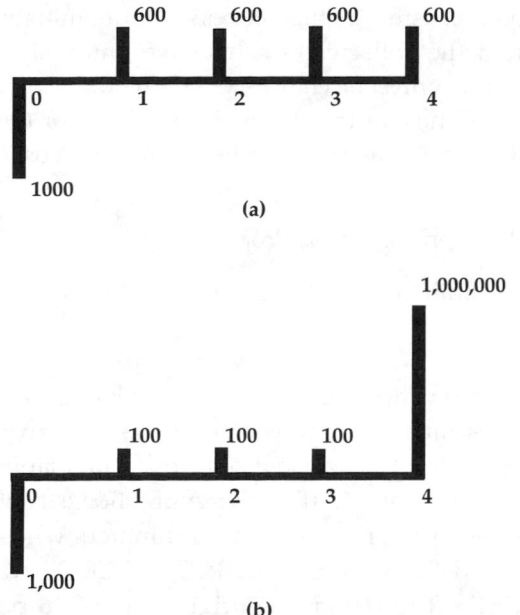

**Figure 4.8 Two investments evaluated using payback period**

certainty. Frequently, although this assumption does not hold exactly, it is not considered restrictive in terms of potential investment decisions. If however the lack of certainty is a significant issue then the analysis is stochastic and the assumptions of certainty are relaxed using probability distributions and statistical techniques to conduct the analysis. The remainder of this section deals with deterministic economic analysis so the assumption of certainty will be assumed to hold. Stochastic techniques are introduced in Section 4.9.5.

### 4.8.2 Deterministic Unconstrained Analysis

Deterministic economic analysis can be further classified into unconstrained deterministic analysis and constrained deterministic analysis. Under unconstrained analysis, all projects within the set available are assumed to be independent. The practical implication of this independence assumption is that an accept/reject decision can be made on each project without regard to the decisions made on other projects. In general this requires that (1) there are sufficient funds available to undertake all proposed projects, (2) there are no mutually exclusive projects, and (3) there are no contingent projects.

A funds restriction creates dependency since, before deciding on a project being evaluated, the evaluator would have to know what decisions had been made on other projects to determine whether sufficient funds were available to undertake the current project. Mutual exclusion creates dependency since acceptance of one of the mutually exclusive projects precludes acceptance of

the others. Contingency creates dependence since prior to accepting a project, all projects on which it is contingent must be accepted.

If none of the above dependency situations are present and the projects are otherwise independent, then the evaluation of the set of projects is done by evaluating each individual project in turn and accepting the set of projects which were individually judged acceptable. This accept or reject judgment can be made using either the PW, AW, IRR, or SIR measure of worth. The unconstrained decision rules for each or these measures of worth are restated below for convenience.

Unconstrained PW Decision Rule: If PW $\geq 0$, then the project is attractive.

Unconstrained AW Decision Rule: If AW $\geq 0$, then the project is attractive.

Unconstrained IRR Decision Rule: If IRR is unique and IRR $\geq$ MARR, then the project is attractive.

Unconstrained SIR Decision Rule: If SIR $\geq 1$, then the project is attractive.

### Example 17

Consider the set of four investment projects whose cash flow diagrams are illustrated in Figure 4.9. If MARR is 12%/yr and the analysis is unconstrained, which projects should be accepted?

Using present worth as the measure of worth:

$$PW_A = -1000+600*(P \mid A,12\%,4) = -1000+600(3.0373) = \$822.38 \Rightarrow \text{Accept A}$$

$$PW_B = -1300+800*(P \mid A,12\%,4) = -1300+800(3.0373) = \$1129.88 \Rightarrow \text{Accept B}$$

$$PW_C = -400+120*(P \mid A,12\%,4) = -400+120(3.0373) = -\$35.52 \Rightarrow \text{Reject C}$$

$$PW_D = -500+290*(P \mid A,12\%,4) = -500+290(3.0373) = \$380.83 \Rightarrow \text{Accept D}$$

Therefore,

Accept Projects A, B, and D and Reject Project C

### 4.8.3 Deterministic Constrained Analysis

Constrained analysis is required any time a dependency relationship exists between any of the projects within the set to be analyzed. In general dependency exists any time (1) there are insufficient funds available to undertake all proposed projects (this is commonly referred to as capital rationing), (2) there are mutually exclusive projects, or (3) there are contingent projects.

Several approaches have been proposed for selecting the best set of projects from a set of potential projects under constraints. Many of these approaches will select the optimal set of acceptable projects under some conditions or will select a set that is near optimal. However, only a few approaches are guaranteed to select the optimal set of projects under all conditions. One of these approaches is presented below by way of a continuation of Example 17.

The first steps in the selection process are to specify the cash flow amounts and cash flow timings for each project in the potential project set. Additionally, a value of MARR to be used in the analysis must be specified. These issues have been addressed in previous sections so further discussion will be omitted here. The next step is to form the set of all possible decision alternatives from the projects. A single decision alternative is a collection of zero, one, or more projects which could be accepted (all others not specified are to be rejected). As an illustration, the possible decision alternatives for the set of projects illustrated in Figure 4.9 are listed in Table 4.7. As a general rule, there will be $2^n$ possible decision alternatives generated from a set of n projects. Thus, for the projects of Figure 4.9, there are $2^4 = 16$ possible decision alternatives. Since this set represents all possible decisions that could be made, one, and only one, will be selected as the best (optimal) decision. The set of decision alternatives developed in this way has the properties of being collectively exhaustive (all possible choices are listed) and mutually exclusive (only one will be selected).

The next step in the process is to eliminate decisions from the collectively exhaustive, mutually exclusive set that represent choices which would violate one (or more) of the constraints on the projects. For the projects of Figure 4.9, assume the following two constraints exist:

Project B is contingent on Project C, and

A budget limit of $1500 exists on capital expenditures at t=0.

Based on these constraints the following decision alternatives must be removed from the collectively exhaustive, mutually exclusive set: any combination that includes B but not C (B only; A&B; B&D; A&B&D), any combination not already eliminated whose t=0 costs exceed $1500 (B&C, A&B&C, A&C&D, B&C&D, A&B&C&D). Thus, from the original set of 16 possible decision alternatives, 9 have been eliminated and need not be evaluated. These results are illustrated in Table 4.8. It is frequently the case in practice that a significant percentage of the original collectively exhaustive, mutually exclusive set will be eliminated before measures of worth are calculated.

The next step is to create the cash flow series for the remaining (feasible) decision alternatives. This is a straight forward process and is accomplished by setting a decision alternative's annual cash flow equal to the sum of the annual cash flows (on a year by year basis) of all projects contained in the decision alternative. Table 4.9 illustrates the results of this process for the feasible decision alternatives from Table 4.8.

The next step is to calculate a measure of worth for each decision alternative. Any of the four consistent measures of worth presented above (PW, AW, IRR,

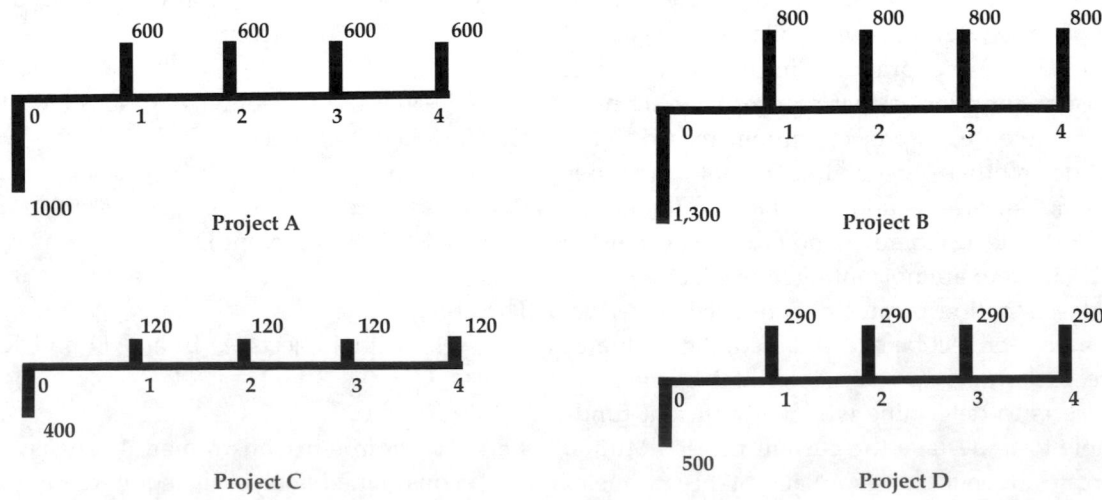

**Figure 4.9 Four investments projects**

or SIR but NOT PBP) can be used. The measures are entirely consistent and will lead to the same decision alternative being selected. For illustrative purposes, PW will be calculated for the decision alternatives of Table 4.9 assuming MARR=12%.

$PW_A =$ -1000 + 600*(P|A,12%,4) = -1000 + 600 (3.0373) = \$822.38

$PW_C =$ -400 + 120*(P|A,12%,4) = -400 + 120 (3.0373) = -\$35.52

$PW_D =$ -500 + 290*(P|A,12%,4) = -500 + 290 (3.0373) = \$380.83

$PW_{A\&C} =$ -1400 + 720*(P|A,12%,4) = -1400 + 720 (3.0373) = \$786.86

$PW_{A\&D} =$ -1500 + 890*(P|A,12%,4) = -1500 + 890 (3.0373) = \$1203.21

$PW_{C\&D} =$ -900 + 410*(P|A,12%,4) = -900 + 410 (3.0373) = \$345.31

$PW_{null} =$ -0 + 0*(P|A,12%,4) = -0 + 0 (3.0373) = \$0.00

**Table 4.7 The decision alternatives from four projects**

| |
|---|
| Accept A only<br>Accept B only<br>Accept C only<br>Accept D only |
| Accept A and B only<br>Accept A and C only<br>Accept A and D only<br>Accept B and C only<br>Accept B and D only<br>Accept C and D only |
| Accept A, B, and C only<br>Accept A, B, and D only<br>Accept A, C, and D only<br>Accept B, C, and D only |
| Accept A, B, C, and D<br>(frequently called the do everything alternative) |
| Accept none<br>(frequently called the do nothing or null alternative) |

**Table 4.8 The decision alternatives with constraints imposed**

| | |
|---|---|
| Accept A only | OK |
| Accept B only | infeasible, B contingent on C |
| Accept C only | OK |
| Accept D only | OK |
| Accept A and B only | infeasible, B contingent on C |
| Accept A and C only | OK |
| Accept A and D only | OK |
| Accept B and C only | infeasible, capital rationing |
| Accept B and D only | infeasible, B contingent on C |
| Accept C and D only | OK |
| Accept A, B, and C only | infeasible, capital rationing |
| Accept A, B, and D only | infeasible, B contingent on C |
| Accept A, C, and D only | infeasible, capital rationing |
| Accept B, C, and D only | infeasible, capital rationing |
| Do Everything | infeasible, capital rationing |
| null | OK |

The decision rules for the various measures of worth under constrained analysis are list below.

Constrained PW Decision Rule: Accept the decision alternative with the highest PW.

Constrained AW Decision Rule: Accept the decision alternative with the highest AW.

Constrained IRR Decision Rule: Accept the decision alternative with the highest IRR.

Constrained SIR Decision Rule: Accept the decision alternative with the highest SIR.

For the example problem, the highest present worth ($1203.21) is associated with accepting projects A and D (rejecting all others). This decision is guaranteed to be optimal (i.e., no feasible combination of projects has a higher PW, AW, IRR, or SIR).

### 4.8.4   Some Interesting Observations Regarding Constrained Analysis

Several interesting observations can be made regarding the approach, measures of worth, and decisions associated with constrained analysis. Detailed development of these observations is omitted here but may be found in many engineering economic analysis texts [White, et al., 1998].

- The present worth of a decision alternative is the sum of the present worths of the projects contained within the alternative. (From above $PW_{A\&D} = PW_A + PW_D$).

- The annual worth of a decision alternative is the sum of the annual worths of the projects contained within the alternative.

- The internal rate of return of a decision alternative is NOT the sum of internal rates of returns of the projects contained within the alternative. The IRR for the decision alternative must be calculated by the trial and error process of finding the value of i that sets the PW of the decision alternative to zero.

- The savings investment ratio of a decision alternative is NOT the sum of the savings investment ratios of the projects contained within the alternative. The SIR for the decision alternative must be calculated from the cash flows of the decision alternative.

- A common, but flawed, procedure for selecting the projects to accept from the set of potential projects involves ranking the projects (not decision alternatives) in preferred order based on a measure of worth calculated for the project (e.g., decreasing project PW) and then accepting projects as far down the list as funds allow. While this procedure will select the optimal set under some conditions (e.g., it works well if the initial investments of all projects are small relative to the capital budget limit), it is not guaranteed to select the optimal set under all conditions. The procedure outlined above will select the optimal set under all conditions.

- Table 4.10 illustrates that the number of decision alternatives in the collectively exhaustive, mutually exclusive set can grow prohibitively large as the number of potential projects increases. The mitigating factor in this combinatorial growth problem is that in most practical situations a high percentage of the possible decision alternatives are infeasible and do not require evaluation.

**Table 4.9 The decision alternatives cash flows**

| yr \ Alt | A only | C only | D only | A&C | A&D | C&D | null |
|----------|--------|--------|--------|------|------|------|------|
| 0 | -1000 | -400 | -500 | -1400 | -1500 | -900 | 0 |
| 1 | 600 | 120 | 290 | 720 | 890 | 410 | 0 |
| 2 | 600 | 120 | 290 | 720 | 890 | 410 | 0 |
| 3 | 600 | 120 | 290 | 720 | 890 | 410 | 0 |
| 4 | 600 | 120 | 290 | 720 | 890 | 410 | 0 |

## Table 4.10 The number of decision alternatives as a function of the number of projects

| Number of Projects | Number of Decision Alternatives |
|---|---|
| 1 | 2 |
| 2 | 4 |
| 3 | 8 |
| 4 | 16 |
| 5 | 32 |
| 6 | 64 |
| 7 | 128 |
| 8 | 256 |
| 9 | 512 |
| 10 | 1,024 |
| 15 | 32,768 |
| 20 | 1,048,576 |
| 25 | 33,554,432 |

### 4.8.5 The Planning Horizon Issue

When comparing projects, it is important to compare the costs and benefits over a common period of time. The intuitive sense of fairness here is based upon the recognition that most consumers expect an investment that generates savings over a longer period of time to cost more than an investment that generates savings over a shorter period of time. To facilitate a fair, comparable evaluation a common period of time over which to conduct the evaluation is required. This period of time is referred to as the planning horizon. The planning horizon issue arises when at least one project has cash flows defined over a life which is greater than or less than the life of at least one other project. This situation did not occur in Example 17 of the previous section since all projects had 4 year lives.

There are four common approaches to establishing a planning horizon for evaluating decision alternatives. These are (1) shortest life, (2) longest life, (3) least common multiple of lives, and (4) standard. The shortest life planning horizon is established by selecting the project with the shortest life and setting this life as the planning horizon. A significant issue in this approach is how to value the remaining cash flows for projects whose lives are truncated. The typical approach to this valuation is to estimate the value of the remaining cash flows as the salvage value (market value) of the investment at that point in its life.

**Example 18**

Determine the shortest life planning horizon for projects A, B, C with lives 3, 5, and 6 years, respectively.

> The shortest life planning horizon is 3 years based on Project A. A salvage value must be established at t=3 for B's cash flows in years 4 and 5. A salvage value must be established at t=3 for C's cash flows in years 4, 5, and 6.

The longest life planning horizon is established by selecting the project with the longest life and setting this life as the planning horizon. The significant issue in this approach is how to handle projects whose cash flows don't extend this long. The typical resolution for this problem is to assume that shorter projects are repeated consecutively (end-to-end) until one of the repetitions extends at least as far as the planning horizon. The assumption of project repeatability deserves careful consideration since in some cases it is reasonable and in others it may be quite unreasonable. The reasonableness of the assumption is largely a function of the type of investment and the rate of innovation occurring within the investment's field (e.g., assuming repeatability of investments in high technology equipment is frequently ill advised since the field is advancing rapidly). If in repeating a project's cash flows, the last repetition's cash flows extend beyond the planning horizon, then the truncated cash flows (those that extend beyond the planning horizon) must be assigned a salvage value as above.

**Example 19**

Determine the longest life planning horizon for projects A, B, C with lives 3, 5, and 6 years, respectively.

> The longest life planning horizon is 6 years based on Project C. Project A must be repeated twice, the second repetition ends at year 6, so no termination of cash flows is required. Project B's second repetition extends to year 10, therefore, a salvage value at t=6 must be established for B's repeated cash flows in years 7, 8, 9, and 10.

An approach that eliminates the truncation salvage value issue from the planning horizon question is the least common multiple approach. The least common multiple planning horizon is set by determining the smallest number of years at which repetitions of all projects would terminate simultaneously. The least common multiple for a set of numbers (lives) can be determined mathematically using algebra. Discussion of this approach is beyond the scope of this chapter. For a small number of projects, the value can be determined by trial and error by examining multiples of the longest life project.

**Example 20**

Determine the least common multiple planning horizon for projects A, B, C with lives 3, 5, and 6 years, respectively.

The least common multiple of 3, 5, and 6 is 30. This can be obtained by trial and error starting with the longest project life (6) as follows:

1st trial: 6*1=6; 6 is a multiple of 3 but not 5; reject 6 and proceed

2nd trial: 6*2=12; 12 is a multiple of 3 but not 5; reject 12 and proceed

3rd trial: 6*3=18; 18 is a multiple of 3 but not 5; reject 18 and proceed

4th trial: 6*4=24; 24 is a multiple of 3 but not 5; reject 24 and proceed

5th trial: 6*5=30; 30 is a multiple of 3 and 5; accept 30 and stop

Under a 30-year planning horizon, A's cash flows are repeated 10 times, B's 6 times, and C's 5 times. No truncation is required.

The standard planning horizon approach uses a planning horizon which is independent of the projects being evaluated. Typically, this type of planning horizon is based on company policies or practices. The standard horizon may require repetition and/or truncation depending upon the set of projects being evaluated.

**Example 21**

Determine the impact of a 5 year standard planning horizon on projects A, B, C with lives 3, 5, and 6 years, respectively.

With a 5-year planning horizon:

Project A must be repeated one time with the second repetition truncated by one year.

Project B is a 5 year project and does not require repetition or truncation.

Project C must be truncated by one year.

There is no single accepted approach to resolving the planning horizon issue. Companies and individuals generally use one of the approaches outlined above. The decision of which to use in a particular analysis is generally a function of company practice and consideration of the reasonableness of the project repeatability assumption and the availability of salvage value estimates at truncation points.

## 4.9    SPECIAL PROBLEMS

### 4.9.1 Introduction

The preceding sections of this chapter outline an approach for conducting deterministic economic analysis of investment opportunities. Adherence to the concepts and methods presented will lead to sound investment decisions with respect to time value of money principles. This section addresses several topics that are of special interest in some analysis situations.

### 4.9.2 Interpolating Interest Tables

All of the examples previously presented in this chapter conveniently used interest rates whose time value of money factors were tabulated in Appendix 4A. How does one proceed if non-tabulated time value of money factors are needed? There are two viable approaches; calculation of the exact values and interpolation. The best and theoretically correct approach is to calculate the exact values of needed factors based on the formulas in Table 4.6.

**Example 22**

Determine the exact value for $(F \mid P, 13\%, 7)$.

From Table 4.6,

$(F \mid P, i, n) = (1+i)^n = (1+.13)^7 = 2.3526$

Interpolation is often used instead of calculation of exact values because, with practice, interpolated values can be calculated quickly. Interpolated values are not "exact" but for most practical problems they are "close enough," particularly if the range of interpolation is kept as narrow as possible. Interpolation of some factors, for instance $(P \mid A, i, n)$, also tends to be less error prone than the exact calculation due simpler mathematical operations.

Interpolation involves determining an unknown time value of money factor using two known values which bracket the value of interest. An assumption is made that the values of the time value of money factor vary linearly between the known values. Ratios are then used to estimate the unknown value. The example below illustrates the process.

**Example 23**
Determine an interpolated value for (F|P,13%,7).

The narrowest range of interest rates which bracket 13% and for which time value of money factor tables are provided in Appendix 4A is 12% to 15%.

The values necessary for this interpolation are

| i values | (F\|P,i%,7) |
|----------|-------------|
| 12% | 2.2107 |
| 13% | (F\|P,13%,7) |
| 15% | 2.6600 |

The interpolation proceeds by setting up ratios and solving for the unknown value, (F|P,13%,7), as follows:

$$\frac{\text{change between rows 2 \& 1 of left column}}{\text{change between rows 3 \& 1 of left column}} =$$

$$\frac{0.13 - 0.12}{0.15 - 0.12} = \frac{(F|P,13\%,7) - 2.2107}{2.6600 - 2.2107}$$

$$\frac{0.01}{0.03} = \frac{(F|P,13\%,7) - 2.2107}{0.4493}$$

$$0.1498 = (F|P,13\%,7) - 2.2107$$

$$(F|P,13\%,7) = 2.3605$$

The interpolated value for (F|P,13%,7), 2.3605, differs from the exact value, 2.3526, by 0.0079. This would imply a $7.90 difference in present worth for every thousand dollars of return at t=7. The relative importance of this interpolation error can be judged only in the context of a specific problem.

### 4.9.3 Non-Annual Interest Compounding

Many practical economic analysis problems involve interest that is not compounded annually. It is common practice to express a non-annually compounded interest rate as follows:

12% per year compounded monthly or 12%/yr/mo.

When expressed in this form, 12%/yr/mo is known as the *nominal* annual interest rate. The techniques covered in this chapter up to this point can not be used directly to solve an economic analysis problem of this type because the interest period (per year) and compounding period (monthly) are not the same. Two approaches can be used to solve problems of this type. One approach involves determining a *period* interest rate, the other involves determining an *effective* interest rate.

To solve this type of problem using a period interest rate approach, we must define the period interest rate:

$$\text{Period Interest Rate} = \frac{\text{Nominal Annual Interest Rate}}{\text{Number of Interest Periods per Year}}$$

In our example,

$$\text{Period Interest Rate} = \frac{12\%/yr/mo}{12\ mo/yr} = 1\%/mo/mo$$

Because the interest period and the compounding period are now the same, the time value of money factors in Appendix 4A can be applied directly. Note however, that the number of interest periods (n) must be adjusted to match the new frequency.

**Example 24**
$2,000 is invested in an account which pays 12% per year compounded monthly. What is the balance in the account after 3 years?

Nominal Annual Interest Rate = 12%/yr/mo

$$\text{Period Interest Rate} = \frac{12\%/yr/mo}{12\ mo/yr} = 1\%/mo/mo$$

Number of Interest Periods = 3 years × 12 mo/yr = 36 interest periods (months)

F = P (F|P,i,n) = $2,000 (F|P,1,36) = $2,000 (1.4308) = $2,861.60

**Example 25**
What are the monthly payments on a 5-year car loan of $12,500 at 6% per year compounded monthly?

Nominal Annual Interest Rate = 6%/yr/mo

$$\text{Period Interest Rate} = \frac{6\%/yr/mo}{12\ mo/yr} = 0.5\%/mo/mo$$

Number of Interest Periods = 5 years × 12 mo/yr = 60 interest periods

A = P (A|P,i,n) = \$12,500 (A|P,0.5,60) = \$12,500 (0.0193) = \$241.25

To solve this type of problem using an effective interest rate approach, we must define the effective interest rate. The effective annual interest rate is the annualized interest rate that would yield results equivalent to the period interest rate as previously calculated. Note however that the effective annual interest rate approach should not be used if the cash flows are more frequent than annual (e.g., monthly). In general, the interest rate for time value of money factors should match the frequency of the cash flows (e.g., if the cash flows are monthly, use the period interest rate approach with monthly periods).

As an example of the calculation of an effective interest rate, assume that the nominal interest rate is 12%/yr/qtr, therefore the period interest rate is 3%/qtr/qtr. One dollar invested for 1 year at 3%/qtr/qtr would have a future worth of:

$$F = P (F|P,i,n) = \$1 (F|P,3,4) = \$1 (1.03)^4$$
$$= \$1 (1.1255) = \$1.1255$$

To get this same value in 1 year with an annual rate the annual rate would have to be of 12.55%/yr/yr. This value is called the effective annual interest rate. The effective annual interest rate is given by $(1.03)^4 - 1 = 0.1255$ or 12.55%.

The general equation for the Effective Annual Interest Rate is:

Effective Annual Interest Rate = $(1 + (r/m))^m - 1$
where:   r =   nominal annual interest rate
             m =   number of interest periods per year

**Example 26**
What is the effective annual interest rate if the nominal rate is 12%/yr compounded monthly?

nominal annual interest rate = 12%/yr/mo
period interest rate = 1%/mo/mo
effective annual interest rate = $(1+0.12/12)^{12} - 1 =$ 0.1268 or 12.68%

**4.9.4 Economic Analysis Under Inflation**
Inflation is characterized by a decrease in the purchasing power of money caused by an increase in general price levels of goods and services without an accompanying increase the value of the goods and services. Inflationary pressure is created when more dollars are put into an economy without an accompanying increase in goods and services. In other words, printing more money without an increase in economic output generates inflation. A complete treatment of inflation is beyond the scope of this chapter. A good summary can be found in Sullivan and Bontadelli [1980].

When consideration of inflation is introduced into economic analysis, future cash flows can be stated in terms of either constant-worth dollars or then-current dollars. Then-current cash flows are expressed in terms of the face amount of dollars (actual number of dollars) that will change hands when the cash flow occurs. Alternatively, constant-worth cash flows are expressed in terms of the purchasing power of dollars relative to a fixed point in time known as the base period.

**Example 27**
For the next 4 years, a family anticipates buying \$1000 worth of groceries each year. If inflation is expected to be 3%/yr what are the then-current cash flows required to purchase the groceries.

To buy the groceries, the family will need to take the following face amount of dollars to the store. We will somewhat artificially assume that the family only shops once per year, buys the same set of items each year, and that the first trip to the store will be one year from today.

Year 1: dollars required \$1000.00*(1.03)=\$1030.00
Year 2: dollars required \$1030.00*(1.03)=\$1060.90
Year 3: dollars required \$1060.90*(1.03)=\$1092.73
Year 4: dollars required \$1092.73*(1.03)=\$1125.51

What are the constant-worth cash flows, if today's dollars are used as the base year.

The constant worth dollars are inflation free dollars, therefore the \$1000 of groceries costs \$1000 each year.

Year 1: \$1000.00
Year 2: \$1000.00
Year 3: \$1000.00
Year 4: \$1000.00

The key to proper economic analysis under inflation is to base the value of MARR on the types of cash flows. If the cash flows contain inflation, then the value of MARR should also be adjusted for inflation. Alternatively, if the cash flows do not contain inflation, then the value of MARR should be inflation free. When MARR does not contain an adjustment for inflation, it is referred to as a real value for MARR. If it contains an inflation adjustment, it is referred to as a combined value for MARR. The relationship between inflation rate, the real value of MARR, and the combined value of MARR is given by:

$$1 + MARR_{COMBINED} = (1 + \text{inflation rate}) * (1 + MARR_{REAL})$$

**Example 28**

If the inflation rate is 3%/yr and the real value of MARR is 15%/yr, what is the combined value of MARR?

$$1 + MARR_{COMBINED} = (1 + \text{inflation rate}) * (1 + MARR_{REAL})$$

$$1 + MARR_{COMBINED} = (1 + 0.03) * (1 + 0.15)$$

$$1 + MARR_{COMBINED} = (1.03) * (1.15)$$

$$1 + MARR_{COMBINED} = 1.1845$$

$$MARR_{COMBINED} = 1.1845 - 1 = 0.1845 = 18.45\%$$

If the cash flows of a project are stated in terms of then-current dollars, the appropriate value of MARR is the combined value of MARR. Analysis done in this way is referred to as *then-current analysis*. If the cash flows of a project are stated in terms of constant-worth dollars, the appropriate value of MARR is the real value of MARR. Analysis done in this way is referred to as then *constant worth analysis*.

**Example 29**

Using the cash flows of Examples 27 and interest rates of Example 28, determine the present worth of the grocery purchases using a constant worth analysis.

Constant worth analysis requires constant worth cash flows and the real value of MARR.

$$PW = 1000 * (P|A,15\%,4)$$
$$= 1000 * (2.8550) = \$2855.00$$

**Example 30**

Using the cash flows of Examples 27 and interest rates of Example 28, determine the present worth of the grocery purchases using a then-current analysis.

Then-current analysis requires then current cash flows and the combined value of MARR.

$$PW = 1030.00 * (P|F,18.45\%,1) + 1060.90 * (P|F,18.45\%,2) + 1092.73 * (P|F,18.45\%,3) +1125.51 * (P|F,18.45\%,4)$$

$$PW = 1030.00 * (0.8442) + 1060.90 * (0.7127) + 1092.73 * (0.6017) +1125.51 * (0.5080)$$

$$PW = 869.53 + 756.10 + 657.50 + 571.76 = 2854.89$$

The notable result of Examples 29 and 30 is that the present worths determined by the constant-worth approach ($2855.00) and the then-current approach ($2854.89) are equal (the $0.11 difference is due to rounding). This result is often unexpected but it is mathematically sound. The important conclusion is that if care is taken to appropriately match the cash flows and value of MARR, the level of general price inflation is not a determining factor in the acceptability of projects. To make this important result hold, inflation must either (1) be included in both the cash flows and MARR (the then-current approach) or (2) be included in neither the cash flows nor MARR (the constant-worth approach).

**4.9.5 Sensitivity Analysis and Risk Analysis**

Often times the certainty assumptions associated with deterministic analysis are questionable. These certainty assumptions include certain knowledge regarding amounts and timing of cash flows as well as certain knowledge of MARR. Relaxing these assumptions requires the use of sensitivity analysis and risk analysis techniques.

Initial sensitivity analyses are usually conducted on the optimal decision alternative (or top two or three) on a single factor basis. Single factor sensitivity analysis involves holding all cost factors except one constant while varying the remaining cost factor through a range of percentage changes. The effect of cost factor changes on the measure of worth is observed to determine whether the alternative remains attractive under the evaluated changes and to determine which cost factor affects the measure of worth the most.

**Example 31**

Conduct a sensitivity analysis of the optimal decision resulting from the constrained analysis of the data in Example 17. The sensitivity analysis should explore the sensitivity of present worth to changes in annual revenue over the range -10% to +10%.

The PW of the optimal decision (Accept A & D only) was determined in Section 4.8.3 to be:

$PW_{A\&D}$ = -1500 + 890*(P|A,12%,4) = -1500 + 890 (3.0373) = \$1203.21

If annual revenue decreases 10%, it becomes 890 - 0.10*890 = 801 and PW becomes

$PW_{A\&D}$ = -1500 + 801*(P|A,12%,4) = -1500 + 801 (3.0373) = \$932.88

If annual revenue increases 10%, it becomes 890 + 0.10*890 = 979 and PW becomes

$PW_{A\&D}$ = -1500 + 979*(P|A,12%,4) = -1500 + 979 (3.0373) = \$1473.52

The sensitivity of PW to changes in annual revenue over the range -10% to +10% is +\$540.64 from \$932.88 to \$1473.52.

**Example 32**

Repeat Example 31 exploring of the sensitivity of present worth to changes in initial cost over the range -10% to +10%.

The PW of the optimal decision (Accept A & D only) was determined in Section 4.8.3 to be:

$PW_{A\&D}$ = -1500 + 890*(P|A,12%,4) = -1500 + 890 (3.0373) = \$1203.21

If initial cost decreases 10% it becomes 1500 - 0.10*1500 = 1350 and PW becomes

$PW_{A\&D}$ = -1350 + 890*(P|A,12%,4) = -1350 + 890 (3.0373) = \$1353.20

If initial cost increases 10% it becomes 1500 + 0.10*1500 = 1650 and PW becomes

$PW_{A\&D}$ = -1650 + 890*(P|A,12%,4) = -1500 + 890 (3.0373) = \$1053.20

The sensitivity of PW to changes in initial cost over the range -10% to +10% is -\$300.00 from \$1353.20 to \$1053.20.

**Example 33**

Repeat Example 31 exploring the sensitivity of the present worth to changes in MARR over the range -10% to +10%.

The PW of the optimal decision (Accept A & D only) was determined in Section 4.8.3 to be:

$PW_{A\&D}$ = -1500 + 890*(P|A,12%,4) = -1500 + 890 (3.0373) = \$1203.21

If MARR decreases 10% it becomes 12% - 0.10*12% = 10.8% and PW becomes

$PW_{A\&D}$ = -1500 + 890*(P|A,10.8%,4) = -1500 + 890 (3.1157) = \$1272.97

If MARR increases 10% it becomes 12% + 0.10*12% = 13.2% and PW becomes

$PW_{A\&D}$ = -1500 + 890*(P|A,13.2%,4) = -1500 + 890 (2.9622) = \$1136.36

The sensitivity of PW to changes in MARR over the range -10% to +10% is -\$136.61 from \$1272.97 to \$1136.36.

The sensitivity data from Examples 31, 32, and 33 are summarized in Table 4.11. A review of the table reveals that the decision alternative A&D remains attractive (PW ≥0) within the range of 10% changes in annual revenues, initial cost, and MARR. An appealing way to summarize single factor sensitivity data is using a "spider" graph. A spider graph plots the PW values determined in the examples and connects them with lines, one line for each factor evaluated. Figure 4.10 illustrates the spider graph for the data of Table 4.11. On this graph, lines with large positive or negative slopes (angle relative to horizontal regardless of whether it is increasing or decreasing) indicate factors to which the present value measure of worth is sensitive. Figure 4.10 shows that PW is least sensitive to changes in MARR (the MARR line is the most nearly horizontal) and most sensitivity to changes in annual revenue (the annual revenue line has the steepest slope). Additional sensitivities could be explored in a similar manner.

**Table 4.11 Sensitivity analysis data table**

| Factor \ Percent Change | - 10% | Base | + 10% |
| --- | --- | --- | --- |
| 1st Cost | 1353.20 | 1203.21 | 1053.20 |
| Annual Revenue | 932.88 | 1203.21 | 1473.52 |
| MARR | 1272.97 | 1203.21 | 1136.36 |

When single factor sensitivity analysis is inadequate to assess the questions which surround the certainty assumptions of a deterministic analysis, risk

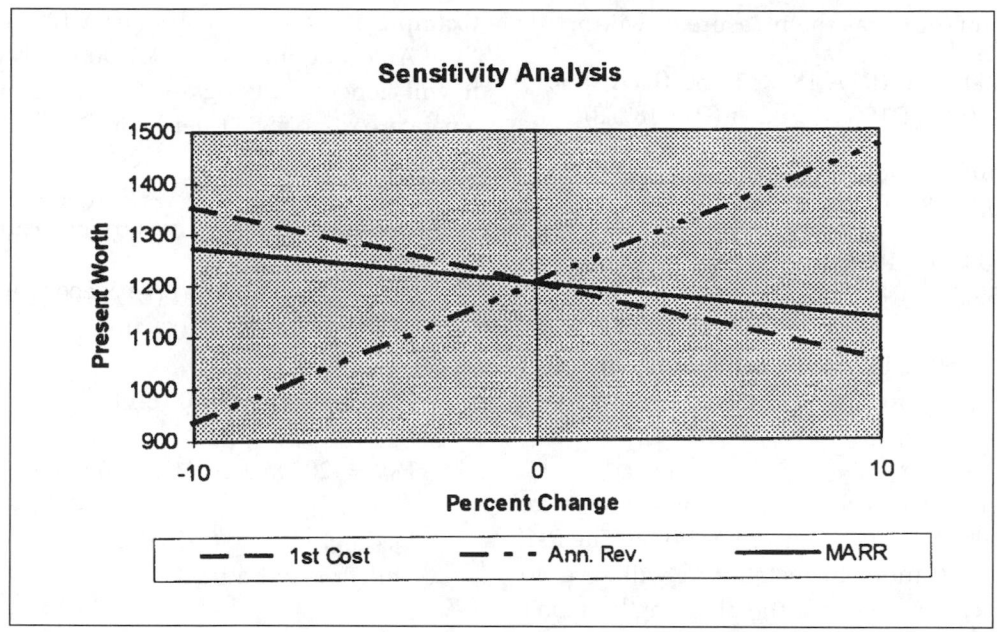

**Figure 4.10. Sensitivity analysis "spider" graph**

analysis techniques can be employed. One approach to risk analysis is the application of probabilistic and statistical concepts to economic analysis. These techniques require information regarding the possible values that uncertain quantities may take on as well as estimates of the probability that the various values will occur. A detailed treatment of this topic is beyond the scope of this chapter. A good discussion of this subject can be found in Park and Sharp-Bette [1990].

A second approach to risk analysis in economic analysis is through the use of simulation techniques and simulation software. Simulation involves using a computer simulation program to sample possible values for the uncertain quantities in an economic analysis and calculating the measure of worth. This process is repeated many times using different samples each time. After many samples have been taken, probability statements regarding the measure of worth may be made. A good discussion of this subject can be found in Park and Sharp-Bette [1990].

## 4.10 SUMMARY AND ADDITIONAL EXAMPLE APPLICATIONS

In this chapter a coherent, consistent approach to economic analysis of capital investments (energy related or other) has been presented. To conclude the chapter, this section provides several additional examples to illustrate the use of time value of money concepts for energy related problems. Additional example applications

as well as a more in depth presentation of conceptual details can be found in the references listed at the end of the chapter. These references are by no means exclusive; many other excellent presentations of the subject matter are also available. Adherence to the concepts and methods presented here and in the references will lead to sound investment decisions with respect to time value of money principles.

### Example 34

In Section 4.3.3 an example involving the evaluation of a baseboard heating and window air conditioner versus a heat pump was introduced to illustrate cash flow diagramming (Figure 4.2). A summary of the differential costs is repeat here for convenience.

- The heat pump costs $1500 more than the baseboard system,

- The heat pump saves $380 annually in electricity costs,

- The heat pump has a $50 higher annual maintenance costs,

- The heat pump has a $150 higher salvage value at the end of 15 years,

- The heat pump requires $200 more in replacement maintenance at the end of year 8.

If MARR is 18%, is the additional investment in the heat pump attractive?

Using present worth as the measure of worth:

PW =  -1500+380*(P I A,18%,15)-50*(P I A,18%,15)
      + 150*(P I F,18%,15) - 200*(P I F,18%,8)

PW =  -1500 + 380*(5.0916) - 50*(5.0916) +
      150*(0.0835) - 200*(0.2660)

PW =  -1500.00 + 1934.81 - 254.58 + 12.53 - 53.20 =
      $139.56

Decision: PW≥0 ($139.56>0.0), therefore the additional investment for the heat pump is attractive.

## Example 35

A homeowner needs to decide whether to install R-11 or R-19 insulation in the attic of her home. The R-19 insulation costs $150 more to install and will save approximately 400 kWh per year. If the planning horizon is 20 years and electricity costs $0.08/kWh is the additional investment attractive at MARR of 10%?

At $0.08/kWh, the annual savings are: 400 kWh * $0.08/kWh = $32.00

Using present worth as the measure of worth:

PW = -150 + 32*(P I A,10%,20)

PW = -150 + 32*(8.5136) = -150 + 272.44 = $122.44

Decision: PW≥0 ($122.44>0.0), therefore the R-19 insulation is attractive.

## Example 36

The homeowner from Example 35 can install R-30 insulation in the attic of her home for $200 more than the R-19 insulation. The R-30 will save approximately 250 kWh per year over the R-19 insulation. Is the additional investment attractive?

Assuming the same MARR, electricity cost, and planning horizon, the additional annual savings are: 250 kWh * $0.08/kWh = $20.00

Using present worth as the measure of worth:

PW = -200 + 20*(P I A,10%,20)

PW = -200 + 20*(8.5136) = -200 + 170.27 = -$29.73

Decision: PW<0 (-$29.73<0.0), therefore the R-30 insulation is not attractive.

## Example 37

An economizer costs $20,000 and will last 10 years. It will generate savings of $3,500 per year with maintenance costs of $500 per year. If MARR is 10%, is the economizer an attractive investment?

Using present worth as the measure of worth:

PW = -20000 + 3500*(P I A,10%,10)
       - 500*(P I A,10%,10)

PW = -20000 + 3500*(6.1446) - 500*(6.1446)

PW = -20000.00 + 21506.10 - 3072.30 = -$1566.20

Decision: PW<0 (-$1566.20<0.0), therefore the economizer is not attractive.

## Example 38

If the economizer from Example 37 has a salvage value of $5000 at the end of 10 years is the investment attractive?

Using present worth as the measure of worth:

PW = -20000 + 3500*(P I A,10%,10)
       - 500*(P I A,10%,10) + 5000*(P I F,10%,10)

PW = -20000 + 3500*(6.1446) - 500*(6.1446)
       + 5000*(0.3855)

PW = -20000.00 + 21506.10 - 3072.30 + 1927.50
       = $361.30

Decision: PW≥0 ($361.30≥0.0), therefore the economizer is now attractive.

## 4.11  References

Brown, R.J. and R.R. Yanuck, 1980, *Life Cycle Costing: A Practical Guide for Energy Managers*, The Fairmont Press, Inc., Atlanta, GA.

Fuller, S.K. and S.R. Petersen, 1994, NISTIR 5165: *Life-Cycle Costing Workshop for Energy Conservation in Buildings: Student Manual*, U.S. Department of Commerce, Office of Applied Economics, Gaithersburg, MD.

Fuller, S.K. and S.R. Petersen, 1995, *NIST Handbook 135: Life-Cycle Costing Manual for the Federal Energy Management Program*, National Technical Information Service, Springfield, VA.

Park, C.S. and G.P. Sharp-Bette, 1990, Advanced Engineering Economics, John Wiley & Sons, New York, NY.

Sullivan, W.G. and J.A. Bontadelli, 1980, "The Industrial Engineer and Inflation," *Industrial Engineering*, Vol. 12, No. 3, 24-33.

Thuesen, G.J. and W.J. Fabrycky, 1993, *Engineering Economy, 8th Edition*, Prentice Hall, Englewood Cliffs, NJ.

White, J.A., K.E. Case, D.B. Pratt, and M.H. Agee, 1998, *Principles of Engineering Economic Analysis*, 3rd Edition, John Wiley & Sons, New York, NY.

# Appendix

Time Value of Money Factors—Discrete Compounding
  i = 1%

| | Single Sums | | Uniform Series | | | | Gradient Series | |
|---|---|---|---|---|---|---|---|---|
| | To Find F Given P (F\|P,i%,n) | To Find P Given F (P\|F,i%,n) | To Find F Given A (F\|A,i%,n) | To Find A Given F (A\|F,i%,n) | To Find P Given A (P\|A,i%,n) | To Find A Given P (A\|P,i%,n) | To Find P Given G (P\|G,i%,n) | To Find A Given G (A\|G,i%,n) |
| n | | | | | | | | |
| 1 | 1.0100 | 0.9901 | 1.0000 | 1.0000 | 0.9901 | 1.0100 | 0.0000 | 0.0000 |
| 2 | 1.0201 | 0.9803 | 2.0100 | 0.4975 | 1.9704 | 0.5075 | 0.9803 | 0.4975 |
| 3 | 1.0303 | 0.9706 | 3.0301 | 0.3300 | 2.9410 | 0.3400 | 2.9215 | 0.9934 |
| 4 | 1.0406 | 0.9610 | 4.0604 | 0.2463 | 3.9020 | 0.2563 | 5.8044 | 1.4876 |
| 5 | 1.0510 | 0.9515 | 5.1010 | 0.1960 | 4.8534 | 0.2060 | 9.6103 | 1.9801 |
| 6 | 1.0615 | 0.9420 | 6.1520 | 0.1625 | 5.7955 | 0.1725 | 14.3205 | 2.4710 |
| 7 | 1.0721 | 0.9327 | 7.2135 | 0.1386 | 6.7282 | 0.1486 | 19.9168 | 2.9602 |
| 8 | 1.0829 | 0.9235 | 8.2857 | 0.1207 | 7.6517 | 0.1307 | 26.3812 | 3.4478 |
| 9 | 1.0937 | 0.9143 | 9.3685 | 0.1067 | 8.5660 | 0.1167 | 33.6959 | 3.9337 |
| 10 | 1.1046 | 0.9053 | 10.4622 | 0.0956 | 9.4713 | 0.1056 | 41.8435 | 4.4179 |
| 11 | 1.1157 | 0.8963 | 11.5668 | 0.0865 | 10.3676 | 0.0965 | 50.8067 | 4.9005 |
| 12 | 1.1268 | 0.8874 | 12.6825 | 0.0788 | 11.2551 | 0.0888 | 60.5687 | 5.3815 |
| 13 | 1.1381 | 0.8787 | 13.8093 | 0.0724 | 12.1337 | 0.0824 | 71.1126 | 5.8607 |
| 14 | 1.1495 | 0.8700 | 14.9474 | 0.0669 | 13.0037 | 0.0769 | 82.4221 | 6.3384 |
| 15 | 1.1610 | 0.8613 | 16.0969 | 0.0621 | 13.8651 | 0.0721 | 94.4810 | 6.8143 |
| 16 | 1.1726 | 0.8528 | 17.2579 | 0.0579 | 14.7179 | 0.0679 | 107.2734 | 7.2886 |
| 17 | 1.1843 | 0.8444 | 18.4304 | 0.0543 | 15.5623 | 0.0643 | 120.7834 | 7.7613 |
| 18 | 1.1961 | 0.8360 | 19.6147 | 0.0510 | 16.3983 | 0.0610 | 134.9957 | 8.2323 |
| 19 | 1.2081 | 0.8277 | 20.8109 | 0.0481 | 17.2260 | 0.0581 | 149.8950 | 8.7017 |
| 20 | 1.2202 | 0.8195 | 22.0190 | 0.0454 | 18.0456 | 0.0554 | 165.4664 | 9.1694 |
| 21 | 1.2324 | 0.8114 | 23.2392 | 0.0430 | 18.8570 | 0.0530 | 181.6950 | 9.6354 |
| 22 | 1.2447 | 0.8034 | 24.4716 | 0.0409 | 19.6604 | 0.0509 | 198.5663 | 10.0998 |
| 23 | 1.2572 | 0.7954 | 25.7163 | 0.0389 | 20.4558 | 0.0489 | 216.0660 | 10.5626 |
| 24 | 1.2697 | 0.7876 | 26.9735 | 0.0371 | 21.2434 | 0.0471 | 234.1800 | 11.0237 |
| 25 | 1.2824 | 0.7798 | 28.2432 | 0.0354 | 22.0232 | 0.0454 | 252.8945 | 11.4831 |
| 26 | 1.2953 | 0.7720 | 29.5256 | 0.0339 | 22.7952 | 0.0439 | 272.1957 | 11.9409 |
| 27 | 1.3082 | 0.7644 | 30.8209 | 0.0324 | 23.5596 | 0.0424 | 292.0702 | 12.3971 |
| 28 | 1.3213 | 0.7568 | 32.1291 | 0.0311 | 24.3164 | 0.0411 | 312.5047 | 12.8516 |
| 29 | 1.3345 | 0.7493 | 33.4504 | 0.0299 | 25.0658 | 0.0399 | 333.4863 | 13.3044 |
| 30 | 1.3478 | 0.7419 | 34.7849 | 0.0287 | 25.8077 | 0.0387 | 355.0021 | 13.7557 |
| 36 | 1.4308 | 0.6989 | 43.0769 | 0.0232 | 30.1075 | 0.0332 | 494.6207 | 16.4285 |
| 42 | 1.5188 | 0.6584 | 51.8790 | 0.0193 | 34.1581 | 0.0293 | 650.4514 | 19.0424 |
| 48 | 1.6122 | 0.6203 | 61.2226 | 0.0163 | 37.9740 | 0.0263 | 820.1460 | 21.5976 |
| 54 | 1.7114 | 0.5843 | 71.1410 | 0.0141 | 41.5687 | 0.0241 | 1.002E+03 | 24.0945 |
| 60 | 1.8167 | 0.5504 | 81.6697 | 0.0122 | 44.9550 | 0.0222 | 1.193E+03 | 26.5333 |
| 66 | 1.9285 | 0.5185 | 92.8460 | 0.0108 | 48.1452 | 0.0208 | 1.392E+03 | 28.9146 |
| 72 | 2.0471 | 0.4885 | 104.7099 | 9.550E-03 | 51.1504 | 0.0196 | 1.598E+03 | 31.2386 |
| 120 | 3.3004 | 0.3030 | 230.0387 | 4.347E-03 | 69.7005 | 0.0143 | 3.334E+03 | 47.8349 |
| 180 | 5.9958 | 0.1668 | 499.5802 | 2.002E-03 | 83.3217 | 0.0120 | 5.330E+03 | 63.9697 |
| 360 | 35.9496 | 0.0278 | 3.495E+03 | 2.861E-04 | 97.2183 | 0.0103 | 8.720E+03 | 89.6995 |

## Time Value of Money Factors—Discrete Compounding
### i = 2%

| n | Single Sums | | Uniform Series | | | | Gradient Series | |
|---|---|---|---|---|---|---|---|---|
| | To Find F Given P (F\|P,i%,n) | To Find P Given F (P\|F,i%,n) | To Find F Given A (F\|A,i%,n) | To Find A Given F (A\|F,i%,n) | To Find P Given A (P\|A,i%,n) | To Find A Given P (A\|P,i%,n) | To Find P Given G (P\|G,i%,n) | To Find A Given G (A\|G,i%,n) |
| 1 | 1.0200 | 0.9804 | 1.0000 | 1.0000 | 0.9804 | 1.0200 | 0.0000 | 0.0000 |
| 2 | 1.0404 | 0.9612 | 2.0200 | 0.4950 | 1.9416 | 0.5150 | 0.9612 | 0.4950 |
| 3 | 1.0612 | 0.9423 | 3.0604 | 0.3268 | 2.8839 | 0.3468 | 2.8458 | 0.9868 |
| 4 | 1.0824 | 0.9238 | 4.1216 | 0.2426 | 3.8077 | 0.2626 | 5.6173 | 1.4752 |
| 5 | 1.1041 | 0.9057 | 5.2040 | 0.1922 | 4.7135 | 0.2122 | 9.2403 | 1.9604 |
| 6 | 1.1262 | 0.8880 | 6.3081 | 0.1585 | 5.6014 | 0.1785 | 13.6801 | 2.4423 |
| 7 | 1.1487 | 0.8706 | 7.4343 | 0.1345 | 6.4720 | 0.1545 | 18.9035 | 2.9208 |
| 8 | 1.1717 | 0.8535 | 8.5830 | 0.1165 | 7.3255 | 0.1365 | 24.8779 | 3.3961 |
| 9 | 1.1951 | 0.8368 | 9.7546 | 0.1025 | 8.1622 | 0.1225 | 31.5720 | 3.8681 |
| 10 | 1.2190 | 0.8203 | 10.9497 | 0.0913 | 8.9826 | 0.1113 | 38.9551 | 4.3367 |
| 11 | 1.2434 | 0.8043 | 12.1687 | 0.0822 | 9.7868 | 0.1022 | 46.9977 | 4.8021 |
| 12 | 1.2682 | 0.7885 | 13.4121 | 0.0746 | 10.5753 | 0.0946 | 55.6712 | 5.2642 |
| 13 | 1.2936 | 0.7730 | 14.6803 | 0.0681 | 11.3484 | 0.0881 | 64.9475 | 5.7231 |
| 14 | 1.3195 | 0.7579 | 15.9739 | 0.0626 | 12.1062 | 0.0826 | 74.7999 | 6.1786 |
| 15 | 1.3459 | 0.7430 | 17.2934 | 0.0578 | 12.8493 | 0.0778 | 85.2021 | 6.6309 |
| 16 | 1.3728 | 0.7284 | 18.6393 | 0.0537 | 13.5777 | 0.0737 | 96.1288 | 7.0799 |
| 17 | 1.4002 | 0.7142 | 20.0121 | 0.0500 | 14.2919 | 0.0700 | 107.5554 | 7.5256 |
| 18 | 1.4282 | 0.7002 | 21.4123 | 0.0467 | 14.9920 | 0.0667 | 119.4581 | 7.9681 |
| 19 | 1.4568 | 0.6864 | 22.8406 | 0.0438 | 15.6785 | 0.0638 | 131.8139 | 8.4073 |
| 20 | 1.4859 | 0.6730 | 24.2974 | 0.0412 | 16.3514 | 0.0612 | 144.6003 | 8.8433 |
| 21 | 1.5157 | 0.6598 | 25.7833 | 0.0388 | 17.0112 | 0.0588 | 157.7959 | 9.2760 |
| 22 | 1.5460 | 0.6468 | 27.2990 | 0.0366 | 17.6580 | 0.0566 | 171.3795 | 9.7055 |
| 23 | 1.5769 | 0.6342 | 28.8450 | 0.0347 | 18.2922 | 0.0547 | 185.3309 | 10.1317 |
| 24 | 1.6084 | 0.6217 | 30.4219 | 0.0329 | 18.9139 | 0.0529 | 199.6305 | 10.5547 |
| 25 | 1.6406 | 0.6095 | 32.0303 | 0.0312 | 19.5235 | 0.0512 | 214.2592 | 10.9745 |
| 26 | 1.6734 | 0.5976 | 33.6709 | 0.0297 | 20.1210 | 0.0497 | 229.1987 | 11.3910 |
| 27 | 1.7069 | 0.5859 | 35.3443 | 0.0283 | 20.7069 | 0.0483 | 244.4311 | 11.8043 |
| 28 | 1.7410 | 0.5744 | 37.0512 | 0.0270 | 21.2813 | 0.0470 | 259.9392 | 12.2145 |
| 29 | 1.7758 | 0.5631 | 38.7922 | 0.0258 | 21.8444 | 0.0458 | 275.7064 | 12.6214 |
| 30 | 1.8114 | 0.5521 | 40.5681 | 0.0246 | 22.3965 | 0.0446 | 291.7164 | 13.0251 |
| 36 | 2.0399 | 0.4902 | 51.9944 | 0.0192 | 25.4888 | 0.0392 | 392.0405 | 15.3809 |
| 42 | 2.2972 | 0.4353 | 64.8622 | 0.0154 | 28.2348 | 0.0354 | 497.6010 | 17.6237 |
| 48 | 2.5871 | 0.3865 | 79.3535 | 0.0126 | 30.6731 | 0.0326 | 605.9657 | 19.7556 |
| 54 | 2.9135 | 0.3432 | 95.6731 | 0.0105 | 32.8383 | 0.0305 | 715.1815 | 21.7789 |
| 60 | 3.2810 | 0.3048 | 114.0515 | 8.768E-03 | 34.7609 | 0.0288 | 823.6975 | 23.6961 |
| 66 | 3.6950 | 0.2706 | 134.7487 | 7.421E-03 | 36.4681 | 0.0274 | 930.3000 | 25.5100 |
| 72 | 4.1611 | 0.2403 | 158.0570 | 6.327E-03 | 37.9841 | 0.0263 | 1.034E+03 | 27.2234 |
| 120 | 10.7652 | 0.0929 | 488.2582 | 2.048E-03 | 45.3554 | 0.0220 | 1.710E+03 | 37.7114 |
| 180 | 35.3208 | 0.0283 | 1.716E+03 | 5.827E-04 | 48.5844 | 0.0206 | 2.174E+03 | 44.7554 |
| 360 | 1.248E+03 | 8.016E-04 | 6.233E+04 | 1.604E-05 | 49.9599 | 0.0200 | 2.484E+03 | 49.7112 |

Time Value of Money Factors—Discrete Compounding
   i = 3%

| | Single Sums | | Uniform Series | | | | Gradient Series | |
|---|---|---|---|---|---|---|---|---|
| | To Find F Given P (F\|P,i%,n) | To Find P Given F (P\|F,i%,n) | To Find F Given A (F\|A,i%,n) | To Find A Given F (A\|F,i%,n) | To Find P Given A (P\|A,i%,n) | To Find A Given P (A\|P,i%,n) | To Find P Given G (P\|G,i%,n) | To Find A Given G (A\|G,i%,n) |
| n | | | | | | | | |
| 1 | 1.0300 | 0.9709 | 1.0000 | 1.0000 | 0.9709 | 1.0300 | 0.0000 | 0.0000 |
| 2 | 1.0609 | 0.9426 | 2.0300 | 0.4926 | 1.9135 | 0.5226 | 0.9426 | 0.4926 |
| 3 | 1.0927 | 0.9151 | 3.0909 | 0.3235 | 2.8286 | 0.3535 | 2.7729 | 0.9803 |
| 4 | 1.1255 | 0.8885 | 4.1836 | 0.2390 | 3.7171 | 0.2690 | 5.4383 | 1.4631 |
| 5 | 1.1593 | 0.8626 | 5.3091 | 0.1884 | 4.5797 | 0.2184 | 8.8888 | 1.9409 |
| 6 | 1.1941 | 0.8375 | 6.4684 | 0.1546 | 5.4172 | 0.1846 | 13.0762 | 2.4138 |
| 7 | 1.2299 | 0.8131 | 7.6625 | 0.1305 | 6.2303 | 0.1605 | 17.9547 | 2.8819 |
| 8 | 1.2668 | 0.7894 | 8.8923 | 0.1125 | 7.0197 | 0.1425 | 23.4806 | 3.3450 |
| 9 | 1.3048 | 0.7664 | 10.1591 | 0.0984 | 7.7861 | 0.1284 | 29.6119 | 3.8032 |
| 10 | 1.3439 | 0.7441 | 11.4639 | 0.0872 | 8.5302 | 0.1172 | 36.3088 | 4.2565 |
| 11 | 1.3842 | 0.7224 | 12.8078 | 0.0781 | 9.2526 | 0.1081 | 43.5330 | 4.7049 |
| 12 | 1.4258 | 0.7014 | 14.1920 | 0.0705 | 9.9540 | 0.1005 | 51.2482 | 5.1485 |
| 13 | 1.4685 | 0.6810 | 15.6178 | 0.0640 | 10.6350 | 0.0940 | 59.4196 | 5.5872 |
| 14 | 1.5126 | 0.6611 | 17.0863 | 0.0585 | 11.2961 | 0.0885 | 68.0141 | 6.0210 |
| 15 | 1.5580 | 0.6419 | 18.5989 | 0.0538 | 11.9379 | 0.0838 | 77.0002 | 6.4500 |
| 16 | 1.6047 | 0.6232 | 20.1569 | 0.0496 | 12.5611 | 0.0796 | 86.3477 | 6.8742 |
| 17 | 1.6528 | 0.6050 | 21.7616 | 0.0460 | 13.1661 | 0.0760 | 96.0280 | 7.2936 |
| 18 | 1.7024 | 0.5874 | 23.4144 | 0.0427 | 13.7535 | 0.0727 | 106.0137 | 7.7081 |
| 19 | 1.7535 | 0.5703 | 25.1169 | 0.0398 | 14.3238 | 0.0698 | 116.2788 | 8.1179 |
| 20 | 1.8061 | 0.5537 | 26.8704 | 0.0372 | 14.8775 | 0.0672 | 126.7987 | 8.5229 |
| 21 | 1.8603 | 0.5375 | 28.6765 | 0.0349 | 15.4150 | 0.0649 | 137.5496 | 8.9231 |
| 22 | 1.9161 | 0.5219 | 30.5368 | 0.0327 | 15.9369 | 0.0627 | 148.5094 | 9.3186 |
| 23 | 1.9736 | 0.5067 | 32.4529 | 0.0308 | 16.4436 | 0.0608 | 159.6566 | 9.7093 |
| 24 | 2.0328 | 0.4919 | 34.4265 | 0.0290 | 16.9355 | 0.0590 | 170.9711 | 10.0954 |
| 25 | 2.0938 | 0.4776 | 36.4593 | 0.0274 | 17.4131 | 0.0574 | 182.4336 | 10.4768 |
| 26 | 2.1566 | 0.4637 | 38.5530 | 0.0259 | 17.8768 | 0.0559 | 194.0260 | 10.8535 |
| 27 | 2.2213 | 0.4502 | 40.7096 | 0.0246 | 18.3270 | 0.0546 | 205.7309 | 11.2255 |
| 28 | 2.2879 | 0.4371 | 42.9309 | 0.0233 | 18.7641 | 0.0533 | 217.5320 | 11.5930 |
| 29 | 2.3566 | 0.4243 | 45.2189 | 0.0221 | 19.1885 | 0.0521 | 229.4137 | 11.9558 |
| 30 | 2.4273 | 0.4120 | 47.5754 | 0.0210 | 19.6004 | 0.0510 | 241.3613 | 12.3141 |
| 36 | 2.8983 | 0.3450 | 63.2759 | 0.0158 | 21.8323 | 0.0458 | 313.7028 | 14.3688 |
| 42 | 3.4607 | 0.2890 | 82.0232 | 0.0122 | 23.7014 | 0.0422 | 385.5024 | 16.2650 |
| 48 | 4.1323 | 0.2420 | 104.4084 | 9.578E-03 | 25.2667 | 0.0396 | 455.0255 | 18.0089 |
| 54 | 4.9341 | 0.2027 | 131.1375 | 7.626E-03 | 26.5777 | 0.0376 | 521.1157 | 19.6073 |
| 60 | 5.8916 | 0.1697 | 163.0534 | 6.133E-03 | 27.6756 | 0.0361 | 583.0526 | 21.0674 |
| 66 | 7.0349 | 0.1421 | 201.1627 | 4.971E-03 | 28.5950 | 0.0350 | 640.4407 | 22.3969 |
| 72 | 8.4000 | 0.1190 | 246.6672 | 4.054E-03 | 29.3651 | 0.0341 | 693.1226 | 23.6036 |
| 120 | 34.7110 | 0.0288 | 1.124E+03 | 8.899E-04 | 32.3730 | 0.0309 | 963.8635 | 29.7737 |
| 180 | 204.5034 | 4.890E-03 | 6.783E+03 | 1.474E-04 | 33.1703 | 0.0301 | 1.076E+03 | 32.4488 |
| 360 | 4.182E+04 | 2.391E-05 | 1.394E+06 | 7.173E-07 | 33.3325 | 0.0300 | 1.111E+03 | 33.3247 |

Time Value of Money Factors—Discrete Compounding
    i = 4%

| | Single Sums | | Uniform Series | | | | Gradient Series | |
|---|---|---|---|---|---|---|---|---|
| | To Find F Given P (F\|P,i%,n) | To Find P Given F (P\|F,i%,n) | To Find F Given A (F\|A,i%,n) | To Find A Given F (A\|F,i%,n) | To Find P Given A (P\|A,i%,n) | To Find A Given P (A\|P,i%,n) | To Find P Given G (P\|G,i%,n) | To Find A Given G (A\|G,i%,n) |
| n | | | | | | | | |
| 1 | 1.0400 | 0.9615 | 1.0000 | 1.0000 | 0.9615 | 1.0400 | 0.0000 | 0.0000 |
| 2 | 1.0816 | 0.9246 | 2.0400 | 0.4902 | 1.8861 | 0.5302 | 0.9246 | 0.4902 |
| 3 | 1.1249 | 0.8890 | 3.1216 | 0.3203 | 2.7751 | 0.3603 | 2.7025 | 0.9739 |
| 4 | 1.1699 | 0.8548 | 4.2465 | 0.2355 | 3.6299 | 0.2755 | 5.2670 | 1.4510 |
| 5 | 1.2167 | 0.8219 | 5.4163 | 0.1846 | 4.4518 | 0.2246 | 8.5547 | 1.9216 |
| 6 | 1.2653 | 0.7903 | 6.6330 | 0.1508 | 5.2421 | 0.1908 | 12.5062 | 2.3857 |
| 7 | 1.3159 | 0.7599 | 7.8983 | 0.1266 | 6.0021 | 0.1666 | 17.0657 | 2.8433 |
| 8 | 1.3686 | 0.7307 | 9.2142 | 0.1085 | 6.7327 | 0.1485 | 22.1806 | 3.2944 |
| 9 | 1.4233 | 0.7026 | 10.5828 | 0.0945 | 7.4353 | 0.1345 | 27.8013 | 3.7391 |
| 10 | 1.4802 | 0.6756 | 12.0061 | 0.0833 | 8.1109 | 0.1233 | 33.8814 | 4.1773 |
| 11 | 1.5395 | 0.6496 | 13.4864 | 0.0741 | 8.7605 | 0.1141 | 40.3772 | 4.6090 |
| 12 | 1.6010 | 0.6246 | 15.0258 | 0.0666 | 9.3851 | 0.1066 | 47.2477 | 5.0343 |
| 13 | 1.6651 | 0.6006 | 16.6268 | 0.0601 | 9.9856 | 0.1001 | 54.4546 | 5.4533 |
| 14 | 1.7317 | 0.5775 | 18.2919 | 0.0547 | 10.5631 | 0.0947 | 61.9618 | 5.8659 |
| 15 | 1.8009 | 0.5553 | 20.0236 | 0.0499 | 11.1184 | 0.0899 | 69.7355 | 6.2721 |
| 16 | 1.8730 | 0.5339 | 21.8245 | 0.0458 | 11.6523 | 0.0858 | 77.7441 | 6.6720 |
| 17 | 1.9479 | 0.5134 | 23.6975 | 0.0422 | 12.1657 | 0.0822 | 85.9581 | 7.0656 |
| 18 | 2.0258 | 0.4936 | 25.6454 | 0.0390 | 12.6593 | 0.0790 | 94.3498 | 7.4530 |
| 19 | 2.1068 | 0.4746 | 27.6712 | 0.0361 | 13.1339 | 0.0761 | 102.8933 | 7.8342 |
| 20 | 2.1911 | 0.4564 | 29.7781 | 0.0336 | 13.5903 | 0.0736 | 111.5647 | 8.2091 |
| 21 | 2.2788 | 0.4388 | 31.9692 | 0.0313 | 14.0292 | 0.0713 | 120.3414 | 8.5779 |
| 22 | 2.3699 | 0.4220 | 34.2480 | 0.0292 | 14.4511 | 0.0692 | 129.2024 | 8.9407 |
| 23 | 2.4647 | 0.4057 | 36.6179 | 0.0273 | 14.8568 | 0.0673 | 138.1284 | 9.2973 |
| 24 | 2.5633 | 0.3901 | 39.0826 | 0.0256 | 15.2470 | 0.0656 | 147.1012 | 9.6479 |
| 25 | 2.6658 | 0.3751 | 41.6459 | 0.0240 | 15.6221 | 0.0640 | 156.1040 | 9.9925 |
| 26 | 2.7725 | 0.3607 | 44.3117 | 0.0226 | 15.9828 | 0.0626 | 165.1212 | 10.3312 |
| 27 | 2.8834 | 0.3468 | 47.0842 | 0.0212 | 16.3296 | 0.0612 | 174.1385 | 10.6640 |
| 28 | 2.9987 | 0.3335 | 49.9676 | 0.0200 | 16.6631 | 0.0600 | 183.1424 | 10.9909 |
| 29 | 3.1187 | 0.3207 | 52.9663 | 0.0189 | 16.9837 | 0.0589 | 192.1206 | 11.3120 |
| 30 | 3.2434 | 0.3083 | 56.0849 | 0.0178 | 17.2920 | 0.0578 | 201.0618 | 11.6274 |
| 36 | 4.1039 | 0.2437 | 77.5983 | 0.0129 | 18.9083 | 0.0529 | 253.4052 | 13.4018 |
| 42 | 5.1928 | 0.1926 | 104.8196 | 9.540E-03 | 20.1856 | 0.0495 | 302.4370 | 14.9828 |
| 48 | 6.5705 | 0.1522 | 139.2632 | 7.181E-03 | 21.1951 | 0.0472 | 347.2446 | 16.3832 |
| 54 | 8.3138 | 0.1203 | 182.8454 | 5.469E-03 | 21.9930 | 0.0455 | 387.4436 | 17.6167 |
| 60 | 10.5196 | 0.0951 | 237.9907 | 4.202E-03 | 22.6235 | 0.0442 | 422.9966 | 18.6972 |
| 66 | 13.3107 | 0.0751 | 307.7671 | 3.249E-03 | 23.1218 | 0.0432 | 454.0847 | 19.6388 |
| 72 | 16.8423 | 0.0594 | 396.0566 | 2.525E-03 | 23.5156 | 0.0425 | 481.0170 | 20.4552 |
| 120 | 110.6626 | 9.036E-03 | 2.742E+03 | 3.648E-04 | 24.7741 | 0.0404 | 592.2428 | 23.9057 |
| 180 | 1.164E+03 | 8.590E-04 | 2.908E+04 | 3.439E-05 | 24.9785 | 0.0400 | 620.5976 | 24.8452 |
| 360 | 1.355E+06 | 7.379E-07 | 3.388E+07 | 2.952E-08 | 25.0000 | 0.0400 | 624.9929 | 24.9997 |

# Time Value of Money Factors—Discrete Compounding
## i = 5%

| n | Single Sums | | Uniform Series | | | | Gradient Series | |
|---|---|---|---|---|---|---|---|---|
| | To Find F Given P (F\|P,i%,n) | To Find P Given F (P\|F,i%,n) | To Find F Given A (F\|A,i%,n) | To Find A Given F (A\|F,i%,n) | To Find P Given A (P\|A,i%,n) | To Find A Given P (A\|P,i%,n) | To Find P Given G (P\|G,i%,n) | To Find A Given G (A\|G,i%,n) |
| 1 | 1.0500 | 0.9524 | 1.0000 | 1.0000 | 0.9524 | 1.0500 | 0.0000 | 0.0000 |
| 2 | 1.1025 | 0.9070 | 2.0500 | 0.4878 | 1.8594 | 0.5378 | 0.9070 | 0.4878 |
| 3 | 1.1576 | 0.8638 | 3.1525 | 0.3172 | 2.7232 | 0.3672 | 2.6347 | 0.9675 |
| 4 | 1.2155 | 0.8227 | 4.3101 | 0.2320 | 3.5460 | 0.2820 | 5.1028 | 1.4391 |
| 5 | 1.2763 | 0.7835 | 5.5256 | 0.1810 | 4.3295 | 0.2310 | 8.2369 | 1.9025 |
| 6 | 1.3401 | 0.7462 | 6.8019 | 0.1470 | 5.0757 | 0.1970 | 11.9680 | 2.3579 |
| 7 | 1.4071 | 0.7107 | 8.1420 | 0.1228 | 5.7864 | 0.1728 | 16.2321 | 2.8052 |
| 8 | 1.4775 | 0.6768 | 9.5491 | 0.1047 | 6.4632 | 0.1547 | 20.9700 | 3.2445 |
| 9 | 1.5513 | 0.6446 | 11.0266 | 0.0907 | 7.1078 | 0.1407 | 26.1268 | 3.6758 |
| 10 | 1.6289 | 0.6139 | 12.5779 | 0.0795 | 7.7217 | 0.1295 | 31.6520 | 4.0991 |
| 11 | 1.7103 | 0.5847 | 14.2068 | 0.0704 | 8.3064 | 0.1204 | 37.4988 | 4.5144 |
| 12 | 1.7959 | 0.5568 | 15.9171 | 0.0628 | 8.8633 | 0.1128 | 43.6241 | 4.9219 |
| 13 | 1.8856 | 0.5303 | 17.7130 | 0.0565 | 9.3936 | 0.1065 | 49.9879 | 5.3215 |
| 14 | 1.9799 | 0.5051 | 19.5986 | 0.0510 | 9.8986 | 0.1010 | 56.5538 | 5.7133 |
| 15 | 2.0789 | 0.4810 | 21.5786 | 0.0463 | 10.3797 | 0.0963 | 63.2880 | 6.0973 |
| 16 | 2.1829 | 0.4581 | 23.6575 | 0.0423 | 10.8378 | 0.0923 | 70.1597 | 6.4736 |
| 17 | 2.2920 | 0.4363 | 25.8404 | 0.0387 | 11.2741 | 0.0887 | 77.1405 | 6.8423 |
| 18 | 2.4066 | 0.4155 | 28.1324 | 0.0355 | 11.6896 | 0.0855 | 84.2043 | 7.2034 |
| 19 | 2.5270 | 0.3957 | 30.5390 | 0.0327 | 12.0853 | 0.0827 | 91.3275 | 7.5569 |
| 20 | 2.6533 | 0.3769 | 33.0660 | 0.0302 | 12.4622 | 0.0802 | 98.4884 | 7.9030 |
| 21 | 2.7860 | 0.3589 | 35.7193 | 0.0280 | 12.8212 | 0.0780 | 105.6673 | 8.2416 |
| 22 | 2.9253 | 0.3418 | 38.5052 | 0.0260 | 13.1630 | 0.0760 | 112.8461 | 8.5730 |
| 23 | 3.0715 | 0.3256 | 41.4305 | 0.0241 | 13.4886 | 0.0741 | 120.0087 | 8.8971 |
| 24 | 3.2251 | 0.3101 | 44.5020 | 0.0225 | 13.7986 | 0.0725 | 127.1402 | 9.2140 |
| 25 | 3.3864 | 0.2953 | 47.7271 | 0.0210 | 14.0939 | 0.0710 | 134.2275 | 9.5238 |
| 26 | 3.5557 | 0.2812 | 51.1135 | 0.0196 | 14.3752 | 0.0696 | 141.2585 | 9.8266 |
| 27 | 3.7335 | 0.2678 | 54.6691 | 0.0183 | 14.6430 | 0.0683 | 148.2226 | 10.1224 |
| 28 | 3.9201 | 0.2551 | 58.4026 | 0.0171 | 14.8981 | 0.0671 | 155.1101 | 10.4114 |
| 29 | 4.1161 | 0.2429 | 62.3227 | 0.0160 | 15.1411 | 0.0660 | 161.9126 | 10.6936 |
| 30 | 4.3219 | 0.2314 | 66.4388 | 0.0151 | 15.3725 | 0.0651 | 168.6226 | 10.9691 |
| 36 | 5.7918 | 0.1727 | 95.8363 | 0.0104 | 16.5469 | 0.0604 | 206.6237 | 12.4872 |
| 42 | 7.7616 | 0.1288 | 135.2318 | 7.395E-03 | 17.4232 | 0.0574 | 240.2389 | 13.7884 |
| 48 | 10.4013 | 0.0961 | 188.0254 | 5.318E-03 | 18.0772 | 0.0553 | 269.2467 | 14.8943 |
| 54 | 13.9387 | 0.0717 | 258.7739 | 3.864E-03 | 18.5651 | 0.0539 | 293.8208 | 15.8265 |
| 60 | 18.6792 | 0.0535 | 353.5837 | 2.828E-03 | 18.9293 | 0.0528 | 314.3432 | 16.6062 |
| 66 | 25.0319 | 0.0399 | 480.6379 | 2.081E-03 | 19.2010 | 0.0521 | 331.2877 | 17.2536 |
| 72 | 33.5451 | 0.0298 | 650.9027 | 1.536E-03 | 19.4038 | 0.0515 | 345.1485 | 17.7877 |
| 120 | 348.9120 | 2.866E-03 | 6.958E+03 | 1.437E-04 | 19.9427 | 0.0501 | 391.9751 | 19.6551 |
| 180 | 6.517E+03 | 1.534E-04 | 1.303E+05 | 7.673E-06 | 19.9969 | 0.0500 | 399.3863 | 19.9724 |
| 360 | 4.248E+07 | 2.354E-08 | 8.495E+08 | 1.177E-09 | 20.0000 | 0.0500 | 399.9998 | 20.0000 |

Time Value of Money Factors—Discrete Compounding
   i = 6%

| | Single Sums | | Uniform Series | | | | Gradient Series | |
|---|---|---|---|---|---|---|---|---|
| | To Find F Given P (F\|P,i%,n) | To Find P Given F (P\|F,i%,n) | To Find F Given A (F\|A,i%,n) | To Find A Given F (A\|F,i%,n) | To Find P Given A (P\|A,i%,n) | To Find A Given P (A\|P,i%,n) | To Find P Given G (P\|G,i%,n) | To Find A Given G (A\|G,i%,n) |
| n | | | | | | | | |
| 1 | 1.0600 | 0.9434 | 1.0000 | 1.0000 | 0.9434 | 1.0600 | 0.0000 | 0.0000 |
| 2 | 1.1236 | 0.8900 | 2.0600 | 0.4854 | 1.8334 | 0.5454 | 0.8900 | 0.4854 |
| 3 | 1.1910 | 0.8396 | 3.1836 | 0.3141 | 2.6730 | 0.3741 | 2.5692 | 0.9612 |
| 4 | 1.2625 | 0.7921 | 4.3746 | 0.2286 | 3.4651 | 0.2886 | 4.9455 | 1.4272 |
| 5 | 1.3382 | 0.7473 | 5.6371 | 0.1774 | 4.2124 | 0.2374 | 7.9345 | 1.8836 |
| 6 | 1.4185 | 0.7050 | 6.9753 | 0.1434 | 4.9173 | 0.2034 | 11.4594 | 2.3304 |
| 7 | 1.5036 | 0.6651 | 8.3938 | 0.1191 | 5.5824 | 0.1791 | 15.4497 | 2.7676 |
| 8 | 1.5938 | 0.6274 | 9.8975 | 0.1010 | 6.2098 | 0.1610 | 19.8416 | 3.1952 |
| 9 | 1.6895 | 0.5919 | 11.4913 | 0.0870 | 6.8017 | 0.1470 | 24.5768 | 3.6133 |
| 10 | 1.7908 | 0.5584 | 13.1808 | 0.0759 | 7.3601 | 0.1359 | 29.6023 | 4.0220 |
| 11 | 1.8983 | 0.5268 | 14.9716 | 0.0668 | 7.8869 | 0.1268 | 34.8702 | 4.4213 |
| 12 | 2.0122 | 0.4970 | 16.8699 | 0.0593 | 8.3838 | 0.1193 | 40.3369 | 4.8113 |
| 13 | 2.1329 | 0.4688 | 18.8821 | 0.0530 | 8.8527 | 0.1130 | 45.9629 | 5.1920 |
| 14 | 2.2609 | 0.4423 | 21.0151 | 0.0476 | 9.2950 | 0.1076 | 51.7128 | 5.5635 |
| 15 | 2.3966 | 0.4173 | 23.2760 | 0.0430 | 9.7122 | 0.1030 | 57.5546 | 5.9260 |
| 16 | 2.5404 | 0.3936 | 25.6725 | 0.0390 | 10.1059 | 0.0990 | 63.4592 | 6.2794 |
| 17 | 2.6928 | 0.3714 | 28.2129 | 0.0354 | 10.4773 | 0.0954 | 69.4011 | 6.6240 |
| 18 | 2.8543 | 0.3503 | 30.9057 | 0.0324 | 10.8276 | 0.0924 | 75.3569 | 6.9597 |
| 19 | 3.0256 | 0.3305 | 33.7600 | 0.0296 | 11.1581 | 0.0896 | 81.3062 | 7.2867 |
| 20 | 3.2071 | 0.3118 | 36.7856 | 0.0272 | 11.4699 | 0.0872 | 87.2304 | 7.6051 |
| 21 | 3.3996 | 0.2942 | 39.9927 | 0.0250 | 11.7641 | 0.0850 | 93.1136 | 7.9151 |
| 22 | 3.6035 | 0.2775 | 43.3923 | 0.0230 | 12.0416 | 0.0830 | 98.9412 | 8.2166 |
| 23 | 3.8197 | 0.2618 | 46.9958 | 0.0213 | 12.3034 | 0.0813 | 104.7007 | 8.5099 |
| 24 | 4.0489 | 0.2470 | 50.8156 | 0.0197 | 12.5504 | 0.0797 | 110.3812 | 8.7951 |
| 25 | 4.2919 | 0.2330 | 54.8645 | 0.0182 | 12.7834 | 0.0782 | 115.9732 | 9.0722 |
| 26 | 4.5494 | 0.2198 | 59.1564 | 0.0169 | 13.0032 | 0.0769 | 121.4684 | 9.3414 |
| 27 | 4.8223 | 0.2074 | 63.7058 | 0.0157 | 13.2105 | 0.0757 | 126.8600 | 9.6029 |
| 28 | 5.1117 | 0.1956 | 68.5281 | 0.0146 | 13.4062 | 0.0746 | 132.1420 | 9.8568 |
| 29 | 5.4184 | 0.1846 | 73.6398 | 0.0136 | 13.5907 | 0.0736 | 137.3096 | 10.1032 |
| 30 | 5.7435 | 0.1741 | 79.0582 | 0.0126 | 13.7648 | 0.0726 | 142.3588 | 10.3422 |
| 36 | 8.1473 | 0.1227 | 119.1209 | 8.395E-03 | 14.6210 | 0.0684 | 170.0387 | 11.6298 |
| 42 | 11.5570 | 0.0865 | 175.9505 | 5.683E-03 | 15.2245 | 0.0657 | 193.1732 | 12.6883 |
| 48 | 16.3939 | 0.0610 | 256.5645 | 3.898E-03 | 15.6500 | 0.0639 | 212.0351 | 13.5485 |
| 54 | 23.2550 | 0.0430 | 370.9170 | 2.696E-03 | 15.9500 | 0.0627 | 227.1316 | 14.2402 |
| 60 | 32.9877 | 0.0303 | 533.1282 | 1.876E-03 | 16.1614 | 0.0619 | 239.0428 | 14.7909 |
| 66 | 46.7937 | 0.0214 | 763.2278 | 1.310E-03 | 16.3105 | 0.0613 | 248.3341 | 15.2254 |
| 72 | 66.3777 | 0.0151 | 1.090E+03 | 9.177E-04 | 16.4156 | 0.0609 | 255.5146 | 15.5654 |
| 120 | 1.088E+03 | 9.190E-04 | 1.812E+04 | 5.519E-05 | 16.6514 | 0.0601 | 275.6846 | 16.5563 |
| 180 | 3.590E+04 | 2.786E-05 | 5.983E+05 | 1.672E-06 | 16.6662 | 0.0600 | 277.6865 | 16.6617 |
| 360 | 1.289E+09 | 7.760E-10 | 2.148E+10 | 4.656E-11 | 16.6667 | 0.0600 | 277.7778 | 16.6667 |

# Time Value of Money Factors—Discrete Compounding
## i = 7%

| | Single Sums | | Uniform Series | | | | Gradient Series | |
|---|---|---|---|---|---|---|---|---|
| | To Find F Given P (F\|P,i%,n) | To Find P Given F (P\|F,i%,n) | To Find F Given A (F\|A,i%,n) | To Find A Given F (A\|F,i%,n) | To Find P Given A (P\|A,i%,n) | To Find A Given P (A\|P,i%,n) | To Find P Given G (P\|G,i%,n) | To Find A Given G (A\|G,i%,n) |
| n | | | | | | | | |
| 1 | 1.0700 | 0.9346 | 1.0000 | 1.0000 | 0.9346 | 1.0700 | 0.0000 | 0.0000 |
| 2 | 1.1449 | 0.8734 | 2.0700 | 0.4831 | 1.8080 | 0.5531 | 0.8734 | 0.4831 |
| 3 | 1.2250 | 0.8163 | 3.2149 | 0.3111 | 2.6243 | 0.3811 | 2.5060 | 0.9549 |
| 4 | 1.3108 | 0.7629 | 4.4399 | 0.2252 | 3.3872 | 0.2952 | 4.7947 | 1.4155 |
| 5 | 1.4026 | 0.7130 | 5.7507 | 0.1739 | 4.1002 | 0.2439 | 7.6467 | 1.8650 |
| 6 | 1.5007 | 0.6663 | 7.1533 | 0.1398 | 4.7665 | 0.2098 | 10.9784 | 2.3032 |
| 7 | 1.6058 | 0.6227 | 8.6540 | 0.1156 | 5.3893 | 0.1856 | 14.7149 | 2.7304 |
| 8 | 1.7182 | 0.5820 | 10.2598 | 0.0975 | 5.9713 | 0.1675 | 18.7889 | 3.1465 |
| 9 | 1.8385 | 0.5439 | 11.9780 | 0.0835 | 6.5152 | 0.1535 | 23.1404 | 3.5517 |
| 10 | 1.9672 | 0.5083 | 13.8164 | 0.0724 | 7.0236 | 0.1424 | 27.7156 | 3.9461 |
| 11 | 2.1049 | 0.4751 | 15.7836 | 0.0634 | 7.4987 | 0.1334 | 32.4665 | 4.3296 |
| 12 | 2.2522 | 0.4440 | 17.8885 | 0.0559 | 7.9427 | 0.1259 | 37.3506 | 4.7025 |
| 13 | 2.4098 | 0.4150 | 20.1406 | 0.0497 | 8.3577 | 0.1197 | 42.3302 | 5.0648 |
| 14 | 2.5785 | 0.3878 | 22.5505 | 0.0443 | 8.7455 | 0.1143 | 47.3718 | 5.4167 |
| 15 | 2.7590 | 0.3624 | 25.1290 | 0.0398 | 9.1079 | 0.1098 | 52.4461 | 5.7583 |
| 16 | 2.9522 | 0.3387 | 27.8881 | 0.0359 | 9.4466 | 0.1059 | 57.5271 | 6.0897 |
| 17 | 3.1588 | 0.3166 | 30.8402 | 0.0324 | 9.7632 | 0.1024 | 62.5923 | 6.4110 |
| 18 | 3.3799 | 0.2959 | 33.9990 | 0.0294 | 10.0591 | 0.0994 | 67.6219 | 6.7225 |
| 19 | 3.6165 | 0.2765 | 37.3790 | 0.0268 | 10.3356 | 0.0968 | 72.5991 | 7.0242 |
| 20 | 3.8697 | 0.2584 | 40.9955 | 0.0244 | 10.5940 | 0.0944 | 77.5091 | 7.3163 |
| 21 | 4.1406 | 0.2415 | 44.8652 | 0.0223 | 10.8355 | 0.0923 | 82.3393 | 7.5990 |
| 22 | 4.4304 | 0.2257 | 49.0057 | 0.0204 | 11.0612 | 0.0904 | 87.0793 | 7.8725 |
| 23 | 4.7405 | 0.2109 | 53.4361 | 0.0187 | 11.2722 | 0.0887 | 91.7201 | 8.1369 |
| 24 | 5.0724 | 0.1971 | 58.1767 | 0.0172 | 11.4693 | 0.0872 | 96.2545 | 8.3923 |
| 25 | 5.4274 | 0.1842 | 63.2490 | 0.0158 | 11.6536 | 0.0858 | 100.6765 | 8.6391 |
| 26 | 5.8074 | 0.1722 | 68.6765 | 0.0146 | 11.8258 | 0.0846 | 104.9814 | 8.8773 |
| 27 | 6.2139 | 0.1609 | 74.4838 | 0.0134 | 11.9867 | 0.0834 | 109.1656 | 9.1072 |
| 28 | 6.6488 | 0.1504 | 80.6977 | 0.0124 | 12.1371 | 0.0824 | 113.2264 | 9.3289 |
| 29 | 7.1143 | 0.1406 | 87.3465 | 0.0114 | 12.2777 | 0.0814 | 117.1622 | 9.5427 |
| 30 | 7.6123 | 0.1314 | 94.4608 | 0.0106 | 12.4090 | 0.0806 | 120.9718 | 9.7487 |
| 36 | 11.4239 | 0.0875 | 148.9135 | 6.715E-03 | 13.0352 | 0.0767 | 141.1990 | 10.8321 |
| 42 | 17.1443 | 0.0583 | 230.6322 | 4.336E-03 | 13.4524 | 0.0743 | 157.1807 | 11.6842 |
| 48 | 25.7289 | 0.0389 | 353.2701 | 2.831E-03 | 13.7305 | 0.0728 | 169.4981 | 12.3447 |
| 54 | 38.6122 | 0.0259 | 537.3164 | 1.861E-03 | 13.9157 | 0.0719 | 178.8173 | 12.8500 |
| 60 | 57.9464 | 0.0173 | 813.5204 | 1.229E-03 | 14.0392 | 0.0712 | 185.7677 | 13.2321 |
| 66 | 86.9620 | 0.0115 | 1.228E+03 | 8.143E-04 | 14.1214 | 0.0708 | 190.8927 | 13.5179 |
| 72 | 130.5065 | 7.662E-03 | 1.850E+03 | 5.405E-04 | 14.1763 | 0.0705 | 194.6365 | 13.7298 |
| 120 | 3.358E+03 | 2.978E-04 | 4.795E+04 | 2.085E-05 | 14.2815 | 0.0700 | 203.5103 | 14.2500 |
| 180 | 1.946E+05 | 5.139E-06 | 2.780E+06 | 3.598E-07 | 14.2856 | 0.0700 | 204.0674 | 14.2848 |
| 360 | 3.786E+10 | 2.641E-11 | 5.408E+11 | 1.849E-12 | 14.2857 | 0.0700 | 204.0816 | 14.2857 |

Time Value of Money Factors—Discrete Compounding
 i = 8%

| n | Single Sums | | Uniform Series | | | | Gradient Series | |
|---|---|---|---|---|---|---|---|---|
| | To Find F Given P (F\|P,i%,n) | To Find P Given F (P\|F,i%,n) | To Find F Given A (F\|A,i%,n) | To Find A Given F (A\|F,i%,n) | To Find P Given A (P\|A,i%,n) | To Find A Given P (A\|P,i%,n) | To Find P Given G (P\|G,i%,n) | To Find A Given G (A\|G,i%,n) |
| 1 | 1.0800 | 0.9259 | 1.0000 | 1.0000 | 0.9259 | 1.0800 | 0.0000 | 0.0000 |
| 2 | 1.1664 | 0.8573 | 2.0800 | 0.4808 | 1.7833 | 0.5608 | 0.8573 | 0.4808 |
| 3 | 1.2597 | 0.7938 | 3.2464 | 0.3080 | 2.5771 | 0.3880 | 2.4450 | 0.9487 |
| 4 | 1.3605 | 0.7350 | 4.5061 | 0.2219 | 3.3121 | 0.3019 | 4.6501 | 1.4040 |
| 5 | 1.4693 | 0.6806 | 5.8666 | 0.1705 | 3.9927 | 0.2505 | 7.3724 | 1.8465 |
| 6 | 1.5869 | 0.6302 | 7.3359 | 0.1363 | 4.6229 | 0.2163 | 10.5233 | 2.2763 |
| 7 | 1.7138 | 0.5835 | 8.9228 | 0.1121 | 5.2064 | 0.1921 | 14.0242 | 2.6937 |
| 8 | 1.8509 | 0.5403 | 10.6366 | 0.0940 | 5.7466 | 0.1740 | 17.8061 | 3.0985 |
| 9 | 1.9990 | 0.5002 | 12.4876 | 0.0801 | 6.2469 | 0.1601 | 21.8081 | 3.4910 |
| 10 | 2.1589 | 0.4632 | 14.4866 | 0.0690 | 6.7101 | 0.1490 | 25.9768 | 3.8713 |
| 11 | 2.3316 | 0.4289 | 16.6455 | 0.0601 | 7.1390 | 0.1401 | 30.2657 | 4.2395 |
| 12 | 2.5182 | 0.3971 | 18.9771 | 0.0527 | 7.5361 | 0.1327 | 34.6339 | 4.5957 |
| 13 | 2.7196 | 0.3677 | 21.4953 | 0.0465 | 7.9038 | 0.1265 | 39.0463 | 4.9402 |
| 14 | 2.9372 | 0.3405 | 24.2149 | 0.0413 | 8.2442 | 0.1213 | 43.4723 | 5.2731 |
| 15 | 3.1722 | 0.3152 | 27.1521 | 0.0368 | 8.5595 | 0.1168 | 47.8857 | 5.5945 |
| 16 | 3.4259 | 0.2919 | 30.3243 | 0.0330 | 8.8514 | 0.1130 | 52.2640 | 5.9046 |
| 17 | 3.7000 | 0.2703 | 33.7502 | 0.0296 | 9.1216 | 0.1096 | 56.5883 | 6.2037 |
| 18 | 3.9960 | 0.2502 | 37.4502 | 0.0267 | 9.3719 | 0.1067 | 60.8426 | 6.4920 |
| 19 | 4.3157 | 0.2317 | 41.4463 | 0.0241 | 9.6036 | 0.1041 | 65.0134 | 6.7697 |
| 20 | 4.6610 | 0.2145 | 45.7620 | 0.0219 | 9.8181 | 0.1019 | 69.0898 | 7.0369 |
| 21 | 5.0338 | 0.1987 | 50.4229 | 0.0198 | 10.0168 | 0.0998 | 73.0629 | 7.2940 |
| 22 | 5.4365 | 0.1839 | 55.4568 | 0.0180 | 10.2007 | 0.0980 | 76.9257 | 7.5412 |
| 23 | 5.8715 | 0.1703 | 60.8933 | 0.0164 | 10.3711 | 0.0964 | 80.6726 | 7.7786 |
| 24 | 6.3412 | 0.1577 | 66.7648 | 0.0150 | 10.5288 | 0.0950 | 84.2997 | 8.0066 |
| 25 | 6.8485 | 0.1460 | 73.1059 | 0.0137 | 10.6748 | 0.0937 | 87.8041 | 8.2254 |
| 26 | 7.3964 | 0.1352 | 79.9544 | 0.0125 | 10.8100 | 0.0925 | 91.1842 | 8.4352 |
| 27 | 7.9881 | 0.1252 | 87.3508 | 0.0114 | 10.9352 | 0.0914 | 94.4390 | 8.6363 |
| 28 | 8.6271 | 0.1159 | 95.3388 | 0.0105 | 11.0511 | 0.0905 | 97.5687 | 8.8289 |
| 29 | 9.3173 | 0.1073 | 103.9659 | 9.619E-03 | 11.1584 | 0.0896 | 100.5738 | 9.0133 |
| 30 | 10.0627 | 0.0994 | 113.2832 | 8.827E-03 | 11.2578 | 0.0888 | 103.4558 | 9.1897 |
| 36 | 15.9682 | 0.0626 | 187.1021 | 5.345E-03 | 11.7172 | 0.0853 | 118.2839 | 10.0949 |
| 42 | 25.3395 | 0.0395 | 304.2435 | 3.287E-03 | 12.0067 | 0.0833 | 129.3651 | 10.7744 |
| 48 | 40.2106 | 0.0249 | 490.1322 | 2.040E-03 | 12.1891 | 0.0820 | 137.4428 | 11.2758 |
| 54 | 63.8091 | 0.0157 | 785.1141 | 1.274E-03 | 12.3041 | 0.0813 | 143.2229 | 11.6403 |
| 60 | 101.2571 | 9.876E-03 | 1.253E+03 | 7.979E-04 | 12.3766 | 0.0808 | 147.3000 | 11.9015 |
| 66 | 160.6822 | 6.223E-03 | 1.996E+03 | 5.010E-04 | 12.4222 | 0.0805 | 150.1432 | 12.0867 |
| 72 | 254.9825 | 3.922E-03 | 3.175E+03 | 3.150E-04 | 12.4510 | 0.0803 | 152.1076 | 12.2165 |
| 120 | 1.025E+04 | 9.753E-05 | 1.281E+05 | 7.803E-06 | 12.4988 | 0.0800 | 156.0885 | 12.4883 |
| 180 | 1.038E+06 | 9.632E-07 | 1.298E+07 | 7.706E-08 | 12.5000 | 0.0800 | 156.2477 | 12.4998 |
| 360 | 1.078E+12 | 9.278E-13 | 1.347E+13 | 7.422E-14 | 12.5000 | 0.0800 | 156.2500 | 12.5000 |

Time Value of Money Factors—Discrete Compounding
i = 9%

| n | Single Sums | | Uniform Series | | | | Gradient Series | |
|---|---|---|---|---|---|---|---|---|
| | To Find F Given P (F\|P,i%,n) | To Find P Given F (P\|F,i%,n) | To Find F Given A (F\|A,i%,n) | To Find A Given F (A\|F,i%,n) | To Find P Given A (P\|A,i%,n) | To Find A Given P (A\|P,i%,n) | To Find P Given G (P\|G,i%,n) | To Find A Given G (A\|G,i%,n) |
| 1 | 1.0900 | 0.9174 | 1.0000 | 1.0000 | 0.9174 | 1.0900 | 0.0000 | 0.0000 |
| 2 | 1.1881 | 0.8417 | 2.0900 | 0.4785 | 1.7591 | 0.5685 | 0.8417 | 0.4785 |
| 3 | 1.2950 | 0.7722 | 3.2781 | 0.3051 | 2.5313 | 0.3951 | 2.3860 | 0.9426 |
| 4 | 1.4116 | 0.7084 | 4.5731 | 0.2187 | 3.2397 | 0.3087 | 4.5113 | 1.3925 |
| 5 | 1.5386 | 0.6499 | 5.9847 | 0.1671 | 3.8897 | 0.2571 | 7.1110 | 1.8282 |
| 6 | 1.6771 | 0.5963 | 7.5233 | 0.1329 | 4.4859 | 0.2229 | 10.0924 | 2.2498 |
| 7 | 1.8280 | 0.5470 | 9.2004 | 0.1087 | 5.0330 | 0.1987 | 13.3746 | 2.6574 |
| 8 | 1.9926 | 0.5019 | 11.0285 | 0.0907 | 5.5348 | 0.1807 | 16.8877 | 3.0512 |
| 9 | 2.1719 | 0.4604 | 13.0210 | 0.0768 | 5.9952 | 0.1668 | 20.5711 | 3.4312 |
| 10 | 2.3674 | 0.4224 | 15.1929 | 0.0658 | 6.4177 | 0.1558 | 24.3728 | 3.7978 |
| 11 | 2.5804 | 0.3875 | 17.5603 | 0.0569 | 6.8052 | 0.1469 | 28.2481 | 4.1510 |
| 12 | 2.8127 | 0.3555 | 20.1407 | 0.0497 | 7.1607 | 0.1397 | 32.1590 | 4.4910 |
| 13 | 3.0658 | 0.3262 | 22.9534 | 0.0436 | 7.4869 | 0.1336 | 36.0731 | 4.8182 |
| 14 | 3.3417 | 0.2992 | 26.0192 | 0.0384 | 7.7862 | 0.1284 | 39.9633 | 5.1326 |
| 15 | 3.6425 | 0.2745 | 29.3609 | 0.0341 | 8.0607 | 0.1241 | 43.8069 | 5.4346 |
| 16 | 3.9703 | 0.2519 | 33.0034 | 0.0303 | 8.3126 | 0.1203 | 47.5849 | 5.7245 |
| 17 | 4.3276 | 0.2311 | 36.9737 | 0.0270 | 8.5436 | 0.1170 | 51.2821 | 6.0024 |
| 18 | 4.7171 | 0.2120 | 41.3013 | 0.0242 | 8.7556 | 0.1142 | 54.8860 | 6.2687 |
| 19 | 5.1417 | 0.1945 | 46.0185 | 0.0217 | 8.9501 | 0.1117 | 58.3868 | 6.5236 |
| 20 | 5.6044 | 0.1784 | 51.1601 | 0.0195 | 9.1285 | 0.1095 | 61.7770 | 6.7674 |
| 21 | 6.1088 | 0.1637 | 56.7645 | 0.0176 | 9.2922 | 0.1076 | 65.0509 | 7.0006 |
| 22 | 6.6586 | 0.1502 | 62.8733 | 0.0159 | 9.4424 | 0.1059 | 68.2048 | 7.2232 |
| 23 | 7.2579 | 0.1378 | 69.5319 | 0.0144 | 9.5802 | 0.1044 | 71.2359 | 7.4357 |
| 24 | 7.9111 | 0.1264 | 76.7898 | 0.0130 | 9.7066 | 0.1030 | 74.1433 | 7.6384 |
| 25 | 8.6231 | 0.1160 | 84.7009 | 0.0118 | 9.8226 | 0.1018 | 76.9265 | 7.8316 |
| 26 | 9.3992 | 0.1064 | 93.3240 | 0.0107 | 9.9290 | 0.1007 | 79.5863 | 8.0156 |
| 27 | 10.2451 | 0.0976 | 102.7231 | 9.735E-03 | 10.0266 | 0.0997 | 82.1241 | 8.1906 |
| 28 | 11.1671 | 0.0895 | 112.9682 | 8.852E-03 | 10.1161 | 0.0989 | 84.5419 | 8.3571 |
| 29 | 12.1722 | 0.0822 | 124.1354 | 8.056E-03 | 10.1983 | 0.0981 | 86.8422 | 8.5154 |
| 30 | 13.2677 | 0.0754 | 136.3075 | 7.336E-03 | 10.2737 | 0.0973 | 89.0280 | 8.6657 |
| 36 | 22.2512 | 0.0449 | 236.1247 | 4.235E-03 | 10.6118 | 0.0942 | 99.9319 | 9.4171 |
| 42 | 37.3175 | 0.0268 | 403.5281 | 2.478E-03 | 10.8134 | 0.0925 | 107.6432 | 9.9546 |
| 48 | 62.5852 | 0.0160 | 684.2804 | 1.461E-03 | 10.9336 | 0.0915 | 112.9625 | 10.3317 |
| 54 | 104.9617 | 9.527E-03 | 1.155E+03 | 8.657E-04 | 11.0053 | 0.0909 | 116.5642 | 10.5917 |
| 60 | 176.0313 | 5.681E-03 | 1.945E+03 | 5.142E-04 | 11.0480 | 0.0905 | 118.9683 | 10.7683 |
| 66 | 295.2221 | 3.387E-03 | 3.269E+03 | 3.059E-04 | 11.0735 | 0.0903 | 120.5546 | 10.8868 |
| 72 | 495.1170 | 2.020E-03 | 5.490E+03 | 1.821E-04 | 11.0887 | 0.0902 | 121.5917 | 10.9654 |
| 120 | 3.099E+04 | 3.227E-05 | 3.443E+05 | 2.905E-06 | 11.1108 | 0.0900 | 123.4098 | 11.1072 |
| 180 | 5.455E+06 | 1.833E-07 | 6.061E+07 | 1.650E-08 | 11.1111 | 0.0900 | 123.4564 | 11.1111 |
| 360 | 2.975E+13 | 3.361E-14 | 3.306E+14 | 3.025E-15 | 11.1111 | 0.0900 | 123.4568 | 11.1111 |

Time Value of Money Factors—Discrete Compounding
    i = 10%

| n | Single Sums | | Uniform Series | | | | Gradient Series | |
|---|---|---|---|---|---|---|---|---|
| | To Find F Given P (F\|P,i%,n) | To Find P Given F (P\|F,i%,n) | To Find F Given A (F\|A,i%,n) | To Find A Given F (A\|F,i%,n) | To Find P Given A (P\|A,i%,n) | To Find A Given P (A\|P,i%,n) | To Find P Given G (P\|G,i%,n) | To Find A Given G (A\|G,i%,n) |
| 1 | 1.1000 | 0.9091 | 1.0000 | 1.0000 | 0.9091 | 1.1000 | 0.0000 | 0.0000 |
| 2 | 1.2100 | 0.8264 | 2.1000 | 0.4762 | 1.7355 | 0.5762 | 0.8264 | 0.4762 |
| 3 | 1.3310 | 0.7513 | 3.3100 | 0.3021 | 2.4869 | 0.4021 | 2.3291 | 0.9366 |
| 4 | 1.4641 | 0.6830 | 4.6410 | 0.2155 | 3.1699 | 0.3155 | 4.3781 | 1.3812 |
| 5 | 1.6105 | 0.6209 | 6.1051 | 0.1638 | 3.7908 | 0.2638 | 6.8618 | 1.8101 |
| 6 | 1.7716 | 0.5645 | 7.7156 | 0.1296 | 4.3553 | 0.2296 | 9.6842 | 2.2236 |
| 7 | 1.9487 | 0.5132 | 9.4872 | 0.1054 | 4.8684 | 0.2054 | 12.7631 | 2.6216 |
| 8 | 2.1436 | 0.4665 | 11.4359 | 0.0874 | 5.3349 | 0.1874 | 16.0287 | 3.0045 |
| 9 | 2.3579 | 0.4241 | 13.5795 | 0.0736 | 5.7590 | 0.1736 | 19.4215 | 3.3724 |
| 10 | 2.5937 | 0.3855 | 15.9374 | 0.0627 | 6.1446 | 0.1627 | 22.8913 | 3.7255 |
| 11 | 2.8531 | 0.3505 | 18.5312 | 0.0540 | 6.4951 | 0.1540 | 26.3963 | 4.0641 |
| 12 | 3.1384 | 0.3186 | 21.3843 | 0.0468 | 6.8137 | 0.1468 | 29.9012 | 4.3884 |
| 13 | 3.4523 | 0.2897 | 24.5227 | 0.0408 | 7.1034 | 0.1408 | 33.3772 | 4.6988 |
| 14 | 3.7975 | 0.2633 | 27.9750 | 0.0357 | 7.3667 | 0.1357 | 36.8005 | 4.9955 |
| 15 | 4.1772 | 0.2394 | 31.7725 | 0.0315 | 7.6061 | 0.1315 | 40.1520 | 5.2789 |
| 16 | 4.5950 | 0.2176 | 35.9497 | 0.0278 | 7.8237 | 0.1278 | 43.4164 | 5.5493 |
| 17 | 5.0545 | 0.1978 | 40.5447 | 0.0247 | 8.0216 | 0.1247 | 46.5819 | 5.8071 |
| 18 | 5.5599 | 0.1799 | 45.5992 | 0.0219 | 8.2014 | 0.1219 | 49.6395 | 6.0526 |
| 19 | 6.1159 | 0.1635 | 51.1591 | 0.0195 | 8.3649 | 0.1195 | 52.5827 | 6.2861 |
| 20 | 6.7275 | 0.1486 | 57.2750 | 0.0175 | 8.5136 | 0.1175 | 55.4069 | 6.5081 |
| 21 | 7.4002 | 0.1351 | 64.0025 | 0.0156 | 8.6487 | 0.1156 | 58.1095 | 6.7189 |
| 22 | 8.1403 | 0.1228 | 71.4027 | 0.0140 | 8.7715 | 0.1140 | 60.6893 | 6.9189 |
| 23 | 8.9543 | 0.1117 | 79.5430 | 0.0126 | 8.8832 | 0.1126 | 63.1462 | 7.1085 |
| 24 | 9.8497 | 0.1015 | 88.4973 | 0.0113 | 8.9847 | 0.1113 | 65.4813 | 7.2881 |
| 25 | 10.8347 | 0.0923 | 98.3471 | 0.0102 | 9.0770 | 0.1102 | 67.6964 | 7.4580 |
| 26 | 11.9182 | 0.0839 | 109.1818 | 9.159E-03 | 9.1609 | 0.1092 | 69.7940 | 7.6186 |
| 27 | 13.1100 | 0.0763 | 121.0999 | 8.258E-03 | 9.2372 | 0.1083 | 71.7773 | 7.7704 |
| 28 | 14.4210 | 0.0693 | 134.2099 | 7.451E-03 | 9.3066 | 0.1075 | 73.6495 | 7.9137 |
| 29 | 15.8631 | 0.0630 | 148.6309 | 6.728E-03 | 9.3696 | 0.1067 | 75.4146 | 8.0489 |
| 30 | 17.4494 | 0.0573 | 164.4940 | 6.079E-03 | 9.4269 | 0.1061 | 77.0766 | 8.1762 |
| 36 | 30.9127 | 0.0323 | 299.1268 | 3.343E-03 | 9.6765 | 0.1033 | 85.1194 | 8.7965 |
| 42 | 54.7637 | 0.0183 | 537.6370 | 1.860E-03 | 9.8174 | 0.1019 | 90.5047 | 9.2188 |
| 48 | 97.0172 | 0.0103 | 960.1723 | 1.041E-03 | 9.8969 | 0.1010 | 94.0217 | 9.5001 |
| 54 | 171.8719 | 5.818E-03 | 1.709E+03 | 5.852E-04 | 9.9418 | 0.1006 | 96.2763 | 9.6840 |
| 60 | 304.4816 | 3.284E-03 | 3.035E+03 | 3.295E-04 | 9.9672 | 0.1003 | 97.7010 | 9.8023 |
| 66 | 539.4078 | 1.854E-03 | 5.384E+03 | 1.857E-04 | 9.9815 | 0.1002 | 98.5910 | 9.8774 |
| 72 | 955.5938 | 1.046E-03 | 9.546E+03 | 1.048E-04 | 9.9895 | 0.1001 | 99.1419 | 9.9246 |
| 120 | 9.271E+04 | 1.079E-05 | 9.271E+05 | 1.079E-06 | 9.9999 | 0.1000 | 99.9860 | 9.9987 |
| 180 | 2.823E+07 | 3.543E-08 | 2.823E+08 | 3.543E-09 | 10.0000 | 0.1000 | 99.9999 | 10.0000 |
| 360 | 7.968E+14 | 1.255E-15 | 7.968E+15 | 1.255E-16 | 10.0000 | 0.1000 | 100.0000 | 10.0000 |

## Time Value of Money Factors—Discrete Compounding
### i = 12%

| n | Single Sums | | Uniform Series | | | | Gradient Series | |
|---|---|---|---|---|---|---|---|---|
| | To Find F Given P (F\|P,i%,n) | To Find P Given F (P\|F,i%,n) | To Find F Given A (F\|A,i%,n) | To Find A Given F (A\|F,i%,n) | To Find P Given A (P\|A,i%,n) | To Find A Given P (A\|P,i%,n) | To Find P Given G (P\|G,i%,n) | To Find A Given G (A\|G,i%,n) |
| 1 | 1.1200 | 0.8929 | 1.0000 | 1.0000 | 0.8929 | 1.1200 | 0.0000 | 0.0000 |
| 2 | 1.2544 | 0.7972 | 2.1200 | 0.4717 | 1.6901 | 0.5917 | 0.7972 | 0.4717 |
| 3 | 1.4049 | 0.7118 | 3.3744 | 0.2963 | 2.4018 | 0.4163 | 2.2208 | 0.9246 |
| 4 | 1.5735 | 0.6355 | 4.7793 | 0.2092 | 3.0373 | 0.3292 | 4.1273 | 1.3589 |
| 5 | 1.7623 | 0.5674 | 6.3528 | 0.1574 | 3.6048 | 0.2774 | 6.3970 | 1.7746 |
| 6 | 1.9738 | 0.5066 | 8.1152 | 0.1232 | 4.1114 | 0.2432 | 8.9302 | 2.1720 |
| 7 | 2.2107 | 0.4523 | 10.0890 | 0.0991 | 4.5638 | 0.2191 | 11.6443 | 2.5515 |
| 8 | 2.4760 | 0.4039 | 12.2997 | 0.0813 | 4.9676 | 0.2013 | 14.4714 | 2.9131 |
| 9 | 2.7731 | 0.3606 | 14.7757 | 0.0677 | 5.3282 | 0.1877 | 17.3563 | 3.2574 |
| 10 | 3.1058 | 0.3220 | 17.5487 | 0.0570 | 5.6502 | 0.1770 | 20.2541 | 3.5847 |
| 11 | 3.4785 | 0.2875 | 20.6546 | 0.0484 | 5.9377 | 0.1684 | 23.1288 | 3.8953 |
| 12 | 3.8960 | 0.2567 | 24.1331 | 0.0414 | 6.1944 | 0.1614 | 25.9523 | 4.1897 |
| 13 | 4.3635 | 0.2292 | 28.0291 | 0.0357 | 6.4235 | 0.1557 | 28.7024 | 4.4683 |
| 14 | 4.8871 | 0.2046 | 32.3926 | 0.0309 | 6.6282 | 0.1509 | 31.3624 | 4.7317 |
| 15 | 5.4736 | 0.1827 | 37.2797 | 0.0268 | 6.8109 | 0.1468 | 33.9202 | 4.9803 |
| 16 | 6.1304 | 0.1631 | 42.7533 | 0.0234 | 6.9740 | 0.1434 | 36.3670 | 5.2147 |
| 17 | 6.8660 | 0.1456 | 48.8837 | 0.0205 | 7.1196 | 0.1405 | 38.6973 | 5.4353 |
| 18 | 7.6900 | 0.1300 | 55.7497 | 0.0179 | 7.2497 | 0.1379 | 40.9080 | 5.6427 |
| 19 | 8.6128 | 0.1161 | 63.4397 | 0.0158 | 7.3658 | 0.1358 | 42.9979 | 5.8375 |
| 20 | 9.6463 | 0.1037 | 72.0524 | 0.0139 | 7.4694 | 0.1339 | 44.9676 | 6.0202 |
| 21 | 10.8038 | 0.0926 | 81.6987 | 0.0122 | 7.5620 | 0.1322 | 46.8188 | 6.1913 |
| 22 | 12.1003 | 0.0826 | 92.5026 | 0.0108 | 7.6446 | 0.1308 | 48.5543 | 6.3514 |
| 23 | 13.5523 | 0.0738 | 104.6029 | 9.560E-03 | 7.7184 | 0.1296 | 50.1776 | 6.5010 |
| 24 | 15.1786 | 0.0659 | 118.1552 | 8.463E-03 | 7.7843 | 0.1285 | 51.6929 | 6.6406 |
| 25 | 17.0001 | 0.0588 | 133.3339 | 7.500E-03 | 7.8431 | 0.1275 | 53.1046 | 6.7708 |
| 26 | 19.0401 | 0.0525 | 150.3339 | 6.652E-03 | 7.8957 | 0.1267 | 54.4177 | 6.8921 |
| 27 | 21.3249 | 0.0469 | 169.3740 | 5.904E-03 | 7.9426 | 0.1259 | 55.6369 | 7.0049 |
| 28 | 23.8839 | 0.0419 | 190.6989 | 5.244E-03 | 7.9844 | 0.1252 | 56.7674 | 7.1098 |
| 29 | 26.7499 | 0.0374 | 214.5828 | 4.660E-03 | 8.0218 | 0.1247 | 57.8141 | 7.2071 |
| 30 | 29.9599 | 0.0334 | 241.3327 | 4.144E-03 | 8.0552 | 0.1241 | 58.7821 | 7.2974 |
| 36 | 59.1356 | 0.0169 | 484.4631 | 2.064E-03 | 8.1924 | 0.1221 | 63.1970 | 7.7141 |
| 42 | 116.7231 | 8.567E-03 | 964.3595 | 1.037E-03 | 8.2619 | 0.1210 | 65.8509 | 7.9704 |
| 48 | 230.3908 | 4.340E-03 | 1.912E+03 | 5.231E-04 | 8.2972 | 0.1205 | 67.4068 | 8.1241 |
| 54 | 454.7505 | 2.199E-03 | 3.781E+03 | 2.645E-04 | 8.3150 | 0.1203 | 68.3022 | 8.2143 |
| 60 | 897.5969 | 1.114E-03 | 7.472E+03 | 1.338E-04 | 8.3240 | 0.1201 | 68.8100 | 8.2664 |
| 66 | 1.772E+03 | 5.644E-04 | 1.476E+04 | 6.777E-05 | 8.3286 | 0.1201 | 69.0948 | 8.2961 |
| 72 | 3.497E+03 | 2.860E-04 | 2.913E+04 | 3.432E-05 | 8.3310 | 0.1200 | 69.2530 | 8.3127 |
| 120 | 8.057E+05 | 1.241E-06 | 6.714E+06 | 1.489E-07 | 8.3333 | 0.1200 | 69.4431 | 8.3332 |
| 180 | 7.232E+08 | 1.383E-09 | 6.026E+09 | 1.659E-10 | 8.3333 | 0.1200 | 69.4444 | 8.3333 |
| 360 | 5.230E+17 | 1.912E-18 | 4.358E+18 | 2.295E-19 | 8.3333 | 0.1200 | 69.4444 | 8.3333 |

Time Value of Money Factors—Discrete Compounding
  i = 15%

| n | Single Sums | | Uniform Series | | | | Gradient Series | |
|---|---|---|---|---|---|---|---|---|
| | To Find F Given P (F\|P,i%,n) | To Find P Given F (P\|F,i%,n) | To Find F Given A (F\|A,i%,n) | To Find A Given F (A\|F,i%,n) | To Find P Given A (P\|A,i%,n) | To Find A Given P (A\|P,i%,n) | To Find P Given G (P\|G,i%,n) | To Find A Given G (A\|G,i%,n) |
| 1 | 1.1500 | 0.8696 | 1.0000 | 1.0000 | 0.8696 | 1.1500 | 0.0000 | 0.0000 |
| 2 | 1.3225 | 0.7561 | 2.1500 | 0.4651 | 1.6257 | 0.6151 | 0.7561 | 0.4651 |
| 3 | 1.5209 | 0.6575 | 3.4725 | 0.2880 | 2.2832 | 0.4380 | 2.0712 | 0.9071 |
| 4 | 1.7490 | 0.5718 | 4.9934 | 0.2003 | 2.8550 | 0.3503 | 3.7864 | 1.3263 |
| 5 | 2.0114 | 0.4972 | 6.7424 | 0.1483 | 3.3522 | 0.2983 | 5.7751 | 1.7228 |
| 6 | 2.3131 | 0.4323 | 8.7537 | 0.1142 | 3.7845 | 0.2642 | 7.9368 | 2.0972 |
| 7 | 2.6600 | 0.3759 | 11.0668 | 0.0904 | 4.1604 | 0.2404 | 10.1924 | 2.4498 |
| 8 | 3.0590 | 0.3269 | 13.7268 | 0.0729 | 4.4873 | 0.2229 | 12.4807 | 2.7813 |
| 9 | 3.5179 | 0.2843 | 16.7858 | 0.0596 | 4.7716 | 0.2096 | 14.7548 | 3.0922 |
| 10 | 4.0456 | 0.2472 | 20.3037 | 0.0493 | 5.0188 | 0.1993 | 16.9795 | 3.3832 |
| 11 | 4.6524 | 0.2149 | 24.3493 | 0.0411 | 5.2337 | 0.1911 | 19.1289 | 3.6549 |
| 12 | 5.3503 | 0.1869 | 29.0017 | 0.0345 | 5.4206 | 0.1845 | 21.1849 | 3.9082 |
| 13 | 6.1528 | 0.1625 | 34.3519 | 0.0291 | 5.5831 | 0.1791 | 23.1352 | 4.1438 |
| 14 | 7.0757 | 0.1413 | 40.5047 | 0.0247 | 5.7245 | 0.1747 | 24.9725 | 4.3624 |
| 15 | 8.1371 | 0.1229 | 47.5804 | 0.0210 | 5.8474 | 0.1710 | 26.6930 | 4.5650 |
| 16 | 9.3576 | 0.1069 | 55.7175 | 0.0179 | 5.9542 | 0.1679 | 28.2960 | 4.7522 |
| 17 | 10.7613 | 0.0929 | 65.0751 | 0.0154 | 6.0472 | 0.1654 | 29.7828 | 4.9251 |
| 18 | 12.3755 | 0.0808 | 75.8364 | 0.0132 | 6.1280 | 0.1632 | 31.1565 | 5.0843 |
| 19 | 14.2318 | 0.0703 | 88.2118 | 0.0113 | 6.1982 | 0.1613 | 32.4213 | 5.2307 |
| 20 | 16.3665 | 0.0611 | 102.4436 | 9.761E-03 | 6.2593 | 0.1598 | 33.5822 | 5.3651 |
| 21 | 18.8215 | 0.0531 | 118.8101 | 8.417E-03 | 6.3125 | 0.1584 | 34.6448 | 5.4883 |
| 22 | 21.6447 | 0.0462 | 137.6316 | 7.266E-03 | 6.3587 | 0.1573 | 35.6150 | 5.6010 |
| 23 | 24.8915 | 0.0402 | 159.2764 | 6.278E-03 | 6.3988 | 0.1563 | 36.4988 | 5.7040 |
| 24 | 28.6252 | 0.0349 | 184.1678 | 5.430E-03 | 6.4338 | 0.1554 | 37.3023 | 5.7979 |
| 25 | 32.9190 | 0.0304 | 212.7930 | 4.699E-03 | 6.4641 | 0.1547 | 38.0314 | 5.8834 |
| 26 | 37.8568 | 0.0264 | 245.7120 | 4.070E-03 | 6.4906 | 0.1541 | 38.6918 | 5.9612 |
| 27 | 43.5353 | 0.0230 | 283.5688 | 3.526E-03 | 6.5135 | 0.1535 | 39.2890 | 6.0319 |
| 28 | 50.0656 | 0.0200 | 327.1041 | 3.057E-03 | 6.5335 | 0.1531 | 39.8283 | 6.0960 |
| 29 | 57.5755 | 0.0174 | 377.1697 | 2.651E-03 | 6.5509 | 0.1527 | 40.3146 | 6.1541 |
| 30 | 66.2118 | 0.0151 | 434.7451 | 2.300E-03 | 6.5660 | 0.1523 | 40.7526 | 6.2066 |
| 36 | 153.1519 | 6.529E-03 | 1.014E+03 | 9.859E-04 | 6.6231 | 0.1510 | 42.5872 | 6.4301 |
| 42 | 354.2495 | 2.823E-03 | 2.355E+03 | 4.246E-04 | 6.6478 | 0.1504 | 43.5286 | 6.5478 |
| 48 | 819.4007 | 1.220E-03 | 5.456E+03 | 1.833E-04 | 6.6585 | 0.1502 | 43.9997 | 6.6080 |
| 54 | 1.895E+03 | 5.276E-04 | 1.263E+04 | 7.918E-05 | 6.6631 | 0.1501 | 44.2311 | 6.6382 |
| 60 | 4.384E+03 | 2.281E-04 | 2.922E+04 | 3.422E-05 | 6.6651 | 0.1500 | 44.3431 | 6.6530 |
| 66 | 1.014E+04 | 9.861E-05 | 6.760E+04 | 1.479E-05 | 6.6660 | 0.1500 | 44.3967 | 6.6602 |
| 72 | 2.346E+04 | 4.263E-05 | 1.564E+05 | 6.395E-06 | 6.6664 | 0.1500 | 44.4221 | 6.6636 |
| 120 | 1.922E+07 | 5.203E-08 | 1.281E+08 | 7.805E-09 | 6.6667 | 0.1500 | 44.4444 | 6.6667 |
| 180 | 8.426E+10 | 1.187E-11 | 5.617E+11 | 1.780E-12 | 6.6667 | 0.1500 | 44.4444 | 6.6667 |
| 360 | 7.099E+21 | 1.409E-22 | 4.733E+22 | 2.113E-23 | 6.6667 | 0.1500 | 44.4444 | 6.6667 |

Time Value of Money Factors—Discrete Compounding
i = 18%

| n | Single Sums | | Uniform Series | | | | Gradient Series | |
|---|---|---|---|---|---|---|---|---|
| | To Find F Given P (F\|P,i%,n) | To Find P Given F (P\|F,i%,n) | To Find F Given A (F\|A,i%,n) | To Find A Given F (A\|F,i%,n) | To Find P Given A (P\|A,i%,n) | To Find A Given P (A\|P,i%,n) | To Find P Given G (P\|G,i%,n) | To Find A Given G (A\|G,i%,n) |
| 1 | 1.1800 | 0.8475 | 1.0000 | 1.0000 | 0.8475 | 1.1800 | 0.0000 | 0.0000 |
| 2 | 1.3924 | 0.7182 | 2.1800 | 0.4587 | 1.5656 | 0.6387 | 0.7182 | 0.4587 |
| 3 | 1.6430 | 0.6086 | 3.5724 | 0.2799 | 2.1743 | 0.4599 | 1.9354 | 0.8902 |
| 4 | 1.9388 | 0.5158 | 5.2154 | 0.1917 | 2.6901 | 0.3717 | 3.4828 | 1.2947 |
| 5 | 2.2878 | 0.4371 | 7.1542 | 0.1398 | 3.1272 | 0.3198 | 5.2312 | 1.6728 |
| 6 | 2.6996 | 0.3704 | 9.4420 | 0.1059 | 3.4976 | 0.2859 | 7.0834 | 2.0252 |
| 7 | 3.1855 | 0.3139 | 12.1415 | 0.0824 | 3.8115 | 0.2624 | 8.9670 | 2.3526 |
| 8 | 3.7589 | 0.2660 | 15.3270 | 0.0652 | 4.0776 | 0.2452 | 10.8292 | 2.6558 |
| 9 | 4.4355 | 0.2255 | 19.0859 | 0.0524 | 4.3030 | 0.2324 | 12.6329 | 2.9358 |
| 10 | 5.2338 | 0.1911 | 23.5213 | 0.0425 | 4.4941 | 0.2225 | 14.3525 | 3.1936 |
| 11 | 6.1759 | 0.1619 | 28.7551 | 0.0348 | 4.6560 | 0.2148 | 15.9716 | 3.4303 |
| 12 | 7.2876 | 0.1372 | 34.9311 | 0.0286 | 4.7932 | 0.2086 | 17.4811 | 3.6470 |
| 13 | 8.5994 | 0.1163 | 42.2187 | 0.0237 | 4.9095 | 0.2037 | 18.8765 | 3.8449 |
| 14 | 10.1472 | 0.0085 | 50.8180 | 0.0197 | 5.0081 | 0.1997 | 20.1576 | 4.0250 |
| 15 | 11.9737 | 0.0835 | 60.9653 | 0.0164 | 5.0916 | 0.1964 | 21.3269 | 4.1887 |
| 16 | 14.1290 | 0.0708 | 72.9390 | 0.0137 | 5.1624 | 0.1937 | 22.3885 | 4.3369 |
| 17 | 16.6722 | 0.0600 | 87.0680 | 0.0115 | 5.2223 | 0.1915 | 23.3482 | 4.4708 |
| 18 | 19.6733 | 0.0508 | 103.7403 | 9.639E-03 | 5.2732 | 0.1896 | 24.2123 | 4.5916 |
| 19 | 23.2144 | 0.0431 | 123.4135 | 8.103E-03 | 5.3162 | 0.1881 | 24.9877 | 4.7003 |
| 20 | 27.3930 | 0.0365 | 146.6280 | 6.820E-03 | 5.3527 | 0.1868 | 25.6813 | 4.7978 |
| 21 | 32.3238 | 0.0309 | 174.0210 | 5.746E-03 | 5.3837 | 0.1857 | 26.3000 | 4.8851 |
| 22 | 38.1421 | 0.0262 | 206.3448 | 4.846E-03 | 5.4099 | 0.1848 | 26.8506 | 4.9632 |
| 23 | 45.0076 | 0.0222 | 244.4868 | 4.090E-03 | 5.4321 | 0.1841 | 27.3394 | 5.0329 |
| 24 | 53.1090 | 0.0188 | 289.4945 | 3.454E-03 | 5.4509 | 0.1835 | 27.7725 | 5.0950 |
| 25 | 62.6686 | 0.0160 | 342.6035 | 2.919E-03 | 5.4669 | 0.1829 | 28.1555 | 5.1502 |
| 26 | 73.9490 | 0.0135 | 405.2721 | 2.467E-03 | 5.4804 | 0.1825 | 28.4935 | 5.1991 |
| 27 | 87.2598 | 0.0115 | 479.2211 | 2.087E-03 | 5.4919 | 0.1821 | 28.7915 | 5.2425 |
| 28 | 102.9666 | 9.712E-03 | 566.4809 | 1.765E-03 | 5.5016 | 0.1818 | 29.0537 | 5.2810 |
| 29 | 121.5005 | 8.230E-03 | 669.4475 | 1.494E-03 | 5.5098 | 0.1815 | 29.2842 | 5.3149 |
| 30 | 143.3706 | 6.975E-03 | 790.9480 | 1.264E-03 | 5.5168 | 0.1813 | 29.4864 | 5.3448 |
| 36 | 387.0368 | 2.584E-03 | 2.145E+03 | 4.663E-04 | 5.5412 | 0.1805 | 30.2677 | 5.4623 |
| 42 | 1.045E+03 | 9.571E-04 | 5.799E+03 | 1.724E-04 | 5.5502 | 0.1802 | 30.6113 | 5.5153 |
| 48 | 2.821E+03 | 3.545E-04 | 1.566E+04 | 6.384E-05 | 5.5536 | 0.1801 | 30.7587 | 5.5385 |
| 54 | 7.614E+03 | 1.313E-04 | 4.230E+04 | 2.364E-05 | 5.5548 | 0.1800 | 30.8207 | 5.5485 |
| 60 | 2.056E+04 | 4.865E-05 | 1.142E+05 | 8.757E-06 | 5.5553 | 0.1800 | 30.8465 | 5.5526 |
| 66 | 5.549E+04 | 1.802E-05 | 3.083E+05 | 3.244E-06 | 5.5555 | 0.1800 | 30.8570 | 5.5544 |
| 72 | 1.498E+05 | 6.676E-06 | 8.322E+05 | 1.202E-06 | 5.5555 | 0.1800 | 30.8613 | 5.5551 |
| 120 | 4.225E+08 | 2.367E-09 | 2.347E+09 | 4.260E-10 | 5.5556 | 0.1800 | 30.8642 | 5.5556 |
| 180 | 8.685E+12 | 1.151E-13 | 4.825E+13 | 2.073E-14 | 5.5556 | 0.1800 | 30.8642 | 5.5556 |
| 360 | 7.543E+25 | 1.326E-26 | 4.190E+26 | 2.386E-27 | 5.5556 | 0.1800 | 30.8642 | 5.5556 |

Time Value of Money Factors—Discrete Compounding
i = 20%

| | Single Sums | | Uniform Series | | | | Gradient Series | |
|---|---|---|---|---|---|---|---|---|
| | To Find F Given P (F\|P,i%,n) | To Find P Given F (P\|F,i%,n) | To Find F Given A (F\|A,i%,n) | To Find A Given F (A\|F,i%,n) | To Find P Given A (P\|A,i%,n) | To Find A Given P (A\|P,i%,n) | To Find P Given G (P\|G,i%,n) | To Find A Given G (A\|G,i%,n) |
| n | | | | | | | | |
| 1 | 1.2000 | 0.8333 | 1.0000 | 1.0000 | 0.8333 | 1.2000 | 0.0000 | 0.0000 |
| 2 | 1.4400 | 0.6944 | 2.2000 | 0.4545 | 1.5278 | 0.6545 | 0.6944 | 0.4545 |
| 3 | 1.7280 | 0.5787 | 3.6400 | 0.2747 | 2.1065 | 0.4747 | 1.8519 | 0.8791 |
| 4 | 2.0736 | 0.4823 | 5.3680 | 0.1863 | 2.5887 | 0.3863 | 3.2986 | 1.2742 |
| 5 | 2.4883 | 0.4019 | 7.4416 | 0.1344 | 2.9906 | 0.3344 | 4.9061 | 1.6405 |
| 6 | 2.9860 | 0.3349 | 9.9299 | 0.1007 | 3.3255 | 0.3007 | 6.5806 | 1.9788 |
| 7 | 3.5832 | 0.2791 | 12.9159 | 0.0774 | 3.6046 | 0.2774 | 8.2551 | 2.2902 |
| 8 | 4.2998 | 0.2326 | 16.4991 | 0.0606 | 3.8372 | 0.2606 | 9.8831 | 2.5756 |
| 9 | 5.1598 | 0.1938 | 20.7989 | 0.0481 | 4.0310 | 0.2481 | 11.4335 | 2.8364 |
| 10 | 6.1917 | 0.1615 | 25.9587 | 0.0385 | 4.1925 | 0.2385 | 12.8871 | 3.0739 |
| 11 | 7.4301 | 0.1346 | 32.1504 | 0.0311 | 4.3271 | 0.2311 | 14.2330 | 3.2893 |
| 12 | 8.9161 | 0.1122 | 39.5805 | 0.0253 | 4.4392 | 0.2253 | 15.4667 | 3.4841 |
| 13 | 10.6993 | 0.0935 | 48.4966 | 0.0206 | 4.5327 | 0.2206 | 16.5883 | 3.6597 |
| 14 | 12.8392 | 0.0779 | 59.1959 | 0.0169 | 4.6106 | 0.2169 | 17.6008 | 3.8175 |
| 15 | 15.4070 | 0.0649 | 72.0351 | 0.0139 | 4.6755 | 0.2139 | 18.5095 | 3.9588 |
| 16 | 18.4884 | 0.0541 | 87.4421 | 0.0114 | 4.7296 | 0.2114 | 19.3208 | 4.0851 |
| 17 | 22.1861 | 0.0451 | 105.9306 | 9.440E-03 | 4.7746 | 0.2094 | 20.0419 | 4.1976 |
| 18 | 26.6233 | 0.0376 | 128.1167 | 7.805E-03 | 4.8122 | 0.2078 | 20.6805 | 4.2975 |
| 19 | 31.9480 | 0.0313 | 154.7400 | 6.462E-03 | 4.8435 | 0.2065 | 21.2439 | 4.3861 |
| 20 | 38.3376 | 0.0261 | 186.6880 | 5.357E-03 | 4.8696 | 0.2054 | 21.7395 | 4.4643 |
| 21 | 46.0051 | 0.0217 | 225.0256 | 4.444E-03 | 4.8913 | 0.2044 | 22.1742 | 4.5334 |
| 22 | 55.2061 | 0.0181 | 271.0307 | 3.690E-03 | 4.9094 | 0.2037 | 22.5546 | 4.5941 |
| 23 | 66.2474 | 0.0151 | 326.2369 | 3.065E-03 | 4.9245 | 0.2031 | 22.8867 | 4.6475 |
| 24 | 79.4968 | 0.0126 | 392.4842 | 2.548E-03 | 4.9371 | 0.2025 | 23.1760 | 4.6943 |
| 25 | 95.3962 | 0.0105 | 471.9811 | 2.119E-03 | 4.9476 | 0.2021 | 23.4276 | 4.7352 |
| 26 | 114.4755 | 8.735E-03 | 567.3773 | 1.762E-03 | 4.9563 | 0.2018 | 23.6460 | 4.7709 |
| 27 | 137.3706 | 7.280E-03 | 681.8528 | 1.467E-03 | 4.9636 | 0.2015 | 23.8353 | 4.8020 |
| 28 | 164.8447 | 6.066E-03 | 819.2233 | 1.221E-03 | 4.9697 | 0.2012 | 23.9991 | 4.8291 |
| 29 | 197.8136 | 5.055E-03 | 984.0680 | 1.016E-03 | 4.9747 | 0.2010 | 24.1406 | 4.8527 |
| 30 | 237.3763 | 4.213E-03 | 1.182E+03 | 8.461E-04 | 4.9789 | 0.2008 | 24.2628 | 4.8731 |
| 36 | 708.8019 | 1.411E-03 | 3.539E+03 | 2.826E-04 | 4.9929 | 0.2003 | 24.7108 | 4.9491 |
| 42 | 2.116E+03 | 4.725E-04 | 1.058E+04 | 9.454E-05 | 4.9976 | 0.2001 | 24.8890 | 4.9801 |
| 48 | 6.320E+03 | 1.582E-04 | 3.159E+04 | 3.165E-05 | 4.9992 | 0.2000 | 24.9581 | 4.9924 |
| 54 | 1.887E+04 | 5.299E-05 | 9.435E+04 | 1.060E-05 | 4.9997 | 0.2000 | 24.9844 | 4.9971 |
| 60 | 5.635E+04 | 1.775E-05 | 2.817E+05 | 3.549E-06 | 4.9999 | 0.2000 | 24.9942 | 4.9989 |
| 66 | 1.683E+05 | 5.943E-06 | 8.413E+05 | 1.189E-06 | 5.0000 | 0.2000 | 24.9979 | 4.9996 |
| 72 | 5.024E+05 | 1.990E-06 | 2.512E+06 | 3.981E-07 | 5.0000 | 0.2000 | 24.9992 | 4.9999 |
| 120 | 3.175E+09 | 3.150E-10 | 1.588E+10 | 6.299E-11 | 5.0000 | 0.2000 | 25.0000 | 5.0000 |
| 180 | 1.789E+14 | 5.590E-15 | 8.945E+14 | 1.118E-15 | 5.0000 | 0.2000 | 25.0000 | 5.0000 |
| 360 | 3.201E+28 | 3.124E-29 | 1.600E+29 | 6.249E-30 | 5.0000 | 0.2000 | 25.0000 | 5.0000 |

Time Value of Money Factors—Discrete Compounding
  i = 25%

| | Single Sums | | Uniform Series | | | | Gradient Series | |
|---|---|---|---|---|---|---|---|---|
| n | To Find F Given P (F\|P,i%,n) | To Find P Given F (P\|F,i%,n) | To Find F Given A (F\|A,i%,n) | To Find A Given F (A\|F,i%,n) | To Find P Given A (P\|A,i%,n) | To Find A Given P (A\|P,i%,n) | To Find P Given G (P\|G,i%,n) | To Find A Given G (A\|G,i%,n) |
| 1 | 1.2500 | 0.8000 | 1.0000 | 1.0000 | 0.8000 | 1.2500 | 0.0000 | 0.0000 |
| 2 | 1.5625 | 0.6400 | 2.2500 | 0.4444 | 1.4400 | 0.6944 | 0.6400 | 0.4444 |
| 3 | 1.9531 | 0.5120 | 3.8125 | 0.2623 | 1.9520 | 0.5123 | 1.6640 | 0.8525 |
| 4 | 2.4414 | 0.4096 | 5.7656 | 0.1734 | 2.3616 | 0.4234 | 2.8928 | 1.2249 |
| 5 | 3.0518 | 0.3277 | 8.2070 | 0.1218 | 2.6893 | 0.3718 | 4.2035 | 1.5631 |
| 6 | 3.8147 | 0.2621 | 11.2588 | 0.0888 | 2.9514 | 0.3388 | 5.5142 | 1.8683 |
| 7 | 4.7684 | 0.2097 | 15.0735 | 0.0663 | 3.1611 | 0.3163 | 6.7725 | 2.1424 |
| 8 | 5.9605 | 0.1678 | 19.8419 | 0.0504 | 3.3289 | 0.3004 | 7.9469 | 2.3872 |
| 9 | 7.4506 | 0.1342 | 25.8023 | 0.0388 | 3.4631 | 0.2888 | 9.0207 | 2.6048 |
| 10 | 9.3132 | 0.1074 | 33.2529 | 0.0301 | 3.5705 | 0.2801 | 9.9870 | 2.7971 |
| 11 | 11.6415 | 0.0859 | 42.5661 | 0.0235 | 3.6564 | 0.2735 | 10.8460 | 2.9663 |
| 12 | 14.5519 | 0.0687 | 54.2077 | 0.0184 | 3.7251 | 0.2684 | 11.6020 | 3.1145 |
| 13 | 18.1899 | 0.0550 | 68.7596 | 0.0145 | 3.7801 | 0.2645 | 12.2617 | 3.2437 |
| 14 | 22.7374 | 0.0440 | 86.9495 | 0.0115 | 3.8241 | 0.2615 | 12.8334 | 3.3559 |
| 15 | 28.4217 | 0.0352 | 109.6868 | 9.117E-03 | 3.8593 | 0.2591 | 13.3200 | 3.4530 |
| 16 | 35.5271 | 0.0281 | 138.1085 | 7.241E-03 | 3.8874 | 0.2572 | 13.7482 | 3.5366 |
| 17 | 44.4089 | 0.0225 | 173.6357 | 5.759E-03 | 3.9099 | 0.2558 | 14.1085 | 3.6084 |
| 18 | 55.5112 | 0.0180 | 218.0446 | 4.586E-03 | 3.9279 | 0.2546 | 14.4147 | 3.6698 |
| 19 | 69.3889 | 0.0144 | 273.5558 | 3.656E-03 | 3.9424 | 0.2537 | 14.6741 | 3.7222 |
| 20 | 86.7362 | 0.0115 | 342.9447 | 2.916E-03 | 3.9539 | 0.2529 | 14.8932 | 3.7667 |
| 21 | 108.4202 | 9.223E-03 | 429.6809 | 2.327E-03 | 3.9631 | 0.2523 | 15.0777 | 3.8045 |
| 22 | 135.5253 | 7.379E-03 | 538.1011 | 1.858E-03 | 3.9705 | 0.2519 | 15.2326 | 3.8365 |
| 23 | 169.4066 | 5.903E-03 | 673.6264 | 1.485E-03 | 3.9764 | 0.2515 | 15.3625 | 3.8634 |
| 24 | 211.7582 | 4.722E-03 | 843.0329 | 1.186E-03 | 3.9811 | 0.2512 | 15.4711 | 3.8861 |
| 25 | 264.6978 | 3.778E-03 | 1.055E+03 | 9.481E-04 | 3.9849 | 0.2509 | 15.5618 | 3.9052 |
| 26 | 330.8722 | 3.022E-03 | 1.319E+03 | 7.579E-04 | 3.9879 | 0.2508 | 15.6373 | 3.9212 |
| 27 | 413.5903 | 2.418E-03 | 1.650E+03 | 6.059E-04 | 3.9903 | 0.2506 | 15.7002 | 3.9346 |
| 28 | 516.9879 | 1.934E-03 | 2.064E+03 | 4.845E-04 | 3.9923 | 0.2505 | 15.7524 | 3.9457 |
| 29 | 646.2349 | 1.547E-03 | 2.581E+03 | 3.875E-04 | 3.9938 | 0.2504 | 15.7957 | 3.9551 |
| 30 | 807.7936 | 1.238E-03 | 3.227E+03 | 3.099E-04 | 3.9950 | 0.2503 | 15.8316 | 3.9628 |
| 36 | 3.081E+03 | 3.245E-04 | 1.232E+04 | 8.116E-05 | 3.9987 | 0.2501 | 15.9481 | 3.9883 |
| 42 | 1.175E+04 | 8.507E-05 | 4.702E+04 | 2.127E-05 | 3.9997 | 0.2500 | 15.9843 | 3.9964 |
| 48 | 4.484E+04 | 2.230E-05 | 1.794E+05 | 5.575E-06 | 3.9999 | 0.2500 | 15.9954 | 3.9989 |
| 54 | 1.711E+05 | 5.846E-06 | 6.842E+05 | 1.462E-06 | 4.0000 | 0.2500 | 15.9986 | 3.9997 |
| 60 | 6.525E+05 | 1.532E-06 | 2.610E+06 | 3.831E-07 | 4.0000 | 0.2500 | 15.9996 | 3.9999 |
| 66 | 2.489E+06 | 4.017E-07 | 9.957E+06 | 1.004E-07 | 4.0000 | 0.2500 | 15.9999 | 4.0000 |
| 72 | 9.496E+06 | 1.053E-07 | 3.798E+07 | 2.633E-08 | 4.0000 | 0.2500 | 16.0000 | 4.0000 |
| 120 | 4.258E+11 | 2.349E-12 | 1.703E+12 | 5.871E-13 | 4.0000 | 0.2500 | 16.0000 | 4.0000 |
| 180 | 2.778E+17 | 3.599E-18 | 1.111E+18 | 8.998E-19 | 4.0000 | 0.2500 | 16.0000 | 4.0000 |
| 360 | 7.720E+34 | 1.295E-35 | 3.088E+35 | 3.238E-36 | 4.0000 | 0.2500 | 16.0000 | 4.0000 |

Time Value of Money Factors—Discrete Compounding
  i = 30%

| | Single Sums | | Uniform Series | | | | Gradient Series | |
|---|---|---|---|---|---|---|---|---|
| | To Find F Given P (F\|P,i%,n) | To Find P Given F (P\|F,i%,n) | To Find F Given A (F\|A,i%,n) | To Find A Given F (A\|F,i%,n) | To Find P Given A (P\|A,i%,n) | To Find A Given P (A\|P,i%,n) | To Find P Given G (P\|G,i%,n) | To Find A Given G (A\|G,i%,n) |
| n | | | | | | | | |
| 1 | 1.3000 | 0.7692 | 1.0000 | 1.0000 | 0.7692 | 1.3000 | 0.0000 | 0.0000 |
| 2 | 1.6900 | 0.5917 | 2.3000 | 0.4348 | 1.3609 | 0.7348 | 0.5917 | 0.4348 |
| 3 | 2.1970 | 0.4552 | 3.9900 | 0.2506 | 1.8161 | 0.5506 | 1.5020 | 0.8271 |
| 4 | 2.8561 | 0.3501 | 6.1870 | 0.1616 | 2.1662 | 0.4616 | 2.5524 | 1.1783 |
| 5 | 3.7129 | 0.2693 | 9.0431 | 0.1106 | 2.4356 | 0.4106 | 3.6297 | 1.4903 |
| 6 | 4.8268 | 0.2072 | 12.7560 | 0.0784 | 2.6427 | 0.3784 | 4.6656 | 1.7654 |
| 7 | 6.2749 | 0.1594 | 17.5828 | 0.0569 | 2.8021 | 0.3569 | 5.6218 | 2.0063 |
| 8 | 8.1573 | 0.1226 | 23.8577 | 0.0419 | 2.9247 | 0.3419 | 6.4800 | 2.2156 |
| 9 | 10.6045 | 0.0943 | 32.0150 | 0.0312 | 3.0190 | 0.3312 | 7.2343 | 2.3963 |
| 10 | 13.7858 | 0.0725 | 42.6195 | 0.0235 | 3.0915 | 0.3235 | 7.8872 | 2.5512 |
| 11 | 17.9216 | 0.0558 | 56.4053 | 0.0177 | 3.1473 | 0.3177 | 8.4452 | 2.6833 |
| 12 | 23.2981 | 0.0429 | 74.3270 | 0.0135 | 3.1903 | 0.3135 | 8.9173 | 2.7952 |
| 13 | 30.2875 | 0.0330 | 97.6250 | 0.0102 | 3.2233 | 0.3102 | 9.3135 | 2.8895 |
| 14 | 39.3738 | 0.0254 | 127.9125 | 7.818E-03 | 3.2487 | 0.3078 | 9.6437 | 2.9685 |
| 15 | 51.1859 | 0.0195 | 167.2863 | 5.978E-03 | 3.2682 | 0.3060 | 9.9172 | 3.0344 |
| 16 | 66.5417 | 0.0150 | 218.4722 | 4.577E-03 | 3.2832 | 0.3046 | 10.1426 | 3.0892 |
| 17 | 86.5042 | 0.0116 | 285.0139 | 3.509E-03 | 3.2948 | 0.3035 | 10.3276 | 3.1345 |
| 18 | 112.4554 | 8.892E-03 | 371.5180 | 2.692E-03 | 3.3037 | 0.3027 | 10.4788 | 3.1718 |
| 19 | 146.1920 | 6.840E-03 | 483.9734 | 2.066E-03 | 3.3105 | 0.3021 | 10.6019 | 3.2025 |
| 20 | 190.0496 | 5.262E-03 | 630.1655 | 1.587E-03 | 3.3158 | 0.3016 | 10.7019 | 3.2275 |
| 21 | 247.0645 | 4.048E-03 | 820.2151 | 1.219E-03 | 3.3198 | 0.3012 | 10.7828 | 3.2480 |
| 22 | 321.1839 | 3.113E-03 | 1.067E+03 | 9.370E-04 | 3.3230 | 0.3009 | 10.8482 | 3.2646 |
| 23 | 417.5391 | 2.395E-03 | 1.388E+03 | 7.202E-04 | 3.3254 | 0.3007 | 10.9009 | 3.2781 |
| 24 | 542.8008 | 1.842E-03 | 1.806E+03 | 5.537E-04 | 3.3272 | 0.3006 | 10.9433 | 3.2890 |
| 25 | 705.6410 | 1.417E-03 | 2.349E+03 | 4.257E-04 | 3.3286 | 0.3004 | 10.9773 | 3.2979 |
| 26 | 917.3333 | 1.090E-03 | 3.054E+03 | 3.274E-04 | 3.3297 | 0.3003 | 11.0045 | 3.3050 |
| 27 | 1.193E+03 | 8.386E-04 | 3.972E+03 | 2.518E-04 | 3.3305 | 0.3003 | 11.0263 | 3.3107 |
| 28 | 1.550E+03 | 6.450E-04 | 5.164E+03 | 1.936E-04 | 3.3312 | 0.3002 | 11.0437 | 3.3153 |
| 29 | 2.015E+03 | 4.962E-04 | 6.715E+03 | 1.489E-04 | 3.3317 | 0.3001 | 11.0576 | 3.3189 |
| 30 | 2.620E+03 | 3.817E-04 | 8.730E+03 | 1.145E-04 | 3.3321 | 0.3001 | 11.0687 | 3.3219 |
| 36 | 1.265E+04 | 7.908E-05 | 4.215E+04 | 2.372E-05 | 3.3331 | 0.3000 | 11.1007 | 3.3305 |
| 42 | 6.104E+04 | 1.638E-05 | 2.035E+05 | 4.915E-06 | 3.3333 | 0.3000 | 11.1086 | 3.3326 |
| 48 | 2.946E+05 | 3.394E-06 | 9.821E+05 | 1.018E-06 | 3.3333 | 0.3000 | 11.1105 | 3.3332 |
| 54 | 1.422E+06 | 7.032E-07 | 4.740E+06 | 2.110E-07 | 3.3333 | 0.3000 | 11.1110 | 3.3333 |
| 60 | 6.864E+06 | 1.457E-07 | 2.288E+07 | 4.370E-08 | 3.3333 | 0.3000 | 11.1111 | 3.3333 |
| 66 | 3.313E+07 | 3.018E-08 | 1.104E+08 | 9.054E-09 | 3.3333 | 0.3000 | 11.1111 | 3.3333 |
| 72 | 1.599E+08 | 6.253E-09 | 5.331E+08 | 1.876E-09 | 3.3333 | 0.3000 | 11.1111 | 3.3333 |
| 120 | 4.712E+13 | 2.122E-14 | 1.571E+14 | 6.367E-15 | 3.3333 | 0.3000 | 11.1111 | 3.3333 |
| 180 | 3.234E+20 | 3.092E-21 | 1.078E+21 | 9.275E-22 | 3.3333 | 0.3000 | 11.1111 | 3.3333 |
| 360 | 1.046E+41 | 9.559E-42 | 3.487E+41 | 2.868E-42 | 3.3333 | 0.3000 | 11.1111 | 3.3333 |

# CHAPTER 5

# BOILERS AND FIRED SYSTEMS

**S.A. PARKER**
Senior Research Engineer, Energy Division
Pacific Northwest National Laboratory
Richland, Washington


**R.B. SCOLLON**
Corporate Manager, Energy Conservation


**R.D. SMITH**
Manager, Energy Generation and Feedstocks
Allied Corporation
Morristown, New Jersey

## 5. 1 INTRODUCTION

Boilers and other fired systems are the most significant energy consumers. Almost two-thirds of the fossil-fuel energy consumed in the United States involves the use of a boiler, furnace, or other fired system. Even most electric energy is produced using fuel-fired boilers. Over 68% of the electricity generated in the United States is produced through the combustion of coal, fuel oil, and natural gas. (The remainder is produced through nuclear, 22%; hydro-electric, 10%; and geothermal and others, <1%.) Unlike many electric systems, boilers and fired systems are not inherently energy efficient.

This chapter and the following chapter on Steam and Condensate Systems examine how energy is consumed, how energy is wasted, and opportunities for reducing energy consumption and costs in the operation of boiler and steam plants. A list of energy and cost reduction measures is presented, categorized as: load reduction, waste heat recovery, efficiency improvement, fuel cost reduction, and other opportunities. Several of the key opportunities for reducing operating costs are presented ranging from changes in operating procedures to capital improvement opportunities. The topics reflect recurring opportunities identified from numerous in-plant audits. Several examples are presented to demonstrate the methodology for estimating the potential energy savings associated with various opportunities. Many of these examples utilize easy to understand nomographs and charts in the solution techniques.

In addition to energy saving opportunities, this chapter also describes some issues relevant to day-to-day operations, maintenance, and troubleshooting. Considerations relative to fuel comparison and selection are also discussed. Developing technologies relative to alternative fuels and types of combustion equipment are also discussed. Some of the technologies discussed hold the potential for significant cost reductions while alleviating environmental problems.

The chapter concludes with a brief discussion of some of the major regulations impacting the operation of boilers and fired systems. It is important to emphasize the need to carefully assess the potential impact of federal, state, and local regulations.

## 5.2 ANALYSIS OF BOILERS AND FIRED SYSTEMS

### 5.2.1 Boiler Energy Consumption

Boiler and other fired systems, such as furnaces and ovens, combust fuel with air for the purpose of releasing the chemical heat energy. The purpose of the heat energy may be to raise the temperature of an industrial product as part of a manufacturing process, it may be to generate high-temperature high-pressure steam in order to power a turbine, or it may simply be to heat a space so the occupants will be comfortable. The energy consumption of boilers, furnaces, and other fire systems can be determined simply as a function of load and efficiency as expressed in the equation:

Energy consumption =
$$\int (load) \times (1/efficiency) \, dt \qquad (5.1)$$

Similarly, the cost of operating a boiler or fired system can be determined as:

Energy cost =
$$\int (load) \times (1/efficiency) \times (fuel \ cost) \, dt \qquad (5.2)$$

As such, the opportunities for reducing the energy consumption or energy cost of a boiler or fired system can be put into a few categories. In order to reduce boiler energy consumption, one can either reduce the load, increase the operating efficiency, reduce the unit fuel energy cost, or combinations thereof.

Of course equations 5.1 and 5.2 are not always that simple because the variables are not always constant. The *load* varies as a function of the process being supported. The *efficiency* varies as a function of the load and other functions, such as time or weather. In addition, the *fuel cost* may also vary as a function of time (such as in seasonal, time-of-use, or spot market rates) or as a function of load (such as declining block or spot market rates.) Therefore, solving the equation for the energy consumption or energy cost may not always be simplistic.

## 5.2.2 Balance Equations

Balance equations are used in an analysis of a process which determines inputs and outputs to a system. There are several types of balance equations which may prove useful in the analysis of a boiler or fired-system. These include a heat balance and mass balance.

### Heat Balance

A heat balance is used to determine where all the heat energy enters and leaves a system. Assuming that energy can neither be created or destroyed, all energy can be accounted for in a system analysis. Energy in equals energy out. Whether through measurement or analysis, all energy entering or leaving a system can be determined. In a simple furnace system, energy enters through the combustion air, fuel, and mixed-air duct. Energy leaves the furnace system through the supply-air duct and the exhaust gases.

In a boiler system, the analysis can become more complex. Energy input comes from the following: condensate return, make-up water, combustion air, fuel, and maybe a few others depending on the complexity of the system. Energy output departs as the following: steam, blowdown, exhaust gases, shell/surface losses, possibly ash, and other discharges depending on the complexity of the system.

### Mass Balance

A mass balance is used to determine where all mass enters and leaves a system. There are several methods in which a mass balance can be performed that can be useful in the analysis of a boiler or other fired system. In the case of a steam boiler, a mass balance can be used in the form of a water balance (steam, condensate return, make-up water, blowdown, and feedwater.) A mass balance can also be used for water quality or chemical balance (total dissolved solids, or other impurity.) The mass balance can also be used in the form of a combustion analysis (fireside mass balance consisting of air and fuel in and combustion gasses and excess air out.) This type of analysis is the foundation for determining combustion efficiency and determining the optimum air-to-fuel ratio.

For analyzing complex systems, the mass and energy balance equations may be used simultaneously such as in solving multiple equations with multiple unknowns. This type of analysis is particularly useful in determining blowdown losses, waste heat recovery potential, and other interdependent opportunities.

### 5.2.3 Efficiency

There are several different measures of efficiency used in boilers and fired systems. While this may lead to some confusion, the different measures are used to convey different information. Therefore, it is important to understand what is being implied by a given efficiency measure.

The basis for testing boilers is the American Society of Mechanical Engineers (ASME) Power Test Code 4.1 (PTC-4.1-1964.) This procedure defines and established two primary methods of determining efficiency: the input-output method and the heat-loss method. Both of these methods result in what is commonly referred to as the gross thermal efficiency. The efficiencies determined by these methods are "gross" efficiencies as apposed to "net" efficiencies which would include the additional energy input of auxiliary equipment such as combustion air fans, fuel pumps, stoker drives, etc. For more information on these methods, see the ASME PTC-4.1-1964 or Taplin 1991.

Another efficiency term commonly used for boilers and other fired systems is combustion efficiency. Combustion efficiency is similar to the heat loss method, but only the heat losses due to the exhaust gases are considered. Combustion efficiency can be measured in the field by analyzing the products of combustion the exhaust gases.

Typically measuring either carbon dioxide ($CO_2$) or oxygen ($O_2$) in the exhaust gas can be used to determine the combustion efficiency as long as there is excess air. Excess air is defined as air in excess of the amount required for stoichiometric conditions. In other words, excess air is the amount of air above that which is theoretically required for complete combustion. In the real world, however, it is not possible to get perfect mixture of air and fuel to achieve complete combustion without some amount of excess air. As excess air is reduced toward the fuel rich side, incomplete combustion begins to occur resulting in the formation of carbon monoxide, carbon, smoke, and in extreme cases, raw unburned fuel. Incomplete combustion is inefficient, expensive, and frequently unsafe. Therefore, some amount of excess air is required to ensure complete and safe combustion.

However, excess air is also inefficient as it results in the excess air being heated from ambient air temperatures to exhaust gas temperatures resulting in a form of heat loss. Therefore while some excess air is required it is also desirable to minimize the amount of excess air.

As illustrated in Figure 5.1, the amount of carbon dioxide, percent by volume, in the exhaust gas reaches a maximum with no excess air stoichiometric conditions. While carbon dioxide can be used as a measure of complete combustion, it can not be used to optimally control the air-to-fuel ratio in a fired system. A drop in the level of carbon dioxide would not be sufficient to inform the control system if it were operating in a condition of excess air or insufficient air. However, measuring oxygen in the exhaust gases is a direct measure of the amount of excess air. Therefore, measuring oxygen in the exhaust gas is a more common and preferred method of controlling the air-to-fuel ratio in a fired system.

### 5.2.4 Energy Conservation Measures

As noted above, energy cost reduction opportunities can generally be placed into one of the following categories: reducing load, increasing efficiency, and reducing unit energy cost. As with most energy conservation and cost reducing measures there are also a few additional opportunities which are not so easily categorized. Table 5.1 lists several energy conservation measures that have

been found to be very cost effective in various boilers and fired-systems.

### 5.3 KEY ELEMENTS FOR MAXIMUM EFFICIENCY

There are several opportunities for maximizing efficiency and reducing operating costs in a boiler or other fired-system as noted earlier in Table 5.1. This section examines in more detail several key opportunities for energy and cost reduction, including excess air, stack temperature, load balancing, boiler blowdown, and condensate return.

### 5.3.1 Excess Air

In combustion processes, excess air is generally defined as air introduced above the stoichiometric or theoretical requirements to effect complete and efficient combustion of the fuel.

There is an optimum level of excess-air operation for each type of burner or furnace design and fuel type. Only enough air should be supplied to ensure complete combustion of the fuel, since more than this amount in-

**Figure 5.1 Theoretical flue gas analysis versus air percentage for natural gas.**

**Table 5.1 Energy Conservation measures
for boilers and fired systems**[a]

**Load Reduction**
Insulation
—steam lines and distribution system
—condensate lines and return system
—heat exchangers
—boiler or furnace
Repair steam leaks
Repair failed steam straps
Return condensate to boiler
Reduce boiler blowdown
Improve feedwater treatment
Improve make-up water treatment
Repair condensate leaks
Shut off steam tracers during the summer
Shut off boilers during long periods of no use
Eliminate hot standby
Reduce flash steam loss
Install stack dampers or heat traps in natural draft boilers
Replace continuous pilots with electronic ignition pilots

**Waste Heat Recovery (a form of load reduction)**
Utilize flash steam
Preheat feedwater with an economizer
Preheat make-up water with an economizer
Preheat combustion air with a recuperator
Recover flue gas heat to supplement other heating system, such as domestic or service hot water, or unit space heater
Recover waste heat from some other system to preheat boiler make-up or feedwater
Install a heat recovery system on incinerator or furnace
Install condensation heat recovery system
—indirect contact heat exchanger
—direct contact heat exchanger

**Efficiency Improvement**
Reduce excess air
Provide sufficient air for complete combustion
Install combustion efficiency control system
—Constant excess air control
—Minimum excess air control
—Optimum excess air and CO control
Optimize loading of multiple boilers
Shut off unnecessary boilers
Install smaller system for part-load operation
—Install small boiler for summer loads
—Install satellite boiler for remote loads
Install low excess air burners
Repair or replace faulty burners
Replace natural draft burners with forced draft burners
Install turbulators in firetube boilers
Install more efficient boiler or furnace system
—high-efficiency, pulse combustion, or condensing boiler or furnace system
Clean heat transfer surfaces to reduce fouling and scale
Improve feedwater treatment to reduce scaling
Improve make-up water treatment to reduce scaling

**Fuel Cost Reduction**
Switch to alternate utility rate schedule
—interruptible rate schedule
Purchase natural gas from alternate source, self procurement of natural gas
Fuel switching
—switch between alternate fuel sources
—install multiple fuel burning capability
—replace electric boiler with a fuel-fired boiler

creases the heat rejected to the stack, resulting in greater fuel consumption for a given process output.

To identify the point of minimum excess-air operation for a particular fired system, curves of combustibles as a function of excess $O_2$ should be constructed similar to that illustrated in Figure 5.2. In the case of a gas-fueled system, the combustible monitored would be carbon monoxide (CO), whereas, in the case of a liquid- or solid fueled system, the combustible monitored would be the Smoke Spot Number (SSN). The curves should be developed for various firing rates as the minimal excess-air operating point will also vary as a function of the firing rate (percent load). Figure 5.2 illustrates two potential curves, one for high-fire and the other for low-fire. The optimal excess-air-control set point should be set at some margin (generally 0.5 to 1%) above the minimum $O_2$ point to allow for response and control variances. It is important to note that some burners may exhibit a gradual or steep CO $O_2$ behavior and this behavior may even change with various firing rates. It is also important to note that some burners may experience potentially unstable operation with small changes in $O_2$ (steep CO-$O_2$ curve behavior).

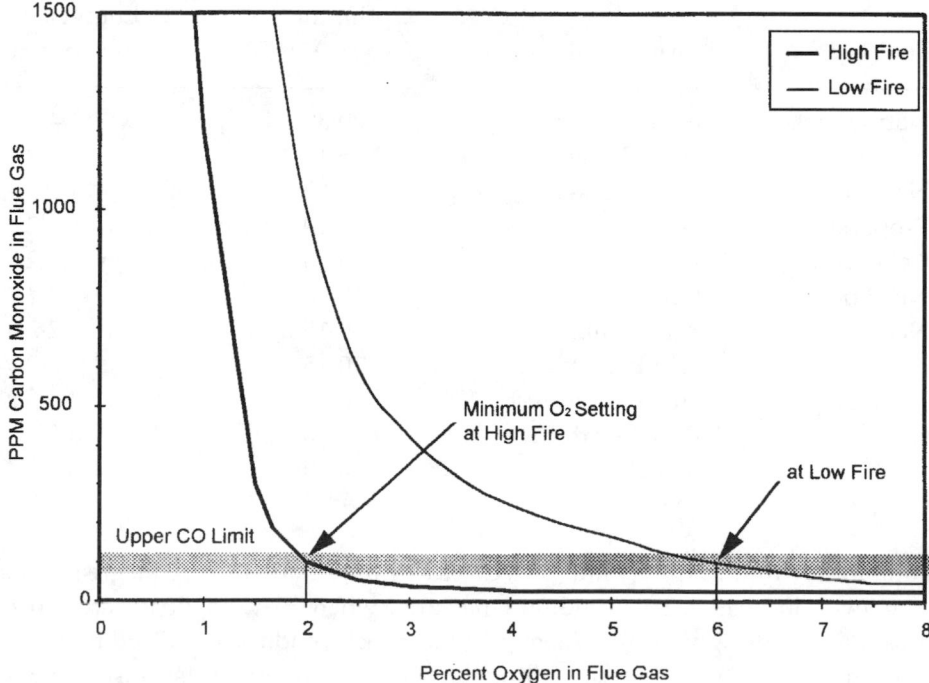

**Figure 5.2 Hypothetical CO-O2 characteristic curve for a gas-fired industrial boiler.**

Upper control limits for carbon monoxide vary depending on the referenced source. Points referenced for gas-fired systems are typically 400 ppm, 200 ppm, or 100 ppm. Today, local environmental regulations may dictate acceptable upper limits. Maximum desirable SSN for liquid fuels is typically SSN=1 for No. 2 fuel oil and SSN=4 for No. 6 fuel oil. Again, local environmental regulations may dictate lower acceptable upper limits.

Typical optimum levels of excess air normally attainable for maximum operating efficiency are indicated in Table 5.2 and classified according to fuel type and firing method.

The amount of excess air (or $O_2$) in the flue gas, unburned combustibles, and the stack temperature rise above the inlet air temperature are significant in defining the efficiency of the combustion process. Excess oxygen ($O_2$) measured in the exhaust stack is the most typical method of controlling the air-to-fuel ratio. However, for more precise control, carbon monoxide (CO) measurements may also be used to control air flow rates in combination with $O_2$ monitoring. Careful attention to furnace operation is required to ensure an optimum level of performance.

Figures 5.3, 5.4, and 5.5 can be used to determine the combustion efficiency of a boiler or other fired system burning natural gas, No. 2 fuel oil, or No. 6 fuel oil respectively so long as the level of unburned combustibles is considered negligible. These figures were derived from H. R. Taplin, Jr., *Combustion Efficiency Tables*, Fairmont Press, Lilburn, GA, 1991. For more information on combustion efficiency including combustion efficiencies using other fuels, see Taplin 1991.

**Where to Look for Conservation Opportunities**

Fossil-fuel-fired steam generators, process fired heaters/furnaces, duct heaters, and separately fired superheaters may benefit from an excess-air-control program. Specialized process equipment, such as rotary kilns, fired calciners, and so on, can also benefit from an air control program.

**How to Test for Relative Efficiency**

To determine relative operating efficiency and to establish energy conservation benefits for an excess-air-control program, you must determine: (1) percent oxygen (by volume) in the flue gas (typically dry), (2) stack temperature rise (the difference between the flue gas temperature and the combustion air inlet temperature), and (3) fuel type.

To accomplish optimal control over avoidable losses, continuous measurement of the excess air is a necessity. There are two types of equipment available to measure

---

### Table 5.2 Typical Optimum Excess Air[a]

| Fuel Type | Firing Method | Optimum Excess Air (%) | Equivalent $O_2$ (by Volume) |
|---|---|---|---|
| Natural gas | Natural draft | 20-30 | 4-5 |
| Natural gas | Forced draft | 5-10 | 1-2 |
| Natural gas | Low excess air | .04-0.2 | 0.1-0.5 |
| Propane | — | 5-10 | 1-2 |
| Coke oven gas | — | 5-10 | 1-2 |
| No. 2 oil | Rotary cup | 15-20 | 3-4 |
| No. 2 oil | Air-atomized | 10-15 | 2-3 |
| No. 2 oil | Steam-atomized | 10-15 | 2-3 |
| No. 6 oil | Steam-atomized | 10-15 | 2-3 |
| Coal | Pulverized | 15-20 | 3-3.5 |
| Coal | Stoker | 20-30 | 3.5-5 |
| Coal | Cyclone | 7-15 | 1.5-3 |

[a]To maintain safe unit output conditions, excess-air requirements may be greater than the optimum levels indicated. This condition may arise when operating loads are substantially less than the design rating. Where possible, check vendors' predicted performance curves. If unavailable, reduce excess-air operation to minimum levels consistent with satisfactory output.

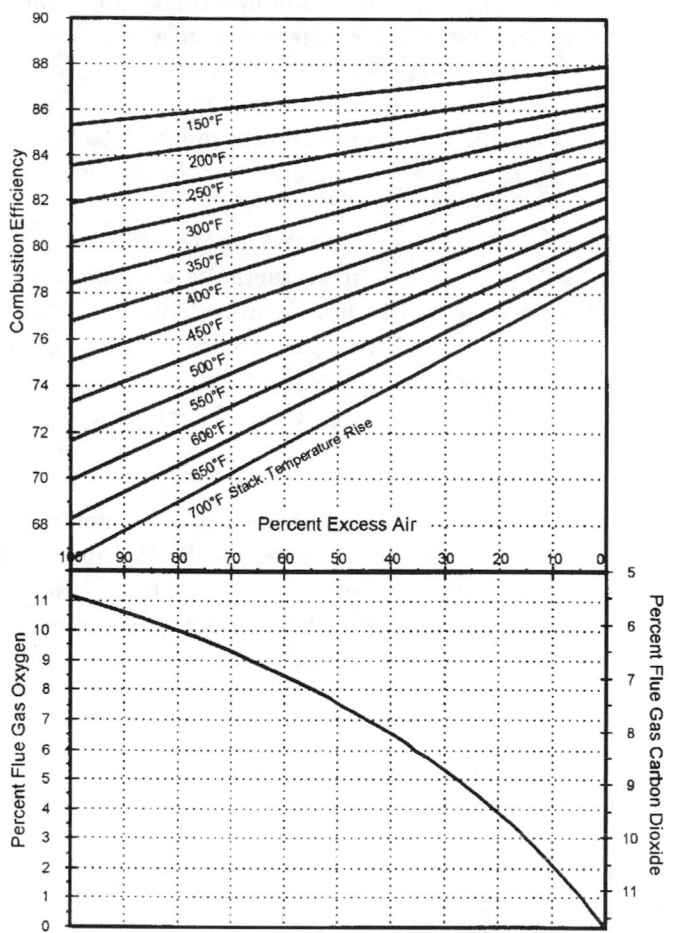

**Figure 5.3 Combustion efficiency chart for natural gas.**

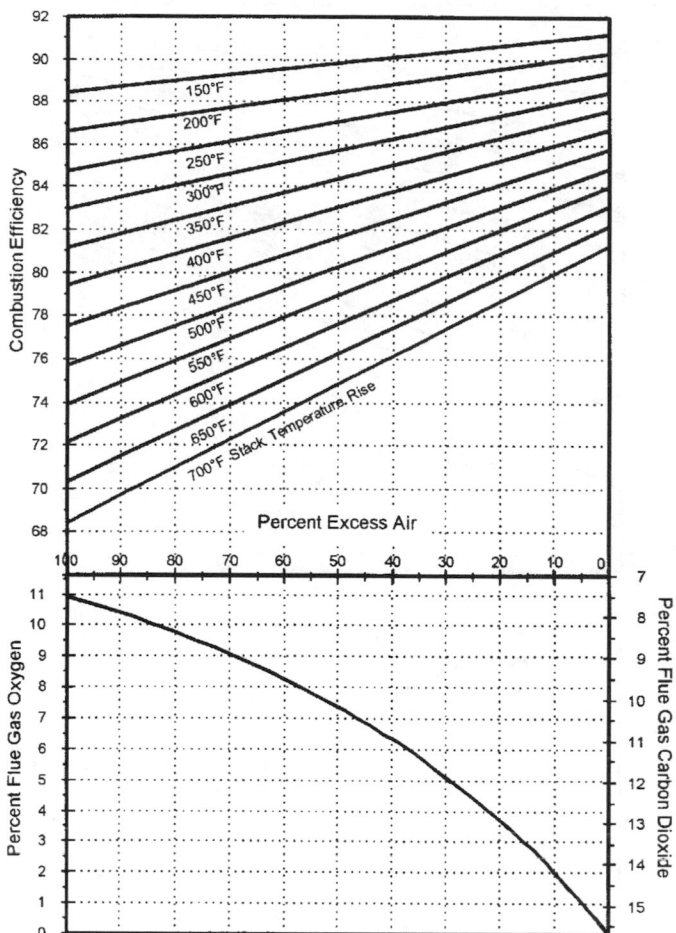

**Figure 5.4 Combustion efficiency chart for number 2 fuel oil.**

flue-gas oxygen and corresponding "excess air": (1) portable equipment such as an Orsat flue-gas analyzer, heat prover, electronic gas analyzer, or equivalent analyzing device; and (2) permanent-type installations probe-type continuous oxygen analyzers (available from various manufacturers), which do not require external gas sampling systems.

The major advantage of permanently mounted equipment is that the on-line indication or recording allows remedial action to be taken frequently to ensure continuous operation at optimum levels. Computerized systems which allow safe control of excess air over the boiler load range have proven economic for many installations. Even carbon monoxide-based monitoring and control systems, which are notably more expensive than simple oxygen-based monitoring and control systems, prove to be cost effective for larger industrial-and utility-sized boiler systems.

Portable equipment only allows performance checking on an intermittent or spot-check basis. Periodic monitoring may be sufficient for smaller boilers or boilers which do not undergo significant change in operating conditions. However, continuous monitoring and control systems have the ability to respond more rapidly to changing conditions, such as load and inlet air conditions.

The stack temperature rise may be obtained with portable thermocouple probes in conjunction with a potentiometer or by installing permanent temperature probes within the exhaust stack and combustion air inlet and providing continuous indication or recording. Each type of equipment provides satisfactory results for the planning and operational results desired.

An analysis to establish performance can be made with the two measurements, percent oxygen and the stack temperature rise, in addition to the particular fuel fired. As an illustration, consider the following example.

**Example**: Determine the potential energy savings associated with reducing the amount of excess air to an optimum level for a natural gas-fired steam boiler.

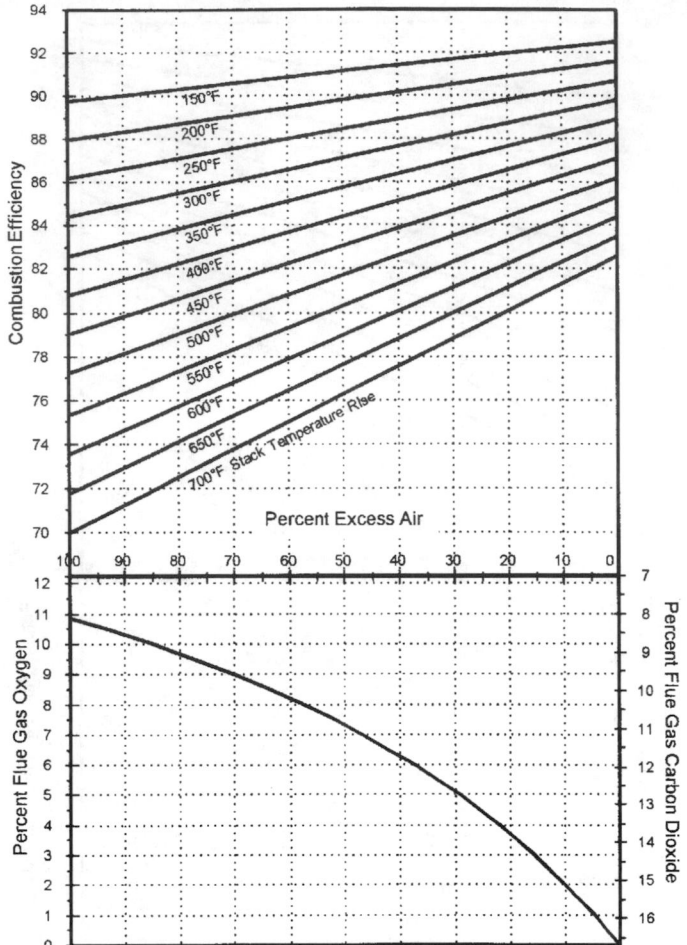

**Figure 5.5 Combustion efficiency chart for number 6 fuel oil.**

**Operating Data.**

| | |
|---|---|
| Current energy consumption | 1,100,000 therms/yr |
| Boiler rated capacity | 600 boiler horsepower |
| Operating hours | 8,500 hr/yr |
| Current stack gas analysis | 9% Oxygen (by volume, dry) |
| | Minimal CO reading |
| Combustion air inlet temperature | 80°F |
| Exhaust gas stack temperature | 580°F |
| Proposed operating condition | 2% Oxygen (by volume, dry) |

**Calculation and Analysis.**

**STEP 1**: Determine current boiler combustion efficiency using Figure 5.6 for natural gas. Note that this is the same figure as Figure 5.3.

A) Determine the current stack temperature rise.
STR = (exhaust stack temperature)
– (combustion air temperature)
STR = 580°F - 80°F = 500°F

B) Enter the chart with an oxygen level of 9% and following a line to the curve, read the percent excess air to be approximately 66%.

C) Continue the line to the curve for a stack temperature rise of 500°F and read the current combustion efficiency to be 76.4%.

**STEP 2:** Determine the proposed boiler combustion efficiency using the same figure.

D) Repeat steps A through C for the proposed combustion efficiency assuming the same stack temperature conditions. Read the proposed combustion efficiency to be 81.4%.

Note that in many cases reducing the amount of excess air will tend to reduce the exhaust stack temperature, resulting in an even more efficient operating condition. Unfortunately, it is difficult to predict the extent of this added benefit.

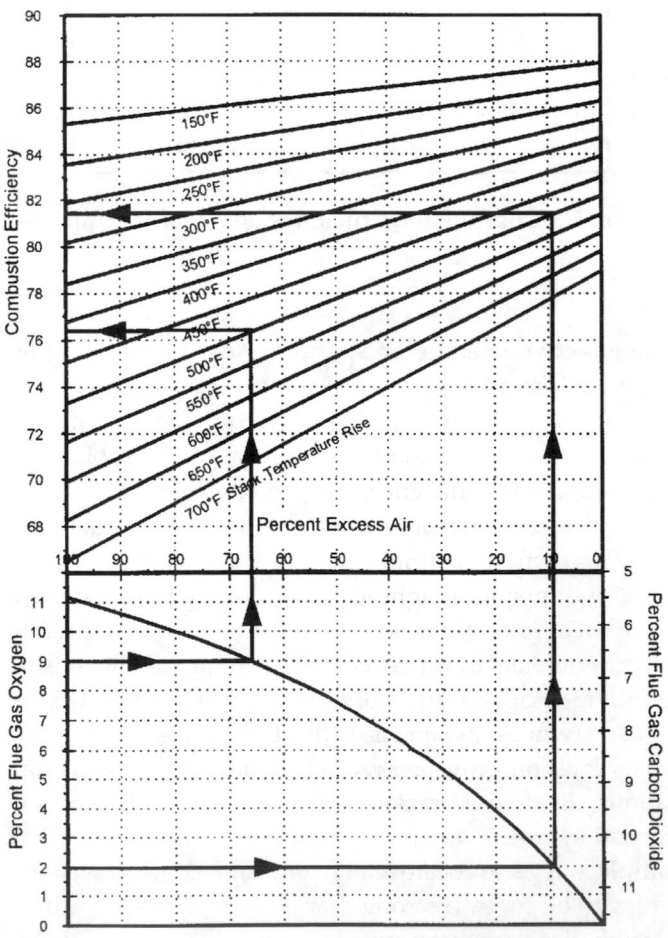

**Figure 5.6. Combustion efficiency curve for reducing excess air example.**

**STEP 3**: Determine the fuel savings.

E) Percent fuel savings = [(new efficiency)
− (old efficiency)]/(new efficiency)
Percent fuel savings = [(81.4%)
− (76.4%)]/(81.4%)
Percent fuel savings = 6.14%

F) Fuel savings =(current fuel consumption)
× (percent fuel savings)
Fuel savings = (1,100,000 therms/yr) × (6.14%)
Fuel savings = 67,540 therms/yr

## Conclusions

This example assumes that the results of the combustion analysis and boiler load are constant. Obviously this is an oversimplification of the issue. Because the air-to-fuel ratio (excess air level) is different for different boiler loads, a more thorough analysis should take this into account. One method to accomplish this would be to perform the analysis at various firing rates, such as high-fire and low-fire. For modulating type boilers which can vary between high- and low-firing rates, a modified bin analysis approach or other bin-type methodology could be employed.

## Requirements to Effect Maximum Economy

To obtain the maximum benefits of an excess-air-control program, the following modifications, additions, checks, or procedures should be considered:

## Key Elements for Maximum Efficiency

1. Ensure that the furnace boundary walls and flue work are airtight and not a source of air infiltration or exfiltration.
   a. Recognized leakage problem areas include (1) test connection for oxygen analyzer or portable Orsat connection; (2) access doors and ash-pit doors; (3) penetration points passing through furnace setting; (4) air seals on soot-blower elements or sight glasses; (5) seals around boiler drums and header expansion joints; (6) cracks or breaks in brick settings or refractory; (7) operation of the furnace at too negative a pressure; (8) burner penetration points; and (9) deterioration of air preheater radial seals or tube-sheet expansion and cracks on tubular air heater applications.
   b. Tests to locate leakage problems: (1) a light test whereby a strong spotlight is placed in the furnace and the unit inspected externally; (2) the use of a pyrometer to obtain a temperature profile on the outer casing. This test generally indicates points where refractory or insulation has deteriorated; (3) a soap-bubble test on suspected penetration points or seal welds; (4) a smoke-bomb test and an external examination for traces of smoke; (5) holding a lighted candle along the casing seams has pinpointed leakage problems on induced- or natural-draft units; (6) operating the forced draft fan on high capacity with the fire out, plus use of liquid chemical smoke producers has helped identify seal leaks; and (7) use of a thermographic device to locate "hot spots" which may indicate faulty insulation or flue-gas leakage.

2. Ensure optimum burner performance.
   a. Table 5.3 lists common burner difficulties that can be rectified through observation and maintenance.
   b. Ascertain integrity of air volume control: (1) the physical condition of fan vanes, dampers, and operators should be in optimum working condition; and (2) positioning air volume controls should be checked for responsiveness and adequacy to maintain optimum air/fuel ratios. Consult operating manual or control manufacturer for test and calibration.
   c. Maintain or purchase high-quality gas analyzing systems: calibrate instrument against a known flue-gas sample.
   d. Purchase or update existing combustion controls to reflect the present state of the art.
   e. Consider adapting "oxygen trim" feature to existing combustion control system.

3. Establish a maintenance program.
   a. Table 5.4 presents a summary of frequent boiler system problems and possible causes.
   b. Perform period maintenance as recommended by the manufacturer.
   c. Keep a boiler operator's log and monitor key parameters.
   d. Perform periodic inspections.

## Guidelines for Day-to-Day Operation

The following steps must be taken to assure peak boiler efficiency and minimum permissible excess-air operation.

1. Check the calibration of the combustion gas analyzer frequently and check the zero point daily.
2. If a sampling system is employed, check to assure proper operation of the sampling system.

3.  The forced-draft damper should be checked for its physical condition to ensure that it is not broken or damaged.
4.  Casing leakage must be detected and stopped.
5.  Routinely check control drives and instruments.
6.  If the combustion gas analyzer is used for monitoring purposes, the excess air must be checked daily. The control may be manually altered to reduce excess air, without shortcutting the safety of operation.
7.  The fuel flow and air flow charts should be carefully checked to ensure that the fuel follows the air on increasing load with proper safety margin and also that the fuel leads the air on decreasing load. This should be compared on a daily shift basis to ensure consistency of safe and efficient operation.
8.  Check the burner flame configuration frequently during each shift and note burner register changes in the operator's log.
9.  Periodically check flue-gas CO levels to ensure complete combustion. If more than a trace amount of CO is present in the flue gas, investigate burner conditions identified on Table 5.3 or fuel supply quality limits such as fuel-oil viscosity/temperature or coal fineness and temperature.

### 5.3.2 Exhaust Stack Temperature

Another primary factor affecting unit efficiency and ultimately fuel consumption is the temperature of combustion gases rejected to the stack. Increased operating efficiency with a corresponding reduction in fuel input can be achieved by rejecting stack gases at the lowest practical temperature consistent with basic design principles. In general, the application of additional heat recovery equipment can realize this energy conservation objective when the measured flue-gas temperature exceeds approximately 250°F. For a more extensive coverage of waste-heat recovery, see Chapter 8.

**Where to Look**

Steam boilers, process fired heaters, and other combustion or heat-transfer furnaces can benefit from a heat-recovery program.

The adaptation of heat-recovery equipment to existing units as discussed in this section will be limited to flue gas/liquid and/or flue gas/air preheat exchangers. Specifically, economizers and air preheaters come under this category. Economizers are used to extract heat energy from the flue gas to heat the incoming liquid process feedstream to the furnace. Flue gas/air preheaters lower the flue-gas temperature by exchanging heat to the incoming combustion air stream.

Planning-quality guidelines will be presented to determine the final sink temperature, as well as comparative economic benefits to be derived by the installation of heat-recovery equipment. Costs to implement this energy conservation opportunity can then be compared against the potential benefits.

### Table 5.3 Malfunctions in Fired Systems

| Malfunction | Fuel | | | Detection | Action |
|---|---|---|---|---|---|
| | Coal | Oil | Gas | | |
| Uneven air distribution to burners | x | x | x | Observe flame patterns | Adjust registers (trial and error) |
| Uneven fuel distribution to burners | x | x | x | Observe fuel pressure gages, or take coal sample and analyze | Consult manufacturer |
| Improperly positioned guns or impellers | x | x | | Observe flame patterns | Adjust guns (trial and error) |
| Plugged or worn burners | x | x | | Visual inspection | Increase frequency of cleaning; install strainers (oil) |
| Damaged burner throats | x | x | x | Visual inspection | Repair |

**Table 5.4 Boiler Performance Troubleshooting**

| System | Problem | Possible Cause |
|---|---|---|
| Heat transfer related | High stack gas temperature | Buildup of gas- or water-side deposits |
| | | Improper water treatment procedure |
| | | Improper soot blower operation |
| Combustion related | High excess air | Improper control system operation |
| | | Low fuel supply pressure |
| | | Change in fuel heating value |
| | | Change in oil viscosity |
| | | Decrease in inlet air temperature |
| | Low excess air | Improper control system operation |
| | | Fan limitations |
| | | Increase in inlet air temperature |
| | High carbon monoxide and combustible emissions | Plugged gas burners |
| | | Unbalanced fuel and air distribution in multiburner furnaces |
| | | Improper air register settings |
| | | Deterioration of burner throat refractory |
| | | Stoker grate condition |
| | | Stoker fuel distribution orientation |
| | | Low fineness on pulverized systems |
| Miscellaneous | Casing leakage | Damaged casing and insulation |
| | Air heater leakage | Worn or improper adjusted seals on rotary heaters |
| | | Tube corrosion |
| | Coal pulverizer power | Pulverizer in poor repair |
| | | Too low classifier setting |
| | Excessive blowdown | Improper operation |
| | Steam leaks | Holes in waterwall tube Valve packing |
| | Missing or loose insulation | Overheating |
| | | Weathering |
| | Excessive soot blower operation | Arbitrary operation schedule that is in excess of requirements |

## How to Test for Heat-Recovery Potential

In assessing overall efficiency and potential for heat recovery, the parameters of significant importance are temperature and fuel type/sulfur content. To obtain a meaningful operating flue-gas temperature measurement and a basis for heat-recovery selection, the unit under consideration should be operating at, or very close to, design and optimum excess-air values as defined on Table 5.2.

Temperature measurements may be made by mercury or bimetallic element thermometers, optical pyrometers, or an appropriate thermocouple probe. The most adaptable device is the thermocouple probe in which an iron or chromel constantan thermocouple is used. Temperature readout is accomplished by connecting the thermocouple leads to a potentiometer. The output of the potentiometer is a voltage reading which may be correlated with the measured temperature for the particular thermocouple element employed.

To obtain a proper and accurate temperature measurement, the following guidelines should be followed:

1.  Locate the probe in an unobstructed flow path and sufficient distance, approximately five diameters downstream or upstream, of any major change of direction in the flow path.

2.  Ensure that the probe entrance connection is relatively leak free.

3.  Take multiple readings by traversing the cross-sectional area of the flue to obtain an average and representative flue-gas temperature.

## Modifications or Additions for Maximum Economy

The installation of economizers and/or flue-gas air preheaters on units not presently equipped with heat-recovery devices and those with minimum heat-recovery equipment are practical ways of reducing stack temperature while recouping flue-gas sensible heat normally rejected to the stack.

There are no "firm" exit-temperature guidelines that cover all fuel types and process designs. However, certain guiding principles will provide direction to the lowest practical temperature level of heat rejection. The elements that must be considered to make this judgment include (1) fuel type, (2) flue-gas dew-point considerations, (3) heat-transfer criteria, (4) type of heat-recovery surface, and (5) relative economics of heat-recovery equipment.

Tables 5.5 and 5.6 may be used for selecting the lowest practical exit-gas temperature achievable with installation of economizers and/or flue-gas air preheaters.

As an illustration of the potential and methodology for recouping flue-gas sensible heat by the addition of heat-recovery equipment, consider the following example.

**Example**: Determine the energy savings associated with installing an economizer or flue-gas air preheater on the boiler from the previous example. Assume that the excess-air control system from the previous example has already been implemented.

### Available Data

| | |
|---|---|
| Current energy consumption | 1,032,460 therms/yr |
| Boiler rated capacity | 600 boiler horsepower |
| Operating hours | 8,500 hr/yr |
| Exhaust stack gas analysis | 2% Oxygen (by volume, dry) |
| | Minimal CO reading |

Current operating conditions:

| | |
|---|---|
| Combustion air inlet temperature | 80°F |
| Exhaust gas stack temperature | 580°F |
| Feedwater temperature | 180°F |
| Operating steam pressure | 110 psia |
| Operating steam temperature | 335°F |

Proposed operating condition:

| | |
|---|---|
| Combustion air inlet temperature | 80°F |
| Exhaust gas stack temperature | 380°F |

### Calculation and Analysis

**STEP 1**: Compare proposed stack temperature against minimum desired stack temperature.

A)  Heat transfer criteria:
$$T_g = T1 + 100°F \text{ (minimum)}$$
$$T_g = 180 + 100°F \text{ (minimum)}$$
$$T_g = 280°F \text{ (minimum)}$$

B)  Flue-gas dew point:
$$T_g = 120°F \text{ (from Figure 5.8)}$$

C)  Proposed stack temperature
$$T_g = 380°F \text{ is acceptable}$$

**STEP 2**: Determine current boiler combustion efficiency using Figure 5.7 for natural gas. Note that this is the same figure as Figure 5.3.

A)  Determine the stack temperature rise.
STR = (exhaust stack temperature)
– (combustion air temperature)
STR = 580°F - 80°F = 500°F

B)  Enter the chart with an oxygen level of 2% and following a line to the curve, read the percent excess air to be approximately 9.3%.

**Table 5.5 Economizers**

|  |  |
|---|---|
| Fuel Type | Test for Determination of Exit Flue-Gas Temperatures |
| Gaseous fuel (minimum percent sulphur) | Heat-transfer criteria: $T_g = T_1 + 100°F$ (minimum): typically the higher of (a) or (b) below. |
| Fuel oils and coal | (a) Heat-transfer criteria: $T_g = T_1 + 100°F$ (min.) (b) Flue-gas dew point (from Figure 5.8 for a particular fuel and percent sulphur by weight |

Where:  $T_g$ = Final stack flue temperature
$T_1$ = Process liquid feed temperature

**Table 5.6 Flue-Gas/Air Preheaters**

|  |  |
|---|---|
| Fuel Type | Test for Determination of Exit Flue-Gas Temperatures |
| Gaseous fuel | Historic economic breakpoint: $T_g$ (min.) = approximately 250°F |
| Fuel oils and coal | Average cold-end considerations; see Figure 5.9 for determination of $T_{ce}$; the exit-gas temperature relationship is $T_g = 2T_{ce} - T_a$ |

Where:  $T_g$  = Final stack flue temperature
$T_{ce}$  = Flue gas air preheater recommended average cold end temperature
$T_a$  = Ambient air temperature

C) Continue the line to the curve for a stack temperature rise of 500°F and read the current combustion efficiency to be 81.4%.

**STEP 3**: Determine the proposed boiler combustion efficiency using the same figure.

D) Repeat steps A through C for the proposed combustion efficiency assuming the new exhaust stack temperature conditions. Read the proposed combustion efficiency to be 85.0%.

**STEP 4**: Determine the fuel savings.

E) Percent fuel savings = [(new efficiency) – (old efficiency)]/(new efficiency)

Percent fuel savings = [(85.0%) - (81.4%)]/(85.0%)
Percent fuel savings = 4.24%

F) Fuel savings =(current fuel consumption) × (percent fuel savings)
Fuel savings = (1,032,460 therms/yr) × (4.24%)
Fuel savings = 43,776 therms/yr

**Conclusion**

As with the earlier example, this analysis methodology assumes that the results of the combustion analysis and boiler load are constant. Obviously this is an oversimplification of the issue. Because the air-to-fuel ratio (excess air level) is different for different boiler loads, a more thorough analysis should take this into account.

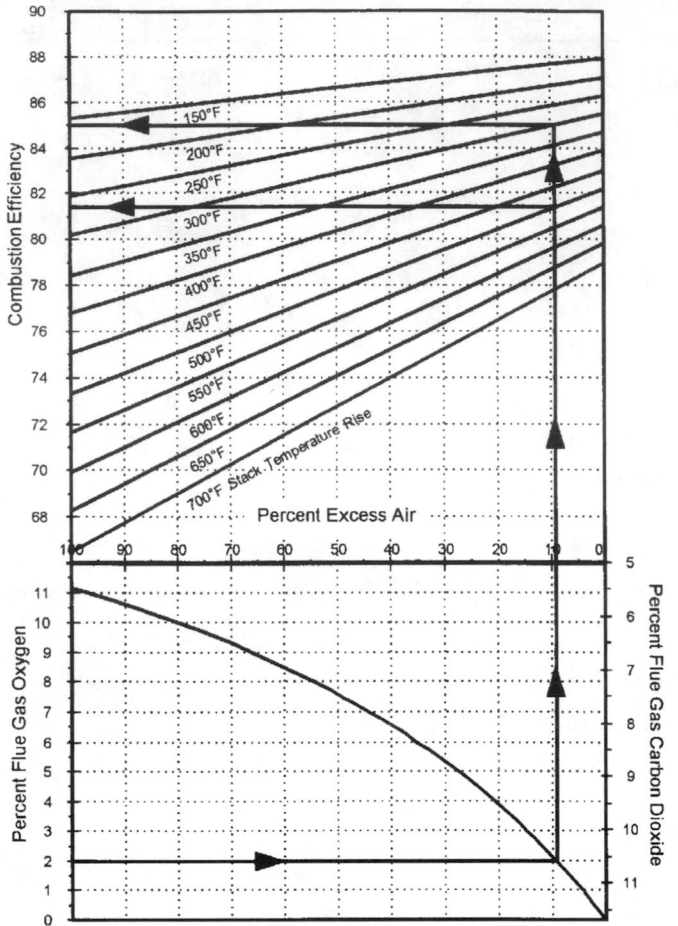

**Figure 5.7 Combustion efficiency curve for stack temperature reduction example.**

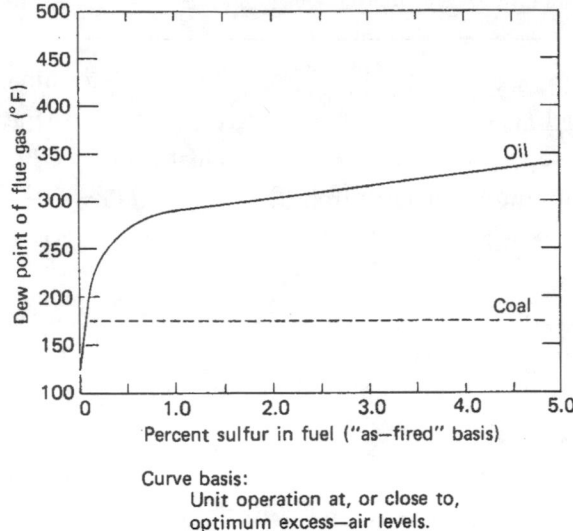

**Figure 5.8 Flue-gas dew point. Based on unit operation *at* or *close to* "optimal" excess-air.**

**Figure 5.9 Guide for selecting flue-gas air preheaters.**

Additional considerations in flue-gas heat recovery include:

1. Space availability to accommodate additional heating surface within furnace boundary walls or adjacent area to stack.

2. Adequacy of forced-draft and/or induced-draft fan capacity to overcome increased resistance of heat-recovery equipment.

3. Adaptability of soot blowers for maintenance of heat-transfer-surface cleanliness when firing ash- and soot-forming fuels.

4. Design considerations to maintain average cold-end temperatures for flue gas/air preheater applications in cold ambient surroundings.

5. Modifications required of flue and duct work and additional insulation needs.

6. The addition of structural steel supports.

7. Adequate pumping head to overcome increased fluid pressure drop for economizer applications.

8. The need for bypass arrangements around economizers or air preheaters.

9. Corrosive properties of gas, which would require special materials.

10. Direct flame impingement on recovery equipment.

## Guidelines for Day-to-Day Operation

1. Maintain operation at goal excess air levels and stack temperature to obtain maximum efficiency and unit thermal performance.

2. Log percent $O_2$ or equivalent excess air, inlet air temperature, and stack temperatures, once per shift or more frequent, noting the unit load and fuel fired.

3. Use oxygen analyzers with recorders for units larger than about $35 \times 10^6$ Btu/hr output.

4. Maintain surface cleanliness by soot blowing at least once per shift for ash- and soot-forming fuels.

5. Establish a more frequent cleaning schedule when heat-exchange performance deteriorates due to firing particularly troublesome fuels.

6. External fouling can also cause high excess air operation and higher stack temperatures than normal to achieve desired unit outputs. External fouling can be detected by use of draft loss gauges or water manometers and periodically (once a week) logging the results.

7. For flue gas/air preheaters, oxygen checks should be taken once a month before and after the heating surface to assess condition of circumferential and radial seals. If $O_2$ between the two readings varies in excess of 1% $O_2$, air heater leakage is excessive to the detriment of operating efficiency and fan horsepower.

8. Check fan damper operation weekly. Adjust fan damper or operator to correspond to desired excess air levels.

9. Institute daily checks on continuous monitoring equipment measuring flue-gas conditions. Check calibration every other week.

10. Establish an experience guideline on optimum time for cleaning and changing oil guns and tips.

11. Receive the "as-fired" fuel analysis on a monthly basis from the supplier. The fuel base may have changed, dictating a different operating regimen.

12. Analyze boiler blowdown every two months for iron. Internal surface cleanliness is as important to maintaining heat-transfer characteristics and performance as external surface cleanliness.

13. When possible, a sample of coal, both raw and pulverized, should be analyzed to determine if operating changes are warranted and if the design coal fineness is being obtained.

### 5.3.3 Waste-Heat-Steam Generation

Plants that have fired heaters and/or low-residence-time process furnaces of the type designed during the era of cheap energy may have potentially significant energy-saving opportunities. This section explores an approach to maximize energy efficiency and provide an analysis to determine overall project viability.

The major problem on older units is to determine a practical and economical approach to utilize the sensible heat in the exhaust flue gas. Typically, many vintage units have exhaust-flue-gas temperatures in the range 1050 to 1600°F. In this temperature range, a conventional flue-gas air preheater normally is not a practical approach because of materials of construction requirements and significant burner front modifications. Additionally, equipping these units with an air preheater could materially alter the inherent radiant characteristics of the furnace, thus adversely affecting process heat transfer. An alternative approach to utilizing the available flue-gas sensible heat and maximizing overall plant energy efficiency is to consider: (1) waste-heat-steam generation: (2) installing an unfired or supplementary fired recirculating hot-oil loop or ethylene glycol loop to effectively utilize transferred heat to a remote location: and (3) installing a process feed economizer.

Because most industrial process industries have a need for steam, the example is for the application of an unfired waste-heat-steam generator.

The hypothetical plant situation is a reformer furnace installed in the plant in 1963 at a time when it was not considered economical to install a waste-heat-steam generator. As a result, the furnace currently vents hot flue gas (1562°F) to the atmosphere after inspiriting ambient air to reduce the exhaust temperature so that standard materials of construction could be utilized.

The flue-gas temperature of 1562°F is predicated on a measured value by thermocouple and is based on a typical average daily process load on the furnace. This induced-draft furnace fires a No. 2 fuel oil and has been optimized for 20% excess air operation. Flue-gas flow is calculated at 32,800 lb/hr. The plant utilizes approximately 180,000 lb/hr of 300-psig saturated steam from three boilers each having a nameplate capacity of 75,000 lb/hr. The plant steam load is shared equally by the three operating boilers, each supplying 60,000 lb/hr. Feedwater to the units is supplied at 220°F from a common water-treating facility. The boilers are fired with low-sulfur (0.1% sulphur by weight) No. 2 fuel oil. Boiler efficiency averages 85% at load. Present fuel costs are $0.76/gal or $5.48/10$^6$ Btu basis of No. 2 fuel oil having a heating value of 138,800 Btu/gal. The basic approach to enhancing plant energy efficiency and minimizing cost is to generate maximum quantities of "waste" heat steam by recouping the sensible heat from the furnace exhaust flue gas.

Certain guidelines would provide a "fix" on the amount of steam that could be reasonably generated. The flue-gas temperature drop could practically be reduced to 65 to 100°F above the boiler feedwater temperature of 220°F. Using an approach temperature of 65°F yields an exit-flue gas temperature of 220 + 65 = 285°F. This assumes that an economizer would be furnished integral with the waste-heat-steam generator.

A heat balance on the flue-gas side (basis of flue-gas temperature drop) would provide the total heat duty available for steam generation. The sensible heat content of the flue gas is derived from Figures 5.10a and 5.10b based on the flue-gas temperature and percent moisture in the flue gas.

Percentage moisture (by weight) in the flue gas is a function of the type of fuel fired and percentage excess-air operation. Typical values of percentage moisture are indicated in Table 5.7 for various fuels and excess air. For No. 2 fuel oil firing at 20% excess air, percent moisture by weight in flue gas is approximately 6.8%.

Therefore, a flue-gas heat balance becomes

| Flue-Gas Temperature Drop (°F) | Sensible Heat in Flue Gas (Btu/lb W.G.) | |
|---|---|---|
| 1562 | 412 | (Fig. 5.15) |
| 285 | 52 | (Fig. 5.14) |
| 1277 | 360 | |

**Figure 5.10a Heat in flue gases vs. percent moisture by weight. (Derived from Keenan and Kayes 1948.)**

**Table 5.7 Percent Moisture by Weight in Flue Gas**

| | Percent Excess Air | | | |
| --- | --- | --- | --- | --- |
| Fuel Type | 10 | 15 | 20 | 25 |
| Natural gas | 12.1 | 11.7 | 11.2 | 10.8 |
| No. 2 fuel oil | 7.3 | 7.0 | 6.8 | 6.6 |
| Coal (varies) | 6.7-5.1 | 6.4-4.9 | 6.3-4.7 | 6.1-4.6 |
| Propane | 10.1 | 9.7 | 9.4 | 9.1 |

The total heat available from the flue gas for steam generation becomes

$$(32,800 \text{ lb.W.G.}) \times (360 \text{ Btu/lb.W.G.}) = (11.8 \times 10^6 \text{ Btu/h})$$

The amount of steam that may be generated is determined by a thermodynamic heat balance on the steam circuit.

Enthalpy of steam at 300 psig saturated
$$h_3 = 1203 \text{ Btu/lb}$$

Enthalpy of saturated liquid at drum pressure of 300 psig

$$h_f = 400 \text{ Btu/lb}$$

Enthalpy corresponding to feedwater temperature of 200°F
$$h_1 = 188 \text{ Btu/lb}$$

For this example, assume that boiler blowdown is 10% of steam flow. Therefore, feedwater flow through the economizer to the boiler drum will be 1.10 times the steam outflow from the boiler drum. Let the steam outflow be designated as x. Equating heat absorbed by the waste-heat-steam generator to the heat available from reducing the flue-gas temperature from 1562°F to 285°F yields the following steam flow:

$$(1.10)(x)(h_f - h_1) + (x)(h_3 - h_f) = 11.8 \times 10^6 \text{ Btu/hr}$$

Therefore,

| | |
| --- | --- |
| steam flow, x | = 11,388 lb/hr |
| feedwater flow | = 1.10(x)= 1.10(11,388)= 12,527 lb/hr |
| boiler blowdown | = 12,527 − 11,388 = 1,139 lb/hr |

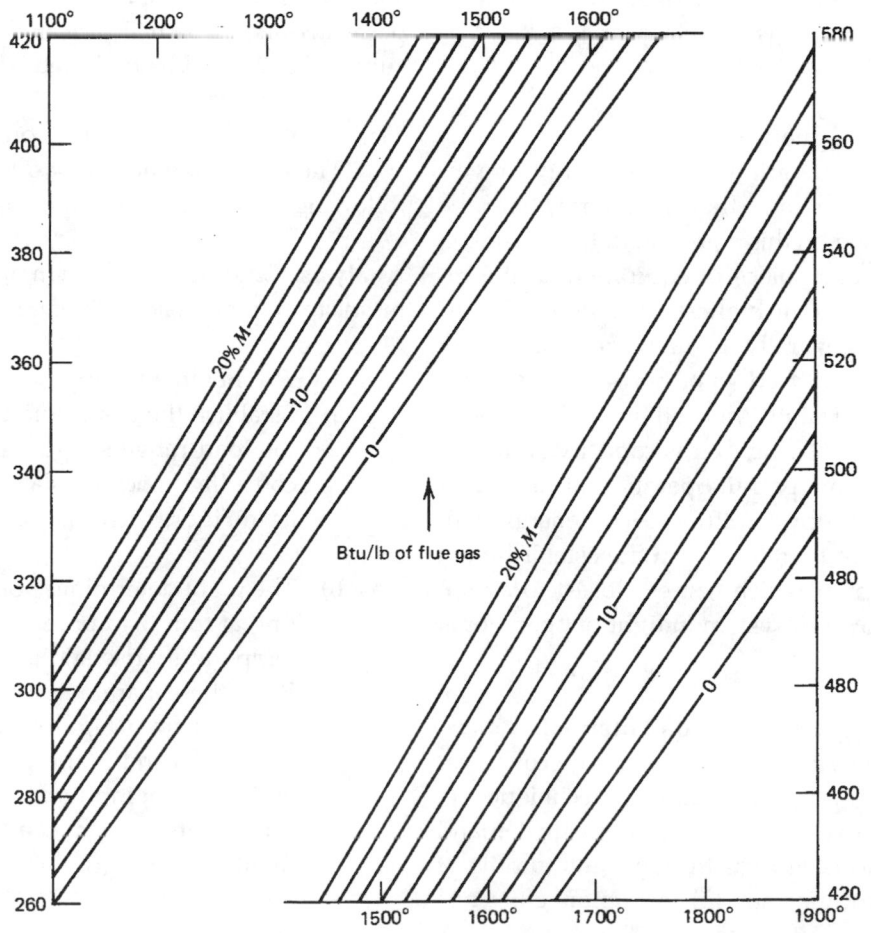

**Figure 5.10a Heat in flue gases vs. percent moisture by weight. (Derived from Keenan and Kayes 1948.)**

Determine the equivalent fuel input in conventional fuel-fired boilers corresponding to the waste heat-steam generator capability. This would be defined as follows:

Fuel input to conventional boilers
= (output)/(boiler efficiency)

Therefore,

Fuel input $= (11.8 \times 10^6 \text{ Btu/h})/(0.85)$
$= 13.88 \times 10^6 \text{ Btu/h}$

This suggests that with the installation of the waste-heat-steam generator utilizing the sensible heat of the reformer furnace flue gas, the equivalent of $13.88 \times 10^6$ Btu/hr of fossil-fuel input energy could be saved in the firing of the conventional boilers while still satisfying the overall plant steam demand.

As with other capital projects, the waste-heat-steam generator must compete for capital, and to be viable, it must be profitable. Therefore, the decision to proceed becomes an economical one. For a project to be considered life-cycle cost effective it must have a net-present value greater than or equal to zero, or an internal rate of return greater than the company's hurdle rate. For a thorough coverage of economic analysis, see Chapter 4.

### 5.3.4 Load Balancing
**Energy Conservation Opportunities**

There is an inherent variation in the energy conversion efficiencies of boilers and their auxiliaries with the operating load imposed on this equipment. It is desirable, therefore, to operate each piece of equipment at the capacity that corresponds to its highest efficiency.

Process plants generally contain multiple boiler units served by common feedwater and condensate return facilities. The constraints imposed by load variations and the requirement of having excess capacity on line to provide reliability seldom permit operation of each piece of equipment at optimum conditions. The energy conservation opportunities therefore lie in the establishment of an operating regimen which comes closest to attaining this goal for the overall system in light of operational constraints.

### How to Test for Energy Conservation Potential

Information needed to determine energy conservation opportunities through load-balancing techniques requires a plant survey to determine (1) total steam demand and duration at various process throughputs (profile of steam load versus runtime), and (2) equipment efficiency characteristics (profile of efficiency versus load).

### Steam Demand

Chart recorders are the best source for this information. Individual boiler steam flowmeters can be totalized for plant output. Demands causing peaks and valleys should be identified and their frequency estimated.

### Equipment Efficiency Characteristics

The efficiency of each boiler should be documented at a minimum of four load points between half and maximum load. A fairly accurate method of obtaining unit efficiencies is by measuring stack temperature rise and percent $O_2$ (or excess air) in the flue gas or by the input/output method defined in the ASME power test codes. Unit efficiencies can be determined with the aid of Figure 5.3, 5.4, or 5.5 for the particular fuel fired. For pump(s) and fan(s) efficiencies, the reader should consult manufacturers' performance curves.

An example of the technique for optimizing boiler loading follows.

**Example**: A plant has a total installed steam-generating capacity of 500,000 lb/hr, and is served by three boilers having a maximum continuous rating of 200,000, 200,000, and 100,000 lb/hr, respectively. Each unit can deliver superheated steam at 620 psig and 700°F with feedwater supplied at 250°F. The fuel fired is natural gas priced into the operation at $3.50/$10^6$ Btu. Total plant steam averages 345,000 lb/hr and is relatively constant.

The boilers are normally operated according to the following loading (top of following page).

**Analysis**. Determine the savings obtainable with optimum steam plant load-balancing conditions.

**STEP 1**. Begin with approach (a) or (b).
a) Establish the characteristics of the boiler(s) over the load range suggested through the use of a consultant and translate the results graphically as in Figures 5.11 and 5.12.

b) The plant determines boiler efficiencies for each unit at four load points by measuring unit stack temperature rise and percent $O_2$ in the flue gas. With these parameters known, efficiencies are obtained from Figures 5.3, 5.4, or 5.5. Tabulate the results and graphically plot unit efficiencies and unit heat inputs as a function of steam load. The results of such an analysis are shown in the tabulation and graphically illustrated in Figures 5.11 and 5.12.

(Unit input) = (unit output)/(efficiency)

| Boiler No. | Size Boiler (103 lb/hr) | Normal Boiler Load (103 lb/hr) | —Measured— | | |
| | | | Stack Temp. (°F) | O2 (%) | Unit Eff. (%) |
|---|---|---|---|---|---|
| 1 | 200 | 140 | 290 | 5 | 85.0 |
| 2 | 200 | 140 | 540 | 6 | 77.4 |
| 3 | 100 | 65 | 540 | 7 | 76.5 |
| Plant steam demand | | 345 | | | |

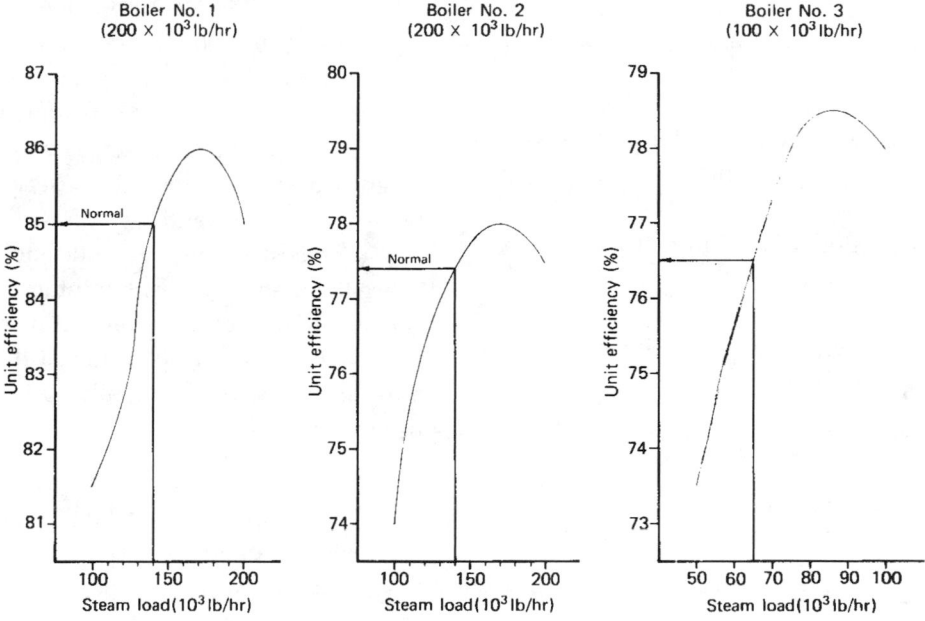

Figure 5.11 Unit efficiency vs. steam load.

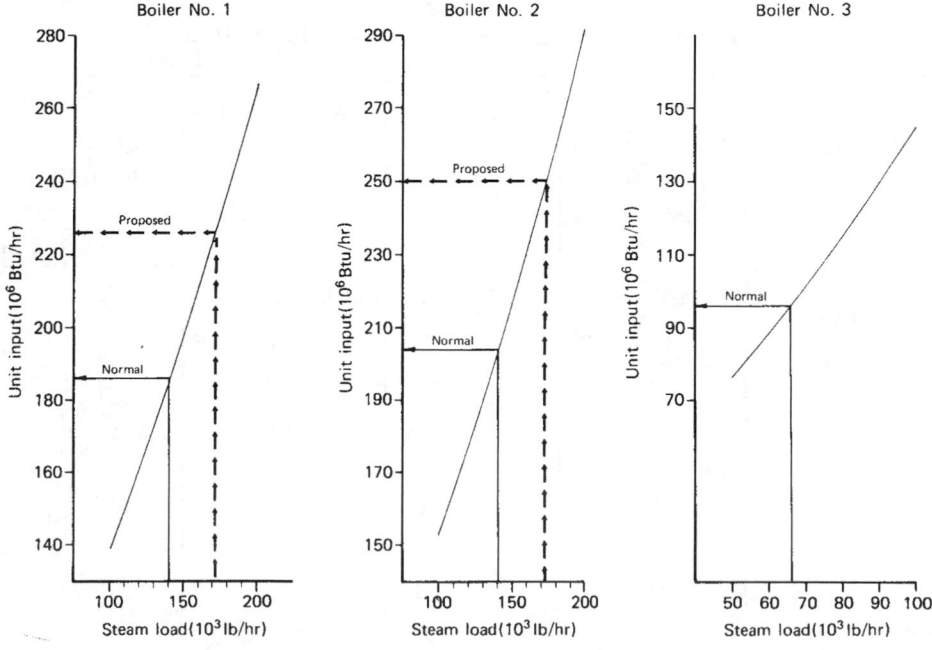

Figure 5.12 Unit input vs. steam load.

**STEP 2**. Sum up the total unit(s) heat input at the present normal operating steam plant load conditions. From Figure 5.12:

| Boiler No. | Steam Load ($10^3$ lb/hr) | Heat Input ($10^6$ Btu/hr) |
|---|---|---|
| 1 | 140 | 186 |
| 2 | 140 | 204 |
| 3 | 65 | 96 |
| Plant totals | 345 | 486 |

**STEP 3**. Optimum steam plant load-balancing conditions are satisfied when the total plant steam demand is met according to Table 5.8.

(Boiler No. 1 input) + (Boiler No. 2 input) + (Boiler No. 3 input) +... = minimum

By trial and error and with the use of Figure 5.12, optimum plant heat input is:

| Boiler No. | Steam Load ($10^3$ lb/hr) | Heat Input ($10^6$ Btu/hr) |
|---|---|---|
| 1 | 173 | 226 |
| 2 | 172 | 250 |
| 3 | (Banked standby) | 476 |
| Plant totals | 345 | |

**STEP 4**. The annual fuel savings realized from optimum load balancing is the difference between the existing boiler input and the optimum boiler input.

Steam plant energy savings
= (existing input) – (optimum input)
= 486 - 476 × $10^6$ Btu/hr
= 10 × $10^6$ Btu/hr
or annually:
= (10 × $10^6$ Btu/hr) × (8500 hr/yr)
× ($3.50/$10^6$ Btu)
= \$297,500/yr

Costs that were not considered in the preceding example are the additional energy savings due to more efficient fan operation and the cost of maintaining the third boiler in banked standby.

The cost savings were possible in this example because the plant had been maintaining a high ratio of total capacity in service to actual steam demand. This results in low-load inefficient operation of the boilers. Other operating modes which generally result in inefficient energy usage are:

1.  Base-loading boilers at full capacity. This can result in operation of the base-loaded boilers and the swing boilers at less than optimum efficiency unnecessarily.

2.  Operation of high-pressure boilers to supply low-pressure steam demands directly via letdown steam.

**Table 5.8 Unit Efficiency and Input Tabulation**

| Boiler No. | Steam Load ($10^3$ lb/hr) | Stack Temperature (°F) | Measured Oxygen (%) | Combustion Efficiency (%) | Output ($10^6$ Btu/hr) | Fuel Input ($10^6$ Btu/hr) |
|---|---|---|---|---|---|---|
| 1 | 200 | 305 | 2 | 85.0 | 226.2 | 266.1 |
|   | 170 | 280 | 2 | 86.0 | 192.3 | 223.6 |
|   | 130 | 300 | 7 | 84.0 | 147.0 | 175.0 |
|   | 100 | 280 | 12 | 81.5 | 113.1 | 138.8 |
| 2 | 200 | 625 | 2 | 77.5 | 226.2 | 291.0 |
|   | 170 | 570 | 4 | 78.0 | 192.3 | 246.5 |
|   | 130 | 520 | 7 | 77.0 | 147.0 | 190.9 |
|   | 100 | 490 | 11 | 74.0 | 113.1 | 152.8 |
| 3 | 100 | 600 | 2 | 78.0 | 113.1 | 145.0 |
|   | 85 | 570 | 2 | 78.5 | 96.1 | 122.5 |
|   | 65 | 540 | 7 | 76.5 | 73.5 | 96.1 |
|   | 50 | 500 | 11 | 73.5 | 56.6 | 76.9 |

3. Operation of an excessive number of auxiliary pumps. This results in throttled, inefficient operation.

## Requirements for Maximum Economy

**Establish a Boiler Loading Schedule.** An optimized loading schedule will allow any plant steam demand to be met with the minimum energy input. Some general points to consider when establishing such a schedule are as follows:

1. Boilers generally operate most efficiently at 65 to 85% full-load rating; centrifugal fans at 80 to 90% design rating. Equipment efficiencies fall off at higher or lower load points, with the decrease most pronounced at low-load conditions.

2. It is usually more efficient to operate a lesser number of boilers at higher loads than a larger number at low loads.

3. Boilers should be put into service in order of decreasing efficiency starting with the most efficient unit.

4. Newer units and units with higher capacity are generally more efficient than are older, smaller units.

5. Generally, steam plant load swings should be taken in the smallest and least efficient unit.

**Optimize the Use of High-Pressure Boilers.** The boilers in a plant that operate at the highest pressure are usually the most efficient. It is, therefore, desirable to supply as much of the plant demand as possible with these units provided that the high-grade energy in the steam can be effectively used. This is most efficiently done by installation of back-pressure turbines providing useful work output, while providing the exhaust steam for low-pressure consumers.

Degrading high-pressure steam through a pressure reducing and desuperheating station is the least efficient method of supplying low-pressure steam demands. Direct generation at the required pressure is usually more efficient by comparison.

**Establish an Auxiliary Loading Schedule.** A schedule for cutting plant auxiliaries common to all boilers in and out of service with rising or falling plant load should be established.

**Establish Procedures for Maintaining Boilers in Standby Mode.** It is generally more economical to run fewer boilers at a higher rating. On the other hand, the integrity of the steam supply must be maintained in the face of forced outage of one of the operating boilers. Both conditions can sometimes be satisfied by maintaining a standby boiler in a "live bank" mode. In this mode the boiler is isolated from the steam system at no load but kept at system operating pressure. The boiler is kept at a pressure by intermittent firing of either the ignitors or a main burner to replace ambient heat losses. Guidelines for live banking of boilers are as follows:

1. Shut all dampers and registers to minimize heat losses from the unit.

2. Establish and follow strict safety procedures for ignitor/burner light-off.

3. For units supplying turbines, take measures to ensure that any condensate which has been formed during banking is not carried through to the turbines. Units with pendant-type superheaters will generally form condensate in these elements.

Operators should familiarize themselves with emergency startup procedures and it should be ascertained that the system pressure decay which will be experienced while bringing the banked boiler(s) up to load can be tolerated.

## Guidelines for Day-to-Day Operation

1. Monitor all boiler efficiencies continuously and immediately correct items that detract from performance. Computerized load balancing may prove beneficial.

2. Ensure that load-balancing schedules are followed.

3. Reassess the boiler loading schedule whenever a major change in the system occurs, such as an increase or decrease in steam demand, derating of boilers, addition/decommissioning of boilers, or addition/removal of heat-recovery equipment.

4. Recheck parameters and validity of established operating mode.

5. Measure and record fuel usage and correlate to steam production and flue-gas analysis for determination of the unit heat input relationship.

6. Keep all monitoring instrumentation calibrated and functioning properly.

7. Optimize excess air operation and minimize boiler blowdown.

**Computerized Systems Available**

There are commercially available direct digital control systems and proprietary sensor devices which accomplish optimal steam/power plant operation, including tie-line purchased power control. These systems control individual boilers to minimum excess air, $SO_2$, $NO_x$, CO (and opacity if desired), and control boiler and cogeneration complexes to reduce and optimize fuel input.

Boiler plant optimization is realized by boiler controls which ensure that the plant's steam demands are met in the most cost-effective manner, continuously recognizing boiler efficiencies that differ with time, load, and fuel quality. Similarly, computer control of cogeneration equipment can be cost effective in satisfying plant electrical and process steam demands.

As with power boiler systems, the efficiencies for electrical generation and extraction steam generation can be determined continuously and, as demand changes occur, loading for optimum overall efficiency is determined.

Fully integrated computer systems can also provide electric tie-line control, whereby the utility tie-line load is continuously monitored and controlled within the electrical contract's limits. For example, loads above the peak demand can automatically be avoided by increasing in-plant power generation, or in the event that the turbines are at full capacity, shedding loads based on previously established priorities.

### 5.3.5 Boiler Blowdown

In the generation of steam, most water impurities are not evaporated with the steam and thus concentrate in the boiler water. The concentration of the impurities is usually regulated by the adjustment of the continuous blowdown valve, which controls the amount of water (and concentrated impurities) purged from the steam drum.

When the amount of blowdown is not properly established and/or maintained, either of the following may happen:

1. If too little blowdown, sludge deposits and carryover will result.

2. If too much blowdown, excessive hot water is removed, resulting in increased boiler fuel requirements, boiler feedwater requirements, and boiler chemical requirements.

Significant energy savings may be realized by utilizing the guides presented in this section for (1) establishing optimum blowdown levels to maintain acceptable boiler-water quality and to minimize hot-water losses, and (2) the recovery of heat from the hot-water blowdown.

**Where to Look For Energy-Saving Opportunities**

The continuous blowdown from any steam-generating equipment has the potential for energy savings whether it is a fired boiler or waste-heat-steam generator. The following items should be carefully considered to maximize savings:

1. Reduce blowdown (BD) by adjustment of the blowdown valve such that the controlling water impurity is held at the maximum allowable level.

2. Maintain blowdown continuously at the minimum acceptable level. This may be achieved by frequent manual adjustments or by the installation of automatic blowdown controls. At current fuel costs, automatic blowdown controls often prove to be economical.

3. Minimize the amount of blowdown required by:
   a. Recovering more clean condensate, which reduces the concentration of impurities coming into the boiler.
   b. Establishing a higher allowable drum solids level than is currently recommended by ABMA standards (see below). This must be done only on recommendation from a reputable water treatment consultant and must be followed up with lab tests for steam purity.
   c. Selecting the raw-water treatment system which has the largest effect on reducing makeup water impurities. This is generally considered applicable only to grass-roots or revamp projects.

4. Recover heat from the hot blowdown water. This is typically accomplished by flashing the water to a low pressure. This produces low-pressure steam (for utilization in an existing steam header) and hot water which may be used to preheat boiler makeup water.

### Tests and Evaluations

**STEP 1**: *Determine Actual Blowdown*. Obtain the following data:

$T$ = ppm of impurities in the makeup water to the deaerator from the treatment plant; obtain average value through lab tests

$B$ = ppm of concentrated impurities in the boiler drum water (blowdown water); obtain average value through lab tests lb/hr MU = lb/hr of makeup water to the deaerator from the water treatment plant; obtain from flow indicator

lb/hr BFW = lb/hr of boiler feedwater to each

lb/hr STM = lb/hr of steam output from each boiler; obtain from flow indicator

lb/hr CR = lb/hr of condensate return

*Note*: percentages for BFW, MU, and CR are determined as a percentage of STM.

Calculate the following:

$$\%MU = lb/hr\ MU \times 100\% / (total\ lb/hr\ BFW)$$
$$= lb/hr\ MU \times 100\% / [(boiler\ no.\ 1\ lb/hr\ BFW) + (boiler\ no.\ 2\ lb/hr\ BFW) + ...]$$

$$\%MU = 100\% - \%CR \qquad (5.3)$$

$$A = ppm\ of\ impurity\ in\ BFW = T \times \%MU \qquad (5.4)$$

Now actual blowdown (BD) may be calculated as a function (percentage) of steam output:

$$\%BD = (A \times 100\%) / (B - A) \qquad (5.5)$$

Converting to lb/hr BD yields

$$lb/hr\ BD = \%\ BD \times lb/hr\ STM \qquad (5.6)$$

*Note:* In using all curves presented in this section. Blowdown must be based on steam output from the boiler as calculated above. Boiler blowdown based on boiler feedwater rate (percent BD BFW) to the boiler should not be used. If blowdown is reported as a percent of the boiler feedwater rate, it may be converted to a percent of steam output using

$$\%BD = \%BD_{BFW} \times (1) / (1 - \%BD_{BFW}) \qquad (5.7)$$

**STEP 2**: *Determine Required Blowdown.* The amount of blowdown required for satisfactory boiler operation is normally based on allowable limits for water impurities as established by the American Boiler Manufacturers Association (ABMA).

These limits are presented in Table 5.9. Modifications to these limits are possible as discussed below. The required blowdown may be calculated using the equations presented above by substituting the ABMA limit for B (concentration of impurity in boiler).

$$\%\ BD_{required} = (A) / (B_{required} - A) \times 100\% \qquad (5.8)$$

$$lb/hr\ BD_{required} = \%\ BD_{required} \times lb/hr\ STM \qquad (5.9)$$

**STEP 3**: *Evaluate the Cost of Excess Blowdown.* The amount of actual boiler blowdown (as calculated in equation 5.4) that is in excess of the amount of required blowdown (as calculated in equation 5.6) is considered as wasting energy since this water has already been heated to the saturation temperature corresponding to the boiler drum pressure. The curves presented in Figure 5.13 provide an easy method of evaluating the cost of excess blowdown as a function of various fuel costs and boiler efficiencies.

As an illustration of the cost of boiler blowdown, consider the following example.

### Table 5.9 Recommended Limits for Boiler-Water Concentrations

| Drum Pressure (psig) | Total Solids | | Alkalinity | | Suspended Solids | | Silica |
|---|---|---|---|---|---|---|---|
| | ABMA | Possible | ABMA | Possible | ABMA | Possible | ABMA |
| 0 to 300 | 3500 | 6000 | 700 | 1000 | 300 | 250 | 125 |
| 301 to 450 | 3000 | 5000 | 600 | 900 | 250 | 200 | 90 |
| 451 to 600 | 2500 | 4000 | 500 | 500 | 150 | 100 | 50 |
| 601 to 750 | 2000 | 2500 | 400 | 400 | 100 | 50 | 35 |
| 751 to 900 | 1500 | — | 300 | 300 | 60 | — | 20 |
| 901 to 1000 | 1250 | — | 250 | 250 | 40 | — | 8 |
| 1001 to 1500 | 1000 | — | 200 | 200 | 20 | — | 2 |

**Example**: Determine the potential energy savings associated with reducing boiler blowdown from 12% to 10% using Figure 5.13.

**Operating Data**

| | |
|---|---|
| Average boiler load | 75,000 lb/hr |
| Steam pressure | 150 psig |
| Make up water temperature | 60°F |
| Operating hours | 8,200 hr/yr |
| Boiler efficiency | 80% |
| Average fuel cost | $2.00/10$^6$ Btu |

**Calculation and Analysis**

Using the curves in Figure 5.13, enter Chart A at 10% blowdown to the curve for 150 psig boiler drum pressure. Follow the line over to chart B and the curve for a unit efficiency of 80%. Then follow the line down to Chart C and the curve for a fuel cost of $2.00/10$^6$ Btu. Read the scale for the equivalent fuel value in blowdown. The cost of the blowdown is estimated at $8.00/hr per 100,000 lb/hr of steam generated. Repeat the procedure for the blowdown rate of 12% and find the cost of the blowdown is

$10.00/hr per 100,000 lb/hr of steam generated. Potential energy savings then is estimated to be

= ($10.00 - 8.00/hr/100,000 lb/hr)
      × (75,000 lb/hr) × (8,200 hr/yr)
= $12,300/yr

**Energy Conservation Methods**

1.  **Minimize Blowdown by Manual Adjustment**. This is accomplished by establishing an operating procedure requiring frequent water quality testing and readjustment of blowdown valves so that water impurities in the boiler are held at the allowable limit. Continuous indicating/recording analyzers may be employed allowing the operator to establish quickly the actual level of water impurity and manually readjust blowdown valves.

2.  **Minimize Blowdown by Automatic Adjustment**. The adjustment of blowdown may be automated by the installation of automatic analyzing equipment and the replacement of manual blowdown valves with control valves (see Figure 5.14). The cost of this equipment is frequently justifiable, particularly

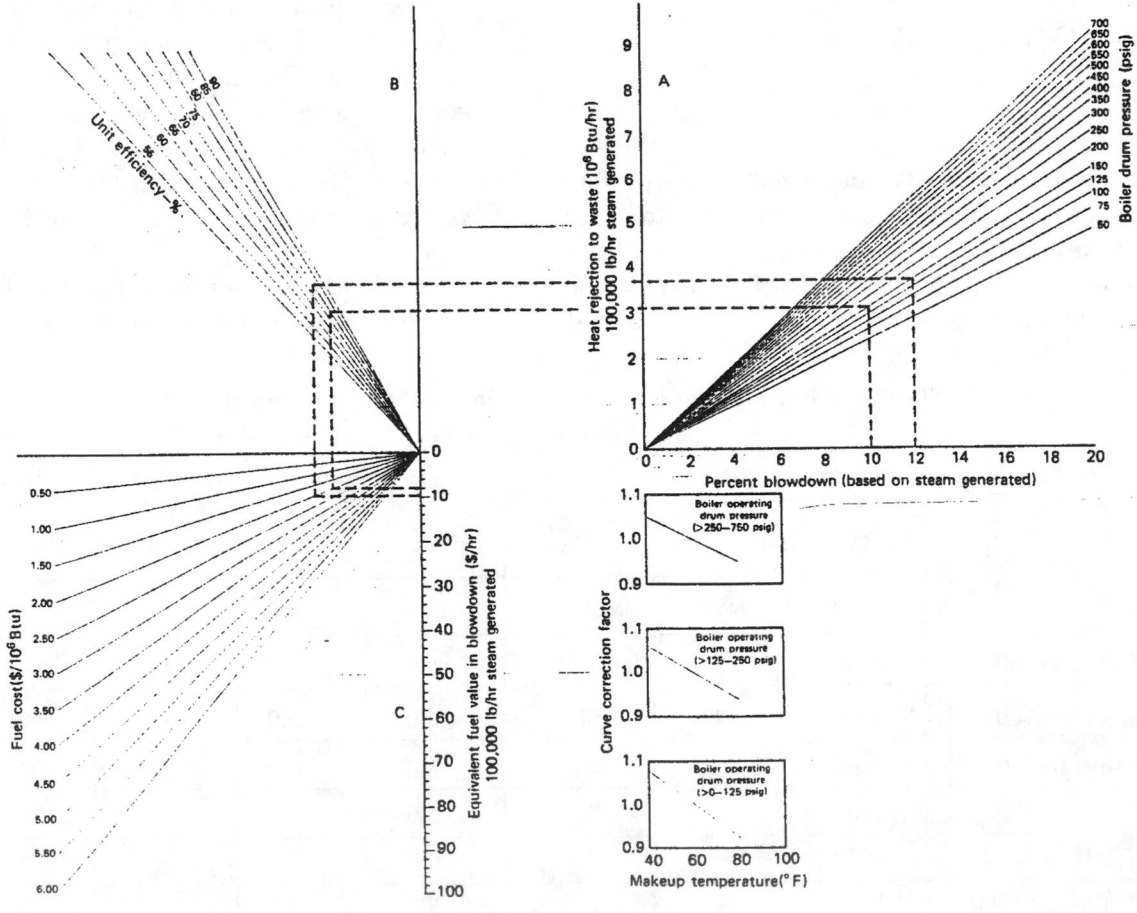

**Figure 5.13 Hourly cost of blowdown.**

when there are frequent load changes on the steam-generating equipment since the automation allows continuous maintenance of the highest allowable level of water contaminants. Literature has approximated that the average boiler plant can save about 20% blowdown by changing from manual control to automatic adjustment.

3. **Decrease Blowdown by Recovering More Condensate**. Since clean condensate may be assumed to be essentially free of water impurities, addition of condensate to the makeup water serves to dilute the concentration of impurities. The change in required blowdown may be calculated using equations 5.3 and 5.5.

**Example**: Determine the effect on boiler blowdown of increasing the rate of condensate return from 50 to 75%:

**Operating Data**
MU water impurities = 10 ppm = T
Maximum allowable limit in drum
= 100 ppm = B

**Calculate**:
For 50% return condensate:
MU = 100% − 50% = 50%
A = 10 × 0.5 = 5 ppm in BFW

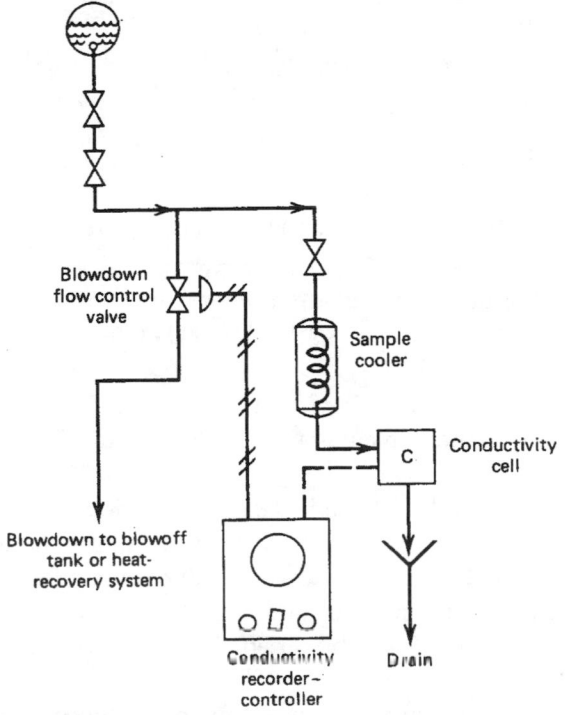

**Figure 5.14 Automated continuous blowdown system.**

% BD = A / (B-A) = 5 / (100-5) = 5.3%

For 75% return condensate:
MU = 100 − 75% = 25%
A = 10 × 0.25 = 2.5 ppm
% BD = 2.5 / (100-2.5) = 2.6%

**Conclusion**
These values may then be used with the curves on Figure 5.13 to approximate the potential energy savings.

4. **Increase Allowable Drum Solids Level.** In some instances it may be possible to increase the maximum allowable impurity limit without adversely affecting the operation of the steam system. However, it must be emphasized that a water treatment consultant should be contacted for recommendation on changes in the limits as given in Table 5.9. The changes must also be followed by lab tests for steam purity to verify that the system is operating as anticipated.

The energy savings may be evaluated by using the foregoing equations for blowdown and the graphs in Figures 5.13 and 5.15. Consider the following example.

**Example**: Determine the blowdown rates as a percentage of steam flow required to maintain boiler drum water impurity concentrations at an average of 3000 ppm and 6,000 ppm.

**Operating Data**
Average makeup water impurity
(measurement)..........................................350 ppm
Condensate return (percent of steam flow)........25%
Assume condensate return free from impurities

**Calculation and Analysis**
Calculate the impurity concentration in the boiler feedwater (BFW):
A = MU impurity × (1.00 - % CR)
A = 350 ppm × (1.00 - 0.25)
A = 262 ppm

*Mathematical solution.*
For drum water impurity level of 3000 ppm:
% BD = A / (B - A)
% BD = 262 / (3000 - 262)
% BD = 9.6%

For drum water impurity level of 6000 ppm:

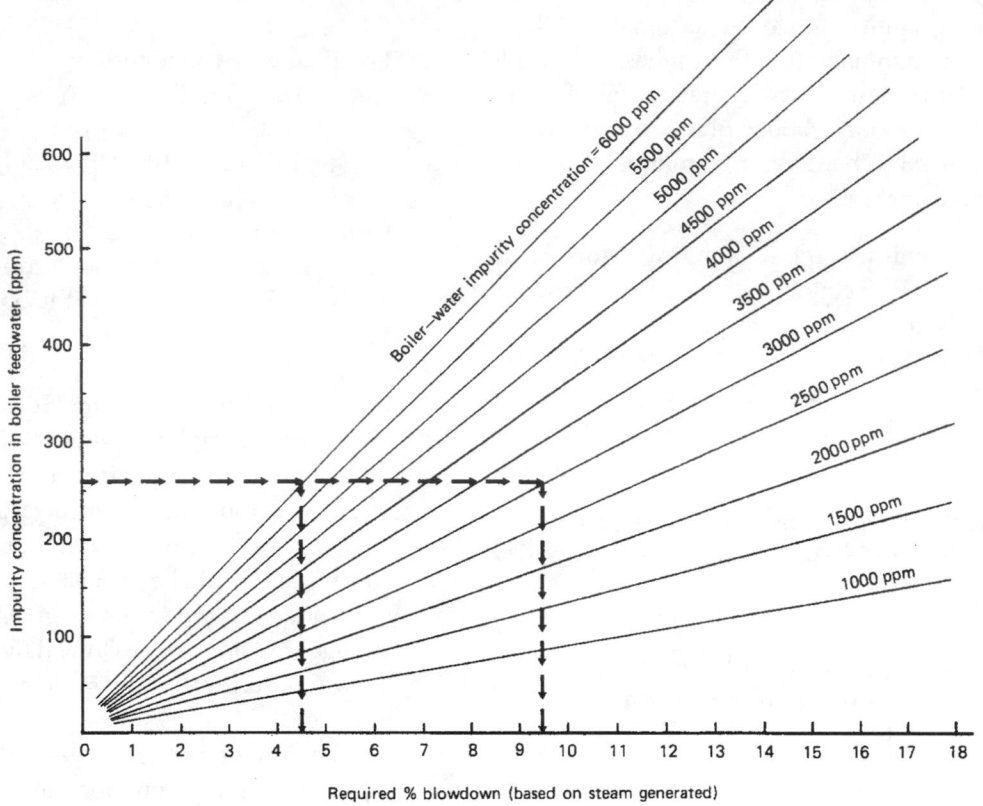

**Figure 5.15 Required percent blowdown. Based on equation 5.5.**

% BD = A/(B - A)
% BD = 262/(6000 - 262)
% BD = 4.6%

*Graphical Solution.* Referring to Figure 5.15
Enter the graph at feedwater impurity level of 262 ppm and follow the line to the curves for 3000 ppm and 6000 ppm boiler drum water impurity level. Then read down to the associated boiler blowdown percentage.

**Conclusion**
The blowdown percentages may not be used in conjunction with Figure 5.13 to determine the annual cost of blowdown and the potential energy cost savings associated with reducing boiler blowdown.

5.  **Select Raw-Water Treatment System for Largest Reduction in Raw-Water Impurities**. Since a large investment would be associated with the installation of new equipment, this energy conservation method is usually applicable to new plants or revamps only. A water treatment consultant should be retained to recommend the type of treatment applicable. An example of how water treatment affects blowdown follows.

**Example**: Determine the effects on blowdown of using a sodium zeolite softener producing a water quality of 350 ppm solids and of using a demineralization unit producing a water quality of 5 ppm solids. The makeup water rate is 30% and the allowable drum solids level is 3000 ppm.

**Solution**:
For sodium zeolite:
% BD = (350 × 0.3 × 100%)/[3000- (350 × 0.3)] = 3.6%

For demineralization unit:
% BD  = (5 × 0.3 × 100%)/[3000 - (5 × 0.3)]
      = 0.6%

Therefore, the percentage blowdown would be reduced by 3.0%.

A secondary benefit derived from increasing feedwater quality is the reduced probability of scale formation in the boiler. Internal scale reduces the effectiveness of heat-transfer surfaces and can result in a reduction of as much as 1 to 2% in boiler efficiency in severe cases.

6.  **Heat Recovery from Blowdown**. Since a certain amount of continuous blowdown must be main-

tained for satisfactory boiler performance, a significant quantity of heat is removed from the boiler. A large amount of the heat in the blowdown is recoverable by using a two-stage heat-recovery system as shown in Figure 5.16 before discharging to the sewer. In this system, blowdown lines from each boiler discharge into a common flash tank. The flashed steam may be tied into an existing header, used directly by process, or used in the deaerator. The remaining hot water may be used to preheat makeup water to the deaerator or preheat other process streams.

The following procedure may be used to calculate the total amount of heat that is recoverable using this system and the associated cost savings.

**STEP 1**. Determine the annual cost of blowdown using the percent blowdown, steam flow rate (lb/hr), unit efficiency, and fuel cost. This can be accomplished in conjunction with Figure 5.13.

**STEP 2**. Determine:

Flash % =    percent of blowdown that is flashed to steam (using Figure 5.17, curve B, at the flash tank pressure or using equation 5.10a or 5.10b)

COND % =    100% - Flash %

$h_{tk}$ =    enthalpy of liquid leaving the flash tank (using Figure 5.17, curve A, at the flash tank pressure)

$h_{ex}$ =    enthalpy of liquid leaving the heat exchanger [using Figure 5.17, curve C; for planning purposes, a 30 to 40°F approach temperature (condensate discharge to makeup water temperature) may be used]

**STEP 3**. Calculate the amount of heat recoverable from the condensate (% QC) using

$$\%QC = [(h_{tk} - h_{ex})/h_{tk}] \times COND \%$$

**STEP 4**. Since all of the heat in the flashed steam is recoverable, the total percent of heat recoverable (% Q) from the flash tank and heat-exchanger system is

$$\% Q = \% QC + Flash \%$$

**Figure 5.16 Typical two-stage blowdown heat-recovery system.**

**STEP 5**. The annual savings from heat recovery may then be determined by using this percent (% Q) with the annual cost of blowdown found in step 1:

annual savings = (% QC/100) × BD cost

To further illustrate this technique consider the following example.

**Example**: Determine the percent of heat recoverable (%Q) from a 150 psig boiler blowdown waste stream, if the stream is sent to a 20 psig flash tank and heat exchanger.

**Available Data**

| | |
|---|---|
| Boiler drum pressure | 150 psig |
| Flash tank pressure | 20 psig |
| Makeup water temperature | 70°F |

Assume a 30°F approach temperature between condensate discharge and makeup water temperature

**Calculation and Analysis**

Referring to Figure 5.17.

Determine Flash % using Chart B:

Entering chart B with a boiler drum pressure of 150 psig and following a live to the curve for a flash tank pressure of 20 psig, read the Steam percentage (Flash %) to be 12.5%.

Determine COND %:
  COND % = 100 - Flash %
  COND % = 100 - 12.5 %
  COND % = 87.5%
Determine $h_{tk}$ using Chart A:
  Entering chart A with a flash tank pressure of 20
  psig and following a line to the curve for saturated liquid, read the enthalpy of the drum water ($h_{tk}$) to be 226 Btu/lb.
Determine $h_{ex}$ using Chart C:
  Assuming a 30°F approach temperature between condensate discharge and makeup water
  temperature, the temperature of the blowdown
  discharge is equal to the makeup water temperature plus the approach temperature which
  equals 100°F (70°F + 30°F).

  Entering chart C with a blowdown heat exchanger rejection temperature of 100°F and following a line to the curve, read the enthalpy of
  the blowdown discharge water to be 68 Btu/lb.
Determine the % QC:
  % QC = [($h_{tk}$ - $h_{ex}$)/$h_{tk}$] × COND %

  % QC = [(226 Btu/lb - 68 Btu/lb)/226 Btu/lb]
          × 87.5%
  % QC = 61.2%
Determine the % Q:
  % Q = % QC + Flash %
  % Q = 61.2% + 12.5%
  % Q = 73.7%

**Conclusion**
Therefore, approximately 73.7% of the heat energy
  can be recovered using this blowdown heat recovery technique.

**More on Flash Steam**
  To determine the amount of flash steam that is generated by high-pressure, high-temperature condensate
being reduced to a lower pressure you can use the following equation:

Flash % = ($h_{HPl}$ - $h_{LPl}$) × 100% / ($h_{LPv}$ − $h_{LPl}$)     (5.10a)
or
Flash % = ($h_{HPl}$ - $h_{LPl}$) × 100% / ($h_{LPevp}$)     (5.10b)

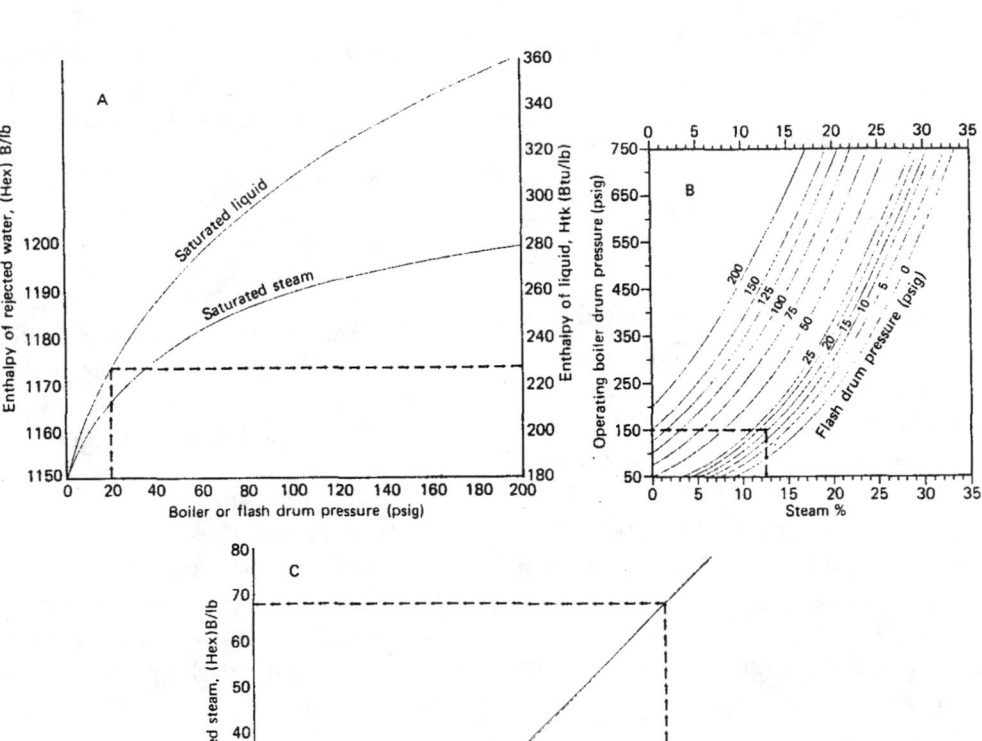

**Figure 5.17 Percent of heat recoverable from blowdown.**

where:   Flash % = amount of flash steam as a percent of
        total mass
  $h_{HPl}$ = enthalpy of the high pressure liquid
  $h_{LPl}$ = enthalpy of the low pressure liquid
  $h_{LPv}$ = enthalpy of the low pressure vapor
$h_{LPevap}$ = evaporation enthalpy of the low pressure
        liquid = $(h_{LPv} - h_{LPl})$

## Guidelines for Day-to-Day Operation

1.   Maintain concentration of impurities in the boiler
     drum at the highest allowable level. Frequent checks
     should be made on water quality and blowdown
     valves adjusted accordingly.

2.   Continuous records of impurity concentration in
     makeup water and boiler drum water will indicate
     trends in deteriorating water quality so that early
     corrective actions may be taken.

3.   Control instruments should be calibrated on a week-
     ly basis.

## 5.3.6 Condensate Return

In today's environment of ever-increasing fuel costs,
the return and utilization of the heat available in clean
steam condensate streams can be a practical and economical energy conservation opportunity. Refer to Chapter 6
for a comprehensive discussion of condensate return. The
information below is presented to summarize briefly and
emphasize the benefits and major considerations pertinent to optimum steam generator operations. Recognized
benefits of return condensate include:

Reduction in steam power plant raw-water makeup and
     associated treatment costs.
Reduction in boiler blowdown requirements resulting in
     direct fuel savings. Refer to section on boiler blow-
     down.
Reduced steam required for boiler feedwater deaeration.
Raw-water and boiler-water chemical cost reduction.
Opportunities for increased useful work output without
     additional energy input.
Reduces objectionable environmental discharges from
     contaminated streams.

## Where to Look

Examine and survey all steam-consuming units
within a plant to determine the present disposition of any
condensate produced or where process modifications can
be made to produce "clean" condensate. Address the following:

1.   Is the condensate clean and being sewered?

2.   Is the stream essentially clean but on occasion be-
     comes contaminated?

3.   If contaminated, can return to the steam system be
     justified by polishing the condensate?

4.   Can raw makeup or treated water be substituted for
     condensate presently consumed?

5.   Is condensate dumped for operating convenience or
     lack of chemical purity?

Results from chemical purity tests, establishing battery limit conditions and analysis of these factors, provide
the basis of obtaining maximum economy.

## Modifications Required for Maximum Economy

Often, the only requirement to gain the benefits of
return condensate is to install the necessary piping and/
or pumping facilities. Other solutions are more complex
and accordingly, require a more in-depth analysis. Chemically "clean" or "contaminated" condensate can be effectively utilized by:

Providing single- or multistage flashing for contaminated streams, and recouping the energy of the flashed
steam. Recovering additional heat from the flash drum
condensate by indirect heat exchange is also a possibility.

Collecting condensate from an atmospheric flash
drum with "automatic" provision to dump on indication
of stream contamination. This concept, when conditions
warrant, allows "normally clean" condensate to be used
within the system.

Installing ion-exchange polishing units for condensate streams which may be contaminated but are significant in quantity and heat value.

Providing a centrally located collection tank and
pump to return the condensate to the steam system. This
avoids a massive and complex network of individual return lines.

Using raw water in lieu of condensate and returning
the condensate to the system. An example is the use of
condensate to regenerate water treatment units.

Changing barometric condensers or other direct-
contact heat exchangers to surface type or indirect exchangers, respectively, and returning the clean condensate to the system.

Collecting condensate from sources normally overlooked, such as space heating, steam tracing, and steam
traps.

Providing flexibility to isolate and sewer individual
return streams to maintain system integrity. Providing

"knockout" or disengaging drum(s) to ensure clean condensate return to the system.

Recovering the heat content from contaminated condensate by indirect heat exchangers. An example is using an exchanger to heat the boiler makeup water.

Returning the contaminated condensate stream to a clarifier or hot lime unit to cleanup for boiler makeup rather than sewering the stream.

Allowing provision for manual water testing of the condensate stream suspected of becoming contaminated.

Using the contaminated condensate for noncritical applications, such as space heating, tank heating, and so on.

### Guidelines for Day-to-Day Operation

1. Maintain the system, including leak detection and insulation repair.

2. Periodically test the return water at its source of entry within the steam system for (a) contamination, (b) corrosion, and (c) acceptable purity.

3. Maintain and calibrate monitoring and analyzing equipment.

4. Ensure that the proper operating regimen is followed; that is, the condensate is returned and not sewered.

## 5.4    FUEL CONSIDERATIONS

The selection and application of fuels to various combustors are becoming increasingly complex. Most existing units have limited flexibility in their ability to fire alternative fuels and new units must be carefully planned to assure the lowest first costs without jeopardizing the future capability to switch to a different fuel. This section presents an overview of the important considerations in boiler and fuel selection. Also refer to Section 5.5.

### 5.4.1 Natural Gas
Natural-gas firing in combustors has traditionally been the most attractive fuel type, because:

1. Gas costs were low until about 2000 when they started to rise dramatically.

2. Only limited fuel-handling equipment typically consisting of pipelines, metering, a liquid knockout drum, and appropriate controls is required.

3. Boiler costs are minimized due to smaller boiler sizes; which result from highly radiant flame charac-

teristics and higher velocities, resulting in enhanced heat transfer and less heating surface.

4. Freedom from capital and operating costs associated with pollution control equipment.

Natural gas, being the cleanest readily available conventional form of fuel, also makes gas-fired units the easiest to operate and maintain.

However, as discussed elsewhere, the continued use of natural gas as fuel to most combustors will probably be limited in the future by government regulations, rising fuel costs, and inadequate supplies. One further disadvantage, which often seems to be overlooked, is the lower boiler efficiency that results from firing gas, particularly when compared to oil or coal.

### 5.4.2 Fuel Oil
**Classifications**
Influential in the storage, handling, and combustion efficiency of a liquid fuel are its physical and chemical characteristics.

Fuel oils are graded as No. 1, No. 2, No. 4, No. 5 (light), No. 5 (heavy), and No. 6. Distillates are Nos. 1 and 2 and residual oils are Nos. 4, 5, and 6. Oils are classified according to physical characteristics by the American Society for Testing and Materials (ASTM) according to Standard D-396.

No. 1 oil is used as domestic heating oil and as a light grade of diesel fuel. Kerosene is generally in a lighter class; however, often both are classified the same. No. 2 oil is suitable for industrial use and home heating. The primary advantage of using a distillate oil rather than a residual oil is that it is easier to handle, requiring no heating to transport and no temperature control to lower the viscosity for proper atomization and combustion. However, there are substantial purchase cost penalties between residual and distillate.

It is worth noting that distillates can be divided into two classes: straight-run and cracked. A straight-run distillate is produced from crude oil by heating it and then condensing the vapors. Refining by cracking involves higher temperatures and pressures or catalysts to produce the required oil from heavier crudes. The difference between these two methods is that cracked oils contain substantially more aromatic and olivinic hydrocarbons which are more difficult to burn than the paraffinic and naphthenic hydrocarbons from the straight-run process. Sometimes a cracked distillate, called industrial No. 2, is used in fuel-burning installations of medium size (small package boiler or ceramic kilns for example) with suitable equipment.

Because of the viscosity range permitted by ASTM, No. 4 and No. 5 oil can be produced in a variety of ways: blending of No. 2 and No. 6, mixture of refinery by-products, through utilization of off-specification products, and so on. Because of the potential variations in characteristics, it is important to monitor combustion performance routinely to obtain optimum results. Burner modifications may be required to switch from, say, a No. 4 that is a blend and a No. 4 that is a distillate.

Light (or cold) No. 5 fuel oil and heavy (or hot) are distinguished primarily by their viscosity ranges: 150 to 300 SUS (Saybolt Universal Seconds) at 100°F and 350 to 750 SUS at 100°F respectively. The classes normally delineate the need for preheating with heavy No. 5 requiring some heating for proper atomization.

No. 6 fuel oil is also referred to as residual, Bunker C, reduced bottoms, or vacuum bottoms. It is a very heavy oil or residue left after most of the light volatiles have been distilled from crude. Because of its high viscosity, 900 to 9000 SUS at 100°F, it can only be used in systems designed with heated storage and sufficient temperature/viscosity at the burner for atomization.

### Heating Value

Fuel oil heating content can be expressed as higher (or gross) heating value and low (or net) heating value. The higher heating value (HHV) includes the water content of the fuel, whereas the lower heating value (LHV) does not. For each gallon of oil burned, approximately 7 to 9 lb of water vapor is produced. This vapor, when condensed to 60°F, releases 1058 Btu. Thus the HHV is about 1000 Btu/lb or 8500 Btu/gal higher than the LHV. While the LHV is representative of the heat produced during combustion, it is seldom used in the United States except for exact combustion calculations.

### Viscosity

Viscosity is a measure of the relative flow characteristics of an oil an important factor in the design and operation of oil-handling and -burning equipment, the efficiency of pumps, temperature requirements, and pipe sizing. Distillates typically have low viscosities and can be handled and burned with relative ease. However, No. 5 and No. 6 oils may have a wide range of viscosities, making design and operation more difficult.

Viscosity indicates the time required in seconds for 60 $cm^3$ of oil to flow through a standard-size orifice at a specific temperature. Viscosity in the United States is normally determined with a Saybolt viscosimeter. The Saybolt viscosimeter has two variations (Universal and Furol) with the only difference being the size of orifice and sample temperature. The Universal has the smallest

opening and is used for lighter oils. When stating an oil's viscosity, the type of instrument and temperature must also be stated.

### Flash Point

Flash point is the temperature at which oil vapors flash when ignited by an external flame. As heating continues above this point, sufficient vapors are driven off to produce continuous combustion. Since flash point is an indication of volatility, it indicates the maximum temperature for safe handling. Distillate oils normally have flash points from 145 to 200°F, whereas the flash point for heavier oils may be up to 250°F. Thus under normal ambient conditions, fuel oils are relatively safe to handle (unless contaminated).

### Pour Point

Pour point is the lowest temperature at which an oil flows under standard conditions. It is 5°F above the oil's solidification temperature. The wax content of the oil significantly influences the pour point (the more wax, the higher the pour point). Knowledge of an oil's pour point will help determine the need for heated storage, storage temperature, and the need for pour-point depressant. Also, since the oil may cool while being transferred, burner preheat temperatures will be influenced and should be watched.

### Sulfur Content

The sulfur content of an oil is dependent upon the source of crude oil. Typically, 70 to 80% of the sulfur in a crude oil is concentrated in the fuel product, unless expensive desulfurization equipment is added to the refining process. Fuel oils normally have sulfur contents of from 0.3 to 3.0% with distillates at the lower end of the range unless processed from a very high sulfur crude. Often, desulfurized light distillates are blended with high-sulfur residual oil to reduce the residual's sulfur content. Sulfur content is an important consideration primarily in meeting environmental regulations.

### Ash

During combustion, impurities in oil produce a metallic oxide ash in the furnace. Over 25 different metals can be found in oil ash, the predominant being nickel, iron, calcium, sodium, aluminum, and vanadium. These impurities are concentrated from the source crude oil during refining and are difficult to remove since they are chemically bound in the oil. Ash contents vary widely: distillates have about 0 to 0.1% ash and heavier oil 0.2 to 1.5%. Although percentages are small, continuous boiler operations can result in considerable accumulations of ash in the firebox.

Problems associated with ash include reduction in heat-transfer rates through boiler tubes, fouling of super-heaters, accelerated corrosion of boiler tubes, and deterioration of refractories. Ashes containing sodium, vanadium, and/or nickel are especially troublesome.

## Other Contaminants

Other fuel-oil contaminants include water, sediment, and sludge. Water in fuel oil comes from condensation, leaks in storage equipment, and/or leaking heating coils. Small amounts of water should not cause problems. However, if large concentrations (such as at tank bottoms) are picked up, erratic and inefficient combustion may result. Sediment comes from dirt carried through with the crude during processing and impurities picked up in storage and transportation. Sediment can cause line and strainer plugging, control problems, and burned/nozzle plugging. More frequent filter cleaning may be required.

Sludge is a mixture of organic compounds that have precipitated after different heavy oils are blended. These are normally in the form of waxes or asphaltenese, which can cause plugging problems.

## Additives

Fuel-oil additives may be used in boilers to improve combustion efficiency, inhibit high-temperature corrosion, and minimize cold-end corrosion. In addition, additives may be useful in controlling plugging, corrosion, and the formation of deposits in fuel-handling systems. However, caution should be used in establishing the need for and application of any additive program. Before selecting an additive, clearly identify the problem requiring correction and the cause of the problem. In many cases, solutions may be found which would obviate the need and expense of additives. Also, be sure to understand clearly both the benefits and the potential debits of the additive under consideration.

Additives to fuel-handling systems may be warranted if corrosion problems persist due to water which cannot be removed mechanically. Additives are also available which help prevent sludge and/or other deposits from accumulating in equipment, which could result in increased loading due to increased pressure drops on pumps and losses in heat-transfer-equipment efficiencies.

Additive vendors claim that excess air can be controlled at lower values when catalysts are used. Although these claims appear to be verifiable, consideration should be given to mechanically controlling $O_2$ to the lowest possible levels. Accurate $O_2$ measurement and control should first be implemented and then modifications to burner assemblies considered. Catalysts, consisting of metallic oxides (typically manganese and barium), have demonstrated the capability of reducing carbon carryover in the flue gas and thus would permit lower $O_2$ levels without smoking. Under steady load conditions, savings can be achieved. However, savings may be negligible under varying loads, which necessitate prevention of fuel-rich mixtures by maintaining air levels higher than optimal.

Other types of combustion additives are available which may be beneficial to specific boiler operating problems. However, these are not discussed here since they are specific in nature and are not necessarily related to improved boiler efficiency. Generally, additives are used when a specific problem exists and when other conventional solutions have been exhausted.

## Atomization

Oil-fired burners, kilns, heat-treating furnaces, ovens, process reactors, and process heaters will realize increased efficiency when fuel oil is effectively atomized. The finer the oil is atomized, the more complete combustion and higher overall combustion efficiency. Obtaining the optimum degree of atomization depends on maintaining a precise differential between the pressure of the oil and pressure of the atomizing agent normally steam or air. The problem usually encountered is that the steam (or air pressure) remains constant while the oil pressure can vary substantially. One solution is the addition of a differential pressure regulator which controls the steam pressure so that the differential pressure to the oil is maintained. Other solutions, including similar arrangements for air-atomized systems, should be reviewed with equipment vendors.

## Fuel-Oil Emulsions

In general, fuel-oil emulsifier systems are designed to produce an oil/water emulsion that can be combusted in a furnace or boiler.

The theory of operation is that micro-size droplets of water are injected and evenly dispersed throughout the oil. As combustion takes place, micro explosions of the water droplets take place which produce very fine oil droplets. Thus more surface area of the fuel is exposed which then allows for a reduction in excess air level and improved efficiency. Unburned particles are also reduced.

Several types of systems are available. One uses a resonant chamber in which shock waves are started in the fluid, causing the water to cavitate and breakdown into small bubbles. Another system produces an emulsion by injecting water into oil. The primary technical difference among the various emulsifier systems currently marketed is the water particle-size distribution that is formed. The mean particle size as well as the particle-size distribution are very important to fuel combustion performance, and these parameters are the primary reasons for differing optimal water content for various systems (from 3% to over 200%). Manufacturers typically claim improvements in boiler efficiency of from 1 to 6% and further savings due to reduced particulate emissions which allow burning of higher-sulfur (lower-cost) fuels. However, recent independent testing seems to indicate that while savings are achieved, these are often at the low end of the range, particularly for units that are already operating near optimum conditions. Greater savings are more probable from emulsifiers that are installed on small, inefficient, and poorly instrumented boilers. Tests have confirmed reductions in particulate and $NO_x$ emissions and the ability to keep boiler internals cleaner. Flames exhibit characteristics of more transparency and higher radiance (similar to gas flames).

Before considering installation of an emulsifier system, existing equipment should be tuned up and put in optimum condition for maximum efficiency. A qualified service technician may be helpful in obtaining best results. After performance tests are completed, meaningful tests with an emulsified oil can be run. It is suggested that case histories of a manufacturer's equipment be reviewed and an agreement be based on satisfaction within a specific time frame.

### 5.4.3 Coal

The following factors have a negative impact on the selection of coal as a fuel:

1. Environmental limitations which necessitate the installation of expensive equipment to control particulates, $SO_2$, and $NO_x$. These requirements, when combined with the low price of oil and gas in the late 1960s and early 1970s, forced many existing industrial and utility coal-fired units to convert to oil or gas. And new units typically were designed with no coal capability at all.

2. Significantly higher capital investments, not only for pollution abatement but also for coal-receiving equipment; raw-coal storage; coal preparation (crushing, conveying, pulverizing, etc.): prepared-coal storage; and ash handling.

3. Space requirements for equipment and coal storage.

4. Higher maintenance costs associated with the installation of more equipment.

5. Concern over uninterrupted availability of coal resulting from strikes.

6. Increasing transportation costs.

The use of coal is likely to escalate even in light of the foregoing factors, owing primarily to its relatively secure availability both on a short-term and long-term basis, and its lower cost. The substantial operating cost savings at current fuel prices of coal over oil or gas can economically justify a great portion (if not all) of the significantly higher capital investments required for coal. In addition, most predictions of future fuel costs indicate that oil and gas costs will escalate much faster than coal.

**Enhancing Coal-Firing Efficiency**

A potentially significant loss from the combustion of coal fuels is the unburned carbon loss. All coal-fired steam generators and coal-fired vessels inherently suffer an efficiency debit attributable to unburned carbon. At a given process output, quantities in excess of acceptable values for the particular unit design and coal rank detract greatly from the unit efficiency, resulting in increased fuel consumption.

It is probable that pulverized coal-fired installations suffer from an inordinately high unburned carbon loss when any of the following conditions are experienced:

A change in the raw-fuel quality from the original design basis.

Deterioration of the fuel burners, burner throats, or burner swirl plates or impellers.

Increased frequency of soot blowing to maintain heat-transfer-surface cleanliness.

Noted increase in stack gas opacity.

Sluggish operation of combustion controls of antiquated design.

Uneven flame patterns characterized by a particularly bright spot in one sector of the flame and a quite dark spot in another.

CO formation as determined from a flue-gas analysis.

Frequent smoking observed in the combustion zone.

Increases in refuse quantities in collection devices.

Lack of continued maintenance and/or replacement of critical pulverized internals and classifier assembly.

High incidence of coal "hang-up" in the distribution piping to the burners.

Frequent manipulation of the air/coal primary and secondary air registers.

**How to Test for Relative Operating Efficiency**

The aforementioned items are general symptoms

that are suspect in causing a high unburned carbon loss from the combustion of coal. However, the magnitude of the problem often goes undetected and remedial action is never taken because of the difficulty in establishing and quantifying this loss while relating it to the overall unit operating efficiency.

In light of this, it is recommended that the boiler manufacturer be consulted to establish the magnitude of this operating loss as well as to review the system equipment and operating methods.

The general test procedure to determine the unburned carbon loss requires manual sampling of the ash (refuse) in the ash pit, boiler hopper(s), air heater hopper, and dust collector hopper and performing a laboratory analysis of the samples. For reference, detailed methods, test procedures, and results are outlined in the ASME Power Test Codes, publications PTC 3.2 (Solid Fuels) and PTC 4.1 (Steam Generating Units), respectively. In addition to the sampling of the ash, a laboratory analysis should be performed on a representative raw coal sample and pulverized coal sample.

Results and analysis of such a testing program will:

1.    Quantify the unburned carbon loss.

2.    Determine if the unburned carbon loss is high for the type of coal fired, unit design, and operating methods.

3.    Reveal a reasonable and attainable value for unburned carbon loss.

4.    Provide guidance for any corrective action.

5.    Allow the plant, with the aid of Figure 5.18, to assess the annual loss in dollars by operating with a high unburned carbon loss.

6.    Suggest an operating mode to reduce the excess air required for combustion.

Figure 5.18 can be used to determine the approximate energy savings resulting from reducing unburned coal fuel loss. The unburned fuel is generally collected with the ash in either the boiler ash hopper(s) or in various collection devices. The quantity of ash collected at various locations is dependent and unique to the system design. The boiler manufacturer will generally specify the proportion of ash normally collected in various ash hoppers furnished on the boiler proper. The balance of ash and unburned fuel is either collected in flue-gas cleanup devices or discharged to the atmosphere. A weighted av-

erage of total percentage combustibles in the ash must be computed to use Figure 5.18. To further illustrate the potential savings from reducing combustible losses in coal-fired systems using Figure 5.18, consider the following example.

**Example**: Determine the benefits of reducing the combustible losses for a coal-fired steam generator having a maximum continuous rating of 145,000 lbs/hr at an operating pressure of 210 psig and a temperature of 475°F at the desuperheater outlet, feedwater is supplied at 250°F.

**Available Data**

| | |
|---|---|
| Average boiler load | 125,000 lb/hr |
| Superheater pressure | 210 psig |
| Superheater temperature | 475°F |
| Fuel type | Coal |
| Measured flue gas oxygen | 3.5% |
| Operating combustibles in ash | 40% |
| Obtainable combustibles in ash | 5% |
| Yearly operating time | 8500 hr/yr |
| Design unit heat output | $150 \times 10^6$ Btu/hr |
| Average unit heat output | $129 \times 10^6$ Btu/hr |
| Average fuel cost | $1.50/$10^6$ Btu |

**Analysis**. Referring to Figure 5.18:
**Chart A.**

| | |
|---|---|
| Operating combustibles | 40% |
| Obtainable combustibles | 5% |

**Chart B.**

| | |
|---|---|
| Design Unit output | $150 \times 10^6$ Btu/hr |

**Chart C.**

| | |
|---|---|
| Average fuel cost | $1.50/$10^6$ Btu |
| Annual fuel savings | $210,000 |

Annual fuel savings corrected to actual operating conditions:

Savings $ = (curve value × [operating heat output)/(design heat output)] × [actual annual operating hours)/(8,760 hr/yr)]

= ($210,000/yr) × [($129 \times 10^6$ Btu/hr)/($150 \times 10^6$ Btu/hr)] × [(8,500 hr/yr)/(8,760 hr/yr)]

= $175,200/yr

**Notes**: If the unit heat output or average fuel cost exceeds the limit of Figure 5.18, use half the particular value and double the savings obtained from Chart C. The annual fuel savings are based on 8,760 hr/yr. For an operating factor other that the curve basis, apply a correction factor to curve results. The application of the charts assumes operating at, or close to, optimum excess air values.

## Modifications for Maximum Economy

It is very difficult to pinpoint and rectify the major problem detracting from efficiency, as there are many interdependent variables and numerous pieces of equipment required for the combustion of coal. Further testing and/or operating manipulations may be required to "zero in" on a solution. Modification(s) that have been instituted with a fair degree of success to reduce high unburned carbon losses and/or high-excess-air operation are:

Modifying or changing the pulverizer internals to increase the coal fineness and thereby enhance combustion characteristics.

Rerouting or modifying air/coal distribution piping to avoid coal hang up and slug flow going to the burners.

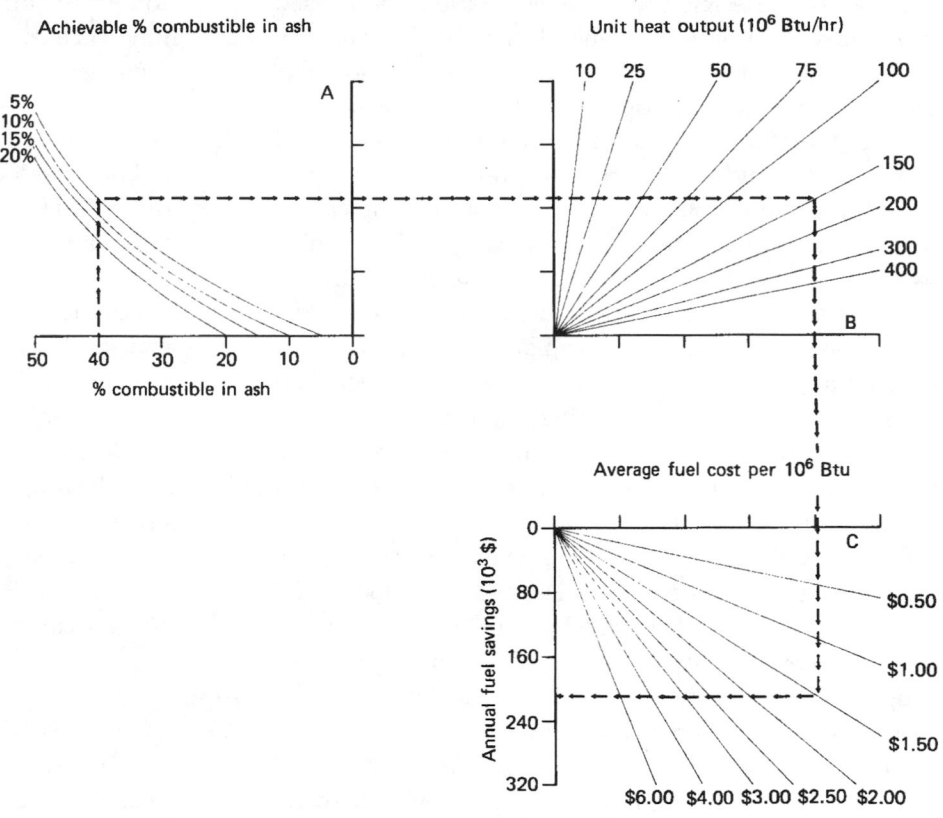

Figure 5.18 Coal annual savings from reducing combustible losses.

Installing additional or new combustion controls to smooth out and maintain consistent performance.

Purchasing new coal feeders compatible with and responsive to unit demand fluctuations.

Calibrating air flow and metering devices to ensure correct air/coal mixtures and velocities at the burner throats.

Installing new classifiers to ensure that proper coal fines reach the burners for combustion. Optimally setting air register positions for proper air/fuel mixing and combustion. Replacing worn and abraded burner impeller plates.

Increasing the air/coal mixture temperature exiting the pulverizers to ensure good ignition without coking.

Cleaning the burning throat areas of deposits.

Installing turning vanes or air foils in the secondary air supply duct or air plenum to ensure even distribution and proper air/fuel mixing at each burner.

Purchasing new, or updating existing combustion controls to reflect the present state of the art.

As can be seen, the solutions can be varied, simple or complex, relatively cheap or quite expensive.

The incentives to correct the problem(s) are offered in Figure 5.18. Compare the expected benefits to the boiler manufacturer's solution and cost to implement and judge the merits.

### Guidelines for Day-to-Day Operation

To a great extent, the items listed which detract from operating performance should be those checked on a day-to-day basis. A checklist should be initiated and developed by the plant to correct those conditions that do not require a shutdown, such as:

1. Maintaining integrity of distribution dampers and air registers with defined positions for operating load.

2. Ensuring pulverizer exit temperature by maintaining air calibration devices and air-moving equipment.

3. Checking feeder speeds and coal hopper valves, ensuring an even and steady coal feed.

4. Frequent blending of the raw coal pile to provide some measure of uniformity.

### Coal Conversion

Converting existing boilers that originally were designed to fire oil and/or natural gas to coal capability is possible for certain types of units provided that significant deratings in steam rate capacities are acceptable to existing operations. Since boiler installations are somewhat custom in design, fuel changes should be addressed on an individual basis. In general, however, conversion of field-erected units to coal firing is technically possible, whereas conversion of shop-assembled boilers is not.

Modifications required to convert shop-assembled units would include installation of ash-handling facilities in the furnace and convection pass, soot-blowing equipment, and modifications to tube spacing each of which is almost physically impossible to do and prohibitive in cost. In addition, even if these modifications could be made, derating of the unit by about two-thirds would be necessary. Alternatives for consideration rather than converting existing shop-assembled units would be to replace with new coal-fired units, or to carefully assess the application of one of an alternative fuel, such as a coal/oil mixture or ultra-fine coal.

Many field-erected boilers can be converted to coal firing without serious compromises being made to good design. The most serious drawback to converting these boilers is the necessity to downrate by 40 to 50%. Although in some instances downrating may be acceptable, most operating locations cannot tolerate losses in steam supply of this magnitude.

Why is downrating necessary? Generally, there are differences between boilers designed for oil and gas and those designed for coal. Burning of coal requires significantly more combustion volume than that needed for oil or gas due to flame characteristics and the need to avoid excessive slagging and fouling of heat-transfer surfaces. This means that a boiler originally designed for oil or gas must be downrated to maintain heat-release rates and firebox exit temperatures within acceptable limits for coal. Other factors influencing downrating are acceptable flue-gas velocities to minimize erosion and tube spacing.

A suggested approach for assuring the viability of coal conversion is:

1. Contact the original boiler manufacturer and determine via a complete inspection the actual modifications that would be required for the boiler.

2. Determine if the site can accommodate coal storage and other associated equipment, including unloading facilities, conveyors, dust abatement, ash disposal, and so on.

3. Select the type of coal to burn, considering cost, availability, and pollution restrictions. Before final selection of a coal, it must be analyzed to determine its acceptability and effect on unit rating. For any installation, final selection must be based on the coal's heating value, moisture content, mineral matter content, ash fusion and chemical characteristics, and/or pulverizers' grindability.

4. Assess over economics.

## Bibliography
AMERICAN SOCIETY OF MECHANICAL ENGINEERS (ASME), ASME Power Test Code (PTC) 4.1-1964; Steam Generating Units, ASME, New York, 1964.

KEENAN, J.H., and J. KAYE, The Gas Tables, Wiley, New York, 1948.

PAYNE, F.W., Efficient Boiler Operations Source Book, 3rd ed., Fairmont, Lilburn, GA., 1991.

SCOLLON, R.B., and R.D. SMITH, Energy Conservation Guidebook, Allied Corporation, Morristown, N.J., 1976

TAPLIN Jr., H.R., Combustion Efficiency Tables, Fairmont, Lilburn, GA., 1991.

# Appendix

DIRECT CONTACT TECHNOLOGY
FOR HOT WATER PRODUCTION

**B. Keith Walker**
QuikWater a Division of Webco Industries
Sand Springs, OK 74063

Direct contact water heating is a heat transfer method in which there are no tubes isolating hot combustion gasses from the fluid to be heated. The exhaust gases are allowed to come into direct contact with water in a totally non-pressurized environment so that all heating occurs at atmospheric pressure.

Direct contact water heaters consist of bodies sometimes made 100% out of stainless steel using a vertical hollow chamber in which water is sprayed at the top of this vertical chamber. The upper portion of this vertical chamber is filled with stainless steel balls (or other shapes) which provide a large heat transfer surface area for the heat of the rising exhaust gasses to be conducted into the water. This "heat transfer zone" is approximately 24" to 36" deep and almost all heat transfer in a direct contact water heater occurs in this area.

In the same vertical column, below the "heat transfer zone" a burner is mounted which provides the energy used to heat the water. Depending on manufacturer, this burner may be substantially removed from the falling water with a dedicated firing chamber or it may reside directly in the path of the water flow with an overhead

metal shield to protect the bulk of the flame from water impingement. The burner is forced draft and is typically fired on natural gas or propane although fuel oils can be used some of the time.

Because the water loses all pressure there must be a pool for water to collect in after passing through a direct contact heater. This pool is either a full sized tank or just a holding area at the base of the vertical heating tower. Water collects in this tank and is then held until the hot water is required at a remote location.

Mechanical pumps must be used to re-pressurize the water in this tank and move the volume required to the needed application. These pumps can provide virtually any pressure that a customer might require and can be used as a pressure booster over the typical water pressure that a facility would see off of a regular city water line.

## ULTRA HIGH EFFICIENCY

Direct contact water heaters are very thermal efficient (hot water energy output ÷ fuel energy in). In all combustion devices water is a by-product of the burning process, because oxygen and hydrogen from the air combine to make $H_2O$. This water is vaporized immediately from liquid to gaseous form in the heat of firing. This phase change of water from liquid to gas is a cooling process and accounts for about a 12% combustion efficiency loss. Direct contact water heaters super cool the combustion gasses below the point where this newly formed water vapor will re-condense back into liquid form. By returning the water back to a liquid state direct contact water heaters are able to reclaim all of the thermal energy that is normally lost out the stack of a traditional boiler. The phase change that steals energy by changing from liquid to gas returns heat energy when the gas is returned back into liquid form. By transferring heat in this way direct contact water heaters are capable of achieving efficiencies that approach 100%.

The thermal efficiency of a direct contact water heater is easily predicted by the exhaust gas temperature.

This can also be translated into determining efficiency by knowing what the inlet water temperature is. If the heating system is in reasonable good order then the exhaust gas temperature should mirror the inlet water temperature by 5 - 10°F. For example if your inlet water is 55°F then your exhaust gas temperature should be about 60°F and the heater efficiency would be about 99%. The typical operating inlet water temperature for a direct contact water heater is between 45°F - 80°F. Which translates to 99.7%-97.7%. If you are recirculating hot water back to the upper portion of a heater then the efficiency of the direct contact water heater will go down for the time frame in which the recirculation occurs. For example if the return water temperature is 160°F then the exhaust gas temperature would be about 165°F making the direct contact heater efficiency about 75% for the time that 160°F is being reheated. These dynamics make direct contact heater system design different from boiler system design. Because in all cases you want the coldest possible return water temperature from a system loop (not the typical 20°F temperature drop for heat exchanger return).

## WATER TEMPERATURE LIMITATIONS

Direct contact water heaters are incapable of producing usable steam and have difficulty in achieving outlet water temperatures higher than 188°F. Direct contact water heaters can achieve water temperatures up to 193°F. However, the efficiency of the heater drops dramatically even with cold water entering the upper portion of the heating tower as large amounts of water are vaporized carrying heat energy away from the heater.

Direct contact water heaters operate with a system pressure just a couple of inches water column above atmospheric pressure. In theory it should be possible under these condition to heat water up to 200°F or even higher in a direct contact water heater. The limiting factor is presented by when the water is heated. The water is heated as the droplet falls through open air in the heat transfer section of the direct contact. Because the heating occurs while the water is moving through the air there are wind aerodynamics that drop the localized pressure around the water droplets to a point that is lower than the available atmospheric pressure. This in turn lowers the temperature at which the liquid water phase changes to a gaseous form. Because if this effect attempting to achieve exit water temperatures higher than 188°F becomes inefficient and unstable. The altitude at which a heater is operating will affect the maximum temperature output capability of the direct contact heater because of the reduction in available atmospheric pressure.

## APPLICATIONS FOR DIRECT CONTACT

Direct contact water heaters can be used in almost any application that has a need for large (sometimes not so large) quantities of hot water. There are three general classifications of water usage. These groups are:

Single Pass Application
Open Loop Recirculation Systems
Special Applications

A short list of applications is given below, though there are many more possible applications.

## WATER QUALITY AND EMISSIONS

Water quality from a direct contact heater is not usually negatively affected after having been heated. There are differences in the designs from the different manufacturers which make the heaters have higher or lower PH drops. After a single pass through the body of a direct contact heating unit the introduction of 7.4 PH water exits the unit between 6.9 and 6.3 PH, but in general the PH change is minimized in heating units that have a dedicated space to complete the combustion of fuels.

Other than $CO_2$ injection into the water stream direct contact heaters do not negatively affect the consistency of the water stream. There are cases in poor heater design where combustion has been quenched early and acids created from incomplete combustion lower the PH an extra amount than is explainable by $CO_2$ injection.

Exhaust emissions from a direct contact water heater is typically lower than the emission rates of a boiler firing with the same burner at the same firing rate. This does not make the units low $NO_x$ emitters, but it does allow certain low $NO_x$ technologies to work even more effectively in direct contact heaters than in comparable boilers.

## OVERALL DIRECT
## CONTACT EFFICIENCY

Thermally, direct contact water heaters are up to 99% efficient, however there are pumps and blowers on the units. An energy calculation of the electricity consumed by these items should be done based on loading, control design, and operating time.

| Single Pass Applications | Open Loop Applications | Special Applications |
|---|---|---|
| Plant Cleanup | Bottle Warming | Hydronic Heating |
| Bird Scalding | Hospital Systems | Caustic Fluids |
| Car Washes | Swimming Pools | Green Houses |
| Concrete Batches | Fruit Cleaning | Hotel Showers |
| Dyeing | Cleanup Systems | Hotel Sinks |
| Parts Cleaning | | Demineralized |
| DA tank feed | | Bio Gas |
| Comm. Laundry | | Multiple Temp. |
| Indust. Laundry | | Institutional |

# CHAPTER 6
# STEAM AND CONDENSATE SYSTEMS

**PHILIP S. SCHMIDT**
Department of Mechanical Engineering
University of Texas at Austin
Austin, Texas

## 6. 1 INTRODUCTION

Nearly half of the energy used by industry goes into the production of process steam, approximately the same total energy usage as that required to heat all the homes and commercial buildings in America. Why is so much of our energy resource expended for the generation of industrial steam?

Steam is one of the most abundant, least expensive, and most effective heat-transfer media obtainable. Water is found everywhere, and requires relatively little modification from its raw state to make it directly usable in process equipment. During boiling and condensation, if the pressure is held constant and both water and steam are present, the temperature also remains constant. Further, the temperature is uniquely fixed by the pressure, and hence by maintaining constant pressure, which is a relatively easy parameter to control, excellent control of process temperature can also be maintained. The conversion of a liquid to a vapor absorbs large quantities of heat in each pound of water. The resulting steam is easy to transport, and because it is so energetic, relatively small quantities of it can move large amounts of heat. This means that relatively inexpensive pumping and piping can be used compared to that needed for other heating media.

Finally, the process of heat transfer by condensation, in the jacket of a steam-heated vessel, for example, is extremely efficient. High rates of heat transfer can be obtained with relatively small equipment, saving both space and capital. For these reasons, steam is widely used as the heating medium in thousands of industries.

Prior to the mid-1970s, many steam systems in industry were relatively energy wasteful. This was not necessarily bad design for the times, because energy was so cheap that it was logical to save first cost, even at the expense of a considerable increase in energy requirements. But things have changed, and today it makes good sense to explore every possibility for improving the energy efficiency of steam systems.

This chapter will introduce some of the language commonly used in dealing with steam systems and will help energy managers define the basic design constraints and the applicability of some of the vast array of manufacturer's data and literature which appears in the marketplace. It will discuss the various factors that produce inefficiency in steam system operations and some of the measures that can be taken to improve this situation. Simple calculation methods will be introduced to estimate the quantities of energy that may be lost and may be partially recoverable by the implementation of energy conservation measures. Some important energy conservation areas pertinent to steam systems are covered in other chapters and will not be repeated here. In particular the reader is referred to Chapters 5 (on boilers), 7 (on cogeneration), and 15 (on industrial insulation).

### 6.1.1 Components of Steam and Condensate Systems

Figure 6.1 shows a schematic of a typical steam system in an industrial plant. The boiler, or steam generator, produces steam at the highest pressure (and therefore the highest temperature) required by the various processes. This steam is carried from the boiler house through large steam mains to the process equipment area. Here, transfer lines distribute the steam to each piece of equipment. If some processes require lower temperatures, the steam may be throttled to lower pressure through a pressure-regulating valve (designated PRV on the diagram) or through a back-pressure turbine. Steam traps located on the equipment allow condensate to drain back into the condensate return line, where it flows back to the condensate receiver. Steam traps also perform other functions, such as venting air from the system on startup, which is discussed in more detail later.

The system shown in Figure 6.1 is, of course, highly idealized. In addition to the components shown, other elements, such as strainers, check valves, and pumping traps may be utilized. In some plants, condensate may be simply released to a drain and not returned; the potential for energy conservation through recovery of condensate will be discussed later. Also, in the system shown, the flash steam produced in the process of throt-

**Fig. 6.1 Typical steam system components.**

tling across the steam traps is vented to the atmosphere. Prevention of this loss represents an excellent opportunity to save Btus and dollars.

### 6.1.2 Energy Conservation Opportunities in Steam Systems

Many opportunities for energy savings exist in steam system operations, ranging from simple operating procedure modifications to major retrofits requiring significant capital expenditures. Table 6.1 shows a checklist of energy conservation opportunities applicable to most steam systems. It is helpful for energy conservation managers to maintain such a running list applicable to their own situations. Ideas are frequently presented in the technical and trade literature, and plant operators often make valuable contributions, since after all, they are the people closest to the problem.

To "sell" such improvements to plant management and operating personnel, it is necessary to demonstrate the dollars-and-cents value of a project or operating change. The following sections discuss the thermal properties of steam, how to determine the steam requirements of plant equipment and to estimate the amount of steam required to make up for system losses, how to assign a dollar value to this steam, and various approaches to alleviating these losses.

## 6.2 THERMAL PROPERTIES OF STEAM

### 6.2.1 Definitions and Terminology

Before discussing numerical calculations of steam properties for various applications, it is necessary to establish an understanding of some terms commonly used in the operation of steam systems.

**British Thermal Unit (Btu).** One Btu is the amount of heat required to raise 1 pound of water 1 degree Fahrenheit in temperature. To get a perspective on this quantity, a cubic foot of natural gas at atmospheric pressure will release about 1000 Btu when burned in a boiler with no losses. This same 1000 Btu will produce a little less than 1 pound of steam at atmospheric conditions, starting from tap water.

**Boiling Point.** The boiling point is the temperature at which water begins to boil at any given pressure. The boiling point of water at sea-level atmospheric pressure is about 212°F. At high altitude, where the atmospheric pressure is lower, the boiling point is also lower. Conversely, the boiling point of water goes up with increasing pressure. In steam systems, we usually refer to the boiling point by another term, "saturation temperature."

**Absolute and Gauge Pressure.** In steam system literature, we frequently see two different pressures used. The "absolute pressure," designated psia, is the true force per unit of area (e.g., pounds per square inch) exerted by the steam on the wall of the pipe or vessel containing it. We usually measure pressures, however, with sensing devices that are exposed to the atmosphere outside, and which therefore register an indication, not of the true force inside the vessel, but of the difference between that force and the force exerted by the outside atmosphere. We call this difference the "gauge pressure," designated psig. Since atmospheric pressure at

**Table 6.1 Checklist of Energy Conservation Opportunities in Steam and Condensate Systems**

## General Operations

1. Review operation of long steam lines to remote single-service applications. Consider relocation or conversion of remote equipment, such as steam-heated storage tanks.
2. Review operation of steam systems used only for occasional services, such as winter-only steam tracing lines. Consider use of automatic controls, such as temperature-controlled valves, to assure that the systems are used only when needed.
3. Implement a regular steam leak survey and repair program.
4. Publicize to operators and plant maintenance personnel the annual cost of steam leaks and unnecessary equipment operations.
5. Establish a regular steam-use monitoring program, normalized to production rate, to track progress in reduction of steam consumption. Publicize on a monthly basis the results of this monitoring effort.
6. Consider revision of the plant-wide steam balance in multipressure systems to eliminate venting of low-pressure steam. For example, provide electrical backup for currently steam-driven pumps or compressors to permit shutoff of turbines when excess low-pressure steam exists.
7. Check actual steam usage in various operations against theoretical or design requirement. Where significant disparities exist, determine the cause and correct it.
8. Review pressure-level requirements of steam-driven mechanical equipment to evaluate feasibility of using lower pressure levels.
9. Review temperature requirements of heated storage vessels and reduce to minimum acceptable temperatures.
10. Evaluate production scheduling of batch operations and revise if possible to minimize startups and shutdowns.

## Steam Trapping

1. Check sizing of all steam traps to assure they are adequately rated to provide proper condensate drainage. Also review types of traps in various services to assure that the most efficient trap is being used for each application.
2. Implement a regular steam trap survey and maintenance program. Train maintenance personnel in techniques for diagnosing trap failure.

## Condensate Recovery

1. Survey condensate sources presently being discharged to waste drains for feasibility of condensate recovery.
2. Consider opportunities for flash steam utilization in low-temperature processes presently using first-generation steam.
3. Consider pressurizing atmospheric condensate return systems to minimize flash losses.

## Mechanical Drive Turbines

1. Review mechanical drive standby turbines presently left in the idling mode and consider the feasibility of shutting down standby turbines.
2. Implement a steam turbine performance testing program and clean turbines on a regular basis to maximize efficiency.
3. Evaluate the potential for cogeneration in multipressure steam systems presently using large pressure-reducing valves.

## Insulation

1. Survey surface temperatures using infrared thermometry or thermography on insulated equipment and piping to locate areas of insulation deterioration. Maintain insulation on a regular basis.
2. Evaluate insulation of all uninsulated lines and fittings previously thought to be uneconomic. Recent rises in energy costs have made insulation of valves, flanges, and small lines desirable in many cases where this was previously unattractive.
3. Survey the economics of retrofitting additional insulation on presently insulated lines, and upgrade insulation if economically feasible.

sea level is usually around 14.7 psi, we can obtain the absolute pressure by simply adding 14.7 to the gauge pressure reading. In tables of steam properties, it is more common to see pressures listed in psia, and hence it is necessary to make the appropriate correction to the pressure indicated on a gauge.

**Saturated and Superheated Steam.** If we put cold water into a boiler and heat it, its temperature will begin to rise until it reaches the boiling point. If we continue to heat the water, rather than continuing to rise in temperature, it begins to boil and produce steam. As long as the pressure remains constant, the temperature will remain at the saturation temperature for the given pressure, and the more heat we add, the more liquid will be converted to steam. We call this boiling liquid a "saturated liquid" and refer to the steam so generated as "saturated vapor." We can continue to add more and more heat, and we will simply generate more saturated vapor (or simply "saturated steam") until the water is completely boiled off. At this point, if we continue to add heat, the steam temperature will begin to rise once more. We call this "superheated steam." This chapter will concentrate on the behavior of saturated steam, because this is the steam condition most commonly encountered in industrial process heating applications. Superheated steam is common in power generation and is often produced in industrial systems when cogeneration of power and process heat is used.

**Sensible and Latent Heat.** Heat input that is directly registered as a change in temperature of a substance is called "sensible heat," for the simple reason that we can, in fact, "sense" it with our sense of touch or with a thermometer. For example, the heating of the water mentioned above before it reaches the boiling point would be sensible heating. When the heat goes into the conversion of a liquid to a vapor in boiling, or vice versa in the process of condensation, it is termed "latent heat." Thus when a pound of steam condenses on a heater surface to produce a pound of saturated liquid at the same temperature, we say that it has released its latent heat. If the condensate cools further, it is releasing sensible heat.

**Enthalpy.** The total energy content of a flowing medium, usually expressed in Btu/lb, is termed its "enthalpy." The enthalpy of steam at any given condition takes into account both latent and sensible heat, and also the "mechanical" energy content reflected in its pressure. Hence steam at 500 psia and 600°F will have a higher enthalpy than steam at the same temperature but at 300 psia. Also, saturated steam at any temperature

and pressure has a higher enthalpy than condensate at the same conditions due to the latent heat content of the steam. Enthalpy, as listed in tables of steam properties, does not include the kinetic energy of motion, but this component is insignificant in most energy conservation applications.

**Specific Volume.** The specific volume of a substance is the amount of space (e.g., cubic feet) occupied by 1 pound of substance. This term will become important in some of our later discussions, because steam normally occupies a much greater volume for a given mass than water (i.e., it has a much greater specific volume), and this must be taken into account when considering the design of condensate return systems.

**Condensate.** Condensate is the liquid produced when steam condenses on a heater surface. As shown later, this condensate still contains a significant fraction of its energy, and can be returned to the boiler to conserve fuel.

**Flash Steam.** When hot condensate at its saturation temperature corresponding to the elevated pressure in a heating vessel rapidly drops in pressure, as, for example, when passing through a steam trap or a valve, it suddenly finds itself at a temperature above the saturation temperature for the new pressure. Steam is thus generated which absorbs sufficient energy to drop the temperature of the condensate to the appropriate saturation level. This is called "flash steam," and the pressure-reduction process is called "flashing." In many condensate return systems, flash steam is simply released to the atmosphere, but it may, in fact, have practical applications in energy conservation.

**Boiler Efficiency.** The boiler efficiency is the percentage of the energy released in the burning of fuel in a boiler which actually goes into the production of steam. The remaining percentage is lost through radiation from the boiler surfaces, blowdown of the boiler water to maintain satisfactory impurity levels, and loss of the hot flue gas up the stack. Although this chapter does go into detail on the subject of boiler efficiency, which is discussed in Chapter 5, it is important to recognize that this parameter relates the energy savings obtainable by conserving steam to the fuel savings obtainable at the boiler, a relation of obvious economic importance. Thus if we save 100 Btu of steam energy and have a boiler with an efficiency of 80%, the actual fuel energy saved would be 100/0.80, or 125 Btu. Because boilers always have an efficiency of less than 100% (and more com-

## Table 6.2 Thermodynamic Properties of Saturated Steam

| (1)<br>Gauge<br>Pressure | (2)<br>Absolute<br>Pressure<br>(psia) | (3)<br>Steam Temp.<br>(°F) | (4)<br>Enthalpy of<br>Sat. Liquid<br>(Btu/lb) | (5)<br>Latent Heat<br>(Btu/lb) | (6)<br>Enthalpy of<br>Steam<br>(Btu/lb) | (7)<br>Specific<br>Volume<br>(ft³/lb) |
|---|---|---|---|---|---|---|
| In. vacuum | | | | | | |
| 29.743 | 0.08854 | 32.00 | 0.00 | 1075.8 | 1075.8 | 3306.00 |
| 29.515 | 0.2 | 53.14 | 21.21 | 1063.8 | 1085.0 | 1526.00 |
| 27.886 | 1.0 | 101.74 | 69.70 | 1036.3 | 1106.0 | 333.60 |
| 19.742 | 5.0 | 162.24 | 130.13 | 1001.0 | 1131.1 | 73.52 |
| 9.562 | 10.0 | 193.21 | 161.17 | 982.1 | 1143.3 | 38.42 |
| 7.536 | 11.0 | 197.75 | 165.73 | 979.3 | 1145.0 | 35.14 |
| 5.490 | 12.0 | 201.96 | 169.96 | 976.6 | 1146.6 | 32.40 |
| 3.454 | 13.0 | 205.88 | 173.91 | 974.2 | 1148.1 | 30.06 |
| 1.418 | 14.0 | 209.56 | 177.61 | 971.9 | 1149.5 | 28.04 |
| Psig | | | | | | |
| 0.0 | 14.696 | 212.00 | 180.07 | 970.3 | 1150.4 | 26.80 |
| 1.3 | 16.0 | 216.32 | 184.42 | 967.6 | 1152.0 | 24.75 |
| 2.3 | 17.0 | 219.44 | 187.56 | 965.5 | 1153.1 | 23.39 |
| 5.3 | 20.0 | 227.96 | 196.16 | 960.1 | 1156.3 | 20.09 |
| 10.3 | 25.0 | 240.07 | 208.42 | 952.1 | 1160.6 | 16.30 |
| 15.3 | 30.0 | 250.33 | 218.82 | 945.3 | 1164.1 | 13.75 |
| 20.3 | 35.0 | 259.28 | 227.91 | 939.2 | 1167.1 | 11.90 |
| 25.3 | 40.0 | 267.25 | 236.03 | 933.7 | 1169.7 | 10.50 |
| 30.3 | 45.0 | 274.44 | 243.36 | 928.6 | 1172.0 | 9.40 |
| 40.3 | 55.0 | 287.07 | 256.30 | 919.6 | 1175.9 | 7.79 |
| 50.3 | 65.0 | 297.97 | 267.50 | 911.6 | 1179.1 | 6.66 |
| 60.3 | 75.0 | 307.60 | 277.43 | 904.5 | 1181.9 | 5.82 |
| 70.3 | 85.0 | 316.25 | 286.39 | 897.8 | 1184.2 | 5.17 |
| 80.3 | 95.0 | 324.12 | 294.56 | 891.7 | 1186.2 | 4.65 |
| 90.3 | 105.0 | 331.36 | 302.10 | 886.0 | 1188.1 | 4.23 |
| 100.0 | 114.7 | 337.90 | 308.80 | 880.0 | 1188.8 | 3.88 |
| 110.3 | 125.0 | 344.33 | 315.68 | 875.4 | 1191.1 | 3.59 |
| 120.3 | 135.0 | 350.21 | 321.85 | 870.6 | 1192.4 | 3.33 |
| 125.3 | 140.0 | 353.02 | 324.82 | 868.2 | 1193.0 | 3.22 |
| 130.3 | 145.0 | 355.76 | 327.70 | 865.8 | 1193.5 | 3.11 |
| 140.3 | 155.0 | 360.50 | 333.24 | 861.3 | 1194.6 | 2.92 |
| 150.3 | 165.0 | 365.99 | 338.53 | 857.1 | 1195.6 | 2.75 |
| 160.3 | 175.0 | 370.75 | 343.57 | 852.8 | 1196.5 | 2.60 |
| 180.3 | 195.0 | 379.67 | 353.10 | 844.9 | 1198.0 | 2.34 |
| 200.3 | 215.0 | 387.89 | 361.91 | 837.4 | 1199.3 | 2.13 |
| 225.3 | 240.0 | 397.37 | 372.12 | 828.5 | 1200.6 | 1.92 |
| 250.3 | 265.0 | 406.11 | 381.60 | 820.1 | 1201.7 | 1.74 |
| | 300.0 | 417.33 | 393.84 | 809.0 | 1202.8 | 1.54 |
| | 400.0 | 444.59 | 424.00 | 780.5 | 1204.5 | 1.16 |
| | 450.0 | 456.28 | 437.20 | 767.4 | 1204.6 | 1.03 |
| | 500.0 | 467.01 | 449.40 | 755.0 | 1204.4 | 0.93 |
| | 600.0 | 486.21 | 471.60 | 731.6 | 1203.2 | 0.77 |
| | 900.0 | 531.98 | 526.60 | 668.8 | 1195.4 | 0.50 |
| | 1200.0 | 567.22 | 571.70 | 611.7 | 1183.4 | 0.36 |
| | 1500.0 | 596.23 | 611.60 | 556.3 | 1167.9 | 0.28 |
| | 1700.0 | 613.15 | 636.30 | 519.6 | 1155.9 | 0.24 |
| | 2000.0 | 635.82 | 671.70 | 463.4 | 1135.1 | 0.19 |
| | 2500.0 | 668.13 | 730.60 | 360.5 | 1091.1 | 0.13 |
| | 2700.0 | 679.55 | 756.20 | 312.1 | 1068.3 | 0.11 |
| | 3206.2 | 705.40 | 902.70 | 0.0 | 902.7 | 0.05 |

What is the temperature inside the line? Coming down column (1) we find a pressure of 150, and moving over to column (3), we note that the corresponding steam temperature at this pressure is about 366°F.

Column (4) lists the enthalpy of the saturated liquid in Btu/lb of water. We can see at the head of the column that this enthalpy is designated as 0 at a temperature [column (3)] of 32°F.

monly around 75 to 80%) there is a built-in "amplifier" on any energy savings effected in the steam system.

### 6.2.2 Properties of Saturated Steam

In calculating the energy savings obtainable through various measures, it is important to understand the quantitative thermal properties of steam and condensate. Table 6.2 shows a typical compilation of the properties of saturated steam.

Columns 1 and 2 list various pressures, either in gauge (psig) or absolute (psia). Note that these two pressures always differ by about 15 psi (14.7 to be more precise). Remember that the former represents the pressure indicated on a normal pressure gauge, while the latter represents the true pressure inside the line. Column 3 shows the saturation temperature corresponding to each of these pressures. Note, for example, that at an absolute pressure of 14.696 (the normal pressure of the atmosphere at sea level) the saturation temperature is 212°F, the figure we are all familiar with. Suppose that we have a pressure of 150 psi indicated on the pressure gauge on a steam line. This is an arbitrary reference point, and therefore the heat indicated at any other temperature tells us the amount of heat added to raise the water from an initial value of 32°F to that temperature. For example, referring back to our 150 psig steam, the water contains about 338.5 Btu/lb: starting from 32°F, 10 lb of water would contain 10 times this number, or about 3385 Btu. We can also subtract one number from another in this column to find the amount of heat necessary to raise the water from one temperature to another. If the water started at 101.74°F, it would contain a heat of 69.7 Btu/lb, and to raise it from this temperature to 366°F would require 338.5—69.7, or about 268.8 Btu for each pound of water. Column (5) shows the latent heat content of a pound of steam for each pressure. For our 150-psig example, we can see that it takes about 857 Btu to convert each pound of saturated water into saturated steam. Note that this is a much larger quantity than the heat content of the water alone, confirming the earlier observation that steam is a very effective carrier of heat; each pound can give up, in this case, 857 Btu when condensed on a surface back to saturated liquid. Column (6), the enthalpy of the saturated steam, represents simply the sum of columns (4) and (5), since each pound of steam contains both the latent heat required to vaporize the water and the sensible heat required to raise the water to the boiling point in the first place.

Column (7) shows the specific volume of the saturated steam at each pressure. Note that as the pressure increases, the steam is compressed; that is, it occupies less space per pound. 150-psig steam occupies only 2.75

ft³/lb; if released to atmospheric pressure (0 psig) it would expand to nearly 10 times this volume. By comparison, saturated liquid at atmospheric pressure has a specific volume of only 0.017 ft³/lb (not shown in the table), and it changes only a few percent over the entire pressure range of interest here. Thus 1 lb of saturated liquid condensate at 212°F will expand more than 1600 times in volume in converting to a vapor. This illustrates that piping systems for the return of condensate from steam-heated equipment must be sized primarily to accommodate the large volume of flashed vapor, and that the volume occupied by the condensate itself is relatively small.

The steam tables can be a valuable tool in estimating energy savings, as illustrated in the following example.

**Example:** A 100-ft run of 6-in. steam piping carries saturated steam at 95 psig. Tables obtained from an insulation manufacturer indicate that the heat loss from this piping run is presently 110,000 Btu/hr. With proper insulation, the manufacturer's tables indicate that this loss could be reduced to 500 Btu/hr. How many pounds per hour of steam savings does this installation represent, and if the boiler is 80% efficient, what would be the resulting fuel savings?

From the insulation manufacturer's data, we can find the reduction in heat loss:

$$\text{heat-loss reduction} = 110{,}000 - 500 = 109{,}500 \text{ Btu/hr}$$

From Table 6.2 at 95 psig (halfway between 90 and 100), the total heat of the steam is about 1188.4 Btu/lb. The steam savings, is therefore

$$\text{steam savings} = \frac{109{,}500 \text{ Btu/hr}}{1188.4 \text{ Btu/lb}}$$

Assume that condensate is returned to the boiler at around 212°F; thus the condensate has a heat content of about 180 Btu/lb. The heat required to generate 95 psig steam from this condensate is 1188.4 − 180.0 or 1008.4 Btu/lb. If the boiler is 80% efficient, then

$$\text{fuel savings} = \frac{1008.4 \text{ Btu/lb} \times 92 \text{ lb/hr}}{0.80}$$

$$= \text{approximately 0.116 million Btu/hr}$$

### 6.2.3 Properties of Superheated Steam

If additional heat is added to saturated steam with no liquid remaining, it begins to superheat and the tem-

## Table 6.3 Thermodynamic Properties of Superheated Steam

| Abs. Press (psi) | | 100 | 200 | 300 | 400 | 500 | 600 | 700 | 800 | 900 | 1000 | 1100 | 1200 | 1300 | 1400 | 1500 |
|---|---|---|---|---|---|---|---|---|---|---|---|---|---|---|---|---|
| 1 | v | 0.0161 | 392.5 | 452.3 | 511.9 | 571.5 | 631.1 | 690.7 | | | | | | | | |
| | h | 68.00 | 1150.2 | 1195.7 | 1241.8 | 1288.6 | 1336.1 | 1384.5 | | | | | | | | |
| 5 | v | 0.0161 | 78.14 | 90.24 | 102.24 | 114.21 | 126.15 | 138.08 | 150.01 | 161.94 | 173.86 | 185.78 | 197.70 | 209.62 | 221.53 | 233.45 |
| | h | 68.01 | 1148.6 | 1194.8 | 1241.3 | 1288.2 | 1335.9 | 1384.3 | 1433.6 | 1483.7 | 1534.7 | 1586.7 | 1639.6 | 1693.3 | 1748.0 | 1803.5 |
| 10 | v | 0.0161 | 38.84 | 44.98 | 51.03 | 57.04 | 63.03 | 69.00 | 74.98 | 80.94 | 86.91 | 92.87 | 98.84 | 104.80 | 110.76 | 116.72 |
| | h | 68.02 | 1146.6 | 1193.7 | 1240.6 | 1287.8 | 1335.5 | 1384.0 | 1433.4 | 1483.5 | 1534.6 | 1586.6 | 1639.5 | 1693.3 | 1747.9 | 1803.4 |
| 15 | v | 0.0161 | 0.0166 | 29.899 | 33.963 | 37.985 | 41.986 | 45.978 | 49.964 | 53.946 | 57.926 | 61.905 | 65.882 | 69.858 | 73.833 | 77.807 |
| | h | 68.04 | 168.09 | 1192.5 | 1239.9 | 1287.3 | 1335.2 | 1383.8 | 1433.2 | 1483.4 | 1534.5 | 1586.5 | 1639.4 | 1693.2 | 1747.8 | 1803.4 |
| 20 | v | 0.0161 | 0.0166 | 22.356 | 25.428 | 28.457 | 31.466 | 34.465 | 37.458 | 40.447 | 43.435 | 46.420 | 49.405 | 52.388 | 55.370 | 58.352 |
| | h | 68.05 | 168.11 | 1191.4 | 1239.2 | 1286.9 | 1334.9 | 1383.5 | 1432.9 | 1483.2 | 1534.3 | 1586.3 | 1639.3 | 1693.1 | 1747.8 | 1803.3 |
| 40 | v | 0.0161 | 0.0166 | 11.036 | 12.624 | 14.165 | 15.685 | 17.195 | 18.699 | 20.199 | 21.697 | 23.194 | 24.689 | 26.183 | 27.676 | 29.168 |
| | h | 68.10 | 168.15 | 1186.6 | 1236.4 | 1285.0 | 1333.6 | 1382.5 | 1432.1 | 1482.5 | 1533.7 | 1585.8 | 1638.8 | 1992.7 | 1747.5 | 1803.0 |
| 60 | v | 0.0161 | 0.0166 | 7.257 | 8.354 | 9.400 | 10.425 | 11.438 | 12.466 | 13.450 | 14.452 | 15.452 | 16.450 | 17.448 | 18.445 | 19.441 |
| | h | 68.15 | 168.20 | 1181.6 | 1233.5 | 1283.2 | 1332.3 | 1381.5 | 1431.3 | 1481.8 | 1533.2 | 1585.3 | 1638.4 | 1692.4 | 1747.1 | 1802.8 |
| 80 | v | 0.0161 | 0.0166 | 0.0175 | 6.218 | 7.018 | 7.794 | 8.560 | 9.319 | 10.075 | 10.829 | 11.581 | 12.331 | 13.081 | 13.829 | 14.577 |
| | h | 68.21 | 168.24 | 269.74 | 1230.5 | 1281.3 | 1330.9 | 1380.5 | 1430.5 | 1481.1 | 1532.6 | 1584.9 | 1638.0 | 1692.0 | 1746.8 | 1802.5 |
| 100 | v | 0.0161 | 0.0166 | 0.0175 | 4.935 | 5.558 | 6.216 | 6.833 | 7.443 | 8.050 | 8.655 | 9.258 | 9.860 | 10.460 | 11.060 | 11.659 |
| | h | 68.26 | 168.29 | 269.77 | 1227.4 | 1279.3 | 1329.6 | 1379.5 | 1429.7 | 1480.4 | 1532.0 | 1584.4 | 1637.6 | 1691.6 | 1746.5 | 1802.2 |
| 120 | v | 0.0161 | 0.0166 | 0.0175 | 4.0786 | 4.6341 | 5.1637 | 5.6831 | 6.1928 | 6.7006 | 7.2060 | 7.7096 | 8.2119 | 8.7130 | 9.2134 | 9.7130 |
| | h | 68.31 | 168.33 | 269.81 | 1224.1 | 1277.4 | 1328.1 | 1378.4 | 1428.8 | 1479.8 | 1531.4 | 1583.9 | 1637.1 | 1691.3 | 1746.2 | 1802.0 |
| 140 | v | 0.0161 | 0.0166 | 0.0175 | 3.4661 | 3.9526 | 4.4119 | 4.8585 | 5.2995 | 5.7364 | 6.1709 | 6.6036 | 7.0349 | 7.4652 | 7.8946 | 8.3233 |
| | h | 68.37 | 168.38 | 269.85 | 1220.8 | 1275.3 | 1326.8 | 1377.4 | 1428.0 | 1479.1 | 1530.8 | 1583.4 | 1636.7 | 1690.9 | 1745.9 | 1801.7 |
| 160 | v | 0.0161 | 0.0166 | 0.0175 | 3.0060 | 3.4413 | 3.8480 | 4.2420 | 4.6295 | 5.0132 | 5.3945 | 5.7741 | 6.1522 | 6.5293 | 6.9055 | 7.2811 |
| | h | 68.42 | 168.42 | 269.89 | 1217.4 | 1273.3 | 1325.4 | 1376.4 | 1427.2 | 1478.4 | 1530.3 | 1582.9 | 1636.3 | 1690.5 | 1745.6 | 1801.4 |
| 180 | v | 0.0161 | 0.0166 | 0.0174 | 2.6474 | 3.0433 | 3.4093 | 3.7621 | 4.1084 | 4.4505 | 4.7907 | 5.1289 | 5.4657 | 5.8014 | 6.1363 | 6.4704 |
| | h | 68.47 | 168.47 | 269.92 | 1213.8 | 1271.2 | 1324.0 | 1375.3 | 1426.3 | 1477.7 | 1529.7 | 1582.4 | 1635.9 | 1690.2 | 1745.3 | 1801.2 |
| 200 | v | 0.0161 | 0.0166 | 0.0174 | 2.3598 | 2.7247 | 3.0583 | 3.3783 | 3.6915 | 4.0008 | 4.3077 | 4.6128 | 4.9165 | 5.2191 | 5.5209 | 5.8219 |
| | h | 68.52 | 168.51 | 269.96 | 1210.1 | 1269.0 | 1322.6 | 1374.3 | 1425.5 | 1477.0 | 1529.1 | 1581.9 | 1635.4 | 1689.8 | 1745.0 | 1800.9 |
| 250 | v | 0.0161 | 0.0166 | 0.0174 | 0.0186 | 2.1504 | 2.4662 | 2.6872 | 2.9410 | 3.1909 | 3.4382 | 3.6837 | 3.9278 | 4.1709 | 4.4131 | 4.6546 |
| | h | 68.66 | 168.63 | 270.05 | 375.10 | 1263.5 | 1319.0 | 1371.6 | 1423.4 | 1475.3 | 1527.6 | 1580.6 | 1634.4 | 1688.9 | 1744.2 | 1800.2 |
| 300 | v | 0.0161 | 0.0166 | 0.0174 | 0.0186 | 1.7665 | 2.0044 | 2.2263 | 2.4407 | 2.6509 | 2.8585 | 3.0643 | 3.2688 | 3.4721 | 3.6746 | 3.8764 |
| | h | 68.79 | 168.74 | 270.14 | 375.15 | 1257.7 | 1315.2 | 1368.9 | 1421.3 | 1473.6 | 1526.2 | 1579.4 | 1633.3 | 1688.0 | 1743.4 | 1799.6 |
| 350 | v | 0.0161 | 0.0166 | 0.0174 | 0.0186 | 1.4913 | 1.7028 | 1.8970 | 2.0832 | 2.2652 | 2.4445 | 2.6219 | 2.7980 | 2.9730 | 3.1471 | 3.3205 |
| | h | 68.92 | 168.85 | 270.24 | 375.21 | 1251.5 | 1311.4 | 1366.2 | 1419.2 | 1471.8 | 1524.7 | 1578.2 | 1632.3 | 1687.1 | 1742.6 | 1798.9 |
| 400 | v | 0.0161 | 0.0166 | 0.0174 | 00162 | 1.2841 | 1.4763 | 1.6499 | 1.8151 | 1.9759 | 2.1339 | 2.2901 | 2.4450 | 2.5987 | 2.7515 | 2.9037 |
| | h | 69.05 | 168.97 | 270.33 | 375.27 | 1245.1 | 1307.4 | 1363.4 | 1417.0 | 1470.1 | 1523.3 | 1576.9 | 1631.2 | 1686.2 | 1741.9 | 1798.2 |
| 500 | v | 0.0161 | 0.0166 | 0.0174 | 0.0186 | 0.9919 | 1.1584 | 1.3037 | 1.4397 | 1.5708 | 1.6992 | 1.8256 | 1.9507 | 2.0746 | 2.1977 | 2.3200 |
| | h | 69.32 | 169.19 | 270.51 | 375.38 | 1231.2 | 1299.1 | 1357.7 | 1412.7 | 1466.6 | 1520.3 | 1574.4 | 1629.1 | 1684.4 | 1740.3 | 1796.9 |
| 600 | v | 0.0161 | 0.0166 | 0.0174 | 0.0186 | 0.7944 | 0.9456 | 1.0726 | 1.1892 | 1.3008 | 1.4093 | 1.5160 | 1.6211 | 1.7252 | 1.8284 | 1.9309 |
| | h | 69.58 | 169.42 | 270.70 | 375.49 | 1215.9 | 1290.3 | 1351.8 | 1408.3 | 1463.0 | 1517.4 | 1571.9 | 1627.0 | 1682.6 | 1738.8 | 1795.6 |
| 700 | v | 0.0161 | 0.0166 | 0.0174 | 0.0186 | 0.0204 | 0.7928 | 0.9072 | 1.010 2 | 1.1078 | 1.2023 | 1.2948 | 1.3858 | 1.4757 | 1.5647 | 1.6530 |
| | h | 69.84 | 169.65 | 270.89 | 375.61 | 487.93 | 1281.0 | 1345.6 | 1403.7 | 1459.4 | 1514.4 | 1569.4 | 1624.8 | 1680.7 | 1737.2 | 1794.3 |
| 800 | v | 0.0161 | 0.0166 | 0.0174 | 0.0186 | 0.0204 | 0.6774 | 0.7828 | 0.8759 | 0.9631 | 1.0470 | 1.1289 | 1.2093 | 1.2885 | 1.3669 | 1.4446 |
| | h | 70.11 | 169.88 | 271.07 | 375.73 | 487.88 | 1271.1 | 1339.2 | 1399.1 | 1455.8 | 1511.4 | 1566.9 | 1622.7 | 1678.9 | 1736.0 | 1792.9 |
| 900 | v | 0.0161 | 0.0166 | 0.0174 | 0.0186 | 0.0204 | 0.5869 | 0.6858 | 0.7713 | 0.8504 | 0.9262 | 0.9998 | 1.0720 | 1.1430 | 1.2131 | 1.2825 |
| | h | 70.37 | 170.10 | 271.26 | 375.84 | 487.83 | 1260.6 | 1332.7 | 1394.4 | 1452.2 | 1508.5 | 1564.4 | 1620.6 | 1677.1 | 1734.1 | 1791.6 |
| 1000 | v | 0.0161 | 0.0166 | 0.0174 | 0.0186 | 0.0204 | 0.5137 | 0.6080 | 0.6875 | 0.7603 | 0.8295 | 0.8966 | 0.9622 | 1.0266 | 1.0901 | 1.1529 |
| | h | 70.63 | 170.33 | 271.44 | 375.96 | 487.79 | 1249.3 | 1325.9 | 1389.6 | 1448.5 | 1504.4 | 1561.9 | 1618.4 | 167~.3 | 1732.5 | 1790.3 |

perature will rise. Table 6.3 shows the thermodynamic properties of superheated steam. Unlike the saturated steam of Table 6.2, in which each pressure had only a single temperature (the saturation temperature) associated with it, superheated steam may exist, for a given pressure, at any temperature above the saturation temperature. Thus the properties must be tabulated as a function of both temperature and pressure, rather than pressure alone. With this exception, the values in the superheated steam table may be used exactly like those in Table 6.2.

**Example:** Suppose, in the preceding example, that the steam line is carrying superheated steam at 250 psia (235 psig) and 500°F (note that both temperature and pressure must be specified for superheated steam). For the same reduction in heat loss (109,500 Btu/hr), how many pounds per hour of steam is saved?

From Table 6.3 the enthalpy of steam at 250 psia and 500°F is 1263.5 Btu/lb. Thus

$$\text{steam savings} = \frac{109{,}500 \text{ Btu/hr}}{1263.5 \text{ Btu/lb}}$$

$$= 86.6 \text{ lb/hr}$$

### Table 6.4 Orders of Magnitude of Convective Conductances

| Heating Process | Order of Magnitude of $h$ (Btu/hr ft$^2$ • F) |
|---|---|
| Free convection, air | 1 |
| Forced convection, air | 5-10 |
| Forced convection, water | 250-1000 |
| Condensation, steam | 5000-10,000 |

### 6.2.4 Heat-Transfer Characteristics of Steam

As mentioned in Section 6.1 steam is one of the most effective heat-transfer media available. The rate of heat transfer from a fluid medium to a solid surface (such as the surface of a heat-exchanger tube or a jacketed heating vessel) can be expressed by Newton's law of cooling:

$$q = h(T_f - T_s)$$

where $q$ is the rate of heat transfer per unit of surface area (e.g., Btu/hr/ft$^2$), $h$ is a proportionality factor called the "convective conductance," $T_f$ is the temperature of the fluid medium, and $T_s$ is the temperature of the surface.

Table 6.4 shows the order of magnitudes of $h$ for several heat-transfer media. Condensation of steam can be several times as effective as the flow of water over a surface for the transfer of heat, and may be 1000 times more effective than a gaseous heating medium, such as air.

In a heat exchanger, the overall effectiveness must take into account the fluid resistances on both sides of the exchanger and the conduction of heat through the tube wall. These effects are generally lumped into a single "overall conductance," $U$, defined by the equation

$$q = U(T_{f1} - T_{f2})$$

where $q$ is defined as before, $U$ is the conductance, and $T_{f1}$ and $T_{f2}$ are the temperatures of the two fluids. In addition, there is a tendency for fluids to deposit "fouling layers" of crystalline, particulate, or organic matter on transfer surfaces, which further impede the flow of heat. This impediment is characterized by a "fouling resistance," which, for design purposes, is usually incorporated as an additional factor in determining the overall conductance.

Table 6.5 illustrates typical values of $U$ (not including fouling) and the fouling resistance for exchangers employing steam on the shell side versus exchangers using a light organic liquid (such as a typical heat-transfer oil). The 30 to 50% higher $U$ values for steam translate directly into a proportionate reduction in required heat-exchanger area for the same fluid temperatures. Furthermore, fouling resistances for the steam-heated exchangers are 50 to 100% lower than for a similar service using an organic heating medium, since pure steam contains no contaminants to deposit on the exchanger surface. From the design standpoint, this means that the additional heating surface incorporated to allow for fouling need not be as great. From the operating viewpoint, it translates into energy conservation, since more heat can be transferred per hour in the exchanger for the same fluid conditions, or the same heating duty can be met with a lower fluid temperature difference if the fouling resistance is lower.

### Table 6.5 Comparison of Steam and Light Organics as Heat-Exchange Media

| Shell-Side Fluid | Tube-Side Fluid | Typical U (Btu/hr ft$^2$ . °F) | Typical Fouling Resistance (hr ft$^2$ . °F/Btu) |
|---|---|---|---|
| Steam | Light organic liquid | 135-190 | 0.001 |
| Steam | Heavy organic liquid | 45-80 | 0.002 |
| Light organic liquid | Light organic liquid | 100-130 | 0.002 |
| Light organic liquid | Heavy organic liquid | 35-70 | 0.003 |

## 6.3 ESTIMATING STEAM USAGE AND ITS VALUE

To properly assess the worth of energy conservation improvements in steam systems, it is first necessary to determine how much steam is actually required to carry out a desired process, how much energy is being wasted through various system losses, and the dollar value of these losses. Such information will be needed to determine the potential gains achievable with insulation, repair or improvement of steam traps, and condensate recovery systems.

### 6.3.1 Determining Steam Requirements

Several approaches can be used to determine process steam requirements. In order of increasing reliability, they include the use of steam consumption tables for typical equipment, detailed system energy balances, and direct measurement of steam and/or condensate flows. The choice of which method is to be used depends on how critical the steam-using process is to the plant's overall energy consumption and how the data are to be used.

For applications in which a high degree of accuracy is not required, such as developing rough estimates of the distribution of energy within a plant, steam consumption tables have been developed for various kinds of process equipment. Table 6.6 shows steam consumption tables for a number of typical industrial and commercial applications. To illustrate, suppose that we wish to estimate the steam usage in a soft-drink bottling plant for washing 2000 bottles/min. From the table we see that, typically, a bottle washer uses about 310 lb of 5-psig steam per hour for each 100 bottles/min of capacity. For a washer with a 2000-bottle/min capacity we would, of course, use about 20 times this value, or 6200 lb/hr of steam. Referring to Table 6.2, we see that 5-psig steam has a total enthalpy of about 1156 Btu/lb. The hourly heat usage of this machine, therefore, would be approximately 1156 × 6200, or a little over 7 million Btu/hr. Remember that this is the heat content of the steam itself, not the fuel heat content required at the boiler, since the boiler efficiency has not yet been taken into account.

Note that most of the entries in Table 6.6 show the steam consumption "in use," not the peak steam consumption during all phases of operation. These figures are fairly reliable for equipment that operates on a more-or-less steady basis; however, they may be quite low for batch-processing operations or operations where the load on the equipment fluctuates significantly during its operation. For this reason, steam equipment manufacturers recommend that estimated steam consumption values be multiplied by a "factor of safety," typically between 2 and 5, to assure that the equipment will operate properly under peak-load conditions. This can be quite important from the standpoint of energy efficiency. For example, if a steam trap is sized for average load conditions only, during startup or heavy-load operations, condensate will tend to back up into the heating vessel, reducing the effective area for condensation and hence reducing its heating capacity. For steam traps and condensate return lines, the incremental cost of slight oversizing is small, and factors of safety are used to assure that the design is adequately conservative to guarantee rapid removal of the condensate. Gross oversizing of steam traps, however, can also cause excessive steam loss. This point is discussed in more detail in the section on steam traps, where appropriate factors of safety for specific applications are given.

A second, and generally more accurate, approach to estimating steam requirements is by direct energy-balance calculations on the process. A comprehensive discussion of energy balances is beyond the scope of this section: analysis of complex equipment should be undertaken by a specialist. It is, however, possible to determine simple energy balances on equipment involving the heating of a single product.

The energy-balance concept simply states that any energy put into a system with steam must be absorbed by the product and/or the equipment itself, dissipated to the environment, carried out with the product, or carried out in the condensate. Recall that the concepts of sensible and latent heat were discussed in Section 6.2 in the context of heat absorption by water in the production of steam. We can extend this concept to consider heat absorption of any material, such as the heating of air in a dryer or the evaporation of water in the process of condensing milk.

The *sensible* heat requirement of any process is defined in terms of the "specific heat" of the material being heated. Table 6.7 gives specific heats for a number of common substances. The specific heat specifies the number of Btu required to raise 1 pound of a substance through a temperature rise of 1°F. Remember that for water, we stated that 1 Btu was, by definition, the amount of heat required to raise 1 lb by 1°F. The specific heat of water, therefore, is exactly 1 (at least near normal ambient conditions). To see how the specific heat can be used to calculate steam requirements for the sensible heating of products, consider the following example.

**Example:** A paint dryer requires about 3000 cfm of 200°F air, which is heated in a steam-coil unit. How

## Table 6.6 Typical Steam Consumption Rates for Industrial and Commercial Equipment

| Type of Installation | Description | Typical Pressure (psig) | Steam Consumption in Use (lb/hr) |
|---|---|---|---|
| Bakeries | Dough-room trough, 8 ft long | 10 | 4 |
| | Oven, white bread, 120-ft$^2$ surface | 10 | 29 |
| Bottle washing | Soft drinks, per 100 bottles/min | 5 | 310 |
| Dairies | Pasteurizer, per 100 gal heated/20 min | 15-75 | 232 (max) |
| Dishwashers | Dishwashing machine | 15-20 | 60-70 |
| Hospitals | Sterilizers, instrument, per 100 in.$^3$, approx. | 40-50 | 3 |
| | Sterilizers, water, per 10 gal, approx. | 40-50 | 6 |
| | Disinfecting ovens, double door, 50-100 ft$^3$, per 10 ft$^3$, approx. | 40-50 | 21 |
| Laundries | Steam irons, each | 100 | 4 |
| | Starch cooker, per 10 gal capacity | 100 | 7 |
| | Laundry presses, per 10-in. length, approx. | 100 | 7 |
| | Tumblers, 40 in., per 10-in. length, approx. | 100 | 38 |
| Plastic molding | Each 12-15 ft$^2$ platen surface | 125 | 29 |
| Paper manufacture | Corrugators, per 1,000 ft$^2$ | 175 | 29 |
| | Wood pulp paper, per 100 lb of paper | 50 | 372 |
| Restaurants | Standard steam tables, per ft of length | 5-20 | 36 |
| | Steamjacketed kettles, 25 gal of stock | 5-20 | 29 |
| | Steamjacketed kettles, 60 gal of stock | 5-20 | 58 |
| | Warming ovens, per 20 ft$^3$ | 5-20 | 29 |
| Silver mirroring | Average steam tables | 5 | 102 |
| Tire shops | Truck molds, large | 100 | 87 |
| | Passenger molds | 100 | 29 |

many pounds of 50-psig steam does this unit require per hour?

The density of air at temperatures of several hundred degrees or below is about 0.075 lb/ft$^3$. The number of pounds of air passing through the dryer is then

$$3000 \text{ ft}^3/\text{min} \times 60 \text{ min/hr} \times 0.075 \text{ lb/ft}^3 = 13,500 \text{ lb/hr}$$

Suppose that the air enters the steam-coil unit at 70°F. Its temperature will then be raised by 200 − 70 = 130°F. From Table 6.7 the specific heat of air is 0.24 Btu/lb °F. The energy required to provide this temperature rise is, therefore,

$$13,500 \text{ lb/hr} \times 0.24 \text{ Btu/lb °F} \times 130°F = 421,200 \text{ Btu/hr}$$

Employing the energy-balance principle, whatever energy is absorbed by the product (air) must be provided by an equal quantity of steam energy, less the energy contained in the condensate.

From Table 6.2 the total enthalpy per pound of 50-psig steam is about 1179.1 Btu, and the condensate (saturated liquid) has an enthalpy of 267.5 Btu/lb. Each pound of steam therefore gives up 1179.1 − 267.5, or 911.6 Btu (i.e., its latent heat) to the air. The steam required is, therefore, just

$$\frac{\text{heat required by air per hour}}{\text{heat released per pound of steam}} = \frac{421,200}{911.6} = 462 \text{ lb/hr}$$

To illustrate how latent heat comes into play when considering the steam requirements of a process, consider another example, this time involving a steam-heat evaporator.

**Example:** A milk evaporator uses a steamjacketed kettle, in which milk is batch-processed at atmospheric pressure. The kettle has a 1500-lb per batch capacity. Milk is heated from a temperature of 80°F to 212°F, where 25% of its mass is then driven off as vapor. Determine the amount of 15-psig steam required per batch, not includ-

## Table 6.7 Specific Heats of Common Materials

| Material | Btu/lb·°F | Material | Btu/lb·°F |
|----------|-----------|----------|-----------|
| *Solids* | | | |
| Aluminum | 0.22 | Iron, cast | 0.49 |
| Asbestos | 0.20 | Lead | 0.03 |
| Cement, dry | 0.37 | Magnesium | 0.25 |
| Clay | 0.22 | Porcelain | 0.26 |
| Concrete, stone | 0.19 | Rubber | 0.48 |
| Concrete, cinder | 0.18 | Silver | 0.06 |
| Copper | 0.09 | Steel | 0.12 |
| Glass, common | 0.20 | Tin | 0.05 |
| Ice, 32°F | 0.49 | Wood | 0.32-0.48 |
| *Liquids* | | | |
| Acetone | 0.51 | Milk | 0.90 |
| Alcohol, methyl, 60-70°F | 0.60 | Naphthalene | 0.41 |
| Ammonia, 104°F | 1.16 | Petroleum | 0.51 |
| Ethylene glycol | 0.53 | Soybean oil | 0.47 |
| Fuel oil, sp. gr. 86 | 0.45 | Tomato juice | 0.95 |
| Glycerine | 0.58 | Water | 1.00 |
| *Gases* | | | |
| *(Constant-Pressure Specific Heats)* | | | |
| Acetone | 0.35 | Carbon dioxide | 0.20 |
| Air, dry, 32-392°F | 0.24 | Methane | 0.59 |
| Alcohol | 0.45 | Nitrogen | 0.24 |
| Ammonia | 0.54 | Oxygen | 0.22 |

ing the heating of the kettle itself.

We must first heat the milk from 80°F to 212°F (sensible heating) and then evaporate off 0.25 × 1500 = 375 lb of water.

From Table 6.7, the specific heat of milk is 0.90 Btu/lb °F. The sensible heat requirement is, therefore,

$$0.90 \times 1500 \text{ lb} \times (212 - 80)°F = 178,200 \text{ Btu/batch}$$

In addition, we must provide the latent heat to vaporize 375 lb of water at 212°F. From Table 6.2, 970.3 Btu/lb is required. The total latent heat is, therefore,

$$375 \times 970.3 = 363,863 \text{ Btu/batch}$$

and the total heat input is

$$363,863 + 178,200 = 542,063 \text{ Btu/batch}$$

This heat must be supplied as the latent heat of 15-psig steam, which, from Table 6.2, is about 945 Btu/lb. The total steam requirement is, then,

$$\frac{542,063 \text{ Btu/batch}}{945 \text{ Btu/lb}} = 574 \text{ lb of 15-psig steam per batch}$$

We could have also determined the startup requirement to heat the steel kettle, if we could estimate its weight, using the specific heat of 0.12 Btu/lb °F for steel, as shown in Table 6.7.

### 6.3.2 Estimating Surface and Leakage Losses

In addition to the steam required to actually carry out a process, heat is lost through the surfaces of pipes, storage tanks, and jacketed heater surfaces, and steam is lost through malfunctioning steam traps, and leaks in flanges, valves, and other fittings. Estimation of these losses is important, because fixing them can often be the most cost-effective energy conservation measure available.

Figure 6.2 illustrates the annual heat loss, based on 24-hr/day, 365-day/yr operation, for bare steam lines at

**Fig. 6.2 Heat loss from bare steam lines.**

**Fig. 6.3 Heat losses from surfaces at elevated temperatures.**

various pressures. The figure shows, for example, that a 100-ft run of 6-in. line operating at 100 psig will lose about 1400 million Btu/yr. The economic return on an insulation retrofit can easily be determined with price data obtained from an insulation contractor.

Figure 6.3 can be used to estimate heat losses from flat surfaces at elevated temperatures, or from already insulated piping runs for which the outside jacket surface temperature is known. The figure shows the heat flow per hour per square foot of exposed surface area as a function of the difference in temperature between the surface and the surrounding air. It will be noted that the nature of the surface significantly affects the magnitude of the heat loss. This is because thermal radiation, which is strongly dependent on the character of the radiating surface, plays an important role in heat loss at elevated temperature, as does convective heat loss to the air.

Another important source of energy loss in steam systems is leakage from components such as loose flanges or malfunctioning steam traps. Figure 6.4 permits estimation of this loss of steam at various pressures leaking through holes. The heat losses are represented in million Btu/yr, based on full-time operation. Using the figure, we can see that a stuck-open steam trap with a 1/8-in. orifice would waste about 600 million Btu/yr of steam energy when leaking from a 100-psig line. This figure can also be used to estimate magnitudes of leakage from other sources of more complicated geometry. It is necessary to first determine an approximate area of leakage (in square inches) and then calculate the equivalent hole diameter represented by that area. The following example illustrates this calculation.

**Example:** A flange on a 200-psig steam line has a leaking gasket. The maintenance crew, looking at the gasket, estimates that it is about 0.020 in. thick and that it is leaking from about 1/8-in. of the periphery of the flange. Estimate the annual heat loss in the steam if the line is operational 8000 hr/yr.

The area of the leak is a rectangle 0.020 in. wide and 1/8-in. in length:

$$\text{leak area} = 0.020 \times 1/8 = 0.0025 \text{ in.}^2$$

An equivalent circle will have an area of $\pi D_2/4$, so if $\pi D_2/4 = 0.0025$, then $D = 0.056$ in. From Figure 6.4, this leak, if occurring year-round (8760 hr), would waste about 200 million Btu/yr of steam energy. For actual operation

$$\text{heat lost} = \frac{8000}{8760} \times 200,000,000 = 182.6 \text{ million Btu/hr}$$

Since the boiler efficiency has not been considered, the actual fuel wastage would be about 25% greater.

### 6.3.3 Measuring Steam and Condensate Rates

In maintaining steam systems at peak efficiency, it is often desirable to monitor the rate of steam flow through the system continuously, particularly at points of major

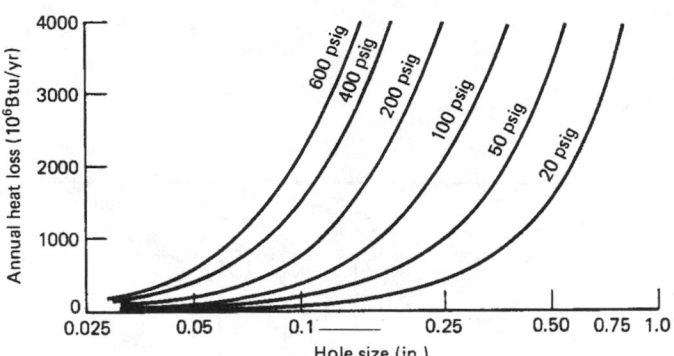

Fig. 6.4  Heat loss from steam leaks.

Fig. 6.5  Orifice flowmeter.

Fig. 6.6  Annular averaging element.

usage, such as steam mains. Figure 6.5 shows one of the most common types of flowmetering device, the calibrated orifice. This is a sharp-edged restriction that causes the steam flow to "neck down" and then re-expand after passing through the orifice. As the steam accelerates to pass through the restriction, its pressure drops, and this pressure drop, if measured, can be easily related to the flow rate. The calibrated orifice is one of a class of devices known as obstruction flowmeters, all of which work on the same principle of restricting the flow and producing a measurable pressure drop. Other types of obstruction meters are the ASME standard nozzle and the venturi. Orifices, although simple to manufacture and relatively easy to install, between flanges for example, are also subject to wear, which causes them, over a period of time, to give unreliable readings. Nozzles and venturis, although more expensive initially, tend to be more resistant to erosion and wear, and also produce less permanent pressure drop once the steam re-expands to fill the pipe. With all of these devices, care must be exercised in the installation, since turbulence and flow irregularities produced by valves, elbows, and fittings immediately upstream of the obstruction will produce erroneous readings.

Figure 6.6 shows another type of flowmetering device used for steam, called an annular averaging element. The annular element principle is somewhat different from the devices discussed above; it averages the pressure produced when steam impacts on the holes facing into the flow direction, and subtracts from this average impact pressure a static pressure sensed by a tube facing downstream. As with obstruction-type flowmeters, the flow must be related to this pressure difference.

A device that does not utilize pressure drop for steam metering applications is the vortex shedding flowmeter, illustrated in Figure 6.7. A solid bar extends through the flow, and as steam flows around the bar, vortices are shed alternately from one side to the other

in its wake, As the vortices shift from side to side, the frequency of shedding can be detected with a thermal or magnetic detector, and this frequency varies directly with the rate of flow. The vortex shedding meter is quite rugged, since the only function of the object extending into the flow stream is to provide an obstruction to generate vortices; hence it can be made of heavy-duty stainless steel. Also, vortex shedding meters tend to be relatively insensitive to variations in the steam properties, since they produce a pulsed output rather than an analog signal.

The target flowmeter, not shown, is also suitable for some steam applications. This type of meter uses a "target," such as a small cylinder, mounted on the end of a metal strut that extends into the flow line. The strut is gauged to measure the force on the target, and if the properties of the fluid are accurately known, this force can be related to flow velocity. The target meter is especially useful when only intermittent measurements are needed, as the unit can be "hot-tapped" in a pipe through a ball valve and withdrawn when not in use. The requirement of accurate property data limits the usefulness of this type of meter in situations where steam conditions vary considerably, especially where high moisture is present.

**Fig. 6.7 Vortex shedding flowmeter with various methods for sensing fluctuations.**

**Fig. 6.8**
**Weigh bucket technique for condensate measurement.**

The devices discussed above are useful in permanent installations where it is desired to continuously or periodically measure steam flow; there is no simple way to directly measure steam flow on a spotcheck basis without cutting into the system. There is, however, a relatively simple indirect method, illustrated in Figure 6.8, for determining the rate of steam usage in systems with unpressurized condensate return lines, or open systems in which condensate is dumped to a drain. If a drain line is installed after the trap, condensate may be caught in a barrel and the weight of a sample measured over a given period of time. Precautions must be taken when using this technique to assure that the flash steam, generated when the condensate drops in pressure as it passes through the trap, does not bubble out of the barrel. This can represent both a safety hazard and an error in the measurements due to the loss of mass in vapor form. The barrel should be partially filled with cold water prior to the test so that flash steam will condense as it bubbles through the water. An energy balance can

also be made on the water at the beginning and end of the test by measuring its temperature, and with proper application of the steam tables, a check can be made to assure that the trap is not blowing through.

Figure 6.9 illustrates another instrument which can be used to monitor condensate flow on a regular basis, the rotameter. A rotameter indicates the flow rate of the liquid by the level of a specially shaped float which rises in a calibrated glass tube, such that its weight exactly balances the drag force of the flowing condensate.

Measurements of this type can be very useful in monitoring system performance, since any unusual change in steam or condensate rate not associated with a corresponding change in production rate would tend to indicate an equipment malfunction producing poor efficiency.

### 6.3.4 Computing the Dollar Value of Steam

In analyzing energy conservation measures for steam systems, it is important to establish a steam value in dollars per pound: this value will depend on the steam pressure, the boiler efficiency, and the price of fuel.

Steam may be valued from two points of view. The more common approach, termed the "enthalpy method," takes into account only the heating capability of the steam, and is most

**Fig. 6.9**
**Liquid rotameter.**

appropriate when steam is used primarily for process heating. The enthalpy method can be illustrated with an example.

**Example:** An oil refinery produces 200-psig saturated steam in a large boiler, some of which is used directly in high-temperature processes, and some of which is let down to 30 psig through regulating valves for use at lower temperatures. The feedwater is added to the boiler at about 160°F. The boiler efficiency has been determined to be 82%, and boiler fuel is priced at $2.20 million Btu. Establish the values of 200-psig and 30-psig steam ($/lb).

Using the enthalpy method, we determine the increased heat content for each steam pressure required from the boiler if feedwater enters at 160°F. From Table 6.2:

total enthalpy of steam at 200 psig
    = 1199.3 Btu/lb
total enthalpy of steam at 30 psig (ignoring superheat
    after PRV)
    = 1172 Btu/lb
enthalpy of feedwater at 160°F
    − approx. 130 Btu/lb (about the same
    as saturated liquid at 160°F)
heat added per lb of 200-psig steam
    = 1199.3 − 130 = 1069.3 Btu/lb
heat added per lb of 30-psig steam
    = 1042 Btu/lb
fuel Btu required per pound at 200 psig
    = 1069.3/0.82 = 1304 Btu
fuel Btu required per lb at 30 psig
    = 1042/0.82 = 1271 Btu

With boiler fuel priced at $2.20/million Btu,
    or .22 per 1000 Btu:

value of 200-psig steam
    = 0.22 × 1.304 =.29/lb or $2.90/1000 lb
value of 30-psig steam
    = 0.22 × 1.271 =.28/lb or $2.80/1000 lb

Another approach to the valuation of steam is termed the "availability" or "entropy" method and takes into account not only the heat content of the steam, but also its power-producing potential if it were expanded through a steam turbine. This method is most applicable in plants where cogeneration (the sequential generation of power and use of steam for process heat) is practiced. The availability method involves some fairly complex thermodynamic reasoning and will not be covered here.

The reader should, however, be aware of its existence in analyzing the economics of cogeneration systems, in which there is an interchangeability between purchased electric power and in-plant power and steam generation.

## 6.4 STEAM TRAPS AND THEIR APPLICATION

### 6.4.1 Functions of Steam Traps

Steam traps are important elements of steam and condensate systems, and may represent a major energy conservation opportunity (or problem, as the case may be). The basic function of a steam trap is to allow condensate formed in the heating process to be drained from the equipment. This must be done speedily to prevent backup of condensate in the system.

Inefficient removal of condensate produces two adverse effects. First, if condensate is allowed to back up in the steam chamber, it cools below the steam temperature as it gives up sensible heat to the process and reduces the effective potential for heat transfer. Since condensing steam is a much more effective heat-transfer medium than stagnant liquid, the area for condensation is reduced, and the efficiency of the heat-transfer process is deteriorated. This results in longer cycle times for batch processes or lower throughput rates in continuous heating processes. In either case, inefficient condensate removal almost always increases the amount of energy required by the process.

A second reason for efficient removal of condensate is the avoidance of "water hammer" in steam systems. This phenomenon occurs when slugs of liquid become trapped between steam packets in a line. The steam, which has a much larger specific volume, can accelerate these slugs to high velocity, and when they impact on an obstruction, such as a valve or an elbow, they produce an impact force not unlike hitting the element with a hammer (hence the term). Water hammer can be extremely damaging to equipment, and proper design of trapping systems to avoid it is necessary.

The second crucial function of a steam trap is to facilitate the removal of air from the steam space. Air can leak into the steam system when it is shut down, and some gas is always liberated from the water in the boiling process and carried through the steam lines. Air mixed with steam occupies some of the volume that would otherwise be filled by the steam itself. Each of these components, air and steam, contributes its share to the total pressure exerted in the system; it is a fundamental thermodynamic principle that, in a mixture of gases, each component contributes to the pressure in the same proportion as its share of the volume of the space. For example, consider a steam system at 100 psia (note that in this case it is necessary to use *absolute* pressures), with 10% of the volume air instead of steam. Therefore, from thermodynamics, 10% of the pressure, or 10 psia, is contributed by the air, and only 90%, or 90 psia, by the steam. Referring to Table 6.2, the corresponding steam temperature is between 316 and 324°F, or approximately 320°F. If the air were not present, the steam pressure would be 100 psia, corresponding to a temperature of about 328°F, so the presence of air in the system reduces the temperature for heat transfer. This means that more steam must be generated to do a given heating job. Table 6.8 indicates the temperature reduction caused by the presence of air in various quantities at given pressures (shown in psig), and shows that the effective temperature may be seriously degraded.

In actual operation the situation is usually even worse than indicated in Table 6.8. We have considered the temperature reduction assuming that the air and steam are uniformly mixed. In fact, on a real heating surface, as air and steam move adjacent to the surface, the steam is condensed out into a liquid, while the air stays behind in the form of vapor. In the region very near the surface, therefore, the air occupies an even larger fraction of the volume than in the steam space as a whole, acting effectively as an insulating blanket on the surface. Suffice it to say that air is an undesirable parasite in steam systems, and its removal is important for proper operation.

Oxygen and carbon dioxide, in particular, have another adverse effect, and this is corrosion in condensate and steam lines. Oxygen in condensate produces pitting or rusting of the surface, which can contaminate the water, making it undesirable as boiler feed, and $CO_2$ in solution with water forms carbonic acid, which is highly corrosive to metallic surfaces. These components must be removed from the system, partially by good steam trapping and partially by proper deaeration of condensate, as is discussed in a subsequent section.

### 6.4.2 Types of Steam Traps and Their Selection

Various types of steam traps are available on the market, and the selection of the best trap for a given application is an important one. Many manufacturers produce several types of traps for specific applications, and manufacturers' representatives should be consulted in arriving at a choice. This section will give a brief introduction to the subject and comment on its relevance to improved energy utilization in steam systems.

Steam traps may be generally classified into three groups: mechanical traps, which work on the basis of the density difference between condensate and steam or air; thermostatic, which use the difference in temperature between steam, which stays close to its saturation temperature, and condensate, which cools rapidly; and thermodynamic, which functions on the difference in flow properties between liquids and vapors.

Figures 6.10 and 6.11 show two types of mechanical traps in common use for industrial applications. Figure 6.10 illustrates the principle of the "bucket trap." In the trap illustrated, an inverted bucket is placed over the inlet line, inside an external chamber. The bucket is attached to a lever arm which opens and closes a valve as the bucket rises and falls in the chamber. As long as condensate flows through the system, the bucket has a negative buoyancy, since liquid is present both inside and outside the bucket. The valve is open and condensate is allowed to drain continuously to the return line. As steam enters the trap it fills the bucket, displacing condensate, and the bucket rises, closing off the valve. Noncondensable gases, such as air and $CO_2$, bubble through a small vent hole and collect at the top of the trap, to be swept out with flash steam the next time the valve opens. Steam may also leak through the vent, but it is condensed on contact with the cool chamber walls, and collects as condensate in the chamber. The vent hole is quite small, so the rate of steam loss through this leakage action is not excessive. As condensate again begins to enter the bucket, it loses buoyancy and begins to drop until the valve opens and again discharges condensate and trapped air.

**Table 6.8 Temperature Reduction Caused by Air in Steam Systems**

| Pressure (psig) | Temp. of Steam, No Air Present | Temp. of Steam Mixed with Various Percentages of Air (by Volume) | | |
|---|---|---|---|---|
| | | 10 | 20 | 30 |
| 10 | 240.1 | 234.3 | 228.0 | 220.9 |
| 25 | 267.3 | 261.0 | 254.1 | 246.4 |
| 50 | 298.0 | 291.0 | 283.5 | 275.1 |
| 75 | 320.3 | 312.9 | 204.8 | 295.9 |
| 100 | 338.1 | 330.3 | 321.8 | 312.4 |

The float-and-thermostatic (F&T) trap, illustrated in Figure 6.11, works on a similar principle. In this case, instead of a bucket, a buoyant float rises and falls in the chamber as condensate enters or is discharged. The float is attached to a valve, similar to the one in

**Fig. 6.10   Inverted bucket trap.**

**Fig. 6.11   Float and thermostatic steam trap.**

**Fig. 6.12   Thermostatic steam trap.**

a bucket trap, which opens and closes as the ball rises and falls. Since there is no natural vent in this trap and the ball cannot distinguish between air and steam, which have similar densities, special provision must be made to remove air and other gases from the system. This is usually done by incorporating a small thermostatically actuated valve in the top of the trap. At low temperature, the valve bellows contracts, opening the vent and allowing air to be discharged to the return line. When steam enters the chamber, the bellows expands, sealing the vent. Some float traps are also available without this thermostatic air-vent feature; external provision must then be provided to permit proper air removal from the system. The F&T-type trap permits continuous discharge of condensate, unlike the bucket trap, which is intermittent. This can be an advantage in certain applications.

Figure 6.12 illustrates a thermostatic steam trap. In this trap, a temperature-sensitive bellows expands and contracts in response to the temperature of the fluid in the chamber surrounding the bellows. When condensate surrounds the bellows, it contracts, opening the drain port. As steam enters the chamber, the elevated temperature causes the bellows to expand and seal the drain. Since air also enters the chamber at a temperature lower than that of steam, the thermostatic trap is naturally self-venting, and it also is a continuous-drain-type trap. The bellows in the trap can be partially filled with a fluid and sealed, such that an internal pressure is produced which counterbalances the external pressure imposed by the steam. This feature makes the bellows-type thermostatic trap somewhat self-compensating for variations in steam pressure. Another type of thermostatic trap, which uses a bimetallic element, is also available. This type of trap is not well suited for applications in which significant variations in steam pressure might be expected, since it is responsive only to temperature changes in the system.

The thermodynamic, or controlled disk steam trap is shown in Figure 6.13. This type of trap is very simple in construction and can be made quite compact and resistant to damage from water hammer. In a thermodynamic trap, a small disk covers the inlet orifice. Condensate or air, moving at relatively low velocity, lifts the disk off its seat and is passed through to the outlet drain. When steam enters the trap, it passes through at high velocity because of its large volume. As the steam passes through the space between the disk and its seat, it impacts on the walls of the control chamber to produce a rise in pressure. This pressure imbalance between the outside of the disk and the side facing the seat causes it to snap shut, sealing off the chamber and preventing the further passage of steam to the outlet. When condensate again enters the inlet side, the disk lifts off the seat and permits its release.

**Fig. 6.13  Disk or thermodynamic steam trap.**

**Fig. 6.14  Drain orifice.**

An alternative to conventional steam traps, the drain orifice is illustrated in Figure 6.14. This device consists simply of an obstruction to the flow of condensate, similar to the orifice flowmeter described in an earlier section but much smaller. This small hole allows the pressure in the steam system to force condensate to drain continuously into the lower-pressure return system. Obviously, if steam enters, rather than condensate, it will also pass through the orifice and be lost. The strategy of using drain orifices is to select an orifice size that permits condensate to drain at such a rate that live steam seldom enters the system. Even if steam does occasionally pass through, the small size of the orifice limits the steam leakage rate to a value much less than would be lost due to a "stuck-open" malfunction of one of the types of traps discussed above. Drain orifices can be successfully applied in systems that have a well-defined and relatively constant condensate load. They are not suited for use where condensate load may vary widely with operating conditions.

As mentioned above, a number of operating requirements must be taken into consideration in selecting the appropriate trap for a given application. Table 6.9 lists these application considerations and presents one manufacturer's ratings on the performance of the various traps discussed above. In selecting a trap for a given application, assistance from manufacturers' representatives should be obtained, since a great body of experience in actual service has been accumulated over the years.

### 6.4.3  Considerations in Steam Trap Sizing

As mentioned earlier in this section, good energy conservation practice demands the efficient removal of condensate from process equipment. It is thus necessary to assure that traps are properly sized for the given condensate load. Grossly oversized traps waste steam by excessive surface heat loss and internal venting, while undersized traps permit accumulation of condensate with resultant loss in equipment heat-transfer effectiveness.

Steam traps are sized based in two specifications, the condensate load (e.g., in lbs/hr or gal/min) and the pressure differential across the trap (in psig). Section 6.3 discussed various methods for estimating condensate loads expected under normal operating conditions.

It is good practice to size the capacity of the trap based on this expected load times a factor of safety to account for peaks at startup and fluctuations in normal operating conditions. It is not unusual for startup condensate loads to be three to four times higher than steady operational loads, and in some applications they may range up to 10 times the steady-state load.

Table 6.10 presents typical factors of safety for condensate capacity recommended by steam trap manufacturers. This indicates typical ranges of factor of safety to consider in various applications. Although there is considerable variation in the recommended values, in both energy and economic terms the cost of oversizing is ordinarily not prohibitive, and conservative safety factors are usually used. The exception to this rule of thumb is in the sizing of disk-type traps, which may not function properly if loaded considerably below design. Drain orifices also must be sized close to normal operating loads. Again, the advice of the manufacturer should be solicited for the specific application in mind.

The other important design specification is the pressure differential over which the trap will operate. Since pressure is the driving force that moves condensate through the trap and on to the receiver, trap capacity will increase, for a given trap size, as the pressure increases. The trap operating-pressure differential is not simply the boiler pressure. On the upstream side of the trap, steam pressure may drop through valves and fittings and through heat-transfer passages in the process equipment. Thus the appropriate upstream pressure is the pressure at the trap inlet, which to a reasonable approximation, can usually be considered to be the process steam pressure at the equipment. Back pressure on the outlet side of the trap must also be considered. This includes the receiver pressure (if the condensate

### Table 6.9 Comparison of Steam Trap Characteristics

| Characteristic | Inverted Bucket | F&T | Disk | Bellows Thermostatic |
|---|---|---|---|---|
| Method of operation | Intermittent | Continuous | Intermittent | Continuous[a] |
| Energy conservation (time in service) | Excellent | Good | Poor | Fair |
| Resistance to wear | Excellent | Good | Poor | Fair |
| Corrosion resistance | Excellent | Good | Excellent | Good |
| Resistance to Hydraulic shock | Excellent | Poor | Excellent | Poor |
| Vents air and $CO_2$ at steam temperature | Yes | No | No | No |
| Ability to vent air at very low pressure (1/4 psig) | Poor | Excellent | NR[b] | Good |
| Ability to handle start-up air loads | Fair | Excellent | Poor | Excellent |
| Operation against back pressure | Excellent | Excellent | Poor | Excellent |
| Resistance to damage from freezing[c] | Good | Poor | Good | Good |
| Ability to purge system | Excellent | Fair | Excellent | Good |
| Performance on very light loads | Excellent | Excellent | Poor | Excellent |
| Responsiveness to slugs of condensate | Immediate | Immediate | Delayed | Delayed |
| Ability to handle dirt | Excellent | Poor | Poor | Fair |
| Comparative physical size | Large[d] | Large | Small | Small |
| Ability to handle "flash steam" | Fair | Poor | Poor | Poor |
| Mechanical failure (open-closed) | Open | Closed | Open[e] | Closed[f] |

[a]Can be intermittent on low load.
[b]Not recommended for low-pressure operations.
[c]Cast iron traps not recommended.
[d]In welded stainless steel construction—medium.
[e]Can fail closed due to dirt.
[f]Can fail open due to wear.

return system is pressurized), the pressure drop associated with flash steam and condensate flow through the return lines, and the head of water associated with risers if the trap is located at a point below the condensate receiver. Condensate return lines are usually sized for a given capacity to maintain a velocity no greater than 5000 ft/min of the flash steam. Table 6.11 shows the expected pressure drop per 100 ft of return line which can be expected under design conditions. Referring to the table, a 60-psig system, for example, returning condensate to an unpressurized receiver (0 psig) through a 2-in. line, would have a return-line pressure drop of just under 1/2 psi/100-ft run, and the condensate capacity of the line would be about 2600 lb/hr. The pressure head produced by a vertical column of water is about 1 psi/2 ft of rise. These components can be summed to estimate the back pressure on the system, and the appropriate pressure for sizing the trap is then the difference between the upstream pressure and the back pressure.

### Table 6.10 Typical Factors of Safety for Steam Traps (Condensate Flow Basis)

| Application | Factor of Safety |
|---|---|
| Autoclaves | 3-4 |
| Blast coils | 3-4 |
| Dry cans | 2-3 |
| Dryers | 3-4 |
| Dry kilns | 3-4 |
| Fan system heating service | 3-4 |
| Greenhouse coils | 3-4 |
| Hospital equipment | 2-3 |
| Hot-water heaters | 4-6 |
| Kitchen equipment | 2-3 |
| Paper machines | 3-4 |
| Pipe coils (in still air) | 3-4 |
| Platen presses | 2-3 |
| Purifiers | 3-4 |
| Separators | 3-4 |
| Steamjacketed kettles | 4-5 |
| Steam mains | 3-4 |
| Submerged surfaces | 5-6 |
| Tracer lines | 2-3 |
| Unit heaters | 3-4 |

## Table 6.11 Condensate Capacities and Pressure Drops for Return Lines[a]

| Supply Press. (psig) | 5 | 15 | | 30 | | | 60 | | | | 100 | | | | | 200 | | | | | |
|---|---|---|---|---|---|---|---|---|---|---|---|---|---|---|---|---|---|---|---|---|---|
| Return Press. (Psig) | 0 | 0 | 5 | 0 | 5 | 10 | 0 | 5 | 10 | 20 | 0 | 5 | 10 | 20 | 30 | 0 | 5 | 10 | 20 | 30 | 50 |
| **Pipe size (in.), schedule 40 pipe** | | | | | | | | | | | | | | | | | | | | | |
| 1/2 | 1,425 | 590 | 1,335 | 360 | 640 | 1,065 | 235 | 370 | 535 | 1,010 | 180 | 270 | 370 | 615 | 955 | 115 | 165 | 215 | 325 | 450 | 760 |
| | 4.0 | 4.0 | 5.3 | 4.0 | 5.3 | 6.5 | 4.0 | 5.3 | 6.5 | 8.9 | 4.0 | 5.3 | 6.5 | 8.9 | 11.3 | 4.0 | 5.3 | 6.5 | 8.9 | 11.3 | 15.9 |
| 3/4 | 2,495 | 1,035 | 2,340 | 635 | 1,125 | 1,855 | 415 | 650 | 940 | 1 770 | 310 | 470 | 645 | 1,085 | 1,675 | 200 | 285 | 375 | 570 | 795 | 1,330 |
| | 2.35 | 2.35 | 3.14 | 2.35 | 3.14 | 3.88 | 2.35 | 3.14 | 3.88 | 5.32 | 2.35 | 3.14 | 3.88 | 5.32 | 6.72 | 2.35 | 3.14 | 3.88 | 5.32 | 6.72 | 9.40 |
| 1 | 4,045 | 1,880 | 3,790 | 1,030 | 1,820 | 3,005 | 670 | 1,055 | 1,520 | 2,865 | 505 | 765 | 1,045 | 1,755 | 2,715 | 325 | 465 | 605 | 925 | 1. 285 | 2,155 |
| | 1.53 | 1.53 | 2.04 | 1.53 | 2.04 | 2.51 | 1.53 | 2.04 | 2.51 | 3.44 | 1.63 | 2.04 | 2.51 | 3.44 | 4.36 | 1.53 | 2.04 | 2.51 | 3.44 | 4.36 | 6.15 |
| 1-1/4 | 7,000 | 2,905 | 6,565 | 1,780 | 3,150 | 5,200 | 1,155 | 1,830 | 2,635 | 4,960 | 875 | 1,320 | 1,810 | 3,035 | 4,695 | 560 | 800 | 1,050 | 1,600 | 2,225 | 3.735 |
| | 0.95 | 0.95 | 1.26 | .95 | 1.26 | 1.55 | .95 | 1.26 | 1.55 | 2.13 | 95 | 1.26 | 1.55 | 2.13 | 2.69 | 0.95 | 1.26 | 1.55 | 2.13 | 2.69 | 3.80 |
| 1-1/2 | 9,530 | 3,955 | 8,935 | 2,425 | 4,290 | 7.080 | 1,575 | 2,490 | 3,585 | 6,750 | 2,190 | 1,795 | 2,465 | 4,135 | 6,395 | 760 | 1,090 | 1,430 | 2,175 | 3,025 | 5,080 |
| | 0.73 | 0.73 | 0.97 | 0.73 | 0.97 | 1.20 | 0.73 | 0.97 | 1.20 | 1.64 | 0.73 | 0.97 | 1.20 | 1.64 | 2.07 | 0.73 | 0.97 | 1.20 | 1.64 | 2.07 | 2.93 |
| 2 | 15,710 | 6,525 | 14,725 | 3,995 | 7,070 | 11,670 | 2,595 | 4,105 | 5,910 | 11,125 | 1,985 | 2.960 | 4,060 | 6,810 | 10,540 | 1,255 | 1,800 | 2,355 | 3,585 | 4,990 | 8,375 |
| | 0.48 | 0.48 | 0.64 | 0.48 | 0.64 | 0.79 | 0.48 | 0.64 | 0.79 | 1.08 | 0.48 | 0.64 | 0.79 | 1.08 | 1.37 | 0.48 | 0.64 | 0.79 | 1.08 | 1.37 | 1.93 |
| 2-1/2 | 22,415 | 9,305 | 21,005 | 5,700 | 10,085 | 16,650 | 3,705 | 5,855 | 8,430 | 15,875 | 2,800 | 4,225 | 5,795 | 9,720 | 15,035 | 1,790 | 2,565 | 3,380 | 5,115 | 7,120 | 11,950 |
| | 0.36 | 0.36 | 0.48 | 0.36 | 0.48 | 0.59 | 0.36 | 0.48 | 0.69 | 0.81 | 0.36 | 0.48 | 0.59 | 0.81 | 1.03 | 0.36 | 0.48 | 0.59 | 0.91 | 1.03 | 1.45 |
| 3 | 34,610 | 14,370 | 32,435 | 8,800 | 15,570 | 25,710 | 5,720 | 9,045 | 13,020 | 24 515 | 4,325 | 6,525 | 8,950 | 15,005 | 23,220 | 2,765 | 3,965 | 5.185 | 7,900 | 10,990 | 18,450 |
| | 0.26 | 0.26 | 0.34 | 0.26 | 0.34 | 0.42 | 0.26 | 0.34 | 0.42 | 0.58 | 0.26 | 0.34 | 0.42 | 0.58 | 0.73 | 0.26 | 0.34 | 0.42 | 0.58 | 0.73 | 1.03 |
| 3-1/2 | 46,285 | 19,220 | 43,380 | 11,765 | 20,825 | 34,385 | 7,650 | 12,095 | 17,410 | 32,785 | 5,785 | 6,725 | 11,970 | 20,070 | 31,050 | 3,695 | 5,300 | 6,940 | 10,565 | 14,700 | 24,675 |
| | 0.21 | 0.21 | 0.27 | 0.21 | 0.27 | 0.34 | 0.21 | 0.27 | 0.34 | 0.46 | 0.21 | 0.27 | 0.34 | 0.46 | 0.59 | 0.21 | 0.27 | 0.314 | 0.46 | 0.59 | 0.83 |
| 4 | 59,595 | 24,745 | 55,855 | 15,150 | 26,815 | 44,275 | 9,850 | 15,575 | 22,415 | 42,210 | 7,450 | 11,235 | 15,410 | 25.840 | 39,960 | 4,780 | 6,825 | 8,935 | 13,600 | 18,925 | 31,770 |
| | 0.17 | 0.17 | 0.23 | 0.17 | 0.23 | 0.28 | 0.17 | 0.23 | 0.28 | 0.38 | 0.17 | 0.23 | 0.28 | 0.38 | 0.49 | 0.17 | 0.23 | 0.28 | 0.36 | 0.49 | 0.25 |
| 5 | 93,655 | 38,890 | 87,780 | 23,810 | 42,140 | 69,580 | 15,480 | 24,475 | 35,230 | 66,335 | 11,705 | 17.660 | 24.220 | 40,610 | 62,830 | 7,475 | 10,725 | 14,040 | 21,375 | 29,745 | 49,930 |
| | 0.12 | 0.12 | 0.16 | 0.12 | 0.16 | 0.20 | 0.12 | 0.16 | 0.20 | 0.05 | 0.12 | 0.16 | 0.20 | 0.05 | 0.17 | 0.12 | 0.16 | 0.20 | 0.05 | 0.17 | 0.11 |
| 6 | 135,245 | 58,160 | 126,760 | 34,385 | 60,855 | 100,480 | 22,350 | 35,345 | 50,875 | 95,795 | 16,905 | 25,500 | 34,975 | 58,645 | 90,735 | 10,800 | 15,490 | 20,270 | 30,865 | 42,950 | 72,105 |
| | 0.10 | 0.10 | 0.13 | 0.10 | 0.13 | 0.04 | 0.10 | 0.13 | 0.04 | 0.05 | 0.10 | 0.13 | 0.04 | 0.05 | 0.01 | 0.10 | 0.13 | 0.04 | 0.05 | 0.01 | 0.01 |
| 8 | 234,195 | 97,245 | 219,505 | 59,540 | 105,380 | 173,995 | 38,705 | 61,205 | 88,095 | 165,880 | 29,270 | 44,160 | 60,565 | 101,650 | 157,115 | 18,700 | 26,820 | 35,105 | 53,450 | 74,175 | 124,855 |
| | 0.02 | 0.02 | 0.02 | 0.02 | 0.02 | 0.01 | 0.02 | 0.02 | 0.01 | 0.01 | 0.02 | 0.02 | 0.01 | 0.01 | 0.01 | 0.02 | 0.02 | 0.01 | 0.01 | 0.01 | 0.01 |

[a]Return-line Capacity (lb/hr) with Pressure Drop (psi) for 100 ft of Pipe at a Velocity of 5000 ft/min.

Table 6.12 shows a typical pressure-capacity table extracted from a manufacturer's catalog. The use of such a table can be illustrated with the following example.

**Example:** A steamjacketed platen press in a plastic lamination operation uses about 500 lb/hr of 30-psig steam in normal operation. A 100-ft run of 1-in. pipe returns condensate to a receiver pressurized to 5 psig: the receiver is located 15 ft above the level of the trap. From the capacity differential pressure specifications in Table 6.12, select a suitable trap for this application.

Using a factor of safety of 3 from Table 6.10, a trap capable of handling 3 × 500 = 1500 lb/hr of condensate will be selected.

To determine the system back pressure, add the receiver pressure, the piping pressure drop, and the hydraulic head due to the elevation of the receiver.

Entering Table 6.11 at 30-psig supply pressure and 5-psig return pressure, the pipe pressure drop for a 1-in. pipe is just slightly over 2 psi at a condensate rate somewhat higher than our 1500 lb/hr; a 2-psi pressure drop is a reasonable estimate.

The hydraulic head due to the 15-ft riser is 2 psi/ft × 15 = 7.5 psi. The total back pressure is, therefore,

## Table 6.12 Typical Pressure-Capacity Specifications for Steam Traps

| | Capacities (lb/hr) | | | |
|---|---|---|---|---|
| Model: | A | B | C | D |
| **Differential pressure (psi)** | | | | |
| 5 | 450 | 830 | 1600 | 2900 |
| 10 | 560 | 950 | 1900 | 3500 |
| 15 | 640 | 1060 | 2100 | 3900 |
| 20 | 680 | 880 | 1800 | 3500 |
| 25 | 460 | 950 | 1900 | 3800 |
| 30 | 500 | 1000 | 2050 | 4000 |
| 40 | 550 | 770 | 1700 | 3800 |
| 50 | 580 | 840 | 1 900 | 4100 |
| 60 | 635 | 900 | 2000 | 4400 |
| 70 | 660 | 950 | 2200 | 3800 |
| 80 | 690 | 800 | 1650 | 4000 |
| 100 | 640 | 860 | 1800 | 3600 |
| 125 | 680 | 950 | 2000 | 3900 |
| 150 | 570 | 810 | 1500 | 3500 |
| 200 | — | 860 | 1600 | 3200 |
| 250 | — | 760 | 1300 | 3500 |
| 300 | — | 510 | 1400 | 2700 |
| 400 | — | 590 | 1120 | 3100 |
| 450 | — | — | 1200 | 3200 |

5 psi (receiver) + 7.5 psi (riser) + 2 psi (pipe) = 14.5 psi

or the differential pressure driving the condensate flow through the trap is 30 − 14.5 = 15.5 psi.

From Table 6.12 we see that a Model C trap will handle 2100 lb/hr at 15-psi differential pressure; this would then be the correct choice.

### 6.4.4 Maintaining Steam Traps for Efficient Operation

Steam traps can and do malfunction in two ways. They may stick in the closed position, causing condensate to back up into the steam system, or they may stick open, allowing live steam to discharge into the condensate system. The former type of malfunction is usually quickly detectable, since flooding of a process heater with condensate will usually so degrade its performance that the failure is soon evidence by a significant change in operating conditions. This type of failure can have disastrous effects on equipment by producing damaging water hammer and causing process streams to back up into other equipment. Because of these potential problems, steam traps are often designed to fail in the open position; for this reason, they are among the biggest energy wasters in an industrial plant. Broad experience in large process plants using thousands of steam traps has shown that, typically, from 15 to 60% of the traps in a plant may be blowing through, wasting enormous amounts of energy. Table 6.13 shows the cost of wasted 100-psig steam (typical of many process plant conditions) for leak diameters characteristic of steam trap orifices. At higher steam pressures, the leakage would be even greater; the loss rate does not go down in direct proportion at lower steam pressures but declines at a rate proportional to the square root of the pressure. For example, a 1/8-in. leak in a system at 60 psig, instead of the 100 psig shown in the table, would still waste

over 75% of the steam rate shown (the square root of 60/100). The cost of wasted steam far outweighs the cost of proper maintenance to repair the malfunctions, and comprehensive steam trap maintenance programs have proven to be among the most attractive energy conservation investments available in large process plants. Most types of steam traps can be repaired, and some have inexpensive replaceable elements for rapid turnaround.

#### Table 6.13 Annual Cost of Steam Leaks

| Leak Diameter (in.) | Steam Wasted per Month (lb)[a] | Cost per Month[b] | Cost per Year[b] |
|---|---|---|---|
| 1/16 | 13,300 | $40 | $480 |
| 1/8 | 52,200 | 156 | 1,890 |
| 1/4 | 209,000 | 626 | 7,800 |
| 1/2 | 833,000 | 2,500 | 30,000 |

[a]Based on 100-psig differential pressure across the orifice.
[b]Based on steam value of $3/1000 lb. Cost will scale in direct proportion for other steam values.

A major problem facing the energy conservation manager is diagnosis of open traps. The fact that a trap is blowing through can often be detected by a rise in temperature at the condensate receiver, and it is quite easy to monitor this simple parameter. There are also several direct methods for checking trap operation. Figure 6.15 shows the simplest approach for open condensate systems where traps drain directly to atmospheric pressure. In proper normal operation, a stream of condensate drains from the line together with a lazy cloud of flash steam, produced as the condensate throttles across the trap. When the trap is blowing through, a

#### Table 6.14 Operating Sounds of Various Types of Steam Traps

| Trap | Proper Operation | Malfunctioning |
|---|---|---|
| Disk type (impulse of thermodynamic) | Opening and snap-closing of disk several times per minute | Rapid chattering of disk as steam blows through |
| Mechanical type (bucket) | Cycling sound of the bucket as it opens and closes | Fails open—sound of steam blowing through<br>Fails closed—no sound |
| Thermostatic type | Sound of periodic discharge if medium to high load; possibly no sound if light load; throttled discharge | Fails closed—no sound |

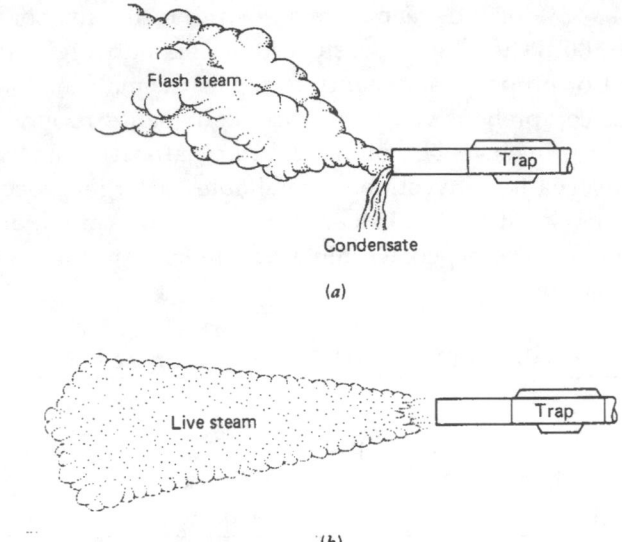

**Fig. 6.15 Visual observation of steam trap operation in open system, (a) Proper operation, (b) Improper operation.**

well-defined jet of live steam will issue from the line with either no condensate, or perhaps a condensate mist associated with steam condensation at the periphery of the jet.

Visual observation is less convenient in a closed condensate system, but can be utilized if a test valve is placed in the return line just downstream of the trap, as shown in Figure 6.16. This system has the added advantage that the test line may be used to actually measure condensate discharge rate as a check on equipment efficiency, as discussed earlier. An alternative in closed condensate return systems is to install a sight glass just downstream of the trap. These are relatively inexpensive and permit quick visual observation of trap operation without interfering with normal production.

Another approach to steam trap testing is to observe the sound of the trap during operation. Table 6.14 describes the sounds made by various types of traps

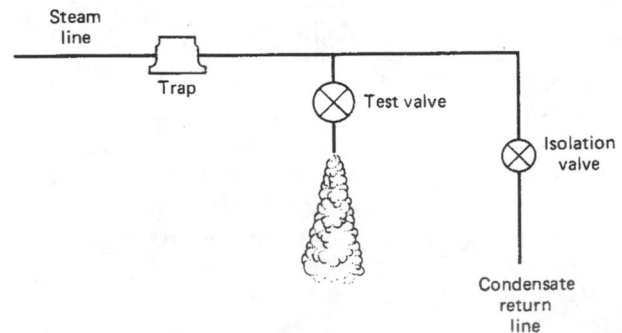

**Fig. 6.16 Visual observation of steam trap operation in closed systems.**

during normal and abnormal operation. This method is most effective with disk-type traps, although it can be used to some extent with the other types as well. An industrial stethoscope can be used to listen to the trap, although under many conditions, the characteristic sound will be masked by noises transmitted from other parts of the system. Ultrasonic detectors may be used effectively in such cases; these devices are, in effect, electronic stethoscopes with acoustic filtering to make them sensitive to sound and vibration only in the very high frequency range. Steam blowing through a trap emits a very high-pitched sound, produced by intense turbulence at the trap orifice, as contrasted with the lower-pitched and lower-intensity sound of liquid flowing through. Ultrasonic methods can, therefore, give a more reliable measure of steam trap performance than conventional "listening" devices.

A third approach to steam trap testing makes use of the drop in saturation temperature associated pressure drop across the trap. Condensate tends to cool rapidly in contact with uninsulated portions of the return line, accentuating the temperature difference. If the temperature on each side of the trap is measured, a sharp temperature drop should be evident. Table 6.15 shows typical temperatures that can be expected on the condensate side for various condensate pressures. In practice, the temperature drop method can be rather uncertain, because of the range of temperatures the condensate may exhibit and because, in blowing through a stuck-open trap, live steam will, itself, undergo some temperature drop. For example, 85-psig saturated steam blowing through an orifice to 15 psig will drop from 328°F to about 300°F, and may then cool further by radiation and convection from uninsulated surfaces. From Table 6.15, the expected condensate-side temperature is about 215 to 238°F for this pressure. Thus although the difference is still substantial, misinterpretation is possible, particularly if accurate measurements of the steam and condensate pressures on each side of the trap are not available.

The most successful programs of steam trap diagnosis utilize a combination of these methods, coupled with a regular maintenance program, to assure that traps are kept in proper operating condition.

This section has discussed the reasons why good steam trap performance can be crucial to successful energy conservation in steam systems. Traps must be properly selected and installed for the given service and appropriately sized to assure efficient removal of condensate and gases. Once in service, expenditures for regular monitoring and maintenance easily pay for themselves in fuel savings.

**Table 6.15 Typical Pipe Surface Temperatures for Various Operating Pressures**

| Operating Pressure (psig) | Typical Line Temperatures (°F) |
|---|---|
| 0 | 190-210 |
| 15 | 215-238 |
| 45 | 248-278 |
| 115 | 295-330 |
| 135 | 304-340 |
| 450 | 395-437 |

## 6.5 CONDENSATE RECOVERY

Condensate from steam systems is wasted, or at least used inefficiently, in many industrial operations. Yet improvements in the condensate system can offer the greatest savings of any of the measures discussed in this chapter. In this section methods are presented for estimating the potential energy and mass savings achievable through good condensate recovery, and the considerations involved in condensate recovery system design are discussed. It is not possible in a brief survey to provide a comprehensive guide to the detailed design of such systems. Condensate return systems can, in fact, be quite complex, and proper design usually requires a careful engineering analysis. The energy manager can, however, define the type of system best suited to the requirements and determine whether sufficient justification exists for a comprehensive design study.

### 6.5.1 Estimation of Heat and Mass Losses in Condensate Systems

The saturated liquid condensate produced when steam condenses on a heating surface still retains a significant fraction of the energy contained in the steam itself. Referring to Table 6.2, for example, it is seen that at a pressure of about 80 psig, each pound of saturated liquid contains about 295 Btu, or nearly 25% of the original energy contained in the steam at the same pressure. In some plants, this condensate is simply discharged to a wastewater system, which is wasteful not only of energy, but also of water and the expense of boiler feedwater treatment. Even if condensate is returned to an atmospheric pressure receiver, a considerable fraction of it is lost in the form of flash steam. Table 6.16 shows the percent of condensate loss due to flashing from systems at the given steam pressure to a flash tank at a lower pressure. For example, if in the 80-psig system discussed above, condensate is returned to a vented receiver instead of discharging it to a drain, nearly 12% is vented to the atmosphere as flash steam. Thus about

**Table 6.16 Percent of Mass Converted to Flash Steam in a Flash Tank**

| Steam Pressure (psig) | Flash Tank Pressure (psig) | | | | | | | | | | |
|---|---|---|---|---|---|---|---|---|---|---|---|
| | 0 | 2 | 5 | 10 | 15 | 20 | 30 | 40 | 60 | 80 | 100 |
| 5 | 1.7 | 1.0 | 0 | | | | | | | | |
| 10 | 2.9 | 2.2 | 1.4 | 0 | | | | | | | |
| 15 | 4.0 | 3.2 | 2.4 | 1.1 | 0 | | | | | | |
| 20 | 4.9 | 4.2 | 3.4 | 2.1 | 1.1 | 0 | | | | | |
| 30 | 6.5 | 5.8 | 5.0 | 3.8 | 2.6 | 1.7 | 0 | | | | |
| 40 | 7.8 | 7.1 | 6.4 | 5.1 | 4.0 | 3.1 | 1.3 | 0 | | | |
| 60 | 10.0 | 9.3 | 8.6 | 7.3 | 6.3 | 5.4 | 3.6 | 2.2 | 0 | | |
| 80 | 11.7 | 11.1 | 10.3 | 9.0 | 8.1 | 7.1 | 5.5 | 4.0 | 1.9 | 0 | |
| 100 | 13.3 | 12.6 | 11.8 | 10.6 | 9.7 | 8.8 | 7.0 | 5.7 | 3.5 | 1.7 | 0 |
| 125 | 14.8 | 14.2 | 13.4 | 12.2 | 11.3 | 10.3 | 8.6 | 7.4 | 5.2 | 3.4 | 1.8 |
| 160 | 16.8 | 16.2 | 15.4 | 14.1 | 13.2 | 12.4 | 10.6 | 9.5 | 7.4 | 5.6 | 4.0 |
| 200 | 18.6 | 18.0 | 17.3 | 16.1 | 15.2 | 14.3 | 12.8 | 11.5 | 9.3 | 7.5 | 5.9 |
| 250 | 20.6 | 20.0 | 19.3 | 18.1 | 17.2 | 16.3 | 14.7 | 13.6 | 11.2 | 9.8 | 8.2 |
| 300 | 22.7 | 21.8 | 21.1 | 19.9 | 19.0 | 18.2 | 16.7 | 15.4 | 13.4 | 11.8 | 10.1 |
| 350 | 24.0 | 23.3 | 22.6 | 21.6 | 20.5 | 19.8 | 18.3 | 17.2 | 15.1 | 13.5 | 11.9 |
| 400 | 25.3 | 24.7 | 24.0 | 22.9 | 22.0 | 21.1 | 19.7 | 18.5 | 16.5 | 15.0 | 13.4 |

3% of the original steam energy (0.12 × 0.25) goes up the vent pipe. The table shows that this loss could be reduced by half by operating the flash tank at a pressure of 30 psig, providing a low-temperature steam source for potential use in other parts of the process. Flash steam recovery is discussed later in more detail.

Even when condensate is fully recovered using one of the methods to be described below, heat losses can still occur from uninsulated or poorly insulated return lines. These losses can be recovered very cost effectively by the proper application of thermal insulation as discussed in Chapter 15.

### 6.5.2 Methods of Condensate Heat Recovery

Several options are available for recovery of condensate, ranging in cost and complexity from simple and inexpensive to elaborate and costly. The choice of which option is best depends on the amount of condensate to be recovered, other uses for its energy, and the potential cost savings relative to other possible investments.

The simplest system, which can be utilized if condensate is presently being discharged, is the installation of a vented flash tank which collects condensate from various points of formation and cools it sufficiently to allow it to be delivered back to the boiler feed tank. Figure 6.17 schematically illustrates such a system. It consists of a series of collection lines tying the points of condensate generation to the flash tank, which allows the liquid to separate from the flash steam; the flash steam is vented to the atmosphere through an open pipe. Condensate may be gravity-drained through a strainer and a trap. To avoid further generation of flash steam, a cooling leg may be incorporated to cool the liquid below its saturation temperature.

Flash tanks must be sized to produce proper separation of the flash steam from the liquid. As condensate is flashed, steam will be generated rather violently, and as vapor bubbles burst at the surface, liquid may be entrained and carried out through the vent. This represents a nuisance, and in some cases a safety hazard, if the vent is located in proximity to personnel or equipment. Table 6.17 permits the estimation of flash tank size required for a given application. Although strictly speaking, flash tanks must be sized on the basis of volume, if a typical length to diameter of about 3:1 is assumed, flash tank dimensions can be represented as the product of diameter times length, which has the units of square feet (area), even though this particular product has no direct physical significance. Consider, for example, the sizing of a vented flash tank for collection of 80-psig condensate at a rate of about 3000 lb/hr. For

a flash tank pressure of 0 psig (atmospheric pressure), the diameter-length product is about 2.5 per 1000 lb. Therefore, a diameter times length of 7.5 ft² would be needed for this application. A tank 1.5 ft in diameter by 5 ft long would be satisfactory. Of course, for flash tanks as with other condensate equipment, conservative design would suggest the use of an appropriate safety factor.

As noted above, venting of flash steam to the atmosphere is a wasteful process, and if significant amounts of condensate are to be recovered, it may be desirable to attempt to utilize this flash steam. Figure 6.18 shows a modification of the simple flash tank system to accomplish this. Rather than venting to the atmosphere, the flash tank is pressurized and flash steam is piped to a low-pressure steam main, where it can be utilized for process purposes. From Table 6.17 it will be noted that the flash tank can be smaller in physical size at elevated pressure, although, of course, it must be properly designed for pressure containment. If in the example above the 80-psig condensate were flashed in a 15-psig tank, only about 2.7 ft² of diameter times length would be required. A tank 1 ft in diameter by 3 ft long could be utilized. Atmospheric vents are usually provided for automatic pressure relief and to allow manual venting if desired.

**Fig. 6.17 Flash tank vented to atmosphere.**

**Fig. 6.18 Pressurized flash tank discharging to low-pressure steam system.**

If pressurized flash steam is to be used, the cost of piping to set up a low-pressure steam system may be significant, particularly if the flash tank is remote from the potential low-pressure steam applications. Thus it is desirable to plan such a system to minimize these piping costs by generating the flash steam near its point of use. Figure 6.19 illustrates such an application. Here an air heater having four sections formerly utilized 100-psi steam in all sections. Because the temperature difference between the steam and the cold incoming air is larger than the difference at the exit end, the condensate load would be unevenly balanced among the four sections, with the heaviest load in the first section; lower-temperature steam could be utilized here. In the revised arrangement shown, 100-psi steam is used in the last three sections; condensate is drained to a 5-psi flash tank, where low-pressure steam is generated and piped to the first section, substantially reducing the overall steam load to the heater. Note that for backup purposes, a pressure-controlled reducing valve has been incorporated to supplement the low-pressure flash steam at light-load conditions. This example shows how flash steam can be used directly without an expensive piping system to distribute it. A similar approach could apply to adjacent pieces of equipment in a multiple-batch operation.

**Table 6.17 Flash Tank Sizing[a]**

| Steam Pressure (psig) | Flash Tank Pressure (psig) | | | | | | | | | | |
|---|---|---|---|---|---|---|---|---|---|---|---|
| | 0 | 2 | 5 | 10 | 15 | 20 | 30 | 40 | 60 | 80 | 100 |
| 400 | 5.41 | 4.70 | 3.89 | 3.01 | 2.44 | 2.03 | 1.49 | 1.15 | 0.77 | 0.56 | 0.42 |
| 350 | 5.14 | 4.45 | 3.66 | 2.84 | 2.28 | 1.91 | 1.38 | 1.07 | 0.70 | 0.51 | 0.37 |
| 300 | 4.86 | 4.15 | 3.42 | 2.62 | 2.11 | 1.75 | 1.26 | 0.96 | 0.62 | 0.44 | 0.31 |
| 250 | 4.41 | 3.82 | 3.12 | 2.39 | 1.91 | 1.56 | 1.11 | 0.85 | 0.52 | 0.37 | 0.25 |
| 200 | 3.98 | 3.40 | 2.80 | 2.12 | 1.68 | 1.37 | 0.97 | 0.72 | 0.43 | 0.28 | 0.18 |
| 175 | 3.75 | 3.20 | 2.61 | 1.95 | 1.57 | 1.26 | 0.87 | 0.64 | 0.38 | 0.23 | 0.15 |
| 160 | 3.60 | 3.08 | 2.50 | 1.86 | 1.46 | 1.19 | 0.80 | 0.59 | 0.34 | 0.21 | 0.12 |
| 150 | 3.48 | 2.98 | 2.41 | 1.80 | 1.40 | 1.14 | 0.77 | 0.56 | 0.31 | 0.19 | 0.10 |
| 140 | 3.36 | 2.86 | 2.31 | 1.72 | 1.35 | 1.08 | 0.72 | 0.52 | 0.29 | 0.16 | 0.08 |
| 130 | 3.24 | 2.76 | 2.23 | 1.65 | 1.29 | 1.02 | 0.67 | 0.49 | 0.26 | 0.14 | 0.07 |
| 120 | 3.12 | 2.65 | 2.15 | 1.57 | 1.22 | 0.97 | 0.64 | 0.44 | 0.23 | 0.12 | 0.04 |
| 110 | 2.99 | 2.52 | 2.05 | 1.50 | 1.15 | 0.91 | 0.58 | 0.40 | 0.20 | 0.09 | 0.02 |
| 100 | 2.85 | 2.41 | 1.92 | 1.40 | 1.07 | 0.85 | 0.53 | 0.36 | 0.16 | 0.06 | |
| 90 | 2.68 | 2.26 | 1.81 | 1.30 | 0.99 | 0.77 | 0.48 | 0.31 | 0.13 | 0.05 | |
| 80 | 2.52 | 2.12 | 1.67 | 1.18 | 0.90 | 0.68 | 0.42 | 0.25 | 0.09 | | |
| 70 | 2.34 | 1.95 | 1.55 | 1.08 | 0.81 | 0.61 | 0.35 | 0.20 | 0.04 | | |
| 60 | 2.14 | 1.77 | 1.39 | 0.96 | 0.70 | 0.52 | 0.27 | 0.14 | | | |
| 50 | 1.94 | 1.59 | 1.22 | 0.81 | 0.58 | 0.41 | 0.20 | 0.08 | | | |
| 40 | 1.68 | 1.36 | 1.02 | 0.67 | 0.44 | 0.30 | 0.11 | | | | |
| 30 | 1.40 | 1.10 | 0.81 | 0.50 | 0.29 | 0.16 | | | | | |
| 20 | 1.06 | 0.81 | 0.55 | 0.28 | 0.12 | | | | | | |
| 12 | 0.75 | 0.48 | 0.28 | | | | | | | | |
| 10 | 0.62 | 0.42 | 0.23 | | | | | | | | |

[a]Flash tank area (ft²) = diameter X length of horizontal tank for discharge of 1000 lb/hr of condensate.

Although the utilization of flash steam in a low-pressure system appears to offer an almost "free" energy source, its practical application involves a number of problems that must be carefully considered. These are all essentially economic in nature.

As mentioned above, the quantity of condensate and its pressure (thus yielding a given quantity of flash steam) must be sufficiently large to provide a significant amount of available energy at the desired pressure. System costs do not go up in simple proportion to capacity. Rather, there is a large initial cost for piping, installation of the flash tank, and other system components, and therefore the overall cost per unit of heat recovered becomes significantly less as the system becomes larger. The nature of the condensate-producing system itself is also important. For example, if condensate is produced at only two or three points from large steam users, the cost of the condensate collection system will be considerably less than that of a system in which there are many small users.

Another important consideration is the potential for application of the flash steam. The availability of 5000 lb/hr of 15-psig steam is meaningless unless there is a need for a heat source of this magnitude in the 250°F temperature range. Thus potential uses must be properly matched to the available supply. Flash steam is most effectively utilized when it can supplement an existing low-pressure steam supply rather than providing the sole source of heat to equipment. Not only must the total average quantity of flash steam match the needs of the process, but the time variations of source and user must be taken into account, since steam cannot be economically stored for use at a later time. Thus flash steam might not be a suitable heat source for sequential batch processes in which the number of operating units is small, such that significant fluctuations in steam demand exist.

When considering the possible conversion to low-pressure steam of an existing piece of equipment presently operating on high-pressure steam, it is important to recognize that steam pressure can have a significant effect on equipment operation. Since a reduction in steam pressure also means a reduction in temperature, a unit may not have adequate heating-surface area to provide the necessary heat capacity to the process at re-

**Fig. 6.19 Flash steam utilization within a process unit.**

duced pressure. Existing steam distribution piping may not be adequate, since steam is lower in density at low pressure than at high pressure. Typically, larger piping is required to transport the low-pressure vapor at acceptable velocities. Although one might expect that the heat losses from the pipe surface might be lower with low-pressure steam because of its lower temperature, in fact, this may not be the case if a large pipe (and hence larger surface area) is needed to handle the lower-pressure vapor. This requirement will also make insulation more expensive.

When flash steam is used in a piece of equipment, the resulting low-pressure condensate must still be returned to a receiver for delivery back to the boiler. Flash steam will again be produced if the receiver is vented, although somewhat less than in the flashing of high-pressure condensate. This flash steam and that produced from the flash tank condensate draining into the receiver will be lost unless some additional provision is made for its recovery, as shown in Figure 6.20. In this system, rather than venting to the atmosphere, the steam rises through a cold-water spray, which condenses it. This spray might be boiler makeup water, for example, and hence the energy of the flash steam is used for makeup preheat. Not only is the heat content of the flash steam saved, but its mass as well, reducing makeup-water requirements and saving the incremental costs of makeup-water treatment. This system has the added advantage that it produces a deaerating effect on the condensate and feedwater. If the cold spray is metered so as to produce a temperature in the tank above about 190°F, dissolved gases in the condensate and feedwater,

**Fig. 6.20  Flash steam recovery in spray tank.**

particularly oxygen and $CO_2$, will come out of solution, and since they are not condensed by the cold-water spray, will be released through the atmospheric vent. As with flash steam systems, this system (usually termed a "barometric condenser" or "spray deaerator") requires careful consideration to assure its proper application. The system must be compatible with the boiler feedwater system, and controls must be provided to coordinate boiler makeup demands with the condensate load.

An alternative approach to the barometric condenser is shown in Figure 6.21. In this system, condensate is cooled by passing it through a submerged coil in the flash tank before it is flashed. This reduces the amount of flash steam generated. Cold-water makeup (possibly boiler feedwater) is regulated by a temperature-controlled valve.

The systems described above have one feature in common. In all cases, the final condensate state is atmospheric pressure, which may be required to permit return of the condensate to the existing boiler-feedwater makeup tank. If condensate can be returned at elevated pressure, a number of advantages may be realized.

Figure 6.22 shows schematics of two pressurized condensate return systems. Condensate is returned, in some cases without the need for a steam trap, to a high-pressure receiver, which routes the condensate directly back to the boiler. The boiler makeup unit and/or deaerator feeds the boiler in parallel with the condensate return unit, and appropriate controls must be incorporated to coordinate the operation of the two units. Systems such as the one shown in Figure 6.22a are available for condensate pressures up to about 15 psig. For higher pressures, the unit can be used in conjunction with a flash tank, as shown in Figure 6.22b. This system would be suitable where an application for 15-psig steam is available. These elevated-pressure systems represent an attractive option in relatively low-

**Fig. 6.21  Flash tank with condensate precooling.**

pressure applications, such as steam-driven absorption chillers. When considering them, care must be exercised to assure that dissolved gases in the boiler makeup are at suitable levels to avoid corrosion, since the natural deaeration effect of atmospheric venting is lost.

One of the key engineering considerations that must be accounted for in the design of all the systems described above is the problem of pumping high-temperature condensate. To understand the nature of the problem, it is necessary to introduce the concept of "net positive suction head" (NPSH) for a pump. This term means the amount of static fluid pressure that must be provided at the inlet side of the pump to assure that no vapor will be formed as the liquid passes through the pump mechanism, a phenomenon known as cavitation. As liquid moves into the pump inlet from an initially static condition, it accelerates and its pressure drops rather suddenly. If the liquid is at or near its saturation temperature in the stationary condition, this sudden drop in pressure will produce boiling and the generation of vapor bubbles. Vapor can also be generated by air coming out of solution at reduced pressure. These bubbles travel through the pump impeller, where the fluid pressure rises, causing the bubbles to collapse. The inrush of liquid into the vapor space produces an impact on the impeller surface which can have an effect comparable to

**Fig. 6.22 Pressurized condensate receiver systems, (a)** *Low-pressure process requirements*, **(b) Flash tank for use with high-pressure systems.**

**Fig. 6.23 Conventional and low-NPSH pumps, (a) Conventional centrifugal pump, (b) Low-NPSH centrifugal pump.**

sandblasting. Clearly, this is deleterious to the impeller and can cause rapid wear. Most equipment operators are familiar with the characteristic "grinding" sound of cavitation in pumps when air is advertently allowed to enter the system, and the same effect can occur due to steam generation in high-temperature condensate pumping.

To avoid cavitation, manufacturers specify a minimum pressure above saturation which must be maintained on the inlet side of the pump, such that, even when the pressure drops through the inlet port, saturation or deaeration conditions will not occur. This minimum-pressure requirement is termed the net positive suction head.

For condensate applications, special low-NPSH pumps have been designed. Figure 6.23 illustrates the difference between a conventional pump and a low-NPSH pump. In the conventional pump (Figure 6.23a) fluid on the suction side is drawn directly into the impeller where the rapid pressure drop occurs in the entry passage. In the low-NPSH pump (Figure 6.23b) a small

"preimpeller" provides an initial pressure boost to the incoming fluid, with relatively little drop in pressure at the entrance. This extra stage of pumping essentially provides a greater head to the entry passage of the main impeller, so that the system pressure at the suction side of the pump can be much closer to saturation conditions than that required for a conventional pump. Low-NPSH pumps are higher in price than conventional centrifugal pumps, but they can greatly simplify the problem of design for high-temperature condensate return, and can, in some cases, actually reduce overall system costs.

An alternative device for the pumping of condensate, called a "pumping trap," utilizes the pressure of the steam itself as the driving medium. Figure 6.24 illustrates the mechanism of a pumping trap. Condensate enters the inlet side and rises in the body until it activates a float-operated valve, which admits steam or compressed air into the chamber. A check valve prevents condensate from being pushed back through the inlet port, and another check valve allows the steam or

**Fig. 6.24 Pumping trap.**

air pressure to drive it out through the exit side. When the condensate level drops to a predetermined position, the steam or air valve is closed, allowing the pumping cycle to start again. Pumping traps have certain inherent advantages over electrically driven pumps for condensate return applications. They have no NPSH requirement, and hence can handle condensate at virtually any temperature without regard to pressure conditions. They are essentially self-regulating, since the condensate level itself determines when the trap pumps; thus no auxiliary electrical controls are required for the system. This has another advantage in environments where explosion-proofing is required, such as refineries and chemical plants. Electrical lines need not be run to the system, since it utilizes steam as a driving force, and the steam line is usually close at hand. Pumping traps operate more efficiently using compressed air, if available, because when steam is introduced to the chamber, some of it condenses before its pressure can drive the condensate out. Thus for the same pressure, more steam is required to give the same pumping capacity as compressed air. The disadvantages of pumping traps are their mechanical complexity, resulting in a susceptibility to maintenance problems, and the fact that they are available only in limited capacities.

The engineering of a complete condensate recovery system from scratch can be a rather involved process, requiring the design of tanks, plumbing, controls, and pumping devices. For large systems, there is little alternative to engineering and fabricating the system to the specific plant requirements. For small- to moderate-capacity applications, however, packaged systems incorporating all of the foregoing components are commercially available. Figures 6.25a and b show examples of two such systems, the former using electrically driven low-NPSH pumps, and the latter utilizing a pumping trap (the lower unit in the figure).

### 6.5.3 Overall Planning Considerations in Condensate Recovery Systems

As mentioned earlier, condensate recovery systems require careful engineering to assure that they are compatible with overall plant operations, that they are safe and reliable, and that they can actually achieve their energy-efficiency potential. In this section a few of the overall planning factors that should be considered are enumerated and discussed.

1. **Availability of Adequate Condensate Sources.** An energy audit should be performed to collect detailed information on the quantity of condensate available from all of the various steam using sources in the plant and the relevant data associated with these sources. Such

(a)

(b)

**Fig. 6.25 Pressurized condensate return systems.**

data include, for example, condensate pressure, quantity, and source location relative to other steam-using equipment and to the boiler room. Certain other information may also be pertinent. For example, if the condensate is contaminated by contact with other process streams, it may be unsuitable for recovery. It is not valid to assume that steam used in the process automatically results in recoverable condensate. Stripping steam used in refining of petroleum and in other separation processes is a good example.

2. **Survey of Possible Flash Steam Applications.** The process should be surveyed in detail to assess what applications presently using first-generation steam might be adaptable to the use of flash steam or to heat recovered from condensate. Temperatures and typical heat loads are necessary but not sufficient. Heat-transfer characteristics of the equipment itself may be important. A skilled heat-transfer engineer, given the present operating characteristics of a process heater can make reasonable estimates of that heater's capability to operate at a lower pressure.

3. **Analysis of Condensate and Boiler Feedwater Chemistry.** To assure that conditions in the boiler are kept in a satisfactory state to avoid scaling and corrosion, a change in feedwater treatment may be required when condensate is recovered and recycled. Water samples from the present condensate drain, from the boiler blow-down, from the incoming water source, and from the outlet of the present feedwater treatment system should be obtained. With this information, a water treatment specialist can analyze the overall water chemical balance and assure that the treatment system is properly configured to maintain good boiler-water conditions.

4. **Piping Systems.** A layout of the present steam and condensate piping system is needed to assess the need for new piping and the adequacy of existing runs. This permits the designer to select the best locations for flash tanks to minimize the need for extensive new piping.

5. **Economic Data. The bottom line on any energy recovery project**, including condensate recovery, is its profitability. In order to analyze the profitability of the system, it is necessary to estimate the quantity of heat recovered, converted into its equivalent fuel usage, the costs and cost savings associated with the water treatment system, and savings in water cost. In addition to the basic capital and installation costs of the condensate

recovery equipment, there may be additional costs associated with modification of existing equipment to make it suitable for flash steam utilization, and there will almost certainly be a cost of lost production during installation and checkout of the new system.

## 6.6 SUMMARY

This chapter has discussed a number of considerations in effecting energy conservation in industrial and commercial steam systems. Good energy management begins by improving the operation of existing systems, and then progresses to evaluation of system modifications to maximize energy efficiency. Methods and data have been presented to assist the energy manager in estimating the potential for savings by improving steam system operations and by implementing system design changes. As with all industrial and commercial projects, expenditure of capital must be justified by reductions in operating expense. With energy costs rising at a rate substantially above the rate of increase of most equipment, it is clear that energy conservation will continue to provide ever more attractive investment opportunities.

## BIBLIOGRAPHY

Armstrong Machine Works, "Steam Conservation Guidelines for Condensate Drainage," *Bulletin M101*, Three Rivers, Mich., 1976.

Babcock and Wilcox Corporation, *Steam: Its Generation and Use*, 38th ed., New York, 1972.

Department of Commerce, National Bureau of Standards, *Energy Conservation Program Guide, for Industry and Commerce*, NBS Handbook 115, Washington, D.C., 1974.

Department of Energy, United Kingdom, *Utilization of Steam, for Process and Heating*, Fuel Efficiency Booklet 3, London, 1977.

Department of Energy, United Kingdom, *Flash Steam and Vapour Recovery*, Fuel Efficiency Booklet 6, London, 1977.

Department of Energy, United Kingdom, *How to Make the Best Use of Condensate*, Fuel Efficiency Booklet 9, London, 1978.

IDA, MASAHIRO, "Utilization of Low Pressure Steam and Condensate Recovery," *TLV Technical Information Bulletin* 5123, TLV Company Ltd., Tokyo, 1975.

ISLES, DAVID L., "Maintaining Steam Traps for Best Efficiency," *Hydrocarbon Processing*, Vol. 56, No. 1 (1977).

JESIONOWSKI, J.M., W.E. DANEKIND, and W.R. RAGER, "Steam Leak Evaluation Technique: Quick and Easy to Use," Oil and Gas Journal, Vol. 74, No. 2 (1976).

KERN, DONALD Q., *Process Heat Transfer*, McGraw-Hill, New York, 1950.

MAY, DONALD, "First Steps in Cutting Steam Costs," *Chemical Engineering*, Vol. 80, No. 26 (1973).

MONROE, E.S., JR., "Select the Right Steam Trap," *Chemical Engineering*, Vol. 83, No. 1 (1976).

MOWER, JOSEPH, "Conserving Energy in Steam Systems," *Plant Engineering*, Vol. 32, No. 8 (1978).

Sarco Company, Inc., *Hook-Up Designs for Steam and Fluid Systems*, 5th ed., Allentown, Pa., 1975.

# CHAPTER 7

# COGENERATION AND DISTRIBUTED GENERATION

**JORGE B. WONG**
General Electric Company
Evansville, Indiana

**JOHN M. KOVACIK**
General Electric Co.
Retired

## 7.1 INTRODUCTION

Cogeneration is broadly defined as the coincident or simultaneous generation of combined heat and power (CHP). In a true cogeneration system a "significant" portion of the generated or recovered heat must be used in a thermal process as steam, hot air, hot water, etc. The cogenerated power is typically in the form of mechanical or electrical energy.

The power may be totally used in the industrial plant that serves as the "host" of the cogeneration system, or may be partially/totally exported to a utility grid. Figure 7.1 illustrates the potential of saving in primary energy when separate generation of heat and electrical energy is substituted by a cogeneration system. This chapter presents an overview of current design, analysis and evaluation procedures.

The combined generation of useful heat and power is not a new concept. The U.S. Department of Energy (1978) reported that in the early 1900s, 58% of the total power produced by on-site industrial plants was cogenerated. However, Pohmeros (1981) stated that by 1950, on-site CHP generation accounted for only 15 percent of total U.S. electrical generation; and by 1974, this figure had dropped to about 5 percent. In Europe, the experience has been very different. The US Department of Energy, US-DOE (1978), reported that "historically, industrial cogeneration has been five to six times more common in some parts of Europe than in the U.S." In 1972, "16% of West Germany's total power was cogenerated by industries; in Italy, 18%; in France, 16%; and in the Netherlands, 10%."

Since the promulgation of the Public Utilities Regulatory Policies Act of 1978 (PURPA), however, U.S. cogeneration design, operation and marketing activities have dramatically increased, and have received a much larger attention from industry, government and academy. As a result of this incentive, newer technologies such as various combined cycles have achieved industrial maturity. Also, new or improved cogeneration-related processes and equipment have been developed and are actively marketed; e.g. heat recovery steam generator and duct burners, gas-engine driven chillers, direct and indirect fired two-stage absorption chiller-heaters, etc.

The impact of cogeneration in the U.S., both as an energy conservation measure and as a means to contribute to the overall electrical power generation capacity cannot be overemphasized. SFA Pacific Inc. (1990) estimated that non-utility generators (NUGs) produce 6-7% of the power generated in the US. And that NUGs will account for about half of the 90,000 MW that will be added during the next decade. More recently, Makansi (1991) reported that around 40,000 MW of independently produced and/or cogenerated (IPP/COGEN) power was put on line since the establishment of PURPA. This constitutes a significant part of new capacity. Makansi also reports that another 60,000 MW of IPP/COGEN power is in construction and in development. This trend is likely to increase since CHP is a supplementary way of increasing the existing U.S. power generation capacity. Thus, distributed generation (DG) is arising as a serious CHP alternative (see Section 7.2.2.4).

PURPA is considered the foremost regulatory instrument in promoting cogeneration, IPPs and/or NUGs. Thus, to promote industrial energy efficiency and resource conservation through cogeneration, PURPA requires electric utilities to purchase cogenerated power at fair rates to both, the utility and the generator. PURPA also orders utilities to provide supplementary and back-up power to qualifying facilities QFs).

The Promulgation of the Energy Policy Act of 1992 EPAct (1992) have brought a "deregulated" and competitive structure to the power industry. EPAct-1992 establishes the framework that allows the possibility of "Retail Wheeling" (anyone can purchase power from any generator, utility or otherwise). This is in addition to the existing Wholesale Wheeling or electricity trade that normally occurs among U.S. utilities. EPAct-1992's main objective is to stimulate competition in the electricity generation sector to and to reduce electricity costs. The impact of EPAct 1992, deregulation, and retailed wheeling on cogeneration can be

**A) Separate generation of heat and electricity in boiler and conventional condensation power plant.**

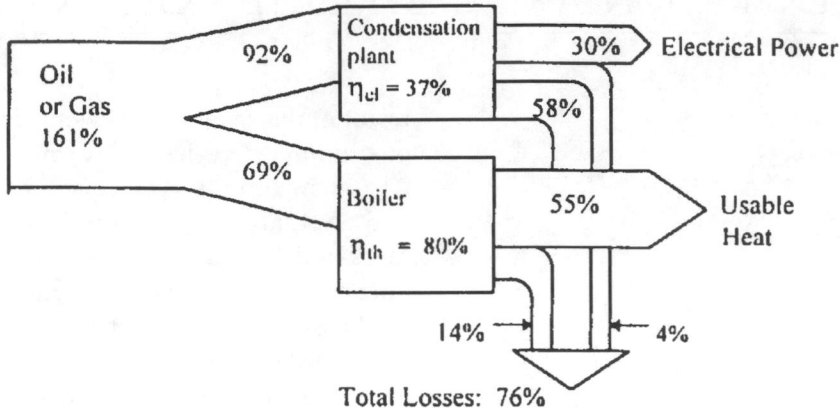

**B) Cogeneration of heat and power.**

Saving in primary energy = (1 - 100% / 161%) - 100 = 38%

**C) Integration of a cogeneration system and an industrial process.**

Figure 7.1 Potential of saving in primary energy when separate generation of heat and electrical energy is substituted by an industrial cogeneration system.

tremendous, since it creates a new class of generating facility called Exempt Wholesale Generators (EWGs). Economies of scale and scope—i.e. the natural higher efficiency of larger CHP plants—could trigger the grow of cogeneration based EWGs. In addition, EPAct-1992 will open the transmission grid to utilities and NUGs by ordering FERC to allow open transmission access to all approved EWGs. However, specific retail wheeling rules have not been legislated at the time of this writing. So far, only a few retail wheeling cases have been started. Thus, the actual impact of deregulation on cogeneration is considered to be uncertain. Nevertheless, some consider that retail wheeling/deregulation and cogeneration could have a synergistic effect and should be able to support each other.

## 7.2.  COGENERATION SYSTEM DESIGN AND ANALYSIS

The process of designing and evaluating a cogeneration system has so many factors that it has been compared to the Rubik's cube—the ingenious game to arrange a multi-colored cube. The change in one of the cube faces will likely affect some other face. The most important faces are: fuel security, regulations, economics, technology, contract negotiation and financing.

### 7.2.1 General Considerations and Definitions

Kovacik (1982) indicates that although cogeneration should be evaluated as a part of any energy management plan, the main prerequisite is that a plant shows a significant and concurrent demand for heat and power. Once this scenario is identified, he states that cogeneration systems can be explored under the following circumstances:

1.   Development of new facilities

2.   Major expansions to existing facilities which increase process heat demands and/or process energy rejection.

3.   When old process and/or power plant equipment is being replaced, offering the opportunity to upgrade the energy supply system.

The following terms and definitions are regularly used in the discussion of CHP systems.

**Industrial Plant:** the facility requiring process heat and electric and/or shaft power. It can be a process plant, a manufacturing facility, a college campus, etc. See Figure 7.1c.

**Process Heat (PH):** the thermal energy required in the industrial plant. This energy is supplied as steam, hot water, hot air, etc.

**Process Returns (PR):** the fluid returned from the industrial plant to the cogeneration system. For systems where the process heat is supplied as steam, the process returns are condensate.

**Net Heat to Process (NHP):** the difference between the thermal energy supplied to the industrial plant and the energy returned to the cogeneration system. Thus, NHP = PH – PR. The NHP may or may not be equal to the actual process heat demand (PH).

**Plant Power Demand (PPD):** the electrical power or load demanded (kW or MW) by the industrial plant. It includes the power required in for industrial processes, air-conditioning, lighting, etc.

**Heat/Power Ratio (H/P):** the heat-to-power ratio of the industrial plant (demand), or the rated heat-to-power ratio of the cogeneration system or cycle (capacity).

**Topping Cycles:** thermal cycles where power is produced prior to the delivery of heat to the industrial plant. One example is the case of heat recovered from a diesel-engine generator to produce steam and hot water. Figure 7.2 shows a diesel engine topping cycle.

**Bottoming Cycle:** power production from the recovery of heat that would "normally" be rejected to a heat sink. Examples include the generation of power using the heat from various exothermic chemical processes and the heat rejected from kilns used in various industries. Figure 7.3 illustrates a bottoming cycle.

**Combined Cycle:** this is a combination of the two cycles described above. Power is produced in a topping cycle—typically a gas-turbine generator. Then, heat exhausted from the turbine is used to produce steam; which is subsequently expanded in a steam turbine to generate more electric or shaft power. Steam can also be extracted from the cycle to be used as process heat. Figure 7.4 depicts a combined cycle.

Figure 7.2 Diesel engine topping cycle.

Figure 7.3 Steam turbine bottoming cycle.

Figure 7.4 Combined cycle: gas-turbine/generator set,
waste heat recovery boiler and steam-turbine/generator set.

**Prime Mover:** a unit of the CHP system that generates electric or shaft power. Typically, it is a gas turbine generator, a steam turbine drive or a diesel-engine generator.

### 7.2.2 Basic Cogeneration Systems

Most cogeneration systems are based on prime movers such as steam turbines, gas turbines, internal combustion engines and packaged cogeneration. Table 7.1 shows typical performance data for various cogeneration systems. Figures in this table (and in this chapter) are based on higher heating values, unless stated otherwise.

### 7.2.2.1 Steam Turbine Systems

Steam turbines are currently used as prime movers in topping, bottoming and combined cycles. There are many types of steam turbines to accommodate various heat/power ratios and loads. For limited expansion (pressure drop) and smaller loads (<4000 HP) lower cost single stage backpressure turbines are used. When several pressure levels are required (and usually for larger loads) multi-stage condensing and non-condensing turbines with induction and/or extraction of steam at intermediate pressures are generally used. Fig. 7.5 shows a variety of condensing and non-condensing turbines.

Four factors must be examined to assure that the maximum amount of power from a CHP steam plant is economically generated based on the process heat required. These factors are: (1) prime-mover size, (2) initial steam conditions, (3) process pressure levels, and (4) feedwater heating cycle.

**1. Prime-Mover Type and Size.** Process heat and plant electric requirements define the type and size of the steam generator. The type of CHP system and its corresponding prime mover are selected by matching the CHP system heat output to the process heat load.

If process heat demands are such that the plant power requirements can be satisfied by cogenerated power, then the size of the prime mover is selected to meet or exceed the "peak" power demand. However, cogeneration may supply only a portion of the total plant power needs. The balance has to be imported through a utility tie. In isolated plants, the balance is generated by

#### Table 7.1 Basic cogeneration systems.

| Cogeneration Systems | Unit Elec. Capacity (kW) | Heat Rate[2] (Btu/kWh) | Electrical[3] Efficiency (%) | Thermal Efficiency (%) | Total Efficiency (%) | Exhaust Temperature (°F) | 125-psig Steam Generation (lbs/hr) |
|---|---|---|---|---|---|---|---|
| Small reciprocating Gas Engines | 1-500 | 25,000 to 10,000 | 14-34 | 52 | 66-86 | 600-1200 | 0-200[1] |
| Large reciprocating Gas Engines | 500-17,000 | 13,000 to 9,500 | 26-36 | 52 | 78-88 | 600-1200 | 100-10,000[1] |
| Diesel Engines | 100-4,000 9,500 | 14,000 to | 24-36 | 50 | 74-86 | 700-1500 | 100-1500[1] |
| Industrial Gas Turbines | 800-10,000 | 14,000 to 11,000 | 24-31 | 50 | 74-81 | 800-1000 | 3,000-30,000 |
| Utility Size Gas Turbines | 10,000-150,000 | 13,000 to 9000 | 26-35 | 50 | 76-85 | 700-800 | 30,000 to 300,000 |
| Steam-Turbine Cycles | 5,000-200,000 | 30,000 to 10,000 | 10-35 | 28 | 38-63 | 350-1000 | 10,000 to 200,000 |

NOTES
[1]Hot water @ 250° is available at 10 times the flow of steam
[2]Heat rate is the fuel heat input (Btu—higher heating value) to the cycle per kWh of electrical output at design (full load) and ISO conditions (60°F ambient temperature and sea level operation)
[3]The electrical generation efficiency in percent of a prime mover can be determined by the formula Efficiency = (3413/Heat Rate)100

Sources: Adapted from Limaye (1985) and Manufacturers Data.

**CONDENSING**

**STRAIGHT FLOW**
For continuous and standby power exhausting to a condenser.

**DUAL-FLOW OPPOSED EXHAUST**
For large steam flow exhausting to condenser to minimize blade stresses and optimize efficiency.

**BACK PRESSURE**

**STRAIGHT FLOW**
The most economic operation where the exhaust steam is used for heating and process purposes.

**NON-AUTOMATIC EXTRACTION**
To provide low flow, up to 15% of throttle flow for heating or process requirements.

**NON-AUTOMATIC EXTRACTION**
All the backpressure benefits plus additional higher pressure bleed for preheating.

**CONTROLLED EXTRACTION**
Used when there is a demand for power and low pressure process steam which may be less than the steam flow required to make the power.

**CONTROLLED EXTRACTION/ INDUCTION**
Where two low pressure steam headers are required.

**CONTROLLED INDUCTION**
Enables the user to produce power from two steam pressure levels with controls which favor the less expensive lower pressure steam flow.

**CONTROLLED EXTRACTION/ INDUCTION**
Used where there is a variable demand for low pressure process steam which at times results in excess low pressure steam available to do work in the turbine.

**Figure 7.5 Various multi-stage steam turbine systems.** *Courtesy of Coppus Engineering/Murray Turbomachinery Corporation.*

additional conventional units. This discussion assumes that both heat and power demands remain constant all times. Hence, the design problem becomes one of specifying two variables: (1) how much power should be cogenerated on-site and (2) how much power should be imported. Thus, given the technological, economical and legal constraints for a particular plant, and assuming the CHP system must be constructed at a minimum overall cost, it becomes a constrained optimization problem.

**2. Initial Steam Conditions.** Many industrial plants do not have adequate process steam demands to generate all the power required. Thus, it is important for the designer to examine those variables over which he has control so he can optimize the amount of power that can

be economically generated. One set of these variables are the initial steam conditions, i.e. the initial pressure and temperature of the steam generated. In general, an increase in initial pressure and/or temperature will increase the amount of energy available for power generation. But the prime mover construction and cost, and the heat demand impose economical limits for the initial steam conditions. Thus, higher initial steam conditions can be economically justified in industrial plants having relatively large process steam demands.

**3. Process Steam Pressure.** For a given set of initial steam conditions, lowering the exhaust pressure also increases the energy available for power generation. However, this pressure is limited—totally, with non-condensing turbines, or partially, with extraction turbines—by the maximum pressure required in the industrial process. For instance, in paper mills, various pressure levels are needed to satisfy various process temperature requirements.

**4. Feedwater Heating.** Feedwater heating through use of steam exhausted and/or extracted from a turbine increases the power that can be generated.

### Calculation of Steam Turbine Power

Given the initial steam conditions (psig, °F) and the exhaust saturated pressure (psig), Theoretical Steam Rates (TSR) specify the amount of steam heat input required to generate a kWh in an ideal turbine. The TSR is defined by

$$TSR \text{ (lb/kWh)} = \frac{3412 \text{ Btu/kWh}}{h_i - h_o \text{ Btu/lb}} \quad (7.1)$$

where $h_i - h_o$ is the difference in enthalpy from the initial steam conditions to the exhaust pressure based on an isentropic (ideal) expansion. These values can be obtained from steam tables or a Mollier chart*. However, they are conveniently tabulated by the American Society of Mechanical Engineers.

The TSR can be converted to the Actual Steam Rate (ASR)

$$ASR \text{ (lb/kWh)} = \frac{TSR}{\eta_{tg}} \quad (7.2)$$

where $\eta_{tg}$ is the turbine-generator overall efficiency, stated or specified at "design" or full-load conditions. Some of the factors that define the overall efficiency of a turbine-generator set are: the inlet volume flow, pressure ratio, speed, geometry of turbine staging, throttling losses, friction losses, generator losses and kinetic losses associated with the turbine exhaust. Most turbine manufacturers provide charts specifying either ASR or $\eta_g$ values. Once the ASR has been established, the net enthalpy of the steam supplied to process (NEP) can be calculated:

NEP (Btu/lb)
$$= Hi - 3500/ASR - Hc(x) - Hm(1-x) \quad (7.3)$$

where Hi = enthalpy at the turbine inlet conditions (Btu/lb)

3500 = conversion from heat to power (Btu/ kWh), including the effect of 2.6% radiation, mechanical and generator losses

Hc = enthalpy of condensate return (Btu/lb)

*See Appendix 1 at the end of the book, "Mollier Diagram for Steam."

Hm = enthalpy of make-up water (Btu/lb)

x = condensate flow fraction in boiler feed-water

(1–x) = make-up water flow fraction in boiler feed water

Hence, assuming a straight flow turbine (See Figure 7.5), the net heat to process (Btu/hr) defined in Section 7.2.1 can be obtained by multiplying equation 7.3 by the flow rate in lb/hr. The analysis of the overall cycle would require the replication of complete heat and mass balance calculations at part-load efficiencies. To expedite these computations, there are a number of commercially available software packages, which also produce mass/ heat balance tables. See Example 8 for a cogeneration software application.

### Selection of Smaller Single-stage Steam Turbines

There exist many applications for smaller units (condensing and noncondensing), especially in mechanical drives or auxiliaries (fans, pumps, etc.); but a typical application is the replacement (or bypass) of a pressure reducing valve (PRV) by a single-stage back-pressure turbine.

After obtaining the TSR from inlet and outlet steam conditions, Figure 7.6 helps in determining the approximate steam rate (ASR) for smaller (<3000HP) single-stage steam turbines. Figure 7.7 is a sample of size and speed ranges available from a turbine manufacturer. It should be noticed that there is an overlap of capacities (e.g. an ET-30 unit can operate in the ET-25 range and in a portion of the ET-15/ET-20 range). Thus, the graph can help in selecting a unit subject to variable steam flows and/or loads.

**Example 1.** A stream of 15,500 lb of saturated steam at 250 psig (406°F) is being expanded through a PRV to obtain process steam at 50 psig. Determine the potential for electricity generation if the steam is expanded using a single-stage back-pressure 360.0 RPM turbine-generator.

Data:

| | |
|---|---|
| Steam flow ($W_s$) | : 15,500 lb/hr |
| Inlet Steam | : 250 psig sat (264.7 psia), 406°F |
| Enthalpy, hi | : 1201.7 Btu/lb (from Mollier chart) |
| Outlet Steam | : 50 psig subcooled (9.67% moisture). |
| Enthalpy, ho | : 1090.8 Btu/lb (from Mollier chart) |
| Turbine Speed | : 3600 RPM |

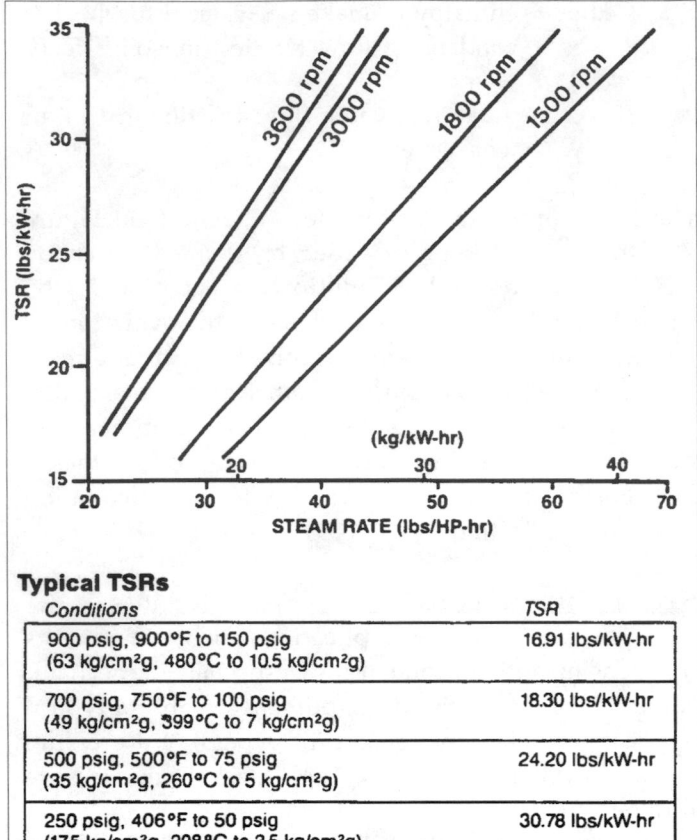

**Typical TSRs**

| Conditions | TSR |
|---|---|
| 900 psig, 900°F to 150 psig<br>(63 kg/cm²g, 480°C to 10.5 kg/cm²g) | 16.91 lbs/kW-hr |
| 700 psig, 750°F to 100 psig<br>(49 kg/cm²g, 399°C to 7 kg/cm²g) | 18.30 lbs/kW-hr |
| 500 psig, 500°F to 75 psig<br>(35 kg/cm²g, 260°C to 5 kg/cm²g) | 24.20 lbs/kW-hr |
| 250 psig, 406°F to 50 psig<br>(17.5 kg/cm²g, 208°C to 3.5 kg/cm²g) | 30.78 lbs/kW-hr |

**Figure 7.6 Estimation of steam rates for smaller (<3,000 HP) single-stage turbines. Courtesy Skinner Engine Co.**

a) Calculate TSR using equation 7.1

$$TSR = \frac{3412 \text{ Btu/kWh}}{h_i - h_o \text{ Btu/lb}}$$

$$= \frac{3412 \text{ Btu/kWh}}{1201.7 - 1090.8 \text{ Btu/lb}}$$

$$= 30.77 \text{ lb/kWh}$$

b) Obtain steam rate from Fig. 7.6, ASR = 38.5 lbs/HP-hr.

c) Calculate potential generation capacity (PGC):

$$PGC = \frac{W_s(0.746 \text{ kW/hp})}{ASR}$$

$$= \frac{(15,500 \text{ lb/hr}) (0.746 \text{ kW/hp})}{38.5 \text{ lbs/hp-hr}} = 300.3 \text{ kW}$$

Figure 7.7 shows that units ET-15 or ET-20 better match the required PGC. Next, the generator would

**Figure 7.7 Power/speed ranges for single stage turbines. Courtesy Skinner Engine Co.**

have to be sized according to a commercially available unit size, e.g. 300 kW.

Selection of Multi-Stage Steam Turbines

Multistage steam turbines provide more flexibility to match various pressure levels and variable flow rates in larger cogeneration applications. Figure 7.5 describes a variety of back-pressure and condensing multi-stage steam turbines.

For turbine selection, a Mollier diagram should be used to explore various multi-stage turbine alternatives. In general, a preliminary analysis should include the following:

— Approximation of actual steam rates.

— Defining number of stages.

— Estimation of stage pressure and temperatures.

— Calculation of full-load and part load steam rates.

— Estimation of induction and/or extraction pressures, temperatures and flow rates for various power outputs.

For larger units (>3000 kW), Figure 7.8 shows a chart to determine the approximate turbine efficiency when the power range, speed and steam conditions are known. Figure 7.9 gives steam rate correction factors for off-design loads and speeds. However, manufacturers of multistage turbines advise that stage selection and other thermodynamic parameters must be evaluated taking into account other important factors such as speed limitations, mechanical stresses, leakage and throttling losses, windage, bearing friction and reheat. Thus, after a preliminary evaluation, it is important to compare notes with the engineers of a turbine manufacturer.

**Example 2.** Steam flow rate must be estimated to design a heat recovery steam generator (a bottoming cycle). The steam is to be used for power generation and is to be expanded in a 5,000 RPM multistage condensing turbine to produce a maximum of 18,500 kW. Steam inlet conditions is 600 psig/750°F and exhaust pressure is 4" HGA (absolute). Additional data are given below.

Data (for a constant entropy steam expansion @ S = 1.61 Btu/lb/°R)

| | | |
|---|---|---|
| PGC | : | 8,500 kW |
| Inlet Steam: | : | 600 psig (615 psia), 750°F |

| | | |
|---|---|---|
| Enthalpy, $h_i$ | : | 1378.9 Btu/lb (from Mollier chart or steam tables) |
| Outlet Steam | : | 4" HG absolute. (2 psia) |
| Enthalpy, $h_o$ | : | 935.0 Btu/lb (from Mollier chart or ASME steam tables) |
| Turbine Speed | : | 5000 RPM |

a) Calculate TSR using equation 7.1 (TSR can also be obtained from ASME Tables or the Mollier chart):

$$TSE \ (lb/kWh) \ = \ \frac{3412 \ Btu/kWh}{h_i - h_o \ Btu/lb}$$

$$= \ \frac{3412 \ Btu/kWh}{1378.9 - 935.0 \ Btu/lb}$$

$$= \ 7.68 \ lb/kWh$$

b) Using the data above, Figure 7.8 gives $\eta_{tg} = 77\%$.

c) Combining equations 7.2 and 7.4 and solving for Ws, the total steam flow required is

$$W_s \ = \ PCG \times TSR/\eta_{tg}$$

$$= \ \frac{18,500 \ kW \times 7.68 \ lb/kWh}{0.77}$$

$$= \ 184,760 \ lb/hr$$

Figure 7.8 Approximate steam turbine efficiency chart for multistage steam turbines (>3,000 kW. *Courtesy Elliot Co.*

Figure 7.9 Steam turbine part-load/speed correction curves. *Courtesy Elliot Co.*

d)   Hence, the waste heat recovery steam generator should be able to generate about 185,000 lb/hr. A unit with a 200,000 lb/hr nominal capacity will likely be specified.

**Example 3.** Find the ASR and steam flow required when the turbine of Example 2 is operated at 14,800 kW and 4,500 RPM.

% Power variation = 14,800 kW/185,500 kW = 80%

% Speed variation = 4500 RPM/5000 RPM = 90%

From Figure 7.9 the power correction factor is 1.04 and the speed correction factor is 1.05. Then, the total correction factor is:

1.04 × 1.05 = 1.09.

Therefore, the part-load ASR is

$$= \frac{7.68 \text{ lb/kWh}}{0.77} \times 1.09 = 10.9 \text{ lb/kWh}$$

The steam flow required (@ 600 psig, 750°F) is

= 10.9 lb/kWh × 14,500 kW

= 158,050 lb/hr.

### 7.2.2.2 Gas Turbine Systems

Gas turbines are extensively applied in industrial plants. Two types of gas turbines are utilized: one is the lighter aircraft derivative turbine and the other is the heavier industrial gas turbine. Both industrial and aircraft engines have demonstrated excellent reliability/availability in the base load service. Due to the nature of the unit designs, aircraft derivative units usually have higher maintenance costs ($/kWh) than the industrial units.

Since gas turbines can burn a variety of liquid and gas fuels and run long times unattended, they are considered to be versatile and reliable. For a fixed capacity, they have the smallest relative foot-print (sq-ft per kW).

### Gas Turbine Based CHP Systems

Exhaust gases from gas turbines (from 600 to 1200°F) offer a large heat recovery potential. The exhaust has been used directly, as in drying processes. Topping cycles have also been developed by using the exhaust gases to generate process steam in heat recovery steam generators (HRSGs). Where larger power loads exist,

high pressure steam is generated to be subsequently expanded in a steam turbine-generator; this constitutes the so called combined cycle (See Fig 7.4).

If the demand for steam and/or power is even higher, the exhaust gases are used (1) as preheated combustion air of a combustion process or (2) are additionally fired by a "duct burner" to increase their heat content and temperature.

Recent developments include combined cycle systems with steam injection or STIG—from the HRSG to the gas turbine (Cheng Cycle)-to augment and modulate the electrical output of the system. The Cheng Cycle allows the cogeneration system to handle a wider range of varying heat and power loads.

All these options present a greater degree of CHP generation flexibility, allowing a gas turbine system to match a wider variety of heat-to-power demand ratios and variable loads.

### Gas Turbine Ratings and Performance

There is a wide range of gas turbine sizes and drives. Available turbines have ratings that vary in discrete sizes from 50 kW to 160,000 kW. Kovacik (1982) and Hay (1988) list the following gas turbine data required for design and off-design conditions:

1.   **Unit Fuel Consumption-Output Characteristics.** These depend on the unit design and manufacturer. The actual specific fuel consumption or efficiency and output also depend on (a) ambient temperature, (b) pressure ratio and (c) part-load operation. Figure 7.11 shows performance data of a gas turbine. Vendors usually provide this kind of information.

2.   **Exhaust Flow Temperature.** This data item allows the development of the exhaust heat recovery system. The most common recovery system are HRSGs which are classified as unfired, supplementary fired and fired units. The amount of steam that can be generated in an unfired or supplementary fired HRSG can be estimated by the following relationship:

$$Ws = \frac{W_g C_p (T_i - T_3) \text{ e L f}}{h_{sh} - h_{sat}} \qquad (7.5)$$

where

$W_s$ =   steam flow rate
$W_g$ =   exhaust flow rate to HRSG
$C_p$ =   specific heat of products of combustion
$T_1$ =   gas temperature-after burner, if applicable

$T_3$ = saturation temperature in steam drum

L = a factor to account radiation and other losses, 0.985

$h_{sh}$ = enthalpy of steam leaving superheater

$h_{sat}$ = saturated liquid enthalpy in the steam drum

e = HRSG effectiveness = $(T_1–T_2)/(T_1–T_3)$, defined by Fig. 7.10.

f = fuel factor, 1.0 for fuel oil, 1.015 for gas.

**3. Parametric Studies for Off-design Conditions.** Varying the amount of primary or supplementary firing will change the gas flow rate or temperature and the HRSG steam output. Thus, according to the varying temperatures, several iterations of equation [7.5] are required to evaluate off-design or part load conditions. When this evaluation is carried over a range of loads, firing rates, and temperatures, it is called a parametric study. Models can be constructed or off-design conditions using gas turbine performance data provided by manufacturers (See Figure 7.11).

**4. Exhaust Pressure Effects on Output and Exhaust Temperature.** Heat recovery systems increase the exhaust backpressure, reducing the turbine output in relation to simple operation (without HRSG). Turbine manufacturers provide test data about inlet and back-pressure effects, as well as elevation effects, on turbine output and efficiency (Fig 7.11).

**Example 4.** In a combined cycle (Fig 7.4), the steam for the turbine of Example 2 must be generated by several HRSGs (See Fig 7.10). To follow variable CHP loads and to optimize overall system reliability, each HRSG will be connected to a dedicated 10-MWe gas turbine-generator (GTG) set. The GTG sets burn fuel-oil and each unit exhausts 140,000 kg/hr of gas at 900°F. Estimate the

total gas flow to the HRSGs, if the system must produce a maximum of 160,000 lb/hr of 615 psia/750°F steam. How many HRSG-gas turbine sets are needed?

DATA (See notation of Equation 7.5 and Figure 7.10)

Gas      $W_g$ = 140,000 kg/hr per gas-turbine unit
Turbines  Cp  = 0.26 Btu/lb/°F, average specific heat
                of gases between $T_1$ and $T_4$
         $T_1$ = 900°F, hot gas temperature
         f   = fuel factor, 1 for fuel oil.

The inlet steam conditions required in the steam turbine of Example 3 are: (615 psia/750°F) and $h_i$ = 1378.9 Btu/lb (from Mollier chart). Thus,

HRSGs    $W_s$ = 160,000 lb/hr from all HRSGs
         T3  = 488.8°F (temp of 615 psia sat steam)
         L   = 0.98
         e   = HRSG effectiveness, 0.9

Therefore,  $h_{Sh} = h_i$ = 1378.9 Btu/lb and,
           $h_{sat}$ = 474.7 Btu/lb (Sat. Water @ 615 psia)

From equation 7.5, the total combustion gas flow required is

$$W_s = \frac{W_g(h_{sh} - h_{sat})}{Cp\,(T_1 - T_3)\,e\,L\,f}$$

$$= \frac{160,000\ lb/hr\ (1381.1 - 474.7\ Btu/lb}{0.26\ Btu/lb/°F\ (900–488.8°F)\ 0.9 \times 0.98 \times 1}$$

$$= 1,537,959.3\ lb/hr\ or\ 427.2\ lb/sec.$$

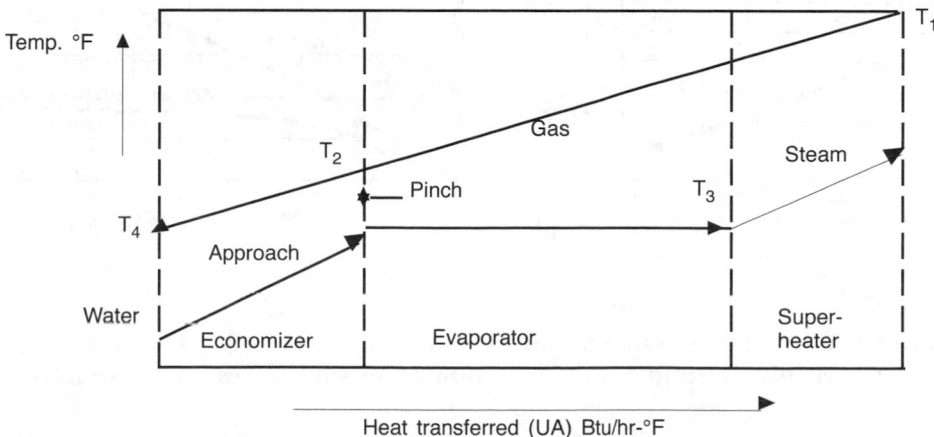

**Figure 7.10 Heat recovery steam generator diagram.**

| FUEL | ISO RATING | POWER kWe | HEAT RATE kJ/kWh (Btu/kWh) | | EXHAUST FLOW kg/hr (lb/hr) | | EXHAUST TEMP °C (°F) | |
|---|---|---|---|---|---|---|---|---|
| Natural Gas | Continuous | 4370 | 12 870 | (12 200) | 75 050 | (165 120) | 497 | (927) |
| | Peak | 4800 | 12 865 | (12 195) | 74 970 | (165 280) | 534 | (994) |
| Distillate | Continuous | 4280 | 13 135 | (12 450) | 75 050 | (165 120) | 497 | (927) |
| | Peak | 4700 | 13 130 | (12 445) | 74 970 | (165 280) | 534 | (994) |

Nominal Performance (gas fuel)

Figure 7.11 Performance ratings and curves of a gas turbine-generator set. Nominal ratings are given at ISO conditions: 59°F, sea level, 60% relative humidity and no external pressure losses. *Courtesy of Solar Turbines, Inc.*

Finally, the number of required gas turbine/HRSG sets is

= [(1,537,959 lb/hr) (1kg/2.21b)/(140,000 kg/hr/set))

= 4.99, i.e. 5 sets.

### 7.2.2.3 Reciprocating Engine Systems

Reciprocating engines include a variety of internally fired, piston driven engines. Their sizes range from 10 bhp to 50,000 bhp. According to Kovacik (1982), the largest unit supplied by a U.S. manufacturer is rated at 13,500 bhp. In larger plants, several units are used to accommodate part load and to provide redundancy and better availability.

In these engines, combustion heat rejected through the jacket water, lube oil and exhaust gases, can be recovered through heat exchangers to generate hot water and/or steam. Fig. 7.13 shows an internal combustion engine cogeneration system.

Exhaust gases have also been used directly. Reciprocating engines are classified by:

— the thermodynamic cycle: Diesel or Otto cycle.

— the rotation speed: high-speed (1200-1800 rpm) medium-speed (500-900 rpm) low speed (450 rpm or less)

— the aspiration type: naturally aspirated or turbocharged

— the operating cycle: two-cycle or four-cycle

— the fuel burned: fuel-oil fired, natural-gas fired.

Reciprocating engines are widely used to move vehicles, generators and a variety of shaft loads. Larger engines are associated to lower speeds, increased torque, and heavier duties. The total heat utilization of CHP systems based on gas-fired or fuel-oil fired engines approach 60-75%. Figure 7.12 shows the CHP balance vs. load of a diesel engine.

**Example 5.** Estimate the amount of 180 F water that can be produced by recovering heat; first from the jacket water and then from the exhaust of a 1200 kW diesel generator. In the average, the engine runs at a 75% load and the inlet water temperature is 70°F. The effectiveness of the jacket water heat exchanger is 90% and exhaust heat exchanger is 80%.

From Figure 7.12, at 75% load, the exhaust heat

and the jacket water heat are 22% and 33%, respectively. Thus, the flow rate of water heated from 70 to 180°F is:

= (1200 kW × 75%) [(90% × 22%) + (80% × 33%)]
  (3412 Btu/kW) (1 lb.°F/Btu)(1 lb/8.33 gal)
  (1 hr/60 min)/(180 − 70°F)

= 25.8 gal per minute.

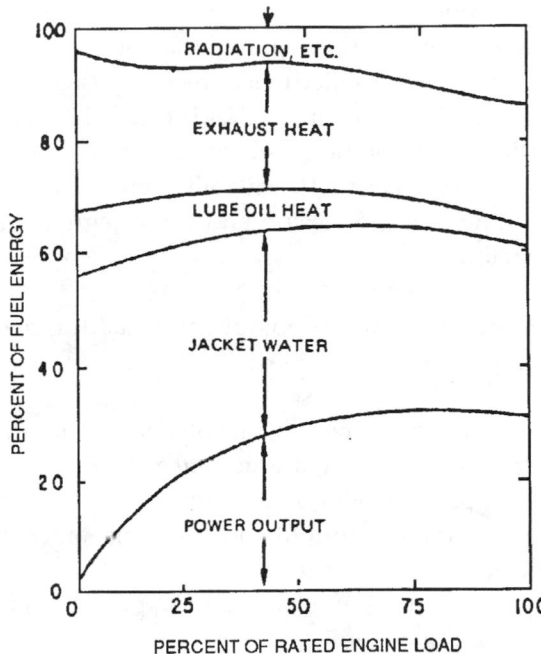

**Fig. 7.12 Diesel engine heat and power balance.**

### 7.2.2.4 Distributed Generation

Distributed Generation (DG) is emerging as the generation of heat and power (CHP) through relatively small distributed units (from 25 MW to a few kilowatts). Typically, these are smaller, self contained, power generation systems located close to, adjacent to, or within the boundaries of a CHP user or consumer facility. Typical fuels include natural gas, liquefied petroleum gas (LPG), kerosene and diesel fuel. DG plants may or may not be interconnected to a utility grid. Earlier DG systems have also been known as packaged cogeneration units.

This section focuses on current developments of the following DG technologies:

- Combustion or Gas Turbines
  (See also section 7.2.2.2)
- Reciprocating Engines
  (See also section 7.2.2.3)
- Fuel Cells
  (See also section 16.5)
- Photovoltaics or solar cells
  (See also section 16.2.6)

- Microturbines
  (See also section 7.2.2.2)

Previous sections in this chapter discuss some of the underlying technologies. Chapter 16, on "Alternative Energy," provides a fundamental and renewable-energy approach to Photovoltaics and Fuel Cells. Next, Table 7.2 shows a comparison of current DG technologies.

## DG Economics

In addition to the economic factors listed in table 7.2, i.e. turn key (installed), heat recovery (installed) and operation and maintenance (O&M) costs, there are other evaluation considerations which are often site specific. These include local fuel availability and cost, size and weight limitations, emission and noise regulations, and other factors.

DG can be used for various purposes and applications with varying economic merit:

1. As a prime mover to supply base-load electrical demand with or without heat recovery.
2. As a peak shaving generator
3. As an uninterruptible power supply, emergency or back-up power unit.
4. As a combination emergency power and peak-shaving generator.

In general, DG as a prime mover for continuous operation does not show feasible economics unless there is significant heat recovery. Thus, for 5000 hrs/yr or more, with heat recovery (60-70% CHP system efficiency), our experience with several projects in the US and the Caribbean shows that for the 200 to 800 kW- range of gas- or diesel fired engine generators, there is an acceptable payback (3 years or less) only if the displaced electricity costs more than $0.10 per kWh and the natural gas fuel is $6.00 per million Btu or less. The evaluation considers a $0.015/ kWh CHP plant operation and maintenance fee, in addition to fuel consumption. See upper part of Table 7.3. The shadowed cells on the upper-right part of the tables list paybacks of three years or less.

Without heat Recovery, DG makes little economic sense for continuous operation, unless an inexpensive and plentiful source of fuel is available; e.g. bio-diesel from waste cooking oil or bio-gas from obtained from agricultural waste, sewer treatment plant and landfills. DG is also feasible in the cases where surplus or pipeline natural or liquefied petroleum gas exists. To have a payback of three years or less, the bottom part of Table 7.3 shows the DG fuel must cost $6/ million Btu or less and the displaced power must cost $0.11/ kWh or more. This sensitivity analysis assumes $0.010/ kWh for operation and maintenance, in addition to fuel cost.

## Table 7.2 Comparison of DG Technologies

| Comparison Factor | Diesel Engine | Gas Engine | Simple Cycle Gas Turbine | Microturbine | Fuel Cells | Photovoltaics |
|---|---|---|---|---|---|---|
| Product Availability | Commercial | Commercial | Commercial | 1999-2005 | 1996-2010 | Commercial |
| Size Range (kW/unit) | 20 - 10,000+ | 50 - 5,000+ | 1,000 -30,000 | 20 - 200 | 50 - 1000+ | 1+ |
| Typical DG Range (kW/unit) | 200 - 2,000 | 300-3,000 | 1,000 - 10,000 | 20-100 | 50 - 200 | 1 - 5 |
| Efficiency (HHV) | 36 - 43% | 28 - 42% | 21 - 40% | 25 - 30% | 35 - 54% | n.a. |
| Genset Package Cost ($/kW) | 125 – 300 | 250 - 600 | 300 - 600 | 350 - 750 | 1500 -3000 | n.a. |
| Turnkey Cost - With no heat recovery ($/kW) | 350 – 500 | 600 - 1000 | 650 - 900 | 600 - 1100 | 1900 - 3500 | 5000 - 10000 |
| Heat Recovery Added Cost ($/kW) | 100- 200 | 75- 150 | 100 - 200 | 75 - 350 | Incl. | n.a. |
| O&M Cost ($/kWh) | 0.005 - 0.010 | 0.007 - 0.015 | 0.003 - 0.008 | 0.005 - 0.010 | 0.005 - 0.010 | 0.001 - 0.004 |

Source: Gas Research Institute (2000): www.gri.org

### Table 7.3. DG Economics and Sensitivity Analysis

| Simple Payback in Years for 5,000 hrs/yr operation and $500/kW installed cost | | | | | | | |
|---|---|---|---|---|---|---|---|
| | DG With Heat Recovery | | Assumes Recovered Heat = Power Output | | | 68% | CHP efficiency |
| Gas Cost ($/MMBTU) | Electrical Cost ($/kWh) | | | | | | |
| | 0.06 | 0.07 | 0.08 | 0.09 | 0.10 | 0.11 | 0.12 |
| 5.00 | 6.13 | 3.80 | 2.75 | 2.16 | 1.78 | 1.51 | 1.31 |
| 6.00 | 9.44 | 4.86 | 3.27 | 2.46 | 1.98 | 1.65 | 1.42 |
| 7.00 | 20.60 | 6.73 | 4.02 | 2.87 | 2.23 | 1.82 | 1.54 |
| 8.00 | N/F | 10.96 | 5.23 | 3.43 | 2.56 | 2.04 | 1.69 |
| 9.00 | N/F | 29.54 | 7.47 | 4.28 | 3.00 | 2.30 | 1.87 |
| 10.00 | N/F | N/F | 13.07 | 5.67 | 3.62 | 2.66 | 2.10 |
| 11.00 | N/F | N/F | 52.22 | 8.39 | 4.56 | 3.13 | 2.39 |
| 12.00 | N/F | N/F | N/F | 16.18 | 6.18 | 3.82 | 2.76 |
| N/F: Not Feasible | DG system generates CHP and includes a heat recovery module ($100/kW installed). | | | | | | |

| Simple Payback in Years for 5,000 hrs/yr operation and $400/kW installed cost | | | | | | | |
|---|---|---|---|---|---|---|---|
| | DG Without Heat Recovery | | | | | 34% | efficiency |
| 5.00 | N/F | 16.00 | 5.33 | 3.20 | 2.29 | 1.78 | 1.45 |
| 6.00 | N/F | N/F | 16.00 | 5.33 | 3.20 | 2.29 | 1.78 |
| 7.00 | N/F | N/F | N/F | 16.00 | 5.33 | 3.20 | 2.29 |
| 8.00 | N/F | N/F | N/F | N/F | 16.00 | 5.33 | 3.20 |
| 9.00 | N/F | N/F | N/F | N/F | N/F | 16.00 | 5.33 |
| 10.00 | N/F | N/F | N/F | N/F | N/F | N/F | 16.00 |
| 11.00 | N/F | N/F | N/F | N/F | N/F | N/F | N/F |
| 12.00 | N/F | N/F | N/F | N/F | N/F | N/F | N/F |
| N/F: Not Feasible | | | | | | | |

## Peak Shaving Optimization Model and Case Study

While DG applications for continuous power generation (with no heat recovery) are rarely economically feasible, DG for peak shaving, in conjunction with back-up or emergency power generation, is generally an attractive business proposition. See Appendix A at the end of this chapter for a detailed case study which shows there is generally a good business case for peak shaving using diesel or gas fired generators. Such case study illustrates an optimization model to estimate the optimal peak shaving generator size (in kW or MW).

## Combustion or Gas Turbines*

Combustion turbine (CT) sizes for distributed generation vary from 1 to 30 MW. CT's are used to power aircraft, marine vessels, gas compressors, utility and industrial generators. In 1998, over 500 CT's were shipped from the US to worldwide facilities totaling 3,500 MW of power capacity. Most of these were sold overseas. The North American market represents only 11% share of the total. The primary application of CT's is as prime mover for continuous power particularly, in combined cycle arrangement, or as a peaking unit to generate during peak demand periods.

Low maintenance, high reliability and high quality exhaust heat make CT's an excellent choice for industrial and commercial CHP applications larger than 3 MW. CT's can burn natural gas, liquid fuels, such as diesel oil, or both gas and liquid (dual-fuel operation). Thus, they contribute to the fuel security of the DG plant. CT emissions can be controlled by using dry low $NO_x$ combustors, water or steam injection, or exhaust treatments such as selective catalytic reduction (SCR). Due to their inherent reliability and remote diagnostic capability, CTs tend to have one of the lowest maintenance costs among DG technologies.

## Reciprocating Engines

Reciprocating internal combustion (IC) engines are a widespread and well-known technology. North American production tops 35 million units per year for automobiles, trucks, construction and mining equipment, lawn care, marine propulsion, and of course, all types of power generation from small portable gen-sets to engines the size of a house, powering generators of several megawatts. Spark ignition engines for power generation use natural gas as the preferred fuel—though they can be set up to run on propane or gasoline. Diesel cycle, compression ignition engines can operate on diesel fuel or heavy oil, or they can be set up in a dual-fuel configuration that

---

*The updates on DG technologies have been obtained from: Fundamentals of Distributed Generation" and "Distributed Generation: A Primer" from the Gas Research Institute web site: *http://www.gri.org/pub/solutions/dg/index.html.*

burns primarily natural gas with a small amount of diesel pilot fuel and can be switched to 100% diesel.

Current generation IC engines offer low first cost, easy start-up, proven reliability when properly maintained, good load-following characteristics, and heat recovery potential. IC engine systems with heat recovery have become a popular form of DG in Europe. Emissions of IC engines have been reduced significantly in the last several years by exhaust catalysts and through better design and control of the combustion process. IC engines are well suited for standby, peaking, and intermediate applications and for combined heat and power (CHP) in commercial and light industrial applications of less than 10 MW.

## Microturbines

Microturbines or turbogenerators are small combustion turbines with outputs of 30-200 kW. Individual units can be packaged together to serve larger loads. Turbogenerator technology has evolved from automotive and truck turbochargers auxiliary power units for airplanes, and small jet engines used for pilot military aircraft.

Recent development of these microturbines has been focused on this technology as the prime mover for hybrid electric vehicles and as a stationary power source for the DG market. In most configurations, the turbine shaft spinning at up to 100,000 rpm drives a high speed generator. This high frequency output is first rectified and then converted to 60 Hz (or 50 Hz). The systems are capable of producing power at around 25 to 30% efficiency by employing a recuperator that transfers heat energy from the exhaust stream back into the incoming air stream. Like larger turbines, these units are capable of operating on a variety of fuels. The systems are air-cooled and some even use air bearings, thereby eliminating both water and oil systems. Low-emission combustion systems are being demonstrated that provide emissions performance comparable to larger CTS. Turbogenerators are appropriately size for commercial buildings or light industrial markets for cogeneration or power-only applications.

## Fuel Cells

Fuel cells (Figure 7.13) produce power electrochemically like a battery rather than like a conventional generating system that converts fuel to heat to shaft-power and finally to electricity. Unlike a storage battery, however, which produces power from stored chemicals, fuel cells produce power when hydrogen fuel is delivered to the negative pole (cathode) of the cell and oxygen in air is delivered to the positive pole (anode). The hydrogen fuel can come from a variety of sources, but the most economic is steam reforming of natural gas—a chemical process that strips the hydrogen from both the fuel and the steam. Several different liquid and solid media can be used to created the fuel cell's electrochemical reaction—phosphoric acid fuel cell (PAFC), molten carbonate fuel cell (MCFC), solid oxide fuel cell (SOFC), and proton exchange membrane (PEM). Each of these media comprises a distinct fuel cell technology with its own performance characteristics and development schedule. PAFCs are in early commercial market development now with 200 kW units delivered to over 120 customers.

The SOFC and MCFC technologies are now in field test or demonstration. PEM units are in early development and testing. Direct electrochemical reactions are generally more efficient than using fuel to drive a heat engine to produce electricity. Fuel cell efficiencies range from 35-40% for the PAFC up to 60% with MCFC and SOFC systems under development. PEM unit efficiencies are as high as 50%. Fuel cells are inherently quiet and extremely clean running. Like a battery, fuel cells produce direct current (DC) that must be run through an inverter to get 60 Hz alternating current (AC). These power electronics components can be integrated with other components as part of a power quality control strategy for sensitive customers. Because of current high costs, fuel cells are best suited to environmentally sensitive areas and customers with power quality concerns. Some fuel cell technology is modular and capable of application in small commercial and even residential markets; other technology utilizes high temperatures in larger sized systems that would be well sited to industrial cogeneration applications.

## Photovoltaics

Photovoltaic power cells use solar energy to produce power. Photovoltaic power is modular and can be sited wherever the sun shines. These systems have been commercially demonstrated in extremely sensitive environmental areas and for remote (grid-isolated) applications. Battery banks are needed to store the energy harnessed during daytime. High costs make these systems a niche technology that is able to compete more on the basis of environmental benefits than on economics. Isolated facilities with need for limited but critical power are typical applications.

### 7.2.3 The Cogeneration Design Process

The following evaluation steps are suggested to carry out cogeneration system design.

1.      Develop the profile of the various process steam (heat) demands at the appropriate steam pressures

**Figure 7.13 Proton Exchange Membrane (PEM ) fuel cells use platinum catalysts to promote the flow of anions (positive ions) through the membrane, thus creating a direct current (DC). An inverter can be used to convert DC to AC.**

for the applications being studied. Also, collect data with regard to condensate returned from the process and its temperature. Data must include daily fluctuations due to normal variations in process needs, as well as seasonal weather effects; including the influence of not-working periods such as weekends, vacation periods, and holidays.

2.  A profile for electric power must be developed in the same manner as the process heat demand profile. These profiles typically include hour-by-hour heat and power demands for "typical" days (or weeks) for each season or month of the year.

3.  Fuel availability and present-day cost as well as projected future costs. The study should also factor process by-product fuels into the development of the energy supply system.

4.  Purchased power availability and its present and expected future cost.

5.  Plant discharge stream data in the same degree of detail as the process heat demand data.

6.  Number and rating of major (demand and generation) equipment items. This evaluation usually establish whether spare capacity and/or supplementary firing should be installed.

7.  Plant, process and CHP system economic lives.

Once this initial data bank has been established, the various alternatives that can satisfy plant heat and power demands can be identified. Subsequently, detailed technical analyses are conducted. Thus, energy balances are made, investment cost estimated, and the economic merit of each alternative evaluated. Some approaches for evaluation are discussed next.

### 7.2.4 Economic Feasibility Evaluation Methods

Cogeneration feasibility evaluation is an iterative process—further evaluations generally require more data. There are a number of evaluation methods using various approaches and different levels of technical detail. Most of them consider seasonal loads and equipment performance characteristics. Some of the most representative methods are discussed as follows.

#### 7.2.4.1  General Approaches For Design and Evaluation

Hay (1988) presents a structured approach for system design and evaluation. It is a sequence of evaluation iterations, each greater than the previous and each producing information whether the costs of the next step is warranted. His suggested design process is based on the following steps.

Step 1: Site Walkthrough and Technical Screening
Step 2: Preliminary Economic Screening
Step 3: Detailed Engineering Design.

Similarly, Butler (1984) considers three steps to perform studies, engineering and construction of cogeneration projects. These are discussed as follow.

**Step 1. Preliminary studies and conceptual engineering.** This is achieved by performing a technical feasibility and economic cost-benefit study to rank and recommend alternatives. The determination of technical feasibility includes a realistic assessment with respect to environmental impact, regulatory compliance, and interface with a utility. Then, an economic analysis-based on the simple payback period-serves as a basis for more refined evaluations.

**Step 2. Engineering and Construction Planning.** Once an alternative has been selected and approved by the owner, preliminary engineering is started to develop the general design criteria. These include specific site information such as process heat and power requirements, fuel availability and pricing, system type definition, modes of operation, system interface, review of alternatives under more detailed load and equipment data, confirm selected alternative and finally size the plant equipment and systems to match the application.

**Step 3. Design Documentation.** This includes the preparation of project flow charts, piping and instrument diagrams, general arrangement drawings, equipment layouts, process interface layouts, building, structural and foundation drawings, electrical diagrams, and specifying an energy management system, if required.

Several methodologies and manuals have been developed to carry out Step 1, i.e. screening analysis and preliminary feasibility studies. Some of them are briefly discussed in the next sections. Steps 2 and 3 usually require ad-hoc approaches according to the characteristics of each particular site. Therefore, a general methodology is not applicable for such activities.

### 7.2.4.2 Preliminary Feasibility Study Approaches

**AGA Manual**—GKCO Consultants (1982) developed a cogeneration feasibility (technical and economical) evaluation manual for the American Gas Association, AGA. It contains a "Cogeneration Conceptual Design Guide" that provides guidelines for the development of plant designs. It specifies the following steps to conduct the site feasibility study:

a)   Select the type of prime mover or cycle (piston engine, gas turbine or steam turbine);

b)   Determine the total installed capacity;

c)   Determine the size and number of prime movers;

d)   Determine the required standby capacity.

According to its authors "the approach taken (in the manual) is to develop the minimal amount of information required for the feasibility analysis, deferring more rigorous and comprehensive analyses to the actual concept study." The approach includes the discussion of the following "Design Options" or design criteria to determine (1) the size and (2) the operation mode of the CHP system.

**Isolated Operation, Electric Load Following**—The facility is independent of the electric utility grid, and is required to produce all power required on-site and to provide all required reserves for scheduled and unscheduled maintenance.

**Baseloaded, Electrically Sized**—The facility is sized for baseloaded operation based on the minimum historic billing demand. Supplemental power is purchased from the utility grid. This facility concept generally results in a shorter payback period than that from the isolated site.

**Baseloaded, Thermally Sized**—The facility is sized to provide most of the site's required thermal energy using recovered heat. The engines operated to follow the thermal demand with supplemental boiler fired as required. The authors point out that: "this option frequently results in the production of more power than is required on-site and this power is sold to the electric utility."

In addition, the AGA manual includes a description of sources of information or processes by which background data can be developed for the specific gas distribution service area. Such information can be used to adapt the feasibility screening procedures to a specific utility.

### 7.2.4.3 Cogeneration System Selection and Sizing.

The selection of a set of "candidate" cogeneration systems entails to tentatively specify the most appropriate prime mover technology, which will be further evaluated in the course of the study. Often, two or more alternative systems that meet the technical requirements are pre-selected for further evaluation. For instance, a plant's CHP requirements can be met by either, a reciprocating engine system or combustion turbine system. Thus, the two system technologies are pre-selected for a more detailed economic analysis.

To evaluate specific technologies, there exist a vast number of technology-specific manuals and references. A representative sample is listed as follows. Mackay (1983) has developed a manual titled "Gas Turbine Cogeneration: Design, Evaluation and Installation." Kovacik (1984) reviews application considerations for both steam turbine and gas turbine cogeneration systems. Limaye (1987) has compiled several case studies on industrial cogeneration applications. Hay (1988) discusses technical and economic considerations for cogeneration application of gas engines, gas turbines, steam engines and packaged systems. Keklhofer (1991) has written a treatise on technical and economic analysis of combined-cycle gas and steam turbine power plants. Ganapathy (1991) has produced a manual on waste heat boilers.

Usually, system selection is assumed to be separate from sizing the cogeneration equipment (kWe). However, since performance, reliability and cost are very dependent on equipment size and number, technology selection and system size are very intertwined evaluation factors. In addition to the system design criteria given by the AGA manual, several approaches for cogeneration system selection and/or sizing are discussed as follows.

### Heat-to-Power Ratio

Canton et al (1987) of The Combustion and Fuels Research Group at Texas A&M University has developed a methodology to select a cogeneration system for a given industrial application using the heat to power ratio (HPR). The methodology includes a series of graphs used 1) to define the load HPR and 2) to compare and match the load HPR to the HPRs of existing equipment. Consideration is then given to either, heat or power load matching and modulation.

### Sizing Procedures

Hay (1987) considers the use of the load duration curve to model variable thermal and electrical loads in system sizing, along with four different scenarios described in Figure 7.14. Each one of these scenarios defines an operating alternative associated to a system size.

Oven (1991) discusses the use of the load duration curve to model variable thermal and electrical loads in system sizing in conjunction with required thermal and electrical load factors. Given the thermal load duration and electrical load duration curves for a particular facility, different sizing alternatives can be defined for various load factors.

Eastey et al. (1984) discusses a model (CO-GENOPT) for sizing cogeneration systems. The basic inputs to the model are a set of thermal and electric profiles, the cost of fuels and electricity, equipment cost and performance for a particular technology. The model calculates the operating costs and the number of units for different system sizes. Then it estimates the net present value for each one of them. Based on the maximum net present value, the "optimum" system is selected. The model includes cost and load escalation.

Wong, Ganesh and Turner (1991) have developed two statistical computer models to optimize cogeneration system size subject to varying capacities/loads and

*Isolated Cogenerator:* User receives all power from cogeneration system *or* electric utility system.

*Electrically Base-Loaded Cogeneration:* User receives power from cogeneration system and purchases power from the electric utility system.

*Thermally Base-Loaded Cogeneration:* User receives power from cogeneration system *and* cogenerator sells power to the electric utility.

*Maximum System:* User receives power from the utility. Cogenerator sells output to the utility grid.

**Figure 7-14. Each operation mode defines a sizing alternative.** *Source: Hay (1987).*

to meet an availability requirement. One model is for internal combustion engines and the other for unfired gas turbine cogeneration systems. Once the user defines a required availability, the models determine the system size or capacity that meets the required availability and maximizes the expected annual worth of its life cycle cost.

## 7.3  COMPUTER PROGRAMS

There are several computer programs-mainly PC based-available for detailed evaluation of cogeneration systems. In opposition to the rather simple methods discussed above, CHP programs are intended for system configuration or detailed design and analysis. For these reasons, they require a vast amount of input data. Below, we examine two of the most well known programs.

### 7.3.1  CELCAP

Lee (1988) reports that the Naval Civil Engineering Laboratory developed a cogeneration analysis computer program known as Civil Engineering Laboratory Cogeneration Program (CELCAP), "for the purpose of evaluating the performance of cogeneration systems on a lifecycle operating cost basis." He states that "selection of a cogeneration energy system for a specific application is a complex task." He points out that the first step in the selection of *cogeneration system* is to make a list of potential candidates. These candidates should include single or multiple combinations of the various types of engine available. The computer program does not specify CHP systems; these must be selected by the designer. Thus, depending on the training and previous experience of the designer, different designers may select different systems of different sizes. After selecting a short-list of candidates, modes of operations are defined for the candidates. So, if there are N candidates and M modes of operation, then NxM alternatives must be evaluated. Lee considers three modes of operation:

1)  Prime movers operating at their full-rated capacity, any excess electricity is sold to the utility and any excess heat is rejected to the environment. Any electricity shortage is made up with imports. Process steam shortages are made-up by an auxiliary boiler.

2)  Prime movers are specified to always meet the entire electrical load of the user. Steam or heat demand is met by the prime mover. An auxiliary boiler is fired to meet any excess heat deficit and excess heat is rejected to the environment.

3)  Prime movers are operated to just meet the steam or heat load. In this mode, power deficits are made up by purchased electricity. Similarly, any excess power is sold back to the utility.

For load analysis, Lee considers that "demand of the user is continuously changing. This requires that data on the electrical and thermal demands of the user be available for at least one year." He further states that "electrical and heat demands of a user vary during the year because of the changing working and weather conditions." However, for evaluation purposes, he assumes that the working conditions of the user-production related CHP load-remain constant and "that the energy-demand pattern does not change significantly from year to year." Thus, to consider working condition variations, Lee classifies the days of the year as working and non-working days. Then, he uses "average" monthly load profiles and "typical" 24-hour load profiles for each class.

"Average" load profiles are based on electric and steam consumption for an average weather condition at the site. A load profile is developed for each month, thus monthly weather and consumption data is required. A best fit of consumption (Btu/month or kWh/month) versus heating and cooling degree days is thus obtained. Then, actual hourly load profiles for working and non-working days for each month of the year are developed. The "best representative" profile is then chosen for the "typical working day" of the month. A similar procedure is done for the non-working days.

Next an energy balance or reconciliation is performed to make sure the consumption of the hourly load profiles agrees with the monthly energy usage. A multiplying factor K is defined to adjust load profiles that do not balance.

$$K_j = E_{mj}/(AE_{wj} + AE_{nwj}) \tag{7.9}$$

where

$K_j$ = multiplying factor for month j

$E_{mj}$ = average consumption (kWh) by the user for the month j selected from the monthly electricity usage versus degree day plot

$AE_{wj}$ = typical working-day electric usage (kWh), i.e. the area under the typical working day electric demand profile for the month j

$AE_{nwj}$ = typical non-working day usage (kWh), i.e. the area under the typical non-working day electric demand profile for the month j.

Lee suggests that each hourly load in the load profiles be multiplied by the K factor to obtain the "correct working and non-working day load profiles for the month." The procedure is repeated for all months of the year for both electric and steam demands. Lee states that "the resulting load profiles represent the load demand for average weather conditions."

Once a number of candidate CHP systems has been selected, equipment performance data and the load profiles are fed into CELCAP to produce the required output. The output can be obtained in a brief or detailed form. In brief form, the output consists of a summary of input data and a life cycle cost analysis including fuel, operation and maintenance and purchased power costs. The detailed printout includes all the information of the brief printout, plus hourly performance data for 2 days in each month of the year. It also includes the maximum hourly CHP output and fuel consumption. The hourly electric demand and supply are plotted, along with the hourly steam demand and supply for each month of the year.

Despite the simplifying assumptions introduced by Lee to generate average monthly and typical daily load profiles, it is evident that still a large amount of data handling and preparation is required before CELCAP is run. By recognizing the fact that CHP loads vary over time, he implicitly justifies the amount of effort in representing the input data through hourly profiles for typical working and non-working days of the month.

If a change occurs in the products, process or equipment that constitute the energy consumers within the industrial plant, a new set of load profiles must be generated. Thus, exploring different conditions requires sensitivity analyses or parametric studies for off-design conditions.

A problem that becomes evident at this point is that, to accurately represent varying loads, a large number of load data points must be estimated for subsequent use in the computer program. Conversely, the preliminary feasibility evaluation methods discussed previously, require very few and only "average" load data. However, criticism of preliminary methods has arisen for not being able to truly reflect seasonal variations in load analysis (and economic analysis) and for lacking the flexibility to represent varying CHP system performance at varying loads.

## 7.3.2 COGENMASTER

Limaye and Balakrishnan (1989) of Synergic Resources Corporation have developed COGENMASTER. It is a computer program to model the technical aspects of alternative cogeneration systems and options, evalu-

ate economic feasibility, and prepare detailed cash flow statements.

COGENMASTER compares the CHP alternatives to a base case system where electricity is purchased from the utility and thermal energy is generated at the site. They extend the concept of an option by referring not only to different technologies and operating strategies but also to different ownership structures and financing arrangements. The program has two main sections: a Technology and a Financial Section. The technology Section includes 5 modules:

- Technology Database Module
- Rates Module
- Load Module
- Sizing Module
- Operating Module

The Financial Section includes 3 modules:
- Financing Module
- Cash Flow Module
- Pricing Module

In COGENMASTER, facility electric and thermal loads may be entered in one of three ways, depending on the available data and the detail required for project evaluation:

— A constant average load for every hour of the year.
— Hourly data for three typical days of the year
— Hourly data for three typical days of each month

Thermal loads may be in the form of hot water or steam; but system outlet conditions must be specified by the user. The sizing and operating modules permit a variety of alternatives and combinations to be considered. The system may be sized for the base or peak, summer or winter, and electric or thermal load. There is also an option for the user to define the size the system in kilowatts. Once the system size is defined, several operation modes may be selected. The system may be operated in the electric following, thermal following or constantly running modes of operation. Thus, N sizing options and M operations modes define a total of NxM cogeneration alternatives, from which the "best" alternative must be selected. The economic analysis is based on simple payback estimates for the CHP candidates versus a base case or do-nothing scenario. Next, depending on the financing options available, different cash flows may be defined and further economic analysis-based on the Net Present Value of the alternatives—may be performed.

## 7.4 U.S. COGENERATION LEGISLATION: PURPA

In 1978 the U.S. Congress amended the Federal Power Act by promulgation of the Public Utilities Regulatory Act (PURPA). The Act recognized the energy saving potential of industrial cogeneration and small power plants, the need for real and significant incentives for development of these facilities and the private sector requirement to remain unregulated.

PURPA of 1978 eliminated several obstacles to cogeneration so cogenerators can count on "fair" treatment by the local electric utility with regard to interconnection, back-up power supplies, and the sale of excess power. PURPA contains the major federal initiatives regarding cogeneration and small power production. These initiatives are stated as rules and regulations pertaining to PURPA Sections 210 and 201; which were issued in final form in February and March of 1980, respectively. These rules and regulations are discussed in the following sections.

Initially, several utilities—especially those with excess capacity-were reticent to buy cogenerated power and have, in the past, contested PURPA. Power (1980) magazine reported several cases in which opposition persisted in some utilities to private cogeneration. But after the Supreme Court ruling in favor of PURPA, more and more utilities are finding that PURPA can work to their advantage. Polsky and Landry (1987) report that some utilities are changing attitudes and are even investing in cogeneration projects.

### 7.4.1 PURPA 201*

Section 201 of PURPA requires the Federal Energy Regulatory Commission (FERC) to define the criteria and procedures by which small power producers (SPPs) and cogeneration facilities can obtain qualifying status to receive the rate benefits and exemptions set forth in Section 210 of PURPA. Some PURPA 201 definitions are stated below.

### Small Power Production Facility

A "Small Power Production Facility" is a facility that uses biomass, waste, or renewable resources, including wind, solar and water, to produce electric power and

*Most of the following sections have been adapted from CFR18 (1990) and Harkins (1980), unless quoted otherwise.

is not greater than 80 megawatts.

Facilities less than 30 MW are exempt from the Public Utility Holding Co. Act and certain state law and regulation. Plants of 30 to 80 MW which use biomass, may be exempted from the above but may not be exempted from certain sections of the Federal Power Act.

### Cogeneration Facility

A "Cogeneration Facility" is a facility which produces electric energy and forms of useful thermal energy (such as heat or steam) used for industrial, commercial, heating or cooling purposes, through the sequential use of energy. A Qualifying Facility (QF) must meet certain minimum efficiency standards as described later. Cogeneration facilities are generally classified as "topping" cycle or "bottoming" cycle facilities.

### 7.4.2 Qualification of a "Cogeneration Facility" or a "Small Power Production Facility" under PURPA

### Cogeneration Facilities

To distinguish new cogeneration facilities which will achieve meaningful energy conservation from those which would be "token" facilities producing trivial amounts of either useful heat or power, the FERC rules establish operating and efficiency standards for both topping-cycle and bottom-cycle NEW cogeneration facilities. No efficiency standards are required for EXISTING cogeneration facilities regardless of energy source or type of facility. The following fuel utilization effectiveness (FUE) values—based on the lower heating value (LHV) of the fuel—are required from QFs.

- For a new topping-cycle facility:

  — No less than 5% of the total annual energy output of the facility must be useful thermal energy.

- For any new topping-cycle facility that uses any natural gas or oil:

  — All the useful electric power and half the useful thermal energy must equal at least 42.5% of the total annual natural gas and oil energy input; and

  — If the useful thermal output of a facility is less than 15% of the total energy output of the facility, the useful power output plus one-half the useful thermal energy output must be no less than 45% of the total energy input of natural gas and oil for the calendar.

For a new bottoming-cycle facility:

- If supplementary firing (heating of water or steam before entering the electricity generation cycle from the thermal energy cycle) is done with oil or gas, the useful power output of the bottoming cycle must, during any calendar year, be no less than 45% of the energy input of natural gas and oil for supplementary firing.

## Small Power Production Facilities

To qualify as a small power production facility under PURPA, the facility must have production capacity of under 80 MW and must get more than 50% of its total energy input from biomass, waste, or renewable resources. Also, use of oil, coal, or natural gas by the facility may not exceed 25% of total annual energy input to the facility.

## Ownership Rules Applying to Cogeneration and Small Power Producers

A qualifying facility may not have more than 50% of the equal interest in the facility held by an electric utility.

### 7.4.3 PURPA 210

Section 210 of PURPA directs the Federal Energy Regulatory Commission (FERC) to establish the rules and regulations requiring electric utilities to purchase electric power from and sell electric power to qualifying cogeneration and small power production facilities and provide for the exemption to qualifying facilities (QF) from certain federal and state regulations.

Thus, FERC issued in 1980 a series of rules to relax obstacles to cogeneration. Such rules implement sections of the 1978 PURPA and include detailed instructions to state utility commissions that all utilities must purchase electricity from cogenerators and small power producers at the utilities' "avoided" cost. In a nutshell, this means that rates paid by utilities for such electricity must reflect the cost savings they realize by being able to avoid capacity additions and fuel usage of their own.

Tuttle (1980) states that prior to PURPA 210, cogeneration facilities wishing to sell their power were faced with three major obstacles:

- Utilities had no obligation to purchase power, and contended that cogeneration facilities were too small and unreliable. As a result, even those co-generators able to sell power had difficulty getting an equitable price.

- Utility rates for backup power were high and often discriminatory

- Cogenerators often were subject to the same strict state and federal regulations as the utility.

PURPA was designed to remove these obstacles, by requiring utilities to develop an equitable program of integrating cogenerated power into their loads.

## Avoided Costs

The costs avoided by a utility when a cogeneration plant displaces generation capacity and/or fuel usage are the basis to set the rates paid by utilities for co-generated power sold back to the utility grid. In some circumstances, the actual rates may be higher or lower than the avoided costs, depending on the need of the utility for additional power and on the outcomes of the negotiations between the parties involved in the cogeneration development process.

All utilities are now required by PURPA to provide data regarding present and future electricity costs on a cent-per-kWh basis during daily, seasonal, peak and off-peak periods for the next five years. This information must also include estimates on planned utility capacity additions and retirements, and cost of new capacity and energy costs.

Tuttle (1980) points out that utilities may agree to pay greater price for power if a cogeneration facility can:

- Furnish information on demonstrated reliability and term of commitment.

- Allow the utility to regulate the power production for better control of its load and demand changes.

- Schedule maintenance outages for low-demand periods.

- Provide energy during utility-system daily and seasonal peaks and emergencies.

- Reduce in-house on-site load usage during emergencies.

- Avoid line losses the utility otherwise would have incurred.

In conclusion, a utility is willing to pay better "buyback" rates for cogenerated power if it is short in capacity, if it can exercise a level of control on the CHP plant and load, and if the cogenerator can provide and/or demonstrate a "high" system availability.

PURPA further states that the utility is not obligated to purchase electricity from a QF during periods that would result in net increases in its operating costs. Thus, low demand periods must be identified by the utility and the cogenerator must be notified in advance. During emergencies (utility outages), the QF is not required to provide more power than its contract requires, but a utility has the right to discontinue power purchases if they contribute to the outage.

### 7.4.4 Other Regulations

Several U.S. regulations are related to cogeneration. For example, among environmental regulations, the Clean Air Act may control emissions from a waste-to-energy power plant. Another example is the regulation of underground storage tanks by the Resource Conservation and Recovery Act (RCRA). This applies to all those cogenerators that store liquid fuels in underground tanks. Thus, to maximize benefits and to avoid costly penalties, cogeneration planners and developers should become savvy in related environmental matters.

There are many other issues that affect the development and operation of a cogeneration project. For further study, the reader is referred to a variety of sources such proceedings from the various World Energy Engineering Congresses organized by the Association of Energy Engineers (Atlanta, GA). Other sources include a general compendium of cogeneration planning considerations given by Orlando (1990), and a manual-developed by Spiewak (1994)—which emphasizes the regulatory, contracting and financing issues of cogeneration.

### 7.5 EVALUATING COGENERATION OPPORTUNITIES: CASE EXAMPLES

The feasibility evaluation of cogeneration opportunities for both, new construction and facility retrofit, require the comparison and ranking of various options using a figure of economic merit. The options are usually combinations of different CHP technologies, operating modes and equipment sizes.

A first step in the evaluation is the determination of the costs of a base-case (or do-nothing) scenario. For new facilities, buying thermal and electrical energy from utility companies is traditionally considered the base case. For retrofits, the present way to buy and/or generate energy is the base case. For many, the base-case scenario is the "actual plant situation" after "basic" energy conservation and management measures have been implemented. That is, cogeneration should be evaluated upon an "efficient" base case plant.

Next, suitable cogeneration alternatives are generated using the methods discussed in sections 7.2 and 7.3. Then, the comparison and ranking of the base case versus the alternative cases is performed using an economic analysis.

Henceforth, this section addresses a basic approach for the economic analysis of cogeneration. Specifically, it discusses the development of the cash flows for each option including the base case. It also discusses some figures of merit such as the gross pay out period (simple payback) and the discounted or internal rate of return. Finally, it describes two case examples of evaluations in industrial plants. The examples are included for illustrative purposes and do not necessarily reflect the latest available performance levels or capital costs.

### 7.5.1 General Considerations

A detailed treatise on engineering economy is presented in Chapter 4. Even so, since economic evaluations play the key role in determining whether cogeneration can be justified, a brief discussion of economic considerations and several evaluation techniques follows.

The economic evaluations are based on examining the incremental increase in the investment cost for the alternative being considered relative to the alternative to which it is being compared and determining whether the savings in annual operating cost justify the increased investment. The parameter used to evaluate the economic merit may be a relatively simple parameter such as the "gross payout period." Or one might use more sophisticated techniques which include the time value of money, such as the "discounted rate of return," on the discretionary investment for the cogeneration systems being evaluated.

Investment cost and operating cost are the expenditure categories involved in an economic evaluation. Operating costs result from the operations of equipment, such as (1) purchased fuel, (2) purchased power, (3) purchased water, (4) operating labor, (5) chemicals, and (6) maintenance. Investment-associated costs are of primary importance when factoring the impact of federal and state income taxes into the economic evaluation. These costs (or credits) include (1) investment tax credits, (2) depreciation, (3) local property taxes, and (4) insurance. The economic evaluation establishes whether the operating and investment cost factors result in sufficient after-tax income to provide the company stockholders an adequate rate of return after the debt obligations with regard to the investment have been satisfied.

When one has many alternatives to evaluate, the

less sophisticated techniques, such as "gross payout," can provide an easy method for quickly ranking alternatives and eliminating alternatives that may be particularly unattractive. However, these techniques are applicable only if annual operating costs do not change significantly with time and additional investments do not have to be made during the study period.

The techniques that include the time value of money permit evaluations where annual savings can change significantly each year. Also, these evaluation procedures permit additional investments at any time during the study period. Thus these techniques truly reflect the profitability of a cogeneration investment or investments.

### 7.5.2 Cogeneration Evaluation Case Examples

The following examples illustrate evaluation procedures used for cogeneration studies. Both examples are based on 1980 investment costs for facilities located in the U.S. Gulf Coast area.

For simplicity, the economic merit of each alternative examined is expressed as the "gross payout period" (GPO). The GPO is equal to the incremental investment for cogeneration divided by the resulting first-year annual operating cost savings. The GPO can be converted to a "discounted rate of return" (DRR) using Figure 7.15. However, this curve is valid only for evaluations involving a single investment with fixed annual operating cost savings with time. In most instances, the annual savings due to cogeneration will increase as fuel costs increase to both utilities and industries in the years ahead. These increased future savings enhance the economics of cogeneration. For example, if we assume that a project has a GPO of three years based on the first-year operating cost savings, Figure 7.15 shows a DRR of 18.7%. However, if the savings due to cogeneration increase 10% annually for the first three operating years of the project and are constant thereafter, the DRR increases to 21.6%; if the sav-

**Fig. 7.15 Discounted rate of return versus gross payout period. Basis: (1) depreciation period, 28 years; (2) sum-of-the-years'-digits depreciation; (3) economic life, 28 years; (4) constant annual savings with time; (5) local property taxes and insurance, 4% of investment cost; (6) state and federal income taxes, 53%; (7) investment tax credit, 10% of investment cost.**

ings increase 10% annually for the first six years, the DRR would be 24.5%; and if the 10% increase was experienced for the first 10 years, the DRR would be 26.6%.

**Example 6**: The energy requirements for a large industrial plant are given in Table 7.3. The alternatives considered include:

**Base case**. Three half-size coal-fired process boilers are installed to supply steam to the plant's 250-psig steam header. All 80-psig steam and steam to the 20-psig deaerating heater is pressure-reduced from the 250-psig steam header. The powerhouse auxiliary power requirements are 3.2 MW. Thus the utility tie must provide 33.2 MW to satisfy the average plant electric power needs.

**Case 1.** This alternative is based on installation of a noncondensing steam turbine generator. The unit initial

### Table 7.3 Plant Energy Supply System Considerations: Example 6

Process steam demands
    Net heat to process at 250 psig. 410°F—317 million Btu/hr avg.
    Net heat to process at 80 psig, 330°F—208 million Btu/hr avg. (peak requirements are 10% greater than
        average values)
Process condensate returns: 50% of steam delivered at 280°F
Makeup water at 80°F
Plant fuel is 3.5% sulfur coal
Coal and limestone for $SO_2$ scrubbing are available at a total cost of $2/million Btu fired
Process area power requirement is 30 MW avg.
Purchased power cost is 3.5 cents/kWh

steam conditions are 1450 psig, 950°F with automatic extraction at 250 psig and 80 psig exhaust pressure. The boiler plant has three half-size units providing the same reliability of steam supply as the Base Case. The feedwater heating system has closed feedwater heaters at 250 psig and 80 psig with a 20 psig deaerating heater. The 20-psig steam is supplied by noncondensing mechanical drive turbines used as powerhouse auxiliary drives. These units are supplied throttle steam from the 250-psig steam header. For this alternative, the utility tie normally provides 4.95 MW. The simplified schematic and energy balance is given in Figure 7.16.

The results of this cogeneration example are tabulated in Table 7.4. Included are the annual energy requirements, the 1980 investment costs for each case, and the annual operating cost summary. The investment cost data presented are for fully operational plants, including offices, stockrooms, machine shop facilities, locker rooms, as well as fire protection and plant security. The cost of land is not included.

The incremental investment cost for Case 1 given in Table 7.4 is $17.2 million. Thus the incremental cost is $609/kW for the 28.25-MW cogeneration system. This illustrates the favorable per unit cost for cogeneration systems compared to coal-fired facilities designed to provide kilowatts only, which cost in excess of $1000/kW.

The impact of fuel and purchased power costs other than Table 7.3 values on the GPO for this example is shown in Figure 7.17. Equivalent DRR values based

**Fig. 7.16 Simplified schematic and energy-balance diagram: Example 6, Case 1. All numbers are flows in $10^3$ lb/hr; Plant requirements given in Table 7.8, gross generation, 30.23 MW; powerhouse auxiliaries, 5.18 MW; net generation, 25.05 MW.**

on first-year annual operating cost savings can be estimated using Figure 7.15.

Sensitivity analyses often evaluate the impact of uncertainties in the installed cost estimates on the profitability of a project. If the incremental investment cost for cogeneration is 10% greater than the Table 7.4 estimate, the GPO would increase from 3.2 to 3.5 years. Thus the DRR would decrease from 17.5% to about 16%, as shown in Figure 7.15.

### Table 7.4 Energy and Economic Summary: Example 6

| Alternative | Base Case | Case 1 |
|---|---|---|
| Energy summary | | |
|     Boiler fuel ($10^6$ Btu/hr HHV) | 599 | 714 |
|     Purchased power (MW) | 33.20 | 4.95 |
| Estimated total installed cost ($10^6$ $) | 57.6 | 74.8 |
| Annual operating costs ($10^6$ $) | | |
|     Fuel and limestone at $2/$10^6$ Btu | 10.1 | 12.0 |
|     Purchased power at 3.5 cents/kWh | 9.8 | 1.5 |
|     Operating labor | 0.8 | 1.1 |
|     Maintenance | 1.4 | 1.9 |
|     Makeup water | 0.3 | 0.5 |
|     Total | 22.4 | 17.0 |
| Annual savings ($10^6$ $) | Base | 5.4 |
| Gross payout period (yrs) | Base | 3.2 |

Basis: (1) boiler efficiency is 87%; (2) operation equivalent to 8400 hr/yr at Table 7-3 conditions; (3) maintenance is 2.5% of the estimated total installed cost; (4) makeup water cost for case 1 is 80 cents/1000 gal *greater than* Base Case water costs; (5) stack gas scrubbing based on limestone system.

**Example 7:** The energy requirements for a chemical plant are presented in Table 7.5. The alternatives considered include:

**Base case.** Three half-size oil-fired packaged process boilers are installed to supply process steam at 150 psig. Each unit is fuel-oil-fired and includes a particulate removal system. The plant has a 60-day fuel-oil-storage capacity. A utility tie provides 30.33 MW average to supply process and boiler plant auxiliary power requirements.

**Case 1.** (Refer to Figure 7.18). This alternative examines the merit of adding a noncondensing steam turbine generator with 850 psig, 825°F initial steam conditions, 150-psig exhaust pressure. Steam is supplied by three half-size packaged boilers. The feedwater heating system is comprised of a 150-psig closed heater and a 20-psig deaerating heater. The steam for the deaerating heater is the exhaust of a mechanical drive turbine (MDT). The MDT is supplied 150-psig steam and drives some of the plant boiler feed pumps. The net generation of this cogeneration system is 6.32 MW when operating at the average 150-psig process heat demand. A utility tie provides the balance of the power required.

**Case 2.** (Refer to Figure 7.19). This alternative is a combined cycle using the 25,000-kW gas turbine generator whose performance is given in Table 7.7. An unfired HRSG system provides steam at both 850 psig, 825°F and 150 psig sat. Plant steam requirements in excess of that available from the two-pressure level unfired HRSG system are generated in an oil fired packaged boiler. The steam supplied to the noncondensing turbine is expanded to the 150-psig steam header. The net generation from the overall system is 26.54 MW. A utility tie provides power requirements in excess of that supplied by the cogeneration system. The plant-installed cost estimates for Case 2 include two half-size package boilers. Thus full steam output can be realized with any steam generator out of service for maintenance.

The energy summary, annual operating costs, and economic results are presented in Table 7.6. The results show that the combined cycle provides a GPO of 2.5 years based on the study fuel and purchased power costs. The incremental cost for Case 2 relative to the Base Case is $395/kW compared to $655/kW for Case 1 relative to the Base Case. This favorable incremental investment cost combined with a FCP of 5510 Btu/kWh contribute to the low CPO.

The influence of fuel and power costs other than those given in Table 7.5 on the GPO for cases 1 and 2 is

**Fig. 7.17 Effect of different fuel and power costs on cogeneration profitability: Example 1.** *Basis:* **Conditions given in Tables 7.3 and 7.4.**

**Table 7.5 Plant Energy Supply System Considerations: Example 7**

Process steam demands
    Net heat to process at 150 psig sat—158.5 million Btu/hr avg. (peak steam requirements are 10% greater than average values)

Process condensate returns: 45% of the steam delivered at 300°F

Makeup water at 80°F

Plant fuel is fuel oil

Fuel cost is $5/million Btu

Process areas require 30 MW

Purchased power cost is 5 cents/kWh

**Fig. 7.18 Simplified schematic and energy-balance diagram: Example 7, Case 1. All numbers are flows in 1000 lb/hr; gross generation, 6.82 MW; powerhouse auxiliaries, 0.50 MW; net generation; 6.32 MW.**

**Fig. 7.19 Simplified schematic and energy-balance diagram: Example 7, Case 2. All numbers are flows in 1000 lb/hr; gross generation, 26.77 MW, powerhouse auxiliaries, 0.23 MW: net generation, 26.54 MW.**

### Table 7.6 Energy and Economic Summary: Example 7

| Alternative | Base Case | Case 1 | Case 2 |
|---|---|---|---|
| Energy summary | | | |
| Fuel ($10^6$ Btu/hr HHV) | | | |
| Boiler | 183 | 209 | 34 |
| Gas turbine | — | 297 | |
| Total | 183 | 209 | 331 |
| Purchased power (MW) | 30.33 | 23.77 | 3.48 |
| Estimated total installed cost ($10^6$ $) | 8.3 | 12.6 | 18.9 |
| Annual operating cost ($10^6$ $) | | | |
| Fuel at $5/M Btu HHV | 7.7 | 8.8 | 13.9 |
| Purchased power at 5 cents/kWh | 12.7 | 10.0 | 1.5 |
| Operating labor | 0.6 | 0.9 | 0.9 |
| Maintenance | 0.2 | 0.3 | 0.5 |
| Makeup water | 0.1 | 0.2 | 0.2 |
| Total | 21.3 | 20.2 | 17.0 |
| Annual savings ($10^6$ $) | Base | 1.1 | 4.3 |
| Gross payout period (yr) | Base | 3.9 | 2.5 |

*Basis*: (1) gas turbine performance per Table 7-7; (2) boiler efficiency, 87%; (3) operation equivalent to 8400 hr/yr at Table 7-5 conditions; (4) maintenance, 2.5% of the estimated total installed costs; (5) incremental makeup water cost for cases 1 and 2 relative to the Base Case. $1 /1000 gal.

shown in Figure 7.20. These GPO values can be translated to DRRs using Figure 7.15.

**Example 8.** A gas-turbine and HRSG cogeneration system is being considered for a brewery to supply baseload electrical power and part of the steam needed for process. An overview of the proposed system is shown in Figure 7.21. This example shows the use of computer tools in cogeneration design and evaluation.

Base Case.: Currently, the plant purchases about 3,500,000 kWh per month at $0.06 per kWh. The brewery uses an average of 24,000 lb/hr of 30 psig saturated steam. Three 300-BHP gas fired boilers produce steam

at 35 psig, to allow for pressure losses. The minimum steam demand is 10,000 lb/hr. The plant operates continuously during ten months or 7,000 hr/year. The base or minimum electrical load during production is 3,200 kW. The rest of the time (winter) the brewery is down for maintenance. The gas costs $3.50/MMBtu.

Case 1: Consider the gas turbine whose ratings are given on Figure 7.11. We will evaluate this turbine in conjunction with an unfired water-tube HRSG to supply part of the brewery's heat and power loads. First, we obtain the ratings and performance data for the selected turbine, which has been sized to meet the electrical base load (3.5 MW). An air washer/evaporative cooler will be installed

**Table 7.7 Steam Generation and Fuel Chargeable to Power: 25,000-kW ISO Gas Turbine and HRSG (Distillate Oil Fuel)**[a]

| Type HRSG | Unfired | | Supplementary Fired | | Fully Fired | |
|---|---|---|---|---|---|---|
| Gas Turbine | | | | | | |
| Fuel ($10^6$ Btu/hr HHV) | 297 | ————————————————————————————————→ | | | | |
| Output (MW) | 21.8 | | 21.6 | | 21.4 | |
| Airflow ($10^3$ lb/hr) | 915 | ————————————————————————————————→ | | | | |
| Exhaust temperature (°F) | 920 | | 922 | | 925 | |
| HRSG fuel ($10^6$ Btu/hr HHV) | NA | | 190 | | 769 | |
| | Steam ($10^3$ lb/hr) | FCP (Btu/kWh HHV) | Steam ($10^3$ lb/hr) | FCP (Btu/kWh HHV) | Steam ($10^3$ lb/hr) | FCP (Btu/kWh HHV) |
| Steam conditions | | | | | | |
| 250 psig sat. | 133 | 6560 | 317 | 5620 | 851 | 4010 |
| 400 psig, 650°F | 110 | 7020 | 279 | 5630 | 751 | |
| 600 psig, 750°F | 101 | 7340 | 268 | 5660 | 722 | |
| 850 psig, 825°F | 93 | 7650 | 261 | 5700 | 703 | |
| 1250 psig, 900°F | — | — | 254 | 5750 | 687 | |
| 1450 psig, 950°F | — | — | 250 | 5750 | 675 | ↓ |

[a]*Basis*: (1) gas turbine performance given for 80°F ambient temperature, sea-level site; (2) HRSG performance based on 3% blowdown, 1-1/2% radiation and unaccounted losses, 228°F feedwater; (3) no HRSG bypass stack loss; (4) gas turbine exhaust pressure loss is 10 in. $H_2O$ with unfired, 14 in. $H_2O$ with supplementary fired, and 20 in. $H_2O$ with fully fired HRSG; (5) fully fired HRSG based on 10% excess air following the firing system and 300°F stack. (6) fuel chargeable to gas turbine power assumes total fuel credited with equivalent 88% boiler fuel required to generate steam; (7) steam conditions are at utilization equipment; a 5% AP and 5°F AT have been assumed from the outlet of the HRSG.

**Fig. 7.20 Effect of different fuel and power cost on co-generation profitability: Example 2. *Basis*: Conditions given in Tables 7.4 and 7.5.**

at the turbine inlet to improve (reduce) the overall heat rate by precooling the inlet air to an average 70°F (80°F or less), during the summer production season Additional operating data are given below.

*Operating Data*

Inlet air pressure losses (filter and air
  pre-cooler): 5" $H_2O$

Exhaust Losses (ducting, by-pass valve,
  HRSG and Stack): 12" $H_2O$

Location Elevation above sea level: 850 ft

Thus, on a preliminary basis, we assume the turbine will constantly run at full capacity, minus the effect of elevation, the inlet air pressure drop and exhaust losses. Since the plant will be located at 850 ft above sea level, from Figure 7.11, the elevation correction factor is 0.90. Hence, the corrected continuous power rating (before deducting pressure losses) when firing natural gas and using 70°F inlet air is:

**Figure 7.21 Gas turbine/HRSG cogeneration application.**

= (Generator Output @ 70°F)
  (Elevation correction @ 850 ft)

= 4,200 kWe × 0.9

= 3,780 kWe

Next, by using the Inlet and Exhaust Power Loss graphs in Figure 7.11, we get the exhaust and inlet losses (@ 3780 kW output): 17 and 7 kW/inch $H_2O$, respectively. So, the total power losses due to inlet and exhaust losses are:

= (17 in)(5 kW/in) + (12 in)(8 kW/in)
= 181 kW

Consequently, the net turbine output after elevation and pressure losses is
    = 3,780 - 181
    = 3,599 kWe

Next, from Figure 7.11 we get the following performance data for 70°F inlet air:

| | |
|---|---|
| Heat rate | : 12,250 Btu/kWh (LHV) |
| Exhaust Temperature | : 935°F |
| Exhaust Flow | : 160,000 lb/hr |

These figures have been used as input data for HGPRO—a prototype HRSG software program developed by V. Ganesh, W.C. Turner and J.B. Wong in 1992 at Oklahoma State University. The program results are shown in Fig. 7.22.

The total installed cost of the complete cogeneration plant including gas turbine, inlet air precooling, HRSG, auxiliary equipment and computer based con-trols is $4,500,000. Fuel for cogeneration is available on a long term contract basis (>5 years) at $2.50/MMBtu. The brewery has a 12% cost of capital. Using a 10-year after tax cash flow analysis with current depreciation and tax rates, <u>should the brewery invest in this cogeneration option</u>? For this evaluation, assume: (1) A 1% inflation for power and non-cogen natural gas; (2) an operation and maintenance (O&M) cost of $0.003/kWh for the first year after the project is installed. Then, the O&M cost should escalate at 3% per year; (3) the plant salvage value is neglected.

**Economic Analysis**

Next, we present system operation assumptions required to conduct a preliminary economic analysis.

1)  The cogeneration system will operate during all the production season (7,000 hrs/year).

2)  The cogeneration system will supply an average of 3.5 MW of electrical power and 24,000 lb of 35 psig steam per hour. The HRSG will be provided with an inlet gas damper control system to modulate and by-pass hot gas flow. This is to allow for variable steam production or steam load-following operation.

3)  The balance of power will be obtained from the existing utility at the current cost ($0.06/kWh)

4)  The existing boilers will remain as back-up units. Any steam deficit (considered to be negligible) will be produced by the existing boiler plant.

5)  The cogeneration fuel (natural gas) will be metered with a dedicated station and will be available at

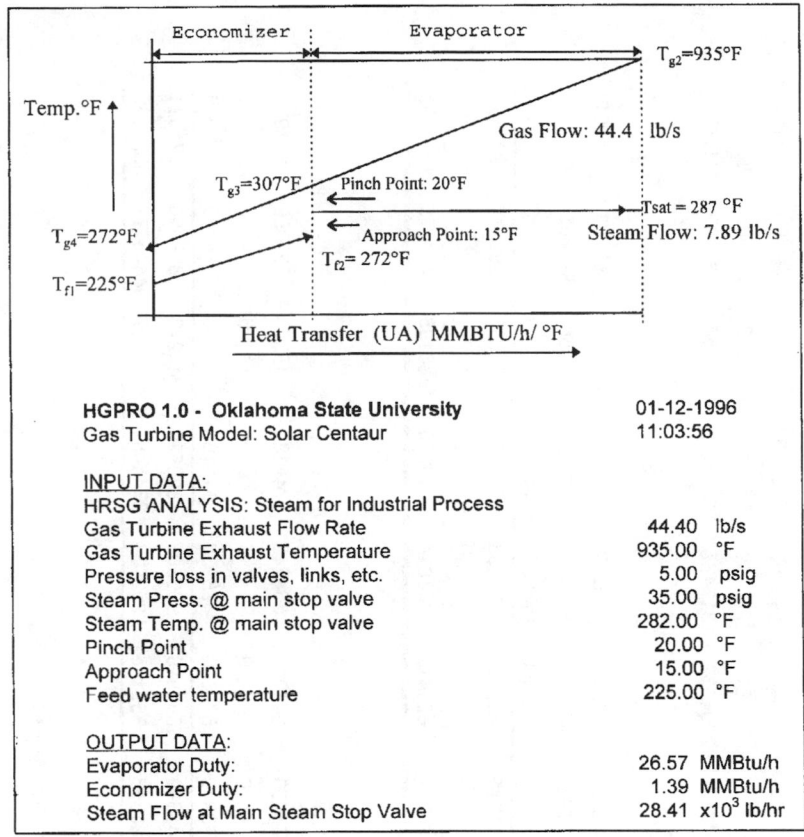

**HGPRO 1.0 - Oklahoma State University**       01-12-1996
Gas Turbine Model: Solar Centaur             11:03:56

INPUT DATA:
HRSG ANALYSIS: Steam for Industrial Process

| | | |
|---|---:|---|
| Gas Turbine Exhaust Flow Rate | 44.40 | lb/s |
| Gas Turbine Exhaust Temperature | 935.00 | °F |
| Pressure loss in valves, links, etc. | 5.00 | psig |
| Steam Press. @ main stop valve | 35.00 | psig |
| Steam Temp. @ main stop valve | 282.00 | °F |
| Pinch Point | 20.00 | °F |
| Approach Point | 15.00 | °F |
| Feed water temperature | 225.00 | °F |

OUTPUT DATA:

| | | |
|---|---:|---|
| Evaporator Duty: | 26.57 | MMBtu/h |
| Economizer Duty: | 1.39 | MMBtu/h |
| Steam Flow at Main Steam Stop Valve | 28.41 | x10³ lb/hr |

**Figure 7.22 Results from HGPRO 1.0, a prototype HRSG software.**

$2.50/MMBtu during the first five years and at $2.75/MMBtu during the next five-year period. Non cogeneration fuel will be available at the current price of $3.50/MMBtu.

The discounted cash flow analysis was carried out using an electronic spreadsheet (Table 7.8). The results of the spreadsheet show a positive net present value. Therefore, when using the data and assumptions given in this case, the cogeneration project appears to be cost effective. The brewery should consider this project for funding and implementation.

*Note:* These numbers ignore breakdowns and possible ratchet clause effects.

## 7.6   CLOSURE

Cogeneration has been used for almost a century to supply both process heat and power in many large industrial plants in the United States. This technology would have been applied to a greater extent if we did not experience a period of plentiful low-cost fuel and reliable low-cost electric power in the 25 years following the end of World War 11. Thus economic rather than technical considerations have limited the application of this energy-saving technology.

The continued increase in the cost of energy is the primary factor contributing to the renewed interest in cogeneration and its potential benefits. This chapter discusses the various prime movers that merit consideration when evaluating this technology. Furthermore, approximate performance levels and techniques for developing effective cogeneration systems are presented.

The cost of all forms of energy is rising sharply. Cogeneration should remain an important factor in effectively using our energy supplies and economically providing goods and services in those base-load applications requiring large quantities of process heat and power.

## 7.6   REFERENCES

1. Butler, C.H., (1984), *Cogeneration: Engineering, Design Financing, and Regulatory Compliance*, McGraw-Hill, Inc., New York, N.Y.

2. Caton, J.A., et al., (1987), *Cogeneration Systems, Texas A&M University*, College Station, TX.

3. CFR-18 (1990): Code of Federal Regulations, Part 292 Regulations Under Sections 201 and 210 of the Public

## INPUT DATA

| | |
|---|---|
| Power & gas cost escalation rate | 1% (year 1-10) |
| Labor, operat. & maint. escalation rat | 3% (year 1-10) |
| Cogen power generation | 3,500 kW |
| Cogen steam production | 24,000 lh/hr |
| Existing boiler efficiency | 80% lh/hr |
| Present Electricity Cost | $0.06 /kWh |
| Present Nat. Gas Cost | $3.50 /MMBtu |
| Turbine plant heat rate, LHV @ 70°F | 12,250 MMBtu/kW    93% LHV/HHV |
| Cogen Nat. Gas Cost | $2.75 /MMBtu |
| Cogen System O&M Cost | $2.50 /MMBtu    $0.003 /kWh    yrs: 6-10 |
| Cogen System Installed Cost | $4,500,000 |
| Operation time: | 7,000 hr/yr. |

## GAS TURBINE / HRSG COGENERATION TEN YEAR ECONOMIC ANALYSIS

| | NPV @ 12.00% | 1996 | 1997 | 1998 | 1999 | 2000 | 2001 | 2002 | 2003 | 2004 | 2005 |
|---|---|---|---|---|---|---|---|---|---|---|---|
| **Savings** | | | | | | | | | | | |
| Power cost savings | 8610742 | 1470000 | 1484700 | 1499547 | 1514542 | 1529688 | 1544985 | 1560435 | 1576039 | 1591799 | 1607717 |
| Nat. Gas to steam savings | 4305371 | 735000 | 742350 | 749774 | 757271 | 764844 | 772492 | 780217 | 788019 | 795900 | 803859 |
| Total Savings | 12916113 | 2205000 | 2227050 | 2249321 | 2271814 | 2294532 | 2317477 | 2340652 | 2364058 | 2387699 | 2411576 |
| **Costs** | | | | | | | | | | | |
| Operating & Maint Costs | -463291 | -73500 | -75705 | -77976 | -80315 | -82725 | -85207 | -87763 | -90396 | -93108 | -95901 |
| Cogen Nat. Gas Cost | -4723554 | -806788 | -806788 | -806788 | -806788 | -806788 | -887466 | -887466 | -887466 | -887466 | -887466 |
| Depreciation: 5-year MACRS | -3520539 | -504000 | -1310400 | -1290240 | -773640 | -579600 | -435960 | -146160 | 0 | 0 | 0 |
| Total Costs | -8707384 | -1384288 | -2192893 | -2175004 | -1660743 | -1469113 | -1408633 | -1121389 | -977862 | -980574 | -983367 |
| **Net Savings** | 4208729 | 820712 | 34157 | 74317 | 611071 | 825419 | 908844 | 1219263 | 1386196 | 1407125 | 1428209 |
| **Cashflows** | | | | | | | | | | | |
| Post-Tax Income/(Loss) | 2569428 | 501045 | 20853 | 45370 | 373059 | 503918 | 554849 | 744360 | 846273 | 859050 | 871921 |
| Investment | -4258929 | -2520000 | -2520000 | 0 | 0 | 0 | 0 | 0 | 0 | 0 | 0 |
| Depreciation Add Back | 3520539 | 504000 | 1310400 | 1290240 | 773640 | 579600 | 435960 | 146160 | 0 | 0 | 0 |
| Total Cashflows | 1831039 | -1514955 | -1188747 | 1335610 | 1146699 | 1083518 | 990809 | 890520 | 846273 | 859050 | 871921 |
| Cumulative Cashflow | | -1514955 | -2703702 | -1368092 | -221393 | 862125 | 1852934 | 2743454 | 3589727 | 4448777 | 5320698 |

**NPV : $1,831,039**

**Table 7.8 After tax discounted cash flow economic analysis.**

Utility Regulatory Policies Act of 1978 With Regard to Small Power Production and Cogeneration, (4-1-90 Edition).

4. Estey P.N., et al., (1984). "A Model for Sizing Cogeneration Systems," Proceedings of the 19th Intersociety Energy Conversion Engineering Conference" Vol. 2 of 4, August, 1984, San Francisco, CA.

5. Ganapathy, V. (1991). *Waste Heat Boiler Deskbook,* The Fairmont Press, Inc., Lilburn, CA.

6. Harkins H.L., (1981), "PURPA New Horizons for Electric Utilities and Industry," IEEE Transactions, Vol. PAS-100, pp 27842789.

7. Hay, N., (1988), *Guide to Natural Gas Cogeneration,* The Fairmont Press, Lilburn, GA.

8. Kehlhofer, R., (1991). *Combined-Cycle Gas & Steam Turbine Power Plants,* The Fairmont Press, Inc. Lilburn, Ga.

9. Kostrzewa, L.J. & Davidson, K.G., (1988). "Packaged Cogeneration," *ASHRAE Journal,* February 1988.

10. Kovacik, J.M., (1982), "Cogeneration," in *Energy Management Handbook,* ed. by W.C. Turner, Wiley, New York, N.Y.

11. Kovacik, J.M., (1985), "Industrial Cogeneration: System Application Consideration," *Planning Cogeneration Systems,* The Fairmont Press, Lilburn, Ga.

12. Lee, R.T.Y., (1988), "Cogeneration System Selection Using the Navy's CELCAP Code," *Energy Engineering, Vol. 85, No. 5, 1988.*

13. Limaye, D.R. and Balakrishnan, S., (1989), "Technical and Economic Assessment of Packaged Cogeneration Systems Using Cogenmaster," *The Cogeneration Journal, Vol. 5, No. 1, Winter 1989-90.*

14. Limaye, D.R., (1985), *Planning Cogeneration Systems,* The Fairmont Press, Atlanta, CA.

15. Limaye, D.R., (1987), *Industrial Cogeneration Applications,* The Fairmont Press, Atlanta, CA.

16. Mackay, R. (1983). "Gas Turbine Cogeneration: Design, Evaluation and Installation." The Garret Corporation, Los Angeles, CA, The Association Of Energy Engineers, Los Angeles CA, February, 1983.

17. Makansi, J., (1991). "Independent Power/Cogeneration, Success Breeds New Obligation-Delivering on Performance," *Power,* October 1991.

18. Mulloney, et. al., (1988). "Packaged Cogeneration Installation Cost Experience," Proceedings of The 11th World Energy Engineering Congress, October 18-21, 1988.

19. Orlando, J.A., (1991). *Cogeneration Planners Handbook,* The Fairmont Press, Atlanta, GA.

20. Oven, M., (1991), "Factors Affecting the Financial Viability Applications of Cogeneration," XII Seminario Nacional Sobre El Uso Racional de La Energia," Mexico City, November, 1991.

21. Polimeros, G., (1981), *Energy Cogeneration Handbook,* Industrial Press Inc., New, York.

22. Power (1980), "FERC Relaxes Obstacles to Cogeneration," *Power,* September 1980, pp 9-10.

23. SFA Pacific Inc. (1990). "Independent Power/Cogeneration, Trends and Technology Update," *Power,* October 1990.

24. Somasundaram, S., et al., (1988), *A Simplified Self-Help Approach To Sizing* of *Small-Scale Cogeneration Systems,* Texas A&M University, College Station, TX

25. Spiewak, S.A. and Weiss L., (1994) *Cogeneration & Small Power Production Manual,* 4th Edition, The Fairmont Press, Inc. Lilburn, CA.

26. Turner, W.C. (1982). *Energy Management Handbook,* John Wiley & Sons, New York, N.Y.

27. Tuttle, D.J., (1980), PURPA 210: New Life for Cogenerators," *Power,* July, 1980.

28. Williams, D. and Good, L., (1994) *Guide to the Energy Policy Act of* 1992, The Fairmont Press, Inc. Lilburn, GA.

29. Wong, J.B., Ganesh, V. and Turner, W.C. (1991), "Sizing Cogeneration Systems Under Variable Loads," 14th World Energy Engineering Congress, Atlanta, GA.

30. Wong, J.B. and Turner W.C. (1993), "Linear Optimization of Combined Heat and Power Systems," Industrial Energy Technology Conference, Houston, March, 1993.

## APPRECIATION

Many thanks to Mr. Lew Gelfand for using and testing over the years the contents of this chapter in the evaluation and development of actual cogeneration opportunities, and to Mr. Scott Blaylock for the information provided on fuel cells and microturbines. Messrs. Gelfand and Blaylock are with DukeEnergy/DukeSolutions.

# Appendix A

*Statistical Modeling of Electric Demand and*
*Peak–Shaving Generator Economic Optimization*
*Jorge B. Wong, Ph.D., PE, CEM*

## ABSTRACT

This paper shows the development a basic electric demand statistical model to obtain the optimal kW–size and the most cost–effective operating time for an electrical peak shaving generator set. This model considers the most general (and simplified) case of a facility with an even monthly demand charge and a uniformly distributed random demand, which corresponds to a linear load–duration curve. A numerical example and computer spreadsheet output illustrate the model.

## INTRODUCTION

Throughout the world, electrical utilities include a hefty charge in a facility's bill for the peak electrical demand incurred during the billing period, usually a month. Such a charge is part of the utility's cost recovery or amortization of newly installed capacity and for operating less efficient power plant capacity during higher load periods.

Demand charge is a good portion of a facility's electrical bill. Typically a demand charge can be as much as 50% of the bill, or more. Thus, to reduce the demand cost, many industrial and commercial facilities try to "manage their loads." One example is by moving some of the electricity–intense operations to "off–peak" hours"—when a facility's electrical load is much smaller and the rates ($/kW) are lower. But, when moving electrical loads to "off–peak" hours is not practical or significant, a facility will likely consider a set of engine–driven or fuel cell generators to run in parallel with the utility grid to supply part or all the electrical load demand during "on–peak" hours. We call these Peak Shaving Generators or PSGs.

While the electric load measurement is instantaneous, the billing demand is typically a 15–to–30– minute average of the instantaneous electrical power demand (kW). To obtain the monthly demand charge, utilities multiply the billing demand by a demand rate. Some utilities charge a flat rate ($/kW–peak per month) for all months of the year. Other utilities have seasonal charges (i.e. different rates for different seasons of the year). Still, others use ratchet clauses to account for the highest "on–peak" season demand of the year.

Thus, the model presented in this paper focuses on the development of a method to obtain the optimal PSG size ($g^*$kW) and PSG operation time (hours per year) for a given facility. This model is for the case of a facility with a constant billing demand rate ($/kW/month) throughout the year. The analysis is based on a linear load–duration curve and uses a simplified life–cycle–cost approach. An example illustrates the underlying approach and optimization method. In addition, the paper shows an EXCEL spreadsheet to implement the optimization model. We call this model PSG–1.

## ELECTRIC DEMAND STATISTICAL MODEL

This section develops the statistical–and–math model for the economical sizing of an electrical peak–shaving generator set (PSG) for a given facility. The fundamental question is: *What is the most economical generator–set size—$g^*$ in kW—for a given site demand profile?* Figure 1 shows a sample record for a facility's electrical demand, which is uniformly distributed between 2000 and 5300 kW. Next, Figure 2 shows the corresponding statistical distributions.

The statistical model of electrical demand is expressed graphically in Figure 2, in terms of two functions:

- The load–duration curve $D(t)$, is the demand as a function of cumulative time $t$ (i.e. the accumulated annual duration $t$ in hrs/year of a given $D(t)$ load in kW), and

- The load frequency distribution $f(D)$ (rectangular shaded area in Figures 1 and 2) is the "uniform" probability density function.

## MODEL ASSUMPTIONS

This statistical model is based upon the following assumptions:

1.  The electrical demand is represented by a linear load–duration curve, as shown in Figure 2. Thus, for a typical year, the facility has a demand $D$ that varies between an upper value $D_u$ (annual maximum) and a lower value $D_1$ (annual minimum). This implies the electrical load is uniformly distributed between the maximum and minimum demands. The facility operates T hours per year.

**Figure 1. Sample record for a uniformly distributed random demand.**

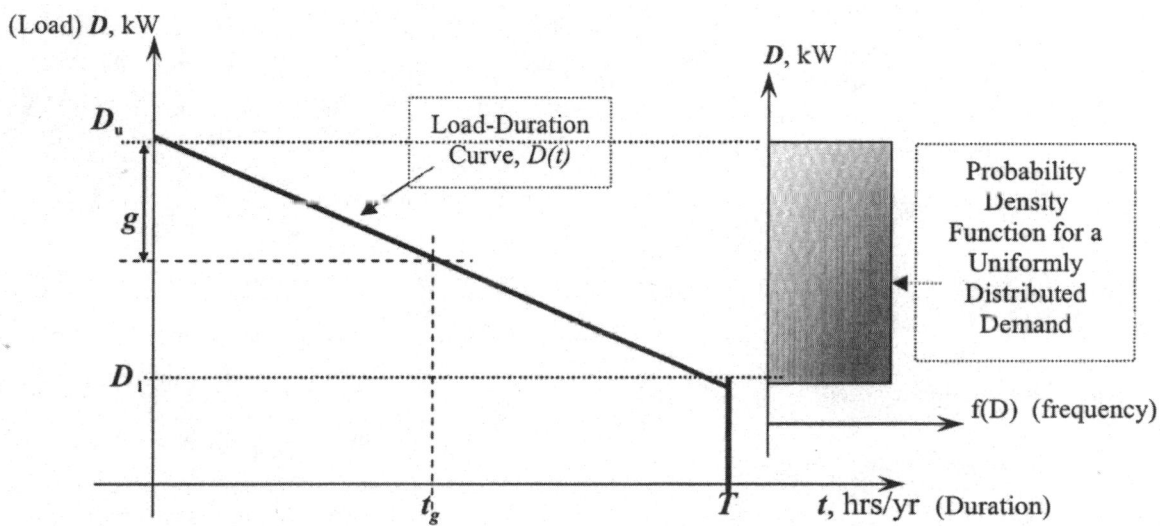

**Figure 2. Load—Duration Curve for uniformly distributed demand.**

2. There is an even energy or consumption rate $Ce$ ($/kWh) throughout the year.

3. There is an even demand rate $Cd$ ($/kW/month) for every month of the year.

4. There is a same demand peak $D_u$ for every month. Demand ratchet clauses are not applicable in this case.

5. The equipment's annual ownership or amortization unit installed cost ($/kW/year) is constant for all sizes of PSGs. The unit ownership or rental cost ($/kW/year) is considered independent of unit size. Ownership, rental or lease annualized costs are denoted by $A_c$.

6. A PSG set is installed to reduce the peak demand by a maximum of $g$ kW, operating $t_g$ hours per year.

## BASE CASE ELECTRICITY ANNUAL COST—WITHOUT PEAK–SHAVING

Consider a facility with the load–duration characteristic shown in Figures 1 and 2. For a unit consumption cost $Ce$, the *annual energy or consumption cost* (without PSG) for the facility is

$$AEC = T \cdot D_1 \cdot Ce + 1/2\ T\ (D_u - D_1)\ Ce$$

Which is equivalent to

$$AEC = T/2 \cdot Ce \ (D_u + D_1) \tag{1}$$

Next, considering a peak demand $D_u$ occurs every month, the *annual demand cost* is defined by

$$ADC = 12 \ D_u \cdot Cd \tag{2}$$

Thus, the total *annual cost* for the facility is

$$TAC = AEC + ADC \tag{3a}$$

Substituting [1] and [2] in equation [3], we have the base case total annual cost:

$$TAC_1 = T/2 \ (D_u + D_1) \ Ce + 12 \ D_u \cdot Cd \tag{3b}$$

## ELECTRICITY ANNUAL COST WITH PEAK–SHAVING

If a peak shaving generator of size $g$ is installed in the facility to run in parallel with the utility grid during peak–load hours, so the maximum load seen by the utility is $(D_u - g)$, then the *electric bill cost* is

$$EBC = T/2 \ (D_u - g + D_1) \ Ce + 12 \ (D_u - g) \cdot Cd$$

In addition, the facility incurs an ownership (amortization) unit cost Ac ($/kW/yr) and operation and maintenance unit cost $O\&M$ ($/kWh). Hence, the *total annual cost* with demand peak shaving is

$$TAC_2 = [T \cdot D_1 + (T + t_g)/2 \ (D_u - g - D_1)]Ce + 12 \ (D_u - g) \ Cd + (Ac + 1/2 \ O\&M \cdot t_g)g \tag{4}$$

## ANNUAL WORTH OF THE PEAK-SHAVING GENERATOR

The *annual worth or net savings* AW ($/yr) of the PSG set are obtained by subtracting equation [4] from equation [3]. That is $AW = TAC_1 - TAC_2$. So,

$$AW = 1/2 \ t_g \cdot g \cdot Ce + 12 \cdot g \cdot Cd - (Ac + 1/2 \cdot O\&M \cdot t_g)g \tag{5}$$

From Figure 2 we obtain $g$: $t_g = (D_u - D_1)$: $T$

So, the expected PSG operating time is

$$t_g = g \cdot T/(D_u - D_1) \tag{6}$$

Substituting the value of $t_g$ in equation [5], we have:

$$AW = g^2 \cdot T/[2(D_u - D_1)] \ Ce + 12 \cdot g \cdot Cd - \{Ac + O\&M \cdot g \cdot T/[2(D_u - D_1)]\} \ g \tag{7}$$

## OPTIMUM CONDITIONS

We next determine the necessary and sufficient conditions for an optimal PSG size $g^*$ and the corresponding maximum AW to exist.

### Necessary Condition

By taking the derivative of *AW*, Equation [7], with respect to $g$ and equating it to zero we obtain the necessary condition for the maximum annual worth or net saving per year. That is:

$$AW' = g \cdot T \cdot Ce/(D_u - D_1) + 12 \ Cd - Ac - g \cdot T \cdot O\&M/(D_u - D_1) = 0 \tag{8}$$

Sufficient Condition. If the second derivative of *AW* with respect to $g$ is negative, i.e. $AW'' < 0$, then *AW* $(g)$ is a strictly convex function of $g$ with a global maximum point. So, by taking the second derivative of *AW* with respect to $g$ and evaluating $AW''$ as an inequality ($<0$) we have:

$$AW'' = T \cdot Ce/(D_u - D_1) - T \cdot O\&M/(D_u - D_1) < 0$$

Multiplying this equation by $(D_u - D_1)/T$ we have the sufficient condition for a maximum *AW* is

$$Ce - O\&M < 0$$
$$\text{or}$$
$$Ce < O\&M$$

Therefore, for a global maximum *AW* to exist, the energy rate $Ce$ must be less than the per unit $O\&M$ cost (including fuel) to operate the peak shaving generator ($/kWh). Since this is the case for most utility rates $Ce$ and commercial PSGs $O\&M$, we can say there is maximum *AW* and an optimal $g^*$ for the typical electrical demand case.

## OPTIMUM PEAK SHAVING GENERATOR SIZE

From equation [6] we can solve for $g$ and find the *optimal PSG size*, $g^*$ (in kW):

$$g^* = (12 \ Cd - Ac) \ (D_u - D_1)/[T \ (O\&M - Ce)] \tag{9}$$

## FOR FURTHER RESEARCH

Further research is underway to develop enhanced models which consider:

- Demand profile flexibility. Other load–duration shapes with different underlying frequency distributions (e.g. triangular, normal and auto–correlated loads).

- Economies of Scale. The fact that larger units have better fuel–to–electricity efficiencies (lower heat rates) and lower per unit installed cost ($/kW).

---

**EXAMPLE.** A manufacturing plant operates 7500 hours per year and has a fairly constant electrical (billing) peak demand every month (See Figure 1). The actual load, however, varies widely between a minimum of 2000 kW and a maximum of 5300 kW (See Figure 2). The demand charge is $10/kW/month and the energy charge is $0.05/kWh. The installed cost of a diesel generator set, the auxiliary electrical switch gear and peak–shaving controls is about $300 per kW. Alternatively, the plant can lease a PSG for $50/kW/yr. The operation and maintenance cost (including diesel fuel) is $0.10/kWh.

Assuming the plant leases the PSG, estimate (1) the optimal PSG size, (2) the annual savings and (3) the PSG annual operation time.

1) The optimal generator size is calculated using equation [9]

$$g^* = \frac{(12. \$10 - \$50/kWh) (5300 - 2000 \text{ kW})}{7500 \text{ h/yr} (\$0.10/kWh - \$0.05/kWh)}$$

$$= \boldsymbol{616 \text{ kW}}$$

2) Using a commercially available PSG of size $g^* = 600$ kW, the potential annual savings are estimated using equation [7]

$$
\begin{aligned}
\mathbf{AW} &= g^2 \cdot T \cdot Ce/[2(D_u - D_1)] + 12 \cdot g \cdot Cd \\
&\quad - \{Ac + O\&M \cdot g \cdot T/[2(D_u - D_1)]\} \, g \\
&= 600^2 \times 7500 \times 0.05/(2(5300 - 2000)) + 12 \times 600 \times 10 \\
&\quad - (\$50 + 0.10 \times 600 \times 7500/(2 \, (5300 - 2000))) \, 600 \\
&= \$20,455 + \$72,000 - \$70,909 \\
&= \$21,546/year
\end{aligned}
$$

3) The expected annual operating time for the PSG is estimated using Equation [6]

| g | AWg |
|---|---|
| (kW) | ($/yr) |
| 0 | 0 |
| 66 | 4373 |
| 132 | 8250 |
| 198 | 11633 |
| 264 | 14520 |
| 330 | 16913 |
| 396 | 18810 |
| 462 | 20213 |
| 528 | 21120 |
| 594 | 21533 |
| 660 | 21450 |
| 726 | 20873 |
| 792 | 19800 |
| 858 | 18233 |
| 924 | 16170 |
| 990 | 13613 |
| 1056 | 10560 |
| 1122 | 7013 |
| 1188 | 2970 |
| 1254 | -1568 |

**Figure 3. PSG-1 Spreadsheet and Chart**

$$t_g = g \cdot T/(D_u - D_1)$$
$$= 600 \times 7500/(5300-2000)$$
$$= 1{,}364 \text{ hours/year}$$

The Excel spreadsheet and chart used to solve this case example is shown in Figure 3.

## CONCLUDING REMARKS

The reader should note that the underlying statistical and optimization model is quite "responsive and robust." That is, the underlying methodology can be used in, or adapted to, a variety of demand profiles and rates, while the results remain relatively valid. A forthcoming paper by this author will show how to adapt the linear load– duration models of Figures 1 and 2 to more complex demand profiles. Thus, for example, one typical case is when the electrical load is represented by a Gauss or normal distribution. Also, we will show how to apply equation [9] to more involved industrial cases with multiple billing seasons and demand rates.

## Appendix References

Beightler, C.S., Phillips, D.T., and Wilde, D.J., *Foundations of Optimization*, Prentice–Hall, Englewood Cliffs, 1979.

Turner, W.C., *Energy Management Handbook*, 4th Edition, the Fairmont Press, Lilburn, GA, 2001.

Hahn & Shapiro, *Statistical Models in Engineering*, John Wiley 1967, Wiley Classics Library, reprinted in 1994.

Witte, L.C., Schmidt, P.S., and Brown, D.R, *Industrial Energy Management and Utilization*, Hemisphere Publishing Co. and Springer–Verlag, Berlin, 1988.

## Appendix Nomenclature

| | |
|---|---|
| $Ac$ | Equipment ownership, lease or rental cost ($/kW/year) |
| $ADC$ | Annual Demand Cost ($/year) |
| $AEC$ | Annual Energy Cost ($/year) |
| $AW$ | Annual Worth ($/year) |
| $Cd$ | Electric demand unit cost ($/kW/month) |
| $Ce$ | Electric energy unit cost ($/kWh) |
| $D$ | Electric demand or load (kW) |
| $D_1$ | Lower bound of a facility's electric demand or minimum load (kW) |
| $D_u$ | Upper bound of a facility's electric demand or maximum load (kW) |
| $EBC$ | Electric bill cost for a facility with PSG, ($/year) |
| $f(D)$ | Frequency of occurrence of a demand, (unit less) |
| $O\&M$ | Operation and Maintenance cost, including fuel cost ($/kWh) |
| $g$ | Peak shaving generator size or rated output capacity (kW) |
| $g^*$ | Optimal peak shaving generator size or output capacity (kW) |
| $t$ | Time, duration of a given load, (hours/year) |
| $t_g$ | Expected time of operation for a PSG, hours/year |
| $T$ | Facility operation time using power(hours/year) |
| $TAC$ | Total annual electric cost |
| $TAC_1$ | Total annual cost, base case w/o PSG ($/year) |
| $TAC_2$ | Total annual cost, with PSG ($/year) |

*Jorge B. Wong, Ph.D., PE, CEM is an energy management advisor and instructor. Jorge helps facility managers and engineers. Contact Jorge: jorgebwong@att.net*

CHAPTER 8

# WASTE-HEAT RECOVERY

**WESLEY M. ROHRER, JR.**
Emeritus Associate Professor of
    Mechanical Engineering
University of Pittsburgh
Pittsburgh, Pennsylvania

## 8. 1 INTRODUCTION

### 8.1.1 Definitions

Waste heat, in the most general sense, is the energy associated with the waste streams of air, exhaust gases, and/or liquids that leave the boundaries of a plant or building and enter the environment. It is implicit that these streams eventually mix with the atmospheric air or the groundwater and that the energy, in these streams, becomes unavailable as useful energy. The absorption of waste energy by the environment is often termed thermal pollution.

In a more restricted definition, and one that will be used in this chapter, waste heat is that energy which is rejected from a process at a temperature high enough above the ambient temperature to permit the economic recovery of some fraction of that energy for useful purposes.

### 8.1.2 Benefits

The principal reason for attempting to recover waste heat is economic. All waste heat that is successfully recovered directly substitutes for purchased energy and therefore reduces the consumption of and the cost of that energy. A second potential benefit is realized when waste-heat substitution results in smaller capacity requirements for energy conversion equipment. Thus the use of waste-heat recovery can reduce capital costs in new installations. A good example is when waste heat is recovered from ventilation exhaust air to preheat the outside air entering a building. The waste-heat recovery reduces the requirement for space-heating energy. This permits a reduction in the capacity of the furnaces or boilers used for heating the plant. The initial cost of the heating equipment will be less and the overhead costs will be reduced. Savings in capital expenditures for the primary conversion devices can be great enough to completely offset the cost of the heat-recovery system. Reduction in capital costs cannot be realized in retrofit installations unless the associated primary energy conversion device has reached the end of their useful lives and are due for replacement.

A third benefit may accrue in a very special case. As an example, when an incinerator is installed to decompose solid, liquid, gaseous or vaporous pollutants, the cost of operation may be significantly reduced through waste-heat recovery from the incinerator exhaust gases.

Finally, in every case of waste-heat recovery, a gratuitous benefit is derived: that of reducing thermal pollution of the environment by an amount exactly equal to the energy recovered, at no direct cost to the recoverer.

### 8.1.3 Potential for Waste-Heat Recovery in Industry

It had been estimated[1] that of the total energy consumed by all sectors of the U.S. economy in 1973, that fully 50% was discharged as waste heat to the environment. Some of this waste is unavoidable. The second law of thermodynamics prohibits 100% efficiency in energy conversion except for limiting cases which are practically and economically unachievable. Ross and Williams,[2] in reporting the results of their second-law analysis of U.S. energy consumption, estimated that in 1975, economical waste-heat recovery could have saved our country 7% of the energy consumed by industry, or $1.82 \times 10^{16}$ Btus (1.82 quads.)

Roger Sant[3] estimated that in 1978 industrial heat recovery could have resulted in a national fuel savings of 0.3%, or $2.65 \times 10^{16}$ Btus (2.65 quads). However, his study included only industrial furnace recuperators.* In terms of individual plants in energy-intensive industries, this percentage can be greater by more than an order of magnitude.

The Annual Energy Review 1991[4] presents data to show that although U.S. manufacturing energy intensity increased by an average of 26.7% during the period 1980 to 1988, the manufacturing sector's energy use efficiency, for all manufacturing, increased by an average of 25.1%. In reviewing the Annual Energy Reviews over the years, it becomes quite clear that during periods of rising fuel

---

*Recuperators are heat exchangers that recover waste heat from the stacks of furnaces to preheat the combustion air. Section 8.4.2 subjects this device to more detailed scrutiny.

prices energy efficiency increases, while in periods of declining fuel prices energy efficiency gains are eroded. Although the average gain in energy use efficiency, in the 7-year period mentioned above, is indeed impressive, several industrial groups accomplished much less than the average or made no improvements at all during that time. As economic conditions change to favor investments in waste-heat recovery there will be further large gains made in energy use efficiency throughout industry.

### 8.1.4 Quantifying Waste Heat

The technical description of waste heat must necessarily include quantification of the following characteristics: (1) quantity, (2) quality, and (3) temporal availability.

The quantity of waste heat available is ordinarily expressed in terms of the enthalpy flow of the waste stream, or

$$\dot{H} = \dot{m}h \qquad\qquad (.1)$$

where

$\dot{H}$ = total enthalpy flow rate of waste stream, Btu/hr

$\dot{m}$ = mass flow rate of waste stream, lb/hr

$h$ = specific enthalpy of waste stream, Btu/lb

The mass flow rate, m, can be calculated from the expression

$$\dot{m} = \rho Q \qquad\qquad (8.2)$$

where    $\rho$  =  density of material, lb/ft$^3$

   $Q$  =  volumetric flow rate, ft$^3$/hr

The potential for economic waste-heat recovery, however, does not depend as much on the quantity available as it does on whether its quality fits the requirements of the potential heating load which must be supplied and whether the waste heat is available at the times when it is required.

The quality of waste heat can be roughly characterized in terms of the temperature of the waste stream. The higher the temperature, the more available the waste heat for substitution for purchased energy. The primary source of energy used in industrial plants are the combustion of fossil fuels and nuclear reaction, both occurring at temperatures approaching 3000°F. Waste heat, of any quantity, is ordinarily of little use at temperatures approaching ambient, although the use of a heat pump can improve the quality of waste heat economically over a limited range of temperatures near

and even below ambient. As an example, a waste-heat stream at 70°F cannot be used directly to heat a fluid stream whose temperature is 100°F. However, a heat pump might conceivably be used to raise the temperature of the waste heat stream to a temperature above 100°F so that a portion of the waste-heat could then be transferred to the fluid stream at 100°F. Whether this is economically feasible depends upon the final temperature required of the fluid to be heated and the cost of owning and operating the heat pump.

### 8.1.5 Matching Loads to Source

It is necessary that the heating load which will absorb the waste heat be available at the same time as the waste heat. Otherwise, the waste heat may be useless, regardless of its quantity and quality. Some examples of synchrony and non-synchrony of waste-heat sources and loads are illustrated in Figure 8.1. Each of the graphs in that figure shows the size and time availability of a waste-heat source and a potential load. In Figure 8.1a the size of the source, indicated by the solid line, is an exhaust stream from an oven operating at 425°F during the second production shift only. One possible load is a water heater for supplying a washing and rinsing line at 135°F. As can be seen by the dashed line, this load is available only during the first shift. The respective quantities and qualities seem to fit satisfactorily, but the time availability of the source could not be worse. If the valuable source is to be used, it will be necessary to (1) reschedule either of the operations to bring them into time correspondence, (2) generate the hot water during the second shift and store it until needed at the beginning of the first shift the next day, or (3) find another heat load which has an overall better fit than the one shown.

In Figure 8.1b we see a waste-heat source (solid line) consisting of the condenser cooling water of an air-conditioning plant which is poorly matched with its load (dashed line)—the ventilating air preheater for the building. The discrepancy in availability is not diurnal as before, but seasonal.

In Figure 8.1c we see an almost perfect fit for source and load, but the total availability over a 24 hour period is small. The good fit occurs because the source, the hot exhaust gases from a heat-treat furnace, is used to preheat combustion air for the furnace burner. However, the total time of availability over a 24-hour period is so small as to cast doubt on the ability to pay off the capital costs of this project.

### 8.1.6 Classifying Waste-Heat Quality

For convenience, the total range of waste-heat temperatures, 80 to 3000°F, is broken down into three

**Figure 8.1 Matching waste-heat sources and loads.**

subranges: high, medium, and low. These classes are designed to match a similar scale which classifies commercial waste-heat-recovery devices. The two systems of classes allow matches to be made between industrial process waste heat and commercially available recovery equipment. Subranges are defined in terms of temperature range as:

| High range | $1100 \leq T \leq 3000$ |
|---|---|
| Medium range | $400 \leq T < 1100$ |
| Low range | $80 \leq T < 400$ |

Waste heat in the high-temperature range is not only the highest quality but is the most useful, and costs less per unit to transfer than lower-quality heat. However, the equipment needed in the highest part of the range requires special engineering and special materials and thus requires a higher level of investment. All of the applications listed in Table 8.1 result from direct-fired processes. The waste heat in the high range is available to do work through the utilization of steam turbines or gas turbines and thus is a good source of energy for cogeneration plants.*

Table 8.2 gives the temperatures of waste gases primarily from direct-fired process equipment in the medium-temperature range. This is still in the temperature range in which work may be economically extracted using gas turbines in the range 15 to 30 psig or steam turbines at almost any desired pressure. It is an economic range for direct substitution of process heat since requirements for equipment are reduced from those in the high-temperature range.

The use of waste heat in the low-temperature range is more problematic. It is ordinarily not practical to extract work directly from the waste-heat source in this temperature range. Practical applications are generally for preheating liquids or gases. At the higher temperatures in this range air preheaters or economizers can be

**Table 8.1 Waste-heat sources in the high-temperature range.**

| Type of Device | Temperature (°F) |
|---|---|
| Nickel refining furnace | 2500-3000 |
| Aluminum refining furnace | 1200-1400 |
| Zinc refining furnace | 1400-2000 |
| Copper refining furnace | 1400-1500 |
| Steel heating furnaces | 1700-1900 |
| Copper reverberatory furnace | 1650-2000 |
| Open hearth furnace | 1200-1300 |
| Cement kiln (dry process) | 1150-1350 |
| Glass melting furnace | 1800-2800 |
| Hydrogen plants | 1200-1800 |
| Solid waste incinerators | 1200-1800 |
| Fume incinerators | 1200-2600 |

*The waste heat generates high-pressure steam in a waste-heat boiler which is used in a steam turbine generator to generate electricity. The turbine exhaust steam at a lower pressure provides process heat. Alternatively, the high-temperature gases may directly drive a gas turbine generator with the exhaust generating low-pressure steam in a waste-heat boiler for process heating.

**Table 8.2 Waste-heat sources in the medium-temperature range.**

| Type of Device | Temperature (°F) |
|---|---|
| Steam boiler exhausts | 450-900 |
| Gas turbine exhausts | 700-1000 |
| Reciprocating engine exhausts | 600-1100 |
| Reciprocating engine exhausts (turbocharged) | 450-700 |
| Heat treating furnaces | 800-1200 |
| Drying and baking ovens | 450-1100 |
| Catalytic crackers | 800-1200 |
| Annealing furnace cooling systems | 800-1200 |
| Selective catalytic reduction systems for $NO_x$ control | 525-750 |

utilized to preheat combustion air or boiler make-up water, respectively. At the lower end of the range heat pumps may be required to raise the source temperature to one that is above the load temperature. An example of an application which need not involve heat pump assistance would be the use of 95°F cooling water from an air compressor to preheat domestic hot water from its ground temperature of 50°F to some intermediate temperature less than 95°F. Electric, gas-fired, or steam heaters could then be utilized to heat the water to the temperature desired. Another application could be the use of 90°F cooling water from a battery of spot welders to preheat the ventilating air for winter space heating. Since machinery cooling can't be interrupted or diminished, the waste-heat recovery system, in this latter case, must be designed to be bypassed or supplemented when seasonal load requirements disappear. Table 8.3 lists some waste-heat sources in the low-temperature range.

### 8.1.7 Storage of Waste Heat

Waste heat can be utilized to adapt otherwise mismatched loads to waste-heat sources. This is possible because of the inherent ability of all materials to absorb energy while undergoing a temperature increase. The absorbed energy is termed stored heat. The quantity that can be stored is dependent upon the temperature rise that can be achieved in the storage material as well as the intrinsic thermal qualities of the material, and can be estimated from the equation

$$Q = \int_{T_1}^{T_2} mC\, dT = \int_{T_1}^{T_2} \rho\, VC\, dT$$

$$= \rho\, VC\, (T - T_0) \text{ for constant specific heat} \quad (8.3)$$

where $m$ = mass of storage material, $lb_m$

$\rho$ = density of storage material, $lb/ft^3$

$V$ = volume of storage material, $ft^3$

$C$ = specific heat of storage material, $Btu/lb_m\ °R$

$T$ = temperature in absolute degrees, °R

The specific heat for solids is a function of temperature which can usually be expressed in the form

$$C_0 = C_0\, [1 + \alpha\, (T - T_0)] \quad (8.4)$$

where  $C_0$ = specific heat at temperature $T_0$

$T_0$ = reference temperature

$\alpha$ = temperature coefficient of specific heat

It is seen from equation 8.3 that storage materials should have the properties of high density and high specific heat in order to gain maximum heat storage for a given temperature rise in a given space. The rate at which heat can be absorbed or given up by the storage material depends upon its thermal conductivity, $k$, which is defined by the equation

$$\frac{\delta Q}{\delta t} = -kA\, \frac{dT}{dx}\bigg|_{x=0} = \dot{Q} \quad (8.5)$$

**Table 8.3 Waste-heat sources in the low-temperature range.**

| Source | Temperature (°F) |
|---|---|
| Process steam condensate | 130-190 |
| Cooling water from: | |
| Furnace doors | 90-130 |
| Bearings | 90-190 |
| Welding machines | 90-190 |
| Injection molding machines | 90-190 |
| Annealing furnaces | 150-450 |
| Forming dies | 80-190 |
| Air compressors | 80-120 |
| Pumps | 80-190 |
| Internal combustion engines | 150-250 |
| Air conditioning and refrigeration condensers | 90-110 |
| Liquid still condensers | 90-190 |
| Drying, baking, and curing ovens | 200-450 |
| Hot-processed liquids | 90-450 |
| Hot-processed solids | 200-450 |

where $t$ = time, hr

$k$ = thermal conductivity, Btu-ft/hr ft$^2$ °F

$A$ = surface area

$$\left.\frac{dT}{dx}\right|_{x=C} = \text{temperature gradient at the surface}$$

Thus additional desirable properties are high thermal conductivity and large surface area per unit mass (specific area). This latter property is inversely proportional to density but can also be manipulated by designing the shape of the solid particles. Other important properties for storage materials are low cost, high melting temperature, and a resistance to spalling and cracking under conditions of thermal cycling. To summarize: the most desirable properties of thermal storage materials are (1) high density, (2) high specific heat, (3) high specific area, (4) high thermal conductivity, (5) high melting temperature, (6) low coefficient of thermal expansion, and (7) low cost.

Table 8.4 lists the thermophysical properties of a number of solids suitable for heat-storage materials.

The response of a storage system to a waste-heat stream is given approximately by the following expression due to Rummel[4]:

$$\frac{Q}{A} = \frac{\Delta T_{l,m}/(\theta + \theta'')}{1/h''\theta + 1/h'\theta' + 1/2.5 C_s \rho_s R_B/k(\theta' + \theta'')} \quad (8.6)$$

where $T_{l,m}$ = logarithmic mean temperature difference based upon the uniform inlet temperature of each stream and the average outlet temperatures

$C_s$ = specific heat of storage material, Btu/lb °F

$\rho_s$ = density of storage material, lb/ft$^3$

$k$ = conductivity of storage material, Btu/hr ft °F

$R_B$ = volume per unit surface area for storage material, ft

$h$ = coefficient of convective heat transfer of gas streams, Btu/hr ft$^2$ °F

$\theta$ = time cycle for gas stream flows, hr

The primed and double-primed values refer, respectively, to the hot and cold entering streams. In cases where the fourth term in the denominator is large compared to the other three terms, this equation should not be

**Table 8.4 Common refractory materials[a,b].**

| Name | Formula | Density (lbm/ft$^3$) | Specific Heat (Btu/lb$_m$) | Mean Thermal Conductivity (Btu/ft hr °F) (to 1000°) | Coefficient of Cubical Expansion (per °F) | Maximum Use Temperature (°F) | Melting Point (°F) |
|---|---|---|---|---|---|---|---|
| Alumina | $Al_2O_3$ | 230 | 0.24 | 2.0 | $8 \times 10^{-6}$ | 3300 | 3700 |
| Beryllium oxide | BeO | 190 | 0.24 | — | $9 \times 10^{-6}$ | 4000 | 4600 |
| Calcium oxide | CaO | 200 | 0.18 | 4.5 | $13 \times 10^{-6}$ | 4200 | 4700 |
| Carbon, graphite | C | 120 | 0.36 | 7 | $3 \times 10^{-6}$ | 4000 | 6500$^c$ |
| Chrome | 40% $Cr_2O_3$ | 200 | 0.20 | 1.0 | $8 \times 10^{-6}$ | 3200 | 3800 |
| Corundum | 90% $Al_2O_3$ | 200 | 0.22 | 1.5 | $7 \times 10^{-6}$ | 3200 | — |
| Forsterite | 2 $MgO$ $SiO_2$ | 160 | 0.23 | 1.2 | $10 \times 10^{-6}$ | 3000 | 3300 |
| Magnesia | MgO | 210 | 0.25 | 2.3 | $11 \times 10^{-6}$ | 4000 | 5000 |
| Magnesium oxide | MgO | 175 | 0.25 | 2.0 | $10 \times 10^{-6}$ | 3500 | 5000 |
| Mullite | 3 $Al_2O_3$ $SiO_2$ | 160 | 0.23 | 1.2 | $5 \times 10^{-6}$ | 3000 | 3350 |
| Silica | $SiO_2$ | 110 | 0.24 | 1.0 | $7 \times 10^{-6}$ | 2800 | 3100 |
| Silicon carbide | SiC | 170 | 0.23 | 8 | $3 \times 10^{-6}$ | 3000 | 4000$^d$ |
| Spinel | $MgO$ $Al_2O_3$ | 220 | 0.23 | 5 | $7 \times 10^{-6}$ | 3300 | — |
| Titanium oxide | $TiO_2$ | 260 | 0.17 | 2.2 | $8 \times 10^{-6}$ | 3000 | 3300 |
| Zircon | $ZrO_2$ $SiO_2$ | 220 | 0.15 | 1.3 | $5 \times 10^{-6}$ | 3500 | 4500 |
| Zirconium oxide | $ZrO_2$ | 360 | 0.13 | 1.3 | $4 \times 10^{-6}$ | 4400 | 4800 |

[a]Most of these materials are available commercially as refractory tile, brick, and mortar. Properties will depend on form, purity, and mixture. Temperatures given should be considered as high limits.

[b]For density in kg/m$^3$, multiply value in lb/ft$^3$ by 16.02. For specific heat in J/kg K, multiply value in Btu/lb$_m$ °F by 4184. For thermal conductivity in W/m K, multiply value in Btu/ft hr °F by 1.73.

[c]Sublimes.

[d]Dissociates.

used. This will occur when the cycle times are short and the thermal resistance to heat transfer is large. In those cases there exists insufficient time for the particles to get heated and cooled. Additional equations for determining the rise and fall in temperatures, and graphs giving temperature histories for the flow streams and the storage material, may be found in Rohsenow and Hartnett.[5]

### 8.1.8 Enhancing Waste Heat with Heat Pumps

Heat pumps offer only limited opportunities for waste-heat recovery simply because the cost of owning and operating the heat pump may exceed the value of the waste heat recovered.

A heat pump is a device that operates cyclically so that energy absorbed at low temperature is transformed through the application of external work to energy at a higher-temperature which can be absorbed by an existing load. The commercial mechanical refrigeration plant can be utilized as a heat pump with small modifications, as indicated in Figure 8.2. The coefficient of performance (COP) of the heat pump cycle is the simple ratio of heat delivered to work required:

$$COP_{HP} = \frac{Q_h}{Q_{net}} = \frac{Q_h}{W_{net}} \tag{8.7}$$

Since the work requirement must be met by a prime mover that is either an electric motor or a liquid-fueled engine, the COP must be considerably greater than 3.0 in order to be an economically attractive energy source. That is true because the efficiency of the prime movers used to drive the heat pump, or to generate the electrical energy for the motor drive, have efficiencies less than 33%. The maximum theoretical COP for an ideal heat pump is given by

**Figure 8.2  Heat pump.**

**Figure 8.3 Theoretical COP vs. load temperature.**

$$COP_H = \frac{1}{1 - T_L/T_H}$$

where  $T_L$  =  temperature of energy source

$T_H$  =  temperature of energy load

The ideal cycle, however, uses an ideal turbine as a vapor expander instead of the usual throttle valve in the expansion line of the mechanical refrigeration plant.

Figure 8.3 is a graph of the theoretical COP versus load temperature for a number of source temperatures. Several factors prevent the actual heat pump from approaching the ideal:

1.   The compressor efficiency is not 100%, but is rather in the range 65 to 85%.

2.   A turbine expander is too expensive to use in any but the largest units. Thus the irreversible throttling process is used instead of an ideal expansion through a turbine. All of the potential turbine work is lost to the cycle.

3.   Losses occur from fluid friction in lines, compressors, and valving.

4.   Higher condenser temperatures and lower evaporator temperatures than the theoretical are required to achieve practical heat flow rates from the source and into the load.

An actual two-stage industrial heat pump installation showed[7] an annual average COP of 3.3 for an average source temperature of 78°F and a load temperature of 190°F. The theoretical COP is 5.8. Except for very carefully designed industrial units, one can expect to achieve actual COP values ranging from 50% to 65% of the theoretical.

An additional constraint on the use of heat pumps is that high-temperature waste heat above 230°F cannot be supplied directly to the heat pump because of the limits imposed by present compressor and refrigerant technology. The development of new refrigerants might raise the limit of heat pump use to 400°F.

### 8.1.9 Dumping Waste Heat

It cannot be emphasized too strongly that the interruption of a waste heat load, either accidentally or intentionally, may impose severe operating conditions on the source system, and might conceivably cause catastrophic failures of that system.

In open system cooling the problem is easier to deal with. Consider the waste-heat recovery from the cooling water from an air compressor. In this case the cooling water is city tap water which flows serially through the water jackets and the intercooler and is then used as makeup water for several heated treatment baths. Should it become necessary to shut off the flow of makeup water to the baths, it would be necessary to valve the cooling water flow to a drain so that the compressor cooling continues with no interruption. Otherwise, the compressor would become overheated and suffer damage.

In a closed cooling system supplying waste heat to a load requires more extensive safeguards and provisions for dumping heat rather than fluid flow. Figure 8.4 is the schematic of a refrigeration plant condenser supplying waste heat for space heating during the winter. Since the heating load varies hourly and daily, and disappears in the summer months, it is necessary to provide an auxiliary heat sink which will accommodate the entire condenser discharge when the waste-heat load disappears. In the installation shown, the auxiliary heat sink is a wet cooling tower which is placed in series with the waste-heat exchanger. The series arrangement is preferable to the alternative parallel arrangement for several reasons. One is that fewer additional controls are needed. Using the parallel arrangement would require that the flows through the two paths be carefully controlled to maintain required condenser temperature and at the same time optimize the waste-heat recovery.

In the above examples the failure to absorb all of the available waste heat had serious consequences on the system supplying the waste heat. A somewhat different waste-heat dumping problem occurs when the effect of excessive waste-heat availability has an adverse affect on the heat sink. An example would be the use of the cooling air stream from an air-cooled screw-type compressor for space heating in the winter months. During the summer months all of the compressor cooling air would have to be dumped to the outdoors in order to prevent overheating of the work space.

### 8.1.10 Open Waste-Heat Exchangers

An open heat exchanger is one where two fluid streams are mixed to form a third exit stream whose energy level (and temperature) is intermediate between the two entering streams. This arrangement has the advantage of extreme simplicity and low fabrication costs with no complex internal parts. The disadvantages are that (1) all flow streams must be at the same pressure, and (2) the contamination of the exit fluids by either of the entrance flows is possible. Several effective applications of open waste-heat exchangers are listed below:

1. The exhaust steam from a turbine-driven feedwater pump in a boiler plant is used to preheat the feedwater in a deaerating feedwater heater.

2. The makeup air for an occupied space is tempered by mixing it with the hot exhaust products from the stack of a gas-fired furnace in a plenum before discharge into the space. This recovery method may be prohibited by codes because of the danger of toxic carbon monoxide; a monitor should be used to test the plenum gases.

**Figure 8.4 System with cold weather condenser pressure control.**

3. The continuous blowdown stream from a boiler plant is used to heat the hot wash and rinse water in a commercial laundry. A steam-heated storage heater serves as the open heater.

**8.1.11 Serial Use of Process Air and Water**

In some applications, waste streams of process air and water can be directly used for heating without prior mixing with other streams. Some practical applications include:

1. Condenser cooling water from batch coolers used directly as wash water in a food-processing plant;

2. steam condensate from wash water heaters added directly to wash water in the bottling section of a brewery;

3. air from the cooling section of a tunnel kiln used as the heating medium in the drying rooms of a refractory;

4. condensate from steam-heated chemical baths returned directly to the baths; and

5. the exhaust gases from a waste-heat boiler used as the heating medium in a lumber kiln.

In all cases, the possibility of contamination from a mixed or a twice-used heat-transport medium must be considered.

**8.1.12 Closed Heat Exchangers**

As opposed to the open heat exchanger, the closed heat exchanger separates the stream containing the heating fluid from the stream containing the heated fluid, but allows the flow of heat across the separating boundaries. The reasons for separating the streams may be:

1. A pressure difference may exist between the two streams of fluid. The rigid boundaries of the heat exchanger are designed to withstand the pressure differences.

2. One stream could contaminate the other if allowed to mix. The impermeable, separating boundaries of the heat exchanger prevents mixing.

3. To permit the use of an intermediate fluid better suited than either of the principal exchange media for transporting waste heat through long distances. While the intermediate fluid is often steam, glycol and water mixtures and other substances can be used to take advantage of their special properties.

Closed heat exchangers fall into the general classification of industrial heat exchangers, however they have many pseudonyms related to their specific form or to their specific application. They can be called recuperators, regenerators, waste-heat boilers, condensers, tube-and-shell heat exchangers, plate-type heat exchangers, feedwater heaters, economizers, and so on. Whatever name is given all perform one basic function: the transfer of heat across rigid and impermeable boundaries. Sections 8.3 and 8.4 provide much more technical detail concerning the theory, application, and commercial availability of heat exchangers.

**8.1.13 Runaround Systems**

Whenever it is necessary to ensure isolation of heating and heated systems, or when it becomes advantageous to use an intermediate transfer medium because of the long distances between the two systems, a runaround heat recovery system is used. Figure 8.5 shows the schematic of a runaround system which recovers heat from the exhaust stream from the heating and ventilating system of a building. The circulating medium is a water-glycol mixture selected for its low freezing point. In winter the exhaust air gives up some energy to the glycol in a heat exchanger located in the exhaust air duct. The glycol is circulated by way of a small pump to a second heat exchanger located in the inlet air duct. The outside air is preheated with recovered waste-heat that substitutes for heat that would otherwise be added in the main heating coils of the building's air handler. During the cooling season the heat exchanger in the exhaust duct heats the exhaust air, and the one in the inlet duct precools the outdoor air prior to its passing through the cooling coils of the air handler. The principal reason for using a runaround system in this application is the long separation distance between the inlet air and the exhaust air ducts. Had these been close together, one air-to-air

**Figure 8.5 Runaround heat-recovery system.**

heat exchanger (with appropriate ducting) could have been more economical.

Figure 8.6 is the schematic diagram of a runaround system used to recover the heat of condensation from a chemical bath steam heater. In this case the bath is a highly corrosive liquid. A leak in the heater coils would cause the condensate to become contaminated and thus do damage to the boiler. The intermediate

**Figure 8.6 Runaround heat-recovery-system process steam source.**

transport fluid isolates the boiler from a potential source of contamination and corrosion. It should be noted that the presence of corrosive chemicals in the bath, which dictated the choice of the runaround system, are also in contact with one side of the condensate heat exchanger. The materials of construction for that heat exchanger should be carefully selected to withstand the corrosion from that chemical.

## 8.2 THE WASTE-HEAT SURVEY

### 8.2.1 How to Conduct the Survey

The survey should be carried out as an integral part of the energy audit of the plant. The survey consists of a systematic study of the sources of waste heat in the plant and of the opportunities for its use. The survey is carried out on three levels. The first step is the identification of every energy-containing nonproduct that flows from the plant. Included are waste streams containing sensible heat (substances at elevated temperatures); examples are hot inert exhaust products from a furnace or cooling water from a compressor. Also to be included are waste streams containing chemical energy (waste fuels), such as carbon monoxide from a heat-treatment furnace or cupola, solvent vapors from a drying oven, or sawdust from a planing mill: Figure 8.7 is an example of a survey form used for listing waste streams leaving the plant.

The second step is to learn more about the original source of the waste-heat stream. Information should be gathered that can lead to a complete heat balance on the equipment or the system that produces it. Since both potential savings and the capital costs tend to be large in waste-heat recovery situations, it is important that the data be correct. Errors in characterization of the energy-containing streams can either make a poor investment look good, or conversely cause a good investment

| Designation | Location | Composition | Flow Rate | Temperature | Exit Heat Rate | Comments |
|---|---|---|---|---|---|---|
|  |  |  |  |  |  |  |

**Figure 8.7 Waste-Heat Source Inventory**

to look bad and be rejected. Because of the high stakes involved, the engineering costs of making the survey may be substantial. Adequate instrumentation must be installed for accurately metering flow streams. The acquired data are used for designing waste-heat-recovery systems. The instruments are then used for monitoring system operation after the installation has been completed. This is to ensure that the equipment is being operated correctly and maintained in optimum condition so that full benefits will be realized from the capital investment. Figure 8.8 is a survey form used to gather information on each individual system or process unit.

### 8.2.2 Measurements

Because waste-heat streams have such variability, it is difficult to list every possible measurement that might be required for its characterization. Generally speaking, the characterization of the quantity, quality, and temporal availability of the waste energy requires that volumetric flow rate, temperature, and flow intervals be measured. Chapter 6 of the NBS Handbook 121[8] is devoted to this topic. A few further generalizations are sufficient for planning the survey operation:

1.  Flow continuity requires that the mass flow rate of any flow stream under steady-state conditions be constant everywhere in the stream; that is,

where Q is the material density, A the cross-sectional flow area, and V the velocity of flow normal to that area. The equation can also be written

$$\rho in Qin = \rho out Qout \qquad (8.9)$$

where Q is the volumetric flow rate. It is safer, more convenient, and usually more accurate to measure low-temperature flows than those at higher temperatures. Thus in many cases the volumetric flow rate of the cold flow and the temperature of the inlet and outlet flows are sufficient to infer the characteristics of the waste-heat stream.

2.  Fuel flows in direct-fired equipment are easily measured with volumetric rate meters. Combustion air and exhaust gas flows are at least an order of magnitude greater than the associated fuel flows. This effectively precludes the use of the volumetric meter because of the expense. However ASME orifice meters, using differential pressure cells are often used for that purpose if the associated pressure drop can be tolerated. It is even cheaper, although less convenient, to determine the volumetric proportions of the flue-gas constituents. Using this data, the air flow quantity and the flue gas flow rate can be calculated from the combustion equation and the law of conservation of mass.

3.  The total energy flux of the fluid streams can be determined from the volumetric flow rate and the temperature using the equation

$$\dot{H} = \rho Q h \qquad (8.10)$$

where   $\dot{H}$  =  enthalpy flux, Btu/hr
        $\rho$  =  density, $lb_m/ft^3$
        $h$  =  specific enthalpy, $Btu/lb_m$
        $Q$  =  volumetric flow rate, $ft^3/hr$

The density, the specific enthalpy, and many other thermal and physical properties as a function of temperature and chemical species is given in tables and graphs found in a number of engineering handbooks[9,10] and other specialized volumes.[11-13]

4.    In order to complete the waste-heat survey for the plant it is not necessary to completely and permanently instrument all systems of interest. One or more gas meters can be temporarily installed and then moved to other locations. In fact, portable instruments can be used for all measurements. While equipment monitoring, should be carried out with permanently installed

### WASTE HEAT SURVEY

SURVEY FORM FOR INDUSTRIAL PROCESS UNITS

NAME OF PROCESS UNIT Tube Reheat Furnace INVENTORY NUMBER ___DE-37___

LOCATION OF PROCESS UNIT, PLANT NAME Plattsburg Works BUILDING ___B___

MANUFACTURER Symonds Eng'ng Ltd MODEL 50-DE-2 SERIAL NUMBER 18031

| | | NAME | FIRING RATE | HHV | TEMPERATURE OF | | | FLUE GAS COMPOSITION % VOLUME | | | | |
|---|---|---|---|---|---|---|---|---|---|---|---|---|
| | | | | | COMB. AIR | FUEL | STACK | CO₂ | O₂ | CO | CH₄ | N₂ |
| FOR DIRECT FIRED UNITS | PRIMARY FUEL | Nat.gas | 4822 | 1030 | 100F | 100F | 2200F | 7.80 | 6.30 | 0.50 | | 85.40 |
| | FIRST ALTERNATIVE | #2 oil | 52.727gph | 131500 | 100F | 100F | 2200F | 9.30 | 7.72 | 0.79 | | 82.19 |
| | SECOND ALTERNAT. | | | | | | | | | | | |

| | | FLOW PATH 1 | FLOW PATH 2 | FLOW PATH 3 | FLOW PATH 4 |
|---|---|---|---|---|---|
| FOR ALL PROCESS UNITS | FLUID COMPOSITION | Steel Tubes | Water | | |
| | FLOW RATE | 50 Tons/hr | 600 gpm | | |
| | INLET TEMPERATURE | 100°F | 80°F | | |
| | OUTLET TEMPERATURE | 2000°F | 131°F | | |
| | DESCRIPTION | Product | Cooling Water | | |

ANNUAL HOURS OPERATION ___5700___ ANNUAL CAPACITY FACTOR, % _____

ANNUAL FUEL COMSUMPTION: PRIMARY FUEL 1.39×10⁶MCF, FIRST ALTERN. 4.43×10⁴ga/ SEC. ALTERN. _____

PRESENT FUEL COST: PRIMARY FUEL $3.21/MCF, FIRST ALTERN. $ 0.535/ga/SEC. ALTERN. _____

ANNUAL ELECTRICAL ENERGY COMSUMPTION, KWHR. 2,047,000

PRESENT ELECTRICAL ENERGY RATE $ 0.0278

**Figure 8.8**

instruments, compromises may be necessary to keep survey costs reasonable. However, permanently installed instruments should become a part of every related capital-improvement program.

5.  For steady-state operations a single temperature can be assigned to each outlet flow stream. But for a process with preprogrammed temperature profiles in time, an average over each cycle must be carefully determined. For a temperature-zoned device, averages of firing rate must be determined carefully over the several burners.

### 8.2.3 Estimation without Measurement

It is risky to base economic predictions used for decisions concerning expensive waste-heat-recovery systems on guesswork. However, when measurements are not possible, it becomes necessary to rely on the best approximations available. The approximations must be made taking full advantage of all relevant data at hand. This should include equipment nameplate data; installation, operating, and maintenance literature; production records; fuel and utility invoices; and equipment logs. The energy auditor must attempt to form a consensus among those most knowledgeable about the system or piece of equipment. This can be done by personally interviewing engineers, managers, equipment operators, and maintenance crews. By providing iterative feedback to these experts, a consensus can be developed. However, estimations are very risky. This author, after taking every possible precaution, using all available data and finding a plausible consensus among the experts, has found his economic projections to be as much as 100% more favorable than actual measurements proved. Those errors would have been avoided by accurate measurements.

### 8.2.4 Constructing the Heat-Balance Diagram

The first law of thermodynamics as applied to a steady flow-steady state system is conveniently written

$$\dot{Q} = \sum_{i=1}^{n} m_1 h_1 + \dot{W} = \sum_{1}^{n} Q_1 h_1 + \qquad (8.11)$$

where

$\dot{Q}$  = net rate of heat loss or gain, Btu/hr

$p_i$  = density of $i$th inflow or outflow, lb/ft$^3$

$Q_i$  = volumetric flow rate of $i$th inflow or outflow, scfh

$h_i$  = specific enthalpy of $i$th inflow or outflow, Btu/lb$_m$

$h'_i$  = specific enthalpy of $i$th inflow or outflow, Btu/scf

$\dot{W}$  = net rate of mechanical or electric work being transferred to or from the system, Btu/hr

$n$  = total number of inlet or outlet paths penetrating system boundaries

Equation 8.11 constitutes the theoretical basis and the mathematical model of the heat-balance diagram shown in Figure 8.9. It corresponds to the data requirements of the survey form shown in Figure 8.8. Using the data taken from that form, we compute the separate terms of the heat balance for a hypothetical furnace as follows:

$$\dot{H}_f \text{ (fuel energy rate)} = \text{firing rate} \times \text{HHV} \qquad (8.12)$$

where HHV is the higher heating value of the fuel n Btu/ft$^3$ or Btu/gal.

$$\dot{H}_{f.1} = 403.8 \times 10^3 \text{ft}^3/\text{hr} \times 1030 \text{ Btu/ft}^3$$

$$= 415.9 \times 10^6 \text{Btu/hr}$$

Because this is a dual-fuel installation, we can also construct a second heat-balance diagram for the alternative fuel:

$$\dot{H}_{f.2} = 3162.6 \text{ gph} \times 131{,}500 \text{ Btu/gal}$$

$$= 415.9 \times 10^6 \text{Btu/hr}$$

Writing the combustion equation on the basis of 100 ft$^3$ of dry flue gas from the flue-gas analysis* for natural gas (fuel no. 1):

$$\alpha \text{Ch}_4 + \text{BO}_2 + \gamma\text{N}_2 = 7.8\text{CO}_2 + 6.3\text{O}_2$$
$$+ 0.5\text{CO} + 85.4\text{N}_2 + 2\alpha\text{H}_2\text{O}$$

Because the chemical atomic species are conserved, we can solve for the relative quantities of air and fuel:

For carbon:  $\alpha = 7.8 + 0.5 = 8.3$

For nitrogen  $\lambda = 85.4$

For oxygen:  $\beta = 7.8 + 6.3 + \dfrac{0.5}{2} + \dfrac{2 \times 8.3}{2} = 22.65$

---

*For simplicity the assumption is made that this natural gas is pure methane. This assumption is often valid for high-Btu fuel gas. For more precision and with other fuel gases, the exact composition should be determined and used in the combustion equation.

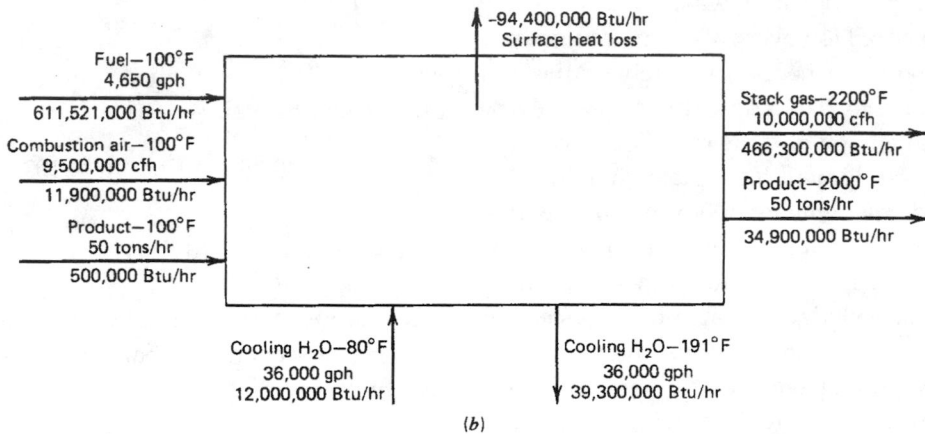

(b)

**Figure 8.9 Heat-balance diagram for reheat furnace(a) Natural gas.(b) No.2 fuel oil.**

$$\frac{A}{F} = \frac{\text{volume of air}}{\text{volume of fuel}} = \frac{22.65 + 85.4}{8.3} = 13.0 \ \frac{\text{ft}^3 \text{air}}{\text{ft}^3 \text{ gas}}$$

$$\dot{H}_{\text{comb. air}} = \frac{A}{F} \times \text{fuel firing rate} \times h'_{\text{air}}$$

$h'_{\text{air}}$ at 100°F is found to be 1.2 Btu/ft$^3$.

$$\dot{H}_{\text{comb.air}} = 13.0 \times 403.8 \ 10^3 \times 1.2$$
$$= 6.3 \times 10^6 \ \text{Btu/hr}$$

The specific enthalpy of each of the flue-gas components at 2200°F is found from Figure 8.10 to be

$$h'_{\text{CO2}} = 69.3 \ \text{Btu/scf}$$
$$h'_{\text{O2}} = 45.4 \ \text{Btu/scf}$$
$$h'_{\text{CO}} = 44.1 \ \text{Btu/scf}$$
$$h'_{\text{N2}} = 43.4 \ \text{Btu/scf}$$
$$h'_{\text{H2O}} = 54.7 \ \text{Btu/scf}$$

$$\dot{H}_{\text{stack gas}} = \text{fuel firing rate} \left( Q_{\text{co}_2} \times h'_{\text{co}_2} \right)$$

$$+ \ Q_{\text{O2}} \times h'_{\text{O2}} + Q_{\text{CO}} \times h'_{\text{CO}} Q_{\text{N}} \times h'_{\text{N2}}$$

$$+ \ Q_{\text{2O}} \times h'_{\text{H2O}}$$

$$= 403.8 \times 10^3 \left( \frac{7.8}{8.3} \times 69.3 + \frac{6.3}{8.3} \times 45.4 \right)$$

$$\left( + \frac{0.5}{8.3} \times 44.1 + \frac{85.4}{8.3} \times 43.4 + \frac{2 \times 8.3}{8.3} \times 54.7 \right) \text{Btu/hr}$$

$$= 265.8 \times 10^6 \text{Btu/hr}$$

Because each fuel has its own chemical composition, the stack-gas composition will be different for each fuel, as will the enthalpy flux: thus the calculation should be repeated for each fuel.

Flow path 1 represents the flow of product through the furnace.

$$\dot{H}_{\text{prod. in}} = m_{\text{prod}} C_{\text{prod}} T_{\text{prod. in}}$$

$$= 50 \text{ tons} \times 2000 \text{ lb/ton} \times 0.115 \text{ Btu/lb} \cdot (100-60)°\text{F}$$

$$= 0.5 \times 10^6 \ \text{Btu/hr}$$

**Figure 8.10 Heat content vs. temperature.**

$\dot{H}_{\text{prod. out}} = m_{\text{prod}} C_{\text{prod}} T_{\text{prod. out}}$

$= 50 \text{ tons} \times 2000 \text{ lb/ton}$

$\times 0.180 \text{ Btu/lb} \cdot {}°F (2000 - 60) \text{ }°F$

$= 34.9 \times 10^6 \text{ Btu/hr}$

Flow path 2 is the cooling-water flow for the conveyor.

$\dot{H}_{\text{CW. in}} = \text{gpm} \times 60 \text{ min/hr} \times 8.33 \text{ lb/gal}$

$\times 1 \text{ Btu/lb} \cdot {}°F \times T_{\text{CW.in}}$

$= 600 \text{ gpm} \times 60 \text{ min/hr} \times 8.33 \text{ lb/gal}$

$\times 1 \text{ Btu/lb} \cdot {}°F (100 - 60) \text{ }°F = 12.0 \times 10^6 \text{ Btu/hr}$

$\dot{H}_{\text{CW. out}} = \text{gpm} \times 60 \text{ min/hr} \times 8.33 \text{ lb/gal}$

$\times 1 \text{ Btu/lb} \cdot {}°F \times T_{\text{CW.out}}$

$= 600 \text{ gpm} \times 60 \text{ min/hr} \times 8.33 \text{ lb/gal}$

$\times 1 \text{ Btu/lb} \cdot {}°F (191 - 60) \text{ }°F = 39.3 \times 10^6 \text{ Btu/hr}$

The heat losses are then estimated from equation 8.11 as

$\dot{Q} = \dot{H}_{\text{stack gas}} - \dot{H}_{f.1} - \dot{H}_{\text{comb. air}} + \dot{H}_{\text{prod. out}}$

$- \dot{H}_{\text{prod. in}} - \dot{H}_{\text{CW. out}} - \dot{H}_{\text{CW. in}}$

$= (265.8 - 415.9 - 6.3 + 34.9 - 0.5 + 39.3$

$- 12.0)10^6 = -94.7 \times 10^6 \text{ Btu/hr}$

where $\dot{Q}$ includes not only the furnace surface losses but any unaccounted for enthalpy flux and all the inaccuracies of measurement and calculation.

Figure 8.9 shows the completed heat-balance diagrams for the reheat furnace. From that diagram we identify three waste-heat streams, as listed in Table 8.5.

### 8.2.5 Constructing Daily Waste-Heat Source and Load Diagrams

The normal daily operating schedule for the furnace analyzed in Section 8.2.4 is plotted in Figure 8.11. It is shown that the present schedule shows furnace operation over two shifts daily, 6 days/week. It is necessary to identify one or more potential loads for each of the two sources.

If high-temperature exhaust stack streams from direct-fired furnaces are recuperated to heat the combustion air stream, then the load and source diagrams are identical. This kind of perfect fit enhances the economics of waste-heat recovery. In order to use the cooling-water stream it will be necessary to find a potential load. Because of the temperature and the quantity of enthalpy flux available, the best fit may be the domestic hot-water load. The hot water is used for wash facilities for the labor force at break times and at the end of each shift. The daily load diagram for the domestic hot-water system is shown in Figure 8.12. Time coincidence for the load and source is for two 10-minute periods prior to lunch and for two 30-minute wash-up periods at the end of the shifts. Because the time coincidence between source and load is for only 1-1/2 hours in each 16 hours, waste-heat recovery will require heat storage.

### 8.2.6 Conceptual Design of the Waste-Heat-Recovery System

Prior to equipment design and before a detailed economic analyses is performed, it is necessary to develop one or more conceptual designs which can serve as a model for the future engineering work. This approach is illustrated by the analyses done in Sections 8.2.4 and 8.2.5 for the two waste-heat streams. An excellent reference text which is useful for the design of waste-heat recovery systems is Hodge's *Analysis and Design of Energy Systems*.[14]

### *Stack-Gas Stream*

Clearly, recuperation is the most promising candidate for heat recovery from high-temperature exhaust gas streams. In the application pictured in Figure 8.13 the hot exhaust gases will be cooled by the incoming combustion air. Because of the temperature of the gases leaving the furnace, the heat exchanger to be selected is a radiation recuperator. This is a concentric tube heat exchanger which replaces the present stack. The incoming combustion air is needed to cool the base of the recuperator and thus parallel flow occurs. Figure 8.13 includes a sketch of the temperature profiles for the two streams. It is seen that in the parallel flow ex-

changer, heat recovery ceases when the two streams approach a common exit temperature. For a well-insulated recuperator the conservation of energy is expressed by the equation

$$Q_{\text{stack gas}} (h'_{\text{stack gas, in}} - h'_{\text{stack gas, out}})$$

$$= Q_{\text{comb. air}} (h'_{\text{comb. air, out}} - h'_{\text{comb.air, in}}) \quad (8.14)$$

Both the right- and left-hand terms represent the heat-recovery rate as well as the decrease in fuel energy required. If the burners or associated equipment have maximum temperature limitations, those temperatures

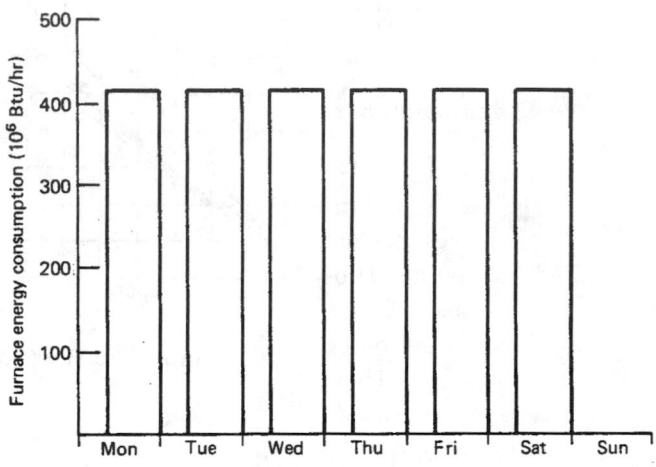

Figure 8.11 Furnace operating schedule.

Figure 8.12 Daily domestic water load.

**Table 8.5 Waste-Heat Streams from Reheat Furnace Fired with Natural Gas**

| Composition | Exhaust Stack: Combustion Products | Cooling Water | Product: Steel Castings |
|---|---|---|---|
| Temperature (°F) | 2200 | 191 | 2000 |
| Flow rate | 57,000,000 cfh | 36,000 gph | 50 tons/hr |
| Enthalpy rate (Btu/hr) | 265,800,000 | 39,300,000 | 34,900,000 |
| Percent fuel energy rate | 64 | 9 | 8 |

**Figure 8.13 Temperature distribution in recuperator.**

also become the high limits for the combustion air, which in turn fixes the maximum allowable enthalpy for the combustion air. Thus the maximum rate of heat recovery is also fixed. Otherwise, the final temperatures of the two stream are based on an optimization of the economic opportunity. This is so because increased heat recovery implies increased recuperator area and thus increased cost. It may also imply higher combustion air temperatures with the resultant increase in fan operating costs and additional investment costs in high-temperature burners, combustion air ducts, and larger fans. In this case we assume that a 100°F temperature difference will occur between the preheated combustion air and the stack gases leaving the recuperator. Equation 8.14 cannot be used directly because the volume rates of fuel and air required are reduced with recuperation. However, if the air/fuel ratio is maintained constant, then $Q_{\text{stack/gas}}/Q_{\text{comb.air}}$ remains almost constant; then equation 8.14 can be written

$$\frac{Q_{\text{stack gas}}}{Q_{\text{comb. air}}} = \text{const.} = \frac{h'_{\text{comb. air, out}} - h'_{\text{comb. air, in}}}{h'_{\text{stack gas, in}} - h'_{\text{stack gas, out}}} \quad (8.15)$$

This equation can be solved with the help of data from Figure 8.9 and the temperature relationship

$$T_{\text{stack gas, out}} = T_{\text{comb. air, out}} + 100°F \quad (8.16)$$

Separate solutions are required for the primary and alternative fuels. The solutions are found by iterating equations 8.15 and 8.16 through a range of temperatures. The value of preheat temperature found for natural gas is 1260°F. For fuel oil the temperature is only slightly

different. The annual heat recovery for each fuel was assumed to be proportional to the total consumption of each fuel, so that the heat recovered was found as

heat recovered

$$= 1.39 \times 10^9 \frac{\text{ft}^3\text{gas}}{\text{yr}} \times 12.95 \frac{\text{ft}^3\text{air}}{\text{ft}^3\text{gas}}$$

$$\times (24.6 - 1.2)\frac{\text{Btu}}{\text{ft}^3} + 4.43 \times 10^6 \frac{\text{gal}}{\text{yr}}$$

$$\times 2040.8 \frac{\text{ft}^3\text{air}}{\text{gal oil}} \times (24.6 - 1.2)\frac{\text{Btu}}{\text{ft}^3}$$

$$= 4.2 \times 10^{11} + 2.1 \times 10^{11} = 6.3 \times 10^{11} \text{Btu/yr}$$

This is an energy savings of

$$\frac{6.3 \times 10^{11}}{1.39 \times 10^9 \times 1030 + 4.43 \times 10^6 \times 131,500} = 0.31 \text{ or } 31\%$$

The predicted cost savings is

| | |
|---|---|
| $1,308,900 | for natural gas |
| 950,200 | for No. 2 fuel oil |
| $2,259,100 | total |

The complete retrofit installation is estimated to cost less than $2,000,000, and the payback period is less than one year. Two points must be emphasized. The entire retrofit installation must be well engineered as a system. This includes the recuperator itself as well as modifications and/or replacement of burners and fans, and the system controls. Only then can the projected system life span be attained and the capital payback actually realized. The cost of lost product must also be factored into the economic analysis if the installation is planned at a time that will cause a plant shutdown. Economics may dictate a delay for the retrofit until the next scheduled or forced maintenance shutdown.

## 8.3 WASTE-HEAT EXCHANGERS

### 8.3.1 Transient Storage Devices

The earliest waste-heat-recovery devices were "regenerators." These consisted of extensive brick work, called "checkerwork," located in the exhaust flues and inlet air flues of high-temperature furnaces in the steel industry. Regenerators are still used to a limited extent in open hearth furnaces and other high-temperature furnaces burning low-grade fuels. It is impossible to achieve steel melt temperature unless regenerators are used to boost the inlet air temperature. In the process vast amounts of waste heat are recovered which would otherwise be supplied by expensive high-Btu fuels. Pairs of regenerators are used alternately to store waste heat from the furnace exhaust gases and then give back that heat to the inlet combustion air. The transfer of exhaust-gas and combustion-air streams from one regenerator to the other is accomplished by using a four-way flapper valve. The design of and estimates of the performance of recuperators follows the principles presented in Section 8.1.7. One disadvantage of this mode of operation is that heat-exchanger effectiveness is maximum only at the beginning of each heating and cooling cycle and falls to almost zero at the end of the cycle. A second disadvantage is that the tremendous mass of the checkerwork and the volume required for its installation raises capital costs above that for the continuous-type air preheaters.

An alternative to the checkerwork regenerator is the heat wheel. This device consists of a permeable flat disk which is placed with its axis parallel to a pair of split ducts and is slowly rotated on an axis parallel to the ducts. The wheel is slowly rotated as it intercepts the gas streams flowing concurrently through the split ducts. Figure 8.14 illustrates those operational features.

As the exhaust-gas stream in the exhaust duct passes through one-half of the disk it gives up some of its heat which is temporarily stored in the disc material. As the disc is turned, the cold incoming air passes through the heated surfaces of the disk and absorbs the energy. The materials used for the disks include metal alloys, ceramics and fiber, depending upon the temperature of the exhaust gases. Heat-exchanger efficiency for the heat wheel has been measured as high as 90% based upon the exhaust stream energy. Further details concerning the heat wheel and its applications are given in Section 8.4.3.

### 8.3.2 Steady-State Heat Exchangers

Section 8.4 treats heat exchangers in some detail. However, several important criteria for selection are listed below.

1. **Flow Arrangements.** These are characterized as:

   Parallel flow     Crossflow
   Counterflow       Mixed flow

   The flow arrangement helps to determine the overall effectiveness, the cost, and the highest achievable temperature in the heated stream. The

**Figure 8.14 Heat wheel.**

latter effect most often dictates the choice of flow arrangement. Figure 8.15 indicates the temperature profiles for the heating and heated streams, respectively. If the waste-heat stream is to be cooled below the cold stream exit, a counterflow heat exchanger must be used.

2. **Character of the Exchange Fluids.** It is necessary to specify the heated and cooled fluids as to:

   Chemical composition
   Physical phase (i.e., gaseous, liquid, solid, or
       multiphase)
   Change of phase, if any, such as evaporating
       or condensing
   These specifications may affect the optimum flow arrangement and/or the materials of construction.

### 8.3.3 Heat-Exchanger Effectiveness

The effectiveness of a heat exchanger is defined as a ratio of the actual heat transferred to the maximum possible heat transfer considering the temperatures of two streams entering the heat exchanger. For a given flow arrangement, the effectiveness of a heat exchanger is directly proportional to the surface area that separates the heated and cooled fluids. The effectiveness of typical heat exchangers is given in Figure 8.16 in terms of the parameter $AU/C_{min}$ where $A$ is the effective heat-transfer area, $U$ the effective overall heat conductance, and $C_{min}$

the mass flow rate times the specific heat of the fluid with minimum $\dot{m}c$. The conductance is the heat rate per unit area per unit temperature difference. Note that as $AU/C_{min}$ increases, a linear relation exists with the effectiveness until the value of $AU/C_{min}$ approaches 1.0. At this point the curve begins to knee over and the increase in effectiveness with $AU$ is drastically reduced. Thus one sees a relatively early onset of the law of diminishing returns for heat-exchanger design. It is implied that one pays heavily for exchangers with high effectiveness.

### 8.3.4 Filtering or Fouling

One of the important heat-exchanger parameters related to surface conditions is termed the fouling factor. The fouling of the surfaces can occur because of film deposits, such as oil films; because of surface scaling due to the precipitation of solid compounds from solution; because of corrosion of the surfaces; or because of the deposit of solids or liquids from two-phase flow streams. The fouling factor increases with increased fouling and causes a drop in heat exchanger effectiveness. If heavy fouling is anticipated, it may call for the filtering of contaminated streams, special materials of construction, or a mechanical design that permits easy access to surfaces for frequent cleaning.

### 8.3.5 Materials and Construction

These topics have been reviewed in previous sections. In summary:

1. High temperatures may require the use of special materials.
2. The chemical and physical properties of exchange fluids may require the use of special materials.
3. Contaminated fluids may require special materials and/or special construction.
4. The additions of tube fins on the outside, grooved surfaces or swaged fins on the inside, and treated or coated surfaces inside or outside may be required to achieve compactness or unusually high effectiveness.

### 8.3.6 Corrosion Control

The standard material of construction for heat exchangers is mild steel. Heat exchangers made of steel are the cheapest to buy because the material is the least expensive of all construction materials and because it is so easy to fabricate. However, when the heat transfer media are corrosive liquids and/or gases, more exotic materials may have to be used. Corrosion tables[15] give the information necessary to estimate the life of the heat exchanger and life-cycle-costing studies allow

**Figure 8.15 Cross-flow heat exchanger.**

valid comparisons of the costs of owning the steel heat exchanger versus one constructed of exotic materials. The problem is whether it will be cheaper to replace the steel heat exchanger at more frequent intervals or to buy a unit made of more expensive materials, but

requiring less frequent replacement. Mechanical designs which permit easy tube replacement lower the cost of rebuilding and favor the use of mild steel heat exchangers.

Corrosion-resisting coatings, such as the TFE plastics, are used to withstand extremely aggressive liquids and gases. However, the high cost of coating and the danger of damaging the coatings during assembly and during subsequent operation limit their use. One disadvantage of using coatings is that they almost invariably decrease the overall conductance of the tube walls and thus necessitate an increase in size of the heat exchanger. The decision to use coatings depends first upon the availability of alternate materials to withstand the corrosion as well as the comparative life-cycle costs, assuming that alternative materials can be found.

Among the most corrosive and widely used materials flowing in heat exchangers are the chlorides such as hydrochloric acid and saltwater. Steel and most steel alloys have extremely short lives in such service. One class of steel alloys that have shown remarkable resistance to chlorides and other corrosive chemicals is called duplex steels[16] and consists of half-and-half ferrite and austenitic microstructures. Because of their high tensile strength, thinner tube walls can be used and this offsets some of the higher cost of the material.

Figure 8.16  Typical heat-exchanger effectiveness.

### 8.3.7 Maintainability

Provisions for gaining access to the internals may be worth the additional cost so that surfaces may be easily cleaned, or tubes replaced when corroded. A shell and tube heat exchanger with flanged and bolted end caps which are easily removed for maintenance is shown in Figure 8.17. Economizers are available with removable panels and multiple one-piece finned, serpentine tube elements, which are connected to the headers with standard compression fittings. The tubes can be removed and replaced on site, in a matter of minutes, using only a crescent wrench.

## 8.4 COMMERCIAL OPTIONS IN WASTE-HEAT-RECOVERY EQUIPMENT

### 8.4.1 Introduction

It is necessary to completely specify all of the operating parameters as well as the heat exchange capacity for the proper design of a heat exchanger, or for the selection of an off-the-shelf item. These specifications will determine the construction parameters and thus the cost of the heat exchanger. The final design will be a compromise among pressure drop (which fixes pump or fan capital and operating costs), maintainability (which strongly affects maintenance costs), heat exchanger effectiveness, and life-cycle cost. Additional features, such as the on-site use of exotic materials or special designs for enhanced maintainability, may add to the initial cost. That design will balance the costs of operation and maintenance with the fixed costs in order to minimize the life-cycle costs. Advice on selection and design of heat exchangers is available from manufacturers and from T.E.M.A.* *Industrial Heat Exchangers*[17] is an excellent guide to heat exchanger selection and includes a directory of heat exchanger manufacturers.

The essential parameters that should be known and specified in order to make an optimum choice of waste-heat recovery devices are:

> Temperature of waste-heat fluid
> Flow rate of waste-heat fluid
> Chemical composition of waste-heat fluid
> Minimum allowable temperature of waste-heat fluid
> Amount and type of contaminants in the waste-heat fluid

**Figure 8.17. Shell and tube heat exchanger.**

> Allowable pressure drop for the waste-heat fluid
> Temperature of heated fluid
> Chemical composition of heated fluid
> Maximum allowable temperature of heated fluid
> Allowable pressure drop in the heated fluid
> Control temperature, if control required

In the remainder of this section, some common types of commercially available waste-heat recovery devices are discussed in detail.

### 8.4.2 Gas-to-Gas Heat Exchangers: Recuperators

Recuperators are used in recovering waste heat to be used for heating gases in the medium- to high-temperature range. Some typical applications are soaking ovens, annealing ovens, melting furnaces, reheat furnaces, afterburners, incinerators, and radiant-heat burners. The simplest configuration for a heat exchanger is the metallic radiation recuperator, which consists of two concentric lengths of metal tubing, as shown in Figure 8.18. This is most often used to extract waste heat from the exhaust gases of a high-temperature furnace for heating the combustion air for the same furnace. The assembly is often designed to replace the exhaust stack.

The inner tube carries the hot exhaust gases while the external annulus carries the combustion air from the atmosphere to the air inlets of the furnace burners. The hot gases are cooled by the incoming combustion air, which then carries additional energy into the combustion chamber. This is energy that does not have to be supplied by the fuel; consequently, less fuel is burned for a given furnace loading. The saving in fuel also means a decrease in combustion air, and therefore stack losses are decreased not only by lowering the stack exit

---
*Tubular Equipment Manufacturers Association, New York, NY

**Figure 8.18 Metallic radiation recuperator.**

temperatures, but also by discharging smaller quantities of exhaust gas. This particular recuperator gets its name from the fact that a substantial portion of the heat transfer from the hot exhaust gases to the surface of the inner tube takes place by radiative heat transfer. The cold air in the annulus, however, is almost transparent to infrared radiation so that only convection heat transfer takes place to the incoming combustion air. As shown in the diagram, the two gas flows are usually parallel, although the configuration would be simpler and the heat transfer more efficient if counterflow were used. The reason for the use of parallel flow is that the cold air often serves the function of cooling the hottest part of the exhaust duct and consequently extends its service life.

The inner tube is often fabricated from high-temperature materials such as high-nickel stainless steels. The large temperature differential at the inlet causes differential expansion, since the outer shell is usually of a different and less expensive material. The mechanical design must take this effect into account. More elaborate designs of radiation recuperators incorporate two sections; the bottom operating in parallel flow, and the upper section using the more efficient counterflow arrangement. Because of the large axial expansions experienced and the difficult stress conditions that can occur at the bottom of the recuperator, the unit is often supported at the top by a freestanding support frame and the bottom is joined to the furnace by way of an expansion joint.

A second common form for recuperators is called the tube-type or convective recuperator. As seen in the schematic diagram of a combined radiation and convective type recuperator in Figure 8.19, the hot gases are carried through a number of small-diameter parallel tubes, while the combustion air enters a shell surrounding the tubes and is heated as it passes over the outside of the tubes one or more times in directions normal to the tubes. If the tubes are baffled as shown so as to allow the air to pass over them twice, the heat exchanger is termed a two-pass convective recuperator; if two baffles are used, a three-pass recuperator; and so on. Although baffling increases the cost of manufacture and also the pressure drop in the air path, it also increases the effectiveness of heat exchange. Tube-type recuperators are generally more compact and have a higher effectiveness than do radiation recuperators, because of the larger effective heat-transfer area made possible through the use of multiple tubes and multiple passes of the air. For maximum effectiveness of heat transfer, combinations of the two types of recuperators are used, with the convection type always following the high-temperature radiation recuperator.

The principal limitation on the heat recovery possible with metal recuperators is the reduced life of the liner at inlet temperatures exceeding 2000°F. This limitation forces the use of parallel flow to protect the bottom of the liner. The temperature problem is compounded when furnace combustion air flow is reduced as the furnace loading is reduced. Thus the cooling of the inner shell is reduced and the resulting temperature rise causes rapid surface deterioration. To counteract this effect, it is necessary to provide an ambient air bypass to reduce the temperature of the exhaust gases. The destruction of a radiation recuperator by overheating is a costly accident. Costs for rebuilding one are about 90% of the cost of a new unit.

To overcome the temperature limitations of metal recuperators, ceramic-tube recuperators have been developed whose materials permit operation to temperatures of 2800°F and on the preheated air side to 2200°F, although practical designs yield air temperatures of 1800°F. Early ceramic recuperators were built of tile and joined with furnace cement. Thermal cycling caused cracking of the joints and early deterioration of the units. Leakage rates as high as 60% were common after short service periods. Later developments featured silicon carbide tubes joined by flexible seals in the air headers. This kind of design, illustrated in Figure 8.20, maintains the seals at a relatively low temperature and the life of seals has been much improved, as evidenced by leakage rates of only a few percent after two years of service.

An alternative design for the convective recuperator is one in which the cold combustion air is heated in a bank of parallel tubes extending into the flue-gas stream normal to the axis of flow. This arrangement is shown in Figure 8.21. The advantages of this configuration are compactness and the ease of replacing individual units. This can be done during full-load operation and minimizes the cost, inconvenience, and possible furnace damage due to a forced shutdown from recuperator failure.

Recuperators are relatively inexpensive and they do reduce fuel consumption. However, their use may require extensive capital improvements. Higher combustion air temperatures may require:

- burner replacement
- larger-diameter air lines with flexible expansion fittings
- cold-air piping for cooling high-temperature burners
- modified combustion controls
- stack dampers
- cold air bleeds
- recuperator protection systems
- larger combustion air fans to overcome the additional pressure drops in the system.

### 8.4.3 Heat Wheels

A rotary regenerator, also called an air preheater or a heat wheel, is used for low- to moderately high-temperature waste-heat recovery. Typical applications are for space heating, curing, drying ovens and heat-treat furnaces. Originally developed as an air preheater for utility steam boilers, it was later adapted, in small sizes, as a regenerator for automotive turbine applications. It has been used for temperatures ranging from 68°F to 2500°F.

Figure 8.22 illustrates the operation of a heat wheel in an air conditioning application. It consists of a porous disk, fabricated of material having a substantial specific heat. The disk is driven to rotates between two side-by-side ducts. One is a cold-gas duct and the other is a hot-gas duct. Although the diagram shows a counterflow configuration, parallel flow can also be used. The axis of the disk is located parallel to and on the plane of the partition between the ducts. As the disk slowly rotates, sensible heat (and in some cases, moisture-containing latent heat) is transferred to the disk by the hot exhaust gas. As the disk moves into

**Figure 8.19. Combined radiation and convective recuperator.**

**Figure 8.20. Silicon-carbide-tube ceramic recuperator.**

the area of the cold duct, the heat is transferred from the disk to the cold air. The overall efficiency of heat transfer (including latent heat) can be as high as 90%.

Heat wheels have been built as large as 70 ft in diameter with air capacities to 40,000 cfm. Multiple units can be used in parallel. This modular approach

**Figure 8.21 Parallel-tube recuperator.**

**Figure 8.22 Rotary regenerator (heat wheel).**

may be used to overcome a mismatch between capacity requirements and the limited number of sizes available in commercial units.

The limitations on the high-temperature range for the heat wheel are primarily due to mechanical difficulties introduced by uneven thermal expansion of the rotating wheel. Uneven expansion can cause excessive deformations of the wheel that result in the loss of adequate gas seals between the ducts and the wheel. The deformation can also result in damage due to the wheel rubbing against its retaining enclosure.

Heat wheels are available in at least four types: 1) A metal frame packed with a core of knitted mesh stainless steel, brass, or aluminum wire, 2) A so-called laminar wheel fabricated from corrugated materials which form many small diameter parallel-flow passages, 3) A laminar wheel constructed from a high-temperature ceramic honeycomb, and 4) A laminar wheel constructed of a fibrous material coated with a hygroscopic so that latent heat can be recovered.

Most gases contain some water vapor since it is a natural component of air and it is also a product of hydrocarbon combustion. Water vapor, as a component of a gas mixture, carries with it its latent heat of evaporation. This latent heat may be a substantial part of the energy contained within the exit-gas streams from air-conditioned spaces or from industrial processes. To recover some of the latent heat in the gas stream, using a heat wheel, the sheet must be coated with a hygroscopic material such as lithium chloride (LiCl) which readily absorbs water vapor to form a hydrate, which in the case of lithium chloride is the hydrate LiCl•$H_2O$. The hydrate consists of one mole of lithium chloride chemically combined with one mole of water vapor. Thus the weight ratio of water to lithium-chloride is 3:7. In a hygroscopic heat wheel, the hot gas stream gives up some

part of its water vapor to the lithium-chloride coating; the gases to be heated are dry and absorb some of the water held in the hydrate. The latent heat in that water vapor adds directly to the total quantity of recovered heat. The efficiency of recovery of the water vapor in the exit stream may be as high as 50%.

Because the pores or passages of heat wheels carry small amounts of gas from the exhaust duct to the intake duct, cross-contamination of the intake gas can occur. If the contamination is undesirable, the carryover of exhaust gas can be partially eliminated by the addition of a purge section located between the intake and exhaust ducts, as shown in Figure 8.23. The purge section allows the passages in the wheel to be cleared of the exhaust gases by introducing clean air which discharges the contaminant to the atmosphere. Note that additional gas seals are required to separate the purge ducts from the intake and exhaust ducts and consequently add to the cost of the heat wheel.

Common practice is to use six air changes of clean air for purging. This results in a reduction of cross-contamination to a value as little as 0.04% for the gas and 0.2% for particulates in laminar wheels, and less than 1.0% total contaminants in packed wheels.

If the heated gas temperatures are to be held constant, regardless of heating loads and exhaust gas temperatures, the heat wheel must be driven at variable speed. This requires a variable-speed drive with a speed-controller with an air temperature sensor as the control element. When operating with outside air in periods of sub-zero temperatures and high humidity, heat wheels may frost up requiring the protection of an air-preheat system. When handling gases containing water-soluble, greasy, or large concentrations of particulates, air filters may be required in the exhaust system upstream from the heat wheel. These features, however, add to the

**Figure 8.23**
**Heat wheel with purge section.**

COOLED WASTE GAS

COLD AIR INLET

HOT AIR TO PROCESS

Figure 8.24 Passive gas to gas regenerator.

HOT WASTE GAS

complexity and the cost of owning and operating the system.

Contaminant buildup on ceramic heat wheels can often be removed by raising the temperature of the exhaust stream to exceed the ignition temperature of the contaminant. However, heat wheels are inherently self-cleaning, because materials entering the wheel from the hot-gas stream tend to be swept out by the reverse flow of the cold-gas stream.

### 8.4.4 Passive Air Preheaters

Passive gas-to-gas regenerators are available for applications where cross-contamination cannot be tolerated. One such type of regenerator, the plate-type, is shown in Figure 8.24. A second type, the heat pipe array is shown in Figure 8.25. Passive air preheaters are used in the low- and medium-temperature applications. Those include drying, curing, and baking ovens; air preheaters in steam boilers; air dryers; waste heat recovery from exhaust steam; secondary recovery from refractory kilns and reverbatory furnaces; and waste heat recovery from conditioned air.

The plate-type regenerator is constructed of alter-nate channels which separate adjacent flows of heated and heating gases by a thin wall of conducting metal. Although their use eliminates cross-contamination, they are bulkier, heavier, and more expensive than a heat wheel of similar heat-recovery and flow capacities. Furthermore, it is difficult to achieve temperature control of the heated gas, while fouling may be a more serious problem.

The heat pipe is a heat-transfer element that is assembled into arrays which are used as compact and efficient passive gas-to-gas heat exchangers. Figure 8.25 shows how the bundle of finned heat pipes extend through the wall separating the inlet and exhaust ducts in a pattern that resembles the conventional finned tube heat exchangers. Each of the separate pipes, however, is a separate sealed element. Each consists of an annular wick on the inside of the full length of the tube, in which an appropriate heat-transfer fluid is absorbed. Figure 8.26 shows how the heat transferred from the hot exhaust gases evaporates the fluid in the wick. This causes the vapor to expand into the center core of the heat pipe. The latent heat of evaporation is carried with the vapor to the cold end of the tube. There it is

Figure 8.25 Heat pipe.

**Figure 8.26 Heat pipe operation.**

removed by transferral to the cold gas as the vapor is recondensed. The condensate is then carried back in the wick to the hot end of the tube. This takes place by capillary action and by gravitational forces if the axis of the tube is tilted from the horizontal. At the hot end of the tube the fluid is then recycled.

The heat pipe is compact and efficient for two reasons. The finned-tube bundle is inherently a good configuration for convective heat transfer between the gases and the outside of the tubes in both ducts. The evaporative-condensing cycle within the heat tubes is a highly efficient method of transferring heat internally. This design is also free of cross-contamination. However, the temperature range over which waste heat can be recovered is severely limited by the thermal and physical properties of the fluids used within the heat pipes. Table 8.6 lists some of the transfer fluids and the temperature ranges in which they are applicable.

### 8.4.5 Gas or Liquid-to-Liquid Regenerators: The Boiler Economizer

The economizer is ordinarily constructed as a bundle of finned tubes, installed in the boiler's breeching. Boiler feedwater flows through the tubes to be heated by the hot exhaust gases. Such an arrangement is shown in Figure 8.27. The tubes are usually connected in a series arrangement, but can also be arranged in series-parallel to control the liquid-side pressure drop. The air-side pressure drop is controlled by the spacing of the tubes and the number of rows of tubes. Economizers are available both prepackaged in modular sizes and designed and fabricated to custom specifications from standard components. Materials for the tubes and fins can be selected to withstand corrosive liquids and/or exhaust gases.

Temperature control of the boiler feedwater is necessary to prevent boiling in the economizer during low-

**Table 8.6 Temperature Ranges for Heat-Transfer Fluids Used in Heat Pipes**

| Fluid | Temperature Range (°F) | | Compatible Metals |
|---|---|---|---|
| Nitrogen | $-300$ | to $-110$ | Stainless steel |
| Ammonia | $-95$ | to $+140$ | Nickel, aluminum, stainless steel |
| Methanol | $-50$ | to $+240$ | Nickel, copper, stainless steel |
| Water | 40 | to 425 | Nickel, copper |
| Mercury | 375 | to 1000 | Stainless steel |
| Sodium | 950 | to 1600 | Nickel, stainless steel |
| Lithium | 1600 | to 2700 | Alloy of niobium and zirconium |
| Silver | 2700 | to 3600 | Alloy of tantalum and tungsten |

Figure 8.27 Boiler economizer.

Figure 8.28 Fuel savings from a gas-fired boiler using economizer.

steam demand or in case of a feedwater pump failure. This is usually obtained by controlling the amount of exhaust gases flowing through the economizer using a damper, which diverts a portion of the gas flow through a bypass duct.

The extent of heat recovery in the economizer may be limited by the lowest allowable exhaust gas temperature in the exhaust stack. The exhaust gases contain water vapor both from the combustion air and from the combustion of the hydrogen that is contained in the fuel. If the exhaust gases are cooled below the dew point of the water vapor, condensation will occur and cause damage to the structural materials. If the fuel also contains sulfur, the sulfur-dioxide will be absorbed by the condensed water to form sulfuric acid. This is very corrosive and will attack the breeching downstream of the economizer and the stack lines. The dew point of the exhaust gases from a natural-gas-fired boiler varies from approximately 138°F for a stoichiometric fuel/air mixture, to 113°F for 100% excess air. Because heat-transmission losses through the stack cause axial temperature gradients from 0.2 to 2°F/ft, and because the stack liner may exist at a temperature 50 to 75°F lower than the gas bulk temperature, it is considered prudent to limit minimum stack temperatures to 300°F, or no lower than 250°F when burning natural gas. When using the fuels containing sulfur, even greater caution is taken. This means that the effectiveness of an economizer is limited unless the exhaust gases from the boiler are relatively hot. Figure 8.28 is a graph of the percent fuel saved plotted against percent excess air for a number of stack gas temperatures using natural gas as a boiler fuel. The plots are based on a 300°F hot-gas temperature leaving the economizer.

### 8.4.6 Shell-and-Tube or Concentric-Tube Heat Exchangers

Shell-and-tube and concentric-tube heat exchangers are used to recover heat in the low and medium range from process liquids, coolants, and condensates of all kinds for heating liquids.

When the medium containing waste heat is either a liquid or a vapor that heats a liquid at a different pressure, a totally exclosed heat exchanger must be used. The two fluid streams must be separated so as to contain their respective pressures. In the shell-and-tube heat exchanger, the shell is a cylinder that contains the tube bundle. Internal baffles may be used to direct the fluid in the shell over the tubes in multiple passes. Because the shell is inherently weaker than the tubes, the higher-pressure fluid is usually circulated in the tubes while the lower-pressure fluid circulates in the shell. However, when the heating fluid is a condensing vapor, it is almost invariably contained within the shell. If the reverse were attempted, the condensation of the vapor within the small-diameter parallel tubes would cause flow instabilities. Shell and tube heat exchangers are produced in a wide range of standard sizes with many combinations of materials for the tubes and the shells. The overall conductance of these heat exchangers range to a maximum of several hundred Btu/hr ft² °F.

A concentric-tube exchanger is used when the fluid pressures are so high that a shell design is uneconomical, or when ease of dissembly is paramount. The hotter fluid is almost invariably contained in the inner tube to minimize surface heat losses. The concentric-tube exchanger may consist of a single straight length, a spiral coil, or a bundle of concentric tubes with hairpin bends.

Shell-and-tube and concentric-tube heat exchangers are used to recover heat in the low and medium range from process liquids, coolants, and condensates of all kinds for heating liquids.

### 8.4.7 Waste-Heat Boilers

Waste-heat boilers are water tube boilers in which hot exhaust gases are used to generate steam. The exhaust gases may be from a gas turbine, an incinerator, a diesel engine, or any other source of medium- to high-temperature waste heat. Figure 8.29 shows a conventional, two-pass waste-heat boiler. When the heat source is in the medium-temperature range, the boiler tends to become bulky. The use of finned tubes extends the heat transfer areas and allows a more compact size. If the quantity of waste heat is insufficient for generating a needed quantity of steam, it is possible to add auxiliary burners to the boiler or an afterburner to the ducting upstream of the boiler. The conventional waste-heat boiler cannot generate super-heated steam so that an external superheater is required if superheat is needed.

A more recently designed waste-heat boiler utilizes a finned-tube bundle for the evaporator, an external drum, and forced recirculation of the feedwater. The design, which is modular, makes for a compact unit with high boiler efficiency. Additional tube bundles can be added for superheating the steam and for preheating the feedwater. The degree of superheat which can be achieved is limited by the waste-heat temperature. The salient features of the boiler are shown on the schematic diagram in Figure 8.30.

Waste-heat boilers are commercially available in capacities from less than 1000 up to 1 million cfm of exhaust gas intake.

### 8.4.8 Input-Output Matrix for Waste-Heat-Recovery Devices

Table 8.8 presents the significant attributes of the most common types of industrial heat exchangers. This matrix is useful in making selections from competing types of heat exchangers for waste-heat recovery.

## 8.5 ECONOMICS OF WASTE-HEAT RECOVERY

### 8.5.1 General

Economic analysis techniques used for analyzing investment potential for waste-heat-recovery systems

are no different from those used for the analysis of any other industrial capital project. These techniques are thoroughly discussed in Chapter 4 of this volume.[18] The economic potential for this class of systems is often limited by factors that are crucial yet overlooked. Although the capital cost of these systems is proportional to the peak rate of heat recovery, the capital recovery depends principally on the annual fuel savings. These savings depend on a number of factors, such as the time distribution of waste-heat source availability, the time distribution of heat-load availability, the availability of waste-heat-recovery equipment that can perform at the specified thermal conditions, and the current and future utility rates and prices of fuel. The inability to accurately predict these factors can make the normal investment decision-making process ineffectual.

There is another important distinction to be made about waste-heat recovery investment. When capital projects involve production-related equipment, the rate

**Figure 8.29 Two-pass waste-heat boiler.**

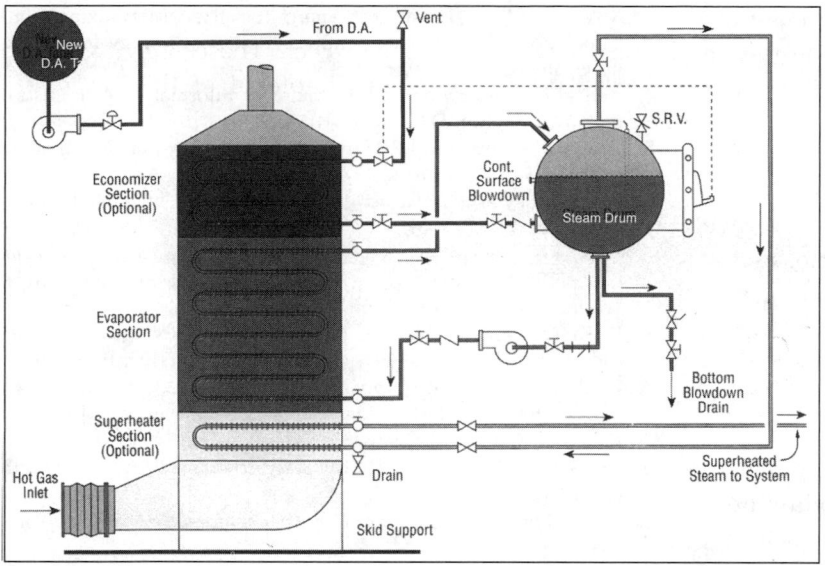

**Figure 8.30 A Recirculation waste-heat boiler.**

**Table 8.8 Operation and Application Characteristics of Industrial Heat Exchangers**

| Commercial Heat-Transfer Equipment | Low temperature: subzero –250°F | Intermediate temp: 250-1200°F | High temperature: 1200-2000°F | Recovers moisture | Large temperature differentials permitted | Packaged units available | Can be retrofit | No cross-contamination | Compact size | Gas-to-gas heat exchange | Gas-to-liquid heat exchanger | Liquid-to-liquid heat exchanger | Corrosive gases permitted with special construction |
|---|---|---|---|---|---|---|---|---|---|---|---|---|---|
| Radiation recuperator | | × | | | × | a | × | × | | × | | | × |
| Convection recuperator | | × | × | | × | × | × | × | | × | | | × |
| Metallic heat wheel | × | × | | b | | × | × | c | × | × | | | × |
| Hygroscopic heat wheel | × | | | × | | × | × | c | × | × | | | |
| Ceramic heat wheel | | × | × | | × | × | | | × | × | | | × |
| Passive regenerator | × | × | | | × | × | × | × | | × | | | × |
| Finned-tube heat exchanger | × | × | | | × | × | × | × | × | | × | | d |
| Tube shell-and-tube exchanger | × | × | | | × | × | × | × | × | | × | × | |
| Waste-heat boilers | × | × | × | | | × | × | × | | | × | | d |
| Heat pipes | × | × | × | | e | × | × | × | × | × | | | × |

[a]Off-the-shelf items available in small capacities only.
[b]Controversial subject. Some authorities claim moisture recovery. Do not advise depending on it.
[c]With a purge section added, cross-contamination can be limited to less than 1% by mass.
[d]Can be constructed of corrosion-resistant materials, but consider possible extensive damage to equipment caused by leaks or tube ruptures.
[e]Allowable temperatures and temperature differential limited by the phase-equilibrium properties of the internal fluid.

of capital recovery can be adjusted by manipulating product selling prices. When dealing with waste-heat-recovery systems, this is generally not an option. The rate of capital recovery is heavily influenced by utility rates and fuel prices which are beyond the influence of the investor. This points out the importance of gathering solid data, and using the best available predictions in preparing analyses for the investment decision process.

### 8.5.2 Effect of Utility Rates and Fuel Prices

Good investment potential exists in an economic climate where energy costs are high relative to equipment costs. However, the costs of goods are sharply influenced by energy costs. This would seem to indicate that the investment potential for waste-heat-recovery projects is not going to improve any faster than energy-cost escalations occur. In one sense this is true of future projects. But because capital projects involve one-time expenditures, which are usually financed by fixed-rate loans, the worth of a present investment will benefit from the rising costs of energy.

### 8.5.3 Effect of Load and Use Factors

The load factor is defined as the ratio of average annual load to rated capacity and the use factor as the fractional part of a year that the equipment is in use. It is clear that the capital recovery rate is directly proportional to these factors.

### 8.5.4 Effects of Reduced System Life

Waste-heat recovery equipment is susceptible to damage from natural and human-made environmental conditions. Damage can result from overheating, freezing, corrosion, collision, erosion, and explosion. Furthermore, capital recovery can never be completed if the equipment fails to achieve its expected life. One must either factor the risks of equipment damage into the economic analysis, or insist that sufficient provision for equipment safety be engineered into the systems.

### References

1. E. Cook, "Energy Flow in Industrial Societies," *Scientific America, Vol.* 225 (Sept. 1971), p. 135.
2. M.H. Ross and R.H. Williams, "The Potential for Fuel Conservation." *Technology Review*, Vol. 79. No. 4 (1977), p. 43.
3. R.W. Sant, "The Least Cost Energy Strategy: Minimizing Consumer Costs through Competition," Report of Energy Productivity Center, CMU, 1979, pp. 35-36.
4. *Annual Energy Review, 1991,* U.S. Energy Information Administration, Washington, D.C.
5. W.M. Rohsenow and J.P. Hartnett, Eds. *Handbook of Heat Transfer,* McGraw-Hill, New York, 1978, pp. 18-73.
6. B.C. Langley, *Heat Pump Technology,* 2nd Ed., Prentice-Hall, Englewood Cliffs, N.J., 1989.
7. K. Kreider and M. McNeil, Eds., *Waste Heat Management Guidebook,* NBS Handbook 121, U.S. Government Printing Office, Washington, D.C., 1976, pp. 121-125.
8. K. Kreider and M. McNeil, Eds., *Waste Heat Management Guidebook,* NBS Handbook 121, U.S. Government Printing Office, Washington, D.C., 1976, pp. 155-184.
9. Baumeister and Marks, Eds., *Standard Handbook for Mechanical Engineers,* 7th ed. McGraw-Hill, New York, 1967.
10. J.K. Salisbury, *Kent's Mechanical Engineers' Handbook, Power,* 12th ed., Wiley, New York, 1967.
11. *Thermodynamic and Transport Properties of Steam,* ASME, New York, 1967.
12. *Selected Values of Chemical Thermodynamic Properties,* NBS Technical Note 270-3, NBS, Washington, D.C., 1968.
13. *Gas Engineers Handbook,* American Gas Association, New York, 19.
14. B.K. Hodge, *Analysis and Design of Energy Systems,* Prentice-Hall, Englewood Cliffs, NJ, 1985.
15. Philip A. Schweitzer, Ed., *Corrosion Resistance Tables, Vols, I and II,* 3rd Edition, Marcel Dekker, New York City, 1991.
16. Weymueller, C.R., Duplex Steels Fight Corrosion, Welding Design and Fabrication, August 1990, pp 30-33, Penton, Cleveland, OH.
17. G. Walker, *Industrial Heat Exchangers,* 2nd Ed., Hemisphere, New York, NY, 1990.
18. K. Kreider and M. McNeil, Eds., *Waste Heat Management Guidebook,* NBS Handbook 121, U.S. Government Printing Office, Washington, D.C., 1976, pp. 25-51.

# CHAPTER 9

# BUILDING ENVELOPE

**KEITH E. ELDER, P.E.**
Iverson Elder Inc.
Bothell, Washington

## 9.1 INTRODUCTION

Building "Envelope" generally refers to those building components that enclose conditioned spaces and through which thermal energy is transferred to or from the outdoor environment. The thermal energy transfer rate is generally referred to as "heat loss" when we are trying to maintain an indoor temperature that is greater than the outdoor temperature. The thermal energy transfer rate is referred to as "heat gain" when we are trying to maintain an indoor temperature that is lower than the outdoor temperature. While many principles to be discussed will apply to both phenomena, the emphasis of this chapter will be upon heat loss.

Ultimately the success of any facility-wide energy management program requires an accurate assessment of the performance of the building envelope. This is true even when no envelope-related improvements are anticipated. Without a good understanding of how the envelope performs, a complete understanding of the interactive relationships of lighting and mechanical systems cannot be obtained.

In addition to a good understanding of basic principles, seasoned engineers and analysts have become aware of additional issues that have a significant impact upon their ability to accurately assess the performance of the building envelope.

1. The actual conditions under which products and components are installed, compared to how they are depicted on architectural drawings.

2. The impact on performance of highly conductive elements within the building envelope; and

3. The extent to which the energy consumption of a building is influenced by the outdoor weather conditions, a characteristic referred to as *thermal mass*.

It is the goal of this chapter to help the reader develop a good qualitative and analytical understanding of the thermal performance of major building envelope components. This understanding will be invaluable in better understanding the overall performance of the facility as well as developing *appropriate* energy management projects to improve performance.

### 9.1.1 Characteristics of Building Energy Consumption

Figure 9.1 below shows superimposed plots of average monthly temperature and fuel consumption for a natural gas-heated facility in the Northwest region of the United States.

Experienced energy analysts recognize the distinc-

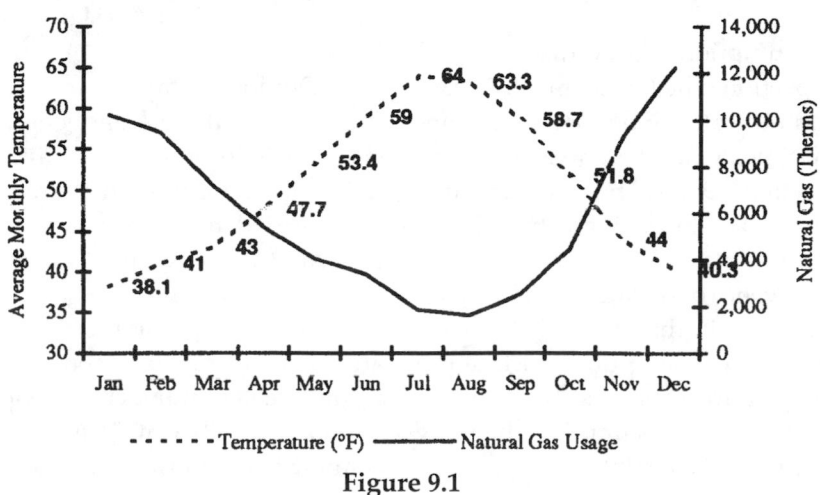

**Figure 9.1**

tive shape of the monthly fuel consumption profile and can often learn quite a bit about the facility just from inspection of this data. For example, the monthly energy consumption is inversely proportional to the average monthly temperature. The lower the average monthly temperature, the more natural gas appears to be consumed.

Figure 9.1 also indicates that there is a period during the summer months when it appears no heating should be required, yet the facility continues to consume some energy. For natural gas, this is most likely that which is consumed for the heating of domestic hot water, but it could also be due to other sources, such as a gas range in a kitchen. This lower threshold of monthly energy consumption is often referred to as the "base," and is characterized by the fact that its magnitude is independent of outdoor weather trends. The monthly fuel consumption which exceeds the "base" is often referred to as the "variable" consumption, and is characterized by the fact that its magnitude is dependent upon the severity of outdoor environmental conditions. The distinction between the base and the variable fuel consumption is depicted below in Figure 9.2.

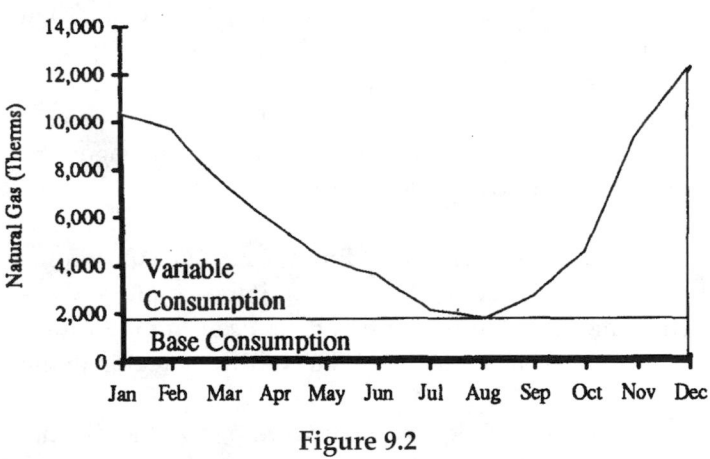

**Figure 9.2**

Often the base consumption can be distinguished from the variable by inspection. The lowest monthly consumption often is a good indicator of the base consumption. However there are times when a more accurate assessment is necessary. Section 9.9 describes one technique for improving the analyst's discrimination between base and variable consumption.

The distinction between base and variable consumption is an important one, in that it is only the variable component of annual fuel consumption which can be saved by building envelope improvements. The more accurate the assessment of the fuel consumption, the more accurate energy savings projections will be.

### 9.1.2 Quantifying Building Envelope Performance

The rate of heat transfer through the building envelope will be found to be related to the following important variables:

1.    Indoor and outdoor temperature;

2.    Conductivity of the individual envelope components; and

3.    The square footage of each of the envelope components.

For a particular building component exposed to a set of indoor and outdoor temperature conditions, these variables are often expressed in equation form by the following:

$$q = UA(T_i - T_o) \qquad (9.1)$$

Where:

$q$ =    the component heat loss, Btu/hr

$U$ =    the overall heat transfer coefficient, Btu/(hr-ft$^2$-°F)

$A$ =    the area of the component, ft$^2$

$T_i$ =    the indoor temperature, °F

$T_o$ =    the outdoor temperature, °F

### 9.1.3 Temperatures for Instantaneous Calculations

The indoor and outdoor temperatures for equation (9.1) are those conditions for which the heat loss needs to be known. Traditionally interest has been in the design heat loss for a building or component, and is determined by using so-called "design-day" temperature assumptions.

Outdoor temperature selection should be made from the ASHRAE Handbook of Fundamentals, for the geographic location of interest. The values published in this chapter are those that are statistically known not to be exceeded more than a prescribed number of hours (such as 2-1/2 %) of the respective heating or cooling season.

For heating conditions in the winter, indoor temperatures maintained between 68 and 72°F result in comfort to the greatest number of people. Indoor temperatures maintained between 74 and 76°F result in the greatest comfort to the most people during the summer (cooling) period.

We will see in later sections that other forms of temperature data collected on an annual basis are useful for evaluating envelope performance on an annual basis.

The next section will take up the topic of how the U-factors for various envelope components are determined.

## 9.2  PRINCIPLES OF ENVELOPE ANALYSIS

The successful evaluation of building envelope performance first requires that the analyst be well-versed in the use of a host of analytical tools which adequately address the unique way heat is transferred through each component. While the heat loss principles are similar, the calculation will vary somewhat from component to component.

### 9.2.1  Heat Loss Through Opaque Envelope Components

We have seen from Equation (9.1) that the heat loss through a component, such as a wall, is proportional to the area of the component, the indoor-outdoor temperature difference, and U, the proportionality constant which describes the temperature-dependent heat transfer through that component. U, currently described as the "U-factor" by ASHRAE[2], is the reciprocal of the total thermal resistance of the component of interest. If the thermal resistance of the component is known, U can be calculated by dividing the total thermal resistance into "1" as shown:

$$U = 1/R_t \tag{9.2}$$

The thermal resistance, $R_t$, is the sum of the individual resistances of the various layers of material which comprise the envelope component. $R_t$ is calculated by adding them up as follows:

$$R_t = R_1 + R_2 + R_n +... \tag{9.3}$$

$R_1$, $R_2$ and $R_n$ represent the thermal resistance of each of the elements in the path of the "heat flow." The thermal resistance of common construction materials can be obtained from the ASHRAE Handbook of Fundamentals. Other physical phenomenon, such as convection and radiation, are typically included as well. For instance, free and forced convection are treated as another form of resistance to heat transfer, and the "resistance" values are tabulated in the ASHRAE Fundamentals Manual for various surface orientations and wind velocities. For example, the outdoor resistance due to forced convection (winter) is usually taken as 0.17 hr/(Btu–ft²–°F) and the indoor resistance due to free convection of a vertical surface is usually taken to be 0.68 hr/(Btu–ft²–°F).

To calculate the overall U-factor, one typically draws a cross-sectional sketch of the building component of interest, assigns resistance values to the various material layers, sums the resistances and uses the reciprocal of that sum to represent U. The total calculation of a wall U-factor is demonstrated in the example below.

**Example**

Calculate the heat loss for 10,000 square feet of wall with 4-inch face brick, R-11 insulation and 5/8-inch sheet rock when the outdoor temperature is 20°F and the indoor temperature is 70°F.

Outside Air Film
4" Brick
R-11 Batt Insulation
5/8" Gypboard
Inside Air Film

**Figure 9.3**

In the ASHRAE Handbook we find that a conservative resistance for brick is 0.10 per inch. Four inches of brick would therefore have a resistance of 0.40. Sheet rock (called gypsum board by ASHRAE) has a resistance of 0.90 per inch, which would be 0.56 for 5/8-inch sheet rock. Batt insulation with a rating of R-11 will have a resistance of 11.0, if expanded to its full rated depth.

The first step is to add all the resistances.

|  | $R_i$ |
| --- | --- |
| Outdoor Air Film (15 m.p.h.) | 0.17 |
| 4-inch Face Brick | 0.40 |
| R-11 Batt Insulation | 11.00 |
| 5/8-inch Gypsum Board | 0.56 |
| Indoor Air Film (still air) | 0.68 |

$$R_t = 12.81 \quad \frac{\text{hr–ft}^2\,\text{F}}{\text{Btu}}$$

The U-factor is the reciprocal of the total resistance

$$U_o = \frac{1}{R_t} = \frac{1}{12.81} = 0.078 \quad \frac{\text{Btu}}{\text{hr–ft}^2\,°\text{F}}$$

The heat loss is calculated by multiplying the U-factor by the area and the indoor-outdoor temperature difference.

$$q = 0.078 \times 10,000 \times (70 - 20)$$

$$= 39,000 \text{ Btu/hr}$$

The accuracy of the previous calculation is dependent on at least two important assumptions. The calculation assumes:

1. The insulation is not compressed; and

2. The layer(s) of insulation has not been compromised by penetrations of more highly conductive building materials.

### 9.2.2 Compression of Insulation

The example above assumed that the insulation is installed according to the manufacturer's instructions. Insulation is always assigned its R-value rating according to a specific standard thickness. If the insulation is compressed into a smaller space than it was rated under, the performance will be less than that published by the manufacturer. For example, R-19 batt insulation installed in a 3-1/2-inch wall might have an effective rating as low as R-13. Table 9.1[7] is a summary of the performance that can be expected from various levels of fiberglass batt insulation types installed in different envelope cavities.

### 9.2.3 Insulation Penetrations

One of the assumptions necessary to justify the use of the one-dimensional heat transfer technique used in equation 9-1 is that the component must be thermally homogeneous. Heat is transferred from the warm side of the component to the colder side and through each individual layer in a series path, much like current flow through simple electrical circuit with the resistances in series. No lateral or sideways heat transfer is assumed to

take place within the layers. For this to be true, the materials in each layer must be continuous and not penetrated by more highly conductive elements.

Unfortunately, there are very few walls in the real world where heat transfer can truly be said to be one-dimensional. Most common construction has wood or metal studs penetrating the insulation, and the presence of these other materials must be taken into consideration.

Traditionally studs are accounted for by performing separate U-factor calculations through both wall sections, the stud and the cavity. These two separate U-factors are then combined in parallel by "weighting" them by their respective wall areas. The following example (Figure 9.4) shows how this would typically be done for a wall whose studs, plates and headers constituted 23% of the total gross wall area.

|  | R-Cavity | R-Frame |
|---|---|---|
| Outdoor Air Film (15 m.p.h.) | 0.17 | 0.17 |
| 4-Inch Face Brick | 0.40 | 0.40 |
| R-11 Batt Insulation | 11.00 | — |
| 3-1/2-Inch Wood Framing | — | 3.59 |
| 5/8-Inch Gypsum Board | 0.56 | 0.56 |
| Indoor Air Film (still air) | 0.6 | 0.68 |

$$R_t = 12.81 \quad 5.40 \quad \frac{\text{hr–ft}^2\text{–°F}}{\text{Btu}}$$

$$U_i = 0.078 \quad 0.185 \quad \frac{\text{hr–ft}^2\text{–°F}}{\text{Btu}}$$

Combining the two U-factors by weighted fractions:

$$U_o = 0.77 \times 0.078 + 023 \times 0.185 = 0.103 \quad \frac{\text{hr–ft}^2\text{–°F}}{\text{Btu}}$$

In the absence of advanced framing construction techniques, wood studs installed 16 inches on-center

| Insulation R-Value at Standard Thickness | | | | | | | | |
|---|---|---|---|---|---|---|---|---|
| R-Value | 38 | 30 | 22 | 21 | 19 | 15 | 13 | 11 |
| Standard Thickness | 12" | 9-½" | 6-¾" | 5-½" | 6-¼" | 3-½" | 3-5/8" | 3-½" |
| Nominal Lumber Sizes, Inches | Actual Depth of Cavity, Inches | Insulation R-Values when Installed in a Confined Cavity | | | | | | |
| 2 x 12 | 11-1/4 | 37 | -- | -- | -- | -- | -- | -- | -- |
| 2 x 10 | 9-1/4 | 32 | 30 | -- | -- | -- | -- | -- | -- |
| 2 x 8 | 7-1/4 | 27 | 26 | -- | -- | -- | -- | -- | -- |
| 2 x 6 | 5-1/2 | -- | 21 | 20 | 21 | 18 | -- | -- | -- |
| 2 x 4 | 3-1/2 | -- | -- | 14 | -- | 13 | 15 | 13 | 11 |

Table 9.1 R-Value of fiberglass batts compressed within various depth cavities.

**Figure 9.4**

- Outside Air Film
- 4" Brick
- R-11 Batt Insulation
- 3-1/2" Fir Framing (23%)
- 5/8" Gypboard
- Inside Air Film

comprise 20-25% of a typical wall (including window framing, sill, etc.). Wood studs installed 24 inches on center comprise approximately 15-20% of the gross wall.

The use of this method is appropriate for situations where the materials in the wall section are sufficiently similar that little or no lateral, or "sideways" heat transfer takes place. Of course a certain amount of lateral heat transfer does take place in every wall, resulting in some error in the above calculation procedure. The amount of error depends on how thermally dissimilar the various elements of the wall are. The application of this procedure to walls whose penetrating members' conductivities deviate from the insulation conductivities by less than an order of magnitude (factor of 10) should provide results that are sufficiently accurate for most analysis of construction materials.

Because the unit R-value for wood is approximately 1.0 per inch and fiberglass batt insulation is R 3.1 per inch, this approach is justified for wood-framed building components.

## 9.3 METAL ELEMENTS IN ENVELOPE COMPONENTS

Most commercial building construction is not wood-framed. Economics as well as the need for fire-rated assemblies has increased the popularity of metal-framing systems over the years. The conductivity of metal framing is significantly more than an order of magnitude greater than the insulation it penetrates. In some instances it is several thousand times greater. However, until recent years, the impact of this type of construction on envelope thermal performance has been ignored by much of the design industry. Yet infrared photography in the field and hot-box tests in the laboratory have demonstrated the severe performance penalty paid for this type

of construction.

The introduction of the metal stud framing system into a wall has the potential to nearly double its heat loss! Just how is this possible, given that a typical metal stud is only about 1/20 of an inch thick? The magnitude of this effect is counter-intuitive to many practicing designers because they have been trained to think of heat transfer through building elements as a one-dimensional phenomenon, (as in the previous examples). But when a highly conductive element such as a metal stud is present in an insulated cavity, two and three-dimensional considerations become extremely important, as illustrated below.

Figure 9.5 shows the temperature distribution centered about a metal stud in a section of insulated wall. The lines, called "isotherms," represent regions with the same temperature. Each line denotes a region that is one degree Fahrenheit different from the adjacent line. These lines are of interest because they help us to visualize the characteristics of the heat flow through the wall section.

**Warm Side**

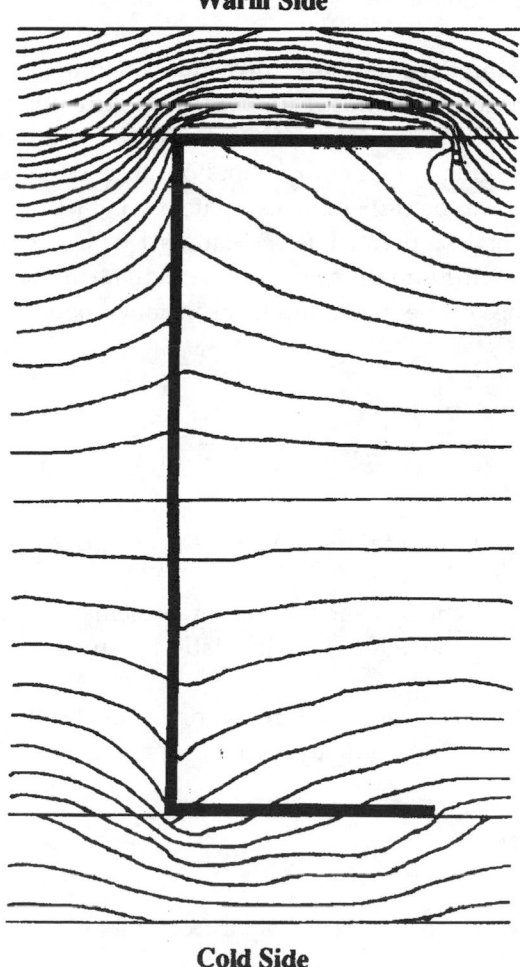

**Cold Side**

**Figure 9.5 Metal stud wall section temperature distribution.**

The direction of heat flow is perpendicular to the lines of constant temperature, and the heat flow is more intense through regions where the isotherms are closer together. It is possible to visualize both the direction and intensity of heat flow through the wall section by observing the isotherms alone.

If the heat flow was indeed one-dimensional and occurring through a thermally homogeneous material, the lines would be horizontal and would show an even and linear temperature-drop progression from the warm side of the wall to the cold side. Notice that the isotherms are far from horizontal in certain regions around the metal stud. Figure 9.5 shows that the heat flow is not parallel, nor does it move directly through the wall section. Also note that the area with the greatest amount of heat flow is not necessarily restricted to the metal part of the assembly. The metal stud has had a negative influence on the insulation in the adjacent region as well. This is the reason for the significant increase in heat loss reported above for metal studs.

Clearly a different approach is required to determine more accurate U-factors for walls with highly conductive elements.

### 9.3.1 Using Parallel-Path Correction Factors

Fortunately, most commercial construction consists of a limited number of metal stud assembly combinations, so the results of laboratory "hot box" tests of typical wall sections can be utilized to estimate the U-factors of walls with metal studs. ASHRAE Standard 90.1 recommends the following equation be used to calculate the equivalent resistance of the insulation layer installed between metal studs:

$$R_t = R_i + R_e \qquad (9.4)$$

Where:

$R_t$ = the total resistance of the envelope assembly

$R_i$ = the resistance of all series elements except the layer of the insulation & studs

$R_e$ = the equivalent resistance of the layer containing the insulation and studs

= $R_{insulation} \times F_c$

Where:

$F_c$ = the correction factor from Table 9.2

The use of the above multipliers in wall U-factor calculation is demonstrated in Figure 9.6.

**Table 9.2 Parallel path correction factors.**

| Size of Members | Framing | Insulation R-Value | Correction Factor, $F_c$ |
|---|---|---|---|
| 2 × 4 | 16 in. O.C. | R-11 | 0.50 |
| 2 × 4 | 24 in. O.C. | R-11 | 0.60 |
| 2 × 6 | 16 in. O.C. | R-19 | 0.40 |
| 2 × 6 | 24 in. O.C. | R-19 | 0.45 |

Outside Air Film
4" Brick
R-11 + Mtl. Stud
5/8" Gypboard
Inside Air Film

**Figure 9.6**

| | $R_i$ |
|---|---|
| Outdoor Air Film (15 m.p.h.) | 0.17 |
| 4-inch Face Brick | 0.40 |
| R-11/Metal Stud 16" O.C. | |
| = 11.0 × 0.50 = | 5.50 |
| 5/8-inch Gypsum Board | 0.56 |
| Indoor Air Film (sill air) | 0.68 |

$$R_t = 7.31 \ \frac{hr - ft^2 - °F}{Btu}$$

$$U_o = \frac{1}{R_t} = \frac{1}{7.31} = 0.137 \ \frac{Btu}{hr - ft^2 - °F}$$

Notice that only one path is calculated through the assembly, rather than two. The insulation layer is simply corrected by the ASHRAE multiplier and the calculation is complete. Also notice that it is only the metal stud/insulation layer that is corrected, not the entire assembly. This does not mean that only this layer is affected by the metal studs, but rather that this is the approach ASHRAE Standard 90.1 intends for the factors to give results consistent with tested performance.

The above example shows that the presence of the

metal stud has increased the wall heat loss by 75%! The impact is even more severe for R-19 walls. The importance of accounting for the impact of metal studs in envelope U-factor calculations cannot be overstated.

### 9.3.3 The ASHRAE "Zone Method"

The ASHRAE Zone Method should be used for U-factor calculation when highly conductive elements are present in the wall that do not fit the geometry or spacing criteria in the above parallel path correction factor table. The Zone Method, described in the ASHRAE Handbook of Fundamentals, is an empirically derived procedure which has been shown to give reasonably accurate answers for simple wall geometries. It is a structured way to calculate the heat transmission through a wall using both series and parallel paths.

The Zone Method takes its name from the fact that the conductive element within the wall influences the heat transmission of a particular region or "zone." The zone of influence is typically denoted as "Zone A," which experiences a significant amount of lateral conduction. The remaining wall section that remains unaffected is called "Zone B." The width of Zone A, which is denoted "W," can be calculated using the following equation:

$$W = m + 2d \qquad (9.5)$$

Where "m" is the width of the metal element at its widest point and "d" is the shortest distance from the metal surface to the outside surface of the entire component. If the metal surface <u>is</u> the outside surface of the component, then "d" is given a minimum value of 0.5 (in English units) to account for the air film.

For example a 3-1/2-inch R-11 insulation is installed between 1.25-inch wide metal studs 16 inches on-center. The assembly is sheathed on both sides with 1/2-inch gypsum board. The variable "m" is the widest part of the stud, which is 1.25 inches. The variable "d" is the distance from the metal element to the outer surface of the wall, which in this case, is the thickness of the gypsum wall board, 0.5 inches. The width of Zone A is:

$$WA = 1.25 + (2 \times 0.5) = \underline{2.25 \text{ inches}}$$

The width of Zone B is:

$$WB = 16 - 2.25 = \underline{13.75 \text{ inches}}$$

Figure 9.7 shows the boundaries of the zones described above.

Actual step-by-step procedures for performing the calculations for Zone A and Zone B can be found in the

**Figure 9.7 Zones A and B.**

ASHRAE Handbook of Fundamentals as well as other publications.[5]

### 9.3.4 Improving the Performance of Envelope Components with Metal Elements

The impact of metal elements can be mitigated by:

1.  Installing the insulation outside of the layer containing metal studs.

2.  Installing interior and exterior finished materials on horizontal "hat" sections, rather than directly on the studs themselves.

3.  Using expanded channel (thermally improved) metal studs.

4.  Using non-conductive thermal breaks at least 0.40 to 0.50 inches thick between the metal element and the inside and outside sheathing. The value of the thermal break conductivity should be at least a factor of 10 less than the metal element in the envelope component.

### 9.3.5 Metal Elements in Metal Building Walls

Many metal building walls are constructed from a corrugated sheet steel exterior skin, a layer of insulation, and sometimes a sheet steel V-rib inner liner. The inner liner can be fastened directly to the steel frame of the building. The exterior cladding is attached to the inner liner and the structural steel of the building through cold formed sheet steel elements called Z-girts. Fiber glass batt insulation is sandwiched between the inner liner and the exterior cladding. Figures 9.8 and 9.9 show details of a typical sheet steel wall.

The steel framing and metal siding materials provide additional opportunities for thermal short circuits to occur. The highly conductive path created by the metal in the girts, purlins and frames connected directly to the metal siding can result in even greater thermal short-cir-

**Figure 9.8 Vertical section view.**

**Figure 9.9 Horizontal section view.**

cuiting than that discussed in Section 9.3.4 for metal studs in insulated walls. Because of this unobstructed high conductivity path, the temperature of the metal element is nearly the same through the entire assembly, rather than varying as heat flows through the assembly.

In laboratory tests, the introduction of 16 gauge Z-Girts spaced 8 feet apart reduced the R-value of a wall with no girts from 23.4 to 16.5 hr–ft$^2$–°F/Btu.[4] Substitution of Z-Girts made of 12-gauge steel reduced the R-value even further, to 15.2 hr–ft$^2$–°F/Btu. The extent of this

heat loss will be dictated by the spacing of the Z-girts, the gauge of metal used and the contact between the metal girts and the metal wall panels.

### 9.3.6 Metal Wall Performance

While the performance of metal building walls will vary significantly, depending on construction features, Tables 9.3-9.5 will provide a starting point in predicting the performance of this type of construction.[6]

### 9.3.7 Strategies for Reducing Heat Loss in Metal Building Walls

#### 9.3.7.1 Maintain Maximum Spacing Between Z-Girts

Tests on a 10-inch wall[8] with a theoretical R-value of 36 hr-ft$^2$-°F/Btu demonstrated a measured R-value of 10.2 hr-ft$^2$-°F/Btu with the girts spaced 2 feet apart. Increasing the spacing to 8 feet between the girts had a corresponding effect of increasing the R-value to 16.5 hr-ft$^2$-°F/Btu. While typical girt spacing in the wall of a metal building is 6 feet on-center, this study does emphasize the desirability of maximizing the girt spacing wherever possible.

#### 9.3.7.2 Use "Thermal" Girts

The thermal performance of metal building walls can be improved by the substitution of thermally improved Z-girts. The use of expanded "thermal" Z-Girts has the potential to increase the effective R-value by 15-20% for a wall with Girts spaced 8 feet apart.[4] A significant amount of the metal has been removed from these thermal girts, which increases the length of the heat flow path and reduces the girt's heat flow area. Figure 9.10 indicates the basic structure of such a girt.

**Figure 9.10 Expanded "thermal" girt.**

**Table 9.3 Wall insulation installed between 16 gauge Z-girts.**

| Unbridged R-Value | Nominal Thickness (Inches) | R-Value 2' O.C. | R-Value 4' O.C. | R-Value 5' O.C. | R-Value 6' O.C. | R-Value 8' O.C. |
|---|---|---|---|---|---|---|
| 6 | 2 | 4.9 | 5.7 | 5.8 | 5.9 | 6.0 |
| 10 | 3 | 6.4 | 7.9 | 8.3 | 8.5 | 8.7 |
| 13 | 4 | 7.4 | 9.4 | 9.9 | 10.2 | 10.6 |
| 19 | 5-1/2 | 9.0 | 12.0 | 12.9 | 13.4 | 14.2 |
| 23 | 7 | 10.2 | 13.8 | 14.9 | 15.6 | 16.5 |
| 36 | 10 | 14.1 | 19.8 | 21.5 | 22.8 | 24.4 |

**Table 9.4 Wall insulation installed between 12 gauge Z-girts.**

| Unbridged R-Value | Nominal Thickness (Inches) | R-Value 2' O.C. | R-Value 4' O.C. | R-Value 5' O.C. | R-Value 6' O.C. | R-Value 8' O.C. |
|---|---|---|---|---|---|---|
| 6 | 2 | 4.8 | 5.6 | 5.8 | 5.9 | 6.0 |
| 10 | 3 | 6.1 | 7.7 | 8.1 | 8.3 | 8.6 |
| 13 | 4 | 6.9 | 9.0 | 9.5 | 9.9 | 10.3 |
| 19 | 5-1/2 | 8.1 | 11.1 | 12.0 | 12.6 | 13.4 |
| 23 | 7 | 8.7 | 12.3 | 13.4 | 14.1 | 15.2 |

Table 9.5 combines test results with calculations to estimate the benefits that may be realized utilizing "thermal" girts.

### 9.3.7.3 Increase Z-Girt-to-Metal Skin Contact Resistance

While the Z-Girt itself does not offer much resistance to heat flow, testing has found that up to half of the resistance of the girt-to-metal skin assembly is attributable to "poor" contact between the girt and the metal skin. Quantitative estimates of the benefit of using less conductive materials to "break" the interface between the two metals are not readily available, but laboratory tests have shown some improvement in thermal performance, just by using half the number of self-tapping screws to fasten the metal skin to the girts.

### 9.3.7.4 Install Thermal Break Between Metal Elements

The impact of this strategy will vary with the flexibility offered by the geometry of the individual wall component, the on-center spacing of the wall girts, thermal break material and contact resistance between the girt and metal wall before consideration of the thermal break. Figure 9.11 illustrates the impact of thermal breaks of varying types and thicknesses installed between a corrugated metal wall and Z-Girts mounted 6 feet on-center.

To be effective, reduction by a factor of 10 or more in thermal conductivity between the thermal break and the metal may be required for a significant improvement in performance. The nominal thickness of the break also plays an important role. Tests of metal panels indicate that, regardless of the insulator, inserts less than 0.40 inches will not normally be sufficient to prevent substantial loss of their insulation value.

### 9.3.8 Calculating U-factors for Metal Building Walls

ASHRAE Standard 90.1 recommends The Johannesson-Vinberg Method[7] for determining the U-factor for sheet metal construction, internally insulated with a metal structure bonded on one or both sides with a metal skin. This method is useful for the prediction of U-factors for metal girt/purlin constructions which can be defined by the geometry described in Figure 9.12.

The dimensions of the elements shown, as well as their conductivities are evaluated in a series of formulations, which are in turn combined in two parallel paths to determine an overall component resistance. The advantage of this method is that metal gauge as well as thermal break thickness and conductivity can be taken into account in the calculation. Caution is advised when using this method to evaluate components with metal-to-metal contact. The method assumes complete thermal contact between all components, which is not always the case in metal building construction. To account for this effect, values for contact resistance can be introduced in the calculation in place of or in addition to the thermal break layer.

**Table 9.5 Wall insulation installed between 16 gauge "thermal" Z-girts.**

| Unbridged R-Value | Nominal Thickness (Inches) | R-Value 2' O.C. | R-Value 4' O.C. | R-Value 5' O.C. | R-Value 6' O.C. | R-Value 8' O.C. |
|---|---|---|---|---|---|---|
| 6 | 2 | 5.6 | 6.2 | 6.3 | 6.4 | 6.4 |
| 10 | 3 | 8.3 | 9.4 | 9.6 | 9.8 | 10.0 |
| 13 | 4 | 10.2 | 11.7 | 12.1 | 12.3 | 12.6 |
| 19 | 5-1/2 | 14.0 | 16.4 | 16.9 | 17.3 | 17.8 |
| 23 | 7 | 16.4 | 19.5 | 20.2 | 20.7 | 21.3 |

Figure 9.11[5] **Impact of thermal break on metal wall R-value corrugated metal with Z-girts 6'0" on-center.**

Figure 9.12 **Metal skinned structure w/internal metal element.**

## 9.4   ROOFS

In many cases the thermal performance of roof structures is similar to that of walls, and calculations can be performed in a similar way to that described in the previous examples. As was the case with walls, metal penetrations through the insulation will exact a penalty. With roofs these penetrations will usually take one of several forms. The first form is the installation of batt insulation between z-purlins, similar to that discussed in Section 9.3. The performance of these assemblies will be similar to that of similarly constructed walls.

### 9.4.1 Insulation Between Structural Trusses

Another common thermal short circuit is that which occurs as a result of installing the insulation between structural trusses with metal components (Figure 9.13).

Table 9.6 is a summary of the derated R-values that might be expected as a result of this type of installation compared to the unbridged R-values published by insulation manufacturers.

### 9.4.2 Insulation Installed "Over-the-Purlin"

One of the most economical methods of insulating the roof of a metal building, shown in Figure 9.14, is to stretch glass fiber blanket insulation over the purlins or trusses, prior to mounting the metal roof panels on top.

The roof purlin is a steel Z-shaped member with a flange width of approximately 2-1/2 inches. The standard purlin spacing for girts supporting roofs in the metal building industry is 5 feet. Using such a method, the insulation is compressed at the purlin/panel interface, resulting in a thermal short circuit. The insulation thickness averages 2-4 inches, but can range up to 6 inches. While thicker blanket insulation can be specified to mitigate this, the thickness is limited by structural considerations. With the reduced insulation thickness at the purlin, there is a diminishing return on the investment in additional insulation. Table 9.7 is a summary of the installed R-values that might be expected using this installation technique.

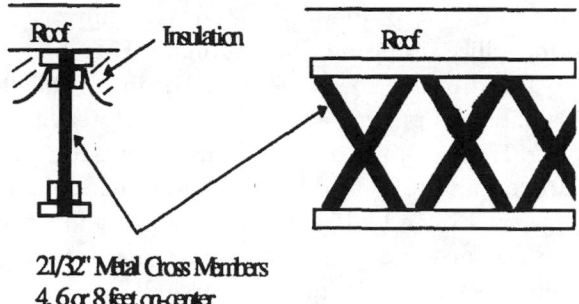

21/32" Metal Cross Members
4, 6 or 8 feet on-center

**Figure 9.13**

**Figure 9.14**

**Table 9.6 Roof insulation installed between metal trusses.[5]**

| Unbridged R-Value | R-Value 4 Ft O.C.[10] | R-Value 6 Ft O.C.[11] | R-Value 8 Ft O.C.[11] |
|---|---|---|---|
| 5 | 4.8 | 4.9 | 4.9 |
| 10 | 9.2 | 9.6 | 9.7 |
| 15 | 13.2 | 14.0 | 14.3 |
| 20 | 17.0 | 18.4 | 18.9 |
| 25 | 20.3 | 22.4 | 23.2 |
| 30 | 23.7 | 26.5 | 27.6 |

#### 9.4.2.1 Calculating Roof Performance for "Over-the-Purlin" Installations

Traditional methods of calculating assembly R-values will not result in accurate estimates of thermal performance for "over-the-purlin" installations. The insulation is not installed uniformly, nor at the thickness required to perform at the rated R-value, and it is penetrated with multiple metal fasteners. The Thermal Insulation Manufacturers Association (TIMA) has developed an empirical formula based on testing of insulation meeting the TIMA 202 specification. Insulation not designed to be laminated (such as filler insulation) will show performance up to 15% less than indicated by the formula below. Assuming the insulation meets the TIMA specification, the U-factor of the completed assembly can be estimated as:

$$U = 0.012 + \frac{0.255}{(0.31 \times R_f + t)} \times \left(1 - \frac{N}{L}\right) + N \times \frac{0.198 + 0.065 \times n}{L} \qquad (9.6)$$

Where:

- $L$ = Length of Building Section, feet
- $N$ = Number of Purlins or Girts in the L Dimension
- $n$ = Fastener Population per Linear Foot of Purlin
- $R_f$ = Sum of Inside and Outside Air Film R-Values, $\frac{hr - ft^2 °F}{Btu}$
- $t$ = Pre-installed insulation thickness (for TIMA 202 type insulation), inches

**Example Calculation Using the TIMA Formula**

Assume a metal building roof structure with the following characteristics:

| | |
|---|---|
| Length of Building Section | = 100 feet |
| Number of Purlins | = 21 purlins |
| Fastener Population per foot of Purlin | = 1 per foot |

Sum of Air Film R-Values = 0.61 + 0.17 = 0.78 $\frac{hr\text{-}ft^2°F}{Btu}$

Insulation thickness = 5 inches

**Table 9.7 Roof insulation installed compressed over-the-purlin[5].**

| Unbridged R-Value | Nominal Thickness (Inches) | R-Value 2' O.C. | R-Value 3' O.C. | R-Value 4' O.C. | R-Value 5' O.C. | R-Value 6' O.C. |
|---|---|---|---|---|---|---|
| 10 | 3 | 5.4 | 6.5 | 7.2 | 7.7 | 8.1 |
| 13 | 4 | 5.7 | 7.1 | 8.0 | 8.7 | 9.3 |
| 16 | 5-1/2 | 5.9 | 7.4 | 8.6 | 9.5 | 10.2 |
| 19 | 6 | 6.0 | 7.7 | 9.0 | 10.0 | 10.9 |

## Example Calculation Using the TIMA Formula

Assume a metal building roof structure with the following characteristics:

Length of Building Section      = 100 feet
Number of Purlins      = 21 purlins
Fastener Population per Foot of Purlin    = 1 per foot

$$\text{Sum of Air Film R-Values} = 0.61 + 0.17 = 0.78 \ \frac{\text{hr-ft}^2{}^\circ\text{F}}{\text{Btu}}$$

Insulation thickness      = 5 inches

$$U = 0.012 + \frac{0.255}{(0.31 \times 0.78 + 6)} \times \left(1 - \frac{21}{105}\right) + 21$$

$$\times \frac{0.198 + 0.065 \times 1}{105} = 0.0973 \ \frac{\text{Btu}}{\text{hr} - \text{ft}^2{}^\circ\text{F}}$$

### 9.4.3 Strategies for Reducing Heat Loss in Metal Building Roofs

Many of the strategies suggested for metal walls are applicable to metal roofs. In addition to strategies that minimize thermal bridging due to metal elements, other features of metal roof construction can have a significant impact on the thermal performance.

Tests have shown significant variation in the thermal performance of insulation depending upon the facing used, even though the permeability of the facing itself has no significant thermal effect. The cause of this difference is in the flexibility of the facing itself. The improvement in permeability rating produces a higher U-factor rating due to the draping characteristics. During installation, the insulation is pulled tightly from eave to eave of the building. The flexibility of vinyl facing allows it to stretch and drape more fully at the purlins than reinforced facing. Figures 9.15 and 9.16 show a generalized illustration of the difference in drape of the vapor retarders and the effect on effective insulation thickness.

The thermal bridging and insulation compression issues discussed above make the following recommendations worth considering.

#### 9.4.3.1 Use the "Roll-Runner" Method of Installing "Over-the-Purlin" Insulation

The "Roll-Runner" installation technique helps to mitigate the effect described above, achieving less insulation compression by improving the draping characteristics of the insulation. This is done using bands or straps to support the suspended batt insulation, as illustrated in Figure 9.17.

Figure 9.15 Vinyl facing.

Figure 9.16 Vinyl facing with glass fiber scrim.

Figure 9.17 Roll Runner Method

Figure 9.18 Insulated purlin with "roll-runner" installation.

### 9.4.3.2 Use Thermal Spacers Between Purlin and Standing Seam Roof Deck

Figure 9.18 depicts a thermal spacer installed between the top of the purlin and the metal roof deck. The impact of this strategy will vary with the geometry of the individual component as well as the location of the spacer.

Table 9.8 summarizes the comparative performance of the two installation techniques discussed above. In all cases, purlins were installed 5 feet on-center. While improvements of 8 to 19 percent in effective R-value are achieved using the Roll-Runner method, performance of 60 to 80 percent can be realized when the purlins were also insulated from direct contact with the metal structure.

### 9.4.3.3 Install Additional Uncompressed Insulation Between Purlins or Bar Joists

This insulation system illustrated in Figure 9.19, is sometimes referred to as "Full Depth/Sealed Cavity," referring to the additional layer of insulation that is installed between purlins or bar joists, and which is not compressed as is the over-the-purlin insulation above it. In this configuration, the main function of the over-the-purlin insulation above is to act as a thermal break.

### 9.4.3.4 Add Rigid Insulation Outside of Purlins

The greatest benefit will be derived where insulation can be added which is neither compressed nor penetrated by conductive elements, as shown in Figure 9.20.

Table 9.9 comparatively summarizes the performance of the above installation for varying levels of insulation installed over purlins with varying on-center spacing.

## 9.5 FLOORS

Floors above grade and exposed to outdoor air can be calculated much the same way as illustrated previously for walls, except that the percentages assumed for floor joists will vary somewhat from that assumed for typical wall constructions.

### 9.5.1 Floors Over Crawl Spaces

The situation is a little different for a floor directly over a crawl space. The problem is that knowledge of the temperature of the crawl space is necessary to perform the calculation. But the temperature of the crawl space is dependent on the number of exposed crawl space surfaces and their U-factors, as well as the impact of crawl space venting, if any. For design-day heat loss calculations, it is usually most expedient to assume a crawl space temperature equal to the outdoor design temperature. This will very nearly be the case for poorly insulated or vented crawl spaces.

When actual, rather than worst-case heat loss is needed, it is necessary to perform a heat balance on the crawl space. The process is described in the ASHRAE Handbook of Fundamentals. Below is a brief summary of the approach.

**Table 9.8 "Roll runner" method**

| Mfgr's Rated R-Value | Nominal Thickness (Inches) | Over the Purlin Method | Roll Runner Method | Roll-Runner with Insulated Purlin |
|---|---|---|---|---|
| 10 | 3 | 7.1 | 7.7 | 12.5 |
| 13 | 4 | 8.3 | 9.1 | 14.3 |
| 19 | 6 | 11.1 | 12.5 | 20.0 |

**Figure 9.19 Insulation suspension system.**

**Figure 9.20 Insulation suspension system.**

**Table 9.9 Insulation "over-the-purlin" w/rigid insulation outside of purlin[14].**

| Uninstalled Unbridged R-Value | Thickness (Inches) | R-Value 2' O.C. | R-Value 3' O.C. | R-Value 4' O.C. | R-Value 5' O.C. | R-Value 6' O.C. |
|---|---|---|---|---|---|---|
| 13 | 2 | 12.2 | 12.9 | 13.3 | 12.5 | 13.8 |
| 17 | 3 | 12.6 | 13.7 | 14.4 | 14.3 | 15.3 |
| 20 | 4 | 12.9 | 14.3 | 15.2 | 15.5 | 16.5 |
| 23 | 5-1/2 | 13.1 | 14.6 | 15.8 | 16.7 | 17.4 |
| 26 | 6 | 13.2 | 14.9 | 16.2 | 18.3 | 18.1 |

Heat loss, $q_{floor}$ from a Floor to a Crawl Space:

$$q_{floor} = q_{perimeter} + q_{ground} + q_{air\ exchange}$$

$$U_f A_f(t_i - t_c) = U_p A_p(t_c - t_o) + U_g A_g(t_c - t_g)$$
$$+ 0.67 \rho c_p V_c(t_c - t_o) \qquad (9.7)$$

Where:

$t_i$ = indoor temperature, °F

$t_o$ = outdoor temperature, °F

$t_g$ = ground temperature, °F

$t_c$ = crawl space temperature, °F

$A_f$ = floor area, ft$^2$

$A_p$ = exposed perimeter area, ft$^2$

$A_g$ = ground area ($A_g = A_f$), ft$^2$

$U_f$ = floor heat transfer coefficient, Btu/(hr–ft$^2$–°F)

$U_g$ = ground coefficient, Btu/hr–ft$^2$°F

$U_p$ = perimeter heat transfer coefficient, Btu/(hr-ft$^2$-°F)

$V_c$ = volume of the crawl space, ft$^3$

$\rho c_p$ = volumetric air heat capacity (0.018 Btu/(ft$^3$–°F)

0.67 = assumed air exchange rate (volume/hour)

The above equation must be solved for $t_c$ the crawl space temperature. Then the heat loss from the space above to the crawl space, using the floor U-factor can be calculated.

### 9.5.2 Floors On Grade

A common construction technique for commercial buildings is to situate the building on a concrete slab right on grade. The actual physics of the situation can be quite complex, but methods have been developed to simplify the problem. In the case of slab-on-grade construction, it has been found that the heat loss is proportional to the perimeter length of the slab, rather than the floor area. Rather than using a U-factor, which is normally associated with a wall or roof *area*, we use an

"F-factor," which is associated with the number of linear *feet* of slab perimeter. The heat loss is given by the equation below:

$$q_{slab} = F \times Perimeter \times (T_{inside} - T_{outside}) \qquad (9.8)$$

F-factors are published by ASHRAE, and are also available in many state energy codes. As an example, the F-factor for an uninsulated slab is 0.73 Btu/(hr-ft-°F). A slab with 24 inches of R-10 insulation installed inside the foundation wall would be 0.54 Btu/(hr-ft-°F).

### 9.5.3 Floors Below Grade

Very little performance information exists on the performance of basement floors. What does exist is more relevant to residential construction than commercial. Fortunately, basement floor loss is usually an extremely small component of the overall envelope performance. For every foot the floor is located below grade, the magnitude of the heat loss is diminished dramatically. Floor heat loss is also affected somewhat by the shortest dimension of the basement. Heat loss is not directly proportional to the outside ambient temperature, as other above grade envelope components. Rather, basement floor loss has been correlated to the temperature of the soil four inches below grade. This temperature is found to vary sinusoidally over the heating season, rather on a daily cycle, as the air temperature does.

A method for determining basement floor and wall heat loss is described, with accompanying calculations, in the ASHRAE Handbook of Fundamentals.

## 9.6  FENESTRATION

The terms "Fenestration," "window," and "glazing" are often used interchangeably. To describe the important aspects of performance in this area requires that terms be defined carefully. "Fenestration" refers to the design and position of windows, doors and other structural openings in a building. When we speak of *windows*, we are actually describing a system of several components. *Glazing* is the transparent component of glass or plastic windows, doors, clerestories, or skylights. The *sash* is a frame in which the glass panes of a window are set. The *Frame* is the complete structural enclosure of the glazing and sash system. *Window* is the term we give to an entire assembly comprised of the sash, glazing, and frame.

Because a window is a thermally nonhomogeneous system of components with varying conductive properties, the thermal performance cannot be accurately approximated by the one-dimensional techniques used to evaluate common opaque building envelope components. The thermal performance of a window system will vary significantly, depending on the following characteristics:

- The number of panes
- The dimension of the space between panes
- The type of gas between the panes
- The emissivity of the glass
- The frame in which the glass is installed
- The type of spacers that separate the panes of glass

### 9.6.1 Multiple Glass Panes

Because of the low resistance provided by the glazing itself, the major contribution to thermal resistance in single pane glazing is from the indoor and outdoor air films. Assuming 0.17 outdoor and 0.68 indoor air film resistances, a single paned glazing unit might be expected to have an overall resistance of better than 0.85 (hr-ft²-°F)/Btu, or a U-factor of 1.18 Btu/(hr-ft²-°F). The addition of a second pane of glass creates an additional space in the assembly, increasing the glazing R-value to 1.85, which results in a U-factor of approximately 0.54 Btu/(hr-ft²-°F). Similarly, the addition of a third pane of glass might increase the overall R-value to 2.85 (hr-ft²°F)/Btu, yielding a U-factor of 0.35 Btu/(hr-ft²-°F). This one-dimensional estimate does not hold true for the entire window unit, but only in the region described by ASHRAE as the "center-of-glass" (see Figure 9.21). Highly conductive framing or spacers will create thermal bridging in much the same fashion as metal studs in the insulated wall evaluated in Section 9.31.

### 9.6.2 Gas Space Between Panes

Most multiple-paned windows are filled with dry air. The thermal performance can be improved by the substitution of gases with lower thermal conductivities. Other gases and gas mixtures used besides dry air are Argon, Krypton, Carbon Dioxide, and Sulfur Hexafluoride. The use of Argon instead of dry air can improve the "center-of-glass" U-factor by 6-9%, depending on the dis-

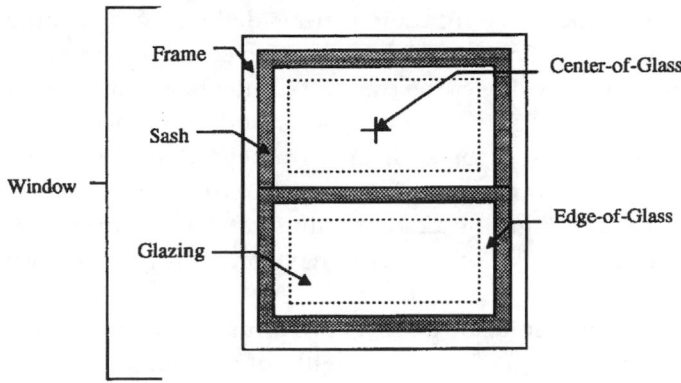

**Figure 9.21 Window components.**

tance between panes, and $CO_2$ filled units achieve similar performance to Argon gas. For spaces up to 0.5 inches, the mixture of Argon and $SF_6$ gas can produce the same performance as Argon, and Krypton can provide superior performance to that of Argon.

### 9.6.3 Emissivity

Emissivity describes the ability of a surface to give off thermal radiation. The lower the emissivity of a warm surface, the less heat loss that it will experience due to radiation. Glass performance can be substantially improved by the application of special low emissivity coatings. The resulting product has come to be known as "Low-E" glass.

Two techniques for applying the Low-E film are sputter and pyrolytic coating. The lowest emittances are achieved with a sputtering process by magnetically depositing silver to the glass inside a vacuum chamber. Sputter coated surfaces must be protected within an insulated glass unit and are often called "soft coat." Pyrolytic coating is a newer method which applies tin oxide to the glass while it is still somewhat molten. The pyrolytic process results in higher emittances than sputter coating, but surfaces are more durable and can be used for single glazed windows. While normal glass has an emissivity of approximately 0.84, pyrolytic coatings can achieve emissivities of approximately 0.40 and sputter coating can achieve emissivities of 0.10 and lower. The emittance of various Low-E glasses will vary considerably between manufacturers.

### 9.6.4 Window Frames

The type of frame used for the window unit will also have a significant impact on the performance. In general, wood or vinyl frames are thermally superior to metal. Metal frame performance can be improved significantly by the incorporation of a "thermal-break." This usually consists of the thermal isolation of the cold side of the frame from the warm side by means of some low-conducting material. Estimation of the performance of a window due to the framing elements is complicated by the variety of configurations and combinations of materials used for sash and frames. Many manufacturers combine materials for structural or aesthetic purposes. Figure 9.22 below illustrates the impact that various framing schemes can have on glass performance schemes.

The center-of-glass curve illustrates the performance of the glazing system without any framing. Notice that single pane glass is almost unaffected by the framing scheme utilized. This is due to the similar order-of-magnitude thermal conductivity between glass and metal. As additional panes are added, emissivities are lowered and low conductivity gases are introduced, the impact of the framing becomes more pronounced, as shown by the increasing performance "spread" toward the right hand side of the chart. Notice that a plain double pane window with a wood or vinyl frame actually has similar performance to that of Low-E glass (hard coating) with Argon gas fill and a metal frame. Also note how flat the curve is for metal-framed windows with no thermal break. It should be clear that first priority should be given to framing systems before consideration is given to Low-E coatings or low conductivity gases.

## 9.6.5 Spacers

Double and triple pane window units usually have continuous members around the glass perimeter to separate the glazing lites and provide an edge seal.

These spacers are often made of metal, which increases the heat transfer between the panes and degrades the performance of the glazing near the perimeter. ASHRAE reports this conductive region to be limited to a 2.5-inch band around the perimeter of the glazing unit, and appropriately describes it as the "edge of glass." Obviously, low conductivity spacers, such as plastic, fiberglass or even glass, are to be preferred over high conductivity spacers, such as metal. The impact of highly conductive spacers is only felt to the extent that other high performance strategies are incorporated into the window. For example, ASHRAE reports no significant performance difference between spacers types when incorporated into window systems with thermally unimproved frames. When thermally improved or thermally broken frames are incorporated, the spacer material becomes a factor in overall window performance.

The best performing windows will incorporate thermally broken frames, non-metallic spacers between the panes, low-emissivity coatings, or low conductivity gases, such as argon or krypton, between the panes.

### 9.6.6 Advanced Window Technologies

#### 9.6.6.1 Suspended Interstitial Films

High performance glazing is available that utilizes one or more Low-E coated mylar or polyester films between the inner and outer glazing of the window unit. High performance is achieved by the addition of a second air space, another Low-E surface, but without the weight of an additional pane of glass.

Figure 9.22 Window system performance comparison.

### 9.6.6.2 *Advanced Spacers*

Significant improvements in spacer design have been made in the past 10-15 years. The "Swiggle Strip" spacer sandwiches a thin piece of corrugated metal between two thicker layers of butyl rubber combined with a desiccant. The conductivity is reported to be one-quarter to one-half of the conductivity of a normal aluminum spacer. Other spacers with conductivities only 6 to 11 percent as great as aluminum have been designed using silicon foam spacers with foil backing or by separating two conventional aluminum spacers with a separate strip of polyurethane foam.

### 9.6.6.3 *Future Window Enhancement Technologies*

Electrochromic glass systems, referred to as "smart windows" sandwich indium tin oxide, amorphous tungsten trioxide, magnesium fluoride and gold between layers of heat-resistant Pyrex. Imposing a small voltage to the window changes its light and heat transmission characteristics, allowing radiation to be reflected in the summer and admitted in the winter.

Windows with R-values of R-16 are theoretically possible if the air or other insulating gas between window panes is replaced with a vacuum. Historically, a permanent vacuum has not been achievable due to the difficulty of forming an airtight seal around the edges of the unit. Work is presently under way on special techniques to create a leakproof seal.

### 9.6.7 Rating the Performance of Window Products

Window energy performance information has not been made available in any consistent form, until recently. Some manufacturers publish R-values that rival many insulation materials. Usually the center-of-glass U-factor is published for glazing, independent of the impact of the framing system that will eventually be used. While the center-of-glass rating alone may be impressive, it does not describe the performance of the entire window. Other manufacturers provide performance data for their framing systems, but they do not reflect how the entire product performs. Even when manufacturers provide ratings for the whole product different methods are used to determine these ratings, both analytical and in laboratory hot box tests. This has been a source of confusion in the industry, requiring some comprehensive standard for the reporting of window thermal performance.

The National Fenestration Rating Council (NFRC), sanctioned by the federal government under the Energy Policy Act of 1992 was established to develop a national performance rating system for fenestration products. The NFRC has established a program where the factors that affect window performance are included in published performance ratings. NFRC maintains a directory of certified products. Currently many local energy codes require that NFRC ratings be used to demonstrate compliance with their window performance standards.

### 9.6.8 Doors

In general, door U-factors can be determined in a similar manner to the walls, roofs and exposed floor demonstrated above. Softwoods have R-values around 1.0 to 1.3 per inch, while hardwoods have R-values that range from 0.80 to 0.95 per inch. A 1-3/8-inch panel door has a U-factor of approximately 0.57 Btu/(Hr-ft$^2$-°F), while a 1-3/4 solid core flush door has a U-factor of approximately 0.40.

As has been shown to be the case with some walls and windows, metal doors are a different story. The same issues affecting windows, such as framing and thermal break apply to metal doors as well. The U-factor of a metal door can vary from 0.20 to 0.60 Btu/(Hr-ft$^2$-°F) depending on the extent to which the metal-to-metal contact can be "broken." In the absence of tested door U-factor data, Table 6 page 24.13 in Chapter 24 of the 1997 ASHRAE Handbook of Fundamentals can be used to estimate the U-factor of typical doors used in residential and commercial construction.

## 9.7   INFILTRATION

Infiltration is the uncontrolled inward air leakage through cracks and interstices in a building element and around windows and doors of a building, caused by the effects of wind pressure and the differences in the outdoor/indoor air density or pressure differentials. The heat loss due to infiltration is described by the following equation:

$$q_{infiltration} = 0.019 \times Q \times (T_{inside} - T_{outside}) \qquad (9.9)$$

Where Q is the infiltration air flow in cubic feet per hour.

The determination of Q is an extremely imprecise undertaking, in that the actual infiltration for similar buildings can vary significantly, even though observable parameters appear to be the same.

### 9.7.1 Estimating Infiltration for Residential Buildings

ASHRAE suggests that the infiltration rate for a residence can be estimated as:

$$Q = L\,[(A(T_i - T_o) + (Bv^2)]^{1/2} \qquad (9.10)$$

Where:

$Q$ = The infiltration rate, $ft^3/hr$

$A$ = Stack Coefficient, $CFM^2/[in^4\text{-}°F]$

$T_i$ = Average indoor Temperature, $°F$

$T_o$ = Average outdoor Temperature, $°F$

$B$ = Wind Coefficient, $CFM^2/[in^4\text{-}(mph)^2]$

$v$ = Average wind speed, mph

$L$ = Crack leakage area, $in^2$

The method is difficult to apply because:

1. L, the total crack area in the building is difficult to determine accurately;

2. The determination of the stack and wind coefficient is subjective;

3. The average wind speed is extremely variable from one micro-climate to another; and

4. Real buildings, built to the same standards, do not experience similar infiltration rates.

ASHRAE reports on the analysis of several hundred public housing units where infiltration varied from 0.5 air changes per hour to 3.5 air changes per hour. If the real buildings experience this much variation, we cannot expect a high degree of accuracy from calculations, unless a significant amount of data are available. However, the studies reported provided some useful guidelines.

In general, older residential buildings without weather-stripping experienced a median infiltration rate of 0.9 air changes per hour. Newer buildings, presumably built to more modern, tighter construction standards, demonstrated median infiltration rates of 0.5 air changes per hour. The structures were unoccupied during tests. It has been estimated that occupants add an estimated 0.10 to 0.15 air changes per hour to the above results.

### 9.7.2 Estimating Infiltration for Complex Commercial Buildings

Infiltration in large commercial buildings is considerably more complex than small commercial buildings or residential buildings. It is affected by both wind speed and "stack effect." Local wind speed is influenced by distance from the reporting meteorological station, elevation, and the shape of the surrounding terrain. The pressure resulting from the wind is influenced by the local wind velocity, the angle of the wind, the aspect ratio of the building, which face of the building is impinged

and the particular location on the building, which in turn is effected by temperature, distance from the building "neutral plane," the geometry of the building exterior envelope elements, and the interior partitions, and all of their relationships to each other. If infiltration due to both wind velocity ($Q_w$) and stack effect ($Q_s$) can be determined, they are combined as follows to determine the overall infiltration rate.

$$Q_{ws} = \sqrt{Q_w^2 + Q_s^2} \qquad (9.11)$$

Where:

$Q_{ws}$ = The combined infiltration rate due to wind and stack effect, $ft^3/hr$

$Q_w$ = The infiltration rate due to wind, $ft^3/hr$

$Q_s$ = The infiltration rate due to stack effect, $ft^3/hr$

While recent research has increased our understanding of the basic physical mechanisms of infiltration, it is all but impossible to accurately calculate anything but a "worst-case" design value for a particular building. Techniques such as the air-change method, and general anecdotal findings from the literature are often the most practical approaches to the evaluation of infiltration for a particular building.

#### 9.7.2.1 Anecdotal Infiltration Findings

General studies have shown that office buildings have air exchange rates ranging from 0.10 to 0.6 air changes per hour with no outdoor intake. To the extent outdoor air is introduced to the building and it is pressurized relative to the local outdoor pressure, the above rates will be reduced.

Infiltration through modern curtain wall construction is a different situation than that through windows "punched" into walls. Studies of office buildings in the United States and Canada have suggested the following approximate leakage rates per unit wall area for conditions of 0.30 inches of pressure (water gauge):

| | |
|---|---|
| Tight Construction | 0.10 CFM/$ft^2$ of curtainwall area |
| Average Construction | 0.30 CFM/$ft^2$ of curtainwall area |
| Leaky Construction | 0.60 CFM/$ft^2$ of curtainwall area |

#### 9.7.2.2 The Air Change Method

While it is difficult to quantitatively predict actual infiltration, it is possible to conservatively predict infiltration that will give us a sense of "worst-case." This requires experience and judgment on the part of the engineer or analyst. ASHRAE Fundamentals Manual published guidelines for estimating infiltration on the basis of the number of "air changes." That is, based on the volume of the space, how many complete changes of air are likely to occur within the space of an hour?

Table 9.10 is a summary of the recommended air changes originally published by ASHRAE in 1972.

These guidelines have been found to be sound over the years and are still widely used by many practicing professionals. The values represent a good starting point for estimating infiltration. They can be modified as required for local conditions, such as wind velocity or excessive building stack effect.

The above values are based on doors and operable windows that are not weather-stripped. ASHRAE originally recommended that the above factors be reduced by 1/3 for weather-stripped windows and doors. This would be a good guideline applicable to modern buildings, which normally have well-sealed windows and doors. Fully conditioned commercial spaces are slightly pressurized and often do not have operable sash. This will also tend to reduce infiltration.

### 9.7.3 An Infiltration Estimate Example

Assume a 20' × 30' room with a 10' ceiling and one exposed perimeter wall with windows. What would the worst-case infiltration rate be?

The total air volume is $20 \times 30 \times 10 = 6,000$ ft$^3$. The table above recommends using 1.0 air changes per hour. However, assuming modern construction techniques, we follow ASHRAE's guideline of using 2/3 of the table values.

$$2/3 \times 6,000 \text{ ft}^3/\text{hr} = 4,000 \text{ ft}^3/\text{hr}$$

The heat loss due to infiltration for this space, if the indoor temperature was 70°F and the outdoor temperature was 25°F would be:

$$q_{\text{infiltration}} = 0.019 \times 4,000 \times (70 - 25)$$
$$= 3,420 \text{ Btu}/\text{hr}$$

## 9.8  SUMMARIZING ENVELOPE PERFORMANCE WITH THE BUILDING LOAD COEFFICIENT

In Section 9.1.2 it was shown that building heat loss is proportional to the indoor-outdoor temperature difference. Our review of different building components, including infiltration has demonstrated that each individual component can be assigned a proportionality constant that describes that particular component's behavior with respect to the temperature difference imposed across it. For a wall or window, the proportionality constant is the U-factor times the component area, or "UA." For a slab-on-grade floor, the proportionality constant is the F-factor times the slab perimeter, or "FP." For infiltration, the proportionality constant is 0.019 time the air flow rate in cubic feet per hour. While not attributed to the building envelope, the effect of ventilation must be accounted for in the building's overall temperature-dependent behavior. The proportionality constant for ventilation is the same for infiltration, except that it is commonly expressed as 1.10 times the ventilation air flow in cubic feet per <u>minute</u>.

The instantaneous temperature-dependent performance of the total building envelope is simply the sum of all the individual component terms. This is sometimes referred to as the *Building Load Coefficient* (BLC). The BLC can be expressed as:

$$BLC = \Sigma UA + \Sigma FP + 0.018\, Q_{INF} + 1.1\, Q_{VENT} \qquad (9.12)$$

Where:

$\Sigma UA$ = the sum of all individual component "UA" products

$\Sigma FP$ = the sum of all individual component "FP" products

$Q_{INF}$ = the building infiltration volume flow rate, in cubic feet per <u>hour</u>

### Table 9.10 Recommended air changes due to infiltration.

| Type of Room | Number of Air Changes Taking Place per Hour |
|---|---|
| Rooms with no windows or exterior doors | 1/2 |
| Rooms with windows or exterior doors on one side | 1 |
| Rooms with windows or exterior doors on two sides | 1-1/2 |
| Rooms with windows or exterior doors on three sides | 2 |
| Entrance Halls | 2 |

$Q_{VENT}$ = the building ventilation volume flow rate, in cubic feet per <u>minute</u>

From the above and Equation (9.1), it follows that the total instantaneous building heat loss can be expressed as:

$$q = BLC(T_{indoor} - T_{outdoor})  \tag{9.13}$$

## 9.9  THERMAL "WEIGHT"

Thermal weight is a qualitative description of the extent to which the building energy consumption occurs in "lock-step" with local weather conditions. Thermal "weight is characterized by the mass and specific heat (heat capacity) of the various components which make up the structure, as well as the unique combination of internal loads and solar exposures which have the potential to offset the temperature-dependent heat loss or heat gain of the facility.

Thermally "Light" Buildings are those whose heating and cooling requirements *are* proportional to the weather. Thermally "Heavy" Buildings are those Buildings whose heating and cooling requirements are *not* proportional to the weather. The "heavier" the building, the less temperature-dependent the building's energy consumption appears to be, and the less accuracy that can be expected from simple temperature-dependent energy consumption calculation schemes.

The concept of thermal weight is an important one when it comes to determining the energy-saving potential of envelope improvements. The "heavier" the building, the less savings per square foot of improved envelope can be expected for the same "UA" improvement. The "lighter" the building, the greater the savings that can be expected. A means for characterizing the thermal "weight" of buildings, as well as analyzing the energy savings potential of "light" and "heavy" buildings will be taken up in Section 9.10.

## 9.10  ENVELOPE ANALYSIS FOR EXISTING BUILDINGS

### 9.10.1 Degree Days

In theory, if one wanted to predict the heat lost by a building over an extended period of time, Equation 9.12 could be solved for each individual hour, taking into account the relevant changes of the variables. This is possible because the change in the value BLC with respect to temperature is not significant and the indoor temperature ($T_i$) is normally controlled to a constant value (such as 70°F in winter). That being the case, the total energy transfer could

be predicted by knowing the summation of the individual deviations of outdoor temperature ($T_o$) from the indoor condition ($T_i$) over an extended period of time.

The summation described has come to be known as "Degree-Days" and annual tabulations of Degree-Days for various climates are published by NOAA, ASHRAE, and various other public and military organizations. Historically an indoor reference point of 65°F has been used to account for the fact that even the most poorly constructed building is capable of maintaining comfort conditions without heating when the temperature is at least 65°F.

Because of the impracticality of obtaining hour-by-hour temperature data for a wide variety of locations, *daily* temperature averages are often used to represent 24-hour blocks of time. The daily averages are calculated by taking the average of daily maximum and minimum temperature recordings, which in turn are converted o Degree Days. Quantitatively this is calculated as:

$$Degree - Days = \frac{(T_{reference} - T_{average})n}{24}  \tag{9.14}$$

Where "n" represents the number of hours in the period for which the Degree Days are being reported, and $T_{reference}$ is a reference temperature at which no heating is assumed to occur. Typically, a reference temperature of 65°F is used. Units of degree hours can be obtained by multiplying the results of any Degree-Day tabulation by 24.

Figure 9.23 shows the monthly natural gas consumption of a metropolitan newspaper building overlaid on a plot of local heating Degree Days.

Because of the clear relationship between heating Degree Days and heating fuel consumption for many buildings, such as the above, the following formula has been used since the 1930's to predict the future heating fuel consumption.

$$E = \frac{q \times DD \times 24}{\Delta t \times H \times Eff}  \tag{9.15}$$

Where:

E  = Fuel consumption, in appropriate units, such as Therms natural gas

q  = The design-day heat loss, Btu/hr

DD = The annual heating Degree-Days (usually referenced to 65°F)

24 = Converts Degree-Days to Degree-Hours

$\Delta t$  = The design day indoor-outdoor temperature difference, °F

H  = Conversion factor for the type of fuel used

Eff = The annualized efficiency of the heating combustion process

**Figure 9.23.**

Note that if he BLC is substituted for $q/\Delta t$, Equation 9.15 becomes:

$$E = \frac{BLC \times DD \times 24}{H \times Eff}$$

(9.16)

Ignoring the conversion terms, we can see that this equation has the potential to describe the relationship between monthly weather trends and the building heating fuel consumption shown in Figure 9.23.

As building construction techniques have improved over the years, and more and more heat-producing equipment has found its way into commercial and even residential buildings, a variety of correction factors have been introduced to accommodate these influences. The Degree-Day method, in its simpler form, is not presently considered a precise method of estimating future building energy consumption. However it is useful for indicating the severity of the heating season for a region and it can prove to be a very powerful tool in the analysis of existing buildings (i.e. ones whose energy consumption is already known).

It is possible to determine an effective Building Load Coefficient for a building by analyzing the monthly fuel consumption and corresponding monthly Degree Days using the technique of linear regression.

### 9.10.2 Analyzing Utility Billings

If we were to plot the monthly natural gas consumption and corresponding Degree-Days for a building, the result might look something like Figure 9.24. It can be seen that the monthly energy consumption appears to follow the weather in an indirect fashion.

If we could draw a line that represented the closest "fit" to the data, it might look like the Figure 9.25.

As a general rule, the simpler the building and its heating (or cooling) system, the better the correlation. Buildings that are not mechanically cooled will show an even better correlation. To put it in terms of our earlier discussion regarding thermal "weight," the "lighter" the building, the better the correlation between building energy consumption and Degree Days. To the extent this line fits the data, it gives us two important pieces of information discussed in Section 9.1.1.

- The Monthly Base Consumption

- The Relationship Between the Weather and Energy Consumption Beyond the Monthly Base, which we have been calling the Building Load Coefficient.

**Figure 9.24.**

Figure 9.25.

### 9.10.3 Using Linear Regression for Envelope Analysis

The basic idea behind linear regression is that any physical relationship where the value of a result is linearly dependent on the value of an independent quantity can be described by the relationship:

$$y = mx + b \qquad (9\text{-}17)$$

where "y" is the dependent variable, x is the independent variable, m is the slope of the line that describes the relationship, and b is the y-intercept, which is the value of y when x = 0.

In the case of monthly building energy consumption, the monthly Degree Days, can be taken as the independent variable, and the monthly fuel consumption as the dependent variable. The slope of the line that relates the monthly consumption to monthly Degree Days is the Building Load Coefficient (BLC). The monthly base fuel consumption is the intercept on the fuel axis (or the monthly fuel consumption when there are no heating Degree Days). The fuel consumption for any month can be determined by multiplying the monthly Degree Days by the Building Load Coefficient and adding it to the monthly base fuel consumption.

$$\text{Fuel Consumption} = \text{BLC} \times \text{DD} + \text{base energy} \qquad (9.18)$$

The value of the above is that once the equation is determined, projected fuel savings can be made by recalculating the monthly fuel consumption using a "conservation-modified" BLC.

The determination of the Building Load Coefficient is useful in the analysis of building envelope for a number of reasons.

• Often the information necessary for the analysis of building envelope components is not available.

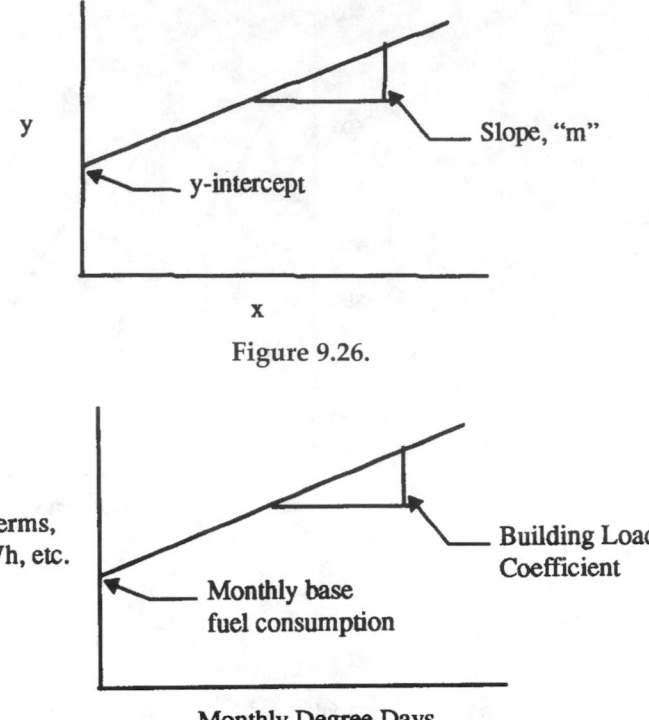

Figure 9.26.

Figure 9.27.

• If envelope information is readily available and the Building Load Coefficient has been determined by analysis of all the individual components, the regression derived BLC gives us a "real-world" check on the calculated BLC.

• Linear regression also gives feedback in terms of how "good" the relationship is between fuel consumption and local climate, giving a good indication as to the thermal "weight" of the building, discussed in Section 9.9.

The following is a brief overview of the regression procedure as it can be applied to corresponding pairs of monthly Degree-Day and fuel consumption data. The Building Load Coefficient which describes the actual performance of the building can be determined with the following relationship:

$$BLC = \frac{n\sum D_i E_i - n\sum D_i \sum E_i}{n\sum D_i^2 - (\sum D_i)^2} \qquad (9.19)$$

Where:

n = the number of degree-day/fuel consumption pairs

$D_i$ = the degree days accumulated for an individual month

$E_i$ = the energy or fuel consumed for an individual month. The monthly base fuel consumption can be calculated as:

$$Base = \frac{\sum E_i}{n} - BLC \frac{\sum D_i}{n} \qquad (9.20)$$

The units of the BLC will be terms of the fuel units per degree day. The BLC will be most valuable for additional analysis if it is converted to units of Btu/(hr-°F) or Watts/°C.

The following example demonstrates how the above might be used to evaluate the potential for envelope improvement in a building.

### Example

A proposal has been made to replace 5,000 ft² of windows in an electrically heated building. The building envelope has been analyzed on a component-by-component basis and found to have an overall analytical BLC = 15,400 Btu/(Hr-°F). Of this total, 6,000 Btu/(Hr-°F) is attributable to the single pane windows, which are assumed to have a U-factor of 1.2 Btu/(hr-ft²-°F). A linear regression is performed on the electric utility data and available monthly Degree-Day data (65°F reference). The BLC is found to be 83.96 kWh/degree-day, which can be converted to more convenient units by the following:

$$BLC = 83.96 \ \frac{kWh}{°F - Days} \times 3413 \ \frac{Btu}{kWh} \times \frac{1 \ Day}{24 \ hours}$$

$$= 11,940 \ Btu/(hr-°F)$$

What is the reason for the discrepancy between the BLC calculated component-by-component and the BLC derived from linear regression of the building's performance?

The explanation comes back to the concept of thermal weight. Remember, the more thermally "heavy" the building, the more independent of outdoor conditions it is. The difference between the value above and the calculated BLC is due in part to internal heat gains in the building that offset some of the heating that would normally be required.

This has significant consequences for envelope retrofit projects. If the potential savings of a window conservation retrofit is calculated on the basis of the direct "UA" improvement, the savings will be overstated in a building such as the above. A more conservative (and realistic) estimate of savings can be made by de-rating the theoretical UA improvement by the ratio of the regression UA to the calculated UA.

### 9.10.4 Evaluating the Usefulness of the Regression Results

The Correlation Coefficient, R, is a useful term that describes how well the derived linear equation accounts for the variation in the monthly fuel consumption of the building. The correlation coefficient is calculated as follows:

$$R = BLC \ \sqrt{\frac{\sum D_i^2 - \frac{\sum D_i^2}{n}}{\sum E_i^2 - \frac{\sum E_i^2}{n}}} \qquad (9.21)$$

The square of the Correlation Coefficient, R,² provides an estimate of the number of independent values (fuel consumption) whose variation is explained by the regression relationship. For example, an R² value of 0.75 tells us that 75% of the monthly fuel consumption data points evaluated can be accounted for by the linear equation given by the regression analysis. As a general rule of thumb, an R² value of 0.80 or above describes a thermally "light" building, and a building whose fuel consumption indicates an R² value significantly less than 0.80 can be considered a thermally "heavy" building.

### 9.10.5 Improving the Accuracy of the Building Load Coefficient Estimate

As discussed elsewhere in Section 9.9, one of the reasons a building might be classified as thermally "heavy" would be the presence of significant internal heat gains, such as lighting, equipment and people. To the extent these gains occur in the perimeter of the building, they will tend to offset the heat loss predicted using the tools described previously in this chapter. The greater the internal heat, relative to the overall building temperature dependence (BLC), the lower the outside temperature has to be before heating is required.

The so-called "building balance temperature" is the theoretical outdoor temperature where the total building heat loss is equal to the internal gain. We discussed earlier that the basis of most published Degree-Day data is a reference (or balance) temperature of 65°F. If data of this sort are used for analysis, the lower the actual building balance temperature than 65°F, the more error will be introduced into the analysis. This accounts for the low R² values encountered in thermally heavy buildings.

While most degree data are published with 65°F as the reference point, it is possible to find compiled sources with 55°F and even 45°F reference temperatures. The use of this monthly data can improve the accuracy of the analysis significantly, if the appropriate data are utilized.

### Example

A regression is performed for a library building with electric resistance heating. The results of the analysis indicate an R² of 0.457. In other words, only 46% of the variation in monthly electricity usage is explained by the regression. The figure below is a plot of the monthly electric consumption versus Degree Days calculated for a 65°F reference.

**Figure 9.28.**

The plot following shows the result of the regression run using monthly Degree Day referenced to 55°F. Notice in the figure that there are less data points showing. This is explained by the observation that all the days with average temperatures greater than 55°F are not included in the data set. The resulting $R^2$ has increased to 0.656.

The next figure shows the result of the regression run using monthly Degree Day referenced to 45°F. The resulting $R^2$ has increased to 0.884. Notice how few data points are left. Normally in statistical analysis great emphasis is placed on the importance of having an adequate data sample for the results to be meaningful. While we would like to have as many points as possible in our analysis, the concern is not so great with this type of analysis. Statistical analysis usually concerns itself with *whether* there is a relationship to be found. The type of analysis we are advocating here assumes that the relationship *does* exist, and we are merely attempting to discover the most accurate form of the relationship. Another way of saying this is that we are using a trial-and-error technique to discover the building balance point.

The final figure shows the result of the regression run with monthly degree days referenced to 40°F. The resulting $R^2$ has *decreased* to 0.535.

We have learned from the above that the balance point of the building analyzed is somewhere between 40° and 50°F, with 45°F probably being pretty close. Additional iterations can be made if more precision is desired, and if the referenced Degree Day data are available.

If the published Degree Day data desired are not available, it is a relatively simple matter to construct it

**Figure 9.29.**

**Figure 9.30.**

**Figure 9.31.**

from the average daily temperatures published for the local climate. If this is done with a spreadsheet, a table can be constructed in such a way that the degree days, BLC, monthly base and $R^2$ values are all linked to an assumed balance temperature. This balance temperature is modified iteratively until $R^2$ is maximized. Of course any energy saving calculations should consistently use both the derived BLC and degree days that accompany the correlation with the highest $R^2$.

## 9.11 ENVELOPE ANALYSIS FOR NEW BUILDINGS

Envelope analysis of new buildings offers a different kind of challenge to the analyst. While new construction offers much greater opportunity for economical improvements to the envelope, the analysis is much more open-ended than that for existing buildings. In other words, we do not know the monthly energy consumption (the answer), and are left totally at the mercy of tools designed to assist us in *predicting* the future energy consumption of an as-yet unconstructed building, the necessary details of which constitute thousands of unknowns.

We say the above to emphasize the extreme difficulty of the task, and the inadvisability of harboring any illusions that this can be done reliably, short of using hour-by-hour computer simulation techniques. Even with the powerful programs currently available, accuracy's no better than 10 to 20 percent should be expected.

Nevertheless, we are often called upon to quantify the benefits of using one envelope strategy in place of another. We are able to quantify the *difference* in annual energy consumption between two options much more reliably than the absolute consumption. A number of techniques have been developed to assist in this process. One of the more popular and useful tools is the *Temperature Bin Method*.

### 9.11.1 The Temperature Bin Method

The Temperature Bin Method requires that instantaneous energy calculations be performed at many different outdoor temperature conditions, with the results multiplied by the number of hours expected at each temperature condition. The "bins" referred to represent the number of hours associated with groups of temperatures, and are compiled in 5°F increments. The hour tabulations are available in annual, monthly and sometimes 8-hour shift totals. All hourly occurrences in a bin are assumed to take place at the bin "center-temperature." For example, it is assumed that 42°F quantitatively represents the 40-44°F bin.

The basic methodology of this method requires calculating the unique heat loss at each bin by multiplying the BLC by the difference between the indoor temperature assumed and the center-temperature for each respective bin. This result is in turn multiplied by the number of hours in the bin. The products of all the

bins are summed to arrive at the total predicted annual heating energy for the building.

The advantage of this "multiple-measure" method over a "single-measure" method, such as the degree day method, is the ability to accommodate other temperature-dependent phenomena in the analysis. For example, the power requirement and capacity of an air-to-air heat pump are extremely temperature dependent. This dependency is easily accommodated by the bin method.

However, just as was the case with the Degree-Day method, energy savings predicted by the bin method may vary significantly from actual, depending on the building balance temperature (building thermal weight). While the balance temperature cannot be predicted for a new building with the techniques previously discussed, it can be estimated with the following equation.

$$T_{balance} = T_{indoor} - \frac{q_{internal}}{BLC} \qquad (9.22)$$

Where:

$T_{balance}$ = The predicted balance temperature, °F

$T_{indoor}$ = The assumed indoor conditioned space temperature, °F

$q_{internal}$ = The assumed internal heat gain in building temperature control zones adjacent to the envelope, Btu/hr

BLC = The Building Load Coefficient, BLC, Btu/(hr-°F)

More accurate energy predictions will result by omitting calculations for bins whose center temperature exceeds the assumed balance temperature. For example, no calculation should be performed for the 50-54°F bin, if the predicted balance temperature is 50°F. The center temperature of 47°F for the 45-49°F bin indicates that a 3°F temperature difference is appropriate for that bin (50- 47°F).

A complete description of the bin method can be found in the ASHRAE Handbook of Fundamentals.

## 9.12 ENVELOPE STANDARDS FOR NEW & EXISTING CONSTRUCTION

The ASHRAE/IESNA Standard 90.1, *Energy Standard for Buildings Except Low-Rise Residential Buildings* is the de-facto standard for commercial building efficiency and is the basis of most local energy codes.

## 9.12.1 ASHRAE 90.1 Compliance Requirements

Compliance with the Standard may be demonstrated utilizing the *Prescriptive Approach* or by performing calculations utilizing the *Building Envelope Tradeoff Option.*

### 9.12.1.1 The Prescriptive Approach

Component criteria necessary for local compliance with the *Prescriptive Approach* are tabulated in tables for 26 separate climate zones, with the appropriate Climate Zone selected based on the local heating and cooling degree-days. One of the features of the new standard is the inclusion of many precalculated building components in the Appendix. A variety of common (and some uncommon) envelope assemblies can be referenced and utilized for demonstrating compliance with the *Prescriptive Approach.* In most cases, this will mean that no calculations will be required by the designer to demonstrate compliance.

### 9.12.1.2 The Building Envelope Tradeoff Option

The *Building Envelope Tradeoff Option* provides more compliance flexibility than might be found in the *Prescriptive Approach.* This option requires the designer to demonstrate that the proposed building envelope results in an *Envelope Performance Factor* that is lower than the *budgeted* one for the project. Because of the impact of the mechanical and lighting systems on heating and cooling energy consumption, the *Building Envelope Tradeoff Option* requires information for the mechanical and lighting systems, as well as for the envelope components. Due to the complexity of this calculation, the ENVSTD Envelope Tradeoff Software, first introduced in 1989, will be available to automate the tradeoff calculations as well as consolidate the required reference data.

## 9.13 SUMMARY

While the above discussion of envelope components has emphasized the information needed to perform rudimentary heat loss calculations, you'll find that the more you understand these basics, the more you begin to understand what makes an efficient building envelope. This same understanding will also guide you in deciding how to prioritize envelope improvement projects in existing buildings.

## 9.14 ADDITIONAL READING

As you can see from this brief introduction, the best source of comprehensive information on building envelope issues is the *ASHRAE Handbook of Fundamentals.* You are encouraged to continue your study of building envelope by reading the following chapters in the 2001 *ASHRAE Handbook of Fundamentals.*

| Chapter | Topic |
|---|---|
| 23,24 | Thermal Insulation and Vapor Retarders |
| 25 | Thermal and Water Vapor Transmission Data |
| 26 | Ventilation and Infiltration |
| 27 | Climatic Design Information |
| 28 | Residential Cooling and Heating Load Calculations |
| 29 | Nonresidential Cooling and Heating Load Calculations |
| 30 | Fenestration |

### 9.14.1 References

1. American Society of Heating, Refrigerating, and Air Conditioning Engineers, Inc., *Heat Transmission Coefficients for Walls, Roofs, Ceilings, and Floors,* 1993.
2. ASHRAE Handbook of Fundamentals, American Society of Heating, Refrigerating and Air Conditioning Engineers, Inc., Atlanta, GA, 1993.
3. ASHRAE Standard 901.-1989, American Society of Heating, Refrigerating, and Air Conditioning Engineers, Inc., Atlanta, GA, 1999.
4. Brown, W.C., *Heat-Transmission Tests on Sheet Steel Walls,* ASHRAE Transactions, 1986, Vol. 92, part 2B.
5. Elder, Keith E., *Metal Buildings: A Thermal Performance Compendium,* Bonneville Power Administration, Electric Ideas Clearinghouse, August 1994.
6. Elder, Keith E., *Metal Elements in the Building Envelope. A Practitioner's Guide,* Bonneville Power Administration, Electric Ideas Clearinghouse, October 1993.
7. Johannesson, Gudni, *Thermal Bridges in Sheet Metal Construction,* Division of Building Technology, Lund Institute of Technology, Lund, Sweden, Report TVHB-3007, 1981 .
8. Loss, W., *Metal Buildings Study: Performance of Materials and Field Validation,* Brookhaven National Laboratory, December 1987.
9. Washington State Energy Code, Washington Association of Building Officials April 1, 1994

# CHAPTER 10

# HVAC SYSTEMS

**ERIC NEIL ANGEVINE, P.E.**
Associate Professor
School of Architecture
Oklahoma State University
Stillwater, Oklahoma

**JENNIFER S. FAIR, P.E.**
Vice President
PSA Consulting Engineers
Oklahoma City, Oklahoma

## 10.1 INTRODUCTION

The mechanical heating or cooling load in a building is dependent upon the various heat gains and losses experienced by the building including solar and internal heat gains and heat gains or losses due to transmission through the building envelope and infiltration (or ventilation) of outside air. The primary purpose of the heating, ventilating, and air-conditioning (HVAC) system in a building is to regulate the dry-bulb air temperature, humidity and air quality by adding or removing heat energy. Due to the nature of the energy forces which play upon the building and the various types of mechanical systems which can be used in non-residential buildings, there is very little relationship between the heating or cooling load and the energy consumed by the HVAC system.

This chapter outlines the reasons why energy is consumed and wasted in HVAC systems for non-residential buildings. These reasons fall into a variety of categories, including energy conversion technologies, system type selection, the use or misuse of outside air, and control strategies. Following a review of the appropriate concerns to be addressed in analyzing an existing HVAC system, the chapter discusses the aspects of human thermal comfort. Succeeding sections deal with HVAC system types, energy conservation opportunities and domestic hot water systems.

## 10.2 SURVEYING EXISTING CONDITIONS

As presented in Chapter 3, the first stage of any effective energy management program is an energy audit of the facility in question. In surveying the HVAC system(s) in a facility, the first step is to find out what you have to work with: what equipment and control systems exist. It is usually beneficial to divide the HVAC systems into two categories: equipment and systems which provide heating and cooling, and equipment and systems which provide ventilation. It is essential to fully document the type and status of all equipment from major components including boilers, chillers, cooling towers and air-handling units to the various control systems, thermostats, valves and gauges, whether automated or manual; in order to later determine what elements can be replaced or improved to realize a saving in energy consumed by the system.

The second step is to determine how the system is operating. This requires that someone measure the operating parameters to determine whether the system actually operates as it was specified to operate. Determine the system efficiency under realistic conditions. This may be significantly different from the theoretical, or full-load efficiency. Determine how the system is operated. What are the hours of operation? Are changes in system controls manual or automatic? Find out how the system is actually operated, which may differ from how the system was designed to be operated. It is best to talk to operators and/or users of the system who know a lot more about how the system operates than the engineers or managers.

If the system is no longer operating at design conditions, it is extremely useful to determine what factors are responsible for the change. Potential causes of operational changes are modifications in the building or system and lapses in maintenance. Have there been structural or architectural changes to the building without corresponding changes to the HVAC system? Have there been changes in building operations? Is the system still properly balanced? Has routine maintenance been performed? Has scheduled preventative maintenance been performed?

Finally, it is useful to determine whether the system can or should be restored to its initial design conditions. If practical, it may be beneficial to carry out the needed maintenance *before* proceeding to analyze the system for further improvements. However, some older systems are so obviously inefficient that bringing them back to original design parameters is not worth the time or expense.

Before continuing with an analysis of the system, it is also useful to determine future plans for the building and the HVAC system which can seriously effect the energy efficiency of system operation. Are there plans to remodel the building or parts of the building? How extensive are proposed changes? Are changes in building operations planned?

Document everything. Only when you have a full record of what the system consists of, how it is operating and how it is operated, and what changes have been made and will be made in the future, can you properly evaluate the benefit of energy conservation techniques which may be applicable to a particular building system.

## 10.3 HUMAN THERMAL COMFORT

The ultimate objective of any heating, cooling and ventilating system is typically to maximize human thermal comfort. Due to the prevalence of simple thermostat control systems for residential and small-scale commercial HVAC systems, it is often believed that human thermal comfort is a function solely, or at least primarily, of air temperature. But this is not the case.

Human thermal comfort is actually maximized by establishing a heat balance between the occupant and his or her environment. Since the body can exchange heat energy with its environment by conduction, convection and radiation, it is necessary to look at the factors which affect these heat transfer processes along with the body's ability to cool itself by the evaporation of perspiration.

All living creatures generate heat by burning food, a process known as metabolism. Only 20 percent of food energy is converted into useful work; the remainder must be dissipated as heat. This helps explain why we remain comfortable in an environment substantially cooler than our internal temperature of nearly 100°F (37°C).

In addition to air temperature, humidity, air motion and the surface temperature of surroundings all have a significant influence on the rate at which the human body can dissipate heat. At temperatures below about 80°F (27°C) most of the body's heat loss is by convection and radiation. Convection is affected mostly by air temperature, but it is also strongly influenced by air velocity. Radiation is primarily a function of the relative surface temperature of the body and its surroundings. Heat transfer by conduction is negligible, since we make minimal physical contact with our surroundings which is not insulated by clothing.

At temperatures above 80°F (27°C) the primary heat loss mechanism is evaporation. The rate of evaporation is dependent on the temperature and humidity of the air, as well as the velocity of air which passes over the body carrying away evaporated moisture.

In addition to these environmental factors, the rate of heat loss by all means is affected by the amount of clothing, which acts as thermal insulation. Similarly, the amount of heat which must be dissipated is strongly influenced by activity level. Thus, the degree of thermal comfort achieved is a function of air temperature, humidity, air velocity, the temperature of surrounding surfaces, the level of activity, and the amount of clothing worn.

In general, when environmental conditions are cool the most important determinant of human thermal comfort is the radiant temperature of the surroundings. In fact, a five degree increase in the mean-radiant temperature of the surroundings can offset a seven degree reduction in air temperature.

When conditions are warm, air velocity and humidity are most important. It is not by accident that the natural response to being too warm is to increase air motion. Similarly, a reduction in humidity will offset an increase in air temperature., although it is usually necessary to limit relative humidity to no more than 70% in summer and no less than 20% in winter.

There is, of course, a human response to air temperature, but it is severely influenced by these other factors. The most noticeable comfort response to air temperature is the reaction to drift, the change of temperature over time. A temperature drift of more than one degree Fahrenheit per hour (0.5°C/hr) will result in discomfort under otherwise comfortable conditions. Temperature stratification can also cause discomfort, and temperature variation within the occupied space of a building should not be allowed to vary by more than 5 degrees F (3°C).

Modern control systems for HVAC systems can respond to more than just the air temperature. One option which has been around for a long time is the humidistat, which senses indoor humidity levels and controls humidification. However, state-of-the-art control systems can measure *operative temperature*, which is the air temperature equivalent to that affected by radiation and convection conditions of an actual environment.* Another useful construct is that of effective temperature, which is a computed temperature that includes the effects of humidity and radiation.†

The location and type of air distribution devices play a role equal in importance to that of effective controls in achieving thermal comfort. The discomfort caused by stratification can be reduced or eliminated by proper distribution of air within the space.

In general terms, thermal comfort can be achieved at air temperatures between about 68°F and 80°F, and relative humidities between 20% and 70%, under varying air velocities and radiant surface temperatures. Figure 10.1 shows the generalized "comfort zone" of dry bulb temperatures and humidities plotted on the psychrometric chart. However it should not be forgotten that human thermal comfort is a complex function of temperature, humidity, air motion, thermal radiation from local surroundings, activity level and amount of clothing.

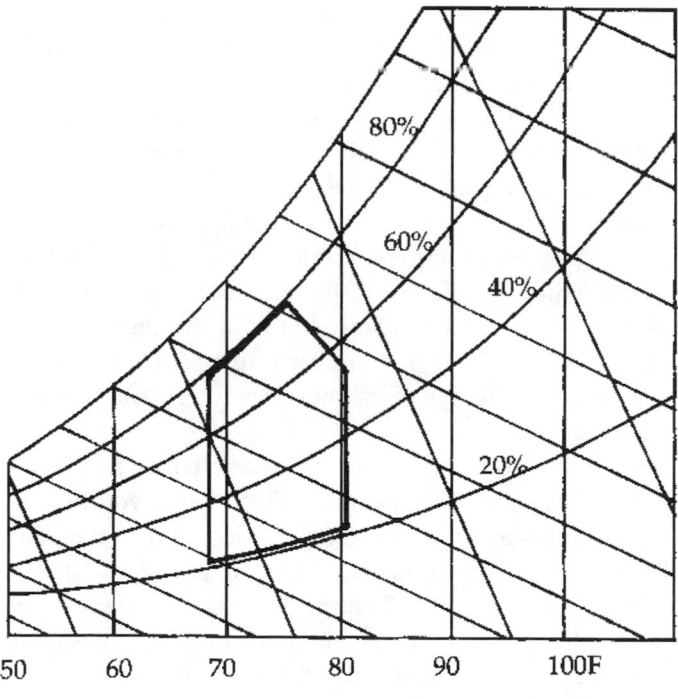

**Figure 10.1 Comfort zone.**

_____

*Operative temperature is technically defined as the uniform temperature of an imaginary enclosure with which an individual experiences the same heat by radiation and convection as in the actual environment.

†Effective temperature is an empirical index which attempts to combine the effect of dry bulb temperature, humidity and air motion into a single figure related to the sensation of thermal comfort at 50% relative humidity in still air.

## 10.4 HVAC SYSTEM TYPES

The energy efficiency of systems used to heat and cool buildings varies widely but is generally a function of the details of the system organization. On the most simplistic level the amount of energy consumed is a function of the source of heating or cooling energy, the amount of energy consumed in distribution, and whether the working fluid is simultaneously heated and cooled. System efficiency is also highly dependent upon the directness of control, which can sometimes overcome system inefficiency.

HVAC system types can be typically classified according to their energy efficiency as highly efficient, moderately efficient or generally inefficient. This terminology indicates only the comparative energy consumption of typical systems when compared to each other. Using these terms, those system types classified as generally inefficient will result in high energy bills for the building in which they are installed, while an equivalent building with a system classified as highly efficient will usually have lower energy bills. However, it is important to recognize that there is a wide range of efficiencies within each category, and that a specific energy-efficient example of a typically inefficient system might have lower energy bills than the least efficient example of a moderately, or even highly efficient type of system.

Figure 10.2 shows the relative efficiency of the more commonly used types of HVAC systems discussed below. The range of actual energy consumption for each system type is a function of other design variables including how the system is configured and installed in a particular building as well as how it is controlled and operated.

To maximize the efficiency of any type of HVAC system, it is important to select efficient equipment, minimize the energy consumed in distribution and avoid simultaneous heating and cooling of the working fluid. It is equally important that the control system directly control the variable parameters of the system.

Most HVAC systems include zones, which are areas within the building which may have different climatic and/or internal thermal loads and for which heat can be supplied or extracted independent of other zones.

The four-volume *ASHRAE Handbook,* published sequentially in a four-year cycle by the American Society of Heating, Refrigerating and Air-Conditioning Engineers, Inc., provides the most comprehensive and authoritative reference on HVAC systems for buildings. The reader is specifically referred to the *ASHRAE*

Least Efficient                          Most Efficient

**Figure 10.2 Relative energy efficiency of air-conditioning systems.**

*Handbook: HVAC Systems and Equipment* for additional information on any of the systems described below.[3] Additional information regarding the most appropriate types of HVAC system for specific applications can be found in the *ASHRAE Handbook: HVAC Applications.*[2]

### 10.4.1 All-Air Systems

The most common types of systems for heating and cooling buildings are those which moderate the air temperature of the occupied space by providing a supply of heated or cooled air from a central source via a network of air ducts. These systems, referred to as all-air systems, increase or decrease the space temperature by altering either the volume or temperature of the air supplied.

Recalling that the most important determinant of thermal comfort in a warm environment is air velocity, most buildings which require cooling employ all-air systems. Consequently, all-air systems are the system of choice when cooling is required. All-air systems also provide the best control of outside fresh air, air quality, and humidity control. An added benefit of forced air systems is that they can often use outside air for cooling interior spaces while providing heating for perimeter spaces. (See §10.5.5; Economizers) The advantages of all-air systems are offset somewhat by the energy consumed in distribution.

All-air systems tend to be selected when comfort

cooling is important and for thermally heavy buildings which have significant internal cooling loads which coincide with heating loads imposed by heat loss through the building envelope.

The components of an all-air HVAC system include an air-handling unit (AHU) which includes a fan, coils which heat and/or cool the air passing through it, filters to clean the air, and often elements to humidify the air. Dehumidification, when required, is accomplished by cooling the air below the dew-point temperature. The conditioned air from the AHU is supplied to the occupied spaces by a network of supply-air ducts and air is returned from conditioned spaces by a parallel network of return-air ducts. (Sometimes the open plenum above a suspended ceiling is used as part of the return-air path.) The AHU and its duct system also includes a duct which supplies fresh outside air to the AHU and one which can exhaust some or all of the return air to the outside. Figure 10.3 depicts the general arrangement of components in an all-air HVAC system.

### Single Duct Systems

The majority of all-air HVAC systems employ a single network of supply air ducts which provide a continuous supply of either warmed or cooled air to the occupied areas of the building.

**Single Zone**—The single duct, single-zone system is the simplest of the all-air HVAC systems. It is one of the most energy-efficient systems as well as one of the least expensive to install. It uses a minimum of distribution energy,* since equipment is typically located within or immediately adjacent to the area which it conditions. The system is directly controlled by a thermostat which turns the AHU on and off as required by the space temperature. The system shown in Figure 10.3 is a single zone system.

Single zone systems can provide either heating or cooling, but provide supply air at the same volume and temperature to the entire zone which they serve. This limits their applicability to large open areas with few windows and uniform heating and cooling loads. Typical applications are department stores, factory spaces, arenas and exhibit halls, and auditoriums.

**Variable Air Volume**—The variable air volume (VAV) HVAC system shown in Figure 10.4 functions much like the single zone system, with the exception that the temperature of individual zones is controlled by a thermostat which regulates the volume of air that is discharged into the space. This arrangement allows a

---

*Distribution energy includes all of the energy used to move heat within the system by fans and pumps. Distribution energy is typically electrical energy.

**Figure 10.3 Elements of an all-air air-conditioning system.**

high degree of local temperature control at a moderate cost. Both installation cost and operating costs are only slightly greater than the single-zone system.

The distribution energy consumed is increased slightly over that of a single-zone system due to the friction losses in VAV control devices, as well as the fact that the fan in the AHU must be regulated to balance the overall air volume requirements of the system. Fan regulation by inlet vanes or outlet dampers forces the fan to operate at less than its optimum efficiency much of the time (see Figure 10.5) Consequently a variable speed fan drive is necessary to regulate output volume of the fan. For the system to function properly, it is necessary that air be supplied at a constant temperature, usually about 55°F (13°C). This requires indirect control of the supply air temperature with an accompanying decrease in control efficiency.

Single-duct VAV systems can often provide limited heating by varying the amount of constant temperature air to the space. By reducing the cooling airflow,

**Figure 10.4 Variable air volume system schematic diagram.**

the space utilizes the lights, people and miscellaneous equipment to maintain the required space temperature. However, if the space requires more heat than can be supplied by internal heat gains, a separate or supplemental heating system must be employed.

Single-duct VAV systems are the most versatile and have become the most widely used of all systems for heating and cooling large buildings. They are appropriate for almost any application except those requiring a high degree of control over humidity or air exchange.

**Reheat systems**—Both the single-zone and single duct VAV systems can be modified into systems which provide simultaneous heating and cooling of multiple zones with the addition of reheat coils for each zone (Figure 10.6). These systems are identical in design to the foregoing systems up to the point where air enters the local ductwork for each zone. In a reheat system supply air passes through a reheat coil which usually contains hot water from a boiler. In a less efficient option, an electrical resistance coil can also be used for reheat. (See comments regarding the efficiency of electric resistance heating in §10.5.6.)

A local thermostat in each zone controls the temperature of the reheat coil, providing excellent control of the zone space temperature. Constant air volume (CAV) reheat systems are typically used in situations which

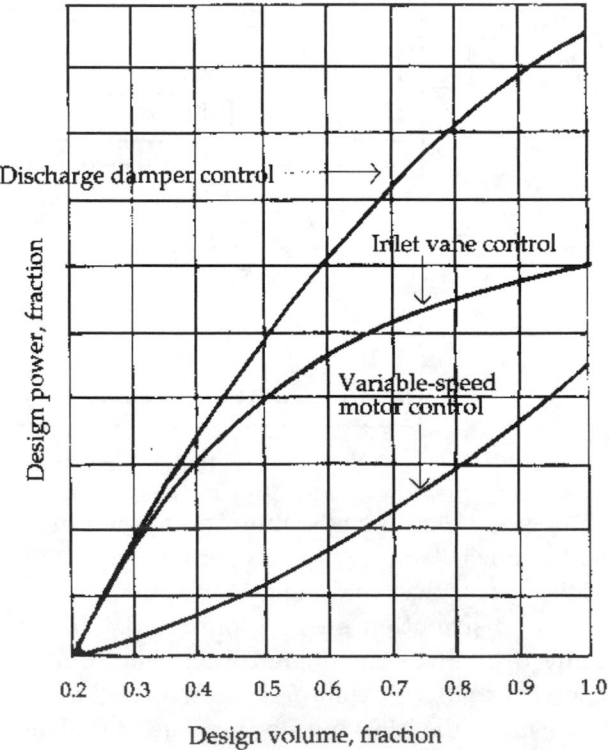

**Figure 10.5 Fan power vs. discharge volume characteristics.**

**Figure 10.6 Reheat system schematic diagram.**

require precise control of room temperature and/or humidity, often with constant airflow requirements, such as laboratories and medical facilities.

Both the CAV and VAV reheat systems are inherently inefficient, representing the highest level of energy consumption of the all-air systems. This is due to the fact that energy is consumed to cool the supply air and then additional energy is consumed to reheat it. In VAV reheat systems, the reheat coil is not activated unless the VAV controls are unable to meet local requirements for *temperature control*, and they are therefore somewhat more *energy efficient* than CAV reheat systems.

Both CAV and VAV reheat systems can also be used with specialized controls to condition spaces with extremely rigid requirements for humidity control, such as museums, printing plants, textile mills and industrial process settings.

**Multizone**—Although commonly misused to indicate any system with thermostatically controlled air-conditioning zones, the multizone system is actually a specific type of HVAC system which is a variation of the single-duct CAV reheat system. In a multizone system, each zone is served by a dedicated supply duct which connects it directly to a central air handling unit (Figure 10.7).

In the most common type of multizone system, the AHU produces warm air at a temperature of about 100°F (38°C) as well as cool air at about 55°F (13°C) which are blended with dampers to adjust the supply air temperature to that called for by zone thermostats. In a variation of this system, a third neutral deck uses outside air as an economizer to replace warm air in the summer or cool air in the winter. In another variation, the AHU produces only cool air which is tempered by reheat coils located in the fan room. In this case, the hot deck may be used as a preheat coil.

Multizone systems are among the least energy efficient, sharing the inherent inefficiency of reheat systems since energy is consumed to simultaneously heat and cool air which is mixed to optimize the supply air temperature. Since a constant volume of air is supplied to each zone, blended conditioned air must be supplied even when no heating or cooling is required.

In addition, multizone systems require a great deal of space for ducts in the proximity of the AHU which restricts the number of zones. They also consume a great deal of energy in distribution, due to the large quantity of constant volume air required to meet space loads. These drawbacks have made multizone systems nearly *obsolete except* in relatively small buildings with only a few zones and short duct runs.

**Figure 10.7 Multizone system schematic diagram.**

## Dual Duct Systems

Dual duct systems are similar to the multizone concept in that both cool supply air and warm supply air are produced by a central AHU. But instead of blending the air in the fan room, separate hot-air ducts and cold-air ducts run parallel throughout the distribution network and air is mixed at terminal mixing boxes in each zone (Figure 10.8). The mixing boxes may include an outlet for delivering air directly to the space, or a duct may connect a branch network with air mixed to a common requirement.

Dual duct systems require the greatest amount of space for distribution ductwork. In order to offset the spatial limitations imposed by this problem, dual duct systems often employ high velocity/high pressure supply ducts, which reduce the size (and cost) of ductwork, as well as the required floor-to-floor height. However this option increases the fan energy required for distribution. Their use is usually limited to buildings with very strict requirements for temperature and or humidity control.

**Constant Volume Dual Duct**—For a long time, the only variation of the dual duct system was a CAV system, which functioned very much like the multizone system. This system exhibits the greatest energy consumption of any all-air system. In addition to the energy required to mix conditioned air even when no heating or cooling is required, it requires a great amount of distribution energy even when normal pressure and low air velocities are used. For these reasons it has become nearly obsolete, being replaced with dual duct VAV or other systems.

**Dual Duct VAV**—Although the dual duct VAV system looks very much like its CAV counterpart, it is far more efficient. Instead of providing a constant volume of supply air at all times, the primary method of responding to thermostatic requirements is through adjusting the volume of either cool or warm supply air.

The properly designed dual duct VAV system functions essentially as two single duct VAV systems operating side by side; one for heating and one for cooling. Except when humidity control is required it is usually possible to provide comfort at all temperatures without actually mixing the two air streams. Even when humidity adjustment is required a good control system can minimize the amount of air mixing required.

The dual-duct VAV system still requires more distribution energy and space than most other systems. The level of indirect control which is necessary to produce heated and cooled air also increases energy consumption. Consequently its use should be restricted to applications which benefit from its ability to provide ex-

**Figure 10.8 Dual duct system schematic diagram.**

ceptional temperature and humidity control and which do not require a constant supply of ventilation air.

### 10.4.2 All-Water Systems

Air is not a convenient medium for transporting heat. A cubic foot of air weighs only about 0.074 pounds (0.34 kg) at standard conditions (70°F; 1 atm.). With a specific heat of about 0.24 Btu/lb°F (0.14 joule/C), one cubic foot can carry less than 0.02 Btu per degree Fahrenheit temperature difference. By comparison, a cubic foot of water weighs 62.4 pounds and can carry 62.4 Btu/ft³.

Water can be used for transporting heat energy in both heating and cooling systems. It can be heated in a boiler to a temperature of 140 to 250°F (60-120°C) or cooled by a chiller to 40 to 50°F (4-10°C), and piped throughout a building to terminal devices which take in or extract heat energy typically through finned coils.

Steam can also be used to transport heat energy. Steam provides most of its energy by releasing the latent heat of vaporization (about 1000 Btu/lb or 2.3 joules/kg). Thus one mass unit of steam provides as much heating as fifty units of water which undergo a 20°F (11°C) temperature change. However, when water vaporizes, it expands in volume more than 1600 times. Consequently liquid water actually carries more energy

per cubic foot than steam and therefore requires the least space for piping.

All-water distribution systems provide flexible zoning for comfort heating and cooling and have a relatively low installed cost when compared to all-air systems. The minimal space required for distribution piping makes them an excellent choice for retrofit installation in existing buildings or in buildings with significant spatial constraints. The disadvantage to these systems is that since no ventilation air is supplied, all-water distribution systems provide little or no control over air quality or humidity and cannot avail themselves of some of the energy conservation approaches of all-air systems.

Water distribution piping systems are described in terms of the number of pipes which are attached to each terminal device:

*One-pipe systems* use the least piping by connecting all of the terminal units in a series loop. Since the water passes through each terminal in the system, its ability to heat or cool becomes progressively less at great distances from the boiler or chiller. Thermal control is poor and system efficiency is low.

*Two-pipe systems* provide a supply pipe and a return pipe to each terminal unit, connected in parallel so that each unit (zone) can draw from the supply as needed. Efficiency and thermal control are both high, but the system cannot provide heating in one zone while cooling another.

*Four-pipe systems* provide a supply and return pipe for both hot water and chilled water, allowing simultaneous heating and cooling along with relatively high efficiency and excellent thermal control. They are, of course, the most expensive to install, but are still inexpensive compared to all-air systems.

*Three-pipe systems* employ separate supply pipes for heating and cooling but provide only a single, common return pipe. Mixing the returned hot water, at perhaps 140°F (60°C), with the chilled water return, at 55°F (13°C), is highly inefficient and wastes energy required to reheat or recool this water. Such systems should be avoided.

## Radiant Heating

Radiant energy is undoubtedly the oldest method of centrally heating buildings, dating to the era of the Roman Empire. Recalling that the most important determinant of thermal comfort when environmental conditions are too cool is the radiant temperature of the physical surroundings, radiant heating systems are among the most economical, so long as the means of producing heat is efficient.

The efficiency of radiant heating is a function primarily of the temperature, area and emissivity of the heat source and the distance between the radiant source and the observer. It is therefore essential that radiant heat sources be located so that they are not obstructed by other objects. Emissivity is an object's ability to absorb and emit thermal radiation, and is primarily related to color. Dark objects absorb and emit radiation better than light colored objects.

There are three categories of radiant heating devices, classified according to the temperature of the source of heat. All may employ electric resistance heating elements, but are more energy-efficient if they employ combustion as a heat source.

Low temperature radiant floors employ the entire floor area as a radiating surface by embedding hot water coils in the floor. The water temperature is typically less than 120°F (50°C). By distributing the heat energy uniformly though the floor, surface temperature is normally below 100°F (40°C).

By increasing the temperature of the radiant surface its area can be reduced. In medium temperature radiant panels, hot water circulates through metal panels, heating them to a temperature of about 140°F (60°C). Consequently the panels must be located out-of-reach, usually on the ceiling or on upper walls.

High temperature *infrared* heaters are typically gas-fired or oil-fired and are discussed below under packaged systems.

Because they are not dependent upon maintaining a static room air temperature, radiant heating systems provide excellent thermal comfort and efficiency in spaces subject to large influxes of outside air, such as factories and warehouses. However they are slow to respond to sudden changes in thermal requirements and malfunctions may be difficult or awkward to correct. Another drawback to radiant systems is that they promote the stratification of room air, concentrating warm air near the ceiling.

## Natural Convection

The simplest all-water system is a system of hydronic (hot-water) convectors. In this system hot water from a boiler or steam-operated hot water converter is circulated through a finned tube, usually mounted horizontally behind a simple metal cover which provides an air inlet opening below the tube and an outlet above. Room air is drawn through the convector by natural convection where it is warmed in passing over the finned tube.

A variation on the horizontal finned-tube hydronic convector is the cabinet convector, which occupies less perimeter space. A cabinet convector would have several finned tubes in order to transfer additional heat to the air passing through it. When this is still insufficient a small electric fan can be added, converting the convector to a *unit heater*. Although an electric resistance element can be used in place of the finned tube, the inefficiency of electric resistance heating should eliminate this option.

Hydronic convectors are among the least expensive heating systems to operate as well as to install. Their use is limited, however, to heating only and they do not provide ventilation, air filtration, nor humidity control.

Hydronic convectors and unit heaters may be used alone in buildings where cooling and mechanical ventilation is not required or to provide heating of perimeter spaces in combination with an all-air cooling system. They are the most suitable type of system for providing heat to control condensation on large expanses of glass on exterior wall systems.

## Fan-coils

A fan-coil terminal is essentially a small air-handling unit which serves a single space without a ducted distribution system; the main difference other than size is that fan coils generally do not have outside air and exhaust provisions. One or more independent terminals are typically located in each room connected to a supply of hot and/or chilled water. At each terminal, a fan in the unit draws room air (sometimes mixed with outside air) through a filter and blows it across a coil of hot water or chilled water and back into the room. Condensate which forms on the cooling, coil must be collected in a drip pan and removed by a drain (Figure 10.9).

Although most fan-coil units are located beneath windows on exterior walls, they may also be mounted horizontally at the ceiling, particularly for installations where cooling is the primary concern.

Technically, a fan-coil unit with an outside air inlet is called a *unit ventilator*. Unit ventilators provide the capability of using cool outside air during cold weather to provide free cooling when internal loads exceed the heat lost through the building envelope. See the discussion of economizers, §10.5.5.

Fan-coil units and unit ventilators are directly controlled by local thermostats, often located within the unit, making this system one of the most energy efficient. Drawbacks to their use is a lack of humidity control and the fact that all maintenance must occur within the occupied space.

Fan-coil units are typically used in buildings which have many zones located primarily along exterior walls, such as schools, hotels, apartments and office buildings. They are also an excellent choice for retrofitting air-conditioning into buildings with low floor-to-floor heights. Although a four-pipe fan-coil system can be used for a thermally heavy building with high internal loads, it

**Figure 10.9 Fan-coil unit.**

suffers the drawback that the cooling of interior zones in warm weather must be carried out through active air-conditioning, since there is no supply of fresh (cool) outside air to provide free cooling separately. They are also utilized to control the space temperature in laboratories where constant temperature make-up air is supplied to all spaces.

**Closed-Loop Heat Pumps**

Individual heat pumps (§10.4.4) have a number of drawbacks in nonresidential buildings. However, closed-loop heat pumps, more accurately called water-to-air heat pumps, offer an efficient option for heating and cooling large buildings. Each room or zone contains a water-source heat pump which can provide heating or cooling, along with air filtration and the dehumidification associated with forced-air air-conditioning.

The water source for all of the heat pumps in the building circulates in a closed piping loop, connected to a cooling tower for summer cooling and a boiler for winter heating. Control valves allow the water to bypass either or both of these elements when they are not needed (Figure 10.10). The primary energy benefit of closed-loop heat pumps is that heat removed from overheated interior spaces is used to provide heat for underheated perimeter spaces during cold weather.

Since the closed-loop heat pump system is an all-water, piped system, distribution energy is low, and since direct, local control is used in each zone, control energy is also minimized, making this system one of the most efficient. Although the typical lack of a fresh-air supply eliminates the potential for an economizer cycle, the heat recovery potential discussed above more than makes up for this drawback.

Heat pump systems are expensive to install and maintenance costs are also high. Careful economic analysis is necessary to be sure that the energy savings will be great enough to offset the added installation and maintenance costs. Closed-loop heat pumps are most applicable to buildings such as hotels which exhibit a wide variety of thermal requirements along with simultaneous heating requirements in perimeter zones and large internal loads or chronically overheated areas such as kitchens and assembly spaces.

**10.4.3 Air & Water systems—Induction**

Once commonly used in large buildings, induction systems employ terminal units installed at the exterior perimeter of the building, usually under windows. A small amount of fresh outside ventilation air is filtered, heated or cooled, and humidified or dehumidified by a central AHU and distributed throughout the building at

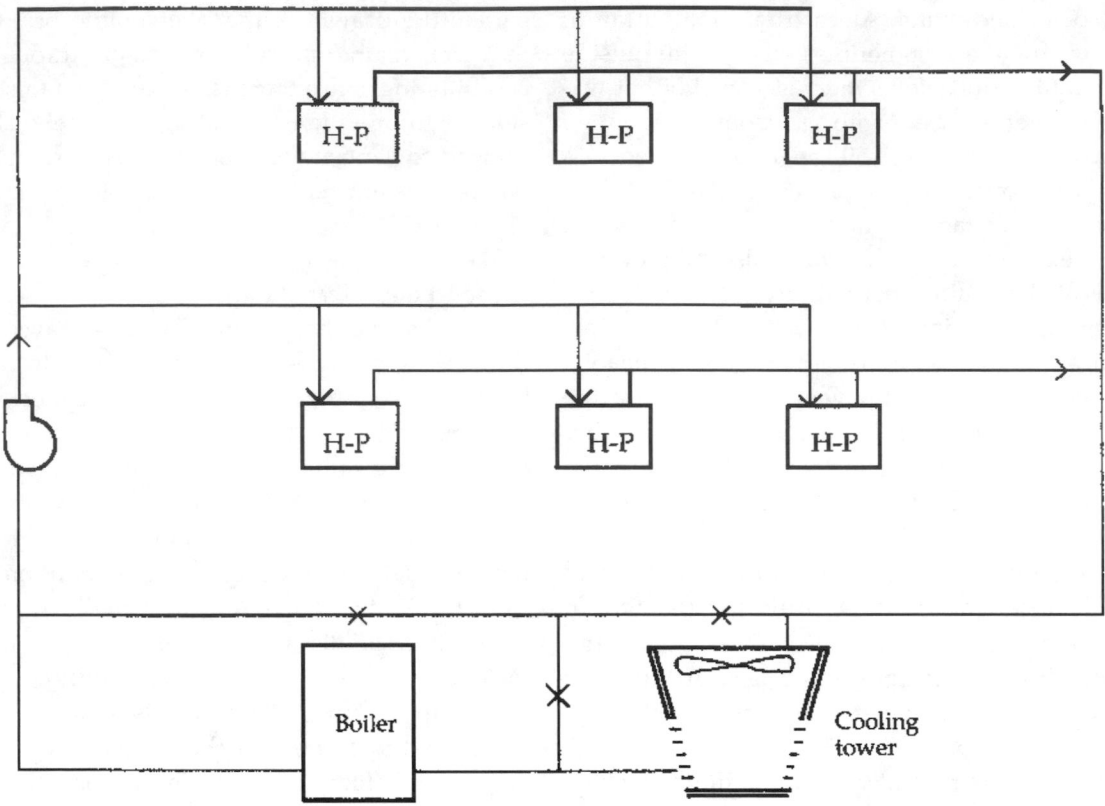

Figure 10.10 Closed-loop heat pump system schematic diagram.

high-velocity by small ducts.

In each terminal unit, this *primary air* is discharged in such a way that it draws in a much larger volume of *secondary air* from the room, which is filtered and passed through a coil for additional heating or cooling (Figure 10.11). The use of primary air as the motive force eliminates the need for a fan in the induction unit. The cooling coil is often deliberately kept at a temperature greater than the dew point temperature of the room air which passes through it, eliminating the need for a condensate drain. Although the standard air-water induction system is a cooling-only system, room terminals can employ reheat coils to heat perimeter zones.

Despite the high pressures and velocities required for the primary air distribution, distribution energy is minimized by the relatively small volume of primary air. But the energy saved in primary air distribution is more than offset by the energy consumed in the indirect control and distribution of cooling water, making air-water induction systems among the least energy efficient.

Air-water induction units tend to be noisy and the system provides negligible control of humidity. The applicability of these systems is limited to buildings with widely varying cooling or heating loads where humidity control is not necessary, such as office buildings. Concerns about indoor air quality limits their use as well.

### 10.4.4 Packaged systems

All of the systems described above may be classified as *central* air-conditioning systems in that they contain certain central elements, typically including a boiler, chiller and cooling tower. Many large buildings

Figure 10.11 Induction unit.

provide heating and cooling with distributed systems of unitary or packaged systems, where each package is a stand-alone system which provides all of the heating and cooling requirements for the area of the building which it serves. Individual units derive their energy from raw energy sources typically limited to electricity and natural gas.

Since large pieces of equipment usually have higher efficiencies than smaller equipment, it might be thought that packaged systems are inherently inefficient when compared to central air-conditioning systems. Yet packaged systems actually use much less energy. There are several reasons for this.

First, there is much less energy used in distribution. Fans are much smaller and pumps are essentially non-existent. In addition, control of the smaller packaged units is local and direct. Typically, the unit is either on or off, which can be a disadvantage when the space use requires that ventilation air not be turned off. However there are some advantages associated with this control flexibility. It allows individual thermal control and accurate metering of use. In addition, if equipment failure occurs it does not affect the entire building.

A third reason for the energy efficiency of packaged systems involves the schedule of operation. While large equipment is more efficient overall, it only operates at this peak efficiency when it is running at full load. Small packaged units, due to their on/off operation, run at full load or not at all. In a central air-conditioning system, the central equipment must run whenever any zone requires heating or cooling, often far from its peak load, optimum efficiency conditions.

A secondary advantage to the use of packaged systems is the advantage of diversity. The design of a large central air-conditioning system sometimes requires that a compromise be made between the ideal type of system for one part of a building and a different type of system for another. When packaged systems are employed, parts of a building with significantly different heating and cooling requirements can be served by different types of equipment. This will always provide improved thermal comfort, and often results in improved efficiency as well.

### Packaged Terminal Air-Conditioners

The most common type of packaged equipment is the packaged terminal air-conditioner, often called a PTAC or incremental unit, due to the fact that increases in equipment can be made incrementally. Examples of PTACs are through-the-wall air-conditioners and single-zone rooftop equipment. Their use is limited to about 500 square feet per unit.

Individual air-to-air (air-source) heat pumps can also be installed as a packaged system. A heat pump is essentially a vapor-compression air-conditioner which can be reversed to extract heat from the outdoor environment and discharge it into the occupied space. A significant drawback to air-source heat pumps is that vapor compression refrigeration becomes inefficient when the evaporator is forced to extract heat from a source whose temperature is 30°F (0°C) or below.

In large systems, heat pumps can utilize a source of circulating water from which to extract heat during cold weather, so that the evaporator temperature never approaches 30°F (0°C). The circulating water would be heated in the coldest weather, and could be cooled by a cooling tower to receive rejected heat during warm weather. These closed-loop heat pumps are discussed under all-water systems above.

### Unit Heaters

Packaged heating-only units typically utilize electricity or natural gas as their primary source of energy. As discussed in a later section, electricity is the most expensive source of heat energy and should be avoided. Natural gas (or liquefied propane) provides a more economical source of heat when used in packaged unit heaters.

Fan-forced unit heaters can disperse heat over a much larger area than packaged air-conditioners. They can distribute heat either vertically or horizontally and respond rapidly to changes in heating requirements.

High temperature infrared radiant heaters utilize a gas flame to produce a high-temperature (over 500°F, 260°C) source of radiant energy. Although they do not respond rapidly to changes in heating requirements, they are essentially immune to massive intrusion of cold outside air. Because they warm room surfaces and physical objects in the space, thermal comfort returns within minutes of an influx of cold air.

HVAC systems may be central or distributed; all-air, all-water, or air-water (induction). Each system type has advantages and disadvantages, not the least important of which is its energy efficiency. An economic analysis should be conducted in selecting an HVAC system type and in evaluating changes in HVAC systems in response to energy concerns.

## 10.5 ENERGY CONSERVATION OPPORTUNITIES

The ultimate objective of any energy management program is the identification of energy conservation opportunities (ECOs) which can be implemented to produce a cost saving. However, it is important to recognize that the fundamental purpose of an HVAC system is to provide human thermal comfort, or the equivalent environmental conditions for some specific process. It is therefore necessary to examine each ECO in the context of its effect on indoor air quality, humidity and thermal comfort standards, air velocity and ventilation requirements, and requirements for air pressurization.

### 10.5.1 Thermal Comfort, Air Quality and Airflow

It is not wise to undertake modifications of an HVAC system to improve energy efficiency without considering the effects on thermal comfort, air quality, and airflow requirements.

**Thermal Comfort**

One of the most serious errors which can be made in modifying HVAC systems is to equate a change in dry bulb air temperature with energy conservation. It is worthwhile to recall that dry bulb air temperature is not the most significant determinant of thermal comfort during either the heating season or cooling season.

Since *thermal comfort in the cooling season is most directly influenced by air motion* cooling energy requirements can be reduced by increasing airflow and/or air motion in occupied spaces without decreasing dry bulb air temperature. During the *heating season thermal comfort is most strongly influenced by radiant heating.* Consequently, changes from forced-air heating to radiant heating can improve thermal comfort while decreasing heat energy requirements.

The total energy, or enthalpy, associated with a change in environmental conditions includes both sensible heat and latent heat. *Sensible* heat is the heat energy required to increase dry bulb temperature. The heat energy associated with a change in moisture content of air is known as *latent* heat. (See Figure 10.12)

Changes in HVAC system design or operation which reduce sensible heating or cooling requirements may increase latent heating or cooling energy which offsets any energy conservation advantage. This is particularly true of economizer cycles (§10.5.5). The use of cool, but humid outside air in an economizer can actually increase energy consumption if dehumidification requirements are increased. For this reason, the most reliable type of economizer control is enthalpy control which prevents the economizer from operating when latent cooling requirements exceed the savings in sensible cooling.

**Air velocity and airflow**

The evaluation of energy conservation opportunities often neglects previously established requirements

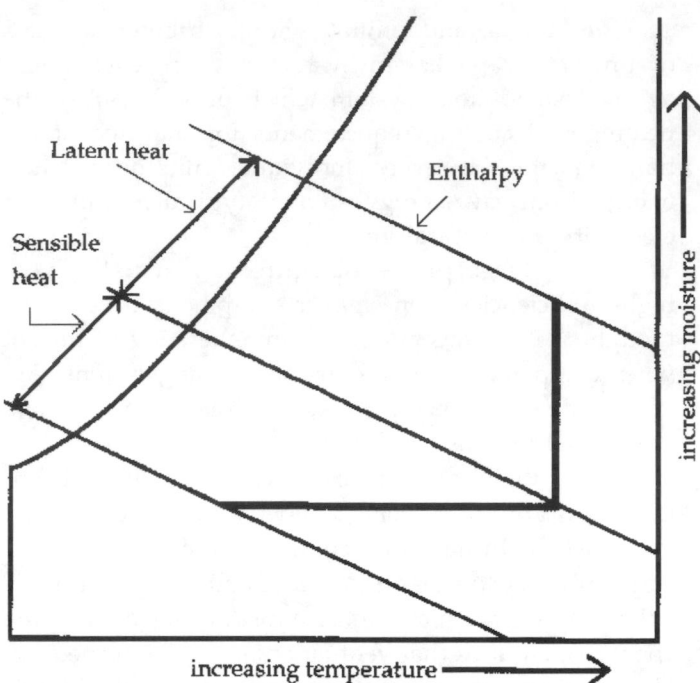

**Figure 10.12 Sensible heat and latent heat.**

for air velocity and airflow. As discussed above, the volume of air supplied and its velocity have a profound influence on human thermal comfort. ECOs which reduce airflow can inadvertently *decrease* thermal comfort. Even more important, airflow cannot be reduced below the volume of outdoor air required by codes for ventilation.

The design of an all-air or air-water HVAC system is much more complex that just providing a supply air duct and thermostat for each space. The completed system must be *balanced* to assure adequate airflow to each space, not only to offset the thermal loads, but also to provide the appropriate *pressurization* of the space.

It is a common practice for supply air to exceed return air in selected spaces to create positive pressurization, which minimizes infiltration and prevents the intrusion of odors and other contaminants from adjacent spaces. Similarly, negative pressurization can be achieved by designing exhaust or return airflow to exceed supply air requirements in order to maintain a sterile field or to force contaminants to be exhausted. Any alterations in air supply or return requirements upset the relationship between supply and return airflows requiring that the system be rebalanced.

**Indoor Air Quality**

Another factor which is often neglected in the application of ECOs is the effect of system changes on indoor air quality. Indoor air quality requirements are most commonly met with ventilation and filtration provided by an all-air or air-water HVAC system. When outdoor air quantities are altered significantly, the effect on indoor air quality is unknown and must be determined.

In a polluted environment, increasing outdoor air volume, for example with an economizer cycle, either increases filtration requirements or results in a deterioration of indoor air quality. On the other hand, reducing airflow to conditioned spaces due, for example, to a conversion to a variable air volume system, reduces the filtration of recycled indoor air which can likewise produce a reduction in air quality. The concerns regarding indoor air quality are discussed in more detail in Chapter 17.

### 10.5.2 Maintenance

The first place to begin in the investigation of any existing HVAC system is to review the condition and maintenance of the system. Inadequate or improper maintenance is a major cause of system inefficiency. The system may have been designed to be energy-efficient and some ECOs may have been previously implemented, but due to a lack of maintenance the system no longer functions in an energy-efficient mode.

A more comprehensive review of maintenance procedures is included in Chapter 14, particularly with regard to boiler maintenance. What follows here a brief summary of some of the things to watch for in inspecting the system for inadequate maintenance.

The control system (see §10.5.6 and Chapter 22) should be performing its function. Thermostats should be properly calibrated. Time clocks, setback devices, and automatic controls should be intact and operating. Bearings and belts should be in good shape and belts should be tight. Bearings and other moving parts should be properly lubricated with the appropriate lubricant. Lubricant selection can actually have a significant effect on system performance and efficiency.

Dirt build-up should be periodically removed especially from heat transfer surfaces, particularly the condenser coils of air-cooled condensers. Filters should be cleaned or replaced regularly rather than waiting for them to appear so dirty as to need cleaning. Dirty filters increase fan horsepower while reducing airflow. If routine and preventative maintenance has not been performed on the system, complete the required maintenance before proceeding to determine other applicable ECOs.

### 10.5.3 Demand Management

Demand management (see also Chapter 11) involves shutting off or deferring the operation of equipment to minimize the peak electrical load which occurs at a given time. This strategy can be extended to shutting off or reducing HVAC equipment requirements when they are not needed. The most common demand management strategies for HVAC systems are scheduled operation and night set-back.

### Operating Schedule

One of the greatest causes of energy waste is unnecessary operation. The most energy-efficient equipment will consume excess energy if it runs when it is not needed. Unnecessary operation tends to be caused by large systems which condition entire buildings or sections of buildings and a lack of control to turn off equipment when it is not needed.

When buildings are conditioned by large central systems, off-hours use of one area may require the operation of the entire system to condition a single space. Sometimes the installation of a local packaged system serving a specialized area can prevent the unnecessary operation of a larger system. For example, a radio station located in a large office building installed a packaged air-conditioner to provide cooling for its studios which operated 24 hours a day while the offices remained on the building system which operated only from 7 a.m. until 5 p.m. Despite the fact that the packaged system was less efficient than the central building system, the energy saved was substantial.

Exhaust fans, dust collectors, and other small equipment which serve specific rooms are often left running continuously due solely to the lack of local control to turn them off. Local switching alone is seldom adequate to save energy since it is difficult to get occupants to turn equipment off and sometimes to turn it on. Passive switching or other automatic devices can eliminate the need to run equipment continuously.

For example an exhaust fan for a seldom-occupied store room can be controlled to turn on with an occupancy sensor or when the lights are turned on and to remain on for a set period of time after the lights are turned off or occupants leave the room. Spring-wound interval timers are useful for controlling small devices which are turned on manually and then left running. Exhaust fans can be interlocked with the central air-conditioning system to shut down when the central system is turned off.

Turning off equipment, particularly central air-conditioning systems, in buildings which are unoccupied at night is another obvious energy saver. Simple time clocks can be used where more sophisticated computer controlled systems are not used, Even when it is necessary to keep the system operating to provide minimal cooling during the night or on weekends, exhaust fans and fresh-air ventilation can be shut down when the building is essentially unoccupied.

Some equipment loads are *deferrable* in that they can be turned off for a short interval and then turned back on with no appreciable impact on operations. Water heating and cooling are two examples, but even space heating can be deferred for periods of up to thirty minutes with no perceptible effect on thermal comfort. During extreme weather outside air can be reduced for short periods in order to improve heating or cooling efficiency. See the discussion of warm-up and cool-down cycles which follows.

Computerized controls can be used to prevent large pieces of equipment from operating simultaneously which increases peak demand and the expense which accompanies increased demand. This is particularly attractive in large systems of distributed packaged equipment.

With adequate storage both heat and "cool" can be stored in order to take advantage of off-peak utility rates and to reduce instantaneous heating or cooling requirements during peak periods. See the discussion of thermal storage in section 10.6.4 and in chapter 19.

**Night Set-back**

The principle of night set-back is to reduce the amount of conditioning provided at night by allowing the interior temperature to drift naturally to a marginal temperature during the night and then to recondition it to normal conditions in the morning. With the possible exception of electric heat pumps, night set-back will always save energy. The energy saved can be computed from the equation:

$$E = [(A_{per} \times U) + q_{vent}] \Delta T \times hr \qquad (10.1)$$

where

| | | |
|---|---|---|
| E | = | |
| $A_{per}$ | = | surface area of perimeter envelope, ft$^2$ |
| U | = | effective U-value of thermal envelope, Btu/hr ft$^2$°F |
| $q_{vent}$ | = | ventilation load, Btu/hr°F |
| $\Delta T$ | = | night setback temperature difference, degrees F |
| hr | = | heating season unoccupied hours |

In a temperate climate this can amount to a saving of 40 percent of the envelope heating energy for a building with little thermal mass which is occupied only ten hours per day. However the amount of energy saved will not always be large and local utility rate structures may be such that the actual cost of the energy saved is less than the cost of the energy to recondition the build-

ing in the morning.

For example, many electric utility companies offer time-of-day rates in which energy used during the night is charged at a rate much lower than that charged for electricity consumed during the day. If an air-conditioned building allowed to heat up during the night must be recooled at the day rate in the morning, night set-back could end up costing more than cooling the building all night.

Similarly, the more energy-efficient the building is the smaller the energy saving will be from night set-back and the less attractive this option will be from a financial standpoint. Night set-back is marginal at best for cooling, and will be only slightly better for heating unless the building has little thermal mass and significant infiltration heat loss.

Where night set-back is a viable option, a secondary question is whether to turn the system off or simply turn the thermostat setpoint down (or up). This is not a question which can be answered simply. The answer is dependent on the length of time the system will be setback and how much temperature variation will occur if the system is turned off. The question of whether it is even advisable to turn the system off is also influenced by humidification and pressurization requirements, air quality concerns, etc. See §10.5.1.

**Warm-up and Cool-down**

When a building is operated with night set-back, it is not necessary to condition the building until the last person leaves in the evening. But the system must be restarted in the morning so that optimum conditions are reached before the beginning of the business day. It is not difficult to determine an appropriate time to let the system begin to drift in the afternoon, but morning warm-up (or cool-down) can require substantial time and energy.

One of the best approaches to re-conditioning a building after night set-back is with the use of an *optimum-time start* device. This microprocessor-based thermostat compares the outside and inside temperatures along with the desired setpoint during the operating cycle. It determines how long it will take to re-condition the building to the setpoint based on previous data and turns the system on at the appropriate time to reach the setpoint temperature just in time. Since time clocks must be set to turn the system on with adequate time to re-condition the building on even the coldest morning, the optimum-time start device can save more than half of the warm-up time on days when the outdoor ambient temperature is moderate.

A second method of saving energy during warm-up or cool down is with a *warm-up/cool-down cycle.*

During the warm-up/cool-down cycle the system recirculates building return air until a temperature within one or two degrees of the setpoint is reached, saving the energy which would be required to heat or cool outside ventilation air. Since outside ventilation air is provided primarily to meet human fresh-air needs there is no harm in reducing or eliminating it during warm-up or cool-down. However modem building codes prohibit eliminating all ventilation air unless the building is truly unoccupied.

Warm-up/cool-down cycles provide a secondary advantage in times of extreme weather. Should the heating and cooling equipment be unable to maintain the setpoint temperature due to a lack of capacity to condition the ventilation air because of its extreme temperature, the warm-up/cool-down cycle could close the outside-air damper and reduce the volume of outdoor air until interior conditions were restored to near the setpoint. Since the system would run continuously in either case, this procedure technically doesn't save energy but it does maintain thermal comfort without expending additional energy.

### 10.5.4 Heat Recovery

Heat recovery systems are used during the heating season to extract waste heat and humidity from exhaust air which is used to preheat cold fresh air from outside. In warm weather, they may be used to extract heat and humidity from warm fresh air and dispel it into the exhaust air stream. Their use is particularly appropriate to buildings with high outside air requirements. Heat recovery systems are discussed in more detail in Chapter 8.

One special type of heat recovery, dubbed bootstrap heating, recovers waste heat from warm condenser water to provide nearly free heat to meet perimeter heating needs. This is most applicable to buildings in which interior zone cooling exceeds the requirements for perimeter heating. Care must be taken in utilizing bootstrap heating to avoid reducing the condenser water temperature below the chiller equipment's minimum entering water temperature.

### 10.5.5 Economizers

One advantage of all-air HVAC systems is that they can utilize outside air to condition interior spaces when it is at an appropriate temperature. *The use of outside air to actively cool interior spaces is referred to as an economizer, or economizer cycle.*

Whenever the outside air is cooler than the cooling setpoint temperature only distribution energy is required to provide cooling with outside air. When the outside air temperature is above the setpoint temperature but less than the return air temperature, it requires less cooling

energy to utilize 100 percent outdoor air for supply air than to condition recycled indoor air. An economizer cycle is simply a control sequence that adjusts outside air and exhaust dampers to utilize 100 percent outside air when its temperature makes it advantageous to do so.

The energy consumed in heating outside air at sea level may be computed from the equation

$$q = 1.08 \, Q \cdot \Delta T \tag{10.2a}$$

where

$q$ = hourly ventilation heating load, Btu/hr
$Q$ = volumetric flow rate of air, cfm
$\Delta T$ = temperature increase, °F

The economizer cycle is most appropriate for thermally massive buildings which have high internal loads and require cooling in interior zones year round. It is ineffective in thermally light buildings and buildings whose heating and cooling loads are dominated by thermal transmission through the envelope. Economizer cycles will provide the greatest benefit in climates having more than 2000 heating degree days per year, since warmer climates will have few days cold enough to permit the use of outside air for cooling.

In theory 100 percent outside air would be utilized for cooling whenever the outside air temperature is below the return air temperature, and this is done in dry climates. But if the outside air has a higher relative humidity than return air and contains significant *latent heat*, the economizer would not be activated at this temperature.

The simplest type of economizer utilizes a *dry-bulb temperature control* which activates the economizer at a predetermined outside dry-bulb temperature, usually the supply air temperature, or about 55°F (13°C). Above 55°F (13°C), minimum outdoor air is supplied for ventilation. Below 55°F (13°C), the quantity of outdoor air is gradually reduced from 100 percent and blended with return air to make 55°F (13°C) supply air.

Eq. 10.2a predicts that the energy required to cool room (return) air 20°F (13°C) from 75°F (24°C) with a supply air temperature of 55°F (13°C) is 1.08 × Q × 20 = 21.6 Q. The energy saved [E, Btu/hr] from the use of a simple economizer which blends outside air with return to provide a supply air temperature, at sea level, is therefore:

$$E = 21.6 \, Q \tag{10.2b}$$

Because it is possible to utilize outdoor air at temperatures above 55°F (13°C) to save cooling energy, a *modified dry-bulb temperature control* can be used. This is identical to the simpler dry-bulb temperature control except that when the outside temperature is between 55°F (13°C) and a preselected higher temperature based on the typical humidity, 100 percent outdoor air is used, but cooled to 55°F for supply air.

The third and most efficient type of control for an economizer cycle is *enthalpy control,* which can instantaneously determine and compare the amount of energy required to cool 100 percent outdoor air with that required to cool the normal blend of return air. It then selects the source which requires the least energy for cooling.

Air-handling units which lack adequate provision for 100 percent outside air can utilize an alternative "wet-side" economizer. This energy-saving technique is a potential retrofit for existing older buildings. Several variations on the wet-side economizer exist, all of which allow the chiller to be shut down. The simplest waterside economizer is a coil in the air-handling unit through which cooling tower water can be circulated to provide free cooling. An alternative arrangement for large chilled water systems interconnects the chilled water circuit directly with the cooling tower, to allow the chiller to be bypassed. To reduce contamination of the chilled water, the cooling tower water should only be coupled to the chilled water circuit through a heat exchanger (Figure 10.13).

The energy saved [E, kWh] due to the use of a wet-side economizer can be determined from a knowledge of the hours [hr] of chiller operation during when the outside temperature is below 40°F (4°C), the capacity of chiller in tons and the kW/ton rating of the chiller:

$$E = hr \times tons \times kW/ton \qquad (10.3)$$

A word of caution is advised. A wet-side economizer uses condenser water at 40-45°F (4-7°C). But the chiller typically operates with condenser water much higher than this. In some cases, the lowest condenser water temperature at which the chiller will operate is 65°F (18°C). This means that when the chiller is turned off there would be no useful coolant until the cooling tower cooled to 45°F (7°C). This "cool down" period is typically no more than 30 minutes. For additional discussion of the wet-side economizer, see §10.6.6.

### 10.5.6 Control Strategies

Control systems play a large part in the energy conservation potential of a HVAC system. To be effective at controlling energy use along with thermal comfort they must be used appropriately, work properly and be set correctly. Overheated or overcooled spaces not only waste energy, they are uncomfortable. An easy way to spot such areas is to walk around the outside of the building and look for open windows which usually indicate overheated (or overcooled) spaces.

The efficiency of control systems is mostly related to the *directness of control.* Simple controls which turn the system off and on when needed provide the most direct control. Systems in which air and/or water temperatures are controlled by parameters other than the actual need for heating and cooling may be said to be indirectly controlled, and are inherently less efficient.

Direct control systems most commonly employ primarily thermostats which turn equipment on or off or adjust the volume of air supplied to a space on the basis of room temperature. Direct thermostat control is most effective in thermally light buildings whose energy requirements are directly proportional to the exterior weather conditions. Where thermostats are an appropriate choice, individual zones must be used for spaces with different heating or cooling requirements.

Many older HVAC systems, particularly heating-only systems, were created as single-zone systems for which the HVAC system is either on throughout the building or entirely off. Self-contained thermostatic valves can be used to modify single-zone convection heating systems and single-zone forced air systems can be readily modified into zoned VAV or reheat air-conditioning systems without replacing the costly central equipment.

The selection of set points, even for directly-controlled systems, is important. Thermostats should be set to maximize thermal comfort rather than relying on previously established design values. The need for humidification varies with outdoor temperature and humidistats should not be set at constant levels and forgotten. The need for humidification is also a function of occupancy and should be adjusted in assembly spaces when they are unused or underutilized. The use of outside air for ventilation when this air must be conditioned likewise can be reduced with proper controls to adjust outside-air dampers. Modem microprocessor-based energy management control systems (EMCS) allow for vastly improved direct control of HVAC systems.

A useful concept made possible by modem electronics is the use of *enthalpy control.* Enthalpy control relates the energy required to cool air at a given dry-bulb temperature to the energy required to remove humidity from moist outdoor air. Enthalpy controls are most commonly used to adjust economizer cycles, determining when it is more efficient to recycle warm indoor air than to condition humid outside air for "free" cooling.

Control systems which are not well-understood or simple to use are subject to misuse by occupants or

Figure 10.13 Wet-side economizer schematic diagram.

building service personnel. For example the controls on a unit ventilator include a room temperature thermostat which controls the valve on the heating or cooling coil, a damper control which adjusts the proportion of fresh air mixed with recirculated room air, and a low-limit thermostat which prevents the temperature of outside air from dropping below a preset temperature (usually 55 to 60°F; 13 to 16°C). A common error of occupants or building custodians in response to a sense that the air supplied by the unit ventilator is too cold is to increase the setpoint on the low-limit thermostat, which prevents

free cooling from outside air or, on systems without a cooling coil, prevents cooling altogether. Controls which are subject to misadjustment by building occupants should be placed so that they cannot be tampered with.

The energy consumption of thermally heavy buildings is less related to either the inside or outside air temperature. Both the heating and cooling loads in thermally heavy buildings are heavily dependent on the heat generated from internal loads and the thermal energy stored in the building mass which may be dissipated at a later time.

In an indirect control system the amount of energy consumed is not a function of human thermal comfort needs, but of other factors such as outdoor temperature, humidity, or enthalpy. Indirect control systems determine the set points for cool air temperature, water temperatures, etc. As a result indirect control systems tend to adjust themselves for peak conditions rather than actual conditions. This leads to overheating or overcooling of spaces with less than peak loads.

One of the most serious threats to the efficiency of any system is the need to heat and cool air or water simultaneously in order to achieve the thermal balance required for adequate conditioning of spaces. Figure 10.14 indicates that 20 percent of the energy consumed in a commercial building might be used to reheat cooled air, offsetting another 6 percent that was used to cool the air which was later reheated. For the example building the energy used to cool reheated air approaches that actually used for space cooling.

Following the 1973 oil embargo federal guidelines encouraged everyone to reduce thermostat settings to 68°F (20°C) in winter and to increase thermostat settings in air-conditioned buildings to 78°F (26°C) in summer. [In 1979, the winter guideline was reduced farther to 65°F (18°C).] The effect of raising the air-conditioning thermostat on a reheat, dual-duct, or multizone system is actually to *increase* energy consumption by increas-

ing the energy required to reheat air which has been mechanically cooled (typically to 55°F; 13°C).

To minimize energy consumption on these types of systems it makes more sense to raise the discharge temperature for the cold-deck to that required to cool perimeter areas to 78°F (26°C) under peak conditions. If the system was designed to cool to 75°F (24°C) on a peak day using 55°F (13°C) air, the cold deck discharge could be increased to 58°F (14.5°C) to maintain space temperatures at no more than 78°F (26°C), saving about $5 per cfm per year. Under less-than-peak conditions these systems would operate *more* efficiently if room temperatures were allowed to fall below 78°F (26°C) than to utilize reheated air to maintain this temperature.

More extensive discussion of energy management control systems may be found in Chapters 12 and 22.

### 10.5.7 HVAC Equipment

The elements which provide heating and cooling to a building can be categorized by their intended function. HVAC equipment is typically classified as heating equipment, including boilers, furnaces and unit heaters; cooling equipment, including chillers, cooling towers and air-conditioning equipment; and air distribution elements, primarily air-handling units (AHUs) and fans. A more lengthy discussion of boilers may be found in Chapter 6, followed by a discussion of steam and condensate systems in Chapter 7. Cooling equipment is discussed in section 10.6, below. What follows here relates mostly to air-handling equipment and distribution systems.

Figure 10.14 depicts the typical energy cost distribution for a large commercial building which employs an all-air reheat-type HVAC system. Excluding the energy costs associated with lighting, kitchen and miscellaneous loads which are typically 25-30 percent of the total, the remaining energy can be divided into two major categories: the energy associated with heating and cooling and the energy consumed in distribution. The total energy consumed for HVAC systems is therefore dependent on the efficiency of individual components, the efficiency of distribution and the ability of the control system to accurately regulate the energy consuming components of the system so that energy is not wasted.

The size (and heating, cooling, or air-moving capacity) of HVAC equipment is determined by the mechanical designer based upon a calculation of the peak internal and envelope loads. Since the peak conditions are arbitrary (albeit well-considered and statistically valid) and it is likely that peak loads will not occur simultaneously throughout a large building or complex

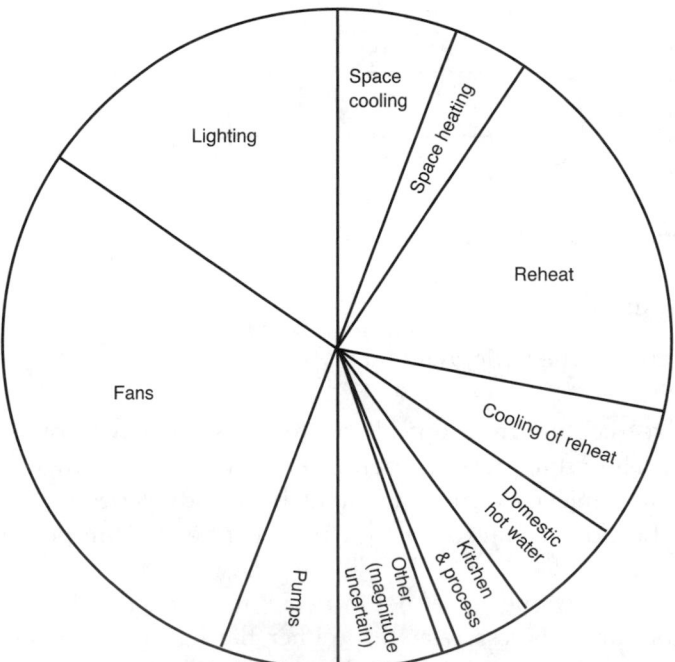

**Figure 10.14 Energy cost distribution for a typical non-residential building using an all-air reheat HVAC system.**

requiring all equipment to operate at its rated capacity, it is common to specify equipment which has a total capacity slightly less than the peak requirement. This diversity factor varies with the function of the space. For example, a hospital or classroom building will use a higher diversity multiplier than an office building.

In sizing heating equipment however, it is not uncommon to provide a total heating capacity from several units which exceeds the design heating load by as much as fifty percent. In this way it is assured that the heating load can be met at any time, even in the event that one unit fails to operate or is under repair.

The selection of several boilers, chillers, or air-handling units whose capacities combine to provide the required heating and cooling capability instead of single large units allows one or more components of the system to be cycled off when loads are less than the maximum.

This technique also allows off-hours use of specific spaces without conditioning an entire building.

### Equipment Efficiency

Efficiency, by definition, is the ratio of the energy output of a piece of equipment to its energy input, in like units to produce a dimensionless ratio. Since no equipment known can produce energy, efficiency will always be a value less than 1.0 (100%).

Heating equipment which utilizes electric resistance appears at first glance to come closest to the ideal of 100 percent efficiency. In fact, every kilowatt of electrical power consumed in a building is ultimately converted to 3413 Btu per hour of heat energy. Since this is a valid unit conversion it can be said that electric resistance heating is 100 percent efficient. What is missing from the analysis however, is the inefficiency of producing electricity, which is most commonly generated using heat energy as a primary energy source.

Electricity generation from heat is typically about 30 percent efficient, meaning that only 30 percent of the heat energy is converted into electricity, the rest being dissipated as heat into the environment. Energy consumed as part of the generation process and energy lost in distribution use up about ten percent of this, leaving only 27 percent of the original energy available for use by the consumer. By comparison, state-of-the-art heating equipment which utilizes natural gas as a fuel is more than eighty percent efficient. Distribution losses in natural gas pipelines account for another 5 percent, making natural gas approximately three times as efficient as a heat energy source than electricity.

The relative efficiency of cooling equipment is usually expressed as a *coefficient of performance* (COP),

which is defined as the ratio of the heat energy extracted to the mechanical energy input in like units. Since the heat energy extracted by modem air conditioning far exceeds the mechanical energy input a COP of up to 6 is possible.

Air-conditioning equipment is also commonly rated by its *energy efficiency ratio* (EER) or seasonal energy efficiency ratio (SEER). EER is defined as the ratio of heat energy extracted (in Btu/hr) to the mechanical energy input in watts. Although it should have dimensions of Btu/hr/watt, it is expressed as a dimensionless ratio and is therefore related to COP by the equation

$$EER = 3.41 \cdot COP \qquad (10.4)$$

Although neither COP nor EER is the efficiency of a chiller or air-conditioner, both are measures which allow the comparison of similar units. The term *air-conditioning efficiency is* commonly understood to indicate the extent to which a given air-conditioner performs to its maximum capacity. As discussed below, most equipment does not operate at its peak efficiency all of the time. For this reason, the *seasonal energy efficiency ratio* (SEER), which takes varying efficiency at partial load into account, is a more accurate measure of air-conditioning efficiency than COP or EER.

In general, equipment efficiency is a function of size. Large equipment has a higher efficiency than small equipment of similar design. But the rated efficiency of this equipment does not tell the whole story. Equipment efficiency varies with the load imposed. All equipment operates at its optimum efficiency when operated at or near its design full-load condition. Both overloading and under-loading of equipment reduces equipment efficiency.

This fact has its greatest impact on system efficiency when large systems are designed to air-condition an entire building or a large segment of a major complex. Since air-conditioning loads vary and since the design heating and cooling loads occur only rarely under the most severe weather or occupancy conditions, most of the time the system must operate under-loaded. When selected parts of a building are utilized for off-hours operation this requires that the entire building be conditioned or that the system operate far from its optimum conditions and thus at far less than its optimum efficiency.

Since most heating and cooling equipment operates at less than its full rated load during most of the year, its part-load efficiency is of great concern. Because of this, most state-of-the-art equipment operates much closer to its full-load efficiency than does older equipment. A knowledge of the actual operating efficiency of existing equipment is important in recognizing economic opportunities to reduce energy consumption through equipment replacement.

## Distribution Energy

Distribution energy is most commonly electrical energy consumed to operate fans and pumps, with fan energy typically being far greater than pump energy except in all-water distribution systems. The performance of similar fans is related by three fan laws which relate fan power, airflow, pressure and efficiency to fan size, speed and air density. The reader is referred to the *ASHRAE Handbook: HVAC Systems and Equipment* for additional information on fans and the application of the fan laws.[3]

Fan energy is a function of the quantity of airflow moved by the fan, the distance over which it is moved, and the velocity of the moving air (which influences the pressure required of the fan). Most HVAC systems, whether central or distributed packaged systems, all-air, all-water, or a combination are typically oversized for the thermal loads that actually occur. Thus the fan is constantly required to move more air than necessary, creating inherent system inefficiency.

One application of the third fan law describes the relationship between fan horsepower (energy consumed) and the airflow produced by the fan:

$$W_1 = W_2 \times (Q_1/Q_2)^3 \qquad (10.5)$$

where

   W = fan power required, hp
   Q = volumetric flow rate, cfm

Because fan horsepower is proportional to the cube of airflow, reducing airflow to 75 percent of existing will result in a reduction in the fan horsepower by the cube of 75 percent, or about 42 percent: $[(0.75)^3 = 0.422]$ Even small increases in airflow result in disproportional increases in fan energy. A ten percent increase in airflow requires 33 percent more horsepower $[1.10^3 = 1.33]$, which suggests that airflow supplied solely for ventilation purposes should be kept to a minimum.

All-air systems which must move air over great distances likewise require disproportionate increases in energy as the second fan law defines the relationship between fan horsepower [W] and pressure [p], which may be considered roughly proportional to the length of ducts connected to the fan:

$$W_1 = W_2 \times (P_1/P_1)^{3/2} \qquad (10.6)$$

The use of supply air at temperatures of less than 55°F (13°C) for primary cooling air permits the use of smaller ducts and fans, reducing space requirements at the same time. This technique requires a complex analysis to determine the economic benefit and is seldom advantageous unless there is an economic benefit associated with space savings.

## System Modifications

In examining HVAC systems for energy conservation opportunities, the less efficient a system is, the greater is the potential for significant conservation to be achieved. There are therefore several "off-the-shelf" opportunities for improving the energy efficiency of selected systems.

**All-air Systems**—Virtually every type of all-air system can benefit from the addition of an economizer cycle, particularly one with enthalpy controls. Systems with substantial outside air requirements can also benefit from heat recovery systems which exchange heat between exhaust air and incoming fresh air. This is a practical retrofit only when the inlet and exhaust ducts are in close proximity to one another.

Single zone systems, which cannot provide sufficient control for varying environmental conditions within the area served can be converted to variable air volume (VAV) systems by adding a VAV terminal and thermostat for each new zone. In addition to improving thermal comfort this will normally produce a substantial saving in energy costs.

VAV systems which utilize fans with inlet vanes to regulate the amount of air supplied can benefit from a change to variable speed or variable frequency fan drives. Fan efficiency drops off rapidly when inlet vanes are used to reduce airflow.

In terminal reheat systems, all air is cooled to the lowest temperature required to overcome the peak cooling load. Modern "discriminating" control systems which compare the temperature requirements in each zone and cool the main airstream only to the temperature required by the zone with the greatest requirements will reduce the energy consumed by these systems. Reheat systems can also be converted to VAV systems which moderate supply air volume instead of supply air temperature, although this is a more expensive alteration than changing controls.

Similarly, dual-duct and multizone systems can benefit from "smart" controls which reduce cooling requirements by increasing supply air temperatures. Hot-deck temperature settings can be controlled so that the temperature of warm supply air is just high enough to meet design heating requirements with 100 percent hot-deck supply air and adjusted down for all other conditions until the hot-deck temperature is at room temperature when outside temperatures exceed 75°F (24°C). Dual duct terminal units can be modified for VAV operation.

An economizer option for multizone systems is the addition of a third "bypass" deck to the multizone air-handling unit. This is not appropriate as a retrofit although an economizer can be utilized to provide cold-deck air as a retrofit.

**All-water systems**—Wet-side economizers are the most attractive common energy conservation measure appropriate to chilled water systems. Hot-water systems benefit most from the installation of self-contained thermostat valves, to create heating zones in spaces formerly operated as single-zone heating systems.

**Air-water Induction**—Induction systems are seldom installed anymore but many still exist in older buildings. The energy-efficiency of induction systems can be improved by the substitution of fan-powered VAV terminals to replace the induction terminals.

## 10.6 COOLING EQUIPMENT

The most common process for producing cooling is vapor-compression refrigeration, which essentially moves heat from a controlled environment to a warmer, uncontrolled environment through the evaporation of a refrigerant which is driven through the refrigeration cycle by a compressor.

Vapor compression refrigeration machines are typically classified according to the method of operation of the compressor. Small air-to-air units most commonly employ a reciprocating or scroll compressor, combined with an air-cooled condenser to form a condensing unit. This is used in conjunction with a direct-expansion (DX) evaporator coil placed within the air-handling unit.

Cooling systems for large non-residential buildings typically employ chilled water as the medium which transfers heat from occupied spaces to the outdoors through the use of chillers and cooling towers.

### 10.6.1 Chillers

The most common type of water chiller for large buildings is the centrifugal chiller which employs a centrifugal compressor to compress the refrigerant, which extracts heat from a closed loop of water which is pumped through coils in air-handling or terminal units within the building. Heat is rejected from the condenser into a second water loop and ultimately rejected to the environment by a cooling tower.

The operating fluid used in these chillers may be either a CFC or HCFC type refrigerant. Many existing centrifugal chillers use CFC-11 refrigerants, the manufacture and use of which is being eliminated under the terms of the Montreal Protocol. New refrigerants HCFC-123 and HCFC-134a are being used to replace the CFC refrigerants but refrigerant modifications to existing equipment will reduce the overall capacity of this equipment by 15 to 25 percent.

Centrifugal chillers can be driven by open or hermetic electric motors or by internal combustion

**Table 10.1 Summary of HVAC System Modifications for Energy Conservation**

| System type | Energy Conservation Opportunities |
|---|---|
| All-air systems (general): | economizer |
| | heat recovery |
| Single zone systems | conversion to VAV |
| Variable air volume (VAV) systems | replace fan inlet vane control with variable frequency drive fan |
| Reheat systems | use of discriminating control systems |
| | conversion to VAV |
| Constant volume dual-duct systems | use of discriminating control systems |
| | conversion to dual duct VAV |
| Multizone systems | use of discriminating control systems |
| | addition of by-pass deck* |
| All-water systems: | |
| hydronic heating systems | addition of thermostatic valves |
| chilled water systems | wet-side economizer |
| Air-water induction systems | replacement with fan-powered VAV terminals |

*Requires replacement of air-handling unit

engines or even by steam or gas turbines. Natural gas engine-driven equipment sized from 50 to 800 tons of refrigeration are available and in some cases are used to replace older CFC-refrigerant centrifugal chillers. These engine-driven chillers are viable when natural gas costs are sufficiently low. Part-load performance modulates both engine speed and compressor speed to match the load profile, maintaining close to the peak efficiency down to 50 percent of rated load. They can also use heat recovery options to take advantage of the engine jacket and exhaust heat.

Turbine-driven compressors are typically used on very large equipment with capacities of 1200 tons or more. The turbine may be used as part of a cogeneration process but this is not required. (For a detailed discussion of cogeneration, see Chapter 7.) If excess steam is available, in industry or a large hospital, a steam turbine can be used to drive the chiller. However the higher load on the cooling tower due to the turbine condenser must be considered in the economic analysis.

Small water chillers, up to about 200 tons of capacity, may utilize reciprocating or screw compressors and are typically air-cooled instead of using cooling towers. An air-cooled chiller uses a single or multiple compressors to operate a DX liquid cooler. Air-cooled chillers are widely used in commercial and large-scale residential buildings.

Other types of refrigeration systems include liquid overfeed systems, flooded coil systems and multi-stage systems. These systems are generally used in large industrial or low-temperature applications.

## 10.6.2 Absorption Chillers

An alternative to vapor-compression refrigeration is absorption refrigeration which uses heat energy to drive a refrigerant cycle, extracting heat from a controlled environment and rejecting it to the environment (Figure 10.15). Thirty years ago absorption refrigeration was known for its low coefficient of performance and high maintenance requirements. Absorption chillers used more energy than centrifugal chillers and were economical only if driven by a source of waste heat.

Today, due primarily to the restriction on the use of CFC and HCFC refrigerants, the absorption chiller is making a comeback. Although new and improved, it still uses heat energy to drive the refrigerant cycle and typically uses aqueous lithium bromide to absorb the refrigerant and water vapor in order to provide a higher coefficient of performance.

The new absorption chillers can use steam as a heat source or be direct-fired. They can provide simultaneous heating and cooling which eliminates the need for a boiler. They do not use CFC or HCFC refrigerants, which may make them even more attractive in years to come. Improved safety and controls and better COP (even at part load) have propelled absorption refrigeration back into the market.

In some cases, the most effective use of refrigeration equipment in a large central-plant scenario is to have some of each type, comprising a hybrid plant. From a mixture of centrifugal and absorption equipment the operator can determine what equipment will provide the lowest operating cost under different con-

**Figure 10.15 Simplified absorption cycle schematic diagram.**

ditions. For example a hospital that utilizes steam year round, but at reduced rates during summer, might use the excess steam to run an absorption chiller or steam-driven turbine centrifugal chiller to reduce its summertime electrical demand charges.

### 10.6.3 Chiller Performance

Most chillers are designed for peak load and then operate at loads less than the peak most of the time. Many chiller manufacturers provide data that identifies a chiller's part-load performance as an aid to evaluating energy costs. Ideally a chiller operates at a desired temperature difference (typically 45-55 degrees F; 25-30 degrees C) at a given flow rate to meet a given load. As the load requirement increases or decreases, the chiller will load or unload to meet the need. A reset schedule that allows the chilled water temperature to be adjusted to meet thermal building loads based on enthalpy provides an ideal method of reducing energy consumption.

Chillers should not be operated at less than 50 percent of rated load if at all possible. This eliminates both surging and the need for hot-gas bypass as well as the potential that the chiller would operate at low efficiency. If there is a regular need to operate a large chiller at less than one-half of the rated load it is economical to install a small chiller to accommodate this load.

### 10.6.4 Thermal Storage

Thermal storage can be another effective way of controlling electrical demand by using stored chilled water or ice to offset peak loads during the peak demand time. A good knowledge of the utility consumption and/or load profile is essential in determining the applicability of thermal storage. See Chapter 19 for a discussion of thermal storage systems.

### 10.6.5 Cooling Towers

Cooling towers use atmospheric air to cool the water from a condenser or coil through evaporation. In general there are three types of cooling tower, named for the relationship between the fan-powered airflow and the flow of water in the tower: counterflow induced draft, crossflow induced draft and counterflow forced draft.

The use of variable-speed, two-speed or three-speed fans is one way to optimize the control of the cooling tower in order to reduce power consumption and provide adequate water cooling capacity. As the required cooling capacity increases or decreases the fans can be sequenced to maintain the approach temperature difference. For most air-conditioning systems this usually varies between 5 and 12 degrees F (3 to 7 degrees C).

When operated in the winter, the quantity of air must be carefully controlled to the point where the water spray is not allowed to freeze. In cold climates it may be necessary to provide a heating element within the tower to prevent freeze-ups. Although electric resistance heaters can be used for this purpose it is far more efficient to utilize hot water or steam as a heat source if available.

### 10.6.6 Wet-side Economizer

The use of "free-cooling" using the cooling tower water to cool supply air or chilled water is referred to as a wet-side economizer. The most common and effective way of interconnecting the cooling tower water to the chilled water loop is through the use of a plate-and-frame heat exchanger which offers a high heat transfer rate and low pressure drop. This method isolates the cooling tower water from the chilled water circuit maintaining the integrity of the closed chilled water loop. Another method is to use a separate circuit and pump that allows cooling tower water to be circulated through a coil located within an air-handling unit.

The introduction of cooling tower water, into the chilled water system, through a so-called strainer cycle, can create maintenance nightmares and should be avoided. The water treatment program required for chilled water is intensive due to the required cleanness of the water in the chilled water loop.

### 10.6.7 Water treatment

A good water treatment program is essential to the maintenance of an efficient chilled water system. Filtering the cooling tower water should be evaluated. In some cases, depending on water quality, this can save the user a great deal of money in chemicals. Pretreating new systems prior to initial start-up will also provide longer equipment life and insure proper system performance.

Chiller performance is based on given design parameters and listed in literature provided by the chiller manufacturer. The performance will vary with building load, chilled water temperature, condenser water temperature and fouling factor. The fouling factor is the resistance caused by dirt, scale, silt, rust and other deposits on the surface of the tubes in the chiller and significantly affects the overall heat transfer of the chiller.

## 10.7 DOMESTIC HOT WATER

The creation of domestic hot water (DHW) represents about 4 percent of the annual energy consumption in typical non-residential buildings. In buildings where sleeping or food preparation occur, including hotels, restaurants, and hospitals, DHW may account for as much as thirty percent of total energy consumption.

Some older lavatory faucets provide a flow of 4 to 6 gal/min (0.25 to 0.38 l/s). Since hand washing is a function more of time than water use, substantial savings can be achieved by reducing water flow. Reduced-flow faucets which produce an adequate spray pattern can reduce water consumption to less than 1 gal/min (0.06 l/s). Flow reducing aerator replacements are also available.

Reducing DHW temperature has also been shown to save energy in non-residential buildings. Since most building users accept water at the available temperature, regardless of what it is, water temperature can be reduced from the prevailing standard of 140°F (60°C) to a 105°F (40°C) utilization temperature saving up to one-half of the energy used to heat the water.

Many large non-commercial buildings employ recirculating DHW distribution systems in order to reduce or eliminate the time required and water wasted in flushing cold water from hot water piping. Recirculating distribution is economically attractive only where DHW use is high and/or the cost of water greatly exceeds the cost of water heating. In most cases the energy required to keep water in recirculating DHW systems hot exceeds the energy used to heat the water actually used.

To overcome this waste of energy there is a trend to convert recirculating DHW systems to localized point-of-use hot water heating, particularly in buildings where plumbing facilities are widely separated. In either case insulation of DHW piping is essential in reducing the waste of energy in distribution. One-inch of insulation on DHW pipes will result in a 50% reduction in the distribution heat loss.

One often-overlooked energy conservation opportunity associated with DHW is the use of solar-heated hot water. Unlike space-heating, the need for DHW is relatively constant throughout the year and peaks during hours of sunshine in non-residential buildings. Year-round use amortizes the cost of initial equipment faster than other active-solar options.

Many of the techniques appropriate for reducing energy waste in DHW systems are also appropriate for energy consumption in heated service water systems for industrial buildings or laboratories.

## 10.8   ESTIMATING HVAC ENERGY CONSUMPTION

The methods for estimating building heating and cooling loads and the consumption of energy by HVAC systems are described in Chapter 9.

### References

1. *ASHRAE Handbook: Fundamentals,* American Society of Heating, Refrigerating and Air-Conditioning Engineers, Inc., Atlanta, 1993.

2. *ASHRAE Handbook: HVAC Applications,* American Society of Heating, Refrigerating and Air-Conditioning Engineers, Inc., Atlanta, 1995.

3. *ASHRAE Handbook: HVAC Systems and Equipment,* American Society of Heating, Refrigerating and Air-Conditioning Engineers, Inc., Atlanta, 1992.

4. *ASHRAE Handbook: Refrigeration,* American Society of Heating, Refrigerating and Air-Conditioning Engineers, Inc., Atlanta, 1994.

# CHAPTER 11

# ELECTRIC ENERGY MANAGEMENT

**K.K. LOBODOVSKY**
BSEE & BSME
Certified Energy Auditor
State of California

## 11.1 INTRODUCTION

Efficient use of electric energy enables commercial, industrial and institutional facilities to minimize operating costs, and increase profits to stay competitive.

The majority of electrical energy in the United States is used to run electric motor driven systems. Generally, systems consist of several components, the electrical power supply, the electric motor, the motor control, and a mechanical transmission system.

There are several ways to improve the systems' efficiency. The cost effective way is to check each component of the system for an opportunity to reduce electrical losses. A qualified individual should oversee the electrical system since poor power distribution within a facility is a common cause of energy losses.

Technology Update Ch. 18[1] lists 20 items to help facility management staff identify opportunities to improve drive system efficiency.

1. Maintain Voltage Levels.
2. Minimize Phase Imbalance.
3. Maintain Power Factor.
4. Maintain Good Power Quality.
5. Select Efficient Transformers.
6. Identify and Fix Distribution System Losses.
7. Minimize Distribution System Resistance.
8. Use Adjustable Speed Drives (ASDs) or 2-Speed Motors Where Appropriate.
9. Consider Load Shedding.
10. Choose Replacement Before a Motor Fails.
11. Choose Energy-Efficient Motors.
12. Match Motor Operating Speeds.
13. Size Motors for Efficiency.
14. Choose 200 Volt Motors for 208 Volt Electrical Systems.
15. Minimize Rewind Losses.
16. Optimize Transmission Efficiency.
17. Perform Periodic Checks.
18. Control Temperatures.
19. Lubricate Correctly.
20. Maintain Motor Records.

Some of these steps require the one-time involvement of an electrical engineer or technician. Some steps can be implemented when motors fail or major capital changes are made in the facility. Others involve development of a motor monitoring and maintenance program.

## 11.2 POWER SUPPLY

Much of this information consists of standards defined by the National Electrical Manufacturers Association (NEMA).

The power supply is one of the major factors affecting selection, installation, operation, and maintenance of an electrical motor driven system. Usual service conditions, defined in NEMA Standard Publication MG1, *Motors and Generators,*[2] include:

- Motors designed for rated voltage, frequency, and number of phases.

- The supply voltage must be known to select the proper motor.

- Motor nameplate voltage will normally be less then nominal power system voltage.

| Nominal Power System Voltage (Volts) | Motor Utilization (Nameplate) Voltage Volts |
|---|---|
| 208 | 200 |
| 240 | 230 |
| 480 | 460 |
| 600 | 575 |
| 2400 | 2300 |
| 4160 | 4000 |
| 6900 | 6600 |
| 13800 | 13200 |

- Operation within tolerance of ±10 percent of the rated voltage.

273

- Operation from a sine wave of voltage source (not to exceed 10 percent deviation factor).

- Operation within a tolerance of ±5 percent of rated frequency.

- Operation within a voltage unbalance of 1 percent or less.

*Operation at other than __usual__ service conditions may result in the consumption of additional energy.*

## 11.3 EFFECTS OF UNBALANCED VOLTAGES ON THE PERFORMANCE OF POLYPHASE SQUIRREL-CAGE INDUCTION MOTORS (MG 1-20.56)

When the line voltages applied to a polyphase induction motor are not equal, unbalanced currents in the stator windings result. A small percentage of voltage unbalance results in a much larger percentage current unbalance. Consequently, the temperature rise of the motor operating at a particular load and percentage voltage unbalance will be greater than for the motor operating under the same conditions with balanced voltages.

Voltages should be evenly balanced as closely as they can be read on a voltmeter. If the voltages are unbalanced, the rated horsepower of polyphase squirrel-cage induction motors should be multiplied by the factor shown in Figure 11.1 to reduce the possibility of damage to the motor. Operation of the motor with more than a 5-percent voltage unbalance is not recommended.

When the derating curve of Figure 11.1 is applied for operation on balanced voltages, the selection and setting of the overload device should take into account the combination of the derating factor applied to the motor and the increase in current resulting from the unbalanced voltages. This is a complex problem involving the variation in motor current as a function of load and voltage unbalance in the addition to the characteristics of the overload device relative to $I_{MAXIMUM}$ or $I_{AVERAGE}$. In the absence of specific information it is recommended that overload devices be selected and/or adjusted at the minimum value that does not result in tripping for the derating factor and voltage unbalance that applies. When the unbalanced voltages are unanticipated, it is recommended that the overload devices be selected so as to be responsive to $I_{MAXIMUM}$ in preference to overload devices responsive to $I_{AVERAGE}$.

**Figure 11.1 Polyphase squirrel-cage induction motors derating factor due to unbalanced voltage.**

## 11.4 EFFECT ON PERFORMANCE— GENERAL (MG 1 20.56.1)

The effect of unbalanced voltages on polyphase induction motors is equivalent to the introduction of a "negative-sequence voltage" having a rotation opposite to that occurring the balanced voltages. This negative-sequence voltage produces an air gap flux rotating against the rotation of the rotor, tending to produce high currents. A small negative-sequence voltage may produce current in the windings considerably in excess of those present under balanced voltage conditions.

### 11.4.1  Unbalanced Defined (MG 1 20.56.2)
The voltage unbalance in percent may be defined as follows:

$$\text{Percent Voltage Unbalance} = 100 \times \frac{\text{Maximum voltage deviation from average voltage}}{\text{average voltage}}$$

Example—With voltages of 220, 215 and 210, the average is 215, the maximum deviation from the average is 5

PERCENT VOLTAGE UNBALANCE
= 100 * 5/215 = 2.3 PERCENT

### 11.4.2  Torque (MG 1 20.56.3)
The locked-rotor torque and breakdown torque are decreased when the voltage is unbalanced. If the voltage unbalance is extremely severe, the torque might not be adequate for the application.

### 11.4.3  Full-Load Speed (MG 1 20.56.4)
The full-load speed is reduced slightly when the motor operates at unbalanced voltages.

### 11.4.4  Currents (MG 1 20.56.5)
The locked-rotor current will be unbalanced but the locked rotor kVA will increase only slightly.

The currents at normal operating speed with unbalanced voltages will be greatly unbalanced in the order of 6 to 10 times the voltage unbalance.

## 11.5 MOTOR

The origin of the electric motor can be traced back to 1831 when Michael Faraday demonstrated the fundamental principles of electromagnetism. The purpose of an electric motor is to convert electrical energy into mechanical energy.

Electric motors are efficient at converting electric energy into mechanical energy. If the efficiency of an electric motor is 80%, it means that 80% of electrical energy delivered to the motor is directly converted to mechanical energy. The portion used by the motor is the difference between the electrical energy input and mechanical energy output.

A major manufacturer estimates that US annual sales exceed 2 million motors. Table 11.1 lists sales volume by motor horsepower. Only 15% of these sales involve high-efficiency motors.[3]

**Table 11.1 Polyphase induction motors annual sales volume.**

| hp | Units |
|---|---|
| 1-5 | 1,300,000 |
| 7.5-20 | 500,000 |
| 25-50 | 140,000 |
| 60-100 | 40,000 |
| 125-200 | 32,000 |
| 250-500 | 11,000 |
| Total | 2,023,000 |

Motor terms are used quite frequently, usually on the assumption that every one knows what they mean or imply. Such is far too often not the case. The following section is a list of motor terms.

## 11.6 GLOSSARY OF FREQUENTLY OCCURRING MOTOR TERMS[4]

**Amps**

*Full Load Amps*

The amount of current the motor can be expected to draw under full load (torque) conditions is called Full Load Amps. It is also known as nameplate amps.

*Locked Rotor Amps*

Also known as starting inrush, this is the amount of current the motor can be expected to draw under starting conditions when full voltage is applied.

*Service Factor Amps*

This is the amount of current the motor will draw when it is subjected to a percentage of overload equal to the service factor on the nameplate of the motor. For example, many motors will have a service factor of 1.15, meaning that the motor can handle a 15% overload. The service factor amperage is the amount of current that the motor will draw under the service factor load condition.

**Code Letter**

The code letter is an indication of the amount of inrush current or locked rotor current that is required by a motor when it is started. Motor code letters usually applied to ratings of motors normally started on full voltage (chart below).

| Code letter | Locked rotor* kVA per horsepower | Horsepower Single-phase | Horsepower Three-phase |
|---|---|---|---|
| F | 5.0 to 5.6 | | 15 up |
| G | 5.6 to 6.3 | 5 | 7.5 to 10 |
| H | 6.3 to 7.1 | 3 | 5 |
| J | 7.1 to 8.0 | 1.5 to 2 | 3 |
| K | 8.0 to 9.0 | 0.75 to 1.00 | 1.5 to 2 |
| L | 9.0 to 10.0 | 0.5 | 1 |

*Locked rotor kVA is equal to the product of the line voltage times motor current divided by 1,000 when the motor is not allowed to rotate; this corresponds to the first power surge required to start the motor. Locked-rotor kVA per horsepower range includes the lower figure up to but not including the higher figure.

## Design

The design letter is an indication of the shape of the torque speed curve. Figure 11.2 shows the typical shape of the most commonly used design letters. They are A, B, C, and D. Design B is the standard industrial duty motor which has reasonable starting torque with moderate starting current and good overall performance for most industrial applications. Design C is used for hard to start loads and is specifically designed to have high starting torque. Design D is the so-called high slip motor which tends to have very high starting torque but has high slip RPM at full load torque. In some respects, this motor can be said to have a 'spongy' characteristic when loads are changing. Design D motors particularly suited for low speed punch press, hoist and elevator applications. Generally, the efficiency of Design D motors at full load is rather poor and thus they are normally used on those applications where the torque characteristics are of primary importance. Design A motors are not commonly specified but specialized motors used on injection molding applications have characteristics similar to Design B. The most important characteristic of Design A is the high pull out torque.

**Figure 11.2**

## Efficiency

Efficiency is the percentage of the input power that is actually converted to work output from the motor shaft. Efficiency is now being stamped on the nameplate of most domestically produced electric motors. See the section 11.14.

$$\text{Efficiency} = \text{EFF} = \frac{746 \times \text{HP Output}}{\text{Watts Input}}$$

## Frame Size

Motors, like suits of clothes, shoes and hats, come in various sizes to match the requirements of the applications. In general, the frame size gets larger with increasing horsepower or with decreasing speeds. In order to promote standardization in the motor industry, NEMA (National Electrical Manufacturers Association) prescribes standard frame sizes for certain horsepower, speed, and enclosure combinations. Frame size pins down the mounting and shaft dimension of standard motors. For example, a motor with a frame size of 56, will always have a shaft height above the base of 3-1/2 inches.

## Frequency

This is the frequency for which the motor is designed. The most commonly occurring frequency in this country is 60 cycles but, internationally, other frequencies such as 25, 40, and 50 cycles can be found.

## Full Load Speed

An indication of the approximate speed that the motor will run when it is putting out full rated output torque or horsepower is called full load speed.

## High Inertia Load

These are loads that have a relatively high fly wheel effect. Large fans, blowers, punch presses, centrifuges, industrial washing machines, and other similar loads can be classified as high inertia loads.

## Insulation Class

The insulation class is a measure of the resistance of the insulating components of a motor to degradation from heat. Four major classifications of insulation are used in motors. They are, in order of increasing thermal capabilities, A, B, F, and H.

| Class of Insulation System | Temperature, Degrees C |
| --- | --- |
| A | 75 |
| B | 905 |
| F | 115 |
| H | 130 |

## Load Types

### Constant Horsepower

The term constant horsepower is used in certain types of loads where the torque requirement is reduced as the speed is increased and vice-versa. The constant horsepower load is usually associated with metal removal applications such as drill presses, lathes, milling machines, and similar types of applications.

### Constant Torque

Constant torque is a term used to define a load characteristic where the amount of torque required to drive the machine is constant regardless of the speed at which it is driven. For example, the torque requirement of most conveyors is constant.

### Variable Torque

Variable torque is found in loads having characteristics requiring low torque at low speeds and increasing values of torque required as the speed is increased. Typically examples of variable torque loads are centrifugal fans and centrifugal pumps.

## Phase

Phase is the indication of the type of power supply for which the motor is designed. Two major categories exist: single phase and three phase. There are some very spotty areas where two phase power is available but this is very insignificant.

## Poles

This is the number of magnetic poles within the motor when power is applied. Poles are always an even number such as 2, 4, 6. In an AC motor, the number of poles work in conjunction with the frequency to determine the synchronous speed of the motor. At 50 and 60 cycles, common arrangements are:

### Synchronous speed

| Poles | 60 Cycles | 50 Cycles |
|-------|-----------|-----------|
| 2     | 3600      | 3000      |
| 4     | 1800      | 1500      |
| 6     | 1200      | 1000      |
| 8     | 900       | 750       |
| 10    | 720       | 600       |

## Power Factor

Percent power factor is a measure of a particular motor's requirements for magnetizing amperage. For more information se section 11.7.

$$\text{Power Factor (3 phase)} = \text{pf} = \frac{\text{Watts Input}}{\text{Volts} \times \text{Amps} \times 1.73}$$

## Service Factor

The service factor is a multiplier that indicates the amount of overload a motor can be expected to handle. For example, a motor with a 1.0 service factor cannot be

278                                                                ENERGY MANAGEMENT HANDBOOK

expected to handle more than its nameplate horsepower on a continuous basis. Similarly, a motor with a 1.15 service factor can be expected to safely handle intermittent loads amounting to 15% beyond its nameplate horsepower.

## Slip

Slip is used in two forms. One is the slip RPM which is the difference between the synchronous speed and the full load speed. When this slip RPM is expressed as a percentage of the synchronous speed, then it is called percent slip or just 'slip.' Most standard motors run with a full loadslip of 2% to 5%.

## Synchronous Speed

This is the speed at which the magnetic field within the motor is rotating. It is also approximately the speed that the motor will run under no load condition. For example, a 4 pole motor running in 60 cycles would have a magnetic field speed of 1800 RPM. The no load speed of that motor shaft would be very close to 1800, probably 1798 or 1799 RPM. The full load speed of the same motor might be 1745 RPM. The difference between the synchronous speed of the full load speed is called the slip RPM of the motor.

## Temperature

*Ambient Temperature.*

Ambient temperature is the maximum safe room temperature surrounding the motor if it is going to be operated continuously at full load. In most cases, the standardized ambient temperature rating is 40°C (104°F). This is a very warm room. Certain types of applications such as on board ships and in boiler rooms, may require motors with a higher ambient temperature capability such as 50°C or 60°C.

*Temperature Rise.*

Temperature rise is the amount of temperature change that can be expected within the winding of the motor from non-operating (cool condition) to its temperature at full load continuous operating condition. Temperature rise is normally expressed in degrees centigrade.

## Time Rating

Most motors are rated in continuous duty which means that they can operate at full load torque continuously without overheating. Motors used on certain types of applications such as waste disposal, valve actuators, hoists, and other types of intermittent loads, will frequently be rated in short term duty such as 5 minutes, 15 minutes, 30 minutes or 1 hour. Just like a human being, a motor can be asked to handle very strenuous work as long as it is not required on a continuous basis.

## Torque

Torque is the twisting force exerted by the shaft or a motor. Torque is measured in inch pounds, foot pounds, and on small motors, in terms of inch ounces.

*Full Load Torque*

Full load torque is the rated continuous torque that the motor can support without overheating within its time rating.

*Peak Torque*

Many types of loads such as reciprocating compressors have cycling torque where the amount of torque required varies depending on the position of the machine. The actual maximum torque requirement at any point is called the peak torque requirement. Peak torque are involved in things such as punch presses and other types of loads where an oscillating torque requirement occurs.

*Pull Out Torque*

Also known as breakdown torque, this is the maximum amount of torque that is available from the motor shaft when the motor is operating at full voltage and is running at full speed. The load is then increased until the maximum point is reached. Refer to Figure 11.3.

*Pull Up Torque*

The lowest point on the torque speed curve for a motor accelerating a load up to full speed is called pull up torque. Some motors are designed to not have a value of pull up torque because the lowest point may occur at the locked rotor point. In this case, pull up torque is the same as locked rotor torque.

**Figure 11.3 Typical speed—torque curve.**

*Starting Torque*

The amount of torque the motor produces when it is energized at full voltage and with the shaft locked in place is called starting torque. This value is also frequently expressed as 'Locked Rotor Torque.' It is the amount of torque available when power is applied to break the load away and start accelerating it up to speed.

**Voltage**

This would be the voltage rating for which the motor is designed. Section 11.2.

## 11.7 POWER FACTOR

### WHAT IS POWER FACTOR (pf)?

It is the mathematical ratio of *ACTIVE POWER* () to *APPARENT POWER* (VA)

$$ pf = \frac{\text{Active power}}{\text{Apparent power}} = W = Cos\ \theta $$

pf angle in degrees = $cos^{-1}\ \theta$

ACTIVE POWER = **W** = "real power" = supplied by the power system to actually turn the motor.

REACTIVE POWER = **VAR** = **(W)tan** $\theta$ = is used strictly to develop a magnetic field within the motor.

or **(VA)² = (W)² + (VAR)²**

NOTE: Power factor may be "leading" or "lagging" depending on the direction of VAR flow.

CAPACITORS can be used to improve the power factor of a circuit with a large inductive load. Current through capacitor LEADS the applied voltage by 90 electrical degrees (VAC), and has the effect of "opposing" the inductive "LAGGING" current on a "one-for-one" (VAR) basis.

### WHY RAISE POWER FACTOR (pf)?

Low (or "unsatisfactory") power factor is caused by the use of inductive (magnetic) devices and can indicate possible low system electrical operating efficiency. These devices are:

*   non-power factor corrected fluorescent and high intensity discharge lighting fixture ballasts (40%-80% pf)

*   arc welders (50%-70% pf)

*   solenoids (20%-50% pf)

*   induction heating equipment (60%-90% pf)

*   lifting magnets (20%-50% pf)

*   small "dry-pack" transformers (30%-95% pf)

*   and most significantly, induction motors (55%-90% pf)

Induction motors are generally the principal cause of low power factor because there are so many in use, and they are usually *not fully loaded*. The correction of the condition of LOW power factor is a problem of vital economic importance in the generation, distribution and utilization of a-c power.

### MAJOR BENEFITS OF POWER FACTOR IMPROVEMENT ARE:

*   increased plant capacity,

*   reduced power factor "penalty" charges for the electric utility,

*   improvement of voltage supply,

*   less power losses in feeders, transformers and distribution equipment.

### WHERE TO CORRECT POWER FACTOR?

Capacitor correction is relatively inexpensive both in material and installation costs. Capacitors can be installed at any point in the electrical system, and will improve the power factor between the point of application and the power source. However, the power factor between the utilization equipment and the capacitor will remain unchanged. Capacitors are usually added at each piece of offending equipment, ahead of groups of small motors (ahead of motor control centers or distribution panels) or at main services. Refer to the National Electrical Code for installation requirements.

The advantages and disadvantages of each type of capacitor installation are listed below:

*Capacitor on each piece of equipment (1,2)*

### ADVANTAGES
* increases load capabilities of distribution system.

* can be switched with equipment; no additional switching is required.

* better voltage regulation because capacitor use follows load.

* capacitor sizing is simplified

* capacitors are coupled with equipment and move with equipment if rearrangements are instituted.

### DISADVANTAGES
* small capacitors cost more per KVAC than larger units (economic break point for individual correction is generally at 10 HP).

*Capacitor with equipment group (3)*

### ADVANTAGES
* increased load capabilities of the service,

* reduced material costs relative to individual correction

* reduced installation costs relative to individual correction

### DISADVANTAGES
* switching means may be required to control amount of capacitance used.

*Capacitor at main service (4,5, & 6)*

### ADVANTAGES

* low material installation costs.

### DISADVANTAGES

* switching will usually be required to control the amount of capacitance used.

* does not improve the load capabilities of the distribution system.

*OTHER CONSIDERATIONS*

Where the loads contributing to power factor are relatively constant, and system load capabilities are not a factor, correcting at the main service could provide a cost advantage. When the low power factor is derived from a few selected pieces of equipment, individual equipment correction would be cost effective. Most capacitors used for power factor correction have built-in fusing; if not, fusing must be provided.

The growing use of ASDs (nonlinear loads) has increased the complexity of system power factor and its corrections. The application of pf correction capacitors without a thorough analysis of the system can aggravate rather than correct the problem, particularly if the fifth and seventh harmonics are present.

*POWER QUALITY REQUIREMENTS[6]*

The electronic circuits used in ASDs may be susceptible to power quality related problems if care is not taken during application, specification and installation. The most common problems include transient overvoltages, voltage sags and harmonic distortion. These power quality problems are usually manifested in the form of nuisance tripping.

TRANSIENT OVERVOLTAGES—Capacitors are devices used in the utility power system to provide power factor correction and voltage stability during periods of heavy loading. Customers may also use capacitors for power factor correction within their facility. When capacitors are energized, a large transient overvoltage may develop causing the ASD to trip.

VOLTAGE SAGS—ASDs are very sensitive to temporary reductions in nominal voltage. Typically, voltage sags are caused by faults on either the customer's or the utility's electrical system.

HARMONIC DISTORTION—ASDs introduce harmonics into the power system due to nonlinear characteristics of power electronics operation. Harmonics are components of current and voltage that are multiples of the normal 60Hz ac sine wave. ASDs produce harmonics which, if severe, can cause motor, transformer and conductor overheating, capacitor failures, misoperation of relays and controls and reduce system efficiencies.

Compliance with IEEE-519 "Recommended Practices and Requirements for Harmonic Control in Electrical Power Systems" is strongly recommended.

## 11.8 HANDY ELECTRICAL FORMULAS & RULES OF THUMB

**Conversion formulas**

| REQUIRED | DIRECT CURRENT | ALTERNATING CURRENT | |
|---|---|---|---|
| | | SINGLE-PHASE | THREE-PHASE |
| AMPERES WHEN H.P. IS KNOWN | $\dfrac{\text{H.P.} \times 746}{\text{E} \times \text{EFF.}}$ | $\dfrac{\text{H.P.} \times 746}{\text{E} \times \text{EFF.} \times \text{P.F.}}$ | $\dfrac{\text{H.P.} \times 746}{1.73 \times \text{E} \times \text{EFF.} \times \text{P.F.}}$ |
| AMPERES WHEN KILOWATTS ARE KNOWN | $\dfrac{\text{KW} \times 1000}{\text{E}}$ | $\dfrac{\text{KW} \times 1000}{\text{E} \times \text{P.F.}}$ | $\dfrac{\text{KW} \times 1000}{1.73 \times \text{E} \times \text{P.F.}}$ |
| AMPERES WHEN KVA IS KNOWN | | $\dfrac{\text{KVA} \times 1000}{\text{E}}$ | $\dfrac{\text{KVA} \times 1000}{1.73 \times \text{E}}$ |
| KILOWATTS | $\dfrac{\text{I} \times \text{E}}{1000}$ | $\dfrac{\text{I} \times \text{E} \times \text{P.F.}}{1000}$ | $\dfrac{\text{I} \times \text{E} \times \text{P.F.} \times 1.73}{1000}$ |
| KVA | | $\dfrac{\text{I} \times \text{E}}{1000}$ | $\dfrac{\text{I} \times \text{E} \times 1.73}{1000}$ |
| HORSEPOWER OUTPUT | $\dfrac{\text{I} \times \text{E} \times \text{EFF.}}{746}$ | $\dfrac{\text{I} \times \text{E} \times \text{EFF.} \times \text{P.F.}}{746}$ | $\dfrac{\text{I} \times \text{E} \times 1.73 \times \text{EFF.} \times \text{P.F.}}{746}$ |

I = AMPERES    H.P. = HORSEPOWER    EFF. = EFFICIENCY (EXPRESSED IN A DECIMAL)
E = VOLTS    P.F. = POWER FACTOR    KW = KILOWATTS     KVA = KILOVOLT-AMPERES

**Rules of thumb.**

At 3600 RPM, a motor develops 1.5 lb.-ft. per HP.
At 1800 RPM, a motor develops 3 lb.-ft. per HP.
At 1200 RPM, a motor develops 4.5 lb.-ft. per HP.

At 550 & 575 Volts, a 3 phase motor draws 1 amp per HP.
At 440 & 460 Volts, a 3 phase motor draws 1.25 amp per HP.
At 220 & 230 Volts, a 3 phase motor draws 2.5 amp per HP.

## 11.9 ELECTRIC MOTOR OPERATING LOADS

Most electric motors are designed to operate at 50 to 100 percent of their rated load. One reason is the motors optimum efficiency is generally 75 percent of the rated load, and the other reason is motors are generally sized for the starting requirements.

Several surveys of installed motors reveal that large portion of motors in use are improperly loaded. Underloaded motors, those loaded below 50 percent of rated load, operate inefficiently and exhibit low power factor. Low power factor increases losses in electrical distribution and utilization equipment, such as wiring, motors, and transformers, and reduces the load-handling capability and voltage regulation of the building's electrical system. Typical part-load efficiency and power factor characteristics are shown in Figure 11.4.

*POWER SURVEY*

Power surveys are conducted to compile meaningful records of energy usage at the service entrance, feeders and individuals loads. These records can be analyzed to prioritize those areas yielding the greatest energy savings. Power surveys also provide information for load scheduling to reduce peak demand and show operational characteristics of loads that may suggest component or system replacement to reduce energy consumption. Only through the measurement of AC power parameters can true cost benefit analysis be performed.[7]

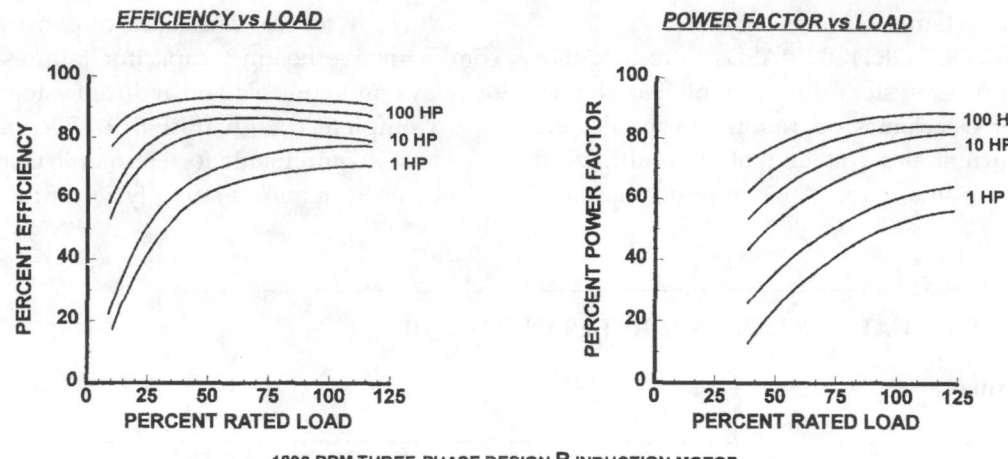

Figure 11.4 Typical part-load efficiency and power factor characteristics

## 11.10 DETERMINING ELECTRIC MOTOR OPERATING LOADS

Determining if electric motors are properly loaded enables a manager to make informed decisions about when to replace them and which replacement to choose. There are several ways to determine motor loads. The best and the simplest way is by direct electrical measurement using a Power Meter. Slip Measurement or Amperage Readings methods can be used to **estimate** the actual load.

## 11.11 POWER METER

To understand the electrical power usage of a facility, load or device, measurements must be taken over a time span to have a profile of the unit's operation. Digital power multimeters, measure Amps, Volts, kWatts, kVars, kVA, Power Factor, Phase Angle and Firing Angle. The GENERAL TEST FORM Figure 11.5 provides a format for documentation with corresponding connection diagrams for various power circuit configurations.

*Such measurements should only be performed by trained personnel*

### Selection of Equipment for Power Measurement or Surveys

When choosing equipment to conduct a power survey, many presentation formats are available including indicating instruments, strip chart recorders and digital devices with numeric printout. For most survey applications, changing loads makes it mandatory for data to be compiled over a period of time. This period may be an hour, day, week or month. Since it is not practical to write down varying readings from an indicating device for a long period of time, a chart recorder or digital device with numeric printout is preferred. If loads vary frequently, an analog trend recording will be easier to analyze than trying to interpret several numeric reports. Digital power survey monitors are typically less expensive than analog recordings systems. Complete microprocessor based power survey systems capable of measuring watts, VARs, kVA, power factor, watt hours, VAR hours and demand including current transformers are available for under $3000. With prices for memory and computers going down, digital devices interfaced to disk or cassette storage will provide a cost effective method for system analysis.[7]

### Loads

When analyzing polyphase motors, it is important to make measurements with equipment suited for the application. Watt measurements or VAR measurements should be taken with a two element device. Power factor should be determined from the readings of both measurements. When variable speed drives are encountered, it is always preferable to take measurements on the line side of the controller. When measurements are required on the load side of the controller, the instrument specifications should be reviewed and if there is a question on the application the manufacturer should be contacted.[7]

## 11.12 SLIP MEASUREMENT

### Conditions

1.  Applied voltage must be within 5% of nameplate rating.

2.  Should not be used on rewound motors.

| TO MEASURE | 1 Phase, 2 Wire | 1 Phase, 3 Wire | 3 Phase, 3 Wire | 3 Phase, 4 Wire | 3 Phase, 4 Wire TAP |
|---|---|---|---|---|---|
| **VOLTAGE** (V) | L - 1 to N | L - 1 to N | A to B Phase | A Phase to N | B Phase to N |
| | | L - 2 to N | C to B Phase | B Phase to N | A Phase to N |
| | | | | C Phase to N | C Phase to N |
| **CURRENT** (A) | L - 1 | L - 1 | A Phase | A Phase | B Phase |
| | | L - 2 | C Phase | B Phase | A Phase |
| | | | | C Phase | C Phase |
| **POWER** (kW) | L - 1 to N | L - 1 to N | | A Phase to N | |
| | | L - 2 to N | | B Phase to N | |
| | | | | C Phase to N | |
| | | Total kW | Total kW | Total kW | Total kW |
| **VOLT-AMPERES REACTIVE** (kVAR) | L - 1 to N | L - 1 to N | | A Phase to N | |
| | | L - 2 to N | | B Phase to N | |
| | | | | C Phase to N | |
| | | Total kVAR | Total kVAR | Total kVAR | Total kVAR |
| **VOLT-AMPERES** (kVA) | L - 1 to N | L - 1 to N | | A Phase to N | |
| | | L - 2 to N | | B Phase to N | |
| | | | | C Phase to N | |
| | | Total kVA | Total kVA | Total kVA | Total kVA |
| **POWER FACTOR** PF % | L - 1 to N | L - 1 to N | | A Phase to N | |
| | | L - 2 to N | | B Phase to N | |
| | | | | C Phase to N | |
| | | Combined PF % | Combined PF % | Combined PF % | Total PF % |

□ VOLTAGE LEAD

○ CURRENT TRANSFORMER with white lead towards load

RE = RED
BE = BLUE
BK = BLACK
WE = WHITE

| SOURCE | SOURCE | SOURCE | SOURCE | SOURCE |
|---|---|---|---|---|
| L1   N | L1   N   L2 | L1   L2   L3 | L1   L2   L3   N | L1   TAP   L2   L3 |
| RD   WE | RD   WE   BE | RD   WE   BE | RD   BE   BK   WE | BE   WE   RD   BK |
| RD | RD   BK   BE | RD   BK   BE | RD   BE   BK | BE   RD   BK |
| LOAD | LOAD | LOAD | LOAD | LOAD |

**Figure 11.5 General test form (for use with power meter).**

3. Motors should be operating under steady load conditions.

4. Should be performed by trained personnel.

Note: Values used in this analysis are subject to rounding errors. For example, full load speed often rounded to the nearest 5 RPM.

**Procedure**

1. Read and record the motors nameplate Full Load Speed. (RPM)

2. Determine Synchronous speed No Load Speed (RPM) (900, 1200, 1800, 3600)

3. Measure and record Operating Load Speed with tachometer. (RPM)

4. Insert the recorded values in the following formula and solve.

$$(\% \text{ Motor load}) = \frac{\text{NLS} - \text{OLS}}{\text{NLS} - \text{FLS}} \times 100$$

*Where:*

NLS = No load or synchronous speed
OLS = Operating load speed
FLS = Full load speed

*Example:*

Consider a 100 HP, 1800 RM Motor
FLS = 1775 RPM, OLS = 1786 RPM

$$(\% \text{ Motor load}) = \frac{1800 - 1786}{1800 - 1775} \times 100 = 56$$

Approximate load on motor = **100 HP × 0.56 = 56 HP**

### 11.13 AMPERAGE READINGS[4]

**Conditions**

1. Applied voltage must be within 5% of nameplate rating.

2. You must be able to disconnect the motor from the load. (By removing V-belts or disconnecting a coupling).

3. Motor must be 7-1/2 HP or larger, 3450, 1725 or 1140 RPM.

4. The indicated line amperage must be below the full load nameplate rating.

**Procedure**

1. Measure and record line amperage with load connected and running.

2. Disconnect motor from load. Measure and record the line amperage when the motor is running without load.

3. Read and record the motors nameplate amperage for the voltage being used.

4. Insert the recorded values in the following formula and solve.

$$(\% \text{ Rated HP}) = \frac{(2 \times \text{LLA}) - \text{NLA}}{(2 \times \text{NPA}) - \text{NLA}} \times 100$$

*Where:*

LLA = Loaded Line Amps
NLA = No Load Line Amps (Motor disconnected from load)
NPA = Nameplate Amperage (For operating voltage)

Please note: This procedure will generally yield reasonably accurate results when motor load is in the 40 to 100% range and deteriorating results at loads below 40%.

*Example:*

• A 20 HP motor driving a pump is operating on 460 volts and has a loaded line amperage of 16.5.

• When the coupling is disconnected and the motor operated at no load the amperage is 9.3.

• The motor nameplate amperage for 460 volts is 24.0.

**Therefore we have:**

Loaded Line Amps     LLA  = 16.5
No Load Amps         NLA  = 9.3
Nameplate Amps       NPA  = 24.0

$$(\% \text{Rated HP}) = \frac{(2 \times 16.5) - 9.3}{(2 \times 24.0) - 9.3} \times 100 = \frac{23.7}{38.7} \times 100 = 61.2\%$$

Approximate load on motor = 20 HP × 0.612 = 12.24 or slightly over 12 HP

### 11.14 ELECTRIC MOTOR EFFICIENCY

The efficiency of a motor is the ratio of the mechanical power output to the electrical power input. It may be expressed as:

$$\text{Efficiency} = \frac{\text{Output}}{\text{Input}} = \frac{\text{Input} - \text{Losses}}{\text{Input}} = \frac{\text{Output}}{\text{Output} + \text{Losses}}$$

Design changes, better materials, and manufacturing improvements reduce motor losses, making premium or energy-efficient motors more efficient than standard motors. Reduced losses mean that an energy-efficient motor produces a given amount of work with less energy input than a standard motor.[3]

In 1989, the National Electrical Manufacturers Association (NEMA) developed a standard definition for energy-efficient motors.[2]

How should we interpret efficiency labels?

**Efficiencies and Different Standards**

| Standard | 7.5 HP motor | 20 HP motor |
|---|---|---|
| International (IEC 34-2) | 82.3% | 89 4% |
| British (BS-269) | 82.3% | 89.4% |
| Japanese (IEC-37) | 85.0% | 90.4% |
| U.S. (IEEE -112 Method B)* | 80.3% | 86.9% |

The critical part of the efficiency comparison calculations is that the efficiencies used **must be comparable**.

The Arthur D Little report contained the following interesting statement: "Reliable and consistent data on motor efficiency is not available to motor appliers. Data published by manufacturers appears to range from very conservative to cavalier."

Recognizing that less than a 10 percent spread in losses is statistically insignificant NEMA has set up efficiency bands. Any motor tested by IEEE - 112, Method B, will carry the nominal efficiency of the highest band for which the average full load efficiency for the model is equal to or above that nominal.

The NEMA nominal efficiency is defined as the average efficiency of a large population of motors of the same design. The spread between nominal efficiency in the table based on increments of 10 percent losses. The spread between the nominal efficiency and the associated minimum is based on an increment of 20 percent losses.

### 11.14.1  The Following is Reprinted From NEMA MG 1-1987 Efficiency (MG 1-12.54)

**Determination of Motor Efficiency and Losses (MG 1-12.54.1)**

Efficiency and losses shall be determined in accordance with IEEE Std 112 Standard Test Procedures for Polyphase Induction Motors and Generators*. The efficiency shall be determined at rated output, voltage, and frequency.

Unless otherwise specified, horizontal polyphase squirrel-cage medium motors rated 1 to 125 horsepower shall be tested by dynamometer (Method B) as described in par. 5.2.2.4 of IEEE Std 112. Motor efficiency shall be calculated using MG 1-12.57 in lieu of Form E of IEEE Std 112. Vertical motors in this horsepower range shall also be tested by Method B if bearing construction permits; otherwise they shall be tested by segregated losses (Method E) as described in par. 5.2.3.1 of IEEE Std 112, including direct measurement of stray-load loss.

The following losses shall be included in determining the efficiency:

1.  Stator $I^2R$.
2.  Rotor $I^2R$.
3.  Core Loss.
4.  Stray load loss.
5.  Friction & windage loss.†
6.  Brush contact loss of wound-rotor machines

Power required for auxiliary items, such as external pumps or fans, that are necessary for the operation of the motor shall be stated separately.

In determining $I^2R$ losses, the resistance of each winding shall be corrected in a temperature equal to an

| Nominal Efficiency | | Minimum Efficiency |
|---|---|---|
| 93.6 ———— 20% Greater Losses → | | 92.4 |
| &#124; | | |
| 10%  Greater Losses | | |
| ↓ | | |
| 93.0 | | |

ambient temperature of 25°C plus the observed rated load temperature rise measured by resistance. When the rated load temperature rise has not been measured, the resistance of the winding shall be corrected to the following temperature:

| Class of Insulation System | Temperature, Degrees C |
|---|---|
| A | 75 |
| B | 95 |
| F | 115 |
| H | 130 |

This reference temperature shall be used for determining $I^2R$ losses at all loads. If the rated temperature rise is specified as that of a lower class of insulation system, the temperature for resistance correction shall be that of the lower insulation class.

NEMA Standard 5-12-1975, revised 6-21-1979; 11-12-1981; 11-20-1986; 1-11-1989.

### Efficiency Of Polyphase Squirrel-cage Medium Motors with Continuous Ratings (MG 1-12.54.2)

The full-load efficiency of Design A and B single-speed polyphase squirrel-cage medium motors in the range of 1 through 125 horsepower for frames assigned in accordance with NEMA Standards Publication No. MG 13, Frame Assignments for Alternating Current Integral-horsepower Induction Motors, (see MG1-1.101) and equivalent Design C ratings shall be identified on the nameplate by a nominal efficiency selected from the Nominal Efficiency column in Table 11.2 (NEMA Table 12.6A) which shall be not greater than the average efficiency of a large population of motors of the same design.

The efficiency shall be identified on the nameplate by the caption "NEMA Nominal Efficiency" or "NEMA Nom. Eff."

The full load efficiency, when operating at rated voltage and frequency, shall be not less than the minimum value indicated in Column B of Table 11.2 (NEMA Table 12.6A) associated with the nominal value in Column A.

Suggested Standard for Future Design 3-16-1977, NEMA Standard 1-17-1980, revised 3-8-1983; 3-14-1991.

The full-load efficiency, when operating at rated voltage and frequency, shall be not less than the minimum value indicated in Column C of Table 11.2 (NEMA Table 12.6A) associated with the nominal value in Column A.

Suggested Standard for Future Design 3-14-1991.

Variations in materials, manufacturing processes, and tests result in motor-to-motor efficiency for a large population of motors of a single design is not a unique efficiency but rather a band of efficiency. Therefore, Table 11.2 (NEMA Table 12.6A) has been established to indicate a logical series of nominal motor efficiencies and the minimum associated with each nominal. The nominal efficiency represents a value which should be used to compute the energy consumption of a motor or group of motors.

Authorized Engineering Information 3-6-1977, revised 1-17-1980;1-11-1989.

### Efficiency Levels of Energy Efficient Polyphase Squirrel-Cage Induction Motors (MG 1-12.55)

The nominal full-load efficiency determined in accordance with MG 1-12.54.1, identified on the nameplate in accordance with MG 1.12.54.2, and having corresponding minimum efficiency in accordance with Column B of Table 11.2 (NEMA Table 12.6A) shall equal or exceed the values listed in Table 11.3 (NEMA Table 12.6B) for the motor to be classified as "energy efficient."

NEMA Std 1-11-1989;3-14-1991.

### Efficiency Levels of Energy Efficient Polyphase Squirrel-Cage Induction Motors (MG 1-12.55A)
(SUGGESTED STANDARD FOR FUTURE DESIGN)

The nominal full-load efficiency determined in accordance with MG 1-12.54.1, identified on the nameplate in accordance with MG 1-12.54.2 and having minimum efficiency in accordance with Column C of Table 11.2 (NEMA Table 12.6A) shall equal or exceed the values listed in Table 11.4 (NEMA Table 12.6C) for the motor to be classified as "energy efficient."

Suggested Standard for future design 9-5-1991.

## 11.15  COMPARING MOTORS

It is essential that motor comparison be done on the same basis as to type, size, load, cost of energy, operating hours and most importantly the efficiency values such as nominal vs. nominal or guaranteed vs. guaranteed.

The following equations are used to compare the two motors.

### For loads not sensitive to motor speed—

Note: Replacing a standard motor with an energy-efficient motor in a centrifugal pump or fan application can result in increased energy consumption if energy-efficient motor operates at a higher RPM.

## Table 11.2 (NEMA Table 12.6A)

| Column A Nominal Efficiency | Column B* Minimum Efficiency based on 20% Loss Difference | Column C† Minimum Efficiency Based on 10% Loss Difference |
|---|---|---|
| 99.0 | 98.8 | 98.9 |
| 98.9 | 98.7 | 98.8 |
| 98.8 | 98.6 | 98.7 |
| 98.7 | 98.5 | 98.6 |
| 98.6 | 98.4 | 98.5 |
| | | |
| 98.5 | 88.2 | 98.4 |
| 98.4 | 98.0 | 98.2 |
| 98.2 | 97.8 | 98.0 |
| 98.0 | 97.6 | 97.8 |
| 97.8 | 97.4 | 97.6 |
| | | |
| 97.6 | 97.1 | 97.4 |
| 97.4 | 96.8 | 97.1 |
| 97.1 | 96.5 | 96.8 |
| 96.8 | 96.2 | 96.5 |
| 96.5 | 95.8 | 96.2 |
| | | |
| 96.2 | 95.4 | 95.8 |
| 95.8 | 95.0 | 95.4 |
| 95.4 | 94.5 | 95.0 |
| 95.0 | 94.1 | 94.5 |
| 94.5 | 93.6 | 94.1 |
| | | |
| 94.1 | 93.0 | 93.6 |
| 93.6 | 92.4 | 93.0 |
| 93.0 | 91.7 | 92.4 |
| 92.4 | 91.0 | 91.7 |
| 91.7 | 90.2 | 91.0 |
| | | |
| 91.0 | 89.5 | 90.2 |
| 90.2 | 88.5 | 89.5 |
| 89.5 | 87.5 | 88.5 |
| 88.5 | 86.5 | 87.5 |
| 87.5 | 85.5 | 86.5 |
| | | |
| 86.5 | 84.0 | 85.5 |
| 85.5 | 82.5 | 84.0 |
| 84.0 | 81.5 | 82.5 |
| 82.5 | 80.0 | 81.5 |
| 81.5 | 78.5 | 80.0 |
| | | |
| 80.0 | 77.0 | 78.5 |
| 78.5 | 75.5 | 77.0 |
| 77.0 | 74.0 | 75.5 |
| 75.5 | 72.0 | 74.0 |
| 74.0 | 70.0 | 72.0 |
| | | |
| 72.0 | 68.0 | 70.0 |
| 70.0 | 66.0 | 68.0 |
| 68.0 | 64.0 | 66.0 |
| 66.0 | 62.0 | 64.0 |
| 64.0 | 59.5 | 62.0 |
| | | |
| 62.0 | 57.5 | 59.5 |
| 59.5 | 55.0 | 57.5 |
| 57.5 | 52.5 | 55.0 |
| 55.0 | 50.5 | 52.5 |
| | | |
| 52.5 | 48.0 | 50.5 |
| | | |
| 50.5 | 46.0 | 48.0 |

*Column B approved as NEMA Standard 3/14/1991
†Column C approved as Suggested Standard for Future Designs 3/14/1991

## Table 11.3 (NEMA Table 12.6B) Full-load efficiencies of energy efficient motors.

### OPEN MOTORS

| HP | 2 POLE (3600) | | 4 POLE (1800) | | 6 POLE (1200) | | 8 POLE (900) | |
|---|---|---|---|---|---|---|---|---|
| | NOMINAL EFFICIENCY | MINIMUM EFFICIENCY | NOMINAL EFFICIENCY | MINIMUM EFFICIENCY | NOMINAL EFFICIENCY | MINIMUM EFFICIENCY | NOMINAL EFFICIENCY | MINIMUM EFFICIENCY |
| 1.0 | ------- | ------- | 82.5 | 80.0 | 77.0 | 74.0 | 72.0 | 68.0 |
| 1.5 | 80.0 | 77.0 | 82.5 | 80.0 | 82.5 | 80.0 | 75.5 | 72.0 |
| 2.0 | 82.5 | 80.0 | 82.5 | 80.0 | 84.0 | 81.5 | 85.5 | 82.5 |
| 3.0 | 82.5 | 80.0 | 86.5 | 84.0 | 85.5 | 82.5 | 86.5 | 84.0 |
| 5.0 | 85.5 | 82.5 | 86.5 | 84.0 | 86.5 | 84.0 | 87.5 | 85.5 |
| 7.5 | 85.5 | 82.5 | 88.5 | 86.5 | 88.5 | 86.5 | 88.5 | 86.5 |
| 10.0 | 87.5 | 85.5 | 88.5 | 86.5 | 90.2 | 88.5 | 89.5 | 87.5 |
| 15.0 | 89.5 | 87.5 | 90.2 | 88.5 | 89.5 | 87.5 | 89.5 | 87.5 |
| 20.0 | 90.2 | 88.5 | 91.0 | 89.5 | 90.2 | 88.5 | 90.2 | 88.5 |
| 25.0 | 91.0 | 89.5 | 91.7 | 90.2 | 91.0 | 89.5 | 90.2 | 88.5 |
| 30.0 | 91.0 | 89.5 | 91.7 | 90.2 | 91.7 | 90.2 | 91.0 | 89.5 |
| 40.0 | 91.7 | 90.2 | 92.4 | 91.0 | 91.7 | 90.2 | 90.2 | 88.5 |
| 50.0 | 91.7 | 90.2 | 92.4 | 91.0 | 91.7 | 90.2 | 91.7 | 90.2 |
| 60.0 | 93.0 | 91.7 | 93.0 | 91.7 | 92.4 | 91.0 | 92.4 | 91.0 |
| 75.0 | 93.0 | 91.7 | 93.6 | 92.4 | 93.0 | 91.7 | 93.6 | 92.4 |
| 100.0 | 93.0 | 91.7 | 93.6 | 92.4 | 93.6 | 92.4 | 93.6 | 92.4 |
| 125.0 | 93.0 | 91.7 | 93.6 | 92.4 | 93.6 | 92.4 | 93.6 | 92.4 |
| 150.0 | 93.6 | 92.4 | 94.1 | 93.0 | 93.6 | 92.4 | 93.6 | 92.4 |
| 200.0 | 93.6 | 92.4 | 94.1 | 93.0 | 94.1 | 93.0 | 93.6 | 92.4 |

### ENCLOSED MOTORS

| HP | 2 POLE (3600) | | 4 POLE (1800) | | 6 POLE (1200) | | 8 POLE (900) | |
|---|---|---|---|---|---|---|---|---|
| | NOMINAL EFFICIENCY | MINIMUM EFFICIENCY | NOMINAL EFFICIENCY | MINIMUM EFFICIENCY | NOMINAL EFFICIENCY | MINIMUM EFFICIENCY | NOMINAL EFFICIENCY | MINIMUM EFFICIENCY |
| 1.0 | ------- | ------- | 80.5 | 77.0 | 75.5 | 72.0 | 72.0 | 68.0 |
| 1.5 | 78.5 | 75.5 | 81.5 | 78.5 | 82.5 | 80.0 | 75.5 | 72.0 |
| 2.0 | 81.5 | 78.5 | 82.5 | 80.0 | 82.5 | 80.0 | 82.5 | 80.0 |
| 3.0 | 82.5 | 80.0 | 84.0 | 81.5 | 84.0 | 81.5 | 81.5 | 78.5 |
| 5.0 | 85.5 | 82.5 | 85.5 | 82.5 | 85.5 | 82.5 | 84.0 | 81.5 |
| 7.5 | 85.5 | 82.5 | 87.5 | 85.5 | 87.5 | 85.5 | 85.5 | 82.5 |
| 10.0 | 87.5 | 85.5 | 87.5 | 85.5 | 87.5 | 85.5 | 87.5 | 85.5 |
| 15.0 | 87.5 | 85.5 | 88.5 | 86.5 | 89.5 | 87.5 | 88.5 | 86.5 |
| 20.0 | 88.5 | 86.5 | 90.2 | 88.5 | 89.5 | 87.5 | 89.5 | 87.5 |
| 25.0 | 89.5 | 87.5 | 91.0 | 89.5 | 90.2 | 88.5 | 89.5 | 87.5 |
| 30.0 | 89.5 | 87.5 | 91.0 | 89.5 | 91.0 | 89.5 | 90.2 | 88.5 |
| 40.0 | 90.2 | 88.5 | 91.7 | 90.2 | 91.7 | 90.2 | 90.2 | 88.5 |
| 50.0 | 90.2 | 88.5 | 92.4 | 91.0 | 91.7 | 90.2 | 91.0 | 89.5 |
| 60.0 | 91.7 | 90.2 | 93.0 | 91.7 | 91.7 | 90.2 | 91.7 | 90.2 |
| 70.0 | 92.4 | 91.0 | 93.0 | 91.7 | 93.0 | 91.7 | 93.0 | 91.7 |
| 100.0 | 93.0 | 91.7 | 93.6 | 92.4 | 93.0 | 91.7 | 93.0 | 91.7 |
| 125.0 | 93.0 | 91.7 | 93.6 | 92.4 | 93.0 | 91.7 | 93.6 | 92.4 |
| 150.0 | 93.0 | 91.7 | 94.1 | 93.0 | 94.1 | 93.0 | 93.6 | 92.4 |
| 200.0 | 94.1 | 93.0 | 94.5 | 93.6 | 94.1 | 93.0 | 94.1 | 93.0 |

## Table 11.4 (NEMA Table 12.6C) (Suggested standard for future design)
## Full-load efficiencies of energy efficient motors.

### OPEN MOTORS

| HP | 2 POLE (3600) | | 4 POLE (1800) | | 6 POLE (1200) | | 8 POLE (900) | |
|---|---|---|---|---|---|---|---|---|
| | NOMINAL EFFICIENCY | MINIMUM EFFICIENCY | NOMINAL EFFICIENCY | MINIMUM EFFICIENCY | NOMINAL EFFICIENCY | MINIMUM EFFICIENCY | NOMINAL EFFICIENCY | MINIMUM EFFICIENCY |
| 1.0 | ------- | ------- | 82.5 | 81.5 | 80.0 | 78.5 | 74.0 | 72.0 |
| 1.5 | 82.5 | 81.5 | 84.0 | 82.5 | 84.0 | 82.5 | 75.5 | 74.0 |
| 2.0 | 84.0 | 82.5 | 84.0 | 82.5 | 85.5 | 84.0 | 85.5 | 84.0 |
| 3.0 | 84.0 | 82.5 | 86.5 | 85.5 | 86.5 | 85.5 | 86.5 | 85.5 |
| 5.0 | 85.5 | 84.0 | 87.5 | 86.5 | 87.5 | 86.5 | 87.5 | 86.5 |
| 7.5 | 87.5 | 86.5 | 88.5 | 87.5 | 88.5 | 87.5 | 88.5 | 87.5 |
| 10.0 | 88.5 | 87.5 | 89.5 | 88.5 | 90.2 | 89.5 | 89.5 | 88.5 |
| 15.0 | 89.5 | 88.5 | 91.0 | 90.2 | 90.2 | 89.5 | 89.5 | 88.5 |
| 20.0 | 90.2 | 89.5 | 91.0 | 90.2 | 91.0 | 90.2 | 90.2 | 89.5 |
| 25.0 | 91.0 | 90.2 | 91.7 | 91.0 | 91.7 | 91.0 | 90.2 | 89.5 |
| 30.0 | 91.0 | 90.2 | 92.4 | 91.7 | 92.4 | 91.7 | 91.0 | 90.2 |
| 40.0 | 91.7 | 91.0 | 93.0 | 92.4 | 93.0 | 92.4 | 91.0 | 90.2 |
| 50.0 | 92.4 | 91.7 | 93.0 | 92.4 | 93.0 | 92.4 | 91.7 | 91.0 |
| 60.0 | 93.0 | 92.4 | 93.6 | 93.0 | 93.6 | 93.0 | 92.4 | 91.7 |
| 75.0 | 93.0 | 92.4 | 94.1 | 93.6 | 93.6 | 93.0 | 93.6 | 93.0 |
| 100.0 | 93.0 | 92.4 | 94.1 | 93.6 | 94.1 | 93.6 | 93.6 | 93.0 |
| 125.0 | 93.6 | 93.0 | 94.5 | 94.1 | 94.1 | 93.6 | 93.6 | 93.0 |
| 150.0 | 93.6 | 93.0 | 95.0 | 94.5 | 94.5 | 94.1 | 93.6 | 93.0 |
| 200.0 | 94.5 | 94.1 | 95.0 | 94.5 | 94.5 | 94.1 | 93.6 | 93.0 |

### ENCLOSED MOTORS

| HP | 2 POLE (3600) | | 4 POLE (1800) | | 6 POLE (1200) | | 8 POLE (900) | |
|---|---|---|---|---|---|---|---|---|
| | NOMINAL EFFICIENCY | MINIMUM EFFICIENCY | NOMINAL EFFICIENCY | MINIMUM EFFICIENCY | NOMINAL EFFICIENCY | MINIMUM EFFICIENCY | NOMINAL EFFICIENCY | MINIMUM EFFICIENCY |
| 1.0 | 75.5 | 74.0 | 82.5 | 81.5 | 80.0 | 78.5 | 74.0 | 72.0 |
| 1.5 | 82.5 | 81.5 | 84.0 | 82.5 | 85.5 | 84.0 | 77.0 | 75.5 |
| 2.0 | 84.0 | 82.5 | 84.0 | 82.5 | 86.5 | 85.5 | 82.5 | 81.5 |
| 3.0 | 85.5 | 84.0 | 87.5 | 86.5 | 87.5 | 86.5 | 84.0 | 82.5 |
| 5.0 | 87.5 | 86.5 | 87.5 | 86.5 | 87.5 | 86.5 | 85.5 | 84.0 |
| 7.5 | 88.5 | 87.5 | 89.5 | 88.5 | 89.5 | 88.5 | 85.5 | 84.0 |
| 10.0 | 89.5 | 88.5 | 89.5 | 88.5 | 89.5 | 88.5 | 88.5 | 87.5 |
| 15.0 | 90.2 | 89.5 | 91.0 | 90.2 | 90.2 | 89.5 | 88.5 | 87.5 |
| 20.0 | 90.2 | 89.5 | 91.0 | 90.2 | 90.2 | 89.5 | 89.5 | 88.5 |
| 25.0 | 91.0 | 90.2 | 92.4 | 91.7 | 91.7 | 91.0 | 89.5 | 88.5 |
| 30.0 | 91.0 | 90.2 | 92.4 | 91.7 | 91.7 | 91.0 | 91.0 | 90.2 |
| 40.0 | 91.7 | 91.0 | 93.0 | 92.4 | 93.0 | 92.4 | 91.0 | 90.2 |
| 50.0 | 92.4 | 91.7 | 93.0 | 92.4 | 93.0 | 92.4 | 91.7 | 91.0 |
| 60.0 | 93.0 | 92.4 | 93.6 | 93.0 | 93.6 | 93.0 | 91.7 | 91.0 |
| 75.0 | 93.0 | 92.4 | 94.1 | 93.6 | 93.6 | 93.0 | 93.0 | 92.4 |
| 100.0 | 93.6 | 93.0 | 94.5 | 94.1 | 94.1 | 93.6 | 93.0 | 92.4 |
| 125.0 | 94.5 | 94.1 | 94.5 | 94.1 | 94.1 | 93.6 | 93.6 | 93.0 |
| 150.0 | 94.5 | 94.1 | 95.0 | 94.5 | 95.0 | 94.5 | 93.6 | 93.0 |
| 200.0 | 95.0 | 94.5 | 95.0 | 94.5 | 95.0 | 94.5 | 94.1 | 93.6 |

Same horsepower—different efficiency.

$$kW_{saved} = hp \times 0.746 \times \left(\frac{100}{E_{STD}} - \frac{100}{E_{EE}}\right)$$

Same horsepower and % load—different efficiency

$$kW_{saved} = hp \times 0.746 \times L \times \left(\frac{100}{E_{STD}} - \frac{100}{E_{EE}}\right)$$

Annual $ savings due to difference in efficiency

$$S = hp \times 0.746 \times L \times C \times N \times \left(\frac{100}{E_{STD}} - \frac{100}{E_{EE}}\right)$$

| | | *Example* |
|---|---|---|
| S | = $ Savings (annual) | 100 |
| hp | = Horsepower | 100 |
| L | = % Load | 100 |
| C | = Energy cost ($/kWh) | 0.08 |
| N | = Operating house (annual) | 4000 |
| $E_{STD}$ | = % Efficiency of standard motor | 91.7 |
| $E_{EE}$ | = % Efficiency of energy eff. motor | 95.0 |
| $RPM_{STD}$ | = Speed of standard motor | 1775 |
| $RPM_{EE}$ | = Speed of energy eff. motor | 1790 |

*For loads sensitive to motor speed*

Above equations should be multiplied by speed ratio correction factor.

*SRCF = Speed Ratio Correction Factor*

$$\left(\frac{RPM_{EE}}{RPM_{STD}}\right)^3$$

*Example:*

S = 100 × 0.746 × 1 × 0.080 × 4000

     × (100/91.7-100/95.0) = $904

S = 100 × 0.746 × 1 × 0.080 × 4000

     × (100/91.7-100/95.0) × (1790/1775)³ = $262

         $ 642 reduction in expected savings.

Relatively minor, 15 RPM, increase in a motor's rotational speed results in a 2.6 percent increase in the load placed upon the motor by the rotating equipment.

## 11.16 SENSITIVITY OF LOAD TO MOTOR RPM

When employing electric motors, for air moving equipment, it is important to remember that the performance of fans and blowers is governed by certain rules of physics. These rules are known as "The Affinity Laws" or "The Fan Laws." There are several parts to it, and are all related to each other in a known manner and when one changes, all others change. For centrifugal loads, even a minor change in the motor's speed translates into significant change in energy consumption and is especially troublesome when the additional air flow is not needed or useful. Awareness of the sensitivity of load and energy requirements to motor speed can help effectively identify motors with specific performance requirements. In most cases we can capture the full energy conservation benefits associated with an energy efficient motor retrofits.

**Terminology of Load to Motor RPM**

CFM   Fan capacity (Cubic Feet per Minute)
       Volume of air moved by the fan per unit of time.

P     Pressure. Pressure produced by the fan that can exist whether the air is in motion or confined in a closed duct.

HP    Horsepower. The power required to drive an air moving device.

RPM   Revolutions Per Minute. The speed at which the shaft of air moving equipment is rotating.

**Affinity Laws or Fan Laws**

$$\text{Law \#1} \quad \frac{CFM_2}{CFM_1} = \frac{RPM_2}{RPM_1}$$

Quantity (CFM) varies as fan speed (RPM)

$$\text{Law \#2} \quad \frac{P_2}{P_1} = \frac{(RPM_2)^2}{(RPM_1)^2}$$

Pressure (P) varies as the <u>square</u> of fan speed (RPM)

$$\text{Law \#3} \quad \frac{HP_2}{HP_1} = \frac{(RPM_2)^3}{(RPM_1)^3}$$

Horsepower (HP) varies as the <u>cube</u> of fan speed (RPM)

*Example*

Fan system <u>32,000</u> CFM

Motor <u>20</u> HP <u>1750</u> RPM (existing)

Motor <u>20</u> HP <u>1790</u> RPM (new EE)

kW = 20 × 0.746 = 14.92 kW
New CFM with new motor = 1790/1750 × 32,000 = 32,731
  or 2.3% increase
New HP = (1790/1750)3 × 20 × = 21.4 HP or 7% increase.
New kW = 21.4 × 0.746 = 15.96 kW 7% increase in kW and
  work performed by motor.

Replacing a standard motor with an energy efficient motor in centrifugal pump or a fan application can result in increased energy consumption if the energy efficient motor operates at a higher RPM. Table 11.5 shows how a 10 RPM increase can negate any savings associated with a high efficiency motor retrofit.

## 11.17 THEORETICAL POWER CONSUMPTION

Figure 11.6 illustrates the energy saving potential of the application of *Adjustable Speed Drive* to an application that traditionally uses throttling control, such as *Discharge Damper*, *Variable Inlet Vane* or *Eddy Current Drive*.

From the standpoint of maximum energy conservation, the most optimal method to reduce fan CFM is to reduce the fan's speed (RPM). This can be accomplished by changing either, the sheaves of the motor, the sheaves of the fan or by varying fan motor speed.

**Applications Involving Extended Periods of Light Load Operation[2]**

A number of methods have been proposed to reduce the voltage applied to the motor in response to the applied load, the purpose of this being to reduce the magnetizing losses during the periods when the full torque capability of the motor is not required. Typical of these devices is the power factor controller. The power factor controller is a device that adjusts the voltage applied to the motor to approximate a preset power factor.

These power factor controllers may, for example, be beneficial for use with small motors operating for extended periods of light loads where the magnetization losses are a relatively high percentage of the total loss. Care must be exercised in the application of these controllers. Savings are achieved only when the controlled motor is operated for extended periods at no load or light load.

Particular care must be taken when considering their use with other than small motors. A typical 10 horsepower motor will have idle losses in the order of 4 or 5 percent of the rated output. In this size range the magnetization losses that can be saved may not be equal to the losses added by the controller plus the additional motor losses caused by the distorted voltage wave form induced by the controller.

**Applications Involving Throttling or By-pass Control.[2]**

Many pump and fan applications involve the control of flow or pressure by means of throttling or bypass devices. Throttling and bypass valves are in effect series and parallel power regulators that perform their function by dissipating the difference between source energy supplied and the desired sink energy.

These losses can be dramatically reduced by controlling the flow rate or pressure by controlling the speed of the pump or fan with a variable speed drive.

Figure 11.6 illustrates the energy saving potential of the application of variable speed drive to an application that traditionally uses throttling control.

A simplistic example will serve to illustrate the savings to be achieved by the use of this powerful energy conservation tool.

Assumptions:
Line Power Required by Fan at full CFM
  without flow control device 100 HP
Full CFM required                1000 hours per year
75% CFM required                 3000 hours per year
50% CFM required                 2000 hours per year
Cost of energy                   $.06 per kWhr

% Power consumption with various flow control methods per Figure 11.6.

**Figure 11.6**

Table 11.5 Hourly operating costs.

# HOURLY OPERATING COSTS
## 100% Load

A MINOR RPM INCREASE IN A MOTOR'S ROTATIONAL SPEED RESULTS IN AN INCREASE IN THE LOAD PLACED UPON THE MOTOR BY ROTATING EQUIPMENT.

| HP | AVERAGE STANDARD MOTOR FL RPM EFF. % | ENERGY EFFICIENT MOTOR SAME RPM EFF. % | ENERGY EFFICIENT MOTOR + 10 RPM EFF. % | $/kWh $0.01 | AVERAGE ELECTRICITY PRICE ( $/kWh) $0.05 | $0.06 | $0.07 | $0.08 | $0.09 | $0.10 | $0.11 |
|---|---|---|---|---|---|---|---|---|---|---|---|
| 5 | 83.8 | | | 0.0445 | $0.22 | $0.27 | $0.31 | $0.36 | $0.40 | $0.45 | $0.49 |
| 5 | | 86.5 | | 0.0431 | $0.22 | $0.26 | $0.30 | $0.34 | $0.39 | $0.43 | $0.47 |
| 5 | | | 86.5 | 0.0439 | $0.22 | $0.26 | $0.31 | $0.35 | $0.39 | $0.44 | $0.48 |
| 7.5 | 85.3 | | | 0.0656 | $0.33 | $0.39 | $0.46 | $0.52 | $0.59 | $0.66 | $0.72 |
| 7.5 | | 88.5 | | 0.0632 | $0.32 | $0.38 | $0.44 | $0.51 | $0.57 | $0.63 | $0.70 |
| 7.5 | | | 88.5 | 0.0643 | $0.32 | $0.39 | $0.45 | $0.51 | $0.58 | $0.64 | $0.71 |
| 10 | 87.2 | | | 0.0856 | $0.43 | $0.51 | $0.60 | $0.68 | $0.77 | $0.86 | $0.94 |
| 10 | | 88.5 | | 0.0843 | $0.42 | $0.51 | $0.59 | $0.67 | $0.76 | $0.84 | $0.93 |
| 10 | | | 88.5 | 0.0857 | $0.43 | $0.51 | $0.60 | $0.69 | $0.77 | $0.86 | $0.94 |
| 15 | 87.6 | | | 0.1277 | $0.64 | $0.77 | $0.89 | $1.02 | $1.15 | $1.28 | $1.41 |
| 15 | | 90.2 | | 0.1241 | $0.62 | $0.74 | $0.87 | $0.99 | $1.12 | $1.24 | $1.36 |
| 15 | | | 90.2 | 0.1262 | $0.63 | $0.76 | $0.88 | $1.01 | $1.14 | $1.26 | $1.39 |
| 20 | 88.4 | | | 0.1688 | $0.84 | $1.01 | $1.18 | $1.35 | $1.52 | $1.69 | $1.86 |
| 20 | | 91.0 | | 0.1640 | $0.82 | $0.98 | $1.15 | $1.31 | $1.48 | $1.64 | $1.80 |
| 20 | | | 91.0 | 0.1667 | $0.83 | $1.00 | $1.17 | $1.33 | $1.50 | $1.67 | $1.83 |
| 25 | 89.2 | | | 0.2091 | $1.05 | $1.25 | $1.46 | $1.67 | $1.88 | $2.09 | $2.30 |
| 25 | | 91.7 | | 0.2034 | $1.02 | $1.22 | $1.42 | $1.63 | $1.83 | $2.03 | $2.24 |
| 25 | | | 91.7 | 0.2068 | $1.03 | $1.24 | $1.45 | $1.65 | $1.86 | $2.07 | $2.28 |
| 30 | 89.2 | | | 0.2509 | $1.25 | $1.51 | $1.76 | $2.01 | $2.26 | $2.51 | $2.76 |
| 30 | | 91.7 | | 0.2441 | $1.22 | $1.46 | $1.71 | $1.95 | $2.20 | $2.44 | $2.68 |
| 30 | | | 91.7 | 0.2482 | $1.24 | $1.49 | $1.74 | $1.99 | $2.23 | $2.48 | $2.73 |
| 40 | 90.2 | | | 0.3308 | $1.65 | $1.98 | $2.32 | $2.65 | $2.98 | $3.31 | $3.64 |
| 40 | | 92.4 | | 0.3229 | $1.61 | $1.94 | $2.26 | $2.58 | $2.91 | $3.23 | $3.55 |
| 40 | | | 92.4 | 0.3284 | $1.64 | $1.97 | $2.30 | $2.63 | $2.96 | $3.28 | $3.61 |
| 50 | 90.1 | | | 0.4140 | $2.07 | $2.48 | $2.90 | $3.31 | $3.73 | $4.14 | $4.55 |
| 50 | | 92.4 | | 0.4037 | $2.02 | $2.42 | $2.83 | $3.23 | $3.63 | $4.04 | $4.44 |
| 50 | | | 92.4 | 0.4105 | $2.05 | $2.46 | $2.87 | $3.28 | $3.69 | $4.11 | $4.52 |
| 60 | 91.0 | | | 0.4919 | $2.46 | $2.95 | $3.44 | $3.93 | $4.43 | $4.92 | $5.41 |
| 60 | | 93.0 | | 0.4813 | $2.41 | $2.89 | $3.37 | $3.85 | $4.33 | $4.81 | $5.29 |
| 60 | | | 93.0 | 0.4895 | $2.45 | $2.94 | $3.43 | $3.92 | $4.41 | $4.89 | $5.38 |
| 75 | 91.9 | | | 0.6088 | $3.04 | $3.65 | $4.26 | $4.87 | $5.48 | $6.09 | $6.70 |
| 75 | | 93.6 | | 0.5978 | $2.99 | $3.59 | $4.18 | $4.78 | $5.38 | $5.98 | $6.58 |
| 75 | | | 93.6 | 0.6079 | $3.04 | $3.65 | $4.26 | $4.86 | $5.47 | $6.08 | $6.69 |
| 100 | 91.7 | | | 0.8135 | $4.07 | $4.88 | $5.69 | $6.51 | $7.32 | $8.14 | $8.95 |
| 100 | | 93.6 | | 0.7970 | $3.99 | $4.78 | $5.58 | $6.38 | $7.17 | $7.97 | $8.77 |
| 100 | | | 93.6 | 0.8106 | $4.05 | $4.86 | $5.67 | $6.48 | $7.30 | $8.11 | $8.92 |
| 125 | 91.7 | | | 1.0169 | $5.08 | $6.10 | $7.12 | $8.14 | $9.15 | $10.17 | $11.19 |
| 125 | | 93.6 | | 0.9963 | $4.98 | $5.98 | $6.97 | $7.97 | $8.97 | $9.96 | $10.96 |
| 125 | | | 93.6 | 1.0132 | $5.07 | $6.08 | $7.09 | $8.11 | $9.12 | $10.13 | $11.15 |
| 150 | 92.9 | | | 1.2045 | $6.02 | $7.23 | $8.43 | $9.64 | $10.84 | $12.05 | $13.25 |
| 150 | | 94.1 | | 1.1892 | $5.95 | $7.13 | $8.32 | $9.51 | $10.70 | $11.89 | $13.08 |
| 150 | | | 94.1 | 1.2094 | $6.05 | $7.26 | $8.47 | $9.68 | $10.88 | $12.09 | $13.30 |
| 200 | 93.1 | | | 1.6026 | $8.01 | $9.62 | $11.22 | $12.82 | $14.42 | $16.03 | $17.63 |
| 200 | | 94.1 | | 1.5855 | $7.93 | $9.51 | $11.10 | $12.68 | $14.27 | $15.86 | $17.44 |
| 200 | | | 94.1 | 1.6125 | $8.06 | $9.68 | $11.29 | $12.90 | $14.51 | $16.13 | $17.74 |

**100 HP MOTOR; 91.7%EFFICIENT;**
**FL SPEED 1775 RPM**
**OPERATES 1000 HOURS PER YEAR;**
**ENERGY COST $0.08 PER kWh**

• COST TO OPERATE THIS MOTOR = $ 6,508

---

**100 HP MOTOR; 91.7%EFFICIENT;**
**FL SPEED 1775 RPM**
**OPERATES 1000 HOURS PER YEAR;**
**ENERGY COST $0.08 PER kWh**

WITH EFFICIENCY IMPROVEMENT
OF 5%
from 91.7% to 96%
COST TO OPERATE THIS MOTOR $ 6,198

$310 SAVINGS

WITH INVESTMENT OF $ ??????

---

**100 HP MOTOR; 91.7%EFFICIENT;**
**FL SPEED 1775 RPM**
**OPERATES 1000 HOURS PER YEAR;**
**ENERGY COST $0.08 PER kWh**

WITH REDUCTION IN OPERATING
HOURS
OF 5%
from 1000 to 950
COST TO OPERATE THIS MOTOR= $6,183
$325 SAVINGS
WITH INVESTMENT OF $ ??????

---

**100 HP MOTOR; 91.7%EFFICIENT;**
**FL SPEED 1775 RPM**
**OPERATES 1000 HOURS PER YEAR;**
**ENERGY COST $0.08 PER kWh**

WITH LOAD SPEED REDUCTION
OF 5%
from 1775 to 1686
COST TO OPERATE THIS MOTOR $ 5,580
$928 SAVINGS
WITH INVESTMENT OF $ ??????

Annual cost of Energy

$$\$ = \frac{\text{hrs x hp x 746 x \% energy consumption x \$/kW}}{1000 \times 100}$$

Annual Cost of Energy Summary:

| Discharge Damper | $24,600 |
| Variable Inlet Vane | $19,600 |
| Eddy Current Drive | $15,900 |
| Adjustable Speed Drive | $13,900 |

## 11.18  MOTOR EFFICIENCY MANAGEMENT

Many think that when one is saying *Motor Efficiency* the logical word to follow is *improvement*. Where are we going? How far can we push manufacturers in our quest for the perfect motor?

During the 102nd Congress, the Markey Bill, H.R. 2451, was introduced. The bill mandated component efficiency standards for such products as lighting, distribution transformers and electric A.C. motors.

This plan was met with opposition by NEMA and other interested groups. They called for a system approach that would recognize the complex nature of the product involved under the plan. The bill passed by the Energy & Power Subcommittee on the theory that the elimination of the least efficient component from the market would ensure that consumers would purchase and use the most efficient products possible.

Although motors tend to be quite efficient in themselves, several factors can contribute to cost-effective replacement or retrofit alternatives to obtain efficiency gain in motors. We are well aware that the electric motor's primary function is to convert electrical energy into mechanical work. It is also important to remember that *good energy management requires a consideration of the total system of which the motor is a part*.

Experience indicates that despite heightened awareness and concern with energy efficiency, the electric motor is either completely neglected or decisions are made on the basis of incomplete information. At this point I would like to quote me.

*"Motors Don't Waste Energy, People Do."*

What this really means is that we must start managing efficiency and not just improving the motor. This is what will improve your corporate bottom line.

## 11.19  MOTORS ARE LIKE PEOPLE

Motors can be managed the same way and with the same skills as people. There are amazing similarities. I have spent years managing both and find there is very little difference between the two.

The expectations are the same for one as for the other. The employee's performance is evaluated to identify improvement opportunities linking them to organizational goals and business objectives. The manager measures the performance of an employee as an individual and as a member of a team. Why then would it not work the same with your motors? Motors employed just as people are and they work as an individual or as a team. They will perform their best if cared for, maintained, evaluated and rewarded.

An on going analysis of motor performance prevents major breakdown. Performance evaluation of a motor should be done as routinely as it is done on an employee. Both the motor and an employee are equally important. Applied motor maintenance will keep the building or plant running smoothly with minimal stress on the system or downtime due to failure.

**MAXIMIZE YOUR EFFECTIVENESS
WITH MOTOR EVALUATION SYSTEM**

## 11.20  MOTOR PERFORMANCE MANAGEMENT PROCESS (MPMP)

The Motor Performance Management Process (MPMP) is designed to be the Motor Manager's primary tool to evaluate, measure and most importantly *manage electric motors*. MPMP focuses on building a stronger relationship between Motor Manager and the electric motor *employed* to perform a task. Specifically, it is a logical, systematic and structured approach to reduce energy waste. Energy waste reduction is fundamental in becoming more efficient in an increasingly competitive market. The implementation of MPMP is more than a good business practice it is an intelligent management resource.

→NEGLECTING YOUR MOTORS
CAN BE A COSTLY MISTAKE←

Motor Managers must understand motor efficiency, how it is achieved and how to conduct an economic evaluation. Timely implementation of MPMP would be an effective way to evaluate existing motor performance in a system and identify improvement opportunities linking them to organizational goals. The following

*Motor Manager* model defines the task of managing and enabling motors to operate in the ways required to achieve business objectives.

$$\frac{Motor}{Manager} \Rightarrow Motor \Rightarrow \frac{Desired}{Operation} \Rightarrow \frac{Business}{Objectives}$$

It is vital to evaluate the differences between motors offered by various manufacturers and only choose those that clearly meet your operating criteria. An investment of 20 to 25% for an Energy Efficient motor, over the cost of a standard motor, can be recovered in a relatively short period of time. Furthermore, with some motors the cost of wasted energy exceeds the original motor price even in the first year of operation.

## 11.21  HOW TO START MPMP

Begin by conscientiously gathering information on each motor used in excess of 2000 hours per year.

Complete a MOTOR RECORD FORM (Figure 11.7) for each motor. (*A detailed profile of your existing Motor.*)

- This form must be established for each motor to serve as a performance record. It will help you to understand WHEN, WHERE, and HOW your motors are used, and identify opportunities to improve drive system efficiency.

- Each item in this form must be addressed, paying particular attention to the following items: *Motor Location, Application, Energy Cost $/kWh, Operating Speed, Operating Load* and *Nameplate information.*

- Motors with special electrical designs or mechanical features should be studied carefully. Some applications require special attention, such as fans, compressors and pumps.

- Motors with a history of repeated repair should be of special interest.

- Examine your completed MOTOR RECORD FORM and select the best candidates for possible retrofit or future replacement.

- Check with your local utility regarding the availability of financial incentives or motor rebates.

Finding the right motor for the application, and calculating its energy and cost savings can be done with the MotorMaster[5] software and WHAT IF motor comparison form Figure 11.8. This form has the capability to compare several motors and analyze potential savings.

| SCENARIO | INPUT | LINE | MOTOR | BELT | FAN | TOTAL |
|---|---|---|---|---|---|---|
| | LOAD % | | | EFFICIENCY % | | |
| 1 | 100.00 | 99.00 | 85.00 | 95.00 | 65.00 | 51.96 |
| 2 | 100.00 | 99.00 | 90.00 | 95.00 | 65.00 | 55.02 |
| 3 | 102.56* | 99.00 | 90.00 | 95.00 | 65.00 | 55.02 |
| | | | KW | | | |
| 1 | 74.60 | 75.35 | 88.65 | 93.32 | 143.56 | 143.56 |
| 2 | 74.60 | 75.35 | 83.73 | 88.13 | 135.59 | 135.59 |
| 3 | 76.51 | 77.28 | 85.87 | 90.39 | 139.06 | 139.06 |
| 1 | $5.97 | $6.03 | $7.09 | $7.47 | $11.49 | $11.49 |
| 2 | $5.97 | $6.03 | $6.70 | $7.05 | $10.85 | $10.85 |
| 3 | $6.12 | $6.18 | $6.87 | $7.23 | $11.12 | $11.12 |
| 1 | $0.00 | $0.06 | $1.06 | $0.37 | $4.02 | $5.52 |
| 2 | $0.00 | $0.06 | $0.67 | $0.35 | $3.80 | $4.88 |
| 3 | $0.00 | $0.06 | $0.69 | $0.36 | $3.89 | $5.00 |

SYSTEM EFFICIENCY IS THE PRODUCT OF INDIVIDUAL EFFICIENCIES

SCENARIO 1
100 HP 1775 RPM 85% EFF

SCENARIO 2
100 HP 1775 RPM 90% EFF

SCENARIO 3
100 HP 1790 RPM 90% EFF

ENERGY COST $ 0.08 /KWH

* FOR LOADS SENSITIVE TO MOTOR SPEED

**System efficiency is the product of individual efficiencies.**

**For example, out of $5.00 total motor losses (Scenario 3) only 69 cents is from the motor itself. The belt and fan account for $3.89 + .36 = $4.25 of the total losses.**

**(The kW amounts shown in the table in the graphic are the sums of the previous components.)**

K.K.LOBODOVSKY 11/1993
(MM_MRF.DOC & XLS)

**MotorManager**

**MOTOR RECORD FORM**

| CUSTOMER NAME: | PREPARED BY: |
|---|---|
| MOTOR LOCATION: | DATE PREPARED: |
| APPLICATION: | ENERGY COST $/KWH |

| MOTOR MANUFACTURER | HP | OPERATING HOURS | % EFF 100% LOAD | % EFF 75% LOAD | % EFF 50% LOAD | % EFF 25% LOAD |
|---|---|---|---|---|---|---|
| MODEL NO. | DESIGN | RPM (FULL LOAD) | VOLTS | AMPS (FULL LOAD) | HZ | SERVICE FACTOR |
| SERIAL NO. | YEAR PURCHASED | FRAME | INSULATION CLASS | TYPE | CODE | PHASE |
| NAME: | YEAR REWOUND | DUTY | POWER FACTOR | | | |

| NO LOAD DATA<br>FROM MOTOR MANUFACTURER | KW | AMPS | VOLTS | CORE LOSS | F. & W. LOSS | STATOR RESIST. |
|---|---|---|---|---|---|---|

| STATOR RESISTANCE DATA | A - B | B - C | C - A | CONDUCTOR | MCM: | AWG : |
|---|---|---|---|---|---|---|
| MEASURED RESISTANCE -> | | | | MATERIAL | CU: | AL: |
| MOTOR SURFACE TEMPERATURE-> | | <-DEG F   DEG C -> | | DISTANCE POINT. TO MOTOR IN FEET -> | | |
| AMBIENT TEMPERATURE-> | | <-DEG F   DEG C -> | | WIRE RESISTANCE IN OHMS / 1000FT-> | | |

| | LOAD TEST | | | NO LOAD TEST | | |
|---|---|---|---|---|---|---|
| TEST DATE | | | | | | |
| CURRENT TRANSFORMER | | | | | | |
| SPEED (RPM) | | | | | | |
| VOLTS A - B | | | | | | |
| VOLTS B - C | | | | | | |
| AMPS A | | | | | | |
| AMPS C | | | | | | |
| KW TOTAL | | | | | | |
| KVAR TOTAL | | | | | | |
| KVA TOTAL | | | | | | |
| PF TOTAL | | | | | | |

**BE = BLUE**
**RD = RED**
**BK = BLACK**
**WE = WHITE**

☐ VOLTAGE LEADS

◯ CURRENT LEADS

WHITE PLUG TOWARDS LOAD

SOURCE

L1   L2   L3

☐RD  ☐WE  ☐BE

◯RD  ◯BK  ◯BE

LOAD

**REMARKS:**

Figure 11.7 Motor record form.

# WHAT IF

$$S = hp \times 0.746 \times L \times C \times N \times \left( \frac{100}{E_{STD}} - \frac{100}{E_{EE}} \right)$$

S=$ SAVINGS  hp= HORSEPOWER  L=%LOAD  C=ENERGY COST  N=OPERATING HRS  E=%EFFICIENCY STD. OR EE.

| | EXISTING | PROPOSAL 1 | PROPOSAL 2 | PROPOSAL 3 | PROPOSAL 4 | PROPOSAL 5 | LINE | FORMULA (L1 = Line 1...) |
|---|---|---|---|---|---|---|---|---|
| ENERGY COST $/KWH | $0.080 | | | | | EXAMPLE | 1 | C |
| OPERATING HRS (PER YEAR) | 4000 | | | | | Baldor | 2 | N |
| MOTOR HP | 100 | | | | | EM4400T-8 | 3 | hp |
| NO-LOAD SPEED RPM | 1800 | | | | | | 4 | NL RPM |
| FULL-LOAD SPEED (SEE Note) | 1775 | | | | | 1780 | 5 | FL RPM |
| Note: FULL-LOAD SPEED information is important for loads sensitive to motor speed. | | | | | | | | |
| EFFICIENCY @ 100% LOAD | 91.7 | | | | | 95.0 | 6 | EFFICIENCY @ 100% LOAD |
| EFFICIENCY @ 75% LOAD | 91.7 | | | | | 95.4 | 7 | EFFICIENCY @ 75% LOAD |
| EFFICIENCY @ 50% LOAD | 91.7 | | | | | 95.5 | 8 | EFFICIENCY @ 50% LOAD |
| INVESTMENT / SALVAGE | $0 | | | | | $7,180 | 9 | INVESTMENT / SALVAGE |
| SLIP | 25 | | | | | 20 | 10 | L4-L5 |
| CALCULATED SPEED @ 75% LOAD | 1,781 | | | | | 1,785 | 11 | L5+(L10*0.25) |
| CALCULATED SPEED @ 50% LOAD | 1,788 | | | | | 1,790 | 12 | L5+(L10*0.5) |
| FOR LOADS NOT SENSITIVE TO MOTOR SPEED | | PROPOSAL 1 | PROPOSAL 2 | PROPOSAL 3 | PROPOSAL 4 | PROPOSAL 5 | LINE | |
| KW @ 100% EFF & LOAD | 81.4 | | | | | 78.5 | 13 | L3*0.746*100/L6 |
| KW @ 75% EFF & LOAD | 61.0 | | | | | 58.6 | 14 | L3*0.746*0.75*100/L7 |
| KW @ 50% EFF & LOAD | 40.7 | | | | | 39.1 | 15 | L3*0.746*0.5*100/L8 |
| KWh @ 100% EFF & LOAD | 325,409 | | | | | 314,105 | 16 | L2*L13 |
| KWh @ 75% EFF & LOAD | 244,057 | | | | | 234,591 | 17 | L2*L14 |
| KWh @ 50% EFF & LOAD | 162,704 | | | | | 156,230 | 18 | L2*L15 |
| OPERATING COST 100%LOAD | $26,033 | | | | | $25,128 | 19 | L1*L16 |
| OPERATING COST 75%LOAD | $19,525 | | | | | $18,767 | 20 | L1*L17 |
| OPERATING COST 50%LOAD | $13,016 | | | | | $12,498 | 21 | L1*L18 |
| FOR LOADS SENSITIVE TO MOTOR SPEED | | PROPOSAL 1 | PROPOSAL 2 | PROPOSAL 3 | PROPOSAL 4 | PROPOSAL 5 | LINE | |
| SPEED RATIO OF        100%LOAD | | | | | | 1.0085 | 22 | (L5prop./L5exist)cubed |
| ( MOTOR B / MOTOR A) CUBE   75%LOAD | | | | | | 1.0063 | 23 | (L11prop./L11exist)cubed |
| (FAN OR AFFINITY LAW)      50%LOAD | | | | | | 1.0042 | 24 | (L12prop./L12exist)cubed |
| KW @ 100% EFF & LOAD | | | | | | 79.2 | 25 | L3*0.746*L22*100/L6 |
| KW @ 75% EFF & LOAD | | | | | | 59.0 | 26 | L3*0.746*0.75*L23*100/L7 |
| KW @ 50% EFF & LOAD | | | | | | 39.2 | 27 | L3*0.746*0.5*L24*100/L8 |
| KWh @ 100% EFF & LOAD | | | | | | 316,767 | 28 | L2*L25 |
| KWh @ 75% EFF & LOAD | | | | | | 236,076 | 29 | L2*L26 |
| KWh @ 50% EFF & LOAD | | | | | | 156,887 | 30 | L2*L27 |
| OPERATING COST 100%LOAD | | | | | | $25,341 | 31 | L1*L28 |
| OPERATING COST 75%LOAD | | | | | | $18,886 | 32 | L1*L29 |
| OPERATING COST 50%LOAD | | | | | | $12,551 | 33 | L1*L30 |
| FOR LOADS NOT SENSITIVE TO MOTOR SPEED | | PROPOSAL 1 | PROPOSAL 2 | PROPOSAL 3 | PROPOSAL 4 | PROPOSAL 5 | LINE | |
| $ SAVINGS (LOSS) @ 100%LOAD | | | | | | $904 | 34 | L19existing - L19proposed |
| $ SAVINGS (LOSS) @ 75%LOAD | | | | | | $757 | 35 | L20existing - L20proposed |
| $ SAVINGS (LOSS) @ 50%LOAD | | | | | | $518 | 36 | L21existing - L21proposed |
| FOR LOADS SENSITIVE TO MOTOR SPEED | | PROPOSAL 1 | PROPOSAL 2 | PROPOSAL 3 | PROPOSAL 4 | PROPOSAL 5 | LINE | |
| $ SAVINGS (LOSS) @ 100%LOAD | | | | | | $691 | 37 | L19existing - L31proposed |
| $ SAVINGS (LOSS) @ 75%LOAD | | | | | | $638 | 38 | L20existing - L32proposed |
| $ SAVINGS (LOSS) @ 50%LOAD | | | | | | $465 | 39 | L21existing - L33proposed |

Figure 11.8 WHAT IF motor comparison form.

## HOW TO GET AROUND IN THE 'WHAT IF' FORM.

| COLUMN | LINE | EXPLANATION |
|---|---|---|
| EXISTING | 1-9 | Information can be taken from the Motor Record Form if previously generated. If not available, generate data. |
| EXISTING | 1 | ENERGY COST $/kWh Self Explanatory |
| EXISTING | 2 | OPERATING HOURS PER YEAR is very important to be as accurate as possible |
| EXISTING | 3 | MOTOR HORSEPOWER Self Explanatory |
| EXISTING | 4 | NO LOAD SPEED RPM (synchronous speed) is usually within 5% of Full Load Speed i.e. 900, 1200, 1800, or 3600 rpm. |
| EXISTING | 5 | FULL LOAD SPEED is found on the nameplate. This information is important for loads sensitive to motor speed. |
| EXISTING | 6 | EFFICIENCY @ 100% LOAD NEMA % efficiency at full load. If motor is relatively new this will be found on the nameplate, if older, it will be necessary to contact the manufacturer or the MotorMaster data base. (WSEO) |
| EXISTING | 7-8 | EFFICIENCY @ 75% AND 50% LOAD This information can be obtained from the manufacturer or the MotorMaster data base. (WSEO) |
| EXISTING | 9 | INVESTMENT/SALVAGE This should include total cost associated with purchase of a motor, such as cost of motor installation, balancing alignment and disposition of existing motor. |
| PROPOSAL 1-5 | 5-9 | Information is acquired from the motor manufacturers catalogs or the MotorMaster data base which contains information on more than 10,000 motors from various manufacturers. |
| EXISTING | 10-21 | The data is calculated using the formulas in the column entitled FORMULA or is automatically calculated if using the What If spreadsheet. |
| PROPOSAL 1-5 | 10-39 | The data is calculated using the formulas in the column entitled FORMULA or is automatically calculated if using the What If spreadsheet. |
| FORMULA | 10-39 | The formulas used for calculating. These formulas may also be used to create a spreadsheet similar to 'What If.' |

## 11.22  NAMEPLATE GLOSSARY

HP—The number of, or fractional part of a horsepower, the motor will produce at rated speed.

RPM—An indication of the approximate speed that the motor will run when it is putting out full rated output torque or horsepower is called full load speed.

VOLTS—Voltage at which the motor may be operated. Generally, this will be 115 Volts, 230 Volts, 115/230 V, or 220/440 V.

AMPS—The amount of current the motor can be expected to draw under full load (torque) conditions is called full load amps. It is also known as nameplate amps.

HZ—Frequency at which the motor is to be operated. This will almost always be 60 Hertz.

SERVICE FACTOR—The service factor is a multiplier that indicates the amount of overload a motor can be expected to handle. For example, a motor with a 1.15 service factor can be expected to safely handle intermittent loads amounting to 15% beyond its nameplate horsepower.

DESIGN—The design letter is an indication of the shape of the torque speed curve. They are A. B. C and D. Design B is the standard industrial duty motor which has reasonable starting torque with moderate starting current and good overall performance for most industrial applications. Design C is used for hard to start loads and is specifically designed to have high starting torque. Design D is the so called high slip motor which tends to have very high starting torque but has high slip rpm at full load torque. Design A motors are not commonly specified but specialized motors used on injection molding applications.

FRAME—Motors, like suits of clothes, shoes and hats, come in various sizes to match the requirements of the application.

INSULATION CLASS—The insulation class is a measure of the resistance of the insulating components of a motor to degradation from heat. Four major classifications of insulation are used in motors. They are, in order of increasing thermal capabilities, A, B, C, and F.

TYPE—Letter code that each manufacturer uses to indicate something about the construction and the power the motor runs on. Codes will indicate split phase, capacitor start, shaded pole, etc.

CODE—This is a NEMA code letter designating the locked rotor kVA per horsepower.

PHASE—The indication of the type of power supply for which the motor is designed. Two major categories exist; single phase and three phase.

AMB. DEG. C.—Ambient temperature is the maximum safe room temperature surrounding the motor if it is going to be operated continuously at full load. In most cases, the standardized ambient temperature rating is 40 degrees C (104 degrees F).

TEMPERATURE RISE—Temperature rise is the amount of temperature change that can be expected within the winding of the motor from non-operating (cool condition) to its temperature at full load continuous operating condition. Temperature rise is normally expressed in degrees centigrade.

DUTY—Most motors are rated for continuous duty which means that they can operate at full load torque continuously without overheating. Motors used on certain types of applications such as waste disposal, valve actuators, hoists, and other types of intermittent loads, will frequently be rated for short term duty such as 5 minutes, 15 minutes, 30 minutes, or 1 hour.

EFF. INDEX OR NEMA %—Efficiency is the percentage of the input power that is actually converted to work output from the motor shaft. Efficiency is now being stamped on the nameplate of most domestically produced electric motors.

## Summary

Over 60 percent of electricity in the United States is consumed by electric motor drive systems. Generally, a motor drive system consists of several components; a power supply, controls, the electric motor and the mechanical transmission system. The function of an electric motor is to convert electric energy into mechanical work. During the conversion the only power consumed by the electric motor is the energy losses within the motor. Since the motor losses are in the range of 5-30% of the input power, it is important to consider the total system of which the motor is a part.

This chapter deals mostly with electric motor drive systems and provides practical methods for managing motors.

One of the "MotorManager" methods is the Motor Performance Management Process (MPMP) which effectively evaluates the performance of existing motors and identifies opportunities to link them to organizational goals. It is a primary tool to evaluate, measure and manage motors and a logical, systematic, structured approach in reducing energy waste, fundamental to efficiency in a competitive market. With minor changes, this process can be used to evaluate other electrical and mechanical equipment.

Considerable attention must be paid to the efficiencies of all electric equipment being purchased today. This is true not only for motors but for transformers of all types and other electrical devices.

To guard against the waste of electrical energy, manufacturers of dry type transformers are designing them with lower than normal conductor and total losses. The reduction in these losses also lowers the temperature rise of the transformer resulting in improved life expectancies as well as a reduction in the air conditioning requirements.

## References

1. Technology Update Produced by the Washington State Energy Office for the Bonneville Power Administration.

2. NEMA Standards publication No. MG 10. *Energy Management Guide for Selection and Use of Polyphase Motors*. National Electrical Manufacturers Association, Washington, D.C. 1989

3. McCoy, G., A. Litman, and J. Douglass. Energy Efficient Electric Motor Selection Handbook. Bonneville Power Admin. 1991

4. Ed Cowern Baldor Electric Motors.

5. **MotorMaster** software developed by the Washington State Energy Office (WSEO) puts information on more than 10,000 three-phase motors at your fingertips. The **MotorMaster** lets you review features and compare efficiency, first cost, and operating cost. The U.S. Department of Energy (DOE) and the Bonneville Power Administration (BPA) funded the development of the software. You may obtain this valuable software or more detailed information by contacting WSEO (360) 956-2000

6. Financing for Energy Efficiency Measures, Virginia Power, POST-AND, REV(0)08-17-93.

7. William E. Lanning and Steven E. Strauss Power Survey Techniques for electrical loads Esterline Angus Instrument Corp.

8. Konstantin Lobodovsky, "Efficient Application of Adjustable Speed Drives," *Motor Manager*, Penn Valley, CA, 1999 (530) 432-2237

# Appendix

## ELECTRONIC ADJUSTABLE SPEED DRIVES: ISSUES AND APPLICATIONS*

**CLINT D. CHRISTENSON**
Oklahoma Industrial Assessment Center
School of Industrial Engineering and Management
322 Engineering North
Oklahoma State University
Stillwater, OK 74078

## INTRODUCTION

Electric motors are used extensively to drive fans, pumps, conveyors, printing presses, and many other processes. A majority of these motors are standard, 3-phase, AC induction motors that operate at a single speed. If the process (fan, pump, etc.) is required to operate at a speed different than the design of the motor, pulleys are applied to adjust the speed of the equipment. If the process requires more than one speed during its operation, various methods have been applied to allow speed variation of a single speed motor. These methods include variable pitch pulley drives, motor-generator sets, inlet or outlet dampers, inlet guide vanes, and Variable Frequency Drives. The following section will briefly discuss each of these speed control technologies.

Changing the pulleys throughout the day to follow demand is not feasible, but the use of a "Reeves" type variable pitch pulleys drive was a common application. These drives utilized a wide belt between two pairs of opposing conical pulleys. As the conical pulleys of the driven shaft were brought together (moved apart) the pulley diameter would increase (decrease) and decrease (increase) the belt speed and the process speed. This type of system is still extensively in use in the food and chemical industries where mixing speeds can dramatically effect product quality. These systems are a mechanical speed adjustment system which has inherent function losses and require routine maintenance.

Motor-Generator sets were used in the past to convert incoming electricity to a form required in the process including changes from AC to DC. The DC output could then be used to synchronize numerous dc motors

---

*Facilities today generally have significant harmonics. Especially when capacitors are used in systems with harmonics and variable frequency drives, professional advice is needed. Also motor capability may be a problem. See (8) page 71 for more information

at the required speed. This was the common type of speed control in printing presses and other "web" type systems. The use of an Eddy-current clutch would vary the output of the generator to the specific needs of the system. The windage and other losses associated with motors are at least doubled with the generator and the efficiency of the system drops drastically at low load situations.

A majority of the commercial and industrial fan and pump speed control techniques employed do not involve speed control at all. These systems utilize inlet dampers, outlet dampers (valves) with or without bypass, or inlet guide vanes to vary the flow to the process. Inlet dampers, used in fan applications, reduce the amount of air supplied to the process by reducing the inlet pressure. Outlet dampers (valves) maintain system pressure (head) seen by the fan (pump) while reducing the actual volume of air (liquid) flowing. Inlet guide vanes are used in fans similar to inlet dampers but the guide vanes are situated such that as air flow is reduced, the circular motion of the fan is imparted upon the incoming air. Each of these control methods operates to reduce the amount of flow with some reduction in energy required.

Variable Frequency Drives (VFD) change the speed of the motor by changing the voltage and frequency of the electricity supplied to the motor based upon system requirements. This is accomplished by converting the AC to DC and then by various switching mechanisms invert the DC to a synthetic AC output with controlled voltage and frequency [Phipps, 1994]. If this process is accomplished properly, the speed of the motor can be controlled over a wide variation in shaft speed (0 rpm through twice name plate) with the proper torque characteristics for the application. The remainder of this paper will discuss the various issues and applications of VFD's. Figure 1 shows the percent power curves versus percent load for simplified centrifugal air handling fan application.

## VARIABLE FREQUENCY DRIVE TYPES

In order to maintain proper power factor and reduce excessive heating of the motor, the name plate volts per hertz ratio must be maintained. This is the main function of the variable frequency drive (VFD). The four main components that make up AC variable frequency drives (VFD's) are the Converter, Inverter, the DC circuit which links the two, and a control unit, shown in Figure 2. The converter contains a rectifier and other circuitry which converts the fixed frequency

**Figure 1. Typical power consumption of various fan control systems. (Source: Moses et. al., 1989)**

AC to DC. The inverter converts the DC to an adjustable frequency, adjustable voltage AC (both must be adjustable to maintain a constant volts to hertz ratio). The DC circuit filters the DC and conducts the DC to the inverter. The control unit controls the output voltage and frequency based upon feedback from the process (e.g. pressure sensor). The three main types of inverter designs are voltage source inverters, current source inverters, and pulse width modulation inverters. Each will be briefly discussed in the next section.

## Inverter Types

The voltage source inverters (VSI) use a silicon controlled rectifier (SCR) to rebuild a pseudosine wave form for delivery to the motor. This is accomplished with a six-step voltage inverter with a voltage source converter and a variable voltage DC bus. As with any SCR system, troublesome harmonics are reflected to the power source. Also the six-step wave form sends current in pulses which can cause the motor to cog at low frequencies, which can damage keyways, couplings, pump impellers, etc. [Phipps, 1994].

The current source inverters (CSI) also use an SCR input from the power source but control the current to the motor rather than the voltage. This is accomplished with a six-step current inverter with a voltage source converter and a variable voltage DC bus. The CSI systems have the same problems with cogging and harmonics as the VSI systems. Many manufacturers only offer VSI or CSI VFD's for larger horsepower sizes (over 300 HP).

The pulse width modulation (PWM) VFD's have become the state of the art concept in the past several years, starting with the smaller hp sizes, and available up to 1500 hp from at least one manufacturer [Phipps, 1994]. PWM uses a simple diode bridge rectifier for power input to a constant voltage DC bus. The PWM inverter develops the output voltage by producing pulses of varying widths which combine to synthesize the desired wave form. The diode bridge significantly reduces the harmonics from the power source. The PWM system produces a current wave form that more closely matches the power line wave form, which reduces adverse heating. The PWM drives also have the advantage of virtually constant power factor at all speeds. Depending on size, it is possible to have power factors over 95% [Phipps, 1994]. Another advantage of the PWM VFD's is that sufficient current frequency (~200+ Hz) is available to operate multiple motors on a single drive, which would be advantageous for the printing press example discussed earlier. The next

**Figure 2. Typical VFD system. (Source: Lobodovsky, 1996)**

section will discuss the types of loads that require adjustable speeds that may be controlled by variable frequency drives.

## VARIABLE SPEED LOADS

The three common types of adjustable speed loads are variable torque, constant torque, and constant horsepower loads. A variable torque load requires much lower torque at low speeds than at high speeds. With this type of load, horsepower varies approximately as the cube of speed and the torque varies approximately as the square of the speed. This type of load is used in applications such as centrifugal fans, pumps, and blowers. A constant torque load requires the same amount of torque at low speed as at high speed. The torque remains constant throughout the speed range, and the horsepower increases or decreases in direct proportion to the speed. A constant torque load is used in applications such as conveyors, positive displacement pumps, some extruders, and for shock loads, overloads, or high inertia loads. A constant horsepower load requires high torque at low speeds, and low torque at high speeds, and therefore constant horsepower at any speed. Constant horsepower loads are encountered in most metal cutting operations, and some extruders [Lobodovsky, 1996]. The savings available from non-centrifugal (constant torque or constant horsepower) loads are based primarily on the VFD's high efficiency (when compared to standard mechanical systems), increased power factor, and reduced maintenance costs. The next section will discuss several applications of VFD's and the issues involved with the application.

## VARIABLE FREQUENCY DRIVE APPLICATIONS

Variable frequency drive systems offer many benefits that result in energy savings through efficient and effective use of electric power. The energy savings are achieved by eliminating throttling, performance, and friction losses associated with other mechanical or electromechanical adjustable speed technologies. Efficiency, quality, and reliability can also be drastically improved with the use of VFD technology. The application of a VFD system is very load dependent and a thorough understanding of the load characteristics is necessary for a successful application. The type of load (i.e. Constant torque, variable torque, constant horsepower) should be known as well as the amount of time that the system operates (or could operate) at less than full speed.

Figures 3 and 4 show the energy savings potential for a variable speed fan and pump, respectively. The VFD pump is compared with a valve control system which would be adjusted to maintain a constant pressure in the system. The VFD fan is compared with both the damper control and the inlet guidevane controls. These curves do not account for system characteristics (i.e., head or static pressure), which would need to be included in an actual design. These figures show that the amount of savings achievable from the VFD is based upon the

Figure 3. Energy savings with VFD fan. (Source: Lobodovsky, 1996)

Figure 4. Savings with VFD pump. (Source: Lobodovsky, 1996)

percent volume flow for both cases. The consumption savings would be determined by the percent of time at a particular load multiplied by the amount of time at that particular load.

## Application Identification

There are many instances where a VFD can be successfully applied. The situation where the existing equipment already utilizes one of the older speed control technologies is the easiest to identify. In the case of the printing press that utilizes an Motor-generator set with a Eddy-current clutch, a single VFD and new AC motors could replace the MG-set and the DC motors. A mixer utilizing variable pitch pulleys ("Reeves" drive) could be replaced with a VFD which would reduce slippage losses and could dramatically improve product quality (through better control) and reliability (solid state versus mechanical).

Equipment (fans or pumps) utilizing dampers or valves to reduce flow is another instance where a VFD may provide a better means of control and energy savings. A variable volume air handler that utilizes damper or inlet guide vane controls could be replaced with a VFD drive controller. This would reduce the amount of energy required to supply the required amount of air to the system. Chiller manufacturers have utilized VFD controllers to replace the standard butterfly inlet valves on centrifugal compressors. This can significantly reduce the part load power requirements of a chiller which occur for a majority of the operating cycle in most applications.

## Application Analysis

Project evaluation methods depend in large part on the size of the project. Smaller and less complex projects may only require reviewing specifications, installation sketches and vendors' quotes. Larger, more complex projects require more detailed engineering drawings and a drive system specialist will need to review the plan [Lobodovsky, 1996]. Once a possible application for a VFD is identified, the load profile (percent load versus time) should be determined. This curve(s) could be developed with the use of demand metering equipment or process knowledge (less desirable). This curve will be used to determine the available savings to justify the project as well as assure that the motor is properly sized. The proper load profile (constant torque, variable torque, etc.) can be compared to that loads corresponding VFD load profile in order to develop savings potential. The motor must be evaluated to assure that the VFD is matched to the motor and load, determine the motor temperature requirements, that the minimum motor speeds are met, among others. Many VFD manufacturers will require that the

existing motor be replaced with a new model to assure that the unit is properly sized and not affected by earlier motor misuse or rewinding. At this stage the expertise of a VFD analyst or sales representative should be brought into the project for further design issues and costs. Phipps includes comprehensive chapters on applying drives to various applications where several check lists are included. The next section includes two case studies of applications of VFD's in industry.

## CASE STUDIES

The following case studies are included as an example of possible VFD applications and the analysis procedures undertaken in the preliminary systems analysis.

### Boiler Combustion Fan

This application involves the use of a VFD to vary the speed of a centrifugal combustion air intake fan (50 nameplate horsepower) on a scotch marine type high pressure steam boiler. The existing system utilizes an actuator to simultaneously vary the amount of gas and air that enters the burner. The air is controlled with the use of inlet dampers (not guide vanes). As the amount of "fire" is reduced, the damper opening is reduced and visa versa. The centrifugal fan and continuous variation in fire rate make this a feasible VFD application. The load profile of the boiler and corresponding motor demand (measured with demand metering equipment) is listed in Table 1. This table also includes the corresponding VFD demand requirements (approximated from Figure 4), kW savings, and kWh savings.

The annual savings for this example totaled 88,000 kWh, which would equate to an annual savings of $4,400 (based upon a cost of energy of $0.05/kWh). Lobodovsky provides an average estimated installed cost of VFD's in this size range at around $350 per horsepower, or an installed cost of $17,500 (50 hp * $350/hp). This would yield a simple payback of around 4 years. This example does not take into account demand savings which may result if the demand reduction corresponds with the plant peak demand. The control of the fan VFD would be able to utilize the same output signal that the existing actuator does.

### Industrial Chiller Plant

A different calculation procedure will be used in the following example. A malting plant in Wisconsin uses seven 550 ton chillers to provide cold water for process cooling. Three of the chillers work all of the time

### Table 1. Boiler combustion fan load profile and VFD Savings

| Operating Time at load (hours) | Percent Full Load Power[2] | Existing Load w/ Damper (kW)[1] | Load with VFD Control (kW)[2] | Kilowatt Savings | Kilowatt-Hour Savings |
|---|---|---|---|---|---|
| 1000 | 100 | 37 | 37 | 0 | 0 |
| 2000 | 105 | 40 | 30 | 10 | 20,000 |
| 3000 | 95 | 35 | 20 | 15 | 45,000 |
| 1000 | 90 | 33 | 10 | 23 | 23,000 |

[1]Measured with a Demand Meter

[2]Approximated using Figure 4

88,000 kWh

and the other four are operated according to the plant's varying demand for cooling. The chillers are currently controlled by variable inlet vanes (VIV).

The typical load diversity and the power input required for one of the chillers under varying load were obtained from the manufacturer. In addition, the data in Table 2 give the power input of a proposed variable frequency drive (also referred to as adjustable speed drive (ASD)) for one chiller that operates about 6,000 hours per year.

A weighted average of the load and duty-cycle fractions gives a load diversity factor (ldf) of 64.2 percent. This means that, on the average, the chiller operates at 64 percent of its full load capacity. A duty-cycle fraction weighted average of the savings attainable by a VFD can be estimated, which is 26.6 percent savings per year. The energy savings due to avoided cost of

electricity usage are computed as follows: Savings = (0.266)(550 ton)(0.7 kW/ton)(0.642)(6,000 hr/yr) = ldf= 394,483 kWh/year The dollar savings at $0.04 per kWh are (394,483 kWh/year)($0.04/kWh) = $15,779/year. For a 500 horsepower variable frequency drive, the installed cost is estimated at $75,000, based upon Lobodovsky average installed cost of $150 per ton for units of this size. Therefore, the payback period is:

$$($75,000)/($15,779/year) = 4.75 \text{ years.}$$

## VFD ATTRIBUTES

These are just a few of the examples to show savings calculations. The advantages of VFD's include other aspects beyond energy savings, including improved productivity, reduced maintenance costs, and improved product quality, among others. The application of VFDs is relatively straight forward but a thorough analysis of the existing system and design of the future system is necessary to assure successful application. The load profile of the existing system is necessary to both determine savings as well as assure the system is properly sized. The specification of the actual VFD should only be done by VFD suppliers and/or experts. The list of references is a good source of more detailed discussions of each of the points discussed in this paper.

**Attached are some additional forms and information on variable speed drives. These are reproduced from a forthcoming book by Mr. Konstantin Lobodovsky.**

### Table 2. Centrifugal chiller load and power input (Source: updated from Moses et Al., 1989)

| Load Fraction (1) | VIV % kW (2) | ASD % kW (3) | Savings % kW (4) | Duty Cycle Fraction (5) |
|---|---|---|---|---|
| 0.3 | 41 | 17 | 24 | 0.08 |
| 0.4 | 53 | 22 | 31 | 0.10 |
| 0.5 | 66 | 33 | 33 | 0.13 |
| 0.6 | 75 | 40 | 35 | 0.18 |
| 0.7 | 82 | 54 | 28 | 0.22 |
| 0.8 | 87 | 63 | 24 | 0.15 |
| 0.9 | 92 | 78 | 14 | 0.09 |
| 1.0 | 100 | 100 | 0 | 0.05 |

| | |
|---|---|
| Column 2. | From typical compressor performance with inlet and vane guide control (York Division of Borg-Warner Corp). |
| Column 3. | From Carrier Corporation's Handbook of Air Conditioning Design: Comparative Performance of Centrifugal Compressor Capacity Control Methods. |
| Column 5. | Actual performance data. |

### References

Lobodovsky, Konstantin K., *Motor Efficiency Management & Applying Adjustable Speed Drives*, April 1996.

Moses, Scott A., Wayne C. Turner, Jorge Wong, and Mark Duffer, "Profit Improvement With Variable Frequency Drives, *Energy Engineering. Vol.* 86, No. 3, 1989.

Phipps, Clarence A. *Variable Speed Drive Fundamentals* Lilburn, GA, The Fairmont Press, 1 994.

**MOTOR MANAGER**

By: Konstantin K. Lobodovsky                7/95                    MM_014    page    4   of   7

# Typical ASD Equipment Costs

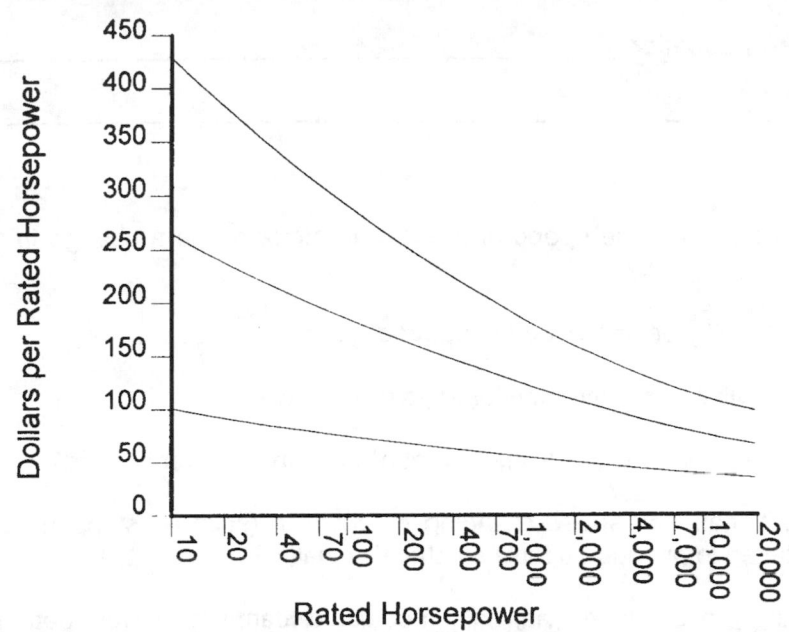

# Typical ASD Installation Costs

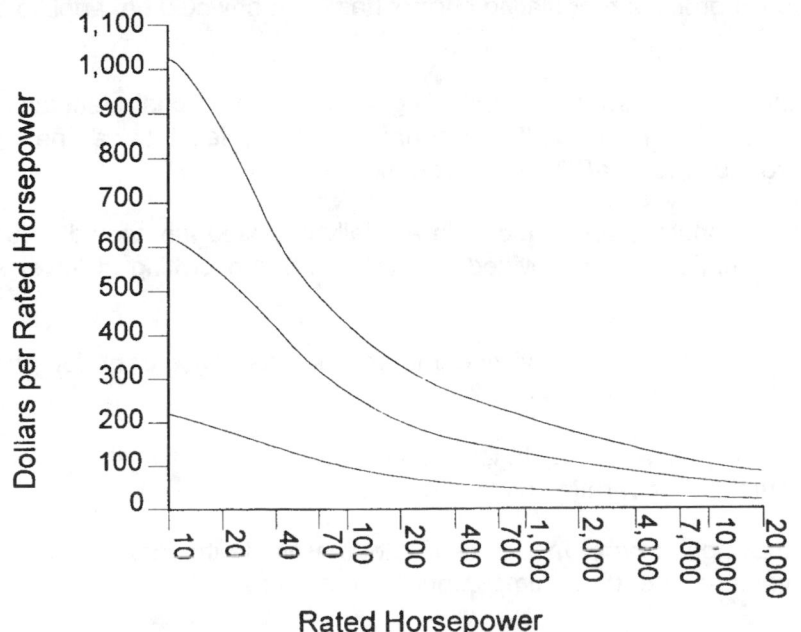

**Typical ASD Equipment and Installation Costs**

**MOTOR MANAGER**  *Resource*

By: Konstantin K. Lobodovsky             7/95                    MM_015    page  4  of  5

# TECHNICAL FEASIBILITY CHECKLIST

**Drive Identification and Location:**_____

_____

1. How would changing the speed of the driving motor cause a change in the process or its rate?

2. Will product quality be improved or impaired?

3. What effects will the improvement or impairment have?

4. In what way can the machinery operate at other than its current speed?

5. In what way can any speed-changing mechanism (such as step pulleys, gears, or fluid drives) be installed to provide suitable electrical signal(s)?

6. If the existing process is mechanically modulated (dampers, valves, gates), how would new sensor(s) be installed to provide suitable electrical signal(s)?

7. If the existing process is electrically modulated (dc motors or wound-rotor induction motors), how would squirrel cage induction motors be adapted to the equipment?

8. Describe how a process modulating control has been or would be, applied to a drive system of this type.

9. Describe the physical space for installing a new or additional electrical motor controller, (Conventional induction motor ASD electronics need at least twice the space of existing starters; synchronous motor ASDs may need more.)

10. If the existing constant speed motor is a totally enclosed fan cooled induction motor, how would additional ventilation be provided if needed, when operating at lower speeds using an ASD?

11. How much derating of the existing motor would be necessary for heating caused by harmonics? Is this derating acceptable?

12. If the machinery is a pump, fan, or compressor, what data sheets or test means are available to estimate the operating characteristics of the unit?

13. Are data, drawings or other means available to estimate the torque requirements at various speeds for machinery other than pumps, fans, or compressors?

14. What drawings or other means available to validate the construction or installation details of the motor and machinery involved?

**MOTOR MANAGER**

*Resource*

By: Konstantin K. Lobodovsky      7/95      MM_015   page   5   of   5

# ECONOMIC FEASIBILITY CHECKLIST

Drive Identification and Location_____

_____

1. How will a change in the speed of the driving motor result in a lower energy requirements?

2. What are your electrical energy costs in terms of your utility bill (consumed kWh, demand charges, etc.) or in terms of product costs?

3. If the existing process is mechanically modulated, what portion of the operating time is at other than maximum flow (or load)?

4. How many hours per week does the equipment operate?

5. How will the use of an ASD improve quality (through better speed control, elimination of waste, product reversion, etc.) and/or ultimately result in lower product costs?

6. What costs associated with drive inefficiencies (friction heat, cooling water, etc.) can be reduced by using an energy-efficient ASD system?

7. What are the costs for maintaining existing mechanical speed-changing equipment (transmissions, etc.)? are they obsolete and in need of replacement?

8. What are the costs of maintaining existing electrical speed-changing equipment (dc and wound-rotor motor, or reduced-voltage starting)? Are they obsolete and in need of replacement?

9. Are there other problems of equipment reliability that cause production delays and higher product costs? Who can they be eliminated by ASDs with self-diagnostic features?

10. What opportunity is there to create additional space by removing large mechanical equipment (transmissions, etc.) with the installation of an ASD controller?

11. Can plant noise be reduced through lessening of the mechanical noise by installing an ASD control, or will the noise of the ASD be excessive?

12. Describe the shutdown arrangements required to provide time for an ASD to be installed.

# ALTERNATIVE METHOD OF VARYING THE SPEED OF THE DRIVEN LOAD

## PAUL DEWEY BOGGS, III
## K.K. LOBODOVSKY

## INTRODUCTION

Variable frequency drives (vfd's), or inverters, have become the universally accepted method of variable speed control of AC motors and their respective loads. This paper addresses an alternative approach to reliable variable speed control of the load and provides a brief look at the developmental history of the magnetic-coupled variable speed drive, the latest advances in the technology, the benefits, and the application of this technology with variable torque loads such as centrifugal fans

## BASIC PRINCIPLE OF OPERATION

The magnetic coupled concept differs vastly from variable frequency drives in that there is no electrical power interruption to the motor. With the motor running continuously, precise speed control is accomplished by varying the magnetic coupling between the motor's output shaft and the load. Most configurations are comprised of two primary elements:

Electromagnet (multi-pole rotor) mounts onto the shaft of the motor.

Steel drum (armature/pulley)

**Drives Basic Parts**

One element (input portion) is affixed to the motors shaft so as to run continuously. The other element (output portion) will have a connection to the driven load. These two elements are separated by an air gap, and have no other mechanical connection other then supportive bearings. By applying current to the coil of the electromagnet rotor, a polarized magnetic field is produced, creating eddy currents on the surface of the drum, magnetically coupling both components and causing the output portion to turn in the same direction as the motor. The speed of the output is dependent on the strength of the magnetic field which is proportionately controlled by the amount or current applied to the electromagnet.

## EVOLUTION OF THE MAGNETIC COUPLED DRIVE

### Foot Mounted Style

In the 1940's and 50's, magnetic coupled devices known as eddy current clutches were effectively used with AC motors and were quickly becoming a popular method of varying the speed of many industrial loads. Although bulky end inefficient, these workhorses were quite reliable and were used in applications such as punch presses, conveyers, winders, end other machine tool situations. These were oversized foot mounted units that initially were designed as a separately housed clutch assembly with an input shaft end output shaft to be coupled in line between the motor and the load. Also offered was motor and clutch combination (one piece) packaged units. In those days, the primary focus was on functionality, performance and maintainability, as energy efficiency was not as important a factor as it is today.

### Shaft-Mounted Styles

In the 1960's some of the first commercially available motor shaft-mounted magnetically coupled drives were offered to the industry. This new design was intended for fractional and small integral horsepower applications, and was novel in that the drive was totally supported by the motor shaft. The product did however have some drawbacks, namely oversized slip rings end problems with brush alignment. It was the user's responsibility to align the brush holder with the slip rings. The intent was to mount the brush holder to the existing motor bolts. Unfortunately, motors supplied by different manufacturers varied significantly and alignment became difficult. Additionally, the slip rings were provided on the drive at the motor shaft entry side, fabricated on a circuit board type material with copper rings racing the motor. Since the diameter of the outer ring was larger than the inner ring, the outer brush would wear out faster. In addition to the uneven end rapid brush wear, the integrity or the circuit board and copper rings were effected by heat, causing separation of the rings from the base material. This design was abandoned soon after initial production.

In the 1980's the problem with brush alignment had been somewhat resolved by a new shaft-mounted design that incorporated a bracket supported by an additional bearing on the drive which maintained reasonable alignment between the brush holder and the slip rings. This design enjoyed some success in the machine tool industry where reduced run-time hours was common this basic design was still flawed, as the slip rings were still located in the motor shaft entry side, causing them to be oversized and progressively larger to accommodate the higher horsepower motor shafts. This created a major headache in the HVAC air handler marketplace, as it was soon discovered that 24-hour duty meant brush changes in some cases as often as once every two to three months on the larger drives. Although an improvement over previous efforts, the location or the bearings being cantilevered to the pulley grooves, caused premature bearing failure in many instances. The additional bearing used to accommodate the brush holder had an unacceptability high failure rate as well. This high maintenance drive has become virtually obsolete in the air handler industry and has been routinely replaced by the more efficient and reliable new brushless designs.

Another design consideration places the pulley grooves out or the outboard side of the drive, a distance away from the motor face. This, by far is the poorest or all approaches because it directly jeopardizes the motor bearings' life expectancy. Since the pulley grooves are not located over the nema shaft extension, applying full rated belt tension exceeds the overhung load rating of the motor in many cases. This outboard pulley design has many documented failures in the field, again compounded by drive hearing failure due to cantilevering effect, and motor bearing damage as well.

**Preferred Design**

The most reliable and field proven design distinguishes itself in many ways from the previous concepts. The rotor/coil assembly rotates constantly with the motor shaft. A one-piece drum/pulley portion is the output-driving member. The pulley grooves are located in-board, closer to the motor face than any other design. The drives' bearings are located directly under the pulley grooves so that maximum belt tension can be applied continuously on all models without harming the drive bearings end yet remains well under the overhung load capacities of the motor. Because the drum is copper lined, the brushless drive runs cooler, and is more efficient than other magnetic-coupled drives. The drive coil requires only one third to one fourth the wattage of other models. It has the fewest parts, weighs less end has the best operating performance or all available designs the unblemished track record approaching five years allows for the longest drive warranty that is available in the industry, matching the full five year motor warranty.

**ENERGY SAVINGS**

Q. How Does The Magnetic-Coupled Drive Save Energy If The Motor Runs Continuously?

Because of the nature of the descending torque load itself, the magnetic-coupled drive takes advantage of the energy savings with variable speed.

The magnetic-coupled drive will take advantage of the affinity laws on variable torque loads such as centrifugal fans and pumps and does so without altering the voltage or interrupting power to the motor. Even with the motor running continuously, the kW required by the motor changes according to the actual load; i.e., loading and unloading of the motor by varying the speed of the load via magnetic coupling between the motor end the load. What you always pay for is the amount of kW used and even with the motor running continuously, the difference in the amount of kW that required of any 3 phase motor from no load to full load is very significant.

As example: 7-1/2 hp motor/3-phase/60hz (typical blower application)

Measured @ full load, max fan speed (typical) = 5.60 kW

Measured @ min load, min fan speed = approximately 0.40 to 0.50 kW

Greater than 10 to 1 kW difference throughout entire speed range, thanks to the affinity laws. Even greater advantages from full load to no load are realized as the motor horsepower size increases. As a result, the power curves are very similar to vfd's on variable torque loads and the kW savings are significant as well. As a general rule, magnetic-coupled drives are more efficient at the top end of the curve, and the vfd's are more efficient further down the curve.

**Q.** Does The Magnetic-Coupled Drive Cause Additional Motor Heating, Even At Very Low Speeds?

The magnetic coupling is electrically isolated from the motor and in effect operates as an infinitely variable; frictionless clutch, allowing tie motors to operate as originally designed at full speed continuously, end with pure uninterrupted AC power. Regardless of the drives' operating speed, the motor never sees any additional heating contributed by the drive. In fact, the drive itself via slip, and not the motor dissipate any additional heat. Since these new drives are efficiently sized to handle the full rated horsepower of this types of loads, the minimum amount of drive heating is effectively dissipated by the drives' own integral fan. Since the motor runs continuously end the drive is simply controllably coupling and uncoupling the load, the effect of the motor loading is no difference in operation than if you had incorporated an infinite number of pulley sizes to provide variable speed to the load.

**Q.** Power Quality Issues, Harmonics?

Since the magnetic-coupled drive does not interrupt the power source to the motor, there are no current harmonics produced nor is there any resultant voltage distortion. There is never any need for filters, reactors, or full rated 3 phase isolation transformers.

**Q.** How Far Can The Controller Be Located From The Motor? Are There Any Limitations?

The distance from the controller to the magnetic drive and motor has worked successfully at distances up to 2000 feet. The only requirement is that the two wires that provide power to the drive coil be sized large enough to allow for any voltage drop (say 14 gauge, typically). No filters or any other devices are required. There is no concern about ever causing any damage to the motor or drive.

**Q.** What About Nuisance Dropouts?

By virtue or the inherently simple design, the drive is always active as long as the motor starter is en-

ergized. Transient over voltages, voltage sags, and harmonic distortion from other sources generally do not effect magnetic-coupled drives unless the duration is long enough to drop out the motor starter circuit.

**Q.** Lightning?

Since eddy-current drives are isolated from the power source, they provide the highest level of immunity to the effects of lightning.

**Q.** Retrofit Existing Motors?

The shaft-mounted magnetic-coupled drive allows for true variable speed retrofit of any existing motor. There is never any need for inverter duty motors. All original motor wiring circuitry can always remain undisturbed.

**Q.** Bypass?

A simple mechanical full speed lock-up feature is standard on all drives. Full speed electrical bypass can be accomplished with one diode.

**Q.** Is There Any Motor Insulation Failure Or Electrical Pitting Of Motor Bearings?

Since die magnetic-Coupled drive does not electrically connect to the motor wires, there is never any possibility or causing any electrically induced harm to the motor windings or the motor bearings.

**Q.** What About Power Factor?

Some utility companies in certain locations may change penalty if the facility's total measured power factor is below an acceptable pre-determined level. In hundreds and hundreds of installations in many varied facilities such as colleges, schools, hospitals, government buildings, and shopping malls, power factor has never been an issue with the new shaft-mounted variable speed drives. However, in the event low power factor requires attention, low cost power factor correction capacitors can be easily installed at each motor location as required.

**Q.** Maintainability, In-House?

This simple technology does not require highly skilled personnel to maintain. The same compact low cost controller is used on all drive sizes 1 through 150 horsepower. Unlike high maintenance brushes and slip rings, the rotary brushless plug-in coupling cartridge can be swapped out in a matter of seconds.

**Q.** For Applications Other Than Variable Torque Loads, How Does The Magnetic-coupled Drive Perform?

In constant torque applications, the drive must be properly sized to allow for additional drive heat dissipation at the lower speed ranges, however there are no added demands on the motor other than the conventional no load to full load conditions, the same as in fixed speed operation considerations. These drives perform very well in many other applicators, such as conveyors, feeders, machine tools, punch-presses, etc., As variable voltage, constant current, or closed loop methods.

**Q.** What's New On The Horizon?

Continued development on the next generation of self-powered variable speed drives is a priority. These new magnetic coupled drives have their own source or power, an integral generator, automatically active when the motor is running. No external power source is required. Installation cost is reduced further because there is no need for the added cost of an electrician. Simply mount the drive to the motor shaft, connect the belts, end then provide a standard 4/20-ma signal to the drive. Loop powered at any horsepower. Imagine that.

## CONCLUSION

The new generation of magnetic-Coupled drive technologies, combined with the ease of installation and the industry's first mega-motor/drive warranty provides an attractively reliable and efficient alternative for variable fan speed control. This rugged product has been successfully used as upgrade replacements for most of the previously mentioned designs and technologies in hundreds of belt driven installations. In critical situations, where maximum reliability and maximum run time is paramount; this approach may well be the preferred alternative.

Www.Payback.Com

E-Mail: info@Coyoteinc.Com

**Typical Power Consumption**

## FAN SHEAVES SELECTION

**Very Important:**

The fan sheaves should always be selected so that when the fan is at its maximum rpm, the payback drive should also be operating as closely as possible to its maximum attainable output speed, typically 1600-1700 rpm.

*See technical data sheet for the specific payback model's sheaves sizes and output speed ranges @ given hp load.

By correctly sizing the fan sheaves, the system will be more efficient, the drive will run cooler, and the drive's bearing life will be optimized. (Selecting too small a driven fan pulley will waste energy and create unnecessary heat dissipation in the drive.)

**For Retrofit Applications:**

There are two easy methods for determining the new fan pulley size to match with the selected model payback drive's sheaves. In both cases, be sure to measure the center to center distance between the existing motor shaft and fan shaft. Observe the belt take-up adjustment on the motor base and allow for a mid-range

**BLOCK DIAGRAM—Variable Frequency Drive (VFD)** Typical required configuration for reduction of harmonic distortion levels induced by Variable Frequency Drives in variable speed applications.

**Block Diagram—Variable Frequency Drive**

**BLOCK DIAGRAM—Magnetic Coupled Drive** Typical configuration for magnetic coupled variable speed drive application. (True Zero Harmonic Distortion)

**Block Diagram—Magnetic Coupled Drive**

take-up measurement to start with for calculating the new belt sizes.

Always take an original rpm reading of the driven (fan) pulley and amp reading of the existing motor at full load and continue to monitor the motor amps after the retrofit. In all correctly sized applications, operation throughout the entire speed range should not exceed the full load amps of the motor.

**Method 1: Pulley Ratio Method**

Step 1: determine the ratio of the existing motor pulley and fan pulley by dividing the driven (fan) pulley diameter by the motor pulley diameter.

(Driven (Fan) Pulley Diameter)/(Motor Pulley Diameter) = (Working Ratio)

Step 2: multiply the listed sheaves diameter of the appropriate model payback drive by the derived working ratio to determine the ideal new driven (fan) sheaves diameter.

(Payback Listed Sheaves Dia.) × (Working Ratio) = (Calculated Fan Sheave Diameter.)

Step 3: select the nearest size sheaves from a sheaves selection book. Using the new selected sheaves diameter in conjunction with the new drive listed sheaves diameter and the center to center measurement, size the belts from a belt selection guide.

Note: taking into account the difference between the motor rpm of 1750 and the drive's maximum attainable output of 1600-1700 rpm, the actual new driven rpm may be slightly less than the original by a factor of this difference rpm, however mechanical lockup, if required in emergency situations should return the driven pulley close to the original rpm.

**Method 2: Rpm Ratio Method**

Step 1: measure the existing driven (fan) pulley rpm with an accurate tachometer instrument.

Step 2: divide the payback drive rated rpm (1600-1700)* by the measured (fan) rpm.

*See applicable model data sheet for max rated output @ given horsepower.

(Payback Max Rated Rpm)/(Existing Driven (Fan) Pulley Rpm) = (Rpm Ratio)

Step 3: multiply the listed sheaves diameter of the appropriate payback drive by the calculated rpm ratio to determine the ideal new driven (fan) sheaves diameter.

(Payback <u>Listed</u> Sheaves Dia.) ¥ (Rpm Ratio) = Calculated Driven Fan Sheave Diameter.

Step 4: select the nearest size sheaves from a sheaves selection book. Using the new selected sheaves diameter

in conjunction with the payback drive listed sheaves diameter and the center to center measurement, size the belts from a belt selection guide.

Note: the new driven pulley maximum rpm will be very closely matched to the original driven pulley rpm, however if mechanical lockup is required for emergency situations, care should be taken that the additional increase in rpm does not cause the system to exceed its maximum capacity or the motor to exceed it's rated motor amps.

**For new installations:** use method 2 for determining the correct sheaves size when the correct model payback drive and desired maximum driven (fan) rpm is already known.

If you need assistance in correctly sizing your tan pulley or have any other questions about your application, please call us at: 817-485-3336 or you may e-mail us at: info@coyoteinc.com.

Pump

Drives basic parts

# Payback Belt Drive Installation Site Survey

By:_____ Date:_____

Contact: _____ Phone: (_____)_____

Address: _____ Fax:  (_____)_____

## AHU#:_____ (Supply FAN) _____ (Return FAN)_____

*EXISTING* MOTOR Pulley DIAMETER _____ Model _____

*EXISTING* FAN Pulley DIAMETER _____ Model _____

*MEASUHED-EXACT* FAN SHAFT DIAMETER_____

FAN BUSHING #:_____ TYPE: (Q.D. Flanged)_____ (Flush Taper-Lock)_____

EXISTING BELT(s) Model and (QTY) _____ (_____)

Center-to-Center distance between Motor and Fan Shafts:

(ALLOWING for MID-RANGE take up Adjustment)_____ INCHES

## Existing MOTOR NAMEPLATE DATA:

HP _____ RPM_____ HZ_____ FRAME#_____

MODEL# _____ (ODP)_____ (TEFC) _____ EFF%_____

AC VOLTS: _____/_____          F.L. AMPS: _____/_____

## IMPORTANT!...Actual Measured Data @ FULL LOAD/RPM:

LINE AC VOLTS_____ AC AMPS_____ FAN RPM_____

## TYPE OF SPEED CONTROL SIGNAL REQUIRED:

Signal Following Type: (4-2Oma)_____ (O-lOvdc)_____ (Other) _____

Pressure (Range /Operating Setpoint) _____-_____/_____ (D/A)__ (R/A)_____

Speed/RPM Meter OPTION (LCD Display & speed sensor) required?_____

Distance Motor / Drive to Controller install location: _____ Feet

Belt Guard Modification / Fabrication required?_____

Submit detailed sketch / layout of AHU, motor, clearance dimensions, and all other pertinent information including belt guard requirements, if applicable.

Fax completed form to Coyote @: (817) 485-9437

CHAPTER 12
# ENERGY MANAGEMENT CONTROL SYSTEMS

**DALE A. GUSTAVSON, C.E.M.**
D.A. Gustavson Company
Orange, California
(714) 639-6100

**TOM LUNNEBERG, P.E.**
Manager, Energy Engineering
CTG Energetics, Inc.
Irvine, CA 92618
(949) 790-0010

**WILLIAM E. CRATTY**
**ALFRED R. WILLIAMS, P.E.**
Energy Conservation Mgmt. Co.
Bethel, Conn.

Competitive economic pressures on owners to reduce building operating expenses are challenging the traditional design and control of heating, ventilating, air conditioning (HVAC) and lighting functions. Facility owners and operators have strong financial incentives to match more closely control, zoning and HVAC equipment sizing to the use of building spaces and outside environmental conditions. This must be done without sacrificing comfort and safety. Energy management systems play a key roll in meeting this challenge.

## 12.1 ENERGY MANAGEMENT SYSTEMS

Energy management is the control of energy consuming devices for the purpose of minimizing energy demand and consumption. Manually toggling on and off devices based upon need is a rudimentary form of energy management. The advent of mechanical devices such as time clocks for automatic toggling and bimetallic strip thermostats to control the output of heating and cooling devices along with pneumatic and electrical transmission systems provided means for developing early energy management systems in the form of automatic temperature controls. The advent of solid state electronic control devices and the increasing power of the microprocessor based personal computer have led to dramatic advances in energy management and what today is termed the energy management control

system (EMCS). The primary difference between early automatic temperature control systems and EMCS is the application of a broad base of variables through programmable logic controllers to optimize the use of energy. While EMCS is used to control building environmental conditions and industrial processes, this chapter will deal only with EMCS applications for control of building HVAC and lighting functions.

### 12.1.1 Direct Digital Control

HVAC building control system manufacturers have greatly enhanced EMCS by incorporating direct digital control (DDC). DDC is defined as a digital computer that measures particular variables, processes this data via control algorithms and controls a terminal device to maintain a given setpoint or the on/off status of an output device. The term "digital" refers only to the fact that input/output information is processed digitally and not that input or output devices are digital. Inputs and outputs relative to a DDC EMCS can be either digital or analog. Typically, most inputs are analog signals converted to digital signals by the computer while the greater portion of outputs are likely to be digital (zero or full voltage).

DDC systems use software to program microprocessors, therefore providing tremendous flexibility for controlling and modifying sophisticated control applications. Changing control sequences by modifying software allows the user to improve performance of control systems throughout a building.

DDC EMCSs can be programmed for customized control of HVAC and lighting systems and perform facility wide energy management routines such as electrical peak demand limiting, ambient condition lighting control, start/stop time optimization, sitewide chilled water and hot water reset, time-of-day scheduling and outdoor air free cooling control. An EMCS using DDC can integrate automatic temperature control functions with energy management functions to ensure that HVAC systems operate in accord with one another for greater energy savings.

The most significant benefits DDC systems offer are: a) the ability to customize the scheduling of equipment and their component devices to react to ever changing conditions of use and weather, and b)

315

additional control modes (integral and derivative) that result in quicker, more accurate control when compared to pneumatic systems. Pneumatic systems inherently offer only proportional (how far is the input from setpoint) control in which the terminal control device linearly varies the output as an input variable changes relative to setpoint. Linear control results in offset or "hunting" by a terminal control device (valve) as it throttles to control to setpoint. By adding the integral (how long has the input been away from setpoint) mode to a proportional controller, offset is minimized as the control point is automatically reset while the controller continually shifts the throttling range. By adding the derivative (how fast is the input approaching/moving away from setpoint) mode the controller can achieve setpoint much quicker and more accurately whenever load varies. Generally, proportional-integral (PI) control is sufficient for building HVAC applications.

The primary objective of an EMCS using DDC is to optimize the control and sequencing of mechanical systems. A DDC EMCS also allows centralized and remote monitoring, supervision and programming maintenance of the HVAC and lighting functions. Additionally, such systems can lead to improved indoor environmental comfort and air quality.

Importantly, EMCS can also perform valuable non-energy related tasks (NERTs). These often overlooked DDC applications, or "extra-standard" functions, can be business specific or facility specific such as follows:

A design/build EMCS specialist for shopping malls adds people counting, vehicle counting, and precipitation logs to normal EMCS monitoring to help mall management lease space and track the success of mall promotions.

During a facility appraisal of a year-old drug store distribution center, a West Coast EMCS consultant/contractor discovered an opportunity to control the facility's conveyor system. The application not only resulted in energy savings, but also streamlined shipping.

An imaginative plant engineer commissioned the writing of custom front-end software for a brand name EMCS to perform noise level control in a factory. This 'stretch' of the EMCS now helps protect the hearing of employees.

DDC EMCSs are being used to control environments in mushroom farms and banana distribution centers, to control pivot-irrigation systems, and even to control snow-making equipment at ski resorts. The limiting of DDC EMCS application strictly to HVAC and lighting control diminishes the financial benefit of this technology to the system owner.

Sometimes a business-specific or facility-specific application will have a greater impact on EMCS justification than does the energy saving component. For example, a 1990 law change in California requires banks to maintain specified levels of lighting near their automatic teller machines (ATMs). The law was passed in response to muggings and killings at poorly lit ATMs. Some banks have responded by randomly installing additional lighting fixtures. This shortsighted approach still begs the question, what happens if the lights go out?

One possible DDC solution would be to measure and log an ATM's light level. When lumens drop below a specified level, that facility's EMCS could simultaneously activate an emergency backup lighting system, notify an alarm monitoring company, and automatically dial out to a central EMCS monitoring station to record and verify that the notification function worked. An alarm light or tone also could be activated at the site to alert the facility personnel to the problem the next morning. All of this by the same microprocessor that controls the HVAC and lighting? Easily! Would it be cost effective? That question might best be asked of the bank's legal department.

For more discussion on NERTs and other benefits of EMCSs, see Section 12.2.

### 12.1.2 Hardware

An EMCS can range from a very simple stand-alone unitary microprocessor based controller with firmware routines (control software logic that the user cannot modify except for setpoints) that provides control of a terminal unit such as a heat pump to a very sophisticated large building DDC EMCS that interfaces with fire and security systems. Although this chapter will generally cite examples from applications in medium to large facilities, there have been numerous documented successes with stand-alone controllers in small buildings. Fast food and dinner houses, auto dealerships, retail stores, bowling centers, super markets, branch banks, small commercial offices, etc., have all benefited from this technology. Typically, EMCS installations at small buildings are 'design/build'. That is, an EMCS manufacturer, distributor, installing contractor, end user, or some combination thereof select the EMCS hardware and decides how it is to be applied.

Much study remains to be done to determine the extent by which firmware routines themselves contribute to the economics in small building EMCS projects. Empirical data suggests that while EMCSs in small facilities do improve the accuracy and response of mechanical system controls, the energy management routines are responsible only partially for the savings which are achieved. In applications where unitary controllers have been treated simply as 'devices', by-in-large savings have proven to be less dramatic than where they have been treated as 'systems' and used as 'tools for saving energy'. Better applied, installed, documented, supported, and maintained EMCS projects have yielded better results than those in which 'black boxes' were hung on walls in broom closets and left alone to work their 'magic'.

What is it that would lead one restaurant chain to bypass or disconnect hundreds of small EMCSs and return to time clock and manual control while at another chain, using nearly identical systems and strategies, a 40% return on investment, better comfort, and more rapid HVAC service response is realized? The opposite results suggest that the "human dimension" is a factor which must be more clearly understood and weighed. This is true for small design/build projects as well as for very large, sophisticated, 'engineered' projects. The "people" factor must not be ignored. Different building managers and occupants have different needs and occupancy habits. Also, no matter how well a system is designed, if the building operators and occupants are not properly taught how to use and maintain it, the EMCS will never live up to its full potential. For more discussion on the importance of training and the "human dimension," see Section 12.3.8.

The selection of EMCS type and sophistication for any given application should balance management and control desires with economics. An all encompassing DDC EMCS will provide the best overall control and management capability, but it is also the most expensive. On the other hand, stand-alone controllers are the least expensive for individual control applications; they also restrict control strategy and management capability. In terms of energy savings, an all encompassing system may not provide significantly greater savings than using stand-alone controllers for all HVAC functions at a given building, but it surely provides for better and easier management of the mechanical systems. However, with appropriate telecommunications software a system of unitary stand-alone controllers connected to modems could prove to be a very effective EMCS for a chain of widely dispersed small facilities.

There are two basic economic opportunities for application of an EMCS. The first is a retrofit of existing buildings with functioning automatic temperature control systems. The second is in new construction. It could be very difficult to justify removing a functioning automatic temperature control system in an existing building to install an all encompassing DDC EMCS solely for energy conservation purposes. For example, an air handling unit is typically managed by a controller that monitors signals for outside air temperature, mixed air temperature, discharge air temperature, and return air temperature, compares the signals to setpoint conditions and in turn transmits a signal to a terminal control device to position the outside air dampers, exhaust dampers, return air dampers, and heating valve or chilled water valve to maintain a predetermined setpoint. If the unit has a variable speed drive, humidifier or dehumidifier there are even more conditions to monitor and control. To remove all of the existing controls and install an all encompassing DDC EMCS could be very costly. An economical alternative is to eliminate the traditional time clock function from the automatic temperature control system and interface a DDC EMCS in a "supervisory" mode over the existing controller by connecting key electric/pneumatic relays (EPs) to DDC outputs and installing strategically placed input sensors. A representative sampling of conditions such as a space temperature or discharge air temperature in conjunction with global inputs and well designed software in most cases would provide adequate management to maintain comfort and save a significant amount of energy.

The first cost of a DDC EMCS installation has decreased dramatically in the past several years, and DDC currently costs about the same or less than a pneumatic control system in most new construction applications. In addition to this first cost advantage, some advantages of DDC over pneumatic controls are: 1) more precise control, 2) unlimited customization of control schemes for energy management and comfort, 3) centralizes and integrates control and monitoring of HVAC and lighting, 4) easier to maintain, and 5) easier to expand and 'grow' with building size and use.

Current day DDC EMCS hardware configuration varies from manufacturer to manufacturer but has a common hierarchical configuration of microprocessor based digital controllers as well as a front-end personal computer. This configuration can be categorized by three levels: 1) terminal equipment level controllers, 2) system level controllers, and 3) the operator interface level. The three levels are networked together via a communications trunk which allows information to be shared between other terminal equipment controllers and all level controllers.

The most important part of a DDC EMCS are the system controllers because they monitor and control most mechanical equipment. A major feature of system level controllers is their ability to handle multiple control loops and functions such as proportional- integral-derivative (PID) control, energy management routines, and alarms while the terminal equipment controllers are single control loop controllers with specific firmware. The operator interface level is generally a personal computer that serves as the primary means to monitor the network for specific data and alarms, customize the control software for downloading to specific system controllers, maintain time-of-day scheduling, and generate management reports. However, it is not uncommon for a single personal computer to serve both as the system controller and operator interface.

The specific method of communications within an EMCS is significant because of the amount of data being processed simultaneously. While methods of data transfer between DDC controllers and the operator interface level vary from manufacturer to manufacturer, communications protocols can be simplified into two distinct categories: 1) poll/response, and 2) peer-to-peer or token-ring-passing network.

A peer-to-peer network does not have a communication master or center point as does the poll/response system because every trunk device, be it a terminal level controller attached to a systems level controller or a system level controller networked to other system level controllers, at some point, has a time slot allowing it to operate as the master in its peer grouping. However, the terminal level controller in a peer-to-peer network usually is a poll/response device. A system using peer-to-peer communication can offer distinct advantages over poll/response communication when redundancy of critical global data is accommodated. These advantages are: 1) communication is not dependent on one device, 2) direct communication between controllers does not require communication through the operator interface level, and 3) global information can be communicated to all controllers quickly and easily.

The speed of system communication in building control usually is not an issue with today's DDC technology. However, the response time for reaction to control parameters can be an issue with the application of a centralized poll/response system where control and monitoring point densities are very high or alarm response time is critical. Generally, peer-to-peer communication distributive systems that network system level controllers and terminal equipment level controllers provide the quickest response time. However, the ever increasing speed of the personal computer is allowing centralized poll/response systems effectively to expand their control and monitoring horizon.

Although quantum leaps in desktop computer processing horsepower lend some advantages to poll/response systems that use a personal computer for centralized control, panelized DDC EMCSs are still by far the most popular approach for providing digital control of building systems. It used to be the case that using a host PC for building system automation allowed more complicated sequences of operation to be implemented, and greater amounts of trend data to be stored. Today, however, reduction in physical size as well as cost for computer memory chips has led to much greater capabilities for stand-alone control units, and as a result there is not much of a performance gap between centralized poll/response and panelized DDC systems.

Be it new construction or a retrofit, when selecting an EMCS the actual needs and requirements of the particular building or campus of buildings must be considered. While facility layout and construction type will influence EMCS configuration, software is the most critical element in any EMCS application.

### 12.1.3 Software

The effectiveness of the software control logic is what provides the building operator with the benefits of an energy management system. While color graphics and other monitoring enhancements are nice features to have, the control logic in the system and terminal controllers is what improves building systems efficiency and produces the energy savings.

Networking DDC devices has provided building operators with the ability to customize the traditional energy management strategies such as time-of-day/holiday scheduling, demand shedding, duty cycling, optimum start/stop and temperature control as well as the ability to implement energy management techniques such as occupied/unoccupied scheduling of discrete building areas, resulting in reduced airflow volumes and unoccupied period setback strategies that greatly reduce operating cost. PID control provides for more accurate, precise and efficient control of building HVAC systems. However, software that provides for adaptive or self tuning control not only enhances savings but also addresses environmental quality. Adaptive control software monitors the performance of a particular control loop and automatically adjusts PID parameters to improve performance. This feature improves control loop response to more complicated and dynamic processes.

If the energy management industry has an Achilles heel it is the specifier who specifies an EMCS by reference to a particular manufacturer's general hardware specifications rather than detailing the specifics of software sequencing logic and system configuration. Specification by manufacturer reference is not a complement to a hardware manufacturer rather it is an indication of EMCS ignorance on the part of the specifier. HVAC mechanical equipment operation sequences are becoming increasingly complex. This is a result of stricter criteria established by national codes (American Society of Heating, Refrigerating and Air Conditioning Engineers' Minimum Outdoor Air Requirements Standard 62 for example) and by building operators who require highly accurate control systems, energy use monitoring and accounting by zone of use. To meet the challenges posed by these stricter codes and building operator demands the EMCS specifier must be thoroughly knowledgeable about all aspects of EMCSs.

As suggested in the earlier discussion of the human dimension phenomenon, it is essential that the EMCS specifier look beyond technology or, as the old saying goes, look beyond the trees to the forest. That is, a specifier must be knowledgeable about not only the building construction and mechanical and lighting systems to be controlled but also the businesses and people which occupy the building. And, of equal importance is the need to assess the level of training and ongoing support that the EMCS owner/operator will need to realize the anticipated benefit.

### 12.1.4 Control Strategies

A DDC EMCS can serve six basic functions: 1) manage the demand or need for energy at any given time, 2) manage the length of time that devices consume energy, 3) set alarms when devices fail or malfunction, 4) facilitate monitoring of HVAC system performance and the functioning of other building systems, 5) assist the building operator to administer equipment maintenance, and 6) provide the building/business owner/operator with non-energy related tasks (NERTs), i.e., extra-standard functions to make the EMCS more effective or advantageous. There are financial benefits for each of these six functions, yet traditionally, only the first two are quantified in an economic analysis. The industry still has work to do to develop new, accurate, reliable models for predicting energy savings by an EMCS. But, to limit the inquiry to just energy savings falls far short of what is necessary for EMCS technology to realize anywhere close to its truly staggering potential.

It is the industry's responsibility to develop models for predicting the value of all EMCS functions. It is too often left to sales and marketing people to explore and, even more importantly, assert the full range of benefits of EMCS technology. The energy and engineering communities tend to stop where science ends and speculation begins. Speculation is a healthy exercise. Here are some questions which might be worth asking:

What is the annual dollar value of an efficient office building 95% leased as compared to a less efficient building 75% leased?

What is the annual dollar value of being able to provide basic comfort troubleshooting by telephone using a personal computer and remote communication software?

What is the annual dollar value of increased productivity in a building, which suffers little or no discomfort?

What is the value of lowering absenteeism by using EMCS to maintain indoor air quality monitoring and preventive action?

What if one of the above benefits is the difference between justifying an EMCS project and not going ahead? The key to making the most of NERTs, or extra-standard EMCS functions is recognizing that: 1) quantified benefits have more impact than unquantified 'hopes' when buying decisions are made, 2) it is the industry's duty to continually observe, track and create new EMCS 'by-products', and 3) only imagination limits the range of extra-standard functions which can be incorporated into the EMCS design. The extent and proficiency of performing these functions depends on the software programming.

The energy management industry has evolved standard time-proven software routines that can be used as a starting point to develop effective programming. These routines are commonly referred to as control strategies. While these routines can be applied in virtually all EMCS installations they must be customized for each building. A description of some of these routines follows.

Daily Scheduling: This routine provides for individual, multiple start and stop schedules for each piece of equipment, for each day of the week.

Holiday Programming: This routine provides for multiple holiday schedules which can be configured up to a year in advance. Typically, each holiday can be programmed for complete shutdown where all zones of control are maintained at setback levels, or for special days requiring the partial shutdown of the facility. In addition, each holiday generally can be designated as a single date or a range of dates.

Yearly Scheduling: Typically, any number of control points can be assigned to special yearly scheduling routines. The system operator usually can enter schedules for yearly scheduled control points for any date during the year. Depending on the particular EMCS software operating system, yearly scheduled dates may be erased once the dates have passed and the schedules have been implemented; or, the scheduled dates may repeat in the following year as do scheduled holiday dates.

Demand Limiting or Load Shedding: Demand limiting can be based on a single electric meter or multiple meters. Generally, loads are assigned to the appropriate meters in a specified order in which equipment is to be shed. Usually, a control point that is allowed to be shed is given a status as a "round-robin" demand point (first off-first on), a "priority" demand point (first off-last on), or a "temperature" demand shed point (load closest to setpoint is shed first). Other parameters per control point include Rated kW, Minimum Shed Time, Maximum Shed Time and Minimum Time Between Shed.

Minimum On-Minimum Off Times: Normally a system turns equipment on and off based on temperatures, schedules, duty cycles, demand limits and other environmental parameters. However, mechanical equipment often is specified to run for a minimum amount of time once started and/or remain off for a minimum amount of time once shut down. Therefore, this routine gives the operator the ability to enter these minimum times for each piece of equipment controlled.

Duty Cycling: This routine provides the ability for a control point to be designated for either temperature compensated (cycle a piece of equipment on and off to maintain a setpoint within a dead band) or straight time dependent (cycle a piece of equipment on and off for distinct time intervals) duty cycling. Control parameters for temperature compensated duty cycling include Total Cycle Lengths, Long and Short Off Cycles and Hi and Low Temperatures. There are separate sets of parameters for both heating and cooling. Time dependent duty cycling uses Total Cycle Length and Off Cycle Lengths. Cycles for each load can be programmed in specific minute increments and based on a selectable offset from the top of the hour.

Optimum Start/Stop: For both heating and cooling seasons DDC systems can provide customized optimum start/stop routines which take into account outside temperature and inside zone temperatures when preparing the building climate for the occupant or shutting the facility down at the end of the day. During unoccupied hours (typically at night) the software tracks the rate of heat loss or gain and then utilizes this data to decide when equipment will be enabled in order to regain desired climatic conditions by the scheduled time of occupation. The same logic is used in reverse for optimum stop.

Night Setback: This routine allows building low temperature limits for nighttime, weekend and holiday hours, as well as parameters and limits for normal occupied operation to be user selectable. Night setback is usually programmed to evaluate outside air temperature in the algorithm.

Hot Water Reset: This routine varies the temperature of the hot water in a loop such that the water temperature is reduced as the heating requirement for the building decreases. The reset temperature for night setback usually is less than the day reset temperature. Typically, reset is accomplished by controlling a three-way valve in the hot water loop; however, depending on boiler type, reset can be accomplished by "floating" the aquastat setpoint. For pneumatic control systems, the primary variable in the reset algorithm is outside air temperature. For newer systems that have DDC controls on air handling units, it is possible to reset the heating hot water temperature based on the "worst case" hot water control valve position. Under this strategy, the heating hot water temperature is reset downwards until the air handling unit with the greatest need for heating has its hot water control valve fully open. For VAV systems that feature DDC control at the zone level, it is also possible to reset the temperature of hot water feeding zonal reheat coils based on the "worst case" zone. Under this strategy, each DDC zone controller reports the position of its hot water reheat control valve, and the temperature it reset downwards until the zone with the greatest reheat requirements has its valve fully open. This strategy can provide a significant reduction in wasteful reheating of conditioned air.

Boiler Optimization: In a facility that has multiple boilers this routine schedules the boilers to maximize plant efficiency by staging the units to give preference to the most efficient boiler, by controlling the burner firing mode when desirable, and by minimizing partial loading.

Chilled Water Reset: This routine varies the temperature of the chilled water in a loop such that the water temperature is increased as the cooling requirement for the building decreases. This is typically done by controlling a three-way valve in the chilled water loop. Chilled water reset also can be accomplished by interfacing the EMCS with the chiller controls to reduce the maximum available cooling capacity such as during demand limiting. For pneumatic control systems, the primary variable in the reset algorithm is outside air temperature. For newer systems that have DDC controls on air handling units, it is possible to reset the chilled water temperature based on the "worst case" chilled water control valve position. Under this strategy, the chilled water temperature is reset upwards until the air handling unit with the greatest need for cooling has its chilled water control valve fully open.

In this way, the chilled water temperature will be reset to as high as possible—which improves chiller capacity and efficiency—without compromising comfort. It must be noted that this strategy must be carefully implemented with variable flow chilled water systems, as it is possible for improvements in chiller efficiency to be eclipsed by increased pumping energy.

Chiller Optimization: In a facility that has multiple chillers this routine schedules the chillers to maximize plant efficiency by staging the units to give preference to the most efficient chiller. Unlike the boiler optimization routine, however, it can be more efficient to operate a chiller plant with units sharing the load rather than fully loading a unit.

Chiller Demand Limiting: This routine limits chiller demand by interfacing the EMCS with the chiller controls to reduce the maximum available cooling capacity in several fixed steps. The primary variable in the demand limiting algorithm is kW demand.

Free Cooling: Free cooling is the use of outside air to augment air conditioning or to ventilate a building when the enthalpy (total heat content) of the outside air is less than the enthalpy of the internal air and there is a desire to cool the building environment. In arid climates, an economizer cycle can work well by measuring dry-bulb temperature only. In climates where humidity is of concern, enthalpy-based controls are preferable as they can provide greater comfort and increased energy savings. An economizer can be categorized as integrated (meaning it can operate in conjunction with mechanical cooling) or non-integrated (meaning that the outside air damper is fixed at its minimum position when mechanical cooling is required).

Recirculation: This routine provides for rapid warm-up during heating and rapid cool-down during cooling by keeping outside air dampers fully closed and return air dampers fully opened during system start-up.

Hot Deck/Cold Deck Temperature Reset: This routine selects the zones or areas with the greater heating and cooling requirements and establishes the minimum hot and cold deck temperature differential which will meet the requirements in order to maximize system efficiency. (NOTE: Multi-zone system)

Motor Speed Control: This routine will vary the speed of fan and pump motors to reduce air and water velocity as loads decrease. Speed control can be accomplished either by controlling a two speed motor with a digital output or a variable speed drive with an analog output.

Manual Override: This routine provides separate manual override schedules to allow for direct control of equipment for specific periods of time.

Duty Logs: DDC systems can track and display various types of information such as last time on, last time off, daily equipment runtimes, temperatures (or other analog inputs) kWh and even remote panel hardware performance. DDC systems can also accumulate and display monthly scheduled and unscheduled runtimes for each piece of equipment controlled or monitored.

Temperature and hardware performance information can be monitored and displayed. Change of state can be logged for all loads and can be reviewed on a daily basis for all loads. In addition, a system can produce line graphics or historical trends of input data and hardware systems information.

Energy consumption logs can be kept separately for every pulse meter point. For electricity these energy logs can record data (by the day) such as daily kilowatt hour consumption, kilowatt demand, time the peak occurred, selected demand limit, time that any load was shed, minimum, maximum and average outside temperatures and degree day information.

Alarm Monitoring and Reporting: DDC systems register and display alarms for conditions such as 1) manual override of machinery at remote locations, 2) equipment failures, 3) high temperatures, 4) low temperatures, invalid temperatures (sensor is being tampered with), and 5) communication problems.

## 12.2   JUSTIFICATION OF EMCSs

Traditionally, a commercial building is constructed by one party (a developer) owned by a second party (investors) and operated by a third party (a facilities management firm). Typically, the building will have one electric meter and a central heating plant. Therefore, total energy cost must be included in the rental structure. Before the 'first energy crisis' and energy was cheap, lessees typically paid a flat rental charge based on square footage rented and nobody paid much attention to operating hours of the HVAC and lighting systems.

When energy prices started their rapid upward spiral building owners and operators implemented measures to pass the cost increase directly on to the lessee. However, now because of the volume of space available and other competitive pressures building owners and operators are being forced to find means of reducing the energy cost component of the base rent in order to reduce operating costs and entice customers with lower base rentals in the lease market.

Depending on the type of HVAC and lighting control systems installed in a building, energy consumption and operating costs can be dramatically different. Therefore, it should go without saying that since operating costs are a key ingredient when establishing competitive rentals, the type of control system installed can have a significant impact on the owner's income. And, to follow this line of reasoning to its conclusion, since rental income is a key ingredient in the appraisal of a commercial facility, the type of control systems installed can have a significant impact on the value placed on the building should the owner want to sell.

A well-designed and properly operated EMCS is an investment which offers the building owner/manager a multitude of benefits, which can oftentimes be bundled and quantified. These benefits include, yet go beyond, impressive payback and return-on-investment figures. In addition to the immediate financial benefits, there are long-range benefits to managing energy and demand as well. Such benefits can be wide-ranging, positively affecting society in general. Consider the following EMCS functions, along with the related benefits:

- **Manage energy consumption and demand**—Lower operating expenses, higher profits and increased competitiveness, keep energy prices affordable in the long term. Less smog, reduced acid rain and less global warming.

- **Optimize operating efficiencies of energy consuming equipment**—Same benefits as above plus extended equipment life which further increases profits and competitiveness.

- **Improve comfort**—Increases occupant productivity and concentration. Helps occupants feel more alert, rested, make fewer mistakes and perform better. Leads to a more vital U.S. economy with better educated and less stressed populace.

- **Improve indoor air quality**—Saves lives, increases productivity, prevents lawsuits, lowers insurance costs and reduces absenteeism. Contributes to better economical and educational systems.

- **Activate alarms when equipment malfunctions**—Provides for less costly disruptions of productivity due to faster response, increased profitability, extended equipment life and increased competitiveness.

- **Assist on- or off-site operator administer service/maintenance**—Results in better records, histories for more efficient, accurate work, less downtime, more productivity and lower cost for service. Keeps polluting vehicles off the street. Leads to higher profits and increased competitiveness.

- **Monitor/log building equipment performance and energy use**—Resulting, accumulated data leads to improved performance of all EMCS functions, thereby increasing magnitude of all benefits itemized above. Also confirms cost avoidance benefit of project.

- **Perform NERTs (non-energy related tasks)**—For every NERT there is a corresponding NEB (non-energy benefit). In fact, the value of NERTs can exceed the value of all the other benefits combined. (Refer back to Section 12.1.1 for specific examples of NERT applications.) Some NEBs can be easily quantified and others cannot, but they should be articulated in your investment justification.

While some of these benefits may seem far-reaching, different building owners and operators have different needs and requirements. Energy management professionals should always be on the alert for specific ways an EMCS can benefit a facility, as each case is unique.

### 12.2.1 An EMCS Opportunity

For purposes of discussion, consider a ten story, multiple tenant, commercial office building configured and occupied as follows as an example of an opportunity for an EMCS application (Figure 12.1).

a. One boiler room in the basement with 2 boilers and 2 redundant sets of circulating pumps (4 in all) to supply hot water to perimeter radiation and heating coils.

b. One domestic hot water system for the entire building.

c. Four mechanical rooms—a north and a south mechanical room on the 5th floor each with 2 air handling units (AHUs) supplying the first 5 floors and a north and a south mechanical room on the 10th floor each with 2 AHUs supplying the top 5 floors. Each AHU has a preheat coil on the inlet side of the supply fan and a chilled water coil and a reheat coil on the discharge side to maintain desired hot deck and cold deck temperatures for a double duct distribution system. The north mechanical room on the 10th floor has a chiller and

**Figure 12.1 Ten-story office building.**

a redundant set of circulators to service the north side AHUs while the south mechanical room on the 10th floor has a chiller and a redundant set of circulators to service the south side AHUs.

d. One roof top mounted cooling tower serving both chillers.

e. The building is zoned by AHU (8 zones) and each room has a mixing box that is controlled by its own thermostat to allow heating and cooling as required.

f. Toilet exhaust is provided by 2 roof mounted exhaust fans that are switched in one of the 10th floor mechanical rooms.

g. Corridor and lobby lighting is switched from a central location while space lighting is operated by a wall switch in each space.

h. The building is normally occupied between 8:00 am and 5:00 p.m. Monday thru Friday and sporadically occupied a various times after normal hours until midnight on work days and occasionally on weekends.

An opportunity for use of DDC EMCS in either retrofits or new construction can be found when considering building use. One of the keys to effective utilization of an EMCS is zoning. Through flexible program scheduling to control the energy consuming systems by zone the building operator can realize substantial energy savings.

Most likely an existing building configured and occupied as the above example building will have manual lighting control and a mechanical time clock based pneumatic system controlling HVAC functions as follows:

a.   Each mechanical room has a time clock activated control panel for occupied/unoccupied switch over of day/night air pressures for space thermostats and start/stop of the mechanical equipment.

b.   Each AHU has a mixed air temperature controller, a cold deck temperature controller and a hot deck temperature controller to maintain supply air temperatures at preset levels.

c.   Outside air dampers are calibrated to maintain desired minimum air intake.

d.   Each zone has a centrally located night thermostat for its respective AHU.

e.   In the boiler room there is a summer/winter change over of the heating and cooling functions.

f.   Corridor and lobby lighting is controlled by a mechanical time clock in the electrical control room.

To accommodate after normal hours occupancy the temperature control system provides for occupied conditions from 7:00 am to midnight on weekdays. On weekends the system provides for occupied conditions from 8:00 am to 5:00 pm. Toilet exhaust fans run continuously. The corridor and lobby lighting controller is set to maintain lights on from 6:00 am to 1:00 am on weekdays and from 7:00 am to 6:00 pm on weekends. Cleaning crews and tenants are responsible for control of lighting in the tenant spaces and toilets.

In multiple tenant, high rise commercial office buildings with HVAC and lighting systems configured and operated in a manner similar to the example facility the DDC EMCS provides an opportunity to save a significant amount of energy and recover costs associated with after normal hours tenant use.

## 12.2.2 The EMCS Retrofit

Without the benefit of a DDC EMCS the operator of the example building must incur after-normal hours operating costs and allocate them over the entire leasing cost. This unfairly spreads the costs generated by a few tenants over the lease costs of all tenants. The overall effect are higher leasing charges for all.

The most cost effective way to retrofit a DDC EMCS in a building controlled by a reasonably maintained pneumatic control system is to interface it with the existing pneumatics in a supervisory mode by eliminating night thermostats, replacing time clock contacts with DDC outputs, direct hard wiring for start/stop control of the equipment not under the influence of the pneumatic system (i.e. toilet exhaust fans, domestic service hot water circulators, and corridor, lobby and exterior lighting); direct hard wiring boilers, chillers and hot water and chilled water circulators for start/stop control, and direct hard wiring to enable/disable the equipment under control of the pneumatic system (i.e. the AHUs and dampers). Enabled equipment is allowed to operate under the control of the pneumatic system. Disabled equipment is shut down (Figure 12.2).

With a supervisory DDC system the building operator can effect a dramatic reduction in energy usage by incorporating a simple timed on/off control strategy for all HVAC and lighting zones that segregates the operation of the equipment into discrete tenant zones. The control strategy would shut down the tenant zones when outside of normal lease times. If a tenant wished to utilize a space after hours, he/she could initial an override (using a simple automated telephone interface from his/her office) to signal the system to preserve his/her working environment. Additionally, on weekends and holidays the entire building can be kept in the unoccupied mode unless a tenant requires extended occupancy of a particular zone. That tenant could initiate an override from any outside telephone to ensure occupied conditions for his/her space upon arrival at the building. Whenever an override is initiated, an operator interface level personal computer could track the energy consumption and/or runtimes of that zone's HVAC equipment for the purpose of billing the tenant an after hours usage charge.

Experience at a multiple tenant, high rise commercial office building on Long Island, New York, has shown that this "tenant strategy" not only results in a dramatic reduction in energy usage but also allows the building operator to reduce the base rental and recover after normal hours operating cost directly from those tenants responsible for the use.

By installing a DDC system as a supervisor of and existing pneumatic control system the building operator

| AI-1 | Boiler #1 water supply temperature | | BO-1 | Boiler #1 S/S |
| AI-2 | Boiler #2 water supply temperature | | BO-2 | Boiler #2 S/S |
| AI-3 | Common boiler water supply temperature | | BO-3 | Circ Pump #1 S/S |
| AI-4 | North loop supply temperature | | BO-4 | Circ Pump #2 S/S |
| AI-5 | South loop supply temperature | | BO-5 | Circ Pump #3 S/S |
| AI-6 | North loop return temperature | | BO-6 | Circ Pump #4 S/S |
| AI-7 | South loop return temperature | | AO-1 | Reset control north |
| AI-8 | Outside air temperature sensor | | AO-2 | Reset control south |
| AI-9 | Ambient light sensor | | BO-7 | Summer winter changeover |
| BI-1 | Circ Pump #1 flow | | BO-8 | Exterior lighting |
| BI-2 | Circ Pump #2 flow | | BO-9 | Domestic hot water |
| BI-3 | Circ Pump #3 flow | | BO-10 | Alarm |
| BI-4 | Circ Pump #4 flow | | | |

SLC

TELC #1    TELC #2    TELC #3    TELC #4

Typical for TELC 2, 3, 4

| AI | ① | Hot deck sensor AHU-1 | | AI-11 | Rm sensor AHU-1 |
| AI | ② | Cold deck sensor AHU-1 | | AI-12 | Rm sensor AHU-1 |
| AI | ③ | Mixed air sensor AHU-1 | | AI-13 | Rm sensor AHU-2 |
| AI | ④ | Return air sensor AHU-1 | | AI-14 | Rm sensor AHU-2 |
| AI | ⑤ | Chiller outlet #1 | | AI-15 | Hot deck sensor AHU-2 |
| AI | ⑥ | Spare | | AI-16 | Cold deck sensor AHU-2 |
| BI | ⑦ | Chilled water Pump #1 flow | | AI-17 | Mixed air sensor AHU-3 |
| BI | ⑧ | Chilled water Pump #2 flow | | AI-18 | Return air sensor AHU-2 |
| BI | ⑨ | Cooling water Pump #1 flow | | AI-19 | Rm sensor AHU-1 |
| BI | ⑩ | Cooling water Pump #2 flow | | AI-20 | Rm sensor AHU-1 |
| | | | | AI-21 | Rm sensor AHU-2 |
| BO | ① | AHU #1 S/S | | AI-22 | Rm sensor AHU-2 |
| BO | ② | AHU #2 S/S | | | |
| BO | ③ | Chiler S/S | | | |
| BO | ④ | Day/night switchover AHU-1 zone | | | |
| BO | 5 | Lighting control | | | |
| BO | 6 | Toilet exhaust | | | |
| BO | 7 | AHU damper control (both units) | | | |
| BO | 8 | Day/night switchover AHU-2 zone | | | |

SLC - System level controller
TELC - Terminal equipment level controller
AI - Analog input
BI - Binary input
AO - Analog output
BO - Binary output
S/S - Stop/start

**Figure 12.2 Office building supervisory EMCS configuration,**

can implement traditional control strategies (see section 12.1.4) to improve building operating efficiency and achieve additional energy cost reductions. In addition to optimizing heating and cooling functions, the toilet exhaust fans, domestic service hot water and corridor and lobby lighting can be interlocked with zone occupied/unoccupied status to shut down these systems when not needed. Pay backs for this type of EMCS retrofit are typically very attractive.

### 12.2.3 New Construction EMCS

If the exampled building were under construction as a new building, the developer would be remiss not to consider a DDC EMCS as part of the original design. In addition to all of the benefits described above, the all encompassing system would provide added benefit in the form of PI control of heating and cooling valves and dampers. The design engineer also would be remiss not to consider improvements in the mechanical design that along with DDC control could significantly improve the energy efficiency of the building.

By combining more aggressive zoning with DDC EMCS in new construction the building designer has the opportunity to better match control of the HVAC with building usage and optimize equipment sizing to better fit loads thus improving equipment operating efficiency. However, the bulk of the EMCS opportunity lies in the thousands of existing buildings which are controlled with pneumatic and electrical/mechanical automatic temperature control systems. Surprisingly enough there are also buildings still being controlled by hand!

## 12.3   SYSTEMS INTEGRATION

The performance of an EMCS is directly proportional to the quality of the designer's systems integration effort. Systems integration can be defined as those activities required to accomplish the precise monitoring and control desired when incorporating an EMCS into a building's HVAC and lighting systems. This definition implies that systems integration includes not only selecting the appropriate EMCS and detailing monitoring points and control interfaces, but also, customizing the software algorithms to meet the user's needs and commissioning the system.

### 12.3.1 Facility Appraisal

An appraisal of the job site is the first step of EMCS systems integration. The appraisal is generally done by interviewing the building users to determine building usage patterns (be careful to differentiate between normally scheduled usage and after hours usage),

to identify any comfort complaints (i.e. are there 'hot' or 'cold' spots in the building, is ventilation adequate, etc.), to assess the attitude that the building maintenance staff has towards an EMCS, and to determine exactly what the building manager expects the system to accomplish.

During the facility appraisal, the following people should be interviewed:

Tenants (by company, department, job function)
Prospective tenants
Delivery services
Security personnel
Service vendors (including janitorial, HVAC, lighting, electrical, life safety)
Building engineers/facility managers
Corporate energy manger, if any
Building owner/CEO/CFO
Property manager and staff
Leasing agents.

Perhaps the most overlooked above are leasing agents. These professionals must understand the benefits of energy efficiency, or they will not be able to sell it to perspective tenants.

### 12.3.2 Equipment List

Next, the designer should develop a complete and accurate equipment list. In a retrofit situation the list should include all of the equipment and devices to be controlled, the present method of control for these items, the operating condition of the items and their controls, and the area each item services. Do not forget to include exterior and interior lighting on the list.

### 12.3.3 Input/Output Point Definition

The third step is to define the input and output points required to achieve the monitoring and control desired, in the most cost-effective or efficient manner. This requires an in-depth appraisal of such items as the equipment list, layout of building spaces, method of building construction, HVAC physical plant makeup and layout, methods of heating and cooling, types of secondary HVAC distribution systems and controls, and the condition of the mechanical equipment and existing controls.

The definition of a control point should include a listing of all inputs that will influence the functioning of that point. The installation specifications should detail this information and indicate the nature of the influence as follows:

| Output Point | Analog Input Point | Notes |
|---|---|---|
| AHU #1 | Outside Air Temperature | |
| | Room #1027 Temperature- - - - - - - - - - - - - - - - - | |
| | Room #935 Temperature | |
| | Room #837 Temperature | Room of greatest |
| | Room #715 Temperature | Demand |
| | Room #625 Temperature- - - - - - - - - - - - - - - - - | |

## 12.3.4 Systems Configuration

Once the inputs and outputs have been defined, the designer needs to define any constraints that might be placed on the EMCS hardware. Building size, building design, building location, building use, design of the physical plant, type and layout of the secondary distribution systems, design of the electrical system, and type and condition of existing controls in retrofits as well as the nature of the personnel operating the building are some of the items to be considered when configuring an EMCS.

Cost of installation can vary significantly depending on the type of system architecture (centralized versus distributive), the method of transmitting information between system controllers and the input/output devices (direct hardwire versus multiplexing versus powerline carrier), the types of system controllers and input/output devices selected (two way versus one way communicators) and the amount of programming required (customized algorithms versus firmware).

Knowing who is going to maintain the software and perform the necessary administrative functions (i.e. inputting schedule changes, monitoring alarms, and producing reports) not only affects how many but also where the operator interface level personal computer is to be located, and the type of monitoring and alarming that will be needed. This will also determine training requirements and the extent of support services needed.

## 12.3.5 Specifics of Software Logic

The most critical phase of systems integration is the software specification. The designer must take care to be as explicit as he/she can when detailing the nature of the algorithms to be installed. This can be done by structuring the software specifications in a manner that clearly defines the algorithms and their interrelationships. The designer who simply includes a manufacturer's general hardware specifications rather than detailing the specifics of software sequencing logic most assuredly dooms the system to failure.

APPENDIX A "EMCS Software Specifications" is a sample of what can be described as a minimally acceptable software specification. Ideally, in addition to a descriptive specification, the designer would leave little doubt as to exactly what he/she has in mind by providing a listing, in easy to understand English format, of all system application programs associated with each piece of equipment. Following is an example of what a listing might look like for control of boilers.

---

## HOT WATER BOILER CONTROL

Where there are multiple boilers in a set the software shall change the lead-lag status of boilers 1 and 2 in the set on a biweekly basis (Note the call to the lead-lag algorithm). This specific algorithm calls the general CENTHEAT algorithm which assumes that the heating circulation pump is stage one of the central heating plant.

For the following control points CONTRACTOR shall apply this HOT WATER BOILER CONTROL algorithm:

1. Boiler 1
2. Boiler 2

VARIABLE DEFINITION:

**User defined variables (operating parameters):**

| | |
|---|---|
| HEATSET: | The heating target temperature (72°F). |
| SETBACKTEMP. | The minimum set back temperature allowed (55°F). |
| MAXOATHEAT: | The maximum outside air temperature for heating to be utilized (65°F). |
| OCCLOWOAT: | The low outside air temperature parameter to be used for hot water reset during occupied time of day (0°F). |
| OCCHIGHOAT: | The high outside air temperature parameter to be used for hot water reset during occupied time of day (60°F). |

OCCLOWSWT:       The low hot water supply temperature parameter to be used for hot water reset during occupied time of day (120°F).

OCCHIGHSWT:      The high hot water supply temperature parameter to be used for hot water reset during occupied time of day (190°F).

UNOCCLOWOAT:     The low outside air temperature parameter to be used for hot water reset during unoccupied time of day (0°F).

UNOCCHIGHOAT:    The high outside air temperature parameter to be used for hot water reset during unoccupied time of day (40°F).

UNOCCLOWSWT:     The low hot water supply temperature parameter to be used for hot water reset during unoccupied time of day (120°F).

UNOCCHIGHSWT:    The high hot water supply temperature parameter to be used for hot water reset during unoccupied time of day (160°F).

## Measured variables:

ACTUALTEMP:      The temperature being controlled by the I/0 point in question. In this case of a central boiler, the ACTUALTEMP shall represent the coolest of all of the zone temperatures. Refer to the Analog Temperature Assignment List to relate the temperature inputs to the control outputs.

OATEMP:          Outside air temperature.

## Calculated variables:

MANOVRFLAG:      Holds the cue for a manual override of normal program logic, 0 = manually override off, 1 manually override on, 2 = no manual override in effect. This is calculated by the MANOVR sub program.

MINFLAG:         Holds the cue for whether a point is currently being controlled by its minimum on or minimum off time, 0 = it is not currently affected by minimum on/off, 1 = it is currently affected by minimum on/off times.

TOD:             Holds a value of 1 if the point is within its time of day schedule, 0 if the point is not within its time of day schedule.

MODETEST:        Holds the cue for the System Mode, 0 = heating season, 1 = non-heating season. Its contents are determined by the MODETEST algorithm that is called from this algorithm.

TURNON:          Holds a value of 1 if the point should be turned on, 0 if the point should be turned off.

## PROGRAM CODE BLOCK

```
        ' identify point characteristics

500     COOLFLAG = 0        ' not a cooling point.
        HEATFLAG = 1        ' heating point.
        TODFLAG = 1         ' time of day point.
        OPTSTFLAG = 1       ' optimum start point.
        TCFLAG = 1          ' temperature control point.
        OALFLAG = 1         ' low outside air temperature point.
        SETBACKFLAG = 1     ' night setback point.
        FREEZEFLAG = 1      ' freeze protection point.

1000    TURNON = 0          ' assume that it can be turned off.
        CALL LEAD-LAG       ' see if it is the lead or lag boiler.
        CALL MANOVR         ' manual override.
            IF MANOVRFLAG = 1 THEN TURNON = 1 : GOTO 3000
        CALL MINTEST        ' minimum on/off test.
            IF MINFLAG = 1 THEN GOTO 3000 ' TURNON is set in MINTEST ' when MINFLAG is = 1
        CALL FREZPROT       ' freeze protection override.
            IF TURNON = 1 THEN GOTO 3000
        CALL MODETEST       ' check to see if in heating or non-heating.
            IF SYSMODE = 1 THEN TURNON = 0 : GOTO 3000

        ' outside air heat lockout.
            IF OATEMP > MAXOATHEAT AND OALFLAG = 1 THEN TURNON = 0 : GOTO 3000
        CALL SNOWDAY — IF SNOWDAYFLAG = 1 THEN TOD = 0 : GOTO 2010
        CALL OPENHOUSE — IF OPHOUSEFLAG = 1 THEN TOD = 1 : GOTO 2000
        CALL ACTIVITY — IF ACTIVITYFLAG = 1 THEN TOD = 1 : GOTO 2000
        CALL TOD            ' see if its scheduled now.

        ' load appropriate hot water parameters, call outside air
        ' reset program to calculate target water temps then call
        ' the appropriate algorithm to set TURNON variable.
2000        IF TOD = 1 THEN
            LOWOAT = OCCLOWOAT : HIGHSWT = OCCHIGHSWT
```

```
          HIGHOAT = OCCHIGHOAT : LOWSWT = OCCLOWSWT
     CALL OARESET

     ' if there is no call for heat, leave boiler off.
          IF ACTUALTEMP > HEATSET THEN TURNON = 0 : GOTO 3000
     CALL CENTHEAT
          END IF

2010      IF TOD = O THEN
          LOWOAT = UNOCCLOWOAT : HIGHSWT = UNOCCHIGHSWT
          HIGHOAT = UNOCCHIGHOAT : LOWSWT = UNOCCLOWSWT
     CALL OARESET

     ' if there is no call for heat, leave boiler off
          IF ACTUALTEMP > SETBACKTEMP THEN TURNON = 0 : GOTO 3000

     CALL CENTHEAT
          END IF

3000  END BLOCK
```

## 12.3.6 Control Point Tie-Ins

To ensure that the EMCS is interfaced to a device to be controlled exactly as desired, the integrator should include sketches detailing the specifics of the interface at the control circuit of that device as exampled in Figure 12.3.

## 12.3.7 Commissioning

The final step of systems integration is commissioning. Commissioning should include both the hardware and software. All of the hardware should be physically inspected for correct installation and the wiring tested for continuity. Inputs should be checked for accuracy against a standard. The operation of all outputs should be demonstrated from the operator interface personal computer or system controller if no personal computer is installed. This inspection should not be taken lightly. The system will not function as designed if any hardware component is incorrectly installed or does not function as desired.

The software should be checked before the system is put on the line. This requires a careful review of all algorithms, setpoints and schedules for compliance with the detailed specifications.

APPENDIX B "EMCS Installation Requirements" provides the basic parameters to formalize specifications to bid the installation of a DDC EMCS. Commissioning procedures are detailed in sections B15, B16 and B17.

## Harmonics

Modern electronic ballasts used in fluorescent lighting systems, as well as other digital switching devices, can generate harmonics in the electrical distribution system. There have been reported cases of such harmonics causing malfunctions in older control systems that communicate using power line carrier control (PLC) technology (such systems use the electrical distribution system as a means for inter-device communication by superimposing their digital "messages" over the 60 Hz power sine wave). Modern PLC systems that use spread spectrum technology are much more tolerant of harmonics, and encounter very few problems.

## Communications Technology

Though most digital control systems rely on hard-wired connections between devices, there are alternative technologies that can be used to simply installation (particularly in retrofit applications), provide a communication link with remote locations, and increase data transmission rates. These include:

- **Power Line Carrier (PLC).** PLC systems use the existing electrical wiring system as a means of achieving inter-device communication. This is accomplished when a device generates a high frequency communication signal that is superimposed over the standard sixty hertz power sine wave. Other devices that are connected to the same electrical system can analyze the spectral content of the incoming power, and extract the superimposed "message" from the sixty hertz carrier wave. PLC systems can provide reduced installation cost because of the reducing wiring requirements, but may be less reliable and slower than hard-wired systems.

- **Wireless Technology**. For locations that are located a significant distance away from a building, there is the ability to send and receive control messages using wireless technology. Some systems are based on

HVAC SCHEMATIC KEY

M       DAMPER MOTOR
T1      MIXED AIR CONTROLLER
T2      OUTSIDE AIR TEMPERATURE LIMIT CONTROLLER
T3      HOT DECK TEMPERATURE CONTROLLER
T4      COLD DECK TEMPERATURE CONTROLLER
T5      INDIVIDUAL ROOM THERMOSTAT
T6      NIGHT THERMOSTAT [REMOVE]
FS      FREEZESTAT
EP-1    MAIN AIR CONTROL
PE-1    PNEUMATIC DAY/NIGHT SWITCHOVER [REMOVE]
SR      SWITCHING RELAY FOR MINIMUM AIR
PRV     MINIMUM AIR PRESSURE REDUCING VALVE
S1      EMCS HOT DECK TEMPERATURE SENSOR
S2      EMCS COLD DECK TEMPERATURE SENSOR
S3      EMCS MIXED AIR TEMPERATURE SENSOR
S4      EMCS RETURN AIR TEMPERATURE SENSOR

Fig. 12.3.5a

simple radio signals, which are relatively low cost, but slow. Other systems are based on cellular telephone technology, which can provide greater range and faster data transmission speed, but which may be more expensive to own and operate.

- **Fiber Optic Communication.** Due to the desire to have ultra fast data transmission capabilities for computer network applications, many buildings and campuses are now installing fiber optic networks. Such networks—which have sufficient bandwidth to provide telephone, computer network, and EMCS communication simultaneously—provide the fastest data transmission rates, but are also expensive to install. Many military bases and college campuses have recently installed fiber optic "backbones" to allow inter-building computer network communication. Such scenarios present an excellent opportunity to replace archaic dial-up modem EMCS systems with modern systems that can provide more elaborate control sequences and better accuracy.

### Industry Standard Communication Protocols

In an effort to achieve true interoperability for multiple pieces of equipment from different manufacturers, there has been significant effort focused on developing standard communication protocols that all manufacturers will adhere to. The advantage of so-called "open" protocols is that, for example, a chiller control panel can share data with a variable speed drive controlling a chilled water pump in order to reduce energy use, with near "plug and play" simplicity. The current market is built around proprietary control systems, making communication between devices from different manufacturers difficult to accomplish in some cases. Open protocols are intended to shift the industry away from proprietary systems that limit control system sophistication and user-friendliness.

Though two standardized protocols are now widely known—BacNET (developed by ASHRAE) and Lonworks (based on the Neuron chip, developed by Echelon Corporation)— and a number of others are being developed and implemented as of this writing, their implementation is still not yet at the scale it needs to be provide true interoperability for a wide range of system types.

For the time being, inter-device communication is typically accomplished using communication "gateways." A gateway serves as a translator between the communication languages of two different pieces of equipment, allowing them to share data and operate synergistically. A gateway can either be a stand-alone piece of hardware, or may be a built-in feature of some DDC panels. Systems integration consultants are often involved on sophisticated control system projects to ensure that all devices are able to communicate with one another.

In the future, it is likely that use of industry standard ("open") communication protocols will achieve critical mass, and as a result nearly all manufacturers will adhere to them. At this point in time, however, their use is still limited.

### 12.3.8 Tackling the Human Dimension in EMCS Specifications

Training should also be considered a part of the commissioning. In fact, training and long-term support are often the most overlooked aspects of an EMCS project. In what form and for whom training should be provided differs for every project. This is because the "people" involved are different every time.

In the end, the performance of an EMCS depends on the operating personnel. Those who are to work with the system must understand it. The most common cause for the failure of an EMCS to achieve intended results is ignorance. If a system is overridden frequently, or if controls are disconnected, or if parameters and schedules are not properly maintained the EMCS becomes useless. APPENDIX B includes training and documentation instructions in sections B10, B11, B12, B13, and B14.

Training is not something that can be adequately described in a few short paragraphs in a specification. In reality, there needs to be Comprehensive, Explicit, and Relevant Training and Support (CERTs). CERTs can be used to prevent misunderstandings and dissatisfaction on the part of the building owner, operators and tenants. CERTs also prevent system disconnections and "vaporwatts" (energy savings that never materialize).

In 12.B11.2, the boiler plate reads, "Contractor shall orient the training specifically to the system installed rather than a general training course." A case can be made for expanding this to include business specific and facility specific instructions. In new construction, these would be determined by peering as far into the future as possible to determine what might be required. The best approach here is to ask copious questions about how the facility will be used, staffed, serviced, and marketed (in the case of an office building).

In retrofit applications, the depth and quality of the questions asked during the facility appraisal (refer back to Section 12.3.8) heavily influence the detail spelled out in the specification. For example, for the EMCS RETROFIT discussed in 12.2.2, a list of facility specific instructions might also include the following:

Program user-friendly, menu-prompted, afterhours access setup routines.

Program user-friendly, menu-prompted, monthly after-hours access bill printing routines. Set up ten (10) test access codes which demonstrate options tenants will have for naming after hours zones. Print out bill examples using hourly rates which will be billed in this building.

Provide one 2-hour training session in after-hours access code programming and bill printing for building engineer.

Provide two 2-hour training sessions including system overview, control strategies, and temperature and status data interpretation, for property manager.

Provide two 2-hour training sessions in after-hours access code logic, benefits, options, and programming for property manager.

Provide one 2-hour training session in after-hours bill printing for property manager and property management secretary.

Provide color copies of access code maps for tenants to easily determine what areas will be illuminated and conditioned when accessed after hours.

Conduct a meeting with all tenants' designated representatives to explain how to request HVAC and/or lighting by telephone and to determine the need for additional access code sub-zones.

Conduct follow-up meeting with all tenants' designated representatives to review first month's billing and deliver access code maps documenting sub-zones requested in first meeting.

Conduct a meeting with janitorial service management personnel to explain the use of access codes (lighting only).

Provide two 2-hour training sessions for outside leasing agents in EMCS benefits (i.e. temperature control, efficiency, improved maintenance and after-hours fairness), why a building with an EMCS is different and better than a building without an EMCS.

Attend up to two lease negotiation meetings during first 12 months after start-up to explain after-hours and other EMCS benefits to prospective tenants.

Act as guest speaker at a tenant meeting to explain EMCS benefits beyond the after-hours fairness already explained in private meetings.

Conduct one 2-hour meeting to explain EMCS operation, provide basic documentation, and establish communication and service-call protocol with all service vendors (i.e. security, HVAC, lighting, and electrical).

The above are only a few of what might be a long list of services which could be included in an EMCS specification for a particular building owner to ensure that projected benefits are realized.

Whether one is a consulting engineer, a contractor doing a design/build project, or a facility director writing a guide specification, every effort should be made to anticipate future needs. EMCS projects priced, negotiated, or bid without the benefit of 'complete' specifications are doomed to less than stellar performance at best and total failure at worse. The human dimension may be the most challenging aspect of all of the factors to weigh and incorporate in an EMCS design.

### 12.3.9 Equipment and Contractor Qualification and Selection

Many EMCS projects have gone astray because decisions as to system manufacturer and installer (or bid list) were overly influenced by the features of EMCS equipment. There is a big difference between 'capabilities' and 'effective application of appropriate control and facility specific strategies'.

Too much emphasis on a 'brand' of EMCS overlooks the vital role of the contractor in the success or failure of a project. Even when protected by the most detailed of specifications, it pays to look beyond the low bid and conduct a thorough investigation of the installers being considered. Several of the major EMCS manufacturers also have contracting divisions. Talent and experience often vary immensely from branch office to branch office. Just because a company has been in business for 50 to 100 years does not automatically mean it is the best choice for a particular project. After all, none of the 'people' who will be involved in the project have worked for that company for 50 to 100 years.

Thorough interviews, inspections of previous job sites, conversations with references, and visits to potential contractors' offices to meet 'the delivery team' are all good ways to increase the chances of quality installation. The delivery team will have a greater impact on the outcome of a project than the EMCS equipment. A thorough investigation of the entire team should be made and only the qualified allowed to compete. The same argument could easily be made for the selection of a consulting engineer.

EMCS technology does not save energy, not on its own at any rate. People save energy by using EMCS technology as a tool. By seeking out the crafts-people and the people most focused on saving energy, the chances of a project paying off are significantly greater.

# Appendix A

## EMCS SOFTWARE SPECIFICATIONS

This Appendix provides a sample of the descriptive portion of an EMCS software specification for a new construction junior high school in Connecticut. The school is constructed of red brick, has a pitched, well insulated roof and comprises 140,000 sq. ft. Entry ways are double doored. Windows comprise a moderate percentage of outside wall area and are of the double pane construction. The building is fully air conditioned. The HVAC systems consists of a central boiler/chiller plant and single duct distribution without static air pressure control. Equipment is as follows: 1) a bank of 10 modular oil fired hot water boilers controlled by the manufacturer's microprocessor controller that stages the units on as needed to maintain a temperature reset schedule as determined by the outside air temperature, 2) 2 two-stage reciprocating chillers connected in parallel, 3) a single cell cooling tower with a two speed fan, 4) one condenser water circulator, 5) a pair of space heating circulators to circulate hot water, 6) a pair of space cooling circulators to circulate chilled water, 7) 17 air handling units—10 of which contain a preheat coil, a supply air fan, an outside air damper and a return air damper supply air to terminal units—7 of the units which contain a cooling coil, a heating coil, a supply air fan, an outside air damper and a return air damper supply air directly to the space, 8) one makeup air unit with an outside air damper, 9) 100 space terminal units each containing a reheat coil, a chilled water coil and a damper, 10) an oil fired domestic service hot water heater and associated circulator to supply heat energy to a heat exchanger in the domestic service hot water storage tank, 11) one circulator to circulate domestic service hot water through the building, 12) 21 attic exhaust fans, 13) 14 cabinet unit heaters, and 14) 27 toilet exhaust fans

all to be controlled by a distributive DDC EMCS with peer-to-peer communication. There are several exhaust fans for ventilating mechanical rooms, fume hoods, science rooms, the auditorium stage and storage rooms. These units are locally controlled on an as needed basis. Occupancy sensors perform space lighting control.

The reader should note that sequencing logic requirements are organized into three sections: 1) administrative, 2) system, and 3) operating.

## 12.A1.0 GENERAL

12.A1.1 The listings in this section are meant to provide a logical guide for the sequence of operations to be used for each point of control. There are many considerations other than those listed in this section which must be incorporated into a proper energy management system control program.

12.A1.2 CONTRACTOR must provide programs in the language that is appropriate for the hardware that is used. The programs written in the primary language of the controller must conform in operation and functionality to the sequences listed in these specifications. CONTRACTOR IS LIABLE FOR PROOF OF COMPLIANCE WITH THESE SPECIFICATIONS. BASIS FOR PROOF OF COMPLIANCE WILL BE COMPARING CONTRACTOR'S WRITTEN PROGRAM CODE WITH THE SEQUENCES PROVIDED IN THESE SPECIFICATIONS AND FIELD TESTING OF ACTUAL SYSTEM OPERATION. EACH CONTROL POINT IN THE SYSTEM MUST BE EXERCISED IN THE PRESENCE OF OWNER'S REPRESENTATIVE TO PROVE PROPER OPERATION.

## 12.A2.0 ADMINISTRATIVE SEQUENCE LOGIC

12.A2.1 BASIC SCHEDULE: Install school occupancy schedules for the building as follows:

a. School Year: Provide capability for the user to easily and conveniently schedule the start and stop dates for the normal school term (School Year = 1). For the current School Year install August 28 as the start date and June 18 as the end date.

b. School Day: Provide capability for the user to easily and conveniently schedule a day of the week during the School Year for normal school activity (School Day = 1). Provide calendar scheduling for the current school year.

c. School Hours: Provide capability for the user to easily and conveniently schedule occupancy hours during the School Day for the normal daily school session (School Hours = 1). For the current year install 7:00 am as the start time and 3:00 pm as the end time.

d. Regular Office Hours: Provide capability for the user to easily and conveniently schedule regular occupancy hours during the School Day for the central administrative office (Regular Office Hours = 1). For the current School Year install 7:00 am as the start time and 5:00 pm as the end time.

e. Vacation Office Hours: Provide capability for the user to easily and conveniently schedule administrative office occupancy hours for days other than a School Day (Vacation Office Hours = 1). For the current year install 8:00 am as the start time and 3:00 pm as the end time.

**12.A2.2** <u>ACTIVITY SCHEDULING</u>: Install capability for the user to easily and conveniently schedule for up to twelve months in advance two (2) occupancy periods per day in addition to the Basic Schedule for an individual area activity (XXX Activity = 1). Program the activities currently scheduled on the current school calendar.

a. Whenever an activity is scheduled, all equipment necessary to maintain normal occupancy conditions for that area at the date and time of that activity shall be enabled for occupancy.

b. The user must be able to easily and conveniently initiate, set time parameters and cancel each of the below listed scheduled activities as a separate function.

1. Clrm East Activity
2. Clrm S.E. Activity
3. Clrm South Activity
4. Clrm S.W. Activity
5. Clrm West Activity
6. Clrm North Activity
7. Media Activity
8. Auditorium Activity
9. Admin Activity
10. Main Gym Activity
11. Aux Gym Activity
12. Music Area Activity
13. Cafeteria Activity
14. Industrial Arts Activity
15. Open House

c. The following are equipment assignments for the above scheduled activity areas.

1. Points affected by CLRM EAST ACTIVITY:
   Boilers—Chillers—AHU-1

2. Points affected by CLRM S.E. ACTIVITY:
   Boilers—Chillers—AHU-2

3. Points affected by CLRM SOUTH ACTIVITY:
   Boilers—Chillers—AHU-3

4. Points affected by CLRM S.W. ACTIVITY:
   Boilers—Chillers—AHU-4

5. Points affected by CLRM WEST ACTIVITY:
   Boilers—Chillers—AHU-5

6. Points affected by CLRM NORTH ACTIVITY:
   Boilers—Chillers—AHU-6

7. Points affected by MEDIA ACTIVITY:
   Boilers—Chillers—AHU-7

8. Points affected by AUDITORIUM ACTIVITY:
   Boilers—Chillers—AHU-8—AHU-15 Domestic Hot Water

9. Points affected by ADMIN ACTIVITY:
   Boilers—Chillers—AHU-9

10. Points affected by MAIN GYM ACTIVITY:
    Boilers—Chillers—AHU-10
    Domestic Hot Water

11. Points affected by AUX GYM ACTIVITY:
    Boilers—Chillers—AHU-11
    Domestic Hot Water

12. Points affected by MUSIC ACTIVITY:
    Boilers—Chillers—AHU-12

13. Points affected by CAFETERIA ACTIVITY:
    Boilers—Chillers—AHU-13—FCU-2
    Domestic Hot Water

14. Points affected by INDUSTRIAL ARTS ACTIVITY:
    Boilers—Chillers—AHU-14—MAU-1

15. Points affected by OPEN HOUSE OVERRIDE:
    All control points

**12.A2.3** INDIVIDUAL OVERRIDES: Install timed overrides for each output point as follows:

a. At any time the user must be able to easily and conveniently Enable/on or Disable/off any output point for a user selected time period.

b. At the end of the override period that point overridden shall return to its normal programmed control and the override period shall be reset to zero.

c. An Individual Override shall have priority over the Snow Day Override.

**12.A2.4** <u>SNOW DAY OVERRIDE</u>: Install an easy and convenient method to override off any scheduled occupancy and keep the entire building in the unoccupied mode (Snow Day =1).

a. A Snow Day override shall automatically terminate at 23:59 hours each day.

**12.A2.5** <u>GLOBAL SOFTWARE POINTS</u>: The user variable software points identified by bold print in this specification shall be referenced in the algorithms such that the user need only to enter a new value for a named software point to effect a value change in all algorithms which include that software variable.

**12.A3.0 SYSTEM SEQUENCE LOGIC**

**12.A3.1** <u>SYSTEM MODE</u>: Install capability for the user to easily and conveniently set and vary annual beginning and ending dates for the heating season to prevent simultaneous heating and cooling. Heating equipment operation will be determined by the time of year (i.e. space heating boilers and circulators will only operate during the heating season And be locked out during the rest of the year regardless of outside air temperature). Install October 15th to May 15th as initial parameters for the heating season. Heating season System Mode = 0, non-heating season System Mode = 1.

**12.A3.2** <u>SUMMER/WINTER</u>: Install user adjustable variables for heating and cooling operations during building occupancy as follows:

a. Whenever System Mode = 0 and the outside air temperature is less than or equal to the user variable **Winter Occupancy Setpoint** (i.e. 65°F), summer/winter shall go to winter. If the outside air temperature is greater than the Winter Occupancy Setpoint <u>and</u> the highest occupied space temperature + greater than some user variable setpoint (i.e. 74°F) <u>and</u> Chiller Plant Operation is Enabled/on then summer/winter shall go to summer to allow the chillers to function. If the outside air temperature is greater than the Winter Occupancy Setpoint and Chiller Plant Operation is Disable/off, summer/winter shall remain in winter to allow maximum ventilation with outside air.

b. Whenever System Mode = 1 and the outside air temperature is greater than or equal to the user variable

Summer Occupancy Setpoint (i.e. 60°F), summer/winter mode shall go to summer to allow the chillers to function. If the outside air temperature is less than the Summer Occupancy Setpoint <u>and</u> the lowest occupied space temperature is greater than some user variable setpoint (i.e. 72°F) <u>and</u> is disabled/off, summer/winter mode shall go to winter to allow maximum ventilation with outside air. If the outside air temperature is less than the Summer Occupancy Setpoint and Chiller Plant Operation is Enabled/on, summer/winter mode shall remain in summer mode.

**12.A3.3** <u>OPTIMUM START</u>: Install an <u>adaptable</u> optimum start of all control points for both heating and cooling functions as follows:

a. During optimum start for both heating and cooling, outside air dampers shall go fully closed; <u>and,</u> classroom and return air dampers shall go fully open; <u>and,</u> all exhaust fans shall be Disabled/off.

b. Optimum start shall be limited to a maximum number of hours and be user adjustable.

**12.A3.4** <u>CHILLER PLANT OPERATION</u>: Install capability for the user to easily and conveniently set and vary annual beginning and ending dates for operation of the chiller plant. Install May 1 to November 10 as initial parameters for Chiller Plant Operation.

**12.A3.5** <u>DESIGNATED SPACE TEMPERATURE SETPOINTS</u>: Install <u>user variable</u> space temperature setpoints as follows:

a. The <u>occupied</u> setpoint for all Classrooms, all Offices, the Media Center, the Music Area, the Cafeteria and Kitchen, The Industrial Arts Area, and the Auditorium shall be 72°F for both heating and cooling functions.

b. The <u>occupied</u> setpoint for the Locker Rooms shall be 74°F for both heating and cooling functions.

c. The <u>occupied</u> setpoint for the Main Gymnasium and the Auxiliary Gymnasium shall be 68°F for heating and 74°F for cooling.

d. The <u>unoccupied</u> heating Setback Temperature shall be 55°F.

e. The <u>unoccupied</u> cooling Setup Temperature shall be 80°F.

**12.A3.6** <u>FREE COOLING</u>: Install free cooling into all algorithms that control air handling unit fans, attic exhaust fans, outside air dampers, return air dampers and space terminal unit dampers as follows:

a.   Free cooling shall be Enabled/on to ventilate the building during <u>unoccupied</u> periods whenever the outside air temperature is greater than the user variable Free Cool Minimum (i.e. 45°F) <u>and</u> the enthalpy of the outside air is less than the enthalpy of the inside <u>air</u> and the average building temperature is greater than the user variable Free Cool Setpoint (i.e. 72°F).

b.   Whenever free cooling is Enabled/on all outside air dampers, return air dampers and space terminal unit dampers shall go to the fully open position; <u>and</u>, all air handling units and attic exhaust fans shall be Enabled/on; <u>and</u>, the chiller plant and boiler plant shall be Disabled/off locked out.

**12.A3.7** <u>MINIMUM ON/OFF</u>: Install a user adjustable Minimum On/Off time period (i.e. 5 minutes) in the control program for each piece of mechanical equipment.

## 12.A4.0 OPERATING SEQUENCE LOGIC

**12.A4.1** <u>AHU START/STOP</u>: Install Enable/Disable of Air Handling Units as follows:

a.   Whenever School Hours = 1 <u>and</u> Snow Day = 0 all Air Handling Units shall be Enabled/on to start the supply air fans.

b.   Whenever Regular Office Hours = 1 or Vacation Office Hours = 1 AHU-9 shall be Enabled/on.

c.   Whenever an individual area activity is scheduled on (i.e. a certain XXX Activity = 1) <u>and</u> Snow Day = 0 the air handling unit serving that area shall be Enabled/on.

d.   Whenever School Hours = 0 <u>and</u> the activity flag for an individual area (i.e. XXX Activity = 0) the air handling unit serving that area shall be Disabled/off.

e.   Whenever Regular Office Hours = 0 or Vacation Office Hours = 0 <u>and</u> Admin Activity = 0 AHU-9 shall be Disabled/off.

**12.A4.2** <u>AUDITORIUM AHU</u>: Install control of the Air Handling Unit for the Auditorium as follows:

a.   Whenever School Day = 1 <u>and</u> Snow Day = 0 <u>and</u> the Auditorium is <u>unoccupied</u> as determined by the Auditorium occupancy sensors AHU-8 shall be duty cycled as a standby load to maintain the Designated Space Temperature for the Auditorium within the user variable Space Dead Band (i.e. 6°F).

b.   If, during the standby mode, the Auditorium should become <u>occupied</u> as determined by the Auditorium occupancy sensors for a user variable time period (i.e. 5 minuets) AHU-8 shall be Enabled/on to maintain the Designated Space Temperature for the Auditorium.

c.   Whenever Auditorium Activity = 1 <u>and</u> Snow Day = 0 AHU-8 shall be Enabled/on to maintain the Designated Space Temperature for the Auditorium.

**12.A4.3** <u>MAIN GYMNASIUM AHU</u>: Install control of the Air Handling Unit for the Main Gymnasium as follows:

a.   Whenever School Day = 1 <u>and</u> Snow Day = 0 AHU-10 shall be duty cycled to maintain the Designated Space Temperature for the Main Gymnasium within the Space Dead Band.

b.   Whenever Main Gym Activity = 1 <u>and</u> Snow Day = 0 AHU-10 shall be Enabled/on to maintain the Designated Space Temperature for the Main Gymnasium.

**12.A4.4** <u>MAKEUP AIR UNIT (MAU -1)</u>: Install Enable/Disable control of MAU-1 such that MAU-1 will be allowed to operate during School Hours and an Industrial Arts Activity. Actual control of on/off while MAU-1 is Enable/on will be by local toggle switch.

**12.A4.5** <u>AHU OUTSIDE AIR & RETURN AIR DAMPERS</u>: Install control of the Air Handling Unit dampers as follows:

a.   Whenever a unit is in normal operation (the supply air fan is functioning and free cool or optimum start is not in effect) <u>and</u> the outside air damper for that unit is not under a lockout the dampers shall be calibrated to maintain the user variable Minimum Fresh Air Intake Volume (i.e. 20%).

b.   Whenever a unit is not operating the outside air damper and return air damper for that unit shall go fully closed.

c. Whenever in System Mode = 0 <u>and</u> the outside air temperature is equal to or greater than the user variable Winter Lockout Setpoint (i.e. 20°F) outside air dampers and return air dampers on operating Air Handling Units shall be Enabled/on and modulate under the control of a user adjustable PID loop to maintain the user variable Mixed Air Temperature (i.e. 55°F) for supply air to the preheat or heating coil.

d. Whenever in System Mode = 1 <u>and</u> the outside air temperature is equal to or less than the user variable Summer Lockout Setpoint (i.e. 85°F) outside air dampers shall be Enabled/on and go to minimum position; <u>and,</u> return air dampers shall go to the fully open position.

e. Whenever the outside air temperature is less than the Winter lockout Setpoint <u>or</u> greater than the Summer Lockout Setpoint outside air dampers shall be Disabled/off and go fully closed; <u>and,</u> return air dampers shall go to the fully open position.

**12.A4.6** <u>AHU PREHEAT COIL CONTROL VALVES</u>: Install control of preheat coil control valves for the Air Handling Units as follows:

a. The AHU preheat coil control valve shall go fully open whenever air flow to the coil stops.

b. Whenever System Mode = 0 <u>and</u> there is air flow to an AHU preheat coil that coil control valve shall modulate under the control of a user adjustable PID loop to maintain the user variable Discharge Air Temperature (i.e. 60°F).

**12.A4.7** <u>AHU SPACE HEATING COIL CONTROL VALVES</u>: Install control of heating coil control valves for the Air Handling Units as follows:

a. The AHU heating coil control valve shall go fully open whenever air flow to the coil stops.

b. Whenever System Mode = 0 <u>and</u> there is air flow to the heating coil that coil control valve shall modulate under the control of a user adjustable PID loop to maintain the designated Space Temperature Setpoint for that space.

**12.A4.8** <u>AHU COOLING COIL CONTROL VALVES</u>: Install control of cooling coil control valves for the Air Handling Units as follows:

a. Whenever System Mode = 0 the cooling coil control valve shall go fully open.

b. Whenever System Mode = 1 <u>and</u> there is air flow to the cooling coil that coil control valve shall modulate under the control of a user adjustable PID loop to maintain the designated Space Temperature Setpoint for that space.

c. Whenever System Mode = 1 <u>and</u> air flow to a cooling coil stops that AHU cooling control valve shall go fully closed.

**12.A4.9** <u>SPACE TERMINAL UNIT DAMPERS & CONTROL VALVES</u>:

a. During an <u>occupancy mode</u> whenever a space is occupied as determined by that space's motion sensor, the damper for the terminal unit serving that space shall go fully open and remain open for the duration of the occupancy; <u>and,</u> the hot water heating valve or chilled water cooling valve, as the season may be, shall modulate under the control of a user adjustable PID loop to maintain the designated Space Temperature Setpoint for that space.

b. During an <u>occupancy mode</u> whenever a space is unoccupied as determined by that space's motion sensor that space's terminal unit damper and heating or chilled water cooling valve, as the season may be, shall go fully closed.

c. If, during an <u>occupancy mode,</u> a space is unoccupied as determined by that space's motion sensor <u>and</u> that space's temperature should exit the user variable Space Dead Band the damper for that space's terminal unit shall go fully open; <u>and,</u> the hot water heating valve or chilled water cooling valve, as the season may be, shall modulate under the control of the PID loop to return that space's temperature to within the Space Dead Band.

d. When a space is in the <u>unoccupied mode and</u> System Mode = 1 <u>and</u> that space's temperature is less than the Setup Temperature that space's terminal unit damper and chilled water cooling valve shall go fully closed.

e. If, during the <u>unoccupied mode and</u> System Mode = 1, the temperature in a space should exceed the Setup Temperature, then the terminal unit damper for that space shall go fully open; <u>and,</u> the chilled water cooling valve for that space shall modulate under the control of the PID loop to cool that space to a temperature that is less than the Setup Temperature by an amount equal to the user variable Setback Differential (i.e. 2°F).

f. If, during the <u>unoccupied mode</u>, the temperature in a space should fall below the Setback Temperature, then the terminal unit damper for that space shall go fully open; <u>and</u>, the hot water heating valve for that space shall modulate under the control of the PID loop to heat that space until the temperature in that space exceeds the Setback Temperature by an amount equal to the Setback Differential.

**12.A4.10** <u>SPACE HEATING CIRCULATORS</u>: Install control of space heating circulators as follows:

a. Whenever the building is <u>occupied</u> during System Mode = 0 <u>and</u> the outside air temperature is less than or equal to the Winter Occupancy Setpoint the lead space heating circulator shall be Enabled/on.

b. Whenever the building is <u>occupied</u> during System Mode = 0 <u>and</u> the outside air temperature is greater than the Winter Occupancy Setpoint the lead space heating circulator shall be Disabled/off.

c. Whenever the building is <u>unoccupied</u> during System Mode = 0 <u>and</u> the outside air temperature is greater than or equal to the user variable Freeze Point (i.e. 35°F) the space heating circulators shall be Disabled/off.

d. Whenever the building is <u>unoccupied</u> during System Mode = 0 <u>and</u> the outside air temperature is greater than or equal to the Freeze Point <u>and</u> the lowest space temperature falls below the Setback Temperature the lead space heating circulator shall be Enabled/on until all space temperatures exceed the Setback Temperature by an amount equal to the Setback Differential.

e. Whenever the outside air temperature is less than the Freeze Point the lead space heating circulator shall be Enabled/on.

f. The circulators will alternate lead/lag every 14 days <u>and</u> the then current lagging circulator will Enable/on in the event that the lead circulator should fail. Circulator failure is to be determined by a flow switch at each circulator discharge.

**12.A4.11** <u>BOILERS</u>: Install control of boiler operation as follows:

a. Unless boiler override is on, boiler operation shall be allowed only if System Mode = 0.

b. Whenever a space heating circulator is Enabled/on the boilers shall be Enabled/on to maintain the appropriate supply water temperature as determined by the boiler reset controller.

c. Whenever space heating circulators are Disabled/off the boilers shall be Disabled/off.

**12.A4.12** <u>CHILLED WATER CIRCULATORS</u>: Install control of chilled water circulators as follows:

a. Whenever the building is <u>occupied and</u> summer/winter is in summer <u>and</u> Chiller Plant Operation is Enabled/on the lead circulator shall be Enabled/on.

b. Whenever the building is <u>unoccupied and</u> the highest space temperature in a cooling area is less than or equal to the Setup Temperature the chilled water circulators shall be Disabled/off.

c. Whenever the building is <u>unoccupied and</u> the highest space temperature in a cooling area is greater than the Setup Temperature the lead chilled water circulator shall be Enabled/on until all space temperatures are less than the Setup Temperature by an amount equal to the Setback Differential.

d. The circulators will alternate lead/lag every 14 days and the then current lagging circulator will Enable/on in the event that the lead circulator should fail. Circulator failure is to be determined by a flow switch at each circulator discharge.

**12.A4.13** <u>CHILLERS</u>: Install control of chillers as follows:

a. Whenever a chilled water circulator is Enabled/on the lead chiller shall be Enabled/on.

b. A chiller shall operate to maintain chilled water discharge (supply to loop) temperature as determined by that chiller's controller.

c. During School Hours only, whenever the lead chiller is Enabled/on <u>and</u> the outside air temperature is greater than the user variable Chiller Maximum Output Point (i.e. 85°F) the lag chiller shall be Enabled/on.

d. Whenever chilled water circulators are Disabled/off the chillers shall be Disabled/off.

e. The chillers will alternate lead/lag every 14 days and the then current lagging chiller will Enable/on in the event that the lead chiller should fail.

**12.A4.14** <u>ATTIC EXHAUST FANS</u>: Install control for operation of attic exhaust fans as follows:

a. Whenever the attic air humidity is greater than the user variable Attic Maximum Humidity Point (i.e. 60%) attic exhaust fans shall be Enabled/on.

**12.A4.15** <u>TOILET EXHAUST FANS</u>: Install control for operation of toilet exhaust fans as follows:

a. Whenever the building is <u>occupied</u> all toilet exhaust fans shall be Enabled/on.

b. Whenever and activity is scheduled on the toilet exhaust fans serving the building area associated with that activity shall be Enabled/on.

c. Whenever the building is <u>unoccupied</u> toilet exhaust fans shall be Disabled/off

**12.A4.16** <u>CABINET UNIT HEATERS</u>: Install control of the cabinet unit heaters as follows:

a. Whenever the building is <u>occupied</u> during System Mode = 0 <u>and</u> the outside air temperature is less than or equal to the Winter Occupancy Setpoint the unit cabinet heaters shall be Enabled/on; and, the hot water heating valve shall modulate under the control of a user adjustable PID loop to maintain space temperature at a user variable setpoint (i.e. 65°F).

b. Whenever the building is <u>occupied</u> during System Mode = 0 <u>and</u> the outside air temperature is greater than the Winter Occupancy Setpoint the unit cabinet heaters shall be Disabled/off.

c. Whenever the building is <u>unoccupied and</u> the outside air temperature is equal to or greater than the Freeze Point the unit cabinet heaters shall be Disabled/off.

d. Whenever the building is <u>unoccupied and</u> the outside air temperature is less than the Freeze Point the unit cabinet heaters shall be Enabled/on; and, the hot water heating valve shall modulate under the control of the PID loop to maintain space temperature at a user variable setpoint (i.e. 55°F).

**12.A4.17** <u>DOMESTIC SERVICE HOT WATER</u>: Install control of the domestic service hot water system as follows:

a. During School Hours <u>and</u> scheduled activities for the Cafeteria, Main Gymnasium and Auxiliary Gymnasium the domestic service hot water heater and loop circulators and burner shall be Enabled/on. The local aquastat will cycle the heater circulator and burner on and off as required to maintain the desired loop temperature.

b. During <u>unoccupied</u> periods the domestic service hot water system shall be Disabled/off.

**12.A4.18** <u>COOLING TOWER</u>: The cooling tower shall be interlocked with the chillers such that whenever a chiller is Enabled/on the condenser water circulator and cooling tower fan are Enable/on.

a. Cooling tower fan speed shall be determined by the cooling tower controller.

**12.A4.19** <u>AHU FREEZE PROTECTION</u>: Install freeze protection for the Air Handling Units and the Make-up Air Unit as follows:

a. Whenever in System Mode = 0 <u>and</u> the air temperature as measured at the face of the discharge side of a preheat or heating coil is less than the user variable Freeze Stat Point (i.e. 45°F) the supply fan for that unit shall be Disabled/off and the outside air damper shall be Disabled/off and go fully closed.

b. If a unit has been Disabled/off because of freeze protection, should the air temperature as measured at the face of the discharge side of a preheat or heating coil increase to a value greater than the Freeze Stat Point plus some user variable differential (i.e. 50°F) that unit shall be returned to its normal programmed control.

# Appendix B

## EMCS INSTALLATION REQUIREMENTS

In addition the Software Specifications and appropriate drawings and sketches necessary to integrate a DDC EMCS into a building, the following can be used as a guide in developing specifications to select a supplier and installer of the desired system.

## 12.B1.0 GENERAL

**12.B1.1** The system within each specific building shall be configured as a distributive (or central as the case may be) control and monitoring system. All computing devices, as designed in accordance with FCC Rules and Regulations Part 15, shall be verified to comply with the requirements for Class A computing devices. The system shall perform all data consolidation, control, timing, alarm, feedback status verification and calculation for the system.

**12.B1.2** The system controller (or controllers as the case may be) shall function as the overall system coordinator, perform automatic energy management functions, control peripheral devices and perform calculations associated with operator interactions, alarm reporting, and event logging.

**12.B1.3** The system shall have at least 48 hours of battery backup to maintain all volatile memory and the real time clock.

**12.B1.4** The system shall have a battery backed uninterruptible real time clock with an accuracy within 10 seconds per day.

**12.B1.5** The system shall operate on 120 VAC, 60 Hz power. Operation shall continue in accordance with the Detailed Specifications from line voltages as low as 95 VAC. Line voltages below the operating range of the system shall be considered outages.

**12.B1.6** The system shall be capable of interfacing with no additional hardware, other than the appropriate cabling, to industry standard terminals/PCs. The interface shall support as a minimum: RS232 EIA port, ASCII character code with baud rate selectable between 300-9600 baud.

**12.B1.7** The system shall have the capability of connecting to a local terminal/PC or printer to perform all programming, modification, monitoring and reporting requirements of the Detailed Specifications.

**12.B1.8** The system shall support a Hayes Smartmodem with full RS232 control signals to communicate with remote terminals and other systems via dial-up voice grade phone lines at a minimum of 1200 baud.

**12.B1.9** The system shall accept control programs in accordance with the Software Specifications, perform automated energy management functions, control peripheral devices and perform all necessary mathematical calculations.

**12.B1.10** The system central processing unit shall be a microcomputer of modular design. The word size shall be 8 bits or larger, with a memory cycle time less than 1 micro second. All chips shall be second sourced.

**12.B1.11** Any control program of a system may affect control of another program if desired. Each program shall have full access to all input/output facilities of the system.

**12.B1.12** The system shall bi-directionally transfer data to input/output units and local control units at a minimum of 720 baud.

**12.B1.13** The system shall provide signal conditioning to insure integrity.

**12.B1.14** The system shall be user friendly and shall allow OWNER to designate appropriate nomenclature for all inputs, outputs programs, routines and menus.

## 12.B2.0 INSTALLATION PARTICULARS

**12.B2.1** CONTRACTOR shall install all equipment, devices, cabling, wiring, etc., in compliance with manufacturer's recommendations.

**12.B2.2** The system shall have key lock protected access to internal circuitry and any output override switches.

**12.B2.3** The system controller(s) shall be properly located to provide easy access, and to ensure the ambient temperature range is +32 degrees F to +95 degrees F, 10% to 95% relative humidity non-condensing. Location and accessibility for removal of covers for repair shall be sufficient to facilitate required access by service personnel. Location shall be approved by OWNER'S REPRESENTATIVE before installation.

**12.B2.4** The load on any output shall not exceed 50% of that output's contact rating.

**12.B2.5** Only one (1) wire pair per input/output point shall terminate within any input/output unit enclosure. All relays, wire junctions, terminal strips, transformers, modems and other devices not an integral part of the input/output unit shall be located external to the input/output unit in a separate suitable enclosure(s) properly mounted.

**12.B2.6** Each processing unit and controller shall have a separate dedicated electric service of the appropriate amperage.

**12.B2.7** All electrical relays, electric/pneumatic relays (EP's) and other control devices shall be mounted in an appropriate enclosure, securely fastened and clearly labeled as to purpose and item controlled. All enclosures shall be of adequate size to allow easy access. A minimum of 70 cubic inches shall be allowed per device and the minimum size enclosure shall be 6" × 6" × 4". All relays shall be base mounted and face out towards their enclosure opening. Relays not applicable to base mounting shall have screw type wire connections.

**12.B2.8** Where stranded wire is used all connections shall be made with solderless connections.

**12.B2.9** All wiring and cabling shall be labeled with a logical numbering system. The labels shall remain consistent from the input/output unit to termination with labels being clearly visible at all terminations and penetrations.

**12.B2.10** For all new work, temperature sensors on piping shall be installed in wells.

**12.B2.11** Where strap-on sensors are used in existing piping, sensors shall be bedded in heat conducting compound with a shield and securely fastened to pipe under insulation with strapping as recommended by manufacturer. If not present, insulation as approved by OWNER'S REPRESENTATIVE shall be installed to extend 12" on both sides of sensor. Sensors shall not be installed on PVC pipe.

**12.B2.12** Sensors in ductwork shall protrude into air stream.

**12.B2.13** Where a system output directly or indirectly controls equipment previously controlled by a time clock, that time clock shall be disconnected or otherwise removed from the control circuit of the equipment.

**12.B2.14** Where a system output directly or indirectly controls equipment which has in its control circuit a Hand On-Off-Auto controller, the Hand On function shall be taken out of the control circuit.

**12.B2.15** All control and sensor wiring in boiler rooms, equipment rooms, electrical rooms and unfinished areas shall be enclosed in metal conduit.

**12.B2.16** All control and sensor wiring in finished areas shall be enclosed in metal wire mold.

**12.B2.17** All control and sensor wiring in non-accessible spaces such as ceilings and tunnels shall be tied in a good workmanship manner acceptable to OWNER'S REPRESENTATIVE (maximum 3' spacing) to a suitable support with approved ties. Said wiring shall not lie on ceiling tiles. All wiring in plenum spaces shall be listed for use in same.

**12.B2.18** Control of boilers, fans, pumps, chillers, compressors or any other equipment which utilizes electric or fossil fuel motors as a power source shall be totally by electrical means.

a. Control points identified as "Electric" control points shall be wired such that the digital signal sent by a control unit is received directly by the device being controlled. Unless otherwise specified, any use of pneumatics to interface an "Electric" control point with the device to be controlled is not acceptable.

**12.B2.19** Use of pneumatics for control of heating/cooling valves, hot/chilled water reset valves, dampers or similar modulating devices is acceptable.

a. In the event that CONTRACTOR should utilize any component of existing pneumatic controls to interface "Pneumatic" control points with the device(s) to be controlled, CONTRACTOR shall be responsible for continuity of the pneumatic signal such that the signal from the control unit is transmitted undiminished and as intended to each device to be controlled.

**12.B2.20** All conduit, boxes, and enclosures shall be metal.

**12.B2.21** All wiring carrying 120 volts or more shall be #12 AWG minimum.

**12.B2.22** All enclosures shall be labeled on the outside stating purpose and relays or devices enclosed. All relays and other devices shall be labeled inside the enclosures to clearly designate their use. Relays are to be labeled at the base or adjacent to the base so that the relay can be replaced without losing the label.

**12.B2.23** Labels shall be of engraved phenolic type riveted to the enclosure. Minimum letter size shall be 1/4". Labels shall be approved by OWNER'S REPRESENTATIVE.

**12.B2.24** All control and sensor wiring shall be a minimum of #18 gauge wire. All wire in plenums shall be plenum rated.

## 12.B3.0 GLOBAL DATA EXCHANGE

**12.B3.1** Each control unit and input/output unit in the system shall be able to exchange information between other control units, input/output units and the system controller(s) each network scan.

**12.B3.2** Any network structure shall be transparent such that each control unit in the system may store and reference any global variable available in the network for use in the controller's calculations or programs.

## 12.B4.0 INPUT/OUTPUT UNITS

**12.B4.1** Input/output units shall be used to connect the data environment to the system and contain all necessary input/output functions to read field sensors and operate controlled equipment based on instructions from the central processing unit.

**12.B4.2** Input/output units shall be fully supervised to detect failures. The input/output units shall report the status of all points in its data environment at the rate of at least once every 60 seconds for temperature control. Slightly longer times may be acceptable for non-critical loads on an exception basis.

**12.B4.3** Upon failure of the input/output unit (including transmission failure), the input/output unit shall automatically force outputs to a predetermined state.

**12.B4.4** For system voltages below 95 VAC the input/output unit shall totally shut down and de-energize its outputs.

## 12.B5.0 INPUTS

**12.B5.1** The system shall be capable of receiving the following input signals:

a. <u>Analog Inputs (AI)</u>: The AI function shall monitor each analog input, perform analog to digital (A/D) conversion, and hold the digital value in a buffer for interrogation.

b. <u>Digital Inputs (DI)</u>: The DI function shall accept dry contact closures.

c. <u>Temperature Inputs</u>: Temperature inputs originating from a thermister, shall be monitored and buffered as an AI, except that automatic conversion to degrees F shall occur without any additional signal conditioning. Cable for temperature input wiring from sensors that incorporate RFI circuitry in the sensor do not require shielded cable. Cable for temperature input wiring that does not incorporate RFI circuitry in the sensor shall be shielded.

d. <u>Pulse Accumulators</u>: The pulse accumulator function shall have the same characteristics as the DI, except that, in addition, a buffer shall be required to totalize pulses between interrogations. The pulse accumulator shall accept rates up to 10 pulse/second.

## 12.B6.0 OUTPUTS

**12.B6.1** The system shall be capable of outputting the following system signals:

a. <u>Digital Outputs (DO)</u>: The DO output function shall provide dry contact closures for momentary and maintained programmable operation of field devices. Any manual overrides shall be reported to the system at each update.

b. <u>Analog Output (AO)</u>: The AO output function may be a true analog output providing a 0-10VDC and/or 0-20ma DC signal; or, accept digital data, perform digital to analog (D/A) conversion and output a pulse within the range of 0.1 to 3276.6 seconds in order to perform pulse width modulation of modulating equipment.

## 12.B7.0 DEVICES

**12.B7.1** <u>Temperature Sensors</u>: Temperature sensors shall have a range suitable to their purpose. For example, typical degree ranges should be as follows: outside air -40°F to +120°F, indoor spaces +40°F to +100°F, hot water +80°F to +260°F, chilled water +20°F to +90°F, cold duct air +35°F to +80°F, and hot duct air +60°F to +200°F. RTDs, thermocouples and sensors based on the National Semiconductor LM 34CZ chip are permitted for sensing temperature. Thermisters of nominal resistance of at least 10K ohms may be used for temperature sensing also. Thermisters shall be encapsulated in epoxy or series 300 stainless steel.

**12.B7.2** <u>Thermowells</u>: Thermowells shall be series 300 stainless steel for use in steam or fuel oil lines, and monel for use in water lines. Thermowells shall be wrought iron for measuring flue gases.

**12.B7.3** <u>Pressure Sensors</u>: Pressure sensors shall withstand up to 150% of rated pressure. Accuracy shall be plus or minus 3% of full scale. Pressure sensors shall be either capsule, diaphragm, bellows or bourdon.

**12.B7.4** <u>Pressure Switches</u>: Pressure switches shall have a repetitive accuracy plus or minus 3% of range and withstand up to 150% of rated pressure. Sensors shall be diaphragm or bourdon tube. Switch operation shall be adjustable over the operating pressure range. The switch shall have an application-rated for C, snapacting wiping contact of platinum alloy, silver alloy or gold plating.

**12.B7.5** <u>Watt-hour transducers</u>: Watt-hour transducers shall have an accuracy of plus or minus 1 percent for kW and kWh and shall be internally selectable without requiring the changing of current or potential transformers.

**12.B7.6** <u>Potential Transformers</u>: Potential transformers shall be in accordance with ANSI C57.13.

**12.B7.7** <u>Current Transformers</u>: Current transformers shall be in accordance with ANSI C57.13.

**12.B7.8** <u>Watt-hour Meters with Diamond Register</u>: Meters shall be in accordance with ANSI C12 and have pulse initiators for remote monitoring of watt-hour consumption and instantaneous demand. Pulse initiators shall utilize light emitting diodes and photo-detectors. Pulse initiators shall consist of form C contacts with a current rating of 2 amps, 10 VA maximum and a life rating of one billion operations.

**12.B7.9** <u>Flow Meters</u>: Flow meter outputs shall be compatible with the system. Accuracy shall be plus or minus 3% of full scale.

**12.B7.10** <u>Control Relays</u>: Control relay contacts shall be rated for the application, with form C contacts, enclosed in a dustproof enclosure. Relays shall have silver cadmium contacts with a minimum life span rating of one million operations. When required, provide relays on H-0-A board with LEDs and relay status feedback.

**12.B7.11** <u>Reed Relays</u>: Reed relays shall have been encapsulated in a glass type container housed in a plastic or epoxy case. Contacts shall be rated for the application. Operating and release times shall be five milliseconds or less. Reed relays shall have a minimum life span rating of 10 million operations.

**12.B7.12** <u>Solid State Relays (SSRs)</u>: Input/output isolation shall be greater than 10E9 ohms with a breakdown voltage of 1500V root mean square or greater at 60 Hz. The contact life shall be 10 x 10E6 operations or greater. The ambient temperature range of SSRs shall be minus 20 to plus 140 degrees F. Input impedance shall not be less than 500 ohms. Relays shall be rated for the application. Operating and release time shall be 100 milliseconds or less. Transient suppression shall be provided as an integral part of the relay.

**12.B7.13** <u>Surge Protection</u>: Surge protection shall meet the following minimum criteria.

    a.  Solid state silicon avalanche junction diode device

    b.  Response time less than or equal to 5ns (nanoseconds)

    c.  Voltage protection level 300 volts maximum

    d.  Service frequency 60 Hertz

    e.  Service voltage 110/120 volts AC

    f.  Life expectancy > 10 years

    g.  Ambient operating temperatures 0 to 40 degrees Celsius

    h.  Suppression power 12,000 watts (1 × 1000 microsecond pulse)

    i.  Instant automatic reset

    j.  UL Listed

## 12.B8.0 OPERATING SYSTEM

**12.B8.1** The operating system shall be real-time operating system requiring no operator interaction to initiate and commence operations. The programming shall include:

    a.  Operations and management of all devices;

    b.  Error detection and recovery from arithmetic and logical functions;

    c.  Editing software to allow the user to develop and alter application programs; and,

    d.  System self-testing.

## 12.B9.0 SOFTWARE

**12.B9.1** The software shall be designed to run twenty four hours a day, three hundred sixty five days a year. It shall continually compile and act upon acquired input and display pertinent information on system operation.

**12.B9.2** The software shall be user friendly, user interactive and provide complete environmental and machine status information at the touch of a key. It shall allow the user to change operation parameters through easily understood display prompts.

**12.B9.3** The software shall be flexible and adaptable. All displays must be clear and concise, using the actual names of the equipment and areas in question rather than employing a cryptic computer code to identify locations and machinery.

**12.B9.4** All operator input shall require a minimum of keystrokes in response to prompted messages clearly identifying the required options for entries.

**12.B9.5** The software modules and data base shall be stored in nonvolatile or battery hacked memory to prevent loss of the operating parameters in the event of a power failure. When the system starts, it shall first reload all the parameters and decide what should be on or off. The system shall then scan the facility, see what equipment is actually running and update all temperatures. If, in fact, the status is no longer what it should be, the system shall stagger start all appropriate equipment.

**12.B9.6** The software shall provide telecommunication capability for remote support at a minimum of 1200 baud. After logging onto the system with a remote terminal, the remote user must have access to all user defined variables (operational parameters) as described hereinafter.

**12.B9.7** The software shall provide the capability to upload and download all data, programs, I/O point databases and user configurable parameters from the remote terminal.

**12.B9.8** The software shall be capable of autodialing a minimum of three different phone numbers, in pulse or dial mode.

**12.B9.9** The software shall update the display of data with the most current data available within 5 seconds of operator request.

**12.B9.10** The software shall allow the user to assign unique identifiers of his choice to each connected point. Identifiers shall have at least eight alpha/numeric characters. All reference to these points in programs, reports and command messages shall be by these identifiers.

**12.B9.11** The software shall provide a simple method for the qualified systems integrator to designate control and monitoring characteristics of each control/monitor point. There shall be a menu selection to print the I/O point listings with current configuration data.

**12.B9.12** The software shall be custom configured for the project and shall conform to the sequences that are explicitly specified in the Software Specifications. All configuration must be completed by CONTRACTOR prior to acceptance. GENERIC SOFTWARE TO BE TOTALLY CONFIGURED BY THE END USER IS UNACCEPTABLE. The operator must be able to easily change operating setpoints and parameters from legible, comprehensive menus, but input/output point definition, control algorithms and machinery interlocks must be custom tailored by CONTRACTOR to conform to the control sequences outlined. These algorithms and interlocks are designed to coordinate the operation of all controlled machinery for optimum energy conservation. Access to input/output point definition and algorithm selection/alternation shall be available to a qualified systems integrator through password protected routines and configuration file manipulation, however, these functions shall be invisible to the system operator.

**12.B9.13** Software shall be modular in nature and easily expandable for both more control points and/or more control features.

**12.B9.14** All input gathered by remote sensors shall be globally accessible to the system controller(s). Software designation rather than hardware configuration shall define the interrelationships of all input/output points.

**12.B9.15** The software shall provide at least 3 password access codes that can be user configurable for leveled access to menu functions. All password protection shall be enabled or disabled through simple menu selections.

**12.B9.16** The execution of control functions shall not be interrupted due to normal user communications including: interrogation, program entry, printout of the program for storage, etc.

**12.B9.17** In the dynamic display mode the software shall have the ability to activate real time logic tracers. These tracers shall display the logical sequences being executed by the program as they occur.

**12.B10.0 ENERGY MANAGEMENT FUNCTIONS**

**12.B10.1** Scheduling: The software shall provide for

up to 5 (user configurable) multiple start and stop schedule(s) for each piece of machinery, for each day of the week. In addition, it shall provide for twenty-six (26) special holiday schedules which can be configured a year in advance. The operator shall have the ability to designate each holiday as either a single date or a range of dates. In addition, each of these 26 holidays shall be configurable as either a complete shutdown day where all zones of control are maintained at setback levels, or a special holiday allowing for the partial shut down of the facility. Each piece of equipment shall have a unique schedule for each of these holidays and the schedules must be identical in format and execution to the normal day-of-week schedules.

**12.B10.2** <u>Demand Shedding (if applicable)</u>: The operator shall be allowed to select the maximum electrical demand limit for each monitored electrical service in the facility that the system will strive to protect. The operator must have the ability to set the order of priority in which equipment is to be shed. The operator must also have the ability to identify any given control point as a "round-robin" shed point (first off-first on), a "priority" demand point (first off-last on) or a "temperature" demand point (load closest to setpoint shed first). Special user selectable demand parameters for each load shall be shed type, shed priority, minimum shed time, maximum shed time, minimum time between shed. There shall be a programmable demand shedding schedule that will determine the time of day during which demand shedding will be active.

**12.B10.3** <u>Minimum on-minimum off times</u>: The operator shall have the ability to enter minimum on and minimum off times for each piece of machinery controlled. This feature shall override all other control parameters to prevent equipment short cycling.

**12.B10.4** <u>Straight Time Duty Cycling</u>: The operator shall be able to establish any given control point as a straight time duty cycled point and to set the cycling parameters. For each cycled load the operator shall be able to designate any two minute segment of the hour as either a cycled off or cycled on period. The program shall always begin the programmed cycle periods at the designated offset from the top of the hour. Cycling parameters shall be able to be copied from point to point, or from one point to a group of points. The parameter copy feature shall automatically stagger the cycle period offset from the top of the hour. There shall be a programmable duty cycling schedule that will determine the time of day during which duty cycling will be active.

**12.B10.5** <u>Temperature Compensated Duty Cycling</u>: The operator shall be able to establish any given control point as a temperature compensated duty cycled point and to set the cycling parameters for both heating and cooling ramps (if applicable) with a temperature Dead Band. For each cycled load the operator shall be able to designate high and low temperatures, long off and short off times for both heating and cooling ramps, and a total cycle length. All time parameters shall be programmable in minutes and the software shall verify data entry to prohibit conflicting time parameters. There shall be a programmable duty cycling schedule that will determine the time of day during which duty cycling will be active.

**12.B10.6** <u>Optimum Start/Stop</u>: For both heating and cooling seasons, the system shall provide customized optimum start/stop routines which take into account outside temperature and inside zone temperatures when preparing the building climate for occupation or shutting the facility down at the end of the day. The software shall track the rate of heat loss or gain in each zone and utilize this data when activating optimum start/stop routines.

**12.B10.7** <u>Temperature Control</u>: Building target temperatures for night time, weekend and holiday hours, as well as parameters and limits on normal occupied operation shall be user selectable. The system will strive to maintain these setpoint temperatures in consideration of other energy management functions such as demand shedding, duty cycling, optimum start/stop, etc. Temperature setpoint parameters for on/off points shall be separate temperatures for heating and cooling (if applicable) with Dead Band. Temperature setpoints for analog control points shall include minimum output position (in %), heating full, heating start, cooling start and cooling full.

**12.B10.8** <u>Air Conditioning Optimization</u>: Air conditioning shall be balanced and controlled considering the following factors:

 a. Date (user selectable start and stop dates);
 b. Time of day;
 c. Temperature (outside and inside zone temperatures);
 d. Electrical load (staggered start ups & cycling);
 e. Short cycle protection (user selectable minimum on and minimum off times); and,
 f. Economizer control of equipment configured to support this type of operation.

**12.B10.9** <u>Freeze Protection</u>: Any control point can be assigned to a special freeze protection menu for use in cold climate areas. Points so assigned shall be programmed to run based on an outside temperature setpoint. This freeze protection algorithm will supersede all other temperature control parameters.

**12.B10.10** <u>Manual Override</u>: Separate software manual overrides shall be available for each controlled load that allow the operator to turn on any load for a specific period of time.

**12.B10.11** <u>Daylight Savings Time</u>: Start and stop dates for Daylight Savings Time shall be programmable and automatically implemented by the system.

**12.B10.12** <u>Alarms</u>: The software shall register and record alarms for the following conditions:

    a.  Failure of controlled equipment to respond to computer commands (where positive feedback is used);
    b.  High analog value (user selectable limits); and,
    c.  Low analog value (user selectable limits).

All alarms must be visible and history shall be maintained for a user selectable number of events with a minimum storage of 10 events. In addition, the system must have the capability to automatically dial out of the building to alert user determined authorities in the event of specific alarms. The operator shall have the ability to enable and disable automatic alarm printing. Alarm logs shall contain the date, time, location and type of alarm registered. Also, the software must record the same data for the time when the alarm cleared.

**12.B10.13** <u>Monitoring</u>: The software shall track and display various types of information on monitored or controlled points including but not limited to:

    a.  Last time on
    b.  Last time off
    c.  Temperature
    d.  Pressure
    e.  Alarms

The software shall have the capability to track analog values on the basis of a user defined time period and store the information for a user configurable number of time periods with a minimum storage of 24 periods.

## 12.B11.0 TRAINING

**12.B11.1** CONTRACTOR shall provide the services of competent instructors who will give full instruction to designated personnel in the operation, maintenance and programming of the system.

**12.B11.2** CONTRACTOR shall orient the training specifically to the system installed rather than a general training course. Instructors shall be thoroughly familiar with the subject matter they are to teach.

**12.B11.3** CONTRACTOR shall provide a training manual for each student which describes in detail the data included on each training program. CONTRACTOR shall provide equipment and material required for classroom training.

**12.B11.4** Training on the functional operation of the system shall include:

    a.  Operation of equipment;
    b.  Programming;
    c.  Diagnostics;
    d.  Failure recovery procedure;
    e.  Alarm formats (where applicable);
    f.  Maintenance and calibration; and,
    g.  Trouble shooting, diagnostics and repair instructions.

## 12.B12.0 SYSTEMS MANUAL

**12.B12.1** CONTRACTOR shall provide three (3) system manuals describing programming and testing, starting with a system overview and proceeding to a detailed description of each software feature. The manual shall instruct the user on programming or reprogramming any portion of the system. This shall include all control programs, algorithms, mathematical equations, variables, setpoints, time periods, messages, and other information necessary to load, alter, test and execute the system. The manual shall include:

**12.B12.2** Complete description of programming language, including commands, editing and writing control programs, algorithms, printouts and logs, mathematical calculations and passwords.

**12.B12.3** Instructions on modifying any control algorithm or parameter, verifying errors, status, changing passwords and initiating or disabling control programs.

**12.B12.4** Complete point identification, including terminal number, symbol, engineering units and control program reference number.

**12.B12.5** Field information including location, device, device type and function, electrical parameters and installation drawing number.

**12.B12.6** Location identification of system control hardware.

**12.B12.7** For each system function, a listing of digital and/or analog hardware required to interface the system to the equipment.

**12.B12.8** Listing of all system application programs associated with each piece of equipment. This listing shall include all control algorithms and mathematical equations. The listing shall be in easy to understand English format. All application programs must be submitted. No proprietary control algorithms will be accepted.

### 12.B13.0 MAINTENANCE MANUAL

**12.B13.1** CONTRACTOR shall provide three (3) maintenance manuals containing descriptions of maintenance on all system components, including sensors and controlled devices. The manual shall cover inspection, periodic preventive maintenance, fault diagnosis and repair or replacement of defective components.

### 12.B14.0 CONTROL DRAWINGS

**12.B14.1** CONTRACTOR shall submit the following Shop Drawings for the new energy management system for approval prior to start of the work.

**12.B14.2** Wiring diagram indicating general system layout for each system.

**12.B14.3** Control drawings including outline of systems being controlled and location of devices. Include floor plans showing locations of EMS panels, wiring runs, junction boxes, EP's, PE's, etc.

### 12.B15.0 CONTRACTOR'S VERIFICATION OF SYSTEM INSTALLATION

**12.B15.1** CONTRACTOR shall provide to OWNER written certification in the form prescribed by OWNER'S REPRESENTATIVE that installation of the system is substantially complete as to all respects of these specifications prior to final checkout pursuant to section 12.B17.0 hereof. This certification shall be submitted to OWNER'S REPRESENTATIVE at least ten (10) calendar days prior to the then current Contract completion date. As part of the certification, CONTRACTOR shall perform as listed below.

**12.B15.2** CONTRACTOR shall verify correct temperature readings and scaling of all temperature inputs by comparing actual space temperature to remote temperature indication as displayed by the system video monitor. Both actual and displayed temperatures shall be recorded in table form and certified as to authenticity by CONTRACTOR. Actual space temperature shall be measured by a device traceable to the National Bureau of Standards. Said device will be noted in the certification.

**12.B15.3** CONTRACTOR shall verify operation of all outputs by operating each output over its full range of operation and observing for correct response at the controlled equipment.

**12.B15.4** CONTRACTOR shall provide to OWNER a list of all existing nonoperational equipment that is interfaced with the system.

**12.B15.5** CONTRACTOR shall perform an inspection of the completed work for compliance with all provisions of these specifications and the Software Specifications.

### 12.B16.0 CONTRACTOR'S VERIFICATION OF SYSTEM PROGRAMMING

**12.B16.1** CONTRACTOR shall provide to OWNER written certification that all system programming is complete as to all respects of these specifications and the Software Specifications prior to final checkout pursuant to section 12.B17.0 hereof. This certification shall be submitted to OWNER'S REPRESENTATIVE at least twenty (20) calendar days prior to the then current Contract completion date. As part of the certification, CONTRACTOR shall perform as listed below.

**12.B16.2** CONTRACTOR shall provide to OWNER a printed copy of all software programming, point designations and schedules installed into the system.

**12.B16.3** CONTRACTOR shall demonstrate to OWNER'S REPRESENTATIVE all software programming as detailed in the Software Specifications. The demonstration shall include but not be limited to verification of the following:

a. Individual equipment control programs;
b. Activity schedule/programs;
c. Holiday schedule;
d. Specific area activity overrides;
e. Freeze protection;
f. Global override;
g. Data logging (trending, last time on, last time off); and,
h. Software interlocks.

## 12.B17.0 FINAL SYSTEMS CHECKOUT

**12.B17.1** CONTRACTOR shall arrange with OWNER'S REPRESENTATIVE a final systems checkout to be performed at least five (5) calendar days prior to the then current Contract completion date.

**12.B17.2** CONTRACTOR shall provide all materials and equipment necessary for final systems checkout.

**12.B17.3** CONTRACTOR shall provide all personnel necessary for final systems checkout. OWNER'S REPRESENTATIVE will act only as an observer for verification and not actively participate in this checkout procedure.

**12.B17.4** CONTRACTOR shall demonstrate operation of the following system capabilities and components in the presence of OWNER'S REPRESENTATIVE:

a. Remote terminal access to the central system controller and all of its peers via telephone modem;

b. Remote uploading and downloading of all programs and data points on the energy management system (to include the central system controller and any peers);

c. Local terminal access to the central system controller and any peers;

d. Operation of each output point as per subparagraph 12.B17.8 below; and,

e. Listing of the status of all input points with verification of dynamic response to changing conditions as per subparagraph 12.B17.7 below.

**12.B17.5** CONTRACTOR shall demonstrate proper operation of each input point to OWNER'S REPRESENTATIVE. Input points will be verified for accuracy, location, scaling and operation. Operation will be verified by including a dynamic change in the ambient condition sensed by input sensor and tracking the sensors response and subsequent return to the static condition.

**12.B17.6** CONTRACTOR shall demonstrate proper operation of each output point to OWNER'S REPRESENTATIVE. Each output point will be verified in four (4) parts as follows:

a. Building automation system to the interface device (i.e. relay, EP, etc.); b. Orientation of interface device (normally open, normally closed, etc.); c. Interface device to the specified equipment; and, d. Actual control/operation of specified equipment.

**12.B17.7** OWNER'S REPRESENTATIVE will conduct an inspection of the physical installation to include, but not be limited to, the following:

a. Mounting of the system controller and peers;
b. Labeling of wires and components;
c. Neat and orderly wiring and terminations;
d. Mounting/location of modem;
e. Mounting/location of auxiliary enclosures;

f. Access to relays and equipment mounted in auxiliary enclosures;

g. Mounting/location of temperature and pressure sensors; and,

h. Completion of associated electrical work including disconnecting all time clocks associated with HVAC equipment and disconnecting "Hand On" position on all associated Hand On-Off-Auto switches.

## 12.B18.0 AS BUILT DRAWINGS

**12.B18.1** CONTRACTOR shall provide to OWNER "as built" drawings in conjunction with CONTRACTOR'S systems installation certification.

**12.B18.2** As built drawings shall include floor layout drawings including but not limited to:

a. Location of all control devices;
b. Location of all controlled equipment;

c. Location of all system controllers and terminal controllers;

d. Location of all wire runs and number of conductors;

e. All spare wires;

f. Location of all circuit and feeds including panel designations and circuit breaker numbers;

g. Room numbers and or space designations;
h. Location of all sensors; and,
i. Wire labeling at terminal points.

**12.B18.3** As built drawings shall include all wiring and

pneumatic schematics showing the systems interfacing for each piece of controlled equipment.

# Appendix C

## THE DDC DICTIONARY

ALGORITHM—An assortment of rules or steps presented for solving a problem.

ANALOG INPUT—A variable input that is sensed, like temperature, humidity or pressure and is sent to a computer for processing.

ANALOG OUTPUT—A variable signal such as voltage, current pneumatic air pressure which would in turn operate a modulating motor which will drive valves, dampers, etc.

ANALOG TO DIGITAL A/D CONVERTER—An integrated circuit that takes data such as a temperature sensor and converts it into digital logic that the microprocessor can recognize. All analog inputs of a DDC system are fed into the microprocessor through an A/D converter.

BAUD RATE—The speed, in bits per second, at which information travels over a communications channel.

BIT—The smallest representation of digital data there is. It has two states; 1 (on) or 0 (off).

BITS OF RESOLUTION—A representation of how finely information can be represented. Eight bits of resolution means information may be divided into 256 different states, ten bits allows division into 1024 states and twelve bits allows division into 4096 states.

BYTE—Eight bits of digital information. Eight bits can represent up to 256 different states of information.

CONTROL LOOP—The strategy that is used to make a piece of equipment operate properly. The loop receives the appropriate inputs and sets the desired condition.

DAISY CHAIN—A wiring scheme where units must follow one another in a specific order.

DEAD BAND—An area around a setpoint where there is no change of state.

DEAD TIME—A delay deliberately placed between two related actions in order to avoid an overlap in operation that could cause equipment problems.

DERIVATIVE CONTROL—A system which changes the output of a controller based on how fast a variable is moving from or to a setpoint.

DIGITAL INPUT—An input where there are only two possibilities, such as on/off, open/closed.

DIGITAL COMMUNICATION BUS—A set of wired, usually a twisted pair for DDC controllers. Information is sent over this bus using a digital value to represent the value of the information.

DIGITAL OUTPUT—An output that has two states, such as on/off, open/closed.

DIGITAL TO ANALOG D/A CONVERTER—An integrated circuit that takes data from the microprocessor and converts it to analog data that is represented by a voltage or current. All analog outputs of a DDC system go through a D/A converter.

DISTRIBUTED CONTROL—A system where all intelligence is not in a single controller. If something fails, other controllers will take over to control a unit.

EPROM—Erasable Programmable Read Only Memory. An integrated circuit that is known as firmware, where software instructions have been "burned" into it.

EEPROM—Electrically Erasable Programmable Read Only Memory. An integrated circuit like the EPROM above except that the information may be altered electronically with the chip installed in a circuit.

FCU—Fan Coil Unit. A type of terminal unit.

FEEDBACK—The signal or signals which are sent back from a controlled process telling the current status.

FIRMWARE—Software that has been programmed into an EPROM.

FLOATING POINT—Numerical data that is displayed or manipulated with automatic positioning of the decimal point.

HP—Heat Pump. A type of terminal unit.

INPUT—Data which is supplied to a computer or control system for processing.

INTEGRAL CONTROL—A system which changes the output of a controller based on how long a variable has been offset from a setpoint. This type of control reduces the setpoint/variable offset.

INTELLIGENT BUILDING—A building that is controlled by a DDC system.

LAG—The delay in response of a system to a controlled change.

LAN—Local Area Network. A communication line through which a computer can transmit and receive information from other computers.

MICROPROCESSOR—The brains of all DDC systems and, for that matter, personal computers. An integrated circuit which has logic and math functions built into it. If fed the proper instructions, it will perform the defined functions.

MODEM—A device which allows the computer to communicate over phone lines. It allows the DDC system to be viewed, operated and programmed through a telephone system.

OFFSET—The difference between a variable and its setpoint.

OPERATOR'S TERMINAL—The computer at which the operator can control a motor or to a relay to start/stop a unit.

OUTPUT—Processed information that is sent by a controller to an actuator to control a motor or to a relay to start/stop a unit.

PI CONTROL—A combination of proportional and integral control. Adequate for almost all HVAC control applications. Control loops using PI control to look at both how far an input is from setpoint and how long it has been away from setpoint.

PID CONTROL—Proportional plus Integral plus Derivative Control. A system which directs control loops to look at how far away an input is from setpoint, how long it has been at setpoint and how fast it is approaching/moving away from setpoint.

PROPORTIONAL CONTROL—A system which linearly varies the output as an input variable changes relative to setpoint. The farther the input is from setpoint, the larger the change in the output.

PULSE WIDTH MODULATION—A means of proportionally modulating an actuator using digital outputs. One output opens the motor, another closes it. Used extensively in VAV boxes for driving the damper motor on the box.

RAM—Random Access Memory. A computer chip that information can be written to and read from. Used to store information needed in control loop calculations.

SETPOINT—A value that has been assigned to a controlled variable. An example would be a cooling setpoint of 76 degrees F.

SOFTWARE—The list of instructions written by an engineer or programmer that makes a controller or computer operate as it does.

TERMINAL UNIT—Part of the mechanical system that serves an individual zone, such as VAV box, hydronic heat pump, fan coil, etc.

TRANSDUCER—A unit that converts one type of signal to another. An electronic signal to a pneumatic signal.

TWISTED PAIR—Two wires in one cable that are twisted the entire length of the cable. DDC systems often use a twisted pair for communication link between controllers.

TWO POSITION—Something that has only two states. A two position valve has two states, either open or closed.

USER INTERFACE—The operator's terminal is usually referred to as the user interface. If allows the operator to interrogate the system and perform desired functions. Most interfaces today are personal computers or minicomputers.

VAV—Variable Air Volume. A terminal unit that varies the amount of air delivered to a space, depending upon the demand for cooling.

WORD—Sixteen bits of digital information. Sixteen bits can represent up to 65,536 states of information.

## EMCS MANUFACTURERS DIRECTORY
**(The following list has not been updated as of this printing, but will be addressed in the next issue.)**

A.E.T. Systems, Inc., 77 Accord Executive Park Drive, Norwell, MA 02061
617-871-4801

Allerton Technologies, 2475 140th Ave. N.E., Bldg. A, Bellevue, WA 98005
206-644-9500

American Auto-Matrix, One Technology Drive, Export, PA 15632
412-733-2000

Andover Controls Corp., 300 Brickstone Square, Andover, MA 01810
508-470-0555

Automated Logic Corp., 1283 Kennestone Circle, Marietta, GA 30066
404-423-7474

Barber-Coleman Co., 1354 Clifford Ave., Rockford, IL 61132
815-877-0241

Broadmoor Electric Co., 1947 Republic Ave., San Leandro, CA 94577
415-483-2000

Carrier Corp., One Carrier Place, Farmington, CT 06034
203-674-3000

Chrontrol Corp., 9707 Candida Street, San Diego, CA 92826
619-566-5656

CSI Control Systems International, Inc., 1625 W. Crosby Road, Carrollton, TX 75006
214-323-1111

D.K. Enterprises, Inc., 7361 Ethel Avenue, Suite 1, North Hollywood, CA 91605
818-764-0819

Danfoss-EMC, Inc., 5650 Enterprise Pkwy., Ft. Myers, FL 33905
813-693-5522

Delta Controls, Inc., 13520 78th Avenue, Surrey, B.C. Canada V5W836
604-590-8184

Dencor, Inc., 1450 W. Evans Avenue, Denver, CO 80223
303-922-1888

EDA Controls Corp., 6645 Singletree Drive, Columbus, OH 43229
614-431-0694

Electronic Systems USA, Inc., 1014 East Broadway, Louisville, KY 40204
502-589-1000

Elemco Building Controls, 1324 Motor Parkway, Hauppauge, NY 11788
516-582-8266

Energy Control Systems, Inc., 2940 Cole Street, Norcross, GA 30071
404-448-0651

Functional Devices, Inc., 310 S. Union Street, Russiaville, IN 46979
317-883-5538

Grasslin Controls Corp., 45 Spear Road, Ramsey, NJ 07446
201-825-9696

Honeywell, Inc., Comm. Bldg. Group, Honeywell Plaza, Minneapolis, MN 55408
612-870-5200

Honeywell, Inc., Res. & Bldg. Controls, 1985 Douglas Drive North, Minneapolis, MN 55422
612-870-5200

Integrated Energy Controls (Enercon Data), 7464 78th St. West, Minneapolis, MN 55435
612-829-1900

ISI Wireless, 3000-D South Highland Drive, Las Vegas, NV
702-733-6500

ISTA Energy Systems Corp., 407 Hope Ave., P.O. Box 618, Roselle, NJ 07203
908-241-8880

Johnson Controls, Inc., 507 E. Michigan Street, P.O. Box 423, Milwaukee, WI 53201
414-274-4128

Kreuter, Mfg., 5000 Peachtree Ind. Blvd., #125, Norcross, GA 30071
404-662-5720

Landis & Gyr Powers, 1000 Deerfield Parkway, Buffalo Grove, IL 60089
708-215-1000

Logic Power, 13500 Wright Circle, Tampa, FL 33626
813-854-1588

Microcontrol Systems, Inc., 6579 N. Sidney Pl., Milwaukee, WI 53209
414-262-3143

North American Technologies, Inc., 1300 N. Florida Mango Road, Ste 32, W. Palm Beach, FL 33409
407-687-3051

Novar Controls Corp., 24 Brown Street, Barberton, OH 44203
216-745-0074

Paragon Electric Company, 606 Parkway Blvd., Two Rivers, WI 54241
414-793-1161

Phonetics, Inc., 901 Tryens Road, Aston, PA 19014
215-558-2700

Process Systems, Inc., 100-A Forsythe Hall Dr, Box 240451, Charlotte, NC 28224
704-588-4660

Robertshaw Controls, 1800 Glenside Drive, Richmond, VA 23261
804-289-4200

Scientific Atlanta, 4300 Northeast Expressway, Atlanta, GA 30340
404-449-2902

Snyder General, Inc., 13600 Industrial Park Blvd., Minneapolis, MN 55440
612-553-5330

Solidyne Corp., 1202 Carnegie Street, Rolling Meadows, IL 60008
619-427-9640

Staefa Control Systems Inc., 8515 Miralani Drive, San Diego, CA 92126
619-530-1000

Teletrol Systems, Inc., 324 Commercial Street, Manchester, NH 03101
603-645-6061

The Watt Stopper, Inc., 296 Brokaw Road, Santa Clara, CA 95050
408-988-5331

Trane Company, The, 20 Yorkton Court, St. Paul, MN 55117
612-490-3900

Triangle Microsystems, Inc., 456 Bacon Street, Dayton, OH 45402
513-223-0373

Unity Systems, Inc., 2606 Spring Street, Redwood City, CA 94063
415-369-3233

Xencom Systems, 12015 Shiloh Road, Suite 155, Dallas, TX 75228
214-991-1643

Zeta Engineering Corp., 797 Industrial Court, Bloomfield Hills, MI 48013
313-332-2828

## References

1. Cratty, William E., "Energy Conservation Design and Services With Ventana," *APEC Journal*, Spring 1990.
2. Crosby, Philip B., *Quality Is Free*, Penguin Books, 1979.
3. Elemco Building Controls, Inc., *Elemco Energy Executive II Control Algorithms*, 1989.
4. Gustavson and Gustavson, *The Business of Energy Management/Doing It Right And Making Money At It*, Gannam/Kubat Publishing, Inc. 1990.
5. Honeywell Inc., *Engineering Manual of Automatic Control For Commercial Buildings*, 1988.
6. Kirkpatrick, Bill, "Computers In EMS/Controls," *Contracting Business*, September 1990.
7. Kirkpatrick, Bill, "Overlay Controls," *Contracting Business*, November 1990.
8. Mehta and Thumann, *Handbook of Energy Engineering*, Fairmont Press, Inc., 1989.
9. Patel, Jayant R., "Case Study: Energy Conservation/A Continual Process At Defense Contractor's Manufacturing Facility," *Energy Engineering*, Vol. 88, No. 4, 1991.
10. Rosenfeld, Shlomo I., "Worker Productivity: Hidden HVAC Cost," *Heating/Piping/Air Conditioning*, September 1991.
11. Rosenfeld, Shlomo I., "Worker Productivity: Hidden HVAC Cost," *Heating/Piping/Air Conditioning*, September 1990.
12. Rosenfeld, Shlomo I., "Worker Productivity: Hidden HVAC Cost," *Heating/Piping/Air Conditioning*, September 1989.
13. Silveria, Kent B., "The New EMCS Generation Has A Lot To Offer," *Consulting-Specifying Engineering*, January 1992.
14. Stern and Aronson., *Energy Use/The Human Dimension*, W.H. Freeman and Company, 1984.
15. Thumann, Albert. *Plant Engineer and Managers Guide To Energy Conservation*, Fairmont Press, Inc. 1989.
16. PECI, "Energy Management Systems... A Practical Guide," 1977.
17. Star, Richard, "Pneumatic Controls In A Digital Age," *Energy and Environmental Magazine*, Fall 1999.
18. Ponds, J. Michael, "NERTS... Non-Energy Relationed Tasks," *Energy and Environmental Magazine*, Summer, 1995.
19. Gustavson, Dale A., "FMCS: The Human Dimension," *Contracting Business*, September 1998.

The authors wish to thank the following organizations for making this directory available:

Electric Power Research Institute, 3412 Hillview Ave., Palo Alto, CA, 94303
Plexus Research, Inc., 289 Great Road, Acton, MA 01720

# CHAPTER 13

# LIGHTING

**ERIC A. WOODROOF, PH.D., CEM**
Johnson Controls, Inc.

**JOHN FETTERS, CEM, CLEP**
Effective Lighting Solutions, Inc.

## 13.1  INTRODUCTION

In today's cost-competitive, market-driven economy, everyone is seeking technologies or methods to reduce energy expenses and environmental impact. Because nearly all buildings have lights, lighting retrofits are very common and generally offer an attractive return on investment. Electricity used to operate lighting systems represents a significant portion of total electricity consumed in the United States. Lighting systems consume approximately 20% of the electricity generated in the United States.[1]

An attractive feature of lighting retrofits is they typically provide savings for both kW and kWh charges. Thus, the potential for dollar savings is increased. Many lighting retrofits can also improve the visual environment and worker productivity. Conversely, if a lighting retrofit reduces lighting quality, worker productivity may drop and the energy savings could be overshadowed by reduced profits. This was the case with the lighting retrofits of the 1970s, when employees were left "in the dark" due to massive de-lamping initiatives. However, due to substantial advances in technologies, today's lighting retrofits can reduce energy expenses while *improving* lighting quality and worker productivity.

This chapter will provide the energy manager with a good understanding of lighting fundamentals, so that he/she can oversee successful lighting upgrades. The Example Section, (near the end of this chapter) contains analyses of a few common lighting retrofits. The Schematics Section contains illustrations of many lamps, ballasts and lighting systems.

## 13.2  LIGHTING FUNDAMENTALS

This section will introduce the important concepts about lighting, and the two objectives of the lighting designer: (1) to provide the right quantity of light, and (2) provide the right quality of light.

### 13.2.1 Lighting Quantity

Lighting quantity is the amount of light provided to a room. Unlike light quality, light quantity is easy to measure and describe.

#### 13.2.1.1 Units

Lighting quantity is primarily expressed in three types of units: watts, lumens and foot-candles (fc). Figure 13.1 shows the relationship between each unit. The watt is the unit for measuring electrical power. It defines the rate of energy consumption by an electrical device when it is in operation. The amount of watts consumed represents the electrical input to the lighting system.

The output of a lamp is measured in lumens. For example, one standard four-foot fluorescent lamp would provide 2,900 lumens in a standard office system. The amount of lumens can also be used to describe the output of an entire fixture (comprising several lamps). Thus, the number of lumens describes how much light is being produced by the lighting system.

The number of foot-candles shows how much light is actually reaching the workplane (or task). Foot-candles are the end result of watts being converted to lumens, the lumens escaping the fixture and traveling through the air to reach the workplane. In an office, the workplane is the desk level. You can measure the amount of foot-candles with a light meter when it is placed on the work surface where tasks are performed. Foot-candle measurements are important because they express the "result" and not the "effort" of a lighting system. The Illuminating Engineering Society (IES) recommends light levels for specific tasks using foot-candles, not lumens or watts.

**Figure 13.1 Units of measurement.**

*Efficacy*

Similar to efficiency, efficacy describes an output/input ratio, the higher the output (while input is kept constant), the greater the efficacy. Efficacy is the amount of lumens per watt from a particular energy source. A common misconception in lighting terminology is that lamps with greater wattage provide more light. However, light sources with high efficacy can provide more light with the same amount of power (watts), when compared to light sources with low efficacy.

Figure 13.2 shows lamp efficacies for various lamp types, based on initial lumen values. These systems will be discussed in greater detail in Section 13.2.3.

### 13.2.1.2 IES Recommended Light Levels

The Illuminating Engineering Society (IES) is the largest organized group of lighting professionals in the United States. Since 1915, IES has prescribed the appropriate light levels for many kinds of visual tasks. Although IES is highly respected, the appropriate amount of light for a given space can be subjective. For many years, lighting professionals applied the philosophy that "more light is better," and light levels recommended by IES generally increased until the 1970s. However, recently it has been shown that occupant comfort decreases when a space has too much light. Numerous experiments have confirmed that the prescribed light levels were excessive and worker productivity was decreasing due to poor visual comfort. Due to these findings, IES revised their Handbook and reduced the recommended light levels for many tasks.

The lighting designer must avoid over-illuminating

a space. Unfortunately, this objective can be difficult because over-illuminated spaces have become the "norm" in many buildings. Although not optimal, the tradition of excessive illumination can be continued simply due to habit, or an organization's reluctance to change. To correct this trend, the first step for a lighting retrofit should be to examine the existing system to determine if it is over-illuminated. Appropriate light levels for all types of visual tasks can be found in the most recent IES Hand-

**Figure 13.2 Lamp Efficacies for Various Lamp Types. (Source: Effective Lighting Solutions, Inc.)**

| Lamp Family | Lamp Type | Watts (nominal) | Efficacy (LPW) |
|---|---|---|---|
| Incandescent | Incandescent | 3 – 1,500 | 4 – 23 |
|  | Halogen | 42 – 1,500 | 14 – 22 |
| CFL | Screw-base | 9 – 85 | 40 – 65 |
|  | Twin-tube Pre-heat (2-pin) | 5 – 13 | 50 – 69 |
|  | Quad Pre-heat (2-pin) | 9 – 26 | 64 – 69 |
|  | Rapid-start (4-pin) | 5 – 57 | 46 – 87 |
| Linear Fluorescent | T2 | 6 – 13 | 52 – 66 |
|  | T5 | 14 – 35 | 96 – 104 |
|  | T5HO | 24 – 80 | 83 – 94 |
|  | T8 4' Standard | 32 | 53 – 88 |
|  | T8 4' Reduced-Wattage | 30 | 94 - 95 |
|  | T8 4' High-Performance | 32 | 94 – 98 |
|  | T10 4' | 40 | 83 |
|  | T12 Reduced-Wattage | 34 | 49 – 85 |
|  | T12 4' Standard | 40 | 53 – 83 |
|  | T12 8' Slimline Reduced-Wattage | 60 | 64 – 97 |
|  | T12 8' Slimline | 75 | 58 – 87 |
|  | T12 8' HO Reduced-Wattage | 95 | 81 – 92 |
|  | T12 8' HO (110-w) | 110 | 55 – 85 |
|  | T12 8' VHO Reduced-Wattage | 195 | 67 |
|  | T12 8' VHO Fluorescent | 215 | 65 – 70 |
| Mercury Vapor | Standard | 50 – 1,000 | 32 – 58 |
|  | Self-ballasted | 160 – 750 | 14 - 19 |
| Metal Halide | Standard Probe-start | 50 – 1,500 | 69 – 115 |
|  | Pulse-start | 50 – 1,000 | 69 – 110 |
|  | Ceramic Arc-tube | 20 – 400 | 83 – 95 |
| High-Pressure Sodium | Standard | 35 – 1,000 | 64 – 150 |
|  | Improved CRI | 70 – 400 | 63 – 94 |
| Low-Pressure Sodium |  | 18 – 180 | 100 – 175 |
| Induction | QL | 55 – 165 | 64 – 73 |
|  | ICETRON™ | 70 – 150 | 70 – 79 |

book. Table 13.1 is a summary of some of the IES recommendations.

It is important to remember that IES light levels correspond to particular visual tasks. In an office, there are many tasks: walking around the office, viewing computer screens, reading and writing on paper. Each task requires a different light level. In the past, lighting designers would identify the task that required the most light and design the lighting system to provide that level of illumination for the entire space. However, as previously stated, these design methods often lead to environments

**Table 13.1 Recommended light levels for visual tasks.**

| Building/Space Type | Guideline Illuminance Range (footcandles) |
|---|---|
| **Commercial interiors** | |
| Art galleries | 30-100 |
| Banks | 50-150 |
| Hotels (rooms and lobbies) | 10-50 |
| Offices | 30-100 |
| -Average reading and writing | 50-75 |
| -Hallways | 10-20 |
| -Rooms with computers | 20-50 |
| Restaurants (dining areas) | 20-50 |
| Stores (general) | 20-50 |
| Merchandise | 100-200 |
| | |
| **Institutional interiors** | |
| Auditoriums/assembly places | 15-30 |
| Hospitals (general areas) | 10-15 |
| Labs/treatment areas | 50-100 |
| Libraries | 30-100 |
| Schools | 30-150 |
| | |
| **Industrial interiors** | |
| Ordinary tasks | 50 |
| Stockroom storage | 30 |
| Loading and unloading | 20 |
| Difficult tasks | 100 |
| Highly difficult tasks | 200 |
| Very difficult tasks | 300-500 |
| Most difficult tasks | 500-1000 |
| | |
| **Exterior** | |
| Building security | 1-5 |
| Floodlighting | |
| (low/high brightness or surroundings) | 5-30 |
| Parking | 1-5 |

with excessive brightness, glare and poor worker productivity.

If the IES tables are applied for each task ("Task Lighting"), a superior lighting system is constructed. For example, in an office with computers there should be up to 30 fc for ambient lighting. Small task lights on desks could provide the additional foot-candles needed to achieve a total illuminance of 50 to 75 fc for reading and writing. Task lighting techniques are discussed in greater detail in Section 13.3.

### 13.2.2 Lighting Quality

Lighting quality can have a dramatic influence on the attitude and performance of occupants. In fact, different "moods" can be created by a lighting system. Consider the behavior of people when they eat in different restaurants. If the restaurant is a fast-food restaurant, the space is usually illuminated by bright white lights, with a significant amount of glare from shiny tables. Occupants rarely spend much time there partly because the space creates an uncomfortable mood and the atmosphere is "fast" (eat and leave). In contrast, consider an elegant restaurant with a candle-lit tables and a "warm" atmosphere. Occupants tend to relax and take more time to eat. Although occupant behavior is also linked to interior design and other factors, lighting quality represents a significant influence. Occupants perceive and react to a space's light color. It is important that the lighting designer be able to recognize and create the subtle aspects of an environment that define the theme of the space. For example, drug and grocery stores use white lights to create a "cool" and "clean" environment. Imagine if these spaces were illuminated by the same color lights as in an elegant restaurant. How would the perception of the store change?

Occupants can be influenced to work more effectively if they are in an environment that promotes a "work-like" atmosphere. The goal of the lighting designer is to provide the appropriate quality of light for a particular task to create the right "mood" for the space.

Employee comfort and performance are worth more than energy savings. Although the cost of energy for lighting ($.50-$1.00/year/ft$^2$) is substantial, it is relatively small compared to the cost of labor ($100-$300/year/ft$^2$). Improvements in lighting quality can yield high dividends for businesses because gains in worker productivity are common when lighting quality is improved. Conversely, if a lighting retrofit reduces lighting quality, occupant performance may decrease, quickly off-setting any savings in energy costs. Good energy managers should remember that buildings were not designed to save energy, they exist to create an environment where people can work efficiently. Occupants should be able to see clearly without being distracted by glare, excessive shadows or other uncomfortable features.

Lighting quality can be divided into four main considerations: Uniformity, Glare, Color Rendering Index and Coordinated Color Temperature.

### 13.2.2.1 Uniformity

The uniformity of illuminance describes how evenly light spreads over an area. Creating uniform illumination requires proper fixture spacing. Non-uniform illuminance creates bright and dark spots, which can cause discomfort for some occupants.

Lighting designers have traditionally specified uniform illumination. This option is least risky because it minimizes the problems associated with non-uniform illumination and provides excellent flexibility for changes in the work environment. Unfortunately, uniform lighting applied over large areas can waste large amounts of energy. For example, in a manufacturing building, 20% of the floor space may require high levels of illumination (100 fc) for a specific visual task. The remaining 80% of the building may only require 40 foot candles. Uniform illumination over the entire space would require 100 fc at any point in the building. Clearly, this is a tremendous waste of energy and money. Although uniform illumination is not needed throughout the entire facility, uniform illumination should be applied on specific tasks. For example, a person assembling small parts on a table should have uniform illumination across the table top.

### 13.2.2.2 Glare

Glare is a sensation caused by relatively bright objects in an occupant's field of view. The key word is *relative*, because glare is most probable when bright objects are located in front of dark environments. For example, a car's high beam headlights cause glare to oncoming drivers at night, yet create little discomfort during the day. *Contrast* is the relationship between the brightness of an object and its background. Although most visual tasks generally become easier with increased contrast, too much brightness causes glare and makes the visual task more difficult. Glare in certain work environments is a serious concern because it usually will cause discomfort and reduce worker productivity.

### Visual Comfort Probability (VCP)

The Visual Comfort Probability is a rating given to a fixture which indicates the percent of people who are comfortable with the glare. Thus, a fixture with a VCP = 80 means that 80% of occupants are comfortable with the amount of glare from that fixture. A minimum VCP of 70

is recommended for general interior spaces. Fixtures with VCPs exceeding 80 are recommended in computer areas and high-profile executive office environments.

To improve a lighting system that has excessive glare, a lighting designer should be consulted. However there are some basic "rules of thumb" which can assist the energy manager. A high-glare environment is characterized by either excessive illumination and reflection, or the existence of very bright areas typically around fixtures. To minimize glare, the energy manager can try to obscure the bare lamp from the occupant's field of view, relocate fixtures or replace the fixtures with ones that have a high VCP.

Reducing glare is commonly achieved by using indirect lighting, using deep cell parabolic troffers, or special lenses. Although these measures will reduce glare, fixture efficiency will be decreased because more light will be "trapped" in the fixture. Alternatively, glare can be minimized by reducing ambient light levels and using task lighting techniques.

### Visual Display Terminals (VDTs)

Today's office environment contains a variety of special visual tasks, including the use of computer monitors or visual display terminals (VDTs). Occupants using VDTs are extremely vulnerable to glare and discomfort. When reflections of ceiling lights are visible on the VDT screen, the occupant has difficulty reading the screen. This phenomena is also called "discomfort glare," and is very common in rooms that are uniformly illuminated by fixtures with low a VCP. Therefore, lighting for VDT environments must be carefully designed, so that occupants remain comfortable. Because the location VDTs can be frequently changed, lighting upgrades should also be designed to be adjustable. Moveable task lights and fixtures with high VCP are very popular for these types of applications. Because each VDT environment is unique, each upgrade must be evaluated on a case-by-case basis.

### 13.2.2.3 Color

Color considerations have an incredible influence on lighting quality. Light sources are specified based on two color-related parameters: the Color Rendering Index (CRI) and the Coordinated Color Temperature (CCT).

### Color Rendering Index (CRI)

In simple terms, the CRI provides an evaluation of how colors appear under a given light source. The index range is from 0 to 100. The higher the number, the easier to distinguish colors. Generally, sources with a CRI > 75 provide excellent color rendition. Sources with a CRI < 55 provide poor color rendition. To provide a "base-case," offices illuminated by most T12 Cool White lamps have a CRI = 62.

It is extremely important that a light source with a high CRI be used with visual tasks that require the occupant to distinguish colors. For example, a room with a color printing press requires illumination with excellent color rendition. In comparison, outdoor security lighting for a building may not need to have a high CRI, but a large quantity of light is desired.

*Coordinated Color Temperature (CCT)*

The Coordinated Color Temperature (CCT) describes the color of the light source. For example, on a clear day, the sun appears yellow. On an over-cast day, the partially obscured sun appears to be gray. These color differences are indicated by a temperature scale. The CCT (measured in degrees Kelvin) is a close representation of the color that an object (black-body) would radiate at a certain temperature. For example, imagine a wire being heated. First it turns red (CCT = 2000K). As it gets hotter, it turns white (CCT = 5000K) and then blue (CCT = 8000K). Although a wire is different from a light source, the principle is similar.

CCT is not related to CRI, but it can influence the atmosphere of a room. Laboratories, hospitals and grocery stores generally use "cool" (blue-white) sources, while expensive restaurants may seek a "warm" (yellow-red) source to produce a candle-lit appearance. Traditionally, office environments have been illuminated by Cool White lamps, which have a CCT = 4100K. However, a more recent trend has been to specify 3500K tri-phosphor lamps, which are considered neutral. Table 13.2 illustrates some common specifications for different visual environments.

### 13.2.3 Lighting System Components

After determining the quantity and quality of illumination required for a particular task, most lighting designers specify the lamp, then the ballast, and finally the fixture to meet the lighting needs. The Schematics Section (near the end of this chapter) contains illustrations of many of the lamps and systems described in this section.

### 13.2.3.1 Lamps

The lamp is the first component to consider in the lighting design process. The lamp choice determines the light quantity, CRI, CCT, relamping time interval and operational costs of the lighting system. This section will only cover the most popular types of lamps. Table 13.3 summarizes the differences between the primary lamps and lighting systems.

**Table 13.2 Sample design considerations for a commercial building.**

| Office Areas | | Light Levels | CRI | Color Temperature | Glare |
|---|---|---|---|---|---|
| Executive | General | 100FC | ≥80 | 3000K | VCP≥70 |
|  | Task | ≥50FC | ≥80 | 3000K | VCP≥70 |
| Private | General | 30-50FC | ≥70 | 3000-3500K | VCP≥70 |
|  | Task | ≥50FC | ≥70 | 3000-3500K | VCP≥70 |
| Open Plan Computers | General | 30-50FC | ≥70 | 3000-3500K | VCP≥90 |
|  | Task | ≥50FC | ≥70 | 3000-3500K | VCP≥90 |
| Hallways | General | 10-20FC | ≥70 | 3000-3500K | VCP≥70 |
|  | Task | 10-20FC | ≥70 | 3000-3500K | VCP≥70 |
| Reception/ Lobby | General | 20-50FC | ≥80 | 3000-5000K | VCP≥90 |
|  | Task | ≥50FC | ≥80 | 3000-3500K | VCP≥90 |
| Conference | General | 10-70FC | ≥80 | 3000-4100K | VCP≥90 |
|  | Task | 10-70FC | ≥80 | 3000-4100K | VCP≥90 |
| Open Plan General | General | 30-70FC | ≥70 | 3500-4100K | VCP≥80 |
|  | Task | 30-70FC | ≥70 | 3500-4100K | VCP≥80 |
| Drafting | General | 70-100FC | ≥70 | 4100-5000K | VCP≥90 |
|  | Task | 100-150FC | ≥780 | 4100-5000K | VCP≥90 |

## Table 13.3 Lamp characteristics

| | Incandescent Including Tungsten Halogen 15-1500 | Fluorescent 15-219 | Compact Fluorescent 4-40 | Mercury Vapor (Self-ballasted) 40-1000 | Metal Halide 175-1000 | High-Pressure Sodium (Improved Color) 70-1000 | Low-Pressure Sodium 35-180 |
|---|---|---|---|---|---|---|---|
| Wattages (lamp only) | | | | | | | |
| Life (hr) | 750-12,000 | 7,500-24,000 | 10,000-20,000 | 16,000-15,000 | 1,500-15,000 | 24,000 (10,000) | 18,000 |
| Efficacy (lumens/W) lamp only | 15-25 | 55-100 | 50-80 | 50-60 (20-25) | 80-100 | 75-140 (67-112) | Up to 180 |
| Lumen maintenance | Fair to excellent | Fair to excellent | Fair | Very good (good) | Good | Excellent | Excellent |
| Color rendition | Excellent | Good to excellent | Good to excellent | Poor to excellent | Very good | Fair | Poor |
| Light direction control | Very good to excellent | Fair | Fair | Very good | Very good | Very good | Fair |
| Relight time | Immediate | Immediate | Imm- 3 seconds | 3-10 min. | 10-20 min. | Less than 1 min. | Immediate |
| Comparative fixture cost | Low: simple | Moderate | Moderate | Higher than fluorescent | Generally higher than mercury | High | High |
| Comparative operating cost | High | Lower than incandescent | Lower than incandescent | Lower than incandescent | Lower than mercury | Lowest of HID types | Low |

*Incandescent*

The oldest electric lighting technology is the incandescent lamp. Incandescent lamps are also the least efficacious (have the lowest lumens per watt) and have the shortest life. They produce light by passing a current through a tungsten filament, causing it to become hot and glow. As the tungsten emits light, it gradually evaporates, eventually causing the filament to break. When this happens, the lamps is said to be "burned-out."

Although incandescent sources are the least efficacious, they are still sold in great quantities because of economies of scale and market barriers. Consumers still purchase incandescent bulbs because they have low initial costs. However, if life-cycle cost analyses are used, incandescent lamps are usually more expensive than other lighting systems with higher efficacies.

*Compact Fluorescent Lamps (CFLs)*

Overview of CFLs:
Compact Fluorescent Lamps (CFLs) are energy efficient, long lasting replacements for some incandescent lamps. CFLs (like all fluorescent lamps) are composed of two parts, the lamp and the ballast. The short tubular lamps can last longer than 8,000 hours. The ballasts (plastic component at the base of tube) usually last longer than 60,000 hours. Some CFLs can be purchased as self-ballasted units, which "screw in" to an existing incandescent socket. For simplicity, this chapter refers to a CFL as a lamp and ballast system. CFLs are available in many styles and sizes.

In most applications, CFLs are excellent replacements for incandescent lamps. CFLs provide similar light quantity and quality while only requiring about 20-30% of the energy of comparable incandescent lamps. In addition, CFLs last 7 to 10 times longer than their incandescent counterparts. In many cases, it is cost-effective to replace an entire incandescent fixture with a fixture specially designed for CFLs.

*The "New Technololgies" Section contains a more thorough explanation of CFLs.*

*Fluorescent*

Fluorescent lamps are the most common light source for commercial interiors in the U.S. They are repeatedly specified because they are relatively efficient, have long lamp lives and are available in a wide variety of styles. For many years, the conventional fluorescent lamp used in offices has been the four-foot F40T12 lamp, which is usually used with a magnetic ballast. However, these lamps are being rapidly replaced by T8 or T5 lamps with electronic ballasts.

The labeling system used by manufacturers may appear complex, however it is actually quite simple. For example, with an F34T12 lamp, the "F" stands for fluorescent, the "34" means 34 watts, and the "T12" refers to the tube thickness. Since tube thickness (diameter) is measured in 1/8 inch increments, a T12 is 12/8 or 1.5 inches in diameter. A T8 lamp is 1 inch in diameter. Some lamp labels include additional information, indicating the CRI and CCT. Usually, CRI is indicated with one digit, like "8" meaning CRI = 80. CCT is indicated by the two digits following, "35" meaning 3500K. For example, a F32T8/841 label indicates a lamp with a CRI = 80 and a CCT = 4100K. Alternatively, the lamp manufacturer might label a lamp with a letter code referring to a specific lamp color. For example, "CW" to mean Cool White lamps with a CCT = 4100K.

Some lamps have "ES," "EE" or "EW" printed on the label. These acronyms attached at the end of a lamp label indicate that the lamp is an energy-saving type. These lamps consume less energy than standard lamps, however they also produce less light.

Tri-phosphor lamps have a coating on the inside of the lamp which improves performance. Tri-phosphor lamps usually provide greater color rendition. A bi-phosphor lamp (T12 Cool White) has a CRI= 62. By upgrading to a tri-phosphor lamp with a CRI = 75, occupants will be able to distinguish colors better. Tri-phosphor lamps are commonly specified with systems using electronic ballasts. Lamp flicker and ballast humming are also significantly reduced with electronically ballasted systems. For these reasons, the visual environment and worker productivity is likely to be improved.

There are many options to consider when choosing fluorescent lamps. Carefully check the manufacturers specifications and be sure to match the lamp and ballast to the application. Table 13.4 shows some of the specifications that vary between different lamp types.

*The "New Technololgies" Section contains a more thorough explanation of the various fluorescent lamp systems available today.*

## High Intensity Discharge (HID)

High-Intensity Discharge (HID) lamps are similar to fluorescent lamps because they produce light by discharging an electric arc through a tube filled with gases. HID lamps generate much more light, heat and pressure within the arc tube than fluorescent lamps, hence the title "high intensity" discharge. Like incandescent lamps, HIDs are physically small light sources, (point sources) which means that reflectors, refractors and light pipes can be effectively used to direct the light. Although originally developed for outdoor and industrial applications, HIDs are also used in office, retail and other indoor applications.

With a few exceptions, HIDs require time to warm up and should not be turned ON and OFF for short intervals. They are not ideal for certain applications because, as point sources of light, they tend to produce more defined shadows than non-point sources such as fluorescent tubes, which emit diffuse light.

Most HIDs have relatively high efficacies and long lamp lives, (5,000 to 24,000+ hours) reducing maintenance re-lamping costs. In addition to reducing maintenance requirements, HIDs have many unique benefits. There are three popular types of HID sources (listed in order of increasing efficacy): Mercury Vapor, Metal Halide and High Pressure Sodium. A fourth source, Low Pressure Sodium, is not technically a HID, but provides similar quantities of illumination and will be referred to as an HID in this chapter. Table 13.3 shows that there are dramatic differences in efficacy, CRI and CCT between each HID source type.

## Mercury Vapor

Mercury Vapor systems were the "first generation" HIDs. Today they are relatively inefficient, provide poor CRI and have the most rapid lumen depreciation rate of all HIDs. Because of these characteristics, other more cost-effective HID sources have replaced mercury vapor

**Table 13.4 Sample fluorescent lamp specifications.**

|  | MANUFACTURERS' INFORMATION | | |
|  | F40T12CW Bi-phosphor | F40T10 Tri-phosphor | F32T8 Tri-phosphor |
| --- | --- | --- | --- |
| CRI | 62 | 83 | 83 |
| CCT (K) | 4,150 | 4,100 or 5,000 | 4,100 or 5,000 |
| Initial lumens | 3,150 | 3,700 | 3,050 |
| Maintained lumens | 2,205 | 2,960 | 2,287 |
| Lumens per watt | 55 | 74 | 71 |
| Rated life (hrs) | 24,000 | 48,000[†] | 20,000 |
| Service life (hrs) | 16,800 | 33,600[†] | 14,000 |

[†]This extended life is available from a specific lamp-ballast combination. Normal T10 lamp lives are approximately 24,000 hours. Service life refers to the typical lamp replacement life.

lamps in nearly all applications. Mercury Vapor lamps provide a white-colored light which turns slightly green over time. A popular lighting upgrade is to replace Mercury Vapor systems with Metal Halide or High Pressure Sodium systems.

*Metal Halide*

Metal Halide lamps are similar to mercury vapor lamps, but contain slightly different metals in the arc tube, providing more lumens per watt with improved color rendition and improved lumen maintenance. With nearly twice the efficacy of Mercury Vapor lamps, Metal Halide lamps provide a white light and are commonly used in industrial facilities, sports arenas and other spaces where good color rendition is required. They are the current best choice for lighting large areas that need good color rendition.

*High Pressure Sodium (HPS)*

With a higher efficacy than Metal Halide lamps, HPS systems are an economical choice for most outdoor and some industrial applications where good color rendition is not required. HPS is common in parking lots and produces a light golden color that allows some color rendition. Although HPS lamps do not provide the best color rendition, (or attractiveness) as "white light" sources, they are adequate for indoor applications at some industrial facilities. The key is to apply HPS in an area where there are no other light source types available for comparison. Because occupants usually prefer "white light," HPS installations can result with some occupant complaints. However, when HPS is installed at a great distance from metal halide lamps or fluorescent systems, the occupant will have no reference "white light" and he/she will accept the HPS as "normal." This technique has allowed HPS to be installed in countless indoor gymnasiums and industrial spaces with minimal complaints.

*Low Pressure Sodium*

Although LPS systems have the highest efficacy of any commercially available HID, this monochromic light source produces the poorest color rendition of all lamp types. With a low CCT, the lamp appears to be "pumpkin orange," and all objects illuminated by its light appear black and white or shades of gray. Applications are limited to security or street lighting. The lamps are physically long (up to 3 feet) and not considered to be point sources. Thus optical control is poor, making LPS less effective for extremely high mounting heights.

LPS has become popular because of its extremely high efficacy. With up to 60% greater efficacy than HPS, LPS is economically attractive. Several cities, such as San Diego, California, have installed LPS systems on streets. Although there are many successful applications, LPS installations must be carefully considered. Often lighting quality can be improved by supplementing the LPS system with other light sources (with a greater CRI).

**13.2.3.2 Ballasts**

With the exception of incandescent systems, nearly all lighting systems (fluorescent and HID) require a ballast. A ballast controls the voltage and current that is supplied to lamps. Because ballasts are an integral component of the lighting system, they have a direct impact on light output. The ballast factor is the ratio of a lamp's light output to a reference ballast. General purpose fluorescent ballasts have a ballast factor that is less than one (typically .88 for most electronic ballasts). Special ballasts may have higher ballast factors to increase light output, or lower ballast factors to reduce light output. As can be expected, a ballast with a high ballast factor also consumes more energy than a general purpose ballast.

*Fluorescent*

Specifying the proper ballast for fluorescent lighting systems has become more complicated than it was 25 years ago, when magnetic ballasts were practically the only option. Electronic ballasts for fluorescent lamps have been available since the early 1980s, and their introduction has resulted in a variety of options.

This section describes the two types of fluorescent ballasts: magnetic and electronic.

*Magnetic*

Magnetic ballasts are available in three primary types.
- Standard core and coil
- High-efficiency core and coil (Energy-Efficient Ballasts)
- Cathode cut-out or Hybrid

Standard core and coil magnetic ballasts are essentially core and coil transformers that are relatively inefficient at operating fluorescent lamps. Although these types of ballasts are no longer sold in the US, they still exist in many facilities. The "high-efficiency" magnetic ballast can replace the "standard ballast," improving the system efficiency by approximately 10%.

"Cathode cut-out" or "hybrid" ballasts are high-efficiency core and coil ballasts that incorporate electronic components that cut off power to the lamp cathodes after the lamps are operating, resulting in an additional 2-watt savings per lamp.

*Electronic*

During the infancy of electronic ballast technology, reliability and harmonic distortion problems hampered their success. However, most electronic ballasts available today have a failure rate of less than one percent, and many distort harmonic current less than their magnetic counterparts. Electronic ballasts are superior to magnetic ballasts because they are typically 30% more energy efficient, they produce less lamp flicker, ballast noise, and waste heat.

In nearly every fluorescent lighting application, electronic ballasts can be used in place of conventional magnetic core and coil ballasts. Electronic ballasts improve fluorescent system efficacy by converting the standard 60 Hz input frequency to a higher frequency, usually 25,000 to 40,000 Hz. Lamps operating on these frequencies produce about the same amount of light, while consuming up to 40% less power than a standard magnetic ballast. Other advantages of electronic ballasts include less audible noise, less weight, virtually no lamp flicker and dimming capabilities.

T12 and T8 ballasts are the most popular types of electronic ballasts. T12 electronic ballasts are designed for use with conventional (T12) fluorescent lighting systems. T8 ballasts offer some distinct advantages over other types of electronic ballasts. They are generally more efficient, have less lumen depreciation, and are available with more options. T8 ballasts can operate one, two, three or four lamps. Most T12 ballasts can only operate one, two or three lamps. Therefore, one T8 ballast can replace two T12 ballasts in a 4 lamp fixture.

Some electronic ballasts are parallel-wired, so that when one lamp burns out, the remaining lamps in the fixture will continue to operate. In a typical magnetic, (series-wired system) when one component fails, all lamps in the fixture shut OFF. Before maintenance personnel can relamp, they must first diagnose which lamp failed. Thus the electronically ballasted system will reduce time to diagnose problems, because maintenance personnel can immediately see which lamp failed.

Parallel-wired ballasts also offer the option of reducing lamps per fixture (after the retrofit) if an area is over-illuminated. This option allows the energy manager to experiment with different configurations of lamps in different areas. However, each ballast operates best when controlling the specified number of lamps.

Due to the advantages of electronically ballasted systems, they are produced by many manufacturers and prices are very competitive. Due to their market penetration, T8 systems (and replacement parts) are more likely to be available, and at lower costs.

*HID*

As with fluorescent systems, High Intensity Discharge lamps also require ballasts to operate. Although there are not nearly as many specification options as with fluorescent ballasts, HID ballasts are available in dimmable and bi-level light outputs. Instant restrike systems are also available.

*Capacitive Switching HID Fixtures*

Capacitive switching or "bi-level" HID fixtures are designed to provide either full or partial light output based on inputs from occupancy sensors, manual switches or scheduling systems. Capacitive-switched dimming can be installed as a retrofit to existing fixtures or as a direct fixture replacement. Capacitive switching HID upgrades can be less expensive than installing a panel-level variable voltage control to dim the lights, especially in circuits with relatively few fixtures.

The most common applications of capacitive switching are athletic facilities, occupancy-sensed dimming in parking lots and warehouse aisles. General purpose transmitters can be used with other control devices such as timers and photosensors to control the bi-level fixtures. Upon detecting motion, the occupancy sensor sends a signal to the bi-level HID ballasts. The system will rapidly bring the light levels from a standby reduced level to about 80 percent of full output, followed by the normal warm-up time between 80 and 100 percent of full light output.

Depending of the lamp type and wattage, the standby lumens are roughly 15-40 percent of full output and the standby wattage is 30-60 percent of full wattage. When the space is unoccupied and the system is dimmed, you can achieve energy savings of 40-70 percent.

### 13.2.3.3 Fixtures (aka Luminaires)

A fixture is a unit consisting of the lamps, ballasts, reflectors, lenses or louvers and housing. The main function is to focus or spread light emanating from the lamp(s). Without fixtures, lighting systems would appear very bright and cause glare.

*Fixture Efficiency*

Fixtures block or reflect some of the light exiting the lamp. The efficiency of a fixture is the percentage of lamp lumens produced that actually exit the fixture in the intended direction. Efficiency varies greatly among different fixture and lamp configurations. For example, using four T8 lamps in a fixture will be more efficient than using four T12 lamps because the T8 lamps are thinner, allowing more light to "escape" between the lamps and out of the fixture. Understanding fixtures is important because a lighting retrofit may involve changing some components of the fixture to improve the efficiency and deliver more light to the task.

The Coefficient of Utilization (CU) is the percent of lumens produced that actually reach the work plane. The CU incorporates the fixture efficiency, mounting height, and reflectances of walls and ceilings. Therefore, improving the fixture efficiency will improve the CU.

*Reflectors*

Installing reflectors in most fixtures can improve its efficiency because light leaving the lamp is more likely to "reflect" off interior walls and exit the fixture. Because lamps block some of the light reflecting off the fixture interior, reflectors perform better when there are less lamps (or smaller lamps) in the fixture. Due to this fact, a common fixture upgrade is to install reflectors and remove some of the lamps in a fixture. Although the fixture efficiency is improved, the overall light output from each fixture is likely to be reduced, which will result in reduced light levels. In addition, reflectors will redistribute light (usually more light is reflected down), which may create bright and dark spots in the room. Altered light levels and different distributions may be acceptable, however these changes need to be considered.

To ensure acceptable performance from reflectors, conduct a trial installation and measure "before" and "after" light levels at various locations in the room. Don't compare an existing system, (which is dirty, old and contains old lamps) against a new fixture with half the lamps and a clean reflector. The light levels may appear to be adequate, or even improved. However, as the new system ages and dirt accumulates on the surfaces, the light levels will drop.

A variety of reflector materials are available: highly reflective white paint, silver film laminate, and anodized aluminum. Silver film laminate usually has the highest reflectance, but is considered less durable. Be sure to evaluate the economic benefits of your options to get the most "bang for your buck."

In addition to installing reflectors within fixtures, light levels can be increased by improving the reflectivity of the room's walls, floors and ceilings. For example, by covering a brown wall with white paint, more light will be reflected back into the workspace, and the Coefficient of Utilization is increased.

*Lenses and Louvers*

Most indoor fixtures use either a lens or louver to prevent occupants from directly seeing the lamps. Light that is emitted in the shielding angle or "glare zone" (angles above 45° from the fixture's vertical axis) can cause glare and visual discomfort, which hinders the occupant's ability to view work surfaces and computer screens. Lenses and louvers are designed to shield the viewer from these uncomfortable, direct beams of light. Lenses and louvers are usually included as part of a fixture when purchased, and they can have a tremendous impact on the VCP of a fixture.

Lenses are sheets of hard plastic (either clear or milky white) that are located on the bottom of a fixture. Clear, prismatic lenses are very efficient because they trap less light within the fixture. Milky-white lenses are called "diffusers" and are the least efficient, trapping a lot of the light within the fixture. Although diffusers have been routinely specified for many office environments, they have one of the lowest VCP ratings.

Louvers provide superior glare control and high VCP when compared to most lenses. As Figure 13.3 shows, a louver is a grid of plastic "shields" which blocks some of the horizontal light exiting the fixture. The most common application of louvers is to reduce the fixture glare in sensitive work environments, such as in rooms with computers. Parabolic louvers usually improve the VCP of a fixture, however efficiency is reduced because more light is blocked by the louver. Generally, the smaller the cell, the greater the VCP and less the efficiency. Deep-cell parabolic louvers offer a better combination of VCP and efficiency, however deep-cell louvers require deep fixtures, which may not fit into the ceiling plenum space.

Table 13.5 shows the efficiency and VCP for various lenses and louvers. VCP is usually inversely related to fixture efficiency. An exception is with the milky-white diffusers, which have low VCP and low efficiency.

*Light Distribution/Mounting Height*

Fixtures are designed to direct light where it is needed. Various light distributions are possible to best suit any visual environment. With "direct lighting," 90-100% of the light is directed downward for maximum use. With "indirect lighting," 90-100% of the light is directed to the ceilings and upper walls. A "semi-indirect" system

**Figure 13.3 Higher shielding angles for improved glare control.**

**Table 13.5 Luminaire efficiency and VCP.**

| Shielding Material | Luminaire Efficiency (%) | Visual Comfort Probability (VCP) |
|---|---|---|
| Clear Prismatic Lens | 60-70 | 50-70 |
| Low Glare Clear Lens | 60-75 | 75-85 |
| Deep-Cell Parabolic Louver | 50-70 | 75-95 |
| Translucent Diffuser | 40-60 | 40-50 |
| Small-Cell Parabolic Louver | 35-45 | 99 |

**Table 13.6 Minimum mounting heights for HIDs**

| Lamp Type | feet above ground |
|---|---|
| 400 W Metal Halide | 16 |
| 1000 W Metal Halide | 20 |
| 200 W High Pressure Sodium | 15 |
| 250 W High Pressure Sodium | 16 |
| 400 W High Pressure Sodium | 18 |
| 1000 W High Pressure Sodium | 26 |

distributes 60-90% down, with the remainder upward. Designing the lighting system should incorporate the different light distributions of different fixtures to maximize comfort and visual quality.

Fixture mounting height and light distribution are presented together since they are interactive. HID systems are preferred for high mounting heights since the lamps are physically small, and reflectors can direct light downward with a high degree of control. Fluorescent lamps are physically long and diffuse sources, with less ability to control light at high mounting heights. Thus fluorescent systems are better for low mounting heights and/or areas that require diffuse light with minimal shadows.

Generally, "high-bay" HID fixtures are designed for mounting heights greater than 20 feet high. "High-bay" fixtures usually have reflectors and focus most of their light downward. "Low-bay" fixtures are designed for mounting heights less than 20 feet and use lenses to direct more light horizontally.

HID sources are potential sources of direct glare since they produce large quantities of light from physically small lamps. The probability of excessive direct glare may be minimized by mounting fixtures at sufficient heights. Table 13.6 shows the minimum mounting height recommended for different types of HID systems.

### 13.2.3.4 Exit Signs

Recent advances in exit sign systems have created attractive opportunities to reduce energy and maintenance costs. Because emergency exit signs should operate 24 hours per day, energy savings quickly recover retrofit costs. There are generally two options, buying a new exit sign, or retrofitting the existing exit sign with new light sources.

Most retrofit kits available today contain adapters that screw into the existing incandescent sockets. Installation is easy, usually requiring only 15 minutes per sign. However, if a sign is severely discolored or damaged, buying a new sign might be required in order to maintain illuminance as required by fire codes.

Basically, there are five upgrade technologies: Compact Fluorescent Lamps (CFLs), incandescent assemblies, Light Emitting Diodes (LED), Electroluminescent panels, and Self Luminous Tubes.

Replacing incandescent sources with compact fluorescent lamps was the "first generation" exit sign upgrade. Most CFL kits must be hard-wired and can not simply screw into an existing incandescent socket. Although CFL kits are a great improvement over incandescent exit signs, more technologically advanced upgrades are available that offer reduced maintenance costs, greater efficacy and flexibility for installation in low (sub-zero) temperature environments.

As Table 13.7 shows, LED upgrades are the most cost-effective because they consume very little energy, and have an extremely long life, practically eliminating maintenance.

**Table 13.7 Exit sign upgrades.**

| Light Source | Watts | Life | Replacement |
|---|---|---|---|
| Incandescent (Long Life) | 40 | 8 months | lamps |
| Compact Fluorescent | 10 | 1.7 years | lamps |
| Incandescent Assembly | 8 | 3 + years | light source |
| Light Emitting Diode (LED) | <4 | >25 | light source |
| Electroluminescent | 1 | 8+ years | panel |
| Self luminous (Tritium) | 0 | 10-20 years | luminous tubes |

Another low-maintenance upgrade is to install a "rope" of incandescent assemblies. These low-voltage "luminous ropes" are an easy retrofit because they can screw into existing sockets like LED retrofit kits. However, the incandescent assemblies create bright spots which are visible through the transparent exit sign and the non-uniform glow is a noticeable change. In addition, the incandescent assemblies don't last nearly as long as LEDs.

Although electroluminescent panels consume less than one watt, light output rapidly depreciates over time. These self-luminous sources are obviously the most energy-efficient, consuming no electricity. However the spent tritium tubes, which illuminate the unit, must be disposed of as a radioactive waste, which will increase over-all costs.

### 13.2.3.5 Lighting Controls

Lighting controls offer the ability for systems to be turned ON and OFF either manually or automatically. There are several control technology upgrades for lighting systems, ranging from simple (installing manual switches in proper locations) to sophisticated (installing occupancy sensors).

#### Switches

The standard manual, single-pole switch was the first energy conservation device. It is also the simplest device and provides the least options. One negative aspect about manual switches is that people often forget to turn them OFF. If switches are far from room exits or are difficult to find, occupants are more likely to leave lights ON when exiting a room.[1] Occupants do not want to walk through darkness to find exits. However, if switches are located in the right locations, with multiple points of control for a single circuit, occupants find it easier to turn systems OFF. Once occupants get in the habit of turning lights OFF upon exit, more complex systems may not be necessary. The point is: switches can be great energy conservation devices as long as they are convenient to use them.

Another opportunity for upgrading controls exists when lighting systems are designed such that all circuits in an area are controlled from one switch, yet not all circuits need to be activated. For example, a college football stadium's lighting system is designed to provide enough light for TV applications. However, this intense amount of light is not needed for regular practice nights or other non-TV events. Because the lights are all controlled from one switch, every time the facility is used all the lights are turned ON. By dividing the circuits and installing one more switch to allow the football stadium to use only 70% of its lights during practice nights, significant energy savings are possible.

Generally, if it is not too difficult to re-circuit a poorly designed lighting system, additional switches can be added to optimize the lighting controls.

#### Time Clocks

Time clocks can be used to control lights when their operation is based on a fixed operating schedule. Time clocks are available in electronic or mechanical styles. However, regular check-ups are needed to ensure that the time clock is controlling the system properly. After a power loss, electronic timers without battery backups can get off schedule—cycling ON and OFF at the wrong times. It requires a great deal of maintenance time to reset isolated time clocks if many are installed.

#### Photocells

For most outdoor lighting applications, photocells (which turn lights ON when it gets dark, and off when sufficient daylight is available) offer a low-maintenance alternative to time clocks. Unlike time clocks, photocells are seasonally self-adjusting and automatically switch ON when light levels are low, such as during rainy days. A photocell is inexpensive and can be installed on each fixture, or can be installed to control numerous fixtures on one circuit. Photocells can also be effectively used indoors, if daylight is available through skylights.

Photocells have worked well in almost any climate, however they should be aimed north (in the northern hemisphere) to "view" the reflected light of the north sky. This way they are not biased by the directionality of east/west exposure or degraded by intense southern exposure. Photocells should also be cleaned when fixtures are re-lamped. Otherwise, dust will accumulate on the photodiode aperture, causing the controls to always perceive it is a cloudy day, and the lights will stay ON.

The least expensive type of photocell uses a cadmium sulfide cell, but these cells lose sensitivity after being in service for a few years by being degraded from their exposure to sunlight. This decreases savings by keeping exterior lighting on longer than required. To avoid this situation, cadmium sulfide cells can be replaced with electronic types that do not lose sensitivity over time. These electronic photocells use solid-state, silicon phototransistors or photodiodes, which last longer as evidenced by their longer warranties—up to 6 years—and can easily pay back before that time with energy and labor savings.

#### Photocells combined with Dimmable Ballasts to allow Daylight Harvesting

Daylight harvesting is a control strategy that can be

applied where diffuse daylight can be used effectively to light interior spaces. There is a widespread misunderstanding that daylighting can only be done in areas where there is a predominance of sunny, clear days, such as California or Arizona. In fact, many places with over 50% cloudy days can cost-effectively use daylight controls.

Daylight harvesting employs strategically located photo-sensors and electronic dimming ballasts. To effectively apply this strategy requires more knowledge than just plugging a sensor into a dimming ballast. Photo-sensors and dimming ballasts form a control system that controls the light level according to the daylight level. The fluorescent lighting is dimmed to maintain a band of light level when there is sufficient daylight present in the space. The output is changed gradually by a fade control so occupants are not disturbed by rapid changes in light level.

### Lumen Depreciation Compensation
### (an additional benefit of a Daylight Harvesting System)

Lighting systems are usually over-designed to compensate for light losses that normally occur during the life time of the system. Alternatively, the "lumen depreciation compensation strategy" allows the design light level to be met without over-designing, thereby providing a more efficient lighting system. The control system works in a way similar to daylight harvesting controls. A photo-sensor detects the actual light level and provides a low-voltage signal to electronic dimming ballasts to adjust the light level. When lamps are new and room surfaces are clean, less power is required to provide the design light level. As lamps depreciate in their light output and as surfaces become dirty, the input power and light level is increased gradually to compensate for these sources of light loss. Some building management systems accomplish this control by using a depreciation algorithm to adjust the output of the electronic ballasts instead of relying on photo-sensors.

#### Table 13.8 Estimated % savings from occupancy sensors.

| Application | Energy Savings |
| --- | --- |
| Offices (Private) | 25-50% |
| Offices (Open Spaces) | 20-25% |
| Rest Rooms | 30-75% |
| Corridors | 30-40% |
| Storage Areas | 45-65% |
| Meeting Rooms | 45-65% |
| Conference Rooms | 45-65% |
| Warehouses | 50-75% |

### Occupancy Sensors

Occupancy sensors save energy by turning off lights in spaces that are unoccupied. When the sensor detects motion, it activates a control device that turns ON a lighting system. If no motion is detected within a specified period, the lights are turned OFF until motion is sensed again. With most sensors, sensitivity (the ability to detect motion) and the time delay (difference in time between when sensor detects no motion and lights go OFF) are adjustable. Occupancy sensors are produced in two primary types: Ultrasonic (US) and Passive Infrared (PIR). Dual-Technology (DT) sensors, that have both ultrasonic and passive infrared detectors, are also available. Table 13.8 shows the estimated percent energy savings from occupancy sensor installation for various locations.

US and PIR sensors are available as wall-switch sensors, or remote sensors such as ceiling mounted or outdoor commercial grade units. With remote sensors, a low-voltage wire connects each sensor to an electrical relay and control module, which operates on common voltages. With wall-switch sensors, the sensor and control module are packaged as one unit. Multiple sensors and/ or lighting circuits can be linked to one control module allowing flexibility for optimum design.

Wall-switch sensors can replace existing manual switches in small areas such as offices, conference rooms, and some classrooms. However, in these applications, a manual override switch should be available so that the lights can be turned OFF for slide presentations and other visual displays. Wall-switch sensors should have an unobstructed coverage pattern (absolutely necessary for PIR sensors) of the room it controls.

Ceiling-mounted units are appropriate in corridors, rest rooms, open office areas with partitions and any space where objects obstruct the line of sight from a wall-mounted sensor location. Commercial grade outdoor units can also be used in indoor warehouses and large aisles. Sensors designed for outdoor use are typically heavy duty, and usually have the adjustable sensitivities and coverage patterns for maximum flexibility. Table 13.9 indicates the appropriate sensors for various applications.

### Ultrasonic Sensors (US)

Ultrasonic sensors transmit and receive high-frequency sound waves above the range of human hearing. The sound waves bounce around the room and return to the sensor. Any motion within the room distorts the sound waves. The sensor detects this distortion and signals the lights to turn ON. When no motion has been detected over a user-specified time, the sensor sends a signal to turn the lights OFF. Because ultrasonic sensors

**Table 13.9 Occupancy sensor applications.**

| Sensor Technology | Private Office | Large Open Office Plan | Partitioned Office Plan | Conference Room | Rest Room | Closets/ Copy Room | Hallways Corridors | Warehouse Aisles Areas |
|---|---|---|---|---|---|---|---|---|
| US Wall Switch | • | | | • | • | • | | |
| US Ceiling Mount | • | • | • | • | • | • | | |
| IR Wall Switch | • | | | • | | • | | |
| IR Ceiling Mount | • | • | • | • | | • | | |
| US Narrow View | | | | | | | • | |
| IR High Mount Narrow View | | | | | | | • | • |
| Corner Mount Wide-View Technology Type | | • | | • | | | | |

need enclosed spaces (for good sound wave echo reflection), they can only be used indoors and perform better if room surfaces are hard, where sound wave absorption is minimized. Ultrasonic sensors are most sensitive to motion toward or away from the sensor. Applications include rooms with objects that obstruct the sensor's line of sight coverage of the room, such as restroom stalls, locker rooms and storage areas.

*Passive Infrared Sensors (PIR)*

Passive Infrared sensors detect differences in infrared energy emanating in the room. When a person moves, the sensor "sees" a heat source move from one zone to the next. PIR sensors require an unobstructed view, and as distance from the sensor increases, larger motions are necessary to trigger the sensor. Applications include open plan offices (without partitions), classrooms and other areas that allow a clear line of sight from the sensor.

*Dual-Technology Sensors (DT)*

Dual-Technology (DT) sensors combine both US and PIR sensing technologies. DT sensors can improve sensor reliability and minimize false switching. However, these types of sensors are still only limited to applications where ultrasonic sensors will work.

*Occupancy Sensor Effect on Lamp Life*

Occupancy Sensors can cause rapid ON/OFF switching which reduces the life of certain fluorescent lamps. Offices without occupancy sensors usually have lights constantly ON for approximately ten hours per day. After occupancy sensors are installed, the lamps may be turned ON and OFF several times per day. Several laboratory tests have shown that some fluorescent lamps lose about 25% of their life if turned OFF and ON every three hours. Although occupancy sensors may cause lamp life to be reduced, the annual burning hours also decreases. Therefore, in most applications, the time period until re-lamp will not increase. However, due to the laboratory results, occupancy sensors should be carefully evaluated if the lights will be turned ON and OFF rapidly. The longer the lights are left OFF, the longer lamps will last.

The frequency at which occupants enter a room makes a difference in the actual percent time savings possible. Occupancy sensors save the most energy when applied in rooms that are not used for long periods of time. If a room is frequently used and occupants re-enter a room before the lights have had a chance to turn OFF, no energy will be saved. Therefore, a room that is occupied once every three hours will be more appropriate for occupancy sensors than a room occupied once every three minutes, even though the percent vacancy time is the same.

*Occupancy Sensors and HIDs*

Although occupancy sensors were not primarily developed for HIDs, some special HID ballasts (bi-level) offer the ability to dim and re-light lamps quickly. Another term for bi-level HID technology is Capacitive Switching HID Fixtures, which are discussed in the HID Ballast Section.

**Lighting Controls via a Facility Management System**

When lighting systems are connected to a Facility Management System (FMS), greater control options can be realized. The FMS could control lights (and other equipment, i.e. HVAC) to turn OFF during non-working hours, except when other sensors indicate that a space is occupied. These sensors include standard occupancy sensors or a card access system, which could indicate which employee is in a particular part of the facility. If the facility is "smart," it will know where the employee works and control the lights and other systems in that area. By wiring all systems to the FMS, there is a greater ability to integrate technologies for maximum performance and savings. *For example, an employee can control lights by entering a code into the telephone system or a computer network.*

Specialized controls for individual work environments (offices or cubicles) are also available. These systems use an occupancy sensor to regulate lights, other electronic systems (and even HVAC systems) in an energy efficient manner. In some systems, remote controls allow the occupant to regulate individual lighting and HVAC systems. These customized systems have allowed some organizations to realize individual productivity gains via more effective and aesthetic work space environments.

## 13.3  PROCESS TO IMPROVE LIGHTING EFFICIENCY

The three basic steps to improving the efficiency of lighting systems:
1. Identify necessary light quantity and quality to perform visual task.
2. Increase light source efficiency if occupancy is frequent.
3. Optimize lighting controls if occupancy is infrequent.

Step 1, identifying the proper lighting quantity and quality is essential to any illuminated space. However, steps 2 & 3 are options that can be explored individually or together. Steps 2 & 3 can both be implemented, but often the two options are economically mutually exclusive. If you can turn OFF a lighting system for the majority of time, the extra expense to upgrade lighting sources is rarely justified. Remember, light source upgrades will only save energy (relative to the existing system) when the lights are ON.

### 13.3.1 Identify necessary light quantities and qualities to perform tasks.

Identifying the necessary light quantities for a task is the first step of a lighting retrofit. Often this step is overlooked because most energy managers try to mimic the illumination of an existing system, even if it is over-illuminated and contains many sources of glare. For many

years, lighting systems were designed with the belief that no space can be over-illuminated. However, the "more light is better" myth has been dispelled and light levels recommended by the IES declined by 15% in hospitals, 17% in schools, 21% in office buildings and 34% in retail buildings.[2] Even with IES's adjustments, there are still many excessively illuminated spaces in use today. Energy managers can reap remarkable savings by simply redesigning a lighting system so that the proper illumination levels are produced.

Although the number of workplane footcandles are important, the occupant needs to have a contrast so that he can perform a task. For example, during the daytime your car headlights don't create enough contrast to be noticeable. However, at night, your headlights provide enough contrast for the task. The same amount of light is provided by the headlights during both periods, but daylight "washes out" the contrast of the headlights.

The same principle applies to offices, and other illuminated spaces. For a task to appear relatively bright, objects surrounding that task must be relatively dark. For example, if ambient light is excessive (150 fc) the occupant's eyes will adjust to it and perceive it as the "norm." However when the occupant wants to focus on something he/she may require an additional light to accent the task (at 200 fc). This excessively illuminated space results in unnecessary energy consumption. The occupant would see better if ambient light was reduced to 30-40 fc and the task light was used to accent the task at 50 fc. As discussed earlier, excessive illumination is not only wasteful, but it can reduce the comfort of the visual environment and decrease worker productivity.

After identifying the proper quantity of light, the proper quality must be chosen. The CRI, CCT and VCP must be specified to suit the space.

### 13.3.2 Increase Source Efficacy

Increasing the source efficacy of a lighting system means replacing or modifying the lamps, ballasts and/or fixtures to become more efficient. In the past, the term "source" has been used to imply only the lamp of a system. However, due to the inter-relationships between components of modern lighting systems, we also consider ballast and fixture retrofits as "source upgrades." Thus increasing the efficacy simply means getting more lumens per watt out of lighting system. For example, to increase the source efficacy of a T12 system with a magnetic ballast, the ballast and lamps could be replaced with T8 lamps and an electronic ballast, which is a more efficacious (efficient) system.

Another retrofit that would increase source efficacy

would be to improve the fixture efficiency by installing reflectors and more efficient lenses. This retrofit would increase the lumens per watt, because with reflectors and efficient lenses, more lumens can escape the fixture, while the power supplied remains constant.

Increasing the efficiency of a light source is one of the most popular types of lighting retrofits because energy savings can almost be guaranteed if the new system consumes less watts than the old system. With reduced lighting load, electrical demand savings are also usually obtained. In addition, lighting quality can be improved by specifying sources with higher CRI and improved performance. These benefits allow capital improvements for lighting systems that pay for themselves through increased profits.

### Task lighting

As a subset of Increasing Source Efficacy, "Task lighting" or "Task/Ambient" lighting techniques involve improving the efficiency of lighting in an entire workplace, by replacing and relocating lighting systems. Task lighting means retrofitting lighting systems to provide appropriate illumination for each task. Usually, this results in a reduction of ambient light levels, while maintaining or increasing the light levels on a particular task. For example, in an office the light level needed on a desk could be 75 fc. The light needed in aisles is only 20 fc. Traditional uniform lighting design would create a workplace where ambient lighting provides 75 fc throughout the entire workspace. Task lighting would create an environment where each desk is illuminated to 75 fc, and the aisles only to 20 fc. Figure 13.4 shows a typical application of Task/Ambient lighting.

Task lighting upgrades are a model of energy efficiency, because they only illuminate what is necessary. Task lighting designs are best suited for office environments with VDTs and/or where modular furniture can incorporate task lighting under shelves. Alternatively,

moveable desk lamps may be used for task illumination. Savings result when the energy saved from reducing ambient light levels exceeds the energy used for task lights.

In most work spaces, a variety of visual tasks are performed, and each employee has lighting preferences. Most workers prefer lighting systems designed with task lighting because it is flexible and allows individual control. For example, older workers may require greater light levels than young workers. Identifying task lighting opportunities may require some creativity, but the potential dollar savings can be enormous.

Task lighting techniques are also applicable in industrial facilities—for example, high intensity task lights can be installed on fork trucks (to supplement headlights) for use in rarely occupied warehouses. With this system, the entire warehouse's lighting can be reduced, saving a large amount of energy.

### 13.3.3 Optimize Lighting Controls

The third step of lighting energy management is to investigate optimizing lighting controls. As shown earlier, improving the efficiency of a lighting system can save a percentage of the energy consumed *while the system is operating*. However, sophisticated controls can turn systems OFF when they are not needed, allowing energy savings to accumulate quickly. The Electric Power Research Institute (EPRI) reports that spaces in an average office building may only be occupied 60-75% of the time, although the lights may be ON for the entire 10 hour day[3]. Lighting controls include switches, time clocks, occupancy sensors and other devices that regulate a lighting system. These systems are discussed in Section 13.2.3, Lighting System Components.

### 13.4 MAINTENANCE

### 13.4.1 Isolated Systems

Most lighting manuals prescribe specialized technologies to efficiently provide light for particular tasks. An example is dimmable ballasts. For areas that have sufficient daylight, dimmable ballasts can be used with integrated circuitry to reduce energy consumption during peak periods. Still, though there may be some shedding of lighting load along the perimeter, these energy cost savings may not represent a great percentage of the building's total lighting load. Further, applications of specialized technologies (such as dimmable ballasts) may be dispersed and isolated in several buildings, which can become a complex maintenance challenge, even if lamp types and locations are recorded properly. If maintenance personnel need to make additional site visits to get the

**Figure 13.4 Task/ambient lighting.**

right equipment to re-lamp or "fine-tune" special systems, the labor costs may exceed the energy cost savings.

In facilities with low potential for energy cost savings, facility managers may not want to spend a great deal of time monitoring and "fine-tuning" a lighting system if other maintenance concerns need attention. If a specialized lighting system malfunctions, repair may require special components, that may be expensive and more difficult to install. If maintenance cannot effectively repair the complex technologies, the systems will fail and occupant complaints will increase. Thus the isolated, complex technology that appeared to be a unique solution to a particular lighting issue is often replaced with a system that is easy to maintain.

In addition to the often eventual replacement of technologies that are difficult to maintain, well intended repairs to the system may accidentally result in "snap-back." "Snap-back" is when a specialized or isolated technology is accidentally replaced with a common technology within the facility. For example, if dimmable ballasts only represent 10% of the building's total ballasts, maintenance personnel might not keep them in stock. When replacement is needed, the maintenance personnel may accidentally install a regular ballast. Thus, the lighting retrofit has "snapped back" to its original condition.

The above arguments are not meant to "shoot down" the application of all new technologies. However, new technologies usually bring new problems. The authors ask that the energy manager carefully consider the maintenance impact when evaluating an isolated technology. Once again, all lighting systems depend on regular maintenance.

### 13.4.2 Maintaining System Performance

As with most manufactured products, lighting systems lose performance over time. This degradation can be the result of Lamp Lumen Depreciation (LLD), Fixture Dirt Depreciation (LDD), Room Surface Dirt Depreciation (RSDD), and many other factors. Several of these factors can be recovered to maintain performance of the lighting system. Figure 13.5 shows the LLD for various types of lighting systems

Lamp Lumen Depreciation occurs because as the lamp ages, its performance degrades. LLD can be accelerated if the lamp is operated in harsh environments, or the system is subjected to conditions for which it was not designed. For example, if a fluorescent system is turned ON and OFF every minute, the lamps and ballasts will not last as long. Light loss due to lamp lumen depreciation can be recovered by re-lamping the fixture.

Fixture Dirt Depreciation and Room Surface Dirt Depreciation block light and can reduce light levels. However, these factors can be minimized by cleaning surfaces and minimizing dust. The magnitude of these factors is dependent on each room, thus recommended cleaning intervals can vary. Generally it is most economical to clean fixtures when re-lamping.

### 13.4.3 Group Re-lamping

Most companies replace lamps when someone notices a lamp is burned out. In a high rise building, this could become a full-time job, running from floor to floor, office to office, disrupting work to open a fixture and replace a lamp. However, in certain cases, it is less costly to group re-lamp on a pre-determined date. Group relamping can

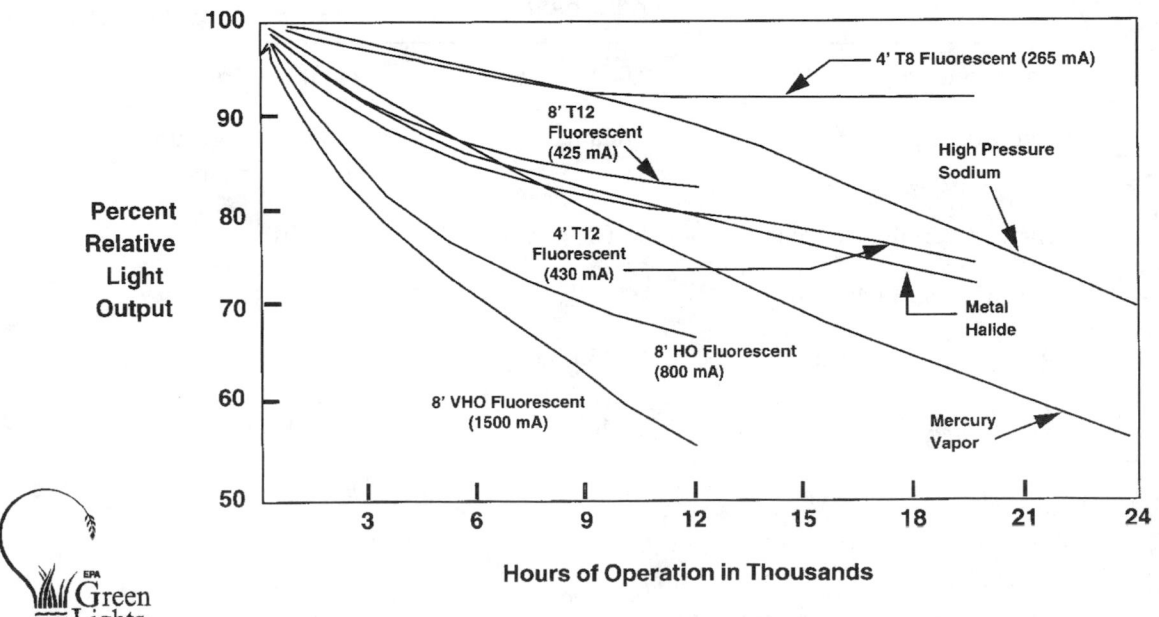

**Figure 13.5 Lamp lumen depreciation (LLD).**

be cost-effective due to economies of scale. Replacing all lamps at one time can be more efficient than relamping "one at a time." In addition, bulk purchasing may also yield savings. The rule of thumb is: group relamp at 50% to 70% of the lamp's rated life. However, depending on site-specific factors and the lumen depreciation of the lighting system, relamping interval may vary.

The facility manager must evaluate their own building, and determine the appropriate relamp interval by observing when lamps start to fail. Due to variations in power voltages (spikes, surges and low power), lamps may have different operating characteristics and lives from one facility to another. It is important to maintain records on lamp and ballast replacements and determine the most appropriate relamping interval. This also helps keep track of maintenance costs, labor needs and budgets.

Group relamping is the least costly method to relamp due to reduced time and labor costs. For example, Table 13.10 shows the benefits of group relamping. As more states adopt legislation requiring special disposal of lighting systems, group relamping in bulk may offer reduced disposal costs due to large volumes of material.

### 13.4.4 Disposal Costs

Disposal costs and regulations for lighting systems vary from state to state. These expenses should be included in an economic analysis of any retrofit. If proper disposal regulations are not followed, the EPA could impose fines and hold the violating company liable for environmental damage in the future.

### 13.5  NEW TECHNOLOGIES & PRODUCTS[1]

The energy efficient lighting market is extremely competitive, forcing manufacturers to develop new products to survive. The development is so rapid, it is challenging to "keep up" with all the latest technologies. This chapter describes the proven technologies, however it is good idea to evaluate the latest developments before implementing a lighting system.

### 13.5.1 Fluorescent Ballasts

Miniaturization of electronic ballasts has been made possible by the use of integrated circuits and surface-mount technologies. The new ballasts are smaller, thinner and lighter.

*Low Profile Housing*

The familiar "brick" shape and weight of ballasts will soon be gone. Reduced parts count and surface mount technology have reduced the size of ballasts as well as improved their reliability. These advances have permitted housings of lower profile and smaller cross-section. Today, some ballasts have a dimensional cross-

**Table 13.10 Group relamping example: 1,000 3-Lamp T8 Lensed troffers**

|  | Spot Relamping (on burn-out) | Group Relamping (@ 70% rated life) |
|---|---|---|
| Relamp cycle | 20,000 hours | 14,000 hours |
| Avg. relamps/year | 525 relamps/yr | 750 relamps/yr (group) 52 relamps/yr (spot) |
| Avg. material cost/year | $1,050/yr | $1,604/yr |
| Lamp disposal @ 0.50 ea. | $236/yr | $375/yr |
| Avg. labor cost/year | $3,150/yr | $1,437/yr |
| TOTAL EXPENSES: | $4,463/yr | $3,416/yr |
| *Assumptions:* *Labor:* | *$6.00/lamp* | *$1.50/lamp* |
| *Material:* | *$2.00/lamp* | *$2.00/lamp* |
| *Operation:* | *3,500 hr/yr* | *3,500 hr/yr* |

[1]The majority of this section was provided by John Fetters, Effective Lighting Solutions. ©Effective Lighting Solutions, Inc.

section of 30 × 30 mm. The advantages of smaller ballast packages include lighter weight, less material, and easier handling and installation. In addition, they fit into the new low-profile fixtures, especially indirect and direct-indirect fixtures.

### Universal Input Voltage

Many facilities have different lighting system voltages in different parts of their buildings. Maintenance personnel are slowed in their ballast replacement task when they don't know the voltage for a particular area of the building. However, ballasts with the universal voltage feature will automatically use any line voltage applied (between 120-277-v). In addition to saving valuable maintenance time when the labor cost of identifying the voltage for each ballast to be replaced, or the expense of distributor restocking of ballasts ordered with the incorrect voltage is included, any cost difference is very affordable. In addition, fewer replacement ballast models need to be stocked.

### Optimizing Ballast Selection

Instant-start ballasts have become the most popular method of starting F32T8/RS rapid-start lamps because of their lower input watts rating compared with rapid-start systems. However, lamp life can be reduced by up to 25% at short burn cycles when lamps are operated instant-start, increasing maintenance costs. In applications where short ON/OFF cycles are common, lamp life increases by using program-start or rapid-start ballasts, instead of instant-start ballasts. Rapid-start operation of rapid-start lamps will ensure normal rated lamp life and program-start ballasts can extend lamp life by up to 50%.

### Dimming electronic ballasts for fluorescent lamps.

Electronic ballasts with dimming functions operate fluorescent lamps at high frequency, just like fixed-output electronic ballasts. Most dimmable ballasts now have separate low-voltage control leads, which can be grouped together to create control zones, which are independent of the power zones. Many dimming ballast designs provide over-voltage protection of the control leads in case line voltage is accidentally applied to the low voltage leads. The control method of choice is 0 to 10vDC, although dimming ballasts are also available, which are designed to accept the AC line phase control signals from incandescent wall-box dimmer controls that dim the fluorescent lamps accordingly.

Dimming ballasts are divided into two categories, based on dimming ranges:

1.  Energy management applications: 100% to 5%
2.  Architectural dimming applications
    100% to 1% (or less)

*Note: Today, dimming ballasts for energy-management applications are also being used in applications that formerly required an architectural dimming ballast, such as conference rooms.*

Dimmable ballasts are available for dimming most linear fluorescent lamps (1-, 2- or 3-lamp versions) including T5 HO lamps. Many of these products start the lamps at any dimmer setting, and do not have to be ramped up to full-light output before they dim. Most of the models available measure less than 15% total harmonic distortion (THD) throughout the dimming range.

Conference and presentation rooms have traditionally been built with two lighting systems. One, an incandescent system, usually uses recessed cans and is dimmed with wall dimmers. The second is usually a fluorescent non-dimming system for general lighting. The incandescent system requires a lot of maintenance, due to the short life of the incandescent lamps. One solution to this situation is to remove the overhead incandescent system and replace the ballasts in the fluorescent system with line-voltage dimming ballasts that can operate from the existing incandescent wall-box dimmer(s). The main benefit of this improvement is lower maintenance cost for a small investment in ballasts and the electrical maintenance staff can make the change.

### Electronic ballasts for compact fluorescent lamps (CFLs)

Several manufacturers have dimming ballasts for rapid-start (4-pin) compact fluorescent lamps (CFLs). Most of these offerings are for the higher-wattage CFLs (26 to 57-w). The lowest dimming limit is 5% and the dimming range varies with the manufacturer. There are designs to accept the AC phase-control signals from incandescent wall-box dimmer controls. This makes upgrading an older incandescent downlight system to an energy-efficient CFL system easy, with no new wiring required.

### 13.5.2 Fluorescent Lamps

*Several smaller, yet brighter fluorescent systems (T2, T5 and T5HO) have flourished in recent years. Smaller systems have been effective in task lighting environments, where less light from a single source is needed. The reduction of unnecessary lighting reduces energy expenses.*

### T2 lamps

These are sub-miniature, 1/4" (0.25") diameter lamps that have side tabs instead of end pins. They are available in standard fluorescent colors of 3000K, 3500K, and 4100K with CRI in the mid-80s. They are rated with a lamp lumen depreciation of 0.95, and only lose 5% of their light output in the first 40% of rated life. T2 lamps have lamp efficacy ratings similar to compact fluorescent in the mid-60s.

Low profile fixtures used for task and under-counter lighting, showcase and decorative lighting have been made possible by these small diameter lamps. Their principal application, however, is for backlighting graphic display panels, which are starting to be done with high-performance light-emitting diodes (LEDs). The use of T2 lamps for this application is not expected to increase.

### T5 lamps

The T5 lamps come in two distinct and different families—standard (high-efficiency) and high output (HO). These recently developed lamps should not be confused with older miniature preheat fluorescent lamps of the same diameter nor with the line of long compact fluorescent lamps of the same diameter.

### Standard (high-efficiency) T5 linear lamps

These 5/8" diameter lamps (Figure 13-6) are equipped with miniature bi-pin bases and are powered by electronic ballasts. All the lamps in this family operate on the same current (170 ma) and have the same surface brightness for all wattages. For cove and cornice applications this is a distinct advantage.

Another reason the T5 lamp is suited for these applications is that they are designed to peak in their lumen rating at 35°C (95°F) vs. 25°C (77°F) for T12 and T8 lamps. This characteristic provides higher light output in confined applications where there is little or no air circulation. In indirect fixtures, this thermal characteristic increases efficiency and gives more usable lumens per watt.

Standard T5 lamps are 12-18% more efficient than T8 lamps (96-106 LPW) and 10-15% more efficient than the T5HO. T5s employ rare-earth phosphors with CRI greater than 80 and lamp lumen maintenance rated at 95%.

There are 4 sizes of standard T5 lamps as shown in Table 13.11, all rated at 20,000 hours (at 3 hours-per-start).

Note that the 28-watt lamp (not quite 4' long) has an initial lumen rating the same as a 4' T8 lamp. However, the millimeter lengths and miniature bi-pin bases preclude their use in standard length linear fluorescent systems and the high bulb-wall brightness limits their use to high ceiling applications because the visible tubes can create too much discomfort glare in low mounting height applications.

### Table 13.11 Standard T5 Lamp Sizes
### (Source: Effective Lighting Solutions, Inc.)

| Nominal Watts | Length mm/(in) | Lumens (initial) | Lumens (maintained) |
|---|---|---|---|
| 14 | 549/(21.6) | 1,350 | 1,283 |
| 21 | 849/(33.4) | 2,100 | 1,995 |
| 28 | 1149/(45.2) | 2,900 | 2,755 |
| 35 | 1449/(57.0) | 3,650 | 3,460 |

### High output T5 linear lamps (SEE TABLE 13.12)

These T5 lamps are physically the same size as standard T5 lamps, but provide higher lumen output. T5HO lamps generate from 1.5 to 2 times the light output of the standard T5 and nearly twice the light output (188%) of T8 and T12 systems with the same number of lamps. One-lamp T5 HO fixtures can replace both lamps of 2-lamp T8 fixtures.

They are approximately 10-15% less efficient than standard T5 lamps (83 to 94 lumens per watt) and can be up to 8% less efficient than standard T8 systems. The surface brightness varies among various wattages and since lamps operate on different currents (see Table 13.12), each lamp wattage requires a unique ballast.

T5 HO lamps are available in the three standard fluorescent color temperatures (cool—4100K, warm—3000K and neutral—3500K) and have a color rendering index greater than 80. Lumen maintenance is rated at 95%. These lamps are now rated at 20,000 hours. Similar to standard T5 lamps, T5HO lamps also peak in their lumen rating at 35°C (95°F) vs. 25°C (77°F) for T12 and T8 lamps. This provides higher light output in confined applications where there is little or no air circulation. In indirect fixtures, this thermal characteristic results in increased efficiency with more usable lumens per watt.

T5HO lamps are being used in designs of slim profile indirect fixtures that take advantage of the smaller lamp. Only 1 lamp per 4-ft section is required, replacing

### Table 13.12 T5HO Lamp Sizes
### (Source: Effective Lighting Solutions, Inc.)

| Nominal Watts | Length mm/(in) | Lamp Current (ma) | Lumens (initial) | Lumens (maintained) |
|---|---|---|---|---|
| 24 | 549/(21.6) | 300 | 2,000 | 1,900 |
| 39 | 849/(33.4) | 340 | 3,500 | 3,325 |
| 54 | 1149/(45.2) | 460 | 5,000 | 4,750 |
| 80 | 1449/(57.0) | 552 | 7,500 | 7,125 |

designs using 2, 4-ft T8 lamps per 4-ft section. The high bulb wall brightness limits their use in direct applications in low ceiling height conditions due to discomfort glare Following the trend to fluorescent, T5HO lamps are being used in high ceiling applications, including high-bay industrial fixtures.

### T8 lamps

*Standard T8 lamps (See Figure 13.6)*

T8 lamps (1" dia) were originally imported from Europe in the early 1980s. The lamps now used in the US are different than their European pre-heat cousins and there are improved models. T8s have been the lamps of choice (along with the high frequency electronic ballasts that drive them) for fluorescent upgrades for several years. T8 lamps are available in 2', 3', 4', and 5' lengths, at 17, 25, 32, and 40-w respectively. These lamps require ballasts that supply 265 ma. There are also two versions of 8' retrofit lamps at 59, or 86-w. U-tubes are available in the new 1 5/8" leg spacing and a retrofit U-tube that has 6" leg spacing that is used to replace 6" leg-spacing T12 U-tubes.

Recent advances in T8 lamps have been in improvements in color rendering and longer life. Extended performance T8 lamps have a life rating of 24,000 (at 3 hours per start)—20% longer than standard T8 lamps. These extended performance lamps operate on the same electronic ballasts designed to operate standard T8 lamps. Lumen maintenance is rated at 0.94 and it levels off after that. Lumen output is slightly higher at 3,000 lumens and CRI is improved to 8. Standard 3000K, 3500K and 4100K colors are provided.

### Reduced-wattage T8 lamps

Sometimes called "energy-saving" T8 lamps, these lamps are available in 28 and 30-w vs. the standard 32-w models. They are designed to replace reduced-wattage, 34-w T12 lamps, when upgrading to electronic ballasts. They are recommended for use only on instant-start elec-

tronic ballasts to provide the higher open-circuit voltage required and they need to be operated above 60°F. In addition, they cost more than standard T8s, but they save about 6% over standard lamps, are TCLP compliant, and have high lumen maintenance (94%).

*High-performance T8 lamps (See Table 13.13)*

These high-lumen lamps (3,100 L) are part of a dedicated lamp/ballast system that can save about 19% over standard T8 systems with the same light output and twice the lamp life (on program-start ballasts) compared to standard instant-start T8 systems. They exhibit high lumen maintenance (95%) and high CRI (86).

**Table 13.13 High-Performance T8 System Watts (Source: Osram-Sylvania)**

| Number of Lamps | 1 | 2 | 3 | 4 |
|---|---|---|---|---|
| Input (system) Watts | 25-w | 48-w | 72-w | 94-w |

### TCLP compliant fluorescent lamps

Over 600 million fluorescent lamp tubes are disposed of every year in the US. Prior to June 1999, the USEPA required that spent fluorescent lamps that did not pass a Toxicity Characteristic Leaching Procedure (TCLP) were to be treated as hazardous waste because they contained more than 0.2 mg/liter (ppm) of mercury.

Standard fluorescent lamps do not pass the TCLP test and were required (prior to the Universal Waste Rule) to be handled as hazardous waste or recycled by using expensive hazardous waste haulers and massive documentation. Fluorescent lamps are now covered by the Universal Waste Rule (as of today). The main result of the inclusion of fluorescent lamps in the universal waste rule is to encourage recycling of spent lamps.

Lamp manufacturers have reduced the mercury content of fluorescent tubes over the past decade to less than half of the original content. In response to public concern for mercury in the environment, the major lamp companies started to produce what they originally called "low-mercury" lamps. Philips Lighting using a proprietary dosing and buffering technology they call ALTO®, produced the first low-mercury fluorescent lamps. 4-foot ALTO fluorescent lamps have less than 10 mg of mercury and therefore will pass the TCLP test. Other lamp companies have followed this trend and now these lamps are called "TCLP compliant" lamps to indicate that the lamps are designed to pass the federal TCLP (Toxic Characteristic Leaching

**Figure 13.6 Some T8 Lamp Sizes (Source: Sylvania)**

Procedure) test.

Low mercury lamps have distinctive colored end caps, usually colored green. Use of TCLP compliant lamps provides users with normal lamp performance, light output, and life and an environmentally friendly option to meet their lighting needs. Although they do not need to be recycled, many end-users are avoiding any liability for their lamp disposal and recycling their spent TCLP compliant lamps.

### 13.5.3 Compact Fluorescent Lamps

Improvements to CFL technologies have been occurring every year since they became commercially available. Products available today provide higher efficacies as well as instant starting, reduced lamp flicker, quiet operation, smaller size and lighter weight. Dimmable CFLs are now available, and it can be expected that their performance will increase with time. The 2700K color (incandescent appearance) has been replaced by the 3000K for commercial applications. "Pre-heat" models start by blinking before they stay ON. Older lamps blink more than new lamps during starting. Rapid-start models start instantly, with no blinking.

*Traditional Problems with CFLs*

CFLs suffer from multiple sensitivities that reduce the light output. They are position-sensitive. Gravity determines where the excess mercury "pools," which affects the mercury vapor pressure that determines the lumen output. Lumen ratings published in lamp catalogs are performed according to ANSI testing standards that require the lamp to be in the vertical, "base up" position. In the base-down position some CFLs produce 20% fewer lumens. In the horizontal position, they produce about 15% less light. Lamp lumen depreciation for CFLs is often more accelerated than for incandescent sources. CFLs are also not recommended for wet applications.

Additional sensitivities include temperature sensitivity that reduces the light output when CFLs are operated above or below their optimum temperature rating. The loss due to temperature is approximately 15-20% and is most noticeable in enclosed fixtures, such as recessed downlights due to self-heating. *However, when the mercury used in a CFL is in the form of an amalgam—an alloy of mercury and other metals, the mercury vapor pressure is reduced without affecting the lamp temperature. This technique makes the lamps less temperature sensitive than conventional CFLs and provides more light at the high and low extremes—above 100°F and below 32°F. Amalgam lamps are not easily identified, but most "triple-tubes" are amalgam products.*

*Screw-base CFLs*

Screw-base CFLs have "Edison" bases and are used to replace incandescent lamps. They have an integral ballast built into the base. Early models had magnetic ballasts built into the base, however most contemporary models have electronic bases, allowing significant size reduction. Some screw-base CFLs are used in commercial applications, but most are used in residential lighting. The newest shape is the spiral or spring shape, shown in Figure 13.7. Higher wattage options are shown in Table 13.14.

*2-pin preheat CFLs (Figure 13.8)*

2-pin CFLs require a separate ballast (which is usually magnetic) located in the fixture. Each lamp has a starter, located in the base, which provides pre-heat starting. Table 13.15 shows twin-tube and quad pre-heat models.

**Figure 13.7 Spiral or Spring Shaped CFLs**

**Table 13.14 Large Screw-base Compact Fluorescent Lamps**
**(Source: www.maxlite.com)**

| Shape | Lumens | Watts | LPW | M.O.D | M.O.L. | Replaces |
|-------|--------|-------|-----|-------|--------|----------|
| Spiral | 3,500 | 55 | 64 | 3.5" | 10" | 200-w incand |
| Quad | 4,200 | 65 | 65 | 3.5" | 11" | 250-w incand |
| Quad | 5,500 | 85 | 65 | 3.5" | 11.8" | 300-w incand |

**Figure 13.8 Twin-tube Pre-heat CFLs (Source: Sylvania)**

**Table 13.15 Pre-Heat Compact Fluorescent Lamps
(Source: Effective Lighting Solutions, Inc.)**

| Description | Lumens | Watts | Lamp Lumens/Watt |
|---|---|---|---|
| Twin-tube Pre-heat | 250 | 5 | 50 |
| | 400 | 7 | 57 |
| | 600 | 9 | 66 |
| | 900 | 13 | 69 |
| Quad Pre-heat | 575 | 9 | 64 |
| | 860 | 13 | 66 |
| | 1200 | 18 | 66 |
| | 1800 | 26 | 69 |

**Figure 13.9 Double Arc-Tube HPS (Source: Sylvania)**

*4-pin rapid-start CFLs*

4-pin rapid-start lamps are available in 16, 18, 24, 26, 28, 32, and 42-watt models. All commercial-grade models are rated at 10,000 hours. The majority of these lamps are T5 (5/8" tube diameter), but some are T4 (1/2" tube diameter). Maximum overall length (MOL) ranges from 3.5" to 5.5." The three primary color temperatures are available—3000K (warm), 3500K (neutral), and 4100K (cool). At least one manufacturer also provides the warmer 2700K color. Rapid-start CFLs are designed for operation on electronic ballasts and can be dimmed when operated on a dimming electronic ballast, designed for the appropriate lamp wattage.

New generation compact fluorescent lamps (CFL) are a significant improvement over the earlier twin and double twin tube types. Instead of using free mercury, these new CFLs use mercury that has been combined with other metals to form an amalgam. The amalgam makes the lamps less sensitive to the effects of temperature and position.

This is an important advantage over standard CFLs and is the reason that many applications using

standard CFLs perform poorly. Amalgam CFLs have stable light output from 23°F to 130°F. Also, amalgam lamps are not position sensitive and exhibit less color shift than conventional CFLs. They do take slightly longer to warm up, but they are at full brightness in less than 3 minutes. Unfortunately, manufacturers do not always clearly identify their amalgam lamps, but most "triple" tubes are amalgam.

CFLs are available in higher wattage for use in high ceiling downlights. A 32-watt triple-tube, amalgam lamp, rated at 2,400 lumens, provides a system replacement for 150-watt incandescent downlights. A 42-watt triple-tube, amalgam lamp, rated at 3,200 lumens, allows its use as a system replacement for high-wattage incandescent downlights and a 57-w rapid start, triple-tube, amalgam lamp, rated at 4,300 lumens, is equivalent to a 200-w incandescent lamp.

At Lightfair International 2003, Philips Lighting unveiled a new multiple burning position, high lumen-output PL-H lamp. These 4-pin, rapid start lamps are used with high frequency, electronic ballasts. They are composed of 6, T5 limbs, joined with bend-and-bridge technology. There are 6 models with wattages ranging from 60 to 120-w. Versatile and powerful they have a lumen output almost double that of other CFLs, up to 9,000 lumens (120-w model), they provide maximum design freedom in many areas, including high ceiling indoor and outdoor applications. In addition, the white light PL-H range promises stable color rendering, long life and high lumen maintenance.

*Dimmable CFLs*

Screw-base dimmable CFLs were introduced in 1996. This lamp is intended to replace incandescent lamps used on wallbox dimming systems. The electronic ballast base in this 1-piece lamp responds to the phase change voltage waveform from most existing dimmers and dims the CFL down to 10% light output.

The dimmable CFL is available in several wattages, the most common being a 23-watt triple-tube amalgam lamp with a lumen rating of 1500 that will replace 90-watt "A" lamps. The major benefit of this lamp is that dimming is accomplished on existing dimming circuits with no additional control wiring required.

### 13.5.4 High Intensity Discharge (HID) Systems:

#### Metal Halide Systems

Metal Halide lamps have become more popular due to technological advancements and consumer preference for "white light." Technologically, the "pulse-start" metal halide systems are a significant improvement in efficiency and performance. Like most electronic ballasts, these operate at high frequency, provide a quicker re-strike time (3-5 minutes) versus standard metal halide systems (6-10 minutes). The pulse-start systems maintain CRI and lumen output better over time.

#### Pulse-start metal halide

Low-wattage metal halide (< 175-w) and high-pressure sodium lamps have used pulse-start technology for many years, using a high voltage pulse starter to ignite the lamps. What is new is the availability of high-wattage, pulse-start metal halide lamps (175-w to 1000-w) that are quickly replacing standard metal halide lamps. There is a new family of arc tubes, called "formed body" that replace the old pinched seal arc tubes and overcome the disadvantages of the old design. The starter electrode, found in standard arc tubes, has been eliminated. The new arc tube design features uniform geometry and higher fill pressures. Improved temperature control is achieved with smaller pinch seals that provide less heat loss, reducing lamp-to-lamp color shift. Formed body arc tube lamps provide a lower ambient temperature limit, -40°F instead of -30°F for standard arc tube lamps. Faster starting and restarting (re-strike) results from the lower mass of the new arc tubes. These changes result in higher lamp efficacy (up to 110 lumens per watt), improved lumen maintenance (up to 80%), consistent lamp-to-lamp color (within 100°K) and 50% faster warm up and re-strike times (three to five minutes vs. eight to 15 minutes).

#### Ceramic metal halide lamps (CMH)

Ceramic arc tube metal halide lamps use the same ceramic material used in high-pressure sodium arc tubes—polycrystalline alumina (PCA). PCA reduces the sodium loss through the more porous glass arc tube used in standard metal halide lamps. This reduces color shift and spectral variation of standard metal halide lamps caused as the sodium is depleted. Metal halide lamps with ceramic arc tubes are designated either CDM (ceramic discharge, metal halide) or CMH (ceramic metal halide) and may also refer to their constant color in their brand name. They are available from 20 to 400-watt with color temperature of either 3000K (warm) or 4000K (cool) and an average rated life from 6,000 to 15,000 hours, depending on the wattage. Ceramic metal halide lamps are started by a pulse starter like PS metal halide lamps and operate best on electronic ballasts. The main advantage of the combination of CMH lamps and electronic ballasts is 10-20% higher lumen output (which also results in a corresponding higher LPW) and the best color stability.

The benefits of CMH lamps include good lamp efficacy (83-95 LPW)—in the same range as older, linear fluorescent lamps; high CRI (83-95); limited color shift (from ± 75K to ± 200K CCT); excellent lamp-to-lamp color consistency; and good lumen maintenance (0.70-0.80).

Applications for these improved color metal halide lamps include high ceilings such as atria, and lobbies of hospitality spaces, downlights, and lighting merchandise—anywhere that the higher CRI and color consistency can be justified. Fade-block models with thin-film coatings on the arc-tube shroud are available for merchandise lighting to help reduce the UV fading of materials. There are also HPS replacement lamps that can be used as an interim solution when converting from a high-pressure sodium system to a white light system.

#### High Pressure Sodium systems

Two new lamp wattages are available to narrow the gap between the 400-w and the 1,000-w standard HPS lamps. Both sizes will probably not survive the market and the 600-watt, 90,000 lumen lamp is not as widely supported by the lighting industry as the 750-w, 105,000 lumen lamp as a good "in between" size. In general, however, all high-pressure sodium lamps are losing ground to white-light sources, such as metal halide or fluorescent.

Several improvements have been introduced in "new-generation" HPS lamps. The major improvement is the elimination of end-of-life cycling that is characteristic of standard high-pressure sodium lamps. However, there are two different design approaches by the three major lamp companies. Two companies have taken the 'notification' approach, in which the lamp turns a distinctive blue color at end of life. A third company simply shuts off the lamp power at end of life.

New HPS lamps have welded bases that replace the old lead soldered bases. Several new models have reduced or zero mercury content, qualifying them as TCLP compliant lamps.

These lamps sacrifice efficacy and life to achieve CRI rating up to 65. Lamp efficacy ranges from 63 to 94 LPW and they have an average rated life of 15,000 hours. They are available in 70, 100, 150, 250, and 400-w models, and have lumen ratings from 4,400 to 37,500.

### Double arc-tube HPS lamps

These HPS lamps are called standby lamps and have two arc tubes, welded together in parallel. However only one arc tube operates when the lamp is ignited. Upon the loss of power, the second arc tube, hot from being in close proximity to the first arc tube, comes ON at about 50% light output. It then comes up to full light output within the strike time of the lamp (~4 min max). Standby lamps are used for safety and security applications and are popular with prison lighting systems as well as roadway systems, with a tested life of 40,000 hours, reducing maintenance time and labor cost.

## 13.5.5 Induction Lighting

### Electrodeless Induction Systems

Since the introduction of the first electric light, a search has been on for long-life lighting. The reason for this search is to reduce the cost associated with changing lighting components at or near their end of rated life—maintenance cost.

The lamps used in induction systems have no electrodes to wear out as other lamps, such as fluorescent and HID lamps do. The lamps can last much longer without electrodes. Long life is the primary advantage of these systems. These systems can provide a good payback where maintenance labor cost is high. When compared with other light sources, electrodeless induction systems will operate 5-8 times longer than fluorescent and metal halide systems and about 4 times longer than HPS systems. In addition, induction lamps come ON relatively quickly and have short re-strike time compared with HID lamps.

Instead of using electrodes to generate electrons as is done in fluorescent lamps, electrodeless systems produce light by means of induction—the use of an electromagnetic field to induce a plasma gas discharge into a tube or bulb that has a phosphor coating. No electrons are needed, since the gas discharge is induced into the bulb or tube by a high-frequency electronic generator that supplies the electromagnetic field. These systems provide white light with a minimum color shift and CRI and lumen depreciation values are similar to fluorescent lamps.

Each of the two primary electrodeless system lamps has a unique size and shape and require new fixtures that are designed to optically match each unique shape. There is no common electronics package for these products since they operate on much different frequencies. The electronics package must be fairly close to the glass envelopes and the maximum mounting distance is restricted to the wire length supplied on the electronics package. These are independent systems and are designed so that both the glass envelopes and the electronics are changed out together at end of life.

### Genura™ Lamp

GE Lighting developed an electrodeless induction lamp labeled Genura™ and introduced it in the US in 1995. Genura™ is a compact R30 reflector lamp with a standard medium base that is intended for use as a retrofit lamp (in place of a 100-watt A lamp, a 75-watt R30 lamp or a 65-watt R30 lamp) in recessed downlights (cans). This product is a lamp and not a system, so it is covered here before the induction systems.

### QL Induction Lighting System (Figure 13.10)

Philips Lighting developed this induction system and introduced it to the European market in 1991. In 1992 the QL was introduced to the US market. The QL system is comprised of three components. 1) a high-frequency generator, 2) a power coupler and 3) the glass bulb. The high-frequency generator is in a separate electronics package that provides the 2.65 MHz current to the power coupler (antenna) through a coax cable. The power coupler sits inside the enclosed glass discharge bulb shaped like a large A lamp. The bulb, which contains an inert gas

**Figure 13.10 QL Lighting System (Source: Philips Lighting)**

and a small amount of mercury is attached to the power coupler by a plastic lamp cap that uses a click system. Like fluorescent lamps, the inside walls of the bulb are coated with a phosphor coating. When the high frequency electromagnetic field is applied to the bulb, the gas is ionized and the lamp produces photons at UV frequency and visible light in the same manner as a fluorescent tube. The photons collide with the phosphor coating and cause the lamp to glow. Full brightness is achieved in 10-15 seconds. The system meets FCC requirements as a low EMI design.

There are three models—55-w, 3,500 lumens, 85-w, 6,000 lumens, and 165-w, 12,000 lumens—with lumen efficacy of 64-73 LPW. The 55-watt model has a maximum overall diameter (MOD) of 85 mm (~3 3/8"), the 85-watt model has a MOD of 111 mm (~ 4 3/8"), and the 165-w model has a MOD of 131 mm (~5 5/32"). QL bulbs are available in two color temperatures—3000K (warm) and 4000K (cool).

The main advantage of the QL system is its long life—average rated life is 100,000 hours. Philips rates life as 20% failures at 60,000 hours. This long life advantage is especially important where maintenance cost is high. The current emphasis in the U.S. is in outdoor lighting systems—street, roadway, and tunnel lighting systems. Several fixture manufacturers have incorporated the QL in their designs.

### ICETRON™ System

The Inductively Coupled Electrodeless system—ICETRON™—was developed by Sylvania. This electrodeless system consists of three parts: 1) a unique rectangular 'donut' shaped bulb—filled with an inert gas and a small amount of mercury, 2) two ring-shaped ferrite core couplers—one at each of the short sides of the bulb, and 3) a separate high-frequency (200-300 KHz) generator. A plug-in connector attaches leads from the couplers to the electronic generator. The driver may be mounted up to 66 feet away from the lamp.

When the high frequency electromagnetic field is applied to the donut-shaped bulb between the ferrite cores at each end, the gas inside the bulb is ionized and produces light by inducing a circulating current in the bulb, which generates photons at UV frequency. These photons collide with the phosphor coating and cause the lamp to glow. ICETRON™ lamps strike and re-strike instantly.

There are three ICETRON™ lamps—70-w, 100-w, and 150-w—and two drivers. Table 13.16 shows the combinations of lamps and drivers and the resulting system performance.

The mercury in the glass envelope is in the form of an amalgam, providing a universal burn situation.

**Table 13.16 ICETRON™ System Performance (Source: Sylvania)**

| Lamp | Driver | System Lumens | System Watts | System LPW |
|------|--------|---------------|--------------|------------|
| 70-w | 100-w | 6,500 | 82 | 79 |
| 100-w | 100-w | 8,000 | 107 | 75 |
| 100-w | 150-w | 11,000 | 157 | 70 |
| 150-w | 150-w | 12,000 | 157 | 76 |

**Figure 13.11 ICETRON™ System (Source: Sylvania)**

Starting temperatures extend down to –40°F, opening up opportunities for low temperature applications such as freezers and coolers. The ICETRON™ bulbs are available in two color temperatures—3500K (neutral) and 4100K (cool) and a CRI of 80. Sylvania rates the lumen maintenance at 70% at 60,000 hours (60% of rated life). This is a departure from the standard method of rating lumen maintenance for other light sources (at 40% rated life). The lumen maintenance curve shows a lumen maintenance value of 75 at 40%. At the rated life of 100,000 hours, the lumen maintenance is about 65%.

The ICETRON™ system meets FCC (non-consumer) requirements and has a low EMI design. The principal advantage of this system is long life—100,000 hours. A comprehensive warranty covers the system for 60 months. Applications where maintenance is difficult and or costly are prime candidates for these long life systems.

### 13.5.6 Remote Source Lighting and Fiber Optics

Remote source lighting systems have the lighting source some distance from the point of delivery. Basically, the light source is connected to a light pipe or fiber optics, which carries the light to the point of application. Remote lighting solutions have become more popular because they fill the needs of projects that have hazardous or underwater environments, walk-in freezers, architectural restrictions or special aesthetic objectives. Remote source lighting systems offer reduced maintenance costs, because lamps can be accessed easily and safely. *For example, light pipes can be effective in gymnasiums or swimming pools. The uniform lighting also can result in a lower glare than single bright fixtures.*

Fiber optics can be used to resolve challenges associated with maintaining aesthetics. Light sources can be installed in rooms outside of a viewing area, with the fiber optics routed through walls (or other obscured spaces—like crown molding) to the application. Like miniature flashlights, the fiber optics can be pointed directly at the needed spot. *For example, gallery or church lighting can be achieved without bulky fixtures getting in the way of the occupant's view.*

## 13.6 SPECIAL CONSIDERATIONS

### 13.6.1 Rules and Regulations
*EPACT-1992*

The National Energy Policy Act of 1992 (EPACT) was designed to dramatically reduce energy consumption via more competitive electricity generation and more efficient buildings, lights and motors. Because lighting is common in nearly all buildings, it is a primary focus of EPACT. The 1992 legislation bans the production of lamps that have low efficacy or CRI. Table 13.17 indicates which lamps are banned and a few options for replacing the banned systems. From left to right, the table shows several options, for each banned system, ranging from the most efficient substitute to the minimum compliance substitute. Generally, the minimum compliance substitute has the lowest initial cost, but after energy costs have been included, the most efficient upgrades have the lowest life-cycle costs.

Often the main expense with a lighting upgrade is the labor cost to install new products; however, the incremental labor cost of installing high-efficiency equipment is minimal. So, it is usually beneficial to install the most efficient technologies because they will have the lowest operational and life-cycle costs. EPACT only eliminates the "bottom of the barrel" in terms of available lighting technology. To keep one step ahead of future lamp bans, it is a good idea to consider upgrades with greater efficiencies than the minimum acceptable substitute.

*Federal Fluorescent Ballast Rule*

An agreement between lighting manufacturers (represented by the National Electrical Manufacturers Association—NEMA) and energy policy advocates (The American Council for an Energy Efficient Economy—ACEEE, The Alliance to Save Energy, and the National Resources Defense Council—NRDC) was finalized on September 2000 and became law as Part 430—Energy Conservation Program for Consumer Products. The new standards are expected to reduce greenhouse gas emissions by 19 million metric tons of carbon and by 60,000 tons of nitrous oxide over the next 20 years—the equivalent of eliminating the emissions of one million cars for 15 years.

The rule promotes T8 electronic systems (without creating efficiency standards for T8 ballasts) by raising the minimum BEF for T12 ballasts to a level that can only be achieved by electronic ballasts. T12 magnetic ballasts are still allowed, but these are a small fraction of a shrinking fluorescent magnetic market. They are less efficient and carry a cost premium, so in actual practice they will not be used.

No magnetic ballasts may be manufactured for the covered lamps (2' U-tubes, 4' rapid-start, 8' instant-start, and 8' HO) after June 30, 2005. Magnetic ballasts for T8 lamps can continue to be manufactured for applications sensitive to infrared (IR) or electromagnetic interference (EMI). Luminaires sold on or after April 1, 2006 that use the covered T12 lamps must incorporate electronic ballasts. An exception is made for magnetic ballasts used for replacement purposes in existing installations, which can be manufactured until June 30, 2010, but must be marked "FOR REPLACEMENT USE ONLY."

There is an implied warning to all fluorescent lamp users that if they have not converted to T8 systems by June 30, 2010, they will have to use T8 ballasts and lamps for spot replacement in their existing T12 systems, which can only result in compatibility problems and a real maintenance headache!

*Electrical Considerations*

Due to the increasingly complex lighting products available today, concern about effects on power distribution systems have risen. In certain situations, lighting retrofits can reduce the power quality of an electrical system. Poor power quality can waste energy and the capacity of an electrical system. In addition, it can harm the electrical distribution system and devices operating on that system.

Electrical concerns peaked when the first generation electronic ballasts for fluorescent lamps caused power quality problems. Due to advances in technology, electronic ballasts available today can *improve* power quality

Table 13.17 EPACT 1992's effect: lamp bans and options.

## F96T12 SLIMLINE (EFFECTIVE MAY 1, 1994)

| NONCOMPLYING LAMPS | MOST EFFICIENT (Ballast change required) | GOOD RETROFIT | MINIMUM COMPLIANCE (See note above) |
|---|---|---|---|
| F96T12/CW (75 W, Cool White) | F96T8/41K-85CRI (4100°K, 85 CRI) F96T8/41K-75CRI (4100°K, 75 CRI) | F96T12/41K-80CRI/ES (60 W, 4100°K, 80 CRI) F96T12/41K-70CRI/ES (60 W, 4100°K, 70 CRI | F96T12/CW/ES (60 W, Cool white, 62 CRI) F96T12/41K-80CRI (75 W, 4100°K, 80 CRI) F96T12/41K-70CRI (75 W, 4100°K, 70 CRI) |
| F96T12/W (75 W, White) | F96T835K-85CRI (3500°K, 85 CRI) F96T8/35K-75CRI (3500°K, 75 CRI) | F96T12/35K-80CRI/ES (60 W, 3500°K, 80 CRI) F96T12/35K-70CRI/ES (60 W, 3500°K, 70 CRI) | F96T12/W/ES (60 W, White, 57 CRI) F96T12/35K-80CRI (75 W, 3500°K, 80 CRI) F96T12/35K-70CRI (75 W, 3500°K, 70 CRI) |
| F96T12/WW (75 W, Warm white) | F96T8/30K-85CRI (3000°K, 85 CRI) F96T8/30K-75CRI (3000°K, 75 CRI) | F96T12/30K-80CRI/ES (60 W, 3000°K, 80 CRI) F96T12/30K-70CRI/ES (60 W, 3000°K, 70 CRI) | F96T12/WW/ES (60 W, Warm white, 52 CRI) F96T12/30K-80CRI (75, W, 3000°K, 80 CRI) F96T12/30K-70CRI (75, W, 3000°K, 70 CRI) |
| F96T12/D (75 W, Daylight) | | F96T12/64-80CRI/ES (60 W, 6400°K, 80 CRI) | F96T12/64K-80CRI (75 W, 6400°K, 80 CRI) |

## F96T12/HO HIGH OUTPUT (EFFECTIVE MAY 1, 1994)

| NONCOMPLYING LAMPS | MOST EFFICIENT (Ballast change required) | GOOD RETROFIT | MINIMUM COMPLIANCE (See note above) |
|---|---|---|---|
| F96T12/CW/HO (110 W, Cool white) | | F96T12/41K-70CRI/HO/ES (95 W, 4100°K, 70 CRI) | F96T12/CW/HO/ES (95, W, Cool white, 62 CRI) F96T12/LW/HO/ES (95 W, Lite white, 48 CRI) F96T12/41K-70CRI/HO (110 W, 4100°K, 70 CRI |
| F96T12/W/HO (110 W, White) | | F96T12/35K-70CRI/HO/ES (95 W, 3500°K, 70 CRI) | F96T12/35K-70CRI/HO (110 W, 3500°K, 70 CRI) |
| F96T12/WW/HO (110 W, Warm white) | | F96T12/30K-70CRI/HO/ES (95 W, 3000°K, 70 CRI) | F96T12/WW/HO/ES (95 W, Warm white, 52 CRI) F96T12/30K-70 CRI/HO (110 W, 3000°K, 70 CRI) |
| F96T12/D/HO (110 W, Daylight) | | F96T12/64K-80CRI/HO/ES (95 W, 6400°K, 80 CRI) | F96T12/64K-80CRI/HO (110 W, 6400°K, 80 CRI) |

*(Continued)*

when replacing magnetically ballasted systems in almost every facility. However, some isolated problems may still occur in electronically sensitive environments such as intensive-care units in hospitals. In these types of areas, special electromagnetic shielding devices are available, and are usually required.

The energy manager should ensure that a new system will improve the power quality of the electrical system.

*Harmonics*

A harmonic is a higher multiple of the primary frequency (usually 60 Hertz) superimposed on the alternating current waveform. A distorted 60 Hz current wave may contain harmonics at 120 Hz, 180 Hz and so on. The harmonic whose frequency is twice that of the fundamental is called the "second-order" harmonic. The harmonic whose frequency is three times the fundamental is the "third-order" harmonic.

Highly distorted current waveforms contain numerous harmonics. The even harmonics (second-order, fourth order, etc.) tend to cancel each other's effects, but

the odd harmonics tend to add in a way that rapidly increases distortion because the peaks and troughs of their waveforms coincide. Lighting products usually indicate a common measurement of distortion percentage: Total Harmonic Distortion (THD). Table 13.18 shows the % THD for various types of lighting and office equipment.

### 13.6.2 HVAC Effects

Nearly all energy consumed by lighting systems is converted to light, heat and noise, which dissipate into the building. Therefore, if the amount of energy consumed by a lighting system is reduced, the amount of heat energy going into the building will also be reduced, and less air-conditioning will be needed. Consequently, the amount of winter-time heating may be increased to compensate for a lighting system that dissipates less heat.

Because most offices use air-conditioning for more months per year than heating, a more efficient lighting system can significantly reduce air-conditioning costs. In addition, air conditioning (usually electric) is much more expensive that heating (usually gas). Therefore, the

### Table 13.17 EPACT 1992's effect: lamp bans and options (*Conclusion*).

#### F40T12 (EFFECTIVE NOVEMBER 1, 1995)

| F40/CW<br>(40 W, Cool white) | F32T8/41K-85CRI/ES<br>(4100°K, 85 CRI)<br>F32T8/41K-75CRI<br>(4100°K, 75 CRI) | F40/41K-80CRI/ES<br>(32 W, 4100°K, 80 CRI)<br>F40/41K-70CRI<br>(34 W, 4100°K, 70 CRI) | F40/CW/ES<br>(34 W, Cool white, 62 CRI)<br>F40/41K-80CRI (40 W, 4100°, 80 CRI)<br>F40/41K-70CRI (40 W, 4100°K, 70 CRI) |
|---|---|---|---|
| F40/W<br>(40 W, White) | F32T8/35K-85CRI<br>(3500°K, 85 CRI)<br>F32T8/35K-75CRI<br>(3500°K, 75 CRI) | F40/35K-80CRI/ES<br>(34 W, 3500°K, 80 CRI<br>F40/35K-70CRI/ES<br>(34 W, 3500°K, 70 CRI) | F40/W/ES (34 W, White, 57 CRI)<br>F40/35K-80CRI (40 W, 3500°K, 80 CRI)<br>F40/35K-70CRI (40 W, 3500°K, 70 CRI) |
| F40/WW<br>(40 W, Warm white)<br>F40/WWX<br>(40 W, Warm white deluxe)<br>F40/WWX/ES<br>(34 W, Warm white deluxe) | F32T8/30K-85CRI<br>(3000°K, 85 CRI)<br>F32T8/30K-75CRI<br>(3000°K, 75 CRI) | F40/30K-80CRI/ES<br>(34 W, 3000°K, 80 CRI)<br>F40/30K-70CRI/ES<br>(34 W, 3000°K, 70 CRI) | F40/WW/ES<br>(34 W, Warm white, 52 CRI)<br>F40/30K-80CRI<br>(40 W, 3000°K, 80 CRI)<br><br>F40/30K-70CRI<br>(40 W, 3000°K, 70 CRI) |
| F40/D (40 W, Daylight)<br>F40/D/ES (34 W, Daylight) | F32T8/50K-75CRI<br>(5000°K, 75 CRI) | F40/64K-80CRI<br>(34 W, 6400°K, 80 CRI) | F40/64K-80CRI<br>(40 W, 6400°K, 80 CRI) |

#### FB40/6 CURVALUME "U-LAMP" (EFFECTIVE NOVEMBER 1, 1995)

| F40/U/6/CW<br>(40 W, Cool white) | F32T8/U/6/41K-85CRI<br>(4100°K, 85 CRI)<br>F32T8/U/6/41K-75CRI<br>(4100°K, 75 CRI) | F40/U/6/41K-70CRI/ES<br>(34 W, 4100°K, 70 CRI) | F40/U/6/CW/ES<br>(34 W, Cool white, 62 CRI)<br>F40/U/6/41K-70CRI<br>(40 W, 4100°K, 70 CRI) |
|---|---|---|---|
| F40/U/6W<br>(40 W, White) | F32T8/U/6/35K-85CRI<br>(3500°K, 85 CRI)<br>F32T8/U/6/35K-75CRI<br>(3500°K, 75 CRI) | F40/U/6/35K-70CRI/ES<br>(34 W, 3500°K, 70 CRI) | F40/U/6/W/ES<br>(34 W, White, 57 CRI)<br>F40/U/6/35K-70CRI<br>(40 W, 3500°K, 70 CRI) |
| F40/U/6/WW<br>(40 W, Warm white)<br>F40/U/6/WWX<br>(40 W, Warm white deluxe) | F32T8/U/6/30K-85CRI<br>(3000°K, 85 CRI)<br>F32T8/U/6/30K-75CRI<br>(3000°K, 75 CRI) | F40/U/6/30K-70CRI/ES<br>(34 W, 3000°K, 70 CRI) | F40/U/6/WW/ES<br>(34 W, Warm white, 52 CRI)<br>F40/U/30K-80CRI (40 W, 3000°K, 80 CRI)<br>F40/U/30K-70CRI (40 W, 3000°K, 70 CRI) |

#### INCANDESCENT REFLECTOR LAMPS (EFFECTIVE NOVEMBER 1, 1995)

| NONCOMPLYING LAMPS | BEST COMPLYING RETROFIT | SUITABLE SUBSTITUTE | EXEMPT |
|---|---|---|---|
| 75PAR38 | 45PAR/Halogen | 50PAR30/Halogen | Colored ty |
| 100PAR38 | 75PAR/Halogen | 75ER30 | Rough ser |
| 150PAR38 | 90PAR/Halogen | 75PAR/Halogen | ER Shape |
| 75/65PAR38 | 45PAR/Halogen | 50PAR30/Halogen | |
| 100/80PAR38 | 75PAR/Halogen | 75ER30 | |
| 150/120PAR38 | 90PAR/Halogen | 75PAR/Halogen | |
| 75R30 | 50PAR30/Longneck/Halogen | 50ER30 | |
| 75R40 | 45PAR/Halogen/Very Wide Flood | 50ER30 | |
| 100R40 | 75PAR/Halogen | 75ER30 | |
| 150R40 | 90PAR/Halogen | 120ER40 | Source: |

savings on air-conditioning electricity are usually worth more dollars than the additional gas cost.

### 13.6.3 The Human Aspect

Regardless of the method selected for achieving energy savings, it is important to consider the human aspect of energy conservation. Buildings and lighting systems should be designed to help occupants work in comfort, safety and enjoyment. Retrofits that improve the lighting quality (and the performance of workers) should be installed, especially when they save money. The recent advances in electronic ballast technology offer an opportunity for energy conservation to actually improve worker productivity. High frequency electronic ballasts and tri-phosphor lamps offer improved CRI, less audible noise and lamp flicker. These benefits have been shown to improve worker productivity and reduce headaches, fatigue and absenteeism.

### Table 13.18 Power quality characteristics for different electric devices.

| | Active Power (W) | Power Factor | Current THD (%) |
|---|---|---|---|
| **Compact fluorescent lighting systems** | | | |
| 13-W quad-tube compact fluorescent lamp w/ NPF magnetic ballast | 16 | 0.54 | 13 |
| 13-W quad-tube compact fluorescent lamp w/ NPF electronic ballast | 13 | 0.50 | 153 |
| 16-W quad-tube compact fluorescent lamp w/ HPF electronic ballast | 16 | 0.91 | 20 |
| **Full-size fluorescent lighting systems (two lamps per ballast)** | | | |
| T12 40-W lamps w/ energy-efficient magnetic ballast for T12 lamps | 87 | 0.98 | 17 |
| T12 34-W lamps w/ energy-efficient magnetic ballasts for T12 lamps | 72 | 0.94 | 22 |
| T10 40-W lamps w/ energy-efficient magnetic ballast for T12 lamps | 93 | 0.98 | 22 |
| T12 40-W lamps w/ electronic ballast for T12 lamps | 72 | 0.99 | 5 |
| T12 34-W lamps w/ electronic ballast for T12 lamps | 62 | 0.99 | 5 |
| T10 40-W lamps w/ electronic ballast for T12 lamps | 75 | 0.99 | 5 |
| T9 34-W lamps w/ electronic ballasts for T12 lamps | 79 | 0 .99 | 5 |
| T9 32-W lamps w/ electronic ballast for T8 lamps | 61 | 0.98 | 6 |
| T8 32-W lamps w/ electronic ballast for T8 lamps | 63 | 0.98 | 6 |
| **High-intensity discharge lighting systems** | | | |
| 400-W high-pressure sodium lamp w/ magnetic transformer | 425 | 0.99 | 14 |
| 400-W metal halide lamp w/ magnetic transformer | 450 | 0.94 | 19 |
| **Incandescent lighting systems** | | | |
| 100-W incandescent A lamp | 101 | 1.00 | 1 |
| 50-W MR16 low-voltage halogen lamp w/ magnetic transformer | 62 | 0.97 | 6 |
| 50-W MR16 low-voltage halogen lamp w/ electronic transformer | 51 | 0.99 | 10 |
| **Office equipment** | | | |
| Desktop computer without monitor | 33 | 0.56 | 139 |
| 13" high-resolution color monitor for desktop computer | 49 | 0.56 | 138 |
| Laser printer while in standby | 29 | 0.40 | 224 |
| Laser printer while printing | 799 | 0.98 | 15 |
| External fax/modem | 5 | 0.73 | 47 |
| Electric pencil sharpener | 85 | 0.41 | 33 |

*NLPIP measured specific products and reported their characteristics. These characteristics may vary substantially for similar products; specifiers should check with product manufacturers for specific information.

NPF = Normal Power Factor

HPF = High Power Factor

*Implementation Tactics*

In addition to utilizing the appropriate lighting products, the implementation method of a lighting upgrade can have a serious impact on its success. To ensure favorable reaction and support from employees, they must be involved in the lighting upgrade. Educat- ing employees and allowing them to participate in the decision process of an upgrade will reduce the resistance of change to a new system. Of critical importance is the maintenance department, because they will have an important role in the future upkeep of the system.

Once the decision has been made to upgrade the lighting in a particular area, and a trial installation has received approval, a complete retrofit should be completed as soon as possible. Due to economies of scale and minimal employee distraction, an all-at-once retrofit is usually optimal. In some cases, an over-night or over-the-weekend installation might be preferred. This method would avoid possible criticisms from side-by-side comparisons of the old and new systems. For example, a task lighting retrofit may appear darker than a uniformly illuminated space adjacent to it. The average worker who believes "more light is better" might protest the retrofit. However, if the upgrade is done over the weekend, the worker may not easily notice the changes.

### 13.6.4 Lighting Waste and the Environment

Upgrading any lighting system will require disposal of lamps and ballasts. Some of this waste may be hazardous and/or require special management. Contact your state to identify the regulations regarding the proper disposal of lighting equipment in your area.

#### *Mercury*

With the exception of incandescent bulbs, nearly all gaseous discharge lamps (fluorescent and HIDs) contain small quantities of mercury that end up in the environment, unless recycled. Mercury is also emitted as a by-product of electricity generation from some fossil-fueled power plants. Although compact fluorescent lamps contain the most mercury per lamp, they save a great deal of energy when compared to incandescent sources. Because they reduce energy consumption, (and avoid power plant emissions) CFLs introduce to the environment less than half the mercury of incandescents.[5] Mercury sealed in glass lamps is also much less available to ecosystems than mercury dispersed throughout the atmosphere. Nevertheless, mercury is not good for our environment and the energy manager should check local disposal codes—*you don't want to break the law.* Mercury in lamps can be recycled, and regulations may soon require it.

#### *PCB Ballasts*

Ballasts produced prior to 1979 may contain Polychlorinated biphenyls (PCBs). Human exposure to these possible carcinogens can cause skin, liver, and reproductive disorders. Older fluorescent and HID ballasts contain high concentrations of PCBs. These chemical compounds were widely used as insulators in electrical equipment such as capacitors, switches and voltage regulators until 1979. The proper method for disposing used PCB ballasts depends on the regulations in the state where the ballasts are removed or discarded. Generators of PCB containing ballast wastes may be subject to notification and liability provisions under the Comprehensive Environmental Response, Compensation and Liability Act of 1980 (CERCLA)—also known as "Superfund."[6]

Generally, the PCB ballast is considered to be a hazardous waste only when the ballast is leaking PCBs. An indication of possible PCB leaking is an oily tar-like substance emanating from the ballast. If the substance contains PCBs, the ballast and all materials it contacts are considered PCB waste, and are subject to state regulations. Leaking PCB ballasts must be incinerated at an EPA approved high-temperature incinerator.

#### *Energy Savings and Reduced Power Plant Emissions*

When appliances use less electricity, power plants don't need to produce as much electricity. Because most power plants use fossil fuels, a reduction in electricity generation results in reduced fossil fuel combustion and airborne emissions. Considering the different types of power plants (and the different fuels used) in different geographic regions, the Environmental Protection Agency has calculated the reduced power plant emissions by saving one kWh. Table 13.19 shows the reduction of $CO_2$, $SO_2$ and $NO_x$ for each kWh per year saved, in different regions of the US.[7]

### 13.7 DAYLIGHTING

Human beings developed with daylight as their primary light source. For thousands of years humans evolved to the frequency of natural diurnal illumination. Daylight is a flicker-free source, generally with the widest spectral power distribution and highest comfort levels. With the twentieth century's trend towards larger buildings and dense urban environments, the development and wide spread acceptance of fluorescent lighting allowed electric light to become the primary source in offices.

Daylighting interior spaces is making a comeback because it can provide good visual comfort, and it can save energy if electric light loads can be reduced. New control technologies and improved daylighting methods allow lighting designers to conserve energy and optimize employee productivity.

There are three primary daylighting techniques available for interior spaces: Utilizing Skylights, Building Perimeter Daylighting and Building Core Daylighting. Skylights are the most primitive and are the most common in industrial buildings. Perimeter daylighting is defined as using natural daylight (when sufficient) such that electric lights can be dimmed or shut off near windows

Table 13.19

| POLLUTION PREVENTION | | | | |
|---|---|---|---|---|
| To estimate pollution prevention of an energy conservation project, use the following formulas and factors. | | | | |
| CO2: | kWh/yr saved | x | emission factor | = | lbs/yr |
| SO2: | kWh/yr saved | x | emission factor | = | g/yr |
| NOx | kWh/yr | x | emission | = | g/yr |

| EPA Regional Emission Factors | | |
|---|---|---|
| **REGION 1:  CT, MA, ME, NH, RI, VT** | | |
| Emission per | CO2 | SO2 | NOx |
| kWh saved: | 1.1 | 4.0 | 1.4 |
| **REGION 2:  NJ, NY, PR, VI** | | |
| Emission per | CO2 | SO2 | NOx |
| kWh saved: | 1.1 | 3.4 | 1.3 |
| **REGION 3:  DC, DE, MD, PA, VA, WV** | | |
| Emission per | CO2 | SO2 | NOx |
| kWh saved: | 1.6 | 8.2 | 2.6 |
| **REGION 4:  AL, FL, GA, KY, MS, NC, SC, TN** | | |
| Emission per | CO2 | SO2 | NOx |
| kWh saved: | 1.5 | 6.9 | 2.5 |
| **REGION 5:  IL, IN, MI, MN, OH, WI** | | |
| Emission per | CO2 | SO2 | NOx |
| kWh saved: | 1.8 | 10.4 | 3.5 |
| **REGION 6:  AR, LA, NM, OK, TX** | | |
| Emission per | CO2 | SO2 | NOx |
| kWh saved: | 1.7 | 2.2 | 2.5 |
| **REGION 7:  IA, KS, MO, NE** | | |
| Emission per | CO2 | SO2 | NOx |
| kWh saved: | 2.0 | 8.5 | 3.9 |
| **REGION 8:  CO, MT, ND, SD, UT, WY** | | |
| Emission per | CO2 | SO2 | NOx |
| kWh saved: | 2.2 | 3.3 | 3.2 |
| **REGION 9:  AZ, CA, HI, NV, Guam, Am Samoa** | | |
| Emission per | CO2 | SO2 | NOx |
| kWh saved: | 1.0 | 1.1 | 1.5 |
| **REGION 10:  AK, ID, OR, WA** | | |
| Emission per | CO2 | SO2 | NOx |
| kWh saved: | 0.1 | 0.5 | 0.3 |

Note:  State pollution emission factors are aggregated by EPA region.

at the building perimeter. Traditionally, the amount of dimming depends on the interior distance from fenestration. However, ongoing research and application of "Core Daylighting Techniques" can stretch daylight penetration distance further into a room. Core daylighting techniques include the use of light shelves, light pipes, active daylighting systems and fiber optics. These technologies will likely become popular in the near future.

Perimeter and core daylighting technologies are technologies that are being further developed to function in the modern office environment. The modern office has many visual tasks that require special considerations to avoid excess illumination or glare. It is important to properly control daylighting in offices, so that excessive glare does not reduce employee comfort or the ability to work on VDTs. A poorly designed or poorly managed daylit space reduces occupant satisfaction and can increase energy use if occupants require additional electric light to balance excessive daylight-induced contrast.

Windows and daylighting typically cause an increased solar heat gain and additional cooling load for HVAC systems. However, development of new glazings and high performance windows has allowed designers to use daylighting without severe heat gain penalties. With dynamic controls, most daylit spaces can now have lower cooling loads than non-daylit spaces with identical fenestration. "The reduction in heat-from-lights due to daylighting can represent a 10% down-sizing in perimeter zone cooling and fans.[8]" However, because there are several parameters, daylighting does not always reduce cooling loads any time it displaces electric light. As window size increases, the maximum necessary daylight may be exceeded, creating additional cooling loads.

Whether interior daylighting techniques can be economically utilized depends on several factors. However the ability to significantly reduce electric lighting loads during utility "peak periods" is extremely attractive.

## 13.8  COMMON RETROFITS

Although there are numerous potential combinations of lamps, ballasts and lighting systems, a few retrofits are very common.

*Offices*

In office applications, popular and profitable retrofits involve installing electronic ballasts and energy efficient lamps, and in some cases, reflectors. Table 13.20 shows how a typical system changes with the addition of reflectors and the removal or substitution of lamps and ballasts. Notice that thin lamps allow more light to exit

the fixture, thereby increasing fixture efficiency. Reflectors improve efficiency by greater amounts when there are less lamps (or thinner lamps) to block exiting light beams.

The expression (Lumens)/(Fixture watt) is an indicator of the overall efficiency of the lighting system. It is similar to the efficacy of a lamp.

*Indoor/Outdoor Industrial*

In nearly all applications with significant annual operating hours, mercury vapor systems can be replaced by metal halide (or high output fluorescent systems). This retrofit will improve CRI, reduce operating and relamping costs. In applications where CRI is not critical, HPS systems (which have a higher efficacy than metal halide systems) can be used.

*Almost Anywhere*

In nearly all applications where incandescent lamps are ON for more than 5 hours per day, switching to CFLs will be cost-effective.

### 13.8.1 Sample Retrofits

This section provides the equations to calculate savings from several different types of retrofits. For each type of retrofit, the calculations shown are based on average conditions and costs, which vary from location to location. For example, annual air conditioning hours will vary from building to building and from state to state. The energy costs used in the following examples were based on $10/kW-month and $.05/kWh. In most industrial settings, demand is also billed. In the following examples demand savings would likely occur in all except for examples # 4 and # 6. To accurately estimate the cost and savings from these types of retrofits, simply insert local values into the equations.

EXAMPLE 1:
UPGRADE T12 LIGHTING SYSTEM TO T8

A hospital had 415 T12 fluorescent fixtures, which operate 24 hours/day, year round. The lamps and ballasts were replaced with T8 lamps and electronic ballasts, which saved about 30% of the energy, and provided higher quality light. Although the T8 lamps cost a little more (resulting in additional lamp replacement costs), the energy savings quickly recovered the expense. In addition, because the T8 system produces less heat, air conditioning requirements during summer months will be reduced. Conversely, heating requirements during winter months will be increased.

Table 13.20 Fluorescent lighting upgrade options.

## ORIGINAL SYSTEM
### Energy Efficient Magnetic Ballast

|                        | 40W lamps (T-12) | 34W lamps (T-12) |
|------------------------|------------------|------------------|
| Number of lamps        | 4                | 4                |
| Total Watts            | 176              | 144              |
| Ballast Factor         | 0.94             | 0.87             |
| Available Lumens       | 12000            | 9700             |
| Luminaire Efficiency   | 0.65             | 0.65             |
| Lumens per Luminaire   | 7800             | 6300             |
| Lumens/Luminaire watt  | 44.3             | 43.8             |

## POTENTIAL RETROFITS

|                        | EE Magnetic Ballast 42W lamps (T-10) | Electronic Rapid Start Ballast 42W lamps (T-10) | Electronic Rapid Start Ballast 32W lamps (T-8) | Electronic Instant Start Ballast 32W lamps(T-8) | Electronic Instant Start Ballast 32W lamps (T-8) |
|------------------------|------|------|------|------|------|
| Number of lamps        | 2    | 2    | 4    | 3    | 2    |
| Total Watts            | 92   | 63   | 112  | 90   | 60   |
| Ballast Factor         | 0.95 | 0.73 | 0.88 | 0.88 | 0.88 |
| Available Lumens       | 7000 | 5400 | 10200| 7700 | 5100 |
| Luminaire Efficiency   | 0.77 | 0.77 | 0.74 | 0.76 | 0.78 |
| Lumens per Luminaire   | 5400 | 4200 | 7500 | 5900 | 4000 |
| Lumens/Luminaire watt  | 58.7 | 66.7 | 67.0 | 65.6 | 66.7 |
| **ADD SILVER REFLECTOR** |    |      |      |      |      |
| Luminaire Efficiency   | 0.83 | 0.83 | 0.78 | 0.81 | 0.84 |
| Lumens/Luminaire       | 5800 | 4500 | 8000 | 6200 | 4300 |
| Lumens/Luminaire watt  | 63.0 | 71.4 | 71.4 | 68.9 | 71.7 |

*References:*
U.S. EPA Green Lights Program, Lighting Upgrade Manual, Apr. 94.
Advance Transformer Specification Guide
Magnetek Specification Guide

NOTES:  40W lamps are rated at 3200 lumens per lamp.
34W lamps are rated at 2800 lumens per lamp.
42W lamps are rated at 3700 lumens per lamp.
32W lamps are rated at 2900 lumens per lamp.
New luminaires may have greater efficiencies, due to highly reflective paints.

## Calculations

### kW Savings

= (# fixtures) [(Present input watts/fixture)—(Proposed input watts/fixture)]

= (415)[(86 watts/T12 fixture)-(60 watts/T8 fixture)]

= 10.8 kW

### kWh Savings

= (kW savings)(Annual Operating Hours)

= (10.8 kW)(8,760 hours/year)

= 94,608 kWh/year

### Air Conditioning Savings

= (kW savings)(Air Conditioning Hours/year)(1/Air Conditioner's COP)

= (10.8 kW)(2000 hours)(1/2.6)

= 8,308 kWh/year

### Additional Gas Cost

= (kW savings)(Heating Hours/year)(.003413 MCF/kWh)(1/Heating Efficiency)(Gas Cost)

= (10.8 kW)(1,500 hours/year)(.003413 MCF/kWh)(1/0.8)($4.00/MCF)

= $ 276/year

### Lamp Replacement Cost

= [(# fixtures)(# lamps/fixture)][((annual operational hours/proposed lamp life)(proposed lamp cost))—((annual hours operation/present lamp life)(present lamp cost))]

= [(415 fixtures)(2 lamps/fixture)][((8,760 hours/20,000 hours)($ 3.00/T8 lamp)) – ((8,760 hours/20,000 hours)($ 1.50/T12 lamp))]

= $ 545/year

### Total Annual Dollar Savings

= (kW Savings)(kW charge)+[(kWh savings)+(Air Conditioning savings)](kWh cost) -(Additional gas cost) – (lamp replacement cost)

= (10.8 kW)($ 120/kW year)+[(94,608 kWh)+(8,308 kWh)]($ 0.05/kWh) -($ 276/year) – ($ 545/year)

= $ 5,621/year

### Implementation Cost

= (# fixtures) (Retrofit cost per fixture)

= (415 fixtures) ($ 45/fixture)

= $ 18,675

### Simple Payback

= (Implementation Cost)/(Total Annual Dollar Savings)

= ($ 18,675)/($ 5,621/year)

= 3.3 years

## EXAMPLE 2: REPLACE INCANDESCENT LIGHTING WITH COMPACT FLUORESCENT LAMPS

A power plant has 111 incandescent fixtures which operate 24 hours/day, year round. The incandescent lamps were replaced with compact fluorescent lamps, which saved over 70% of the energy, and last over ten times as long. Because the lamp life is so much longer, there is a maintenance relamping labor savings. Air conditioning savings or heating costs were not included because these fixtures are located in a high-bay building which is not heated or air-conditioned.

## Calculations

### Watts Saved Per Fixture

= (Present input watts/fixture) – (Proposed input watts/fixture)

= (150 watts/fixture) – (30 watts/fixture)

= 120 watts saved/fixture

### kW Savings

= (# fixtures)(watts saved/fixture)(1 kW/1000 watts)

= (111 fixtures)(120 watts/fixtures)(1/1000)

= 13.3 kW

### kWh Savings

= (Demand savings)(annual operating hours)

= (13.3 kW)(8,760 hours/year)

= 116,683 kWh/year

### Lamp Replacement Cost

= [(Number of Fixtures)(cost per CFL Lamp)(operating hours/lamp life)] – [(Number of existing incandescent bulbs)(cost per bulb)(operating hours/lamp life)]

= [(111 Fixtures)($10/CFL lamp)(8,760 hours/10,000 hours)] – [(111 bulbs)($1.93/type "A" lamp)(8,760 hours/750 hours)]

= $ – 1,530/year§      §Negative cost indicates savings.

### Maintenance Relamping Labor Savings

= [(# fixtures)(maintenance relamping cost per fixture)] [((annual hours operation/present lamp life))-((annual hours operation/proposed lamp life))]

= [(111 fixtures)($1.7/fixture)][((8,760/750))-((8,760/10,000))]

= $ 2,039/year

### Total Annual Dollar Savings

= (kWh savings)(kWh cost)+ (kW savings)(kW cost) – (lamp replacement cost) + (maintenance relamping labor savings)

= (116,683 kWh)($.05/kWh)+(13.3)($120/kW year) -(-1,530/year) + (2,039/year)

= $ 10,999/year

Total Implementation Cost

= [(# fixtures)(cost/CFL ballast and lamp)] + (retrofit labor cost)]

= (111 fixtures)($45/fixture)

= $ 4,995

Simple Payback

= (Total Implementation Cost)/(Total Annual Dollar Savings)

= ($ 4,995)/(10,999/year)

= 0.5 years

## EXAMPLE 3: INSTALL OCCUPANCY SENSORS

In this example, an office building has many individual offices that are only used during portions of the day. After mounting wall-switch occupancy sensors, the sensitivity and time delay settings were adjusted to optimize the system. The following analysis is based on an average time savings of 35% per room. Air conditioning costs and demand charges would likely be reduced, however these savings are not included.

Calculations

kWh Savings

= (# rooms)(# fixtures/room)(input watts/fixture) (1 kW/1000 watts) (Total annual operating hours)(estimated % time saved/100)

= (50 rooms)(4 fixtures/room)(144 watts/fixture)(1/1000) (4,000 hours/year)(.35)

= 40,320 kWh/year

Total Annual Dollar Savings ($/Year)

= (kWh savings/year)(kWh cost)

= (40,320 kWh/year)($.05/kWh)

= $ 2,016/year

Implementation Cost

= (# occupancy sensors needed)[(cost of occupancy sensor)+ (installation time/room)(labor cost)]

= (50)[($ 75)+(1 hour/sensor)($20/hour)]

= $ 4,750

Simple Payback

= (Implementation Cost)/(Total Annual Dollar Savings)

= ($ 4,750)/($ 2,016/year)

= 2.4 years

## EXAMPLE 4: RETROFIT EXIT SIGNS WITH L.E.D.s

An office building had 117 exit signs, which used incandescent bulbs. The exit signs were retrofitted with LED exit kits, which saved 90% of the energy. Even though the existing incandescent bulbs were "long-life" models, (which are expensive) material and maintenance savings were significant. Basically, the hospital should not have to relamp exit signs for 25 years!

Calculations

Input Wattage – Incandescent Signs

= (Watt/fixture) (number of fixtures)

= (40 Watts/fix) (117 fix)

= 4.68 kW

Input Wattage – LED Signs

= (Watt/fixture) (number of fixtures)

= (3.6 Watts/fix) (117 fix)

= .421 kW

kW Savings

= (Incandescent Wattage) – (LED Wattage)

= (4.68 kW) – (.421 kW)

= 4.26 kW

kWh Savings

= (kW Savings)(operating hours)

= (4.26 kW)(8,760 hours)

= 37,318 kWh/yr

Lamp Replacement Cost

= [(Number of LED Exit Fixtures)(cost per LED Fixture)(operating hours/Fixture life)] – [(Number of existing Exit lamps)(cost per Exit lamp)(operating hours/lamp life)]

= [(117 Fixtures)($ 60/lamp kit)(8,760 hours/219,000 hours)] – [(234 Exit lamps)($5.00/lamp)(8,760 hours/8,760 hours)]

= -$ 889/year[§]          [§]Negative cost indicates savings.

Maintenance Relamping Labor Savings

= (# signs)(Number of times each fixture is relamped/yr)(time to relamp one fixture)(Labor Cost)

= (117 signs)(1 relamp/yr)(.25 hours/sign)($20/hour)

= $585/year

Annual Dollar Savings

= [(kWh savings)(electrical consumption cost)] + [(kW savings)(kW cost)] + [Maintenance Cost Savings]-[lamp replacement cost]

= [(37,318 kWh)($.05/kWh)] + [(4.26 kW)($120/kW yr)]+ [$585/yr] – [– $889/yr]

= $ 3,851/year

Implementation Cost
= [# Proposed Fixtures][(Cost/fixture + Installation Cost/fixture)]
= [117][$60/fixture + $5/fixture]
= $ 7,605

Simple Payback
= (Implementation Cost)/(Annual Dollar Savings)
= ($7,605)/($3,851/yr)
= 2 years.

## EXAMPLE 5: REPLACE OUTSIDE MERCURY VAPOR LIGHTING SYSTEM WITH HIGH AND LOW PRESSURE SODIUM LIGHTING SYSTEM

A parking lot is illuminated by mercury vapor lamps, which are relatively inefficient. The existing fixtures were replaced with a combination of High Pressure Sodium (HPS) and Low Pressure Sodium (LPS) lamps. The LPS provides the lowest-cost illumination, while the HPS provides enough color rendering ability to distinguish the colors of cars. By replacing the fifty 400 watt Mercury Vapor lamps with ten 250 watt HPS and forty 135 watt LPS fixtures, the company saved approximately $ 2,750/year with an installed cost of $12,500 and a payback of 4.6 years.

## EXAMPLE 6:
## REPLACE "U" LAMPS WITH STRAIGHT T8 TUBES

The existing fixtures were 2' by 2' Lay-In Troffers with two F40T12CW "U" lamps, with a standard ballast consuming 96 watts per fixture. The retrofit was to remove the "U" lamps and install three F017T8 lamps with an electronic ballast, which had only 47 watts per fixture.

## 13.9    SUMMARY

In summary, this chapter will help the energy manager make informed decisions about lighting. The following "recipe" reviews some of the main points that influence the effectiveness of lighting retrofits.

### A Recipe for Successful Lighting Retrofits
1.  Identify visual task—Distinguish between tasks that involve walking and tasks that involve reading small print.

2.  Identify lighting needs for each task-Use IES tables to determine target light levels.

3.  Research available products and lighting techniques—Talk to lighting manufacturers about your objectives, let them help you select the products. Perhaps they will offer a demonstration or trial installation. Be aware of the relative costs, especially the costs associated with specialized technologies.

4.  Identify lamps to fulfill lighting needs—Pick the lamp that has the proper CRI, CCT, lamp life and lumen output.

5.  Identify ballasts and fixtures to fulfill lighting needs. Select the proper ballast factor, % THD, voltage, fixture light distribution, lenses or baffles, fixture efficiency.

6.  Identify the optimal control technology: Decide whether to use IR, US or DT Occupancy Sensors. Know when to use time clocks or install switches.

7.  Consider system variations to optimize:
    Employee performance—Incorporate the importance of lighting quality into the retrofit process.
    Energy savings-Pick the most efficient technologies that are cost-effective.
    Maintenance—Installing common systems for simple maintenance, group re-lamping, maintenance training.
    Ancillary effects—Consider effects on the HVAC system, security, safety, etc.

8.  Publicize results—As with any energy management program, your job depends on demonstrating progress. By making energy cost savings known to the employees and upper-level management, all people contributing to the program will know that there is a benefit to their efforts.

9.  Continually look for more opportunities—The lighting industry is constantly developing new products that could improve profitability for your company. Keep in-touch with new technologies and methods to avoid "missing the boat" on a good opportunity. Table 13.21 offers a more complete listing of energy saving ideas for lighting retrofits.

## 13.10 SCHEMATICS

### Lamps

#### Fluorescent

**Bulb Shapes** (Not Actual Sizes)

The size and shape of a bulb is designated by a letter or letters followed by a number. The letter indicates the shape of the bulb while the number indicates the diameter of the bulb in eighths of an inch. For example, "T-12" indicates a tubular shaped bulb having a diameter of $^{12}/_8$ or $1^1/_2$ inches. The following illustrations show some of the more popular bulb shapes and sizes.

T-5 Miniature Bipin

T-8 Medium Bipin

T-10 Medium Bipin

T-12 Medium Bipin

T-12 Mogul Bipin

T-12 Recessed Double Contact

T-12 Recessed Double Contact (Jacketed)

T-8 Single Pin Slimline

T-12 Single Pin Slimline

T-9 4-Pin Circline

T-8 Medium Bipin U-Bent Lamp

**SL*Lamps**

SL*17    SL*18    SL*18/R40

**SLS Lamps**

SLS 9,11    SLS 15, 20, 23

**PL Lamps**

PL-S 5/SYS    PL-C 13/SYS    PL-S 5    PL-C 13    PL-L

PL Adapter Systems    PL Replacement Bulbs

## Fluorescent (continued)

### Biax Lamps

### Base Types

### Lamp and Ballast System

**HID**

### Mercury Vapor

### Metal Halide

### High Pressure Sodium

### Low Pressure Sodium

## Exit Signs

### LED

- **1.8-3.6 Input Watts/Fixture. (Replaces standard 20-25 watt lamps.)**
- Convert existing incandescent EXIT signs to use energy efficient LED liehg strips.
- Each kit contains two LED light strips and a reflective backing to provide even light distribution and a new red lens for the fixture.
- Estimated life is 25 years.
- Complies with OSHA and NFPA requirements.
- Available in four base styles to fit existing sockets or as a hard wire kit.
- LED light strips emit a bright red light and are not recommended for use with green signs.
- In addition to DGSC standard warranty, manufacturer's 25-year warranty applies.
- UL approved.

### CFL

Quick connecting adapter screws into existing incandescent socket. For use with medium screw base societs.

Two lamp EXIT sign retrofit system, backup lamp will take over if the primary lamp fails.

UL approved.
- Lamp: 9 watt, twin tube compact fluorescent
- Lumens: 600
- Lamp Avg Life: 10,000 hours
- Ballast Losses: 2 watts/ballast
- System Input Watts: 11 watts
- Minimum Starting Temperature: 0°F

**Unit Dimensions: 8"L X 4-5/8"H X 1"H**

## Occupancy Sensors

### Ceiling Mounted

### Wall Mounted

## Table 13.21 Energy saving checklist.

**Lighting Needs**

* Visual tasks: specification        Identify specific visual tasks and locations to determine recommended illuminances for tasks and for surrounding areas.

* Safety and aesthetics             Review lighting requirements for given applications to satisfy safety and aesthetic criteria.

* Over-illuminated                   In existing spaces, identify applications where maintained illumination is greater than recom-
  application                        mended. Reduce energy by adjusting illuminance to meet recommended levels.

* Groupings: similar                 Group visual tasks having the same illuminance requirements, and avoid widely separated
  visual tasks                       workstations.

* Task lighting                      Illuminate work surfaces with luminaires properly located in or on furniture; provide lower ambient levels.

* Luminance ratios                   Use wall-washing and lighting of decorative objects to balance brightness.

**Space Design and Utilization**

* Space plan                         When possible, arrange for occupants working after hours to work in close proximity to one another.

* Room surfaces                      Use light colors for walls, floors, ceilings and furniture to increase utilization of light, and re-
                                     duce connected lighting power to achieve required illuminances. Avoid glossy finishes on room
                                     and work surfaces to limit reflected glare.

* Space utilization                  Use modular branch circuit wiring to allow for flexibility in moving, relocating or adding
  *branch circuit wiring*            luminaires to suit changing space configurations.

* Space utilization:                 Light building for occupied periods only, and when required for security or cleaning pur-
                                     poses
  occupancy                          (see chapter 31, Lighting Controls).

**Daylighting**

* Daylight compensation              If daylighting can be used to replace some electric lighting near fenestration during substan-
                                     tial periods of the day, lighting in those areas should be circuited so that it may be controlled
                                     manually or automatically by switching or dimming.

* Daylight sensing                   Daylight sensors and dimming systems can reduce electric lighting energy.

* Daylight control                   Maximize the effectiveness of existing fenestration-shading controls (interior and exterior) or
                                     automatically by switching or dimming.

* Space utilization                  Use daylighting in transition zones, in lounge and recreational areas, and for functions where
                                     the variation in color, intensity and direction may be desirable. Consider applications where
                                     daylight can be utilized as ambient lighting, supplemented by local task lights.

**Lighting Sources: Lamps and Ballasts**

* Source efficacy                    Install lamps with the highest efficacies to provide the desired light source color and distribu-
                                     tion requirements.

*(Continued)*

* Fluorescent lamps — Use T8 fluorescent and high-wattage compact fluorescent systems for improved source efficacy and color quality.

* Ballasts — Use electronic or energy efficient ballasts with fluorescent lamps.

* HID — Use high-efficacy metal halide and high-pressure sodium light sources for exterior floodlighting.

* Incandescent — Where incandescent sources are necessary, use reflector halogen lamps for increased efficacy.

* Compact fluorescent — Use compact fluorescent lamps, where possible, to replace incandescent sources.

* Lamp wattage reduced-wattage lamps — Use reduced-wattage lamps where illuminance is too high.

* Control compatibility — If a control system is used, check compatibility of lamps and ballasts with the control device.

* System change — Substitute metal halide and high-pressure sodium systems for existing mercury vapor lighting systems.

## Luminaires

* Maintained efficiency — Select luminaires which do not collect dirt rapidly and which can be easily cleaned.

* Improved maintenance — Improved maintenance procedures may enable a lighting system with reduced wattage to provide adequate illumination throughout systems or component life.

* Luminaire efficiency replacement or relocation — Check luminaire effectiveness for task lighting and for overall efficiency; if ineffective or inefficient, consider replacement or relocation.

* Heat removal — When luminaire temperatures exceed optimal system operating temperatures, consider using special luminaires to improve lamp performance and reduce heat gain to the space.

* Maintained efficiency — Select a lamp replacement schedule for all light sources, to more accurately predict light loss factors and possibly decrease the number of luminaires required.

## Lighting controls

* Switching; local control — Install switches for local and convenient control of lighting by occupants. This should be in combination with a building-wide system to turn lights off when the building is unoccupied.

* Selective switching — Install selective switching of luminaires according to groupings of working tasks and different working hours.

* Low-voltage switching systems — Use low-voltage switching systems to obtain maximum switching capability.

* Master control system — Use a programmable low-voltage master switching system for the entire building to turn lights on and off automatically as needed, with overrides at individual areas.

* Multipurpose spaces — Install multi-circuit switching or preset dimming controls to provide flexibility when spaces are used for multiple purposes and require different ranges of illuminance for various activities. Clearly label the control cover plates.

* "Tuning" illuminance — Use switching and dimming systems as a means of adjusting illuminance for variable lighting requirements.

*(Continued)*

\*  Scheduling                   Operate lighting according to a predetermined schedule, based on occupancy.

\*  Occupant/motion sensors      Use occupant/motion sensors for unpredictable patterns of occupancy.

\*  Lumen maintenance            Fluorescent dimming systems may be utilized to maintain illuminance throughout lamp life, thereby saving energy by compensating for lamp-lumen depreciation and other light loss factors.

\*  Ballast switching            Use multilevel ballasts and local inboard-outboard lamp switching where a reduction in illuminances is sometimes desired.

### Operation and Maintenance

\*  Education                    Analyze lighting used during working and building cleaning periods, and institute an education program to have personnel turn off incandescent lamps promptly when the space is not in use, fluorescent lamps if the space will not be used for 10 min. or longer, and HID lamps (mercury, metal halide, high-pressure sodium) if the space will not be used for 30 min. or longer.

\*  Parking                      Restrict parking after hours to specific lots so lighting can be reduced to minimum security requirements in unused parking areas.

\*  Custodial service            Schedule routine building cleaning during occupied hours.

\*  Reduced illuminance           Reduce illuminance during building cleaning periods.

\*  Cleaning schedules            Adjust cleaning schedules to minimize time of operation, by concentrating cleaning activities in fewer spaces at the same time and by turning off lights in unoccupied areas.

\*  Program evaluation            Evaluate the present lighting maintenance program, and revise it as necessary to provide the most efficient use of the lighting system.

\*  Cleaning and maintenance      Clean luminaires and replace lamps on a regular maintenance schedule to ensure proper illuminance levels are maintained.

\*  Regular system checks         Check to see if all components are in good working condition. Transmitting or diffusing media should be examined, and badly discolored or deteriorated media replaced to improve efficiency.

\*  Renovation of luminaries      Replace outdated or damaged luminaires with modern ones which have good cleaning capabilities and which use lamps with higher efficacy and good lumen maintenance characteristics.

\*  Area maintenance              Trim trees and bushes that may be obstructing outdoor luminaire distribution and creating unwanted shadow.

## 13.11 GLOSSARY[9]

**AMPERE:** The standard unit of measurement for electric current that is equal to one coulomb per second. It defines the quantity of electrons moving past a given point in a circuit during a specific period. Amp is an abbreviation.

**ANSI:** Abbreviation for American National Standards Institute.

**ARC TUBE:** A tube enclosed by the outer glass envelope of a HID lamp and made of clear quartz or ceramic that contains the arc stream.

**ASHRAE:** American Society of Heating, Refrigerating and Air-Conditioning Engineers.

**AVERAGE RATED LIFE:** The number of hours at which half of a large group of product samples have failed.

**BAFFLE:** A single opaque or translucent element used to control light distribution at certain angles.

**BALLAST:** A device used to operate fluorescent and HID lamps. The ballast provides the necessary starting voltage, while limiting and regulating the lamp current during operation.

**BALLAST CYCLING:** Undesirable condition under which the ballast turns lamps ON and OFF (cycles) due to the overheating of the thermal switch inside the ballast. This may be due to incorrect lamps, improper voltage being supplied, high ambient temperature around the fixture, or the early stage of ballast failure.

**BALLAST EFFICIENCY FACTOR:** The ballast efficiency factor (BEF) is the ballast factor (see below) divided by the input power of the ballast. The higher the BEF—within the same lamp ballast type—the more efficient the ballast.

**BALLAST FACTOR:** The ballast factor (BF) for a specific lamp-ballast combination represents the percentage of the rated lamp lumens that will be produced by the combination.

**CANDELA:** Unit of luminous intensity, describing the intensity of a light source in a specific direction.

**CANDELA DISTRIBUTION:** A curve, often on polar coordinates, illustrating the variation of luminous intensity of a lamp or fixture in a plane through the light center.

**CANDLEPOWER:** A measure of luminous intensity of a light source in a specific direction, measured in candelas (see above).

**COEFFICIENT OF UTILIZATION:** The ratio of lumens from a fixture received on the work plane to the lumens produced by the lamps alone. (Also called "CU")

**COLOR RENDERING INDEX (CRI):** A scale of the effect of a light source on the color appearance of an object compared to its color appearance under a reference light source. Expressed on a scale of 1 to 100, where 100 indicates no color shift. A low CRI rating suggests that the colors of objects will appear unnatural under that particular light source.

**COLOR TEMPERATURE:** The color temperature is a specification of the color appearance of a light source, relating the color to a reference source heated to a particular temperature, measured by the thermal unit Kelvin. The measurement can also be described as the "warmth" or "coolness" of a light source. Generally, sources below 3200K are considered "warm;" while those above 4000K are considered "cool" sources.

**COMPACT FLUORESCENT:** A small fluorescent lamp that is often used as an alternative to incandescent lighting. The lamp life is about 10 times longer than incandescent lamps and is 3-4 times more efficacious. Also called PL, Twin-Tube, CFL, or BIAX lamps.

**CONTRAST:** The relationship between the luminance of an object and its background.

**DIFFUSE:** Term describing dispersed light distribution. Refers to the scattering or softening of light.

**DIFFUSER:** A translucent piece of glass or plastic sheet that shields the light source in a fixture. The light transmitted throughout the diffuser will be directed and scattered.

**DIRECT GLARE:** Glare produced by a direct view of light sources. Often the result of insufficiently shielded light sources. (SEE GLARE)

**DOWNLIGHT:** A type of ceiling fixture, usually fully recessed, where most of the light is directed downward. May feature an open reflector and/or shielding device.

**EFFICACY:** A metric used to compare light output to energy consumption. Efficacy is measured in lumens per watt. Efficacy is similar to efficiency, but is expressed in dissimilar units. For example, if a 100-watt source produces 9000 lumens, then the efficacy is 90 lumens per watt.

**ELECTRONIC BALLAST:** A ballast that uses semi-conductor components to increase the frequency of fluorescent lamp operation-typically, in the 20-40 kHz range. Smaller inductive components provide the lamp current control. Fluorescent system efficiency is increased due to high frequency lamp operation.

**ENERGY-SAVING BALLAST:** A type of magnetic ballast designed so that the components operate more efficiently, cooler and longer than a "standard magnetic" ballast. By US law, standard magnetic ballasts can no longer be manufactured.

**ENERGY-SAVING LAMP:** A lower wattage lamp, generally producing fewer lumens.

**FLUORESCENT LAMP:** A light source consisting of a tube filled with argon, along with krypton or other inert gas. When electrical current is applied, the resulting arc emits ultraviolet radiation that excites the phosphors inside the lamp wall, causing them to radiate visible light.

**FOOTCANDLE (FC):** The English unit of measurement of the illuminance (or light level) on a surface. One footcandle is equal to one lumen per square foot.

**FOOTLAMBERT:** English unit of luminance. One footlambert is equal to 1/p candelas per square foot.

**GLARE:** The effect of brightness or differences in brightness within the visual field sufficiently high to cause annoyance, discomfort or loss of visual performance.

**HARMONIC:** For a distorted waveform, a component of the wave with a frequency that is an integer multiple of the fundamental.

**HID:** Abbreviation for high intensity discharge. Generic term describing mercury vapor, metal halide, high pressure sodium, and (informally) low pressure sodium light sources and fixtures.

**HIGH-BAY:** Pertains to the type of lighting in an industrial application where the ceiling is 20 feet or higher. Also describes the application itself.

**HIGH OUTPUT (HO):** A lamp or ballast designed to operate at higher currents (800 mA) and produce more light.

**HIGH PRESSURE SODIUM LAMP:** A high intensity discharge (HID) lamp whose light is produced by radiation from sodium vapor (and mercury).

**HVAC:** Heating, ventilating and air conditioning systems.

**ILLUMINANCE:** A photometric term that quantifies light incident on a surface or plane. Illuminance is commonly called light level. It is expressed as lumens per square foot (footcandles), or lumens per square meter (lux).

**INDIRECT GLARE:** Glare produced from a reflective surface.

**INSTANT START:** A fluorescent circuit that ignites the lamp instantly with a very high starting voltage from the ballast. Instant start lamps have single-pin bases.

**LAMP LUMEN DEPRECIATION FACTOR (LLD):** A factor that represents the reduction of lumen output over time. The factor is commonly used as a multiplier to the initial lumen rating in illuminance calculations, which compensates for the lumen depreciation. The LLD factor is a dimensionless value between 0 and 1.

**LAY-IN-TROFFER:** A fluorescent fixture; usually a 2' x 4' fixture that sets or "lays" into a specific ceiling grid.

**LED:** Abbreviation for light emitting diode. An illumination technology used for exit signs. Consumes low wattage and has a rated life of greater than 80 years.

**LENS:** Transparent or translucent medium that alters the directional characteristics of light passing through it. Usually made of glass or acrylic.

**LIGHT LOSS FACTOR (LLF):** Factors that allow for a lighting system's operation at less than initial conditions. These factors are used to calculate maintained light levels. LLFs are divided into two categories, recoverable and non-recoverable. Examples are lamp lumen depreciation and fixture surface depreciation.

**LOUVER:** Grid type of optical assembly used to control light distribution from a fixture. Can range from small-cell plastic to the large-cell anodized aluminum louvers used in parabolic fluorescent fixtures.

**LOW-PRESSURE SODIUM:** A low-pressure discharge lamp in which light is produced by radiation from sodium vapor. Considered a monochromatic light source (most colors are rendered as gray).

**LUMEN:** A unit of light flow, or luminous flux. The lumen rating of a lamp is a measure of the total light output of the lamp.

**FIXTURE EFFICIENCY:** The ratio of total lumen output of a fixture and the lumen output of the lamps, expressed as a percentage. For example, if two fixtures use the same lamps, more light will be emitted from the fixture with the higher efficiency.

**FIXTURE:** A complete lighting unit consisting of a lamp or lamps, along with the parts designed to distribute the light, hold the lamps, and connect the lamps to a power source. Also called a fixture.

**MEAN LIGHT OUTPUT:** Light output in lumens at 40% of the rated life.

**MERCURY VAPOR LAMP:** A type of high intensity discharge (HID) lamp in which most of the light is produced by radiation from mercury vapor. Emits a blue-green cast of light. Available in clear and phosphor-coated lamps.

**METAL HALIDE:** A type of high intensity discharge (HID) lamp in which most of the light is produced by radiation of metal halide and mercury vapors in the arc tube. Available in clear and phosphor-coated lamps.

**OCCUPANCY SENSOR:** Control device that turns lights OFF after the space becomes unoccupied. May be ultrasonic, infrared or other type.

**PHOTOCELL:** A light sensing device used to control fixtures and dimmers in response to detected light levels.

**POWER FACTOR:** Power factor is a measure of how effectively a device converts input current and voltage into useful electric power. Power factor is the ratio of kW/kVA.

**RAPID START (RS):** The most popular fluorescent lamp/ballast combination used today. This ballast quickly and efficiently preheats lamp cathodes to start the lamp. Uses a "bi-pin" base.

**REACTIVE POWER:** Power that creates no useful work; it results when current is not in phase with voltage. Calculated using the equation:

$$\text{reactive power} = V \times A \times \sin \phi$$

where $\phi$ is the phase displacement angle.

**RECESSED:** The term used to describe the fixture that is flush mounted into a ceiling.

**RETROFIT:** Refers to upgrading a fixture, room, or building by installing new parts or equipment.

**ROOT-MEAN-SQUARE (rms):** The effective average value of a periodic quantity such as an alternating current or voltage wave, calculated by averaging the squared values of the amplitude over one period, and taking the square root of that average.

**SPACE CRITERION:** A maximum distance that interior fixtures may be spaced that ensures uniform illumination on the work plane. The fixture height above the work plane multiplied by the spacing criterion equals the center-to-center fixture spacing.

**SPECULAR:** Mirrored or polished surface. The angle of reflection is equal to the angle of incidence. This word describes the finish of the material used in some louvers and reflectors.

**T12 LAMP:** Industry standard for a fluorescent lamp that is 12 one-eighths (1.5 inches) in diameter.

**TANDEM WIRING:** A wiring option in which a ballast is shared by two or more fixtures. This reduces labor, materials, and energy costs. Also called "master-slave" wiring.

**VCP:** Abbreviation for visual comfort provability. A rating system for evaluating direct discomfort glare. This method is a subjective evaluation of visual comfort expressed as the percent of occupants of a space who will be bothered by direct glare. VCP allows for several factors: fixture luminances at different angles of view, fixture size, room size, fixture mounting height, illuminance, and room surface reflectivity. VCP tables are often provided as part of photometric reports for specific fixtures.

**TOTAL HARMONIC DISTORTION (THD):** For current or voltage, the ratio of a wave's harmonic content to its fundamental component, expressed as a percentage. Also called "harmonic factor," it is a measure of the extent to which a waveform is distorted by harmonic content.

**VERY HIGH OUTPUT (VHO):** A fluorescent lamp that operates at a "very high" current (1500mA), producing more light output than a "high output" lamp (800 mA) or standard output lamp (430mA).

**WATT (W):** The unit for measuring electrical power. It defines the rate of energy consumption by an electrical device when it is in operation. The energy cost of operating an electrical device is calculated as its wattage times the hours of use. In single phase circuits, it is related to volts and amps by the formula: Volts x Amps x PF = Watts. (Note: For AC circuits, PF must be included.)

**WORK PLANE:** The level at which work is done and at which illuminance is specified and measured. For office applications, this is typically a horizontal plane 30 inches above the floor (desk height).

**References**

1. Mills, E., Piette, M., 1993, "Advanced Energy-Efficient Lighting Systems: Progress and Potential," *Energy—The International Journal*, 18 (2) pp. 75.
2. Woodroof, E., 1995, *A LECO Screening Procedure for Surveyors*, Oklahoma State University Press—Thesis, p. 54.
3. Bartlett, S., 1993, *Energy—The International Journal*, 18 p.99.
4. Flagello, V, 1995, "Evaluating Cost-Efficient P.I.R. Occupancy Sensors: A Guide to Proper Application," *Strategic Planning for Energy and the Environment*, 14 (3) p. 52.
5. Mills, E., Piette, M., 1993, "Advanced Energy-Efficient Lighting Systems: Progress and Potential," *Energy—The International Journal*, 18 (2), p.90.
6. United States EPA Green Lights Program, 1995, *Lighting Upgrade Manual—Lighting Waste Disposal*, p.2.
7. United States EPA Green Lights Program, 1995, "Green Lights Update" p.2, March 1995.
8. R. Rundquist, 1991, "Daylighting controls: Orphan of HVAC design," *ASHRAE Journal,*, p. 30, November 1991.
9. United States EPA Green Lights Program, 1995, *Lighting Upgrade Manual—Lighting Fundamentals*.

# CHAPTER 14
# ENERGY SYSTEMS MAINTENANCE

**W.J. KENNEDY, P.E.**
Department of Industrial Engineering
Clemson University
Clemson, SC 29634-0920

Energy systems maintenance is the maintenance of all systems that use or affect the use of energy. These systems are found in every kind of organization that uses energy, whether a hospital, a church, a store, a university, a warehouse, or a factory. Energy systems maintenance includes such routine maintenance tasks as lubrication, examination, and cleaning of electrical contacts and calibrating thermostats, and such non-routine tasks as repainting walls to increase the effective lighting, cleaning fins on condensers, and cleaning damper blades and linkages.

A good energy systems maintenance program can save a company substantial amounts of money in wasted steam and wasted electricity and in the lost production and additional expense caused by preventable equipment breakdowns. Other benefits include general cleanliness, improved employee morale, and increased safety. In a good maintenance program, planning, scheduling, and monitoring are all carried out in a predictable and well-organized manner. This chapter is designed to assist the reader to develop such a program. In addition, a special section is included on some recent developments and on some of the special problems and opportunities associated with material handling systems.

The maintenance part of this chapter is divided into three sections. The first section gives a four-step procedure for developing a maintenance program for energy management. This plan relies on a detailed knowledge of the systems within any plant and the components of these systems. The systems are described and the main components of each system are listed. The second section provides a more detailed description of the maintenance of these components and constitutes the main body of this chapter. The final section describes some of the instruments useful in the maintenance associated with energy management.

## 14.1 DEVELOPING THE MAINTENANCE PROGRAM

There are four steps in the development of a maintenance program. Step 1 is to determine the present condition of the existing facility. This step includes a detailed examination of each of the major energy-consuming systems. The output from this examination is a list of the motors, lights, transformers, and other components that make up each system, together with a report of the condition of each. Step 2 is the preparation of a list of routine maintenance tasks with an estimate of the number of times that each task must be performed. This list should also include, for each task, the craft, the needed material, the time required, and the appropriate equipment. These data are then incorporated in step 3 into a regular schedule for the accomplishment of the desired maintenance. Step 4 is the monitoring necessary to keep the program in force once it has been initiated. These four steps are discussed in more detail in the following sections.

### 14.1.1 Determine the Present Condition of the Facility

The purpose of this step is to create a starting point. The questions to be answered are: (1) What equipment and systems are in the building? (2) What material is available to describe each system and/or its components? (3) What needs to be done to get the energy-related systems into working condition and to keep them that way?

This step should incorporate (1) vendor data and operating specifications for as much of the installed equipment as possible, kept in a notebook, a file cabinet, or a computer data base (depending upon the size and complexity of the facility); (2) a diagram of each major system, showing the location of all important equipment and the direction of all fluid flows; (3) a complete list of all the equipment in the building, showing the name, location, and condition of each item; and (4) a comprehensive list of maintenance tasks required for each piece of equipment. This information constitutes an equipment reference source unique to the equipment in your facility. It should be kept current, with each addition

401

initialed and dated, and its location should be known to all maintenance personnel and their supervisors.

The preparation of the notebook and the compilation of the other data is made much easier if separate energy-related systems are defined within the plant so that the systems can be examined in turn. A suggested classification of these systems would constitute (1) the building envelope, including all surfaces of the facility exposed to the outside; (2) the boiler and steam distribution system; (3) the heating, ventilating, and air-conditioning system, together with its controls; (4) the electrical system; (5) the lights, windows, and adjacent reflective walls, ceilings, and floors; (6) the hot-water distribution system; (7) the compressed-air distribution system; and (8) the manufacturing system, consisting of motors and specialized energy-consuming equipment used in the creation of products. Each of these systems should be inspected, the condition of each part noted, and a diagram drawn if appropriate. The following descriptions and tables have been prepared to assist you in this inspection.

### Building Envelope

This consists of all parts of the facility that can leak air into or out of any building. Its components and appropriate initial maintenance measures are given in Table 14.1. The envelope should be described by a blueprint of the building, showing locations and compositions of all outside walls, windows, ceilings, and floors, and the locations of all outside doors. The primary malfunction of this system is leakage of air, and this leakage can often be detected by sight—looking for cracks— or by noting the presence of a draft. Infrared scanning from the outside can also be helpful. The benefits from maintaining this system are (1) a reduction in the amount of air that must be heated or cooled, and (2) increased comfort due to decreased drafts.

### Boilers and Steam Distribution Systems

(These systems are described in greater detail in Chapters 5 and 6). A boiler is often the largest consumer of fuel in a factory or building. Any improvements that maintenance can make in its operation are therefore immediately reflected in decreased energy consumption and decreased energy cost. If the steam distribution system has leaks or is not properly insulated, these faults cause the boiler to generate more steam than is needed; eliminating these problems saves money. But boilers can malfunction, and steam leaks can cause severe burns if maintenance is performed by untrained personnel. The first step in proper boiler maintenance is usually to get the boiler and the steam distribution system inspected by a licensed professional. It is possible, however, to examine your own boiler and determine whether your system has some of the more conspicuous problems. To estimate the value of repairs to this system, assume that any boiler that has not been adjusted for two years can have its efficiency increased by 25% by a suitable adjustment, with a corresponding decrease in fuel consumption and in projected fuel costs. In steam distribution, a defective steam trap can typically waste 50,000,000 Btu/yr, at a cost of between $100 and $1000, depending upon the source of fuel. The savings from boiler and steam distribution system maintenance measures are thus worth pursuing. The system components and an appropriate set of initial maintenance measures are given in Table 14.2.

### Heating, Ventilating, and Air-Conditioning Systems

(These systems are described in more detail in Chapter 10.) The purposes of these systems are to supply enough air of the right temperature to keep people comfortable and to exhaust harmful or unpleasant air contaminants. A complete description of this system should include a blueprint with the location of all damp-

**Table 14.1 Problems and solutions: the building envelope.**

| System Component | Problem | Initial Maintenance Action |
|---|---|---|
| Door | Loose fitting | Weatherstripping, new threshold, or frame repair |
| | Does not close | If problem is caused by air pressure, balance intake and exhaust air; correct door fit; check door-closing hardware |
| Window | Air leakage | Weatherstrip, caulk, or add storm window |
| | Broken | Replace broken panes |
| Wall | Drafts from wall openings | Caulk or seal openings on outside of wall |
| | Cracks | Patch or seal if air is entering or leaving |
| Ceiling | Drafts around exhaust piping | Caulk or repair flashing |
| Roof | Holes | Patch or cover |

**Table 14.2 Problems and solutions: the boiler and steam distribution system.**

| System Component | Problem | Initial Maintenance Action |
|---|---|---|
| Boiler | Inoperable gauges | Overhaul boiler controls as soon as possible |
| | Most recent boiler adjustment at least two years ago | Have boiler adjusted for most efficient firing |
| | Scale deposits on water side of shutdown boiler | Remove scale; check water-softening system |
| | Boiler stack temperature more than 150°F above steam or water temperature | Clean tubes and adjust fuel burner (Ref. 1) |
| | Fuel valves leak | Repair |
| | Stack shows black smoke or haze when boiler is operating | Check combustion controls |
| | Rust in water gauge | Check return line for evidence of corrosion (Ref. 2) |
| | Safety valves not checked or tagged | Have inspection performed immediately |
| Steam trap | Leaks | Have inspection performed; repair or replace |
| Steam valve | Leaks | Repair |
| Steam line | Lines uninsulated | Have insulation installed |
| | Water hammer noted | Fix steam trap |
| Condensate return | Uninsulated | Insulate if hot to touch |
| Condensate tank | Steam plumes at tank vents | Check and repair leaking steam traps |
| | No insulation | Install insulation |
| Condensate pumps | Excessive noise | Check and repair |
| | Leaks | Replace packing: overhaul or replace pump if necessary |

ers, fans, and ducts, a complete diagram of the control system showing the location of all gauges, thermostats, valves, and other components, and a list of the correct operating ranges for each dial and gauge. The descriptive material should also include any vendor-supplied manuals and the engineering diagrams and reports prepared when each system was installed or modified. Since many systems have been modified since their installation, it is also desirable to have someone prepare a diagram of your system as it is now.

Determining the present condition of the system includes preparing a diagram of the existing system based on an actual survey of the system; placing labels on each valve, gauge, and piece of equipment; and examining each system component to see if it is working. Table 14.3 gives a list of some of the more expensive troubles that may be encountered.

Significant amounts of money may be saved by proper maintenance of this system. These savings come from three sources: reducing the energy used by the system and its associated cost, decreasing the amount of unanticipated repair that is necessary in the absence

of good maintenance, and reducing downtime caused when the system does not work and conditions become either uncomfortable or unsafe. Since energy maintenance on this system usually includes maintaining the temperature established by company policy, and since company policies have increasingly favored a high temperature threshold for cooling and a low threshold for heating, energy maintenance is directly responsible for realizing the considerable savings made possible by these policies. It is not uncommon for a good maintenance policy to cause a decrease of 50% in the energy consumption of a building. In addition to the energy cost savings, a good energy maintenance program can spot deterioration of equipment and can enable repair to be scheduled at a time that will not cause extensive disruption of work. Finally, there have been cases when a buildup of poisonous gas was caused when exhaust fans ceased to function. Such problems can often be avoided with a good maintenance program aimed at the ventilation system.

The controls for a heating, ventilating, and air-conditioning system can range from a simple thermostat

**Table 14.3 Problems and solutions: the heating, ventilating, and air-conditioning system.**

| Component | Problem | Initial Maintenance Action |
|---|---|---|
| Filter | Excessively dirty | Replace or clean |
| Damper | Blocked open or linkage disconnected | Check damper controls |
| | Leaks badly | Clean and overhaul |
| Ductwork | Open joints | Repair |
| | Loose insulation in duct work | Replace and attach firmly |
| | Water leakage or rust spots | Repair |
| | Crushed | Replace |
| Grillwork | Air flow impossible due to dirt | Remove and clean |
| | Blocked by equipment | Remove equipment |
| Fan | Motor not hooked up to fan | Disconnect motor or install fan belts |
| | Excess noise | Check bearings, belt tension mountings, dirty blades |
| | Insufficient ventilation | Check fan and surrounding duct work and grill work |
| | Belt too tight or too loose | Adjust motor mount |
| | Pulleys misaligned | Correct alignment |
| Pump | Hot-water pump is cold (or vice versa) | Inspect valving; check direction of flow |
| Blower | Not moving air in acceptable quantities | Check direction of rotation and change wiring if needed; clean if dirty |
| | Excessive noise | Check bearings, coupler |
| | Rotation wrong direction | Check wiring |
| | Shaft does not turn freely by hand | Check lubricant; repair |
| Chiller | Leaks | Repair |
| Cooling tower | Scaling on spray nozzles | Remove by chipping or by chemical means |
| | Leaks | Repair |
| | Cold water too warm | Check pumps, fans, and fill; clean louvers and fill |
| | Excessive water drift | Check drift elimination, metering orifices, and basins (for leaks); check for overpumping |
| Compressor | | See Table 14.7 |
| Thermostat | Temperature reading not accurate | Calibrate |
| | Leaks water or oil from mounting | Check pneumatic control lines |

to a modern computer-controlled network. The amount of troubleshooting that you can do will be directly proportional to your knowledge of the system. If you know a lot about it, you may be able to accomplish a great deal, even initially. If your knowledge is limited, get advice from someone who has installed such systems, obtain all the operating manuals that pertain to your system, and get estimates of the cost needed to bring your control system up to its designed operation. You can save larger amounts of money by simply adjusting controls—for example, by using a lower temperature on weekends and when no one is in the facility—than you can by any other comparable expense of effort. It pays to get your control system in operating condition and to keep it that way.

*Electrical System*

Many industrial electrical distribution systems are being used in ways not foreseen by the designers. These changes in use can cause some problems in energy consumption and in safety. If a motor is operating at a lower voltage than it was designed for, it is probably using more amperage than was intended and is causing unnecessary losses in transmission lines. If the wires are too small for the load, line losses can be large, and fire hazards increase significantly. Other problems that can create unnecessary energy loss are voltage imbalance in three-phase motors and leaks from voltage sources to ground.[1] Most of these problems are safety hazards as well as expensive in energy costs, and it is imperative that they be checked.

It is desirable to have a qualified electrician or electrical contractor examine your facility to find safety problems. At the same time, a load survey should be performed to determine whether your wiring, transformers, and switch gear are appropriately sized for the load they are currently carrying. To supplement this formal examination, Table 14.4 gives a list of trouble indicators you should look for as you are ascertaining the current condition of your facility. This list should be used in conjunction with the formal survey.

Another problem that may be costing money is a low power factor. The power factor is the ratio between the resistive component of ac power and the total (resistive and inductive) supplied. Because it costs more to provide electricity with a low power factor, it is common for electrical utilities to make an additional charge if the power factor is less than some value, typically 75%. If your electrical bill includes a charge for a low power factor, it may be worthwhile to have a power-factor meter installed and to focus management attention on the problem of increasing the power factor. Equipment that contributes to a low power factor includes welding machines, induction motors, power transformers, electric arc furnaces, and fluorescent light ballasts. A low power factor can be corrected with the installation of capacitors or with a separate power supply for machinery causing the problem. The maintenance of this equipment then becomes another item on the list of scheduled energy management maintenance. Power-factor management is covered in Chapter 11.

### Lights, Windows, and Reflective Surfaces

Maintenance of the lighting adjacent to reflective walls, ceilings, and floors serves several purposes in energy management. The energy consumed by lights is significant, as is the energy used by the heating, ventilating, and air-conditioning system to remove the heat put into a building by the lights. But the psychological value may have a greater impact. If lights are conspicuously absent from corridors and, where not needed, from management offices, people tend to look upon energy conservation as a program that is being taken seriously, and they begin to take it more seriously themselves. Similarly, a plant where energy management has been encouraged but where no attention has been paid to lighting is often seen as a plant where energy management is not taken seriously. (For a more complete discussion of lighting, see Chapter 13.)

Many factors modify the effectiveness of the lighting system. Of particular importance are the condition of the lights, the cleanliness of the luminaires, and the cleanliness of the walls, ceiling, and floors. When determining the condition of this system, a light meter should be carried as standard equipment. This will show the rooms where the IES (Illuminating Engineering Society) or other lighting standards have been exceeded or where not enough light is present. There may also be some additional problems, which are described in Table 14.5. Windows can also be a source of light. If they are used this way in your facility, they should be cleaned. If they are not used as a light source, consider boarding them up to avoid having to remove the heat that they allow in from the sun. The lighting systems in a facility are important, both in energy cost and as a morale factor. They deserve your attention.

### Hot-Water Distribution System

Hot water can have several uses within a facility. It can be used for sterilization, for industrial cleaning, as a source of process heat, or for washing hands. The maintenance principles for all of these uses are the same, but the temperatures that must be maintained may differ. The purpose of energy systems maintenance in this area is to keep the temperatures as low as possible, to prevent leaks, to keep insulation in repair, and to keep heat-transfer surfaces clean. To put this into perspective,

**Table 14.4 Problems and solutions: the electrical system.**

| Component | Problem | Initial Maintenance Action |
|---|---|---|
| Transformer | Leaking oil | Have electric company check at once |
| | Not ventilated | Install ventilation or provide for natural ventilation |
| | Dirt or grease in transformer and control room | Install air filtering system to insure clean contacts |
| | Water on control room floor | Install drainage or stop leaks into control room |
| Contact | Burned spots | Indicates shorting; repair immediately (Ref. 4) |
| | Frayed wire | May cause shorting; use tape to secure frayed ends |
| Switch | Sound of arcing, lights flicker | Replace |

**Table 14.5 Problems and solutions: lights, windows, and reflective surfaces.**

| Component | Problem | Initial Maintenance Action |
|---|---|---|
| Light | Illuminate unused space | Remove light and store for later use elsewhere |
| | Flickers (fluorescents) | Replace quickly |
| | Too little light | Increase lighting to acceptable levels |
| | Ballasts buzz | Adjust voltage or change ballast types |
| | Smoking | Replace ballast; check contacts and electrical wiring; do not use until condition is remedied |
| Wall | Dirty or greasy | Clean |
| | Painted with dark paint | On next painting use brighter paint |
| Floor | Hard to keep clean | Examine possibility of changing floor surface |
| Window | Dirty | Clean, if used for light; otherwise, consider boarding over to prevent solar gain and heat loss |

a leak of 1 cup/min of water that has been heated from 55°F to 180°F uses about 30 million Btu/yr. This is approximately equivalent to 30,000 ft$^3$ of natural gas.

Part of the description of this system should include hot-water temperatures throughout the facility. Hot-water temperatures should be kept at a low level unless there is some good reason for keeping them high. It is also unlikely that water needs to be kept hot during weekends. A further area for improved maintenance is in the insulation of hot-water tanks and lines. If a line or tank is hot to the touch, it should probably be insulated, both for the energy saved and for safety. Adding 1 in. of insulation to a water main carrying 150°F water can save as much as $1.60 per foot per year, depending upon the fuel used as a heat source. Heat-transfer surfaces should also be examined—any fins or radiators that are plugged up with debris or dirt are causing the hot-water heater to consume more energy. Table 14.6 gives some additional

troubleshooting suggestions for the initial maintenance.

*Air Compressors and the Air Distribution System*

Compressed air usually serves one of three functions: as a control medium, for cleaning, or as a source of energy for tools or machines. As a control medium, it serves to regulate various parts of heating, ventilating, and air-conditioning systems. In cleaning, compressed air can be used to dry materials or blow away various kinds of dirt. And it can be a convenient source of energy for tools or for various kinds of hydraulic equipment. All three uses are affected badly by line leaks and by poor compressor performance. When a pneumatic control system for building temperature develops leaks, the usual result is that only hot air comes into a room. Thus this kind of leak creates two kinds of energy waste: excess running of the compressor and excess heating. If the air is used for cleaning, the affect of a moderate leak

**Table 14.6 Problems and solutions: hot-water distribution system.**

| Component | Problem | Initial Maintenance Action |
|---|---|---|
| Faucet | Leaks | Fix; replace with spring-actuated units |
| | Water too hot for washing | Turn down thermostat |
| Piping | Hot to touch | Lower water temperature or add insulation |
| | Leaking | Replace |
| | Scale buildup | Consider installation of water-softening unit |
| Water storage tank | Hot to touch | Insulate or lower water temperature |
| | Leaks | Repair or replace |
| Electric boiler | Scale buildup | Install water-softening unit or start regular flushing operation |
| | On during periods when no people are in facility | Install time clocks to regulate use |
| Radiator | Finned surface badly fouled | Clean with soap, water, and a brush |
| | Obstructed | Remove obstructions |

**Table 14.7 Problems and solutions: air compressors and the air distribution system.**

| Component | Problem | Initial Maintenance Action |
|---|---|---|
| Compressor (Ref. 5) | Low suction pressures | Look for leaks on low-pressure side; overhaul if necessary |
| | Gauges do not work | Repair or replace |
| | Excess vibration | Check mountings and initial installation instructions |
| | Cold crankcase heater during compressor operation | Check for lubrication problems |
| | Loose wiring or frayed wires | Repair |
| | Leaks on high-pressure side | Examine compressor closely; check gaskets, connections, etc., and replace if necessary |
| Air line | Leaks | Repair |
| | Water in line | Look for leak at compressor |
| | Oil in line | Look for lubrication leak in compressor |

may only be that more air is used; this depends upon the individual application. If air is used as an energy source for motors or tools, the speed of the motor or tool usually depends upon the air pressure, and any decrease in this pressure will affect the performance of the motor. Another thing—controls are particularly affected by oil, water, and dust particles in control air, so it is of utmost importance that the compressor be checked regularly to see that the air cleaner is working as intended. More detail on compressors can be found in Chapter 12 of Ref. 5. Table 14.7 gives a guide to the initial maintenance procedures for air compressors and air distribution systems.

*The Manufacturing Equipment System*

This section would not be complete without a discussion of some of the more common equipment used in manufacturing. Some of this equipment has unique maintenance requirements, and these must be obtained from the manufacturer. Much equipment, however, is common to many companies or facilities, and this includes motors, ovens, and time clocks.

**Motors.** Motors can consume excess amounts of energy if they are improperly mounted, if they are not hooked up to their load, or, in the case of three-phase motors, if the voltages in the opposing legs are different. In this last case, the *ASHRAE Equipment Handbook* states:

*With three-phase motors, it is essential that the phase voltages be balanced. If not, a small voltage imbalance can produce a greater current unbalance and a much greater temperature rise which can result in nuisance overload trips or motor failures. Motors should not be operated where the voltage unbalance is greater than 1% without first consulting the manufacturer.*[5]

**Bearings.** Bearing wear is one of the more significant failure types in a motor, and this can be avoided by a procedure devised by Harold Tornberg, a plant manager and former industrial engineer for Safeway Stores, Inc. The Tornberg procedure is to connect a stethoscope to a decibel meter and to take readings at both ends of a motor. The reading on the driving end of the motor will usually be 2 to 3 dB above the reading on the inactive end. If there is no difference, the bearings on the inactive end probably need to be lubricated or replaced. If the difference is 5 to 6 dB, the bearings on the active end need to be replaced. If the difference is 7 to 9 dB, the bearing is turning inside the housing, and the motor should be overhauled as soon as possible. If the difference in readings is more than 9 dB, the motor is on the verge of failure and should be replaced immediately. These procedures were developed on motors in use at Safeway Stores and are in use throughout the Safeway system. To adapt these procedures to your facility, buy or make an instrument that allows you to determine the noise level at a particular point. Then survey the motors in your facility using the criteria discussed above. Take one of the motors that the test has indicated as needing repair into your shop, and tear it down. If you find that the failure was worse than these standards indicated, lower the decibel limit needed to indicate a problem. If you find that the failure was not as bad as you thought, raise the standards. In any case, this procedure is one that can be adapted to your needs to indicate bearing failures before they occur.

**Ovens.** Ovens use much energy, and many standard operating practices are available to help decrease associated waste.[6] In describing the initial maintenance actions that must be performed, check the seals, controls, refractory, and insulation. The possible problems are shown in Table 14.8. Correcting these will save energy and improve operating efficiency at the same time. Remember also that heat lost from an oven must be removed by the air-conditioning system.

**Time Clocks.** Time clocks can be used to significant advantage in the control of equipment that can be turned off at regular intervals. But two problems may occur to eliminate any savings that might otherwise be generated. First, people can wire around time clocks or otherwise obstruct their operation. Or maintenance personnel can forget to reset them after power outages. If either of these happens, all the potential energy savings possible with their use will have been lost. These items are included in Table 14.8.

### 14.1.2 Prepare a List of Routine Maintenance Actions with Time Estimates, Materials, and Frequency for Each

This is the second step in the four-step procedure for developing a maintenance program. The first step was to determine the present condition of each system in the facility, using Tables 14.1 to 14.8 to help locate trouble spots. The products of this step are: (1) a list of all the equipment in the facility by system; (2) a list of the major one-time maintenance problems associated with this equipment; and (3) a notebook with these lists and with the diagrams for each of the major systems. The next major step is to augment the notebook by a list of preventive energy maintenance actions for each system together with an estimate of the materials needed, the time required, and the maintenance frequency for each piece of equipment. Table 14.9 gives representative inspection intervals to help you develop your own procedures. For convenience, the items of equipment are arranged in alphabetical order, and the maintenance actions are described only briefly. A more detailed description of the required maintenance for some items is given

in Section 14.2 and necessary instruments are discussed in Section 14.5.

Incidentally, the lists described above should be maintained in a form that maintenance personnel can easily use. If your personnel are familiar with database searching, or if you have a computer system that you have used, understand, and like, then by all means use a computer system. Such systems have information retrieval capabilities that can be most helpful, they offer fast access to information, and they are easy to update. They also don't take up much room. On the other hand, they can be a barrier to getting work done. If the software is unfamiliar, if the maintenance people don't like to use it, or if keeping it up to date is difficult, then a manual system is better. The objective is to get the maintenance done, and done right, not to demonstrate another use for a computer.

**Time Standards.** Accurate time standards for maintenance are difficult to obtain except where the maintenance actions are the same whenever they are repeated. In this case, predetermined time standard systems such as MTM and MOST are claimed to work, as are some detailed standards such as those developed by the Navy.[7] But most maintenance actions include troubleshooting. Troubleshooting depends upon the condition of the equipment, the maintenance history, and the skill of the maintenance personnel. In general, hiring journeymen rather than apprentices is an investment that is well worthwhile. To estimate the amount of maintenance time that will be needed, consult equipment manufacturers first, then other users of the same kind of equipment. Modify these estimates to include the experience of your personnel, the present condition of the equipment, and the availability of necessary repair parts. Then record your time estimates, compare these estimates with actual experience, and revise the estimates to conform with your actual experience.

**Table 14.8 Motors, ovens, and time clocks.**

| Component | Problem | Initial Maintenance Action |
| --- | --- | --- |
| Motor | Noisy | Check bearings |
| | Too hot | Check voltage on both legs of three-phase input |
| | Vibrates | Check mounting |
| Oven | Door gaskets worn | Replace |
| | Insulation or refractory brick missing | Replace |
| | Thermostats out of calibration | Replace or recalibrate |
| Time clock | Does not work | Repair or replace |
| | Time incorrect | Adjust clock to correct time |

**Table 14.9 Table of preventive maintenance actions.**

|  | Frequency |
|---|---|
| Air lines | |
| 1. Check for leaks | |
| Blowers | Annually |
| 1. Inspect belts for tension and alignment | |
| 2. Inspect pulley wheels | |
| 3. Inspect for dirt and grease | |
| Boilers (see Section 14.2.1) | |
| 1. Check temperature and pressure | Daily |
| 2. Clean tubes and other heating surfaces | As needed |
| 3. Check water gauge glass | Daily |
| 4. Remove scale | Annually |
| 5. Perform flue-gas analysis | Monthly |
| 6. Calibrate controls | Annually |
| Chillers (Ref. 5) | Annually |
| 1. Clean condenser and oil cooler | |
| 2. Calibrate controls | |
| 3. Check electrical connections | |
| 4. Inspect valves and bearings | |
| Condensate return system | |
| 1. Check valves, pumps, and lines | Annually |
| Cooling coils | |
| 1. Brush and wash with soap | Quarterly or when needed |
| 2. Clean drip pan drain | When needed |
| Compressors | |
| 1. Check oil levels | Monthly |
| 2. Check wiring | Annually |
| 3. Visual check for leaks | Monthly |
| 4. Log oil temperature and pressure | Monthly |
| 5. Remove rust with wire brush | Annually |
| 6. Replace all drive belts | Annually |
| Condenser | |
| 1. Clean fan | Annually |
| 2. Brush off coil | Monthly |
| Controls | |
| 1. Calibrate thermostats | Semiannually (Ref. 4) |
| 2. Get professional inspection of control system | According to equipment specifications |
| 3. Check gauges to see that readings are in correct range | Monthly |
| 4. Examine control tubing for leaks | Monthly |
| Cooling towers (Ref. 3) | |
| 1. Inspect for clogging and unusual noise | Daily |
| 2. Check gear reducer oil | Level—weekly. For sludge andwater—  monthly |
| 3. Inspect for leakage | Semiannually |
| 4. Tighten loose bolts | Semiannually |
| 5. Clean suction screen | Weekly |
| Dampers | |
| 1. Check closure | Every 6 weeks |
| 2. Clean with brush | Semiannually |

**Table 14.9 (*Continued*)**

| | Frequency |
|---|---|
| **Ductwork** | Semiannually |
|   1.  Inspect and refasten loose insulation | |
|   2.  Check and repair leaks | |
|   3.  Inspect for crushed or punctured ducts; repair | |
| **Electrical system** | |
|   1.  Inspect equipment for frayed or burned wiring | Semiannually |
|   2.  Perform electrical load analysis | Whenever major equipment changes occur, or every 2 years |
| **Fans** | |
|   1.  Check fan blades and clean if necessary | Semiannually |
|   2.  Check fan belts for proper tension and wear | Monthly |
|   3.  Check pulleys for wear and alignment | Semiannually |
|   4.  Check for drive noise, loose belts, and excessive vibration | Semiannually |
| **Faucets** | |
|   1.  Check for leaks; replace washers if needed | Annually |
| **Filters, air** | |
|   1.  Replace | When dirty, or monthly |
|   2.  Check for gaps around filters | When replacing |
|   3.  Inspect electrical power equipment (for roll filters; from Ref. 6) | Monthly |
| **Filter, oil** | |
|   1.  Clean and oil | Whenever compressor oil is changed |
| **Gauges** | |
|   1.  Check calibration | Annually |
|   2.  Check readings | As needed |
| **Grillwork** | |
|   1.  Remove dirt, grease, bugs | Monthly |
|   2.  Remove obstructions from in front of grill | Monthly |
|   3.  Check air direction | Monthly |
|   4.  Recaulk seams | As required |
| **Leaks** | |
|   1.  Check for refrigerant leaks | Annually |
| **Lights, inside** | |
|   1.  Perform group relamping | See Chapter 13 |
|   2.  Perform survey of lighting in actual use | Semiannually |
|   3.  Clean luminaires | Office area: every 6 months; laboratories: every 2 months; maintenance shops: every 6 months; heavy manufacturing areas: every 3 months; warehouses: every 12 months |
|   4.  Replace flickering lights | Immediately |
| **Lights, outside** | See Chapter 13 |
| **Motors** | |
|   1.  Lubricate bearings | When needed |
|   2.  Check alignment | Semiannually |
|   3.  Check mountings | Semiannually |
| **Ovens** | Semiannually |
|   1.  Check insulation | |

**Table 14.9 (*Continued*)**

|  | Frequency |
|---|---|
| 2. Check controls | Annually |
| 3. Check firebrick and insulation | Annually |
| Piping | |
| 1. Check mountings | Annually |
| 2. Check for leaks | Semiannually |
| Pumps | |
| 1. Check lubrication | Monthly |
| 2. Examine for leaks | Semiannually |
| Steam traps, (Ref. 8), high pressure (250 psig or more) | |
| 1. Test | Daily-weekly |
| Steam traps, medium pressure (30-250 psig) | |
| 1. Test | Weekly-monthly |
| Steam traps, low pressure | |
| 1. Test | Monthly-annually |
| Steam traps, all (see Section 14.2.3) | |
| 1. Take apart and check for dirt and corrosion | If test shows problems |
| Thermostats | |
| 1. Check calibration | Annually |
| Time clocks | |
| 1. Reset | After every power outage |
| 2. Check and clean | Annually |
| Transformers (Ref. 6) | |
| 1. Inspect gauges and record readings | Monthly |
| 2. Remove debris | Monthly |
| 3. Sample and test dielectric | Annually |
| 4. Remove cover and check for water | Every 10 years |
| Walls and windows | |
| 1. Check for air infiltration | Semiannually |
| 2. Clean | Regularly, when dirty |

### 14.1.3 Prepare a Maintenance Schedule

Section 14.1.2 gave a list of maintenance actions needed for many equipment items, together with approximate maintenance frequencies. To develop your own maintenance schedule, first list the equipment you have in your facility. Then, using the table, estimate the maintenance or inspection frequencies for each unit. A section of such a table is shown in Table 14.10. This table should go into the workbook with manufacturers' manuals described earlier. This master table and the workbook are used to develop two files: an equipment file and a tickler file.

The equipment file is a list of cards describing the maintenance for each unit. A unit may have more than one card, depending upon the complexity of its maintenance. These cards contain troubleshooting directions, a description of how to perform specific main-

tenance on each piece of equipment, and a record of individual equipment maintenance. The cards should be updated every time maintenance is performed or whenever a new troubleshooting procedure is devised. Figure 14.1 shows such a card. The tickler file has a separate card for each day and gives the maintenance to be performed on that day. This file has two parts. The first part is for actions to be done monthly, quarterly, semiannually, or annually and is arranged by months. The second part of the file is for actions to be done weekly and is arranged by day of the week. In preparing this tickler file, items that require the same skills should be grouped together where possible to save time and to take advantage of any economies of scale. (An alternative is to do all tasks in the same area at the same time.) Figure 14.2 shows an example of one card from the tickler file. Ideally, all of these files should be on a computer, but see the note in Section 14.1.2.

The final step is to use the tickler and equipment files to set up an operating schedule by assigning each task to a particular person or crew.

### 14.1.4 Follow-up and Monitoring

Management action is necessary to see that the prescribed maintenance policies are being followed. This step, the fourth in the development of an energy maintenance program, has three objectives. First, to ensure that maintenance is being performed as scheduled. This objective can be accomplished by periodic inspections of the records and of the action recorded. The second objective is for the management to get an update on the condition of the facility so as to anticipate and plan for capital expenditures relating to maintenance. This planning requires knowledge of the current condition of the plant. The third objective is to update all the files.

The energy equipment notebook should be changed every time a new kind of equipment is installed or whenever a manufacturer announces improved maintenance procedures for equipment already on hand. The equipment card file should be changed at the same time, with abbreviated versions of any new maintenance instructions. The equipment file should also be changed whenever someone discovers an improved way of maintaining an item on a card. The tickler file should be changed to reflect the actual needs of the facility. If, for example, the original standard on filter changing was 0.5 hr per change, and a new method is developed that enables these to be changed in 0.3 hr, the new method should be noted on the equipment card file and the new standard on the tickler card files. Similarly, if the frequency of filter changes was originally listed as every month, and experience shows that 3 months is

a better interval, the three months should be noted in the tickler file. By following this procedure, the time standards and the maintenance frequencies can be kept current, and the maintenance plan can be adapted to each individual facility.

**Table 14.10 Example of equipment maintenance and frequencies.**

| Equipment | Location | Action | Standard, Frequency |
|---|---|---|---|
| Air-handling units | Bldg 211 | Clean filters | 0.7 hr, monthly |
| | | Clean fan blades | 1.2 hr, annually |
| | | Check ducts | 1.4 hr, annually |
| | | Replace fan belts | 0.6 hr, annually |
| | | Check belt tension | 0.1 hr, monthly |

**Fig. 14.1 Equipment record.** (Courtesy of Safeway Stores, Inc.)

## 14.2 DETAILED MAINTENANCE PROCEDURES

Section 14.1 gave an outline for the development of a maintenance program for use in energy management. To maintain some of the more complex equipment properly, it is necessary to have more detail than could be presented in Table 14.9. This section is therefore devoted to a more complete discussion of boilers, pumps, and steam traps.

### 14.2.1 Boilers

Money spent in proper boiler maintenance is one of the best investments a company can make. The benefits are substantial. Suppose, for example, that your gas bill is $1,500,000 per year and that 90% of this, or $1,350,000, is for your central heating plant. Improving your boiler efficiency 5% saves $67,500 per year, in addition to any benefits that you may realize through decreased downtime, decreased repair costs, and other nontrivial expenses. To gain this benefit requires proper boiler maintenance, either by your own personnel or by a boiler service company hired to provide this service on a regular basis. The following section is designed to help you decide what must be done, how often, and why. Two warnings must be observed. First, safety precautions must be known and observed at all times. *Make sure that all safety interlocks are functioning before doing any work on a boiler.* Second, nothing in the following section should be construed as superseding manufacturers' instructions or local safety and environmental regulations. Keeping these warnings in mind, consider the following problems, their effects, and the best means of correcting them.

The basic reference for the following material is *Guidelines for Industrial Boiler Performance Improvement*, published by the Environmental Protection Agency.[9]

*Excess $O_2$ in Stack Gas*
The amount of oxygen

mixed with fuel is directly related to the amount of excess air introduced to the boiler, as shown in Figure 14.3. The problem of excess $O_2$ is simply this: If there is too little air available to combine with fuel, incomplete combustion takes place, the boiler smokes, and hazards of boiler malfunction increase dramatically (in an oil- or coal-fired boiler) or the carbon monoxide (CO) concentration builds up (in a gas-fired boiler). If too much air is introduced, a great deal of the energy in the fuel is used to heat up the excess air, and the efficiency of the boiler decreases. These inefficiencies are shown in Figures 14.4 and 14.5. The two curves in each figure reflect the range within which the curve of a typical boiler would be expected to fall. The exact curve for your boiler should be established by a careful test at each of the firing rates you anticipate.

Maintenance Department

Weekly Lubrication Schedule

Day of week  Tuesday, 11/13/79 _____ Plant

Date _____ 19____ By _____

| Alto Roll Slicer | | | Tail Off Conveyor | | |
|---|---|---|---|---|---|
| Clean and oil all moving parts; clean and | | | Clean and oil drive chain, #1 oil; check | | |
| oil all drive chains, #1 oil; check all gear- | | | gearmotor oil level, #8 oil. | | |
| motors oil level, #8 oil; check variable- | | | | | |
| speed internal drive belt for wear; | | | All Cooling Conveyors | | |
| check out machine. | | | Oil all transfer roller bearings, #30 oil. | | |
| | | | | | |
| United Roll Bagger | | | Model-K-Roll Machine | | |
| Clean and oil all moving parts; clean and | | | Check head oilers; oil sifter linkeage, | | |
| oil all drive chains, #1 oil; check gear | | | sifter bushings, chain take up idler | | |
| reducer oil level, #8 oil; check drive belts | | | sprocket, #1 oil; check vacuum com- | | |
| for wear; check out, wipe off machine. | | | pressor oil level, compressor oil. | | |
| | | | | | |
| Kwik-Lok | | | Pan-O-Mat | | |
| Clean and oil all moving parts, #1 oil; | | | Oil all sifter linkeage, sifter bushings, vari- | | |
| check out, wipe off machine. | | | able speed adjusting pulley, chain take up | | |
| | | | idler sprocket, control arm linkeage, cam | | |
| | | | rollers, gate bushings, tail off conveyor | | |
| Brush Unit | | | pulleys, #1 oil. | | |
| Remove side cover; clean and oil all | | | | | |
| chains, #1 oil; check out, wipe off | | | | | |
| | | | | | |
| | | | Total time | | |
| The following items need attention: | | | | | |
| machine. | | | | | |

**Fig. 14.2 Tickler file card.** (Courtesy of Safeway Stores, Inc.)

Fig. 14.3 **Relationship between boiler excess air and stack-gas concentration of excess oxygen.** (Adapted from Ref. 9.)

Fig. 14.4 **Typical smoke-$O_2$ characteristic curves for coal- or oil-fired industrial boilers. Curve 1, gradual smoke/$O_2$ characteristic; curve 2, steep smoke/$O_2$ characteristic.** (Adapted from Ref. 9.)

The combustion controls on the boiler should be used to decrease the amount of excess air to a point where the amount of excess $O_2$ represents a compromise between the optimal minimum and the point at which actual waste of energy occurs. The effect of reducing excess oxygen toward the optimum can be dramatic; if a reduction in excess air from 80% to 20% can be accompanied by a reduction in stack-gas temperature from 500°F to 400°F, the resulting improvement in boiler efficiency can be 6 to 8% or more. On an annual gas bill of $1,000,000, this is a reduction of $60,000 to $80,000 in gas alone.

But it is not usually desirable to operate at the minimum indicated on Figures 14.4 and 14.5, for several reasons. The consequences of exceeding the minimum O2 percentage by a small amount are not as damaging as those of being less than the minimum by a corresponding amount. If the minimum is exceeded slightly, fuel is wasted. If, however, there is insufficient excess air and thus insufficient excess oxygen, smoking or CO buildup can occur, the tubes can become fouled, and the boiler can malfunction. Since there is usually some play in the combustion control system, a setting greater than the minimum helps to guard against these problems. Typical values are shown in Table 14.11.

The amount of O2 in stack gas can be measured continuously, by a recording gas analyzer, or periodically, by an Orsat analyzer. The advantages of a continuous, mounted unit are that readings are always available to boiler operating personnel for daily boiler maintenance and for troubleshooting. The cost of an installed gas

Fig. 14.5 **Typical CO-$O_2$, characteristic curves for gas-fired industrial boilers. Curve 1, gradual CO/$O_2$ characteristic: curve 2, steep CO/$O_2$ characteristic.** (Adapted from Ref. 9.)

analyzing unit will vary from application to application, but typical costs are $200 to $1000. The readings should be logged every hour.

Sudden changes in the percentage of excess oxygen from hour to hour are unlikely but should be analyzed immediately if they occur. Some of the more common

## Table 14.11 Typical values for minimum excess $O_2$.

| Typical Fuel Type | Minimum Values of Excess $O_2$ at High Firing Rates (%) |
|---|---|
| Natural gas | 0.5-3.0 |
| Oil fuels | 2.0-4.0 |
| Pulverized coal | 3.0-6.0 |
| Coal stoker | 4.0-8.0 |

*Source:* Ref. 9.

causes for changes in excess $O_2$ are tube fouling, changes in fuel or in atmospheric conditions, or damage to the control system. The base for comparison should be established by the boiler service personnel or the boiler contractor at the time of initial installation of the boiler, and he or she should be consulted immediately if any unusual conditions are observed.

### Excessive Stack-Gas Temperature

This problem can be caused by water-side or fire-side fouling. If fire-side fouling (i.e., a buildup of soot, ash, or other particles) is taking place, the rate of heat transfer from the boiler to the water or steam it is heating is impeded, and the stack gas is correspondingly hotter. The effect of soot on fuel consumption is noticeable, according to the American Boiler Manufacturers Association (ABMA).[10] Efficiency losses due to soot are approximately as given in Table 14.12. For this reason, the ABMA recommends that tubes be cleaned once every shift where practical. Where such cleaning is not practical, it is recommended that tubes be cleaned whenever the stack temperature rises 75°F.

### Table 14.12 Typical fuel losses due to soot.

| Soot Layer on Heating Surfaces (in.) | Increase in Fuel Consumption (%) |
|---|---|
| 1/32 | 2.5 |
| 1/16 | 4.4 |
| 1/8 | 8.5 |

*Source:* Ref. 10.

If the fouling is on the water side because of scale buildup, accumulation of mud or slime, or for some other reason, heat transfer will be impeded as described above, and increased stack temperature will result. The effect of this scaling is seen in Table 14.13. The detrimental effect can be severe.

As with excess oxygen, it is advisable to monitor

## Table 14.13 Effect of water-side scaling.

| Thickness of Scale (in.) | Loss of Heat (%) | | |
|---|---|---|---|
| | Soft Carbonate | Hard Carbonate | Hard Sulfate |
| 1/50 | 3.5 | 5.2 | 3.0 |
| 1/32 | 7.0 | 8.3 | 6.0 |
| 1/25 | 8.0 | 9.9 | 9 0 |
| 1/20 | 10.0 | 11.2 | 11.0 |
| 1/16 | 12.5 | 12.6 | 12.6 |
| 1/11 | 15.0 | 14.3 | 14.3 |
| 1/9 | n.a. | 16.0 | 16.0 |

*Source:* Ref. 10.

stack temperature on an hourly basis and to record the readings. By graphing the readings, it is possible to determine how much variation is caused by shifting load, when given temperature indicates that something unusual has happened, and when a gradual temperature rise indicates that it is time to clean the water side or the fire side of the boiler. If soot blowers are being used, the stack temperature should drop immediately after the tubes are cleaned. If the temperature does not drop, the soot blowers may not be working or the thermometer may have become fouled—in any case, something is wrong.

### Smoking or Excess CO

Excess CO, in the case of natural gas, or smoking for coal or oil fuels, gives an indication that something has changed. Changes in the fuel composition or wear in some component of the burner can cause these problems, or there may be a change in the air supply. In any case, the problem should be corrected immediately. A stack-gas analyzer can be used to monitor the percentage of CO on a continuous basis.

### Flame Appearance

The appearance of the flame can give some valuable information. If the pattern is unusual, there may have been changes in the burner tips or in other parts of the burner, or there may be a malfunction in a related part of the boiler. Also, examining the flame pattern can show if part of the boiler is getting overheated. At the same time the flame is being examined, the inside of the boiler can be examined as far as possible to see that everything is in order—the stoker (if it is a coal-fired boiler), the refractory, the burners, and so on. The flame check is quick and should not replace the other inspections noted above, but it can provide additional information. Such a check should be performed every hour.

*Record Keeping*

The Environmental Protection Agency guidelines recommend that a log be kept on each boiler with the information shown in Table 14.14.[9] In addition to recording this information, plotting it on a graph can give early indications of unusual trends or cycles in the data; these patterns can then be incorporated into the general system used to indicate when something is about to go wrong.

*Maintenance Actions and Frequencies*

Table 14.15 gives a list of the most common boiler maintenance actions that need to be performed annually. Table 14.16 gives a checklist of other routine maintenance items. If your staff is trained in boiler operation and maintenance this table can be used to help define a pattern of boiler maintenance. If you decide to hire boiler maintenance done by an outside firm, the table can help you to determine what they should do and to determine whether you would gain more from doing this work with your own personnel.

Ultimately, your own maintenance personnel will do much of the routine maintenance. To be sure they know what to expect and how to perform these tasks, have a boiler service representative or contractor train each person in the more routine kinds of boiler operation, such as reading a sight glass, inspecting the flame, and blowing down the boiler. In particular, they should learn enough about the boiler that they know how to operate it safely. Keep in mind that many maintenance people are undertrained as well as underpaid and that they are not familiar with procedures for handling live steam. Without proper training, they may also let sight glasses go dry, and they may miss such facts as that the low-water cutoff valve is not working. These maintenance people can be the most expensive ones you hire—if their lack of training causes you to lose a boiler.

Another note—the given maintenance intervals are averages. If your boiler uses a great deal of makeup water, more sludge can develop, and this will have to be removed by blowing down more often than indicated. If your boiler is a closed system with very few leaks, it is possible that water quality will not be a problem. In that case, the blowdowns can be much less frequent. You must adapt the procedures to your own boiler, preferably with the help of a local professional and the vendor.

### 14.2.2 Package Boilers

The foregoing discussion applies to the operator of a large boiler with complete controls and one or more

**Table 14.14 Boiler information to be logged.**

General data to establish unit output
    Steam flow, pressure
    Superheated steam temperature (if applicable)
    Feedwater temperature
Firing system data
    Fuel type (in multifuel boilers)
    Fuel flow rate
    Oil or gas supply pressure
    Pressure at burners
    Fuel temperature
    Burner damper settings
    Windbox-to-furnace air pressure differential
    Other special system data unique to particular installation
Air flow indication
    Air preheater inlet gas $O_2$
    Stack gas $O_2$
    Optional: air flow pen, forced-draft fan damper position, forced-draft fan amperes
Flue-gas and air temperature
    Boiler outlet gas
    Economizer or air heater outlet gas
    Air temperature to air heater
Unburned combustion indication
    CO measurement
    Stack appearance
    Flame appearance
Air and flue-gas pressures
    Forced-draft fan discharge
    Furnace boiler outlet
    Economizer differential
    Air heater and gas-side differential
Unusual conditions
    Steam leaks
    Abnormal vibration or noise
    Equipment malfunctions
    Excessive makeup water
Blowdown operation
Soot-blower operation

*Source:* Ref. 9.

persons directly responsible for the boiler operation. In many cases, however, needs are met by a package boiler—a self-contained unit that generally requires little maintenance. Depending upon the quantity of fresh water used annually, water treatment may or may not be critically important. Most of the important maintenance procedures are covered in the operating manual that comes with the boiler, and this manual should be kept available and up to date. Some of the more important

**Table 14.15 Boiler checklist for annual maintenance.**

| Check | Examine for: |
| --- | --- |
| Safety interlocks | Operability—must work |
| Boiler trip circuits | Operability |
| Burners | |
|   1. Oil tip openings | Erosion or deposits |
|   2. Oil temperatures | Must meet manufacturer's specifications |
|   3. Atomizing steam pressure | Must meet manufacturer's specifications |
|   4. Burner diffusers | Burned or broken, properly located in burner throat |
|   5. Oil strainers | In place, clean |
|   6. Throat refractory | In good condition |
| Gas injection system | |
|   1. Orifices | Unobstructed |
|   2. Filters and moisture traps | In place, clean, and operating |
|   3. Burner parts | Missing or damaged |
| Coal burners | |
|   1. Burner components | Working properly |
|   2. Coal | Fires within operating specifications |
|   3. Grates | Excessive wear |
|   4. Stokers | Location and operation |
|   5. Air dampers | Unobstructed, working |
|   6. Cinder reinjection system | Working, unobstructed |
| Combustion Controls | |
|   1. Fuel valves | Move readily, clean |
|   2. Control linkages and dampers | Excessive "play" |
|   3. Fuel supply inlet pressures on atomizing | Meet manufacturer's specifications steam or air systems |
|   4. Controls | Smooth response to varying loads |
|   5. Gauges | Functioning and calibrated |
| Furnace | |
|   1. Fire-side tube surfaces | Soot and fouling |
|   2. Soot blowers | Operating properly |
|   3. Baffling | Damaged; gas leaks |
|   4. Refractory and insulation | Cracks, missing insulation |
|   5. Inspection ports | Clean |
| Water treatment | |
|   1. Gauges | Working properly |
|   2. Blowdown valves | Working properly |
|   3. Water tanks | Sludge |
|   4. Water acidity | Within specifications |

*Source:* Ref. 9.

**Table 14.16 Boiler checklist for routine maintenance.**

| Action | Frequency |
| --- | --- |
| Check safety controls | Daily |
| Check stack-gas analysis | Daily or more often |
| Blow down water in gauges | Weekly |
| Blow down sludge from condensate tanks | Whenever needed; frequency depends upon amount of makeup water used |
| Have water chemistry checked | Quarterly |
| Perform combustion efficiency check; log results | Daily or weekly |
| Check and record pressures and readings from boiler gauges | Daily or weekly |

procedures for keeping a package boiler running are described below.

1. *Safety and relief valves* should be checked occasionally to see that they work and that they will reseat properly. They should be checked carefully to avoid excessive steam loss and scaling. If any such valve fails to work properly, the fact should be noted and the valve fixed by a boiler service representative at the first opportunity.

2. *Air supply* should be kept open so that the boiler can have enough combustion air at all times. Restricting combustion air by blocking the boiler air openings or by blocking all air openings into the boiler room can create a buildup of carbon monoxide and/or cause the boiler to operate inefficiently.

3. *Low-water and high-water gauges and controls* should be flushed periodically to remove sludge. If sludge builds up in a float-operated valve, the valve can fail to operate, with expensive consequences to the boiler. The gauges should be flushed periodically and should not be allowed to get rusty or clogged.

4. *Combustion controls* should be inspected regularly and adjusted if sooting or burner wear is taking place. Sooting causes a great drop in boiler efficiency (see Table 14.12), and burner wear can cause irregular firewall wear, inefficient combustion, and the need for increased boiler maintenance. It is generally worthwhile to have your combustion controls inspected at least annually.

5. *Tubes* can be cleaned if scaling occurs by removing head plates at either end and running a special brush through each tube. For detailed procedures and necessary safety precautions, see your local boiler service representative. Tubes can be replaced if necessary, but the job calls for special skills and should be done by someone trained for this job.

### 14.2.3 Steam Traps[8]

A steam trap is a mechanical device used to remove air, carbon dioxide, and condensed steam and to prevent steam from flowing freely into the condensate return system. Steam traps are necessary for several reasons. If air is not removed from steam, the oxygen can dissolve in low-temperature steam condensate and help cause corrosion of the valves, pipes, and coils in the steam distribution system. If carbon dioxide is not removed, it can combine with steam condensate to form carbonic acid, another major source of corrosion in the steam distribution system. Air and carbon dioxide also act as insulators to impede heat transfer from

the steam; their presence creates a partial pressure that lowers the steam temperature and the heat-transfer rate.

Perhaps the main function of steam traps, however, is to permit the removal of steam condensate from a system while simultaneously preventing the free escape of steam. In this last function, the energy of the steam is kept within the system, and the amount of the live steam within the facility is controlled.

Steam traps occur as parts of nearly every steam distribution system. They are often not maintained, and this lack of maintenance can create a hidden cost that is significant. The cost of a steam trap failure is dependent upon the failure mode, with a strong dependence upon the original design. The two main failure modes are failing open and failing shut, and the design consideration of most maintenance interest is proper drainage. Consider these problems in turn.

### Problems if Trap Fails Open

If a trap fails open, live steam flows directly from the steam system through the trap, a pressure buildup is caused in the condensate return system and the condensate return lines are heated unnecessarily, or the steam is released directly to the air. In the first case, the back pressure in the condensate lines may cause other steam traps to fail, and the condensate may not be removed from the steam distribution system. In the second case, steam discharging to air costs as much as any other kind of steam leak and creates pressure losses. These costs can be estimated using Table 14.17 and the formula

$$C = \text{waste} \times \frac{1100 \text{ Btu/lb}}{\text{boiler eff.}} \times \frac{\text{energy cost/million Btu}}{1,000,000}$$

where

$C$ = energy cost per month due to steam loss

waste = number of pounds of steam wasted per month, from Table 14.17

1100 Btu/lb = approximate value for total heat in saturated steam at 100 psi

boiler eff. = boiler efficiency, usually 60 to 80%

energy cost/million Btu = cost that can be obtained from your fuel supplier

These figures are close estimates for steam at 100-psi pressure; for other pressures between 50 and 200 psi, the cost is nearly proportional to the pressure, using 100 psi and Table 14.17.

**Table 14.17 Cost of Steam Leaks at 100 psi Assuming a Boiler Efficiency of 80% and Input Energy Cost of $2 per Million Btu**

| Size of Orifice (in.) | Steam Wasted per Month (lb) | Total Cost per Month | Total Cost per Year |
|---|---|---|---|
| 1/2 | 835,000 | $2,480 | $29,760 |
| 7/16 | 637,000 | 1,892 | 22,704 |
| 3/8 | 470,000 | 1,396 | 16,752 |
| 5/16 | 325,000 | 965 | 11,580 |
| 1/4 | 210,000 | 624 | 7,488 |
| 3/16 | 117,000 | 347 | 4,164 |
| 1/8 | 52,500 | 156 | 1,872 |

### Problems if Trap Fails Shut

If a steam trap fails shut, it acts like a plug in that part of the steam system. Condensate builds up and cools, heat transfer stops, and noncondensable gases dissolve in the water and create corrosion in the pipes and in the equipment served by the trap. If a blocked trap causes condensation to build up in a line containing active steam, the steam may push water ahead of it to form a water hammer. To visualize the effects of water hammer, think of water at 90 miles/hr colliding with the inside of a pipe. Severe damage can result.

### Problems of Improper Drainage

If steam condensate is not drained properly, weight accumulates, heat transfer stops, and freezing may result. Water weighs 62.4 lb/ft³, steam at 100 psi about 0.26 lb/ft³. If a 6-in. pipe is filled with condensate rather than steam, it is carrying 12.25 lb per linear foot rather than 0.05 lb. The heat transfer problem has been presented earlier. Another problem may be the freezing of pipes, particularly if the system depends upon proper operation of the steam trap to prevent freezing.

There are four basic types of steam traps, and each requires a different troubleshooting procedure. The following sections show how each trap type operates, what can go wrong with it, and what troubleshooting procedures are recommended.

### Open Bucket Traps

In the trap shown in Figure 14.6, the condensate enters through the top. Since the entering hole is at the side of the bucket, condensate flows into the body of the trap, raising the bucket and causing the valve to plug the condensate relief hole. When the bucket has raised as far as the trap frame will allow, condensate fills up the trap and finally overflows into the bucket, decreasing its buoyancy. When enough water has flowed into the bucket, it drops, opening the condensate return valve. Steam pressure then forces the condensate into the condensate return system. Because of the design of the bucket, some of the condensate within the bucket is also emptied at this time, but the bucket is never completely empty, thus maintaining a water seal over the outlet and preventing steam from blowing through the trap. The cycle then repeats.

The advantages of this kind of trap are its reliability and its general ease of servicing. It can develop leaks in the bucket, and pivot wear or the accumulation of dirt can make it unworkable.

**Fig. 14.6 Open bucket steam trap.** (Courtesy of Armstrong Machine Works.)

**Open Bucket Traps, Troubleshooting and Maintenance.** The operation of an open bucket trap is characterized by its regular opening and closing. This creates a noise which can be detected with an industrial stethoscope. When a trap is first installed, the frequency of clicking should be noted on the body of the trap. If this frequency changes drastically without a corresponding change in steam pressure, the trap should be taken apart (after isolating it from live steam !) and overhauled. If the trap is not making noise and is cold, determine whether condensate is getting to the trap. If no condensate or steam is reaching the trap, check strainers, valves, and lines upstream of the trap to discover where the blockage is occurring. If condensate is reaching the trap, check the pressure of the steam coming into the trap. If this pressure is too great, the trap will not work. If the pressure is correct and the trap is not working, valve off the trap so that no live steam is coming from either the steam distribution system or the condensate return and take off the cover or remove the trap from the line. Look for dirt, worn parts, or worn or plugged orifices. It is also possible that some fluctuations in steam pressure caused the trap to lose the small amount of water in the bucket necessary for its operation (the *priming*). This condition can be detected if the trap is blowing live steam. Close the inlet valve for a few minutes to let condensate build up, then let this condensate into the trap and it should work.

Another problem that this trap can have is caused by the need to have the priming water in the trap at all times. The trap is automatically vulnerable to freezing if the external temperature gets cold enough and if the supply of steam is shut off. If such freezing occurs, many traps have provisions for the admission of live steam to thaw the ice.

*Note:* There is a difference between *flash steam* and *live steam*. Flash steam is formed when condensate at a high temperature is exposed to air at a lower temperature, and part of it "flashes" to steam. Flash steam has very little pressure and gives an irregular, undirected flow pattern. Live steam however, is steam under pressure and has a well-defined, consistent pattern. The difference is significant, since the presence of flash steam is expected, whereas live steam gives an indication that something is wrong.

### Inverted Bucket Traps

The inverted bucket trap (Figure 14.7) works like the open bucket trap described above, with two exceptions. First, the trap is opened and closed by steam lifting a bucket rather than by condensate lowering it.

**Fig. 14.7 Inverted bucket steam trap.** (Courtesy of Armstrong Machine Works)

Second, the design of the trap is self-cleaning—dirt is scoured from the bottom of the trap automatically as a result of the trap design. Thus there is less likelihood that the trap will become plugged up with dirt, although a large dirt particle could still keep the bucket lifted up or the valve jammed open. Both of these problems can be alleviated by installing and maintaining a strainer upstream of the trap. This trap is maintained using the procedures outlined for the maintenance of the bucket trap described above. This trap can also freeze or lose its prime.

### Float and Thermostatic Traps

The float and thermostatic (F&T) steam trap is often used where steam pressures vary over a wide range and where gravity may be the only force available to remove condensate. This trap, shown in Figure 14.8, has a float-controlled valve to let condensate out and a thermostatically controlled valve, shut at condensate temperatures but open at lower temperatures to let out air and other nondissolved gases. It can be used where continuous condensate flow is desirable, unlike the bucket traps, which yield only intermittent flow.

**Float and Thermostatic Traps, Troubleshooting and Maintenance Procedures.** In case of no steam flow, the float may have developed a leak or have collapsed, or the air valve may be plugged. The mechanism may be worn and in need of replacement. If the trap is blowing live steam, the air hole may be blocked open because of wear or by a piece of scale, or the trap may be filled with dirt. To check any of these, valve off the trap and examine or remove the trap. Freezing can also be a problem.

**Fig. 14.8 Float and thermostatic steam trap.** (Courtesy of Armstrong Machine Works.)

## Disk Traps

Another common type of steam trap is the disk trap, one example of which is shown in Figure 14.9. When condensate or air enters from the left, it pushes up the disk and flows out through the passage on the right. When steam enters the trap, it passes around the edge of the disk and creates a downward pressure on the top of the trap. When this pressure is great enough, the disk drops and the trap closes. When the trap closes, the disk is no longer heated by the steam, and the steam above the trap cools, giving lower pressure against the top of the disk. When this pressure is lower than the pressure from underneath, the trap opens and the process repeats itself.

**Disk Traps, Troubleshooting and Maintenance Procedures.** The contact area between the disk and its seat can become corroded or blocked, and this can cause a problem. Other problems may include installation backward to the proper direction of steam flow. To maintain this trap, first check the installation to see if the manufacturer's directions have been followed. Then, if installation is correct, valve off the trap from live steam, remove the top, and check on the condition of the disk and its contact surface. Unlike the bucket traps, this trap has no prime, so losing its prime is not a problem.

## Thermostatic Traps

This trap operates on the general principles illustrated in Figure 14.10. In this trap, the bellows or bimetallic strips are extended when hot and contracted when cold. Since condensate is cold (relative to steam), its pressure causes the tap to contract, opening the outlet and allowing condensate and noncondensable gases such as air and carbon dioxide to escape. When these have left the trap, steam comes in and heats up the bellows or bimetallic strip. This expands, closing the outlet. As steam condenses, the quantity of condensate builds up, the valve opens again, and the cycle repeats. This trap can be used to give a continuous flow of condensate if desired.

**Thermostatic Traps, Troubleshooting and Maintenance.** Depending upon the mounting of the trap, dirt and scale may not be a problem in the outlet. If the bellows leaks and fills with condensate, the trap may fail instead. If a bimetallic strip is used instead of bellows, corrosion can be a problem. Like most other traps, failure can be caused in this trap if scale or dirt gets between the valve and its seat. Also, the dirt screen can get plugged up, causing the trap to fail closed.

## Steam Distribution System, Maintenance Frequency

The steam distribution system, including piping, steam-heated equipment, steam traps, and the condensate return system, should be checked regularly for leaks and to ensure correct operation. The frequency of inspection should vary with the age of the installation—in a new system, dirt and other foreign matter in the pipes will be deposited in strainers, steam traps, and the bottom of steam-heated equipment. After the installation of a new system, it thus makes sense to check steam traps, strainers, and drains frequently, perhaps every 3 months. As the system gets older, scale can become a problem if the water is not properly treated. If scaling is a problem, the steam distribution system will have to be maintained more often than without scaling—semiannually rather than annually, for example, on steam trap inspection. The entire steam system should be inspected carefully at least once each year for leaks and to keep steam traps and steam-heated equipment operating correctly. When this inspection and maintenance procedure is followed regularly, the amount of steam needed will be reduced, and the cost savings will be substantial.

## 14.3 MATERIALS HANDLING MAINTENANCE

Much energy is used in materials handling. Proper maintenance can reduce this amount of energy in the operation of lift trucks, conveyors, and the ancillary bat-

**Fig. 14.9 Disk-type steam trap.** (Courtesy of Armstrong Machine Works.)

**Fig. 14.10 Thermostatic steam trap (bimetal).**

teries, motors, fans, and tires. The proper maintenance actions are described in great detail in the operating and maintenance manuals of equipment manufacturers, and much of this material is condensed from those publications. For detailed instructions on the maintenance of a particular equipment item, refer to the manual for that item. The material in this chapter can, however, be used to develop check-off lists and to give a good idea of the types and amounts of maintenance required.

### 14.3.1 Electric Lift Trucks[11,12]

The general maintenance principles that apply to an electric lift truck are (1) keep the battery charged, (2) make sure that all fluid levels are where they should be, (3) check the brakes and keep them working correctly, (4) inspect the electrical equipment and keep it running, and (5) lubricate the truck according to manufacturer's specifications as follows.

#### Battery

The battery should be inspected at least weekly with specific attention paid to the specific gravity, the contacts, and the electrolyte level. The specific gravity should be between 1.250 and 1.275 and should not vary more than 0.020 between cells. Special care must be taken to see that the charger is connected up with correct polarity and the correct voltage, and smoking should be prohibited near the charging so as not to ignite the hydrogen gas liberated during charging. If the battery fails to hold a charge, a test with a hydrometer can often detect whether a particular cell is at fault; if no cell is clearly bad, have a load test performed by someone who knows the necessary safety precautions.

The battery contacts should be inspected daily for corrosion. At the same time, the wires leading to the contacts should also be inspected for cracked insulation and for other signs of wear. The level of electrolyte should be inspected daily and any deficiencies made up using distilled water.

#### Hydraulic Components

The level of hydraulic oil should be inspected daily and filled if it is below the lower mark on the dipstick. It should be changed at least every 2000 hours. As the daily hydraulic level inspection is being performed, the operator should also check all hydraulic lines for leaks and fix them (or have them fixed) immediately.

The brakes should also be checked daily to see that the shoes are not dragging and that the brake pedal has the right amount of play and resistance to foot pressure. Brake shoes should be inspected and adjusted at least once a year.

#### Tires

If the tires on a particular lift truck are pneumatic, it is important that they be kept inflated to the pressures recommended in the operating manual and that the inflation be checked at least weekly Underinflation can lead to excessive wear. Underinflation also increases electrical consumption. If the tires are made of hard rubber, they should be inspected daily so that the operator knows their condition. If a tire has large chunks out of it, it may constitute a safety hazard, and the irregular ride that this problem generates may be damaging to the material being transported. If tires show irregular wear, the front end may be out of line—this should be checked and fixed.

#### General Lubrication

General lubrication should follow the schedule and procedures laid out in the operation manual for the particular lift truck being maintained. This lubrication should also include inspection, adjustment, and oiling of the hoist chain.

### 14.3.2 Lift Trucks, Non-electric

The main points to maintain on a lift truck powered by diesel, gasoline, or LPG are the engine, the hydraulic system, the transmission, and the braking system. Batteries are important, but their care is not as central to the maintenance of these lift trucks as it is to the maintenance of electric-powered trucks.

#### Engine

Special attention should be paid to the oil level, the radiator, and fan belts. The oil level should be inspected daily, and the oil and oil filter should be changed every 150 hours (unless the environment is very dirty or muddy, in which case the oil should be changed more often). The oil level on the dipstick should be between the high and low marks—if it is too low, more should be added. At the same time this check is being made, the operator should look for any oil leaks. The operator should also try to keep track of the frequency of adding oil as warning of unseen leaks. The coolant level in the radiator should be checked before each day's work, either by looking at the coolant level in the recovery bottle or by carefully removing the radiator cap to check the level there. Fan belts should be checked periodically. If they are too tight, bearing wear will be excessive; if they are too loose, insufficient cooling and/or generator power will take place. More details on the maintenance of diesel engines are given in Section 14.4.

## The Hydraulic System

If the hydraulic system fails on a lift truck, people can get hurt. Any hydraulic pumps that are run without fluid can burn out, causing a significant and usually avoidable expense. The way that such an expense can be avoided is by careful checking of the hydraulic levels daily and by daily inspection of all hydraulic connections for leaks.

## Transmission

The fluid level in the transmission should also be checked daily. This minor inconvenience can save the cost of a new transmission and the loss of the vehicle during the repair time.

## Brakes

Part of the daily check should be a quick check of the foot brake and the parking brake. The footbrake pedal should move a short distance—1/4 to 1/2 in., depending upon the truck model—before the brake engages, and the brake should respond firmly to pressure. The parking brake should be checked to see if it provides the braking necessary to act as a backup for the foot brakes. Air brakes should have the drain cock opened until no more water escapes with the air.

### 14.3.3 Conveyor Systems

Conveyor belts, in-floor towlines, and similar chain- or belt-driven equipment have three important areas where proper maintenance can save energy and money. These are in controls, drive motors and gear, and the actual moving belt or line.

## Conveyor Controls

One way to save energy on controls is to install and maintain controls that cause the conveyor to move only when there is material to be moved. This measure has saved significant amounts of energy in coal mines and in pneumatic conveying systems, for example, where in one case the total energy consumed was reduced by 90%. The additional wear and tear on starting motors was not found to be significant when these motors were maintained according to the original procedures specified at the time of purchase of the equipment. Adding a load-related starting control saves on conveyor wear and on each part of the equipment that is no longer operating continuously.

## Drive Motors and Gears

Drive motors should be checked regularly using the Tornberg procedure for motors described in Section 14.1.1. When the motor is being inspected, the rest of the driving gear should also be inspected, with the details being dependent upon the power transmission system being used by the conveyor.

## Conveyor Belts or Lines

The belt, towline, or other equipment used to move the material should be checked regularly. The kinds of problem to include in the inspection are worn belts; noisy rollers, indicating possible bearing problems; belts that are too tight or too loose; and any place that shows conspicuous wear. Such an inspection should be performed regularly and coupled with regular lubrication of appropriate points. Manufacturing specifications should be consulted as a source for specific lubrication directions for all conveying equipment.

## 14.4 TRUCK OPERATION AND MAINTENANCE[13]

Most organizations use trucks in some form, whether for coal haulage, parcel delivery, or earth moving. Wherever trucks are used, there are opportunities for saving energy and money. The three general rules for achieving these savings are: (1) match the equipment to the load, (2) keep the equipment tuned up and in good repair, and (3) turn it off if you do not need to have it running. The following section gives more details on these rules as they relate to general operation and to the operation of diesel and gasoline engines. The information presented here has been distilled from manuals produced by Cummins, Caterpillar, Detroit Diesel, and Hyster.

### 14.4.1 Management Decisions

The person supervising a company with a vehicle fleet can have a large impact upon the costs of running that fleet in at least five different areas: (1) matching the vehicle to the job, (2) maintenance of the vehicle yard, (3) instituting a system of fuel accountability, (4) taking advantage of unavoidable delays to create more efficient operation, and (5) keeping all vehicles tuned up.

## Matching the Vehicle to the Job

This takes two forms. First is the policy that assigns small passenger or pickups to personnel on one-person trips; delivery vans for local deliveries, where the demand calls for more capacity than a pickup provides; and semi-trucks only where justified by the amount of material to be moved. Within each category the same rationale is applied—use the larger vehicles only when their use is necessitated by the load. Once the vehicles have been chosen or assigned based on their intended use, the second rule can come into play—if you do not need all the power of the vehicle now, see if it can be modified so that the maximum power used is lowered. This can happen where large earth-moving or transporting equipment has been designed for a 10% grade and where actual conditions do not have more than a 7% grade. If the actual requirement is less than the equipment was designed for, the governor or the fuel pump can be modified to limit the horsepower available to the operator.

### Maintaining the Vehicle Yard

Mud resists vehicle travel and makes maintenance into a chore. The effect of mud, whether it is on a haul road or in a vehicle yard, can be seen in Figure 14.11. If a significant part of your fleet travel is along roads under your maintenance or in a yard you control, this figure can help you decide whether to surface the road or yard or how often to grade the road. Particularly where your vehicles have 30% or more of their travel off the road, the savings can be substantial.

### Instituting Fuel Accountability

If your company does not have a system accounting for the use of every gallon of fuel, you probably could benefit from instituting such a program. The program has two benefits: It enables you to determine which of your vehicles is using the most fuel, and it decreases losses to employee vehicles. If you can locate the vehicles that are least efficient, you can start the process to find out whether something is wrong with the vehicle, whether some operating procedures need to be changed, or whether the particular design is inefficient. In any case, you have a better idea of where your vehicle fuel is going.

### Taking Advantage of Bottlenecks

If your facility has certain inherent bottlenecks, such as a power shovel that has less capacity than the trucks serving it, it may be possible to have these work to your advantage. In the example of the shovel, fuel can be saved by slowing down on the return trip from the dumping area, using the time that would be wasted waiting for the shovel in the additional time along the haul road. Make adjustments to the vehicle engine if operators are unwilling to conform to these practices.

### Keep Vehicles Tuned Up

When a 1% improvement in vehicle efficiency can create a savings of $1 per day for each vehicle, it makes sense to keep vehicles tuned up. Keeping a vehicle fleet in tune saves three ways: in energy consumption, in frequency of downtime and the repairs involved, and in the cost of repairs that do have to be made. The tune-up procedure for each vehicle is prescribed by the manufacturer.

### 14.4.2 Daily Maintenance Practices

In addition to the general maintenance practices described above, daily procedures performed by the operators can create substantial savings in fuel and operating costs. In addition to the mandatory inspection of such safety items as brakes, the horn, and lights (if appropriate), a daily inspection should be made of tires, batteries, air hoses, fuel and lubricant levels, and air filters.

### Tires

Proper inflation of tires reduces rolling resistance and thus reduces energy consumption. Figure 14.11 shows a typical curve relating inflation and fuel efficiency. In addition to the fuel savings, proper inflation increases tire life and decreases the total cost per mile of running each vehicle.

### Batteries

If one cell of a battery is weaker than the other cells, the generator uses more power than if all cells were in good condition. This additional power cuts into fuel efficiency. To avoid this problem, the water level in each cell should be checked daily and filled with distilled water whenever the level is below the prescribed fill mark, and the battery voltages should be checked frequently.

### Air Hoses

These should be checked for leaks as soon as they are under pressure. Air leaks cause

**Fig. 14.11 Gasoline consumption as a function of rolling resistance.**

the air compressor to use more power than necessary and create another waste of fuel.

### Fuel and Lubricant Levels

It is generally recommended that fuel tanks not be filled to more than 95% of the tank capacity. The extra 5% is to allow for expansion of the fuel as it heats up. If the tanks are filled to the top, this expansion can cause an overflow of fuel, a waste and possible safety hazard. Lubricant levels should be checked routinely to detect leaks and to keep proper lubrication in the engine, transmission, and other systems.

### Air Filters

An engine uses many gallons of air for every gallon of fuel burned, so keeping the air clean is critical. If the air filter gets clogged, the engine must use additional amounts of energy to overcome the resistance and thus extra fuel is consumed; if the air filter gets too dirty, the engine will not run. Also, particles of the filter or dirt particles may be drawn into various parts of the engine, causing unnecessary wear.

### 14.4.3 Operating Practices

In addition to the daily maintenance check described above, each operator can do several other things to decrease the fuel consumption of his or her vehicle. These measures include eliminating unnecessary idling, warming the engine correctly, choosing the most economical speeds for a given load, keeping the hydraulic system above stall speed, and keeping the cooling temperature within specified limits. The following paragraphs discuss these actions in more detail.

### Eliminating Unnecessary Idling

The three largest makers of diesels are unanimous in condemning unnecessary idling as a waste of energy and as hard on the engine, although some idling is necessary.

If equipment will be idled for more than 10 to 15 minutes, the cost for shutdown and restarting the engine will generally be less than the cost of fuel for idling. The engine should not be shut down immediately after a hard run, however, but should be allowed to idle for a few minutes to allow engine temperatures to drop. Otherwise, a surge of hot coolant to the heads can cause damage. Also, turbocharger damage can occur if coolant oil and air flow are shut off too abruptly.

### Warming Up the Engine

When starting, the engine should be warmed so that the coolant is in the recommended operating range before putting a load on the vehicle. This is recommended for diesel engines. For gasoline engines, check

with a reputable dealer to find whether this warm-up period is necessary.

### Choosing the Most Economical Speed

Every vehicle operator should know the most economical rpm range to achieve good fuel economy while delivering the desired horsepower. The best rpm range can be determined from performance curves, such as Figure 14.12, and from the fuel consumption map, such as Figure 14.13. If these curves describe the engine operated by a particular operator, that person should know that operating at 1700 rpm rather than 2300 rpm saves approximately 8% in fuel. Similarly, mechanical drive equipment should be operated in the highest gear practical. Since the economical operation of a vehicle fleet may enable the company to stay in business, saving this amount of fuel should be important to any operator.

### Operating the Hydraulic System at the Right Speed

Care should be taken to operate the hydraulic system with enough power that the pumps do not stall. Stalling a hydraulic pump creates heat but does not get work done. Similarly, if operating a converter, the operator should downshift to get more power rather than allow the converter to stall.

### Maintaining Recommended Coolant Temperatures

The operator should keep the coolant temperature in the range recommended by the manufacturers in order to achieve maximum fuel economy.

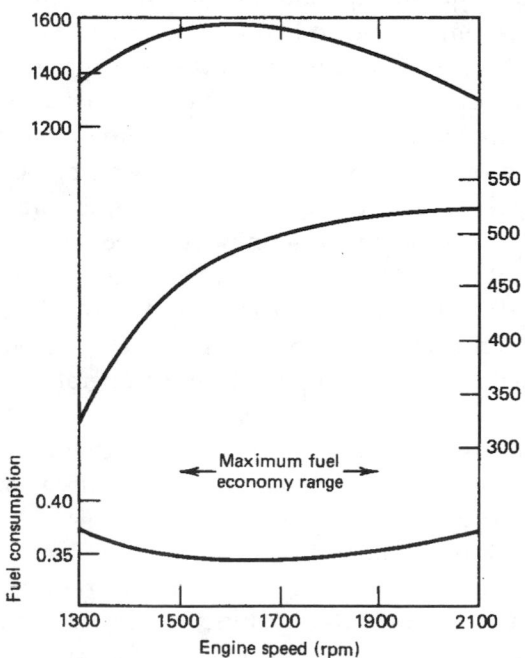

**Fig. 14.12 Typical diesel performance curve.** (Courtesy of Cummins Engine Company.)

**Fig. 14.13 Fuel consumption map.** (Courtesy of Cummins Engine Company.)

## 14.5 MEASURING INSTRUMENTS

Instruments serve two functions in maintenance—checking equipment condition and on-line monitoring. It is often impossible to know whether some equipment is functioning properly without measuring a temperature, pressure, or other parameter associated with the operation of the equipment. For example, the only indication that a steam trap is not working properly may be the absence of a regular clicking as the trap opens and closes, and this noise may be detectable only with a stethoscope. A similar use of instruments is checking on the effectiveness of repairs. Another function of instrumentation is to observe major system operating parameters so that equipment can be operated with the minimum effective use of energy. Instrumentation for this purpose is generally mounted permanently, with gauges readable either at the equipment or at some central location.

Measuring instruments used in maintenance can be classified by type or by function—by type according to what they measure, by function according to what system they serve. A classification of instruments into these categories is shown in Table 14.18, followed by descriptions of each instrument in the table.

### 14.5.1 Electrical Measurements

Electrical measurements can be used to assess the condition of individual items of equipment and to analyze the energy consumption patterns of the entire facility. Troubleshooting and monitoring of individual units can be accomplished with portable wattmeters, multimeters, and power-factor meters. Many of these instruments are available in recording models also for permanent installations for monitoring changes in overall consumption. Safety precautions must be known and carefully observed in the installation and use of all equipment, particularly that involving electricity.

*Multimeters*

A multimeter is capable of measuring voltage, current, and resistance. These meters come in many ranges. One of the more common types is the clip-on meter capable of measuring 0 to 300 A, 0 to 600 V, and 0 to 1000 Ω. This meter typically sells for approximately $50 and is easily carried in a belt holster. It can be used to check many parts of the electrical system and to make electrical checks on motors.

*Power-Factor Meter*

This meter is used to determine the ratio between the resistive component and the total (resistive and inductive) electric power supplied. It can be used to check the power factor on equipment and to determine whether a power-factor-improvement program is working.

*Recording Ammeter*

This records the electrical consumption over a prescribed period. Its most important function is to enable someone to determine the magnitude and timing of peak load demand and the magnitude of the underlying base load (the equipment that is on all the time). A recording ammeter can also help determine the effectiveness of an ongoing energy management program. It can be used to assist operating personnel in shifting peak loads and in determining the magnitude and timing of any secondary peaks. Such a meter, installed, costs $200 to $1000 and can provide a payback period of one to two months. Similar meters are available to record watts, volts, and power factor.

*Wattmeter*

A wattmeter is used to determine directly the amount of power used by a piece of equipment or by a facility. It can be used to analyze electrical consumption by enabling maintenance personnel to first determine the total power being used in an area, then find what

**Table 14.18 Instruments for use in energy management.**

| System | Instrument | Portable | Mounted | Parameter Measured |
|---|---|:---:|:---:|---|
| Building envelope | Infrared photography | × | | Heat loss |
| Boilers | See Chapter 5 and Section 14.5 | | | |
| Steam traps | Thermometer (sensitive) | × | | Temperature differences between inlet and outlet |
| | Stethoscope | × | | Opening and closing noise |
| Heating, ventilating, and air conditioning | Flow hood | × | | Air flow rates |
| | Pitot tube | × | | Air flow rates |
| | Inclined-tube manometer | | × | Pressure differential between two points |
| | Bourdon tube | × | × | Pressure |
| | Pocket thermometer | × | | Temperature of rooms, pipes, etc. |
| | Orifice flowmeters | | × | Flow rates of air or steam |
| | Psychrometer | × | × | Humidity |
| Electrical | Ammeter, recording | | × | Electricity usage, peaks |
| | Ammeter, clip-on, with probes | × | | Voltage, current, resistance |
| | Wattmeter, recording | | × | Power consumption |
| | Power-factor meter | × | × | Electrical power factor |
| Lighting | Industrial lightmeter | × | | Light levels |
| Hot water | Thermometer | × | × | Temperature |
| Air compressors | Pressure gauges | | × | Oil pressure, air pressure |
| | Stethoscope with gauge | × | | Bearing wear on motors |
| | Multimeters | × | | Voltage balance on three-phase motors |
| | Strobe light | × | | Motor vibration |
| | Infrared film and camera | × | | Bearing wear |

equipment is using the power, then turn off machines and shift loads from peak periods to off-peak periods.

### 14.5.2 Light Measurements

Reducing unnecessary lighting has psychological impacts that can far exceed any direct impact on energy consumption. To convince people that their areas are overlit or underlit, however, it is necessary to have well-established lighting standards and a way to compare existing conditions against those standards. The standards have been provided by the Illuminating Engineering Society in the *IES Handbook* and are discussed in Chapter 13 of this book.

#### Industrial Light Meter

This instrument is typically designed to measure incident light directly and can be used to measure reflected light and light transmittance as well. An industrial light meter differs from a photographic light meter in two ways: (1) it reads directly in footcandles

in scales that are approximately linear rather than the logarithmic scales used in photographic light meters; (2) it is designed to receive light from wide angles rather than focusing on a particular part of the viewing field. Such a meter is small, light, and inexpensive, and sufficiently useful that it should be part of the equipment of every maintenance supervisor.

### 14.5.3 Pressure Measurements

If equipment is not operating in its proper pressure range, it can be damaged or the equipment it serves can be damaged. This is true whether the pressure being measured comes from oil (compressors or lift trucks), fuel (boilers), steam (boilers), or other sources. Measuring pressure using portable equipment generally requires that the equipment to be measured be equipped with a fitting specifically designed for pressure-sensing equipment. With such fittings in place, it is possible to use a Bourdon tube or a diaphragm gauge for routine inspections. Where information is to be available at any time, or when the readings are to be taken so often that portable pressure-sensing equipment becomes a nuisance, it may be desirable to permanently install a Bourdon tube or a diaphragm gauge, or to rely on a manometer.

### Bourdon Gauge

This common pressure gauge consists of a curved tube closed at one end with the other end connected to the pressure to be measured. When the pressure inside the tube is greater than the pressure outside the tube, the tube tends to straighten, and the amount of change in length or curvature can be translated directly into a gauge reading. Such gauges are available in many pressure ranges and accuracies.

### Diaphragm Gauge

If the pressure inside a bellows or on one side of a diaphragm is greater than the pressure outside the bellows or on the other side of the diaphragm, the bellows or the diaphragm will move. The amount of movement is related to the pressure between the inside and outside the bellows (or on the different sides of the diaphragm). These pressure gauges are also very common.

### Manometer

If a glass tube has liquid in it, and if one end is open to the air and the other end is exposed to a pressure other than air pressure, the end with the higher pressure will have a lower liquid level. This kind of gauge can be mounted across a filter bank to indicate when a clogged filter is causing pressure to build up, or it can be mounted in a place where a much higher pressure is to be measured. If the tube is inclined, a smaller pressure difference is detected, as in the inclined-tube manometer. These gauges are easy to read, easy to maintain (the glass must be kept clean, and the inlet and outlet holes must be kept clear of debris), and inexpensive.

### 14.5.4 Stack-Gas Analysis

As explained in Section 14.2.1, proper maintenance of boilers depends heavily on knowledge of stack-gas composition. If there is too much molecular oxygen, the boiler is operating inefficiently; if there is too much carbon monoxide or too much smoke, the boiler is operating inefficiently and creating an operating hazard. Thus it is important to keep track of $O_2$, CO, and smoke. Three types of monitoring equipment meet these needs: Orsat kits, permanently mounted meters, and smoke detectors.

### Orsat Apparatus

The Orsat analysis kit consists of three tubes filled with potassium hydroxide, potassium pyrogallate, and cuprous chloride, respectively. Flue gas is introduced into all three tubes, and the amount of carbon dioxide, oxygen, and carbon monoxide removed in the first, second, and third tubes, respectively, indicates the proportion of those gases in the flue gas. Any gas remaining is assumed to be nitrogen. This apparatus is capable of accurate readings when representative samples of flue gas have been taken from different points in the flue stream, when the apparatus does not leak, and when the operator is well trained in its use.[14]

### Smoke Detectors

Smoke detectors work by comparing the amount of light going through a sample of smoke with some standard shades. If a Ringlemann scale is used, the smoke number is between 1 and 4, if the Bacharach scale, between 1 and 9. Smoke detectors can be either portable or permanently mounted.

### 14.5.5 Temperature Measurement

Temperature measurements are essential to energy management and proper maintenance in at least four situations: for comfort, to determine where heat is leaking from a building, to define abnormally hot areas in a machine, and to use in the analysis of boiler operations and of industrial operations using process heat. Temperatures in these situations call for instruments such as a pocket thermometer, infrared photography, and permanently mounted devices possibly using thermocouples.

### Pocket Thermometers

Every person who must set thermostats in a building and listen to the complaints of uncomfortable people needs one of these. It is possible to get a rugged thermometer, small enough to fit into a shirt pocket and accurate to within ± 1/2°F between 0 and 220°F, for $12 to $16. Such a thermometer can be used to calibrate thermostats, and to check complaints. By placing the heat-sensing end on a hot pipe, it is possible to estimate the temperature of the material in the pipe or to determine that it is beyond the temperature range of the thermometer. This thermometer can also be used to check the temperature of water coming from various points of a culinary hot-water system. The thermometer is small and inexpensive, and its value as a tool in energy management makes it indispensable.

## Infrared Equipment

This equipment works by sensing infrared radiation, a kind of radiation given off by warm materials. Infrared equipment, whether photographic or using a television-like device, senses differences in temperature. It can therefore detect heat leaks that are not revealed by a visual inspection. Infrared examinations have been useful in detecting areas of major heat loss in buildings or between buildings where steam leaks were occurring. Such inspections have also proven valuable in detecting areas of unseen friction in motors and thus finding problem areas before the problems have become major.

## Thermocouples[15]

To create a thermocouple, two wires or strips of different material are joined at both ends. Then one end is kept at a constant temperature and the other end is used in the temperature probe. If the ends are at different temperatures, a voltage difference will occur between the ends. Measuring this voltage provides a way to estimate the temperature difference between the two ends. Instruments using this principle are found in many places, particularly where a permanent meter is desired for temperature in a remote or inaccessible place.

## 14.5.6 Velocity and Flow-Rate Measurement

In many situations, it is necessary to know the flow rate of some substance, such as air or steam, to determine where energy is being used. For example, a factory with several buildings and a central steam system may not be metered so that the steam consumption of each building can be determined. As another example, it is generally difficult to estimate the amount of air being moved by a fan in an air-conditioning system without making some measurement of air velocity. Three types of measuring instruments used are flow hoods, pitot tubes, and orifice plates.

## Flow Hoods

A flow hood resembles an inverted pyramid with the top replaced by a small cube. The inverted pyramid is made of cloth treated to minimize air leaks. The small cube is a turbine that generates current which is measured by an attached meter. In practice, the opening of the hood is placed over the grill emitting the air. Air is forced to the base of the pyramid, turns the turbine, and generates electricity. The meter is calibrated in ft/min (or m/sec). Since the cross-sectional area is known at the point where the velocity is being measured, the number of ft$^3$/min (or m$^3$/sec) can be immediately calculated.

## Pitot Tubes

The pitot tube operates on the principle that air flow across the end of an open tube creates a pressure drop, and a measurement of this pressure drop can be converted into a measurement of the air velocity at the end of the tube. Pitot tubes have been used in applications ranging from air flow rates from a duct to determining the air speed of an airplane. They can be used to estimate the air flow velocity across a duct and may be more convenient than flow hoods for some applications. Care must be taken, however, that enough readings are made to give a representative velocity profile—one reading is not enough.

## Orifice Plates[15]

An orifice plate consists of a disk with a hole of known diameter mounted in a pipe or duct with a manometer attached to the pipe upstream and downstream from the orifice plate. Since the hole in the orifice plate is always smaller than the inside diameter of the pipe, the pressure downstream of the orifice plate is smaller than the pressure upstream of the plate. This pressure difference can be used together with the diameter of the orifice and the inside diameter of the pipe and the density of the material to give a value for the flow rate.

## 14.5.7 Vibration Measurement

Vibration is found in most mechanical devices that move. Sometimes this noise is helpful, as in the case of the noise caused by a bucket steam trap opening and closing. Often, however, an increase in vibration of a machine is an indication that something is going wrong. Among the many instruments that can be used to check vibrations, two are of particular value in the maintenance associated with energy management: the stethoscope and the stroboscope.

## Stethoscope

A stethoscope brings noise from a particular place to the ears of the observer and by isolating the source seems to amplify the noise. Two valuable uses for this are in the Tornberg procedure for monitoring motor bearings, described in Section 14.1, 1, and in the procedure for monitoring steam trap operation of steam traps described in Section 14.2.3.

## Stroboscope

A stroboscope is a flashing light whose flashes occur at regular intervals that can be changed at the will of the operator. When the flashes occur at the same time that a particular part of rotating machinery passes one point, the flashes make the machinery appear to stand still. Thus any lateral vibrations of the equipment appear in slow motion and can be examined in detail. This procedure also shows cracks that open up only when the machinery is in motion. By detecting incipient trouble before it causes the equipment to be shut down, the stroboscope enables maintenance personnel to order replacement parts and to schedule corrective maintenance at convenient times rather than at the times dictated by equipment breakdowns.

## 14.6 SAVING ENERGY DOLLARS IN MATERIALS HANDLING AND STORAGE

The earlier parts of this chapter have dealt with various topics in energy-efficient maintenance of equipment and facilities. Two other topics of interest not covered elsewhere in this handbook are energy management in materials handling and new devices for energy cost savings. This section covers the first of those topics.

Section 14.3 discussed the maintenance of certain kinds of materials handling equipment, specifically lift trucks and conveyor systems. In addition to proper maintenance, however, there are many operating changes that together can save a large fraction of the materials handling energy cost. These cost savings can be realized if a systematic approach is followed. One approach is to answer these four questions:

1.  What is energy being used for now, and is all of this use productive?
2.  How much electricity, gas, and oil is used now, and can this use or its cost be reduced by modifying equipment usage?
3.  Can the working hours or the locations of people be changed to reduce energy requirements without adverse impacts?
4.  How can the energy management program be monitored? These questions are discussed in Sections 14.6.1 to 14.6.4.

### 14.6.1 Analyzing Present Energy Usage

Energy is used in three ways in materials handling and storage: to move material, to condition spaces for material, and to condition spaces for people who are moving or storing material. Energy used to move material includes fuel or electricity for lift trucks, diesel or electricity for conveyor power, or cranes and electricity for automatic storage and retrieval systems. Examples of energy to condition spaces for material includes refrigeration for vegetables, controlled tempera-

ture storage for electric components, and materials and ventilation of areas used to store volatile liquids. The energy used to condition spaces for people includes heating, cooling, and ventilating a warehouse to make it comfortable for persons working there, air conditioning a cab of a large crane, and providing a comfortable working area for the secretaries supporting the materials handling and storage functions. These three functions also interact—if gasoline is used in a warehouse lift truck, the amount of ventilation air required for personnel is 8000 ft$^3$/min per lift truck, a requirement that is not present for electric lift trucks. (Nonelectric trucks may, however, have offsetting advantages for individual applications.) Energy used for conditioning material storage spaces can also interact in other ways with energy used to condition space for people—for example, waste heat from a refrigerator has been used to heat office space.

### 14.6.2 Walk-Through Audit

The first step in analyzing energy usage in your materials handling and storage system is to find out what equipment is being used to move material and, when it is turned on, to check the conditions that are being maintained for material. Then determine where the people are located in the system and what conditions are being maintained for their benefit. To do this, three walk-through audits are recommended. The first audit is performed during working hours in an attempt to find practices that can be improved. For this audit, you should be equipped with an industrial light meter (about $20 to $30) and a pocket thermometer ($10 to $15). As you walk around the facility, you should write or record potential improvements for later analysis. These improvements can be found in answer to questions such as these:

1.  Does equipment need to be idling when no one is using it, and, if so, what is a reasonable maximum for an idling period?
2.  Is the temperature being maintained unnecessarily high or low—is it necessary to air-condition this space in the summer or to heat it in the winter?
3.  Is all the lighting necessary? Standard guidelines for warehouse lighting in the *IES Handbook*[16] range from 5 to 50 footcandles, depending upon whether the storage is in active use and upon the amount of visual effort needed to distinguish one item from another.
4.  Is a great deal of conditioned air lost whenever trucks unload at the facility?

The second walk-through audit is performed when

office personnel have left for the day. The intent of this audit is to discover equipment and lights that are on but are not needed for the security of the building or its contents. This audit should focus on equipment, lights, and unnecessary heating and cooling of office spaces. The use of a night setback—setting the thermostat to 55°F at night—can save as much as 20% of the bill for office heating and cooling.

The third audit is the 2 a.m. audit, performed sometime after most people have left the building. In this audit, look for motors that are on unnecessarily, for lights that are on but are not needed for security purposes, and for temperatures that are higher or lower than they need to be.

Completion of the three walk-through audits gives a qualitative survey of the equipment being used, lighting levels and temperatures being maintained, and some operating improvements that can be instituted. This is a good start. More information is necessary, however, before an energy management program can achieve its potential, and this is provided by the next step.

## 14.6.3 Finding and Analyzing Improvements that Cost Money

Walk-through audits can uncover operating practices that can be improved, but a different kind of analysis is needed to discover possible capital-intensive improvements. This analysis has three parts: examination of past energy bills, use of a checklist, and economic justification of the most promising improvements. (The economic analysis is discussed in Chapter 4.)

### Examining Past Bills

One purpose in examining energy bills is to determine the total amount that can be saved. If your bill for fuel for vehicles is $25,000 per year, then $25,000 is the upper annual limit on savings. Another purpose for examining bills is to find what factors are significant in the billing for your facility. If, for example, you are being charged for a low power factor on your electrical bill, power-factor improvement may be worthwhile. If your electrical bill shows a factor labeled "power" or "demand," you are being charged for the peak power you use. To find whether peak electrical demand presents an opportunity for cost savings, compute the load factor by the formula

$$\text{load factor} = \frac{\text{kWh used in billing period}}{\text{peak demand} \times \text{hours in billing period}}$$

(The billing period in days is given on the bill.) If the load factor is less than 70 to 80%, there are significant

periods of high electrical usage. To determine when these occur, have your electrical utility install a recording ammeter or install one yourself. When you find the peak usage times, examine your equipment to find what causes the peaks and see if some of this can be rescheduled at off-peak times.

A third purpose for examining past bills is to establish a base for comparison for your energy management program. For your continued economic health it is essential that you manage your energy consumption. In order to find the measures that have worked for your facility, however, you need to be able to show that these measures have caused you to fall below previous energy usage. If measures you institute do not cause the consumption to change significantly, you know that you need to do more.

### Possible Areas for Energy Consumption Improvement

As soon as your bills have been examined, you are in a position to examine possible areas where money invested can have large returns in energy cost containment. (Much of this material was taken, with permission, from Ref. 17.) These areas, in abbreviated form, are presented in Tables 14.19 to 14.22.

### 14.6.4 Monitoring

When the energy consumption base has been established and the energy management plan has begun to take effect, energy consumption must be monitored. Monitoring serves three functions: to evaluate progress, to show which measures work, and to show which measures do not work.

The first step in a monitoring program is to choose some measure of energy consumption to monitor. Each kind of energy used in the facility should be monitored, with separate graphs for gasoline, propane, gas, and electric usage (kWh) and electric demand (kW). These should be plotted for the 12 months previous to the installation of the energy management plan, where possible, to establish a base against which to judge improvement. If your facility does not have a separate electric meter or separate accounting procedures for fuel, it is probably worthwhile to install them so you can know exactly what effect your program is having on your energy consumption.

Once the energy management plan has gone into effect, the next step is to graph the consumption, month by month, to see whether the program has helped and, if so, how much. If the program has not decreased energy consumption, you can use the monitoring equipment to analyze your energy use in more detail—to find what areas or pieces of equipment are using the energy. Then

**Table 14.19 Materials handling energy savings: doors.**

| Problem | Solution |
|---|---|
| People use truck bay doors to enter and leave building | Install personnel doors<br>Benefits include less loss of heated or cooled air |
| Open passages between heated and cooled areas | Use strip curtain doors if lift trucks use the passageway, unless there is a pressure difference between the areas |
| Truck bays open to outside air | Install adjustable dock cushions |

*Source:* Ref. 17.

**Table 14.20 Materials handling energy savings: industrial trucks.**

| Problem | Solution |
|---|---|
| Ventilation is used primarily to get rid of truck fumes | 1. Modify trucks to reduce ventilation requirements if possible<br>2. Replace diesel or propane trucks by electric trucks<br>3. Replace with more efficient trucks |
| Excessive demand charge for electric power | 1. Have electric trucks charged during off-peak hours<br>2. Install drop-in dc generators |
| Excessive fuel consumption | 1. Reduce truck idle time to manufacturer's specifications<br>2. Use smaller trucks where possible<br>3. Use a communication system between dispatcher and trucks to reduce unloaded travel time<br>4. Use improved ignition and carburetion devices<br>5. Use more energy-efficient truck tires |

*Source:* Ref. 17.

**Table 14.21 Materials handling energy savings: hoists and cranes.**

| Problem | Solution |
|---|---|
| Cranes powered by compressed air | Use electrical hoists and cranes (5 hp on compressor = 1 hp on hoist) |
| Idling | Use switches to turn crane off when not in use for a preset time (5-10 min) |
| Too many moves | Package in larger unit loads |
| Most loads too small for full use of hoist or crane capacity | Size the crane or hoist to load moved most often |
| Cranes running during off hours | Turn off as part of regular closing check list |

*Source:* Ref. 17.

**Table 14.22 Materials handling energy savings: conveyors.**

| Problem | Solution |
|---|---|
| Excessive idling | 1. Install controls to keep conveyors running only when loaded<br>2. Wire conveyors to light switch |
| Motors running at less than top load | 1. Downsize motors and install slow-start controls |
| Motors burn out | 1. Institute preventive maintenance program for motors (see Section 14.1.1)<br>2. Replace single-phase with three-phase motors |

concentrate on these, and start again.

When this monitoring has shown that a particular measure is particularly effective, find out why, and copy its best features elsewhere. If a measure did not work, find out why and make sure that the bad features of it are not being duplicated elsewhere.

## 14.7 RECENT DEVELOPMENTS

Two recent developments that have had dramatic impacts on energy management are the maturing of computer-based energy maintenance management systems and the increased availability of low-cost remotely accessible sensors.

### 14.7.1 Energy Maintenance Management Systems

An ideal maintenance management system would have these attributes:

The most recent energy costs and production data would be available, broken down as much as possible into costs by areas or by product, depending upon the amount of metering that is installed. Variance reports should be available on request to use in comparing actual and projected energy costs.

A complete file of maintenance tasks should be available, with the following information for each:

- Identification number of equipment to be maintained
- Maintenance tasks to be performed, with instructions and check lists for each task
- Frequency of maintenance
- Expected maintenance duration
- Skills needed
- Equipment needed for fault diagnosis and for repair
- Priority and justification if appropriate
- Repair parts needed

This file can then be used to generate daily maintenance schedules and to project needs for repair parts. With sufficient additional cost and production data, it can also be used to estimate the total cost and benefit of the maintenance function as performed, and it can be used for planning.

It should be possible to incorporate a daily log of preventive and repair maintenance actions performed. This file would include, for each unit of equipment maintained, the following:

- Identification number of equipment to be maintained
- Date and time
- Person in charge
- Equipment condition
- Maintenance tasks performed
- Time needed for repair
- Repair parts used
- Additional relevant notes

These characteristics should be built into a management information system in such a way that reports and graphs are easily composed and retrieved. Such systems have existed for some time, and, when properly designed, they can provide a great deal of information quickly.

But the problem with such a system lies in making it usable by maintenance technicians, and in keeping it up to date. (Note: the following description comes from *Guide to Energy Management*, by Capehart, Turner, and Kennedy; Fairmont Press, 2000).

"Most maintenance people don't like spread sheets. The information is important, but it will not be collected if collecting it is more trouble than it appears to be worth. There is a solution—the hand held computer. One company (and by now probably three or four more) has developed a system with the following characteristics:

- It is easy to use by the technician in the field. The technician can record the conditions he/she has found and the actions taken to remedy the problem, without writing anything down.

- A permanent record is kept for each machine or area maintained, and this record can be accessed in the field with a minimum of effort.

- The equipment with the records is lightweight and pocket-portable.

- Data from monitoring equipment can be incorporated into the equipment database easily.

- All equipment and labor data are available for more sophisticated analysis at a central site, and the analysis results can be immediately available for use in the field.

The technology for this system includes

1. A computer capable of handling a large database and many requests for service quickly. Such computers are readily available and not expensive.

2. A portable hand-held transmitter/receiver capable of displaying information sent from the main computer and of sending information back to it in a form that the

main computer could understand and analyze. Such equipment is also available now and is compatible with most large maintenance management systems such as those by SAP.

3.    The software and hardware to tie the computer and the field units together.

Such a system has been demonstrated by Datastream (Greenville, SC) (web address: www.dstm.com), Field Data Specialists, Inc. (www.trapbase.com) and others. The advantages of such a system are many. First, the information is likely to be more accurate and more complete than information from paper-based systems. Second, it is possible to tailor the analysis of a particular machine problem to the particular machine, knowing the history of repairs on that machine. Third, good data can be kept for use in spare parts inventory calculations so that this element of repair delay is eliminated. Finally, it is easy for the technicians in the field to use, so its use is more likely than for a more cumbersome system or one based on paper."

## 14.7.2 New Sensors and Monitoring Equipment

*   Data loggers: Recently, inexpensive battery-powered instruments have become available that record measurements (temperature, relative humidity, light intensity, on/off, open/closed, voltage and events) over time. These data loggers (the name used by the Onset Computer Corporation) are small, battery-powered devices that are equipped with a microprocessor, data storage and sensor. Most data loggers utilize turn-key software on a personal computer to initiate the logger and view the collected data. Such equipment is available from the Onset Computer Corporation, at www.onsetcomp.com. The data loggers can keep track of the condition of equipment and, when used with appropriate software, can project times when preventive maintenance should be performed.

*   Ultrasonic detectors for steam trap testing. One classical way to check steam traps was the "screwdriver method," where a technician put one end of the screwdriver on the steam trap and the other end in his ear. He then listened for noises of steam and condensate flow and for unusual trap noises. This method has been made more sophisticated by the use of ultrasonic detectors—stethoscopes with amplifiers and monitors, with software to translate all noises into the audible range. They usually have headphones and frequently have meters to indicate the frequencies being detected. Ultrasonic detectors can be used to monitor the operation of steam traps

and to detect and localize steam leaks. The cost of steam saved by fixing one steam trap or one leak can frequently pay for the entire cost of the detectors. Information on ultrasonic systems can be found at www.enerchecksystems.com (Enercheck Systems, Inc.) or from other vendors.

## 14.7.3 Monitoring Systems for Energy Maintenance Management

In addition to the sensors and monitoring equipment described above, several companies specialize in systems designed to show an operator the condition of equipment in the boiler room and related areas. Many of these units are available off-the-shelf. These include:

*   Hot water reset and boiler cutout controls (www.mge.com). These adjust the heating water temperature to a lower temperature when the ambient outside air is warm, and they turn off the heating system when it is no longer necessary.

*   Boiler controls. These have been around for a long time, with the displays continually advancing in the information and timeliness they display. Some of the vendors for these are Foster-Wheeler, Honeywell, Warrick, McDonnell Miller, Tekmar, Yokogawa, Omega Engineering, Primary Flow Signal, Inc., and others. Parameters controlled and displayed typically include water levels, steam pressures within a boiler and in the steam distribution system, steam and condensate flows, and temperatures. In addition, it is possible to get monitors for pH, conductivity, and flue gas emissions including $NO_x$, CO, and $O_2$. The sensor information from these systems can be transmitted to displays at the operating stations through hard wired circuits or by the use of radio frequency transmitters and receivers. These displays assist significantly in managing the boiler and steam distribution system, and they provide continuous data for use in boiler maintenance.

*   Boiler sequencing controls. The question of which boilers to turn on for the most efficient operation is addressed with the use of boiler sequencing controls.

*   Steam traps. Maintaining steam traps has been a cumbersome process, in part because it has not been easy to determine when a steam trap is working. Armstrong has tackled this problem with their Steam Eye™ system where traps are tested continuously and failed traps are reported by a radio transmitter.

By having a probe in the steam trap, such a unit can signal if the trap is operating normally, if steam is blowing through the trap without any condensing, or if the trap is flooded with condensate and inoperative. The results of monitoring are displayed in three files: monitored traps, non-critical trap faults, and new critical trap faults. Since this monitoring system saves much of the time needed to determine the location and nature of steam trap problems, it appears to be a significant advance over the time when each trap needed to be examined by either a stethoscope or by an ear held on one end of a long screwdriver. The Armstrong company web site is www.armstrong-intl.com.

Most of the large boiler companies supply the customer with custom-tailored display and data communication systems. Most companies that design and build boilers pay particular attention to boiler control systems. A recent brochure put out by The McBurney Corporation, whose areas of expertise include waste and by-product boiler systems, has displays for the Boiler Fuel Delivery system, the Boiler Steam and Feed water treatment system, and separate modules for the turbine generator, fuel delivery, boiler burner management, and combustion control.(www.mcburney.com) The units in each of these modules are controlled by programmable logic controllers that are linked to the main control display.

## References

1. National Electrical Contractors Association and National Electrical Manufacturers Association, *Total Energy Management: a Practical Handbook on Energy Conservation and Management*, National Electrical Contractors Association, 1976.

2. *Operation and Maintenance Manual for Steel Boilers* (Including Water Treatment), The Hydronics Institute, 34 Russo Place, Berkeley Heights, N.J. 07922, 1964.

3. "Operation and Maintenance Instructions, Class 600 Industrial Cross-Flow Cooling Towers," The Marley Co., 5800 Foxridge Drive, Mission, Kans. 66202.

4. New *HVAC Energy Saving Developments*, Air Distribution Associates, Inc., 955 Lively Boulevard, Wood Dale, Ill. 60191.

5. *ASHRAE Handbook and Product Directory, 1979, Equipment*, American Society of Heating, Refrigerating and Air Conditioning Engineers, Inc., New York, 1979.

6. Thomas F. Sack, *A Complete Guide to Building and Plant Maintenance*, Prentice-Hall, Englewood Cliffs, N.J., 1971.

7. *Unit Price Standards Handbook*, Depts. of the Navy, the Army, and the Air Force, NAVFAC P-716.

8. *Steam Conservation Guidelines for Condensate Drainage*, Armstrong Machine Works, Three Rivers, Mich. 49093.

9. Dale E. Shore and Michael W. McElroy, *Guidelines for Industrial Boiler Performance Improvement*, EPA Report EPA-600/8-77-003a, Jan. 1977, U.S. Environmental Protection Agency, Industrial Environmental Research Laboratory, Research Triangle Park, N.C.

10. W.H. Axtman, *Boiler Fuel Management and Energy Conservation*, American Boiler Manufacturers Association, Suite 700, 1500 Wilson Boulevard, Arlington, VA 22209.

11. *Owner's and Operator's Guide, E-Series Sit Drive*, Hyster Company, Danville, IL 61832, 1976.

12. *Owner's and Operator's Guide, Challenger*, Hyster Company, Danville, IL 61832, 1975.

13. *Off-Highway Fuel Savings Guide*, Cummins Engine Co., Inc., Columbus, Ind., 1977.

14. *Automative Marketing Bulletin 78-25*, Cummins Engine Co., Inc., Columbus, IN.

15. *ASHRAE Handbook and Product Directory, 1977 Fundamentals*, American Society of Heating, Refrigerating and Air Conditioning Engineers, Inc., New York, 1978.

16. "IES Recommended Levels of Illumination," *IES Lighting Handbook*, 5th ed., Illuminating Engineering Society of North America, New York, 1977.

17. Wayne C. Turner, "The Impact of Energy on Material Handling," presented at the 1980 Materials Handling Institute Seminar, St. Louis, MO.

18. The MP5 Enterprise System, by Datastream Systems, Inc., 50 Datastream Plaza, Greenville, SC 29605.

# CHAPTER 15
# INDUSTRIAL INSULATION

**JAVIER A. MONT, PH.D., CEM**
Johnson Controls, Inc.
Industrial Global Accounts
Chesterfield, Missouri

**MICHAEL R. HARRISON**
Manager, Engineering and Technical Services
Johns-Manville Sales Corp.
Denver, Colorado

Thermal insulation is a mature technology that has changed significantly in the last few years. What has not changed is the fact that it still plays a key role in the overall energy management picture. In fact, the use of insulation is mandatory for the efficient operation of any hot or cold system. It is interesting to consider that by using insulation, the entire energy requirements of a system are reduced. Most insulation systems reduce the unwanted heat transfer, either loss or gain, by at least 90% as compared to bare surfaces.

Since the insulation system is so vital to energy-efficient operations, the proper selection and application of that system is very important. This chapter describes the various insulation materials commonly used in industrial applications and explores the criteria used in selecting the proper products. In addition, methods for determining the proper insulation thickness are developed, taking into account the economic trade-offs between insulation costs and energy savings.

## 15.1 FUNDAMENTALS OF THERMAL INSULATION DESIGN THEORY

The basic function of thermal insulation is to retard the flow of unwanted heat energy either to or from a specific location. To accomplish this, insulation products are specifically designed to minimize the three modes of heat transfer. The efficiency of an insulation is measured by an overall property called thermal conductivity.

### 15.1.1 Thermal Conductivity

The thermal conductivity, or $k$ value, is a measure of the amount of heat that passes through 1 square foot of 1-inch-thick material in 1 hour when there is a temperature difference of 1°F across the insulation thickness. Therefore, the units are Btu-in./hr ft² °F. This property relates only to homogeneous materials and has nothing to do with the surfaces of the material. Obviously, the lower the $k$ value, the more efficient the insulation. Since products are often compared by this property, the measurement of thermal conductivity is very critical. The American Society of Testing Materials (ASTM) has developed sophisticated test methods that are the standards in the industry. These methods allow for consistent evaluation and comparison of materials and are frequently used at manufacturing locations in quality control procedures.

### Conduction

Heat transfer in this mode results from atomic or molecular motion. Heated molecules are excited and this energy is physically transferred to cooler molecules by vibration. It occurs in both fluids (gas and liquid) and solids, with gas conduction and solid conduction being the primary factors in insulation technology.

Solid conduction can be controlled in two ways: by utilizing a solid material that is less conductive and by utilizing less of the material. For example, glass conducts heat less readily than steel and a fibrous structure has much less through-conduction than does a solid mass. Gas conduction does not lend itself to simple modification. Reduction can be achieved by either reducing the gas pressure by evacuation or by replacing the air with a heavy-density gas such as Freon®. In both cases, the insulation must be adequately sealed to prevent reentry of air into the modified system. However, since gas conduction is a major component of the total thermal conductivity, applications requiring very low heat transfer often employ such gas-modified products.

### Convection

Heat transfer by convection is a result of hot fluid rising in a system and being replaced by a colder, heavier fluid. This fluid heats, rises, and carries more heat away from the heat source. Convective heat transfer is minimized by the creation of small cells within which the temperature gradients are small. Most thermal insulations are porous structures with enough density to block radiation and provide structural integrity. As such, convection is virtually eliminated within the insulation except for applications where forced convection is being driven through the insulation structure.

### Radiation

Electromagnetic radiation is responsible for much of the heat transferred through an insulation and increases in its significance as temperatures increase. The radiant energy will flow even in a vacuum and is governed by the emittance and temperature of the surfaces involved. Radiation can be controlled by utilizing surfaces with low emittance and by inserting absorbers or reflectors within the body of the insulation. The core density of the material is a major factor, with radiation being reduced by increased density. The interplay between the various heat-transfer mechanisms is very important in insulation design. High density reduces radiation but increases solid conduction and material costs. Gas conduction is very significant, but to alter it requires permanent sealing at additional cost. In addition, the temperature in which the insulation is operating changes the relative importance of each mechanism. Figure 15.1 shows the contribution of air conduction, fiber conduction, and radiation in a glass fiber insulation at various densities and mean temperatures.

### 15.1.2 Heat Transfer

There are many texts dedicated to the physics of heat transfer, some of which are listed in the references. In its simplest form, however, the basic law of energy flow can be stated as follows:

*A steady flow of energy through any medium of transmission is directly proportional to the force causing the flow and inversely proportional to the resistance to that force.*

In dealing with heat energy, the forcing function is the temperature difference and the resistance comes from whatever material is located between the two temperatures.

$$\text{heat flow} = \frac{\text{temperature difference}}{\text{resistance to heat flow}}$$

This is the fundamental equation upon which all heat-transfer calculations are based.

### Temperature Difference

By definition, heat transfer will continue to occur until all portions of the system are in thermal equilibrium (i.e., no temperature difference exists). In other words, no amount of insulation is able to provide enough resistance to totally stop the flow of heat as long as a temperature difference exists. For most insulation applications, the two temperatures involved are the operating temperature of the piping or equipment and the surrounding ambient air temperature.

### Thermal Resistance

Heat flow is reduced by increasing the thermal resistance of the system. The two types of resistances commonly encountered are mass and surface resistances. Most insulations are homogeneous and as such have a thermal conductivity or $k$ value. Here the insulation resistance, $R_I = tk/k$, where tk represents the thickness of the insulation. In cases of nonhomogeneous products such as multifoil metallic insulations, the thermal properties of the products at their actual finished thicknesses are expressed as conductances rather than conductivities based on a 1-in. thickness. In this case the resistance $R_I = 1/C$, where $C$ represents the measured conductance.

The other component of insulation resistance is the surface resistance, $R_s = 1/f$, where $f$ represents the surface film coefficient. These values are dependent on the emittance of

**Fig. 15.1 Contribution of each mode of heat transfer. (From Ref. 15.)**

the surface and the temperature difference between the surface and the temperature difference between the surface and the surrounding environment.

Thermal resistances are additive and as such are the most convenient terms to deal with. Following are several expressions for the heat-transfer equation, showing the relationships between the commonly used $R$, $C$, and $U$ values. For a single insulation with an outer film:

$$Q = \frac{\Delta t}{R_I + R_s}$$

$$= \frac{\Delta t}{tk/k + 1/f}$$

$$= \frac{\Delta t}{1/C_I + 1/f} \quad \text{Where the insulation is not homogeneous}$$

$$= U \, \Delta t \quad \text{where } U = \frac{1}{R_{\text{total}}} = \frac{1}{R_I + R_s}$$

$U$ is termed the overall coefficient of heat transmission of the insulation system.

## 15.2 INSULATION MATERIALS

Marketplace needs, in conjunction with active research and development programs by manufacturers, are responsible for a continuing change in insulation materials available to industry. Some products have been used for decades, whereas others are relatively new and are still being evaluated. The following sections describe the primary insulation materials available today, but first, the important physical properties will be discussed.

### 15.2.1 Important Properties

Each insulation application has a unique set of requirements as it relates to the important insulation properties. However, certain properties emerge as being the most useful for comparing different products and evaluating their fitness for a particular application. Table 15.1 lists the insulation types and product properties that are discussed in detail below. One area that will not be discussed is industrial noise control. Thermal insulations are often used as acoustical insulations for their absorption or attenuation properties. Many texts are available for reference in this area.

### Table 15.1 Industrial insulation types and properties.

| Insulation Type and Forms(a) | Temp. Range (°F) | Density (lb/cu.ft) | Thermal Conductivity [Btu-in/hr-ft2-°F at Tmean (°F)] | | | | | | | | | | | | Compressive Strength (psi) at % Deformation | Fire Hazard Classification Flame-Spread- Smoke Developed | Cell Structure (Permeability and Moisture Absorption) |
|---|---|---|---|---|---|---|---|---|---|---|---|---|---|---|---|---|---|
| | | | -300 | -100 | 0 | 75 | 100 | 200 | 300 | 500 | 700 | 800 | 900 | 1400 | | | |
| Calcium silicate blocks, shapes and P/C | to 1200 | 11-15 | | | | | 0.38 | 0.41 | 0.44 | 0.52 | 0.62 | | 0.72 | | 100-250 at 5% | Non-combustible | Open cell |
| Glass fiber blankets | to 1000 | 1 | | | | 0.24 | 0.25 | 0.34 | 0.46 | 0.78 | | | | | | | Open cell |
| | | 2 | | | | 0.22 | 0.22 | 0.30 | 0.36 | 0.57 | | | | | 0.02-3.5 at 10% | Non-combustible to 25/50 | |
| Glass fiber boards | to 850 | 3 | | | | 0.22 | 0.23 | 0.27 | 0.32 | 0.49 | | | | | | | |
| Glass fiber P/C | -20 to 850 | 3 | | | | 0.22 | 0.23 | 0.31 | 0.39 | 0.62 | | | | | | | |
| Mineral fiber blocks, boards and P/C | to 1800 | 15-24 | | | | | 0.32 | 0.37 | 0.42 | 0.52 | 0.62 | | 0.74 | | 1-18 at 10% | Non-combustible to 25/50 | Open cell |
| Cellular glass blocks, boards and P/C | to 900 | 8 | 0.18 | 0.24 | 0.29 | 0.33 | 0.34 | 0.41 | 0.49 | 0.70 | | | | | 100 at 5% | Non-combustible | Closed cell |
| Expanded perlite blocks, shapes and P/C | to 1500 | 13-15 | | | | | 0.4 | 0.45 | 0.5 | 0.6 | 0.71 | | 0.83 | | 90 at 5% | Non-combustible | Open cell |
| Urethane foam blocks and P/C | (-100 to -450) to 225 | to 1.5 | | | | 0.16-0.18 | | | | | | | | | 16-75 at 10% | 25-75 to 140-400 | 95% closed cell |
| Polyisocyanurate foam blocks and P/C | to 250 | 2 | | | | 0.14 | 0.15 | | | | | | | | 17-25 at 10% | 25-55 to 100 | 85-90% closed cell |
| Phenolic foam P/C | -40 to 250 | 2-3 | | | | 0.22 | 0.26 | | | | | | | | 13-22 at 10% | 25/50 | Open cell |
| Elastomeric closed cell sheets and P/C | to 400 | 8.5-9.5 | | | | 0.29 | 0.32 | | | | | | | | 40 at 10% | 25-75 to 115-490 | Closed cell |
| Ceramic fiber blankets | to 2600 | 6-8 | | | | | | | | 0.47-0.50 | | 0.70-0.60 | | 1.20-1.80 | 0.5-1 at 10% | Non-combustible | Open cell |

(a) P/C means pipe covering.

Sources: Refs. 17, 18 and manufacturers' literature.

### Temperature-Use Range

Since all products have a point at which they become thermally unstable, the upper temperature limit of an insulation is usually quite important. In some cases the physical degradation is gradual and measured by properties such as high-temperature shrinkage or cracking. In such cases, a level is set for the particular property and the product is rated to a temperature at which that performance level is not exceeded. Occasionally, the performance levels are established by industry standards, but frequently, the manufacturers establish their own acceptance levels based on their own research and application knowledge.

In other cases, thermal instability is very rapid rather than gradual. For example, a product containing an organic binder may have a certain temperature at which an exothermic reaction takes place due to a too-rapid binder burnout. Since this type of reaction can be catastrophic, the temperature limit for such a product may be set well below the level at which the problem would occur.

Low-end temperature limits are usually not specified unless the product becomes too brittle or stiff and, as such, unusable at low temperatures. The most serious problem with low-temperature applications is usually vapor transmission, and this is most often related to the vapor-barrier jacket or coating rather than to the insulation. In general, products are eliminated from low-temperature service by a combination of thermal efficiency and cost.

### Thermal Conductivity

This property is very important in evaluating insulations since it is the basic measure of thermal efficiency, as discussed in Section 15.1.1. However, a few points must be emphasized. Since the $k$ value changes with temperature, it is important that the insulation *mean* temperature be used rather than the operating temperature. The mean temperature is the average temperature within the insulation and is calculated by summing the hot and cold surface temperatures and dividing by 2: $(t_h + t_s)/2$. Thermal conductivity data are always published per mean temperature, but many users incorrectly make comparisons at operating temperatures.

A second concern relates to products which have $k$ values that change with time. In particular, foam products often utilize an agent that fills the cells with a gas heavier than air. Shortly after manufacture, some of this gas migrates out, causing an increase in thermal conductivity. This new value is referred to as an "aged $k$" and is more realistic for design purposes.

### Compressive Strength

This property is important for applications where the insulation will see a physical load. It may be a full-time load, such as in buried lines or insulation support saddles, or it may be incidental loading from foot traffic. In either case, this property gives an indication of how much deformation will occur under load. When comparing products it is important to identify the percent compression at which the compressive strength is reported. Five and 10% are the most common, and products should be compared at the same level.

### Fire Hazard Classification

Insulation materials are involved with fire in two ways: fire hazard and fire protection. Fire protection refers to the ability of a product to withstand fire exposure long enough to protect the column, pipe, or vessel it is covering. This topic is discussed in Section 15.3.1.

Fire hazard relates to the product's contribution to a fire by either flame spread or smoke development. The ASTM E-84 tunnel test is the standard method for rating fire hazard and compares the FS/SD (flame spread/smoke developed) to that of red oak, which has a 100/100 rating. Typically, a 25/50 FHC is specified where fire safety is an important concern. Certain concealed applications allow higher ratings, while the most stringent requirements require a noncombustible classification.

### Cell Structure

The internal cell structure of an insulation is a primary factor in determining the amount of moisture the product will absorb as well as the ease in which vapor will pass through the material. Closed-cell structures tend to resist both actions, but the thickness of the cell walls as well as the base material will also influence the long-term performance of a closed-cell product. In mild design conditions such as chilled-water lines in a reasonable ambient, closed-cell products can be used without an additional vapor barrier. However, in severe conditions or colder operating temperatures, an additional vapor barrier is suggested for proper performance.

### Available Forms

An insulation material may be just right for a specific application, but if it is not manufactured in a form compatible with the application, it cannot be used. Insulation is available in different types (Ref. 19).

**Loose-fill insulation and insulating cements.** Loose-fill insulation consists of fibers, powders, granules or nodules which are poured or blown into walls or other irregular spaces. Insulating cements are mixtures of a loose material with water or other binder which are blown on a surface and dried in place.

**Flexible, semirigid and rigid insulation.** Flexible and semirigid insulation, which are available in sheets or rolls, are used to insulate pipes and ducts. Rigid insulation is available in rectangular blocks, boards or sheets and are also used to insulate pipes and other surfaces. The most common forms of insulation are flexible blankets, rigid boards and blocks, pipe insulation half-sections and full-round pipe sections.

**Formed-in-place insulation.** This type of insulation can be poured, frothed or sprayed in place to form rigid or semirigid foam insulation. They are available as liquid components, expandable pellets or fibrous materials mixed with binders.

**Removable-reusable insulation covers.** Used to insulate components that require routine maintenance (like valves, flanges, expansion joints, etc.). These covers use belts, Velcro or stainless steel hooks to reduce the installation time.

### Other Properties

For certain applications and thermal calculations, other properties are important. The pH of a material is occasionally important if a potential for chemical reaction exists. Density is important for calculating loads on support structures and occasionally has significance with respect to the ease of installation of the product. The specific heat is used together with density in calculating the amount of heat stored in the insulation system, primarily of concern in transient heat-up or cool-down cycles.

### 15.2.2 Material Description

#### Calcium Silicate

These products are formed from a mixture of lime and silica and various reinforcing fibers. In general, they contain no organic binders, so they maintain their physical integrity at high temperatures. The calcium silicate products are known for exceptional strength and durability in both intermediate- and high-temperature applications where physical abuse is a problem. In addition, their thermal performance is superior to other products at the higher operating temperatures.

#### Glass Fiber

Fiberglass insulations are supplied in more forms, sizes, and temperature limits than are other industrial insulations. All of the products are silica-based and range in density from 0.6 to 12 lb/ft$^3$. The binder systems employed include low-temperature organic binders, high-temperature organic/antipunk binders, and needled mats with no binders at all. The resulting products include flexible blankets, semirigid boards, and preformed one-piece pipe covering for a very wide range of applications from cryogenic to high temperature. In general, the fiberglass products are not considered load bearing.

Most of the organic binders used begin to oxidize (burn out) in the range 400 to 500°F. The loss of binder somewhat reduces the strength of the product in that area, but the fiber matrix composed of long glass fibers still gives the product good integrity. As a result, many fiberglass products are rated for service above the binder temperature, and successful experience indicates that they are completely suitable for numerous applications.

#### Mineral Fiber/Rock Wool

These products are distinguished from glass fiber in that the fibers are formed from molten rock or slag rather than silica. Most of the products employ organic binders similar to fiberglass but the very high temperature, high-density blocks use inorganic clay-type binders. The mineral wool fibers are more refractory (heat resistant) than glass fibers, so the products can be used to higher temperatures. However, the mineral wool fiber lengths are much shorter than glass and the products do contain a high percentage of unfiberized material. As a result, after binder burnout, the products do not retain their physical integrity very well and long-term vibration or physical abuse will take its toll.

#### Cellular Glass

This product is composed of millions of completely sealed glass cells, resulting in a rigid insulation that is totally inorganic. Since the product is closed cell, it will not absorb liquids or vapors and thus adds security to cryogenic or buried applications, where moisture is always a problem. Cellular glass is load bearing, but also somewhat brittle, making installation more difficult and causing problems in vibrating or flexing applications. At high temperatures, thermal-shock cracking can be a problem, so a cemented multilayer construction is used. The thermal conductivity of cellular glass is higher than for most other products, but it has unique features that make it the best product for certain applications.

#### Expanded Perlite

These products are made from a naturally occurring mineral, perlite, that has been expanded at a high temperature to form a structure of tiny air cells surrounded by vitrified product. Organic and inorganic binders together with reinforcing fibers are used to hold the structure together. As produced, the perlite materials have low moisture absorption, but after heating and oxidizing the organic material, the absorption increases dramatically. The products are rigid and load bearing but have lower compressive strengths and higher thermal conductivities than the calcium silicate products and are also much more brittle.

### Plastic Foams

There are three foam types finding some use in industrial applications, primarily for cold service. They are all produced by foaming various plastic resins.

**Polyurethane/Isocyanurate Foams.** These two types are rigid and offer the lowest thermal conductivity since they are expanded with fluorocarbon blowing agents. However, sealing is still required to resist the migration of air and water vapor back into the foam cells, particularly under severe conditions with large differentials in vapor pressure. The history of urethanes is plagued with problems of dimensional stability and fire safety. The isocyanurates were developed to improve both conditions, but they still have not achieved the 25/50 FHC (fire hazard classification) for a full range of thickness. As a result, many industrial users will not allow their use except in protected or isolated areas or when covered with another fire-resistant insulation. The advantage of these foam products is their low thermal conductivity, which allows less insulation thickness to be used, of particular importance in very cold service.

**Phenolic Foam.** These products have achieved the required level of fire safety, but do not offer k values much different from fiberglass. They are rigid enough to eliminate the need for special pipe saddle supports on small lines. However, the present temperature limits are so restrictive that the products are primarily limited to plumbing and refrigeration applications.

### Polyimide Foams

Polyimide foams are used as thermal and acoustical insulation. This material is fire resistant (FS/SD) of 10/10) and lightweight, so it requires fewer mechanical fastening devices. Thermal insulation is available in open-cell structure. Temperature stability limits its application to chilled water lines and systems up to 100°F.

### Elastomeric Cellular Plastic

These products combine foamed resins with elastomers to produce a flexible, closed-cell material. Plumbing and refrigeration piping and vessels are the most common applications, and additional vapor-barrier protection is not required for most cold service conditions. Smoke generation has been the biggest problem with the elastomeric products and has restricted their use in 25/50 FHC areas. To reduce installation costs, elastomeric pipe insulation is available in 6-ft long, presplit tubular sections with a factory-applied adhesive along the longitudinal joint.

### Refractories

Insulating refractories consist primarily of two types, fiber and brick.

**Ceramic Fiber.** These alumina-silica products are available in two basic forms, needled and organically bonded. The needled blankets contain no binders and retain their strength and flexibility to very high temperatures. The organically bonded felts utilize various resins which provide good cold strength and allow the felts to be press cured up to 18 lb/cu.ft. density. However, after the binder burns out, the strength of the felt is substantially reduced. The bulk ceramic fibers are also used in vacuum forming operations where specialty parts are molded to specific shapes.

**Insulating Firebrick.** These products are manufactured from high-purity refractory clays with alumina also being added to the higher temperature grades. A finely graded organic filler which burns out during manufacture provides the end product with a well-designed pore structure, adding to the product's insulating efficiency. Insulating firebricks are lighter and therefore store less heat than the dense refractories and are superior in terms of thermal efficiency.

### Protective Coatings and Jackets

Any insulation system must employ the proper covering to protect the insulation and ensure long-term performance. Weather barriers, vapor barriers, rigid and soft jackets, and a multitude of coatings exist for all types of applications. It is best to consult literature and representatives of the various coating manufacturers to establish the proper material for a specific application. Jackets with reflective surfaces (like aluminum and stainless steel jackets) have low emissivity ($\varepsilon$). For this reason, reflective jackets have lower heat loss than plain or fabric jackets (high emissivity). In hot applications, this will result in higher surface temperatures and increase the risk of burning personnel. In cold applications, surface temperatures will be lower, which could cause moisture condensation. Regarding jacketing material, existing environment and abuse conditions and desired esthetics usually dictate the proper material. Section 15.3.3 will discuss jacketing systems typically used in industrial work.

## 15.3 INSULATION SELECTION

The design of a proper insulation system is a two-fold process. First, the most appropriate material must be selected from the many products available. Second, the proper thickness of material to use must be determined. There is a link between these two decisions in that one product with superior thermal performance may require less thickness than another material, and the thickness reduction may reduce the cost. In many cases, however, the thermal values are so close that the same thickness is specified for all the candidate materials. This section deals with the process of material selection. Section 15.4 addresses thickness determination.

### 15.3.1 Application Requirements

Section 15.2.1 discussed the insulation properties that are of most significance. However, each application will have specific requirements that are used to weigh the importance of the various properties. There are three items that must always be considered to determine which insulations are suitable for service. They are operating temperature, location or ambient environment, and form required.

#### Operating Temperature

This parameter refers to the hot or cold service condition that the insulation will be exposed to. In the event of operating design temperatures that may be exceeded during overrun conditions, the potential temperature extremes should be used to assure the insulation's performance.

**Cryogenic** (–455 to –150°F). Cryogenic service conditions are very critical and require a well-designed insulation system. This is due to the fact that if the system allows water vapor to enter, it will not only condense to a liquid but will subsequently expand and destroy the insulation. Proper vapor barrier design is critical in this temperature range. Closed-cell products are often used since they provide additional vapor resistance in the event that the exterior barrier is damaged or inadequately sealed.

For the lowest temperatures where the maximum thermal resistance is required, vacuum insulations are often employed. These insulations are specially designed to reduce all the modes of heat transfer. Multiple foil sheets (reduced radiation) are separated by a thin mat filler of fiberglass (reduced solid conduction) and are then evacuated (reduced convection and gas conduction). These "super insulations" are very efficient as long as the vacuum is maintained, but if a vacuum failure occurs, the added gas conduction drastically reduces the efficiency.

Finely divided powders are also used for bulk, cavity-fill insulation around cryogenic equipment. With these materials, only a moderate vacuum is required, and in the event that the vacuum fails, the powder still acts as an insulation. It is, however, very important to keep moisture away from the powders, as they are highly absorbent and the ingress of moisture will destroy the system.

Some plastic foams are suitable for cryogenic service, whereas others become too brittle to use. They must all have additional vapor sealing since high vapor pressures can cause moisture penetration of the cell walls. Closed-cell foamed glass (cellular glass) is quite suitable for this service in all areas except those requiring great thermal efficiency. Since it is not evacuated and has solid structure, the thermal conductivity is relatively high.

Because of the critical nature of much cryogenic work, it is very common to have the insulation system specifically designed for the job. The increased use of liquefied gases (natural and propane) together with cryogenic fluids in manufacturing processes will require continued use and improvement of these systems.

**Low Temperature (–150 to 212°F).** This temperature range includes the plumbing, HVAC, and refrigeration systems used in all industries from residential to aerospace. There are many products available in this range, and the cost of the installed thermal efficiency is a large factor. Products typically used are glass fiber, plastic foams, phenolic foam, elastomeric materials, and cellular glass. In below-ambient conditions, a vapor barrier is still required, even though as the service approaches ambient temperature, the necessary vapor resistance becomes less. Above-ambient conditions require little special attention, with the exception of plastic foams, which approach their temperature limits around 200°F.

Because of the widespread requirement for the plumbing and HVAC services within residential and commercial buildings, the insulations are subject to a variety of fire codes. Many codes require a flame spread rating less than 25 for exposed material and smoke ratings from 50 to 400, depending on location. A composite rating of 25/50 FHC is suitable for virtually all applications, with a few applications requiring non-combustibility.

**Intermediate Temperature (212 to 1000°F).** The great majority of steam and hot process applications fall within this operating range. Refineries, power plants, chemical plants, and manufacturing operations all require insulation for piping and equipment at these temperatures. The products generally used are calcium silicate, glass fiber, mineral wool, and expanded perlite. Most of the fiberglass products reach their temperature limit somewhere in this range, with common break-points at 450, 650, 850, and 1000°F for various products.

There are two significant elements to insulation selection at these temperatures. First, the thermal conductivity values change dramatically over the range of mean temperatures, especially for light-density products under 18lb/cu.ft. This means, for example, that fiberglass pipe covering will be more efficient than calcium silicate for the lower temperatures, with the calcium silicate having an advantage at the higher temperatures. A thermal conductivity comparison is of value in making sure that the insulation mean temperature is used rather than the operating temperature.

The second item relates to products that use organic binders in their manufacture. All the organics will burn out somewhere within this temperature range, usually between 400 and 500°F. Many products are designed to be used above that temperature, whereas others are not. This is mentioned here only to call attention to the fact that some structural strength is usually lost with organic binder.

**High Temperature (1000 to 1600°F).** Superheated steam, boiler exhaust ducting, and some process operations deal with temperatures at this level. Calcium silicate, mineral wool, and expanded perlite products are commonly used together with the lower-limit ceramic fibers. Except for a few clay-bonded mineral wool materials, these products reach their temperature limits in this range. Thermal instability, as shown by excessive shrinkage and cracking, is usually the limiting factor.

**Refractory (1600 to 3600°F).** Furnaces and kilns in steel mills, heat treating and forging shops, as well as in brick and tile ceramic operations, operate in this range. Many types of ceramic fiber are used, with alumina-silica fibers being the most common. Insulating firebrick, castables, and bulk-fill materials are all necessary for meeting the wide variety of conditions that exist in refractory applications. Again, thermal instability is the controlling factor in determining the upper temperature limits of the many products employed.

### Location

The second item to consider in insulation selection is the location of the system. Location includes many factors that are critical to choosing the most cost-effective product for the life of the application. Material selection based on initial price only without regard to location can be not only inefficient, but dangerous under certain conditions.

**Surrounding Environment.** For an insulation to remain effective, it must maintain its thickness and thermal conductivity over time. Therefore, the system must either be protected from or able to withstand the rigors of the environment. An outdoor system needs to keep water from entering the insulation, and in most areas, the jacketing must hold up under radiant solar load. Indoor applications are generally less demanding with regard to weather resistance, but there are washdown areas that see a great deal of moisture. Also, chemical fumes, atmospheres, or spillage may seriously affect certain jacketing materials and should be evaluated prior to specifications. Direct burial applications are normally severe, owing to soil loading, corrosiveness, and moisture. It is imperative that the barrier material be sealed from groundwater and resistant to corrosion. Also, the insulation must have a compressive strength sufficient to support the combined weight of the pipe, fluid, soil backfill, and potential wheel loads from ground traffic.

Another concern is insulation application on austenetic stainless steel, a material subject to chloride stress-corrosion cracking. There are two specifications most frequently used to qualify insulations for use on these stainless steels: MIL-1-24244 and Nuclear Regulatory Commission NRC Reg. Guide 1.36. The specifications require, first, a stress-corrosion qualification test on actual steel samples; then, on each manufacturing lot to be certified, a chemical analysis must be performed to determine the amount of chlorides, fluorides, sodium, and silicates present in the product. The specific amounts of sodium and silicates required to neutralize the chlorides and fluorides are stated in the specifications.

There are many applications where vibration conditions are severe, such as in gas turbine exhaust stacks. In general, rigid insulations such as calcium silicate withstand this service better than do fibrous materials, especially at elevated temperatures. If the temperature is high enough to oxidize the organic binder, the fibrous products lose much of their compressive strength and resiliency. On horizontal piping, the result can be an oval-shaped pipe insulation which is reduced in thickness on the top of the pipe and sags below the underside of the pipe, thus reducing the thermal efficiency of the system. On vertical piping and equipment with pinned-on insulation, the problem of sag is reduced, but the vibration can still tend to degrade the integrity of the insulation.

Location in a fire-prone area can affect the insulation selection in two ways. First, the insulation system cannot be allowed to carry the fire to another area; this is fire hazard. Second, the insulation can be selected and designed to help protect the piping or equipment from the fire. There are many products available for just fireproofing such areas as structural steel columns, but in general they are not very efficient thermal insulations. When an application requires both insulation during operation and protection during a fire, calcium silicate is probably the best selection. This is due to the water of hydration in the product, which must be driven off before the system will rise above the steam temperature. Other high-density, high-temperature products are used as well. With all the products it is important that the jacketing system be designed with stainless steel bands and/or jacketing since the insulation must be maintained on the piping in order to protect it. Figure 15.2 shows fire test results for three materials per the ASTM E-119 fire curve and indicates the relative level of fire protection provided by each material.

A final concern deals with the transport of volatile fluids through piping systems. When leaks occur around flanges or valves, these fluids can seep into the insulation. Depending on the internal insulation structure, the surface area may be increased significantly, thus reducing the fluid's flash point. If this critical temperature drops below the operating temperature of the system, autoignition can occur, thus creating a fire hazard. In areas where leaks are a problem, either a leakage drain must be provided to remove the fluid or else a closed-cell material such as cellular glass should be used, since it will not absorb the fluid.

The previous discussion is intended to draw attention to specific application requirements, not necessarily to determine the correct insulation to be used. Each situation should be evaluated for its own requirements, and in areas of special concern (auto-ignition, fire, etc.) the manufacturer's representative should be called upon to answer questions specific to the product.

**Resistance to Physical Abuse.** Although this issue is related to location, it is so important that it needs its own discussion. In commercial construction and many light industrial facilities, the pipe and equipment insulation is either hidden or isolated from any significant abuse. In such cases, little attention need be given to this issue. However, in most heavy industrial applications, physical abuse and the problems caused by it are matters of great concern.

Perhaps by definition, physical abuse differs from physical loading in that loading is planned and designed for, whereas abuse is not. For example, with cold piping, pipe support saddles are often located external to the insulation and vapor barrier. This puts the combined weight of the pipe and fluid onto the lower portion of the insulation. This is a designed situation, and a rigid material is inserted between the pipe and the saddle

**Fig. 15.2 Fire resistance test data for pipe insulation.** (Used by permission from Ref. 6.)

to carry the load. However, if a worker decides to use the insulated pipe as a scaffold support, a walkway, or a hoist support, the insulation may not be designed to support such a load and damage will occur. A quick walk-through of any industrial facility will show much evidence of "unusual" or "unanticipated" abuse. In point of fact, many users have seen so much of this that they now design for the abuse, having determined that it is "usual" for their facility.

The effects of abuse are threefold. First, dented and creased aluminum jacketing is unsightly and lowers the overall appearance of the plant. Nonmetal jackets may become punctured and torn. The second point is that wherever the jacketing is deformed, the material under it is compressed and as a result is a much poorer insulation since the thickness has been reduced. Finally, on outside lines some deformation will undoubtedly occur at the jacketing overlap. This allows for water to enter the system, further degrading the insulation and reducing the thermal efficiency.

In an effort to deal with the physical abuse problem, some specifications call for all horizontal piping to be insulated with rigid material while allowing a fibrous option on the vertical lines. Others modify this specification by requiring rigid insulation to a height of 6 to 10 ft on vertical lines to protect against lateral abuse. Still, in facilities that have a history of a rough environment, it is most common to specify the rigid material for all piping and equipment except that which is totally enclosed or isolated.

As previously mentioned, the primary insulation material choice is between rigid and nonrigid materials. Calcium silicate, cellular glass, and expanded perlite products fit the rigid category, whereas most mineral wool and fiberglass products are nonrigid. Over the years, the calcium silicate products have become the standard for rigorous services, combining good thermal efficiency with exceptional compressive strength and abuse resistance. The maintenance activities associated with rigid insulations are significantly less than the maintenance and replacement needs of the softer insulations in abuse areas. The costs associated with this are discussed in a later section. However, it is also recognized that often, maintenance activities are lacking, which results in a deteriorated insulation system operating at reduced efficiency for a long period of time.

### Form Required

The third general category to consider is the insulation form required for the application. Obviously, pipe insulation and flat sheets are manufactured for specific purposes, and the lack of a specific form eliminates that product from consideration. However, there are subtle differences between form that can make a significant difference in installation costs and system efficiency.

On flat panels, the two significant factors are panel size and the single-layer thickness available. A fibrous 4 × 8 ft sheet is applied much more rapidly than four 2 × 4 ft sheets, and it is possible that the number of pins required might be reduced. In regard to thickness, if one material can be supplied 4 in. thick as a single layer as opposed to two 2-in.-thick layers, the first option will result in significant labor savings. The same holds true for 18-in.- vs. 12-in.-wide rigid block installation.

Fibrous pipe insulations have three typical forms: one-piece hinged snap-on, two piece halfsections, and flexible blanket wraparound. For most pipe sizes, the one-piece material is the fastest to install and may not require banding if the jacket is attached to the insulation and secured to itself. Two-piece products must be wired in place and then subsequently jacketed in a separate operation. Wrap-around blankets are becoming more popular, especially for large-diameter pipe and small vessels. They come in standard roll lengths and are cut to length on the job site.

Rigid insulations also have different forms, which vary with the manufacturer. The two-piece half-section pipe insulation is standard. However, these sections can be supplied prejacketed with aluminum, which in effect gives a one-piece hinged section that does not need a separate application of insulation and jacket. Also, thicknesses up to 6 in. are available, eliminating the need for double-layer applications where they are not required for expansion reasons. The greatest diversity comes in the large-diameter pipe sizes. Quads (quarter sections)

are available and are both quicker to install and thermally more efficient than is scored block bent around the pipe. Similarly, curved radius blocks are available for sizes above quads and provide a better fit than does flat beveled block or scored block.

The important point is that the available forms of insulation may well affect the decision as to which material to select and which manufacturer to purchase that material from. It is unwise to assume that all manufacturers offer the same sizes and forms or that the cost to install the product is not affected by its form.

### 15.3.2 Cost Factors

Section 15.3.1 dealt with the process of selecting the materials best suited for a specific application. In some cases the requirements are so stringent that only one specific material is acceptable. For most situations, however, more than one insulation material is suitable, even though they may be rank-ordered by anticipated performance. In these cases, several cost factors should be considered to determine which specific material (and/or manufacturer) should be selected to provide the best system for the lowest cost.

### Initial Cost

In new construction, the owner is usually interested primarily in the installed cost of the insulation system. As long as the various material options provide similar thermal performance, they can be compared on an equal basis. The contractor, on the other hand, is much more concerned about the insulation form and its effect on installation time. It may be of substantial benefit to the contractor to utilize a more costly material that can be installed more efficiently for reduced labor costs. In a highly competitive market, these savings need to be passed through to the owner for the contractor to secure the job. The point is that the lowest-cost material does not necessarily become the lowest-cost installed material, so all acceptable alternatives should be evaluated.

### Maintenance Cost

To keep their performance and appearance at acceptable levels, all insulation systems must be maintained. This means, for example, that outdoor weather protection must be replaced when damaged to prevent deterioration of the insulation. If left unattended, the entire system may need to be replaced. In a high-abuse area, a nonrigid insulation may need to be replaced quite frequently in order to maintain performance. Aesthetics often play an important part in maintenance activities, depending on the type of operation and its location. In such cases there is benefit in utilizing an insulation that maintains its form and if possible aids the jacketing or coating in resisting abuse.

The trade-off comes between initial cost and maintenance cost in that a less costly system may well require greater maintenance. Unfortunately, the authorities for initial construction and ongoing maintenance are often split, so the owner may not be aware of the future consequences of the initial system selection. It is imperative that both aspects be viewed together.

### Lost Heat Cost

If the various suitable insulations are properly evaluated, a more thermally efficient product should require less thickness to meet the design parameters However, if a common thickness is specified for all products, there can be a substantial difference in heat loss or gain between the systems In such a case, the more efficient product should receive financial credit for transferring less heat, and this should be considered in the overall cost calculations

Referring to the previous discussion on maintenance costs, there was an underlying assumption that maintenance would be performed to the extent that the original thermal efficiency would be maintained. In reality, maintenance is usually not performed until the situation is significantly deteriorated and sometimes not even then. The result of this is reduced thermal efficiency for much of the life of a maintenance-intensive system. It is very difficult to assign a figure to the amount of additional heat transfer due to deterioration. In a wet climate, for example, a torn jacket will allow moisture into the system and drastically affect the performance. Conversely, in a dry area, the insulation might maintain its performance for quite some time. Still, when dealing with maintenance costs, it is a valid concern that systems in need of maintenance generally are transferring more heat at greater cost than are systems requiring less maintenance.

### Design Life

The anticipated project life is the foundation upon which all costs are compared. Since there are trade-offs between initial cost and ongoing maintenance and heat-loss costs, the design life is important in determining the total level of the ongoing costs. To illustrate, consider the difference between designing a 40-year power plant and a two-year experimental process. Assuming that the insulation in the experimental process will be scrapped at the close of the project, it makes no sense to use a more costly insulation that has lower maintenance requirements, since those future benefits will never be realized. Similarly, utilizing a less costly but maintenance-intensive system when the design life is 40 years makes little sense, since the additional front-end costs could be regained in only a few years of reduced maintenance costs.

### 15.3.3 Typical Applications

This section is designed to give a brief overview of commonly used materials and application techniques. For a detailed study of application, techniques, and recommendations, see Ref. 7, as well as the guide specifications supplied by most insulation manufacturers.

### The Heat Plant

*Boilers* are typically insulated with fiberglass or mineral wool boards, with some usage of calcium silicate block when extra durability is desired. Powerhouse boilers are normally insulated on-site with the fibrous insulation being impaled on pins welded to the boiler. Box-rib aluminum is then fastened to the stiffeners or buckstays as a covering for the insulation. In most commercial and light industrial complexes, package boilers are normally used. These are insulated at the factory, usually with fiberglass or mineral wool.

*Breechings* and other high-temperature duct work are insulated with calcium silicate (especially where traffic patterns exist), mineral wool, and high-temperature fiberglass. On very large breechings, prefabricated panels are used, as discussed in the following paragraph. H-bar systems supporting the fibrous materials are common, with the aluminum lagging fastened to the outside of the H-bar members. Also, many installations utilize roadmesh over the duct stiffeners, creating an air space, and then wire the insulation to the mesh substrate. Indoors, a finish coat of cement may be used rather than metal lagging.

*Precipitators* are typically insulated with prefabricated panels filled with mineral wool or fiberglass blankets. For large, flat areas, such panels provide very efficient installations, as the panels are simply secured to the existing structure with self-tapping screws. H-bar and Z-bar systems are also used to contain the fibrous boards.

*Steam piping* insulation varies with temperature and location, as discussed earlier. Calcium silicate wired in place and then jacketed with corrugated or plain aluminum is very widely used. The jacketing is either screwed at the overlap or banded in place. Fiberglass is used extensively in low-pressure steam work in areas of limited abuse. Mineral wool and expanded perlite can also be used for higher-temperature steam, but calcium silicate is the standard.

## Process Work

*Hot process piping and vessels* are typically insulated with calcium silicate, mineral wool, or high-temperature fiberglass. Horizontal applications are generally subject to more abuse than vertical and as such have a higher usage of calcium silicate. Many vessels have the insulation banded in place and then the metal lagging banded in place separately. Most vessel heads have a cement finish and may or may not be subsequently covered with metal. Recent product developments have provided a fiberglass wraparound product for large-diameter piping and vessels. This flexible material conforms to the curvature and need only be pinned at the bottom of a horizontal vessel. Banding is then used to secure both the insulation and the jacketing. In areas of chemical contamination, stainless steel jacketing is frequently used.

*Cold process vessels and piping* also use a variety of insulations, depending on the minimum temperature and the thermal efficiency required. Cellular glass is widely used in areas where the closed-cell structure is an added safeguard (the material is still applied with a vapor-barrier jacket or coating). It is also used wherever there is a combined need for closed-cell structure and high compressive strength. However, the polyurethane materials are much more efficient thermally, and in cryogenic work, maximum thermal resistance is often required. Multiple vapor barriers are used with the urethanes to prevent the migration of moisture throughout the entire system. In all cold work, the workmanship, particularly on the outer vapor barrier, is extremely critical. There are many other specially engineered systems for cryogenic work, as discussed in Section 15.3.1.

*Fluid storage tanks* located outdoors are typically insulated with fiberglass insulation. Prefabricated panels are either installed on studs or banded in place. Also, the jacketing can be banded on separately over the insulation. A row of cellular glass is placed along the base of the tank to prevent moisture from wicking up into the fiberglass. Sprayed urethane is also used on tanks that will not exceed 200°F but a trade-off exists between cost efficiency and the long-term durability of the system.

*Tank roofs* are a problem because of the need for a rigid walking surface as well as a lagging system that will shed water. Many tops are left bare for this reason, whereas others utilize a spray coating of cork-filled mastic, which provides only minimal insulation. Rigid fiberglass systems can be made to work with a well-designed covering system that drains properly. The most secure system is to use a built-up roofing system similar to those used on flat-top buildings. The installation is generally more costly, but acceptable long-term performance is much more probable.

## HVAC System

*Duct work* constructed of sheet metal is usually wrapped with light-density fiberglass with a preapplied foil and kraft facing. The blanket is overlapped and then stapled, with tape or mastic being applied if the duct flow is cold and a vapor barrier is required. Support pins are required to prevent sag on the bottom of horizontal ducts. Fiberglass duct liner is used inside sheet-metal ducts to provide better sound attenuation along the duct; this provides a thermal benefit as well. For exposed duct work, a heavier-density fiberglass board may be used as a wrap to provide a more acceptable appearance. In all cases, the joints in the sheet metal ducts should be sealed with tape or caulking to minimize air leakage and allow the transport of air to the desired location, rather than losing much of it along the run.

*Rigid fiberglass duct board and round duct* are also used in many low-pressure applications. These products form the duct itself as well as providing the thermal, acoustical, and vapor-barrier requirements most often needed. The closure system used to join the duct sections also acts to seal the system for minimum air leakage.

*Chillers and chilled water expansion tanks* are usually insulated with closed-cell elastomeric sheet to prevent condensation on the equipment. The joints are sealed and a finish may or may not be applied to the outside, depending on location.

*Piping* for both hot and cold service is normally insulated with fiberglass pipe insulation. On cold work, the vapor-barrier jacket is sealed at the overlap with either an adhesive or a factory-applied self-seal lap. If staples are used, they should be dabbed with mastic to secure the vapor resistance. Aluminum jacketing is often used on outside work with a vapor barrier applied beneath if it is cold service. Domestic hot- and cold-water plumbing and rain leaders are also commonly insulated with fiberglass. Insulation around piping supports takes many forms, depending on the nature of the hanger or support. On cold work, the use of a clevis hanger on the outside of the insulation requires a high-density insert to support the weight of the piping. This system eliminates the problem of adequately sealing around penetrations of the vapor barrier.

## 15.4 INSULATION THICKNESS DETERMINATION

This section presents formulas and graphical procedures for calculating heat loss, surface temperature, temperature drop, and proper insulation thickness. Over the last years, computer programs that perform these calculations are more readily available to customers (see section 15.5.4). But still, it is important to understand the basics for their use. Although the overall objective is to determine the right amount of insulation that should be used, some of the equations use thickness as an input variable rather than solving for it. However, all the calculations are simply manipulations or further refinements of the equation in Section 15.1.2:

$$Q = \frac{\Delta t}{R_I + R_s}$$

Following is a list of symbols, definitions, and units to be used in the heat-transfer calculations.

$t_a$ = ambient temperature, °F

$t_s$ = surface temperature of insulation next to ambient, °F

$t_h$ = hot surface temperature, normally operating temperature (cold surface temperature in cold applications), °F

$k$ = thermal conductivity of insulation always determined at mean temperature, Btu-In./hr ft² °F

$t_m$ = $(t_h + t_s)/2$ = mean temperature of insulation, °F

$t_h$ = $(t_{in} + t_{out})/2$ = average hot temperature when fluid enters at one temperature and leaves at another, °F

tk = thickness of insulation, in.

$r_1$ = actual outer radius of steel pipe or tubing, in.

$r_2$ = $(r_1 + tk)$ = radius to outside of insulation on piping, in.

Eq tk = $r_2 \operatorname{Ln}(r_2/r_1)$ = equivalent thickness of insulation on a pipe, in.

$f$ = surface air film coefficient, Btu/hr ft² °F

$R_s$ = $1/f$ = surface resistance, hr ft² °F/Btu

$R_I$ = tk/k = thermal resistance of insulation, hr ft² °F/Btu

$Q_F$ = heat flux through a flat surface, Btu/hr ft²

$Q_p$ = $Q_f\left(\frac{2\pi_2}{12}\right)$ = heat flux through a pipe, Btu/hr lin. ft

$A$ = area of insulation surface, ft²

$L$ = length of piping, lin. ft

$Q_{T-}$ = $Q_F \times A$ or $Q_p \times L$ = total heat loss, Btu/hr

$H$ = time, hr

$C_p$ = specific heat of material. Btu/lb. °F

$\rho$ = density, lb/ft³

$M$ = mass *flow* rate of a material, lb/hr

$\Delta$ = difference by subtraction, unit less

RH = relative humidity, %

DP = dew-point temperature, °F

### 15.4.1 Thermal Design Objective

The first step in determining how much insulation to use is to define what the objective is. There are many reasons for using insulation, and the amount to be used will definitely vary based on the objective chosen. The four broad categories, which include most applications, are (1) personnel protection, (2) condensation control, (3) process control, and (4) economics. Each of these is discussed in detail, with sample problems leading through the calculation sequence.

### 15.4.2 Fundamental Concepts

*Thermal Equilibrium*

A very important law in heat transfer is that under steady-state conditions, the heat *flow* through any portion of the insulation system is the same as the heat *flow* through any other part of the system. Specifically, the heat *flow* through the insulation equals the heat *flow* from the surface to the ambient, so the temperature difference for each section is proportional to the resistance for each section:

$$\text{heat flow} = \frac{\text{temperature difference}}{\text{resistance to heat flow}}$$

$$Q = \frac{t_h - t_a}{R_I + R_s} = \frac{t_h - t_s}{R_I} = \frac{t_s - t_a}{R_s}$$

Because all of the heat flows $Q$ are equal, this relationship is used to check surface temperature or other interface temperatures. For an analysis concerned with the inner surface film coefficient, the same reasoning applies.

$$Q = \frac{t_h - t_a}{R_{s1} + R_1 + R_{s2}} = \frac{t_h - t_{s1}}{R_{s1}} = \frac{t_{s1} - t_{s2}}{R_1} = \frac{t_{s2} - t_a}{R_{s2}}$$

Or for a system with two insulation materials involved, the interface temperature $t_{if}$ between the materials is involved.

$$Q = \frac{t_h - t_a}{R_{I1} + R_{I2} + R_s} = \frac{t_h - t_{if}}{R_{I2}} = \frac{t_{if} - t_s}{R_{I2}} = \frac{t_s - t_a}{R_s} = \frac{t_{if} - t_a}{R_{I2} + R_s}$$

It should be apparent that the heat flow $Q$ is also equal for any combination of $\Delta t$ and $R$ values, as shown by the last equivalency above, which utilized two parts of the system instead of just one.

Finally, it is of critical importance to calculate the $R_I$ values using the insulation mean temperature, not the operating temperature. The mean temperature is the sum of the temperatures on either side of the insulation divided by 2. Again for the last set of equivalencies:

$$t_m \text{ for } R_{I1} = \frac{t_h + t_{if}}{2}$$

$$t_m \text{ for } R_{I2} = \frac{t_{if} - t_s}{2}$$

### Pipe vs. Flat Calculations—Equivalent Thickness

Because the radial heat flows in a path from a smaller-diameter pipe, through the insulation, and then off a larger-diameter surface, a phenomenon termed "equivalent thickness" (Eq tk) occurs. Because of the geometry and the dispersion of the heat to a greater area, the pipe really "sees" more insulation than is actually there. When the adjustment is made to enter a greater insulation thickness into the calculation, the standard flat geometry formulas can be used by substituting Eq tk for tk into the equations.

The formula for equivalent thickness is

$$\text{Eq tk} = r_2 \ln\frac{r_2}{r_1}$$

where $r_1$ and $r_2$ are the inner and outer radii of the insulation system. For example, an 8-in. IPS with 3-in. insulation would lead to an equivalent thickness as follows (8-in. IPS has 8.625 in. actual outside diameter):

$$r_1 = \frac{8.625}{2} = 4.31 \text{ (Table 15.2)}$$

$$r_2 = r_1 + \text{tk} = 4.31 + 3 = 7.31$$

$$\text{Eq tk} = 7.31 \ln\frac{7.31}{4.31} = 7.31 \text{ In } 1.70$$

$$= 3.86 \qquad \textit{actual outside diameter}$$

This Eq tk can hen be used in the flat geometry equation by substituting Eq tk for tk.

$$Q = \frac{t_h - t_k}{\text{Eq tk}/k + R_s}$$

The example above used an even insulation thickness of 3 in. Some products are manufactured to such even thicknesses, and Table 15.2 lists the Eq tk for such products. However, many products are manufactured to "simplified" thicknesses, which allow a proper fit when nesting double-layer materials. ASTM-C-5859 lists these standard dimensions, and Table 15.3 shows Eq tk values for the simplified thicknesses. Figure 15.3 also shows the conversion for any thickness desired and will be used later in the reverse fashion.

### Surface Resistance

There is always diversity of opinion when it comes to selecting the proper values for the surface resistance $R_s$. The surface resistance is affected by surface emittance, surface air velocity, and the surrounding environment. Heat-transfer texts have developed procedures for calculating $R_s$ values, but they are all based on speculated values of emittance and air velocity. In actuality, the emittance of a surface often changes with time, temperature, and surface contamination, such as dust. As a result, it is unnecessary to labor over calculating specific $R_s$ values, when the conditions are estimates at best.

Table 15.4 lists a series of $R_s$ values based on three different surface conditions and the temperature difference between the surface and ambient air. Also included are single-point $R_s$ values for three different surface air velocities. See the note at the bottom of Table 15.4 relating to the effect of $R_s$ on heat-transfer calculations.

### 15.4.3 Personnel Protection

Workers need to be protected from high-temperature piping and equipment in order to prevent skin burns. Before energy conservation analyses became commonplace, many insulation systems were designed simply to maintain a "safe-touch" temperature on the outer jacket. Now, with energy costs so high, personnel protection calculations are generally limited to temporary installations or waste-heat systems, where the energy being transferred will not be further utilized.

Normally, safe-touch temperatures are specified in the range 130 to 150°F, with 140°F being used most often. It is important to remember that the surface temperature is directly related to the surface resistance $R_s$, which in turn depends on the emittance of the surface. As a result, an aluminum jacket will be hotter than a dull mastic coating over the same amount of insulation. This is demonstrated below.

### Calculation

The objective is to calculate the amount of insulation required to attain a specific surface temperature. As noted earlier,

$$\frac{t_h - t_s}{R_1} = \frac{t_s - t_a}{R_s}$$

Therefore,

$$R_1 = R_s\left(\frac{t_h - t_s}{t_s - t_a}\right) = \frac{tk}{k}(flat) = \frac{Eq\,tk}{k}(pipe)$$

Therefore,

$$tk \text{ or } Eq\,tk = kR_s\left(\frac{t_h - t_s}{t_s - t_a}\right)$$

**Fig. 15.3 Equivalent thickness chart. (From Ref. 16.)**

**Table 1**          **5.2 Equivalent thickness values for even insulation thicknesses.**

| Nominal Pipe Size (in.) | $r_1$ | Actual Thickness (in.) | | | | | | |
|---|---|---|---|---|---|---|---|---|
| | | 1 | 1-1/2 | 2 | 2-1/2 | 3 | 3-1/2 | 4 |
| 1/2 | 0.420 | 1.730 | 2.918 | 4.238 | 5.662 | 7.172 | 8.755 | 10.402 |
| 3/4 | 0.525 | 1.626 | 2.734 | 3.966 | 5.297 | 6.712 | 8.199 | 9.747 |
| 1 | 0.658 | 1.532 | 2.563 | 3.711 | 4.953 | 6.275 | 7.665 | 9.117 |
| 1-1/4 | 0.830 | 1.447 | 2.405 | 3.472 | 4.626 | 5.856 | 7.153 | 8.507 |
| 1-1/2 | 0.950 | 1.403 | 2.321 | 3.342 | 4.449 | 5.629 | 6.872 | 8.171 |
| 2 | 1.188 | 1.337 | 2.195 | 3.148 | 4.177 | 5.276 | 6.436 | 7.648 |
| 2-1/2 | 1.438 | 1.287 | 2.099 | 2.997 | 3.968 | 5.001 | 6.093 | 7.234 |
| 3 | 1.750 | 1.242 | 2.012 | 2.858 | 3.771 | 4.742 | 5.768 | 6.840 |
| 3-1/2 | 2.000 | 1.217 | 1.959 | 2.772 | 3.649 | 4.582 | 5.564 | 6.592 |
| 4 | 2.250 | 1.194 | 1.916 | 2.704 | 3.549 | 4.448 | 5.396 | 6.386 |
| 4-1/2 | 2.500 | 1.178 | 1.880 | 2.645 | 3.464 | 4.337 | 5.253 | 6.211 |
| 5 | 2.781 | 1.163 | 1.846 | 2.590 | 3.388 | 4.231 | 5.118 | 6.043 |
| 6 | 3.313 | 1.138 | 1.799 | 2.510 | 3.270 | 4.071 | 4.911 | 5.790 |
| 7 | 3.813 | 1.120 | 1.761 | 2.453 | 3.184 | 3.956 | 4.759 | 5.604 |
| 8 | 4.313 | 1.108 | 1.737 | 2.407 | 3.116 | 3.863 | 4.644 | 5.452 |
| 9 | 4.813 | 1.097 | 1.714 | 2.369 | 3.056 | 3.783 | 4.541 | 5.330 |
| 10 | 5.375 | 1.088 | 1.693 | 2.333 | 3.007 | 3.714 | 4.450 | 5.214 |
| 11 | 5.875 | 1.079 | 1.675 | 2.305 | 2.972 | 3.663 | 4.383 | 5.123 |
| 12 | 6.375 | 1.076 | 1.662 | 2.286 | 2.936 | 3.619 | 4.321 | 5.048 |
| 14 | 7.000 | 1.069 | 1.647 | 2.265 | 2.900 | 3.569 | 4.258 | 4.969 |
| 16 | 8.000 | 1.059 | 1.639 | 2.231 | 2.858 | 3.504 | 4.178 | 4.866 |
| 18 | 9.000 | 1.053 | 1.622 | 2.206 | 2.822 | 3.449 | 4.110 | 4.776 |
| 201 | 0.000 | 1.048 | 1.608 | 2.188 | 2.789 | 3.411 | 4.051 | 4.711 |
| 24 | 12.000 | 1.040 | 1.589 | 2.163 | 2.736 | 3.347 | 3.971 | 4.598 |
| 30 | 15.000 | 1.032 | 1.572 | 2.122 | 2.704 | 3.281 | 3.874 | 4.497 |

*Source:* Ref. 1          6.

**Table 15.3 Equivalent thickness values for simplified insulation thicknesses.**

| Nominal Pipe Size (in.) | Actual Thickness (in.) | | | | | | | |
|---|---|---|---|---|---|---|---|---|
| | $r1$ | 1 | 1-1/2 | 2 | 2-1/2 | 3 | 3-1/2 | 4 |
| 1/2 | 0.420 | 1.730 | 3.053 | 4.406 | 6.787 | 8.253 | 9.972 | 12.712 |
| 3/4 | 0.523 | 1.435 | 2.660 | 3.885 | 5.996 | 7.447 | 8.965 | 10.642 |
| 1 | 0.638 | 1.715 | 2.770 | 4.013 | 5.358 | 6.702 | 8.112 | 9.581 |
| 1-1/4 | 0.830 | 1.281 | 2.727 | 3.333 | 4.552 | 5.777 | 7.070 | 8.420 |
| 1-1/2 | 0.950 | 1.457 | 2.382 | 4.025 | 5.253 | 6.476 | 7.759 | 9.179 |
| 2 | 1.188 | 1.438 | 2.367 | 3.398 | 4.446 | 5.561 | 6.733 | 8.027 |
| 2-1/2 | 1.438 | 1.383 | 2.765 | 3.657 | 4.737 | 5.815 | 7.015 | 8.195 |
| 3 | 1.750 | 1.286 | 2.114 | 2.968 | 3.889 | 4.868 | 5.965 | 7.046 |
| 3-1/2 | 2.000 | 1.625 | 2.459 | 3.258 | 4.166 | 5.251 | 6.266 | 7.256 |
| 4 | 2.230 | 1.281 | 2.010 | 2.806 | 3.659 | 4.059 | 5.577 | 6 543 |
| 4-1/2 | 2.300 | 1.564 | 2.351 | 3.152 | 4.905 | 4.962 | 5.907 | 7 080 |
| 5 | 2.781 | 1.202 | 1.893 | 2.639 | 3.489 | 4.339 | 5.230 | 6.461 |
| 6 | 3.313 | 1.138 | 1.799 | 2.555 | 3.317 | 4.122 | 5.237 | 6.015 |
| 7 | 3.813 | | 1.804 | 2.495 | 3.230 | 4.153 | 4.969 | 5.821 |
| 8 | 4.313 | | 1.776 | 2.445 | 3.391 | 4.010 | 4.842 | 5.768 |
| 9 | 4.813 | | 1.752 | 2.579 | 3.232 | 3.971 | 4.786 | 5.583 |
| 10 | 5.375 | | 1.810 | 2.457 | 3.108 | 3.850 | 4.591 | 5.361 |
| 11 | 5.875 | | 1.793 | 2.428 | 3.140 | 3.793 | 4.519 | 5.271 |
| 12 | 6.375 | | 1.777 | 2.405 | 3.103 | 3.745 | 4.456 | 5.241 |
| 14 | 7.000 | | 1.647 | 2.265 | 2.900 | 3.569 | 4.258 | 4.969 |
| 16 | 8.000 | | 1.639 | 2.231 | 2.858 | 3.504 | 4.178 | 4.866 |
| 18 | 9.000 | | 1.622 | 2.206 | 2.822 | 3.449 | 4.110 | 4.776 |
| 20 | 10.000 | | 1.608 | 2.188 | 2.789 | 3.411 | 4.051 | 4.711 |
| 24 | 12.000 | | 1.589 | 2.163 | 2.736 | 3.347 | 3.971 | 4.598 |
| 30 | 15.000 | | 1.572 | 2.122 | 2.704 | 3.281 | 3.874 | 4.497 |

*Source:* Ref. 16.

**Table 15.4 $R_s$ Values[a] (hr · ft² °F/Btu).**

| $t_s - t_a$ (°F) | Still Air Plain, Fabric, Dull Metal: $\varepsilon = 0.95$ | Aluminum: $\varepsilon = 0.2$ | Stainless Steel: $\varepsilon = 0.4$ |
|---|---|---|---|
| 10 | 0.53 | 0.90 | 0.81 |
| 25 | 0.52 | 0.88 | 0.79 |
| 50 | 0.50 | 0.86 | 0.76 |
| 75 | 0.48 | 0.84 | 0 75 |
| 100 | 0.46 | 0.80 | 0 72 |
| **With Wind Velocities** Wind Velocity (mph) | | | |
| 5 | 0.35 | 0.41 | 0.40 |
| 10 | 0.30 | 0.35 | 0.34 |
| 20 | 0.24 | 0.28 | 0.27 |

*Source:* Courtesy of Johns-Manville, Ref. 16.

[a]For heat-loss calculations, the effect of $R_s$ is small compared to $R_l$, so the accuracy of $R_s$ is not critical. For surface temperature calculations, $R_s$ is the controlling factor and is therefore quite critical. The values presented in Table 15.4 are commonly used values for piping and flat surfaces. More precise values based on surface emittance and wind velocity can be found in the references.

**Example.** For a 4-in. pipe operating at 700°F in an 85°F ambient temperature with aluminum jacketing over the insulation, determine the thickness of calcium silicate that will keep the surface temperature below 140°F. Since this is a pipe, the equivalent thickness must first be calculated and then converted to actual thickness.

**STEP 1.** Determine $k$ at $t_m = (700 + 140)/2 = 420°F$. $k = 0.49$ from Table 15.1 or appendix Figure 15.A1 for calcium silicate.

**STEP 2.** Determine $R_s$ from Table 15.4 for aluminum. $t_s - t_a = 140 - 85 = 55$. So $R_s = 0.85$.

**STEP 3.** Calculate Eq tk:

$$Eq\ tk\ =\ (0.49)(0.85)\ \frac{700 - 140}{140 - 85}$$

$$=\ 4.24\ in.$$

**STEP 4.** Determine the actual thickness from Table 15.2. The effect of 4.24 in. on a 4-in. pipe can be accomplished by using 3 in. of insulation.

*Note:* Thickness recommendations are always increased to the next 1-in. increment. If a surface temperature calculation happens to fall precisely on an even increment (such as 3 in.), it is advisable to be conservative and increase to the next increment (such as 3-1/2 in.). This reduces the criticality of the $Rs$ number used. In the preceding example, it would not be unreasonable to recommend 3-1/2 in. of insulation, since it was found to be so close to 3 in.

To illustrate the effect of surface type, consider he same example with a mastic coating.

**Example.** From Table 15.4, $R_s = 0.50$, so

$$Eq\ tk\ =\ (0.49)(0.50)\ \frac{700 - 140}{140 - 85}$$

$$=\ 2.49\ in.$$

This corresponds to an actual thickness requirement on a 4-in. pipe of 2 in. This compares with 3 in. required for an aluminum-jacketed system. It is of interest to note that even though the aluminum system has a higher surface temperature, the actual heat loss is less because of the higher surface resistance value.

### Graphical Method

The calculations illustrated above can also be carried out using graphs which set the heat loss through the insulation equal to the heat loss off the surface, following the discussion in Section 15.4.2.

Figure 15.4 will be used for several different calculations. The following example gives the four-step procedure for achieving the desired surface temperature for personnel protection. The accompanying diagram outlines this procedure.

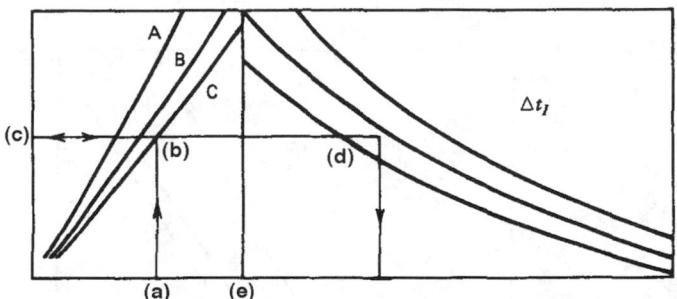

**Example.** We follow the procedure of the first example, again using aluminum jacketing.

**STEP 1.** Determine $t_s - t_a$, $140 - 85 = 55°F$.

**STEP 2.** In the diagram, proceed vertically from (a) of $\Delta t = 55$ to the curve for aluminum jacketing (b).

**STEP 2a.** Although not required, read the heat loss $Q = 65$ Btu/hr ft$^2$) (c).

**STEP 3.** Proceed to the right to (d), the appropriate curve for $t_h - t_s = 700 - 140 = 560°F$. Interpolate between lines as necessary.

**STEP 4.** Proceed down to read the required insulation resistance $R_t = 8.6$ at (e). Since $R = tk/k$ or Eq tk/$k$,

$$tk\ or\ Eq\ tk = R_t k$$

$$t_m = \frac{700 + 140}{2} = 420°F$$

$k = 0.49$ from appendix Figure 15.A1 and

$$tk\ or\ Eq\ tk = (8.6)\ (0.49) = 4.21\ in.$$

which compares well with the 4.24 in. from the earlier calculation.

The conversion of Eq tk to actual thickness required for pipe insulation is done in the same manner, using Figure 15.3.

A better understanding of the procedure involved in utilizing this quick graphical method will be obtained after working through the remainder of the calculations in this section.

### 15.4.4 Condensation Control

On cold systems, either piping or equipment, insulation must be employed to prevent moisture in the warmer surrounding air from condensing on the colder surfaces. The insulation must be of sufficient thickness to keep the insulation surface temperature above the dew point of the surrounding air. Essentially, the calculation procedures are identical to those for personnel protection except that the dew-point temperature is substituted for the desired surface temperature. (*Note:* The surface temperature should be kept 1 or 2° above the dew point to prevent condensation at that temperature.)

### *Dew-Point Determination*

The condensation (saturation) temperature, or dew point, is dependent on the ambient dry-bulb and wet-bulb temperatures. With these two values and the use of a psychrometric chart, the dew point can be determined. However, for most applications, the relative humidity is more readily attainable, so the dew point is determined using dry-bulb temperature and relative humidity rather than wet-bulb. Table 15.5 is used to find the proper dew-point temperature.

### *Calculation*

This equation is identical to the previous surface-temperature problem except that the surface temperature $t_s$ now takes on the value of the dew point of the ambient air. Also, $t_h$ now represents the cold operating temperature.

$$\text{tk or Eq tk} = kR_s \left( \frac{t_h - t_s}{t_s - t_a} \right)$$

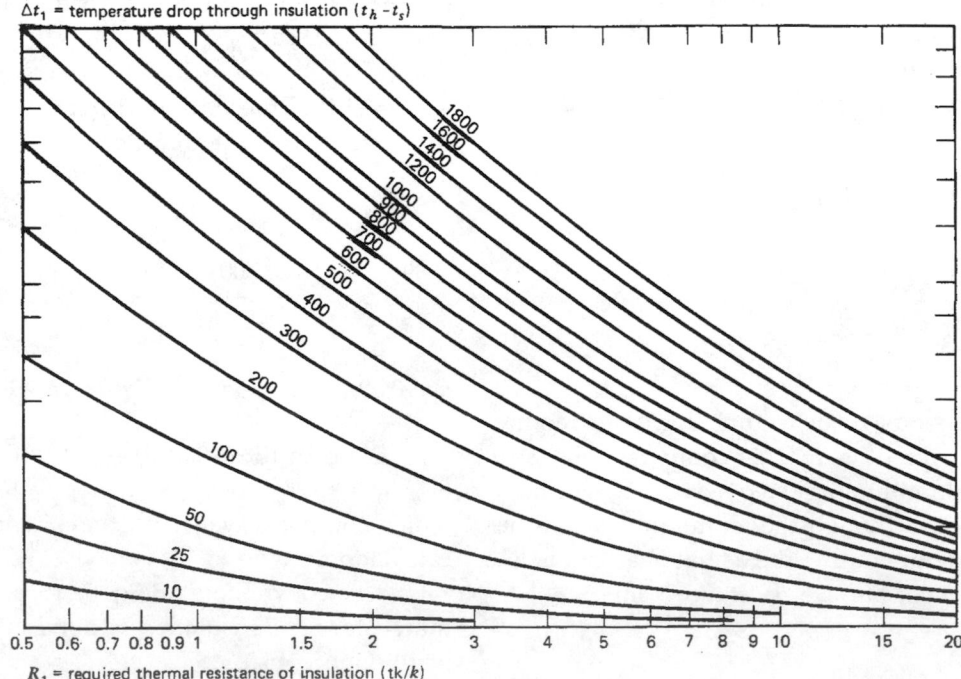

Fig. 15.4 Heat loss and surface temperature graphical method. (From Ref. 16.)

**Example.** For a 6-in.-diameter chilled-water line operating at 35°F in an ambient of 90°F and 85% RH, determine the thickness of fiberglass pipe insulation with a composite kraft paper jacket required to prevent condensation.

**STEP 1.** Determine the dew point (DP) using either a psychrometric chart or Table 15.5. DP at 90°F and 85% RH = 85°F. (In Step 5, the thickness is rounded up, which yields a higher temp.)

**STEP 2.** Determine $k$ at $t_m = (35 + 85)/2 = 60°F$. $k$ at 60°F = 0.23, from Table 15.1 or appendix Figure 15.A2.

**STEP 3.** Determine $Rs$ from Table 15.4. $\Delta t$ here is $(t_{a'} - t_s)$ rater than $(t_s - t_a)$, $t_a - t_s = 90 - 85 = 5°F$, $R_s = 0.54$.

**STEP 4.** Calculate Eq tk.

$$Eq\ tk = (0.23)(0.54)\frac{35 - 85}{85 - 90}$$

$$= 1.24\ in.$$

**STEP 5.** Determine the actual thickness from Figure 15.2 for 6-in. pipe, 1.24 in. Eq tk. The actual thickness is 1.5 in.

### Graphical Method

The graphical procedures are as described in Section 15.4.3. As the applications become colder, it is apparent that the required insulation thicknesses will become larger, with $R_I$ values toward the right side of Figure 15.4. It is suggested that the graphical procedure not be used when the resulting $R_I$ values must be determined from a very flat portion of the $(t_h - t_s)$ curve (anytime the numbers are to the far right of Figure 15.4). It is difficult to read the graph with sufficient accuracy, particularly in light of the simplicity of the mathematical calculation.

### Thickness Chart for Fiberglass Pipe Insulation

Table 15.6 gives the thickness requirements for fiberglass pipe insulation with a white, all-purpose jacket in still air. The calculations are based on the lowest

**Table 15.5 Dew-point temperature.**

| Dry-Bulb Temp. (°F) | Percent Relative Humidity | | | | | | | | | | | | | | | | | | |
|---|---|---|---|---|---|---|---|---|---|---|---|---|---|---|---|---|---|---|---|
| | 10 | 15 | 20 | 25 | 30 | 35 | 40 | 45 | 50 | 55 | 60 | 65 | 70 | 75 | 80 | 85 | 90 | 95 | 100 |
| 5 | −35 | −30 | −25 | −21 | −17 | −14 | −12 | −10 | −8 | −6 | −5 | −4 | −2 | −1 | 1 | 2 | 3 | 4 | 5 |
| 10 | −31 | −25 | −20 | −16 | −13 | −10 | −7 | −5 | −3 | −2 | 0 | 2 | 3 | 4 | 5 | 7 | 8 | 9 | 10 |
| 15 | −28 | −21 | −16 | −12 | −8 | −5 | −3 | −1 | 1 | 3 | 5 | 6 | 8 | 9 | 10 | 12 | 13 | 14 | 15 |
| 20 | −24 | −16 | −8 | −4 | −2 | 2 | 4 | 6 | 8 | 10 | 11 | 13 | 14 | 15 | 16 | 18 | 19 | 20 | |
| 25 | −20 | −15 | −8 | −4 | 0 | 3 | 6 | 8 | 10 | 12 | 15 | 16 | 18 | 19 | 20 | 21 | 23 | 24 | 25 |
| 30 | −15 | −9 | −3 | 2 | 5 | 8 | 11 | 13 | 15 | 17 | 20 | 22 | 23 | 24 | 25 | 27 | 28 | 29 | 30 |
| 35 | −12 | −5 | 1 | 5 | 9 | 12 | 15 | 18 | 20 | 22 | 24 | 26 | 27 | 28 | 30 | 32 | 33 | 34 | 35 |
| 40 | −7 | 0 | 5 | 9 | 14 | 16 | 19 | 22 | 24 | 26 | 28 | 29 | 31 | 33 | 35 | 36 | 38 | 39 | 40 |
| 45 | −4 | 3 | 9 | 13 | 17 | 20 | 23 | 25 | 28 | 30 | 32 | 34 | 36 | 38 | 39 | 41 | 43 | 44 | 45 |
| 50 | −1 | 7 | 13 | 17 | 21 | 24 | 27 | 30 | 32 | 34 | 37 | 39 | 41 | 42 | 44 | 45 | 47 | 49 | 50 |
| 55 | 3 | 11 | 16 | 21 | 25 | 28 | 32 | 34 | 37 | 39 | 41 | 43 | 45 | 47 | 49 | 50 | 52 | 53 | 55 |
| 60 | 6 | 14 | 20 | 25 | 29 | 32 | 35 | 39 | 42 | 44 | 46 | 48 | 50 | 52 | 54 | 55 | 57 | 59 | 60 |
| 65 | 10 | 18 | 24 | 28 | 33 | 38 | 40 | 43 | 46 | 49 | 51 | 53 | 55 | 57 | 59 | 60 | 62 | 63 | 65 |
| 70 | 13 | 21 | 28 | 33 | 37 | 41 | 45 | 48 | 50 | 53 | 55 | 57 | 60 | 62 | 64 | 65 | 67 | 68 | 70 |
| 75 | 17 | 25 | 32 | 37 | 42 | 46 | 49 | 52 | 55 | 57 | 60 | 62 | 64 | 66 | 69 | 70 | 72 | 74 | 75 |
| 80 | 20 | 29 | 35 | 41 | 46 | 50 | 54 | 57 | 60 | 62 | 65 | 67 | 69 | 72 | 74 | 75 | 77 | 78 | 80 |
| 85 | 23 | 32 | 40 | 45 | 50 | 54 | 58 | 61 | 64 | 67 | 69 | 72 | 74 | 76 | 78 | 80 | 82 | 83 | 85 |
| 90 | 27 | 36 | 44 | 49 | 54 | 58 | 62 | 66 | 69 | 72 | 74 | 77 | 79 | 81 | 83 | 85 | 87 | 89 | 90 |
| 95 | 30 | 40 | 48 | 54 | 59 | 63 | 67 | 70 | 73 | 76 | 79 | 82 | 84 | 86 | 88 | 90 | 91 | 93 | 95 |
| 100 | 34 | 44 | 52 | 58 | 63 | 68 | 71 | 75 | 78 | 81 | 84 | 86 | 88 | 91 | 92 | 94 | 96 | 98 | 100 |
| 105 | 38 | 48 | 56 | 62 | 67 | 72 | 76 | 79 | 82 | 85 | 88 | 90 | 93 | 95 | 97 | 99 | 101 | 103 | 105 |
| 110 | 41 | 52 | 60 | 66 | 71 | 77 | 80 | 84 | 87 | 90 | 92 | 95 | 98 | 100 | 102 | 104 | 106 | 108 | 110 |
| 115 | 45 | 56 | 64 | 70 | 75 | 80 | 84 | 88 | 91 | 94 | 97 | 100 | 102 | 105 | 107 | 109 | 111 | 113 | 115 |
| 120 | 48 | 60 | 68 | 74 | 79 | 85 | 88 | 92 | 96 | 99 | 102 | 105 | 107 | 109 | 112 | 114 | 116 | 118 | 120 |
| 125 | 52 | 63 | 72 | 78 | 84 | 89 | 93 | 97 | 100 | 104 | 107 | 109 | 111 | 114 | 117 | 119 | 121 | 123 | 125 |

temperature in each temperature range. Three temperature/humidity conditions are depicted.

### 15.4.5 Process Control

Included under this heading will be all the calculations other than those for surface temperature and economics. It is often necessary to calculate the heat flow through a given insulation thickness, or conversely, to calculate the thickness required to achieve a certain heat flow rate. The final situation to be addressed deals with temperature drop in both stagnant and flowing systems.

*Heat Flow for a Specified Thickness*

**Calculation Equations.** Again, the basic equation for a single insulation material is

$$Q_F = \frac{t_h - t_a}{R_I + R_s}$$

**Example.** For an 850°F boiler operating indoors in an 80°F ambient temperature insulated with 4 in. of calcium silicate covered with 0.016 in. aluminum jacketing, determine the heat loss per square foot of boiler surface and the surface temperature.

**STEP 1.** Find $k$ for calcium silicate at $t_m$. Assume that $t_s = 140°F$. Then $t_m = (850 + 140)/2 = 495°F$, k at 495°F = 0.53, from Table 15-1 or appendix Figure 15.A1.

**STEP 2.** Determine $R_s$ for aluminum from Table 15.4. $t_s - t_a = 140 - 0 = 60°F$, so $R_s = 0.85$.

**STEP 3.** Calculate $R_I = 4/0.53 = 7.5$.

**STEP 4.** Calculate

$$QF = \frac{850 - 80}{7.5 + 0.85} = 92 \text{ Btu/hr} \bullet \text{ft2}$$

**STEP 5.** Calculate the surface temperature *ts*, as follows:

$$R_s \times Q_F = t_s - t_a$$
$$(R_s \times Q_F) + ta = ts$$
$$t_s = (0.85 \times 92) + 80$$
$$= 158°F$$

**STEP 6.** Calculate $t_m$ to check assumption and to check the $k$ value used.

$$t_m = \frac{850 - 80}{2}$$
$$= 504°F$$

Since $k$ at 504°F = $k$ at 495°F (assumed) = 0.53, the assumption is okay. A check on $R_s$ can also be made based on the calculated surface temperature.

**STEP 7.** If the assumption is not okay, recalculate using a new $k$ value based on the new $t_m$.

The $Q_F$ used above is for flat surfaces. In determining heat flow from a pipe, the same equations are used with Eq tk substituted for tk in the $R_I$ calculation as discussed in Section 15.4.2. Often, it is desired to express pipe heat losses in terms of Btu/hr-lin.-ft rather than

**Table 15.6 Fiberglass pipe insulation: minimum thickness to prevent condensation[a].**

| Operating Pipe Temperature (°F) | 80°F and 90% RH | | 80°F and 70% RH | | 80°F and 50% RH | |
|---|---|---|---|---|---|---|
| | Pipe Size (in.) | Thickness (in.) | Pipe Size (in.) | Thickness (in.) | Pipe Size (in.) | Thickness (in.) |
| 0-34 | Up to 1 | 2 | Up to 8 | 1 | Up to 8 | 1/2 |
| | 1-1/4 to 2 | 2-1/2 | 9-30 | 1-1/2 | 9-30 | 1 |
| | 2-1/2 to 8 | 3 | | | | |
| | 9-30 | 3-1/2 | | | | |
| 35-49 | Up to 1-1/2 | 1-1/2 | Up to 4 | 1/2 | Up to 30 | 1/2 |
| | 2-8 | 2 | 412-30 | | | |
| | 9-30 | 3 | | | | |
| 50-70 | Up to 3 | 1-1/2 | Up to 30 | 1/2 | Up to 30 | 1/2 |
| | 3-1/2 to 20 | 2 | | | | |
| | 21-30 | 2-1/2 | | | | |

*Source:* Courtesy of Johns-Manville, Ref. 16.
[a]Based on still air and AP Jacket.

Btu/hr ft². This is termed $Q_P$, with

$$Q_P - Q_F\left(\frac{2\pi r_2}{12}\right)$$

**Graphical Method.** Figure 15.4 may again be used in lieu of calculations. The main difference from the previous chart usage is that surface temperature is now an unknown, and must be determined such that thermal equilibrium exists.

**Example.** Determine the heat loss from the side walls of a vessel operating at 300°F in an 80°F ambient temperature. Two inches of 3-lb/ft³ fiberglass is used with aluminum lagging.

**STEP 1.** Assume a surface temperature $t_s$ = 120°F.

**STEP 2.** Calculate

$$t_m = \frac{t_h = t_s}{2} = \frac{300 + 120}{2} = 210°F$$

Determine $k$ from appendix Figure 15.A3 at 210°F. $k$ = 0.27.

**STEP 3.** Calculate $R_I$ = tk/$k$ = 2/0.27 = 7.41.

**STEP 4.** Go to position (a) on the chart shown for $RI$ = 7.41 and read vertically to (b), where $t_h - t_s$ = 180°F.

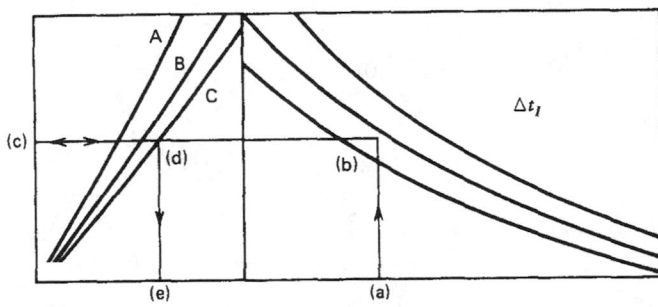

**STEP 5.** Read to the left to (c) for heat loss $Q$ = 24 Btu/hr ft².

**STEP 6.** Read down from the proper surface curve from (d) to (e), which represents $ts - ta$, to check the surface-temperature assumption. For aluminum, $ts - ta$ (chart) is 21°F, compared with the 120 − 80 = 40°F assumption.

**STEP 7.** Calculate a new surface temperature 80 + 21 = 101°F; then calculate a new $tm$, = (300 + 101)/2 =

200.5°F. Then find a new $k$ = 0.26, which gives a new $RI$ = 2/126 = 7.69.

**STEP 8.** Return to step 4 with the new $RI$ and proceed. This example shows the insensitivity of heat loss to changes in surface temperature since the new $Q$ = 22 Btu/hr ft².

For pipe insulation, the same procedure is followed except that $RI$ is calculated using the equivalent thickness. Also, conversion to heat loss per linear foot must be done separately after the square-foot loss is determined.

### Thickness for a Specified Heat Loss

Again, a surface temperature $t_s$ must first be assumed ad then checked for accuracy at the end of the calculation.

From Section 15.4.2,

$$Q = \frac{t_h - t_s}{R_I} = \frac{t_h - t_s}{tk/k}$$

$$tk \ (\text{or Eq th}) = k\left(\frac{t_h - t_s}{Q}\right)$$

where $k$ is determined at $t_m = (t_h + t_s)/2$.

**Example.** How much calcium silicate insulation is required on a 650°F duct in an 80°F ambient temperature if the maximum heat loss is 50 Btu/hr ft²? The insulation will be finished with a mastic coating.

**STEP 1.** Assume that $t_s$ = 105°F. So $t_m$ = (650 + 105)/2 = 377°F. $k$ from Table 15-1 or appendix Figure 15.A1 at 377°F = 0.46.

**STEP 2.** Find tk as follows:

$$tk = k\left(\frac{t_h - t_s}{Q}\right) = 0.46\left(\frac{650 - 105}{50}\right)$$

$$= 5.01 \text{ in.}$$

**STEP 3.** Check surface temperature assumption by $t_s = (Q \times R_s) + t_a$ using $R_s$ = 0.52. From Table 15.4 for a mastic finish,

$$t_s = 50(0.52) + 80$$
$$= 106°F$$

(Note that this in turn changes the $t_s - t_a$ from 40 to 25, which changes $R_s$ from 0.49 to 0.51, which is insignificant.)

For a graphical solution to this problem, Figure 15.4 is again used. It is simply a matter of reading across the desired $Q$ level and adjusting the $t_s$ and $R_I$ values to reach equilibrium. Thickness is then determined by $tk = kR_I$.

### Temperature Drop in a System

The following discussion is quite simplified and is not intended to replace the service of the process design engineer. The material is presented to illustrate how insulation ties into the process design decision.

**Temperature Drop in Stationary Media over Time.** The procedure calls for standard heat-flow calculations now tied into the heat content of the fluid. To illustrate, consider the following example.

**Example.** A water storage tank is calculated to have a surface area of 400 ft² and a volume of 790 ft³. How much will the temperature drop in a 72-hr period with an ambient temperature of 0°F, assuming that the initial water temperature is 50°F? The tank is insulated with 2-in. fiberglass with a mastic coating.

Before proceeding, realize that the maximum heat transfer will occur when the water is at 50°F. As it drops in temperature, the heat-transfer rate is reduced due to a smaller temperature difference. As a first approximation, it is reasonable to use the maximum heat transfer based on 50°F. Then if the temperature drop is significant, an average water temperature can be used in the second iteration.

**STEP 1.** Assume a surface temperature, calculate the mean temperature, find the $k$ factor from Table 15.1 or appendix Figure 15A.3, and determine $R_s$ from Table 15.4. With $t_s = 10°F$, $_m = 30°F$, $k = 0.22$, and $R_s = 0.53$.

**STEP 2.** Calculate heat loss with fluid at 50°F.

$$Q_F = \frac{50 - 0}{(2/0.22) + 0.53} = 5.2 \text{ Btu/hr} \cdot \text{ft}^2$$

$$Q_F = Q_F \times A = 5.2 \,(400) = 2080 \text{ Btu/hr}$$

**STEP 3.** Calculate the amount of heat that must be lost for the entire volume of water to drop 1°F.

Available heat per °F
$$= \text{volume} \times \text{density} \times \text{specific heat}$$
$$= 790 \text{ ft}^3 \times 62.4 \text{ lb/ft}^3 \times 1 \text{ Btu/lb F}$$
$$= 49,296 \text{ Btu/°F}$$

**STEP 4.** Calculate the temperature drop in 72 hr by determining the total heat flow over the period: $Q$

= 2080 × 72 = 149,760 Btu. Divide this by the available heat per 1°F drop:

$$\frac{149,760 \text{ Btu}}{49,296 \text{ Btu/°F}} = 3.04°F \text{ drop}$$

This procedure may also be used for fluid lying stationary in a pipeline. In this case it is easiest to do all the calculations for 1 linear foot rather than for the entire length of pipe.

One conservative aspect of this calculation is that the heat capacity of the metal tank or pipe is not included in the calculation. Since the container will have to decrease in temperature with the fluid, there is actually more heat available than was used above.

**Temperature Drop in Flowing Media.** There are two common situations in this category, the first involving flue gases and the second involving water or other fluids with a thickening or freezing point. This section discusses the flue-gas problem and the following section, freeze protection.

A problem is encountered with flue gases that have fairly high condensation temperatures. Along the length of a duct run, the temperature will drop, so insulation is added to control the temperature drop. This calculation is actually a heat balance between the mass flow rate of energy input and the heat loss energy outlet.

For a round duct of radius $r_1$ and length $L$, gas enters at $t_h$, and must not drop below $t_{min}$ (the dew point). The flow rate is $M$ lb/hr and the gas has a specific heat of $C_p$ Btu/lb °F. Therefore, the maximum allowable heat loss in Btu/hr is

$$Q_t = MC_p\Delta t = MC_p(t_h - t_{min})$$

Also,

$$Q_T = Q_P \times L = \frac{t_h - t_a}{R_I - R_s} \times \frac{2\pi r_2}{12} \times L$$

where

$$t_h = \left(\frac{t_{in} + t_{out}}{2}\right)$$

(A conservative simplification would be to set $t_h = t_{in}$ since the higher temperature, $t_{in}$, will cause a greater heat loss.)

To simplify on large ducts, assume that $r_1 = r_2$ (ignore the insulation-thickness addition to the surface area). Therefore,

$$\frac{t_h - t_a}{R_I - R_s} \times \frac{2\pi r_1}{12} \times L = MC_p(t_h - t_{min})$$

and

$$R_I + R_s = \left( \frac{t_h - t_a}{t_h - t_{min}} \right) \times \frac{2\pi r_1}{12} \times \frac{L}{MC_P}$$

Therefore,

$$R_I = \left[ \frac{t_h - t_a}{MC_P(t_h - t_{min})} \times \frac{2\pi r_1}{12} \times L \right] - R_s$$

$$= \frac{tk}{k} \therefore tk = R_I \times k$$

**Example.** A 48-in.-diameter duct 90 ft long in a 60°F ambient temperature has gas entering at 575°F and 15,000 cfm. The gas density standard conditions is 0.178 lb/ft$^3$ and the gas outlet must not be below 555°F. $Cp$ = 0.18 Btu/lb °F. Determine the thickness of calcium silicate required to keep the outlet temperature above 565°F, giving a 10°F buffer to account for the interior film coefficient. A more sophisticated approach calculates an interior film resistance $R_s$ (interior) instead of using a 10°F or larger buffer. The resulting equation for $Q_p$ would be

$$Q_p = \frac{t_h - t_a}{R_s(\text{inferior}) + R_I + R_s} \times \frac{2\pi r_2}{12}$$

This equation, however, will not be used.

**STEP 1.** Determine $t_h$ the average gas temperature, = (575 + 565)/2 = 570°F. (A logarithmic mean could be calculated for more accuracy, but it is usually not necessary.)

**STEP 2.** Determine M lb/hr. The flow rate is 15,000 cfm of hot gas (570°F). At standard conditions 1 atm, 70°F), the flow rate must be determined by the absolute temperature ratio:

$$\frac{t_h + 460}{70 + 460} = \frac{15,000}{\text{std. flow}}$$

$$\text{Std. flow} = 15,000 \left( \frac{70 + 460}{570 + 460} \right)$$

$$= 6262 \text{ cfm std. gas (or scfm)}$$

$$M = 6262 \text{ cfm} \times 0.178 \text{ lb/ft}^3 \times 60 \text{ min/hr}$$

$$= 66,878 \text{ lb/hr}$$

**STEP 3.** Determine $Rs$ from Table 15.4 assuming $t_s$ = 80°F and a dull surface $R_s$ = 0.5.

**STEP 4.** Calculate $R_I$.

$$R_I = \frac{570 - 60}{(66,878) \ (0.18) \ (575–565)} \times \frac{2\pi 24}{12} \times 90 - 0.52$$

$$= 4.79 - 0.52$$

$$= 4.27$$

**STEP 5.** Calculate the thickness. Assume that $t_s$ = 80°F.

$$t_m = \frac{570 + 80}{2} = 325°F$$

$k$ at 325°F = 0.45 for calcium silicate from Appendix Figure 15.A1.

$$tk = R_I \times k = 4.27 \times 0.45$$
$$= 1.93 \text{ in.}$$

**STEP 6.** The thickness required for this application is 2 in. of calcium silicate. Again, a more conservative recommendation would be 2-1/2 in.

*Note:* The foregoing calculation is quite complex. It is, however, the basis for many process control and freeze-prevention calculations. The two equations for $Q$, can be manipulated to solve for the following:

Temperature drop, based on a given thickness and flow rate.

Minimum flow rate, based on given thickness and temperature drop.

Minimum length, based on thickness, flow rate, and temperature drop.

**Freeze Protection.** Four different calculations can be performed with regard to water-line freezing (or the unacceptable thickening of any fluid).

1.  Determine the time required for a stagnant, insulated water line to reach 32°F.

2.  Determine the amount of heat tracing required to prevent freezing.

3.  Determine the flow rate required to prevent freezing of an insulated line.

4.  Determine the insulation required to prevent freezing of a line with a given flow rate.

Calculations 1 and 2 relate to Section 15.4.5, where we dealt with stationary media. To apply the same principles to the freeze problems, the following modifications should be made.

a.  In calculation 1, the heat transfer should be based on the average water temperature between the starting temperature and freezing:

$$t_h = \frac{t_{start} + 32}{2}$$

b. Rather than solving for temperature drop, given the number of hours, the hours are determined based on

$$\text{hours to freeze} = \frac{\text{available heat}}{\text{heat loss/hr}} \frac{\text{Btu}}{\text{Btu/hr}}$$

where available heat is $WCp\ \Delta t$, with

$W=$ lb of water
$Cp=$ specific heat of water (1 Btu/lb °F)
$\Delta t= t_{start} - 32$

c. In calculation 2, the heat-loss value should be calculated based upon the minimum temperature at which the system should stay, for example, 35°F. The heat tracing should provide enough heat to the system to offset the naturally occurring losses of the pipe. Heat-trace calculations are quite complex and many variables are involved. References 8 and 10 should be consulted for this type of work.

Calculations 3 and 4 relate to Section 4.5.3, dealing with flows. In the case of water, the minimum temperature can beset at 32°F and the heat-transfer rate is again on an operating average temperature

$$t_h = \frac{t_{start} + 32}{2}$$

The equations given can be manipulated to solve for flow rate or insulation thickness.

As an aid in estimating the amount of insulation for freeze protection, Table 15.7 shows both the hours to freezing and the minimum flow rate to prevent freezing based on different insulation thicknesses. These figures are based on an initial water temperature of 42°F, an ambient temperature of – 10°F, a surface resistance of 0.54, and a thermal conductivity for fiberglass pipe insulation of $k = 0.23$.

### 15.4.6 Operating Conditions

Like all other calculations, heat-transfer equations yield results that are only as accurate as the input variables used. The operating conditions chosen for the heat-transfer calculations are critical to the result, and very misleading conclusions can be drawn if improper conditions are selected.

The term "operating conditions" refers to the environment surrounding the insulation system. Some of the variable conditions are operating temperature, ambient temperature, relative humidity, wind velocity, fluid type, mass flow rate, line length, material volume, and others. Since many of these variables are constantly changing, the selection of a proper value must be made on some logical basis. Following are three suggested methods for determining the appropriate variable values.

1. **Worst Case.** If a severe failure might occur with insufficient insulation, a worst-case approach is probably warranted. For example, freeze protection should obviously be based on the historical temperature extremes rather than on yearly averages. Similarly, exterior condensation control should be based on both ambient temperature and humidity extremes in addition to the lowest operating temperature. The *ASHRAE Handbook of Fundamentals* as well as U.S. Weather Bureau data give proper design conditions for most locales. In process areas, an appropriate example involves flue-gas condensation. Here the minimum flow rate is the most critical and should be used in the calculation.

As a general rule, worst-case conditions will result in greater insulation thickness than will average conditions. In some cases the difference is very substantial, so it is important to determine initially if a worst-case calculation is required.

2. **Worst Season Average.** When a heating or cooling process is only operating part of a year, it is sensible to consider the average conditions only during that period of time. However, in year-round operations, a seasonal average is also justified in many cases. For example, personnel protection requires a maximum surface temperature that is dependent on the ambient air temperature. Taking the average summer daily maximum temperature is more practical than taking the absolute maximum ambient that could occur. The following example illustrates this.

**Example.** Consider an 8-in.-diameter, 600°F waste-heat line operating indoors with an average daily high of 80°F (but occasionally it will be 105°F). To maintain the surface below 135°F, 2 in. of calcium silicate is required with the 80°F ambient, whereas 3-1/2 in. is required with the 105°F ambient. The difference is significant and must be weighed against the benefit of the additional insulation in terms of worker safety.

3. **Yearly Average.** Economic calculations for continuously operating equipment should be based

**Table 15.7 Hours to freeze and flow rate required to prevent freezing[a].**

| Nominal Pipe Size (in.) | 1 in | | 2 in. | | 3 in | |
|---|---|---|---|---|---|---|
| | Hours to Freeze | gpm/100 ft | Hours to Freeze | gpm/100 ft | Hours to Freeze | gpm/100 ft |
| 1/2 | 0.30 | 0.087 | 0.42 | 0.282 | 0.50 | 0.053 |
| 3/4 | 0.47 | 0.098 | 0.66 | 0.070 | 0.79 | 0.058 |
| 1 | 0.66 | 0.113 | 0.96 | 0.078 | 1.16 | 0.065 |
| 1-1/2 | 0.90 | 0.144 | 1.35 | 0.096 | 1.67 | 0.078 |
| 2 | 1.72 | 0.169 | 2.64 | 0.110 | 3.31 | 0.088 |
| 2-1/2 | 2.13 | 0.195 | 3.33 | 0.124 | 4.24 | 0.098 |
| 3 | 2.81 | 0.228 | 4.50 | 0.142 | 5.80 | 0.110 |
| 4 | 3.95 | 0.279 | 6.49 | 0.170 | 8.49 | 0.130 |
| 5 | 5.21 | 0.332 | 8.69 | 0.199 | 11.54 | 0.150 |
| 6 | 6.48 | 0.386 | 10.98 | 0.228 | 14.71 | 0.170 |
| 7 | 7.66 | 0.437 | 13.14 | 0.255 | 17.75 | 0.189 |
| 8 | 8.89 | 0.487 | 15.37 | 0.282 | 20.89 | 0.207 |

*Source:* Ref. 16.
[a]Calculations based on fiberglass pipe insulation with k = 0.23, initial water temperature of 42°F, and ambient air temperature of −10°F. Flow rate represents the gallons per minute required in a 100-ft pipe and may be prorated for longer or shorter

on yearly average operating conditions rather than on worst-case design conditions. Since the intent is to maximize the owner's financial return, an average condition will not overstate the savings as the worst case or worst season might. A good approach to process work is to calculate the economic thickness based on yearly averages and then check the sufficiency of that thickness under the worst-case design conditions. That way, both criteria are met.

### 15.4.7 Bare-Surface Heat Loss

It is often desirable to determine if any insulation is required and also to compare bare surface losses with those using insulation. Table 15.8 gives bare-surface losses based on the temperature difference between the surface and ambient air. Actual temperature conditions between those listed can be arrived at by interpolation. To illustrate, consider a bare, 8-in.-diameter pipe operating at 250°F in an 80°F ambient temperature. $\Delta t = 250 - 80 = 170$°F. $Q$ for $\Delta t$ of 150°F = 812.5 Btu/ hr-lin.-ft; $Q$ for $\Delta t$ of 200°F = 1203 Btu/hr lin. ft. Interpolating between 150 and 200°F gives

$$Q_{170} = Q_{150} + (2/5)(Q_{200} - Q_{150})$$

$$= 812.5 + 0.4(1203 - 812.5)$$

$$= 968.7 \text{ Btu/hr lin. ft}$$

### 15.5 INSULATION ECONOMICS

Thermal insulation is a valuable tool in achieving energy conservation. However, to strive for maximum energy conservation without regard for economics is not acceptable. There are many ways to manipulate the cost and savings numbers, and this section explains the various approaches and the pros and cons of each.

#### 15.5.1 Cost Considerations

Simply stated, if the cost of insulation can be recouped by a reduction in total energy costs, the insulation investment is justified. Similarly, if the cost of additional insulation can be recouped by the additional energy-cost reduction, the expenditure is justified. There is a significant difference between the "full thickness" justification and the "incremental" justification. This is discussed in detail in Section 15.5.3. The following discussions will generally use the incremental approach to economic evaluation.

*Insulation Costs*

The insulation costs should include everything that it takes to apply the material to the pipe or vessel and to properly cover it to finished form. Certainly, it is more costly to install insulation 100 ft in the air than it is from ground level, and metal jackets are more costly than all-purpose indoor jackets. Anticipated maintenance costs

## Table 15.8 Heat loss from bare surfaces[a].

| Normal Pipe Size (in.) | Temperature Difference (°F) | | | | | | | | | | | | | | | |
|---|---|---|---|---|---|---|---|---|---|---|---|---|---|---|---|---|
| | 50 | 100 | 150 | 200 | 250 | 300 | 350 | 400 | 450 | 500 | 550 | 600 | 700 | 800 | 900 | 1000 |
| 1/2 | 22 | 47 | 79 | 117 | 162 | 215 | 279 | 355 | 442 | 541 | 650 | 772 | 1,047 | 1,364 | 1,723 | 2,123 |
| 3/4 | 27 | 59 | 99 | 147 | 203 | 269 | 349 | 444 | 552 | 677 | 812 | 965 | 1,309 | 1,705 | 2,153 | 2,654 |
| 1 | 34 | 75 | 124 | 183 | 254 | 336 | 437 | 555 | 691 | 846 | 1,016 | 1,207 | 1,637 | 2,133 | 2,694 | 3,320 |
| 1-1/4 | 42 | 94 | 157 | 232 | 321 | 425 | 552 | 702 | 873 | 1,070 | 1,285 | 1,527 | 2,071 | 2,697 | 3,406 | 4,198 |
| 1-1/2 | 49 | 107 | 179 | 265 | 367 | 487 | 632 | 804 | 1,000 | 1,225 | 1,471 | 1,748 | 2,371 | 3,088 | 3,899 | 4,806 |
| 2 | 61 | 134 | 224 | 332 | 459 | 608 | 790 | 1,004 | 1,249 | 1,530 | 1,837 | 2,183 | 2,961 | 3,856 | 4,870 | 6,002 |
| 2-1/2 | 74 | 162 | 271 | 401 | 556 | 736 | 956 | 1,215 | 1,512 | 1,852 | 2,224 | 2,643 | 3,584 | 4,669 | 5,896 | 7,267 |
| 3 | 89 | 197 | 330 | 489 | 677 | 897 | 1,164 | 1,480 | 1,841 | 2,256 | 2,708 | 3,219 | 4,365 | 5,685 | 7,180 | 8,849 |
| 3-1/2 | 102 | 225 | 377 | 558 | 773 | 1,024 | 1,329 | 1,690 | 2,102 | 2,576 | 3,092 | 3,675 | 4,984 | 6,491 | 8,198 | 10,100 |
| 4 | 115 | 254 | 424 | 628 | 869 | 1,152 | 1,496 | 1,901 | 2,365 | 2,898 | 3,479 | 4,135 | 5,607 | 7,304 | 9,224 | 11,370 |
| 4-1/2 | 128 | 282 | 471 | 698 | 965 | 1,280 | 1,662 | 2,113 | 2,628 | 3,220 | 3,866 | 4,595 | 6,231 | 8,116 | 10,250 | 12,630 |
| 5 | 142 | 313 | 524 | 776 | 1,074 | 1,424 | 1,848 | 2,350 | 2,923 | 3,582 | 4,300 | 5,111 | 6,931 | 9,027 | 11,400 | 14,050 |
| 6 | 169 | 373 | 624 | 924 | 1,279 | 1,696 | 2,201 | 2,799 | 3,481 | 4,266 | 5,121 | 6,086 | 8,254 | 10,750 | 13,580 | 16,730 |
| 7 | 195 | 430 | 719 | 1,064 | 1,473 | 1,952 | 2,534 | 3,222 | 4,007 | 4,910 | 5,894 | 7,006 | 9,501 | 12,380 | 15,630 | 19,260 |
| 8 | 220 | 486 | 813 | 1,203 | 1,665 | 2,207 | 2,865 | 3,643 | 4,531 | 5,552 | 6,666 | 7,922 | 10,740 | 13,990 | 17,670 | 21,780 |
| 9 | 246 | 542 | 907 | 1,343 | 1,859 | 2,464 | 3,198 | 4,066 | 5,057 | 6,197 | 7,440 | 8,842 | 11,990 | 15,620 | 19,720 | 24,310 |
| 10 | 275 | 606 | 1,014 | 1,502 | 2,078 | 2,755 | 3,576 | 4,547 | 5,655 | 6,930 | 8,320 | 9,888 | 13,410 | 17,470 | 22,060 | 27,180 |
| 11 | 300 | 661 | 1,106 | 1,638 | 2,267 | 3,005 | 3,901 | 4,960 | 6,169 | 7,560 | 9,076 | 10,790 | 14,630 | 19,050 | 24,060 | 29,660 |
| 12 | 326 | 718 | 1,202 | 1,779 | 2,463 | 3,265 | 4,238 | 5,338 | 6,701 | 8,212 | 9,859 | 11,720 | 15,890 | 20,700 | 26,140 | 32,210 |
| 14 | 357 | 783 | 1,319 | 1,952 | 2,703 | 3,582 | 4,650 | 5,912 | 7,354 | 9,011 | 10,820 | 12,860 | 17,440 | 22,710 | 28,680 | 35,350 |
| 16 | 408 | 901 | 1,508 | 2,232 | 3,090 | 4,096 | 5,317 | 6,759 | 8,407 | 10,300 | 12,370 | 14,700 | 19,940 | 25,970 | 32,790 | 40,410 |
| 18 | 460 | 1,015 | 1,698 | 2,514 | 3,480 | 4,612 | 5,987 | 7,612 | 9,467 | 11,600 | 13,930 | 16,550 | 22,450 | 29,240 | 36,930 | 45,510 |
| 20 | 510 | 1,127 | 1,885 | 2,790 | 3,862 | 5,120 | 6,646 | 8,449 | 10,510 | 12,880 | 15,460 | 18,380 | 24,920 | 32,460 | 40,990 | 50,520 |
| 24 | 613 | 1,353 | 2,263 | 3,350 | 4,638 | 6,148 | 7,980 | 10,150 | 12,620 | 15,460 | 18,570 | 22,060 | 29,920 | 38,970 | 49,220 | 60,660 |
| 30 | 766 | 1,690 | 2,827 | 4,186 | 5,795 | 7,681 | 9,971 | 12,680 | 15,770 | 19,320 | 23,200 | 27,570 | 37,390 | 48,700 | 61,500 | 75,790 |
| Flat | 98 | 215 | 360 | 533 | 738 | 978 | 1,270 | 1,614 | 2,008 | 2,460 | 2,954 | 3,510 | 4,760 | 6,200 | 7,830 | 9,650 |

Source: Ref. 16.

[a]Losses given in Btu/hr lin. ft of bare pipe at various temperature differences and Btu/hr-ft$^2$ for flat surfaces. Heat losses were calculated for still air and $\varepsilon = 0.95$ (plain, fabric or dull metals).

should also be included based on the material and application involved. The variations in labor costs due to both time and base rate should be evaluated for each particular insulation system design and locale. In other words, insulation costs tend to be job specific as well as being differentiated by product.

### Lost Heat Costs

Reducing the amount of unwanted heat loss is the function of insulation, and the measurement of this is in Btu. The key to economic analyses rests in the dollar value assigned to each Btu that is wasted. At the very least, the energy cost must include the raw-fuel cost, modified by the conversion efficiency of the equipment. For example, if natural gas costs $2.50/million Btu and it is being converted to heat at 70% efficiency, the effective cost of the Btu is 2.50/0.70 = $3.57/million Btu.

The cost of the heat plant is always a point of discussion. Many calculations ignore this capital cost on the basis that a heat plant will be required whether insulation is used or not. On the other hand, the only purpose of the heat plant is to generate usable Btus. So the cost of each Btu should reflect the capital plant cost ammortized over the life of the plant. The recent trend that seems most reasonable is to assign an incremental cost to increases in capital expenditures. This cost is stated as dollars per 1000 Btu per hour. This gives credit to a well-insulated system that requires less Btu/hr capacity.

### Other Costs

As the economic calculations become more sophisticated, other costs must be included in the analysis. The major additions are the cost of money and the tax effect of the project. Involving the cost of money recognizes the real fact that many projects are competing for each investment dollar spent.

Therefore, the money used to finance an insulation project must generate a sufficient after-tax return or the money will be invested elsewhere to achieve such a return. This topic, together with an explanation of the use of discount factors, is discussed in detail in Chapter 4.

The effect of taxes can also be included in the analysis as it relates to fuel expense and depreciation. Since both of these items are expensed annually, the after-tax cost is significantly reduced. The final example in Section 15.5.3 illustrates this.

### 15.5.2. Energy Savings Calculations

The following procedure shows how to estimate the energy cost savings resulting from installing thermal insulation.

*Procedure*

**STEP 1. Calculate present heat losses ($Q_{Tpres}$).** You can use one of the following methods to calculate the heat losses of the present system:

- Heat flow equations. These equations are in Section 15.4.2.

- Graphical method. Consists of Steps 1, 2 and 2a of the graphical method presented in Section 15.4.3.

- Table values. Table 15.8 presents heat losses values for bare surfaces (dull metals).

**STEP 2. Determine insulation thickness (tk).** Using Section 15.4, you can determine the insulation thickness according to your specific needs. Depending on the pipe diameter and temperature, the first inch of insulation can reduce bare surface heat losses by approximately 85-95% (Ref. 20). Then, for a preliminary economic evaluation, you can use tk = 1-in. If the evaluation is not favorable, you will not be able to justify a thicker insulation. On the other hand, if the evaluation is favorable, you will need to determine the appropriate insulation thickness and reevaluate the investment.

**STEP 3. Calculate heat losses with insulation ($Q_{Tins}$).** Use the equations from Section 15.4.5.

**STEP 4. Determine heat loss savings ($Q_{Tsavings}$).** Subtract the heat losses with insulation from the present heat losses ($Q_{Tsavings} = Q_{Tpres} - Q_{Tins}$)

**STEP 5. Estimate fuel cost savings.** Estimate the amount of fuel used to generate each Btu wasted and use this value to calculate the energy cost savings. With this savings, you can evaluate the insulation investment using any appropriate financial analysis method (see Section 15.5.3).

**Example.** For the example presented in section 15.4.3, determine the fuel cost savings resulting from insulating the pipe with 3-1/2 in. of calcium silicate.

*Data*

- Pipe data: 4-in pipe operating at 700°F in an 85°F ambient temperature.

- Jacket type: Aluminum.

- Pipe length: 100-ft.

- Operating hours: 4,160 hr/yr

- Fuel data: Natural gas, burned to heat the fluid in the pipe at $3/MCF. Efficiency of combustion is approximately 80%

**STEP 1.** Determine present heat loss. From Table 15.8 (4-in. pipe, temperature difference = $t_s - t_a$ = 700 – 85 = 615°F), heat loss = 4,356 Btu/hr-lin.ft. Then,

$$Q_{Tpres} = \text{(heat loss/lin.ft)(length)}$$
$$= \text{(4,356 Btu/hr-lin.ft.)(100 ft)}$$
$$= 435,600 \text{ Btu/hr}$$

**STEP 2.** Determine insulation thickness. In this example, the surface temperature has to be below 140°F, which is accomplished with an insulation thickness = tk = 3.5-in.

**STEP 3.** Determine heat losses with insulation. For this example, we need to calculate the heat losses for tk = 3.5-in following the procedure outlined in Section 15.4.5.

1) From the example in Section 15.4.3, $t_s$ = 140 °F, $k$ = 0.49 and $R_s$ = 0.85.

2) From Table 15.2, Eq tk for 3-1/2-in insulation n a 4-in. pipe = 5.396 in. Then,

$$R_I = \text{Eq tk}/k = 5.396/0.49 = 11$$

3) Calculate heat loss $Q_F$:

$$Q_F = \frac{700 - 85}{11 + 0.85} = 52 \text{ Btu/hr ft}^2$$

4) Calculate surface temperature $t_s$:
$$t_s = t_a + R_s \times Q_F = 85 + (0.85 \times 52) = 129°F$$

5) Calculate $t_m$ = (700+129)/2 = 415°F. The insulation thermal conductivity at 415°F is 0.49, which is close

enough to the assumed value (see Appendix 15.1). Then, $Q_F$ = 52 Btu/hr ft$^2$.

6) Determine the outside area of insulated pipe. From Table 15.2, pipe radius = $rl$ = 2.25-in., then, outside insulated area (ft$^2$):

$$= 2\pi \ (rl+tk)(length)/(12 \ in./ft)$$

$$= 2\pi \ (2.25 \ in+3.5 \ in.)(100 \ ft)/(12 \ in./ft)$$

$$= 301 \ ft^2$$

7) Calculate heat losses with insulation:

$$Q_{Tins} = (Q_F)(outside \ area)$$

$$= (52 \ Btu/hr \ ft^2)(301 \ ft^2)$$

$$= 15,652 \ Btu/hr$$

**STEP 4.** Determine heat losses savings $Q_{Tsavings}$:

$$Q_{Tsavings} = (Q_{Tpres} - Q_{Tins})(hr/yr)$$

$$= (435,600-15,652 \ Btu/hr)(4,160 \ hr/yr)$$
$$(1 \ MMBtu/10^6 \ Btu)$$

$$= 1,747 \ MMBtu/yr$$

**STEP 5.** Determine fuel cost savings. Assuming 1 MCF = 1 MMBtu:

Fuel savings

$$= \ (Q_{Tsavings})(conversion \ factor)/$$

$$(combustion \ efficiency)$$

$$= \ (1,747 \ MMBtu/yr)(1 \ MCF/MMBtu)/(0.8)$$

$$= \ 2,184 \ MCF/yr$$

Then,

Fuel cost savings  = (fuel savings) (fuel cost)

$$= (2,184 \ MCF/yr) \ (\$3/MCF)$$

$$= \$6,552/yr$$

### 15.5.3 Financial Analysis Methods— Sample Calculations

Chapter 4 offers a complete discussion of the various types of financial analyses commonly used in industry. A review of that material is suggested here, as the methods discussed below rely on this basic understanding.

To select the proper financial analysis requires an understanding of the degree of sophistication required

by the decision maker. In some cases, a quick estimate of profitability is all that is required. At other times, a very detailed cash flow analysis is in order. The important point is to determine what level of analysis is desired and then seek to communicate at that level. Following is an abbreviated discussion of four primary methods of evaluating an insulation investment: (1) simple payback; (2) discounted payback; (3) minimum annual cost using a level annual equivalent; and (4) present-value cost analysis using discounted cash flows.

### Economic Calculations

Basically, a simple payback period is the time required to repay the initial capital investment with the operating savings attributed to that investment. For example, consider the possibility of upgrading a present insulation thickness standard.

|  | Current Standard | Thickness Upgraded Thickness | Difference |
|---|---|---|---|
| Insulation investment ($) | 225,000 | 275,000 | 50,000 |
| Annual fuel cost ($) | 40,000 | 30,000 | 10,000 |

$$Simple \ payback = \frac{investment \ difference}{annual \ fuel \ saving} = \frac{50,000}{10,000} = 5.0 \ years$$

This calculation represents the incremental approach, which determines the amount of time to recover the additional $50,000 of investment.

In the following table, the full thickness analysis is similar except that the upgraded thickness numbers are now compared to an uninsulated system with zero insulation investment.

|  | Uninsulated System | Upgraded Thickness | Difference |
|---|---|---|---|
| Insulation investment ($) | 0 | 275,000 | 275,000 |
| Annual fuel cost ($) | 340,000 | 30,000 | 310,000 |

$$Simple \ payback = \frac{275,000}{310,000} = 0.89 \ year$$

The magnitude of the difference points out the danger in talking about payback without a proper definition of terms. If in the second example, management had a payback requirement of 3 years, the full insulation investment easily complies, whereas the incremental investment does not. Therefore, it is very important to understand the intent and meaning behind the payback requirement.

Although simple payback is the easiest financial calculation to make, its use is normally limited to rough estimating and the determination of a level of financial risk for a certain investment. The main drawback with this simple analysis is that it does not take into account the time value of money, a very important financial consideration.

### Time Value of Money

Again, see Chapter 4. The significance of the cost of money is often ignored or underestimated by those who are not involved in their company's financial mainstream. The following methods of financial analysis are all predicated on the use of discount factors that reflect the cost of money to the firm. Table 15.9 is an abbreviated table of present-value factors for a steady income stream over a number of years. Complete tables are found in Chapter 4.

### Discounted Payback

Although similar to simple payback, the utilization of the discount factor makes the savings in future years worth less in present-value terms. For discounted payback, then, the annual savings times the discount factor must now equal the investment to achieve payback in present-value dollars. Using the same example:

| | Thickness | | |
| | Current Standard | Upgraded Thickness | Difference |
|---|---|---|---|
| Insulation investment ($) | 225,000 | 275,000 | 50,000 |
| Annual fuel cost ($) | 40,000 | 30,000 | 10,000 |

Now, payback occurs when:

investment = discount factor × annual savings

50,000 = (discount factor) × 10,000

so solving for the discount factor,

$$\text{discount factor} = \frac{\text{investment}}{\text{annual savings}} = \frac{50,000}{10,000} = 5.0$$

For a 15% cost of money, read down the 15% column of Table 15.9 to find a discount factor close to 5. The corresponding number of years is then read to the left, approximately 10 years in this case. For a cost of

money of only 5%, the payback is achieved in about 6 years. Obviously, a 0% cost of money would be the same as the simple payback calculation of 5 years.

### Minimum Annual Cost Analysis

As previously discussed, an insulation investment must involve a lump-sum cost for insulation as well as a stream of fuel costs over the many years. One method of putting these two sets of costs into the same terms is to spread out the insulation investment over the life of the project. This is done by dividing the initial investment by the appropriate discount factor in Table 15.9. This produces a "level annual equivalent" of the investment for each year which can then be added to the annual fuel cost to arrive at a total annual cost.

Utilizing the same example with a 20-year project life and 10% cost of money:

| | Thickness | |
| | Current Standard | Upgraded Thickness |
|---|---|---|
| Insulation investment ($) | 225,000 | 275,000 |

For 20 years at 10%, the discount factor is 8.514 (Table 15.9), so

| | | Thickness | |
|---|---|---|---|
| Equivalent annual insulation costs | | $\dfrac{225,000}{8.514}$ | $\dfrac{275,000}{8.514}$ |
| | = | 26,427 | 32,300 |
| Annual fuel cost ($) | | 40,000 | 30,000 |
| Total annual cost ($) | | 66,427 | 62,300 |

Therefore, on an annual cost basis, the upgraded thickness is a worthwhile investment because it reduces the

**Table 15.9 Present-Value Discount Factors for an Income of $1 Per Year for the Next *n* Years**

| | Cost of Money at: | | | | |
|---|---|---|---|---|---|
| Years | 5% | 10% | 15% | 20% | 25% |
| 1 | .952 | .909 | .870 | .833 | .800 |
| 2 | 1.859 | 1.736 | 1.626 | 1.528 | 1.440 |
| 3 | 2.723 | 2.487 | 2.283 | 2.106 | 1.952 |
| 4 | 3.546 | 3.170 | 2.855 | 2.589 | 2.362 |
| 5 | 4.329 | 3.791 | 3.352 | 2.991 | 2.689 |
| 10 | 7.722 | 6.145 | 5.019 | 4.192 | 3.571 |
| 15 | 10.380 | 7.606 | 5.847 | 4.675 | 3.859 |
| 20 | 12.460 | 8.514 | 6.259 | 4.870 | 3.954 |

annual costs by $4127.

Now, to illustrate again the importance of using a proper cost of money, change the 10% to 20% and recompute the annual cost. The 20% discount factor is 4.870.

| | Thickness Current Standard | Upgraded Thickness |
|---|---|---|
| Equivalent annual insulation cost ($) | 225,000 | 275,000 |
| | 4.870 | 4.870 |
| = | 46,201 | 56,468 |
| Add the annual fuel cost ($) | 40,000 | 30,000 |
| Total annual cost | 86,201 | 86,468 |

In this case, the higher cost of money causes the upgraded annual cost to be greater than the current cost, so the project is not justified.

### Present-Value Cost Analysis

The other method of comparing project costs is to bring all the future costs (i.e., fuel expenditures) back to today's dollars by discounting and then adding this to the initial investment. This provides the total present-value cost of the project over its entire life cycle, and projects can be chosen based on the minimum present-value cost. This discounted cash flow (DCF) technique is used regularly by many companies because it allows the analyst to view a project's total cost rather than just the annual cost and assists in prioritizing among many projects.

| | Thickness Current Standard | Upgraded Thickness |
|---|---|---|
| Annual fuel cost ($) | 40,000 | 30,000 |

For 20 years at 10% the discount factor is 8.514 (Table 15.9), so

| | | |
|---|---|---|
| Present value of fuel cost over 20 years | 40,000 × 8.514 | 30,000 × 8.514 |
| | =340,560 | 255,420 |
| Insulation investment ($) | 225,000 | 275,000 |
| Total present-value cost of insulation project | $565,460 | $530,420 |

Again, the lower total project cost with the upgraded thickness option justifies that project.

So far, the effect of taxes and depreciation has been ignored so as to concentrate on the fundamentals. However, the tax effects are very significant on the cash flow to the company and should not be ignored. In the case

where the insulation investment is capitalized utilizing a 20-year straight-line depreciation schedule and a 48% tax rate, the following effects are seen (see table at top of next page).

This illustrates the significant impact of both taxes and depreciation. In the preceding analysis, the PV benefit of upgrading was (565,460 − 530,420) = $35,040. In this case, the cash flow benefit is reduced to (356,116 − 351,626) = $4490.

The final area of concern relates to future increases in fuel costs. So far, all the analyses have assumed a constant stream of fuel costs, implying no increase in the base cost of fuel. This assumption allows the use of the PV factor in Table 15.9. To accommodate annual fuel-price increases, either an average fuel cost over the project life is used or each year's fuel cost is discounted separately to PV terms. Computerized calculations permit this, whereas a manual approach would be extremely laborious.

### 15.5.4 Economic Thickness (ETI) Calculations

Section 15.5.3 developed the financial analyses often used in evaluating a specific insulation investment. As presented, however, the methods evaluate only two options rather than a series of thickness options. Economic thickness calculations are designed to evaluate each 1/2-in. increment and sum the insulation and operating costs for each increment. Then the option with the lowest total annual cost is selected as the economic thickness. Figure 15.5 graphically illustrates the optimization method. In addition, it shows the effect of additional labor required for double- and triple-layer insulation applications.

Mathematically, the lowest point on the total-cost curve is reached when the incremental insulation cost equals the incremental reduction in energy cost. By definition, the economic thickness is:

*that thickness of insulation at which the cost of the next increment is just offset by the energy savings due to that increment over the life of the project.*

### Historical development

A problem with the McMillan approach was the large number of charts that were needed to deal with all the operating and financial variables. In 1949, Union Carbide Corp. in a cooperation with West Virginia University established a committee headed by W.C. Turner to establish practical limits for the many variables and to develop a manual for performing the calculations. This was done, and in 1961, the manual was published by the National Insulation Manufacturers Association (previously called TIMA and

| | Thickness Current Standard | Upgraded Thickness |
|---|---|---|
| 1. Annual energy cost ($) | 40,000 | 30,000 |
| 2. After-tax energy cost ($) ((1) × (1.0 − 0.48)) | 20,800 | 15,600 |
| 3. Insulation depreciation ($ tax benefit) | | |
| (225,000/20 yr)(0.48) | 5,400 | |
| (275,000/20 yr)(0.48) | | 6,600 |
| 4. Net annual cash costs [$; (2) − (3)] | 15,400 | 9,000 |
| 5. Present-value factor for 20 years at 10% = 8.514 | | |
| 6. Present value of annual cash flows [$; (4) × (5)] | 131,116 | 76,626 |
| 7. Present value of cash flow for insulation purchase ($) | 225,000 | 275,000 |
| 8. Present-value cost of project [$; (6) + (7)] | 356,116 | 351,626 |

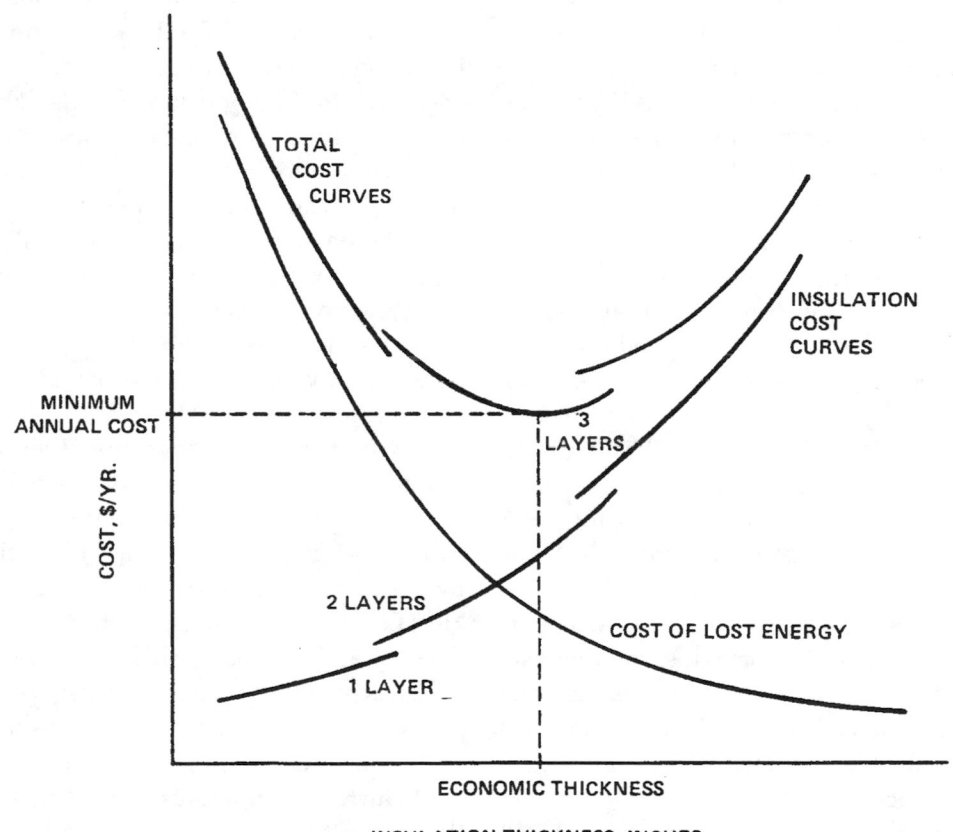

**Fig. 15.5 Economic thickness of insulation (ETI) concept.**

now NAIMA, North American Insulation Manufacturers Association). The manual was entitled *How to Determine Economic Thickness of Insulation* and employed a number of nomographs and charts for manually performing the calculations.

Since that time, the use of computers has greatly changed the method of ETI calculations. In 1973, TIMA released several programs to aid the design engineer in selecting the proper amount of insulation. Then in 1976, the Federal Energy Administration (FEA) published a no-

mograph manual entitled *Economic Thickness of Industrial Insulation* (Conservation Paper #46). In 1980, these manual methods were computerized into the "Economic Thickness of Industrial Insulation for Hot and Cold Surfaces." Through the years, NAIMA developed a version for personal computers; the newest program was renamed 3EPLUS and calculates the ETI thickness of insulation.

Perhaps the most significant change occurring is that most large owners and consulting engineers are developing and using their own economic analysis programs, specifically tailored to their needs. As both heat-transfer and financial calculations become more sophisticated, these programs will continue to be upgraded and their usefulness in the design phase will increase.

### Nomograph Methods

A nomograph methods is not presented here, but the interested reader can review the following references:

- FEA manual (Ref. 12). This manual provides a fairly complete but time-consuming nomograph method.

- 1972 ASHRAE Handbook of Fundamentals, Chapter 17 (Ref. 13) which provides a simplified, one-page nomograph. This approach is satisfactory for a quick determination, but it lacks the versatility of the more complex approach. The nomograph has been eliminated in the latest edition and reference is made to the computer analyses and the FEA manual.

### Computer Programs

Several insulation manufacturers offer to run the analysis for their customers. Also, computer programs such as the 3EPLUS are available for customers who want to run the analysis on their own. The 3EPLUS software is an ETI program developed by the North American Insulation Manufacturers Association and the Steam Challenge Program. The program, available for free download (Ref. 14), calculates heat losses, energy and cost savings, thickness for maximum surface temperature and optimum thickness of insulation.

All the insulation owning costs are expressed on an equivalent uniform annual cost basis. This program uses the ASTM C680 method for calculating the heat loss and surface temperatures. Each commercially available thickness is analyzed, and the thickness with the lowest annual cost is the economic thickness (ETI).

Figure 15.6 shows the output generated by the NAIMA 3EPLUS program. The first several lines are a readout of the input data. The different variables used in the program allow to simulate virtually any job condition. The same program can be used for retrofit analyses and bare-surface calculations. There are two areas of input data that are not fully explained in the output. The first is the installed insulation cost. The user has the option of entering the installed cost for each particular thickness or using an estimating procedure developed by the FEA (now DOE).

The second area that needs explanation is the insulation choice, which relates to the thermal conductivity of the material. The example in Figure 15.6 shows the insulation as Glass Fiber Blanket. The program includes the thermal conductivity equations of several generic types of thermal insulation, which were derived from ASTM materials specifications. The user has the option of supplying thermal conductivity data for other materials.

The lower portion of the output supplies seven columns of information. The first and second columns are input data, while the others are calculated output. The program also calculates the reduction in $CO_2$ emissions by insulating to economic thickness. The meaning of columns two to seven of the output are explained below.

**Annual Cost ($/yr).** This is the annual operating cost including both energy cost and the amortized insulation cost. Tax effects are included. This value is the one that determines the economic thickness. As stated under the columns, the lowest annual cost occurs with 2.50 in. of insulation which is the economic thickness.

**Payback period (yr).** This value represents the discounted payback period of the specific thickness as compared to the reference thickness. In this example, the reference thickness variable was input as zero, so the payback is compared to the uninsulated condition.

**Present Value of Heat Saved ($/ft).** This gives the energy cost savings in discounted terms as compared to the uninsulated condition. As discussed earlier, the first increment has the most impact on energy savings, but the further incremental savings are still justified, as evidenced by the reduction of annual cost to the 2.50-in. thickness. Heat Loss (Btu/ft). This calculation allows the user to check the expected heat loss with that required for a specific process. It is possible that under certain conditions a thickness greater than the economic thickness may be required to achieve a necessary process requirement.

**Surface Temperature (°F).** This final output allows the user to check the resulting surface temperature to assure that the level is within the safe-touch range. The ETI program is very sophisticated. It employs sound methods of both thermal and financial analysis and provides output that is relevant and useful to the design engineer and owner. NAIMA makes this program available to those desiring to have it on their own computer systems. In addition, several of the insulation manufacturers offer to run the analysis for their customers and send them a program output.

## Figure 15.6 NAIMA 3E computer program output.

| | |
|---|---|
| Project Name = | Date = 11-13-1995 |
| Project Number = | Engineer = |
| System = | Contact = |
| Location = | Phone = |

| | |
|---|---|
| Fuel Type = | Gas |
| First Year Price = | 3.36 $ per mcf |
| Heating Value = | 1000 Btu per cf |
| Efficiency = | 80.0% |
| Annual Fuel Inflation Rate = | 6.0% |
| Annual hours of operation = | 8320 hours |

ECONOMIC DATA

| | |
|---|---|
| Interest rate or Return on investment = | 10.0% |
| Effective Income Tax Rate = | 30% |
| Physical Plant Depreciation Period = | 7 years |
| New Insulation Depreciation Period = | 7 years |
| Incremental Equipment Investment Rate = | 3.47 $/MMBtu/hr |
| Percent of New Insulation Cost for | |
| Annual Insulation Maintenance = | 2% |
| Percent of Annual Fuel Bill for | |
| Physical Plant Maintenance = | 1% |
| Ambient temperature = | 75 F |
| Emittance of outer jacketing = | 0.10 |
| Wind speed = | 0 mph |
| Emittance of existing surface = | 0.80 |
| Reference thickness for payback calculations = | 0.0 inches |

Insulation material = GLASS FIBER BLANKET
Horizontal Pipe

| | |
|---|---|
| Pipe Size = | 5 inch |
| Average Installation Complexity factor = | 1.20 |
| Performance Service factor = | 1.00 |

Insulation costs estimated by FEA method

| | |
|---|---|
| Labor rate = | 38.35 $/hr |
| Productivity factor = | 100 |
| Price of 2x2 pipe insulation = | 4.97 $/ln ft |
| Price of 2 inch block = | 1.71 $/sqft |

Operating Temperature 450 F

| Insulation Thick Inches | Cost $/ft | Annual Cost $/ft | Payback Period Years | Pres Value Heat Saved $/ft | Heat Loss Btu/ft | Surf Temp F |
|---|---|---|---|---|---|---|
| Bare | | 57.37 | | | 1834 | 450 |
| 1.0 | 10.18 | 9.25 | 0.2 | 1133.90 | 226 | 193 |
| 1.5 | 12.10 | 8.07 | 0.2 | 1170.00 | 174 | 163 |
| 2.0 | 14.96 | 7.76 | 0.3 | 1190.76 | 145 | 146 |
| 2.5 | 17.27 | 7.60 | 0.4 | 1205.83 | 124 | 133 |
| 3.0 | 19.68 | 7.71 | 0.4 | 1214.97 | 111 | 125 |
| 4.0 | 25.45 | 8.36 | 0.6 | 1228.33 | 92 | 113 |
| Double layer | | | | | | |
| 3.0 | 22.30 | 8.27 | 0.5 | 1214.97 | 111 | 125 |
| 4.0 | 29.22 | 9.18 | 0.7 | 1228.33 | 92 | 113 |
| 5.0 | 36.14 | 10.34 | 0.9 | 1235.90 | 81 | 106 |
| 6.0 | 43.07 | 11.63 | 1.1 | 1240.32 | 75 | 102 |
| Triple layer | | | | | | |
| 6.0 | 67.48 | 16.91 | 1.7 | 1240.32 | 75 | 102 |
| 7.0 | 77.39 | 18.86 | 2.0 | 1244.51 | 69 | 99 |
| 8.0 | 87.00 | 20.79 | 2.3 | 1247.76 | 64 | 96 |

The Economic Thickness is single layer 2.5 inches.

The savings for the economic thickness is 49.77 $/ln ft/yr and the reduction in Carbon Dioxide emissions is 1608 lbs/lnft/yr.

## APPENDIX 15.1

### Typical Thermal Conductivity Curves Used in Sample Calculations*

**Fig. 15.A1 Calcium silicate.**

**Fig. 15.A2 Fiberglass pipe insulation.**

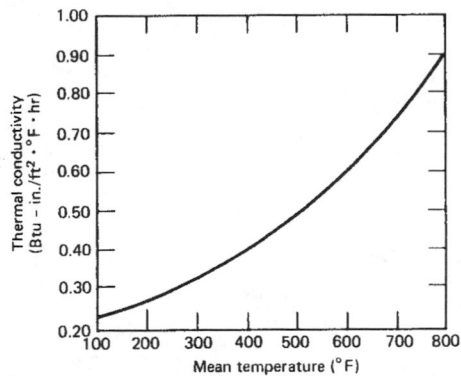

**Fig. 15.A3 Fiberglass board, 3 lb/ft$^3$.**

*Current manufacturers' data should always be used for calculations.

## References

1. American Society for Testing and Materials, Annual Book of ASTM Standards: Part 18—Thermal and Cryogenic Insulating Materials; Building Seals and Sealants; Fire Test; Building Constructions; Environmental Acoustics; Part 17—Refractories, Glass and Other Ceramic Materials; Manufactured Carbon and Graphite Products.

2. W.H. McADAMS, Heat Transmission, McGraw-Hill, New York, 1954.

3. E.M. SPARROW and R.D. CESS, Radiation Heat Transfer, McGraw-Hill, New York, 1978.

4. L.L. BERANEK, Ed., Noise and Vibration Control, McGraw-Hill, New York, 1971.

5. F.A. WHITE, Our Acoustic Environment, Wiley, New York, 1975.

6. M. KANAKIA, W. HERRERA, and F. HUTTO, JR., "Fire Resistance Tests for Thermal Insulation," Journal of Thermal Insulation, Apr. 1978, Technomic, Westport, Conn.

7. Commercial and Industrial Insulation Standards, Midwest Insulation Contractors Association, Inc., Omaha, Neb., 1979.

8. J.F. MALLOY, Thermal Insulation, Reinhold, New York, 1969.

9. American Society for Testing and Materials, Annual Book of ASTM Standards, Part 18, STD C-585.

10. J.F. MALLOY, Thermal Insulation, Reinhold, New York, 1969, pp. 72-77, from Thermon Manufacturing Co. technical data.

11. L.B. McMlLLAN, "Heat Transfer through Insulation in the Moderate and High Temperature Fields: A Statement of Existing Data," No. 2034, The American Society of Mechanical Engineers, New York, 1934.

12. Economic Thickness of Industrial Insulation, Conservation Paper No. 46, Federal Energy Administration, Washington, D.C., 1976. Available from Superintendent of Documents, U.S. Government Printing Office, Washington, D.C. 20402 (Stock No. 041-018-00115-8).

13. ASHRAE Handbook of Fundamentals, American Society of Heating, Refrigerating and Air Conditioning Engineers, Inc., New York, 1972, p. 298.

14. NAIMA 3 E's Insulation Thickness Computer Program, North American Insulation Manufacturers Association, 44 Canal Center Plaza, Suite 310, Alexandria, VA 22314.

15. P. Greebler, "Thermal Properties and Applications of High Temperature Aircraft Insulation," American Rocket Society, 1954. Reprinted in Jet Propulsion, Nov.-Dec. 1954.

16. Johns-Manville Sales Corporation, Industrial Products Division, Denver, Colo., Technical Data Sheets.

17. ASHRAE Handbook of Fundamentals, American Society of Heating, Refrigerating and Air Conditioning Engineers, Inc., Atlanta, GA, 1992, p.22.16.

18. W.C. Turner and J.F. Malloy, Thermal Insulation Handbook, Robert E. Krieger Publishing Co. And McGraw Hill, 1981.

19. Ahuja, A., "Thermal Insulation: A Key to Conservation," Consulting-Specifying Engineer, January 1995, p. 100-108.

20. U.S. Department of Energy, "Industrial Insulation for Systems Operating Above Ambient Temperature," Office of Industrial Technologies, Bulletin ORNL/M-4678, Washington, D.C., September 1995.

# CHAPTER 16
# USE OF ALTERNATIVE ENERGY

**JERALD D. PARKER**
Professor Emeritus, Oklahoma State University
Stillwater, Oklahoma
Professor Emeritus, Oklahoma Christian University
Oklahoma City, Oklahoma

## 16.1 INTRODUCTION

Any energy source that is classified as an "alternative energy source" is that because, at one time it was not selected as the best choice. If the original choice of an energy source was a proper one the use of an alternative energy source would make sense only if some condition has changed. This might be:

1.   Present or impending nonavailability of the present energy source
2.   Change in the relative cost of the present and the alternative energy
3.   Improved reliability of the alternative energy source
4.   Environmental or legal considerations

To some, an alternative energy source is a nondepleting or renewable energy source, and, for many it is this characteristic that creates much of the appeal. Although the terms "alternative energy source" and "renewable energy sources" are not intended by this writer to be synonymous, it will be noted that some of the alternative energy sources discussed in this section are renewable.

It is also interesting that what we now think of as alternative energy sources, for example solar and wind, were at one time important conventional sources of energy. Conversely, natural gas, coal, and oil were, at some time in history, alternative energy sources. Changes in the four conditions listed above, primarily conditions 2 and 3, have led us full circle from the use of solar and wind, to the use of natural gas, coal, and oil, and back again in some situations to a serious consideration of solar and wind.

In a strict sense, technical feasibility is not a limitation in the use of the alternative energy sources that will be discussed. Solar energy can be collected at any

reasonable temperature level, stored, and utilized in a variety of ways. Wind energy conversion systems are now functioning and have been for many years. Refuse-derived fuel has also been used for many years. What is important to one who must manage energy systems are the factors of economics, reliability, and in some cases, the nonmonetary benefits, such as public relations.

Government funding for R&D as well as tax incentives in the alternative energy area dropped sharply during the decade of the eighties and early nineties. This caused many companies with alternative energy products to go out of business, and for others to cut back on production or to change into another product or technology line. Solar thermal energy has been hit particularly hard in this respect, but solar powered photovoltaic cells have had continued growth both in space and in terrestrial applications. Wind energy systems have continued to be installed throughout the world and show promise of continued growth. The burning of refuse has met with some environmental concerns and strict regulations. Recycling of some refuse materials such as paper and plastics has given an alternative to burning. Fuel cells continue to increase in popularity in a wide variety of applications including transportation, space vehicles, electric utilities and uninterruptible power supplies.

Surviving participants in the alternative energy business have in some cases continued to grow and to improve their products and their competitiveness. As some or all of the four conditions listed above change, we will see rising or falling interest on the part of the government, industry and private individuals in particular alternative energy systems.

## 16.2 SOLAR ENERGY

### 16.2. 1 Availability

"Solar energy is free!" states a brochure intended to sell persons on the idea of buying their solar products. "There's no such thing as a free lunch" should come to mind at this point. With a few exceptions, one must invest capital in a solar energy system in order to reap the benefits of this alternative energy source. In addition to the cost of the initial capital investment, one is usually faced with additional periodic or random costs due to operation and maintenance. Provided that the solar system does its expected task in a reasonably reliable manner, and presuming that the conventional energy source is available and satisfactory, the important question usually is: Did it save money compared to the conventional system? Obviously, the cost of money, the cost of conventional fuel, and the cost and performance of the solar system are all important factors. As a first step in looking at the feasibility of solar energy, we will consider its availability.

Solar energy arrives at the outer edge of the earth's atmosphere at a rate of about 428 Btu/hr ft² (1353 W/m²). This value is referred to as the solar constant. Part of this radiation is reflected back to space, part is absorbed by the atmosphere and re-emitted, and part is scattered by atmospheric particles. As a result, only about two-thirds of the sun's energy reaches the surface of the earth. At 40° north latitude, for example, the noontime radiation rate on a flat surface normal to the sun's rays is about 300 Btu/hr · ft² on a clear day. This would be the approximate maximum rate at which solar energy could be collected at that latitude. A solar collector tracking the sun so as to always be normal to the sun's rays could gather approximately $3.6 \times 10^3$ Btu/ft²-day as an absolute upper limit. To gather 1 million Btu/day, for example, would require about 278 ft² (26 m²) of movable collectors, collecting all the sunlight that would strike them on a clear day.

Since no collector is perfect and might collect only 70% of the energy striking it, and since the percent sunshine might also be about 70%, a more realistic area would be about 567 ft² (53 m²) to provide 1 million Btu of energy per day. In the simplest terms, would the cost of constructing, operating, and maintaining a solar system consisting of 567 ft² of tracking solar collectors justify a reduction in conventional energy usage of 1 million Btu/day? Fixed collectors might be expected to deliver approximately 250,000 Btu/yr for each square foot of surface.

A most important consideration which was ignored in the discussion above was that of the system's ability to use the solar energy when it is available. A space-heating system, for example, cannot use solar energy in the summer. In industrial systems, energy demand will rarely correlate with solar energy availability. In some cases, the energy can be stored until needed, but in most systems, there will be some available solar energy that will not be collected. Because of this factor, particular types of solar energy systems are most likely to be economically viable. Laundries, car washes, motels, and restaurants, for example, need large quantities of hot water almost every day of the year. A solar water-heating system seems like a natural match for such cases. On the other hand, a solar system that furnishes heat only during the winter, as for space heating, may often be a poor economic investment.

The amount of solar energy available to collect in a system depends upon whether the collectors move to follow or partially follow the sun or whether they are fixed. In the case of fixed collectors, the tilt from horizontal and the orientation of the collectors may be significant. The remainder of this section considers the energy available to fixed solar collecting systems.

Massive amounts of solar insolation data have been collected over the years by various government and private agencies. The majority of these data are hourly or daily solar insolation values on a horizontal surface, and the data vary considerably in reliability. Fixed solar collectors are usually tilted at some angle from the horizontal so as to provide a maximum amount of total solar energy collected over the year, or to provide a maximum amount during a particular season of the year. What one needs in preliminary economic studies is the rate of solar insolations on tilted surfaces.

Figure 16.1 shows the procedure for the conversion of horizontal insolation to insolation on a tilted surface. The measured insolation data on a horizontal surface consist of direct radiation from the sun and diffuse radiation from the sky. The total radiation must be split into these two components (step A) and each component analyzed separately (steps B and C). In addition, the solar energy reflected from the ground and other surroundings must be added into the total (step D). Procedures for doing this are given in Refs. 1 to 4.

**Figure 16.1 Conversion of horizontal insolation to insolation on tilted surface.**

A very useful table of insolation values for 122 cities in the United States and Canada is given in Ref. 5. These data were developed from measured weather data using the methods of Refs. 2 and 3 and are only as reliable as the original weather data, perhaps ± 10%. A summary of the data for several cities is given in Table 16.1.

One of the more exhaustive compilations of U.S. solar radiation data is that compiled by the National Climatic Center in Asheville, North Carolina, for the Department of Energy. Data from 26 sites were rehabilitated and then used to estimate data for 222 stations, shown in Figure 16.2. A summary of these data is tabulated in a textbook by Lunde.[6] It should be remembered that measured data from the past do not predict what will happen in the future. Insolation in any month can be quite variable from year to year at a given location.

Another approach is commonly used to predict insolation on a specified surface at a given location. This method is to first calculate the clear-day insolation, using knowledge of the sun's location in the sky at the given time. The clear-day insolation is then corrected by use of factors describing the clearness of the sky at a given location and the average percent of possible sunshine.

The clear-sky insolation on a given surface is readily found in references such as the *ASHRAE Handbook of Fundamentals*. A table of percent possible sunshine for several cities is given in Table 16.2.

## 16.2.2 Solar Collectors

A wide variety of devices may be used to collect solar energy. A general classification of types is given in Figure 16.3. Tracking-type collectors are usually used where relatively high temperatures (above 250°F) are required. These types of collectors are discussed at the end of this section. The more common fixed, flat-plate collector will be discussed first, followed by a discussion of tube-type or mildly concentrating collectors.

The flat-plate collector is a device, usually faced to the south (in the northern parts of the globe) and usually at some fixed angle of tilt from the horizontal. Its purpose is to use the solar radiation that falls upon it to raise the temperature of some fluid to a level above the ambient conditions. That heated fluid, in turn, may be used to provide hot water or space heat, to drive an engine or a refrigerating device, or perhaps to remove moisture from a substance. A typical glazed flat-plate solar collector of the liquid type is shown in Figure 16.4.

The sun's radiation has a short wavelength and easily passes through the glazing (or glazings), with only about 10 to 15% of the energy typically reflected and absorbed in each glazing. The sunlight that passes through is almost completely absorbed by the absorber surface and raises the absorber temperature. Heat loss out the back from the absorber plate is minimized by the use of insulation. Heat loss out the front is decreased somewhat by the glazing, since air motion is restricted. The heated

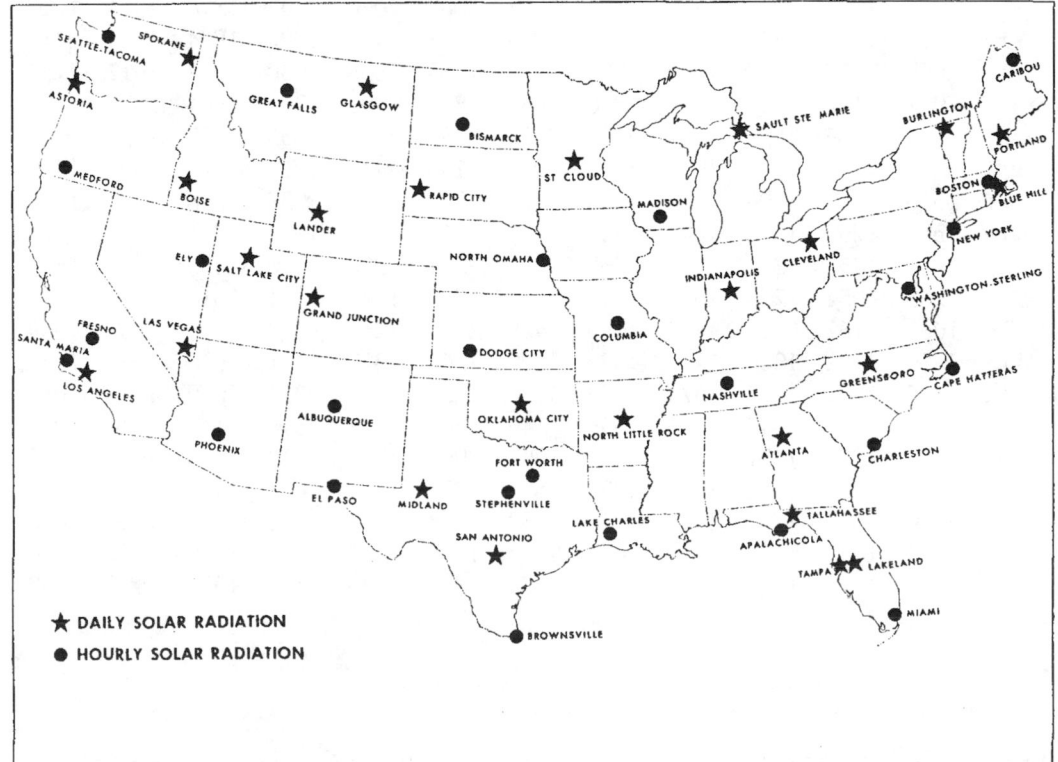

**Figure 16.2 Weather stations for which rehabilitated measured (asterisks) and derived data have been collected. [From *SOLMET, Volume 1*, and *Input Data for Solar Systems*, Nov. 1978, prepared by NOAA for DOE, Interagency agreement E (49-26)-1041. Some data are given in Ref. 6.]**

★ DAILY SOLAR RADIATION
● HOURLY SOLAR RADIATION

**Table 16.1 Average Daily Radiation on Tilted Surfaces for Selected Cities**

Average Daily Radiation (Btu/day ft²).

| City | Slope | Jan. | Feb. | Mar. | Apr. | May | June | July | Aug. | Sept. | Oct. | Nov. | Dec. |
|---|---|---|---|---|---|---|---|---|---|---|---|---|---|
| Albuquerque, NM | hor. | 1134 | 1436 | 1885 | 2319 | 2533 | 2721 | 2540 | 2342 | 2084 | 1646 | 1244 | 1034 |
| | 30 | 1872 | 2041 | 2295 | 2411 | 2346 | 2390 | 2289 | 2318 | 2387 | 2251 | 1994 | 1780 |
| | 40 | 2027 | 2144 | 2319 | 2325 | 2181 | 2182 | 2109 | 2194 | 2369 | 2341 | 2146 | 1942 |
| | 50 | 2127 | 2190 | 2283 | 2183 | 1972 | 1932 | 1889 | 2028 | 2291 | 2369 | 2240 | 2052 |
| | vert. | 1950 | 1815 | 1599 | 1182 | 868 | 754 | 795 | 1011 | 1455 | 1878 | 2011 | 1927 |
| Atlanta, GA | hor. | 839 | 1045 | 1388 | 1782 | 1970 | 2040 | 1981 | 1848 | 1517 | 1288 | 975 | 740 |
| | 30 | 1232 | 1359 | 1594 | 1805 | 1814 | 1801 | 1782 | 1795 | 1656 | 1638 | 1415 | 1113 |
| | 40 | 1308 | 1403 | 1591 | 1732 | 1689 | 1653 | 1647 | 1701 | 1627 | 1679 | 1496 | 1188 |
| | 50 | 1351 | 1413 | 1551 | 1622 | 1532 | 1478 | 1482 | 1571 | 1562 | 1679 | 1540 | 1233 |
| | vert. | 1189 | 1130 | 1068 | 899 | 725 | 659 | 680 | 811 | 990 | 1292 | 1332 | 1107 |
| Boston, MA | hor. | 511 | 729 | 1078 | 1340 | 1738 | 1837 | 1826 | 1565 | 1255 | 876 | 533 | 438 |
| | 30 | 830 | 1021 | 1313 | 1414 | 1677 | 1701 | 1722 | 1593 | 1449 | 1184 | 818 | 736 |
| | 40 | 900 | 1074 | 1333 | 1379 | 1592 | 1595 | 1623 | 1536 | 1450 | 1234 | 878 | 803 |
| | 50 | 947 | 1101 | 1322 | 1316 | 1477 | 1461 | 1494 | 1448 | 1417 | 1254 | 916 | 850 |
| | vert. | 895 | 950 | 996 | 831 | 810 | 759 | 791 | 857 | 993 | 1044 | 842 | 820 |
| Chicago, IL | hor. | 353 | 541 | 836 | 1220 | 1563 | 1688 | 1743 | 1485 | 1153 | 763 | 442 | 280 |
| | 30 | 492 | 693 | 970 | 1273 | 1502 | 1561 | 1639 | 1503 | 1311 | 990 | 626 | 384 |
| | 40 | 519 | 716 | 975 | 1239 | 1425 | 1563 | 1544 | 1447 | 1307 | 1024 | 662 | 403 |
| | 50 | 535 | 723 | 959 | 1180 | 1322 | 1341 | 1421 | 1363 | 1274 | 1034 | 682 | 415 |
| | vert. | 479 | 602 | 712 | 746 | 734 | 707 | 754 | 806 | 887 | 846 | 610 | 373 |
| Ft. Worth, TX | hor. | 927 | 1182 | 1565 | 1078 | 2065 | 2364 | 2253 | 2165 | 1841 | 1450 | 1097 | 898 |
| | 30 | 1368 | 1550 | 1807 | 1065 | 1891 | 2060 | 2007 | 2097 | 2029 | 1859 | 1604 | 1388 |
| | 40 | 1452 | 1601 | 1803 | 1020 | 1755 | 1878 | 1845 | 1979 | 1995 | 1907 | 1698 | 1488 |
| | 50 | 1500 | 1614 | 1758 | 957 | 1586 | 1663 | 1648 | 1820 | 1914 | 1908 | 1749 | 1549 |
| | vert. | 1315 | 1286 | 1196 | 569 | 728 | 679 | 705 | 890 | 1185 | 1459 | 1509 | 1396 |
| Lincoln, NB | hor. | 629 | 950 | 1340 | 1752 | 2121 | 2286 | 2268 | 2054 | 1808 | 1329 | 865 | 629 |
| | 30 | 958 | 1304 | 1605 | 1829 | 2004 | 2063 | 2088 | 2060 | 2092 | 1818 | 1351 | 1027 |
| | 40 | 1026 | 1363 | 1620 | 1774 | 1882 | 1909 | 1944 | 1971 | 2087 | 1894 | 1450 | 1113 |
| | 50 | 1068 | 1389 | 1597 | 1679 | 1724 | 1720 | 1763 | 1838 | 2030 | 1922 | 1512 | 1170 |
| | vert. | 972 | 1162 | 1156 | 989 | 856 | 788 | 828 | 992 | 1350 | 1561 | 1371 | 1100 |
| Los Angeles, CA | hor. | 946 | 1266 | 1690 | 1907 | 2121 | 2272 | 2389 | 2168 | 1855 | 1355 | 1078 | 905 |
| | 30 | 1434 | 1709 | 1990 | 1940 | 1952 | 1997 | 2138 | 2115 | 2066 | 1741 | 1605 | 1439 |
| | 40 | 1530 | 1776 | 1996 | 1862 | 1816 | 1828 | 1966 | 2002 | 2037 | 1788 | 1706 | 1550 |
| | 50 | 1587 | 1799 | 1953 | 1744 | 1644 | 1628 | 1758 | 1845 | 1959 | 1791 | 1762 | 1620 |
| | vert. | 1411 | 1455 | 1344 | 958 | 760 | 692 | 744 | 918 | 1230 | 1383 | 1537 | 1479 |
| New Orleans, LA | hor. | 788 | 954 | 1235 | 1518 | 1655 | 1633 | 1537 | 1533 | 1411 | 1316 | 1024 | 729 |
| | 30 | 1061 | 1162 | 1356 | 1495 | 1499 | 1428 | 1369 | 1456 | 1490 | 1604 | 1402 | 1009 |
| | 40 | 1106 | 1182 | 1339 | 1424 | 1389 | 1309 | 1263 | 1371 | 1451 | 1626 | 1464 | 1058 |
| | 50 | 1125 | 1174 | 1292 | 1324 | 1256 | 1170 | 1137 | 1259 | 1381 | 1610 | 1490 | 1082 |
| | vert. | 944 | 899 | 847 | 719 | 599 | 546 | 548 | 647 | 843 | 1189 | 1240 | 929 |
| Portland, OR | hor. | 578 | 872 | 1321 | 1495 | 1889 | 1992 | 2065 | 1774 | 1410 | 1005 | 578 | 508 |
| | 30 | 1015 | 1308 | 1684 | 1602 | 1836 | 1853 | 1959 | 1830 | 1670 | 1427 | 941 | 941 |
| | 40 | 1114 | 1393 | 1727 | 1569 | 1746 | 1739 | 1848 | 1771 | 1680 | 1502 | 1020 | 1042 |
| | 50 | 1184 | 1442 | 1727 | 1502 | 1622 | 1594 | 1702 | 1673 | 1651 | 1539 | 1073 | 1116 |
| | vert. | 1149 | 1279 | 1326 | 953 | 889 | 824 | 890 | 989 | 1172 | 1309 | 1010 | 1109 |

*Source: Ref. 5.*

**Table 16.2 Mean percentage of possible sunshine for selected U.S. cities.**

| Station | Jan. | Feb. | Mar. | Apr. | May | June | July | Aug. | Sept. | Oct. | Nov. | Dec. | Annual |
|---|---|---|---|---|---|---|---|---|---|---|---|---|---|
| Albuquerque, NM | 70 | 72 | 72 | 76 | 79 | 84 | 76 | 75 | 81 | 80 | 79 | 70 | 76 |
| Atlanta, GA | 48 | 53 | 57 | 65 | 68 | 68 | 62 | 63 | 64 | 67 | 60 | 47 | 60 |
| Boston, MA | 47 | 56 | 57 | 56 | 59 | 62 | 64 | 63 | 61 | 58 | 48 | 48 | 57 |
| Chicago, IL | 44 | 49 | 53 | 56 | 63 | 69 | 73 | 70 | 65 | 61 | 47 | 41 | 59 |
| Ft. Worth, TX | 56 | 57 | 65 | 66 | 67 | 75 | 78 | 78 | 74 | 70 | 63 | 58 | 68 |
| Lincoln, NB | 57 | 59 | 60 | 60 | 63 | 69 | 76 | 71 | 67 | 66 | 59 | 55 | 64 |
| Los Angeles, CA | 70 | 69 | 70 | 67 | 68 | 69 | 80 | 81 | 80 | 76 | 79 | 72 | 73 |
| New Orleans, LA | 49 | 50 | 57 | 63 | 66 | 64 | 58 | 60 | 64 | 70 | 60 | 46 | 59 |
| Portland, OR | 27 | 34 | 41 | 49 | 52 | 55 | 70 | 65 | 55 | 42 | 28 | 23 | 48 |

*Source:* Ref. 7.

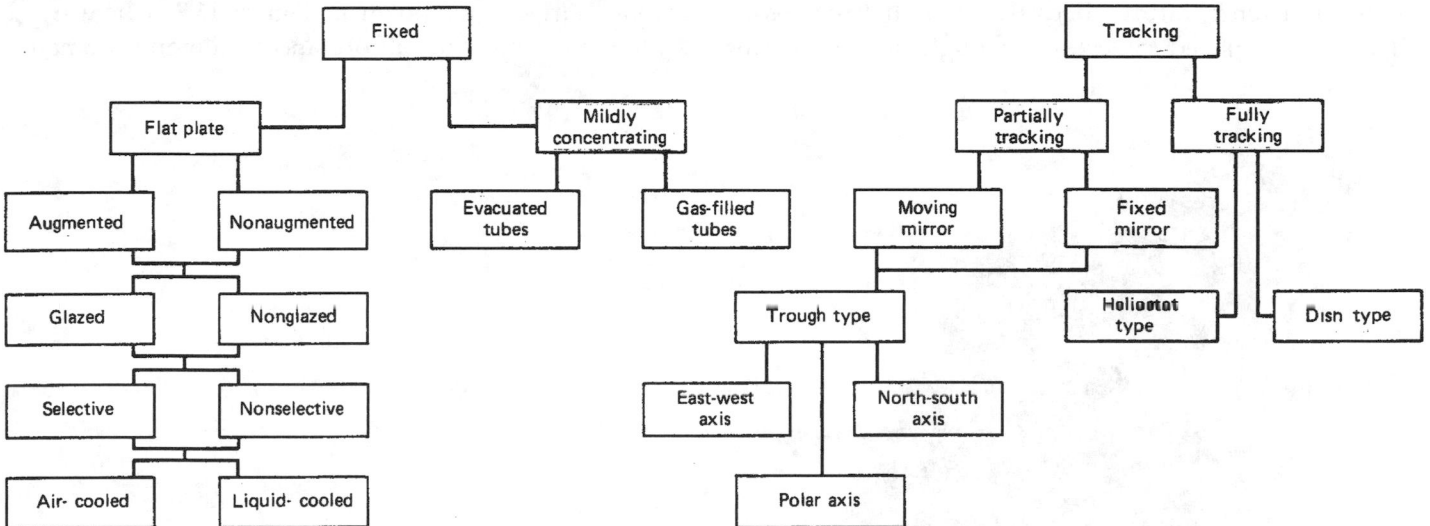

**Figure 16.3 Types of solar collectors.**

absorber plate also radiates energy back toward the sky, but this radiation is longer-wavelength radiation and most of this radiation not reflected back to the absorber by the glazing is absorbed by the glazing. The heated glazing, in turn, converts some of the absorbed energy back to the air space between it and the absorber plate. The trapping of sunlight by the glazing and the consequent heating is known as the "greenhouse effect."

Energy is removed from the collector by the coolant fluid. A steady condition would be reached when the absorber temperature is such that losses to the coolant and to the surroundings equal the energy gain from the solar input. When no energy is being removed from the collectors by the coolant, the collectors are said to be at *stagnation.* For a well-designed solar collector, that stagnation temperature may be well above 300°F. This must be considered in the design of solar collectors and solar systems, since loss of coolant pumping power might be

expected to occur sometime during the system lifetime. A typical coolant flow rate for flat-plate collectors is about 0.02 gpm/ft² of collector surface (for a 20°F rise).

The fraction of the incident sunlight that is collected by the solar collector for useful purposes is called the collector efficiency. This efficiency depends upon several variables, which might change for a fixed absorber plate design and fixed amount of back and side insulation. These are:

1. Rate of insolation
2. Number and type of glazing
3. Ambient air temperature
4. Average (or entering) coolant fluid temperature

A typical single-glazed flat-plate solar collector efficiency curve is given in Figure 16.5. The measured performance can be approximated by a straight line. The left intercept is related to the product τα, where τ is the transmittance of the glazing and α is the absorptance of the absorber plate. The slope of the line is related to the magnitude of the heat losses from the collector, a flatter line representing a collector with reduced heat-loss characteristics.

A comparison of collector efficiencies for unglazed, single-glazed, and double-glazed flat-plate collectors is shown in Figure 16.6. Because of the lack of glazing reflections, the unglazed collector has the highest efficiencies at the lower collector temperatures. This factor, combined with its lower cost, makes it useful for swimming pool heating. The single-glazed collector also performs well at lower collector temperatures, but like the unglazed collector, its efficiency drops off at higher collection temperatures because of high front losses. The double-glazed collector, although not performing too well at lower temperatures, is superior at the higher temperatures and might be used for space heating and/or cooling applications. The efficiency of an evacuated tube collector is also shown in Figure 16.6. It can be seen that it performs very poorly at low temperatures, but because of small heat losses, does very well at higher temperatures.

A very important characteristic of a solar collector surface is its selectivity, the ratio of its absorptance $\alpha_s$ for sunlight to its emittance $\varepsilon$ for long-wavelength radiation. A collector surface with a high value of $\alpha_s/\varepsilon$ is called a selective surface. Since these surfaces are usually formed by a coating process, they are sometimes called *selective coatings*. The most common commercial selective coating is *black chrome*. The characteristics of a typical black chrome surface are shown in Figure 16.7, where $\alpha_\lambda = \varepsilon_\lambda$, the monochromatic absorptance and monochromatic

**Figure 16.4 Typical double-glazed flat-plate collector, liquid type, internally manifolded. (Courtesy LOF.)**

**Figure 16.5 Efficiency of a typical liquid-type solar collector panel.**

emittance of the surface. Note that at short wavelengths (~ 0.5 μ), typical of sunlight, the absorptance is high. At the longer wavelengths (~2 μ and above), where the absorber plate will emit most of its energy, the emittance is high. Selective surfaces will generally perform better than ordinary blackened surfaces. The performance of a flat black collector and a selective coating collector are compared in Figure 16.8. The single-glazed selective collector performance is very similar to the double-glazed nonselective collector. Economic considerations usually lead one to pick a single-glazed, selective or a double-glazed, nonselective collector over a double-glazed, selective collector, although this decision depends heavily upon quoted or bid prices.

Air-type collectors are particularly useful where hot air is the desired end product. Air collectors have distinct advantages over liquid-type collectors:

1. Freezing is not a concern.
2. Leaks, although undesirable, are not as detrimental as in liquid systems.
3. Corrosion is less likely to occur.

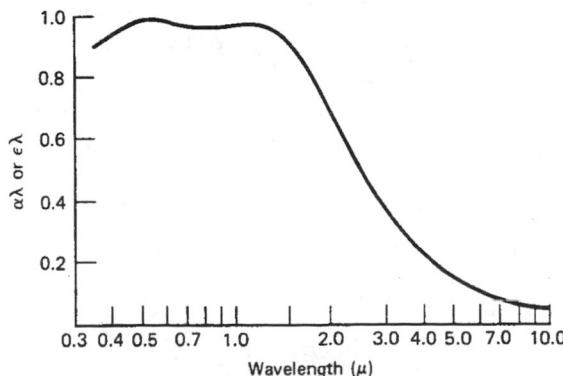

**Figure 16.7 Characteristics of a typical selective (black chrome) collector surface.**

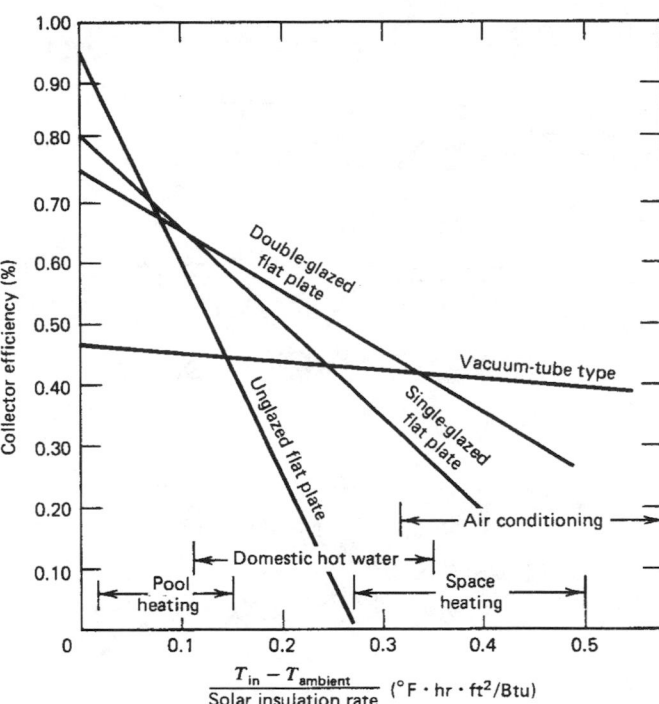

**Figure 16.6 Comparison of collector efficiencies for various liquid type collectors.**

Air systems may require large expenditures of fan power if the distances involved are large or if the delivery ducts are too small. Heat-transfer rates to air are typically lower than those to liquids, so care must be taken in air collectors and in air heat exchangers to provide sufficient heat-transfer surface. This very often involves the use of extended surfaces or fins on the sides of the surface, where air is to be heated or cooled. Typical air collector designs are shown in Figure 16.9.

Flat-plate collectors usually come in modules

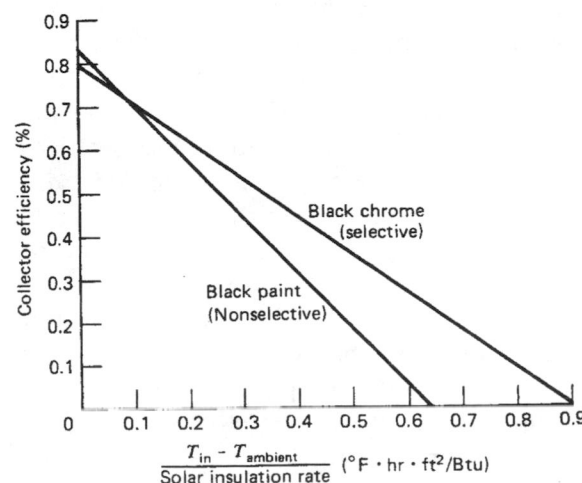

**Figure 16.8 Comparison of the efficiencies of selective and nonselective collectors.**

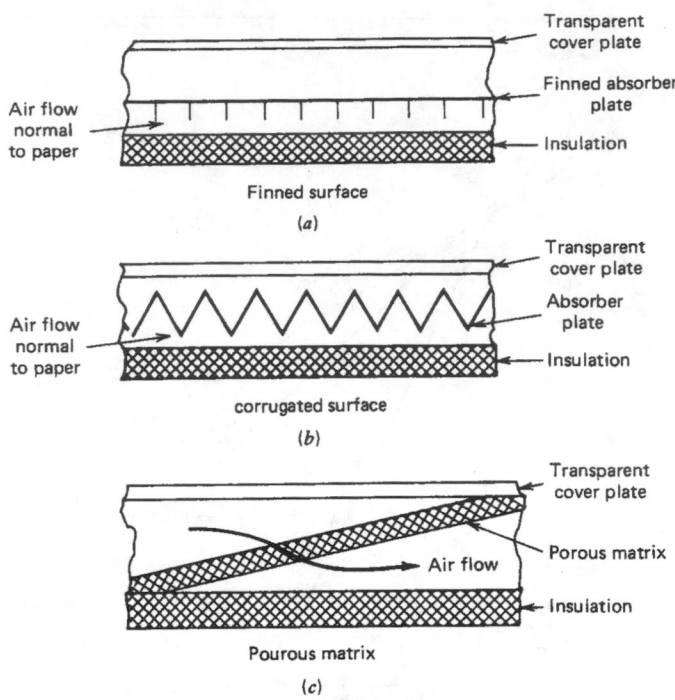

**Figure 16.9 Typical air collector designs. (a) Finned surface. (b) Corrugated surface. (c) Porous matrix.**

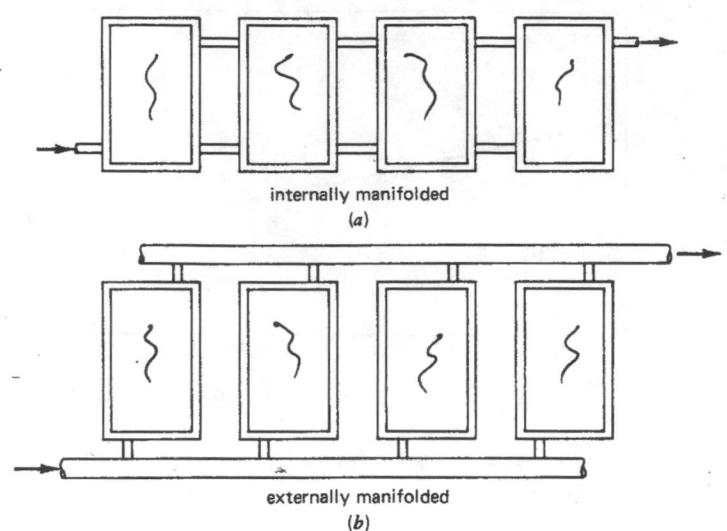

**Figure 16.10 Examples of collectors hooked in parallel. (a) Internally manifolded. (b) Externally manifolded.**

about 3 ft wide by 7 ft tall, although there is no standard size. Collectors may have internal manifolds or they may be manifolded externally to form collector arrays (Figure 16.10). Internally manifolded collectors are easily connected together, but only a small number can be hooked together in a single array and still have good flow distribution. Small arrays (5 to 15) are often piped together with similar arrays in various series and parallel arrangements to give the best compromise between nearly uniform flow rates in each collector, and as small a pressure drop and total temperature rise as can be attained. Externally manifolded collectors are easily connected in balanced arrays if the external manifold is properly designed. However, these types of arrays require more field connections, however, have more exposed piping to insulate, and are not as neat looking.

The overall performance of a collector array, measured in terms of the collector array efficiency, may be quite a bit less than the collector efficiency of the individual collectors. This is due primarily to unequal flow distribution between collectors, larger temperature rises in series connections than in single collectors, and heat losses from the connecting piping. A good array design will minimize these factors together with the pumping requirements for the array.

Concentrating collectors provide relatively high temperatures for applications such as air conditioning, power generation, and the furnishing of industrial or

process heat above 250°F (121°C). They generally cannot use the diffuse or scattered radiation from the sky and must track so that the sun's direct rays will be concentrated on the receiver. The theory is simple. By concentrating the sun's rays on a very small surface, heat losses are reduced at the high temperature desired. An important point to make is that concentrating collectors do not increase the amount of energy above that which falls on the mirrored surfaces; the energy is merely concentrated to a smaller receiver surface.

A typical parabolic trough-type solar collector array is shown in Figure 16.11. Here the concentrating surface or mirror is moved, to keep the sun's rays concentrated as much as possible on the receiver, in this case a tube through which the coolant flows. In some systems the tube moves and the mirrored surfaces remain fixed.

This type collector can be mounted on an east-west axis and track the sun by tilting the mirror or receiver in a north-south direction (Figure 16.12a). An alternative is to mount the collectors on a north-south axis and track the sun by rotating in an east-west direction (Figure 16.12b). A third scheme is to use a polar mount, aligning the trough and receiver parallel to the earth's pole, or inclined at some angle to the pole, and tracking east to west (Figure 16.12c). Each has its advantages and disadvantages and the selection depends upon the application. A good discussion of concentrating collectors is given in Ref. 8.

Fully tracking collectors may be a parabolic disk with a "point source" or may use a field of individual nearly flat moving mirrors or heliostats, concentrating their energy on a single source, such as might be installed

**Figure 16.11 Typical parabolic trough-type solar collector array (Suntec, Inc.).**

(a)

(b)

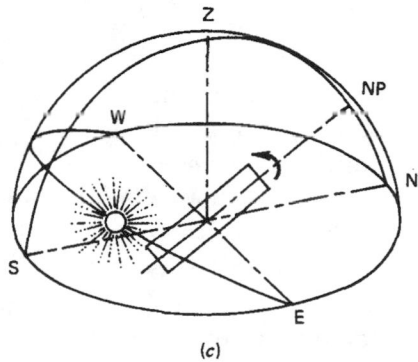

(c)

**Figure 16.12 Trough-type collector arrangements for sun tracking. (a) N-S horizontal E-W tracking. (b) E-W horizontal N-S tracking. (c) Polar axis E-W tracking.**

on a tower (a power tower). Computers usually control the heliostat motion. Some trough-type collectors are also fully tracking, but this is not too common. All partial and fully tracking collectors must have some device to locate the sun in the sky, either by sensing or by prediction. Tracking motors, and in some cases flexible or movable line connections, are additional features of tracking systems. Wind loads can be a serious problem for any solar collector array that is designed to track. Ability to withstand heavy windloads is perhaps the biggest single advantage of the flat-plate, fixed collector array.

### 16.2.3 Thermal Storage Systems

Because energy demand is almost never tied to solar energy availability, a storage system is usually a part of the solar heating or cooling system. The type of storage may or may not depend upon the type of collectors used. With air-type collectors, however, a rock-bed type of storage is sometimes used (Figure 16.13). The rocks are usually in the size range 3/4 to 2 in. in diameter to give the best combination of surface area and pressure drop. Air flow must be down for storing and up for removal if this type system is to perform properly. Horizontal air flow through a storage bed should normally be avoided. An air flow rate of about 2 cfm/ft$^2$ of collector is recommended. The amount of storage required in any solar heating system is tied closely to the amount of collector surface area installed, with the optimum amount being determined by a computer calculation. As a rule of thumb, for rough estimates one should use about 75 lb of rock per square foot of air-type collectors. If the storage is too large, the system will not be able to attain sufficiently high temperatures, and in addition, heat losses will be high. If the storage is too small, the system will overheat at times and may not collect and store a large enough fraction of the energy available.

The most common solar thermal storage system is one that uses water, usually in tanks. As a rule the water storage tank should contain about 1.8 gal/ft$^2$ of collector surface. Water has the highest thermal storage capabil-

**Figure 16.13 Rock-bed-type storage system.**

**Figure 16.14 External heat exchanger between collectors and main storage.**

ity of any common single-phase material per unit mass or per unit volume. It is inexpensive, stable, nontoxic, and easily replaced. Its main disadvantage is its high vapor pressure at high temperatures. This means that high pressures must be used to prevent boiling at high temperatures.

Water also freezes, and therefore in most climates, the system must either (1) drain all of the collector fluid back into the storage tank, or (2) use antifreeze in the collectors and separate the collector fluid from the storage fluid by use of a heat exchanger.

Drain-down systems must be used cautiously because one failure to function properly can cause severe damage to the collectors and piping. It is the more usual practice in large systems to use a common type of heat exchanger, such as a shell-and-tube exchanger, placed external to the storage tank, as shown in Figure 16.14. Another method, more common to small solar systems, is to use coils of tubing around the tank or inside the tank, as shown in Figure 16.15.

In any installation using heat exchangers between the collectors and storage, the exchanger must have sufficient surface for heat transfer to prevent impairment of system performance. Too small a surface area in the exchanger causes the collector operating temperature to be higher relative to the storage tank temperature, and the collector array efficiency decreases. As a rough rule of thumb, the exchanger should be sized so as to give an effectiveness of at least 0.60, where the effectiveness is the actual temperature decrease of the collector fluid passing through the exchanger to the maximum possible temperature change. The maximum possible would be the difference between the design temperature of the collector fluid entering the exchanger and the temperature entering from the storage tank.

Stratification normally occurs in water storage

**Figure 16.15 Internal heat exchanger between collector and storage medium.**

systems, with the warmest water at the top of the tank. Usually, this is an advantage, and flow inlets to the tank should be designed so as not to destroy this stratification. The colder water at the bottom of the tank is usually pumped to the external heat exchanger and the warmer, returning water is placed at the top or near the center of the tank. Hot water for use is usually removed from the top of the tank.

Phase-change materials (PCMs) have been studied extensively as storage materials for solar systems. They depend on the ability of a material to store thermal energy during a phase change at constant temperature. This is called latent storage, in contrast to the sensible storage of rock and water systems. In PCM systems large quantities of energy can be stored with little or no change in temperature. The most common PCMs are the eutectic salts. Commercial PCMs are relatively expensive and, to a certain extent, not completely proven as to lifetime and reliability. They offer distinct advantages, however, particularly in regard to insulation and space requirements, and will no doubt continue to be given attention.

### 16.2.4 Control Systems

Solar systems should operate automatically with little attention from operating personnel. A good control system will optimize the performance of the system with reliability and at a reasonable cost. The heart of any solar thermal collecting system is a device to turn on the collector fluid circulating pump (and other necessary devices) when the sun is providing sufficient insolation so that energy can be collected and stored, or used. With flat-plate collectors it is common to use a differential temperature controller (Figure 16.16), a device with two temperature sensors. One sensor is normally located on the collector fluid outlet and the other in the storage tank near the outlet to the heat exchanger (or at the level of the internal heat exchanger). When the sun is out, the fluid in the collector is heated. When a prescribed temperature difference (about 20°F) exists between the two sensors, the controller turns on the collector pump and other necessary devices. If the temperature difference drops below some other prescribed difference (about 3 to 5°F), the controller turns off the necessary devices. Thus clouds or sundown will cause the system to shut down and prevent not only the unnecessary loss of heat to the collectors but also the unnecessary use of electric-

ity. The distinct temperature difference to start and to stop is to prevent excessive cycling.

Differential temperature controllers are available with adjustable temperature difference settings and can also be obtained to modulate the flow of the collector fluid, depending upon the solar energy available.

Controllers for high-temperature collectors, such as evacuated tubes and tracking concentrators, sometimes use a light meter to sense the level of sunlight and turn on the pumps. Some concentrating collectors are inverted for protection when light levels go below a predetermined value.

In some systems the storage fluid must be kept above some minimum value (e.g., to prevent freezing). In such cases a *low-temperature controller* is needed to turn on auxiliary heaters if necessary. A *high-temperature controller* may also be needed to bypass the collector fluid or to turn off the system so that the storage fluid is not overheated.

Figure 16.17 shows a control diagram for a solar-heated asphalt storage system (Figure 16.18) in which the fluid must be kept between two specified temperatures. Solar heat is used whenever it is available (collector pump on). If the storage temperature drops below the specified minimum, the pump *and* an electric heater are turned on to circulate electrically heated fluid to the tank. If the tank fluid gets too warm, the system shuts off. Almost any required control pattern can be developed for solar systems using the proper arrangement of a differential temperature controller, high- and low-temperature controllers, relays, and electrically operated valves.

### 16.2.5 Sizing and Economics

An article on how to identify cost-effective solar-thermal applications is given in the *ASHRAE Journal*[9]. In almost any solar energy system the largest single expense are the solar collector panels and support structure. For this reason the system is usually "sized" in terms of collector panel area. Pumps, piping, heat exchangers, and storage tanks are then selected to match.

Very rarely can a solar thermal system provide 100% of the energy requirements for a given application. The optimum-size solar system is the one that is the most economical on some chosen basis. The computations may be based on (1) lowest life-cycle cost, (2) quickest payout, (3) best rate of return on investment, and, (4) largest annual savings. All of these computations involve the initial installed cost, the operating and maintenance costs, the life of the equipment, the cost of money, the cost of fuel, and the fuel escalation rate, in ad-

**Figure 16.16 Installation of a differential temperature controller in a liquid heating system.**

**Figure 16.17 Control system for the solar-heated asphalt storage tank of Figure 16.**

**Figure 16.18 Flow schematic of a solar-heated asphalt storage**

dition to computations involving the amount of energy furnished by the solar system.

A typical set of calculations might lead to the results shown in Figure 16.19, the net annual savings per year versus the collector area, with the present cost of fuel as a parameter.[10] Curve "a" represents a low fuel cost, the net savings is negative, and the system would cost rather than save money. Curve "b" represents a slightly higher fuel cost where a system of about 800 ft$^2$ of collectors would break even.

Curves "c" and "d," representing even higher fuel costs, show a net savings, with optimum savings occurring at about 1200 and 2000 ft$^2$, respectively.

High interest rates tend to reduce the economic viability of solar systems. High fuel costs obviously have the opposite effect, as does a longer life of the equipment. Federal and state tax credits would also have an important effect on the economics of solar energy as an

alternative energy source. Technical improvements and lower first costs can obviously have an important effect on the economics, but contributions of these two factors have been relatively slow in coming.

### 16.2.6  Solar Cells

Solar cells use the electronic properties of semiconductor material to convert sunlight directly into electricity. They are widely used today in space vehicles and satellites, and in terrestrial applications requiring electricity at remote locations. Since the conversion is direct, solar cells are not limited in efficiency by the Carnot principle. A wide variety of text under titles such as solar cells, photovoltaics, solar electricity, and semiconductor technology are available to give details of the operating principles, technology and system applications of solar cells.

Most solar cells are very large area p-n junction

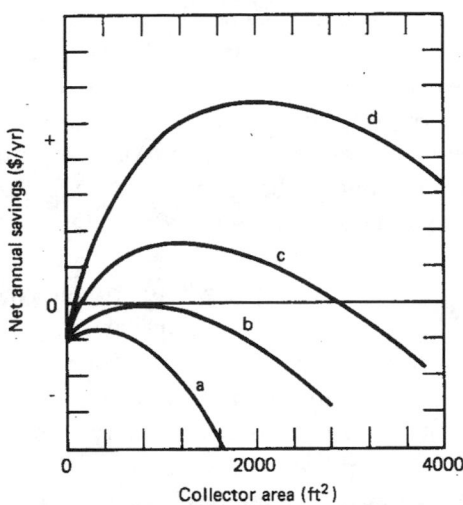

**Figure 16.19 Collector area optimization curves for a typical solar heating system. Ref. 10.**

diodes. Figure 16.20a. A p-n junction has electronic asymmetry. The n-type regions have large electron densities but small hole densities. Electrons flow readily through the material but holes find it very difficult. P-type material has the opposite characteristic. Excess electron-hole pairs are generated throughout the p-type material when it is illuminated. Electrons flow from the p-type region to the n-type and a flow of holes occurs in the opposite direction. If the illuminated p-n junction is electrically short circuited a current will flow in the short-circuiting lead. The normal rectifying current-voltage characteristic of the diode is shown in Figure 16.20b. When illuminated (insulated) the current generated by the illumination is superimposed to give a characteristic where power can be extracted.

The characteristic voltage and current parameters of importance to utilizing solar cells are shown in Figure 16.20b. The short-circuit current $I_{sc}$ is, ideally, equal to the light generated current $I_L$. The open-circuit voltage $V_{oc}$ is determined by the properties of the semiconductor. The particular point on the operating curve where the power is maximum, the rectangle defined by $V_{mp}$ and $I_{mp}$ will have the greatest area. The fill factor FF is a measure of how "square" the output characteristics are. It is given by:

$$FF = \frac{V_{mp}I_{mp}}{V_{oc}I_{sc}}$$

Ideally FF is a function only of the open-circuit voltage and in cells of reasonable efficiency has a value in the range of 0.7 to 0.85.

Most solar cells are made by doping silicon, the second most abundant element in the earth's crust. Sand is reduced to metallurgical-grade silicon, which is then further purified and converted to single-crystal silicon wafers. The wafers are processed into solar cells which are then encapsulated into weatherproof modules. Boron is used to produce p-type wafers and phosphorus is the most common material used for the n-type impurity. Other types of solar cells include CdS, CdTe, and GaAs. Theoretical maximum conversion efficiencies of these simple cells are around 21 to 23 percent.

Major factors which, when present in real solar cells, prohibit the attainment of theoretical efficiencies include reflection losses, incomplete absorption, only partial utilization of the energy, incomplete collection of electron-hole pairs, a voltage factor, a curve factor and internal series resistance.

Thin-film solar cells have shown promise in reducing the cost of manufacturing and vertical junction

(a)

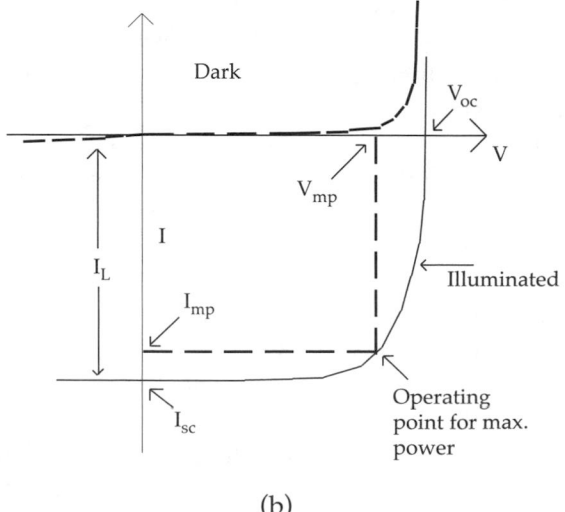

(b)

**Figure 16.20 Nomenclature of solar cells.**

cells have been shown to have high end-of-life efficiencies. Solar cells are subject to weathering and radiation damage. Care must be taken in solar arrays to avoid poor interconnection between cells, and increased series resistance due to deterioration of contacts.

Solar cells are arranged in a variety of series and parallel arrangements to give the voltage-current characteristics desired and to assure reliability in case of individual cell failure. Fixed arrays are placed at some optimal slope and usually faced due south in the northern hemisphere. Large arrays are usually placed on a structure allowing tracking of the sun similar to those used for concentrating solar thermal collectors. In some arrays the sunlight is concentrated before it is allowed to impinge on the solar cells. Provision must be made for thermal energy removal since the solar cell typically converts only a small fraction of the incident sunlight into electrical power. Increasing temperature of the cell has a dominant effect on the open circuit voltage, causing the power output and efficiency to decrease. For silicon cells the power output decreases by 0.4 to 0.5 % per degree Kelvin increase.

Provisions must usually be made for converting the direct current generated by the array into the more useful alternating current at suitable frequency and voltage. In many systems where 24 hour/day electricity is needed some type of storage must be provided.

## 16.3 WIND ENERGY

Wind energy to generate electricity is most feasible at sites where wind velocities are consistently high and reasonably steady. Ideally these sites should be remote from densely populated areas, since noise generation, safety, and disruption of TV images may be problems. On the other hand the generators must be close enough to a consumer that the energy produced can be utilized without lengthy transmission. An article in the EPRI journal (11) gives a good update on wind energy in the electric utility industry as of 1999. Another very useful source of information about wind energy is available from the American Wind Energy Association (12) and from its web site. This group publishes the AWEA Wind Energy Weekly and maintains an archive of back issues. According to Awea 3,600 megawatts of new wind energy capacity were installed in 1999 worldwide, bringing total installed capacity of 13,400 MW. In the United States 895 MW of new generating capacity was added between July 1998 and June 1999. In addition more than 180 MW of equipment was installed in repowering (replacing) older wind equipment. Some of the growth has

been due to supportive government policies at both the state and federal levels, some due to the technology's steadily improving economics, and some due to electric utilities developing "green" policies for customers preferring nonpolluting sources. Early growth was in the mountain passes of California. More recently rapid growth in wind energy has occurred in Minnesota and Iowa as a result of legislative mandates. Other states are expected to follow.

The seacoast of Europe, where strong winds blow consistently, continues to be a popular siting for wind turbines. European manufacturers account for 90 percent of the turbines installed worldwide.

Cost of wind-powered electricity has fallen by about 80% since the early 1980's and is expected to continue to fall as the technology develops. The average cost in 2000 was in the 5 cents/kWh range. To be competitive with conventional sources this cost will have to be cut approximately in half.

### 16.3. 1 Availability

A panel of experts from NSF and NASA estimated that the power potentially available across the continental United States, including offshore sites and the Aleutian arc, is equivalent to approximately $10^5$ GW of electricity.[13] This was about 100 times the electrical generating capacity of the United States. Figure 16.21 shows the areas in the United States where the average wind velocities exceed 18 miles/hr (6 meters/sec) at 150 ft (45.7 m) above ground level. As an approximation, the wind velocity varies approximately as the 1/7 power of distance from the ground.

The power that is contained in a moving air stream per unit area normal to the flow is proportional to the cube of the wind velocity. Thus small changes in wind velocity lead to much larger changes in power available. The equation for calculating the power density of the wind is

$$\frac{P}{A} = \frac{1}{2} \rho V^3 \tag{16.1}$$

where  $P$  = power contained in the wind

$A$  = area normal to the wind velocity

$\rho$  = density of air (about 0.07654 lbm/ft³ or 1.23 kg/m³)

$V$  = velocity of the air stream

Consistent units should be selected for use in equation 16.1. It is convenient to rewrite equation 16.1 as

$$\frac{P}{A} = KV^3 \tag{16.2}$$

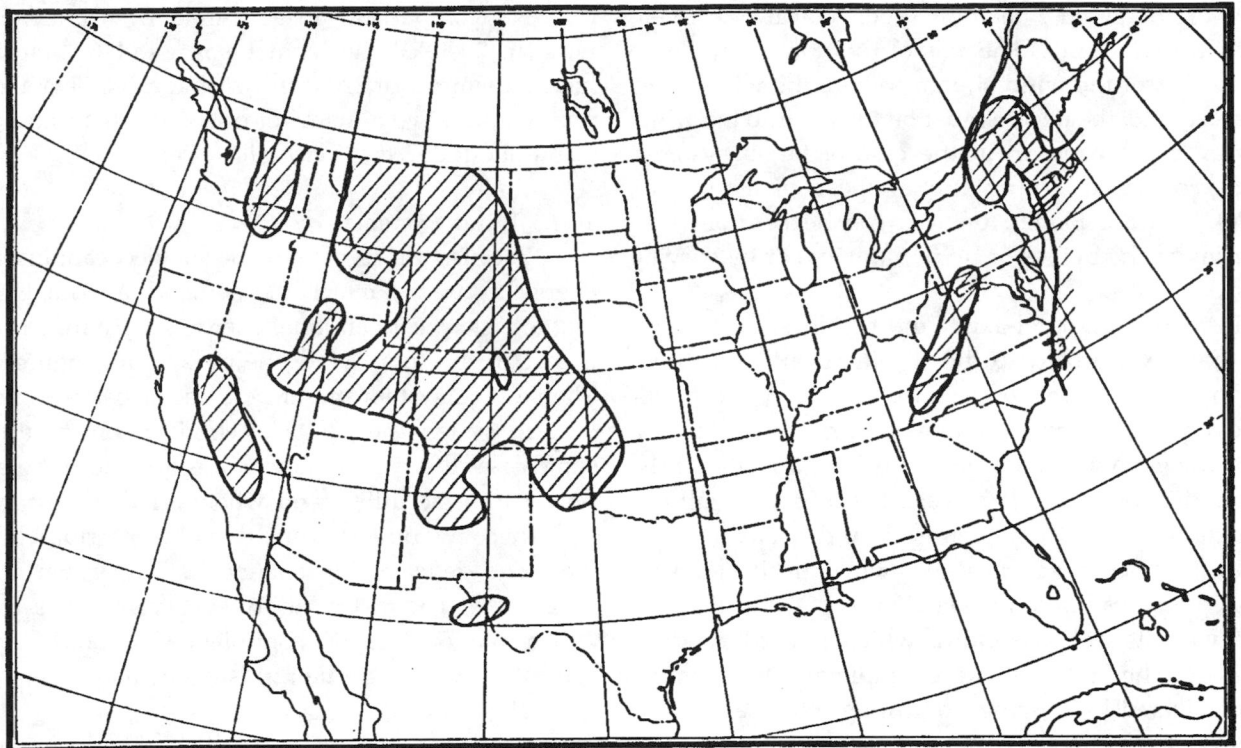

**Figure 16.21 Areas in the United States where average wind speeds exceed 18 miles/hr (8 miles/sec) at 150 ft (45.7 m) elevation above ground level. (From Ref. 13.)**

If the power density $P/A$ is desired in the units $W/ft^2$, then the value of $K$ depends upon the units selected for the velocity $V$. Values of $K$ for various units of velocity are given in Table 16.3.

The fraction of the power in a wind stream that is converted to mechanical shaft power by a wind device is given by the power coefficient $Cp$.

It can be shown that only 16/27 or 0.5926 of the power in a wind stream can be extracted by a wind machine, since there must be some flow velocity downstream from the device for the air to move out of the way. This upper limit is called the Betz coefficient (or Glauert's limit). No wind device can extract this theoretical maximum. More typically, a device might extract some fraction, such as 70%, of the theoretical limit. Thus a real device might extract approximately $(0.5926)(0.70) = 41\%$ of the power available. Such a device would have an aerodynamic efficiency of 0.70 and a power coefficient of 0.41. The power conversion capability of such a device could be determined by using equation 16.2 and Table 16.3. Assume a 20-mile/hr wind. Then

$$\left(\frac{P}{A}\right)_{actual} = \left(5.08 \times 10^{-3}\right)(20)^3(0.41) = 16.7 \ W/ft^2$$

Notice that for a 30-mile/hr wind the power conversion

**Table 16.3 Values of K to Give P/A (W/ ft²) in Equation 16.2[a].**

| Units of $V$ | $K$ |
| --- | --- |
| ft/sec | $1.61 \times 10^{-3}$ |
| miles/hr | $5.08 \times 10^{-3}$ |
| km/hr | $1.22 \times 10^{-3}$ |
| m/sec | $5.69 \times 10^{-2}$ |
| knots | $7.74 \times 10^{-3}$ |

[a]To convert w/ft² to W/m², multiply by 10.76.

capability would be 56.2 W/ft², or more than three times as much.

Because the power conversion capability of a wind device varies as the cube of the wind velocity, one cannot predict the annual energy production from a wind device using mean wind velocity. Such a prediction would tend to underestimate the actual energy available.

### 16.3.2 Wind Devices

Wind conversion devices have been proposed and built in a very wide variety of types. The most general types are shown in Figure 16.22. The most common type is the horizontal-axis, head-on type, typical of conventional farm windmills. The axis of rotation is parallel to the direction of the wind stream. Where the wind direction is variable, the device must be turned into the wind, either by a tail vane or, in the case of larger systems, by a servo device. The rotational speed of the single-, double-, or three-bladed devices can be controlled by feathering of the blades or by flap devices or by varying the load.

In most horizontal-axis wind turbines, the generator is directly coupled to the turbine shaft, sometimes through a gear drive. In the case of the bicycle multi-bladed type, the generator may be belt driven off the rim, or the generator hub may be driven directly off the rim by friction. In the later case there is no rotational speed control except that imposed by the load.

In the case of a vertical-axis wind turbine (VAWT) such as the Savonius or the Darrieus types, the direction of the wind is not important, which is a tremendous advantage. The system is more simple and there are no stresses created by yawing or turning into the wind as occurs on horizontal-axis devices. The VAWT are also lighter in weight, require only a short tower base, and can have the generator near the ground. VAWT enthusiasts claim much lower costs than for comparable horizontal-axis systems.

The side wind loads on a VAWT are accommodated by guy wires or cables stretched from the ground to the upper bearing fixture.

The Darrieus-type VAWT can have one, two, three, or more blades, but two or three are most common. The curved blades have an airfoil cross section with very low starting torque and a high tip-to-wind speed.

The Savonius-type turbine has a very high starting torque but a relatively low tip-to-wind speed. It is primarily a drag-type device, whereas the Darrieus type is primarily a lift-type device. The Savonius and the Darrieus types are sometimes combined in a single turbine to give good starting torque and yet maintain good performance at high rotational speeds.

Figure 16.23 shows the variation of the power coefficient $Cp$ as the ratio of blade tip speed to wind speed varies for different types of wind devices. It can be seen that two-blade types operating at relatively high speed ratios have the highest value of $Cp$, in the range of 0.45, which is fairly close to the limiting value of the Betz coefficient (0.593). The Darrieus rotor is seen to have a slightly lower maximum value, but like the two-blade type, performs best at high rotational speeds. The American (bicycle) multi-blade type is seen to perform best at lower ratios of tip to wind speed, as does the Savonius.

For comparison, in a 17-mile/hr (7.6-meters/sec) wind, a 2000-kW horizontal-axis wind turbine would have a diameter of 220 ft (67 m) and a 2000-kW Darrieus type would have a diameter of 256 ft (78 m) and would stand about 312 ft (95 m) tall.[12]

### 16.3.3 Wind Systems

Because the typical wind device cannot furnish energy to exactly match the demand, a storage system and a backup conventional energy source may be made a part of the total wind energy system (Figure 16.24). The storage system might be a set of batteries and the backup system might be electricity from a utility. In some cases the system may be designed to put electrical power into the utility grid whenever there is a surplus and to draw power from the utility grid whenever there is a deficiency of energy. Such a system must be synchronized with the utility system and this requires either rotational speed control or electronic frequency control such as might be furnished by a field-modulated generator.

Economics favors the system that feeds surplus power into the utility grid over the system with storage, but the former does require reversible metering devices and a consenting utility. Some states have and others probably will pass laws that require public utilities to accept such power transfers.

### 16.3.4 Wind Characteristics—Siting

The wind characteristics given in Figure 16.21 are simple average values. The wind is almost always quite variable in both speed and direction. Gusting is a rapid up-and-down change in wind speed and/or direction. An important characteristic of the wind is the number of hours that the wind exceeds a particular speed. This information can be expressed as speed-duration curves, such as those shown in Figure 16.25 for three sites in the United States. These curves are similar to the load-duration curves used by electric utilities.

Because the power density of the wind depends on the cube of the wind speed, the distribution of annual average energy density of winds of various speeds will be quite different for two sites with different average wind speeds. A comparison between sites having average velocities of 13 and 24 miles/hr (5.8 and 10.7 meters/sec) is given in Figure 16.25. The area under the curve is indicative of the total energy available per unit area per year for each case.

Sites should be selected where the wind speed is as high and steady as possible. Rough terrain and the presence of trees or building should be avoided. The crest of a well-rounded hill is ideal in most cases, whereas

**Figure 16.22 Types of wind-conversion devices. (From Ref. 13.)**

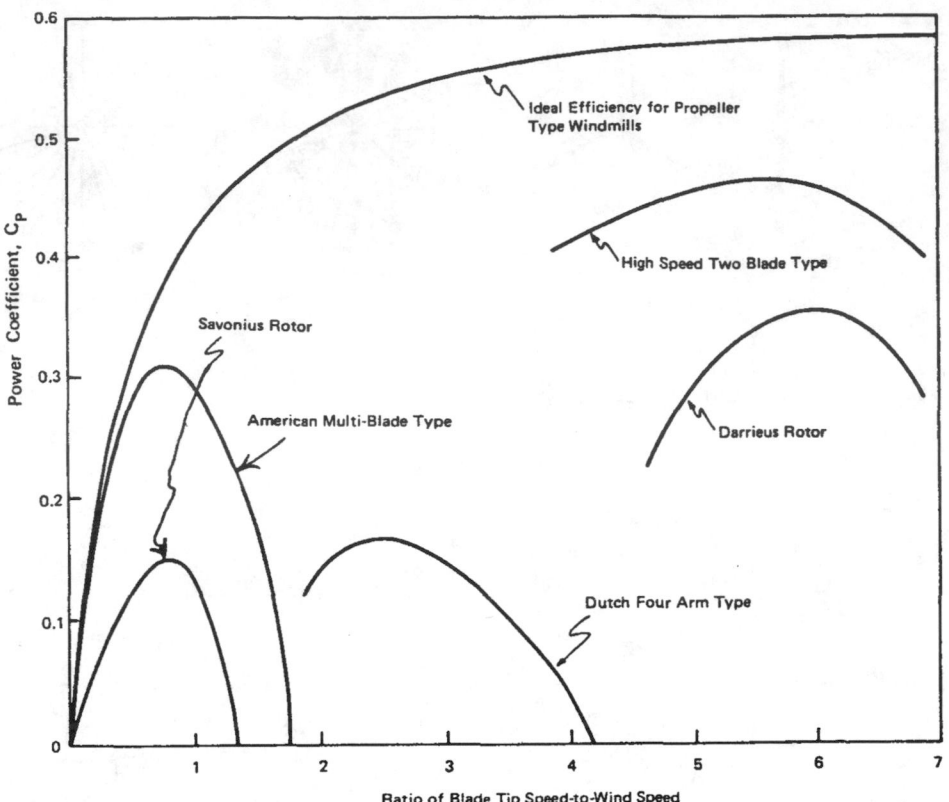

**Figure 16.23 Typical pressure coefficients of several wind turbine devices. (From Ref. 13.)**

**Figure 16.24 Typical WECS with storage. (From Ref. 13.)**

a peak with sharp, abrupt sides might be very unsatisfactory, because of flow reversals near the ground. Mountain gaps that might produce a funneling effect could be most suitable.

### 16.3.5 Performance of Turbines and Systems

There are three important wind speeds that might be selected in designing a wind energy conversion system (WECS). They are (1) cut-in wind speed, (2) rated wind speed, and (3) cut-off wind speed. The names are descriptive in each case. The wind turbine is kept from turning at all by some type of brake as long as the wind speed is below the cut-in value. The wind turbine is shut off-completely at the cut-off wind speed to prevent damage to the turbine. The rated wind speed is the lowest speed at which the system can generate its rated power. If frequency control were not important, a wind turbine would be permitted to rotate at a variable speed as the wind speed changed. In practice, however, since frequency control must be maintained, the wind turbine rotational speed might be controlled by varying the load on the generator when the wind speed is between the

**Figure 16.25 Annual average speed-duration curves for three sites. (From Ref. 13.)**

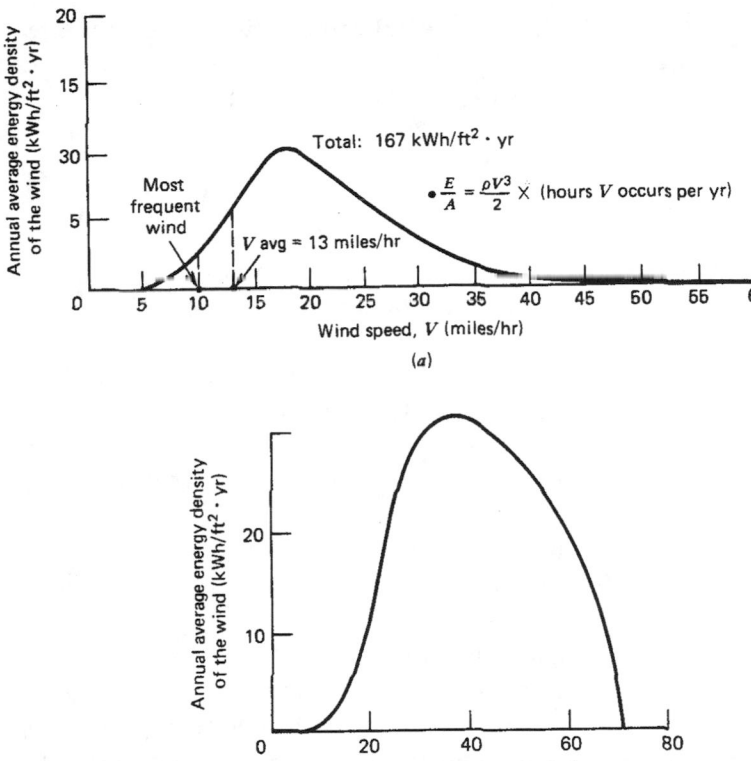

**Figure 16.26 Comparison of distribution of annual average energy density at two sites. (From Ref. 13.) (a) $V_{avg}$ = 13 miles/hr. (b) $V_{avg}$ = 24 miles/hr.**

content of the wind at that site is low for speeds below the cut-in speed and somewhat above the rated speed.

Another useful curve is the actual annual power density output of a WECS (Figure 16.28). The curve shows the hours that the device would actually operate and the hours of operation at full rated power. The curve is for a system with a rated wind speed of 30 miles/hr (13.4 meters/sec), a cut-in velocity of 15 miles/hr (6.7 meters/sec) and a cut-off velocity of 60 miles/hr (26.8 meters/sec) with constant output above 30 miles/hr.

### 16.3.6 Loadings and Acoustics

Blades on wind turbine devices have a variety of extraneous loads imposed upon them. Rotor blades may be subject to lead-lag motions, flapping, and pitching. These motions and some of their causes are shown in Figure 16.29. These loads can have a serious effect on the system performance, reliability, and lifetime.

Acoustics can be a serious problem with wind devices, especially in populated areas. The DOE/ NASA device at Boone, North Carolina, caused some very serious low-frequency (~1 Hz) noises and was taken out of service.

The most promising wind systems from an economic standpoint appear to be mid-size, propeller-type systems, located in large numbers at one site and controlled from a central terminal. Most systems will likely be owned by an electric utility or sell their power to a utility.

### 16.4 REFUSE-DERIVED FUEL

#### 16.4.1 Process Wastes

Typical composition of solid waste is shown in Table 16.4. It can be seen that more than 70% by weight is combustible. More important, more than 90% of the volume of typical solid waste can be eliminated by combustion. Burning waste as fuel has the advantage of not only replacing scarce fossil fuels but also greatly reducing the problem of waste disposal.

Solid wastes affect public health, the environment, and also present an opportunity for reuse or recycling of the material. Managing this in an optimum way is sometimes called *integrated solid waste management*. A textbook on that subject (15) provides much more detail than can be furnished in this brief handbook discussion. That reference estimates that between 2500 and 7750 pounds of waste is generated per person

cut-in and rated speed. When the wind speed is greater than the rated speed but less than cut-out speed, the spin can be controlled by changing the blade pitch on the turbine. This is shown in Figure 16.27 for the 100-kW DOE/NASA system at Sandusky, Ohio. A system such as that shown in Figure 16.27 does not result in large losses of available wind power if the average energy

Figure 16.27 Power output of a 100-kW WECS at various wind speeds. (From Ref. 13.)

Figure 16.28 Actual annual power density output of a WECS. (From Ref. 13.)

each year in the United States. Included in this is typically 2225 pounds of municipal waste, 750 pounds of industrial waste, and between 250 and 3000 pounds of agricultural waste per person each year.

The total mass of solid wastes in the United States reached more than 41 billion tons in 1971[14] and is probably more than double that amount today. It was estimated that each person in the United States consumed 660 Ib of packaging material in 1976, and uses over 5 lb/day of waste products.

The heating value of the refuse would be an important consideration in any refuse-derived fuel (RDF) application. Typical heating values of solid waste refuse components are given in Table 16.5. Other values are given in Ref. 16.

## Table 16.4 Typical composition of solid waste.

Food wastes—12% by weight
  Garbage (10%)
  Fats (2%)
Noncombustibles—24% by weight
  Ashes (10%)
  Metals (8%): cans, wire, and foil
  Glass and ceramics (6%): bottles primarily
Rubbish—64% by weight
  Paper (42%): various types, some with fillers
  Leaves (5%) Grass (4%)
  Street sweepings (3%)
  Wood (2.4%): packaging, furniture, logs, twigs
  Brush (1.5%)
  Greens (1.5%)
  Dirt (1%)
  Oil, paints (0.8%)
  Plastics (0.7%): polyvinyl chloride, polyethylene, styrene, etc., as found in packaging, housewares, furniture, toys, and nonwoven synthetics
  Rubber (0.6%): shoes, tires, toys, etc.
  Rags (0.6%): cellulose, protein, and woven synthetics
  Leather (0.3%): shoes, tires, toys, etc.
  Unclassified (0.6%)

*Source:* Ref. 14.

### 16.4.2 Refuse Preparation

There are several routes by which waste can be used to generate steam or electricity. The possible paths for municipal wastes are shown in Figure 16.30. In the past the most common method was for the refuse to

**Figure 16.29 Rotor blade motions and their causes. (From Ref. 13.)**

ROTOR BLADE MOTIONS

CAUSES OF EXTRANEOUS SYSTEM LOADS

**Figure 16.30 Possible paths to convert municipal refuse to fuel. (From Ref. 15.)**

be burned unprepared in a waterwall steam generator. This technology is simple and well developed and costs can be accurately predicted. Another approach is for the refuse to be placed in a landfill and gas formed from decomposition of the organic material recovered and burned. This is not efficient and requires land for use in the landfill. Refuse may be given some treatment, such as shredding and separation, and then burned in a waterwall steam generator.

The more sophisticated methods involve some type of treatment after shredding to change the refuse into a more desirable fuel form. This may involve converting the shredded refuse into a gas or liquid or into solid pellets.

**Table 16.5 Typical Heating Values of Solid Waste Refuse Components**[a]

|                                      | MJ/kg        | Btu/lb        |
|--------------------------------------|--------------|---------------|
| Domestic refuse                      |              |               |
|   Garbage                  | 4.23         | 1,820         |
|   Grass                    | 8.88         | 3,820         |
|   Leaves                   | It 39        | 4,900         |
|   Rags (cotton, linen)     | 14.97        | 6,440         |
|   Brush, branches          | 16.60        | 7,140         |
|   Paper, cardboard, cartons, bags | 17.81 | 7,660         |
|   Wood, crates, boxes, scrap | 18.19      | 7,825         |
| Industrial scrap and plastic refuse  |              |               |
|   Boot, shoe trim, and scrap | 19.76      | 8,500         |
|   Leather scrap            | 23.24        | 10,000        |
|   Cellophane               | 27.89        | 12,000        |
|   Waxed paper              | 27.89        | 12,000        |
|   Rubber                   | 28.45        | 12,240        |
|   Polyvinyl chloride       | 40.68        | 17,500        |
|   Tires                    | 41.84        | 18,000        |
|   Oil, waste, fuel-oil residue | 41.84    | 18,000        |
|   Polyethylene             | 46.12        | 19,840        |
| Agricultural                         |              |               |
|   Bagasse                  | 8.37-15.11   | 3,600-6,500   |
|   Bark                     | 10.46-12.09  | 4,500-5,200   |
|   Rice hulls               | 12.15-15.11  | 5,225-6,500   |
|   Corncobs                 | 18.60-19.29  | 8,000-8,300   |
| Composite                            |              |               |
|   Municipal                | 10.46-15.11  | 4,500-6,500   |
|   Industrial               | 15.34-16.97  | 6,600-7,300   |
|   Agricultural             | 6.97-13.95   | 3,000-6,000   |

*Source:* Ref. 14.

[a] Calorific value in MJ/kg (Btu/lb) as fired.

The shredding, which can be done wet or dry (as received), converts the refuse into a relatively homogeneous mixture. The shredding is usually done by hammermills or crushers. This shredding operation is costly both in terms of energy and maintenance. One reference gives 1977 maintenance costs of 60 cents/ton of waste.[16]

Problems with fire, explosions, vibrations, and noise are common in the shredding operation.

Density separation increases the fuel's heating value, minimizes wear on transporting and boiler heat-transfer surfaces, and makes the ash more usable. Resource recovery can be an important by-product, with separation of metals and glass for resale.

### 16.4.3 Pyrolysis and Other Processes

Pyrolysis is the thermal decomposition of material in the absence of oxygen. The product can be a liquid or a gas suitable for use as a fuel. There are many pyrolysis projects in the research and development stage. An apparently successful pyrolytic heat-recovery system is described in Ref. 17. The new plant saved $53,000 the first year while disposing of 90% of the firm's waste. It

was expected that the system would pay out in approximately three years. Emissions were said to be below standards set by EPA. A second pyrolytic heat recovery is being installed to take care of expanding needs.

Anaerobic digestion processes, similar to those used in wastewater treatment facilities, can also be used to convert the shredded, separated waste into a fuel. About 3 scf of methane can be produced from about 1 lb of refuse. In this process the shredded organic material is mixed with nutrients in an aqueous slurry, heated to about 140°F, and circulated through a digester for several days. The off-gas has a heating value of about 600 Btu/scf but can be upgraded to nearly pure methane.

Solid fuel pellets can also be prepared from refuse which are low in inorganics and moisture and with heating values around 7500 to 8000 Btu/lbm (17,000 to 19,000 kJ/kg). Some pellets have been found to be too fibrous to be ground in the low- and medium-speed pulverizers that might normally be found in coal-fired plants.

### 16.4.4 Refuse Combustion

The major problems in firing refuse in steam generators seems to be fouling of heat-transfer surfaces and corrosion. Fouling is caused by slag and fly-ash deposition. It is reduced by proper sizing of the furnace, by proper arrangement of heat-transfer surfaces, and by proper use of boiler cleaning equipment.[17]

Corrosion in RDF systems is usually due to:

1. Reducing environment caused by stratification or improper distribution of fuel and air.

2. Halogen corrosion caused by presence of polyvinyl chloride (PVC) in the refuse.

3. Low-temperature corrosion caused when some surface in contact with the combustion gases is below the dew-point temperature of the gas.

It appears that many existing coal-fired boilers can be modified to use suitably prepared refuse as a fuel.

One type of refuse that is both abundant and readily usable in certain types of furnaces and steam generators is tires. Paper mills, cement plants and electric utilities are among those who have discovered this abundant and readily available fuel, with a higher heating value than coal and with less production of pollutants such as nitrous oxides and ash.[18]

Old tires have been a problem in landfills because their shape permits the trapping of water and methane. Tire piles which catch fire are difficult to extinguish and may burn for days, spewing black smoke into the air and oozing oil into the ground. Burning tires as fuel solves the difficult problems created by the U.S. inventory of old tires, estimated to be about 850 million, and the 250 million tires added to the piles each year. Added incentives using old tires are offered by most state governments and tire collectors typically charge a tipping fee for taking a used tire. The company or utility operating the furnace can assume responsibility for direct collection or hire that done by a vendor.

Most of the experience in using tires as a fuel in the U.S. electric utility industry has been with cyclone-fired boilers, which make up less than 10 percent of utility capacity. Little modification is required in these type boilers except for the conveyer system bringing the fuel to the combustion chambers. The metal bead wire around the rim of the tires must be removed prior to combustion and any other metal left after burning must be removed magnetically. The tires are typically burned in cyclone-fired boilers as chips about 1 inch square and as a small percentage of the total fuel. Slightly larger sizes have been successfully burned in stoker-fired units where the fuel sits on a moving grate near the bottom of the boiler.

Tires are more difficult to burn successfully in pulverized-coal boilers except as whole tires in wet-bottom furnaces operating at high temperatures, around 3200°F. In one project whole tires were fed into the boiler at 10 second intervals by a conveyor and lock hopper system. There appears to be good promise of burning tires in fluidized-bed combustion systems, particularly if they are designed in advance to handle such fuel. In the near future it is likely that all tires being discarded will find their way directly into use in some material manufacturing, such as road surfacing, or will be used as a fuel. The split between the various uses will be determined primarily by the economics.

## 16.5  FUEL CELLS

All energy conversion processes that utilize the concept of a heat engine are limited in their thermal efficiency by the Carnot principle. The fuel cell, since it is a direct conversion device, has the advantage of not being limited by that principle. The fuel cell is an electrochemical device in which the chemical energy of a conventional fuel is converted directly and efficiently into low voltage direct-current electrical energy. It can be thought of as a primary battery in which the fuel and oxidizer are stored external to the battery and are fed to it as needed.

In the last decade the fuel cell has emerged as one of the most promising alternative energy technologies. Because of their modular capabilities fuel cells can be built in a wide range of sizes from 200 kW units to power an individual building to 100 megawatt plants to add baseload capacity to an electric utility system. Small plants can operate with efficiencies similar to those of large plants. They can produce high grade waste heat for use in cogeneration or in space heating applications, yielding total energy efficiencies approaching 85 percent. Their reliability makes them useful as uninterruptible power systems for hospitals, communication and computer companies, and hotels. They make little noise, pose little danger to those around and are generally acceptable in close quarters. They can use a variety of fuels and change out between fuels can be accomplished rapidly. They can provide VAR control, they have a quick ramp rate and they can be remotely controlled in unattended operation.

It is estimated that by the year 2010 there will be approximately 130 gigawatts of new electrical generating capacity (26). The fuel cells will have opportunity to be a part of this expanding capacity in all of the areas of repowering, new central power plants , industrial generation and in commercial/ residential generation. Many improvements have been brought forth for use in commercial fuel cells. In the demonstration systems there do not seem to be any serious technical barriers. A major effort will be directed to making the cells more economically competitive and reliable with suitable lifetimes of operation.

Fundamentals of fuel cell operation are described in texts on direct energy conversion, such as the one by Angrist (19) and in the Fuel Cell Handbook (20). A schematic of a fuel cell is given in Figure 16.31. The fuel, which is in gaseous form, diffuses through the anode and is oxidized. This releases electrons to the external circuit. The oxidizer gas diffuses through the cathode where it is reduced by the electrons coming from the anode through the external circuit. The resulting oxidation products are carried away. In contrast to the combustion in a heat engine, where electrons pass directly from the fuel molecules to the oxidizer molecules, the fuel cell keeps the fuel molecules from mixing with the oxidizer molecules. The transfer of electrons takes place through the path containing the load.

The theoretical efficiency of a simple fuel cell is the maximum useful work we can obtain (the change in Gibbs free energy) divided by the heat of reaction. Since the change in Gibbs free energy is given by $\Delta G = \Delta H - T\Delta S$, the efficiency of the simple cell is:

**Figure 16.31 Simple schematic of a fuel cell.**

$$\eta = \frac{G}{H} = 1 - \frac{T\ S}{H}$$

If any heat is being rejected by the cell T$\Delta$S is not zero and the cell efficiency will be less than zero. Other factors that must be considered in real fuel cells are losses associated with attendant accessories, undesirable reactions taking place in the cell, some hindrance of the reaction at the anode or cathode, a concentration gradient in the electrolyte, and Joule heating in the electrolyte. The difference between the maximum useful work and the actual work must appear as rejected heat, in keeping with the first law of thermodynamics. A voltage efficiency is defined by:

$$\eta_v = \frac{V_{ac}}{V}$$

The Faradaic or current efficiency $\eta_F$ is also defined for the fuel cell, and is the fraction of the reaction which is occurring electrochemically to give current. The part of the chemical free energy that actually results in electrical energy is the product of $\eta_v\ \eta_F$.

The ideal fuel would be hydrogen, which if used with pure oxygen as the oxidizer, would give pure water as the product of combustion.

In non-space (utility) operation the four most promising types of fuel cells may be classified by their electrolytes and the corresponding temperatures at which they operate. These are:

| Electrolyte | Temperature |
|---|---|
| phosphoric acid fuel cell (PAFC) | 200°C |
| proton exchange membrane (PEM) | 80°C |
| molten carbonate fuel cells (MCFC) | 650°C |
| solid oxide fuel cell (SOFC) | 1000°C |

Operating temperature has a very significant effect on materials of construction, fabrication methods, and the way in which a unit may be applied.

The phosphoric acid (PAFC) is the only type of fuel cell that is in commercial production. The other three are in various stages of research and development and demonstration. The MCFC and SOFCs are of interest since they promise higher efficiencies and lower first costs than the PAFC. The PEMs are primarily suited for residential/business and transportation application.

The most common PAFC applications today are cogeneration units placed on site and fired with natural gas. Promising fuels for the near future applications include methanol and coal.

The proton exchange membrane (PEM) fuel cells are a recent development, and have a solid polymer membrane electrolyte that can be operated at 80°C. One major developer has been Ballard Power Systems of Canada. Ballard is utilizing plastic for the gas manifolds and other components in an attempt to reduce costs. The Chicago Transit Authority have been using this type of cell in some of their buses.

The molten carbonate fuel cells (MCFC) operate at 650°C, high enough to allow the cell's rejected heat to be used in a thermal steam cycle. Another advantage of the temperature is that reforming can take place within the cell, using a reforming catalyst, thus improving the cell efficiency. The cell operation does not require the use of precious metal catalysts and major cell-stack components can be stamped from less expensive metals. Corrosion prevention and selection of materials for construction are the major technical challenges.

Full-scale demonstrations are now being tested and show promising results both in performance and cost. Earlier demonstration tests pinpointed problems that seem to be solvable by improved control strategies, operating procedure and equipment design.

The solid oxide fuel cell (SOFC) can be fabricated in a variety of shapes due to the use of a solid ceramic electrolyte. Corrosion problems are alleviated and natural gas can be used directly without external reforming. The high temperatures do lead to problems in thermal expansion mismatches and sealing between cells.

Developments on SOFCs have taken place continuously for a longer period of time than any of the other cell types mentioned above. Westinghouse has been a primary developer and has shown their ability to operate these cells for long periods of time. State-of-the-art fuel cell technology is being demonstrated in a 160-kilowatt plant. These plants allow more flexibility in the choice of fuels and show excellent performance in combined cycle operation. The SOFCs can easily follow changing load requirements by adjusting air and fuel flows.

In March of 2000 the U.S. Department of Energy announced that FuelCell Energy, Inc. (FCE) had completed one year of commercial design validation and endurance testing of a 250 kilowatt-class Direct FuelCell™ (DFC™) power plant in Connecticut. Internal conversion of fuel gas to hydrogen occurs directly instead of externally in a separate unit. This promises reduces costs and makes more efficient use of what would have been wasted excess heat. The demonstration involved a record running time of 8600 hours for a carbonate fuel cell fuel cell stack. The stack was the largest of its type ever assembled up to this time. The next step toward commercialization will be field trials of the packaged submegawatt product.

Several organizations have formed to promote fuel cell research, development and commercialization. The World Fuel Cell Council (21) was founded as a nonprofit association in 1991 by a number of fuel cell manufacturers and material suppliers. Its objective is to promote the most rapid commercialization of fuel cell technology worldwide. Members of the Council include companies involved in the development and use of a variety of fuel cell technologies for both stationary and mobile applications.

The U.S. Fuel Cell Council (22) is an industry association dedicated to fostering the commercialization of fuel cells in the United States. It provides technical advice, conducts outreach activities to inform user industries, works to raise public awareness, conducts education programs, provides networking opportunities, and establishes links to comparable activities in the U.S. and elsewhere.

The Fuel Cell Commercialization Group (FCCG) has a mission to commercialize carbonate fuel cells for power generation (23). FCCG members are electric and gas utilities and other energy users that have recognized the opportunity and the value of early involvement in the development and commercialization of this very promising technology. The FCCG is working to design a multi-megawatt carbonate fuel cell power plant that meets utilities' needs for a commercial product, available in the year 2002. FCCG members provide product definition, information exchange, and other market feedback critical to the commercialization process, and will purchase the first fuel cell power plants produced by FCE. The Alliance to Commercialize Carbonate Technology (24), a membership alliance of utilities and industry, was created to help bring the next generation of fuel cell technology, known as molten carbonate fuel cells (MCFC), into commercial markets.

The mission of the Nation Fuel Cell Research Center (NFCRC) is to promote and support the genesis of a fuel cell industry by providing technological leadership within a vigorous program of research, development and demonstration (25). NFCRC's goal is to become a focal point or "technological incubator" for advancing fuel cell technology.

A large amount of support for fuel cell research and development is provided by the United States Department of Energy (26). The department maintains an informative web site, which can lead interested parties into the wide range of activities in this area. (http://www.fe. doe. gov)

## References

1. S.E. Fuller, "The Relationship between Diffuse, Total and Extraterrestrial Solar Radiation," Journal *of Solar Energy and Technology, Vol.* 18, No. 3 (1976), pp. 259263.
2. B.V.H. Liu and R.C. Jordan, "Availability of Solar Energy for Flat-Plate Solar Heat Collectors," Appli*cation of Solar Energy for Heating and Cooling of Buildings,* ASHRAE GRP-170, American Society of Heating, Refrigerating and Air Conditioning Engineers, Inc., New York, 1977.
3. S.A. Klein, "Calculations of Monthly Average Insolation on Tilted Surfaces," Vol. 1, *Proceedings of the ISES and SESC Joint Conference,* Winnipeg, 1976.
4. D.W. Ruth and R.E. Chant, "The Relationship of Diffuse Radiation to Total Radiation in Canada," *Journal of Solar Energy and Technology, Vol.* 18, No. 2 (1976), pp. 153-154.
5. "Introduction to Solar Heating and Cooling Designing and Sizing," DOE/CS-0011 UC-59a, b, c, Aug. 1978.
6. Peter J. Lunde, *Solar Thermal Engineering,* Wiley, New York, 1980.
7. *Climatic Atlas of the United States, U.S.* Government Printing Office, Washington, D.C., 1968.
8. A.B. Meinel and M.P. Meinel, *Applied Solar Energy, an Introduction,* Addison Wesley, Reading, Mass., 1976.
9. Maria Telkes, "Solar Energy Storage," *ASHRAE Journal,* Sept. 1974, pp. 38-44.
10. *ERDA Facilities Solar Design Handbook,* ERDA 77-65, Aug. 1977.
11. Taylor Moore, "Wind Power: Gaining Momentum", *EPRI journal,* Winter 1999, EPRI, Palo Alto, CA.
12. American Wind Energy Association, 122 C Street NW, Washington, DC, 20001, e-mail windmail@awea.org.
13. F.R. Eldridge, *Wind Machines,* NSF-RA-N-75-051, prepared for NSF by the Mitre Corporation, Oct. 1975, U.S. Government Printing Office, Washington, D.C. (Stock No. 038000-00272-4).
14. R.F. Rolsten et al., "Solid Waste as Refuse Derived Fuel," *Nuclear Technology, Vol.* 36 (mid-Dec. 1977), pp. 314-327.
15. G. Tchobanoglous, H. Theisen, and S.A. Vigil, *Integrated Solid Waste Management, McGraw-Hill,* New York, 1993.
16. "Power from Wastes-A Special Report," *Power,* Feb. 1975.
17. A.W. Morgenroth, "Solid Waste as a Replacement Energy Source," *ASHRAE journal,* June 1978.
18. J.H. Hirschenhofer, D.B. Stauffer, R.P. Engleman, M.G. Klett, *Fuel Cell Handbook, Business /* Technology Books, November 1998.
19. World Fuel Cell Council, Kroegerstrasse 5, D-60313 Frankfurt/M, Germany .
20. U.S. Fuel Cell Council, 1625 K. Street NW Suite 75, Washington, DC, 2000
21. Fuel Cell Commercialization Group, 1800 M Street, NW, Suite 300, Washington, DC 20036-5802.
22. The Alliance to Commercialize Carbonate Technology, 1110 Capitol Way S., Suite 307, Olympia, Washington 98501.
23. National Fuel Cell Research Center, University of California, Irvine, Engineering Laboratory Facility, Irvine, California 92697-3550.
24. Fuel Cell Systems, U.S. Department of Energy, Federal Energy Technology Center, 3610 Collins Ferry Road, Morgantown, WV 26507-0880

# CHAPTER 17

# INDOOR AIR QUALITY

H.E. BARNEY BURROUGHS
Atlanta, Ga.

## 17.1 INTRODUCTION AND BACKGROUND

IAQ and Energy Management are in an undeniably linked relationship. The purpose of this chapter is to provide the energy manager with a general and overall understanding of the IAQ issue. From this understanding will come the knowledge of the delicate balance between IAQ and energy management and, especially, the confidence that they can work together synergistically and positively to create buildings that operate both healthfully and efficiently.

Many authors and experts focus on the mid-70s with its energy crisis as the spawning ground of today's IAQ problems. Skyrocketing energy costs led to tight building construction which resulted in drastic reductions in ventilation air and infiltration. This yielded the early expression "Tight Building Syndrome." Unfortunately, this label was not only misleading, it was inadequate in providing a full explanation for the issue. It resulted in the focus of blame on *ventilation* and its inherent *energy* cost as being the "cause all"—"fix all"— of indoor air quality. This is simply not the case, then or now as we enter the new millennium.

There were a number of other trends and changes that were happening at the same time. These combined and intermingled to cause today's situation of many existing or potential problem buildings. Let's examine these issues so that we have a better understanding of how the IAQ situation developed.

### 17.1.1 Impact of "Least Cost" Design and Construction

*"Value Engineering"* has lost its once positive connotation. Now we know it really means devalued and *"cheapened."* This "least cost" mentality led to many design/construction decisions and tactics that save on first cost. However, both the owner and the occupants too often pay a deferred price in increased life/cycle costs, discomfort, illness and health costs, and low productivity. In the extreme, first cost savings return their dubi-ous "value" multifold in the form of lost lease income, expensive remediation, and even litigation.

### 17.1.2 Technical Advances in Office Equipment and Processes

What modern office could survive without the high speed/volume copier, the fax, the laser printer, and the desktop computer terminal? In addition to their heat load, these bring ozone, Volatile Organic Compounds (VOCs), and submicron respirable particles from the print toner into the indoor environment.

### 17.1.3 Changes in Office Layout and Furnishings

High density office layout is now possible through "office systems." In addition to the VOC contaminant outgassing of their fabric and fiberboard components, work stations can add unpredicted barriers to air flow and ventilation effectiveness. Further, they facilitate increased occupancy density which can exceed original shell building design parameters.

### 17.1.4 Shift of Materials used in Construction

Polymers and plastics are now widely used in construction components as glues, binders, and soiling retardants. Typical are formaldehyde (HCHO) and other aldehydes which are ubiquitous components of furniture, fabric, particle board, and plastic surfaces. There is a widening usage of fabrics and fleecy materials in the space. Polymeric glues and high solvent-loaded adhesives were also widely used to attach them. The usage of carpeting is now widespread in schools, for example, where hard surfaces previously prevailed. Porous wall dividers and wall coverings add to this fabric burden and contribute to both outgassing and "sinks." An example is 4PC (4-phenyl-cyclo-hexane) which was a component in the carpet backing. Exposure to high concentration peaks or prolonged exposure over time to these chemicals can trigger a health response risk known as MCS (Multiple-Chemical-Sensitivity).

### 17.1.5 Focus on Energy Conservation in HVAC Systems

Many well-meaning energy managers and engineers designated energy conservation above comfort (remember the "Building Emergency Thermostat Settings" [BETS] of the Carter era). This "feeding frenzy" of energy saving brought on VAV (Variable Air Volume) distribution that was designed to *completely* shutdown supply air to zones with low demand. This resulted in lowered or nonexistent ventilation effectiveness. This meant that conditioned dilution air was not being delivered to the occupied zone. This allows air within the space to become "aged" or stale and stuffy and the system to go out of proper air balance. Unventilated space also becomes a "sink" for contaminants. Contaminants remaining in the space deposit onto and into fleecy and porous surfaces. Then, they re-emerge over time by outgassing due to changes in temperature, humidity, and airflow.

### 17.1.6 Focus on Ventilation

Because of the energy concern and its resulting predominance, ventilation, and the initial label of "Tight Building Syndrome" received undue attention. This is not to say that ventilation is not integrally involved in many IAQ problems. However, this early focus overshadowed the roles of source control and filtration as control mechanisms. It also minimized the many other aspects of good IAQ such as maintenance, moisture control, housekeeping, commissioning, and other facility management practices.

### 17.1.7 Prevailing National Priorities

The early focus on energy and the overall environment lessened under the Reagan Administration with the de-emphasis on regulation. However, the Federal regulators focused on Radon and Asbestos as being serious indoor environmental threats to the general public health and welfare. Heavy regulations were created regarding asbestos in public buildings, such as schools and commercial buildings. The early focus of the regulations was mitigation through removal. This created tremendous financial burden on both the private and public building management segment. Of course, the regulations and their interpretation have since softened. Their impact, however, was to financially burden facility and plant management teams and to sensitize decision makers against environmental concerns.

### 17.1.8 "No Maintenance" Mentality

The traditional "if it ain't broke, don't fix it" attitude matched with the "least cost" perspective has contributed heavily to the current status of building environmental quality. Inadequate, deferred, or nonexistent maintenance can destroy the finest of designs. Most IAQ experts agree that buildings that do not

work properly due to poor upkeep are the most likely candidates for Sick Building Syndrome. EPA researchers claim over half of the complaint buildings they visit are plagued with poor maintenance.

### 17.1.9 Time Marches On

Few energy managers today remember the OPEC crisis that began the energy panics in 1974. Buildings constructed in the early stages of this period have considerable age. Systems, controls, decorations, roofs, elevators, and HVAC systems, have been subjected to decades of use and abuse. Poor maintenance, poor filtration, low ventilation rates, water leaks and least cost purchasing decisions have taken their toll. It was revealed, for example, in the trial concerning the EPA headquarters building in Washington, D.C. that the ductwork had only 1/32" of uniform dirt buildup. Yet, this equated to *tons* of particulate contamination that had to be cleaned out of the building distribution system. This had accumulated over years of operating with low efficiency blanket media particulate filters. The aging process alone is a major factor to deal with when considering IAQ problems.

### 17.1.10 Low Filtration is No Filtration

Low efficiency disposable type furnace filters or blanket media were and are widely used in commercial buildings. They represent the least cost, however, they also represent least performance in providing contaminant control and protection of the systems, the building, and the occupants. The resulting poor filtration allowed contamination accumulation on cooling coils, ductwork, and the facility. This robs efficiency and energy because of operating with fouled inefficient systems, like dirty coils. Poor filtration also provides nutrients for microbial growth as well as a source of odor and VOC sinks. Furthermore, the conditioned and occupied space is not protected from the pollutants generated in the outdoor environment.

### 17.1.11 "It's No Problem Boss!"

The natural response by Facility Managers to IAQ complaints was "frustration mixed with skepticism" and a "tendency to downplay complaints"[1]. Handling of IAQ problems was, therefore, slow or nonexistent. Thus, the real problems are complicated by the perception of lack of concern, lack of response, lack of communications and just poor human relations. This inappropriate response adds human anger to the health effects equation and it usually equals litigation.

### 17.1.12 Public Awareness prevails

IAQ issues have been in all the pertinent trade

publications, most network TV news shows and many hometown newspapers. The general public was conditioned to environmental issues with the focus on radon, asbestos, PCB, insecticides, fertilizers, ozone and lead. Problem buildings like EPA buildings and numerous high profile public courthouses make great headlines.

Thus, your tenants, your employees, even your family and children know about air quality in their space. The general public is therefore suspicious of "places that make them sick" and "chemical smells" and "black stuff that comes out of the ductwork or sticks to walls." This awareness has raised the expectation of occupants regarding the quality of the air in their personal space.

### 17.1.13 Summary of Trends

It should be obvious that there are a number of issues and related trends that worked together to bring us the IAQ issue. The interesting point is that many of these influences are totally under the control of the facility and plant management team. Like the philosopher of the Okefenokee (Walt Kelly's Pogo) once stated. *"We has found the enemy—and he is us."*

## 17.2 WHAT IS THE CURRENT SITUATION?

Have we learned anything? This discussion is intended to brief the reader on our current knowledge of the IAQ issue—now called IEQ (Indoor Environmental Quality).

### 17.2.1 Symptoms of IEQ

We still lack a full understanding of the health effect, linkages and consequences of indoor air contaminant exposure. This is especially true of the more controversial ailments, such as MCS (Multiple Chemical Sensitivity). However, we do have a good working understanding of the health related symptoms that are reported in a sick building. Listed in Table 17.1[2], are the health effects that are typical of occupant complaints. What makes them symptoms of "Sick Building Syndrome" is that they occur only when the individual is exposed to the building environment. When no longer in the exposure environment, the symptoms tend to go away or lessen dramatically. If the symptoms persist even after leaving the space and they fit an identifiable medical diagnostic pattern, then the ailment is likely to be a "Building Related Illness" such as Legionnaire's Disease or Hypersensitivity Pneumonitis. Unfortunately, some of these symptoms are vague and are easily confused with other causes, such as the common cold,

preexisting allergies or the prevailing "flu bug." This fuels the response of facility managers to treat the occupant complaints as "walk-in" problems rather than caused by the building space.

**Table 17.1 Typical IEQ Symptoms**

| |
|---|
| Burning eyes |
| Headache & sinusitis |
| Runny nose |
| Tightness in chest |
| Cough & hoarseness |
| Lethargy & fatigue |
| General malaise |
| Sore throat |
| Itching skin |
| Various combinations of the above |

### 17.2.2 Causes of IEQ Problems

The more common and obvious causes of IEQ problems are known and are listed in Table 17.2. However, they seldom represent a *single* causative element because most problem buildings exhibit multiple and intermingled causes. Further, all of the involved causes have to be brought under control to assure that the building returns to a nonproblem status. We still lack clear linkages between specific cause and the resulting health effect. Yet, we know that the ultimate and overriding cause of IEQ problems is the unusual or elevated concentration of components in the air supplied to the space. The typical air components noted in Table 17.3 are prevalent in most all indoor air and exposure to them in routine concentrations does not bring about health effects or discomfort. However, it is the elevated concentrations of these same components that contaminate the space and bring on health complaints. Thus, it is *contaminant amplification* that is the primary IEQ cause. Therefore, the search for the specific cause and effect in a specific building site should focus on those elements that impact the contaminant amplification and retention, such as water and dirt as amplifiers of fungal growth. When we fully understand these factors, we can then diagnose building related complaints. They also provide insight regarding what makes one building "sick" and another "healthy."

### 17.2.3 No Clear Definition of "Acceptable" Air

We still lack a clearly stated and measurable definition of "acceptable indoor air quality." The primary current parameter is tied to an outdoor air ventilation rate prescribed by the ASHRAE Ventilation Standard 62[4]. This rate is, in turn, tied to the human occupancy

### Table 17.2 Typical IEQ Causes

Low ventilation rates
Inadequate ventilation efficiency
Poor quality makeup air
Poor filtration
Inadequate humidity control
Water incursion
Poor maintenance
Microbiological accumulation and growth
Internal pollutants
Ground gas
Space usage and activity
Concurrent stressors

### Table 17.3 Typical Air Contaminants

Respirable particles
Viable particles
Allergens
Spores
Odors
Corrosive gases
VOCs & Formaldehyde
Hydrocarbons
Reactive inorganic compounds
Oxidants

level of the building and is stated in CFM *per occupant.* This has proven to be an inadequate yardstick because it ignores the contaminant load of the building, its furnishings, and the activities within the space. Further, it promotes the philosophy that air quality and comfort is solely a function of the ventilation rate. This is definitely not the case as has been discussed previously. Thus, designers, building owners, and facility managers need a better target guideline than furnished by Standard 62 for defining and determining acceptable indoor air quality. The Standard specifically defines it as "air in which there are no known contaminants at harmful concentrations as determined by cognizant authorities and with which a substantial majority (80% or more) of the people exposed do not express dissatisfaction." Thus, the issue becomes a *complaint* based and a *comfort* based issue. The 20% dissatisfaction factor also becomes irrelevant when we realize that just one highly vocal or highly affected individual can trigger an IAQ incident. The current ASHRAE Standing Standard Project Committee (SSPC) is revising Standard 62 through addenda that are issued under the continuos maintenance process. They are wrestling with the rationale for establishing

more meaningful and appropriate ventilation rates. In the meantime, it is very important for the IEQ manager to be aware of actual ventilation rates and related occupancy populations within the facility.

### 17.2.4 Litigation as the Driver

With the failure of repeated legislative initiatives and the sidetrack of proposed OSHA regulations, the remaining driver of IEQ is the risk of litigation. Lawsuits in both the worker's compensation and civil courts abound. Often, they are highly publicized and involve millions of dollars in liability and mitigation. The specter of litigation is, therefore, a powerful and destruction force at work in the IEQ arena. A legacy of the proposed OSHA regulation[5] persists, however, in the almost universal response of the building management community to the smoking issue. After only the announcement of a proposed ruling that would ban smoking in the workplace, the voluntary control of tobacco smoking in public space was widespread. This provides some indication of the power of regulation and the influence that it could impose on the indoor air quality issue. This powerful influence provides some guidance to the potential outcome of the ASHRAE Standard 62 revision process as it issues addenda for public review. This may be particularly true for this standard; addenda are being drafted in mandatory code language. Managers of energy and IEQ should track standards and regulations pertaining to IEQ very closely and carefully.

## 17.3 SOLUTIONS AND PREVENTION OF IEQ PROBLEMS

The following discussions provide both practical and philosophical aspects to the prevention and mitigation of the causes of IEQ complaints.

### 17.3.1 Understand the Issue!

Some experts in the field have re-expressed the issue in the broader term "IEQ" (Indoor Environmental Quality) and this is now used in this text. This recognizes that the perceived quality of the conditioned space also includes the broader issues of lighting, sound, ergonomics, and psycho/social issues.

Even as a complex and multidisciplined issue, IEQ control and prevention reduces to simple principles. These principles are inherent in the definition of air *conditioning* as "the control of temperature, humidity, distribution, and cleanliness of air supplied to occupied space." When a building HVAC system fails to address any aspect of this definition, then complaints, discomfort, and health effects can result. Cleanliness of the air is probably the most overlooked but perhaps the most crucial aspect in commercial buildings. Thus, an overriding cause of complaints and adverse health effects from IEQ is an abnormal or excessive exposure to accumulated airborne contaminants.

The tactics to control excessive contaminant buildup in conditioned space are: dilution, extraction, and source control.[4]

*Dilution* is the reduction of contaminant load by *addition* of clean or at least lesser contaminated air, i.e., *ventilation.*

*Extraction is* the reduction of contaminant load by *subtraction* of contaminants by means of *filtration.*

*Source Control is* the reduction of contaminant load by *elimination* by means of exhaust or selection of low emitting components and practices.

**Most of the management, control, or prevention suggestions reflect attention to all three of these control tactics. Of these, however, the most cost effective is source control and the least cost effective is ventilation. Thus, filtration and source control should receive priority when analyzing mitigation alternatives from a cost effectiveness standpoint.**

### 17.3.2 Treat the Issue as Reality!

An IEQ complaint may be real or it may be imagined—but it is *real* to the individual reporting the complaint. Thus, potential IEQ complaints require prompt and positive response from the facility management team. This sends a proactive message to employees and tenants that their comfort, health, and feelings are important to you. This avoids frustration and anger experienced by occupants and avoids a legitimate charge of negligence on your part. Complaints should be logged and follow-up documented which aids in providing information that is helpful in the diagnostics process. Frequency, nature, location, timing and affected individuals provide guidance in determining the causal factors of the incidents. This also helps in the determination of the severity and urgency of the situation.

### 17.3.3 Put Someone in Charge and Make Them Smart!

Appoint an experienced member of the facility management team as your IEQ Manager. Many times this is the Energy Manager because of the close relationship between IEQ and energy management issues. This is consistent with the recommendation in the EPA Building Owners Manual. This individual should be the clearing house for IEQ complaints, investigation, problem resolutions, and record keeping. This also becomes the individual who receives training to become the in-house "guru." The IEQ manager should become very familiar with causes, prevention, and mitigation tactics. ASHRAE, IFMA, BOMA, EPA, AIHA, and AWMA are associations or organizations that are sources of information and training on IEQ matters. In addition, the AEE (Association of Energy Engineers) offers a course on IEQ Fundamentals that leads to a certification (Certified Indoor Air Quality Professional—CIAQP).

### 17.3.4 Establish Proactive Monitoring!

I now hear building managers say "Sure, we respond to IEQ complaints"! That is not proactive but *"reactive response."* Proactive response takes effort, time, and money just like preventative maintenance. Periodic facility inspections and assessments will provide baseline air quality performance data. It can also reveal imminent problems that can be proactively mitigated before occupant complaints occur. This entails a physical inspection walk-through of the entire facility to identify potential IEQ causes and practices before they manifest into complaints. It also may be advisable to perform baseline air diagnostics on key environmental parameters such as temperature and relative humidity, Carbon Dioxide, Carbon Monoxide, Total Volatile Organics, Airborne Microbials, Airborne Particles, and Formaldehyde. This will enable the Facility Manager to establish "norms" for the facility that can be used for comparative purposes if complaints occur and performance levels alter.

### 17.3.5 Use ASHRAE Standard 62

Really use it, which means to go beyond the ventilation tables. Table 17.4[4] numerically designates outdoor air levels for ventilation using the Ventilation Rate Procedure of Standard 62-1999. Most designers stop at this table and never go beyond. Thus, they don't realize that the Standard also covers such issues as outdoor air quality, ventilation effectiveness, monitoring, maintenance, filtration, humidity control, and source control. The Indoor Air Quality alternative procedure allows for consideration of energy management concerns through outdoor air reduction and air cleaning. Because of this broad coverage, Standard 62 becomes an excellent overall protocol for addressing many of the issues discussed above, including energy management.

### 17.3.6 Test, Adjust and Balance Again!

Unfortunately, "TAB" work is usually considered a one-time event—done at system start-up. Then the report is filed away. However, a building is a dynamic model, changing as it matures through its life cycle of changing layouts, tenants, turnover and usage. TAB becomes an effective tool to assure ongoing maximized system performance and ventilation effectiveness.

### 17.3.7 Exploit Filtration!

Perhaps the most under-utilized of the prevention tactics is filtration, both gas phase and particulate. High efficiency filtration addresses contaminant control by extraction, can cost-effectively substitute for ventilation to save energy, and can act as source control. More often than not, urban outdoor air does not even meet the ambient air quality levels prescribed by Standard 62. Though Standard 62 is largely silent on filtration, a new addendum will incorporate a minimum particulate filter efficiency. This will be set at the particle extraction performance of not less than the ASHRAE medium efficiency filter. This is the minimum efficiency value that defines a good pleated filter. Filtration can also cleanse the outdoor dilution air to eliminate it as a contaminant source and enhance its effectiveness as a space dilutant. Further, self-contained air cleaners can provide localized air quality improvement using both gaseous and par-

**Table 17.4 Excerpts from Table II - ASHRAE Standard 62-1999**

| Application | Estimated Maximum Occupancy P/1000 ft$^2$ or 100 m$^2$ | Outdoor Air Requirements cfm person | cfm/ft$^2$ | Comments |
|---|---|---|---|---|
| **Offices** | | | | |
| Office space | 7 | 20 | | Some office equipment may require local exhaust. |
| Reception areas | 60 | 15 | | |
| Telecommunication centers and data entry areas | 60 | 20 | | |
| Conference rooms | 50 | 20 | | Supplementary smoke-removal equipment may be required. |
| **Public Spaces** | | | | |
| Corridors and utilities | | | 0.05 | |
| Public restrooms, cfm/wc or urinal | | 50 | | Mechanical exhaust with no recirculation is recommended. Normally supplied by transfer air, local mechanical exhaust: with no recirculation recommended |
| Locker and dressing rooms | | | 0.5 | |
| Smoking lounge | 70 | 60 | | |
| Elevators | | | 1.00 | Normally supplied by transfer air. |
| **Retail Stores, Sales Floors, and Show Room Floors** | | | | |
| Basement and street | 30 | | 0.30 | |
| Upper floors | 20 | | 0.20 | |

ticulate filter components. Tables 17.5 and 17.6[3] provide additional usages and benefits from more extensive usage of filtration.

### Table 17.5 Usages of Filtration

Protect mechanical equipment such as heat exchange coils

Protect systems such as air distribution ductwork, outlets, and porous components

Protect occupied space such as wall and ceiling surfaces

Protect occupants from contaminant exposure

Protect processes such as pharmaceuticals and electronic chip fabrication

Provide clean makeup air to use as high quality dilution air

Protect environment from contaminated exhaust

Provide source control to avoid reintrainment into the return public air

Augment ventilation air through treatment of the return air

### Table 17.6 Benefits of Filtration

Increased system efficiency from clean components
Increase system life
Lower maintenance costs
Lower housekeeping costs
Avoidance of product failure
Enhanced energy management
Lower health risk and costs
Increased productivity and reduced absenteeism
Avoiding efficiency losses from the insulating effect of dirt build-up on heat transfer surfaces—HVAC coils

### 17.3.8 Use Lifelong Commissioning!

Thought by many to be just system start-up, commissioning is emerging as an essential component of the successful quality construction or renovation process. The commissioning process really commences with the first vision of a construction process, and then proceeds throughout the life of the building. (See ASHRAE Guideline #1, Commissioning).[7] There is a very strong role for the commissioning agent to assure that air quality is built into and retained within the building construction process.

### 17.3.9 Leverage the Use of Source Control!

Controlling the contaminant, whether it be an odor or a toxic substance, is most effectively and efficiently done at the source. This can mean localized

exhaust of contaminants to atmosphere. It can mean localized spot filtration using air cleaners. It can mean a rigorous review of all MSDS records and restricting introduction of purposeful, but harmful, products and chemicals to the space. This includes pesticides and routine housekeeping chemicals. It can also mean a review of construction and buildout components for outgassing properties and/or potential. Specific control tactics are scheduled in Table 17.7[9] and are quoted from *Managing Indoor Air Quality*, an excellent reference book for the energy manager.

### 17.3.10 Look at Controls!

As the nerve center of the building, controls play a major role in effective and efficient building management... getting the right air to the right place at the right time. They both monitor and control response to the varying conditions within and without the building envelope. They can impact air quality in the way that outdoor air levels are managed, distribution systems respond, and mechanical equipment reacts to both the changing dynamics of the building and the needs of the occupants.

### 17.3.11 Diligent Maintenance!

One of the most frequently cited deficiencies in air quality, proper operation and maintenance can be an effective deterrent to IEQ problems. Cleanliness and avoidance of water in the systems are critical to maintaining high air quality. Poor maintenance can unfortunately negate the finest of designs and construction efforts. Likewise, housekeeping and janitorial protocols should be examined. Cleaning chemicals, vacuuming techniques, and other purposeful practices such as pest control, should all be scrutinized to make sure that they are lowering and not aggravating contaminant levels.

### 17.3.12 Perform Life Cycle Costing!

As opposed to "least cost" thinking, (see Chapter 4) life cycle costing must re-emerge as the real cost decision model. For example, high efficiency filters are more expensive first cost than cheap furnace filters. Yet, they are more cost effective when the energy impact, health impact, and long life cycle issues are considered. Recognize the real "life" cycle and the ecological and environmental cost of a product from its cradle to its grave. And the "Green Building" folks would like us to think in terms of cradle to cradle—including the cost and ecological impact of recycling. For example, in some urban communities, furnace filters have to be demanufactured at great cost

## Table 17.7 Typical Control Tactics

| Contaminant | Sources | Control Techniques |
| --- | --- | --- |
| Asbestos | Furnace, pipe, wall ceiling insulation fireproofing, acoustical and floor tiles | 1. Enclose: shield<br>2. Encapsulate; seal<br>3. Remove<br>4. Label ACM<br>5. Use precautions against breathing when disturbed |
| Bioaerosols | Wet insulation, carpet, ceiling tile, wall coverings, furniture, air conditioners, dehumidifiers, cooling towers, drip pans, and cooling coils. People, pets, plants, insects, and soil | Use effective filters. Check and clean any areas with standing water. Be sure condensate pans drain and are clean. Treat with algicides. Maintain humidifiers and dehumidifiers. Check & clean duct linings. Keep surfaces clean. |
| Carbon Monoxide (CO) | Vehicle exhaust, esp. attached garages; unvented kerosene heaters and gas appliances; tobacco smoke; malfunctioning furnaces | Check and repair furnaces, flues, heat exchangers, etc. for leaks. Use only vented combustion appliances. Be sure exhaust from garage does not enter air intake. |
| Combustion Products ($NO_x$ and CO) | Incomplete combustion process | Use vented appliances and heaters. Avoid air from loading docks and garages entering air intake. Check HVAC for leaks regularly, repair promptly. Use plants as air cleaners. Filter particulates. |
| Environmental Tobacco Smoke (ETS) | Passive smoking; sidestream and mainstream smoke | Eliminate smoking; confine smokers to designated areas; or isolate smokers with direct outside exhaust. Increase ventilation. Filter contaminants. |
| Formaldehyde (HCHO) | Building products; i.e., paneling,, materials with lower particleboard, plywood urea-formaldehyde insulation<br><br>as well as fabrics and furnishings | Selective purchasing of materials with lower formaldehyde emissions. Barrier coatings and sealants. New construction commissioning. |
| Radon | Soil around basements and slab on grade | Ventilate crawl spaces. Ventilate sub-slab. Seal cracks, holes, around drain pipes. Positive pressure in tight basements.<br>NOTE: Increased ventilation does not necessarily reduce radon levels. |
| Volatile Organic Compounds (VOCs) | Solvents in adhesives cleaning agents, paints, fabrics, tobacco smoke,<br><br>linoleum, pesticides, gasoline, photocopying materials, refrigerants, building material | Avoid use of solvents and pesticides indoors. If done, employ time of use and excessive ventilations. Localized exhaust near source when feasible. Selective purchasing. Increase ventilation. |

to be disposed in solid waste handling sites. Likewise, air quality concerns and their resulting impact on the human asset base must be considered when design/construct decisions are made. Shortcuts made for reasons of budget, time, or convenience, will demand payment in due course.

### 17.3.13 Prepare for Regulation and Litigation Now!

The proactive Facility Team will prepare for this eventuality by bringing your facilities and your practices into compliance with current "state of the art" and to document your efforts.

It is never too soon to prepare for the eventuality of litigation. The best defense against either litigation or regulation is to proactively follow the recommendations in this chapter. Then document, document, document! Document your design path! Document your decision path!! Document your baseline performance path using proactive response to the IEQ issue!!!

### References

1. *Building Operating Management Magazine,* August, 1994
2. Burroughs, H.E. *The Role of Risk Assessment as a Mitigation Protocol for IAQ Causes.* World Energy Engineering Congress, Association of Energy Engineers Conference, 1991.
3. Burroughs, H.E. *Air Filtration. A New Look At An Existing IAQ Asset.* Carrier Global Engineering Conference, 1994.
4. ASHRAE Standard 62-1999, Ventilation for Acceptable Indoor Air Quality. American Society of Heating, Refrigerating, and Air-Conditioning Engineers, Inc., 1791 Tullie Circle, NE, Atlanta, GA. 30329
5. *Department of Labor. Federal Register (OSHA),* Occupational Safety and Health Administration, Pages 1 5969 -1 6039.
6. *Building Air Quality: A Guide For Building Owners And Facility Managers.* US EPA, 401 M St., Washington, DC 20460.
7. ASHRAE Guideline 1-1997, Guideline for Commissioning of HVAC Systems. American Society of Heating, Refrigerating, and Air-Conditioning Engineers, Inc. 1791 Tullie Circle, NE, Atlanta, GA. 30329.
8. ASHRAE 52.2-1999, Methods of Testing General Ventilation Air-Cleaning Devices for Removal Efficiency by Particle Size.
9. Burroughs, H.E. and Hansen, Shirley J., *Managing Indoor Air Quality,* Second Edition, 1998. The Fairmont Press, Lilburn, GA.

# CHAPTER 18

# ELECTRIC AND GAS UTILITY RATES
# FOR COMMERCIAL AND INDUSTRIAL CUSTOMERS

LYNDA J. WHITE
RICHARD A. WAKEFIELD
JAIRO A. GUTIERREZ
CSA Energy Consultants
Arlington, VA

## 18.1 INTRODUCTION

### Purpose and Limitations

The main focus of this chapter on rates is to provide information on how an average commercial or industrial customer can identify potential rate-related ways of reducing its energy costs. The basic costs incurred by electric and gas utilities are described and discussed. How these costs are reflected in the final rates to commercial and industrial customers is illustrated. Some examples of gas and electric rates and how they are applied are included. In addition, this chapter identifies some innovative rates that were developed by electric and gas utilities as a response to the increasing pressure for the development of a more competitive industry.

Because of the breadth and complexity of the subject matter, the descriptions, discussions and explanations presented in this chapter can not cover every specific situation. Energy consumption patterns are often unique to a particular commercial and/or industrial activity and therefore case by case evaluations are strongly suggested. The purpose is to present some general cost background and guidelines to better understand how to identify potential energy cost savings measures.

### 18.1.2 General Information

Historically, electric and gas utility rate structures were developed by the utilities themselves within a much less complex regulatory environment, by simply considering market factors (demand) as well as cost factors (supply). Today the increasing pressures to develop more competitive markets have forced utilities to reconsider their traditional pricing procedures. Other factors affecting today's electric and gas markets include rising fuel prices, environmental concerns, and energy conservation mandates. These factors and pressures have affected gas and electric utility costs and hence their rates to their final customers.

In general, electric and gas rates differ in structure according to the type and class of consumption. Differences in rates may be due to actual differences in the costs incurred by a utility to serve one specific customer vs. another. Utility costs also vary according to the time when the service is used. Customers using service at off-peak hours are less expensive to serve than on-peak users. Since electricity cannot be stored, and since a utility must provide instantaneous and continuous service, the size of a generation plant is determined by the aggregate amount of service taken by all its customers at any particular time.

The main cost elements generally included in rate-making activities are: energy costs, customer costs, and demand costs. Each of these is discussed in the next section.

## 18.2 UTILITY COSTS

Utilities perform their activities in a manner similar to that of any other privately-owned company. The utility obtains a large portion of its capital in the competitive money market to build its system. It sells a service to the public. It must generate enough revenues to cover its operating expenses and some profit to stay in business and to attract capital for future expansions of its system.

In general there are two broad types of costs incurred by a utility in providing its service. First, there are the fixed capital costs associated with the investment in the facilities needed to produce (or purchase) and deliver the service. Some of the expenses associated with fixed capital costs include interest on debts, depreciation, insurance, and taxes. Second, there are the expenses associated with the operation and maintenance of those same facilities. These expenses include such things as salaries and benefits, spare parts, and the purchasing, handling, preparing, and transporting of energy resources. The rates paid by utility customers are designed to generate the necessary revenues to recover both types of costs. Both capital and operation and maintenance costs are allocated between the major cost elements incurred by a utility.

### 18.2.1 Cost Components

The major costs to a utility can be separated into three components. These include customer costs, energy/commodity costs, and demand costs. These cost components are briefly described below.

*Customer Costs*

Customer costs are those costs incurred in the connection between customer and utility. They vary with the number of customers, not with the amount of use by the customer. These costs include the operating and capital costs associated with metering (original cost and on-going meter-reading costs), billing, and maintenance of service connections.

*Energy/Commodity Costs*

Energy and commodity costs consist of costs that vary with changes in consumption of kilowatt-hours (kWh) of electricity or of cubic feet of gas. These are the capital and operating costs that change only with the consumption of energy, such as fuel costs and production supplies. They are not affected by the number of customers or overall system demand.

*Demand Costs*

Electric utilities must be able to meet the peak demand—the period when the greatest number of customers are simultaneously using service. Gas utilities must be responsive to daily or hourly peak use of gas. In either case, the utility will need to generate or purchase enough power to cover its firm customers' needs at all times. Demand-related costs are dependent upon overall system requirements. Demand costs can be allocated in many different ways, but utilities tend to allocate on-peak load. Included in these costs are the capital and operating costs for production, transmission, and storage (in the case of gas utilities) that vary with demand requirements.

### 18.2.2 Allocation of Costs

Once all costs are identified, the utility must decide how to allocate these costs to its various customer classes. How much of each cost component is directly attributable to serving a residential, a lighting, or a manufacturing customer? In answer to this question, each utility performs a cost-of-service study to devise a set of allocation factors that will allow them to equitably divide these costs to the various users. After the costs are allocated, the utility devises a rate structure designed to collect sufficient revenue to cover all its costs, plus a fair rate of

return (currently, this is running between 10 and 14% of the owners' equity.)

## 18.3  RATE STRUCTURES

### 18.3.1 Basic Rate Structure

The rate tariff structure generally follows the major cost component structure. The rates themselves usually consist of a customer charge, an energy charge, and a demand charge. Each type of charge may consist of several individual charges and may be varied by the time or season of use.

*Customer Charge*

This is generally a flat fee per customer ranging from zero to $25 for a residential customer to several thousand dollars for a large industrial customer. Some utilities base the customer charge to large industrial customers on the level of maximum annual use.

*Energy Charge*

This is a charge for the use of energy, and is measured in dollars per kilowatt-hour for electricity, or in dollars per therm or cubic foot of gas. The energy charge often includes a fuel adjustment factor that allows the utility to change the price allocated for fuel cost recovery on a monthly, quarterly, or annual basis without resorting to a formal rate hearing. This passes the burden of variable fuel costs (either increases or decreases) directly to the consumer. Energy charges are direct charges for the actual use of energy.

*Demand Charge*

The demand charge is usually not applied to residential or small commercial customers, though it is not always limited to large users. The customer's demand is generally measured with a demand meter that registers the maximum demand or maximum average demand in any 15-, 30-, or 60-minute period in the billing month. For customers who do not have a demand meter, an approximation may be made based on the number of kilowatt hours consumed. Gas demand is determined over an hour or a day and is usually the greatest total use in the stated time period.

Another type of demand charge that may be included is a reactive power factor charge; a charge for kilovolt-amp reactive demand (kVAR). This is a method used to charge for the power lost due to a mismatch between the line and load impedance. Where the power-factor charge is significant, corrective action can be taken, for example by adding capacitance to electric motors.

Demand may be "ratcheted" back to a period of greater use in order to provide the utility with revenues to maintain the production capabilities to fulfill the greater-use requirement. In other words, if a customer uses a maximum demand of 100 kW or 100 MMBtu one month, then uses 60 kW or 60 MMBtu for the next six months, he/she may have to pay for 100 kW or MMBtu each month until the ratchet period (generally 12 months) is over.

### 18.3.2 Variations

Utilities use a number of methods to tailor their rates to the needs of their customers. Some of the different structures used to accomplish this include: seasonal pricing; block pricing; riders; discounts; and innovative rates.

#### Seasonal Pricing

Costs usually vary by season for most utilities. These variations may be reflected in their rates through different demand and energy charges in the winter and summer. When electric utilities have a seasonal variation in their charges, usually the summer rates are higher than the winter rates, due to high air conditioning use. Gas utilities will generally have winter rates that are higher than summer rates, reflecting increased space-heating use.

#### Block Pricing

Energy and demand charges may be structured in one of three ways: 1) a declining block structure; 2) an inverted block structure; or 3) a flat rate structure. An inverted block pricing structure increases the rate as the consumption increases. A declining block pricing method decreases the rate as the user's consumption increases. When a rate does not vary with consumption levels it is a "flat" structure. With the declining or inverted block structures, the number of kWh, MMBtu, or therms used is broken into blocks. The unit cost (cents per kWh or cents per MMBtu or therm) is lower or higher for each succeeding block.

A declining block reflects the fact that most utilities can generate additional electricity or provide additional gas for lower and lower costs—up to a point. The capital costs of operation are spread over more usage. The inverted block structure reflects the fact that the incremental cost of production exceeds the average cost of energy. Hence, use of more energy will cause a greater cost to the utility.

Most utilities offer rates with more than one block pricing structure. A utility may offer some combination of inverted, declining and flat block rates, often reflecting seasonal energy cost differentials as well as use differen-

tials. For example, a gas utility may use an inverted block pricing structure in the winter that reflects the higher energy costs in that period, but use either a flat or declining block pricing structure in the summer when energy costs are lower.

#### Riders

A "Rider" modifies the structure of a rate based on specific qualifications of the customer. For example, a customer may be on a general service rate and subscribe to a rider that reduces summer energy charges where the utility is granted physical control of the customers air conditioning load.

#### Discounts

The discount most often available is the voltage discount offered by electric utilities. A voltage discount provides for a reduction in the charge for energy and/or demand if the customer receives service at voltages above the standard voltage. This may require the customer to install, operate and maintain the equipment necessary to reduce the line voltage to the appropriate service voltage. Each customer must evaluate the economics of the discounts against the cost of the equipment they will have to provide.

#### Innovative Rates

Increased emphasis on integrated resource planning, demand-side management and the move to a more competitive energy marketplace has focused utility attention on innovative rates. Those rates designed to change customer load use, help customers maintain or increase market share, or provide the utility with a more efficient operating arena are innovative. Most rates offered today fit into the innovative category.

## 18.4 INNOVATIVE RATE TYPES

Utilities have designed a variety of rate types to accomplish different goals. Some influence the customer to use more or less energy or use energy at times that are helpful to the utility. Others are designed to retain or attract customers. Still others are designed to encourage efficient use of energy. The following are some of the innovative rate types that customers should know about.

**Time-of-Use Rates** are used for the pricing of electricity only. The primary purpose of the time-of-use (TOU) rate is to send the proper pricing signals to the consumer regarding the cost of energy during specific times of the day. Generally, a utility's daytime load is higher than its nighttime load, resulting in higher daytime production costs. Proper TOU price signals will encourage customers to defer energy use until costs are lower. TOU rates are usually offered as options to customers, though some utilities have mandatory TOU rates.

**End-Use Rates**, these rates include air-conditioning, all-electric, compressed natural gas, multi-family, space-heating, thermal energy storage, vehicle fuel and water-heating rates. These rates are all intended to encourage customers to use energy for a specific end-use.

**Financial Incentive Rates** include rates such as residential assistance, displacement, economic development, and surplus power rates. Assistance rates provide discounts to residential customers who meet specific low-income levels, are senior citizens or suffer from some physical disability. Displacement rates are offered by electric utilities to customers who are capable of generating their own electricity. The price offered to these customers for utility-provided power is intended to induce the customer to "displace" its own generated electricity with utility-provided electricity. Economic development rates are generally offered by utilities to provide economic incentives for businesses to remain, locate, or expand into areas which are economically distressed. This type of rate is an attempt to attract new customers into the area and to get existing customers to expand until the area is revitalized. Surplus power rates are offered to large commercial and industrial customers. They are offered gas or electric capacity at greatly reduced prices when the utility has an excess available for sale.

**Interruptible Rates** generally apply to commercial and industrial customers. The utilities often offer several options with respect to the customer's ability to interrupt. Prices vary based on the amount of capacity that is interruptible, the length of the interruption, and the notification time before interruption. Such interruptions are generally, but not always, customer controlled. In addition, the total number of interruptions and the maximum annual hours of interruption may be limited.

## 18.5  CALCULATION OF A MONTHLY BILL

Following is the basic formula used for calculating the monthly bill under a utility rate. The sum of these components will result in the monthly bill.

1) Customer Charge
   - Customer charge = fixed monthly charge
2) Energy Charge
   - Energy charge = dollars × energy use
   - Energy/Fuel Cost Adjustment = dollars × energy use

3) Demand Charge
   - Demand charge = dollars × demand
   - Reactive Demand Charges (electric only) = dollars × measured kilovolt-ampere reactive demand
4) Tax/Surcharge
   - Tax/surcharge = either sum of one or more of items 1-3 above multiplied by tax percentage, dollars × energy use, or dollars × demand

Some examples illustrating the calculation of a monthly bill follow. These examples are actual rate schedules used by the utilities shown. The information regarding the consumption of electricity used in the sample calculations are based on the typical figures shown in Table 18.1.

### 18.5.1  Commercial General Service with Demand Component

The commercial general service rate often involves the use of demand charges. Table 18.2 provides rate data from Public Service Electric & Gas Company. A sample bill is calculated using the data from Table 18.1 for a convenience store.

Energy Usage – 17,588 kWh; Billing Demand – 30 kW; Season – Summer (June)

| | |
|---|---:|
| Customer Charge: | $3.74 |
| Energy Charge: | $1,661.01 |
| 17,588 kWh ×.09444 | |
| Energy Cost Adjustment: | –$302.37 |
| 17,588 kWh × <.017192> | |
| Demand Charge: | $262.34 |
| (1 kW × 4.53) + (29 kW × 8.89) | |
| | |
| Total Monthly Charge: | **$1,624.72** |

*Energy Usage Measured in kWh per kW Demand*

The standard measure of electric energy blocks is in kilowatt hours. An alternative measure used by electric utilities for commercial and industrial rates is energy per unit of demand (e.g., kWh per kW). This block measurement is illustrated by Virginia Power's Schedule GS-2 Intermediate rate in Table 18.3. We are calculating the bill for October use. According to Table 18.1, the convenience store used 16,003 kWh in October and had a billing demand of 32 kW. The bill for this use is as follows.

| | |
|---|---:|
| Customer Charge: | $21.17 |
| Energy Charge: | |
| Block 1 - (150 kWh × 32 kW), | |
| or 4,800 kWh @ $0.04840 = | $232.32 |

**Table 18.1 Typical usage patterns.**

|  | Convenience Store | | Office Building | | Shopping Center | |
|---|---|---|---|---|---|---|
|  | kW | kWh | kW | kWh | kW | kWh |
| JAN | 28 | 12,049 | 660 | 194,500 | 1,935 | 892,000 |
| FEB | 30 | 13,097 | 668 | 215,500 | 1,905 | 795,000 |
| MAR | 31 | 15,001 | 668 | 223,500 | 1,740 | 719,000 |
| APR | 32 | 13,102 | 600 | 177,500 | 1,515 | 633,000 |
| MAY | 30 | 14,698 | 345 | 101,000 | 1,425 | 571,000 |
| JUN | 30 | 17,588 | 293 | 95,000 | 1,455 | 680,000 |
| JUL | 32 | 17,739 | 375 | 118,000 | 1,440 | 661,000 |
| AUG | 35 | 17,437 | 420 | 156,500 | 1,395 | 667,000 |
| SEP | 31 | 18,963 | 555 | 130,000 | 1,455 | 733,000 |
| OCT | 32 | 16,003 | 540 | 207,000 | 1,455 | 646,000 |
| NOV | 30 | 17,490 | 570 | 169,500 | 1,545 | 675,000 |
| DEC | 29 | 12,684 | 600 | 172,500 | 1,740 | 768,000 |

**Table 18.2 Commercial general service with demand component.**

| Company: | Public Service Electric & Gas Company |
|---|---|
| Rate Class: | Commercial |
| Rate Type: | General Service |
| Rate Name: | Schedule GLP |
| Service: | Electric |
| Effective Date: | 01/01/94 |

Qualifications: General purposes where demand is less then 150 kW.

|  | | Winter Oct-May | | Summer Jun-Sep |
|---|---|---|---|---|
| Customer Charge: | | $3.74 | | $3.74 |
| Minimum Charge: | | $3.74 | | $3.74 |
| Energy Cost Adjustment: | | -$0.019037 | | –$0.017192 |
| Tax Rate: | | $0.00 | | $0.00 |
| Surcharge: | | $0.00 | | $0.00 |
| # of Energy Blocks: | | 1 | | 1 |
| Block 1 Size: | > | 0 | > | 0 |
| Block 1 Energy Charge: | | $0.09444 | | $0.09444 |
| No. of Demand Blocks: | | 2 | | 2 |
| Block 1 Size: | | 1 | | 1 |
| Block 2 Size: | > | 1 | > | 1 |
| Block 1 Demand Charge: | | $4.53 | | $4.53 |
| Block 2 Demand Charge: | | $7.84 | | $8.89 |

BILLING DEMAND: Estimated by dividing kWh usage by 100 or, if metered, greatest average 30-minute demand in the month.

Block 2 - (150 kWh × 32 kW),
     or 4,800 kWh @ $0.02712 =                    $130.18
Block 3 - (all remaining kWh),
     or 6,403 kWh @ $0.01173 =                     $75.11
Energy Cost Adjustment: (16,003 × $0.01418)       $226.92
Demand Charge: (32 kW × $5.023)                   $160.74
Subtotal:                                         $846.44
Tax Rate:
     25% of first $50 of subtotal                  $12.50
     12% on remainder of subtotal                  $95.57
Total Monthly Charge:                             **$954.51**

*Time-of-Use Rates*

Time-of-use rates are calculated very differently from general service rates. The customer's use must be recorded on a time-of-use meter, so that billing can be calculated on the use in each time period. In the following sample calculation from Long Island Lighting Company (Table 18.4), we assume the customer has responded to the price signal and has relatively little on-peak usage.

Energy Usage - 1,200 kWh; Season - Summer; On-Peak Period - 10:00 AM-8:00 PM, Monday-Friday; On-Peak Usage - 12.5%

Customer Charge:                                   $9.79
Energy Charge:                                    $176.53
     150 on-peak kWh @ $0.3739 = $56.09
     1050 off-peak kWh @ $0.1147 = $120.44
Energy Cost Adjustment:                            $2.88
Tax:                                              $10.02
Total Monthly Charge:                            **$199.22**

## 18.5.2 Gas General Service Rates

Gas general service rates are calculated in a similar manner to the electric general service rates. Gas rates are priced in dollars per MMBtu, or in dollars per MCF, depending upon the individual utility's unit of measurement.

**Table 18.3 Energy usage measured in kWh per kW of demand.**

| Company: | Virginia Electric & Power Company | |
|---|---|---|
| Rate Class: | Commercial | |
| Rate Type: | General Service | |
| Rate Name: | Schedule GS-2 - Intermediate | |
| Effective Date: | 1/01/94 | |
| Qualifications: | Non-residential use with at least three billing demands => 30 kW in the current and previous 11 billing months; but not more than two billing months of 500 kW or more. | |

| | OCT-MAY | JUN-SEP |
|---|---|---|
| Customer Charge: | $21.17 | $21.17 |
| Minimum Charge: | See Note | See Note |
| Energy Cost Adjustment: | $0.01418 | $0.01418 |
| Tax Rate: | See Note | See Note |
| No. of Energy Blocks: | 3 | 3 |
| Block 1 Size: | 150 kWh/kW demand | 150 kWh/kW demand |
| Block 2 Size: | 150 kWh/kW demand | 150 kWh/kW demand |
| Block 3 Size: | > 300 kWh/kW demand | > 300 kWh/kW demand |
| Block 1 Energy Charge: | $0.0484 | $0.0484 |
| Block 2 Energy Charge: | $0.02712 | $0.02712 |
| Block 3 Energy Charge: | $0.01173 | $0.01173 |
| No. of Demand Blocks: | 1 | 1 |
| Block 1 Size: | > 0 | > 0 |
| Block 1 Demand Charge: | $5.023 | $6.531 |

MINIMUM CHARGE: Greater of: 1) contract amount; or 2) sum of customer charge, energy charge and adjustments, plus $1.604 times the maximum average 30-minute demand measured in the month. TAX: Tax is 25% of the first $50, and 12% of the excess.

BILLING DEMAND: Maximum average 30-minute demand measured in the month, but not less than the maximum demand determined in the current and previous 11 months when measured demand has reached 500 kW or more.

**Table 18.4 Electric time-of-use rate.**

| Company: | Long Island Lighting Company |
|---|---|
| Rate Class: | Residential |
| Rate Type: | Time-of-Use |
| Rate Name: | Schedule SC 1-VMRP, Rate 2 |
| Effective Date: | 04/11/95 |
| Qualifications: | |

Use for all residential purposes where consumption is 39,000 kWh or less for year ending September 30, or under 12,600 kWh for June through September.

| | OCT-MAY | JUN-SEP |
|---|---|---|
| Customer Charge: | $9.79 | $9.79 |
| Minimum Charge: | $16.53 | $16.53 |
| Energy Cost Adjustment: | $0.0024 | $0.0024 |
| Tax Rate: | 5.29389% | 5.29389% |
| No. of Energy Blocks: | 1 | 1 |
| Block 1 Size: | > 0 | > 0 |
| Block 1 Energy Charge: | | |
| On-Peak: | $0.1519 | $0.3739 |
| Off-Peak: | $0.0978 | $0.1147 |

TAX: applied to total bill

PEAK PERIOD:

> On-Peak Hours: 10 a.m.-8 p.m., MON-FRI.
> Off-Peak Hours: All remaining hours.

## Commercial General Service Rate with Demand Component

Gas rates for the commercial or industrial customer may involve a demand charge. The method for calculating this type of bill is done in the same manner as an electric rate with a demand charge. The example in Table 18.5 comes from Northern Illinois Gas Company.

Energy Usage - 11,500 MMBtu; Billing Demand - 400 MMBtu; Season - Winter

| | |
|---|---|
| Customer Charge: | $325.00 |
| Energy Charge: | $4,899.00 |
| Purchased Gas Adjustment (Energy Cost Adjustment): | $28,221.00 |
| Demand Charge: | $2,828.00 |
| Tax: | $1,849.92 |
| Total Charge: | **$38,122.92** |

## 18.6 CONDUCTING A LOAD STUDY

Once a customer understands how utility rates are implemented, he/she can perform a simple load study to make use of this information. A load study will help the energy user to identify his load patterns, amount and time of occurrence of maximum load, and the load factor. This information can be used to modify use in ways that can lower electric or gas bills. It can also help the customer to determine the most appropriate rate to use.

The basic steps of a utility load study are shown in Table 18.6.

The first step is to collect historical load data. Past bills are one source for this information. One year of data is necessary to identify seasonal patterns; two or more years of data is preferable. Select a study period that is fairly representative of normal consumption conditions.

The next step is to organize the data so that use patterns are evident. One way to analyze the data is to plot the kWh usage, the maximum demand, and the load factor. The load factor is the ratio of the average demand to the maximum demand. The average demand is determined by the usage in kWh divided by the total number of hours (24 × number of days) in the billing period. The number of days in the billing period may vary depending on how often the meter is read. (See Table 18.7.)

**Table 18.5 Commercial gas rate with demand component.**

| Company: | Northern Illinois Gas Company |
| Rate Class: | Commercial/Industrial Large |
| Rate Type: | General Service |
| Rate Name: | Schedule 6 |
| Effective Date: | 09/01/89 |
| Qualifications: | General commercial use. All charges are shown per MMBtu. |

|  | Winter | | Summer |
| --- | --- | --- | --- |
| Customer Charge: | $325.00 | | $325.00 |
| Minimum Charge: | $3,000.00 | | $3,000.00 |
| Purchased Gas Adjustment: | $2.454 | | $2.61 |
| Tax Rate: | 5.1% | | 5.1% |
| No. of Energy Blocks: | 1 | | 1 |
| Block 1 Size: | > 0 | > | 0 |
| Block 1 Energy Charge: | $0.426 | | $0.426 |
| No. of Demand Blocks: | 1 | | 1 |
| Block 1 Size: | > 0 | > | 0 |
| Block 1 Demand Charge: | $7.07 | | $5.388 |

TAX: applied to the total bill and is the lower of 5% of revenues or $0.24 per MMBtu, plus 0.10%.
BILLING DEMAND: per MMBtu of customer's Maximum Daily Contract Quantity.
The PGA and Demand charges shown here change monthly. Winter charges applied in January 1994; and Summer charges applied in July 1993.

---

**Table 18.6 Basic steps for conducting a load analysis.**

1) Collect historical load data
   - compile data for at least one year
2) Organize data by month for
   - kWh consumption
   - maximum kW demand
   - load factor
3) Review data for
   - seasonal patterns of use
   - peak demands
4) Determine what demand or use can be eliminated or reduced
5) Review load data with utility

---

**Table 18.7 Load factor calculation.**

$$\text{Load Factor} = \frac{\text{average demand (kW)}}{\text{maximum demand (kW)}}, \quad \text{where}$$

$$\text{Average Demand} = \frac{\text{kWh usage}}{(24) \times (\text{number of days})}$$

Example: December office building load from Table 18.1

$$\text{Average Demand} = \frac{172,500 \text{ kWh}}{(24) \times (31)} = 231.9 \text{ kW}$$

$$\text{Load Factor} = \frac{231.9 \text{ kW}}{600 \text{ kW}} = 0.39$$

---

Next, review the data. Seasonal variations will be easily pinpointed. For example, most buildings will show a seasonal trend with two peaks. One will occur in winter and another during summer, reflecting seasonal heating and cooling periods. There may be other peaks due to some aspect of some industrial process, such as a cannery where crops are processed when they are harvested.

In Figure 18.1, kWh, maximum kW, and average kW are plotted from the data in Table 18.1 for the shopping center and the office building.

Note that the shopping center has a dominant winter peak—it is in the winter that the maximum kWh and kW are used for the year. The load factor ranges between 0.54 and 0.70. Overall, the average demand for this customer is about 60% of the peak demand.

The office building shows a different pattern. The load factor ranges between 0.35 and 0.52. This reflects the fact that the office building is really used less than half

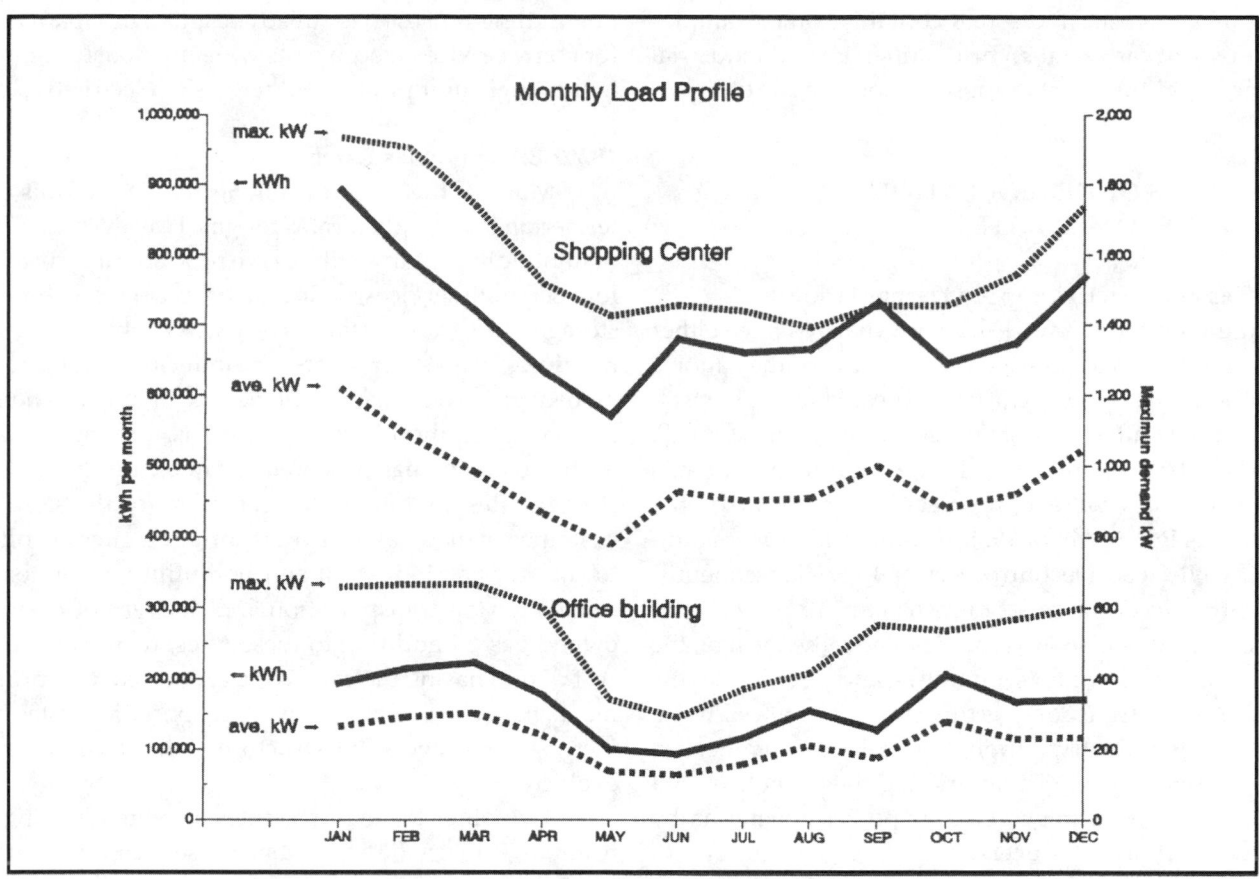

**Figure 18.1 Monthly load profile.**

the time. Generally, working hours span from 8 a.m. to 6 p.m. Although some electrical load continues during the night hours, it is not as intense as during the normal office hours.

If the load factor were 1, this would imply uniform levels of use—in effect, a system that was turned on and left running continuously. This may be the case with some manufacturing processes such as steel mills and refineries.

The fourth step is to determine what demand or use can be eliminated, reduced, or redirected. How can the shopping center reduce its energy costs? By reducing or shifting the peak demand it can shave demand costs. Although overall consumption is not necessarily reduced, the demand charge is reduced. Where demand ratchets are in place, shaving peak demand may result in savings over a period of several months, not just the month of use. One way to shift peak demand is to install thermal storage units for space cooling purposes; this will shift day time load to night time, giving the customer an overall higher load factor. This may qualify the customer for special rates from the utility as well.

Where there may not be much that can be done about the peak demand, (in a high load factor situation)

more emphasis should be placed on methods to reduce usage. Some examples: turn up the thermostat at night during the summer, down during the winter; install motion detectors to turn off unnecessary lights; turn off other equipment that is not in use.

Where the customer is charged for electric service on a time-of-use basis, a more sophisticated load study should be performed. The data collected should consist of hourly load data over at least one year. This data can be obtained through the use of recording meters. Once acquired, the data should be organized to show use patterns on a monthly basis with Monday through Friday (or Saturday, depending on the customer's uses) use plotted separately from weekend use. Review of this data should show where shaving or shifting energy or demand can lower overall electricity bills.

Once the customer has obtained a better understanding of his energy usage patterns, he can discuss with his utility how to best benefit from them. The utility most likely will be interested because it will also receive some benefits. The customer can consider implementing certain specific measures to better fit in the utility's load pattern, and at the same time improve his energy use. The customer's benefit will generally be associated with less energy-related costs. Table 18.8 contains some examples of options that can be taken by commercial and industrial customers and the effect of those options on the utility.

## 18.7 EFFECTS OF DEREGULATION ON CUSTOMER RATES

### 18.7.1 Gas and Electric Supply Deregulation

In the period since 1980, many changes have either occurred or begun to occur in the structure of the nation's electric and gas supply industries. These changes have already begun to affect the rate types and structures for U.S. gas and electricity consumers. In the natural gas industry, well-head prices were deregulated as a result of the Natural Gas Policy Act of 1978 and the subsequent Natural Gas Well-Head Decontrol Act of 1989. Subsequently, FERC introduced a number of restructuring rules (Order Nos. 436, 500, and 636) that dramatically change the regulation of the nation's pipelines and provide access for end-users to transport gas purchased at the well-head. In the electric industry, supply deregulation commenced with passage of the Public Utility Regulatory Policies Act of 1978, which encouraged electric power generation by certain non-utility producers. The Energy Policy Act of 1992 further deregulated production and mandated open transmission access for wholesale transfers of electricity between qualified suppliers and wholesale customers.

These legal and regulatory changes will have a significant and lasting effect on the rate types and rate structures experienced by end-users. In the past, most gas and electric customers paid a single bundled rate that reflected all costs for capacity and energy, storage, delivery, and administration. Once customers are given the opportunity to purchase their gas and electric resources directly from producers, it then becomes necessary to unbundle the costs associated with production from the costs associated with transportation and delivery to end users. This unbundling process has already resulted in separate rates for many services whose costs were previously combined in the single unit price for either gas or electricity.

### 18.7.2 Effect on Gas Rates

Much of the discussion in Section 18.3 of this chapter pertains to bundled rates for gas. However, as a result of unbundling, many utilities are now offering customers four separate services, including balancing, procurement, storage, and transportation of gas. Gas balancing rates provide charges for over- or under-use of customer-owned gas over a specified period of time. When the customer has the utility procure gas for transportation to the customer, gas procurement rates are charged. Gas storage rates are offered to customers for the storage of customer-owned gas. Gas transportation rates are offered to commercial, industrial and non-utility generator customers for the transportation and delivery of customer-owned gas. In addition to these rates, there is the actual cost of purchasing the gas to be transported. Gas procurement, balancing, storage and transportation rates have increased in usage as the structure of the gas industry has evolved.

Two other types of gas rates also are evolving as a result of industry deregulation. These include negotiated gas rates and variable gas rates. The former refers to rates that are negotiated between individual customers and the utility. Such rates are often subject to market conditions. The latter, variable gas service rates, refer to rates that vary from month to month. A review of all of the gas service rates collected by the Gas Research Institute (GRI) in 1994 indicated that 52% of the gas utilities surveyed offered at least one type of variable pricing. Such rates are often indexed to an outside factor, such as the price of gasoline or the price of an alternative fuel, and they usually vary between established floor and ceiling prices. The most common types of variable rates are those offered for transportation services.

**Table 18.8 Customer options and their effects on utility.**

| OPTIONS | | |
|---|---|---|
| *Commercial* | *Industrial* | *Utility Effect* |
| Accept direct control of water heaters | Subscribe to interruptible rates | Reduction of load during peak periods |
| Store hot water to increase space heating | Add nighttime operations | Builds load during off-peak periods |

### 18.7.3 Effect on Electric Rates

In the past, most U.S. electric customers have paid a single bundled rate for electricity. Many of these customers purchased from a utility that produced, transmitted, and delivered the electricity to their premises. In other cases, customers purchased from a distribution utility that had itself purchased the electricity at wholesale from a generating and transmitting utility. In both of these cases, the customer paid for electricity at a single rate that did not distinguish between the various services required to produce and deliver the power. In the future, as a result of the deregulation process already underway, there is a far greater likelihood that initially large customers, and later many smaller customers, will have the ability to select among a number of different suppliers. In most of these cases, however, the transmission and delivery of the purchased electricity will continue to be a regulated monopoly service. Consequently, future electricity consumers are likely to receive separate bills for:

- electric capacity and energy;
- transmission; and
- distribution.

In some cases, a separate charge may also be made for system control and administrative services, depending on exact industry structure in the given locality. For each such charge, a separate rate structure will apply. At present, it appears likely that there will be significant regional and local differences in the way these rates evolve and are implemented.

### GLOSSARY

There are a few terms that the user of this document needs to be familiar with. Below is a listing of common terms and their definitions.

**Billing Demand**: The billing demand is the demand that is billed to the customer. The electric billing demand is generally the maximum demand or maximum average measured demand in any 15-, 30-, or 60-minute period in the billing month. The gas billing demand is determined over an hour or a day and is usually the greatest total use in the stated time period.

**British Thermal Unit (Btu)**: Quantity of heat needed to bring one pound of water from 58.5 to 59.5 degrees Fahrenheit under standard pressure of 30 inches of mercury.

**Btu Value**: The heat content of natural gas is in Btu per cubic foot. Conversion factors for natural gas are:

- Therm = 100,000 Btu;
- 1 MMBtu = 1,000,000 Btu = 1 Decatherm.

**Contract Demand**: The demand level specified in a contractual agreement between the customer and the utility. This level of demand is often the minimum demand on which bills will be determined.

**Controllable Demand**: A portion or all of the customer's demand that is subject to curtailment or interruption directly by the utility.

**Cubic Foot**: Common unit of measurement of gas volume; the amount of gas required to fill one cubic foot.

**Curtailable Demand**: A portion of the customer's demand that may be reduced at the utility's direction. The customer, not the utility, normally implements the reduction.

**Customer Charge**: The monthly charge to a customer for the provision of the connection to the utility and the metering of energy and/or demand usage.

**Demand Charge**: The charge levied by a utility for metered demand of the customer. The measurement of demand may be either in kW or kVA.

**Dual-Fuel Capability**: Some interruptible gas rates require the customer to have the ability to use a fuel other than gas to operate their equipment.

**Energy Blocks**: Energy block sizes for gas utilities are either in MCFs or in MMBtus. The standard measures of energy block sizes for electric utilities are kWhs. However, several electric utilities also use an energy block size based on the customers' demand level (i.e. kWh per kW). Additionally, some electric utilities combine the standard kWh value with the kWh per kW value.

**Energy Cost Adjustment (ECA)**: A fuel cost factor charged for energy usage. This charge usually varies on a periodic basis, such as monthly or quarterly. It reflects the utilities' need to recover energy related costs in a volatile market. It is often referred to as the fuel cost adjustment, purchased power adjustment or purchased gas adjustment.

**Excess or Non-Coincidental Demand**: Some utilities charge for demands in addition to the on- or off-peak demands in time-of-use rates. An excess demand is demand used in off-peak time periods that exceeds usage during on-peak hours. Non-coincidental demand is the maximum demand measured any time in a billing period. This charge is usually in addition to the on- or off-peak demand charges.

**Firm Demand**: The demand level that the customer can rely on for uninterrupted use.

**Interruptible Demand**: All of the customer's demand may be completely interrupted at the utility's direction. Either the customer or the utility may implement the interruption.

**MCF**: Thousand (1000) cubic feet.

**MMCF**: Million (1,000,000) cubic feet.

**Minimum Charge**: The minimum monthly bill that will be charged to a customer. This generally is equal to the customer charge, but may include a minimum demand charge as well.

**Off-Peak Demand**: Greatest demand measured in the off-peak time period.

**On-Peak Demand**: Greatest demand measured in the on-peak time period.

**Ratchet**: A ratchet clause sets a minimum billing demand that applies during peak and/or non-peak months. It is usually applied as a percentage of the peak demand for the preceding season or year.

**Reactive Demand**: In electric service, some utilities have a special charge for the demand level in kilovolt-amperes reactive (kVAR) that is added to the standard demand charge. This value is a measure of the customer's power factor.

**Surcharge**: A charge levied by utilities to recover fees or imposts other than taxes.

**Therm**: A unit of heating value equal to 100,000 Btu.

**Transportation Rates**: Rates for the transportation of customer-owned gas. These rates do not include purchase or procurement of gas.

**Voltage Discounts**: Most electric utilities offer discounted rates to customers who will take service at voltages other than the general distribution voltages. The voltages for which discounts are generally offered are Secondary, Primary, Sub-transmission and Transmission. The actual voltage of each of these levels vary from utility to utility.

## References

1. Acton, J.P., Gelbard, E.H., Hosek, J.R., & Mckay, D.J. (1980, February). *British Industrial Response to the Peak-Load Pricing of Electricity*. The Rand Corporation, R-2508-DOE/DWP.

2. David, A.K., & Li, Y.Z. (1991, November). A Comparison of System Response For Different Types of Real-Time Pricing. IEEE International Conference on Advances In Power System Control, *Operation and Management*. Hong Kong, p. 385-390.

3. Anonymous (1997, August). *Energy User News*. Chilton Co., p. 32.

4. EPRI (1980, October). Industrial Response To Time Of Day Pricing—A Technical and Economic Assessment Of Specific Load Management Strategies. Gordian Associates, EA-1573, Research Project 1212-2.

5. Hanser, P., Wharton, J., & Fox-Penner, P. (March 1, 1997). Real-time Pricing—Restructuring's Big Bang? *Public Utilities Fortnightly*, 135 (5), p. 22-30.

6. Kirsch, L. D., Sullivan, R.L., & Flaim, T.A. (1988, August). Developing Marginal Costs For Real-Time Pricing. IEEE *Transactions on Power Systems*, 3 (3), p. 1133-1138.

7. Mykytyn Consulting group, Inc. (1997). Electric Utilities and Tariffs. PowerRates [Online]. Available: http://www.mcgi.com/pr/samples/utility_list.html [November 4, 1997].

8. O'Sheasy, M. (MIKE.T.OSHEASY@gpc.com). (1998, March 12). RTP. E-mail to Mont, J. (mjavier@okstate.edu).

9. Tabors, R.D., Schweppe, F.C., & Caraminis, M.C. (1989, May). Utility Experience with Real-Time Rates. *Transactions on Power Systems*, 4(2), p. 463-471.

10. Tolley, D.L. (1988, January). Industrial Electricity Tariffs. *Power Engineering Journal*. p. 27-34.

# CHAPTER 19

# THERMAL ENERGY STORAGE

**CLINT CHRISTENSON**
Johnson Controls, Inc.

## 19.1 INTRODUCTION

A majority of the technology developed for energy management has dealt with the more efficient *consumption* of electricity, rather than timing the demand for it. Variable frequency drives, energy efficient lights, electronic ballasts and energy efficient motors are a few of these consumption management devices. These techniques often only impact a small portion of the facilities demand (when compared to say the mechanical cooling equipment), which is normally a major portion of the facilities overall annual electric bill. The management of demand charges deals very little with conservation of energy, but mainly with the ability of a generator to supply power *when* needed. It is this timing of consumption that is the basis of demand management and the focus of thermal energy storage (TES).

Experts agree that demand management is actually not a form of energy *conservation* but a form of cost *management*. Throughout the 1980's and most of the 1990's, Demand Side Management (DSM) was done by utilities in order to manage generating capacity and costs by promoting demand reduction though incentives (financial rewards) and disincentives (rate structures). Most of the incentive programs have ceased due to surplus generation capacity and the approach of retail electrical deregulation. A deregulated market place will surely impact the cost of energy for many customers and if the commodity pricing experiments of the recent past are any indication of the future, demand costs, and thus demand management, will remain as an important cost control strategy for utilities and the energy users.

Utilities often charge more for energy and demand during certain periods in the form of on-peak rates and ratchet clauses. The process of managing the generation capacity that a particular utility has "on-line" involves the utilization of those generating units that produce power most efficiently first since these units would have the lowest avoided costs (ultimately the actual cost of energy). When the loads are approaching the connected generation capacity of the utility, additional generating

**Figure 19.1 Typical office building chiller consumption profile.**

units must be brought on line. Each additional unit has an incrementally higher avoided cost since these "peaking units" units are less efficient and used less often. This has prompted many organizations to implement some form of demand management.

Thermal energy storage (TES) is the concept of generating and storing energy in the form of heat or cold for use during peak periods. For the profile in Figure 19.1, a cooling storage system could be implemented to reduce or eliminate the need to run the chillers during the on-peak rate period. *By running the chillers during off-peak hours and storing this capacity for use during the on-peak hours, a reduction in energy costs can be realized.* If this type of system is implemented during new construction or when equipment is being replaced, smaller capacity chillers can be installed, since the chiller can spread the production of the total load over the entire day, rather than being sized for peak loads.

Thermal energy storage has been used for centuries, but only recently have large electrical users taken advantage of the technique for cost management. The process involves storing Btu's (or lack of Btu's) for use when either a heat source or a heat sink is required. The use of eaves, root cellars, ground coupled heat pump systems, and adobe type thermal mass could all be considered forms of thermal storage. Today, the ability to take advantage of a source of inexpensive energy (whether waste heat source or time based rate structure) for use during a later time of more expensive energy has extended the applications of TES. For this particular chapter, the focus of discussion will concentrate on the storage of cooling capacity and the storage of heat will not be considered. The two main driving forces behind the storage of cooling capacity, rate structure and cooling system management, will be discussed in the following paragraphs.

Often the chiller load and efficiency follow the chiller consumption profile, in that the chiller is running at high load, i.e. high efficiency, only a small portion of the day. This is due to the HVAC system having to produce cooling when it is needed as well as to be able to handle instantaneous peak loads. With smaller chiller systems designed to handle the base and peak loads during off-peak hours, the chillers can run at higher average loads and thus higher efficiencies. Appendix A following this chapter lists several manufacturers of thermal energy storage systems.

Thermal energy storage also has the ability to balance the daily loads on a cooling system. Conventional air conditioning system must employ a chiller large enough to handle the peak cooling demand as it occurs. This mandates that the cooling system will be required to operate in a load following mode, varying the output of the system in response to changes in the cooling requirements. In systems that operate within a one or two shift operation or those that are much more climatically based, can benefit from the smoothing characteristics of TES. A school for example that adds a new wing, could utilize the existing refrigeration system during the evening to generate cooling capacity to be stored for use during the day. Although additional piping and pumping capacity would need to be added to the addition, new chiller capacity may not have to be added. A new construction project that would have similar single shift cooling demand profile could utilize a smaller chiller in combination with storage to better balance the chiller operation. This could significantly reduce the capital cost of the renovation in addition to any rate based savings as discussed above.

Companies often control the demand of electricity by utilizing some of the techniques listed above and other consumption management actions which also reduce

demand. More recently the ability to shift the *time* when electricity is needed has provided a means of balancing or shifting the demand for electricity to "off-peak" hours. This technique is often called demand balancing or demand shifting. This demand balancing may best be seen with the use of an example 24-hour chiller consumption plot during the peak day, Figure 19.1 and Table 19.1. This facility exhibits a typical single shift building load profile. Note that the load listed in this table for the end of hour 1 identifies the average load between midnight and 1:00 a.m., and that for end of hour 2 is the average load between 1 and 2 a.m., and so on. This example will employ a utility rate schedule with a summer on-peak demand period from 10 am to 5:59 p.m., an 8-hour period. Moving load from the on-peak rate period to the off-peak period can both balance the demand and reduce residual ratch-

**Table 19.1 Example chiller consumption profile**

| Chiller Consumption Profile | | |
|---|---|---|
| | Chiller Load | |
| End of Hour | (Tons) | Rate |
| 1 | 100 | Reg |
| 2 | 120 | Reg |
| 3 | 125 | Reg |
| 4 | 130 | Reg |
| 5 | 130 | Reg |
| 6 | 153 | Reg |
| 7 | 165 | Reg |
| 8 | 230 | Reg |
| 9 | 270 | Reg |
| 10 | 290 | Reg |
| 11 | 340 | On-Peak |
| 12 | 380 | On-Peak |
| 13 | 450 | On-Peak |
| 14 | 490 | On-Peak |
| 15 | 510 | On-Peak |
| 16 | 480 | On-Peak |
| 17 | 410 | On-Peak |
| 18 | 360 | On-Peak |
| 19 | 250 | Reg |
| 20 | 210 | Reg |
| 21 | 160 | Reg |
| 22 | 130 | Reg |
| 23 | 125 | Reg |
| 24 | 115 | Reg |

| | | |
|---|---|---|
| Daily Total | 6123 | Ton-Hrs |
| Daily Avg. | 255.13 | Tons |
| Peak Total | 3420 | Ton-Hrs |
| Peak Demand | 510 | Tons |

eted peak charges. Thermal energy storage is one method available to accomplish just that.

## 19.2 STORAGE SYSTEMS

There are two general types of storage systems, ones that shut the chiller down during on-peak times and run completely off the storage system during that time are known as "full storage systems." Those designed to have the chiller run during the on-peak period supplementing the storage system are known as "partial storage systems." The full storage systems have a higher first cost since the chiller is off during peaking times and the cooling load must be satisfied by a larger chiller running fewer hours and a larger storage system storing the excess. The full storage systems do realize greater savings than the partial system since the chillers are completely turned off during on-peak periods. Full storage systems are often implemented in retrofit projects since a large chiller system may already be in place.

A partial storage system provides attractive savings with less initial cost and size requirements. New construction projects will often implement a partial storage system so that the size of both the chiller and the storage system can be reduced. Figures 19.2 and 19.3 and Tables 19.2 and 19.3 demonstrate the chiller load required to satisfy the cooling needs of the office building presented in Figure 19.1 for the full and partial systems, respectively. Column 2 in these tables represents the building cooling load each hour, and column 3 represents the chiller output for each hour. Discussion of the actual calculations that are required for sizing these different systems is included in a subsequent section. For simplicity sake, these numbers do not provide for any system losses, which will also be discussed in a later section.

The full storage system has been designed so that the total daily chiller load is produced during the off-peak hours. This eliminates the need to run the chillers during the on-peak hours, saving the increased rates for demand charges during this period and as well as any future penalties due to ratchet clauses. The partial storage system produces 255.13 tons per hour during the entire day, storing excess capacity for use when the building demand exceeds the chiller production. This provides the ability to control the chiller load, limit the peak chiller demand to 255.13 kW,[*] and still take advantage of the off-peak rates for a portion of the on-peak chiller load.

---

[*]assuming COP = 3.5, then $kW/ton = \dfrac{3.517}{COP} = 1.0\ kW/ton$

## Table 19.2 Full storage chiller consumption profile.

Chiller Consumption Profile—Full Storage System

| 1 | 2 | 3 | 4 |
|---|---|---|---|
| End of Hour (Tons) | Cooling Load (Tons) | Chiller Load[2] | Rate |
| 1 | 100 | 382.69 | Reg |
| 2 | 120 | 382.69 | Reg |
| 3 | 125 | 382.69 | Reg |
| 4 | 130 | 382.69 | Reg |
| 5 | 130 | 382.69 | Reg |
| 6 | 153 | 382.69 | Reg |
| 7 | 165 | 382.69 | Reg |
| 8 | 230 | 382.69 | Reg |
| 9 | 270 | 382.69 | Reg |
| 10 | 290 | 382.69 | Reg |
| 11 | 340 | 0 | On-Peak |
| 12 | 380 | 0 | On-Peak |
| 13 | 450 | 0 | On-Peak |
| 14 | 490 | 0 | On-Peak |
| 15 | 510 | 0 | On-Peak |
| 16 | 480 | 0 | On-Peak |
| 17 | 410 | 0 | On-Peak |
| 18 | 360 | 0 | On-Peak |
| 19 | 250 | 382.69 | Reg |
| 20 | 210 | 382.69 | Reg |
| 21 | 160 | 382.69 | Reg |
| 22 | 130 | 382.69 | Reg |
| 23 | 125 | 382.69 | Reg |
| 24 | 115 | 382.69 | Reg |

| | Without storage | With storage |
|---|---|---|
| Daily Total (Ton-Hrs) | 6123 | 6123 |
| Daily Avg (Tons): | 255.13[1] | 255.13 |
| Peak Total (Ton-Hrs) | 3420[3] | 0[4] |
| Peak Demand (Tons) | 510[3] | 0[4] |

[1] $\dfrac{6123\ Ton\text{-}Hr}{24\ Hours} = 255.13\ Avg\ Tons$

[2] $\dfrac{6123\ Ton\text{-}Hr}{16\ Hours} = 382.69\ Avg\ Tons$

[3]This peak load is supplied by the TES, not the chiller
[4]This is the chiller load and peak during on-peak periods

An advantage of partial load systems is that they can provide a means of improving the performance of a system that can handle the cumulative cooling load, but not the instantaneous peak demands of the building. In such a system, the chiller could be run nearer optimal load continuously throughout the day, with the excess cooling tonnage being stored for use during the peak

**Figure 19.2 Full storage chiller consumption profile.**

**Figure 19.3 Partial storage chiller consumption profile.**

**Table 19.3 Partial storage chiller consumption profile.**

Chiller Consumption Profile
Partial Storage System

| 1 | 2 | 3 | 4 |
|---|---|---|---|
| Hour of Day | Cooling Load (Tons) | Chiller Load (Tons)[1] | Rate |
| 1 | 100 | 255.13 | Reg |
| 2 | 120 | 255.13 | Reg |
| 3 | 125 | 255.13 | Reg |
| 4 | 130 | 255.13 | Reg |
| 5 | 130 | 255.13 | Reg |
| 6 | 153 | 255.13 | Reg |
| 7 | 165 | 255.13 | Reg |
| 8 | 230 | 255.13 | Reg |
| 8 | 270 | 255.13 | Reg |
| 10 | 290 | 255.13 | Reg |
| 11 | 340 | 255.13 | On-Peak |
| 12 | 380 | 255.13 | On-Peak |
| 13 | 450 | 255.13 | On-Peak |
| 14 | 490 | 255.13 | On-Peak |
| 15 | 510 | 255.13 | On-Peak |
| 16 | 480 | 255.13 | On-Peak |
| 17 | 410 | 255.13 | On-Peak |
| 18 | 360 | 255.13 | On-Peak |
| 19 | 250 | 255.13 | Reg |
| 20 | 210 | 255.13 | Reg |
| 21 | 160 | 255.13 | Reg |
| 22 | 130 | 255.13 | Reg |
| 23 | 125 | 255.13 | Reg |
| 24 | 115 | 255.13 | Reg |

|  | Without storage | With storage |
|---|---|---|
| Daily Total (Ton-Hrs) | 6123 | 6123 |
| Daily Avg (Tons): | 255.13 | 255.13 |
| Peak Total (Ton-Hrs): | 3420[2] | 2041[3] |
| Peak Demand (Tons): | 510[2] | 255.13[3] |

[1] $\dfrac{6123 \text{ Ton-Hr}}{24 \text{ Hours}} = 255.13$ Avg Tons

[2]This peak load is supplied by the TES, supplemented by the chiller
[3]This is the chiller load and peak during on-peak period.

periods. An optional method for utilizing partial storage is a system that already utilizes two chillers. The daily cooling load could be satisfied by running both chillers during the off-peak hours, storing any excess cooling capacity, and running only one chiller during the on-peak period, to supplement the discharge of the storage system. This also has the important advantage of offering a reserve chiller during peak load times. Figure 19.4 shows the chiller consumption profile for this optional partial storage arrangement and Table 19.4 lists the consumption values. Early and late in the cooling season, the partial load system could approach the full load system characteristics. As the cooling loads and peaks begin to decline, the storage system will be able to handle more of the on-peak requirement, and eventually the on-peak chiller could also be turned off. A system such as this can be designed to run the chillers at optimum load, increasing efficiency of the system.

## 19.3  STORAGE MEDIUMS

There are several methods currently in use to store cold in thermal energy storage systems. These are water, ice, and phase change materials. The water systems simply store chilled water for use during on-peak periods. Ice systems produce ice that can be used to cool the actual chilling water, utilizing the high latent heat of fusion. Phase change materials are those materials that exhibit properties, melting points for example, that lend themselves to thermal energy storage. Figure 19.5a represents the configuration of the cooling system with either a water or phase change material thermal storage system and Figure 19.5b represents a general configuration of a TES utilizing ice as the storage medium The next few sections will discuss these different mediums.

### 19.3.1  Chilled Water Storage

Chilled water storage is simply a method of storing chilled water generated during off-peak periods in a large tank or series of tanks. These tanks are the most commonly used method of thermal storage. One factor to this popularity is the ease to which these water tanks can be interfaced with the existing HVAC system. The chillers are not required to produce chilled water any colder than presently used in the system so the system efficiency is not sacrificed. The chiller system draws warmer water from one end of the system and this is replaced with chilled water in the other. During the off-peak charge cycle, the temperature of the water in the storage will decline until the output temperature of the chiller system is approached or reached This chilled water is then withdrawn during the on-peak discharge cycle, supplementing or replacing the chiller(s) output.

Facilities that have a system size constraint such as lack of space often install a series of small insulated tanks that are plumbed in series. Other facilities have installed a single, large volume tank either above or below ground. The material and shape of these tanks vary greatly from installation to installation. These large tanks are often de-

**Figure 19.4  Optional partial storage chiller profile.**

signed very similar to municipal water storage tanks. The main performance factors in the design of these tank systems, either large or multiple, is location and insulation. An Electric Power Research Institute's (EPRI) Commercial Cool Storage Field Performance Monitoring Project (RP-2732-05) Report states that the storage efficiencies of tanks significantly decrease if tank walls were exposed to sunlight and outdoor ambient conditions and/or had long hold times prior to discharging[7]. To minimize heat gain, tanks should be out of the direct sun whenever possible. The storage efficiency of these tanks is also decreased significantly if the water is stored for extended periods.

One advantage to using a single large tank rather than the series of smaller ones is that the temperature differential between the warm water intake and the chilled water outlet can be maintained. This is achieved utilizing the property of thermal stratification where the warmer water will migrate to the top of the tank and the colder to the bottom. Proper thermal stratification can only be maintained if the intake and outlet diffusers are located at the top and bottom of the tank and the flow rates of the water during charge and discharge cycles is kept low This will reduce a majority of the mixing of the two temperature waters. Another method used to assure that the two temperature flows remain separated is the use of a movable bladder, creating a physical partition. One top/bottom diffuser tank studied in the EPRI study used

**Table 19.4 Partial storage chiller consumption profile.**

| 1 | 2 | 3 | 4 |
|---|---|---|---|
| Hour of Day | Cooling Load (Tons) | Chiller Load[1,2] (Tons) | Rate |
| 1 | 100 | 306 | Reg |
| 2 | 120 | 306 | Reg |
| 3 | 125 | 306 | Reg |
| 4 | 130 | 306 | Reg |
| 5 | 130 | 306 | Reg |
| 6 | 153 | 306 | Reg |
| 7 | 165 | 306 | Reg |
| 8 | 230 | 306 | Reg |
| 9 | 270 | 306 | Reg |
| 10 | 290 | 306 | Reg |
| 11 | 340 | 153 | On-Peak |
| 12 | 380 | 153 | On-Peak |
| 13 | 450 | 153 | On-Peak |
| 14 | 490 | 153 | On-Peak |
| 15 | 510 | 153 | On-Peak |
| 16 | 480 | 153 | On-Peak |
| 17 | 410 | 153 | On-Peak |
| 18 | 360 | 153 | On-Peak |
| 19 | 250 | 306 | Reg |
| 20 | 210 | 306 | Reg |
| 21 | 160 | 306 | Reg |
| 22 | 130 | 306 | Reg |
| 23 | 125 | 306 | Reg |
| 24 | 115 | 306 | Reg |

|  | Without storage | With storage |
|---|---|---|
| Daily Total (Ton-Hrs): | 6123 | 6123 |
| Daily Avg (Tons): | 255.13 | 255.13 |
| On-Peak (Ton-Hrs): | 3420[3] | 1225[4] |
| Peak Demand (Tons): | 510[3] | 153[4] |

[1]
$$\frac{(6123 \text{ Ton-Hr}) (2 \text{ Chillers Operating})}{(16 \text{ Hours})(2 \text{ Chillers}) + (8 \text{ Hours})(1 \text{ Chiller})} = 306 \text{ Tons}$$

[2]
$$\frac{(6123 \text{ Ton-Hr}) (1 \text{ Chiller Operating})}{(16 \text{ Hours})(2 \text{ Chillers}) + (8 \text{ Hours})(1 \text{ Chiller})} = 153 \text{ Tons}$$

[3]This peak load is supplied by the TES, supplemented by the chiller.
[4]This is the chiller load and peak during the on-peak period.

a thermocouple array, installed to measure the chilled water temperature at one foot intervals from top to bottom of the tank. This tank had a capacity of 550,000 gallons and was 20 feet deep but had only a 2.5 foot blend zone over which the temperature differential was almost 20 degrees[7].

The advantages of using water as the thermal storage medium are:

**Figure 19.5a Water & eutectic storage system configuration.**

**Figure 19.5b Ice storage system configuration.**

1. Retrofitting the storage system with the existing HVAC system is very easy,

2. Water systems utilize normal evaporator temperatures,

3. With proper design, the water tanks have good thermal storage efficiencies,

4. Full thermal stratification maintains chilled water temperature differential, maintaining chiller loading and efficiencies, and

5. Water systems have lower auxiliary energy consumption than both ice and phase change materials since the water has unrestricted flow through the storage system.

### 19.3.2 Ice Storage

Ice storage utilizes water's high latent heat of fusion to store cooling energy. One pound of ice stores 144 Btu's of cooling energy while chilled water only contains 1 Btu per pound $-°F$[7,8]. This reduces the required storage volume approximately 75%[7] if ice systems are used rather than water. Ice storage systems form ice with the chiller system during off-peak periods and this ice is used to generate chilled water during on-peak periods.

There are two main methods in use to utilize ice for on-peak cooling. The first is considered a static system in which serpentine expansion coils are fitted within a insulated tank of cooling water. During the charging cycle, the cooling water forms ice around the direct expansion coil as the cold gases pass through it (see Figure 19.5b). The thickness of the ice varies with the ice building time (charge time) and heat transfer area. During the discharge cycle, the cooling water contained in the tank is used to cool the building and the warmer water returned from the building is circulated through the tank, melting the ice, and using its latent heat of fusion for cooling.

The second major category of thermal energy storage systems utilizing ice can be considered a dynamic system. This system has also been labeled a plate ice maker or ice harvester. During the charging cycle the cooling water is pumped over evaporator "plates" where ice is actually produced. These thin sheets of ice are fed into the cooling water tank, dropping the temperature. During on-peak periods, this chilled water is circulated through the building for cooling. This technology is considered dynamic due to the fact that the ice is removed from the evaporator rather than simply remaining on it.

Static ice storage systems are currently available in factory-assembled packaged units which provide ease of installation and can provide a lower initial capital cost. When compared to water storage systems, the size and weight reduction associated with ice systems makes them very attractive to facilities with space constraints One main disadvantage to ice systems is the fact that the evaporator must be cold enough to produce ice. These evaporator temperatures usually range from 10° to 25° while most chiller evaporator temperatures range from 42° to 47°[9]. This required decrease in evaporator temperature results in a higher energy demand per ton causing some penalty in cooling efficiency. The EPRI Project reported that chillers operating in chilled water or eutectic salt (phase change material) used approximately 20% less energy than chillers operating in ice systems (0.9 vs. 1.1 kW/ton)[7]. The advantages of using ice as the thermal storage medium are:

1. Retrofitting the storage system with the existing HVAC chilled water system is feasible,

2. Ice systems require less space than that required by the water systems,

3. Ice systems have higher storage but lower refrigeration efficiencies than those of water, and

4. Ice systems are available in packaged units, due to smaller size requirements

### 19.3.3 Phase Change Materials

The benefit of capturing latent heat of fusion while maintaining evaporating temperatures of existing chiller systems can be realized with the use of phase change materials. There are materials that have melting points higher than that of water that have been successfully used in thermal energy storage systems. Several of these materials fall into the general category called "eutectic salts" and are salt hydrates which are mixtures of inorganic salts and water. Some eutectic salts have melting (solidifying) points of 47°[7], providing the opportunity for a direct retrofit using the existing chiller system since this is at or above the existing evaporator temperatures. In a thermal storage system, these salts are placed in plastic containers, which are immersed within an insulated chilled water tank. During the charging cycle, the chilled water flows through the gaps between the containers, freezing the salts within them. During the on-peak discharge, the warmer building return water circulates through the tank, melting the salts and utilizing the latent heat of fusion to cool the building. These salt solutions have latent heat of fusion around 40 Btu/lb[9].

This additional latent heat reduces the storage volume by 66% of that required for an equivalent capacity water storage system[9]. Another obvious benefit of using eutectic salts is that the efficiency of the chillers is not sacrificed, as stated earlier, since the phase change occurs around normal evaporator suction temperatures. One problem with the eutectic salt systems is that the auxiliary energy consumption is higher since the chilled water must be pumped through the array of eutectic blocks. The auxiliary energy consumption of the ice systems is higher than both the water and eutectic salt systems since the chilled water must be pumped through the ice system coils, nozzles, and heat exchangers. The EPRI study found that the chilled water systems had an average auxiliary energy use of 0.43 kWh/Ton-Hr compared to the phase change systems (eutectic and ice) average auxiliary energy use of 0.56 kWh/Ton-Hr[5]. The advantages of using eutectic salts as the thermal storage medium are that they:

1. can utilize the existing chiller system for generating storage due to evaporator temperature similarity,

2. require less space than that required by the water systems, and

3. have higher storage and equivalent refrigeration efficiencies to those of water.

## 19.4 SYSTEM CAPACITY

The performance of thermal storage systems depends upon proper design. If it is sized too small or too large, the entire system performance will suffer. The following section will explain this sizing procedure for the example office building presented earlier. The facility has a maximum load of 510 tons, a total cooling requirement of 6,123 Ton-Hours, and a on-peak cooling requirement of 3,420 Ton-Hours. This information will be analyzed to size a conventional chiller system, a partial storage system, a full storage system, and the optional partial storage system. These results will then be used to determine the actual capacity needed to satisfy the cooling requirements utilizing either a chilled water, a eutectic salt, or an ice thermal storage system. Obviously some greatly simplifying assumptions are made.

### 19.4.1 Chiller System Capacity

The conventional system would need to be able to handle the peak load independently, as seen in Figure 19.1. A chiller or series of chillers would be needed to produce the peak cooling load of 510 tons. Unfortunately,

packaged chiller units usually are available in increments that mandate excess capacity but for simplicity one 600-ton chiller will be used for this comparison. The conventional chiller system will provide cooling as it is needed and will follow the load presented in Figure 19.1 and Table 19.1.

To determine the chiller system requirement of a cooling system utilizing partial load storage, further analysis is needed. Table 19.1 showed that the average cooling load of the office building was 255.13 tons per hour. The ideal partial load storage system will run at this load (see Figure 19.3 and Table 19.3). The chiller system would need to be sized to supply the 255.13 tons per hour, so one 300-ton chiller will be used for comparison purposes. Table 19.5 shows how the chiller system would operate at 255.13 tons per hour, providing cooling required for the building directly and charging the storage system with the excess. Although the storage system supplements the cooling system for 2 hours before the peak period, the cooling load is always satisfied.

Comparing the peak demand from the bottoms of columns 2 and 3 of Table 19.5 shows that the partial storage system reduced this peak load almost 50% (510 – 255.13 = 254.87 Tons). Column 4 shows the tonnage that is supplied to the storage system and column 5 shows the amount of cooling contained in the storage system at the end of each hour of operation. This system was design so that there would be zero capacity remaining in the thermal storage tanks after the on-peak period. The values contained at the bottom of Table 19.5 are the total storage required to assure that there is no capacity remaining and the maximum output required from storage. These values will be utilized in the next section to determine the storage capacity required for each of the different storage mediums.

The full storage system also requires some calculations to determine the chiller system size. Since the chillers will not be used during the on-peak period, the entire daily cooling requirement must be generated during the off-peak periods. Table 19.1 listed the total cooling load as 6,123 Ton-Hours for the peak day. Dividing this load over the 16 off-peak hours yields that the chillers must generate 383 tons of cooling per hour (6,123 Ton-Hours/16 hours). A 450-ton chiller will be utilized in this situation for comparison purposes. Table 19.6 shows how the chiller system would operate at 383 tons per hour, providing cooling required for the building directly and charging the storage system with the excess.

Comparing the peak demand with and without storage in Table 19.6 shows that the full storage system eliminated all load from the on-peak period. Column 4 shows the tonnage that is supplied to the storage system

**Table 19.5 Partial storage operation profile.**

Thermal Storage Operation Profile
Partial Storage System

| 1<br>End of<br>Hour<br>(Tons) | 2<br>Cooling<br>Load<br>(Tons) | 3<br>Chiller<br>Load<br>(Ton-Hrs) | 4<br>Capacity to<br>Storage<br>(Ton-Hrs) | 5<br>Capacity<br>In Storage | 6<br>Storage<br>Cycle |
|---|---|---|---|---|---|
| 1 | 100 | 255.13 | 155 | 696 | Charge |
| 2 | 120 | 255.13 | 135 | 831 | Charge |
| 3 | 125 | 255.13 | 130 | 961 | Charge |
| 4 | 130 | 255.13 | 125 | 1086 | Charge |
| 5 | 130 | 255.13 | 125 | 1211 | Charge |
| 6 | 153 | 255.13 | 102 | 1314 | Charge |
| 7 | 165 | 255.13 | 90 | 1404 | Charge |
| 8 | 230 | 255.13 | 25 | 1429 | Charge |
| 9 | 270 | 255.13 | -15 | 1414 | Discharge |
| 10 | 290 | 255.13 | -35 | 1379 | Discharge |
| 11 | 340 | 255.13 | -85 | 1294 | Discharge |
| 12 | 380 | 255.13 | -125 | 1169 | Discharge |
| 13 | 450 | 255.13 | -195 | 974 | Discharge |
| 14 | 490 | 255.13 | -235 | 740 | Discharge |
| 15 | 510 | 255.13 | -255 | 485 | Discharge |
| 16 | 480 | 255.13 | -225 | 260 | Discharge |
| 17 | 410 | 255.13 | -155 | 105 | Discharge |
| 18 | 360 | 255.13 | -105 | 0 | Discharge |
| 19 | 250 | 255.13 | 5 | 5 | Charge |
| 20 | 210 | 255.13 | 45 | 50 | Charge |
| 21 | 160 | 255.13 | 95 | 145 | Charge |
| 22 | 130 | 255.13 | 125 | 271 | Charge |
| 23 | 125 | 255.13 | 130 | 401 | Charge |
| 24 | 115 | 255.13 | 140 | 541 | Charge |

|  | Without storage | With storage | | |
|---|---|---|---|---|
| Daily Total (Ton-Hrs): | 6123 | 6123 | | |
| Daily Avg (Tons): | 255.13 | 255.13 | | |
| Peak Total (Ton-Hrs): | 3420 | 2041 | Storage Total = | 1429 |
| Peak Demand (Tons): | 510 | 255.13 | Peak Storage Output = | 255 |

Column 4 = Column 3 – Column 2
Column 5(n) = Column 5(n–1) + Column 4(n)

and column 5 shows the amount of cooling contained in the storage system at the end of each hour of operation. This system was designed so that there would be 0 capacity remaining in the thermal storage tanks after the on-peak period, as shown at the bottom of Table 19.6. The values in Table 19.6 will be utilized in the next section to determine the storage capacity required for each of the different storage mediums.

The optional partial storage system is a blend of the two systems presented earlier. Values given in Table 19.7 and Figure 19.4 are one combination of several possibili-

ties that would drop the consumption and peak demand during the on-peak period. Once again this system has been designed to run both chillers during off-peak hours and run only one during on-peak hours. Benefits of this arrangement are that the current chiller system could be used in combination with the storage system and that the storage system does not require as much capacity as the full storage system. Also, a reserve chiller is available during peak-load times.

Comparing the peak demand with and without storage in Table 19.7 shows that the optional partial stor-

**Table 19.6 Full storage operation profile.**

Thermal Storage Operation Profile
Full Storage System

| 1 Hour of Day | 2 Cooling Load (Tons) | 3 Chiller Load (Tons) | 4 Capacity to Storage (Ton-Hrs) | 5 Capacity In Storage (Ton-Hrs) | 6 Storage Cycle |
|---|---|---|---|---|---|
| 1 | 100 | 383 | 283 | 1589 | Charge |
| 2 | 120 | 383 | 263 | 1852 | Charge |
| 3 | 125 | 383 | 258 | 2109 | Charge |
| 4 | 130 | 383 | 253 | 2362 | Charge |
| 5 | 130 | 383 | 253 | 2615 | Charge |
| 6 | 153 | 383 | 230 | 2844 | Charge |
| 7 | 165 | 383 | 218 | 3062 | Charge |
| 8 | 230 | 383 | 153 | 3215 | Charge |
| 9 | 270 | 383 | 113 | 3327 | Charge |
| 10 | 290 | 383 | 93 | 3420 | Charge |
| 11 | 340 | 0 | -340 | 3080 | Discharge |
| 12 | 380 | 0 | -380 | 2700 | Discharge |
| 13 | 450 | 0 | -450 | 2250 | Discharge |
| 14 | 490 | 0 | -490 | 1760 | Discharge |
| 15 | 510 | 0 | -510 | 1250 | Discharge |
| 16 | 480 | 0 | -480 | 770 | Discharge |
| 17 | 410 | 0 | -410 | 360 | Discharge |
| 18 | 360 | 0 | -360 | 0 | Discharge |
| 19 | 250 | 383 | 133 | 133 | Charge |
| 20 | 210 | 383 | 173 | 305 | Charge |
| 21 | 160 | 383 | 223 | 528 | Charge |
| 22 | 130 | 383 | 253 | 781 | Charge |
| 23 | 125 | 383 | 258 | 1038 | Charge |
| 24 | 115 | 383 | 268 | 1306 | Charge |

|  | Without storage | With storage | | |
|---|---|---|---|---|
| Daily Total (Ton-Hrs): | 6123 | 6123 | | |
| Daily Avg (Tons): | 255.13 | 255.13 | | |
| Peak Total (Ton-Hrs): | 3420 | 0 | Storage Total = | 3420 |
| Peak Demand (Tons): | 510 | 0 | Peak Storage Output = | 510 |

Column 4 = Column 3 – Column 2
Column 5(n) = Column 5(n–1) + Column 4(n)

age system reduces the peak load from 510 tons to 153 tons, or approximately 70% during the on-peak period. Column 4 shows the tonnage that is supplied to the storage system and column 5 shows the amount of cooling contained in the storage system at the end of each hour of operation. This system was designed so that there would be zero capacity remaining in the thermal storage tanks after the on-peak period. The values contained at the bottom of Table 19.7 are the total storage capacity required and the maximum output required from storage. These values will be utilized in the next section to determine the storage capacity required for each of the different storage mediums. Table 19.8 summarizes the performance parameters for the three configurations discussed above. The next section summarizes the procedure used to determine the size of the storage systems required to handle the office building.

**19.4.2 Storage System Capacity**

Each of the storage mediums has different size requirements to satisfy the needs of the cooling load. This section will describe the procedure to find the actual

## Table 19.7 Optional partial storage operation profile.

### Thermal Storage Operation Profile—Optional Partial Storage System

| 1<br>End of<br>Hour | 2<br>Cooling<br>Load<br>(Tons) | 3<br>Chiller<br>Load<br>(Tons) | 4<br>Capacity to<br>Storage<br>(Ton-Hrs) | 5<br>Capacity<br>In Storage<br>(Ton-Hrs) | 6<br>Storage<br>Cycle |
|---|---|---|---|---|---|
| 1 | 100 | 306 | 206 | 1053 | Charge |
| 2 | 120 | 306 | 186 | 1239 | Charge |
| 3 | 125 | 306 | 181 | 1420 | Charge |
| 4 | 130 | 306 | 176 | 1597 | Charge |
| 5 | 130 | 306 | 176 | 1773 | Charge |
| 6 | 153 | 306 | 153 | 1926 | Charge |
| 7 | 165 | 306 | 141 | 2067 | Charge |
| 8 | 230 | 306 | 76 | 2143 | Charge |
| 9 | 270 | 306 | 36 | 2179 | Charge |
| 10 | 290 | 306 | 16 | 2195 | Charge |
| 11 | 340 | 153 | -187 | 2008 | Discharge |
| 12 | 380 | 153 | -227 | 1782 | Discharge |
| 13 | 450 | 153 | -297 | 1485 | Discharge |
| 14 | 490 | 153 | -337 | 1148 | Discharge |
| 15 | 510 | 153 | -357 | 791 | Discharge |
| 16 | 480 | 153 | -327 | 464 | Discharge |
| 17 | 410 | 153 | -257 | 207 | Discharge |
| 18 | 360 | 153 | -207 | 0 | Discharge |
| 19 | 250 | 306 | 56 | 56 | Charge |
| 20 | 210 | 306 | 96 | 152 | Charge |
| 21 | 160 | 306 | 146 | 298 | Charge |
| 22 | 130 | 306 | 176 | 475 | Charge |
| 23 | 125 | 306 | 181 | 656 | Charge |
| 24 | 115 | 306 | 191 | 847 | Charge |

|  | Without storage | With storage |  |  |
|---|---|---|---|---|
| Daily Total (Ton-Hrs): | 6123 | 6123 | | |
| Daily Avg (Tons): | 255.13 | 255.12 | | |
| Peak Total (Ton-Hrs): | 3420 | 1225 | Storage Total = | 2195 |
| Peak Demand (Tons): | 510 | 153 | Peak Storage Output = | 357 |

Column 4 = Column 3 - Column 2
Column 5(n) = Column 5(n-1) + Column 4(n)

volume or size of the storage system for the partial load system for each of the different storage mediums. The design of the chiller and thermal storage system must provide enough chilled water to the system to satisfy the peak load, so particular attention should be paid to the pumping and piping. Table 19.9 summarizes the size requirement of each of the three different storage options.

To calculate the capacity of the partial load storage system, the relationship between capacity (C), mass (M), specific heat of material (Cp), and the coil temperature differential $(T_2-T_1)$ shown in Figure 19.5a will be used:

$$C = M\,Cp\,(T_2-T_1)$$

where:
$M = lbm$
$Cp = Btu/lbm\,°R$
$(T_2-T_1) = °R$

The partial load system required that 1,429 Ton-Hrs be stored to supplement the output of the chiller during on-peak periods. This value does not allow for any thermal loss which normally occurs. For this discussion, a conservative value of 20% is used, which is an average suggested in the EPRI report[7]. This will increase the storage

**Table 19.8 System performance comparison.**

| PERFORMANCE PARAMETERS | SYSTEM | | | |
| --- | --- | --- | --- | --- |
| | Conventional No Storage | Partial Storage | Full Storage | Optional Partial |
| Overall Peak Demand (Tons) | 510 | 255.13 | 383 | 306 |
| On-Peak, Peak Demand (Tons) | 510 | 255.13 | 0 | 153 |
| On-Peak Chiller Consumption (Ton-Hrs) | 3,420 | 2,041 | 0 | 1,225 |
| Required Storage Capacity[1] (Ton-Hrs) | — | 1,379 | 3,420 | 2,195 |
| MAXIMUM STORAGE OUTPUT[1] (Tons) | — | 255 | 510 | 357 |

[1]Values from Table 19.5, 19.6, and 19.7. Represent the capacity required to be supplied by the TES.

requirements to 1,715 Ton-Hrs and chilled water storage systems in this size range cost approximately \$200/Ton-Hr including piping and installation[5]. Assuming that there are 12,000 Btu's per Ton-Hr, this yields:

$$C = (1,715 \text{ Ton-Hrs})*(12,000 \text{ Btu}/\text{Ton-Hr})$$
$$= 20.58 \times 10^6 \text{ Btu's.}$$

Assuming $(T_2 - T_1) = 12°$ and $Cp = 1 \text{ Btu}/\text{lb}_m \text{ °R}$, the relation becomes:

$$M = \frac{C}{Cp(T_2 - T_1)} = \frac{20.58 \times 10^6 \text{ Btus}}{1 \text{ Btu}/\text{lb}_m -°R)(12°R)} = 1.72 \times 10^6 \text{ lbm H}_2\text{O}$$

$$\text{Volume of Water} = \text{Mass}/\text{Density} = \frac{1.72 \times 10^6 \text{ lb}_m}{62.5 \text{ lb}_m/\text{Ft}^3}$$

$$= 27,520 \text{ Ft}^3 \text{ or}$$
$$\frac{1.72 \times 10^6 \text{ lb}_m}{8.34 \text{ lb}_m/\text{gal}} = 206,235 \text{ gal.}$$

Sizing the storage system utilizing ice is completed in a very similar fashion. The EPRI study states that the ice storage tanks had average daily heat gains 3.5 times greater than the chilled water and eutectic systems due to the higher coil temperature differential $(T_2 - T_1)$ To allow for these heat gains a conservative value of 50% will be added to the actual storage capacity, which is an average suggested in the EPRI report[7] This will increase the storage requirements to 2,144 Ton-Hrs. Assuming that there are 12,000 Btu's per Ton-Hr, this yields: (2,144 Ton-Hrs)*(12,000 Btu's/Ton-Hr) = 25.73 × 10⁶ Btu's. The ice systems utilize the latent heat of fusion so the $C_1$ now becomes

$$C_1 = \text{Latent Heat} = 144 \text{ Btu}/\text{lb}_m.$$

Because the latent heat of fusion, which occurs at 32°F, is so large compared to the sensible heat, the sensible heat (Cp) is not included in the calculation. The mass of water required to be frozen becomes:

$$M = C/C_1 = \frac{25.73 \times 10^6 \text{ Btus}}{(144 \text{ Btu}/\text{lb}_m)} = 1.79 \times 10^5 \text{ lb}_m \text{ H}_2\text{O}$$

$$\text{Volume of Ice} = \frac{\text{Mass}}{\text{Density}} = \frac{1.79 \times 10^5 \text{ lb}_m}{62.5 \text{ lb}_m/\text{Ft}^3}$$
$$= 2,864 \text{ Ft}^3$$

This figure is conservative since the sensible heat has been ignored but calculates the volume of ice needed to be generated. The actual volume of ice needed will vary and the total amount of water contained in the tank around the ice coils will vary greatly. The ability to purchase pre-packaged ice storage systems makes their sizing quite easy For this situation, two 1,080 Ton-Hr ice storage units will be purchased for approximately \$150/

## Table 19.9 Complete system comparison.

| Performance Parameters | SYSTEM | | | |
|---|---|---|---|---|
| | Conventional No Storage | Partial Storage | Full Storage | Optional Partial |
| **CHILLER** | | | | |
| SIZE (# and Tons) | 1 @ 600 | 1 @ 300 | 1 @ 450 | 2 @ 175 |
| COST($) | 180,000 | 90,000 | 135,000 | 105,000 |
| **WATER STORAGE** | | | | |
| Capacity (Ton-Hrs) | — | 1,715 | 4,104 | 2,634 |
| Volume (cubic feet) | — | 27,484 | 65,769 | 42,212 |
| Volume (gallons) | — | 205,635 | 492,086 | 315,827 |
| Cost per Ton-Hr ($) | — | 200 | 135 | 165 |
| Storage cost ($) | — | 343,000 | 554,040 | 434,610 |
| **ICE STORAGE** | | | | |
| Capacity (Ton-Hrs) | | 2,144 | 5,130 | 3,293 |
| # and size (Ton-Hrs) | — | 2 @ 1,080 | 4 @ 1,440 | 3 @ 1,220 |
| Ice volume (cubic feet) | — | 2,859 | 6,840 | 4,391 |
| Cost per Ton-Hr (S) | — | 150 | 150 | 150 |
| Storage cost ($)[1] | — | 324,000 | 864,000 | 549,000 |
| **EUTECTIC STORAGE** | | | | |
| Capacity (Ton-Hrs) | | 1,715 | 4,104 | 2,634 |
| Eutectic vol (cubic feet) | — | 8,232 | 19,699 | 12,643 |
| Cost per Ton-Hr ($) | — | 250 | 200 | 230 |
| Storage cost ($) | — | 428,750 | 820,000 | 605,820 |

[1](2 units)(1,080 Ton-Hrs/units)($150/Ton-Hr) = $324,000
Note: The values in this table vary slightly from those in the text from additional significant digits.

Ton-Hr including piping and installation[4] (note that this provides 2,160 Ton-Hrs compared to the needed 2,144 Ton-Hrs).

Sizing the storage system utilizing the phase change materials or eutectic salts is completed just as the ice storage system. The EPRI study states that the eutectic salt storage tanks had average daily heat gains approximately the same as that of the chilled water systems. To allow for these heat gains a conservative value of 20% is added to the actual storage capacity[5]. This increases storage requirements to 1,715 Ton-Hrs. Assuming there are 12,000 Btu's per Ton-Hr, this yields: (1,715 Ton-Hrs)*(12,000 Btu's/Ton-Hr) = 20.58 × 10^6 Btu's.

The eutectic system also utilizes the latent heat of fusion like the ice system and the temperature differential shown in Figure 19.5a is not used in the calculation. The $C_1$ now becomes:

$$C_1 = \text{Latent Heat} = 40 \text{ Btu/lbm}$$

$$M = C/C_1 = \frac{20.58 \times 10^6 \text{ Btus}}{(40 \text{ Btu/lmb})} = 5.15 \times 10^5 \text{ lbm}$$

$$\frac{\text{Volume of}}{\substack{\text{Eutectic Salts (assuming} \\ \text{density = water)}}} = \frac{\text{Mass}}{\text{Density}} = \frac{5.15 \times 10^5 \text{ lbm}}{62.5 \text{ lbm/ft}^3}$$

$$= 8,232 \text{ ft}^3$$

The actual volume of eutectic salts needed would need to be adjusted for density differences in the various combinations of the salts. Eutectic systems have not been studied in great detail and factory sized units are not yet readily available. The EPRI report[7] studied a system that required 1,600 Ton-Hrs of storage which utilized approximately 45,000 eutectic "bricks" contained in an

80,600 gallon tank of water. For this situation, a similar eutectic storage unit will be purchased for approximately $250/Ton-Hr including piping and installation. The ratio of Ton-Hrs required for partial storage and the required tank size will be utilized for sizing the full and optional partial storage systems.

Table 19.9 summarizes the sizes and costs of the different storage systems and the actual chiller systems for each of the three storage arrangements. The values presented in this example are for a specific case and each application should be analyzed thoroughly. The cost per ton hour of a water system dropped significantly as the size of the tanks rises as will the eutectic systems since the engineering and installation costs are spread over more capacity. Also we ignored the sensible heat of the ice and eutectic systems.

## 19.5 ECONOMIC SUMMARY

Table 19.9 covered the approximate costs of each of the three system configurations utilizing each of the three different storage mediums. Table 19.8 listed the various peak day performance parameters of each of the systems presented. To this point, the peak day chiller consumption has been used to size the system. To analyze the savings potential of the thermal storage systems, much more information is needed to determine daily cooling and chiller loads and the respective storage system performance. To calculate the savings accurately, a daily chiller consumption plot is needed for at least the summer peak period. These values can then be used to determine the chiller load required to satisfy the cooling demands. Only the summer months may be used since most of the cooling takes place and a majority of the utilities "time of use" charges (on-peak rates) are in effect during that time. There are several methods available to estimate or simulate building cooling load. Some of these methods are available in a computer simulation format or can also be calculated by hand.

For the office building presented earlier, an alternative method will be used to estimate cooling savings. An estimate of a monthly, average day cooling load will be used to compare the operating costs of the respective cooling configurations. For simplicity, it is assumed that the peak month is July and that the average cooling day is 90% of the cooling load of the peak day. The average cooling day for each of the months that make up the summer cooling period are estimated based upon July's average cooling load. These factors are presented in Table 19.10 for June through October[11]. These factors are applied to the hourly chiller load of the average July day to determine the season chiller/TES operation loads. The monthly average day, hourly chiller loads for each of the three systems are presented in Table 19.11. The first column for each month in Table 19.11 lists the hourly cooling demand. The chiller consumption required to satisfy this load utilizing each of the storage systems is also listed. This table does not account for the thermal efficiencies used to size the systems but for simplicity, these values will be used to determine the rate and demand savings that will be achieved after implementing the system. The formulas presented for the peak day thermal storage systems operations have been used for simplicity. These chiller loads do not represent the optimum chiller load since some of partial systems approach full storage systems during the early and late cooling months. The bottom of the table contains the totals for the chiller systems. These totaled average day values will now be used to calculate the savings. The difference between the actual cooling load and the chiller load is the approximate daily savings for each day of that month.

A hypothetical southwest utility rate schedule will be used to apply economic terms to these savings. The electricity consumption rate is $0.04/kWh and the de-

### Table 19.10 Average summer day cooling load factors.

| MONTH | kW FACTOR[1] | PEAK TONS[2] | kWh FACTOR[1] | Ton-Hrs/day[3] |
|-------|--------------|--------------|---------------|----------------|
| JUNE | 0.8 | 360 | 0.8 | 4,322 |
| JULY | 1 | 450 | 1 | 5,403 |
| AUGUST | 0.9 | 405 | 0.9 | 4,863 |
| SEPT | 0.7 | 315 | 0.7 | 3,782 |
| OCT | 0.5 | 225 | 0.5 | 2,702 |

[1]kW and kWh factors were estimated to determine utility cost savings.
[2]The average day peak load is estimated to be 90% of the peak day. The kW factor for each month is multiplied by the peak months average tonnage. For JUNE: PEAK TONS = (0.8)*(450) = 360
[3]The average day consumption is estimated to be 90% of the peak day. The kWh factor for each month is multiplied by the peak months average consumption. For JUNE: CONSUMPTION = (0.8)*(5,403) = 4,322

### Table 19.11 Monthly average day chiller load profiles.

| END OF HOUR | JUNE (in tons) | | | | JULY (in tons) | | | | AUGUST (in tons) | | | | SEPTEMBER (in tons) | | | | OCTOBER (in tons) | | | |
|---|---|---|---|---|---|---|---|---|---|---|---|---|---|---|---|---|---|---|---|---|
| | Actual | partial | full | optional | Actual | partial | full | optional | Actual | partial | full | optional | Actual | partial | full | optional | Actual | partial | full | optional |
| 1 | 71 | 180 | 270 | 216 | 88 | 225 | 338 | 270 | 79 | 203 | 304 | 243 | 62 | 158 | 236 | 189 | 44 | 113 | 169 | 135 |
| 2 | 85 | 180 | 270 | 216 | 106 | 225 | 338 | 270 | 95 | 203 | 304 | 243 | 74 | 158 | 236 | 189 | 53 | 113 | 169 | 135 |
| 3 | 88 | 180 | 270 | 216 | 110 | 225 | 338 | 270 | 99 | 203 | 304 | 243 | 77 | 158 | 236 | 189 | 55 | 113 | 169 | 135 |
| 4 | 92 | 180 | 270 | 216 | 115 | 225 | 338 | 270 | 103 | 203 | 304 | 243 | 80 | 158 | 236 | 189 | 57 | 113 | 169 | 135 |
| 5 | 92 | 180 | 270 | 216 | 115 | 225 | 338 | 270 | 103 | 203 | 304 | 243 | 80 | 158 | 236 | 189 | 57 | 113 | 169 | 135 |
| 6 | 108 | 180 | 270 | 216 | 135 | 225 | 338 | 270 | 122 | 203 | 304 | 243 | 95 | 158 | 236 | 189 | 68 | 113 | 169 | 135 |
| 7 | 116 | 180 | 270 | 216 | 146 | 225 | 338 | 270 | 131 | 203 | 304 | 243 | 102 | 158 | 236 | 189 | 73 | 113 | 169 | 135 |
| 8 | 162 | 180 | 270 | 216 | 203 | 225 | 338 | 270 | 183 | 203 | 304 | 243 | 142 | 158 | 236 | 189 | 101 | 113 | 169 | 135 |
| 9 | 191 | 180 | 270 | 216 | 238 | 225 | 338 | 270 | 214 | 203 | 304 | 243 | 167 | 158 | 236 | 189 | 119 | 113 | 169 | 135 |
| 10 | 205 | 180 | 270 | 216 | 256 | 225 | 338 | 270 | 230 | 203 | 304 | 243 | 179 | 158 | 236 | 189 | 128 | 113 | 169 | 135 |
| 11 | 240 | 180 | 0 | 108 | 300 | 225 | 0 | 135 | 270 | 203 | 0 | 122 | 210 | 158 | 0 | 95 | 150 | 113 | 0 | 88 |
| 12 | 268 | 180 | 0 | 108 | 335 | 225 | 0 | 135 | 302 | 203 | 0 | 122 | 235 | 158 | 0 | 95 | 168 | 113 | 0 | 68 |
| 13 | 318 | 180 | 0 | 108 | 397 | 225 | 0 | 135 | 357 | 203 | 0 | 122 | 278 | 158 | 0 | 95 | 199 | 113 | 0 | 68 |
| 14 | 346 | 180 | 0 | 108 | 432 | 225 | 0 | 135 | 389 | 203 | 0 | 122 | 303 | 158 | 0 | 95 | 216 | 113 | 0 | 68 |
| 15 | 360 | 180 | 0 | 108 | 450 | 225 | 0 | 135 | 405 | 203 | 0 | 122 | 315 | 158 | 0 | 95 | 225 | 113 | 0 | 68 |
| 16 | 339 | 180 | 0 | 108 | 424 | 225 | 0 | 135 | 381 | 203 | 0 | 122 | 296 | 158 | 0 | 95 | 212 | 113 | 0 | 68 |
| 17 | 289 | 180 | 0 | 108 | 362 | 225 | 0 | 135 | 326 | 203 | 0 | 122 | 253 | 158 | 0 | 95 | 181 | 113 | 0 | 68 |
| 18 | 254 | 180 | 0 | 108 | 318 | 225 | 0 | 135 | 286 | 203 | 0 | 122 | 222 | 158 | 0 | 95 | 159 | 113 | 0 | 68 |
| 19 | 176 | 180 | 270 | 216 | 221 | 225 | 338 | 270 | 199 | 203 | 304 | 243 | 154 | 158 | 236 | 189 | 110 | 113 | 169 | 135 |
| 20 | 148 | 180 | 270 | 216 | 185 | 225 | 338 | 270 | 167 | 203 | 304 | 243 | 130 | 158 | 236 | 189 | 93 | 113 | 169 | 135 |
| 21 | 113 | 180 | 270 | 216 | 141 | 225 | 338 | 270 | 127 | 203 | 304 | 243 | 99 | 158 | 236 | 189 | 71 | 113 | 169 | 135 |
| 22 | 92 | 180 | 270 | 216 | 115 | 225 | 338 | 270 | 103 | 203 | 304 | 243 | 80 | 158 | 236 | 189 | 57 | 113 | 169 | 135 |
| 23 | 88 | 180 | 270 | 216 | 110 | 225 | 338 | 270 | 99 | 203 | 304 | 243 | 77 | 158 | 236 | 189 | 55 | 113 | 169 | 135 |
| 24 | 81 | 180 | 270 | 216 | 101 | 225 | 338 | 270 | 91 | 203 | 304 | 243 | 71 | 158 | 236 | 189 | 51 | 113 | 169 | 135 |

| TOTALS: | JUNE | partial | full | optional | JULY | partial | full | optional | AUG | partial | full | optional | SEPT | partial | full | optional | OCT | partial | full | optional |
|---|---|---|---|---|---|---|---|---|---|---|---|---|---|---|---|---|---|---|---|---|
| TOTAL: (ton-hrs) | 4,322 | 4,322 | 4,322 | 4,322 | 5,403 | 5,403 | 5,403 | 5,403 | 4,862 | 4,862 | 4,862 | 4,862 | 3,782 | 3,782 | 3,782 | 3,782 | 2,701 | 2,701 | 2,701 | 2,701 |
| AVG: (tons) | 180 | 180 | 180 | 180 | 225 | 225 | 225 | 225 | 203 | 203 | 203 | 203 | 158 | 158 | 158 | 158 | 113 | 113 | 113 | 113 |
| OFF-PEAK MAX: (tons) | 205 | 180 | 270 | 216 | 256 | 225 | 338 | 270 | 230 | 203 | 304 | 243 | 179 | 158 | 236 | 189 | 128 | 113 | 169 | 135 |
| ON-PEAK MAX: (tons) | 360 | 180 | 0 | 108 | 450 | 225 | 0 | 135 | 405 | 203 | 0 | 122 | 315 | 158 | 0 | 95 | 225 | 113 | 0 | 68 |
| ON-PEAK CONSUMP: (ton-hrs) | 2,414 | 1,441 | 0 | 864 | 3,018 | 1,801 | 0 | 1,081 | 2,716 | 1,621 | 0 | 972 | 2,112 | 1,261 | 0 | 756 | 1,509 | 900 | 0 | 540 |

For June Partial Storage: (4,322 Ton–Hrs)/(24 Hrs) = 180 Tons
For June Full Storage: (4,322 Ton–Hrs)/(16 Hrs) = 270 Tons

mand rate during the summer is $3.50/kW per month for the peak demand during the off-peak hours and $5.00/kW per month for the peak demand during the on-peak hours. These summer demand rates are in effect from June through October. This rate schedule only provides savings from balancing the demand, although utilities often have cheaper off-peak consumption rates. It can be seen that the off-peak demand charge assures that the demand is leveled and not merely shifted. This rate schedule will be applied to the total values in Table 19.11 and multiplied by the number of days in each month to determine the summer savings. These savings are contained in Table 19.12. The monthly average day loads in Table 19.11 are assumed to be 90% of the actual monthly peak billing demand, and are adjusted accordingly in Table 19.12. The total monthly savings for each of the chiller/TES systems is determined at the bottom of each monthly column.

These cost savings are not the only monetary justification for implementing TES systems. Utilities often ex-

tend rebates and incentives to companies installing thermal energy storage systems to shorten their respective payback period. This helps the utility reduce the need to build new generation plants. The southwest utility serving the office building studied here offers $200 per design day peak kW shifted to off-peak hours up to $200,000.

## 19.6 CONCLUSIONS

Thermal energy storage will play a large role in the future of demand side management programs of both private organizations and utilities. An organization that wishes to employ a system wide energy management strategy will need to be able to track, predict and control their load profile in order to minimize utility costs. This management strategy will only become more critical as electricity costs become more variable in a deregulated market. Real time pricing and multi-facility contracts will further enhance the savings potential of demand management, within which thermal energy storage should become a valuable tool.

The success of the thermal storage system and the HVAC system as a whole depend on many factors:

- The chiller load profile,

- The utility rate schedules and incentive programs,

- The condition of the current chiller system,

- The space available for the various systems,

- The selection of the proper storage medium, and

- The proper design of the system and integration of this system into the current system.

Thermal storage is a very attractive method for an organization to reduce electric costs and improve system management. New installation projects can utilize storage to reduce the initial costs of the chiller system as well as savings in operation. Storage systems will become easier to justify in the future with increased mass production, technical advances, and as more companies switch to storage.

**References**

1. Cottone, Anthony M., "Featured Performer: Thermal Storage," in *Heating Piping and Air Conditioning*, August 1990, pp. 51-55.
2. Hopkins, Kenneth J., and James W. Schettler, "Thermal Storage Enhances Heat Recovery," in *Heating Piping and Air Conditioning,*, March 1990, pp. 45-50.
3. Keeler, Russell M., "Scrap DX for CW with Ice Storage," in *Heating Piping and Air Conditioning,*, August 1990, pp. 59-62.
4. Lindemann, Russell, Baltimore Aircoil Company, Personnel Phone Interview, January 7, 1992.
5. Mankivsky, Daniel K., Chicago Bridge and Iron Company, Personnel Phone Interview, January 7, 1992.
6. Pandya, Dilip A., "Retrofit Unitary Cool Storage System," in *Heating Piping and Air Conditioning,*, July 1990, pp. 35-37.
7. Science Applications International Corporation, *Operation Performance of Commercial Cool Storage Systems* Vols. 1 & 2, Electric Power Research Institute (EPRI) Palo Alto, September 1989.
8. Tamblyn, Robert T., "Optimizing Storage Savings," in *Heating Piping and Air Conditioning,*, August 1990 pp. 43-46.

**Table 19.12 Summer monthly system utility costs and TES savings.**

| | JUNE (30 days) | | | | JULY (31 days) | | | | AUGUST (31 days) | | | | SEPTEMBER (30 days) | | | | OCTOBER (31 days) | | | |
|---|---|---|---|---|---|---|---|---|---|---|---|---|---|---|---|---|---|---|---|---|
| | Actual | Partial | Full | Optional | Actual | Partial | Full | Optional | Actual | Partial | Full | Optional | Actual | Partial | Full | Optional | Actual | Partial | Full | Optional |
| ON-PEAK, PEAK (kW)[1] | 400 | 200 | 0 | 120 | 500 | 250 | 0 | 150 | 450 | 226 | 0 | 136 | 350 | 176 | 0 | 106 | 250 | 126 | 0 | 76 |
| OFF-PEAK, PEAK (kW)[1] | 228 | 200 | 300 | 240 | 284 | 250 | 376 | 300 | 256 | 226 | 338 | 270 | 199 | 176 | 262 | 210 | 142 | 126 | 188 | 150 |
| CONSUMPTION (kW-Hr)[1] | 4,322 | 4,322 | 4,322 | 4,322 | 5,403 | 5,403 | 5,403 | 5,403 | 4,862 | 4,862 | 4,862 | 4,862 | 3,782 | 3,782 | 3,782 | 3,782 | 2,701 | 2,701 | 2,701 | 2,701 |
| DEMAND COST ($)[2] | 2,797 | 1,700 | 1,050 | 1,440 | 3,496 | 2,125 | 1,314 | 1,800 | 3,144 | 1,917 | 1,182 | 1,623 | 2,446 | 1,492 | 918 | 1,263 | 1,748 | 1,067 | 657 | 903 |
| CONSUMPTION COST ($)[3] | 5,186 | 5,186 | 5,186 | 5,186 | 6,700 | 6,700 | 6,700 | 6,700 | 6,029 | 6,029 | 6,029 | 6,029 | 4,538 | 4,538 | 4,538 | 4,538 | 3,349 | 3,349 | 3,349 | 3,349 |
| TOTAL COST ($)[4] | 7,984 | 6,886 | 6,236 | 6,626 | 10,195 | 8,825 | 8,014 | 8,500 | 9,173 | 7,946 | 7,211 | 7,652 | 6,985 | 6,031 | 5,456 | 5,801 | 5,097 | 4,416 | 4,006 | 4,252 |
| SAVINGS ($)[5] | | 1,097 | 1,747 | 1,357 | | 1,371 | 2,181 | 1,696 | | 1,227 | 1,962 | 1,522 | | 954 | 1,528 | 1,183 | | 681 | 1,091 | 845 |

**Table 19.13 Available demand management incentives.**

| Performance Parameters | System | | | |
|---|---|---|---|---|
| | Conventional No Storage | Partial Storage | Full Storage | Optional Partial |
| Actual On-Peak Demand[1] (kW) | 510 | 255 | 0 | 153 |
| On-Peak Demand Shifted[2] (kW) | | 255 | 510 | 357 |
| Utility Subsidy[3] ($) | | 51,000 | 102,000 | 71,400 |

[1]Yearly design peak demand from Table 19.8.
[2]Demand shifted from design day on-peak period. For partial: 510 kW - 255 kW    = 255 kW.
[3]Based upon $200/kW shifted from design day on-peak period. For partial: 255 kW * $200/kW = $51,000.

# Appendix 19-A

Partial list of manufacturers of thermal storage systems. Source: *Energy User News*, Vol. 22, No. 12, December 1997.

| Manufacturer | Storage Type | Capacity (ton-hours) |
|---|---|---|
| Applied Thermal Technologies | Ice, Ice coil | 450 |
| Baltimore Aircoil Co. | Ice, gylcol solid ice, ice coil | 237-761 |
| Berg Chilling Systems Ltd. | ice coil | 100-10,000+ |
| Calmac Manufacturing Corp | ice, glycol solid ice, eutectic | 570 |
| CBI Walker Inc. | water, hot water | 2,000 -120,000+ |
| Chester-Jensen Co. Inc. | ice coil | 12-1,200 |
| Chicago Bridge & Iron Co. | water | 500+ |
| Cryogel | ice, encapsulated ice | 100-40,000 |
| Delta-Therm Corp. | | unlimited |
| Dunham-Bush Inc. | ice | 120, 180, 240 |
| FAFCO Inc. | Ice, gylcol solid ice | 125, 250, 375, 500 |
| Group Thermo Inc. | water, hot water | to 1,800 GPH |
| Henry Vogt Machine Co. | ice | unlimited |
| Morris & Associates | ice | 50 - 200 tons/day |
| Nalgun Corp. | Ice, water | 2,000 and up |
| Paul Mueller Co. | ice slurry | 3-1000+ |
| Perma Pipe | water | |
| Phoenix Thermal Storage | ice, hot water | 3 to 5 |
| Precision Parts Corp. | hot water | 440-17,800 gallons |
| Reaction Thermal Systems | ice | 242-4,244 |
| Steffes ETS Inc. | ceramic brick | 1.32-9 kW |
| Steibel Eltron Inc. | ceramic brick | |
| Store-More | ice coil ice, water, encapsulated ice, | 270-20,000 |
| The Trane Co. | glycol solid ice, ice coil | 60-145,000 |
| Turbo Refrigeration | ice | 10-340 |
| Vogt Tube Ice | ice | unlimited |
| York international Corp. | ice | 200 - 60,000 |

9.  Thumann, Albert, *Optimizing HVAC Systems* The Fairmont Press, Inc., 1988.
10. Thumann, Albert and D. Paul Mehta, *Handbook of Energy Engineering*, The Fairmont Press, Inc., 1991.
11. Wong, Jorge-Kcomt, Dr. Wayne C. Turner, Hemanta Agarwala, and Alpesh Dharia, *A Feasibility Study to Evaluate Different Options for Installation of a New Chiller With/Without Thermal Energy Storage System*, study conducted for the Oklahoma State Office Buildings Energy Cost Reduction Project, Revised 1990.

# Appendix 19-B

Partial list of Utility Cash Incentive Programs.
Source: Dan Mankivsky, Chicago Bridge & Iron, August 1991.

| STATE | CASH INCENTIVE | |
| --- | --- | --- |
| - Electric Utility | $/kW Shifted | Maximum |
| **ARIZONA** | | |
| - Arizona Public Service | 75-125 | no limit |
| - Salt River Project | 60-250 | no limit |
| **CALIFORNIA** | | |
| - American Public Utilities Dept. | 60 | 50,000 |
| - L.A. Dept of Water & Power | 250 | 40% cost |
| - Pacific Gas & Electric | 300 | 50%-70% |
| - Pasadena Public Utility | 300 | no limit |
| - Riverside Public Utility | 200 | no limit |
| - Sacramento Municipal Util Dist. | 200 | no limit |
| - San Diego Gas & Electric | 50-200 | no limit |
| - Southern California Edison | 100 | 300,000 |
| **DISTRICT OF COLUMBIA** | | |
| - Patomac Electric Power Co. | 200-250 | no limit |
| **FLORIDA** | | |
| - Florida Power & Light Co. | 250/ton | no limit |
| - Florida Power Corp. | 160-180 | 25% |
| - Tampa Electric Co. | 200 | no limit |
| **INDIANA** | | |
| - Indianapolis Power & Light | 200 | no limit |
| - Northern Indiana Public Service | 200/ton | |
| **MARYLAND** | | |
| - Baltimore Gas & Electric | 200 | no limit |
| - Patomac Electric Power Co. | 200-250 | no limit |
| **MINNESOTA** | | |
| - Northern States Power | 400/ton | no limit |
| **NEVADA** | | |
| - Nevada Power | 100-150 | no limit |
| **NEW JERSEY** | | |
| - Atlantic Electric | 150 | 200,000 |
| - Jersey Central Power & Light | 300 | 250,000 |
| - Orange & Rockland Utilities | 250 | no limit |
| - Public Service Electric & Gas | 125-250 | no limit |

*(Continued)*

| STATE<br>- Electric Utility | CASH INCENTIVE | |
| --- | --- | --- |
| | $/kW Shifted | Maximum |
| **NEW YORK** | | |
| - Central Edison Gas & Electric | 25/Ton-Hr | equip cost |
| - Consolidated Edison Co. | 600 | no limit |
| - Long Island Lighting Co. | 300-500 | no limit |
| - New York State Electric & Gas | 113 | no limit |
| - Orange & Rockland Utilities | 250 | no limit |
| - Rochester Gas & Electric | 200-300 | 70,000 |
| **NORTH DAKOTA** | | |
| - Northern States Power | 400/ton | no limit |
| **OHIO** | | |
| - Cincinnati Gas L Electric | 150 | no limit |
| - Toledo Edison | 200-250 | — |
| **OKLAHOMA** | | |
| - Oklahoma Gas & Electric | 125-200 | 225,000 |
| **PENNSYLVANIA** | | |
| - Metropolitan Edison | 100-250 | 40,000 |
| - Orange & Rockland Utilities | 250 | no limit |
| - Pennsylvania Electric | 250 | no limit |
| - Pennsylvania Power & Light | 100 | no limit |
| - Philadelphia Electric | 100-200 | 25,000 |
| **SOUTH DAKOTA** | | |
| - Northern States Power | 400/ton | no limit |
| **TEXAS** | | |
| - Austin Electric Department | 300 | 150,000 |
| - El Paso Electric Company | 200 | no limit |
| - Gulf States Utilities | 250 | — |
| - Houston Lighting & Power | 350 | — |
| - Texas Utilities (Dallas Power, Texas Electric Service, and Texas Power & Light) | 125-250 | no limit |
| **WISCONSIN** | | |
| - Madison Gas & Electric | 60 - 80 | no limit |
| - Northern States Power | 175 | no limit |
| - Wisconsin Electric Power | 350 | no limit |

*Note: Some states have additional programs not listed here and some of the listed programs have additional limitations.

# CHAPTER 20

# CODES, STANDARDS & LEGISLATION

**ALBERT THUMANN, P.E., CEM**
Association of Energy Engineers
Atlanta, GA

**CLINT CHRISTENSON**
Johnson Controls, Inc.

**Editor's Note:** EPACT-2005 was just being made federal law as this edition was being finalized. Look for details of EPACT-2005 legislation in future editions.

---

This chapter presents an historical perspective on key codes, standards, and regulations which have impacted energy policy and are still playing a major role in shaping energy usage. The Energy Policy Act of 1992 is far reaching and its implementation is impacting electric power deregulation, building codes and new energy efficient products. Sometimes policy makers do not see the far reaching impact of their legislation. The Energy Policy Act for example has created an environment for retail competition. Electric utilities will drastically change the way they operate in order to provide power and lowest cost. This in turn will drastically reduce utility sponsored incentive and rebate programs which have influenced energy conservation adoption.

## 20.1 THE ENERGY POLICY ACT OF 1992

This comprehensive legislation is far reaching and impacts energy conservation, power generation, and alternative fuel vehicles as well as energy production. The federal as well as private sectors are impacted by this comprehensive energy act. Highlights are described below:

### Energy Efficiency Provisions
Buildings
- Requires states to establish minimum commercial building energy codes and to consider minimum residential codes based on current voluntary codes.

Utilities
- Requires states to consider new regulatory standards that would: require utilities to undertake integrated resource planning; allow efficiency programs to be at least as profitable as new supply options; and encourage improvements in supply system efficiency.

Equipment Standards
- Establishes efficiency standards for: commercial heating and air-conditioning equipment; electric motors; and lamps.

- Gives the private sector an opportunity to establish voluntary efficiency information/labeling programs for windows, office equipment and luminaires, or the Dept. of Energy will establish such programs.

Renewable Energy
- Establishes a program for providing federal support on a competitive basis for renewable energy technologies. Expands program to promote export of these renewable energy technologies to emerging markets in developing countries.

Alternative Fuels
- Gives Dept. of Energy authority to require a private and municipal alternative fuel fleet program starting in 1998. Provides a federal alternative fuel fleet program with phased-in acquisition schedule; also provides state fleet program for large fleets in large cities.

Electric Vehicles
- Establishes comprehensive program for the research and development, infrastructure promotion, and vehicle demonstration for electric motor vehicles.

Electricity
- Removes obstacles to wholesale power competition in the Public Utilities Holding Company Act by allowing both utilities and non-utilities to form exempt wholesale generators without triggering the PUHCA restrictions.

Global Climate Change
- Directs the Energy Information Administration to establish a baseline inventory of greenhouse gas emissions and establishes a program for the voluntary reporting of those emissions. Directs the Dept. of Energy to prepare a report analyzing the strategies

for mitigating global climate change and to develop a least-cost energy strategy for reducing the generation of greenhouse gases.

### Research and Development

*   Directs the Dept. of Energy to undertake research and development on a wide range of energy technologies, including: energy efficiency technologies, natural gas end-use products, renewable energy resources, heating and cooling products, and electric vehicles.

## 20.2 STATE CODES

The Energy Policy Act of 1992 called for states to establish minimum commercial building energy codes and to consider the same for residential codes. Prior to this regulation, many states had some level of energy efficiency included in building codes (ASHRAE 90-80, CA Title 24, etc.), but most did not address the advances in equipment, materials or designs that would impact energy usage. A 1991 study by the Alliance to Save Energy found that most states employed codes that were very outdated, which may have initiated that portion of EPACT-1992.

The development of efficiency standards normally is undertaken by a consortium of interested parties in order to assure that the performance level is economically attainable. The groups for building efficiency standards are made up of building designers, equipment suppliers, construction professionals, efficiency experts, and others. There are several trade groups and research institutions that have developed standards as well as some states that developed their own. The approved standards are merely words on paper until a state or local agency adopts these standards into a particular building code. Once this occurs, officials (state or local) have the authority to inspect and assure that the applicable codes are enforced during design and construction.

The main organization responsible for developing building systems and equipment standards, at least in the commercial sector is the American Society of Heating, Refrigeration, and Air-conditioning Engineers (ASHRAE).

More than three quarters of the states have adopted ASHRAE Standard 90-80 as a basis for their energy efficiency standard for new building design. The ASHRAE Standard 90-80 is essentially "prescriptive" in nature. For example, the energy engineer using this standard would compute the average conductive value for the building walls and compare it against the value in the standard. If the computed value is above the recommendation, the amount of glass or building construction materials would need to be changed to meet the standard.

Most states have initiated "Model Energy Codes" for efficiency standards in lighting and HVAC. Probably one of the most comprehensive building efficiency standards is California Title 24. Title 24 established lighting and HVAC efficiency standards for new construction, alterations and additions of commercial and non-commercial buildings.

ASHRAE Standard 90-80 has been updated into two new standards:

ASHRAE 90.1-1999 Energy Efficient Design of New Buildings Except New Low-Rise Residential Buildings

ASHRAE 90.2-1993 Energy Efficient Design of New Low Rise Residential Building

The purposes of ASHRAE Standard 90.1-1999 are:

(a) set minimum requirement for the energy efficient design of new buildings so that they may be constructed, operated, and maintained in a manner that minimizes the use of energy without constraining the building function nor the comfort or productivity of the occupants.

(b) provide criteria for energy efficient design and methods for determining compliance with these criteria.

(c) provide sound guidance for energy efficient design.

In addition to recognizing advances in the performance of various components and equipment, the Standard encourages innovative energy conserving designs. This has been accomplished by allowing the building designer to take into consideration the dynamics that exist between the many components of a building through use of the System Performance Method or the Building Energy Cost Budget Method compliance paths. The standard, which is cosponsored by the Illuminating Engineering Society of North America, includes an extensive section on lighting efficiency, utilizing the Unit Power Allowance Method.

The standard also addresses the design of the following building systems:

- Electrical power,
- Auxiliary systems including elevators and retail refrigeration,
- Building envelope,
- HVAC systems,
- HVAC equipment,
- Service water heating and equipment, and
- Energy Management.

ASHRAE has placed 90.1 and 90.2 under continuous maintenance procedures by a Standing Standard Project Committee, which allows corrections and interpretations to be adopted through addenda.

## 20.3 MODEL ENERGY CODE

In 1994, the nation's model code organizations, Council of American Building Officials (CABO), Building Officials and Code Administrators International (BOCA), International Conference of Building Officials (ICBO), and Southern Building Codes Congress International (SBCCI), created the International Code Council (ICC). The purpose of the new coalition was to develop a single set of comprehensive building codes for new residential and commercial buildings, and additions to such buildings. The 2000 International Energy Conservation Code (IECC) was published in February of 2000 along with ten other codes, collectively creating the 2000 Family of International codes. These codes are the successor to the 1998 IECC and the 1995 Model Energy Code (MEC) as well as all of the previous MECs.

The IECC establishes minimum design and construction parameters for energy-efficient buildings through the use of prescriptive and performance based provisions. The 2000 IECC has been refined and simplified in response to the needs of the numerous users of the model energy code. It establishes minimum thermal performance requirements for building ceilings, walls, floors/foundations, and windows, and sets minimum efficiencies for lighting, mechanical and power systems in buildings. Currently EPACT-1992 references MEC 95 as the recommended building efficiency code. The Department of Energy is considering certifying the 2000 IECC as the most cost-effective residential energy-efficiency standard available. Once this determination is announced, EPACT-1992 requires states to determine the appropriateness of revising their residential energy codes to meet or exceed the 2000 IECC.

The publication of the 2000 IECC offers states and local jurisdictions the opportunity to apply for financial and technical assistance offered by DOE's Building Standards and Guidelines Program. If the standards are codified by these entities, their code enforcement agencies will have opportunities to utilize the support infrastructure already established by the national model code organizations. More information can be obtained at the building Codes Assistance Projects web site: www. crest. org/efficiency/.

## 20.4 FEDERAL ENERGY EFFICIENCY REQUIREMENTS

The federal sector is a very large consumer of energy in the United States. There are actually over 500,000 federal buildings with a combined energy cost of $10 billion per year. Managers and operators of these installations (mostly Department of Defense and Postal Service) have very little incentive to conserve energy or improve efficiency. Any work that is accomplished toward these goals would have normally been kept in the coffers and consumed by other functions as unencumbered funds. The OPEC oil embargo brought into focus the impact of energy costs and the US dependence on foreign sources of energy upon our economy. In 1975, the Energy Policy and Conservation Act directed the federal government to develop mandatory standards for agency procurement policies with respect to energy efficiency; and, develop and implement a 10-year plan for energy conservation in federal buildings, including mandatory lighting, thermal and insulation standards. This act was formalized with the Energy Conservation and Production Act in 1977, which established a 10% savings goal by 1985 over a 1975 baseline. The National Energy Conservation Policy Act of 1978 further defined the federal energy initiative with the following stipulations:

- Establishes the use of Life-cycle-cost (LCC) method of project analysis,

- Establishes publication of Energy Performance Targets,

- Requires LCC audits and retrofits of federal buildings by 1990,

- Establishes Federal Photovoltaic Program,

- Buildings exceeding 1000 square feet are subject to energy audits, and

- Establishes a Federal Solar Program.

In 1988 the Federal Energy Management Implementation Act (FEMIA 1988) amended the Federal Energy Initiative by removing the requirements to perform the LCC audits by 1990 and extended the deadline of 10 percent savings goals to 1995. FEMIA also allowed the Secretary of Energy to set the discount rate used in LCC analysis and directed the various federal agencies to establish incentive for energy conservation. The National Defense Authorization Acts for FY 89, 90, and 91 added the following provisions:

- Establishes incentive for shared energy savings contracts in DOD, allowing half of first year savings to be used for welfare, morale, and recreation activities at the facility. The other half to be used for additional conservation measures.

- Expands DOD's shared energy savings incentive to include half of first 5 years of savings.

- Requires the Secretary of Defense to develop plan for maximizing Cost effective energy savings, develop simplified contracting method for shared energy savings, and report annually to congress on progress.

- Expands DOD incentives to participate in utility rebate programs and to retain two-thirds of funds saved.

The President has power to invoke their own standards, in the form of Executive Orders, under which, agencies of the federal government must adhere. Presidents Bush and Clinton have both further increased and extended the efficiency improvements required to be undertaken by the federal sector. On June 3, 1999, President Clinton signed the order titled, "Greening the Government Through Efficient Energy Management." The order required federal agencies to achieve by 2010:

- 35% greater energy efficiency in buildings relative to 1985 levels, and

- 30% cut in greenhouse gas emissions from building-related energy use relative to 1990.

The order also directs agencies to maximize the use of energy savings performance contracts and utility contracts, in which private companies make energy improvements on federal facilities at their own expense and receive a portion of the resulting savings. Life cycle cost analysis must be used so agencies see the long term savings from energy investments rather than merely the low bidder selection criteria. The order requires that everything from light bulbs to boilers be energy efficient as well as the use of renewable energy technologies and sources such as solar, wind, geothermal and biomass. This order also mandated that the DOE, DOD and GSA provide relevant training or training materials for those programs that they make available to all federal agencies relating to energy management strategies contained in this order. A complete text of E.O. 13123 can be found on the FEMP Web site www.greeninginterior.doi.gov/13123.html.

## 20.5 INDOOR AIR QUALITY (IAQ) STANDARDS[1]

Indoor Air Quality (IAQ) is an emerging issue of concern to building managers, operators, and designers. Recent research has shown that indoor air is often less clean than outdoor air and federal legislation has been proposed to establish programs to deal with this issue on a national level. This, like the asbestos issue, will have an impact on building design and operations. Americans today spend long hours inside buildings, and building operators, managers and designers must be aware of potential IAQ problems and how they can be avoided.

IAQ problems, sometimes termed "Sick Building Syndrome," have become an acknowledged health and comfort problem. Buildings are characterized as sick when occupants complain of acute symptoms such as headache, eye, nose and throat irritation, dizziness, nausea, sensitivity to odors and difficulty in concentrating. The complaints may become more clinically defined so that an occupant may develop an actual building-related illness that is believed to be related to

[1]Source: Indoor Air Quality: Problems & Cures, M. Black & W. Robertson, Presented at 13th World Energy Engineering Congress.

IAQ problems.

The most effective means to deal with an IAQ problem is to remove or minimize the pollutant source, when feasible. If not, dilution and filtration may be effective.

Dilution (increased ventilation) is to admit more outside air to the building, ASHRAE's 1981 standard recommended 5 CFM/person outside air in an office environment. The new ASHRAE ventilation standard, 62-1989, now requires 20 CFM/person for offices if the prescriptive approach is used. Incidentally, it was the energy cost of treating outside air that led to the 1981 standard. The superseded 1973 standard recommended 15-25 CFM/person.

Increased ventilation will have an impact on building energy consumption However, this cost need not be severe. If an airside economizer cycle is employed and the HVAC system is controlled to respond to IAQ loads as well as thermal loads, 20 CFM/person need not be adhered to and the economizer hours will help attain air quality goals with energy savings at the same time.

In the fall of 1999 ASHRAE Standard 62-1999 was issued, "Ventilation for Acceptable Indoor Air Quality." The new standard contained the entire 1989 version, which remains unchanged, along with four new addenda. The reference in the 89 standard that the ventilation levels could accommodate a moderate amount of smoking was removed, due to troubles with secondhand tobacco smoke. The new standard also removed reference to thermal comfort, which is covered by other ASHRAE Standards. Attempts were made to clarify the confusion concerning how carbon dioxide can be used to determine air contamination. A statement was also added to assure that designers understand that merely following the prescribed ventilation rates does not ensure acceptable indoor air quality. The Standard was added to the continuous review process, which will mandate firms keep up with the perpetual changes, corrections and clarifications. There are many issues that are still under review as addendums to the standard. The types of buildings that are covered were limited to commercial and institutional, and the methods of calculation of the occupancy levels have been clarified. ASHRAE offers a subscription service that updates all addendum and interpretations. One of the main issues that should be considered during design of HVAC systems is that the outdoor air ventilation is required to be delivered cfm, which may be impacted with new variable volume air handling systems.

Energy savings can be realized by the use of improved filtration in lieu of the prescriptive 20 CFM/ person approach. Improved filtration can occur at the air handler, in the supply and return ductwork, or in the spaces via self-contained units. Improved filtration can include enhancements such as ionization devices to neutralize airborne biological matter and to electrically charge fine particles, causing them to agglomerate and be more easily filtered.

The Occupational Safety and Health Administration (OSHA) announced a proposed rule on March 25, 1994 that would regulate indoor air quality (IAQ) in workplaces across the nation. The proposed rule addresses all indoor contaminants but a significant step would ban all smoking in the workplace or restrict it to specially designed lounges exhausted directly to the outside. The smoking rule would apply to all workplaces while the IAQ provisions would impact "non-industrial" indoor facilities.

There is growing consensus that the most promising way to achieve good indoor air quality is through contaminant source control. Source control is more cost effective than trying to remove a contaminant once it has disseminated into the environment. Source control options include chemical substitution or product reformulating, product substitution, product encapsulation, banning some substances or implementing material emission standards. Source control methods except emission standards are incorporated in the proposed rule.

## 20.6 REGULATIONS & STANDARDS IMPACTING CFCs

For years, chlorofluorocarbons (CFCs) were used in air-conditioning and refrigeration systems. However, because CFCs are implicated in the depletion of the earth's ozone layer, regulations required the complete phaseout of the production of new CFCs by the turn of the century. Many companies, like DuPont, developed alternative refrigerants to replace CFCs. The need for alternatives will become even greater as regulatory cutbacks cause continuing CFC shortages.

Air-conditioning and refrigeration systems designed to operate with CFCs will need to be retrofitted (where possible) to operate with alternative refrigerants so that these systems can remain in use for their intended service life.

DuPont and other companies are commercializing their series of alternatives—hydrochlorofluorocarbon (HCFC) and hydrofluorocarbon (HFC) compounds. See Table 20.1.

The Montreal Protocol which is being implemented by the United Nations Environment Program

**Table 20.1 Candidate Alternatives for CFCs in Existing Cooling Systems**

| CFC | Alternative | Potential Retrofit Applications |
|---|---|---|
| CFC-11 | HCFC-123 | Water and brine chillers; process cooling |
| CFC-12 | HFC-134a or Ternary Blends | Auto air conditioning; medium temperature commercial food display and transportation equipment; refrigerators/freezers; dehumidifiers; ice makers; water fountains |
| CFC-114 | HCFC-124 | Water and brine chillers |
| R-502 | HFC-125 | Low-temperature commercial food equipment |

(UNEP) is a worldwide approach to the phaseout of CFCs. A major revision to the Montreal Protocol was implemented at the 1992 meeting in Copenhagen which accelerated the phaseout schedule.

The reader is advised to carefully consider both the "alternate" refrigerants entering the market place *and* the alternate technologies available. Alternate refrigerants come in the form of HCFCs and HFCs. HFCs have the attractive attribute of having no impact on the ozone layer (and correspondingly are not named in the Clean Air Act). Alternative technologies include absorption and ammonia refrigeration (established technologies since the early 1900's), as well as desiccant cooling.

Taxes on CFCs originally took effect January 1, 1990. The Energy Policy Act of 1992 revised and further increased the excise tax effective January 1, 1993.

Another factor to consider in ASHRAE Guidelines 3-1990—Reducing Emission of Fully Halogenated Chlorofluorocarbon (CFC) Refrigerants in Refrigeration and Air-Conditioning Equipment and Applications:

The purpose of this guideline is to recommend practices and procedures that will reduce inadvertent release of fully halogenated chlorofluorocarbon (CFC) refrigerants during manufacture, installation, testing, operation, maintenance, and disposal of refrigeration and air-conditioning equipment and systems.

The guideline is divided into 13 sections. Highlights are as follows:

The Design Section deals with air-conditioning and refrigeration systems and components and identifies possible sources of loss of refrigerants to atmosphere. Another section outlines refrigerant recovery reuse and disposal. The Alternative Refrigerant section discusses replacing R11, R12, R113, R114, R115 and azeotropic mixtures R500 and R502 with HCFCs such as R22.

## 20.7 REGULATORY AND LEGISLATIVE ISSUES IMPACTING AIR QUALITY

### 20.7.1 Clean Air Act Amendment

On November 15, 1990, the new Clean Air Act (CAA) was signed by President Bush. The legislation includes a section entitled Stratospheric Ozone Protection (Title VI). This section contains extraordinarily comprehensive regulations for the production and use of CFCs, halons, carbon tetrachloride, methyl chloroform, and HCFC and HFC substitutes. These regulations will be phased in over the next 40 years, and they will impact every industry that currently uses CFCs.

The seriousness of the ozone depletion is such that as new findings are obtained, there is tremendous political and scientific pressure placed on CFC end-users to phase out use of CFCs. This has resulted in the U.S., under the signature of President Bush in February 1992, to have accelerated the phaseout of CFCs.

### 20.7.2 Kyoto Protocol

The United States ratified the United Nations' Framework Convention on Climate Change, which is also known as the Climate Change Convention, on December, 4, 1992. The treaty is the first binding international legal instrument to deal directly with climate change. The goal is to stabilize green house gases in the atmosphere that would prevent human impact on global climate change, The nations that signed the treaty come together to make decisions at meetings call Conferences of the Parties. The 38 parties are grouped into two groups, developed industrialized nations (Annex I countries) and developing countries (Annex 11). The Kyoto Protocol, an international agreement reached in Kyoto in 1997 by the third Conference of the Parties (COP-3), aims to lower emissions from two groups of three greenhouse gases: Carbon dioxide, methane, and nitrous oxide and the second group of hydrofluorocarbon (HFCs), sulfur hexafluoride and perfluorocarbons. Emissions are meant to be reduced and limited to levels found in 1990 or 1995, depending upon the gases considered. The requirements will impact future clean air amendments, particularly for point sources. These requirements will further impact the implementation of distributed generation sources, which are discussed in the following section.

## 20.8 REGULATORY AND LEGISLATIVE ISSUES IMPACTING COGENERATION & INDEPENDENT POWER PRODUCTION[2]

Federal, state and local regulations must be addressed when considering any cogeneration project. This section will provide an overview of the federal regulations that have the most significant impact on cogeneration facilities.

### 20.8.1 Federal Power Act

The Federal Power Act asserts the federal government's policy toward competition and anti-competitive activities in the electric power industry. It identifies the Federal Energy Regulatory Commission (FERC) as the agency with primary jurisdiction to prevent undesirable anti-competitive behavior with respect to electric power generation. Also, it provides cogenerators and small power producers with a judicial means to overcome obstacles put in place by electric utilities.

### 20.8.2 Public Utility Regulatory Policies Act (PURPA)

This legislation was part of the 1978 National Energy Act and has had perhaps the most significant effect on the development of cogeneration and other forms of alternative energy production in the past decade. Certain provisions of PURPA also apply to the exchange of electric power between utilities and cogenerators.

PURPA provides a number of benefits to those cogenerators who can become Qualifying Facilities (QFs) under the act. Specifically, PURPA

- Requires utilities to purchase the power made available by cogenerators at reasonable buy-back rates.

These rates are typically based on the utilities' cost.

- Guarantees the cogenerator or small power producer interconnection with the electric grid and the availability of backup service from the utility.

- Dictates that supplemental power requirements of cogenerator must be provided at a reasonable cost.

- Exempts cogenerators and small power producers from federal and state utility regulations and associated reporting requirements of these bodies.

In order to assure a facility the benefits of PURPA, a cogenerator must become a Qualifying Facility. To achieve Qualifying Status, a cogenerator must generate electricity and useful thermal energy from a single fuel source. In addition, a cogeneration facility must be less than 50% owned by an electric utility or an electric utility holding company. Finally, the plant must meet the minimum annual operating efficiency standard established by FERC when using oil or natural gas as the principal fuel source. The standard is that the useful electric power output plus one half of the useful thermal output of the facility must be no less than 42.5% of the total oil or natural gas energy input. The minimum efficiency standard increases to 45% if the useful thermal energy is less than 15% of the total energy output of the plant.

### 20.8.3 Natural Gas Policy Act (NGPA)

The major objective of this legislation was to create a deregulated national market for natural gas. It provides for incremental pricing of higher cost natural gas supplies to industrial customers who use gas, and it allows the cost of natural gas to fluctuate with the cost of fuel oil. Cogenerators classified as Qualifying Facilities under PURPA are exempt from the incremental pricing schedule established for industrial customers.

### 20.8.4 Resource Conservation and Recovery Act of 1976 (RCRA)

This act requires that disposal of non-hazardous solid waste be handled in a sanitary landfill instead of an open dump. It affects only cogenerators with biomass and coal-fired plants. This legislation has had little, if any, impact on oil and natural gas cogeneration projects.

### 20.8.5 Public Utility Holding Company Act of 1935

The Public Utility Holding Company Act of 1935 (the 35 Act) authorizes the Securities and Exchange Commission (SEC) to regulate certain utility

[2]Source: *Georgia Cogeneration Handbook*, published by the Governor's Office of Energy Resources.

"holding companies" and their subsidiaries in a wide range of corporate transactions.

The Energy Policy Act of 1992 creates a new class of wholesale-only electric generators—"exempt wholesale generators" (EWGs)—which are exempt from the Public Utility Holding Company Act (PUHCA). The Act dramatically enhances competition in U.S. wholesale electric generation markets, including broader participation by subsidiaries of electric utilities and holding companies. It also opens up foreign markets by exempting companies from PUHCA with respect to retail sales as well as wholesale sales.

### 20.8.6 Moving towards a deregulated electric power marketplace

EPACT-1992 set into motion a widespread movement for utilities to become more competitive. Retail wheeling proposals were set into motion in states such as California, Wisconsin, Michigan, New Mexico, Illinois and New Jersey. There are many issues involved in a deregulated power marketplace and public service commission rulings and litigation will certainly play a major role in the power marketplace of the future. Deregulation has already brought about several important developments:

- Utilities will need to become more competitive. Downsizing and minimization of costs including elimination of rebates are the current trend. This translates into lower costs for consumers. For example Southern California Edison announced that the system average price will be reduced from 10.7 cents/kWh to lower than 10 cents by the year 2000. This would be a 25% reduction after adjusting for inflation.

- Utilities will merge to gain a bigger market share. For example, Wisconsin Electric Power Company merged with Northern States Power; this merger of two utilities resulted in a savings of $2 billion over 10 years.

- Utilities are forming new companies to broaden their services. Energy service companies, financial loan programs and purchasing of related companies are all part of the new utility strategy.

- In 1995 one hundred power marketing companies have submitted applicants to FERC. Power marketing companies will play a key role in brokering power between end users and utilities in different states and in purchasing of new power generation facilities.

- Utilities will need to restructure to take advantage of deregulation. Generation Companies

may be split away from other operating divisions such as transmission and distribution. Vertical disintegration will be part of the new utility structure.

- Utilities will weigh the cost of repowering and upgrading existing plants against purchasing power from a third party.

Chapter 24 discusses many more issues on the topic of electrical deregulation.

## 20.9 OPPORTUNITIES IN THE SPOT MARKET[3]

### Basics of the Spot Market

A whole new method of contracting has emerged in the natural gas industry through the spot market. The market has developed because the Natural Gas Policy Act of 1978 (NGPA) guaranteed some rights for end-users and marketers in the purchasing and transporting of natural gas. It also put natural gas supplies into a more competitive position with deregulation of several categories.

The Federal Energy Regulatory Commission (FERC) provided additional rulings that facilitated the growth of the spot market. These rulings included provisions for the Special Marketing Programs in 1983 (Order 2346) and Order 436 in 1985, which encouraged the natural gas pipelines to transport gas for end-users through blanket certificates.

The change in the structure of markets in the natural gas industry has been immense in terms of both volumes and the participants in the market. By year-end 1986, almost 40% of the interstate gas supply was being transported on a carriage basis. Not only were end-users participating in contract carriage, but local distribution companies (LDCs) were accounting for about one half of the spot volumes on interstate pipelines.

The "spot market" or "direct purchase" market refers to the purchase of gas supplies directly from the producer by a marketer, end-user or LDC. (The term "spot gas" is often used synonymously with "best efforts gas," "interruptible gas," "direct purchase gas" and "self-help gas.") This type of arrangement cannot be called new because the pipelines have always sold some supplies directly to end-users.

The new market differs from the past arrangements in terms of the frequency in contracting and the volumes involved in such contracts. Another characteristic of the spot market is that contracts are short-term, usually only 30 days, and on an interruptible basis. The interruptible nature of spot market supplies is an important key to understanding the operation of the spot market and the costs of dealing in it. On both the production and transportation sides, all activities in transportation or purchasing supplies are on a "best efforts" basis. This means that when a cold snap comes the direct purchaser may not get delivery on his contracts because of producer shutdowns, pipeline capacity and operational problems or a combination of these problems. The "best efforts" approach to dealing can also lead to problems in transporting supplies when demand is high and capacity limited.

## FERC's Order No. 436

The impetus for interstate pipeline carriage came with FERC's Order No. 436, later slightly changed and renumbered No. 500, which provided more flexibility in pricing and transporting natural gas. In passing the 1986 ruling, FERC was attempting to get out of the day-to-day operations of the market and into more generic rule making. More significantly, FERC was trying to get interstate pipelines out of the merchant business into the transportation business—a step requiring a major restructuring of contracting in the gas industry.

FERC has expressed an intent to create a more competitive market so that prices would signal adjustments in the markets. The belief is that direct sales ties between producers and end-users will facilitate market adjustments without regulatory requirements clouding the market. As more gas is deregulated, FERC reasoned that natural gas prices will respond to the demand: Lower prices would assist in clearing excess supplies; then as markets tightened, prices would rise drawing further investment into supply development.

## FERC Order No. 636

Order 636 required significant "Restructuring" in interstate pipeline services, starting in the fall of 1993. The original Order 636:

- Separates (unbundles) pipeline gas sales from transportation

- Provides open access to pipeline storage

- Allows for "no notice" transportation service

- Requires access to upstream pipeline capacity

- Uses bulletin boards to disseminate information

- Provides for a "capacity release" program to temporarily sell firm transportation capacity

- Pregrants a pipeline the right to abandon gas sales

- Bases rates on straight fixed variable (SFV) design

- Passes through 100% of transition costs in fixed monthly charges to firm transport customers

## FERC Order No. 636A

Order 636A makes several relatively minor changes in the original order and provides a great deal of written defense of the original order's terms. The key changes are:

- Concessions on transport and sales rates for a pipeline's traditional "small sales" customers (like municipalities).

- The option to "release" (sell) firm capacity for less than one month—*without posting it on a bulletin board system or bidding.*

- Greater flexibility in designing special transportation rates (i.e., off-peak service) while still requiring overall adherence to the straight fixed variable rate design.

- Recovery of 10% of the transition costs from the interruptible transportation customers (Part 284).

Court action is still likely on the Order. Further, each pipeline will submit its own unique tariff to comply with the Order. As a result, additional changes and variations are likely to occur.

## 20.10 THE CLIMATIC CHANGE ACTION PLAN

The Climatic Change Action Plan was established April 21, 1993 and includes the following:

- Returns U.S. greenhouse gas emissions to 1990 levels by the year 2000 with cost effective domestic actions.

- Includes measures to reduce all significant greenhouse gases, carbon dioxide, methane,

# CHAPTER 21

# NATURAL GAS PURCHASING

**CAROL FREEDENTHAL**
JOFREEnergy Consulting
Houston, Texas

## 21.1 PREFACE

This is the second full revision for this chapter, Natural Gas Purchasing. Chapter 21 was originally written when the book was published in 1993. Rewrite for the first revision was a completely new effort done in 1996. As of 2000, the industry has continued to change and is still in the conversion from a federally regulated, price-controlled business to an economically dynamic, open industry, and this is a completely revised writing. Changes are continuing to shape the industry differently, especially when coupled with the changes coming from the potential decontrol of the electric power industry. To make even more changes, the impact of ECommerce business-to-business is beginning to play a role in this industry. When this revision was started, only one company offered the web for gas marketing. As of 2000, five additional companies had launched ECommerce business-to-business natural gas trading.

The old natural gas business is really a new business. Its structure goes back 150 years but it is more like a new industry. It has the typical growth and turmoil of a new business. Energy products, especially natural gas and electricity, are new businesses as the country goes into the new millennium. Newly "reformed" companies, new marketing organizations, new systems affecting gas marketing, and even, a new industry structure makes it necessary to start from scratch in writing the revision for this chapter.

Like the new millennium, the natural gas industry and equally as important, the total energy business is going through its own transition. Change will continue as companies and businesses try different strategies. ECommerce will play a major role in the industry's transition. This phase of the transition is amorphous and makes it difficult to predict the exact course of events for the future. Things that appeared far-out years ago are becoming closer to reality. The newest buzzwords, "distributive electricity" includes the use of fuel cells and small dual cycle turbine driven generators by residential and small commercial users. Both of these are becoming economically feasible. The impact on the gas and electric industries is unknown. This is a time of change for the new energy business. Marketing and supplying energy products like natural gas and electricity will go through many changes before optimum conditions are found.

A few things are for sure. Natural gas is becoming the major fuel for stationary power uses in the United States. Long dominated by oil products for this use, now gas is becoming the leader. Coal continues as a major fuel source for electric generation. Consumption of coal for power generation has reached record levels in recent years but environmental concerns and the required high capital for new coal burning generating plants will reduce coal's market share. The public's dislike of nuclear power and the high costs to build plants with the safety desired means no growth for this industry. A new philosophy will have to be developed by society recognizing safety and environmental benefits of atomic power before new nuclear facilities will be built.

The natural gas industry, just like the power industry, which is going through its own decontrol activities, change will be a way of life always. Companies in the energy field and in associated areas such as communications, financial, systems, etc. will continue to merge, acquire, spin off, and change their structure and goals. As the country goes into the new millennium, these are industries in transition and will change along with the growth industries in cyberspace. A big difference from the old, staid and conservative electric and gas utilities of the prior century! Change and growth are the way.

Regardless of this, one factor continues to dominate. The profit motive is still the driving force of the industry today. It will not change but will continue into the future. Economics will govern change and be the basis for decision making. All the transformations—buying and selling of companies, new marketing companies, new systems for handling the merged assets, etc. will all be subject to one metric; is it profitable? Already, some acquisitions made by large electric and gas companies to bring together various parts of the energy industry have come apart because the final economics did not pass muster.

The purpose of this chapter is to give the fuel buyer, for any operations or industry, the knowledge and information needed to buy natural gas for fuel. The buyer may be in a large petrochemical plant where natural gas is a major raw material or may be the commercial user having hundreds of apartments needing gas for heat and hot water or plant operator where the gas is used for process steam. It might be a first time experience or an on-going job for the buyer. This chapter will give the background and information to find natural gas supplies for any need at the lowest cost and highest service and security. The chapter will include information on history of the industry, sources of supply, transportation, distribution, storage, contracts, regulatory, and financial considerations needed to buy natural gas.

## 21.2 INTRODUCTION

Natural gas is predominately the compound "methane," $CH_4$. It has the chemical structure of one carbon atom and four hydrogen atoms. It is the simplest of the carbon based chemicals and has been a fuel for industry, for illumination, and for heating and some cooling of homes, offices, schools, and factories. Natural gas is also, a major fuel for generating electricity. In addition to fuel uses, gas is a major feedstock for the chemical industry in making such products and their derivatives as ammonia and methanol. Natural gas is used in refining and chemical plants as a source for hydrogen needed by these processing businesses. Through the reforming process, hydrogen is stripped from the methane leaving carbon dioxide, which has its usefulness in chemical manufacturing or use, by itself in cooling, carbonated drinks, or crude oil recovery.

The term "natural gas industry" includes the people, equipment, and systems starting in the fields where the wells are located and the natural gas is produced. It includes other field tasks as gathering, treating, and processing. Transportation to storage or to interstate or intrastate pipelines for further transportation to the market area storage, or to the distribution system for delivery to the consumer and the burner tip, are part of the system. The burner tip might be in a boiler, hot water heater, combustion engine, or a chemical reactor to name a few of the many uses for natural gas.

Natural gas is produced in the field by drilling into the earth's crust anywhere from a couple of thousand feet to five miles in depth. Once the gas is found and the well completed to bring the gas to the earth's surface, it is treated if necessary to remove acid impurities and again, if necessary, processed to take out liquid hydrocarbons of longer carbon chains than methane, which has a single carbon atom. After processing, the gas is transported in pipelines to consuming areas where distribution companies handle the delivery to the specific consumer.

In addition to the people and companies directly involved in the production, transportation, storage, and marketing of natural gas, there are countless other businesses and people involved in assisting the gas industry to complete its tasks. There are systems companies, regulatory and legal professionals, financial houses, banks, and a host of other businesses assisting the natural gas industry. Figure 21.1 shows the many parts of the industry as it is known today. The money flowing through the major sections of the industry are shown in Figure 21.2. The $85 billion industry shown in the diagram only represents the functions in getting natural gas, the commodity, to market and consumption. Not included in the overall industrial revenues are the moneys generated by the sales and resale of gas before its consumption, the processing and marketing of natural gas liquids coming from the gas, and the financial markets where gas futures and other financial instruments are sold and traded. These are big businesses also. Estimates are that the physical gas is traded three to four times before consumption. In the financial markets, gas volumes 10 to 12 times the amount of gas consumed on an average day are traded daily.

Figure 21.3 should be of most interest to the natural gas buyer as it depicts the various sales points, stages, and handling the gas goes through in getting from the wellhead to the burner tip—from the wellhead to the consumer. As one can see in the diagram, there are many alternate paths the gas can travel before coming to its end use as a fuel or feedstock for chemical manufacturing. Each one of the stages on the flowsheet represents an added value point in the travel to consumption. Raw gas coming from the wellhead many times has sufficient quality to go directly into a transporting pipeline for delivery to the consuming area. Sometimes the gas needs field treating and/or processing to meet pipeline specifications for acceptance into the pipeline.

The gas industry is the oldest utility except for water and sanitation. In the middle of the 19th century, many large cities used a synthetic gas made from passing steam over coal to light downtown areas and provide central heating systems. Big cities like Baltimore, New York, Boston, and many more cities and municipalities used gas for illumination. Many utilities from that period exist today and are still gas and electric suppliers to the areas they serve.

In the early days of the gas business, there was no natural gas, as known today. Instead, these utilities produced a synthetic gas for both the illumination and the central heating systems. The synthetic gas, sometimes called "water gas" because of the method of producing it,

Figure 21.1 Natural Gas Industry Flowsheet.

**TOTAL U.S. NATURAL GAS REVENUES BY YEARS**

SOURCE: EIA, NGM
MAY 2000

| YEAR | TOTAL, $B | WELLHEAD, $/MMBtu |
|------|-----------|-------------------|
| 1995 | 76.60 | 1.55 |
| 1996 | 89.17 | 2.17 |
| 1997 | 96.67 | 2.32 |
| 1998 | 84.74 | 1.94 |
| 1999 | 84.70 | 2.07 |

JOFREE HOUSTON, TX 77002 CF 8JUL2000

Figure 21.2 Gas Industry Money Flow for Business Activities.

**Figure 21.3 Wellhead to Consumer Flowsheet**

had bad attributes—it contained a high content of hydrogen and carbon monoxide, two bad actors for a gas used in homes, businesses, and factories. People died when exposed to it because of the carbon monoxide, and buildings blew-up because of the hydrogen when free gas from leaks or pipe ruptures was ignited. When natural gas came on the scene in the early 1900s, where it was available, it quickly replaced the old manufactured gas. About the same time, advances were made in electricity so that cities and municipalities changed to electricity for lighting and illumination. Natural gas quickly lost its market for municipal lighting.

Natural gas was originally an unwanted by-product from the oil fields. Problem was getting rid of it. Flaring was used, but this was a waste of good natural resources. Around the beginning of the century, associated gas from Ohio oil fields was shipped to Cleveland in wooden pipes to replace the then used synthetic gas. In the early days of the industry, the limitations to greater uses of natural gas were that gas was produced in only certain parts of the country and transportation was available for only very short distances. Market penetration was thwarted by the ability to ship it. There were no long distance pipelines in the early days of the industry. Natural gas made a great replacement for the synthetic counterpart—methane is essentially safe as far as toxicity and is much safer as far as explosion. Gas' growth was dependent on building long distance pipelines. Not until the 1930s did the industry have the capability of making strong enough, large steel piping needed for the long-distance pipelines. Completion of major interstate pipelines to carry gas from producing regions to consumers was the highlight of the 1930s to the start of World War II in the early 1940s.

Pipeline construction came to a halt and was dormant until the war's end. Construction went full force

after the war to insure delivering the most economical and easiest fuel to America's homes, commercial facilities and industrial players. Even today with the start of the new millennium, some areas of the U.S. still do not have a fully developed natural gas distribution and delivery system. Areas in the West where population is sparse, parts of the Northeast where oil prices were too competitive to delivered gas prices, and other parts of the country lacking distribution systems for the same reasons are still without natural gas. Many of these use what is called "bottled gas," a mixture of propane and butane or propane only for home heating and other critical uses. Just recently, new supplies and pipelines were developed to bring natural gas to the Northeast U.S. from Canada. Additional distribution systems will bring more gas to more customers through the country from the tip of Florida to the North Central and West Northern states.

Ever since natural gas became available for fuel, it was under some form of government economic control. Through the tariff mechanism for pricing natural gas, the government had the power to make gas prices more or less attractive to competing fuels. Further, with the government controlling wellhead prices and slow to make changes in prices as conditions changed, it became difficult and economically undesirable to expand natural gas production. Government price controls hampered the growth of the U.S. natural gas business. The gas shortages of the mid-1970s are an example of government control stifling expansion and growth. There was no shortage of gas reserves, only a shortage of incentives for producers to develop and supply the gas. The free market builds its own controls to foster competition and growth.

Congress passed the Natural Gas Policy Act of 1978 to change the policy of government economic control. A few years of transition were needed before significant changes began in the industry. Real impact started in 1985. Even today, the industry is still in transition. The federal decontrol changed interstate marketing and movement of natural gas. Gas at the local levels where the state Public Utility Commission or similar local government has control, is still heavily regulated. Decontrol at the federal level is slowly filtering down to local agencies. As of 2002, some states began moving to "open transportation" rules. A current obstacle to the swifter implementation of rules at the state and local levels is the tie of gas and electricity as utilities within state regulatory control. With the electric industry going through its own "decontrol," many wanted to see the much larger electric industry work out the utility problems. Then gas could follow with less negotiating and discussion. The electric timetable is now years behind its planned evolution and this has slowed gas local control further.

With the price of gas changing each year, the total industry value changes. The industry in nominal annual terms is roughly a $100 Billion business. Electricity is around $230 Billion. Many electric companies that were both gas and electricity utilities even before deregulation, have bought major natural gas pipelines or gas distributors. Large electric companies bought into the natural gas industry whether they purchased transporters, distributors, or marketing companies. Interestingly, in a relatively few years, some of these combinations have come apart because of poor profitability.

Electric and gas utility companies have gone after transportation and marketing companies. Surprisingly, none of the expanding companies have sought to buy, at the beginning of the gas business, the oil and gas exploration and production companies (E&P companies). These are the companies looking for natural gas and then producing it. While all of the transporting companies, whether long distance or distribution in nature and, further, whether electric and/or natural gas in business, have shied away from the production companies, other E&P companies have merged or acquired smaller operations to add to the total capability of the company. The significant changes during the 1990's saw major E&P companies acquire even major and independent E&P assets.

## 21.3  NATURAL GAS AS A FUEL

Why has natural gas grown in popularity? What makes it a fuel of choice in so many industries as the new millennium begins? What shortcomings does it have? Figure 21.4 shows the change in basic fuels mix used in the U.S. in 1985 and 1999. Nuclear, which started in 1960, enjoyed a period of rapid growth. The high costs for all the safety engineered into the plants had made it an uneconomical system towards the end of the century. There are no nuclear plants scheduled for construction. Even some of those completed and running and some with the initial construction still in progress were shut down or converted into natural gas fired units. The only change that will be seen in nuclear generation of electricity is plant efficiencies will be improved for the units continuing to operate.

Coal usage in the U.S. has grown in recent years with record coal production in the late 1990s. Coal is by far the major fuel used for electric generation, commanding a 56% market share. It has many negative properties like the need for railroads for transportation, high pollution from the burner after-products, and poor handling characteristics including being dirty, losses on storage, and the difficulties of moving a solid material, including the disposal of the remaining ash. Still, coal has a number of things going for it which will keep coal in use for many years to come. The ready availability and abundance are major merits. The stability of coal prices will always give coal a place in the market. Figure 21.5 shows the comparison in prices among coal, natural gas, and oil products for the period 1985 through 1999. Coal at about a dollar per million British units (MMBtu) is not only much cheaper per unit of energy, but also has the advantages of availability and abundance. Coal will slowly lose position because of its disadvantages of pollution and higher costs to meet changing standards and high capital costs for building new generating plants.

Petroleum products have lost market share in the later years because of their costs and the dependence of the U.S. on foreign suppliers for crude and crude oil products. Oil products used for electric generation include distillate fuel oil, a relatively lightweight oil, which during the refining process can have most of the sulfur removed during that process. Low sulfur fuels are desirable to keep emissions low for environmental reasons. The other major oil product used is residual fuel oil, the bottom of the barrel from the refining process. This is a heavy, hard to transport fuel with many undesirable ingredients that become environmental problems after combustion. Many states have put costly tariffs on using residual fuel oil because of its environmental harm when used.

**1985 - 74 Quads**

**1999 - 93 Quads**

**Figure 21.4 U.S. Basic Fuels 1985 & 1999 (Quadrillion Btu)**

Natural gas is the nation's second largest source of fuel and a major source of feedstock for chemical production. Plentiful supplies at economically satisfactory prices, a well developed delivery system of pipelines to bring gas from the wellhead to the consumer, and its environmental attractiveness has made natural gas the choice of fuel for many applications. Going into the new millennium, natural gas will be a popular fuel. As a fuel for industry for heating and generating electricity and as a feedstock for chemicals, natural gas is very attractive. For residential and commercial applications, the security of supply and efficiency in supplying makes it an ideal fuel. Even though natural gas is a fossil fuel, it has the lowest ratio of combustion-produced carbon dioxide to energy released. Carbon dioxide is the biggest culprit in the concern for global warming.

Natural gas consumption data are followed in four major areas by the Federal Energy Information Agency in addition to its listing the data for natural gas used in the fields for lease and plant fuel and as fuel for natural gas pipelines; residential, commercial, industrial, and electric generation. Natural gas demand has always, in modern times led the amount of gas produced except for the mid-1970s when the country experienced a severe natural gas supply shortage. In those years, while there were more than sufficient reserves in the ground to meet demand, the control of gas pricing by the federal government stymied the initiative of producers to meet demand. Potential supply was available but the lack of profit incentive prevented meeting demand in those years. Demand increased because of changes and shortages in crude oil supplies. Early 1970s were the start of the change in crude pricing and the country was faced with decreased supplies from foreign producers. Crude prices doubled almost overnight, but because natural gas was price controlled and could not meet the rising prices, supplies in the interstate market suffered.

The major market for natural gas is the industrial sector. Residential is next and commercial and electric generating take about the same amount. Figure 21.6 graphically depicts the share each market took for 1999. The residential market is basically for home heating and hot water fuel. The commercial market is for space heating. Use of natural gas in industrial plants when used for space heating is included in this category. The industrial category covers all other uses of natural gas in industry and includes gas used by industrial locations for power generation until earlier 2000. All power generation is now included in the category of electric generation. The major demand factor in all categories is weather. Residential and commercial consumption are most affected by weather since these two categories reflect space heating. Electric generation is weather sensitive also since the summer electric load is responsive to the air conditioning load needed for the hot weather. Even though the industrial load is not as sensitive to weather as are other categories, it does reflect the additional heating load needed for the process industries when temperatures fall and raw materials including process air and/or water are much colder.

Natural gas has tremendous potential to gain even greater use in the generation of electricity in several ways. First, it could be the choice fuel to replace aging nuclear plants that will not be re-certified as they age. Further

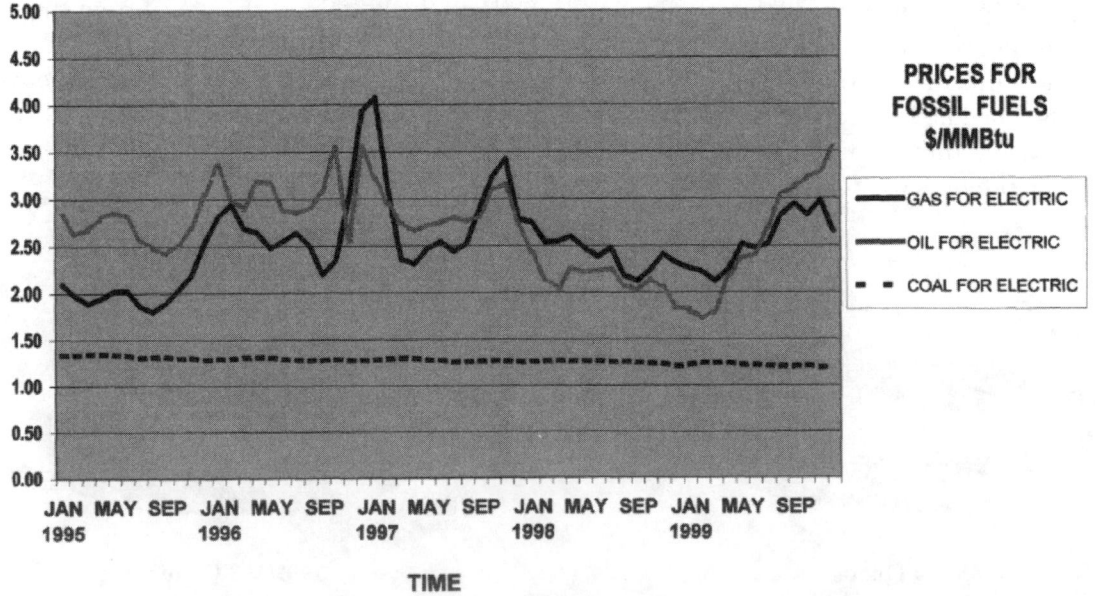

**Figure 21.5 Fuel Prices for Generating Electricity 1995-1999**

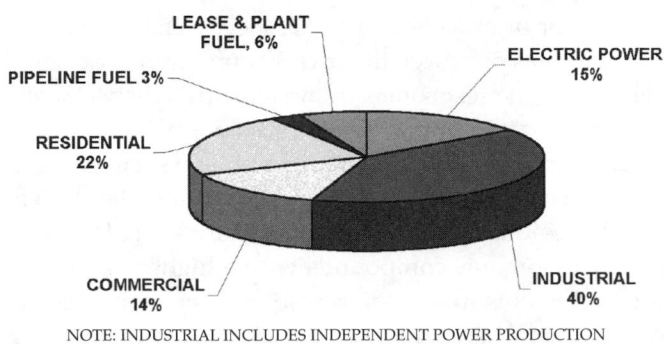

NOTE: INDUSTRIAL INCLUDES INDEPENDENT POWER PRODUCTION

JOFREE HOUSTON, TX 77002 CF 20 JUN 2000

**Figure 21.6 Natural Gas Markets 1999**

as coal plants age and need replacement or need to be replaced because of environmental causes, natural gas is the ideal fuel. It is easier to get to the plant and handle in the plant, the environmental needs are much smaller, and the capital required for the generating plant and facilities is much lower. Natural gas is the fuel of choice among the fuels currently available.

Even if the electric systems in effect now were to change to more "distributive" in nature, such as fuel cells or small, dual cycle gas turbines, natural gas would be the ideal fuel. Some planners see fuel cells or turbines being used by residential units so that each household could have its own source of electricity. When houses needed additional power, they would draw it from the utility lines. When the fuel cell produces more than needed, the utility would take the excess. Most fuel cell work today involves hydrogen and oxygen as the combined fuels for operation. Natural gas could be the source of hydrogen. Since many homes already have natural gas piped to the house, it would be easy to handle this new fuel to make electric power locally. In addition, distributive power generation could use small, gas turbines for power supply. Commercial users would be possible users of these systems also.

### 21.3.1 Supply

Natural gas is a product coming from the earth. As discussed previously, the major component of natural gas is the chemical compound methane, CH4. Methane is the product formed when organic matter like trees and foliage decays without sufficient oxygen available to completely transform the carbon in these materials to carbon dioxide. Theory is natural gas deep in the ground is a product of decaying material from past millions of years of the earth's history. Chemical elements available as the matter decayed gives the methane such contaminants as hydrogen sulfide, carbon dioxide, nitrogen, and many more compounds and elements. Natural gas comes from shallow depths as little as a few thousand feet into the

earth and as deep as 20 to 25 thousand feet. Natural gas wells are drilled on dry land and on water covered land. Current drilling in the Gulf of Mexico deep waters is in water depths up to around 3,000 feet.

Natural gas quantities are measured in two sets of units. The volume of the gas at standard conditions is one measure. Basically, at standard conditions of temperature and pressure, the amount of natural gas in a volume of a cubic foot is a standard measure. Since a cubic foot is a relatively small volume when talking of natural gas, the usual term is a thousand cubic feet (Mcf). Still as a volume measurement, the next largest unit would be a million cubic feet (MMcf) which is a thousand, thousand cubic feet. A billion cubic feet is expressed as Bcf and a trillion is Tcf.

Since natural gas is not a pure compound but a mixture of many products formed from the decaying organic matter, the energy content of each cubic foot at standard conditions is another method of measuring natural gas quantities. The energy units used in the U.S. are British thermal units (Btu), the amount of heat needed to raise a pound of water one degree Fahrenheit at standard conditions of pressure and at 60 degrees Fahrenheit. A typical cubic foot of gas, if of pure methane, would have about 1000 Btu per cubic foot (Btu/cf). Gas coming from wells can range from very low heat contents (200 to 300 Btu/cf) because of non-combustible contaminants like oxygen, carbon dioxide, nitrogen, water, etc. to energy contents of 1500 to 1800 Btu/cf. The additional heat comes from liquid hydrocarbons of higher carbon contents entrained in the gas. The higher carbon content molecules are known as "natural gas liquids" (NGLs). Also, other combustible gases like hydrogen sulfide contained in natural gas can raise the heat content of the gas produced.

Data from the Federal Energy Information Agency (EIA) show an "average" cubic foot of gas produced in the U.S. as dry natural gas in recent years would have had an average of 1,028 Btu/cf. Gas is treated and/or processed to remove the contaminants lowering or raising the Btu quantity per cubic foot to meet pipeline specifications for handling and shipping and the gas. Pipeline quality natural gas is 950 to 1150 Btu per cubic foot.

A frequently used term to describe the energy content of natural gas when sold at the local distribution level, such as residential, commercial or small industrial users, is the "therm." A therm is equivalent to 100,000 Btu. Ten therms would make a "dekatherm" (Dt) and would be equivalent to a million Btu (MMBtu). The therm makes it easier when discussing smaller quantities of natural gas.

When exploration and production companies search for gas in the ground, they refer to the quantities located as reserves. This is a measure of the gas the companies expect to be able to produce from the fields where the gas was found. Through various exploration methods—basic geophysical studies of the ground and surrounding areas to the final steps of development wells are used for more accurately pin-pointing reserve volumes. Reserves are the inventory these companies hold and from which gas is produced to fill market needs. In 1997, the U.S. government's Department of Energy showed U.S. natural gas reserves in the order of magnitude of 170 trillion cubic feet (Tcf) of economically recoverable reserves. Without any replacement, this would be a 5- to 7-year life of existing reserves at current consumption rates. U.S. exploration and production companies are continuously looking for new reserves to replace the gas taken from the ground for current consumption. From 1994 to 1997, producers found reserves equal or more in volume to gas produced during that year. The reserve volumes are from areas where gas is already being produced and represent a very secure number for the amount of gas thought to be in the ground and economically feasible to produce. These are called recoverable reserves based on produced and flowing gas.

The next level of measuring reserves is gas held behind these recoverable producing reserves. A little less secure and a little more speculative but, still a good chance of producing as designated. Using this category, just for the U.S., there are enough gas reserves for 25 to 35 years depending on the amount consumed each year. There are abundant gas reserves in North America to assure a steady supply for the near term and future. In addition to the U.S. reserves, gas in Mexico and Canada are considered a part of the U.S. supply or the total North American supply. Mexico contributes very little to the US supply at this time because its gas production and transportation systems are limited. As gas demand and prices increase, Mexico could play an important role as an U.S. supplier. As already noted, considerable amounts of gas come from Canada.

In addition to these two levels of gas reserves, there are additional categories "possible" or potential of reserves. These become more speculative but are still an important potential supply for the future. Some of these may become more important sooner than expected. A good example is the gas supplies coming from coal seam sources. Considerable gas is produced in New Mexico from these sources which were not expected to be such large suppliers until much later in time. Additional potential supplies but with long lead times for further development is gas from hydrates and gas from sources deeper in the Gulf Coast.

Natural gas produced from wells where crude oil is the major product is termed "associated gas." Roughly 40% of the gas produced in the U.S. comes from associated wells while the rest comes from wells drilled specifically for natural gas. Only differences between the gas produced from the two types of wells are the associated wells gas might contain greater amounts of what has been mentioned previously as "natural gas liquids" (NGL). These liquids are organic compounds with a higher number of carbon elements in each of the molecules making up that compound and are entrained in the gas as minute liquid droplets. Methane, which is the predominant compound in natural gas, has one carbon and four hydrogens in the molecule. The two-carbon molecule is called ethane, the three-carbon molecule is propane, four-carbon molecule is butane, and the fifth, is pentane. All molecules with more than five carbons are collected with the pentanes and the product is called "pentane plus." It is also known as "natural gasoline" which must be further refined before it can be used as motor fuel. The NGL are removed by physical means either through absorption in an organic solvent or through cryogenically cooling the gas stream so that the liquids can be separated from the methane and each other.

There are markets for the individual NGL products. The ethane is used by the chemical industry for making plastics. Propane is also used in the chemical industry but finds a significant market as fuel. Butanes go to the chemical and fuels market and the pentanes plus are basically feedstock for the motor fuels production from refineries. The overall NGL market is about a $10 to 15 Billion a year business depending on the product prices. Prices for NGL vary as the demand varies for each of the specific products and bear little relationship to the price paid for the natural gas. When gas prices are high and NGL prices are low, profitability on the NGL is very poor. At the times, when the profitability is poor, the ethane will be re-injected back into the natural gas stream and sold with the gas to boost the heat content of the gas.

A second difference between associated and gas well gas is strictly of a regulatory nature. Gas from associated wells is produced with no quantity regulations so that the maximum amount of crude oil can be produced from the well. Gas from "gas only" wells depending on the state where produced, may be subject to production restrictions because of market, conservation, or other conditions. Major natural gas producing areas in the U.S. are Texas, Louisiana, Oklahoma, and New Mexico. These states, including the offshore areas along the Gulf Coast stretching from Alabama to the southern tip of Texas, account for over 80% of the gas produced in the country. Figure 21.7, North American Gas Producing Areas, shows the gas producing states in the United States and the import locations for Canadian gas and for LNG receiving terminals. Other states with significant gas production are California, Wyoming, Colorado, New York, Pennsylvania, Alabama, Mississippi, and Michigan. A total of 18 states supply commercial quantities of natural gas according to the Federal EIA.

A major supplier of gas for the U.S. is Canada. While imports do come from other countries, Canada by far, is the major supplier to the lower 48 states. Natural gas coming from Canada is transported by pipeline into the U.S. The small amounts of gas coming from Mexico also travel by pipeline. Imports from other countries into the U.S. are transported as liquefied natural gas (LNG). Here natural gas at the producing country is cooled and compressed until it is liquid. The reduction in volume is roughly 20 times the original volume. The liquefied gas with its reduced volume is now economically sized for shipping. The liquefied gas is transported between

countries in large vessels, which are essentially very large cryogenically insulated, floating containers. The LNG is received at terminals in the U.S. where it is re-vaporized to gas. During this step, large quantities of refrigeration are available from the expanding liquid to gas. The cooling "energy" is sold and used in commercial applications to recoup some of the costs in making the gas into LNG. There are currently four terminals in the U.S. for receiving and handling LNG. These are in Boston, Lake Charles, LA, Baltimore, MD, and off the coast of Georgia at Elba Island. The Baltimore and Georgia locations were shut down years ago when natural gas prices would not justify LNG sales. Current plans are to reopen both facilities shortly.

Overall imports into the U.S. have grown considerably since the mid-1980s when only 843 Bcf were imported in 1985. Natural gas imports in 1999 increased for the 13th consecutive year to 3,548 Bcf, 16.0 percent of

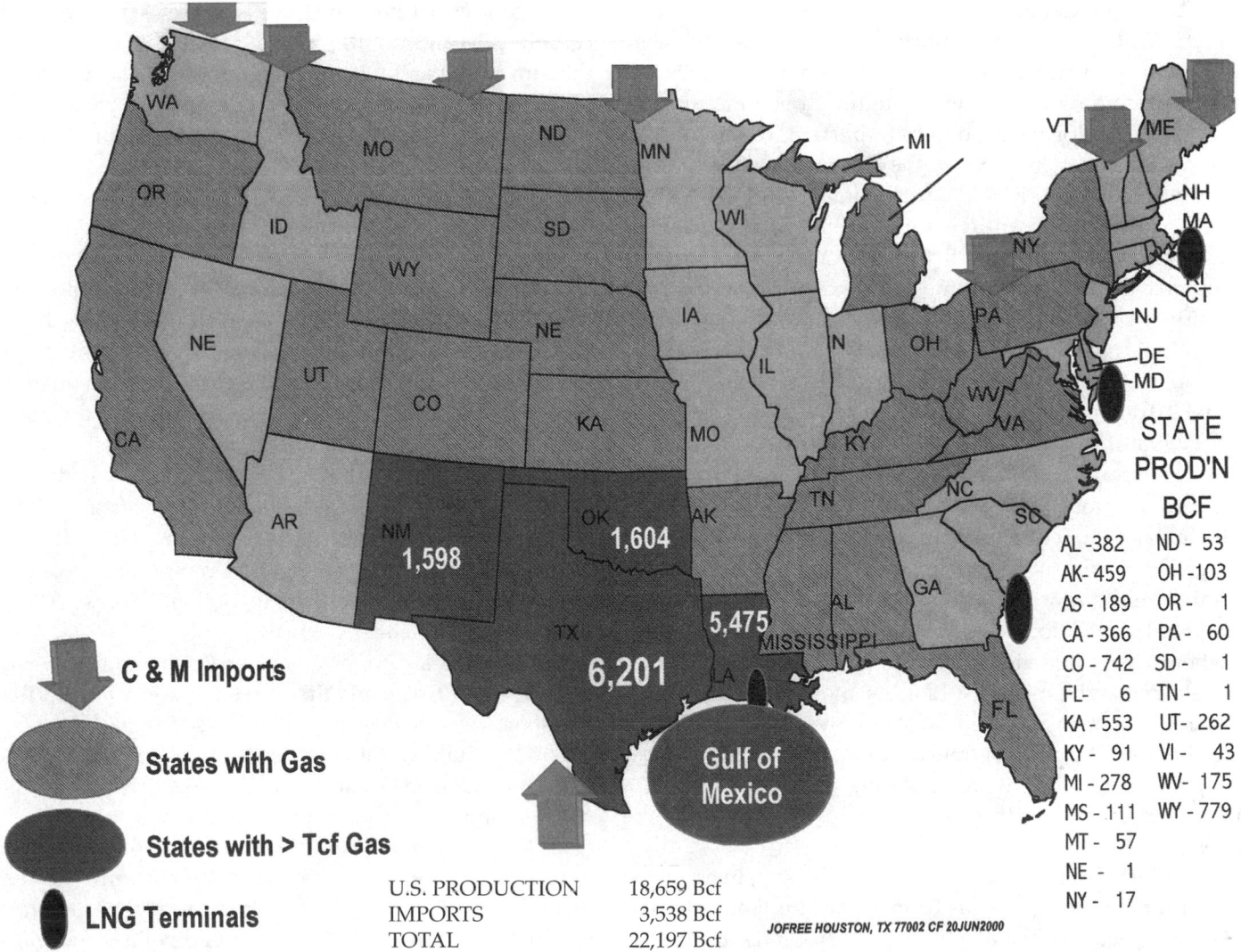

Figure 21.7  North American Gas Producing Areas in 1999

total U.S. gas supply. Canada supplied 93.9 percent of the total imports in 1999. Of the total imports, only 4.5 percent were received as LNG. Canada did much in the late 1990s to expand the pipeline systems bringing gas to the U.S. Additional pipelines are scheduled for completion early in the new millennium. Most Canadian production is in the provinces of Alberta and British Columbia. New production did come on from the Eastern Coast late in the last century and was imported into the U.S. from the Maritime Provinces. Since 1985, Canadian imports have more than quadrupled and Canada plays a major role in the expected additional supply needed to meet the demand for the years to come. Estimates are Canadian gas volumes will increase insuring the supply of gas for U.S. demand in future years. The Alliance Pipeline was completed in late 2000 and added 1.3 Bcf/day of supply to the U.S. Already, Canadian gas makes up a significant portion of the gas going to the U.S. Northeast. Figure 21.7 shows the major importing locations for gas coming into the U.S. from Canada.

While natural gas imports into the U.S. as LNG were small in comparison to the total gas imported in 1999, the amount coming in 1999 was roughly three times that received the prior year. Equally important, the number of countries supplying LNG to the U.S. increased from three to six. Algeria continued to be the major supplier with 75 Bcf in 1999 but recently completed production facilities in Trinidad supplied 49 Bcf in the same year. Plans are to make all the terminals in the U.S. operative so that additional LNG supplies can be expected. Locations of all terminals are shown in Figure 21.7

### 21.3.2 Transportation

Natural gas in the United States is transported almost exclusively by pipeline. From the time the natural gas leaves the wellhead, whatever route it takes in getting to the burner tip, it is through a pipe! Short or long distance, regardless, natural gas is transported in pipe. Only exceptions are the few times compressed natural gas is transported by truck for short distances. And, in some locations where gas is liquefied (LNG) for storage for use during peak demand times, the LNG is moved by truck also. Movement of gas through these two means is insignificant in the overall picture of transporting natural gas.

When talking of transporting natural gas through pipelines, there are three main groups of pipelines to be considered:

**Gathering System**: These are the pipelines in the field for collecting the gas from the individual wells and bringing it to either a central point for pick up by the long-haul pipeline or to a central treating and/or processing facility.

**Long-haul transportation**: This is the pipeline picking up the gas at the gathering point, or if a highly productive well near a pipeline, from the well itself and moving the gas to a city-gate for delivery to the distribution company or to a sales point for a large user where the gas is delivered directly to the consumer. The long-haul pipeline can be either an interstate pipeline that crosses from one state into another or an intrastate pipeline where the transportation is only within the state where the gas was produced. The interstate pipelines are economically controlled by the Federal Energy Regulatory Commission (FERC). The operating regulations fall under the Department of Transportation (DOT). The Environmental Regulatory Agency has jurisdiction regardless of the type of pipeline in regard to environmental matters. The interstate pipelines are still economically regulated by the Federal Energy Regulatory Commission (FERC) since these are utilities engaged in interstate commerce.

Intrastate pipelines are economically regulated by state agencies. Utilities are granted a license to operate in certain areas and are allowed to make a rate of return on their invested capital. This is different from the non-regulated businesses where they compete to make profits from the operations. As utilities, the rates for transportation are set through regulatory procedures. The pipeline makes a rate case for presentation to the FERC for authorization to charge the rates shown in the case. The pipeline is allowed to recover all of its costs for transporting the gas and make a return on the invested capital of the pipeline. Natural gas pipelines offer essentially two basic types of rates for transporting natural gas: firm and interruptible. With firm transportation, the transportation buyer is guaranteed a certain volume capacity daily for the gas it wants transported. The buyer is obligated to pay a portion of the transportation charge regardless whether its uses the volume or not on a daily basis. This is called a "demand charge" and is a part of the transportation tariff. The second part of the tariff is the commodity charge and is a variable charge depending on how much gas is transported by the pipeline.

Pipelines also offer an "interruptible" tariff where space is on a "first come-first served" basis. Interruptible transportation carries no guarantee to the party buying the transportation that space in the pipeline will be available when needed. The tariff here is usually very close to the commodity rate under the firm transportation.

The methodology of the ratemaking procedure used to recover the pipeline's costs and rate of return is such that when a pipeline sells all of its firm transportation, it will make its allowed rate of return. A pipeline can legally exceed its accepted rate of return based on its handling of the firm and interruptible transportation. Typically, the pipeline has about 80% of its volume contracted in firm transportation. When a firm transporter does not use its full capacity, the pipeline can mitigate the costs to that pipeline by selling its firm transportation to another transporter as interruptible transportation.

Many of the transportation contracts for firm transportation are terminating in the 2000 period. With the changes in the marketing system and the shift in the merchant role, some pipelines may have difficulty in filling their firm transportation sufficiently, This may bring some reduction in transportation costs which the gas buyer may be able to exploit. Further, the gas buyer at times can use what is called "back hauling" to get a lower rate for gas transportation. An example of this might be gas coming from Canada through the North Central U.S. area such as Chicago. A buyer for this gas might be located in the Southwest, say in Texas. Rather than ship gas from Chicago to Texas and have to pay the full tariff, a shipper might exchange gas in Texas for the gas to come from Chicago to Texas. In turn, the gas coming from Canada would be sold in the Chicago area as "Texas" gas. Here the shipper would pay the much lower fee for the "paper transportation" of the gas volumes. This would be a back haul arrangement.

The interstate pipeline community is relatively small. Many of the pipelines have merged or were acquired by other utilities since the regulatory changes in the industry took the merchant function from them and made them strictly transporters. There are 25 major interstate pipelines moving gas from the production areas of the country to the consumer. These are owned or controlled by only 13 companies. Table 21.1, U.S. Interstate Natural Gas Pipelines, lists the major U.S. interstate pipelines, and the parent company having ownership. In all likelihood, even more mergers and acquisitions will occur to bring the number of separate companies even lower.

Intrastate pipeline companies are within the state where the gas is produced. Many of these have miles of pipeline comparable to the interstate systems but, do not cross state lines. Within the state, these pipelines serve the same mission as the interstate pipelines; bringing the gas from the field whether the well or gathering point to the city gate for distribution by the local distributor or directly to a large consumer. Some of the larger ones for the gas producing states are listed in Table 21.2. While the pipelines themselves are no longer sellers of natural gas, the buyer should review the pipelines' systems to see if there is a close connection possible so a direct supply might be made from the pipeline to the consumer. In cases where a pipeline is close to a plant or other large user, a marketer or the buyer itself can make arrangements for the short-

haul pipeline to bring gas from the transporting pipeline to the facility.

Pipeline transportation might include more than one pipeline to complete the shipment from well to burner tip. Who pays for the transportation at each step is open to negotiation between the gas supplier and the buyer. Usually, the producers are responsible for the gathering and field costs of getting the gas to the transportation pipeline's inlet, which may be on the pipeline or at a terminal point, sometimes designated as a "hub." Many times when the transporting pipeline goes through a producing field, the producer will only be responsible for gathering charges to get the gas from the wellhead to the field's central point for discharge into the pipeline's inlet. The gathering and field charges along with the transportation to the transporting pipeline inlet is what makes the difference between wellhead gas prices and "into pipe" gas prices.

## Table 21.1 U.S. Major Interstate Natural Gas Pipelines

| PIPELINE | PARENT COMPANY | PIPELINE HEADQUARTERS |
|---|---|---|
| Panhandle Pipeline | CMS Energy | Houston, TX |
| Trunkline Pipeline | | Houston, TX |
| ANR Pipeline | Coastal Corp. | Houston, TX |
| | | Detroit, MI |
| CIG Pipeline | | Colorado Springs, CO |
| Columbia Gas Tran'n | Columbia Energy Co. | Charleston, WV |
| Columbia Gulf Trans'n | | Houston, TX |
| CNG Pipeline | Dominican Energy | Pittsburgh, PA |
| Algonquin Gas Trans'n | Duke Energy | Boston, MA |
| Texas Eastern Pipeline | | Houston, TX |
| El Paso Pipeline | El Paso Energy | Houston, TX |
| Sonat Gas | | Houston, TX |
| Tennessee Pipeline | | Houston, TX |
| Florida Gas (50%) | Enron Corporation | Houston, TX |
| Northern Natural Gas | | Omaha, NE |
| Transwestern Pipeline | | Houston, TX |
| NGPL | Kinder Morgan | Houston, TX |
| Gateway United | Koch Industries | Houston, TX |
| Wiliston Basin | MDU Resources | Bismarck, ND |
| National Fuel Gas | National Fuel Gas | Buffalo, NY |
| Northern Border | Northern Border | Omaha, NE |
| PCT | Pacific Gas & Electric | San Francisco, CA |
| Questar Pipeline | Questar Energy | Salt Lake City, UT |
| Mississippi River | Reliant Industries | St. Louis, MO |
| Noram Pipeline | | Houston, TX |
| Northwest Pipeline | Williams Companies | Salt Lake City, UT |
| Texas Gas Pipeline | | Owensboro, KY |
| Transco Pipeline | | Houston, TX |
| Williams Gas Pipeline | | Tulsa, OK |

Who pays for the transportation charges from the transporting pipeline's pick-up to the city gate or distribution company's inlet, even if it includes more than one transporting pipeline, is negotiable between the seller or marketing company and the buyer. The marketing company selling the gas might quote a delivered price to the buyer, especially, if the seller is holding transportation rights with the pipeline handling the transportation. If the buyer has transportation rights, he might take the gas FOB (Free on Board, the point where title transfers and where transportation charges to that point are included in the sales price) at the transportation pipeline's inlet. These are all part of the marketing and negotiating in moving gas from the field to the city gate and/or the consumer.

What are typical prices for transporting natural gas from producing area to consumers in various parts of the country where there is no intrastate gas? The buyer can get detailed information from the pipeline tariffs which can be gotten from the FERC and other sources like trade letters and magazines.

Pipeline rates or tariffs are set by the regulatory agencies involved. There is some negotiation possible. Still, the gas in different locations will have a value based on market conditions regardless of transportation rates. This is called "basis differential." Some typical basis differentials between hubs and major markets are shown in Figure 21.8. These were developed from published prices given in trade publications for a several month period to get representative values.

For natural gas to be carried in transportation pipelines, it must meet certain conditions of quality and composition. This was previously referred to as "pipeline specifications." These standards include the heating content of the gas per unit volume; i.e. British thermal units per cubic feet (Btu/cf). Typically, pipeline quality gas will

**Table 21.2 U.S. Major Natural Gas Intrastate Pipelines—Summer 2000**

| STATE | PIPELINE | PARENT | HEADQUARTERS |
|---|---|---|---|
| ALABAMA | Southeast Alabama Gas | Southeast Alabama Gas | Andalusia, Al |
| CALIFORNIA | Pacific Gas Trans'n | Pacific Gas & Electric Co. | San Francisco, CA |
| | Southern California Gas | Sempr Energy | Los Angeles, CA |
| LOUISIANA | Chandeleur Pipeline Co. | Chandeleur Pipeline Co. | Woodlands, TX |
| | Louisiana Interstate Pln | AEP Corp. | Alexandria, LA |
| | Mid Louisiana. Gas Co. | Midcoast Energy Resources | Houston, TX |
| NEW MEXICO | Gas Company of New Mexico | Public Service Co. of New Mexico | Albuquerque, NM |
| OKLAHOMA | Enogex, Inc. | Enogex, Inc. | Oklahoma City, OK |
| | Oneoak Gas Tran'n | Oneoak Inc. | Tulsa, OK |
| TEXAS | Aquilia Gas Pipeline | Utilicorp | Omaha, NE |
| | Ferguson-Burleson County Gas | Mitchell Energy & Dev't Corp. | Woodlands, TX |
| | Houston Gas Pipeline | Enron Energy | Houston, TX |
| | Lone Star Gas Pipeline | Ensearch | Dallas, TX |
| | Midcon Texas Pipeline | Midcon Texas | Houston, TX |
| | PG&E Texas Pipeline | PG&E | Houston, TX |
| | Westar Transmission | Kinder-Morgan | Houston, TX |
| | Winnie Pipeline Co. | Mitchell Energy & Dev't Corp. | Woodlands, TX |

be around 1,000 Btu/cf. Gas coming out of the well, can range from very low values to over 1,500 to 1,600 Btu/cf. The lower values come from gas having contaminants like carbon dioxide or nitrogen in the stream while the higher values come from the gas containing entrained liquid hydrocarbons or hydrogen sulfide. The contaminants are removed in treating, for the hydrogen sulfide and other acid impurities, and processing facilities for the liquid hydrocarbons such as ethane, propane, etc. Typically, pipeline quality gas will run around 1,000 Btu/cf with a range of from 950 to 1150 Btu/cf. The exact amount is measured in the stream as the gas is sold on a Btu basis. Typical other specifications for pipeline transmission of natural gas are given in Table 21.3.

**Distribution**: Once the natural gas is moved from the producing area it can travel from a few miles to thousands of miles in getting to its destination. The usual terminating point for the gas is at a city gate where the local distribution company (LDC) delivers it to the individual user whether it is a commercial, residential, or industrial consumer. In some cases where the consumer is a large industrial or an electric generating plant, the gas might go

**Table 21.3 Natural Gas Interstate Pipeline Specifications.**

Contaminants may not exceed the following levels:

20 grains of elemental sulfur per 100 cubic feet

1 grain of hydrogen sulfide per 100 cubic feet

7 pounds of water per million cubic feet

3 percent of carbon dioxide by volume

Other impurity (i.e. oxygen, nitrogen, dirt, gum, etc.) if their levels *exceed amounts that the buyer must incur costs to make the gas meet pipeline specifications.*

Source: *Handbook on Gas Contracts*, Thomas G. Johnson, IED Press, Inc. Oklahoma City, OK. 1982, page 63

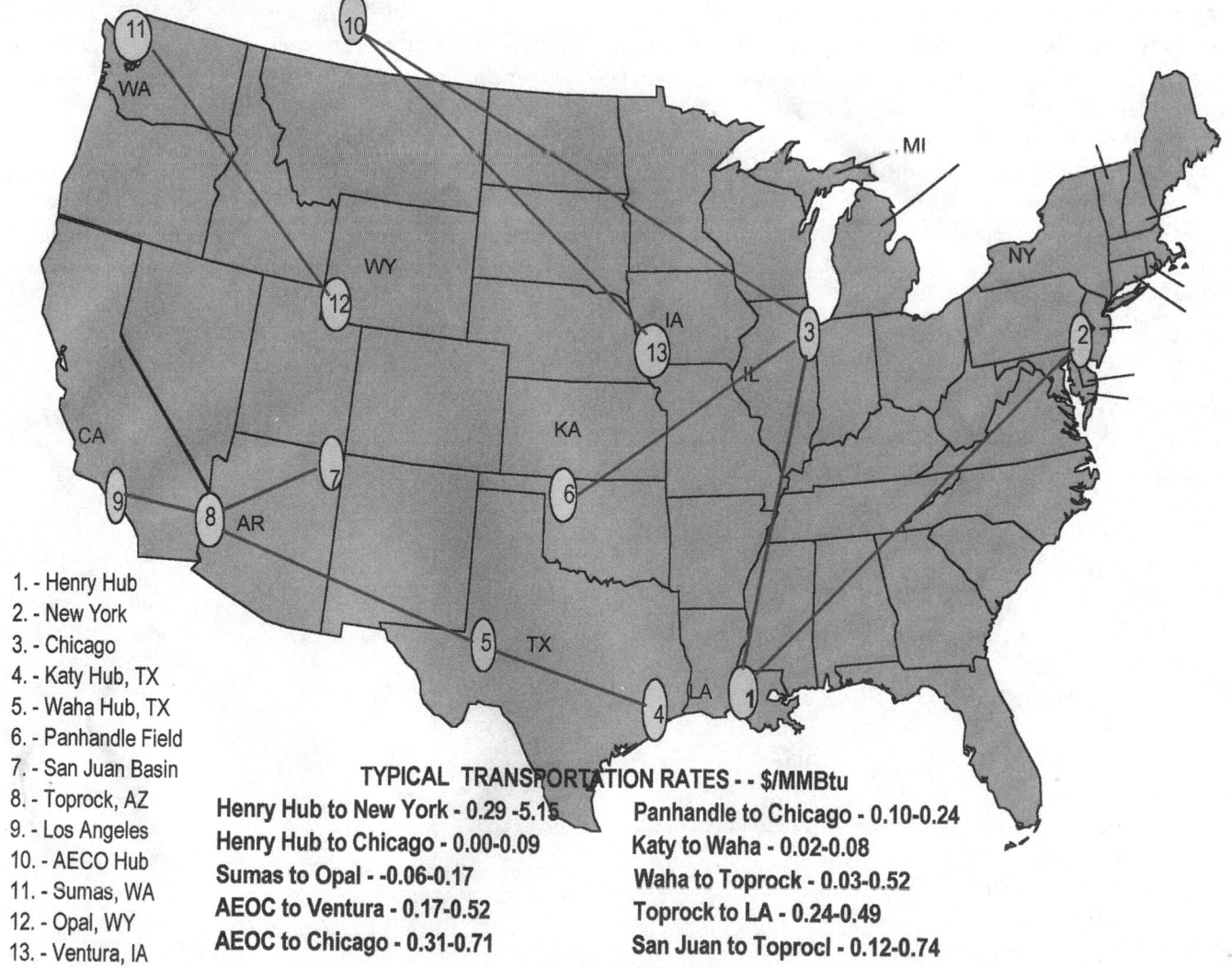

1. - Henry Hub
2. - New York
3. - Chicago
4. - Katy Hub, TX
5. - Waha Hub, TX
6. - Panhandle Field
7. - San Juan Basin
8. - Toprock, AZ
9. - Los Angeles
10. - AECO Hub
11. - Sumas, WA
12. - Opal, WY
13. - Ventura, IA

**TYPICAL TRANSPORTATION RATES - - $/MMBtu**

Henry Hub to New York - 0.29 -5.15
Henry Hub to Chicago - 0.00-0.09
Sumas to Opal - -0.06-0.17
AEOC to Ventura - 0.17-0.52
AEOC to Chicago - 0.31-0.71

Panhandle to Chicago - 0.10-0.24
Katy to Waha - 0.02-0.08
Waha to Toprock - 0.03-0.52
Toprock to LA - 0.24-0.49
San Juan to Toprocl - 0.12-0.74

Figure 21.8 Typical Natural Gas Basis Differentials between Hubs and Major Market Points.

directly from the long haul transporter to the consumer. There are hundreds of distribution companies in the country. Some are investor owned utilities while many are municipality owned and operated. Some are co-ops formed for distributing the gas.

The trade association representing this group of gas companies almost exclusively is the American Gas Association, headquartered outside of Washington, DC. Information and data on the industry as a whole, and on distribution companies can be obtained from this organization. Its address and web site are listed in Table 21.4.

The local distribution company is usually regulated by the state regulatory agency such as the Public Service Commission. It may also be under local regulation by the city or municipality it serves. This group of natural gas transporters is yet to be deregulated throughout the country. Some states, Georgia the most notable, have passed new regulations much like the decontrol of the national pipelines. In these locations, the transporter is strictly a mover of gas and has no merchant function. It may have a subsidiary or affiliated company doing the merchant function or marketing of the gas. The eventual result of deregulation at this level will be for local distribution companies to offer open access to their transportation facilities. Each state will have to make its decision as to whether the LDC is freed from the merchant role or retains it if only in part along with offering open transportation for other merchants to move gas to the final consumer.

The odorizing of natural gas so that its presence can be detected easily since natural gas as such is an odorless gas, is usually done by the local distribution company before distributing the gas. The odorant is a sulfur containing hydrocarbon with an obnoxious odor that can be detected by human smell even when used in very small, minute quantities in the gas. While it is commonly thought all natural gas must be odorized when it is sold to the user, this is not necessarily correct. Gas going to industrial uses where the sulfur containing material giving the odor could be harmful to the process need not be odorized. There are both federal and state regulations governing the odorization. In buying natural gas, the buyer should insure the contract includes provision for adding the odorant and whose responsibility it is for proper addition and monitoring.

### 21.3.3 Economics

Natural gas prices were originally set by the federal regulatory agency having jurisdiction over natural gas. The original methodology for price setting was much like the rate of return methodology for pipeline transportation tariffs. This was a direct function of the believed costs of finding, developing and producing natural gas. As discussed previously, the low prices paid at the wellhead prevented the natural gas industry from maintaining the necessary supply and caused the dire gas shortages of the mid-1970s. After natural gas prices were decontrolled, and natural gas became a true commodity, prices are a

**Table 21.4  Federal Agencies & Natural Gas Trade Associations**

| ORGANIZATION | INFORMATION & SERVICES | WEBSITE |
|---|---|---|
| **FEDERAL & MAJOR STATE AGENCIES FOR NATURAL GAS & ENERGY REGULATION** | | |
| 1  Department of Commerce | Information on offshore production of gas and oil | www.doc.gov. |
| 2  Department of Energy (DOE) Energy Information Agency | Information on energy products; supply, demand, consumption, prices, | www.eia.doe.gov |
| 3  Department of Transportation | Regulates the safety of pipelines used in transporting natural gas. | www.dot.gov |
| 4  Federal Energy Regulatory Commission | Regulates natural gas pipeline tariffs and facilities. | www.ferc.fed.us |
| 5  Louisiana Office of Conservation | Regulatory board for Louisiana natural gas operations. | www.dnr. state.la.us |
| 6  New Mexico Public Regulation Commission | Regulatory board for New Mexico. | www.nmprc.state.nm.us |
| 7  Oklahoma Conservation Commission | Regulatory board for Oklahoma. | www. okcc. state. ok. us |
| 8  Texas Railroad Commission | Regulatory board for Texas. | www.rrc.state.tx.us |

*(Continued)*

| ORGANIZATION | INFORMATION & SERVICES | WEBSITE |
|---|---|---|
| **NATURAL GAS & RELATED ENERGY TRADE ASSOCIATIONS** | | |
| 1 American Gas Association (AGA) | Trade organization on natural gas; major source of information on gas local distribution companies. | www.aga.org |
| 2 American Petroleum Institute (API) | Represents the nations oil and gas industries | www.api.org |
| 3 Association of Energy Engineers | Organization supplying information and services for energy efficiency, energy services, deregulation, facilities, management, etc. | www.aeecenter.org |
| 4 Canadian Energy Research Institute (CERI) | Responsible for Canadian energy research. | www.ceri.ca |
| 5 Edison Electric Institute | Represents electric industry; major area is investor -owned electric companies. | www.eei.org |
| 6 Gas Industry Standards Board (GISB) | Industry forum for development of natural gas measurement methods and standards for gas transmission. | www.gisb.org |
| 7 Gas Processors Association | Trade association for natural gas processors, a group of companies extracting gas liquids from natural gas streams and marketing the products. | www.gasprocessors.com |
| 8 GasMart | Annual marketing meeting for natural gas suppliers, transporters, customers, and marketers. | www.gasmart.com |
| 9 Gas Research Institute | Develops technical solutions for natural gas and related energy markets. | www.gri.org |
| 10 Interstate Natural Gas Association of America | Voice of the interstate natural gas system including the pipelines and companies supplying natural gas. | www.ingaa.org |
| 11 National Energy Marketers Association | National non-profit trade association representing all facets of the energy business. | www.energymarketers.com |
| 12 Natural Gas Information & Education Resources | Sweb-site dedicated to natural gas education and history. | www.naturalgas.org |
| 13 Natural Gas Supply Association (NGSA) | Represents independent and integrated producers and marketers of natural gas. | www.ngsa.org |
| 14 Southern Gas Association (SFA ) | Links people, ideas, and information for transmission, distribution, and marketing of natural gas to all customers served by member companies. | www.southerngas.org |

reflection of the normal economic factors impacting commodity pricing.

The price for natural gas at the burner tip is dependent on many things—market conditions, supply/demand balances, economic conditions, and many more including the activity of natural gas financial markets, prices for competing fuels, etc. In the early stages of the industry, because natural gas was considered a burdensome by product of the crude oil industry, it was sold for very low prices. When crude oil was around $2/barrel (B) or about 30 cents per million British thermal units (MMBtu), natural gas under federal price control sold for a penny or two per thousand cubic feet or roughly the same per MMBtu. In actual heating value, a thousand cubic feet of natural gas has close to a million Btu. A barrel of fuel oil is 42 gallons of oil and about six million Btu depending on the grade of fuel oil. On an economic basis of energy content, natural gas prices for a thousand cubic feet compared to a 42 gallon barrel of oil, should be close to one-sixth the value of the oil, i.e. an $18 barrel of oil would be equivalent to $3/Mcf or $3/MMBtu natural gas. Very seldom has the price, even after decontrol, reached this ratio. Instead, the value of gas has run about half or about one-twelfth or one-tenth the value of the oil product. In 2000, oil prices (West Texas Intermediate, WTI) were around $35/B. At the same time natural gas prices were over $5.00/MMBtu. Gas prices were about one-sixth the value of oil in dollars per barrel, very close to the energy equivalent of competing fuel, residual fuel oil. For the first time gas at the wellhead was at parity with crude oil in $/Btu.

The fear in 2000 was gas supplies would not meet demand. Gas demand is increasing as more and more power plants are being completed that will use natural gas as fuel. This will do much to balance the gas demand between summer and winter. In summer the gas will go as fuel for electric generation to meet the hot weather requirements for power for air-conditioning, and in winter it will go as fuel for heating. Many buyers fear the high summer demand will prevent storage filling believed necessary for the winter heating season.

Pricing is not a logical phenomenon. Data and basic considerations can help in predicting prices but the final price is very dependent on perception—market perception at the time. Too many of the variables are unknown precisely enough for pricing to be a scientific conclusion. Forecasting prices is art. Perception of the value based on supply/demand parameters sets the price. The market itself will do a lot to raise or lower the price. Further, the large financial market compared to the physical market for natural gas has an immense impact on the prevailing price. Gas prices can "spike" for many reasons—real or perceived. Hurricanes, hot weather spells, changes in the economy, etc. can make prices go up or down quickly and significantly. Short-term changes are always a possibility. Seasonality at times has little bearing on the current price. Natural gas prices have dropped precipitously in the middle of January and reached highs for the year in "shoulder months." Eventually, prices come back to reality but in the time they are moving large dollar gains or losses can occur.

In looking at gas prices, it is necessary to know where the gas is sold as prices vary according to where the sale is made in the wellhead to consumer path. Unlike crude oil, which is transformed into various commercial products, each with its own value, natural gas is essentially the same once it enters the transportation portion of the journey to the burner tip. Its value does increase as it moves through the system going from the wellhead to the consumer because of the added value of the transportation and services bringing the gas to market.

The simplest place for pricing natural gas is gas sold at the wellhead. Gas priced on a Btu value at the wellhead will accurately reflect the value of the gas further down the chain even though wellhead gas might need to be gathered, treated, and/or processed. Once the gas is pipeline quality, its price reflects where in the transportation line from sales point to ending sales point it is at the time. Anywhere in the chain, the wellhead price can be determined by net backing the price to the wellhead by subtracting the additional costs to get to the point of pricing. The value of the gas, since it is a commodity, is open to supply/demand forces on the pricing. In addition, since at times pipeline capacity for moving the gas is susceptible to supply/demand restraints for capacity, the price differentials based on location can be affected also. This is how the basis differential of gas prices between different locations occurs. This is reflected in Figure 21.8. Gas purchased at the wellhead is done so on a wellhead price. Gas purchased further downstream might be termed "into pipe price" or "hub price" if coming from a central point where gas supplies come together for distribution to pipelines for long-haul transportation.

The New York Mercantile in making a futures market for natural gas, has its main contract, at the Henry Hub, in Louisiana because of the hub's central locality and easy accessibility. The difference between prices for major hubs or selling locations is termed the "basis" pricing and can vary much depending on the current supply/demand factors.

The first of the typical major pricing points for natural gas would be the wellhead or field price. This might include actual sales at the wellhead, at a central location after the gas is gathered in the field, or at the tailgate of a treating and/or processing plant depending on the plant location in relation to the transporting pipeline.

The next major pricing point would be the "into pipe" price where the gas goes into a pipeline for transportation to the marketing area or to a "hub" for further redirection and transportation to the consuming area. The hub has the ability to dispatch the gas coming in on one pipeline to another in the variety of pipelines coming into and leaving the hub.

From the hub pricing, gas would then be priced at the city gate where it is transferred to a local distribution company for delivery to the consumer. The pricing for the consumer would be based on the "sales point" price, which would be a total price for the gas including all the transportation and services required to get the gas to the user's receipt point.

The individual price paid by the buyer is dependent on many factors starting from the wellhead pricing to the price at the meter coming onto the buyer's property. In generalities, the government and other reporting services report the prices at the major pricing points and at the consumers' location. The major consuming sectors where prices are reported are the residential, commercial, industrial, and electric generating markets. Since the progression from each of the stages from production to market carries a cost factor, it is important to know where in the delivery chain the price quoted applies. Figure 21.9 is a comparison of prices at each of these major market points for 1998.

### 21.3.4 Environmental

Environmentally, natural gas is the preferred fuel. Even though it is a fossil fuel, the amount of carbon dioxide released is the lowest per unit of energy received of the major fossil fuels. Natural gas is ideal for its handling and transportation qualities. Its environmental advantages makes it the most popular fuel and fuel of choice for many applications. It presents no unique environmental concerns to the user and as long as the supply is pipeline quality, the fuel source is of no concern in regard to environmental purposes.

### 21.3.5 Regulatory Changes

To the average gas buyer, the new natural gas industry presents few regulatory problems or concerns other than those imposed by local or state authorities. The federal regulations from prior years on natural gas have been reduced. While natural gas pricing is no longer under federal regulations, it is still tied to some of the original federal natural gas laws. In today's markets, these are essentially of no interference to commerce. It does mean that under certain extreme conditions, federal regulations could again be imposed on natural gas and certain uses could be restricted.

For the current conditions, the buyer mainly has to be concerned with local and state rules and regulations. Transportation, storage and handling regulations are again local and state but here, federal agencies do play a role. The Department of Transportation and the Environmental Protection Agency have jurisdiction in the areas of pipeline safety and environment, respectively. Buyers should insure in their negotiations and contracts with sellers, transporters, and providers that all regulations are covered and the responsibility for meeting these rules are a part of the transaction. The contracts for buying and transporting should speak directly to whose responsibility meeting the requirements will fall and which parties will be responsible for the consequences if failure occurs.

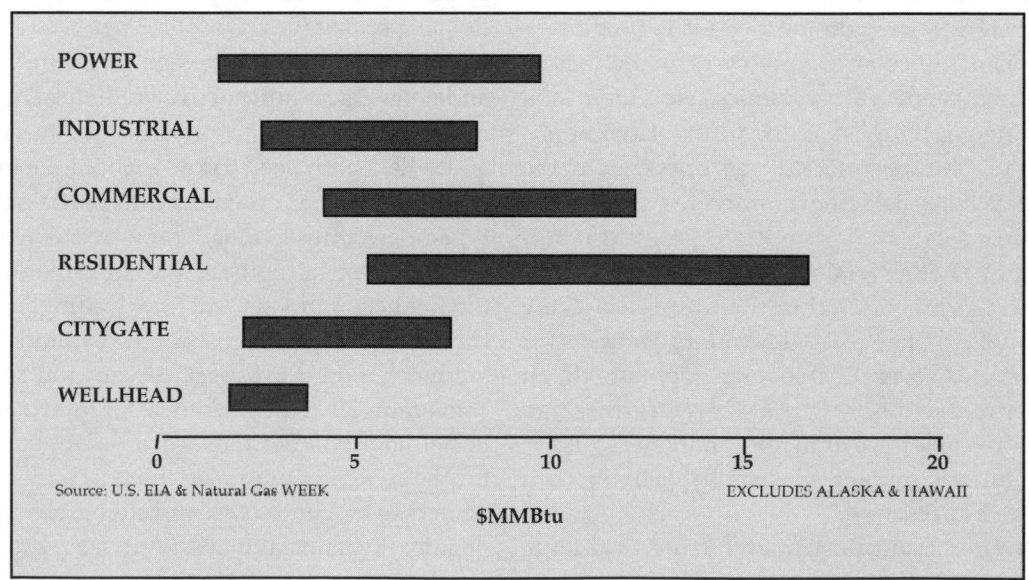

Figure 21.9  Natural Gas Prices by Sales Points for 1998

Agencies having responsibility for natural gas regulations at the federal and state levels are easily accessible. Table 21.4 lists the major federal agencies including the web sites. State Public Utility Commissions (PUC) can easily be located if information is necessary. Further, many law firms and consultants specialize in the regulatory aspects and should be contacted if necessary.

## 21.4 BUYING NATURAL GAS

To buy natural gas for either small or large operations, a thorough knowledge of the structure of the natural gas marketing system is essential. Again, this is the big change from the days when the industry was under price regulation by the federal government. Gas sales to consumers were through only one route—producers to pipeline transporter and merchant to local distributor to consumer. In the beginning of the transition to open marketing, this was referred to as "system gas." In localities where LDC are still the merchants, this is still the case. In a very small number of occasions, the chain was shortened to producer to pipeline to major consumer. Now—even with states in general still having control over the local distribution, the chain can be as short as producers to consumer or more generally, producers to marketers to consumers for relatively large users and producers to marketers to distributors for residential and most commercial and small industrial applications. This is the free market for natural gas. Buyers are free to pick any marketer or seller to supply gas. Open transportation is available to everyone—at least it should be!

### 21.4.1 Physical and Financial Markets

Since natural gas is a commodity—it is fungible—and its supply is at times at the mercy of many factors including weather, demand, economics, etc., there is a market for buying gas supplies in the future. Commonly, this is called the "futures market" as opposed to the physical market where the actual commodity goes to the buyer either for resale or consumption. Many users of natural gas buy or "hedge" on the commodity market to take advantage of prices offered in the future. The New York Commodity Exchange (NYMEX) offers contracts for up to 36 months and several banks and operators do an over the counter market offering prices even further out. The consumers or sellers (producers, marketers, users, etc.) using the futures market are usually hedging as a means of price risk protection.

As an example, a fertilizer manufacturer is a large user of natural gas for making ammonia and derivatives for use in fertilizers and industrial chemicals. If it takes

the ammonia manufacturer an average of 60 to 90 days from the time he buys the raw material natural gas to be ready to sell it as ammonia, he has to worry about the price of both the ammonia and the price of the replacement natural gas changing during the period. If he uses $2 per MMBtu gas for the ammonia and then after selling the ammonia has to buy $3 per MMBtu gas to make new ammonia, he could be at a price disadvantage in the ammonia market. To "hedge" against these kinds of price changes, the manufacturer can buy "futures" when he starts making ammonia with the $2 raw material. He can protect his future-buying price for the raw material, which represent 70-80 percent of the manufactured cost of the ammonia, by hedging his future purchases.

Since the prices on the futures market move constantly, almost daily for the near term market and less as time goes out, the futures market makes an ideal medium of wagering what the price will be in the future. The "speculators" who come into the market have no need for the commodity nor will they most likely ever take actual physical ownership of it. Their purpose is strictly to wager on where the price will be on a certain date. It can be either up or down from the price on the day they buy "futures." This is not a small market but one in billions of dollars. In 1999, it was estimated that for every billion cubic feet of gas consumed in an average day, 10 to 12 billion cubic feet were traded on the NYMEX exchange and other markets. Of course, some of this 12-fold excess of consumption went to hedging, but roughly speculators traded 90 percent. The average amount of gas consumed per day in 1999 without regard to seasonality was about 60 billion cubic feet. Using the wellhead price of around $3.00/MMBtu, about $180 million was traded each average day for the consumption of gas. In the financial trading markets, almost two billion dollars per day were traded!

Other than to have mentioned the financial market and show its significance in the natural gas industry, this chapter is devoted to physical gas buying. The buyers and sellers both need to know about the financial markets and evaluate their own need to participate or not in this type of gas transactions. There are many marketing companies, financial houses, and consultants well versed in the financial markets and how trading in these can lower the over all purchase costs of the commodity. Many books are written on this aspect. Buyers and sellers should become familiar with all sources of information in this area in helping to either maximize the return for the product for the sellers or minimize the purchase costs for the consumers. The comments on buying gas for use does not negate the financial market but, leaves it to other sources for the users to learn how to work within the financial framework including its benefits and risks. Knowledge of the financial markets are necessary because of the impact the financial market has on the physical market and prices for natural gas.

## 21.4.2 Actually Buying the Gas

So—how does the gas user get down to the basics of buying natural gas? Do they call the local distributor, if the consuming facility is in an area served by the local distributor, or does the buyer shop around for the best price and service? Again, information and knowledge are the secret to success. The buyer must know what is needed to determine what path to follow in buying natural gas. If the buyer is looking for a source of gas for a new operations, one never before using natural gas as the fuel, then they must estimate the necessary parameters to know how much is needed to fill the requirements of that operation. If the buyer is replacing an expiring contract or having to change vendors, then they have the historical record to help in knowing what is needed to renew the supply sources. They can use the existing information and records to predict with greater accuracy what volume of gas will be needed, the changes on a daily or other time basis that will be needed and what were the prior costs for the gas supply. With this information, the fuel buyer can look for new sources to meet the needs more efficiently and cheaply.

The very first question to be answered is how much natural gas will be consumed on a daily basis and what will be the range of use on a daily, weekly, monthly and annual basis. The information could even be a question of an hourly rate as to how large a swing does the user anticipate. These are the big questions to answer in making the first step in trying to select a supplier or seller. Knowing the quantity and conditions of where the rate will vary are crucial to starting the buying process. Whether the consumer is a large or small user of gas will play a major role in what selections are open to it for purchasing gas. The physical conditions prevailing in the area of the location using the gas will play a role because of regulations of the area and the actual physical availability of pipes for transporting the gas to the consumer.

Typically, the break from a small user to a large one is a rate of about one thousand cubic feet per day or in energy units, about a million Btu per day. Most local gas distribution companies will talk in "therms" and "dekatherms" rather than Btu or cubic feet. The dekatherm is ten therms. Each therm is 100,000 Btu. Each dekatherm is one million Btu. The line between large and small users is not rigid. Applications coming close to this approximation may still meet the criteria for going the large user route. If the user is on the small side, depending on the state or location of the use, it still may have an alternative of buying from the local distribution company or using the LDC for transportation only and buying the commodity from a marketing entity. Making contact with marketing companies, which will be discussed later, and getting information on the local regulatory rules will help in making this decision. Many local distribution companies have set up their own non-regulated marketing companies to help consumers buy gas at the lowest price with the required service criteria of their own operations. One should not forget the potential of ECommerce, the newest way to buy and sell natural gas. A smart buyer will look at all possible sources for meeting his requirements at the lowest price but with reliability and service.

In buying a commodity like natural gas, price alone should not be the only criteria. Service (security of service, emergency additional supplies, etc.) equally impacts the buyer's bottom line as does price in meeting fuel requirements. Having a cheap supply of gas where its availability is so uncertain as to disrupt plant or business operations is really an expensive supply when looked at in the total picture. Security of supply or additional supplies, etc. is a valuable consideration to be included in pricing natural gas sources.

The large users—those over the thousand cubic feet level or close to it, should investigate all possible sources for supply and transportation. Their sources may go all the way back to the wellhead or producer marketing companies. Depending on how large a supply is needed at a given location, the buyer may include dealing with pipelines and distribution companies for transportation and delivery of the gas. Once the buyer knows in general which direction to go, the big issues then become finding a marketer, transportation, and contracts for the services and commodity.

## 21.4.3 Natural Gas Marketers

Marketers come in varying forms, sizes, and descriptions. One can look at it much like purchasing gasoline at the local filling station—"Full-Service" or "Self-Serve." To add a little more variety or confusion, gas buying and selling is moving to ECommerce and the business-to-business Internet capabilities. When the start of marketing companies began in the mid-1980s to take the place of the merchant function performed by the pipelines, it was almost anyone with a telephone and a pencil could be called a natural gas marketer. Through the years, with a number of the marketing companies taking on added scope and abilities, the "fly-by-night," less reliable marketers were pushed out of the business. Even some of the more reputable, better financed groups have gone because of the inability to be profitable in a fast moving, sometimes, irrational market place. With financial trading exceeding high volumes of trading each day, risk becomes an even more important element of consideration.

Marketing natural gas is more than just selling and delivering gas to the consumer. The gas business is big business running into revenues of around $100 Billion per year depending on the exact price for the commodity that year. The $100 Billion is only a measure of the actual commodity trading on an idealized basis of direct trades from producers to marketers to consumers. Actually, an average cubic foot of gas most likely gets traded three to four times before coming to the consumer, the entity with the burner tip that will consume the gas and put it out of the market. This is only for the physical side of the trading—the place where the commodity actually is moved to a final destination for consumption. The total natural gas consumption in 1999 was 22 trillion cubic feet (Tcf).

This pales in comparison to the financial markets where 10 to 12 time the volume traded each day in the physical market of consumed gas is traded in the financial sector. The money moved in this arena is beyond the $100 Billion discussed previously. At times, the market is responding more to the financial than to the physical drives. The speculators are doing more to move the market than the actual users who need the natural gas for fuel or feedstock. Like all commodities, natural gas makes an ideal medium for financial trading. There are those who need to make a play in the market for the protection or risk-adverse properties the market gives. Those who produce the gas and those using large quantities can buy some protection of the future price by buying futures. This is "hedging." The futures buyer is taking a position for a given month in the future where the price he pays will be the price for the quantity of gas he purchased futures for on that given month. He/she has locked in the price for gas anywhere from a month forward to 36 months forward. Whether buying or selling gas, hedging is a tool to relieve some of the risk in buying or selling a relatively volatile commodity.

The volatility of natural gas prices (no pun intended) makes it an ideal commodity for speculators to make a market in it for the sheer purpose of making money. The speculator is betting the price will be higher or lower on a given date and is willing to take a position by buying the commodity for trading at that time. Much of the trading in natural gas is for speculation and this can only add to the volatility of the market place. While most of the hedgers bring a relatively simple mentality to the market place based on supply/demand parameters, the economy, and other pertinent factors, the speculators have a "statistics" of their own for playing the market. Basically, the speculators are "market technicians" and play a statistical analysis of the market itself for buying and selling the commodities. The mentality of the speculator is basically, "who needs to know all the details of the commodity, the market place itself shows the results and following the market place with its own statistical tools is the way to go." Of course, many of these speculators are very large in the amount of money they control. When the signals show its time to buy or sell, very large sums of money can come into or leave the market. Easy to see how this can make the price of the commodity very volatile. Figure 21.5 shows the prices for natural gas, coal, and crude oil for the last five years to give a comparison among these three major fuels for electric generation as to the market volatility of each one.

### 21.4.4 Finding the Seller

Now, to whom one goes for buying natural gas is a question of the degree of service expected and needed. A large user wanting to hedge prices to insure a stronger control in the price paid for the commodity might go to a "full-service" marketer while someone needing a relatively small amount of gas at a reasonable price can call the local distribution company or a more "self-service" marketing company. The selection is difficult because there are so many choices. There are roughly 30 major marketing companies handling natural gas and any where up to a couple of hundred smaller groups. There are the local distribution companies in the area. Most of them, in addition to selling "system gas" will have an affiliate or subsidiary selling market sensitive priced natural gas also. System gas is natural gas the LDC has purchased for resale to its local customers. Since this customer base includes residential and commercial customers as well as the industrial sector, the average price will be higher usually. Most local distribution companies have made available open access transportation so that large industrial user can bring in its own gas supply and let the local company transport it to the buyer's facility. As part of its tariff, the LDC will set a minimum amount of gas the buyer uses as a criteria for allowing the buyer to purchase its own gas and use the LDC for transportation. The tariff will set the cost for transportation by the LDC. In addition to the transportation costs, rate of return, and other pipeline costs in the charge, in many tariffs a provision includes any local taxes or fees made by local governments for transporting natural gas.

The LDC or pipeline affiliate will only sell the commodity. The buyer might also have a choice of buying from other marketers and can "shop" its purchase needs to get the best package of prices, services, and other options. The local distribution company would most likely be the transporter for the customer. In some locations, this may not be the case depending on the location of the buyer and accessibility to other pipelines for transportation.

Remember, price alone should not be the only consideration in purchasing natural gas. Dependability and service have a definite value. One should always keep this in mind when buying gas supplies. It might be wonderful to buy the cheapest gas only to be unable to get it when weather or other problems make the supply scarce!

After the prospective buyer has determined what its needs are and what alternatives it can live with, then the buyer should find the best source for buying the supply that meets those necessities. Looking for a good supplier can be a big part of the purchasing decision. But, it is an important element in the total economics that the buyer succeeds in selecting the right source or sources. The buying could include the transportation of the gas to the consumer or, again, depending on the sophistication of the buyers, they might purchase transportation separately from buying the commodity. These are all part of the difficulties of purchasing a gas supply. Because of the many parameters to be covered and the need to know the players and the system, many companies seek consultants to help either initially or continually to make better decisions in gas purchases. The difficulty is unless the buyer is in the buying sector almost continuously, he or she will be at a distinct disadvantage in seeking the optimum natural gas sources. The expense of using a marketing company or a consultant can be a very small price in finding the most effective and efficient source of supply. Table 21.5 lists the largest marketing companies whether "full-service" or "self-service" in style.

Looking at Table 21.5, one can see that many of the major marketing companies listed are a subsidiary or partner of natural gas pipeline companies, either the long-distance movers or the local distribution ones. Other big marketing companies are a subsidiary of the natural gas producer. Companies like Shell Oil (subsidiary Coral Energy), Texaco, and Exxon-Mobil all have marketing companies. Most of these are more to the "Self-Serve" type where selling the gas they produce or gas from their associates or partners in the field is the purpose for their marketing operations. The buyer should sample a large enough group of marketers to insure getting the best price, reliability, and service. Selecting a gas supply source is not an overnight task. The work needed should be in proportion to the value of the gas to the operation. If large supplies of gas are needed, differences of a penny or two or reliability are very important.

The new area touched upon only lightly because of its newness is buying gas using ECommerce and the business-to-business Internet. Many of the major gas

### Table 21.5 Natural Gas Marketing Companies

## VOLUME OF GAS SOLD FOR 1ST QTR 2000 ESTIMATED BY JOFREE

| RANK | Company | Parent Company | Estimated 1st Qtr Average Volume Bcfd | Type Service | E-Commerce Busines-to-Business | Main Office | Web Site |
|---|---|---|---|---|---|---|---|
| 1 | Enron Capital & Trade | Enron Corporation | 20.6 | Full | Yes | Houston, TX | www.enrononline.com |
| 2 | Duke Energy Trading & Marketing | Duke Energy & Exxon- Mobil | 12.0 | Full | Yes | Houston, TX | www.duke-energy.com |
| 3 | Aquila Energy | UtiliCorp | 10.7 | Full | Yes | Houston, TX | www.aquilaenergy.com |
| 4 | Coral Energy Resources, L. P. | Shell USA & Shell Canada | 10.8 | Full | Yes | Houston, TX | www.coralconnect.com |
| 5 | NorAm | Reliant Energy | 9.9 | Full | No | Houston, TX | www.reliantenergy.com |
| 6 | Dynegy Trading | Dynegy Corporation | 9.6 | Full | Yes | Houston, TX | www.dynegy.com |
| 7 | PG&E Energy Marketing | PG&E | 7.7 | Full | No | Houston, TX | www.pgees.com |
| 8 | Southern Company Energy Marketing | Vastar & Southern Company | 6.9 | Full | Yes | Atlanta, GA | www.southernco.com |
| 9 | Sempra Energy Marketing | Sempra Energy | 8.5 | Full | Yes | Houston, TX | www.sempra.com |
| 10 | TransCanada Gas Services | TransCanada Pipeline | 6.9 | Full | No | Houston, TX | www.transcanada.com |
| 11 | AEP Energy Services | American Electric Power Corporation | 3.3 | Full | No | Columbus, OH | www.aep.com |
| 12 | El Paso Energy Marketing | El Paso Energy | 5.6 | Full | Yes | Houston, TX | www.epenergy.com |
| 13 | Koch Energy Marketing | Koch Corporation | 5.6 | Full | No | Houston, TX | www.koch.com |
| 14 | Amoco Gas Marketing | BP Amoco PLC | 5.8 | Full | No | Houston, TX | www.bpamoco.com |
| 15 | Engage Energy | Coastal Corporation & Westcoast Energy | 5.7 | Full | No | Houston, TX | www.engageenergy.com |
| 16 | Williams Energy Services Co. | The Williams Companies | 3.8 | Full | No | Tulsa, OK | www.williams.com |
| 17 | Exxon Mobil | Exxon-Mobil | 4.4 | Self | No | Houston, TX | www.exxon.com |
| 18 | TXU | Texas Utilities | 4.0 | Full | No | Dallas, TX | www.txu.com |
| 19 | Conoco | Conoco | 3.4 | Self | No | Houston, TX | www.conoco.com |
| 20 | Texaco Natural Gas | Texaco | 3.4 | Self | No | Houston, TX | www.texaco.com |

*JOFREE* HOUSTON, TX 77002 CF6/29/00S 24JUN2000

marketers, either singularly or joining into a consortium, are making markets buying and selling gas through the Internet. Table 21.5 lists the marketing companies currently, summer of 2000, ready to trade using cyberspace and Table 21.6 lists all ECommerce companies doing business in the energy sector. The list will grow and players will change with time. The Internet marketers make it easy for the knowledgeable trader to buy or sell gas without having to use a marketer or broker. How much additional effort the buyer will need to complete the sale and transportation will depend on how this system of marketing grows and prospers. After only a short time of this method of marketing being in existence, large enough volumes have been traded to see the value and potential for ECommerce business in the natural gas industry.

### 21.4.5 Natural Gas Pricing

Natural gas is a commodity. There are many suppliers and the commodity is fungible. Its price is a function of its availability. Simple, but true. When supply is perceived sufficient, gas prices based on current conditions would be in the $2-3/MMBtu range. Over-supply will see the price drop significantly sometimes coming down 40 to 50 percent of this price. Tight supplies can do the same with the cap sometimes on a short-term basis, being almost unlimited. Early summer 2000 prices teetered around the $4/MMBtu range and went above $5/MMBtu by early fall because of fears of short supplies caused by the summer electric generating high demand, gas storage filling requirements, and winter gas supply. While the movement is based on supply/demand parameters, the problem is two-fold: no one knows the supply/demand picture with accuracy and secondly, fact and perception play unequal roles. In the end, each buyer and seller must make its own decision on where the price will go in the short and long term futures.

Historically, natural gas prices in the beginning were cents per thousand cubic feet. After crude oil prices became market sensitive in the early 1970s, it was not until natural gas prices were decontrolled that gas in interstate commerce came up to realistic prices ranging from over a dollar to $5-6/MMBtu. Gas prices during the 1970s, before decontrol, for the intrastate market quickly came to market sensitive levels of $3 to $6/MMBtu. The Natural Gas Policy Act of 1978 ended the difference between interstate and intrastate pricing. The high price for natural gas after decontrol was an affect of the legislation, which set up about a dozen pricing categories. When the gas surpluses of the mid-1980s started, where the legislation had set the "maximum lawful price," it did nothing for a minimum price. The gas merchants of that time, the pipelines, brought the prices down to the $2/MMBtu

range quickly. Since 1985, natural gas prices have varied from around a $1/MMBtu to current highs above $5/MMBtu. Figure 21.5 shows natural gas price history during the period 1995 through 1999.

There are tools to help in price analysis and forecasting. In addition to the sources for tracking the current gas prices, there are tools for helping in projections of future prices. Services that can supply forecasts based on their interpretation of the future are available. Many of the financial houses making a market in natural gas and other energy futures have current material on their analysis of gas markets. The federal government has many publications and resources for tracking and estimating gas supply, demand, and pricing. And, of course, there are many consultants offering pricing, supply, and demand forecasts. Many of the sources are free. Two things to keep in mind regardless of the source of information. First, forecasting is an art. There are statistical methods and programs to help in making predictions but many of the assumptions are based on the forecasters' ability and experience. It is still art not fact or science. Who can predict with accuracy and precision the weather for a week or six months out? Hurricanes come, blizzards come and sometimes little is known before hand. There is even a difference if the extremely cold weather comes during the week or only on the weekend. During the week, schools and business facilities need gas for heating; weekends they are closed. The second point is simple. If the forecaster has an ax to grind, be careful of the conclusions! Since it is an art, the forecaster who has a specific purpose can be prejudiced whether conscious of it or not. Some of the trade sources for natural gas price reporting and forecasting are listed in Table 21.7.

Since gas is a commodity and depends more on the factors of supply/demand for pricing than actual costs, gas prices vary significantly over a short time period. Each month, some of the gas trade publications (see Table 21.7) give what is called the "gas price index." This number is based on the price sellers and buyers are using at the end of the month and it becomes the index for the next month. The index changes each month and many contracts use the index from a given publication as the price gas will be bought or sold for that month. Sometimes the contract will call for a penny or two per million Btu above or below the index. The major index used at this time is based on natural gas sold at the Henry Hub in Louisiana, a very common place for gas sales and trades. There are many more places where gas is traded and each of these will have an index of its own or a "basis price," a method for converting from the Henry Hub price to that location's price. It is usually based on the value of the gas at that location versus Henry Hub and the added cost

**Table 21.6 Commerce & Energy**

| Number | Internet Address | Activities & Purpose |
|--------|-----------------|---------------------|
| 1 | icapenergy.com | ICAP Energy, Inc. is an over-the-counter broker of natural gas, electricity and weather derivatives. |
| 2 | www.apx.com | Is International. APX leverages the Internet to trade all electricity products—the electricity commodity itself, the transmission rights needed for delivery, and the ancillary services that support and reliability. All integrated. |
| 3 | www.buyenergyonline.com | Buy and sell energy—Great Britain. |
| 4 | www.capacitycenter.com | Alerts to find capacity on natural gas pipelines. |
| 5 | www.coralconnect.com | Lots of information. Allows one to buy and trade gas and electricity with Coral Energy. |
| 6 | www.energy.com | Deregulation Status, Consumer Education, Links to energy suppliers, etc. |
| 7 | www.energycrossroads.com | "The e-partner of choice for small to mid-sized utilities." |
| 8 | www.energyon.com | Buy energy on retail basis. |
| 9 | www.greenmountain.com | Green power focus. |
| 10 | www.houstonstreet.com | Electronic trading. |
| 11 | www.oilexchange.com | Oil property sales. |
| 12 | www.onlinechoice.com. | Gives customers access to buying pools. Also allows you to state your needs and receive a bid back. |
| 13 | www.powerchoice.com | Pepco offering to help in gas and electricity choice and to provide other electricity services such as power surge protection. |
| 14 | www. redmeteor.com | Energy Trading system. |
| 15 | www.scana.com | SCANA Online is an Internet-based energy auction. |
| 16 | www.itron.com | Provides interactive energy e-business solutions for optimizing energy usage and energy procurement processes. |
| 17 | www.tradecapture.com | TradeCapture.com is an innovation which will give you one stop shopping for multiple commodities and locations in the physical and financial commodity markets. Has offices in Canada, Mexico and Great Britain. |
| 18 | www.trueadvantage.com | TrueAdvantage is a leading provider of private-label sales leads and RFP services to the online B2B marketplace including energy. |
| 19 | www.truequote.com | Provides Price Discovery. |
| 20 | www.utilisave.com | Utility cost management, cost recovery, and e-commerce procurement solutions. |

of transportation between the two locations. The basis does not always vary as the value of gas transportation changes. When gas prices are rising, the basis value can increase and vice-versa. Examples of these differences can be found in the trade publications listing natural gas prices, see Table 21.7.

### 21.4.6 Contracts for Purchasing Natural Gas

As has been said previously, the major change to buying natural gas in the new millennium is the ability to buy from many sources. This can mean buying from the actual producer regardless of where the consumer is located, to buying from local or national marketers or the local distribution company. A consumer might buy from the local distribution company in its area either directly from the utility or from a non-regulated marketing subsidiary of the utility. Another major difference today is the consumer can buy the commodity and the transportation separately or together depending on the source of the gas, the quantity, the service required, and/or the location of the consumer

**Table 21.7 Natural Gas Price Reporting Sources**

|   | SOURCE | MEDIA | TIMING | TELEPHONE | WEBSITE |
|---|--------|-------|--------|-----------|---------|
| 1 | ACEO/NGX | Internet | Same day & near month | | www.naturalgas.com |
| 2 | Bloomberg Energy | Internet | Spot Market Current less 20 | A/C 609/279-4261 | www.bloomberg.com/energy |
| 3 | CNN The Financial Network | Internet | minutes | | www.CNNfn.com |
| 4 | ENERFAX | Internet | Daily | | www.enerfax.com |
| 5 | GAS DAILY | Printed, Fax, & Electronic | Daily | 800/424-2908 Crutchfield Energy Data | www.ftenergyusa.com/gasdaily |
| 6 | GASearch | Internet | Market intelligence | 972/247-2968 | www.gasearch.com |
| 7 | Platts | Printed, Fax, & Electronic | Biweekly & Current | A/C 800/752-8878 | www.mhenergy.com |
| 8 | Natural Gas | Internet | Various | | www.naturalgas.com |
| 9 | NATURAL GAS MONTH, EIA | Internet & Printed | Monthly | | www.eia.doe.gov/oil_gas |
| 10 | NATURAL GAS WEEK | Printed, Fax, & Electronic | Weekly & Current | 800/621-0050 | www.energyintel.com |
| 11 | NATURAL GAS INTELLIGENCE | Printed, Fax, & Electronic | Current futures Weekly & Current | A/C 703/318-8848 | www.intelligencepress.com |
| 12 | NYMEX | Internet | delayed 30 min | | www.nymex.com |
| 13 | Reuters | Fax & Electronic | Current delayed | A/C 800/438-6992 | www.reuters.com |

both for physical and/or regulatory reasons.

These changes in how gas can be purchased have brought changes in how contracts are written between the supplier and the consumer. If the buyer is responsible for its own transportation, it would mean having to contract for this transportation as well as contracting for the gas supply. It also opens up new considerations. The buyer wants to make sure they are well protected in getting the gas they pay for from the vendor and if the case is such, the transporter as well. In addition, the buyer must be concerned he is protected from any liability that might occur because of damage caused by the gas in the sale and delivery to the user. Contracts are legal documents covering these elements and need to be clear and accurate. After something happens—such as being charged for gas not received or for someone hurt in an accident involving the gas in question—is not the time to start looking at the contract. Who is responsible, or what limits there are for the difference between paying for a volume of gas and receiving a smaller amount, and any other conditions and situations differing from what was expected should be stated in the contract. Recourse and responsibilities should be spelled out in the contract.

With contracts being legal documents, the expense

and time to insure proper legal resources are used in negotiating and drawing up the contracts for buying and delivering natural gas are well worth the effort. Even in very short times of delivery or for very small quantities of gas to be purchased and delivered, contracts must accurately and legally cover protection of all parties involved in the transaction. This is where an "ounce of prevention is worth a pound of cure!"

A contract or contracts between the two or more parties will spell out the details of the transactions needed to purchase and deliver the gas from the source to the consumer. Many of today's gas deals are done over the telephone based upon agreed-to basic terms. Some are being done through computer and cyberspace. Whenever there is an on going relationship of supplying natural gas over a period of time regardless of the length of the time of delivery, there will be a contract or contracts covering buying and selling conditions including transportation, delivery, metering, payment, ownership, etc. There might be a contract to supply natural gas for as little as an hour or as long as a year or two and up to as long as 10 to 20 years. The long-term contracts of the controlled period when the transporting company marketed gas are no longer in use. Typically today, regardless of the term of the contract, provisions are included for price adjustments and for security of supply. Also, typically today, contracts longer than six months or a year are considered long-term contracts. Contracts up to three to four years can be for a fixed price or a market based price depending on the whims of the buyer and seller. Most fixed priced contracts today will be based on the financial market for futures contracts to protect the buyer and seller from catastrophe due to sudden market peaks where the seller would be obligated to supply gas at a fixed price when prices are rising for the commodity. Further, the contract will protect the seller from the buyer ending the contract prematurely. Likewise, the buyer will want protection should prices drop significantly to reopen the pricing provisions.

A contract is an allocation of risks between the buyer and seller. It is the same between the party buying the transportation and the transporter. Every business deal involves risks and the contract sets the responsibilities of the parties so that there should be little argument if something does not go according to plan. The seller is taking the risk of supplying gas and the risk of getting its payment for the gas and services supplied. The buyer is taking the risk of having a reliable, secure source of gas for its business needs. These are the major risks each party is taking and the contract is a written document to insure both are protected as much as possible. Contracts are written documents to help in allocating these risks. But, even the best contract, written by the best lawyers and negotiators, is really, no better than the people offering the commodity and services and the people buying the services and commodity. No contract will help if the party involved is not honorable, trustworthy and capable of doing what it claims it can do in the contract. Further, signing a contract and then planning on going to court to enforce it is a waste of good time and assets of either party. Contracts are like locks on doors—they are for the benefit of good people to insure no one gets confused or forgets the details of the arrangement. Contracts do little to protect from dishonest or untrustworthy business associates. Of course, even with good contracts and good intentions of the parties, things go wrong and contract disputes arise. These disputes can involve large loses of management time and company assets. Well-written and negotiated contracts can keep the disputes to a minimum in happening and to minimum losses when the unexpected does occur.

Since one or more contracts may be needed to purchase and deliver natural gas, the buyer should be careful in his actions. Contracts for the purchase of natural gas will usually have the following major areas of consideration as listed below. Many of these will apply to the

transportation contracts as well unless the purchase of the gas includes the transportation. Since today, sellers and buyers will vary considerably in their position in the respective industries, the contract needs to be tailored specifically between the two or more parties involved in the transactions. A contract for buying natural gas from a local distribution company will be different in many aspects from the contract between a marketing company and the buyer. The local distribution company is a regulated entity and many elements that will be in a contract are part of the regulatory aspect. Most of the specific items the utility will have to abide by are given in its tariffs, which are filed with local or state regulatory agencies. Some of the major elements of the gas purchase contracts are outlined in the following:

1. **Purpose & Scope of the Agreement**—What is to be accomplished by the contract. Who will be supplying gas, how will it be transported, and who will receive the gas. Additional comments as to the potential use, whether a sole supplier, etc. might also be included in this section.

2. **Definitions**—Lists the standard and special terms used in the contract. Especially important in natural gas contracts because of the uniqueness of the commodity, the industry ways of doing business, and the specific parameters of the operations the gas is being purchased for by this contract.

3. **Term of the Agreement**—A statement giving how long the agreement will be in force and what conditions will terminate the agreement. Some contracts will include information on methods and options of extending the contract past the initial terms of the agreement.

4. **Quantity**—Here, the details of the total quantity of natural gas to be sold and delivered by the seller and received by the buyer will be stated. Information on the daily contract quantity (DCQ) or even hourly contract quantity will be stated. Any specific deviations from the regular amount such as swing quantities needed during high production or other causes are listed here. Penalties the seller is willing to accept for the buyer's failure to take the quantity of gas set in the contract will be listed in this section. Also, the converse, penalties the buyer is willing to accept for the seller's non-performance according to the contract will be stated in this section. If there is any take-or-pay language, this is the section for it. Take-or-pay

is an agreement for the buyer to pay for gas if contracted but not taken. The Buyer usually has a period to make up the deficiency. The section will also state whether the gas is being sold on a firm basis with the buyer and seller obligated as stated to perform or if the gas is being made available or will be taken on a "best efforts" basis. Very important in this section could be the ways the buyer "nominates" takes for certain periods. The section will include means for balancing the account and give additional penalties for under- or over-quantities of gas taken by the Buyer. Other subjects that play a role in the quantity of gas to be supplied such as well or reserve measurements if buying directly from a producer or other supply considerations if buying from a marketer can be in this section.

5. **Price**—Price to be paid for the natural gas to be delivered by this contract as well as any statements regarding price escalators and/or means to renegotiate the price will be stated in this section. Omissions of statements to this effect can be construed as a statement of the contract so care must be exercised on what is said and what is left out. Any language needed for agreement on price indexing or other means of adjusting the price to current market conditions will be included. The writing should include provisions for both price increases as well as decreases if this is the desired purpose of the statements regarding changes for market or other conditions. The Price Section will cover any additional expenses or costs the buyer is willing to undertake in addition to the direct cost of the gas. If the contract is with the producer or an interest owner in a gas well, the section will state who is responsible for any gathering, treating, or processing costs. Again, for a contract with the seller being a producer, provisions will be in this section for who has responsibility for severance and other taxes, royalties or other charges the producer might be liable for payment. Pricing units most commonly used today will be energy units such as British thermal units (Btu). The dollar value per thousand Btu is typical such as gas at $0.003/MBtu. Since a typical volume measurement of natural gas is a thousand cubic feet (Mcf) and typical pipeline quality gas of this volume would have about a million Btu in energy units, the typical unit for gas sales is a million Btu or MMBtu. The above example would be $ 3.00/ MMBtu. The Pricing Section will also include

language in today's contracts protecting both the buyer and the seller from the vagaries of the natural gas markets today. While these in effect reduce the coverage of the contract and change some of the allocations of risks set by the contract, often both parties are willing to have a contract with legal means of changing the pricing conditions of the contract. The long-term, fixed price contracts went out with decontrol. There are still fixed price contracts but, the seller will protect its position by going to the financial market and buying futures to protect his position of supplying long-term, fixed price natural gas. Since the seller is taking steps to insure supplying the gas at a fixed price by buying futures, the seller will protect his actions by putting clauses in the contract to protect this position should the buyer fail to take the gas as contracted.

6. **Transportation**—Transportation details as to who has responsibility, the transporter, costs, etc. to deliver the gas to the buyer must be included in the contract. Crucial items are who is responsible for arranging the transportation, who will pay for the transportation, and whether the transportation will be on a firm or interruptible basis. The transportation must cover the full course of bringing the natural gas from the source to the buyer's location including bringing it to the accepted delivery point(s) as stated in the next paragraph. The buyer must insure they are covered in case transportation is unobtainable or ceases after delivery has started. The contract will include any special conditions on either the buyer's or seller's part to take into account any special situations either party could have to interfere with transporting the gas from the source to the delivery point(s). Further, any regulatory matters pertaining to the gas transportation should be referred to in this section and in more detail in the regulatory section discussed further in this chapter. This section should cover who has responsibility for overages and balancing of the account, measurement, disagreements on quantities, and payments of transportation and associated charges.

7. **Delivery Point(s)**—Since the delivery point or points are different in each situation, the contract needs to state delivery or alternate delivery points in very specific language. This clause can become a very crucial one in times where there is a dispute over quantities of gas sold or received. There is

also a slightly different interpretation of this clause in light of the new sales methods where there is a separate contract for the sale of the natural gas and one for the transportation. To the gas buyer, the only real delivery point is when the natural gas crosses the meter and valve where the gas comes directly into the buyer's system. The buyer wants to be responsible only for the gas received in his system. What gas is presumed sold or dispatched at some other location such as where the gas might come off of an interstate pipeline into the pipe of the local pipeline delivering the gas to the consumer is really not the concern of the buyer. This is an argument between the two pipelines or between the pipeline and the seller depending on the contract for transportation between the significant parties. Delivery point is also crucial in assessing responsibility for problems that might arise from the gas in question. An explosion or fire resulting from the improper handling of the gas and the ensuing legal action by one or more party could be influenced by the delivery point as to who had responsibility for the gas at the time and location of the accident. At times, delivery points may need to be changed to meet requirements of either parties and the need to change should be included in this section to insure that changing the locations according to the contract do not in any way negate the contract or the terms in the contract.

8. **Measurement & Quality**—Methods, conditions, timing, and authority for the measurement of the gas volume and quality are given in this section. Usually, a trade association or other organization's methods and requirements for measurement are called for in this section to insure the proper measurement of the quantity of gas sold or bought and the quality of the gas under the contract. Remedies or alternatives should be included in this section for those cases where the gas fails to meet the quality requirements of the contract whether on a short-time, unexpected rare or single occurrence or a continuing failure to meet the specifications.

9. **Billing**—Terms for the billing, who is responsible for payment, manner and methods of payment, etc. are included in this section.

10. **Force Majeure**—This the clause in the contract to protect both parties affected by a totally unforeseeable occurrence which is beyond the control of the party seeking protection from the responsibilities of the contract. Many times, this is referred to as an "act of God" and includes severe weather, acts of war or insurrection, strikes, etc.

11. **Warranty of Title**—The clause guarantees the seller has title to the gas and can sell it. Included are allowances for the buyer to recover damages if there is a failure of title should another party make claim to ownership of the gas.

12. **Regulatory**—All necessary permits, licenses, etc. must be obtained according to FERC regulations and any state or local authorities having jurisdiction over the selling and transportation of the natural gas. The party or parties having the responsibility for obtaining these and payments required should be fully covered.

13. **Assignments**—Specifications for the transfer of rights under the contract are covered in this section. This could be an important item in light of the various changes occurring in the gas industry today. The buyer should insure coverage includes changes that might impact gas transportation as well as the commodity if the seller was responsible for transportation as well as for the natural gas.

## 21.5 NEW FRONTIERS FOR THE GAS INDUSTRY

A number of challenges face the natural gas industry going into the new millennium. Each of these will have an impact on the future buying and trading of the commodity. A summary of these follows:

1. Complete the natural gas decontrol to the final level—local markets in each state.

2. Complete the development of an energy industry that incorporates other energy sources like power, fuel oil, etc.

3. Develop the delivery system to insure secure supply of the larger quantities of natural gas demand forecast for future years.

4. Develop new supply sources to meet the forecasted demand through the 2015 period forward.

Each one of these will play an important role in the gas industry of the future. More important, each one will require capital flows into the industry to make them possible.

# CHAPTER 22

# CONTROL SYSTEMS

STEVE DOTY, PE, CEM

## 22.1 INTRODUCTION

Control systems are an integral part of many energy related processes. Control systems can be as simple as a residential thermostat, to very complex computer controlled systems for multiple buildings, to industrial process control. Their diligence and repeatability can also serve to maintain the savings of the project improvements for years, further justifying their existence by providing economic return to the customer. This chapter will introduce the reader to some concepts of automatic control theory, followed by practical applications useful to the field of the Energy Professional. Upon completing this chapter the reader will gain a basic understanding of common terms, control technology and control mode categories, basic input and output instrumentation, and the practical need to temper 'things possible' with the skill level of the operators who will inherit it. The importance of system controllability and user-friendliness, as primary design parameters, will be stressed. Basic control strategies will be discussed. Estimating savings from the use of Automatic Controls will be discussed. Finally, there will be an introduction to some complex optimization methods, and suggested topics for further study. Examples are used throughout.

**The intent of this chapter is to focus on the application of Automatic Controls as a tool to achieve energy savings.** Some background information on controls is obviously a necessary prerequisite; however this chapter does not attempt to cover all aspects of automatic control theory or application or to make the reader a controls expert. Less emphasis will be placed on hardware and theory, and more emphasis will be placed on practical applications and tangible benefits. A basic background in energy-related subject matter, common energy units, common energy-consuming machinery and systems, HVAC concepts, and the general field of energy engineering is assumed. The field of Automatic Controls is a busy technology with lots of jargon, hardware and software variations and details, and the sheer volume of it can create an air of mystery and awe. If this chapter

is successful, the reader will be able to separate the fundamental control concepts from the technical details, and effectively apply Automatic Controls to achieve energy savings.

Many of the examples given are for commercial building HVAC and lighting systems, since these examples are common and should be familiar to the reader. Similar concepts and considerations apply equally to other fields of endeavor where energy savings are a driving force.

A Glossary of Terms provides clarification of common terms used in the field of Automatic Controls.

## 22.2 WHY AUTOMATIC CONTROL?

- **Regulation:** Many things need attention and adjustment to compensate for changing conditions, or varying demands. Examples of this are common in living organisms, such as body temperature, blood pressure, etc. Process control regulation is really just emulating the concepts of such natural processes. The field of automatic control is similar in that we *continually adjust some device to cause a particular measured variable to remain at a desired state.'*
  Examples:
  The need to throttle heating and cooling equipment sized for maximum load that is effectively over-sized at part load conditions.
  Varying occupancy, and systems attendant to the occupants (lighting, ventilation).
  Varying product throughput rate through manufacturing facilities.
  Varying demands, and the need to maintain level or full state for water or fuel reservoirs, feed or coal bins, etc.
- **Coordination:** Organizing or sequencing multiple processes in a logical and efficient manner is an important aspect of automatic control applications.
- **Automation:** Human beings can make very good manual controllers because we can think on our feet and consider many variables together, but most control tasks are repetitive and suitable for mechanization. Automatic operation allows people to provide oversight of system operations and more

effectively utilize their time.

- *Consistency:* Manual control by people can be effective, although we are not all that repeatable and are sometimes forgetful. Using machinery for automatic control adds the improvement of consistent, repeatable operations. The repeatability and consistency feature of automatic control is very important in manufacturing.

- *Conservation:* Supplemental enhancement control routines can be incorporated to reduce energy use while still maintaining good control. It is important to note that control systems do not necessarily reduce energy consumption, unless specifically applied and designed for that purpose.

## 22.3  WHY OPTIMIZATION?

The *80-20 Rule* reminds us that we can usually hope to achieve 80 percent of the measure's potential with 20 percent of the difficulty, but the remaining portion requires much more effort. For this text, *Optimizing* refers to reducing energy consumption as much as possible, often approaching the barrier of diminishing returns. Optimization can be characterized as taking over where the basic controls left off and working on the remaining opportunities—the ones that aren't as easy to attain. The appropriate use of optimization depends upon the customer's priorities, and these should be tested before the decision to *optimize* is made. Of course, from an energy conservation or ecology standpoint, we should all press for that last 20 percent. But if maximum simplicity controls that require only basic skills are a main focus of the customer, optimization may not be a good application. Similarly, projects where reliability is priority #1 may be better served with basic control routines, allowing the extra 20 percent potential to slip away to gain the advantages of simplicity. Economics always comes into play, and some optimization projects (chasing the last 20 percent) may not have the attractive payback periods of their 80 percent counterparts. Most projects represent some balance of these interests, depending upon the needs of the customer. It is important to understand that optimization for maximum benefit will not be for everyone.

A case in point for optimization is the subject of fixed set points, which are often a matter of convenience or approximation, and usually represent a compromise in optimal energy use. The more factors we can take into account, the closer to optimal will be the result, as stated by Liptak: "*...multivariable optimization is the approach of common sense. It is the control technique applied by*

*nature, and frequently it is also the simplest and most elegant method of control.*" [1: pp xi]

To summarize, *the desires for maximum simplicity and maximum efficiency are at odds with each other.* A system that is perceived as being too complex will likely fall into disrepair and be bypassed or unplugged. If the customer is committed to squeezing their energy costs through optimization, they will need to also embrace the technology and be willing to adapt and change along with the process. It is almost a given that pushing the envelope of optimization requires the operations personnel to accept additional complication and raise the bar of required operational skill. This concept should be discussed in advance to be sure the project isn't set up to fail by being unacceptably complex.

## 22.4  TECHNOLOGY CLASSIFICATIONS

### 22.4.1 Introduction

The following is a very important first statement before any discussion about control hardware: "*The type of hardware used in optimization is less important than the understanding of the process and of the control concepts that are to be implemented.*" [1: pp 42]

The main goal should be to become clear about the process fundamentals and what should happen—then the parts and pieces are just details. This discussion of different available hardware types is a familiar but sometimes laborious and dull part of any controls text. Remember that Automatic Controls are really nothing more than machines that do for us what we would do ourselves if we had nothing better to do; they do work for us like any other tool, and are only as clever as the people who craft them.

### 22.4.2  Conventional Electric
###         (also On-Off or Two-State)

Electricity is used as the power source. Control is discrete (on-off, high-low, cut-in/cut-out, etc.). Contact closures are used to implement control logic. This principle of control has widespread use, varying from simple and familiar control to complex interlocks and Boolean Logic.

Examples:
- A basic home heating thermostat *cuts in* at 67 degrees F and *cuts out* at 69 degrees F, thereby maintaining a room temperature of approximately 68 degrees F.
- A well pump controller *cuts in* when the storage tank pressure is 50 psig and *cuts out* when the storage tank pressure is 70 psig, thereby

maintaining a system pressure of approximately 60 psig.

- An interlock circuit that prevents an exhaust fan from starting until the associated make-up air damper is first proven open.

The differentiator between Conventional Electric controls and Analog Electronic Controls is the discrete (two-state) nature of the inputs and outputs; analog controls have varying rate inputs and outputs.

### 22.4.2.1 Floating Control.

A variation of this two-state control is "floating control." Technically still "on-off," this unique control method has a system control action similar to analog (modulating) control. With floating control, whenever the signal is sufficiently off set point, the motor actuator (moving a valve or damper) is energized, providing correction to the process. What makes floating control different is that once the feedback measurement indicates the process has returned to set point, the actuator is de-energized but the device (valve or damper) holds its last position—which may be 1/4 open, 1/2 open, or any midrange point. Because of this, floating control can achieve tighter control than simple on-off control, approaching that of true modulating control. Floating control is less expensive to implement than modulating control, and is used for small terminal HVAC control valves and dampers, as well as large machinery like chiller inlet vanes. See also Control Modes for a diagram that illustrates floating control action.

### 22.4.3 Analog Electronic

Electricity is used as the power source. The key difference between analog and conventional electric control is *modulation*. Analog controls have variable inputs and outputs, not just two states. Minor changes in output positioning of controlled devices make tighter control possible than with two-state (on-off) electric controls.

The hardware for analog electronic controls may include resistors, rheostats, Wheatstone bridges, operational amplifiers, or may use solid state components to measure the process and modulate the output devices.

Some considerations of analog electronic controls:

- Unless a control dead band is built into the controller, it is likely that the controlled device will be activated by even the most minute deviation from set point. The action may be nearly imperceptible, but if this occurs the repeated hovering about the ideal set point may over work the actuators and cause reduced device lifespan.

- Analog electronic technology *"connectibility"* options are generally limited to remote adjustments, remote alarm panels, etc. This, and modest cost, make them a popular choice for basic HVAC manufactured systems that come with factory-installed controls.

### 22.4.4 Pneumatic

This is the general term for controls that use compressed air as the motive force for control inputs and output, instead of electricity. Analog pneumatic pressures are alternately coded and de-coded into control units; for example: 3-15 psig = 0-100 degF. Discrete pneumatic pressures are also coded; for example, 0 psig = off, 20 psig = on. Pneumatic devices can be two-state or modulating, but are most commonly modulating. Pneumatic controls often have interface devices that communicate pneumatic signals to and from their electric counterparts, such as Pressure-to-Electric switches (PE switches), and various Electric-to-Pneumatic solenoids and transducers.

Some considerations for pneumatic controls:

- Air supply quality is critical. No oil or water allowed at the instruments!

- Contamination in the main air system represents a potential single point of failure for many or all pneumatic controls.

### 22.4.5 Digital Control (also called Direct Digital Control or DDC)

This control technology uses microprocessors to provide control. A big advantage to DDC controls is the fact that changes to the system are often made with software and do not automatically require physical changes and cost like other technologies. Discrete (on-off) information is readily absorbed as a "1" or "0." For analog processing, interface equipment called digital-to-analog converters (D/A) and analog-to-digital converters (A/D) are used. The higher the resolution of the A/D and D/A conversion, the closer the digital signals resemble true analog signals, allowing smoother control.

One of the great enhancements of digital controls over the last 30 years has been the concept of *distributed control*. This technology shift occurred in response to customer complaints of excessive dependence on single hardware or software points, and widespread loss of control after a single item failure. Current best practice uses multiple smaller controllers at the points of control. These communicate upstream to a supervisory operator workstation, but each single failure point now affects a much smaller area, increasing overall process reliability.

- **DDC Graphics**: A Graphical User Interface (GUI) is an overlay onto most DDC control systems. While labor-intensive to create, these are useful for gaining acceptance of the control system, and can reduce the skill level needed for digital control system navigation. The GUI uses easily recognized icons and symbols, color-coded messages and alarms, and other visual methods, to increase the user-friendliness of the system. The GUI is analogous to the labeled display boards of traditional pneumatics.

- **Field User Interface**: For effective operations and maintenance, the user interface should not be limited to operator workstation graphics. Like the pressure gages or test ports of a pneumatic system, some form of user interface should be provided at the point of control to invite personnel to interact. This may be a fixed display-adjusting device that can scroll through a few points, to a portable operator workstation that can be connected at will. Without some provision for user interface in the field, the DDC system can introduce a frustration to the operations personnel who cannot "see" what is going on at the equipment locations.

- **DDC Controller Hardware**: Today's digital controls come in a variety of styles, to suit the application. For common applications, mass produced and low cost *definite purpose controllers* are available. These use factory programming, also called *'canned' software,* which speeds application and start-up, but may limit the opportunities to create special instructions or for optimization, unless sufficient programming is available from the supervisory controller to achieve the intended result. Multi-purpose *generic controllers* are also available that utilize *custom programming*. These generally come with a fixed number of input and output points, but the *point capacity* can often be expanded with multiplexers or other point expander cards. Since the programming is custom, these controllers are the natural choice for optimization routines that require dynamic calculations and adaptability. For control processes with a large number of points, point expander boards are usually preferable to multiple linked controllers because having a single processor brain eliminates some unpredictable operations modes that come from partial failures, e.g. if just one of the controllers fails. Conversely, too many expander boards can turn this application into a large-effect single point of failure.

- **Proprietary vs. Open DDC Controls**: The topic of proprietary controls, captive customers, open

protocols, hybrid systems, gateways, etc. is a hot topic in the control industry today. While the concept is not difficult to grasp, details and implications are very numerous and complex, and beyond the scope of this text. The basic argument FOR proprietary systems is the one-stop-shopping ease of purchasing, and security of all-encompassing technical resources for system maintenance and system integration; for large or complex systems this can be an important benefit. The basic argument AGAINST proprietary systems is the *proprietary lock* on the customer, and how the lack of competition almost certainly results higher replacement costs that escalate during the ownership life cycle.

Current state of the industry includes both proprietary systems and *"open protocol"* systems, and both are viable, requiring a customer choice where new systems are proposed. Gateways and translator offerings are hardware go-between solutions that offer limited connectivity between proprietary systems. These are common solutions, but can create new problems even as they solve others—and many retain the proprietary nature they are intended to solve.

- **DDC Information Technology (IT) System Maintenance:** Systems that share communication infrastructure with other, more sensitive, systems require careful attention to information configuration and maintenance. The ideal approach is to have computer Information Technology (IT) capabilities on staff; devoting their time to control system integration activities as an in-house expense; however this may be cost prohibitive for all but the largest systems. The more common 'drop-in' approach is to select an industry partner to provide this technical service, although this can gravitate back toward proprietary solutions unless the person maintains an independent status from any equipment manufacturer.

## 22.5 CONTROL MODES

### 22.5.1 Introduction

Deciding which *control mode* to apply is important, regardless of the technology used. It is important to understand that these modes can be implemented using many of the available technology types. In many cases, simple on-off control is adequate, and very appropriate. In other cases, the desired effect can only be achieved with modulating controls. The following are basic control

| TECHNOLOGY | PROS | CONS |
|---|---|---|
| *Conventional Electric* | • Simple<br>• Proven<br>• Easy to understand and troubleshoot<br>• Inexpensive<br>• Accommodates "and-or" logic with simple series-parallel circuits (relay logic).<br>• Floating control variation approaches analog control quality with reduced cost. | • Definite purpose controls are not flexible without hardware change.<br>• Limited capabilities for optimization. |
| *Analog Electronic* | • Lowest cost stand-alone modulating control option.<br>• Often standard equipment on packaged HVAC equipment. | • Long term drift unless properly maintained.<br>• Short-lived components.<br>• Definite purpose controls are not flexible without hardware change.<br>• Often not user-friendly.<br>• Limited capabilities for optimization.<br>• Input/Output hardware items, especially actuators, are more expensive than conventional on-off devices. |
| *Pneumatic* | • Simple<br>• Proven<br>• End devices extremely durable<br>• Can be long lived if designed and maintained properly. | • Long term drift unless properly maintained.<br>• Temperature dependent drift.<br>• Definite purpose controls are not flexible without hardware change.<br>• Limited capabilities for optimization.<br>• Susceptible to system-wide failure if compressed air system is contaminated.<br>• Complicated adjustments for pneumatic controllers<br>• First cost includes infrastructure cost (tubing).<br>• Future operator work force may have a reduction in skill level for this technology. |
| *Digital Control (DDC)* | • Excellent flexibility, using software changes with existing hardware.<br>• Best option for optimization due to computing power.<br>• Opportunities for communication linkage to other, compatible equipment for large data gathering/workstation benefit.<br>• Convenient method of recording key measurements, baseline information, historical data, trends, run-times, etc.<br>• Convenient remote monitoring. | • Highest first cost.<br>• First cost includes infrastructure cost (cabling).<br>• Often outdated before physically worn out, due to rapid technology advances.<br>• Tendency for proprietary equipment manufacturers to create captive customers and high life cycle costs for repair parts and software upgrades. |

**Figure 22.1 Pros and Cons of Different Control Technologies**

modes. The accompanying diagrams will illustrate typical system performance.

The term *"system capacitance"* refers to the rate of response of a system to a stimulus. Systems with a large capacitance tend to resist change and the effects of control are felt more slowly than systems with smaller capacitance. Comparing the effect to a flywheel or relative mass is a good way to describe this concept. Another useful example to illustrate system capacitance is an instantaneous electric water heater (small volume of water) compared to a standard residential tank-type water heater—upon energizing the heater elements, the water temperature in the tank unit changes much more slowly because it has more mass, and we say it has greater system capacitance.

The term *"gain"* is a control term synonymous with sensitivity, and is usually an adjustable value used to tune the controls. If a quicker response is desired for a small input change, the gain is increased. This makes the input change more noticeable, and results in a stronger output reaction from the controller.

### 22.5.2 On-Off Control

Also called *"two position control"* this rudimentary

mode is used with equipment that is either *on* or *off*. A nominal set point exists but is rarely actually achieved, except in passing. A *range* of control values must be tolerated to avoid short-cycling the equipment, and temperature ranges are often fairly wide for this reason. In the case of equipment that cannot be modulated, this is often the only choice. The smoothness of control depends strongly upon the system capacitance; systems with very low capacitance can experience short cycling problems using two position control.

### 22.5.3 Floating Control

A hybrid combination of on-off control and modulating control, also called *incremental* control. As with on-off control, there is a control range (cut-in/cut-out). However, unlike on-off control, floating control systems have the ability to maintain a mid-position of the controlled device, instead of full-on or full-off. Between the cut-in and cut-out thresholds the controlled device merely holds its last position. The process variable is not actually under control within this range, and is seen to *float* with the load until it crosses a threshold to get another incremental nudge in the correcting direction. This control is not as tight as true modulating control, but is inexpensive and reliable. Equipment items from terminal units to 1000 HP Water Chillers are controlled in this manner with good success. Note that floating control is limited to processes that change slowly and do not require very tight control.

### 22.5.4 Proportional-Only Control (P)

This is the basic modulating control, and what most commercial pneumatic and analog electronic systems utilize. It is essentially an error-sensing device with a fixed gain or amplification. A control

output is issued to regulate a process, and the magnitude of the output is directly *proportional* to the size of the error. This type of control is economical and reliable. A characteristic offset (residual error) is natural with this type of controller, and the size of the offset will increase with load. This offset is due to the inherent fixed-gain of

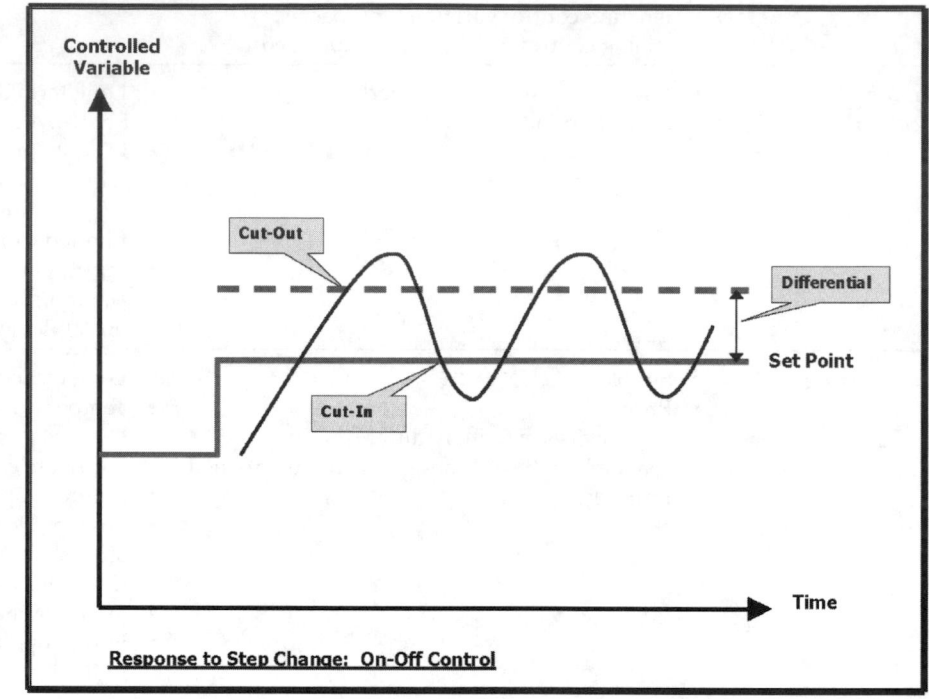

**Figure 22.2 On-Off Control Mode Diagram**

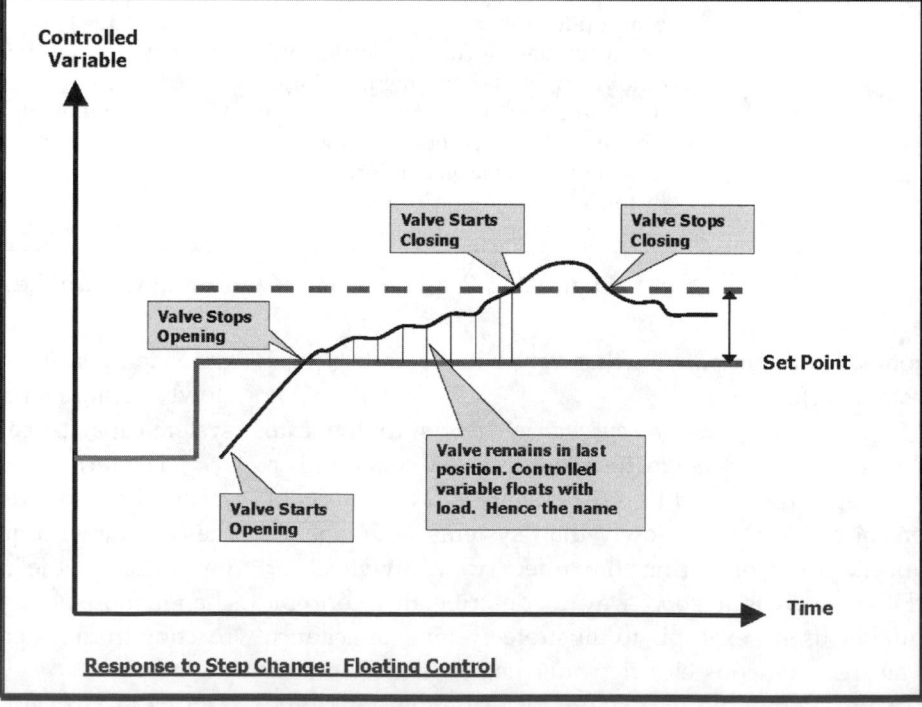

**Figure 22.3 Floating Control Mode Diagram**

the controller.

If the proportional control action is too sensitive (gain set too high), the controller's response will be excessive, and oscillation or *hunting* will occur. When this occurs, the controller output (and the equipment connected to it) oscillate up-and-down, open-and-closed, etc. and don't settle out.

### 22.5.5 Proportional-Plus-Integral Control (PI)

An *integral* function is added to a proportional-only controller to eliminate the residual error. This control action adjusts the gain to a stronger and stronger value until the error is eliminated. In theory, the integral controller will not rest as long as any error exists, however it is common to allow a small acceptable error band around the nominal set point to prevent incessant low-level hunting as the controller seeks the perfect "zero" error condition. In practice this has the appearance of slowly but surely building up the output to taper off the error. Since the "wind up" effect is slow, but also relentless, integral control problems can occur in processes that change rapidly. Also, if the controller is left active for long periods while the controlled equipment has been turned off, the integral controller will "wind up" to a 100% output. Upon start-up after a long period of wind-up, the integral function can be strong enough to "stick" at full output for long periods of time with complete loss of control. Therefore, whenever integral control is used, some form of hardware or software interlock

must be used as an *anti-wind-up* measure. The most common approach to preventing wind-up is to simply turn off the controller when the process is stopped.

### 22.5.6 Proportional-Integral–Derivative Control (PID)

PID Control is used to accommodate rapid changes in process and minimize overshoot. This is done by reacting to the rate of change of error (derivative) instead of the

**Figure 22.4 Proportional Control Mode Diagram**

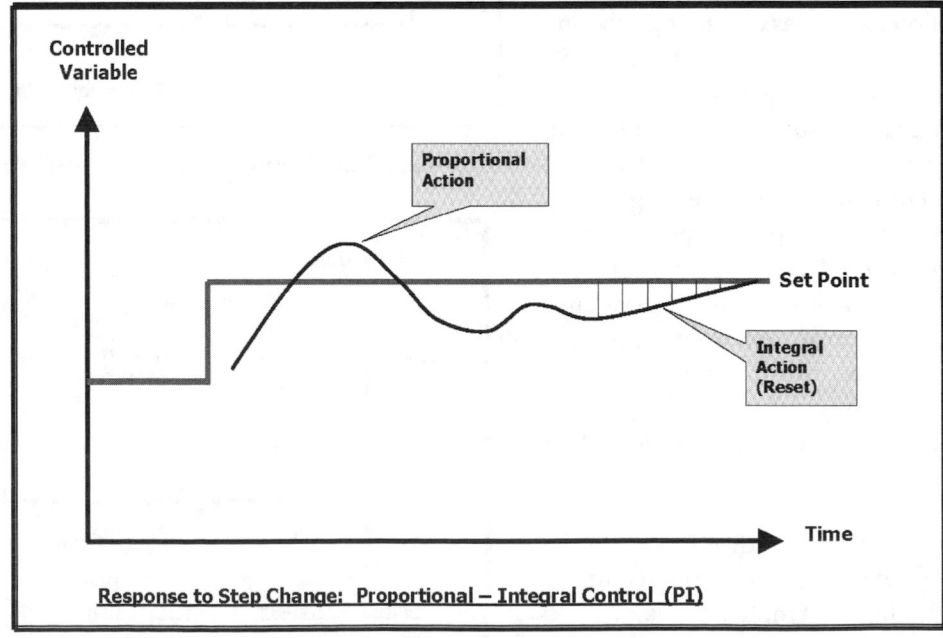

**Figure 22.5 Proportional-integral Control Mode Diagram**

magnitude of the error (proportional) or the duration of the error (integral). In reviewing the characteristic response curves, PID control looks like the absolute best and, in fact, does provide the tightest control of all the control modes. However, the derivative gain is very touchy to set up properly and can easily cause instability of control, especially at the beginning of a batch process or after a large step change. In HVAC work, the derivative term is seldom used to avoid the potential for instability, and since most HVAC processes are relatively slow acting and are tolerant of temporary overshoot of PI control. In many process control applications, PID control is essential since close control is often tied to product quality.

### 22.5.7 Sequencing

While not actually a separate control mode, this topic deserves special attention since it has great potential for energy savings. A significant blight in many industry processes is the overlapping of adjacent and opposing processes. A common example is HVAC *reheat*, where air that has just been cooled with an energy source is now being heated. This is analogous to driving around with the accelerator fixed and controlling the vehicle speed with the brakes; even if the detriment to the brake system is neglected, the effect upon vehicle fuel economy is obvious. There are many examples of heating/cooling overlap, some deliberate but most inadvertent—but all make overall energy consumption higher than necessary. One very effective tool in combating or correcting this is effective equipment sequencing. Ignoring some of the imperfections of real-world control systems, some control systems that appear sequenced will actually overlap occasionally, or over time. Some things that contribute to inadvertent overlap are:

- Instrument drift.
- System influence upon actuator range of motion.
- Normal overshoot from on-off controls.

- Normal offset from proportional control.
- Normal time lag from Integral control.
- Independent, but sequenced, modulating controllers.
- Adjacent processes, such as two comfort zones in an open bay with widely different user set points.

Where the process does not require absolute tight control, building in some blank spaces or *dead band* between sequenced elements is a simple way to achieving and sustaining energy savings.

## 22.6 INPUT/OUTPUT DEVICES

### 22.6.1 Introduction

There are many available input and output devices,

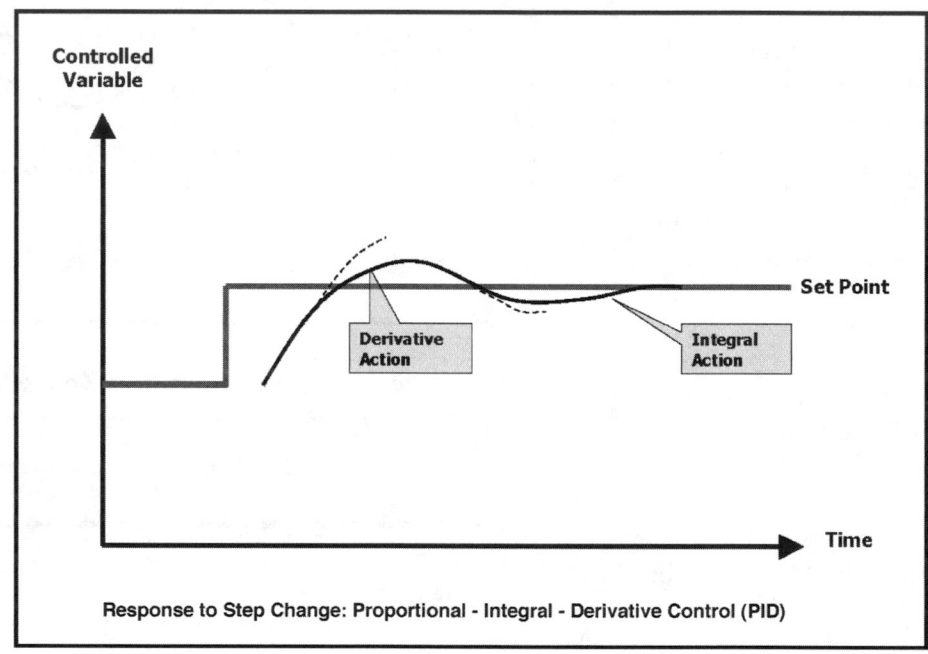

**Response to Step Change: Proportional - Integral - Derivative Control (PID)**

**Figure 22.6 Proportional-Integral-Derivative Control Mode Diagram**

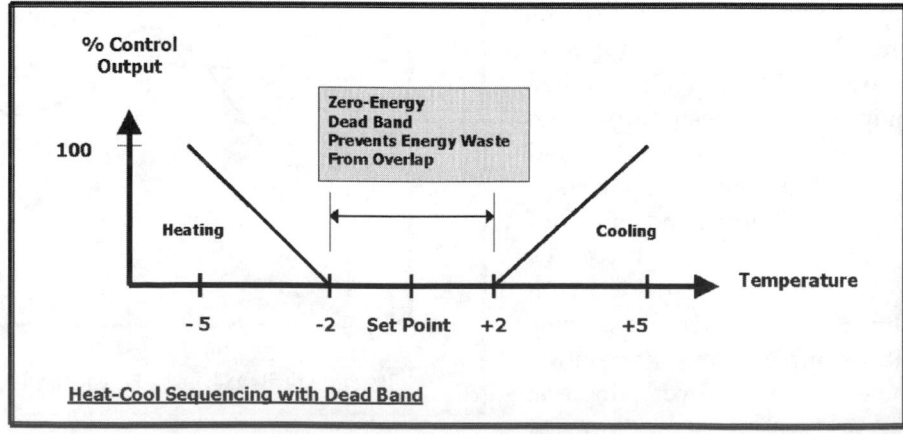

**Heat-Cool Sequencing with Dead Band**

**Figure 22.7 Sequencing with Dead Band**

serving many basic and specific needs, along with ranges of quality and other features as required for the job, and the Controls Application Engineer quickly becomes familiar with many of these in great detail. The Energy Professional may choose to delve into the sea of products, but can also be very effective by providing only performance-based generic requirements and managing the project from a more macro view.

For input and output devices, there are basic distinctions between *Transducers, Switches, Sensors,* and *Transmitters* which are useful to understand.

- **Transducers:** These are the core of any instrument, and are used to convert the basic physical phenomena of interest into a form more useful to the instrument.
  Examples:
  — Temperature: bimetal coil, two-phase gas bellows, thermistor, RTD, etc.
  — Pressure: Bourdon tube, diaphragm, strain gage, etc.
  The term 'transducer' is also commonly applied to output signal form changing equipment, such as converting an analog electronic signal to a corresponding pneumatic signal for use by a pneumatic actuator.

- **Switches:** Devices that can have two states (on-off, open-close) etc., used for regulating on-off electrical circuits as inputs or to actuate other equipment or devices in a two-position manner. Solenoid valves, relays, etc. are in this class of instruments.

- **Sensors:** These are "passive" input devices that can be read directly by the controller with no signal conditioning. They are usually limited to short cable lengths and further limited to transducers with inherent linear outputs.
  Examples:
  — RTD (resistance temperature detector) that can be wired and read directly to the controller.

- **Transmitters:** These are the typical analog input instrument. A transducer is coupled with some form of pneumatic or electronic signal conditioning. The transmitter output is a linear, standard value easily decoded as an input. Often, the signal conditioning allows for stable signal over a large distance to allow remote location of the device, hence the term 'transmitter.'
  Examples:

| | |
|---|---|
| — 0-100 degF | → 3-15 psig |
| — 0-100 degF | → 4-20 mA (milliamps) |
| — 0-100 degF | → 0-10 VDC (Volts DC) |
| — 4-20mA output | → 3-15 psig |

### 22.6.2 Conventional Devices and Wiring

Conventional wiring architecture consists of devices in the field; each wired back to the input or output terminals of the controller, with a dedicated set of wires (often a pair of wires). Thus, for 100 instruments there would be 100 sets of wires finding their way home to the controller. This type of wiring is very straight forward and is referred to as *home run* wiring. At the control panel the number of wires is the greatest and the system of wires fans out and disperses, the further it gets from the central point.

### 22.6.3 Addressable ('Smart') Devices and Wiring

Also called *'smart'* devices, these devices can be controlled directly by a digital control system, by giving the device a unique identification code or "address" to differentiate it from all other similar devices. Usually each *addressable* device has a means of setting up its own unique address or name, to permit uniqueness in communication. *Addressable* devices relay their information or accept their commands digitally, and do not rely on transmitters or the like. Addressable devices have the distinct advantage of reduced wiring, since a single loop of communication trunk wiring can be shared by multiple devices; on systems with large number of points, the difference can be remarkable. However, addressable devices cost more than conventional (non-addressable) devices, since they include on-board communication hardware, as well as A/D and D/A converters for addressable analog devices. Since they cost more, their application requires balancing the cost vs. the benefits. Of the devices listed below, the ones currently most popular are the simple contact closure input/output devices used in large quantities for security, fire alarm, and lighting control systems, where the economies of scale show an advantage. The single loop communication can present new failure modes compared to conventional *home run* wiring methods; i.e. if a single cable is cut how many devices are affected?

Common Addressable Devices:
- Fire alarm devices—input and output
- Lighting control devices—input and output
- Security devices—input and output

Other Addressable Devices:
- Lighting ballast

- Actuators
- Transmitters and sensors
- Thermostats
- Relays

### 22.6.4 Linearization

**"For simplicity of design, a linear relationship between input and output is highly desirable." [3: pp 28]**

Linearity is a high priority in instrumentation, for both inputs and outputs. This is because the controller's algorithms or ratios are arithmetic in nature, comparing the input to a standard and producing a derived output. Linearization makes the effect of the controller's decisions predictable and manageable. Understanding how linearization can affect the control system is important to assure project success.

Input/Output instruments are considered linear when an incremental change of input value produces an equal increment of output value regardless of the value of the input or location of the device's range. For example, a change of 1 degree F might produce a change of 0.16mA in transmitter output. If truly linear, the device would produce this 0.16mA change in current if reading 0 to 1 degF, or if reading 100 to 101 degF. **To the extent that non-linearities exist in the control loop, errors and unpredictability will also exist and so should be identified and minimized.**

Many natural phenomena are linear, and many are not. Linear Examples include metal resistance with respect to temperature changes, static pressure with respect to depth of liquid, and volume of a vertical cylinder with respect to level. Non-Linear Examples include the volume of a conical container or a horizontal cylinder with respect to level, the change in flow with respect to butterfly valve or single blade damper position, and a heat exchanger heat transfer rate with respect to flow rate. Some natural phenomena behave in predictable, but non-linear fashion. These include thermocouples and differential pressure flow meters (head loss devices). Thermocouples are linearized by a look-up table or mathematical expression that defines the non-linearity, while head-loss meters are linearized with a square root extractor. Most common input measurements have already been linearized by the instrument manufacturers. Instead of hardware characterization, it is possible to linearize inputs and outputs through software. This "software linearization" is sometimes used with industrial controls, but seldom with commercial controls.

A common control issue with linearization is control valves and dampers. These are notorious for having non-linear response with respect to travel position. Control valves are generally characterized as either *linear, quick opening* or *equal percentage*. For valves, the flow "character" is achieved by specially contouring the valve plug, to influence the flow rates at different valve stem positions. Characterized ball valves are also available, to greatly improve the inherent quick-opening flow pattern of these valve trim shapes. Control dampers and butterfly valves are either flat blade or flat disk shapes and do not have selectable characterized flow patterns like control valves do. In the case of control dampers, about all that can be done to reduce non-linear air flow through the damper is to down-size it to create a high pressure drop, which may create other complications or costs, especially in outside air ducts and large ducts without room for transitions. Where linear control of an air stream is important, *air valves* can be used which are available with characterized flow patterns.

Modulation of a heat exchange process is a common automatic control application. The following example applies for most types of heat exchanges including shell and tube, tube and fin, etc., and includes all HVAC air coils. The heat transfer characteristics of a heat exchanger can be likened to a 'quick opening valve' since the incremental change in heat transfer for the first fraction of fluid input is much higher than the last fraction. Controlling flow linearly through a heat exchanger will yield a non-linear output with associated control problems, especially tuning issues. In this case, the non-linear flow characteristic of a control valve is deliberately used to improve process control. The standard approach to correct this is to use a control valve with a flow characteristic that is a mirror image of the inherent heat exchanger performance (*equal percentage type*), canceling this inherent phenomenon so the overall control effect is nearly linear. This example not only illustrates how linearity is important to a control system, but also points out that *for heat exchanger applications, the control valve selected should almost always be the equal percentage type.*

## 22.7 VALVES AND DAMPERS

### 22.7.1 Valve and Damper Selection

Selection criteria usually includes a minimum pressure drop for proper *authority* and best linearity over the process.

#### 22.7.1.1 Valve Sizing

The typical sizing procedure for a hydronic control valve is 5 psig wide open pressure drop. The following explains the reason for this. The underlying principle for reasonable control of a heat transfer coil is for the control

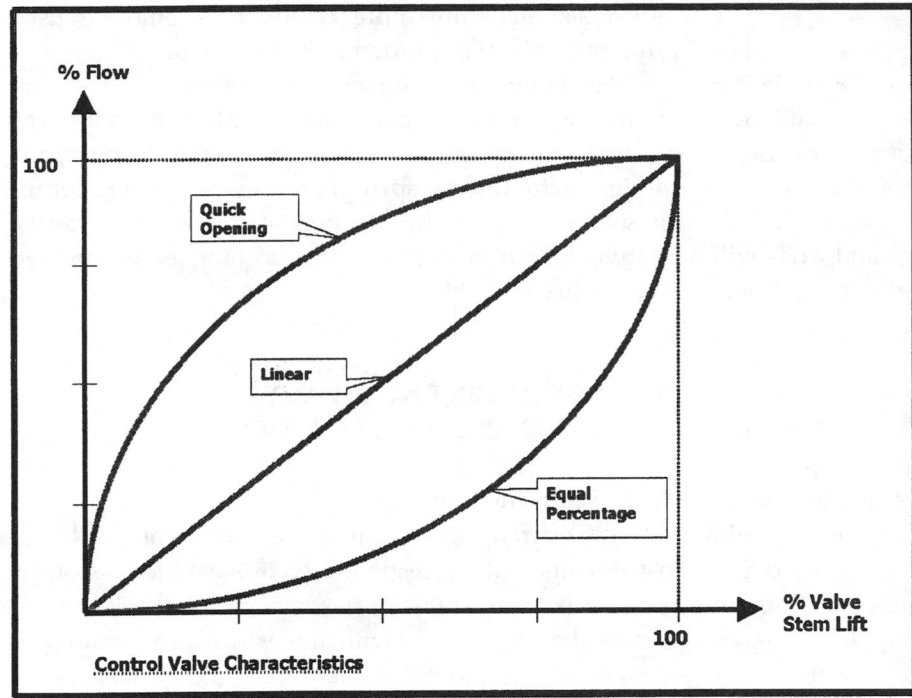

**Figure 22.8 Control Valve Characteristics**

valve wide open pressure drop to be at least as high as the coil it controls. So, if an air handler coil is sized for a 5 psig pressure drop at full flow, then the control valve at 5 psig full flow pressure drop would be appropriate. It is common practice to select HVAC water coils at 10 ft. w.c. or so wide open pressure drop, which equates to 10/2.31 = 4.3 psig. Thus the 5-psi valve pressure drop convention is a reflection of the coil sizing convention. An extension of this logic would be that if the coil were selected at 1 psig wide open pressure drop, then the control valve sizing criteria could also be reduced. This is in fact the case, although seldom done in practice since coils selected at extremely low pressure drop have other, new issues, namely reduced velocity and laminar flow heat transfer degradation at reduced flows.

> _Editorial Comment:_ _The practice of adding system resistance to achieve good control is counterintuitive and definitely an opportunity for improvement in the industry, because adding circuit resistance to any fluid handling system increases the system energy requirements._

### 22.7.1.2 Leak-By and Close Off

When specifying control actuators, specifications and close attention are important to achieve reliable close-off performance. This is true for both valves and dampers, especially those with marginal actuator close-off ratings, excessive system pressure, large damper sections, metal-

seated valves, undercut butterfly valves, and actuators without a positive seating mechanism to impart a residual tight seating force. Spring-return pneumatic actuators are forgiving in this sense because they inherently provide residual seating force. Some electronic actuators have mechanisms to provide ample seating force, but others rely on simple travel adjustments that define the open and closed positions—these are very undesirable since the opportunity for internal leak by, with subsequent heat/cool overlap, is high. Ball valves can be used for modulating service if they have characterized seats, and are inherently tight seating.

Actuators that are only marginally strong enough to close off against system flow can rob the system of efficiency over time as system pressures change, valve stems bind, damper axles stiffen, etc. Determining system needs and requiring close off ratings well in excess of this (e.g. at least 50% more) is good practice for sustainable operation.

### 22.7.1.3 Damper Sizing

Just like valves, down-sizing dampers will improve control at the expense of raising system pressure. In practice, dampers are often left duct-sized, even though resulting control is poor. There are several reasons for this:

- Dampers are relatively cheap, compared to the transition costs for a duct fabricator.

- Dampers are often large, and transitions take up space that may not be available.

- At outside air intakes, a down-sized damper can cause rain or snow entrainment.

For air flows control by dampers, other than HVAC, proper sizing will yield more linear control and is recommended where practical. For control purposes, the _opposed blade_ damper is normally used, since its aperture size and overall resistance varies the most directly with travel. Conversely, "parallel blade" or round dampers are highly non-linear in nature and hard to control unless drastically down-sized, or used for two position control only.

### 22.7.1.4 Other Damper Considerations

Large damper sections are often problematic. For cost reasons, there is often a desire to use fewer, larger actuators and link the dampers together so an adjacent actuator is driven by another damper, not an actuator. In practice, this can easily result in one end of a long section substantially open even as the actuator-end is closed. This is due to the fact that the damper blades and axles will twist and stretch. Methods to prevent this undesirable condition include multiple actuators or jack-shafting.

### 22.7.2  Valve and Damper Actuators

Like other instruments, actuators come in a variety of styles and quality levels, and each has its pros and cons. Significant differences exist between manufacturers that make generalizations and rules of thumb difficult. One thing is certain about actuators: they are a moving part, with mechanical components, and will require maintenance; therefore consideration should be given to life cycle cost and maintainability. Some of the lower cost actuators are not intended to be serviced. For each of the types listed there are *serviceable* and *throwaway* variations, as well as spring-return/non-spring return types. Types of actuators include:

- Pneumatic spring return
- Pneumatic air-to-open/air-to-close
- Electric motor/gear reduction in oil bath
- Electric motor/gear reduction-open air
- Electric hysterisis/stalling motor
- Hydraulic
- Wax motors (thermal expansion)
- System powered actuators/using air or water system pressure as the motive force
- Self-powered actuators/using a capillary bulb and bellows

### 22.7.2.1 Actuator "Normal" (spring return) Position

For valves and dampers, the phrases *normally open* or (N.O.) and *normally closed* (N.C.) refer to the device position with no power applied, where a spring-returning mechanism exists to drive it to one position or another. Valves required to have a "fail position" are necessary in many applications to provide an increased measure of reliability if control power is lost. In the case of comfort heating and cooling, the choice is made by asking *"upon a loss of power, control signal or air pressure, would I rather have full heat, full cool, or don't care?"* In other cases, there are other operational issues like overheating, over-humidifying, etc. that should be considered. Without the spring return feature, the actuator will simply remain at its last position prior to the power interruption. Spring-

return actuators are more costly, so should be used prudently, where the added cost is justified.

For large pneumatic cylinder actuators, a measure of fail-safe control can be provided without the expense of a spring system; using an *air-to-open/air-to-close* actuator (no spring) and a small spring return air solenoid valve, the position of the actuator can be relatively assured on power loss, as long as compressed air remains available.

## 22.8  INSTRUMENT ACCURACY, REPEATABILITY, AND DRIFT

### 22.8.1  Introduction

Like anything else, there are different grades of instruments with corresponding costs, so the task of the specifier is to separate the needs from the wants, and to balance the performance with the costs. With an awareness of some of the basic considerations and of instrument grades and selection criteria, good decisions are usually evident. Leaving the instrument selection entirely up to the vendor may or may not be the best approach. To the extent that the 'standard offering' instrument portfolio has good performance, this can save money, however a review of the proposed instruments is advised just to be sure. When reviewing product literature for instrumentation, like any other equipment, *it is often as important what is not said, as what is said on a component specification sheet*. For example, if long term drift is not mentioned, ask yourself "why is that?"

### 22.8.2  Accuracy

In the realm of instrument calibration and accuracy specifications, there are two important terms: **"percent of reading"** and **"percent of span."** It is not enough to say "+/- 5%" when specifying a calibration tolerance or instrument accuracy rating. The tolerance should be stated as either +/- xx degF, +/- xx psi, etc., +/- xx% of reading, or +/- xx% of span. Usually the instruments that are rated in terms of +/-% of reading are the higher quality instruments.

The following example illustrates the important difference between *"% of reading"* and *"% of span"* concepts. Consider a 0-100 psig (span) transmitter that is indicating 10 psig (reading).

Accuracy Spec "+/- 5% of *reading,*"
    Acceptable Range 9.5-10.5 psig

Accuracy Spec "+/- 5% of *span,*"
    Acceptable Range 5-15 psig

| Measurement | Instrument Tolerance<br>Note:<br>Industrial instrument tolerances are much tighter. |
|---|---|
| *Temperature (space or room)* | +/- 1.0 degF |
| *Temperature (duct)* | +/- 0.5 degF |
| *Pressure (duct static)* | +/- 0.5 in. w.c. |
| *Pressure (other)* | +/- 5.0 pct of reading |
| *Flow rates for air* | +/- 5.0 pct of reading |
| *Flow rates for water* | +/- 5.0 pct of reading |
| *Relative Humidity* | +/- 5.0 pct of reading |
| *Oxygen or CO2 monitor* | +/- 1.0 pct of span |

**Figure 22.9. Suggested Tolerances For Commercial Instruments.**

### 22.8.3   Repeatability

Repeatability is self-defining; it is the ability of an instrument or process to faithfully repeat itself, given identical conditions. For instrumentation, this is synonymous with precision, and the mark of better grade instruments. Regarding instrumentation, a general statement is that *accuracy can be adjusted, but repeatability is a function of the instrument quality and cannot be changed*. Repeatability is determined, in large part, by the stability of the output in the face of environmental changes, such as ambient temperature effect, voltage fluctuation, pressure fluctuation, etc. Instruments whose readings are susceptible to changes in ambient temperature can be very problematic if located in areas where the temperature is expected to change, however this specification (and the cost to mitigate it with higher quality instruments) is of much less concern if located in areas of constant temperature. Commercial grade pneumatic instruments are usually susceptible to problems from ambient temperature changes, since thermal expansion changes the volume of tubing, size of orifices, etc. Some lower grade electronic components are also affected by ambient temperature changes.

### 22.8.4   Drift

Drift is an undesirable but inevitable quality for any instrument. Long-term drift is attributable to many factors including normal degradation from age. A general rule of thumb is that the long term drift, from all sources combined, leave the instrument reading within reason for a period of five (5) years, to reduce the need for O/M activities and constant maintenance of the device, leaving it to serve instead of being served by the facility.

Auto-Zero Feature: Some devices, such as stack gas sensors and very-low range differential pressure transmitters, periodically re-establish the zero point of their output span by simultaneously providing a zero input condition and automatically adjusting the output to a zero value.

### 22.9   BASIC CONTROL BLOCK DIAGRAMS

### 22.9.1   Introduction

Besides control technology and control mode choices, a basic consideration of control strategy is whether it is open or closed loop.

### 22.9.2   Closed Loop Control

The controlled system impacts the measured variable, and process measurement provides feedback to the controller.

Examples of Closed Loop Control:

- Room thermostat controls a heating water valve, to regulate heat to that room.

- Leaving water temperature sensor for a heat exchanger controls the steam inlet valve to that heat exchanger.

### 22.9.3   Open Loop Control

An open-loop control system is characterized as one whose output has no impact on the measured variable, and so any form of process feedback is impossible.

Examples of Open Loop Control:

- **Turning off the building heating boilers in April and leaving them off until September.** A common control strategy, this follows the common sense notion that boilers should not be needed in summer, but does not address spring and fall vary well and can lead to both discomfort and energy loss when weather is unseasonably cold or warm, usually near the calendar cutoff dates. This is open-loop because the outside air temperature is unaffected by whether or not the boiler is on. This is also open-loop to the building itself, because *call for heat* feedback will not be heard.

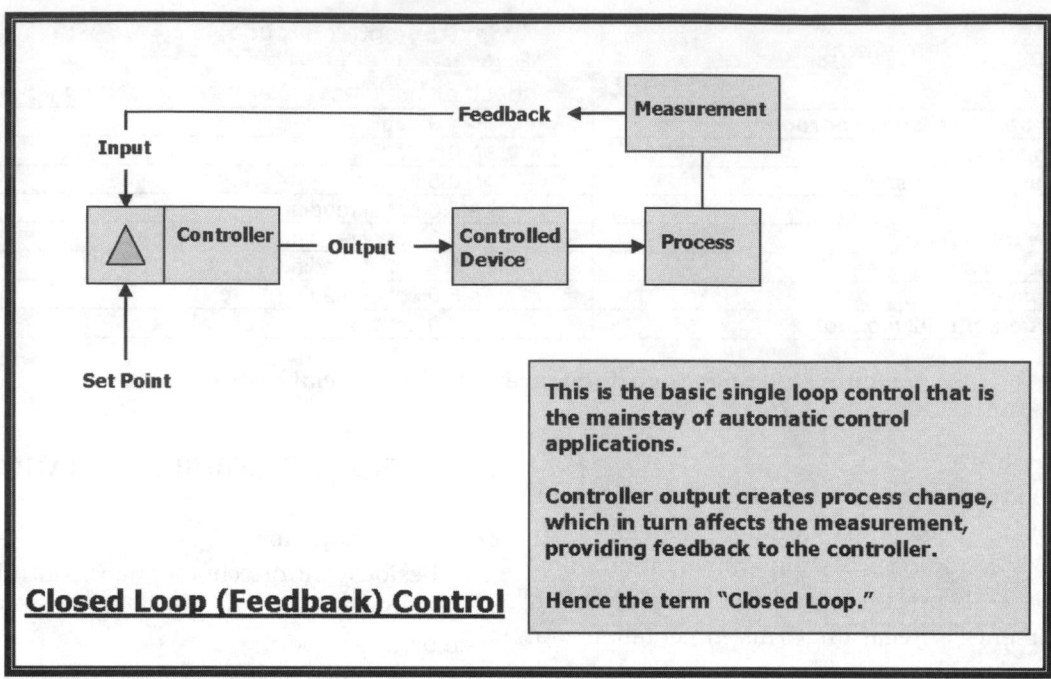

Figure 22.10 Closed Loop Control

---

- **Automatic reset of hot water temperature from outside air.** A common control strategy, this provides general compensation based on the common sense notion that '*the colder it gets outside the more heat we'll need,*' however is open-loop since the outside air temperature is unaffected by water temperature.
- **Starting the building HVAC system one hour before occupancy.** A common strategy, this follows the common sense notion that the building will need some time to warm up (or cool down) after being off all night, or all weekend. It provides general compensation for the thermal lag in the building mass, but is open-loop because variations in actual time required in different seasons is not considered. For example it may take (4) hours after a long and cold weekend but only a half-hour after a single night in spring, but the controller is blind to these facts.
- **Thermostat in room 1 controls the hot water control valve serving room 2.** In this example, closed loop feedback control is changed into open loop control due to a design or installation error.

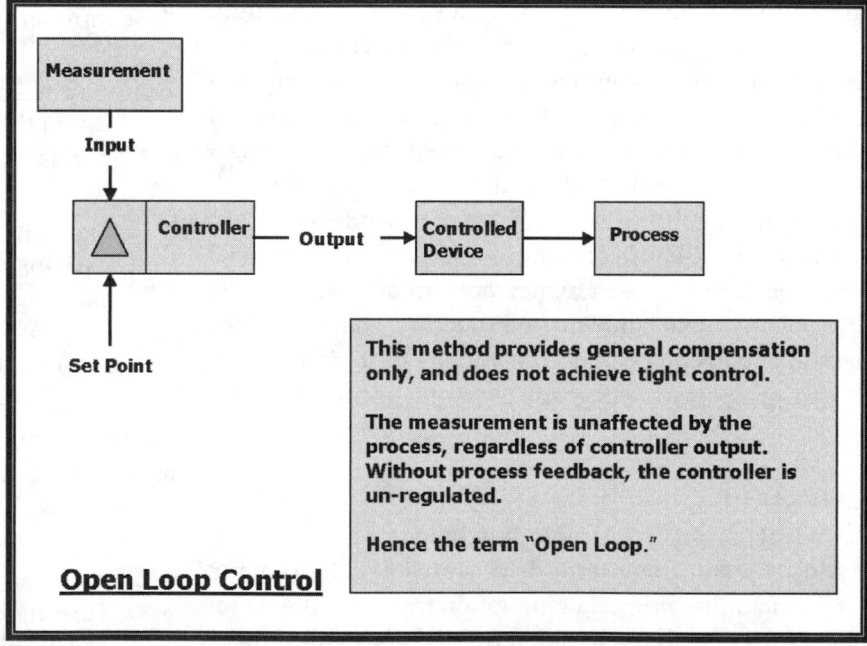

Figure 22.11 Open Loop Control

## 22.10 KEY ELEMENTS OF SUCCESSFULLY APPLIED AUTOMATIC CONTROLS

### 22.10.1 Examples of Good and Bad Control Applications
Example.
- BAD. Two thermostats in an exterior room; one serving the perimeter heat and the other serving

| CONTROL SUCCESS ITEM | REMARKS |
|---|---|
| **Have a good understanding of the process intended to be controlled.** | Understanding the system to be put under control is essential to a properly applied control system.<br><br>See Examples below for some good and bad control applications. |
| **Understand reasonable and unreasonable expectations of the controls.** | Maintain proper perspective of controls in the overall list of project priorities:<br><br>**Project priorities, in ascending order. [4]**<br>    Priority 1.    Safety<br>    Priority 2.    Reliability<br>    Priority 3.    Control and Regulation<br>    Priority 4.    Energy Efficiency<br><br>Use controls to automate and optimize, but not to solve inherent, systemic problems.<br><br>Understand that controls can break, require maintenance, don't last forever.<br><br>Energy efficiency should only be a focus after the higher priorities have been satisfied, and none of the energy efficiency work should compromise any higher priority. |
| **Verify the system is controllable and lends itself to control.** | Without this condition met, the control application will not be successful.<br><br>See Examples below for good and bad controllability. |
| **Know what should be accomplished, be able to define clearly and in detail what is to be controlled, and how.** | The clear and thorough Sequence of Operation in plain English is a very good sanity check for the control application. This exercise is a great check of the proposed control system. If it can be accurately described in words or with diagrams, it can probably be built. |
| **Verify that the basis for savings is sound.** | Understand the underlying principles of that which is to be controlled, including how the result constitutes an improvement. Understand the control behavior and energy use as-is, and compare that to the proposed control method. Verify that there will be a savings benefit to justify the work. |
| **Take care not to solve one problem and create a new one at the same time.** | Consider each control system for _what could happen if_.... and try to ferret out as many of those as practical.<br>Examples:<br>  •  If the control system quits, will all the lights be OFF or will they be ON by default?<br>  •  If power is lost, will anything freeze up?<br>  •  With all controls in their default states, would it be better to have full heating or full cooling, or don't care?<br>  •  If the actuator fails in its last commanded state (half way), how will anyone know that has happened?<br>  •  If the sensing tube falls off and the fan ramps up to full speed, will there be any duct damage?<br><br>Provide a responsible review to be sure the controls do not create new issues, especially those that could create safety issues or equipment damage.<br><br>Peer review is always good to double check for pitfalls and opportunities for improvement. |
| **Make sure the controls will provide economic return.** | Identify approximate costs and savings attributable to the controls. As a sanity check, almost any control system proposal that costs more than the equipment it controls or has a payback of greater than 5-years is suspect.<br><br>Prioritize control system expenditures with need. For example, if 2/3 of the savings can be achieved with a very simple and economical system, it may be appropriate to stop there, depending upon the owner's staff or project ambitions for savings.<br><br>Avoid Bells and Whistles. Be critical of each proposed 'point'. Ask how this will benefit the owner operationally or financially; and which points the customer can do without.<br><br>Separate the core control features from the 'gee whiz' points that have minimal value. Monitor-only points add project cost and extend payback periods, so use these add-on's with discretion. |
| **Understand the user so that the new control application is a facility-fit.** | The control system must survive if it is to remain viable over the long run. Consider the user's point of view:<br>  •  Identify the level of sophistication the operators will accept.<br>  •  Make provisions for the system to be user-friendly.<br>  •  Accommodate Owner preferences for materials, system supplier, man-machine interface, etc.<br>  •  Identify needs for additional staff training.<br><br>See Examples below for user-friendly design features. |
| **Have sufficient knowledge of control theory and terminology to be conversant with the Owner and a Controls Contractor.** | Communication using this trade language is helpful, but remember not to sweat the details. With the basic control expectations made clear, the process fundamentals sound, and the project economically viable, many of the details can be left to the specialty contractor. |

**Figure 22.12 Controls Application Checklist**

the VAV cooling box overhead air distribution.
- GOOD. One thermostat in a perimeter room; sequencing the VAV box and perimeter heat, with a dead band in between to prevent overlap.

### 22.10.2 Examples of Good and Bad System Controllability
Example.
- BAD. A single step controller on a 50-ton HVAC package unit that has one compressor attempting to maintain a constant 55 degree discharge temperature. Excessive equipment cycling will result, regardless of controls, and the consequence of poor control and likely premature equipment failure.
- GOOD. A four-step controller on a 50-ton HVAC package unit that has four compressors, where 12.5 tons represents a typical part load day for the building. Cut-in and cut-out set wider than the temperature change of each stage, to prevent short-cycling. Minimum on-off times further safeguard equipment from short-cycle damage.

### 22.10.3 Examples of User-Friendly Control Design Features
- Match the system complexity to the user's level of sophistication, or their willingness to learn. If it's too complex, they'll probably just unplug it 6-months after start-up. Training will help to raise skill levels and avoid over-simplification. Be patient, especially for anything complex. Any type of measurement that allows the user to see the savings created by the new control system will help spark interest and encourage ownership and buy-in.
- Control diagrams at the control panel as a handy resource.
- User adjustment for comfort control applications.
- Devices located in accessible locations for repairs and calibration.
- Redundant Visual Indicators by key inputs and outputs.
- Valve position indicators that are visible from the floor.
- Temperature and pressure gages adjacent to key measurement points.
- Man machine interface provision at the process location.
- Override Provisions for key output points to accommodate emergencies without disconnecting things.
- Standard parts that are readily available.
- Sensors and Transmitters with long-term stability for minimal need for calibration.
- Arrange control wiring/tubing with trouble-shooting in mind.
- Power on-off switches and air shut-off valves, in key locations to assist the service person.
- 3-valve manifolds at flow meters to allow zero-calibration.
- Drain and Vent valves at fluid transmitters for routine calibration.
- Minimum on-times and off-times to prevent short cycling.
- Make normal-use operator parameters adjustable, and critical parameters non adjustable, to allow user interaction and reduce inadvertent changes for the worse.
- Plug-In relays for easy replacement.
- Relays with LED lights to indicate at a glance when the relay is on or off.
- Separate control loops that are interrelated in a process designed with coordinated and linked set points, to prevent adjustment of one controller setting from inadvertently impacting another one.
- Control system logic design to identify faulty sensor readings, to prevent faulty control action.
- In general, try to provide a control design that will be accepted, and will last.

## 22.11 OPERATIONS AND MAINTENANCE

For the control design to stand the test of time (e.g. to be sustainable), it must be a good fit for the Owner.

## 22.12 EXPECTED LIFE OF CONTROL EQUIPMENT

Be realistic about how long things will last. Nothing lasts forever. For example, a control project based on a 10-year life will need to include cost of repairs to be realistic. Note for the table below that life spans shown vary by manufacturer and grade. This information is offered as a prompt for realistic life cycle cost estimating when controls are applied.

| USER-FRIENDLY ITEM | REMARKS |
|---|---|
| *Be Clear !* | Without explicitly clear diagrams and operating instructions, the controls can be perceived as a complicated contraption, and quickly fall into disrepair, eroding the intended savings. Clear and thorough documentation is helpful for sustained operation of these systems. Making control intentions clear from the beginning helps ensure Owner buy-in and support. |
| *Man-Machine Interface (MMI)* | This can take many forms, but has the common theme of convenient communication of key parameters to operations personnel. Consider the following as appropriate:<br>• Gages and thermometers by key instruments.<br>• Test wells adjacent to sensors for calibration or portable readings.<br>• Local display-adjust panels (fixed or portable) as a window into the DDC controller.<br>• In some cases, a single MMI can serve several equipment control items in a room, to reduce cost. |
| *Training* | Essential for operations staff buy-in. The owner needs to understand it and inherit it – and be proud of it. Involvement during the start-up and functional testing phase is a great way to gain familiarity and start training. |
| *Tuning* | Each control loop should be rigorously tested by large step changes to verify stability. For DDC controls, time-trends are very useful for verifying that the control loops are tuned to be responsive, without hunting. Be sure the tuning constants are backed up if DDC controls are used. Be realistic for how tight the controls need to be; striving for absolute zero-error control can cause the system to oscillate around the perfect set point, thereby causing excess end device activity and premature failures. Allowing a small variance around the nominal set point can allow the actuators to last a lot longer. |
| *Calibration* | Don't accept factory calibration. Check and verify each new instrument and repeat the exercise periodically – e.g. each two years. Instruments drift, especially commercial grade ones. Industrial grade instruments hold tolerance much better but are much more expensive. Some commercial instruments have no calibration adjustment, so the calibration would be more like verification; if not within specified tolerances, adjust or replace it. |
| *Control Diagrams* | As much detail as possible. DON'T LOSE THESE! As the years go by, they will become difficult or impossible to replace. If possible, keep the control diagrams and sequences with each control panel, near the equipment it controls. There are almost always changes in the original shop drawings as the installed system molds itself to the project conditions. The as-built version of the shop drawings is more valuable than the original or submittal diagrams, and should be part of the close-out documents provided to the Owner. As-Builts must be accurate to be effective. |
| *Sequence of Operation* | Sequences should be painstakingly clear. This cannot be overemphasized. The sequence should serve more than the installing technician; it should provide the basis for functional testing, Owner training including new hires, and troubleshooting over the years. A good sequence will explain each of the system's functions in detail, conveying the intent of the sequence and not just stages of operation. |
| *Start-up and Check Out.* | Insist on a good QC plan for start-up and check out of the controls. Understand that Quality Control is not governed by any industry standards – some suppliers apply significant effort to this and other so not. For large and complicated systems, requesting a written plan and final report is a good idea. This includes tuning.<br>Examples:<br>• Point-to-point testing (ringing) for all inputs and outputs from the hardware device all the way into the graphic screen. This verifies all points in between and is strongly recommended.<br>• Computer trends and step changes are a very good way to verify control stability and proper tuning.<br>• Points list printout<br>• VAV min/max settings printout<br>In all cases, avoid the QC method of *attention by exception* which is to say the QC is a reactive process in response to complaints. |
| *Functional Testing* | *"Show Me"* time. Don't be bashful at this stage. Take it through its paces from top to bottom, and don't be afraid to kick the tires a little bit. Include all pertinent operational modes and related safety controls. The purpose of this step is to see that the system is meeting the project's intentions for functionality, energy saving, and generally if it is doing what it is supposed to. Functional testing also verifies that the Owner is getting what they have paid for, and is very appropriate as a final payment contingency item. It's also a great way for the Owner to learn, and can be an effective part of Owner training. A written functional test plan is useful to follow and get to the point of being done (for payment), and also can be the basis of repeat verification testing over time. |
| *Periodic Exercising* | For controls that only act periodically instead of continuously this can make sense.<br>Examples:<br>• The change-over valves for a two pipe system need to move quickly and seat firmly to avoid leak by between the cooling and heating systems. A physical check of these once a year makes sense.<br>• Equipment that has been drained or shut off for a season, as part of the normal start-up work to bring it back on line. |
| *Standardized parts* | For additions to facilities, especially large facilities, inventories and familiarization of parts are important issues. Often, there are preferences for or against certain types of parts, features, etc. which should be incorporated to provide a more friendly new system. |
| *Minimum Documents for Owner's Use* | • Sequence of Operation.<br>• Control drawings, as-installed.<br>• Functional testing procedures and results.<br>• Tuning and calibration parameters.<br>• Copy of training materials, or video taping, for re-training and new hires.<br>• Operations and Maintenance manuals, trouble shooting, and emergency procedures.<br>• Installed device catalog data |

**Figure 22.13 User-Friendly Control System Checklist**

| DEVICE | TYPICAL LIFE |
|---|---|
| Pneumatic Cylinder Actuator | 20 years |
| Electronic Actuator – oil-filled motor style | 10-15 years |
| Electronic Actuator – mini hydraulic | 5 years |
| Open-air direct coupled electric actuators | 5-7 years |
| Residential style electric "clock motor" valve actuator with stalling motor | 5 years |
| VAV box controller with built-in actuator | 5-7 years |
| Low range differential air pressure sensor, analog | 5 years |
| Specialty sensors (humidity, CO2) | 5 years |
| Economizer dampers | 10 years |
| Globe control valve with adjustable packing | 15 years |
| Analog electronic controls, in general | 7-10 years |
| DDC   controls, before technology advances render them obsolete, working or not | 7-10 years |

Figure 22.14 Typical Control Equipment Life Spans

## 22.13  BASIC ENERGY-SAVING CONTROL APPLICATIONS

Common themes throughout the following list of applications:

- Strive to satisfy most of the people most of the time, but not all the people all the time (80-20 Rule)
- Try not to run any equipment continuously, or off-season
- Avoid heating, cooling, and lighting areas that are unoccupied
- "Just enough" air pressure, water pressure, etc. to satisfy the point of use
- "Just enough" heating and cooling, as determined by the point of use
- "Just enough" ventilation air for the people and to make up the exhaust needs

- Eliminate simultaneous heating and cooling wherever possible; minimize where unavoidable

## 22.14   ADVANCED ENERGY-SAVING CONTROL APPLICATIONS

(See Figures 22.17 and 22.18.)

## 22.15  FACILITIES OPERATIONS CONTROL APPLICATIONS

Automatic Controls, especially DDC controls, can be a very valuable tool for facilities personnel. Computerized maintenance management, early detection, remote servicing, automatic notification, trends and logs, and other features can be used to improve facility operational quality.

| LIGHTING CONTROL- BASIC | REMARKS |
|---|---|
| *Programmed Start-Stop* | Timed coincident with occupant schedules.<br><br>Two features are important for acceptance of these systems:<br>• A warning that the lights are about to go out<br>• Local overrides that allow people to keep the lights on |
| *Photo Cell Control of Outdoor Lights* | Outdoor lights off during day-lit hours<br><br>Time switch to shut off some site lighting after dark, but security lighting stays on. |
| *Occupancy Sensor* | Use for individual rooms, multi-purpose rooms, warehouses, etc., that are only occasionally occupied. For warehouses, it is good practice to leave a few lights on continuously for safety and security.  Not practical for restrooms, since they can leave people in the dark. |
| *Timed Tenant Override* | When after-hours tenants invoke the override, limit the amount of time this will remain active before the system reverts to unoccupied mode again (e.g. 2 hours), and what actually gets overridden. |

Figure 22.15 Basic Lighting Control Applications

| HVAC CONTROL - BASIC | REMARKS |
|---|---|
| *Programmed Start-Stop* | Any equipment left to run continuously should be controlled for automatic start-stop based on occupancy schedules. Can be Energy Management System (EMS) control or can be a simple time switch. Outside air dampers and exhaust fans are normally off during unoccupied times. |
| *Optimal Start* | Use to delay HVAC system start-up as long as possible. During the warm-up mode, the building is normally unoccupied so the outside air damper can remain closed and exhaust fans off until occupied. |
| *Optimal Stop* | Use to turn off primary heating and cooling equipment (Chillers / Boilers) shortly before the end of the occupied period, utilizing the thermal lag from the circulating fluid, air, building and furniture mass, and the fact that most people will not detect (or complain about) a temperature change of less than 2 degrees F. Ideally, the comfort conditions will have slipped just 2 degrees F just as the unoccupied event begins. |
| *Occupied/Unoccupied Mode* | Use to set-up and set-back space temperatures, and to reduce or deactivate outside air intake during unoccupied times. Typical unoccupied temperatures are 55 degF heating and 85 degF cooling. |
| *Timed Tenant Override* | When after-hours tenants invoke the override for comfort, limit the amount of time this will remain active before the system reverts to unoccupied mode again (e.g. 2 hours), and what equipment needs to operate. |
| *Dead band Thermostats* | Use or adjust thermostats to allow for a zero-energy dead band, a range where no action is taken by the HVAC system. A simple, but effective routine that should be part of every non-critical comfort control application. The dead band separation of heating and cooling controlled devices provides a positive separation to prevent the energy waste associated with heat/cool overlap. |
| *Restrict Tenant Adjustment Limits* | Remote lock-out of tenant adjustment for space temperature, or limiting the adjustment to +/- 2 degrees F at most. |
| *VAV Box Control Dead Band* | Independent VAV heating and cooling set points with a minimum 5 degree zero-energy dead band in between, in lieu of a single setting with a proportional band. This will increase the zero-energy dead band and reduce simultaneous heating and cooling.<br><br>Note: For normal VAV systems where the ventilation air comes mixed with the supply air, there is a *minimum stop* setting, for minimum ventilation, that creates some heating-cooling overlap. The dead band application for VAV boxes minimizes, but does not eliminate this overlap. |
| *Air Handler Control* | Independent preheat, mixed air, and cooling coil control set points, with a minimum 5 degree zero-energy dead band in between, in lieu of a single setting with a proportional band. This will reduce simultaneous heating and cooling. |
| *Outside Air for Morning Building Cool-Down* | Use outside air for pre-occupancy cool-down cycle in the summer, in conjunction with scheduled off time and/or optimal start routines. This routine takes advantage of cool nights and flushes the warm air out of the building that has accumulated and avoids or reduces demand on the primary cooling equipment.<br><br>Normally limited to dry climates. In humid climates or whenever the outside air humidity is higher than building humidity, this has the potential to add load if subsequent de-humidification (and energy expenditure) occurs as a result. |
| *Extended Air Economizer Operation* | Extend air-side economizer operation as far as practical, with the use of enthalpy comparison sensors, in lieu of the standard practice of limiting the economizer operation to 55 degrees F or less. With this method, whenever the outside air has less energy in it than the return air, cooling costs will be reduced by using outside air instead of return air. On days when the outside air is significantly drier than inside air, the free cooling range can be extended and outside air temperatures above return air temperatures can be utilized. When the outside air is humid, 55 degrees F is a smart cutoff point. |
| *Boiler Lockout from Outside Air* | Generally, prohibit boiler operation above 60 degrees F. Suggested control would be to cut-in at 55 and cut-out at 60 degrees F outside air. This is open loop control, but saves energy compared to "start based on demand" routines, since all it takes is one strong demand point to run the large machine for extended periods. |

**Figure 22.16 Basic HVAC Control Applications**

| HVAC CONTROL - BASIC | REMARKS |
|---|---|
| *Boiler Lockout from Outside Air* | Generally, prohibit boiler operation above 60 degrees F. Suggested control would be to cut-in at 55 and cut-out at 60 degrees F outside air. This is open loop control, but saves energy compared to "start based on demand" routines, since all it takes is one strong demand point to run the large machine for extended periods. |
| *Chiller Lockout from Outside Air* | Generally, prohibit chiller operation below 50 degrees F, and coordinate with air and water economizer settings. Suggested control would be to 'cut-in' at 55 and 'cut-out' at 50 degrees F outside air. This is open loop control, but saves energy compared to "start based on demand" routines, since all it takes is one strong demand point to run the large machine for extended periods. |
| *Load Limiting* | Reducing demand for primary electrical HVAC equipment can save money on demand costs. This has application for chillers, large variable speed fans, and large variable speed pumps. By load limiting to 90%, this keeps 90% of design capacity available but will reduce electric demand by 20-25%. Since 30-40% of the electric costs for large facilities are often from demand, this will create savings. |
| *Outside Air Reset of Hot Water Converters* | Compared to a constant temperature setting, this will reduce standby losses. In the cooling season, these losses are also unnecessary cooling loads. |
| *Mid-Range Vestibule Temperature* | For vestibule spaces, temper the space to a mid-range temperature, half way between indoor and outdoor temperatures. |
| *Operable Window Interlock* | Where operable windows are used, provide automatic control interlock to disable the HVAC serving that room, to avoid heating and cooling the great outdoors. |
| *Roll-Up Door Interlock* | Provide automatic control interlock to disable the HVAC serving interior area, to avoid heating and cooling the great outdoors when the door is up. This will encourage people to keep the door closed. |
| *Staggered Heating and Cooing System Start-Up*<br><br>*(Buildings with demand charges)* | After extended 'off' periods, such as night set back, normal automatic control response will be to drive the heating and cooling equipment in order to reach the occupied set point. To avoid setting the utility *maximum electrical demand* value and subsequent higher demand charges coming out of unoccupied periods, bring large electrically driven heating and cooling equipment loads on in <u>segments</u> for 30 minutes or until the process has "caught up" and is no longer at full load. . NOTE: this is not in reference to inrush currents for starting motors.<br><br>For <u>example</u>, consider a building heated by electric resistance heat. By default, coming out of unoccupied periods during winter will result in 100% of the heat being energized simultaneously. Controlling in 2 or more zones and allowing ample *pull-up* time for zone 1 to get to temperature and begin cycling normally before starting zone 2 will reduce the maximum demand for that day and, if done diligently through automation, will reduce the overall seasonal demand and demand charges.<br><br><u>Examples:</u><br>• Electric resistance heat (winter)<br>• Packaged HVAC cooling equipment (summer) |

Figure 22.16 Basic HVAC Control Applications (*Concluded*)

| LIGHTING CONTROL-ADVANCED | REMARKS |
|---|---|
| *Daylight Harvesting* | Control perimeter lighting on/off or modulate, in response to daytime sunlight entering the building.<br><br>In-board / out-board switching for interior lighting. |
| *Programmable Ballast* | Special "addressable" fluorescent ballast are available that can be controlled individually, without conventional relays. Remote control, custom scheduling by room, by area, and *by light fixture* are possible with this system to minimize lighting use. |

Figure 22.17 Advanced Lighting Control Applications

| HVAC CONTROL-ADVANCED | REMARKS |
|---|---|
| | Note: Many optimization routines rely on end-use polling of demand or valve/damper positions such as "most open valve" routines. This can be done cost effectively without actual measurement of position – by polling the individual "percent commanded output". This is referred to as "implied position" and is acceptable n most cases in lieu of actual position.<br><br>Note also that routines that use polling have the potential to be inefficiently operated if one errant measurement exists. For polling space temperatures, for example, limiting the user adjustment is strongly recommended in conjunction with demand polling of space controls. Additionally, it may make sense to 'discard' the high and low values from such polling to prevent errant operation. |
| *VAV Box Optimal Minimum Setting* | The usual method of calculating VAV box minimums is to assume full load conditions and the outside air percent of the total supply air at those conditions. With constant outside air control, the mixture is richer in outside air at supply air flows that are below maximum. Therefore, the VAV minimum stops can be reset automatically based on the actual OA mix as loads change.<br>Example<br>• A 20% VAV minimum stop is required based on 25% OA at design conditions. This same VAV box could have a minimum position of 10% when the supply fan is at 50%, since there is now twice as much OA in that air stream. The effect of this is to reduce over-cooling and reheating when the VAV box is at minimum while the zone thermostat continues to call for less air. |
| *Optimal Supply Air Static Pressure Setting – Variable Air Volume (VAV) Systems* | This requires polling of individual VAV boxes for air valve position and reduces system pressure until at least one box is 90% open, thereby providing the optimal system duct pressure (just enough pressure). This reduces fan horsepower. |
| *Optimal Supply Water Pressure Setting – Variable Pumping Systems* | This requires polling individual air handler for control valve position and reduces system pressure until at least one control valve is 90% open, thereby providing the optimal system water pressure (just enough pressure). This reduces pump horsepower. |
| *Optimal Evaporative Cooling Setting* | This uses outdoor air wet bulb temperature, which can be measured directly, but is usually calculated from temperature and humidity. The evaporative process can get close to, but never reach or exceed, the wet bulb temperature. By knowing the wet bulb temperature, the control system will know its boundaries and won't try to achieve something it cannot. For water-cooled refrigeration equipment, the low limit for the reset is normally around 55 or 60 degrees F and the chiller mfg needs to be consulted to confirm. Resetting the cooling water temperature down in this way, in lieu of a constant temperature setting, will reduce kW/ton energy use and demand. NOTE: achieving colder condenser water is a trade off between improved chiller kW/ton and increased cooling tower fan kW/ton, so evaluation for diminishing returns is required. |

**Figure 22.18 Advanced HVAC Control Applications**

*(Continued)*

| HVAC CONTROL-ADVANCED | REMARKS |
|---|---|
| ***Optimal Supply Air Temp Set Point for VAV Air Systems*** | This requires polling individual VAV box air valve position and reheat valve position. This utilizes a fixed temperature for cooling (e.g. 52 degrees F) with no reset at all until polling of individual boxes indicates that most of the boxes are in heating – only then is the SA temperature allowed to gradually be reset upward to 62 degrees F max. The reset is accomplished by polling VAV reheat valve positions and the air temperature is gradually reset upwards until at least one VAV box reheat valve is 90% open, thereby providing optimal air temperature (just warm enough). Simultaneously, VAV damper positions are polled to be sure enough cooling is being provided for any zone still calling for cooling, and cooling will prevail if the two are at odds.<br><br>A similar, but less accurate, method uses the supply fan VAV output percent command along with outside air to estimate when most of the VAV boxes are at minimum and in the heating mode, signaling the time to begin resetting supply air temperature. An important feature of this reset schedule is that NO reset occurs during summer months, preserving the basis of VAV savings. A simple reset schedule is used for this method as follows:<br><br><table><tr><td>SA Fan % Capacity</td><td>Outside Air Temp</td><td>Supply Air Temp Set point</td></tr><tr><td>100%</td><td>Any</td><td>55</td></tr><tr><td>40%*</td><td>40**</td><td>55</td></tr><tr><td>20%</td><td>10</td><td>62</td></tr></table><br>* The percent of the supply fan capacity at which the reset begins should correspond to the aggregate VAV box minimum settings. For example, if the aggregate (weighed average) VAV box setting is 30%, then a supply fan value of 40% indicates that most, but not all, VAV boxes are at minimum.<br><br>** The temperature at which the reset begins should correspond to the building thermal **break-even** point, which is the point below which cooling is not required, other than some interior zones. |
| ***$CO_2$ Demand Controlled Ventilation (DCV)*** | Has application for large open assembly areas, characterized by a single point of $CO_2$ control in the occupied space of the single zone unit serving that area. Applications include a ballroom, theatre or open plan office area. A separate $CO_2$ point of control would be required for each dividable meeting area, each assembly area, and each group area, so that areas of great ventilation demand are served appropriately and do not get overlooked by sensor averaging.<br><br>In essence, this control method uses a $CO_2$ sensor(s) as a people counter, thereby optimizing the use of outside air, and the energy required to condition it. |
| ***Optimal Ventilation by People Count*** | A combination of number of people and the duration at each location would provide the necessary information to calculate the exact amount of ventilation air needed. This method puts the welfare of the occupant first, but also minimizes the energy use by providing the optimal ventilation quantity (just enough ventilation) at any given time.<br><br>Note: This routine is conceptual only, but is presented due to the large potential for energy savings. To the author's knowledge, the friendly "people counter" instrument, other than a turnstile, hasn't been developed yet.<br><br>This may be possible, manually or automatically, based on ticket sales for assembly occupancies. |
| ***Optimal Ventilation Effectiveness by Season*** | Adjust ventilation rates based on improved "**e**" (ventilation effectiveness) in summer (value 1.0) compared to winter (value 0.8), to save outside air conditioning energy costs. |

**Figure 22.18 Advanced HVAC Control Applications**

*(Continued)*

| HVAC CONTROL-ADVANCED | REMARKS |
|---|---|
| *Optimal Sequencing of Multiple Chillers / Boilers* | Strategically selecting cut-in and cut-out points to keep the primary equipment operating in its most efficient range. Typically, maintaining this equipment between 50-90% load achieves good efficiency, but verifying the actual best-efficiency points or range for each system, from manufacturer's load profile data, is recommended. |
| *Optimal Reset of Hot Water Converters* | This requires polling of individual hot water points of use, for control valve position, such that at least one valve is open 90%. By resetting based on demand, this routine will provide the optimal water temperature (just hot enough). |
| *Multi-Zone System: Reset of Hot Deck and Cold Deck temperatures based on zone demand* | This requires polling of individual multi-zone (MZ) mixing dampers to determine the greatest cooling and heating demands, ideally so that at least one MZ damper in cooling and heating is fully open. By resetting hot and cold deck from space demand, optimal heating and cooling (just enough of each) will be provided in the hot and cold decks. Since these systems inherently mix cooled and warmed air, this reduces simultaneous heating and cooling. |
| *Induction Units* | Reset secondary coil temperature based on zone demand: This requires polling or individual induction unit coil control valve position, so that at least one secondary coil control valve is mostly open. |
| *Dual Duct Terminal Units* | Unless constant volume is critical to the process service, splitting the two dampers for independent control allows optimization. With the dampers split (separate actuators), and ventilation air in one or the other air streams (usually the cold duct), control like a VAV box with a dead band. During a call for cooling, the hot duct damper would remain closed while the cold duct damper modulates to maintain temperature. As cooling demand decreases, the cold duct damper throttles toward closed and finally reaches its minimum position (for ventilation). Here it will float within the dead band. If space temperature falls sufficiently to require heating, the hot duct damper would begin to throttle open with the cold duct damper at minimum. This sequence minimizes heating-cooling overlap and significantly reduces the energy use of the dual duct system that otherwise has simultaneous heating and cooling waste built into it. |
| *District Heating and Cooling Delta-T Control* | By actively controlling differential temperatures through buildings points of use to be as high has practical, flow can be reduced and pumping costs minimized. This can be touchy and, if too aggressive, can lead to comfort issues, but worthwhile.<br><br>Many, if not all, of these distribution systems pump more water than they need to. There are a variety of reasons for this, which are beyond the scope of this report, but many of them can be mitigated by imposing a control limit of *minimum delta T* (differential temperature) across the coil, building, or segment of the distribution system. This has the effect of requiring the point of use to extract all the available energy out of the circulated fluid before returning it.<br><br>Control is implemented by some manner of throttling device, either a control valve or a pump speed, and supervised by differential temperature measurement and/or flow measurement.<br><br>Applications vary, but the common theme is to wring the heating or cooling energy out of the water and reduce flow when possible, while still maintaining comfort. |

**Figure 22.18 Advanced HVAC Control Applications**

*(Concluded)*

| FACILITIES OPERATIONS APPLICATION EXAMPLES | REMARKS |
|---|---|
| *Just in Time Scheduled Filter Changes (predictive maintenance)* | Air mover efficiency is lost if filter replacement is postponed. Filter costs are higher than necessary when filters are changed too early. Identify filter change-out points.<br><br>For constant volume air systems this is a straight forward application of a filter differential pressure (dP) switch, set to announce the need for service<br><br>VAV systems will require special scheduled off-hour events to open the VAV boxes and speed up the fan to a fixed high air flow rate, in order to establish a baseline and determine actual filter loading. |
| *Monitor Heat Exchanger Approach Temperatures (predictive maintenance)* | Efficiency is lost when heat transfer surfaces become fouled. On large equipment, it is practical to monitor the approach temperature and perform maintenance as required to keep equipment running optimally. After 'clean condition' benchmarks are determined, the control system can monitor actual approach and provide alert messages when fouling is indicated.<br>Examples:<br><br>• Boiler stack temperature vs. process temperature.<br>• Chiller condenser temperature vs. refrigerant temperature.<br>• Heat exchanger inlet and outlet<br>• Air-cooling-coil coldest air to coldest water differential |
| *VAV Box Controls Health / Sanity Check* | VAV box controllers and actuators are inexpensive and have a short life span. It is very common to find systems left to their own accord for a number of years to have 25% of the VAV boxes unresponsive and stuck in some last position, with obvious control and energy ramifications. Using the front end, it is relatively easy to set up an automated polling routine to sequentially go from box-to-box and command incremental increases in air flow and heating coil flow, and log whether or not a process change occurred (e.g. is the box alive or not), thereby generating a list of boxes needing attention. Doing this annually is a good idea. |
| *EMS Monitoring for Energy Use Benchmarking* | Using the Energy Management System to monitor the primary energy sources is useful to establish benchmark usage patterns for the building.<br><br>**Suggested primary energy metering points for a typical building:**<br>Gas<br>• Whole Building Consumption<br>Electric<br>• Whole Building Demand<br>• Whole Building Consumption<br>• Lighting Loads sub-meter<br>• Mechanical Loads sub-meter<br>• Other / Plug Loads sub-meter<br><br>Additional sub-metering may be used, as appropriate and trended and logged by the Energy Management System. For example, sub-metering of Data Centers or other high-use areas, or large sub-let areas, may make sense for both energy and business considerations.<br><br>Overall Energy Use Index (EUI) values, in kBtu/SF-yr, kBtu/SF-degree day-mo, or similar normalized units, are also very useful as a building energy use watchdog, to provide early detection of things going awry. This provides a reminder mechanism to prompt the operations personnel to keep the building energy use on track.<br><br>Once the benchmarks are set, the Energy Management System can easily be set with energy use thresholds to alert the Owner if usage creeps up over time. This early detection mechanism serves to prompt investigation and corrective measures. This energy use feedback will help sustain the intended building energy use over time. |
| *EMS for Measurement and Verification* | The trending and reporting features of Energy Management Systems are well suited for Measurement and Verification activities. The differential *before and after* energy use can be identified using definite purpose energy meters (kWh, kW, flow meters, Btu meters, etc) or calculated values. |

**Figure 22.19 Facilities Operations Control Applications**

## 22.16 CONTROL SYSTEM APPLICATION PITFALLS TO AVOID

The following are some common pitfalls to avoid in applying Automatic Controls.

## 22.17 COSTS AND BENEFITS OF AUTOMATIC CONTROL

Automatic Controls are unique in that they often provide both tangible and intangible costs and benefits. Tangible Benefits, like other energy-related projects, include energy savings and demand savings. Intangible benefits are those that are difficult to quantify or predict. Some of these only apply to Digital Control systems.

## 22.18 ESTIMATING SAVINGS FROM APPLIED AUTOMATIC CONTROL SYSTEMS

### 22.18.1 Introduction

Economic barriers are among the greatest obstacle for the Energy Professional, and Automatic Controls are no exception. Acceptance of the cost of new technology almost always requires economic justification, and rightly so. Finding ways to predict and then demonstrate the savings is necessary should be the goal of any Automatic Controls project, since this will build momentum for subsequent projects.

While it is often easy to visualize that savings will occur from Automatic Control improvements, estimating them can be a very daunting and intimidating challenge. This presents a dilemma to the Energy Professional since cost justification is almost always an expectation. Projects that have merit but defy quantification may be overlooked.

Done accurately, the cost saving calculations can be laborious and expensive, posing a barrier to otherwise viable projects. One approach is to produce reasonable estimates using abbreviated estimating methods or 'rules of thumb' where possible. As with all estimates, being conservative is a key to success so that project performance is seen to *under-sell and over-deliver*. Even when not a contract requirement to guarantee savings, there is always an expectation that the estimated costs and savings come to pass; further, the credibility of the energy professional is built largely on the accuracy of these estimates. Bearing in mind that there will always be uncertainties and uncontrolled variables at work, it is usually good practice to *de-rate the calculated savings*. By artificially reducing savings (de-rating) and artificially increasing projected costs (contingency allowance), two things happen.

**Effect of de-rating project estimated savings:**
- The odds of project performance exceeding expectations increases.
- Projects with marginal Return on Investment (ROI) look worse and may be eliminated.

The applied de-rate will depend on the level of uncertainty. A high degree of uncertainty suggests the need for a higher de-rate. A value between 20-30% is

| PITFALL | REMARKS |
|---|---|
| *Too Complex* | It will only get unplugged if overly complex. Often simple solutions will work well and stay working longer. Find the balance between complexity and optimization that fits the customer. |
| *Use of Experimental or Beta Site Products* | BETA stands for<br><br>**B**rand new item, and we aren't<br>**E**xactly sure it will work, so please be our<br>**T**est site,<br>**A**nd let us know what does/doesn't work.<br><br>It doesn't actually stand for that, but that is what it means to the customer. As a rule, don't volunteer the customer to be a guinea pig. Use proven solutions. |
| *Introducing New Single Points of Failure* | Evaluate systems for weaknesses. Avoid bottlenecks and central hub issues where possible, and have contingency plans where they are unavoidable. |
| *Creating a Captive Customer – Proprietary Systems* | The Owner needs to make this decision, and there is no 'right' answer. There are pros and cons to both proprietary and open systems, and both ways can achieve good control results. |

**Figure 22.20 Control System Pitfalls**

| BENEFITS | | COSTS | |
|---|---|---|---|
| TANGIBLE | INTANGIBLE | TANGIBLE | INTANGIBLE |
| Energy Savings | Reliability improvements through increased monitoring. | First Cost. | Higher skill level required for maintenance personnel (DDC). |
| Demand Savings | Automated equipment operating logs and ease of archiving. | Control System Repair Costs.<br><br>Often overlooked, these should be budgeted for like any other building system. A rule of thumb is 5% of initial cost per year as an allowance for Operations and Maintenance. | Increased training costs (DDC). |
| Labor Savings. A touchy subject. Whether job elimination is used as project justification is an individual decision. | Reliability improvements from expedited / remote troubleshooting, early detection and other pro-active benefits from Trending, Reporting, and Alarm features of DDC controls. | Labor costs for the IT (Information Technology) work on the computer-side of the control system, especially when interfacing with other systems or sharing common cabling. Costs may be in-house special personnel or contracted. | Shorter technology obsolescence cycle for DDC controls means earlier system replacement, and increased life cycle costs. |
| | Predictive / planned maintenance. | | Increased time spent in attendance of the controls and operator workstations (DDC). |
| | Improved comfort and occupant productivity.<br><br>NOTE: Some attempts have been made to identify associated cost savings from productivity, but these are usually in reference to corrective actions made to areas where very poor conditions existed previously. So, smaller improvements in comfort probably would not yield proportional results. | | Higher maintenance costs if proprietary DDC system chosen |
| | Avoided cost of replacement of existing controls that need replacing anyway. | | Electrical consumption of controls and instrumentation. |

**Figure 22.21 Control System Costs and Benefits**

suggested for most applications, although there are cases where no de-rate is needed.

Examples:

- Easy to Quantify. A lighting replacement project with 24-7 operation need not be de-rated at all since there are few, if any, uncertainties. This is easy to quantify, since only one parameter has changed (light fixture energy efficiency).

- Hard to Quantify. An Automatic Control project that includes multiple control system improvements implemented at the same time, such as variable pumping, free cooling modes, supply air reset, boiler lock-out, morning warm-up, and new quarter-turn (no leak by) terminal unit control valves. This adds uncertainty for what savings comes from individual measures.

From a pure simplicity standpoint, a project consisting of several control improvements could be implemented one at a time, with six months of post-project Measurement and Verification (M&V), therefore changing only one thing at a time. This would have the clear advantage of knowing what savings occurred from what measure, but would probably not be done in practice due to the protracted length of time for the entire project to be implemented, as well as lost savings from delayed implementation.

## 22.18.2 Replacement Costs

A common practice for existing facilities is to propose equipment replacement using energy efficiency as justification. In general, it is easy to justify the differential cost of upgrades to higher efficiency equipment, but often impossible to justify the entire replacement project on energy savings alone. Burdening the project cost with unrelated expenses, such as equipment replacement that was due anyway, makes the payback look worse and creates an unfair perception of long paybacks. Whenever possible, energy improvement expenses should be fairly separated from normal replacement project costs. Equipment that is near the end of its useful life should be a planned replacement expense, regardless of the desire to reduce energy costs. If replaced early, the remaining value of the equipment may be appropriately 'charged' to the energy project, but not the entire project cost since this would need to be done anyway.

## 22.18.3  Barriers to quantifying savings
   from controls changes

- Control parameters are almost always user-adjustable and will usually be fine-tuned during the life of the project, including the post-project measurement period.
- Control algorithms often include multiple variables which do not act independently. Consequently the effects on energy consumption often defy isolating and quantifying separately. An uncertainty is whether one measure will interact with another. In many cases, the aggregate savings may be less than the sum of the parts, in which case the savings would be over-stated if calculated independently.
- Control system improvements that improve quality, comfort, ventilation, or other features from some deficient state may actually increase energy use in some ways, eroding the overall savings of the measure itself.

### 22.18.4  Methodology

Establish realistic baselines. Identify variables and their interactions. Identify competing or complementing processes that will subtract from measure savings. Reduce the number variables to simplify analysis. Treat uncertainties with contingencies, by either inflating the baseline or de-rating the savings. With the baseline established, use experience to evaluate calculated savings as percent of total expenses. Use all these indicators as sanity checks to avoid over-stating savings. After project completion, get after-the-fact measurements where possible to compare actual to estimated savings; collecting this real data will allow improved estimating and reduced contingencies over time.

### 22.18.5  Quantified Savings Examples.

The following are some examples of how to generally quantify savings from Automatic Controls measures. Many of the methods use rules of thumb, and assumptions to simplify the work. More accurate results can be had by using a rigorous computer model, but since time is money the luxury of detailed models is not always available. Without an excessive investment in time, the methods shown below will yield results close enough to identify probable savings and to tell if the measure is worthwhile or not. While this is by no means a complete listing, it is hoped to convey the general method of abbreviated energy accounting for multi-faceted processes when improvements are proposed. Some of the examples show straight forward benefits, and others show benefits with parasitic losses or competing processes. The first example below is Condenser Water Reset, and is discussed in detail. Other abbreviated solutions are provided in the table that follows.

### 22.18.6  Detailed Example: Condenser Water Reset.

Note: The cooling tower term **"approach"** is the difference between leaving condenser water temperature and ambient wet bulb temperature.

**The variables:**

The chiller kW/ton varies according to load, and the cooling tower kW/ton varies according to wet bulb temperature and approach. The cooling tower fan energy varies according to chiller ton hours, tower capacity, and condenser temperature set point with respect to ambient wet bulb. Overall savings varies with annual ton-hours, chiller efficiency, the ratio of chiller-to-tower efficiency (kW/ton), chiller low limit in accepting colder condenser water, and coincident wet bulb temperatures. The wet bulb variable means that any rules of thumb developed for annual savings would be area/climate specific.

Although the chiller energy is reduced directly as the condenser temperature is lowered (1-1.5% per degree), the cooling tower efficiency decreases at lower approach temperatures, and eventually a point is reached where the savings in chiller energy are met by the added cooling tower horsepower. This break-even point varies with the cooling tower sizing, and towers with high hp/ton (small box, big fan) will hit the wall sooner than more generously sized cooling towers. The cooling tower energy penalty varies depending upon the approach temperature and rises sharply for each degree below 7 degrees F approach. Some common values of the energy penalty, in percent increase per degree lowered, are as follows:

| 12 degF Approach | 7.3% |
| 11 degF Approach | 7.8% |
| 10 degF Approach | 8.3% |
| 9 degF Approach | 8.9% |
| 8 degF Approach | 9.7% |
| 7 degF Approach | 10.6% |

*8.9% Average cooling tower fan
energy penalty per degree lowered.*

The chillers place a physical limit on this control process; some can take colder condenser water than others. Centrifugal chillers can accept from 55 to 70 degrees F entering condenser water, depending upon the manufacturer. Screw chillers are generally limited to 70 degF entering condenser water. Many reciprocating chillers are limited to 70 degrees F, but some can operate at lower temperatures. **IMPORTANT: For each application, the limits of the machinery need to be identified and the control system must stay within those limits to assure no damage or detriment is done to the chillers.**

Refer to the table below. For a chiller with 0.5kW/ton efficiency, paired with a 0.07 kW/ton cooling tower, a proposed 5 degree reduction (from 12 to 7 degrees F approach), would yield *worse* energy consumption than leaving it at 12 degrees F, due to the high cooling tower fan energy penalty. This same example, with all things equal except a 0.04 kW/ton cooling tower, would save 0.35% per degree lower overall cooling energy. This example shows that **during operation at or near design wet bulb conditions, most or all of the theoretical chiller savings from condenser water reset will usually be negated by the added cooling tower energy use, unless the cooling tower is very efficient (0.04 kW/ton or less).**

While the energy savings near design conditions may be marginal, most of the chiller hours will be at wet bulb conditions that are more favorable, and this is where the energy savings are attained. The control strategy capitalizes on this by continuously adjusting the operating set point based on ambient wet bulb. For those chiller operating hours when the wet bulb is significantly lower than design and the cooling tower can easily produce colder water with little tower energy penalty, the savings will be much more pronounced and closer to the theoretical 1-1.5% per degree. Of course, the chiller load during these shoulder seasons or overnight periods are usually lower than maximum, so **knowing the chiller load profile and coincident wet bulb is key to quantifying savings.**

**Chillers that can accept very cold condenser water and have cooling needs coincident with reduced wet bulb temperatures such as in the Southwestern US, when it is easy to achieve colder condenser water temperatures, can achieve 10-15% annual overall cooling energy savings using condenser water reset, compared to a fixed temperature set point of 70 degrees F.**

The following diagram illustrates the pronounced effect the system cooling tower has upon the overall cooling savings of the condenser water reset control routine. Remember, without the cooling tower fan energy expense, the savings from the chiller would be 1-1.5% per degree lowered.

The following diagram shows how the relative hp rating of the cooling tower (kW/ton) affects the optimal cooling tower leaving water temperature set point. The "approach" value is a parameter of the optimized condenser temperature calculation, and should be selected based on the cooling tower in use for best economy.

The "approach" value is a parameter of the optimized condenser temperature calculation, and should be selected based on the cooling tower in use for best economy. Suggested values of cooling tower approach for best overall cooling efficiency (kW/ton) are shown below. By using the "Ratio" instead of specific combinations,

| Nominal Cooling Tower kW/ton | Nominal Ratio of kW/ton Chiller / Tower | Overall Cooling Energy Savings % per degF lowered |
|---|---|---|
| 0.03 | 16.7 | 0.63% |
| 0.04 | 12.5 | 0.35% |
| 0.05 | 10.0 | 0.08% |
| 0.06 | 8.3 | -0.18% |
| 0.07 | 7.1 | -0.42% |
| 0.08 | 6.3 | -0.66% |
| 0.09 | 5.6 | -0.88% |
| 0.10 | 5.0 | -1.10% |

**Effect of Cooling Tower Selection on Overall Energy Savings From Condenser Water Reset**
(Lowered from 12 to 7 deg F Approach, Chiller 0.5 kW/ton)

**Figure 22.22 Cooling Tower Effect on Condenser Water Reset Savings**

| Chiller kW/ton | Tower kW/ton | Ratio of Chiller / Tower kW/ton | Typical Lowest Economical Tower Approach For Condenser Water Reset, degF |
|---|---|---|---|
| 0.5 | 0.1 | 5:1 | 15 |
| 0.5 | 0.085 | 6:1 | 13.5 |
| 0.5 | 0.07 | 7:1 | 12 |
| 0.5 | 0.06 | 8:1 | 11 |
| 0.5 | 0.05 | 10:1 | 10 |
| 0.5 | 0.04 | 12.5:1 | 8.5 |
| 0.5 | 0.03 | 17:1 | 7 |

**Figure 22.23 Cooling Tower Effect on Optimum Condenser Water Temperature**

this information can be applied to any combination of chiller and cooling tower. This is the value inserted in the sequence "*...optimum cooling tower set point shall be equal to the calculated wet bulb temperature plus approach...,*" and provides further evidence of the importance of amply sized, low HP cooling towers.

## EXAMPLES OF QUANTIFYING SAVINGS FROM AUTOMATIC CONTROL MEASURES

**(See Figure 22.24.)**

## 22.19 CONCLUSION AND FURTHER STUDY

Automatic controls are useful for basic regulation, and quality control of processes and environments. They can also be leveraged for energy savings through optimization. Properly applied, these systems are reliable and cost effective.

Let us now return to the Chapter Intent, so see if the objectives were met. The stated purpose of this chapter was **to focus on the application of Automatic Controls as a tool to achieve energy savings.** The reader should review the titles of each section, reflect on the key topics they have taken away, and decide if the stated objective was met. If you have gained insight into how automatic controls can help you achieve your energy goals, have increased your comfort level with the subject matter, know important questions to ask the vendors, and are eager to put these systems to work for you, then this chapter has been worthwhile. If you do not feel the objective was met for you, consider re-reading the material, seeking out supplemental material, or contacting the author for further discussion. Very often a second text on the same subject will amplify the effect of the first text, by reinforcing concepts with different examples and descriptions.

This chapter was necessarily made brief, and does not pretend to be a complete treatment of the subject. The following advanced topics were not addressed, and are listed for the interested reader to pursue through additional study.

- Addressable I/O devices, network communications, and equipment interfacing technologies.
- Boolean Logic.
- Cascade Control.
- Computerized Maintenance Management Systems (CMMS).
- Control Loop Interaction.
- Ergonomic Considerations. Trends, logs, reports, alarms, graphics.
- Failure Modes, Mitigation, and Fault Tolerance.
- Feed Forward Control.
- Fuzzy Logic.
- IT Security.
- Ladder Logic.
- Loop Tuning. Step changes, dead time, stability, response time.
- Measurement and Verification.
- Open Protocols, XML, Gateways, and Translators for DDC networking.
- SAMA logic diagrams, multi-element industrial control methods.
- SCADA systems.
- Self-Powered Controls.
- Wireless Controls.

| MEASURE | MEASURE SAVINGS | OTHER PROCESS OR MEASURE THAT SUBTRACTS FROM SAVINGS | REMARKS |
|---|---|---|---|
| *Programmed HVAC Air System Start-Stop* | Rule of Thumb: HVAC Air Systems left on continuously without setback instead of being turned off each night have been found to increase annual energy use by 15%. | | Values stated were estimated on one building from envelope losses from not setting back temperatures in winter, plus air horsepower for fans that never stop.<br><br>Calculated savings were conservatively reduced by 30%. |
| *Chiller - Condenser Water Reset* | Rule of thumb: 1-1.5% chiller energy (kW/ton) reduction per degree lowered. | Additional cooling tower fan horsepower subtracts from measure savings, proportional to the additional cooling tower hours of operation.<br><br>Increased tower fan energy can consume from 40% to 100% of the chiller theoretical savings at or near design wet bulb conditions. | This is a VERY dynamic calculation.<br><br>Ignoring cooling tower fan horsepower penalty will over-state savings.<br><br>Knowing the chiller load profile and coincident wet bulb is essential to quantifying savings.<br><br>See detailed example above. |
| *Chiller – Constant Volume Chilled Water Reset* | Rule of thumb: 1-1.5% chiller energy (kW/ton) reduction per degree the chilled water temperature is raised.<br><br>This can be applied to the annual aggregate load (ton-hours) with little error, and does not need a load profile. | | This can also cause loss of humidity control in many climates, and the risk of mold damage may overshadow the potential for energy savings.<br><br>If done in humid climates, layering this control with indoor and ambient wet bulb temperatures would be suggested, to restrict the reset activity to times of low humidity. |
| *Chiller – Variable Volume Chilled Water Reset* | Rule of thumb: 1-1.5% chiller energy (kW/ton) reduction per degree the chilled water temperature is raised.<br><br>This can be applied to the annual aggregate load (ton-hours) with little error, and does not need a load profile. | Additional pump horsepower subtracts from measure savings, according to affinity laws:<br><br>$HP2 = HP1 \left(\frac{Q2}{Q1}\right)^3$<br><br>Energy transport efficiency is high, so pumping costs are usually a small fraction of chiller costs. This is not a linear relationship, so a load profile is required. Make two identical load profiles in a spreadsheet, and insert volume and static pressure for the existing and proposed systems.<br><br>$Pump\ bhp = \frac{gpm * head * sp.\ Grav}{3960 * eff}$ | Since pump energy is usually small compared to chiller energy, per ton, there is a net savings, but the added flows negate most (e.g. 2/3) of the chiller savings.<br><br>See humidity concerns above in "Chiller – Constant Volume Chilled Water Reset" |
| *Air Handler – Constant Volume Supply Air Reset in cooling mode* | Rule of thumb: 1-1.5% chiller energy reduction per degree the supply air temperature is raised, for hours when mechanical cooling is used.<br><br>This can be applied to the annual aggregate air handler mechanical cooling load with little error, and does not need a load profile. | Hours when air-economizer is used do not count toward these savings. | See humidity concerns above in "Chiller – Constant Volume Chilled Water Reset" |

**Figure 22.24 Quantifying Control System Savings**

| MEASURE | MEASURE SAVINGS | OTHER PROCESS OR MEASURE THAT SUBTRACTS FROM SAVINGS | REMARKS |
|---|---|---|---|
| *Air Handler – VAV Supply Air Reset in cooling mode* | Rule of thumb is **1-1.5%** chiller energy reduction per degree the supply air temperature is raised, for hours when mechanical cooling is used.<br><br>This can be applied to the annual aggregate air handler mechanical cooling load with little error, and does not need a load profile. | Hours when air-economizer is used do not count toward these savings.<br><br>Additional fan horsepower subtracts from measure savings, according to affinity laws:<br>$$HP2 = HP1 \left(^{Q2}/_{Q1}\right)^3$$<br><br>Energy transport efficiency is <u>low</u>, so air moving costs are usually a large fraction of chiller costs. This is not a linear relationship, so a load profile is required. Make two identical load profiles in a spreadsheet, and insert volume and static pressure for the existing and proposed systems.<br><br>Fan bhp =<br>$$^{cfm \, * \, tsp}/_{6356 \, * \, eff}$$ | Usually a net loss, unless amply sized ducts exist such that air friction losses are low.<br><br>See humidity concerns above in "Chiller – Constant Volume Chilled Water Reset" |
| *VAV Supply air reset, once most of the VAV boxes are confirmed to be at minimum positions and in heating mode.* | Elevating the supply air when most boxes are at minimum positions reduces the reheat coil burden.<br><br>This is because the VAV reheat system must first heat the primary air to room temperature, before any building heating occurs. This is known as the inherent VAV reheat penalty. | Additional fan horsepower subtracts slightly from measure savings.<br><br>Energy transport efficiency is low, but at minimum flows the fan laws have reduced baseline horsepower substantially.<br><u>Example:</u><br>• If at 40% of max, fan HP will be 15% or less of full load.<br><br>Heat from the extra fan energy occurs only in heating mode and provides some beneficial heating, but at efficiencies equal to electric resistance heating. The amount subtracted from the measure savings is the added fan Hp, expressed in Btu, less the corresponding HVAC heating input needed for the same amount of heat. | This should provide a net savings.<br><br>If evaluated at minimum air flow conditions (winter), the extra fan hp is a low magnitude penalty.<br><br>Once the primary air flow is at minimum (with 80-90% of the boxes at minimum), then raising supply air temperature produces savings that outweigh the added fan energy. The aggregate VAV box minimum air flow settings establish the winter supply air flow rate for the fan horsepower penalty calculation. |

Figure 22.24 Quantifying Control System Savings (*Continued*)

| MEASURE | MEASURE SAVINGS | OTHER PROCESS OR MEASURE THAT SUBTRACTS FROM SAVINGS | REMARKS |
|---|---|---|---|
| **Boiler lock-out above 60 degrees F, in lieu of continuous operation.** | Use weather data to determine hours above 60 degrees F when the boiler will now be off, and estimate savings from boiler standby losses for these times.<br><br>Thermal losses from distribution piping will add to cooling load if boiler runs in summer.<br><br>Rule of Thumb: Boilers left idling all summer instead of being turned off in warm weather have been found to increase gas use by 30%.  Values were taken from an Hourly Analysis Program for one building Climate Zone 2. Calculated savings were conservatively reduced by 30%. | | Boiler efficiency drops rapidly below 75% load, due to standby losses. Efficiency loss is roughly 1% thermal efficiency less for each 6% capacity reduction, for operation below 50% load.<br><br>Losses will be from boiler casing/stack losses, circulating pumps, and piping thermal loss for these hours. |
| **Chiller lock-out below 55 degrees F, in lieu of continuous operation.** | Use weather data to determine hours below 55 degrees F when the chiller will now be off, and cooling loads during these times (if any).  Estimate savings from chiller efficiency at reduced load for the hours it should have been off. | | Chiller efficiency drops rapidly below 50% load.  Efficiency loss is roughly 0.1kW/ton worse for each 6% capacity reduction, for operation below 50% load (water-cooled chiller).<br><br>For air-cooled chillers, the efficiency may not drop much, due to the reduced ambient temperature beneficial effect.<br><br>Load for the chiller will be from control valves that don't close, pumping, and heat absorbed into the distribution piping.<br><br>If air handler winter pre-heat is set to 55 degrees F (sequenced with mixing dampers), simultaneous heating and cooling can exist between the preheat coil and cooling coil if they are each set to the same 55 degrees F.  This loss can be estimated at 5% of the preheat coil load. |

**Figure 22.24 Quantifying Control System Savings (*Continued*)**

| MEASURE | MEASURE SAVINGS | OTHER PROCESS OR MEASURE THAT SUBTRACTS FROM SAVINGS | REMARKS |
|---|---|---|---|
| **Constant Volume Hot water converter reset** | Reduced distribution system piping heat losses.<br><br>Estimate from hours and heat loss tables, using average systems temperatures, and take-offs of pipe sizes and lengths.<br><br>If the heating system runs during cooling season, these losses (now credits) are amplified as new cooling loads, increasing benefit. | Evaluate <u>when</u> (winter or summer?) and <u>where</u> the piping heat losses occur. Wherever the thermal losses are beneficial, they subtract from measure savings.<br><br>If boiler lockout is used, losses from summer hours must be subtracted from measure savings. | |
| **Variable Flow Hot water converter reset** | Reduced distribution system piping heat losses.<br><br>Estimate from average systems temperatures and hours, using heat loss tables.<br><br>If the heating system runs during cooling season, these losses (now credits) are amplified as new cooling loads, increasing benefit. | Variable speed pumping will ramp up as temperature is lowered, increasing pumping costs and tending to offset the savings in standby piping losses.<br><br>Heat from the extra pump energy occurs only in heating mode and provides some beneficial heating, but at the efficiency of electric resistance heating. This beneficial heat subtracts from measure savings.<br><br>Evaluate <u>when</u> (winter or summer?) and <u>where</u> the piping heat losses occur. Wherever the thermal losses are beneficial, they subtract from measure savings.<br><br>If boiler lockout is used, losses from summer hours must be subtracted from measure savings. | For annual hours with the boiler active, the piping losses can be estimated. From these, subtract added pumping costs.<br><br>Energy transport energy penalty for hydronic systems is low, so this may still produce good savings. |
| **Control off or reduce lighting in a thermally <u>heavy</u> building** | Electric savings is from the hours the lights are now off, compared to existing hours of operation (or reduced wattage if de-lamped)<br><br>In summer, excess heat from inefficient lighting is amplified since it is also an extra cooling load. | | This measure assumes the building is almost always in cooling, so the extra lighting energy is almost never a benefit at all. |

**Figure 22.24 Quantifying Control System Savings (*Continued*)**

| MEASURE | MEASURE SAVINGS | OTHER PROCESS OR MEASURE THAT SUBTRACTS FROM SAVINGS | REMARKS |
|---|---|---|---|
| *Control off or reduce lighting in a thermally light building* | Electric savings is from the hours the lights are now off, compared to existing hours of operation (or reduced wattage if de-lamped).<br><br>Excess heat from inefficient lighting also adds cooling load during cooling season. | Excess heat from inefficient lighting is actually part of the building heating but at the efficiency of electric resistance heating. This loss of beneficial waste heat in winter subtracts from measure savings. The winter energy loss is equal to the heat output saved from lighting.<br><br>The amount subtracted from the measure savings is the avoided lighting waste heat, less the corresponding HVAC heating input needed for the same amount of heat. | So, the reduced heat loss from better lights means the HVAC heating system must increase output to compensate. Unless the building heating system is electric resistance, the HVAC heating system will be more efficient at producing heat than waste heat from lighting.<br><br>Note that for buildings with electric resistance heat this would result in no overall savings during heating season.<br><br>**If heating and cooling seasons are reasonably balanced, it may be acceptable to ignore both effects, as counteracting.** |
| *Use night-time unoccupied mode and morning warm-up in lieu of constant conditioning* | During unoccupied periods when the indoor temperature drifts between occupied and ambient temperatures, skin loads reduce proportional to the reduced dT.<br><br><u>Rule of thumb:</u><br>1% savings per degree of setback, if kept there for at least 8 hours. | If electric heat is used, a pitfall can exist in the morning warm-up by increasing electric demand, reducing or negating the measure savings, especially if there is a utility ratchet clause.<br><br>Avoid this by bringing the building up in segments, allowing sufficient time between stages to prevent large demand events. | |
| *Change indoor cooling temperature set points from 74 cooling / 72 heating to 76 cooling and 70 heating.* | Similar effect as night set-back, except use 2% per degree setback for continuous operation instead of 1% for setback. | If local adjustment is available to the tenants, some adjacent areas may have sufficiently different local set points to cause simultaneous heating and cooling, reducing measure savings. | Limiting the local user adjustment to +/- 2 degrees F from nominal set point will help ensure these savings are achieved. |

**Figure 22.24 Quantifying Control System Savings (*Continued*)**

| MEASURE | MEASURE SAVINGS | OTHER PROCESS OR MEASURE THAT SUBTRACTS FROM SAVINGS | REMARKS |
|---|---|---|---|
| *Air-side economizer* | Use weather data and building load characteristics (thermal break even point) to determine when the building needs cooling indoors and free cooling is simultaneously available.<br><br>Both the number of hours of economizer cooling and the magnitude of displaced cooling load at those hours are required to make this determination. | The actual cooling load will be a mix of envelope loads and internal loads, and will decline at lower temperatures – the slope of the load decline curve will depend on the relative magnitudes of the thermal and internal load constituents.<br><br>For thermally light buildings, where the thermal break even outdoor temperature is higher than 55 or 60 degrees F, savings will be near zero. | For thermally heavy buildings, e.g. those with high internal heat gains that need cooling all the time, the air economizer can be very beneficial since both the hours of benefit and the displaced load (ton-hours) will be high.<br><br>Rules of thumb for savings are climate dependent, and will vary by region.<br><br>In dry climates, this can be estimated with fair accuracy using outdoor dry bulb temperature profiles, but in other regions, the coincident OA humidity or wet bulb profile will need to be over-layed as well to establish cooling loads.<br><br>For thermally heavy buildings cooling benefit is viable whenever OA temp is 55 degF or below, even in humid climates or when raining. |
| *Replace terminal reheat valves with quarter turn 'no leak-by' valves.* | Savings are from hot water use when not needed, over-heating spaces from leakage.<br><br>For heating systems that run in the summer, cooling savings ADD to the hot water savings.<br><br>Where cooling is available at the same time, the cooling load will increase to compensate for the unintended heat load from leaking control valves. | | For low cost light commercial control valves – those with marginal shutoff ratings and characteristic short life spans – leakage is very common.<br><br>For rooms with only heating the classic symptom is overheating in the spaces during winter. In extreme cases, the occupant will control indoor temperature by opening the windows in winter.<br><br>For rooms with cooling too, such as a VAV box with reheat, this may go un-noticed for years.<br><br>In some facilities with large terminal reheat loads (e.g. hospital) this can represent 20% of summer cooling load, and for most systems with conventional globe-style heating valves that are over 10-years old, assuming a loss equal to 10% of summer cooling load is reasonable. |

**Figure 22.24 Quantifying Control System Savings (*Continued*)**

| MEASURE | MEASURE SAVINGS | OTHER PROCESS OR MEASURE THAT SUBTRACTS FROM SAVINGS | REMARKS |
|---|---|---|---|
| *Reduce VAV Supply Air Duct Static Pressure according to Demand* <br><br> *Or* <br><br> *Reduce Variable Flow Hydronic System Differential Pressure according to Demand* | Basis for savings is shown in the remarks column. Load profile is needed to determine pct flow needed in the building, and then compare incremental flows at the two static pressures. <br><br> This calculation uses the difference in downstream duct/pipe pressure (SP1 and SP2), not the total pressure, so the "tsp" or "head" are the difference between the two. <br><br> For summer operation, the extra fan / pump horsepower also creates new cooling load and ADDS to the measure savings. | This approach assumes a constant supply temperature. Anything that raises supply temperatures will increase mass flows (air flow or water flow) and reduce measure savings. | $HP2 = HP1 \times \left(^{SP2}/_{SP1}\right)^{y1.5}$ <br><br> Example: a 50 bhp fan load at 1.3 in.w.c. static pressure is reduced to 1.0 in. w.c. by adjusting the control set point. New motor load = $50 \text{ bhp} \times \left(^{1.0}/_{1.3}\right)^{y1.5} = 33.7$ bhp and savings would be 50-33.7 = 16.3 bhp <br><br> NOTE: For a variety of reasons, the entire theoretical savings of VFD applications do not always come to pass. To avoid savings expectations that aren't met, de-rating the theoretical savings is suggested. The author uses a conservative factor of 0.7 in such estimates. <br><br> Circulating fan and pump motor loads will change throughout the day and by season, so a load profile should be used to quantify savings. Make two identical load profiles in a spreadsheet, using the same volumetric flow rates and different static pressures (in. w.c.) for the existing and proposed systems. <br><br> Fan bhp = $^{cfm * tsp}/_{6356 * eff}$ <br><br> Pump bhp = $^{gpm * head * sp. Grav}/_{3960 * eff}$ |
| *Demand Controlled Ventilation* | Savings are from avoided ventilation due to reduced occupancy. <br><br> Calculate the summer and winter ventilation cost per person, using the OA CFM per person and seasonal weather data profiles. With an occupancy profile, compare costs for constant vs. variable ventilation. | Building exhaust requirements and pressurization requirements will limit how far down the ventilation rate can go, regardless of the number of people. | Not all local building departments allow this. <br><br> Savings are climate dependent. In high humidity areas, ventilation cost is very high. In moderate and dry climates, it is much lower and may only be a cost during winter. |

**Figure 22.24 Quantifying Control System Savings (*Continued*)**

| MEASURE | MEASURE SAVINGS | OTHER PROCESS OR MEASURE THAT SUBTRACTS FROM SAVINGS | REMARKS |
|---|---|---|---|
| *Cost Benefit from Operations and Maintenance Efficiency Improvements* | (DDC) Reduced labor costs from the potential to reduce service calls to individual sites when a condition can be analyzed and a setting can be adjusted remotely, and organizational advantages of a computerized maintenance system use (CMMS).<br><br>Reduced repair costs by avoiding equipment shutdown or damage from early detection (pre-alarms) for out-of-norm conditions. | Difficult to quantify.<br><br>Benchmark data not available | A good control system can be an excellent tool for Operations and Maintenance personnel. Overall efficiency gains, as well as general quality of operations, are benefits of using the DDC system to its potential.<br><br>Parameters / alarms to configure and monitor may include equipment that is off when it should be on, lights that are on when they should be off, prompting t clear user overrides, process temperatures above or below set point indicating stuck valves or other problems, etc. |

**Figure 22.24 Quantifying Control System Savings (*Continued*)**

| MEASURE | MEASURE SAVINGS | OTHER PROCESS OR MEASURE THAT SUBTRACTS FROM SAVINGS | REMARKS |
|---|---|---|---|
| *Cost Benefit from Improved Comfort and Indoor Air Quality* | Limited data. One study conservatively estimated the following productivity losses: [5: pp 4-9]<br><br>Comfort:<br>("Impaired Air and Thermal Quality")<br>1.5%<br><br>Indoor Air Quality:<br>3.5% for an unhealthy building.<br>("Conditions similar to an SBS building,…but with a lower percentage of employees affected.")<br><br>Indoor Air Quality:<br>6% for SBS / BRI<br>[Sick Building Syndrome / Building-Related Illness]<br><br>To identify a dollar value of improvements to comfort or IAQ, us the appropriate percentage productivity increase with the total productivity benchmark data, expressed in $/SF (1995 dollars).<br><br>For example, if a 1% productivity increase is expected for a 25,000 SF office building with a total productivity rate of $97 per SF-yr, the savings (with a 30% de-rate) would be $0.01*97*25,000*0.7 or approximately $17,000 per year. | These can be real savings, but are generally intangible, so can always be argued. | Use this data with caution and understand its limitations to avoid over-stating savings.<br><br>It is suggested that this be used **only** in extreme cases where detriment from poor comfort or poor indoor air quality is a core issue and project driver.<br><br>A de-rate of 30% is suggested, to make sure such claims are conservative. Remember it is better to under-sell and over-deliver than the reverse.<br><br>**Total worker productivity costs in terms of $/SF**<br>**[5: pp 4-8, 4-11]**<br>Values below were derived from the tables in the referenced document, in 1995 dollars.<br><br>$13    **Assembly**<br>$45    **Education**<br>$19    **Food Service**<br>$110   **Healthcare**<br>$19    **Lodging**<br>$23    **Mercantile and Services**<br>$97    **Office** |

**Figure 22.24 Quantifying Control System Savings (*Continued*)**

| MEASURE | MEASURE SAVINGS | OTHER PROCESS OR MEASURE THAT SUBTRACTS FROM SAVINGS | REMARKS |
|---|---|---|---|
| *Cost Benefit from Improved Operator / Technician Training* | Limited data.  One study concluded:<br><br>*'Improved HVAC preventive maintenance resulting directly from technician training has a significant energy savings potential. From a study of nine Community Colleges in California, savings potential estimates ranged from 6% to 19% of total annual campus energy costs, or $0.09-0.26/SF-yr.'* [6]) | These can be real savings, but are generally intangible, so can always be argued. | Use this data with caution and understand its limitations to avoid over-stating savings.  It is suggested to use **only** in extreme cases where detriment from <u>poor maintenance</u> is a core issue and project driver.<br><br>A de-rate of 30% is suggested, to make sure such claims are conservative. Remember it is better to under-sell and over-deliver than the reverse. |
| *Cost Benefit from Calibration* | Tough to estimate.  The question is which direction are the calibrations off?<br><br>Field experience has shown a rule of thumb of 10% energy waste from badly neglected system calibration is reasonable. Note that the overlap could easily be twice as much if the primary heating and cooling systems run continuously. | Worst case is two adjacent sequenced measurements that drift together and overlap.  This can increase cooling load by 20% from false loading, plus the heating energy spent.  Note that these same two measurements could also drift apart, thereby creating a nice dead band.<br><br>Likewise, if space temperatures drift down in winter, indicating they are colder than they really are, the heating system will use more energy to heat the space than designed – this same drift (if it is the same thermostat) will reduce cooling load in summer.<br><br>If one air handler static pressure sensor drifts down, the fan will work harder to achieve a false value.  But the neighboring air handler static sensor may have drifted the other way. | Allowing the instruments to get out of calibration will almost surely cause inefficient operation and higher energy costs, mostly due to heating-cooling overlap or other bucking processes.<br><br>A suggested approach is to calibrate all instruments every two years as a matter of course.  That way system confidence will be enhanced from good faith in the instruments.<br><br>A de-rate of 30% is suggested, to make sure such claims are conservative. Remember it is better to under-sell and over-deliver than the reverse. |

**Figure 22.24 Quantifying Control System Savings (*Concluded*)**

## 22.20 GLOSSARY OF TERMS

**Accuracy:** A performance measurement of an instrument to produce a value equal to the actual value. Regarding instrumentation, a general statement is that accuracy can be adjusted, but repeatability is a function of the instrument.

**Addressable Devices:** Also called *smart* devices. A term used to describe equipment that can be given a unique identifying *address* and controlled directly by a digital control system, often in reference to commodity-type equipment that exists in quantity in a facility. By having unique addresses, multiple devices can share a common communication bus and reduce the volume of wiring.

**Analog:** Variable (input or output), contrasting to discrete.

**Bi-Metal:** A basic temperature transducer formed by joining two metals with different thermal expansion properties. Common shapes are coiled ribbons (that turn upon a temperature change), bars (that bend upon a temperature change), and disks (that warp and 'snap' upon a temperature change). Movement is predictable and repeatable, and is integral to many temperature control instruments.

**Binary:** Synonymous with *on-off* or *discrete* (input or output), contrasting to analog.

**Boolean Logic:** A logical technique used to formulate precise queries using true-false connectors or 'operators' between concepts. The primary operators are AND, OR, and NOT. Words or concepts joined with these operators, and parentheses are used to organize the sequence groups of concepts. The true-false nature of Boolean Logic makes it compatible with binary logic used in digital computers, since a TRUE or FALSE result can be easily represented by a 0 or 1. Named after George Boole.

**Building Automation System (BAS):**

BAS is the term given to a computerized automatic control system when the primary focus: "…is on automating as much as possible to save labor." **[7: pp 38.10]**. Building Automation System (BAS) is currently the most common term used to describe computerized control in buildings that provide one or more of the functions:

> Energy Management System (EMS)
> Facility Management System (FMS)
> Energy Monitoring and Control (EMCS)

**Cascade Control:** Also called Master-Sub Master. This is a combination of two controllers in series with a related measured variable. The output of the first controller becomes an input to the second controller which can then amplify it. Often, the output of the first (master) becomes the set point of the second controller (sub-master). Useful in improving modulating control of systems with very slow process response.

**Cavitation:** A problem phenomenon with fluid flow, often with control valves or pumps, where the local pressure drop is sufficiently high to cause temporary boiling of the fluid. When the pressure is sufficiently regained, the vapor bubbles created by boiling collapse and create a shock. If this occurs within the vicinity of fluid handling apparatus it can damage the equipment.

**Closed Loop:** A control system which includes related process feedback input(s) and controlled device output(s), collectively forming a regulating process. The defining characteristic is the feedback input point that senses changes in the process and the effect of the controlled device, to close the loop of communication.

**Control Loop:** The general term for a collection of control system components used to regulate a process that includes as a minimum a Controller, Set Point, Control Element (Controlled Device), Associated Process, Process Measurement (Measured Variable).

**Controlled Device:** The manipulated device responding to the controller output, that then impacts the process itself. The item being manipulated by the controller.

**Controller:** A device that compares set point to measured values, and determines an appropriate output response.

**Cut In:** A parameter of two-position control. The cut-**in** value is where the controlled variable is sufficiently beyond the set point for the controlled equipment to be turned on (cut-in).

**Cut Out:** A parameter of two-position control. The cut-**out** value is where the controlled variable is sufficiently beyond the set point for the controlled equipment to be turned off (cut-out).

**Dead Band:** Also called zero energy dead band. Refers to a deliberate gap in control span between sequenced, usually conflicting, processes, to avoid control overlap and subsequent operational or energy consumption issues.

Example:

A single temperature controller that sequences both heating and cooling equipment to a common duct or space may use a Dead Band to prevent simultaneous heating and cooling.

**Dead Time:** The time between a change in the process input and when that change is felt in the downstream process measurement. Often a function of the time it takes material to flow from one point to another.

**De-Bouncing:** A type of discrete input signal conditioning used for digital control systems. Mechanical snap-acting contacts actually open and close many times and can trigger control output problems when monitored by a high speed digital circuit. The de-bouncing signal conditioning has, as its purpose, to filter out the bouncing 'noise' so the controller sees a simple open-or-closed state.

**Demand Limiting:** A control strategy with the purpose of reducing electrical demand, not necessarily energy. This is applied to facilities with utility demand charges, and usually is designed to reduce peak (max) demand thereby reducing utility demand charges.

**Derivative Control Mode:** Also called rate or anticipatory control, and is used to increase response time after a disturbance or step change. For a given rate of change of error, there is a unique value of controller output. This control mode reacts to the rate of change of error, not the magnitude of error. This mode cannot be used alone since there is no output when the error is zero.

**Direct Acting:** Control Action that increases its output as the measured variable increases above set point.

**Discrete:** Synonymous with on-off or binary (input or output). Characterized by having two possible states. Typical <u>example</u> is an open-close contact input, or relay output.

**Dry Contact:** A discrete signal, output or input, that is made with a contact closure such as a mechanical relay or switch, that has no voltage at the terminals and depends upon voltage from the initiating circuit to read its two-state resistance. Contrasting, when the device has a voltage at the terminals (as if to light a light bulb when the contacts close) it is termed the wetting voltage.

**Duty Cycling:** A demand-limiting strategy where multiple electric points of use, often motors, which otherwise run continuously, are controlled to run some fraction of time, e.g. 40 minutes on and 20 minutes off. By coordinating the run and off times of multiple items, the aggregate utility electric demand can be reduced, to generate demand savings.

**Energy Monitoring and Control System (EMCS):** See Building Automation System

**Energy Management System (EMS):** EMS is the term given to a computerized automatic control system when the primary focus: "…is on saving energy by specific automatic control programs." **[7: pp 38.10,11].** See also Building Automation System.

**Facility Management System (FMS):** FMS is the term given to a computerized automatic control system when the primary focus: "…goes beyond HVAC controls and/or beyond a single building, such as including fire, security, or manufacturing systems." **[7: pp 38.11].** See also See Building Automation System.

**Feedback:** The measurement of the process output that is returned to the process controller that is also influenced by the controlled device. Feedback is the differentiating factor for a closed-loop control system, contrasted to an open-loop control system.

**Floating Control Mode:** A form of discrete (on-off) control with a null position whereby the controller holds the last output when set point is achieved, rather than returning to zero output.

**Gain:** A tuning constant that multiplies an input or output parameter, usually to increase sensitivity and improve control.

**GUI:** Graphical User Interface. The software that overlays the machine coding, adding the user-friendliness look and feel to a digital control system work station.

**Hunting:** Chronic, repeating oscillation (overshoot and undershoot) of a modulating control system, usually indicating a poorly tuned control loop.

**Hysteresis:** A physical phenomenon best explained as inertia or a body's tendency to stay at rest. In control systems this can affect input and output instruments, and controlled devices (valves and dampers). The hysteresis effect will require a different value of control to cause a change or movement, depending upon whether approaching the value from higher or lower values. Generally, hysteresis is an issue for heavier controlled devices or with minute I/O changes.

**Indicating Transmitter:** A transmitter with an auxiliary display or meter to provide local indication of the measurement and/or output signal.

**Input/Output Points (I/O):** The general term for the identifiable instruments (points) used to relay the control system information into (input points) and out from (output points) the controller, linking the controller to the actual process.

**Integral Control Mode:** Also called reset control, and is used to return the process to zero error by producing an output any time the error is other than zero. Significant is that the output increases over time and is useful to return the process to a zero error state.

**Interlock:** A control strategy that requires one process (discrete or analog) to depend upon the on-off state of a separate but related process.
<u>Examples:</u>
— A boiler firing interlocked with combustion air

damper requires the damper to be proven open prior to firing. This interlock would be for safety and is normally "hard-wired," so as not to depend upon any software intervention.

— A cooling coil control routine interlocked with supply fan operation requires the supply fan to be proven running prior to engaging in control of the cooling coil. This interlock would be for normal control since the feedback measurement is downstream of the cooling coil and would not be sensed without air flow, and is usually a software interlock.

**Ladder Diagram:** Also called an elementary wiring diagram. This is an electrical circuit representation where the high and low voltage terminals (120VAC, 24VAC, 24VDC, etc) are shown as vertical lines on opposite sides of the page. Each circuit sharing this power source, with its control contacts and load, becomes a "rung" on the ladder.

**M&V:** Measurement and Verification

**Master-Sub Master Control:** See "Cascade Control"

**Measured Variable:** The measurement representative of the process that is the basis for controller action, relative to set point.

**Measurement and Verification (M&V):** The after-the-fact activities that are used to verify whether and to what extent energy savings have occurred, to gage the actual energy savings and cost savings benefit of a project. Often, these are compared to estimated savings. In some cases, M&V is part of a contract stipulation for guaranteed savings and may form the basis of payment from one party to another. Depth and rigor of these activities varies depending upon need for accuracy and available funds.

**Minimum Position:** An adjustment parameter of controlled devices, referring to the lowest level of control action allowed, regardless of further reduction in controller signal.
Examples:
Minimum fan speed
Minimum damper position

**Modulating:** See also "Analog." Varying position in response to need, contrasting to two-position.

**Most-open Valve:** A control algorithm that senses damper or valve positions at the various points of use to determine demand. This strategy serves the most demanding area, but strives to provide just enough process media to satisfy it, thereby reducing overall system energy use. Similar control routine applies to "most open damper."

**Open Loop Control:** The controlled system has no impact on the measured variable. No amount of

change in the controlled variable or in the output of the controller will cause a change in the measured variable.

**Optimization:** Actively monitoring and controlling each of the pertinent parameters of a process to maximize productivity and/or to minimize energy use. Often requires control of multiple dynamic variables. Optimization can apply to single equipment items or large systems.

**Oscillation:** Repeated over and over-shooting about the desired set point. For two-position control, this is normal. For modulating control, this is abnormal. See "Hunting."

**Overshoot:** A measure of control system response to changing loads. After a corrective action is taken, and before settling out at the new level, the controlled variable is usually driven beyond the set point as it tries to recover. The extent to which it temporarily goes beyond the mark is the overshoot.

**Point - Controls context:** Usually used with Digital Control systems, each unique identifiable input or output item. The number of system points represents the size of the system, e.g. the number of connected instruments that can be individually controlled. Most commonly used in reference to hardware items, but also applies to software points.

**Proportional Band:** The range of error (+/-) that will cause the proportional controller output to vary between 0-100%. As the proportional gain increases, the proportional band decreases.

**Proportional Control Mode:** Produces a linear proportional output that is a direct response to the measured error, configured to counteract the error and provide basic regulation. A residual offset error is characteristic of this control mode, e.g. a true zero-error state can never be achieved using proportional-only control.

**Proportional Offset:** The amount of residual error remaining in the process using proportional-only control.

**Pulse Width Modulation:** An output signal conditioning strategy that converts a modulating 0-100% output to a percentage of on time output, to impose the proportional control onto a discrete device. Often used in the control of electric resistance heating or incremental control electric motor actuators.

**Real Time:** (general context): A computer system that responds to inputs without delay. (Computer science context): A computer system that updates information at the same rate it receives information.

**Relay:** A discrete output interface device, used to control a large current (via heavy contacts) with a small current (the relay coil current) and/or to isolate controller low voltage power (DC) from the controlled equipment.

**Relay Logic:** The term given to discrete logical control (if-then, and, or) accomplished by arranging relay contact in series and/or parallel electrical circuits.

**Repeatability:** A performance parameter that measures the ability of a process or instrument to produce the same value of output during repeated trials. Instrument repeatability is synonymous with instrument precision.

**Reset or Reset Schedule:** A control algorithm whereby one set point is varied, usually linearly, between two values, as a function of another analog value. Example: Resetting hot water temperature set point (HW) from outside air temperature (OA):

Reset Schedule

| OA | HW |
|----|-----|
| 70 | 120 |
| 30 | 180 |

**Resolution:** The minimum measurable value of an input variable or the minimum incremental change of the output variable. In digital control systems, this is often a measurement criterion of the A/D or D/A transformation, e.g. the number of steps provided and how closely it emulates true analog control.

**Reverse Acting:** Control Action that decreases its output as the measured variable increases above set point.

**RTD:** Resistance Temperature Detector: A transducer that leverages the physical properties of certain materials that predictably change resistance as a function of temperature. RTD response is characterized as being nearly linear and stable over time.

**SCADA:** Supervisory Control and Data Acquisition. These are industrial control systems with very large point counts, lots going on, often widespread and remote, and often critical in nature. Thousands of point systems are common. Example: Utility energy and water metering and control.

**SCR:** Silicon Controlled Rectifier. A device used to proportionally control an electric load, often a resistance heater or motor. The rectifier is an electronic switch so is technically an on-off device. However, SCR's are normally controlled at very high speeds in time-proportioning fashion to create a modulating effect.

**Sensor:** The general name for a device whose purpose is to sense some media and translate the measurement into a convenient and predictable form, often electrical, pneumatic.

**Set Point:** The desired steady state of a controlled process. The goal of the control activity.

**Settling Time:** "…the time required to for the process-control loop to bring the dynamic variable back to within the allowable range,…" [3: pp 10]

**Short Cycling:** A dysfunction of some two-position control applications where the cut-in and cut-out settings are too narrow or the process rate is to fast, with the result being in rapid cycling of equipment, often to the detriment of the equipment.

**Soft Start:** Any of several methods or reducing start-up demand upon a motor, usually by reduced voltage or mechanically unloading the driven load.

**Step Change:** Also known as a disturbance or bump, this describes a sudden and significant event process output change. In the context of control loop tuning, a step change is a valuable testing function that will demonstrate the ability of the tuned components to react properly to sudden process changes without undue loss of control, excess drift from set point, hunting, or other control anomalies.

**System Capacitance:** A general term that describes the relative strength or capacity of the controlled equipment to effect a change to the process measured variable, and a good indicator of system controllability and control mode choice. In lay terms, a low system capacitance acts like the equipment is over-sized with the tendency to short-cycle as a result. In technical terms this is measured by the pct process change resulting from a step change in output. Systems with high capacitance can usually be controlled with any control mode. Systems with low capacitance are often troublesome. Systems with low capacitance may experience short-cycling and instability using on-off control unless excessively wide range is used. Modulated systems with low system capacitance may require careful tuning and complex proportional control (PI, PID) to maintain stable control.

**Thermistor:** A semi-conductor device used to measure temperature. Lightweight and inexpensive, thermistors are popular for low cost or non-critical applications such as residential or light commercial temperature control, where minor errors and long term drift are acceptable. Thermistor outputs are not linear with respect to changes in temperature, and usually require some form of signal conditioning to avoid errors.

**Thermocouple:** A basic transducer for temperature measurement that uses the physical principal of galvanic action between dissimilar materials. The galvanic response is a function of temperature and so, with proper signal conditioning, can be used to measure temperature.

**Throttling Range:** "...the amount of change in the controlled variable that causes the controlled device to move from one extreme to the other, from full-open to full-closed." **[2: pp 1:21]**

**Transducer:** A device that performs the initial input or output conversion of a dynamic variable into proportional electrical or pneumatic information, often a very low level signal requiring further signal conditioning.

Examples include: Thermocouple, thermistor, RTD, variable capacitance dP cell, strain gage vibration transducer, flow element (orifice plate, etc)

**Transmitter:** The general name for an input device that provides signal conditioning from a transducer or other basic measurement signal, transforming it into some proportional information in a useful form, often with the ability to send the information over long distance without loss of accuracy (transmit).

**Tuning:** Adjustment of gains and various parameters of a dynamic control system to achieve acceptable and stable control.

**Two-position Control Mode:** Also called On-Off control, this is the most elementary control mode. Output is either 100% if sufficiently below set point, or 0% if sufficiently above set point. Over and undershoot is normal for this control mode. The least costly of all control methods. Regarding two position control: "Generally, the two-position control mode is best adapted to large scale systems with relatively slow process rates." **[3: pp 290]**.

**Wax Motor:** A self-powered actuator whose motive force is thermal expansion of a wax pellet/bellows assembly. The expansion is translated into a linear movement to vary the position of another device. Common uses include small thermally compensated devices such as shower mixing valves, and thermally compensated air diffusers.

## Sources

1.  *Optimization of Unit Operations*, Bela Liptak, pub. Chilton Book Company, 1987.
2.  "Fundamentals of HVAC Control Systems," Steven Taylor, PE, ASHRAE, Atlanta Georgia, 2004.
3.  *Process Control Instrumentation Technology*, second ed, Curtis D. Johnson, pub. John Wiley & Sons, 1982.
4.  "Hierarchy of HVAC Design Needs," David Schwaller, PE, *ASHRAE Journal* August 2003.
5.  "Productivity Benefits Due to Improved Indoor Air Quality," NEMI, National Energy Management Institute, August 1995.
6.  "Quantifying The Energy Benefits Of HVAC Maintenance Training and Preventive Maintenance," AEE, *Energy Engineering*; Vol. 96; Issue 2; 1999.
7.  "Computer Applications," *1999 ASHRAE Applications Handbook*, ASHRAE, Atlanta Georgia, 1999.

# CHAPTER 23

# ENERGY SECURITY AND RELIABILITY

**BRADLEY L. BRACHER, P.E., C.E.M.**
**Great Plains Energy Consulting, Inc.**
Oklahoma City, OK

## 23.1 INTRODUCTION

Reliable utility services are vital to all industrial, commercial and military installations. Loss of electricity, thermal fuels, water, environmental control, or communications systems can bring many operations to an immediate halt resulting in significant economic loss due to unscheduled downtime, loss of life, or threat to national security. These services are delivered by vast, complex networks with many components. A small number of damaged components is often sufficient to disable portions of these networks or halt operation of the entire system. These component failures can be caused by equipment failure, natural disaster, accidents, and sabotage.

The need for security continues to grow in all areas. It has become increasingly apparent in recent years that law enforcement agencies cannot provide the needed resources and personnel to protect citizens, corporations, and private property. Theft and vandalism are ever present. Terrorism, shootings, and bombings continue at a level pace and increase in intensity. The transition from an industrial economy to an information economy is bound to be accompanied by political, social, and economic turmoil. Law enforcement does what it can to detect and prevent crime in advance but will of necessity be forever relegated to a primary role of investigating after the fact. Responsibility for security now, as it always has, falls upon individuals and private corporations to secure their own well-being.

Many companies raised their awareness of energy security issues while preparing for anticipated problems associated with Year 2000 (Y2K) computer problems. Managers feared widespread utility outages initiated by computer malfunctions. Fortunately, business and industry took this threat seriously and acted in advance to prevent major problems. Not knowing the extent of problems that might occur on January 1, 2000, many facility managers critically examined the utility supply systems they rely on for the first time, installed back-up systems,

and developed contingency plans. With Y2K behind them without major incident, many managers have now directed their attention elsewhere. However, the reliability of utility systems is not a dead issue.

Many analysts are watching the pending deregulation of the electric utility industry to see how that will affect system reliability. Some utilities have curtailed routine maintenance in anticipation of mandatory divestiture of assets. Work like tree trimming, system expansion, and replacement of aging equipment is deferred. This permits an increase of current profit and reduces future financial risk that could occur if regulators do not permit utilities to fully recover stranded system costs when they unbundle services. Wise managers are anticipating the risks now and preparing to take action to improve the energy security of their facilities. Only time will reveal the true extent of these threats.

Energy security is the process of assessing the risk of loss from unscheduled utility outages and developing cost effective solutions to mitigate or minimize that risk. This chapter will explore the vulnerable network nature of utility systems and the events or threats that disrupt utility services. Methods will be presented to assess these threats and identify those most likely to result in serious problems. Actions to counter these threats will be introduced along with a methodology to evaluate the cost effectiveness of proposed actions. Finally, the links between energy security and energy management will be discussed.

### 23.1.1 Principles of Security

General physical security involves four areas of concern: deterrence, delay, detection, and intervention. Deterrence is a way of making a potential target unattractive to those wishing to engage in mischief or mayhem. Properly implemented deterrence leads the would be attacker to believe their actions would not result in the desired outcome, require a level of effort not commensurate with the objective, or entail a high likelihood of capture. Examples of physical security include area lighting, physical barriers, visible intrusion detection systems, and the presence of guards. It is not cost effective to post guards at most utility installations like substations due to the large number of sites. Roving security patrols on an unpredictable schedule can be effective.

Delay mechanisms increase the amount of time and effort required to accomplish an unauthorized entry and execute a criminal task. The increased effort requires additional planning and manpower on the part of the criminal and thus serves as a deterrent. Examples include fences, barbed wire, razor wire, secure doors and windows, locks, and channels that restrict the flow of people and vehicles. A delay mechanism must add enough time to the criminal's task to permit security forces to arrive and intercept the intruder and is, therefore, most effective when combined with detection systems.

Detection systems are intrusion alarms and video monitoring in both the visible and infrared spectrums. These systems alert security forces to the presence of intruders and allow them to respond in a manner that brings the intrusion to an end with a minimum of loss or damage. The time required to penetrate the facility, carry out the theft or assault, and exit the facility must be greater than the longest probable response time of security forces. The response time dictates the design and selection of delay mechanisms.

Intervention is the final line of defense in physical security. It consists primarily of guards or security forces. Intruders must ultimately be confronted face-to-face if aggression is to be halted. Their physical presence serves as a deterrent. They can be permanently placed for critical facilities. They can be deployed on a full time basis to less critical facilities when intelligence information indicates a high threat environment. Roving patrols can be used for multiple facilities of a less critical nature. Security forces should be able to respond in mass to a detected intrusion to abort crimes in progress.

All of these physical security fundamentals can be used to reduce the vulnerability of utility systems. Additional countermeasures specific to utility systems should be implemented. Actions such as improving component reliability, installing redundant systems, preparing for rapid recovery, and contingency planning are discussed at length later.

### 23.1.2 Utility Systems As Interdependent Networks

Modern industrial, institutional, and commercial facilities are dependent on many utility systems. The utility services supporting modern facilities include electricity, thermal fuels (natural gas, fuel oil, coal), water, steam, chilled water, compressed air, sanitary sewer, industrial waste, communication systems, and transportation systems. Some of these systems are restricted to the site. Others are a maze of distribution paths and processing stations connecting the site to distant generation or production facilities. The individual components of these systems form networks.

Any failed component in a network has the potential to disrupt or degrade the performance of the entire system. Some networks contain only a single path linking the site to the source of its utility supply. The loss of that single path results in total disruption at the site. Other networks have redundant paths. Alternate routes are available to serve the site at full or partial capacity should one of the routes be damaged. In a redundant system, loads normally carried by the damage portion of the system will be shifted to an operational part of the network. This places extra stress on the remaining portion of the system. Marginal components sometimes fail under this additional load, causing additional segments of the network to fail.

Utility networks do not operate in isolation; they are linked to each other in a web of dependence. Each provides a vital service to the other. The impact of one failed utility network is felt in many other systems. The effects spread like ripples on a pond. More systems will fail if redundant or back-up systems are not in place. Often, the easiest way for a saboteur to disable a utility system is to damage another system on which it depends.

**Example.** Figure 23.1 illustrates the mutual dependence utility systems share with each other at one facility. This facility generates some of its own electricity with natural gas engines, pumps a portion of its water from wells using electric pumps, utilizes steam absorption chillers, and produces compressed air from a mixture of electric and gas engine compressors. A disruption of natural gas will immediately halt the on-site generation of electricity and steam and curtail the production of compressed air since these systems depend directly on natural gas as their prime mover. Failure of one utility will cascade to other systems. The absorption chillers would also experience a forced outage since they are powered by steam which is experiencing a forced outage due to the lack of natural gas. All process equipment that requires any of these disrupted utilities will be idle until full utility service is restored.

### 23.1.3 Threats to Utility Systems

The individual components of a utility system can be damaged or disabled in many ways. The network itself may be wholly or partially disrupted depending on the criticality of the component or components damaged. The threats to network components can be segregated into four categories: equipment failure, natural disaster, accidents, and sabotage.

Equipment failure is the normal loss of system components as they reach the end of their expected life. The reliability of most electrical and mechanical components

Figure 23.1 Interdependence of utility networks.

follows a "bath-tub curve" illustrated in Figures 23.2 and 23.3. These curves begin with a high rate of failure early in the life cycle. This failure rate rapidly decreases until it reaches a minimum value that remains relatively stable throughout the normal operating period. The failure rate increases again once the component enters the wear-out phase.

Natural disasters are the most common cause of widespread network failure. They include earthquake, hurricane, tornado, wind, lightning, fire, flood, ice, and animal damage. Utility companies are well versed in dealing with these situations and generally have the means to effect repairs rapidly. However, widespread damage can leave some customers without service for days resulting in substantial economic loss.

Accidents are unintentional human actions such as traffic accidents, operator error, fires, and improper design or modification of the system. Operator error played a significant role in the failure of the Chernoble and Three Mile Island nuclear plants. It also contributed to both of the New York City blackouts. Once a failure sequence has begun, the system is in an abnormal state. Operators may lack the experience to know how the system will respond to a given corrective action or might fail to respond quickly enough to a dynamic and rapidly changing situation.

Sabotage covers the realm of intentional human action. It includes the entire spectrum from a single disgruntled employee to acts of terrorism to military action. Terrorism has been a fact of life in many countries for a long time. Its use in the United States has increased great-

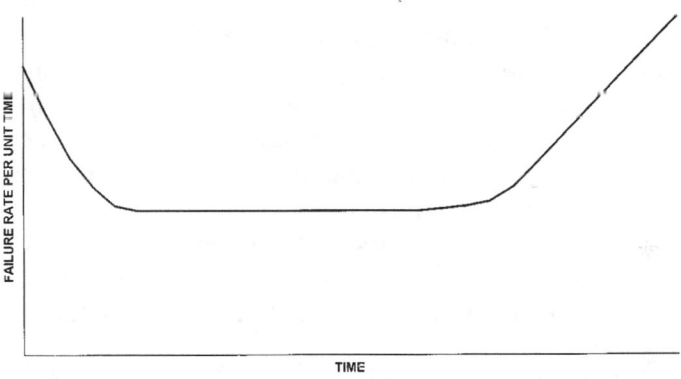

Figure 23.2 Typical "bath-tub curve" for electronic equipment showing failure rate per unit time.

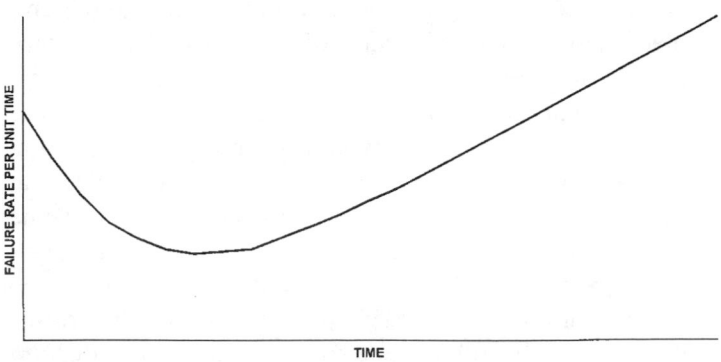

Figure 23.3 Typical "bath-tub curve" for mechanical equipment showing failure rate per unit time.

ly in recent years. Most terrorist organizations have social or political motivations that could lead them to target industrial or commercial operations. Military actions are beyond the scope of what most industrial or commercial facilities can handle. These types of threats are of interest only to the military itself.

## 23.2 RISK ANALYSIS METHODS

Risk analysis is a necessary first step in the process of minimizing the losses associated with unscheduled utility outages. The purpose of risk analysis is to understand the system under study and establish a knowledge base from which resources can be optimally allocated to counter known threats. This understanding comes from a systematic evaluation that collects and organizes information about the failure modes of a utility system. Once these failure modes are understood decisions can be made that reduce the likelihood of a failure, facilitate operation under duress conditions, and facilitate rapid restoration to normal operating conditions.

Two broad categories of risk analysis methods are available. The inductive methods make assumptions about the state of specific system components or some initiating event and then determine the impact on the entire system. They can be used to examine a system to any level of detail desired, but are generally only used to provide an overview. It is impractical and often unnecessary to examine every possible failure or combination of failures in a system. When the complexity or importance of a system merits more detailed analysis a deductive method is used.

Deductive analysis makes an assumption about the condition of the entire system and then determines the state of specific components that lead to the assumed condition. Fault tree analysis is the most common and most useful deductive technique. The deductive method is preferred because it imposes a framework of order and objectivity in place of what is often a subjective and haphazard process.

Probabilities can be incorporated into all of these methods to estimate the overall reliability of the system. Probabilistic techniques are best used when analyzing equipment failures but have also been used with some success in the evaluation of human error or accident. They are of less value when evaluating natural disasters and meaningless when applied to sabotage or other forms of organized hostility. The results would not only depend on the probability of a sabotage attempt occurring but also on the probability that all actions necessary to disable the system were successfully accomplished during the

attack. This type of data is extremely difficult to obtain and is highly speculative. Therefore, the analysis is best conducted under the assumption that an adverse situation will definitely occur and proceed to determine what specific scenarios will result in system failure[1].

### 23.2.1 Inductive Methods

A number of analysis tools are available to examine the effects of single component failures of a system. Three common and well developed techniques are Failure Mode and Effect Analysis (FMEA), Failure Mode Effect and Criticality Analysis (FMECA) and Fault Hazard Analysis (FHA). These methods are very similar and build upon the previous technique by gradually increasing in scope. Most utilize failure probabilities of individual components to estimate the reliability of the entire system. Preprinted forms are generally used to help collect and organize the information.

FMEA recognizes that components can fail in more than one way. All of the failure modes for each component are listed along with the probability of failure. These failure modes are then sorted into critical and non-critical failures. The non-critical failures are typically ignored in the name of economy. Any failure modes with unknown consequences should be considered critical. Figure 23.4 is a typical data sheet used in FMEA.

Failure Mode Effect and Criticality Analysis (FMECA) is very similar to FMEA except that the criticality of the failure is analyzed in greater detail and assurances and controls are described for limiting the likelihood of such failures[2]. FMECA has four steps. The first identifies the faulted conditions. The second step explores the potential effects of the fault. Next, existing corrective actions or countermeasures are listed that minimize the risk of failure or mitigate the impact of a failure. Finally, the situation is evaluated to determine if adequate precautions have been made and if not to identify additional action items. This technique is of particular value when working with the system operators. It can lead to an excellent understanding of how a system is actually operated as opposed to how the designer intended for it to be operated. Figure 23.5 is a data sheet used with FMECA.

Fault Hazard Analysis (FHA) is another permutation of Failure Mode and Effect Analysis (FMEA). Its value lies in an ability to detect faults that cross organizational boundaries. Figure 23.6 is a data collection form for use with FHA. It is the same data form used with FMEA with three addition columns. Faults in column five are traced up to an organizational boundary. Column six is added to list upstream components that could cause the fault in question. Factors that cause secondary failures are listed in column seven. These are things like operational conditions or environmental variables known to affect the component. A remarks column is generally included to summarize the situation. FHA is an excellent starting point if more detailed examination is anticipated since the data are collected in a format that is readily used in Fault Tree Analysis.

These techniques concern themselves with the effects of single failures. Systems with single point failures tend to be highly vulnerable if special precautions are not taken and these methods will highlight those vital components. However, critical utility systems are normally designed to be redundant. The majority of forced outages in a redundant system will be caused by the simultaneous failure of multiple components. Techniques that only consider single point failures will significantly underestimate the vulnerability of a system. No matter how unlikely an event or combination of events may seem, experience has proven that improbable events do occur.

The Double Failure Matrix (DFM) examines the effects of two simultaneous failures. A square grid is laid out with every failure mode of every component listed along the columns and also along the rows. The intersec-

tion of any two failure modes in the matrix represents a double failure in the system. The criticality of that double failure is listed at the intersection. Critical and catastrophic failures are explored further to identify corrective actions or alternative designs. The diagonal along the grid is the intersection of a component failure with itself. This is the set of single mode failures. It can be used as a first cut analysis and later expanded to encompass the entire set of two component failures if desired. This method becomes quite cumbersome or prohibitively difficult with large or complex systems.

Event Tree Analysis is an exhaustive methodology that considers every possible combination of failed components. It is known as a tree because the pictorial illustration branches like a tree every time another component is included. The tree begins with a fully operational system that represents the trunk of the tree. The first component is added and the tree branches in two directions. One branch represents the normal operational state of the component and the other represents the failed state. The second component is considered next. A branch representing the operational and failed states is

| FAILURE MODE AND EFFECTS ANALYSIS | | | | | |
|---|---|---|---|---|---|
| COMPONENT | FAILURE PROBABILITY | FAILURE MODE | % FAILURES BY MODE | EFFECTS | |
| | | | | CRITICAL | NON-CRITICAL |
| | | | | | |
| | | | | | |
| | | | | | |
| | | | | | |
| | | | | | |
| | | | | | |
| | | | | | |
| | | | | | |
| | | | | | |
| | | | | | |
| | | | | | |
| | | | | | |
| | | | | | |

Figure 23.4 Data collection sheet for Failure Mode and Effect Analysis (FMEA).

| FAILURE MODE EFFECT AND CRITICALITY ANALYSIS | | | |
|---|---|---|---|
| FAULT IDENTIFICATION | POTENTIAL EFFECTS OF FAULT | COMPENSATION OR CONTROL | ACTION REQUIRED |
| | | | |
| | | | |
| | | | |
| | | | |
| | | | |
| | | | |
| | | | |
| | | | |
| | | | |
| | | | |
| | | | |
| | | | |
| | | | |

Figure 23.5 Data collection sheet for Failure Mode Effect and Criticality Analysis (FMECA).

## FAULT HAZARD ANALYSIS

| COMPONENT | FAILURE PROBABILITY | FAILURE MODE | % FAILURES BY MODE | EFFECTS | | UPSTREAM COMPONENTS THAT INITIATE FAULT | FACTORS THAT CAUSE SECONDARY FAILURES | COMMENTS |
|---|---|---|---|---|---|---|---|---|
| | | | | CRITICAL | NON-CRITICAL | | | |
| | | | | | | | | |
| | | | | | | | | |
| | | | | | | | | |
| | | | | | | | | |
| | | | | | | | | |
| | | | | | | | | |
| | | | | | | | | |
| | | | | | | | | |
| | | | | | | | | |
| | | | | | | | | |
| | | | | | | | | |
| | | | | | | | | |

**Figure 23.6 Data collection sheet for Failure Hazard Analysis (FHA).**

added to each of the existing branches, resulting in a total of four branches. Additional components are added in a like manner until the entire system has been included. The paths from the trunk to the tip of each branch are then evaluated to determine the state of the entire system for every combination of failed components. Complex systems can be in a state of total success, total failure, or some variant of partial success or failure.

**Example.** Figure 23.7 is a simplified one-line diagram of an electric distribution system for a facility with critical loads. The main switch gear at the facility is a double-ended substation fed from two different commercial power sources. All critical loads are isolated on a single buss which can receive power from either commercial source or an on-site emergency generator. The facility engineer conducted a Failure Mode Effect and Criticality Analysis. His findings are listed in Figure 23.8. Three single point failures were identified that result in a forced outage of the critical loads if any one of these components fails: automatic transfer switch, main breaker at the critical load switch

gear, and the buss in the critical load switch gear. Action items were listed for all components to improve system reliability. Most of the breakers require on-site spares for maximum reliability, however, a number of these can be shared to minimize the expense. For example, the main breakers at switch gear "A" and "B" can share a spare since they are of the same size. The most important action items are those for the single point failures.

FMECA identifies only single point failures and can significantly underestimate the risk of a forced outage since improbable events like double failures do occur. A Double Failure Matrix (DFM) was constructed in Figure 23.9 to identify the combinations of two failed components that would deprive the critical loads of electricity. Rules were defined to gauge the criticality of each double failure. Loss of any one of the three sources of electricity is a level 2 failure with marginal consequences. Loss of any two of the three sources is a level 3 failure with critical consequences and loss of all three sources is a level 4 failure with catastrophic consequences since all critical

Figure 23.7 **Simplified electric distribution system for example.**

loads are in an unscheduled forced outage. The upper portion of the matrix is left blank in this example since it is a mirror image of the lower portion.

The highlighted diagonal is the intersection of a component with itself and represents the single point failures. The results of the FMECA are verified since the three components identified as single point failures are now shown to be level 4 failures along the diagonal. Since these components are single point failures their rows and columns are filled with 4's denoting a catastrophic loss of power when they and any other component are simultaneously failed. Six additional level 4 failures appear on the chart that represent true double component failures. All of these involve the inability to transfer commercial power to the critical load buss and a simultaneous failure of the emergency generator or its support equipment. This could lead the facility manager to investigate alternate systems configurations that have more than one way of transferring commercial power to the critical loads. Additionally, the threats that can reasonably be expected to damage the components involved in level 4 failures should be carefully explored and countermeasures implemented to reduce the

likelihood of damage. The level 3 failures that occur along the diagonal also merit special attention.

**23.2.2 Deductive Method**

Fault Tree Analysis is the most useful of the deductive methods and is preferred by the author above all the inductive methods. Benefits include an understanding of all system failure modes, identification of the most critical components in a complex network, and the ability to objectively compare alternate system configurations. Fault trees use a logic that is essentially the reverse of that used in event trees. In this method a particular failure condition is considered and a logic tree is constructed that identifies the various combinations and sequence of other failures that lead to the failure being considered. This method is frequently used as a qualitative evaluation method in order to assist the designer, planner or operator in deciding how a system may fail and what remedies may be used to overcome the cause of failure[3].

The fault tree is a graphical model of the various combinations of faults that will result in a predefined undesired condition. Examples of this undesired condi-

## FAILURE MODE EFFECT AND CRITICALITY ANALYSIS

| FAULT IDENTIFICATION | POTENTIAL EFFECTS OF FAULT | COMPENSATION OR CONTROL | ACTION REQUIRED |
|---|---|---|---|
| Commercial Power "A" | No commercial power from source "A" | Close Tie and feed from Comm Power "B" | Investigate security & reliability of system supplying Comm Power "A" |
| Main Breaker Swgr "A" | No commercial power from source "A" | Close Tie and feed from Comm Power "B" | Conduct regular inspections & preventive maintenance & keep spare on-site |
| Buss Swgr "A" | No commercial power to critical loads or non-critical loads on Swgr "A" | ATS transfers critical loads to Emergency Generator Repair or replace buss | Conduct regular inspections & preventive maintenance Prevent animal & water damage |
| Commercial Power "B" | No commercial power from source "B" | Critical loads fed by preferred comm power source "A" | Investigate security & reliability of system supplying Comm Power "B" |
| Main Breaker Swgr "B" | No commercial power from source "B" | Critical loads fed by preferred comm power source "A" | Conduct regular inspections & preventive maintenance & keep spare on-site |
| Buss Swgr "B" | No commercial power from source "B" to crit loads & No power to non-crit loads "B" | Critical loads fed by preferred comm power source "A" Repair or replace Buss | Conduct regular inspections & preventive maintenance Prevent animal & water damage |
| Tie Between Swgr "A" and Swgr "B" | Reduced reliability Additional failures could lead to forced outage | Repair or replace tie switch | Conduct regular inspections & preventive maintenance |
| Critical Load Breaker Swgr "A" | No commercial power to critical loads | ATS transfers critical loads to Emergency Generator | Conduct regular inspections & preventive maintenance & keep spare on-site |
| Automatic Transfer Switch | Forced outage - No power to critical loads | Repair or replace ATS | Conduct regular inspections & preventive maintenance & test ATS monthly |
| Main Breaker Critical Load Swgr | Forced outage - No power to critical loads | Repair or replace breaker | Conduct regular inspections & preventive maintenance & keep spare on site |
| Buss Critical Load Swgr | Forced outage - No power to critical loads | Repair or replace buss | Conduct regular inspections & preventive maintenance Prevent animal & water damage |
| Emergency Generator | Reduced reliability - No on-site back-up to commercial power | Critical loads fed by preferred comm power source "A" - Repair gen. | Conduct regular inspections & preventive maintenance & test generator monthly |
| Breaker Emergency Generator | Reduced reliability - No on-site back-up to commercial power | Critical loads fed by preferred comm power source "A" Repair or replace breaker | Conduct regular inspections & preventive maintenance & keep spare on-site |
| Control System Emergency Generator | Reduced reliability - No on-site back-up to commercial power | Critical loads fed by preferred comm power source "A" - repair controls | Conduct regular inspections & preventive maintenance & test monthly |

Figure 23.8 Failure Mode Effect and Criticality Analysis (FMECA) for example.

## DOUBLE FAILURE MATRIX

KEY:
1 Negligible
2 Marginal
3 Critical
4 Catastrophic

| | Commercial Power "A" | Main Breaker Swgr "A" | Buss Swgr "A" | Commercial Power "B" | Main Breaker Swgr "B" | Buss Swgr "B" | Tie Between Swgr "A" and Swgr "B" | Critical Load Breaker Swgr "A" | Automatic Transfer Switch | Main Breaker Critical Load Swgr | Buss Critical Load Swgr | Emergency Generator | Breaker Emergency Generator | Control System Emergency Generator |
|---|---|---|---|---|---|---|---|---|---|---|---|---|---|---|
| Commercial Power "A" | 2 | | | | | | | | | | | | | |
| Main Breaker Swgr "A" | 2 | 2 | | | | | | | | | | | | |
| Buss Swgr "A" | 3 | 3 | 3 | | | | | | | | | | | |
| Commercial Power "B" | 3 | 3 | 3 | 2 | | | | | | | | | | |
| Main Breaker Swgr "B" | 3 | 3 | 3 | 2 | 2 | | | | | | | | | |
| Buss Swgr "B" | 3 | 3 | 3 | 2 | 2 | 2 | | | | | | | | |
| Tie Between Swgr "A" and Swgr "B" | 3 | 3 | 3 | 2 | 2 | 2 | 2 | | | | | | | |
| Critical Load Breaker Swgr "A" | 3 | 3 | 3 | 3 | 3 | 3 | 3 | 3 | | | | | | |
| Automatic Transfer Switch | 4 | 4 | 4 | 4 | 4 | 4 | 4 | 4 | 4 | | | | | |
| Main Breaker Critical Load Swgr | 4 | 4 | 4 | 4 | 4 | 4 | 4 | 4 | 4 | 4 | | | | |
| Buss Critical Load Swgr | 4 | 4 | 4 | 4 | 4 | 4 | 4 | 4 | 4 | 4 | 4 | | | |
| Emergency Generator | 3 | 3 | 4 | 3 | 3 | 3 | 3 | 4 | 4 | 4 | 4 | 2 | | |
| Breaker Emergency Generator | 3 | 3 | 4 | 3 | 3 | 3 | 3 | 4 | 4 | 4 | 4 | 2 | 2 | |
| Control System Emergency Generator | 3 | 3 | 4 | 3 | 3 | 3 | 3 | 4 | 4 | 4 | 4 | 2 | 2 | 2 |

**Figure 23.9 Double Failure Matrix (DFM) for electric distribution system in example.**

tion are:
a) Total loss of electricity to surgical suite.
b) Total loss of chilled water to computer facility.
c) Steam boiler unable to generate steam.
d) Loss of natural gas feedstock to fertilizer plant.
e) Water supply to major metropolitan area curtailed to half of minimum requirement.
f) Environmental conditioning of controlled experiment is interrupted for more than thirty minutes.

The faults can be initiated by sabotage actions, software failures, component hardware failures, human errors, or other pertinent events. The relationship between these events is depicted with logic gates.

The most important event and logic symbols are shown in Figure 23.10. A basic event is an initiating fault that requires no further development or explanation. The basic event is normally associated with a specific component or subsystem failure. The undeveloped event is a failure that is not considered in further detail because it is not significant or sufficient information is not available.

Logic gates are used to depict the relationship between two or more events and some higher level failure of the system. This higher level failure is known as the output of the gate. These higher level failures are combined using logic gates until they culminate in the top event of the tree, which is the previously defined undesired event. A simple fault tree is illustrated in Figure 23.11.

The "OR" gate shows that the higher level failure will occur if at least one of the input events occurs. An "OR" gate could be used to model two circuit breakers in series on a radial underground feeder. If either circuit breaker is opened the circuit path will be broken and all loads served by that feeder will be deprived of electricity.

The output event of an "AND" gate occurs only if all of the input events occur simultaneously. It would be used to model two redundant pumps in parallel. If one pump fails the other will continue to circulate fluid through the system and no higher order failure will occur. If both pumps fail at the same time for any reason the entire pumping subsystem fails.

The logic of the fault tree is analyzed using boolean algebra to identify the minimal cut sets. A minimal cut set is a collection of system components which, when failed, cause the failure of the system. The system is not in a failed state if any one of the components in this set has not failed or is restored to operation. In fault tree terminology, a cut set is a combination of basic events that will result in the undesired top event of the tree. A computer is normally used to automate this tedious and error prone mathematical procedure.

Cut sets are utilized because they directly correspond to the modes of system failure. In a simple case, the cut sets do not provide any insights that are not already quite obvious. In more complex systems, where the system failure modes are not so obvious, the minimal cut set computation provides the analyst with a thorough and systematic method to identify the combinations of component failures which culminate in the top event. Once an exhaustive list of cut sets is assembled, they can be analyzed to determine which components occur in failure modes with the highest frequency. These, along with the single point failures, are the most critical components of the entire system and merit special attention to keep them out of harm's way.

## 23.3  COUNTERMEASURES

Corrective actions, or countermeasures, are implemented to reduce the risk of an unscheduled outage. These measures should initially be focused on single component failures and those with the most catastrophic consequences when failed. The second priority are those components that occur in the largest number of cut sets. Countermeasures fall into three broad categories: protective measures, redundant systems, and rapid recovery.

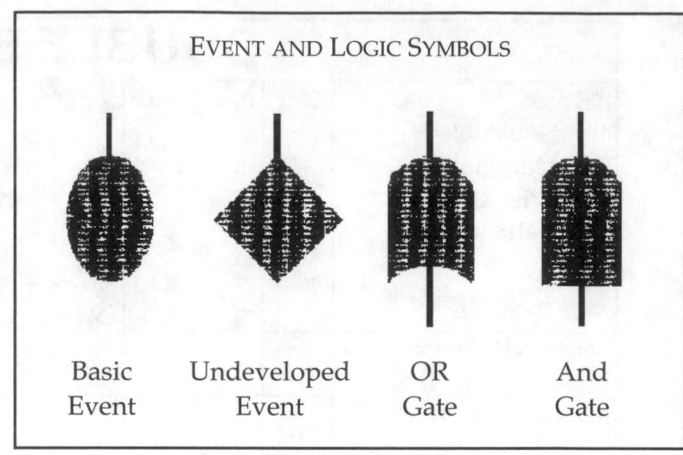

**Figure 23.10 Event and logic symbols**

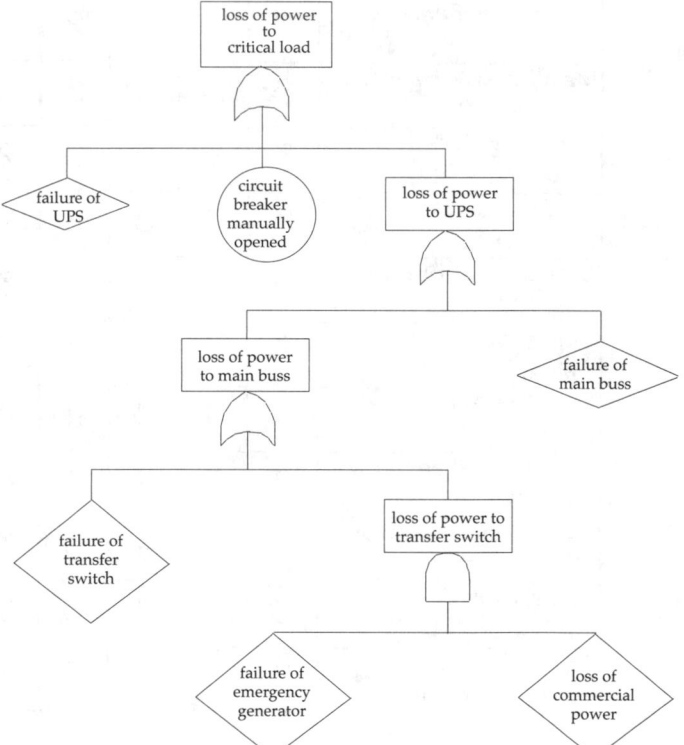

**Figure 23.11 Sample fault tree for electric distribution system with uninterruptible power supply and emergency generator.**

Risk analysis contributes vital information to the countermeasure process by imparting an understanding of the system failure modes. By knowing how the system can fail and what components contribute the those failures, the facility manager can make informed decisions that allocate resources to those countermeasures that mitigate the most significant risks.

Facilities and the utility systems that support them are complex. No single countermeasure is sufficient to mitigate all risk. A multi-faceted approach is required. Just as a three-legged stool will not stand on one leg, neither will a facility be secure against disruption of utility support without a comprehensive approach.

### 23.3.1 Physical Security

Protective countermeasures are actions taken before a crisis occurs to safeguard the components of a utility system against mechanical or electrical damage. Systems are normally constructed to withstand vandalism, severe weather, and other foreseeable events. Critical systems are normally designed with high reliability components.

Physical barriers are the first protective action that should be considered. They establish a physical and psychological deterrent to unauthorized access. Their purpose is to define boundaries for both security and safety, detect entry, and delay and impede unauthorized entry. Fences are the most common barrier. They are used to secure the perimeter of a site. In high value sites, fencing should be supplemented with barbed wire or razor wire. Walls are another excellent barrier. They are routinely used to isolate electrical vaults and mechanical rooms within a building for safety reasons. Proper access control makes these areas more secure as well as safe. It should be remembered that physical barriers are not sufficient to deter a determined adversary intent on causing damage. When the situation warrants, physical barriers must be supplemented with surveillance systems and guards. Area lighting is used in combination with physical barriers to further deter intrusion and aid security personnel in the detection of unauthorized entry.

Many components of utility systems are built in exposed locations that make them vulnerable to traffic accidents or tampering. Pipeline components such as valves and regulators can be buried in pits. Berms can protect many components like metering stations and regulators by sheltering them from vehicular traffic, deflecting blasts, and obscuring the line of sight necessary for firearm damage. Bollards offer excellent protection against delivery vehicles and lawn mowers. Creek and river crossing should be designed to withstand the full force of flooding. This includes any large debris that might be carried by the current.

Site planning is an important aspect of energy security. Vital components should be located away from perimeter fences and shielded from view when possible. Dispersal of resources is another excellent strategy. When redundant system components like electrical transformers are co-located they can all be disabled by a single event. Hardening of components and the structures that house them is appropriate in some instances. Hardening makes a system tolerant to the blast, shock, and vibration caused by explosions or earthquakes.

### 23.3.2 Component Reliability

Reliability improvement of individual components can significantly reduce outages related to equipment failures. This is especially applicable to single point failures. Equipment brands and models known to have high a mean time between failure (MTBF) should be specified. Newly constructed systems should be thoroughly tested at load to insure they are beyond the known infant mortality period. An ongoing preventive maintenance program should be implemented for existing facilities to keep them in a state of high reliability. Such a program should include through inspection, adjusting, lubrication, and replacement of failed redundant components. Aging components should be replaced before they enter the wear out region. Standby systems, such as emergency electric generators, should be periodically tested at full load. These procedures require trained personnel, test equipment, and meticulous record keeping.

### 23.3.3 Redundant Systems

Many actions can be taken to facilitate continued operation when a portion of the utility infrastructure is crippled or otherwise unavailable to the facility. The most important of these is redundancy. Critical facilities should be supported by multiple independent supply routes. Electrical feeders should follow different geographic routes and, ideally, come from different substations. Telephones lines can sometimes to routed to different switches. Backup generators and uninterruptible power supplies guard against disturbances and disruptions that occur off-site in the electric grid. Fuels subject to curtailment, such as natural gas, can be supplemented with alternate fuels such as fuel oil or propane-air mixing systems that can be stored on-site. Sites that burn coal should maintain a stockpile to guard against strikes or transportation uncertainties. Redundant systems should be sized with sufficient excess capacity to carry all critical loads and all support functions, such as lighting and environmental conditioning, that are required for continuous operation.

### 23.3.4 Rapid Recovery

Once a crisis has developed, the overriding goal is to get the system operating in a normal state as rapidly as possible and resume full scale operations. Stockpiling critical or hard to get parts will reduce the recovery time more than any other action. These components should have been identified during the risk analysis. They must be stored separately from the components they are intended to replace to preclude the possibility of a single threat damaging both the primary component and the spare. Emergency response teams must be available to effect the repairs. If damage is expected to exceed the capabilities of in-house personnel, additional parts suppliers and alternate repair crews should be identified and contracts in place to facilitate their rapid deployment. Simulation of recovery actions is an excellent training tool to prepare crews for operation under adverse conditions. Realistic simulation also helps identify unexpected obstacles like limited communication capability, electronic locks that won't open without electricity, and vehicles that can't be refueled.

Any number of things can and will go wrong once an emergency has been initiated. Operating with the lights off, both literally and figuratively, is a demanding task even for the prepared. Portable generators and lights are needed for twenty-four hour repair operations. Self contained battery-powered light carts should be available for use inside buildings. Clear lines of authority must be defined in advance. Crisis coordinators must be able to contact response teams even if the telephones don't work. Crew members need to know where to report and to whom. These things must be thought out in advance and documented in contingency plans.

### 23.3.5 Contingency Planning

Contingency planning is a vital follow-up to the countermeasure process. Plans give order to the chaos surrounding a catastrophic event. A manager's thinking is not always clear in the fog of a crisis. Lines of communication break down. A plan provides the necessary framework of authority to implement restoration actions. The plan identifies the resources necessary to effect repairs and sets priorities to follow if the damage is extensive. Occasionally the planning process identifies critical components not previously identified or highlights bottlenecks such as communication systems that may become overloaded during the crisis.

The goal of contingency planning is rapid recovery to a normal operating state. Risk analysis plays an important role in planning. It identifies those high-risk components or sub-systems that are most vulnerable or result in catastrophic situations when failed. Contingency plans must be developed to complement the implementation of countermeasures that reduce the likelihood of damage occurring. Risk analysis also identifies those components that only yield moderate consequences when damaged. These components are frequently not vital enough to merit the expense of constructing physical barriers or redundant systems. A solid contingency plan identifying sources of parts and labor to effect repairs may be the least cost option.

The most important task performed in the plan is the designation of an emergency coordinator and delineation of authorities and responsibilities. A single coordinator is necessary to insure priorities are followed and to resolve conflicts that may develop. Each member of the response team should have a well defined role. Proper planning will insure no vital tasks "fall through the cracks" unnoticed. No plan can anticipate every contingency that may arise, but a well written plan provides a framework that can be adapted to any situation. An energy contingency plan should, as a minimum, contain the following items:

- Definition of specific authorities and responsibilities
- Priorities for plant functions and customers
- Priorities for protection and restoration of resources
- Curtailment actions
- Recall of personnel (in-house, contract, and mutual aid)
- Location of spares
- Location and contacts for repair equipment
- Equipment suppliers

Each critical component identified during risk analysis should be addressed in detail.

Contingency plans should be exercised on a regular basis. This familiarizes the staff with the specific roles they should assume during an emergency. Practicing these tasks allows them to become proficient and provides an opportunity to identify deficiencies in the plan. Exercising also creates an awareness that a crisis can occur. This gets personnel thinking about actions that reduce risk in existing and new systems. Exercises can range from a paper game or role playing to a full scale simulation that actually interrupts the power to portions of the facility. While full scale simulation is expensive in terms of both labor and productivity, many valuable lessons can be learned that would not otherwise be apparent until an actual disruption occurred.

The most valuable product of repetitive gaming, for all of the people that truly participate, is the unfolding of a process of how to react to a real energy emergency, a process including a communications net, recall procedures, and information systems of all sorts. Repetitive gaming tends to fine-tune the contingency plan, exposing short comings, honing it down to an effective and working document. It is impossible to learn true energy emergency management in a totally calm, non-emergency atmosphere.[4]

### 23.4 ECONOMICS OF ENERGY SECURITY AND RELIABILITY

The consequences of utility outages cover the spectrum from minor annoyances to catastrophic loss of life or revenue. Financial losses can be grouped in several convenient categories. Loss of work in progress covers many things. A few examples include parts damaged during machining operations or by being stranded in cleaning vats, spoiled meats or produces in cold storage, interruption of time sensitive chemical process, and lost or corrupted data. Lost business opportunities are common in retail outlets that depend on computerized registers. Businesses that require continuous contact with customers such as reservation centers, data processing facilities, and communications infrastructure also experience lost business opportunities when utilities are disrupted. Many data processing and reservation centers estimate the cost of interruptions in the six figure range per hour of downtime.

The cost of implementing countermeasures should be commensurate with the cost of an unscheduled outage. This requires a knowledge of the financial impact of an outage. If a probabilistic risk analysis technique was utilized, the frequency and duration of outages can be estimated using conventional reliability theory. Costs associated with recovery operations, damaged work in progress, and lost business opportunities are calculated for each type of outage under analysis. The expected annual loss can then be estimated for each scenario by multiplying the duration by the hourly cost and adding the cost of repairs and recovery operations. Countermeasure funding can be prioritized on the basis of expected financial loss, probability of occurrence, payback time, or other relevant measure.

For non-probabilistic techniques, a more subjective approach must be utilized. The duration of most outages can be estimated based on known repair times. The accuracy of these estimates becomes questionable when widespread damage requires repair crews to service multiple sites or stocks of spare parts are exhausted. The absence of hard data on the frequency of such outages increases the subjective nature of the estimate. One technique is to simply assume that a crisis will occur and allocate resources to mitigate the risk on the basis of financial loss per occurrence or a cost/benefit ratio.

Payback period and cost/benefit ratio are the most common economic analysis tools used to rank countermeasures competing for budget dollars. They can also be used to determine if a particular measure should be funded at all. Payback is calculated by dividing the cost to implement a countermeasure by the expected annual loss derived from a probabilistic risk analysis. The payback period can be evaluated using normal corporate policy. A payback of two years or less is usually sufficient to justify the expenditure. The cost/benefit ratio is calculated by dividing the cost to implement a

countermeasure by the losses associated with a single forced outage. If the ratio is about one the expenditure will be paid back after a single incident. If the ratio is about one-half, the expenditure will be paid back after two incidents.

Insurance should be considered to shift the risk of financial loss to another party. Business interruption policies are available to compensate firms for lost opportunities and help cover the cost of recovery operations. The cost of insurance policies should be treated as an alternate countermeasure and compared using payback or cost/benefit ratio.

## 23.5  LINKS TO ENERGY MANAGEMENT

Many energy security projects can piggy-back on energy conservation projects at a nominal incremental cost. Purchased utility costs can be reduced while reliability is improved. This can dramatically alter the economics of implementing certain countermeasures. Fuel switching strategies let installations take advantage of interruptible natural gas tariffs or transportation contracts. Peak shaving with existing or proposed generators will reduce electric demand charges. Time-of-use rates and real-time pricing make self-generation very attractive during certain seasons or at particular times of the day.

Many industrial facilities purchase utilities on interruptible supply contracts to reduce purchased utility costs. Curtailments are usually of short duration and are often contractually limited to a specified number of hours per year. Alternate fuel sources are normally installed prior to entering into an interruptible supply contract. An example would be to replace the natural gas supply to a steam plant with fuel oil or propane. This gives the facility the option of burning the least expensive fuel. Caution must be exercised when designing a dual-fuel system. Utility service is often curtailed at a time when it is most needed. Natural gas is typically curtailed during the coldest days of winter when demand is highest and supplies become limited due to freezing of wells. An extended curtailment could outlast the on-site supply of an alternate fuel. The same cold weather which caused the primary fuel to be curtailed may also result in shortages of the alternate fuel. An unscheduled plant shutdown could occur if the storage system is undersized or if proper alternate arrangement were not made in advance.

Thermal energy storage systems are installed to reduce on-peak electric demand charges. They can also play a key role in facility reliability. Chillers are frequently considered non-critical loads and are not connected to emergency generators. When a disruption occurs, critical equipment that requires cooling must be shut-down until electricity is restored and the chilled water system is returned to operation. The time required to restart the chilled water system and critical equipment can last many hours longer than the electrical disruption that initiated the event. When a storage system is in place, only the chilled water circulating pumps need to be treated as critical loads powered by the generator. Chilled water can be continuously circulated from the storage system to the equipment, thus precluding the need for a shut-down in all but the longest of outages. When the chillers are removed from the emergency generator additional process equipment can be connected in its place and kept operating during outages or a smaller generator can be installed at a lower first cost.

Energy conservation also has a direct impact on energy security. Conservation projects reduce the amount of energy required to perform at full capacity. By consuming less energy, an installation with a fixed quantity of alternate fuels stored on-site can remain in operation longer under adverse conditions. More efficient operations can also reduce the size of stand-by generators, uninterruptible power supplies, and similar systems. Smaller equipment usually means reduced construction costs and improved economics.

## 23.6  IMPACT OF UTILITY DEREGULATION

Deregulation of the utility industry has the potential to impact energy security and reliability in a way the could greatly exceed the consequences of Year 2000 problems. Y2K problems such as forced outages of power plants or widespread blackouts failed to materialize even in countries that did little to prepare for the much anticipated century date rollover. This absence of Y2K problems has led many to conclude the threats never existed at all, which is not a correct perception. The potential for disruption did exist, it was just of unknown magnitude. Many energy managers will conclude that deregulation also carries with it no risk and will do nothing to prepare.

The risk associated with utility deregulation is also of unknown magnitude but is of a different nature than Y2K. Deregulation is not a one-time event. It is a permanent structural change in an entire industry that will have long term effects which evolve over time. Some of the issues that can be predicted now include reduced reliability of transmission and distribution systems, reduced reliability of generation systems, higher potential

for contract default, and market price risk.

Many utilities have reduced scheduled maintenance of transmission and distribution systems and generation systems. This decision has two root causes. The first is the wave of downsizing that swept the industry during the 1990s. These attempts to reduce operating costs and improve the bottom line have resulted in extended maintenance cycles, fewer spare parts, and smaller crews responding to forced outages and natural disasters. While many of these companies are showing temporary profit increases, they are also reducing system reliability and extending mean time to repair. The long term consequences of these decisions are yet to be fully felt.

A secondary motivation for deferred maintenance is the uncertainty associated with deregulation. Regulators will require utilities to unbundle the services offered. Integrated companies offering generation, transmission and distribution, billing, and other customer services will evolve into multiple, sometimes competing, companies in much the same way the telephone industry was broken up. Regulators may not permit full recovery of capital investments in infrastructure. The fear of not recovering these "stranded costs" has lead some utilities to reduce their investment in replacement or upgraded equipment. The consequences are the same: reduced reliability.

While transmission and distribution is expected to remain a monopoly industry, many questions remain to be answered regarding system operation and reliability. Generation will become a competitive industry. Old, inefficient plants with high heat rates will not be able to compete on a cost basis in the new order but must continue to operate and fill the demand for electricity. New merchant plants and distributed generators with high efficiencies are being planned and constructed now. These, along with the old utility plants, will be unregulated plants competing for customers and operated for profit.

Under the old, regulated structure, utilities had multiple generators that improved reliability. If one plant failed, the system had sufficient spinning reserve to immediately compensate. Many unregulated operators have only a single plant in the region with no spinning reserve to insure reliability. If a generation provider experiences a forced outage the customer may have no choice but to purchase power from a default provider or on the short term spot market at greatly increased costs. The only alternative would be an immediate curtailment of all operations. Recent system disruptions in the midwest and pacific northwest have seen spot market prices rise several orders of magnitude until failed plants were returned to service.

Experience with deregulated natural gas provides an indicator of future problems that might occur with electricity. Gas producers contract to provide gas. If the spot market price of gas increases before the completion of the contract term, then the producer's profit is reduced. Unethical operators have defaulted on contract obligations by refusing to allocate the gas to the original contracted customer. They instead sell the gas on the spot market to the highest bidder. This forces the original customer to curtail load or obtain supplies on the spot market at current prices. The customer's attempt to obtain stable commodity prices and minimize market risk through contractual instruments is nullified by the unethical acts of a supplier. Similar situations are bound to occur as the electric industry deregulates.

## 23.7  SUMMARY

All facilities require a continuous and adequate supply of utility support to function. Energy security is the process of evaluating utility systems and implementing actions that minimize the impact of unscheduled outages that prevent a facility from operating at full capacity. Utility systems are networks with many components. Loss of a few, or in some cases one, critical components is sufficient to disable a network or leave it operating at partial capacity. Additionally, utility networks are not independent. They support each other in a symbiotic manner. The collapse of one network can lead to a domino effect that causes other networks to fail.

The critical components of a utility network can be identified by using risk analysis techniques. The inductive methods make assumptions about the status of individual components and then determine what impact is felt on the entire system. Deductive methods, notably fault tree analysis, use an opposite approach. Fault tree analysis assumes the system is in some undesired condition and proceeds to determine what combinations of failed components will result in that condition. The combinations of failed components are called cut sets. The component failures can be caused by equipment failure, natural disaster, accident, or sabotage.

Countermeasures are actions that prevent or minimize the impact of utility disruptions. Three countermeasures that should not be neglected by any facility manager are physical protection, redundancy, and stockpiling of critical spare parts. Coupled with contingency planning, these counter measures will greatly enhance the energy security of any installation. Contingency plans should be exercised for maximum effectiveness.

Utility disruptions can cause a wide range of impacts on the affected facility. In most cases, this impact can be expressed as a dollar value. This financial impact is useful in evaluating the economics of implementing countermeasures. The most common economic analysis tools are payback and cost/benefit ratio. Many countermeasures can be made more cost effective by linking them to energy conservation projects.

### References

1.   Bracher, Bradley L., Utility Risk Analysis, Proceedings of the 8th World Energy Engineering Congress, 1989, p. 601.
2.   U.S. Nuclear Regulatory Commission, *Fault Tree Handbook*, p. II-4.
3.   Billinton and Allan, *Reliability Evaluation of Engineering Systems: Concepts and Techniques*, p. 113.
4.   Broussard, Peter A., *Energy Security for Industrial Facilities - Contingency Planning for Power Disruption*, PennWell Publishing Company, 1994, pg. 111.

# CHAPTER 24

# UTILITY DEREGULATION AND ENERGY SYSTEM OUTSOURCING

**GEORGE R. OWENS, P.E. C.E.M.**
Energy and Engineering Solutions, Inc.

## 24.0 INTRODUCTION

"Utility Deregulation," "Customer Choice," "Unbundled Rates," "Re-regulation," "Universal Service Charge," "Off Tariff Gas," "Stranded Costs," "Competitive Transition Charge (CTC)," "Caps and Floors," "Load Profiles" and on and on are the new energy buzzwords. They are all the jargon are being used as customers, utilities and the new energy service suppliers become proficient in doing the business of utility deregulation.

Add to that the California energy shortages and rolling blackouts, the Northeast and Midwest outages of 2003, scandal, rising energy prices, loss of price protection in deregulated states and you can see why utility deregulation is increasingly on the mind of utility customers throughout the United States and abroad.

With individual state actions on deregulating natural gas in the late 80's and then the passage of the Energy Policy Act (EPACT) of 1992, the process of deregulating the gas and electric industry was begun. Because of this historic change toward a competitive arena, the utilities, their customers, and the new energy service providers have begun to reexamine their relationships.

How will utility customers, each with varying degrees of sophistication, choose their suppliers of these services? Who will supply them? What will it cost? How will it impact comfort, production, tenants and occupants? How will the successful new players bring forward the right product to the marketplace to stay profitable? And how will more and better energy purchases improve the bottom line?

This chapter reviews the historic relationships between utilities, their customers, and the new energy service providers, and the tremendous possibilities for doing business in new and different ways.

The following figure portrays how power is generated and how it is ultimately delivered to the end customer.

1. Generator – Undergoing deregulation
2. Generator Substation – See 1

**The Power Flow Diagram**

3. Transmission System – Continues to be regulated by the Federal Energy Regulatory Commission (FERC) for interstate and by the individual states for in-state systems
4. Distribution Substation – Continues to be regulated by individual states
5. Distribution Lines – See 4
6. End Use Customer – As a result of deregulation, will be able to purchase power from a number of generators. Will still be served by the local "wires" distribution utility which is regulated by the state.

## 24.1 AN HISTORICAL PERSPECTIVE OF THE ELECTRIC POWER INDUSTRY

At the turn of the century, vertically integrated electric utilities produced approximately two-fifths of the nation's electricity. At the time, many businesses (nonutilities) generated their own electricity. When utilities began to install larger and more efficient generators and more transmission lines, the associated increase in convenience and economical service prompted many industrial consumers to shift to the utilities for their electricity needs. With the invention of the electric motor came the inevitable use of more and more home appliances. Consumption of electricity skyrocketed along with the utility share of the nation's generation.

The early structure of the electric utility industry was predicated on the concept that a central source of power supplied by efficient, low-cost utility generation, transmission, and distribution was a natural monopoly. In addition to its intrinsic design to protect consumers, regulation generally provided reliability and a fair rate of return to the utility. The result was traditional rate base regulation.

For decades, utilities were able to meet increasing demand at decreasing prices. Economies of scale were achieved through capacity additions, technological advances, and declining costs, even during periods when the economy was suffering. Of course, the monopolistic environment in which they operated left them virtually unhindered by the worries that would have been created by competitors. This overall trend continued until the late 1960s, when the electric utility industry saw decreasing unit costs and rapid growth give way to increasing unit costs and slower growth.

The passage of EPACT-1992 began the process of drastically changing the way that utilities, their customers, and the energy services sector deal (or do not deal) with each other. Regulated monopolies are out and customer choice is in. The future will require knowledge, flexibility, and maybe even size to parlay this changing environment into profit and cost saving opportunities.

One of the provisions of EPACT-1992 mandates open access on the transmission system to "wholesale" customers. It also provides for open access to "exempt wholesale generators" to provide power in direct competition with the regulated utilities. This provision fostered bilateral contracts (those directly between a generator and a customer) in the wholesale power market. The regulated utilities then continue to transport the power over the transmission grid and ultimately, through the distribution grid, directly to the customer.

What EPACT-1992 did not do was to allow for "retail" open access. Unless you are a wholesale customer, power can only be purchased from the regulated utility. However, EPACT-1992 made provisions for the states to investigate retail wheeling ("wheeling" and "open access" are other terms used to describe deregulation). Many states have held or are currently holding hearings. Several states either have or will soon have pilot programs for retail wheeling. The model being used is that the electric generation component (typically 60-70% of the total bill), will be deregulated and subject to full competition. The transmission and distribution systems will remain regulated and subject to FERC and state Public Service Commission (PSC) control.

A new comprehensive energy bill, EPACT-2005, was signed into law in 2005, just as this edition was being finalized. Look for expanded discussion of EPACT-2005 in future editions of this chapter. This bill affects energy production, including renewables, energy conservation, regulations on the country's transmission grids, utility deregulation as well as other energy sectors. Tax incentives to spur change are key facets of EPACT-2005.

## ELECTRIC INDUSTRY DEREGULATION TIME LINE

1992 - Passage of EPACT and the start of the debate.

1995 & 1996 - The first pilot projects and the start of special deals. Examples are: The automakers in Detroit, New Hampshire programs for direct purchase including industrial, commercial and residential, and large user pilots in Illinois and Massachusetts.

1997 - Continuation of more pilots in many states and almost every state has deregulation on the legislative and regulatory commission agenda.

1998 - Full deregulation in a few states for large users (i.e., California and Massachusetts). Many states have converged upon 1/1/98 as the start of their deregulation efforts with more pilots and the first 5% roll-in of users, such as Pennsylvania and New York.

2000 - Deregulation of electricity became common for most industrial and commercial users and began to penetrate the residential market in several states. These included Maryland, New Jersey, New York, and Pennsylvania among others. See figure 24.1.

2002/3- Customers have always had a "backstop" of regulated pricing. Now that the transition periods are nearing their end, customers are faced with the option of buying electricity on the open market without a regulated default price.

2003 - During the summer, parts of the northeast and upper Midwest experience a massive blackout that shuts down businesses and residential customers. The adequacy of the transmission system is blamed.

2005 - EPACT-2005 becomes law

## 24.2 THE TRANSMISSION SYSTEM AND THE FEDERAL ENERGY REGULATORY COMMISSION'S (FERC) ROLE IN PROMOTING COMPETITION IN WHOLESALE POWER

Even before the passage of EPACT in 1992, FERC played a critical role in the competitive transformation of wholesale power generation in the electric power industry. Specific initiatives include notices of proposed rulemaking that proposed steps toward the expansion of competitive wholesale electricity markets. FERC's Order 888, which was issued in 1996, required public utilities that own, operate, or control transmission lines to file tariffs that were non-discriminatory at rates that are no higher than what the utility charges itself. These actions essentially opened up the national transmission grid to non-discretionary access on the wholesale level (public utilities, municipalities and rural cooperatives). This order did not give access to the transmission grid to retail customers.

In an effort to ensure that the transmission grid is opened to competition on a non-discriminatory basis, Independent System Operators (ISO's) are being formed in many regions of the country. An ISO is an independent operator of the transmission grid and is primarily responsible for reliability, maintenance (even if the day-to-day maintenance is performed by others) and security. In addition, ISO's generally provide the following functions: congestion management, administering transmission and ancillary pricing, making transmission information publicly available, etc.

## 24.3 STRANDED COSTS

Stranded costs are generally described as legitimate, prudent and verifiable costs incurred by a public utility or a transmitting utility to provide a service to a customer that subsequently are no longer used. Since the asset or capacity is generally paid for through rates, ceasing to use the service leaves the asset, and its cost, stranded. In the case of de-regulation, stranded costs are created when the utility service or asset is provided, in whole or in part, to a deregulated customer of another public utility or transmitting utility. Stranded costs emerge because new generating capacity can currently be built and operated at costs that are lower than many utilities' embedded costs. Wholesale and retail customers have, therefore, an incentive to turn to lower cost producers. Such actions make it difficult for utilities to recover all their prudently incurred costs in generating facilities.

Stranded costs can occur during the transition to a fully competitive wholesale power market as some wholesale customers leave a utility's system to buy power from other sources. This may idle the utility's existing generating plants, imperil its fuel contracts, and inhibit its capability to undertake planned system expansion leading to the creation of "stranded costs." During the transition to a fully competitive wholesale power market, some utilities may incur stranded costs as customers switch to other suppliers. If power previously sold to a departing customer cannot be sold to an alternative buyer, or if other means of mitigating the stranded costs cannot be found, the options for recovering stranded costs are limited.

The issue of stranded costs has become contentious in the state proceedings on electric deregulation. Utilities

Retail access is either currently available to all or some customers or will soon be available. Those states are Arizona, Connecticut, Delaware, District of Columbia, Illinois, Maine, Maryland, Massachusetts, Michigan, New Hampshire, New Jersey, New York, Ohio, Oregon, Pennsylvania, Rhode Island, Texas, and Virginia. In Oregon, no customers are currently participating in the State's retail access program, but the law allows nonresidential customers access. Yellow colored states are not actively pursuing restructuring. Those states are Alabama, Alaska, Colorado, Florida, Georgia, Hawaii, Idaho, Indiana, Iowa, Kansas, Kentucky, Louisiana, Minnesota, Mississippi, Missouri, Nebraska, North Carolina, North Dakota, South Carolina, South Dakota, Tennessee, Utah, Vermont, Washington, West Virginia, Wisconsin, and Wyoming. In West Virginia, the Legislature and Governor have not approved the Public Service Commission's restructuring plan, authorized by HB 4277. The Legislature has not passed a resolution resolving the tax issues of the PSC's plan, and no activity has occurred since early in 2001. A green colored state signifies a delay in the restructuring process or the implementation of retail access. Those states are Arkansas, Montana, Nevada, New Mexico, and Oklahoma. California is the only blue colored state because direct retail access has been suspended.
**\*As of January 30, 2003, Department of Energy, Energy Information Administration**

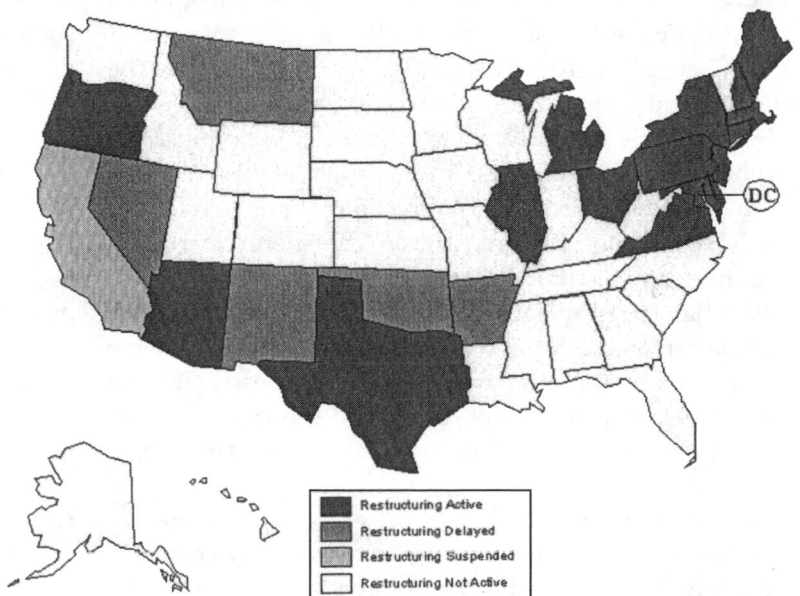

Restructuring Active
Restructuring Delayed
Restructuring Suspended
Restructuring Not Active

**Figure 24.1 Status of State Electric Industry Restructuring Activity\***

have argued vehemently that they are justified in recovering their stranded costs. Customer advocacy groups, on the other hand, have argued that the stranded costs proposed by the utilities are excessive. This is being worked out in the state utility commissions. Often, in exchange for recovering stranded costs, utilities are joining in settlement agreements that offer guaranteed rate reductions and opening up their territories to deregulation.

## 24.4  STATUS OF STATE ELECTRIC INDUSTRY RESTRUCTURING ACTIVITY

Electric deregulation on the retail level is determined by state activity. Many states have or are in the process of enacting legislation and/or conducting proceedings. See Figure 24.1.

## 24.5  TRADING ENERGY - MARKETERS AND BROKERS

With the opening of retail electricity markets in several states, new suppliers of electricity have developed beyond the traditional vertically integrated electric utility. Energy marketers and brokers are the new companies that are being formed to fill this need. An energy marketer is one that buys electricity or gas commodity and transmission services from traditional utilities or other suppliers, then resells these products. An energy broker, like a real estate broker, arranges for sales but does not take title to the product. There are independent energy marketers and brokers as well as unregulated subsidiaries of the regulated utility.

According to The Edison Electric Institute, the energy and energy services market was $360 billion in 1996 and was expected to grow to $425 billion in 2000. To help put these numbers in perspective, this market is over six times the telecommunications marketplace. As more states open for competition, the energy marketers and brokers are anticipating strong growth. Energy suppliers have been in a merger and consolidation mode for the past few years. This will probably continue at the same pace as the energy industry redefines itself even further. Guidance on how to choose the right supplier for your business or clients will be offered later on in this chapter

The trading of electricity on the commodities market is a rather new phenomenon. It has been recognized that the marketers, brokers, utilities and end users need to have vehicles that are available for the managing of risk in the sometimes-volatile electricity

market. The New York Mercantile Exchange (NYMEX) has instituted the trading of electricity along with its more traditional commodities. A standard model for an electricity futures contract has been established and is traded for delivery at several points around the country. As these contracts become more actively traded, their usefulness will increase as a means to mitigate risk. An example of a risk management play would be when a power supplier locks in a future price via a futures or options contract to protect its position at that point in time. Then if the prices rise dramatically, the supplier's price will be protected.

## 24.6  THE IMPACT OF DEREGULATION

Historically, electricity prices have varied by a factor of two to one or greater, depending upon where in the county the power is purchased. See Figure 24.2. These major differences even occur in utility jurisdictions that are joined. The cost of power has varied because of several factors, some of which are under the utilities control and some that are not, such as:

- Decisions on projected load growth
- The type of generation
- Fuel selections
- Cost of labor and taxes
- The regulatory climate

All of these factors contribute to the range of pricing. Customers have been clamoring for the right to choose the supplier and gain access to cheaper power for quite some time. This has driven regulators to impose utility deregulation, often with opposition from the incumbent utilities.

Many believe that electric deregulation will even out this difference *and* bring down the total average price through competition. There are others that do not share that opinion. Most utilities are already taking actions to reduce costs. Consolidations, layoffs, and mergers are occurring with increased frequency. As part of the transition to deregulation, many utilities are requesting and receiving rate freezes and reductions in exchange for stranded costs.

One factor has remained a constant until the early 2000's. Customers have always had a "backstop" of regulated pricing until recently. Now that the transition periods are nearing their end, customers are faced with the option of buying electricity on the open market without a regulated default price. The risks to customers have increased dramatically. And, energy consultants

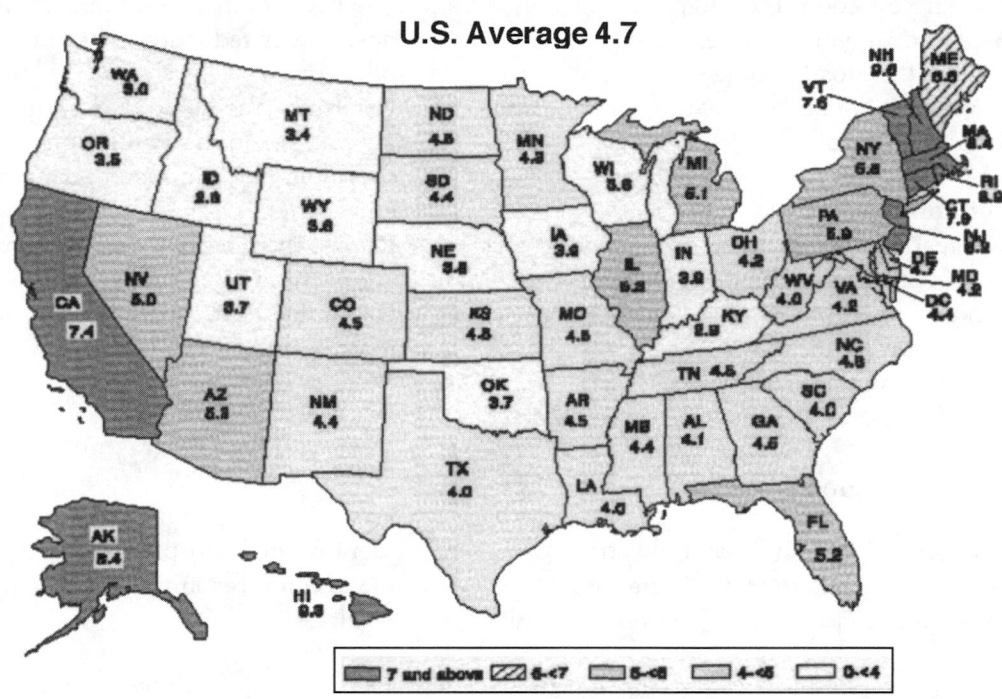

**U.S. Average 4.7**

Source: Energy Information Administration, Form EIA-861, "Annual Electric Utility Report," (1995).

**Figure 24.2 Electricity Cost by State**

Average Revenue from Electric Sales to Industrial Consumers by State, 1995 (Cents per Kilowatt-hour)

and ESCOs are having a difficult time predicting the direction of electricity costs.

All of this provides for interesting background and statistics, but what does it mean to energy managers interested in providing and procuring utilities, commissioning, O&M (operations and maintenance), and the other energy services required to build and operate buildings effectively? Just as almost every business enterprise has experienced changes in the way that they operate in the 90's and 2000 and beyond, the electric utilities, their customers and the energy service sector must also transform. Only well-prepared companies will be in a position to take advantage of the opportunities that will present themselves after deregulation. Building owners and managers need to be in a position to actively participate in the early opening states. The following questions will have to be answered by each and every company if they are to be prepared:

- Will they participate in the deregulated electric market?
- Is it better to do a national account style supply arrangement or divide the properties by region and/or by building type?
- How will electric deregulation affect their relationships with tenants in commercial, governmental

and institutional properties?
- Would there be a benefit for multi-site facilities to partake in purchasing power on their own?
- Should the analysis and operation of electric deregulation efforts be performed in-house or by consultants or a combination?
- What criteria should be used to select the energy suppliers when the future is uncertain?

## 24.7 THE TEN-STEP PROGRAM TO SUCCESSFUL UTILITY DEREGULATION

In order for the building sector to get ready for the new order and answer the questions raised above, this ten-step program has been developed to ease the transition and take advantage of the new opportunities. This Ten Step program is ideally suited to building owners and managers as well as energy engineers that are in the process of developing their utility deregulation program.

Step #1 - Know Thyself
- When do you use the power
- Distinguish between summer vs. winter, night vs. day

- What load can you control/change
- What $$$ goal does your business have
- What is your 24 hr. load profile
- What are your in-house engineering, monitoring and financial strengths

Step #2 - Keep Informed
- Read, read, read—network, network, network
- Interact with your professional organizations
- Talk to vendors, consultants, and contractors
- Subscribe to trade publications
- Attend seminars and conferences
- Utilize internet resources—news groups, WWW, E-mail
- Investigate buyer's groups

Step #3 - Talk to Your Utilities (all energy types)
- Recognize customer relations are improving
- Discuss alternate contract terms or other energy services
- Find out if they are "for" or "agin" deregulation
- Obtain improved service items (i.e., reliability)
- Tell them your position and what you want. Now is not the time to be bashful
- Renegotiate existing contracts

Step #4 - Talk to Your Future Utility(ies)
- See Step #3
- Find out who is actively pursuing your market
- Check the neighborhood, check the region, look nationally
- Develop your future relationships
- Partner with Energy Service Companies (ESCOs), power marketing, financial, vendor and other partners for your energy services needs

Step #5 - Explore Energy Services Now
(Why wait for deregulation?)
- Implement "standard" energy projects such as lighting, HVAC, etc.
- Investigate district cooling/heating
- Explore selling your central plant
- Calculate square foot pricing
- Buy comfort, Btus or GPMs; not kWhs
- Outsource your Operations and Maintenance
- Consider other work on the customer side of the meter

Step #6 - Understand the Risks
- Realize that times will be more complicated in the future
- Consider the length of a contract term in uncertain times

- Identify whether you want immediate reductions now, larger reductions later or prices tied to some other index
- Determine the value of a flat price for utilities
- Be wary of losing control of your destiny-turning over some of the operational controls of your energy systems
- Realize the possibility some companies will not be around in a few years
- Determine how much risk you are willing to take in order to achieve higher rewards

Step #7 - Solicit Proposals
- Meet with the bidders prior to issuing the Request For Proposal (RFP)
- Prepare the RFP for the services you need
- Identify qualified players
- Make commissioning a requirement to achieve the results

Step #8 - Evaluate Options
- Enlist the aid of internal resources and outside consultants
- Narrow the playing field and interview the finalists prior to awarding
- Prepare a financial analysis of the results over the life of the project—Return on Investment (ROI) and Net Present Value (NPV)
- Remember that the least first cost may or may not be the best value
- Pick someone that has the financial and technical strengths for the long term
- Evaluate financial options such as leasing or shared

Step #9 - Negotiate Contracts
Remember the following guidelines when negotiating a contract:
- The longer the contract, the more important the escalation clauses due to compounding
- Since you may be losing some control, the contract document is your only protection
- The supplying of energy is not regulated like the supplying of kWhs are now
- The clauses that identify the party taking responsibility for an action, or "Who Struck John" clauses, are often the most difficult to negotiate
- Include monitoring and evaluation of results
- Understand how the contract can be terminated and what the penalties for early termination are

Step #10 - Sit Back and Reap the Rewards
- Monitor, measure, and compare

- Don't forget Operations & Maintenance for the long term
- Keep looking, there are more opportunities out there
- Get off your duff and go to Step #1 for the next round of reductions

## 24.8 AGGREGATION

Aggregation is the grouping of utility customers to jointly purchase commodities and/or other energy services. There are many aggregators already formed or being formed in the states where utility deregulation is occurring. There are two basic forms of aggregation:

1.  Similar Customers with Similar Needs
    Similar customers may be better served via aggregation even if they have the same load profiles
    - Pricing and risk can be tailored to similar customers needs
    - Similar billing needs can be met
    - Cross subsidization would be eliminated
    - Trust in the aggregator; i.e. BOMA for office building managers membership

2.  Complementary Customers that May Enhance the Total
    Different load profiles can benefit the aggregated group by combining different load profiles.
    - Match a manufacturing facility with a flat or inverted load profile to an office building that has a peaky load profile, etc.
    - Combining of load profiles is more attractive to a supplier than either would be individually

### Why Aggregate?
Some potential advantages to aggregating are:
- Reduction of internal administration expense
- Shared consulting expenses
- More supplier attention resulting from a larger bid
- Lower rates may be the result of a larger bid
- Lower average rates resulting from combining dissimilar user profiles

### Why Not Aggregate?
Some potential disadvantages from aggregating are:
- If you are big enough, you are your own aggregation
- Good load factor customers may subsidize poor load factor customers
- The average price of an aggregation may be lower than your unique price
- An aggregation cannot meet "unique" customer requirements

### Factors that affect the decision on joining an aggregation
Determine if an aggregation is right for your situation by considering the following factors. An understanding of how these factors apply to your operation will result in an informed decision.
- Size of load
- Load profile
- Risk tolerance
- Internal abilities (or via consulting)
- Contract length flexibility
- Contract terms and conditions flexibility
- Regulatory restrictions

## 24.9 IN-HOUSE VS. OUTSOURCING ENERGY SERVICES

The end user sector has always used a combination of in-house and outsourced energy services. Many large managers and owners have a talented and capable staff to analyze energy costs, develop capital programs, and operate and maintain the in-place energy systems. Others (particularly the smaller players who cannot justify an in-house staff) have outsourced these functions to a team of consultants, contractors, and utilities. These relationships have evolved recently due to downsizing and returning to the core businesses. In the new era of deregulation, the complexion of how energy services are delivered will evolve further.

Customers and energy services companies are already getting into the utility business of generating and delivering power. Utilities are also getting into the act by going beyond the meter and supplying chilled/hot water, conditioned air, and comfort. In doing so, many utilities are setting up unregulated subsidiaries to provide commissioning, O&M, and many other energy services to customers located within their territory, and nationwide as well.

A variety of terms are often used: Performance Contracting, Energy System Outsourcing, Utility Plant Outsourcing, Guaranteed Savings, Shared Savings, Sell/Leaseback of the central plant, Chauffage (used in Europe), Energy Services Performance Contract (ESPC), etc. Definitions are as follows:
- Performance Contracting
  Is the process of providing a specific improvement

such as a lighting retrofit or a chiller change-out, usually using the contractor's capital and then paying for the project via the savings over a specific period of time. Often the contractor guarantees a level of savings. The contractor supplies capital, engineering, equipment, installation, commissioning and often the maintenance and repair.

- Energy System Outsourcing
Is the process of divesting of the responsibilities and often the assets of the energy systems to a third party. The third party then supplies the commodity, whether it be chilled water, steam, hot water, electricity, etc., at a per unit cost. The third party supplier then is responsible for the improvement capital and operations and maintenance of the energy system for the duration of the contract.

**Advantages**

The advantages of a performance contract or an energy system outsourcing project revolves around four major areas:

1. Core Business Issues
Many industries and corporations have been re-examining all of their non-core functions to determine if they would be better served by outsourcing these functions. Performance contracting or outsourcing can make sense if someone can be found that can do it better and cheaper than what can be managed by an in-house staff. Then the building managers can oversee the contractor and not the complete operation. This may allow the building to devote additional time and resources to other core business issues such as increasing revenues and reducing health care costs.

2. Monetization
One of the unique features of a performance contract or an energy system outsourcing project is the opportunity to obtain an up front payment. There is an extreme amount of flexibility available depending upon the needs. The amount available can range from zero dollars to the approximate current value of the installation. The more value placed on the up front payment will necessarily cause the monthly payments to increase as well as the total amount of interest paid.

3. Deferred Capital Costs
Many electrical and HVAC energy systems are at an age or state of repair that would necessitate

the infusion of a major capital investment in the near future. These investments are often required to address end-of-life, regulatory and efficiency issues. Either the building owner or manager could provide the capital or a third party could supply it and then include the repayment in a commodity charge plus interest; ("there are no free lunches").

4. Operating Costs
The biggest incentive to a performance contract or an energy system outsourcing project is that if the right supplier is chosen with the right incentives, then the total cost to own and operate the central plant can be less. The supplier, having expertise and volume in their core area of energy services, brings this to reality. With this expertise and volume, the supplier should be able to purchase supplies at less cost, provide better-trained personnel and implement energy and maintenance saving programs. These programs can range from capital investment of energy saving equipment to optimizing operations, maintenance and control programs.

**Disadvantages**

Potentially, there are several disadvantages to undertaking a performance contract or an outsourcing project. The items identified in this section need to be recognized and mitigated as indicated here and in the Risk Management section.

1. Loss of Control
As with any service, if it is outsourced, the service is more difficult to control. The building is left with depending upon the skill, reliability and dedication of the service supplier and the contract to obtain satisfactory results. Even with a solid contract; if the supplier does not perform or goes out of business, the customer will suffer (see the Risk Management section). Close coordination between the building and the supplier will be necessary over the long term of the contract to adjust to changing conditions.

2. Loss of Flexibility
Unless addressed adequately in the contract, changes that the building wants or needs to make can cause the economics of the project to be adversely affected. Some examples are:
- Changes in hours of operation
- New systems that require additional cooling or heating, such as an expansion or renova-

tion, conversion of office or storage space to other uses, additional equipment requiring additional cooling, etc.

- Scheduling outages for maintenance or repairs
- Using in house technicians for other services throughout the building. If this situation occurs in current operation, provisions for additional building staff or having the supplier make the technician available needs to be arranged. If additional costs are indicated, they should be included in the financial analysis.

3. Cost Increases
This only becomes a disadvantage if the contract does not adequately foresee and cover every contingency and changing situation adequately. To protect themselves, the suppliers will try to put as much cost risk onto the customer as possible. It is the customer and the customer's consultants and attorneys responsibility to define the risks and include provisions in the contract.

**Financial Issues**

The basis for success of a performance contract or an energy system outsourcing project is divided between the technical issues, contract terms, supplier's performance and how the project will be financed. These types of projects are as much (if not more) about the financial deal than the actual supplying of a commodity or a service. (See Chapter 4 -Economic Analysis and Life Cycle Costing) The answers to some basic questions will help guide the decision making process.

- *Is capital required during the term of the project?*
The question of the need for capital is one of the major driving factors of a performance contract or an energy outsourcing project. Capital invested into the HVAC and electrical systems for efficiency upgrades, end of life replacements, increased reliability or capacity and environmental improvements can be financed through the program.

- *Who will supply the capital and at what rate?*
The answer to the question of who will be supplying the capital should be made based upon your ability to supply capital from internal operations, capital improvement funds, borrowing ability and any special financing options such as tax free bonds or other low interest sources. If capital is needed for other uses such as expansions and other revenue generating or cost reduction measures,

then energy system outsourcing may be a good choice.

- *Is there a desire to obtain a payment up front?*
As stated previously, a performance contract or energy system outsourcing project presents the opportunity to obtain a payment up front for the assets of the HVAC and electrical systems. However, any up-front payment increases the monthly payment over the term of the contract and should be considered similar to a loan.

- *Does the capital infusion and better operations generate enough cash flow to pay the debt?*
This is the sixty-four dollar question. Only by performing a long-term evaluation of the economics of the project with a comparison to the in house plan can the financial benefits be fairly compared. A Net Present Value and Cash Flow analysis should be used for the evaluation of a performance contract or energy system outsourcing project. It shows the capital and operating impact of the owner continuing to own and operate a HVAC and electrical systems. This is compared to a third party outsourced option. The analysis should be for a long enough period to incorporate the effect of a major capital investment. This is often done for a 20-year period. This type of analysis would allow the building owner or manager to evaluate the financial impact of the project over the term of the contract. Included in the analysis should be a risk sensitivity assessment that would bracket and define the range of results based upon changing assumptions.

**Other Issues**

1. Management and Personnel Issues
- Management - Usually, an in-house manager will need to be assigned to manage the supplier and the contract and to verify the accuracy of the billing. An in-house technical person or an outside consultant should have the responsibility to periodically review the condition of the equipment to protect the long-term value of the central plant.

- Personnel - Existing employees need to be considered. This may or may not have a monetary consequence due to severance or other policies. If there is an impact, it needs to be reflected in the analysis. It would usually be to the building's benefit if the years of knowledge and experience represented by the current engineers could be transferred to the new supplier. Another personnel concern is the effect on the moral of the employees due to their fear of losing their jobs.

2. Which services to outsource?

   Where there are other services located in the central plant that are not outsourced, these need to be identified in the documents. These could include compressed air for controls, domestic water, hot water, etc. A method of allocating costs for shared services will need to be established and managed through the duration of the contract.

3. Product specifications

   The properties of the supplied service need to be adequately described to judge if the supplier is meeting the terms of the contract. Quantities like temperature, water treatment values, pressure, etc. needs to be well defined.

4. Early Termination

   There should be several options in the contract for early termination. The most obvious is for lack of performance. In this case, lack of performance can range from total disruption of service to not meeting the defined values of the commodity to letting the equipment deteriorate. There should also be the ability to have the building owner terminate the contract if the building owner decides that they want to take the central plant in-house or find another contractor. If the supplier is in default, then a "make whole" payment would be required of the building to terminate the contract in this case.

**Risk Management**

   As with any long-term commitment, the most important task is to identify all of the potential risks, evaluate their consequences and probability and then to formulate strategies that will mitigate the risks. This could be in the form of the contract document language or other financial instruments for protection. One of the most important areas of risk management mitigation is to choose a supplier that will deliver what is promised over the entire contract period.

1. How to Choose a Supplier

   In addition to price, the following factors are important to the success of a project and should be evaluated before selecting a supplier.
   - Track record
   - Knowledge of your business, priorities and risk tolerance
   - Size

   - Financial backing
   - Customer service and reporting
   - "Staying Power"

2. Long Term Contracts

   Because the potential supplier will be investing capital for increased life, reliability and efficiency, the contract needs to be long enough to recover the costs and provide a positive cash flow. The length of the project can vary from three to five years for a simple, small-scale project up to ten to twenty years for one of increased complexity. Cost impacts at the termination of the contract needs to be adequately addressed, such as:
   - Renewals
   - Buyouts
   - Equipment leases
   - Equipment condition at the end of the contract

3. Changing Assumptions
   - Interest rates
   - Utility rates
   - Maintenance and repair costs
   - Areas served (i.e., expansions/renovations/contractions)
   - Regulations; building specific, environmental, OSHA, local codes, etc.
   - Utility deregulation

4. Other Risks
   - The impact of planned or unplanned outages of the central plant
   - The consequences of the supplier not being able to maintain chilled water temperature or steam pressure
   - "Take or Pay" This provision of a contract requires the customer to pay a certain amount even if they do not use the commodity
   - Defaults and Remedies

**24.10 SUMMARY**

   This chapter presented information on the changing world of the utility industry in the new millennium. Starting in the 80's with gas deregulation and the passage of the Energy Policy Act of 1992 for electricity, the method of providing and purchasing energy was changed forever. Utilities began a slow change from vertically integrated monopolies to providers of regulated wires and transmission services. Some utilities continued to supply generation services, through their unregulated enterprises and by independent power producers in the deregulated markets while others sold their generation assets and became "wires" companies. Customers became confused in the early stages of deregulation, but by the end of the 1990's some became more knowledgeable and successful in buying deregulated natural gas and electricity.

In the early 2000's, difficulties have developed in the deregulated utility arena. California rescinded deregulation (except for existing contracts) after shortages, rolling blackouts and price increases sent the utilities into a tailspin. The great blackout of 2003 raises concerns about the reliability of the transmission system. And the loss of regulated rates provides more challenges to customers and their consultants. However, many customers continue to participate in the deregulated markets to obtain reduced (or stable) prices, reduce their risk of big price swings and incorporate energy reduction programs with energy procurement programs.

Another result of deregulation has been a re-examination by customers of outsourcing their energy needs. Some customers have "sold" their energy systems to energy suppliers and are now purchasing Btus instead of kWhs. The energy industry responded with energy service business units to meet this new demand for outsourcing. Performance contracting and energy system outsourcing can be advantageous when the organization does not have internal expertise to execute these projects and when other sources of capital are needed. However, performance contracting and energy system outsourcing is not without peril if the risks are not understood and mitigated. Before undertaking a performance contract or energy system outsourcing project, the owner or manager first needs to define the financial, technical, legal and operational issues of importance. Next, the proper resources, whether internal or outsourced, need to be marshaled to define the project, prepare the Request for Proposal, evaluate the suppliers and bids, negotiate a contract and monitor the results, often over a long period. If these factors are properly considered and executed, the performance contract or energy system outsourcing often produce results that could not be obtained via other project methods.

## BIBLIOGRAPHY

*Power Shopping* and *Power Shopping II*, A publication of the Building Owners and Managers Association (BOMA) International, 1201 New York Avenue, N.W., N.W., Suite 300, Washington, DC 20005.

*The Changing Structure of the Electric Power Industry: Historical Overview*, United States Department of Energy, Energy Information Administration, Washington, DC.

*The Ten Step Program to Successful Utility Deregulation for Building Owners and Managers*, George R Owens PE CEM, President Energy and Engineering Solutions, Inc. (EESI), 9449 Penfield Ct., Columbia, MD 21045.

*Performance Contracting and Energy System Outsourcing*, George R Owens PE CEM, President Energy and Engineering Solutions, Inc. (EESI), 9449 Penfield Ct., Columbia, MD 21045.

*Generating Power and Getting It to The Consumer*, Edison Electric Institute, 701 Pennsylvania Ave NW, Washington, DC, 20004.

*The Changing Structure of the Electric Power Industry: An Update*, US Department of Energy, Energy Information Administration, DOE/EIA-0562(96)

*PJM Electricity Futures*, New York Mercantile Exchange (NYMEX) web page, www.nymex.com

## SOME USEFUL INTERNET RESOURCES

10 Step paper - www.eesienergy.com

State activities - www.eia.doe.gov/cneaf/electricity/chg_str/

State regulatory commissions www.naruc.org

Utilities - www.utilityconnection.com

Maillist - AESP-NET@AESP.org

# CHAPTER 25

# FINANCING ENERGY MANAGEMENT PROJECTS

**ERIC A. WOODRUFF, PH.D., CEM, CEP, CLEP**
Johnson Controls, Inc.

## 25.1 INTRODUCTION

Financing can be a key success factor for projects. This chapter's purpose is to help facility managers understand and apply the financial arrangements available to them. Hopefully, this approach will increase the implementation rate of good energy management projects, which would have otherwise been cancelled or postponed due to lack of funds.

Most facility managers agree that energy management projects (EMPs) are good investments. Generally, EMPs reduce operational costs, have a low risk/reward ratio, usually improve productivity and even have been shown to improve a firm's stock price.[1] Despite these benefits, many cost-effective EMPs are not implemented due to financial constraints. A study of manufacturing facilities revealed that first-cost and capital constraints represented over 35% of the reasons cost-effective EMPs were not implemented.[2] Often, the facility manager does not have enough cash to allocate funding, or can not get budget approval to cover initial costs. Financial arrangements can mitigate a facility's funding constraints,[3] allowing additional energy savings to be reaped.

Alternative finance arrangements can overcome the "initial cost" obstacle, allowing firms to implement more EMPs. However, many facility managers are either unaware or have difficulty understanding the variety of financial arrangements available to them. Most facility managers use simple payback analyses to evaluate projects, which do not reveal the added value of after-tax benefits.[4] Sometimes facility managers do not implement an EMP because financial terminology and contractual details intimidate them.[5]

To meet the growing demand, there has been a dramatic increase in the number of finance companies specializing in EMPs. At a recent World Energy Engineering Congress, finance companies represented the most common exhibitor type. These financiers are introducing new payment arrangements to implement EMPs. Often, the financier's innovation will satisfy the unique customer needs of a large facility. This is a great service

however, most financiers are not attracted to small facilities with EMPs requiring less than $100,000. Thus, many facility managers remain unaware or confused about the common financial arrangements that could help them implement EMPs.

Numerous papers and government programs have been developed to show facility managers how to use quantitative (economic) analysis to evaluate financial arrangements.[4,5,6] (Refer to Chapter 4 of this book.) *Quantitative analysis includes computing the simple payback, net present value (NPV), internal rate of return (IRR), or life-cycle cost of a project with or without financing.* Although these books and programs show how to evaluate the economic aspects of projects, they do not incorporate qualitative factors like strategic company objectives, (which can impact the financial arrangement selection). Without incorporating a facility manager's qualitative objectives, it is hard to select an arrangement that meets all of the facility's needs. A recent paper showed that qualitative objectives can be at least as important as quantitative objectives.[9]

This chapter hopes to provide some valuable information, which can be used to overcome the previously mentioned issues. The chapter is divided into several sections to accomplish three objectives. Sections 2 and 3 *introduce the basic financial arrangements* via a simple example. In sections 4 and 5, financial terminology is defined and each arrangement is explained in greater detail while applied to a case study. The remaining sections show *how to match financial arrangements to different projects and facilities.*

## 25.2 FINANCIAL ARRANGEMENTS: A SIMPLE EXAMPLE

Consider a small company "PizzaCo" that makes frozen pizzas, and distributes them regionally. PizzaCo uses an old delivery truck that breaks down frequently and is inefficient. Assume the old truck has no salvage value and is fully depreciated. PizzaCo's management would like to obtain a new and more efficient truck to reduce expenses and improve reliability. However, they do not have the cash on hand to purchase the truck. Thus, they consider their financing options.

### 25.2.1 Purchase the Truck with a Loan or Bond

Just like most car purchases, PizzaCo borrows money from a lender (a bank) and agrees to a monthly re-payment plan. Figure 25.1 shows PizzaCo's annual cash flows for a loan. The solid arrows represent the financing cash flows between PizzaCo and the bank. Each year, PizzaCo makes payments (on the principal, plus interest based on the unpaid balance), until the balance owed is zero. The payments are the negative cash flows. Thus, at time zero when PizzaCo borrows the money, they receive a large sum of money from the bank, which is a positive cash flow (which will be used to purchase the truck).

The *dashed* arrows represent the truck purchase as well as savings cash flows. Thus, at time zero, PizzaCo purchases the truck (a negative cash flow) with the money from the bank. Due to the new truck's greater efficiency, PizzaCo's annual expenses are reduced (which is a savings). The annual savings are the positive cash flows. The remaining cash flow diagrams in this chapter utilize the same format.

PizzaCo could also purchase the truck by selling a bond. This arrangement is similar to a loan, except investors (not a bank) give PizzaCo a large sum of money (called the bond's "par value"). Periodically, PizzaCo would pay the investors only the interest accumulated. As Figure 25.2 shows, when the bond reaches maturity, PizzaCo returns the par value to the investors. The equipment purchase and savings cash flows are the same as with the loan.

### 25.2.2 Sell Stock to Purchase the Truck

In this arrangement, PizzaCo sells its stock to raise money to purchase the truck. In return, PizzaCo is expected to pay dividends back to shareholders. Selling stock has a similar cash flow pattern as a bond, with a few subtle differences. Instead of interest payments to bondholders, PizzaCo would pay dividends to share-

holders until some future date when PizzaCo could buy the stock back. However, these dividend payments are not mandatory, and if PizzaCo is experiencing financial strain, it does not need to distribute dividends. On the other hand, if PizzaCo's profits increase, this wealth will be shared with the new stockholders, because they now own a part of the company.

### 25.2.3 Rent the Truck

Just like renting a car, PizzaCo could rent a truck for an annual fee. This would be equivalent to a true lease. The rental company (lessor) owns and maintains the truck for PizzaCo (the lessee). PizzaCo pays the rental fees (lease payments) which are considered tax-deductible business expenses.

Figure 25.3 shows that the lease payments (solid arrows) start as soon as the equipment is leased (year zero) to account for lease payments paid in advance. Lease payments "in arrears" (starting at the end of the first year) could also be arranged. However, the leasing company may require a security deposit as collateral. Notice that the savings cash flows are essentially the same as the previous arrangements, except there is no equipment purchase, which is a large negative cash flow at year zero.

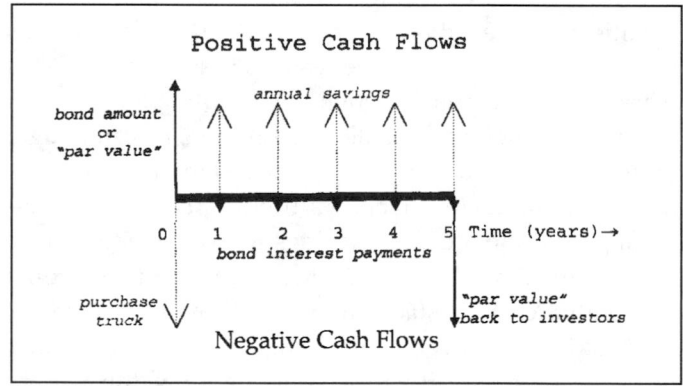

**Figure 25.2 PizzaCo's Cash Flows for a Bond.**

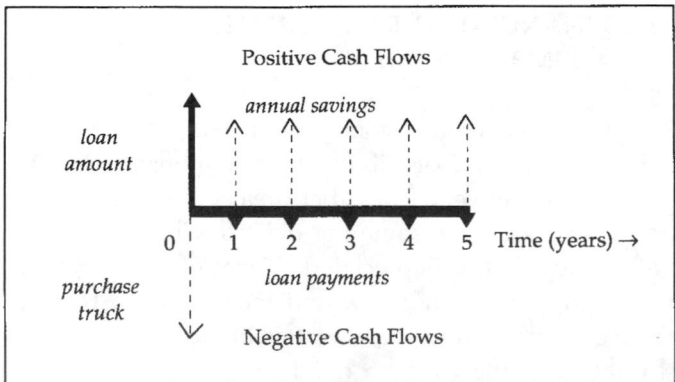

**Figure 25.1 PizzaCo's Cash Flows for a Loan.**

**Figure 25.3. PizzaCo's Cash Flows for a True Lease.**

In a true lease, the contract period should be shorter than the equipment's useful life. The lease is cancelable because the truck can be leased easily to someone else. At the end of the lease, PizzaCo can either return the truck or renew the lease. In a separate transaction, PizzaCo could also negotiate to buy the truck at the fair market value.

If PizzaCo wanted to secure the option to buy the truck (for a bargain price) at the end of the lease, then they would use a capital lease. A capital lease can be structured like an installment loan, however ownership is not transferred until the end of the lease. The lessor retains ownership as security in case the lessee (PizzaCo) defaults on payments. Because the entire cost of the truck is eventually paid, the lease payments are larger than the payments in a true lease, (assuming similar lease periods). Figure 25.4 shows the cash flows for a capital lease with advance payments and a bargain purchase option at the end of year five.

There are some additional scenarios for lease arrangements. A "vendor-financed" agreement is when the lessor (or lender) is the equipment manufacturer. Alternatively, a third party could serve as a financing source. With "third party financing," a finance company would purchase a new truck and lease it to PizzaCo. In either case, there are two primary ways to repay the lessor.

1.  With a "fixed payment plan"; where payments are due whether or not the new truck actually saves money.

2.  With a "flexible payment plan"; where the savings from the new truck are shared with the third party, until the truck's purchase cost is recouped with interest. This is basically a "shared savings" arrangement.

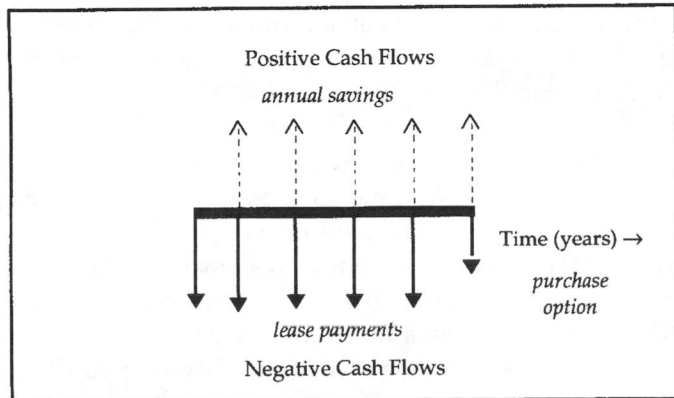

**Figure 25.4 PizzaCo's Cash Flows for a Capital Lease.**

### 25.2.4 Subcontract Pizza Delivery to a Third Party

Since PizzaCo's primary business is not delivery, it could subcontract that responsibility to another company. Let's say that a delivery service company would provide a truck and deliver the pizzas at a reduced cost. Each month, PizzaCo would pay the delivery service company a fee. However, this fee is guaranteed to be less than what PizzaCo would have spent on delivery. Thus, PizzaCo would obtain savings without investing any money or risk in a new truck. This arrangement is analogous to a performance contract.

This arrangement is very similar to a third-party lease and a shared savings agreement. However with a performance contract, the contractor assumes most of the risk, (because they supply the equipment, with little or no investment from PizzaCo). The contractor also is responsible for ensuring that the delivery fee is less than what PizzaCo would have spent. For the PizzaCo example, the arrangement would designed under the conditions below.

*   The delivery company owns and maintains the truck. It also is responsible for all operations related to delivering the pizzas.

*   The monthly fee is related to the number of pizzas delivered. This is the performance aspect of the contract; if PizzaCo doesn't sell many pizzas, the fee is reduced. *A minimum amount of pizzas may be required by the delivery company (performance contractor) to cover costs.* Thus, the delivery company assumes these risks:
    1.  PizzaCo will remain solvent, and
    2.  PizzaCo will sell enough pizzas to cover costs, and
    3.  the new truck will operate as expected and will actually reduce expenses per pizza, and
    4.  the external financial risk, such as inflation and interest rate changes, are acceptable.

*   Because the delivery company is financially strong and experienced, it can usually obtain loans at low interest rates.

*   The delivery company is an expert in delivery; it has specially skilled personnel and uses efficient equipment. Thus, the delivery company can deliver the pizzas at a lower cost (even after adding a profit) than PizzaCo.

Figure 25.5 shows the net cash flows according to PizzaCo. Since the delivery company simply reduces

PizzaCo's operational expenses, there is only a net savings. There are no negative financing cash flows. Unlike the other arrangements, the delivery company's fee is a less expensive substitute for PizzaCo's in-house delivery expenses. With the other arrangements, PizzaCo had to pay a specific financing cost (loan, bond or lease payments, or dividends) associated with the truck, whether or not the truck actually saved money. In addition, PizzaCo would have to spend time maintaining the truck, which would detract from its core focus: making pizzas. With a performance contract, the delivery company is paid from the operational savings it generates. Because the savings are greater than the fee, there is a net savings. Often, the contractor guarantees the savings.

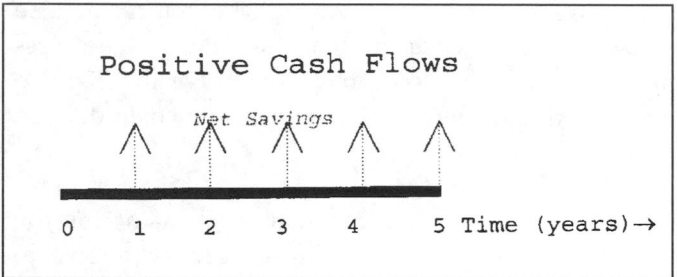

**Figure 25.5 PizzaCo's Cash Flows for a Performance Contract.**

*Supplementary Note: Combinations of the basic finance arrangements are possible. For example, a shared savings arrangement can be structured within a performance contract. Also, performance contracts are often designed so that the facility owner (PizzaCo) would own the asset at the end of the contract.*

## 25.3 FINANCIAL ARRANGEMENTS: DETAILS AND TERMINOLOGY

To explain the basic financial arrangements in more detail, each one is applied to an energy management-related case study. To understand the economics behind each arrangement, some finance terminology is presented below.

### 25.3.1 Finance Terminology

Equipment can be purchased with cash on-hand (officially labeled "retained earnings"), a loan, a bond, a capital lease or by selling stock. Alternatively, equipment can be utilized with a true lease or with a performance contract.

Note that with performance contracting, the building owner is not paying for the equipment itself, but the benefits provided by the equipment. *In the Simple Example, the benefit was the pizza delivery. PizzaCo was not concerned with what type of truck was used.*

The decision to purchase or utilize equipment is partly dependent on the company's strategic focus. If a company wants to delegate some or all of the responsibility of managing a project, it should use a true lease, or a performance contact.[10] However, if the company wants to be intricately involved with the EMP, purchasing and self-managing the equipment could yield the greatest profits. When the building owner purchases equipment, he/she usually maintains the equipment, and lists it as an asset on the balance sheet so it can be depreciated.

Financing for purchases has two categories:

1.  *Debt Financing*, which is borrowing money from someone else, or another firm. (using loans, bonds and capital leases)

2.  *Equity Financing*, which is using money from your company, or your stockholders. (using retained earnings, or issuing common stock)

In all cases, the borrower will pay an interest charge to borrow money. The interest rate is called the "cost of capital." The cost of capital is essentially dependent on three factors: (1) the borrower's credit rating, (2) project risk and (3) external risk. External risk can include energy price volatility, industry-specific economic performance as well as global economic conditions and trends. The cost of capital (or "cost of borrowing") influences the return on investment. If the cost of capital increases, then the return on investment decreases.

The "minimum attractive rate of return" (MARR) is a company's "hurdle rate" for projects. *Because many organizations have numerous projects "competing" for funding, the MARR can be much higher than interest earned from a bank, or other risk-free investment.* Only projects with a return on investment greater than the MARR should be accepted. The MARR is also used as the discount rate to determine the "net present value" (NPV).

### 25.3.2 Explanation of Figures and Tables

Throughout this chapter's case study, figures are presented to illustrate the transactions of each arrangement. Tables are also presented to show how to perform the economic analyses of the different arrangements. The NPV is calculated for each arrangement.

It is important to note that the NPV of a particular arrangement can change significantly if the cost of capital, MARR, equipment residual value, or project life is adjusted. Thus, the examples within this chapter are provided only to illustrate how to perform the analyses. The cash flows and interest rates are estimates, which can vary from project to project. To keep the calculations simple, end-of-year cash flows are used throughout this chapter.

Within the tables, the following abbreviations and equations are used:

EOY = End of Year
Savings = pre-Tax Cash Flow
Depr. = Depreciation
Taxable Income = Savings - Depreciation - Interest Payment
Tax = (Taxable Income)*(Tax Rate)
ATCF = After Tax Cash Flow = Savings – Total Payments – Taxes

Table 25.1 shows the basic equations that are used to calculate the values under each column heading within the economic analysis tables.

*been fully depreciated, he/she can claim the book value as a tax-deduction.**

## 25.4 APPLYING FINANCIAL ARRANGEMENTS: A CASE STUDY

Suppose PizzaCo (*the "host" facility*) needs a new chilled water system for a specific process in its manufacturing plant. The installed cost of the new system is $2.5 million. The expected equipment life is 15 years, however the process will only be needed for 5 years, after which the chilled water system will be sold at an estimated market value of $1,200,000 (book value at year five = $669,375). The chilled water system should save PizzaCo about $1 million/year in energy savings. PizzaCo's tax rate is 34%. The equipment's annual maintenance and insurance cost is $50,000. PizzaCo's MARR is 18%. Since at the end of year 5, PizzaCo expects to sell the asset for an amount greater than its book value, the

#### Table 25.1 Table of Sample Equations used in Economic Analyses.

| A | B | C | D | E | F | G | H | I | J |
|---|---|---|---|---|---|---|---|---|---|
| | | | Payments | | | Principal | Taxable | | |
| EOY | Savings | Depreciation | Principal | Interest | Total | Outstanding | Income | Tax | ATCF |
| n | | | | | | | | | |
| n+1 | | = (MACRS %)* | | | =(D) +(E) | =(G at year n) | =(B)–(C)–(E) | =(H)*(tax rate) | =(B)–(F)–(I) |
| n+2 | | (Purchase Price) | | | | –(D at year n+1) | | | |

Regarding depreciation, the "modified accelerated cost recovery system" (MACRS) is used in the economic analyses. This system indicates the percent depreciation claimable year-by-year after the equipment is purchased. Table 25.2 shows the MACRS percentages for seven-year property. *For example, after the first year, an owner could depreciate 14.29% of an equipment's value. The equipment's "book value" equals the remaining unrecovered depreciation. Thus, after the first year, the book value would be 100%-14.29%, which equals 85.71% of the original value. If the owner sells the property before it has*

#### Table 25.2 MACRS Depreciation Percentages.

| EOY | MACRS Depreciation Percentages for 7-Year Property |
|---|---|
| 0 | 0 |
| 1 | 14.29% |
| 2 | 24.49% |
| 3 | 17.49% |
| 4 | 12.49% |
| 5 | 8.93% |
| 6 | 8.92% |
| 7 | 8.93% |
| 8 | 4.46% |

*To be precise, the IRS uses a "half-year convention" for equipment that is sold before it has been completely depreciated. In the tax year that the equipment is sold, (say year "x") the owner claims only of the MACRS depreciation percent for that year. (This is because the owner has only used the equipment for a fraction of the final year.) Then on a separate line entry, (in the year "x*"), the remaining unclaimed depreciation is claimed as "book value." The x* year is presented as a separate line item to show the book value treatment, however x* entries occur in the same tax year as "x."

additional revenues are called a "capital gain," (which equals the market value – book value) and are taxed. If PizzaCo sells the asset for less than its book value, PizzaCo incurs a "capital loss."

PizzaCo does not have $2.5 million to pay for the new system, thus it considers its finance options. PizzaCo is a small company with an average credit rating, which means that it will pay a higher cost of capital than a larger company with an excellent credit rating. As with any borrowing arrangement, if investors believe that an investment is risky, they will demand a higher interest rate.

### 25.4.1 Purchase Equipment with Retained Earnings (Cash)

If PizzaCo did have enough retained earnings (cash on-hand) available, it could purchase the equipment without external financing. Although external finance expenses would be zero, the benefit of tax-deductions (from interest expenses) is also zero. Also, any cash used to purchase the equipment would carry an "opportunity cost," because that cash could have been used to earn a return somewhere else. This opportunity cost rate is usually set equal to the MARR. In other words, the company lost the opportunity to invest the cash and gain at least the MARR from another investment.

Of all the arrangements described in this chapter, purchasing equipment with retained earnings is probably the simplest to understand. For this reason, it will serve as a brief example and introduction to the economic analysis tables that are used throughout this chapter.

#### 25.4.1.1 Application to the Case Study

Figure 25.6 illustrates the resource flows between the parties. In this arrangement, PizzaCo purchases the chilled water system directly from the equipment manufacturer.

Once the equipment is installed, PizzaCo recovers the full $1 million/year in savings for the entire five

years, but must spend $50,000/year on maintenance and insurance. At the end of the five-year project, PizzaCo expects to sell the equipment for its market value of $1,200,000. Assume MARR is 18%, and the equipment is classified as 7-year property for MACRS depreciation. Table 25.3 shows the economic analysis for purchasing the equipment with retained earnings.

Reading Table 25.3 from left to right, and top to bottom, at EOY 0, the single payment is entered into the table. Each year thereafter, the savings as well as the depreciation (which equals the equipment purchase price multiplied by the appropriate MACRS % for each year) are entered into the table. Year by year, the taxable income = savings – depreciation. The taxable income is then taxed at 34% to obtain the tax for each year. The after-tax cash flow = savings - tax for each year.

At EOY 5, the equipment is sold before the entire value was depreciated. EOY 5* shows how the equipment sale and book value are claimed. In summary, the NPV of all the ATCFs would be $320,675.

### 25.4.2 Loans

Loans have been the traditional financial arrangement for many types of equipment purchases. A bank's willingness to loan depends on the borrower's financial health, experience in energy management and number of years in business. Obtaining a bank loan can be difficult if the loan officer is unfamiliar with EMPs. Loan officers and financiers may not understand energy-related terminology (demand charges, kVAR, etc.). In addition, facility managers may not be comfortable with the financier's language. Thus, to save time, a bank that can understand EMPs should be chosen.

Most banks will require a down payment and collateral to secure a loan. However, securing assets can be difficult with EMPs because the equipment often becomes part of the real estate of the plant. *For example, it would be very difficult for a bank to repossess lighting fixtures from a retrofit.* In these scenarios, lenders may be willing to secure other assets as collateral.

**Figure 25.6 Resource Flows for Using Retained Earnings**

**25.7 Resource Flow Diagram for a Loan.**

**Table 25.3 Economic Analysis for Using Retained Earnings.**

| EOY | Savings | Depr. | Payments | | | Principal | Taxable | Tax | ATCF |
| | | | Principal | Interest | Total | Outstanding | Income | | |
|---|---|---|---|---|---|---|---|---|---|
| 0 | | | | | 2,500,000 | | | | -2,500,000 |
| 1 | 950,000 | 357,250 | | | | | 592,750 | 201,535 | 748,465 |
| 2 | 950,000 | 612,250 | | | | | 337,750 | 114,835 | 835,165 |
| 3 | 950,000 | 437,250 | | | | | 512,750 | 174,335 | 775,665 |
| 4 | 950,000 | 312,250 | | | | | 637,750 | 216,835 | 733,165 |
| 5 | 950,000 | 111,625 | | | | | 838,375 | 285,048 | 664,953 |
| 5* | 1,200,000 | 669,375 | | | | | 530,625 | 180,413 | 1,019,588 |
| | | 2,500,000 | | | | | | | |

Net Present Value at 18%: $320,675

Notes:
| | | | |
|---|---|---|---|
| Loan Amount: | 0 | | |
| Loan Finance Rate: | 0% | MARR | 18% |
| | | Tax Rate | 34% |

MACRS Depreciation for 7-Year Property, with half-year convention at EOY 5
Accounting Book Value at end of year 5: 669,375
Estimated Market Value at end of year 5: 1,200,000
EOY 5* illustrates the Equipment Sale and *Book* Value
Taxable Income: =(Market Value - Book Value)
=(1,200,000 - 669,375) = $530,625

---

### 25.4.2.1 Application to the Case Study

Figure 25.7 illustrates the resource flows between the parties. In this arrangement, PizzaCo purchases the chilled water system with a loan from a bank. PizzaCo makes equal payments (principal + interest) to the bank for five years to retire the debt. Due to PizzaCo's small size, credibility, and inexperience in managing chilled water systems, PizzaCo is likely to pay a relatively high cost of capital. For example, let's assume 15%.

PizzaCo recovers the full $1 million/year in savings for the entire five years, but must spend $50,000/year on maintenance and insurance. At the end of the five-year project, PizzaCo expects to sell the equipment for its market value of $1,200,000. Tables 25.4 and 25.5 show the economic analysis for loans with a zero down payment and a 20% down payment, respectively. Assume that the bank reduces the interest rate to 14% for the loan with the 20% down payment. Since the asset is listed on PizzaCo's balance sheet, PizzaCo can use depreciation benefits to reduce the after-tax cost. In addition, all loan interest expenses are tax-deductible.

### 25.4.3 Bonds

Bonds are very similar to loans; a sum of money is borrowed and repaid with interest over a period of time. The primary difference is that with a bond, the issuer

(PizzaCo) periodically pays the investors only the interest earned. This periodic payment is called the "coupon interest payment." *For example, a $1,000 bond with a 10% coupon will pay $100 per year. When the bond matures, the issuer returns the face value ($1,000) to the investors.*

Bonds are issued by corporations and government entities. Government bonds generate tax-free income for investors, thus these bonds can be issued at lower rates than corporate bonds. This benefit provides government facilities an economic advantage to use bonds to finance projects.

### 25.4.3.1 Application to the Case Study

Although PizzaCo (a private company) would not be able to obtain the low rates of a government bond, they could issue bonds with coupon interest rates competitive with the loan interest rate of 15%.

In this arrangement, PizzaCo receives the investors' cash (bond par value) and purchases the equipment. PizzaCo uses part of the energy savings to pay the coupon interest payments to the investors. When the bond matures, PizzaCo must then return the par value to the investors. See Figure 25.8.

As with a loan, PizzaCo owns, maintains and depreciates the equipment throughout the project's life. All coupon interest payments are tax-deductible. At the end

### Table 25.4 Economic Analysis for a Loan with No Down Payment.

| EOY | Savings | Depr. | Principal | Payments Interest | Total | Principal Outstanding | Taxable Income | Tax | ATCF |
|-----|---------|-------|-----------|--------|-------|--------------|---------|-----|------|
| 0 | | | | 2,500,000 | | | | | |
| 1 | 950,000 | 357,250 | 370,789 | 375,000 | 745,789 | 2,129,211 | 217,750 | 74,035 | 130,176 |
| 2 | 950,000 | 612,250 | 426,407 | 319,382 | 745,789 | 1,702,804 | 18,368 | 6,245 | 197,966 |
| 3 | 950,000 | 437,250 | 490,368 | 255,421 | 745,789 | 1,212,435 | 257,329 | 187,492 | 116,719 |
| 4 | 950,000 | 312,250 | 563,924 | 181,865 | 745,789 | 648,511 | 455,885 | 55,001 | 49,210 |
| 5 | 950,000 | 111,625 | 648,511 | 97,277 | 745,789 | 0 | 741,098 | 251,973 | -47,761 |
| 5* | 1,200,000 | 669,375 | | | 530,625 | 180,413 | 1,019,588 | | |
| | | 2,500,000 | | | | | | | |

Net Present Value at 18%: $757,121

Notes: Loan Amount: 2,500,000 (used to purchase equipment at year 0)

Loan Finance Rate: 15%  MARR 18%

Tax Rate 34%

MACRS Depreciation for 7-Year Property, with half-year convention at EOY 5

Accounting Book Value at end of year 5: 669,375

Estimated Market Value at end of year 5: 1,200,000

EOY 5* illustrates the Equipment Sale and *Book* Value

Taxable Income: =(Market Value - Book Value)

=(1,200,000 - 669,375) = $530,625

### Table 25.5 Economic Analysis for a Loan with a 20% Down-Payment,

| EOY | Savings | Depr. | Principal | Payments Interest | Total | Principal Outstanding | Taxable Income | Tax | ATCF |
|-----|---------|-------|-----------|--------|-------|--------------|---------|-----|------|
| 0 | | | | | 500,000 | 2,000,000 | | | -500,000 |
| 1 | 950,000 | 357,250 | 302,567 | 280,000 | 582,567 | 1,697,433 | 312,750 | 106,335 | 261,098 |
| 2 | 950,000 | 612,250 | 344,926 | 237,641 | 582,567 | 1,352,507 | 100,109 | 34,037 | 333,396 |
| 3 | 950,000 | 437,250 | 393,216 | 189,351 | 582,567 | 959,291 | 323,399 | 109,956 | 257,477 |
| 4 | 950,000 | 312,250 | 448,266 | 134,301 | 582,567 | 511,0241 | 503,449 | 171,173 | 196,260 |
| 5 | 950,000 | 111,625 | 511,024 | 71,543 | 582,567 | 0 | 766,832 | 260,723 | 106,710 |
| 5* | 1,200,000 | 669,375 | | | | | 530,625 | 180,413 | 1,019,588 |
| | | 2,500,000 | | | | | | | |

Net Present Value at 18%: $710,962

Notes: Loan Amount: 2,000,000 (used to purchase equipment at year 0)

Loan Finance Rate: 14%  MARR 18%

500,000  Tax Rate 34%

MACRS Depreciation for 7-Year Property, with half-year convention at EOY 5

Accounting Book Value at end of year 5: 669,375

Estimated Market Value at end of year 5: 1,200,000

EOY 5* illustrates the Equipment Sale and *Book* Value

Taxable Income: =(Market Value - Book Value)

=(1,200,000 - 669,375) = $530,625

**Figure 25.8 Resource Flow Diagram for a Bond.**

of the five-year project, PizzaCo expects to sell the equipment for its market value of $1,200,000. Table 25.6 shows the economic analysis of this finance arrangement.

### 25.4.4 Selling Stock

Although less popular, selling company stock is an equity financing option which can raise capital for projects. For the host, selling stock offers a flexible repayment schedule, because dividend payments to shareholders aren't absolutely mandatory. Selling stock is also often used to help a company attain its desired capital structure. However, selling new shares of stock dilutes the power of existing shares and may send an inaccurate "signal" to investors about the company's financial strength. If the company is selling stock, investors may think that it is desperate for cash and in a poor financial

condition. Under this belief, the company's stock price could decrease. However, recent research indicates that when a firm announces an EMP, investors react favorably.[11] On average, stock prices were shown to increase abnormally by 21.33%.

By definition, the cost of capital (rate) for selling stock is:

$$\text{cost of capital}_{\text{selling stock}} = D/P$$
$$\text{where} \quad D = \text{annual dividend payment}$$
$$P = \text{company stock price}$$

However, in most cases, the after-tax cost of capital for selling stock is higher than the after-tax cost of debt financing (using loans, bonds and capital leases). This is because interest expenses (on debt) are tax deductible, but dividend payments to shareholders are not.

In addition to tax considerations, there are other reasons why the cost of debt financing is less than the financing cost of selling stock. Lenders and bond buyers (creditors) will accept a lower rate of return because they are in a less risky position due to the reasons below.

- Creditors have a contract to receive money at a certain time and future value (stockholders have no such guarantee with dividends).

- Creditors have first claim on earnings (interest is paid before shareholder dividends are allocated).

**Table 25.6 Economic Analysis for a Bond.**

| EOY | Savings | Depr. | Principal | Payments Interest | Total | Principal Outstanding | Taxable Income | Tax | ATCF |
|-----|---------|-------|-----------|-------------------|-------|-----------------------|----------------|-----|------|
| 0 | | | | | | 2,500,000 | | | |
| 1 | 950,000 | 357,250 | | 375,000 | 375,000 | 2,500,000 | 217,750 | 74,035 | 500,965 |
| 2 | 950,000 | 612,250 | | 375,000 | 375,000 | 2,500,000 | -37,250 | -12,665 | 587,665 |
| 3 | 950,000 | 437,250 | | 375,000 | 375,000 | 2,500,000 | 137,750 | 46,835 | 528,165 |
| 4 | 950,0 0 | 312,250 | | 375,000 | 375,000 | 2,500,000 | 262,750 | 89,335 | 485,665 |
| 5 | 950,000 | 111,625 | 2,500,000 | 375,000 | 2,875,000 | 0 | 463,375 | 157,548 | -2,082,548 |
| 5* | 1,200,000 | 669,375 | | | | | 530,625 | 180,413 | 1,019,588 |
| | | 2,500,000 | | | | | | | |

Net Present Value at 18%: 953,927

Notes: Loan Amount: 2,500,000 (used to purchase equipment at year 0)

Loan Finance Rate: 0%   MARR   18%

Tax Rate   34%

MACRS Depreciation for 7-Year Property, with half-year convention at EOY 5

Accounting Book Value at end of year 5:   669,375

Estimated Market Value at end of year 5:   1,200,000

EOY 5* illustrates the Equipment Sale and *Book* Value

Taxable Income:   =(Market Value - Book Value)
=(1,200,000 - 669,375) = $530,625

- Creditors usually have secured assets as collateral and have first claim on assets in the event of bankruptcy.

Despite the high cost of capital, selling stock does have some advantages. This arrangement does not bind the host to a rigid payment plan (like debt financing agreements) because dividend payments are not mandatory. The host has control over when it will pay dividends. Thus, when selling stock, the host receives greater payment flexibility, but at a higher cost of capital.

### 25.4.4.1 Application to the Case Study

As Figure 25.9 shows, the financial arrangement is very similar to a bond, at year zero the firm receives $2.5 million, except the funds come from the sale of stock. Instead of coupon interest payments, the firm distributes dividends. At the end of year five, PizzaCo repurchases the stock. Alternatively, PizzaCo could capitalize the dividend payments, which means setting aside enough money so that the dividends could be paid with the interest generated.

Table 25.7 shows the economic analysis for issuing stock at a 16% cost of equity capital, and repurchasing the stock at the end of year five. (For consistency of comparison to the other arrangements, the stock price does not change during the contract.) Like a loan or

bond, PizzaCo owns and maintains the asset. Thus, the annual savings are only $950,000. PizzaCo pays annual dividends worth $400,000. At the end of year 5, PizzaCo expects to sell the asset for $1,200,000.

Note that Table 25.7 is slightly different from the other tables in this chapter:

Taxable Income = Savings – Depreciation, and

ATCF = Savings – Stock Repurchases - Dividends - Tax

### 25.4.5 Leases

Firms generally own assets, however it is the use of these assets that is important, not the ownership. Leasing is another way of obtaining the use of assets. There are numerous types of leasing arrangements, ranging

**Figure 25.9 Resource Flow Diagram for Selling Stock.**

**Table 25.7 Economic Analysis of Selling Stock.**

| EOY | Savings | Depr. | Stock Transactions | | Dividend | Taxable | Tax | ATCF |
| | | | Sale of Stock | Repurchase | Payments | Income | | |
|---|---|---|---|---|---|---|---|---|
| 0 | | $2,500,000 from Stock Sale is used to purchase equipment, thur ATCF = 0 | | | | | | |
| 1 | 950,000 | 357,250 | | | 400,000 | 592,750 | 201,535 | 348,465 |
| 2 | 950,000 | 612,250 | | | 400,000 | 337,750 | 114,835 | 435,165 |
| 3 | 950,000 | 437,250 | | | 400,000 | 512,750 | 174,335 | 375,665 |
| 4 | 950,000 | 312,250 | | | 400,000 | 637,750 | 216,835 | 333,165 |
| 5 | 950,000 | 111,625 | | 2,500,000 | 400,000 | 838,375 | 285,048 | 2,235,048 |
| 5* | 1,200,000 | 669,375 | | | | 530,625 | 180,413 | 1,019,588 |
| | | 2,500,000 | | | | | | |

Net Present Value at 18%:                                    477,033

Notes:  Value of Stock Sold (which is repurchased after year 5 2,500,000 (used to purchase equipment at year 0)
Cost of Capital = Annual Dividend Rate:                16%            MARR =         18%
Tax Rate =      34%

MACRS Depreciation for 7-Year Property, with half-year convention at EOY 5
Accounting Book Value at end of year 5:           669,375
Estimated Market Value at end of year 5:           1,200,000
*EOY 5\* illustrates the Equipment Sale and Book Value*
              *Taxable Income*:        = (Market Value - Book Value)
                                      = (1,200,000 - 669,375) = $530,625

from basic rental agreements to extended payment plans for purchases. Leasing is used for nearly one-third of all equipment utilization.[12] Leases can be structured and approved very quickly, even within 48 hours. Table 25.8 lists some additional reasons why leasing can be an attractive arrangement for the lessee.

**Table 25.8 Good Reasons to Lease.**

*Financial Reasons*
- With some leases, the entire lease payment is tax-deductible.
- Some leases allow "off-balance sheet" financing, preserving credit lines

*Risk Sharing*
- Leasing is good for short-term asset use, and reduces the risk of getting stuck with obsolete equipment
- Leasing offers less risk and responsibility

Basically, there are two types of leases; the "true lease" (a.k.a. "operating" or "guideline lease") and the "capital lease." One of the primary differences between a true lease and a capital lease is the tax treatment. In a true lease, the lessor owns the equipment and receives the depreciation benefits. However, the lessee can claim the entire lease payment as a tax-deductible business expense. In a capital lease, the lessee (PizzaCo) owns and depreciates the equipment. However, only the interest portion of the lease payment is tax-deductible. In general, a true lease is effective for a short-term project, where the company does not plan to use the equipment when the project ends. A capital lease is effective for long-term equipment.

### 25.4.5.1 The True Lease

Figure 25.10 illustrates the legal differences between a true lease and a capital lease.[13] A true lease (or operating lease) is strictly a rental agreement. The word "strict" is appropriate because the Internal Revenue Service will only recognize a true lease if it satisfies the following criteria:

1. the lease period must be less than 80% of the equipment's life, and

2. the equipment's estimated residual value must be ≥20% of its value at the beginning of the lease, and

3. there is no "bargain purchase option," and

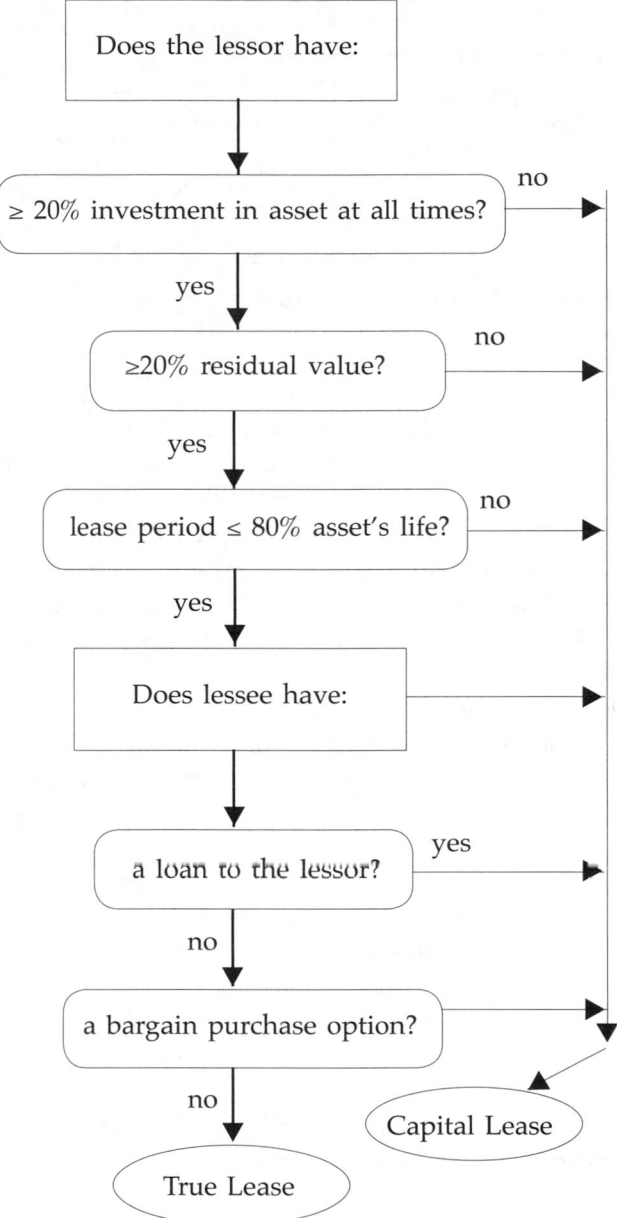

**Figure 25.10 Classification for a True Lease.**

4. there is no planned transfer of ownership, and

5. the equipment must not be custom-made and only useful in a particular facility.

### 25.4.5.2 Application to the Case Study

It is unlikely that PizzaCo could find a lessor that would be willing to lease a sophisticated chilled water system and after five years, move the system to another facility. Thus, obtaining a true lease would be unlikely. However, Figure 25.11 shows the basic relationship between the lessor and lessee in a true lease. A third-party leasing company could also be involved by purchasing

**Figure 25.11 Resource Flow Diagram for a True Lease.**

the equipment and leasing to PizzaCo. Such a resource flow diagram is shown for the capital lease.

Table 25.9 shows the economic analysis for a true lease. Notice that the lessor pays the maintenance and insurance costs, so PizzaCo saves the full $1 million per year. PizzaCo can deduct the entire lease payment of $400,000 as a business expense. However PizzaCo does not obtain ownership, so it can't depreciate the asset.

### 25.4.5.3 The Capital Lease

The capital lease has a much broader definition than a true lease. A capital lease fulfills any one of the following criteria:

1.  the lease term ≥80% of the equipment's life;
2.  the present value of the lease payments ≥80% of the initial value of the equipment;
3.  the lease transfers ownership;
4.  the lease contains a "bargain purchase option," which is negotiated at the inception of the lease.

Most capital leases are basically extended payment plans, except ownership is usually not transferred until the end of the contract. This arrangement is common for large EMPs because the equipment (such as a chilled water system) is usually difficult to reuse at another facility. With this arrangement, the lessee eventually pays for the entire asset (plus interest). In most capital leases, the lessee pays the maintenance and insurance costs.

The capital lease has some interesting tax implications because the lessee must list the asset on its balance sheet from the beginning of the contract. Thus, like a loan, the lessee gets to depreciate the asset and only the interest portion of the lease payment is tax deductible.

### 25.4.5.4 Application to the Case Study

Figure 25.12 shows the basic third-party financing relationship between the equipment manufacturer, lessor and lessee in a capital lease. The finance company (lessor) is shown as a third party, although it also could be a division of the equipment manufacturer. Because the finance company (with excellent credit) is involved, a lower cost of capital (12%) is possible due to reduced risk of payment default.

Like an installment loan, PizzaCo's lease payments cover the entire equipment cost. However, the lease payments are made in advance. Because PizzaCo is considered the owner, it pays the $50,000 annual maintenance expenses, which reduces the annual savings to $950,000. PizzaCo receives the benefits of depreciation and tax-deductible interest payments. To be consistent with the analyses of the other arrangements, PizzaCo would sell the equipment at the end of the lease for its market value. Table 25.10 shows the economic analysis for a capital lease.

### Table 25-9 Economic Analysis for a True Lease

| EOY | Savings | Depr. | Lease Payments | Total | Principal Outstanding | Taxable Income | Tax | ATCF |
|---|---|---|---|---|---|---|---|---|
| 0 | | | 400,000 | 400,000 | | -400,000 | | -400,000 |
| 1 | 1,000,000 | 0 | 400,000 | 400,000 | | 600,000 | 204,000 | 396,000 |
| 2 | 1,000,000 | 0 | 400,000 | 400,000 | | 600,000 | 204,000 | 396,000 |
| 3 | 1,000,000 | 0 | 400,000 | 400,000 | | 600,000 | 204,000 | 396,000 |
| 4 | 1,000,000 | 0 | 400,000 | 400,000 | | 600,000 | 204,000 | 396,000 |
| 5 | 1,000,000 | 0 | | | | 1,000,000 | 340,000 | 660,000 |

Net Present Value at 18%: $953,757

Notes: Annual Lease Payment: 400,000  
MARR = 18%  
Tax Rate 34%

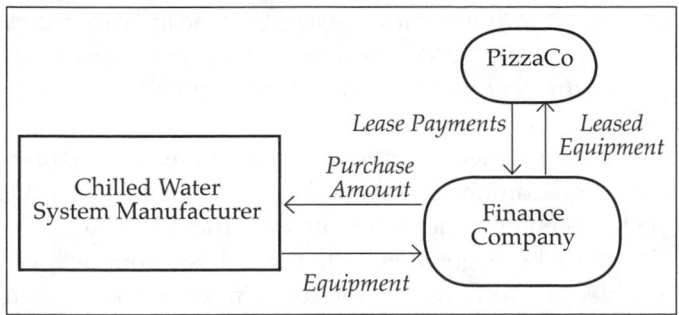

**Figure 25.12 Resource Flow Diagram for a Capital Lease.**

### 25.4.5.5 The Synthetic Lease

A synthetic lease is a "hybrid" lease that combines aspects of a true lease and a capital lease. Through careful structuring and planning, the synthetic lease appears as an operating lease for accounting purposes (enables the Host to have off-balance sheet financing), yet also appears as a capital lease for tax purposes (to obtain depreciation for tax benefits). Consult your local financing expert to learn more about synthetic leases; they must be carefully structured to maintain compliance with the associated tax laws.

With most types of leases, loans and bonds the monthly payments are fixed, regardless of the equip-

ment's utilization, or performance. However, shared savings agreements can be incorporated into certain types of leases.

### 25.4.6 Performance Contracting

Performance contracting is a unique arrangement that allows the building owner to make necessary improvements while investing very little money upfront. The contractor usually assumes responsibility for purchasing and installing the equipment, as well as maintenance throughout the contract. But the unique aspect of performance contracting is that the contractor is paid based on the performance of the installed equipment. Only after the installed equipment actually reduces expenses does the contractor get paid. Energy service companies (ESCOs) typically serve as contractors within this line of business.

Unlike most loans, leases and other fixed payment arrangements, the ESCO is paid based on the performance of the equipment. In other words, if the finished product doesn't save energy or operational costs, the host doesn't pay. This aspect removes the incentive to "cut corners" on construction or other phases of the project, as with bid/spec contracting. In fact, often there is an incentive to exceed savings estimates. For this reason, performance contracting usually entails a more "facility-

**Table 25.10 Economic Analysis for a Capital Lease.**

| EOY | Savings | Depr. | Payments | | | Principal Outstanding | Taxable Income | Tax | ATCF |
|-----|---------|-------|----------|----------|-------|----------------------|----------------|-----|------|
| | | | Principal | Interest | Total | | | | |
| 0 | | | 619,218 | 0 | 619,218 | 1,880,782 | | -619,218 | |
| 1 | 950,000 | 357,250 | 393,524 | 225,694 | 619,218 | 1,487,258 | 367,056 | 124,799 | 205,983 |
| 2 | 950,000 | 612,250 | 440,747 | 178,471 | 619,218 | 1,046,511 | 159,279 | 54,155 | 276,627 |
| 3 | 950,000 | 437,250 | 493,637 | 125,581 | 619,218 | 552,874 | 387,169 | 131,637 | 199,145 |
| 4 | 950,000 | 312,250 | 552,874 | 66,345 | 619,218 | 0 | 571,405 | 194,278 | 136,503 |
| 5 | 950,000 | 111,625 | | | | | 838,375 | 285,048 | 664,953 |
| 5* | 1,200,000 | 669,375 | | | | | 530,625 | 180,413 | 1,019,588 |
| | | 2,500,000 | | | | | | | |
| | | | | | | Net Present Value at 18%: | | | $681,953 |

Notes:    Total Lease Amount:      2,500,000
However, Since the payments are in advance, the first payment is analogous to a Down-Payment
Thus the actual amount borrowed is only = 2,500,000 - 619,218 = 1,880,782
Lease Finance Rate:    12%                MARR  18%
Tax Rate                34%
MACRS Depreciation for 7-Year Property, with half-year convention at EOY 5
Accounting Book Value at end of year 5:    669,375
Estimated Market Value at end of year 5:   1,200,000
EOY 5* illustrates the Equipment Sale and Book Value
                        Taxable Income:    = (Market Value - Book Value)
                                          = (1,200,000 - 669,375) = $530,625

wide" scope of work (to find extra energy savings), than loans or leases on particular pieces of equipment.

With a facility-wide scope, many improvements can occur at the same time. For example, lighting and air conditioning systems can be upgraded at the same time. In addition, the indoor air quality can be improved. With a comprehensive facility management approach, a "domino-effect" on cost reduction is possible. For example, if facility improvements create a safer and higher quality environment for workers, productivity could increase. As a result of decreased employee absenteeism, the workman's compensation cost could also be reduced. These are additional benefits to the facility.

Depending on the host's capability to manage the risks (equipment performance, financing, etc.) the host will delegate some of these responsibilities to the ESCO. In general, the amount of risk assigned to the ESCO is directly related to the percent savings that must be shared with the ESCO.

For facilities that are not in a good position to manage the risks of an energy project, performance contracting may be the only economically feasible implementation method. *For example, the US Federal Government used performance contracting to upgrade facilities when budgets were being dramatically cut. In essence, they "sold" some of their future energy savings to an ESCO, in return for receiving new equipment and efficiency benefits.*

In general, performance contracting may be the best option for facilities that:

- are severely constrained by their cash flows;

- have a high cost of capital;

- don't have sufficient resources, such as a lack of in-house energy management expertise or an inadequate maintenance capacity*;

- are seeking to reduce in-house responsibilities and focus more on their core business objectives; or

- are attempting a complex project with uncertain reliability or if the host is not fully capable of managing the project. *For example, a lighting retrofit has a high probability of producing the expected cash flows, whereas a completely new process does not have*

*the same "time-tested" reliability. If the in-house energy management team cannot manage this risk, performance contracting may be an attractive alternative.*

Performance contracting does have some drawbacks. In addition to sharing the savings with an ESCO, the tax benefits of depreciation and other economic benefits must be negotiated. Whenever large contracts are involved, there is reason for concern. One study found that 11% of customers who were considering EMPs felt that dealing with an ESCO was too confusing or complicated.[14] Another reference claims, "with complex contracts, there may be more options and more room for error."[15] Therefore, it is critical to choose an ESCO with a good reputation and experience within the types of facilities that are involved.

There are a few common types of contracts. The ESCO will usually offer the following options:

- guaranteed fixed dollar savings;
- guaranteed fixed energy (MMBtu) savings;
- a percent of energy savings; or
- a combination of the above.

Obviously, facility managers would prefer the options with "guaranteed savings." However this extra security (and risk to the ESCO) usually costs more. The primary difference between the two guaranteed options is that guaranteed fixed dollar savings contracts ensure dollar savings, even if energy prices fall. *For example, if energy prices drop and the equipment does not save as much money as predicted, the ESCO must pay (out of its own pocket) the contracted savings to the host.*

Percent energy savings contracts are agreements that basically share energy savings between the host and the ESCO. The more energy saved, the higher the revenues to both parties. However, the host has less predictable savings and must also periodically negotiate with the ESCO to determine "who saved what" when sharing savings. There are numerous hybrid contracts available that combine the positive aspects of the above options.

### 25.4.6.1 Application to the Case Study

PizzaCo would enter into a hybrid contract; *percent energy savings/guaranteed arrangement.* The ESCO would purchase, install and operate a highly efficient chilled water system. The ESCO would guarantee that PizzaCo would save the $1,000,000 per year, but PizzaCo would pay the ESCO 80% of the savings. In this way, PizzaCo would not need to invest any money, and would simply collect the net savings of $200,000 each year. To avoid periodic negotiations associated with shared savings

---

*Maintenance capacity represents the ability that the maintenance personnel will be able to maintain the new system. It has been shown that systems fail and are replaced when maintenance concerns are not incorporated into the planning process. See Woodroof, E. (1997) "Lighting Retrofits: Don't Forget About Maintenance," Energy Engineering, 94(1) pp. 59-68.

agreements, the contract could be worded such that the ESCO will provide guaranteed energy savings worth $200,000 each year.

With this arrangement, there are no depreciation, interest payments or tax-benefits for PizzaCo. However, PizzaCo receives a positive cash flow with no investment and little risk. At the end of the contract, the ESCO removes the equipment. At the end of most performance contracts, the host usually acquires or purchases the equipment for fair market value. However, for this case study, the equipment was removed to make a consistent comparison with the other financial arrangements.

Figure 25.13 illustrates the transactions between the parties. Table 25.11 presents the economic analysis for performance contracting.

Note that Table 25.11 is slightly different from the other tables in this chapter: Taxable Income = Savings – Depreciation – ESCO Payments.

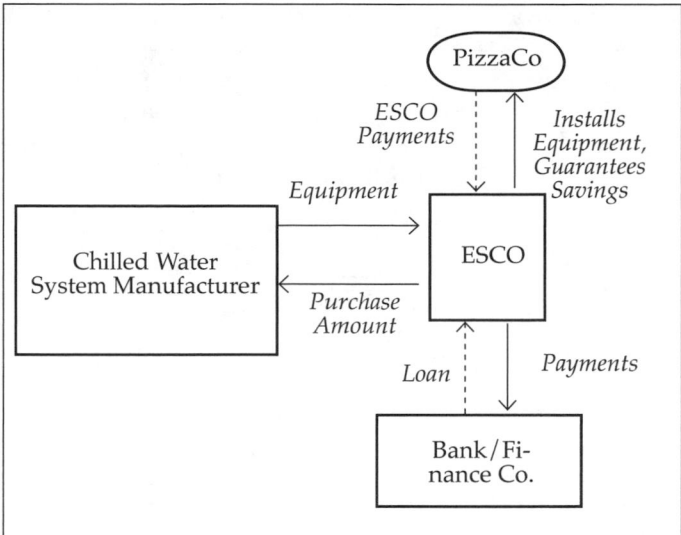

**Figure 25.13 Transactions for a Performance Contract.**

### 25.4.7 Summary Of Tax Benefits

Table 25.12 summarizes the tax benefits of each financial arrangement presented in this chapter.

### 25.4.8 Additional Options

Combinations of the basic financial arrangements can be created to enhance the value of a project. A sample of the possible combinations are described below.

• Third party financiers often cooperate with performance contracting firms to implement EMPs.

• Utility rebates and government programs may provide additional benefits for particular projects.

• Tax-exempt leases are available to government facilities.

• Insurance can be purchased to protect against risks relating to equipment performance, energy savings, etc.

• Some financial arrangements can be structured as non-recourse to the host. Thus, the ESCO or lessor would assume the risks of payment default. However, as mentioned before, profit sharing increases with risk sharing.

Attempting to identify the absolute best financial arrangement is a rewarding goal, unless it takes too long. As every minute passes, potential dollar savings are lost forever. When considering special grant funds, rebate programs or other unique opportunities, it is important to consider the lost savings due to delay.

**Table 25.11 Economic Analysis of a Performance Contract.**

| EOY | Savings | Depr. | ESCO Payments | Total | Principal Outstanding | Taxable Income | Tax | ATCF |
|---|---|---|---|---|---|---|---|---|
| 0 | | | | | | | | |
| 1 | 1,000,000 | 0 | 800,000 | 800,000 | | 200,000 | 68,000 | 132,000 |
| 2 | 1,000,000 | 0 | 800,000 | 800,000 | | 200,000 | 68,000 | 132,000 |
| 3 | 1,000,000 | 0 | 800,000 | 800,000 | | 200,000 | 68,000 | 132,000 |
| 4 | 1,000,000 | 0 | 800,000 | 800,000 | | 200,000 | 68,000 | 132,000 |
| 5 | 1,000,000 | 0 | 800,000 | 800,000 | | 200,000 | 68,000 | 132,000 |
| | | | | Net Present Value at 18%: | | | | $412,787 |

Notes: ESCO purchases/operates equipment. Host pays ESCO 80% of the savings = $800,000.
The contract could also be designed so that PizzaCo can buy the equipment at the end of year 5.

**Table 25.12 Host's Tax Benefits for each Arrangement.**

| ARRANGEMENT | Depreciation Benefits | Interest Payments are Tax-Deductible | Total Payments are Tax-Deductible |
|---|---|---|---|
| Retained Earnings | X | | |
| Loan | X | X | |
| Bond | X | X | |
| Sell Stock | X | | |
| Capital Lease | X | X | |
| True Lease | | | X |
| Performance Contract | | | X |

## 25.5 "PROS" & "CONS" OF EACH FINANCIAL ARRANGEMENT

This section presents a brief summary of the "Pros" and "Cons" of each financial arrangement from the host's perspective.

### Loan

"Pros":
- host keeps all savings,
- depreciation & interest payments are tax-deductible,
- host owns the equipment, and
- the arrangement is good for long-term use of equipment

"Cons":
- host takes all the risk, and must install and manage project

### Bond

Has the same Pros/Cons as loan, and "Pro":
- good for government facilities, because they can offer a tax-free rate (that is lower, but considered favorable by investors)

### Sell Stock

Has the same Pros/Cons as loan, and "Pro":
- selling stock could help the host achieve its target capital structure

"Con":
- dividend payments (unlike interest payments) are not tax-deductible, and
- dilutes company control

### Use Retained Earnings

Has the same Pros/Cons as loan, and "Pro":
- host pays no external interest charges. However retained earnings do carry an opportunity cost, because such funds could be invested somewhere at the MARR.

"Con":
- host loses tax-deductible benefits of interest charges

### Capital Lease

Has the same Pros/Cons as loan, and "Pro":
- Greater flexibility in financing, possible lower cost of capital with third-party participation

### True Lease

"Pros":
- allows use of equipment, without ownership risks,
- reduced risk of poor performance, service, equipment obsolescence, etc.,
- good for short-term use of equipment, an
- entire lease payment is tax-deductible

"Cons":
- no ownership at end of lease contract, and
- no depreciation tax benefits

### Performance Contract

"Pros":
- allows use of equipment, with reduced installment/operational risks, and
- reduced risk of poor performance, service, equipment obsolescence, etc., and

- allows host to focus on its core business objectives

"Cons":
- potentially binding contracts, legal expenses, and increased administrative costs, and
- host must share project savings

### 25.5.1 Rules of Thumb

When investigating financing options, consider the following generalities:

Loans, bonds and other host-managed arrangements should be used when a customer has the resources (experience, financial support, and time) to handle the risks. Performance contracting (ESCO assumes most of the risk) is usually best when a customer doesn't have the resources to properly manage the project. Remember that with any arrangement where the host delegates risk to another firm, the host must also share the savings.

Leases are the "middle ground" between owning and delegating risks. Leases are very popular due to their tax benefits.

True leases tend to be preferred when:
- the equipment is needed on a short-term basis;
- the equipment has unusual service problems that cannot be handled by the host;
- technological advances cause equipment to become obsolete quickly; or
- depreciation benefits are not useful to the lessee.

Capital Leases are preferred when:
- the installation and removal of equipment is costly;
- the equipment is needed for a long time; or
- the equipment user desires to secure a "bargain purchase option."

## 25.6 CHARACTERISTICS THAT INFLUENCE WHICH FINANCIAL ARRANGEMENT IS BEST

There are at least three types of characteristics that can influence which financial arrangement should be used for a particular EMP. These include facility characteristics, project characteristics and financial arrangement characteristics. In this section, quantitative characteristics are bulleted with this symbol: $. The qualitative characteristics are bulleted with this symbol: ☺ Note that qualitative characteristics are generally "strategic" and are not associated with an exact dollar value.

A few of the Facility Characteristics include:

☺ The long-term plans of facility. For example, is the facility trying to focus on core business objectives and outsourcing other tasks, such as EMPs?

$ The facility's current financial condition. Credit ratings and ability to obtain loans can determine whether certain financial arrangements are feasible.

☺ The experience and technical capabilities of in-house personnel. Will additional resources (personnel, consultants, technologies, etc.) be needed to successfully implement the project?

$ The facility's ability to obtain rebates from the government, utilities, or other organizations. For example, there are Dept. of Energy subsidies available for DOE facilities.

$ The facility's ability to obtain tax benefits. For example, government facilities can offer tax-exempt interest rates on bonds.

A few of the Project Characteristics include:

$ The project's economic benefits. Net Present Value, Internal Rate of Return and Simple Payback.

☺ The project's complexity and overall risk. For example, a complex project that has never been done before has a different level of risk than a standard lighting retrofit.

☺ The project's alignment with the facility's long-term objectives. Will this project's equipment be needed for long-term goals?

☺ The project's cash flow schedule and the variance between cash flows. For example, there may be significant differences in the acceptability of a project based on when revenues are received.

A few of the Financial Arrangement Characteristics include:

$ The economic benefit of a project using a particular financial arrangement. The Net Present Value and Internal Rate of Return can be influenced by the financial arrangement selected.

☺  The impact on the corporate capital structure. For example, will additional debt be required to finance the project? Will additional liabilities appear on the firm's balance sheet and impact the image of the company to investors?

☺  The flexibility of the financial arrangement. For example, can the facility manager alter the contract and payment terms in the event of revenue shortfall or changes in operational hours?

## 25.7 INCORPORATING STRATEGIC ISSUES WHEN SELECTING FINANCIAL ARRANGEMENTS

Because strategic issues can be important when selecting financial arrangements, the facility manager should include them in the selection process. The following questions can help assess a facility manager's needs.

- Does the facility manager want to manage projects or outsource?
- Are net positive cash flows required?
- Will the equipment be needed for long-term needs?
- Is the facility government or private?
- If private, does the facility manager want the project's assets on or off the balance sheet?
- Will operations be changing?

From the research experience, a Strategic Issues Financing Decision Tree was developed to guide facility managers to the financial arrangement which is most likely optimal. Figure 25.14 illustrates the decision tree, which is by no means a rule, but it embodies some general observations from the industry.

Working the tree from the top to bottom, the facility manager should assess the project and facility characteristics to decide whether it is strategic to manage the project or outsource. If outsourced, the "performance contract" would be the logical choice.* If the facility manager wants to manage the project, the next step (moving down the tree) is to evaluate whether the project's equipment will be needed for long or short-term purposes. If short-term, the "true lease" is logical. If it is a long-term project, in a government facility, the "bond" is likely to be the best option. If the facility is

in the private sector, the facility manager should decide whether the project should be on or off the balance sheet. An off-balance sheet preference would lead back to the "true lease." If the facility manager wants the project's assets on the balance sheet, the Net Present Value (or other economic benefit indicator) can help determine which "host-managed" arrangement (loan, capital lease or cash) would be most lucrative.

## 25.8  CHAPTER SUMMARY

It is clear that knowing the strategic needs of the facility manager is critical to selecting the best arrangement. There are practically an infinite number of financial alternatives to consider. This chapter has provided some information on the basic financial arrangements. Combining these arrangements to construct the best contract for your facility is only limited by your creativity.

## 25.9  GLOSSARY

**Capitalize**
To convert a schedule of cash flows into a principal amount, called capitalized value, by dividing by a rate of interest. In other words, to set aside an amount large enough to generate (via interest) the desired cash flows forever.

**Capital or Financial Lease**
Lease that under Statement 13 of the Financial Accounting Standards Board must be reflected on a company's balance sheet as an asset and corresponding liability. Generally, this applies to leases where the lessee acquires essentially all of the economic benefits and risks or the leased property.

**Depreciation**
The amortization of fixed assets, such as plant and equipment, so as to allocate the cost over their depreciable life. Depreciation reduces taxable income, but is not an actual cash flow.

**Energy Service Company (ESCO)**
Company that provides energy services (and possibly financial services) to an energy consumer.

**Host**
The building owner or facility that uses the equipment.

**Lender**
Individual or firm that extends money to a borrower

---

*It should be noted that a performance contract could be structured using leases and bonds.

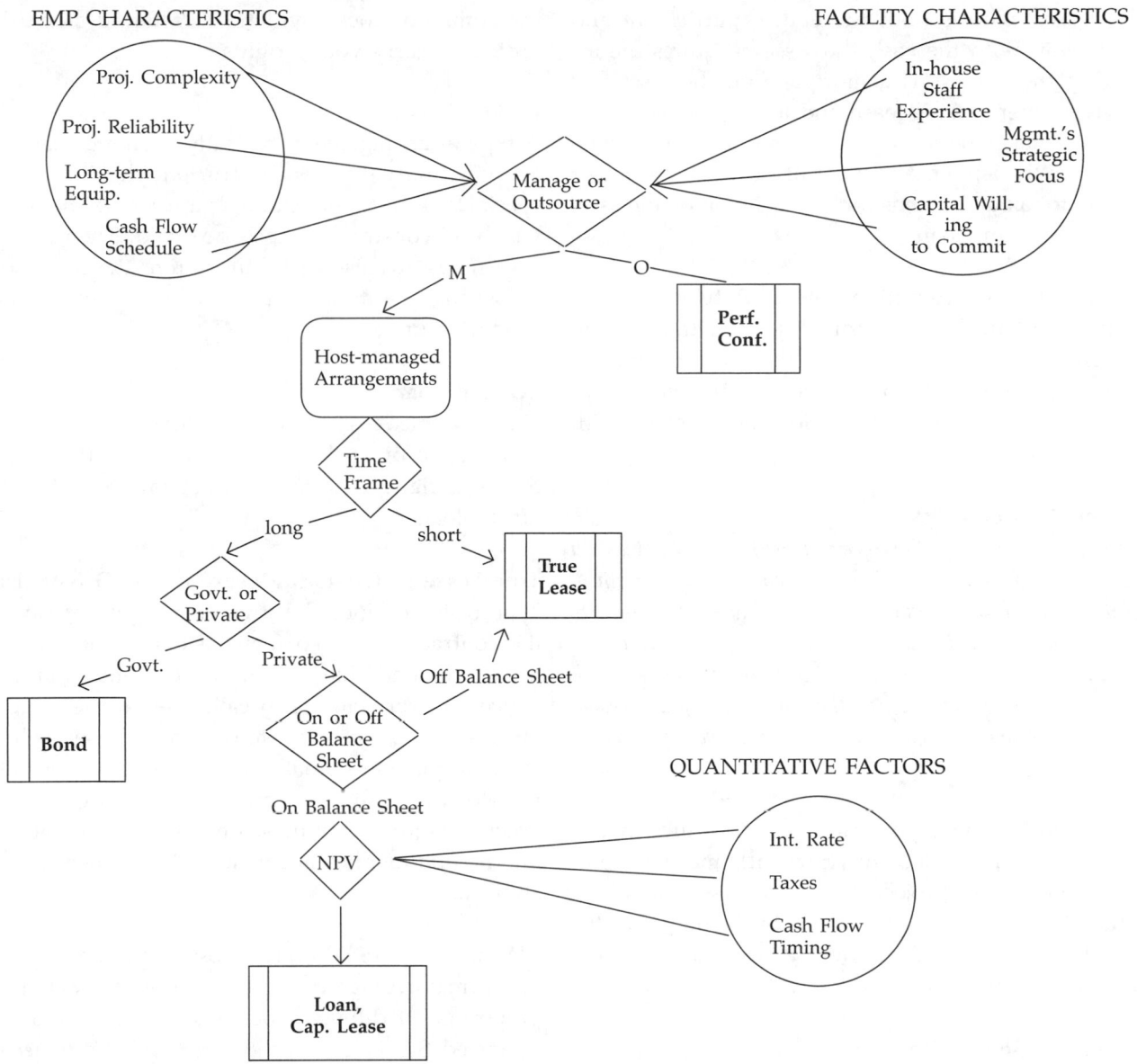

EMP CHARACTERISTICS

FACILITY CHARACTERISTICS

QUANTITATIVE FACTORS

**Figure 25.14 Strategic Issues Financing Decision Tree.**

with the expectation of being repaid, usually with interest. Lenders create debt in the form of loans or bonds. If the borrower is liquidated, the lender is paid off before stockholders receive distributions.

**Lessee**
The renter. The party that buys the right to use equipment by making lease payments to the lessor.

**Lessor**
The owner of the leased equipment.

**Line of Credit**
An informal agreement between a bank and a borrower indicating the maximum credit the bank will extend.

A line of credit is popular because it allows numerous borrowing transactions to be approved without the re-application paperwork.

**Liquidity**
Ability of a company to convert assets into cash or cash equivalents without significant loss. For example, investments in money market funds are much more liquid than investments in real estate.

**Leveraged Lease**
Lease that involves a lender in addition to the lessor and lessee. The lender, usually a bank or insurance company, puts up a percentage of the cash required to purchase the asset, usually more than half. The balance is put up

by the lessor, who is both the equity participant and the borrower. With the cash the lessor acquires the asset, giving the lender (1) a mortgage on the asset and (2) an assignment of the lease and lease payments. The lessee then makes periodic payments to the lessor, who in turn pays the lender. As owner of the asset, the lessor is entitled to tax deductions for depreciation on the asset and interest on the loan.

### MARR (Minimum Attractive Rate of Return)

MARR is the "hurdle rate" for projects within a company. MARR is used to determine the NPV; the annual after-tax cash flow is discounted at MARR (which represents the rate the company could have received with a different project).

### Net Present Value (NPV)

*As the saying goes, "a dollar received next year is not worth as much as a dollar today." The NPV converts the worth of that future dollar into what is worth today. NPV converts future cash flows by using a given discount rate. For example, at 10%, $1,000 dollars received one year from now is worth only $909.09 dollars today. In other words, if you invested $909.09 dollars today at 10%, in one year it would be worth $1,000.*

NPV is useful because you can convert future savings cash flows back to "time zero" (present), and then compare to the cost of a project. If the NPV is positive, the investment is acceptable. In capital budgeting, the discount rate used is called the hurdle rate and is usually equal to the incremental cost of capital.

### "Off-Balance Sheet" Financing

Typically refers to a True Lease, because the assets are not listed on the balance sheet. Because the liability is not on the balance sheet, the Host appears to be financially stronger. However, most large leases must be listed in the footnotes of financial statements, which reveals the "hidden assets."

### Par Value or Face Value

Equals the value of the bond at maturity. For example, a bond with a $1,000 dollar par value will pay $1,000 to the issuer at the maturity date.

### Preferred Stock

A hybrid type of stock that pays dividends at a specified rate (like a bond), and has preference over common stock in the payment of dividends and liquidation of assets. However, if the firm is financially strained, it can avoid paying the preferred dividend as it would

the common stock dividends. Preferred stock doesn't ordinarily carry voting rights.

### Project Financing

A type of arrangement, typically meaning that a Single Purpose Entity (SPE) is constructed. The SPE serves as a special bank account. All funds are sent to the SPE, from which all construction costs are paid. Then all savings cash flows are also distributed from the SPE. The SPE is essentially a mini-company, with the sole purpose of funding a project.

### Secured Loan

Loan that pledges assets as collateral. Thus, in the event that the borrower defaults on payments, the lender has the legal right to seize the collateral and sell it to pay off the loan.

### True Lease or Operating Lease or Tax-Oriented Lease

Type of lease, normally involving equipment, whereby the contract is written for considerably less time than the equipment's life and the lessor handles all maintenance and servicing; also called service lease. Operating leases are the opposite of capital leases, where the lessee acquires essentially all the economic benefits and risks of ownership. Common examples of equipment financed with operating leases are office copiers, computers, automobiles and trucks. Most operating leases are cancelable.

### WACC (Weighted Average Cost of Capital)

The firm's average cost of capital, as a function of the proportion of different sources of capital: Equity, Debt, Preferred Stock, etc. *For example, a firm's target capital structure is:*

| Capital Source | Weight ($w_i$) |
|----------------|----------------|
| Debt | 30% |
| Common Equity | 60% |
| Preferred Stock | 10% |

*and the firm's costs of capital are:*

before tax cost of debt $= k_d = 10\%$
cost of common equity $= k_s = 15\%$
cost of preferred stock $= k_{ps} = 12\%$

*Then the weighted average cost of capital will be:*
$$\text{WACC} = w_d k_d (1-T) + w_s k_s + w_{ps} k_{ps}$$

where  $w_i$ = weight of Capital Source$_i$
T = tax rate = 34%
After-tax cost of debt = $k_d(1-T)$

Thus,

$$WACC = (.3)(.1)(1-.34) + (.6)(.15) + (.1)(.12)$$

$$WACC = 12.18\%$$

## References

1. Wingender, J. and Woodroof, E., (1997) "When Firms Publicize Energy Management Projects Their Stock Prices Go Up: How High?—As Much as 21.33% within 150 days of an Announcement," *Strategic Planning for Energy and the Environment*, Vol. 17(1), pp. 38-51.

2. U.S. Department of Energy, (1996) "Analysis of Energy-Efficiency Investment Decisions by Small and Medium-Sized Manufacturers," U.S. DOE, Office of Policy and Office of Energy Efficiency and Renewable Energy, pp. 37-38.

3. Woodroof, E. and Turner, W. (1998), "Financial Arrangements for Energy Management Projects," *Energy Engineering* 95(3) pp. 23-71.

4. Sullivan, A. and Smith, K. (1993) "Investment Justification for U.S. Factory Automation Projects," *Journal of the Midwest Finance Association*, Vol. 22, p. 24.

5. Fretty, J. (1996), "Financing Energy-Efficient Upgraded Equipment," Proceedings of the 1996 International Energy and Environmental Congress, Chapter 10, Association of Energy Engineers.

6. Pennsylvania Energy Office, (1987) The Pennsylvania Life Cycle Costing Manual.

7. United States Environmental Protection Agency (1994). ProjectKalc, Green Lights Program, Washington DC

8. Tellus Institute, (1996), P2/Finance version 3.0 for Microsoft Excel Version 5, Boston MA.

9. Woodroof, E. And Turner, W. (1999) "Best Ways to Finance Your Energy Management Projects," *Strategic Planning for Energy and the Environment*, Summer 1999, Vol. 19(1) pp. 65-79.

10. Cooke, G.W., and Bomeli, E.C., (1967), *Business Financial Management*, Houghton Mifflin Co., New York.

11. Wingender, J. and Woodroof, E., (1997) "When Firms Publicize Energy Management Projects: Their Stock Prices Go Up," *Strategic Planning for Energy and the Environment*, 17 (1) pp. 38-51.

12. Sharpe, S. and Nguyen, H. (1995) "Capital Market Imperfections and the Incentive to Lease," *Journal of Financial Economics*, 39(2), p. 271-294.

13. Schallheim, J. (1994), *Lease or Buy?*, Harvard Business School Press, Boston, p. 45.

14. Hines, V. (1996),"EUN Survey: 32% of Users Have Signed ESCO Contracts," *Energy User News* 21(11), p.26.

15. Coates, D.F. and DelPonti, J.D. (1996), "Performance Contracting: a Financial Perspective" *Energy Business and Technology Sourcebook*, Proceedings of the 1996 World Energy Engineering Congress, Atlanta. p. 539-543.

# CHAPTER 26

# COMMISSIONING FOR ENERGY MANAGEMENT

**DAVID E. CLARIDGE**
Professor
Mechanical Engineering Department
Texas A&M University

**MINGSHENG LIU**
Associate Professor
Architectural Engineering
University of Nebraska - Lincoln

**W.D. TURNER**
Professor
Mechanical Engineering Department
Texas A&M University

## 26.1 INTRODUCTION TO COMMISSIONING FOR ENERGY MANAGEMENT

Commissioning an existing building has been shown to be a key energy management activity over the last decade, often resulting in energy savings of 10%, 20% or sometimes 30% without significant capital investment. Commissioning is more often applied to new buildings today than to existing buildings, but the energy manager will have more opportunities to apply the process to an existing building as part of the overall energy management program. Hence, this chapter emphasizes commissioning applied to existing buildings, but also provides some commissioning guidance for the energy manager who is involved in a construction project.

Commissioning an existing building provides several benefits in addition to being an extremely effective energy management strategy. It typically provides an energy payback of one to three years. In addition, building comfort is improved, systems operate better and maintenance cost is reduced. Commissioning measures typically require no capital investment, though the process often identifies maintenance that is required before the commissioning can be completed. Potential capital upgrades or retrofits are often identified during the commissioning activities, and knowledge gained during the process permits more accurate quantification of benefits than is possible with a typical audit. Involve-

ment of facilities personnel in the process can also lead to improved staff technical skills.

This chapter is intended to provide the energy manager with the information needed to make the decision to conduct an in-house commissioning program or to select and work with an outside commissioning provider. There is no single definition of commissioning for an existing building, or for new buildings, so several widely used commissioning definitions are given. The commissioning process used by the authors in existing buildings is described in some detail, and common commissioning measures and commissioning resources are described so the energy manager can choose how to implement a commissioning program. Monitoring and verification is very important to a successful commissioning program. Some commissioning specific M&V issues are discussed, particularly the role of M&V in identifying the need for follow-up commissioning activities. Commissioning a new building is described from the perspective of the energy manager. Three case studies illustrate different applications of the commissioning process as part of the overall energy management program.

## 26.2 COMMISSIONING DEFINITIONS

To commission a navy ship refers to the order or process that makes it completely ready for active duty. Over the last two decades, the term has come to refer to the process that makes a building or some of its systems completely ready for use. In the case of existing buildings, it generally refers to a restoration or improvement in the operation or function of the building systems.

### 26.2.1 New Building Commissioning

ASHRAE defines building commissioning as: "the process of ensuring systems are designed, installed, functionally tested, and operated in conformance with the design intent. Commissioning begins with planning and includes design, construction, start-up, acceptance, and training and can be applied throughout the life of the building. Furthermore, the commissioning process encompasses and coordinates the traditionally separate functions of systems documentation, equipment start-up, control system calibration, testing and balancing, and performance testing."[1]

This guideline was restricted to new buildings, but it later became evident that while initial start-up problems were not an issue in older buildings, most of the other problems that commissioning resolved were even more prevalent in older systems.

### 26.2.2  Recommissioning

Recommissioning refers to commissioning a building that has already been commissioned at least once. After a building has been commissioned during the construction process, recommissioning ensures that the building continues to operate effectively and efficiently. Buildings, even if perfectly commissioned, will normally drift away from optimum performance over time, due to system degradation, usage changes, or failure to correctly diagnose the root cause of comfort complaints. Therefore, recommissioning normally reapplies the original commissioning procedures in order to keep the building operating according to design intent or it may modify them for current operating needs.

Optimally, recommissioning becomes part of a facility's continuing O&M program. There is not yet a consensus on recommissioning frequency, but some consider that it should occur every 3 to 5 years. If there are frequent build-outs or changes in building use, recommissioning should be applied more often.[2]

### 26.2.3  Retrocommissioning

Retrocommissioning is the first-time commissioning of an existing building. Many of the steps in the retrocommissioning process are similar to those for commissioning. Retrocommissioning, however, occurs after construction, as an independent process, and its focus is usually on energy-using equipment such as mechanical equipment and related controls. Retrocommissioning may or may not bring the building back to its original design intent, since the usage may have changed or the original design documentation may no longer exist.[3]

### 26.2.4  Continuous Commissioning®[45]

Continuous Commissioning (CC$^{SM}$) is an ongoing process to resolve operating problems, improve comfort, optimize energy use, and identify retrofits for existing commercial and institutional buildings and central plant facilities. CC focuses on improving overall system control and operations for the building, as it is currently utilized, and on meeting existing facility needs. CC is much more than an operations and maintenance program. It is not intended to ensure that a building's systems function as originally designed, but it ensures

that the building and its systems operate optimally to meet the current uses of the building. As part of the CC process, a comprehensive engineering evaluation is conducted for both building functionality and system functions. Optimal operational parameters and schedules are developed based on actual building conditions and current occupancy requirements.

## 26.3  THE COMMISSIONING PROCESS IN EXISTING BUILDINGS

There are multiple terms that describe the commissioning process for existing buildings as noted in the previous section. Likewise, there are many adaptations of the process itself. The same practitioner will implement the process differently in different buildings, based on the budget and the owner requirements. The process described here is the process used by the chapter authors when the owner wants a thorough commissioning job. The terminology used will refer to the continuous commissioning process, but many of the steps are the same for retrocommissioning or recommissioning. The model described assumes that a commissioning provider is involved, since that is normally the case. Some (or all) of the steps may be implemented by the facility staff if they have the expertise and adequate staffing levels to take on the work.

CC focuses on improving overall system control and operations for the building, as it is currently utilized, and on meeting existing facility needs. It does not ensure that the systems function as originally designed, but ensures that the building and systems operate optimally to meet the current requirements. During the CC process, a comprehensive engineering evaluation is conducted for both building functionality and system functions. The optimal operational parameters and schedules are developed based on actual building conditions and current occupancy requirements. An integrated approach is used to implement these optimal schedules to ensure practical local and global system optimization and persistence of the improved operation schedules.

### 26.3.1 Commissioning Team

The CC team consists of a project manager, one or more CC engineers and CC technicians, and one or more designated members of the facility operating team. The primary responsibilities of the team members are shown in Table 26.1. The project manager can be an owner representative or a CC provider representative. It is essential that the engineers have the qualifications and experience to perform the work specified in the table. The designated facility team members generally include at least one lead HVAC technician and an EMCS operator or engineer. It is essential that the designated members of the facility operating team actively participate in the process and be convinced of the value of the measures proposed and implemented, or operation will rapidly revert to old practices.

### 26.3.2 CC Process

The CC process consists of two phases. The first phase is the project development phase that identifies the buildings and facilities to be included in the project and develops the project scope. At the end of this phase, the CC scope is clearly defined and a CC contract is signed as described in Section 26.3.2.1. The second phase implements CC and verifies project performance through the six steps outlined in Figure 26.1 and described in Section 26.3.2.2.

### 26.3.2.1 Phase 1: Project Development
**Step 1: Identify Buildings or Facilities**

**Objective**: Screen potential CC candidates with minimal effort to identify buildings or facilities that will receive a CC audit. The CC candidate can be a building, an entire facility, or a piece of equipment. If the building is part of a complex or campus, it is desirable to select the entire facility as the CC candidate since one mechanical problem may be rooted in another part of the building or facility.

**Approach**: The CC candidates can be selected based on one or more of the following criteria:
- The candidate provides poor thermal comfort
- The candidate consumes excessive energy, and/or
- The design features of the facility HVAC systems are not fully used.

If one or more of the above criteria fits the description of the facility, it is likely to be a good candidate for CC. CC can be effectively implemented in buildings that have received energy efficiency retrofits, in newer buildings, and in existing buildings that have not received energy efficiency upgrades. In other words, virtually any building can be a potential CC candidate.

The CC candidates can be selected by the building owner or the CC provider. However, the building owner is usually in the best position to select the most promising candidates because of his or her knowledge of the facility operation and costs. The CC provider

**Table 26.1 Commissioning team members and their primary responsibilities.**

| Team Member(s) | Primary Responsibilities |
|---|---|
| Project Manager | 1. Coordinate the activities of building personnel and the commissioning team<br>2. Schedule project activities |
| CC Engineer(s) | 1. Develop metering and field measurement plans<br>2. Develop improved operational and control schedules<br>3. Work with building staff to develop mutually acceptable implementation plans<br>4. Make necessary programming changes to the building automation system<br>5. Supervise technicians implementing mechanical systems changes<br>6. Project potential performance changes and energy savings<br>7. Conduct an engineering analysis of the system changes<br>8. Write the project report |
| Designated Facility Staff | 1. Participate in the initial facility survey<br>2. Provide information about problems with facility operation<br>3. Suggest commissioning measures for evaluation<br>4. Approve all CC measures before implementation<br>5. Actively participate in the implementation process |
| CC Technicians | 1. Conduct field measurements<br>2. Implement mechanical, electrical, and control system program modifications and changes, under the direction of the project engineer |

**Figure 26.1 Outline of Phase II of the CC Process: Implementation & Verification**

should then perform a preliminary analysis to check the feasibility of using the CC process on candidate facilities before performing a CC audit.

The following information is needed for the preliminary assessment:

- Monthly utility bills (both electricity and gas) for at least 12 months (actual bills preferable to a table of historic energy and demand data because meter reading dates are needed)
- General building information: size, function, major equipment, and occupancy schedules
- O&M records, if available
- Description of any problems in the building, such as thermal comfort, indoor air quality, moisture, or mildew.

An experienced engineer should review this information and determine the potential of the CC process to improve comfort and reduce energy cost. The CC projects often improve building comfort and reduce building energy consumption at the same time. However, some of the CC measures may increase building energy consumption in order to satisfy room comfort and indoor air requirements. For example, providing building minimum outside air will certainly increase the cooling energy consumption during summer and heating consumption during winter compared to operating the building with no outside air. If the potential justifies a CC audit, a list of preliminary commissioning measures for evaluation in a CC audit should be developed. If the owner is interested in proceeding at this point, a CC audit may be performed.

**Step 2: Perform CC Audit and Develop Project Scope**

**Objectives**: The objectives of this step are to:
- Define owner's requirements
- Check the availability of in-house technical support such as CC technicians
- Identify major CC measures
- Estimate potential savings from CC measures and cost to implement

**Approach**: The owner's representative, the CC project manager and the CC project engineer will meet. The expectations and interest of the building owner in comfort improvements, utility cost reductions, and maintenance cost reductions will be discussed and documented. The availability and technical skills of in-house technicians will be discussed. After this discussion, a walkthrough must be conducted to identify the feasibility of the owner expectations for comfort performance and improved energy performance. During the walkthrough, the CC engineer and project manager will identify major CC measures applicable to the building. An in-house technician should participate in this walk-through to provide a local operational perspective and input. The project engineer estimates the potential savings and the commissioning cost and together with the project manager prepares the CC audit report that documents these findings as well as the owner expectations.

**Special Considerations:**
- A complete set of mechanical and control system design documentation is needed

- The CC engineer and technician will make preliminary measurements of key equipment operating parameters during the walk-through

- Any available measured whole building level or sub-metered energy consumption data from stand-alone meters or the building automation system should be utilized while preparing the report.

A CC audit report must be completed that lists and describes preliminary CC measures, the estimated energy savings from implementation, and the cost of carrying out the CC process on the building(s) evaluated in the CC audit.

There may be more than one iteration or variation at each step described here, but once a contract is signed, the process moves to Phase 2 as detailed below.

### 26.3.2.2 Phase 2: CC Implementation and Verification
### Step 1: Develop CC plan and form the project team

**Objectives:**
- Develop a detailed work plan

- Identify the entire project team

- Clarify the duties of each team member

**Approach:** The CC project manager and project engineer develop a detailed work plan for the project, that includes major tasks, their sequence, time requirements, and technical requirements. The work plan is then presented to the building owner or representative(s) at a meeting attended by any additional CC engineers and/or technicians on the project team. During the meeting, the owner contact personnel and in-house technicians who will work on the project should be identified. If in-house technicians are going to conduct measurements and system adjustments, additional time should be included in the schedule unless they are going to be dedicated full time to the CC project. Typically, in-house technicians must continue their existing duties and cannot devote full time to the CC effort, which results in project delays. In-house staff may also require additional training. The work plan may need to be modified, depending on the availability and skill levels of in-house staff utilized.

**Special Issues:**
- Availability of funding to replace/repair parts found broken
- Time commitment of in-house staff
- Training needs of in-house staff

**Deliverable:** CC Report Part I—CC Plan that includes project scope and schedule, project team, and task duties of each team member.

### Step 2: Develop performance baselines

**Objectives:**
- Document existing comfort conditions
- Document existing system conditions
- Document existing energy performance

**Approach:** Document all known comfort problems in individual rooms resulting from too much heating, cooling, noise, humidity, odors (especially from mold or mildew), or lack of outside air. Also, identify and document any HVAC system problems including:
- Valve and damper hunting
- Disabled systems or components
- Operational problems
- Frequently replaced parts

An interview and walk-through may be required although most of this information is collected during the CC audit and Step 1. Room comfort problems should be quantified using hand held meters or portable data loggers. System and/or component problems should be documented based on interviews with occupants and technical staff in combination with field observations and measurements.

Baseline energy models of building performance are necessary to document the energy savings after commissioning. The baseline energy models can be developed using one or more of the following types of data:
- Short-term measured data obtained from data loggers or the EMCS system
- Long-term hourly or 15-minute whole building energy data, such as whole-building electricity, cooling and heating consumption, and/or
- Utility bills for electricity, gas, and/or chilled or hot water

The whole-building energy baseline models normally include whole building electricity, cooling energy, and heating energy models. These models are generally expressed as functions of outside air temperature since both heating and cooling energy are normally weather dependent.

Any component baseline models should be represented using the most relevant physical parameter(s) as the independent variable(s). For example, the fan motor power should be correlated with the fan airflow and pump motor energy consumption should be correlated with water flow.

Short-term measured data are often the most cost effective and accurate if the potential savings of CC measures are independent of the weather. For example, a single true power measurement can be used to develop the baseline fan energy consumption if the pulley is to be changed in a constant air volume system. Short-term data are useful for determining the baseline for specific pieces of equipment, but are not reliable for baselining overall building energy use. They may be used with calibrated simulation to obtain plausible baselines when no longer term data is available.

Long-term measurements are normally required since potential savings of CC measures are weather dependent. These measurements provide the most convincing evidence of the impact of CC projects. Long-term data also help in continuing to diagnose system faults during ongoing CC. Although more costly than short-term measured data, long-term data often produces additional savings making them the preferred data type. For example, unusual energy consumption patterns can be easily identified using long-term short-interval measured data. "Fixing" these unusual patterns can result in significant energy savings. Generally speaking, long-term interval data for electricity, gas, and thermal usage are preferred.

Utility bills may be used to develop the energy use baselines if the CC process will result in energy savings that are a significant fraction (>15%) of baseline use and if the building functions and use patterns will remain the same throughout the project.

The CC engineers should provide the metering options(s) that meet the project requirements to the building owner or representative. A metering method should be selected from the options presented by the CC engineer, and a detailed metering implementation plan developed. It may be necessary to hire a metering subcontractor if an energy information system is installed prior to implementation of the CC measures. More detailed information on savings determination is in contained the measurement and verification chapter of this handbook (Chapter 27).

**Special Considerations:**
- Use the maintenance log to help identify major system problems
- Select a metering plan that suits the CC goals and the facility needs
- Always consider and measure or obtain weather data as part of the metering plan
- Keep meters calibrated. When the EMCS system is used for metering, both sensors and transmitters should be calibrated using field measurements

**Deliverables:** CC Report Part II: Report on Current Building Performance, that includes current energy performance, current comfort and system problems, and metering plans if new meters are to be installed. Alternatively, if utility bills are used to develop the baseline energy models, the report should include baseline energy models.

**Step 3: Conduct system measurements and develop proposed CC measures**

**Objectives:**
- Identify current operational schedules and problems
- Develop solutions to existing problems
- Develop improved operation and control schedules and setpoints
- Identify potential cost effective energy retrofit measures

**Approach:** The CC engineer should develop a detailed measurement cut-sheet for each major system. The cut-sheet should list all the parameters to be measured, and all mechanical and electrical parts to be checked. The CC engineer should also provide measurement training to the technician if a local technician is used to perform system measurements. The CC technicians should follow the procedures on the cut-sheets to obtain the measurements using appropriate equipment.

The CC engineer conducts an engineering analysis to develop solutions for the existing problems; and develops improved operation and control schedules and setpoints for terminal boxes, air handling units (AHUs), exhaust systems, water and steam distribution systems, heat exchangers, chillers, boilers, and other components or systems as appropriate. Cost effective energy retrofit measures can also be identified and documented during this step, if desired by the building owner.

**Special Considerations:**
- Trend main operational parameters using the EMCS and compare with the measurements from hand meters
- Print out EMCS control sequences and schedules
- Verify system operation in the building and compare to EMCS schedules

**Deliverable:** CC Report Part III: Existing System Conditions. This report includes:
- Existing control sequences and setpoints for all major equipment, such as AHU supply air temperature, AHU supply static pressures, terminal box minimum airflow and maximum airflow values, water loop differential pressure setpoints, and

equipment on/off schedules
- List of disabled control sequences and schedules
- List of malfunctioning equipment and control devices
- Engineering solutions to the existing problems and a list of repairs required
- Proposed improved control and operation sequences and schedules

**Step 4: Implement CC measures**

**Objectives:**
- Obtain approval for each CC measure from the building owner's representative prior to implementation
- Implement solutions to existing operational and comfort problems
- Implement and refine improved operation and control schedules

**Approach:** The CC project manager and project engineer should present the engineering solutions to existing problems and the improved operational and control schedules to the building owner's representative in one or more meetings. The in-house operating staff should be invited to this meeting(s). All critical questions should be answered. It is important at this point to get "buy-in" and approval from both the building owner's representative and the operating staff. The meeting(s) will decide the following issues:
- Approval, modification or disapproval of each CC measure
- Implementation sequence of CC measures
- Implementation schedules

CC implementation should start by solving existing problems. The existing comfort and difficult control problems are the first priority of the occupants, operating staff, and facility owner. Solving these problems improves occupant comfort and increases productivity. The economic benefits from comfort improvements are sometimes higher than the energy cost savings, though less easily quantified. The successful resolution of existing problems can also earn trust in the CC engineer from facility operating staff, facility management and the occupants.

Implementation of the improved operation and control schedules should start at the end of the comfort delivery system, such as at the terminal boxes, and end with the central plant. This procedure provides benefits to the building occupants as quickly as possible. It also reduces the overall load on the system. If the process is reversed, the chiller plant is commissioned first. The chiller sequences are developed based on the current load. After the rest of the commissioning is complete, the chiller load may decrease by 30%, resulting in a need to revise the chiller operating schedules.

The CC engineers should develop a detailed implementation plan that lists each major activity. The CC technician should follow this plan in implementing the measures. The CC engineers should closely supervise the implementation and refine the operational and control schedules as necessary. The CC engineers should also be responsible for the key software changes as necessary.

Following implementation, the new operation and control sequences must be documented in a way that helps the building staff understand why they were implemented. Maintenance procedures for these measures should be provided. If any measures have not been implemented due to temporary impediments such as an out of stock part, recommendations for their future implementation should be included.

**Special Considerations:**
- Ensure that the owner's technical representative understands each major measure
- Encourage in-house technician involvement in the implementation and/or have them implement as many measures as possible
- Document improvements in a timely manner.

**Deliverable:** CC Report Part IV: CC Implementation. This report includes detailed documentation of implemented operation and control sequences, maintenance procedures for these measures, and recommendations for measures to be implemented in the future.

**Step 5: Document comfort improvements and preliminary energy savings**

**Objectives:**
- Document improved comfort conditions
- Document improved system conditions
- Document preliminary energy savings

**Approach:** The comfort measurements taken in Step 2 (Phase 2) should be repeated at the same locations under comparable conditions to determine impact on room conditions. The measured parameters, such as temperature and humidity, should be compared with the measurements from Step 2.

The M&V procedures adopted in Step 2 should be used to determine the early post-CC energy performance.

Energy performance should be compared under the same occupancy conditions and weather normalized.

**Special Considerations:**
* Savings analyses should follow accepted measurement and verification protocols such as the IPMVP
* Comfort conditions should conform to appropriate guidelines/design documents such as ASHRAE Standard 55.

**Deliverable:** CC Report, Part 5: Preliminary Measurement and Verification Report. This report includes results of detailed measurements of room conditions and energy consumption after CC activities, and any retrofit recommendations that may be provided. The room conditions should be compared with those from the pre-CC period. The projected annual energy savings should be determined according to the M&V approach adopted in Step 2.

**Step 6: Keep the commissioning continuous**

**Objectives:**
* Maintain improved comfort and energy performance
* Provide measured annual energy savings.

**Approach:** The CC engineers should review the system operation periodically to identify any operating problems and develop improved operation and control schedules as described below.

The CC engineers should provide follow-up phone consultation to the operating staff as needed, supplemented by site visits. This will allow the operating staff to make wise decisions and maintain the savings and comfort in years to come. If long term measured data are available, the CC engineers should review the energy data quarterly to evaluate the need for a site visit. If the building energy consumption has increased, the CC engineers determine possible reasons and verify with facility operating staff. Once the problem(s) is identified, the CC engineer should visit the site, develop measures to restore the building performance, and supervise the facility staff in implementing the measures. If the CC engineer can remotely log onto the EMCS system, the CC engineer can check the existing system operation quarterly using the EMCS system. When a large number of operation and control measures are disabled, a site visit is necessary. If the CC engineer cannot evaluate the facility using long-term measured energy data and EMCS system information, the CC engineer should visit the facility semi-annually.

One year after CC implementation is complete, the CC engineer should write a project follow-up report that documents the first-year savings, recommendations or changes resulting from any consultation or site visits provided, and any recommendations to further improve building operations.

**Special Considerations:** Operating personnel often have a high turnover rate, and it is important to train new staff members in the CC process and make sure they are aware of the reasons the CC measures were implemented
Ongoing follow-up is essential if the savings are to be maintained at high levels over time.

**Deliverable:** Special CC Report, that documents measured first-year energy savings, results from first-year follow-up, recommendations for ongoing staff training, and a schedule of follow-up CC activities.

### 26.3.3 Uses of Commissioning in the Energy Management Process

Commissioning can be used as a part of the energy management program in several different ways. It can be used;
* As a stand-alone measure
* As a follow-up to the retrofit process
* As an Energy Conservation Measure (ECM) in a retrofit program
* To ensure that a new building meets or exceeds its energy performance goals

#### 26.3.3.1 A stand-alone measure

Commissioning is probably most often implemented in existing buildings because it is the most cost-effective step the owner can take to increase the energy efficiency of the building, generally offering a pay-back under three years, and often 1-2 years[6]. The CC process also provides a high level of understanding of the building and its operation, enabling retrofit recommendations developed as part of the CC process to be made with a high level of certainty. The load reductions resulting from implementation of the CC process may also, for example, enable use of a smaller high efficiency chiller.

#### 26.3.3.2 A follow-up to the retrofit process

CC has often been used to provide additional savings after a successful retrofit[7,8] and an illustrative case study is provided in Section 26.4.3. It has also been used numerous times to make an under-performing retrofit meet or exceed the original expectations. The process was initially developed for these purposes as part of the Texas LoanSTAR program.

### 26.3.3.3 As an Energy Conservation Measure (ECM) in a retrofit program

The rapid pay-back that generally results from CC may be used to lower the pay-back of a package of measures to enable inclusion of a desired equipment replacement that has a longer pay-back in a retrofit package[9]. This is illustrated by a case study in Section 26.3.4. In this approach the CC engineers conduct the CC audit in parallel with the retrofit audit conducted by the design engineering firm. Because the two approaches are different and look at different opportunities, it is very important to closely coordinate these two audits. For example, the CC engineer may determine a need for a variable frequency drive on a chilled water pump. This is a retrofit opportunity for the audit engineer and should be written up as a retrofit ECM. Similarly, the audit engineer may encounter a CC opportunity during the building walk-through audit, which should be reported to the CC engineer.

### 26.3.3.4 To ensure that a new building meets or exceeds its energy performance goals

Commissioning is generally used for a new building to ensure that the systems work and provide comfort for the occupants with minimum start-up problems. It also has been found to reduce expensive change orders and other construction problems. It may also be used to significantly improve the efficiency of a new building by optimizing operation to meet its actual loads and uses instead of working to design assumptions[10,11].

The commissioning process has been described using an outside provider. It is certainly possible to perform commissioning using internal personnel when the needed skills are available on staff and these engineers and technicians can be assigned to the commissioning process. This is directly analogous to the retrofit process. Most energy audits and retrofit designs are performed by external consultants, but they can and are provided by internal personnel on occasion.

### 26.3.4 Case Study With CC As An ECM[12]

Prairie View A&M University is a 1.7 million square foot campus, with most buildings served by a central thermal plant. Electricity is purchased from a local electric co-op.

University staff identified the need for a major plant equipment replacements on campus. They wished to use the Texas LoanSTAR program to finance the project. The LoanSTAR program finances energy efficiency upgrades for public buildings, requiring that the aggregate energy payback of all energy conservation measures (ECMs) financed be ten years or less. The program requires that participating state agencies meter all buildings/plants receiving the ECMs and implement a comprehensive M&V program. The cost of the detailed investment grade audit and the mandatory M&V can be rolled into the loan, but the simple payback must still meet the ten-year criterion. This typically means that the aggregate payback of the ECMs must be 8 to 8-1/2 years (without the audit and M&V costs included). Replacement of items such as chillers, cooling towers, and building automation systems typically have paybacks of considerably more than 10 years. Hence, they can only be included in a loan if packaged with low payback measures that bring the aggregate payback below 10 years.

The university administration wanted to maximize the loan amount to get as much equipment replacement as possible. They also wanted to ensure that the retrofits work properly after they are installed. To maximize their loan dollars, they chose to include CC as an ECM. They also chose to include the audit and M&V costs in the loan to minimize up front costs.

The LoanSTAR Program provides a brief walk-through audit of the candidate buildings and plants. This audit is performed to determine whether there is sufficient retrofit potential to justify a more thorough investment grade audit.

The CC audit is conducted in parallel with the retrofit audit conducted by the engineering design firm, when CC is to be included as an ECM. The two approaches look at different opportunities, but there can be some overlap, so it is very important to closely coordinate both audits. For example, the CC engineer may determine a need for a variable frequency drive on a chilled water pump. This is a retrofit opportunity for the audit engineer and should be written up as a retrofit ECM. Similarly, the audit engineer may encounter a CC opportunity during the building audit, which should be reported to the CC engineer. It is particularly important that the savings estimated by the audit team are not "double counted." The area of greatest overlap in this case was the building automation system. Considerable care was taken not to mix improved EMCS operation with operational improvements determined by the CC engineer, so both measures received proper credit.

The same design engineering firm conducted both the initial walk-through audit and the detailed, investment grade audit. ESL CC engineers likewise conducted an initial CC walk-through audit as well as the detailed CC audit.

The CC measures identified included:

- hot and cold deck temperature resets
- extensive EMCS programming to avoid simultaneous heating and cooling
- air and water balancing
- duct static pressure resets
- sensor calibration/repair
- improved start/stop/warm-up/shutdown schedules

The CC engineers took the measurements required and collected data on building operation during the CC audit to perform a calibrated simulation on the major buildings. Available metered data and building EMCS data were also used. The CC energy savings were then written as an ECM and discussed with the design engineer. Any potential overlaps were removed. The combined ECMs were then listed, and the total savings determined.

Table 26.2 summarizes the ECMs identified from the two audits.

The CC savings were calculated to be $204,563, as determined by conducting calibrated simulation of 16 campus buildings and by engineering calculations of savings from improved loop pumping. No CC savings were claimed for central plant optimization. Those savings were all applied to ECM #7, although it seems likely that additional CC savings will accrue from this measure. The simple payback from CC is slightly under three years, making it by far the most cost effective of the ECMs to be implemented. The CC savings represent nearly 30% of the total project savings.

Perhaps more importantly, CC accounted for 2/3 of the "surplus" savings dollars available to buy down the payback of the chillers and EMCS upgrade. Without CC as an ECM, the University would have had to choose which ECMs to delete, one chiller and the EMCS upgrades; or some combination of chillers and limited building EMCS upgrades to meet the ten-year payback criteria. With CC, however, the university was able to include all these hardware items, and still meet the ten-year payback.

## 26.4. COMMISSIONING MEASURES

CC measures can be placed in two basic categories. The first category includes a number of long-time energy management measures that eliminate operation when it isn't needed, or simply "shut it off if it isn't needed." A number of these measures are a bit more complex than simply turning it off, but all are widely recognized and practiced. However, opportunities to implement some of these measures are often found, even in well-run facilities. These are discussed in some detail since most facility personnel can implement these measures. The second category of measures can broadly be categorized as implementing control practices that are optimized to the facility. These measures require a relatively high level of knowledge and skill to analyze the operation of a building, develop the optimal control sequences, and then implement them. These measures are presented in less detail, but references are provided for the reader who wishes to learn more. Some measures include the implementation of retrofits or new hardware in ways that are relatively new and innovative to provide rapid pay-back comparable to the other CC measures. These measures are sometimes considered as a separate category, but are not discussed to that level of detail in this chapter.

### 26.4.1   Eliminating Unnecessary Operation

Commissioning begins with simple measures that are included in any good energy management program. Simple rules like shut off any system that isn't needed are the beginning point of a good commissioning program as well as a good energy management program.

#### 26.4.1.1 Remove Foot Heaters and Turn Off Desk Fans

The presence of foot heaters and desk fans indicates an unsuitable working environment and wastes energy as well. To turn off foot heaters and desk fans, the following actions should be taken:

- Adjust the individual zone temperature setpoint according to the occupant's desires;
- Balance zone airflow if foot heaters are used in a portion of the zone;
- Adjust AHU supply air temperature and static pressure if the entire building is too cold or too hot
- Repair existing mechanical and control problems, such as replacing diffusers of the wrong type and relocate return air grilles, to maintain a comfortable zone temperature

Different people require different temperatures to feel comfortable. Some organizations, however, mandate the zone temperature setpoint for both summer and winter. This often leads to comfort complaints and negatively impacts productivity. The operating staff must place comfort as a priority and adjust the room temperature setpoint as necessary. Workers should be asked to dress appropriately during the summer and winter to maintain their individual comfort if setpoints are centrally mandated for a facility. Most complaints can be eliminated when the room temperature is within the range of ASHRAE's recommended comfort zone.

### Table 26.2: Summary of Energy Cost Measures (ECMs)[13]

| ECM # | ECM | Annual Savings | | | | Cost to Implement | Simple Payback |
|-------|-----|----------------|---|---|---|-------------------|----------------|
| | | Electric kWh/yr | Electric Demand kW/yr | Gas MCF/yr | Cost Savings | | |
| #1 | Lighting | 1,565,342 | 5,221 | (820) | $94,669 | $ 561,301 | 6.0 |
| #2 | Replace Chiller #3 | 596,891 | 1,250 | -0- | $33,707 | $ 668,549 | 19.8 |
| #3 | Repair Steam System | -0- | -0- | 13,251 | $58,616 | $ 422,693 | 7.2 |
| #4 | Install Motion Sensors | 81,616 | -0- | (44.6) | $ 3,567 | $ 26,087 | 7.3 |
| #5 | Add 2 Bldgs. to CW Loop | 557,676 | 7,050 | -0- | $ 60,903 | $ 508,565 | 8.4 |
| #6 | Add Chiller #4 | 599,891 | 1,250 | -0- | $33,707 | $ 668,549 | 19.8 |
| #7 | Primary/Secondary Pumping | 1,070,207 | -0- | -0- | $49,230 | $ 441,880 | 9.0 |
| #8 | Replace DX Systems | 38,237 | 233 | -0- | $ 2,923 | $ 37,929 | 13.0 |
| #9 | Replace DDC/EMCS | 2,969,962 | 670 | 2,736 | $151,488 | $2,071,932 | 13.7 |
| #10 | Continuous Commissioning | 2,129,855 | -0- | 25,318 | $204,563 | $ 605,000 | 3.0 |
| | Assessment Reports | | | | | $ 102,775 | |
| | Metering | | | | | $ 157,700 | |
| | M&V | | | | | $ 197,500 | |
| | | 9,606,677 | 15,674 | 40,440 | $693,373 | $6,470,460 | 9.3 |

### 26.4.1.2 Turn off Heating Systems During Summer

Heating is not needed for most buildings during the summer. When the heating system is on, hot water or steam often leaks through control valves, causes thermal comfort problems, and consumes excessive cooling and heating energy. To improve building comfort and decrease heating and cooling energy consumption, the following actions should be taken:

- Turn off boilers or heat exchangers if the entire building does not need heating
- Manually valve off heating and preheating coils if the heating system has to be on for other systems
- Reset differential pressure of the hot water loop to a lower value to prevent excessive pressure on control valves during the summer
- Trouble-shoot individual zones or systems, that have too many cold complaints
- Do not turn heating off too early in the spring to avoid having to turn the system back on repeatedly

This measure may be applied in constant air volume systems in dry climates. When the reheat system is shut off, room comfort may be maintained by increasing supply air temperature. This measure is not suitable for other climates where the cooling coil leaving air temperature has to be controlled below 57°F to control room humidity levels.

This simple measure results in significant energy savings as well as improved comfort in most buildings. Figure 26.2 compares the measured heating energy consumption before and after manually shutting off AHU heating valves in a building in Austin, Texas.

This building has a floor area of 147,000 square feet with two dual duct VAV systems. Before closing the heating coil manual valves, the average daily steam consumption varied from a low of 0.2 MMBtu/hr to a high of about 0.28 MMBtu/hr. After the manual valves were closed, steam leakage was eliminated through the heating coil. The steam consumption immediately dropped to slightly above 0.10 MMBtu/hr. Since the manual valves in this building can stay closed for

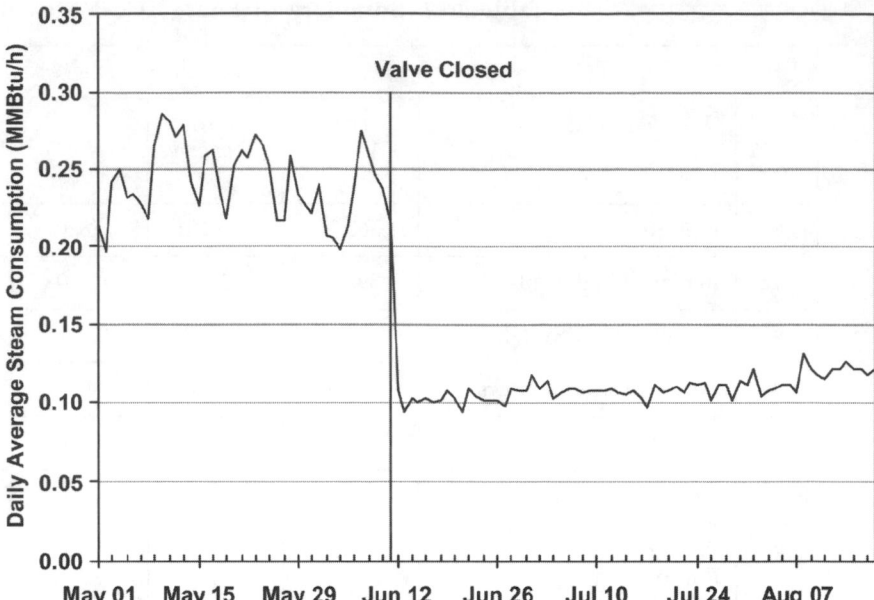

**Figure 26.2 Comparison of measured daily average steam consumption before and after manually shutting off heating coil valves in the Business Administration Building at the University of Texas at Austin[14].**

more than seven months, the annual steam savings are 756 MMBtu/yr. The same amount of chilled water will also be saved if the building remains at the same temperature, so the cooling energy savings will be 756 MMBtu/yr as well. The annual energy cost savings is $7,560 at an energy price of $5/MMBtu. This savings is not particularly large, but the only action required was shutting two manual valves.

### 26.4.1.3 Turn Off Systems During Unoccupied Hours

If a building is not occupied at nights or on weekends, the HVAC system may often be turned off completely during these periods. With a properly designed warm-up/cool-down, building comfort can be maintained with significant energy savings. In a commercial or institutional building, office equipment and lighting make up a large portion (often 50% or more) of the electrical requirements. However, a significant portion of a building (15% or more) is normally unoccupied during office hours due to travel, meetings, vacations, and sick leave. Turning off systems during unoccupied hours results in significant energy savings without degrading occupant comfort. This measure can be achieved by the following actions:

- Turn off lights, computers, printers, fax machines, desk fans, and other office equipment when leaving the office
- Turn off lights and set back room thermostats after cleaning
- Turn off AHUs at nights and on weekends. A schedule needs to be developed for each zone or air handling unit. *Turning off the system too early in the evening or turning the system on too late in*

*the morning may cause comfort problems. Conversely turning off a system too late in the evening and turning the system on too early in the morning may lose considerable savings*

- Turn off the boiler hot water pump at night during the summer when AHUs are turned off
- Turn off chillers and chilled water pumps when free cooling is available or when AHUs are turned off

Figure 26.3 presents the measured building electricity consumption, excluding chiller consumption, before and after implementation of AHU and office equipment turn-off on nights and weekends in the Stephen F. Austin Building (SFA) in Austin, Texas.

The Stephen F. Austin Building has 470,000 square feet of floor area with 22 dual duct AHUs. During the first phase of implementation, 16 AHUs were turned off from midnight to 4 a.m. weekdays and weekends. During the second phase, 22 AHUs were turned off from 11:00 p.m. to 5 a.m. during weekdays and weekends. In addition, during the second phase, all occupants were asked to turn off office equipment when they leave their office.

The measured results show that the nighttime whole building electricity use decreased from 1,250 kW to 900 kW during the first phase. During the second phase, the nighttime minimum electricity decreased to 800 kW.

It was observed that the daily peak electricity consumption after night shutdowns began is significantly lower than the base peak. For example, the lowest peak during the second phase is 1,833 kW, which is 8% lower

**Figure 26.3 Hourly whole building electricity consumption at the SFA Building before and after night shut down of AHUs was initiated[15].**

than the base peak. The lower electricity peak indicates that some office equipment remained off during the daytime or the employees were more conscientious in turning off lights and equipment when they left the office. The annual energy cost savings, including electricity, heating, and cooling, were determined to be $100,000/yr using measured hourly data.

### 26.4.1.4 Slow Down Systems During Unoccupied/Lightly-Occupied Hours

Most large buildings are never completely unoccupied. It is not uncommon to find a few people working regardless of the time of day. The zones that may be used during the weekends or at night, are also unpredictable. System shut down often results in complaints. Substantial savings can be achieved while maintaining comfort conditions in a building by an appropriate combination of the following actions:

- Reset outside air intake to a lower level (0.05 cfm/ft²) during these hours during hot summer and cold winter weather. Outside air can be reduced since there will be very few people in the building. Check outside and exhaust air balance to maintain positive building pressure
- Reset the minimum airflow to a lower value, possibly zero, for VAV terminal boxes
- Program constant volume terminal boxes as VAV boxes, and reset the minimum flow from the maximum to a lower value, possibly zero during unoccupied hours
- Reset AHU static pressure and water loop differential pressure to lower values
- Set supply air fan at a lower speed

These measures maintain building comfort while minimizing energy consumption. The savings are often comparable with the shut down option. Figure 26.4 presents the measured hourly fan energy consumption in the Education Building at the University of Texas at Austin.

The Education Building has 251,000 ft² of floor area with eight 50-hp AHUs that are operated on VFDs. Prior to the introduction of this measure, the motor control center (MCC) energy consumption was almost constant. The CC measure implemented was to set the fan speed at 30% at night and on weekends. The nighttime slow roll decreased the fan power from approximately 50 kW to approximately 25 kW while maintaining building comfort.

### 26.4.1.5 Limit Fan Speed During Warm Up and Cool Down Periods

If nighttime shut down is implemented, warm-up is necessary during the winter and cool down is required during the summer. During warm-up and cool down periods, fan systems are often run at maximum speed since all terminal boxes require either maximum heating or maximum cooling. A simple fan speed limit can reduce fan power significantly. This principle may also be used in other systems, such as pumps. The following actions should be taken to achieve the fan energy savings:

- Determine the optimal start up time using 80% (adjustable) fan capacity if automatic optimal start up is used
- Set the fan speed limit at 80% (adjustable) manually and extend the warm up or cool down period

**Figure 26.4 Measured Post-CC hourly supply fan electricity consumption in the Education Building**[16].

by 25%. If the speed limit is set at another fractional value (x), determine the warm up period using the following equation:

$$T_n = \frac{T_{exist}}{x}$$

- Keep outside air damper closed during warm-up and cool-down periods

The fan energy savings increase significantly as the fan speed limit decreases. Figure 26.5 shows the theoretical fan power savings. When the fan speed limit is 50% of design fan speed, the potential fan energy savings are 75% of the fan energy even if the fan runs twice as long. The theoretical model did not consider the variable speed drive loss. The actual energy savings will normally be somewhat lower than the model projected value.

Note that if the outside air damper cannot be closed tightly, extra thermal energy may be required to cool or warm up outside air that leaks through the damper. This factor should be considered when this measure is used.

### 26.4.2 Operational Efficiency Measures for AHU Systems

Air handler systems normally condition and distribute air inside buildings. A typical AHU system consists of some combination of heating and cooling coils, supply and return air fans, filters, humidifiers, dampers, ductwork, terminal boxes, and associated safety and control devices, and may include an economizer. As the building load changes, AHUs change one or more of the following parameters to maintain building com-

fort: outside air intake, total airflow, static pressure, and supply air temperature and humidity. Both operating schedules and initial system set up, such as total airflow and outside airflow, significantly impact building energy consumption and comfort. The following ten major CC measures should be used to optimize AHU operation and control schedules:

- Adjust total airflow for constant air volume systems
- Set minimum outside air intake correctly
- Improve static pressure set-point and schedule
- Optimize supply air temperatures
- Improve economizer operation and control
- Improve coupled control AHU operation
- Valve off hot air flow for dual duct AHUs during summer
- Install VFD on constant air volume systems
- Install airflow control for VAV systems
- Improve terminal box operation

#### 26.4.2.1. Adjust Total Air Flow and Fan Head for Constant Air Volume Systems

Air flow rates are significantly higher than required in most buildings primarily due to system over-sizing. In some large systems, an oversized fan causes over-pressurization in terminal boxes. This excessive pressurization is the primary cause of room noise. The excessive airflow often causes excessive fan energy consumption, excessive heating and cooling energy consumption, humidity control problems, and excessive noise in terminal boxes[18].

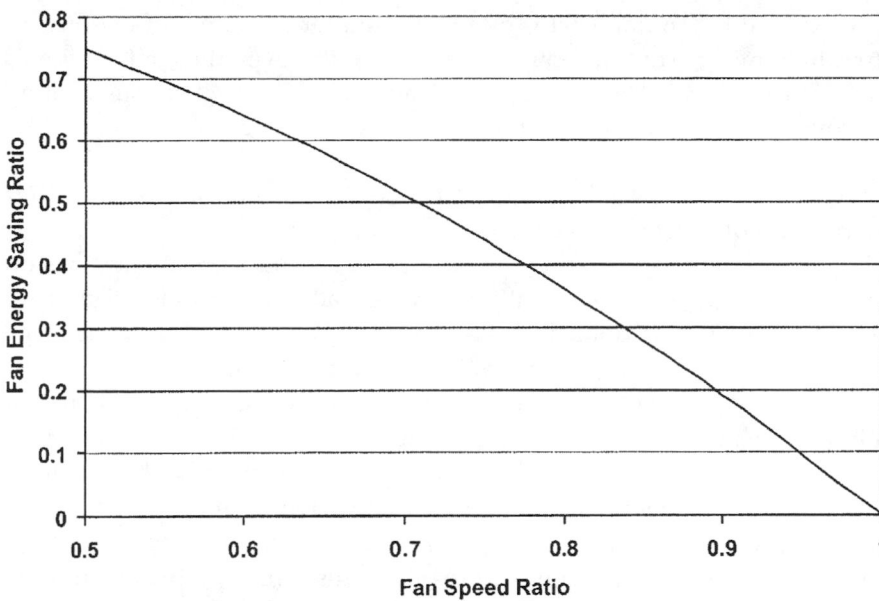

**Figure 26.5 Potential fan energy savings using fan speed limiting[16].**

### 26.4.2.2. Set Minimum Outside Air Intake Correctly

Outside air intake rates are often significantly higher than design values in existing buildings due to lack of accurate measurements, incorrect design calculations and balancing, and operation and maintenance problems. Excessive outside air intake is caused by the mixed air chamber pressure being lower than the design value, by significant outside air leakage through the maximum outside air damper on systems with an economizer, by the minimum outside air intake being set to use minimum total airflow for a VAV system, or by lower than expected/designed occupancy.

### 26.4.2.3. Improve Static Pressure Setpoint and Schedule

The supply air static pressure is often used to control fan speed and ensure adequate airflow to each zone. If the static pressure setpoint is lower than required, some zones may experience comfort problems due to lack of airflow. If the static pressure setpoint is too high, fan power will be excessive. In most existing terminal boxes, proportional controllers are used to maintain the airflow setpoint. When the static pressure is too high, the actual airflow is higher than its setpoint. The additional airflow depends on the setting of the control band. Field measurements[19] have found that the excessive airflow can be as high as 20%. Excessive airflow can also occur when terminal box controllers are malfunctioning. For pressure dependent terminal boxes, high static pressure causes significant excessive airflow. Consequently, high static pressure often causes unnecessary heating and cooling energy consumption. A higher than necessary static pressure setpoint is also the primary reason for noise problems in buildings.

### 26.4.2.4. Optimize Supply Air Temperatures

Supply air temperatures (cooling coil discharge air temperature for single duct systems; cold deck and hot deck temperatures for dual duct systems) are the most important operation and control parameters for AHUs. If the cold air supply temperature is too low, the AHU may remove excessive moisture during the summer using mechanical cooling. The terminal boxes must then warm the over-cooled air before sending it to each individual diffuser for a single duct AHU. More hot air is required in dual duct air handlers. The lower air temperature consumes more thermal energy in either system. If the cold air supply temperature is too high, the building may lose comfort control. The fan must supply more air to the building during the cooling season, so fan power will be higher than necessary. The goal of optimal supply air temperature schedules is to minimize combined fan power and thermal energy consumption or cost. Although developing optimal reset schedules requires a comprehensive engineering analysis, improved (near optimal) schedules can be developed based on several simple rules. Guidelines for developing improved supply air temperature reset schedules are available for four major types of AHU systems[20].

### 26.4.2.5. Improve Economizer Operation and Control

An economizer is designed to eliminate mechanical cooling when the outside air temperature is lower than the supply air temperature setpoint and decrease mechanical cooling when the outside air temperature is between the cold deck temperature and a high temperature limit, which is typically less than 70°F. Economizer control is often implemented so it controls mixed air temperature at the cold deck temperature or simply 55°F. This control algorithm is far from optimum. It may, in fact, actually increase the building energy consumption. The economizer operation can be improved using the following steps:

1. Integrate economizer control with optimal cold deck temperature reset. It is tempting to ignore cold deck reset when the economizer is operating, since the cooling is free. However, cold deck reset normally saves significant heating.

2. For a draw-through AHU, set the mixed air temperature 1°F lower than the cold deck temperature setpoint. For a blow-through unit, set the mixed air temperature at least 2°F lower than the supply air temperature setpoint. This will eliminate chilled water valve hunting and unnecessary mechanical cooling.

3. For a dual duct AHU, the economizer should be disabled if the hot air flow is higher than the cold air flow since the heating energy penalty is then typically higher than cooling energy savings.

4. Set the economizer operating range as wide as possible. For dry climates, the economizer should be activated when the outside air temperature is between 30°F and 75°F, between 30°F and 65°F for normal climates, and between 30°F and 60°F for humid climates. When proper return and outside air mixing can be achieved, the economizer can be activated even when the outside air temperature is below 30°F.

5. Measure the true mixed air temperature. Most mixing chambers do not achieve complete mixing of the return air and outside air before reaching the cooling coil. It is particularly important that mixed air temperature be measured accurately when an economizer is being used. An averaging temperature sensor should be used for the mixed air temperature measurement.

### 26.4.2.6. Improve Coupled Control AHU Operation

Coupled control is often used in single-zone single-duct, constant volume systems. Conceptually, this system provides cooling or heating as needed to maintain the setpoint temperature in the zone and uses simultaneous heating and cooling only when the humidistat indicates that additional cooling (followed by reheat) is needed to provide humidity control. However, the humidistat is often disabled for a number of reasons. To control room relative humidity level, the control signals or spring ranges are overlapped. Simultaneous heating and cooling often occurs almost continuously.

### 26.4.2.7. Valve Off Hot Air Flow for Dual Duct AHUs During Summer

During the summer, most commercial buildings do not need heating. Theoretically, hot air should be zero for dual duct VAV systems. However, hot air leakage through terminal boxes is often significant due to excessive static pressure on the hot air damper. For constant air volume systems, hot air flow is often up to 30% of the total airflow. During summer months, hot air temperatures as high as 140°F have been observed due to hot water leakage through valves. The excessively high hot air temperature often causes hot complaints in some locations. Eliminating this hot air flow can improve building thermal comfort, reduce fan power, cooling consumption, and heating consumption.

### 26.4.2.8. Install VFD on Constant Air Volume Systems

The building heating load and cooling load varies significantly with both weather and internal occupancy conditions. In constant air volume systems, a significant amount of energy is consumed unnecessarily due to humidity control requirements. Most of this energy waste can be avoided by simply installing a VFD on the fan without a major retrofit effort. Guidelines for VFD installation are available for dual duct, multi-zone, and single duct systems[21].

### 26.4.2.9. Airflow Control for VAV Systems

Airflow control of VAV systems has been an important design and research subject since the VAV system was introduced. An airflow control method should: (1) ensure sufficient airflow to each space or zone; (2) control outside air intake properly; and (3) maintain a positive building pressure. These goals can be achieved using the variable speed drive volume tracking (VS-DVT) method[22,23].

### 26.4.2.10. Improve Terminal Box Operation

The terminal box is the end device of the AHU system. It directly controls room temperature and airflow. Improving the set up and operation are critical for room comfort and energy efficiency. The following measures are suggested:
- Set minimum air damper position properly for pressure dependent terminal boxes.
- Use VAV control algorithm for constant air volume terminal boxes.

- Use airflow reset.
- Integrate lighting and terminal box control.
- Integrate airflow and temperature reset.
- Improve terminal box control performance.

### 26.4.3 Case Study—AHU CC in a Previously Retrofit Building[24]

#### 26.4.3.1 Case Study Building Description

The building studied in this paper is the 295,000 gross ft$^2$ (226,000 ft$^2$ net) Zachry Engineering Center (ZEC), located on the Texas A&M University campus (30°N, 96°W) where the average January temperature is 50°F and average July temperature is 84°F, and pictured in Figure 26.6. The building has four-floors plus an unconditioned basement parking level. It was constructed in the early 1970s and is a heavy structure with 6 in. concrete floors and insulated exterior walls made of precast concrete and porcelain-plated steel panels. About 12% of the exterior wall area is covered with single-pane, bronze-tinted glazing. The windows are recessed approximately 24 in. from the exterior walls, which provides some shading. Approximately 2835 ft$^2$ of northeast-facing clerestory windows admit daylight into the core of the building.

The ZEC includes offices, classrooms, laboratories and computer rooms and is open 24 hours per day, 365 days per year with heaviest occupancy during normal working hours between 8 a.m. and 6 p.m. on weekdays. Occupancy, electrical consumption and chilled water consumption show marked weekday/weekend differences with peak weekend electrical consumption less than 10% above the nightly minimum; weekday holiday occupancy is similar to weekend usage with intermediate usage on weekdays between semesters when class

**Figure 26.6 The Zachry Engineering Center on the Texas A&M campus[25].**

rooms are not in use, but laboratories and offices are occupied.

HVAC Systems. Twelve identical dual-duct systems with 40 hp fans rated at 35,000 cfm and eight smaller air handlers (3 hp average) supply air to the zones in the building. Supply and return air ducts are located around the perimeter of the building. These were operated with a constant outdoor air intake at a nominal value of 10% of design flow. The large dual-duct constant air volume (DDCAV) systems were converted to dual-duct variable-air volume (DDVAV) systems accompanied by connection to the campus energy management and control system (EMCS) in 1991. This retrofit successfully reduced fan power consumption by 44%, cooling consumption by 23%, and heating consumption by 84%.

Monitoring of energy use. In the engineering center about 50 channels of hourly data have been recorded and collected each week since May 1989. The sensors are scanned every 4 seconds and the values are integrated to give hourly totals or averages as appropriate. The important channels for savings measurement are those for air handler electricity consumption and whole-building heating and cooling energy use. Air handler electricity consumption is measured at the building's motor control center (MCC) and represents all of the air-handling units and most of the heating, ventilating, and air-conditioning (HVAC)-related pumps in the building. Cooling and heating energy use are determined by a Btu meter which integrates the monitored fluid flow rate and temperature difference across the supply and return lines of the chilled- and hot-water supply to the building. The majority of the 50 channels of monitored information come from one air handler that was highly instrumented.

### 4.3.2 Continuous Commissioning of the Zachry Engineering Center

The Continuous Commissioning process was applied to the Zachry Engineering Center in 1996-97. In this case, the initial survey and specification of monitoring portions of the CC process were not performed since the university president had decided to implement CC on campus based on its success in numerous other locations rather than on the results of individual building surveys. Metering had been installed much earlier in the Zachry Engineering Center as part of the retrofit process, so performance baselines were already available.

### Conduct system measurements and develop proposed CC measures

The facility survey found that the building control system set-up was far from optimum and found numerous other problems in the building as well. The basic

**1990, 1994 & 1997 MCC Consumption vs. Temperature**

**Figure 26.7 ZEC daily motor control center (MCC) electric consumption in 1990 before the retrofit, in 1994 after the retrofit, and in 1997 after CC[26].**

control strategies found in the building are summarized in Table 26.3 in the column labeled "Pre-CC Control Practice." The ranges shown for constant parameters reflect different constant values for different individual air handlers.

The control practices shown in the table are all widely used, but none are optimal for this building. The campus controls engineer worked closely with the CC engineers during the survey. The items shown in Table 26.3 could all be determined by examination of the control system in the building, but the facility survey also examines a great deal of the equipment throughout the building and found numerous cases of valves that let too much hot or cold water flow, control settings that caused continuous motion and unnecessary wear on valves, air ducts that had blown off of the terminal boxes, kinks in air ducts that led to rooms that could not be properly heated or cooled, etc. Following the survey, the building performance was analyzed and CC measures including optimum control schedules were developed for the building in cooperation with the campus controls engineer.

**Implement CC measures**

Following the survey, the building performance was analyzed and optimum control schedules were developed for the building in cooperation with the campus control engineer. The air handlers, pumps and terminal

boxes had major control parameters changed to values shown in the "Post-CC" column of Table 26.3. Most of the control parameters were optimized to vary as a function of outside air temperature, $T_{oa}$, as indicated.

In addition to optimizing the control settings for the heating and cooling systems in the building, numerous problems specific to individual rooms, ducts, or terminal boxes were diagnosed and resolved. These included items like damper motors that were disconnected, bent air ducts that could not supply enough air to properly control room temperature, leaking air dampers, dampers that indicated open when only partly open, etc.

Problems of this sort often had led to occupant complaints that were partially resolved without fixing the real problem. For example, if a duct was constricted so inadequate flow reached a room, the pressure in the air handler might be increased to get additional flow into the room. "Fixes" like this typically improve room comfort, but sometimes lead to additional heating and cooling consumption in every other room on the same air handler.

**Document energy savings**

Implementation of these measures resulted in significant additional savings beyond the original savings from the VAV retrofit and controls upgrade as shown in Figures 26.7, 26.8 and 26.9. Figure 26.7 shows the motor control center power consumption as a func-

**Table 26.3. Major control settings in the
Zachry Engineering Center before and after implementation of CC[27].**

| Parameter | Pre-CC Control Practice | Post-CC Control Practice |
|---|---|---|
| Pressure in air ducts | Constant at 2.5 - 3.5 in $H_2O$ | 1- 2 in$H_2O$ as $T_{oa}$ increases |
| Cold air temperature | Constant at 50°F - 55°F | 60°F - 55°F as $T_{oa}$ increases |
| Hot air temperature | Constant at 110°F - 120°F | 90°F - 70°F as $T_{oa}$ increases |
| Air flow to rooms | Variable - but inefficient | Optimized min/max flow and damper operation |
| Heating pump control | Operated continuously | On when $T_{oa}>55°F$ |
| Cooling pump control | Variable speed with shut-off | Pressure depends on flow |

tion of ambient temperature for 1990, 1994 and 1997. It is evident that the minimum fan power has been cut in half and there has been some reduction even at summer design conditions. Figure 26.9 shows the hot water consumption for 1990, 1994 and 1997, again as a function of daily average temperature. The retrofit reduced the annual hot water (HW) consumption for heating to only 16% of the baseline, so there is little room for further reduction. However, it can be seen that the CC measures further reduced HW consumption, particularly at low temperatures. The largest savings from the CC measures are seen in the chilled water consumption as shown in Figure 26.8. The largest fractional savings occur at low ambient temperatures, but the largest absolute savings occur at the highest ambient temperatures.

The annualized consumption values for the baseline, post-retrofit and post-CC conditions are shown in Table 26.4. The MCC consumption for 1997 was 1,209,918 kWh, 74% of the 1994 consumption and only 41% of the 1990 consumption. On an annual basis, the post-CC HW consumption normalized to 1994 weather was 1940 MMBtu, a reduction to only 10% of baseline consumption and a reduction of 34% from the 1994 consumption. The CC measures reduced the post-CC chilled water (CHW) consumption to 17,400 MMBtu, a reduction of 17,800 MMBtu which is noticeably larger than the 13,900 MMBtu savings produced by the retrofit. The CHW savings accounted for the largest portion of the CC savings in this cooling-dominated climate.

**Generalized application of case study and conclusions**

The major energy savings from the CC activities in the case study building resulted from five items.

1. Optimization of duct static pressures at lower levels. This reduces fan power and also reduces damper leakage that increases both heating and cooling consumption.

2. Optimization of cold air temperatures. Most buildings in our experience use a constant set-point for

**Table 26.4. Consumption at the Zachry Engineering Center before and after retrofit and after implementation of CC measures[28].**

| | Baseline Consumption | Post-retrofit | Post-Retrofit | Post-CC | Post-CC |
|---|---|---|---|---|---|
| Fan Power | 2,950,000 kWh | 1,640,000 kWh | 56% | 1,210,000 kWh | 41% |
| Chilled Water | 45,800 MMBtu | 35,300 MMBtu | 77% | 17,400 MMBtu | 37% |
| Heating Water | 18,800 MMBtu | 2940 MMBtu | 16% | 1940 MMBtu | 10% |

the cold air temperature which is very inefficient. Even if this set-point is changed, it is seldom optimized.

3. Optimization of hot air temperatures. We find that most buildings modulate hot air temperature according to outside air temperature, but it is normally significantly higher than necessary.

4. Optimize settings on VAV boxes. Most VAV terminal boxes have minimum flow set-points at night that are too high.

5. Optimize pump control. Static pressure set-points on chilled water and hot water pumps are generally set higher than necessary.

The CC process has been illustrated by application to a large building which had earlier had a major retrofit performed. The post-CC consumption values represent 41% of the pre-retrofit fan and pump consumption, 10% of the pre-retrofit hot water consumption, and 38% of the pre-retrofit chilled water consumption. Using the baseline energy prices, the post-CC consumption reflects an HVAC energy cost that is only 36% of the baseline HVAC cost and is only 65% of the HVAC cost after the

retrofit. The measures implemented in the case study building are quite typical of CC measures implemented in other buildings. These results are better than average for the process, but are not one-of-a-kind.

### 26.4.4 CC Measures for Water/Steam Distribution Systems

Distribution systems include central chilled water, hot water, and steam systems, that deliver thermal energy from central plants to buildings. In turn, the system distributes the chilled water, hot water and steam to AHU coils and terminal boxes. Distribution systems mainly consist of pumps, pipes, control valves, and variable speed pumping devices. This section focuses on the CC measures for optimal pressure control, water flow control, and general optimization.

#### 26.4.4.1 Improve Building Chilled Water Pump Operation

Most building chilled water pumping systems are equipped with variable speed devices (VSDs). If a VSD is not installed, retrofit of a VSD is generally recommended. The discussion here is limited to systems where a VSD is installed. The goal of pumping optimization is to avoid excessive differential pressures across the control valves, while providing enough water to

**1990, 1994 and 1997 CHW vs. Temperature**

°1990 CHW   ◇1994 CHW   ▲1997 CHW

**Temperature (°F)**

Figure 26.8. ZEC daily chilled water consumption for 1990 before the retrofit, 1994 after the retrofit, and 1997 after CC[29].

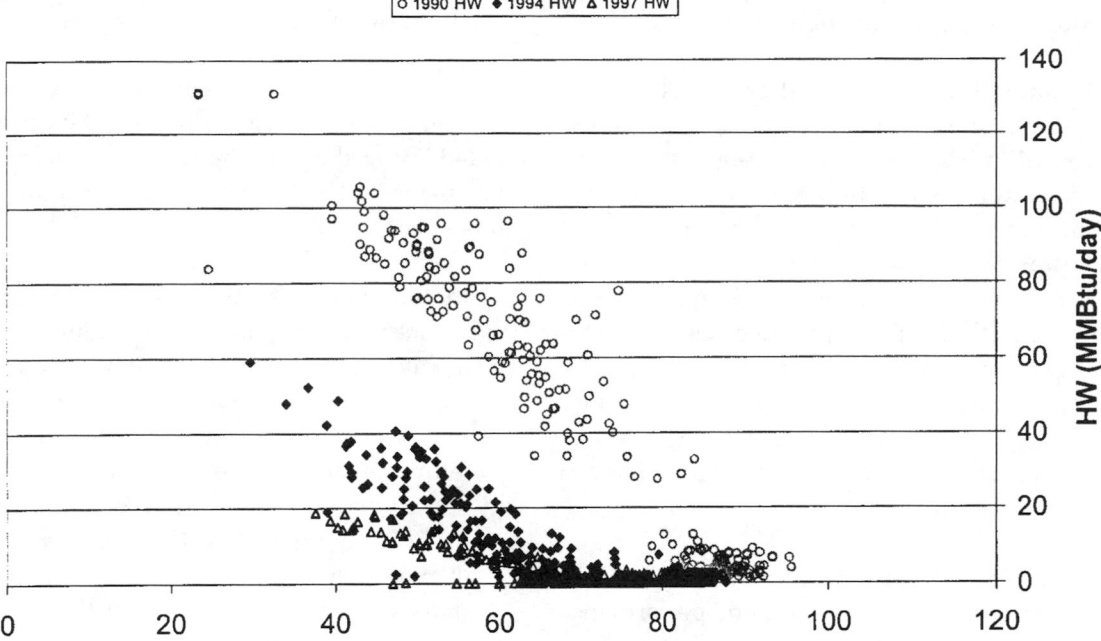

**Figure 26.9. ZEC daily hot water consumption for 1990 before the retrofit, 1994 after the retrofit, and 1997 after CC[30].**

each building, coil, or other end use. An optimal pump differential pressure schedule should be developed that provides adequate pressure across the hydraulically most remote coil in the system under all operating conditions, but does not provide excess head.

### 26.4.4.2 Improve Secondary Loop Operation

For buildings supplied by a secondary loop from a central plant, building loop optimization should be performed before the secondary loop optimization.

*Source Distributed Systems*: If there are no building pumps, the secondary pumps must provide the pressure head required to overcome both the secondary loop and the building loop pressure losses. In this case, the secondary loop is called a source distributed system. The secondary loop pumps should be controlled to provide enough pressure head for the most remote coil. If VFDs are installed, the differential pressure can be controlled by modulating pump speed. Otherwise, the differential pressure can be modulated by changing the number of pumps in operation.

*Source Distributed Systems With Building Pumps*: In most campus settings, each building has a pump. The optimal differential pressure setpoint should then be determined by optimizing the secondary loop pressure

setpoint so the combined secondary pump and building pumping power is minimized. This can be done by developing a pressure reset schedule that requires maximum building pump power at the most hydraulically remote building on the secondary loop. This may occur with a negative differential pressure across the most remote building.

### 26.4.4.3 Improving Central Plant Water Loop Operation

The central plant loop optimization should be performed after secondary loop optimization.

*Single Loop Systems*: For most heating distribution systems and some chilled water systems, a single loop is used instead of primary and secondary systems. Under partial load conditions, fewer pumps can be used for both chillers and heat exchangers. This can result in less pump power consumption.

*Primary-secondary Loop Systems*: Primary-secondary systems are the most common chilled water distribution systems used with central chiller plants. This design is based on the assumption that the chilled water flow through the chiller must be maintained at the design level. This is seldom needed. Due to this incorrect assumption, a significant amount of pumping power is wasted in numerous central plants. Design engineers may or may not include an isolation valve on the by-pass line of the primary loop. Procedures are available to optimize pump operation for both cases[31,32,33,34].

### 26.4.5 CC Measures for Central Chiller Plants

The central chiller plant includes chillers, cooling towers, a chilled water distribution system, and the condenser water distribution system. Although a secondary pumping system may be physically located inside the central plant, commissioning issues dealing with secondary loops are discussed in the previous section. The central chiller plant produces chilled water using electricity, steam, hot water, or gas. The detailed commissioning measures vary with the type of chiller and this section gives general commissioning measures that apply to a typical central cooling plant and that can produce significant energy savings.

*Use the Most Efficient Chillers*: Most central chiller plants have several chillers with different performance factors or efficiencies. The differences in performance may be due to the design, to performance degradation, age, or operational problems. One chiller may have a higher efficiency at a high load ratio while another may have a higher efficiency at a lower load ratio. Running chillers with the highest performance can result in significant energy savings because you will be providing the greatest output for the least input.

*Reset the Supply Water Temperature*: Increasing the chilled water supply temperature can decrease chiller electricity consumption significantly. The general rule-of-thumb is that a one degree Fahrenheit increase corresponds to a decrease in compressor electricity consumption of 1.7%. The chilled water supply temperature can be reset based on either cooling load or ambient conditions.

*Reset Condenser Return Water Temperature*: Decreasing cooling tower return water temperature has the same effect as increasing the chilled water supply temperature. The cooling tower return temperature should be reset based on weather conditions. The following provides general guidelines:
- The cooling tower return water temperature setpoint should be at least 5°F (adjustable based on tower design "approach") higher than the ambient wet bulb temperature. This prevents excessive cooling tower fan power consumption.
- The cooling tower water return temperature should not be lower than 65°F for chillers made before 1999, and should not be lower than 55°F

for newer chillers. It is also recommended that you consult the chiller manufacturer for specific limits on allowable condenser water temperature.

The cooling tower return water temperature reset can be implemented using the BAS. If it cannot be implemented using the BAS, operators can reset the setpoint daily using the daily maximum wet bulb or dry bulb temperature.

Decreasing the cooling tower return temperature may increase fan power consumption. However, fan power may not necessarily increase with lower cooling tower return water temperature. The following tips can help.
- Use all towers. For example, use all three towers when one of the three chillers is used. This may eliminate fan power consumption entirely. The pump power may actually stay the same. Be sure to keep the other two tower pumps off.
- Never turn on the cooling tower fan before the by-pass valve is totally closed. If the by-pass valve is not totally closed, the additional cooling provided by the fan is not needed and will not be used. Save the fan power!
- Balance the water distribution to the towers and within the towers. Towers are often seen where water is flowing down only one side of the tower, or one tower may have twice the flow of another. This significantly increases the water return temperature from the towers.

*Increase Chilled Water Return Temperature*: Increasing chilled water return temperature has the same effect as increasing chilled water supply temperature. It can also significantly decrease the secondary pump power since the higher the return water temperature (for a given supply temperature), the lower the differential temperature, and the lower the chilled water flow. Maximizing chilled water return temperature is much more important than optimizing supply water temperature since it often provides much more savings potential. It is hard to increase supply temperature 5°F above the design setpoint. It is often easy to increase the return water temperature as much as 7°F by conducting water balance and shutting off by-pass of three-way valves.

*Use Variable Flow under Partial Load Conditions*: Typical central plants use primary-secondary loops. A constant speed primary pump is often dedicated to a particular chiller. When the chiller is turned on, the pump is on. Chilled water flow through each chiller is maintained at the design flow rate by this operating schedule. When the building-loop flow is less than the chiller loop flow, part of the chiller flow by-passes the building and returns to the chiller. This practice causes excessive primary pump power consumption and excessively low entering water temperature to the chiller, which increases the compressor power consumption.

It is generally perceived that chilled water flows must remain constant for chiller operational safety. Actually, most new chillers allow chilled water flow as low as 30% of the design value. The chilled water flow can be decreased to be as low as 50% for most existing chillers if the proper procedures are followed[35].

Varying chilled water flow through a chiller can result in significant pump power savings. Although the primary pumps are kept on all the time, the secondary pump power consumption is decreased significantly when compared to the conventional primary and secondary system operation. Varying chilled water flow through the chillers will also increase the chiller efficiency when compared to constant water flow with chilled water by-pass. More information can be found in a paper by Liu[36].

*Optimize Chiller Staging*: For most chillers, the kW/ton decreases (COP increases) as the load ratio increases from 40% to 80%. When the load ratio is too low, the capacity modulation device in the chiller lowers the chiller efficiency. When the chiller has a moderate load, the capacity modulation device has reasonable efficiency. The condenser and evaporator are oversized for the load under this condition so the chiller efficiency is higher. When the chiller is at maximum load, the evaporator and condenser are marginally sized, reducing the chiller efficiency below its maximum value. Running chillers in the high efficiency range can result in significant electrical energy savings and can improve the reliability of plant operation. Optimal chiller staging should be developed using the following procedures:

- Determine and understand the optimal load range for each chiller. This information should be available from the chiller manufacturer. For example, chiller kW/ton typically has a minimum value (best efficiency) when the chiller load is somewhere between 50% and 70% of the design value.
- Turn on the most efficient chiller first. Optimize the pump and fan operation accordingly.
- Turn on more chillers to maintain the load ratio (chiller load over the design load) within the optimal efficiency range for each chiller. This assumes that the building by-pass is closed. If the building by-pass cannot be closed, the minimum chiller load ratio should be maintained at 50% or higher to limit primary pumping power increases

*Maintain Good Operating Practices*: The operating procedures recommended by the manufacturer should be followed. It is important to calibrate the temperature, pressure, and current sensors and the flow switches periodically. The temperature sensors are especially important for maintaining efficient operation. Control parameters must be set properly, particularly the time delay relay.

### 26.4.6 CC Measures for Central Heating Plants

Central heating plants produce hot water, steam, or both, typically using either natural gas, coal or oil as fuel. Steam, hot water, or both are distributed to buildings for HVAC systems and other end uses, such as cooking, cleaning, sterilization and experiments. Boiler plant operation involves complex chemical, mechanical and control processes. Energy performance and operational reliability can be improved through numerous measures. However, the CC measures discussed in this section are limited to those that can be implemented by an operating technician, operating engineers, and CC engineers.

#### 26.4.6.1 Optimize Supply Water Temperature and Steam Pressure

Steam pressure and hot water temperature are the most important safety parameters for a central heating plant. Reducing the boiler steam pressure and hot water temperature has the following advantages:

- Improves plant safety
- Increases boiler efficiency and decreases fuel consumption
- Increases condensate return from buildings and improves building automation system performance
- Reduces hot water and steam leakage through malfunctioning valves

#### 26.4.6.2 Optimize Feed Water Pump Operation

The feed water pump is sized based on boiler design pressure. Since most boilers operate below the design pressure, the feed water pump head is often significantly higher than required. This excessive pump head is often dropped across pressure reducing valves and manual valves. Installing a VSD on the feed water pump in such cases can decrease pump power consumption and improve control performance. Trimming the impeller or changing feed water pumps may also be feasible, and the cost may be lower. However, the VSD provides more flexibility, and it can be adjusted to any level. Consequently, it maximizes the savings and can be adjusted to future changes as well.

### 26.4.6.3 Optimize Airside Operation

The key issues are excessive airflow and flu gas temperature control. Some excess airflow is required to improve the combustion efficiency and avoid having insufficient combustion air during fluctuations in airflow. However, excessive airflow will consume more thermal energy since it has to be heated from the outside air temperature to the flue gas temperature. The boiler efficiency goes down as excessive airflow increases. The flue gas temperature should be controlled properly. If the flue gas temperature is too low, acid condensation can occur in the flue. If the flue gas temperature is too high, it carries out too much thermal energy. The airside optimization starts with a combustion analysis, that determines the combustion efficiency based on the flu gas composition, flu gas temperature, and fuel composition. The typical combustion efficiency should be higher than 80%. If the combustion efficiency is lower than this value, available procedures[37,38] should be used to determine the reasons.

### 26.4.6.4 Optimize Boiler Staging

Most central plants have more than one boiler. Using optimal staging can improve plant energy efficiency and reduce maintenance cost. The optimal staging should be developed using the following guidelines:
- Measure boiler efficiency.
- Run the higher efficiency boiler as the primary system and run the lower efficiency boiler as the back up system.
- Avoid running any boiler at a load ratio less than 40% or higher than 90%.
- If two boilers are running at average load ratios less than 60%, no stand-by boiler is necessary. If three boilers are running at loads of less than 80%, no stand by boiler is necessary.

Boiler staging involves boiler shut off, start up, and standby. Realizing the large thermal inertial and the temperature changes between shut off, standby, and normal operation, precautions must be taken to prevent corrosion damage and expansion damage. Generally speaking, short-term (monthly) turn on/off should be avoided for steam boilers. Hot water boilers are sometimes operated to provide water temperatures as low as 80°F. This improves distribution efficiency, but may lead to acid condensate in the flue. The hot water temperature must be kept high enough to prevent this condensation.

### 26.4.6.5 Improve Multiple Heat Exchanger Operation

Heat exchangers are often used in central plants or buildings to convert steam to hot water or high temperature hot water to lower temperature hot water. If more than one heat exchanger is installed, use as many heat exchangers as possible provided the average load ratio is 30% or higher. This approach provides the following benefits:

- Lower pumping power. For example, if two heat exchangers are used instead of one under 100% load, the pressure loss through the heat exchanger system will be decreased by 75%. The pumping power will also be decreased by 75%.

- Lower leaving temperature on the heat source. The condensate should be super-cooled when the heat exchangers are operated at low load ratio. The exit hot water temperature will be lower than the design value under the partial load condition. This will result in less water or steam flow and more energy extracted from each pound of water or steam. For example, the condensate water may be sub-cooled from 215°F to 150°F under low heat exchanger loads. Compared with leaving the heat exchanger at 215°F, each pound of steam delivers 65 Btu more thermal energy to the heat exchanger.

Using more heat exchangers will result in more surface heat loss. If the load ratio is higher than 30%, the benefits mentioned above normally outweigh the heat loss. More information can be found in a paper by Liu et al.[39]

### 26.4.6.6 Maintain Good Operating Practices:

Central plant operation involves both energy efficiency and safety issues. Proper safety and maintenance guidelines should be followed. The following maintenance issues should be carefully addressed:

- Blowdown: check blowdown setup if a boiler is operating at partial load most of the time. The purpose of blowdown is to remove the mineral deposits in the drum. This deposit is proportional to the cumulative make-up water flow, which is then proportional to the steam or hot water production. The blowdown can often be set back significantly. If the load ratio is 40% or higher, the blowdown can be reset proportional to the load ratio. If the load ratio is less than 40%, keep the blowdown rate at 40% of the design blowdown rate.

- Steam traps: check steam traps frequently. Steam traps have a tendency to fail, failed position is usually open, and leakage costs can be significant. A steam trap maintenance program is recommended. Consult the manufacturer and other manuals for proper procedures and methods.

- Condensate return: inspect the condensate return frequently. Make sure you are returning as much condensate as possible. This is very expensive water. It has high energy content and is treated water. When you lose condensate, you have to pay for the make-up water, chemicals, fuel, and, in some cases, sewage costs.

### 26.4.7 Continuous Commissioning Guidelines

Guidelines can be used to assist in carrying the CC process out in an orderly manner. An abbreviated sample set is provided in Appendix A that provides a basic check list, procedures, and documentation requirements.

## 26.5 ENSURING OPTIMUM BUILDING PERFORMANCE

The CC activities described in the previous sections will optimize building system operation and reduce energy consumption. To ensure excellent long-term performance, the following activities should be conducted.

1. Document CC activities,
2. Measure energy and maintenance cost savings,
3. Train operating and maintenance staff,
4. Measure energy data, and continuously measure energy performance, and
5. Obtain on-going assistance from CC engineers.

This section discusses guidelines to perform these tasks.

### 26.5.1 Document the CC Project

The documentation should be brief and accurate. *The operating sequences should be documented accurately and carefully*. This documentation should not repeat the existing building documentation. It should describe the procedures implemented, including control algorithms, and briefly give the reasons behind these procedures. The emphasis is on accurate and usable documentation. The documentation should be easily used by operating staff. For example, operating staff should be able to create operating manuals and procedures from the document.

The CC project report should include accurate documentation of current energy performance, building data, AHUs and terminal boxes, water loops and pumps, control system, and performance improvements.

### 26.5.2 Measure Energy Savings

Most building owners expect the CC project to pay for itself through energy savings. Measurement of en-

ergy savings is one of the most important issues for CC projects. The measurements should follow the procedures described in Chapter 27 (Measurement and Verification) of this handbook. Chapter 27 describes procedures from the *International Performance Measurement and Verification Protocol*[40] (IPMVP). This section will provide a very brief description of these procedures, emphasizing issues that are important in M&V for CC projects.

The process for determining savings as adopted in the IPMVP defines energy savings, $E_{save}$, as:

$$E_{Save} = E_{base} - E_{post}$$

where $E_{base}$ is the "baseline" energy consumption before the CC measures were implemented, and $E_{post}$ is the measured consumption following implementation of the CC measures.

Figure 26.10 shows the daily electricity consumption of the air handlers in a large building in which the HVAC systems were converted from constant volume systems to VAV systems using variable frequency drives. Consumption is shown for slightly over a year before the VFDs were installed (Pre), for about three months of construction and for about two years after installation (Post). In this case, the base daily electricity consumption is 8,300 kWh/day. The post-retrofit electricity consumption is 4,000 kWh/day, corresponding to electricity savings of 4,300 kWh/day. During the construction period, the savings are slightly lower.

However, in most cases, consumption shows more variation from day to day and month to month than that shown by the fan power for these constant speed fans. Hence, determination of the baseline must consider a number of factors, including weather changes, changes in occupancy schedule, changes in number of occupants, remodeling of the spaces, equipment changes, etc.

In the IPMVP, the baseline energy use, $E_{base}$, is determined either from direct measurements, utility bills, or from a model of the building operation before the retrofit (or commissioning). The post-installation energy use is generally simply the measured energy use, but it may be determined from a model if measured data are not available. Before and after comparisons are often normalized with factors such as weather, occupancy, etc.

The IPMVP includes four different M&V techniques or options. These options, may be summarized as Option A—some measurements, but mostly stipulated savings, Option B: measurement at the system or device level, Option C—measurement at the whole-building or facility level, and Option D—determination from calibrated simulation. Each option has its advantages for some special applications.

The cost savings must also consider changes in

**Figure 26.10 Daily electricity consumption for approximately one year before a retrofit and two years after the retrofit[41].**

utility rates. It is generally recommended that the utility rates in place before the retrofit or CC measures were implemented be used if any savings projections were made, since those projections were made based on the rates in effect at that time.

In general, the least expensive M&V method that will provide the accuracy required should be used. Utility billing data will be the least expensive data whenever it is available. It will sometimes provide the required accuracy for CC projects, but has the disadvantage that it may take considerable time before the improved performance is clearly evident. The discussion below generally assumes that higher frequency data is being used.

### 26.5.2.1 Option A - Stipulated Savings
### (Involving some measurements)

The stipulated option determines savings by measuring the capacity or the efficiency of a system before and after retrofit or commissioning, and then multiplies the difference by an agreed upon or "stipulated" factor such as the hours of operation, or the load on the system. This option focuses on a physical determination of equipment changes to ensure that the installation meets contract specifications. Key performance factors (e.g. lighting wattage) are determined with spot or short-

term measurements and operational factors (e.g. lighting operating hours) are stipulated based on historical data or spot measurement. Performance factors are measured or checked yearly. This method provides reliable savings estimation for applications where the energy savings are independent of weather and occupancy conditions.

For example, during the CC process, the fan pulley was decreased from 18" to 16" for a constant volume AHU. The fan power savings can be determined using the following method:

- Measure the fan power consumption before changing the pulley and the power consumption after changing the pulley.
- Determine the number of hours in operation.
- Determine the fan power savings as the product of the hourly fan power energy savings and the number of hours in operation.

If the energy consumption varies with occupancy and weather conditions, this option should not be used. For example, the minimum airflow setting was adjusted from 50% to 0% for 100 VAV terminal boxes at night and during weekends. Since the airflow depends on both internal and external loads, and the airflow may not be 0% even if the minimum flow setting is 0%, this method cannot be used to determine savings.

If Option A can provide the required accuracy, it will generally be the second least expensive method, after utility data.

### 26.5.2.2 Option B - Device/System Level Measurement

Within Option B, savings are determined by continuous measurements taken throughout the project term at the device or system level. Individual loads or end-uses are monitored continuously to determine performance and the long-term persistence of the measures installed. The base line model can be developed using the measured energy consumption and other parameters. The energy savings can be determined as the difference between baseline energy consumption and the measured energy consumption. This method provides the best saving estimation for an individual device or system.

The data collected can also be used to improve or optimize the system operation, and as such is particularly valuable for continuous commissioning projects. Since measurements are taken throughout the project term, the cost is higher than option A.

### 26.5.2.3 Option C - Whole Building Level Measurement

Determines savings by analyzing "whole-building" or facility level data measured during the baseline period and the post-installation period. This option is required when it is desired to measure interaction effects, e.g. the impact of a lighting retrofit on the cooling consumption as well as savings in lighting energy. The data used may be utility data, or sub-metered data.

The minimum number of measurement channels recommended for performance assurance or savings measurement will be the number needed to separate heating, cooling and other electric uses. The actual number of channels will vary, depending on whether pulses are taken from utility meters, or if two or three current transformers are installed to measure the three phase power going into a chiller. Other channels may need to be added, depending on the specific measures that are being evaluated.

Option C requires that installation of the proper systems/equipment and proper operating practices must be confirmed. It determines savings from metered data taken throughout the project term. The major limitation in the use of Option C for savings determination is that the size of the savings must be larger than the error in the baseline model. The major challenge is accounting for changes other than those associated with the ECMs or the commissioning changes implemented.

Accurate determination of savings using Option C normally requires 12 months of continuous data before a retrofit and continuous data after retrofit. However, for commissioning applications, a shorter period of data during which daily average ambient conditions cover a large fraction of normal yearly variation is often adequate.

### 26.5.2.4 Option D - Calibrated Simulation

Savings are determined through simulation of the facility components and/or the whole facility. The most powerful application of this approach calibrates a simulation model to baseline consumption data. For commissioning applications, it is recommended that calibration be to daily or hourly data. This type of calibration may be carried out most rapidly if simulated data is compared to measured data as a function of ambient temperature. Wei et al.[42] have developed "energy signatures" that greatly aid this process. More information can be found in Liu and Claridge[43] and a manual providing instructions on use of the method is available[44].

Just as for the other options, the implementation of proper operating practices should be confirmed. It is particularly important that personnel experienced in the use of the particular simulation tool conduct the analysis. The simulation analysis needs to be well documented, with electronic and hard copies of the simulation program input and output preserved.

### 26.5.2.5 Data Used to Determine Savings

Note that monthly bills may be used to estimate the energy savings. This method is one version of Option C described above. It is typically the least expensive method of verification. It will work fine if the following conditions are met:

1. Significant savings are expected at the utility meter level
2. Savings are too small to cost-justify more data
3. There will be no changes in
   a. Equipment
   b. Schedules
   c. Occupancy
   d. Space utilization

The case shown in Figure 26.11 is an example where monthly bills clearly show the savings. The savings were large and consistent following the retrofit until June. At this point, a major deviation occurred. The presence of other metering at this site showed that the utility bill was incorrect. Further investigation showed that the utility meter had been changed and this had not been considered in the bill sent. The consumption included in this bill was greater than the site would have used if it had used the peak demand recorded on the utility

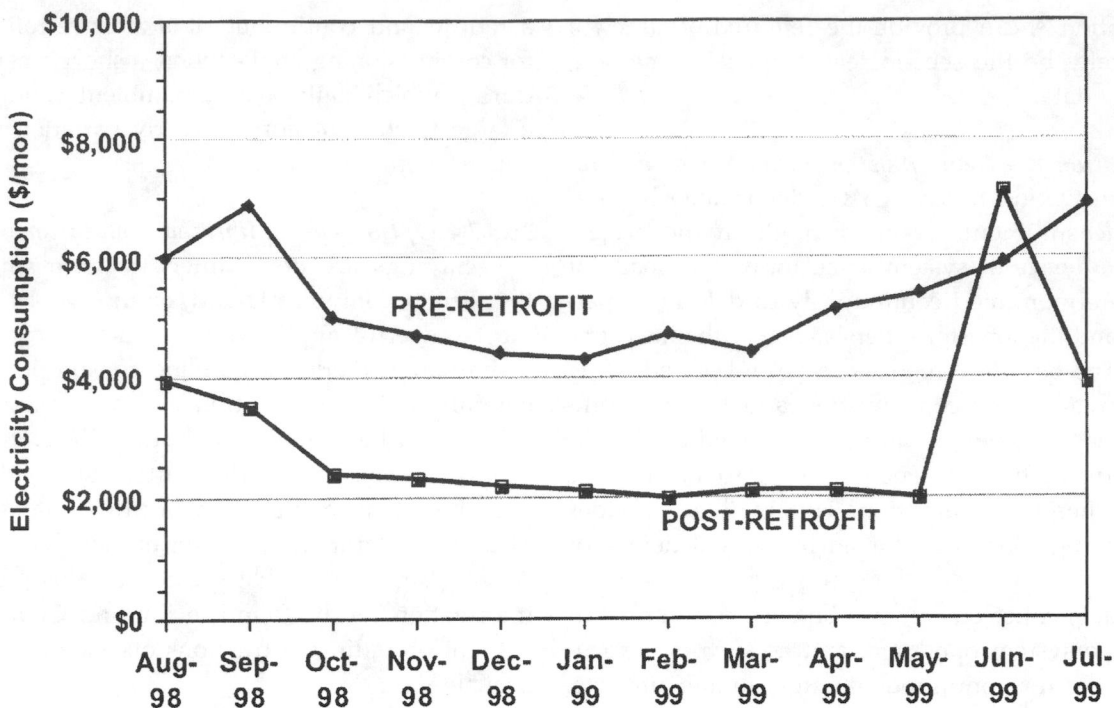

**Figure 26.11 Comparison of monthly utility bills before (top line) and after (bottom line) a retrofit[45].**

meter for every hour of the billing period!

However, daily or hourly data will show the results of commissioning measures much more quickly and are an extremely valuable diagnostic tool when problems arise as described in Section 26.5.4. Hence it is recommended that such data be used for savings determination and follow-up tracking whenever possible.

### 26.5.3 Trained Operating and Maintenance Staff

Efficient building operation begins with a qualified and committed staff. Since the CC process generally makes changes in the way a building is operated to improve comfort and efficiency, it is essential that the operators be a part of the commissioning team. They need to work with the CC engineers, propose CC measures and implement or help implement them. In addition to actively participating in the CC process, formal technical training should be provided to ensure that the operating staff understands the procedures implemented so they can perform trouble-shooting properly.

### 26.5.4 Continuously Measure Energy Performance

The measurement of energy consumption data is very important to maintain building performance and maintain CC savings. The metered data can be used to:
1. *Identify and solve problems.* Metered consumption data is needed to be sure that the building is still operating properly. If there is a component failure

or an operating change that makes such a small change in comfort or operating efficiency that it is not visible in metered consumption data, it generally isn't worth worrying about. If it does show up as even a marginal increase in consumption, trouble-shooting should be initiated.

2. *Trend/measure energy consumption data.* This continuing activity is the first line of defense against declining performance. The same procedures used to establish a pre-CC baseline can be used to establish a baseline for post-CC performance, and this post-CC baseline can be used as a standard against which future performance is compared. Consumption that exceeds this baseline for a few days, or even a month may not be significant, but if it persists much more than a month, trouble-shooting should be used to find out what has led to the increase. If it is the result of a malfunctioning valve, you can fix it. If it is the result of 100 new computers added to the building, you will adjust the baseline accordingly.

3. *Trend and check major operating parameters.* Parameters such as cold-deck temperatures, zone supply temperatures, etc. should be trended periodically for comparison with historic levels. This can be extremely valuable when trouble-shooting and

when investigating consumption above the post-CC baseline.

4. *Find the real problems when the system needs to be repaired or fixed.* It is essential that the same fundamental approach used to find and fix problems while the CC process is initiated be used whenever new hot calls or cold calls are received.

### 26.5.5 Utilize Expert Support as Needed

It is inevitable that a problem will come up which, even after careful trouble-shooting points toward a problem with one or more of the CC measures that have been implemented. Ask the CC providers for help in solving such problems before undoing an implemented CC measure. Sometimes it will be necessary to modify a measure that has been implemented. The CC engineers will often be able to help with finding the most efficient solution, and they will sometimes be able to help you find another explanation, so the problem can be remedied without changing the measure.

Ask help from the CC providers when you run into a new problem or situation. Problems occasionally crop up that defy logical explanation. These are the problems that generally get resolved by trying one of three things that seem like possible solutions, and playing

with system settings until the problem goes away. This is one of the most important situations in which expert help is needed. Without the expert's help, these kinds of problems—and the trial and error solutions—can undo the measure's savings potential.

### 26.5.6 How Well Do Commissioning Savings Persist?

The Energy Systems Laboratory at Texas A&M has conducted a study of 10 buildings on the Texas A&M campus that had CC measures implemented in 1996-97[46,47]. Table 26.5 shows the baseline cost of combined heating, cooling and electricity use of each building and the commissioning savings for 1998 and 2000. The baseline consumption and savings for each year were normalized to remove any differences due to weather.

Looking at the totals for the group of 10 buildings, heating and cooling consumption increased by $207,258 (12.1%) from 1998 to 2000, but savings from the earlier commissioning work were still $985,626. However, it may also be observed that almost three-fourths of this consumption increase occurred in two buildings, the Kleberg Building, and G. Rollie White Coliseum. The increased consumption of the Kleberg Building was due to a combination of component failures and control problems as described in the case study in Section 26.5.6.1. The increased consumption in G. Rollie White Coliseum

Table 26.5. Commissioning savings in 1998 and 2000 for 10 buildings on the Texas A&M campus[48].

| Building | Baseline Use ($/Yr) | 1998 Savings ($/Yr) | 2000 Savings($/yr) |
|---|---|---|---|
| Kleberg Building | $484,899 | $313,958 | $247,415 |
| G.R. White Coliseum | $229,881 | $154,973 | $71,809 |
| Blocker Building | $283,407 | $76,003 | $56,738 |
| Eller O&M Building | $315,404 | $120,339 | $89,934 |
| Harrington Tower | $ 145,420 | $64,498 | $48,816 |
| Koldus Building | $ 192,019 | $57,076 | $61,540 |
| Richardson Petroleum Building | $273,687 | $120,745 | $120,666 |
| Veterinary Medical Center Addition | $324,624 | $87,059 | $92,942 |
| Welmer Business Building | $224,481 | $47,834 | $68,145 |
| Zachry Engineering Center | $436,265 | $150,400 | $127,620 |
| Totals | $2,910,087 | $ 1,192,884 | $985,626 |

was due to different specific failures and changes, but was qualitatively similar to Kleberg since it resulted from a combination of component failures and control changes. The five buildings that showed consumption changes of more than 5% from 1998 to 2000 were all found to have different control settings that appear to account for the changed consumption (including the decrease in the Wehner Business Building).

These data suggest that commissioning savings generally persist, but tracking can subsequently uncover problems that did not cause comfort problems, but have increased consumption by $10,000-$100,000 per year in large buildings.

### 26.5.6.1 Commissioning persistence case study - Kleberg Building[49]

The Kleberg Building is a teaching/research facility on the Texas A&M campus consisting of classrooms, offices and laboratories, with a total floor area of approximately 165,030 ft[2]. Ninety percent of the building is heated and cooled by two (2) single duct variable air volume (VAV) air handling units (AHU) each having a pre-heat coil, a cooling coil, one supply air fan (100 hp), and a return air fan (25 hp). Two smaller constant volume units handle the teaching/lecture rooms in the building. The campus plant provides chilled water and hot water to the building. The two (2) parallel chilled water pumps (2 × 20 hp) have variable frequency drive control. There are 120 fan-powered VAV boxes with terminal reheat in 12 laboratory zones and 100 fan-powered VAV boxes with terminal reheat in the offices. There are six (6) exhaust fans (10-20 hp, total 90 hp) for fume hoods and laboratory general exhaust. The air handling units, chilled water pumps and 12 laboratory zones are controlled by a direct digital control (DDC) system. DDC

controllers modulate dampers to control exhaust airflow from fume hoods and laboratory general exhaust.

A CC investigation was initiated in the summer of 1996 due to the extremely high level of simultaneous heating and cooling observed in the building (Abbas, 1996). Figures 26.12 and 26.13 show daily heating and cooling consumption (expressed in average kBtu/hr) as functions of daily average temperature. The Pre-CC heating consumption data given in Figure 26.13 shows very little temperature dependence as indicted by the regression line derived from the data. Data values were typically between 5 and 6 MMBtu/hr with occasional lower values. The cooling data (Figure 26.12) shows more temperature dependence and the regression line indicates that average consumption on a design day would exceed 10 MMBtu/hr. This corresponds to only 198 sq.ft./ton based on average load, which is a high cooling density.

It was soon found that the preheat was operating continuously, heating the mixed air entering the cooling coil to approximately 105°F, instituted in response to a humidity problem in the building. The preheat was turned off and heating and cooling consumption both dropped by about 2 MMBtu/hour as shown by the middle clouds of data in Figures 26.12 and 26.13. Subsequently, the building was thoroughly examined and a comprehensive list of commissioning measures was developed and implemented. The principal measures implemented that led to reduced heating and cooling consumption were:

- Preheat to 105°F was changed to preheat to 40°F
- Cold deck schedule changed from 55°F fixed to vary from 62°F to 57°F as ambient temperature varies from 40°F to 60°F

**Figure 26.12 Pre-CC and post-CC heating water consumption at the Kleberg Building vs. daily average outdoor temperature[50].**

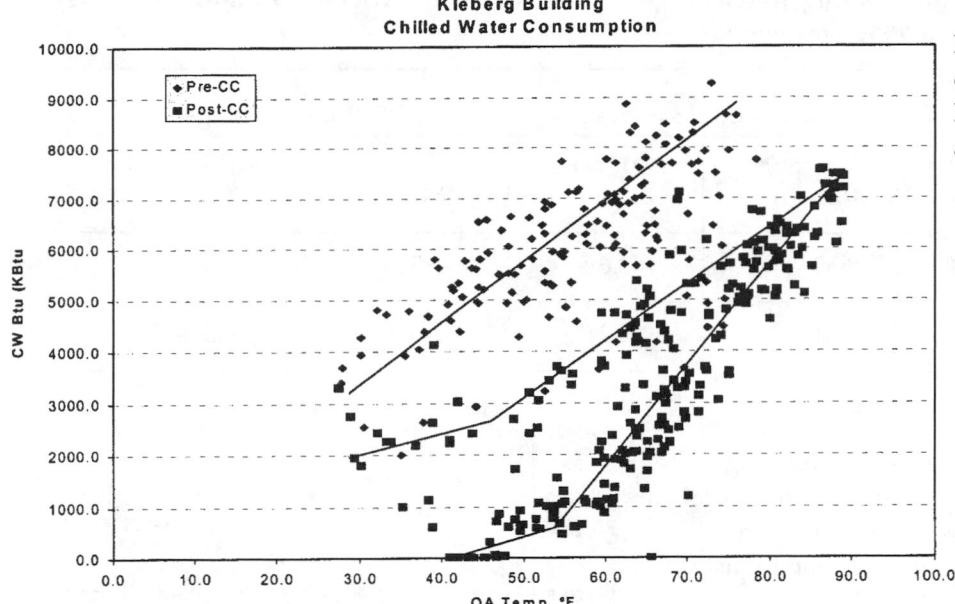

**Figure 26.13 Pre-CC and post-CC chilled water consumption at the Kleberg Building vs. daily average outdoor temperature[51].**

- Economizer—set to maintain mixed air at 57°F whenever outside air below 60°F
- Static pressure control—reduced from 1.5 inH$_2$O to 1.0 inH$_2$O and implemented night-time set back to 0.5 inH$_2$O
- Replaced or repaired a number of broken VFD boxes
- Chilled water pump VFDs were turned on.

Additional measures implemented included changes in CHW pump control—changed so one pump modulates to full speed before the second pump comes on instead of operating both pumps in parallel at all times, building static pressure was reduced from 0.05 in. w.c. to 0.02 in. w.c., and control changes were made to eliminate hunting in several valves. It was also observed that there was a vibration at a particular frequency in the pump VFDs that influenced the operators to place these VFDs in the manual mode, so it was recommended that the mountings be modified to solve this problem.

These changes further reduced chilled water and heating hot water use as shown in Figures 26.12 and 26.13 for a total annualized reduction of 63% in chilled water use and 84% in hot water use. Additional follow-up conducted from June 1998 through April 1999 focused on air balance in the 12 laboratory zones, general exhaust system rescheduling, VAV terminal box calibration, adjusting the actuators and dampers, and calibrating fume hoods and return bypass devices to remote DDC control[52] (Lewis, et al. 1999). These changes reduced electricity consumption by about 7% or 30,000 kWh/mo.

In 2001 it was observed that chilled water savings for 2000 had declined to 38% and hot water savings to

62% as shown in Table 26.6. Chilled water data for 2001 and the first three months of 2002 are shown in Figure 26.14. The two lines shown are the regression fits to the chilled water data before CC implementation and after implementation of CC measures in 1996 as shown in Figure 26.13. It is evident that consumption during 2001 is generally appreciably higher than immediately following implementation of CC measures. The CC group performed field tests and analyses that soon focused on two single-duct VAV AHU systems, two chilled water pumps, and the Energy Management Control System (EMCS) control algorithms as described in Chen et al.[53] Several problems were observed as noted below.

**Problems Identified**
- The majority of the VFDs were running at a constant speed near 100% speed.

- VFD control on two chilled water pumps was again bypassed to run at full speed.

- Two chilled water control valves were leaking badly. Combined with a failed electronic to pneumatic switch and the high water pressure noted above, this resulted in discharge air temperatures of 50°F and lower and activated preheat continuously.

- A failed pressure sensor and two failed CO$_2$ sensors put all outside air dampers to the full open position.

- The damper actuators were leaking and unable to maintain pressure in some of the VAV boxes. This

**Table 26.6. Chilled water and heating water usage and saving in the Kleberg Building for three different years normalized to 1995 weather[54].**

| Type | Pre-CC | Post-CC Use/Savings | | 2000 Use/Savings | |
|------|--------|------|------|------|------|
| | Baseline (MMBtu/yr) | Use (MMBtu/yr) | Savings % | Use (MMBtu/yr) | Savings (%) |
| CHW | 72935 | 26537 | 63.6% | 45431 | 37.7% |
| HW | 43296 | 6841 | 84.2% | 16351 | 62.2% |

caused cold air to flow through the boxes even when they were in the heating mode, resulting in simultaneous heating and cooling. Furthermore some of the reheat valves were malfunctioning. This caused the reheat to remain on continuously in some cases.

- Additional problems identified from the field survey included the following: 1) high air resistance from the filters and coils, 2) errors in a temperature sensor and static pressure sensor, 3) high static pressure setpoints in AHU1 and AHU2.

This combination of equipment failure compounded by control changes that returned several pumps and fans to constant speed operation had the consequence of increasing chilled water use by 18,894 MMBtu and hot water use by 9,510 MMBtu. This amounted to an increase of 71% in chilled water use and more than doubled hot water use from two years earlier

These problems have now been corrected and building performance has returned to previously low levels as illustrated by the data for April-June 2002 in Figure 26.4. These data are all below the lower of the two regression lines and is comparable to the level achieved after additional CC measures were implemented in 1998-99.

## 26.6 COMMISSIONING NEW BUILDINGS FOR ENERGY MANAGEMENT

The energy manager's effort is generally directed toward the improving the efficiency of existing buildings. However, whenever the organization initiates design and construction of a new building that will become part of the energy manager's portfolio of buildings, it is extremely important that the energy manager become an active part of the design and construction team to ensure that the building incorporates all appropriate energy efficiency technologies. It is just as

**Figure 26.14 ChW data for the Kleberg Building for January 2001- June 2002[55].**

important that the perspective of operational personnel be included in the design process so it will be possible to effectively and efficiently operate the building. One of the best ways to accomplish these objectives is to commission the building as it is designed and built.

The primary motivation for commissioning HVAC systems is generally to achieve HVAC systems that work properly to provide comfort to building occupants at low cost. In principle, all building systems should be designed, installed, documented, tested, and building staff trained in their use. In practice, competitive pressures and fee structures, and financial pressures to occupy new buildings as quickly as possible generally result in buildings that are handed over to the owners with minimal contact between designers and operators, minimal functional testing of systems, documentation that largely consists of manufacturers system component manuals, and little or no training for operators. This in turn leads to problems such as mold growth in walls of new buildings, rooms that never cool properly, air quality problems, etc. Such experiences were doubtless the motivation for the facility manager for a large university medical center who stated a few years ago that he didn't want to get any new buildings. He only wanted three-year old buildings in which the problems had been fixed.

Although commissioning provides higher quality buildings and results in fewer initial and subsequent operational problems, most owners will include commissioning in the design and construction process only if they believe they will benefit financially from commissioning. It is much more difficult to document the energy cost savings from commissioning a new building than an existing building. There is no historical use pattern to use as a baseline. However, it has been estimated that new building commissioning will save 8% in energy cost alone compared with the average building which is not commissioned[56]. This offers a payback for the cost of commissioning in just over four years from the energy savings alone and also provides improved comfort and air quality.

Commissioning is often considered to be a punch-list process that ensures that the systems in a building function before the building is turned over to the owner. However, the process outlined in Table 26.7 shows the process beginning in the pre-design phase. It is most effective if allowed to influence both design and construction. It is essential that the energy manager be involved in the commissioning process on the owner's team no later than the design phase of the construction process. This permits input into the design process that can have major impact on the efficiency of the building as built and can lead to a building that has far fewer operational problems.

## 26.7 SUMMARY

Commissioning of existing buildings is emerging as one of the most cost effective ways for an energy manager to lower operating costs, and typically does so with no capital investment, or with a very minimal amount. It has been successfully implemented in several hundred buildings and provides typical paybacks of one to three years.

It is much more than the typical O&M program. It does not ensure that the systems function as originally designed, but focuses on improving overall system control and operations for the building as it is currently utilized and on meeting existing facility needs. During the CC process, a comprehensive engineering evaluation is conducted for both building functionality and system functions. The optimal operational parameters and schedules are developed based on actual building conditions. An integrated approach is used to implement these optimal schedules to ensure practical local and

**Table 26.7. The commissioning process for new buildings[57].**

1. **Conception or pre-design phase**
   (a) Develop commissioning objectives
   (b) Hire commissioning provider
   (c) Develop design phase commissioning requirements
   (d) Choose the design team

2. **Design phase**
   (a) Commissioning review of design intent
   (b) Write commissioning specifications for bid documents
   (c) Award job to contractor
   (d) Develop commissioning plan

3. **Construction/Installation phase**
   (a) Gather and review documentation
   (b) Hold commissioning scoping meeting and finalize plan
   (c) Develop pre-test checklists
   (d) Start up equipment or perform pre-test checklists to ensure readiness for functional testing during acceptance

4. **Acceptance phase**
   (a) Execute functional test and diagnostics
   (b) Fix deficiencies
   (c) Retest and monitor as needed
   (d) Verify operator training
   (e) Review O&M manuals
   (f) Building accepted by owner

5. **Post-acceptance phase**
   (a) Prepare and submit final report
   (b) Perform deferred tests (if needed)
   (c) Develop recommissioning plan/schedule

global system optimization and to ensure persistence of the improved operational schedules.

The approach presented in this chapter begins by conducting a thorough examination of all problem areas or operating problems in the building, diagnoses these problems, and develops solutions that solve these problems while almost always reducing operating costs at the same time. Equipment upgrades or retrofits may be implemented as well, but have not been a factor in the case studies presented, except where the commissioning was used to finance equipment upgrades. This is in sharp contrast to the more usual approach to improving the efficiency of HVAC systems and cutting operating costs that primarily emphasizes system upgrades or retrofits to improve efficiency.

Commissioning of new buildings is also an important option for the energy manager, offering an opportunity to help ensure that new buildings have the energy efficiency and operational features that are most needed.

## 26.8. FOR ADDITIONAL INFORMATION

Two major sources of information on commissioning existing buildings are the *Continuous Commissioning*SM Guidebook: Maximizing Building Energy Efficiency and Comfort (Liu, M., Claridge, D.E. and Turner, W.D., Federal Energy Management Program, U.S. Dept. of Energy, 144 pp., 2002) and *A Practical Guide for Commissioning Existing Buildings* (Haasl, T. and Sharp, T., Portland Energy Conservation, Inc. and Oak Ridge National Laboratory for U.S. DOE, ORNL/TM-1999/34, 69 pp. + App., 1999). Much of this chapter has been abridged and adapted from the CC Guidebook.

There are a much wider range of materials available that treat commissioning of new buildings. Two documents that provide a good starting point are *ASHRAE Guideline 1-1996: The HVAC Commissioning Process* (American Society of Heating, Refrigerating and Air-Conditioning Engineers, Atlanta, GA, 1996) and the *Building Commissioning Guide—Version 2.2* (U.S. GSA and U.S. DOE, 1998 by McNeil Technologies, Inc. and Enviro-Management & Research, Inc. available at www.emrinc.com).

The case studies in this chapter have been largely abridged and adapted from the following three papers:

Claridge, D.E., Liu, M., Deng, S., Turner, W.D., Haberl, J.S., Lee, S.U., Abbas, M., Bruner, H., and Veteto, B., "Cutting Heating and Cooling Use Almost in Half Without Capital Expenditure in a Previously Retrofit Building," *Proc. of 2001 ECEEE Summer Study*, Mandeliu, France, Vol. 2, pp. 74-85, June 11-16, 2001.

Turner, W.D., Claridge, D.E., Deng, S. and Wei, G., "The Use of Continuous Commissioning SM As An Energy Conservation Measure (ECM) for Energy Efficiency Retrofits," Proc. of 11th National Conference on Building Commissioning, Palm Springs, CA, CD, May 20-22, 2003.

Claridge, D.E., Turner, W.D., Liu, M., Deng, S., Wei, G., Culp, C., Chen, H. and Cho, S.Y., "Is Commissioning Once Enough?," *Solutions for Energy Security & Facility Management Challenges: Proc. of the 25th WEEC*, Atlanta, GA, pp. 29-36, Oct. 9-11, 2002.

## References

1. ASHRAE, *ASHRAE Guideline 1-1996: The HVAC Commissioning Process*, American Society of Heating, Refrigerating and Air-Conditioning Engineers, Atlanta, GA, 1996.
2. U.S. Department of Energy, Building Commissioning: The Key to Quality Assurance, Washington, DC. 1999.
3. ibid.
4. Continuous Commissioning is a registered trademark and CC is a service mark of the Texas Engineering Experiment Station (TEES). Contact TEES for further information.
5. Liu, M., Claridge, D.E. and Turner, W.D., *Continuous Commissioning SM Guidebook: Maximizing Building Energy Efficiency and Comfort*, Federal Energy Management Program, U.S. Dept. of Energy, 144 *pp., 2002.*
6. Claridge, D.E., Haberl, J., Liu, M., Houcek, J., and Athar, A., "Can You Achieve 150% of Predicted Retrofit Savings: Is It Time for Recommissioning?" *A CEEE 1994 Summer Study on Energy Efficiency In Buildings Proceedings: Commissioning, Operation and Maintenance, Vol. 5,* American Council for an Energy Efficient Economy, Washington, D.C., pp. 73-87, 1994.
7. Claridge, D.E., Liu, M., Deng, S., Turner, W.D., Haberl, J.S., Lee, S.U., Abbas, M., Bruner, H., and Veteto, B., "Cutting Heating and Cooling Use Almost in Half Without Capital Expenditure in a Previously Retrofit Building," *Proc. of 2001 ECEEE Summer Study*, Mandeliu, France, Vol. 2, pp. 74-85, June 11-16, 2001.
8. Turner, W. D., Claridge, D. E., Deng, S. and Wei, G., "The Use of Continuous Commissioning SM As An Energy Conservation Measure (ECM) for Energy Efficiency Retrofits," Proc. of 11th National Conference on Building Commissioning, Palm Springs, CA, CD, May 20-22, 2003.
9. Zhu, Y., Liu, M., Claridge, D.E., Feary, D. and Smith, T., "A Continuous Commissioning Case Study of a State-of-the-Art Building," *Proceedings of the 5th National Commissioning Conference*, Huntington Beach, CA, pp. 13.1 - 13. 10, April, 1997.
10. Liu, M., Zhu, Y., Powell, T., and Claridge, D.E., "System Optimization Saves $195,000/yr. in a New Medical Facility," *Proceedings of the 6th National Conference on Building Commissioning*, Lake Buena Vista, FL, pp. 14.2.1-14.2.11, May 18-20, 1998.
11. Turner, et al., op. cit.
12. Source: Adapted from Turner et al. op. cit.
13. Source: Liu, Claridge, and Turner, op. cit.
14. Source: ibid.
15. Source: ibid.
16. Source: ibid.
17. Liu, M., Zhu, Y., Park, B.Y., Claridge, D.E., Feary, D.K. and Gain, J., "Airflow Reduction to Improve Building Comfort and Reduce Building Energy Consumption - A Case Study," *ASHRAE Transactions-Research, Vol.* 105, Part 1, pp. 3 84 - 3 90, 1999.
18. Liu, M., Zhu, Y., Claridge, D. and White, E., "Impacts of Static Pressure Set Level on the HVAC Energy Consumption and Indoor Conditions," *ASHRAE Transactions-Research.* Volume 103, Part 2, pp. 221-228, 1997.
19. Liu, Claridge, and Turner, op. cit.
20. ibid.
21. ibid.
22. Liu, M., "Variable Speed Drive Volumetric Tracking (VSDVT) for Airflow Control in Variable Air Volume (VAV) Systems," Proceedings of Thirteenth Symposium on Improving Building Systems in Hot and Humid Climates, San Antonio, TX, pp. 195-198, May 15-16, 2002.
23. Claridge, et al. 2001, op. cit.
24. Source: ibid.
25. Source: ibid.
26. Source: ibid.
27. Source: ibid.
28. Source: ibid.
29. Source: ibid.

30. Liu, Claridge, and Turner, op. cit.

31. Liu, M., "Variable Water Flow Pumping for Central Chilled Water Systems," *ASME Journal of Solar Energy Engineering, Vol.* 124, pp. 300-304, 2002.

32. Deng, S., Turner, W.D., Batten, T., and Liu, M., "Continuous Conumriissioning[SM] of a Central Chilled Water and Heating Hot Water System," Proc. Twelfth Symposium on Improving Building Systems in Hot and Humid Climates, San Antonio, TX, pp. 199-206, May 15-16, 2000.

33. Deng, S. Turner, W. D., Claridge, D. E., Liu, M., Bruner, H., Chen, H. and Wei, G. "Retrocommissioning of Central Chilled/ Hot Water Systems," *ASHRAE Transactions-Research, Vol.* 108, Part 2, pp. 75-81, 2002.

34. Liu, Claridge, and Turner, op. cit.

35. Liu, M., op. cit.

36. Liu, Claridge, and Turner, op. cit.

37. Wei, G., Liu, M., and Claridge, D.E., "In-situ Calibration of Boiler Instrumentation Using Analytic Redundancy," *International Journal of Energy Research, Vol.* 25, pp. 375-387, 2001.

38. Liu, M. et al., 1998, op. cit.

39. IpMVp Committee, *International Performance Measurement & Verification Protocol.- Concepts and Options for Determining Energy and Water Savings, Vol. 1, U.S.* Dept. of Energy, DOE/GO-102001-1187, January, 2001, 86 pp

40. Source: Liu, Claridge, and Turner, op. cit.

41. Wei, G., Liu, M., and Claridge, D.E., "Signatures of Heating and Cooling Energy Consumption for Typical AHUs," *The Eleventh Symposium on Improving Building Systems in Hot and Humid Climates Proceedings,* Fort Worth, Texas, pp. 387-402, June 1-2, 1998.

42. Liu, M. and Claridge, D.E., "Use of Calibrated HVAC System Models to Optimize System Operation," *ASME Journal of Solar Energy Engineering, Vol.* 120, pp. 131-138, 1998.

43. Claridge, D.E., Bensouda, N., Lee, S.U., Wei, G., Heinemeier, K., and Liu, M., *Manual of Procedures for Calibrating Simulations of Building Systems,* submitted to the Lawrence Berkeley National Laboratory under Sponsored Contract #66503346 as part of the California Energy Commission Public Interest Energy Research Program, 94 pp., October, 2003.

44. Source: Liu, Claridge, and Turner, op. cit.

45. Turner, W.D., Claridge, D.E., Deng, S., Cho, S., Liu, M., Hagge, T., Darnell, C., Jr., and Bruner, H., Jr., "Persistence of Savings Obtained from Continuous Conunissioning[SM]," *Proc. of 9th National Conference on Building Commissioning,* Cherry Hill, NJ, pp. 20-1.1 - 20-1.13, May 9-11, 2001.

46. Claridge, D.E., Turner, W.D., Liu, M., Deng, S., Wei, G., Culp, C., Chen, H. and Cho, S.Y., "Is Commissioning Once Enough?," *Solutions for Energy Security & Facility Management Challenges: Proc. of the 25 h WEEC,* Atlanta, GA, pp. 29-36, Oct. 9-11, 2002.

47. Source: ibid.

48. ibid.

49. Source: ibid.

50. Source: ibid.

51. Lewis, T., H. Chen, and M. Abbas, "CC summary for the Kleberg Building," Internal ESL Report, July, 1999.

52. Chen, H., Deng, S., Bruner, H., Claridge, D. and Turner, W.D., "Continuous Commissioning[SM] Results Verification And Follow-Up For An Institutional Building - A Case Study," *Proc. 13th Symposium on Improving Building Systems in Hot and Humid Climates,* Houston, TX, pp. 87-95, May 20-23, 2002.

53. Source: Claridge, D.E. et al. 2002, op. cit.

54. Source: ibid.

55. PECI, "National Strategy for Building Commissioning," Portland Energy Conservation, Inc., Portland, OR, 1999.

56. Adapted from Haasl, T. and Sharp, T., *A Practical Guide for Commissioning Existing Buildings,* Portland Energy Conservation, Inc. and Oak Ridge National Laboratory for U.S. DOE, ORNL/TM-1999/34, 69 pp. + App., 1999.

# CHAPTER 27

# MEASUREMENT AND VERIFICATION OF ENERGY SAVINGS

**JEFF S. HABERL, PH.D., P.E.**
**CHARLES C. CULP, PH.D., P.E.**
Energy Systems Laboratory
Texas A&M University

## 27.1 INTRODUCTION

### 27.1.1 M&V Method Selection

Measurement and verification (M&V) has a dual role. First, M&V quantifies the savings being obtained. This applies to the initial savings and the long-term savings. Since the persistence of savings has been shown to decrease with time,[1] long term M&V provides data to make these savings sustainable. Second, M&V must be cost effective so that the cost of measurement and the analysis does not consume the savings.[2,3] The 1997 International Performance Measurement and Verification Protocol (IPMVP) set the target costs for M&V to be in the range of 1% to 10%, depending upon the Option selected, of the construction cost for the life of the ECM. Most approaches fall in the recommended range of 3% to 10% of the construction cost. The IPVMP 2001 removed this guidance on the recommended costs for M&V. Currently, a goal of about 5% of the savings per year has evolved as a preferred criterion for costing M&V, since the cost justification directly results from the calculation.

A general procedure for selecting an approach can be summarized by the following five steps:

a)  *In general one wants to try to...* Perform Monthly Utility Bill Before/After Analysis.
b)  *And if this does not work, then...* Perform Daily or Hourly Before/After Analysis.
c)  *And if this does not work, then...* Perform Component Isolation Analysis.
d)  *And if this does not work, then...* Perform Calibrated Simulation Analysis.
e)  *Then...* Report savings and Finish Analysis.

### 27.1.2 History of M&V
#### 27.1.2.1 History of Building Energy Measurement.

The history of the measurement of building energy use can be traced back to the 19th century for electricity, and earlier for fuels such as coal and wood, which were used to heat buildings.[4,5,6,7] By the 1890s, although electricity was common in many new commercial build-ings, its use was primarily for incandescent lighting and, to a lesser extent, for the electric motors associated with ventilating buildings since most of the work in office buildings was carried out during daylight hours. The metering of electricity closely paralleled the spread of electricity into cities as its inventors needed to recover the cost of its production through the collection of payments from electric utility customers.[8,9] Commercial meters for the measurement of flowing liquids in pipes can be traced back to the same period, beginning with the invention of the first commercial flow meter by Clemens Herschel in 1887, which used principles based on the pitot tube and venturi flow meter, invented in 1732 and 1797 by their respective namesakes.[10] Commercial meters for the measurement of natural gas can likewise be traced to the sale and distribution of natural gas, which paralleled the development of the electric meters.

#### 27.1.2.2 History of M&V in the U.S.
##### 27.1.2.2.1 History of M&V

The history of the measurement and verification of building energy use parallels the development and use of computerized energy calculations in the 1960s, with a much accelerated awareness in 1973, when the embargo on Mideast oil made energy a front page issue.[11,12] During the 1950s and 1960s most engineering calculations were performed using slide rules, engineering tables and desktop calculators that could only add, subtract, multiply and divide. Since the public was led to believe energy was cheap and abundant,[13] the measurement and verification of the energy use in a building was limited for the most part to simple, unadjusted comparisons of monthly utility bills.

In the 1960s several efforts were initiated to formulate and codify equations that could predict dynamic heating and cooling loads, including efforts at the National Bureau of Standards to predict temperatures in fallout shelters,[14] and the 1967 HCC program developed by a group of mechanical engineering consultants,[15] which used the Total Equivalent Temperature Difference/Time Averaging (TETD/TA) method. The popularity of this program prompted the American Society of Heating, Refrigeration and Air-Conditioning Engineers (ASHRAE) to embark on a series of efforts that eventually delivered today's modern, general purpose simulation programs[16] (i.e., DOE-2, BLAST, EnergyPlus, etc.), which utilize thermal response factors,[17,18] as well as algorithms for simulation of the quasi-steady-state performance of primary and secondary equipment.[19] One of these efforts was to validate the hourly calculations with field measurements at the Legal Aid Building on the Ohio State University campus,[20] which is probably the first application of a calibrated, building energy simulation program.

Some of the earliest efforts to develop standardized methods for the M&V of building energy use began with efforts to normalize residential heating energy use in single-family and multi-family buildings,[21] which include the Princeton Scorekeeping Method (PRISM),[22] a forerunner to ASHRAE's Variable-based Degree Day (VBDD) calculation method. In commercial buildings, numerous methods have been reported over the years[23,24,25] varying from weather normalization using monthly utility billing data,[26,27,28] daily and hourly methods,[29] and even dynamic inverse models using resistance-capacitance (RC) networks.[30] Procedures and methodologies to baseline energy use in commercial buildings began to appear in several publications in the 1980s[31,32,33] and the early 1990s.[34,35,36] Modeling toolkits and software have been developed that are useful in developing performance metrics for buildings, as well HVAC system components. These efforts include

the Princeton Scorekeeping Software (PRISM),[37] which is useful for developing variable-based degree day models of monthly or daily data, ASHRAE's HVAC01 software for modeling primary HVAC systems such as boilers, and chillers,[38] ASHRAE's HVAC02 software for modeling secondary HVAC systems including air-handlers, blowers, cooling coils and terminal boxes,[39] and ASHRAE's Research Projects 827-RP for *in-situ* measurement of chillers, pumps, and blowers,[40] Research Project 1004-RP for *in-situ* measurement of thermal storage systems,[41] Research Project 1050-RP Toolkit for Calculating Linear, Change-point Linear and Multiple-Linear Inverse Building Energy Analysis Models, and Research Project 1093-RP toolkit for calculating diversity factors for energy and cooling loads.[43]

In 1989, a report by Oak Ridge National Laboratory[44] classified the diverse commercial building analysis methods into five categories, including: annual total energy and energy intensity comparisons, linear regression and component models, multiple linear regression, building simulation, and dynamic (inverse) thermal performance models. In 1997 a reorganized and expanded version of this classification appeared in the *ASHRAE Handbook of Fundamentals*, and is shown in Tables 27.1 and 27.2. In Table 27.1 different methods of analyzing building energy are presented, which have been clas-

Table 27.1 ASHRAE's 1997 Classification of Methods for the Thermal Analysis of Buildings.[57]

| METHOD: | FORWARD | INVERSE | HYBRID | COMMENTS: |
|---|---|---|---|---|
| Steady State Methods | | | | |
| Simple linear regression | | X | | One dependent parameter, one independent parameter. May have slope and y-intercept. |
| Multiple linear regression | | X | X | One dependent parameter, multiple independent parameters. |
| Modified degree day method | X | | | Based on fixed reference temperature of 65F. |
| Variable base degree day method | X | | | Variable reference temperatures. |
| Traditional ASHRAE bin method and inverse bin method | X | X | X | Hours in temperature bin times load for that bin. |
| Change point models: 3-parameter (PRISM CO,HO), 4-parameter, 5-parameter (PRISM HC). | | X | X | Uses daily or monthly utility billing data and average period temperatures. |
| ASHRAE TC 4.7 modified bin method | X | | X | Modified bin method with cooling load factors. |
| Dynamic Methods | | | | |
| Thermal network | X | X | X | Uses equivalent thermal parameters (inverse mode). |
| Response factors | X | | | Tabulated or as used in simulation programs. |
| Fourier Analysis | X | X | X | Frequency domain analysis convertible to time domain. |
| ARMA Model | | X | | Autoregressive Moving Average model. |
| ARMA Model | | X | | Multiple-input autoregressive moving average model. |
| BEVA, PSTAR | X | X | X | Combination of ARMA and Fourier series, includes loads in time domain. |
| Modal analysis | X | X | X | Bldg. described by diagonalized differential equation using nodes. |
| Differential equation | | X | | Analytical linear differential equation. |
| Computer simulation (DOE-2, BLAST) | X | | X | Hourly simulation programs with system models. |
| Computer emulation (HVACSIM+, TRNSYS) | X | | X | Sub-hourly simulation programs. |
| Artificial Neural Networks | | X | X | Connectionist models. |

**Table 27.2 ASHRAE's 1997 Decision Diagram for Selection of Model.**[58]

| METHOD: | USAGE: (1) | DIFFICULTY: | TIME SCALE: (2) | CALC. TIME: | VARIABLES: | ACCURACY: |
|---|---|---|---|---|---|---|
| Simple linear regression | ES | Simple | D,M | Very Fast | T | Low |
| Multiple linear regression | D,ES | Moderate | D,M | Fast | T,H,S,W,t | Medium |
| Inverse ASHRAE bin method | ES | Moderate | H | Fast | T | Medium |
| Change point models. | D,ES | Moderate | H,D,M | Fast | T | Medium |
| ASHRAE TC 4.7 modified bin method | ES, De | Moderate | H | Medium | T,S,tm | Medium |
| Thermal network | D,ES,C | Complex | S,H | Fast | T,S,tm | High |
| Fourier Series Analysis | D,ES,C | Complex | S,H | Medium | T,H,S,W,t,tm | High |
| ARMA Model | D,ES,C | Complex | H | Medium | T,H,S,W,t,tm | High |
| Modal analysis | D,ES,C | Complex | H | Medium | T,H,S,W,t,tm | High |
| Differential equation | D,ES,C | Very Complex | S,H | Fast | T,H,S,W,t,tm | High |
| Computer Simulation (Component-based) | D,ES,C, De | Very Complex | S,H | Slow | T,H,S,W,t,tm | Medium |
| Computer simulation (Fixed schematic) | D,ES,De | Very Complex | H | Slow | T,H,S,W,t,tm | Medium |
| Computer emulation | D,C | Very Complex | S,H | Very Slow | T,H,S,W,t,tm | High |
| Artificial Neural Networks | D,ES,C | Complex | S,H | Medium | T,H,S,W,t,tm | High |

NOTE: (1) Usage shown includes: diagnostics (D), energy savings calculations (ES), design (De), and control (C).

(2) Time scales shown are hourly (H), daily (D), monthly (M), and sub-hourly (S).

(3) Variables include: temperature (T), humidity (H), solar (S), wind (W), time (t), thermal mass (TM).

sified according to model type, including: forward, inverse, and hybrid models.[45]

In the first method, forward modeling, a thermodynamic model is created of a building using fundamental engineering principles to predict the hypothetical energy use of a building for 8,760 hours of the year given the location, and weather conditions. This requires a complete description of the building, system, or component of interest, as well as the physical description of the building geometry, geographical location, system type, wall insulation value, etc. Forward models are normally used to design and size HVAC systems, and have begun to be used to model existing building, using a technique referred to as calibrated simulation.

In the second method, inverse modeling, an empirical analysis is conducted on the behavior of the building as it relates to one or more driving forces or parameters. This approach is referred to as a system identification, parameter identification or inverse modeling. To develop an inverse model, one must assume a physical configuration of the building or system, and then identify the parameter of interest using statistical analysis.[46] Two primary types of inverse models have been reported in the literature, including: steady state inverse models and dynamic inverse models. A third category, hybrid models, consists of models that have characteristics of both forward and inverse models.[47]

The simplest steady-state inverse model regresses monthly utility consumption data against average bill-

ing period temperatures. More robust methods include multiple linear regression, change-point linear regression, and Variable-Based Degree Day regressions as indicated in Table 27.1. The advantage of steady-state inverse models is that their use can be automated and applied to large datasets where monthly utility billing data and average daily temperatures for the billing periods are available. Steady-state inverse models can also be applied to daily data which allows one to compensate for differences in weekday and weekend use.[48] Unfortunately, steady state inverse models are insensitive to dynamic effects (i.e., thermal mass), and other variables (for example humidity and solar gain), and are difficult to apply to certain building types, for example buildings that have strong on/off schedule dependent loads, or buildings that display multiple change-points.

Dynamic inverse models include: equivalent thermal network analysis,[49] ARMA models,[50,51] Fourier series models,[52,53] machine learning,[54] and artificial neural networks.[55,56] Unlike steady-state, inverse models, dynamic models are capable of capturing dynamic effects such as thermal mass which traditionally have required the solution of a set of differential equations. The disadvantages of dynamic inverse models are that they are increasingly complex, and need more detailed measurements to "tune" the model.

Hybrid models are models that contain forward and inverse properties. For example, when a traditional fixed-schematic simulation program such as DOE-2 or

BLAST (or even a component-based simplified model) is used to simulate the energy use of an existing building then one has a forward analysis method that is being used in an inverse application, i.e., the forward simulation model is being calibrated or fit to the actual energy consumption data from a building in much the same way that one fits a linear regression of energy use to temperature.

Table 27.2 presents information that is useful for selecting an inverse model where usage of the model (diagnostics - D, energy savings calculations - ES, design De, and control - C), degree of difficulty in understanding and applying the model, time scale for the data used by the model (hourly - H, daily - D, monthly - M, and sub-hourly - S), calculation time, and input variables used by the models (temperature - T, humidity - H, solar - S, wind - W, time - t, thermal mass - tm), and accuracy are used to determine the choice of a particular model.

*27.1.2.2.2 History of M&V Protocols in the United States*

The history of measurement and verification protocols in the United States can be traced to independent M&V efforts in different regions of the country as shown in Table 27.3, with states such as New Jersey, California, and Texas developing protocols that contained varying procedures for measuring the energy and demand savings from retrofits to existing buildings. These efforts

culminated in the development of the USDOE's 1996 North American Energy Measurement and Verification Protocol (NEMVP),[59] which was accompanied by the USDOE's 1996 FEMP Guidelines,[60] both relying on analysis methods developed in the Texas LoanSTAR program.[61] In 1997 the NEMVP was updated and republished as the International Performance Measurement and Verification Protocols (IPMVP). The IPMVP was then expanded in 2001 into two volumes: Volume I covering Energy and Water Savings,[63] and Volume II covering Indoor Environmental Quality.[64] In 2003 Volume III of the IPMVP was published, which covers protocols for new construction.[65] Finally, in 2002 American Society of Heating Refrigeration Air-conditioning Engineers (ASHRAE) released Guideline 14-2002: Measurement of Energy and Demand Savings,[66] which is intended to serve as the technical document for the IPMVP.

### 27.1.3 Performance Contracts

In order to reduce costs and improve the HVAC and lighting systems in its buildings, the U.S. federal government has turned to the private energy efficiency sector to develop methods to finance and deliver energy efficiency to the government. One of these arrangements, the performance contract, often includes a guarantee of performance, which benefits from accurate, reliable measurement and verification. In such a contract

### Table 27.3: History of M&V Protocols

2003 - IPMVP - 2003 Volume III (new construction)
2002 - ASHRAE Guideline 14 - 2002
2001 - IPMVP - 2001 Volume I & II (revised and expanded IPMVP)
1998 - Texas State Performance Contracting Guidelines
1997 - IPMVP (revised NEMVP)
1996 - FEMP Guidelines
1996 - NEMVP
1995 - ASHRAE Handbook - Ch. 37 "Building Energy Monitoring"
1994 - PG&E Power Saving Partner "Blue Book"
1993 - NAESCO M&V Protocols
1993 - New England AEE M&V Protocols
1992 - California CPUC M&V Protocols
1989 - Texas LoanSTAR Program
1988 - New Jersey M&V Protocols
1985 - First Utility Sponsored Large Scale Programs to Include M&V
1985 - ORNLs "Field Data Acq. For Bld & Eqp Energy Use Monitoring"
1983 - International Energy Agency "Guiding Principles for Measurement"
1980s - USDOE funds the End - Use Load and Consumer Assessment Program (ELCAP)
1980s - First Utility Sponsored Large Scale Programs to Include M&V
1970s - First Validation of Simulations
1960s - First Building Energy Simulations on Mainframe Computers

all costs of the project (i.e., administration, measurement and verification, overhead and profit) are paid for by the energy saved from the energy or water conservation projects. In principle, this is a very attractive option for the government, since it avoids paying the initial costs of the retrofits, which would have to come from shrinking taxpayer revenues. Instead, the costs are paid over a series of years because the government agrees to pay the Energy Service Company (ESCO) an annual fee that equals the annual normalized costs savings of the retrofit (plus other charges). This allows the government to finance the retrofits by paying a pre-determined annual utility bill over a series of years, which equals the utility bill during the base year had the retrofit not occurred. In reality, because the building has received an energy conservation retrofit, the actual utility bill is reduced, which allows funding the annual fee to the ESCO without realizing any increase in the total annual utility costs (i.e., utility costs plus the ESCO fee). Once the performance contract is paid off, the total annual utility bills for the government are reduced and the government receives the full savings amount of the retrofit.

### 27.1.3.1 Definitions, Roles and Participants

There are many different types of performance contracts, which vary according to risk, and financing, including Guaranteed Savings, and Shared Savings.[67] In a Guaranteed Savings performance contract a fixed payment is established that repays the ESCO's debt financing of the energy conservation retrofit, and any fees associated with the project. In return, the ESCO guarantees that the energy savings will cover the fixed payment to the ESCO. Hence, in a Guaranteed Savings contract the ESCO is responsible for the majority of the project risks. In a Shared Savings performance contract payments to the ESCO are based on an agreed-upon portion of the estimated savings generated by the retrofit. In such contracts the M&V methods selected determine the level of risk, and the responsibilities of the ESCO, and building owner. In both types of contracts the measurement and verification of the energy savings plays a crucial role in determining payment amounts.

## 27.2 OVERVIEW OF MEASUREMENT AND VERIFICATION METHODS

Nationally recognized protocols for measurement and verification have evolved since the publication of the 1996 NEMVP as shown in Table 27.4. This evolution reflects the consensus process that the Department of Energy has chosen as a basis for the protocols. This process was chosen to produce methods that all parties agree can be used by the industry to determine savings from performance contracts, varying in accuracy and cost from partial stipulation to complete measurement. In 1996 three M&V methods were included in the NEMVP: Option A: measured capacity with stipulated consumption; Option B: end-use retrofits, which utilized measured capacity and measured consumption; and Option C: whole-facility or main meter measurements, which utilize before after regression models.

In 1997, Options A, B and C were modified and relabeled, and Option D: calibrated simulation was added. Also included in the 1997 IPMVP was a chapter on measuring the performance of new construction, which primarily utilized calibrated simulation, and a discussion of the measurement of savings due to water conservation efforts. In 2001 the IPMVP was published in two volumes: Volume I, which covers Options A, B, and C, which were redefined and relabeled from the 1997 IPMVP, and Volume II, which covers indoor environmental quality (IEQ), and includes five M&V approaches for IEQ, including: no IEQ M&V, M&V based on modeling, short-term measurements, long-term measurements, and a method based on occupant perceptions of IEQ. In 2003 the IPMVP released Volume III, which contains four M&V methods: Option A: partially measured Energy Conservation Measure (ECM) isolation, Option B: ECM isolation, Option C: whole-building comparisons, and Option D: whole-building calibrated simulation.

In 2002 ASHRAE released Guideline 14-2002: Measurement of Energy and Demand Savings, which is intended to serve as the technical document for the IPMVP. As the name implies, Guideline 14 contains approaches for measuring energy and demand savings from energy conservation retrofits to buildings. This includes three methods: a retrofit isolation approach, which parallels Option B of the IPMVP, a whole-building approach, which parallels Option C of the IPMVP, and a whole-building calibrated simulation approach, which parallels Option D of the 1997 and 2001 IPMVP. ASHRAE's Guideline 14 does not explicitly contain an approach that parallels Option A in the IPMVP, although several of the retrofit isolation approaches use partial measurement procedures, as will be discussed in a following section.

### 27.2.1 M&V Methods: Existing Buildings

In general, a common theme between the NEMVP, IPMVP and ASHRAE's Guideline 14-2002, is that M&V methods for measuring energy and demand savings in existing building are best represented by the following three approaches: retrofit isolation approach, a whole-building approach, and a whole-building calibrated simulation approach. Similarly, the measurement of the performance of new construction, renewables, and water use utilize one or more of these same methods.

### 27.2.1.1 Retrofit Isolation Approach

The retrofit isolation approach is best used when end use capacity, demand or power can be measured during the baseline period, and after the retrofit for short-term period(s) or continuously over the life of the project. This approach can use continuous measurement of energy use both before and after the retrofit. Likewise, periodic, short-term measurements can be used during the baseline and after the retrofit to determine the retrofit savings. Often such short-term measurements are accompanied by periodic inspections of the equipment to assure that the equipment is operating as specified. In most cases energy use is calculated by developing representative models of the isolated component load (i.e., the kW or Btu/hr) and energy end-use (i.e., the kWh or Btu).

#### 27.2.1.1.1 Classifications of Retrofits

According to ASHRAE's Guideline 14-2002 retrofit isolation approach, components or end-uses can be classified according to the following definitions:[68]

*   Constant Load, Constant use. Constant load, constant use systems consist of systems where the energy used by the system is constant (i.e., varies by less than 5%) and the use of the system is constant (i.e., varies by less than 5%) through both the baseline and post-retrofit period.

*   Constant Load, Variable use. Constant load, variable use systems consist of systems where the energy used by the system is constant (i.e., varies by less than 5%) but the use of the system is variable (i.e., varies by more than 5%) through either the baseline or post-retrofit period.

*   Variable Load, Constant use. Variable Load, constant use systems consist of systems where the energy used by the system is variable (i.e., varies by more than 5%) but the use of the system is constant (i.e., varies by less than 5%) through either the baseline or post-retrofit period.

*   Variable Load, Variable use. Variable Load, variable use systems consist of systems where the energy used by the system is variable (i.e., varies by more than 5%) and the use of the system is variable (i.e., varies by more than 5%) through either the baseline or post-retrofit period.

Use of these classifications then allows for a simplified decision table (Table 27.5) to be used in determining which type of retrofit-isolation procedure to use. For example, in the first row (i.e., a CL/TS-pre-retrofit to CL/TS-post-retrofit), if a constant load with a known or timed schedule is replaced with a new device that has a reduced constant load, and a known or constant schedule, then the pre-retrofit and post-retrofit metering can be performed with one-time load measurement(s). Contrast this with the last row (i.e., a VL/VS-pre-retrofit to VL/VS-post-retrofit), if a variable load, with a timed or variable schedule is replaced with

**Table 27.4: Evolution of M&V Protocols in the United States.**

| 1996 NEMVP | 1997 IPMVP | 2001/2003 IPMVP | 2002 ASHRAE GUIDELINE 14 |
|---|---|---|---|
| OPTION A: Measured Capacity Stipulated Consumption | OPTION A: End-use Retrofits: Measured Capacity, Stipulated Consumption | VOLUME I: OPTION A: Partially Measured Retrofit Isolation | |
| OPTION B: End-use Retrofits: Measured Capacity, Measured Consumption | OPTION B: End-use Retrofits: Measured Capacity, Measured Consumption | VOLUME I: OPTION B: Retrofit Isolation | RETROFIT ISOLATION APPROACH |
| OPTION C: Whole-facility or Main Meter Measurement | OPTION C: Whole-facility or Main Meter Measurement | VOLUME I: OPTION C: Whole-building | WHOLE-BUILDING APPROACH |
| | OPTION D: Calibrated Simulation | VOLUME I: OPTION D: Calibrated Simulation | WHOLE-BUILDING CALIBRATED SIMULATION APPROACH |
| | | VOLUME II: IEQ M&V 5 Approaches | |
| | Measurement and Verification of New Buildings | VOLUME III: New Construction | |
| | EXAMPLE: Water Projects | | |

a new device that has a reduced variable load, and a variable schedule, then the pre-retrofit and post-retrofit metering should use continuous or short-term measurement that are sufficient in length to allow for the characterization of the performance of the component to be accomplished with a model (e.g., regression, or engineering model).

### 27.2.1.1.2 Detailed Retrofit Isolation Measurement and Verification Procedures

Appendix E of ASHRAE's Guideline 14-2002 contains detailed retrofit isolation procedures for the measurement and verification of savings, including: pumps, fans, chillers, boilers and furnaces, lighting, and large and unitary HVAC systems. In general, the procedures were drawn from the previous literature, including ASHRAE's Research Project 827-RP[70] (i.e., pumps, fans, chillers), various published procedures for boilers and furnaces,[71,72,73,74,75,76,77] lighting procedures, and calibrated HVAC calibration simulations.[78,79] A review of these procedures, which vary from simple one-time measurements, to complex, calibrated air-side psychrometric models, is described it the following sections.

**Table 27.5: Metering Requirements to Calculate Energy and Demand Savings from the ASHRAE Guideline 14-2002.[69]**

| Pre-Retrofit | Retrofit changes | Required metering | |
|---|---|---|---|
| | | Retrofit | Post-retrofit |
| CL/TS | Load but still CL | One time load measurement | One time load measurement |
| CL/TS | Load to VL | One time load measurement | Sufficient load measurements to characterize load |
| CL/TS | Schedule but still TS | One time load measurement (either pre- or post-retrofit) | |
| CL/TS | Schedule to VS | One time load measurement (either pre- or post-retrofit) | Sufficient measurement of runtime |
| CL/TS | Load but still CL and schedule but still TS | One time load measurement | One time load measurement |
| CL/TS | Load to VL and schedule but still TS | One time load measurement | Sufficient load measurements to characterize load |
| CL/TS | Load but still CL and schedule to VS | One time load measurement | One time load measurement and sufficient measurement of runtime |
| CL/TS | Load to VL and schedule to VS | One time load measurement | Sufficient load measurements to characterize load |
| CL/VS | Load but still CL | One time load measurement and sufficient measurement of runtime | One time load measurement and sufficient measurement of runtime |
| CL/VS | Load to VL | One time load measurement and sufficient measurement of runtime | Sufficient load measurements to characterize load |
| CL/VS | Schedule to TS | One time load measurement (either pre- or post-retrofit) and sufficient measurement of runtime | |
| CL/VS | Schedule but still VS | One time load measurement (either pre- or post-retrofit) and sufficient measurement of runtime | Sufficient measurement of runtime |
| CL/VS | Load but still CL and schedule to TS | One time load measurement and sufficient measurement of runtime | One time load measurement |
| CL/VS | Load to VL and schedule but still TS | One time load measurement and sufficient measurement of runtime | Sufficient load measurements to characterize load |
| CL/VS | Load but still CL and schedule to VS | One time load measurement and sufficient measurement of runtime | One time load measurement and sufficient measurement of runtime |
| CL/VS | Load to VL and schedule but still VS | One time load measurement and sufficient measurement of runtime | Sufficient load measurements to characterize load |
| VL/TS or VS | Load to CL | Sufficient load measurements to characterize load | One time load measurement and sufficient measurement of runtime |
| VL/TS or VS | Load but still VL | Sufficient load measurements to characterize load | Sufficient load measurements to characterize load |
| VL/TS or VS | Schedule still or to TS | Sufficient load measurements to characterize load | Sufficient load measurements to characterize load |
| VL/TS or VS | Schedule to or still VS | Sufficient load measurements to characterize load | Sufficient load measurements to characterize load |
| VL/TS or VS | Load to CL and schedule still or to TS | Sufficient load measurements to characterize load | One time load measurement |
| VL/TS or VS | Load but still VL and schedule still or to TS | Sufficient load measurements to characterize load | Sufficient load measurements to characterize load |
| VL/TS or VS | Load to CL and schedule to or still VS | Sufficient load measurements to characterize load | One time load measurement and sufficient measurement of runtime |
| VL/TS or VS | Load but still VL and schedule to or still VS | Sufficient load measurements to characterize load | Sufficient load measurements to characterize load |

CL = constant load
TS = timed (known) schedule
VL = variable load
VS = variable (unknown) schedule

## A. Pumps

Most large HVAC systems utilize electric pumps for moving heating/cooling water from the building's primary systems (i.e., boiler or chiller) to the building's secondary systems (i.e., air-handling units, radiators, etc.) where it can condition the building's interior. Such pumping systems use different types of pumps, varying control strategies, and piping layouts. Therefore, the characterization of pumping electric power depends on the system design and control method used. Pumping systems can be characterized by the three categories shown in Table 27.6.[80] Table 27.7 shows the six pump testing methods, including the required measurements, applications and procedures steps.

ASHRAE's Research Project 827-RP[81] developed six *in-situ* methods for measuring the performance of pumps of varying types and controls. To select a method the user needs to determine the pump system type and control, and the desired level of uncertainty, cost, and degree of intrusion. The user also needs to record

### Table 27.6: Applicability of Test Methods to Common Pumping Systems from the ASHRAE Guideline 14-2002.[82]

| Test Method: | Pumping System: | | |
|---|---|---|---|
| | Constant Speed, Constant Volume | Constant Speed, Variable Volume | Variable Speed, Variable Volume |
| 1.    Single Point | ✓ | | |
| 2.    Single Point with Manufacturer's Pump Curve | | ✓ | |
| 3.    Multiple Point with Imposed Loads at Pump | | ✓ | |
| 4.    Multiple Point with Imposed Loads at Zone | | ✓ | ✓ |
| 5.    Multiple Point through Short Term Monitoring | | ✓ | ✓ |
| 6.    No-Flow Test for Pump Characteristics | ✓ | ✓ | ✓ |

### Table 27.7: Pump Testing Methods from ASHRAE Guideline 14-2002.[83]

| METHOD | PUMPS |
|---|---|
| Method #1: Single Point Test | Measure: i) volumetric flow rate, ii) coincident RMS power, iii) differential pressure, iv) and rotational speed while the pump is at typical operating conditions.<br><br>Applications: Constant volume constant speed pumping systems. Used to confirm design operating conditions and pump and system curves.<br><br>Steps:<br>• Operate pump at typical existing operating conditions for the system.<br>• Measure pump suction and discharge pressure, or differential pressure.<br>• Measure pump capacity.<br>• Measure motor RMS power input.<br>• Measure speed.<br>• Calculate pump and energy characteristics. |
| Method #2: Single Point Test with Manufacturer's Curve | Measure: i) volumetric flow rate, ii) coincident RMS power, iii) differential pressure, iv) rotational speed while the pump is at typical operating conditions.<br><br>Applications: Variable volume constant speed pumping systems. Used with manufacturer's data on the pump, motor, and drive system to determine power at other operating points.<br><br>Steps:<br>• Obtain manufacturer's pump performance curves.<br>• Operate pump at typical existing operating conditions.<br>• Measure pump suction and discharge pressure, or differential pressure.<br>• Measure pump capacity.<br>• Measure motor RMS power input.<br>• Measure speed.<br>• Calculate pump and energy characteristics. |
| Method #3: Multiple Point Test with Imposed Loads at Pump or Fan | Measure: i) volumetric flow rate and ii) coincident RMS power while the pump is at operated at a range of flow load conditions as prescribed in the test procedures.<br><br>Applications: Variable volume constant speed pumping systems. The loads are imposed downstream of the pump with existing control valves. Pump operation follows the pump curve. Pump differential pressure and rotational speed may also be measured for more complete pump system evaluation.<br><br>Steps: |

**Table 27.7 (*Continued*)**

| | |
|---|---|
| | • Operate pump with system configuration set for maximum flow.<br>• Measure pump capacity.<br>• Measure motor RMS power input.<br>• Change system configuration to reduce flow and repeat measurement steps 2 and 3.<br>• Calculate pump and energy characteristics. |
| Method #4: Multiple Point Test with Imposed Loads at Zone | Measure: i) volumetric flow rate and ii) coincident RMS power for a range of building or zone thermal loads as prescribed in the test procedures.<br><br>Applications: Variable volume systems. The loads are imposed on the building or zones such that the system will experience a broad range of flow rates. The existing pump variable speed control strategy is allowed to operate. Pump differential pressure and rotational speed may also be measured for more complete pump system evaluation.<br><br>Steps:<br>• Operate pump with system configured for maximum flow rate.<br>• Measure pump capacity.<br>• Measure motor RMS power input.<br>• Change system configuration and repeat measurement steps 2 and 3.<br>• Calculate pump and energy characteristics. |
| Method #5: Multiple Point Test through Short Term Monitoring | Measure: i) volumetric flow rate and ii) coincident RMS power for a range of building or zone thermal loads as prescribed in the test procedures.<br><br>Applications: Variable volume variable speed systems. A monitoring period must be selected such that the system will experience a broad range of loads and pump flow rates. Pump differential pressure and rotational speed may also be measured for more complete pump system evaluation.<br><br>Steps:<br>• Choose appropriate time period for test.<br>• Monitor pump operation and record data values for pump capacity and motor RMS power input.<br>• Calculate pump and energy characteristics. |
| Method #6: No-Flow Test for Pump Characteristics | Measure: i) differential pressure at zero flow conditions (shut-off head) and compare to manufacturer's pump curves to determine impeller size.<br><br>Applications: All types of centrifugal pumps (not recommended for use on positive displacement pumps)<br><br>Steps:<br>• Run pump at design operating conditions and close discharge valve completely.<br>• Measure pump suction and discharge pressure, or differential pressure.<br>• Measure speed.<br>• Calculate shut-off head.<br>• Compare shut-off head with manufacturer's pump performance curve to determine and/or verify impeller diameter. |

the pump and motor data (i.e., manufacturer, model and serial number), fluid characteristics and operating conditions. The first two methods (i.e., single-point and single-point with a manufacturer's curve) involve testing at a single operating point. The third and fourth procedures involve testing at multiple operating points under imposed system loading. The fifth method also involves multiple operating points, in this case obtained through short term monitoring of the system without imposed loading. The sixth procedure operates the pump with the fluid flow path completely blocked. While the sixth procedure is not useful for generating a power versus load relationship, it can be used to confirm manufacturer's data or to identify pump impeller diameter. A summary of the methods is provided below. Additional details can be found by consulting ASHRAE's Guideline 14-2002.

## A-1. Constant Speed and Constant Volume Pumps

Constant volume pumping systems use three way valves and bypass loops at the end-use or at the pump. As the load varies in the system, pump pressure and flow are held relatively constant, and the pump input power remains nearly constant. Because pump motor speed is constant, constant volume pumping systems have a single operating point. Therefore, measuring the power use at the operating point (i.e., a single point measurement) and the total operating hours are enough to determine annual energy use.

## A-2. Constant Speed and Variable Volume Pumps

Variable pumping systems with constant speed pumps use two-way control valves to modulate flow to the end-use as required. In constant speed variable volume pumping systems, the flow varies along the pump curve as the system pressure drop changes in response to the load. In some cases, a bypass valve may be modulated if system differential pressure becomes too large. Such systems have a single possible operating point for any given flow, as determined by the pump curve at that flow rate. In such systems the second and third testing methods can be used to characterize the pumps energy use at varying conditions. In the second procedure, measurements of *in-situ* power use is performed at one flow rate and manufacturer's data on the pump, motor, and drive system are used to create a part load power use curve. In the third testing method *in-situ* measurements are made of the electricity use of the pump with varying loads imposed on the system using existing control, discharge, or balancing valves. The fourth and fifth methods can also be used to characterize the pump electricity use. Using one of these methods the part load power use curve and a representative flow load frequency distribution are used to determine annual energy use.

### A-3. Variable Speed and Variable Volume Pumps

Like the constant speed variable volume system, flow to the zone loads is typically modulated using two-way control valves. However, in variable speed variable volume pumping systems, a static pressure controller is used to adjust pump speed to match the flow load requirements. In such systems the operating point cannot be determined solely from the pump curve and flow load because a given flow can be provided at various pressures or speeds. Furthermore, the system design and control strategy place constraints on either the pressure or flow. Such systems have a range of system curves which call for the same flow rate, depending on the pumping load. 827-RP provides two options (i.e., multiple point with imposed loads and short term monitoring) for accurately determining the *in-situ* part load power use. In both cases, the characteristics of the *in-situ* test include the pump and piping system (piping, valves, and controllers), therefore the control strategy is included within the data set. In the fourth method (i.e., multiple point with imposed load at the zone), the pump power use is measured at a range of imposed loads. These imposed loads are done at the zone level to account for the *in-situ* control strategy and system design. In the fifth method (i.e., multiple point through short term monitoring), the pump system is monitored as the building experiences a range of thermal loads, with no artificial imposition of loads. If the monitored loads reflect the full range of loads, then an accurate part load power curve can be developed that represents the full range of annual load characteristics. For methods #4 and #5, the measured part load power use curves and flow load frequency distribution are used to determine annual energy use.

### A-4. Calculation of Annual Energy Use

Once the pump performance has been measured the annual energy use can be calculated using the following procedures, depending upon whether the system is a constant volume or variable volume pumping system. Savings are then calculated by comparing the annual energy use of the baseline with the annual energy use of the post-retrofit period.

*Constant Volume Constant Speed Pumping Systems.* In a constant volume constant speed pumping systems the volume of the water moving through the pump is almost constant, and therefore the power load of the pump is virtually constant. The annual energy calculation is therefore a constant times the frequency of the operating hours of the pump.

$$E_{annual} = T * P$$

where:

$T$ = annual operating hours
$P$ = equipment power input

*Variable Volume Pumping Systems.* For variable volume pumping systems the volume of water moving through the pump varies over time, hence the power demand of the pump and motor varies. The annual energy use then becomes a frequency distribution of the load times the power associated with each of the bins of operating hours. *In-situ* testing is used to determine the power associated with the part load power use.

$$E_{annual} = \sum_i \left( T_i * P_i \right)$$

where:

$i$ = bin index, as defined by the load frequency distribution
$T_i$ = number of hours in bin $i$
$P_i$ = equipment power use at load bin $I$

### B. Fans

Most large HVAC systems utilize fans or air-handling units to deliver heating and cooling to the building's interior. Such air-handling systems use different types of fans, varying control strategies, and duct layouts. Therefore, the characterization of fan electric power depends on the system design and control method used. Fan systems can be characterized by the three categories shown in Table 27.8.[84]

In a similar fashion as pumping systems, ASHRAE's Research Project 827-RP developed five *in-situ* methods for measuring the performance of fans of varying types and controls. To select a method the user needs to determine the system type and control, and the desired level of uncertainty, cost, and degree of intrusion. The user also needs to record the fan and motor data (i.e., manufacturer, model and serial number), as well as the operating conditions (i.e., temperature, pressure and humidity of the air stream). The first two methods (i.e., single-point and single-point with a manufacturer's curve) involve testing at a single operating point. The third and fourth procedures involve testing at multiple operating points under imposed system loading. The fifth method also involves multiple operating points, in this case obtained through short term monitoring of the system without imposed loading. Additional details about fan testing procedures can be found by consulting ASHRAE's Guideline 14-2002.

### B-1. Calculation of Annual Energy Use

Once the fan performance has been measured the annual energy use can be calculated using the following procedures, depending upon whether the system is a constant volume or variable volume system. Savings are then calculated by comparing the annual energy use of the baseline with the annual energy use of the post-retrofit period.

*Constant Volume Fan Systems.* In a constant volume system the volume of the air moving across the fan is almost constant, and therefore the power load of the fan is virtually constant. The annual energy calculation is therefore a constant times the frequency of the operating hours of the fan.

$$E_{annual} = T * P$$

where:

$T$ = annual operating hours
$P$ = equipment power input

*Variable Volume Systems.* For variable volume systems the volume of the air being moved by the fan varies over time, hence the power demand of the fan and motor varies. The annual energy use then becomes a frequency distribution of the load times the power associated with each of the bins of operating hours. *In-situ* testing is used to determine the power associated with the part load power use.

**Table 27.8: Applicability of Test Methods to Common Fan Systems from the ASHRAE Guideline 14-2002.**[85]

| Test Method: | Fan System: | | |
|---|---|---|---|
| | Constant Speed, Constant Volume | Constant Speed, Variable Volume | Variable Speed, Variable Volume |
| 1.  Single Point Test | ✓ | | |
| 2.  Single Point with Manufacturer's Fan Data | | ✓ | |
| 3.  Multiple Point with Imposed Loads at Fan | | ✓ | |
| 4.  Multiple Point with Imposed Loads at Zone | | ✓ | ✓ |
| 5.  Multiple Point through Short Term Monitoring | | ✓ | ✓ |

**Table 27.9: Fan Testing Methods from ASHRAE Guideline 14-2002.**[86]

| METHOD | FANS |
|---|---|
| Method #1: Single Point Test | Measure: i) volumetric flow rate, ii) coincident RMS power use, iii) fan differential pressure, and iiii) fan rotational speed while the fan is at typical operating conditions.<br><br>Applications: Constant volume fan systems. Data are used to confirm design operating conditions and fan and system curves.<br><br>Steps:<br>• Operate fan at typical existing operating conditions for the system.<br>• Measure fan inlet and discharge pressure or (preferably) differential pressure.<br>• Measure fan flow capacity.<br>• Measure motor RMS power input.<br>• Measure fan speed.<br>• Calculate fan and energy characteristics. |
| Method #2: Single Point Test with Manufacturer's Curve | Measure: i) volumetric flow rate, ii) coincident RMS power use, iii) fan differential pressure, and iv) fan rotational speed while the fan is at typical operating conditions.<br><br>Applications: Variable volume systems without fan control. Data are used with manufacturer's data on the fan, motor, and drive system and engineering principles to determine power at other operating points. |

*(Continued)*

**Table 27.9: (Continued)**

| | |
|---|---|
| | Steps:<br>• Obtain manufacturer's fan performance curves.<br>• Operate fan at typical existing operating conditions.<br>• Measure fan inlet and discharge pressure, or differential pressure.<br>• Measure fan flow capacity.<br>• Measure motor RMS power input.<br>• Measure fan speed.<br>• Calculate fan and energy characteristics. |
| Method #3: Multiple Point Test with Imposed Loads at Pump or Fan | Measure: i) volumetric flow rate and ii) coincident RMS power while the fan is at operated at a range of flow rate conditions as prescribed in the test procedures.<br><br>Applications: Variable volume systems without fan control. The loads are imposed downstream of the fan with existing dampers. Fan operation follows the fan curve. Fan differential pressure and rotational speed may also be measured for more complete fan system evaluation.<br><br>Steps:<br>• Operate fan with system configuration set for maximum flow.<br>• Measure fan flow capacity.<br>• Measure motor RMS power input.<br>• Change system configuration to reduce flow and repeat measurement steps 2 and 3.<br>• Calculate fan and energy characteristics. |
| Method #4: Multiple Point Test with Imposed Loads at Zone | Measure: i) volumetric flow rate and ii) coincident RMS power use while the fan is operated at a range of flow rate conditions as prescribed in the test procedures.<br><br>Applications: Variable volume systems. Thermal loads are imposed at the building or zone level such that the system will experience a broad range of flow rates. The existing fan variable speed control strategy is allowed to operate. Fan differential pressure and rotational speed may also be measured for more complete fan system evaluation.<br><br>Steps:<br>• Operate fan with system configured for maximum flow rate.<br>• Measure fan capacity.<br>• Measure motor RMS power input.<br>• Change system configuration and repeat measurement steps 2 and 3.<br>• Calculate fan and energy characteristics. |
| Method #5: Multiple Point Test through Short Term Monitoring | Measure: i) volumetric flow rate and ii) coincident RMS power while the fan operates at a range of flow rates.<br><br>Applications: Variable volume systems. The range of flow rates will depend on the building or zones experiencing a wide range of thermal loads. A time period must be selected such that the system will experience a broad range of loads and fan flow rates. The existing fan variable speed control strategy is allowed to operate. Fan differential pressure and rotational speed may also be measured for more complete fan system evaluation.<br><br>Steps:<br>• Choose appropriate time period for test.<br>• Monitor fan operation and record data values for fan capacity and motor RMS power input.<br>• Calculate fan and energy characteristics. |

$$E_{annual} = \sum_i \left( T_i * P_i \right)$$

where:

$i$ = bin index, as defined by the load frequency distribution

$T_i$ = number of hours in bin i

$P_i$ = equipment power use at load bin $I$

### C. Chillers

In a similar fashions as pumps and fans, *in-situ* chiller performance measurements have been also been developed as part of ASHRAE Research Project 827-RP. These models provide useful performance testing methods to evaluate annual energy use and peak demand characteristics for installed water-cooled chillers and selected air-cooled chillers. These procedures require short-term testing of the part load performance of an installed chiller system over a range of building thermal loads and coincident ambient conditions. The test methods determine chiller power use at varying thermal loads using thermodynamic models or statistical models with inputs from direct measurements, or manufacturer's data. With these models annual energy use can be determined using the resultant part load power use curve with a load frequency distribution. Such models are capable of calculating the chiller power use as a function of the building thermal load, evaporator and condenser flow rates, entering and leaving chilled water temperatures, entering condenser water temperatures, and internal chiller controls. ASHRAE's Guideline 14-2002 describes two models for calculating the power input of a chiller, including simple and temperature dependent thermodynamic models.[87,88,89] A third method, which uses a tri-quadratic regression model such as those found in the DOE-2 simulation program,[90,91,92,93,94] also provides acceptable performance models, provided that measurements are made over the full operating range.

C-1. Simple Thermodynamic Model

Both the simple thermodynamic model and the temperature-dependent thermodynamic model express chiller efficiency as $1/COP$ because it has a linear relationship with $1/$(evaporator load). The simpler version of the chiller model developed predicts a linear relationship between $1/COP$ and $1/Q_{evap}$, which is independent of the evaporator supply temperature or condenser temperature returning to the chiller. The full form of simple thermodynamic model is shown in the equation below.

$$\frac{1}{COP} = -1 + \left(T_{cwRT}/T_{chwST}\right)$$

$$+ \left(\frac{1}{Q_{evap}}\right)\left(\frac{q_{evap}T_{cwRT}}{T_{chwST}} - q_{cond}\right) + f\,HX$$

where:

$T_{cwRT}$ = Entering (return) condenser water temperature (Kelvin)

$T_{chwST}$ = Leaving (supply) evaporator water temperature (Kelvin)

$Q_{evap}$ = Evaporator load

$q_{evap}$ = rate of internal losses in evaporator

$q_{cond}$ = rate of internal losses in condenser

$f_{HX}$ = dimensionless term.[95]

This equation reduces to a simple form that allows for the determination of two coefficients using linear regression, which is shown in the following equation.

$$\frac{1}{COP} = -C_1\left(\frac{1}{Q_{evap}}\right) + C_0$$

In this simplified form the coefficient $c_1$ characterizes the internal chiller losses, while the coefficient $c_0$ combines the other terms of the simple model. The COP figure of merit can be converted into conventional efficiency measures of COP or kW per ton using the following relationships:

Coefficient of Performance (COP):

$$COP = \frac{kW\ refrigeration\ effect}{kW\ input}$$

Energy Efficiency Ratio (EER):

$$EER = \frac{Btu/hr\ refrigeration\ effect}{Watts\ input} = 3.412\ COP$$

Power per Ton (kW/ton):

$$kW/ton = \frac{kW\ input}{tons\ refrigeration\ effect} = 12/EER$$

The simple thermodynamic model can be determined with relatively few measurements of the chiller load (evaporator flow rate, entering and leaving chilled water temperatures) and coincident RMS power use. Unfortunately, variations in the chilled water supply (i.e., the temperature of the chilled water leaving the evaporator) and the condenser water return temperature are not considered. Hence, this model is best used with chiller systems that maintain constant temperature control of evaporator and condenser temperatures. In systems with varying temperatures a temperature-dependent thermodynamic model, or tri-quadratic model yield a more accurate performance prediction.

C-2. Temperature-dependent Thermodynamic Model

The temperature dependent thermodynamic model includes the losses in the heat exchangers of the evaporator and condenser, which are expressed as a function of the chilled water supply and condenser water return temperatures. The resulting expression uses three coefficients ($A_0$, $A_1$, $A_2$), which are found with linear regression, as shown in the equation that follows.

$$\frac{1}{COP} = -1 + \left(T_{cwRT}/T_{chwST}\right)$$

$$+ \frac{-A_0 + A_1(T_{cwRT}) - A_2(T_{cwRT}/T_{chwST})}{Q_{evap}}$$

Use of this temperature dependent thermodynamic model requires the measurement of the chiller load (i.e., evaporator flow rate, entering and leaving chilled water temperatures), coincident RMS power use, and condenser water return temperature. Since this model is sensitive to varying temperatures it is applicable to a wider range of chiller systems.

To use the temperature dependent model, measured chiller thermal load, coincident RMS power use, chilled water supply temperature, and condenser water return temperatures are used to calculate the three coefficients ($A_0$, $A_1$, $A_2$).

To determine $A_2$ the following equation is plotted against $T_{cwRT}/T_{chwST}$ (Kelvin), with value of $A_2$ being determined from the regression lines, which should resemble a series of straight parallel lines, one for each condenser temperature setting.

$$\alpha = \left(\frac{1}{COP} + 1 - (T_{cwRT}/T_{chwST})\right)Q_{evap}$$

The coefficients $A_0$ and $A_1$ are determined by plotting $\beta$ from the next equation, using the already determined value of $A_2$, versus the condenser water return temperature $T_{cwRT}$ (Kelvin). This should result in a group of data points forming a single straight line. The slope of the regression line determines the value of coefficient $A_1$ while the intercept determines the value of coefficient $A_0$.

$$\beta = \left(\frac{1}{COP} + 1 - (T_{cwRT}/T_{chwST})\right)Q_{evap}$$

$$+ A_2\left(T_{cwRT}/T_{chwST}\right)$$

After $A_0$, $A_1$, and $A_2$ have been determined using $\alpha$ and $\beta$ and from the equations above, the 1/COP can be calculated and used to determine the chiller performance over a wide range of measured input parameters of chiller load, chilled water supply temperature, and condenser water return temperature.

## C-3. Quadratic Chiller Models

Chiller performance models can also be calculated with quadratic models, which can include models that express the chiller power use as a function of the chiller load (quadratic), as a function of the chiller load and chilled water supply temperature (bi-quadratic), or as a function of the chiller load, evaporator supply temperature and condenser return temperature (tri-quadratic). Such models use the quadratic functional form used in the DOE-2 energy simulation program to model part-load equipment and plant performance characteristics. Two examples of quadratic models are shown below, one for a monitoring project where chiller electricity use, chilled water production, chilled water supply temperature, and condenser water temperature returning to the chiller were available, which uses a tri-quadratic model as follows:

$$\begin{aligned} kW/ton = {} & a + b \times Tons + c \times T_{cond} + d \times T_{evap} + e \times Tons^2 \\ & + f \times T_{cond}^2 \\ & + g \times T_{evap}^2 + h \times Tons \times T_{cond} \\ & + I \times T_{evap} \times Tons \\ & + j \times T_{cond} \times T_{evap} + k \times Tons \times T_{cond} \times T_{evap} \end{aligned}$$

In a second example, chiller electricity use is modeled with a bi-quadratic model that includes only the chilled water production, and chilled water supply temperature, which reduces to the following form. Either model can easily be calculated from field data in a spreadsheet using multiple linearized regression.

$$\begin{aligned} kW/ton = {} & a + b \times Tons + c \times Tk_{evap} + d \times Tons^2 \\ & + e \times T_{evap}^2 + f \times T_{evap} \times Tons \end{aligned}$$

## C-4. Example: Quadratic Chiller Models

An example of a quadratic chiller performance analysis model is provided from hourly measurements that were taken at to determine the baseline model of a cooling plant at an Army base in Texas. Figure 27.1 shows the time series data that were recorded during June and August of 2002. The upper trace is the chiller thermal load (tons), and the lower trace is the ambient temperature during this period. Figure 27.2 shows a time series plot of the recorded temperatures of the condenser water returning to the chiller (upper trace), and the chilled water supply temperatures (lower trace). In Figures 27.3 and 27.4 the performance of the chiller is shown as the chiller efficiency (i.e., kW/ton) versus the chiller load (tons). In Figure 27.3 linear ($R^2 = 34.3\%$) and quadratic ($R^2 = 53.4\%$) models of the chiller are superimposed over the measured data from the chiller to illustrate how a quadratic model fits the chiller data. In Figure 27.4 a tri-quadratic model ($R^2 = 83.7\%$) is shown superimposed over the measured data.

A quick inspection of the $R^2$ goodness-of-fit indicators for the linear, quadratic and tri-quadratic models begins to shed some light on how well the models are fitting the data. However, one must also inspect how well the model is predicting the chiller performance at the intended operation points. For example, although a linear model has an inferior $R^2$ when compared to a quadratic model, for this particular chiller, it gives similar performance values for cooling loads ranging from 200 to 450 tons. Choosing the quadratic model improves the prediction of the chiller performance for values below 200 tons. However, it significantly under predicts the kW/ton at 350 tons and over-predicts the kW/ton at values over 500 tons. Hence, both models should be used with caution.

The tri-quadratic model has an improved $R^2$ of 83.7% and does not seem to contain any ranges where the model's bias is significant from the measured data (excluding the few stray points which are caused by transient data). Therefore, in the case of this chiller, the additional effort to gather and analyze the chiller load against the chilled water supply temperature and condenser water return temperature is well justified.

## C-5. Calculation of Annual Energy Use

Once the chiller performance has been determined the annual energy use can be calculated using the simple or temperature dependent models to determine the power demand of the chiller at each bin of the cooling load distribution. For chillers with varying temperatures, a load frequency distribution, which contains the two water temperatures, provides the operating hours of the chiller at each bin level. The energy use $E_i$, and power level $P_i$ are given by the equations below. The total annual energy use is then the sum of the product of the number of hours in each bin times the chiller power associated with that bin. Savings are then calculated by comparing the annual energy use of the baseline with the annual energy use of the post-retrofit period.

$$E_i = T_i * P_i$$
$$P_i = (1 / Eff_i) \ (Q_{evap,i})$$
$$E_{annual} = \sum_i (T_i * P_i)$$

where:

$i$ = bin index, as defined by load frequency distribution

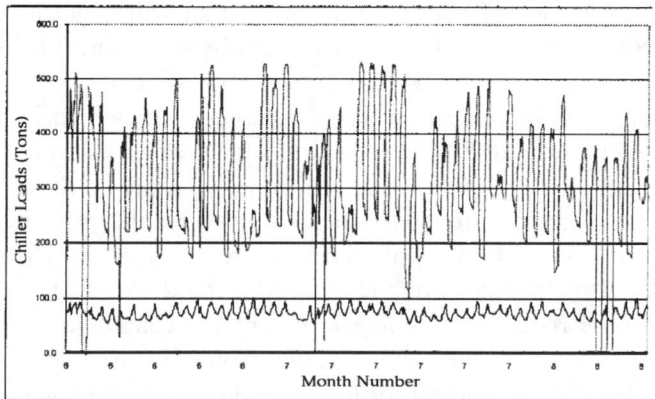

Figure 27.1: Example chiller analysis. Time series plot of chiller load (upper trace, tons) and ambient temperature (lower trace, degrees F).

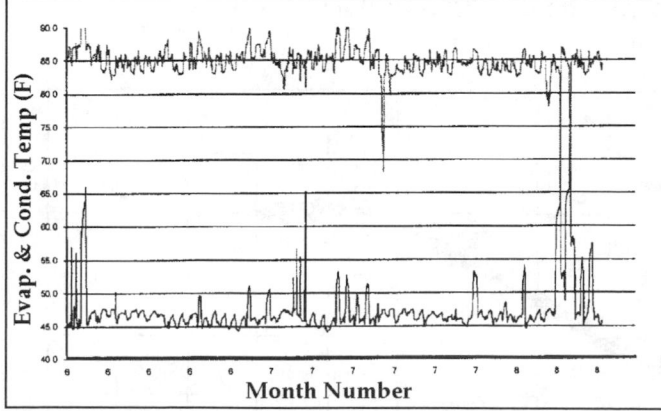

Figure 27.2: Example chiller analysis. Time series plot of condenser water return temperature (upper trace, degrees F) and chilled water supply temperature (lower trace, degrees F).

Figure 27.3: Example chiller analysis. Chiller performance plot of chiller efficiency (kW/ton) versus the chiller cooling load. Comparisons of linear ($R^2$ = 34.3%) and quadratic ($R^2$ = 53.4.3%) chiller models are shown.

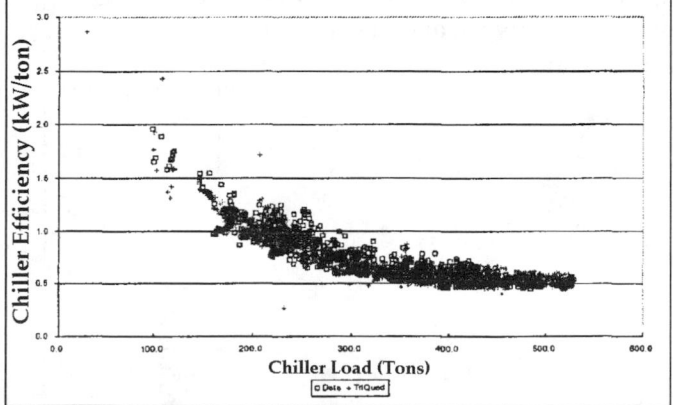

Figure 27.4: Example chiller analysis. Chiller performance plot of chiller efficiency (kW/ton) versus the chiller cooling load. In this figure a tri-quadratic chiller model ($R^2$ = 83.7%) is shown.

$T_i$ = number of hours in bin $i$

$P_i$ = equipment power use at load bin $i$

$Eff_i$ = chiller 1/COP in bin $i$

$Q_{evap,i}$ = chiller load in bin $i$

### D. Boilers and Furnaces

*In-situ* boiler and furnace performance measurements, for non-reheat boilers and furnaces, are listed in Appendix E of ASHRAE Guideline 14-2002. These procedures, which were obtained from the previously noted published literature on performance measurements of boilers and furnaces,[96,97,98,99,100,101,102] are grouped into four methods (i.e., single-point, single-point with manufacturer's data, multiple point with imposed loads, and multiple point tests using short term monitoring) that use three measurement techniques (i.e., direct method, direct heat loss method, and indirect combustion method), for a total of twelve methods.

The choice of method depends on boiler type (i.e., constant fire boiler, or variable-fire boilers), and availability of measurements (i.e., fuel meters, steam meters, etc.). For constant fire boilers, the boiler load is virtually constant. Therefore, a single measurement or series of measurements a full load will characterize the boiler or furnace efficiency at a given set of ambient conditions. For variable fire boilers the fuel use and output of the boiler varies. Therefore, the efficiency of the boiler will vary depending upon the load of the boiler as described by the manufacturer's efficiency curve. Figure 27.5 shows an example of the measured performance of a variable-fire, low pressure steam boiler installed at an army base in Texas.[103]

### D-1. Boiler Efficiency Measurements

There are three principal methods for determining boiler efficiency, the direct method (i.e., Input-Output method), the direct heat loss method, also known as the indirect method, and the indirect combustion efficiency method. The first two are recognized by the American Society of Mechanical Engineers (ASME) and are mathematically equivalent. They give identical results if all the heat balance factors are considered and the boiler measurements performed without error. ASME has formed committees from members of the industry and developed the performance test codes[105] that detail procedures of determining boiler efficiency by the first two methods mentioned above. The accuracy of boiler performance calculations is dependent on the quantities measured and the method used to determine the efficiency. In the direct efficiency method, these quantities are directly related to the overall efficiency. For example, if the measured boiler efficiency is 80%, then an error of 1% in one of the quantities measured will result in a 0.8% error in the efficiency. Conversely, in the direct heat loss method, the measured parameters are related to the boiler losses. Therefore, for the same boiler which had an efficiency of 80%, a measurement error of 1% in any quantity affects the overall efficiency by only 0.2% (i.e., 1% of the measured losses of 20%). As a result, the direct heat loss method is inherently more accurate than the direct method for boilers. However, the direct heat loss method requires more measurement and calculations. In general, boiler efficiencies range from 75% to 95% for utility boilers, and for industrial and commercial boilers, the average efficiency ranges from 76% to 83% on gas, 78% to 89% on oil and 85% to 88% for coal.[106,107]

**Figure 27.5: Example Boiler Performance Curve from Short Term Monitoring.**[104]

- **Direct Method**

  The direct method (i.e., the input-output method) is the simplest method to determine boiler efficiency. In this method, the heat supplied to the boiler and the heat absorbed by the water in the boiler in a given time period are directly measured. Using the direct method, the efficiency of a non-reheat boiler is given by:[108]

$$\eta_b = \frac{Q_a}{Q_i} \times 100$$

where

| | | |
|---|---|---|
| $Q_a$ | = | heat absorbed (Btu/hr) = $\Sigma m_o h_o - \Sigma m_i h_i$ |
| $m_o h_o$ | = | mass flow-enthalpy products of working fluid streams leaving boiler envelope, including main steam, blowdown, soot blowing steam, etc. |
| $m_i h_i$ | = | mass flow-enthalpy products of working fluid streams entering boiler envelope, including feedwater, desuperheating sprays, etc. |
| $Q_i$ | = | heat inputs (Btu/hr) = $V_{fuel} \times HHV + Q_c$ |
| $V_{fuel}$ | = | volumetric flow of fuel into boiler (SCF/hr) |
| HHV | = | fuel higher heating value (Btu/SCF), and |
| $Q_c$ | = | heat credits (Btu/hr). Heat credits are defined as the heat added to the envelope of the steam generating unit other than the chemical heat in the fuel "as fired." These credits include quantities such as sensible heat in the fuel, the entering air and the atomizing steam. Other credits include heat from power conversion in the pulverizer or crusher, circulating pump, primary air fan and recirculating gas fan. |

- **Direct Heat Loss Method**

  In the direct heat loss method the boiler efficiency equals 100% minus the boiler losses. The direct heat loss method tends to be more accurate than the direct method because the direct heat loss method focuses on determining the heat lost from the boiler, rather than on the heat absorbed by the working fluid. The direct heat loss method determines efficiency using the following:[109]

$$\eta_b = \frac{Q_a}{Q_i} \times 100 = \frac{Q_i - Q_{loss}}{?} \times 100$$

$$= 100 - L_{df} - L_{fh} - L_{am} - L_{rad} - L_{conv} - L_{bd} - L_{inc} - L_{unacct}$$

where

| | | |
|---|---|---|
| $Q_{loss}$ | = | heat losses (Btu/hr), |
| $L_{df}$ | = | dry flue gas heat loss (%), |
| $L_{fh}$ | = | fuel hydrogen heat loss (%), |
| $L_{am}$ | = | combustion air moisture heat loss (%), |
| $L_{rad}$ | = | radiation heat loss (%), |
| $L_{conv}$ | = | convection heat loss (%), |
| $L_{inc}$ | = | incombusted fuel loss (%), |
| $L_{bd}$ | = | blowdown heat loss (%), |
| $L_{unacct}$ | = | unaccounted for heat losses (%). |

Using this method the flue gas losses (sensible and latent heat) due to radiation and convection, incomplete combustion, and blowdown are accounted for. In most boilers the flue gas loss is the largest loss, which can be determined by a flue gas analysis. Flue gas losses vary with flue gas exit temperature, fuel composition and type of firing.[110] Radiation and convection loss can be obtained from the standard curves.[111] Unaccounted for heat losses can also be obtained from published industry sources,[112] which cite losses of 1.5% for solid fuels, and 1% for gaseous or liquid fuel boilers. Losses from boiler blowdown should also be measured. Typical values can be found in various sources.[113,114]

- **Indirect Combustion Method**

  The indirect combustion method can also be used to measure boiler efficiency. The combustion efficiency is the measure of the fraction of fuel-air energy available during the combustion process, calculated from the folowing:[115,116]

$$\eta_c = \frac{|h_p| - |h_f + h_a|}{Q_i} \times 100$$

where

| | | |
|---|---|---|
| $\eta_c$ | = | combustion efficiency (%) |
| $h_p$ | = | enthalpy of products (Btu/lb) |
| $h_f$ | = | enthalpy of fuel (Btu/lb) |
| $h_a$ | = | enthalpy of combustion air, Btu/lb |
| $Q_i$ | = | heat inputs (Btu/hr) = Vfuel * HHV + Qc |

Indirect combustion efficiency can be related to direct efficiency or direct heat loss efficiency measurements using the following:[117,118]

$$\eta_b = \eta_c - L_{rad} - L_{conv} - L_{unacct}$$

On the right side of the equation the loss terms are usually small for well insulated boilers. These terms must be accounted for when boilers are poorly insulated or operated poorly (i.e., excessive blowdown control, etc.).

Table 27-10 provides a summary of the performance measurement methods (i.e., single-point, single-point with manufacturer's data, multiple point with imposed loads, and multiple point tests using short term monitoring), which use three efficiency measurement techniques (i.e., direct method, direct heat loss method, and indirect combustion method) that are listed in Appendix E of ASHRAE Guideline 14-2002. For each method the pertinent measurements are listed along with the steps that should be taken to calculate the efficiency of the boiler or furnace being measured.

### D-2. Calculation of Annual Energy Use

Once the boiler performance has been measured the annual energy use can be calculated using the following procedures, depending upon whether the system is a constant fire boiler or variable fire boiler. Savings are then calculated by comparing the annual energy use of the baseline with the annual energy use of the post-retrofit period.

- **Constant Fire Boilers**

  In constant fire boilers the method assumes the load and fuel use are constant when the boiler is operating. Therefore, the annual fuel input is simply the full-load operating hours of the boiler times the fuel input. The total annual energy use is given by:

  $$E_{annual} = T * P$$

  where: T = annual operating hours under full load
  P = equipment power use

- **Variable Fire Boilers**

  For variable fire boilers the output of the boiler and fuel input vary according to load. Hence a frequency distribution of the load is needed that provides the operating hours of the boiler at each bin level. *In-situ* testing is then used to determine the efficiency of the boiler or furnace for each bin. The total annual energy use for variable fire boilers is given by:

$$E_{annual} = \sum_i \left( T_i * P_i \right)$$

where:

i = bin index, as defined by load variable frequency distribution

$T_i$ = number of hours in bin i

$P_i$ = equipment fuel input (& efficiency) at load bin (i)

### E. Lighting

One of the most common retrofits to commercial buildings is to replace inefficient T-12 fluorescents and magnetic ballasts with T-8 fluorescents and electronic ballasts. This type of retrofit saves electricity associated with the use of the more efficient lighting, and depending on system type, can reduce cooling energy use because of reduced internal loads from the removal of the inefficient lighting. In certain climates, depending on system type, this can also mean an increase in heating loads, which are required to offset the heat from the inefficient lighting. Previously published studies show the cooling interaction can increase savings by 10 to 20%. The increased heating requirements can reduce savings by 5 to 20%.[120] Therefore, where the costs can be justified, accurate measurement of total energy savings can involve before/after measurements of the lighting loads, cooling loads, and heating loads.

### E-1. Lighting Methods

ASHRAE Guideline 14-2002 (see Table 27.11) provides six measurement methods to account for the electricity and thermal savings, varying from methods that utilize sampled before/after measurements to methods

**Table 27.10: Boiler and Furnace Performance Testing Methods from ASHRAE Guideline 14-2002.[119]**

| METHOD | DESCRIPTION OF THE METHOD |
|---|---|
| Method #1a: Single Point Test (direct method) | Measure: i) mass flow and enthalpy of fluid streams leaving the boiler (main steam, blowdown, etc.) ii) mass flow and enthalpy of fluid streams entering the boiler (feedwater, desuperheating sprays, etc.), iii) heat inputs.<br><br>Applications: Non-reheat boilers and furnaces.<br><br>Steps:<br>• Operate boiler at typical existing operating conditions for the system.<br>• Measure mass flow and enthalpy of fluid streams leaving the boiler (main steam, blowdown, etc.).<br>• Measure mass flow and enthalpy of fluid streams entering the boiler (feedwater desuperheating sprays, etc.).<br>• Measure heat inputs.<br>• Calculate efficiency using the direct efficiency method.<br>• Calculate boiler and efficiency characteristics. |

## Table 27.10 (*Continued*)

| | |
|---|---|
| Method #1b: Single Point Test (direct heat loss method) | Measure: i) all boiler losses (dry flue gas loss, fuel hydrogen heat loss, combustion air moisture heat loss, radiation heat loss, convection heat loss, unburned fuel loss, blowdown loss & unaccounted for losses), ii) heat inputs.<br><br>Applications: Non-reheat boilers and furnaces.<br><br>Steps:<br>• Operate boiler at typical existing operating conditions for the system.<br>• Measure all boiler losses (dry flue gas loss, fuel hydrogen heat loss, combustion air moisture heat loss, radiation heat loss, convection heat loss, unburned fuel loss, blowdown loss & unaccounted for losses).<br>• Measure heat inputs.<br>• Calculate efficiency using direct heat loss method.<br>• Calculate boiler and efficiency characteristics. |
| Method #1c: Single Point Test (indirect combustion method) | Measure: i) enthalpy of all combustion products, ii) enthalpy of fuel, iii) enthalpy of combustion air, iv) heat inputs.<br><br>Applications: Non-reheat boilers and furnaces.<br><br>Steps:<br>• Operate boiler at typical existing operating conditions for system.<br>• Measure enthalpy of all combustion products, the enthalpy of the fuel, the enthalpy of combustion air.<br>• Measure heat inputs.<br>• Calculate efficiency using the indirect combustion method.<br>• Calculate boiler and efficiency characteristics. |
| Method #2a: Single Point Test with Manufacturer's Data (direct method) | Measure: i) mass flow and enthalpy of fluid streams leaving the boiler (main steam, blowdown, etc.) ii) mass flow and enthalpy of fluid streams entering the boiler (feedwater, desuperheating sprays, etc.), iii) heat inputs.<br><br>Applications: Non-reheat boilers and furnaces. Data are used with manufacturer's published boiler efficiency curves and engineering principles to determine efficiency at other operating points. Boiler efficiency at other operating points is assumed to follow the manufacturer's curve.<br><br>Steps:<br>• Operate boiler at typical existing operating conditions for the system<br>• Obtain manufacturer's boiler efficiency curve.<br>• Measure mass flow and enthalpy of fluid streams leaving the boiler (main steam, blowdown, etc.).<br>• Measure mass flow and enthalpy of fluid streams entering the boiler (feedwater desuperheating sprays, etc.).<br>• Measure heat inputs.<br>• Calculate efficiency using the direct efficiency method for the single point and compare to manufacturer's curve.<br>• Calculate boiler and efficiency characteristics. |
| Method #2b: Single Point Test with Manufacturer's Data (direct heat loss method) | Measure: i) all boiler losses (dry flue gas loss, fuel hydrogen heat loss, combustion air moisture heat loss, radiation heat loss, convection heat loss, unburned fuel loss, blowdown loss & unaccounted for losses), ii) heat inputs.<br><br>Applications: Non-reheat boilers and furnaces. Data are used with manufacturer's published boiler efficiency curves and engineering principles to determine efficiency at other operating points. Boiler efficiency at other operating points is assumed to follow the manufacturer's curve. If single point does not confirm manufacturer's curve within 5% another boiler efficiency method will need to be used.<br><br>Steps:<br>• Operate boiler at typical existing operating conditions for the system.<br>• Obtain manufacturer's boiler efficiency curve.<br>• Measure all boiler losses (dry flue gas loss, fuel hydrogen heat loss, combustion air moisture heat loss, radiation heat loss, convection heat loss, unburned fuel loss, blowdown loss & unaccounted for losses).<br>• Measure heat inputs.<br>• Calculate efficiency using direct heat loss method for a single point and compare to manufacturer's curve.<br>• Calculate boiler and efficiency characteristics. |
| Method #2c: Single Point Test with Manufacturer's Data (indirect combustion method) | Measure: i) enthalpy of all combustion products, ii) enthalpy of fuel, iii) enthalpy of combustion air, iv) heat inputs.<br><br>Applications: Non-reheat boilers and furnaces. Data are used with manufacturer's published boiler efficiency curves and engineering principles to determine efficiency at other operating points. Boiler efficiency at other operating points is assumed to follow the manufacturer's curve. If single point does not confirm manufacturer's curve within 5% another boiler efficiency method will need to be used.<br><br>Steps:<br>• Operate boiler at typical existing operating conditions for system.<br>• Obtain manufacturer's boiler efficiency curve.<br>• Measure enthalpy of all combustion products, the enthalpy of the fuel, the enthalpy of combustion air.<br>• Measure heat inputs.<br>• Calculate efficiency using the indirect combustion method, compare to manf. curve.. |

## Table 27.10 (*Continued*)

| | • Calculate boiler and efficiency characteristics. |
|---|---|
| Method #3a: Multiple Point Test with Imposed Loads (direct method) | Measure over a range of operating conditions: i) mass flow and enthalpy of fluid streams leaving the boiler (main steam, blowdown, etc.) ii) mass flow and enthalpy of fluid streams entering the boiler (feedwater, desuperheating sprays, etc.), iii) heat inputs.  Different loads are imposed on the boiler and measurements repeated.  Boiler operation is assumed to follow manufacturer's efficiency curve.<br><br>Applications:  Non-reheat boilers and furnaces.<br><br>Steps:<br>• Obtain manufacturer's efficiency curves.<br>• Operate boiler at a given load.<br>• Measure mass flow and enthalpy of fluid streams leaving the boiler (main steam, blowdown, etc.).<br>• Measure mass flow and enthalpy of fluid streams entering the boiler (feedwater desuperheating sprays, etc.).<br>• Measure heat inputs.<br>• Calculate efficiency using the direct efficiency method.<br>• Change load on boiler and repeat steps 2 through 6.<br>• Calculate boiler and efficiency characteristics. |
| Method #3b: Multiple Point Test with Imposed Loads (direct heat loss method) | Measure over a range of operating conditions: i) all boiler losses (dry flue gas loss, fuel hydrogen heat loss, combustion air moisture heat loss, radiation heat loss, convection heat loss, unburned fuel loss, blowdown loss & unaccounted for losses), ii) heat inputs.  Different loads are imposed on the boiler and measurements repeated.  Boiler operation is assumed to follow manufacturer's efficiency curve.<br><br>Applications: Non-reheat boilers and furnaces.<br><br>Steps:<br>• Obtain manufacturer's boiler efficiency curve.<br>• Operate boiler at a given load.<br>• Measure all boiler losses (dry flue gas loss, fuel hydrogen heat loss, combustion air moisture heat loss, radiation heat loss, convection heat loss, unburned fuel loss, blowdown loss & unaccounted for losses).<br>• Measure heat inputs.<br>• Calculate efficiency using direct heat loss method for a single point and compare to manufacturer's curve.<br>• Change load on boiler and repeat steps 2 through 5.<br>• Calculate boiler and efficiency characteristics. |
| Method #3c: Multiple Point Test with Imposed Loads (indirect combustion method) | Measure over a range of operating conditions: i) enthalpy of all combustion products, ii) enthalpy of fuel, iii) enthalpy of combustion air, iv) heat inputs.  Different loads are imposed on the boiler and measurements are repeated.  Boiler operation is assumed to follow the manufacturer's efficiency curve.<br><br>Applications: Non-reheat boilers and furnaces.<br><br>Steps:<br>• Obtain manufacturer's boiler efficiency curve.<br>• Operate boiler at a given load.<br>• Measure enthalpy of all combustion products, the enthalpy of the fuel, the enthalpy of combustion air.<br>• Measure heat inputs.<br>• Calculate efficiency using the indirect combustion method and compare to manufacturer's curve.<br>• Change load on boiler and repeat steps 2 through 5.<br>• Calculate boiler and efficiency characteristics. |
| Method #4a: Multiple Point Test through Short Term Monitoring (direct method) | Monitor over a range of operating conditions: i) mass flow and enthalpy of fluid streams leaving the boiler (main steam, blowdown, etc.) ii) mass flow and enthalpy of fluid streams entering the boiler (feedwater, desuperheating sprays, etc.), iii) heat inputs.  The range of boiler loads should cover the normally expected loads that the boiler will experience (low and high).<br><br>Applications:  Non-reheat boilers and furnaces.<br><br>Steps:<br>• Choose appropriate time period for test.<br>• Monitor boiler operation and record data values for mass flow and enthalpy of fluid streams leaving the boiler (main steam, blowdown, etc.), mass flow and enthalpy of fluid streams entering the boiler (feedwater desuperheating sprays, etc.), and heat inputs.<br>• Calculate efficiency using the direct efficiency method.<br>• Calculate boiler and efficiency characteristics. |
| Method #4b: Multiple Point Test through Short Term Monitoring (direct heat loss method) | Monitor over a range of operating conditions: i) all boiler losses (dry flue gas loss, fuel hydrogen heat loss, combustion air moisture heat loss, radiation heat loss, convection heat loss, unburned fuel loss, blowdown loss & unaccounted for losses), ii) heat inputs.  The range of boiler loads should cover the normally expected loads that the boiler will experience (low and high).<br><br>Applications: Non-reheat boilers and furnaces.<br><br>Steps:<br>• Choose appropriate period for the test.<br>• Monitor all boiler losses (dry flue gas loss, fuel hydrogen heat loss, combustion air moisture heat loss, |

**Table 27.10 (*Continued*)**

| | |
|---|---|
| | radiation heat loss, convection heat loss, unburned fuel loss, blowdown loss & unaccounted for losses), and monitor heat inputs.<br>• Calculate efficiency using direct heat loss method for a single point and compare to manufacturer's curve.<br>• Calculate boiler and efficiency characteristics. |
| Method #4c: Multiple Point Test through Short Term Monitoring (indirect combustion efficiency method) | Monitor over a range of operating conditions: i) enthalpy of all combustion products, ii) enthalpy of fuel, iii) enthalpy of combustion air, iv) heat inputs. The range of boiler loads should cover the normally expected loads that the boiler will experience (low and high).<br><br>Applications: Non-reheat boilers and furnaces.<br><br>Steps:<br>• Choose appropriate time period for test.<br>• Monitor: enthalpy of all combustion products, the enthalpy of the fuel, the enthalpy of combustion air, and monitor heat inputs.<br>• Calculate efficiency using the indirect combustion method and compare to manufacturer's curve.<br>• Calculate boiler and efficiency characteristics. |

that use sub-metered before-after lighting measurement with measurements of increases or decreases to the heating and cooling systems from the removal of the internal lighting load. In general, the calculation of savings from lighting retrofits involves ascertaining the wattage or power reduction associated with the new fixtures, which is then multiplied times the hours per day (i.e., lighting usage profiles) that the lights are used. The lighting usage profiles can be calculated based on appropriate estimates of use, measured at the electrical distribution panel, or sampled with lighting loggers. Figure 27.6 shows an example of weekday-weekend profiles calculated with ASHRAE's Diversity Factor Toolkit.[121]

Some lighting retrofits involve the installation of daylighting sensors to dim fixtures near the perimeter of the building or below skylights when lighting levels can be maintained with daylighting, thus reducing the electricity used for supplemental lighting. Measuring the savings from such daylighting retrofits usually involves before-after measurements of electrical power and lighting usage profiles.

Any lighting retrofit should include an assessment of the existing lighting levels, which is measured during daytime and nighttime conditions. All lighting retrofits should achieve and maintain lighting levels recommended by the Illuminating Engineering Society of North America (IESNA).[122] Any pre-retrofit lighting levels not maintaining IESNA lighting levels should be brought to the attention of the building owner or administrator. In the following section the six methods, which are described in the ASHRAE Guideline 14-2002 are summarized. Table 27.11 contains the lighting performance measurement methods from ASHRAE's Guideline 14-2002.

• **Method #1:** Baseline and post-retrofit measured lighting power levels and stipulated diversity profiles.

In Method #1 before-after lighting power levels for a representative sample of lighting fixtures are measured using a Wattmeter, yielding an average Watt/fixture measurement for the pre-retrofit fixtures and post-retrofit fixtures. Lighting usage profiles are estimated or stipulated using the best available information, which represents the lighting usage profiles for the fixtures. This method works best for exterior lighting fixtures or lighting fixtures controlled by a timer or photocell. Lighting fixtures located in hallways, or any interior lighting fixtures that is operated 24 hours per day, 7 days per week or controlled by a timer is also suitable for this method. Savings benefits or penalties from thermal interactions are not included in this method.

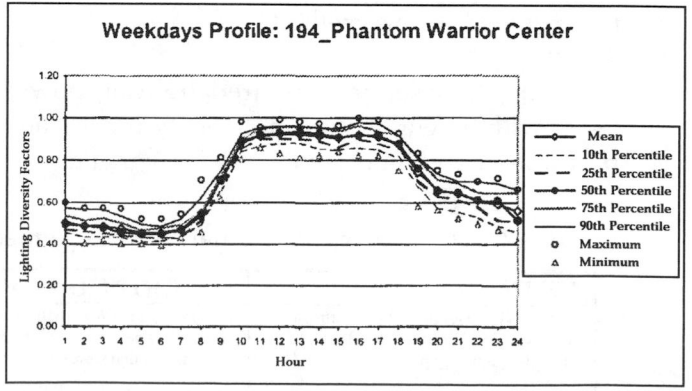

**Figure 27.6: Example Weekday-Weekend Lighting Profiles.**

- Method #2: Baseline and post-retrofit measured lighting power levels and sampled baseline and post-retrofit diversity profiles.

    In Method #2 before-after lighting power levels for a representative sample of lighting fixtures are measured using a Wattmeter, yielding an average Watt/fixture measurement for the pre-retrofit fixtures and post-retrofit fixtures. Lighting usage profiles are measured with portable lighting loggers, or portable current meters attached to lighting circuits to determine the lighting usage profiles for the fixtures. This method is appropriate for any interior or exterior lighting circuit that has predictable usage profiles. Savings benefits or penalties from thermal interactions are not included in this method.

- Method #3: Baseline measured lighting power levels with baseline sampled diversity profiles and post-retrofit power levels with post-retrofit continuous diversity profile measurements.

    In Method #3 pre-retrofit lighting power levels for a representative sample of lighting fixtures are measured using a Wattmeter, yielding an average Watt/fixture measurement for the pre-retrofit fixtures. Pre-retrofit lighting usage profiles are measured with portable lighting loggers, or portable current meters attached to lighting circuits to determine the lighting usage profiles for the fixtures. Post-retrofit lighting usage is measured continuously using either sub-metered lighting electricity measurements, or post-retrofit lighting power levels for a representative sample of lighting fixtures times a continuously measured diversity profile (i.e., using lighting loggers or current measurements on lighting circuits). This method is appropriate for any interior or exterior lighting circuit that has predictable usage profiles. Savings benefits or penalties from thermal interactions are not included in this method.

- Method #4: Baseline measured lighting power levels with baseline sampled diversity profiles and post-retrofit continuous sub-metered lighting.

In Method #4 pre-retrofit lighting power levels for a representative sample of lighting fixtures are measured using a Wattmeter, yielding an average Watt/fixture measurement for the pre-retrofit fixtures. Pre-retrofit lighting usage profiles are measured with portable lighting loggers, or portable current meters attached to lighting circuits to determine the lighting usage profiles for the fixtures. Post-retrofit lighting usage is measured continuously using sub-metered lighting electricity measurements. This method is appropriate for any interior or exterior lighting circuit that has predictable usage profiles. Savings benefits or penalties from thermal interactions are not included in this method.

- Method #5: Includes methods #1, #2, or #3 with measured thermal effect (heating & cooling).

    In Method #5 pre-retrofit and post-retrofit lighting electricity use is measured with Methods #1, #2, #3 or #4, and the thermal effect is measured using the component isolation method for the cooling or heating system.

    This method is appropriate for any interior lighting circuit that has predictable usage profiles. Savings benefits or penalties from thermal interactions are included in this method.

- Method #6: Baseline and post-retrofit sub-metered lighting measurements and thermal measurements.

    In Method #6 pre-retrofit and post-retrofit lighting electricity use is measured continuously using sub-metering, and the thermal effect is measured using whole-building cooling and heating sub-metered measurements. This method is appropriate for any interior lighting circuit. Savings benefits or penalties from thermal interactions are included in this method.

E-2. Calculation of Annual Energy Use

The calculation of annual energy use varies according to lighting calculation method as shown in Table

**Table 27.11: Lighting Performance Measurement Methods from ASHRAE Guideline 14-2002.[123]**

| METHOD | DESCRIPTION OF THE METHOD |
|---|---|
| Method #1: Baseline and post-retrofit measured lighting power levels and stipulated diversity profiles. | Description: i) Obtain before-after lighting power levels using RMS watt/fixture measurements for the pre-retrofit fixtures and the post-retrofit fixtures, ii) stipulate the lighting usage profiles using the best available information that represents lighting usage profiles for the facility. |
| | Application: Exterior lighting on a timer or photocell. |
| | Interior hallway lighting or any interior lighting used continuously or on a timer. |
| | Steps: |

## Table 27.11 (*Continued*)

|  |  |
|---|---|
|  | <ul><li>Obtain measured RMS watt/fixture data for pre-retrofit and post-retrofit fixtures.</li><li>Count the fixtures associated with each functional area in the building (e,g,, areas that have different usage profiles).</li><li>Define the lighting usage profiles for each functional area using the appropriate information that represents lighting usage profiles (e.g., continuously on, on during evening hours, etc.).</li><li>Calculate lighting energy usage characteristics.</li></ul> |
| Method #2: Before/after measured lighting power levels with sampled before/after diversity profiles. | Description: i) Measure lighting power levels using RMS watt meter for a sample of the pre-retrofit fixtures and the post-retrofit fixtures, ii) measure the lighting usage profiles using light loggers or portable metering attached to the lighting circuits.<br><br>Application: Any exterior lighting or interior lighting with predictable usage profiles.<br><br>Steps:<ul><li>Measure watt/fixture using RMS watt meter for pre-retrofit and post-retrofit fixtures.</li><li>Count the fixtures associated with each functional area in the building (i.e., areas that have different usage profiles).</li><li>Sample lighting usage profiles for each functional area using lighting loggers and/or portable submetered RMS watt meters on lighting circuits.</li><li>Calculate lighting energy usage characteristics.</li></ul> |
| Method #3: Baseline measured lighting power levels with baseline sampled diversity profiles and post-retrofit power levels with continuous diversity profile measurements. | Description: i) Obtain lighting power levels using RMS watt/fixture measurements for the pre-retrofit fixtures and the post-retrofit fixtures, ii) sample the baseline lighting usage profiles using light loggers or RMS watt measurements on submetered lighting circuits, iii) continuously measure the post-retrofit lighting usage profiles using light loggers or RMS watt measurements on submetered lighting circuits.<br><br>Application: Any exterior lighting or interior lighting.<br><br>Steps:<ul><li>Obtain lighting power levels using RMS watt/fixture measurements for the pre-retrofit fixtures and the post-retrofit fixtures.</li><li>Sample the baseline lighting usage profiles using light loggers or RMS watt measurements on submetered lighting circuits.</li><li>Continuously measure the post-retrofit lighting usage profiles using light loggers or RMS watt measurements on submetered lighting circuits.</li><li>Calculate lighting energy usage characteristics.</li></ul> |
| Method #4: Baseline measured lighting power levels with baseline sampled diversity profiles and post-retrofit continuous sub-metered lighting. | Description: i) Obtain lighting power levels using RMS watt/fixture measurements for the pre-retrofit fixtures, ii) sample the baseline lighting usage profiles using light loggers or RMS watt measurements on submetered lighting circuits, iii) continuously measure the post-retrofit lighting power usage using RMS watt measurements on submetered lighting circuits.<br><br>Application: Any exterior lighting or interior lighting.<br><br>Steps:<ul><li>Obtain lighting power levels using RMS watt/fixture measurements for the pre-retrofit fixtures.</li><li>Sample the baseline lighting usage profiles using light loggers or RMS watt measurements on submetered lighting circuits.</li><li>Continuously measure the post-retrofit lighting usage using RMS watt measurements on submetered lighting circuits.</li><li>Calculate lighting energy usage characteristics.</li></ul> |
| Method #5: Method #1, #2, or #3 with stipulated thermal effects. | Description: i) Obtain lighting power profiles and usage using Method(s) #1, #2, #3, or #4 ii) Calculate the heating or cooling system efficiency using HVAC component isolation methods described in this document, iii) Calculate decrease in cooling load and increase in heating load.<br><br>Application: Any interior lighting.<br><br>Steps:<ul><li>Obtain lighting power profiles and usage using Method(s) #1, #2, or #3,</li><li>Calculate the heating or cooling system efficiency using HVAC component isolation methods described in this document,</li><li>Calculate lighting energy usage characteristics.</li><li>Calculate decrease in cooling load and increase in heating load.</li></ul> |
| Method #6: Before/after sub-metered lighting and thermal measurements. | Description: i) Obtain lighting energy usage by measuring RMS lighting use continuously at the sub-metered level for pre-retrofit and post-retrofit conditions, ii) Obtain thermal energy use data by measuring sub-metered cooling or heating energy use for pre-retrofit and post-retrofit conditions, iii) develop representative lighting usage profiles from the sub-metered lighting data.<br><br>Application: Any interior lighting projects.<br>Any exterior lighting projects (no thermal interaction).<br><br>Steps:<ul><li>Obtain measured sub-metered lighting data for pre-retrofit and post-retrofit periods.</li><li>Develop representative lighting usage profiles from the sub-metered lighting data.</li><li>Calculate lighting energy usage characteristics.</li><li>Calculate decrease in cooling load and increase in heating load.</li></ul> |

27.12. The savings are determined by comparing the annual lighting energy use during the baseline period to the annual lighting energy use during the post-retrofit period. In Methods #5 and #6 the thermal energy effect can either be calculated using the component efficiency methods or it can be measured using whole-building, before-after cooling and heating measurements. Electric demand savings can be calculated using Methods #5 and #6 using diversity factor profiles from the pre-retrofit period and continuous measurement in the post-retrofit period. Peak electric demand reductions attributable to reduced chiller loads can be calculated using the component efficiency tests for the chillers. Savings are then calculated by comparing the annual energy use of the baseline with the annual energy use of the post-retrofit period.

## F. HVAC Systems

As mentioned previously, during the 1950s and 1960s most engineering calculations were performed using slide rules, engineering tables and desktop calculators that could only add, subtract, multiply and divide. In the 1960s efforts were initiated to formulate and codify equations that could predict dynamic heating and cooling loads, including efforts to simulate HVAC systems. In 1965 ASHRAE recognized that there was a need to develop public-domain procedures for calculating the energy use of HVAC equipment and formed the Presidential Committee on Energy Consumption, which became the Task Group on Energy Requirements (TGER) for Heating and Cooling in 1969.[125] TGER commissioned two reports that detailed the public domain procedures for calculating the dynamic heat transfer through the building envelopes,[126] and procedures for simulating the performance and energy use of HVAC systems.[127] These procedures became the basis for today's public-domain building energy simulation programs such as BLAST, DOE-2, and EnergyPlus.[128,129]

In addition, ASHRAE has produced several additional efforts to assist with the analysis of building energy use, including a modified bin method,[130] the HVAC-01[131] and HVAC-02[132] toolkits, and HVAC simulation accuracy tests[133] which contain detailed algorithms and computer source code for simulating secondary and primary HVAC equipment. Studies have also demonstrated that properly calibrated simplified HVAC system models can be used for measuring the performance of commercial HVAC systems.[134,135,136,137]

**Table 27.12: Lighting Calculations Methods from ASHRAE Guideline 14-2002.[124]**

| TYPE OF MEASUREMENT | PRE-RETROFIT ELECTRICITY USAGE CALCULATIONS | POST-RETROFIT ELECTRICITY USAGE CALCULATIONS | THERMAL ENERGY USAGE CALCULATIONS |
|---|---|---|---|
| Method #1: Baseline and post-retrofit measured lighting power levels and stipulated diversity profiles. | For each lighting circuit:<br><br>Annual energy use = (Power levels) x (24-hr *stipulated* profiles) x (number of days assigned to each profile) | For each lighting circuit:<br><br>Annual energy use = (Power levels) x (24-hr *stipulated* profiles) x (number of days assigned to each profile) | None. |
| Method #2: Before/after measured lighting power levels with sampled before/after diversity profiles. | For each lighting circuit:<br><br>Annual energy use = (Power levels) x (24-hr *sampled* profiles) x (number of days assigned to each profile) | For each lighting circuit:<br><br>Annual energy use = (Power levels) x (24-hr *sampled* profiles) x (number of days assigned to each profile) | None. |
| Method #3: Baseline measured lighting power levels with baseline sampled diversity profiles and post-retrofit power levels with continuous diversity profile measurements. | For each lighting circuit:<br><br>Annual energy use = (Power levels) x (24-hr *sampled* profiles) x (number of days assigned to each profile) | For each lighting circuit:<br><br>Annual energy use = (Power levels) x (continuous diversity profile measurements) | None. |
| Method #4: Baseline measured lighting power levels with baseline sampled diversity profiles and post-retrofit continuous sub-metered lighting. | For each lighting circuit:<br><br>Annual energy use = (Power levels) x (24-hr *sampled* profiles) x (number of days assigned to each profile) | For each lighting circuit:<br><br>Annual use = sub-metered lighting energy use. | None. |
| Method #5: Method #1, #2, #3, or #4 with calculated thermal effect. | Annual use = method #1, #2, #3, or #4 as appropriate. | Annual use = method #1, #2, #3, or #4 as appropriate. | Pre and post thermal load from the lighting is calculated using the component efficiency measurement methods for HVAC systems. |
| Method #6: Before/after sub-metered lighting and thermal measurements. | For each lighting circuit:<br><br>Annual use = sub-metered lighting energy use. | | Pre and post thermal load is calculated using before-after whole-building cooling and heating sub-metered measurements. |

## F-1. HVAC System Types

In order to facilitate the description of measurement methods that are applicable to a wide range of HVAC systems, it is necessary to categorize HVAC systems into groups, such as single zone, steady state systems to the more complex systems such a multi-zone systems with simultaneous heating and cooling. To accomplish this two layers of classification are proposed, in the first layer, systems are classified into two categories: systems that provide heating or cooling under separate thermostatic control, and systems that provide heating and cooling under a combined control. In the second classification, systems are grouped according to: systems that provide constant heating rates, systems that provide varying heating rates, systems that provide constant cooling rates, systems that provide varying cooling rates.

- HVAC systems that provide heating or cooling at a constant rate include: single zone, 2-pipe fan coil units, ventilating and heating units, window air conditioners, evaporative cooling. Systems that provide heating or cooling at a constant rate can be measured using: single-point tests, multi-point tests, short-term monitoring techniques, or *in-situ* measurement combined with calibrated, simplified simulation.

- HVAC systems that provide heating or cooling at varying rates include: 2-pipe induction units, single zone with variable speed fan and/or compressors, variable speed ventilating and heating units, variable speed, and selected window air conditioners. Systems that provide heating or cooling at varying rates can be measured using: single-point tests, multi-point tests, short-term monitoring techniques, or short-term monitoring combined with calibrated, simplified simulation.

- HVAC systems that provide simultaneous heating and cooling include: multi-zone, dual duct constant volume dual duct variable volume, single duct constant volume w/reheat, single duct variable volume w/reheat, dual path systems (i.e., with main and preconditioning coils), 4-pipe fan coil units, and 4-pipe induction units. Such systems can be measured using: *in-situ* measurement combined with calibrated, simplified simulation.

## F-2. HVAC System Testing Methods

In this section four methods are described for the *in-situ* performance testing of HVAC systems as shown in Table 27.14, including: a single point method that uses manufacturer's performance data, a multiple point method that includes manufacturer's performance data, a multiple point that uses short-term data and manufacturer's performance data, and a short-term calibrated simulation. Each of these methods is explained in the sections that follow.

- Method #1: Single point with manufacturer's performance data

  In this method the efficiency of the HVAC system is measured with a single-point (or a series) of field measurements at steady operating conditions. On-site measurements include: the energy input to system (e.g., electricity, natural gas, hot water or steam), the thermal output of system, and the temperature of surrounding environment. The efficiency is calculated as the measured output/input. This method can be used in the following constant systems: single zone systems, 2-pipe fan coil units, ventilating and heating units, single speed window air conditioners, and evaporative coolers.

**Table 27.13: Relationship of HVAC Test Methods to Type of System.**

| Test Method | HVAC System | | |
| --- | --- | --- | --- |
| | Constant Heating or Cooling | Variable Heating or Cooling | Simultaneous Heating and Cooling |
| 1. Single Point with Manufacturer's Performance Data | ✓ | | |
| 2. Multiple Point with Manufacturer's Performance Data. | ✓ | ✓ | |
| 3. Multiple Point through Short Term Monitoring with Manufacturer's Data. | ✓ | ✓ | |
| 4. Short Term Monitoring and Calibrated, Simplified Simulation. | ✓ | ✓ | ✓ |

- Method #2: Multiple point with manufacturer's performance data

  In this method the efficiency of the HVAC system is measured with multiple points on the manufacturer's performance curve. On-site measurements include: the energy input to system (e.g., electricity, natural gas, hot water or steam), the thermal output of system, the system temperatures, and the temperature of surrounding environment. The efficiency is calculated as the measured output/input, which varies according to the manufacturer's performance curve. This method can be used in the following systems: single zone (constant or varying), 2-pipe fan coil units, ventilating and heating units (constant or varying), window air conditioners (constant or varying), evaporative cooling (constant or varying) 2-pipe induction units (varying), single zone with variable speed fan and/or compressors, variable speed ventilating and heating units, and variable speed window air conditioners.

- Method #3: Multiple point using short-term data and manufacturer's performance data

  In this method the efficiency of the HVAC system is measured continuously over a short-term period, with data covering the manufacturer's performance curve. On-site measurements include: the energy input to system (e.g., electricity, natural gas, hot water or steam), the thermal output of system, the system temperatures, and the temperature of surrounding environment. The efficiency is calculated as the measured output/input, which varies according to the manufacturer's performance curve. This method can be used in the following systems: single zone (constant or varying), 2-pipe fan coil units, ventilating and heating units (constant or varying), window air conditioners (constant or varying), evaporative cooling (constant or varying) 2-pipe induction units (varying), single zone with variable speed fan and/or compressors, variable speed ventilating and heating units, and variable speed window air conditioners.

- Method #4: Short-term monitoring and calibrated, simplified simulation

  In this method the efficiency of the HVAC system is measured continuously over a short-term period, with data covering the manufacturer's performance curve. On-site measurements include: the energy input to system (e.g., electricity, natural gas, hot water or steam), the thermal output of system, the system temperatures, and the temperature of surrounding environment. The efficiency is calculated using a calibrated air-side simulation of the system, which can include manufacturer's performance curves for various components. Similar measurements are repeated after the retrofit. This method can be used in the following systems: single zone (constant or varying), 2-pipe fan coil units, ventilating and heating units (constant or varying), window air conditioners (constant or varying), evaporative cooling (constant or varying), 2-pipe induction units (varying), single zone with variable speed fan and/or compressors, variable speed ventilating and heating units, variable speed window air conditioners, multi-zone, dual duct constant volume, dual duct variable volume, single duct constant volume w/reheat, single duct variable volume w/reheat, dual path systems (i.e., with main and preconditioning coils), 4-pipe fan coil units, 4-pipe induction units

### F-3. Calculation of Annual Energy Use

The calculation of annual energy use varies according to HVAC calculation method as shown in Table 27.15. The savings are determined by comparing the annual HVAC energy use and demand during the baseline period to the annual HVAC energy use and demand during the post-retrofit period.

### Whole-building or Main-meter Approach
*Overview*

The whole-building approach, also called the main-meter approach, includes procedures that measure the performance of retrofits for those projects where whole-building pre-retrofit and post-retrofit data are

**Table 27.14: HVAC System Testing Methods.**[138,139]

| METHOD | DESCRIPTION OF THE METHOD |
|---|---|
| Method #1: Single point with manufacturer's performance data | Measure: i) energy input to system (e.g., electricity, natural gas, hot water or steam), ii) thermal output of system, iii) temperature of surrounding environment (adjust for differences in efficiency with manufacturer's data).<br><br>Applications:<br>• single zone (constant mode)<br>• 2-pipe fan coil units (constant mode)<br>• ventilating and heating units,<br>• single speed window air conditioners |

## Table 27.14 (*Continued*)

| | |
|---|---|
| | •   evaporative coolers<br><br>Steps:<br>•   Measure energy input to system (i.e, electricity, natural gas, hot water or steam).<br>•   Measure thermal output of system.<br>•   Measure temperature of space where system is operating.<br>•   Calculate efficiency as output/input. If efficiency does not agree to within 5% of manufacturer's performance data then use method #2, #3 or #4.<br>•   Adjust efficiency, using manufacturer's data if surrounding environmental conditions vary significantly over the year (i.e., ambient temperatures that the HVAC unit is exposed to).<br>•   Calculate savings by comparing differences in before-after component efficiency calculations applied to continuously measured post-retrofit thermal output energy requirements or sampled thermal load profiles. |
| Method #2: Multiple point with manufacturer's performance data | Measure at multiple points: i) energy input to system (e.g., electricity, natural gas, hot water or steam), ii) thermal output of system, iii) temperature of surrounding environment (adjust for differences in efficiency with manufacturer's data).<br><br>Applications:<br>•   single zone (constant or varying)<br>•   2-pipe fan coil units<br>•   ventilating and heating units (constant or varying)<br>•   window air conditioners (constant or varying)<br>•   evaporative cooling (constant or varying)<br>•   2-pipe induction units (varying)<br>•   single zone with variable speed fan and/or compressors<br>•   variable speed ventilating and heating units<br>•   variable speed window air conditioners<br><br>Steps:<br>•   Measure energy input to system at multiple points (i.e., electricity, natural gas, hot water or steam).<br>•   Measure corresponding thermal output of system at multiple points.<br>•   Measure temperature of space where system is operating during all tests.<br>•   Calculate efficiency as output/input. If efficiency does not agree to within 5% of manufacturer's performance data then use method #3 or #4.<br>•   Adjust efficiency, using manufacturer's data if surrounding environmental conditions vary significantly over the year (i.e., ambient temperatures that the HVAC unit is exposed to).<br>•   Calculate savings by comparing differences in before-after component efficiency calculations applied to continuously measured post-retrofit thermal output energy requirements or sampled thermal load profiles. |
| Method #3: Multiple point using short-term data and manufacturer's performance data | Continuously measure over a short-term period: i) energy input to system (e.g., electricity, natural gas, hot water or steam), ii) thermal output of system, iii) temperature of surrounding environment (adjust for differences in efficiency with manufacturer's data).<br><br>Applications:<br>•   single zone (constant or varying)<br>•   2-pipe fan coil units<br>•   ventilating and heating units (constant or varying)<br>•   window air conditioners (constant or varying)<br>•   evaporative cooling (constant or varying)<br>•   2-pipe induction units (varying)<br>•   single zone with variable speed fan and/or compressors<br>•   variable speed ventilating and heating units<br>•   variable speed window air conditioners<br><br>Steps:<br>•   Continuously measure over a short-term period energy input to system (e.g., electricity, natural gas, hot water or steam).<br>•   Measure corresponding thermal output of system at multiple points.<br>•   Measure temperature of space where system is operating during all tests.<br>•   Calculate efficiency as output/input. If efficiency does not agree to within 5% of manufacturer's performance data then use method #4.<br>•   Adjust efficiency, using manufacturer's data if surrounding environmental conditions vary significantly over the year (i.e., ambient temperatures that the HVAC unit is exposed to).<br>•   Calculate savings by comparing differences in before-after component efficiency calculations applied to continuously measured post-retrofit thermal output energy requirements or sampled thermal load profiles. |
| Method #4: Short-term monitoring and calibrated, simplified simulation | Measure over a representative range: i) thermal and electric energy input to system (e.g., electricity, natural gas, hot water or steam), ii) thermal output of system, iii) temperature of surrounding environment (may be used to adjust for losses to space), iv) develop an air-side simulation model that is representative of the system (Knebel 1983), v) calibrate the air-side model to the measured data for both pre-retrofit and post-retrofit conditions.<br><br>Applications:<br>•   single zone (constant or varying)<br>•   2-pipe fan coil units<br>•   ventilating and heating units (constant or varying)<br>•   window air conditioners (constant or varying)<br>•   evaporative cooling (constant or varying)<br>•   2-pipe induction units (varying)<br>•   single zone with variable speed fan and/or compressors |

**Table 27.14 (*Continued*)**

- variable speed ventilating and heating units
- variable speed window air conditioners
- multi-zone
- dual duct constant volume
- dual duct variable volume
- single duct constant volume w/reheat
- single duct variable volume w/reheat
- dual path systems (i.e., with main and preconditioning coils)
- 4-pipe fan coil units
- 4-pipe induction units

Steps:
- Measure thermal input to system over a representative range (e.g., electricity, natural gas, hot water or steam).
- Measure thermal output of system.
- Measure temperature of space where system is operating.
- Measure important system operation characteristics (e.g., cooling coil setpoint, heating coil setpoint, mixture of outside air to returning air, etc.)
- For systems using chilled water, calculate efficiency as output/input over a range of conditions representative of operating conditions.
- For systems using direct expansion of refrigerant, calculate efficiency as output/input over a range of varying cooling supply temperatures and heat rejection supply temperatures (i.e., to capture the efficiency of the A/C unit over varying outside conditions).
- Develop an airside simulation model that is representative of the system (Knebel 1983) and calibrate the air-side model to the measured data for both pre-retrofit and post-retrofit conditions.
- Calculate savings by applying the calibrated simulation models for the baseline and post-retrofit system to continuously measured post-retrofit cooling requirements or sampled cooling load profiles.

**Table 27.15: HVAC Performance Measurement Methods from ASHRAE Guideline 14-2002.[140]**

| TYPE OF MEASUREMENT | PRE-RETROFIT ENERGY USAGE AND ELECTRIC DEMAND CALCULATIONS | POST-RETROFIT ENERGY USAGE AND ELECTRIC DEMAND CALCULATIONS |
|---|---|---|
| Method #1: Single point with manufacturer's performance data | For each HVAC system:<br><br>Annual energy use = (Measured Energy Use) x (Full Load Runtime Hours)<br><br>Monthly demand use = (Measured Peak Electric Demand Use for System) x (System on/off Status for Each Month) | For each HVAC system:<br><br>Annual energy use = (Measured Energy Use) x (Full Load Runtime Hours)<br><br>Monthly demand use = (Measured Peak Electric Demand Use for System) x (System on/off Status for Each Month) |
| Method #2: Multiple point with manufacturer's performance data | For each HVAC system:<br><br>Annual energy use = (Measured Energy Use for Each Operating Point) x (Full Load Runtime Hours for Each Operating Point)<br><br>Monthly demand use = (Measured Peak Electric Demand Use for System for Each Operating Point) x (Maximum Operating Point for Each Month) | For each HVAC system:<br><br>Annual energy use = (Measured Energy Use for Each Operating Point) x (Full Load Runtime Hours for Each Operating Point)<br><br>Monthly demand use = (Measured Peak Electric Demand Use for System for Each Operating Point) x (Maximum Operating Point for Each Month) |
| Method #3: Multiple point using short-term data and manufacturer's performance data | For each HVAC system:<br><br>Annual energy use = (Measured Energy Use for Each Operating Point) x (Full Load Runtime Hours for Each Operating Point)<br><br>Monthly demand use = (Measured Peak Electric Demand Use for System for Each Operating Point) x (Maximum Operating Point for Each Month) | For each HVAC system:<br><br>Annual energy use = (Measured Energy Use for Each Operating Point) x (Full Load Runtime Hours for Each Operating Point)<br><br>Monthly demand use = (Measured Peak Electric Demand Use for System for Each Operating Point) x (Maximum Operating Point for Each Month) |
| Method #4: Short-term monitoring and calibrated, simplified simulation | For each HVAC system:<br><br>Annual energy use = (Simulated Energy Use Using Calibrated Air-side Model) x (Binned Weather Data)<br><br>Monthly demand use = (Simulated Peak Electric Demand Use for System) x (Maximum Bin Temperature for Month) | For each HVAC system:<br><br>Annual energy use = (Simulated Energy Use Using Calibrated Air-side Model) x (Binned Weather Data)<br><br>Monthly demand use = (Simulated Peak Electric Demand Use for System) x (Maximum Bin Temperature for Month) |

available to determine the savings, and where the savings are expected to be significant enough that the difference between pre-retrofit and post-retrofit usage can be measured using a whole-building approach. Whole-building methods can use monthly utility billing data (i.e., demand or usage), or continuous measurements of the whole-building energy use after the retrofit on a more detailed measurement level (weekly, daily or hourly). Sub-metering measurements can also be used to develop the whole-building models, providing that the measurements are available for the pre-retrofit and post-retrofit period, and that meter(s) measures that portion of the building where the retrofit was applied. Each sub-metered measurement then requires a separate model. Whole-building measurements can also be used on stored energy sources, such as oil or coal inventories. In such cases, the energy used during a period needs to be calculated (i.e., any deliveries during the period minus measured reductions in stored fuel).

In most cases, the energy use and/or electric demand are dependent on one or more independent variables. The most common independent variable is outdoor temperature, which affects the building's heating and cooling energy use. Other independent variables can also affect a building's energy use and peak electric demand, including: the building's occupancy (i.e., often expressed as weekday or weekend models), parking or exterior lighting loads, special events (i.e., Friday night football games), etc.

*Whole-building Energy Use Models*

Whole-building models usually involve the use of a regression model that relates the energy use and peak demand to one or more independent variables. The most widely accepted technique uses linear or change-point linear regression to correlate energy use or peak demand as the dependent variable with weather data and/or other independent variables. In most cases the whole-building model has the form:

$$E = C + B_1 V_1 + B_2 V_2 + B_3 V_3 + \dots$$

where

$E$ = the energy use or demand estimated by the equation,

$C$ = a constant term in energy units/day or demand units/billing period,

$B_n$ = the regression coefficient of an independent variable $V_n$,

$V_n$ = the independent driving variable.

In general, when creating a whole-building model for a number of different regression models are tried for a particular building and the results are compared and the best model selected using $R^2$ and CV (RMSE). Table 27.16 and Figure 27.7 contain models listed in ASHRAE's Guideline 14-2002, which include steady-state constant or mean models, models adjusted for the days in the billing period, two-parameter models, three-parameter models or variable-based degree-day models, four-parameter models, five-parameter models, and multivariate models. All of these models can be calculated with ASHRAE Inverse Model Toolkit (IMT), which was developed from Research Project 1050-RP.[141]

The steady-state, linear, change-point linear, variable-based degree-day and multivariate inverse models contained in ASHRAE's IMT have advantages over other types of models. First, since the models are simple, and their use with a given dataset requires no human intervention, the application of the models can be on can be automated and applied to large numbers of build-

**Table 27.16: Sample Models for the Whole-Building Approach from ASHRAE Guideline 14-2002.[152]**

| Name | | Independent Variable(s) | Form | Examples |
|------|---|-------------------------|------|----------|
| No Adjustment /Constant Model | | None | $E = E_b$ | Non weather sensitive demand |
| Day Adjusted Model | | None | $E = E_b \times \dfrac{day_b}{day_c}$ | Non weather sensitive use (fuel in summer, electricity in summer) |
| Two-Parameter Model | | Temperature | $E = C + B_1 (T)$ | |
| Three-Parameter Models | | Degree days/Temperature | $E = C + B_1 (DD_{BT})$ <br> $E = C + B_1 (B_2 - T)^+$ <br> $E = C + B_1 (T - B_2)^+$ | Seasonal weather sensitive use (fuel in winter, electricity in summer for cooling) Seasonal weather sensitive demand |
| Four-Parameter, Change Point Model | | Temperature | $E = C + B_1 (B_3 - T)^+ - B_2 (T - B_3)^+$ <br> $E = C - B_1 (B_3 - T)^+ + B_2 (T - B_3)^+$ | |
| Five-Parameter Models | | Degree days/Temperature | $E = C - B_1 (DD_{TH}) + B_2 (DD_{TC})$ <br> $E = C + B_1 (B_3 - T)^+ + B_2 (T - B_4)^+$ | Heating and cooling supplied by same meter. |
| Multi-Variate Models | | Degree days/Temperature, other independent variables | Combination form | Energy use dependent non-temperature based variables (occupancy, production, etc.). |

ings, such as those contained in utility databases. Such a procedure can assist a utility, or an owner of a large number of buildings, identify which buildings have abnormally high energy use. Second, several studies have shown that linear and change-point linear model coefficients have physical significance to operation of heating and cooling equipment that is controlled by a thermostat.[142,143,144,145] Finally, numerous studies have reported the successful use of these models on a variety of different buildings.[146,147,148,149,150,151]

Steady-state models have disadvantages, including: an insensitivity to dynamic effects (e.g., thermal mass), insensitivity to variables other than temperature (e.g., humidity and solar), and inappropriateness for certain building types, for example building that have strong on/off schedule dependent loads, or buildings that display multiple change-points. If whole-building models are required in such applications, alternative models will need to be developed.

## A. One-parameter or Constant Model

One-parameter, or constant models are models where the energy use is constant over a given period. Such models are appropriate for modeling buildings that consume electricity in a way that is independent of the outside weather conditions. For example, such models are appropriate for modeling electricity use in buildings which are on district heating and cooling systems, since the electricity use can be well represented by a constant weekday-weekend model. Constant models are often used to model sub-metered data on lighting use that is controlled by a predictable schedule.

## B. Day-adjusted Model

Day-adjusted models are similar to one-parameter constant models, with the exception that the final coefficient of the model is expressed as an energy use per day, which is then multiplied by the number of days in the billing period to adjust for variations in the utility billing cycle. Such day-adjusted models are often used with one, two, three, four and five-parameter linear or change-point linear monthly utility models, where the energy use per period is divided by the days in the billing period before the linear or change-point linear regression is performed.

## C. Two-parameter Model

Two-parameter models are appropriate for modeling building heating or cooling energy use in extreme climates where a building is exposed to heating or cooling year-around, and the building has an HVAC system with constant controls that operates continu-

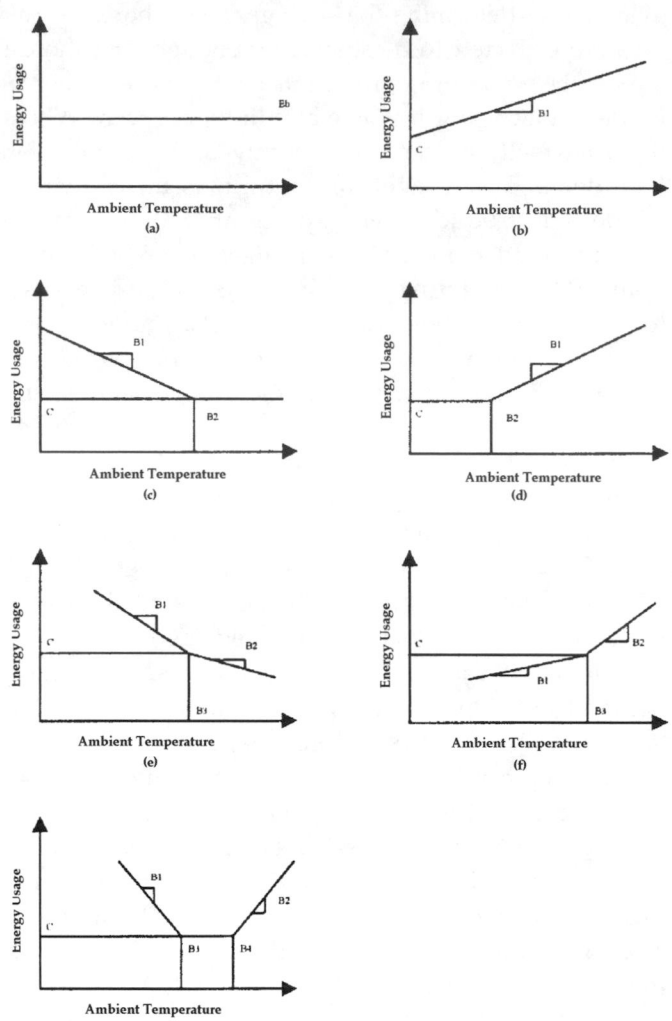

**Figure 27.7: Sample Models for the Whole-building Approach. Included in this figure is: (a) mean or one-parameter model, (b) two-parameter model, (c) three-parameter heating model (similar to a variable based degree-day model (VBDD) for heating), (d) three-parameter cooling model (VBDD for cooling), (e) four-parameter heating model, (f) four-parameter cooling model, and (g) five-parameter model.[153]**

ously. Examples include outside air pre-heating systems in arctic conditions, or outside air pre-cooling systems in near-tropical climates. Dual-duct, single-fan, constant-volume systems, without economizers can also be modeled with two-parameter regression models. Constant use, domestic water heating loads can also be modeled with two-parameter models, which are based on the water supply temperature.

## D. Three-parameter Model

Three-parameter models, which include change-point linear models or variable-based, degree day models, can be used on a wide range of building types, including residential heating and cooling loads, small commercial buildings, and models that describe the gas used by boiler thermal plants that serve one or more buildings. In Table 27.16, three-parameter models have several formats, depending upon whether or not the model is a variable based degree-day model or three-parameter, change-point linear models for heating or cooling. The variable-based degree day model is defined as:

$$E = C + B_1 (DD_{BT})$$

where

    $C$ = the constant energy use below (or above) the change point, and

    $B_1$ = the coefficient or slope that describes the linear dependency on degree-days,

    $DD_{BT}$ = the heating or cooling degree-days (or degree hours), which are based on the balance-point temperature.

The three-parameter change-point linear model for heating is described by[154]

$$E - C + B_1 (B_2 - T)^+$$

where

    $C$ = the constant energy use above the change point,

    $B_1$ = the coefficient or slope that describes the linear dependency on temperature,

    $B_2$ = the heating change point temperature,

    $T$ = the ambient temperature for the period corresponding to the energy use,

    + = positive values only inside the parenthesis.

The three-parameter change-point linear model for cooling is described by

$$E = C + B_1 (T - B_2)^+$$

where

    $C$ = the constant energy use below the change point,

    $B_1$ = the coefficient or slope that describes the linear dependency on temperature,

    $B_2$ = the cooling change point temperature,

    $T$ = the ambient temperature for the period corresponding to the energy use,

    + = positive values only for the parenthetical expression.

## E. Four-parameter Model

The four-parameter change-point linear heating model is typically applicable to heating usage in buildings with HVAC systems that have variable-air volume, or whose output varies with the ambient temperature. Four-parameter models have also been shown to be useful for modeling the whole-building electricity use of grocery stores that have large refrigeration loads, and significant cooling loads during the cooling season. Two types of four-parameter models are listed in Table 27.16, including a heating model and a cooling model. The four-parameter change-point linear heating model is given by

$$E = C + B_1 (B_3 - T)^+ - B_2 (T - B_3)^+$$

where

    $C$ = the energy use at the change point,

    $B_1$ = the coefficient or slope that describes the linear dependency on temperature below the change point,

    $B_2$ = the coefficient or slope that describes the linear dependency on temperature above the change point

    $B_3$ = the change-point temperature,

    $T$ = the temperature for the period of interest,

    + = positive values only for the parenthetical expression.

The four-parameter change-point linear cooling model is given by

$$E = C - B_1 (B_3 - T)^+ + B_2 (T - B_3)^+$$

where

    $C$ = the energy use at the change point,

    $B_1$ = the coefficient or slope that describes the linear dependency on temperature below the change point,

    $B_2$ = the coefficient or slope that describes the linear dependency on temperature above the change point

    $B_3$ = the change-point temperature,

    $T$ = the temperature for the period of interest,

    + = positive values only for the parenthetical expression.

## F. Five-parameter Model

Five-parameter change-point linear models are useful for modeling the whole-building energy use in buildings that contain air conditioning and electric heating. Such models are also useful for modeling the weather dependent performance of the electricity consumption of variable air volume air-handling units. The basic form for the weather dependency of either case is shown in Figure 27.7f, where there is an increase in electricity use below the change point associated with heating, an increase in the energy use above the change point associated with cooling, and constant energy use between the heating and cooling change points. Five-parameter change-point linear models can be described using variable-based degree day models, or a five-parameter model. The equation for describing the energy use with variable-based degree days is

$$E = C - B_1 \ (DD_{TH}) + B_2 \ (DD_{TC})$$

where

$C$ = the constant energy use between the heating and cooling change points,

$B_1$ = the coefficient or slope that describes the linear dependency on heating degree-days,

$B_2$ = the coefficient or slope that describes the linear dependency on cooling degree-days,

$DD_{TH}$ = the heating degree-days (or degree hours), which are based on the balance-point temperature.

$DD_{TC}$ = the cooling degree-days (or degree hours), which are based on the balance-point temperature.

The five-parameter change-point linear model that is based on temperature is

$$E = C + B_1 \ (B_3 - T)^+ + B_2 \ (T - B_4)^+$$

where

$C$ = the energy use between the heating and cooling change points,

$B_1$ = the coefficient or slope that describes the linear dependency on temperature below the heating change point,

$B_2$ = the coefficient or slope that describes the linear dependency on temperature above the cooling change point

$B_3$ = the heating change-point temperature,

$B_4$ = the cooling change-point temperature,

$T$ = the temperature for the period of interest,

$+$ = positive values only for the parenthetical expression.

## G. Whole-building Peak Demand Models

Whole-building peak electric demand models differ from whole-building energy use models in several respects. First, the models are not adjusted for the days in the billing period since the model is meant to represent the peak electric demand. Second, the models are usually analyzed against the maximum ambient temperature during the billing period. Models for whole-building peak electric demand can be classified according to weather-dependent and weather-independent models.

### G-1. Weather-dependent
#### Whole-building Peak Demand Models

Weather-dependent, whole-building peak demand models can be used to model the peak electricity use of a facility. Such models can be calculated with linear and change-point linear models regressed against maximum temperatures for the billing period, or calculated with an inverse bin model.[155,156]

### G-2. Weather-independent
#### Whole-building Peak Demand Models

Weather-independent, whole-building peak demand models are used to measure the peak electric use in buildings or sub-metered data that do not show significant weather dependencies. ASHRAE has developed a diversity factor toolkit for calculating weather-independent whole-building peak demand models as part of Research Project 1093-RP. This toolkit calculates the 24-hour diversity factors using a quartile analysis. An example of the application of this approach is given in the following section.

*Example: Whole-building energy use models*

Figure 27.8 presents an example of the typical data requirements for a whole-building analysis, including one year of daily average ambient temperatures and twelve months of utility billing data. In this example of a residence, the daily average ambient temperatures were obtained from the National Weather Service (i.e., the average of the published min/max data), and the utility bill readings represent the actual readings from the customer's utility bill. To analyze these data several calculations need to be performed. First, the monthly electricity use (kWh/month) needs to be divided by the days in the billing period to obtain the average daily electricity use for that month (kWh/day). Second, the average daily temperatures need to be calculated from the published NWS min/max data. From these average daily temperatures the average billing period temperature need to be calculated for each monthly utility bill.

The data set containing average billing period temperatures and average daily electricity use is then analyzed with ASHRAE's Inverse Model Toolkit (IMT)[157] to determine a weather normalized consumption as shown in Figures 27.9 and 27.10. In Figure 27.9 the twelve monthly utility bills (kWh/period) are shown plotted against the average billing period temperature along with a three-parameter change-point model calculated with the IMT. In Figure 27.10 the twelve monthly utility bills, which were adjusted for days in the billing period (i.e., kWh/day) are shown plotted against the average billing period temperature along with a three-parameter change-point model calculated with the IMT. In the analysis for this house, the use of an average daily model improved the accuracy of the unadjusted model (i.e., Figure 27.9) from an $R^2$ of 0.78 and CV (RMSE) of 24.0% to an $R^2$ of 0.83 and a CV (RMSE) of 19.5% for the adjusted model (i.e., Figure 27.10), which indicates a significant improvement in the model.

In another example the hourly steam use (Figure 27.11) and hourly electricity use (Figure 27.13) for the U.S. DOE Forrestal Building is modeled with a daily weekday-weekend three-parameter, change-point model for the steam use (Figure 27.12), and an hourly weekday-weekend demand model for the electricity use (Figure 27.14). To develop the weather-normalized model for the steam use the hourly steam data and hourly weather data were first converted into average daily data, then a three-parameter, weekday-weekend model was calculated using the EModel software,[158] which contains similar algorithms as ASHRAE's IMT. The resultant model, which is shown in Figure 27.12 along with the daily steam, is well described with an $R^2$ of 0.87 an RMSE of 50,085.95 kBtu/day and a CV (RMSE) of 37.1%.

In Figure 27.14 hourly weather-independent 24-hour weekday-weekend profiles have been created for

**Figure 27.8: Example Data for Monthly Whole-building Analysis (upper trace, daily average temperature, F, lower points, monthly electricity use, kWh/day).**

**Figure 27.9 Example Unadjusted Monthly Whole-building Analysis (3P Model) for kWh/period ($R^2$ = 0.78, CV (RMSE) = 24.0%).**

**Figure 27.10. Example Adjusted Whole-building Analysis (3P Model) for kWh/day ($R^2$ = 0.83, CV (RMSE) = 19.5%).**

the whole-building electricity use using ASHRAE's 1093-RP Diversity Factor Toolkit.[159] These profiles can be used to calculate the baseline whole-building electricity use (i.e., using the mean hourly use) by multiplying times the expected weekdays and weekends in the year. The profiles can also be used to calculate the peak electricity use (i.e., using the 90th percentile).

### Calculation of Annual Energy Use

Once the appropriate whole-building model has been chosen and applied to the baseline data, the annual energy use for the baseline period and the post-retrofit period are then calculated. Savings are then calculated by comparing the annual energy use of the baseline with the annual energy use of the post-retrofit period.

### Whole-building Calibrated Simulation Approach

Whole-building calibrated simulation normally requires the hourly simulation of an entire building, including the thermal envelope, interior and occupant loads, secondary HVAC systems (i.e., air handling units), and the primary HVAC systems (i.e., chillers, boilers). This is usually accomplished with a general purpose simulation program such as BLAST, DOE-2 or EnergyPlus, or similar proprietary programs. Such programs require an hourly weather input file for the location in which the building is being simulated. Calibrating the simulation refers to the process whereby selected outputs from the simulation are compared and eventually matched with measurements taken from an actual building. A number of papers in the literature have addressed techniques for accomplishing these calibrations, and include results from case study buildings where calibrated simulations have been developed for various purposes. [160, 161, 162, 163, 164, 165, 166, 167, 168, 169, 170, 171,172,173,174,175]

### Applications of Calibrated Whole-building Simulation.

Calibrated whole-building simulation can be a useful approach for measuring the savings from energy conservation retrofits to buildings. However, it is generally more expensive than other methods, and therefore it is best reserved for applications where other, less costly approaches cannot be used. For example, calibrated simulation is useful in projects where either pre-retrofit or post-retrofit whole-building metered electrical data are not available (i.e., new buildings or buildings without meters such as many college campuses with central facilities). Calibrated simulation is desired in projects where there are significant interactions between retrofits, for example lighting retrofits combined with changes to HVAC systems, or chiller retrofits. In such cases the whole-building simulation program can account for the interactions, and in certain cases, actually isolate interactions to allow for end-use energy allocations. It is useful in projects where there are significant changes in the facility's energy use during or after a retrofit has been installed, where it may be necessary to account for additions to a building that add or subtract thermal loads from the HVAC system. In other cases, demand may change over time, where the changes are not related to the energy conservation measures. Therefore, adjustments to account for these changes will be also be needed. Finally, in many newer buildings, as-built design simulations are being delivered as a part of the building's final documents. In cases where such simulations are properly documented they can be calibrated to the baseline conditions and then used to calculate and measure retrofit savings.

Unfortunately, calibrated, whole-building simulation is not useful in all buildings. For example, if a building cannot be readily simulated with available simulation programs, significant costs may be incurred in

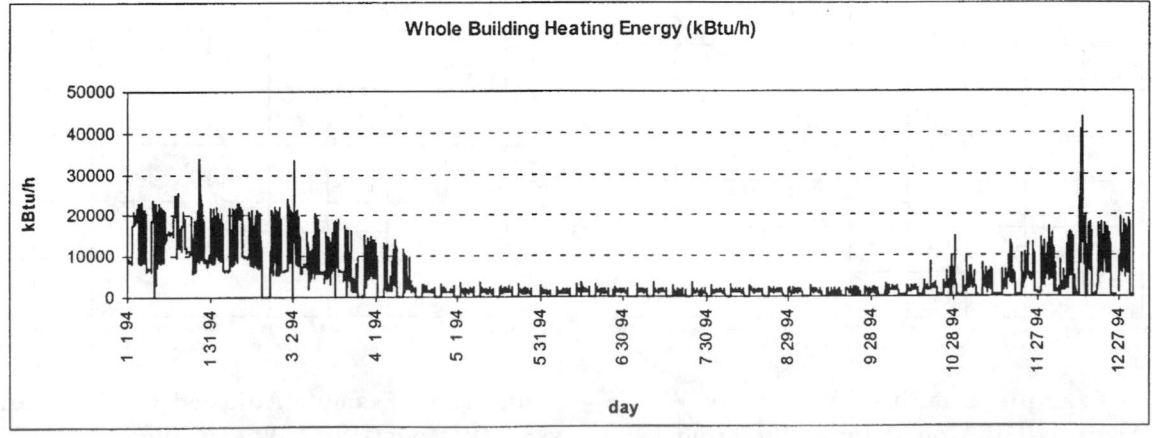

**Figure 27.11: Example Heating Data for Daily Whole-building Analysis.**

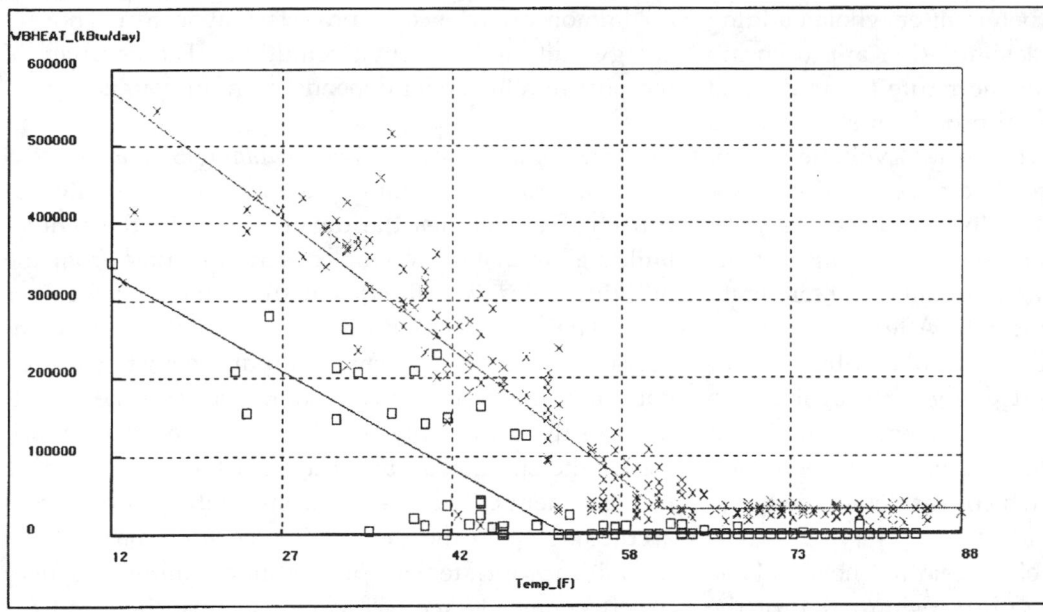

Figure 27.12: Example Daily Weekday-weekend Whole-building Analysis (3P Model) for Steam Use (kBtu/day, $R^2$ = 0.87, RMSE = 50,085.95, CV (RMSE) = 37.1%). Weekday use (x), weekend use (□).

Figure 27.13: Example Electricity Data for Hourly Whole-building Demand Analysis.

Figure 27.14: Example Weekday-weekend Hourly Whole-building Demand Analysis (1093-RP Model) for Electricity Use.

modifying a program or developing a new program to simulate only one building (e.g., atriums, underground buildings, buildings with complex HVAC systems that are not included in a simulation program's system library). Additional information about calibrated, whole-building simulation can be found in ASHRAE's Guideline 14-2002.

Figure 27.15 provides an example of the use of calibrated simulation to measure retrofit savings in a project where pre-retrofit measurements were not available. In this figure both the before-after whole-building approach and the calibrated simulation approach are illustrated. On the left side of the figure the traditional whole-building, before-after approach is shown for a building that had a dual-duct, constant volume system (DDCV) replaced with a variable air volume (VAV) system. In such a case where baseline data are available, the energy use for the building is regressed against the coincident weather conditions to obtain the representative baseline regression coefficients. After the retrofit is installed, the energy savings are calculated by comparing the projected pre-retrofit energy use against the measured post-retrofit energy use, where the projected pre-retrofit energy use calculated with the regression model (or empirical model), which was determined with the facility's baseline DDCV data.

In cases where the baseline data are not available (i.e., the right side of the figure), a simulation of the building can be developed and calibrated to the post-retrofit conditions (i.e., the VAV system). Then, using the calibrated simulation program, the pre-retrofit energy use (i.e., DDCV system) can be calculated for conditions in the post-retrofit period, and the savings calculated by comparing the simulated pre-retrofit energy use against the measured post-retrofit energy use. In such a case the calibrated post-retrofit simulation can also be used to fill-in any missing post-retrofit energy use, which is

a common occurrence in projects that measure hourly energy and environmental conditions. The accuracy of the post-retrofit model depends on numerous factors.

*Methodology for Calibrated Whole-building Simulation*

Calibrated simulation requires a systematic approach that includes the development of the whole-building simulation model, collection of data from the building being retrofitted and the coincident weather data. The calibration process then involves the comparison of selected simulation outputs against measured data from the systems being simulated, and the adjustment of the simulation model to improve the comparison of the simulated output against the corresponding measurements. The choice of simulation program is a critical step in the process, which must balance the model appropriateness, algorithmic complexity, user expertise, and degree of accuracy against the resources available to perform the modeling.

Data collection from the building includes the collection of data from the baseline and post-retrofit periods, which can cover several years of time. Building data to be gathered includes such information as the building location, building geometry, materials characteristics, equipment nameplate data, operations schedules, temperature settings, and at a minimum whole-building utility billing data. If the budget allows, hourly whole-

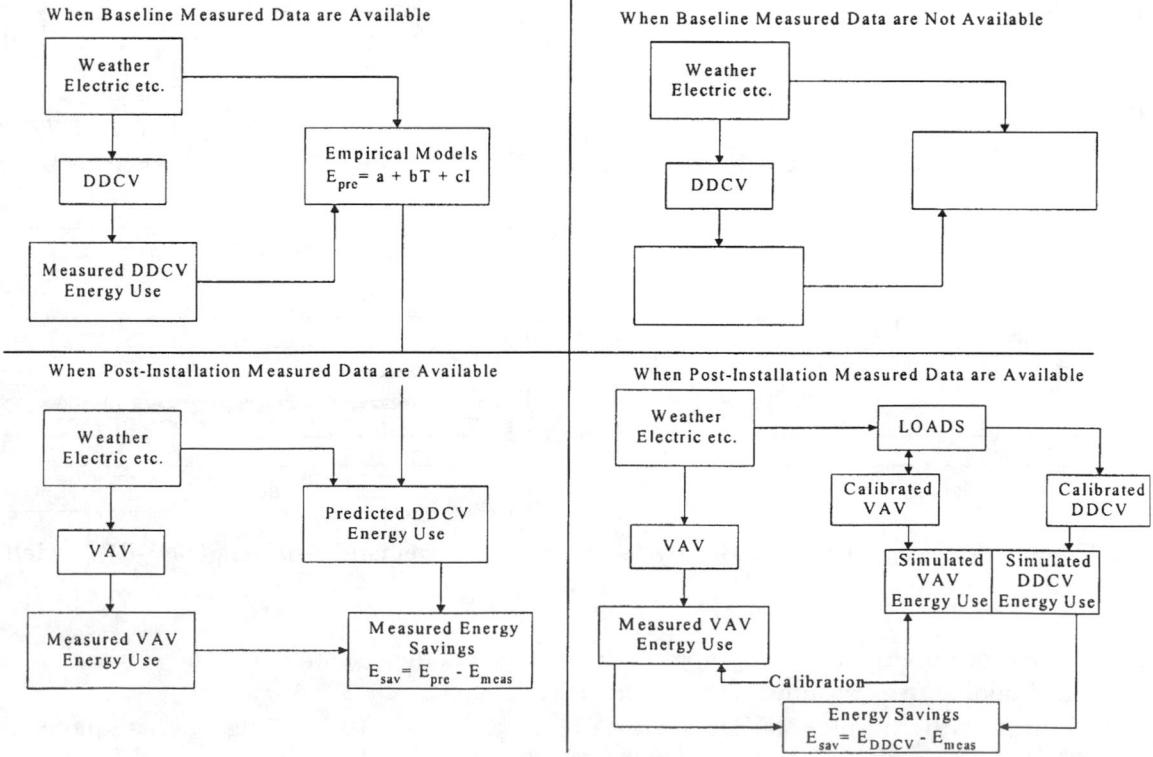

**Figure 27.15: Flow Diagram for Calibrated Simulation Analysis of Air-Side HVAC System.**[176]

building energy use and environmental data can be gathered to improve the calibration process, which can be done over short-term, or long-term period.

Figure 27.16 provides an illustration of a calibration process that used hourly graphical and statistical comparisons of the simulated versus measured energy use and environmental conditions. In this example, the site-specific information was gathered and used to develop a simulation input file, including the use of measured weather data, which was then used by the DOE-2 program to simulate the case study building. Hourly data from the simulation program was then extracted and used in a series of special-purpose graphical plots to help guide the calibration process (i.e., time series, bin and 3-D plots). After changes were made to the input file, DOE-2 was then run again, and the output compared against the measured data for a specific period. This process was then repeated until the desired level of calibration was reached, at which point the simulation was proclaimed to be "calibrated." The calibrated model was then used to evaluate how the new building was performing compared to the design intent.

A number of different calibration tools have been reported by various investigators, ranging from simple X-Y scatter plots to more elaborate statistical plots and indices. Figures 27.17, 27.18 and 27.19 provide examples of several of these calibration tools. In Figure 27.17 an example of an architectural rendering tool is shown that assists the simulator with viewing the exact placement of surfaces in the building, as well as shading from nearby buildings, and north-south orientation. In Figure 27.18 temperature binned calibration plots are shown comparing the weather dependency of an hourly simulation against measured data. In this figure the upper plots show the data as scatter plots against temperature. The lower plots are statistical, temperature-binned box-whisker-mean plots, which include the super positioning of measured mean line onto the simulated mean line to facilitate a detailed evaluation. In Figure 27.19 comparative three-dimensional plots are shown that show measured data (top plot), simulated data (second plot from the top), simulated minus measured data (second plot from the bottom, and measured minus simulated data (bottom plot). In these plots the day-of-the-year is the scale across the page (y axis), the hour-of-the-day is the scale projecting into the page (x axis), and the hourly

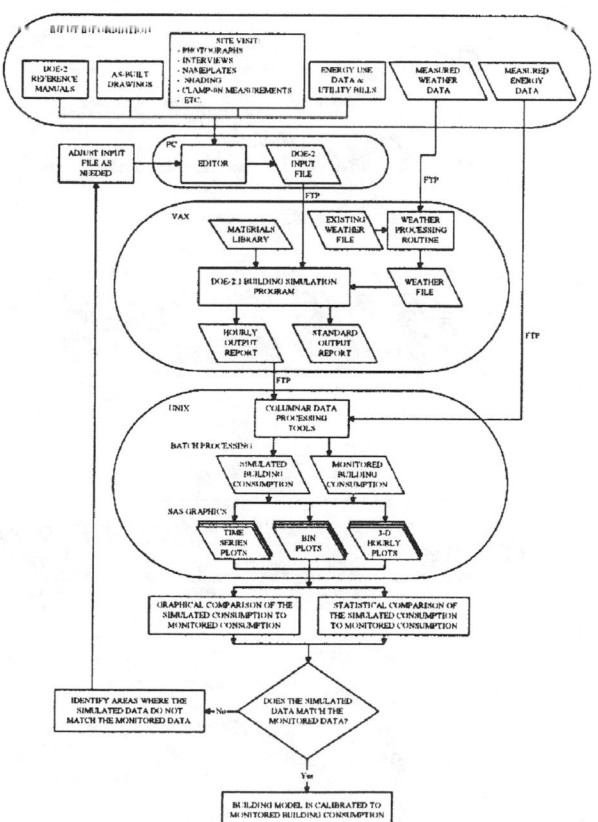

Figure 27.16: Calibration Flowchart. This figure shows the sequence of processing routines that were used to develop graphical calibration procedures.[178]

Figure 27.17: Example Architecture Rendering of the Robert E. Johnson Building, Austin, Texas.[179,180]

electricity use is the vertical scale of the surface above the x-y plane. These plots are useful for determining how well the hourly schedules of the simulation match the schedules of the real building, and can be used to identify other certain schedule-related features. For example, in the front of plot (b) the saw-toothed feature is indicating on/off cycling of the HVAC system, which is not occurring in the actual building.

Table 27.17 contains a summary of the procedures used for developing a calibrated, whole-building simulation program, as defined in ASHRAE's Guideline 14-2002. In general, to develop a calibrated simulation, detailed information is required for a building, including information about the building's thermal envelope (i.e., the walls, windows, roof, etc.), information about the building's operation, including temperature settings, HVAC systems, and heating-cooling equipment that existed both during the baseline and post-retrofit period. This information is input into two simulation files, one for the baseline and one for the post-retrofit conditions. Savings are then calculated by comparing the two simulations of the same building, one that represents the baseline building, and one that represents the building's operations during the post-retrofit period.

### 27.2.2 Role of M&V

Each Energy Conservation Measure (ECM) presents particular requirements. These can be grouped in functional sections as shown in Table 27.18. Unfortu-

nately, in most projects, numerous variables exist so the assessments can be easily disputed. In general, the low risk (L)—reasonable payback ECMs exhibit steady performance characteristics that tend not to degrade or become easily noticed when savings degradation occurs. These include lighting, constant speed motors, two-speed motors and IR radiant heating. The high risk (H)—reasonable payback ECMs include EMCSs, variable speed drives and control retrofits. The savings from these ECMs can be overridden by building operators and not be noticed until years later. Most other ECMs fall in the category of "it depends." The attention that the operations and maintenance directs at these dramatically impacts the sustainability of the operation and the savings. With an EMCS, operators can set up trend reports to measure and track occupancy schedule overrides, the various reset schedule overrides, variable speed drive controls and even monitor critical parameters which track mechanical systems performance. illustrates a "most likely" range of ratings for the various categories.[183]

Often, building envelope or mechanical systems need to be replaced. Building systems have finite life-

Figure 27.19: Comparative Three-dimensional Plots. (a) Measured Data. (b) Simulated Data. (c) Simulated-Measured Data. (d) Measured-Simulated Data.

Figure 27.18: Temperature Bin Calibration Plots. This figure shows the measured and simulated hourly weekday data as scatter plots against temperature in the upper plots and as statistical binned box-whisker-mean plots in the lower plots.[181]

**Table 27.17: Calibrated, whole-building Simulation Procedures from ASHRAE Guideline 14-2002.[177]**

| STEP | PROCEDURES | DATA OR INFORMATION REQUIRED |
|---|---|---|
| Step 1: Calibrated Simulation Plan | • Develop Baseline Scenario<br>• Develop Post-retrofit Scenario<br>• Select simulation package<br>• Select calibration tools | • Information about existing conditions, building location, surroundings, etc.<br>• Information about post-retrofit conditions. |
| Step 2: Data Collection From Existing Building | • Visit site and collect data about baseline and post-retrofit conditions, characteristics, plans,<br>• Collect energy use data, and information about building operations, and systems<br>• Determine internal heat loads<br>• Prepare data for input into simulation | • Utility billing data for baseline and for post-retrofit periods<br>• Interval energy use and environmental data from building<br>• Equipment schedules, thermostat settings, HVAC system types, performance measurements |
| Step 3: Data Input Into Simulation Program | • Develop whole-building simulation or building, including envelope, loads, systems, plant<br>• Use architectural rendering to confirm inputs | • All building characteristics, system types, schedules, setpoints, etc., needed for developing simulation. |
| Step 4: Baseline Simulation Run and Comparison | • Run simulation.<br>• Compare with measured data<br>• Repeat process until simulation matches actual building | • All building characteristics, system types, schedules, setpoints, etc., needed for developing simulation.<br>• On-sites measurements of energy and environmental data |
| Step 5: Post-retrofit Simulation Run and Comparison. | • Run simulation.<br>• Compare with measured data<br>• Repeat process until simulation matches actual building | • All building characteristics, system types, schedules, setpoints, etc., needed for developing simulation.<br>• On-sites measurements of energy and environmental data |
| Step 6: Calculate Savings | • Compare baseline and post-retrofit simulations | |
| Step 7: Report findings | • Provide baseline & post-retrofit building descriptions.<br>• Document building measurements<br>• Include simulation plan<br>• Present results<br>• Provide input/output files from simulation | |

times, ranging from two to five years for most light bulbs to 10 to 20+ years for chillers and boilers. Building envelope replacements like insulation, siding, roof, windows and doors can have lifetimes from 10 to 50 years. In these instances, life cycle costing should be done to compare the total cost of upgrading to more efficient technology. Also, the cost of M&V should be considered when determining how to sustain the savings and performance of the replacement. In many cases, the upgraded efficiency will have a payback of less than 10 years when compared to the current efficiency of the existing equipment. Current technology high efficiency upgrades normally use controls to acquire the high efficiency. These controls often connect to standard interfaces so that they communicate with today's state of the art Energy Management and Control Systems (EMCSs).

### 27.2.3 Cost/Benefit Analysis

The target for work for the USAF has been 5% of the savings.[184] The cost of the M&V can exceed 5%

if the risk of losing savings exceeds predefined limits. The Variable Speed Drive ECM illustrates these opportunities and risks. VSD equipment exhibits high reliability. Equipment type of failures normally happen when connection breaks occur with the control input, the remote sensor. Operator induced failures occur then the operator sets the unit to 100% speed and does not re-enable the control. Setting the unit to 100% can occur for legitimate reasons. These reasons include running a test, overriding a control program that does not provide adequate speed under specific, and typically unusual, circumstances, or requiring 100% operation for a limited time. The savings disappear if the VSD remains at 100% operating speed.

For example, consider a VSD ECM with ten (10) motors with each motor on a different air handling unit. Each motor has fifty (50) horsepower (HP). The base case measured these motors running 8760 hours per year at full speed. Assume that the loads on the motors matched the nameplate 50 HP at peak loads.

**Table 27.18: Overview of Risks and Costs for ECMs.**

| ECM Strategies | ECM | | | M&V | | |
| --- | --- | --- | --- | --- | --- | --- |
| | Implement Cost | Payback | Savings | Implement Cost | Short Term Savings Risk | Long Term Savings Risk |
| Boiler Replace/Rebuild | M | M | M | H | M | H |
| Building Envelope Upgrades | H | L | M | H | L | M |
| Central Heating Plant Decentralization | VH | M | M | | H | M |
| Chiller Plant Decentralization | VH | L | TBD | | | |
| Chiller Replace/Rebuild | H | L | M | | | |
| Constant Speed Motors | L | L | L | | L | L |
| Cooling Tower Replace/Rebuild | H | L | M | | | |
| DDC Controls | M-H | M | H | | | |
| EMCS | H | L-M | H | | | |
| Ground Source HP | M | M | L-M | | | |
| HP Replace | L | M | L | | | |
| IR Radiant Heating | M-H | M | H | | | |
| Lighting | M | H | TBD | | | |
| Commissioning | H | VH | H | | | |
| Propane Air Plant | H | L-M | | | | |
| Steam Traps | L | TBD | | | | |
| Thermal Storage | H | L | | | | |
| Tower Free Cooling | M-H | M | | | | |
| Varaiable Speed Motors | L | H | | | | |
| **(VH)** Very High | >$1M | <3 years | >$100K | >$100K | >15%/yr | >50% |
| **(H)** High | >$100K | <5 years | >$10K | >$10K | >10%/yr | >35% |
| **(M)** Medium | >$10K | <7 years | >$1K | >$1K | >5%/yr | >25% |
| **(L)** Low | >$1K | > 8 years | <$1K | <$1K | <4%/yr | <10% |

Although the actual load on a AHU fan varies with the state of the terminal boxes, assume that the load average equates to 80% of the full load since the duct pressure will rise as the terminal boxes reduce flow at the higher speed. Table 27.19 contains the remaining assumptions. To correctly determine the average power load, the average power must either be integrated over the period of consumption or the bin method must be used. For the purposes of this example, the 14.4% value will be used.

The equation below shows the relationship between the fan speed and the power consumed. The exponent has been observed to vary between 2.8 (at high flow) and 2.7 (at reduced flow) for most duct systems. This includes the loss term from pressure increases at a given fan speed. Changing the exponent from 2.8 to 2.7 reduces the savings by less than 5%.

$$\text{Pwr} = \text{Pwr}_0 \times \left( \frac{\% \text{ Speed}}{\text{Full Speed}} \right)^{2.8}$$

Demand savings will not be considered in this example. Demand savings will likely be very low if the utility has a 12-month ratchet clause and the summer load requires some full speed operation during peak times. Assuming a $12.00/kW per month demand charge, demand savings could be high for off-season months if the demand billing resets monthly. Without a ratchet clause, rough estimates have yearly demand savings ranging up to $17,000 if the fan speed stays under 70% for 6 months per year. Yearly demand savings jump to over $20,000 if the fan speed stays under 60% for 6 months per year.

**Table 27.19: VSD Example Assumptions.**

| | Pre-ECM | Post-ECM | Comments |
| --- | --- | --- | --- |
| Hours / Yr | 8760 | 8760 | |
| $ / kWh | $0.06 | $0.06 | |
| $ / kW / month | $12.00 | $12.00 | |
| Average Speed | 100% | 50% | |
| PowerCurveExponent | 2.8 | 2.8 | Contains duct loss impact |
| Average Power Load | 80% | 14.4% | See Equation XXX |
| Total Motor HP | 500 | 500 | |
| Total kW | 373 | 384 | At full speed, ~3% loss in inverter |

Figure 27.20 illustrates the savings expected from the VSD ECM by hours of use per year. The 5% and 10% of Savings lines define the amount available for M&V expenditures at these levels. In this example, the ECM savings exceeds $253,000 per year. Five percent (5%) of savings over a 20-year project life makes $253K available for M&V and ten percent (10%) of the savings makes $506K available over the 20-year period. If the motors run less frequently than continuous, savings decrease as shown in Figure 27.20. Setting up the M&V program to monitor the VSDs on an hourly basis and report savings on a monthly report requires monitoring the VSD inverter with an EMCS to poll the data and create reports.

To provide the impact of the potential losses from losing the savings, assume the savings degrades at a loss of 10% of the total yearly savings per year. Studies have shown that control ECMs like the VSD example can expect to see 20% to 30% degradation in savings in 2 to 3 years. Figure 27.21 illustrates what happens to the

savings in 20 years with 10% of the savings spent on M&V. Note that the losses exceed the M&V cost during the first year, resulting in a net loss of almost $3,000,000 over the 20-year period. Figure 27.22 shows the savings per year with a 10% loss of savings. M&V costs remain at 10% of savings. At the end of the 20-year period, the savings drop to almost $30,000 per year out of a potential savings of over $250,000 per year.

This example shows the cumulative impact of losing savings on a year by year basis. The actual savings amounts will vary depending upon the specific factors in an ECM and can be scaled to reflect a specific application. Increasing the M&V cost to reduce the loss of savings often makes sense and must be carefully thought through.

### 27.2.4 Cost Reduction Strategies

M&V strategies can be cost reduced by lowering the requirements for M&V or by statistical sampling. Reducing requirements involves performing trade-offs with the risks and benefits of having reliable numbers to determine the savings and the costs for these measurements.

#### 27.2.4.1 Constant Load ECMs

Lighting ECMs can save 30% of the pre-ECM energy and have a payback in the range of 3 to 6 years. Assuming that the lighting ECM was designed and implemented per the specifications and the savings were verified to be occurring, just verifying that the storeroom has the correct ballasts and lamps may constitute acceptable M&V on a yearly basis. This costs far less than performing a yearly set of measurements, analyzing them and then creating reports. In this case, other safeguards should be implemented to assure that the bulb and ballast replacement occurs and meets the

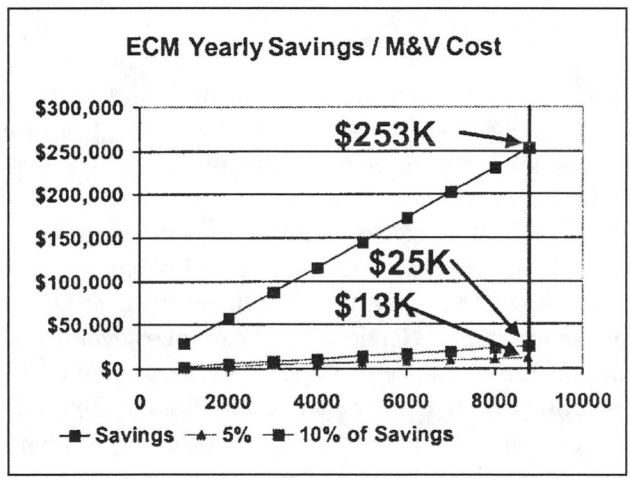

**Figure 27.20: Example VSD EMC Yearly Savings/M&V Cost.**

**Figure 27.21: Yearly Impact of Ongoing Losses.**

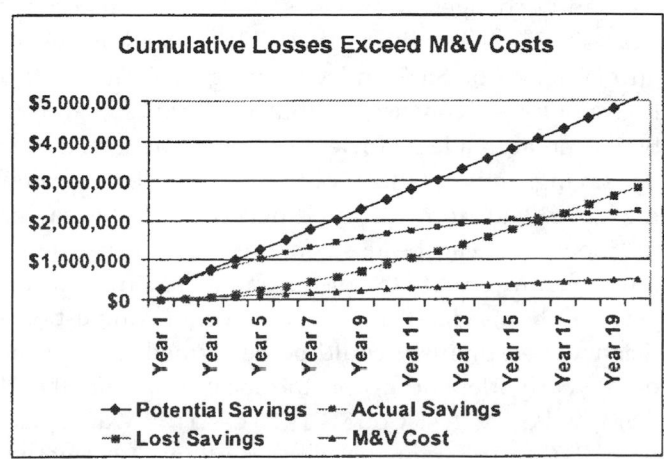

**Figure 27.22: Cumulative Impact of Savings Loss.**

requirements specified.

High efficiency motor replacements provide another example of constant load ECMs. The key short term risks with motor replacements involve installing the right motor with all mechanical linkages and electrical components installed correctly. Once verified, the long term risks for maintaining savings occur when the motor fails. The replacement motor must be the correct motor or savings can be lost. A sampled inspection reduces this risk. Make sure to inspect all motors at least once every five (5) years.

### 27.2.4.2 Major Mechanical Systems

Boilers, chillers, air handler units, cooling towers comprise the category of manor mechanical equipment in buildings. They need to be considered separately as each carry their own set of short-term and long-term risks. In general, measurements provide necessary risk reduction. The question becomes: What measurements reduce the risk of savings loss by an acceptable amount?

First a risk assessment needs to be performed. The short-term risks for boilers involve installing the wrong size or installing the boiler improperly (not to specifications). Long-term savings sustainability risks tend to focus on the water side and the fire side. Water deposits ($K^+$, $Ca^{++}$, $Mg^+$) will form on the inside of the tubes and add a thermal barrier to the heat flow. The fire side can add a layer of soot if the $O_2$ level drops too low. Either of these reduce the efficiency of the boiler over the long haul. Generally this can take several years to impact the efficiency if regular tune-ups and water treatment occurs.

Boilers come in a wide variety of shapes and sizes. Boiler size can be used as a defining criterion for measurements. Assume that natural gas or other boiler fuels cost about $5.00 per MMBtu. Although fuel price constantly changes, it provides a reference point for this analysis. Thus a boiler with 1MMBtu per hour output, an efficiency of 80% and operating at 50% load 3500 hours per year, consumes about $11,000 per year. If this boiler replaced a less efficient boiler, say at 65%, then the net savings amounts to about $2,500 per year, assuming the same load from the building. At 5% of the savings, $125 per year can be used for M&V. This does not allow much M&V. At 10% of the annual savings, $250 per year can be used. At this level of cost, a combustion efficiency measurement could be performed, either yearly or bi-yearly, depending on the local costs. In 2003 the ASME's Power Test Code 4.1 (PTC-4.1)[185] was replaced with PTC4. Either of these codes allows two methods to measuring boiler efficiency. The first method uses

the energy in equals energy out—using the first law of thermodynamics. This requires measuring the Btu input via the gas flow and the Btu output via the steam (or water) flow and temperatures. The second method measures the energy loss due to the content and temperature of the exhausted gases, radiated energy from the shell and piping and other loss terms (like blowdown). The energy loss method can be performed in less than a couple of hours. The technician performing these measurements must be skilled or significant errors will result in the calculated efficiency. The equation below shows the calculations required.

$$Efficiency = 100\% - Losses + Credits$$

The losses term includes the temperature of the exhaust gas and a measure of the unburned hydrocarbons by measuring $CO_2$ or $O_2$ levels, the loss due to excess CO and a radiated term. Credits seldom occur but could arise from solar heating the makeup water or similar contributions. The Greek letter "$\eta$" usually denotes efficiency.

As with boilers, a risk assessment needs to be performed for chilers. The short term risks for chillers involve sizing or improper installation. Long term savings sustainability risks focus on the condenser water system, as circulation occurs in an open system. Water deposits ($K^+$, $Ca^{++}$, $Mg^+$, organics) will form on the inside of the condenser tubes and add a barrier to the thermal flow. These reduce the efficiency of the chiller over the long haul. Generally this can take several years to impact the efficiency if proper water treatment occurs. Depending on the environmental conditions, the quality of the makeup water and the water treatment, condenser tube fouling should be checked every year or at least every other year.

Chillers consume electricity in the case of most centrifugal, screw, scroll and reciprocating compressors. Direct-fired absorbers and engine driven compressors use a petroleum based fuel. As with boilers, chiller size and application sets the basic energy consumption levels. Assume, for the purpose of this example, that electricity provides the chiller energy. Older chillers with water towers often operate at the 0.8 to 1.3 kW per ton level of efficiency. New chillers with water towers can operate in the 0.55 to 0.7 range of efficiency. Note that the efficiency of any chiller depends upon the specific operating conditions. Also assume the following: 500 Tons centrifugal chiller with the specifications shown in Table 27.20. Under these conditions the chiller produces 400 Tons of chilled water and requires an expenditure of $ 38,000 per year, considering both energy use and demand charges. Some utilities only charge demand charges on the transmission and delivery (T&D) parts of the rate structure. In that case, the cost at $0.06/kWh would be closer to $28,000. Using the 5% (10%) guideline for M&V costs as a percentage of savings leaves almost $1,100 ($2,200) per year to spend on M&V. This creates an allowable expenditure over a 20-year project of $22,000 ($44,000) for M&V. If the utility has a ratchet clause in the rate structure, the amount for M&V increases to $1,700 ($3,400) per year. At $1,100 per year, trade-offs will need to be made to stay within that "budget." The risks need to be weighed and decisions made as to what level of M&V costs will be allowed.

To determine the actual efficiency of a chiller requires accurate measurements of the chilled water flow, the difference between the chilled water supply and return temperatures and the electrical power provided to the chiller. Costs can be reduced using an EMCS if only temperature, flow and power sensors need to be installed.

sustainability risks. When an operator overrides a strategy and forgets to re-enable it, the savings disappear. A common EMCS ECM requires the installation of equipment and programs used to set back temperatures or turn off equipment. Short term risks involve setting up the controls so that performance enhances, or at least does not degrade, the comfort of the occupants. When discomfort occurs, either occupants set up "portable electric reheat units" or operators override the control program. For example, when the night set-back control does not get the space to comfort by occupancy, operators typically override instead of adjusting the parameters in the program. These actions tend to occur during peak loading times and then not get re-enabled during milder times. Long term risks cover the same area as short term risks. A new operator or a failure in remote equipment that does not get fixed will likely cause the loss of savings. Estimating the savings cost for various projects can be done when the specifics are known.

### Table 27.20: Example of Savings with a 500 Ton Chiller.

|  | No Rachet-4 Mo/Yr | | 13 Month Rachet | |  |
|  | Pre-ECM | Post-ECM | Pre-ECM | Post-ECM | Comments |
|---|---|---|---|---|---|
| Hours / Yr | 2000 | 2000 | 2000 | 2000 |  |
| Tons Cooling | 400 | 400 | 400 | 400 |  |
| $ / kWh | $0.06 | $0.06 | $0.06 | $0.06 |  |
| $ / kW / month | $12.00 | $12.00 | $12.00 | $12.00 |  |
| Average Efficiency | 0.90 | 0.57 | 0.90 | 0.57 | kW / Ton |
| Yearly kWh | 720000 | 456000 | 720000 | 456000 |  |
| Yearly kWh Cost | $43,200 | $27,360 | $43,200 | $27,360 | Energy use cost |
| Yearly kW Cost | $17,280 | $10,944 | $51,840 | $32,832 | Demand cost |
| Yearly Cost | $60,480 | $38,304 | $95,040 | $60,192 | Total |
| Savings |  | $22,176 |  | $34,848 |  |

Cooling tower replacement requires knowledge of the risks and costs involved. As with boilers and chillers, the primary risks involve the water treatment. Controls can be used to improve the efficiency of a chiller/tower combination by as much as 15% to 20%. As has been previously stated, control ECMs often get overridden and the savings disappear.

#### 27.2.4.3 Control Systems

Control ECMs encompass a wide spectrum of capabilities and costs. Upgrading a pneumatic control system and installing EP (electronic to pneumatic) transducers involves the simple end. The complex side could span installing a complete EMCS with sophisticated controls, with various reset, pressurization and control strategies. Generally, EMCSs function as basic controls and do not get widely used in sophisticated applications.

Savings due to EMCS controls bear high

### Table 27.21: Sampling Requirements.

| Precision | 20% | 20% | 10% |
|---|---|---|---|
| Confidence | 80% | 90% | 90% |
|  |  |  |  |
| Population Size, N | Sample Size, n | | |
| 4 | 3 | 4 | 4 |
| 12 | 6 | 8 | 11 |
| 20 | 8 | 10 | 16 |
| 30 | 9 | 11 | 21 |
| 40 | 9 | 12 | 26 |
| 50 | 10 | 13 | 29 |
| 60 | 10 | 14 | 32 |
| 70 | 10 | 15 | 35 |
| 80 | 10 | 15 | 37 |
| 90 | 10 | 15 | 39 |
| 100 | 10 | 15 | 41 |
| 200 | 11 | 16 | 51 |
| 300 | 11 | 17 | 56 |
| 400 | 11 | 17 | 59 |
| 500 | 11 | 17 | 60 |
| Infinite | 11 | 17 | 68 |

Risk abatement can be as simple as requiring a trend report weekly or at least monthly. M&V costs can generally be easily held under 5% when using an EMCS and creating trend reports.

### 27.2.5 M&V Sampling Strategies

M&V can be made significantly lower cost by sampling. Sampling also reduces the timeliness of obtaining specific data on specific equipment. The benefits of sampling arise when the population of items increases. Table 27.21 (M&V Guidelines: Measurement and Verification for Federal Energy Projects, Version 2.2, Appendix D) illustrates how confidence and precision impact the number of samples required in a given population of items.

Lighting ECMs may involve thousands of fixtures. For example, to obtain a savings estimate for 1,000 or more fixtures, with a confidence of 80% and a precision of 20%, 11 fixtures would need to be sampled. If the requirements increased to a confidence of 90% and a precision of 10%, 68 fixtures would need to be sampled.

The boiler ECM also represents opportunity for M&V cost reduction using sampling. Assume that the ECM included replacing 50 boilers. If a confidence of 80% and a precision of 20% satisfy the requirements, 10 boilers would need to be sampled. The cost is then reduced to 20% of the cost of measuring all boilers, a significant savings. A random sampling to select the sample set can easily be implemented.

### References

1. Claridge D.E., Turner, W.D., Liu, M., Deng, S., Wei, G., Culp, C., Chen, H., and Cho, S. 2002. "Is Commissioning Once Enough?" Solutions for Energy Security and Facility Management Challenges: Proceedings of the 25th WEEC, Atlanta, GA, October 19-11, 2002, pp. 29-36.
2. Haberl, J., Lynn, B., Underwood, D., Reasoner, J., Rury, K. 2003. "Development an M&V Plan and Baseline for the Ft. Hood ESPC Project," ASHRAE Seminar Presentation, (June).
3. C. Culp, K.Q. Hart, B. Turner, S. Berry-Lewis, 2003. "Energy Consumption Baseline: Fairchild AFB's Major Boiler Retrofit," ASHRAE Seminar (January).
4. Arnold, D., 1999. "The Evolution of Modem Office Buildings and Air Conditioning," ASHRAE Journal, American Society of Heating Refrigeration Air-conditioning Engineers, Atlanta, GA, pp. 40-54, (June).
5. Donaldson, B., Nagengast. 1994. Heat and Cold: Mastering the Great Indoors. American Society of Heating Refrigeration Air-conditioning Engineers, Atlanta, GA.
6. Cheney, M., Uth, R. 1999. Tesla: Master of Lightning. Barnes and Noble Books, New York, N,Y.
7. Will, H. 1999. The First Century of Air Conditioning. American Society of Heating Refrigeration Air-conditioning Engineers, Atlanta, GA.
8. Israel, P. 1998. Edison: A Life of Invention. John Wiley and Sons, New York, N.Y.
9. EEI 1981. Handbook for Electricity Metering, 8th Edition with Appendix, Edition Electric Institute, Washington D.C.
10. Miller, R. 1989. Flow Measurement Engineering Handbook. McGraw Hill, New York, N.Y.
11. American Institute of Physics, 1975. Efficient Use of Energy: The APS Studies on the Technical Aspects of the More Efficient Use of Energy, American Physical Society, New York, N.Y., (A report on the 1973 summer study at Princeton University).
12. National Geographic, February 1981. Special Report on Energy: Facing up to the Problem, Getting Down to Solutions, National Geographic Society, Washington, D.C.
13. Scientific American 1971. Energy and Power. W.H. Freeman and Company, San Francisco, CA. (A reprint of eleven articles that appeared in the September 1971 Scientific American).
14. Kusuda, T. 1999. "Early History and Future Prospects of Building System Simulation," Proceedings of the Sixth International Building Performance Simulation Association (IBPSA BS' 99), Kyoto, Japan, (September).
15. APEC 1967. HCC-heating/cooling load calculations program. Dayton, Ohio, Automated Procedures for Engineering Consultants.
16. Ayres, M., Stamper, E. 1995. "Historical Development of Building Energy Calculations," ASHRAE Transactions, Vol. 101, Pt. 1. American Society of Heating Refrigeration Air-conditioning Engineers, Atlanta, GA.
17. Stephenson, D., and Mitalas, G. 1967. "Cooling Load Calculations by Thermal Response Factor Method," ASHRAE Transactions, Vol. 73, pt. 1.
18. Mitalas, G. and Stephenson, D. 1967, "Room Thermal Response Factors," ASHRAE Transactions, Vol. 73, pt. 2.
19. Stoecker, W. 1971. Proposed Procedures for Simulating the Performance of Components and Systems for Energy Calculations, 2 "d Edition, American Society of Heating Refrigeration Air-conditioning Engineers, Atlanta, GA.
20. Sepsy, C. 1969. "Energy Requirements for Heating, and Cooling Buildings (ASHRAE RP 66-OS), Ohio State University.
21. Socolow, R. 1978. Saving Energy in the Home: Princeton's Experiments at Twin Rivers, Ballinger Publishing Company, Cambridge, Massachusetts, (This book contains a collection of papers that were also published in Energy and Buildings, Vol. 1, No. 3., (April)).
22. Fels, M. 1986. Special Issue Devoted to Measuring Energy Savings: The Scorekeeping Approach, Energy and Buildings, Vol. 9, Nos. 1 & 2, Elsevier Press, Lausanne, Switzerland, (February/May).
23. DOE 1985. Proceedings of the DOE/ORNL Data Acquisition Workshop, Oak Ridge National Laboratory, Oak Ridge, TN, (October).
24. Lyberg, M. 1987. Source Book for Energy Auditors: Vols. 1 & 2, International Energy Agency, Stockholm, Sweden, (Report on IEA Task XI).
25. IEA 1990. Field Monitoring For a Purpose. International Energy Agency Workshop, Chalmers University, Gothenburg, Sweden, (April).
26. Oninicomp 1984. Faser Software, Omnicomp, Inc., State College, PA, (monthly accounting software with VBDD capability).
27. Eto, J. 1988. "On Using Degree-days to Account for the Effects of Weather on Annual Energy Use in Office Buildings," Energy and Buildings, Vol. 12, No. 2, pp. 113-127.
28. SRC Systems 1996. Metrix: Utility Accounting System, Berkeley, CA, (monthly accounting software with combined VBDD/multiple regression capabilities).
29. Haberl, J. and Vajda. E. 1988. "Use of Metered Data Analysis to Improve Building Operation and Maintenance: Early Results From Two Federal Complexes," Proceedings of the ACEEE 1988 Summer Study on Energy Efficient Buildings, Pacific Grove, CA, pp. 3.98 - 3.111, (August).
30. Sonderegger, R. 1977. Dynamic Models of House Heating Based on Equivalent Thermal Parameters, Ph.D. Thesis, Center

for Energy and Environmental Studies, Report No. 57, Princeton University.

31. DOE 1985. op.cit.

32. Lyberg, M. 1987. op.cit.

33. IEA 1990. op.cit.

34. ASHRAE 199 1. Handbook of HVAC Applications, Chapter 37: Building Energy Monitoring, American Society of Heating Refrigeration Air-conditioning Engineers, Atlanta, GA.

35. Haberl, J., and Lopez, R. 1992. "LoanSTAR Monitoring Workbook: Workbook and Software for Monitoring Energy in Buildings," submitted to the Texas Governor's Energy Office, Energy Systems Laboratory, Texas A&M University, (August).

36. Claridge, D., Haberl, J., O'Neal, D., Heffington, W., Turner, D., Tombari, C., Roberts, M., Jaeger, S. 1991. "Improving Energy Conservation Retrofits with Measured Savings." ASHRAE Journal, Volume 33, Number 10, pp. 14-22, (October).

37. Fels, M., Kissock, K., Marean, M., and Reynolds, C. 1995. PRISM, Advanced Version 1.0 User's Guide, Center for Energy and Envirom-nental Studies, Princeton University, Princeton, N.J., (January).

38. ASHRAE 1999, I-IVACO 1 Toolkit: A Toolkit for Primary HVAC System Energy Calculation, ASHRAE Research Project -RP 665, Lebrun, J., Bourdouxhe, J-P, and Grodent, M., American Society of Heating Refrigeration Air-conditioning Engineers, Atlanta, GA.

39. ASHRAE 1993. HVAC02 Toolkit: Algorithms and Subroutines for Secondary HVAC System Energy Calculations, ASHRAE Research Project - 827-RP, Authors: Brandemuehl, M., Gabel, S.,Andresen, American Society of Heating Refrigeration Air-conditioning Engineers, Atlanta, GA.

40. Brandemuehl, M., Krarti, M., Phelan, J. 1996, "827-RP Final Report: Methodology Development to Measure In-Situ Chiller, Fan, and Pump Performance," ASHRAE Research, ASHRAE, Atlanta, GA, (March).

41. Haberl, J., Reddy, A., and Elleson, J. 2000a. "Determining Long-Term performance Of Cool Storage Systems From Short-Term Tests, Final Report," submitted to ASHRAE under Research Project 1004-RP, Energy Systems Laboratory Report ESL-TR-00/08-01, Texas A&M University, 163 pages, (August).

42. Kissock, K., Haberl, J., and Claridge, D. 2001. "Development of a Toolkit for Calculating Linear, Changepoint Linear and Multiple-Linear Inverse Building Energy Analysis Models: Final Report," submitted to ASHRAE under Research Project 1050-RP, University of Dayton and Energy Systems Laboratory, (December).

43. Abushakra, B., Haberl, J., Claridge, D., and Sreshthaputra, A. 2001 "Compilation Of Diversity Factors And Schedules For Energy And Cooling Load Calculations; ASHRAE Research Project 1093: Final Report," submitted to ASHRAE under Research Project 1093-RP, Energy Systems Lab Report ESL-TR-00/06-01, Texas A&M University, 150 pages, (June).

44. MacDonald, J. and Wasserman, D. 1989. Investigation of Metered Data Analysis Methods for Commercial and Related Buildings, Oak Ridge National Laboratory Report No. ORNL/CON-279, (May).

45. Rabl, A. 1988. "Parameter Estimation in Buildings: Methods for Dynamic Analysis of Measured Energy Use," Journal of Solar Energy Engineering, Vol. 110, pp. 52-66.

46. Rabl, A., Riahle, A. 1992. "Energy Signature Model for Commercial Buildings: Test With Measured Data and Interpretation," Energy and Buildings, Vol. 19, pp. 143-154.

47. Gordon, J.M. and Ng, K.C. 1994. "Thermodynamic Modeling of Reciprocating Chillers," Journal of Applied Physics, Volume 75, No. 6, March 15, 1994, pp. 2769-2774.

48. Claridge, D.E., Haberl, J. S., Sparks, R., Lopez, R., Kissock, K. 1992. "Monitored Conu-nercial Building Energy Data: Reporting the Results." 1992 ASHRAE Transactions Symposium Paper, Vol. 98, Part 1, pp. 636-652.

49. Sonderegger, R. 1977, op.cit.

50. Subbarao, K., Burch, J., Hancock, C.E, 1990. "How to accurately measure the load coefficient of a residential building," Journal of Solar Energy Engineering, in preparation.

51. Reddy, A. 1989. "Application of Dynamic Building Inverse Models to Three Occupied Residences Monitored Non-intrusively," Proceedings of the Thermal Performance of Exterior Envelopes of Buildings IV, ASHRAE/DOE/BTECC/CIBSE.

52. Shurcliff, W.A. 1984. "Frequency Method of Analyzing a Building's Dynamic Thermal Performance, W.A. Shurcliff, 19 Appleton St., Cambridge, MA.

53. Dhar, A. 1995, "Development of Fourier Series and Artificial Neural Networks Approaches to Model Hourly Energy Use in Commercial Buildings," Ph.D. Dissertation, Mechanical Engineering Department, Texas A&M University, May.

54. Miller, R., and Seem, J. 199 1. "Comparison of Artificial Neural Networks with Traditional Methods of Predicting Return Time from Night Setback," ASHRAE Transactions, Vol. 97, Pt.2, pp. 500-508.

55. J.F. Kreider and X.A. Wang, (199 1). "Artificial Neural Networks Demonstration for Automated Generation of Energy Use Predictors for Commercial Buildings." ASHRAE Transactions, Vol. 97, part 1.

56. Kreider, J. and Haberl, J. 1994. "Predicting Hourly Building Energy Usage: The Great Energy Predictor Shootout: Overview and Discussion of Results," ASHRAE Transactions-Research, Volume 100, Part 2, pp. 1104 - 1118, (June).

57. ASHRAE 1997. Handbook of Fundamentals, Chapter 30: Energy Estimating and Modeling Methods, American Society of Heating Refrigeration Air-conditioning Engineers, Atlanta, GA., p. 30.27 (Copied with perinission).

58. ASHRAE 1997. op.cit., p. 30.28 (Copied with permission).

59. USDOE 1996. North American Energy Measurement and Verification Protocol (NEMVP), United States Department of Energy DOE/EE-0081, (March).

60. FEMP 1996. Standard Procedures and Guidelines for Verification of Energy Savings Obtained Under Federal Savings Performance Contracting Programs, USDOE Federal Energy Management Program (FEMP).

61. Haberl, J., Claridge, D., Turner, D., O'Neal, D., Heffington, W., Verdict, M. 2002. "LoanSTAR After 11 Years: A Report on the Successes and Lessons Learned From the LoanSTAR Program," Proceedings of the 2nd International Conference for Enhanced Building Operation," Richardson, Texas, pp. 131-138, (October).

62. USDOE 1997. International Performance Measurement and Verification Protocol (IPMVP), United States Department of Energy DOE/EE-0157, (December).

63. USDOE 2001. International Performance Measurement and Verification Protocol (IPMVP): Volume 1: Concepts and Options for Determining Energy and Water Savings, United States Department of Energy DOE/GO-102001-1187 (January).

64. USDOE 2001. International Performance Measurement and Verification Protocol (IPMVP): Volume 11: Concepts and Practices for Improved Indoor Environrnental Quality, United States Department of Energy DOE/GO-102001-1188 (January).

65. USDOE 2003. International Performance Measurement and Verification Protocol (IPMVP): Volume 11: Concepts and Practices for Improved Indoor Environmental Quality, United States Department of Energy DOE/GO-102001-1188 (January).

66. ASHRAE 2002.Guideline 14: Measurement of Energy and Demand Savings, American Society of Heating Refrigeration Air-conditioning Engineers, Atlanta, GA (September).

67. Hansen, S. 1993. Performance Contracting for Energy and Environmental Systems, Fairmont Press, Lilbum, GA, pp. 99-100.

68. ASHRAE 2002. op.cit., pp. 27-30.

69. Ibid, p. 30 (Copied with permission).
70. Brandemuehl et al. 1996. op.cit.
71. Wei, G. 1997. "A Methodology for In-situ Calibration of Steam Boiler Instrumentation," MS Thesis, Mechanical Engineering Department, Texas A&M University, August.
72. Dukelow, S.G. 1991. The Control of Boilers. Research Triangle Park, NC: Instrument Society of America.
73. Dyer, F.D. and Maples, G. 1981. Boiler Efficiency Improvement. Boiler Efficiency Institute. Auburn: AL.
74. Garcia-Borras, T. 1983. Manual for Improving Boiler and Furnace Performance. Houston, TX: Gulf Publishing Company.
75. Aschner, F.S. 1977. Planning Fundamentals of Thermal Power Plants. Jerusalem, Israel: Israel Universities Press.
76. ASME 1974 Performance Test for Steam Units — PTC 4. 1 a 1974.
77. Babcock and Wilcox. 1992. "Steam: Its generation and Use," Babcock and Wilcox, Barberton, Ohio, ISBN 0-9634570-0-4.
78. Katipamula, S., and Claridge, D. 1992. "Monitored Air Handler Performance and Comparison with a Simplified System Model," ASHRAE Transactions, Vol. 98, Pt 2., pp. 341-35 1.
79. Liu, M., and Claridge, D. 1995 "Application of Calibrated HVAC Systems to Identify Component Malfunctions and to Optimize the Operation and Control Schedules," ASME/JSME International Solar Energy Conference, pp. 209-217.
80. ASHRAE 2002. op.cit., p. 144.
81. Brandemuehl, et al. 1996. op. cit.
82. ASHRAE 2002, op cit., p. 144, (Copied with permission).
83. ASHRAE 2002. op.cit., pp. 144-147, (Copied with permission).
84. ASHRAE 2002. op.cit., p. 144.
85. Ibid, p. 148, (Copied with permission).
86. ASHRAE 2002. op.cit., pp. 147-149, (Copied with permission).
87. Gordon, J.M. and Ng, K.C. 1994. op.cit.
88. Gordon, J.M. and Ng, K.C., 1995. "Predictive and diagnostic aspects of a universal thermodynamic model for chillers," International Journal of Heat Mass Transfer, 38(5), p.807.
89. Gordon, J.M., Ng, K.C., and Chua, H.T., 1995. "Centrifugal chillers: thermodynamic modeling and a diagnostic case study," International Journal of Refrigeration, 18(4), p.253.
90. LBL. 1980. DOE-2 User Guide, Ver. 2. 1. Lawrence Berkeley Laboratory and Los Alamos National Laboratory, Rpt No. LBL-8689 Rev. 2; DOE-2 User Coordination Office, LBL, Berkeley, CA.
91. LBL. 198 1. DOE-2 Engineers Manual, Ver. 2. 1 A, Lawrence Berkeley Laboratory and Los Alamos National Laboratory, Rpt No. LBL-1 1353; DOE-2 User Coordination Office, LBL, Berkeley, CA.
92. LBL. 1982. DOE-2.1 Reference Manual Rev. 2.1A. Lawrence Berkeley Laboratory and Los Alamos National Laboratory, Rpt No. LBL-8706 Rev. 2; DOE-2 User Coordination Office, LBL, Berkeley, CA. LBL.
93. LBL. 1989. DOE-2 Supplement, Ver 2. 1 D. Lawrence Berkeley Laboratory, Rpt No. LBL-8706 Rev. 5 Supplement. DOE-2 User Coordination Office, LBL, Berkeley, CA.
94. Haberl, J. S., Reddy, T. A., Figueroa, I., Medina, M. 1997. "Overview of LoanSTAR Chiller Monitoring and Analysis of In-Situ Chiller Diagnostics Using ASHRAE RP827 Test Method," Proceedings of the PG&E Cool Sense National Integrated Chiller Retrofit Forum (September).
95. According to Gordon et al. 1995, fHx is a dimensionless term that is norrrially negligible.
96. Wei, G. 1997. op.cit.
97. Dukelow, S.G. 1991. op.cit.
9.8 Dyer, F.D. and Maples, G. 198 1. op.cit.
99. Garcia-Borras, T. 1983. op.cit.
100. Aschner, F.S. 1977. op.cit.
101. ASME 1974. op.cit.
102. Babcock and Wilcox. 1992. op.cit.
103. Haberl, J., Lynn, B., Underwood, D., Reasoner, J., Rury, K. 2003. op.cit.
104. Haberl, et al. 2003. ibid.
105. ASME, 1974. Power Test Codes (PTC) 4. la, Steam Generating Units. New York: ASME.
106. Stallard, G.S. and Jonas, T.S. 1996. Power Plant Engineering: Combustion Processes. New York: Chapman & Hall.
107. Payne, F.W. 1985. Efficient Boiler Operations Sourcebook. Atlanta, GA: The Fairmont Press.
108. Wei 1997. op.cit.
109. Wei 1997. ibid.
110. Aschner 1977. op.cit.
111. Babcock and Wilcox 1992. op.cit.
112. Dukelow 1991. op.cit.
113. Witte, L.C., Schmidt, P.S., and Brown, DR. 1988. Industrial Energy Management and Utilization. New York: Hemisphere Publishing Corporation.
114. Aschner 1977. op.cit.
115. Thumann, A. 1988. Guide to Improving Efficiency of Combustion Systems. Lilburn, GA: The Fairmont Press.
116. Wei 1997. op.cit.
117. Garcia-Borras, T. 1983. Manual for Improving Boiler and Furnace Performance. Houston, TX: Gulf Publishing Company.
118. Wei 1997. op.cit.
119. ASHRAE 2002, op.cit., pp. 154-156, (Copied with permission).
120. Bou Saada, T., Haberl, J., Vajda, J., and Harris, L. 1996. "Total Utility Savings From the 37,000 Fixture Lighting Retrofit to the USDOE Forrestal Building," Proceedings of the 1996 ACEEE Summery Study, (August).
121. Abushakra, B., Sreshthaputra, A., Haberl, J., and Claridge, D. 2001. "Compilation of Diversity Factors and Schedules for Energy and Cooling Load Calculations - Final Report," submitted to ASHRAE under Research Project 1093-RP, Energy Systems Lab Report ESL-TR-01/04-01, Texas A&M University, (April).
122. IESNA 2003. Lighting Handbook, 9h Edition, Illuminating Engineering Society of North America, New York, N.Y.
123. ASHRAE 2002, op.cit., pp. 156-159, (Copied with permission).
124. ASHRAE 2002, op. cit., p. 160, (Copied with permission).
125. Ayres, M., Stamper, E. 1995, op.cit.
126. ASHRAE 1969. Procedures for Determining Heating and Cooling Loads for Computerized Energy Calculations: Algorithms for Building Heat Transfer Sub-routines. M. Lokmanhekim, Editor, American Society of Heating Refrigeration Air-conditioning Engineers, Atlanta, GA.
127. ASHRAE 1971. Procedures for Simulating the Performance of Components and Systems for Energy Calculations. Stoecker, W.F. Stoecker, editor, 2 nd edition, American Society of Heating Refrigeration Airconditioning Engineers, Atlanta, GA
128. BLAST. 1993. BLAST Users Manual. BLAST Support Office, University of Illinois Urbana-Champaign.
129. LBL 1980, 1981, 1982, 1989, op.cit.
130. Kriebel, D.E., 1983. Simplified Energy Analysis Using the Modified Bin Method, American Society of Heating, Refrigerating and Air-Conditioning Engineers, Inc., Atlanta, Georgia.
131. ASHRAE 1999. op.cit.
132. ASHRAE 1993. op.cit.
133. Yuill, G., K., Haberl, J.S. 2002. Development of Accuracy Tests For Mechanical System Simulation. Final Report for ASHRAE Research Project 865-RP, The University of Nebraska at Lincoln, (July).
134. Katipamula, S. and Claridge, D.E., 1993. " Use of Simplified Systems Models to Measure Retrofit Savings," ASME Journal of Solar Energy Engineering, Vol. 115, pp. 57-68, May.
135. Liu, M. and Claridge , D.E., 1995. "Application of Calibrated HVAC System Models to Identify Component Malfunctions

and to Optimize the Operation and Control Schedules," Solar Engineering 1995, W.B. Stine, T. Tanaka and D.E. Claridge (Eds.), ASME/JSME/JSES International Solar Energy Conference, Maui, Hawaii, March.

136. Liu, M. and Claridge, D.E., 1998. "Use of Calibrated HVAC System Models to Optimize System Operation," Journal of Solar Energy Engineering, May 1998, Vol. 120.

137. Liu, M., Wei, G., Claridge, D., E., 1998, "Calibrating AHU Models Using Whole Building Cooling and Heating Energy Consumption Data," Proceedings of 1998 ACEEE Surrurier Study on Energy Efficiency in Buildings. Vol. 3.

138. Haberl, J., Claridge, D., Turner. D. 2000b. "Workshop on Energy Measurement, Verification and Analysis Technology," Energy Conservation Task Force, Federal Reserve Bank, Dallas, Texas (April).

139. This table contains material adapted from proposed HVAC System Testing Methods for ASHRAE Guideline 14-2002, which were not included in the published ASHRAE Guideline 14-2002.

140. Haberl et al. 2000b, op. cit.

141. Kissock et al. 2001. op.cit.

142. Fels 1986. op.cit.

143. Rabl 1988. op.cit.

144. Rabl and Raihle 1992. op.cit.

145. Claridge et al. 1992. op.cit.

146. Reddy, T.A., Haberl, J.S., Saman, N.F., Turner, W.D., Claridge, D.E., Chalifoux, A.T. 1997. "Baselining Methodology for Facility-Level Monthly Energy Use - Part 1: Theoretical Aspects," ASHRAE Transactions-Research, Volume 103, Part 2, pp. 336-347, (June).

147. Reddy, T.A., Haberl, J.S., Saman, N.F., Turner, W.D., Claridge, D.E., Chalifoux, A.T. 1997 "Baselining Methodology for Facility-Level Monthly Energy Use - Part 2: Application to Eight Army Installations," ASHRAE Transactions-Research, Volume 103, Part 2, pp. 348-359, (June).

148. Haberl, J., Thamilseran, S., Reddy, A., Claridge, D., O'Neal, D., Turner, D. 1998. "Baseline Calculations for Measuring and Verification of Energy and Demand Savings in a Revolving Loan Program in Texas," ASHRAE Transactions-Research, Volume 104, Part 2, pp. 841-858, (June).

149. Turner, D., Claridge, D., O'Neal, D., Haberl, J., Heffington, W., Taylor, D., Sifuentes, T. 2000. "Program Overview: The Texas LoanSTAR Program: 1989 - 1999 A 10-year Experience," Proceedings of the 2000 ACEEE Summery Study on Energy Efficiency in Buildings, Volume 4, pp. 4.365-4.376, (August).

150. Haberl, J., Sreshthatputra, A., Claridge, D., Turner, D. 2001. "Measured Energy Indices for 27 Office Buildings," Proceedings of the 1st International Conference for Enhanced Building Operation," Austin, Texas, pp. 185-200, (July).

151. Beasley, R., Haberl, J. 2002. "Development of a Methodology for Baselining The Energy Use of Large Multi-building Central Plants," ASHRAE Transactions-Research, Volume 108, Part 1, pp. 251-259, (January).

152. ASHRAE 2002. op.cit. p. 25, (Copied with permission).

153. Haberl et al. 2000b, op. cit.

154. Temperatures below zero are calculated as positive increases away from the change point temperature.

155. Thamilseran, S., Haberl, J. 1995. "A Bin Method for Calculating Energy Conservation Retrofits Savings in Commercial Buildings," Proceedings of the 1995 ASME/JSME/JSES International Solar Energy Conference, Lahaina, Maui, Hawaii, pp. 111- 124 (March). .

156. Thamilseran, S. 1999. "An Inverse Bin Methodology to Measure the Savings from Energy Conservation Retrofits in Commercial Buildings," Ph.D. Thesis, Mechanical Engineering Department, Texas A&M University, (May).

157. Kissock et al. 2001. op.cit.

158. Kissock, J.K, Xun,W., Sparks, R., Claridge, D., Mahoney, J. and

Haberl, J., 1994. "EModel Version 1.4de," Texas A&M University, Energy Systems Laboratory, Department of Mechanical Engineering, Texas A&M University, College Station, TX, December.

159. Abushakra et al. 2001. op.cit.

160. Haberl, J., Bou-Saada, T. 1998. "Procedures for Calibrating Hourly Simulation Models to Measured Building Energy and Environmental Data," ASME Journal of Solar Energy Engineering, Volume 120, pp. 193-204, (August).

161. Clarke, J.A, Strachan, P.A. and Pemot, C.. 1993. An Approach to the Calibration of Building Energy Simulation Models. ASHRAE Transactions. 99(2): 917-927.

162. Diamond, S.C. and Hunn, B.D.. 1981. Comparison of DOE-2 Computer Program Simulations to Metered Data for Seven Commercial Buildings. ASHRAE Transactions. 87(l): 1222-123 1.

163. Haberl, J., Bronson, D., Hinchey, S. and O'Neal, D. 1993. "Graphical Tools to Help Calibrate the DOE-2 Simulation Program to Non-weather Dependent Measured Loads," 1993 ASHRAE Journal, Vol. 35, No. 1, pp. 27-32, (January).

164. Haberl, J., Bronson, D. and O'Neal, D. 1995. "An Evaluation of the Impact of Using Measured Weather Data Versus TMY Weather Data in a DOE-2 Simulation of an Existing Building in Central Texas." ASHRAE Transactions Technical Paper no. 3921, Vol. 101, Pt. 2, (June).

165. Hinchey, S.B. 1991. Influence of Thermal Zone Assumptions on DOE-2 Energy Use Estimations of a Commercial Building. M.S. Thesis, Energy Systems Report No. ESL-TH-91/09-06, Texas A&M University, College Station, TX

166. Hsieh, E.S. 1988, Calibrated Computer Models of Commercial Buildings and Their Role in Building Design and Operation. M.S. Thesis, PU/CEES Report No. 230, Princeton University, Princeton, NJ.

167. Hunn, B.D., Banks, J.A. and Reddy, S.N. 1992. Energy Analysis of the Texas Capitol Restoration. The DOE-2 User News. 13 (4): 2- 10.

168. Kaplan, M.B., Jones, B. and Jansen, J. 1990a. DOE-2. I C Model Calibration with Monitored End-use Data. Proceedings from the ACEEE 1990 Summer Study on Energy Efficiency in Buildings, Vol. 10, pp. 10. 11510.125.

169. Kaplan, M.B., Caner, P. and Vincent, G.W. 1992. Guidelines for Energy Simulation of Commercial Buildings. Proceedings from the ACEEE 1992 Summer Study on Energy Efficiency in Buildings, Vol. 1, pp. 1.137-1.147.

170. Katipamula, S. and Claridge, D.E., 1993. " Use of Simplified Systems Models to Measure Retrofit Savings," ASME Journal of Solar Energy Engineering, Vol. 115, pp.57-68, May.

171. Liu, M. and Claridge, D.E., 1995. "Application of Calibrated HVAC System Models to Identify Component Malfunctions and to Optimize the Operation and Control Schedules," Solar Engineering 1995, W.B. Stine, T. Tanaka and D.E. Claridge (Eds.), ASME/JSME/JSES International Solar Energy Conference, Maui, Hawaii, March.

172. Liu, M. and Claridge, D. E., 1998. "Use of Calibrated HVAC System Models to Optimize System Operation," Journal of Solar Energy Engineering, May 1998, Vol. 120.

173. Liu, M., Wei, G., Claridge, D. E., 1998, "Calibrating AHU Models Using Whole Building Cooling and Heating Energy Consumption Data," Proceedings of 1998 ACEEE Summer Study on Energy Efficiency in Buildings. Vol. 3.

174. Manke, J., Hittle, D. and Hancock 1996. "Calibrating Building Energy Analysis Models Using Short Term Test Data," Proceedings of the 1996 International ASME Solar Energy Conference, p. 369, San Antonio, TX

175. McLain, H.A., Leigh, S.B., and MacDonald, J.M1993. Analysis of Savings Due to Multiple Energy Retrofits in a Large Office Building. Oak Ridge National Laboratory, ORNL Report No. ORNL/CON-363, Oak Ridge, TN.

176. Haberl et al. 2000b, op. cit.

177. ASHRAE 2002. op.cit. p. 35-43, (Copied with permission).

178. Bou-Saada, T. 1994. An Improved Procedure for Developing A Calibrated Hourly Simulation Model of an Electrically Heated and Cooled Commecial Building, Master's Thesis, Mechanical Engineering Department, Texas A&M University, (December), p. 54.

179. Sylvester, K., Song, S., Haberl, J., and Turner, D. 2002. Case Study: Energy Savings Assessment for the Robert E. Johnson State Office Building in Austin, Texas," IBPSA Newsletter, Vol. 12, Number 2, pp. 22-28, (Summer).

180. Huang & Associates. 1993. DrawBDL user's guide. 6720 Potrero Ave., El Cerrito, California, 94530.

181. Bou Saada, T. 1994. "An Improved Procedure for Developing A Calibrated Hourly Simulation Model of an Electrically Heated and Cooled Commercial Building," Master's Thesis, Mechanical Engineering Department, Texas A&M University, (December), p. 150.

182. Bou-Saada 1994. op. cit. p. 144.

183. C. Culp, K. Q. Hart, B. Turner, S. Berry-Lewis, 2003. "Cost Effective Measurement and Verification at Fairchild AFB, International Conference on Enhance Building Operation," Energy Systems Laboratory Report, Texas A&M University, (October).

184. Culp et al. 2003, ibid.

185. ASME 1974 op cit.

# CHAPTER 28

# GROUND-SOURCE HEAT PUMPS
# APPLIED TO COMMERCIAL BUILDINGS

**STEVEN A. PARKER, P.E., C.E.M.**
**DONALD L. HADLEY**
Energy Science and Technology Directorate
Pacific Northwest National Laboratory[1]
Richland, WA

## 28.1 ABSTRACT

Ground-source heat pumps can provide an energy-efficient, cost-effective way to heat and cool commercial facilities. While ground-source heat pumps are well established in the residential sector, their application in larger, commercial-style, facilities is lagging, in part because of limited experience with the technology by those in decision-making positions. Through the use of a ground-coupling system, a conventional water-source heat pump design is transformed to a unique means of utilizing thermodynamic properties of earth and groundwater for efficient operation throughout the year in most climates. In essence, the ground (or groundwater) serves as a heat source during winter operation and a heat sink for summer cooling. Many varieties in design are available, so the technology can be adapted to almost any site. Ground-source heat pump systems can be used widely in commercial-building applications and, with proper installation, offer great potential for the commercial sector, where increased efficiency and reduced heating and cooling costs are important. Ground-source heat pump systems require less refrigerant than conventional air-source heat pumps or air-conditioning systems, with the exception of direct-expansion-type ground-source heat pump systems.

Installation costs are relatively high but are offset by low maintenance and operating expenses and efficient energy use. The greatest barrier to effective use is improper design and installation; well-trained, experienced, and responsible designers and installers is of critical importance.

This chapter provides information and procedures that an energy manager can use to evaluate most ground-source heat pump applications. Ground-source heat pump operation, system types, design variations, energy savings, and other benefits are explained. Guidelines are provided for appropriate application and installation. Two case studies are presented to give the reader a sense of the actual costs and energy savings. A list of manufacturers and references for further reading are included for prospective users who have specific or highly technical questions not fully addressed in this chapter. Sample case spreadsheets are also provided.

## 28.2 BACKGROUND

This chapter is based on a Federal Technology Alert sponsored by the U.S. Department of Energy (DOE), Federal Energy Management Program (FEMP). The original Federal Technology Alert was published in 1995 and updated in 2001. The material was updated in 2005 to develop this chapter.

### 28.2.1 The DOE Federal Energy Management Program

The federal government is the largest energy consumer in the nation. Annually, in its 500,000 buildings and 8,000 locations worldwide, it uses nearly 1.4 quadrillion Btu (quads) of energy, costing approximately $8 billion. This represents 1.5% of all primary energy consumption in the United States[2]. The DOE Federal Energy Management Program was established in 1974 to provide direction, guidance, and assistance to federal agencies in planning and implementing energy management programs that will improve the energy efficiency and fuel flexibility of the federal infrastructure.

Over the years several federal laws and Executive Orders have shaped FEMP's mission. These include the Energy Policy and Conservation Act of 1975; the National Energy Conservation and Policy Act of 1978; the Federal

---

[1]Pacific Northwest National Laboratory is operated for the U.S. Department of Energy by Battelle Memorial Institute under contract DE-AC05-76RL01830.

[2]DOE. 2001. Annual Report to Congress on Federal Government Energy Management and Conservation programs, Fiscal Year 1999. DOE/EE-0252. U.S. Department of Energy. Washington, DC.

Energy Management Improvement Act of 1988; Executive Order 12759 in 1991; the National Energy Policy Act of 1992 (EPAct 1992); Executive Order 12902 in 1994; Executive Order 13123 in 1999; and the Energy Policy Act of 2005 (EPAct 2005).

The DOE Federal Energy Management Program is currently involved in a wide range of energy-assessment activities, including conducting new technology demonstrations, to hasten the penetration of energy-efficient technologies into the federal marketplace.

### 28.2.2 The FEMP New Technology Demonstrations Activity

The Energy Policy Act of 1992, and subsequent Executive Orders, mandated that energy consumption in federal buildings be reduced by 35% from 1985 levels by the year 2010. The Energy Policy Act of 2005 calls for even more energy reduction. To achieve this goal, the DOE Federal Energy Management Program sponsors a series of program activities to reduce energy consumption at federal installations nationwide. One of these program activities, new technology demonstrations, is tasked to accelerate the introduction of energy-efficient and renewable technologies into the federal sector and to improve the rate of technology transfer.

In addition to technology demonstrations, FEMP sponsors a series of publications that are designed to disseminate information on new and emerging technologies. These publications include:

**Federal Technology Alerts**—longer summary reports that provide details on energy-efficient, water-conserving, and renewable-energy technologies that have been selected for further study for possible implementation in the Federal sector. Additional information on Federal Technology Alerts is provided below.

**Technology Installation Reviews**— concise reports describing a new technology and providing case study results, typically from another demonstration program or pilot project.

**Technology Focuses**—brief information on new, energy-efficient, environmentally friendly technologies of potential interest to the Federal sector.

### 28.2.3 More on Federal Technology Alerts

Federal Technology Alerts provide summary information on candidate energy-saving technologies developed and manufactured in the United States. The technologies featured in the Federal Technology Alerts have already entered the market and have some experience but are not in general use in the Federal sector.

The goal of the Federal Technology Alerts is to improve the rate of technology transfer of new energy-saving technologies within the Federal sector by providing the right people in the field with accurate, up-to-date information on the new technologies so that they can make informed decisions on whether the technologies are suitable for their sites. The information in the Federal Technology Alerts typically includes a description of the candidate technology; a description of its performance, applications and field experience to date; a list of manufacturers; and important sources for additional information. Appendixes provide supplemental information and example worksheets for the technology.

FEMP sponsors publication of the Federal Technology Alerts to facilitate information sharing between manufacturers and government staff. While the technology featured promises potential Federal sector energy savings, the Federal Technology Alerts do not constitute FEMP's endorsement of a particular product, because FEMP has not independently verified performance data provided by manufacturers. Readers should note the publication date and consider the Federal Technology Alerts as an accurate picture of the technology and its performance at the time of publication. Product innovations and the entrance of new manufacturers or suppliers should be anticipated since the date of publication. FEMP encourages interested energy and facility managers to contact the manufacturers and other sites directly, and to use the worksheets in the Federal Technology Alerts to aid in their purchasing decisions.

## 28.3  INTRODUCTION TO GROUND-SOURCE HEAT PUMPS

Ground-source heat pumps are known by a variety of names: geoexchange heat pumps, ground-coupled heat pumps, geothermal heat pumps, earth-coupled heat pumps, ground-source systems, groundwater source heat pumps, well water heat pumps, solar energy heat pumps, and a few other variations. Some names are used to describe more accurately the specific application; however, most are the result of marketing efforts and the need to associate (or disassociate) the heat pump systems from other systems. This chapter refers to them as ground-source heat pumps except when it is necessary to distinguish a specific design or application of the technology. A typical ground-source heat pump system design applied to a commercial facility is illustrated in Figure 28.1.

It is important to remember that the primary equipment used for ground-source heat pumps are water-source heat pumps. What makes a ground-source heat pump different (unique, efficient, and usually more expensive to install) is the ground-coupling system. In

**Figure 28.1. Typical ground-source heat pump system applied to a commercial facility**

addition, most manufacturers have developed extended-range water-source heat pumps for use as ground-source heat pumps.[3]

A conventionally designed water-source heat pump system would incorporate a boiler as a heat source during the winter heating operation and a cooling tower to reject heat (heat sink) during the summer cooling operation. This system type is also sometimes called a boiler/tower water-loop heat pump system. The water loop circulates to all the water-source heat pumps connected to the system. The boiler (for winter operation) and the cooling tower (for summer operation) provide a fairly constant water-loop temperature, which allows the water-source heat pumps to operate at high efficiency.

A conventional air-source heat pump uses the outdoor ambient air as a heat source during the winter heating operation and as a heat sink during the summer cooling operation. Air-source heat pumps are subject to higher temperature fluctuations of the heat source and heat sink. They become much less effective—and less efficient—at extreme ambient air temperatures. This is particularly true at low temperatures. In addition, heat transfer using air as a transfer medium is not as effective as water systems because of air's lower thermal mass.

A ground-source heat pump uses the ground (or in some cases groundwater) as the heat source during the winter heating operation and as the heat sink during the summer cooling operation. Ground-source heat pumps may be subject to higher temperature fluctuations than conventional water-source heat pumps but not as high as air-source heat pumps. Consequently, most manufacturers have developed extended-range systems. The extended-range systems operate more efficiently while subject to the extended-temperature range of the water loop. Like water-source heat pumps, ground-source heat pumps use a water loop between the heat pumps and the heat source/heat sink (the earth). The primary exception is the direct-expansion ground-source heat pump, which is described in more detail later in this chapter.

Ground-source heat pumps take advantage of the thermodynamic properties of the earth and groundwater. Temperatures below the ground surface do not fluctuate significantly through the day or the year as do ambient air temperatures. Ground temperatures a few feet below the surface stay relatively constant throughout the year. For this reason, ground-source heat pumps remain extremely efficient throughout the year in virtually any climate.

## 28.4 ABOUT THE TECHNOLOGY

In 1999, an estimated 400,000 ground-source heat pumps were operating in residential and commercial ap-

---

[3]The extended-range designation is important. Conventional water-source heat pumps are designed to operate with a water-loop as a heat sink that maintains a narrow temperature range. Ground-source heat pumps, however, are typically required to operate with a water-loop heat sink under a wider range of temperatures.

plications, up from 100,000 in 1990. In 1985, it was estimated that only around 14,000 ground-source heat pump systems were installed in the United States. Annual sales of approximately 45,000 units were reported in 1997. With a projected annual growth rate of 10%, 120,000 new units would be installed in 2010, for a total of 1.5 million units in 2010 (Lund and Boyd 2000). In Europe, the estimated total number of installed ground-source heat pumps at the end of 1998 was 100,000 to 120,000 (Rybach and Sanner 2000). Nearly 10,000 ground-source heat pumps have been installed in U.S. Federal buildings, over 400 schools and thousands of low-income houses and apartments (ORNL/SERDP, no date).

Although ground-source heat pumps are used throughout the United States, the majority of new ground-source heat pump installations in the United States are in the southern and mid-western states (from North Dakota to Florida). Oklahoma, Texas, and the East Coast have been particularly active with new ground-source heat pump installations. Environmental concerns, particularly from the potential for groundwater contamination with a leaking ground loop, and a general lack of understanding of the technology by HVAC companies and installers have limited installations in the West (Lund and Boyd 2000). Usually the technology does well in an area where it has been actively promoted by a local utility or the manufacturer.

Ground-source heat pumps are not a new idea. Patents on the technology date back to 1912 in Switzerland (Calm 1987). One of the oldest ground-source heat pump systems, in the United Illuminating headquarters building in New Haven, Connecticut, has been operating since the 1930s (Pratsch 1990). Although ground-source heat pump systems are probably better established today in rural and suburban residential areas because of the land area available for the ground loop, the market has expanded to urban and commercial applications.

The vast majority of ground-source heat pump installations utilize unitary equipment consisting of multiple water-source heat pumps connected to a common ground-coupled loop. Most individual units range from 1 to 10 tons (3.5 to 35.2 kW), but some equipment is available in sizes up to 50 tons (176 kW). Large-tonnage commercial systems are achieved by using several unitary water-source heat pumps, each responsible for an individual control zone.

One of the largest commercial ground-source heat pump systems is at Stockton College in Pomona, New Jersey, where 63 ground-source heat pumps totaling 1,655 tons (5,825 kW) are connected to a ground-coupled loop consisting of 400 wells, each 425 feet (129 m) deep (Gahran 1993).

Public schools are another good application for the ground-source heat pump technology with over 400 installations nationwide. In 1995, the Lincoln, Nebraska, Public School District built four new 70,000 square foot elementary schools. Space conditioning loads are met by 54 ground-source heat pumps ranging in size from 1.4 to 15 tons, with a total cooling capacity of 204 tons. Gas-fired boilers provided hot water for pre-heating of the outside air and for terminal re-heating. Compared with other similar new schools, these four ground-source heat pump conditioned facilities used approximately 26% less source energy per square foot of floor area (Shonder et al. 1999).

Multiple unitary systems are not the only arrangement suitable for large commercial applications. It is also possible to design large centralized heat-pump system consisting of reciprocating and centrifugal compressors (up to 19.5 million Btu/h) and to use these systems to support central-air-handling units, variable air-volume systems, or distributive two-pipe fan coil units.

### 28.4.1 How the Technology Works

Heat normally flows from a warmer medium to a colder one. This basic physical law can only be reversed with the addition of energy. A heat pump is a device that does so by essentially "pumping" heat up the temperature scale, then transferring it from a cold material to a warmer one by adding energy, usually in the form of electricity. A heat pump functions by using a refrigerant cycle similar to the household refrigerator. In the heating mode, a heat pump removes the heat from a low temperature source, such as the ground or air, and supplies that heat to a higher temperature sink, such as the heated interior of a building. In the cooling mode, the process is reversed and the heat is extracted from the cooler inside air and rejected to the warmer outdoor air or other heat sink. For space conditioning of buildings, heat pumps that remove heat from outdoor air in the heating mode and reject it to outdoor air in the cooling mode are common. These are normally called air-source or air-to-air heat pumps. Air-source heat pumps have the disadvantage that the greatest requirement for building heating or cooling is necessarily coincident with the times when the outdoor air is least effective as a heat source or sink. Below about 37°F (2.8°C), supplemental heating is required to meet the heating load. For this reason, air-source heat pumps are essentially unfeasible in cold climates with outdoor temperatures below 37°F (2.8°C) for extended periods of time.

The efficiency of any heat pump is inversely proportional to the temperature difference between the conditioned space and the heat source (heating mode) or heat sink (cooling mode) as can be easily shown by a simple thermodynamic analysis (Reynolds and Perkins 1977). For this reason, air-source heat pumps are less efficient and have a lower heating capacity in the heating mode at low outdoor air temperatures. Conversely, air-source heat pumps are also less efficient and have a lower cooling capacity in the cooling mode at high outdoor air temperatures. Ground-source heat pumps, however, are not impacted directly by outdoor air temperatures. Ground-source heat pumps use the ground, groundwater, or surface water, which are more thermally stable and not subjected to large annual swings of temperature, as a heat source or sink.

### 28.4.2 Other benefits

The primary benefit of ground-source heat pumps is the increase in operating efficiency, which translates to a reduction in heating and cooling costs, but there are additional advantages. One notable benefit is that ground-source heat pumps, although electrically driven, are classified as a renewable-energy technology. The justification for this classification is that the ground acts as an effective collector of solar energy. The renewable-energy classification can affect federal goals and potential federal funding opportunities.

An environmental benefit is that ground-source heat pumps typically use 25% less refrigerant than split-system air-source heat pumps or air-conditioning systems. Ground-source heat pumps generally do not require tampering with the refrigerant during installation. Systems are generally sealed at the factory, reducing the potential for leaking refrigerant in the field during assembly.

Ground-source heat pumps also require less space than conventional heating and cooling systems. While the requirements for the indoor unit are about the same as conventional systems, the exterior system (the ground coil) is underground, and there are no space requirements for cooling towers or air-cooled condensers. In addition, the ground-coupling system does not necessarily limit future use of the land area over the ground loop, with the exception of siting a building. Interior space requirements are also reduced. There are no floor space requirements for boilers or furnaces, just the unitary systems and circulation pumps. Furthermore, many distributed ground-source heat pump systems are designed to fit in ceiling plenums, reducing the floor space requirement of central mechanical rooms.

Compared with air-source heat pumps that use outdoor air coils, ground-source heat pumps do not require defrost cycles or crankcase heaters and there is virtually no concern for coil freezing. Cooling tower systems require electric resistance heaters to prevent freezing in the tower basin, also not necessary with ground-source heat pumps.

It is generally accepted that maintenance requirements are also reduced, although research continues directed toward verifying this claim. It is clear, however, that ground-source heat pumps eliminate the exterior fin-coil condensers of air-cooled refrigeration systems and eliminate the need for cooling towers and their associated maintenance and chemical requirements. This is a primary benefit cited by facilities in highly corrosive areas such as near the ocean, where salt spray can significantly reduce outdoor equipment life.

Ground-source heat pump technology offers further benefits: less need for supplemental resistance heaters, no exterior coil freezing (requiring defrost cycles) such as that associated with air-source heat pumps, improved comfort during the heating season (compared with air-source heat pumps, the supply air temperature does not drop when recovering from the defrost cycle), significantly reduced fire hazard over that associated with fossil fuel-fired systems, reduced space requirements and hazards by eliminating fossil-fuel storage, and reduced local emissions from those associated with other fossil fuel-fired heating systems.

Another benefit is quieter operation, because ground-source heat pumps have no outside air fans. Finally, ground-source heat pumps are reliable and long-lived, because the heat pumps are generally installed in climate-controlled environments and therefore are not subject to the stresses of extreme temperatures. Because of the materials and joining techniques, the ground-coupling systems are also typically reliable and long-lived. For these reasons, ground-source heat pumps are expected to have a longer life and require less maintenance than alternative (more conventional) technologies.

### 28.4.3 Ground-Coupled System Types

The ground-coupling systems used in ground-source heat pumps fall under three main categories: closed-loop, open-loop and direct-expansion. These are illustrated in Figure 28.2 and discussed in the following sections. The type of ground coupling employed will affect heat pump system performance (therefore the heat pump energy consumption), auxiliary pumping energy requirements, and installation costs. Choice of the most appropriate type of ground coupling for a site is usually a function of specific geography, available land area, and life-cycle cost economics.

#### Closed-loop Systems

Closed-loop systems consist of an underground network of sealed, high-strength plastic pipe[4] acting as a heat exchanger. The loop is filled with a heat transfer fluid, typically water or a water-antifreeze[5] solution, although other heat transfer fluids may be used.[6] When cooling requirements cause the closed-loop liquid temperature to rise, heat is transferred to the cooler earth. Conversely, when heating requirements cause the closed-loop fluid temperature to drop, heat is absorbed from the warmer earth. Closed-loop systems use pumps to circulate the heat transfer fluid between the heat pump and the ground loop. Because the loops are closed and sealed, the heat pump heat exchanger is not subject to mineral buildup and there is no direct interaction (mixing) with groundwater.

There are several varieties of closed-loop configurations including horizontal, spiral, vertical, and submerged.

**Horizontal Loops**

Horizontal loops, illustrated in Figure 28.2a, are often considered when adequate land surface is available. The pipes are placed in trenches, typically at a depth of 4 to 10 feet (1.2 to 3.0 m). Depending on the specific design, from one to six pipes may be installed in each trench. Although requiring more linear feet of pipe, multiple-pipe configurations conserve land space, require less trenching, and therefore frequently cost less to install than single-pipe configurations. Trench lengths can range from 100 to 400 feet per system cooling ton (8.7 to 34.6 m/kW), depending on soil characteristics and moisture content, and the number of pipes in the trench. Trenches are usually spaced from 6 to 12 feet (1.8 to 3.7 m) apart.

These systems are common in residential applications but are not frequently applied to large-tonnage commercial applications because of the significant land area required for adequate heat transfer. The horizontal-loop systems can be buried beneath lawns, landscaping, and parking lots. Horizontal systems tend to be more popular where there is ample land area with a high water table.

---

[4]Acceptable piping includes high quality polyethylene or polybutylene. PVC is not acceptable in either heat transfer characteristics or strength.
[5]Common heat transfer fluids include water or water mixed with an antifreeze, such as: sodium chloride, calcium chloride, potassium carbonate, potassium acetate, ethylene glycol, propylene gycol, methyl alcohol, or ethyl alcohol.
[6]Note that various heat transfer fluids have different densities and thermodynamic properties. Therefore, the heat transfer fluid selected will affect the required pumping power and the amount of heat transfer pipe. Furthermore, some local regulations may limit the selection and use of certain antifreeze solutions.

Figure 28.2. Ground-coupling system types

- **Advantages**: Trenching costs typically lower than well-drilling costs; flexible installation options.

- **Disadvantages**: Large ground area required; ground temperature subject to seasonal variance at shallow depths; thermal properties of soil fluctuate with season, rainfall, and burial depth; soil dryness must be properly accounted for in designing the required pipe length, especially in sandy soils and on hilltops that may dry out during the summer; pipe system could be damaged during backfill process; longer pipe lengths are required than for vertical wells; antifreeze solution viscosity increases pumping energy, decreases the heat transfer rate, and thus reduces overall efficiency; lower system efficiencies.

**Spiral Loops**

A variation on the multiple pipe horizontal-loop configuration is the spiral loop, commonly referred to as the "slinky." The spiral loop, illustrated in Figure 28.2b, consists of pipe unrolled in circular loops in trenches; the horizontal configuration is shown.

Another variation of the spiral-loop system involves placing the loops upright in narrow vertical trenches. The spiral-loop configuration generally requires more piping, typically 500 to 1,000 feet per system cooling ton (43.3 to 86.6 m/kW) but less total trenching than the multiple horizontal-loop systems described above. For the horizontal spiral-loop layout, trenches are generally 3 to 6 feet (0.9 to 1.8 m) wide; multiple trenches are typically spaced about 12 feet (3.7 m) apart. For the vertical spiral-loop layout, trenches are generally 6 inches (15.2 cm) wide; the pipe loops stand vertically in the narrow trenches. In cases where trenching is a large component of the overall installation costs, spiral-loop systems are a means of reducing the installation cost. As noted with horizontal systems, slinky systems are also generally associated with lower-tonnage systems where land area requirements are not a limiting factor.

- **Advantages**: Requires less ground area and less trenching than other horizontal loop designs; installation costs sometimes less than other horizontal loop designs.

- **Disadvantages**: Requires more total pipe length than other ground-coupled designs; relatively large ground area required; ground temperature subject to seasonal variance; larger pumping energy requirements than other horizontal loops defined above; backfilling the trench can be difficult with certain soil types and the pipe system could be damaged during backfill process.

## Vertical Loops

Vertical loops, illustrated in Figure 28.2c, are generally considered when land surface is limited. Wells are bored to depths that typically range from 75 to 300 feet (22.9 to 91.4 m) deep. The closed-loop pipes are inserted into the vertical well. Typical piping requirements range from 200 to 600 feet per system cooling ton (17.4 to 52.2 m/kW), depending on soil and temperature conditions. Multiple wells are typically required with well spacing not less than 15 feet (4.6 m) in the northern climates and not less than 20 feet (6.1 m) in southern climates to achieve the total heat transfer requirements. A 300- to 500-ton capacity system can be installed on one acre of land, depending on soil conditions and ground temperature.

There are three basic types of vertical-system heat exchangers: U-tube, divided-tube, and concentric-tube (pipe-in-pipe) system configurations.

- **Advantages**: Requires less total pipe length than most closed-loop designs; requires the least pumping energy of closed-loop systems; requires least amount of surface ground area; ground temperature typically not subject to seasonal variation.

- **Disadvantage**: Requires drilling equipment; drilling costs frequently higher than horizontal trenching costs; some potential for long-term heat buildup underground with inadequately spaced bore holes.

## Submerged Loops

If a moderately sized pond or lake is available, the closed-loop piping system can be submerged, as illustrated in Figure 28.2d. Some companies have installed ponds on facility grounds to act as ground-coupled systems; ponds also serve to improve facility aesthetics. Submerged-loop applications require some special considerations, and it is best to discuss these directly with an engineer experienced in the design applications. This type of system requires adequate surface area and depth to function adequately in response to heating or cooling requirements under local weather conditions. In general, the submerged piping system is installed in loops attached to concrete anchors. Typical installations require around 300 feet of heat transfer piping per system cooling ton (26.0 m/kW) and around 3,000 square feet of pond surface area per ton (79.2 m$^2$/kW) with a recommended minimum one-half acre total surface area. The concrete anchors act to secure the piping, restricting movement, but also hold the piping 9 to 18 inches (22.9 to 45.7 cm) above the pond floor, allowing for good convective flow of water around the heat transfer surface area. It is also recommended that the heat-transfer loops be at least 6 to 8 feet (1.8 to 2.4 m) below the pond surface, preferably deeper. This maintains adequate thermal mass even in times of extended drought or other low-water conditions. Rivers are typically not used because they are subject to drought and flooding, both of which may damage the system.

- **Advantages**: Can require the least total pipe length of closed-loop designs; can be less expensive than other closed-loop designs if body of water available.

- **Disadvantage**: Requires a large body of water and may restrict lake use (i.e., boat anchors).

## Open-Loop Systems

Open-loop systems use local groundwater or surface water (i.e., lakes) as a direct heat transfer medium instead of the heat transfer fluid described for the closed-loop systems. These systems are sometimes referred to specifically as "ground-water-source heat pumps" to distinguish them from other ground-source heat pumps. Open-loop systems consist primarily of extraction wells, extraction and reinjection wells, or surface water systems. These three types are illustrated in Figures 28.2e, 28.2f, and 28.2g, respectively.

A variation on the extraction well system is the standing column well. This system reinjects the majority of the return water back into the source well, minimizing the need for a reinjection well and the amount of surface discharge water.

There are several special factors to consider in open-loop systems. One major factor is water quality. In open-loop systems, the primary heat exchanger between the refrigerant and the groundwater is subject to fouling, corrosion, and blockage. A second major factor is the adequacy of available water. The required flow rate through the primary heat exchanger between the refrigerant and the groundwater is typically between 1.5 and 3.0 gallons per minute per system cooling ton (0.027 and 0.054 L/s-kW). This can add up to a significant amount of water and can be affected by local water resource regulations. A third major factor is what to do with the discharge stream. The groundwater must either be re-injected into the ground by separate wells or discharged to a surface system such as a river or lake. Local codes and regulations may affect the feasibility of open-loop systems.

Depending on the well configuration, open-loop systems can have the highest pumping load requirements of any of the ground-coupled configurations. In ideal conditions, however, an open-loop application can be the most economical type of ground-coupling system.

- **Advantages**: Simple design; lower drilling requirements than closed-loop designs; subject to better thermodynamic performance than closed-loop systems because well(s) are used to deliver groundwater at ground temperature rather than as a heat exchanger delivering heat transfer fluid at temperatures other than ground temperature; typically lowest cost; can be combined with potable water supply well; low operating cost if water already pumped for other purposes, such as irrigation.

- **Disadvantages**: Subject to various local, state, and Federal clean water and surface water codes and regulations; large water flow requirements; water availability may be limited or not always available; heat pump heat exchanger subject to suspended matter, corrosive agents, scaling, and bacterial contents; typically subject to highest pumping power requirements; pumping energy may be excessive if the pump is oversized or poorly controlled; may

require well permits or be restricted for extraction; water disposal can limit or preclude some installations; high cost if reinjection well required.

**Direct-Expansion Systems**

Each of the ground-coupling systems described above uses an intermediate heat transfer fluid to transfer heat between the earth and the refrigerant. Use of an intermediate heat transfer fluid necessitates a higher compression ratio in the heat pump to achieve sufficient temperature differences in the heat transfer chain (refrigerant to fluid to earth). Each also requires a pump to circulate water between the heat pump and the ground-couple. Direct-expansion systems, illustrated in Figure 28.2h, remove the need for an intermediate heat transfer fluid, the fluid-refrigerant heat exchanger, and the circulation pump. Copper coils are installed underground for a direct exchange of heat between refrigerant and earth. The result is improved heat transfer characteristics and thermodynamic performance.

The coils can be buried either in deep vertical trenches or wide horizontal excavations. Vertical trenches typically require from 100 to 150 square feet of land surface area per system cooling ton (2.6 to 4.0 m$^2$/kW) and are typically 9 to 12 feet (2.7 to 3.7 m) deep. Horizontal installations typically require from 450 to 550 square feet of land area per system cooling ton (11.9 to 14.5 m$^2$/kW) and are typically 5 to 10 feet (1.5 to 3.0 m) deep. Vertical trenching is not recommended in sandy, clay or dry soils because of the poor heat transfer.

Because the ground coil is metal, it is subject to corrosion (the pH level of the soil should be between 5.5 and 10, although this is normally not a problem). If the ground is subject to stray electric currents and/or galvanic action, a cathodic protection system may be required. Because the ground is subject to larger temperature extremes from the direct-expansion system, there are additional design considerations. In winter heating operation, the lower ground coil temperature may cause the ground moisture to freeze. Expansion of the ice buildup may cause the ground to buckle. Also, because of the freezing potential, the ground coil should not be located near water lines. In the summer cooling operation, the higher coil temperatures may drive moisture from the soil. Low moisture content will change soil heat transfer characteristics.

At the time this chapter was initially drafted (1995), only one U.S. manufacturer offered direct-expansion ground-source heat pump systems. However, new companies have released similar direct-expansion systems. In November 2005, the Geothermal Heat Pump Consortium web site identified four manufacturers of direct exchange systems. Systems were available from 16,000 to 83,000 Btu/h (heating/cooling capacity) (4.7 to 24.3 kW). Larger commercial applications would require multiple units with individual ground coils.

- **Advantages**: Higher system efficiency; no circulation pump required.

- **Disadvantages**: Large trenching requirements for effective heat transfer area; ground around the coil subject to freezing (may cause surface ground to buckle and can freeze nearby water pipes); copper coil should not be buried near large trees where root system may damage the coil; compressor oil return can be complicated, particularly for vertical heat exchanger coils or when used for both heating and cooling; leaks can be catastrophic; higher skilled installation required; installation costs typically higher; this system type requires more refrigerant than most other systems; smaller infrastructure in the industry.

### 28.4.4 Variables Affecting Design and Performance

Among the variables that have a major impact on the sizing and effectiveness of a ground-coupling system, the importance of underground soil temperatures and soil type deserves special mention.

### Underground Soil Temperature

The soil temperature is of major importance in the design and operation of a ground-source heat pump. In an open-loop system, the temperature of groundwater entering the heat pump has a direct impact on the efficiency of the system. In a closed-loop system and in the direct-expansion system, the underground temperature will affect the size of the required ground-coupling system and the resulting operational effectiveness of the underground heat exchanger. Therefore, it is important to determine the underground soil temperature before selecting a system design.

Annual air temperatures, moisture content, soil type, and ground cover all have an impact on underground soil temperature. In addition, underground temperature varies annually as a function of the ambient surface air temperature swing, soil type, depth, and time lag. Figure 28.3 contains a map of the United States indicating mean annual underground soil temperatures and amplitudes of annual surface ground temperature swings. Figure 28.4, though for a specific location, illustrates how the annual soil temperature varies with depth, soil type, and season. For vertical ground-loop systems, the mean annual earth temperature (Figure 28.3a) is an important factor in the

ground-loop design. With horizontal ground-loop systems, the ground surface annual temperature variation (Figure 28.3b and Figure 28.4b) becomes an important design consideration.

### Soil and Rock Classification

The most important factor in the design and successful operation of a closed-loop ground-source heat pump system is the rate of heat transfer between the closed-loop ground-coupling system and the surrounding soil and rock. The thermal conductivity of the soil and rock is the critical value that determines the length of pipe required. The pipe length, in turn, affects the installation cost as well as the operational effectiveness, which in turn affects the operating cost. Because of local variations in soil type and moisture conditions, economic designs may vary by location. Soil classifications include coarse-grained sands and gravels, fine-grained silts and clays, and loam (equal mixtures of sand, silt, and clay). Rock classifications are broken down into nine different petrologic groups. Thermal conductivity values vary significantly within each of the nine groups. Each of these classifications plays a role in determining the thermal conductivity and thereby affects the design of the ground-coupling system. For more information on the thermal properties of soils and rocks and how to identify the different types of soils and rocks, see Soil and Rock Classification for the Design of Ground-Coupled Heat Pump Systems (STS Consultants 1989).

### Series versus Parallel Flow

Closed-loop ground-coupled heat exchangers may be designed in series, parallel, or a combination of both. In series systems, the heat transfer fluid can take only one path through the loop, whereas in parallel systems the fluid can take two or more paths through the circuit. The selection will affect performance, pumping requirements, and cost. Small-scale ground-coupling systems can use either series or parallel-flow design, but most large ground-coupling systems use parallel-flow systems. The advantages and disadvantages of series and parallel systems are summarized below. In large systems, pressure drop and pumping costs need to be carefully considered or they will be very high. Variable-speed drives can be used to reduce pumping energy and costs during part-load conditions. Total life-cycle cost and design limitations should be used to design a specific system.

- **Series-System Advantages**: Single path flow and pipe size; easier air removal from the system; slightly higher thermal performance per linear foot of

(a) Mean earth temperature, $T_M$ (°F)

**Figure 28.4. Soil temperature variation Source: OSU (1988)**

(b) Earth surface temperature amplitude, $A_S$ (°F)

**Figure 28.3. Mean annual soil temperatures. Source: OSU (1988)**

pipe because larger pipe size required in the series system.

- **Series-System Disadvantages**: Larger fluid volume of larger pipe in series requires greater antifreeze volumes; higher pipe cost per unit of performance; increased installed labor cost; limited capacity (length) caused by fluid pressure drop characteristics; larger pressure drop resulting in larger pumping load; requires larger purge system to remove air from the piping network.

- **Parallel-System Advantages**: Smaller pipe diameter has lower unit cost; lower volume requires less antifreeze; smaller pressure drop resulting in smaller pumping load; lower installation labor cost.

- **Parallel-System Disadvantages**: Special attention required to ensure air removal and flow balancing between each parallel path to result in equal length loops.

### 28.4.5 Variations

The ground-coupling system is what makes the ground-source heat pump unique among heating and air-conditioning systems and, as described above, there are several types of ground-coupling systems. In addition, variations to ground-source heat pump design and installation can save additional energy or reduce installation costs. Three notable variations are described below.

## Cooling-Tower-Supplemented System

The ground-coupling system is typically the largest component of the total installation cost of a ground-source heat pump. In southern climates or in thermally heavy commercial applications where the cooling load is the driving design factor, supplementing the system with a cooling tower or other supplemental heat rejection system can reduce the required size of a closed-loop ground-coupling system. The supplemental heat rejection system is installed in the loop by means of a heat exchanger (typically a plate and frame heat exchanger) between the facility load and the ground couple. A cooling tower system is illustrated in Figure 28.5. The cooling tower acts to pre-cool the loop's heat transfer fluid upstream of the ground couple, which lowers the cooling-load requirement on the ground-coupling system. By significantly reducing the required size of the ground-coupling system, using a cooling tower can lower the overall installation cost. This type of system is operating successfully in several commercial facilities, including some mission-critical facilities at Fort Polk in Louisiana.

## Solar-Assisted System

In northern climates where the heating load is the driving design factor, supplementing the system with solar heat can reduce the required size of a closed-loop ground-coupling system. Solar panels, designed to heat water, can be installed into the ground-coupled loop (by means of a heat exchanger or directly), as illustrated in Figure 28.6. The panels provide additional heat to the heat transfer fluid. This type of variation can reduce the required size of the ground-coupled system and increase heat pump efficiency by providing a higher temperature heat transfer fluid.

## Hot Water Recovery/Desuperheating

The use of heat pumps to provide hot water is becoming common. Because of their high efficiency, this practice makes economic sense. Most manufacturers offer an option to include desuperheating heat exchangers to provide hot water from a heat pump. These dual-wall heat exchangers are installed in the refrigerant loop to recover high temperature heat from the superheated refrigerant gas. Hot-water recovery systems can supplement, or sometimes replace, conventional facility water-heating systems. With the heat pump in cooling mode, hot-water recovery systems increase system operating efficiency while acting as a waste-heat-recovery device—and provide essentially free hot water. When the load is increased during the heating mode, the heat pump still provides heating and hot water more efficiently and less expensively than other systems.

### 28.4.6 System Design and Installation

More is becoming known about the design and installation of ground-source heat pumps. Design-day cooling and heating loads are determined through traditional design practices such as those documented by the American Society of Heating, Refrigerating, and Air-Conditioning Engineers (ASHRAE). Systems are also zoned using commonly accepted design practices.

The key issue that makes ground-source heat pumps unique is the design of the ground-coupling system. Most operational problems with ground-source heat pumps

**Figure 28.5. Cooling-tower-supplemented system for cooling-dominated loads**

**Figure 28.6. Solar-assisted system for heating-dominated loads**

stem from the performance of the ground-coupling system. Today, software tools are available to support the design of the ground-coupling systems that meet the needs of designers and installers. These tools are available from several sources, including the International Ground-Source Heat Pump Association (IGSHPA). In addition, several manufacturers have designed their own proprietary tools more closely tuned to their particular system requirements.

Ground loops can be placed just about anywhere—under landscaping, parking lots, or ponds. Selection of a particular ground-coupling system (vertical, horizontal, spiral, etc.) should be based on life-cycle cost of the entire system, in addition to practical constraints. Horizontal closed-loop ground-coupling systems can be installed using a chain-type trenching machine, horizontal boring machine, backhoe, bulldozer, or other earth-moving heavy equipment. Vertical applications (for both open and closed systems) require a drilling rig and qualified operators. Most applications of ground-source heat pumps to large facilities use vertical closed-loop ground-coupling systems primarily because of land constraints. Submerged-loop applications require some special considerations and, as noted earlier, it is best to discuss these directly with an experienced design engineer.

It is important to assign overall responsibility for the entire ground-source heat pump system to a single individual or contractor. Installation of the system, however, will involve several trades and contractors, many of whom may not have worked together in previous efforts. In addition to refrigeration/air-conditioning and sheet metal contractors, installation involves plumbers, and (in the case of vertical systems) well drillers. Designating a singular responsible party and coordinating activities will significantly reduce the potential for problems with installation, startup, and proper operation.

In heating-dominated climates, a mixture of antifreeze and water must be used in the ground-coupling loops if loop temperatures are expected to fall below about 41°F (5°C). A study by Heinonen (1997) establishes the important considerations for antifreeze solutions for ground-source heat pump systems and provides guidance on selection.

One note of caution to the designer: some regulations, installation manuals, and/or local practices call for partial or full grouting of the borehole. The thermal conductivity of materials normally used for grouting is very low compared with the thermal conductivity of most native soil formations. Thus, grouting tends to act as insulation and hinders heat transfer to the ground. Some experimental work by Spilker (1998) has confirmed the negative impact of grout on borehole heat transfer. Under

heat rejection loading, average water temperature was nearly 11°F (6°C) higher for a 6.5-in. (16.5-cm) diameter borehole backfilled with standard bentonite grout than for a 4.75-in. (12.1-cm) diameter borehole backfilled with thermally enhanced bentonite grout. Using fine sand as backfill in a 6.5-in. (16.5-cm) diameter borehole lowered the average water temperature over 14°F (8°C) compared with the same-diameter bore backfilled with standard bentonite grout. For a typical system (Spilker 1998) with a 6.5-in. (16.5-cm) diameter borehole, the use of standard bentonite grout would increase the required bore length by 49% over fine sand backfill in the same borehole. By using thermally enhanced grout in a smaller 4.75-in. (12.1-cm) borehole, the bore length is increased by only 10% over fine sand backfill in the larger 6.5-in. (16.5-cm) diameter borehole. Thus, the results of this study (Spilker 1998) suggest three steps that may be taken to reduce the impact of grout on vertical borehole system performance:

- Reduce the amount of grout used to the bare minimum. Sand or cuttings may be used where allowed, but take care to ensure that the entire interstitial space between the piping and the borehole diameter is filled.

- Use thermally enhanced grout wherever possible. For information on thermally enhanced grout consult ASHRAE (1997) and Spilker (1998).

- Reduce the borehole diameter as much as possible to mitigate the effects of the grout or backfill used. The regulatory requirements for vertical boreholes used for ground-coupling heat exchangers vary widely by state. Current state and federal regulations, as well as related building codes, are summarized at the Geothermal Heat Pump Consortium web site (www.geoexchange.org/publications/regs.htm).

### 28.4.7 Summary of Ground-Loop Design Software

Because of the diversity in loads in multi-zone buildings, the design of the ground-coupling heat exchanger (the ground loop) must be based on peak block load rather than the installed capacity. This is of paramount importance because ground coupling is usually a major portion of the total ground-source heat pump system cost, and over-sizing will render a project economically unattractive.

In the residential sector, many systems have been designed using rules-of-thumb and local experience, but for commercial-scale systems such practices are ill advised. For all but the most northern climates, commercial-scale buildings will have significantly more heat rejection than extraction. This imbalance in heat rejection/extraction can cause heat buildup in the ground to the point where heat pump performance is adversely affected and hence system efficiency and possibly occupant comfort suffer. (This is an important consideration in producintg accurate life-cycle cost estimates of energy use.) Proper design for commercial-scale systems almost always benefits from the use of design software. Software for commercial-scale ground-source heat pump system design should consider the interaction of adjacent loops and predict the potential for long-term heat buildup in the soil. Some sources of PC-based design software packages that address this need are:

- **GchpCalc**, Version 3.1, Energy Information Services, Tel: (205) 799-4591. This program includes built-in tables for heat pump equipment from most manufacturers. Input is in the form of heat loss/gain during a design day and the approximate equivalent full-load heating hours and equivalent full-load cooling hours. Primary output from the program is the ground loop length required. This program will also calculate the optimal size for a supplemental fluid cooler for hybrid systems, as discussed later.

- **GLHEPRO**, International Ground Source Heat Pump Association (IGSHPA), Tel: (800) 626-4747. Input required is monthly heating/cooling loads on heat pumps and monthly peak loads either entered directly by user or read from BLAST or Trane System Analyzer and Trane Trace output files. Output includes long-term soil temperature effect from rejection/extraction imbalance. The current configuration of the program has some constraints on selection of borehole spacing, depth, and overall layout that will be removed from a future version now being prepared.

- **GS2000**, Version 2.0, Caneta Research Inc., Tel: (905) 542-2890, email: caneta@compuserve.com. Heating/cooling loads are input as monthly totals on heat pumps or, alternatively, monthly loads on the ground loop may be input. Equipment performance is input at ARI/ISO rating conditions. For operating conditions other than the rating conditions, the equipment performance is adjusted based on generic heat pump performance relationships.

Each of these programs requires input about the soil thermal properties, borehole resistance, type of piping and borehole arrangement, fluid to be used, and other design parameters. Many of the required inputs will be available from tables of default values. The designer should be careful to ensure that the values chosen are representative of the actual conditions to be encountered to ensure efficient and cost-effective designs. Test borings and *in situ* thermal conductivity analysis to determine the type of soil formations and aquifer locations will substantially improve design accuracy and may help reduce costs. Even with the information from test borings, some uncertainty will remain with respect to the soil thermal properties. These programs make it possible to vary design parameters easily within the range of anticipated values and determine the sensitivity of the design to a particular parameter (OTL 1999). In some instances, particularly very large projects, it may be advisable to obtain specific information on ground-loop performance by thermal testing of a sample borehole (Shonder and Beck 2000).

## 28.5 APPLICATION

This section addresses technical aspects of applying ground-source heat pumps. The range of applications and climates in which the technology has been installed are discussed. The advantages, limitations, and benefits are enumerated. Design and integration considerations for ground-source heat pumps are highlighted, including energy savings estimates, equipment warranties, relevant codes and standards, equipment and installation costs, and utility incentives.

### 28.5.1 Application Screening

A ground-source heat pump system is one of the most efficient technologies available for heating and cooling. It can be applied in virtually any climate or building category. Although local site conditions may dictate the type of ground-coupling system employed, the high first cost and its impact on the overall life-cycle cost are typically the constraining factors.

The operating efficiency of ground-source heat pumps is very dependent on the entering water temperature, which, in turn, depends on ground temperature, system load, and size of ground loop. As with any HVAC system, the system load is a function of the facility, internal activities, and the local weather. Furthermore, with ground-source heat pumps, the load on the ground-coupling system may impact the underground temperature. Therefore, energy consumption will be closely tied to the relationship between the annual load distribution and the annual ground loop-temperature distribution (e.g., their joint frequency distribution).

There are several techniques to estimate the annual energy consumption of ground-source heat pump systems. The most accurate methodologies use computer simulation, and several software systems now support the analysis of ground-source heat pumps. These methods, while more accurate than hand techniques, are also difficult and expensive to employ and are therefore more appropriate when additional detail is required rather than as an initial screening tool.

The bin method is another analytical tool for screening technology applications. In general, a bin method is a simple computational procedure that is readily adaptable to a spreadsheet-type analysis and can be used to estimate the energy consumption of a given application and climate. Bin methods rely on load and ambient wet and dry bulb temperature distributions. This methodology is used in the case study presented later in this chapter.

### 28.5.2 Where to Apply Ground-Source Heat Pumps

Ground-source heat pumps are generally applied to air-conditioning and heating systems, but may also be used in any refrigeration application. The decision whether to use a ground-source heat pump system is driven primarily by economics. Almost any HVAC system can be designed using a ground-source heat pump. The primary technical limitation is a suitable location for the ground-coupling system. The following list identifies some of the best applications of ground-source heat pumps.

- Ground-source heat pumps are probably least cost-prohibitive in new construction; the technology is relatively easy to incorporate.

- Ground-source heat pumps can also be cost effective to replace an existing system at the end of its useful life, or as a retrofit, particularly if existing ductwork can be reused with minimal modification.

- In climates with either cold winters or hot summers, ground-source heat pumps can operate much more efficiently than air-source heat pumps or other air-conditioning systems. Ground-source heat pumps are also considerably more efficient than other electric heating systems and, depending on the heating fuel cost, may be less expensive to operate than other heating systems.

- In climates with by high daily temperature swings, ground-source heat pumps show superior efficiency. In addition, in climates characterized by large daily temperature swings, the ground-coupling system also offers some thermal storage capability, which

may benefit the operational coefficient of performance.

- In areas where natural gas is not available or where the cost of natural gas or other fuel is high compared with electricity, ground-source heat pumps are economical. They operate with a heating coefficient of performance in the range of 3.0 to 4.5, compared with conventional heating efficiencies in the range of 80% to 97%. Therefore, when the cost of electricity (per Btu) is less than 3.5 times that of conventional heating fuels (per Btu), ground-source heat pumps have lower energy costs.

- Areas of high natural gas (or fuel oil) costs will favor ground-source heat pumps over conventional gas (or fuel oil) heating systems. High electricity costs will favor ground-source heat pumps over air-source heat pumps.

- In facilities where multiple temperature control zones or individual load control is beneficial, ground-source heat pumps provide tremendous capability for individual zone temperature control because they are primarily designed using multiple unitary systems.

- In areas where drilling costs are low, vertical-loop systems may be especially attractive. The initial cost of the ground-source heat pump system is one of the prime barriers to the economics. In locations with a significant ground-source heat pump industry infrastructure (such as Oklahoma, Louisiana, Florida, Texas, and Indiana), installation costs may be lower and the contractors more experienced. This, however, is changing as the market for ground-source heat pumps grows.

### 28.5.3 What to Avoid

The following precautions should be followed when the application of ground-source heat pump technology is considered:

- Avoid threaded plastic pipe connections in the ground loop. Specify thermal fusion welding. Unlike conventional water-source heat pump systems where the water loop temperature ranges from 60° to 90°F (15.6° to 32.2°C), ground-source systems are subject to wider temperature ranges (20° to 110°F [-6.7° to 43.3°C]), and the resulting expansion and contraction may result in leaks at the threaded connection. It is also generally recommended to specify piping and joining methods approved by International Ground-Source Heat Pump Association (IGSHPA).

- Check local water and well regulations. Regulations affecting open-loop systems are common, and local regulations can vary significantly. Some local regulations may require reinjection wells rather than surface drainage. Some states require permits to use even private ponds as a heat source/sink.

- Have the ground-source heat pump system installed as a complete and balanced assemblage of components, each of which must be properly designed, sized, and installed (Giddings 1988). Also, have the system installed under the responsibility of a single party. If the entire system is installed by three different professionals, none of whom understands or appreciates the other two parts of the system, then the system may not perform satisfactorily.

- One of the most frequent problems cited is improper sizing of the heat pump or the ground-coupling system. Approved calculation procedures should be used in the sizing process—as is the case with any heating or air-conditioning system regardless of technology. ASHRAE has established one of the most widely known and accepted standards for the determination of design heating and cooling loads. Sizing the ground-coupling system is just as critical. Because of the uncertainty of soil conditions, a site analysis to determine the thermal conductivity and other heat transfer properties of the local soil may be required. This should be the responsibility of the designing contractor because it can significantly affect the final design.

- Avoid inexperienced designers and installers (see above). Check on the previous experience of potential designers and installers. It is also generally recommended to specify IGSHPA certified installers.

### 28.5.4 Design and Equipment Integration

The purpose of this chapter is to familiarize the energy manager and facility engineer with the benefits and liabilities of ground-source heat pumps in their application to commercial buildings. It is beyond the scope of this chapter to fully explain the design requirements of a ground-source heat pump system. It is, however, important that the reader know the basic steps in the design process.

The design of a ground-source heat pump system will generally follow the following sequence:

1. Determine local design conditions, climatic and soil thermal characteristics.

2. Determine local water, well, and grouting requirements.

3. Determine building heating and cooling loads at design conditions.

4. Select the alternative HVAC system components, including the indoor air-distribution system type; size the alternatives as required; and select equipment that will meet the demands calculated in Step 2 (using the preliminary estimate of the entering water temperatures to determine the heat pump's heating and cooling capacities and efficiencies).

5. Determine the monthly and annual building heating and cooling energy requirements.

6. Make preliminary selection of a ground-coupling system type.

7. Determine a preliminary design of the ground-coupling system.

8. Determine the thermal resistance of the ground-coupling system.

9. Determine the required length of the ground-coupling system; recalculate the entering and exiting water temperatures on the basis of system loads and the ground-coupling system design.

10. Redesign the ground-coupling system, as required, to balance the requirements of the system load (heating and cooling) with the effectiveness of the ground-coupling system. Note that designing and sizing the ground-coupling system for one season (such as cooling) will impact its effectiveness and ability to meet system load requirements during the other season (such as heating).

11. Perform life-cycle cost analysis on the system design (or system design alternatives).

Although the design procedure for the ground-coupling system is an iterative and sometimes difficult process, several sources are available to simplify the task. First, an experienced designer should be assigned responsibility for the heat pump and ground-coupling system designs. Several manufacturers of ground-source heat pump equipment have their own software tools to support the design of large, commercial-type systems. However, for those who typically design systems in-house, there are support tools available. Software programs are available to support the design of ground-source heat pump HVAC systems and the ground-coupling system. Several software tools are available through the IGSHPA, including an Earth-Coupled Analysis Program and a Ground-Loop Heat Exchanger Design Program. In addition, several technical design manuals also are available through IGSHPA, ASHRAE, and equipment manufacturers (refer to earlier section for an introduction to ground-loop design software).

There are several different approaches for incorporating ground-source heat pumps into the HVAC design. However, most applications in large facilities involve multiple smaller heat pump units (<20 ton) applied in a modular zone control system and connected to a common water loop and associated ground-coupling system. Although some agencies are experimenting with larger equipment sizes, most manufacturers are supporting the development of efficient smaller systems (1/2 ton to 15 tons [1.8 to 52.8 kW]).

### 28.5.5 Equipment Warranties

The prospective user should ask potential suppliers, contractors, and installers about equipment warranties. The heat pump equipment is typically guaranteed free from manufacturer defects from 1 to 5 years. Some manufacturers offer extended warranties up to 10 years. Residential applications have been found to have longer warranties than commercial applications.

Warranties should also be requested for the ground-coupling system, which is less common. Some installers and pipe manufacturers have offered limited ground-

coupling system warranties as long as 50 years. Quality control in the installation of the ground-coupling system has been a concern in the industry. The IGSHPA now offers training and certification programs for installers and designers. In addition, the IGSHPA also administers a registration program for those organizations in the industry. These services have gone far to improve quality control and customer satisfaction.

The track records reported in the literature for the ground-coupling systems are good. In cases where pipe joints were thermally welded (fused) and followed the standards recommended by IGSHPA, systems have proven reliable and resistant to system leaks.

### 28.5.6 Energy Codes and Standards

Applications of ground-source heat pumps are subject to building and facility energy codes and standards. In addition, the equipment used is subject to commercial equipment energy codes and standards. Most energy regulations that impact commercial buildings derive from ASHRAE standards, specifically ASHRAE Standard 90.1. Minimum equipment efficiency standards, as identified in ASHRAE Standard 90.1-2004, for commercial equipment relative to ground-source heat pumps are shown in Table 28.1.

Although codes and standards identify minimum efficiencies, such as those identified above, they do not fully communicate the energy efficiency that is achieved by today's heat pumps. A review of manufacturer's literature on commercially available equipment indicates that cooling efficiencies[7] (EERs) of 13.4 to 20 Btu/W-h and heating efficiencies (COPs) of 3.1 to 4.3 are readily available.[8] When

---

[7]Rating based on ANSI/ARI/ASHRAE ISO Standard 13256-1-2005 for closed-loop ground-source heat pumps.
[8]For more information on the various terms used to define efficiency in HVAC systems (see Appendix).

**Table 28.1. Minimum Commercial Equipment Efficiency Ratings Standards (ASHRAE 90.1-2004)**

| Technology Application | Category (capacity) | Rating Condition* (entering water temperature) | Minimum Performance (ASHRAE 90.1-2004) | |
|---|---|---|---|---|
| | | | Cooling | Heating |
| Water-source heat pumps | < 17 kBtuh | 86°F (30°C) | 11.2 EER | - - - |
| | | 68°F (20°C) | - - - | 4.2 COP |
| | ≥ 17 kBtuh and < 65 kBtuh | 86°F (30°C) | 12.0 EER | - - - |
| | | 68°F (20°C) | - - - | 4.2 COP |
| | ≥ 65 kBtuh and < 135 kBtuh | 86°F (30°C) | 12.0 EER | - - - |
| | | 68°F (20°C) | - - - | 4.2 COP |
| Groundwater-source heat pumps | < 135 kBtuh | 59°F (15°C) | 16.2 EER | - - - |
| | | 50°F (10°C) | - - - | 3.6 COP |
| Ground-source heat pumps | < 135 kBtuh | 77°F (25°C) | 13.4 EER | - - - |
| | | 32°F (0°C) | - - - | 3.1 COP |

*Reference Standard: ANSI/ARI/ASHRAE ISO Standard 13256-1: Water-source Heat Pumps – Testing and Rating for Performance – Part 1: Water-to-Air and Brine-to-Air Heat Pumps

comparing equipment efficiencies, it is important to make an appropriate comparison. The efficiency of any heating or cooling equipment varies with application, load, and related heat-source and heat-sink temperatures.

As partially illustrated in Table 28.1, ISO Standard 13256-1 provides separate equipment rating conditions for water-source heat pumps, ground-water-source heat pumps, and closed-loop ground-source heat pumps. The difference among these rating conditions is in the application, not necessarily the equipment. Therefore, a given heat pump rated under closed-loop ground-source heat pump conditions would have a different EER (cooling) and COP (heating) than if the same heat pump were rated under water-source heat pump rating conditions and different still if the equipment were rated under the ground-water-source heat pump rating conditions.

Also, standard ratings are for specific temperatures and operating conditions. Standard ratings do not necessarily reflect the efficiency of systems under true seasonal operating conditions. In determining the efficiency of heating or cooling equipment for estimating energy consumption or potential energy savings, the efficiency should be corrected for the appropriate operating conditions.

### 28.5.7 Utility Incentives and Support

Many utilities are promoters of ground-source heat pumps, and many offer incentive programs and support. The Geothermal Heat Pump Consortium web site (www.geoexchange.org/incentives/incentives.htm) identifies federal and state incentives for ground-source heat pump systems for both residential and commercial applications. Incentives identified include rebates (by ton, by unit, and/or by kWh energy saved), low interest loans, sales tax exemptions, tax credits, and/or tax deductions. Rebates for ground-source heat pumps identified ranged from $150 to $600 per ton. For commercial facilities, some incentives ran into several thousand dollars. In addition to the technology-specific rebate programs, some utilities also offer custom rebate programs or programs that are not technology specific but are based on the energy savings regardless of the technology employed. Readers are encouraged to contact their local utility to find out more about what programs, services, and rebates may be offered by the utility to promote energy management and new energy-efficient technologies.

### 28.6  TECHNOLOGY PERFORMANCE

In 1999, an estimated 400,000 ground-source heat pumps were operating in residential and commercial applications, up from 100,000 in 1990. With a projected annual growth rate of 10%, 120,000 new units would be installed in 2010, for a total of 1.5 million units in 2010 (Lund and Boyd 2000). The majority of new ground-source heat pump installations in the United States are for residential applications in the southern and mid-western states. Environmental concerns and a general lack of understanding of the technology by HVAC companies and installers have limited installations in the west.

### 28.6.1 Field Experience

Observations about field performance of ground-source heat pumps obtained from federal and private-sector users are summarized in this section.

The large numbers of reported installations testify to the stability of this technology. Most sites contacted report satisfaction with the overall performance (energy efficiency, maintenance, and comfort) of the technology. One of the world's largest installations of ground-source heat pumps is at Fort Polk, Louisiana, where over 4000 units were installed in family housing as part of a major retrofit project (Hughes and Shonder 1998). They also have systems installed in several larger facilities, including some cooling-tower hybrid systems (discussed earlier in this chapter). Personnel at Fort Polk are obviously pleased with the performance of their ground-source heat pumps, and the energy manager reports that the systems have performed better than expected and they plan to use the technology for all heating and cooling requirements on the base.

The maintenance engineer at the hospital in Love County, Oklahoma, reported that no problems have occurred since nine units were installed in their facility in the early 1990s. They perform routine maintenance on the system. They particularly like the technology because it takes up less space and gives better heating and cooling control than former systems.

An interview with the energy manager primarily responsible for the ground-source heat pump installation at the Oklahoma State Capitol Building revealed no problems with the heat pumps or the ground loop; however, the building was poorly zoned, which impeded effective temperature control. Approval from the local Water Resource Board for drilling the wells proved difficult but was primarily a matter of educating the board about the technology and the closed-loop system. The Capitol building system, which is a cooling-tower-supplemented design, contains 855-cooling tons (3,009.6 kW) connected to a common ground-loop system consisting of 372 250-foot (76.2 m) vertical wells.

One facility manager did report problems with the ground-source heat pump system installed in a facility at the Dugway Proving Grounds, Utah. The system uses

22.5-ton (8.8-kW) units. Although the actual problem had not been identified, the system is not performing up to expectations. The belief was that the ground loop was undersized. Sizing of the ground loop is one of the most important design factors and is a function of system load, ground temperature, and local soil conditions. As noted earlier in this chapter, the type of soil and rock in contact with the pipe loop has serious implications for the length of ground loop required for adequate heat transfer. Inexperienced designers or installers are more likely to undersize the systems, resulting in inadequate system performance, or to oversize the systems, resulting in increased installation costs. The importance of good design, documentation, installation, and commissioning cannot be overemphasized.

In early 1995, at the National Training Center in Fort Irwin, California, 220 new single-family houses were built with an innovative ground-source heat pump system. The thermal sink/source for the heat pumps is geothermally heated 75°F (23.9°C) groundwater. The groundwater is pumped into a storage reservoir adjacent to the housing project. The reservoir water is then circulated through a double-walled heat exchanger, transferring thermal energy to/from a secondary closed circulation loop connected to individual heat pumps in each residence. Operating conditions and problems with maintaining correct water flow through the central facility heat exchanger contributed to lower than expected energy savings during the first year of operation. There was no indication that the heat pumps themselves were not performing as designed. Limited anecdotal information from the occupants indicates that overall satisfaction with the level of thermal comfort from the system was high.[9]

Vertical-bore, ground-coupled heat pump systems were installed in four new elementary schools in Lincoln, Nebraska, in 1995. Each school required 54 heat pumps ranging in size from 1.4 to 15 tons (204-tons total cooling capacity). With the heat pump systems, utility costs for these four schools are nearly half of that of other schools in the district, and the systems are providing a comfortable, complaint-free environment (Shonder et al. 1999).

Additional case studies are identified on the Geothermal Heat Pump Consortium web page (http://www.geoexchange.org/).

## 28.6.2 Energy Savings

The most important reason to consider the application of ground-source heat pumps to commercial buildings is the potential energy savings and its impact on overall life-cycle cost of the heating and cooling system. Ground-source heat pumps save energy and money because the equipment operates more efficiently than conventional systems, the maintenance costs are lower (see next section for maintenance benefits), and the equipment has a longer life expectancy than conventional unitary equipment. In addition, a ground-source heat pump does not require a defrost cycle, or, in most situations, backup electric resistance heat, as do air-source heat pumps.

By comparison, the average cooling efficiency at commercial-type facilities is estimated to be an EER of 8.0 (2.33 COP) for existing facilities and an EER of 10.0 (2.93 COP) for new facilities. Ground-source heat pump systems have the potential to reduce consumption of cooling energy by 30% to 50% and to reduce heating energy by 20% to 40% compared with typical air-source heat pumps.

A review of manufacture's literature on commercially available systems indicates that cooling efficiencies (EERs) of 13.4 to 20 Btu/W-h and heating efficiencies (COPs) of 3.1 to 4.3 are readily available.[10] A study prepared by DOE estimates energy-saving ranges of 17% to 42% comparing ground-source heat pumps to air-source heat pumps, depending on region (Calm 1987). Figure 28.7 illustrates the range of efficiencies typical of various heating and cooling equipment.

## 28.6.3 Maintenance

The ground-source heat pump technology is mature and reliable. Systems have standard warranties ranging from 1 to 5 years. The heat pump units are self-contained, and maintenance requirements are relatively straightforward; no new maintenance skills are necessary. Because heat pump equipment is not exposed to outdoor elements, the units actually require less maintenance than typical air-source heat pumps. One site reported a problem with cottonwood trees clogging outside condenser units of the previous air-conditioning systems. This maintenance problem was eliminated with the application of ground-source heat pumps.

In closed-loop systems, the ground loop is virtually maintenance free. The circulating pump(s) requires routine maintenance, as with any pump and motor system, and the water loop (a closed system) should be routinely monitored for temperature, pressure, flow, and antifreeze concentration. Unless there is a leak, no action is generally required.

---

[9]D.L. Hadley and L. Lkevgard. August 1997. *Family Housing Energy Savings Verification, National Training Center, Fort Irwin, CA. Preliminary Findings: Cooling Season Energy Savings.* Letter Report, Pacific Northwest National Laboratory, Richland, Washington.

[10]Rating based on ANSI/ARI/ASHRAE ISO Standard 13256-1-2005 for closed-loop ground-source heat pumps.

**Figure 28.7. Typical heating and cooling equipment efficiencies**

In open-loop systems, the well requires maintenance similar to any water well. The system should be routinely monitored for temperature, pressure, and flow. Because groundwater is being supplied to the heat pump, the heat exchangers should be routinely inspected for potential fouling and scale buildup.

Maintenance costs for ground-source heat pumps are about the same or less than conventional equipment. For example, maintenance costs for four schools in the Lincoln, Nebraska, equipped with ground-source heat pump systems have been well documented (Martin et al. 1999, 2000). The results of these studies are summarized in Table 28.2.

**Table 28.2 Maintenance Costs for Public Schools in Lincoln, Nebraska**

| HVAC System Type | Preventive Maintenance (cents/yr;ft$^2$) | Repair (cents/hr-ft$^2$) | Total (cents/hr-ft$^2$) |
|---|---|---|---|
| GSHP System | 7.12.1 | 9.2 | |
| Conventional system | 5.9 to 12.6 | 2.9 to 6.1 | 8.8 to 18.7 |

Source: Martin et al. 1999, 2000

Another analysis, this one by the Geothermal Heat Pump Consortium, found that the average total preventive and corrective maintenance costs for 25 ground-source-heat-pump-equipped buildings were approximately 11 cents/ft$^2$, compared to 30 to 40 cents/ft$^2$ for conventional systems (Cane 1998).

### 28.6.4 Installation Costs

Application-specific parameters, such as equipment capacity, type, refrigerant, air-distribution system, control system, plumbing configuration, and ground-coupling system type, significantly affect the total cost of the overall system. While equipment costs are competitive, installation costs vary significantly.

To illustrate the potential range of installation costs, the following examples have been reported.

- Stockton State College in Pomona, New Jersey, retrofit a system that totaled 1,655 tons (5,826 kW) at a total cost of $5,246,000 (Gahran 1993). The facility received grants and rebates reducing the capital outlay to $135,000. The system included 63 rooftop units, 500 variable-air-volume (VAV) boxes, and a 3,500- point energy management control system (EMCS). The unit cost was $3,170/ton in 1993 dollars.

- WaterFurnace, a ground-source heat pump manufacturer, designed the technology using a submerged pond closed-loop system into its new office building located in Fort Wayne, Indiana. The system totaled 134 tons (471.7 kW) at a cost of $239,800 (WaterFurnace WF639). The unit cost was $1,790/ton in 1991 dollars.

- Salem Community College in Carney's Point, New Jersey, retrofit a system that totaled 160 tons at a total cost of $284,000 (Gahran 1994). The system included 32 heat pumps. The unit cost was $1,775/ton in 1993 dollars.

- Paint Lick Elementary School in Garrard County, Kentucky, designed the technology into the new school building. The system totaled 123 tons (433 kW) at a total cost of $380,010 (WaterFurnace WF666). The unit cost was $3,090/ton in 1992 dollars.

- Maywood Elementary School in Hammond, Indiana, designed the technology into a new school building. The unit totaled 250 tons (880 kW) at a total cost of $1,277,190 (WaterFurnace WF925). The system consisted of 74 heat pumps and a ground-coupling system consisting of 244 vertical wells. The unit cost was $5,110/ton in 1994 dollars.

- The Lincoln, Nebraska, school district installed ground-source heat pump systems in four new elementary schools. In each school, the system consisted of 54 heat pumps ranging in size from 1.4 tons to 15 tons, with a total cooling capacity of 180 tons (630 kW). Four gas-fired boilers with a capacity of 330,000 Btuh each provide hot water for preheat and terminal reheat. The total heat pump cost per school was approximately $657,000 ($3,650/ton) in 1995 dollars (Shonder et al. 1999).

- Kavanaugh (1995) concluded that the average cost of ground-source heat pumps (including unit, loop, duct, and installation) ranged from $2360/ton for a 5-ton horizontal loop to $3000/ton for a 3-ton vertical loop system. Compared to a 3-ton conventional system, the added cost was $1250 to $1550 per ton.

Installation costs are expected to drop as the ground-source heat pump industry infrastructure grows and designers and installers become more experienced. Reducing installation costs is one of the prime goals of the International Ground-Source Heat Pump Association and the Geothermal Heat Pump Consortium.

### 28.6.5 Other Impacts

There are no significant negative environmental impacts associated with ground-source heat pumps. There is, however, the potential for systems to be affected by some local codes and regulations. The most likely source of conflict, if any, lies with the installation of the ground-coupling system. Working with an experienced installer is the best advice. However, local electric utilities and other local sites with existing ground-source heat pump installations are other sources of information about local permit and regulation issues.

With the application of any electrotechnology, there is a potential environmental benefit. Installing a ground-source heat pump system in lieu of a fossil-fuel heating

system will reduce local emissions. Furthermore, installing a more efficient electrotechnology such as a ground-source heat pump system for cooling will reduce source emissions at the utility power plant. Typical emission reductions per MWh of energy conserved are 0.3 pounds (0.14 kg) of particulates, 3.3 pounds (1.5 kg) of sulfur oxides, 5.3 pounds (2.4 kg) of nitrogen oxides, and 1,720 pounds (780 kg) of carbon dioxide. These numbers vary with time and region, depending on the power generation fuel mix (EPA 1994; Nemeth 1993).

## 28.7  HYPOTHETICAL CASE STUDIES

The purpose of these hypothetical case studies is to assist the energy manager or facility engineer in estimating the energy consumption and costs associated with the construction and operation of ground-source heat pump systems and comparing them with those for conventional HVAC technologies. The goal is to estimate energy consumption and savings, not to design systems.

There are several methods for estimating energy consumption of HVAC technologies, from simplistic degree-day calculations to sophisticated hour-by-hour energy modeling and simulation systems supported by computer programs. The examples used in this chapter are based on an outdoor temperature bin method. This method is described in more detail in Closed-Loop/ Closed-Source Heat Pump Systems: Installation Guide (OSU 1988).

Two case studies were developed for this chapter. In both, estimates of the potential energy consumption and life-cycle costs of the ground-source heat pump technology are compared with conventional HVAC technologies. In each study, the facility is a hypothetical administrative building of typical single-story construction. The facilities in each example operate continuously (no night setback temperature control). In the first example, the building is located in upstate New York; in the second example, the same building is located in central Oklahoma.

### 28.7.1 Example 1: Upstate New York Facility

The hypothetical administrative office facility is in upstate New York. Mean local weather conditions are 7,331 (base 65°F) heating degree-days (4,073 heating Celsius degree-days) and 472 (base 65°F) cooling degree-days (262 cooling Celsius degree-days). The 97.5% heating design temperature is -5°F (-20.6°C) and the 2.5% cooling design temperature is 85°F (29.4°C). The mean annual earth temperature is 49°F (9.4°C).

The electric rate schedule for this example consists of a demand charge of $5.51/kW-month and an energy charge of $0.05/kWh. The gas rate schedule consists of an energy charge of $0.54/therm.

The heating load for this facility during the design day is estimated to be 3,309 kBtu/h (970.6 kW). The design cooling load is 1,273 kBtu/h (373.4 kW). The balance temperature for the facility is 60°F (15.6°C). The balance temperature is the temperature at which neither heating nor cooling is required to meet the comfort requirements of the facility.

**Technology Description**

Three technologies are compared in this example. The conventional technology is a rooftop air-conditioning system with integral natural gas furnace heaters. For this example, the capacity and input power are based on a commercially available rooftop unit with a rated heating capacity of 270 kBtu/h (79.2 kW) and a rated cooling capacity of 162 kBtu/h (47.5 kW) at ARI conditions. The equipment selected has a cooling efficiency of 8.2 EER and 10.1 IPLV[11] at ARI conditions. The heating efficiency is rated at 80%. The indoor air fan is rated at 2.50 kW input and the outdoor air fans are rated at 2.67 kW input. To meet the heating requirements during the design day, 13 rooftop units are required. This number also meets the cooling load requirements. The estimated equipment life of this alternative, according to ASHRAE, is 15 years (ASHRAE HVAC Applications Handbook, 2005, pp 36.3).

The second alternative is a rooftop air-source heat pump system with electric resistance supplemental heaters. For this example, the capacity and input power are based on a commercially available air-source heat pump. The rooftop units have a rated cooling capacity of 162 kBtu/h (47.5 kW) and a rated heating capacity of 166 kBtu/h (48.7 kW) (high temperature) and 102 kBtu/h (29.9 kW) (low temperature) at ARI Standard 340 test conditions. The equipment selected has a rated cooling efficiency of 8.5 EER and a rated heating efficiency of 3.0 COP (high temperature) and 2.1 COP (low temperature). The indoor and outdoor air fan input power is included in the power input loads. To meet the cooling requirements during the design day, seven rooftop units are required. To meet the added load during the heating design day, each rooftop unit is equipped with 125-kW supplemental heaters. The estimated equipment life of this alternative, according to ASHRAE, is 15 years (ASHRAE HVAC Applications Handbook, 2005, pp 36.3).

The third alternative is a ground-source heat pump system using extended-range water-source heat pumps and a vertical closed-loop ground-coupling system. For

this example, the capacity and input power are based on a commercially available ground-source heat pump. The heat pumps have a rated cooling capacity of 57.5 kBtu/h (16.9 kW) and a rated heating capacity of 60.0 kBtu/h (17.6 kW) at ARI 330 conditions. The equipment selected has a rated cooling efficiency of 14.49 EER and a rated heating efficiency of 4.0 COP. The indoor air fan input power is included in the power input loads. To meet the cooling requirements during the design day, 22 units are required. To meet the added load during the heating design day, each unit is equipped with a 33-kW supplemental heater. It is also assumed that the water-loop system employs a variable-speed drive for added energy savings. The estimated equipment life of this alternative, according to ASHRAE, is 19 years (ASHRAE HVAC Applications Handbook, 2005, pp 36.3).

**Savings Potential**

Energy consumption for the three alternatives is estimated by an outdoor temperature bin method. The calculations are performed in a spreadsheet using 5°F bin data because this is a form in which data are readily available (TM 5-785).

The main assumption underlying this type of analysis is that the facility is "thermally light." This implies that the heating and cooling loads on the building are proportional to the outside air temperature. A "thermally heavy" facility would be a building in which the cooling load on the building is not very proportional to the outside air temperature because of significant internal heat loads or a thermally massive structure, which tends to "hold" the heat or cold. In the case of a "thermally heavy" facility, the building heating and cooling loads would have to be calculated using a methodology other than the one utilized in the following analysis.

The analysis for the ground-source heat pump option shown in Table 28.3 begins with an estimate of the building load for each bin. Column 1 indicates the midpoint of each temperature bin, e.g., the bin from 95° to 99°F has a midpoint temperature of 97°F. Column 2 gives the number of hours that occur in each bin over a "typical" year for that location. Column 3 estimates the corresponding building load. This is calculated by linear interpolation between no heating or cooling load at the facility balance temperature and the design heating or cooling load at the design heating or cooling temperature. The actual equations used in this analysis are listed in Table 28.4

The entering water temperature (EWT) is estimated in Column 4 of Table 28.3. This is the temperature of the water entering the heat pump from the ground-coupling system. For initial estimating, the entering water temperature is a linear interpolation between two points for

---

[11]EER and IPLV have units of Btu/W-h in this example.

**Table 28.3. Upper New York Facility Ground-Source Heat Pump Energy Consumption Bin Method Analysis Example**

| Avg. Bin Temp. (°F) | Bin Hours (h/yr) | Bldg. Load (kBtuh) | EWT (°F) | Net GSHP Capacity (kBtuh) | Theor. Run Time (%) | Partial Load Factor | Actual Run Time (%) | GSHP Eff'y (EER or COP) | GSHP Input Power (kW) | Suppl. Heater Power (kW) | Pump Flow Rate (gpm) | GSHP Electric Energy (kWh) | Suppl. Heater Energy (kWh) | Circ. Pump Energy (kWh) | Circ. Pump Energy (kWh) | Total Electric Energy (kWh) |
|---|---|---|---|---|---|---|---|---|---|---|---|---|---|---|---|---|
| (1) | (2) | (3) | (4) | (5) | (6) | (7) | (8) | (9) | (10) | (11) | (12) | (13) | (15) | (16cv) | (16vsd) | (17) |
| Cooling Mode: | | | | | | | | | | | | | | | | |
| 102 | 0 | 2,139 | 108 | 1,124.2 | 1.00 | 1.00 | 1.00 | 10.70 | 105.02 | n/a | 310 | 0 | 0 | 0 | 0 | 0 |
| 97 | 4 | 1,884 | 101 | 1,167.1 | 1.00 | 1.00 | 1.00 | 11.70 | 99.76 | n/a | 310 | 399 | 0 | 36 | 36 | 435 |
| 92 | 15 | 1,629 | 94 | 1,210.0 | 1.00 | 1.00 | 1.00 | 12.80 | 94353 | n/a | 310 | 1,418 | 0 | 134 | 134 | 1,551 |
| 87 | 88 | 1,375 | 87 | 1,252.9 | 1.00 | 1.00 | 1.00 | 14.04 | 89.25 | n/a | 310 | 7,854 | 0 | 785 | 785 | 8,639 |
| 82 | 206 | 1,120 | 80 | 1,295.9 | 0.86 | 0.97 | 0.89 | 15.43 | 83.99 | n/a | 278 | 15,482 | 0 | 1,838 | 1,472 | 16,954 |
| 77 | 360 | 866 | 73 | 1,338.8 | 0.65 | 0.91 | 0.71 | 17.00 | 78.74 | n/a | 220 | 20,104 | 0 | 3,213 | 1,616 | 21,720 |
| 72 | 540 | 611 | 66 | 1,381.7 | 0.44 | 0.86 | 0.51 | 18.80 | 73.48 | n/a | 159 | 20,391 | 0 | 4,819 | 1,273 | 21,664 |
| 67 | 690 | 356 | 59 | 1,424.6 | 0.25 | 0.81 | 0.31 | 20.88 | 68.22 | n/a | 96 | 14,495 | 0 | 6,158 | 584 | 15,079 |
| 62 | 704 | 102 | 52 | 1,467.5 | 0.07 | 0.77 | 0.09 | 23.31 | 62.97 | n/a | 28 | 4,009 | 0 | 6,283 | 51 | 4,060 |
| Heating Mode: | | | | | | | | | | | | | | | | |
| 57 | 680 | 153 | 48 | 1,024.1 | 0.15 | 0.79 | 0.19 | 3.59 | 83.56 | 0.00 | 59 | 10,763 | 0 | 6,068 | 218 | 10,981 |
| 52 | 671 | 407 | 48 | 1,010.8 | 0.40 | 0.85 | 0.47 | 3.57 | 82.98 | 0.00 | 147 | 26,371 | 0 | 5,988 | 1,343 | 27,714 |
| 47 | 618 | 622 | 47 | 997.6 | 0.66 | 0.92 | 0.72 | 3.55 | 82.41 | 0.00 | 225 | 36,891 | 0 | 5,515 | 2,894 | 39,785 |
| 42 | 631 | 916 | 46 | 984.3 | 0.93 | 0.98 | 0.95 | 3.53 | 81.83 | 0.00 | 294 | 48,915 | 0 | 5,631 | 5,053 | 53,968 |
| 37 | 754 | 1,171 | 45 | 971.0 | 1.00 | 1.00 | 1.00 | 3.50 | 81.25 | 58.58 | 310 | 61,265 | 44,168 | 6,729 | 6,729 | 112,162 |
| 32 | 821 | 1,425 | 44 | 957.7 | 1.00 | 1.00 | 1.00 | 3.48 | 80.68 | 137.07 | 310 | 66,235 | 112,534 | 7,327 | 7,327 | 186,096 |
| 27 | 580 | 1,680 | 43 | 944.5 | 1.00 | 1.00 | 1.00 | 3.46 | 80.10 | 215.56 | 310 | 46,458 | 125,025 | 5,176 | 5,176 | 176,658 |
| 22 | 441 | 1,934 | 43 | 931.2 | 1.00 | 1.00 | 1.00 | 3.43 | 79.52 | 294.05 | 310 | 35,069 | 129,676 | 3,936 | 3,936 | 168,681 |
| 17 | 326 | 2,189 | 42 | 917.9 | 1.00 | 1.00 | 1.00 | 3.41 | 78.95 | 372.54 | 310 | 25,736 | 121,449 | 2,909 | 2,909 | 150,094 |
| 12 | 242 | 2,444 | 41 | 904.7 | 1.00 | 1.00 | 1.00 | 3.38 | 78.37 | 451.03 | 310 | 18,965 | 109,150 | 2,160 | 2,160 | 130,275 |
| 7 | 165 | 2,698 | 40 | 891.4 | 1.00 | 1.00 | 1.00 | 3.36 | 77.79 | 529.52 | 310 | 12,836 | 87,371 | 1,472 | 1,472 | 101,679 |
| 2 | 101 | 2,953 | 39 | 878.1 | 1.00 | 1.00 | 1.00 | 3.33 | 77.21 | 608.01 | 310 | 7,799 | 61,409 | 901 | 901 | 70,109 |
| -3 | 62 | 3,207 | 38 | 864.8 | 1.00 | 1.00 | 1.00 | 3.31 | 76.64 | 726.00 | 310 | 4,751 | 45,012 | 553 | 553 | 50,317 |
| -8 | 33 | 3,462 | 37 | 851.6 | 1.00 | 1.00 | 1.00 | 3.28 | 76.06 | 726.00 | 310 | 2,510 | 23,958 | 294 | 294 | 26,762 |
| -13 | 14 | 3,716 | 37 | 838.3 | 1.00 | 1.00 | 1.00 | 3.25 | 75.48 | 726.00 | 310 | 1,057 | 10,164 | 125 | 125 | 11,346 |
| -18 | 6 | 3,971 | 36 | 825.0 | 1.00 | 1.00 | 1.00 | 3.23 | 74.91 | 726.00 | 310 | 449 | 4,356 | 54 | 54 | 4,859 |
| -23 | 2 | 4,225 | 35 | 811.7 | 1.00 | 1.00 | 1.00 | 3.20 | 74.33 | 726.00 | 310 | 149 | 1,452 | 18 | 18 | 1,619 |
| Cooling | | | | | | | | | | | | 84,152 | 0 | 23,265 | 5,951 | 90,103 |
| Heating | | | | | | | | | | | | 406,219 | 875,724 | 54,856 | 41,162 | 1,323,105 |
| Total | | | | | | | | | | | | 490,371 | 875,724 | 78,122 | 47,113 | 1,413,207 |

the heating cycle and again for the cooling cycle. During zero building load conditions, which occurs at the facility balance temperature, the entering water temperature is assumed to be equal to the ground temperature. During the peak load, the entering EWT is set to the maximum temperature desired during the peak cooling season load; similarly, it is set to the minimum temperature desired during the peak heating season load. Publications by the International Ground-Source Heat Pump Association recommend for initial calculations these temperatures be set to 100°F maximum (37.8°C) and 37°F (2.8°C) minimum, although in this example, the minimum temperature was set to 35°F (1.7°C). Today, most commercial systems are commonly designed at 90°F (32.2°C) maximum EWT. The estimated entering water temperatures are refined later in the design process, as the ground-coupling system design is finalized.

Column 5 in Table 28.3 is the net capacity of the ground-source heat pump equipment. This is taken from equipment specifications based on the entering water temperature. Columns 6, 7, and 8 are used to determine the loss in efficiency caused by part-load operating characteristics of equipment, which result in increased run time on the units. The part-load factor is dependent on

the equipment "degradation factor." The degradation factor is typically assumed to be 0.25 unless the equipment manufacturer has tested the unit and determined a lower value.

Column 10 is the input power of the ground-source heat pump equipment. Like the net capacity, this is taken from equipment specifications based on the entering water temperature. Column 9 is the efficiency of the ground-source heat pump alone and is determined by dividing the net capacity by the input power and correcting for the appropriate units of measure. The cooling efficiency is typically expressed as the energy-efficiency ratio (EER) and has units of Btu/watt-h. Heating efficiency is typically expressed as the coefficient of performance (COP) and is a dimensionless measure.

Column 11 is the average input power required by the supplemental heaters to meet the estimated building heating load. It is estimated by taking the difference between the building heating load and the capacity of the heat pumps. When the ground-source heat pumps can meet the building load, there is no load on the supplemental heaters. The maximum load does not exceed the installed capacity of the supplemental heaters. Not included in this simplified example is an air-to-air heat

## Table 28.4. Summary of Equations for Bin Method Energy Consumption Analysis

| Name | Column | Season | Description/Equation |
|---|---|---|---|
| Average temp (°F) | (1) | Both | Average bin temperature<br>= Midpoint of temperature bin from weather data |
| Bin hours (hr/yr) | (2) | Both | Bin hours<br>= Number of hours in temperature bin from weather data |
| Building load | (3) | Cooling<br><br>Heating | Building load<br>= (avg. bin temp. - fac. balance temp.) * [(fac. design load) / (fac. design temp. - fac. balance temp.)]<br>= (fac. balance temp. - avg. bin temp.) * [(fac. design load) / (fac. balance temp. - fac. design temp.)] |
| Entering water temperature | (4) | Cooling<br><br>Heating | Entering water temperature<br>= (ground temp.) + [(max. entering water temp. – ground temp.) / (max. bin temp. - fac. balance temp<br>(avg. bin temp. - fac. balance temp.)<br>= (min. entering water temp.) + [ (ground temp. - min. entering water temp.) / (fac. balance temp. - m<br>temp.)] * (avg. bin temp. - min. bin temp.) |
| Net equipment capacity | (5) | Both | Net equipment capacity<br>= base on equipment specifications corrected for temperature |
| Theoretical run time | (6) | Both | Theoretical run time<br>= (column 3) / (column 5);  maximum = 1.00 |
| Partial load factor | (7) | Both | Partial load factor<br>= 1.00 - { degradation factor * [ 1.00 - (column 6)] }; maximum = 1.00 |
| Actual run time | (8) | Both | Actual run time<br>= (column 6) / (column 7) |
| Efficiency | (9) | Cooling<br>Heating | Efficiency<br>= (column 5) / (column 10)<br>= (column 5) / [ (column 10) * 3.412] |
| HVAC equipment input power | (10) | Both | HVAC equipment input power<br>= based on equipment specifications corrected for temperature, may also include indoor and outdoor |
| Supplemental heater power | (11) | Heating | Supplemental heater power<br>= [ (column 3) - (column 5) ] /3.412; minimum = 0, maximum = capacity |
| Average water-loop pump flow rate | (12) | Both | Average water-loop pump flow rate<br>= [ no. of units * flow rate per unit * (column 8) ] |
| HVAC equipment energy | (13) | Both | HVAC equipment energy<br>= (column 2) * (column 8) * (column 10) |
| Fan energy | (14) | Cooling<br><br><br>Heating | Fan energy<br>= (column 2) * (column 8) * (total indoor air fan load + total outdoor air fan load); ignore if included<br>column 10<br>= (column 2) * (column 8) * (total indoor air fan load); ignore if included in column 10 |
| Supplemental heater energy | (15) | Heating | Supplemental heater energy = (column 2) * (column 11) |
| Constant-volume water-loop pump energy | (16cv) | Both | Constant-volume water-loop pump energy<br>= [no. of units * flow rate per unit * pump pressure rise in p.s.i. * 0.746 * (column 2)] / [1715 * pum<br>motor eff.] |
| Variable-speed drive water-loop pump energy | (16vsd) | Both | Variable-speed drive water-loop pump energy<br>= (column 16cv) * [ (column 8)$^2$] |
| Total electric energy | (17) | Both | Total electric energy<br>= (column 13) + (column 14) + (column 15) + (column 16) |
| Total electric gas furnace energy | (18) | Heating | Total natural gas furnace energy<br>= (column 2) * (column 5) * (column 8) / [AFUE * 100] |

recovery system that would reduce the need for supplemental heating. These recovery systems are common in cold region applications and can be specified to be part of the basic rooftop unit.

The average pump flow rate in Column 12 is estimated based on the number of heat pumps, the actual run time (Column 8), and the rated flow per unit. Most ground-source heat pumps are rated at around 3 gpm per ton of cooling (0.054 L/s-kW), although this varies by manufacturer.

The ground-source heat pump energy consumption, Column 13, is the result of the input power multiplied by the percent run time multiplied by the number of hours in each bin (Column 10 x Column 8 x Column 2). The supplemental heater energy consumption is similarly

estimated as the result of the supplemental heater power multiplied by the run time multiplied by the number of hours in each bin (Column 11 x Column 8 x Column 2).

The energy consumption of the ground-coupling loop is dependent on the configuration of the ground-coupling and control systems. For this example, it is assumed the net pressure rise of the circulation pump at full design flow is 48 psi and, as a result of the control system, the variable-speed drive reduces the energy requirement as a square function of the flow rate (assumes automatic water shut-off valves at each heat pump that close when the unit is off). The total energy consumption (Column 16) is the sum of the component energy consumptions (Column 13 + Column 14 + Column 15). For the ground-source heat pump system in this example, the annual energy consumption is estimated to be 1,413,207 kWh/yr. The monthly-billed demand is estimated by examining the corresponding temperature bin during the time the monthly peak demand is typically set for the facility. For this example, it is estimated that the demand is 4,355 kW-month/yr.

Not included in the sample equipment design is an air-to-air heat recovery system that would reduce the need for supplemental electric resistance heating. Heat pump manufacturers make these units available as part of the rooftop unit for installations in colder northern climates.

Similar spreadsheets were developed for each of the other technology alternatives using a standard bin method. Table 28.5 shows the analysis for a conventional air-conditioning with natural-gas furnace option. The analysis for an air-source heat pump option is shown in Table 28.6.

### Life-Cycle Cost

The total installation cost, including material, labor, overhead, and profit, for the ground-source heat pump option is $329,350 compared with the conventional system at $454,100 and the air-source heat pump system at $213,000. The operations and maintenance cost for the ground-source heat pump system is estimated to be $3,700/yr compared with $8,775/yr for the conventional system and $3,300/yr for the air-source heat pump system. Using the Building Life-Cycle Cost[12] software available from the National Institute for Standards and Technology (NIST), the total life-cycle costs for the three alternatives are estimated at $1,502,942 for the ground-source heat pump system, $1,516,482 for the air-source heat pump system, and $1,639,262 for the conventional system. A common life-cycle of 15 years was used in this analysis. A summary of the energy and cost factors and the results of the life-cycle cost analysis for the three alternatives are shown in Table 28.7.

---

[12]BLCC is available at no charge from the DOE Federal Energy Management Program web site at http://www.eere.energy.gov/femp/. As of November 2005, BLCC version 5.3-05 was the most current version available.

### Table 28.5. Upper New York Facility Conventional Air-Conditioning and Gas Furnace System Bin Method Analysis Example

| Avg. Bin Temp (°F) | Bin Hours (h/yr) | Bldg. Load (kBtuh) | Net Equip. Capacity (kBtuh) | Theor. Run Time (%) | Partial Load Factor | Actual Run Time (%) | A/C Eff'y (EER) | A/C Input Power (kW) | A/C Electric Energy (kWh) | Fan Energy (kWh) | Total Electric Energy (kWh) | Nat. Gas Furnace Energy (therms) |
|---|---|---|---|---|---|---|---|---|---|---|---|---|
| (1) | (2) | (3) | (5) | (6) | (7) | (8) | (9) | (10) | (13) | (14) | (17) | (18) |
| Cooling Mode: | | | | | | | | | | | | |
| 102 | 0 | 2,139 | 2,308.8 | 0.93 | 0.98 | 0.94 | 9.95 | 232.05 | 0 | 0 | 0 | n/a |
| 97 | 4 | 1,884 | 2,386.8 | 0.79 | 0.95 | 0.83 | 10.74 | 222.30 | 741 | 224 | 965 | n/a |
| 92 | 15 | 1,629 | 2,453.1 | 0.66 | 0.92 | 0.13 | 11.29 | 217.23 | 2,363 | 131 | 3,094 | n/a |
| 87 | 88 | 1,375 | 2,511.6 | 0.55 | 0.89 | 0.62 | 11.67 | 215.28 | 11,693 | 3,651 | 15,344 | n/a |
| 82 | 206 | 1,120 | 2,570.1 | 0.44 | 0.86 | 0.51 | 12.05 | 213.33 | 22.300 | 7,026 | 29,326 | n/a |
| 77 | 360 | 866 | 2,628.6 | 0.33 | 0.83 | 0.40 | 12.44 | 211.38 | 30.108 | 9,513 | 39,681 | n/a |
| 72 | 540 | 611 | 2,687.1 | 0.23 | 0.81 | 0.28 | 12.83 | 209.43 | 31,813 | 10,229 | 42,102 | n/a |
| 67 | 690 | 356 | 2,745.6 | 0.13 | 0.78 | 0.17 | 13.23 | 207.48 | 23,753 | 7,694 | 31,447 | n/a |
| 62 | 704 | 102 | 2,804.1 | 0.04 | 0.76 | 0.05 | 13.64 | 205.53 | 6,923 | 2,264 | 9,187 | n/a |
| Heating Mode: | | | | | | | | | | | | |
| 57 | 680 | 153 | 3,510.0 | 0.04 | 0.76 | 0.06 | n/a | n/a | n/a | 1,264 | 1,264 | 1,106 |
| 52 | 671 | 407 | 3,510.0 | 0.12 | 0.78 | 0.15 | n/a | n/a | n/a | 3,248 | 3,248 | 4,385 |
| 47 | 618 | 622 | 3,510.0 | 0.19 | 0.80 | 0.24 | n/a | n/a | n/a | 4,751 | 4,751 | 6,413 |
| 42 | 631 | 916 | 3,510.0 | 0.26 | 0.82 | 0.32 | n/a | n/a | n/a | 6,567 | 6,567 | 8,865 |
| 37 | 754 | 1,171 | 3,510.0 | 0.33 | 0.83 | 0.40 | n/a | n/a | n/a | 9,809 | 9,809 | 13,242 |
| 32 | 821 | 1,425 | 3,510.0 | 0.41 | 0.85 | 0.48 | n/a | n/a | n/a | 12,125 | 12,125 | 17,179 |
| 27 | 580 | 1,680 | 3,510.0 | 0.48 | 0.87 | 0.55 | n/a | n/a | n/a | 10,374 | 10,374 | 14,005 |
| 22 | 441 | 1,934 | 3,510.0 | 0.55 | 0.89 | 0.62 | n/a | n/a | n/a | 8,898 | 8,898 | 12,012 |
| 17 | 326 | 2,189 | 3,510.0 | 0.62 | 0.91 | 0.69 | n/a | n/a | n/a | 7,294 | 7,294 | 9,847 |
| 12 | 242 | 2,444 | 3,510.0 | 0.70 | 0.92 | 0.75 | n/a | n/a | n/a | 5,925 | 5,925 | 7,999 |
| 7 | 165 | 2,698 | 3,510.0 | 0.77 | 0.94 | 0.82 | n/a | n/a | n/a | 4,375 | 4,375 | 5,906 |
| 2 | 101 | 2,953 | 3,510.0 | 0.84 | 0.96 | 0.88 | n/a | n/a | n/a | 2,875 | 2,875 | 3,882 |
| -3 | 62 | 3,207 | 3,510.0 | 0.91 | 0.98 | 0.93 | n/a | n/a | n/a | 1,882 | 1,882 | 2,540 |
| -8 | 33 | 3,462 | 3,510.0 | 0.99 | 1.00 | 0.99 | n/a | n/a | n/a | 1,061 | 1,061 | 1,433 |
| -13 | 14 | 3,716 | 3,510.0 | 1.00 | 1.00 | 1.00 | n/a | n/a | n/a | 455 | 455 | 614 |
| -18 | 6 | 3,971 | 3,510.0 | 1.00 | 1.00 | 1.00 | n/a | n/a | n/a | 195 | 195 | 263 |
| -23 | 2 | 4,225 | 3,510.0 | 1.00 | 1.00 | 1.00 | n/a | n/a | n/a | 65 | 65 | 88 |
| Cooling | | | | | | | | | 129,754 | 41,391 | 171,145 | 0 |
| Heating | | | | | | | | | 0 | 81,763 | 81,763 | 110,380 |
| Total | | | | | | | | | 129,754 | 123,154 | 252,908 | 110,380 |

**Table 28.6. Upper New York Facility Air-Source Heat Pump System Bin Method Analysis Example**

| Avg. Bin Temp. (°F) | Bin Hours (h/yr) | Bldg. Load (kBtuh) | Net ASHP Capacity (kBtuh) | Theor. Run Time (%) | Partial Load Factor | Actual Run Time (%) | ASHP Eff'y (EER or COP) | ASHP Input Power (kW) | Suppl. Heater Power (kW) | ASHP Electric Energy (kWh) | Suppl. Heater Energy (kWh) | Total Electric Energy (kWh) |
|---|---|---|---|---|---|---|---|---|---|---|---|---|
| (1) | (2) | (3) | (5) | (6) | (7) | (8) | (9) | (10) | (11) | (13) | (15) | (17) |
| Cooling Mode: | | | | | | | | | | | | |
| 102 | 0 | 2,139 | 1,230.0 | 1.00 | 1.00 | 1.00 | 10.74 | 114.50 | n/a | 0 | 0 | 0 |
| 97 | 4 | 1,884 | 1,268.3 | 1.00 | 1.00 | 1.00 | 11.44 | 110.88 | n/a | 444 | 0 | 444 |
| 92 | 15 | 1,629 | 1,306.5 | 1.00 | 1.00 | 1.00 | 12.18 | 107.26 | n/a | 1,609 | 0 | 1,609 |
| 87 | 88 | 1,375 | 1,344.8 | 1.00 | 1.00 | 1.00 | 12.97 | 103.65 | n/a | 9,121 | 0 | 9,121 |
| 82 | 206 | 1,120 | 1,383.1 | 0.81 | 0.95 | 0.85 | 13.83 | 100.03 | n/a | 17,523 | 0 | 17,523 |
| 77 | 360 | 866 | 1,421.3 | 0.61 | 0.90 | 0.68 | 14.74 | 96.41 | n/a | 23,429 | 0 | 23,429 |
| 72 | 540 | 611 | 1,459.6 | 0.42 | 0.85 | 0.49 | 15.73 | 92.80 | n/a | 24,545 | 0 | 24,545 |
| 67 | 690 | 356 | 1,497.9 | 0.24 | 0.81 | 0.29 | 16.80 | 89.18 | n/a | 18,089 | 0 | 18,089 |
| 62 | 704 | 102 | 1,536.1 | 0.07 | 0.77 | 0.09 | 17.95 | 85.56 | n/a | 5,210 | 0 | 5,210 |
| Heating Mode: | | | | | | | | | | | | |
| 57 | 680 | 153 | 1,413.3 | 0.11 | 0.78 | 0.14 | 3.11 | 133.14 | 0.00 | 12,591 | 0 | 12,591 |
| 52 | 671 | 407 | 1,318.8 | 0.31 | 0.83 | 0.37 | 3.03 | 127.54 | 0.00 | 31,949 | 0 | 31,949 |
| 47 | 618 | 622 | 1,226.4 | 0.54 | 0.88 | 0.61 | 2.93 | 122.57 | 0.00 | 46,192 | 0 | 46,192 |
| 42 | 631 | 916 | 1,135.4 | 0.81 | 0.95 | 0.85 | 2.82 | 118.02 | 0.00 | 63,148 | 0 | 63,148 |
| 37 | 754 | 1,171 | 1,012.9 | 1.00 | 1.00 | 1.00 | 2.61 | 113.89 | 46.30 | 85,873 | 34,910 | 120,784 |
| 32 | 821 | 1,425 | 869.4 | 1.00 | 1.00 | 1.00 | 2.32 | 110.04 | 162.96 | 90,343 | 133,789 | 224,132 |
| 27 | 580 | 1,680 | 782.6 | 1.00 | 1.00 | 1.00 | 2.14 | 107.03 | 263.00 | 62,077 | 152,540 | 214,617 |
| 22 | 441 | 1,934 | 733.6 | 1.00 | 1.00 | 1.00 | 2.06 | 104.58 | 351.96 | 46,120 | 155,215 | 201,335 |
| 17 | 326 | 2,189 | 682.5 | 1.00 | 1.00 | 1.00 | 1.96 | 101.92 | 441.54 | 33,226 | 143,942 | 177,168 |
| 12 | 242 | 2,444 | 630.0 | 1.00 | 1.00 | 1.00 | 1.86 | 99.12 | 531.53 | 23,987 | 128,629 | 152,617 |
| 7 | 165 | 2,698 | 567.0 | 1.00 | 1.00 | 1.00 | 1.76 | 94.43 | 624.59 | 15,581 | 103,058 | 118,636 |
| 2 | 101 | 2,953 | 497.0 | 1.00 | 1.00 | 1.00 | 1.65 | 88.48 | 719.71 | 8,936 | 72,691 | 81,627 |
| -3 | 62 | 3,207 | 433.3 | 1.00 | 1.00 | 1.00 | 1.52 | 83.58 | 875.00 | 5,182 | 54,250 | 59,432 |
| -8 | 33 | 3,462 | 373.4 | 1.00 | 1.00 | 1.00 | 1.38 | 79.38 | 875.00 | 2,620 | 28,875 | 31,495 |
| -13 | 14 | 3,716 | 337.4 | 1.00 | 1.00 | 1.00 | 1.32 | 75.18 | 875.00 | 1,053 | 12,250 | 13,303 |
| -18 | 6 | 3,971 | 316.4 | 1.00 | 1.00 | 1.00 | 1.31 | 70.98 | 875.00 | 426 | 5,250 | 5,676 |
| -23 | 2 | 4,225 | 295.4 | 1.00 | 1.00 | 1.00 | 1.30 | 66.78 | 875.00 | 134 | 1,750 | 1,884 |
| Cooling | | | | | | | | | | 99,969 | 0 | 99,969 |
| Heating | | | | | | | | | | 529,437 | 1,027,149 | 1,556,586 |
| Total | | | | | | | | | | 629,406 | 1,027,149 | 1,656,555 |

## 28.7.2 Example 2: Oklahoma City Facility

This example places the same building in a new environment. The location of the facility is in Oklahoma City, Oklahoma. Typical local weather conditions are 3,588°F-day/yr (base 65°F) heating degree-days (1,993 heating Celsius degree-days) and 2,068 (base 65°F) cooling degree-days (1,149 cooling Celsius degree-days). The 97.5% heating design temperature is 13°F (-10.6°C) and the 2.5% cooling design temperature is 96°F (35.6°C). The mean annual earth temperature is 62°F (16.7°C).

The electric rate schedule for this example consists of a summer demand charge of $12.35/kW-month, a winter demand charge of $4.48/kW-month, and an energy charge of $0.0236/kWh. The gas rate schedule consists of an energy charge of $3.40/mcf ($0.34/therm).

The corresponding heating load for this facility during the design day is estimated to be 2,392 kBtu/h (701.7 kW). The design cooling load is 1,832 kBtu/h (537.4 kW). The same balance temperature of 60°F (15.6°C) is estimated for the facility.

### Technology Description

The same three technologies are compared in this example, but the number of units has changed because the building loads are now different. The conventional technology is a rooftop air-conditioning system with integral natural gas furnace heaters. To meet the cooling requirements during the actual cooling design day, 10 rooftop units are required. This number also meets the heating load requirements.

The second alternative is a rooftop air-source heat pump system with electric resistance supplemental heaters. To meet the cooling requirements during the actual cooling design day, 10 rooftop units are required. To meet the added load during the heating design day, each rooftop unit is equipped with a 50-kW supplemental heater.

The third alternative is a ground-source heat pump system using extended range water-source heat pumps and a vertical closed-loop ground-coupling system. To meet the cooling requirements during the actual cooling design day, 35 units are required. To meet the added load during the heating design day, each unit is equipped with a 10-kW supplemental heater. The water-loop system employs a variable-speed drive for added energy savings.

### Savings Potential

Energy consumption is estimated with the same spreadsheet temperature bin model as the previous case.

## Table 28.7. Example 1: Summary of Results

| Upstate New York Facility | Conventional system | Air-Source Heat Pump | Ground-Source Heat Pump |
|---|---|---|---|
| Number of units | 13 | 7 | 22 |
| Nominal capacity (tons) | | | |
| Each | 13.5 | 13.5 | 4.8 |
| Total | 175.5 | 94.5 | 105.6 |
| Supplemental heaters (kW) | | | |
| Each | n/a | 125 | 33 |
| Total | n/a | 875 | 726 |
| Equipment capacity (kBtuh) (at design conditions) | | | |
| Summer | 2,535.0 | 1,360.1 | 1,270.1 |
| Winter | 3,510.0 | 3,395.0 | 3,336.6 |
| Energy consumption (/yr) | | | |
| Electricity (kWh) | 252,908 | 1,656,555 | 1,413,207 |
| Demand* (kW-mo) | 1,481 | 4,200 | 4,355 |
| Natural gas (therm) | 110,380 | 0 | 0 |
| Total energy ($10^6$ Btu) | 11,901 | 5,562 | 4,822 |
| Energy Cost ($/yr) | | | |
| Electricity | 12,645 | 82,828 | 70,660 |
| Demand | 8,160 | 23,142 | 23,996 |
| Natural gas | 59,605 | 0 | 0 |
| Total energy | 80,411 | 105,970 | 94,656 |
| O&M Costs ($/yr) | 8,775 | 3,300 | 3,700 |
| Installed Costs (S) | 454,100 | 212,500 | 329,300 |
| Equipment life (yr) | 15 | 15 | 15 |
| Total life-cycle cost ($) | 1,639,262 | 1,516,482 | 1,502,942 |

*Assumptions are required to estimate the monthly billing demand. In this case, the authors matched a temperature bin to when the peak demand window was likely to occur. A sensitivity analysis could be performed by changing the assumptions.

The results for the ground-source heat pump option are shown in Table 28.8. Table 28.9 shows the analysis for the conventional air-conditioning and natural-gas furnace option. Table 28.10 shows the analysis for the air-source heat pump option.

### Life-Cycle Cost

The total installation cost for the ground-source heat pump option (including material, labor, overhead, and profit) is estimated to be $451,500, compared with the conventional system at $349,300 and the air-source heat pump system at $236,000. The operations and maintenance cost for the ground-source heat pump system is estimated to be $5,880/yr, compared with $6,750/yr for the conventional system and $4,725/yr for the air-source heat

pump system. Through BLCC, the total life-cycle costs for the three alternatives were estimated to be $938,282 for the ground-source heat pump system, $800,065 for the air-source heat pump system, and $974,924 for the conventional system. A summary of the energy and cost factors and the results of the life-cycle cost analysis for each of the three options are shown in Table 28.11.

The fact that the ground-source heat pump system did not have the lowest life-cycle cost in this evaluation should not imply that ground-source heat pumps are never cost effective in this region. On the contrary, the purpose of this example is to illustrate that the cost effectiveness of ground-source heat pumps is a function of system heating and cooling load requirements, local utility costs, and most importantly, local installation costs.

**Table 28.8. Oklahoma City Facility Ground-Source Heat Pump Energy Consumption Bin Method Analysis Example**

| Avg. Bin Temp. (°F) | Bin Hours (h/yr) | Bldg. Load (kBtuh) | EWT (°F) | Net GSHP Capacity (kBtuh) | Theor. Run Time (%) | Partial Load Factor | Actual Run Time (%) | GSHP Eff'y (EER or COP) | GSHP Input Power (kW) | Suppl. Heater Power (kW) | Pump Flow Rate (gpm) | GSHP Electric Energy (kWh) | Suppl. Heater Energy (kWh) | Circ. Pump Energy (kWh) | Circ. Pump Energy (kWh) | Total Electric Energy (kWh) |
|---|---|---|---|---|---|---|---|---|---|---|---|---|---|---|---|---|
| (1) | (2) | (3) | (4) | (5) | (6) | (7) | (8) | (9) | (10) | (11) | (12) | (13) | (15) | (16cv) | (16vsd) | (17) |
| **Cooling Mode:** | | | | | | | | | | | | | | | | |
| 112 | 0 | 2,646 | 117 | 1,699.8 | 1.00 | 1.00 | 1.00 | 9.55 | 177.93 | n/a | 494 | 0 | 0 | 0 | 0 | 0 |
| 107 | 1 | 2,392 | 112 | 1,751.3 | 1.00 | 1.00 | 1.00 | 10.20 | 171.63 | n/a | 494 | 172 | 0 | 14 | 14 | 186 |
| 102 | 18 | 2,137 | 106 | 1,802.8 | 1.00 | 1.00 | 1.00 | 10.90 | 165.32 | n/a | 494 | 2,976 | 0 | 256 | 256 | 3,231 |
| 97 | 94 | 1,883 | 101 | 1,854.3 | 1.00 | 1.00 | 1.00 | 11.66 | 159.02 | n/a | 494 | 14,948 | 0 | 1,335 | 1,335 | 16,282 |
| 92 | 240 | 1,628 | 96 | 1,905.8 | 0.85 | 0.96 | 0.89 | 12.48 | 152.71 | n/a | 438 | 32,500 | 0 | 3,407 | 2,679 | 35,180 |
| 87 | 393 | 1,374 | 91 | 1,957.2 | 0.70 | 0.93 | 0.76 | 13.37 | 146.41 | n/a | 374 | 43,644 | 0 | 5,580 | 3,210 | 46,854 |
| 82 | 603 | 1,120 | 85 | 2,008.7 | 0.56 | 0.86 | 0.63 | 14.34 | 140.10 | n/a | 309 | 52,945 | 0 | 8,561 | 3,362 | 56,308 |
| 77 | 829 | 829 | 80 | 2,060.2 | 0.42 | 0.85 | 0.49 | 15.40 | 133.80 | n/a | 242 | 54,477 | 0 | 11,770 | 2,839 | 57,316 |
| 72 | 948 | 611 | 75 | 2,111.7 | 0.29 | 0.82 | 0.35 | 16.56 | 127.49 | n/a | 174 | 42,505 | 0 | 13,459 | 1,665 | 44,170 |
| 67 | 819 | 356 | 69 | 2,163.1 | 0.16 | 0.79 | 0.21 | 17.85 | 121.19 | n/a | 103 | 20,659 | 0 | 11,628 | 504 | 21,163 |
| 62 | 729 | 102 | 64 | 2,214.6 | 0.05 | 0.76 | 0.06 | 19.28 | 114.88 | n/a | 30 | 5,054 | 0 | 10,350 | 38 | 5,092 |
| **Heating Mode:** | | | | | | | | | | | | | | | | |
| 57 | 654 | 153 | 61 | 1,931.3 | 0.08 | 0.77 | 0.10 | 3.88 | 146.07 | 0.00 | 51 | 9,811 | 0 | 9,285 | 98 | 9,909 |
| 52 | 627 | 407 | 58 | 1,872.9 | 0.22 | 0.80 | 0.27 | 3.82 | 143.53 | 0.00 | 133 | 24,323 | 0 | 8,902 | 650 | 24,973 |
| 47 | 608 | 622 | 56 | 1,814.5 | 0.36 | 0.84 | 0.43 | 3.77 | 140.99 | 0.00 | 214 | 37,160 | 0 | 8,632 | 1,622 | 38,782 |
| 42 | 592 | 916 | 54 | 1,756.1 | 0.52 | 0.88 | 0.59 | 3.72 | 138.45 | 0.00 | 292 | 48,566 | 0 | 8,405 | 2,951 | 51,516 |
| 37 | 537 | 1,171 | 51 | 1,697.7 | 0.69 | 0.92 | 0.75 | 3.66 | 135.91 | 0.00 | 369 | 54,559 | 0 | 7,624 | 4,260 | 58,820 |
| 32 | 446 | 1,425 | 49 | 1,639.3 | 0.87 | 0.97 | 0.90 | 3.60 | 133.37 | 0.00 | 443 | 53,457 | 0 | 6,332 | 5,114 | 58,571 |
| 27 | 282 | 1,679 | 47 | 1,580.9 | 1.00 | 1.00 | 1.00 | 3.54 | 130.84 | 28.91 | 494 | 36,896 | 8,152 | 4,004 | 4,004 | 49,051 |
| 22 | 174 | 1,934 | 44 | 1,522.5 | 1.00 | 1.00 | 1.00 | 3.48 | 128.30 | 120.60 | 494 | 22,323 | 20,985 | 2,470 | 2,470 | 45,779 |
| 17 | 85 | 2,188 | 42 | 1,464.1 | 1.00 | 1.00 | 1.00 | 3.41 | 125.76 | 212.30 | 494 | 10,689 | 18,046 | 1,207 | 1,207 | 29,942 |
| 12 | 49 | 2,443 | 40 | 1,405.6 | 1.00 | 1.00 | 1.00 | 3.34 | 123.22 | 304.00 | 494 | 6,038 | 14,896 | 696 | 696 | 21,629 |
| 7 | 16 | 2,697 | 37 | 1,347.2 | 1.00 | 1.00 | 1.00 | 3.27 | 120.68 | 350.00 | 494 | 1,931 | 5,600 | 227 | 227 | 7,758 |
| 2 | 9 | 2,952 | 35 | 1,288.8 | 1.00 | 1.00 | 1.00 | 3.20 | 118.14 | 350.00 | 494 | 1,063 | 3,150 | 128 | 128 | 4,341 |
| -3 | 1 | 3,206 | 33 | 1,230.4 | 1.00 | 1.00 | 1.00 | 3.12 | 115.60 | 350.00 | 494 | 116 | 350 | 14 | 14 | 480 |
| -8 | 0 | 3,461 | 30 | | 1.00 | 1.00 | 1.00 | 3.04 | 113.06 | 350.00 | 494 | 0 | 0 | 0 | 0 | 0 |
| Cooling | | | | | | | | | | | | 269,880 | 0 | 66,359 | 15,902 | 285,782 |
| Heating | | | | | | | | | | | | 306,932 | 71,179 | 57,926 | 23,441 | 401,551 |
| Total | | | | | | | | | | | | 576,812 | 71,179 | 124,284 | 39,342 | 687,333 |

The reader is encouraged to perform similar calculations for a specific site.

The above examples assume that the facility will responsible for all costs associated with the construction and operation of the installed ground-source heat pump system. However, if the equipment were installed under an energy-savings performance contract (ESPC) program, the installation cost could be zero and the economic justification for the project would be significantly different.

## 28.8 THE TECHNOLOGY IN PERSPECTIVE

The future of ground-source heat pump technology looks good because there are many potential commercial applications. Although installation costs are typically higher for ground-source heat pumps than for other technologies, the decision criteria should be based on life-cycle costs rather than first costs; then, a ground-source heat pump system can be the most cost-effective alternative. According to the EPA study, *Space Conditioning: The Next Frontier*, ground-source heat pumps are consistently the most energy-efficient, least polluting of all space conditioning technologies throughout the country (EPA 1993).

The limited industry volume has been one of the factors holding back the development of a broader contractor base. The technology has not enjoyed broad national promotion. According to the results of one survey on the barriers to ground-source heat pumps, HVAC, plumbing, and Architectural/Engineering contractors are conservative, and are therefore reluctant to commit to what they regard as innovative and (to some) unproven technology and equipment (Technical Marketing Associates 1988). Installation of ground-source heat pump systems requires skills beyond those of most HVAC or plumbing contractors. Until recently, only a small number of contractors have performed enough installations to develop an extensive base of experience and expertise in ground-source heat pumps.

### 28.8.1 The Technology's Development

The ground-source heat pump technology has been shown through laboratory testing, field testing, and theoretical analysis to be technically valid and economically attractive in many applications. Energy savings have been verified in a large number of field tests over the past 30 years. In several installations, reductions in maintenance

**Table 28.9. Oklahoma City Facility Conventional Air-Conditioning and Gas Furnace System Bin Method Analysis Example**

| Avg. Bin Temp (°F) | Bin Hours (h/yr) | Bldg. Load (kBtuh) | Net Equip. Capacity (kBtuh) | Theor. Run Time (%) | Partial Load Factor | Actual Run Time (%) | A/C Eff'y (EER) | A/C Input Power (kW) | A/C Electric Energy (kWh) | Fan Energy (kWh) | Total Electric Energy (kWh) | Nat. Gas Furnace Energy (therms) |
|---|---|---|---|---|---|---|---|---|---|---|---|---|
| (1) | (2) | (3) | (5) | (6) | (7) | (8) | (9) | (10) | (13) | (14) | (17) | (18) |
| Cooling Mode: | | | | | | | | | | | | |
| 112 | 0 | 2,646 | 1,621.0 | 1.00 | 1.00 | 1.00 | 8.63 | 187.90 | 0 | 0 | 0 | 0 |
| 107 | 1 | 2,392 | 1,706.0 | 1.00 | 1.00 | 1.00 | 9.25 | 184.40 | 184 | 52 | 236 | 0 |
| 102 | 18 | 2,137 | 1,776.0 | 1.00 | 1.00 | 1.00 | 9.95 | 178.50 | 3,213 | 931 | 4,144 | 0 |
| 97 | 94 | 1,883 | 1,836.0 | 1.00 | 1.00 | 1.00 | 10.74 | 171.00 | 16,074 | 4,860 | 20,934 | 0 |
| 92 | 240 | 1,628 | 1,887.0 | 0.86 | 0.97 | 0.89 | 11.29 | 167.10 | 35,837 | 11,088 | 46,924 | 0 |
| 87 | 393 | 1,374 | 1,932.0 | 0.71 | 0.93 | 0.77 | 11.67 | 165.60 | 49,886 | 15,574 | 65,461 | 0 |
| 82 | 603 | 1,120 | 1,977.0 | 0.57 | 0.89 | 0.64 | 12.05 | 164.10 | 62,850 | 19,801 | 82,652 | 0 |
| 77 | 829 | 829 | 2,022.0 | 0.43 | 0.86 | 0.50 | 12.44 | 162.10 | 67,298 | 21,398 | 88,696 | 0 |
| 72 | 948 | 611 | 2,067.0 | 0.30 | 0.82 | 0.36 | 12.83 | 161.10 | 54,766 | 17,576 | 72,342 | 0 |
| 67 | 819 | 356 | 2,112.0 | 0.17 | 0.79 | 0.21 | 13.23 | 159.60 | 27,831 | 9,015 | 36,846 | 0 |
| 62 | 729 | 102 | 2,157.0 | 0.05 | 0.76 | 0.06 | 13.64 | 158.10 | 7,139 | 2,334 | 9,473 | 0 |
| Heating Mode: | | | | | | | | | | | | |
| 57 | 654 | 153 | 2,700.0 | 0.06 | 0.76 | 0.07 | n/a | n/a | 0 | 1,210 | 1,210 | 1,633 |
| 52 | 627 | 407 | 2,700.0 | 0.15 | 0.79 | 0.19 | n/a | n/a | 0 | 3,001 | 3,001 | 4,051 |
| 47 | 608 | 622 | 2,700.0 | 0.25 | 0.81 | 0.30 | n/a | n/a | 0 | 4,591 | 4,591 | 6,198 |
| 42 | 592 | 916 | 2,700.0 | 0.34 | 0.83 | 0.41 | n/a | n/a | 0 | 6,015 | 6,015 | 8,120 |
| 37 | 537 | 1,171 | 2,700.0 | 0.43 | 0.86 | 0.51 | n/a | n/a | 0 | 6,780 | 6,780 | 9,154 |
| 32 | 446 | 1,425 | 2,700.0 | 0.53 | 0.88 | 0.60 | n/a | n/a | 0 | 6,673 | 6,673 | 9,008 |
| 27 | 282 | 1,679 | 2,700.0 | 0.62 | 0.91 | 0.69 | n/a | n/a | 0 | 4,843 | 4,843 | 6,538 |
| 22 | 174 | 1,934 | 2,700.0 | 0.72 | 0.93 | 0.77 | n/a | n/a | 0 | 3,354 | 3,354 | 4,527 |
| 17 | 85 | 2,188 | 2,700.0 | 0.81 | 0.95 | 0.85 | n/a | n/a | 0 | 1,808 | 1,808 | 2,441 |
| 12 | 49 | 2,443 | 2,700.0 | 0.90 | 0.98 | 0.93 | n/a | n/a | 0 | 1,135 | 1,135 | 1,533 |
| 7 | 16 | 2,697 | 2,700.0 | 1.00 | 1.00 | 1.00 | n/a | n/a | 0 | 400 | 400 | 540 |
| 2 | 9 | 2,952 | 2,700.0 | 1.00 | 1.00 | 1.00 | n/a | n/a | 0 | 225 | 225 | 304 |
| -3 | 1 | 3,206 | 2,700.0 | 1.00 | 1.00 | 1.00 | n/a | n/a | 0 | 25 | 25 | 34 |
| -8 | 0 | 3,461 | 2,700.0 | 1.00 | 1.00 | 1.00 | n/a | n/a | 0 | 0 | 0 | 0 |
| Cooling | | | | | | | | | 325,079 | 102,629 | 427,708 | 0 |
| Heating | | | | | | | | | 0 | 40,060 | 40,060 | 54,081 |
| Total | | | | | | | | | 325,079 | 142,688 | 467,768 | 54,081 |

costs have also been verified. The technology has gained rapid acceptance from users.

The remaining barriers to rapid implementation include (1) acceptance from new users and engineers unfamiliar with the technology, (2) out-of-date cost estimating guides for construction and maintenance, (3) general lack of engineers able to design ground-source heat pump systems, and (4) lack of contractors to install and service ground-source heat pump systems. This chapter is intended to address some of these concerns by reporting on the collective experience of ground-source heat pump users and evaluators and by providing application guidance. For actual design guidance, readers should refer to any of the many publications available, some of which are listed at the end of this chapter.

In addition to the U.S. Department of Energy and the Environmental Protection Agency, two associations are working to address the barriers to increased utilization of ground-source heat pumps: the International Ground Source Heat Pump Association (www.igshpa. okstate.edu) and the Geothermal Heat Pump Consortium

(www.geoexchange.org).

The International Ground Source Heat Pump Association (IGSHPA) is a non-profit, member-driven organization established in 1987 to advance ground-source heat pump technology on local, state, national and international levels. IGSHPA conducts ground-source heat pump system installation training and geothermal research. The mission of the International Ground Source Heat Pump Association and its membership is to promote the use of ground-source heat pump technology worldwide through education and communication.

The government-industry-utility consortium, Geothermal Heat Pump Consortium, Inc., was formed in 1994 to create a self-sustaining ground-source heat pump market. The Geothermal Heat Pump Consortium is a nonprofit organization working to raise awareness and increase the use of ground-source heat pump technology throughout the United States. As a cooperative venture, the Consortium counts among its partners electric utilities, equipment manufacturers, architects, designers, engineers, contractors, builders, drillers, energy

### Table 28.10. Oklahoma City Facility Air-Source Heat Pump System Bin Method Analysis Example

| Avg. Bin Temp (°F) | Bin Hours (h/yr) | Bldg. Load (kBtuh) | Net ASHP Capacity (kBtuh) | Theor. Run Time (%) | Partial Load Factor | Actual Run Time (%) | ASHP Eff'y (EER or COP) | ASHP Input Power (kW) | Suppl. Heater Power (kW) | ASHP Electric Energy (kWh) | Suppl. Heater Energy (kWh) | Total Electric Energy (kWh) |
|---|---|---|---|---|---|---|---|---|---|---|---|---|
| (1) | (2) | (3) | (5) | (6) | (7) | (8) | (9) | (10) | (11) | (13) | (15) | (17) |
| Cooling Mode: | | | | | | | | | | | | |
| 112 | 0 | 2,646 | 1,647.8 | 1.00 | 1.00 | 1.00 | 9.48 | 173.90 | n/a | 0 | 0 | 0 |
| 107 | 1 | 2,392 | 1,702.5 | 1.00 | 1.00 | 1.00 | 10.09 | 168.73 | n/a | 169 | 0 | 169 |
| 102 | 18 | 2,137 | 1,757.1 | 1.00 | 1.00 | 1.00 | 10.74 | 163.57 | n/a | 2,944 | 0 | 2,944 |
| 97 | 94 | 1,883 | 1,811.8 | 1.00 | 1.00 | 1.00 | 11.44 | 158.40 | n/a | 14,890 | 0 | 14,890 |
| 92 | 240 | 1,628 | 1,866.5 | 0.87 | 0.97 | 0.90 | 12.18 | 153.23 | n/a | 33,143 | 0 | 33,143 |
| 87 | 393 | 1,374 | 1,921.1 | 0.72 | 0.93 | 0.77 | 12.97 | 148.07 | n/a | 44,808 | 0 | 44,808 |
| 82 | 603 | 1,120 | 1,975.8 | 0.57 | 0.89 | 0.64 | 13.83 | 142.90 | n/a | 54,759 | 0 | 54,759 |
| 77 | 829 | 829 | 2,030.5 | 0.43 | 0.86 | 0.50 | 14.74 | 137.73 | n/a | 56,798 | 0 | 56,798 |
| 72 | 948 | 611 | 2,085.1 | 0.29 | 0.82 | 0.36 | 15.73 | 132.57 | n/a | 44,709 | 0 | 44,709 |
| 67 | 819 | 356 | 2,139.8 | 0.17 | 0.79 | 0.21 | 16.80 | 127.40 | n/a | 21,942 | 0 | 21,942 |
| 62 | 729 | 102 | 2,194.5 | 0.05 | 0.76 | 0.06 | 17.95 | 122.23 | n/a | 5,426 | 0 | 5,426 |
| Heating Mode: | | | | | | | | | | | | |
| 57 | 654 | 153 | 2,019.0 | 0.08 | 0.77 | 0.10 | 3.11 | 190.20 | 0.00 | 12,234 | 0 | 12,234 |
| 52 | 627 | 407 | 1,884.0 | 0.22 | 0.80 | 0.27 | 3.03 | 182.20 | 0.00 | 30,706 | 0 | 30,706 |
| 47 | 608 | 622 | 1,752.0 | 0.38 | 0.84 | 0.45 | 2.93 | 175.10 | 0.00 | 47,611 | 0 | 47,611 |
| 42 | 592 | 916 | 1,622.0 | 0.56 | 0.89 | 0.63 | 2.82 | 168.60 | 0.00 | 63,254 | 0 | 63,254 |
| 37 | 537 | 1,171 | 1,447.0 | 0.81 | 0.95 | 0.85 | 2.61 | 162.70 | 0.00 | 74,223 | 0 | 74,223 |
| 32 | 446 | 1,425 | 1,242.0 | 1.00 | 1.00 | 1.00 | 2.32 | 157.20 | 53.64 | 70,111 | 23,924 | 94,035 |
| 27 | 282 | 1,679 | 1,118.0 | 1.00 | 1.00 | 1.00 | 2.14 | 152.90 | 164.56 | 43,118 | 46,407 | 89,525 |
| 22 | 174 | 1,934 | 1,048.0 | 1.00 | 1.00 | 1.00 | 2.06 | 149.40 | 259.66 | 25,996 | 45,181 | 71,176 |
| 17 | 85 | 2,188 | 975.0 | 1.00 | 1.00 | 1.00 | 1.96 | 145.40 | 355.63 | 12,376 | 30,229 | 42,605 |
| 12 | 49 | 2,443 | 900.0 | 1.00 | 1.00 | 1.00 | 1.86 | 141.60 | 452.20 | 6,938 | 22,158 | 29,096 |
| 7 | 16 | 2,697 | 810.0 | 1.00 | 1.00 | 1.00 | 1.76 | 134.90 | 500.00 | 2,158 | 8,000 | 10,158 |
| 2 | 9 | 2,952 | 710.0 | 1.00 | 1.00 | 1.00 | 1.65 | 126.40 | 500.00 | 1,138 | 4,500 | 5,638 |
| -3 | 1 | 3,206 | 619.0 | 1.00 | 1.00 | 1.00 | 1.52 | 119.40 | 500.00 | 119 | 500 | 619 |
| -8 | 0 | 3,461 | 534.0 | 1.00 | 1.00 | 1.00 | 1.38 | 113.40 | 500.00 | 0 | 0 | 0 |
| Cooling | | | | | | | | | | 279,589 | 0 | 279,589 |
| Heating | | | | | | | | | | 389,983 | 180,898 | 570,880 |
| Total | | | | | | | | | | 669,571 | 180,898 | 850,469 |

service companies, and other private sector companies that operate in the ground-source heat pump market as well as national, state and local organizations and public agencies.

### 28.8.2 Technology Outlook

The outlook for ground-source heat pumps is bright. In 1999, an estimated 400,000 ground-source heat pumps were operating in residential and commercial applications, up from 100,000 in 1990. In 1985, it was estimated that only around 14,000 ground-source heat pump systems were installed in the United States. Annual sales of approximately 45,000 units were reported in 1997. With a projected annual growth rate of 10%, 120,000 new units would be installed in 2010, for a total of 1.5 million units in 2010.

In Europe, the estimated total number of installed ground-source heat pumps at the end of 1998 was 100,000 to 120,000. Nearly 10,000 ground-source heat pumps have been installed in U.S. federal buildings, over 400 schools, and thousands of low-income houses and apartments.

### 28.9 MANUFACTURERS

There are a number of manufacturers of ground-source heat pumps and water-source heat pump that can be ground coupled. The following is list of U.S. manufacturers identified on the Geothermal Heat Pump Consortium web site[13].

Advanced Geothermal Technology
P.O. Box 6469
Reading, PA 19610
Tel: (610) 796-1450
Fax: (610) 736-0571

American Geothermal, Inc.
1037 Old Salem Road
Murfreesboro, TN 37129
Tel: (615) 890-6985; Tel: (800) 776-8039
Fax: (615) 890-6926
www.amgeo.com

[13]List obtained from the Geothermal Heat Pump Consortium, Inc. web site, www.geoexchange.org/local/manufacturers.htm (last accessed 1 November 2005).

### Table 28.11. Example 2: Summary of Results

| Oklahoma City Facility | Conventional system | Air-Source Heat Pump | Ground-Source Heat Pump |
|---|---|---|---|
| Number of units | 10 | 10 | 35 |
| Nominal capacity (tons) | | | |
| Each | 13.5 | 13.5 | 4.8 |
| Total | 135 | 135 | 168 |
| Supplemental heaters (kW) | | | |
| Each | n/a | 50 | 10 |
| Total | n/a | 500 | 350 |
| Equipment capacity (kBtuh) (at design conditions) | | | |
| Summer | 1,848.0 | 1,822.7 | 1,864.6 |
| Winter | 2,700.0 | 2,621.0 | 2,611.5 |
| Energy consumption (/yr) | | | |
| Electricity (kWh) | 467,708 | 850,469 | 687,333 |
| Demand* (kW-mo) | | | |
| - Winter season | 455 | 2,748 | 1,931 |
| - Summer season | 842 | 770 | 70 |
| Natural gas (therm) | 54,081 | 0 | 0 |
| Total energy ($10^6$ Btu) | 7,004 | 2,902 | 2.345 |
| Energy Cost ($/yr) Electricity | 11,038 | 20,071 | 16,221 |
| Demand | 12,437 | 21,821 | 18,160 |
| Natural gas | 18,388 | 0 | 0 |
| Total energy | 41,863 | 41,892 | 34,381 |
| O&M Costs ($/yr) | 6,750 | 4,725 | 5,880 |
| Installed Costs (S) | 349,300 | 236,000 | 451,500 |
| Equipment life (yr) | 15 | 15 | 15 |
| Total life-cycle cost ($) | 974,924 | 800,065 | 938,282 |

*Assumptions are required to estimate the monthly billing demand. In this case, the authors matched a temperature bin to when the peak demand window was likely to occur. A sensitivity analysis could be performed by changing the assumptions.

ClimateMaster
7300 SW 44th Street
P.O. Box 25788
Oklahoma City, OK 73125
Tel: (405) 745-6000
Fax: (405) 745-3629
www.climatemaster.com

Earth To Air Systems, Inc.
123 SE Parkway Court, Ste 160
Franklin, TN 37064 - USA
Ph: (615) 595-2888
Fax: (615) 595-1080
www.earthtoair.com

Econar Energy Systems Corp.
19230 Evans Street
Elk River, MN 55330
Tel: (612) 241-3110
Tel: (800) 4-ECONAR
Fax: (612) 241-3111
www.econar.com

ECR Technologies
3536 DMG Drive
Lakeland, FL 33811
Tel: (863) 701-0096
Fax: (863) 701-7796
www.ecrtech.com

Enertran Technologies, Inc.
517 McCormick Blvd.
London, Ontario N5W 4C8
Canada
Tel: (519) 452-7049
Fax: (519) 452-0106
www.enertran.ca

FHP Manufacturing
601 NW 65th Court
Ft. Lauderdale, FL 33309
Tel: (954) 776-5471
Fax: (954) 776-5529
www.fhp-mfg.com

HydroDelta Corporation
Tel: (724) 728-4230
Tel: (412) 373-5800
1000 Rico Road
Monroeville, PA 15146
www.hydrodelta.com

Hydro-Temp Corporation
P.O. Box 566
Pocahontas, AR 72455
Tel: (870) 892-8343
Tel: (800) 382-3113
Fax: (870) 892-8323
http://www.hydro-temp.com

McQuay International
4900 Technology Park Blvd.
Auburn, NY 13021
Tel: (315) 282-6243
Fax: (315) 282-6417
www.mcquay.com

Millbrook Industries - Hydronic Division
41659 256th Street
RR #3, Box 265
Mitchell, SD 57301
Tel: (605) 995-0241
Fax: (605) 996-8837
www.hydronmodule.com

The Trane Company
182 Cotton Belt Parkway
McGregor, TX 76657
Tel: (254) 299-6329
Fax: (817) 299-6671
www.trane.com

WaterFurnace International
9000 Conservation Way
Fort Wayne, IN 46809
Tel: (260) 478-5667 ext. 3221
Tel: (800) 222-5667
Fax: (260) 478-3029
www.waterfurnace.com

## 28.10  FOR FURTHER INFORMATION

### Trade Associations

American Society of Heating, Refrigerating, Refrigerating
and Air Conditioning Engineers (ASHRAE)
1791 Tullie Circle, NE
Atlanta, GA 30329
Tel: (404) 636-8400
Fax: (404) 321-5478
www.ashrae.org

Canadian Earth Energy Association
130 Slater Street, Suite 1050
Ottawa, ON Canada K1P6E2
Tel: (613) 230-2332
Fax: (613) 237-1480 Fax
www.earthenergy.ca

Electric Power Research Institute (EPRI)
P.O. Box 50490
3412 Hillview Ave.
Palo Alto, CA 94303
www.epri.com

Geothermal Heat Pump Consortium, Inc. (Geoexchange)
1050 Connecticut Ave, NW, Suite 1000
Washington, D.C.
Tel: (202) 558-7175
Fax: (202) 558-6759
www.geoexchange.org

Geothermal Resource Council
P.O. Box 1350
Davis, CA 95617
www.geothermal.org

International Ground-Source Heat Pump Association
(IGSHPA)
498 Cordell South
Oklahoma State University
Stillwater, OK 74078
Tel: (800) 626-4747
Fax: (405) 744-5283
www.igshpa.okstate.edu

National Rural Electric Cooperative Assn. (NRECA)
1800 Massachusetts Ave., NW
Washington, D.C. 20036
www.nreca.org

National Ground Water Association
601 Dempsey Road
Westerville, OH 43801
Tel: (800) 551-7379
www.ngwa.org

Michigan Geothermal Energy Association
2859 West Jolly Road
Okemos, MI 48864
Tel: (800) 417-5555
www.earthcomfort.com

## Newsletters

"Heat Pump News Exchange," Electric Power Research Institute (EPRI), published
quarterly and distributed by EPRI, Palo Alto, California.

"The Source," International Ground Source Heat Pump Association, published quarterly and distributed by Oklahoma State University Ground Source Heat Pump Publications, Stillwater, Oklahoma.

"Geo-Heat Center Quarterly Bulletin," Geo-Heat Center, published quarterly and distributed by Geo-Heat Center, Oregon Institute of Technology, Klamath Falls, OR.

"Down to Earth," published by the Michigan Geothermal Energy Association, Okemos, MI.

"Outside the Loop," published quarterly by the University of Alabama, Tuscaloosa, Alabama (www.oit. edu/~geoheat/otl/index.htm).

## User and Third-Party Field and Lab Test Reports[14]

Collie, M.J. (editor) 1979. "Comparison of Water-Source and Air-Source Heat Pumps in Northern Environment," a chapter in Heat Pump Technology for Saving Energy. Noyes Data Corporation, Park Ridge, New Jersey.

*Hughes, P.J. and J.A. Shonder. 1998. The Evaluation of a 4000-Home Geothermal Heat Pump Retrofit at Fort Polk, Louisiana: Final Report. ORNL/CON-460, Oak Ridge National Laboratory, Oak Ridge, Tennessee.

Phetteplace, G., H. Ueda, and D. Carbee. 1992. "Performance of Ground-Coupled Heat Pumps in Military Family Housing Units." Solar Engineering. G0656A-1992, American Society of Mechanical Engineers, New York, NY.

Phetteplace, G. 1995. Ground-Coupled Heat Pumps for Family Housing Units. FEAP-UG-CRREL-95/01, U.S. Army Center for Public Works, Alexandria, Virginia.

Svec, O.J. 1987. "Potential of Ground Heat Source Systems." International Journal of Energy Research, Vol. 11, pp. 573-581.

## Design and Installation Guides

*ASHRAE. 1995. "Chapter 29: Geothermal Energy." ASHRAE Handbook: Heating, Ventilating, and Air-Conditioning Applications. American Society of Heating, Refrigerating, and Air- Conditioning Engineers, Inc. Atlanta, Georgia.

Bose, J.E., J.D. Parker, and F.C. McQuiston. 1985. Design/Data Manual for Closed-Loop Ground-Coupled Heat Pump Systems. American Society of Heating, Refrigerating, and Air-Conditioning Engineers, Inc., Atlanta, Georgia.

Caneta Research, Inc. 1995. Commercial/Institutional Ground-Source Heat Pump Engineering Manual. American Society of Heating, Refrigerating, and Air-Conditioning Engineers, Inc., Atlanta, Georgia.

*International Ground-Source Heat Pump Association (not dated). Ground Source Systems: Design and Installation Standards. Oklahoma State University, Stillwater, Oklahoma.

*Oklahoma State University. 1988. Closed-Loop/Ground-Source Heat Pump Systems: Installation Guide. International Ground-Source Heat Pump Association, Stillwater, Oklahoma.

*STS Consultants. 1989. Soil and Rock Classification for the Design of Ground-Coupled Heat Pump Systems: Field Manual. CU-6600, Electric Power Research Institute, Palo Alto, California.

*WaterFurnace. October 1991. Polyethylene Pond/Lake Loop, Mat Design. WF390, WaterFurnace International, Fort Wayne, Indiana.

---

[14]*Denotes literature cited in the technical body of this chapter.

## Manufacturer's Application Notes

Command-Aire. 1988. Earth Energy Heat Pumps: Commercial Water Source Heat Pump Systems; Application, Installation, Operation Manual. Bulletin C-A10-0191.

Guarino, D.S. March 1993. "WaterFurnace: From the Ground Up." Reprint from Contracting Business.

## Utility, Information Service, or
## Government Agency Tech-Transfer Literature

Bas, E. February 13, 1995. "Geothermal Market Receives Hefty Boost from Consortium." The Air Conditioning, Heating, and Refrigeration News, p.1.

*Calm, J.M. September 1987. Proceedings of the Workshop on Ground-Source Heat Pumps. From the workshop held in Albany, New York, October 27 through November 1, 1986. HPC-WR-2. International Energy Agency Heat Pump Center.

Goldfish, L.H., and R.A. Simonelli. 1988. Ground-Source and Hydronic Heat Pump Market Study. EM-6062, Electric Power Research Institute, Palo Alto, California.

Ground Source Systems: Educational and Marketing Material Catalog. International Ground Source Heat Pump Association, Stillwater, Oklahoma.

GS-Systems: An Answer to U.S. Energy and Environmental Concerns: A comprehensive report on ground-source heat pumps and their benefits (not dated). International Ground-Source Heat Pump Association, Oklahoma State University, Stillwater, Oklahoma.

*Pratsch, L.W. 1990. "Geothermal Heat Pumps: A Major Opportunity for the Utility Industry." Presented at the Geothermal Energy Conference, Columbus, Ohio, November 14, 1990.

Scofield, M., and P. Joyner. September 1991. "Heat Pumps for Northern Climates." EPRI Journal 16(6), pp. 28-33.

## Case Studies

Duffy, G. March 22, 1993. "120-ton geothermal system replaces 30-year-old relics." The Air Conditioning, Heating, and Refrigeration News.

*Gahran, A. September 1993. "Grants, Util. Rebate Pay 97% of Ground-Source Heat Pump Project Cost." Energy User News 18(9) pp. 10, 51, 72.

*Gahran, A. January 1994. "New Ground-Source Heat Pumps Cut Energy, Maintenance Costs $68K." Energy User News 19(1) p. 16.

Greenleaf, J. July-September 1982. "College invests in an energy-efficient future." Brown and Caldwell Quarterly. KCPL. 1994. A Model of Efficiency from the Ground Up. Commercial case study 2M/3-135, Kansas City Power and Light, Kansas City, Missouri.

KCPL. 1994. Innovative Heating and Cooling System Allows Increased Efficiencies to Surface. Commercial case study 4-132, Kansas City Power and Light, Kansas City, Missouri.

KCPL. 1994. Station's Ground Source Heat Pump System Finds Efficiency in Unusual Places. Commercial case study 4-118, Kansas City Power and Light, Kansas City, Missouri.

PSO. "Facility Type: Family Fitness Center, Technology Application: Ground- Source Heat Pump." Power Profile. Number 1, Public Service Company of Oklahoma, Tulsa, Oklahoma.

Randazzo, M. November 1994. "Retrofit Cut Elec. Use by 32MMkWh/Yr." Energy User News 19(11) pp. 1,14.

WaterFurnace. Dallas' "Hope of the Sun" goes underground for affordable, efficient, environmental housing. Case Study 5, WF915, WaterFurnace International, Fort Wayne, Indiana.

*WaterFurnace. Elementary school teaches lesson in efficiency. Case Study 2, WF666, WaterFurnace International, Fort Wayne, Indiana.

WaterFurnace. Geothermal comfort goes low-rise in Kisilano, British Columbia. Case Study 8, WF923, WaterFurnace International, Fort Wayne, Indiana.

WaterFurnace. John Deere dealership comforts its clientele from the floor up. Case Study 11, WF922, WaterFurnace International, Fort Wayne, Indiana.

WaterFurnace. Last century's structure retrofitted for the next. Case Study 7, WF924, WaterFurnace International, Fort Wayne, Indiana.

WaterFurnace. Patients treated to geothermal comfort at Dupont Medical Center. Case Study 12, WF942, WaterFurnace International, Fort Wayne, Indiana.

WaterFurnace. Salem Community college enhances environment and education. Case Study 4, WF907, WaterFurnace International, Fort Wayne, Indiana.

*WaterFurnace. Three Indiana schools compare HVAC notes. Case Study 9, WF925, WaterFurnace International, Fort Wayne, Indiana.

*WaterFurnace. WaterFurnace builds state-of-the-art facility. Case Study 1, WF639, WaterFurnace International, Fort Wayne, Indiana.

WaterFurnace. WaterFurnace goes down under, down under. Case Study 10, WF921, WaterFurnace International, Fort Wayne, Indiana.

WaterFurnace. Whitehawk Ranch looks earthward for heating and cooling. Case Study 6, WF916, WaterFurnace International, Fort Wayne, Indiana.

WaterFurnace. York County contributes to a healthier environment. Case Study 3, WaterFurnace International, Fort Wayne, Indiana.

## Codes and Standards

*Energy-Efficient Design of New Buildings Except Low-Rise Residential Buildings. 2004. ASHRAE/IESNA Standard 90.1-2004, American Society of Heating, Refrigerating, and Air- Conditioning Engineers, Inc., Atlanta, Georgia.

*Water-source Heat Pumps – Testing and Rating for Performance – Part 1: Water-to-air and Brine-to-air Heat Pumps. 2005. ANSI/ARI/ASHRAE ISO Standard 13256-1-2005, American Society of Heating, Refrigerating, and Air- Conditioning Engineers, Inc., Atlanta.

*Ground Source Closed-Loop Heat Pumps. 1998. ANSI/ARI Standard 330-98, Air-Conditioning and Refrigeration Institute, Inc., Arlington, Virginia.

*Ground Water-Source Heat Pumps. 1998. ARI Standard 325-98. Air-Conditioning and Refrigeration Institute, Inc., Arlington, Virginia.

*Water-Source Heat Pumps. 1998. ANSI/ARI Standard 320-98. Air-Conditioning and Refrigeration Institute, Inc., Arlington, Virginia.

## Other Sources

American Society of Heating, Refrigerating, and Air-Conditioning Engineers (ASHRAE). 1997. Ground Source Heat Pumps — Design of Geothermal Systems for Commercial and Institutional Buildings. ASHRAE, Atlanta, GA.

Cane, D., A. Morrison, and C. Ireland. 1998. "Maintenance and Service Costs of Commercial Building Ground-Source Heat Pump Systems." ASHRAE Transactions, Vol 104(2): 699-706.

Duffy, G. July/August 1989. "Commercial Earth-Coupled Heat Pumps Are Ready To Roll." Engineered Systems, pp. 40-47.

*Electric Power Research Institute (EPRI). May 1993. 1992 Survey of Utility Demand-Side Management Programs. EPRI-TR 102193, Palo Alto, California.

*EPA. 1993. Space Conditioning: The Next Frontier. Office of Air and Radiation, 430-R-93-004 (4/93), EPA, Washington, D.C.

*EPA. 1994. Voluntary Reporting Formats for Greenhouse Gas Emissions and Carbon Sequestration. (Public Review Draft, May 1994). EPA, Washington, D.C.

*Geothermal Heat Pumps." February 1990. Custom Builder. 5(2) pp. 32-35.

*Giddings, T. February 1988. "Ground Water Heat Pumps: Why Doesn't Everybody Want One?" Well Water Journal 42(2) pp. 32-33.

Gilmore, V.E. June 1988. "Neo-geo heat pump." Popular Science 232(6) pp. 88-89,112.

*Heinonen, E.W., R.E. Tapscott, M.W. Wildin, and A.N. Beall. February 1997. "Assessment of Antifreeze Solutions for Ground-Source Heat Pump Systems." Report 908RP, American Society of Heating, Refrigerating, and Air-Conditioning Engineers (ASHRAE), Atlanta.

Hurlburt, S. February 1988. "Ground Water Heat Pumps - The Second Wave." Well Water Journal 42(2) pp. 34-40.

Kavanaugh, S., C Gilbreath, J. Kilpatrick. 1995. Cost Containment for Ground-Source Heat Pumps, Final Report. University of Alabama, Tuscaloosa.

Kavanaugh, S. September 1992. "Ground-coupled heat pumps for commercial buildings." ASHRAE Journal 34(9), 30-37.

*Lund, J.W. and T. L. Boyd. 2000. "Geothermal Direct-Use in the United States in 2000." Geo-Heat Center Quarterly Bulletin, Vol. 21, No. 1 (March), Klamath Falls, Oregon, pp. 1-5.

Martin, M.A., D.J. Durfee, and P.J. Hughes. 1999. "Comparing Maintenance Costs of Geothermal Heat Pump Systems with Other HVAC Systems in Lincoln Public Schools: Repair, Service, and Corrective Action." ASHRAE Transactions, Vol. 105(2): 1208-1216.

Martin, M.A., M.G. Madgett, and P.J. Hughes. 2000. "Comparing Maintenance Costs of Geothermal Heat Pump Systems with Other HVAC Systems: Preventive Maintenance Actions and Total Maintenance Costs." ASHRAE Transactions, Vol. 106(1): 1-15.

*Nemeth, R.J., D.E. Fornier, and L.A. Edgar. 1993. Renewables and Energy- Efficiency Planning. U.S. Army Construction Engineering Research Laboratory, Champaign, Illinois.

Oak Ridge National Laboratory (not dated). "Big Savings from the World's Largest Installation of Geothermal Heat Pumps at Fort Polk, Louisiana." Oak Ridge National Laboratory, Oak Ridge, Tennessee.

Outside the Loop. 1999. "Impact of Conductivity Error on Design Results." Outside the Loop - Volume 2, Number 3, University of Alabama, Tuscaloosa, Alabama.

Phetteplace, G., H. Ueda, and D. Carbee. April 1992. "Performance of Ground-Coupled Heat Pumps in Military Family Housing Units." Proceedings of the 1992 ASME International Solar Energy Conference. Maui, Hawaii, 4-8 April, 1992, pp. 377-383.

Reynolds, William C., and Herry C. Perkins. 1977. Engineering Thermodynamics, pp. 287-289. McGraw-Hill Book Co., New York.

Rybach, L., and B. Sanner. March 2000. "Ground-Source Heat Pump Systems the European Experience." Geo-Heat Center Quarterly Bulletin 21(1):16-26, Klamath Falls, OR.

*Shonder J.A., D. Durfee, M.A. Martin, P.J. Hughes, and T.R. Sharp. 1999. "Benchmark for Performance: Geothermal Applications in Lincoln Public Schools." ASHRAE Transactions 105(2): 1199-1207.

Shonder, J. A., M. A. Martin, T. R. Sharp, D. J. Durfee, and P. J. Hughes. 1999. "Benchmark for Performance: Geothermal Applications in Lincoln Public Schools." ASHRAE SE-99-20-03, Atlanta.

Shonder, J. A., V. Baxter, J. Thorton, and P. Hughes. June 1999. "A New Comparison of Vertical Ground Heat Exchanger Design Methods for Residential Applications." ASHRAE Transactions.

Shonder, J. A., V. Baxter, P. Hughes, and J. Thorton (draft - no date). "A Comparison of Vertical Ground Heat Exchanger Design Software for Commercial Applications." ASHRAE Transactions.

Shonder, J.A. and J.V. Beck. 2000. "Field Test of a New Method for Determining Soil Formation Thermal Conductivity and Borehole Resistance from In Situ Measurements." ASHRAE Transactions.

*Spilker, E. H. January 1998. "Ground-Coupled Heat Pump Loop Design Using Thermal Conductivity Testing and the Effect of Different Backfill Materials on Vertical Bore Length." Proceedings of the American Society of Heating, Refrigerating, and Air-Conditioning Engineers (ASHRAE). January 1998 meeting, San Francisco, paper SF98-1-3.

Swanson, G.L. August 1986. "Heating and Cooling with GWHPs." Well Water Journal 40(8) pp. 44-46.

*Technical Marketing Associates, Inc. 1988. Ground-Source and Hydronic Heat Pump Market Study. EM-6062, Electric Power Research Institute, Palo Alto, California.

*TM 5-785. 1978. Engineering Weather Data. Department of the Army Technical Manual. TM 5-785, U.S. Government Printing Office, Washington, D.C.

# Appendix

### PERFORMANCE AND EFFICIENCY TERMINOLOGY

There are many terms used in the heating, air-conditioning, and refrigeration industry that convey performance and efficiency. Many of these terms are synonymous, others are not. When comparing various systems, it is important to understand what the assorted terms are, how they are determined, and their relationship. The following provides a brief description of many (but not all) of the terms used to convey efficiency. Use of the terms is also summarized in Table 28.A.1.

**Annual Fuel Utilization Efficiency (AFUE):** For fuel-fired systems such as boilers and furnaces, AFUE is defined as the ratio of annual output energy to annual input energy. This term is generally applied to systems <=300,000 Btu/ h input. AFUE is a weighted average efficiency under standard rated conditions at various part-load conditions and also includes any non-heating-season pilot input losses. AFUE is a unitless term generally expressed as a percentage.

**Coefficient of Performance (COP):** The COP is the basic parameter used to report the efficiency of refrigerant-based systems. It is a unitless term. This term is universal in its use but not in its meaning. COP can be used to define both cooling efficiency or heating efficiency, such as for a heat pump. For cooling, COP is defined as the ratio of the rate of heat removal to the rate of energy input to the compressor, in consistent units. For heating, COP is defined as the ratio of rate of heat delivered to the rate of energy input to the compressor, in consistent units. COP can be used to define the efficiency at a single (standard or nonstandard) rated condition or a weighted average (seasonal) condition. Depending on its use, the term may or may not include the energy consumption of auxiliary systems such as indoor or outdoor fans, chilled water pumps, or cooling tower systems. For purposes of comparison, the higher the COP the more efficient the system. For mathematical purposes, COP can be treated as an efficiency. (COP of 2.00 = 200% efficient) For unitary heat pumps, ratings at two standard (outdoor) temperatures (47°F and 17°F [8.3°C and -8.3°C]) may be reported based on old ARI standards.

**Combustion Efficiency ($n_c$ or $E_c$):** For fuel-fired systems, this efficiency term is defined as the ratio of the fuel energy input minus the flue gas losses (dry flue gas, incomplete combustion and moisture formed by combustion of hydrogen) to the fuel energy input. In the U.S., fuel-fired efficiencies are reported based on the higher heating value of the fuel. Other countries report fuel-fired efficiencies based on the lower heating value of the fuel. The combustion efficiency is calculated by determining the fuel gas losses as a percent of fuel burned. [$E_c = 1 -$ (flue gas losses)]

**Energy Efficiency Ratio (EER):** The EER is a term generally used to define the cooling efficiency of unitary air-conditioning and heat pump systems. The term implies that the efficiency is determined at a single rated condition specified by the appropriate equipment standard and is defined as the ratio of net cooling capacity (heat removed in Btu/h) to the total input rate of electric energy required (watt). The units of EER are Btu/w-h. It is important to note that this efficiency term typically includes the energy requirement of auxiliary systems such as the indoor and outdoor fans. For purposes of comparison, the higher the EER the more efficient the system. To convert to a COP, divide the EER by 3.412. [COP = EER/3.412]

**Heating Seasonal Performance Factor (HSPF):** The term HSPF is similar to the term SEER, except it is used to signify the seasonal heating efficiency of heat pumps. The HSPF is a weighted average efficiency over a range of outside air conditions following a specific standard test method. The term is generally applied to heat pump systems less than 60,000 Btu/h (rated cooling capacity.) The units of HSPF are Btu/w-h. It is important to note that this efficiency term typically includes the energy requirement of auxiliary systems such as the indoor and outdoor fans. For purposes of comparison, the higher the HSPF the more efficient the system.

**Integrated Part-Load Value (IPLV):** The term IPLV is used to signify the cooling efficiency related to a typical (hypothetical) season rather than a single rated condition. The IPLV is calculated by determining the weighted average efficiency at part-load capacities specified by an accepted standard. It is also important to note that IPLVs are typically calculated using the same condensing temperature for each part-load condition and IPLVs do not include cycling or load/unload losses. The units of IPLV are not consistent in the literature; therefore, it is important to confirm which units are implied when the term IPLV is used. ASHRAE Standard 90.1 (using ARI reference standards) uses the term IPLV to report seasonal cooling efficiencies for both seasonal COPs (unitless) and seasonal EERs (Btu/w-h), depending on the equipment capacity category; and most chiller manufacturers report seasonal efficiencies for large chillers as IPLV using units of kW/ton. Depending on how a cooling system loads and unloads (or cycles), the IPLV can be between 5 and 50% higher than the EER at the standard rated condition.

**kW/ton:** The term kW/ton is generally used for large commercial and industrial air-conditioning, heat pump, and refrigeration systems. The term is defined as the ratio of the rate of energy consumption (kW) to the rate of heat removal (ton) at a rated condition. As the term suggests, the units are kW/ton. Because refrigeration systems of this capacity are typically custom designed, the reported kW/ton generally implies only the compressor and does not include the auxiliaries. However, for specific references, auxiliaries can be added to report the overall system efficiency using this term. It is important to note that this term is inverse to the other performance and efficiency terminology. Therefore, for purposes of comparison, the lower the kW/ton the more efficient the system. To convert to a COP, divide 12,000 by the product of 3,412 multiplied by the kW/ton. [COP = 12,000/{(3,412) * (kW/ton)}]

**Seasonal Energy Efficiency Ratio (SEER):** The term SEER is used to define the average annual cooling efficiency of an air conditioning or heat pump system. The term SEER is similar to the term EER but is related to a typical (hypothetical) season rather than for a single rated condition. The SEER is a weighted average of EERs over a range of rated outside air conditions following a specific standard test method. The term is generally applied to systems less than 60,000 Btu/h. The units of SEER are Btu/W-h. It is important to note that this efficiency term typically includes the energy requirements of auxiliary systems such as the indoor and outdoor fans. For purposes of comparison, the higher the SEER the more efficient the system. Although SEERs and EERs cannot be directly compared, the SEERs usually range from 0.5 to 1.0 higher than corresponding EERs.

**Thermal Efficiency ($n_t$ or $E_t$):** This efficiency term is generally defined as the ratio of the heat absorbed by the water (or the water and steam) to the heat value of the energy consumed. The combustion efficiency of a fuel-fired system will be higher than its thermal efficiency. See ASME Power Test Code 4.1 for more details on determining the thermal efficiency of boilers and other fuel-fired systems. In the U.S., fuel-fired efficiencies are typically reported based on the higher heating value of the fuel. Other countries typically report fuel-fired efficiencies based on the fuel's lower heating value. The difference between a fuel's higher heating value and its lower heating value is the latent energy contained in the water vapor (in the exhaust gas) which results when hydrogen (from the fuel) is burned. The efficiency of a system based on a fuel's lower heating value can be 10 to 15% higher than its efficiency based on a fuel's higher heating value.

**Table 28.A.1.**
**Summary of Performance and Efficiency Terminology**

| Operating Mode | Design Rated Conditions | Seasonal Average Conditions |
|---|---|---|
| Cooling | COP | COP |
| | EER | SEER |
| | kW/ton | IPLV |
| Heating | COP | AFUE |
| | $E_c$ | COP |
| | $E_t$ | HSPF |

# CHAPTER 29

# SUSTAINABILITY AND HIGH PERFORMANCE GREEN BUILDINGS

## LEED FOR NEW CONSTRUCTION AND EXISTING BUILDINGS

**NICK STECKY**
NJS Associates, LLC

## 29.1 BEGINNINGS

The publication of Rachel Carson's *Silent Spring* in 1962 alerted the general public to the dangers of pesticides, in particular the dangers to humans. This helped precipitate the rise of an environmental movement, politics and laws in the United States during the sixties and seventies. New laws were passed to protect the environment. These included:

- The National Environmental Policy Act of 1969—this created the Federal Environmental Protection Agency.
- The Clean Air Act was passed in 1970. This greatly expanded the protection of two previous laws, the Air Pollution Control Act of 1955 and the first Clean Air Act of 1963.
- The Water Pollution Control Act of 1972.
- The Endangered Species Act of 1973.
- The formation of the Federal Department of Energy in the late seventies.

But a problem with environmentalism was beginning to brew. There came a tension, an apparent conflict between the need to preserve the environment and the need to grow and expand the economy and jobs. Environmentalists began to be seen as opponents of growth and industry. There appeared to be a contradiction between business and protection of the environment. Environmentalism began to be seen as just another "special interest" group which simply added cost to running a business with very little added value.

In addition to legislation, other events were occurring during the seventies, eighties, and nineties which encouraged the development of sustainability:

- Earth Day, launched April 22, 1970.
- Nuclear power suffered major setbacks with the incidents at Three Mile Island and Chernobyl.
- Major oceanic oil spills, including the EXXON Valdez.
- The OPEC oil crises of the mid-seventies and early eighties.
- The mid-seventies natural gas shortages that caused many plants to close during the winter to preserve gas for home heating.
- Discovery that ozone depleting compounds, such as CFC refrigerants, were destroying the ozone layer of the atmosphere.

During the eighties, in reaction to the forces of high energy costs, inadequate energy supplies, environmentalism, and pollution control, a new approach to designing, building and operating buildings began to develop. It was recognized that buildings consume significant percentages of our resources, open space and energy. However, some of the new approaches were not without problems. For example, architects experimented with half or even fully buried homes. But many times these homes had problems such as humidity, which led to mold. As a result of the desire to save energy on ventilation air, ASHRAE modified Standard 62 and reduced ventilation rates to five cubic feet per minute (CFM) for offices. This contributed to what was later called sick building syndrome, or SBS. SBS was a catchall phrase for any building that provided an uncomfortable, irritating, and possibly unhealthy indoor environment for the occupants. Employees in these buildings generally had higher rates of absenteeism, lower productivity, and higher incidences of lawsuits against building owners and employers.

### 29.1.1 The Sustainability Movement

While buildings and occupants were suffering through difficulties as described above, the nineties gave rise to the sustainability movement. It was in 1999 that the book *Natural Capitalism* by P. Hawken, A. Lovins, and L. H. Lovins was released. The authors present the connection between economics and environmentalism, and argued that they are mutually supportive, not mutually exclusive. *Natural Capitalism* points out that as in its title, nature itself is capital. For example, what is the value of a clean lake? It is drinking water that needs less expensive water treatment before being potable. It is the recreational value and income for some, through swimmers and boaters. If the lake were destroyed through pollution, water treatment costs would rise and lake use revenues go down. Seems simple enough, but it has mostly been ignored until now.

Historically, the environmental movement consisted primarily of regulation, legislation, and mandates. However, this regulatory approach has often been seen by the business community as an obstacle to growth. But a new form of green revolution was emerging. This one may succeed where traditional legislative environmentalism has had limited success. This new form is a larger view, taking economic, community, and technological considerations into account, as well as environmental. This new sustainability recognizes the need to consider the cost-benefit analysis in evaluating and/or promoting various programs. And the themes presented in *Natural Capitalism* help us identify and quantify the complete costs and benefits associated with sustainability.

### 29.1.2 Sustainability Defined

Some of the many definitions of sustainability that exist include:

### Design Ecology Project
Sustainability is a state or process that can be maintained indefinitely. The principles of sustainability integrate three closely intertwined elements—the environment, the economy, and the social system—into a system that can be maintained in a healthy state indefinitely.

### Brundtland Commission of the UN
Development is sustainable "if it meets the needs of the present without compromising the ability of future generations to meet their own needs."

### ASHRAE defines sustainability as
"providing for the needs of the present without detracting from the ability to fulfill the needs of the future."

Note the commonality of theme. The concept is rooted in maintaining our current standards of living without jeopardizing future generation's standards of living.

## 29.2 SUSTAINABILITY GIVES RISE TO THE GREEN BUILDING MOVEMENT

The late eighties and early nineties were a crucial developmental period for the green design movement. Leaders of green design included William McDonough, Paul Hawken, John Picard, Bill Browning, and David Gottfried, who later went on to be one of the co-founders of the USGBC. The movement acknowledged that buildings represent a very significant usage of resources, land, and energy, and that improvements to the ways in which we design, construct, operate, and decommission buildings could make significant contributions to improvement of the environment and overall sustainability.

Today, owners, occupants, and communities are beginning to hold buildings to higher standards. Industry leaders are responding by creating physical assets that save energy and resources, and are more satisfying and productive, economical as well as environmentally accountable. Building owners who want the greatest return on investment can take a path that is green both economically and environmentally. How are they doing it? They are doing it through integrated solutions for the design, construction, maintenance, and operations, as well as the ultimate disposal of a building.

An integrated, or whole building approach, may mean that up-front costs may be no more than conventional construction, but life cycle costs over the life of the asset will be lower. By designing, building, and operating in an integrated way, owners can expect high performance buildings that offer:

- Increased efficiencies of systems and use of resources and energy.
- Quality indoor environments that are healthy, secure, pleasing and productive for occupants and operators.
- Optimal economic and environmental performance.
- Wise use of building sites, assets and materials.
- Landscaping, material use and recycling efforts inspired by the natural environment.
- Lessened human impact upon the natural environment.

### 29.2.1 Formation of the Unites States Green Building Council, USGBC

In 1993, David Gottfried, a developer, Mike Italiano, an environmental attorney, and Rick Fedrizzi of Carrier Corporation got together to form the United States Green Building Council, USGBC. They had become concerned about the fragmentation of the building industry as it relates to sustainable design. At the time, there was no clear consensus among industry professionals as to what constituted a "green design." The USGBC was initially formed to create an educational organization that would bring building professionals together to promote sustainable design. In 1997, the USGBC was awarded a $200,000 grant from the Federal Department of Energy, and the USGBC was off and running. The USGBC has made a large impact on the design and building industry. There are more than 5,500 members consisting of individuals, large and leading corporations, governmental entities, universities, educational entities, consultants, product manufacturers, trade associations, and more.

### 29.2.2 Making It All Come Together: The USGBC

The U.S. Green Building Council, (USGBC.org), a balanced consensus coalition representing every sector of the building industry, spent five years developing, testing, and refining the LEED Green Building Rating System. LEED stands for "Leadership in Energy & Environmental Design," and when adopted from the start of a project helps facilitate integration throughout the design, construction, and operation of buildings.

*MISSION STATEMENT*
*The U.S. Green Building Council is the nation's foremost coalition of leaders from across the building industry working to promote buildings that are environmentally responsible, profitable, and healthy places to live and work.*

The mission of this unprecedented coalition is to accelerate the adoption of green building practices, technologies, policies, and standards. The USGBC is a committee-based organization endeavoring to move the green building industry forward with market-based solutions. Another vital function of the council is linking industry and government. The council has formed effective relationships and priority programs with key federal agencies, including the U.S. DOE, EPA, and GSA.

The council's membership is open and balanced. It is comprised of leading and visionary representation from all segments of the building industry including product manufacturers, environmental groups, building owners, building professionals, utilities, city government, research institutions, professional societies and universities. This type of representation provides a unique, integrated platform for carrying out important programs and activities.

### LEED™ Overview

The LEED Green Building Rating System™ is a priority program of the US Green Building Council. It is a

voluntary, consensus-based, market-driven building rating system based on existing proven technology. It evaluates environmental performance from an integrated or "whole building" perspective over a building's life cycle, providing a definitive standard for what constitutes a "green building."

LEED™ is based on accepted energy and environmental principles and strikes a balance between known effective practices and emerging concepts. Unlike other rating systems currently in existence, the development of LEED Green Building Rating System™ was initiated by the US Green Council Membership, representing all segments of the building industry. It also has been open to public scrutiny.

LEED™ is an assessment system that incorporates third party verification and is designed for rating new and existing commercial, institutional, and high-rise residential buildings. It is a feature-oriented system where credits, also called points, are earned for satisfying each criterion. Different levels of green building certification are awarded based on the total credits earned. The system is designed to be comprehensive in scope, yet simple in operation.

### 29.2.3 General Introduction & Discussion

There are several key points in this section:

- LEED is becoming nationally accepted at local, state and federal levels.
- The engineering community such as ASHRAE, AEE, IESNA has the knowledge and skills that can add significant value to a LEED design team.
- LEED has value to the engineering community, both as owners/operators of facilities or as design team members.
- It is a WIN-WIN proposition for us all.
- Green buildings are a process, not a collection of technologies.
- Encourage engineering participation/membership in USGBC and its local chapters.
- Encourage professional accreditation and attendance at LEED workshops.
- Encourage building owners/operators to register and certify projects.

There are many different terms for sustainable buildings, but basically they all convey the same message:

- Sustainable Design
- High Performance Buildings
- High Efficiency Buildings
- Integrated Building Design
- Green Buildings

Sustainable buildings may be our goal, but the most common term used for these high-performance buildings is green buildings. For some, the term green buildings may sound too environmentally focused, but the way it is used here, it represents the all-inclusive idea of sustainable buildings.

**Characteristics of Sustainable Green Buildings**
- Optimal environmental and economic performance.
- Increased efficiencies saving energy and resources.
- Satisfying, productive, quality indoor spaces.
- Whole building design, construction and operation over the entire life cycle.
- A fully integrated approach—teams, processes, systems.

### 29.2.4 Green Buildings

Green buildings are designed and constructed in accordance with practices that significantly reduce or eliminate the negative impact of buildings on the environment and its occupants. This includes design, construction, operations, AND, ultimately, demolition. Five fundamental categories constitute the USGBC green building designation. They are:

- Sustainable Site Planning
- Safeguarding Water and Water Efficiency
- Energy Efficiency and Renewable Energy
- Conservation of Materials and Resources
- Indoor Environmental Quality

All relate back to the previous definitions and discussions of sustainability.

All of these are contained in the LEED standard, Leadership in Energy and Environmental Design. This is the trademark rating system developed by the United States Green Building Council, USGBC.

Besides the LEED rating system to define and describe sustainable buildings, there are others such as:

- The British Research Establishment Environmental Assessment Method (BREEAM) was launched in 1990 and is increasing in its use.
- Canada's Building Environmental Performance Assessment Criteria (BEPAC) began in 1994, but was never fully implemented due to its complexity. Canada has now licensed use of LEED from the USGBC.
- The Hong Kong Building Environmental Assessment Method (HK-BEAM) is currently in pilot form.

- The USGBC LEED family of programs.
- State and regional guides include high performance building guidelines in NY and PA's guidelines for creating high performance buildings, as well as California's programs for school construction.
- Green Globes is a web-based self-assessment program that guides the integration of green principles into a building's design.

LEED strives to encompass a wide band of sustainability that includes:
- Society and Community—recognizes that buildings exist to serve the needs of the community, but that they must also minimize their impacts.
- Environment—again striving to minimize negative impacts on the environment
- Economics—recognizes that the adoption of these sustainability initiatives by business will require economic benefits that can be delivered by green buildings.
- Energy—recognizes that energy plays a key role in building operating costs as well as a sustainable energy future.

### 29.2.5 Benefits of LEED Buildings
#### The Environment

At the onset, LEED was created to standardize the concept of building green to offer the building industry a universal program that provided concrete guidelines for the design and construction of sustainable buildings for a livable future. As such, it is firmly rooted in the conservation of our world's resources. Each credit point awarded through the rating system reduces our demand and footprint upon the natural environment.

#### Economics

Conventional wisdom says that the construction of an environmentally friendly, energy efficient building brings with it a substantial price tag and extended timetables. This need not be the case. Breakthroughs in building materials, operating systems and integrated technologies have made building green not only a timely, cost effective alternative, but a preferred method of construction among the nation's leading professionals.

#### From the USGBC

"Smart business people recognize that high performance green buildings produce more than just a cleaner, healthier environment. They also positively impact the bottom line. Benefits include: better use of building materials, significant operational savings, and increased workplace productivity."

In many instances, green alternatives to conventional building methods are less expensive to purchase and install. An even larger number provide tremendous operational savings. The USGBC and LEED offer the building industry a fiscally sound platform on which to build their case for whole-building design and construction. Green buildings can show a positive return on investment for owners and builders.

Economic benefits also can include:

- Improved occupant performance—employee productivity rises, students' grades improve. California schools have analyses that show children in high performance facilities have improved test scores.
- Absenteeism is reduced.
- Retail stores have observed measurable sales improvements in stores with daylighting.

The savings associated with productivity gains can be the single largest category of savings in an office building. For example, salaries per square foot can be in the range of ten times or more than the cost of energy per square foot in a typical office building. So while the energy engineering community agonizes over extracting each cent in energy costs, there are greater savings potentials by leveraging the increased employee productivity through high performance buildings. As described earlier, these buildings provide a superior indoor environment with regard to lighting, noise control, temperature and humidity control, ventilation, and fresh air.

Building green also enhances asset value. According to organizations such as the International Facilities Management Association (IFMA) and the Building Owners and Managers Association (BOMA), the asset value of a property rises at a rate of ten times the value of the operational savings. For example, if green building efficiency reduces operating costs by $1/sq ft per year, the asset value of that property rises by ten times that amount, or $10. For a 300,000 sq ft office complex, annual savings could be $300,000, and the asset value increase would be approximately $3,000,000. It pays to be efficient.

**Societal Implications**

In addition to the environmental and economic benefits of building green, there are also significant societal benefits. These include increased productivity, a healthy work environment, comfort, and satisfaction, to name a few. The good news about green buildings can then be leveraged with local and trade media through press releases, ceremonies, or events. By calling attention to the building and its certification status, owners speak volumes about themselves. The USGBC writes:

"Like a strong prospectus, building green sends the right message about a company or organization: it's well run, responsible, and committed to the future"

This, too, has a direct effect on the bottom line.

### 29.2.6 Benefits to the Architectural and Engineering Community

All too often, market forces drive a building design team to focus on minimum first cost regardless of what the overall life cycle costs of this minimum first cost design may be. The end result is that all elements of society, from owner, to occupants, and to the community end up with a building which is less than what it could have or should have been. It used too many resources in construction, its energy and operating costs are high, and it does not provide the optimum indoor environment for employee productivity.

However, installation of the LEED process of forming an integrated design team on conceptual design day one promotes the formation of a creative solution to the particular buildings needs being planned for. Within this creative roundtable of equals, the team is able to maximize use of the collective wisdom of the members and develop a design concept that can be both green and economic. For example, increasing the use of natural light to displace some artificial light can result in a reduced load on air conditioning systems. These AC systems can then be downsized, resulting in reduced equipment first costs and reduction in electricity use, both through the artificial lighting reduction and reduced cooling loads. Small savings multiply and reverberate throughout the design, ultimately having significant impacts on overall first costs and operating costs.

When building "green," the sum is larger than the individual parts only when all components are integrated into a single unified system. Integration draws upon every aspect of the building to realize efficiencies, cost savings, and continuous returns on investment.

The most significant benefit to the architectural/engineering (A/E) design community is that LEED promotes and rewards creative solutions. Sustainable and green designs are not yet commodity skills that all firms can lay claim to. For those A/E firms seeking to provide more value to their clients, LEED is a way to achieve this. LEED can become a standard for design excellence, and provides an A/E with a brand differentiator versus the competition. LEED can become an outstanding competitive edge in the marketplace for firms seeking a leadership position of excellence.

**LEED Acceptance**

There's a groundswell of acceptance taking place.

When LEED was first introduced in 2000, there was reluctance to accept it. It represented a new way of looking at the design, construction, operations, and disposal of facilities. It called for raising the bar, which many have been slow to accept. There were many questions about higher first costs, paybacks, overall benefits, ability of the design community to deliver, doubts about the technologies involved, etc. The perception was that green buildings were too futuristic, unobtainable, and required too many tradeoffs to be able to deliver a practical, efficient facility where people could live, work, and play.

Some elements of these doubts remain, especially as to the financial benefits question. However, as we gain experience with green buildings, we are developing the experience and the data needed to resolve these doubts.

**The First Cost "Premium" of LEED**

Conventional wisdom says that green buildings cost more. However, as the industry becomes more experienced with the actual delivery of green buildings, green is becoming more cost-neutral. In an article titled "The Costs and Financial Benefits of High Performance Buildings," Greg Kats of Capital E analyzed 40 California LEED buildings for the "cost premium" of LEED. The study consisted of 32 office buildings and eight schools:

- The eight LEED-certified buildings (the basic level of LEED certification) cost an average of 0.7% more.
- The twenty-one Silver rated buildings cost an average 1.9% more.
- The nine Gold rated buildings cost average of 2.2% more.
- The two Platinum rated buildings cost an average of 6.8% more.

**29.2.7 LEED Described**

*Why was LEED created?*
- LEED Developed as a way to define and quantify what constitutes a sustainable green design.
- Defines "green" by providing a standard for measurement.
- Addresses the "greenwashing" issue, such as false or exaggerated claims.
- Facilitates positive results for the environment, occupant health, and financial return.
- Use as a design guideline.
- Recognizes leaders.
- Stimulates green competition.
- Establishes market value with recognizable national "brand."
- Raises consumer awareness.
- Transforms the marketplace.

- Promotes a whole-building, integrated design process.

*What is LEED*
- Consists of performance-based and prescriptive-based criteria.
- Focuses on the whole building system instead of the components.
- Life cycle based, not first cost.
- Promotes architectural and engineering innovation, i.e., innovation LEED credits.
- Provides a third-party verification process to ensure quality and compliance.

The growing family of LEED building rating systems includes:

LEED NC for new construction,
LEED EB for existing buildings,
LEED CI for commercial interiors,
LEED CS for commercial core & shell,
LEED H for homes, currently in pilot.

Other LEED programs are being developed.

**29.3  INTRODUCING THE LEED NC RATING SYSTEM: A TECHNICAL REVIEW**

The LEED format for rating a green building consists of two categories:
- Prerequisites—These are mandatory requirements and all must be satisfied before a building can be certified.
- Credits—Each credit is optional, with each contributing to the overall total of credits. This will determine the level a building will be rated at, i.e. Certified, Silver, Gold or Platinum.

**29.3.1 Sustainable Sites—14 Possible Points**
Prerequisite:
- Erosion & Sedimentation Control—Control erosion to reduce negative impacts on water and air quality by complying with the EPA storm water management requirements for construction activities.

Credits:
- Site Selection—Avoid development of inappropriate sites and reduce the environmental impact from the location of a building on a site. One point.
- Urban Redevelopment—Channel development to urban areas with existing infrastructures, protecting

green fields and preserving habitat and natural resources. One point.

- Brownfield Redevelopment—Rehabilitate damaged sites where development is complicated by real or perceived environmental contamination, thereby reducing pressure on undeveloped land. One point.
- Alternative Transportation—Four points available: one each for public transportation access, bicycle storage, alternative fuel vehicles, and parking capacity.
- Reduced Site Disturbance—Conserve existing natural areas and restore damaged areas to provide habitat and promote biodiversity. Two credits available: one each to protect or restore open space, and/or reduce the development footprint.
- Storm Water Management—Limit disruption of natural water flows by minimizing storm water runoff, increasing on-site infiltration and reducing contaminants. Two credits available: one each for no increase in the rate or quantity of runoff and/ or treatment systems designed to remove total dissolved solids, phosphorous, and comply with EPA guidelines.
- Heat Island Effect—Reduce heat islands, which are thermal gradient differences between developed and underdeveloped areas. There is one point each for roof and non-roof applications, up to two points.
- Light Pollution Reduction—one point. The intent is to eliminate light trespass from the building site, improve night sky access, and reduce development impact on nocturnal environments.

## 29.3.2 Water Efficiency—5 possible points
No prerequisites in this category.
Credits:

- Water Efficient Landscaping—Limit or eliminate the use of potable water for landscape irrigation. Up to two points available. One each for use of high efficiency irrigation technology, and/or using captured rain or recycled water for irrigation.
- Innovative Wastewater Technologies—one point. Reduce the generation of wastewater and potable water demand, while increasing the local aquifer recharge.
- Water Use Reduction—20% reduction is one point, 30% reduction is two points. Maximize water efficiency within the building to reduce the burden on municipal water supply and wastewater systems.

## 29.3.3 Energy & Atmosphere—17 possible points
Three prerequisites:

- Fundamental Building Systems Commission-

ing—Verify and ensure that fundamental building elements and systems such as HVAC are designed, installed, and calibrated to operate as intended.
- Minimum Energy Performance—Establish the minimum level of energy for the base building systems which is compliance with ASHRAE/IES 90.1-1999.
- CFC Reduction in HVAC&R Equipment—This requires zero use of CFC based refrigerants in new buildings such as R11 and R12.

Credits:

- Optimize Energy Efficiency—possible ten points. Achieve increasing levels of energy performance above the prerequisite standard, ASHRAE 90.1-1999, to reduce environmental impacts associated with excessive energy use. 20% better is two points, 30% is four points, and on up to 60% better is worth ten points.
- Renewable Energy—three points possible, one point each for 5%, 10%, 20% of total energy. Encourage and recognize increasing levels of self-supply through renewable technologies to reduce environmental impacts associated with fossil fuel energy use.
- Additional Commissioning—one point. Verify and ensure that the *entire* building, including the building envelope, is designed, constructed, and calibrated to operate as intended. This as opposed to the prerequisite which called for only *fundamental* systems.
- Elimination of HCFC s and HALONS—one point. Reduce ozone depletion and support early compliance with the Montreal Protocol. This applies to refrigerants such as R22 and R123.
- Measurement & Verification—one point. Provide for the ongoing accountability and optimization of building energy and water consumption performance over time.
- Green Power—one point. Encourage the development and use of grid-source energy technologies on a net zero pollution basis by the purchase of green power that meets the Center for Resource Solutions Green-E products.

## 29.3.4 Materials & Resources—13 possible points
Prerequisite:

- Storage & Collection of Recyclables—Facilitate the reduction of waste generated by building occupants that is hauled to and disposed on in landfills.

Credits:

- Building Reuse—two points possible at 75% and 100% reuse of building shell and non-shell. A third point is available by maintaining 100% of an existing building's structure and 50% of the non-shell such as walls, floor coverings, and ceilings. Purpose is to extend the life cycle of existing building stock.

- Construction Waste Management—two points available. This is to divert construction, demolition, and land clearing debris from landfill disposal. Redirect recyclable material back into the manufacturing process. 50% recycled or salvaged materials by weight gets one point, 75% earns one more point.

- Resource Reuse—two possible points at 5% or 10% of using salvaged or refurbished materials. The intent is to extend the life cycle of targeted building materials by reducing environmental impacts related to materials manufacturing and transport.

- Recycled Content—two possible points. One point specifying a minimum of 25% building materials that contain post consumer recycled materials. An additional point is available if an additional 25% is recycled content.

- Local/Regional Materials—two possible points either manufactured or harvested locally. The purpose is to increase the demand for building products that are manufactured locally, less than 500 miles, thereby reducing the environmental impacts resulting from long-distance transportation.

- Rapidly Renewable Materials—one point. Reduce the use and depletion of finite raw and long cycle renewable materials by replacing them with renewables.

- Forest Stewardship Council, FSC, certified wood. One point

Note: regarding "certified products." This does not mean "LEED certified" but certified by other entities such as the Forest Stewardship Council FSC. Beware of manufactures claiming "LEED certified" products. The USGBC and LEED do not certify products. What they do is adopt industry standards as applicable, such as the FSC certified wood.

### 29.3.5  Indoor Environmental Quality (IEQ)— 15 possible points

Two prerequisites:

- Minimum IAQ Performance—Comply with ASHRAE 62-2004 indoor air quality standard to prevent the development of air quality problems.

- Environmental Tobacco Smoke (ETS) Control—Prevent exposure of building occupants and systems to ETS.

Credits:
- Carbon Dioxide Monitoring—one point. Provide an

HVAC system which can monitor and ventilate a building based upon $CO_2$ levels.

- Ventilation Effectiveness—one point. Provide for the effective delivery of mixed and outdoor air to support health, safety, and comfort of occupants. Uses ASHRAE 129 methodology.

- Construction IAQ Management Plan—two possible points. Develop an IAQ management plan for during construction and before occupancy, one point. An additional point is available by conducting a two-week building flushout prior to occupancy using 100% outside air.

- Low Emitting Materials—four possible points. One point for low VOC adhesives. One point for low VOC paints. One point for carpet exceeding the Carpet & Rug Institute Green Label IAQ Program. One point for composite wood and agrifiber products containing no additional urea formaldehyde resins.

- Indoor Chemical & Pollutant Source Control—one point. This is to avoid exposing occupants to potentially hazardous chemicals that adversely affect IAQ.

- Controllability of Systems—two possible points, one each for perimeter and non-perimeter. This is to provide a high level of individual occupant control of thermal, ventilation, and lighting systems.

- Thermal Comfort—comply with ASHRAE Standard 55—one point. This relates to providing a thermally comfortable environment that supports healthy and productive performance of occupants.

- Permanent Monitoring System—one point. These are permanent systems to monitor temperature and humidity which also allow occupants to have partial control over these parameters.

- Daylighting & Views—two points available. This is to provide a connection between indoor spaces and outdoor environments through the introduction of sunlight and views into the occupied areas of the building. For one point, achieve a daylight factor of 2% in 75% of all space occupied for critical visual tasks, but not including the likes of laundry rooms, copying rooms etc. For an additional point, achieve the 2% rating in 90% of spaces.

### 29.3.6 Innovation & Design Process—5 possible points
- Use of LEED-accredited professional—one point
- Innovation in design—four possible points

### 29.3.7 Discussion
Note: energy and engineering skills are applicable to as many as 58% of total available points on water, energy, and IAQ. Energy related credits are the largest category

of available credits.

May '04 *Energy User News* article by Peter D'Antonio, entitled "The LEEDing Way."

The article analyzes the activity in energy & atmosphere, E&A, and indoor environmental quality, IEQ, for the first 53 LEED-certified buildings:

- Regarding E&A, average points earned is only 5.3 out of the possible 17!
- This is the lowest % achieved in any of the five categories! So although E & A is the largest plum, few appear to be taking advantage of it.
- Renewable energy points are earned in fewer than 10% of the certified buildings.
- Regarding IEQ points, ventilation effectiveness and controllability points are achieved in less than one third of buildings.

This EUN article provides support for the proposition that engineers, the AEE, ASHRAE, the IESNA, and others are not maximizing the potential contributions to LEED buildings. Hence, there is a significant opportunity to take on a larger role in the design, construction, and operations of green buildings.

Other factors for the design team to consider are the forces that drive the LEED points on a project. Many times, for the design team of an LEED project, it boils down to, "How many points can we get?" This becomes especially the case for the mechanical, electrical, and plumbing (MEP) team members. They control or influence approximately 75% the total LEED credits on a job. Commonly called "pointchasing," it is an effort by the design team to achieve the maximum available points at the minimum cost and effort. And although it is a rather ugly approach to green building design, it has become a matter of fact that teams will focus on points. It is even possible that the acquisition of points is one of the elements in the design contract. Possibly a bonus is linked to points achieved?

But this can and should be "managed" by the owner. It gets back to the integrated design process and the setting of goals during the design charette. Does the owner want very high energy efficiency, or does he want to make an environmental statement with a green roof?

## 29.4 LEED FOR EXISTING BUILDINGS RATING SYSTEM (LEED-EB) ADOPTED IN 2004

### LEED for existing buildings

Addresses:
- Operation and upgrades of existing buildings.
- Pilot: 2002—2004.
- Initial certification and ongoing re-certification.

- Achievements:
  —More than 95 registered buildings.
  —4 certified buildings.
- Range of Users:
  —Federal, state, and local government; schools, colleges and universities, commercial buildings.

LEED-EB approval completed and released for use, Oct. 2004.

### Why LEED-EB Is So Important

Drawing on similarities to LEED NC, LEED EB has a larger potential impact and resultant benefits to society simply because there are many times more existing buildings than new construction. LEED EB focuses on where the greatest impact potential is.

### 29.4.1 LEED for Existing Buildings Rating System
### LEED-EB Rating System Goals

- Help building owners upgrade and operate their buildings in a sustainable way over the long term. Avoid the "saw tooth" approach (upgrade, decline, upgrade, decline).
- Support high productivity by building occupants.
- Operations
  —Helps building owners upgrade and operate their buildings in a sustainable way over the long term.
  —Reduces building operating costs.
  —Solves building operating problems.
  —Improves indoor environment.
  —Supports higher productivity of building occupants.

- Communications
  —Helps building managers, operators, and service providers communicate the importance of effective ongoing building operation and maintenance to decision makers in their organization.
  —Helps building managers and operators make sustainability part of the culture of their organization.
  —Helps CEOs and CFOs make sustainability part of the culture of their organization.
  —Helps communicate organization's sustainability commitments and achievements to its customers and the community.

### Prerequisites and Credits

Same categories as for other LEED Rating Systems:
- Sustainable Sites
- Water Efficiency
- Energy and Atmosphere

- Materials and Resources
- Indoor Environmental Quality
- Innovation and Accredited Professional

**LEED-EB Rating System**

Four Levels of Certification:

- LEED-EB Certified　　32-39 points
- Silver Level　　　　　40-47 points
- Gold Level　　　　　　48-63 points
- Platinum Level　　　　64-85 points

### 29.4.2 LEED for EB Technical Review

Similar to LEED NC, all prerequisites must be satisfied, and the credits are optional depending upon the final points and certification level desired.

#### 29.4.2.1 Sustainable Sites—14 possible points

Two prerequisites:

- Erosion and Sedimentation Control—Control erosion to reduce negative impacts on water and air quality.
- Age of Building—two years old or more.

Credits:

- Plan for Green Site & Bldg Exterior Management—up to two points. Encourage grounds/site/building exterior management practices that have the lowest environmental impact possible and preserve ecological integrity, enhance diversity, and protect wildlife while supporting building performance.
- Hi Development Density Building & Area—one point. Channel development to urban areas with existing infrastructure, protect greenfields, and preserve habitat and natural resources.
- Environmentally Preferable Alt Transportation—up to four points available. One point each for: public transportation access, bicycle storage and changing rooms, alternative fueled vehicles, car pooling, and telecommuting.
- Reduced Site Disturbance—up to two points. One point for protecting or restoring 50% of site area. An additional point to protect or restore open space at 75% of the site area.
- Storm water Management—up to two points. One point for measures that mitigate at least 25% of the annual storm water falling on the site. An additional point for mitigation of at least 50% of storm water.
- Reduce Heat Islands Effect (roof and non-roof)—up to two points. One point for reduction of heat islands. An additional point is available for an ENERGYSTAR-compliant roof.

- Light Pollution Reduction—one point. Eliminate light trespass from the building and site, improve the night sky, and reduce development impact on nocturnal environments.

#### 29.4.2.2 Water Use and Water Efficiency— 5 possible points

Two Prerequisites:

- Minimum Water Efficiency—maximize fixture water efficiency within buildings to reduce the burden on potable water supply and wastewater systems.
- Discharge Water Compliance—protect natural habitat, waterways, and water supply from pollutants carried by building discharge water.

Credits:

- Water Efficient Landscaping—up to two points. Requires the use of water efficient irrigation technologies or captured rain and recycled water to reduce potable water consumption for irrigation. The first point is based on a 50% reduction, and an additional point is available for 95% reduction in potable water use.
- Innovative Wastewater Technology—one point. Reduce the generation of wastewater and potable water demand, while increasing the local aquifer recharge.
- Water Use Reduction—up to two points. Maximize fixture potable water efficiency to reduce burdens on potable and wastewater municipal systems. The first point is for a 10% reduction and an additional point is available for a 20% reduction.

#### 29.4.2.3 Energy and Atmosphere—23 possible points

Three Prerequisites

- Existing Building Commissioning—Verify that fundamental buildings systems are performing as intended.
- Minimum Energy Performance—Satisfy the minimum level of energy efficiency using the ENERGYSTAR portfolio manager. Needs a rating of 60 or more.
- Ozone Protection—Reduce ozone depletion potentials by not using CFC refrigerants such as R11 and R12.

Credits:

- Optimize Energy Performance—up to ten points available. Achieve increasing levels of energy efficiency above the ENERGYSTAR prerequisite score of 60. Thus a score 63 earns one point, 79 earns five points, up to a maximum of ten points for a 99 rating.

- On-site & Off-site Renewable Energy—up to four points. The first point is for 5% on-site renewable OR 25% off-site renewables up to a maximum of four points for 30% on-site renewable energy OR 100% off-site renewable energy.
- Building Operations & Maintenance—up to three points. One point each for maintenance staff education, building systems maintenance, building systems monitoring.
- Additional Ozone Protection—one point. Reduce ozone depletion potential in compliance with Montreal Protocol. Thus HCFC refrigerants such as R22 and R123 are not used.
- Performance Measurement—up to four points. Have in place a continuous metering system for a number of facilities functions as follows: lighting systems, electric and gas metering, cooling load, chilled water system efficiency, irrigation water metering, boiler efficiencies, HVAC systems such as economizers, variable speed pumps and fans, air distribution, and emissions monitoring. Note these all can be incorporated into the building automation system (BAS).
- Documenting Sustainable Building Cost Impacts— one point. Document overall building operating costs for previous five years and track changes in the overall operating costs.

### 29.4.2.4 Materials and Resources -16 possible points
Two prerequisites:
- Source Reduction and Waste Management—establish minimum source reduction and recycling program elements.

Toxic Material Source Reduction—reduced mercury in lamps.

Credits:
- Construction, Demolition and Renovation Waste Management—up to two points. First point for diverting 50% or more of construction, demolition and land clearing waste from landfills. An additional point if 75% or more is diverted.
- Optimize Use of Alternative Materials—up to five points available. Maintain a sustainable purchasing program covering at least office paper, office equipment, furniture, furnishings, and building materials. One point is awarded for each 10% of total purchases that achieve criteria such as 70% salvaged materials, 10% post consumer recycled, 50% rapidly renewables, FSC-certified wood, and materials manufactured within 500 miles of the site.

- Optimize Use of IAQ Compliant Products—up to two points. These relate to the purchase of low emitting materials such as for carpets, sealants, paints and coatings, composite materials and agrifiber products with no added urea formaldehyde.
- Sustainable Cleaning Products and Materials—up to three points. Points accumulate based upon quantities of products that meet the Green Seal GS-37 or comply with the California Code of Regulations for VOCs. Disposable janitorial paper products and trash bags meet the requirements of the EPA comprehensive procurement guidelines.
- Occupant Recycling—up to three points. Set up divert/recycle programs for occupants. 30% is one point, 20% another, and the third point if 30% of total waste stream is diverted or recycled.
- Additional Toxic Material Source Reduction—one point. Establish a program to reduce the potential amounts of mercury brought into the building through lamps.

### 29.4.2.5 Indoor Environmental Quality—
### 22 points available
Four prerequisites:
- Outside Air Introduction and Exhaust Systems— Satisfy ASHRAE 62-2004 for IAQ.
- Environmental Tobacco Smoke (ETS) Control—Prevent or minimize occupant exposure to ETS.
- Asbestos Removal or Encapsulation—Establish an asbestos remediation and control management plan.
- PCB Removal—Establish a PCB management plan including a facility survey for PCBs.

Credits:
- Outside Air Delivery Monitoring—one point. Provide permanent monitoring systems on ventilation system performance measuring outdoor air and $CO_2$.
- Increased Ventilation—one point. Increase ventilation rates to exceed ASHRAE 62-2004 by 30%.
- Construction IAQ Management Plan—one point. Prevent any IAQ problems from arising due to construction/renovation work. Isolate occupied areas from dust, noise and other irritants.
- Documenting Productivity Impacts—up to two points. Document the history of absenteeism, productivity, and health care costs and submit to the USGBC.
- Indoor Chemical & Pollution Source Control—up to two points. Reduce the exposure of occupants to dusts and particulates by using filters of effective-

ness of MERV 13 or greater. An additional point is earned by reducing occupants' exposures to contaminants that may arise from operations such as copying, faxing, etc.

- Controllability of Systems—up to two points. Points available for occupant control of lighting systems and another for HVAC and temperature control.
- Thermal Comfort—ASHRAE Standard 55- up to two points. First point for compliance with the standard and an additional point is available for a permanent monitoring system to ensure compliance.
- Daylighting and Views—up to four points. Provide a connection between indoor spaces and the outdoor environment through the introduction of sunlight and views. Points available for 50% and 75% of spaces that have a 2% daylight factor. And two more points available for 45% of spaces (1 point) and 90% of spaces (1 point) that have direct line of sight vision to the outdoors.
- Contemporary IAQ Practice—one point. Enhance IAQ performance by optimizing practices and developing procedures to prevent the development of IAQ problems.
- Green Cleaning—up to six points. Points available for cleaning entryway systems, isolation of janitorial closets, low environmental impact cleaning policy, low environmental impact pest management policy, and low environmental impact cleaning equipment policy.

### 29.4.2.6 Innovation and Accredited Professional
5 possible points
- LEED EB Innovation in Operation, Upgrades and Maintenance—up to four points.
- LEED-accredited Professional—one point.

## 29.5 SUMMARY DISCUSSION OF TWO NEW LEED PROGRAMS:

### LEED-CI for Commercial Interiors and LEED-CS for Core and Shell

LEED-CI addresses the specifics of tenant spaces primarily in office, retail and industrial buildings. It was formally adopted in the fall of 2004. A companion rating is LEED for core & shell which is currently under development and in its pilot phase. Adoption is expected the fall of 2005. Together, LEED-CI and LEED-CS will establish green building criteria for commercial office real estate for use by both developers and tenants.

LEED-CI serves building owners and occupants as well as the interior designers and architects who design building interiors and the teams of professionals who install them. It addresses performance areas including water efficiency, energy efficiency, HVAC systems & equipment, resource utilization, furnishings, and indoor environmental quality.

### 29.5.1 LEED for Commercial Interiors (CI)
- Addresses the design and construction of interiors in existing buildings and tenant fit-outs in new core and shell buildings.
- Pilot: 2002—2004.
- Achievements: More that 45 projects in pilot.
- LEED CI Adopted in fall 2004.

### 29.5.1.1 LEED-CI Point Distribution
The same five basic categories as the other LEED rating systems are used.

**Possible Points**

| | |
|---|---|
| Sustainable Sites | 7 |
| Water Efficiency | 2 |
| Energy & Atmosphere | 12 |
| Materials & Resources | 14 |
| Indoor Environmental Quality | 17 |
| Innovation & Design Process | 4 |
| LEED Accredited Professional | 1 |
| Total Points Available | 57 |

**4 Levels of Certification**

| | |
|---|---|
| Certified | 21-26 |
| Silver | 27-31 |
| Gold 2-41 | |
| Platinum | 42-57 |

Because of its nature, i.e., the interior parts of a building, the energy engineer has less opportunity to aid in earning points in this program. This is the only program to date wherein the E & A credits are not the largest category. However, engineers in general do have the potential to aid points especially in the indoor environmental quality category.

### 29.5.2 LEED for Core & Shell, CS
Based upon the LEED NC rating system for new construction and major renovation, LEED CS was developed in recognition of the unique nature of core and shell developments. In particular, there is the lack of developer control over key aspects such as interior finishes, lighting, and HVAC distribution. Thus the scope of CS is limited to those elements of the project under the direct control of

the developer.

With its standards for CS and CI, the USGBC addresses the entire building—core, shell, and interiors. The responsibilities for particular sections, however are assigned to those parties having direct control over them. LEED CS is still in its pilot phase as of early 2006.

### 29.5.2.1 LEED CS Credit Categories

Below is a summary of where the points will be for LEED CS. Note its similarities to LEED NC and that energy & atmosphere is the largest points category.

**Possible Points**

| | |
|---|---|
| Sustainable Sites | 15 |
| Water Efficiency | 5 |
| Energy & Atmosphere | 16 |
| Materials & Resources | 11 |
| Indoor Environmental Quality | 13 |
| Innovation and Design Process | 4 |
| LEED Accredited Professional | 1 |
| Total Points Available | 65 |

**4 Levels of Certification**

| | |
|---|---|
| Certified | 24-29 points |
| Silver | 30-35 points |
| Gold | 36-47 points |
| Platinum | 48-64 points |

You can download all four of the LEED rating systems. The rating systems download are free. However, other tools and workbooks such as the reference guides do have a fee associated with them, with discounts given to members.

- Visit U.S. Green Building Council Web Site at www.usgbc.org/leed.
- Choose rating system.
- Click on rating system you would like to download.

### 29.6 THE LEED PROCESS

- Design Team Integration
- Project Registration
- Project Certification
- Documentation

**The LEED Design Process:**

When does a Green Design Begin?

*"It Begins in the Beginning"*

Critical to success is the integration of the design TEAM on Day 1 of design. LEED is a marketplace transformer. It is a paradigm shift away from top down, minimum first cost emphasis. The hierarchical old-fashioned way was design, bid, build.

But the LEED design process is one of integrated, holistic building design, construction, operations and maintenance. All participate as equals on a construction roundtable:

- Owner Operations Personnel
- Owner
- Architect
- Engineer
- Construction Manager
- Contractors & Subcontractors
- Equipment Suppliers & Manufacturers
- Commissioning Authority—Watchdog Role

During the very early stages of a green building's development, a design charette should be held. This refers to meetings that are held over the course of a day or two, wherein the entire team, the construction roundtable group, gets together to develop the roadmap to successful green building.

- The ENTIRE team joins in—all stakeholders, including the owner, designers, commissioning authority, and operations.
- Gain buy-in and consensus.
- Explore environmental issues.
- Propose alternatives.
- Identify modeling and resource allocation.
- Use the LEED checklists as guide for level of green desired.
- Use an outside facilitator who specializes in integrated design.
- Present examples of resources and ways to trace costs and benefits of modeling.
- Establish a task/responsible team to track and manage compliance with the process.
- Determine an LEED Leader who will be the "watchdog" over points. The commissioning authority can be a good choice for this.

### 29.6.1 The Energy Engineer's Goal: Get Invited!

It is during the design charette which occurs at the earliest moments of a project, that the energy engineer can provide the maximum overall benefit to the project. It is during this time that key choices are made about the lighting, HVAC, and building envelope. The energy engineer can help guide the team to the most appropriate energy efficient design strategies based upon the team's energy goals and the available energy sources.

On an LEED project, a major concern of the design team is simply, "How do I get the points?" Commonly called "points chasing," it is an effort by the design team to achieve the maximum available points at the minimum cost and effort. However, this can be ameliorated or driven by the owner. During the design charette, it is the owner's responsibility to clearly identify the goals and objectives of the project. If it is simply, "get me the most points," the team will point chase. If the goals are, for example, to have the highest energy efficiency possible, or the most daylighting possible, then other choices, alternatives, and evaluations may be examined.

### 29.6.2 Example and Discussion on Obtaining LEED Points on a Project

The following is an example of how in an integrated green building design, a "simple" decision such as to have a computerized building automation system (BAS), can have substantial overall impact on acquiring points.

Knowing that LEED buildings will generally require sophisticated controls, the BAS can have a major role in obtaining and facilitating points. The following is a summary of the impact a BAS may have on LEED points. The design team can use this example to guide them in the same process for other techniques and/or technologies. Again, the emphasis is on integrated design. We are designing an integrated building, not a collection of parts and systems.

#### 29.6.2.1 Some Examples of BAS Influence on LEED Credits:

- Sustainable Sites—Light pollution reduction through use of controls.
- Water Efficiency—use of metering to document water consumption—although not a credit itself, it can facilitate credits in this category.

**Energy & Atmosphere—many credits and much influence here:**

- Energy Prerequisite and Optimized Energy Performance—ten credits are in play here. The BAS as an integral part of the energy consuming systems including lighting, HVAC, load management, etc., and helps earn credits through performance improvements that will be quantified in the building energy simulations required by ASHRAE 90.1

- Commissioning—BAS aids the commissioning authority in performing their duties, a time saver.

- Measurement and Verification—one prescriptive credit—provide for accountability & optimization of energy and water consumption over time.

- Optimize Energy Performance—in LEED CI, up to four prescriptive credits for lighting and power controls. Other credits in energy performance, an additional four.

- Energy Submetering—in LEED CI, measure for energy accountability. Up to two prescriptive credits.

- Building Operations & Maintenance—three prescriptive credits, relating to staff education, building systems maintenance, and building systems monitoring.

- Performance Measurement—enhanced metering and emission reduction reporting—up to four credits prescriptive.

**Indoor Environmental Quality**
- $CO_2$ Monitoring—one prescriptive credit.
- Increase Ventilation Effectiveness—BAS can aid in earning this credit.
- Controllability of Systems—up to two prescriptive credits.
- Thermal Comfort—up to two prescriptive credits—comply with ASHRAE Standard 55 and permanently monitor temperature and humidity.
- Outdoor Air Delivery Monitoring—one prescriptive credit.

The conclusion is that a BAS can directly add to points accumulation, but indirectly has a great deal of influence on other points. The intent of these examples is to demonstrate how design decisions flow through the entire integrated design, having direct effects on some credits and indirect effects on others.

### 29.6.3 Marketing LEED and Sustainability to the Community, Owners, and Designers

LEED offers a great deal of value to various members of the community. The owners benefit by having high performance buildings that are cost efficient and provide for better employee productivity. The design community benefits by now having a way to craft a stronger value message for superior architecture and design. LEED provides a way for designers to qualify and quantify their competitive advantage over other non-green designers. The community benefits by having a program such as LEED that promotes urban and brownfield development, reduces demands on infrastructure such as roads and waste disposal, improves the environment, and provides for a healthy living style.

Elaborating as to why a design team should be promoting LEED design, let us examine the overall life cycle cost of a typical commercial office building.

Ownership Cost Breakdown—40 Year Life Cycle Costs

Construction or First Cost is 11%
Financing is 14%
Alterations are 25%
Operations are 50%

It is interesting that the cost that most design teams grapple with, first costs, and keeping them low, is actually the least significant cost element in the overall life cycle costs of a building. Thus, those decisions to keep first costs low by specifying cheaper designs and equipment can have a serious negative impact on the overall life cycle performance of a facility.

Additionally, considering that first cost is only 11% of the total life cycle cost and that A/E fees are only 6% to 8% of that 11%, or .88% of the total costs, could it be beneficial to the owner to pay more for superior architecture and engineering? This is because the designs and selections made by the design team have a great deal of leverage on the total costs of life cycle ownership costs.

### 29.6.4 Credits that Engineers can help in acquiring

Many times it is believed that the architectural profession has the most potential to aid in acquiring LEED credits. However it is the engineering profession that in fact influences the most points. For example:

- Energy & atmosphere and indoor environments are 40% of available points.
- Minimum energy performance is a prerequisite!
- Energy modeling is required.
- Energy measurement and verification is a credit.
- Commissioning is required.

### Other benefits engineers bring to the design team

- Requires creativity vs. CAD commodity design. Creativity is desired and rewarded.
- Promotes investment in A/E design $ to value engineer before, not after.
- Promotes the collective wisdom of the integrated design team.
- This is the interplay between professions that occurs during the design charette.
- Catalyst for the design team to do the high performance job that it is capable of.
- Provides a value message of premium engineering and design to the owner—this may be the path to higher fees and/or more work?

### 29.6.5 Impediments to Green Acceptance

Typically, the first and possibly most serious impediment to the wide-scale adoption of LEED is the perception that it costs more. The facts are that it may add cost, from 1% to 5%, depending upon the level of green the ownership team has identified in that design charette. But it does not necessarily cost more if the design team is clever about making design decisions and using all available resources that may be at hand.

Hint: If trying to promote LEED, look to identify market conditions in the project's locale that supports LEED. Many states have various programs to incentivize energy efficiency and other marketplace conditions which can affect the viability of a green project. The following uses New Jersey as an example of "market conditions" that can drive LEED adoption:

### The Case for LEED in New Jersey:

First cost is less of an issue because of high efficiency equipment incentives. There is a program through the NJ Board of Public Utilities called NJ Smart Start Buildings, which provides rebates for high efficiency equipment such as lighting, HVAC, boilers, and chillers, as well as commissioning and design team meetings. Essentially much of the cost differential between cheap inefficient equipment and high efficiency equipment is offset by the rebates.

Renewable energy sources are promoted by statewide programs such as the Clean Energy Program. Similar to Smart Start, renewables such as wind, solar PV, and biomass projects are rebated up to 50% of the initial installed cost.

ASHRAE Std 90.1-1999 is the State Energy Code as well as the prerequisite for energy and atmosphere LEED credits. So there is no additional cost for NJ buildings to comply with this. In other states that may not have this code requirement, compliance with ASHRAE 90.1 would add cost.

Many brownfields are available for development with incentives from the NJ Economic Development Administration, NJEDA. This can be a simple prescriptive credit.

Mass transportation is generally adjacent to the brownfields, which aids in acquiring more points. NJ, as the most densely populated state in the nation, has many former industrial sites which are in inner city areas close to mass transit and part of urban renewal. Thus a brownfield site can facilitate a number of other credits.

Environmental—voluntary LEED adoption decreases need for additional regulation. NJ is one of the most regulated and legislated states, but if we have more adoption of LEED, many of the goals of environmental legislation can be achieved voluntarily.

High energy costs in NJ promote equipment and operational cost efficiencies. Paybacks on high efficiency equipment are quicker than in other states, which helps drive the recognized value of energy efficiency versus low first cost equipment.

Other states and regions may have similar incentives and programs. New York and California are two that come to mind. In addition, the Energy Policy Act of 2005, promises to encourage energy efficiency through various programs.

## 29.7  ASHRAE GUIDES DEVELOPED TO SUPPORT LEED

### Introduction to ASHRAE's GreenGuide and the Advanced Energy Design Guide

Besides the market conditions which may favor LEED adoption, organizations, such as ASHRAE, are developing tools to help the design teams accomplish their green design goals.

During the year 2000, the USGBC released its first Green Building Rating System. Called "Leadership in Energy and Environmental Design for New Construction," LEED 2.0, later revised and reissued as LEED 2.1 NC. Although the LEED rating system incorporated many ASHRAE standards into the rating system, the architectural community, not the engineering community, was in the vanguard of LEED. ASHRAE recognized the potential value of LEED to the community and the building industry and the value that ASHRAE could bring to the LEED program. Under then president-elect Bill Coad, Tech Committee 1.10 Energy Resources was tasked with developing a handbook or guide for sustainable engineering design, specifically targeted for ASHRAE members. In addition, during this time, after dialog between both organizations, the USGBC and ASHRAE entered into a partnering agreement. The result of these efforts was the development of the ASHRAE *GreenGuide*, released in December 2003, to assist the USGBC in its efforts at promoting sustainable design. Additionally, in 2005, ASHRAE released the *Advanced Energy Design Guide for Small Office Buildings* (less than 20,000 square feet). This guide provides a prescriptive description and requirements for various building components and systems that would be energy efficient.

### 29.7.1 ASHRAE GreenGuide

The ASHRAE *GreenGuide* aids designers of sustainable, high performance green facilities. It offers various "green tips" to aid the integrated design team in developing a green building. In 2002, the American Society of Refrigerating and Air-Conditioning Engineers, ASHRAE,

and the United States Green Building Council, USGBC, entered into a partnering agreement to team together to promote green buildings. This *GreenGuide* was developed by ASHRAE to assist the USGBC in their efforts at promoting sustainable design.

The guide was developed to provide guidance on how to apply green design techniques. Its purpose is to help the designer of a "green design" with the question of "What do I do next?" It is organized to be relevant to the audience, useful, and practical, and to encourage innovative ideas from the design team. A key component of the guide is the "green tips," which will be covered in some detail later in this section.

However, the guide is not a consensus document, and one does not have to agree with all elements of the guide for it to be helpful. It was not developed to motivate the use of green design.

### 29.7.1.1 Green Design, Sustainability and Good Design

"Green" has become one of those words that can have too many possible meanings. One of the USGBC's initial goals was to provide a definition of green through the development and release of the LEED rating system. It was here that we had a measurable, quantifiable way of determining how green a building was. It also addresses the "greenwashing" issue wherein all types of green technologies and techniques could be employed, some valid and others questionable, all in the effort to be able to label a building green. The conclusion is that green buildings are LEED buildings. This message is almost universally accepted, in the USA as well as internationally. But do be aware that there are other strong green rating systems that have been developed in Canada and Europe. For example Canada has the Green Building Challenge, GBC, and Britain has the Building Research Establishment Environmental Assessment Method, BREEAM. According to the guide, the consensus on green buildings is that they achieve a high level of performance over the full life cycle in the following areas:

- Minimal consumption of nonrenewable, depletable resources such as water, energy, land, and materials.
- Minimal atmospheric emissions that have negative environmental effects.
- Minimal discharge of harmful liquid and solid materials, including demolition debris at the end of a building's life.
- Minimal negative impact on site ecosystems.
- Maximum quality of indoor environment including lighting, air quality, temperature, and humidity.

### 29.7.1.2 "Good" Design

ASHRAE asks whether good design is intrinsically green design. The *GreenGuide* authors make the distinction between green and good. Good design includes:

- Meeting the purpose and needs of the building's owners and occupants.
- Meets the requirements of health and safety.
- Achieves a good indoor environment consisting of thermal comfort, indoor air quality, acoustical comfort, and is compatible with the surrounding buildings.
- Creates the intended emotional impact on building's occupants.

But this is not green design in the sense that it does not address energy conservation, environmental impact, low impact emissions, and waste disposal. So integrating good and green design, such as the LEED rating system, helps us achieve the optimum building for the owners' needs as well as the needs of society. Hence the authors of *GreenGuide* "strongly advocate that buildings should strive to achieve both."

The *GreenGuide* emphasizes the design process. This process is the first crucial element in producing a green building. There needs to be an integrated design team created in the beginning. This team should include the owner, project manager, representative of the end user, architect, mechanical engineer, plumbing and fire protection engineer, electrical engineer, lighting designer, structural engineer, landscaping specialist, civil engineer, energy analyst, environmental consultant, commissioning authority, construction manager, cost estimator, building operator, and code enforcement representative.

Each individual profession works together in a team environment to establish the building's goals and the manner in which these goals will be achieved. Each profession must be able to recognize the impact of one another on others designs and process.

For example, during conceptual design, the architect will determine the size and number of floors of the building. The building envelope will determine the size of HVAC equipment, as well as the types of systems being evaluated. The energy analyst will advise the team on the energy cost implications of their selections. Everyone is interdependent upon the others. What should come at the end of this iterative process is a single integrated design that functions a unit and not as a collection of parts.

Integration and interdependency of the design team professionals are the keys to a successful green design.

### 29.7.2 Conceptual Engineering Design

The principle intent of the *GreenGuide* is to assist the design-engineering professionals in integrating their skills and bringing value into the green design. The guide discusses a number of design responsibilities and suggests a number "green tips" to the design team. You may find that these suggestions are not new to you, and many of these concepts have been in use or at least in consideration for years. However, with the advent of the desire to build green, which requires high performance systems, these techniques have a much better chance of being incorporated into the green building than in the past when first cost was possibly the primary concern of on owner and the design team. The tips are arranged so that they are listed after the design section responsibility to which it is most closely linked.

#### 29.7.2.1 Load Determination

Loads are determined by summing up internal and external gains and losses. The HVAC engineer can assist the architect in determining necessary characteristics of the building envelope. Working together with the energy analyst and others, they will aid the architect in selecting building orientation, insulation, fenestration, roofing, lighting, day lighting, systems sizes, and efficiencies, etc. A key element to strive for is the initial reduction of loads resulting from efficient building envelope, lighting loads, power loads and A/C tonnages.

**Green Tip #1: Night Precooling**

This involves the circulation of cool air during the nighttime hours during the cooling season with the intent of cooling the structure. There are two variations on this theme. First is the use of fans only to bring in cool ventilation air; this is a somewhat passive technique. The other is to actively run the HVAC plants to precool the facility, potentially below daytime occupied temperatures to take advantage of the building's thermal mass. Parameters to consider when evaluating this strategy are local diurnal temperature variations, humidity levels, and thermal coupling of the circulated air to the building's thermal mass.

#### 29.7.2.2 Space Thermal/Comfort Delivery Systems

Occupant comfort and health are important and the quality of the indoor environment promotes satisfied, productive workers within the building. Green buildings provide a more productive workplace environment.

Thermal comfort is primarily concerned with satisfying ASHRAE Standard 55 for temperature and humidity requirements.

Indoor air quality is primarily concerned with ASHRAE Standard 62 for fresh air and ventilation. However, both of these can negatively impact energy consumption and thus the following green tips are offered:

### Green Tip #2 Air-to-Air Heat Recovery—
### Heat Exchange Enthalpy Wheels

This is a rotary cylinder filled with an air-permeable medium with a large internal surface area. Intake and exhaust air streams pass through opposite ends of the wheel in a reverse flow configuration. Latent and sensible heat are then transferred from exhaust to inlet air, thereby recovering some of the conditioning energy that was invested in the exhaust air.

### Green Tip #3 Air-to-Air Heat Recovery—
### Heat Pipe Systems

These are passive devices, usually configured as tubes with fins for maximum surface area. They contain a thermal fluid that transports sensible heat only between exhausts and inlet air streams.

### Green Tip #4 Air-to-Air Heat Recovery—
### Run Around Systems

This consists of energy recovery coils in the exhaust and inlet air streams and a circulating loop containing a heat transfer fluid. These systems do not transfer latent energy. An option added to this system and the heat pipe system is the use of an indirect evaporative water process that can reduce cooling loads in addition to the heat recovery process.

### Green Tip #5 Displacement Ventilation

This technique supplies conditioned air at a temperature slightly lower than the desired room temperature at low velocities horizontally at the floor level. Returns are located in the ceiling. This supply air rises by convection, picks up the room load and exits through the ceiling returns.

### Green Tip #6 Dedicated Outdoor Air Systems (DOAS)

This uses a dedicated, separate air handler to condition the outdoor air before delivering it directly to the occupants. The air delivered should be conditioned sufficiently to not adversely affect the thermal comfort of the occupants. The only absolute with this system is that the ventilation air must be delivered directly to the space from a separate system. Control strategy, energy recovery, and leaving air conditions are all variables that can be fixed by the designer.

### Green Tip #7 Ventilation Demand Control Using $CO_2$

$CO_2$ concentrations are measured in a space and ventilation rates are automatically adjusted by the BAS to maintain $CO_2$ concentrations within predetermined limits. This system is best used in buildings and spaces with variable occupancies such as public spaces, theaters, meeting rooms etc.

### Green Tip #8 Hybrid Ventilation

This allows the controlled introduction of outside air ventilation into a building by both mechanical and passive means. It is sometimes called "mixed mode ventilation." It has built-in strategies to allow the mechanical and passive portions to work with one another so as to not cause additional ventilation loads as would occur using mechanical ventilation alone. This is a non- conventional technique that has the promise of reducing operating expenses as well as providing a healthier stimulating environment.

### 29.7.2.3 Energy Distribution Systems

These provide the heating, cooling, lighting, and electric power throughout the building. The most common media to distribute energy are steam, hydronics (water), air and electricity. Steam supply, because of its pressure characteristics, does not need to be pumped although generally the steam condensate is pumped back to the boilers. Water and air are the principle media that require mechanical intervention for distribution, and hence can be major consumers of electric power for pumps and fans.

### Green Tip #9 Variable Flow/Variable
### Speed Pumping Systems

Pumps and fans can be significant users of electric power. In a conventional application, the pumps and fans operate at a fixed rate based on maximum design conditions, regardless of actual loads. Adding variability to the pumping and fans to allow modulation of flows based upon actual systems needs as opposed to design conditions can provide significant electrical savings.

### 29.7.2.4 Energy Conservation Systems

This section focuses on the equipment that generates electricity, steam, hot water, and chilled water. These include distributed electrical generation, boilers, furnaces, electrically driven water chillers and thermally driven absorption chillers.

### Green Tip #10 Low-$NO_x$ Burners

These are natural gas burners that improve energy efficiency and lower emissions of oxides of nitrogen, $NO_x$. They can be purchased as an option for new equipment or retrofitted to existing equipment.

### Green Tip #11 Combustion Air Preheating

This tip preheats combustion air by using waste

heat from the exhaust stack, increasing energy efficiency of equipment such as boilers and furnaces. The principle is to introduce preheated hot air into the combustor instead of cold air, thereby reducing energy consumption.

### Green Tip #12 Combustion Space/Water Heaters

These consist of a storage water heater, a heat delivery system such as fan coils or baseboards, and associated pumps and controls. The single unit does dual duty, as a water heater for domestic hot water heater and as a source of hot water for the hydronic heating system.

### Green Tip #13 Ground Source Heat Pumps, GSHP

These extract heat stored in the ground when in the heating mode, and rejects heat removed from the building into the ground in the cooling season. They consist of a loop of piping or a well in the ground and indoor units consisting of evaporators and condensers connected into the ground water loop. GSHPs can reduce the energy required for space heating, cooling, and service water in commercial/institutional buildings by as much as 50%.

### Green Tip #14 Water Loop Heat Pump Systems

Consists of multiple water source heat pumps within a building and tied into a neutral temperature loop that serves as a heat source and a heat sink. The loop temperature in turn is maintained at this neutral point by supplementing with heat from a boiler or cooling from a cooling tower.

### Green Tip #15 Thermal Energy Storage for Cooling

This is an active storage system that uses the building's cooling equipment to remove heat, usually during the night and off-peak periods, to take advantage of lower-cost electricity during those periods and make ice slurry or chilled water. This enables a number of control and operational strategies. For example, smaller chillers can be purchased and the building peak loads are satisfied with ice made during the off- peak periods.

### Green Tip #16 Double Effect Absorption Chillers

Chilled water for facility cooling can be driven by electricity or thermal energy such as steam. In absorption chillers, thermal energy such as steam is used to drive a process using water and a salt solution such as lithium bromide in a vacuum-sealed shell to produce chilled water. There are no ozone depleting refrigerants used in this process. Absorbers come in single and double effect types, the double effect having a COP of about 1.2 versus the single effect COP of about .8.

### Green Tip #17 Gas Engine Driven Chillers

Chilled water systems that use energy sources besides electricity can help offset high electricity costs. Generally, these are engines run on natural gas and are coupled to a chiller compressor section. Essentially, it is an engine replacing the electric motor of an electrically driven chiller. Gasoline and diesel fuel can also be used depending upon engine types selected.

### Green Tip #18 Gas-Fired Chiller/Heater

Gas-fired absorption chillers are a special type of absorption chillers wherein the thermal energy source is a direct burner typically firing natural gas, although other fuels could be used. This is as opposed to the conventional absorber operated with steam. The chiller/heater can do double duty both to make cold and hot water simultaneously.

### Green Tip #19 Desiccant Cooling and Dehumidification

Rotary desiccant dehumidifiers use solid desiccants such as silica gel to attract water from the air. Humid air is dehumidified in one part of the desiccant bed while a different part of the bed is dried for reuse by a second air stream.

### Green Tip #20 Evaporative Cooling

This technique can be used to reduce the amount of energy consumed by mechanical cooling equipment. There are two types of evaporation coolers, direct and indirect. Direct introduces water directly into the air stream, and the water evaporates, reducing the dry bulb temperature of the air while raising the relative humidity. Indirect systems spray water onto a coil and, through evaporation from the fins of the coils, reduces the dry bulb temperature also. However, no water is added to the air stream with this method.

#### 29.7.2.5 Energy and Water Sources

The designer may not have much option in the selections of energy sources for the building being designed. Typically, energy is provided by the area utilities in the form of gas and electric. However, the designer can consider options to supplement the conventional energy sources with renewables such as wind, solar photovoltaics, PV, and hydro as the site permits. PV is generally the most applicable renewable energy source for buildings. The others are too site specific.

### Green Tip #21 Solar Energy System—Photovoltaics

Sunlight shines on solid state crystals of silicon and generates low voltage direct current electricity. Applications simply require full access to the sun and sufficient space, typically on the roof, to generate useful amounts of electricity. The low voltage DC can be inverted and voltage boosted to be supplied directly into the building's electrical distribution system. Thus there is no need to locate a specific low voltage DC load for the power produced.

### 29.7.2.6 Lighting Systems

The *GreenGuide* section here is designed to familiarize the HVAC&R engineer with lighting systems and their potential impact on the equipment sizes and designs. However, it states that it is best to engage a lighting designer who will design according to IESNA standards (Illuminating Engineering Society-North America). However, the guide does make one suggestion for lighting.

**Green Tip #22 Light Conveyor**

A light conveyor is a large pipe or duct with reflective sides that transmits artificial or natural light along its length. There are occasions wherein designers have used light tracking to enhance the performance of the light conveyors.

### 29.7.2.7 Plumbing and Fire Protection Systems

Although not usually within the purview of the HVAC designer's expertise, both must be able to work together in developing a fully integrated design. There are several tips that include aspects of both skill sets.

**Green Tip #23 Water Conserving Plumbing Fixtures**

Water conserving strategies can save owners both consumption and demand charges. Options to be considered for water conservation are:
- Infrared faucet sensors
- Delayed action shut off valves
- Low flow toilets
- Faucets with flow restrictors
- Metering faucets
- Water efficient dishwashers
- Waterless urinals
- Closed cooling towers to eliminate drift and filters for cleaning tower water

**Green Tip #24 Graywater Systems**

Graywater is defined as wastewater coming from operations such as showers, bathtubs, washing machines and sinks. This is separate from blackwater, which is wastewater from toilets and sinks that contain organic or toxic materials. Where allowed by code, graywater can be filtered, treated, stored, and later used for nonpotable uses such as irrigation of landscaping and flushing of toilets. Distribution would be through a piping system clearly separated from all others.

**Green Tip #25 Point-of-Use Domestic Hot Water Heaters**

These provide small quantities of hot water at the point of use, without a tie into a central hot water source. Generally electrically heated, savings are obtained through the avoidance of large amounts of hot water storage, and thermal losses of hot water distribution piping.

**Green Tip #26 Direct Contact Water Heaters**

This consists of a heat exchanger in which flue gases are in direct contact with the water to be heated. Cold water enters the top of the heat exchanger, natural gas is burned and flows up through the heat exchanger wherein the water is cascading down, acquiring the heat of the burned gas. Although there is direct contact between exhaust gases and the water, there is very little contamination of the water. These are suitable for dairy and food processing uses, as well as many other processes.

**Green Tip #27 Rainwater Harvesting**

Although this is not a new concept, it has been around for thousands of years, and is a simple and effective technology to apply. Rainwater is channeled into cisterns for nonpotable uses as needed for irrigation, toilet flushing etc.

### 29.7.2.8 Controls

These can be thought of as the "nervous systems" of the building's mechanical and electrical infrastructure. Controls can be used for temperature and humidity control, ventilation control, energy management and analysis, etc. (See also Chapter 22.)

**Green Tip #28 Mixed Air Temperature Reset**

This refers to the mix of outside and return air that exist on an operating system supply air handling unit prior to any "new" energy being applied to it. The concept is to reset the mixed air temperature (MAT) to a temperature that just satisfies the lowest cold air demand. Reset controls involve raising the setpoint of the MAT controls based on input that indicates the demand of the zone needing the coldest air, limited by the minimum amount of outside air for IAQ purposes.

**Green Tip #29 Cold Deck Temperature Reset, CDT, with Humidity Override**

CDT is similar to MAT, but it applies to air leaving a cooling coil. The concept here is to allow the discharge temperature from the cooling coil to go to higher temperatures when demand for cooling is low. However, this strategy can create high humidity conditions indoors on occasion. Thus humidity sensors are located in the spaces to override the rising cold deck temperatures and drive temperatures back down to extract more moisture based upon demands for temperature and humidity control.

Other elements of green, good design covered by the *GreenGuide* but with no specific green tips suggested are:

- Expressing and Testing Concepts
- Completing Design and Documentation for Construction
- Post Design—Construction to Demolition
- Builder/Contractor Selection
- Construction
- Commissioning
- Operation/Maintenance/Performance Evaluation

### 29.7.3 ASHRAE Advanced Energy Design Guide

This ASHRAE manual for small office buildings is intended for buildings up to 20,000 square feet in size. It is a prescriptive description of how designers and contractors can achieve up to 30% reduction in energy consumption in comparison to ASHRAE Standard 90.1-1999. This would be worth four credits in LEED NC. The manual, although primarily of use for new construction (NC) design, can also be used as a planning tool for LEED EB because the measures given are a form of best energy practices for new or existing buildings.

The guide is divided into three basic chapters:

(I) Integrated Process to Achieve Energy Savings
1. Pre-Design Phase—Prioritize Goals
2. Design Phase
3. Construction
4. Acceptance
5. Occupancy
6. Operation

(II) The guide provides recommendations by climate. It divides the USA into 8 zones, with #1 the warmest zone (the southern tip of Florida) to #8 for Alaska.

(III) For each climate zone there is a table providing prescriptive recommendations for the building components listed below:
1. Roofs
2. Walls
3. Floors
4. Slabs
5. Doors
6. Vertical Glazing
7. Skylighting
8. Interior Lighting
9. HVAC category includes air-conditioning, heating, air side economizers, ventilation and air handling ducts.
10. Service Water Heating

**References:**
ASHRAE *GreenGuide*, 2003—Editor David L. Grumman
ASHRAE Advanced Energy Design Guide
ASHRAE Website *www.ASHRAE.ORG*
*Energy User News*, May 2004 by Peter D'Antonio, entitled "The LEED-ing Way."
The United States Green Building Council, www.USGBC.ORG

# THERMAL SCIENCES REVIEW

**L.C. WITTE**
Professor of Mechanical Engineering
University of Houston
Houston, Texas

## I.1. INTRODUCTION

Many technical aspects of energy management involve the relationships that result from the thermal sciences: classical thermodynamics, heat transfer, and fluid mechanics. For the convenience of the user of this handbook, brief reviews of some applicable topics in the thermal sciences are presented. Derivation of equations are omitted; for readers needing those details, references to readily available literature are given.

## I.2. THERMODYNAMICS

Classical thermodynamics represents our understanding of the relationships of energy transport to properties and characteristics of various types of systems. This science allows us to describe the global behavior of energy-sensitive devices. The relationships that can be developed will find application in both fluid-flow and heat-transfer systems.

A thermodynamic system is a region in space that we select for analysis. The boundary of the system must be defined. Usually, the system boundary will coincide with the physical shell of a piece of hardware. A closed system is one where no mass may cross the boundary, whereas an open system, sometimes called a control volume, will generally have mass flowing through it.

We generally divide energy into two categories: stored and transient types of energy. The stored forms are potential, kinetic, internal, chemical, and nuclear. These terms are fairly self-descriptive in that they relate to ways in which the energy is stored. Chemical and nuclear energy represent the energy tied up in the structure of the molecular and atomic compounds themselves. These two types of stored energy presently form the prime energy sources for most industrial and utility applications and thus are of great importance to us.

Potential, kinetic, and internal energy forms generally are nonchemical and nonnuclear in nature. They relate to the position, velocity, and state of material in a thermodynamic system. More detailed representations of these energy forms will be shown later.

### I.2.1 Properties and States

Thermodynamic systems are a practical necessity for the calculation of energy transformations. But to do this, certain characteristics of the system must be defined in a quantifiable way. These characteristics are usually called properties of the system. The properties form the basis of the state of the system. A state is the overall nature of the system defined by a unique relationship between properties.

Properties are described in terms of four fundamental quantities: length, mass, time, and temperature. Mass, length, and time are related to a force through Newton's second law.

In addition to the fundamental quantities of a system there are other properties of thermodynamic importance. They are pressure, volume, internal energy, enthalpy, and entropy—$P$, $V$, $U$, $H$, and $S$, respectively.

**Equation of State.** Returning to the concept of a state, we use an equation of state to relate the pertinent properties of a system. Generally, we use $P = P(m, V, T)$ as the functional equation of state. The most familiar form is the ideal gas equation of state written as

$$PV = mRT \tag{I.1}$$

or

$$PV = n\bar{R}T \tag{I.2}$$

Equation I.1 is based on the mass in a system, whereas equation I.2 is molal-based. $R$ is called the gas constant and is unique to a particular gas. $\bar{R}$, on the other hand, is called the universal gas constant and retains the same value regardless of the gas (i.e., $\bar{R} = 1545$ ft $lb_f/lb_m \cdot mol \cdot °R = 1.9865$ Btu/$lb_m \cdot mol \cdot °R$. It can be easily shown that $R$ is simply $\bar{R}/M$, where $M$ is the molecular weight of the gas.

The ideal gas equation is useful for superheated but not for saturated vapors. A vexing question that often arises is what is the range of application of the ideal gas equation. Figure I.1 shows the limits of appli-

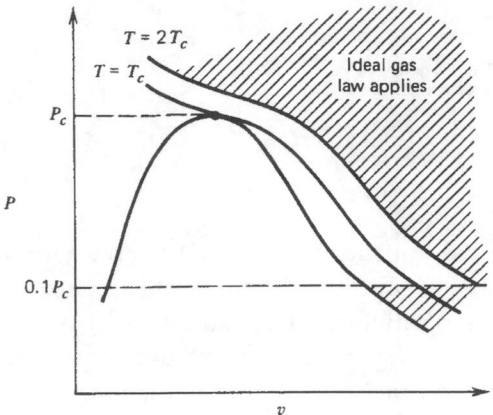

**Fig. I.1 Applicability of ideal gas equation of state.**

cability. The shaded areas demonstrate where the ideal gas equation applies to within 10% accuracy.

If high pressures or vapors near saturation are involved, other means of representing the equation of state are available. The compressibility factor $Z$ for example $Z = PV/RT$, is a means of accounting for nonideal gas conditions. Several other techniques are available; see Ref. 1 for example.

Changes of state for materials that do not behave in an ideal way can be calculated by use of generalized charts for property changes. For example, changes in enthalpy and entropy can be presented in terms of reduced pressure and temperature, $P_r$ and $T_r$. The reduced properties are the ratio of the actual to the critical properties. These charts (see Appendix II) can be used to calculate the property changes for any change in state for those substances whose thermophysical properties are well documented.

**Ideal Gas Mixtures.** Where several gases are mixed, a way to conveniently represent the properties of the mixture is needed. The simplest way is to treat the system as an ideal gas mixture. In combustion systems where fuel vapor and air are mixed, and in the atmosphere where oxygen nitrogen and water vapor form the essential elements of air, the concept of the ideal gas mixture is very useful.

There are two ways to represent gas mixtures. One is to base properties on mass, called the gravimetric approach. The second is based on the number of moles in a system, called a molal analysis. This leads to the definition of a mass fraction, $x_i = m_i/m_t$, where $m_i$ is the mass of the $i$th component in the mixture and $m_t$ is the total mass of the system, and the mole fraction, $y_i = n_i/n_t$, where $n$ represents the number of moles.

Commonly, the equation of state for an ideal mixture involves the use of Dalton's law of additive pressures. This uses the concept that the volume of a system is occupied by all the components. Using a two-component system as an example, we write

$$P_1V_1 = n_1\bar{R}T_1$$

and $\qquad P_2V_2 = n_2\bar{R}T_2$

Also $\qquad P_1V_1 = n_1\bar{R}T_2$

Since $V$, $\bar{R}$, and $T$ are the same for all three equations above, we see that

$$P_t = P_1 + P_2$$

and $\qquad (P_1 P_2)V = (n_1 n_1)\bar{R}T$

for a mixture. Each gas in the mixture of ideal gases then behaves in all respects as though it exists alone at the volume and temperature of the total system.

### I.2.2 Thermodynamic Processes

A transformation of a system from one state to another is called a process. A cycle is a set of processes by which a system is returned to its initial state. Thermodynamically, it is required that a process be quantifiable by relations between properties if an analysis is to be possible.

A process is said to be reversible if a system can be returned to its initial state along a reversed process line with no change in the surroundings of the system. In actual practice a reversible process is not possible. All processes contain effects that render them irreversible. For example, friction, nonelastic deformation, turbulence, mixing, and heat transfer are all effects that cause a process to be irreversible. The reversible process, although impossible, is valuable, because it serves as a reference value. That is, we know that the ideal process is a theoretical limit toward which we can strive by minimizing the irreversible effects listed above.

Many processes can be described by a phrase which indicates that one of its properties or characteristics remains constant during the process. Table I.1 shows the more common of these processes together with expressions for work, heat transfer, and entropy change for ideal gases.

### I.2.3 Thermodynamic Laws

Thermodynamic laws are relationships between mass and energy quantities for both open and closed systems. In classical form they are based on the conservation of mass for a system with no relativistic effects. Table I.2 shows the conservation-of-mass relations for

## Table I.1 Ideal Gas Processes[a]

| Process | Describing Equations | $_1W_2$ | $_1Q_2$ | $S_2 - S_1$ |
|---|---|---|---|---|
| Isometric or constant volume | $V = c,^{(1)}\ v = c, \dfrac{v}{t} = c$ | 0 | $U_2 - U_1 = m\,(u_2 - u_1)$<br>$= mc_v\,(T_2 - T_1)^{(2)}$ | $m\left(s_2^0 - s_1^0 - R\ln\dfrac{\rho_2}{\rho_1}\right)^{(3)}$<br>$= mc_v\,\ln\dfrac{T_2}{T_1}$ |
| Isobaric, isopiestic, or constant pressure | $P = c,\ V/T = c, \dfrac{v}{t} = c$ | $p\,(V_2 - V_1)$<br>$= mR(T_2 - T_1)$ | $H_2 - H_1 = m\,(h_2 - h_1)$<br>$= mc_p\,(T_2 - T_1)$ | $m\left(s_2^0 - s_1^0\right)$<br>$= mc_p\ \ln\dfrac{T_2}{T_1}$ |
| Isothermal or constant temperature | $T = c,\ pV = pv = c$ | $p_1V_1\ \ln\ r^{(4)} = (p_2 - V_2)\ \ln\ r$<br>$= mRT\ \ln\ r$ | $_1W_2$ | $mR\ \ln\ r$ |
| Isentropic or reversible adiabatic | $s = c$<br><br>$pV^{k\,(5)} = c,\ TV^{k-1}$<br>$= c,\ p^{k-1}T^{k-1} = c$ | $U_1 - U_2 = m\,(u_1 - u_2)$<br>$\dfrac{P_1V_1 - P_2V_2}{k-1} = \dfrac{mR\,(T_1 - T_2)}{k-1}$ | 0 | 0 |
| Polytropic | $pV^{k\,(6)} = c,\ TV^{n-1}$<br>$= c,\ p^{n-1}T^{n-1} = c$ | $\dfrac{P_1V_1 - P_2V_2}{n-1} = \dfrac{mR\,(T_1T_2)}{n-1}$ | Use first law | $m\left(s_2^0 - s_1^0 - R\ \ln\dfrac{\rho_2}{\rho_1}\right)$<br>$= \dfrac{k-n}{k-1}\ \ln\ r$ |

[a](1) $c$ stands for an unspecified constant; (2) the second line of each entry applies when $c_p$ and $c_v$ are independent of temperature; (3) $s_2 - s_1 = s^0(T_2) - s^0(T_1) - R\ln(p_2/p_1)$; (4) $r$ = volume ratio or compression ratio = $V_2/V_1$; (5) $k = c_p/c_v > 1$; (6) $n$ = polytropic exponent.

various types of systems. In energy conservation, the first and second laws of thermodynamics form the basis of most technical analysis.

For open systems two approaches to analysis can be taken, depending upon the nature of the process. For steady systems, the steady-state steady-flow assumption

### Table I.2 Law of Conservation of Mass

Closed system, any process: m = constant

Open system, SSSF:
$$\sum_{in} \dot{m} = \sum_{out} \dot{m}$$

Open system, USUF:
$$(m_2 - m_2)_{c.v.} = \left[\sum_{in} \dot{m} = \sum_{out} \dot{m}\right] t$$

Open system, general case:
$$\frac{d}{dt}\,m_{c.v.} = \sum_{in} \dot{m} = \sum_{in} \dot{m}\ \text{or}\ \frac{d}{dt}\int_v \rho\,dV = \int_A \rho\,dV_{rn}\,dA$$

(SSSF) is adequate. This approach assumes that the state of the material is constant at any point in the system. For transient processes, the uniform flow uniform state (UFUS) assumption fits most situations. This involves the assumption that at points where mass crosses the system boundary, its state is constant with time. Also, the state of the mass in the system may vary with time but is uniform at any time.

Tables I.3 and I.4 give listings of the conservation-of-energy (first-law) and the second-law relations for various systems.

The first law simply gives a balance of energy during a process. The second law, however, extends the utility of thermodynamics to the prediction of both the possibility of a proposed process or the direction of a system change following a perturbation of the system. Although the first law is perhaps more directly valuable in energy conservation, the implications of the second law can be equally illuminating.

## Table I.3 First Law of Thermodynamics

Closed system, cyclic process:

$$\oint dQ = \oint dW$$

Closed system, state 1 to state 2:

$$_1Q_2 = E_2 - E_1 + {_1W_2}$$

$$E = internal\ energy + kinetic\ energy + potential\ energy = m(u + V^2/2 + gz)$$

Open system:

$$\dot{Q}_{c.v.} = \frac{d}{dt}\int_v \rho\left(u + V^2/2 + gz\right)dV + \int_A \rho\left(h + V^2/2 + gz\right)V_{rn}dA + \dot{W}_{c.v.}$$

where enthalpy per unit mass $h = u + pv$;

alternative form:

$$\dot{Q}_{c.v.} + \sum_{in}\dot{m}\left(h + V^2/2 + gz\right) = \dot{W}_{c.v.} + \sum_{out}\dot{m}\left(h + V^2/2 + gz\right) + \dot{E}_{c.v.}$$

where

$$\dot{E}_{c.v.} = \frac{d}{dt}\int_v \rho\left(u + V^2/2 + gz\right)dV$$

Open system, steady state steady flow (SSSF):

$$\dot{Q}_{c.v.} + \sum_{in}\dot{m}\left(h + V^2/2 + gz\right) = \dot{W}_{c.v.} + \sum_{out}\dot{m}\left(h + V^2/2 + gz\right)$$

Open system, uniform state uniform flow (USUF):

$$_1Q_{c.v.} + \sum_{in}m\left(h + V^2/2 + gz\right) = {_1W_{2c.v.}} + \sum_{out}m\left(h + V^2/2 + gz\right) + \left[m\left(u + V^2/2 + gz\right)\right]_{1c.v.}^{2c.v.}$$

There are two statements of the second law. The two, although appearing to be different, actually can be shown to be equivalent. Therefore, we state only one of them, the Kelvin-Planck version:

*It is impossible for any device to operate in a cycle and produce work while exchanging heat only with bodies at a single fixed temperature.*

The other statement is called the Clausius statement.

The implications of the second law are many. For example, it allows us to (1) determine the maximum possible efficiency of a heat engine, (2) determine the maximum coefficient of performance for a refrigerator, (3) determine the feasibility of a proposed process, (4) predict the direction of a chemical or other type of process, and (5) correlate physical properties. So we see that the second law is quite valuable.

### I.2.4 Efficiency

Efficiency is a concept used to describe the effectiveness of energy conversion devices that operate in cycles as well as in individual system components that operate in processes. Thermodynamic efficiency, $\eta$, and coefficient of performance, COP, are used for devices that operate in cycles. The following definitions apply:

$$\eta = \frac{W_{net}}{Q_H} \qquad COP = \frac{Q_L}{W}$$

where $Q_L$ and $Q_H$ represent heat transferred from cold and hot regions, respectively, $W_{net}$ is useful work pro-

## Table I.4 Second Law of Thermodynamics

Closed system, cyclic process:

$$\oint \frac{dQ}{T} \leq 0$$

Closed system, state 1 to state 2:

$$\oint_1^2 \frac{dQ}{T} \leq S_2 - S_1 = m\,(S_2 - S_1)$$

Open system:

$$\frac{d}{dt}\int_V \rho\, s\, dV + \sum_{out} \dot{m}s \geq \sum_{in} \dot{m}s + \int_A \frac{\dot{Q}}{T}\, dA$$

or

$$\frac{d}{dt}\int_V \rho\, s\, dV + \int_V \rho\, s\, V_{rn} dA \geq \int_A \frac{\dot{Q}}{T}\, dA$$

Open system, SSSF:

$$\sum_{out} \dot{m}s + \sum_{in} \dot{m}s \int_A \frac{\dot{Q}}{T}\, dA$$

Open system, USUF:

$$(m_2 s_2 - m_1 s_1)_{c.v.} + \sum_{out} \dot{m}s \geq \sum_{in} \dot{m}s \int_A \frac{\dot{Q}_{c.v.}}{T}\, dA$$

duced in a heat engine, and W is the work required to drive the refrigerator. A heat engine produces useful work, while a refrigerator uses work to transfer heat from a cold to a hot region. There is an ideal cycle, called the Carnot cycle, which yields the maximum efficiency for heat engines and refrigerators. It is composed of four ideal reversible processes; the efficiency of this cycle is

$$\eta_c = 1 - \frac{T_L}{T_H}$$

and the COP is

$$[COP]_c = \frac{TL/TH}{1 - TL/TH}$$

These represent the best possible performance of cyclic energy conversion devices operating between temperature extremes, $T_H$ and $T_L$. The thermodynamic efficiency

should not be confused with efficiencies applied to devices that operate along a process line. This efficiency is defined as

$$\eta_{device} = \frac{actual\ energy\ transfer}{ideal\ energy\ transfer}$$

for a work-producing device and

$$\eta_{device} = \frac{ideal\ energy\ transfer}{actual\ energy\ transfer}$$

for a work-consuming device. Note these definitions are such that $\eta < 1$. These efficiencies are convenience factors in that the actual performance can be calculated from an ideal process line and the efficiency, which generally must be experimentally determined. Table I.5 shows the most commonly encountered versions of efficiencies.

### I.2.5 Power and Refrigeration Cycles

Many cycles have been devised to convert heat into work, and vice versa. Several of these take advantage of the phase change of the working fluid: for example, the Rankine, the vapor compression, and the absorption cycles. Others involve approximations of thermodynamic processes to mechanical processes and are called air-standard cycles.

## Table I.5 Thermodynamic Efficiency

Heat engines and refrigerators:

Engine efficiency $\eta \equiv W/Q_H \leq \eta_{Carnot} = (T_H - T_L)/T_H < 1$

Heat pump c.o.p. $\beta' \equiv Q_H/W \leq \beta'_{Carnot} = T_H/(T_H - T_L) > 1$

Refrigerator c.o.p. $\beta \equiv Q_L/W \leq \beta_{Carnot} = T_L/(T_H - T_L)$,

$\quad 0 < \beta < \infty,\ (Q_H/Q_L)_{Carnot} = T_H - T_L$

Process efficiencies

$\quad \eta_{ad,\ turbine} = w_{actual,\ adiabatic}/w_{isentropic}$

$\quad \eta_{ad,\ compressor} = w_{isentropic}/w_{actual,\ adiabatic}$

$\quad \eta_{ad,\ nozzle} = K.E._{actual,\ adiabatic}/K.E._{isentropic}$

$$\eta_{nozzle} = \frac{V_a^2/2gc}{V_s^2/2gc}$$

$\eta_{cooled\ nozzle} = w_{isentropic\ rev.}/w_{actual}$

The Rankine cycle is probably the most frequently encountered cycle in thermodynamics. It is used in almost all large-scale electric generation plants, regardless of the energy source (gas, coal, oil, or nuclear). Many modern steam-electric power plants operate at super-critical pressures and temperatures during the boiler heat addition process. This leads to the necessity of reheating between high- and lower-pressure turbines to prevent excess moisture in the latter stages of turbine expansion (prevents blade erosion). Feedwater heating is also extensively used to increase the efficiency of the basic Rankine cycle (see Ref. 1 for details).

The vapor compression cycle is almost a reversed Rankine cycle. The major difference is that a simple expansion valve is used to reduce the pressure between the condensor and the evaporator rather than being a work-producing device. The reliability of operation of the expansion valve is a valuable trade-off compared to the small amount of work that could be reclaimed. The vapor compression cycle can be used for refrigeration or heating (heat pump).

In the energy conservation area, applications of the heat pump are taking on added emphasis. The device is useful for heating from an electrical source (compressor) in situations where direct combustion is not available. Additionally, the device can be used to upgrade the temperature level of waste heat recovered at a lower temperature.

Air-standard cycles, useful both for power generation and heating/cooling applications, are the thermodynamic approximations to the processes occurring in the actual devices. In the actual cases, a thermodynamic

cycle is not completed, necessitating the approximations. Air-standard cycles are analyzed by using the following approximations:

1.   Air is the working fluid and behaves as an ideal gas.

2.   Combustion and exhaust processes are replaced by heat exchangers.

Other devices must be analyzed component by component using property data for the working fluids (see Appendix II). Figure I.2 gives a listing of various power systems with their corresponding thermodynamic cycle and other pertinent information.

### I.2.6 Combustion Processes

The combustion process continues to be the most prevalent means of energy conversion. Natural and manufactured gases, coal, liquid fuel/air mixtures, even wood and peat are examples of energy sources requiring combustion.

There are two overriding principles of importance in analyzing combustion processes. They are the combustion equation and the first law for the combustion chamber. The combustion equation is simply a mass balance between reactants and products of the chemical reaction combustion process. The first law is the energy balance for the same process using the results of the combustion equation as input.

In practice, we can restrict our discussion to hydrocarbon fuels, meaning that the combustion equation (chemical balance) is written as

### Table I.6 Characteristics of Some of the Hydrocarbon Families

| Family | Formula | Structure | Saturated |
|---|---|---|---|
| Paraffin | $C_nH_{2n+2}$ | Chain | Yes |
| Olefin | $C_nH_{2n}$ | Chain | No |
| Diolefin | $C_nH_{2n-2}$ | Chain | No |
| Naphthene | $C_nH_{2n}$ | Ring | Yes |
| Aromatic Benzene | $C_nH_{2n-6}$ | Ring | No |
| Aromatic Naphthalene | $C_nH_{2n-12}$ | Ring | No |

Molecular structure of some hydrocarbon fuels:

Chain structure,
saturated

Chain structure,
unsaturated

Ring structure,
saturated

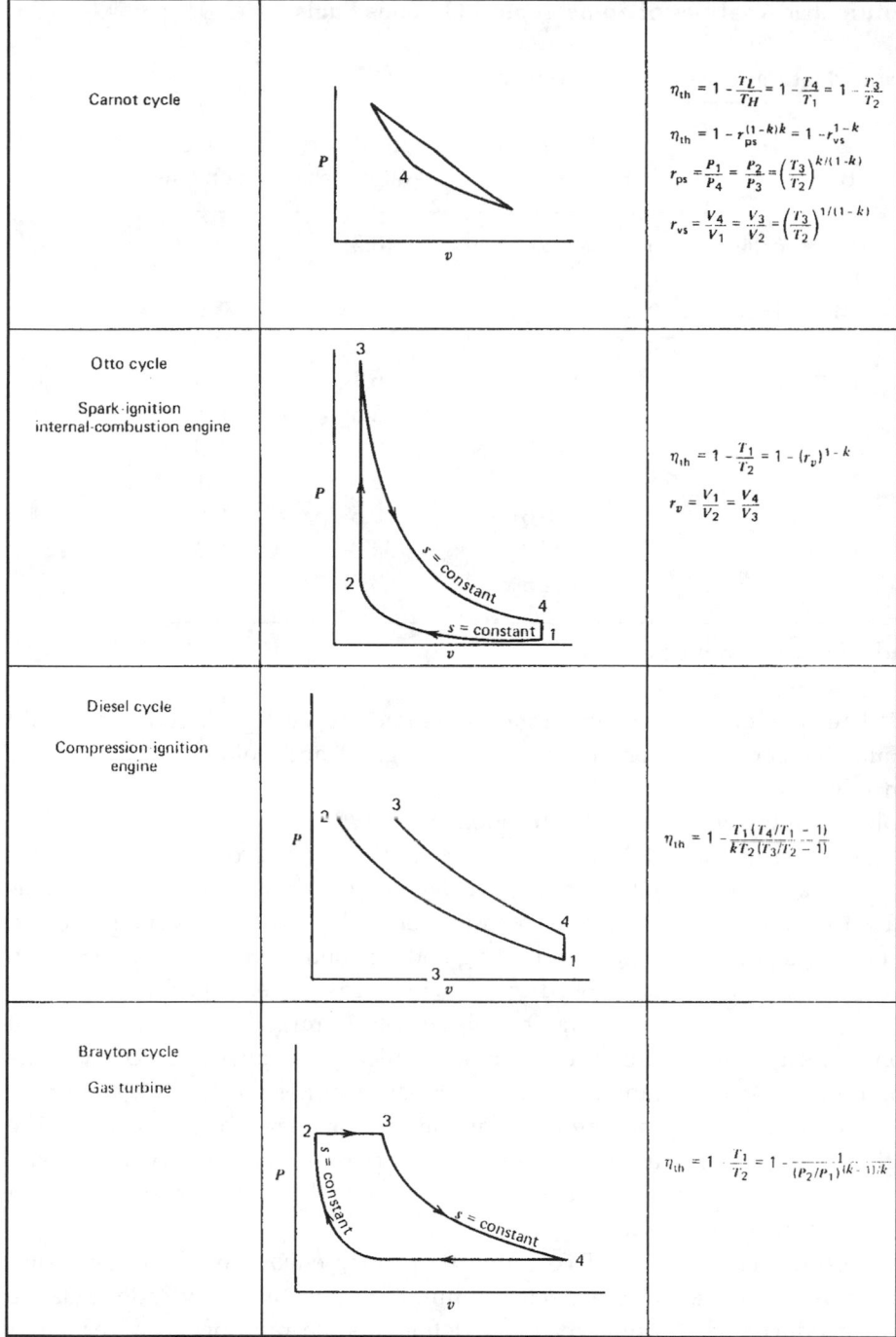

| Carnot cycle | | $\eta_{th} = 1 - \frac{T_L}{T_H} = 1 - \frac{T_4}{T_1} = 1 - \frac{T_3}{T_2}$ <br> $\eta_{th} = 1 - r_{ps}^{(1-k)k} = 1 - r_{vs}^{1-k}$ <br> $r_{ps} = \frac{P_1}{P_4} = \frac{P_2}{P_3} = \left(\frac{T_3}{T_2}\right)^{k/(1-k)}$ <br> $r_{vs} = \frac{V_4}{V_1} = \frac{V_3}{V_2} = \left(\frac{T_3}{T_2}\right)^{1/(1-k)}$ |
| Otto cycle <br><br> Spark-ignition internal-combustion engine | | $\eta_{th} = 1 - \frac{T_1}{T_2} = 1 - (r_v)^{1-k}$ <br> $r_v = \frac{V_1}{V_2} = \frac{V_4}{V_3}$ |
| Diesel cycle <br><br> Compression ignition engine | | $\eta_{th} = 1 - \frac{T_1(T_4/T_1 - 1)}{kT_2(T_3/T_2 - 1)}$ |
| Brayton cycle <br><br> Gas turbine | | $\eta_{th} = 1 - \frac{T_1}{T_2} = 1 - \frac{1}{(P_2/P_1)^{(k-1)/k}}$ |

**Fig. I.2 Air standard cycles.**

$$C_xH_y + \alpha(O_2 + 3.76N_2) \rightarrow b\ CO_2 + c\ CO_2$$

$$+ e\ H_2O + d\ O_2 + 3.76a\ N_2$$

This equation neglects the minor components of air; that is, air is assumed to be 1 mol of $O_2$ mixed with 3.76 mol of $N_2$. The balance is based on 1 mol of fuel $C_xH_y$. The unknowns are determined for each par-

ticular application. Table I.6 gives the characteristics of some of the hydrocarbons. Table I.7 shows the volumetric analyses of several gaseous fuels.

Once a combustion process is decided upon (i.e., the fuel to be used and the heat transfer/combustion chamber are selected), the relative amount of fuel and air become of prime importance. This is because the air/fuel ratio (AF) controls the temperature of the combustion zone and the energy available to be transferred to a working fluid or converted to work. Stoichiometric air is that quantity of air required such that no oxygen would appear in the products. Excess air occurs when more than enough air is provided to the combustion process. Ideal combustion implies perfect mixing and complete reactions. In this case theoretical air (TA) would yield no free oxygen in the products. Excess air then is actual air less theoretical air.

Most industrial combustion processes conform closely to a steady-state, steady-flow case. The first law for an open control volume surrounding the combustion zone can then be written. If we assume that $Q$ and $W$ are zero and that $\Delta$K.E. and $\Delta$P.E. are negligible, then the following equation results:

$$\sum_{\text{products}} (H_e - H_{\text{ref}}) =$$

$$\sum_{\text{reactants}} (H_i - H_{\text{ref}}) + H_{\text{comb}} \quad (I.3)$$

Subscripts $i$ and $e$ refer to inlet and exit conditions, respectively. $H_{\text{ref}}$ is the enthalpy of each component at some reference temperature. $\Delta H$comb represents the heat of combustion for the fuel and, in general, carries a negative value, meaning that heat would have to be transferred out of the system to maintain inlet and exit temperatures at the same level.

The adiabatic flame temperature occurs when the

**Table I.7 Volumetric Analyses of Some Typical Gaseous Fuels**

| Constituent | Various Natural Gases | | | | Producer Gas from Bituminous Coal | Carbureted Water Gas | Coke Oven Gas |
|---|---|---|---|---|---|---|---|
| | A | B | C | D | | | |
| Methane | 93.9 | 60.1 | 67.4 | 54.3 | 3.0 | 10.2 | 32.1 |
| Ethane | 3.6 | 14.8 | 16.8 | 16. | | | |
| Propane | 1.2 | 13.4 | 15.8 | 16.2 | | | |
| Butanes plus[a] | 1.3 | 4.2 | | 7.4 | | | |
| Ethene | | | | | | 6.1 | 3.5 |
| Benzene | | | | | | 2.8 | 0.5 |
| Hydrogen | | | | | 14.0 | 40.5 | 46.5 |
| Nitrogen | | 7.5 | | 5.8 | 50.9 | 2.9 | 8.1 |
| Oxygen | | | | | 0.6 | 0.5 | 0.8 |
| Carbon monoxide | | | | | 27.0 | 34.0 | 6.3 |
| Carbon dioxide | | | | | 4.5 | 3.0 | 2.2 |

[a]This includes butane and all heavier hydrocarbons.

combustion zone is perfectly insulated. The solution of equation I.3 would give the adiabatic flame temperature for any particular case. The maximum adiabatic flame temperature would occur when complete combustion occurs with a minimum of excess $O_2$ appearing in the products.

Appendix II gives tabulated values for the important thermophysical properties of substances important in combustion.

**Gas Analysis.** During combustion in heaters and boilers, the information required for control of the burner settings is the amount of excess air in the fuel gas. This percentage can be a direct reflection of the efficiency of combustion.

The most accurate technique for determining the volumetric makeup of combustion by-products is the Orsat analyzer. The Orsat analysis depends upon the fact that for hydrocarbon combustion the products may contain $CO_2$, $O_2$, CO, $N_2$, and water vapor. If enough excess air is used to obtain complete combustion, no CO will be present. Further, if the water vapor is removed, only $CO_2$, $O_2$, and $N_2$ remain.

The Orsat analyzer operates on the following principles. A sample of fuel gas is first passed over a desiccant to remove the moisture. (The amount of water vapor can be found later from the combustion equation.) Then the sample is exposed in turn to materials that absorb first the $CO_2$, then the $O_2$, and finally the CO (if present). After each absorption the volumetric change is carefully measured in a graduated pipette system. The

remaining gas is assumed to be $N_2$. Of course, it could contain some trace of gases and pollutants.

### I.2.7 Psychrometry

Psychrometry is the science of air/water vapor mixtures. Knowledge of the behavior of such systems is important, both in meteorology and industrial processes, especially heating and air conditioning. The concepts can be applied to other ideal gas/water vapor mixtures.

Air and water vapor mixed together at a total pressure of 1 atm is called atmospheric air. Usually, the amount of water vapor in atmospheric air is so minute that the vapor and air can be treated as an ideal gas. The air existing in the mixture is often called dry air, indicating that it is separate from the water vapor coexisting with it.

Two terms frequently encountered in psychrometry are relative humidity and humidity ratio. Relative humidity, $\phi$, is defined as the ratio of the water vapor pressure to the saturated vapor pressure at the temperature of the mixture. Figure I.3 shows the relation between points on the T–s diagram that yield the relative humidity. Relative humidity cannot be greater than unity or 100%, as is normally stated.

The humidity ratio, $\omega$, on the other hand, is defined as the ratio of the mass of water vapor to the mass of dry air in atmospheric air, $\omega = m_v/m_a$. This can be shown to be $\omega = v_a/v_v$ and a relationship between $\omega$ and $\phi$ exists, $\omega = (va/v_g)\phi$, where $v_g$ refers to the specific volume of saturated water vapor at the temperature of the mixture.

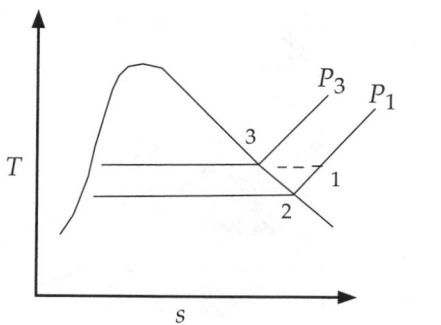

**Fig. I.3 Behavior of water in air:** $\phi = P_1/P_3$; $T_2$ = dew point.

A convenient way of describing the condition of atmospheric air is to define four temperatures: dry-bulb, wet-bulb, dew-point, and adiabatic saturation temperatures. The dry-bulb temperature is simply that temperature which would be measured by any of several types of ordinary thermometers placed in atmospheric air.

The dew-point temperature (point 2 on Figure I.3) is the saturation temperature of the water vapor at its existing partial pressure. In physical terms it is the mixture temperature where water vapor would begin to condense if cooled at constant pressure. If the relative humidity is 100% the dew-point and dry-bulb temperatures are identical.

In atmospheric air with relative humidity less than 100%, the water vapor exists at a pressure lower than saturation pressure. Therefore, if the air is placed in contact with liquid water, some of the water would be evaporated into the mixture and the vapor pressure would be increased. If this evaporation were done in an insulated container, the air temperature would decrease, since part of the energy to vaporize the water must come from the sensible energy in the air. If the air is brought to the saturated condition, it is at the adiabatic saturation temperature.

A psychrometric chart is a plot of the properties of atmospheric air at a fixed total pressure, usually 14.7 psia. The chart can be used to quickly determine the properties of atmospheric air in terms of two independent properties, for example, dry-bulb temperature and relative humidity. Also, certain types of processes can be described on the chart. Appendix II contains a psychrometric chart for 14.7-psia atmospheric air. Psychrometric charts can also be constructed for pressures other than 14.7 psia.

## I.3  HEAT TRANSFER

Heat transfer is the branch of engineering science that deals with the prediction of energy transport caused by temperature differences. Generally, the field is broken down into three basic categories: conduction, convection, and radiation heat transfer.

*Conduction* is characterized by energy transfer by internal microscopic motion such as lattice vibration and electron movement. Conduction will occur in any region where mass is contained and across which a temperature difference exists.

*Convection* is characterized by motion of a fluid region. In general, the effect of the convective motion is to augment the conductive effect caused by the existing temperature difference.

*Radiation* is an electromagnetic wave transport phenomenon and requires no medium for transport. In fact, radiative transport is generally more effective in a vacuum, since there is attenuation in a medium.

### I.3.1  Conduction Heat Transfer

The basic tenet of conduction is called Fourier's law,

$$\dot{Q} = -kA\frac{dT}{dx}$$

The heat flux is dependent upon the area across which energy flows and the temperature gradient at that plane. The coefficient of proportionality is a material property, called thermal conductivity $k$. This relationship always applies, both for steady and transient cases. If the gradient can be found at any point and time, the heat flux density, $\dot{Q}/A$, can be calculated.

**Conduction Equation.** The control volume approach from thermodynamics can be applied to give an energy balance which we call the conduction equation. For brevity we omit the details of this development; see Refs. 2 and 3 for these derivations. The result is

$$G + K\nabla^2 T = -\rho C \frac{\partial T}{\partial \tau} \tag{I.4}$$

This equation gives the temperature distribution in space and time, G is a heat-generation term, caused by chemical, electrical, or nuclear effects in the control volume. Equation I.4 can be written

$$\nabla^2 T + \frac{G}{K} = \frac{\rho C}{k}\frac{\partial T}{\partial \tau}$$

The ratio $k/\rho C$ is also a material property called thermal diffusivity u. Appendix II gives thermophysical properties of many common engineering materials.

For steady, one-dimensional conduction with no heat generation,

$$\frac{D^2T}{dx^2} = 0$$

This will give $T = ax + b$, a simple linear relationship between temperature and distance. Then the application of Fourier's law gives

$$\dot{Q} = kA\frac{T}{x}$$

a simple expression for heat transfer across the $\Delta x$ distance. If we apply this concept to insulation for example, we get the concept of the R value. $R$ is just the resistance to conduction heat transfer per inch of insulation thickness (i.e., $R = 1/k$).

**Multilayered, One-Dimensional Systems.** In practical applications, there are many systems that can be treated as one-dimensional, but they are composed of layers of materials with different conductivities. For example, building walls and pipes with outer insulation fit this category. This leads to the concept of overall heat-transfer coefficient, $U$. This concept is based on the definition of a convective heat-transfer coefficient,

$$\dot{Q} = hA \ T$$

This is a simplified way of handling convection at a boundary between solid and fluid regions. The heat-transfer coefficient $h$ represents the influence of flow conditions, geometry, and thermophysical properties on the heat transfer at a solid-fluid boundary. Further discussion of the concept of the $h$ factor will be presented later.

Figure I.4 represents a typical one-dimensional, multilayered application. We define an overall heat-transfer coefficient $U$ as

$$Q = UA \ (T_i - T_0)$$

We find that the expression for $U$ must be

$$U = \frac{1}{\dfrac{1}{h_1} + \dfrac{x_1}{k_1} + \dfrac{x_2}{k_2} + \dfrac{x_3}{k_3} + \dfrac{1}{h_0}}$$

This expression results from the application of the conduction equation across the wall components and the convection equation at the wall boundaries. Then, by noting that in steady state each expression for heat must be equal, we can write the expression for $U$, which contains both convection and conduction effects. The $U$ factor is extremely useful to engineers and architects in a wide variety of applications.

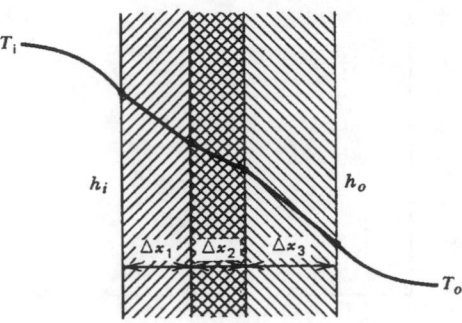

**Fig. I.4 Multilayered wall with convection at the inner and outer surfaces.**

The $U$ factor for a multilayered tube with convection at the inside and outside surfaces can be developed in the same manner as for the plane wall. The result is

$$U = \frac{1}{\dfrac{1}{h_0} + \sum_j \dfrac{r_0 \ln\left(r_j + 1/r_j\right)}{k_j} + \dfrac{1r_0}{h_i r_i}}$$

where $r_i$ and $r_0$ are inside and outside radii.

*Caution:* The value of $U$ depends upon which radius you choose (i.e., the inner or outer surface).

If the inner surface were chosen, we would get

$$U = \frac{1}{\dfrac{1r_i}{h_0 r_0} + \sum_j \dfrac{r_i \ln\left(r_j + 1/r_j\right)}{k_j} + \dfrac{1}{h_i}}$$

However, there is no difference in heat-transfer rate; that is,

$$Q_0 = U_i A_i T_{overall} = U_0 A_0 T_{overall}$$

so it is apparent that

$$U_i A_i = U_0 A_0$$

for cylindrical systems.

**Finned Surfaces.** Many heat-exchange surfaces experience inadequate heat transfer because of low heat-transfer coefficients between the surface and the adjacent fluid. A remedy for this is to add material to the surface. The added material in some cases resembles a fish "fin," thereby giving rise to the expression "a finned surface." The performance of fins and arrays of fins is an important item in the analysis of many heat-exchange devices. Figure I.5 shows some possible shapes for fins.

The analysis of fins is based on a simple energy balance between one-dimensional conduction down the length of the fin and the heat convected from the exposed surface to the surrounding fluid. The basic equation that applies to most fins is

$$\frac{d_2\theta}{dx^2} + \frac{1}{A}\frac{dA}{dx}\frac{d\theta}{dx} - \frac{h}{k}\frac{1}{A}\frac{dS}{dx}\theta = 0 \qquad (I.5)$$

when $\theta$ is $(T - T_\infty)$, the temperature difference between fin and fluid at any point; $A$ is the cross-sectional area of the fin; $S$ is the exposed area; and $x$ is the distance along the fin. Chapman[2] gives an excellent discussion of the development of this equation.

The application of equation I.5 to the myriad of possible fin shapes could consume a volume in itself. Several shapes are relatively easy to analyze; for example, fins of uniform cross section and annular fins can be treated so that the temperature distribution in the fin and the heat rate from the fin can be written. Of more utility, especially for fin arrays, are the concepts of fin efficiency and fin surface effectiveness (see Holman[3]).

Fin efficiency $\eta_f$ is defined as the ratio of actual heat loss from the fin to the ideal heat loss that would occur if the fin were isothermal at the base temperature. Using this concept, we could write

$$\dot{Q}_{fin} = A_{fin}^h (T_b - T_\infty)\eta_f$$

$\eta_f$ is the factor that is required for each case. Figure I.6 shows the fin efficiency for several cases.

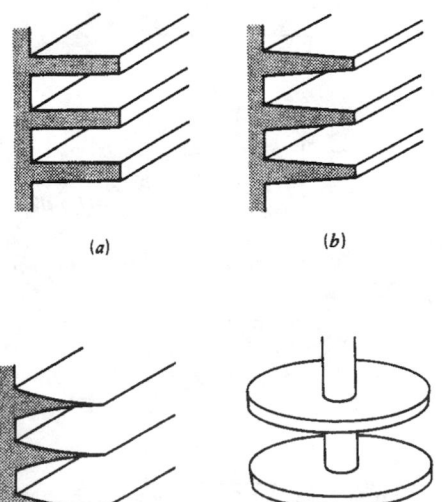

**Fig. I.5 Fins of various shapes. (a) Rectangular, (b) Trapezoidal, (c) Arbitrary profile, (d) Circumferential.**

Surface effectiveness K is defined as the actual heat transfer from a finned surface to that which would occur if the surface were isothermal at the base temperature. Taking advantage of fin efficiency, we can write

$$K = \frac{(A - A_f)h\,\theta_0 + \eta_f A_f\theta_0}{A_h\theta_0} \qquad (I.6)$$

Equation I.6 reduces to

$$K = 1 \quad \frac{A_f}{A} \quad (1 - \eta_f)$$

which is a function only of geometry and single fin efficiency. To get the heat rate from a fin array, we write

$$Q_{array} = Kh\,(T_b - T_\infty)\,A$$

where $A$ is the total area exposed.

**Transient Conduction.** Heating and cooling problems involve the solution of the time-dependent conduction equation. Most problems of industrial significance occur when a body at a known initial temperature is suddenly exposed to a fluid at a different temperature. The temperature behavior for such unsteady problems can be characterized by two dimensionless quantities, the Biot number, Bi = $hL/k$, and the Fourier modulus, Fo = $\alpha\tau/L^2$. The Biot number is a measure of the effectiveness of conduction within the body. The Fourier modulus is simply a dimensionless time.

If Bi is a small, say Bi $\leq$ 0.1, the body undergoing the temperature change can be assumed to be at a uniform temperature at any time. For this case,

$$\frac{T - T_f}{T_i - T_f} = \exp\left[-\left(\frac{hA}{\rho CV}\right)\tau\right]$$

where $T_f$ and $T_i$ are the fluid temperature and initial body temperature, respectively. The term $(\rho CV/hA)$ takes on the characteristics of a time constant.

If Bi $\geq$ 0.1, the conduction equation must be solved in terms of position and time. Heisler[4] solved the equation for infinite slabs, infinite cylinders, and spheres. For convenience he plotted the results so that the temperature at any point within the body and the amount of heat transferred can be quickly found in terms of Bi and Fo. Figures I.7 to I.10 show the Heisler charts for slabs and cylinders. These can be used if $h$ and the properties of the material are constant.

### I.3.2 Convection Heat Transfer

Convective heat transfer is considerably more complicated than conduction because motion of the medium is involved. In contrast to conduction, where many geometrical configurations can be solved analytically, there are only limited cases where theory alone will give convective heat-transfer relationships. Consequently, convection is largely what we call a semi-empirical science. That is, actual equations for heat transfer are based strongly on the results of experimentation.

**Convection Modes.** Convection can be split into several subcategories. For example, forced convection refers to the case where the velocity of the fluid is completely independent of the temperature of the fluid. On the other hand, natural (or free) convection occurs when the temperature field actually causes the fluid motion through buoyancy effects.

We can further separate convection by geometry into external and internal flows. Internal refers to channel, duct, and pipe flow and external refers to unbounded fluid flow cases. There are other specialized forms of convection, for example the change-of-phase phenomena: boiling, condensation, melting, freezing, and so on. Change-of-phase heat transfer is difficult to predict analytically. Tongs[5] gives many of the correlations for boiling and two-phase flow.

**Dimensional Heat-Transfer Parameters.** Because experimentation has been required to develop appropriate correlations for convective heat transfer, the use of generalized dimensionless quantities in these correlations is preferred. In this way, the applicability of experimental data covers a wider range of conditions and fluids. Some of these parameters, which we generally call "numbers," are given below:

$$\text{Nusselt number: Nu} = \frac{hL}{k}$$

where $k$ is the fluid conductivity and L is measured along the appropriate boundary between liquid and solid; the Nu is a nondimensional heat-transfer coefficient.

$$\text{Reynolds number: Re} = \frac{Lu}{\upsilon}$$

defined in Section I.4: it controls the character of the flow

$$\text{Prandtl number: Pr} = \frac{C\mu}{k}$$

ratio of momentum transport to heat-transport characteristics for a fluid: it is important in all convective cases, and is a material property

$$\text{Grashof number: Gr} = \frac{g\,\beta(T - T_\infty)L^3}{\upsilon^2}$$

serves in natural convection the same role as Re in forced convection: that is, it controls the character of the flow

$$\text{Stanton number: St} = \frac{h}{\rho\,uC_p}$$

(a)

(b)

**Fig. I.6 (a) Efficiencies of rectangular and triangular fins, (b) Efficiencies of circumferential fins of rectangular profile.**

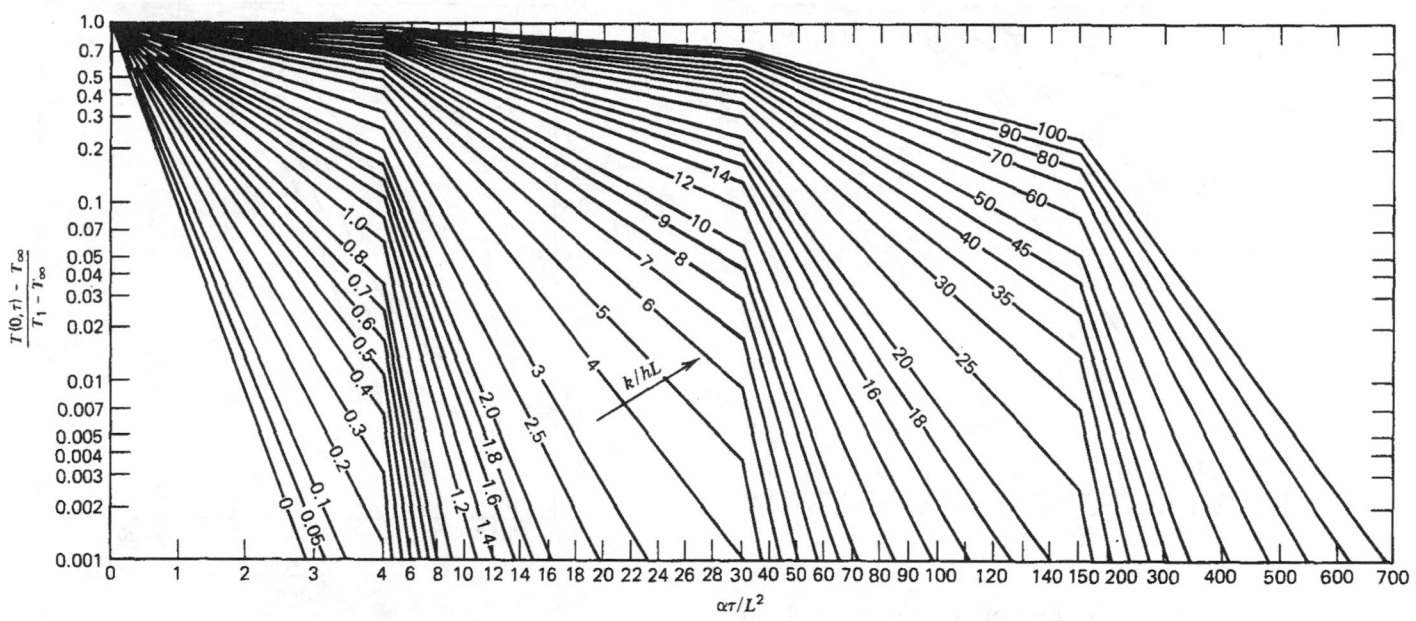

**Fig. I.7 Midplane temperature for an infinite plate of thickness 2L. (From Ref. 4.)**

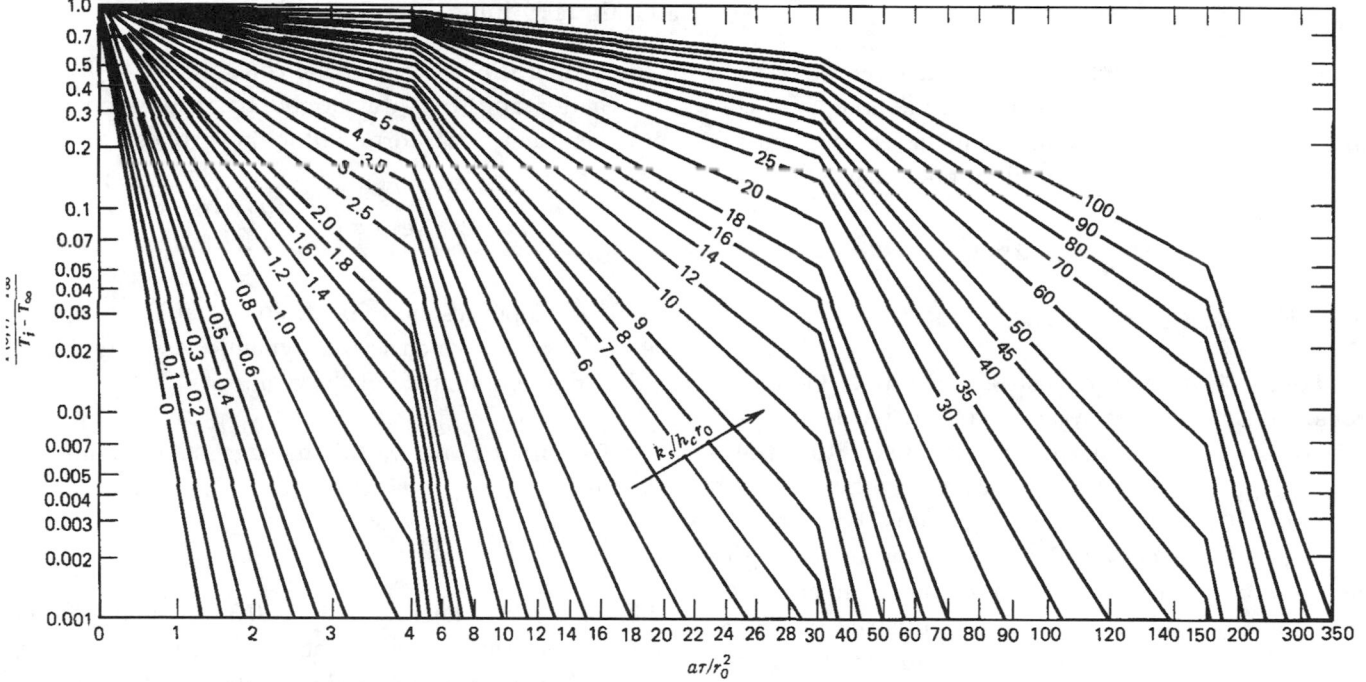

**Fig. I.8 Axis temperature for an infinite cylinder of radius $r_o$. (From Ref. 4.)**

also a nondimensional heat-transfer coefficient: it is very useful in pipe flow heat transfer.

In general, we attempt to correlate data by using relationships between dimensionless numbers: for example, in many convection cases, we could write Nu = Nu(Re, Pr) as a functional relationship. Then it is possible either from analysis, experimentation, or both, to write an equation that can be used for design calculations. These are generally called working formulas.

**Forced Convection Past Plane Surfaces.** The average heat-transfer coefficient for a plate of length $L$ may be calculated from

$$Nu_L = 0.664 \, (Re_L)^{1/2} (Pr)^{1/3}$$

if the flow is laminar (i.e., if $Re_L \leq 4{,}000$). For this case the fluid properties should be evaluated at the mean film temperature $T_m$, which is simply the arithmetic

**Fig. I.9 Temperature as a function of center temperature in an infinite plate of thickness 2L. (From Ref. 4.)**

**Fig. I.10 Temperature as a function of axis temperature in an infinite cylinder of radius $r_o$. (From Ref. 4.)**

average of the fluid and the surface temperature.

For turbulent flow, there are several acceptable correlations. Perhaps the most useful includes both laminar leading edge effects and turbulent effects. It is

$$Nu = 0.0036 \, (Pr)^{1/3} \, [(Re_L)^{0.8} - 18.700]$$

where the transition Re is 4,000.

**Forced Convection Inside Cylindrical Pipes or Tubes.** This particular type of convective heat transfer is of special engineering significance. Fluid flows through pipes, tubes, and ducts are very prevalent, both in laminar and turbulent flow situations. For example, most heat exchangers involve the cooling or heating of fluids in tubes. Single pipes and/or tubes are also used to transport hot or cold liquids in industrial processes. Most of the formulas listed here are for the $0.5 \leq Pr \leq 100$ range.

**Laminar Flow.** For the case where $Re_D < 2300$, Nusselt showed that $Nu_D = 3.66$ for long tubes at a constant tube-wall temperature. For forced convection cases (laminar and turbulent) the fluid properties are evaluated at the bulk temperature $T_b$. This temperature, also called the mixing-cup temperature, is defined by

$$T_b = \frac{\int_0^R uTr \, dr}{\int_0^R ur \, dr}$$

if the properties of the flow are constant.

Sieder and Tate developed the following more convenient empirical formula for short tubes:

$$Nu_D = 1.86 \, (Re_D)^{1/3} (Pr)^{1/3} \left(\frac{D}{L}\right)^{1/3} \left(\frac{\mu}{\mu_s}\right)^{0.14}$$

The fluid properties are to be evaluated at $T_b$ except for the quantity $\mu_s$, which is the dynamic viscosity evaluated at the temperature of the wall.

**Turbulent Flow.** McAdams suggests the empirical relation

$$Nu_D = 0.023 \, (Pr_D)^{0.8}(Pr)^n \qquad (I.7)$$

where $n = 0.4$ for heating and $n = 0.3$ for cooling. Equation I.7 applies as long as the difference between the pipe surface temperature and the bulk fluid temperature is not greater than 10°F for liquids or 100°F for gases.

For temperature differences greater then the limits specified for equation I.7 or for fluids more viscous than water, the following expression from Sieder and Tate will give better results:

$$NU_D = 0.027 \, (Pr_D)^{0.8}(Pr)^{1/3} \left(\frac{\mu}{\mu_s}\right)^{0.14}$$

Note that the McAdams equation requires only a knowledge of the bulk temperature, whereas the Sieder-Tate expression also requires the wall temperature. Many people prefer equation I.7 for that reason.

Nusselt found that short tubes could be represented by the expression

$$\text{Nu}_D = 0.036 \, (\text{Pe}_D)^{0.8}(\text{Pr})^{1/3}\left(\frac{\mu}{\mu_s}\right)^{0.14}\left(\frac{D}{L}\right)^{1/18}$$

For noncircular ducts, the concept of equivalent diameter can be employed, so that all the correlations for circular systems can be used.

**Forced Convection in Flow Normal to Single Tubes and Banks.** This circumstance is encountered frequently, for example air flow over a tube or pipe carrying hot or cold fluid. Correlations of this phenomenon are called semi-empirical and take the form $\text{Nu}_D = C(\text{Re}_D)^m$. Hilpert, for example, recommends the values given in Table I.8. These values have been in use for many years and are considered accurate.

Flows across arrays of tubes (tube banks) may be even more prevalent than single tubes. Care must be exercised in selecting the appropriate expression for the tube bank. For example, a staggered array and an in-line array could have considerably different heat-transfer characteristics. Kays and London[6] have documented many of these cases for heat-exchanger applications. For a general estimate of order-of-magnitude heat-transfer coefficients, Colburn's equation

$$\text{Nu}_D = 0.33 \, (\text{Re}_D)^{0.6} \, (\text{Pr})^{1/3}$$

is acceptable.

**Free Convection Around Plates and Cylinders.** In free convection phenomena, the basic relationships take on the functional form $\text{Nu} = f(\text{Gr}, \text{Pr})$. The Grashof number replaces the Reynolds number as the driving function for flow.

In all free convection correlations it is customary to evaluate the fluid properties at the mean film temperature $T_m$, except for the coefficient of volume expansion $\beta$, which is normally evaluated at the temperature of the undisturbed fluid far removed from the surface—namely, $T_f$. Unless otherwise noted, this convention should be used in the application of all relations quoted here.

Table I.9 gives the recommended constants and exponents for correlations of natural convection for vertical plates and horizontal cylinders of the form $\text{Nu} = C \cdot \text{Ra}^m$. The product $\text{Gr} \cdot \text{Pr}$ is called the Rayleigh number (Ra) and is clearly a dimensionless quantity associated with any specific free convective situation.

### I.3.3 Radiation Heat Transfer

Radiation heat transfer is the most mathematically complicated type of heat transfer. This is caused primarily by the electromagnetic wave nature of thermal radiation. However, in certain applications, primarily high-temperature, radiation is the dominant mode of heat transfer. So it is imperative that a basic understanding of radiative heat transport be available. Heat transfer in boiler and fired-heater enclosures is highly dependent upon the radiative characteristics of the surface and the hot combustion gases. It is known that for a body radiating to its surroundings, the heat rate is

$$\dot{Q} = \varepsilon \sigma A\left(T^4 - T_s^4\right)$$

where $\varepsilon$ is the emissivity of the surface, $\sigma$ is the Stefan-Boltzmann constant, $\sigma = 0.1713 \times 10^{-8}$ Btu/hr ft$^2 \cdot$ R$^4$. Temperature must be in absolute units, R or K. If $\varepsilon = 1$ for a surface, it is called a "blackbody," a perfect emitter of thermal energy. Radiative properties of various surfaces are given in Appendix II. In many cases, the heat exchange between bodies when all the radiation emitted by one does not strike the other is of interest. In this case we employ a shape factor $F_{ij}$ to modify the basic transport equation. For two blackbodies we would write

$$\dot{Q}_{12} = F_{12}\sigma A\left(T_1^4 - T_2^4\right)$$

**Table I.8 Values of C and m for Hilpert's Equation**

| Range of $N_{\text{ReD}}$ | $C$ | $m$ |
|---|---|---|
| 1–4 | 0.891 | 0.330 |
| 4–40 | 0.821 | 0.385 |
| 40–4000 | 0.615 | 0.466 |
| 4000–40,000 | 0.175 | 0.618 |
| 40,000–250,000 | 0.0239 | 0.805 |

**Table I.9 Constants and Exponents for Natural Convection Correlations**

| Ra | Vertical Plate[a] | | Horizontal Cylinders[b] | |
|---|---|---|---|---|
| | $c$ | $m$ | $c$ | $m$ |
| $10^4 < \text{Ra} < 10^9$ | 0.59 | 1/4 | 0.525 | 1/4 |
| $10^9 < \text{Ra} < 10^{12}$ | 0.129 | 1/3 | 0.129 | 1/3 |

[a]Nu and Ra based on vertical height $L$.
[b]Nu and Ra based on diameter $D$.

for the heat transport from body 1 to body 2. Figures I.11 to I.14 show the shape factors for some commonly encountered cases. Note that the shape factor is a function of geometry only.

Gaseous radiation that occurs in luminous combustion zones is difficult to treat theoretically. It is too complex to be treated here and the interested reader is referred to Siegel and Howell[7] for a detailed discussion.

## I.4    FLUID MECHANICS

In industrial processes we deal with materials that can be made to flow in a conduit of some sort. The laws that govern the flow of materials form the science that is called fluid mechanics. The behavior of the flowing fluid controls pressure drop (pumping power), mixing efficiency, and in some cases the efficiency of heat transfer. So it is an integral portion of an energy conservation program.

### I.4.1 Fluid Dynamics

When a fluid is caused to flow, certain governing laws must be used. For example, mass flows in and out of control volumes must always be balanced. In other words, conservation of mass must be satisfied.

In its most basic form the continuity equation (conservation of mass) is

$$\iint_{\text{C.S.}} \rho(\bar{v}\cdot\bar{n})\, dA + \frac{\partial}{\partial_t} \iiint_{\text{C.V.}} \rho\, dV = 0$$

In words, this is simply a balance between mass entering and leaving a control volume and the rate of mass storage. The $\rho(\bar{v}\cdot\bar{n})$ terms are integrated over the control surface, whereas the $\rho\, dV$ term is dependent upon an integration over the control volume.

For a steady flow in a constant-area duct, the continuity equation simplifies to

$$\dot{m} = \rho_f A_c \bar{u} = \text{constant}$$

That is, the mass flow rate $\dot{m}$ is constant and is equal to the product of the fluid density $\rho_f$, the duct cross section $A_c$, and the average fluid velocity $\bar{u}$.

If the fluid is compressible and the flow is steady, one gets

$$\frac{\dot{m}}{\rho_f} = \text{constant} = (\bar{u}A_c)(\bar{u}A_c)_2$$

where 1 and 2 refer to different points in a variable area duct.

### I.4.2 First Law—Fluid Dynamics

The first law of thermodynamics can be directly applied to fluid dynamical systems, such as duct flows. If there is no heat transfer or chemical reaction and if the internal energy of the fluid stream remains unchanged, the first law is

$$\frac{V_i^2 - V_e^2}{2g_c} + \frac{z_i - z_e}{g_c}g + \frac{p_i - p_e}{\rho} + \left(w_p - w_f\right) = 0 \tag{I.8}$$

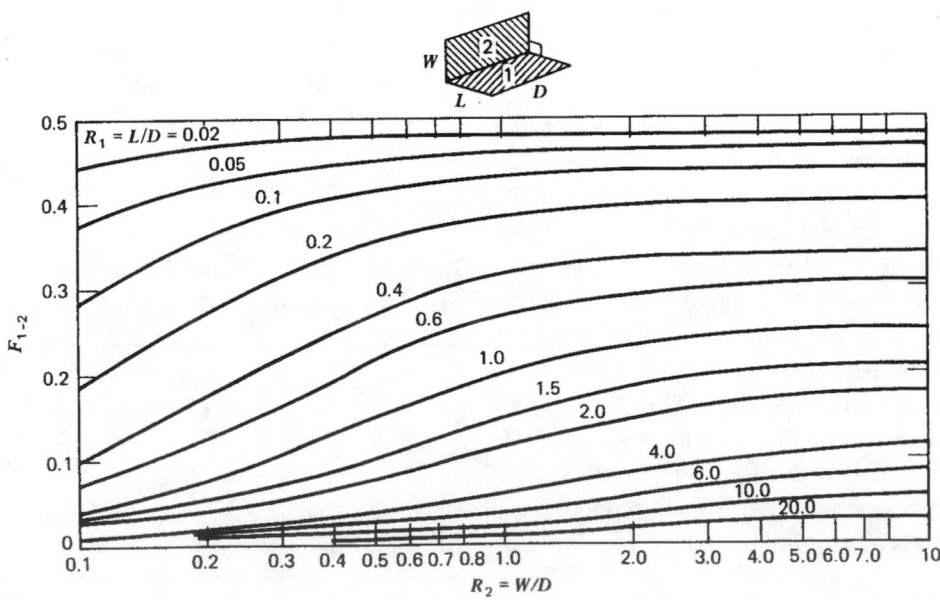

Fig. I.11 Radiation shape factor for perpendicular rectangles with a common edge.

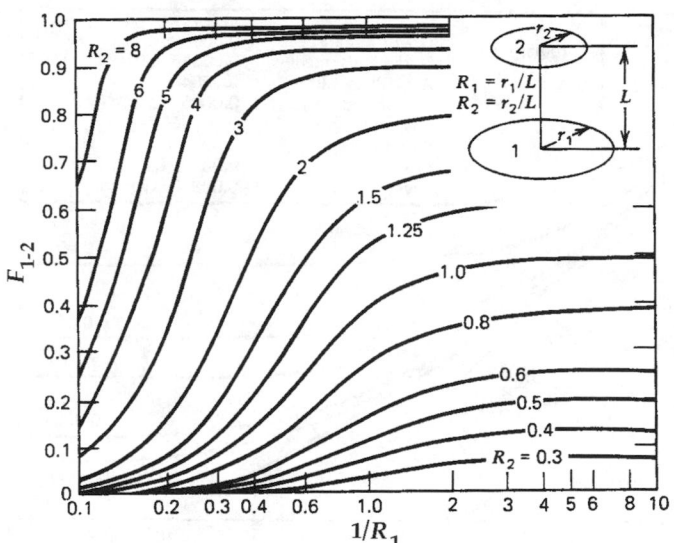

Fig. I.12 Radiation shape factor for parallel, concentric disks.

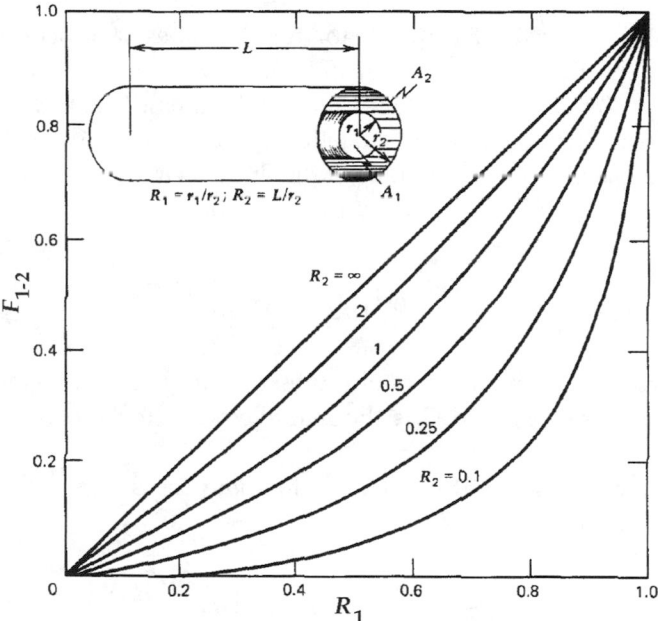

Fig. I.13 Radiation shape factor for concentric cylinders of finite length.

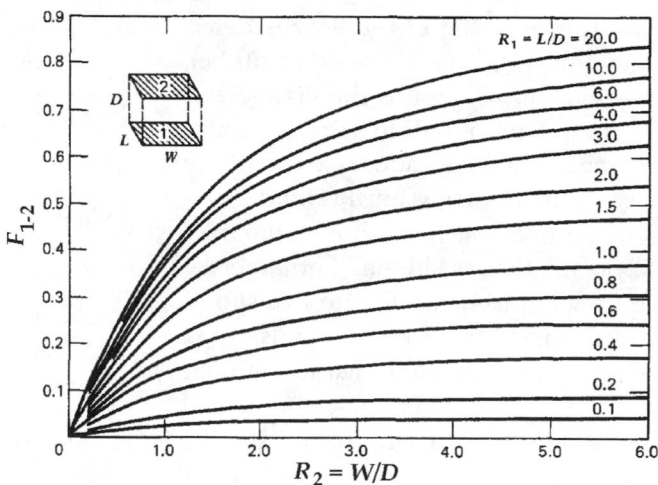

Fig. I.14 Radiation shape factor for parallel, directly opposed rectangles.

Fig. I.15 The first law applied to adiabatic flow system.

where the subscripts $i$ and $e$ refer to inlet and exit conditions and $w_p$ and $w_f$ are pump work and work required to overcome friction in the duct. Figure I.15 shows schematically a system illustrating this equation.

Any term in equation I.8 can be converted to a rate expression by simply multiplying by , the mass flow rate. Take, for example, the pump horsepower,

$$W\left(\frac{\text{energy}}{\text{time}}\right) = \dot{m}w_p\left(\frac{\text{mass}}{\text{time}}\right)\left(\frac{\text{energy}}{\text{mass}}\right)$$

In the English system, horsepower is

$$\text{hp} = \dot{m}\left(\frac{\text{lb}_m}{\text{sec}}\right)w_p = \left(\frac{\text{ft} \cdot \text{lb}_f}{\text{lb}_m}\right) \times \left(\frac{1\ \text{hp}-\text{sec}}{500\ \text{ft}-\text{lb}}\right) = \left(\frac{\dot{m}w_p}{550}\right)$$

Referring back to equation I.8, the most difficult term to determine is usually the frictional work term $w_f$. This is a term that depends upon the fluid viscosity, the flow conditions, and the duct geometry. For simplicity, $w_f$ is generally represented as

$$w_f = \frac{p_f}{\rho}$$

when $\Delta p_f$ is the frictional pressure drop in the duct. Further, we say that

$$\frac{p_f}{\rho} = \frac{2\,f\bar{u}^2 L}{g_c\,D}$$

in a duct of length $L$ and diameter $D$. The friction factor $f$ is a convenient way to represent the differing influence of laminar and turbulent flows on the friction pressure drop.

The character of the flow is determined through the Reynolds number, $Re = \rho u D/\mu$, where $\mu$ is the viscosity of the fluid. This nondimensional grouping represents the ratio of dynamic to viscous forces acting on the fluid.

Experiments have shown that if $Re \leq 2300$, the flow is laminar. For larger $Re$ the flow is turbulent. Figure I.16 shows how the friction factor depends upon the $Re$ of the flow. Note that for laminar flow the $f$ vs. $Re$ curve is single-valued and is simply equal to $16/Re$. In the turbulent regime, the wall roughness $e$ can affect the friction factor because of its effect on the velocity profile near the duct surface.

If a duct is not circular, the equivalent diameter $D_e$ can be used so that all the relationships developed for circular systems can still be used. $D_e$ is defined as

$$D_e = \frac{4A_c}{P}$$

$P$ is the "wetted" perimeter, that part of the flow cross section that touches the duct surfaces. For a circular system $D_e = 4(\pi D^2/4\pi D) = D$, as it should. For an annular duct, we get

$$D_e = \frac{\left(\pi D_o^2/4 - \pi D_i^2/4\right)4}{\pi D_o + \pi D_i} = \frac{\pi(D_o + D_i)(D_o + D_i)}{\pi D_o + \pi D_i}$$

$$= D_o + D_i$$

**Pressure Drop in Ducts.** In practical applications, the essential need is to predict pressure drops in piping and duct networks. The friction factor approach is adequate for straight runs of constant area ducts. But valves nozzles, elbows, and many other types of fittings are necessarily included in a network. This can be accounted for by defining an equivalent length $L_e$ for the fitting. Table I.10 shows $L_e/D$ values for many different fittings.

**Pressure Drop across Tube Banks.** Another commonly encountered application of fluid dynamics is the pressure drop caused by transverse flow across arrays of heat-transfer tubes. One technique to calculate this effect is to find the velocity head loss through the tube bank:

$$N_v = fNF_d$$

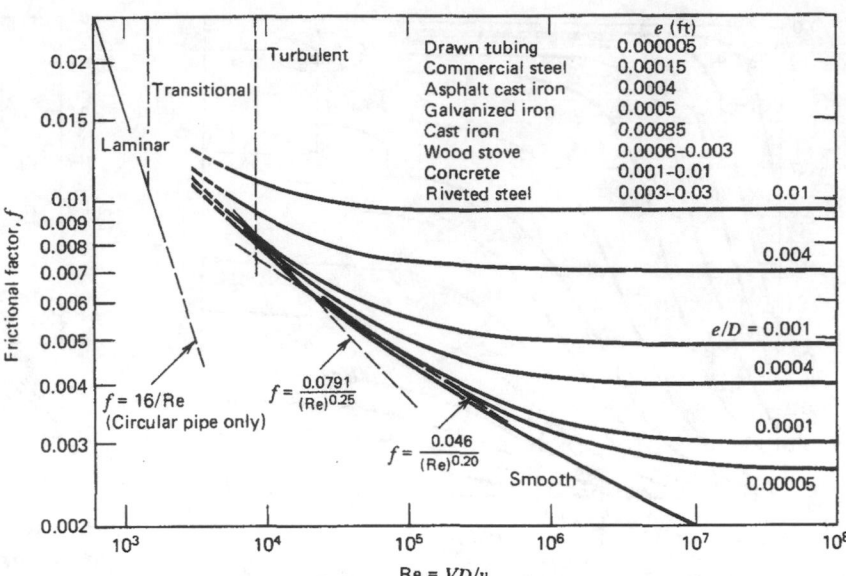

**Fig. I.16 Friction factors for straight pipes.**

where $f$ is the friction factor for the tubes (a function of the $Re$), $N$ the number of tube rows crossed by the flow, and $F_d$ is the "depth factor." Figures I.17 and I.18 show the $f$ factor and $F_d$ relationship that can be used in pressure-drop calculations. If the fluid is air, the pressure drop can be calculated by the equation

$$p = N\left(\frac{30}{B}\right)\frac{T}{1.73 \times 10^5}\left(\frac{G}{10^3}\right)^2$$

where $B$ is the atmospheric pressure (in. Hg), $T$ is temperature (°R), and $G$ is the mass velocity (lbm/ft² hr).

**Bernoulli's Equation.** There are some cases where the equation

$$\frac{p}{\rho} + \frac{u^2}{2} + gz = \text{constant}$$

which is called Bernoulli's equation, is useful. Strictly speaking, this equation applies for inviscid, incompressible, steady flow along a streamline. However, even in pipe flow where the flow is viscous, the equation can be applied because of the confined nature of the flow. That is, the flow is forced to behave in a streamlined manner. Note that the first law equation (I.8) yields Bernoulli's equation if the friction drop exactly equals the pump work.

### I.4.3 Fluid-Handling Equipment

For industrial processes, another prime application of fluid dynamics lies in fluid-handling equipment.

## Table I.10 $L_e/D$ for Screwed Fittings, Turbulent Flow Only[a]

| Fitting | $L_e/D$ |
| --- | --- |
| 45° elbow | 15 |
| 90° elbow, standard radius | 31 |
| 90° elbow, medium radius | 26 |
| 90° elbow, long sweep | 20 |
| 90° square elbow | 65 |
| 180° close return bend | 75 |
| Swing check valve, open | 77 |
| Tee (as el, entering run) | 65 |
| Tee (as el, entering branch) | 90 |
| Couplings, unions | Negligible |
| Gate valve, open | 7 |
| Gate valve, 1/4 closed | 40 |
| Gate valve, 1/2 closed | 190 |
| Gate valve, 3/4 closed | 840 |
| Globe valve, open | 340 |
| Angle valve, open | 170 |

[a]Calculated from Crane Co. Tech. Paper 409, May 1942.

Pumps, compressors, fans, and blowers are extensively used to move gases and liquids through the process network and over heat-exchanger surfaces. The general constraint in equipment selection is a matching of fluid handler capacity to pressure drop in the circuit connected to the fluid handler.

Pumps are used to transport liquids, whereas compressors, fans, and blowers apply to gases. There are features of performance common to all of them. For purposes of illustration, a centrifugal pump will be used to discuss performance characteristics.

**Centrifugal Machines.** Centrifugal machines operate on the principle of centrifugal acceleration of a fluid element in a rotating impeller/housing system to achieve a pressure gain and circulation.

The characteristics that are important are flow rate (capacity), head, efficiency, and durability. $Q_f$ (capacity), $h_p$ (head), and $\eta_p$ (efficiency) are related quantities, dependent basically on the fluid behavior in the pump and the flow circuit. Durability is related to the wear, corrosion, and other factors that bear on a pump's reliability and lifetime.

Figure I.19 shows the relation between flow rate and related characteristics for a centrifugal pump at constant speed. Graphs of this type are called performance curves; fhp and bhp are fluid and brake horsepower, respectively. The primary design constraint is a matching

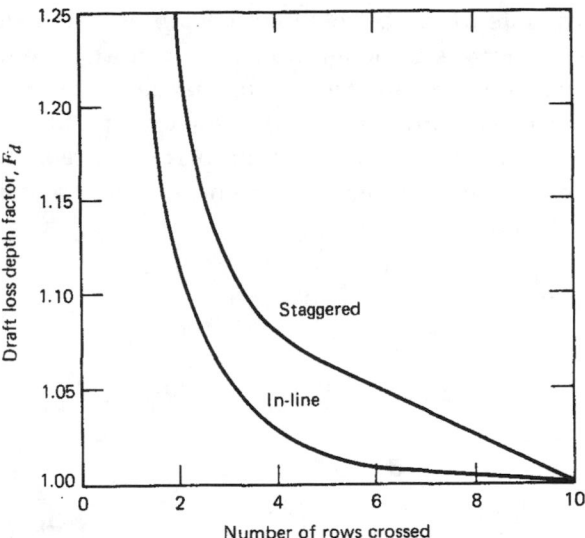

**Fig. I.17 Depth factor for number of tube rows crossed in convection banks.**

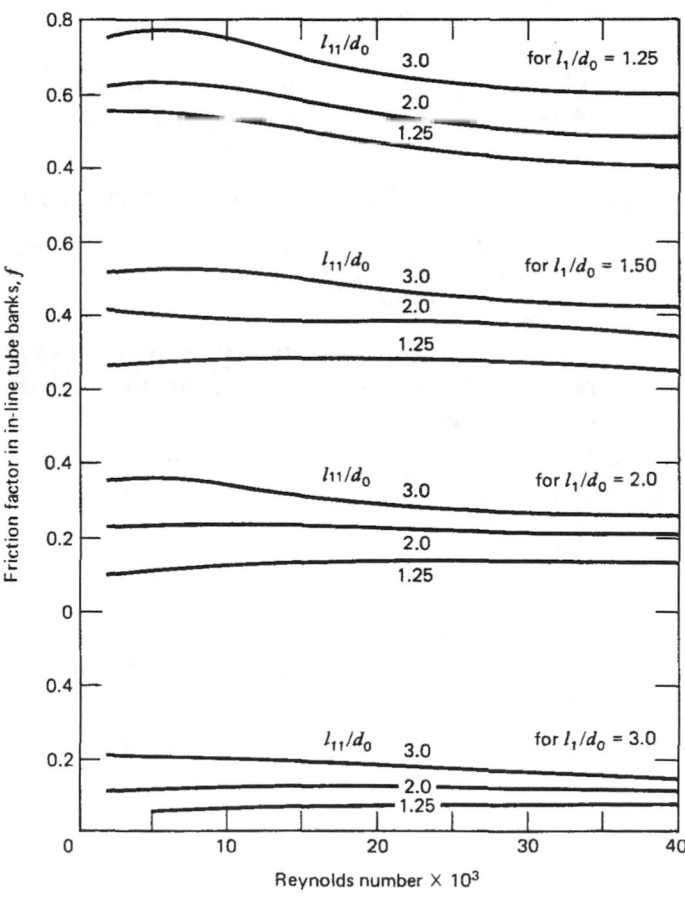

**Fig. I.18 Friction factor $f$ as affected by Reynolds number for various in-line tube patterns, crossflow gas or air, $d_o$, tube diameter; $l_\perp$, gap distance perpendicular to the flow; $l_\parallel$, gap distance parallel to the flow.**

of flow rate to head. Note that as the flow-rate requirement is increased, the allowable head must be reduced if other pump parameters are unchanged.

Analysis and experience has shown that there are scaling laws for centrifugal pump performance that give the trends for a change in certain performance parameters. Basically, they are:

Efficiency:

$$\eta_p = f_1\left(\frac{Q_f}{D^3 n}\right)$$

Dimensionless head:

$$\frac{h_p g}{D^2 n^2} = f_2\left(\frac{Q_f}{D^3 n}\right)$$

Dimensionless brake horsepower:

$$\frac{\text{bhp} \cdot \text{g}}{\gamma D^2 n^3} = f_3\left(\frac{Q_f}{D^3 n}\right)$$

where $D$ is the impeller diameter, $n$ is the rotary impeller speed, $g$ is gravity, and $\gamma$ is the specific weight of fluid.

The basic relationships yield specific proportionalities such as $Q_f \propto n$ (rpm), $h_p \propto n^2$, $fhp \propto n^3$, $Q_f \propto \dfrac{1}{D^2}$, $h_p \propto \dfrac{1}{D^4}$, and $fh_p \propto \dfrac{1}{D^4}$.

For pumps, density variations are generally negligible since liquids are incompressible. But for gas-handling equipment, density changes are very important. The scaling laws will give the following rules for changing density:

$$h_p \propto \rho$$
$$fh_p \propto \rho \qquad\qquad (Q_f, n \text{ constant})$$

$$\left.\begin{array}{c} n \\ fhp \\ Q_f \end{array}\right\} \propto \rho^{-1/2} \qquad ! $$
$$\qquad\qquad (h_p \text{ constant}$$

$$\left.\begin{array}{c} n \\ Q_f \\ h_p \end{array}\right\} \propto \frac{1}{\rho}$$
$$\qquad\qquad (\dot{m} \text{ constant})$$

$$fhp \propto \frac{1}{\rho^2}$$

For centrifugal pumps, the following equations hold:

Fig. I.19 Performance curve for a centrifugal pump.

$$fhp = \frac{Q_f \rho g h_p}{550 g_c}$$

$$\eta_p p = \frac{Q_f \rho g h_p 550 g_c}{\text{bhp}} = \frac{fhp}{\text{bhp}}$$

system efficiency $\eta_s = \eta_p \times \eta_m$     (motor efficiency)

It is important to select the motor and pump so that at nominal operating conditions, the pump and motor operate at near their maximum efficiency.

For systems where two or more pumps are present, the following rules are helpful. To analyze pumps in parallel, add capacities at the same head. For pumps in series, simply add heads at the same capacity.

There is one notable difference between blowers and pump performance. This is shown in Figure I.20. Note that the bhp continues to increase as permissible head goes to zero, in contrast to the pump curve when bhp approaches zero. This is because the kinetic energy imparted to the fluid at high flow rates is quite significant for blowers.

Manufactures of fluid-handling equipment provide excellent performance data for all types of equipment. Anyone considering replacement or a new installation should take full advantage of these data.

Fluid-handling equipment that operates on a principle other than centrifugal does not follow the centrifugal scaling laws. Evans[8] gives a thorough treatment of most types of equipment that would be encountered in industrial application.

**Fig. I.20 Variation of head and bhp with flow rate for a typical blower at constant speed.**

## References

1. G.J. Van Wylen and R.E. Sonntag, *Fundamentals of Classical Thermodynamics*, 2nd ed., Wiley, New York, 1973.
2. A.S. Chapman, *Heat Transfer*, 3rd ed., Macmillan, New York, 1974.
3. J.P. Holman, *Heat Transfer*, 4th ed., McGraw-Hill, New York, 1976.
4. M.P. Heisler, *Trans. ASME*, Vol. 69 (1947), p. 227.
5. L.S. Tong, *Boiling Heat Transfer and Two-Phase Flow*, Wiley, New York, 1965.
6. W.M. Kays and A.L. London, *Compact Heat Exchangers*, 2nd ed., McGraw-Hill, New York, 1963.
7. R. Siegel and J.R. Howell, *Thermal Radiation Heat Transfer*, McGraw-Hill, New York, 1972.
8. FRANK L. Evans, JR., *Equipment Design Handbook for Refineries and Chemical Plants*, Vols. 1 and 2, Gulf Publishing, Houston, Tex., 1974.

## SYMBOLS

### Thermodynamics

| | |
|---|---|
| AF | air/fuel ratio |
| $C_p$ | constant-pressure specific heat |
| $C_v$ | constant-volume specific heat |
| $C_{p0}$ | zero-pressure constant-pressure specific heat |
| $C_{v0}$ | zero-pressure constant-volume specufic heat |
| $e, E$ | specific energy and total energy |
| $g$ | acceleration due to gravity |
| $g, G$ | specific Gibbs function and total Gibbs function |
| $g_e$ | a constant that relates force, mass, length, and time |
| $h, H$ | specific enthalpy and total enthalpy |
| k | specific heat ratio: $C_p/C_v$ |

| | |
|---|---|
| K.E. | kinetic energy |
| $lb_f$ | pound force |
| $lb_m$ | pound mass |
| lb mol | pound mole |
| $m$ | mass |
| $\dot{m}$ | mass rate of flow |
| $M$ | molecular weight |
| $n$ | number of moles |
| $n$ | polytropic exponent |
| $P$ | pressure |
| $P_i$ | partial pressure of component i in a mixture |
| P.E. | potential energy |
| $P_r$ | relative pressure as used in gas tables |
| $q, Q$ | heat transfer per unit mass and total heat transfer |
| $\dot{Q}$ | rate of heat transfer |
| $QH, QL$ | heat transfer from high- and low-temperature bodies |
| $R$ | gas constant |
| $\bar{R}$ | universal gas constant |
| $s, S$ | specific entropy and total entropy |
| $t$ | time |
| $T$ | temperature |
| $u, U$ | specific internal energy and total internal energy |
| $v, V$ | specific volume and total volume |
| V | velocity |
| $V_r$ | relative velocity |
| $w, W$ | work per unit mass and total work |
| $W$ | rate of work, or power |
| $w_{rev}$ | reversible work between two states assuming heat transfer with surroundings |
| $x$ | mass fraction |
| $Z$ | elevation |
| $Z$ | compressibility factor |

### Greek Letters

| | |
|---|---|
| $\beta$ | coefficient of performance for a refrigerator |
| $\beta'$ | coefficient of performance for a heat pump |
| $\eta$ | efficiency |
| $\rho$ | density |
| $\phi$ | relative humidity |
| $\omega$ | humidity ratio or specific humidity |

### Subscripts

| | |
|---|---|
| $c$ | property at the critical point |
| c.v. | control volume |
| $e$ | state of a substance leaving a control volume |
| $f$ | formation |
| $f$ | property of saturated liquid |
| $fg$ | difference in property for saturated vapor |

|   | and saturated liquid |
| $g$ | property of saturated vapor |
| $r$ | reduced property |
| $s$ | isentropic process |

**Superscripts**

| - | bar over symbol denotes property on a molal basis (over $V$, $H$, $S$, $U$, $A$, $G$, the bar denotes partial molal property) |
| ° | property at standard-state condition |
| * | ideal gas |
| $L$ | liquid phase |
| $S$ | solid phase |
| $V$ | vapor phase |

**Heat Transfer—Fluid Flow**

| $A$ | surface area |
| $A_m$ | profile area for a fin |
| Bi | Biot number, $(hL/k)$ |
| $c_p$ | specific heat at constant pressure |
| $c$ | specific heat |
| $D$ | diameter |
| $D_e$ | hydraulic diameter |
| $F_{i\text{-}j}$ | shape factor of area $i$ with respect to area $j$ |
| $f$ | friction factor |
| Gr | Grashof number, $g\,\beta\Delta T L_c^3/v^2$ |
| $g$ | acceleration due to gravity |
| $g_c$ | gravitational constant |
| $h$ | convective heat-transfer coefficient |
| $k$ | thermal conductivity |
| $m$ | mass |
| $\dot{m}$ | mass rate of flow |
| $N$ | number of rows |
| Nu | Nusselt number, $hL/k$ |
| Pr | Prandtl number, $\mu C_p/k$ |
| $p$ | pressure |
| $Q$ | volumetric flow rate |
| $Q$ | rate of heat flow |
| Ra | Rayleigh number, $g\,\beta\Delta T L_c^3/v\,\alpha$ |

| Re | Reynolds number, $\rho u_{av} L_c/\mu$ |
| $r$ | radius |
| St | Stanton number, $h/Cp\,\rho u_\infty$ |
| $T$ | temperature |
| $U$ | overall heat-transfer coefficient |
| $u$ | velocity |
| $ux$ | free-stream velocity |
| $V$ | volume |
| $V$ | velocity |
| $W$ | rate of work done |

**Greek Symbols**

| $\alpha$ | thermal diffusivity |
| $\beta$ | coefficient of thermal expansion |
| $\Delta$ | difference, change |
| $\varepsilon$ | surface emissivity |
| $\eta_f$ | fin effectiveness |
| $\mu$ | viscosity |
| $v$ | kinematic viscosity |
| $\rho$ | density |
| $\sigma$ | Stefan-Boltzmann constant |
| $\tau$ | time |

**Subscripts**

| $b$ | bulk conditions |
| cr | critical condition |
| $c$ | convection |
| cond | conduction |
| conv | convection |
| $e$ | entrance, effective |
| $f$ | fin, fluid |
| $i$ | inlet conditions |
| $o$ | exterior condition |
| $0$ | centerline conditions in a tube at $r = 0$ |
| $o$ | outlet condition |
| $p$ | pipe, pump |
| $s$ | surface condition |
| $\infty$ | free-stream condition |

# APPENDIX II
# CONVERSION FACTORS AND PROPERTY TABLES

Compiled by
**L.C. WITTE**

Professor of Mechanical Engineering
University of Houston
Houston, Texas

## Table II.1 Conversion Factors

| To Obtain: | Multiply: | By: |
|---|---|---|
| Acres | Sq miles | 640.0 |
| Atmospheres | Cm of Hg @ 0 deg C | 0.013158 |
| Atmospheres | Ft of $H_2O$ @ 39.2 F. | 0.029499 |
| Atmospheres | Grams/sq cm | 0.00096784 |
| Atmospheres | In. Hg @ 32 F | 0.033421 |
| Atmospheres | In. $H_2O$ @ 39.2 F | 0.0024583 |
| Atmospheres | Pounds/sq ft | 0.00047254 |
| Atmospheres | Pounds/sq in. | 0.068046 |
| Btu | Ft-lb | 0.0012854 |
| Btu | Hp-hr | 2545.1 |
| Btu | Kg-cal. | 3.9685 |
| Btu | kW-hr | 3413 |
| Btu | Watt-hr | 3.4130 |
| Btu/(cu ft) (hr) | kW/liter | 96,650.6 |
| Btu/hr | Mech. hp | 2545.1 |
| Btu/hr | kW | 3413 |
| Btu/hr | Tons of refrigeration | 12,000 |
| Btu/hr | Watts | 3.4127 |
| Btu/kW hr | Kg cal/kW hr | 3.9685 |
| Btu/(hr) (ft) (deg F) | Cal/(sec) (cm) (deg C) | 241.90 |
| Btu/(hr) (ft) (deg F) | Joules/(sec) (cm) (deg C) | 57.803 |
| Btu/(hr) (ft) (deg F) | Watts/(cm) (deg C) | 57.803 |
| Btu/(hr) (sq ft) | Cal/(sec) (sq cm) | 13,273.0 |
| Btu/min | Ft-lb/min | 0.0012854 |
| Btu/min | Mech. hp | 42.418 |
| Btu/min | kW | 56.896 |
| Btu/lb | Cal/gram | 1.8 |
| Btu/lb | Kg cal/kg | 1.8 |
| Btu/(lb) (deg F) | Cal/(gram) (deg C) | 1.0 |
| Btu/(lb) (deg F) | Joules/(gram) (deg C) | 0.23889 |
| Btu/sec | Mech. hp | 0.70696 |
| Btu/sec | Mech. hp (metric) | 0.6971 |
| Btu/sec | Kg-cal/hr | 0.0011024 |
| Btu/sec | kW | 0.94827 |
| Btu/sq ft | Kg-cal/sq meter | 0.36867 |

**Table II.1 Continued**

| To Obtain: | Multiply: | By: |
|---|---|---|
| Calories | Ft-lb | 0.32389 |
| Calories | Joules | 0.23889 |
| Calories | Watt-hr | 860.01 |
| Cal/(cu cm) (sec) | kW/liter | 0.23888 |
| Cal/gram | Btu/lb | 0.55556 |
| Cal/(gram) (deg C) | Btu/(lb) (deg F) | 1.0 |
| Cal/(sec) (cm) (deg C) | Btu/(hr) (ft) (deg F) | 0.0041336 |
| Cal/(sec) (sq cm) | Btu/(hr) (sq ft) | 0.000075341 |
| Cal/(sec) (sq cm) (deg C) | Btu/(hr) (sq ft) (deg F) | 0.0001355 |
| Centimeters | Inches | 2.540 |
| Centimeters | Microns | 0.0001 |
| Centimeters | Mils | 0.002540 |
| Cm of Hg @ 0 deg C | Atmospheres | 76.0 |
| Cm of Hg @ 0 deg C | Ft of $H_2O$ @ 39.2 F | 2.242 |
| Cm of Hg @ 0 deg C | Grams/sq cm | 0.07356 |
| Cm of Hg @ 0 deg C | In. of $H_2O$ @ 4 C | 0.1868 |
| Cm of Hg @ 0 deg C | Lb/sq in. | 5.1715 |
| Cm of Hg @ 0 deg C | Lb/sq ft | 0.035913 |
| Cm/deg C | In./deg F | 4.5720 |
| Cm/sec | Ft/min | 0.508 |
| Cm/sec | Ft/sec | 30.48 |
| Cm/(sec) (sec) | Gravity | 980.665 |
| Cm of $H_2O$ @39.2 F | Atmospheres | 1033.24 |
| Cm of $H_2O$ @39.2 F | Lb/sq in. | 70.31 |
| Centipoises | Centistokes | Density |
| Centistokes | Centipoises | l/density |
| Cu cm | Cu ft | 28,317 |
| Cu cm | Cu in. | 16.387 |
| Cu cm | Gal. (USA, liq.) | 3785.43 |
| Cu cm | Liters | 1000 03 |
| Cu cm | Ounces (USA, liq.) | 29.573730 |
| Cu cm | Quarts (USA, liq.) | 946.358 |
| Cu cm/sec | Cu ft/min | 472.0 |
| Cu ft | Cords (wood) | 128.0 |
| Cu ft | Cu meters | 35.314 |
| Cu ft | Cu yards | 27.0 |
| Cu ft | Gal. (USA, liq.) | 0.13368 |
| Cu ft | Liters | 0.03532 |
| Cu ft/min | Cu meters/sec | 2118.9 |
| Cu ft/min | Gal. (USA, liq./sec) | 8.0192 |
| Cu ft/lb | Cu meters/kg | 16.02 |
| Cu ft/lb | Liters/kg | 0.01602 |
| Cu ft/sec | Cu meters/min | 0.5886 |
| Cu ft/sec | Gal. (USA, liq.)/min | 0.0022280 |
| Cu ft/sec | Liters/min | 0.0005886 |
| Cu in. | Cu centimeters | 0.061023 |
| Cu in. | Gal. (USA, liq.) | 231.0 |
| Cu in. | Liters | 61.03 |
| Cu in. | Ounces (USA. liq.) | 1.805 |

**Table II.1 Continued**

| To Obtain: | Multiply: | By: |
|---|---|---|
| Cu meters | Cu ft | 0.028317 |
| Cu meters | Cu yards | 0.7646 |
| Cu meters | Gal. (USA. liq.) | 0.0037854 |
| Cu meters | Liters | 0.001000028 |
| Cu meters/hr | Gal./min | 0.22712 |
| Cu meters/kg | Cu ft/lb | 0.062428 |
| Cu meters/min | Cu ft/min | 0.02832 |
| Cu meters/min | Gal./sec | 0.22712 |
| Cu meters/sec | Gal./min | 0.000063088 |
| Cu yards | Cu meters | 1.3079 |
| Dynes | Grams | 980.66 |
| Dynes | Pounds (avoir.) | 444820.0 |
| Dyne-centimeters | Ft-lb | 13,558,000 |
| Dynes/sq cm | Lb/sq in. | 68947 |
| Ergs | Joules | 10,000,000 |
| Feet | Meters | 3.281 |
| Ft of $H_2O$ @ 39.2 F | Atmospheres | 33.899 |
| Ft of $H_2O$ @ 39.2 F | Cm of Hg @ 0 deg C | 0.44604 |
| Ft of $H_2O$ @ 39.2 F | In. of Hg @ 32 deg F | 1.1330 |
| Ft of $H_2O$ @ 39.2 F | Lb/sq ft | 0.016018 |
| Ft of $H_2O$ @ 39.2 F | Lb/sq in. | 2.3066 |
| Ft/min | Cm/sec | 1.9685 |
| Ft/min | Miles (USA. statute)/hr | 88.0 |
| Ft/sec | Knots | 1.6889 |
| Ft/sec | Meters/sec | 3.2808 |
| Ft/sec | Miles (USA, statute)/hr | 1.4667 |
| Ft/(sec) (sec) | Gravity (sea level) | 32.174 |
| Ft/(sec) (sec) | Meters/(sec) (sec) | 3.2808 |
| Ft-lb | Btu | 778.0 |
| Ft-lb | Joules | 0.73756 |
| Ft-lb | Kg-calories | 3087.4 |
| Ft-lb | kW-hr | 2,655,200 |
| Ft-lb | Mech. hp-hr | 1,980,000 |
| Ft-lb/min | Btu/min | 778.0 |
| Ft-lb/min | Kg cal/min | 3087.4 |
| Ft-lb/min | kW | 44,254.0 |
| Ft-lb/min | Mech. hp | 33,000 |
| Ft-lb/sec | Btu/min | 12.96 |
| Ft-lb/sec | kW | 737.56 |
| Ft-lb/sec | Mech. hp | 550.0 |
| Gal. (Imperial, liq.) | Gal. (USA. Liq.) | 0.83268 |
| Gal. (USA, liq.) | Barrels (petroleum, USA) | 42 |
| Gal. (USA. liq.) | Cu ft | 7.4805 |
| Gal. (USA. liq.) | Cu meters | 264.173 |
| Gal. (USA. liq.) | Cu yards | 202.2 |
| Gal. (USA. liq.) | Gal. (Imperial, liq.) | 1.2010 |
| Gal. (USA. liq.) | Liters | 0.2642 |
| Gal. (USA, liq.)/min | Cu ft/sec | 448.83 |
| Gal. (USA, liq.)/min | Cu meters/hr | 4.4029 |

**Table II.1 Continued**

| To Obtain: | Multiply: | By: |
|---|---|---|
| Gal. (USA. liq.)/sec | Cu ft/min | 0.12468 |
| Gal. (USA. liq.)/sec | Liters/min | 0.0044028 |
| Grains | Grams | 15.432 |
| Grains | Ounces (avoir.) | 437.5 |
| Grains | Pounds (avoir.) | 7000 |
| Grains/gal. (USA. liq.) | Parts/million | 0.0584 |
| Grams | Grains | 0.0648 |
| Grams | Ounces (avoir.) | 28.350 |
| Grams | Pounds (avoir.) | 453.5924 |
| Grams/cm | Pounds/in. | 178.579 |
| Grams/(cm) (sec) | Centipoises | 0.01 |
| Grams/cu cm | Lb/cu ft | 0 .016018 |
| Grams/cu cm | Lb/cu in. | 27.680 |
| Grams/cu cm | Lb/gal. | 0.119826 |
| Gravity (at sea level) | Ft/(sec) (sec) | 0.03108 |
| Inches | Centimeters | 0.3937 |
| Inches | Microns | 0.00003937 |
| Inches of Hg @ 32 F | Atmospheres | 29.921 |
| Inches of Hg @ 32 F | Ft of $H_2O$ @ 39.2 F | 0.88265 |
| Inches of Hg @ 32 F | Lb/sq in. | 2.0360 |
| Inches of Hg @ 32 F | In. of $H_2O$ @ 4 C | 0.07355 |
| Inches of $H_2O$@ 4 C | In. of Hg @ 32 F | 13.60 |
| Inches of H2O @ 39.2 F | Lb/sq in. | 27.673 |
| Inches/deg F | Cm/deg C | 0.21872 |
| Joules | Btu | 1054.8 |
| Joules | Calories | 4.186 |
| Joules | Ft-lb | 1.35582 |
| Joules | Kg-meters | 9.807 |
| Joules | kW-hr | 3,600,000 |
| Joules | Mech. hp-hr | 2,684,500 |
| Kg | Pounds (avoir.) | 0.45359 |
| Kg-cal | Btu | 0.2520 |
| Kg-cal | Ft-lb | 0.00032389 |
| Kg-cal | Joules | 0.0002389 |
| Kg-cal | kW-hr | 860.01 |
| Kg-cal | Mech. hp-hr | 641.3 |
| Kg-cal/kg | Btu/lb | 0.5556 |
| Kg-cal/kW hr | Btu/kW hr | 0.2520 |
| Kg-cal/min | Ft-lb/min | 0.0003239 |
| Kg-cal/min | kW | 14,33 |
| Kg-cal/min | Mech. hp | 10.70 |
| Kg-cal/sq meter | Btu/sq ft | 2.712 |
| Kg/cu meter | Lb/cu ft | 16.018 |
| Kg/(hr) (meter) | Centipoises | 3.60 |
| Kg/liter | Lb/gal. (USA, liq.) | 0.11983 |
| Kg/meter | Lbm | 1.488 |
| Kg/sq cm | Atmospheres | 1.0332 |
| Kg sq cm | Lb/sq in . | 0.0703 |
| Kg/sq meter | Lb/sq ft | 4.8824 |

**Table II.1 Continued**

| To Obtain: | Multiply: | By: |
| --- | --- | --- |
| Kg/sq meter | Lb/sq in. | 703.07 |
| Km | Miles (USA, statute) | 1.6093 |
| kW | Btu/min | 0.01758 |
| kW | Ft-lb/min | 0.00002259 |
| kW | Ft-lb/sec | 0.00135582 |
| kW | Kg-cal/hr | 0.0011628 |
| kW | Kg-cal/min | 0.069767 |
| kW | Mech. hp | 0.7457 |
| kW-hr | Btu | 0.000293 |
| kW-hr | Ft-lb | 0.0000003766 |
| kW-hr | Kg-cal | 0.0011628 |
| kW-hr | Mech. hp-hr | 0.7457 |
| Knots | Ft/sec | 0.5921 |
| Knots | Miles/hr | 0.8684 |
| Liters | Cu ft | 28 . 316 |
| Liters | Cu in. | 0.01639 |
| Liters | Cu centimeters | 999.973 |
| Liters | Gal. (Imperial. liq.) | 4.546 |
| Liters | Gal. (USA, liq.) | 3.78533 |
| Liters/kg | Cu ft/lb | 62.42621 |
| Liters/min | Cu ft/sec | 1699.3 |
| Liters/min | Gal. (USA. liq.)/min | 3.785 |
| Liters/sec | Cu ft/min | 0.47193 |
| Liters/sec | Gal./min | 0.063088 |
| Mech. hp | Btu/hr | 0.0003929 |
| Mech. hp | Btu/min | 0.023575 |
| Mech. hp | Ft-lb/sec | 0.0018182 |
| Mech. hp | Kg-cal/min | 0.093557 |
| Mech. hp | kW | 1.3410 |
| Mech. hp-hr | Btu | 0.00039292 |
| Mech. hp-hr | Ft-lb | 0.00000050505 |
| Mech. hp-hr | Kg-calories | 0.0015593 |
| Mech. hp-hr | kW-hr | 1.3410 |
| Meters | Feet | 0.3048 |
| Meters | Inches | 0.0254 |
| Meters | Miles (Int., nautical) | 1852.0 |
| Meters | Miles (USA, statute) | 1609.344 |
| Meters/min | Ft/min | 0.3048 |
| Meters/min | Miles (USA. statute)/hr | 26.82 |
| Meters/sec | Ft/sec | 0.3048 |
| Meters/sec | Km/hr | 0.2778 |
| Meters/sec | Knots | 0.5148 |
| Meters/sec | Miles (USA, statute)/hr | 0.44704 |
| Meters/(sec) (sec) | Ft/(sec) (sec) | 0.3048 |
| Microns | Inches | 25,400 |
| Microns | Mils | 25.4 |
| Miles (Int., nautical) | Km | 0.54 |
| Miles (Int., nautical) | Miles (USA, statute) | 0.8690 |
| Miles (Int., nautical)/hr | Knots | 1.0 |

**Table II.1 Continued**

| To Obtain: | Multiply: | By: |
|---|---|---|
| Miles (USA, statute) | Km | 0.6214 |
| Miles (USA, statute) | Meters | 0.0006214 |
| Miles (USA, statute) | Miles (Int., nautical) | 1.151 |
| Miles (USA, statute)/hr | Knots | 1.151 |
| Miles (USA, statute)/hr | Ft/min | 0.011364 |
| Miles (USA, statute)/hr | Ft/sec | 0.68182 |
| Miles (USA, statute)/hr | Meters/min | 0.03728 |
| Miles (USA, statute)/hr | Meters/sec | 2.2369 |
| Milliliters/gram | Cu ft/lb | 62.42621 |
| Millimeters | Microns | 0.001 |
| Mils | Centimeters | 393.7 |
| Mils | Inches | 1000 |
| Mils | Microns | 0.03937 |
| Minutes | Radians | 3437.75 |
| Ounces (avoir. ) | Grains (avoir. ) | 0.0022857 |
| Ounces (avoir.) | Grams | 0.035274 |
| Ounces (USA, liq.) | Gal. (USA, liq.) | 128.0 |
| Parts/million | Gr/gal. (USA, liq.) | 17.118 |
| Percent grade | Ft/100 ft | 1.0 |
| Pounds (avoir.) | Grains | 0.0001429 |
| Pounds (avoir.) | Grams | 0.0022046 |
| Pounds (avoir.) | Kg | 2.2046 |
| Pounds (avoir.) | Tons, long | 2240 |
| Pounds (avoir.) | Tons, metric | 2204.6 |
| Pounds (avoir.) | Tons, short | 2000 |
| Pounds/cu ft | Grams/cu cm | 62.428 |
| Pounds/cu ft | Kg/cu meter | 0.062428 |
| Pounds/cu ft | Pounds/gal. | 7.48 |
| Pounds/cu in . | Grams/cu cm | 0.036127 |
| Pounds/ft | Kg/meter | 0.67197 |
| Pounds/hr | Kg/min | 132.28 |
| Pounds/(hr) (ft) | Centipoises | 2.42 |
| Pounds/inch | Grams/cm | 0.0056 |
| Pounds/(sec) (ft) | Centipoises | 0.000672 |
| Pounds/sq inch | Atmospheres | 14.696 |
| Pounds/sq inch | Cm of Hg @ 0 deg C | 0.19337 |
| Pounds/sq inch | Ft of $H_2O$ @ 39.2 F | 0.43352 |
| Pounds/sq inch | In. Hg @ l 32 F | 0.491 |
| Pounds/sq inch | In. $H_2O$ @ 39.2 F | 0.0361 |
| Pounds/sq inch | Kg/sq cm | 14 . 223 |
| Pounds/sq inch | Kg/sq meter | 0.0014223 |
| Pounds/gal. (USA, liq.) | Kg/liter | 8.3452 |
| Pounds/gal. (USA, liq.) | Pounds/cu ft | 0.1337 |
| Pounds/gal. (USA, liq.) | Pounds/cu inch | 231 |
| Quarts (USA, liq.) | Cu cm | 0.0010567 |
| Quarts (USA, liq.) | Cu in. | 0.01732 |
| Quarts (USA, liq.) | Liters | 1.057 |
| Sq centimeters | Sq ft | 929.0 |
| Sq centimeters | Sq inches | 6.4516 |

## Table II.1 Continued

| To Obtain: | Multiply: | By: |
|---|---|---|
| Sq ft | Acres | 43,560 |
| Sq ft | Sq meters | 10.764 |
| Sq inches | Sq centimeters | 0.155 |
| Sq meters | Acres | 4046.9 |
| Sq meters | Sq ft | 0.0929 |
| Sq mlles (USA. statute) | Acres | 0.001562 |
| Sq mils | Sq cm | 155.000 |
| Sq mils | Sq inches | 1,000.000 |
| Tons (metric ) | Tons (short) | 0.9072 |
| Tons (short) | Tons (metric) | 1.1023 |
| Watts | Btu/sec | 1054.8 |
| Yards | Meters | 1.0936 |

### Table II.2-1  Saturated Steam

| Temp. $t$ (°F) | Abs. Press. $p$ (psi) | Specific Volume Sat. Liquid $v_f$ | Specific Volume Evap $v_{fg}$ | Specific Volume Sat. Vapor $v_g$ | Enthalpy Sat. Liquid $h_f$ | Enthalpy Evap $h_{fg}$ | Enthalpy Sat. Vapor $h_g$ | Entropy Sat. Liquid $s_f$ | Entropy Evap $s_{fg}$ | Entropy Sat. Vapor $s_g$ | Temp. $t$ (°F) |
|---|---|---|---|---|---|---|---|---|---|---|---|
| | | | | | *Temperature Table* | | | | | | |
| 32.0" | 0.08859 | 0.016022 | 3304.7 | 3304.7 | − 0.0179 | 1075.5 | 1075.5 | 0.0000 | 2.1873 | 2.1873 | 32.0" |
| 34.0 | 0.09600 | 0.016021 | 3061.9 | 3061.9 | 1.996 | 1074.4 | 1076.4 | 0.0041 | 2.1762 | 2.1802 | 34.0 |
| 36.0 | 0.10395 | 0.016020 | 2839.0 | 2839.0 | 4.008 | 1073.2 | 1077.2 | 0.0081 | 2.1651 | 2.1732 | 36.0 |
| 38.0 | 0.11249 | 0.016019 | 2634.1 | 2634.2 | 6.018 | 1072.1 | 1078.1 | 0.0122 | 2.1541 | 2.1663 | 38.0 |
| 40.0 | 0.12163 | 0.016019 | 2445.8 | 2445.8 | 8.027 | 1071.0 | 1079.0 | 0.0162 | 2.1432 | 2.1594 | 40.0 |
| 42.0 | 0.13143 | 0.016019 | 2272.4 | 2272.4 | 10.035 | 1069.8 | 1079.9 | 0.0202 | 2.1325 | 2.1527 | 42.0 |
| 44.0 | 0.14192 | 0.016019 | 2112.8 | 2112.8 | 12.041 | 1068.7 | 1080.7 | 0.0242 | 2.1217 | 2.1459 | 44.0 |
| 46.0 | 0.15314 | 0.016020 | 1965.7 | 1965.7 | 14.047 | 1067.6 | 1081.6 | 0.0282 | 2.1111 | 2.1393 | 46.0 |
| 48.0 | 0.16514 | 0.016021 | 1830.0 | 1830.0 | 16.051 | 1066.4 | 1082.5 | 0.0321 | 2.1006 | 2.1327 | 48.0 |
| 50.0 | 0.17796 | 0.016023 | 1704.8 | 1704.8 | 18.054 | 1065.3 | 1083.4 | 0.0361 | 2.0901 | 2.1262 | 50.0 |
| 52.0 | 0.19165 | 0.016024 | 1589.2 | 1589.2 | 20.057 | 1064.2 | 1084.2 | 0.0400 | 2.0798 | 2.1197 | 52.0 |
| 54.0 | 0.20625 | 0.016026 | 1482.4 | 1482.4 | 22.058 | 1063.1 | 1085.1 | 0.0439 | 2.0695 | 2.1134 | 54.0 |
| 56.0 | 0.22183 | 0.016028 | 1383.6 | 1383.6 | 24.059 | 1061.9 | 1086.0 | 0.0478 | 2.0593 | 2.1070 | 56.0 |
| 58.0 | 0.23843 | 0.016031 | 1292.2 | 1292.2 | 26.060 | 1060.8 | 1086.9 | 0.0516 | 2.0491 | 2.1008 | 58.0 |
| 60.0 | 0.25611 | 0.016033 | 1207.6 | 1207.6 | 28.060 | 1059.7 | 1087.7 | 0.0555 | 2.0391 | 2.0946 | 60.0 |
| 62.0 | 0.27494 | 0.016036 | 1129.2 | 1129.2 | 30.059 | 1058.5 | 1088.6 | 0.0593 | 2.0291 | 2.0885 | 62.0 |
| 64.0 | 0.29497 | 0.016039 | 1056.5 | 1056.5 | 32.058 | 1057.4 | 1089.5 | 0.0632 | 2.0192 | 2.0824 | 64.0 |
| 66.0 | 0.31626 | 0.016043 | 989.0 | 989.1 | 34.056 | 1056.3 | 1090.4 | 0.0670 | 2.0094 | 2.0764 | 66.0 |
| 68.0 | 0.33889 | 0.016046 | 926.5 | 926.5 | 36.054 | 1055.2 | 1091.2 | 0.0708 | 1.9996 | 2.0704 | 68.0 |
| 70.0 | 0.36292 | 0.016050 | 868.3 | 868.4 | 38.052 | 1054.0 | 1092.1 | 0.0745 | 1.9900 | 2.0645 | 70.0 |
| 72.0 | 0.38844 | 0.016054 | 814.3 | 814.3 | 40.049 | 1052.9 | 1093.0 | 0.0783 | 1.9804 | 2.0587 | 72.0 |
| 74.0 | 0.41550 | 0.016058 | 764.1 | 764.1 | 42.046 | 1051.8 | 1093.8 | 0.0821 | 1.9708 | 2.0529 | 74.0 |
| 76.0 | 0.44420 | 0.016063 | 717.4 | 717.4 | 44.043 | 1050.7 | 1094.7 | 0.0858 | 1.9614 | 2.0472 | 76.0 |
| 78.0 | 0.47461 | 0.016067 | 673.8 | 673.9 | 46.040 | 1049.5 | 1095.6 | 0.0895 | 1.9520 | 2.0415 | 78.0 |
| 80.0 | 0.50683 | 0.016072 | 633.3 | 633.3 | 48.037 | 1048.4 | 1096.4 | 0.0932 | 1.9426 | 2.0359 | 80.0 |
| 82.0 | 0.54093 | 0.016077 | 595.5 | 595.5 | 50.033 | 1047.3 | 1097.3 | 0.0969 | 1.9334 | 2.0303 | 82.0 |
| 84.0 | 0.57702 | 0.016082 | 560.3 | 560.3 | 52.029 | 1046.1 | 1098.2 | 0.1006 | 1.9242 | 2.0248 | 84.0 |
| 86.0 | 0.61518 | 0.016087 | 227.5 | 527.5 | 54.026 | 1045.0 | 1099.0 | 0.1043 | 1.9151 | 2.0193 | 86.0 |
| 88.0 | 0.65551 | 0.016093 | 496.8 | 496.8 | 56.022 | 1043.9 | 1099.9 | 0.1079 | 1.9060 | 2.0139 | 88.0 |

**Table II.2-1    Continued**

| Temp. t (°F) | Abs. Press. p (psi) | Specific Volume | | | Enthalpy | | | Entropy | | | Temp. t (°F) |
|---|---|---|---|---|---|---|---|---|---|---|---|
| | | Sat. Liquid $v_f$ | Evap $v_{fg}$ | Sat. Vapor $v_g$ | Sat. Liquid $h_f$ | Evap $h_{fg}$ | Sat. Vapor $h_g$ | Sat. Liquid $s_f$ | Evap $s_{fg}$ | Sat. Vapor $s_g$ | |
| 90.0 | 0.69813 | 0.016099 | 468.1 | 468.1 | 58.018 | 1042.7 | 1100.8 | 0.1115 | 1.8970 | 2.0086 | 90.0 |
| 92.0 | 0.74313 | 0.016105 | 441.3 | 441.3 | 60.014 | 1041.6 | 1101.6 | 0.1152 | 1.8881 | 2.0033 | 92.0 |
| 94.0 | 0.79062 | 0.016111 | 416.3 | 416.3 | 62.010 | 1040.5 | 1102.5 | 0.1188 | 1.8792 | 1.9980 | 94.0 |
| 96.0 | 0.84072 | 0.016117 | 392.8 | 392.9 | 64.006 | 1039.3 | 1103.3 | 0.1224 | 1.8704 | 1.9928 | 96.0 |
| 98.0 | 0.89356 | 0.016123 | 370.9 | 370.9 | 66.003 | 1038.2 | 1104.2 | 0.1260 | 1.8617 | 1.9876 | 98.0 |
| 100.0 | 0.94924 | 0.016130 | 350.4 | 350.4 | 67.999 | 1037.1 | 1105.1 | 0.1295 | 1.8530 | 1.9825 | 100.0 |
| 102.0 | 1.00789 | 0.016137 | 331.1 | 331.1 | 69.995 | 1035.9 | 1105.9 | 0.1331 | 1.8444 | 1.9775 | 102.0 |
| 104.0 | 1.06965 | 0.016144 | 313.1 | 313.1 | 71.992 | 1034.8 | 1106.8 | 0.1366 | 1.8358 | 1.9725 | 104.0 |
| 106.0 | 1.1347 | 0.016151 | 296.16 | 296.18 | 73.99 | 1033.6 | 1107.6 | 0.1402 | 1.8273 | 1.9675 | 106.0 |
| 108.0 | 1.2030 | 0.016158 | 280.28 | 280.30 | 75.98 | 1032.5 | 1108.5 | 0.1437 | 1.8188 | 1.9626 | 108.0 |
| 110.0 | 1.2750 | 0.016165 | 265.37 | 265.39 | 77.98 | 1031.4 | 1109.3 | 0.1472 | 1.8105 | 1.9577 | 110.0 |
| 112.0 | 1.3505 | 0.016173 | 251.37 | 251.38 | 79.98 | 1030.2 | 1110.2 | 0.1507 | 1.8021 | 1.9528 | 112.0 |
| 114.0 | 1.4299 | 0.016180 | 238.21 | 238.22 | 81.97 | 1029.1 | 1111.0 | 0.1542 | 1.7938 | 1.9480 | 114.0 |
| 116.0 | 1.5133 | 0.016188 | 225.84 | 225.85 | 83.97 | 1027.9 | 1111.9 | 0.1577 | 1.7856 | 1.9433 | 116.0 |
| 118.0 | 1.6009 | 0.016196 | 214.20 | 214.21 | 85.97 | 1026.8 | 1112.7 | 0.1611 | 1.7774 | 1.9386 | 118.0 |
| 120.0 | 1.6927 | 0.016204 | 203.25 | 203.26 | 87.97 | 1025.6 | 1113.6 | 0.1646 | 1.7693 | 1.9339 | 120.0 |
| 122.0 | 1.7891 | 0.016213 | 192.94 | 192.95 | 89.96 | 1024.5 | 1114.4 | 0.1680 | 1.7613 | 1.9293 | 122.0 |
| 124.0 | 1.8901 | 0.016221 | 183.23 | 183.24 | 91.96 | 1023.3 | 1115.3 | 0.1715 | 1.7533 | 1.9247 | 124.0 |
| 126.0 | 1.9959 | 0.016229 | 174.08 | 174.09 | 93.96 | 1022.2 | 1116.1 | 0.1749 | 1.7453 | 1.9202 | 126.0 |
| 128.0 | 2.1068 | 0.016238 | 165.45 | 165.47 | 95.96 | 1021.0 | 1117.0 | 0.1783 | 1.7374 | 1.9157 | 128.0 |
| 130.0 | 2.2230 | 0.016247 | 157.32 | 157.33 | 97.96 | 1019.8 | 1117.8 | 0.1817 | 1.7295 | 1.9112 | 130.0 |
| 132.0 | 2.3445 | 0.016265 | 149.64 | 149.66 | 99.95 | 1018.7 | 1118.6 | 0.1851 | 1.7217 | 1.9068 | 132.0 |
| 134.0 | 2.4717 | 0.016265 | 142.40 | 142.41 | 101.95 | 1017.5 | 1119.5 | 0.1884 | 1.7140 | 1.9024 | 134.0 |
| 136.0 | 2.6047 | 0.016274 | 135.55 | 135.57 | 103.95 | 1016.4 | 1120.3 | 0.1918 | 1.7063 | 1.8980 | 136.0 |
| 138.0 | 2.7438 | 0.016284 | 129.09 | 129.11 | 105.95 | 1015.2 | 1121.1 | 0.1951 | 1.6986 | 1.8937 | 138.0 |
| 140.0 | 2.8892 | 0.016293 | 122.98 | 123.00 | 107.95 | 1014.0 | 1122.0 | 0.1985 | 1.6910 | 1.8895 | 140.0 |
| 142.0 | 3.0411 | 0.016303 | 117.21 | 117.22 | 109.95 | 1012.9 | 1122.8 | 0.2018 | 1.6534 | 1.8852 | 142.0 |
| 144.0 | 3.1997 | 0.016312 | 111.74 | 111.76 | 111.95 | 1011.7 | 1123.6 | 0.2051 | 1.6759 | 1.8810 | 144.0 |
| 146.0 | 3.3653 | 0.016322 | 106.58 | 106.59 | 113.95 | 1010.5 | 1124.5 | 0.2084 | 1.6684 | 1.8769 | 146.0 |
| 148.0 | 3.5381 | 0.016332 | 101.68 | 101.70 | 115.95 | 1009.3 | 1125.3 | 0.2117 | 1.6610 | 1.8727 | 148.0 |
| 150.0 | 3.7184 | 0.016343 | 97.05 | 97.07 | 117.95 | 1008.2 | 1126.1 | 0.2150 | 1.6536 | 1.8686 | 150.0 |
| 152.0 | 3.9065 | 0.016353 | 92.66 | 92.68 | 119.95 | 1007.0 | 1126.9 | 0.2183 | 1.6463 | 1.8646 | 152.0 |
| 154.0 | 4.1025 | 0.016363 | 88.50 | 88.52 | 121.95 | 1005.8 | 1127.7 | 0.2216 | 1.6390 | 1.8606 | 154.0 |
| 156.0 | 4.3068 | 0.016374 | 84.56 | 84.57 | 123.95 | 1004.6 | 1128.6 | 0.2248 | 1.6318 | 1.8566 | 156.0 |
| 158.0 | 4.5197 | 0.016384 | 80.82 | 80.83 | 125.96 | 1003.4 | 1129.4 | 0.2281 | 1.6245 | 1.8526 | 158.0 |
| 160.0 | 4.7414 | 0.016395 | 77.27 | 77.29 | 127.96 | 1002.2 | 1130.2 | 0.2313 | 1.6174 | 1.8487 | 160.0 |
| 162.0 | 4.9722 | 0.016406 | 73.90 | 73.92 | 129.96 | 1001.0 | 1131.0 | 0.2345 | 1.6103 | 1.8448 | 162.0 |
| 164.0 | 5.2124 | 0.016417 | 70.70 | 70.72 | 131.96 | 999.8 | 1131.8 | 0.2377 | 1.6032 | 1.8409 | 164.0 |
| 166.0 | 5.4623 | 0.016428 | 67.67 | 67.68 | 133.97 | 998.6 | 1132.6 | 0.2409 | 1.5961 | 1.8371 | 166.0 |
| 168.0 | 5.7223 | 0.016440 | 64.78 | 64.80 | 135.97 | 997.4 | 1133.4 | 0.2441 | 1.5892 | 1.8333 | 168.0 |
| 170.0 | 5.9926 | 0.016451 | 62.04 | 62.06 | 137.97 | 996.2 | 1134.2 | 0.2473 | 1.5822 | 1.8295 | 170.0 |
| 172.0 | 6.2736 | 0.016463 | 59.43 | 59.45 | 139.98 | 995.0 | 1135.0 | 0.2505 | 1.5753 | 1.8258 | 172.0 |
| 174.0 | 6.5656 | 0.016474 | 56.95 | 56.97 | 141.98 | 993.8 | 1135.8 | 0.2537 | 1.5684 | 1.8221 | 174.0 |
| 176.0 | 6.8690 | 0.016486 | 54.59 | 54.61 | 143.99 | 992.6 | 1136.6 | 0.2568 | 1.5616 | 1.8184 | 176.0 |
| 178.0 | 7.1840 | 0.016498 | 52.35 | 52.36 | 145.99 | 991.4 | 1137.4 | 0.2600 | 1.5548 | 1.8147 | 178.0 |
| 180.0 | 7.5110 | 0.016510 | 50.21 | 50.22 | 148.00 | 990.2 | 1138.2 | 0.2631 | 1.5480 | 1.8111 | 180.0 |
| 182.0 | 7.850 | 0.016522 | 48.172 | 18.189 | 150.01 | 989.0 | 1139.0 | 0.2662 | 1.5413 | 1.8075 | 182.0 |
| 184.0 | 8.203 | 0.016534 | 46.232 | 46.249 | 152.01 | 987.8 | 1139.8 | 0.2694 | 1.5346 | 1.8040 | 184.0 |
| 186.0 | 8.568 | 0.016547 | 44.383 | 44.400 | 154.02 | 986.5 | 1140.5 | 0.2725 | 1.5279 | 1.8004 | 186.0 |
| 188.0 | 8.947 | 0.016559 | 42.621 | 42.638 | 156.03 | 985.3 | 1141.3 | 0.2756 | 1.5213 | 1.7969 | 188.0 |
| 190.0 | 9.340 | 0.016572 | 40.941 | 40.957 | 158.04 | 984.1 | 1142.1 | 0.2787 | 1.5148 | 1.7934 | 190.0 |
| 192.0 | 9.747 | 0.016585 | 39.337 | 39.354 | 160.05 | 982.8 | 1142.9 | 0.2818 | 1.5082 | 1.7900 | 192.0 |
| 194.0 | 10.168 | 0.016598 | 37.808 | 37.824 | 162.05 | 981.6 | 1143.7 | 0.2848 | 1.5017 | 1.7865 | 194.0 |
| 196.0 | 10.605 | 0.016611 | 36.348 | 36.364 | 164.06 | 980.4 | 1144.4 | 0.2879 | 1.4952 | 1.7831 | 196.0 |
| 198.0 | 11.058 | 0.016624 | 34.954 | 34.970 | 166.08 | 979.1 | 1145.2 | 0.2910 | 1.4888 | 1.7798 | 198.0 |
| 200.0 | 11.526 | 0.016637 | 33.622 | 33.639 | 168.09 | 977.9 | 1146.0 | 0.2940 | 1.4824 | 1.7764 | 200.0 |
| 204.0 | 12.512 | 0.016664 | 31.135 | 31.151 | 172.11 | 975.4 | 1147.5 | 0.3001 | 1.4697 | 1.7698 | 204.0 |
| 208.0 | 13.568 | 0.016691 | 28.862 | 28.878 | 176.14 | 972.8 | 1149.0 | 0.3061 | 1.4571 | 1.7632 | 208.0 |
| 212.0 | 14.696 | 0.016719 | 26.782 | 26.799 | 180.17 | 970.3 | 1150.5 | 0.3121 | 1.4447 | 1.7568 | 212.0 |
| 216.0 | 15.901 | 0.016747 | 24.878 | 24.894 | 184.20 | 967.8 | 1152.0 | 0.3181 | 1.4323 | 1.7505 | 216.0 |
| 220.0 | 17.186 | 0.016775 | 23.131 | 23.148 | 188.23 | 965.2 | 1153.4 | 0.3241 | 1.4201 | 1.7442 | 220.0 |
| 224.0 | 18.556 | 0.016805 | 21.529 | 21.545 | 192.27 | 962.6 | 1154.9 | 0.3300 | 1.4081 | 1.7380 | 224.0 |
| 228.0 | 20.015 | 0.016834 | 20.056 | 20.073 | 196.31 | 960.0 | 1156.3 | 0.3359 | 1.3961 | 1.7320 | 228.0 |
| 232.0 | 21.567 | 0.016864 | 18.701 | 18.718 | 200.35 | 957.4 | 1157.8 | 0.3417 | 1.3842 | 1.7260 | 232.0 |
| 236.0 | 23.216 | 0.016895 | 17.454 | 17.471 | 204.40 | 954.8 | 1159.2 | 0.3476 | 1.3725 | 1.7201 | 236.0 |
| 240.0 | 24.968 | 0.016926 | 16.304 | 16.321 | 208.45 | 952.1 | 1160.6 | 0.3533 | 1.3609 | 1.7142 | 240.0 |
| 244.0 | 26.826 | 0.016958 | 15.243 | 15.260 | 212.50 | 949.5 | 1162.0 | 0.3591 | 1.3494 | 1.7085 | 244.0 |

**Table II.2-1 Continued**

| Temp. t (°F) | Abs. Press. p (psi) | Specific Volume | | | Enthalpy | | | Entropy | | | Temp. t (°F) |
|---|---|---|---|---|---|---|---|---|---|---|---|
| | | Sat. Liquid $v_f$ | Evap $v_{fg}$ | Sat. Vapor $v_g$ | Sat. Liquid $h_f$ | Evap $h_{fg}$ | Sat. Vapor $h_g$ | Sat. Liquid $s_f$ | Evap $s_{fg}$ | Sat. Vapor $s_g$ | |
| 248.0 | 28.796 | 0.016990 | 14.264 | 14.281 | 216.56 | 946.8 | 1163.4 | 0.3649 | 1.3379 | 1.7028 | 248.0 |
| 252.0 | 30.883 | 0.017022 | 13.358 | 13.375 | 220.62 | 944.1 | 1164.7 | 0.3706 | 1.3266 | 1.6972 | 252.0 |
| 256.0 | 33.091 | 0.017055 | 12.520 | 12.538 | 224.69 | 941.4 | 1166.1 | 0.3763 | 1.3154 | 1.6917 | 256.0 |
| 260.0 | 35.427 | 0.017089 | 11.745 | 11.762 | 228.76 | 938.6 | 1167.4 | 0.3819 | 1.3043 | 1.6862 | 260.0 |
| 264.0 | 37.894 | 0.017123 | 11.025 | 11.042 | 232.83 | 935.9 | 1168.7 | 0.3876 | 1.2933 | 1.6808 | 264.0 |
| 268.0 | 40.500 | 0.017157 | 10.358 | 10.375 | 236.91 | 933.1 | 1170.0 | 0.3932 | 1.2823 | 1.6755 | 268.0 |
| 272.0 | 43.249 | 0.017193 | 9.738 | 9.755 | 240.99 | 930.3 | 1171.3 | 0.3987 | 1.2715 | 1.6702 | 272.0 |
| 276.0 | 46.147 | 0.017228 | 9.162 | 9.180 | 245.08 | 927.5 | 1172.5 | 0.4043 | 1.2607 | 1.6650 | 276.0 |
| 280.0 | 49.200 | 0.017264 | 8.627 | 8.644 | 249.17 | 924.6 | 1173.8 | 0.4098 | 1.2501 | 1.6599 | 280.0 |
| 284.0 | 52.414 | 0.01730 | 8.1280 | 8.1453 | 253.3 | 921.7 | 1175.0 | 0.4154 | 1.2395 | 1.6548 | 284.0 |
| 288.0 | 55.795 | 0.01734 | 7.6634 | 7.6807 | 257.4 | 918.8 | 1176.2 | 0.4208 | 1.2290 | 1.6498 | 288.0 |
| 292.0 | 59.350 | 0.01738 | 7.2301 | 7.2475 | 261.5 | 915.9 | 1177.4 | 0.4263 | 1.2186 | 1.6449 | 292.0 |
| 296.0 | 63.084 | 0.01741 | 6.8259 | 6.8433 | 265.6 | 913.0 | 1178.6 | 0.4317 | 1.2082 | 1.6400 | 296.0 |
| 300.0 | 67.005 | 0.01745 | 6.4483 | 6.4658 | 269.7 | 910.0 | 1179.7 | 0.4372 | 1.1979 | 1.6351 | 300.0 |
| 304.0 | 71.119 | 0.01749 | 6.0955 | 6.1130 | 273.8 | 907.0 | 1180.9 | 0.4426 | 1.1877 | 1.6303 | 304.0 |
| 308.0 | 75.433 | 0.01753 | 5.7655 | 5.7830 | 278.0 | 904.0 | 1182.0 | 0.4479 | 1.1776 | 1.6256 | 308.0 |
| 312.0 | 79.953 | 0.01757 | 5.4566 | 5.4742 | 282.1 | 901.0 | 1183.1 | 0.4533 | 1.1676 | 1.6209 | 312.0 |
| 316.0 | 84.688 | 0.01761 | 5.1673 | 5.1849 | 286.3 | 897.9 | 1184.1 | 0.4586 | 1.1576 | 1.6162 | 316.0 |
| 320.0 | 89.643 | 0.01766 | 4.8961 | 4.9138 | 290.4 | 894.8 | 1185.2 | 0.4640 | 1.1477 | 1.6116 | 320.0 |
| 324.0 | 94.826 | 0.01770 | 4.6418 | 4.6595 | 294.6 | 891.6 | 1186.2 | 0.4692 | 1.1378 | 1.6071 | 324.0 |
| 328.0 | 100.245 | 0.01774 | 4.4030 | 4.4208 | 298.7 | 888.5 | 1187.2 | 0.4745 | 1.1280 | 1.6025 | 328.0 |
| 332.0 | 105.907 | 0.01779 | 4.1788 | 4.1966 | 302.9 | 885.3 | 1188.2 | 0.4798 | 1.1183 | 1.5981 | 332.0 |
| 336.0 | 111.820 | 0.01783 | 3.9681 | 3.9859 | 307.1 | 882.1 | 1189.1 | 0.4850 | 1.1086 | 1.5936 | 336.0 |
| 340.0 | 117.992 | 0.01787 | 3.7699 | 3.7878 | 311.3 | 878.8 | 1190.1 | 0.4902 | 1.0990 | 1.5892 | 340.0 |
| 344.0 | 124.430 | 0.01792 | 3.5834 | 3.6013 | 315.5 | 875.5 | 1191.0 | 0.4954 | 1.0894 | 1.5849 | 344.0 |
| 348.0 | 131.142 | 0.01797 | 3.4078 | 3.4258 | 319.7 | 872.2 | 1191.1 | 0.5006 | 1.0799 | 1.5806 | 348.0 |
| 352.0 | 138.138 | 0.01801 | 3.2423 | 3.2603 | 323.9 | 868.9 | 1192.7 | 0.5058 | 1.0705 | 1.5763 | 352.0 |
| 356.0 | 145.424 | 0.01806 | 3.0863 | 3.1044 | 328.1 | 865.5 | 1193.6 | 0.5110 | 1.0611 | 1.5721 | 356.0 |
| 360.0 | 153.010 | 0.01811 | 2.9392 | 2.9573 | 332.3 | 862.1 | 1194.4 | 0.5161 | 1.0517 | 1.5678 | 360.0 |
| 364.0 | 160.903 | 0.01816 | 2.8002 | 2.8184 | 336.5 | 858.6 | 1195.2 | 0.5212 | 1.0424 | 1.5637 | 364.0 |
| 368.0 | 169.113 | 0.01821 | 2.6691 | 2.6873 | 340.8 | 855.1 | 1195.9 | 0.5263 | 1.0332 | 1.5595 | 368.0 |
| 372.0 | 177.648 | 0.01826 | 2.5451 | 2.5633 | 345.0 | 851.6 | 1196.7 | 0.5314 | 1.0240 | 1.5554 | 372.0 |
| 376.0 | 186.517 | 0.01831 | 2.4279 | 2.4462 | 349.3 | 848.1 | 1197.4 | 0.5365 | 1.0148 | 1.5513 | 376.0 |
| 380.0 | 195.729 | 0.01836 | 2.3170 | 2.3353 | 353.6 | 844.5 | 1198.0 | 0.5416 | 1.0057 | 1.5473 | 380.0 |
| 384.0 | 205.294 | 0.01842 | 2.2120 | 2.2304 | 357.9 | 840.8 | 1198.7 | 0.5466 | 0.9966 | 1.5432 | 384.0 |
| 388.0 | 215.220 | 0.01847 | 2.1126 | 2.1311 | 362.2 | 837.2 | 1199.3 | 0.5516 | 0.9876 | 1.5392 | 388.0 |
| 392.0 | 225.516 | 0.01853 | 2.0184 | 2.0369 | 366.5 | 833.4 | 1199.9 | 0.5567 | 0.9786 | 1.5352 | 392.0 |
| 396.0 | 236.193 | 0.01858 | 1.9291 | 1.9477 | 370.8 | 829.7 | 1200.4 | 0.5617 | 0.9696 | 1.5313 | 396.0 |
| 400.0 | 247.259 | 0.01864 | 1.8444 | 1.8630 | 375.1 | 825.9 | 1201.0 | 0.5667 | 0.9607 | 1.5274 | 400.0 |
| 404.0 | 258.725 | 0.01870 | 1.7640 | 1.7827 | 379.4 | 822.0 | 1201.5 | 0.5717 | 0.9518 | 1.5234 | 404.0 |
| 408.0 | 270.600 | 0.01875 | 1.6877 | 1.7064 | 383.8 | 818.2 | 1201.9 | 0.5766 | 0.9429 | 1.5195 | 408.0 |
| 412.0 | 282.894 | 0.01881 | 1.6152 | 1.6340 | 388.1 | 814.2 | 1202.4 | 0.5816 | 0.9341 | 1.5157 | 412.0 |
| 416.0 | 295.617 | 0.01887 | 1.5463 | 1.5651 | 392.5 | 810.2 | 1202.8 | 0.5866 | 0.9253 | 1.5118 | 416.0 |
| 420.0 | 308.780 | 0.01894 | 1.4808 | 1.4997 | 396.9 | 806.2 | 1203.1 | 0.5915 | 0.9165 | 1.5080 | 420.0 |
| 424.0 | 322.391 | 0.01900 | 1.4184 | 1.4374 | 401.3 | 802.2 | 1203.5 | 0.5964 | 0.9077 | 1.5042 | 424.0 |
| 428.0 | 336.463 | 0.01906 | 1.3591 | 1.3782 | 405.7 | 798.0 | 1203.7 | 0.6014 | 0.8990 | 1.5004 | 428.0 |
| 432.0 | 351.00 | 0.01913 | 1.30266 | 1.32179 | 410.1 | 793.9 | 1204.0 | 0.6063 | 0.8903 | 1.4966 | 432.0 |
| 436.0 | 366.03 | 0.01919 | 1.24887 | 1.26806 | 414.6 | 789.7 | 1204.2 | 0.6112 | 0.8816 | 1.4928 | 436.0 |
| 440.0 | 381.54 | 0.01926 | 1.19761 | 1.21687 | 419.0 | 785.4 | 1204.4 | 0.6161 | 0.8729 | 1.4890 | 440.0 |
| 444.0 | 397.56 | 0.01933 | 1.14874 | 1.16806 | 423.5 | 781.1 | 1204.6 | 0.6210 | 0.8643 | 1.4853 | 444.0 |
| 448.0 | 414.09 | 0.01940 | 1.10212 | 1.12152 | 428.0 | 776.7 | 1204.7 | 0.6259 | 0.8557 | 1.4815 | 448.0 |
| 452.0 | 431.14 | 0.01947 | 1.05764 | 1.07711 | 432.5 | 772.3 | 1204.8 | 0.6308 | 0.8471 | 1.4778 | 452.0 |
| 456.0 | 448.73 | 0.01954 | 1.01518 | 1.03472 | 437.0 | 767.8 | 1204.8 | 0.6356 | 0.8385 | 1.4741 | 456.0 |
| 460.0 | 466.87 | 0.01961 | 0.97463 | 0.99424 | 441.5 | 763.2 | 1204.8 | 0.6405 | 0.8299 | 1.4704 | 460.0 |
| 464.0 | 485.56 | 0.01969 | 0.93588 | 0.95557 | 446.1 | 758.6 | 1204.7 | 0.6454 | 0.8213 | 1.4667 | 464.0 |
| 468.0 | 504.83 | 0.01976 | 0.89885 | 0.91862 | 450.7 | 754.0 | 1204.6 | 0.6502 | 0.8127 | 1.4629 | 468.0 |
| 472.0 | 524.67 | 0.01984 | 0.86345 | 0.88329 | 455.2 | 749.3 | 1204.5 | 0.6551 | 0.8042 | 1.4592 | 472.0 |
| 476.0 | 545.11 | 0.01992 | 0.82958 | 0.84950 | 459.9 | 744.5 | 1204.3 | 0.6599 | 0.7956 | 1.4555 | 476.0 |
| 480.0 | 566.15 | 0.02000 | 0.79716 | 0.81717 | 464.5 | 739.6 | 1204.1 | 0.6648 | 0.7871 | 1.4518 | 480.0 |
| 484.0 | 587.81 | 0.02009 | 0.76613 | 0.78622 | 469.1 | 734.7 | 1203.8 | 0.6696 | 0.7785 | 1.4481 | 484.0 |
| 488.0 | 610.10 | 0.02017 | 0.73641 | 0.75658 | 473.8 | 729.7 | 1203.5 | 0.6745 | 0.7700 | 1.4444 | 488.0 |
| 492.0 | 633.03 | 0.02026 | 0.70794 | 0.72820 | 478.5 | 724.6 | 1203.1 | 0.6793 | 0.7614 | 1.4407 | 492.0 |
| 496.0 | 656.61 | 0.02034 | 0.68065 | 0.70100 | 483.2 | 719.5 | 1202.7 | 0.6842 | 0.7528 | 1.4370 | 496.0 |
| 500.0 | 680.86 | 0.02043 | 0.65448 | 0.67492 | 487.9 | 714.3 | 1202.2 | 0.6890 | 0.7443 | 1.4333 | 500.0 |
| 504.0 | 705.78 | 0.02053 | 0.62938 | 0.64991 | 492.7 | 709.0 | 1201.7 | 0.6939 | 0.7357 | 1.4296 | 504.0 |

**Table II.2-1    Continued**

| Abs. Press. $p$ (psi) | Temp. $t$ (°F) | Specific Volume | | | Enthalpy | | | Entropy | | | Temp. $t$ (°F) |
|---|---|---|---|---|---|---|---|---|---|---|---|
| | | Sat. Liquid $v_f$ | Evap $v_{fg}$ | Sat. Vapor $v_g$ | Sat. Liquid $h_f$ | Evap $h_{fg}$ | Sat. Vapor $h_g$ | Sat. Liquid $s_f$ | Evap $s_{fg}$ | Sat. Vapor $s_g$ | |
| 508.0 | 731.40 | 0.02062 | 0.60530 | 0.62592 | 497.5 | 703.7 | 1201.1 | 0.6987 | 0.7271 | 1.4258 | 508.0 |
| 512.0 | 757.72 | 0.02072 | 0.58218 | 0.60289 | 502.3 | 698.2 | 1200.5 | 0.7036 | 0.7185 | 1.4221 | 512.0 |
| 516.0 | 784.76 | 0.02081 | 0.55997 | 0.58079 | 507.1 | 692.7 | 1199.8 | 0.7085 | 0.7099 | 1.4183 | 516.0 |
| 520.0 | 812.53 | 0.02091 | 0.53864 | 0.55956 | 512.0 | 687.0 | 1199.0 | 0.7133 | 0.7013 | 1.4146 | 520.0 |
| 524.0 | 841.04 | 0.02102 | 0.51814 | 0.53916 | 516.9 | 681.3 | 1198.2 | 0.7182 | 0.6926 | 1.4108 | 524.0 |
| 528.0 | 870.31 | 0.02112 | 0.49843 | 0.51955 | 521.8 | 675.5 | 1197.3 | 0.7231 | 0.6839 | 1.4070 | 528.0 |
| 532.0 | 900.34 | 0.02123 | 0.47947 | 0.50070 | 526.8 | 669.6 | 1196.4 | 0.7280 | 0.6752 | 1.4032 | 532.0 |
| 536.0 | 931.17 | 0.02134 | 0.46123 | 0.48257 | 531.7 | 663.6 | 1195.4 | 0.7329 | 0.6665 | 1.3993 | 536.0 |
| 540.0 | 962.79 | 0.02146 | 0.44367 | 0.46513 | 536.8 | 657.5 | 1194.3 | 0.7378 | 0.6577 | 1.3954 | 540.0 |
| 544.0 | 995.22 | 0.02157 | 0.42677 | 0.44834 | 541.8 | 651.3 | 1193.1 | 0.7427 | 0.6489 | 1.3915 | 544.0 |
| 548.0 | 1028.49 | 0.02169 | 0.41048 | 0.43217 | 546.9 | 645.0 | 1191.9 | 0.7476 | 0.6400 | 1.3876 | 548.0 |
| 552.0 | 1062.59 | 0.02182 | 0.39479 | 0.41660 | 552.0 | 638.5 | 1190.6 | 0.7525 | 0.6311 | 1.3837 | 552.0 |
| 556.0 | 1097.55 | 0.02194 | 0.37966 | 0.40160 | 557.2 | 632.0 | 1189.2 | 0.7575 | 0.6222 | 1.3797 | 556.0 |
| 560.0 | 1133.38 | 0.02207 | 0.36507 | 0.38714 | 562.4 | 625.3 | 1187.7 | 0.7625 | 0.6132 | 1.3757 | 560.0 |
| 564.0 | 1170.10 | 0.02221 | 0.35099 | 0.37320 | 567.6 | 618.5 | 1186.1 | 0.7674 | 0.6041 | 1.3716 | 564.0 |
| 568.0 | 1207.72 | 0.02235 | 0.33741 | 0.35975 | 572.9 | 611.5 | 1184.5 | 0.7725 | 0.5950 | 1.3675 | 568.0 |
| 572.0 | 1246.26 | 0.02249 | 0.32429 | 0.34678 | 578.3 | 604.5 | 1182.7 | 0.7775 | 0.5859 | 1.3634 | 572.0 |
| 576.0 | 1285.74 | 0.02264 | 0.31162 | 0.33426 | 583.7 | 597.2 | 1180.9 | 0.7825 | 0.5766 | 1.3592 | 576.0 |
| 580.0 | 1326.17 | 0.02279 | 0.29937 | 0.32216 | 589.1 | 589.9 | 1179.0 | 0.7876 | 0.5673 | 1.3550 | 580.0 |
| 584.0 | 1367.7 | 0.02295 | 0.28753 | 0.31048 | 594.6 | 582.4 | 1176.9 | 0.7927 | 0.5580 | 1.3507 | 584.0 |
| 588.0 | 1410.0 | 0.02311 | 0.27608 | 0.29919 | 600.1 | 574.7 | 1174.8 | 0.7978 | 0.5485 | 1.3464 | 588.0 |
| 592.0 | 1453.3 | 0.02328 | 0.26499 | 0.28827 | 605.7 | 566.8 | 1172.6 | 0.8030 | 0.5390 | 1.3420 | 592.0 |
| 596.0 | 1497.8 | 0.02345 | 0.25425 | 0.27770 | 611.4 | 558.8 | 1170.2 | 0.8082 | 0.5293 | 1.3375 | 596.0 |
| 600.0 | 1543.2 | 0.02364 | 0.24384 | 0.26747 | 617.1 | 550.6 | 1167.7 | 0.8134 | 0.5196 | 1.3330 | 600.0 |
| 604.0 | 1589.7 | 0.02382 | 0.23374 | 0.25757 | 622.9 | 542.2 | 1165.1 | 0.8187 | 0.5097 | 1.3284 | 604.0 |
| 608.0 | 1637.3 | 0.02402 | 0.22394 | 0.24796 | 628.8 | 533.6 | 1162.4 | 0.8240 | 0.4997 | 1.3238 | 608.0 |
| 612.0 | 1686.1 | 0.02422 | 0.21442 | 0.23865 | 634.8 | 524.7 | 1159.5 | 0.8294 | 0.4896 | 1.3190 | 612.0 |
| 616.0 | 1735.9 | 0.02444 | 0.20516 | 0.22960 | 640.8 | 515.6 | 1156.4 | 0.8348 | 0.4794 | 1.3141 | 616.0 |
| 620.0 | 1786.9 | 0.02466 | 0.19615 | 0.22081 | 646.9 | 506.3 | 1153.2 | 0.8403 | 0.4689 | 1.3092 | 620.0 |
| 624.0 | 1839.0 | 0.02489 | 0.18737 | 0.21226 | 653.1 | 496.6 | 1149.8 | 0.8458 | 0.4583 | 1.3041 | 624.0 |
| 628.0 | 1892.4 | 0.02514 | 0.17880 | 0.20394 | 659.5 | 486.7 | 1146.1 | 0.8514 | 0.4474 | 1.2988 | 628.0 |
| 632.0 | 1947.0 | 0.02539 | 0.17044 | 0.19583 | 665.9 | 476.4 | 1142.2 | 0.8571 | 0.4364 | 1.2934 | 632.0 |
| 636.0 | 2002.8 | 0.02566 | 0.16226 | 0.18792 | 672.4 | 465.7 | 1138.1 | 0.8628 | 0.4251 | 1.2879 | 636.0 |
| 640.0 | 2059.9 | 0.02595 | 0.15427 | 0.18021 | 679.1 | 454.6 | 1133.7 | 0.8686 | 0.4134 | 1.2821 | 640.0 |
| 644.0 | 2118.3 | 0.02625 | 0.14644 | 0.17269 | 685.9 | 443.1 | 1129.0 | 0.8746 | 0.4015 | 1.2761 | 644.0 |
| 648.0 | 2178.1 | 0.02657 | 0.13876 | 0.16534 | 692.9 | 431.1 | 1124.0 | 0.8806 | 0.3893 | 1.2699 | 648.0 |
| 652.0 | 2239.2 | 0.02691 | 0.13124 | 0.15816 | 700.0 | 418.7 | 1118.7 | 0.8868 | 0.3767 | 1.2634 | 652.0 |
| 656.0 | 2301.7 | 0.02728 | 0.12387 | 0.15115 | 707.4 | 405.7 | 1113.1 | 0.8931 | 0.3637 | 1.2567 | 656.0 |
| 660.0 | 2365.7 | 0.02768 | 0.11663 | 0.14431 | 714.9 | 392.1 | 1107.0 | 0.8995 | 0.3502 | 1.2498 | 660.0 |
| 664.0 | 2431.1 | 0.02811 | 0.10947 | 0.13757 | 722.9 | 377.7 | 1100.6 | 0.9064 | 0.3361 | 1.2425 | 664.0 |
| 668.0 | 2498.1 | 0.02858 | 0.10229 | 0.13087 | 731.5 | 362.1 | 1093.5 | 0.9137 | 0.3210 | 1.2347 | 668.0 |
| 672.0 | 2566.6 | 0.02911 | 0.09514 | 0.12424 | 740.2 | 345.7 | 1085.9 | 0.9212 | 0.3054 | 1.2266 | 672.0 |
| 676.0 | 2636.8 | 0.02970 | 0.08799 | 0.11769 | 749.2 | 328.5 | 1077.6 | 0.9287 | 0.2892 | 1.2179 | 676.0 |
| 680.0 | 2708.6 | 0.03037 | 0.08080 | 0.11117 | 758.5 | 310.1 | 1068.5 | 0.9365 | 0.2720 | 1.2086 | 680.0 |
| 684.0 | 2782.1 | 0.03114 | 0.07349 | 0.10463 | 768.2 | 290.2 | 1058.4 | 0.9447 | 0.2537 | 1.1984 | 684.0 |
| 688.0 | 2857.4 | 0.03204 | 0.06595 | 0.09799 | 778.8 | 268.2 | 1047.0 | 0.9535 | 0.2337 | 1.1872 | 688.0 |
| 692.0 | 2934.5 | 0.03313 | 0.05797 | 0.09110 | 790.5 | 243.1 | 1033.6 | 0.9634 | 0.2110 | 1.1744 | 692.0 |
| 696.0 | 3013.4 | 0.03455 | 0.04916 | 0.08371 | 804.4 | 212.8 | 1017.2 | 0.9749 | 0.1841 | 1.1591 | 696.0 |
| 700.0 | 3094.3 | 0.03662 | 0.03857 | 0.07519 | 822.4 | 172.7 | 995.2 | 0.9901 | 0.1490 | 1.1390 | 700.0 |
| 702.0 | 3135.5 | 0.03824 | 0.03173 | 0.06997 | 835.0 | 144.7 | 979.7 | 1.0006 | 0.1246 | 1.1252 | 702.0 |
| 704.0 | 3177.2 | 0.04108 | 0.02192 | 0.06300 | 854.2 | 102.0 | 956.2 | 1.0169 | 0.0876 | 1.1046 | 704.0 |
| 705.0 | 3198.3 | 0.04427 | 0.01304 | 0.05730 | 873.0 | 61.4 | 934.4 | 1.0329 | 0.0527 | 1.0856 | 705.0 |
| 705.47[b] | 3208.2 | 0.05078 | 0.00000 | 0.05078 | 906.0 | 0.0 | 906.0 | 1.0612 | 0.0000 | 1.0612 | 705.47[b] |

**Table 11.2-1    Continued**

| Abs. Press. $p$ (psi) | Temp. $t$ (°F) | Specific Volume | | | Enthalpy | | | Entropy | | | Temp. $t$ (°F) |
|---|---|---|---|---|---|---|---|---|---|---|---|
| | | Sat. Liquid $V_f$ | Evap $v_{fx}$ | Sat. Vapor $v_x$ | Sat. Liquid $h_f$ | Evap $h_{fx}$ | Sat. Vapor $h_x$ | Sat. Liquid $s_f$ | Evap $s_{fx}$ | Sat. Vapor $s_f$ | |

*Pressure Table*

| Abs. Press. | Temp. | Sat. Liquid | Evap | Sat. Vapor | Sat. Liquid | Evap | Sat. Vapor | Sat. Liquid | Evap | Sat. Vapor | Temp. |
|---|---|---|---|---|---|---|---|---|---|---|---|
| 0.08865 | 32.018 | 0.016022 | 3302.4 | 3302.4 | 0.0003 | 1075.5 | 1075.5 | 0.0000 | 2.1872 | 2.1872 | 0.08865 |
| 0.25 | 59.323 | 0.016032 | 1235.5 | 1235.5 | 27.382 | 1060.1 | 1087.4 | 0.0542 | 2.0425 | 2.0967 | 0.25 |
| 0.50 | 79.586 | 0.016071 | 641.5 | 641.5 | 47.623 | 1048.6 | 1096.3 | 0.0925 | 1.9446 | 2.0370 | 0.50 |
| 1.0 | 101.74 | 0.016136 | 333.59 | 333.60 | 69.73 | 1036.1 | 1105.8 | 0.1326 | 1.8455 | 1.9781 | 1.0 |
| 5.0 | 162.24 | 0.016407 | 73.515 | 73.532 | 130.20 | 1000.9 | 1131.1 | 0.2349 | 1.6094 | 1.8443 | 5.0 |
| 10.0 | 193.21 | 0.016592 | 38.404 | 38.420 | 161.26 | 982.1 | 1143.3 | 0.2836 | 1.5043 | 1.7879 | 10.0 |
| 14.696 | 212.00 | 0.016719 | 26.782 | 26.799 | 180.17 | 970.3 | 1150.5 | 0.3121 | 1.4447 | 1.7568 | 14.696 |
| 15.0 | 213.03 | 0.016726 | 26.274 | 26.290 | 181.21 | 969.7 | 1150.9 | 0.3137 | 1.4415 | 1.7552 | 15.0 |
| 20.0 | 227.96 | 0.016834 | 20.070 | 20.087 | 196.27 | 960.1 | 1156.3 | 0.3358 | 1.3962 | 1.7320 | 20.0 |
| 30.0 | 250.34 | 0.017009 | 13.7266 | 13.7436 | 218.9 | 945.2 | 1164.1 | 0.3682 | 1.3313 | 1.6995 | 30.0 |
| 40.0 | 267.25 | 0.017151 | 10.4794 | 10.4965 | 236.1 | 933.6 | 1169.8 | 0.3921 | 1.2844 | 1.6765 | 40.0 |
| 50.0 | 281.02 | 0.017274 | 8.4967 | 8.5140 | 250.2 | 923.9 | 1174.1 | 0.4112 | 1.2474 | 1.6586 | 50.0 |
| 60.0 | 292.71 | 0.017383 | 7.1562 | 7.1736 | 262.2 | 915.4 | 1177.6 | 0.4273 | 1.2167 | 1.6440 | 60.0 |
| 70.0 | 302.93 | 0.017482 | 6.1875 | 6.2050 | 272.7 | 907.8 | 1180.6 | 0.4411 | 1.1905 | 1.6316 | 70.0 |
| 80.0 | 312.04 | 0.017573 | 5.4536 | 5.4711 | 282.1 | 900.9 | 1183.1 | 0.4534 | 1.1675 | 1.6208 | 80.0 |
| 90.0 | 320.28 | 0.017659 | 4.8779 | 4.8953 | 290.7 | 894.6 | 1185.3 | 0.4643 | 1.1470 | 1.6113 | 90.0 |
| 100.0 | 327.82 | 0.017740 | 4.4133 | 4.4310 | 298.5 | 888.6 | 1187.2 | 0.4743 | 1.1284 | 1.6027 | 100.0 |
| 110.0 | 334.79 | 0.01782 | 4.0306 | 4.0484 | 305.8 | 883.1 | 1188.9 | 0.4834 | 1.1115 | 1.5950 | 110.0 |
| 120.0 | 341.27 | 0.01789 | 3.7097 | 3.7275 | 312.6 | 877.8 | 1190.4 | 0.4919 | 1.0960 | 1.5879 | 120.0 |
| 130.0 | 347.33 | 0.01796 | 3.4364 | 3.4544 | 319.0 | 872.8 | 1191.7 | 0.4998 | 1.0815 | 1.5813 | 130.0 |
| 140.0 | 353.04 | 0.01803 | 3.2010 | 3.2190 | 325.0 | 868.0 | 1193.0 | 0.5071 | 1.0681 | 1.5752 | 140.0 |
| 150.0 | 358.43 | 0.01809 | 2.9958 | 3.0139 | 330.6 | 863.4 | 1194.1 | 0.5141 | 1.0554 | 1.5695 | 150.0 |
| 160.0 | 363.55 | 0.01815 | 2.8155 | 2.8336 | 336.1 | 859.0 | 1195.1 | 0.5206 | 1.0435 | 1.5641 | 160.0 |
| 170.0 | 368.42 | 0.01821 | 2.6556 | 2.6738 | 341.2 | 854.8 | 1196.0 | 0.5269 | 1.0322 | 1.5591 | 170.0 |
| 180.0 | 373.08 | 0.01827 | 2.5129 | 2.5312 | 346.2 | 850.7 | 1196.9 | 0.5328 | 1.0215 | 1.5543 | 180.0 |
| 190.0 | 377.53 | 0.01833 | 2.3847 | 2.4030 | 350.9 | 846.7 | 1197.6 | 0.5384 | 1.0113 | 1.5498 | 190.0 |
| 200.0 | 381.80 | 0.01839 | 2.2689 | 2.2873 | 355.5 | 842.8 | 1198.3 | 0.5438 | 1.0016 | 1.5454 | 200.0 |
| 210.0 | 385.91 | 0.01844 | 2.16373 | 2.18217 | 359.9 | 839.1 | 1199.0 | 0.5490 | 0.9923 | 1.5413 | 210.0 |
| 220.0 | 389.88 | 0.01850 | 2.06779 | 2.08629 | 364.2 | 835.4 | 1199.6 | 0.5540 | 0.9834 | 1.5374 | 220.0 |
| 230.0 | 393.70 | 0.01855 | 1.97991 | 1.99846 | 368.3 | 831.8 | 1200.1 | 0.5588 | 0.9748 | 1.5336 | 230.0 |
| 240.0 | 397.39 | 0.01860 | 1.89909 | 1.91769 | 372.3 | 828.4 | 1200.6 | 0.5634 | 0.9665 | 1.5299 | 240.0 |
| 250.0 | 400.97 | 0.01865 | 1.82452 | 1.84317 | 376.1 | 825.0 | 1201.1 | 0.5679 | 0.9585 | 1.5264 | 250.0 |
| 260.0 | 404.44 | 0.01870 | 1.75548 | 1.77418 | 379.9 | 821.6 | 1201.5 | 0.5722 | 0.9508 | 1.5230 | 260.0 |
| 270.0 | 407.80 | 0.01875 | 1.69137 | 1.71013 | 383.6 | 818.3 | 1201.9 | 0.5764 | 0.9433 | 1.5197 | 270.0 |
| 280.0 | 411.07 | 0.01880 | 1.63169 | 1.65049 | 387.1 | 815.1 | 1202.3 | 0.5805 | 0.9361 | 1.5166 | 280.0 |
| 290.0 | 414.25 | 0.01885 | 1.57597 | 1.59482 | 390.6 | 812.0 | 1202.6 | 0.5844 | 0.9291 | -1.5135 | 290.0 |
| 300.0 | 417.35 | 0.01889 | 1.52384 | 1.54274 | 394.0 | 808.9 | 1202.9 | 0.5882 | 0.9223 | 1.5105 | 300.0 |
| 350.0 | 431.73 | 0.01912 | 1.30642 | 1.32554 | 409.8 | 794.2 | 1204.0 | 0.6059 | 0.8909 | 1.4968 | 350.0 |
| 400.0 | 444.60 | 0.01934 | 1.14162 | 1.16095 | 424.2 | 780.4 | 1204.6 | 0.6217 | 0.8630 | 1.4847 | 400.0 |
| 450.0 | 456.28 | 0.01954 | 1.01224 | 1.03179 | 437.3 | 767.5 | 1204.8 | 0.6360 | 0.8378 | 1.4738 | 450.0 |
| 500.0 | 467.01 | 0.01975 | 0.90787 | 0.92762 | 449.5 | 755.1 | 1204.7 | 0.6490 | 0.8148 | 1.4639 | 500.0 |
| 550.0 | 476.94 | 0.01994 | 0.82183 | 0.84177 | 460.9 | 743.3 | 1204.3 | 0.6611 | 0.7936 | 1.4547 | 550.0 |
| 600.0 | 486.20 | 0.02013 | 0.74962 | 0.76975 | 471.7 | 732.0 | 1203.7 | 0.6723 | 0.7738 | 1.4461 | 600.0 |
| 650.0 | 494.89 | 0.02032 | 0.68811 | 0.70843 | 481.9 | 720.9 | 1202.8 | 0.6828 | 0.7552 | 1.4381 | 650.0 |
| 700.0 | 503.08 | 0.02050 | 0.63505 | 0.65556 | 491.6 | 710.2 | 1201.8 | 0.6928 | 0.7377 | 1.4304 | 700.0 |
| 750.0 | 510.84 | 0.02069 | 0.58880 | 0.60949 | 500.9 | 699.8 | 1200.7 | 0.7022 | 0.7210 | 1.4232 | 750.0 |
| 800.0 | 518.21 | 0.02087 | 0.54809 | 0.56896 | 509.8 | 689.6 | 1199.4 | 0.7111 | 0.7051 | 1.4163 | 800.0 |
| 850.0 | 525.24 | 0.02105 | 0.51197 | 0.53302 | 518.4 | 679.5 | 1198.0 | 0.7197 | 0.6899 | 1.4096 | 850.0 |
| 900.0 | 531.95 | 0.02123 | 0.47968 | 0.50091 | 526.7 | 669.7 | 1196.4 | 0.7279 | 0.6753 | 1.4032 | 900.0 |
| 950.0 | 538.39 | 0.02141 | 0.45064 | 0.47205 | 534.7 | 660.0 | 1194.7 | 0.7358 | 0.6612 | 1.3970 | 950.0 |
| 1000.0 | 544.58 | 0.02159 | 0.42436 | 0.44596 | 542.6 | 650.4 | 1192.9 | 0.7434 | 0.6476 | 1.3910 | 1000.0 |
| 1050.0 | 550.53 | 0.02177 | 0.40047 | 0.42224 | 550.1 | 640.9 | 1191.0 | 0.7507 | 0.6344 | 1.3851 | 1050.0 |
| 1100.0 | 556.28 | 0.02195 | 0.37863 | 0.40058 | 557.5 | 631.5 | 1189.1 | 0.7578 | 0.6216 | 1.3794 | 1100.0 |
| 1150.0 | 561.82 | 0.02214 | 0.35859 | 0.38073 | 564.8 | 622.2 | 1187.0 | 0.7647 | 0.6091 | 1.3738 | 1150.0 |
| 1200.0 | 567.19 | 0.02232 | 0.34013 | 0.36245 | 571.9 | 613.0 | 1184.8 | 0.7714 | 0.5969 | 1.3683 | 1200.0 |

**Table 11.2-1 Continued**

| Abs. Press. $p$ (psi) | Temp. $t$ (°F) | Specific Volume | | | Enthalpy | | | Entropy | | | Temp. $t$ (°F) |
|---|---|---|---|---|---|---|---|---|---|---|---|
| | | Sat. Liquid $v_f$ | Evap $v_{fx}$ | Sat. Vapor $v_x$ | Sat. Liquid $h_f$ | Evap $h_{fx}$ | Sat. Vapor $h_x$ | Sat. Liquid $s_f$ | Evap $s_{fx}$ | Sat. Vapor $s_f$ | |
| 1250.0 | 572.38 | 0.02250 | 0.32306 | 0.34556 | 578.8 | 603.8 | 1182.6 | 0.7780 | 0.5850 | 1.3630 | 1250.0 |
| 1300.0 | 577.42 | 0.02269 | 0.30722 | 0.32991 | 585.6 | 594.6 | 1180.2 | 0.7843 | 0.5733 | 1.3577 | 1300.0 |
| 1350.0 | 582.32 | 0.02288 | 0.29250 | 0.31537 | 592.3 | 585.4 | 1177.8 | 0.7906 | 0.5620 | 1.3525 | 1350.0 |
| 1400.0 | 587.07 | 0.02307 | 0.27871 | 0.30178 | 598.8 | 576.5 | 1175.3 | 0.7966 | 0.5507 | 1.3474 | 1400.0 |
| 1450.0 | 591.70 | 0.02327 | 0.26584 | 0.28911 | 605.3 | 567.4 | 1172.8 | 0.8026 | 0.5397 | 1.3423 | 1450.0 |
| 1500.0 | 596.20 | 0.02346 | 0.25372 | 0.27719 | 611.7 | 558.4 | 1170.1 | 0.8085 | 0.5288 | 1.3373 | 1500.0 |
| 1550.0 | 600.59 | 0.02366 | 0.24235 | 0.26601 | 618.0 | 549.4 | 1167.4 | 0.8142 | 0.5182 | 1.3324 | 1550.0 |
| 1600.0 | 604.87 | 0.02387 | 0.23159 | 0.25545 | 624.2 | 540.3 | 1164.5 | 0.8199 | 0.5076 | 1.3274 | 1600.0 |
| 1650.0 | 609.05 | 0.02407 | 0.22143 | 0.24551 | 630.4 | 531.3 | 1161.6 | 0.8254 | 0.4971 | 1.3225 | 1650.0 |
| 1700.0 | 613.13 | 0.02428 | 0.21178 | 0.23607 | 636.5 | 522.2 | 1158.6 | 0.8309 | 0.4867 | 1.3176 | 1700.0 |
| 1750.0 | 617.12 | 0.02450 | 0.20263 | 0.22713 | 642.5 | 513.1 | 1155.6 | 0.8363 | 0.4765 | 1.3128 | 1750.0 |
| 1800.0 | 621.02 | 0.02472 | 0.19390 | 0.21861 | 648.5 | 503.8 | 1152.3 | 0.8417 | 0.4662 | 1.3079 | 1800.0 |
| 1850.0 | 624.83 | 0.02495 | 0.18558 | 0.21052 | 654.5 | 494.6 | 1149.0 | 0.8470 | 0.4561 | 1.3030 | 1850.0 |
| 1900.0 | 628.56 | 0.02517 | 0.17761 | 0.20278 | 660.4 | 485.2 | 1145.6 | 0.8522 | 0.4459 | 1.2981 | 1900.0 |
| 1950.0 | 632.22 | 0.02541 | 0.16999 | 0.19540 | 666.3 | 475.8 | 1142.0 | 0.8574 | 0.4358 | 1.2931 | 1950.0 |
| 2000.0 | 635.80 | 0.02565 | 0.16266 | 0.18831 | 672.1 | 466.2 | 1138.3 | 0.8625 | 0.4256 | 1.2881 | 2000.0 |
| 2100.0 | 642.76 | 0.02615 | 0.14885 | 0.17501 | 683.8 | 446.7 | 1130.5 | 0.8727 | 0.4053 | 1.2780 | 2100.0 |
| 2200.0 | 649.45 | 0.02669 | 0.13603 | 0.16272 | 695.5 | 426.7 | 1122.2 | 0.8828 | 0.3848 | 1.2676 | 2200.0 |
| 2300.0 | 655.89 | 0.02727 | 0.12406 | 0.15133 | 707.2 | 406.0 | 1113.2 | 0.8929 | 0.3640 | 1.2569 | 2300.0 |
| 2400.0 | 662.11 | 0.02790 | 0.11287 | 0.14076 | 719.0 | 384.8 | 1103.7 | 0.9031 | 0.3430 | 1.2460 | 2400.0 |
| 2500.0 | 668.11 | 0.02859 | 0.10209 | 0.13068 | 731.7 | 361.6 | 1093.3 | 0.9139 | 0.3206 | 1.2345 | 2500.0 |
| 2600.0 | 673.91 | 0.02938 | 0.09172 | 0.12110 | 744.5 | 337.6 | 1082.0 | 0.9247 | 0.2977 | 1.2225 | 2600.0 |
| 2700.0 | 679.53 | 0.03029 | 0.08165 | 0.11194 | 757.3 | 312.3 | 1069.7 | 0.9356 | 0.2741 | 1.2097 | 2700.0 |
| 2800.0 | 684.96 | 0.03134 | 0.07171 | 0.10305 | 770.7 | 285.1 | 1055.8 | 0.9468 | 0.2491 | 1.1958 | 2800.0 |
| 2900.0 | 690.22 | 0.03262 | 0.06158 | 0.09420 | 785.1 | 254.7 | 1039.8 | 0.9588 | 0.2215 | 1.1803 | 2900.0 |
| 3000.0 | 695.33 | 0.03428 | 0.05073 | 0.08500 | 801.8 | 218.4 | 1020.3 | 0.9728 | 0.1891 | 1.1619 | 3000.0 |
| 3100.0 | 700.28 | 0.03681 | 0.03771 | 0.07452 | 824.0 | 169.3 | 993.3 | 0.9914 | 0.1460 | 1.1373 | 3100.0 |
| 3200.0 | 705.08 | 0.04472 | 0.01191 | 0.05663 | 875.5 | 56.1 | 931.6 | 1.0351 | 0.0482 | 1.0832 | 3200.0 |
| 3208.2ᶜ | 705.47 | 0.05078 | 0.00000 | 0.05078 | 906.0 | 0.0 | 906.0 | 1.0612 | 0.0000 | 1.0612 | 3208.2ᶜ |

ᵃ The states shown are metastable.
ᵇ Critical temperature.
ᶜ Critical pressure.
*Source*: Copyright 1967 ASME (Abridged); reprinted by permission.

## Table II.2-2 Superheated Steam[a]

Temperature (°F)

| Abs. Press. (psi) (Sat. Temp.) | | Sat. Water | Sat. Steam | 200 | 250 | 300 | 350 | 400 | 450 | 500 | 600 | 700 | 800 | 900 | 1000 | 1100 | 1200 |
|---|---|---|---|---|---|---|---|---|---|---|---|---|---|---|---|---|---|
| 1 (101.74) | Sh | | | 98.26 | 148.26 | 198.26 | 248.26 | 298.26 | 348.26 | 398.26 | 498.26 | 598.26 | 698.26 | 798.26 | 898.26 | 998.26 | 1098.26 |
| | v | 0.01614 | 333.6 | 392.5 | 422.4 | 452.3 | 482.1 | 511.9 | 541.7 | 571.5 | 631.1 | 690.7 | 750.3 | 809.8 | 869.4 | 929.0 | 988.6 |
| | h | 69.73 | 1105.8 | 1150.2 | 1172.9 | 1195.7 | 1218.7 | 1241.8 | 1265.1 | 1288.6 | 1336.1 | 1384.5 | 1433.7 | 1483.8 | 1534.9 | 1586.8 | 1639.7 |
| | s | 0.1326 | 1.9781 | 2.0509 | 2.0841 | 2.1152 | 2.1445 | 2.1722 | 2.1985 | 2.2237 | 2.2708 | 2.3144 | 2.3551 | 2.3934 | 2.4296 | 2.4640 | 2.4969 |
| 5 (162.24) | Sh | | | 37.76 | 87.76 | 137.76 | 187.76 | 237.76 | 287.76 | 337.76 | 437.76 | 537.76 | 637.76 | 737.76 | 837.76 | 937.76 | 1037.76 |
| | v | 0.01641 | 73.53 | 78.14 | 84.21 | 90.24 | 96.25 | 102.24 | 108.23 | 114.21 | 126.15 | 138.08 | 150.01 | 161.94 | 173.86 | 185.78 | 197.70 |
| | h | 130.20 | 1131.1 | 1148.6 | 1171.7 | 1194.8 | 1218.0 | 1241.3 | 1264.7 | 1288.2 | 1335.9 | 1384.3 | 1433.6 | 1483.7 | 1534.7 | 1586.7 | 1639.6 |
| | s | 0.2349 | 1.8443 | 1.8716 | 1.9054 | 1.9369 | 1.9664 | 1.9943 | 2.0208 | 2.0460 | 2.0932 | 2.1369 | 2.1776 | 2.2159 | 2.2521 | 2.2866 | 2.3194 |
| 10 (193.21) | Sh | | | 6.79 | 56.79 | 106.79 | 156.79 | 206.79 | 256.79 | 306.79 | 406.79 | 506.79 | 606.79 | 706.79 | 806.79 | 906.79 | 1006.79 |
| | v | 0.01659 | 38.42 | 38.84 | 41.93 | 44.98 | 48.02 | 51.03 | 54.04 | 57.04 | 63.03 | 69.00 | 74.98 | 80.94 | 86.91 | 92.87 | 98.84 |
| | h | 161.26 | 1143.3 | 1146.6 | 1170.2 | 1193.7 | 1217.1 | 1240.6 | 1264.1 | 1287.8 | 1335.5 | 1384.0 | 1433.4 | 1483.5 | 1534.6 | 1586.6 | 1639.5 |
| | s | 0.2836 | 1.7879 | 1.7928 | 1.8273 | 1.8593 | 1.8892 | 1.9173 | 1.9439 | 1.9692 | 2.0166 | 2.0603 | 2.1011 | 2.1394 | 2.1757 | 2.2101 | 2.2430 |
| 14.696 (212.00) | Sh | | | | 38.00 | 88.00 | 138.00 | 188.00 | 238.00 | 288.00 | 388.00 | 488.00 | 588.00 | 688.00 | 788.00 | 888.00 | 988.00 |
| | v | .0167 | 26.799 | | 28.42 | 30.52 | 32.60 | 34.67 | 36.72 | 38.77 | 42.86 | 46.93 | 51.00 | 55.06 | 59.13 | 63.19 | 67.25 |
| | h | 180.17 | 1150.5 | | 1168.8 | 1192.6 | 1216.3 | 1239.9 | 1263.6 | 1287.4 | 1335.2 | 1383.8 | 1433.2 | 1483.4 | 1534.5 | 1586.5 | 1639.4 |
| | s | .3121 | 1.7568 | | 1.7833 | 1.8158 | 1.8459 | 1.8743 | 1.9010 | 1.9265 | 1.9739 | 2.0177 | 2.0585 | 2.0969 | 2.1332 | 2.1676 | 2.2005 |
| 15 (213.03) | Sh | | | | 36.97 | 86.97 | 136.97 | 186.97 | 236.97 | 286.97 | 386.97 | 486.97 | 586.97 | 686.97 | 786.97 | 886.97 | 986.97 |
| | v | 0.01673 | 26.290 | | 27.837 | 29.899 | 31.939 | 33.963 | 35.977 | 37.985 | 41.986 | 45.978 | 49.964 | 53.946 | 57.926 | 61.905 | 65.882 |
| | h | 181.21 | 1150.9 | | 1168.7 | 1192.5 | 1216.2 | 1239.9 | 1263.6 | 1287.3 | 1335.2 | 1383.8 | 1433.2 | 1483.4 | 1534.5 | 1586.5 | 1639.4 |
| | s | 0.3137 | 1.7552 | | 1.7809 | 1.8134 | 1.8437 | 1.8720 | 1.8988 | 1.9242 | 1.9717 | 2.0155 | 2.0563 | 2.0946 | 2.1309 | 2.1653 | 2.1982 |
| 20 (227.96) | Sh | | | | 22.04 | 72.04 | 122.04 | 172.04 | 222.04 | 272.04 | 372.04 | 472.04 | 572.04 | 672.04 | 772.04 | 872.04 | 972.04 |
| | v | 0.01683 | 20.087 | | 20.788 | 22.356 | 23.900 | 25.428 | 26.946 | 28.457 | 31.466 | 34.465 | 37.458 | 40.447 | 43.435 | 46.420 | 49.405 |
| | h | 196.27 | 1156.3 | | 1167.1 | 1191.4 | 1215.4 | 1239.2 | 1263.0 | 1286.9 | 1334.9 | 1383.5 | 1432.9 | 1483.2 | 1534.3 | 1586.3 | 1639.3 |
| | s | 0.3358 | 1.7320 | | 1.7475 | 1.7805 | 1.8111 | 1.8397 | 1.8666 | 1.8921 | 1.9397 | 1.9836 | 2.0244 | 2.0628 | 2.0991 | 2.1336 | 2.1982 |
| 25 (240.07) | Sh | | | | 9.93 | 59.93 | 109.93 | 159.93 | 209.93 | 259.93 | 359.93 | 459.93 | 559.93 | 659.93 | 759.93 | 859.93 | 959.93 |
| | v | 0.01693 | 16.301 | | 16.558 | 17.829 | 19.076 | 20.307 | 21.527 | 22.740 | 25.153 | 27.557 | 29.954 | 32.348 | 34.740 | 37.130 | 39.518 |
| | h | 208.52 | 1160.6 | | 1165.6 | 1190.2 | 1214.5 | 1238.5 | 1262.5 | 1286.4 | 1334.6 | 1383.3 | 1432.7 | 1483.0 | 1534.2 | 1586.2 | 1639.2 |
| | s | 0.3535 | 1.7141 | | 1.7212 | 1.7547 | 1.7856 | 1.8145 | 1.8415 | 1.8672 | 1.9149 | 1.9588 | 1.9997 | 2.0381 | 2.0744 | 2.1089 | 2.1418 |
| 30 (250.34) | Sh | | | | | 49.66 | 99.66 | 149.66 | 199.66 | 249.66 | 349.66 | 449.66 | 549.66 | 649.66 | 749.66 | 849.66 | 949.66 |
| | v | 0.01701 | 13.744 | | | 14.810 | 15.859 | 16.892 | 17.914 | 18.929 | 20.945 | 22.951 | 24.952 | 26.949 | 28.943 | 30.936 | 32.927 |
| | h | 218.93 | 1164.1 | | | 1189.0 | 1213.6 | 1237.8 | 1261.9 | 1286.0 | 1334.2 | 1383.0 | 1432.5 | 1482.8 | 1534.0 | 1586.1 | 1639.0 |
| | s | 0.3682 | 1.6995 | | | 1.7334 | 1.7647 | 1.7937 | 1.8210 | 1.8467 | 1.8946 | 1.9386 | 1.9795 | 2.0179 | 2.0543 | 2.0888 | 2.1217 |
| 35 (259.29) | Sh | | | | | 40.71 | 90.71 | 140.71 | 190.71 | 240.71 | 340.71 | 440.71 | 540.71 | 640.71 | 740.71 | 840.71 | 940.71 |
| | v | 0.01708 | 11.896 | | | 12.654 | 13.562 | 14.453 | 15.334 | 16.207 | 17.939 | 19.662 | 21.379 | 23.092 | 24.803 | 26.512 | 28.220 |
| | h | 228.03 | 1167.1 | | | 1187.8 | 1212.7 | 1237.1 | 1261.3 | 1285.5 | 1333.9 | 1382.8 | 1432.3 | 1482.7 | 1533.9 | 1586.0 | 1638.9 |
| | s | 0.3809 | 1.6872 | | | 1.7152 | 1.7468 | 1.7761 | 1.8035 | 1.8294 | 1.8774 | 1.9214 | 1.9624 | 2.0009 | 2.0372 | 2.0717 | 2.1046 |
| 40 (267.25) | Sh | | | | | 32.75 | 82.75 | 132.75 | 182.75 | 232.75 | 332.75 | 432.75 | 532.75 | 632.75 | 732.75 | 832.75 | 932.75 |
| | v | 0.01715 | 10.497 | | | 11.036 | 11.838 | 12.624 | 13.398 | 14.165 | 15.685 | 17.195 | 18.699 | 20.199 | 21.697 | 23.194 | 24.689 |
| | h | 236.14 | 1169.8 | | | 1186.6 | 1211.7 | 1236.4 | 1260.8 | 1285.0 | 1333.6 | 1382.5 | 1432.1 | 1482.5 | 1533.7 | 1585.8 | 1638.8 |
| | s | 0.3921 | 1.6765 | | | 1.6992 | 1.7312 | 1.7608 | 1.7883 | 1.8143 | 1.8624 | 1.9065 | 1.9476 | 1.9860 | 2.0224 | 2.0569 | 2.0899 |
| 45 (274.44) | Sh | | | | | 25.56 | 75.56 | 125.56 | 175.56 | 225.56 | 325.56 | 425.56 | 525.56 | 625.56 | 725.56 | 825.56 | 925.56 |
| | v | 0.01721 | 9.399 | | | 9.777 | 10.497 | 11.201 | 11.892 | 12.577 | 13.932 | 15.276 | 16.614 | 17.950 | 19.282 | 20.613 | 21.943 |
| | h | 243.49 | 1172.1 | | | 1185.4 | 1210.4 | 1235.7 | 1260.2 | 1284.6 | 1333.3 | 1382.3 | 1431.9 | 1482.3 | 1533.6 | 1585.7 | 1638.7 |
| | s | 0.4021 | 1.6671 | | | 1.6849 | 1.7173 | 1.7471 | 1.7748 | 1.8010 | 1.8492 | 1.8934 | 1.9345 | 1.9730 | 2.0093 | 2.0439 | 2.0768 |
| 50 (281.02) | Sh | | | | | 18.98 | 68.98 | 118.98 | 168.98 | 218.98 | 318.98 | 418.98 | 518.98 | 618.98 | 718.98 | 818.98 | 918.98 |
| | v | 0.01727 | 8.514 | | | 8.769 | 9.424 | 10.062 | 10.688 | 11.306 | 12.529 | 13.741 | 14.947 | 16.150 | 17.350 | 18.549 | 19.746 |
| | h | 250.21 | 1174.1 | | | 1184.1 | 1209.9 | 1234.9 | 1259.6 | 1284.1 | 1332.9 | 1382.0 | 1431.7 | 1482.2 | 1533.4 | 1585.6 | 1638.6 |
| | s | 0.4112 | 1.6586 | | | 1.6720 | 1.7048 | 1.7349 | 1.7628 | 1.7890 | 1.8374 | 1.8816 | 1.9227 | 1.9613 | 1.9977 | 2.0322 | 2.0652 |
| 55 (287.07) | Sh | | | | | 12.93 | 62.93 | 112.93 | 162.93 | 212.93 | 312.93 | 412.93 | 512.93 | 612.93 | 712.93 | 812.93 | 912.93 |
| | v | 0.01733 | 7.787 | | | 7.945 | 8.546 | 9.130 | 9.702 | 10.267 | 11.381 | 12.485 | 13.583 | 14.677 | 15.769 | 16.859 | 17.948 |
| | h | 256.43 | 1175.9 | | | 1182.9 | 1208.9 | 1234.2 | 1259.1 | 1283.6 | 1332.6 | 1381.8 | 1431.5 | 1482.0 | 1533.3 | 1585.5 | 1638.5 |
| | s | 0.4196 | 1.6509 | | | 1.6601 | 1.6933 | 1.7237 | 1.7518 | 1.7781 | 1.8266 | 1.8710 | 1.9121 | 1.9507 | 1.987 | 2.022 | 2.055 |
| 60 (292.71) | Sh | | | | | 7.29 | 57.29 | 107.29 | 157.29 | 207.29 | 307.29 | 407.29 | 507.29 | 607.29 | 707.29 | 807.29 | 907.29 |
| | v | 0.01738 | 7.174 | | | 7.257 | 7.815 | 8.354 | 8.881 | 9.400 | 10.425 | 11.438 | 12.446 | 13.450 | 14.452 | 15.452 | 16.450 |
| | h | 262.21 | 1177.6 | | | 1181.6 | 1208.0 | 1233.5 | 1258.5 | 1283.2 | 1332.3 | 1381.5 | 1431.3 | 1481.8 | 1533.2 | 1585.3 | 1638.4 |
| | s | 0.4273 | 1.6440 | | | 1.6492 | 1.6934 | 1.7134 | 1.7417 | 1.7681 | 1.8168 | 1.8612 | 1.9024 | 1.9410 | 1.9774 | 2.0120 | 2.0450 |
| 65 (297.98) | Sh | | | | | 2.02 | 52.02 | 102.02 | 152.02 | 202.02 | 302.02 | 402.02 | 502.02 | 602.02 | 702.02 | 802.02 | 902.02 |
| | v | 0.01743 | 6.653 | | | 6.675 | 7.195 | 7.697 | 8.186 | 8.667 | 9.615 | 10.552 | 11.484 | 12.412 | 13.337 | 14.261 | 15.183 |
| | h | 267.63 | 1179.1 | | | 1180.3 | 1207.0 | 1232.7 | 1257.9 | 1282.7 | 1331.9 | 1381.3 | 1431.1 | 1481.6 | 1533.0 | 1585.2 | 1638.3 |
| | s | 0.4344 | 1.6375 | | | 1.6390 | 1.6731 | 1.7040 | 1.7324 | 1.7590 | 1.8077 | 1.8522 | 1.8935 | 1.9321 | 1.9685 | 2.0031 | 2.0361 |
| 70 (302.93) | Sh | | | | | | 47.07 | 97.07 | 147.07 | 197.07 | 297.07 | 397.07 | 497.07 | 597.07 | 697.07 | 797.07 | 897.07 |
| | v | 0.01748 | 6.205 | | | | 6.664 | 7.133 | 7.590 | 8.039 | 8.922 | 9.793 | 10.659 | 11.522 | 12.382 | 13.240 | 14.097 |
| | h | 272.74 | 1180.6 | | | | 1206.0 | 1232.0 | 1257.3 | 1282.2 | 1331.6 | 1381.0 | 1430.9 | 1481.5 | 1532.9 | 1585.1 | 1638.2 |
| | s | 0.4411 | 1.6316 | | | | 1.6640 | 1.6951 | 1.7237 | 1.7504 | 1.7993 | 1.8439 | 1.8852 | 1.9238 | 1.9603 | 1.9949 | 2.0279 |
| 75 (307.61) | Sh | | | | | | 42.39 | 92.39 | 142.39 | 192.39 | 292.39 | 392.39 | 492.39 | 592.39 | 692.39 | 792.39 | 892.39 |
| | v | 0.01753 | 5.814 | | | | 6.204 | 6.645 | 7.074 | 7.494 | 8.320 | 9.135 | 9.945 | 10.750 | 11.553 | 12.355 | 13.155 |
| | h | 277.56 | 1181.9 | | | | 1205.0 | 1231.2 | 1256.7 | 1281.7 | 1331.3 | 1380.7 | 1430.7 | 1481.3 | 1532.7 | 1585.0 | 1638.1 |
| | s | 0.4474 | 1.6260 | | | | 1.6554 | 1.6868 | 1.7156 | 1.7424 | 1.7915 | 1.8361 | 1.8774 | 1.9161 | 1.9526 | 1.9872 | 2.0202 |

**Table II.2-2   Continued**

| Abs. Press. (psi) (Sat. Temp.) | | Sat. Water | Sat. Steam | 350 | 400 | 450 | 500 | 550 | 600 | 700 | 800 | 900 | 1000 | 1100 | 1200 | 1300 | 1400 |
|---|---|---|---|---|---|---|---|---|---|---|---|---|---|---|---|---|---|
| **80** (312.04) | Sh | | | 37.96 | 87.96 | 137.96 | 187.96 | 237.96 | 287.96 | 387.96 | 487.96 | 587.96 | 687.96 | 787.96 | 887.96 | 987.96 | 1087.96 |
| | v | 0.01757 | 5.471 | 5.801 | 6.218 | 6.622 | 7.018 | 7.408 | 7.794 | 8.560 | 9.319 | 10.075 | 10.829 | 11.581 | 12.331 | 13.081 | 13.829 |
| | h | 282.15 | 1183.1 | 1204.0 | 1230.5 | 1256.1 | 1281.3 | 1306.2 | 1330.9 | 1380.5 | 1430.5 | 1481.1 | 1532.6 | 1584.9 | 1638.0 | 1692.0 | 1746.8 |
| | s | 0.4534 | 1.6208 | 1.6473 | 1.6790 | 1.7080 | 1.7349 | 1.7602 | 1.7842 | 1.8289 | 1.8702 | 1.9089 | 1.9454 | 1.9800 | 2.0131 | 2.0446 | 2.0750 |
| **85** (316.26) | Sh | | | 33.74 | 83.74 | 133.74 | 183.74 | 233.74 | 283.74 | 383.74 | 483.74 | 583.74 | 683.74 | 783.74 | 883.74 | 983.74 | 1083.74 |
| | v | 0.01762 | 5.167 | 5.445 | 5.840 | 6.223 | 6.597 | 6.966 | 7.330 | 8.052 | 8.768 | 9.480 | 10.190 | 10.898 | 11.604 | 12.310 | 13.014 |
| | h | 286.52 | 1184.2 | 1203.0 | 1229.7 | 1255.5 | 1280.8 | 1305.8 | 1330.6 | 1380.2 | 1430.3 | 1481.0 | 1532.4 | 1584.7 | 1637.9 | 1691.9 | 1746.8 |
| | s | 0.4590 | 1.6159 | 1.6396 | 1.6716 | 1.7008 | 1.7279 | 1.7532 | 1.7772 | 1.8220 | 1.8634 | 1.9021 | 1.9386 | 1.9733 | 2.0063 | 2.0379 | 2.0682 |
| **90** (320.28) | Sh | | | 29.72 | 79.72 | 129.72 | 179.72 | 229.72 | 279.72 | 379.72 | 479.72 | 579.72 | 679.72 | 779.72 | 879.72 | 979.72 | 1079.72 |
| | v | 0.01766 | 4.895 | 5.128 | 5.505 | 5.869 | 6.223 | 6.572 | 6.917 | 7.600 | 8.277 | 8.950 | 9.621 | 10.290 | 10.958 | 11.625 | 12.290 |
| | h | 290.69 | 1185.3 | 1202.0 | 1228.9 | 1254.9 | 1280.3 | 1305.4 | 1330.2 | 1380.0 | 1430.1 | 1480.8 | 1532.3 | 1584.6 | 1637.8 | 1691.8 | 1746.7 |
| | s | 0.4643 | 1.6113 | 1.6323 | 1.6646 | 1.6940 | 1.7212 | 1.7467 | 1.7707 | 1.8156 | 1.8570 | 1.8957 | 1.9323 | 1.9669 | 2.0000 | 2.0316 | 2.0619 |
| **95** (324.13) | Sh | | | 25.87 | 75.87 | 125.87 | 175.87 | 225.87 | 275.87 | 375.87 | 475.87 | 575.87 | 675.87 | 775.87 | 875.87 | 975.87 | 1075.87 |
| | v | 0.01770 | 4.651 | 4.845 | 5.205 | 5.551 | 5.889 | 6.221 | 6.548 | 7.196 | 7.838 | 8.477 | 9.113 | 9.747 | 10.380 | 11.012 | 11.643 |
| | h | 294.70 | 1186.2 | 1200.9 | 1228.1 | 1254.3 | 1279.8 | 1305.0 | 1329.9 | 1379.7 | 1429.9 | 1480.6 | 1532.1 | 1584.5 | 1637.7 | 1691.7 | 1746.6 |
| | s | 0.4694 | 1.6069 | 1.6253 | 1.6580 | 1.6876 | 1.7149 | 1.7404 | 1.7645 | 1.8094 | 1.8509 | 1.8897 | 1.9262 | 1.9609 | 1.9940 | 2.0256 | 2.0559 |
| **100** (327.82) | Sh | | | 22.18 | 72.18 | 122.18 | 172.18 | 222.18 | 272.18 | 372.18 | 472.18 | 572.18 | 672.18 | 772.18 | 872.18 | 972.18 | 1072.18 |
| | v | 0.01774 | 4.431 | 4.590 | 4.935 | 5.266 | 5.588 | 5.904 | 6.216 | 6.833 | 7.443 | 8.050 | 8.655 | 9.258 | 9.860 | 10.460 | 11.060 |
| | h | 298.54 | 1187.2 | 1199.9 | 1227.4 | 1253.7 | 1279.3 | 1304.6 | 1329.6 | 1379.5 | 1429.7 | 1480.4 | 1532.0 | 1584.4 | 1637.6 | 1691.6 | 1746.5 |
| | s | 0.4743 | 1.6027 | 1.6187 | 1.6516 | 1.6814 | 1.7088 | 1.7344 | 1.7586 | 1.8036 | 1.8451 | 1.8839 | 1.9205 | 1.9552 | 1.9883 | 2.0199 | 2.0502 |
| **105** (331.37) | Sh | | | 18.63 | 68.63 | 118.63 | 168.63 | 218.63 | 268.63 | 368.63 | 468.63 | 568.63 | 668.63 | 768.63 | 868.63 | 968.63 | 1068.63 |
| | v | 0.01778 | 4.231 | 4.359 | 4.690 | 5.007 | 5.315 | 5.617 | 5.915 | 6.504 | 7.086 | 7.665 | 8.241 | 8.816 | 9.389 | 9.961 | 10.532 |
| | h | 302.24 | 1188.0 | 1198.8 | 1226.6 | 1253.1 | 1278.8 | 1304.2 | 1329.2 | 1379.2 | 1429.4 | 1480.3 | 1531.8 | 1584.2 | 1637.5 | 1691.5 | 1746.4 |
| | s | 0.4790 | 1.5988 | 1.6122 | 1.6455 | 1.6755 | 1.7031 | 1.7288 | 1.7530 | 1.7981 | 1.8396 | 1.8785 | 1.9151 | 1.9498 | 1.9828 | 2.0145 | 2.0448 |
| **110** (334.79) | Sh | | | 15.21 | 65.21 | 115.21 | 165.21 | 215.21 | 265.21 | 365.21 | 465.21 | 565.21 | 665.21 | 765.21 | 865.21 | 965.21 | 1065.21 |
| | v | 0.01782 | 4.048 | 4.149 | 4.468 | 4.772 | 5.068 | 5.357 | 5.642 | 6.205 | 6.761 | 7.314 | 7.865 | 8.413 | 8.961 | 9.507 | 10.053 |
| | h | 305.80 | 1188.9 | 1197.7 | 1225.8 | 1252.5 | 1278.3 | 1303.8 | 1328.9 | 1379.0 | 1429.2 | 1480.1 | 1531.7 | 1584.1 | 1637.4 | 1691.4 | 1746.4 |
| | s | 0.4834 | 1.5950 | 1.6061 | 1.6396 | 1.6698 | 1.6975 | 1.7233 | 1.7476 | 1.7928 | 1.8344 | 1.8732 | 1.9099 | 1.9446 | 1.9777 | 2.0093 | 2.0397 |
| **115** (338.08) | Sh | | | 11.92 | 61.92 | 111.92 | 161.92 | 211.92 | 261.92 | 361.92 | 461.92 | 561.92 | 661.92 | 761.92 | 861.92 | 961.92 | 1061.92 |
| | v | 0.01785 | 3.881 | 3.957 | 4.265 | 4.558 | 4.841 | 5.119 | 5.392 | 5.932 | 6.465 | 6.994 | 7.521 | 8.046 | 8.570 | 9.093 | 9.615 |
| | h | 309.25 | 1189.6 | 1196.7 | 1225.0 | 1251.8 | 1277.9 | 1303.3 | 1328.6 | 1378.7 | 1429.0 | 1479.9 | 1531.6 | 1584.0 | 1637.2 | 1691.4 | 1746.3 |
| | s | 0.4877 | 1.5913 | 1.6001 | 1.6340 | 1.6644 | 1.6922 | 1.7181 | 1.7425 | 1.7877 | 1.8294 | 1.8682 | 1.9049 | 1.9396 | 1.9727 | 2.0044 | 2.0347 |
| **120** (341.27) | Sh | | | 8.73 | 58.73 | 108.73 | 158.73 | 208.73 | 258.73 | 358.73 | 458.73 | 558.73 | 658.73 | 758.73 | 858.73 | 958.73 | 1058.73 |
| | v | 0.01789 | 3.7275 | 3.7815 | 4.0786 | 4.3610 | 4.6341 | 4.9009 | 5.1637 | 5.6813 | 6.1928 | 6.7006 | 7.2060 | 7.7096 | 8.2119 | 8.7130 | 9.2134 |
| | h | 312.58 | 1190.4 | 1195.6 | 1224.1 | 1251.2 | 1277.4 | 1302.9 | 1328.2 | 1378.4 | 1428.8 | 1479.8 | 1531.4 | 1583.9 | 1637.1 | 1691.3 | 1746.2 |
| | s | 0.4919 | 1.5879 | 1.5943 | 1.6286 | 1.6592 | 1.6872 | 1.7132 | 1.7376 | 1.7829 | 1.8246 | 1.8635 | 1.9001 | 1.9349 | 1.9680 | 1.9996 | 2.0300 |
| **130** (347.33) | Sh | | | 2.67 | 52.67 | 102.67 | 152.67 | 202.67 | 252.67 | 352.67 | 452.67 | 552.67 | 652.67 | 752.67 | 852.67 | 952.67 | 1052.67 |
| | v | 0.01796 | 3.4544 | 3.4699 | 3.7489 | 4.0129 | 4.2672 | 4.5151 | 4.7589 | 5.2384 | 5.7118 | 6.1814 | 6.6486 | 7.1140 | 7.5781 | 8.0411 | 8.5033 |
| | h | 318.95 | 1191.7 | 1193.4 | 1222.5 | 1249.9 | 1276.4 | 1302.1 | 1327.5 | 1377.9 | 1428.4 | 1479.4 | 1531.1 | 1583.6 | 1636.9 | 1691.1 | 1746.1 |
| | s | 0.4998 | 1.5813 | 1.5833 | 1.6182 | 1.6493 | 1.6775 | 1.7037 | 1.7283 | 1.7737 | 1.8155 | 1.8545 | 1.8911 | 1.9259 | 1.9591 | 1.9907 | 2.0211 |
| **140** (353.04) | Sh | | | | 46.96 | 96.96 | 146.96 | 196.96 | 246.96 | 346.96 | 446.96 | 546.96 | 646.96 | 746.96 | 846.96 | 946.96 | 1046.96 |
| | v | 0.01803 | 3.2190 | | 3.4661 | 3.7143 | 3.9526 | 4.1844 | 4.4119 | 4.8588 | 5.2995 | 5.7364 | 6.1709 | 6.6036 | 7.0349 | 7.4652 | 7.8946 |
| | h | 324.96 | 1193.0 | | 1220.8 | 1248.7 | 1275.3 | 1301.3 | 1326.8 | 1377.4 | 1428.0 | 1479.1 | 1530.8 | 1583.4 | 1636.7 | 1690.9 | 1745.9 |
| | s | 0.5071 | 1.5752 | | 1.6085 | 1.6400 | 1.6686 | 1.6949 | 1.7196 | 1.7652 | 1.8071 | 1.8461 | 1.8828 | 1.9176 | 1.9508 | 1.9825 | 2.0129 |
| **150** (358.43) | Sh | | | | 41.57 | 91.57 | 141.57 | 191.57 | 241.57 | 341.57 | 441.57 | 541.57 | 641.57 | 741.57 | 841.57 | 941.57 | 1041.57 |
| | v | 0.01809 | 3.0139 | | 3.2208 | 3.4555 | 3.6799 | 3.8978 | 4.1112 | 4.5298 | 4.9421 | 5.3507 | 5.7568 | 6.1612 | 6.5642 | 6.9661 | 7.3671 |
| | h | 330.65 | 1194.1 | | 1219.1 | 1247.4 | 1274.3 | 1300.5 | 1326.1 | 1376.9 | 1427.6 | 1478.7 | 1530.5 | 1583.1 | 1636.5 | 1690.7 | 1745.7 |
| | s | 0.5141 | 1.5695 | | 1.5993 | 1.6313 | 1.6602 | 1.6867 | 1.7115 | 1.7573 | 1.7992 | 1.8383 | 1.8751 | 1.9099 | 1.9431 | 1.9748 | 2.0052 |
| **160** (363.55) | Sh | | | | 36.45 | 86.45 | 136.45 | 186.45 | 236.45 | 336.45 | 436.45 | 536.45 | 636.45 | 736.45 | 836.45 | 936.45 | 1036.45 |
| | v | 0.01815 | 2.8336 | | 3.0060 | 3.2288 | 3.4413 | 3.6469 | 3.8480 | 4.2420 | 4.6295 | 5.0132 | 5.3945 | 5.7741 | 6.1522 | 6.5293 | 6.9055 |
| | h | 336.07 | 1195.1 | | 1217.6 | 1246.0 | 1273.3 | 1299.6 | 1325.4 | 1376.4 | 1427.2 | 1478.4 | 1530.3 | 1582.9 | 1636.3 | 1690.5 | 1745.6 |
| | s | 0.5206 | 1.5641 | | 1.5906 | 1.6231 | 1.6522 | 1.6790 | 1.7039 | 1.7499 | 1.7919 | 1.8310 | 1.8678 | 1.9027 | 1.9359 | 1.9676 | 1.9980 |
| **170** (368.42) | Sh | | | | 31.58 | 81.58 | 131.58 | 181.58 | 231.58 | 331.58 | 431.58 | 531.58 | 631.58 | 731.58 | 831.58 | 931.58 | 1031.58 |
| | v | 0.01821 | 2.6738 | | 2.8162 | 3.0288 | 3.2306 | 3.4255 | 3.6158 | 3.9879 | 4.3536 | 4.7155 | 5.0749 | 5.4325 | 5.7888 | 6.1440 | 6.4983 |
| | h | 341.24 | 1196.0 | | 1215.6 | 1244.7 | 1272.2 | 1298.8 | 1324.7 | 1375.8 | 1426.8 | 1478.0 | 1530.0 | 1582.6 | 1636.1 | 1690.4 | 1745.4 |
| | s | 0.5269 | 1.5591 | | 1.5823 | 1.6152 | 1.6447 | 1.6717 | 1.6968 | 1.7428 | 1.7850 | 1.8241 | 1.8610 | 1.8959 | 1.9291 | 1.9608 | 1.9913 |
| **180** (373.08) | Sh | | | | 26.92 | 76.92 | 126.92 | 176.92 | 226.92 | 326.92 | 426.92 | 526.92 | 626.92 | 726.92 | 826.92 | 926.92 | 1026.92 |
| | v | 0.01827 | 2.5312 | | 2.6474 | 2.8508 | 3.0433 | 3.2286 | 3.4093 | 3.7621 | 4.1084 | 4.4508 | 4.7907 | 5.1289 | 5.4657 | 5.8014 | 6.1363 |
| | h | 346.19 | 1196.9 | | 1213.8 | 1243.4 | 1271.2 | 1297.9 | 1324.0 | 1375.3 | 1426.3 | 1477.7 | 1529.7 | 1582.4 | 1635.9 | 1690.2 | 1745.3 |
| | s | 0.5328 | 1.5543 | | 1.5743 | 1.6078 | 1.6376 | 1.6647 | 1.6900 | 1.7362 | 1.7784 | 1.8176 | 1.8545 | 1.8894 | 1.9227 | 1.9545 | 1.9849 |
| **190** (377.53) | Sh | | | | 22.47 | 72.47 | 122.47 | 172.47 | 222.47 | 322.47 | 422.47 | 522.47 | 622.47 | 722.47 | 822.47 | 922.47 | 1022.47 |
| | v | 0.01833 | 2.4030 | | 2.4961 | 2.6915 | 2.8756 | 3.0525 | 3.2246 | 3.5601 | 3.8889 | 4.2140 | 4.5365 | 4.8572 | 5.1766 | 5.4949 | 5.8124 |
| | h | 350.94 | 1197.6 | | 1212.0 | 1242.0 | 1270.1 | 1297.1 | 1323.3 | 1374.8 | 1425.9 | 1477.4 | 1529.4 | 1582.1 | 1635.7 | 1690.0 | 1745.1 |
| | s | 0.5384 | 1.5498 | | 1.5667 | 1.6006 | 1.6307 | 1.6581 | 1.6835 | 1.7299 | 1.7722 | 1.8115 | 1.8484 | 1.8834 | 1.9166 | 1.9484 | 1.9789 |
| **200** (381.80) | Sh | | | | 18.20 | 68.20 | 118.20 | 168.20 | 218.20 | 318.20 | 418.20 | 518.20 | 618.20 | 748.20 | 818.20 | 918.20 | 1018.20 |
| | v | 0.01839 | 2.2873 | | 2.3598 | 2.5480 | 2.7247 | 2.8939 | 3.0583 | 3.3783 | 3.6915 | 4.0008 | 4.3077 | 4.6128 | 4.9165 | 5.2191 | 5.5209 |
| | h | 355.51 | 1198.3 | | 1210.1 | 1240.6 | 1269.0 | 1296.2 | 1322.6 | 1374.3 | 1425.5 | 1477.0 | 1529.1 | 1581.9 | 1635.4 | 1689.8 | 1745.0 |
| | s | 0.5438 | 1.5454 | | 1.5593 | 1.5938 | 1.6242 | 1.6518 | 1.6773 | 1.7239 | 1.7663 | 1.8057 | 1.8426 | 1.8776 | 1.9109 | 1.9427 | 1.9732 |

**Table II.2-2  Continued**

| Abs. Press. (psi) (Sat. Temp.) | | Sat. Water | Sat. Steam | Temperature (°F) 400 | 450 | 500 | 550 | 600 | 700 | 800 | 900 | 1000 | 1100 | 1200 | 1300 | 1400 | 1500 |
|---|---|---|---|---|---|---|---|---|---|---|---|---|---|---|---|---|---|
| 210 (385.91) | Sh | | | 14.09 | 64.09 | 114.09 | 164.09 | 214.09 | 314.09 | 414.09 | 514.09 | 614.09 | 714.09 | 814.09 | 914.09 | 1014.09 | 1114.09 |
| | v | 0.01844 | 2.1822 | 2.2364 | 2.4181 | 2.5880 | 2.7504 | 2.9078 | 3.2137 | 3.5128 | 3.8080 | 4.1007 | 4.3915 | 4.6811 | 4.9695 | 5.2571 | 5.5440 |
| | h | 359.91 | 1199.0 | 1208.02 | 1239.2 | 1268.0 | 1295.3 | 1321.9 | 1373.7 | 1425.1 | 1476.7 | 1528.8 | 1581.6 | 1635.2 | 1689.6 | 1744.8 | 1800.8 |
| | s | 0.5490 | 1.5413 | 1.5522 | 1.5872 | 1.6180 | 1.6458 | 1.6715 | 1.7182 | 1.7607 | 1.8001 | 1.8371 | 1.8721 | 1.9054 | 1.9372 | 1.9677 | 1.9970 |
| 220 (389.88) | Sh | | | 10.12 | 60.12 | 110.12 | 160.12 | 210.12 | 310.12 | 410.12 | 510.12 | 610.12 | 710.12 | 810.12 | 910.12 | 1010.12 | 1110.12 |
| | v | 0.01850 | 2.0863 | 2.1240 | 2.2999 | 2.4638 | 2.6199 | 2.7710 | 3.0642 | 3.3504 | 3.6327 | 3.9125 | 4.1905 | 4.4671 | 4.7426 | 5.0173 | 5.2913 |
| | h | 364.17 | 1199.6 | 1206.3 | 1237.8 | 1266.9 | 1294.5 | 1321.2 | 1373.2 | 1424.7 | 1476.3 | 1528.5 | 1581.4 | 1635.0 | 1689.4 | 1744.7 | 1800.6 |
| | s | 0.5540 | 1.5374 | 1.5453 | 1.5808 | 1.6120 | 1.6400 | 1.6658 | 1.7128 | 1.7553 | 1.7948 | 1.8318 | 1.8668 | 1.9002 | 1.9320 | 1.9625 | 1.9919 |
| 230 (393.70) | Sh | | | 6.30 | 56.30 | 106.30 | 156.30 | 206.30 | 306.30 | 406.30 | 506.30 | 606.30 | 706.30 | 806.30 | 906.30 | 1006.30 | 1106.30 |
| | v | 0.01855 | 1.9985 | 2.0212 | 2.1919 | 2.3503 | 2.5008 | 2.6461 | 2.9276 | 3.2020 | 3.4726 | 3.7406 | 4.0068 | 4.2717 | 4.5355 | 4.7984 | 5.0606 |
| | h | 368.28 | 1200.1 | 1204.4 | 1236.3 | 1265.7 | 1293.6 | 1320.4 | 1372.7 | 1424.2 | 1476.0 | 1528.2 | 1581.1 | 1634.8 | 1689.3 | 1744.5 | 1800.5 |
| | s | 0.5588 | 1.5336 | 1.5385 | 1.5747 | 1.6062 | 1.6344 | 1.6604 | 1.7075 | 1.7502 | 1.7897 | 1.8268 | 1.8618 | 1.8952 | 1.9270 | 1.9576 | 1.9869 |
| 240 (397.39) | Sh | | | 2.61 | 52.61 | 102.61 | 152.61 | 202.61 | 302.61 | 402.61 | 502.61 | 602.61 | 702.61 | 802.61 | 902.61 | 1002.61 | 1102.61 |
| | v | 0.01860 | 1.9177 | 1.9268 | 2.0928 | 2.2462 | 2.3915 | 2.5316 | 2.8024 | 3.0661 | 3.3259 | 3.5831 | 3.8385 | 4.0926 | 4.3456 | 4.5977 | 4.8492 |
| | h | 372.27 | 1200.6 | 1202.4 | 1234.9 | 1264.6 | 1292.7 | 1319.7 | 1372.1 | 1423.8 | 1475.6 | 1527.9 | 1580.9 | 1634.6 | 1689.1 | 1744.3 | 1800.4 |
| | s | 0.5634 | 1.5299 | 1.5320 | 1.5687 | 1.6006 | 1.6291 | 1.6552 | 1.7025 | 1.7452 | 1.7848 | 1.8219 | 1.8570 | 1.8904 | 1.9223 | 1.9528 | 1.9822 |
| 250 (400.97) | Sh | | | | 49.03 | 99.03 | 149.03 | 199.03 | 299.03 | 399.03 | 499.03 | 599.03 | 699.03 | 799.03 | 899.03 | 999.03 | 1099.03 |
| | v | 0.01865 | 1.8432 | | 2.0016 | 2.1504 | 2.2909 | 2.4262 | 2.6872 | 2.9410 | 3.1909 | 3.4382 | 3.6837 | 3.9278 | 4.1709 | 4.4131 | 4.6546 |
| | h | 376.14 | 1201.1 | | 1233.4 | 1263.5 | 1291.8 | 1319.0 | 1371.6 | 1423.4 | 1475.3 | 1527.6 | 1580.6 | 1634.4 | 1688.9 | 1744.2 | 1800.2 |
| | s | 0.5679 | 1.5264 | | 1.5629 | 1.5951 | 1.6239 | 1.6502 | 1.6976 | 1.7405 | 1.7801 | 1.8173 | 1.8524 | 1.8858 | 1.9177 | 1.9482 | 1.9776 |
| 260 (404.44) | Sh | | | | 45.56 | 95.56 | 145.56 | 195.56 | 295.56 | 395.56 | 495.56 | 595.56 | 695.56 | 795.56 | 895.56 | 995.56 | 1095.56 |
| | v | 0.01870 | 1.7742 | | 1.9173 | 2.0619 | 2.1981 | 2.3289 | 2.5808 | 2.8256 | 3.0663 | 3.3044 | 3.5408 | 3.7758 | 4.0097 | 4.2427 | 4.4750 |
| | h | 379.90 | 1201.5 | | 1231.9 | 1262.4 | 1290.9 | 1318.2 | 1371.1 | 1423.0 | 1474.9 | 1527.3 | 1580.4 | 1634.2 | 1688.7 | 1744.0 | 1800.1 |
| | s | 0.5722 | 1.5230 | | 1.5573 | 1.5899 | 1.6189 | 1.6453 | 1.6930 | 1.7359 | 1.7756 | 1.8128 | 1.8480 | 1.8814 | 1.9133 | 1.9439 | 1.9732 |
| 270 (407.80) | Sh | | | | 42.20 | 92.20 | 142.20 | 192.20 | 292.20 | 392.20 | 492.20 | 592.20 | 692.20 | 792.20 | 892.20 | 992.20 | 1092.20 |
| | v | 0.01875 | 1.7101 | | 1.8391 | 1.9799 | 2.1121 | 2.2388 | 2.4824 | 2.7186 | 2.9509 | 3.1806 | 3.4084 | 3.6349 | 3.8603 | 4.0849 | 4.3087 |
| | h | 383.56 | 1201.9 | | 1230.4 | 1261.2 | 1290.0 | 1317.5 | 1370.5 | 1422.6 | 1474.6 | 1527.1 | 1580.1 | 1634.0 | 1688.5 | 1743.9 | 1800.0 |
| | s | 0.5764 | 1.5197 | | 1.5518 | 1.5848 | 1.6140 | 1.6406 | 1.6885 | 1.7315 | 1.7713 | 1.8085 | 1.8437 | 1.8771 | 1.9090 | 1.9396 | 1.9690 |
| 280 (411.07) | Sh | | | | 38.93 | 88.93 | 138.93 | 188.93 | 288.93 | 388.93 | 488.93 | 588.93 | 688.93 | 788.93 | 888.93 | 988.93 | 1088.93 |
| | v | 0.01880 | 1.6505 | | 1.7665 | 1.9037 | 2.0322 | 2.1551 | 2.3909 | 2.6194 | 2.8437 | 3.0655 | 3.2855 | 3.5042 | 3.7217 | 3.9384 | 4.1543 |
| | h | 387.12 | 1202.3 | | 1228.8 | 1260.0 | 1289.1 | 1316.8 | 1370.0 | 1422.1 | 1474.2 | 1526.8 | 1579.9 | 1633.8 | 1688.4 | 1743.7 | 1799.8 |
| | s | 0.5805 | 1.5166 | | 1.5464 | 1.5798 | 1.6093 | 1.6361 | 1.6841 | 1.7273 | 1.7671 | 1.8043 | 1.8395 | 1.8730 | 1.9050 | 1.9356 | 1.9649 |
| 290 (414.25) | Sh | | | | 35.75 | 85.75 | 135.75 | 185.75 | 285.75 | 385.75 | 485.75 | 585.75 | 685.75 | 785.75 | 885.75 | 985.75 | 1085.75 |
| | v | 0.01885 | 1.5948 | | 1.6988 | 1.8327 | 1.9578 | 2.0772 | 2.3058 | 2.5269 | 2.7440 | 2.9585 | 3.1711 | 3.3824 | 3.5926 | 3.8019 | 4.0106 |
| | h | 390.60 | 1202.6 | | 1227.3 | 1258.9 | 1288.1 | 1316.0 | 1369.5 | 1421.7 | 1473.9 | 1526.5 | 1579.6 | 1633.5 | 1688.2 | 1743.6 | 1799.7 |
| | s | 0.5844 | 1.5135 | | 1.5412 | 1.5750 | 1.6048 | 1.6317 | 1.6799 | 1.7232 | 1.7630 | 1.8003 | 1.8356 | 1.8690 | 1.9010 | 1.9316 | 1.9610 |
| 300 (417.35) | Sh | | | | 32.65 | 82.65 | 132.65 | 182.65 | 282.65 | 382.65 | 482.65 | 582.65 | 682.65 | 782.65 | 882.65 | 982.65 | 1082.65 |
| | v | 0.01889 | 1.5427 | | 1.6356 | 1.7665 | 1.8883 | 2.0044 | 2.2263 | 2.4407 | 2.6509 | 2.8585 | 3.0643 | 3.2688 | 3.4721 | 3.6746 | 3.8764 |
| | h | 393.99 | 1202.9 | | 1225.7 | 1257.7 | 1287.2 | 1315.2 | 1368.9 | 1421.3 | 1473.6 | 1526.2 | 1579.4 | 1633.3 | 1688.0 | 1743.4 | 1799.6 |
| | s | 0.5882 | 1.5105 | | 1.5361 | 1.5703 | 1.6003 | 1.6274 | 1.6758 | 1.7192 | 1.7591 | 1.7964 | 1.8317 | 1.8652 | 1.8972 | 1.9278 | 1.9572 |
| 310 (420.36) | Sh | | | | 29.64 | 79.64 | 129.64 | 179.64 | 279.64 | 379.64 | 479.64 | 579.64 | 679.64 | 779.64 | 879.64 | 979.64 | 1079.64 |
| | v | 0.01894 | 1.4939 | | 1.5763 | 1.7044 | 1.8233 | 1.9363 | 2.1520 | 2.3600 | 2.5638 | 2.7650 | 2.9644 | 3.1625 | 3.3594 | 3.5555 | 3.7509 |
| | h | 397.30 | 1203.2 | | 1224.1 | 1256.5 | 1286.3 | 1314.5 | 1368.4 | 1420.9 | 1473.2 | 1525.9 | 1579.2 | 1633.1 | 1687.8 | 1743.3 | 1799.4 |
| | s | 0.5920 | 1.5076 | | 1.5311 | 1.5657 | 1.5960 | 1.6233 | 1.6719 | 1.7153 | 1.7553 | 1.7927 | 1.8280 | 1.8615 | 1.8935 | 1.9241 | 1.9536 |
| 320 (423.31) | Sh | | | | 26.69 | 76.69 | 126.69 | 176.69 | 276.69 | 376.69 | 476.69 | 576.69 | 676.69 | 776.69 | 876.69 | 976.69 | 1076.69 |
| | v | 0.01899 | 1.4480 | | 1.5207 | 1.6462 | 1.7623 | 1.8725 | 2.0823 | 2.2843 | 2.4821 | 2.6774 | 2.8708 | 3.0628 | 3.2538 | 3.4438 | 3.6332 |
| | h | 400.53 | 1203.4 | | 1222.5 | 1255.2 | 1285.3 | 1313.7 | 1367.8 | 1420.5 | 1472.9 | 1525.6 | 1578.9 | 1632.9 | 1687.6 | 1743.1 | 1799.3 |
| | s | 0.5956 | 1.5048 | | 1.5261 | 1.5612 | 1.5918 | 1.6192 | 1.6680 | 1.7116 | 1.7516 | 1.7890 | 1.8243 | 1.8579 | 1.8899 | 1.9206 | 1.9500 |
| 330 (426.18) | Sh | | | | 23.82 | 73.82 | 123.82 | 173.82 | 273.82 | 373.82 | 473.82 | 573.82 | 673.82 | 773.82 | 873.82 | 973.82 | 1073.82 |
| | v | 0.01903 | 1.4048 | | 1.4684 | 1.5915 | 1.7050 | 1.8125 | 2.0168 | 2.2132 | 2.4054 | 2.5950 | 2.7828 | 2.9692 | 3.1545 | 3.3389 | 3.5227 |
| | h | 403.70 | 1203.6 | | 1220.9 | 1254.0 | 1284.4 | 1313.0 | 1367.3 | 1420.0 | 1472.5 | 1525.3 | 1578.7 | 1632.7 | 1687.5 | 1742.9 | 1799.2 |
| | s | 0.5991 | 1.5021 | | 1.5213 | 1.5568 | 1.5876 | 1.6153 | 1.6643 | 1.7079 | 1.7480 | 1.7855 | 1.8208 | 1.8544 | 1.8864 | 1.9171 | 1.9466 |
| 340 (428.99) | Sh | | | | 21.01 | 71.01 | 121.01 | 171.01 | 271.01 | 371.01 | 471.01 | 571.01 | 671.01 | 771.01 | 871.01 | 971.01 | 1071.01 |
| | v | 0.01908 | 1.3640 | | 1.4191 | 1.5399 | 1.6511 | 1.7561 | 1.9552 | 2.1463 | 2.3333 | 2.5175 | 2.7000 | 2.8811 | 3.0611 | 3.2402 | 3.4186 |
| | h | 406.80 | 1203.8 | | 1219.2 | 1252.8 | 1283.4 | 1312.2 | 1366.7 | 1419.6 | 1472.2 | 1525.0 | 1578.4 | 1632.5 | 1687.3 | 1742.8 | 1799.0 |
| | s | 0.6026 | 1.4994 | | 1.5165 | 1.5525 | 1.5836 | 1.6114 | 1.6606 | 1.7044 | 1.7445 | 1.7820 | 1.8174 | 1.8510 | 1.8831 | 1.9138 | 1.9432 |
| 350 (431.73) | Sh | | | | 18.27 | 68.27 | 118.27 | 168.27 | 268.27 | 368.27 | 468.27 | 568.27 | 668.27 | 768.27 | 868.27 | 968.27 | 1068.27 |
| | v | 0.01912 | 1.3255 | | 1.3725 | 1.4913 | 1.6002 | 1.7028 | 1.8970 | 2.0832 | 2.2652 | 2.4445 | 2.6219 | 2.7980 | 2.9730 | 3.1471 | 3.3205 |
| | h | 409.83 | 1204.0 | | 1217.5 | 1251.5 | 1282.4 | 1311.4 | 1366.2 | 1419.2 | 1471.8 | 1524.7 | 1578.2 | 1632.3 | 1687.1 | 1742.6 | 1798.9 |
| | s | 0.6059 | 1.4968 | | 1.5119 | 1.5483 | 1.5797 | 1.6077 | 1.6571 | 1.7009 | 1.7411 | 1.7787 | 1.8141 | 1.8477 | 1.8798 | 1.9105 | 1.9400 |
| 360 (434.41) | Sh | | | | 15.59 | 65.59 | 115.59 | 165.59 | 265.59 | 365.59 | 465.59 | 565.59 | 665.59 | 765.59 | 865.59 | 965.59 | 1065.59 |
| | v | 0.01917 | 1.2891 | | 1.3285 | 1.4454 | 1.5521 | 1.6525 | 1.8421 | 2.0237 | 2.2009 | 2.3755 | 2.5482 | 2.7196 | 2.8898 | 3.0592 | 3.2279 |
| | h | 412.81 | 1204.1 | | 1215.8 | 1250.3 | 1281.5 | 1310.6 | 1365.6 | 1418.7 | 1471.5 | 1524.4 | 1577.9 | 1632.1 | 1686.9 | 1742.5 | 1798.8 |
| | s | 0.6092 | 1.4943 | | 1.5073 | 1.5441 | 1.5758 | 1.6040 | 1.6536 | 1.6976 | 1.7379 | 1.7754 | 1.8109 | 1.8445 | 1.8766 | 1.9073 | 1.9368 |
| 380 (439.61) | Sh | | | | 10.39 | 60.39 | 110.39 | 160.39 | 260.39 | 360.39 | 460.39 | 560.39 | 660.39 | 760.39 | 860.39 | 960.39 | 1060.39 |
| | v | 0.01925 | 1.2218 | | 1.2472 | 1.3606 | 1.4635 | 1.5598 | 1.7410 | 1.9139 | 2.0825 | 2.2484 | 2.4124 | 2.5750 | 2.7366 | 2.8973 | 3.0572 |
| | h | 418.59 | 1204.4 | | 1212.4 | 1247.7 | 1279.5 | 1309.0 | 1364.5 | 1417.9 | 1470.8 | 1523.8 | 1577.4 | 1631.6 | 1686.5 | 1742.2 | 1798.5 |
| | s | 0.6156 | 1.4894 | | 1.4982 | 1.5360 | 1.5683 | 1.5969 | 1.6470 | 1.6911 | 1.7315 | 1.7692 | 1.8047 | 1.8384 | 1.8705 | 1.9012 | 1.9307 |

**Table II.2-2   Continued**

| Abs. Press. (psi) (Sat. Temp.) | | Sat. Water | Sat. Steam | Temperature (°F) 450 | 500 | 550 | 600 | 650 | 700 | 800 | 900 | 1000 | 1100 | 1200 | 1300 | 1400 | 1500 |
|---|---|---|---|---|---|---|---|---|---|---|---|---|---|---|---|---|---|
| **400** (444.60) | Sh | | | 5.40 | 55.40 | 105.40 | 155.40 | 205.40 | 255.40 | 355.40 | 455.40 | 555.40 | 655.40 | 755.40 | 855.40 | 955.40 | 1055.40 |
| | $v$ | 0.01934 | 1.1610 | 1.1738 | 1.2841 | 1.3836 | 1.4763 | 1.5646 | 1.6499 | 1.8151 | 1.9759 | 2.1339 | 2.2901 | 2.4450 | 2.5987 | 2.7515 | 2.9037 |
| | $h$ | 424.17 | 1204.6 | 1208.8 | 1245.1 | 1277.5 | 1307.4 | 1335.9 | 1363.4 | 1417.0 | 1470.1 | 1523.3 | 1576.9 | 1631.2 | 1686.2 | 1741.9 | 1798.2 |
| | $s$ | 0.6217 | 1.4847 | 1.4894 | 1.5282 | 1.5611 | 1.5901 | 1.6163 | 1.6406 | 1.6850 | 1.7255 | 1.7632 | 1.7988 | 1.8325 | 1.8647 | 1.8955 | 1.9250 |
| **420** (449.40) | Sh | | | .60 | 50.60 | 100.60 | 150.60 | 200.60 | 250.60 | 350.60 | 450.60 | 550.60 | 650.60 | 750.60 | 850.60 | 950.60 | 1050.60 |
| | $v$ | 0.01942 | 1.1057 | 1.1071 | 1.2148 | 1.3113 | 1.4007 | 1.4856 | 1.5676 | 1.7258 | 1.8795 | 2.0304 | 2.1795 | 2.3273 | 2.4739 | 2.6196 | 2.7647 |
| | $h$ | 429.56 | 1204.7 | 1205.2 | 1242.4 | 1275.4 | 1305.8 | 1334.5 | 1362.3 | 1416.2 | 1469.4 | 1522.7 | 1576.4 | 1630.8 | 1685.8 | 1741.6 | 1798.0 |
| | $s$ | 0.6276 | 1.4802 | 1.4808 | 1.5206 | 1.5542 | 1.5835 | 1.6100 | 1.6345 | 1.6791 | 1.7197 | 1.7575 | 1.7932 | 1.8269 | 1.8591 | 1.8899 | 1.9195 |
| **440** (454.03) | Sh | | | | 45.97 | 95.97 | 145.97 | 195.97 | 245.97 | 345.97 | 445.97 | 545.97 | 645.97 | 745.97 | 845.97 | 945.97 | 1045.97 |
| | $v$ | 0.01950 | 1.0554 | | 1.1517 | 1.2454 | 1.3319 | 1.4138 | 1.4926 | 1.6445 | 1.7918 | 1.9363 | 2.0790 | 2.2203 | 2.3605 | 2.4998 | 2.6384 |
| | $h$ | 434.77 | 1204.8 | | 1239.7 | 1273.4 | 1304.2 | 1333.2 | 1361.1 | 1415.3 | 1468.7 | 1522.1 | 1575.9 | 1630.4 | 1685.5 | 1741.2 | 1797.7 |
| | $s$ | 0.6332 | 1.4759 | | 1.5132 | 1.5474 | 1.5772 | 1.6040 | 1.6286 | 1.6734 | 1.7142 | 1.7521 | 1.7878 | 1.8216 | 1.8538 | 1.8847 | 1.9143 |
| **460** (458.50) | Sh | | | | 41.50 | 91.50 | 141.50 | 191.50 | 241.50 | 341.50 | 441.50 | 541.50 | 641.50 | 741.50 | 841.50 | 941.50 | 1041.50 |
| | $v$ | 0.01959 | 1.0092 | | 1.0939 | 1.1852 | 1.2691 | 1.3482 | 1.4242 | 1.5703 | 1.7117 | 1.8504 | 1.9872 | 2.1226 | 2.2569 | 2.3903 | 2.5230 |
| | $h$ | 439.83 | 1204.8 | | 1236.9 | 1271.3 | 1302.5 | 1331.8 | 1360.0 | 1414.4 | 1468.0 | 1521.5 | 1575.4 | 1629.9 | 1685.1 | 1740.9 | 1797.4 |
| | $s$ | 0.6387 | 1.4718 | | 1.5060 | 1.5409 | 1.5711 | 1.5982 | 1.6230 | 1.6680 | 1.7089 | 1.7469 | 1.7826 | 1.8165 | 1.8488 | 1.8797 | 1.9093 |
| **480** (462.82) | Sh | | | | 37.18 | 87.18 | 137.18 | 187.18 | 237.18 | 337.18 | 437.18 | 537.18 | 637.18 | 737.18 | 837.18 | 937.18 | 1037.18 |
| | $v$ | 0.01967 | 0.9668 | | 1.0409 | 1.1300 | 1.2115 | 1.2881 | 1.3615 | 1.5023 | 1.6384 | 1.7716 | 1.9030 | 2.0330 | 2.1619 | 2.2900 | 2.4173 |
| | $h$ | 444.75 | 1204.8 | | 1234.1 | 1269.1 | 1300.8 | 1330.5 | 1358.8 | 1413.6 | 1467.3 | 1520.9 | 1574.9 | 1629.5 | 1684.7 | 1740.6 | 1797.2 |
| | $s$ | 0.6439 | 1.4677 | | 1.4990 | 1.5346 | 1.5652 | 1.5925 | 1.6176 | 1.6628 | 1.7038 | 1.7419 | 1.7777 | 1.8116 | 1.8439 | 1.8748 | 1.9045 |
| **500** (467.01) | Sh | | | | 32.99 | 82.99 | 132.99 | 182.99 | 232.99 | 332.99 | 432.99 | 532.99 | 632.99 | 732.99 | 832.99 | 932.99 | 1032.99 |
| | $v$ | 0.01975 | 0.9276 | | 0.9919 | 1.0791 | 1.1584 | 1.2327 | 1.3037 | 1.4397 | 1.5708 | 1.6992 | 1.8256 | 1.9507 | 2.0746 | 2.1977 | 2.3200 |
| | $h$ | 449.52 | 1204.7 | | 1231.2 | 1267.0 | 1299.1 | 1329.1 | 1357.7 | 1412.7 | 1466.6 | 1520.3 | 1574.4 | 1629.1 | 1684.4 | 1740.3 | 1796.9 |
| | $s$ | 0.6490 | 1.4639 | | 1.4921 | 1.5284 | 1.5595 | 1.5871 | 1.6123 | 1.6578 | 1.6990 | 1.7371 | 1.7730 | 1.8069 | 1.8393 | 1.8702 | 1.8998 |
| **520** (471.07) | Sh | | | | 28.93 | 78.93 | 128.93 | 178.93 | 228.93 | 328.93 | 428.93 | 528.93 | 628.93 | 728.93 | 828.93 | 928.93 | 1028.93 |
| | $v$ | 0.01982 | 0.8914 | | 0.9466 | 1.0321 | 1.1094 | 1.1816 | 1.2504 | 1.3819 | 1.5085 | 1.6323 | 1.7542 | 1.8746 | 1.9940 | 2.1125 | 2.2302 |
| | $h$ | 454.18 | 1204.5 | | 1228.3 | 1264.8 | 1297.4 | 1327.7 | 1356.5 | 1411.8 | 1465.9 | 1519.7 | 1573.9 | 1628.7 | 1684.0 | 1740.0 | 1796.7 |
| | $s$ | 0.6540 | 1.4601 | | 1.4853 | 1.5223 | 1.5539 | 1.5818 | 1.6072 | 1.6530 | 1.6943 | 1.7325 | 1.7684 | 1.8024 | 1.8348 | 1.8657 | 1.8954 |
| **540** (475.01) | Sh | | | | 24.99 | 74.99 | 124.99 | 174.99 | 224.99 | 324.99 | 424.99 | 524.99 | 624.99 | 724.99 | 824.99 | 924.99 | 1024.99 |
| | $v$ | 0.01990 | 0.8577 | | 0.9045 | 0.9884 | 1.0640 | 1.1342 | 1.2010 | 1.3284 | 1.4508 | 1.5704 | 1.6880 | 1.8042 | 1.9193 | 2.0336 | 2.1471 |
| | $h$ | 458.71 | 1204.4 | | 1225.3 | 1262.5 | 1295.7 | 1326.3 | 1355.3 | 1410.9 | 1465.1 | 1519.1 | 1573.4 | 1628.2 | 1683.6 | 1739.7 | 1796.4 |
| | $s$ | 0.6587 | 1.4565 | | 1.4786 | 1.5164 | 1.5485 | 1.5767 | 1.6023 | 1.6483 | 1.6897 | 1.7280 | 1.7640 | 1.7981 | 1.8305 | 1.8615 | 1.8911 |
| **560** (478.84) | Sh | | | | 21.16 | 71.16 | 121.16 | 171.16 | 221.16 | 321.16 | 421.16 | 521.16 | 621.16 | 721.16 | 821.16 | 921.16 | 1021.16 |
| | $v$ | 0.01998 | 0.8264 | | 0.8653 | 0.9479 | 1.0217 | 1.0902 | 1.1552 | 1.2787 | 1.3972 | 1.5129 | 1.6266 | 1.7388 | 1.8500 | 1.9603 | 2.0699 |
| | $h$ | 463.14 | 1204.2 | | 1222.2 | 1260.3 | 1293.9 | 1324.9 | 1354.2 | 1410.0 | 1464.4 | 1518.6 | 1572.9 | 1627.8 | 1683.3 | 1739.4 | 1796.1 |
| | $s$ | 0.6634 | 1.4529 | | 1.4720 | 1.5106 | 1.5431 | 1.5717 | 1.5975 | 1.6438 | 1.6853 | 1.7237 | 1.7598 | 1.7939 | 1.8263 | 1.8573 | 1.8870 |
| **580** (482.57) | Sh | | | | 17.43 | 67.43 | 117.43 | 167.43 | 217.43 | 317.43 | 417.43 | 517.43 | 617.43 | 717.43 | 817.43 | 917.43 | 1017.43 |
| | $v$ | 0.02006 | 0.7971 | | 0.8287 | 0.9100 | 0.9824 | 1.0492 | 1.1125 | 1.2324 | 1.3473 | 1.4593 | 1.5693 | 1.6780 | 1.7855 | 1.8921 | 1.9980 |
| | $h$ | 467.47 | 1203.9 | | 1219.1 | 1258.0 | 1292.1 | 1323.4 | 1353.0 | 1409.2 | 1463.7 | 1518.0 | 1572.4 | 1627.4 | 1682.9 | 1739.1 | 1795.9 |
| | $s$ | 0.6679 | 1.4495 | | 1.4654 | 1.5049 | 1.5380 | 1.5668 | 1.5929 | 1.6394 | 1.6811 | 1.7196 | 1.7556 | 1.7898 | 1.8223 | 1.8533 | 1.8831 |
| **600** (486.20) | Sh | | | | 13.80 | 63.80 | 113.80 | 163.80 | 213.80 | 313.80 | 413.80 | 513.80 | 613.80 | 713.80 | 813.80 | 913.80 | 1013.80 |
| | $v$ | 0.02013 | 0.7697 | | 0.7944 | 0.8746 | 0.9456 | 1.0109 | 1.0726 | 1.1892 | 1.3008 | 1.4093 | 1.5160 | 1.6211 | 1.7252 | 1.8284 | 1.9309 |
| | $h$ | 471.70 | 1203.7 | | 1215.9 | 1255.6 | 1290.3 | 1322.0 | 1351.8 | 1408.3 | 1463.0 | 1517.4 | 1571.9 | 1627.0 | 1682.6 | 1738.8 | 1795.6 |
| | $s$ | 0.6723 | 1.4461 | | 1.4590 | 1.4993 | 1.5329 | 1.5621 | 1.5884 | 1.6351 | 1.6769 | 1.7155 | 1.7517 | 1.7859 | 1.8184 | 1.8494 | 1.8792 |
| **650** (494.89) | Sh | | | | 5.11 | 55.11 | 105.11 | 155.11 | 205.11 | 305.11 | 405.11 | 505.11 | 605.11 | 705.11 | 805.11 | 905.11 | 1005.11 |
| | $v$ | 0.02032 | 0.7084 | | 0.7173 | 0.7954 | 0.8634 | 0.9254 | 0.9835 | 1.0929 | 1.1969 | 1.2979 | 1.3969 | 1.4944 | 1.5909 | 1.6864 | 1.7813 |
| | $h$ | 481.89 | 1202.8 | | 1207.6 | 1249.6 | 1285.7 | 1318.3 | 1348.7 | 1406.0 | 1461.2 | 1515.9 | 1570.7 | 1625.9 | 1681.6 | 1738.0 | 1794.9 |
| | $s$ | 1.6828 | 1.4381 | | 1.4430 | 1.4858 | 1.5207 | 1.5507 | 1.5775 | 1.6249 | 1.6671 | 1.7059 | 1.7422 | 1.7765 | 1.8092 | 1.8403 | 1.8701 |
| **700** (503.08) | Sh | | | | | 46.92 | 96.92 | 146.92 | 196.92 | 296.92 | 396.92 | 496.92 | 596.92 | 696.92 | 796.92 | 896.92 | 996.92 |
| | $v$ | 0.02050 | 0.6556 | | | 0.7271 | 0.7928 | 0.8520 | 0.9072 | 1.0102 | 1.1078 | 1.2023 | 1.2948 | 1.3858 | 1.4757 | 1.5647 | 1.6530 |
| | $h$ | 491.60 | 1201.8 | | | 1243.4 | 1281.0 | 1314.6 | 1345.6 | 1403.7 | 1459.4 | 1514.4 | 1569.4 | 1624.8 | 1680.7 | 1737.2 | 1794.3 |
| | $s$ | 1.6928 | 1.4304 | | | 1.4726 | 1.5090 | 1.5399 | 1.5673 | 1.6154 | 1.6580 | 1.6970 | 1.7335 | 1.7679 | 1.8006 | 1.8318 | 1.8617 |
| **750** (510.84) | Sh | | | | | 39.16 | 89.16 | 139.16 | 189.16 | 289.16 | 389.16 | 489.16 | 589.16 | 689.16 | 789.16 | 889.16 | 989.16 |
| | $v$ | 0.02069 | 0.6095 | | | 0.6676 | 0.7313 | 0.7882 | 0.8409 | 0.9386 | 1.0306 | 1.1195 | 1.2063 | 1.2916 | 1.3759 | 1.4592 | 1.5419 |
| | $h$ | 500.89 | 1200.7 | | | 1236.9 | 1276.1 | 1310.7 | 1342.5 | 1401.5 | 1457.6 | 1512.9 | 1568.2 | 1623.8 | 1679.8 | 1736.4 | 1793.6 |
| | $s$ | 0.7022 | 1.4232 | | | 1.4598 | 1.4977 | 1.5296 | 1.5577 | 1.6065 | 1.6494 | 1.6886 | 1.7252 | 1.7598 | 1.7926 | 1.8239 | 1.8538 |
| **800** (518.21) | Sh | | | | | 31.79 | 81.79 | 131.79 | 181.79 | 281.79 | 381.79 | 481.79 | 581.79 | 681.79 | 781.79 | 881.79 | 981.79 |
| | $v$ | 0.02087 | 0.5690 | | | 0.6151 | 0.6774 | 0.7323 | 0.7828 | 0.8759 | 0.9631 | 1.0470 | 1.1289 | 1.2093 | 1.2885 | 1.3669 | 1.4446 |
| | $h$ | 509.81 | 1199.4 | | | 1230.1 | 1271.1 | 1306.8 | 1339.3 | 1399.1 | 1455.8 | 1511.4 | 1566.9 | 1622.7 | 1678.9 | 1735.7 | 1792.9 |
| | $s$ | 0.7111 | 1.4163 | | | 1.4472 | 1.4869 | 1.5198 | 1.5484 | 1.5980 | 1.6413 | 1.6807 | 1.7175 | 1.7522 | 1.7851 | 1.8164 | 1.8464 |
| **850** (525.24) | Sh | | | | | 24.76 | 74.76 | 124.76 | 174.76 | 274.76 | 374.76 | 474.76 | 574.76 | 674.76 | 774.76 | 874.76 | 974.76 |
| | $v$ | 0.02105 | 0.5330 | | | 0.5683 | 0.6296 | 0.6829 | 0.7315 | 0.8205 | 0.9034 | 0.9830 | 1.0606 | 1.1366 | 1.2115 | 1.2855 | 1.3588 |
| | $h$ | 518.40 | 1198.0 | | | 1223.0 | 1265.9 | 1302.8 | 1336.0 | 1396.8 | 1454.0 | 1510.0 | 1565.7 | 1621.6 | 1678.0 | 1734.9 | 1792.3 |
| | $s$ | 0.7197 | 1.4096 | | | 1.4347 | 1.4763 | 1.5102 | 1.5396 | 1.5899 | 1.6336 | 1.6733 | 1.7102 | 1.7450 | 1.7780 | 1.8094 | 1.8395 |
| **900** (531.95) | Sh | | | | | 18.05 | 68.05 | 118.05 | 168.05 | 268.05 | 368.05 | 468.05 | 568.05 | 668.05 | 768.05 | 868.05 | 968.05 |
| | $v$ | 0.02123 | 0.5009 | | | 0.5263 | 0.5869 | 0.6388 | 0.6858 | 0.7713 | 0.8504 | 0.9262 | 0.9998 | 1.0720 | 1.1430 | 1.2131 | 1.2825 |
| | $h$ | 526.70 | 1196.4 | | | 1215.5 | 1260.6 | 1298.6 | 1332.7 | 1394.4 | 1452.2 | 1508.5 | 1564.4 | 1620.6 | 1677.1 | 1734.1 | 1791.6 |
| | $s$ | 0.7279 | 1.4032 | | | 1.4223 | 1.4659 | 1.5010 | 1.5311 | 1.5822 | 1.6263 | 1.6662 | 1.7033 | 1.7382 | 1.7713 | 1.8028 | 1.8329 |

**Table II.2-2　Continued**

| Abs. Press. (psi) (Sat. Temp.) | | Sat. Water | Sat. Steam | 550 | 600 | 650 | 700 | 750 | 800 | 850 | 900 | 1000 | 1100 | 1200 | 1300 | 1400 | 1500 |
|---|---|---|---|---|---|---|---|---|---|---|---|---|---|---|---|---|---|
| **950** (538.39) | Sh | | | 11.61 | 61.61 | 111.61 | 161.61 | 211.61 | 261.61 | 311.61 | 361.61 | 461.61 | 561.61 | 661.61 | 761.61 | 861.61 | 961.61 |
| | v | 0.02141 | 0.4721 | 0.4883 | 0.5485 | 0.5993 | 0.6449 | 0.6871 | 0.7272 | 0.7656 | 0.8030 | 0.8753 | 0.9455 | 1.0142 | 1.0817 | 1.1484 | 1.2143 |
| | h | 534.74 | 1194.7 | 1207.6 | 1255.1 | 1294.4 | 1329.3 | 1361.5 | 1392.0 | 1421.5 | 1450.3 | 1507.0 | 1563.2 | 1619.5 | 1676.2 | 1733.3 | 1791.0 |
| | s | 0.7358 | 1.3970 | 1.4098 | 1.4557 | 1.4921 | 1.5228 | 1.5500 | 1.5748 | 1.5977 | 1.6193 | 1.6595 | 1.6967 | 1.7317 | 1.7649 | 1.7965 | 1.8267 |
| **1000** (544.58) | Sh | | | 5.42 | 55.42 | 105.42 | 155.42 | 205.42 | 255.42 | 305.42 | 355.42 | 455.42 | 555.42 | 655.42 | 755.42 | 855.42 | 955.42 |
| | v | 0.02159 | 0.4460 | 0.4535 | 0.5137 | 0.5636 | 0.6080 | 0.6489 | 0.6875 | 0.7245 | 0.7603 | 0.8295 | 0.8966 | 0.9622 | 1.0266 | 1.0901 | 1.1529 |
| | h | 542.55 | 1192.9 | 1199.3 | 1249.3 | 1290.1 | 1325.9 | 1358.7 | 1389.6 | 1419.4 | 1448.5 | 1505.4 | 1561.9 | 1618.4 | 1675.3 | 1732.5 | 1790.3 |
| | s | 0.7434 | 1.3910 | 1.3973 | 1.4457 | 1.4833 | 1.5149 | 1.5426 | 1.5677 | 1.5908 | 1.6126 | 1.6530 | 1.6905 | 1.7256 | 1.7589 | 1.7905 | 1.8207 |
| **1050** (550.53) | Sh | | | | 49.47 | 99.47 | 149.47 | 199.47 | 249.47 | 299.47 | 349.47 | 449.47 | 549.47 | 649.47 | 749.47 | 849.47 | 949.47 |
| | v | 0.02177 | 0.4222 | | 0.4821 | 0.5312 | 0.5745 | 0.6142 | 0.6515 | 0.6872 | 0.7216 | 0.7881 | 0.8524 | 0.9151 | 0.9767 | 1.0373 | 1.0973 |
| | h | 550.15 | 1191.0 | | 1243.4 | 1285.7 | 1322.4 | 1355.8 | 1387.2 | 1417.3 | 1446.6 | 1503.9 | 1560.7 | 1617.4 | 1674.4 | 1731.8 | 1789.6 |
| | s | 0.7507 | 1.3851 | | 1.4358 | 1.4748 | 1.5072 | 1.5354 | 1.5608 | 1.5842 | 1.6062 | 1.6469 | 1.6845 | 1.7197 | 1.7531 | 1.7848 | 1.8151 |
| **1100** (556.28) | Sh | | | | 43.72 | 93.72 | 143.72 | 193.72 | 243.72 | 293.72 | 343.72 | 443.72 | 543.72 | 643.72 | 743.72 | 843.72 | 943.72 |
| | v | 0.02195 | 0.4006 | | 0.4531 | 0.5017 | 0.5440 | 0.5826 | 0.6188 | 0.6533 | 0.6865 | 0.7505 | 0.8121 | 0.8723 | 0.9313 | 0.9894 | 1.0468 |
| | h | 557.55 | 1189.1 | | 1237.3 | 1281.2 | 1318.8 | 1352.9 | 1384.7 | 1415.2 | 1444.7 | 1502.4 | 1559.4 | 1616.3 | 1673.5 | 1731.0 | 1789.0 |
| | s | 0.7578 | 1.3794 | | 1.4259 | 1.4664 | 1.4996 | 1.5284 | 1.5542 | 1.5779 | 1.6000 | 1.6410 | 1.6787 | 1.7141 | 1.7475 | 1.7793 | 1.8097 |
| **1150** (561.82) | Sh | | | | 39.18 | 89.18 | 139.18 | 189.18 | 239.18 | 289.18 | 339.18 | 439.18 | 539.18 | 639.18 | 739.18 | 839.18 | 939.18 |
| | v | 0.02214 | 0.3807 | | 0.4263 | 0.4746 | 0.5162 | 0.5538 | 0.5889 | 0.6223 | 0.6544 | 0.7161 | 0.7754 | 0.8332 | 0.8899 | 0.9456 | 1.0007 |
| | h | 564.78 | 1187.0 | | 1230.9 | 1276.6 | 1315.2 | 1349.9 | 1382.2 | 1413.0 | 1442.8 | 1500.9 | 1558.1 | 1615.2 | 1672.6 | 1730.2 | 1788.3 |
| | s | 0.7647 | 1.3738 | | 1.4160 | 1.4582 | 1.4923 | 1.5216 | 1.5478 | 1.5717 | 1.5941 | 1.6353 | 1.6732 | 1.7087 | 1.7422 | 1.7741 | 1.8045 |
| **1200** (567.19) | Sh | | | | 32.81 | 82.81 | 132.81 | 182.81 | 232.81 | 282.81 | 332.81 | 432.81 | 532.81 | 632.81 | 732.81 | 832.81 | 932.81 |
| | v | 0.02232 | 0.3624 | | 0.4016 | 0.4497 | 0.4905 | 0.5273 | 0.5615 | 0.5939 | 0.6250 | 0.6845 | 0.7418 | 0.7974 | 0.8519 | 0.9055 | 0.9584 |
| | h | 571.85 | 1184.8 | | 1224.2 | 1271.8 | 1311.5 | 1346.9 | 1379.7 | 1410.8 | 1440.9 | 1499.4 | 1556.9 | 1614.2 | 1671.6 | 1729.4 | 1787.6 |
| | s | 0.7714 | 1.3683 | | 1.4061 | 1.4501 | 1.4851 | 1.5150 | 1.5415 | 1.5658 | 1.5883 | 1.6298 | 1.6679 | 1.7035 | 1.7371 | 1.7691 | 1.7996 |
| **1300** (577.42) | Sh | | | | 22.58 | 72.58 | 122.58 | 172.58 | 222.58 | 272.58 | 322.58 | 422.58 | 522.58 | 622.58 | 722.58 | 822.58 | 922.58 |
| | v | 0.02269 | 0.3299 | | 0.3570 | 0.4052 | 0.4451 | 0.4804 | 0.5129 | 0.5436 | 0.5729 | 0.6287 | 0.6822 | 0.7341 | 0.7847 | 0.8345 | 0.8836 |
| | h | 585.58 | 1180.2 | | 1209.9 | 1261.9 | 1303.9 | 1340.8 | 1374.6 | 1406.4 | 1437.1 | 1496.3 | 1554.3 | 1612.0 | 1669.8 | 1727.9 | 1786.3 |
| | s | 0.7843 | 1.3577 | | 1.3860 | 1.4340 | 1.4711 | 1.5022 | 1.5296 | 1.5544 | 1.5773 | 1.6194 | 1.6578 | 1.6937 | 1.7275 | 1.7596 | 1.7902 |
| **1400** (587.07) | Sh | | | | 12.93 | 62.93 | 112.93 | 162.93 | 212.93 | 262.93 | 312.93 | 412.93 | 512.93 | 612.93 | 712.93 | 812.93 | 912.93 |
| | v | 0.02307 | 0.3018 | | 0.3176 | 0.3667 | 0.4059 | 0.4400 | 0.4712 | 0.5004 | 0.5282 | 0.5809 | 0.6311 | 0.6798 | 0.7272 | 0.7737 | 0.8195 |
| | h | 598.83 | 1175.3 | | 1194.1 | 1251.4 | 1296.1 | 1334.5 | 1369.3 | 1402.0 | 1433.2 | 1493.2 | 1551.8 | 1609.9 | 1668.0 | 1726.3 | 1785.0 |
| | s | 0.7966 | 1.3474 | | 1.3652 | 1.4181 | 1.4575 | 1.4900 | 1.5182 | 1.5436 | 1.5670 | 1.6096 | 1.6484 | 1.6845 | 1.7185 | 1.7508 | 1.7815 |
| **1500** (596.20) | Sh | | | | 3.80 | 53.80 | 103.80 | 153.80 | 203.80 | 253.80 | 303.80 | 403.80 | 503.80 | 603.80 | 703.80 | 803.80 | 903.80 |
| | v | 0.02346 | 0.2772 | | 0.2820 | 0.3328 | 0.3717 | 0.4049 | 0.4350 | 0.4629 | 0.4894 | 0.5394 | 0.5869 | 0.6327 | 0.6773 | 0.7210 | 0.7639 |
| | h | 611.68 | 1170.1 | | 1176.3 | 1240.2 | 1287.9 | 1328.0 | 1364.0 | 1397.4 | 1429.2 | 1490.1 | 1549.2 | 1607.7 | 1666.2 | 1724.8 | 1783.7 |
| | s | 0.8085 | 1.3373 | | 1.3431 | 1.4022 | 1.4443 | 1.4782 | 1.5073 | 1.5333 | 1.5572 | 1.6004 | 1.6395 | 1.6759 | 1.7101 | 1.7425 | 1.7734 |
| **1600** (604.87) | Sh | | | | | 45.13 | 95.13 | 145.13 | 195.13 | 245.13 | 295.13 | 395.13 | 495.13 | 595.13 | 695.13 | 795.13 | 895.13 |
| | v | 0.02387 | 0.2555 | | | 0.3026 | 0.3415 | 0.3741 | 0.4032 | 0.4301 | 0.4555 | 0.5031 | 0.5482 | 0.5915 | 0.6336 | 0.6748 | 0.7153 |
| | h | 624.20 | 1164.5 | | | 1228.3 | 1279.4 | 1321.4 | 1358.5 | 1392.8 | 1425.2 | 1486.9 | 1546.6 | 1605.6 | 1664.3 | 1723.2 | 1782.3 |
| | s | 0.8199 | 1.3274 | | | 1.3861 | 1.4312 | 1.4667 | 1.4968 | 1.5235 | 1.5478 | 1.5916 | 1.6312 | 1.6678 | 1.7022 | 1.7347 | 1.7657 |
| **1700** (613.13) | Sh | | | | | 36.87 | 86.87 | 136.87 | 186.87 | 236.87 | 286.87 | 386.87 | 486.87 | 586.87 | 686.87 | 786.87 | 886.87 |
| | v | 0.02428 | 0.2361 | | | 0.2754 | 0.3147 | 0.3468 | 0.3751 | 0.4011 | 0.4255 | 0.4711 | 0.5140 | 0.5552 | 0.5951 | 0.6341 | 0.6724 |
| | h | 636.45 | 1158.6 | | | 1215.3 | 1270.5 | 1314.5 | 1352.9 | 1388.1 | 1421.2 | 1483.8 | 1544.0 | 1603.4 | 1662.5 | 1721.7 | 1781.0 |
| | s | 0.8309 | 1.3176 | | | 1.3697 | 1.4183 | 1.4555 | 1.4867 | 1.5140 | 1.5388 | 1.5833 | 1.6232 | 1.6601 | 1.6947 | 1.7274 | 1.7585 |
| **1800** (621.02) | Sh | | | | | 28.98 | 78.98 | 128.98 | 178.98 | 228.98 | 278.98 | 378.98 | 478.98 | 578.98 | 678.98 | 778.98 | 878.98 |
| | v | 0.02472 | 0.2186 | | | 0.2505 | 0.2906 | 0.3223 | 0.3500 | 0.3752 | 0.3988 | 0.4426 | 0.4836 | 0.5229 | 0.5609 | 0.5980 | 0.6343 |
| | h | 648.49 | 1152.3 | | | 1201.2 | 1261.1 | 1307.4 | 1347.2 | 1383.3 | 1417.1 | 1480.6 | 1541.4 | 1601.2 | 1660.7 | 1720.1 | 1779.7 |
| | s | 0.8417 | 1.3079 | | | 1.3526 | 1.4054 | 1.4446 | 1.4768 | 1.5049 | 1.5302 | 1.5753 | 1.6156 | 1.6528 | 1.6876 | 1.7204 | 1.7516 |
| **1900** (628.56) | Sh | | | | | 21.44 | 71.44 | 121.44 | 171.44 | 221.44 | 271.44 | 371.44 | 471.44 | 571.44 | 671.44 | 771.44 | 871.44 |
| | v | 0.02517 | 0.2028 | | | 0.2274 | 0.2687 | 0.3004 | 0.3275 | 0.3521 | 0.3749 | 0.4171 | 0.4565 | 0.4940 | 0.5303 | 0.5656 | 0.6002 |
| | h | 660.36 | 1145.6 | | | 1185.7 | 1251.3 | 1300.2 | 1341.4 | 1378.4 | 1412.9 | 1477.4 | 1538.8 | 1599.1 | 1658.8 | 1718.6 | 1778.4 |
| | s | 0.8522 | 1.2981 | | | 1.3346 | 1.3925 | 1.4338 | 1.4672 | 1.4960 | 1.5219 | 1.5677 | 1.6084 | 1.6458 | 1.6808 | 1.7138 | 1.7451 |
| **2000** (635.80) | Sh | | | | | 14.20 | 64.20 | 114.20 | 164.20 | 214.20 | 264.20 | 364.20 | 464.20 | 564.20 | 664.20 | 764.20 | 864.20 |
| | v | 0.02565 | 0.1883 | | | 0.2056 | 0.2488 | 0.2805 | 0.3072 | 0.3312 | 0.3534 | 0.3942 | 0.4320 | 0.4680 | 0.5027 | 0.5365 | 0.5695 |
| | h | 672.11 | 1138.3 | | | 1168.3 | 1240.9 | 1292.6 | 1335.4 | 1373.5 | 1408.7 | 1474.1 | 1536.2 | 1596.9 | 1657.0 | 1717.0 | 1771.1 |
| | s | 0.8625 | 1.2881 | | | 1.3154 | 1.3794 | 1.4231 | 1.4578 | 1.4874 | 1.5138 | 1.5603 | 1.6014 | 1.6391 | 1.6743 | 1.7075 | 1.7389 |
| **2100** (642.76) | Sh | | | | | 7.24 | 57.24 | 107.24 | 157.24 | 207.24 | 257.24 | 357.24 | 457.24 | 557.24 | 657.24 | 757.24 | 857.24 |
| | v | 0.02615 | 0.1750 | | | 0.1847 | 0.2304 | 0.2624 | 0.2888 | 0.3123 | 0.3339 | 0.3734 | 0.4099 | 0.4445 | 0.4778 | 0.5101 | 0.5418 |
| | h | 683.79 | 1130.5 | | | 1148.5 | 1229.8 | 1284.9 | 1329.3 | 1368.4 | 1404.4 | 1470.9 | 1533.6 | 1594.7 | 1655.2 | 1715.4 | 1775.7 |
| | s | 0.8727 | 1.2780 | | | 1.2942 | 1.3661 | 1.4125 | 1.4486 | 1.4790 | 1.5060 | 1.5532 | 1.5948 | 1.6327 | 1.6681 | 1.7014 | 1.7330 |
| **2200** (649.45) | Sh | | | | | .55 | 50.55 | 100.55 | 150.55 | 200.55 | 250.55 | 350.55 | 450.55 | 550.55 | 650.55 | 750.55 | 850.55 |
| | v | 0.02669 | 0.1627 | | | 0.1636 | 0.2134 | 0.2458 | 0.2720 | 0.2950 | 0.3161 | 0.3545 | 0.3897 | 0.4231 | 0.4551 | 0.4862 | 0.5165 |
| | h | 695.46 | 1122.2 | | | 1123.9 | 1218.0 | 1276.8 | 1323.1 | 1363.3 | 1400.0 | 1467.6 | 1530.9 | 1592.5 | 1653.3 | 1713.9 | 1774.4 |
| | s | 0.8828 | 1.2676 | | | 1.2691 | 1.3523 | 1.4020 | 1.4395 | 1.4708 | 1.4984 | 1.5463 | 1.5883 | 1.6266 | 1.6622 | 1.6956 | 1.7273 |
| **2300** (655.89) | Sh | | | | | | 44.11 | 94.11 | 144.11 | 194.11 | 244.11 | 344.11 | 444.11 | 544.11 | 644.11 | 744.11 | 844.11 |
| | v | 0.02727 | 0.1513 | | | | 0.1975 | 0.2305 | 0.2566 | 0.2793 | 0.2999 | 0.3372 | 0.3714 | 0.4035 | 0.4344 | 0.4643 | 0.4935 |
| | h | 707.18 | 1113.2 | | | | 1205.3 | 1268.4 | 1316.7 | 1358.1 | 1395.7 | 1464.2 | 1528.3 | 1590.3 | 1651.5 | 1712.3 | 1773.1 |
| | s | 0.8929 | 1.2569 | | | | 1.3381 | 1.3914 | 1.4305 | 1.4628 | 1.4910 | 1.5397 | 1.5821 | 1.6207 | 1.6565 | 1.6901 | 1.7219 |

**Table II.2-2  Continued**

| Abs. Press. (psi) (Sat. Temp.) | | Sat. Water | Sat. Steam | 700 | 750 | 800 | 850 | 900 | 950 | 1000 | 1050 | 1100 | 1150 | 1200 | 1300 | 1400 | 1500 |
|---|---|---|---|---|---|---|---|---|---|---|---|---|---|---|---|---|---|
| **2400** (662.11) | Sh | | | 37.89 | 87.89 | 137.89 | 187.89 | 237.89 | 287.89 | 337.89 | 387.89 | 437.89 | 487.89 | 537.89 | 637.89 | 737.89 | 837.89 |
| | v | 0.02790 | 0.1408 | 0.1824 | 0.2164 | 0.2424 | 0.2648 | 0.2850 | 0.3037 | 0.3214 | 0.3382 | 0.3545 | 0.3703 | 0.3856 | 0.4155 | 0.4443 | 0.4724 |
| | h | 718.95 | 1103.7 | 1191.6 | 1259.7 | 1310.1 | 1352.8 | 1391.2 | 1426.9 | 1460.9 | 1493.7 | 1525.6 | 1557.0 | 1588.1 | 1649.6 | 1710.8 | 1771.8 |
| | s | 0.9031 | 1.2460 | 1.3232 | 1.3808 | 1.4217 | 1.4549 | 1.4837 | 1.5095 | 1.5332 | 1.5553 | 1.5761 | 1.5959 | 1.6149 | 1.6509 | 1.6847 | 1.7167 |
| **2500** (668.11) | Sh | | | 31.89 | 81.89 | 131.89 | 181.89 | 231.89 | 281.89 | 331.89 | 381.89 | 431.89 | 481.89 | 531.89 | 631.89 | 731.89 | 831.89 |
| | v | 0.02859 | 0.1307 | 0.1681 | 0.2032 | 0.2293 | 0.2514 | 0.2712 | 0.2896 | 0.3068 | 0.3232 | 0.3390 | 0.3543 | 0.3692 | 0.3980 | 0.4259 | 0.4529 |
| | h | 731.71 | 1093.3 | 1176.7 | 1250.6 | 1303.4 | 1347.4 | 1386.7 | 1423.1 | 1457.5 | 1490.7 | 1522.9 | 1554.6 | 1585.9 | 1647.8 | 1709.2 | 1770.4 |
| | s | 0.9139 | 1.2345 | 1.3076 | 1.3701 | 1.4129 | 1.4472 | 1.4766 | 1.5029 | 1.5269 | 1.5492 | 1.5703 | 1.5903 | 1.6094 | 1.6456 | 1.6796 | 1.7116 |
| **2600** (673.91) | Sh | | | 26.09 | 76.09 | 126.09 | 176.09 | 226.09 | 276.09 | 326.09 | 376.09 | 426.09 | 476.09 | 526.09 | 626.09 | 726.09 | 826.09 |
| | v | 0.02938 | 0.1211 | 0.1544 | 0.1909 | 0.2171 | 0.2390 | 0.2585 | 0.2765 | 0.2933 | 0.3093 | 0.3247 | 0.3395 | 0.3540 | 0.3819 | 0.4088 | 0.4350 |
| | h | 744.47 | 1082.0 | 1160.2 | 1241.1 | 1296.5 | 1341.9 | 1382.1 | 1419.2 | 1454.1 | 1487.7 | 1520.2 | 1552.2 | 1583.7 | 1646.0 | 1707.7 | 1769.1 |
| | s | 0.9247 | 1.2225 | 1.2908 | 1.3592 | 1.4042 | 1.4395 | 1.4696 | 1.4964 | 1.5208 | 1.5434 | 1.5646 | 1.5848 | 1.6040 | 1.6405 | 1.6746 | 1.7068 |
| **2700** (679.53) | Sh | · | | 20.47 | 70.47 | 120.47 | 170.47 | 220.47 | 270.47 | 320.47 | 370.47 | 420.47 | 470.47 | 520.47 | 620.47 | 720.47 | 820.47 |
| | v | 0.03029 | 0.1119 | 0.1411 | 0.1794 | 0.2058 | 0.2275 | 0.2468 | 0.2644 | 0.2809 | 0.2965 | 0.3114 | 0.3259 | 0.3399 | 0.3670 | 0.3931 | 0.4184 |
| | h | 757.34 | 1069.7 | 1142.0 | 1231.1 | 1289.5 | 1336.3 | 1377.5 | 1415.2 | 1450.7 | 1484.6 | 1517.5 | 1549.8 | 1581.5 | 1644.1 | 1706.1 | 1767.8 |
| | s | 0.9356 | 1.2097 | 1.2727 | 1.3481 | 1.3954 | 1.4319 | 1.4628 | 1.4900 | 1.5148 | 1.5376 | 1.5591 | 1.5794 | 1.5988 | 1.6355 | 1.6697 | 1.7021 |
| **2800** (684.96) | Sh | | | 15.04 | 65.04 | 115.04 | 165.04 | 215.04 | 265.04 | 315.04 | 365.04 | 415.04 | 465.04 | 515.04 | 615.04 | 715.04 | 815.04 |
| | v | 0.03134 | 0.1030 | 0.1278 | 0.1685 | 0.1952 | 0.2168 | 0.2358 | 0.2531 | 0.2693 | 0.2845 | 0.2991 | 0.3132 | 0.3268 | 0.3532 | 0.3785 | 0.4030 |
| | h | 770.69 | 1055.8 | 1121.2 | 1220.6 | 1282.2 | 1330.7 | 1372.8 | 1411.2 | 1447.2 | 1481.6 | 1514.8 | 1547.3 | 1579.3 | 1642.2 | 1704.5 | 1766.5 |
| | s | 0.9468 | 1.1958 | 1.2527 | 1.3368 | 1.3867 | 1.4245 | 1.4561 | 1.4838 | 1.5089 | 1.5321 | 1.5537 | 1.5742 | 1.5938 | 1.6306 | 1.6651 | 1.6975 |
| **2900** (690.22) | Sh | | | 9.78 | 59.78 | 109.78 | 159.78 | 209.78 | 259.78 | 309.78 | 359.78 | 409.78 | 459.78 | 509.78 | 609.78 | 709.78 | 809.78 |
| | v | 0.03262 | 0.0942 | 0.1138 | 0.1581 | 0.1853 | 0.2068 | 0.2256 | 0.2427 | 0.2585 | 0.2734 | 0.2877 | 0.3014 | 0.3147 | 0.3403 | 0.3649 | 0.3887 |
| | h | 785.13 | 1039.8 | 1095.3 | 1209.6 | 1274.7 | 1324.9 | 1368.0 | 1407.2 | 1443.7 | 1478.5 | 1512.1 | 1544.9 | 1577.0 | 1640.4 | 1703.0 | 1765.2 |
| | s | 0.9588 | 1.1803 | 1.2283 | 1.3251 | 1.3780 | 1.4171 | 1.4494 | 1.4777 | 1.5032 | 1.5266 | 1.5485 | 1.5692 | 1.5889 | 1.6259 | 1.6605 | 1.6931 |
| **3000** (695.33) | Sh | | | 4.67 | 54.67 | 104.67 | 154.67 | 204.67 | 254.67 | 304.67 | 354.67 | 404.67 | 454.67 | 504.67 | 604.67 | 704.67 | 804.67 |
| | v | 0.03428 | 0.0850 | 0.0982 | 0.1483 | 0.1759 | 0.1975 | 0.2161 | 0.2329 | 0.2484 | 0.2630 | 0.2770 | 0.2904 | 0.3033 | 0.3282 | 0.3522 | 0.3753 |
| | h | 801.84 | 1020.3 | 1060.5 | 1197.9 | 1267.0 | 1319.0 | 1363.2 | 1403.1 | 1440.2 | 1475.4 | 1509.4 | 1542.4 | 1574.8 | 1638.5 | 1701.4 | 1763.8 |
| | s | 0.9728 | 1.1619 | 1.1966 | 1.3131 | 1.3692 | 1.4097 | 1.4429 | 1.4717 | 1.4976 | 1.5213 | 1.5434 | 1.5642 | 1.5841 | 1.6214 | 1.6561 | 1.6888 |
| **3100** (700.28) | Sh | | | | 49.72 | 99.72 | 149.72 | 199.72 | 249.72 | 299.72 | 349.72 | 399.72 | 449.72 | 499.72 | 599.72 | 699.72 | 799.72 |
| | v | 0.03681 | 0.0745 | | 0.1389 | 0.1671 | 0.1887 | 0.2071 | 0.2237 | 0.2390 | 0.2533 | 0.2670 | 0.2800 | 0.2927 | 0.3170 | 0.3403 | 0.3628 |
| | h | 823.97 | 993.3 | | 1185.4 | 1259.1 | 1313.0 | 1358.4 | 1399.0 | 1436.7 | 1472.3 | 1506.6 | 1539.9 | 1572.6 | 1636.7 | 1699.8 | 1762.5 |
| | s | 0.9914 | 1.1373 | | 1.3007 | 1.3604 | 1.4024 | 1.4364 | 1.4658 | 1.4920 | 1.5161 | 1.5384 | 1.5594 | 1.5794 | 1.6169 | 1.6518 | 1.6847 |
| **3200** (705.08) | Sh | | | | 44.92 | 94.92 | 144.92 | 194.92 | 244.92 | 294.92 | 344.92 | 394.92 | 444.92 | 494.92 | 594.92 | 694.92 | 794.92 |
| | v | 0.04472 | 0.0566 | | 0.1300 | 0.1588 | 0.1804 | 0.1987 | 0.2151 | 0.2301 | 0.2442 | 0.2576 | 0.2704 | 0.2827 | 0.3065 | 0.3291 | 0.3510 |
| | h | 875.54 | 931.6 | | 1172.3 | 1250.9 | 1306.9 | 1353.4 | 1394.9 | 1433.1 | 1469.2 | 1503.8 | 1537.4 | 1570.3 | 1634.8 | 1698.3 | 1761.2 |
| | s | 1.0351 | 1.0832 | | 1.2877 | 1.3515 | 1.3951 | 1.4300 | 1.4600 | 1.4866 | 1.5110 | 1.5335 | 1.5547 | 1.5749 | 1.6126 | 1.6477 | 1.6806 |
| **3300** | Sh | | | | | | | | | | | | | | | | |
| | v | | | | 0.1213 | 0.1510 | 0.1727 | 0.1908 | 0.2070 | 0.2218 | 0.2357 | 0.2488 | 0.2613 | 0.2734 | 0.2966 | 0.3187 | 0.3400 |
| | h | | | | 1158.2 | 1242.5 | 1300.7 | 1348.4 | 1390.7 | 1429.5 | 1466.1 | 1501.0 | 1534.9 | 1568.1 | 1632.9 | 1696.7 | 1759.9 |
| | s | | | | 1.2742 | 1.3425 | 1.3879 | 1.4237 | 1.4542 | 1.4813 | 1.5059 | 1.5287 | 1.5501 | 1.5704 | 1.6084 | 1.6436 | 1.6767 |
| **3400** | Sh | | | | | | | | | | | | | | | | |
| | v | | | | 0.1129 | 0.1435 | 0.1653 | 0.1834 | 0.1994 | 0.2140 | 0.2276 | 0.2405 | 0.2528 | 0.2646 | 0.2872 | 0.3088 | 0.3296 |
| | h | | | | 1143.2 | 1233.7 | 1294.3 | 1343.4 | 1386.4 | 1425.9 | 1462.9 | 1498.3 | 1532.4 | 1565.8 | 1631.1 | 1695.1 | 1758.2 |
| | s | | | | 1.2600 | 1.3334 | 1.3807 | 1.4174 | 1.4486 | 1.4761 | 1.5010 | 1.5240 | 1.5456 | 1.5660 | 1.6042 | 1.6396 | 1.6728 |
| **3500** | Sh | | | | | | | | | | | | | | | | |
| | v | | | | 0.1048 | 0.1364 | 0.1583 | 0.1764 | 0.1922 | 0.2066 | 0.2200 | 0.2326 | 0.2447 | 0.2563 | 0.2784 | 0.2995 | 0.3198 |
| | h | | | | 1127.1 | 1224.6 | 1287.8 | 1338.2 | 1382.2 | 1422.2 | 1459.7 | 1495.5 | 1529.9 | 1563.6 | 1629.2 | 1693.6 | 1757.2 |
| | s | | | | 1.2450 | 1.3242 | 1.3734 | 1.4112 | 1.4430 | 1.4709 | 1.4962 | 1.5194 | 1.5412 | 1.5618 | 1.6002 | 1.6358 | 1.6691 |
| **3600** | Sh | | | | | | | | | | | | | | | | |
| | v | | | | 0.0966 | 0.1296 | 0.1517 | 0.1697 | 0.1854 | 0.1996 | 0.2128 | 0.2252 | 0.2371 | 0.2485 | 0.2702 | 0.2908 | 0.3106 |
| | h. | | | | 1108.6 | 1215.3 | 1281.2 | 1333.0 | 1377.9 | 1418.6 | 1456.5 | 1492.6 | 1527.4 | 1561.3 | 1627.3 | 1692.0 | 1755.9 |
| | s | | | | 1.2281 | 1.3148 | 1.3662 | 1.4050 | 1.4374 | 1.4658 | 1.4914 | 1.5149 | 1.5369 | 1.5576 | 1.5962 | 1.6320 | 1.6654 |
| **3800** | Sh | | | | | | | | | | | | | | | | |
| | v | | | | 0.0799 | 0.1169 | 0.1395 | 0.1574 | 0.1729 | 0.1868 | 0.1996 | 0.2116 | 0.2231 | 0.2340 | 0.2549 | 0.2746 | 0.2936 |
| | h | | | | 1064.2 | 1195.5 | 1267.6 | 1322.4 | 1369.1 | 1411.2 | 1450.1 | 1487.0 | 1522.4 | 1556.8 | 1623.6 | 1688.9 | 1753.2 |
| | s | | | | 1.1888 | 1.2955 | 1.3517 | 1.3928 | 1.4265 | 1.4558 | 1.4821 | 1.5061 | 1.5284 | 1.5495 | 1.5886 | 1.6247 | 1.6584 |
| **4000** | Sh | | | | | | | | | | | | | | | | |
| | v | | | | 0.0631 | 0.1052 | 0.1284 | 0.1463 | 0.1616 | 0.1752 | 0.1877 | 0.1994 | 0.2105 | 0.2210 | 0.2411 | 0.2601 | 0.2783 |
| | h | | | | 1007.4 | 1174.3 | 1253.4 | 1311.6 | 1360.2 | 1403.6 | 1443.6 | 1481.3 | 1517.3 | 1552.2 | 1619.8 | 1685.7 | 1750.6 |
| | s | | | | 1.1396 | 1.2754 | 1.3371 | 1.3807 | 1.4158 | 1.4461 | 1.4730 | 1.4976 | 1.5203 | 1.5417 | 1.5812 | 1.6177 | 1.6516 |
| **4200** | Sh | | | | | | | | | | | | | | | | |
| | v | | | | 0.0498 | 0.0945 | 0.1183 | 0.1362 | 0.1513 | 0.1647 | 0.1769 | 0.1883 | 0.1991 | 0.2093 | 0.2287 | 0.2470 | 0.2645 |
| | h | | | | 950.1 | 1151.6 | 1238.6 | 1300.4 | 1351.2 | 1396.0 | 1437.1 | 1475.5 | 1512.2 | 1547.6 | 1616.1 | 1682.6 | 1748.0 |
| | s | | | | 1.0905 | 1.2544 | 1.3223 | 1.3686 | 1.4053 | 1.4366 | 1.4642 | 1.4893 | 1.5124 | 1.5341 | 1.5742 | 1.6109 | 1.6452 |
| **4400** | Sh | | | | | | | | | | | | | | | | |
| | v | | | | 0.0421 | 0.0846 | 0.1090 | 0.1270 | 0.1420 | 0.1552 | 0.1671 | 0.1782 | 0.1887 | 0.1986 | 0.2174 | 0.2351 | 0.2519 |
| | h | | | | 909.5 | 1127.3 | 1223.3 | 1289.0 | 1342.0 | 1388.3 | 1430.4 | 1469.7 | 1507.1 | 1543.0 | 1612.3 | 1679.4 | 1745.3 |
| | s | | | | 1.0556 | 1.2325 | 1.3073 | 1.3566 | 1.3949 | 1.4272 | 1.4556 | 1.4812 | 1.5048 | 1.5268 | 1.5673 | 1.6044 | 1.6389 |

**Table II.2-2  Continued**

| Abs. Press. (psi) (Sat. Temp.) | Sat. Water | Sat. Steam | Temperature (°F) | | | | | | | | | | | | | |
|---|---|---|---|---|---|---|---|---|---|---|---|---|---|---|---|---|
| | | | 750 | 800 | 850 | 900 | 950 | 1000 | 1050 | 1100 | 1150 | 1200 | 1250 | 1300 | 1400 | 1500 |
| 4600 | Sh | | | | | | | | | | | | | | | |
| | v | | 0.0380 | 0.0751 | 0.1005 | 0.1186 | 0.1335 | 0.1465 | 0.1582 | 0.1691 | 0.1792 | 0.1889 | 0.1982 | 0.2071 | 0.2242 | 0.2404 |
| | h | | 883.8 | 1100.0 | 1207.3 | 1277.2 | 1332.6 | 1380.5 | 1423.7 | 1463.9 | 1501.9 | 1538.4 | 1573.8 | 1608.5 | 1676.3 | 1742.7 |
| | s | | 1.0331 | 1.2084 | 1.2922 | 1.3446 | 1.3847 | 1.4181 | 1.4472 | 1.4734 | 1.4974 | 1.5197 | 1.5407 | 1.5607 | 1.5982 | 1.6330 |
| 4800 | Sh | | | | | | | | | | | | | | | |
| | v | | 0.0355 | 0.0665 | 0.0927 | 0.1109 | 0.1257 | 0.1385 | 0.1500 | 0.1606 | 0.1706 | 0.1800 | 0.1890 | 0.1977 | 0.2142 | 0.2299 |
| | h | | 866.9 | 1071.2 | 1190.7 | 1265.2 | 1323.1 | 1372.6 | 1417.0 | 1458.0 | 1496.7 | 1533.8 | 1569.7 | 1604.7 | 1673.1 | 1740.0 |
| | s | | 1.0180 | 1.1835 | 1.2768 | 1.3327 | 1.3745 | 1.4090 | 1.4390 | 1.4657 | 1.4901 | 1.5128 | 1.5341 | 1.5543 | 1.5921 | 1.6272 |
| 5000 | Sh | | | | | | | | | | | | | | | |
| | v | | 0.0338 | 0.591 | 0.0855 | 0.1038 | 0.1185 | 0.1312 | 0.1425 | 0.1529 | 0.1626 | 0.1718 | 0.1806 | 0.1890 | 0.2050 | 0.2203 |
| | h | | 854.9 | 1042.9 | 1173.6 | 1252.9 | 1313.5 | 1364.6 | 1410.2 | 1452.1 | 1491.5 | 1529.1 | 1565.5 | 1600.9 | 1670.0 | 1737.4 |
| | s | | 1.0070 | 1.1593 | 1.2612 | 1.3207 | 1.3645 | 1.4001 | 1.4309 | 1.4582 | 1.4831 | 1.5061 | 1.5277 | 1.5481 | 1.5863 | 1.6216 |
| 5200 | Sh | | | | | | | | | | | | | | | |
| | v | | 0.0326 | 0.0531 | 0.0789 | 0.0973 | 0.1119 | 0.1244 | 0.1356 | 0.1458 | 0.1553 | 0.1642 | 0.1728 | 0.1810 | 0.1966 | 0.2114 |
| | h | | 845.8 | 1016.9 | 1156.0 | 1240.4 | 1303.7 | 1356.6 | 1403.4 | 1446.2 | 1486.3 | 1524.5 | 1561.3 | 1597.2 | 1666.8 | 1734.7 |
| | s | | 0.9985 | 1.1370 | 1.2455 | 1.3088 | 1.3545 | 1.3914 | 1.4229 | 1.4509 | 1.4762 | 1.4995 | 1.5214 | 1.5420 | 1.5806 | 1.6161 |
| 5400 | Sh | | | | | | | | | | | | | | | |
| | v | | 0.0317 | 0.0483 | 0.0728 | 0.0912 | 0.1058 | 0.1182 | 0.1292 | 0.1392 | 0.1485 | 0.1572 | 0.1656 | 0.1736 | 0.1888 | 0.2031 |
| | h | | 838.5 | 994.3 | 1138.1 | 1227.7 | 1293.7 | 1348.4 | 1396.5 | 1440.3 | 1481.1 | 1519.8 | 1557.1 | 1593.4 | 1663.7 | 1732.1 |
| | s | | 0.9915 | 1.1175 | 1.2296 | 1.2969 | 1.3446 | 1.3827 | 1.4151 | 1.4437 | 1.4694 | 1.4931 | 1.5153 | 1.5362 | 1.5750 | 1.6109 |
| 5600 | Sh | | | | | | | | | | | | | | | |
| | v | | 0.0309 | 0.0447 | 0.0672 | 0.0856 | 0.1001 | 0.1124 | 0.1232 | 0.1331 | 0.1422 | 0.1508 | 0.1589 | 0.1667 | 0.1815 | 0.1954 |
| | h | | 832.4 | 975.0 | 1119.9 | 1214.8 | 1283.7 | 1340.2 | 1389.6 | 1434.3 | 1475.9 | 1515.2 | 1552.9 | 1589.6 | 1660.5 | 1729.5 |
| | s | | 0.9855 | 1.1008 | 1.2137 | 1.2850 | 1.3348 | 1.3742 | 1.4075 | 1.4366 | 1.4628 | 1.4869 | 1.5093 | 1.5304 | 1.5697 | 1.6058 |
| 5800 | Sh | | | | | | | | | | | | | | | |
| | v | | 0.0303 | 0.0419 | 0.0622 | 0.0805 | 0.0949 | 0.1070 | 0.1177 | 0.1274 | 0.1363 | 0.1447 | 0.1527 | 0.1603 | 0.1747 | 0.1883 |
| | h | | 827.3 | 958.8 | 1101.8 | 1201.8 | 1273.6 | 1332.0 | 1382.6 | 1428.3 | 1470.6 | 1510.5 | 1548.7 | 1585.8 | 1657.4 | 1726.8 |
| | s | | 0.9803 | 1.0867 | 1.1981 | 1.2732 | 1.3250 | 1.3658 | 1.3999 | 1.4297 | 1.4564 | 1.4808 | 1.5035 | 1.5248 | 1.5644 | 1.6008 |
| 6000 | Sh | | | | | | | | | | | | | | | |
| | v | | 0.0298 | 0.0397 | 0.0579 | 0.0757 | 0.0900 | 0.1020 | 0.1126 | 0.1221 | 0.1309 | 0.1391 | 0.1469 | 0.1544 | 0.1684 | 0.1817 |
| | h | | 822.9 | 945.1 | 1084.6 | 1188.8 | 1263.4 | 1323.6 | 1375.7 | 1422.3 | 1465.4 | 1505.9 | 1544.6 | 1582.0 | 1654.2 | 1724.2 |
| | s | | 0.9758 | 1.0746 | 1.1833 | 1.2615 | 1.3154 | 1.3574 | 1.3925 | 1.4229 | 1.4500 | 1.4748 | 1.4978 | 1.5194 | 1.5593 | 1.5960 |
| 6500 | Sh | | | | | | | | | | | | | | | |
| | v | | 0.0287 | 0.0358 | 0.0495 | 0.0655 | 0.0793 | 0.0909 | 0.1012 | 0.1104 | 0.1188 | 0.1266 | 0.1340 | 0.1411 | 0.1544 | 0.1669 |
| | h | | 813.9 | 919.5 | 1046.7 | 1156.3 | 1237.8 | 1302.7 | 1358.1 | 1407.3 | 1452.2 | 1494.2 | 1534.1 | 1572.5 | 1646.4 | 1717.6 |
| | s | | 0.9661 | 1.0515 | 1.1506 | 1.2328 | 1.2917 | 1.3370 | 1.3743 | 1.4064 | 1.4347 | 1.4604 | 1.4841 | 1.5062 | 1.5471 | 1.5844 |
| 7000 | Sh | | | | | | | | | | | | | | | |
| | v | | 0.0279 | 0.0334 | 0.0438 | 0.0573 | 0.0704 | 0.0816 | 0.0915 | 0.1004 | 0.1085 | 0.1160 | 0.1231 | 0.1298 | 0.1424 | 0.1542 |
| | h | | 806.9 | 901.8 | 1016.5 | 1124.9 | 1212.6 | 1281.7 | 1340.5 | 1392.2 | 1439.1 | 1482.6 | 1523.7 | 1563.1 | 1638.6 | 1711.1 |
| | s | | 0.9582 | 1.0350 | 1.1243 | 1.2055 | 1.2689 | 1.3171 | 1.3567 | 1.3904 | 1.4200 | 1.4466 | 1.4710 | 1.4938 | 1.5355 | 1.5735 |
| 7500 | Sh | | | | | | | | | | | | | | | |
| | v | | 0.0272 | 0.0318 | 0.0399 | 0.0512 | 0.0631 | 0.0737 | 0.0833 | 0.0918 | 0.0996 | 0.1068 | 0.1136 | 0.1200 | 0.1321 | 0.1433 |
| | h | | 801.3 | 889.0 | 992.9 | 1097.7 | 1188.3 | 1261.0 | 1322.9 | 1377.2 | 1426.0 | 1471.0 | 1513.3 | 1553.7 | 1630.8 | 1704.6 |
| | s | | 0.9514 | 1.0224 | 1.1033 | 1.1818 | 1.2473 | 1.2980 | 1.3397 | 1.3751 | 1.4059 | 1.4335 | 1.4586 | 1.4819 | 1.5245 | 1.5632 |
| 8000 | Sh | | | | | | | | | | | | | | | |
| | v | | 0.0267 | 0.0306 | 0.0371 | 0.0465 | 0.0571 | 0.0671 | 0.0762 | 0.0845 | 0.0920 | 0.0989 | 0.1054 | 0.1115 | 0.1230 | 0.1338 |
| | h | | 796.6 | 879.1 | 974.4 | 1074.3 | 1165.4 | 1241.0 | 1305.5 | 1362.2 | 1413.0 | 1459.6 | 1503.1 | 1544.5 | 1623.1 | 1698.1 |
| | s | | 0.9455 | 1.0122 | 1.0864 | 1.1613 | 1.2271 | 1.2798 | 1.3233 | 1.3603 | 1.3924 | 1.4208 | 1.4467 | 1.4705 | 1.5140 | 1.5533 |
| 8500 | Sh | | | | | | | | | | | | | | | |
| | v | | 0.0262 | 0.0296 | 0.0350 | 0.0429 | 0.0522 | 0.0615 | 0.0701 | 0.0780 | 0.0853 | 0.0919 | 0.0982 | 0.1041 | 0.1151 | 0.1254 |
| | h | | 792.7 | 871.2 | 959.8 | 1054.5 | 1144.0 | 1221.9 | 1288.5 | 1347.5 | 1400.2 | 1448.2 | 1492.9 | 1535.3 | 1615.4 | 1691.7 |
| | s | | 0.9402 | 1.0037 | 1.0727 | 1.1437 | 1.2084 | 1.2627 | 1.3076 | 1.3460 | 1.3793 | 1.4087 | 1.4352 | 1.4597 | 1.5040 | 1.5439 |
| 9000 | Sh | | | | | | | | | | | | | | | |
| | v | | 0.0258 | 0.0288 | 0.0335 | 0.0402 | 0.0483 | 0.0568 | 0.0649 | 0.0724 | 0.0794 | 0.0858 | 0.0918 | 0.0975 | 0.1081 | 0.1179 |
| | h | | 789.3 | 864.7 | 948.0 | 1037.6 | 1125.4 | 1204.1 | 1272.1 | 1333.0 | 1387.5 | 1437.1 | 1482.9 | 1526.3 | 1607.9 | 1685.3 |
| | s | | 0.9354 | 0.9964 | 1.0613 | 1.1285 | 1.1918 | 1.2468 | 1.2926 | 1.3323 | 1.3667 | 1.3970 | 1.4243 | 1.4492 | 1.4944 | 1.5349 |
| 9500 | Sh | | | | | | | | | | | | | | | |
| | v | | 0.0254 | 0.0282 | 0.0322 | 0.0380 | 0.0451 | 0.0528 | 0.0603 | 0.0675 | 0.0742 | 0.0804 | 0.0862 | 0.0917 | 0.1019 | 0.1113 |
| | h | | 786.4 | 859.2 | 938.3 | 1023.4 | 1108.9 | 1187.7 | 1256.6 | 1318.9 | 1375.1 | 1426.1 | 1473.1 | 1517.3 | 1600.4 | 1679.0 |
| | s | | 0.9310 | 0.9900 | 1.0516 | 1.1153 | 1.1771 | 1.2320 | 1.2785 | 1.3191 | 1.3546 | 1.3858 | 1.4137 | 1.4392 | 1.4851 | 1.5263 |
| 10000 | Sh | | | | | | | | | | | | | | | |
| | v | | 0.0251 | 0.0276 | 0.0312 | 0.0362 | 0.0425 | 0.0495 | 0.0565 | 0.0633 | 0.0697 | 0.0757 | 0.0812 | 0.0865 | 0.0963 | 0.1054 |
| | h | | 783.8 | 854.5 | 930.2 | 1011.3 | 1094.2 | 1172.6 | 1242.0 | 1305.3 | 1362.9 | 1415.3 | 1463.4 | 1508.6 | 1593.1 | 1672.8 |
| | s | | 0.9270 | 0.9842 | 1.0432 | 1.1039 | 1.1638 | 1.2185 | 1.2652 | 1.3065 | 1.3429 | 1.3749 | 1.4035 | 1.4295 | 1.4763 | 1.5180 |
| 10500 | Sh | | | | | | | | | | | | | | | |
| | v | | 0.0248 | 0.0271 | 0.0303 | 0.0347 | 0.0404 | 0.0467 | 0.0532 | 0.0595 | 0.0656 | 0.0714 | 0.0768 | 0.0818 | 0.0913 | 0.1001 |
| | h | | 781.5 | 850.5 | 923.4 | 1001.0 | 1081.3 | 1158.9 | 1228.4 | 1292.4 | 1351.1 | 1404.7 | 1453.9 | 1500.0 | 1585.8 | 1666.7 |
| | s | | 0.9232 | 0.9790 | 1.0358 | 1.0939 | 1.1519 | 1.2060 | 1.2529 | 1.2946 | 1.3371 | 1.3644 | 1.3937 | 1.4202 | 1.4677 | 1.5100 |

**Table II.2-2   Continued**

| Abs. Press. (psi) (Sat. Temp.) | | Sat. Water | Sat. Steam | 750 | 800 | 850 | 900 | 950 | 1000 | 1050 | 1100 | 1150 | 1200 | 1250 | 1300 | 1400 | 1500 |
|---|---|---|---|---|---|---|---|---|---|---|---|---|---|---|---|---|---|
| | | | | | | | | | | | | Temperature (°F) | | | | | |
| 11000 | $v$ | | | 0.0245 | 0.0267 | 0.0296 | 0.0335 | 0.0386 | 0.0443 | 0.0503 | 0.0562 | 0.0620 | 0.0676 | 0.0727 | 0.0776 | 0.0868 | 0.0952 |
| | $h$ | | | 779.5 | 846.9 | 917.5 | 992.1 | 1069.9 | 1146.3 | 1215.9 | 1280.2 | 1339.7 | 1394.4 | 1444.6 | 1491.5 | 1578.7 | 1660.6 |
| | $s$ | | | 0.9196 | 0.9742 | 1.0292 | 1.0851 | 1.1412 | 1.1945 | 1.2414 | 1.2833 | 1.3209 | 1.3544 | 1.3842 | 1.4112 | 1.4595 | 1.5023 |
| 11500 | $v$ | | | 0.0243 | 0.0263 | 0.0290 | 0.0325 | 0.0370 | 0.0423 | 0.0478 | 0.0534 | 0.0588 | 0.0641 | 0.0691 | 0.0739 | 0.0827 | 0.0909 |
| | $h$ | | | 777.7 | 843.8 | 912.4 | 984.5 | 1059.8 | 1134.9 | 1204.3 | 1268.7 | 1328.8 | 1384.4 | 1435.5 | 1483.2 | 1571.8 | 1654.7 |
| | $s$ | | | 0.9163 | 0.9698 | 1.0232 | 1.0772 | 1.1316 | 1.1840 | 1.2308 | 1.2727 | 1.3107 | 1.3446 | 1.3750 | 1.4025 | 1.4515 | 1.4949 |
| 12000 | $v$ | | | 0.0241 | 0.0260 | 0.0284 | 0.0317 | 0.0357 | 0.0405 | 0.0456 | 0.0508 | 0.0560 | 0.0610 | 0.0659 | 0.0704 | 0.0790 | 0.0869 |
| | $h$ | | | 776.1 | 841.0 | 907.9 | 977.8 | 1050.9 | 1124.5 | 1193.7 | 1258.0 | 1318.5 | 1374.7 | 1426.6 | 1475.1 | 1564.9 | 1648.8 |
| | $s$ | | | 0.9131 | 0.9657 | 1.0177 | 1.0701 | 1.1229 | 1.1742 | 1.2209 | 1.2627 | 1.3010 | 1.3353 | 1.3662 | 1.3941 | 1.4438 | 1.4877 |
| 12500 | $v$ | | | 0.0238 | 0.0256 | 0.0279 | 0.0309 | 0.0346 | 0.0390 | 0.0437 | 0.0486 | 0.0535 | 0.0583 | 0.0629 | 0.0673 | 0.0756 | 0.0832 |
| | $h$ | | | 774.7 | 838.6 | 903.9 | 971.9 | 1043.1 | 1115.2 | 1184.1 | 1247.9 | 1308.8 | 1365.4 | 1418.0 | 1467.2 | 1558.2 | 1643.1 |
| | $s$ | | | 0.9101 | 0.9618 | 1.0127 | 1.0637 | 1.1151 | 1.1653 | 1.2117 | 1.2534 | 1.2918 | 1.3264 | 1.3576 | 1.3860 | 1.4363 | 1.4808 |
| 13000 | $v$ | | | 0.0236 | 0.0253 | 0.0275 | 0.0302 | 0.0336 | 0.0376 | 0.0420 | 0.0466 | 0.0512 | 0.0558 | 0.0602 | 0.0645 | 0.0725 | 0.0799 |
| | $h$ | | | 773.5 | 836.3 | 900.4 | 966.8 | 1036.2 | 1106.7 | 1174.8 | 1238.5 | 1299.6 | 1356.5 | 1409.6 | 1459.4 | 1551.6 | 1637.4 |
| | $s$ | | | 0.9073 | 0.9582 | 1.0080 | 1.0578 | 1.1079 | 1.1571 | 1.2030 | 1.2445 | 1.2831 | 1.3179 | 1.3494 | 1.3781 | 1.4291 | 1.4741 |
| 13500 | $v$ | | | 0.0235 | 0.0251 | 0.0271 | 0.0297 | 0.0328 | 0.0364 | 0.0405 | 0.0448 | 0.0492 | 0.0535 | 0.0577 | 0.0619 | 0.0696 | 0.0768 |
| | $h$ | | | 772.3 | 834.4 | 897.2 | 962.2 | 1030.0 | 1099.1 | 1166.3 | 1229.7 | 1291.0 | 1348.1 | 1401.5 | 1451.8 | 1545.2 | 1631.9 |
| | $s$ | | | 0.9045 | 0.9548 | 1.0037 | 1.0524 | 1.1014 | 1.1495 | 1.1948 | 1.2361 | 1.2749 | 1.3098 | 1.3415 | 1.3705 | 1.4221 | 1.4675 |
| 14000 | $v$ | | | 0.0233 | 0.0248 | 0.0267 | 0.0291 | 0.0320 | 0.0354 | 0.0392 | 0.0432 | 0.0474 | 0.0515 | 0.0555 | 0.0595 | 0.0670 | 0.0740 |
| | $h$ | | | 771.3 | 832.6 | 894.3 | 958.0 | 1024.5 | 1092.3 | 1158.5 | 1221.4 | 1283.0 | 1340.2 | 1393.8 | 1444.4 | 1538.8 | 1626.5 |
| | $s$ | | | 0.9019 | 0.9515 | 0.9996 | 1.0473 | 1.0953 | 1.1426 | 1.1872 | 1.2282 | 1.2671 | 1.3021 | 1.3339 | 1.3631 | 1.4153 | 1.4612 |
| 14500 | $v$ | | | 0.0231 | 0.0246 | 0.0264 | 0.0287 | 0.0314 | 0.0345 | 0.0380 | 0.0418 | 0.0458 | 0.0496 | 0.0534 | 0.0573 | 0.0646 | 0.0714 |
| | $h$ | | | 770.4 | 831.0 | 891.7 | 954.3 | 1019.6 | 1086.2 | 1151.4 | 1213.8 | 1275.4 | 1332.9 | 1386.4 | 1437.3 | 1532.6 | 1621.1 |
| | $s$ | | | 0.8994 | 0.9484 | 0.9957 | 1.0426 | 1.0897 | 1.1362 | 1.1801 | 1.2208 | 1.2597 | 1.2949 | 1.3266 | 1.3560 | 1.4087 | 1.4551 |
| 15000 | $v$ | | | 0.0230 | 0.0244 | 0.0261 | 0.0282 | 0.0308 | 0.0337 | 0.0369 | 0.0405 | 0.0443 | 0.0479 | 0.0516 | 0.0552 | 0.0624 | 0.0690 |
| | $h$ | | | 769.6 | 829.5 | 889.3 | 950.9 | 1015.1 | 1080.6 | 1144.9 | 1206.8 | 1268.1 | 1326.0 | 1379.4 | 1430.3 | 1526.4 | 1615.9 |
| | $s$ | | | 0.8970 | 0.9455 | 0.9920 | 1.0382 | 1.0846 | 1.1302 | 1.1735 | 1.2139 | 1.2525 | 1.2880 | 1.3197 | 1.3491 | 1.4022 | 1.4491 |
| 15500 | $v$ | | | 0.0228 | 0.0242 | 0.0258 | 0.0278 | 0.0302 | 0.0329 | 0.0360 | 0.0393 | 0.0429 | 0.0464 | 0.0499 | 0.0534 | 0.0603 | 0.0668 |
| | $h$ | | | 768.9 | 828.2 | 887.2 | 947.8 | 1011.1 | 1075.7 | 1139.0 | 1200.3 | 1261.1 | 1319.6 | 1372.8 | 1423.6 | 1520.4 | 1610.8 |
| | $s$ | | | 0.8946 | 0.9427 | 0.9886 | 1.0340 | 1.0797 | 1.1247 | 1.1674 | 1.2073 | 1.2457 | 1.2815 | 1.3131 | 1.3424 | 1.3959 | 1.4433 |

ᵃ Sh-superheat, °F; $v$ specific volume, ft³/lb; $h$ enthalpy, Btu/lb; $s$ entropy, Btu/°F · lb.
*Source*: Copyright 1967 ASME (Abridged); reprinted by permission.

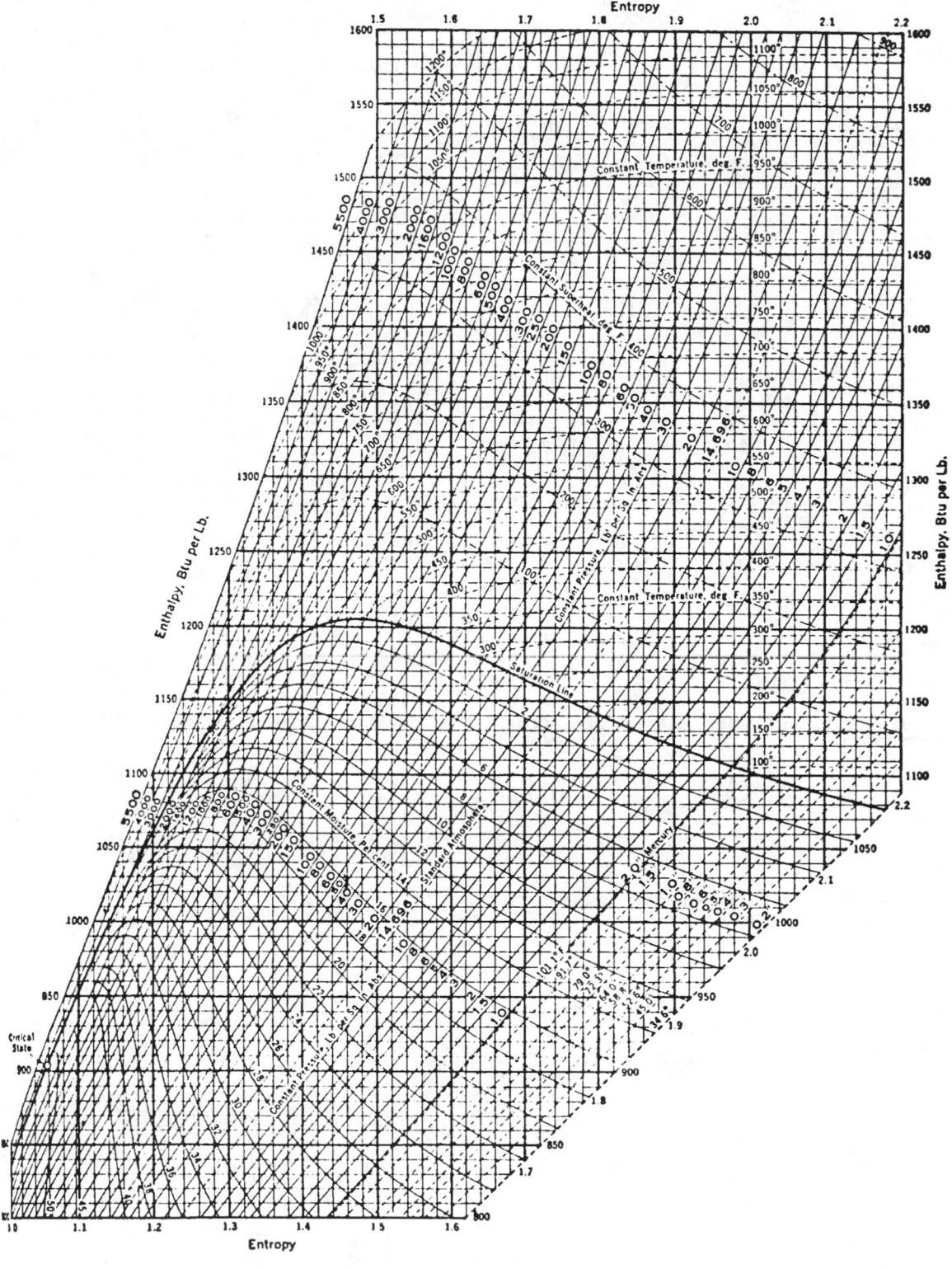

## Mollier Diagram for Steam

*Source:* Modified and greatly reduced from J.H. Keenan and F.G. Keyes, *Thermodynamic Properties of Steam*, John Wiley & Sons Inc., New York, 1936; reproduced by permission of the publishers.

**Table II.2-4   Thermodynamic Properties of Saturated Freon-12**

| Temp. $T$ (°F) | Abs. Press. $P$ (lb$_f$/in.$^2$) | Specific Volume (ft$^3$/lb$_m$) Sat. Liquid $v_f$ | Evap. $v_{fg}$ | Sat. Vapor $v_g$ | Enthalpy (Btu/lb$_m$) Sat. Liquid $h_f$ | Evap. $h_{fg}$ | Sat. Vapor $h_g$ | Entropy (Btu/lb$_m$ · °R) Sat. Liquid $s_f$ | Evap. $s_{fg}$ | Sat. Vapor $s_g$ |
|---|---|---|---|---|---|---|---|---|---|---|
| −130 | 0.41224 | 0.009736 | 70.7203 | 70.730 | −18.609 | 81.577 | 62.968 | −0.04983 | 0.24743 | 0.19760 |
| −120 | 0.64190 | 0.009816 | 46.7312 | 46.741 | −16.565 | 80.617 | 64.052 | −0.04372 | 0.23731 | 0.19359 |
| −110 | 0.97034 | 0.009899 | 31.7671 | 31.777 | −14.518 | 79.663 | 65.145 | −0.03779 | 0.22780 | 0.19002 |
| −100 | 1.4280 | 0.009985 | 21.1541 | 22.164 | −12.466 | 78.714 | 66.248 | −0.03200 | 0.21883 | 0.18683 |
| −90 | 2.0509 | 0.010073 | 15.8109 | 15.821 | −10.409 | 77.764 | 67.355 | −0.02637 | 0.21034 | 0.18398 |
| −80 | 2.8807 | 0.010164 | 11.5228 | 11.533 | −8.3451 | 76.812 | 68.467 | −0.02086 | 0.20229 | 0.18143 |
| −70 | 3.9651 | 0.010259 | 8.5584 | 8.5687 | −6.2730 | 75.853 | 69.580 | −0.01548 | 0.19464 | 0.17916 |
| −60 | 5.3575 | 0.010357 | 6.4670 | 6.4774 | −4.1919 | 74.885 | 70.693 | −0.01021 | 0.18716 | 0.17714 |
| −50 | 7.1168 | 0.010459 | 4.9637 | 4.9742 | −2.1011 | 73.906 | 71.805 | −0.00506 | 0.18038 | 0.17533 |
| −40 | 9.3076 | 0.010564 | 3.8644 | 3.8750 | 0 | 72.913 | 72.913 | 0 | 0.17373 | 0.17373 |
| −30 | 11.999 | 0.010674 | 3.0478 | 3.0585 | 2.1120 | 71.903 | 74.015 | 0.00496 | 0.16733 | 0.17229 |
| −20 | 15.267 | 0.010788 | 2.4321 | 2.4429 | 4.2357 | 70.874 | 75.110 | 0.00983 | 0.16119 | 0.17102 |
| −10 | 19.189 | 0.010906 | 1.9628 | 1.9727 | 6.3716 | 69.824 | 76.196 | 0.01462 | 0.15527 | 0.16989 |
| 0 | 23.849 | 0.011030 | 1.5979 | 1.6089 | 8.5207 | 68.750 | 77.271 | 0.01932 | 0.14956 | 0.16888 |
| 10 | 29.335 | 0.011160 | 1.3129 | 1.3241 | 10.684 | 67.651 | 78.335 | 0.02395 | 0.14403 | 0.16798 |
| 20 | 35.736 | 0.011296 | 1.0875 | 1.0988 | 12.863 | 66.522 | 79.385 | 0.02852 | 0.13867 | 0.16719 |
| 30 | 43.148 | 0.011438 | 0.90736 | 0.91880 | 15.058 | 65.361 | 80.419 | 0.03301 | 0.13347 | 0.16648 |
| 40 | 51.667 | 0.011588 | 0.76198 | 0.77357 | 17.273 | 64.163 | 81.436 | 0.03745 | 0.12841 | 0.16586 |
| 50 | 61.394 | 0.011746 | 0.64362 | 0.65537 | 19.507 | 62.926 | 82.433 | 0.04184 | 0.12346 | 0.16530 |
| 60 | 72.433 | 0.011913 | 0.54648 | 0.55839 | 21.766 | 61.643 | 83.409 | 0.04618 | 0.11861 | 0.16479 |
| 70 | 84.888 | 0.012089 | 0.46609 | 0.47818 | 24.050 | 60.309 | 84.359 | 0.05048 | 0.11386 | 0.16434 |
| 80 | 98.870 | 0.012277 | 0.39907 | 0.41135 | 26.365 | 58.917 | 85.282 | 0.05475 | 0.10917 | 0.16392 |
| 90 | 114.49 | 0.012478 | 0.34281 | 0.35529 | 28.713 | 57.461 | 86.174 | 0.05900 | 0.10453 | 0.16353 |
| 100 | 131.86 | 0.012693 | 0.29525 | 0.30794 | 31.100 | 55.929 | 87.029 | 0.06323 | 0.09992 | 0.16315 |
| 110 | 151.11 | 0.012924 | 0.25577 | 0.26769 | 33.531 | 54.313 | 87.844 | 0.06745 | 0.09534 | 0.16279 |
| 120 | 172.35 | 0.013174 | 0.22019 | 0.23326 | 36.013 | 52.597 | 88.610 | 0.07168 | 0.09073 | 0.16241 |
| 130 | 195.71 | 0.013447 | 0.19019 | 0.20364 | 38.553 | 50.768 | 89.321 | 0.07583 | 0.08609 | 0.16202 |
| 140 | 221.32 | 0.013746 | 0.16424 | 0.17799 | 41.162 | 48.805 | 89.967 | 0.08021 | 0.08138 | 0.16159 |
| 150 | 249.31 | 0.014078 | 0.14156 | 0.15564 | 43.850 | 46.684 | 90.534 | 0.08453 | 0.07657 | 0.16110 |
| 160 | 279.82 | 0.014449 | 0.12159 | 0.13604 | 46.633 | 44.373 | 91.006 | 0.08893 | 0.07260 | 0.16053 |
| 170 | 313.00 | 0.014871 | 0.10386 | 0.11873 | 49.529 | 41.830 | 91.359 | 0.09342 | 0.06643 | 0.15985 |
| 180 | 349.00 | 0.015360 | 0.08794 | 0.10330 | 52.562 | 38.999 | 91.561 | 0.09804 | 0.06096 | 0.15900 |
| 190 | 387.98 | 0.015942 | 0.073476 | 0.089418 | 55.769 | 35.792 | 91.561 | 0.10284 | 0.05511 | 0.15793 |
| 200 | 430.09 | 0.016659 | 0.060069 | 0.076728 | 59.203 | 32.075 | 91.278 | 0.10789 | 0.04862 | 0.15651 |
| 210 | 475.52 | 0.017601 | 0.047242 | 0.064843 | 62.959 | 27.599 | 90.558 | 0.11332 | 0.03921 | 0.15453 |
| 220 | 524.43 | 0.018986 | 0.035154 | 0.053140 | 67.246 | 21.790 | 89.036 | 0.11943 | 0.03206 | 0.15149 |
| 230 | 577.03 | 0.021854 | 0.017581 | 0.039435 | 72.893 | 12.229 | 85.122 | 0.12739 | 0.01773 | 0.14512 |
| 233.6 (critical) | 596.9 | 0.02870 | 0 | 0.02870 | 78.86 | 0 | 78.86 | 0.1359 | 0 | 0.1359 |

*Source*: Copyright 1955 and 1956 E. I. du Pont de Nemours & Company, Inc. reprinted by permission.

**Table II.2-5   Superheated Freon-12**

| Temp. (°F) | $v$ | $h$ | $s$ | $v$ | $h$ | $s$ | $v$ | $h$ | $s$ |
|---|---|---|---|---|---|---|---|---|---|
| | 5 lb$_f$/in.$^2$ | | | 10 lb$_f$/in.$^2$ | | | 15 lb$_f$/in.$^2$ | | |
| 0 | 8.0611 | 78.852 | 0.19663 | 3.9809 | 78.246 | 0.18471 | 2.6201 | 77.902 | 0.17751 |
| 20 | 8.4265 | 81.309 | 0.20244 | 4.1691 | 81.014 | 0.19061 | 2.7494 | 80.712 | 0.18349 |
| 40 | 8.7903 | 84.090 | 0.20812 | 4.3556 | 83.828 | 0.19635 | 2.8770 | 83.561 | 0.18931 |
| 60 | 9.1528 | 86.922 | 0.21367 | 4.5408 | 86.689 | 0.20197 | 3.0031 | 86.451 | 0.19498 |
| 80 | 9.5142 | 89.806 | 0.21912 | 4.7248 | 89.596 | 0.20746 | 3.1281 | 89.383 | 0.20051 |
| 100 | 9.8747 | 92.738 | 0.22445 | 4.9079 | 92.548 | 0.21283 | 3.2521 | 92.357 | 0.20593 |
| 120 | 10.234 | 95.717 | 0.22968 | 5.0903 | 95.546 | 0.21809 | 3.3754 | 95.373 | 0.21122 |
| 140 | 10.594 | 98.743 | 0.23481 | 5.2720 | 98.586 | 0.22325 | 3.4981 | 98.429 | 0.21640 |
| 160 | 10.952 | 101.812 | 0.23985 | 5.4533 | 101.669 | 0.22830 | 3.6202 | 101.525 | 0.22148 |
| 180 | 11.311 | 104.925 | 0.24479 | 5.6341 | 104.793 | 0.23326 | 3.7419 | 104.661 | 0.22646 |
| 200 | 11.668 | 108.079 | 0.24964 | 5.8145 | 107.957 | 0.23813 | 3.8632 | 107.835 | 0.23135 |
| 220 | 12.026 | 111.272 | 0.25441 | 5.9946 | 111.159 | 0.24291 | 3.9841 | 111.046 | 0.23614 |

**Table II.2-5 Continued**

| Temp. (°F) | $v$ | $h$ | $s$ | $v$ | $h$ | $s$ | $v$ | $h$ | $s$ |
|---|---|---|---|---|---|---|---|---|---|
| | 20 lb$_f$/in.$^2$ | | | 25 lb$_f$/in.$^2$ | | | 30 lb$_f$/in.$^2$ | | |
| 20 | 2.0391 | 80.403 | 0.17829 | 1.6125 | 80.088 | 0.17414 | 1.3278 | 79.765 | 0.17065 |
| 40 | 2.1373 | 83.289 | 0.18419 | 1.6932 | 83.012 | 0.18012 | 1.3969 | 82.730 | 0.17671 |
| 60 | 2.2340 | 86.210 | 0.18992 | 1.7723 | 85.965 | 0.18591 | 1.4644 | 85.716 | 0.18257 |
| 80 | 2.3295 | 89.168 | 0.19550 | 1.8502 | 88.950 | 0.19155 | 1.5306 | 88.729 | 0.18826 |
| 100 | 2.4241 | 92.164 | 0.20095 | 1.9271 | 91.968 | 0.19704 | 1.5957 | 91.770 | 0.19379 |
| 120 | 2.5179 | 95.198 | 0.20628 | 2.0032 | 95.021 | 0.20240 | 1.6600 | 94.843 | 0.19918 |
| 140 | 2.6110 | 98.270 | 0.21149 | 2.0786 | 98.110 | 0.20763 | 1.7237 | 97.948 | 0.20445 |
| 160 | 2.7036 | 101.380 | 0.21659 | 2.1535 | 101.234 | 0.21276 | 1.7868 | 101.086 | 0.20960 |
| 180 | 2.7957 | 104.528 | 0.22159 | 2.2279 | 104.393 | 0.21778 | 1.8494 | 104.258 | 0.21463 |
| 200 | 2.8874 | 107.712 | 0.22649 | 2.3019 | 107.588 | 0.22269 | 1.9116 | 107.464 | 0.21957 |
| 220 | 2.9789 | 110.932 | 0.23130 | 2.3756 | 110.817 | 0.22752 | 1.9735 | 110.702 | 0.22440 |
| 240 | 3.0700 | 114.186 | 0.23602 | 2.4491 | 114.080 | 0.23225 | 2.0351 | 113.973 | 0.22915 |
| | 35 lb$_f$/in.$^2$ | | | 40 lb$_f$/in.$^2$ | | | 50 lb$_f$/in.$^2$ | | |
| 40 | 1.1850 | 82.442 | 0.17375 | 1.0258 | 82.148 | 0.17112 | 0.80248 | 81.540 | 0.16655 |
| 60 | 1.2442 | 85.463 | 0.17968 | 1.0789 | 85.206 | 0.17712 | 0.84713 | 84.676 | 0.17271 |
| 80 | 1.3021 | 88.504 | 0.18542 | 1.1306 | 88.277 | 0.18292 | 0.89025 | 87.811 | 0.17862 |
| 100 | 1.3589 | 91.570 | 0.19100 | 1.1812 | 91.367 | 0.18854 | 0.93216 | 90.953 | 0.18434 |
| 120 | 1.4148 | 94.663 | 0.19643 | 1.2309 | 94.480 | 0.19401 | 0.97313 | 94.110 | 0.18988 |
| 140 | 1.4701 | 97.785 | 0.20172 | 1.2798 | 97.620 | 0.19933 | 1.0133 | 97.286 | 0.19527 |
| 160 | 1.5248 | 100.938 | 0.20689 | 1.3282 | 100.788 | 0.20453 | 1.0529 | 100.485 | 0.20051 |
| 180 | 1.5789 | 104.122 | 0.21195 | 1.3761 | 103.985 | 0.20961 | 1.0920 | 103.708 | 0.20563 |
| 200 | 1.6327 | 107.338 | 0.21690 | 1.4236 | 107.212 | 0.21457 | 1.1307 | 106.958 | 0.21064 |
| 220 | 1.6862 | 110.586 | 0.22175 | 1.4707 | 110.469 | 0.21944 | 1.1690 | 110.235 | 0.21553 |
| 240 | 1.7394 | 113.865 | 0.22651 | 1.5176 | 113.757 | 0.22420 | 1.2070 | 113.539 | 0.22032 |
| 260 | 1.7923 | 117.175 | 0.23117 | 1.5642 | 117.074 | 0.22888 | 1.2447 | 116.871 | 0.22502 |
| | 60 lb$_f$/in.$^2$ | | | 70 lb$_f$/in.$^2$ | | | 80 lb$_f$/in.$^2$ | | |
| 60 | 0.69210 | 84.126 | 0.16892 | 0.58088 | 83.552 | 0.16556 | ... | ... | ... |
| 80 | 0.72964 | 87.330 | 0.17497 | 0.61458 | 86.832 | 0.17175 | 0.52795 | 86.316 | 0.16885 |
| 100 | 0.76588 | 90.528 | 0.18079 | 0.64685 | 90.091 | 0.17768 | 0.55734 | 89.640 | 0.17489 |
| 120 | 0.80110 | 93.731 | 0.18641 | 0.67803 | 93.343 | 0.18339 | 0.58556 | 92.945 | 0.18070 |
| 140 | 0.83551 | 96.945 | 0.19186 | 0.70836 | 96.597 | 0.18891 | 0.61286 | 96.242 | 0.18629 |
| 160 | 0.86928 | 100.776 | 0.19716 | 0.73800 | 99.862 | 0.19427 | 0.63943 | 99.542 | 0.19170 |
| 180 | 0.90252 | 103.427 | 0.20233 | 0.76708 | 103.141 | 0.19948 | 0.66543 | 102.851 | 0.19696 |
| 200 | 0.93531 | 106.700 | 0.20736 | 0.79571 | 106.439 | 0.20455 | 0.69095 | 106.174 | 0.20207 |
| 220 | 0.96775 | 109.997 | 0.21229 | 0.82397 | 109.756 | 0.20951 | 0.71609 | 109.513 | 0.20706 |
| 240 | 0.99988 | 113.319 | 0.21710 | 0.85191 | 113.096 | 0.21435 | 0.74090 | 112.872 | 0.21193 |
| 260 | 1.0318 | 116.666 | 0.22182 | 0.87959 | 116.459 | 0.21909 | 0.76544 | 116.251 | 0.21669 |
| 280 | 1.0634 | 120.039 | 0.22644 | 0.90705 | 119.846 | 0.22373 | 0.78975 | 119.652 | 0.22135 |
| | 90 lb$_f$/in.$^2$ | | | 100 lb$_f$/in.$^2$ | | | 125 lb$_f$/in.$^2$ | | |
| 100 | 0.48749 | 89.175 | 0.17234 | 0.43138 | 88.694 | 0.16996 | 0.32943 | 87.407 | 0.16455 |
| 120 | 0.51346 | 92.536 | 0.17824 | 0.45562 | 92.116 | 0.17597 | 0.35086 | 91.008 | 0.17087 |
| 140 | 0.53845 | 95.879 | 0.18391 | 0.47881 | 95.507 | 0.18172 | 0.37098 | 94.537 | 0.17686 |
| 160 | 0.56268 | 99.216 | 0.18938 | 0.50118 | 98.884 | 0.18726 | 0.39015 | 98.023 | 0.18258 |
| 180 | 0.58629 | 102.557 | 0.19469 | 0.52291 | 102.257 | 0.19262 | 0.40857 | 101.484 | 0.18807 |
| 200 | 0.60941 | 105.905 | 0.19984 | 0.54413 | 105.633 | 0.19782 | 0.42642 | 104.934 | 0.19338 |
| 220 | 0.63213 | 109.267 | 0.20486 | 0.56492 | 109.018 | 0.20287 | 0.44380 | 108.380 | 0.19853 |
| 240 | 0.65451 | 112.644 | 0.20976 | 0.58538 | 112.415 | 0.20780 | 0.46081 | 111.829 | 0.20353 |
| 260 | 0.67662 | 116.040 | 0.21455 | 0.60554 | 115.828 | 0.21261 | 0.47750 | 115.287 | 0.20840 |
| 280 | 0.69849 | 119.456 | 0.21923 | 0.62546 | 119.258 | 0.21731 | 0.49394 | 118.756 | 0.21316 |
| 300 | 0.72016 | 122.892 | 0.22381 | 0.64518 | 122.707 | 0.22191 | 0.51016 | 122.238 | 0.21780 |
| 320 | 0.74166 | 126.349 | 0.22830 | 0.66472 | 126.176 | 0.22641 | 0.52619 | 125.737 | 0.22235 |
| | 150 lb$_f$/in.$^2$ | | | 175 lb$_f$/in.$^2$ | | | 200 lb$_f$/in.$^2$ | | |
| 120 | 0.28007 | 89.800 | 0.16629 | ... | ... | ... | ... | ... | ... |
| 140 | 0.29845 | 93.498 | 0.17256 | 0.24595 | 92.373 | 0.16859 | 0.20579 | 91.137 | 0.16480 |
| 160 | 0.31566 | 97.112 | 0.17849 | 0.26198 | 96.142 | 0.17478 | 0.22121 | 95.100 | 0.17130 |
| 180 | 0.33200 | 100.675 | 0.18415 | 0.27697 | 99.823 | 0.18062 | 0.23535 | 98.921 | 0.17737 |
| 200 | 0.34769 | 104.206 | 0.18958 | 0.29120 | 103.447 | 0.18620 | 0.24860 | 102.652 | 0.18311 |
| 220 | 0.36285 | 107.720 | 0.19483 | 0.30485 | 107.036 | 0.19156 | 0.26117 | 106.325 | 0.18860 |
| 240 | 0.37761 | 111.226 | 0.19992 | 0.31804 | 110.605 | 0.19674 | 0.27323 | 109.962 | 0.19387 |
| 260 | 0.39203 | 114.732 | 0.20485 | 0.33087 | 114.162 | 0.20175 | 0.28489 | 113.576 | 0.19896 |
| 280 | 0.40617 | 118.242 | 0.20967 | 0.34339 | 117.717 | 0.20662 | 0.29623 | 117.178 | 0.20390 |
| 300 | 0.42008 | 121.761 | 0.21436 | 0.35567 | 121.273 | 0.21137 | 0.30730 | 120.775 | 0.20870 |
| 320 | 0.43379 | 125.290 | 0.21894 | 0.36773 | 124.835 | 0.21599 | 0.31815 | 124.373 | 0.21337 |
| 340 | 0.44733 | 128.833 | 0.22343 | 0.37963 | 128.407 | 0.22052 | 0.32881 | 127.974 | 0.21793 |

**Table II.2-5   Continued**

| Temp. (°F) | $v$ | $h$ | $s$ | $v$ | $h$ | $s$ | $v$ | $h$ | $s$ |
|---|---|---|---|---|---|---|---|---|---|
| | 250 lb$_f$/in.$^2$ | | | 300 lb$_f$/in.$^2$ | | | 400 lb$_f$/in.$^2$ | | |
| 160 | 0.16249 | 92.717 | 0.16462 | ... | ... | ... | ... | ... | ... |
| 180 | 0.17605 | 96.925 | 0.17130 | 0.13482 | 94.556 | 0.16537 | 0.091005 | 93.718 | 0.16092 |
| 200 | 0.18824 | 100.930 | 0.17747 | 0.14697 | 98.975 | 0.17217 | 0.10316 | 99.046 | 0.16888 |
| 220 | 0.19952 | 104.809 | 0.18326 | 0.15774 | 103.136 | 0.17838 | 0.11300 | 103.735 | 0.17568 |
| 240 | 0.21014 | 108.607 | 0.18877 | 0.16761 | 107.140 | 0.18419 | 0.11300 | 103.735 | 0.17568 |
| 260 | 0.22027 | 112.351 | 0.19404 | 0.17685 | 111.043 | 0.18969 | 0.12163 | 108.105 | 0.18183 |
| 280 | 0.23001 | 116.060 | 0.19913 | 0.18562 | 114.879 | 0.19495 | 0.12949 | 112.286 | 0.18756 |
| 300 | 0.23944 | 119.747 | 0.20405 | 0.19402 | 118.670 | 0.20000 | 0.13680 | 116.343 | 0.19298 |
| 320 | 0.24862 | 123.420 | 0.20882 | 0.20214 | 122.430 | 0.20489 | 0.14372 | 120.318 | 0.19814 |
| 340 | 0.25759 | 127.088 | 0.21346 | 0.21002 | 126.171 | 0.20963 | 0.15032 | 124.235 | 0.20310 |
| 360 | 0.26639 | 130.754 | 0.21799 | 0.21770 | 129.900 | 0.21423 | 0.15668 | 128.112 | 0.20789 |
| 380 | 0.27504 | 134.423 | 0.22241 | 0.22522 | 133.624 | 0.21872 | 0.16285 | 131.961 | 0.21258 |
| | 500 lb$_f$/in.$^2$ | | | 600 lb$_f$/in.$^2$ | | | | | |
| 220 | 0.064207 | 92.397 | 0.15683 | ... | ... | ... | | | |
| 240 | 0.077620 | 99.218 | 0.16672 | 0.047488 | 91.024 | 0.15335 | | | |
| 260 | 0.087054 | 104.526 | 0.17421 | 0.061922 | 99.741 | 0.16566 | | | |
| 280 | 0.094923 | 109.277 | 0.18072 | 0.070859 | 105.637 | 0.17374 | | | |
| 300 | 0.10190 | 113.729 | 0.18666 | 0.078059 | 110.729 | 0.18053 | | | |
| 320 | 0.10829 | 117.997 | 0.19221 | 0.084333 | 115.420 | 0.18663 | | | |
| 340 | 0.11426 | 122.143 | 0.19746 | 0.90017 | 119.871 | 0.19227 | | | |
| 360 | 0.11992 | 126.205 | 0.20247 | 0.95289 | 124.167 | 0.19757 | | | |
| 380 | 0.12533 | 130.207 | 0.20730 | 0.10025 | 128.355 | 0.20262 | | | |
| 400 | 0.13054 | 134.166 | 0.21196 | 0.10498 | 132.466 | 0.20746 | | | |
| 420 | 0.13559 | 138.096 | 0.21648 | 0.10952 | 136.523 | 0.21213 | | | |
| 440 | 0.14051 | 142.004 | 0.22087 | 0.11391 | 140.539 | 0.21664 | | | |

*Source*: Copyright 1955 and 1956 E. I. du Pont de Nemours & Company, Inc.; reprinted by permission.

**Table II.2-6   Thermodynamic Properties of Saturated Ammonia**

| Temp. (°F) | Abs. Press. $P$ (lb$_f$/in.$^2$) | Specific Volume (ft$^3$/lb$_m$) | | | Enthalpy (Btu/lb$_m$) | | | Entropy (Btu/lb$_m$ · °R) | | |
|---|---|---|---|---|---|---|---|---|---|---|
| | | Sat. Liquid $v_f$ | Evap. $v_{fg}$ | Sat. Vapor $v_g$ | Sat. Liquid $h_f$ | Evap. $h_{fg}$ | Sat. Vapor $h_g$ | Sat. Liquid $s_f$ | Evap. $s_{fg}$ | Sat. Vapor $s_g$ |
| −60 | 5.55 | 0.0228 | 44.707 | 44.73 | −21.2 | 610.8 | 589.6 | −0.0517 | 1.5286 | 1.4769 |
| −55 | 6.54 | 0.0229 | 38.357 | 38.38 | −15.9 | 607.5 | 591.6 | −0.0386 | 1.5017 | 1.4631 |
| −50 | 7.67 | 0.0230 | 33.057 | 33.08 | −10.6 | 604.3 | 593.7 | −0.0256 | 1.4753 | 1.4497 |
| −45 | 8.95 | 0.0231 | 28.597 | 28.62 | −5.3 | 600.9 | 595.6 | −0.0127 | 1.4495 | 1.4368 |
| −40 | 10.41 | 0.02322 | 24.837 | 24.86 | 0 | 597.6 | 597.6 | 0.000 | 1.4242 | 1.4242 |
| −35 | 12.05 | 0.02333 | 21.657 | 21.68 | 5.3 | 594.2 | 599.5 | 0.0126 | 1.3994 | 1.4120 |
| −30 | 13.90 | 0.0235 | 18.947 | 18.97 | 10.7 | 590.7 | 601.4 | 0.0250 | 1.3751 | 1.4001 |
| −25 | 15.98 | 0.0236 | 16.636 | 16.66 | 16.0 | 587.2 | 603.2 | 0.0374 | 1.3512 | 1.3886 |
| −20 | 18.30 | 0.0237 | 14.656 | 14.68 | 21.4 | 583.6 | 605.0 | 0.0497 | 1.3277 | 1.3774 |
| −15 | 20.88 | 0.02381 | 12.946 | 12.97 | 26.7 | 580.0 | 606.7 | 0.0618 | 1.3044 | 1.3664 |
| −10 | 23.74 | 0.02393 | 11.476 | 11.50 | 32.1 | 576.4 | 608.5 | 0.0738 | 1.2820 | 1.3558 |
| −5 | 26.92 | 0.02406 | 10.206 | 10.23 | 37.5 | 572.6 | 610.1 | 0.0857 | 1.2597 | 1.3454 |
| 0 | 30.42 | 0.02419 | 9.092 | 9.116 | 42.9 | 568.9 | 611.8 | 0.0975 | 1.2377 | 1.3352 |
| 5 | 34.27 | 0.02432 | 8.1257 | 8.150 | 48.3 | 565.0 | 613.3 | 0.1092 | 1.2161 | 1.3253 |
| 10 | 38.51 | 0.02446 | 7.2795 | 7.304 | 53.8 | 561.1 | 614.9 | 0.1208 | 1.1949 | 1.3157 |
| 15 | 43.14 | 0.02460 | 6.5374 | 6.562 | 59.2 | 557.1 | 616.3 | 0.1323 | 1.1739 | 1.3062 |
| 20 | 48.21 | 0.02474 | 5.8853 | 5.910 | 64.7 | 553.1 | 617.8 | 0.1437 | 1.1532 | 1.2969 |
| 25 | 53.73 | 0.02488 | 5.3091 | 5.334 | 70.2 | 548.9 | 619.1 | 0.1551 | 1.1328 | 1.2879 |
| 30 | 59.74 | 0.02503 | 4.8000 | 4.825 | 75.7 | 544.8 | 620.5 | 0.1663 | 1.1127 | 1.2790 |
| 35 | 66.26 | 0.02518 | 4.3478 | 4.373 | 81.2 | 540.5 | 621.7 | 0.1775 | 1.0929 | 1.2704 |
| 40 | 73.32 | 0.02533 | 3.9457 | 3.971 | 86.8 | 536.2 | 623.0 | 0.1885 | 1.0733 | 1.2618 |
| 45 | 80.96 | 0.02548 | 3.5885 | 3.614 | 92.3 | 531.8 | 624.1 | 0.1996 | 1.0539 | 1.2535 |
| 50 | 89.19 | 0.02564 | 3.2684 | 3.294 | 97.9 | 527.3 | 625.2 | 0.2105 | 1.0348 | 1.2453 |
| 55 | 98.06 | 0.02581 | 2.9822 | 3.008 | 103.5 | 522.8 | 626.3 | 0.2214 | 1.0159 | 1.2373 |
| 60 | 107.6 | 0.02597 | 2.7250 | 2.751 | 109.2 | 518.1 | 627.3 | 0.2322 | 0.9972 | 1.2294 |
| 65 | 117.8 | 0.02614 | 2.4939 | 2.520 | 114.8 | 513.4 | 628.2 | 0.2430 | 0.9786 | 1.2216 |
| 70 | 128.8 | 0.02632 | 2.2857 | 2.312 | 120.5 | 508.6 | 629.1 | 0.2537 | 0.9603 | 1.2140 |
| 75 | 140.5 | 0.02650 | 2.0985 | 2.125 | 126.2 | 503.7 | 629.9 | 0.2643 | 0.9422 | 1.2065 |
| 80 | 153.0 | 0.02668 | 1.9283 | 1.955 | 132.0 | 498.7 | 630.7 | 0.2749 | 0.9242 | 1.1991 |
| 85 | 166.4 | 0.02687 | 1.7741 | 1.801 | 137.8 | 493.6 | 631.4 | 0.2854 | 0.9064 | 1.1918 |
| 90 | 180.6 | 0.02707 | 1.6339 | 1.661 | 143.5 | 488.5 | 632.0 | 0.2958 | 0.8888 | 1.1846 |
| 95 | 195.8 | 0.02727 | 1.5067 | 1.534 | 149.4 | 483.2 | 632.6 | 0.3062 | 0.8713 | 1.1775 |
| 100 | 211.9 | 0.02747 | 1.3915 | 1.419 | 155.2 | 477.8 | 633.0 | 0.3166 | 0.8539 | 1.1705 |
| 105 | 228.9 | 0.02769 | 1.2853 | 1.313 | 161.1 | 472.3 | 633.4 | 0.3269 | 0.8366 | 1.1635 |
| 110 | 247.0 | 0.02790 | 1.1891 | 1.217 | 167.0 | 466.7 | 633.7 | 0.3372 | 0.8194 | 1.1566 |
| 115 | 266.2 | 0.02813 | 1.0999 | 1.128 | 173.0 | 460.9 | 633.9 | 0.3474 | 0.8023 | 1.1497 |
| 120 | 286.4 | 0.02836 | 1.0186 | 1.047 | 179.0 | 455.0 | 634.0 | 0.3576 | 0.7851 | 1.1427 |
| 125 | 307.8 | 0.02860 | 0.9444 | 0.973 | 185.1 | 448.9 | 634.0 | 0.3679 | 0.7679 | 1.1358 |

*Source*: National Bureau of Standards Circular No. 142, *Tables of Thermodynamic Properties of Ammonia*.

**Table II.2-7  Thermodynamic Properties of Superheated Ammonia**

| Abs. Press. (Sat. Temp.) (lbf/in.²) | | 0 | 20 | 40 | 60 | 80 | 100 | 120 | 140 | 160 | 180 | 200 | 220 |
|---|---|---|---|---|---|---|---|---|---|---|---|---|---|
| | | | | | | | | Temperature (°F) | | | | | |
| 10 | $v$ | 28.58 | 29.90 | 31.20 | 32.49 | 33.78 | 35.07 | 36.35 | 37.62 | 38.90 | 40.17 | 41.45 | |
| (−41.34) | $h$ | 618.9 | 629.1 | 639.3 | 649.5 | 659.7 | 670.0 | 680.3 | 690.6 | 701.1 | 711.6 | 722.2 | |
| | $s$ | 1.477 | 1.499 | 1.520 | 1.540 | 1.559 | 1.578 | 1.596 | 1.614 | 1.631 | 1.647 | 1.664 | |
| 15 | $v$ | 18.92 | 19.82 | 20.70 | 21.58 | 22.44 | 23.31 | 24.17 | 25.03 | 25.88 | 26.74 | 27.59 | |
| (−27.29) | $h$ | 617.2 | 627.8 | 638.2 | 648.5 | 658.9 | 669.2 | 679.6 | 690.0 | 700.5 | 711.1 | 721.7 | |
| | $s$ | 1.427 | 1.450 | 1.471 | 1.491 | 1.511 | 1.529 | 1.548 | 1.566 | 1.583 | 1.599 | 1.616 | |
| 20 | $v$ | 14.09 | 14.78 | 15.45 | 16.12 | 16.78 | 17.43 | 18.08 | 18.73 | 19.37 | 20.02 | 20.66 | 21.3 |
| (−16.64) | $h$ | 615.5 | 626.4 | 637.0 | 647.5 | 658.0 | 668.5 | 678.9 | 689.4 | 700.0 | 710.6 | 721.2 | 732.0 |
| | $s$ | 1.391 | 1.414 | 1.436 | 1.456 | 1.476 | 1.495 | 1.513 | 1.531 | 1.549 | 1.565 | 1.582 | 1.598 |
| 25 | $v$ | 11.19 | 11.75 | 12.30 | 12.84 | 13.37 | 13.90 | 14.43 | 14.95 | 15.47 | 15.99 | 16.50 | 17.02 |
| (−7.96) | $h$ | 613.8 | 625.0 | 635.8 | 646.5 | 657.1 | 667.7 | 678.2 | 688.8 | 699.4 | 710.1 | 720.8 | 731.6 |
| | $s$ | 1.362 | 1.386 | 1.408 | 1.429 | 1.449 | 1.468 | 1.486 | 1.504 | 1.522 | 1.539 | 1.555 | 1.571 |
| 30 | $v$ | 9.25 | 9.731 | 10.20 | 10.65 | 11.10 | 11.55 | 11.99 | 12.43 | 12.87 | 13.30 | 13.73 | 14.16 |
| (−.57) | $h$ | 611.9 | 623.5 | 634.6 | 645.5 | 656.2 | 666.9 | 677.5 | 688.2 | 698.8 | 709.6 | 720.3 | 731.1 |
| | $s$ | 1.337 | 1.362 | 1.385 | 1.406 | 1.426 | 1.446 | 1.464 | 1.482 | 1.500 | 1.517 | 1.533 | 1.550 |
| 35 | $v$ | | 8.287 | 8.695 | 9.093 | 9.484 | 9.869 | 10.25 | 10.63 | 11.00 | 11.38 | 11.75 | 12.12 |
| (5.89) | $h$ | | 622.0 | 633.4 | 644.4 | 655.3 | 666.1 | 676.8 | 687.6 | 698.3 | 709.1 | 719.9 | 730.7 |
| | $s$ | | 1.341 | 1.365 | 1.386 | 1.407 | 1.427 | 1.445 | 1.464 | 1.481 | 1.498 | 1.515 | 1.531 |
| 40 | $v$ | | 7.203 | 7.568 | 7.922 | 8.268 | 8.609 | 8.945 | 9.278 | 9.609 | 9.938 | 10.27 | 10.59 |
| (11.66) | $h$ | | 620.4 | 632.1 | 643.4 | 654.4 | 665.3 | 676.1 | 686.9 | 697.7 | 708.5 | 719.4 | 730.3 |
| | $s$ | | 1.323 | 1.347 | 1.369 | 1.390 | 1.410 | 1.429 | 1.447 | 1.465 | 1.482 | 1.499 | 1.515 |
| 45 | $v$ | | 6.363 | 6.694 | 7.014 | 7.326 | 7.632 | 7.934 | 8.232 | 8.528 | 8.822 | 9.115 | 9.406 |
| (16.87) | $h$ | | 618.8 | 630.8 | 642.3 | 653.5 | 664.6 | 675.5 | 686.3 | 697.2 | 708.0 | 718.9 | 729.9 |
| | $s$ | | 1.307 | 1.331 | 1.354 | 1.375 | 1.395 | 1.414 | 1.433 | 1.450 | 1.468 | 1.485 | 1.501 |
| 50 | $v$ | | | 5.988 | 6.280 | 6.564 | 6.843 | 7.117 | 7.387 | 7.655 | 7.921 | 8.185 | 8.448 |
| (21.67) | $h$ | | | 629.5 | 641.2 | 652.6 | 663.7 | 674.7 | 685.7 | 696.6 | 707.5 | 718.5 | 729.4 |
| | $s$ | | | 1.317 | 1.340 | 1.361 | 1.382 | 1.401 | 1.420 | 1.437 | 1.455 | 1.472 | 1.488 |
| 60 | $v$ | | | 4.933 | 5.184 | 5.428 | 5.665 | 5.897 | 6.126 | 6.352 | 6.576 | 6.798 | 7.019 |
| (30.21) | $h$ | | | 626.8 | 639.0 | 650.7 | 662.1 | 673.3 | 684.4 | 695.5 | 706.5 | 717.5 | 728.6 |
| | $s$ | | | 1.291 | 1.315 | 1.337 | 1.358 | 1.378 | 1.397 | 1.415 | 1.432 | 1.449 | 1.466 |
| 70 | $v$ | 4.401 | 4.615 | 4.822 | 5.025 | 5.224 | 5.420 | 5.615 | 5.807 | 6.187 | 6.563 | | |
| (37.7) | $h$ | 636.6 | 648.7 | 660.4 | 671.8 | 683.1 | 694.3 | 705.5 | 716.6 | 738.9 | 761.4 | | |
| | $s$ | 1.294 | 1.317 | 1.338 | 1.358 | 1.377 | 1.395 | 1.413 | 1.430 | 1.463 | 1.494 | | |
| 80 | $v$ | 3.812 | 4.005 | 4.190 | 4.371 | 4.548 | 4.722 | 4.893 | 5.063 | 5.398 | 5.73 | | |
| (44.4) | $h$ | 634.3 | 646.7 | 658.7 | 670.4 | 681.8 | 693.2 | 704.4 | 715.6 | 738.1 | 760.7 | | |
| | $s$ | 1.275 | 1.298 | 1.320 | 1.340 | 1.360 | 1.378 | 1.396 | 1.414 | 1.447 | 1.478 | | |
| 90 | $v$ | 3.353 | 3.529 | 3.698 | 3.862 | 4.021 | 4.178 | 4.332 | 4.484 | 4.785 | 5.081 | | |
| (50.47) | $h$ | 631.8 | 644.7 | 657.0 | 668.9 | 680.5 | 692.0 | 703.4 | 714.7 | 737.3 | 760.0 | | |
| | $s$ | 1.257 | 1.281 | 1.304 | 1.325 | 1.344 | 1.363 | 1.381 | 1.400 | 1.432 | 1.464 | | |
| 100 | $v$ | 2.985 | 3.149 | 3.304 | 3.454 | 3.600 | 3.743 | 3.883 | 4.021 | 4.294 | 4.562 | | |
| (56.05) | $h$ | 629.3 | 642.6 | 655.2 | 667.3 | 679.2 | 690.8 | 702.3 | 713.7 | 736.5 | 759.4 | | |
| | $s$ | 1.241 | 1.266 | 1.289 | 1.310 | 1.331 | 1.349 | 1.368 | 1.385 | 1.419 | 1.451 | | |
| 140 | $v$ | | 2.166 | 2.288 | 2.404 | 2.515 | 2.622 | 2.727 | 2.830 | 3.030 | 3.227 | 3.420 | |
| (74.79) | $h$ | | 633.8 | 647.8 | 661.1 | 673.7 | 686.0 | 698.0 | 709.9 | 733.3 | 756.7 | 780.0 | |
| | $s$ | | 1.214 | 1.240 | 1.263 | 1.284 | 1.305 | 1.324 | 1.342 | 1.376 | 1.409 | 1.440 | |
| 180 | $v$ | | | 1.720 | 1.818 | 1.910 | 1.999 | 2.084 | 2.167 | 2.328 | 2.484 | 2.637 | |
| (89.78) | $h$ | | | 639.9 | 654.4 | 668.0 | 681.0 | 693.6 | 705.9 | 730.1 | 753.9 | 777.7 | |
| | $s$ | | | 1.199 | 1.225 | 1.248 | 1.269 | 1.289 | 1.308 | 1.344 | 1.377 | 1.408 | |
| 220 | $v$ | | | | 1.443 | 1.525 | 1.601 | 1.675 | 1.745 | 1.881 | 2.012 | 2.140 | 2.265 |
| (102.42) | $h$ | | | | 647.3 | 662.0 | 675.8 | 689.1 | 701.9 | 726.8 | 751.1 | 775.3 | 799.5 |
| | $s$ | | | | 1.192 | 1.217 | 1.239 | 1.260 | 1.280 | 1.317 | 1.351 | 1.383 | 1.413 |
| 240 | $v$ | | | | 1.302 | 1.380 | 1.452 | 1.521 | 1.587 | 1.714 | 1.835 | 1.954 | 2.069 |
| (108.09) | $h$ | | | | 643.5 | 658.8 | 673.1 | 686.7 | 699.8 | 725.1 | 749.8 | 774.1 | 798.4 |
| | $s$ | | | | 1.176 | 1.203 | 1.226 | 1.248 | 1.268 | 1.305 | 1.339 | 1.371 | 1.402 |
| 260 | $v$ | | | | 1.182 | 1.257 | 1.326 | 1.391 | 1.453 | 1.572 | 1.686 | 1.796 | 1.904 |
| (113.42) | $h$ | | | | 639.5 | 655.6 | 670.4 | 684.4 | 697.7 | 723.4 | 748.4 | 772.9 | 797.4 |
| | $s$ | | | | 1.162 | 1.189 | 1.213 | 1.235 | 1.256 | 1.294 | 1.329 | 1.361 | 1.391 |
| 280 | $v$ | | | | 1.078 | 1.151 | 1.217 | 1.279 | 1.339 | 1.451 | 1.558 | 1.661 | 1.762 |
| (118.45) | $h$ | | | | 635.4 | 652.2 | 667.6 | 681.9 | 695.6 | 721.8 | 747.0 | 771.7 | 796.3 |
| | $s$ | | | | 1.147 | 1.176 | 1.201 | 1.224 | 1.245 | 1.283 | 1.318 | 1.351 | 1.382 |

*Source*: National Bureau of Standards Circular No. 142, *Tables of Thermodynamic Properties of Ammonia*.

**Table II.2-8   Thermodynamic Properties of Saturated Nitrogen**

| Temp. (°R) | Abs. Press. $P$ (lb$_f$/in.$^2$) | Specific Volume (ft$^3$/lb$_m$) Sat. Liquid $v_f$ | Evap. $v_{fg}$ | Sat. Vapor $v_g$ | Enthalpy (Btu lb$_m$) Sat. Liquid $h_f$ | Evap. $h_{fg}$ | Sat. Vapor $h_g$ | Entropy (Btu/lb$_m$ · °R) Sat. Liquid $s_f$ | Evap. $s_{fg}$ | Sat. Vapor $s_g$ |
|---|---|---|---|---|---|---|---|---|---|---|
| 113.670 | 1.813 | 0.01845 | 23.793 | 23.812 | 0.000 | 92.891 | 92.891 | 0.00000 | 0.81720 | 0.81720 |
| 120.000 | 3.337 | 0.01875 | 13.570 | 13.589 | 3.113 | 91.224 | 94.337 | 0.02661 | 0.76020 | 0.78681 |
| 130.000 | 7.654 | 0.01929 | 6.3208 | 6.3401 | 8.062 | 88.432 | 96.494 | 0.06610 | 0.68025 | 0.74634 |
| 139.255 | 14.696 | 0.01984 | 3.4592 | 3.4791 | 12.639 | 85.668 | 98.306 | 0.09992 | 0.61518 | 0.71510 |
| 140.000 | 15.425 | 0.01989 | 3.3072 | 3.3271 | 13.006 | 85.436 | 98.443 | 0.10253 | 0.61026 | 0.71279 |
| 150.000 | 28.120 | 0.02056 | 1.8865 | 1.9071 | 17.945 | 82.179 | 100.124 | 0.13628 | 0.54786 | 0.68414 |
| 160.000 | 47.383 | 0.02132 | 1.1469 | 1.1682 | 22.928 | 78.458 | 101.476 | 0.16795 | 0.49093 | 0.65888 |
| 170.000 | 74.991 | 0.02219 | 0.7299 | 0.7521 | 28.045 | 74.383 | 102.427 | 0.19829 | 0.43754 | 0.63584 |
| 180.000 | 112.808 | 0.02323 | 0.4789 | 0.5021 | 33.411 | 69.478 | 102.889 | 0.22805 | 0.38599 | 0.61404 |
| 190.000 | 162.761 | 0.02449 | 0.3190 | 0.3435 | 39.153 | 63.582 | 102.735 | 0.25789 | 0.33464 | 0.59254 |
| 200.000 | 226.853 | 0.02613 | 0.2119 | 0.2380 | 45.283 | 56.474 | 101.757 | 0.28780 | 0.28237 | 0.57017 |
| 210.000 | 307.276 | 0.02845 | 0.1354 | 0.1639 | 52.061 | 47.474 | 99.536 | 0.31894 | 0.22607 | 0.54501 |
| 220.000 | 406.739 | 0.03249 | 0.0750 | 0.1075 | 60.336 | 34.536 | 94.872 | 0.35494 | 0.15698 | 0.51192 |
| 226.000 | 477.104 | 0.03806 | 0.0374 | 0.0755 | 68.123 | 20.423 | 88.546 | 0.38789 | 0.09037 | 0.47826 |

*Source*: Abstracted from National Bureau of Standards Technical Note 129A, *The Thermodynamic Properties of Nitrogen from 114 to 540 R between 1.0 and 3000 psia, Supplement A (British Units)*, by Thomas R. Strobridge.

**Table II.2-9   Thermodynamic Properties of Superheated Nitrogen**

| Temp. (°R) | $v$ (ft$^3$/lb$_m$) | $h$ (Btu/lb$_m$) | $s$ (Btu/lb$_m$ · °R) | $v$ (ft$^3$/lb$_m$) | $h$ (Btu/lb$_m$) | $s$ (Btu/lb$_m$ · °R) | $v$ (ft$^3$/lb$_m$) | $h$ (Btu/lb$_m$) | $s$ (Btu/lb$_m$ · °R) |
|---|---|---|---|---|---|---|---|---|---|
| | 14.7 lb$_f$/in.$^2$ | | | 20 lb$_f$/in.$^2$ | | | 50 lb$_f$/in.$^2$ | | |
| 150 | 3.7782 | 101.086 | 0.7343 | 2.7395 | 100.715 | 0.7109 | | | |
| 200 | 5.1366 | 113.849 | 0.8078 | 3.7538 | 113.625 | 0.7852 | 1.4534 | 112.315 | 0.7159 |
| 250 | 6.4680 | 126.443 | 0.8640 | 4.7397 | 126.293 | 0.8418 | 1.8663 | 125.432 | 0.7744 |
| 300 | 7.7876 | 138.958 | 0.9096 | 5.7138 | 138.850 | 0.8875 | 2.2662 | 138.239 | 0.8212 |
| 350 | 9.1015 | 151.432 | 0.9481 | 6.6820 | 151.351 | 0.9261 | 2.6599 | 150.896 | 0.8602 |
| 400 | 10.412 | 163.882 | 0.9814 | 7.6469 | 163.821 | 0.9594 | 3.0502 | 163.471 | 0.8938 |
| 450 | 11.721 | 176.319 | 1.0107 | 8.6098 | 176.271 | 0.9887 | 3.4385 | 175.997 | 0.9233 |
| 500 | 13.028 | 188.748 | 1.0368 | 9.5714 | 188.710 | 1.0149 | 3.8255 | 188.492 | 0.9496 |
| 540 | 14.073 | 198.690 | 1.0560 | 10.340 | 198.657 | 1.0341 | 4.1344 | 198.474 | 0.9688 |
| | 100 lb$_f$/in.$^2$ | | | 200 lb$_f$/in.$^2$ | | | 500 lb$_f$/in.$^2$ | | |
| 200 | 0.6834 | 109.931 | 0.6585 | 0.2884 | 103.911 | 0.5875 | | | |
| 250 | 0.9078 | 123.948 | 0.7212 | 0.4272 | 120.763 | 0.6631 | 0.1321 | 108.378 | 0.5608 |
| 300 | 1.1169 | 137.205 | 0.7696 | 0.5420 | 135.076 | 0.7153 | 0.1966 | 128.168 | 0.6335 |
| 350 | 1.3192 | 150.133 | 0.8094 | 0.6490 | 148.589 | 0.7570 | 0.2473 | 143.838 | 0.6819 |
| 400 | 1.5181 | 162.888 | 0.8435 | 0.7522 | 161.718 | 0.7921 | 0.2932 | 158.205 | 0.7202 |
| 450 | 1.7149 | 175.540 | 0.8733 | 0.8532 | 174.630 | 0.8225 | 0.3368 | 171.933 | 0.7526 |
| 500 | 1.9103 | 188.129 | 0.8998 | 0.9529 | 187.408 | 0.8494 | 0.3790 | 185.292 | 0.7807 |
| 540 | 2.0660 | 198.170 | 0.9192 | 1.0319 | 197.567 | 0.8690 | 0.4120 | 195.807 | 0.8010 |
| | 1000 lb$_f$/in.$^2$ | | | 2000 lb$_f$/in.$^2$ | | | 3000 lb$_f$/in.$^2$ | | |
| 250 | 0.0384 | 78.126 | 0.4145 | 0.0286 | 70.290 | 0.3596 | 0.0261 | 69.719 | 0.3371 |
| 300 | 0.0828 | 115.224 | 0.5514 | 0.0398 | 97.820 | 0.4599 | 0.0321 | 93.216 | 0.4228 |
| 350 | 0.1150 | 135.789 | 0.6150 | 0.0552 | 122.614 | 0.5366 | 0.0403 | 116.066 | 0.4933 |
| 400 | 0.1417 | 152.487 | 0.6597 | 0.0699 | 142.869 | 0.5908 | 0.0493 | 136.883 | 0.5490 |
| 450 | 0.1659 | 167.637 | 0.6954 | 0.0833 | 160.406 | 0.6321 | 0.0582 | 155.522 | 0.5930 |
| 500 | 0.1887 | 181.969 | 0.7256 | 0.0958 | 176.411 | 0.6659 | 0.0667 | 172.551 | 0.6289 |
| 540 | 0.2063 | 193.069 | 0.7470 | 0.1053 | 188.526 | 0.6892 | 0.0732 | 185.361 | 0.6535 |

*Source*: Abstracted from National Bureau of Standards Technical Note 129A, *The Thermodynamic Properties of Nitrogen from 114 to 540 R between 1.0 and 3000 psia, Supplement A (British Units)*, by Thomas R. Strobridge.

## Table II.2-10  Pressure/Enthalpy Diagram. FREON-22

**Table II.2-11   Pressure/Enthalpy Diagram, FREON-11**

**Table II.2-12  Thermodynamic Properties of Air at Low Pressure**

| $T$ (°R) | $h$ (Btu/lb$_m$) | $u$ (Btu/lb$_m$) | $s°$ (Btu/lb$_m$ · °R) | $T$ (°R) | $h$ (Btu/lb$_m$) | $u$ (Btu/lb$_m$) | $s°$ (Btu/lb$_m$ · °R) |
|---|---|---|---|---|---|---|---|
| 200 | 47.67 | 33.96 | 0.36303 | 1200 | 291.30 | 209.05 | 0.79628 |
| 220 | 52.46 | 37.38 | 0.38584 | 1220 | 296.41 | 212.78 | 0.80050 |
| 240 | 57.25 | 40.80 | 0.40666 | 1240 | 301.52 | 216.53 | 0.80466 |
| 260 | 62.03 | 44.21 | 0.42582 | 1260 | 306.65 | 220.28 | 0.80876 |
| 280 | 66.82 | 47.63 | 0.44356 | 1280 | 311.79 | 224.05 | 0.81280 |
| 300 | 71.61 | 51.04 | 0.46007 | 1300 | 316.94 | 227.83 | 0.81680 |
| 320 | 76.40 | 54.46 | 0.47550 | 1320 | 322.11 | 231.63 | 0.82075 |
| 340 | 81.18 | 57.87 | 0.49002 | 1340 | 327.29 | 235.43 | 0.82464 |
| 360 | 85.97 | 61.29 | 0.50369 | 1360 | 332.48 | 239.25 | 0.82848 |
| 380 | 90.75 | 64.70 | 0.51663 | 1380 | 337.68 | 243.08 | 0.83229 |
| 400 | 95.53 | 68.11 | 0.52890 | 1400 | 342.90 | 246.93 | 0.83604 |
| 420 | 100.32 | 71.52 | 0.54058 | 1420 | 348.14 | 250.79 | 0.83975 |
| 440 | 105.11 | 74.93 | 0.55172 | 1440 | 353.37 | 254.66 | 0.84341 |
| 460 | 109.90 | 78.36 | 0.56235 | 1460 | 358.63 | 258.54 | 0.84704 |
| 480 | 114.69 | 81.77 | 0.57255 | 1480 | 363.89 | 262.44 | 0.85062 |
| 500 | 119.48 | 85.20 | 0.58233 | 1500 | 369.17 | 266.34 | 0.85416 |
| 520 | 124.27 | 88.62 | 0.59173 | 1520 | 374.47 | 270.26 | 0.85767 |
| 540 | 129.06 | 92.04 | 0.60078 | 1540 | 379.77 | 274.20 | 0.86113 |
| 560 | 133.86 | 95.47 | 0.60950 | 1560 | 385.08 | 278.13 | 0.86456 |
| 580 | 138.66 | 98.90 | 0.61793 | 1580 | 390.40 | 282.09 | 0.86794 |
| 600 | 143.47 | 102.34 | 0.62607 | 1600 | 395.74 | 286.06 | 0.87130 |
| 620 | 148.28 | 105.78 | 0.63395 | 1620 | 401.09 | 290.04 | 0.87462 |
| 640 | 153.09 | 109.21 | 0.64159 | 1640 | 406.45 | 294.03 | 0.87791 |
| 660 | 157.92 | 112.67 | 0.64902 | 1660 | 411.82 | 298.02 | 0.88116 |
| 680 | 162.73 | 116.12 | 0.65621 | 1680 | 417.20 | 302.04 | 0.88439 |
| 700 | 167.56 | 119.58 | 0.66321 | 1700 | 422.59 | 306.06 | 0.88758 |
| 720 | 172.39 | 123.04 | 0.67002 | 1720 | 428.00 | 310.09 | 0.89074 |
| 740 | 177.23 | 126.51 | 0.67665 | 1740 | 433.41 | 314.13 | 0.89387 |
| 760 | 182.08 | 129.99 | 0.68312 | 1760 | 438.83 | 318.18 | 0.89697 |
| 780 | 186.94 | 133.47 | 0.68942 | 1780 | 444.26 | 322.24 | 0.90003 |
| 800 | 191.81 | 136.97 | 0.69558 | 1800 | 449.71 | 326.32 | 0.90308 |
| 820 | 196.69 | 140.47 | 0.70160 | 1820 | 455.17 | 330.40 | 0.90609 |
| 840 | 201.56 | 143.98 | 0.70747 | 1840 | 460.63 | 334.50 | 0.90908 |
| 860 | 206.46 | 147.50 | 0.71323 | 1860 | 466.12 | 338.61 | 0.91203 |
| 880 | 211.35 | 151.02 | 0.71886 | 1880 | 471.60 | 342.73 | 0.91497 |
| 900 | 216.26 | 154.57 | 0.72438 | 1900 | 477.09 | 346.85 | 0.91788 |
| 920 | 221.18 | 158.12 | 0.72979 | 1920 | 482.60 | 350.98 | 0.92076 |
| 940 | 226.11 | 161.68 | 0.73509 | 1940 | 488.12 | 355.12 | 0.92362 |
| 960 | 231.06 | 165.26 | 0.74030 | 1960 | 493.64 | 359.28 | 0.92645 |
| 980 | 236.02 | 168.83 | 0.74540 | 1980 | 499.17 | 363.43 | 0.92926 |
| 1000 | 240.98 | 172.43 | 0.75042 | 2000 | 504.71 | 367.61 | 0.93205 |
| 1020 | 245.97 | 176.04 | 0.75536 | 2020 | 510.26 | 371.79 | 0.93481 |
| 1040 | 250.95 | 179.66 | 0.76019 | 2040 | 515.82 | 375.98 | 0.93756 |
| 1060 | 255.96 | 183.29 | 0.76496 | 2060 | 521.39 | 380.18 | 0.94026 |
| 1080 | 260.97 | 186.93 | 0.76964 | 2080 | 526.97 | 384.39 | 0.94296 |
| 1100 | 265.99 | 190.58 | 0.77426 | 2100 | 532.55 | 388.60 | 0.94564 |
| 1120 | 271.03 | 194.25 | 0.77880 | 2120 | 538.15 | 392.83 | 0.94829 |
| 1140 | 276.08 | 197.94 | 0.78326 | 2140 | 543.74 | 397.05 | 0.95092 |
| 1160 | 281.14 | 201.63 | 0.78767 | 2160 | 549.35 | 401.29 | 0.95352 |
| 1180 | 286.21 | 205.33 | 0.79201 | 2200 | 560.59 | 409.78 | 0.95868 |
|  |  |  |  | 2300 | 588.82 | 431.16 | 0.97123 |
|  |  |  |  | 2400 | 617.22 | 452.70 | 0.98331 |

*Source*: Abridged from Table 1 in Joseph H. Keenan and Joseph Kaye, *Gas Tables*, John Wiley & Sons, Inc., New York; copyright 1948.

**Table II.2-14   Generalized Entropy Correction Chart**

**Table II.2-13.   Generalized Enthalpy Correction Chart**

**Table II.3  Critical Constants**

| Substance | Formula | Molecular Weight | Temperature | | Pressure | | Volume (ft³/lb mol) |
|---|---|---|---|---|---|---|---|
| | | | °K | °R | atm | lb$_f$/in.² | |
| Ammonia | NH₃ | 17.03 | 405.5 | 729.8 | 111.3 | 1636 | 1.16 |
| Argon | Ar | 39.944 | 151 | 272 | 48.0 | 705 | 1.20 |
| Bromine | Br₂ | 159.832 | 584 | 1052 | 102 | 1500 | 2.17 |
| Carbon dioxide | CO₂ | 44.01 | 304.2 | 547.5 | 72.9 | 1071 | 1.51 |
| Carbon monoxide | CO | 28.01 | 133 | 240 | 34.5 | 507 | 1.49 |
| Chlorine | Cl₂ | 70.914 | 417 | 751 | 76.1 | 1120 | 1.99 |
| Deuterium (normal) | D₂ | 4.00 | 38.4 | 69.1 | 16.4 | 241 | — |
| Helium | He | 4.003 | 5.3 | 9.5 | 2.26 | 33.2 | 0.926 |
| Helium | He | 3.00 | 3.34 | 6.01 | 1.15 | 16.9 | — |
| Hydrogen (normal) | H₂ | 2.016 | 33.3 | 59.9 | 12.8 | 188.1 | 1.04 |
| Krypton | Kr | 83.7 | 209.4 | 376.9 | 54.3 | 798 | 1.48 |
| Neon | Ne | 20.183 | 44.5 | 80.1 | 26.9 | 395 | 0.668 |
| Nitrogen | N₂ | 28.016 | 126.2 | 227.1 | 33.5 | 492 | 1.44 |
| Nitrous oxide | N₂O | 44.02 | 309.7 | 557.1 | 71.7 | 1054 | 1.54 |
| Oxygen | O₂ | 32.00 | 154.8 | 278.6 | 50.4 | 736 | 1.25 |
| Sulfur dioxide | So₂ | 64.06 | 430.7 | 775.2 | 77.8 | 1143 | 1.95 |
| Water | H₂O | 18.016 | 647.4 | 1165.3 | 218.3 | 3204 | 0.90 |
| Xenon | Xe | 131.3 | 289.75 | 521.55 | 58.0 | 852 | 1.90 |
| Benzene | C₆H₆ | 78.11 | 562 | 1012 | 48.6 | 714 | 4.17 |
| n-Butane | C₄H₁₀ | 58.120 | 425.2 | 765.2 | 37.5 | 551 | 4.08 |
| Carbon tetrachloride | CCl₄ | 153.81 | 556.4 | 1001.5 | 45.0 | 661 | 4.42 |
| Chloroform | CHCl₃ | 119.39 | 536.6 | 965.8 | 54.0 | 794 | 3.85 |
| Dichlorodifluoromethane | CCl₂F₂ | 120.92 | 384.7 | 692.4 | 39.6 | 582 | 3.49 |
| Dichlorofluoromethane | CHCl₂F | 102.93 | 451.7 | 813.0 | 51.0 | 749 | 3.16 |
| Ethane | C₂H₆ | 30.068 | 305.5 | 549.8 | 48.2 | 708 | 2.37 |
| Ethyl alcohol | C₂H₅OH | 46.07 | 516.0 | 929.0 | 63.0 | 926 | 2.68 |
| Ethylene | C₂H₄ | 28.052 | 282.4 | 508.3 | 50.5 | 742 | 1.99 |
| n-Hexane | C₆H₁₄ | 86.172 | 507.9 | 914.2 | 29.9 | 439 | 5.89 |
| Methane | CH₄ | 16.012 | 191.1 | 343.9 | 45.8 | 673 | 1.59 |
| Methyl alcohol | CH₄OH | 32.04 | 513.2 | 923.7 | 78.5 | 1154 | 1.89 |
| Methyl chloride | CH₃Cl | 50.49 | 416.3 | 749.3 | 65.9 | 968 | 2.29 |
| Propane | C₃H₈ | 44.094 | 370.0 | 665.9 | 42.0 | 617 | 3.20 |
| Propene | C₃H₆ | 42.078 | 365.0 | 656.9 | 45.6 | 670 | 2.90 |
| Propyne | C₃H₄ | 40.062 | 401 | 722 | 52.8 | 776 | — |
| Trichlorofluoromethane | CCl₃F | 137.38 | 471.2 | 848.1 | 43.2 | 635 | 3.97 |

*Source*: K. A. Kobe and R. E. Lynn, Jr., *Chemical Reviews*, Vol. 52 (1953), pp. 117–236.

**Table II.4-1  Enthalpy of Formation, Gibbs Function of Formation, and Absolute Entropy of Various Substances at 77 °F (25°C) and 1 Atm Pressure**

| Substance | Formula | M | State | $\bar{h}_f^\circ$ | | $\bar{g}_f^\circ$ | | $\bar{s}^\circ$ |
|---|---|---|---|---|---|---|---|---|
| | | | | cal/g mol | Btu/lb mol | cal/g mol | Btu/lb mol | cal/g mol · °K, Btu/ lb mol · °R |
| Carbon monoxide[a] | CO | 28.011 | Gas | −26,417 | −47,551 | −32,783 | −59,009 | 47.214 |
| Carbon dioxide[a] | CO₂ | 44.011 | Gas | −94,054 | −169,297 | −94,265 | −169,677 | 51.072 |
| Water[a,b] | H₂O | 18.016 | Gas | −57,798 | −104,036 | −54,636 | −98,345 | 45.106 |
| Water[b] | H₂O | 18.016 | Liquid | −68,317 | −122,971 | −56,690 | −102,042 | 16.716 |
| Methane[a] | CH₄ | 16.043 | Gas | −17,895 | −32,211 | −12,145 | −21,861 | 44.490 |
| Acetylene[a] | C₂H₂ | 26.038 | Gas | 54,190 | 97,542 | 49,993 | 89,987 | 48.004 |
| Ethene[a] | C₂H₄ | 28.054 | Gas | 12,496 | 22,493 | 16,281 | 29,306 | 52.447 |
| Ethane[c] | C₂H₆ | 30.070 | Gas | −20,236 | −36,425 | −7,860 | −14,148 | 54.85 |
| Propane[c] | C₃H₈ | 44.097 | Gas | −24,820 | −44,676 | −5,614 | −10,105 | 64.51 |
| Butane[c] | C₄H₁₀ | 58.124 | Gas | −30,150 | −54,270 | −4,100 | −7,380 | 74.12 |
| Octane[c] | C₈H₁₈ | 114.23 | Gas | −49,820 | −89,680 | 3,950 | 7,110 | 111.55 |
| Octane[c] | C₈H₁₈ | 114.23 | Liquid | −59,740 | −107,532 | 1,580 | 2,844 | 86.23 |
| Carbon[a] (graphite) | C | 12.011 | Solid | 0 | 0 | 0 | 0 | 1.359 |

[a] From JANAF *Thermochemical Tables*, Thermal Research Laboratory, The Dow Chemical Company, Midland, Mich.
[b] From *Circular 500*, National Bureau of Standards, Washington D.C.
[c] From F. D. Rossini et al., *API Research Project 44*.

**Table II.4-2    Enthalpy of Combustion of Some Hydrocarbons at 25°C (77°F)**

| Hydrocarbon | Formula | Liquid $H_2O$ in Products (Negative of Higher Heating Value) | | Vapor $H_2O$ in Products (Negative of Lower Heating Value) | |
|---|---|---|---|---|---|
| | | Liquid Hydrocarbon (Btu/lb$_m$ fuel) | Gaseous Hydrocarbon (Btu/lb$_m$ fuel) | Liquid Hydrocarbon (Btu/lb$_m$ fuel) | Gaseous Hydrocarbon (Btu/lb$_m$ fuel) |
| **Parafin family** | | | | | |
| Methane | $CH_4$ | | −23,861 | | −21,502 |
| Ethane | $C_2H_6$ | | −22,304 | | −20,416 |
| Propane | $C_3H_8$ | −21,490 | −21,649 | −19,773 | −19,929 |
| Butane | $C_4H_{10}$ | −21,134 | −21,293 | −19,506 | −19,665 |
| Pentane | $C_5H_{12}$ | −20,914 | −21,072 | −19,340 | −19,499 |
| Hexane | $C_6H_{14}$ | −20,772 | −20,930 | −19,233 | −19,391 |
| Heptane | $C_7H_{16}$ | −20,668 | −20,825 | −19,157 | −19,314 |
| Octane | $C_8H_{18}$ | −20,591 | −20,747 | −19,100 | −19,256 |
| Decane | $C_{10}H_{22}$ | −20,484 | −20,638 | −19,020 | −19,175 |
| Dodecane | $C_{12}H_{26}$ | −20,410 | −20,564 | −18,964 | −19,118 |
| **Olefin family** | | | | | |
| Ethene | $C_2H_4$ | | −21,626 | | −20,276 |
| Propene | $C_3H_6$ | | −21,033 | | −19,683 |
| Butene | $C_4H_8$ | | −20,833 | | −19,483 |
| Pentene | $C_5H_{10}$ | | −20,696 | | −19,346 |
| Hexene | $C_6H_{12}$ | | −20,612 | | −19,262 |
| Heptene | $C_7H_{14}$ | | −20,552 | | −19,202 |
| Octene | $C_8H_{16}$ | | −20,507 | | −19,157 |
| Nonene | $C_9H_{18}$ | | −20,472 | | −19,122 |
| Decene | $C_{10}H_{20}$ | | −20,444 | | −19,094 |
| **Alkylbenzene family** | | | | | |
| Benzene | $C_6H_6$ | −17,985 | −18,172 | −17,259 | −17,446 |
| Methylbenzene | $C_7H_8$ | −18,247 | −18,423 | −17,424 | −17,601 |
| Ethylbenzene | $C_8H_{10}$ | −18,488 | −18,659 | −17,596 | −17,767 |
| Propylbenzene | $C_9H_{12}$ | −18,667 | −18,832 | −17,722 | −17,887 |
| Butylbenzene | $C_{10}H_{14}$ | −18,809 | −18,970 | −17,823 | −17,984 |

**Table II.4-3    Enthalpy of Formation at 25°C, Ideal Gas Enthalpy, and Absolute Entropy at 1 Atm Pressure**

| Temperature | | Nitrogen, Diatomic ($N_2$) (Sept. 30, 1965) $(\bar{h}^{\circ}_f)_{298} = 0$ cal/g mol = 0 Btu/lb mol, $M = 28.016$ | | | Nitrogen Monatomic (N) (Mar. 31, 1961) $(\bar{h}^{\circ}_f)_{298} = 112,965$ cal/g mol = 203,337 Btu/lb mol, $M = 14.008$ | | |
|---|---|---|---|---|---|---|---|
| °K | °R | $\bar{h}^{\circ} - \bar{h}^{\circ}_{298}$ (cal/g mol) | $\bar{s}^{\circ}$ (cal/g mol · °K, Btu/lb mol · °R) | $\bar{h}^{\circ} - \bar{h}^{\circ}_{537}$ (Btu/lb mol) | $\bar{h}^{\circ} - \bar{h}^{\circ}_{298}$ (cal/g mol) | $\bar{s}^{\circ}$ (cal/g mol · °K, Btu/lb mol · °R) | $\bar{h}^{\circ} - \bar{h}^{\circ}_{537}$ (Btu/lb mol) |
| 0 | 0 | −2,072 | 0 | −3,730 | −1,481 | 0 | −2,666 |
| 100 | 180 | −1,379 | 38.170 | −2,483 | −984 | 31.187 | −1,771 |
| 200 | 360 | −683 | 42.992 | −1,229 | −488 | 34.631 | −878 |
| 298 | 537 | 0 | 45.770 | 0 | 0 | 36.614 | 0 |
| 300 | 540 | 13 | 45.813 | 23 | 9 | 36.645 | 16 |
| 400 | 720 | 710 | 47.818 | 1,278 | 506 | 38.074 | 911 |
| 500 | 900 | 1,413 | 49.386 | 2,543 | 1,003 | 39.183 | 1,805 |
| 600 | 1,080 | 2,125 | 50.685 | 3,825 | 1,500 | 40.089 | 2,700 |
| 700 | 1,260 | 2,853 | 51.806 | 5,135 | 1,996 | 40.855 | 3,593 |
| 800 | 1,440 | 3,596 | 52.798 | 6,473 | 2,493 | 41.518 | 4,487 |
| 900 | 1,620 | 4,335 | 53.692 | 7,839 | 2,990 | 42.103 | 5,382 |
| 1000 | 1,800 | 5,129 | 54.507 | 9,232 | 3,487 | 42.627 | 6,277 |
| 1100 | 1,980 | 5,917 | 55.258 | 10,651 | 3,984 | 43.100 | 7,171 |
| 1200 | 2,160 | 6,718 | 55.955 | 12,092 | 4,481 | 43.532 | 8,066 |
| 1300 | 2,340 | 7,529 | 56.604 | 13,552 | 4,977 | 43.930 | 8,959 |
| 1400 | 2,520 | 8,350 | 57.212 | 15,030 | 5,474 | 44.298 | 9,853 |
| 1500 | 2,700 | 9,179 | 57.784 | 16,522 | 5,971 | 44.641 | 10,748 |
| 1600 | 2,880 | 10,015 | 58.324 | 18,027 | 6,468 | 44.962 | 11,642 |
| 1700 | 3,060 | 10,858 | 58.835 | 19,544 | 6,965 | 45.263 | 12,537 |

**Table II.4-3 Continued**

| Temperature | | Oxygen, Diatomic (O₂) (Sept. 30, 1965) | | | Oxygen, Monatomic (O) (June 30, 1962) | | |
|---|---|---|---|---|---|---|---|
| | | $(\bar{h}_f^\circ)_{298}$ = 0 cal/g mol = 0 Btu/lb mol, $M$ = 32.00 | | | $(\bar{h}_f^\circ)_{298}$ = 59,559 cal/g mol = 107,206 Btu/lb mol, $M$ = 16.00 | | |
| °K | °R | $\bar{h}^\circ - \bar{h}_{298}^\circ$ (cal/g mol) | $\bar{s}^\circ$ (cal/g mol · °K, Btu/lb mol · °R) | $\bar{h}^\circ - \bar{h}_{537}^\circ$ (Btu/lb mol) | $\bar{h}^\circ - \bar{h}_{298}^\circ$ (cal/g mol) | $\bar{s}^\circ$ (cal/g mol · °K, Btu/lb mol · °R) | $\bar{h}^\circ - \bar{h}_{537}^\circ$ (Btu/lb mol) |
| 1800 | 3,240 | 11,707 | 59.320 | 21,073 | 7,461 | 45.547 | 13,430 |
| 1900 | 3,420 | 12,560 | 59.782 | 22,608 | 7,958 | 45.815 | 14,324 |
| 2000 | 3,600 | 13,418 | 60.222 | 24,152 | 8,455 | 46.070 | 15,219 |
| 2100 | 3,780 | 14,280 | 60.642 | 25,704 | 8,952 | 46.313 | 16,114 |
| 2200 | 3,960 | 15,146 | 61.045 | 27,263 | 9,449 | 46.544 | 17,903 |
| 2300 | 4,140 | 16,015 | 61.431 | 28,827 | 9,946 | 46.765 | 17,008 |
| 2400 | 4,320 | 16,886 | 61.802 | 30,395 | 10,444 | 46.977 | 18,799 |
| 2500 | 4,500 | 17,761 | 62.159 | 31,970 | 10,941 | 47.180 | 19,694 |
| 2600 | 4,680 | 18,638 | 62.503 | 33,548 | 11,439 | 47.375 | 20,590 |
| 2700 | 4,860 | 19,517 | 62.835 | 35,131 | 11,938 | 47.563 | 21,488 |
| 2800 | 5,040 | 20,398 | 63.155 | 36,716 | 12,437 | 47.745 | 22,387 |
| 2900 | 5,220 | 21,280 | 63.465 | 38,304 | 12,936 | 47.920 | 23,285 |
| 3000 | 5,400 | 22,165 | 63.765 | 39,897 | 13,437 | 48.090 | 24,187 |
| 3200 | 5,760 | 23,939 | 64.337 | 43,090 | 14,441 | 48.414 | 25,994 |
| 3400 | 6,120 | 25,719 | 64.877 | 46,294 | 15,451 | 48.720 | 27,812 |
| 3600 | 6,480 | 27,505 | 65.387 | 49,509 | 16,469 | 49.011 | 29,644 |
| 3800 | 6,840 | 29,295 | 65.871 | 52,731 | 17,495 | 49.288 | 31,491 |
| 4000 | 7,200 | 31,089 | 66.331 | 55,960 | 18,531 | 49.554 | 33,356 |
| 4200 | 7,560 | 32,888 | 66.770 | 59,198 | 19,580 | 49.810 | 35,244 |
| 4400 | 7,920 | 34,690 | 67.189 | 62,442 | 20,643 | 50.057 | 37,157 |
| 4600 | 8,280 | 36,496 | 67.591 | 65,693 | 21,721 | 50.297 | 39,098 |
| 4800 | 8,640 | 38,306 | 67.976 | 68,951 | 22,816 | 50.530 | 41,069 |
| 5000 | 9,000 | 40,119 | 68.346 | 72,214 | 23,928 | 50.757 | 43,070 |
| 5200 | 9,360 | 41,935 | 68.702 | 75,483 | 25,059 | 50.978 | 45,106 |
| 5400 | 9,720 | 43,755 | 69.045 | 78,759 | 26,210 | 51.195 | 47,178 |
| 5600 | 10,180 | 45,579 | 69.377 | 82,042 | 27,380 | 51.408 | 49,284 |
| 5800 | 10,540 | 47,406 | 69.698 | 85,331 | 28,570 | 51.617 | 51,426 |
| 6000 | 10,800 | 49,237 | 70.008 | 88,627 | 29,780 | 51.822 | 53,604 |
| 0 | 0 | −2,075 | 0 | −3,735 | −1,608 | 0 | −2,894 |
| 100 | 180 | −1,381 | 41.395 | −2,486 | −1,080 | 32.466 | −1,944 |
| 200 | 360 | −685 | 46.218 | −1,233 | −523 | 36.340 | −941 |
| 298 | 537 | 0 | 49.004 | 0 | 0 | 38.468 | 0 |
| 300 | 540 | 13 | 49.047 | 23 | 10 | 38.501 | 18 |
| 400 | 720 | 724 | 51.091 | 1,303 | 528 | 39.991 | 950 |
| 500 | 900 | 1,455 | 52.722 | 2,619 | 1,038 | 41.131 | 1,868 |
| 600 | 1,080 | 2,210 | 54.098 | 3,978 | 1,544 | 42.054 | 2,779 |
| 700 | 1,260 | 2,988 | 55.297 | 5,378 | 2,048 | 42.831 | 3,686 |
| 800 | 1,440 | 3,786 | 56.361 | 6,815 | 2,550 | 43.501 | 4,590 |
| 900 | 1,620 | 4,600 | 57.320 | 8,280 | 3,052 | 44.092 | 5,494 |
| 1000 | 1,800 | 5,427 | 58.192 | 9,769 | 3,552 | 44.619 | 6,394 |
| 1100 | 1,980 | 6,266 | 58.991 | 11,279 | 4,051 | 45.095 | 7,292 |
| 1200 | 2,160 | 7,114 | 59.729 | 12,805 | 4,551 | 45.529 | 8,192 |
| 1300 | 2,340 | 7,971 | 60.415 | 14,348 | 5,049 | 45.928 | 9,088 |
| 1400 | 2,520 | 8,835 | 61.055 | 15,903 | 5,548 | 46.298 | 9,986 |
| 1500 | 2,700 | 9,706 | 61.656 | 17,471 | 6,046 | 46.642 | 10,883 |
| 1600 | 2,880 | 10,583 | 62.222 | 19,049 | 6,544 | 46.963 | 11,779 |
| 1700 | 3,060 | 11,465 | 62.757 | 20,637 | 7,042 | 47.265 | 12,676 |
| 1800 | 3,240 | 12,354 | 63.265 | 22,237 | 7,540 | 47.550 | 13,572 |
| 1900 | 3,420 | 13,249 | 63.749 | 23,848 | 8,038 | 47.819 | 14,468 |
| 2000 | 3,600 | 14,149 | 64.210 | 25,468 | 8,536 | 48.074 | 15,365 |
| 2100 | 3,780 | 15,054 | 64.652 | 27,097 | 9,034 | 48.317 | 16,261 |
| 2200 | 3,960 | 15,966 | 65.076 | 28,739 | 9,532 | 48.549 | 17,158 |
| 2300 | 4,140 | 16,882 | 65.483 | 30,388 | 10,029 | 48.770 | 18,052 |
| 2400 | 4,320 | 17,804 | 65.876 | 32,047 | 10,527 | 48.982 | 18,949 |
| 2500 | 4,500 | 18,732 | 66.254 | 33,718 | 11,026 | 49.185 | 19,847 |
| 2600 | 4,680 | 19,664 | 66.620 | 35,395 | 11,524 | 49.381 | 20,743 |
| 2700 | 4,860 | 20,602 | 66.974 | 37,084 | 12,023 | 49.569 | 21,641 |
| 2800 | 5,040 | 21,545 | 67.317 | 38,781 | 12,522 | 49.751 | 22,540 |
| 2900 | 5,220 | 22,493 | 67.650 | 40,487 | 13,022 | 49.926 | 23,440 |
| 3000 | 5,400 | 23,446 | 67.973 | 42,203 | 13,522 | 50.096 | 24,340 |
| 3200 | 5,760 | 25,365 | 68.592 | 45,657 | 14,524 | 50.419 | 26,143 |
| 3400 | 6,120 | 27,302 | 69.179 | 49,144 | 15,529 | 50.724 | 27,952 |
| 3600 | 6,480 | 29,254 | 69.737 | 52,657 | 16,537 | 51.012 | 29,767 |
| 3800 | 6,840 | 31,221 | 70.269 | 56,198 | 17,549 | 51.285 | 31,588 |

**Table II.4-3   Continued**

| Temperature | | Carbon Dioxide (CO$_2$) (Sept. 30, 1965) $(\bar{h}_f^\circ)_{298} = -94,054$ cal/g mol $= -169,297$ Btu/lb mol, $M = 44.011$ | | | Carbon Monoxide (CO) (Sept 30, 1965) $(\bar{h}_f^\circ)_{298} = -26,417$ cal/g mol $= -47,551$ Btu/lb mol, $M = 28.011$ | | |
|---|---|---|---|---|---|---|---|
| °K | °R | $\bar{h}^\circ - \bar{h}_{298}^\circ$ (cal/g mol) | $\bar{s}^\circ$ (cal/g mol · °K, Btu/lb mol · °R) | $\bar{h}^\circ - \bar{h}_{537}^\circ$ (Btu/lb mol) | $\bar{h}^\circ - \bar{h}_{298}^\circ$ (cal/g mol) | $\bar{s}^\circ$ (cal/g mol · °K, Btu/lb mol · °R) | $\bar{h}^\circ - \bar{h}_{537}^\circ$ (Btu/lb mol) |
| 4000 | 7,200 | 33,201 | 70.776 | 59,762 | 18,565 | 51.546 | 33,417 |
| 4200 | 7,560 | 35,193 | 71.262 | 63,347 | 19,586 | 51.795 | 35,255 |
| 4400 | 7,920 | 37,196 | 71.728 | 66,953 | 20,611 | 52.033 | 37,100 |
| 4600 | 8,280 | 39,208 | 72.176 | 70,574 | 21,641 | 52.262 | 38,954 |
| 4800 | 8,640 | 41,229 | 72.606 | 74,212 | 22,676 | 52.482 | 40,817 |
| 5000 | 9,000 | 43,257 | 73.019 | 77,863 | 23,715 | 52.695 | 42,687 |
| 5200 | 9,360 | 45,292 | 73.418 | 81,526 | 24,760 | 52.899 | 44,568 |
| 5400 | 9,720 | 47,332 | 73.803 | 85,198 | 25,809 | 53.097 | 46,456 |
| 5600 | 10,180 | 49,377 | 74.175 | 88,879 | 26,863 | 53.289 | 48,353 |
| 5800 | 10,540 | 51,426 | 74.535 | 92,567 | 27,921 | 53.475 | 50,258 |
| 6000 | 10,800 | 53,479 | 74.883 | 96,262 | 28,984 | 53.655 | 25,171 |
| 0 | 0 | −2,238 | 0 | −4,028 | −2,072 | 0 | −3,730 |
| 100 | 180 | −1,543 | 42.758 | −2,777 | −1,379 | 39.613 | −2,483 |
| 200 | 360 | −816 | 47.769 | −1,469 | −683 | 44.435 | −1,229 |
| 298 | 537 | 0 | 51.072 | 0 | 0 | 47.214 | 0 |
| 300 | 540 | 16 | 51.127 | 29 | 13 | 47.257 | 23 |
| 400 | 720 | 958 | 53.830 | 1,724 | 711 | 49.265 | 1,280 |
| 500 | 900 | 1,987 | 56.122 | 3,577 | 1,417 | 50.841 | 2,551 |
| 600 | 1,080 | 3,087 | 58.126 | 5,557 | 2,137 | 52.152 | 3,847 |
| 700 | 1,260 | 4,245 | 59.910 | 7,641 | 2,873 | 53.287 | 5,171 |
| 800 | 1,440 | 5,453 | 61.522 | 9,815 | 3,627 | 54.293 | 6,529 |
| 900 | 1,620 | 6,702 | 62.992 | 12,064 | 4,397 | 55.200 | 7,915 |
| 1000 | 1,800 | 7,984 | 64.344 | 14,371 | 5,183 | 56.028 | 9,329 |
| 1100 | 1,980 | 9,296 | 65.594 | 16,733 | 5,983 | 56.790 | 10,769 |
| 1200 | 2,160 | 10,632 | 66.756 | 19,138 | 6,794 | 57.496 | 12,229 |
| 1300 | 2,340 | 11,988 | 67.841 | 21,578 | 7,616 | 58.154 | 13,709 |
| 1400 | 2,520 | 13,362 | 68.859 | 24,052 | 8,446 | 58.769 | 15,203 |
| 1500 | 2,700 | 14,750 | 69.817 | 26,550 | 9,285 | 59.348 | 16,713 |
| 1600 | 2,880 | 16,152 | 70.722 | 29,074 | 10,130 | 59.893 | 18,234 |
| 1700 | 3,060 | 17,565 | 71.578 | 31,617 | 10,980 | 60.409 | 19,764 |
| 1800 | 3,240 | 18,987 | 72.391 | 34,177 | 11,836 | 60.898 | 21,305 |
| 1900 | 3,420 | 20,418 | 73.165 | 36,752 | 12,697 | 61.363 | 22,855 |
| 2000 | 3,600 | 21,857 | 73.903 | 39,343 | 13,561 | 61.807 | 24,410 |
| 2100 | 3,780 | 23,303 | 74.608 | 41,945 | 14,430 | 62.230 | 25,974 |
| 2200 | 3,960 | 24,755 | 75.284 | 44,559 | 15,301 | 62.635 | 27,542 |
| 2300 | 4,140 | 26,212 | 75.931 | 47,182 | 16,175 | 63.024 | 29,115 |
| 2400 | 4,320 | 27,674 | 76.554 | 49,813 | 17,052 | 63.397 | 30,694 |
| 2500 | 4,500 | 29,141 | 77.153 | 52,454 | 17,931 | 63.756 | 32,276 |
| 2600 | 4,680 | 30,613 | 77.730 | 55,103 | 18,813 | 64.102 | 33,863 |
| 2700 | 4,860 | 32,088 | 78.286 | 57,758 | 19,696 | 64.435 | 35,453 |
| 2800 | 5,040 | 33,567 | 78.824 | 60,421 | 20,582 | 64.757 | 37,048 |
| 2900 | 5,220 | 35,049 | 79.344 | 63,088 | 21,469 | 65.069 | 38,644 |
| 3000 | 5,400 | 36,535 | 79.848 | 65,763 | 22,357 | 65.370 | 40,243 |
| 3200 | 5,760 | 39,515 | 80.810 | 71,127 | 24,139 | 65.945 | 43,450 |
| 3400 | 6,120 | 42,507 | 81.717 | 76,513 | 25,927 | 66.487 | 46,669 |
| 3600 | 6,480 | 45,508 | 82.574 | 81,914 | 27,719 | 66.999 | 49,894 |
| 3800 | 6,840 | 48,518 | 83.388 | 87,332 | 29,516 | 67.485 | 53,129 |
| 4000 | 7,200 | 51,538 | 84.162 | 92,768 | 31,316 | 67.946 | 56,369 |
| 4200 | 7,560 | 54,566 | 84.901 | 98,219 | 33,121 | 68.387 | 59,618 |
| 4400 | 7,920 | 57,601 | 85.607 | 103,682 | 34,930 | 68.807 | 62,874 |
| 4600 | 8,280 | 60,644 | 86.284 | 109,159 | 36,741 | 69.210 | 66,134 |
| 4800 | 8,640 | 63,695 | 86.933 | 114,651 | 38,557 | 69.596 | 69,403 |
| 5000 | 9,000 | 66,753 | 87.557 | 120,155 | 40,375 | 69.967 | 72,675 |
| 5200 | 9,360 | 69,819 | 88.158 | 125,674 | 42,196 | 70.325 | 75,953 |
| 5400 | 9,720 | 72,893 | 88.738 | 131,207 | 44,021 | 70.669 | 79,238 |
| 5600 | 10,180 | 75,976 | 89.299 | 136,757 | 45,849 | 71.001 | 82,528 |
| 5800 | 10,540 | 79,068 | 89.841 | 142,322 | 47,679 | 71.332 | 85,822 |
| 6000 | 10,800 | 82,168 | 90.367 | 147,902 | 49,513 | 71.663 | 89,123 |

**Table II.4-3    Continued**

| Temperature | | Water ($H_2O$) (Mar. 31, 1961) $(\bar{h}_f^\circ)_{298} = -57{,}798$ cal/g mol $= -104{,}036$ Btu/lb mol, $M = 18.016$ | | | Hydroxyl (OH) (Mar. 31, 1966) $(\bar{h}_f^\circ)_{298} = 9432$ cal/g mol $= 16{,}978$ Btu/lb mol, $M = 17.008$ | | |
|---|---|---|---|---|---|---|---|
| °K | °R | $\bar{h}^\circ - \bar{h}_{298}^\circ$ (cal/g mol) | $\bar{s}^\circ$ (cal/g mol · °K, Btu/lb mol · °R) | $\bar{h}^\circ - \bar{h}_{537}^\circ$ (Btu/lb mol) | $\bar{h}^\circ - \bar{h}_{298}^\circ$ (cal/g mol) | $\bar{s}^\circ$ (cal/g mol · °K, Btu/lb mol · °R) | $\bar{h}^\circ - \bar{h}_{537}^\circ$ (Btu/lb mol) |
| 0 | 0 | −2,367 | 0 | −4,261 | −2,192 | 0 | −3,946 |
| 100 | 180 | −1,581 | 36.396 | −2,846 | −1,467 | 35.726 | −2,641 |
| 200 | 360 | −784 | 41.916 | −1,411 | −711 | 40.985 | −1,280 |
| 298 | 537 | 0 | 45.106 | 0 | 0 | 43.880 | 0 |
| 300 | 540 | 15 | 45.155 | 27 | 13 | 43.925 | 23 |
| 400 | 720 | 825 | 47.484 | 1,485 | 725 | 45.974 | 1,305 |
| 500 | 900 | 1,654 | 49.334 | 2,977 | 1,432 | 47.551 | 2,578 |
| 600 | 1,080 | 2,509 | 50.891 | 4,516 | 2,137 | 48.837 | 3,847 |
| 700 | 1,260 | 3,390 | 52.249 | 6,102 | 2,845 | 49.927 | 5,121 |
| 800 | 1,440 | 4,300 | 53.464 | 7,740 | 3,556 | 50.877 | 6,401 |
| 900 | 1,620 | 5,240 | 54.570 | 9,432 | 4,275 | 51.724 | 7,695 |
| 1000 | 1,800 | 6,209 | 55.592 | 11,176 | 5,003 | 52.491 | 9,005 |
| 1100 | 1,980 | 7,210 | 56.545 | 12,978 | 5,742 | 53.195 | 10,336 |
| 1200 | 2,160 | 8,240 | 57.441 | 14,832 | 6,491 | 53.847 | 11,684 |
| 1300 | 2,340 | 9,298 | 58.288 | 16,736 | 7,252 | 54.455 | 13,054 |
| 1400 | 2,520 | 10,384 | 59.092 | 18,691 | 8,023 | 55.027 | 14,441 |
| 1500 | 2,700 | 11,495 | 59.859 | 20,691 | 8,805 | 55.566 | 15,849 |
| 1600 | 2,880 | 12,630 | 60.591 | 22,734 | 9,596 | 56.077 | 17,273 |
| 1700 | 3,060 | 13,787 | 61.293 | 24,817 | 10,397 | 56.563 | 18,715 |
| 1800 | 3,240 | 14,964 | 61.965 | 26,935 | 11,207 | 57.025 | 20,173 |
| 1900 | 3,420 | 16,160 | 62.612 | 29,088 | 12,024 | 57.467 | 21,643 |
| 2000 | 3,600 | 17,373 | 63.234 | 31,271 | 12,849 | 57.891 | 23,128 |
| 2100 | 3,780 | 18,602 | 63.834 | 33,484 | 13,681 | 58.296 | 24,626 |
| 2200 | 3,960 | 19,846 | 64.412 | 35,723 | 14,520 | 58.686 | 26,136 |
| 2300 | 4,140 | 21,103 | 64.971 | 37,985 | 15,364 | 59.062 | 27,655 |
| 2400 | 4,320 | 22,372 | 65.511 | 40,270 | 16,214 | 59.424 | 29,185 |
| 2500 | 4,500 | 23,653 | 66.034 | 42,575 | 17,069 | 59.773 | 30,724 |
| 2600 | 4,680 | 24,945 | 66.541 | 44,901 | 17,929 | 60.110 | 32,272 |
| 2700 | 4,860 | 26,246 | 67.032 | 47,243 | 18,794 | 60.436 | 33,829 |
| 2800 | 5,040 | 27,556 | 67.508 | 49,601 | 19,662 | 60.752 | 35,392 |
| 2900 | 5,220 | 28,875 | 67.971 | 51,975 | 20,535 | 61.058 | 36,963 |
| 3000 | 5,400 | 30,201 | 68.421 | 54,362 | 21,411 | 61.355 | 38,540 |
| 3200 | 5,760 | 32,876 | 69.284 | 59,177 | 23,174 | 61.924 | 41,713 |
| 3400 | 6,120 | 35,577 | 70.102 | 64,039 | 24,949 | 62.462 | 44,908 |
| 3600 | 6,480 | 38,300 | 70.881 | 68,940 | 26,735 | 62.973 | 48,123 |
| 3800 | 6,840 | 41,043 | 71.622 | 73,877 | 28,532 | 63.458 | 51,358 |
| 4000 | 7,200 | 43,805 | 72.331 | 78,849 | 30,338 | 63.922 | 54,608 |
| 4200 | 7,560 | 46,583 | 73.008 | 83,849 | 32,153 | 64.364 | 57,875 |
| 4400 | 7,920 | 49,375 | 73.658 | 88,875 | 33,976 | 64.788 | 61,157 |
| 4600 | 8,280 | 52,181 | 74.281 | 93,926 | 35,807 | 65.195 | 64,453 |
| 4800 | 8,640 | 55,000 | 74.881 | 99,000 | 37,644 | 65.586 | 67,759 |
| 5000 | 9,000 | 57,829 | 75.459 | 104,092 | 39,489 | 65.963 | 71,080 |
| 5200 | 9,360 | 60,669 | 76.016 | 109,204 | 41,340 | 66.326 | 74,412 |
| 5400 | 9,720 | 63,520 | 76.553 | 114,336 | 43,197 | 66.676 | 77,755 |
| 5600 | 10,180 | 66,381 | 77.074 | 119,486 | 45,060 | 67.015 | 81,108 |
| 5800 | 10,540 | 69,251 | 77.577 | 124,652 | 46,929 | 67.343 | 84,472 |
| 6000 | 10,800 | 72,131 | 78.065 | 129,836 | 48,803 | 67.661 | 87,845 |

**Table II.4-3   Continued**

| Temperature | | Hydrogen, Diatomic ($H_2$) (Mar. 31, 1961) $(\bar{h}^\circ_f)_{298} = 0$ cal/g mol $= 0$ Btu/lb mol, $M = 2.016$ | | | Hydrogen, Monatomic (H) (Sept. 30, 1965) $(\bar{h}^\circ_f) = 52,100$ cal/g mol $= 93,780$ Btu/lb mol, $M = 1.008$ | | |
|---|---|---|---|---|---|---|---|
| °K | °R | $\bar{h}^\circ - \bar{h}^\circ_{298}$ (cal/g mol) | $\bar{s}^\circ$ (cal/g mol · °K, Btu/lb mol · °R) | $\bar{h}^\circ - \bar{h}^\circ_{537}$ (Btu/lb mol) | $\bar{h}^\circ - \bar{h}^\circ_{298}$ (cal/g mol) | $\bar{s}^\circ$ (cal/g mol · °K, Btu/lb mol · °R) | $\bar{h}^\circ - \bar{h}^\circ_{537}$ (Btu/lb mol) |
| 0 | 0 | −2,024 | 0 | −3,643 | −1,481 | 0 | −2,666 |
| 100 | 180 | −1,265 | 24.387 | −2,277 | −984 | 21.965 | −1,771 |
| 200 | 360 | −662 | 28.520 | −1,192 | −488 | 25.408 | −878 |
| 298 | 537 | 0 | 31.208 | 0 | 0 | 27.392 | 0 |
| 300 | 540 | 13 | 31.251 | 23 | 9 | 27.423 | 16 |
| 400 | 720 | 707 | 33.247 | 1,273 | 506 | 28.852 | 911 |
| 500 | 900 | 1,406 | 34.806 | 2,531 | 1,003 | 29.961 | 1,805 |
| 600 | 1,080 | 2,106 | 36.082 | 3,791 | 1,500 | 30.867 | 2,700 |
| 700 | 1,260 | 2,808 | 37.165 | 5,054 | 1,996 | 31.632 | 3,593 |
| 800 | 1,440 | 3,514 | 38.107 | 6,325 | 2,493 | 32.296 | 4,487 |
| 900 | 1,620 | 4,226 | 38.946 | 7,607 | 2,990 | 32.881 | 5,382 |
| 1000 | 1,800 | 4,944 | 39.702 | 8,899 | 3,487 | 33.404 | 6,277 |
| 1100 | 1,980 | 5,670 | 40.394 | 10,206 | 3,984 | 33.878 | 7,171 |
| 1200 | 2,160 | 6,404 | 41.033 | 11,527 | 4,481 | 34.310 | 8,066 |
| 1300 | 2,340 | 7,148 | 41.628 | 12,866 | 4,977 | 34.708 | 8,959 |
| 1400 | 2,520 | 7,902 | 42.187 | 14,224 | 5,474 | 35.076 | 9,853 |
| 1500 | 2,700 | 8,668 | 42.716 | 15,602 | 5,971 | 35.419 | 10,748 |
| 1600 | 2,880 | 9,446 | 43.217 | 17,003 | 6,468 | 35.739 | 11,642 |
| 1700 | 3,060 | 10,233 | 43.695 | 18,419 | 6,965 | 36.041 | 12,537 |
| 1800 | 3,240 | 11,030 | 44.150 | 19,854 | 7,461 | 36.325 | 13,430 |
| 1900 | 3,420 | 11,836 | 44.586 | 21,305 | 7,958 | 36.593 | 14,324 |
| 2000 | 3,600 | 12,651 | 45.004 | 22,772 | 8,455 | 36.848 | 15,219 |
| 2100 | 3,780 | 13,475 | 45.406 | 24,255 | 8,952 | 37.090 | 16,114 |
| 2200 | 3,960 | 14,307 | 45.793 | 25,753 | 9,449 | 37.322 | 17,088 |
| 2300 | 4,140 | 15,146 | 46.166 | 27,263 | 9,945 | 37.542 | 17,901 |
| 2400 | 4,320 | 15,993 | 46.527 | 28,787 | 10,442 | 37.754 | 18,796 |
| 2500 | 4,500 | 16,848 | 46.875 | 30,326 | 10,939 | 37.957 | 19,690 |
| 2600 | 4,680 | 17,708 | 47.213 | 31,874 | 11,436 | 38.152 | 20,585 |
| 2700 | 4,860 | 18,575 | 47.540 | 33,435 | 11,933 | 38.339 | 21,479 |
| 2800 | 5,040 | 19,448 | 47.857 | 35,006 | 12,430 | 38.520 | 22,374 |
| 2900 | 5,220 | 20,326 | 48.166 | 36,587 | 12,926 | 38.694 | 23,267 |
| 3000 | 5,400 | 21,210 | 48.465 | 38,178 | 13,423 | 38.862 | 24,161 |
| 3200 | 5,760 | 22,992 | 49.040 | 41,386 | 14,417 | 39.183 | 25,951 |
| 3400 | 6,120 | 24,794 | 49.586 | 44,629 | 15,410 | 39.484 | 27,738 |
| 3600 | 6,400 | 26,616 | 50.107 | 47,909 | 16,404 | 39.768 | 29,527 |
| 3800 | 6,840 | 28,457 | 50.605 | 51,223 | 17,398 | 40.037 | 31,316 |
| 4000 | 7,200 | 30,317 | 51.082 | 54,571 | 18,391 | 40.292 | 33,104 |
| 4200 | 7,560 | 32,194 | 51.540 | 57,949 | 19,385 | 40.534 | 34,893 |
| 4400 | 7,920 | 34,088 | 51.980 | 61,358 | 20,379 | 40.765 | 36,682 |
| 4600 | 8,280 | 35,999 | 52.405 | 64,798 | 21,372 | 40.986 | 38,470 |
| 4800 | 8,640 | 37,926 | 52.815 | 68,267 | 22,366 | 41.198 | 40,259 |
| 5000 | 9,000 | 39,868 | 53.211 | 71,762 | 23,359 | 41.400 | 42,046 |
| 5200 | 9,360 | 41,825 | 53.595 | 75,285 | 24,353 | 41.595 | 43,835 |
| 5400 | 9,720 | 43,797 | 53.967 | 78,835 | 25,347 | 41.783 | 45,625 |
| 5600 | 10,180 | 45,783 | 54.328 | 82,409 | 26,340 | 41.963 | 47,412 |
| 5800 | 10,540 | 47,783 | 54.679 | 86,009 | 27,334 | 42.138 | 49,201 |
| 6000 | 10,800 | 49,796 | 55.020 | 89,633 | 28,328 | 42.306 | 50,990 |

*Source*: Thermochemical data are from the JANAF *Thermochemical Tables*, Thermal Research Laboratory, The Dow Chemical Company, Midland Mich. The date each table was issued is indicated.

## Table II.5-1 Thermal Properties of Metals

| Metal | Properties at 68°F $\rho$ (lbm/ft³) | $C_p$ (Btu/lbm·°F) | $k$ (Btu/hr·ft·°F) | $\alpha$ (ft²/hr) | $k$(Btu/hr·ft·°F) −148°F −100°C | 32°F 0°C | 212°F 100°C | 392°F 200°C | 572°F 300°C | 752°F 400°C | 1112°F 600°C | 1472°F 800°C | 1832°F 1000°C | 2192°F 1200°C |
|---|---|---|---|---|---|---|---|---|---|---|---|---|---|---|
| **Aluminum** | | | | | | | | | | | | | | |
| Pure | 169 | 0.214 | 132 | 3.665 | 134 | 132 | 132 | 132 | 132 | | | | | |
| Al-Cu (Duralumin): 94–96 Al, 3–5 Cu, trace Mg | 174 | 0.211 | 95 | 2.580 | 73 | 92 | 105 | 112 | | | | | | |
| Al-Mg (Hydronalium): 91–95 Al, 5–9 Mg | 163 | 0.216 | 65 | 1.860 | 54 | 63 | 73 | 82 | | | | | | |
| Al-Si (Silumin): 87 Al, 13 Si | 166 | 0.208 | 95 | 2.773 | 86 | 94 | 101 | 107 | | | | | | |
| Al-Si (Silumin, copper bearing): 86.5 Al, 12.5 Si, 1 Cu | 166 | 0.207 | 79 | 2.311 | 69 | 79 | 83 | 88 | 93 | | | | | |
| Al-Si (Alusil): 78–80 Al, 20–22 Si | 164 | 0.204 | 93 | 2.762 | 83 | 91 | 97 | 101 | 103 | | | | | |
| Al-Mg-Si: 97 Al, 1 Mg, 1 Si, 1 Mn | 169 | 0.213 | 102 | 2.859 | — | 101 | 109 | 118 | | | | | | |
| **Lead** | 710 | 0.031 | 20 | 0.924 | 21.3 | 20.3 | 19.3 | 18.2 | 17.2 | | | | | |
| **Iron** | | | | | | | | | | | | | | |
| Pure | 493 | 0.108 | 42 | 0.785 | 50 | 42 | 39 | 36 | 32 | 28 | 23 | 21 | 20 | 21 |
| Wrought iron (C < 0.50%) | 490 | 0.11 | 34 | 0.634 | — | 34 | 33 | 30 | 28 | 26 | 21 | 19 | 19 | 19 |
| Cast iron (C ≈ 4%) | 454 | 0.10 | 30 | 0.666 | | | | | | | | | | |
| Steel (C$_{max}$ ≈ 1.5%) | | | | | | | | | | | | | | |
| Carbon steel (C ≈ 0.5%) | 489 | 0.111 | 31 | 0.570 | — | 32 | 30 | 28 | 26 | 24 | 20 | 17 | 17 | 18 |
| 1.0% | 487 | 0.113 | 25 | 0.452 | — | 25 | 25 | 24 | 23 | 21 | 19 | 17 | 16 | 17 |
| 1.5% | 484 | 0.116 | 21 | 0.376 | — | 21 | 21 | 21 | 20 | 19 | 18 | 16 | 16 | 17 |
| Nickel steel (Ni ≈ 0%) | 493 | 0.108 | 42 | 0.785 | | | | | | | | | | |
| 10% | 496 | 0.11 | 15 | 0.279 | | | | | | | | | | |
| 20% | 499 | 0.11 | 11 | 0.204 | | | | | | | | | | |
| 30% | 504 | 0.11 | 7 | 0.118 | | | | | | | | | | |
| 40% | 510 | 0.11 | 6 | 0.108 | | | | | | | | | | |
| 50% | 516 | 0.11 | 8 | 0.140 | | | | | | | | | | |
| 60% | 523 | 0.11 | 11 | 0.182 | | | | | | | | | | |
| 70% | 531 | 0.11 | 15 | 0.258 | | | | | | | | | | |
| 80% | 538 | 0.11 | 20 | 0.344 | | | | | | | | | | |
| 90% | 547 | 0.11 | 27 | 0.452 | | | | | | | | | | |
| 100% | 556 | 0.106 | 52 | 0.892 | | | | | | | | | | |
| Invar (Ni ≈ 36%) | 508 | 0.11 | 6.2 | 0.108 | | | | | | | | | | |
| Chrome steel (Cr = 0%) | 493 | 0.108 | 42 | 0.785 | 50 | 42 | 39 | 36 | 32 | 28 | 23 | 21 | 20 | 21 |
| 1% | 491 | 0.11 | 35 | 0.645 | — | 36 | 32 | 30 | 27 | 24 | 21 | 19 | 19 | |
| 2% | 491 | 0.11 | 30 | 0.559 | — | 31 | 28 | 26 | 24 | 22 | 19 | 18 | 18 | |
| 5% | 489 | 0.11 | 23 | 0.430 | — | 23 | 22 | 21 | 21 | 19 | 17 | 17 | 17 | 17 |
| 10% | 486 | 0.11 | 18 | 0.344 | — | 18 | 18 | 18 | 17 | 17 | 16 | 16 | 17 | |
| 20% | 480 | 0.11 | 13 | 0.258 | — | 13 | 13 | 13 | 13 | 14 | 14 | 15 | 17 | |
| 30% | 476 | 0.11 | 11 | 0.204 | | | | | | | | | | |
| Cr-Ni (chrome-nickel) | | | | | | | | | | | | | | |
| 15 Cr, 10 Ni | 491 | 0.11 | 11 | 0.204 | | | | | | | | | | |
| 18 Cr, 8 Ni (V2A) | 488 | 0.11 | 9.4 | 0.172 | — | 9.4 | 10 | 10 | 11 | 11 | 13 | 15 | 18 | |
| 20 Cr, 15 Ni | 489 | 0.11 | 8.7 | 0.161 | | | | | | | | | | |
| 25 Cr, 20 Ni | 491 | 0.11 | 7.4 | 0.140 | | | | | | | | | | |
| Ni-Cr (nickel-chrome) | | | | | | | | | | | | | | |
| 80 Ni, 15 Cr | 532 | 0.11 | 10 | 0.172 | | | | | | | | | | |
| 60 Ni, 15 Cr | 516 | 0.11 | 7.4 | 0.129 | | | | | | | | | | |
| 40 Ni, 15 Cr | 504 | 0.11 | 6.7 | 0.118 | | | | | | | | | | |
| 20 Ni, 15 Cr | 491 | 0.11 | 8.1 | 0.151 | — | 8.1 | 8.7 | 8.7 | 9.4 | 10 | 11 | 13 | | |
| Cr-Ni-Al: 6 Cr, 1.5 Al, 0.5 Si (Sicromal 8) | 482 | 0.117 | 13 | 0.237 | | | | | | | | | | |
| 24 Cr, 2.5 Al, 0.5 Si (Sicromal 12) | 479 | 0.118 | 11 | 0.194 | | | | | | | | | | |
| Manganese steel (Ma = 0%) | 493 | 0.118 | 42 | 0.784 | | | | | | | | | | |
| 1% | 491 | 0.11 | 29 | 0.538 | | | | | | | | | | |
| 2% | 491 | 0.11 | 22 | 0.376 | — | 22 | 21 | 21 | 21 | 20 | 19 | | | |
| 5% | 490 | 0.11 | 13 | 0.247 | | | | | | | | | | |
| 10% | 487 | 0.11 | 10 | 0.194 | | | | | | | | | | |
| Tungsten steel (W = 0%) | 493 | 0.108 | 42 | 0.785 | | | | | | | | | | |
| 1% | 494 | 0.107 | 38 | 0.720 | | | | | | | | | | |
| 2% | 497 | 0.106 | 36 | 0.677 | — | 36 | 34 | 31 | 28 | 26 | 21 | | | |
| 5% | 504 | 0.104 | 31 | 0.591 | | | | | | | | | | |
| 10% | 519 | 0.100 | 28 | 0.527 | | | | | | | | | | |
| 20% | 551 | 0.093 | 25 | 0.484 | | | | | | | | | | |
| Silicon steel (Si = 0%) | 493 | 0.108 | 42 | 0.785 | | | | | | | | | | |
| 1% | 485 | 0.11 | 24 | 0.451 | | | | | | | | | | |
| 2% | 479 | 0.11 | 18 | 0.344 | | | | | | | | | | |
| 5% | 463 | 0.11 | 11 | 0.215 | | | | | | | | | | |

## Table II.5-1, Continued

| Metal | Properties at 68°F | | | | | | | | | | | | | |
|---|---|---|---|---|---|---|---|---|---|---|---|---|---|---|
| | | | | | | | | | $k$(Btu/hr-ft-.F) | | | | | |
| | $\rho$ (lb$_m$/ft$^3$) | $C_p$ (Btu/lb$_m$•°F) | $k$ (Btu/hr • ft • °F) | $\alpha$ (ft$^2$/hr) | −148°F −100°C | 32°F 0°C | 212°F 100°C | 392°F 200°C | 572°F 300°C | 752°F 400°C | 1112°F 600°C | 1472°F 800°C | 1832°F 1000°C | 2192°F 1200°C |
| **Copper** | | | | | | | | | | | | | | |
| Pure | 559 | 0.0915 | 223 | 4.353 | 235 | 223 | 219 | 216 | — | 210 | 204 | | | |
| Aluminum bronze: 95 Cu, 5 Al | 541 | 0.098 | 48 | 0.903 | | | | | | | | | | |
| Bronze: 75 Cu. 25 Sn | 541 | 0.082 | 15 | 0.333 | | | | | | | | | | |
| Red brass: 85 Cu. 9 Sn. 6 Zn | 544 | 0.092 | 35 | 0.699 | — | 34 | 41 | | | | | | | |
| Brass: 70 Cu. 30 Zn | 532 | 0.092 | 64 | 1.322 | 51 | — | 74 | 83 | 85 | 85 | | | | |
| German silver 62 Cu, 15 Ni. 22 Zn | 538 | 0.094 | 14.4 | 0.290 | 11.1 | — | 18 | 23 | 26 | 28 | | | | |
| Constantan: 60 Cu, 40 Ni | 557 | 0.098 | 13.1 | 0.237 | 12 | — | 12.8 | 15 | | | | | | |
| **Magnesium** | | | | | | | | | | | | | | |
| Pure | 109 | 0.242 | 99 | 3.762 | 103 | 99 | 97 | 94 | 91 | | | | | |
| Mg-Al (electrolytic) 6-8% Al, 1-2% Zn | 113 | 0.24 | 38 | 1.397 | — | 30 | 36 | 43 | 48 | | | | | |
| Mg-Mn: 2% Mn | 111 | 0.24 | 66 | 2.473 | 54 | 64 | 72 | 75 | | | | | | |
| **Molybdenum** | 638 | 0.060 | 79 | 2.074 | 80 | 79 | 79 | | | | | | | |
| **Nickel** | | | | | | | | | | | | | | |
| Pure (99.9%) | 556 | 0.1065 | 52 | 0.882 | 60 | 54 | 48 | 42 | 37 | 34 | | | | |
| Impure (99.2%) | 556 | 0.106 | 40 | 0.677 | — | 40 | 37 | 34 | 32 | 30 | 32 | 36 | 39 | 40 |
| Ni-Cr: 90 Ni. 10 Cr | 541 | 0.106 | 10 | 0.172 | — | 9.9 | 10.9 | 12.1 | 13.2 | 14.2 | | | | |
| 80 Ni, 20 Cr | 519 | 0.106 | 7.3 | 0.129 | — | 7.1 | 8.0 | 9.0 | 9.9 | 10.9 | 13.0 | | | |
| **Silver** | | | | | | | | | | | | | | |
| Purest | 657 | 0.0559 | 242 | 6.601 | 242 | 241 | 240 | 238 | | | | | | |
| Pure (99.9%) | 657 | 0.0559 | 235 | 6.418 | 242 | 237 | 240 | 216 | 209 | 208 | | | | |
| **Tungsten** | 1208 | 0.0321 | 94 | 2.430 | — | 96 | 87 | 82 | 77 | 73. | 65 | 44 | | |
| **Zinc. Pure** | 446 | 0.0918 | 64.8 | 1.591 | 66 | 65 | 63 | 61 | 58 | 54 | | | | |
| **Tin. pure** | 456 | 0.0541 | 37 | 1.505 | 43 | 38.1 | 34 | 33 | | | | | | |

*Source:* From E.R.G. Eckert and R.M. Drake, *Heat and Mass Transfer-,* copyright 1959 McGraw-Hill; used with the permission of McGraw-Hill Book Company.

## Table II.5-2 Thermal Properties of Some Nonmetals

| Substance | $C_p$ (Btu/lb$_m$ • °F) | $\rho$ lb$_m$/°F$^3$ | $t$ (°F) | $k$ (Btu/hr • ft$^2$ • °F) | $\alpha$ (ft$^2$/hr) |
|---|---|---|---|---|---|
| **Structural** | | | | | |
| Asphalt | | | 68 | 0.43 | a |
| Bakelite 0.38 | b | 79.5 | 68 | 0.134 | b   0.0044 |
| **Bricks** | | | | | |
| Common | 0.20 | d   100 | d | 68 | 0.40   a   0.02 |
| Face | | 128   d | 68 | 0.76 | a |
| Carborundum brick | | | 1110 | 10.7 | a |
| | | | 2550 | 6.4 | a |
| Chrome brick | 0.20 | d   188 | 392   d | 1.34 | a   0.036 |
| | | | 1022 | 1.43 | a   0.038 |
| | | | 1652 | 1.15 | a   0.031 |
| Diatomaceous earth (fired) | | | 400 | 0.14 | a |
| | | | 1600 | 0.18 | a |
| Fire clay brick (burnt 2426 F) | 0.23 | d   128 | 932   d | 0.60 | a   0.020 |
| | | | 1472 | 0.62 | a   0.021 |
| | | | 2012 | 0.63 | a   0.021 |
| Fire clay brick (burnt 2642 F) | 0.23 | d   145 | 912   d | 0.74 | a   0.022 |
| | | | 1472 | 0.79 | a   0.024 |
| | | | 2012 | 0.81 | a   0.024 |
| Fire clay brick (Missouri) | 0.23 | d   165 | 392   f | 0.58 | a   0.015 |
| | | | 1112 | 0.85 | a   0.022 |
| | | | 2552 | 1.02 | a   0.027 |
| Magnesite | 0.27 | d | 400 | 2.2 | a |
| | | | 1200 | 1.6 | a |
| | | | 2200 | 1.1 | a |
| Cement, portland | | 94 | | 0.17 | a |
| Cement, mortar | | | 75 | 0.67 | a |
| Concrete 0.21 | b | 119-144 | 68   b | 0.47-0.81 | b   0.019-0.027 |
| Concrete, cinder | | | 75 | 0.44 | a |

**Table II.5-2  Continued**

| Substance | $C_p$ (Btu/lb$_m \cdot$°F) | | $\rho$ (lb$_m$/ft$^3$) | | $t$ (°F) | $k$ (Btu/hr$\cdot$ft$^2 \cdot$°F) | | $\alpha$ (ft$^2$/hr) |
|---|---|---|---|---|---|---|---|---|
| Glass, plate | 0.2 | b | 169 | b | 68 | 0.44 | b | 0.013 |
| Glass, borosilicate | | | 139 | b | 86 | 0.63 | b | |
| Plaster, gypsum | 0.2 | d | 90 | d | 70 | 0.28 | a | 0.016 |
| Plaster, metal lath | | | | | 70 | 0.27 | a | |
| Plaster, wood lath | | | | | 70 | 0.16 | a | |
| Stone | | | | | | | | |
|   Granite | 0.195 | d | 165 | d | | 1.0–2.3 | a | 0.031–0.071 |
|   Limestone | 0.217 | d | 155 | d | 210–570 | 0.73–0.77 | a | 0.022–0.023 |
|   Marble | 0.193 | b | 156–169 | b | 68 | 1.6 | b | 0.054 |
|   Sandstone | 0.17 | b | 135–144 | b | 68 | 0.94–1.2 | b | 0.041–0.049 |
| Wood, cross grain | | | | | | | | |
|   Balsa | | | 8.8 | a | 86 | 0.032 | a | |
|   Cypress | | | 29 | d | 86 | 0.056 | a | |
|   Fir | 0.65 | d | 26.0 | b | 75 | 0.063 | a | 0.0037 |
|   Oak | 0.57 | d | 38–30 | b | 86 | 0.096 | a | 0.0049 |
|   Yellow pine | 0.67 | d | 40 | d | 75 | 0.085 | a | 0.0032 |
|   White pine | | | 27 | d | 86 | 0.065 | a | |
| Wood, radial | | | | | | | | |
|   Oak | 0.57 | b | 38–30 | b | 68 | 0.10–0.12 | b | { 0.0043– \\ 0.0047 |
|   Fir | 0.65 | b | 26.0–26.3 | b | 68 | 0.08 | b | 0.0048 |
| Insulating | | | | | | | | |
|   Asbestos | | | 29.3 | b | { −328 \\ 32 | 0.043 \\ 0.090 | b \\ b | |
| | | | 36.0 | b | { 32 \\ 212 \\ 392 \\ 752 | 0.087 \\ 0.111 \\ 0.120 \\ 0.129 | b \\ b \\ b \\ b | |
| | | | 43.5 | b | { −328 \\ 32 | 0.09 \\ 0.135 | b \\ b | |
|   Asbestos cement | | | | | | 1.2 | a | |
|   Asbestos cement board | | | | | 68 | 0.43 | a | |
|   Asbestos sheet | | | | | 124 | 0.096 | a | |
|   Asbestos felt (40 laminations per inch) | | | | | { 100 \\ 300 \\ 500 | 0.033 \\ 0.040 \\ 0.048 | a \\ a \\ a | |
|   Asbestos felt (20 laminations per inch) | | | | | { 100 \\ 300 \\ 500 | 0.045 \\ 0.055 \\ 0.065 | a \\ a \\ a | |
|   Asbestos, corrugated (4 plies per inch) | | | | | { 100 \\ 200 \\ 300 | 0.05 \\ 0.058 \\ 0.069 | a \\ a \\ a | |
|   Balsam wool | | | 2.2 | a | 90 | 0.023 | a | |
|   Cardboard, corrugated | | | | | | 0.037 | a | |
|   Celotex | | | | | 90 | 0.028 | a | |
|   Corkboard | | | 10 | b | 86 | 0.025 | b | |
|   Cork, expanded scrap | 0.45 | b | 2.8–7.4 | b | 68 | 0.021 | b | 0.006–0.017 |
|   Cork, ground | | | 9.4 | b | 86 | 0.025 | b | |
| Insulating | | | | | | | | |
|   Diatomaceous earth (powdered) | | | 10 | e | { 200 \\ 400 \\ 600 | 0.029 \\ 0.038 \\ 0.048 | e \\ e \\ e | |
|   Diatomaceous earth (powdered) | | | 14 | e | { 200 \\ 400 \\ 600 | 0.033 \\ 0.039 \\ 0.046 | e \\ e \\ e | |
|   Diatomaceous earth (powdered) | | | 18 | e | { 200 \\ 400 \\ 600 | 0.040 \\ 0.045 \\ 0.049 | e \\ e \\ e | |
|   Felt, hair | | | 8.2 | c | { 20 \\ 100 \\ 200 | 0.0237 \\ 0.0269 \\ 0.0310 | c \\ c \\ c | |

**Table II.5-2    Continued**

| Substance | $C_p$ (Btu/lb$_m$·°F) | | $\rho$ (lb$_m$/ft$^3$) | | $t$ (°F) | $k$ (Btu/hr·ft$^2$·°F) | | $\alpha$ (ft$^2$/hr) |
|---|---|---|---|---|---|---|---|---|
| Felt, hair | | | 11.4 | c | 20<br>100<br>200 | 0.0212<br>0.0254<br>0.0299 | c<br>c<br>c | |
| Felt, hair | | | 12.8 | c | 20<br>100<br>200 | 0.0233<br>0.0262<br>0.0295 | c<br>c<br>c | |
| Fiber insulating board | | | 14.8 | b | 70 | 0.028 | b | |
| Glass wool | | | 1.5 | c | 20<br>100<br>200 | 0.0217<br>0.0313<br>0.0435 | c<br>c<br>c | |
| Glass wool | | | 4.0 | c | 20<br>100<br>200 | 0.0179<br>0.0239<br>0.0317 | c<br>c<br>c | |
| Glass wool | | | 6.0 | c | 20<br>100<br>200 | 0.0163<br>0.0218<br>0.0288 | c<br>c<br>c | |
| Kapok | | | | | 86 | 0.020 | a | |
| Magnesia, 85% | | | 16.9 | c | 100<br>200<br>300<br>400 | 0.039<br>0.041<br>0.043<br>0.046 | a<br>a<br>a<br>a | |
| Rock wool | | | 4.0 | c | 100<br>200 | 0.0224<br>0.0317 | c<br>c | |
| Rock wool | | | 8.0 | c | 20<br>100<br>200 | 0.0171<br>0.0228<br>0.0299 | c<br>c<br>c | |
| Rock wool | | | 12.0 | c | 20<br>100<br>200 | 0.0183<br>0.0226<br>0.0281 | c<br>c<br>c | |
| **Miscellaneous** | | | | | | | | |
| Aerogel, silica | | | 8.5 | b | 248 | 0.013 | b | |
| Clay | 0.21 | b | 91.0 | b | 68 | 0.739 | b | 0.039 |
| Coal, anthracite | 0.30 | b | 75–94 | b | 68 | 0.15 | b | 0.005–0.006 |
| Coal, powdered | 0.31 | b | 46 | b | 86 | 0.067 | b | 0.005 |
| Cotton | 0.31 | b | 5 | b | 68 | 0.034 | b | 0.075 |
| Earth, coarse | 0.44 | b | 128 | b | 68 | 0.30 | b | 0.0054 |
| Ice | 0.46 | b | 57 | b | 32 | 1.28 | b | 0.048 |
| Rubber, hard | | | 74.8 | b | 32 | 0.087 | b | |
| Sawdust | | | | | 75 | 0.034 | a | |
| Silk | 0.33 | b | 3.6 | b | 68 | 0.021 | b | 0.017 |

*Source:* Adapted from (a) A. I. Brown and S. M. Marco, *Introduction to Heat Transfer*, 3rd ed., McGraw-Hill, New York, 1958; (b) E. R. G. Eckert, *Introduction to the Transfer of Heat and Mass*, McGraw-Hill, New York, 1950; (c) R. H. Heilman, *Industrial and Engineering Chemistry*, Vol. 28 (1936), p. 782; (d) L. S. Marks, *Mechanical Engineers' Handbook*, 6th ed., McGraw-Hill, New York, 1958; (e) R. Calvert, *Diatomaceous Earth*, Chemical Catalog Company, Inc., 1930; (f) H. F. Norton, *Journal of the American Ceramic Society*, Vol. 10 (1957), p. 30.

**Table II.5-3  Property Values of Fluids in Saturated State**

| $t$ (°F) | $\rho$ (lb/ft$^3$) | $C_p$ (Btu/ lb·°F) | $\nu$ (ft$^2$/sec) | $k$ (Btu/ hr·ft·°F) | $\alpha$ (ft$^2$/hr) | Pr | $\beta$ (1/R) |
|---|---|---|---|---|---|---|---|
| \multicolumn{8}{c}{*Water (H$_2$O)*} |

**Water (H$_2$O)**

| $t$ (°F) | $\rho$ (lb/ft$^3$) | $C_p$ (Btu/lb·°F) | $\nu$ (ft$^2$/sec) | $k$ (Btu/hr·ft·°F) | $\alpha$ (ft$^2$/hr) | Pr | $\beta$ (1/R) |
|---|---|---|---|---|---|---|---|
| 32 | 62.57 | 1.0074 | $1.925 \times 10^{-5}$ | 0.319 | $5.07 \times 10^{-3}$ | 13.6 | |
| 68 | 62.46 | 0.9988 | 1.083 | 0.345 | 5.54 | 7.02 | $0.10 \times 10^{-3}$ |
| 104 | 62.09 | 0.9980 | 0.708 | 0.363 | 5.86 | 4.34 | |
| 140 | 61.52 | 0.9994 | 0.514 | 0.376 | 6.02 | 3.02 | |
| 176 | 60.81 | 1.0023 | 0.392 | 0.386 | 6.34 | 2.22 | |
| 212 | 59.97 | 1.0070 | 0.316 | 0.393 | 6.51 | 1.74 | |
| 248 | 59.01 | 1.015 | 0.266 | 0.396 | 6.62 | 1.446 | |
| 284 | 57.95 | 1.023 | 0.230 | 0.395 | 6.68 | 1.241 | |
| 320 | 56.79 | 1.037 | 0.204 | 0.393 | 6.70 | 1.099 | |
| 356 | 55.50 | 1.055 | 0.186 | 0.390 | 6.68 | 1.004 | |
| 392 | 54.11 | 1.076 | 0.172 | 0.384 | 6.61 | 0.937 | |
| 428 | 52.59 | 1.101 | 0.161 | 0.377 | 6.51 | 0.891 | |
| 464 | 50.92 | 1.136 | 0.154 | 0.367 | 6.35 | 0.871 | |
| 500 | 49.06 | 1.182 | 0.148 | 0.353 | 6.11 | 0.874 | |
| 537 | 46.98 | 1.244 | 0.145 | 0.335 | 5.74 | 0.910 | |
| 572 | 44.59 | 1.368 | 0.145 | 0.312 | 5.13 | 1.019 | |

**Ammonia (NH$_3$)**

| $t$ (°F) | $\rho$ | $C_p$ | $\nu$ | $k$ | $\alpha$ | Pr | $\beta$ |
|---|---|---|---|---|---|---|---|
| −58 | 43.93 | 1.066 | $0.468 \times 10^{-5}$ | 0.316 | $6.75 \times 10^{-3}$ | 2.60 | |
| −40 | 43.18 | 1.067 | 0.437 | 0.316 | 6.88 | 2.28 | |
| −22 | 42.41 | 1.069 | 0.417 | 0.317 | 6.98 | 2.15 | |
| −4 | 41.62 | 1.077 | 0.410 | 0.316 | 7.05 | 2.09 | |
| 14 | 40.80 | 1.090 | 0.407 | 0.314 | 7.07 | 2.07 | |
| 32 | 39.96 | 1.107 | 0.402 | 0.312 | 7.05 | 2.05 | |
| 50 | 39.09 | 1.126 | 0.396 | 0.307 | 6.98 | 2.04 | |
| 68 | 38.19 | 1.146 | 0.386 | 0.301 | 6.88 | 2.02 | $1.36 \times 10^{-3}$ |
| 86 | 37.23 | 1.168 | 0.376 | 0.293 | 6.75 | 2.01 | |
| 104 | 36.27 | 1.194 | 0.366 | 0.285 | 6.59 | 2.00 | |
| 122 | 35.23 | 1.222 | 0.355 | 0.275 | 6.41 | 1.99 | |

**Carbon Dioxide (CO$_2$)**

| $t$ (°F) | $\rho$ | $C_p$ | $\nu$ | $k$ | $\alpha$ | Pr | $\beta$ |
|---|---|---|---|---|---|---|---|
| −58 | 72.19 | 0.44 | $0.128 \times 10^{-5}$ | 0.0494 | $1.558 \times 10^{-3}$ | 2.96 | |
| −40 | 69.78 | 0.45 | 0.127 | 0.0584 | 1.864 | 2.46 | |
| −22 | 67.22 | 0.47 | 0.126 | 0.0645 | 2.043 | 2.22 | |
| −4 | 64.45 | 0.49 | 0.124 | 0.0665 | 2.110 | 2.12 | |
| 14 | 61.39 | 0.52 | 0.122 | 0.0635 | 1.989 | 2.20 | |

**Eutectic Calcium Chloride Solution (29.9% CaCl$_2$)**

| $t$ (°F) | $\rho$ | $C_p$ | $\nu$ | $k$ | $\alpha$ | Pr | $\beta$ |
|---|---|---|---|---|---|---|---|
| −58 | 82.39 | 0.623 | $39.13 \times 10^{-5}$ | 0.232 | $4.52 \times 10^{-3}$ | 312 | |
| −40 | 82.09 | 0.6295 | 26.88 | 0.240 | 4.65 | 208 | |
| −22 | 81.79 | 0.6356 | 18.49 | 0.248 | 4.78 | 139 | |
| −4 | 81.50 | 0.642 | 11.88 | 0.257 | 4.91 | 87.1 | |
| 14 | 81.20 | 0.648 | 7.49 | 0.265 | 5.04 | 53.6 | |
| 32 | 80.91 | 0.654 | 4.73 | 0.273 | 5.16 | 33.0 | |
| 50 | 80.62 | 0.660 | 3.61 | 0.280 | 5.28 | 24.6 | |
| 68 | 80.32 | 0.666 | 2.93 | 0.288 | 5.40 | 19.6 | |
| 86 | 80.03 | 0.672 | 2.44 | 0.295 | 5.50 | 16.0 | |
| 104 | 79.73 | 0.678 | 2.07 | 0.302 | 5.60 | 13.3 | |
| 122 | 79.44 | 0.685 | 1.78 | 0.309 | 5.69 | 11.3 | |

**Glycerin [C$_3$H$_5$(OH)$_3$]**

| $t$ (°F) | $\rho$ | $C_p$ | $\nu$ | $k$ | $\alpha$ | Pr | $\beta$ |
|---|---|---|---|---|---|---|---|
| 32 | 79.66 | 0.540 | 0.0895 | 0.163 | $3.81 \times 10^{-3}$ | $84.7 \times 10^3$ | |
| 50 | 79.29 | 0.554 | 0.0323 | 0.164 | 3.74 | 31.0 | |
| 68 | 78.91 | 0.570 | 0.0127 | 0.165 | 3.67 | 12.5 | $0.28 \times 10^{-3}$ |
| 86 | 78.54 | 0.584 | 0.0054 | 0.165 | 3.60 | 5.38 | |
| 104 | 78.16 | 0.600 | 0.0024 | 0.165 | 3.54 | 2.45 | |
| 122 | 77.72 | 0.617 | 0.0016 | 0.166 | 3.46 | 1.63 | |

**Table II.5-3  Continued**

| $t$ (°F) | $\rho$ (lb/ft³) | $C_p$ (Btu/ lb·°F) | $\nu$ (ft²/sec) | $k$ (Btu/ hr·ft·°F) | $\alpha$ (ft²/hr) | Pr | $\beta$ (1/R) |
|---|---|---|---|---|---|---|---|
| | | | *Ethylene Glycol* $[C_2H_4(OH_2)]$ | | | | |
| 32 | 70.59 | 0.548 | $61.92 \times 10^{-5}$ | 0.140 | $3.62 \times 10^{-3}$ | 615 | |
| 68 | 69.71 | 0.569 | 20.64 | 0.144 | 3.64 | 204 | $0.36 \times 10^{-3}$ |
| 104 | 68.76 | 0.591 | 9.35 | 0.148 | 3.64 | 93 | |
| 140 | 67.90 | 0.612 | 5.11 | 0.150 | 3.61 | 51 | |
| 176 | 67.27 | 0.633 | 3.21 | 0.151 | 3.57 | 32.4 | |
| 212 | 66.08 | 0.655 | 2.18 | 0.152 | 3.52 | 22.4 | |
| | | | *Engine Oil (Unused)* | | | | |
| 32 | 56.13 | 0.429 | 0.0461 | 0.085 | $3.53 \times 10^{-3}$ | 47100 | |
| 68 | 55.45 | 0.449 | 0.0097 | 0.084 | 3.38 | 10400 | $0.39 \times 10^{-3}$ |
| 104 | 54.69 | 0.469 | 0.0026 | 0.083 | 3.23 | 2870 | |
| 140 | 53.94 | 0.489 | $0.903 \times 10^{-3}$ | 0.081 | 3.10 | 1050 | |
| 176 | 53.19 | 0.509 | 0.404 | 0.080 | 2.98 | 490 | |
| 212 | 52.44 | 0.530 | 0.219 | 0.079 | 2.86 | 276 | |
| 248 | 51.75 | 0.551 | 0.133 | 0.078 | 2.75 | 175 | |
| 284 | 51.00 | 0.572 | 0.086 | 0.077 | 2.66 | 116 | |
| 320 | 50.31 | 0.593 | 0.060 | 0.076 | 2.57 | 84 | |
| | | | *Mercury (Hg)* | | | | |
| 32 | 850.78 | 0.0335 | $0.133 \times 10^{-5}$ | 4.74 | $166.6 \times 10^{-3}$ | 0.0288 | |
| 68 | 847.71 | 0.0333 | 0.123 | 5.02 | 178.5 | 0.0249 | $1.01 \times 10^{-4}$ |
| 122 | 843.14 | 0.0331 | 0.112 | 5.43 | 194.6 | 0.0207 | |
| 212 | 835.57 | 0.0328 | 0.0999 | 6.07 | 221.5 | 0.0162 | |
| 302 | 828.06 | 0.0326 | 0.0918 | 6.64 | 246.2 | 0.0134 | |
| 392 | 820.61 | 0.0375 | 0.0863 | 7.13 | 267.7 | 0.0116 | |
| 482 | 813.16 | 0.0324 | 0.0823 | 7.55 | 287.0 | 0.0103 | |
| 600 | 802 | 0.032 | 0.0724 | 8.10 | 316 | 0.0083 | |
| | | | *Carbon Dioxide* $(CO_2)$ | | | | |
| 32 | 57.87 | 0.59 | 0.117 | 0.0604 | 1.774 | 2.38 | |
| 50 | 53.69 | 0.75 | 0.109 | 0.0561 | 1.398 | 2.80 | |
| 68 | 48.23 | 1.2 | 0.098 | 0.0504 | 0.860 | 4.10 | $7.78 \times 10^{-3}$ |
| 86 | 37.32 | 8.7 | 0.086 | 0.0406 | 0.108 | 28.7 | |
| | | | *Sulfur Dioxide* $(SO_2)$ | | | | |
| −58 | 97.44 | 0.3247 | $0.521 \times 10^{5}$ | 0.140 | $4.42 \times 10^{-3}$ | 4.24 | |
| −40 | 95.94 | 0.3250 | 0.456 | 0.136 | 4.38 | 3.74 | |
| −22 | 94.43 | 0.3252 | 0.399 | 0.133 | 4.33 | 3.31 | |
| −4 | 92.93 | 0.3254 | 0.349 | 0.130 | 4.29 | 2.93 | |
| 14 | 91.37 | 0.3255 | 0.310 | 0.126 | 4.25 | 2.62 | |
| 32 | 89.80 | 0.3257 | 0.277 | 0.122 | 4.19 | 2.38 | |
| 50 | 88.18 | 0.3259 | 0.250 | 0.118 | 4.13 | 2.18 | |
| 68 | 86.55 | 0.3261 | 0.226 | 0.115 | 4.07 | 2.00 | $1.08 \times 10^{-3}$ |
| 86 | 84.86 | 0.3263 | 0.204 | 0.111 | 4.01 | 1.83 | |
| 104 | 82.98 | 0.3266 | 0.186 | 0.107 | 3.95 | 1.70 | |
| 122 | 81.10 | 0.3268 | 0.174 | 0.102 | 3.87 | 1.61 | |
| | | | *Methyl Chloride* $(CH_3Cl)$ | | | | |
| −58 | 65.71 | 0.3525 | $0.344 \times 10^{5}$ | 0.124 | $5.38 \times 10^{-3}$ | 2.31 | |
| −40 | 64.51 | 0.3541 | 0.342 | 0.121 | 5.30 | 2.32 | |
| −22 | 63.46 | 0.3561 | 0.338 | 0.117 | 5.18 | 2.35 | |
| −4 | 62.39 | 0.3593 | 0.333 | 0.113 | 5.04 | 2.38 | |
| 14 | 61.27 | 0.3629 | 0.329 | 0.108 | 4.87 | 2.43 | |
| 32 | 60.08 | 0.3673 | 0.325 | 0.103 | 4.70 | 2.49 | |
| 50 | 58.83 | 0.3726 | 0.320 | 0.099 | 4.52 | 2.55 | |
| 68 | 57.64 | 0.3788 | 0.315 | 0.094 | 4.31 | 2.63 | |
| 86 | 56.38 | 0.3860 | 0.310 | 0.089 | 4.10 | 2.72 | |
| 104 | 55.13 | 0.3942 | 0.303 | 0.083 | 3.86 | 2.83 | |
| 122 | 53.76 | 0.4034 | 0.295 | 0.077 | 3.57 | 2.97 | |

*Source:* From E. R. G. Eckert and R. M. Drake, *Heat and Mass Transfer*; copyright 1959 McGraw-Hill; Used with the permission of McGraw-Hill Book Company.

**Table II.5-4  Property Values of Gases at Atmospheric Pressure**

| $T$ (°F) | $\rho$ (lb/ft$^3$) | $C_p$ (Btu/ lb·°F) | $\mu$ (lb/sec·ft) | $\nu$ (ft$^2$/sec) | $k$ (Btu/ hr·ft·°F) | $\alpha$ (ft$^2$/hr) | Pr |
|---|---|---|---|---|---|---|---|
| | | | | *Air* | | | |
| −280 | 0.2248 | 0.2452 | $0.4653 \times 10^{-5}$ | $2.070 \times 10^{-5}$ | 0.005342 | 0.09691 | 0.770 |
| −190 | 0.1478 | 0.2412 | 0.6910 | 4.675 | 0.007936 | 0.2226 | 0.753 |
| −100 | 0.1104 | 0.2403 | 0.8930 | 8.062 | 0.01045 | 0.3939 | 0.739 |
| −10 | 0.0882 | 0.2401 | 1.074 | 10.22 | 0.01287 | 0.5100 | 0.722 |
| 80 | 0.0735 | 0.2402 | 1.241 | 16.88 | 0.01516 | 0.8587 | 0.708 |
| 170 | 0.0623 | 0.2410 | 1.394 | 22.38 | 0.01735 | 1.156 | 0.697 |
| 260 | 0.0551 | 0.2422 | 1.536 | 27.88 | 0.01944 | 1.457 | 0.689 |
| 350 | 0.0489 | 0.2438 | 1.669 | 31.06 | 0.02142 | 1.636 | 0.683 |
| 440 | 0.0440 | 0.2459 | 1.795 | 40.80 | 0.02333 | 2.156 | 0.680 |
| 530 | 0.0401 | 0.2482 | 1.914 | 47.73 | 0.02519 | 2.531 | 0.680 |
| 620 | 0.0367 | 0.2520 | 2.028 | 55.26 | 0.02692 | 2.911 | 0.680 |
| 710 | 0.0339 | 0.2540 | 2.135 | 62.98 | 0.02862 | 3.324 | 0.682 |
| 800 | 0.0314 | 0.2568 | 2.239 | 71.31 | 0.03022 | 3.748 | 0.684 |
| 890 | 0.0294 | 0.2593 | 2.339 | 79.56 | 0.03183 | 4.175 | 0.686 |
| 980 | 0.0275 | 0.2622 | 2.436 | 88.58 | 0.03339 | 4.631 | 0.689 |
| 1070 | 0.0259 | 0.2650 | 2.530 | 97.68 | 0.03483 | 5.075 | 0.692 |
| 1160 | 0.0245 | 0.2678 | 2.620 | 106.9 | 0.03628 | 5.530 | 0.696 |
| 1250 | 0.0232 | 0.2704 | 2.703 | 116.5 | 0.03770 | 6.010 | 0.699 |
| 1340 | 0.0220 | 0.2727 | 2.790 | 126.8 | 0.03901 | 6.502 | 0.702 |
| 1520 | 0.0200 | 0.2772 | 2.955 | 147.8 | 0.04178 | 7.536 | 0.706 |
| 1700 | 0.0184 | 0.2815 | 3.109 | 169.0 | 0.04410 | 8.514 | 0.714 |
| 1880 | 0.0169 | 0.2860 | 3.258 | 192.8 | 0.04641 | 9.602 | 0.722 |
| 2060 | 0.0157 | 0.2900 | 3.398 | 216.4 | 0.04880 | 10.72 | 0.726 |
| 2240 | 0.0147 | 0.2939 | 3.533 | 240.3 | 0.05098 | 11.80 | 0.734 |
| 2420 | 0.0138 | 0.2982 | 3.668 | 265.8 | 0.05348 | 12.88 | 0.741 |
| 2600 | 0.0130 | 0.3028 | 3.792 | 291.7 | 0.05550 | 14.00 | 0.749 |
| 2780 | 0.0123 | 0.3075 | 3.915 | 318.3 | 0.05750 | 15.09 | 0.759 |
| 2960 | 0.0116 | 0.3128 | 4.029 | 347.1 | 0.0591 | 16.40 | 0.767 |
| 3140 | 0.0110 | 0.3196 | 4.168 | 378.8 | 0.0612 | 17.41 | 0.783 |
| 3320 | 0.0105 | 0.3278 | 4.301 | 409.9 | 0.0632 | 18.36 | 0.803 |
| 3500 | 0.0100 | 0.3390 | 4.398 | 439.8 | 0.0646 | 19.05 | 0.831 |
| 3680 | 0.0096 | 0.3541 | 4.513 | 470.1 | 0.0663 | 19.61 | 0.863 |
| 3860 | 0.0091 | 0.3759 | 4.611 | 506.9 | 0.0681 | 19.92 | 0.916 |
| 4160 | 0.0087 | 0.4031 | 4.750 | 546.0 | 0.0709 | 20.21 | 0.972 |
| | | | | *Helium* | | | |
| −456 | | 1.242 | $5.66 \times 10^{-7}$ | | 0.0061 | | |
| −400 | 0.0915 | 1.242 | 33.7 | $3.68 \times 10^{-5}$ | 0.0204 | 0.1792 | 0.74 |
| −200 | 0.211 | 1.242 | 84.3 | 39.95 | 0.0536 | 2.044 | 0.70 |
| −100 | 0.0152 | 1.242 | 105.2 | 69.30 | 0.0680 | 3.599 | 0.694 |
| 0 | 0.0119 | 1.242 | 122.1 | 102.8 | 0.0784 | 5.299 | 0.70 |
| 200 | 0.00829 | 1.242 | 154.9 | 186.9 | 0.0977 | 9.490 | 0.71 |
| 400 | 0.00637 | 1.242 | 184.8 | 289.9 | 0.114 | 14.40 | 0.72 |
| 600 | 0.00517 | 1.242 | 209.2 | 404.5 | 0.130 | 20.21 | 0.72 |
| 800 | 0.00439 | 1.242 | 233.5 | 531.9 | 0.145 | 25.81 | 0.72 |
| 1000 | 0.00376 | 1.242 | 256.5 | 682.5 | 0.159 | 34.00 | 0.72 |
| 1200 | 0.00330 | 1.242 | 277.9 | 841.0 | 0.172 | 41.98 | 0.72 |
| | | | | *Hydrogen* | | | |
| −406 | 0.05289 | 2.589 | $1.079 \times 10^{-6}$ | $2.040 \times 10^{-5}$ | 0.0132 | 0.0966 | 0.759 |
| −370 | 0.03181 | 2.508 | 1.691 | 5.253 | 0.0209 | 0.262 | 0.721 |
| −280 | 0.01534 | 2.682 | 2.830 | 18.45 | 0.0384 | 0.933 | 0.712 |
| −190 | 0.01022 | 3.010 | 3.760 | 36.79 | 0.0567 | 1.84 | 0.718 |
| −100 | 0.00766 | 3.234 | 4.578 | 59.77 | 0.0741 | 2.99 | 0.719 |
| −10 | 0.00613 | 3.358 | 5.321 | 86.80 | 0.0902 | 4.38 | 0.713 |
| 80 | 0.00511 | 3.419 | 6.023 | 117.9 | 0.105 | 6.02 | 0.706 |
| 170 | 0.00438 | 3.448 | 6.689 | 152.7 | 0.119 | 7.87 | 0.697 |
| 260 | 0.00383 | 3.461 | 7.300 | 190.6 | 0.132 | 9.95 | 0.690 |
| 350 | 0.00341 | 3.463 | 7.915 | 232.1 | 0.145 | 12.26 | 0.682 |
| 440 | 0.00307 | 3.465 | 8.491 | 276.6 | 0.157 | 14.79 | 0.675 |
| 530 | 0.00279 | 3.471 | 9.055 | 324.6 | 0.169 | 17.50 | 0.668 |
| 620 | 0.00255 | 3.472 | 9.599 | 376.4 | 0.182 | 20.56 | 0.664 |
| 800 | 0.00218 | 3.481 | 10.68 | 489.9 | 0.203 | 26.75 | 0.659 |
| 980 | 0.00191 | 3.505 | 11.69 | 612 | 0.222 | 33.18 | 0.664 |
| 1160 | 0.00170 | 3.540 | 12.62 | 743 | 0.238 | 39.59 | 0.676 |
| 1340 | 0.00153 | 3.575 | 13.55 | 885 | 0.254 | 46.49 | 0.686 |
| 1520 | 0.00139 | 3.622 | 14.42 | 1039 | 0.268 | 53.19 | 0.703 |
| 1700 | 0.00128 | 3.670 | 15.29 | 1192 | 0.282 | 60.00 | 0.715 |
| 1880 | 0.00118 | 3.720 | 16.18 | 1370 | 0.296 | 67.40 | 0.733 |
| 1940 | 0.00115 | 3.735 | 16.42 | 1429 | 0.300 | 69.80 | 0.736 |

**Table II.5-4  Continued**

| $T$ (°F) | $\rho$ (lb/ft$^3$) | $C_p$ (Btu/ lb·°F) | $\mu$ (lb/sec·ft) | $\nu$ (ft$^2$/sec) | $k$ (Btu/ hr·ft·°F) | $\alpha$ (ft$^2$/hr) | Pr |
|---|---|---|---|---|---|---|---|
| | *Oxygen* | | | | | | |
| −280 | 0.2492 | 0.2264 | 5.220 × 10$^{-6}$ | 2.095 × 10$^{-5}$ | 0.00522 | 0.09252 | 0.815 |
| −190 | 0.1635 | 0.2192 | 7.721 | 4.722 | 0.00790 | 0.2204 | 0.773 |
| −100 | 0.1221 | 0.2181 | 9.979 | 8.173 | 0.01054 | 0.3958 | 0.745 |
| −10 | 0.0975 | 0.2187 | 12.01 | 12.32 | 0.01305 | 0.6120 | 0.725 |
| 80 | 0.0812 | 0.2198 | 13.86 | 17.07 | 0.01546 | 0.8662 | 0.709 |
| 170 | 0.0695 | 0.2219 | 15.56 | 22.39 | 0.01774 | 1.150 | 0.702 |
| 260 | 0.0609 | 0.2250 | 17.16 | 28.18 | 0.02000 | 1.460 | 0.695 |
| 350 | 0.0542 | 0.2285 | 18.66 | 34.43 | 0.02212 | 1.786 | 0.694 |
| 440 | 0.0487 | 0.2322 | 20.10 | 41.27 | 0.02411 | 2.132 | 0.697 |
| 530 | 0.0443 | 0.2360 | 21.48 | 48.49 | 0.02610 | 2.496 | 0.700 |
| 620 | 0.0406 | 0.2399 | 22.79 | 56.13 | 0.02792 | 2.867 | 0.704 |
| | *Nitrogen* | | | | | | |
| −280 | 0.2173 | 0.2561 | 4.611 × 10$^{-6}$ | 2.122 × 10$^{-5}$ | 0.005460 | 0.09811 | 0.786 |
| −100 | 0.1068 | 0.2491 | 8.700 | 8.146 | 0.01054 | 0.3962 | 0.747 |
| 80 | 0.0713 | 0.2486 | 11.99 | 16.82 | 0.01514 | 0.8542 | 0.713 |
| 260 | 0.0533 | 0.2498 | 14.77 | 27.71 | 0.01927 | 1.447 | 0.691 |
| 440 | 0.0426 | 0.2521 | 17.27 | 40.54 | 0.02302 | 2.143 | 0.684 |
| 620 | 0.0355 | 0.2569 | 19.56 | 55.10 | 0.02646 | 2.901 | 0.686 |
| 800 | 0.0308 | 0.2620 | 21.59 | 70.10 | 0.02960 | 3.668 | 0.691 |
| 980 | 0.:267 | 0.2681 | 23.41 | 87.68 | 0.03241 | 4.528 | 0.700 |
| 1160 | 0.0237 | 0.2738 | 25.19 | 98.02 | 0.03507 | 5.404 | 0.711 |
| 1340 | 0.0213 | 0.2789 | 26.88· | 126.2 | 0.03741 | 6.297 | 0.724 |
| 1520 | 0.0194 | 0.2832 | 28.41 | 146.4 | 0.03958 | 7.204 | 0.736 |
| 1700 | 0.0178 | 0.2875 | 29.90 | 168.0 | 0.04151 | 8.111 | 0.748 |
| | *Carbon Dioxide* | | | | | | |
| −64 | 0.1544 | 0.187 | 7.462 × 10$^{-6}$ | 4.833 × 10$^{-5}$ | 0.006243 | 0.2294 | 0.818 |
| −10 | 0.1352 | 0.192 | 8.460 | 6.257 | 0.007444 | 0.2868 | 0.793 |
| 80 | 0.1122 | 0.208 | 10.051 | 8.957 | 0.009575 | 0.4103 | 0.770 |
| 170 | 0.0959 | 0.215 | 11.561 | 12.05 | 0.01183 | 0.5738 | 0.755 |
| 260 | 0.0838 | 0.225 | 12.98 | 15.49 | 0.01422 | 0.7542 | 0.738 |
| 350 | 0.0744 | 0.234 | 14.34 | 19.27 | 0.01674 | 0.9615 | 0.721 |
| 440 | 0.0670 | 0.242 | 15.63 | 23.33 | 0.01937 | 1.195 | 0.702 |
| 530 | 0.0608 | 0.250 | 16.85 | 27.71 | 0.02208 | 1.453 | 0.685 |
| 620 | 0.0558 | 0.257 | 18.03 | 32.31 | 0.02491 | 1.737 | 0.668 |
| | *Carbon Monoxide* | | | | | | |
| −64 | 0.09699 | 0.2491 | 9.295 × 10$^{-6}$ | 9.583 × 10$^{-5}$ | 0.01101 | 0.4557 | 0.758 |
| −10 | 0.0525 | 0.2490 | 10.35 | 12.14 | 0.01239 | 0.5837 | 0.750 |
| 80 | 0.07109 | 0.2489 | 11.990 | 16.87 | 0.01459 | 0.8246 | 0.737 |
| 170 | 0.06082 | 0.2492 | 13.50 | 22.20 | 0.01666 | 1.099 | 0.728 |
| 260 | 0.05329 | 0.2504 | 14.91 | 27.98 | 0.01864 | 1.397 | 0.722 |
| 350 | 0.04735 | 0.2520 | 16.25 | 34.32 | 0.0252 | 1.720 | 0.718 |
| 440 | 0.04259 | 0.2540 | 17.51 | 41.11 | 0.02232 | 2.063 | 0.718 |
| 530 | 0.03872 | 0.2569 | 18.74 | 48.40 | 0.02405 | 2.418 | 0.721 |
| 620 | 0.03549 | 0.2598 | 19.89 | 56.04 | 0.02569 | 2.786 | 0.724 |
| | *Ammonia (NH$_3$)* | | | | | | |
| −58 | 0.0239 | 0.525 | 4.875 × 10$^{-6}$ | 2.04 × 10$^{-4}$ | 0.0099 | 0.796 | 0.93 |
| 32 | 0.0495 | 0.520 | 6.285 | 1.27 | 0.0127 | 0.507 | 0.90 |
| 122 | 0.0405 | 0.520 | 7.415 | 1.83 | 0.0156 | 0.744 | 0.88 |
| 212 | 0.0349 | 0.534 | 8.659 | 2.48 | 0.0189 | 1.015 | 0.87 |
| 302 | 0.0308 | 0.553 | 9.859 | 3.20 | 0.0226 | 1.330 | 0.87 |
| 392 | 0.0275 | 0.572 | 11.08 | 4.03 | 0.0270 | 1.713 | 0.84 |
| | *Steam (H$_2$O Vapor)* | | | | | | |
| 224 | 0.0366 | 0.492 | 8.54 × 10$^{-6}$ | 2.33 × 10$^{-4}$ | 0.0142 | 0.789 | 1.060 |
| 260 | 0.0346 | 0.481 | 9.03 | 2.61 | 0.0151 | 0.906 | 1.040 |
| 350 | 0.0306 | 0.473 | 10.25 | 3.35 | 0.0173 | 1.19 | 1.010 |
| 440 | 0.0275 | 0.474 | 11.45 | 4.16 | 0.0196 | 1.50 | 0.996 |
| 530 | 0.0250 | 0.477 | 12.66 | 5.06 | 0.0219 | 1.84 | 0.991 |
| 620 | 0.0228 | 0.484 | 13.89 | 6.09 | 0.0244 | 2.22 | 0.986 |
| 710 | 0.0211 | 0.491 | 15.10 | 7.15 | 0.0268 | 2.58 | 0.995 |
| 800 | 0.0196 | 0.498 | 16.30 | 8.31 | 0.0292 | 2.99 | 1.000 |
| 890 | 0.0183 | 0.506 | 17.50 | 9.56 | 0.0317 | 3.42 | 1.005 |
| 980 | 0.0171 | 0.514 | 18.72 | 10.98 | 0.0342 | 3.88 | 1.010 |
| 1070 | 0.0161 | 0.522 | 19.95 | 12.40 | 0.0368 | 4.38 | 1.019 |

*Source:* From E. R. G. Eckert and R. M. Drake, *Heat and Mass Transfer*; copyright 1959 by McGraw-Hill; used with permission of McGraw-Hill Book Company.

**Table II.5-5  Zero-Pressure Properties of Gases**

| Gas | Chemical Formula | Molecular Weight | $R$ (ft lb$_f$/lb$_m \cdot$ °R) | $C_{p0}$ (Btu/lb$_m \cdot$ °R)[a] | $C_{v0}$ (Btu/lb$_m \cdot$ °R)[a] | $k$[a] |
|---|---|---|---|---|---|---|
| Air | — | 28.97 | 53.34 | 0.240 | 0.171 | 1.400 |
| Argon | Ar | 39.94 | 38.66 | 0.1253 | 0.0756 | 1.667 |
| Carbon dioxide | $CO_2$ | 44.01 | 35.10 | 0.203 | 0.158 | 1.285 |
| Carbon monoxide | CO | 28.01 | 55.16 | 0.249 | 0.178 | 1.399 |
| Helium | He | 4.003 | 386.0 | 1.25 | 0.753 | 1.667 |
| Hydrogen | $H_2$ | 2.016 | 766.4 | 3.43 | 2.44 | 1.404 |
| Methane | $CH_4$ | 16.04 | 96.35 | 0.532 | 0.403 | 1.32 |
| Nitrogen | $N_2$ | 28.016 | 55.15 | 0.248 | 0.177 | 1.400 |
| Oxygen | $O_2$ | 32.000 | 48.28 | 0.219 | 0.157 | 1.395 |
| Steam | $H_2O$ | 18.016 | 85.76 | 0.445 | 0.335 | 1.329 |

[a] $C_{p0}$, $C_{v0}$, and $k$ are at 80°F.

**Table II.5-6  Constant-Pressure Specific Heats of Various Substances at Zero Pressure**

| Gas or Vapor | Equation: $\bar{C}_{p0}$ in Btu/lb mol · °R, $T$ in °R | Range (°R) | Max. Error (%) |
|---|---|---|---|
| $O_2$ | $\bar{C}_{p0} = 11.515 - \dfrac{172}{\sqrt{T}} + \dfrac{1530}{T}$ | 540–5000 | 1.1 |
| | $= 11.515 - \dfrac{172}{\sqrt{T}} + \dfrac{1530}{T}$ $+ \dfrac{0.05}{1000}(T - 4000)$ | 5000–9000 | 0.3 |
| $N_2$ | $\bar{C}_{p0} = 9.47 - \dfrac{3.47 \times 10^3}{T} + \dfrac{1.16 \times 10^6}{T^2}$ | 540–9000 | 1.7 |
| CO | $\bar{C}_{p0} = 9.46 - \dfrac{3.29 \times 10^3}{T} + \dfrac{1.07 \times 10^6}{T^2}$ | 540–9000 | 1.1 |
| $H_2$ | $\bar{C}_{p0} = 5.76 + \dfrac{0.578}{1000}T + \dfrac{20}{\sqrt{T}}$ | 540–4000 | 0.8 |
| | $= 5.76 + \dfrac{0.578}{1000}T + \dfrac{20}{\sqrt{T}}$ $- \dfrac{0.33}{1000}(T - 4000)$ | 4000–9000 | 1.4 |
| $H_2O$ | $\bar{C}_{p0} = 19.86 \dfrac{597}{\sqrt{T}} + \dfrac{7500}{T}$ | 540–5400 | 1.8 |
| $CO_2$ | $\bar{C}_{p0} = 16.2 - \dfrac{6.53 \times 10^3}{T} + \dfrac{1.41 \times 10^6}{T^2}$ | 540–6300 | 0.8 |
| $CH_4$ | $\bar{C}_{p0} = 4.52 + 0.00737T$ | 540–1500 | 1.2 |
| $C_2H_4$ | $\bar{C}_{p0} = 4.23 + 0.01177T$ | 350–1100 | 1.5 |
| $C_2H_6$ | $\bar{C}_{p0} = 4.01 + 0.01636T$ | 400–1100 | 1.5 |
| $C_3H_8$ | $\bar{C}_{p0} = 2.258 + 0.0320T - 5.43 \times 10^{-6}T^2$ | 415–2700 | 1.8 |
| $C_4H_{10}$ | $\bar{C}_{p0} = 4.36 + 0.0403T - 6.83 \times 10^{-6}T^2$ | 540–2700 | 1.7 |
| $C_8H_{18}$ | $\bar{C}_{p0} = 7.92 + 0.0601T$ | 400–1100 | est. 4 |
| $C_{12}H_{26}$ | $\bar{C}_{p0} = 8.68 + 0.0889T$ | 400–1100 | est. 4 |

*Source:* From R. L. Sweigert and M. W. Beardsley, *Bulletin No. 2*, Georgia School of Technology, 1938, except $C_3H_8$ and $C_4H_{10}$, which are from H. M. Spencer, *Journal of the American Chemical Society*, Vol. 67 (1945), p. 1859.

**Table II.6  Psychrometric Chart for Air/Steam Mixtures**

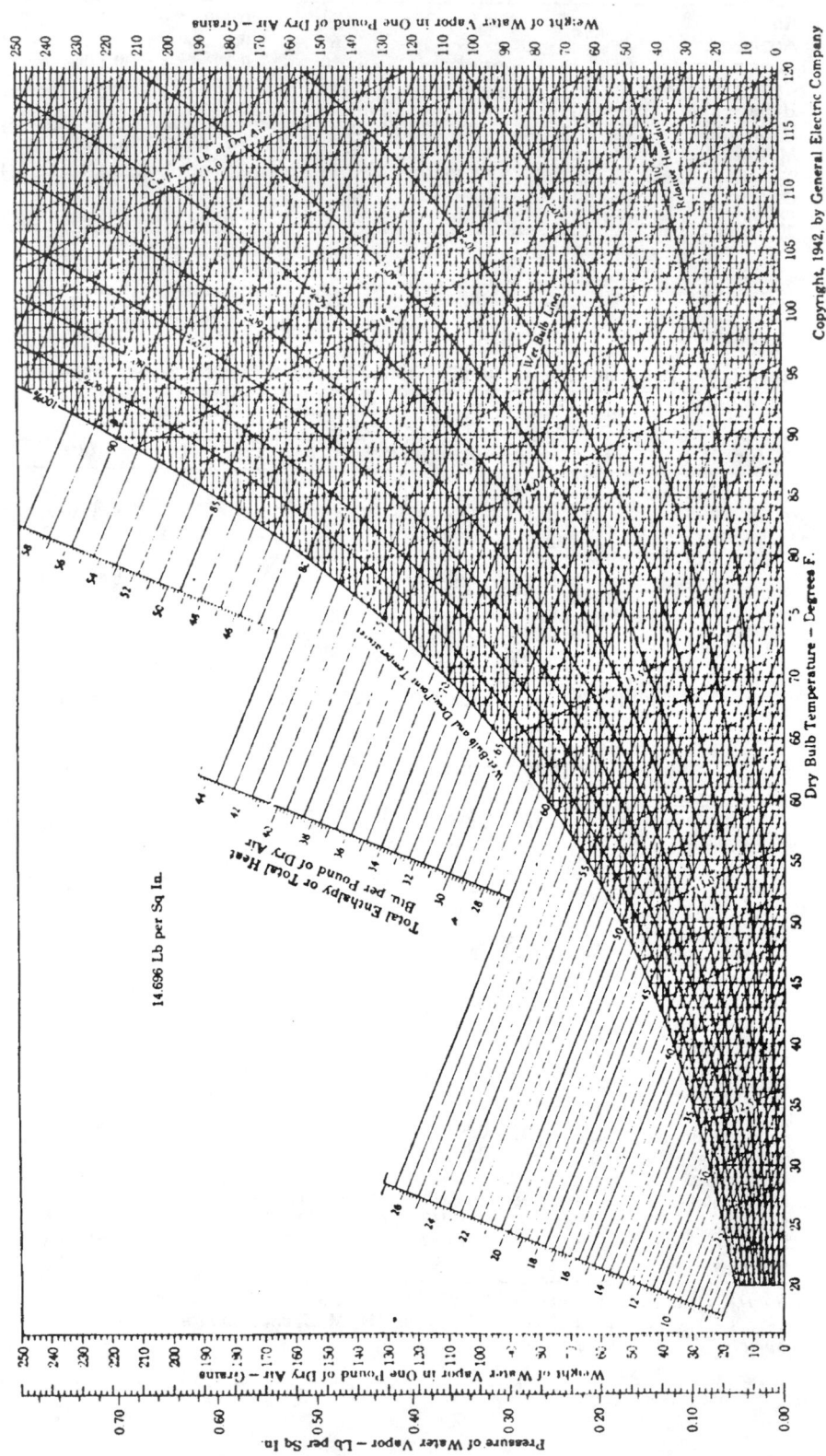

**Table II.7  Total Emissivity Data**

| Surface | °C | °F | $\epsilon$ |
|---|---|---|---|
| *Metals* | | | |
| **Aluminum** | | | |
| Polished, 98% pure | 200–600 | 400–1100 | 0.04–0.06 |
| Commercial sheet | 100 | 200 | 0.09 |
| Rough plate | 40 | 100 | 0.07 |
| Heavily oxidized | 100–550 | 200–1000 | 0.20–0.33 |
| **Antimony** | | | |
| Polished | 40–250 | 100–500 | 0.28–0.31 |
| **Bismuth** | | | |
| Bright | 100 | 200 | 0.34 |
| **Brass** | | | |
| Highly polished | 250 | 500 | 0.03 |
| Polished | 40 | 100 | 0.07 |
| Dull plate | 40–250 | 100–500 | 0.22 |
| Oxidized | 40–250 | 100–500 | 0.46–0.56 |
| **Chromium** | | | |
| Polished sheet | 40–550 | 100–1000 | 0.08–0.27 |
| **Cobalt** | | | |
| Unoxidized | 250–550 | 500–1000 | 0.13–0.23 |
| **Copper** | | | |
| Highly polished electrolytic | 100 | 200 | 0.02 |
| Polished | 40 | 100 | 0.04 |
| Slightly polished | 40 | 100 | 0.12 |
| Polished, lightly tarnished | 40 | 100 | 0.05 |
| Dull | 40 | 100 | 0.15 |
| Black oxidized | 40 | 100 | 0.76 |
| **Gold** | | | |
| Pure, highly polished | 100–600 | 200–1100 | 0.02–0.035 |
| **Inconel** | | | |
| X, stably oxidized | 230–900 | 450–1600 | 0.55–0.78 |
| B, stably oxidized | 230–1000 | 450–1750 | 0.32–0.55 |
| X and B, polished | 150–300 | 300–600 | 0.20 |
| **Iron and Steel** | | | |
| Mild steel, polished | 150–500 | 300–900 | 0.14–0.32 |
| Steel, polished | 40–250 | 100–500 | 0.07–0.10 |
| Sheet steel, ground | 1000 | 1700 | 0.55 |
| Sheet steel, rolled | 40 | 100 | 0.66 |
| Sheet steel, strong rough oxide | 40 | 100 | 0.80 |
| Steel, oxidized at 1100°F | 250 | 500 | 0.79 |
| Cast iron, with skin | 40 | 100 | 0.70–0.80 |
| Cast iron, newly turned | 40 | 100 | 0.44 |
| Cast iron, polished | 200 | 400 | 0.21 |
| Cast iron, oxidized | 40–250 | 100–500 | 0.57–0.66 |
| Iron, red rusted | 40 | 100 | 0.61 |
| Iron, heavily rusted | 40 | 100 | 0.85 |
| Wrought iron, smooth | 40 | 100 | 0.35 |
| Wrought iron, dull oxidized | 20–360 | 70–680 | 0.94 |
| Stainless, polished | 40 | 100 | 0.07–0.17 |
| Stainless, after repeated heating and cooling | 230–930 | 450–1650 | 0.50–0.70 |
| **Lead** | | | |
| Polished | 40–250 | 100–500 | 0.05–0.08 |
| Gray, oxidized | 40 | 100 | 0.28 |
| Oxidized at 390°F | 200 | 400 | 0.63 |
| Oxidized at 1100°F | 40 | 100 | 0.63 |
| **Magnesium** | | | |
| Polished | 40–250 | 100–500 | 0.07–0.13 |
| **Manganin** | | | |
| Bright rolled | 100 | 200 | 0.05 |
| **Mercury** | | | |
| Pure, clean | 40–100 | 100–200 | 0.10–0.12 |

**Table II.7  Continued**

| Surface | °C | °F | ε |
|---|---|---|---|
| Molybdenum | | | |
|   Polished | 40–250 | 100–500 | 0.06–0.08 |
|   Polished | 550–1100 | 1000–2000 | 0.11–0.18 |
|   Filament | 550–2800 | 1000–5000 | 0.08–0.29 |
| Monel | | | |
|   After repeated heating and cooling | 230–930 | 450–1650 | 0.45–0.70 |
|   Oxidized at 1100°F | 200–600 | 400–1100 | 0.41–0.46 |
|   Polished | 40 | 100 | 0.17 |
| Nickel | | | |
|   Polished | 40–250 | 100–500 | 0.05–0.07 |
|   Oxidized | 40–250 | 100–500 | 0.35–0.49 |
|   Wire | 250–1100 | 500–2000 | 0.10–0.19 |
| Platinum | | | |
|   Pure, polished plate | 200–600 | 400–1100 | 0.05–0.10 |
|   Oxidized at 1100°F | 250–550 | 500–1000 | 0.07–0.11 |
|   Electrolytic | 250–550 | 500–1000 | 0.06–0.10 |
|   Strip | 550–1100 | 1000–2000 | 0.12–0.14 |
|   Filament | 40–1100 | 100–2000 | 0.04–0.19 |
|   Wire | 200–1370 | 400–2500 | 0.07–0.18 |
| Silver | | | |
|   Polished or deposited | 40–550 | 100–1000 | 0.01–0.03 |
|   Oxidized | 40–550 | 100–1000 | 0.02–0.04 |
|   German silver," polished | 250–550 | 500–1000 | 0.07–0.09 |
| Tin | | | |
|   Bright tinned iron | 40 | 100 | 0.04–0.06 |
|   Bright | 40 | 100 | 0.06 |
|   Polished sheet | 100 | 200 | 0.05 |
| Tungsten | | | |
|   Filament | 550–1100 | 1000–2000 | 0.11–0.16 |
|   Filament | 2800 | 5000 | 0.39 |
|   Filament, aged | 40–3300 | 100–6000 | 0.03–0.35 |
|   Polished | 40–550 | 100–1000 | 0.04–0.08 |
| Zinc | | | |
|   Pure polished | 40–250 | 100–500 | 0.02–0.03 |
|   Oxidized at 750°F | 400 | 750 | 0.11 |
|   Galvanized, gray | 40 | 100 | 0.28 |
|   Galvanized, fairly bright | 40 | 100 | 0.23 |
|   Dull | 40–250 | 100–500 | 0.21 |
| *Nonmetals* | | | |
| Asbestos | | | |
|   Board | 40 | 100 | 0.96 |
|   Cement | 40 | 100 | 0.96 |
|   Paper | 40 | 100 | 0.93–0.95 |
|   Slate | 40 | 100 | 0.97 |
| Brick | | | |
|   Red, rough | 40 | 100 | 0.93 |
|   Silica | 1000 | 1800 | 0.80–0.85 |
|   Fireclay | 1000 | 1800 | 0.75 |
|   Ordinary refractory | 1100 | 2000 | 0.59 |
|   Magnesite refractory | 1000 | 1800 | 0.38 |
|   White refractory | 1100 | 2000 | 0.29 |
|   Gray, glazed | 1100 | 2000 | 0.75 |
| Carbon | | | |
|   Filament | 1050–1420 | 1900–2600 | 0.53 |
|   Lampsoot | 40 | 100 | 0.95 |
| Clay | | | |
|   Fired | 100 | 200 | 0.91 |
| Concrete | | | |
|   Rough | 40 | 100 | 0.94 |

**Table II.7  Continued**

| Surface | °C | °F | ε |
|---|---|---|---|
| Corundum | | | |
|   Emery rough | 100 | 200 | 0.86 |
| Glass | | | |
|   Smooth | 40 | 100 | 0.94 |
|   Quartz glass (2 mm) | 250–550 | 500–1000 | 0.96–0.66 |
|   Pyrex | 250–550 | 500–1000 | 0.94–0.75 |
|   Gypsum | 40 | 100 | 0.80–0.90 |
| Ice | | | |
|   Smooth | 0 | 32 | 0.97 |
|   Rough crystals | 0 | 32 | 0.99 |
|   Hoarfrost | −18 | 0 | 0.99 |
| Limestone | 40–250 | 100–500 | 0.95–0.83 |
| Marble | | | |
|   Light gray, polished | 40 | 100 | 0.93 |
|   White | 40 | 100 | 0.95 |
| Mica | 40 | 100 | 0.75 |
| Paints | | | |
|   Aluminum, various ages and compositions | 100 | 200 | 0.27–0.62 |
|   Black gloss | 40 | 100 | 0.90 |
|   Black lacquer | 40 | 100 | 0.80–0.93 |
|   White paint | 40 | 100 | 0.89–0.97 |
|   White lacquer | 40 | 100 | 0.80–0.95 |
|   Various oil paints | 40 | 100 | 0.92–0.96 |
|   Red lead | 100 | 200 | 0.93 |
| Paper | | | |
|   White | 40 | 100 | 0.95 |
|   Writing paper | 40 | 100 | 0.98 |
|   Any color | 40 | 100 | 0.92–0.94 |
|   Roofing | 40 | 100 | 0.91 |
| Plaster | | | |
|   Lime, rough | 40–250 | 100–500 | 0.92 |
| Porcelain | | | |
|   Glazed | 40 | 100 | 0.93 |
| Quartz | 40–550 | 100–1000 | 0.89–0.58 |
| Rubber | | | |
|   Hard | 40 | 100 | 0.94 |
|   Soft, gray rough | 40 | 100 | 0.86 |
| Sandstone | 40–250 | 100–500 | 0.83–0.90 |
| Snow | (−12) − (−6) | 10–20 | 0.82 |
| Water | | | |
|   0.1 mm or more thick | 40 | 100 | 0.96 |
| Wood | | | |
|   Oak, planed | 40 | 100 | 0.90 |
|   Walnut, sanded | 40 | 100 | 0.83 |
|   Spruce, sanded | 40 | 100 | 0.82 |
|   Beech | 40 | 100 | 0.94 |
|   Planed | 40 | 100 | 0.78 |
|   Various | 40 | 100 | 0.80–0.90 |
|   Sawdust | 40 | 100 | 0.75 |

*Source:* From E. M. Sparrow and R. D. Cess, *Radiation Heat Transfer*, rev. ed.; copyright © 1970 by Wadsworth Publishing Company, Inc.; reprinted by permission of the publisher, Brooks/ Cole Publishing Company, Monterey, Calif.
[a] German silver is actually an alloy of copper, nickel, and zinc.

# APPENDIX III
# REVIEW OF ELECTRICAL SCIENCE

**RUSSELL L. HEISERMAN, Ed.D.**

School of Technology
Oklahoma State University
Stillwater, Oklahoma

## III. 1 INTRODUCTION

This brief review of electrical science is intended for those readers who may use electrical principles only on occasion and is intended to be supportive of the material found in those chapters of the handbook based on electrical science. The review consists of selected topics in basic ac circuit theory presented at a nominal analytical level. Much of the material deals with power in ac circuits and principles of power-factor improvement.

## III.2 REVIEW OF VECTOR ALGEBRA

Vector algebra is the mathematics most appropriate for ac circuit problems. Most often electric quantities, voltage and current, are not in phase in ac circuits, so phase relationships as well as magnitude have to be considered. This brief review will cover the basic idea of a vector quantity and then refresh the process of adding, subtracting, multiplying, and dividing vectors.

### III.2.1 Review

A vector is a quantity having both direction and magnitude. Familiar vector quantities are velocity and force. Other familiar quantities, such as speed, volume, area, and mass, have magnitude only.

A vector quantity is expressed as having both magnitude and direction, such as

$$Ae^{\pm j\theta}$$

where $A$ is the magnitude and $e^{\pm j\theta}$ expresses the direction in the complex plane (Figure III.1).

The important feature of this vector notation is to note that the angle of displacement is in fact an exponent. This feature is significant, since it will allow the use of the law of exponents when multiplying, dividing, or raising to a power.

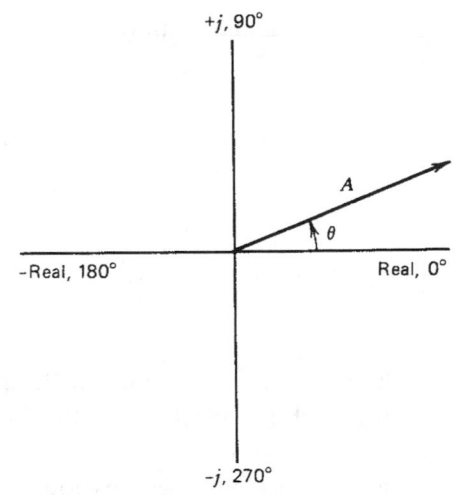

**Fig. III.1 The generalized vector $Ae^{j\theta}$ shown in the complex plane. If $j$ is positive, it is referenced to the positive real axis with a counterclockwise displacement. If $j$ is negative, it is referenced to the positive real axis with a clockwise displacement.**

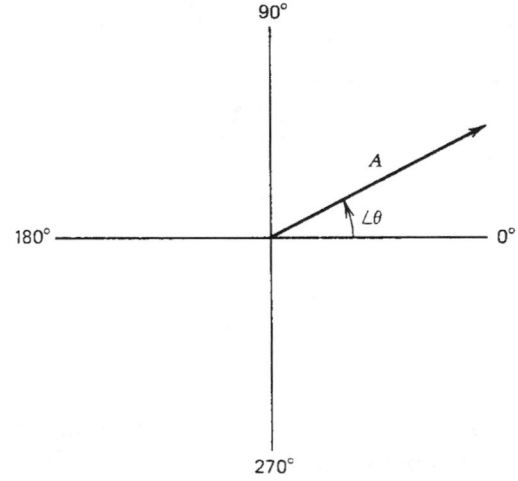

**Fig. III.2 The vector A $\angle\theta$ shown in the polar coordinate system. A vector expressed as A $\angle\theta$ is said to be in polar form.**

Common practice has created a shorthand for expressing vectors. This method is quicker to write and for many, more clearly expresses the idea of a vector:

$$A \angle \theta$$

This shorthand is read as a vector magnitude $A$ operating or pointing in the direction $\theta$. It is termed the polar representation of a vector as shown in Figure III.2.

Now the function $e^{j\theta}$ may be expressed or resolved into its horizontal and vertical components in the complex plane:

$$e^{j\theta} = \cos\theta + j\sin\theta$$

The vector has been resolved and expressed in rectangular form. Using the shorthand notation

$$A\angle\theta = A\cos\theta + jA\sin\theta$$

where $A\cos\theta$ is the vector projection on the real axis and $jA\sin\theta$ is the vector projection on the imaginary axis, as shown in Figure III.3.

Both rectangular and polar expressions of a vector quantity are useful when performing mathematical operations.

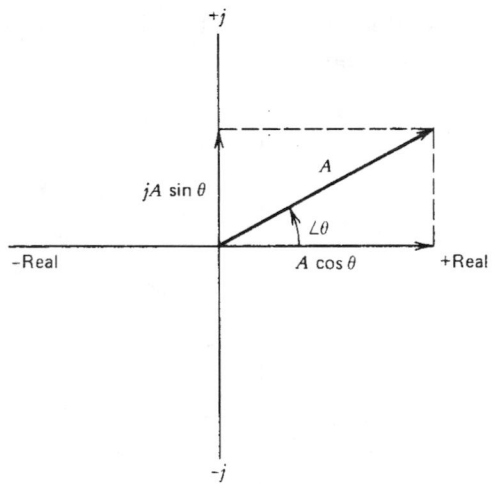

Fig. III.3 The vector A $\angle\theta$ shown together with its rectangular components.

### III.2.2 Addition and Subtraction of Vectors

When adding or subtracting vectors, it is most convenient to use the rectangular form. This is best demonstrated through an example. Suppose that we have two vectors, $20\angle30°$ and $25\angle-45°$, and these vectors are to be added. The quickest way to accomplish this is to resolve each vector into its rectangular components, add the real components, then add the imaginary components, and, if needed, express the results in polar form:

$$20\angle30° = 20\cos30° + j20\sin30°$$

$$= 17.3 + j10$$

$$25\angle-45° = 25\cos45° - j25\sin45°$$

$$= 17.7 - j17.7$$

A calculator is a handy tool for resolving vectors. Many calculators have automatic programs for converting vectors from one form to another.

Now adding we obtain

$$
\begin{array}{r}
17.3 + j10 \\
(+)\quad \underline{17.7 - j17.7} \\
35.0 - j7.7
\end{array}
$$

By inspection, this vector is seen to be slightly greater in magnitude than 35.0 and at a small angle below the positive real axis. Again using a calculator to express the vector in polar form: $35.8\angle-12.4°$, an answer in agreement with what was anticipated. Figure III.4 shows roughly the same result using a graphical technique. Subtraction is accomplished in much the same way. Suppose that the vector $25\angle-45°$ is to be subtracted from the vector $20\angle30°$.

$$20\angle30° \quad = 17.3 + j10$$

$$25\angle-45° = 17.7 - j17.7$$

To subtract

$$
\begin{array}{r}
17.3 + j10 \\
\underline{17.7 - j17.7}
\end{array}
$$

first change sign of the subtrahend and then add:

$$
\begin{array}{r}
17.3 + j10 \\
\underline{-17.7 + j17.7} \\
-0.4 + j27.7
\end{array}
$$

The effect of changing the sign of the subtrahend is to push the vector back through the origin. as shown in Figure III.5.

The resulting vector appears to be about 28 units long and barely in the second quadrant. The calculator gives $27.7\angle90.8°$.

### III.2.3 Multiplication and Division of Vectors

Vectors are expressed in polar form for multiplication and division. The magnitudes are multiplied or divided and the angles follow the rules governing

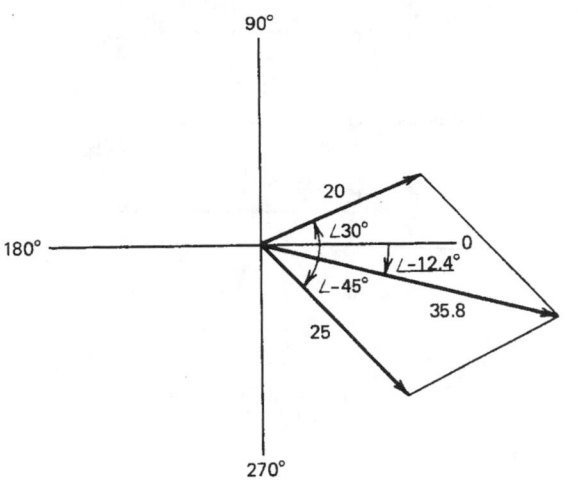

**Fig. III.4 Use of the graphical parallelogram method for adding two vectors. The result or sum is the diagonal originating at the origin of the coordinate system.**

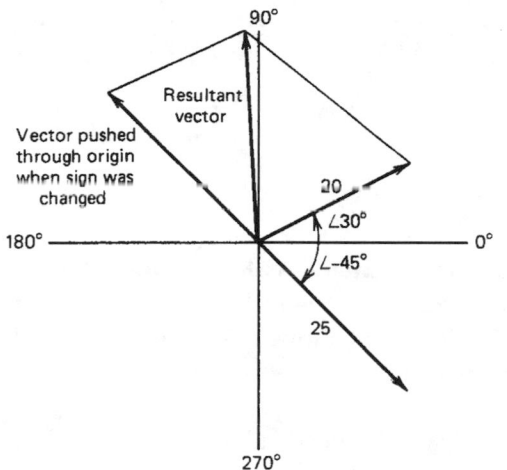

**Fig. III.5 Graphical solution to subtraction of vectors.**

exponents, added when multiplying, subtracted when dividing. Consider

$$20\angle 30° \times 25\angle{-45}° = 500\angle{-15}°$$

The magnitudes are multiplied and the angles are added. Consider

$$\frac{20 \angle 30°}{25\angle{-45}°} = 0.8 \angle 75°$$

The magnitudes are divided and the angle of the divisor is subtracted from the angle of dividend.

Raising to powers is a special case of multiplication. The magnitude is raised to the power and the angle is multiplied by the power. Consider

$$(20\angle 30)^3 = 8000\angle 90°$$

or consider

$$(20\angle 30°)^{1/2} = 4.47 / 15°$$

### III.2.4 Summary

Vector manipulation is straightforward and easy to do. This presentation is intended to refresh those techniques most commonly used by those working at a practical level with ac electrical circuits. It has been the author's intent to exclude material on dot and cross products in favor of techniques that tend to allow the user more of a feeling for what is going on.

## III.3 RESISTANCE, INDUCTANCE, AND CAPACITANCE

The three types of electric circuit elements having distinct characteristics are resistance, inductance, and capacitance. This brief review will focus on the characteristics of these circuit elements in ac circuits to support later discussions on circuit impedance and power-factor-improvement principles.

### III.3. 1 Resistance

Resistance $R$ in an ac circuit is the name given to circuit elements that consume real power in the form of heat, light. mechanical work, and so on. Resistance is a physical property of the wire used in a distribution system that results in power loss commonly called $I^2R$ loss. Resistance can be thought of as a name given that portion of a circuit load that performs real work, that is, the portion of the power fed to a motor that results in measurable mechanical work being accomplished.

If resistance is the only circuit element in an ac circuit, the physical properties of that circuit are easily summarized, as shown in Figure III.6. The important property is that the voltage and current are in phase.

Since the current and voltage are in phase and the ac source is a sine wave, the power used by the resistor is easily computed from root-mean-square (rms) (effective) voltage and current readings taken with a typical multimeter. The power is computed by taking the product of the measured voltage in volts or kilovolts and the measured current in amperes:

$$P(\text{watts}) = V(\text{volts}) \times I \text{ (amperes)}$$

where $V$ is the voltage measured in volts and $I$ is the current measured in amps. Both quantities are measured with an rms reading meter.

In many industrial settings the voltage may be

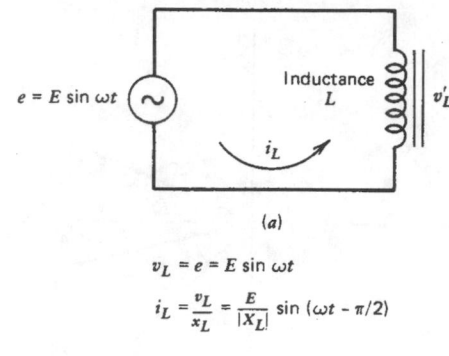

$$v_R = e = E \sin \omega t$$

$$i_R = \frac{v_R}{R} = \frac{E}{R} \sin \omega t$$

where　$\omega = 2\pi f$
　　　　$f$ = frequency in hertz

**Fig. III.6 Circuit showing an ac source with radian frequency $\omega$. The current through the resistor is in phase with the voltage across the resistor.**

measured in kilovolts and current in amperes. The power is computed as the product of current and voltage and expressed as kilowatts:

$$P \text{ (kilowatts)} = V \text{ (kilovolts)} \times I \text{ (amps)}$$

If it is unhandy to measure both voltage and current, one can compute power using only voltage or current if the resistance $R$ is known:

$$P \text{ (watts)} = I^2 \text{(amps)} \times R \text{ (ohms)}$$

or

$$P(\text{watts}) = \frac{V^2(\text{volts})}{R(\text{ohms})}$$

### III.3.2 Inductance

Inductance $L$ in an ac circuit is usually formed as coils of wire, such as those found in motor windings, solenoids, or inductors. In a real circuit it is impossible to have only pure inductance, but for purposes of establishing background we will take the theoretical case of a pure inductance so that its circuit properties can be isolated and presented.

An inductor is a circuit element that uses no real power; it simply stores energy in the form of a magnetic field and will give up this stored energy, alternately storing energy and giving it up every half-cycle. The result of this storing and giving up energy when an inductor is driven by a sine-wave source is to put the measured magnetizing current ($i_c$) 90° out of phase with the driving voltage. The magnetizing current lags behind the driving voltage by 90°. If pure inductance were the load of a sine-wave generator, we could summarize its characteristics as in Figure III.7.

An inductor limits the current flowing through it by reacting with the voltage change across it. This property is called inductive reactance $X_L$. The inductive

(a)

$$v_L = e = E \sin \omega t$$

$$i_L = \frac{v_L}{x_L} = \frac{E}{|X_L|} \sin (\omega t - \pi/2)$$

where $\pi/2$ is 90°; this expresses
the idea that the current lags
the voltage by 90°

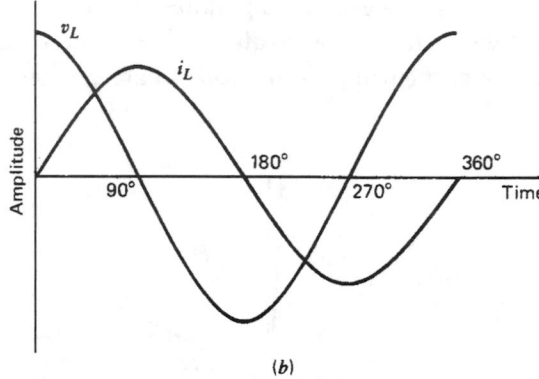

(b)

**Fig. III.7 (a) Ac circuit with pure inductance. (b) Plot of the voltage across the inductor $v_L$, and the current $i_L$ through it. The plot shows a 90° displacement between the current and the voltage.**

reactance of a coil whose inductance is known in henrys (H) may be computed using the expression

$$X_L = 2\pi f L$$

where $f$ is the frequency in hertz and $L$ is the coil's inductance in henrys.

### III.3.3 Capacitance

Capacitance $C$, like inductance, only stores and gives up energy. However, the voltage and current phasing is exactly opposite that of a inductor in an ac circuit. The current in an ac circuit containing only capacitance leads the voltage by $\angle 90°$. Figure III.8 summarizes the characteristics of an ac circuit with a pure capacity load. A capacitor also reacts to changes. This property is called capacitive reactance $X_c$. The capacity reactance may be computed by using the expression

$$X_c = \frac{1}{2\pi f C}$$

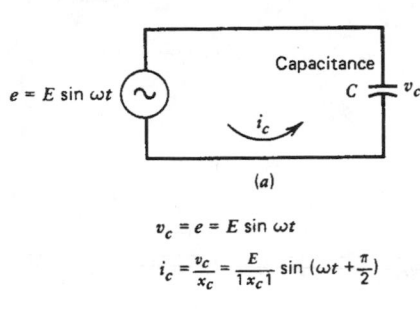

$$v_c = e = E \sin \omega t$$

$$i_c = \frac{v_c}{x_c} = \frac{E}{1x_c 1} \sin (\omega t + \frac{\pi}{2})$$

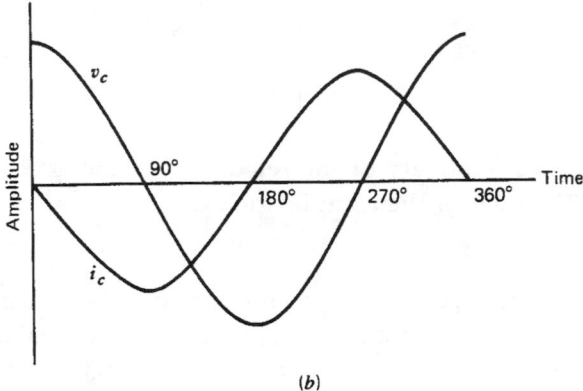

**Fig. III.8 (a) Ac circuit with pure capacitance. (b) Plot of the voltage across the capacitor $v_c$, and the current $i_c$ through it. The plot shows a 90° displacement between the current and the voltage.**

where $f$ is the frequency in hertz and $C$ is the capacity in farads.

### III.3.4 Summary

Circuit elements are resistance that consumes real power and two reactive elements that only store and give up energy. These two reactive elements, capacitors and inductors, have opposite effects on the phase displacement between the current and voltage in ac circuits. These opposite effects are the key to adding capacitors in an otherwise inductive circuit for purposes of reducing the current-voltage phase displacement. Reducing the phase displacement improves the power factor of the circuit. (Power factor is defined and discussed later.)

### III.4 IMPEDANCE

In the preceding section it was mentioned that pure inductance does not occur in a real-world circuit. This is because the wire that is used to form the most carefully made coil still has resistance. This section considers circuits containing resistance and inductive reactance and circuits containing resistance and capacity reactance. At-

tention will be given to the notation used to describe such circuits since vector algebra must be used exclusively.

### III.4.1 Circuits with Resistance and Inductive Reactance

Figure III.9 shows a circuit that has both resistive and inductive elements. Such a circuit might represent a real inductor with the resistance representing the wire resistance, or such a circuit might be a simple model of a motor, with the inductance reflecting the inductive characteristics of the motor's windings and the resistance representing both the wire resistance and the real power consumed and converted to mechanical work performed by the motor.

In Figure III.9 the current is common to both circuit elements. Recall that the voltage across the resistor is in phase with this current while the voltage across the inductor leads the current. This idea is shown by plotting these quantities in the complex plane. Since $i$ is the reference, it is plotted on the positive real axis as shown in Figure III.10.

The voltage across the resistor is in phase with the current, so it is also on the positive real axis. whereas the voltage across the inductor is on the positive $j$ axis

**Fig. III.9 Circuit with both resistance and inductance. The circuit current i is common to both elements.**

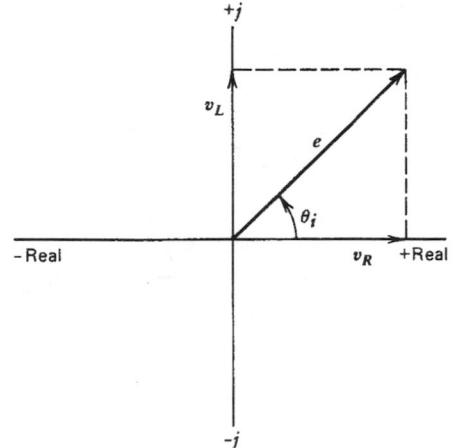

**Fig. III.10 Circuit voltages and current plotted in the complex plane. Both $i$ and $V_R$ are on the positive real axis since they are in phase, $V_L$ is on the positive $j$ axis since it leads the current by 90°.**

since it leads the current by 90°. However, the sum of the voltages must be the source voltage $e$. Figure III.10 shows that the two voltages must be added as vectors:

$$e = v_r + jv_L$$

$$e = i_R + jiX_L$$

If we call the ratio of voltage to current the circuit impedance, then

$$Z = \frac{e}{i} = R + jX_L$$

$Z$, the circuit impedance, is a complex quantity and may be expressed in either polar or rectangular form:

$$Z = R + jX_L$$

or

$$Z = |Z| \underline{/\theta}$$

In circuits with resistance and inductance the complex impedance will have a positive phase angle and if $R$ and $X_L$ are plotted in the complex plane, $X_L$ is plotted on the positive j axis, as shown in Figure III.11.

### III.4.2 Circuits with Resistance and Capacity Reactance

Circuits containing resistance and capacitance are approached about the same way. Going through a similar analysis and looking at the relationship among $R$, $X_c$, and $Z$ would show that $X_c$ is plotted on the negative $j$ axis, as shown in Figure III.12.

### III.4.3 Summary

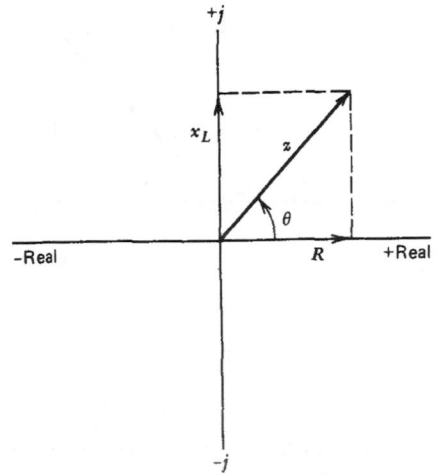

**Fig. III.11 Plot in the complex plane showing the complex relationship of $R$, $X_L$, and $Z$.**

**Fig. III.12 Summary of the relationship among $R$, $X_c$, and $Z$ shown in the complex plane.**

In circuits containing both resistive and reactive elements, the resistance is plotted on the positive real axis while the reactances are plotted on the imaginary axis. The fact that inductive and capacitive reactance causes opposite phase displacements (has opposite effects in ac circuits) is further emphasized by plotting their reactance effects in opposite directions on the imaginary axis of the complex plane. The case is building for why capacitors might be used in an ac circuit with inductive loading to improve the circuit's power factor.

## III.5 POWER IN AC CIRCUITS

This section considers three aspects of power in ac circuits. First, the case of a circuit containing resistance and inductance is discussed, followed by the introduction of the power triangle for circuits containing resistance and inductance. Finally, power-factor improvement by the use of capacitors is presented.

### III.5.1 Power in a Circuit Containing Both Resistance and Inductance

Figure III.13 reviews this situation through a circuit drawing and the voltages and currents shown in the complex plane. Meters are in place that read the effective or rms voltage $V$ across the complex load and the effective or rms line current $I$.

Power is usually thought of as the product of voltage and the current in a circuit. The question is: The current $I$ times *which* voltage will yield the correct or true power? This is an important question, since Figure III.13b shows three voltages in the complex plane.

Each of the three products may be taken, and each

(a)

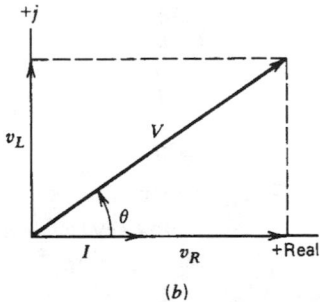

(b)

**Fig. III.13 (a) Circuit having resistance and inductance; meters are in place to measure the line current *I* and the voltage V. (b) Relationship between the various voltages and the line current for this circuit.**

has a name and a meaning. Taking the ammeter reading *I* times the voltmeter reading yields the *apparent power*. The apparent power is the load current-load voltage product without regard to the phase relationship of the current and voltage. This figure by itself is meaningless:

$$P(\text{apparent}) = IV$$

If the voltmeter could be connected across the resistor only, to measure $v_R$, then the line current-voltage product would yield the *true power*, since the current and voltage are in phase.

$$P(\text{true}) = Iv_R$$

Usually, this connection cannot be made, so the true power of a load is measured with a special meter called a wattmeter that automatically performs the following calculation.

$$P(\text{true}) = IV \cos \theta$$

Note that in Figure III.13b, the circuit voltage *V* and the resistance voltage $V_R$ are related through the cosine of θ. The third product that could be taken is called *imaginary power* or VAR, the voltampere reactive product.

$$P(\text{imaginary}) = Iv_L$$

This is the power that is alternately stored and given up by the inductor to maintain its magnetic field. None of this reactive power is actually used.

If the voltages in the foregoing examples were measured in kilovolts, the three values computed would be the more familiar:

$$P(\text{apparent}) = kVA$$

$$P(\text{real}) = kW$$

$$P(\text{imaginary}) = kVAR$$

This discussion, together with Figure III.13b, leads to the power triangle.

### III.5.2 The Power Triangle

The power triangle consists of three values, kVA, kW, and kVAR, arranged in a right triangle. The angle between the line current and voltage, Θ, becomes an important factor in this triangle. Figure III.14 shows the power triangle.

To emphasize the relationship between these three quantities, an example may be helpful. Suppose that we have a circuit with inductive characteristics and using a voltmeter, ammeter, and wattmeter the following values are measured:

$$\text{watts} = 1.5 \text{ kW}$$

$$\text{line current} = 10 \text{ A}$$

$$\text{line voltage} = 240 \text{ V}$$

From this information we should be able to determine the kVA, Θ, and the kVAR.

The kVA can be computed directly from the voltmeter and ammeter readings:

$$\text{kVA} = (10 \text{ A})(0.24) \text{ kV} = 2.4 \text{ kVA}$$

**Fig. III.14 Power triangle for an inductive load. The angle θ is the angle of displacement between the line voltage and the line current.**

Looking at the triangle in Figure III.14 and recalling some basic trigonometry, we have

and θ is the angle whose cosine equals 0.625. This can be looked up in a table or calculated using a hand calculator that computes trig functions:

$$\theta = \cos^{-1} 0.625 = 51.3°$$

Again referring to the power triangle and a little trig, we see that

$$kVAR = kVA \sin \theta$$

$$= 2.4 \text{ kVA} \sin 51.3°$$

$$= 1.87 \text{ kVAR}$$

Figure III.15 puts all these measured and calculated data together in a power triangle.

Of particular interest is the ratio kW/kVA. This ratio is called the *power factor* (PF) of the circuit. So the power factor is the ratio of true power to apparent power in a circuit. This is also the cosine of the angle θ, the angle of displacement between the line voltage and the line current. To improve the power factor, the angle θ must be reduced. This could be accomplished by reducing the kVAR side of the triangle.

**Fig. III.15 Organization of the measured and computed data of the example into a power triangle.**

### III.5.3 Power-Factor Improvement

Recall that inductive reactance and capacity reactance are plotted in opposite directions on the imaginary axis, *j*. Thus it should be no surprise to consider that kVAR produced by a capacitive load behave in an opposite way to kVAR produced by inductive loads. This is the case and is the reason capacitors are commonly added to circuits having inductive loads to improve power factor (reduce the angle θ).

Suppose in the example being considered that enough capacity is added across the load to offset the effects of 90% of the inductive load. That is, we will try

(a)

(b)

**Fig. III.16 (a) Inductive circuit with capacity added to correct power factor. (b) Power vectors showing the relationship among kW, kVAR inductive, and kVAR capacitive.**

to improve the power factor by better than 90%. Figure III.16 shows the circuit arrangement with the kW and kVAR vectors drawn to show their relationship.

Following the example through, consider Figure III.17, where 90% of the kVAR inductive load has been neutralized by adding the capacitor.

Working with the modified triangle in Figure III.17, we can compute the new θ, call it $\theta_2$.

$$\theta_2 = \tan^{-1} \frac{0.19}{1.5}$$
$$= 7.2$$

Again, a calculator comes in handy.

Since the new power factor is the cosine of $\theta_2$, we compute

$$PF \text{ new} = \text{cosine } 7.2 = 0.99$$

certainly an improvement.

Recall that the power factor can be expressed as a ratio of kW to kVA. From this idea we can compute a new kVA value:

$$PF = 0.99 = \frac{1.5 \text{ kW}}{kVA \text{ new}}$$

or

Fig. III.17 Resulting net power triangle when the capacitor is added. A new kVA can be calculated as well as a new θ.

$$\text{kVA new} = \frac{1.5 \text{ kW}}{0.99}$$

$$= 1.52 \text{ kVA}$$

The line voltage did not change, so the line current must be lower.

$$1.52 \text{ kVA new} = 0.24 \text{ kV} \times I \text{ new}$$

$$I \text{ new} = \frac{1.52 \text{ kVA}}{0.24} = 6.3 \text{ A}$$

Comparing the original circuit to the circuit after adding capacity, we have:

|  | Inductive Circuit | Improved Circuit |
| --- | --- | --- |
| Line voltage | 240 V | 240 V |
| Line current | 10 A | 6.3 A |
| PF | 62.5% | 99% |
| kVA | 2.4 kVA | 1.52 kVA |
| kW | 1.5 kW | 1.5 kW |
| kVAR | 1.87 kVAR | 0.19 kVAR |

The big improvement noted is the reduction of line current by 37% with no decrease in real power, kW, used by the load. Also note the big change in kVA; less generating capacity is used to meet the same real power demand (generator input power is determined by KVA output).

### III.5.4 Summary

Through an example it has been demonstrated how the addition of a capacitor across an inductive load can improve power factor, reduce line current, and reduce the amount of generating capacity required to supply the load. The way this comes about is by having the capacitor supply the inductive magnetizing current locally. Since inductive and capacitive elements store and release power at different times in each cycle, this reactive current simply flows back and forth between the capacitor and inductor of the load. This idea is

Fig. III.18 (a) Pictorial showing the inductive load of the example in this section. (b) The load with a capacitor added. With the exchange of the kVAR current between the capacitor and inductive load, very little kVAR current is supplied by the generator.

reinforced by Figure III.18. Adding capacitors to inductive loads can free generating capacity, reduce line loss, improve power factor, and in general be cost effective in controlling energy bills.

## III.6 THREE-PHASE POWER

Three-phase power is the form of power most often distributed to industrial users. This form of transmission has three advantages over single-phase systems: (1) less copper is required to supply a given power at given voltage; (2) if the load of each phase of the three-phase source is identical, the instantaneous output of the alternator is constant; and (3) a three-phase system produces a magnetic field of constant density that rotates at the line frequency—this greatly reduces the complexity of motor construction.

The author realizes that both delta systems and wye systems exist, but will concentrate on four-wire wye systems as being representative of internal distribution systems. This type of internal distribution system allows the customer both single-phase and three-phase service. Our focus will be on measuring power and determining power factor in four-wire three-phase wye-connected systems.

### III.6.1 The Four-Wire Wye-Connected System

Figure III.19 shows a generalized four-wire wye-connected system. The coils represent the secondary windings of the transformers at the site substation while the generalized loads represent phase loads that are the

**Table III.1 How to Select Capacitor Ratings for Induction Motors/*Source: 1.***

**Reference No. 1:** For motor designs pre-dating TRI-CLAD 700® and CUSTOM 8000® Motors
(See GED-6063-02, Reference No. 2, for TRI-CLAD and CUSTOM 8000 motor designs)

Now it's easy to choose the right capacitors for your induction motors. Just refer to the following tables to find the kvar required by your particular motors. Locate motors by horsepower, rpm, and number of poles. All ratings are based on General Electric motor designs.

Tables I and V are also applicable to standard wound-rotor, open-type, three-phase, 60-cycle motors, provided the kvar values in the table are multiplied by a factor of 1.1, and the reduction in line current is increased by multiplying the values in the table by 1.05.

When selecting and installing capacitors, keep in mind the following: A capacitor located at the motor releases the maximum system capacity and is most effective in reducing system losses. Also, for a motor that runs continuously, or nearly so, it is usually most economical to locate the capacitor right at the motor terminals and switch it with the motor.

**TABLE II—220-, 440-, 550-VOLT MOTOR, TOTALLY-ENCLOSED, FAN-COOLED, GENERAL ELECTRIC TYPE K (NEMA DESIGN "B"), NORMAL STARTING TORQUE, NORMAL STARTING CURRENT**

| Induction Motor Horsepower Rating | 3600 / 2 Kvar | % AR | 1800 / 4 Kvar | % AR | 1200 / 6 Kvar | % AR | 900 / 8 Kvar | % AR | 720 / 10 Kvar | % AR | 600 / 12 Kvar | % AR |
|---|---|---|---|---|---|---|---|---|---|---|---|---|
| 2 | 1 | 17 | 1 | 20 | 1 | 23 | 1 | 24 | ... | ... | ... | ... |
| 3 | 1 | 11 | 1 | 19 | 1 | 19 | 2 | 24 | ... | ... | ... | ... |
| 5 | 1 | 9 | 2 | 15 | 2 | 19 | 2 | 20 | ... | ... | ... | ... |
| 7½ | 1 | 6 | 2 | 13 | 4 | 19 | 4 | 20 | ... | ... | ... | ... |
| 10 | 2 | 6 | 2 | 11 | 4 | 16 | 5 | 15 | 7.5 | 20 | 7.5 | 27 |
| 15 | 4 | 6 | 4 | 11 | 4 | 13 | 5 | 15 | 7.5 | 20 | 7.5 | 24 |
| 20 | 4 | 6 | 5 | 11 | 5 | 13 | 7.5 | 15 | 10 | 20 | 10 | 24 |
| 25 | 5 | 6 | 5 | 8 | 5 | 9 | 7.5 | 15 | 10 | 17 | 10 | 18 |
| 30 | 5 | 6 | 7.5 | 8 | 7.5 | 9 | 10 | 15 | 10 | 15 | 10 | 18 |
| 40 | 7.5 | 6 | 10 | 8 | 10 | 9 | 10 | 15 | 10 | 15 | 15 | 17 |
| 50 | 7.5 | 6 | 10 | 8 | 10 | 9 | 15 | 15 | 15 | 12 | 20 | 17 |
| 60 | 10 | 6 | 10 | 8 | 10 | 9 | 15 | 12 | 20 | 12 | 25 | 17 |
| 75 | 15 | 6 | 15 | 8 | 15 | 9 | 20 | 11 | 25 | 12 | 35 | 17 |
| 100 | 15 | 6 | 20 | 8 | 25 | 9 | 25 | 11 | 40 | 11 | 45 | 17 |
| 125 | 20 | 6 | 25 | 7 | 30 | 9 | 30 | 11 | 45 | 12 | 45 | 15 |
| 150 | 25 | 6 | 30 | 7 | 30 | 9 | 40 | 11 | 45 | 12 | 50 | 15 |
| 200 | 35 | 6 | 40 | 7 | 60 | 9 | 60 | 11 | 55 | 11 | 60 | 13 |
| 250 | 40 | 5 | 50 | 6 | 60 | 9 | 80 | 11 | 60 | 11 | 100 | 13 |
| 300 | 50 | 5 | 45 | 6 | 80 | 8 | 80 | 10 | 80 | 10 | 125 | 13 |
| 350 | 60 | 5 | 70 | 6 | 80 | 8 | 80 | 9 | ... | ... | ... | ... |
| 400 | 60 | 5 | 80 | 6 | 80 | 6 | 160 | ... | ... | ... | ... | ... |
| 450 | 70 | 5 | 100 | 6 | ... | ... | ... | ... | ... | ... | ... | ... |
| 500 | 70 | 5 | ... | ... | ... | ... | ... | ... | ... | ... | ... | ... |

**TABLE I—220-, 440-, AND 550-VOLT MOTORS, ENCLOSURE OPEN—INCLUDING DRIPPROOF AND SPLASHPROOF, GENERAL ELECTRIC TYPE K (NEMA DESIGN "B"), NORMAL STARTING TORQUE AND CURRENT**

| Induction Motor Horsepower Rating | 3600 / 2 Kvar | % AR | 1800 / 4 Kvar | % AR | 1200 / 6 Kvar | % AR | 900 / 8 Kvar | % AR | 720 / 10 Kvar | % AR | 600 / 12 Kvar | % AR |
|---|---|---|---|---|---|---|---|---|---|---|---|---|
| 2 | 1 | 16 | 1 | 20 | 1 | 22 | 1 | 24 | ... | ... | ... | ... |
| 3 | 1 | 10 | 1 | 16 | 2 | 21 | 2 | 24 | ... | ... | ... | ... |
| 5 | 1 | 9 | 2 | 16 | 2 | 21 | 2 | 21 | ... | ... | ... | ... |
| 7½ | 1 | 8 | 2 | 13 | 2 | 15 | 4 | 21 | 5 | 29 | ... | ... |
| 10 | 2 | 8 | 2 | 13 | 4 | 15 | 5 | 21 | 5 | 25 | 7.5 | 34 |
| 15 | 4 | 8 | 4 | 13 | 5 | 15 | 7.5 | 15 | 7.5 | 23 | 7.5 | 25 |
| 20 | 4 | 7 | 5 | 9 | 5 | 12 | 7.5 | 15 | 10 | 23 | 10 | 24 |
| 25 | 4 | 7 | 5 | 9 | 5 | 11 | 7.5 | 12 | 10 | 23 | 10 | 23 |
| 30 | 5 | 7 | 7.5 | 9 | 5 | 11 | 10 | 12 | 10 | 15 | 10 | 19 |
| 40 | 5 | 5 | 7.5 | 9 | 10 | 11 | 10 | 12 | 10 | 15 | 15 | 19 |
| 50 | 7.5 | 5 | 10 | 7 | 10 | 9 | 15 | 12 | 20 | 15 | 25 | 19 |
| 60 | 7.5 | 5 | 10 | 7 | 10 | 9 | 15 | 11 | 20 | 15 | 30 | 19 |
| 75 | 10 | 5 | 10 | 7 | 15 | 9 | 15 | 10 | 30 | 15 | 40 | 19 |
| 100 | 15 | 5 | 20 | 7 | 25 | 9 | 30 | 10 | 40 | 15 | 45 | 17 |
| 125 | 15 | 5 | 20 | 7 | 30 | 9 | 35 | 10 | 45 | 13 | 60 | 17 |
| 150 | 15 | 5 | 25 | 6 | 30 | 9 | 40 | 9 | 50 | 13 | 60 | 17 |
| 200 | 40 | 5 | 40 | 6 | 45 | 8 | 50 | 9 | 70 | 13 | 75 | 17 |
| 250 | 45 | 5 | 50 | 6 | 50 | 8 | 70 | 9 | 75 | 12 | 90 | 17 |
| 300 | 50 | 5 | 50 | 6 | 70 | 8 | 75 | 9 | 75 | 11 | 105 | 17 |
| 350 | 50 | 5 | 50 | 5 | 75 | 8 | 80 | 9 | 80 | 11 | 105 | 17 |
| 400 | 60 | 5 | 60 | 5 | 75 | 8 | 100 | 9 | 100 | 11 | 110 | 17 |
| 450 | 60 | 5 | 75 | 5 | 75 | 6 | 100 | 9 | 100 | 11 | 110 | 17 |
| 500 | 70 | 5 | 90 | 5 | 90 | 5 | 110 | 9 | 120 | 11 | 120 | 17 |

**TABLE III—220-, 440-, 550-VOLT MOTORS, ENCLOSURE OPEN—INCLUDING DRIPPROOF AND SPLASHPROOF, GENERAL ELECTRIC TYPE KG (NEMA DESIGN "C"), HIGH STARTING TORQUE AND NORMAL STARTING CURRENT**

| Induction Motor Horsepower Rating | 1800 / 4 Kvar | % AR | 1200 / 6 Kvar | % AR | 900 / 8 Kvar | % AR | 720 / 10 Kvar | % AR |
|---|---|---|---|---|---|---|---|---|
| 3 | 1 | 15 | 2 | 19 | 2 | 30 | ... | ... |
| 5 | 2 | 15 | 2 | 15 | 4 | 26 | ... | ... |
| 7½ | 2 | 15 | 2 | 15 | 4 | 22 | ... | ... |
| 10 | 2 | 12 | 4 | 12 | 5 | 17 | ... | ... |
| 15 | 4 | 12 | 4 | 12 | 5 | 17 | ... | ... |
| 20 | 4 | 12 | 5 | 12 | 7.5 | 17 | ... | ... |
| 25 | 7.5 | 9 | 7.5 | 11 | 10 | 17 | ... | ... |
| 30 | 7.5 | 9 | 7.5 | 11 | 10 | 15 | ... | ... |
| 40 | 10 | 9 | 10 | 11 | 10 | 13 | ... | ... |
| 50 | 10 | 9 | 10 | 9 | 15 | 13 | ... | ... |
| 60 | 10 | 8 | 15 | 9 | 20 | 13 | 20 | 18 |
| 75 | 15 | 8 | 15 | 9 | 20 | 12 | 30 | 18 |
| 100 | 20 | 8 | 25 | 9 | 30 | 12 | 40 | 18 |
| 125 | 20 | 7 | 30 | 8 | 30 | 12 | 50 | 18 |
| 150 | 35 | 7 | 35 | 8 | 45 | 12 | ... | ... |
| 200 | 40 | 7 | 50 | 8 | 70 | 12 | ... | ... |
| 250 | 50 | 7 | 50 | 8 | 80 | 12 | ... | ... |
| 300 | 60 | 7 | 80 | 8 | 80 | 10 | ... | ... |
| 350 | 60 | 7 | 80 | 8 | 80 | 9 | ... | ... |

**NOTE:** A capacitor located on the motor side of the overload relay reduces current through the relay, and therefore, a smaller relay may be necessary. The motor-overload relay should be selected on the basis of the motor full-load nameplate current reduced by the percent reduction in line current (% AR) due to capacitors.

Fig. III.19 Generalized four-wire wye-connected system. The coils A, B, and C represent the three transformer secondaries at the site substation; while $Z_A$, ZB. and $Z_C$ are the generalized loads seen by each phase.

**TABLE IV—220-, 440-, 550-VOLT MOTORS, TOTALLY-ENCLOSED, FAN-COOLED, GENERAL ELECTRIC TYPE KG (NEMA DESIGN "C"), HIGH STARTING TORQUE, NORMAL STARTING CURRENT**

| Induction Motor Horse-power Rating | 1800 4 | | 1200 6 | | 900 8 | | 720 10 | | 600 12 | |
|---|---|---|---|---|---|---|---|---|---|---|
| | Kvar | % AR | Kvar | % AR | Kvar | % AR | Kvar | % AR | Kvar | % AR |
| 3 | .... | ... | 2 | 20 | 2 | 27 | .... | ... | .... | ... |
| 5 | 2 | 10 | 2 | 19 | 2 | 21 | .... | ... | .... | ... |
| 7½ | 2 | 10 | 4 | 17 | 4 | 21 | .... | ... | .... | ... |
| 10 | 2 | 9 | 4 | 17 | 5 | 21 | .... | ... | .... | ... |
| 15 | 4 | 9 | 5 | 14 | 5 | 16 | .... | ... | .... | ... |
| 20 | 5 | 9 | 5 | 11 | 7.5 | 16 | .... | ... | .... | ... |
| 25 | 7.5 | 9 | 5 | 10 | 10 | 10 | 15 | 10 | 15 | 10 | 18 |
| 30 | 7.5 | 9 | 10 | 10 | 10 | 10 | 15 | 10 | 15 | 10 | 18 |
| 40 | 10 | 9 | 10 | 9 | 10 | 12 | 15 | 15 | 15 | 18 |
| 50 | 10 | 9 | 10 | 9 | 15 | 12 | 15 | 15 | 25 | 18 |
| 60 | 15 | 9 | 15 | 9 | 20 | 12 | 30 | 15 | 30 | 18 |
| 75 | 15 | 7 | 15 | 8 | 20 | 11 | 30 | 15 | 40 | 16 |
| 100 | 20 | 7 | 25 | 8 | 25 | 11 | 40 | 15 | 45 | 16 |
| 125 | 25 | 7 | 35 | 8 | 35 | 11 | 50 | 15 | 50 | 15 |
| 150 | 35 | 7 | 35 | 7 | 50 | 11 | 60 | 14 | 50 | 13 |
| 200 | 40 | 7 | 45 | 7 | 50 | 10 | 70 | 14 | 70 | 13 |

**TABLE V—2300- AND 4000-VOLT MOTORS, ENCLOSURE OPEN—INCLUDING DRIPPROOF AND SPLASHPROOF, GENERAL ELECTRIC TYPE K (NEMA DESIGN "B"), NORMAL STARTING TORQUE AND CURRENT**

| Induction Motor Horse-power Rating | 3600 2 | | 1800 4 | | 1200 6 | | 900 8 | | 720 10 | | 600 12 | |
|---|---|---|---|---|---|---|---|---|---|---|---|---|
| | Kvar | % AR | Kvar | % AR | Kvar | % AR | Kvar | % AR | Kvar | % AR | Kvar | % AR |
| 100 | .... | 7 | 25 | 10 | 25 | 9 | 25 | 11 | 25 | 11 | 25 | 11 |
| 125 | 25 | 7 | 25 | 8 | 25 | 9 | 25 | 9 | 25 | 10 | 50 | 15 |
| 150 | 25 | 6 | 25 | 7 | 25 | 7 | 25 | 9 | 50 | 11 | 50 | 14 |
| 200 | 25 | 6 | 25 | 6 | 50 | 8 | 50 | 9 | 50 | 11 | 75 | 14 |
| 250 | 50 | 7 | 50 | 6 | 50 | 8 | 50 | 8 | 75 | 11 | 100 | 14 |
| 300 | 50 | 6 | 50 | 6 | 75 | 8 | 75 | 9 | 75 | 9 | 100 | 13 |
| 350 | 50 | 5 | 50 | 5 | 75 | 8 | 75 | 8 | 75 | 9 | 100 | 14 |
| 400 | 50 | 5 | 50 | 5 | 75 | 5 | 100 | 9 | 100 | 9 | 125 | 12 |
| 450 | 75 | 5 | 75 | 5 | 75 | 5 | 100 | 6 | 100 | 8 | 100 | 8 |
| 500 | 75 | 6 | 75 | 5 | 100 | 5 | 125 | 6 | 125 | 8 | 125 | 8 |
| 600 | 75 | 5 | 100 | 5 | 125 | 6 | 125 | 7 | 125 | 8 | 150 | 8 |
| 700 | 100 | 5 | 100 | 5 | 100 | 5 | 125 | 7 | 150 | 8 | 150 | 8 |
| 800 | 100 | 5 | 125 | 5 | 125 | 5 | 150 | 7 | 150 | 8 | 150 | 8 |

**TABLE VI—2300- AND 4000-VOLT MOTORS, TOTALLY-ENCLOSED, FAN-COOLED, GENERAL ELECTRIC TYPE K (NEMA DESIGN "B"), NORMAL STARTING TORQUE, NORMAL STARTING CURRENT**

| Induction Motor Horse-power Rating | 3600 2 | | 1800 4 | | 1200 6 | | 900 8 | | 720 10 | | 600 12 | |
|---|---|---|---|---|---|---|---|---|---|---|---|---|
| | Kvar | % AR | Kvar | % AR | Kvar | % AR | Kvar | % AR | Kvar | % AR | Kvar | % AR |
| 100 | .... | | 25 | 8 | 25 | 8 | 25 | 9 | 25 | 10 | 25 | 12 |
| 125 | 25 | 5 | 25 | 7 | 25 | 8 | 25 | 9 | 25 | 9 | 50 | 14 |
| 150 | 25 | 6 | 25 | 6 | 25 | 8 | 50 | 10 | 50 | 10 | 50 | 14 |
| 200 | 25 | 5 | 25 | 6 | 50 | 9 | 50 | 11 | 50 | 10 | 75 | 14 |
| 250 | 25 | 4 | 50 | 7 | 50 | 7 | 75 | 10 | 75 | 12 | 75 | 13 |
| 300 | 50 | 5 | 50 | 5 | 50 | 7 | 75 | 10 | 75 | 10 | 100 | 14 |
| 350 | 50 | 5 | 50 | 5 | 75 | 8 | 75 | 9 | 100 | 11 | 125 | 15 |
| 400 | 75 | 5 | 75 | 7 | 125 | 9 | 100 | 9 | 100 | 10 | 125 | 14 |
| 450 | 50 | 4 | 100 | 8 | 100 | 8 | 100 | 8 | 100 | 10 | 125 | 13 |
| 500 | 75 | 5 | 125 | 8 | 125 | 8 | 125 | 8 | 125 | 10 | 150 | 13 |

**TABLE VII—2300- AND 4000-VOLT MOTORS, ENCLOSURE OPEN—INCLUDING DRIPPROOF AND SPLASHPROOF, GENERAL ELECTRIC TYPE KG (NEMA DESIGN "C"), HIGH STARTING TORQUE AND NORMAL STARTING CURRENT**

| Induction Motor Horse-power Rating | 1800 4 | | 1200 6 | | 900 8 | | 720 10 | |
|---|---|---|---|---|---|---|---|---|
| | Kvar | % AR | Kvar | % AR | Kvar | % AR | Kvar | % AR |
| 100 | 25 | 7 | 25 | 10 | 25 | 10 | 25 | 10 |
| 125 | 25 | 7 | 25 | 9 | 25 | 9 | 25 | 10 |
| 150 | 25 | 8 | 25 | 9 | 25 | 9 | .... | ... |
| 200 | 50 | 9 | 50 | 9 | 50 | 10 | .... | ... |
| 250 | 50 | 9 | 50 | 8 | 50 | 9 | .... | ... |
| 300 | 50 | 7 | 75 | 9 | 75 | 10 | .... | ... |
| 350 | 50 | 6 | 75 | 8 | 75 | 9 | .... | ... |

**TABLE VIII—2300- AND 4000-VOLT MOTORS, TOTALLY-ENCLOSED, FAN-COOLED, GENERAL ELECTRIC TYPE KG (NEMA DESIGN "C"), HIGH-STARTING TORQUE, NORMAL STARTING CURRENT**

| Induction Motor Horse-power Rating | 1200 6 | | 900 8 | | 720 10 | | 600 12 | |
|---|---|---|---|---|---|---|---|---|
| | Kvar | % AR | Kvar | % AR | Kvar | % AR | Kvar | % AR |
| 75 | .... | ... | .... | ... | 25 | 16 | 25 | 12 |
| 100 | 25 | 9 | 25 | 9 | 25 | 12 | 50 | 12 |
| 125 | 25 | 9 | 25 | 8 | 50 | 18 | 50 | 17 |
| 150 | 25 | 8 | 50 | 14 | 50 | 15 | 50 | 15 |
| 200 | 25 | 7 | 50 | 10 | 75 | 15 | 74 | 14 |

**TABLE IX—440-VOLT OPEN, DRIPPROOF, OIL-FIELD MOTORS, 1200-RPM, GENERAL ELECTRIC TYPE KG, KR AND KOF**

| Induction Motor Horsepower Rating | High Starting Torque—Low Starting Current | | | | | |
|---|---|---|---|---|---|---|
| | Type KG Motor | | Type KR Motor | | Type KOF Motor | |
| | Kvar | % AR | Kvar | % AR | Kvar | % AR |
| 5 | 3 | 25½ | 3 | 14½ | 4 | 24 |
| 7½ | 5 | 25½ | 4 | 20½ | 6 | 24 |
| 10 | 5 | 13½ | 4 | 12½ | 6 | 19 |
| 15 | 6 | 17½ | 6 | 14 | 6 | 13 |
| 20 | 8 | 17 | 10 | 20½ | 10 | 16 |
| 25 | 10 | 15 | 10 | 14½ | 10 | 15 |
| 30 | 10 | 16 | 10 | 14½ | 10 | 13 |
| 40 | 10 | 13½ | 10 | 13 | 10 | 10 |
| 50 | 10 | 10½ | 10 | 7½ | 15 | 10 |
| 60 | 10 | 9 | 15 | 9½ | 15 | 9 |
| 75 | 15 | 10½ | 15 | 8½ | .... | .... |
| 100 | 30 | 13 | .... | .... | .... | .... |

**CAPACITORS BENEFIT YOUR DISTRIBUTION SYSTEM BY**

- Reducing power costs
- Releasing system capacity
- Improving voltage levels
- Reducing system losses

INDUSTRIAL AND POWER CAPACITOR PRODUCTS DEPT. ● GENERAL ELECTRIC CO. ● HUDSON FALLS, N.Y.

**GENERAL ⓖⓔ ELECTRIC**

GED-6063.01A
12-72 (10M) 6200

sum loads on each phase. These loads may be composites of single-phase services and three-phase motors being fed by the distribution system. N is the neutral or return.

To determine the power and power factor of any phase A, B, or C, consider that phase as if it were a single-phase system. Measure the real power, kW, delivered by the phase by use of a wattmeter and measure and compute the volt-ampere product, apparent power, kVA, using a voltmeter and ammeter.

The power factor of the phase can then be determined and corrected as needed. Each phase can be treated independently in turn. The only caution to note is to make the measurements during nominal load periods, this will allow power-factor correction for the most common loading.

If heavy motors are subject to intermittent duty, additional power and power-factor information can be gathered while they are operating. Capacitors used to correct power factor for these intermittent loads should be connected to relays so that they are across the motors and on phase only when the motor is on; otherwise, overcorrection can occur.

**Table III.2 How to Select Capacitor Ratings for Induction Motors/*Source: 2.***

**how to select capacitor ratings for induction motors**

Reference No. 2: For TRI-CLAD 700® and CUSTOM 8000® motors only.
(See GED-6063-01, Reference No. 1, for motor designs pre-dating TRI-CLAD 700 and CUSTOM 8000 Line)

Now it's easy to choose the right capacitors for your induction motors. Just refer to the following tables to find the kvar required by your particular motors. Locate motors by horsepower, rpm, and number of poles. All ratings are based on General Electric motor designs.

Tables I and V are also applicable to standard, wound-rotor, open-type, three-phase, 60-cycle motors, provided the kvar values in the table are multiplied by a factor of 1.1, and the reduction in line current is increased by multiplying the values in the table by 1.05.

When selecting and installing capacitors, keep in mind the following: A capacitor located at the motor releases the maximum system capacity and is most effective in reducing system losses. Also, for a motor that runs continuously, or nearly so, it is usually most economical to locate the capacitor right at the motor terminals and switch it with the motor.

**TABLE I—230-, 460-, AND 575-VOLT MOTORS, ENCLOSURE OPEN—INCLUDING DRIPPROOF GENERAL ELECTRIC TYPE K (NEMA DESIGN "B"), NORMAL STARTING TORQUE AND CURRENT**

| Induction Motor Horsepower Rating | Nominal Motor Speed in Rpm and Number of Poles | | | | | | | | | | | |
| | 3600 2 | | 1800 4 | | 1200 6 | | 900 8 | | 720 10 | | 600 12 | |
| | Kvar | %AR | Kvar | %AR | Kvar | %AR | Kvar | %AR | Kvar | %AR | Kvar | %AR |
|---|---|---|---|---|---|---|---|---|---|---|---|---|
| 2 | 1 | 14 | 1 | 24 | 1 | 28 | 2 | 42 | ... | .. | 3 | 50 |
| 3 | 1 | 14 | 2 | 24 | 2 | 26 | 4 | 42 | 3 | 40 | 4 | 49 |
| 5 | 2 | 14 | 2 | 21 | 3 | 26 | 4 | 31 | 4 | 40 | 5 | 49 |
| 7½ | 2 | 14 | 4 | 21 | 4 | 21 | 4 | 26 | 7.5 | 40 | 10 | 49 |
| 10 | 4 | 14 | 4 | 17 | 5 | 21 | 5 | 26 | 7.5 | 36 | 10 | 41 |
| 15 | 5 | 12 | 5 | 17 | 5 | 19 | 10 | 26 | 7.5 | 31 | 10 | 34 |
| 20 | 5 | 12 | 7.5 | 17 | 7.5 | 19 | 10 | 23 | 10 | 29 | 15 | 34 |
| 25 | 5 | 12 | 7.5 | 17 | 7.5 | 19 | 10 | 23 | 15 | 24 | 15 | 34 |
| 30 | 7.5 | 11 | 7.5 | 17 | 10 | 19 | 15 | 23 | 15 | 24 | 25 | 32 |
| 40 | 7.5 | 10 | 15 | 17 | 15 | 19 | 20 | 23 | 20 | 24 | 30 | 32 |
| 50 | 10 | 10 | 15 | 17 | 20 | 19 | 25 | 23 | 24 | 24 | 35 | 32 |
| 60 | 10 | 10 | 20 | 17 | 30 | 19 | 30 | 23 | 30 | 22 | 45 | 32 |
| 75 | 15 | 10 | 25 | 14 | 30 | 16 | 30 | 17 | 35 | 21 | 40 | 19 |
| 100 | 15 | 10 | 30 | 14 | 30 | 12 | 35 | 16 | 40 | 15 | 45 | 17 |
| 125 | 30 | 10 | 35 | 12 | 30 | 12 | 50 | 14 | 45 | 15 | 50 | 17 |
| 150 | 30 | 10 | 35 | 11 | 35 | 12 | 50 | 14 | 50 | 14 | 60 | 17 |
| 200 | 35 | 10 | 50 | 11 | 50 | 12 | 70 | 14 | 70 | 13 | 90 | 17 |
| 250 | 35 | 10 | 55 | 9 | 70 | 12 | 85 | 14 | 90 | 13 | 100 | 17 |
| 300 | 35 | 10 | 65 | 9 | 75 | 12 | 95 | 14 | 100 | 13 | 110 | 17 |
| 350 | 40 | 10 | 80 | 9 | 85 | 12 | 125 | 14 | 120 | 13 | 150 | 17 |
| 400 | 100 | 10 | 80 | 8 | 100 | 12 | 140 | 14 | 150 | 13 | 150 | 17 |
| 450 | 100 | 10 | 90 | 8 | 140 | 12 | 150 | 13 | 150 | 13 | 175 | 17 |
| 500 | 100 | 8 | 115 | 8 | 150 | 12 | 150 | 12 | 175 | 13 | 175 | 17 |

**TABLE II—230-, 460-, 575-VOLT MOTOR, TOTALLY ENCLOSED, FAN-COOLED GENERAL ELECTRIC TYPE K (NEMA DESIGN "B"), NORMAL STARTING TORQUE, NORMAL STARTING CURRENT**

| Induction Motor Horsepower Rating | Nominal Motor Speed in Rpm and Number of Poles | | | | | | | | | | | |
| | 3600 | | 1800 | | 1200 | | 900 | | 720 | | 600 | |
| | Kvar | %AR | Kvar | %AR | Kvar | %AR | Kvar | %AR | Kvar | %AR | Kvar | %AR |
|---|---|---|---|---|---|---|---|---|---|---|---|---|
| 2 | 1 | 15 | 1 | 24 | 1 | 28 | 2 | 42 | ... | ... | 3 | 50 |
| 3 | 1 | 15 | 2 | 24 | 2 | 28 | 4 | 32 | 3 | 41 | 4 | 49 |
| 5 | 2 | 15 | 2 | 20 | 3 | 26 | 4 | 29 | 4 | 35 | 5 | 49 |
| 7½ | 3 | 15 | 4 | 20 | 4 | 22 | 4 | 29 | 7.5 | 35 | 10 | 49 |
| 10 | 3 | 11 | 4 | 17 | 5 | 22 | 5 | 29 | 7.5 | 24 | 10 | 34 |
| 15 | 4 | 10 | 5 | 17 | 7.5 | 22 | 10 | 29 | 7.5 | 24 | 10 | 34 |
| 20 | 5 | 10 | 7.5 | 17 | 10 | 21 | 10 | 22 | 15 | 23 | 15 | 34 |
| 25 | 5 | 10 | 7.5 | 17 | 10 | 21 | 10 | 22 | 15 | 20 | 20 | 31 |
| 30 | 5 | 10 | 7.5 | 17 | 10 | 21 | 15 | 22 | 15 | 20 | 20 | 31 |
| 40 | 10 | 10 | 10 | 17 | 15 | 20 | 20 | 22 | 20 | 20 | 30 | 31 |
| 50 | 10 | 10 | 15 | 12 | 25 | 21 | 23 | 21 | 25 | 22 | 35 | 31 |
| 60 | 12 | 9 | 15 | 12 | 25 | 19 | 25 | 19 | 30 | 22 | 40 | 30 |
| 75 | 15 | 9 | 20 | 12 | 25 | 15 | 30 | 19 | 35 | 21 | 40 | 30 |
| 100 | 20 | 9 | 25 | 12 | 30 | 13 | 40 | 19 | 40 | 12 | 50 | 30 |
| 125 | 20 | 9 | 35 | 12 | 30 | 13 | 45 | 18 | 50 | 12 | 60 | 30 |
| 150 | 25 | 9 | 40 | 11 | 25 | 13 | 55 | 18 | 50 | 12 | 70 | 30 |
| 200 | 30 | 9 | 40 | 8 | 60 | 13 | 60 | 18 | 70 | 12 | 75 | 30 |
| 250 | 60 | 9 | 50 | 8 | 60 | 13 | 115 | 18 | 100 | 12 | 125 | 30 |
| 300 | 65 | 7 | 50 | 8 | 60 | 13 | 140 | 18 | 125 | 12 | 150 | 30 |
| 350 | 70 | 7 | 55 | 8 | 80 | 13 | 150 | 18 | 150 | 12 | 150 | 30 |
| 400 | 70 | 7 | 40 | 8 | 130 | 13 | 160 | 17 | 175 | 12 | 175 | 30 |
| 450 | 90 | 7 | 95 | 8 | 145 | 13 | 160 | 17 | 175 | 12 | 200 | 30 |
| 500 | 100 | 7 | 110 | 7 | 170 | 13 | 210 | 17 | ... | | ... | .. |

**TABLE III—230-, 460-, 575-VOLT MOTORS, ENCLOSURE OPEN—INCLUDING DRIPPROOF, GENERAL ELECTRIC TYPE KG (NEMA DESIGN "C") HIGH STARTING TORQUE AND NORMAL STARTING CURRENT**

| Induction Motor Horsepower Rating | Nominal Motor Speed in Rpm and Number of Poles | | | | | | | | | |
| | 1800 4 | | 1200 6 | | 900 8 | | 720 10 | | 600 12 | |
| | Kvar | %AR | Kvar | %AR | Kvar | %AR | Kvar | %AR | Kvar | %AR |
|---|---|---|---|---|---|---|---|---|---|---|
| 3 | ... | .. | 2 | 28 | 4 | 42 | ... | ... | ... | ... |
| 5 | 2 | 21 | 3 | 26 | 4 | 32 | ... | ... | ... | ... |
| 7.5 | 4 | 21 | 4 | 22 | 4 | 29 | ... | ... | ... | ... |
| 10 | 4 | 17 | 5 | 22 | 5 | 29 | ... | ... | 20 | 40 |
| 15 | 5 | 17 | 7.5 | 22 | 10 | 29 | ... | ... | ... | ... |
| 20 | 5 | 17 | 7.5 | 21 | 10 | 25 | ... | ... | 30 | 40 |
| 25 | 7.5 | 17 | 7.5 | 21 | 10 | 25 | ... | ... | 35 | 39 |
| 30 | 7.5 | 17 | 10 | 21 | 15 | 23 | 20 | 26 | 45 | 39 |
| 40 | 10 | 17 | 15 | 21 | 20 | 23 | ... | ... | ... | ... |
| 50 | 20 | 17 | 15 | 21 | 25 | 23 | 30 | 28 | ... | ... |
| 60 | 20 | 15 | 30 | 21 | 25 | 23 | 35 | 28 | ... | ... |
| 75 | 25 | 14 | 30 | 17 | 40 | 23 | 45 | 15 | 45 | 17 |
| 100 | 30 | 13 | 30 | 14 | 50 | 23 | 40 | 15 | 50 | 17 |
| 125 | 35 | 12 | 40 | 14 | 50 | 16 | 45 | 15 | 50 | 17 |
| 150 | 35 | 10 | 45 | 13 | 50 | 14 | 50 | 13 | 60 | 17 |
| 200 | 50 | 10 | 55 | 11 | 50 | 14 | 90 | 13 | 90 | 17 |
| 250 | 55 | 9 | 70 | 11 | 85 | 14 | 90 | 13 | 100 | 17 |
| 300 | 65 | 9 | 75 | 10 | 95 | 14 | 100 | 13 | 110 | 17 |
| 350 | 80 | 9 | 85 | 9 | 125 | 14 | 120 | 13 | 150 | 17 |

**NOTE:** A capacitor located on the motor side of the overload relay reduces current through the relay, and therefore, a smaller relay may be necessary. The motor-overload relay should be selected on the basis of the motor full-load nameplate current reduced by the percent reduction in line current (% AR) due to capacitors.

In the special case of a four-wire wye-connected system with balanced loading, two wattmeters may be used to monitor the power consumed on the service and also allow computation of the power factor from the two wattmeter readings.

### III.6.1.1 Balanced Four-Wire Wye-Connected System

Figure III.20 shows a balanced system containing two wattmeters. The sum of these two wattmeter readings are the total real power being used by the service:

$$P_T = P_1 + P_2$$

Further, the angle of displacement between each line current and voltage can be computed from $P_1$ and $P_2$:

$$\theta = \tan^{-1} \sqrt{3}\,\frac{P_2 - P_1}{P_2 + P_1}$$

and the power factor PF = cos θ.

**TABLE IV—230-, 460-, 575-VOLT MOTORS, TOTALLY-ENCLOSED, FAN-COOLED, GENERAL ELECTRIC TYPE KG (NEMA DESIGN "C"), HIGH STARTING TORQUE, NORMAL STARTING CURRENT**

| Induction Motor Horsepower Rating | 1800 / 4 | | 1200 / 6 | | 900 / 8 | | 720 / 10 | | 600 / 12 | |
|---|---|---|---|---|---|---|---|---|---|---|
| | Kvar | %AR | Kvar | %AR | Kvar | %AR | Kvar | %AR | Kvar | %AR |
| 3 | ... | ... | 2 | 28 | 4 | 42 | ... | ... | ... | ... |
| 5 | 2 | 20 | 3 | 26 | 4 | 32 | ... | ... | ... | ... |
| 7.5 | 4 | 19 | 4 | 22 | 4 | 30 | ... | ... | ... | ... |
| 10 | 4 | 19 | 5 | 22 | 5 | 30 | ... | ... | ... | ... |
| 15 | 5 | 19 | 7.5 | 22 | 10 | 30 | ... | ... | ... | ... |
| 20 | 7.5 | 19 | 10 | 21 | 10 | 29 | ... | ... | 20 | 40 |
| 25 | 7.5 | 19 | 10 | 19 | 10 | 29 | ... | ... | | |
| 30 | 7.5 | 19 | 10 | 20 | 10 | 29 | 25 | 31 | 30 | 40 |
| 40 | 15 | 19 | 13 | 19 | 20 | 24 | ... | ... | 35 | 38 |
| 50 | 15 | 19 | 20 | 19 | 20 | 22 | 30 | 28 | 40 | 37 |
| 60 | 15 | 19 | 25 | 18 | 25 | 22 | 35 | 28 | 40 | 37 |
| 75 | 25 | 13 | 25 | 15 | 35 | 22 | 45 | 28 | 40 | 37 |
| 100 | 30 | 13 | 25 | 12 | 45 | 20 | 40 | 28 | 30 | 37 |
| 125 | 30 | 13 | 30 | 12 | 45 | 20 | 50 | 28 | 50 | 37 |
| 150 | 40 | 11 | 40 | 12 | 50 | 20 | 50 | 28 | 70 | 37 |
| 200 | 40 | 8 | 60 | 12 | 60 | 20 | 70 | 28 | 75 | 38 |

**TABLE V—2400- AND 4160-VOLT MOTORS, ENCLOSURE OPEN—INCLUDING DRIPPROOF AND SPLASHPROOF, GENERAL ELECTRIC TYPE K (NEMA DESIGN "B") NORMAL STARTING TORQUE AND CURRENT**

| Induction Motor Horsepower Rating | 3600 / 2 | | 1800 / 4 | | 1200 / 6 | | 900 / 8 | | 720 / 10 | | 600 / 12 | |
|---|---|---|---|---|---|---|---|---|---|---|---|---|
| | Kvar | %AR | Kvar | %AR | Kvar | %AR | Kvar | %AR | Kvar | %AR | Kvar | %AR |
| 100 | ... | .. | 25 | 11 | 25 | 12 | 50 | 24 | 25 | 14 | 25 | 20 |
| 125 | ... | .. | 25 | 9 | 25 | 12 | 50 | 13 | 25 | 14 | 50 | 20 |
| 150 | 25 | 9 | 25 | 9 | 25 | 12 | 50 | 13 | 50 | 14 | 75 | 20 |
| 200 | 25 | 9 | 25 | 9 | 50 | 12 | 50 | 13 | 75 | 14 | 100 | 20 |
| 250 | 25 | 9 | 50 | 8 | 50 | 12 | 75 | 13 | 75 | 14 | 100 | 20 |
| 300 | 50 | 9 | 50 | 8 | 75 | 12 | 100 | 13 | 100 | 14 | 125 | 20 |
| 350 | 75 | 9 | 50 | 8 | 75 | 12 | 100 | 12 | 100 | 14 | 125 | 19 |
| 400 | 75 | 9 | 50 | 8 | 100 | 12 | 100 | 12 | 125 | 14 | 150 | 19 |
| 450 | 100 | 9 | 75 | 8 | 100 | 12 | 125 | 12 | 125 | 14 | 150 | 19 |
| 500 | 100 | 8 | 100 | 8 | 125 | 12 | 125 | 12 | 150 | 14 | 200 | 19 |
| 600 | 125 | 8 | 125 | 8 | 175 | 12 | 150 | 11 | 150 | 14 | 200 | 17 |
| 700 | 150 | 8 | 150 | 6 | 200 | 11 | 150 | 10 | 200 | 14 | 200 | 15 |
| 800 | 175 | 8 | 150 | 7 | 175 | 10 | 175 | 10 | 225 | 13 | 250 | 15 |

**TABLE VI—2400- AND 4160-VOLT MOTORS, TOTALLY-ENCLOSED, FAN-COOLED, GENERAL ELECTRIC TYPE K (NEMA DESIGN "B"), NORMAL STARTING TORQUE, NORMAL STARTING CURRENT**

| Induction Motor Horsepower Rating | 3600 / 2 | | 1800 / 4 | | 1200 / 6 | | 900 / 8 | | 720 / 10 | | 600 / 12 | |
|---|---|---|---|---|---|---|---|---|---|---|---|---|
| | Kvar | %AR | Kvar | %AR | Kvar | %AR | Kvar | %AR | Kvar | %AR | Kvar | %AR |
| 100 | ... | .. | 25 | 17 | ... | .. | 50 | 22 | 25 | 12 | 50 | 15 |
| 125 | ... | .. | 50 | 17 | 25 | 15 | 50 | 17 | 50 | 12 | 50 | 15 |
| 150 | 25 | 6 | 25 | 12 | 50 | 15 | 50 | 17 | 50 | 12 | 75 | 15 |
| 200 | 25 | 6 | 50 | 12 | 75 | 15 | 50 | 17 | 50 | 12 | 100 | 15 |
| 250 | 25 | 6 | 50 | 11 | 75 | 15 | 75 | 17 | 75 | 12 | 100 | 15 |
| 300 | 50 | 6 | 50 | 11 | 75 | 13 | 125 | 17 | 100 | 12 | 125 | 15 |
| 350 | 50 | 6 | 50 | 11 | 75 | 13 | 125 | 17 | 125 | 12 | 150 | 15 |
| 400 | 75 | 6 | 125 | 11 | 125 | 13 | 150 | 17 | 150 | 12 | 200 | 15 |
| 450 | 75 | 6 | 125 | 10 | 150 | 13 | 175 | 17 | 200 | 12 | 225 | 15 |
| 500 | 75 | 6 | 125 | 8 | 175 | 13 | 225 | 17 | 225 | 12 | 225 | 15 |

**TABLE VII—2400- AND 4160-VOLT MOTORS, ENCLOSURE OPEN—INCLUDING DRIPPROOF AND SPLASHPROOF, GENERAL ELECTRIC TYPE KG (NEMA DESIGN "C"), HIGH STARTING TORQUE AND NORMAL STARTING CURRENT**

| Induction Motor Horsepower Rating | 1800 / 4 | | 1200 / 6 | | 900 / 8 | | 720 / 10 | |
|---|---|---|---|---|---|---|---|---|
| | Kvar | %AR | Kvar | %AR | Kvar | %AR | Kvar | %AR |
| 100 | 25 | 10 | 25 | 11 | 25 | 13 | 25 | 14 |
| 125 | 25 | 8 | 25 | 9 | 25 | 13 | 25 | 14 |
| 150 | 25 | 7 | 50 | 12 | 50 | 13 | 50 | 14 |
| 200 | 25 | 8 | 50 | 12 | 50 | 13 | 75 | 14 |
| 250 | 25 | 8 | 50 | 12 | 75 | 13 | 75 | 14 |
| 300 | 50 | 8 | 75 | 12 | 100 | 13 | 100 | 14 |
| 350 | 50 | 8 | 75 | 12 | 100 | 12 | 100 | 14 |

**TABLE VIII—2400- AND 4160-VOLT MOTORS, TOTALLY-ENCLOSED, FAN-COOLED, GENERAL ELECTRIC TYPE KG (NEMA DESIGN "C"), HIGH STARTING TORQUE, NORMAL STARTING CURRENT**

| Induction Motor Horsepower Rating | 1200 / 6 | | 900 / 8 | | 720 / 10 | | 600 / 12 | |
|---|---|---|---|---|---|---|---|---|
| | Kvar | %AR | Kvar | %AR | Kvar | %AR | Kvar | %AR |
| 75 | ... | .. | ... | .. | 25 | 12 | 50 | 15 |
| 100 | ... | .. | ... | .. | 25 | 12 | 50 | 15 |
| 125 | 25 | 10 | 50 | 17 | 25 | 12 | 75 | 15 |
| 150 | ... | .. | 50 | 17 | 50 | 12 | 75 | 15 |
| 200 | 75 | 15 | 50 | 17 | 50 | 12 | 100 | 15 |

**CAPACITORS BENEFIT YOUR DISTRIBUTION SYSTEM BY**

- Reducing power costs
- Releasing system capacity
- Improving voltage levels
- Reducing system losses

This quick method for monitoring power and power factor is useful in determining both fixed capacitors to be tied across each phase for the nominal load, and the capacitors that are switched in only when intermittent loads come on-line.

The two-wattmeter method is useful for determining real power consumed in either wye- or delta-connected systems with or without balanced loads:

$$P_T = P_1 + P_2$$

However, the use of these readings for determining phase power factor as well is restricted to the case of balanced loads.

**III.6.2 Summary**

This brief coverage of power and power-factor determination in three-phase systems covers only the very basic ideas in this important area. It is the aim of this brief coverage to recall or refresh ideas once learned but seldom used.

Tables III.1 and III.2 were supplied by General Electric. who gave permission for the reproduction of their materials in this handbook.

**Fig. III.20 Four-wire wye-connected system with wattmeter connections detailed. Solid circle voltage connections to wattmeter; open circle, current connections to wattmeter.**

# INDEX